DRUG DISCOVERY HANDBOOK

DRUG DISCOVERY HANDBOOK

Edited by

SHAYNE COX GAD, PH.D., D.A.B.T.
Gad Consulting Services
Cary, North Carolina

⊛WILEY-INTERSCIENCE

A JOHN WILEY & SONS, INC., PUBLICATION

Copyright © 2005 by John Wiley & Sons, Inc. All rights reserved

Published by John Wiley & Sons, Inc., Hoboken, New Jersey
Published simultaneously in Canada

No part of this publication may be reproduced, stored in a retrieval system, or transmitted in any form or by any means, electronic, mechanical, photocopying, recording, scanning, or otherwise, except as permitted under Section 107 or 108 of the 1976 United States Copyright Act, without either the prior written permission of the Publisher, or authorization through payment of the appropriate per-copy fee to the Copyright Clearance Center, Inc., 222 Rosewood Drive, Danvers, MA 01923, (978) 750-8400, fax (978) 750-4470, or on the web at www.copyright.com. Requests to the Publisher for permission should be addressed to the Permissions Department, John Wiley & Sons, Inc., 111 River Street, Hoboken, NJ 07030, (201) 748-6011, fax (201) 748-6008, or online at http://www.wiley.com/go/permission.

Limit of Liability/Disclaimer of Warranty: While the publisher and author have used their best efforts in preparing this book, they make no representations or warranties with respect to the accuracy or completeness of the contents of this book and specifically disclaim any implied warranties of merchantability or fitness for a particular purpose. No warranty may be created or extended by sales representatives or written sales materials. The advice and strategies contained herein may not be suitable for your situation. You should consult with a professional where appropriate. Neither the publisher nor author shall be liable for any loss of profit or any other commercial damages, including but not limited to special, incidental, consequential, or other damages.

For general information on our other products and services or for technical support, please contact our Customer Care Department within the United States at (800) 762-2974, outside the United States at (317) 572-3993 or fax (317) 572-4002.

Wiley also publishes its books in a variety of electronic formats. Some content that appears in print may not be available in electronic formats. For more information about Wiley products, visit our web site at www.wiley.com.

Library of Congress Cataloging-in-Publication Data:

Drug discovery handbook / edited by Shayne Gad.
    p. cm.
    Includes index.
    ISBN-13 978-0-471-21384-0 (cloth)
    ISBN-10 0-471-21384-5 (cloth)
    1. Drugs—Research—Handbooks, manuals, etc. 2. Drug development—Handbooks, manuals, etc. 3. Drugs—Design—Handbooks, manuals, etc. 4. Pharmaceutical chemistry—Handbooks, manuals, etc. I. Gad, Shayne C., 1948–

RM301.25.D784 2005
615'.19—dc22

2004027077

Printed in the United States of America
10 9 8 7 6 5 4 3 2

CONTRIBUTORS

Dmitri Artemov, JHU ICMIC Program, Department of Radiology, The Johns Hopkins University School of Medicine, Baltimore, Maryland, *Novel Imaging Agents for Molecular MR Imaging of Cancer*

Kadir Aslan, Institute of Fluorescence, University of Maryland Biotechnology Institute, Baltimore, Maryland, *Metal-Enhanced Fluorescence: Application to High-Throughput Screening and Drug Discovery*

Jürgen Bajorath, Department of Life Science Informatics, B-IT, International Center for Information Technology, Rheinische Friedrich-Wilhelms-University Bonn, Bonn, Germany, *Molecular Similarity Methods and QSAR Models as Tools for Virtual Screening*

Martyn N. Banks, Lead Discovery and Profiling, Applied Biotechnology Division, Bristol Myers Squibb Co., Pharmaceutical Research Institute, Wallingford, Connecticut, *High-Throughput Screening: Evolution of Technology and Methods*

Richard D. Beger, Division of Systems Toxicology, National Center for Toxicological Research, Food and Drug Administration, Jefferson, Arkansas, *Combining NMR Spectral Information with Associated Structural Features to Form Computationally Nonintensive, Rugged, and Objective Models of Biological Activity*

Oren E. Beske, Vitra Bioscience, Inc., Mountain View, California, *Simultaneous Screening of Multiple Cell Lines Using the CellCard System*

Zaver M. Bhujwalla, JHU ICMIC Program, Department of Radiology, The Johns Hopkins University School of Medicine, Baltimore, Maryland, *Novel Imaging Agents for Molecular MR Imaging of Cancer*

Sean M. Biggs, Department of Pathology, Department of Cell Biology and Physiology, and Cancer Research and Treatment Center, University of New Mexico School of Medicine, University of New Mexico Health Sciences Center, Albuquerque, New Mexico, *High-Throughput Flow Cytometry*

George E. Billman, Department of Physiology and Cell Biology, The Ohio State University, Columbus, Ohio, *Cardiac Sarcolemmal ATP-Sensitive Potassium Channel Antagonists: Novel Ischemia-Selective Antiarrhythmic Agents*

Cristian Bologa, Department of Biochemistry and Molecular Biology and Office of Biocomputing, University of New Mexico School of Medicine, University of New Mexico Health Sciences Center, Albuquerque, New Mexico, *High-Throughput Flow Cytometry*

Hans Bräuner-Osborne, Department of Medicinal Chemistry, Danish University of Pharmaceutical Sciences, Copenhagen, Denmark, *GABA and Glutamate Receptor Ligands and Their Therapeutic Potential in CNS Disorders*

Andrej Bugrim, GeneGo, St. Joseph, Michigan, *Systems Biology: Applications in Drug Discovery*

Dan A. Buzatu, Division of Systems Toxicology, National Center for Toxicological Research, Food and Drug Administration, Jefferson, Arkansas, *Combining NMR Spectral Information with Associated Structural Features to Form Computationally Nonintensive, Rugged, and Objective Models of Biological Activity*

Angela M. Cacace, Lead Discovery and Profiling, Applied Biotechnology Division, Bristol Myers Squibb Co., Pharmaceutical Research Institute, Wallingford, Connecticut, *High-Throughput Screening: Evolution of Technology and Methods*

Zehui Cao, Department of Chemistry, University of Florida, Gainesville, Florida, *Cancer Cell Proteomics Using Molecular Aptamers*

Hui Chen, Department of Chemistry, University of Florida, Gainesville, Florida, *Cancer Cell Proteomics Using Molecular Aptamers*

CONTRIBUTORS

Amy K. Chesterfield, Discovery-Neurosciences Research, Eli Lilly and Company, Indianapolis, Indiana, *Methods for the Design and Analysis of Replicate-Experiment Studies to Establish Assay Reproducibility and the Equivalence of Two Potency Assays*

C.H. Cho, Department of Pharmacology, Faculty of Medicine, The University of Hong Kong, Hong Kong, China, *Herbal Medicines and Animal Models of Gastrointestinal Diseases*

Michael J. Corey, Chromos, Seattle, Washington, *Coupled Luminescent Methods in Drug Discovery: 3-Min Assays for Cytotoxicity and Phosphatase Activity*

Xizhong Cui, Critical Care Medicine Department, Clinical Center, National Institutes of Health, Bethesda, Maryland, *Factors Influencing the Efficacy of Mediator-Specific Anti-Inflammatory, Glucocorticoid, and Anticoagulant Therapies for Sepsis*

Katherine J. Deans, Critical Care Medicine Department, Clinical Center, National Institutes of Health, Bethesda, Maryland; Department of Chemistry, Massachusetts General Hospital, Boston, Massachusetts, *Factors Influencing the Efficacy of Mediator-Specific Anti-Inflammatory, Glucocorticoid, and Anticoagulant Therapies for Sepsis*

Erik de Clercq, Rega Institute for Medical Research, Katholieke Universiteit Leuven, Leuven, Belgium, *Strategies in the Design of Antiviral Drugs*

Brian J. Eastwood, Statistics and Information Sciences, Eli Lilly and Company, Indianapolis, Indiana, *Methods for the Design and Analysis of Replicate-Experiment Studies to Establish Assay Reproducibility and the Equivalence of Two Potency Assays*

Bruce S. Edwards, Department of Pathology and Cancer Research and Treatment Center, University of New Mexico School of Medicine, University of New Mexico Health Sciences Center, Albuquerque, New Mexico, *High-Throughput Flow Cytometry*

Peter Q. Eichacker, Critical Care Medicine Department, Clinical Center, National Institutes of Health, Bethesda, Maryland, *Factors Influencing the Efficacy of Mediator-Specific Anti-Inflammatory, Glucocorticoid, and Anticoagulant Therapies for Sepsis*

Sean Ekins, GeneGo, St. Joseph, Michigan, *Systems Biology: Applications in Drug Discovery*

Christian C. Felder, Discovery-Neurosciences Research, Eli Lilly and Company, Indianapolis, Indiana, *Methods for the Design and Analysis of Replicate-Experiment Studies to Establish Assay Reproducibility and the Equivalence of Two Potency Assays*

Michael D. Feese, deCODE biostructures, Bainbridge Island, Washington, *Protein X-Ray Crystallography in Drug Discovery*

Bente Frølund, Department of Medicinal Chemistry, Danish University of Pharmaceutical Sciences, Copenhagen, Denmark, *GABA and Glutamate Receptor Ligands and Their Therapeutic Potential in CNS Disorders*

Brian L. Furman, Department of Physiology and Pharmacology, University of Strathclyde, Strathclyde Institute of Biomedical Sciences, Glasgow, Scotland, *Endocrine and Metabolic Agents*

Chris D. Geddes, Institute of Fluorescence, University of Maryland Biotechnology Institute, Baltimore, Maryland; Center for Fluorescence Spectroscopy, University of Maryland School of Medicine, Baltimore, Maryland, *Metal-Enhanced Fluorescence: Application to High-Throughput Screening and Drug Discovery*

Jeremy R. Greenwood, Department of Medicinal Chemistry, Danish University of Pharmaceutical Sciences, Copenhagen, Denmark, *GABA and Glutamate Receptor Ligands and Their Therapeutic Potential in CNS Disorders*

Ignacy Gryczynski, Center for Fluorescence Spectroscopy, University of Maryland School of Medicine, Baltimore, Maryland, *Metal-Enhanced Fluorescence: Application to High-Throughput Screening and Drug Discovery*

Michael Haley, Critical Care Medicine Department, Clinical Center, National Institutes of Health, Bethesda, Maryland, *Factors Influencing the Efficacy of Mediator-Specific Anti-Inflammatory, Glucocorticoid, and Anticoagulant Therapies for Sepsis*

Piet Herdewijn, Rega Institute for Medical Research, Katholieke Universiteit Leuven, Leuven, Belgium, *Strategies in the Design of Antiviral Drugs*

John G. Houston, Applied Biotechnology Division, Bristol Myers Squibb Co., Pharmaceutical Research Institute, Wallingford, Connecticut, *High-Throughput Screening: Evolution of Technology and Methods*

CONTRIBUTORS

Tomi Järvinen, Department of Pharmaceutical Chemistry, University of Kuopio, Kuopio, Finland, *Design and Pharmaceutical Applications of Prodrugs*

Tommy N. Johansen, Department of Medicinal Chemistry, Danish University of Pharmaceutical Sciences, Copenhagen, Denmark, *GABA and Glutamate Receptor Ligands and Their Therapeutic Potential in CNS Disorders*

Hidong Kim, deCODE biostructures, Bainbridge Island, Washington, *Protein X-Ray Crystallography in Drug Discovery*

Robert J. Kinders, *Coupled Luminescent Methods in Drug Discovery: 3-Min Assays for Cytotoxicity and Phosphatase Activity*

J.K.S. Ko, School of Chinese Medicine, The Hong Kong Baptist University, Hong Kong, China, *Herbal Medicines and Animal Models of Gastrointestinal Diseases*

Povl Krogsgaard-Larsen, Department of Medicinal Chemistry, Danish University of Pharmaceutical Sciences, Copenhagen, Denmark, *GABA and Glutamate Receptor Ligands and Their Therapeutic Potential in CNS Disorders*

Duane B. Lakings, Drug Safety Evaluation Consulting, Inc., Elgin, Texas, *Biological and Chemistry Assays Available During Drug Discovery and Developability Assessment*

Joseph R. Lakowicz, Center for Fluorescence Spectroscopy, University of Maryland School of Medicine, Baltimore, Maryland, *Metal-Enhanced Fluorescence: Application to High-Throughput Screening and Drug Discovery*

Ying Li, Department of Chemistry, University of Florida, Gainesville, Florida, *Cancer Cell Proteomics Using Molecular Aptamers*

Tommy Liljefors, Department of Medicinal Chemistry, Danish University of Pharmaceutical Sciences, Copenhagen, Denmark, *GABA and Glutamate Receptor Ligands and Their Therapeutic Potential in CNS Disorders*

Thorsteinn Loftsson, Faculty of Pharmacy, University of Iceland, Reykjavik, Iceland, *Design and Pharmaceutical Applications of Prodrugs*

Arianna Loregian, Department of Histology, Microbiology, and Medical Biotechnologies, University of Padova, Padova, Italy, *Strategies and Methods in Monitoring and Targeting Protein–Protein Interactions*

Ulf Madsen, Department of Medicinal Chemistry, Danish University of Pharmaceutical Sciences, Copenhagen, Denmark, *GABA and Glutamate Receptor Ligands and Their Therapeutic Potential in CNS Disorders*

Joanna Malicka, Center for Fluorescence Spectroscopy, University of Maryland School of Medicine, Baltimore, Maryland, *Metal-Enhanced Fluorescence: Application to High-Throughput Screening and Drug Discovery*

Prabodhika Mallikratchy, Department of Chemistry, University of Florida, Gainesville, Florida, *Cancer Cell Proteomics Using Molecular Aptamers*

Mar Masson, Faculty of Pharmacy, University of Iceland, Reykjavik, Iceland, *Design and Pharmaceutical Applications of Prodrugs*

Brian R. McNaughton, Department of Chemistry, University of Rochester, Rochester, New York, *Combinatorial Chemistry in the Drug Discovery Process*

Benjamin L. Miller, Department of Biochemistry and Biophysics, Department of Dermatology, University of Rochester, Rochester, New York, *Combinatorial Chemistry in the Drug Discovery Process*

Peter C. Minneci, Critical Care Medicine Department, Clinical Center, National Institutes of Health, Bethesda, Maryland; Department of Chemistry, Massachusetts General Hospital, Boston, Massachusetts, *Factors Influencing the Efficacy of Mediator-Specific Anti-Inflammatory, Glucocorticoid, and Anticoagulant Therapies for Sepsis*

Susan L. Mooberry, Department of Physiology and Medicine, Southwest Foundation for Biomedical Research, San Antonio, Texas, *Targets and Approaches for Cancer Drug Discovery*

Charles Natanson, Critical Care Medicine Department, Clinical Center, National Institutes of Health, Bethesda, Maryland, *Factors Influencing the Efficacy of Mediator-Specific Anti-Inflammatory, Glucocorticoid, and Anticoagulant Therapies for Sepsis*

William Neil, Bristol-Myers Squibb, New Brunswick, New Jersey, *Using Microsoft Excel as a Laboratory Data Management Tool*

Mogens Nielsen, Department of Medicinal Chemistry, Danish University of Pharmaceutical Sciences, Copenhagen, Denmark, *GABA and Glutamate Receptor Ligands and Their Therapeutic Potential in CNS Disorders*

CONTRIBUTORS

Tatiana Nikolskaya, GeneGo, St. Joseph, Michigan, *Systems Biology: Applications in Drug Discovery*

Yuri Nikolsky, GeneGo, St. Joseph, Michigan, *Systems Biology: Applications in Drug Discovery*

Peter Nollert, deCODE biostructures, Bainbridge Island, Washington, *Protein X-Ray Crystallography in Drug Discovery*

Jonathan O'connell, Lead Discovery and Profiling, Applied Biotechnology Division, Bristol Myers Squibb Co., Pharmaceutical Research Institute, Wallingford, Connecticut, *High-Throughput Screening: Evolution of Technology and Methods*

Marius Olah, Department of Biochemistry and Molecular Biology and Office of Biocomputing, University of New Mexico School of Medicine, University of New Mexico Health Sciences Center, Albuquerque, New Mexico, *High-Throughput Flow Cytometry*

Paul D. Olivo, Apath, LLC, St. Louis, Missouri, *Respiratory Viruses*

Tudor I. Oprea, Department of Biochemistry and Molecular Biology and Office of Biocomputing, University of New Mexico School of Medicine, University of New Mexico Health Sciences Center, Albuquerque, New Mexico, *High-Throughput Flow Cytometry*

Giorgio Palù, Department of Histology, Microbiology, and Medical Biotechnologies, University of Padova, Padova, Italy, *Strategies and Methods in Monitoring and Targeting Protein–Protein Interactions*

Keykavous Parang, Department of Biomedical and Pharmaceutical Sciences, University of Rhode Island, Kingston, Rhode Island, *Protein Kinase Inhibitors in Drug Discovery*

Steve Pascolo, CureVac, GmbH, Tübingen, Germany, *RNA-Based Therapies*

Eric R. Prossnitz, Department of Cell Biology and Physiology and Cancer Research and Treatment Center, University of New Mexico School of Medicine, University of New Mexico Health Sciences Center, Albuquerque, New Mexico, *High-Throughput Flow Cytometry*

Jarkko Rautio, Department of Pharmaceutical Chemistry, University of Kuopio, Kuopio, Finland, *Design and Pharmaceutical Applications of Prodrugs*

Nathan T. Ross, Department of Biochemistry and Biophysics, University of Rochester, Rochester, New York, *Combinatorial Chemistry in the Drug Discovery Process*

A. Erik Rubin, Bristol-Myers Squibb, New Brunswick, New Jersey, *Using Microsoft Excel as a Laboratory Data Management Tool*

Mark F. Russo, Bristol-Myers Squibb, New Brunswick, New Jersey, *Using Microsoft Excel as a Laboratory Data Management Tool*

Dihua Shangguan, Department of Chemistry, University of Florida, Gainesville, Florida, *Cancer Cell Proteomics Using Molecular Aptamers*

Peter C. Simons, Department of Pathology and Cancer Research and Treatment Center, University of New Mexico School of Medicine, University of New Mexico Health Sciences Center, Albuquerque, New Mexico, *High-Throughput Flow Cytometry*

Larry A. Sklar, Department of Pathology and Cancer Research and Treatment Center, University of New Mexico School of Medicine, University of New Mexico Health Sciences Center, Albuquerque, New Mexico, *High-Throughput Flow Cytometry*

Charles B. Spainhour, Clark's Summit, Pennsylvania, *Natural Products*

Bart L. Staker, deCODE biostructures, Bainbridge Island, Washington, *Protein X-Ray Crystallography in Drug Discovery*

Gongqin Sun, Department of Cell and Molecular Biology, University of Rhode Island, Kingston, Rhode Island, *Protein Kinase Inhibitors in Drug Discovery*

Weihong Tan, Department of Chemistry, University of Florida, Gainesville, Florida, *Cancer Cell Proteomics Using Molecular Aptamers*

Zhiwen Tang, Department of Chemistry, University of Florida, Gainesville, Florida, *Cancer Cell Proteomics Using Molecular Aptamers*

Anna Waller, Department of Pathology and Cancer Research and Treatment Center, University of New Mexico School of Medicine, University of New Mexico Health Sciences Center, Albuquerque, New Mexico, *High-Throughput Flow Cytometry*

Sandy Weinberg, Fast Trak BioDefense, GE Healthcare, Atlanta, Georgia, *Age of Regulation*

Gerald J. Whartenby, Regulatory Consultant for GE Healthcare, Philadelphia, Pennsylvania, *Age of Regulation*

Jon G. Wilkes, Division of Systems Toxicology, National Center for Toxicological Research, Food and Drug Administration, Jefferson, Arkansas, *Combining NMR Spectral Information with Associated Structural Features to Form Computationally Nonintensive, Rugged, and Objective Models of Biological Activity*

Mary C. Wolff, Discovery-Neurosciences Research, Eli Lilly and Company, Indianapolis, Indiana, *Methods for the Design and Analysis of Replicate-Experiment Studies to Establish Assay Reproducibility and the Equivalence of Two Potency Assays*

Susan M. Young, Department of Pathology and Cancer Research and Treatment Center, University of New Mexico School of Medicine, University of New Mexico Health Sciences Center, Albuquerque, New Mexico, *High-Throughput Flow Cytometry*

CONTENTS

Preface   xix

Introduction: Drug Discovery in the 21st Century   1

1 Natural Products   11
Charles B. Spainhour

2 Cancer Cell Proteomics Using Molecular Aptamers   73
Weihong Tan, Zehui Cao, Dihua Shangguan, Ying Li, Zhiwen Tang, Prabodhika Mallikratchy, and Hui Chen

3 Molecular Similarity Methods and QSAR Models as Tools for Virtual Screening   87
Jürgen Bajorath

4 Systems Biology: Applications in Drug Discovery   123
Sean Ekins, Andrej Bugrim, Yuri Nikolsky, and Tatiana Nikolskaya

5 High-Throughput Flow Cytometry   185
Larry A. Sklar, Peter C. Simons, Anna Waller, Sean M. Biggs, Susan M. Young, Marius Olah, Cristian Bologa, Tudor I. Oprea, Eric R. Prossnitz, and Bruce S. Edwards

6	**Combining NMR Spectral Information with Associated Structural Features to Form Computationally Nonintensive, Rugged, and Objective Models of Biological Activity** Richard D. Beger, Dan A. Buzatu, and Jon G. Wilkes	227
7	**Using Microsoft Excel® as a Laboratory Data Management Tool** A. Erik Rubin, Mark F. Russo, and William Neil	287
8	**Age of Regulation** Sandy Weinberg and Gerald J. Whartenby	337
9	**Simultaneous Screening of Multiple Cell Lines Using the CellCard System** Oren E. Beske	353
10	**Protein X-ray Crystallography in Drug Discovery** Peter Nollert, Michael D. Feese, Bart L. Staker, and Hidong Kim	373
11	**Biological and Chemistry Assays Available During Drug Discovery and Developability Assessment** Duane B. Lakings	457
12	**Strategies and Methods in Monitoring and Targeting Protein–Protein Interactions** Arianna Loregian and Giorgio Palù	483
13	**High-Throughput Screening: Evolution of Technology and Methods** Martyn N. Banks, Angela M. Cacace, Jonathan O'Connell, and John G. Houston	559
14	**Metal-Enhanced Fluorescence: Application to High-Throughput Screening and Drug Discovery** Kadir Aslan, Ignacy Gryczynski, Joanna Malicka, Joseph R. Lakowicz, and Chris D. Geddes	603
15	**Methods for the Design and Analysis of Replicate-Experiment Studies to Establish Assay Reproducibility and the Equivalence of Two Potency Assays** Brian J. Eastwood, Amy K. Chesterfield, Mary C. Wolff, and Christian C. Felder	667

CONTENTS

16 Coupled Luminescent Methods in Drug Discovery: 3-Min Assays for Cytotoxicity and Phosphatase Activity 689
Michael J. Corey and Robert J. Kinders

17 Design and Pharmaceutical Applications of Prodrugs 733
Tomi Järvinen, Jarkko Rautio, Mar Masson, and Thorsteinn Loftsson

18 GABA and Glutamate Receptor Ligands and Their Therapeutic Potential in CNS Disorders 797
Ulf Madsen, Hans Bräuner-Osborne, Jeremy R. Greenwood, Tommy N. Johansen, Povl Krogsgaard-Larsen, Tommy Liljefors, Mogens Nielsen, and Bente Frølund

19 Cardiac Sarcolemmal ATP-Sensitive Potassium Channel Antagonists: Novel Ischemia-Selective Antiarrhythmic Agents 909
George E. Billman

20 Factors Influencing the Efficacy of Mediator-Specific Anti-Inflammatory, Glucocorticoid, and Anticoagulant Therapies for Sepsis 937
Peter C. Minneci, Katherine J. Deans, Michael Haley, Xizhong Cui, Charles Natanson, and Peter Q. Eichacker

21 Combinatorial Chemistry in the Drug Discovery Process 961
Nathan T. Ross, Brian R. McNaughton, and Benjamin L. Miller

22 Herbal Medicines and Animal Models of Gastrointestinal Diseases 1013
C.H. Cho and J.K.S. Ko

23 Endocrine and Metabolic Agents 1037
Brian L. Furman

24 Respiratory Viruses 1105
Paul D. Olivo

25 Strategies in the Design of Antiviral Drugs 1135
Erik De Clercq and Piet Herdewijn

26 Protein Kinase Inhibitors in Drug Discovery 1191
Keykavous Parang and Gongqin Sun

27 RNA-Based Therapies 1259
Steve Pascolo

28	**Novel Imaging Agents for Molecular MR Imaging of Cancer** *Dmitri Artemov and Zaver M. Bhujwalla*	**1309**
29	**Targets and Approches for Cancer Drug Discovery** *Susan L. Mooberry*	**1343**
Index		**1375**

PREFACE

This *Drug Discovery Handbook* represents a unique attempt to survey the different approaches to discovering potential new therapeutic moieties. Such moieties are the backbone of both the pharmaceutical industry and the prime axis for the advancement of medical science.

The volume is unique in that it seeks to cover possible approaches to drug discovery as broadly as possible while not just doing so in a superficial manner. The 29 chapters cover all the major approaches to the problem of identifying potential drugs and were written by leading representatives from each of these approaches.

I hope that this banquet is satisfying and useful to my colleagues in the field.

Select figures of this title are available in full color at ftp://ftp.wiley.com/public/sci_tech_med/drug_discovery/.

S.C. GAD

INTRODUCTION: DRUG DISCOVERY IN THE 21st CENTURY

SHAYNE COX GAD

I.1 INTRODUCTION

The discovery, development, and registration of a pharmaceutical is an immensely expensive operation and represents a rather unique challenge. For every 9000 to 10,000 compounds specifically synthesized or isolated as potential therapeutics, one (on average) will actually reach the market. This process is illustrated diagrammatically in Figure 1. Each successive stage in the process is more expensive, making it of great interest to identify as early as possible those agents that are likely not to go the entire distance, allowing a concentration of effort on the compounds that have the highest probability of reaching the market. Compounds "drop out" of the process primarily for three reasons:

1. Toxicity or (lack of) tolerance
2. (Lack of) efficacy
3. (Lack of) bioavailability of the active moiety in humans

Early identification of poor or noncompetitive candidates in each of these three categories is thus extremely important [1], forming the basis for the use of screening in pharmaceutical discovery and development. How much and which resources to invest in screening and each successive step in support of the development of a potential drug are matters of strategy and phasing that are detailed later in this Introduction. In vitro methods are increasingly

---

Drug Discovery Handbook, by Shayne Cox Gad
Copyright © 2005 by John Wiley & Sons, Inc.

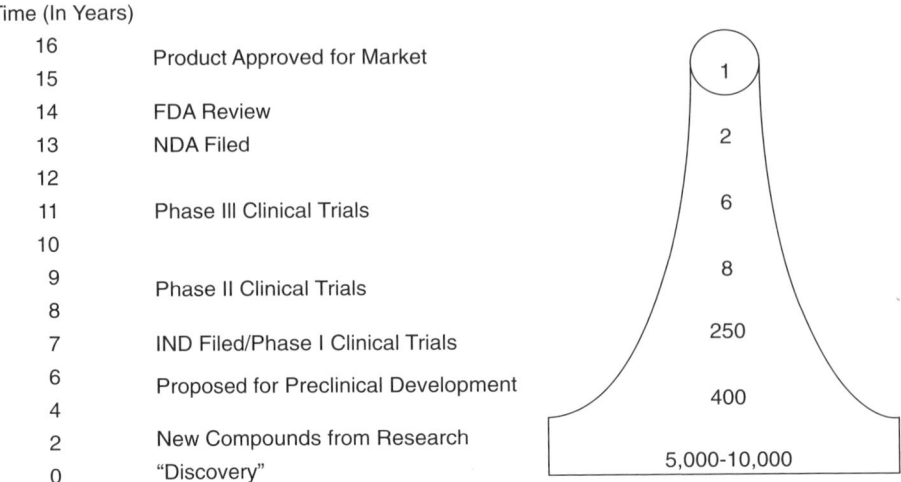

Figure I.1 Attrition during the development of new molecules with a promise of therapeutic potential. Over the course of taking a new molecular entity through scale-up, safety, and efficacy testing, and, finally, to market, typically only 1 out of every 9000 to 10,000 will go to the marketplace.

providing new tools for use in both early screening and the understanding of mechanisms of observed toxicity in preclinical and clinical studies [2, 3], particularly with the growing capabilities and influence of genomic and proteomic technologies. This is increasingly important as the societal concern over drug prices has grown [4]. Additionally, the marketplace for new drugs is exceedingly competitive. The rewards for being either early (first or second) into the marketplace or achieving a significant therapeutic advantage are enormous in terms of eventual market share. Additionally, the first drug approved sets agency expectations for those drugs that follow. In mid-2004, there were 263 pharmaceutical products awaiting approval (93 of these biotech products)—the "oldest" having been in review 10 years) and some 2300 additional agents in the IND stage. Not all of these (particularly the oldest) will be economically successful).

The usual way in which transition (or "flow") of new molecules between the different phases is handled in drug discovery/development is to use a tiered screening or testing approach. Each tier generates more specific data (and costs more to do so) and draws on the information generated in earlier tiers to refine the design of new studies. Different tiers are keyed to the support of successive decision points (go/no-go points) in the development process, with the intent of reducing risks (as to efficacy bioavailability and safety) as early as possible.

The first real critical decisions concerning the potential advancement of a compound to evaluation in clinical trials are the most difficult. They require an understanding of how well particular in vitro or in vivo work in predicting adverse effects in humans (usually very well, but there are notable lapses; for example, giving false positives and false negatives) and an understanding of what initial clinical trials are intended to do. Though an "approved" IND grants one entry into limited evaluations of drug effects in humans, flexibility in the execution and analysis of these studies offers a significant opportunity to also investigate efficacy [5].

Once past the discovery and initial lead or candidate selection stages, each aspect of development becomes more tightly connected with the other aspects of the development of a compound, particularly the potential clinical aspects. These interconnections are coordinated by project management systems. Many times during the early years of the development process, biological evaluation of efficacy and safety constitutes the rate-limiting step—it is, in the language of project management, on the critical path.

Another way in which pharmaceutical development varies from toxicology as practiced in other industries is that it is a much more multidisciplinary and integrated process. This particularly stands out in the incorporation of the evaluation of ADME (absorption, distribution, metabolism, and excretion) aspects in the safety evaluation process. These pharmacokinetic/metabolism (PKM) aspects are evaluated for each of the animal model species utilized to predict the safety of a potential drug prior to evaluation in humans. Frequently, in vitro characterizations of metabolism for model (or potential model) species and humans are performed to allow optimal model selection and understanding of findings. This allows for an early appreciation of both the potential bioavailability of active drug moieties and the relative predictive values of the various biological models. Such data early on are also very useful (in fact, sometimes essential) in setting does levels for later animal studies and in projecting safe dose levels for clinical use. Unlike the case in most other realms of development of biologically active molecules, one is not limited to extrapolating the relationships between administered dose and systemic effects. Rather, one has significant information on systemic levels of the therapeutic moiety—typically, total area under the curve (AUC), peak plasma levels (C_{max}), and plasma half-lives, at a minimum.

I.2 SCREENS: THEIR USE AND INTERPRETATION IN DRUG DISCOVERY

Much (perhaps even most) of what is performed in safety assessment can be considered screening—trying to determine if some effect is or is not (to an acceptable level of confidence) present [6]. The general concepts of such screens are familiar to toxicologists in the pharmaceutical industry because the approach is a major part of the activities of the pharmacologists involved in the discovery of new compounds. But the principles underlying screening are not generally well recognized or understood. And such understanding is essential to the proper use, design, and analysis of screens [7, 8]. Screens are the biological equivalent of exploratory data analysis, or EDA [9]. Each test or assay has an associated activity criterion, that is, a level above which the activity of interest is judged to be present. If the result for a particular test compound meets this criterion, the compound may pass to the next stage. This criterion could be based on statistical significance (e.g., all compounds with observed activities significantly greater than the control at the 5 percent level could be tagged). However, for early screens, such a formal criterion may be too strict, resulting in few compounds begin identified as "active."

A useful indicator of the efficacy of an assay series is the frequency of discovery of truly active compounds. The frequency is related to the probability of discovery and to the degree of risk (hazard to health) associated with an active compound passing a screen undetected. These two factors in turn depend on the distribution of activities in the series of compounds being tested and the chances of rejecting or accepting compounds with given activities at each stage.

Statistical modeling of the assay system may lead to the improvement of the design of the system by reducing the interval between discoveries of active compounds. The objectives behind a screen and considerations of (1) costs for producing compounds and testing and (2) the degree of uncertainly about test performance will determine desired performance characteristics of specific cases. In the most common case of early toxicity screens performed to remove possible problem compounds, preliminary results suggest that it may be beneficial to increase the number of compounds tested, decrease the numbers of animals per group, and increase the range and number of doses. The result will be less information on more structure, but there will be an overall increase in the frequency of discovery of active compounds (assuming that truly active compounds are entering the system at a steady rate).

The methods described here are well suited to analyzing screening data when the interest is truly in detecting the absence of an effect with little chance of false negatives. There are many forms of graphical analysis methods available, including some newer forms that are particularly well suited to multivariate data (the types that are common in more complicated screening test designs). It is intended that these aspects of analysis will be focused on in a later publication.

The design of each assay and the choice of the activity criterion should, therefore, be adjusted, bearing in mind the relative costs of retaining false positives and rejecting false negatives. Decreasing the group sizes in the early assays reduces the chance of obtaining significance at any particular level (such as 5 percent), so that the activity criterion must be relaxed, in a statistical sense, to allow more compounds through. At some stage, however, it becomes too expensive to continue screening many false positives, and the criteria must be tightened accordingly. Where the criteria are set depends on what acceptable noise levels are in a screening system.

Characteristics of Screens

An excellent introduction to the characteristics of screens in Redman's [10] interesting approach, which identifies four characteristics of an assay. Redman assumes that a compound is either active or inactive and that the proportion of activities in a compound can be estimated from past experience. After testing, a compound will be classified as positive or negative (i.e., possessing or lacking activity). It is then possible to design the assay so as to optimize the following characteristics:

1. Sensitivity: the ratio of true positives to total activities
2. Specificity: the ratio of true negatives to total inactives
3. Positive accuracy: the ratio of true to observed positives
4. Negative accuracy: the ratio of true to observed negatives
5. Capacity: the number of compounds that can be evaluated
6. Reproducibility: the probability that a screen will produce the same result at another time (and, perhaps, in some other lab)

An advantage of testing many compounds is that it gives the opportunity to average activity evidence over structural classes or to study quantitative structure–activity relationships (QSARs). Quantitative structure–activity relationships can be used to predict the activity of new compounds and thus reduce the chance of in vivo testing on negative compounds. The use of QSARs can increase the proportion of truly active compounds passing through the system.

To simplify this presentation, data sets drawn only from neuromuscular screening activity were used. However, the evaluation and approaches should be valid for all similar screening data sets, regardless of source. The methods are not sensitive to the biases introduced by the degree of interdependence found in many screening batteries that use multiple measures (such as the neurobehavioral screen).

1. Screens almost always focus on detecting a single end point of effect (such as mutagenicity, lethality, neurotoxicity, or development toxicity) and have a particular set of operating characteristics in common.
2. A large number of compounds are evaluated, so ease and speed of performance (which may also be considered efficiency) are very desirable characteristics.
3. The screen must be very sensitive in its detection of potential effective agents. An absolute minimum of active agents should escape detection; that is, there should be very few false negatives (in other words, the type II error rate or β level should be low). Stated yet another way, the signal gain should be way up.

4. It is desirable that the number of false positives be small (i.e., there should be a low type I error rate of α level).
5. Items 2 to 4, which are all to some degree contradictory, require the involved researchers to agree on a set of compromises, starting with the acceptance of a relatively high α level (0.10 or more), that is, an increased noise level.
6. In an effort to better serve item 2, safety assessment screens are frequently performed in batteries so that multiple end points are measured in the same operation. Additionally, such measurements may be repeated over a period of time in each model as a means of supporting item 3.
7. This screen should use small amounts of compound to make item 1 possible and should allow evaluation of materials that have limited availability (such as novel compounds) early on in development.
8. Any screening system should be validate initially using a set of blind (positive and negative) controls. These blind controls should also be evaluated in the screening system on a regular basis to ensure continuing proper operation of the screen. As such, the analysis techniques used here can then be used to ensure the quality or modify the performance of a screening system.
9. The more that is known about the activity of interest, the more specific the form of screen that can be employed. As specificity increases, so should sensitivity.
10. Sample (group) sizes are generally small.
11. The data tend to be imprecisely gathered (often because researchers are unsure of what they are looking for) and therefore possess extreme within-group variability. Control and historical data are not used to adjust for variability or modify test performance.
12. Proper dose selection is essential for effective and efficient screen design and conduct. If insufficient data are available, a suitably broad range of doses must be evaluated (however, this technique is undesirable on multiple grounds, as has already been pointed out).

It should be kept in mind that there are a number of common mistakes (in both the design and conduct of studies and in how information from studies is used) that have led to unfortunate results, ranging from losses in time and money to the discarding of perfectly good potential drugs. Such outcomes are indeed the great disasters in drug discovery—especially since many of them are avoidable if attention is paid to a few basic principles.

It is quite possible to design a study for failure. Common shortfalls include:

1. Using the wrong animal model.
2. Using the wrong route or dosing regimen.

3. Using the wrong vehicle or formulation of test material.
4. Using the wrong dose level. In studies where several dose levels are studied, the worst outcome is to have an effect at the lowest dose level tested (i.e., the safe dosage in animals remains unknown). The next worst outcome is to have no effect at the highest dose tested (generally meaning that the signs of toxicity remain unknown, invalidating the study in the eyes of many regulatory agencies).
5. Making leaps of faith. An example is to set dosage levels based on others' data and to then dose all test animals. At the end of the day, all animals in all dose levels are dead. The study is over; the problem remains.
6. Using the wrong concentration of test materials in a screen. Many effects are very concentration dependent.

The design and conduct of discovery screens and programs also require an understanding of some basic concepts:

1. The studies are performed to establish or deny a specific activity of a compound, rather than to characterize the toxicity of a compound.
2. Because pharmaceuticals are intended to affect the functioning of biological systems and safety assessment characterizes the effects of higher-than-therapeutic doses of compounds, it is essential that one be able to differentiate between hyperpharmacology and true (undesirable) adverse effects.
3. Focus of the development process for a new pharmaceutical is an essential aspect of success but is also difficult to maintain. Clinical research units generally desire to pursue as many or as broad claims as possible for a new agent and frequently also apply pressure for the development of multiple forms for administration by different routes.

This volume will present a wide variety of approaches to the discovery and identification of potential new drugs. In assembling this volume, these approaches were derived thinking of four large categories. This approach is a traditional one, focusing on using some accepted (at least to the researcher) screens for a specific therapeutic activity to identify active or promising structures.

Therapeutic Area Approach: Diseases Seeking Drugs

This approach is perhaps the most traditional one, focusing on using some accepted (at least to the researcher) screen for a specific therapeutic activity to identify active or promising structure. Most drug discovery in the past started with such mass screening of selected molecules.

Mechanism Approach: Drugs Seeking Diseases

This approach is still currently popular. One or more compounds possessing a mechanism of activity are evaluated for activity in a specific disease. Or several: The challenge is the plausibility of suitable therapeutic activity and the degree of validity of one by hypothesis for a disease process.

Medicinal Chemistry Approach

This is also a traditional approach starting with molecular structures of known properties and seeking to modify structure to improve and optimize desirable features. The medicinal chemist utilizes knowledge of what past structural modifications have meant in terms of functional differences.

Technique-Based Approaches

Since the early 1990s, the most popular approaches have utilized new technologies based on expanded knowledge of how to characterize genetic and molecular level processes in humans and other species. Regrettably, as of late 2004, these methods have not yet borne the expected fruit. But perhaps the necessary period of learning how to use such tools is almost over.

Genomics The field of genomics, particularly high-throughput sequencing and characterization of expressed human genes, has created new opportunities for drug discovery. Knowledge of all the human genes and their functions may allow effective preventive measures and change drug research strategy and drug discovery development processes. Pharmacogenomics (dating back to the 1970s) is the application of genomic technologies such as gene sequencing, statistical genetics, and gene expression analysis to drugs in clinical development and on the market. It applies the large-scale systematic approaches of genomics to speed the discovery of drug response markers, whether they act at the level of the drug target, drug metabolism, or disease pathways. The potential implication of genomics and pharmacogenomics in drug discovery and development is that treatments for diseases could be identified according to genetic and specific individual markers, selecting medications and dosages that are optimized for individual patients.

Combinatorial Chemistry By reacting a set of starting chemicals in every possible combination, combinatorial chemistry gives life scientists the ability to create molecules in huge numbers and to test them for sought-after properties. Those abilities have attracted the attention of pharmaceutical companies. In recent years, virtually every large drug company has set up a combinatorial chemistry group.

The groups have revolutionized drug search programs. Before the mid-1990s, a single chemist could make perhaps 5 compounds per week. Now, he or she can create up to 100 new compounds in the same period of time. The work helps to find leads for potential drugs and significantly speeds up optimization of those leads.

Proteomics Information can be used to identify proteins associated with a disease, which computer software can then use as targets for new drugs. For example, if a certain protein is implicated in a disease, the three-dimensional structure of that protein provides the information a computer program needs to design drugs to interfere with the action of the protein. A molecule that fits the active site of an enzyme but cannot be released by the enzyme will inactivate the enzyme. This is the basis of new drug discovery tools, which aim to find new drugs to inactivate proteins involved in disease. As genetic differences among individuals are found, researchers will use these same techniques to develop personalized drugs that are more effective for the individual.

REFERENCES

1. Fishlock, D. (1990, April 24). Survival of the fittest drugs. *Financial Times*, pp. 16–17.
2. Gad, S. C. (1998b). A tier testing strategy incorporating *in vitro* testing methods for pharmaceutical safety assessment. *Humane Innovations Alternatives Animal Experimentation, 3*, 75–79.
3. Gad, S. C. (2001). *In Vitro Toxicology*, 2nd ed. Taylor and Francis, Philadelphia, PA.
4. Littlehales, C. (1999). The Price of a Pill, *Modern Drug Discovery*, Jan/Feb 1999, pp. 21–30.
5. O'Grady, J., Linet, O. I. (1990). *Early Phase Drug Evaluation in Man*. CRC Press, Boca Raton, FL.
6. Zbinden, G., Elsner, J., Boelsterli, U. A. (1984). Toxicological screening. *Reg. Toxicol. Pharamcol. 4*, 275–286.
7. Gad, S. C. (1988a). An approach to the design and analysis of screening studies in toxicology, *J. Am. Coll. Of Toxicol. 7*(2), 127–138.
8. Gad, S. C. (1989a). Principles of screening in toxicology: With special emphasis on applications to neurotoxicology, *J. Am. Coll. Toxicol. 8*(1), 21–27.
9. Tukey, J. W. (1977). *Exploratory Data Analysis*. Addison-Wesley, Reading, MA.
10. Redman, C. (1981). Screening compounds for clinically active drugs. In C. R. Buncher, J. Tsay (Eds), *Statistics in the Pharmaceutical Industry*. Marcel Dekker, New York, pp. 19–42.

BIBLIOGRAPHY

Alder, S., Zbinden, G. (1988). *National and International Drug Safety Guidelines*. MTC Verlag, Zollikon, Switzerland.

Angell, M. (2004). *The Truth About the Drug Companies*. Random House, New York.

Beyer, K. (1978). Discovery, development and delivery of new drugs. *Monographs in Pharmacology and Physiology*, No. 12. Spectrum, New York.

FDA (Food and Drug Administration) (1987a). Good laboratory practice regulations: Final rule. 21 CFR Part 58, *Fed. Reg.*, September 4, 1987.

FDA (Food and Drug Administration) (1987b). New drug, antibiotic, and biologic drug produce regulations. 21 CFR Parts 312, 314, 511, and 514, *Federal Register*, *52*(53), 8798–8857.

French, S. (1982). *Sequencing and Scheduling.* Halsted, New York.

Gad, S. C. (2000). *Product Safety Evaluation Handbook*, 2nd ed. Marcel Dekker, New York.

Goozner, M. (2004). *The $800 Million Pill.* University of California Press, Los Angeles.

Guarino, R. A. (1987). *New Drug Approval Process.* Marcel Dekker, New York.

Hamner, C. E. (1982). *Drug Development.* CRC Press, Boca Raton, FL, pp. 53–80.

Kliem, R. L. (1986). *The Secrets of Successful Project Management.* Wiley, New York.

Knutson, J. R. (1980). *How to Be a Successful Project Manager.* American Management Associations, New York.

Leber, P. (1987). FDA: The federal regulations of drug development. In H. Y. Meltzer (Ed.), *Psychopharmacology: The Third Generation of Progress.* Raven, New York, pp. 1675–1683.

Levy, A. E., Simon, R. C., Beerman, T. H., Fold, R. M. (1977). Scheduling of toxicology protocol studies. *Comput. Biomed. Res. 10*, 139–151.

Matoren, G. M. (1984). *The Clinical Process in the Pharmaceutical Industry.* Marcel Dekker, New York, pp. 273–284.

Sneader, W. (1986). *Drug Development: From Laboratory to Clinic.* Wiley, New York.

Spector, R., Park, G. D., Johnson, G. F., Vessell, E. S. (1988). Therapeutic drug monitoring. *Clin. Pharmacol. Therapeu. 43*, 345–353.

Spilker, B. (1994). *Multinational Drug Companies*, 2nd ed. Raven, New York.

Zbinden, G. (1992). *The Source of the River Po.* Haag and Herschen, Frankfurt, Germany.

1

NATURAL PRODUCTS

CHARLES B. SPAINHOUR

1.1	INTRODUCTION	12
1.2	HISTORY AND BACKGROUND OF THE USE OF NATURAL PRODUCTS AS THERAPEUTIC AGENTS	12
1.3	NATURAL PRODUCTS RESEARCH AND DEVELOPMENT—AN UPDATE	14
1.4	DISCOVERY OF NATURAL PRODUCTS	22
	Literature Sources	22
	Environmental Sources	23
1.5	ESSENTIAL PHARMACODYNAMICS	33
	Protein Targets of Drug Action	33
	General Principles of Drug Action	34
	Molecular Aspects of Drug Action—Receptors	37
	Molecular Aspects of Drug Action—Ion Channels	39
	Molecular Aspects of Drug Action—G-Protein-Coupled Systems	40
	Molecular Aspects of Drug Action—Receptors as Enzymes	40
	Molecular Aspects of Drug Action—Transcription Factors	41
	Molecular Aspects of Drug Action—Other Targets	41
	Drug Tolerance	42
1.6	SCREENING FOR NATURAL PRODUCT ACTIVITY	43
1.7	ISOLATION AND PURIFICATION OF NATURAL PRODUCTS	47
1.8	STRUCTURE IDENTIFICATION OF NATURAL PRODUCTS	51
1.9	SYNTHESIS OF NATURAL PRODUCTS	53
1.10	DEVELOPMENT OF NATURAL PRODUCTS	55
	Regulatory Guidelines and Nonclinical Development	55
	Learning from the Mistakes of the Past in the Development of Natural Products	58
1.11	FUTURE OF NATURAL PRODUCTS	60
	References	63

Drug Discovery Handbook, by Shayne Cox Gad
Copyright © 2005 by John Wiley & Sons, Inc.

1.1 INTRODUCTION

By definition, the word *natural* is an adjective referring to something that is present in or produced by nature and not artificial or man-made. When the word *natural* is used in verbiage or written, many times it is assumed that the definition is something good or pure. However, many effective poisons are natural products [145]. The term *natural products* today is quite commonly understood to refer to herbs, herbal concoctions, dietary supplements, traditional Chinese medicine, or alternative medicine [72]. That will not be the case in this chapter. The information presented here will be restricted to the discovery and development of modern drugs that have been isolated or derived from natural sources. While in some cases, such discovery and development may have been based on herbs, folklore, or traditional or alternative medicine, the research and discovery of, along with the development of, herbal remedies or dietary supplements typically present different challenges with different goals [93, 152]. So while the stories of herbs and drugs are very much intertwined, it needs to be fully appreciated that the use of herbs as natural product therapy is different than the use of herbs as a platform for drug discovery and further development.

1.2 HISTORY AND BACKGROUND OF THE USE OF NATURAL PRODUCTS AS THERAPEUTIC AGENTS

Natural products are generally either of prebiotic origin or originate from microbes, plants, or animal sources [115, 116]. As chemicals, natural products include such classes of compounds as terpenoids, polyketides, amino acids, peptides, proteins, carbohydrates, lipids, nucleic acid bases, ribonucleic acid (RNA), deoxyribonucleic acid (DNA), and so forth. Natural products are not just accidents or products of convenience of nature. More than likely they are a natural expression of the increase in complexity of organisms [76]. Interest in natural sources to provide treatments for pain, palliatives, or curatives for a variety of maladies or recreational use reaches back to the earliest points of history.

Nature has provided many things for humankind over the years, including the tools for the first attempts at therapeutic intervention [115, 116]. Neanderthal remains have been found to contain the remnants of medicinal herbs [72]. The *Nei Ching* is one of the earliest health science anthologies ever produced and dates back to the thirtieth century BC [115, 116]. Some of the first records on the use of natural products in medicine were written in cuneiform in Mesopotamia on clay tablets and date to approximately 2600 BC [29, 30, 115, 116]. Indeed, many of these agents continue to exist in one form or another to this day as treatments for inflammation, influenza, coughing, and parasitic infestation. Chinese herb guides document the use of herbaceous plants as far back in time as 2000 BC [72]. In fact, *The Chinese Materia Medica* has been repeatedly documented over centuries starting at about 1100 BC [29, 30]. Egyptians have been found to have documented uses of various herbs in 1500 BC [29, 30, 72]. The best known of these documents is the Ebers Papyrus, which documents nearly 1000 different substances and formulations, most of which are plant-based medicines [115, 116]. Asclepius (in 1500 BC) was a physician in ancient Greece who achieved fame in part because of his use of plants in medicine [72]. A collection of Ayurvedic hymns in India from 1000 BC and earlier describes the uses of over 1000 different herbs. This work served as the basis for *Tibetan Medicine* translated from Sanskrit during the eighth century [29, 30]. Theophrastus, a philosopher and natural scientist in approximately 300 BC, wrote a *History of Plants* in which he addressed the medicinal qualities of herbs and the ability to cultivate them. The Greek botanist Pedanious Dioscorides in approximately AD 100 produced a work entitled *De Materia Medica*, which today is still a very well-known European document on the use of herbs in medicine. Galen (AD 130–200), practiced and taught pharmacy and medicine in Rome and published over two dozen books on his areas of interest. Galen was well-known for his complex formulations containing numerous and multiple ingredients. Monks in monasteries in the Middle Ages (fifth to the twelfth centuries) copied manuscripts about herbs and their uses [29, 30, 72]. However, it should not go unrecognized that it was the Arabs who were responsible for maintaining the documentation of much of the Greek and Roman knowledge of herbs and natural products and expanding that information with their own knowledge of Chinese and Indian herbal medicine [29, 30]. The Persian philosopher and physician Avicenna produced a work entitled *Canon Medicinae*, which is considered to be the definitive summarization of Greek and Roman medicine. Li Shih-Chen produced a Chinese drug encyclopedia during the Ming Dynasty entitled *Pen-ts'as kang mu* in AD 1596, which records 1898 herbal drugs and 8160 prescriptions [115, 116]. John Wesley, the founder of Methodism, had a profoundly negative view on the status of physicians within society and in 1747 wrote a book entitled *Primitive Physic*, which was a popular reference book of the time detailing numerous natural cures [72]. When the colonists originally came to America, they lacked trained physicians and so turned to the Native Americans for advice in healing practices. Such a lack of conventional medicine and physicians in early America spawned the production of various types of almanacs and other publications that contained various natural product-based recipes and assorted tidbits of medical information. Indeed, in an effort to curry favor with commoners, physicians themselves turned to the production of self-treatment guides for the general public. Various types of societies and botanical clubs held meetings and published different types of communiqués to educate the public with regard to the availability of natural products and how they could be helpful to an individual's health. Samuel Thompson's *Thompson's New Guide to Health* was one very popular publication. For a variety of different reasons, the interest in natural products continues to this very day [6, 8, 17, 39, 72, 81, 88, 90, 104]. The first commercial pure natural product introduced for therapeutic use is generally considered to be the narcotic morphine, marketed by Merck in 1826 [118]. The first semisynthetic pure drug based on a natural product, aspirin, was introduced by Bayer in 1899.

1.3 NATURAL PRODUCT RESEARCH AND DEVELOPMENT—AN UPDATE

The World Health Organization estimates that approximately 80 percent of the world's population relies primarily on traditional medicines as sources for their primary health care [44]. Over 100 chemical substances that are considered to be important drugs that are either currently in use or have been widely used in one or more countries in the world have been derived from a little under 100 different plants. Approximately 75 percent of these substances were discovered as a direct result of chemical studies focused on the isolation of active substances from plants used in traditional medicine [29, 30]. The number of medicinal herbs used in China in 1979 has been estimated to be numbered at 5267 [115, 116]. More current statistics based on prescription data from 1993 in the United States show that over 50 percent of the most prescribed drugs had a natural product either as the drug or as the starting point in the synthesis or design of the actual end chemical substance [118]. Thirty-nine percent of the 520 new drugs approved during the period 1983 through 1994 were either natural products or derivatives of natural products [65]. Indeed, if one looks at new drugs from an indication perspective over the same period of time, over 60 percent of antibacterials and antineoplastics were again either natural products themselves or based on structures of natural products. Of the 20 top-selling drugs on the market in the year 2000 that are not proteins, 7 of these were either derived from natural products or developed from leads generated from natural products. This select group of drugs generates over 20 billion U.S. dollars of revenue on an annual basis [60, 65].

Drug development over the years has relied only on a small number of molecular prototypes to produce new medicines [65]. Indeed, only approximately 250 discrete chemical structure prototypes have been used up to 1995, but most of these chemical platforms have been derived from natural sources.

While recombinant proteins and peptides are gaining market share, low-molecular-weight compounds still remain the predominant pharmacologic choice for therapeutic intervention [60]. Just a small sampling of the many available examples of the commercialization of modern drugs from natural products along with their year of introduction, indication, and company are: Orlistat, 1999, obesity, Roche; Miglitol, 1996, antidiabetic (Type II), Bayer; Topotecan, 1996, antineoplastic, SmithKline Beecham; Docetaxel, 1995, antineoplastic, Rhône-Poulenc Rorer; Tacrolimus, 1993, immunosuppressant, Fujisawa; Paclitaxel, 1993, antineoplastic, Bristol-Myers Squibb.

The overwhelming concern today in the pharmaceutical industry is to improve the ability to find new drugs and to accelerate the speed with which new drugs are discovered and developed. This will only be successfully accomplished if the procedures for drug target elucidation and lead compound identification and optimization are themselves optimized. Analysis of the human genome will provide access to a myriad number of potential targets that will need to be evaluated [60, 65]. The process of high-throughput screening enables the testing of increased numbers of targets and samples to the extent that approximately 100,000 assay points per day are able to be generated. However, the ability to accelerate the identification of pertinent lead compounds will only be achieved with the implementation of new ideas to generate varieties of structurally diverse test samples [60, 65, 66]. Experience has persistently and repeatedly demonstrated that nature has evolved over thousands of years a diverse chemical library of compounds that are not accessible by commonly recognized and frequently used synthetic approaches. Natural products have revealed the ways to new therapeutic approaches, contributed to the understanding of numerous biochemical pathways and have established their worth as valuable tools in biological chemistry and molecular and cellular biology. Just a few examples of some natural products that are currently being evaluated as potential drugs are (natural product, source, target, indication, status): manoalide, marine sponge, phospholipage-A_2 Ca^{2+}-release, anti-inflammatory, clincial trials; dolastatin 10, sea hare, microtubules, antineoplastic, nonclinical; staurosporine, streptomyces, protein kinase C, antineoplastic, clinical trials; epothilone, myxobacterium, microtubules, antineoplastic, research; calanolide A, B, tree, DNA polymerase action on reverse transcriptase, acquired immunodeficiency syndrome (AIDS), clinical trials; huperzine A, moss, cholinesterase, alzheimer's disease, clinical trials [60].

The costs of drug discovery and drug development continue to increase at astronomical rates, yet despite these expenditures, there is a decrease in the number of new medicines introduced into the world market. Despite the successes that have been achieved over the years with natural products, the interest in natural products as a platform for drug discovery has waxed and waned in popularity with various pharmaceutical companies. Natural products today are most likely going to continue to exist and grow to become even more valuable as sources of new drug leads. This is because the degree of chemical diversity found in natural products is broader than that from any other source, and the degree of novelty of molecular structure found in natural products is greater than that determined from any other source [31, 65, 142].

Where are these opportunities? Well, research into the use of plant-derived natural products alone in just the field of medicine covers a broad spectrum of activities [35, 67, 166, 168, 169]. Examples of such biological activity profiles would include, but are not limited to, nootropics, psychoactive agents, dependence attenuators, anticonvulsants, sedatives, analgesics, anti-inflammatory agents, antipyretics, neurotransmission modulators, autonomic activity modulators, autacoid activity modulators, anticoagulants, hyoplipidemics, antihypertensive agents, cardioprotectants, positive ionotropes, antitussives, antiasthmatics, pulmonary function enhancers, antiallergens, hypoglycemic agents, antifertility agents, fertility-enhancing agents, wound healing agents, dermal healing agents, bone healing agents, compounds useful in the prevention of urinary calculi as well as their dissolution, gastrointestinal motility modulators, gastric ulcer protectants, immunomodulators, hepato-protective agents, myelo-protective agents, pancreato-protective agents, oculo-protective agents, membrane stabilizers, hemato-protective agents, antioxidants, agents protective against oxidative stress, antineoplastics, antimicrobials, antifungal agents, antiprotozoal agents, antihelminthics, and nutraceuticals [35]. Many frontiers remain within the field of natural products that can provide opportunities to improve our quality of life.

Fungal disease has historically been a difficult clinical entity with which to effectively deal. Fungal diseases can include more than just a mycosis and can also include allergic reactions to fungal proteins and toxic reactions to fungal toxins. Mycoses as a group include diseases that are significantly more serious and life-threatening than nail infestations, athletes foot, or "jock-itch." Indeed, increasing numbers of overtly healthy individuals are becoming victims of the complications of fungal infestation. The reasons for this are that increasing numbers of people are receiving immunomodulatory treatment for an organ transplant or some underlying chronic systemic pathology, antineoplastic chemotherapy for cancer, or have been the recipients of proper or improper use of powerful antibiotics. Additionally there are a number of individuals within society that are infected with the human immunodeficiency virus (HIV). The available drugs to treat mycoses have been limited [5]. Furthermore, in this armamentarium, there are problems with dose-limiting nephrotoxicity, the rapid development of resistance, drug–drug interactions of concern, and a fungistatic mechanism of action. Thus there is an urgent need for the development of more efficacious antifungal agents with fewer limitations and less side effects. Ideally such compounds should possess good distribution characteristics, a novel mechanism of action, and a broad-spectrum cidal antifungal activity. The discovery and isolation of an echinocandin-type lipopeptide (FR901379) and lipopeptidolactone (FR901469) from microbes has been a significant achievement. These compounds are water soluble and inhibit the synthesis of 1,3-β-glycan, a key component of the fungal cell wall. Furthermore, since the cell wall is a feature particular to fungi and is not present in eukaryotic cells, such inhibitors certainly have the potential to demonstrate selective toxicity against the fungi and not against the animal or human host. The ultimate modifications of the lipopeptide and lipopeptidolactone referenced above led to the discovery of micafungin (FK463), which is currently in phase III clinical trials. This work along with the relatively recent approval of caspofungin (Merck) as a therapeutic agent for the treatment of disseminated aspergillosis are significant achievements in that they demonstrate that a melding of the proper research to identify and develop appropriate targets with the chemical and biological diversity found in natural products can be very rewarding.

Much ado has been made over recent years about endocrine disruptors and their effects on humans [33]. It needs to be recognized that endocrine disruptors are not just synthetic chemicals but can also be natural products. The use of natural product endocrine disruptors may provide significant insight into our understanding of the mechanisms by which the evolution of the genome can protect transactivation of the sex hormone receptors and aid in the development of drugs, which can protect the embryo during its development from hormone disruptive effects.

Diabetes is a multisystemic affliction, having impact on nearly every body organ. As a disease, it kills more individuals on a per annum basis than AIDS and breast cancer combined [148]. The impact on the quality of life of an individual suffering with diabetes is profound. A number of natural products currently exist that demonstrate hypoglycemic activity. Indeed, depending upon the source that one might use, there are approximately 800 to 1200 plants that exhibit hypoglycemic activity. While research and development efforts in this particular area thus far are largely restricted to traditional medicine uses, future research may well identify a potent antidiabetic agent.

The incidences of neuropsychiatric disorders are steadily increasing as our population increases in size and age. Such disorders include, but are not limited to, seizure disorders, schizophrenia, dementia, mania, aggression, memory loss, psychoses, age-related cognitive decline, depression, anxiety states, mood disorders, substance abuse, and substance dependence. There is a large body of data available that suggests the use of many natural products as potential treatments for these conditions and other neuropsychiatric disorders [18, 91, 92]. Indeed, a number of plant extracts have been associated with the treatment of various categories of mental symptoms and various types of receptor selectivity [18]. A very controversial potential psychotherapeutic agent is *Gingko biloba* [52]. A lack of understanding of mechanism of action, misidentification of materials, contamination of materials, intrinsic toxicity, and absence of standardization all contribute to this controversy. Further fractionation, isolation, and characterization of active components of these and other plants will undoubtedly lead to the discovery of novel neuropsychiatric agents as well as the debunking of other alleged therapies.

There are numerous blood-based diseases that afflict humans. These would include, but are not limited to, anemia, blood group incompatibility, blood protein disorders, bone marrow diseases, hemoglobinopathies, hemorrhagic diatheses, leukemia, disorders of leukocyte dysfunction, platelet disorders, and erythrocyte aggregation disorders. A number of natural products have been reported in the literature to be of value in the treatment of Epstein-Barr virus infection, leukemia, thrombosis and coagulopathy, malaria, anemia, and bone marrow diseases [113]. Extracts from the fungus *Trichothecium roseum*, the sea cucumber *Cucumaria japonica*, the legume *Amorpha fruitcosa*, the tree *Magnolia officinalis*, and others may be useful in the therapeutic management of Epstein-Barr virus infection. Extracts from the basidiomycetes *Mycena pura* and *Nidula candida* may be useful in the treatment of leukemia. Compounds isolated from *Streptomyces platensis* may be useful in the treatment of thrombocytopenia. Compounds obtained from the marine sponge *Aplysina archeri* have been reported to inhibit the growth of the feline leukemia virus. Scalarane-type bishomo-sesterterpenes isolated from the marine sponge *Phyllospongia foliascens* have been reported to exhibit cytotoxic, antithrombocytic, and vasodilation activities. It should be noted that a number of natural products are based on the coumarin nucleus and as such may exhibit antithrombotic and antiplatelet activities. A number of blood-sucking animals have small, low-molecular-weight proteins in their salivas that interfere with the clotting of blood and therefore might be of value as potential anticoagulants. *Streptomyces hygroscopicus ascomyceticus* manufactures a macrolide that has been reported to have immunosuppressant activity and may prove to be beneficial in preventing transplant rejection in humans. It is entirely possible that these compounds and others offer sufficient structural diversity, range of biological activities, and differing mechanisms of action that new, safer, and more efficacious drugs to treat blood-based disorders could well burgeon from this library.

A wide variety of natural products are claimed to possess immunosuppressant activity, but it is often difficult to dissect this activity away from associated cytotoxicity [101]. Since the first heart transplant in the late 1960s, medicine has progressed to the point where most organ transplants have become relatively routine procedures. The survival of individuals with transplants is owed in large part to the discovery of the fungal metabolite cyclosporine A in 1970 and its widespread use starting in 1978. Indeed, cyclosporine A has achieved such success that it is currently being evaluated for value in the treatment of Crohn's disease, systemic lupus erythematosus, and rheumatoid arthritis. Research efforts abound in the area of natural products and immunosuppression. A methyl analog of oligomycin F isolated from *Streptomyces ostreogriseus* has been reported to quite effectively suppress B-cell activation and T-cell activation in the presence of mitogens at concentrations comparable to that of cyclosporine A. Concanamycin F first isolated from *Streptomyces diastatochromogenes* in 1992 has been found to possess a wide array of biological activities including immunosuppressive and antiviral activities. The experimental immunosuppressant (+)-discodermolide isolated from the marine sponge *Discodermia dissoluta* exhibits relatively nonspecific immunosuppression, causing the cell cycle to arrest during G_2 and M phases. This compound's current primary interest is as a potential antineoplastic agent since it stabilizes microtubules and prevents depolymerization, effectively causing cell cyclic arrest during the metaphase to anaphase transition. This same mode of activity is shared with Taxol (Paclitaxel), the epothilones, eleutherobin, and the sarcodictyins. The didemnins, cyclic peptides, were first isolated from the marine tunicate *Trididemnum solidum* and exhibit immunosuppressive activity through a generalized cytotoxicity mediated by inhibition of progression through the G_1 phase of the cell cycle by an unknown mechanism. The trichopolyns I to V from the fungus *Trichoderma polysporum* are lipopeptides that suppress the proliferation of lymphocytes in the murine allogeneic mixed lymphocyte response assay. Triptolide from the plant *Tripterygium winfordii* demonstrates immunosuppressant activity through the inhibition of IL-2 receptor expression and signal transduction. The novel heteroaromatic compound lymphostin, obtained from *Streptomyces* KY11783 has demonstrated immunosuppressant activity through its potent inhibition of the lymphocyte kinase $p56^{lck}$. Over the last decade, research activities on immunosuppressants of natural product origin have focused on the mechanisms of inhibition of T-cell activation and proliferation. This approach has been fruitful, leading to the generation of significant information about signaling pathways between T cells, greater detail about the roles of T cells in immune function, and the discovery of Tacrolimus (Prograf) from the soil fungus *Streptomyces tsukubaensis*. As immunological research progresses, increasingly more potential targets will be elucidated for immunomodulatory therapeutic intervention. Natural products will undoubtedly provide a sound platform for the delivery of natural-product-based therapeutic agent candidates.

Natural-products-based anticancer drug discovery continues to be an active area of research throughout the world [34, 102, 112, 147]. While cancer incidences and the frequencies of types of cancer may vary from country to country, the most common sites for the development of neoplasia are generally considered to be the breast, colon/rectum, prostate, cervix/uterus, esophagus/stomach, pancreas, liver, lung, urinary bladder, kidney, ovary, oral cavity, and blood (leukemia and non-Hodgkin lymphoma) [147]. Currently, the chemotherapeutic management of these tumors involves a variety of different plant-based chemicals that are either currently in use or in clinical trials and include such drug classes as the vinca alkaloids, lignans, taxanes, stilbenes, flavones, cephalotaxanes, camptothecins, and taxanes. Despite the wide range of organ structure, type, and function, great similarities exist between the organs with regard to the pathogenesis of cancer. As more and more details of the molecular biology of cancer are revealed, more targets will present themselves for possible therapeutic chemical intervention in the growth and development of neoplasms. A somewhat new approach is that of cancer chemoprevention, where chemoprevention is defined as the prevention, delay, or reversal of carcinogenesis [112]. A few of the more promising cancer chemopreventive agents are (compound, plant source, target): brusatol, *Brucea javanica*, differentiation; zapotin, *Casimiroa edulis*, differentiation and apoptosis; apigenin, *Mezoneuron cacullatum*, antimutagenesis; deguelin, *Mundelea sericea*, inhibitor of ornithine decarboxylase; brassinin, *Brassica* spp., inducer of quinine reductase; and resveratrol, *Cassia quinquangulata*, cyclooxygenase inhibitor. A final note with regard to this approach is that it is important to appreciate that the distinction between chemopreventive agent and chemotherapeutic agent can become quite blurred.

A recurrent theme in neoplasia is the alteration of cell cycle control. One therapeutic approach to the treatment of neoplasia is the development of a treatment that would return to normal the altered cell cycle [143]. Cyclin-dependent kinases (CDKs) control the progression of a cell through its growth cycle. CDKs are regulated through a series of site-specific complex mechanisms, and the components of such mechanisms include activating cyclins and endogenous CDK inhibitors. Processes of such mechanisms involve regulatory phosphorylation. There are natural products such as butyrolactone and staurosporine that are currently known to be able to provide such activity. These compounds and others generated from their platform are adenosine 5'-triphosphate (ATP) site-directed inhibitors and directly antagonize the activity of CDKs. Further research should more fully elucidate the most efficacious endpoint of CDK inhibition and lead to the control of neoplastic growth and possibly even bring about cytostasis or apoptosis.

The introduction of active agents derived from natural sources into the anticancer weaponry has already significantly changed the futures of many individuals afflicted with cancer of many different types. Continued research into natural sources will continue to deliver newer and more promising chemicals and chemical classes of anticancer agents with novel mechanisms of action that will improve survival rates to even higher degrees.

Human immunodeficiency virus infection is a devastating, globally widespread disease that consumes significant health-care dollars in the due course of management of patients [79]. Most of the currently useful anti-HIV agents are nucleosides and are limited in use due to severe toxicity and emerging drug resistance. Natural products, with their broad chemical structural diversity, provide an excellent opportunity to deliver significant therapeutic advances in the treatment of HIV [167]. Many natural products with novel structures have been identified as having anti-HIV activities [79, 167]. Betulinic acid, a triterpenoid isolated from *Syzigium claviflorum*, has been found to contain anti-HIV activity in lymphocytes. The quassinoside glycoside isolated from *Allanthus altissima* has been found to inhibit HIV replication. Artemisinin, isolated from *Artemisia anuua*, is a sesquiterpene lactone that is of special interest because of its novel structure, potent antimalarial activity, and activity against *Pneumocystis carinii*. A novel phorbol ester isolated from *Excoecaria agallocba* has been reported to be a potent inhibitor of HIV-1 reverse transcriptase. Indeed, most of the natural product chemicals that are attracting interest in this area of research are secondary metabolites such as terpenes, phenolics, peptides, alkaloids, and carbohydrates and are also inhibitors of HIV reverse transcriptase. Other target opportunities in the life cycle of the human immunodeficiency virus available for exploitation are: (1) attachment of virus to cell surface, (2) penetration and fusion of the virus with the cell membrane, (3) reverse transcription via reverse transcriptase, (4) integration into the host genome, (5) synthesis of viral proteins including zinc fingers, and (6) processing of viral polypeptide with HIV protease and assembly of viral proteins and DNA into a viral particle, maturation, and extrusion of the mature virus [167].

Infectious viral diseases remain a worldwide problem. Viruses have been resistant to therapy or treatment longer than most other forms of life because their nature is to depend on the cells that they infect for their multiplication and survival [41]. Such a characteristic has made the development of effective antiviral chemotherapeutic agents difficult. Today there are few effective antivirals available for use. In order to confidently wage the war against viruses, research efforts are now turning to the molecular diversity available from natural products. For the period 1983 to 1994, seven out of 10 synthetic agents approved by the Food and Drug Administration (FDA) for use as antivirals were based on a natural product. These drugs are famciclovir, stavudine, zidovudine, zalcitabine, ganciclovir, sorivudine, and didanosine. The viral genome can be composed of either RNA or DNA and HIV, which was discussed earlier is an RNA containing virus. The general potential targets of antiviral chemotherapy are: (1) attachment of virus to host cell, (2) penetration of the host cell by the virus, (3) viral particle uncoating, release, and transport of viral nucleic acid and transport proteins, (4) nucleic acid polymerase release/activation, (5) translation of mRNA (messenger RNA) to polypeptides (early proteins), (6) transcription of mRNA, (7) replication of nucleic acids, (8) protein synthesis (late proteins), (9) viral polypeptide cleavage into polypeptides necessary for maturation, (10) assembly of viral capsids and precursors, (11) encapsidation of nucleic acid, (12) envelopment, and (13) release. Early antiviral research focused on compounds that inhibited viral DNA synthesis, purine, and pyrimidine nucleoside analogs. Today most current antiviral agents target RNA-based viruses and the inhibition of reverse transcriptase in order to block the transcription of the RNA genome to DNA. Such inhibition would prevent the synthesis of viral mRNA and proteins. Protease inhibitors affect the synthesis of late viral proteins and viral packaging activity. There are no currently available drugs that target early viral protein synthesis. Antiviral compound research has included alkaloids, carbohydrates, chromones, coumarins, flavonoids, lignans, phenolics, quinines, xanthones, phenylpropanoids, tannins, terpenes, steroids, iridoids, thiopenes, polyacetylenes, lactones, butenolides, phospholipids, proteins, peptides, and lectins. While plants have been a common hunting ground, many other sources are now starting to be explored, especially the marine environment. The use of natural products in the field of antiviral research appears to be limited only by the imagination of the researcher.

This review has demonstrated that natural products are indeed viable sources and resources for drug discovery and development [3]. Indeed, without natural products, medicine would be lacking in therapeutic tools in several important clinical areas such as neurodegenerative disease, cardiovascular disease, solid tumors treatment, and immunoinflammatory disease [4, 64, 122]. Furthermore, the continual emergence of new natural product chemical structure skeletons, with interesting biological activities along with the potential for chemical modification and synthesis bode well for the utility of natural products. Finally, the uses of natural products need to be by no means restricted to pharmaceuticals but can also be expanded to agrochemicals. For example, the use of pyrethrins obtained from *Chrysanthemum* spp. as insecticides has been very popular over the years and persists today. Research continues into the use of natural products as pesticides. While the success stories have not been as numerous or spectacular for herbicides as they have been for drugs and pesticides, there have been victories along the way and the future holds strong potential for this field also [40, 99].

1.4 DISCOVERY OF NATURAL PRODUCTS

Literature Sources

Natural products can come from anywhere. People most commonly think of plants first when talking about natural products, but trees and shrubs can also provide excellent sources of material that could provide the basis of a new therapeutic agent. Animals too, whether highly developed or poorly developed, whether they live on land, sea, or in the air can be excellent sources of natural products. Bacteria, smuts, rusts, yeasts, molds, fungi, and many other forms of what we consider to be primitive life can provide compounds or leads to compounds that can potentially be very useful therapeutic agents. Suffice it to say that natural products can come from any point or level on the phylogenetic tree. When searching for natural products, one should never feel that a form of life is too low, simple, or grotesque to provide a compound of interest. However, before one goes marching out into the woods, sailing out into the sea, climbing the highest mountains, or descending into the deepest caves, it is appropriate to perform a little bit of research, and hence a visit to the library becomes the first step in any search for a natural product. Remember that the use of a natural product as a therapeutic agent requires that one match some particular characteristic of the compound with a disease or condition. This matching process involves a two-tiered process. The first tier can be comprised of a thorough evaluation of the pathophysiologic condition of interest including any pertinent history, etiology, clinical manifestations, biochemistry, clinical chemistry, hematology, physiology, pathology, and therapeutics. With the therapeutic target in mind and a complete understanding of the pathophysiology of the condition, one can then begin a search for a natural product that has some particular characteristic that might suggest that it has utility as a potential therapeutic agent. Alternatively, one could take the approach of observing or finding a particular characteristic of a natural product and then searching for a useful disease or condition to treat with the material. The choice of which path to follow is a personal one, with either selection being equally useful. Admittedly, in the past and not infrequently the search for and investigation of a natural product arose out of serendipity.

There has been an explosion of information in the biomedical sciences over the last 25 years, and attempting to find information about natural products can be a challenging task [39]. Ideally, it is important to know the history, folklore, origin of use, source, chemical structure, availability, method of preparation, pharmacology, toxicology, and therapeutics of any natural product. However, the reality is that many times even for compounds or preparations that have been used for centuries, there are significant gaps in this portfolio of desired information. Nonetheless, a trip to the library can be very useful. It is amazing what research projects individuals have become engaged in over the years. Excellent sources of information on natural products can be readily found on the Internet. Some of these sites available on the Internet are shown in Table 1.1. Information retrieval services or search engines are also quite useful tools. These, however, can be expensive and require the aid of someone skilled in their use. Some of the more common sources are shown in Table 1.2. A number of helpful books have been published on natural products over the years, but the preponderance of them dwells on the topic of plants. The more common of these references are shown in Table 1.3. Finally, periodicals probably represent the most current and timely sources of research and information on natural products. Some of the better and more prominent journals are shown in Table 1.4.

Environmental Sources

Myriad opportunities abound throughout nature that can provide natural products with significant therapeutic potential [20, 29, 30, 61, 77, 83]. These opportunities can present themselves from almost any niche of nature and most likely some that have not even yet been discovered.

TABLE 1.1 Internet Sites

American Botanical Council http://www.herbalgram.org	American Herbalists Guild http://www.americanherbalistsguild.com/top.htm
Complementary & Alternative Methods http://www.cancer.org/eprise/main/docroot/eto/eto_5?sitearea=eto	Dr. Duke's Phytochemical and Ethnobotanical Databases http://www.ars-grin.gov/duke
Herb Research Foundation http://www.herbs.org	Herbal Education Services http://www.botanicalmedicine.org
Herbal Medicine: Internet Resources: Alternative Medicine http://www.pitt.edu/~cbw/herb.html	HerbMed http://www.herbmed.org/
International Herb Association http://www.iherb.org	MEDLINEplus: Alternative Medicine http://www.nlm.nih.gov/medlineplus/alternativemedicine.html
National Center for Complementary and Alternative Medicine http://nccam.nih.gov/	World Health Organization Publications http://www.who.int/dsa/cat98/trad8.htm

Adapted from DerMarderosian and Beutler [39].

TABLE 1.2 Information Retrieval Services

BIOSIS http://www.biosis.org	Chemical Abstracts Service http://www.cas.org/
Current Contents and Science 　Citation Index http://www.isinet.com/isi	*Excerpta Botanica. Section A, Taxonomica 　et Chorologica*, International Association 　for Plant Taxonomy, G. Fischer, Stuttgart, 　New York.
The Herb Research Foundation http://www.herbs.org/	*IPA (International Pharmaceutical Abstracts)*, http://info.cas.org/ONLINE/DBSS/ipass.html. Database contains international coverage of 　pharmacy and health-related literature.
Lynn Index, Massachusetts 　College of Pharmacy	*Medicinal and Aromatic Plants Abstracts*, 　Publications and Information Directorate, 　Council of Scientific and Industrial Research 　(CSIR), New Delhi, India.
MEDLINE, MEDLARS http://www.nlm.nih.gov/	NAPRALERT (NAtural PRoducts ALERT) http://www.aq.uiuc.edu/~ffh/napra.html
NAPRONET http://ccl.net/chemistry/ resources/ tips/list/ NAPRONET/index.shtml	Poisindex System infor@mdx.com or http://www.micromedex.com/products/ poisindex
Toxicology Information 　Response Center (TIRC) http://www.ornl.gov/ TechResources/tirc/hmepg.html	TOXLINE http://toxnet.nlm.nih.gov/cgi-bin/sis/htmlgen? TOXLINE

Adapted from DerMarderosian and Beutler [39].

Microbes The observation of the effects of microbial secondary metabolites on pathogenic fungi and bacteria spawned the antibiotic era [29, 30, 38]. Since its inception, humankind has grown to take for granted the wonders of antibiotics. Indeed, results of antibiotic use were so impressive that compounds of this general type were for the most part the only chemicals used against pathogenic microorganisms. Due to escalating research and development costs, the difficulties in identifying novel structures and the problems in finding new mechanisms of action, the golden era of antibiotics appeared to be meeting its own demise. Many individuals even professed that the use of antibiotics might even be passé, with modern medicine choosing more modern techniques for treatment. However, this same library of antibiotics had, over the same time, also been found to exhibit other biological properties that might be beneficial to humankind. Accordingly, research into the complete biological activity profiles of antibiotics began with the intent of identifying the utility of these compounds for various pharmacological or agrochemical applications. This shift in focus expanded the search for new natural products from microbes, where microbial metabolites might be used to treat diseases other than those caused

TABLE 1.3 Books

L. Aikman. *Nature's Healing Arts: From Folk Medicine to Modern Drugs.* Washington, DC: National Geographic Society, 1977.

J. L. Beal and E. Reinhard, eds. *Natural Products as Medicinal Agents: Plenary Lectures of the International Research Congress on Medicinal Plant Research*, Strasbourg, July 1980. Stuttgart: Hippokrates, c1981.

M. Blumenthal, ed. *Herbal Medicine: Expanded Commission E Monographs.* Austin, TX: American Botanical Council, 2000.

M. Bricklin. *The Practical Encyclopedia of Natural Healing*, new rev. ed. Emmaus, PA: Rodale Press, 1983.

W. Bucherl, E. E. Buckley, and V. Deulofeu, eds. *Venomous Animals and Their Venoms*, 3 vols. New York: Academic Press, 1968–71.

F. Densmore, *How Indians Use Wild Plants for Food, Medicine, and Crafts.* Washington, DC: Government Printing Office, 1928. Reprint, New York: Dover, 1974.

J. A. Duke. *CRC Handbook of Medicinal Herbs.* Boca Raton, FL: CRC Press, 1985.

C. Facciola. *Cornucopia: A Source Book of Edible Plants.* Vista, CA: Kampong Publications, 1990.

B. W. Halstead. *Poisonous and Venomous Marine Animals of the World*, 2nd rev. ed. Princeton, NJ: Darwin Press, 1988.

D. Hoffman. *The Herbal Handbook: A User's Guide to Medical Herbalism.* Rochester, VT: Healing Arts Press, 1998.

H. Baslow. *Marine Pharmacology: A Study of Toxins and Other Biologically Active Substances of Marine Origin.* Baltimore: Williams & Wilkins, 1969.

W. H. Blackwell. *Poisonous and Medicinal Plants* Englewood Cliffs, NJ: Prentice Hall, 1990.

M. Blumenthal, ed. *The Complete German Commission E Monographs: Therapeutic Guide to Herbal Medicines.* Austin, TX: American Botanical Council, 1998.

British Herbal Pharmacopoeia. Great Britain: British Herbal Medicine Association, 1996.

M. Castleman. *The Healing Herbs: The Ultimate Guide to the Curative Power of Nature's Medicines.* Emmaus, PA: Rodale Press, 1991.

A. H. Der Marderosian and L. E. Liberti. *Natural Products Medicine: A Scientific Guide to Foods, Drugs, Cosmetics.* Philadelphia: G.F. Stickley, 1988.

W. C. Evans, *Trease and Evans' Pharmacognosy*, 14th ed. London: WB Saunders, 1996.

Foster S. *Tyler's Honest Herbal: A Sensible Guide to the Use of Herbs and Related Remedies.* 4th ed. New York: Haworth Herbal Press, 1998.

G. Henslow. *The Plants of the Bible: Their Ancient and Mediaeval History Popularly Described.* London: Masters, 1906.

D. Hoffman. ed. *The Information Sourcebook of Herbal Medicine.* Freedom, CA: Crossing Press, 1994.

TABLE 1.3 *Continued*

R. W. Kerr. *Herbalism through the Ages*, 7th ed. San Jose, CA: Supreme Grand Lodge of AMORC, 1980.

J. M. Kingsbury. *Poisonous Plants of the United States and Canada.* Englewood Cliffs, NJ: Prentice-Hall, 1964.

K. F. Lampe and M. A. McCann. *AMA Handbook of Poisonous and Injurious Plants.* Chicago: American Medical Association, 1985.

W. H. Lewis, and M. P. F. Elvin-Lewis. *Medical Botany: Plants Affecting Man's Health.* New York: Wiley, 1977.

D. J. Mabberly. *The Plant-Book* 2nd ed. Cambridge: Cambridge University Press, 1997.

J. F. Morton. *Atlas of Medicinal Plants of Middle America: Bahamas to Yucaton.* Springfield, IL: C. C. Thomas, 1981.

P. Ody. *The Complete Medicinal Herbal.* New York: Dorling Kindersley, 1993.

G. Penso. *Inventory of Medicinal Plants Used in the Different Countries.* Geneva: World Health Organization, 1980.

F. Rosengarten. *The Book of Spices.* Wynnewood, PA: Livingston, 1969.

J. E. Simon. *Herbs, An Indexed Bibliography, 1971–1980: The Scientific Literature on Selected Herbs, and Aromatic and Medicinal Plants of the Temperate Zone.* Hamden, CT: Shoe String Press, 1984.

J. M. Kingsbury. *Deadly Harvest: A Guide to Common Poisonous Plants.* New York: Holt, Rinehart and Winston, 1965.

P. Krogsgaard-Larsen, S. B. Christensen, H. Kofod, eds. *Natural Products and Drug Development: Proceedings of the Alfred Benzon Symposium 20 Held at the Premises of the Royal Danish Academy of Sciences and Letters, Copenhagen, 7–11 August 1983.* Copenhagen: Munksgaard, 1984.

A. Y. Leung. *Encyclopedia of Common Natural Ingredients Used in Food, Drugs, and Cosmetics*, 2nd ed. New York: Wiley, 1996.

I. E. Liener. ed. *Toxic Constituents of Plant Foodstuffs*, 2nd ed. New York: Academic Press, 1980.

J. E. Meyer. *The Herbalist*, rev ed. Glenwood, IL: Meyerbooks, 1986.

J. F. Morton. *Major Medicinal Plants: Botany, Culture, and Uses.* Springfield, IL: Thomas, 1977.

A. Osol and R. Pratt, eds. *The United States Dispensatory*, 27th ed. Philadelphia: Lippincott, 1973.

T. Robinson. *The Organic Constituents of Higher Plants: Their Chemistry and Interrelationships*, 6th ed. North Amherst, MA: Cordus Press, 1991.

P. Schauenberg. *Guide to Medicinal Plants.* New Canaan, CT: Keats, 1977.

D. G. Spoerke. *Herbal Medications.* Santa Barbara, CA: Woodbridge Press, 1990.

TABLE 1.3 *Continued*

R. P. Steiner, ed. *Folk Medicine: The Art and the Science*. Washington, DC: American Chemical Society, 1986.	T. Swain, ed. *Plants in the Development of Modern Medicine*. Cambridge, MA: Harvard University Press, 1972.
M. Sweet. *Common Edible and Useful Plants of the East and Midwest*. Healdsburg, CA: Naturegraph Publishers, 1975.	M. Sweet. *Common Edible and Useful Plants of the West*. Healdsburg, CA: Naturegraph Publishers, 1976.
V. E. Tyler, L. R. Brady, and J. E. Robbers. *Pharmacognosy*, 9th ed. Philadelphia: Lea and Febiger, 1988.	World Health Organization. *WHO Monographs on Selected Medicinal Plants*, Vol. 1. Geneva: World Health Organization, 1999.
H. W. Youngken and J .S. Karas. *Common Poisonous Plants of New England*. U.S. Public Health Service Pub. No. 1220. Washington, DC: Government Printing Office, 1964.	M. Zohary. *Plants of the Bible*. New York: Cambridge University Press, 1982.

Adapted from DerMarderosian and Beutler [39].

by bacteria and fungi. Microorganisms have proven to be an excellent source of novel natural products including polyketide and peptide antibiotics as well as classes of other biologically active compounds [123]. Today, microbial metabolites are used as antineoplastic agents (e.g., mitomycin), immunosuppressive agents (e.g., rapamycin), hypocholesterolemic agents (e.g., pravastatin), enzyme inhibitors (e.g., desferal), antimigraine agents (e.g., ergot alkaloids), herbicides (e.g., bialaphos), antiparasitic agents (e.g., salinomycin), bioinsecticides (e.g., tetranactin), and ruminant growth promoters (e.g., monensin) [38]. It is noteworthy that some of these compounds when originally discovered failed in their development for their original uses as either antibiotics or as agricultural fungicides. Bacteriocins are ribosomally produced antibiotic peptides and proteins that can be subdivided into different categories, lantibiotics, and microcins. Lantibiotics are produced by Gram-positive bacteria and microcins are produced by Gram-negative bacteria. Both lantibiotics and microcins possess an ability to form pores or punch holes in membranes of susceptible microorganisms. This property is of interest to the food industry, as bacteriocins are produced by *Lactococcus* spp., which are used in the preservation of various foodstuffs [123].

The cyanobacterium *Nostoc ellipsosporum* has been found to produce a novel protein (CV-N), which has generated interest because of its viricidal activity and apparent potential as an anti-HIV therapeutic agent. The antiviral activity of this chemical is reported to be mediated through specific interactions with the HIV envelope glycoproteins gp120 and possibly gp41.

TABLE 1.4 Periodicals

American Journal of Natural Medicine http://www.impakt.com	Botanical Review http://www.nybg.org/bsci/spub/botr/frntpg3b.html
Bulletin on Narcotics http://www.odccp.org/bulletin_on_narcotics.html	Canadian Journal of Botany http://www.nrc.ca/cgi-bin/cisti/journals/rp/rp2_desc_e?cjb
Canadian Journal of Herbalism http://www.herbalists.on.ca/journal/	Economic Botany http://www.econbot.org
European Journal of Herbal Medicine http://www.ejhm.co.uk/	Herb Companion Press http://www.interweave.com/
Herb Quarterly http://www.herbquarterly.com	International Journal of Aromatherapy http://www.harcourt-international.com/journals/ijar
Journal of Ethnopharmacology http://www.elsevier.com/locate/jethpharm	Journal of Natural Products http://pubs.acs.org/journals/jnprdf/index.html
Medical Anthropology: Cross Cultural Studies in Health and Illness http://www.sfu.ca/medanth	Medical Herbalism: A Journal for the Clinical Practitioner http://medherb.com/MHHOME.SHTML
Natural Health http://www.naturalhealth1.com	Natural Product Letters http://www.tandf.co.uk/journals/titles/10575634.html
Natural Product Reports http://www.rsc.org/is/journals/current/npr/nprpub.htm	Pharmaceutical Biology http://www.szp.swets.nl/szp/frameset.htm
Phytochemistry: The International Journal of Plant Biochemistry and Molecular Biology, The Journal of the Phytochemical Society of Europe and the Phytochemical Society of North America http://www.elsevier.nl/locate/inca/273	Phytomedicine: International Journal of Phytotherapy and Phytopharmacology http://www.urbanfischer.de/journals
Phytotherapy Research http://www3.interscience.wiley.com/cgi-bin/jtoc?ID=12567	Plant Foods for Human Nutrition http://www.wkap.nl/jrnltoc.htm/0921-9668/contents
Planta Medica: Natural Products and Medicinal Plant Research http://www.thieme.de/plantamedica/fr_inhalt.html	Toxicon: Official Journal of The International Society on Toxicology http://www.elsevier.com/locate/toxicon
Veterinary and Human Toxicology, American Academy of Veterinary and Comparative Toxicology, Comparative Toxicology Laboratories	Z Naturforsch http://www.ncbi.nlm.nih.gov/entrez/query.fcgi

Adapted from DerMarderosian and Beutler [39].

Research has further revealed that CV-N is a new class of antiviral agent because of its unique interaction with envelope glycoproteins.

Biologically active proteins produced from fungi should not be ignored. *Trichoderma viride* produces a polypeptide, alamethicin, that has demonstrated ion-gating activity. Some fungal ribotoxins such as mitogillin have been found to act as specific ribonucleases. The edible mushroom *Rozites caperata* has been found to produce a compound, RC-183, which inhibits herpes simplex virus (HSV-1 and HSV-2) in a murine animal model.

Various compounds isolated from microbes and fungi have gained favor in their utility as tools to investigate such activities as nerve growth and mechanisms of microbial infiltration and pathogenesis. The fruiting body of *Pleurotus ostreatus* has been reported to produce a lectin that demonstrates potent antitumor activity in mice. Similarly, another lectin from *Volvariella volvacea* shows antiproliferative activity against various tumor cell lines via the mediation of a concentration-dependent stimulation of the expression of cyclin kinase inhibitors resulting in cell cycle arrest in the G_2/M phase. Many other fungal lectins have been isolated, but complete determinations of their biological activity profiles remain to be completed.

The ingenious application of molecular and cellular biology to detect receptor agonistic and antagonistic activities of compounds has tremendously aided drug research activity. In addition to the many opportunities that exist for microbial secondary metabolites with regard to medicine and agrochemicals, time has also demonstrated the increased need for the development of novel antibiotics because of the development of resistant strains of pathogens, the appearance of new diseases, and the inadequacy or toxicity of current drugs [37]. The diversity of microorganisms is of a staggering quantity, and only an extremely small proportion of bacteria and fungi have been examined for the production of potentially useful secondary metabolites.

Plants The higher plants produce a variety of different types of compounds, including biologically active proteins. Some of these types of compounds are even shared with other organisms, and they include such chemical families as lectins, defensins, cyclotides, and ribosome-inactivating proteins [123]. Ribosome-inactivating proteins are a group of proteins exhibiting a wide spectrum of biological activities, including a ribonucleolytic activity for which the group is named. These compounds can be obtained from *Panax ginseng* and other plants and have been reported to demonstrate antifungal and antiviral activities. Ribosome-inactivating proteins from *Phytolacca americana* have been reported to be active against HIV and from *Saponaria officinalis* to possess antineoplastic activity. Plant antimicrobial peptides comprise another large group of biologically active compounds. This group of compounds can be further subdivided into thionins, defensins, cyclotides, and lectins. Thionins are small proteins that selectively form disulfide bridges with other proteins or form ion channels in membranes. This ability to make membranes more permeable suggests the potential for antimicrobial activity. Defensins are cysteine-rich peptides that also permeabilize membranes but appear to be very specific in their activity, targeting fungal cell membranes and not mammalian or bacterial cell membranes. Cyclotides are a protein family, whose mechanism of action has not yet been elucidated, but have demonstrated inhibitory activity against HIV-1. Finally, the lectins are proteins that have a noncatalytic domain that binds reversibly to specific carbohydrates. This activity encompasses potentially a wide spectrum of biological activities including antineoplastic activity, immunostimulation activity, immunosuppression activity, antimicrobial activity, and antimicrobial activity.

Higher plants have been over time an extremely popular source of natural products [23, 95, 123, 127, 130]. Since 1961 approximately nine different compounds derived from plants have been approved in the United States as antineoplastic agents [95]. These drugs include vinblastine, vincristine, vinorelbine, etoposide, teniposide, paclitaxel, docetaxel, topotecan, and irinotecan. The mechanisms of action of these compounds ranges from that of tubulin inhibition to the inhibition of the essential DNA enzymes topoisomerase I and topoisomerase II or both toposimoerases I and II.

Podophyllum peltatum reportedly has curative properties on the venereal wart, *Condyloma acuminatum*, along with other therapeutic benefits [130]. The active glycoside component of extracts from *P. peltatum* has been found to inhibit mitosis in vitro. Furthermore, derivatives of this compound have been found to be capable of arresting cells in either late S phase or early G_2 phase, without inhibiting microtubule assembly. The latex from the plant *Euphorbia lateriflora* has been used both as a purgative and a cure for ringworm [23]. The Pacific yew tree yielded the antineoplastic, paclitaxel [29, 30]. *Newboutonia vellutina*, a Euphorbiaceae, has been reported to have utility both as a parasiticide as well as a treatment for gastric distension [23]. Compounds isolated and identified from this source will undoubtedly continue to make strong contributions to modern therapeutics.

Insects Various polypeptides of interest have been isolated from the venoms of arachnids and arthropods that prey on insects [123]. Indeed a variety of reviews have been published on ion channel toxins from scorpions and specific venom proteins and neurotoxins from arthropod venoms and their effects on the cardiovascular system. The caterpillar *Lonomia achelous* has been reported to be the source of a biologically active protein that causes a coagulopathy that is mediated via specific interactions with Factor V in the coagulation cascade. While this discovery requires further work to evaluate its therapeutic potential, the discovery may open new opportunities in thrombosis research. Insect peptides have been the subject of research into the immune defense system of insects but have not yet been investigated for effects and potential benefit in humans. Compounds from this peptide group include such sources as the termite (*Pseudacanthotermes spiniger*), the mosquito (*Anopheles gambiae*), the moth (*Heliothis viriscens*), and the beetle (*Oryctes rhinoc-*

eros). Insect-derived natural products offer another strong potential avenue for the development of future drugs.

Vertebrates Research into a variety of antimicrobial peptides, such as magainins, defensins, cathelicidins, and protegrins generated by vertebrates has over recent time become very popular. Cathelicidin-type peptides are a broad range of antimicrobial proteins that have been isolated from rabbits, mice, sheep, and humans. Cathelicidins are composed of two different domains, the cathelin and antimicrobial domains. The cathelin domain becomes bactericidal after cleavage from the antimicrobial domain. Purportedly, these materials bind to lipopolysaccharide and neutralize its activity. The pit viper, *Bothrops jaracaca*, produces a compound that spurred the synthesis of the angiotensin-converting enzyme (ACE) inhibitors captropril and enalpril [29, 30]. The skin of the poisonous frog, *Epipedobates tricolor*, produces epibatidine. This substance has ultimately led to the creation of a new class of analgesics. Other vertebrates produce compounds that have been the subjects of substantial research, which offer potential opportunities to identify useful compounds in the areas of cardiovascular function, immune function, and central nervous system function.

Marine Organisms The marine environment, arguably the original source of all life, is a rich source of bioactive compounds [14, 45–48, 84, 89, 96, 144, 157]. More than 70 percent of our planet's surface is covered by the oceans, and some experts feel that the potentially available biodiversity on the deep seafloor or coral reefs is greater than that existing in the rainforests [62].

Consider the fact that many marine organisms have soft bodies and lead a sedentary lifestyle, making a chemical system of defense almost essential for survival. Marine organisms have evolved the ability to synthesize such toxic compounds or extract or convert pertinent compounds from other marine microorganisms. Natural products from marine organisms are released into the water and therefore are rapidly diluted, accordingly they must be very potent materials to have the desired end effect. The richly available marine biodiversity that is available to us has to this point only been explored to an extremely limited extent. Furthermore, the primary chemical diversity available from marine organisms is most likely capable of delivering an even greater abundance of secondary metabolites for research use. For all of these reasons it is believed that the natural products that are available from the seas and oceans provide a tremendous opportunity for the discovery of novel therapeutic agents [62].

The first discovery of a marine-based biologically active compound of therapeutic interest was really quite by accident approximately 10 years after the end of the World War II [29, 30]. The C-nucleosides isolated from the Caribbean sponge *Cryptotheca crypta* were found to possess antiviral activity. This discovery eventually led to the development of cytosine arabinoside, a useful antineoplastic agent. Biologically active marine proteins derived from the venom of marine snails of the *Conus* genus have attracted a significant level of research over the years [123]. These conotoxin peptides interact in a unique fashion with voltage-gated ion channels to induce a wide spectrum of pharmacological effects. Such effects include anesthesia, analgesia, and anticonvulsant activity. The conotoxin ziconotide is currently under review in the United States for use in the treatment of chronic, opiate-resistant pain. According to some estimates, there are most likely approximately 1000 different *Conus* snails. Each snail produces up to approximately 200 different venoms. The broad spectrum of biological activities manifested by each of these venom components multiplied by the number of snails and venom components available suggests significant opportunity for new drugs from the snail alone. The mussel *Mytilis edulis* has been reported to produce antibacterial peptides and cytotoxic lectins. Horseshoe crabs produce a variety of different antibacterial peptides and proteins. Indeed, *Limulus polyphemus* produces an interesting group of antimicrobial peptides referred to as polyphemusins, and a synthetic peptide based on the sequence of polyphemusin II has been reported to strongly inhibit the cytopathic effect of infection with HIV.

It has been reported that the tunicate *Styela clava* produces α-helical antimicrobial peptides called clavanins that are homologous to the magainins produced by certain types of frogs. The marine worm *Cerebratulus lacteus* produces neurotoxic polypeptides, which have been found to contain the ability to make membranes more permeable. Sponges have been found to produce a wide variety of interesting secondary metabolites. For example, lectins isolated from *Chondrilla nucula* have achieved utility in the histochemical labeling of melanoma and breast and thyroid carcinomas. Another compound, the protein mapacalcine, produced by *Cliona vastifica*, has been reported to specifically block non-L-type calcium channels in murine duodenal myocytes, while at the same time not exhibiting any affect on T-type calcium flux or potassium or chloride currents. Various proteins from sponges have been reported to selectively kill human tumor cells. For example, a protein from the sponge *Tethya ingalli* lyses ovarian cancer cells; a glycoprotein from *Pachymatisma johnstonii* has been discovered to inhibit cell growth at the G_0/G_1 phase via a unique mechanism in a non-small-cell-bronchopulmonary carcinoma line. Interestingly, this same chemical has been found to have potential as an antiparasitic in the treatment of leishmaniasis, by demonstrating cytotoxic activity against the parasite in the promastigote and amastigote stages of the life cycle. The protein niphatevirin isolated from the sponge *Niphates erecta* has been discovered to inhibit HIV-induced cytopathic effects, cell-to-cell fusion, and syncytium formation via interacting directly with the CD4 cellular receptor. Finally, a compound isolated from the sponge *Microciona prolifera* has been reported to bind to gp120 and resultantly protect T-lymphoblastoid cells from infection with HIV.

The so-called mining of the sea for potential drugs did not start in earnest until the mid-1970s because of a technical inability to effectively gain access to the biodiversity that exists within the seas and oceans. Despite this slow start, the potential for contribution of new drugs is staggering, a vision that undoubtedly will become reality over the next decade. Examples of such research activity and fulfillment of this future hope are (chemical, source, chemical class, chemical target, therapeutic Indication): AM336 (AMRAD), cone snail, peptide, ion channels, chronic pain; GTS21 (Taiho), nemertine worm, anabaseine-derivative, ion channels, Alzheimer's disease and schizophrenia; LAF389 (Novartis), sponge, amino acid derivative, methionine aminopeptidase inhibitor, cancer; OAS1000 (OsteoArthritis Sciences), soft coral, diterpene-pentoseglycoside, PLA_2 inhibitor, wound healing and inflammation; ILX651 (Ilex Oncology), sea slug, peptide, microtubule-interference, cancer; Cematodin (Knoll), sea slug, peptide, microtubule interference, cancer; Yondelis, sea squirt, isoquinolone, DNA-interactive agent, cancer; Alipidin, sea squirt, cyclic depsipeptide, oxidative stress inducer, cancer; Kahalalide F, sea slug/alga, cyclic depsipeptide, lysosomotropic compound, cancer; KRN7000 (Kirin), sponge, α-galactosylceramide, immunostimulatory agent, cancer; squalamine lactate, shark, aminosteroid, calcium-binding protein antagonist, cancer; IPL512602 Inflazyme/Aventis], sponge, steroid, unknown, inflammation [62].

1.5 ESSENTIAL PHARMACODYNAMICS

Modern science has provided a detailed understanding of the interaction of many therapeutic drugs with biological systems at a biochemical or molecular biological level. The proper use of such information can provide a wealth of tools for the discovery of new drug opportunities by providing a framework to permit comparison of the mechanisms of action, biological activities, therapeutic indices, and the therapeutic potentials of drugs along with the ability to forecast possible problems [7, 87, 121, 131, 134, 137, 159].

Protein Targets for Drug Action

The actions of drugs can be divided into those occurring at specific sites and those that are nonspecific. Nonspecific effects are typically mediated through a generalized effect in many organs, and the response observed depends upon the distribution of the drug. The response is usually associated with the organ or organs having the highest concentrations of the drug. Specific effects are produced by an interaction of the drug with a specific site or sites either on the cell membrane or inside of the cell. The protein targets for drug action on mammalian cells can be broadly divided into receptors, ion channels, enzymes, transcription factors, and other nonspecific sites of action. The term *target* can also be broadly defined to include such things as microorganisms and cancer cells, but this discussion will be limited to normal mammalian systems or cells. It should be noted that while there are other types of protein that are known to function as drug targets, it must be appreciated that there exist many drugs whose sites of action have not yet been elucidated in detail. Furthermore, many drugs are known to bind to plasma proteins as well as to various cellular constituents, without producing any obvious physiological effect. Nevertheless the above generalization that most drugs act on one of the five types of protein targets listed above is a reasonable initial working classification system.

General Principles of Drug Action

There are a variety of different types of drug action. Accordingly, drugs can be classified into specific categories such as agonists, antagonists, partial agonists, inverse agonists, allosteric modulators, enzyme inhibitors or activators, and those having nonspecific action.

Agonists bind to a receptor or site of action and produce a conformational change in the receptor or that site, which mimics the action of the normal physiologic binding ligand. At low concentrations, the activity of the drug is additive with the natural ligand. Drugs can differ in both their affinity or strength of binding and the rate of association and dissociation from the receptor or binding site. The affinity or strength of binding of the drug to the receptor ultimately determines the concentration necessary to produce a response and therefore is directly related to the potency of the drug. For some compounds a maximal response may of necessity invoke the contribution of all available receptors, but for most drugs a maximal response is produced while some receptors remain unoccupied. The presence of spare receptors becomes an important point when considering changes in the numbers of available receptors resulting from adaptive responses occurring in response to either chronic exposure to a drug or the irreversible binding of an antagonist. The rate of binding or dissociation of a drug to a receptor or site of action is generally of little importance in determining the rate of onset or termination of a drug's elicited effect in vivo because such behavior depends mostly on the rate of delivery to and removal from the target organ, in other words, the rates of absorption and elimination of the drug from the body. The effect of changes in the numbers of receptors on the dose–response curve for an agonist depends on the potency of the agonist, receptor occupancy by the agonist, and the efficacy of the agonist.

Antagonists bind to a receptor but do not elicit the necessary conformational change required to produce the normal response–effect. These types of compounds will block access to the receptor or binding site by the normal physiologic ligand. It is important to keep in mind that antagonist-induced effects may only be observable when the normal agonist or ligand is present. The binding of most clinically useful antagonists is both reversible and competitive. Consequently, typically a receptor or site of action blockage can be overcome by increasing the concentration of the natural ligand or another agonist drug. Most antagonists shift the dose–response curve to the right but do not alter the magnitude of the maximum possible response.

Partial agonists demonstrate both agonist and antagonist activities. For these types of drugs, the activity demonstrated by the drug is a function of the concentration of the natural ligand or agonist. Maximal binding of the partial agonist to a receptor will produce only a submaximal response. This is most likely the result of incomplete amplification of the receptor signal via G proteins. A partial agonist will demonstrate agonist activity at low concentrations of the natural ligand, but the dose–response will not attain maximal activity even when all of the receptors are occupied. Alternatively, at high concentrations of the natural ligand a partial agonist will behave as an antagonist because it will prevent the access of the natural ligand to the receptor, thereby preventing a maximal response.

Inverse agonists act in such a fashion on receptors as to produce a change opposite to that caused by an agonist. While not totally understood, the discovery of this phenomenon has given rise to the theory that receptors exist in equilibrium between active and inactive forms in the absence of an agonist ligand. The presence of an agonist will increase the proportion of active receptors, but the presence of an inverse agonist will shift the balance toward the more inactive receptors, thereby reducing the level of basal activity. Antagonists, when bound to a receptor will block the activity not only of agonists but also of inverse agonists. The mechanism of action of inverse agonists is not well characterized but may involve the destabilization of receptor–G protein coupling. Inverse agonists may preferentially bind to the inactivated form of the receptor, shifting the equilibrium away from the active form. An additional complication to this whole concept is the fact that compounds that are normal antagonists for some tissue receptors act as inverse agonists for the same receptor in a different tissue. The concept of inverse antagonism and its use in the therapeutic utility of drugs remains to be fully understood and better characterized.

Allosteric modulators do not act directly on a ligand/receptor site but may bind elsewhere on the receptor to enhance or decrease the binding of the natural physiologic ligand to the receptor. Some drugs have a target site of action that is an enzyme, and these compounds can effect their action on either the catalytic site or elsewhere on the molecule at an allosteric site. Finally some compounds merely elicit a broad and generalized effect that causes the desired therapeutic outcome. Examples of such activity might be that of osmotic diuretics and their ability to induce diuresis through the general action of osmosis or general anesthetics, which modulate neuronal cell membrane fluidity.

Drug action also involves the demonstration of a number of different important properties. These are specificity, selectivity, potency, and efficacy. Specificity refers to the fact that many drugs act only at one type of receptor (e.g., cholinergic versus noradrenergic). Drugs that are not specific in their action for one type of receptor display a wide variety of not necessarily desirable side effects.

The site of action of a drug may in some cases involve one or more members of a family or group of receptors. In such a situation, where the drug may act preferentially with one member of the group, it is said to be selective in its action. Alternatively, the drug may show a similar affinity for more than one member of a group or family of receptors and is therefore referred to as being nonselective. Drugs may demonstrate a predilection for a particular receptor type or subtype and may bind to a different receptor type or subtype to different degrees. In these situations, it is possible to determine the individual dose–response relationships for each receptor type or subtype. The selectivity of a drug is a measure of the degree of separation of the individual dose–response curves for different receptor types or subtypes. Ultimately, the expression of selectivity is dependent upon the dose and the concentration of the drug at the different receptors. High concentrations of an agonist drug will typically generate a maximal occupancy of all of the involved and available receptor types or subtypes with little to no discrimination or selectivity. High concentrations of an agonist drug can even create a blockade of the receptor. One should always be careful to include the appropriate qualifications when referring to selectivity and nonselectivity because populations of receptors can be related or unrelated to each other. For example, one can speak of the selectivity of a drug for α-adrenoreceptors as compared to β-adrenoceptors or β_1-adrenoceptors as compared to β_2-adrenoceptors.

The in vitro potency of a drug is determined by the strength of its binding to a receptor. This is also referred to as the affinity of the drug for the receptor. The more potent a drug is, the lower will be the concentration required to affect its binding to a receptor and to provide a response for an agonist or to block a response for the case of an antagonist. Potencies of different drugs are generally compared or contrasted using ratios of the different doses required to either elicit or block an equivalent response. Dose–response curves are typically S-shaped. If the mechanism of action is identical or very similar for the drugs being compared, the linear or midportions between the lower and upper plateaus of the dose–response curve are commonly found to be parallel. To make comparisons between different points on distinct dose–response curves more obvious and facile, a linearizing transform can be utilized for each sigmoid curve. A popular technique for linearizing data is the logit plot. Potencies are generally not determined by comparing responses at identical or similar doses. This is because with the selection of a single administered dose, inherent differences between individual drugs will cause too large of a difference in drug responses. Rather, relative potencies are determined by an evaluation of the ratio of the doses for different drugs that produce equivalent responses on the respective drug dose–response curves. For example, the doses that cause 50 percent inhibition of the target activity on various dose–response curves could be compared for different drugs. Drugs that demonstrate in vitro the highest affinities for a receptor are generally the most potent. However, the in vivo dose–response relationship will be a function of the delivery of the drug to the site of action. Such delivery is related to the absorption, distribution, and elimination (pharmacokinetics) of a drug. Therefore, it is important to keep in mind that for a series of related drugs in vivo behavior may not always reflect in vitro receptor binding properties.

The efficacy of a drug is its ability to produce the maximum possible response. For example, drugs can be classified as full agonists or partial agonists. Full agonists produce an increase in response with increase in concentration up to that point at which the maximum possible response is elicited. A partial agonist will produce an increase in response with an increase in concentration, but it cannot produce a maximal possible response.

Molecular Aspects of Drug Action—Receptors

The activities of most cellular processes are highly controlled in order for the cell to exist under optimum homeostatic conditions not only when at rest but also in response to the myriad of physiological and metabolic demands that are placed upon it. These reflex actions require specific responses appropriate for a given cell type to signals elicited as a result of a change in physiological state. These signals most commonly are specific chemicals that are released into the general circulation or are locally released but then are recognized by a targeted or specific cell. Ultimately the condition of homeostasis is based upon (a) the generation of a chemical signal, (b) the recognition of a chemical signal, and (c) the generation of an appropriate response or cellular change(s).

A chemical signal eventually binds to a specific type of macromolecule of a cell. This binding in and of itself then triggers a cellular response. The chemical signal is referred to as the ligand. The cellular macromolecule is referred to as the receptor. Receptors can be located in different places. They may be located in the cell membrane in order to accept extracellular ligands that cannot cross or only cross with difficulty the cell membrane. Alternatively, they can be located in the cytoplasm, where they will react with lipid-soluble ligands that can pass through the cell membrane. The binding produces a receptor–ligand complex that will then generate an appropriate cellular response. There are three general types of responses: direct, indirect, and second messenger. A direct response might be the inhibition of a specific process. An indirect response might involve the interaction of the receptor–ligand complex with another moiety (e.g., macromolecule) to continue or complete the process. A second-messenger response would involve the production of an additional chemical signal (second messenger), which would ultimately control a cellular process.

There are different structural and functional classifications of receptors, but generally speaking there are just a few functional families whose members share both common mechanisms of action and similarities in molecular structure. There are at least four main types of receptors: types 1 through 4. The classification of receptors is based on molecular structure and the nature of the receptor–effector linkage. Type 1 receptors are typically located in a membrane and are directly coupled to an ion channel. Type 2 receptors are located in a membrane and are coupled by a G protein to an enzyme or channel. Type 3 receptors are located in membranes and are directly coupled to an enzyme. Finally, type 4 receptors are located in the nucleus or cytosol and are coupled via DNA to gene transcription.

Although there are some exceptions, ligands typically bind to receptors in a reversible fashion. Accordingly, the intensity and duration of the response generated from the binding of the ligand to the receptor is a function of the lifetime of the ligand–receptor complex. The interaction(s) between a receptor and a ligand most commonly do not involve the formation of permanent covalent bonds, but rather do involve the formation of weaker and reversible forces, such as ionic bonding, hydrogen bonding, van der Waals forces, and hydrophobic interactions.

Within a physiologic entity, there are myriad possible extracellular and intracellular chemical signals that are produced that can affect multiple different processes. Subsequently, a very important property of a receptor is its specificity or the extent to which a receptor can recognize, discriminate, and respond to only one signal. Some receptors demonstrate a very high degree of specificity and will bind only a single endogenous ligand, while other receptors are less specific and will bind a number of different endogenous ligands. This ability of receptors to recognize, discriminate, and bind to a given ligand depends on the degree and type of interaction between the receptor molecule and specific chemical structural characteristics of the ligand. Chemical structural differences between ligands may be very subtle in nature. However, receptor specificity occurs because the generation of a fully functional receptor–ligand complex requires the formation of reversible binding interaction between various different molecular sites on the ligand and on the receptor in a special three-dimensional spatial relationship. Receptors themselves are proteins folded into a three-dimensional configuration so that the specific arrangement of reversible interaction sites on the receptor itself is consolidated into a very small volume, referred to as the receptor binding site.

As receptors and ligands each have three-dimensional configurations, the ligand must be presented to the receptor in a very specific three-dimensional configuration to bind and produce a functional receptor–ligand complex. Many drugs exist in different stereoisomeric forms, and each of these various stereoisomers may well exhibit different receptor binding behavior and resulting response. Different isomers can be equally active or some active, some partially active and some even toxic. Furthermore, it is important to realize that if a particular drug exists in equal concentrations of two different isomeric forms, only one of which is active and the other of which is inactive, then only 50 percent of the drug mixture is therapeutic. This has led over the years to the recognition of the value of the separation and purification of individual isomers of drugs for development as potential therapeutic agents.

Different types of receptors recognize and bind to different ligands. Furthermore, there may be different subtypes of a given receptor, each of which recognizes or binds to the same specific ligand but generates different intracellular responses. Various receptor subtypes are often found in different tissues, organs, or distribution patterns in an organ and hence can produce different end effects. The existence of receptor subtypes creates the opportunity for specific drugs to produce extremely selective actions with fewer unwanted side effects.

At any given time, the number of receptors in a cell is not static but rather is a dynamic. There is a high turnover of receptors as they are continuously removed and replaced. Drug treatment itself can either increase the number of receptors, a phenomenon that is called upregulation, or decrease the number of receptors, a phenomenon referred to as downregulation. This potential change in numbers of receptors can be an important aspect of drug treatment and management of a clinical case to achieve a desired therapeutic response.

Molecular Aspects of Drug Action—Ion Channels

Receptors for several neurotransmitters exist as agonist-regulated, ion-selective channels in the plasma membrane of a cell and are referred to as ligand-gated ion channels. These receptors send their signals by altering a cell's membrane potential or its ionic composition. This group of receptors includes nicotinic cholinergic receptors, γ-aminobutyric acid receptors (subtype A), glycine receptors, aspartate receptors, glutamate receptors, nicotinic cholinergic receptors, and 5-hydroxytryptamine (subtype 3) receptors. These receptors are all multiple subunit proteins. Each protein subunit spans the cell membrane, and the subunits are arranged symmetrically in such a fashion as to form a channel. This channel opens and closes upon proper stimulation as a result of specifically induced molecular structural changes.

While some ion channels are linked to a receptor and open only when the receptor is occupied by an agonist or ligand, there are other types of ion channels, which themselves serve as targets for drug action. This type of interaction can be indirect, involving other intermediates, but the interaction can also be direct with the behavior of the channel being modulated by the binding of a drug directly to a part or parts of the channel protein. The simplest type of ion channel blocking involves a physical chemical barricade of the channel opening by the drug itself. More complex types of direct ion channel blocking involve drug–channel protein interactions. Normally, an ion channel opens in response to membrane depolarization. In these situations, a drug, by binding with ion channel protein alters or modulates the probability of the channel opening. The degree of inhibition or facilitation of the opening of the ion channel is a function of the chemical structure of the drug itself. In this latter case, a drug actually affects the gating of the channel, while in the former case the drug actually blocks penetration of the channel. The modulation of ion channels by drugs is a very important mechanism through which pharmacologic effects are mediated at the cellular level.

Molecular Aspects of Drug Action—G-Protein-Coupled Systems

There is a large family of receptors that utilize heterotrimeric guanosine 5'-triphosphate (GTP)-binding regulatory proteins. These regulatory proteins, known as G proteins, behave as transducers to send signals to their effector proteins. Ligands for G-protein-coupled receptors include eicosanoids, a variety of lipid signaling molecules, various peptides, different proteins, and a host of biogenic amines. Effectors for G-protein-regulated receptors include adenyl cyclase, phospholipase C_β, Ca^{2+} currents, rho GTP exchange catalysts, inward-rectifying K^+ currents, and phosphatidyl inositol-3-kinase. G-protein receptors occur widely in nature and are widely used drug targets.

G-protein-coupled receptors span the cell membrane and exist as a bundle of seven α helices. A cleft exists in the three-dimensional configuration of these seven α helices, which binds agonists on the extracellular face of the receptor. Alternatively, agonists can also bind to a globular ligand-binding domain located on the extracellular face of the G-protein-coupled receptor. The G proteins themselves bind to the cytoplasmic face of the receptor. These receptors respond to the binding of an agonist by promoting the binding of GTP to the G protein. The binding of the GTP to the G protein activates the G protein, which in turn activates the effector protein. The G protein remains in an activated state until it hydrolyzes the bound GTP to guanosine 5'-diphosphate (GDP). G proteins are composed of an α-GTP-binding subunit, which is specifically recognized by the receptor and an associated dimmer of β and γ subunits. G proteins are defined by α-subunit composition, such as α_s, α_i, α_o, α_q, and α_{13}. The activation of the G_α subunit by GTP permits the regulation of an effector protein and promotes the release of $G_{\beta\gamma}$ subunits, which in turn regulate their own group of effectors.

There are many different G-protein-coupled receptor types in a cell. Each type is specific for one of many different G proteins. Each G_α subunit can in turn regulate one or more effectors. Therefore, receptors that bind to multiple different ligands can focus their signals through a single G protein. Alternatively, a G-protein-coupled receptor is also capable of sending multiple signals through activation of more than one G protein species. It should be appreciated that receptor G-protein effector systems are intricately versatile and complex networks.

Molecular Aspects of Drug Action—Receptors as Enzymes

Receptors with inherent enzymatic activity are most commonly cell surface protein kinases. These receptors demonstrate their regulatory activity by phosphorylating various effector proteins at the inner face of the cell membrane. The biochemical reaction of phosphorylation changes the molecular structure, biological properties, and hence the biological activities of an effector or its interactions with other protein molecules. Catalytic activities include tyrosine kinase, tyrosine phosphatase, guanyl cyclase, serine kinase, and threonine kinase. However, tyrosine residues are the most common substrate. Basic receptor protein kinases are composed of an agonistic-binding domain on the extracellular membrane surface, a single element that spans the cell membrane and a protein kinase domain on the cytoplasmic face of the cell membrane. Other variations exist of this basic model and include oligomerization and/or the elaboration of other regulatory or protein binding domains to the cytoplasmic portion of the receptor. Some protein kinase receptors lack the attached intracellular enzymic domain, and in response to agonist binding, attract, attach, or link to and activate a distinct soluble and mobile protein kinase on the cytoplasmic side of the cell membrane.

There can be further variation on the above structural theme for receptor enzymes. For example, tyrosine phosphatases possess an extracellular enzymatic domain. In other receptor enzymes, the intracellular domain is not a protein kinase, but rather a guanyl cyclase, which synthesizes a second messenger, adenosine-3′,5′-cyclic monophosphate (cAMP). Other variations of this receptor type may also be possible.

Before leaving this category of receptor, it is appropriate to mention second messengers. A variety of signals within a cell are sent via second-messenger pathways. While there are few cytoplasmic second messengers, their release and presence can affect many different activities. Recognized second messengers include cAMP, nitric oxide, diacylglycerol, Ca^{2+}, inositol phosphates, and guanosine-3,5′-cyclic monophosphate (cGMP). Second messengers can exert control on each other either directly or indirectly. Directly, they can alter the metabolism of other second messengers. Indirectly, they can share intracellular targets. Such a complex pattern of regulatory activity permits a cell to respond to the presence of an agonist with an integrated and concerted expression of second messengers and responses. Cyclic AMP is the best known of the second messengers and is synthesized by adenyl cyclase under the control of different G-protein-coupled receptors.

Molecular Aspects of Drug Action—Transcription Factors

There are receptors for steroid hormones, thyroid hormones, retinoids, vitamin D, and other molecules that are soluble DNA binding proteins that regulate the transcription of specific genes. These receptors are part of a larger family of transcription factors that are regulated by phosphorylation, by association with other proteins, binding to metabolites, or binding to regulatory ligands. These receptors exist as homo- and heterodimers, and their actions and activity can also be regulated by higher order oligomerization with other regulatory molecules.

Molecular Aspects of Drug Action—Other Targets

Enzymes can also serve as targets for drugs. Typically, the drug is a substrate that is structurally related to the normal substrate and acts as a competitive inhibitor of the enzyme. Another type of enzyme–drug interaction occurs when a drug is a false substrate. In this type of situation, the drug undergoes the enzyme-mediated chemical transformation to form an atypical product, which then undermines the normal metabolic pathway. This altered function of the metabolic pathway occurs as a result of the creation of a substrate that is not usable for the next step in the metabolic process or ultimately generates a partially functional or nonfunctional end product. It should be mentioned that drugs can also be metabolized by enzymes in such a fashion as to either convert the drug into a toxic reactive intermediate causing significant cellular damage or convert the drug from a totally inert moiety (prodrug) into an active compound with therapeutic benefit.

In addition to the sites and mechanisms of action discussed above, drugs may also bind to and either activate or inhibit other specific sites. These could include specific types of cells, cellular organelles, transport proteins, specific ion pump, and the like. A discussion of all of these additional targets is beyond the scope of this chapter.

Drug Tolerance

An observed decrease in drug response with repeated doses is commonly referred to as the development of tolerance. Tolerance may occur as a result of a decrease in the concentration of the drug at the receptor site or through a decrease in response of the receptor to the same concentration of the drug. Some drugs can stimulate their own metabolism, are therefore eliminated more rapidly with repeated dosage, and accordingly less drug is available to elicit a response. However, the most clinically important and relevant examples of tolerance result from changes in numbers of receptors and subsequent quantitative changes in concentration–response relationships.

Desensitization is a term that is used to describe changes in the dose–response relationship, irrespective of time, arising from a decrease in response of the receptor. Desensitization can result from decreased G-protein coupling, decreased receptor binding affinity, downstream modulation of the initial signal, or the downregulation (decreased numbers) of receptors. This latter process is slow, and takes hours to days to become effective. However, extracellular receptors coupled to G proteins can show rapid desensitization within minutes during continued activation that can occur through two different mechanisms. The first of these mechanisms is homologous desensitization, in which enzymes are activated as a result of ligand binding to a receptor–G protein complex. These enzymes include G-protein-coupled receptor kinases (GRKs), which interact with the βγ subunit of the G protein and inactivate the occupied receptor protein by phosphorylation. The second of these mechanisms is heterologous desensitization, in which the receptor, whether occupied or not, is inactivated through phosphorylation by a cAMP-dependent kinase, which can be switched on by a variety of signals that increase cAMP. Both α_1-adrenoceptors and muscarinic receptors, which are linked to phospholipase C, may also undergo desensitization via receptor phosphorylation, which uncouples the G protein from the receptor. Regardless of the mechanism, ultimately the phosphorylated receptor protein is eventually internalized and subjected to intracellular dephosphorylation before reentering the cytoplasmic membrane. Finally, the downstream modulation of a signal may occur through either a feedback mechanism or through simple depletion of an essential cofactor.

1.6 SCREENING FOR NATURAL PRODUCT ACTVITY

What is a screen? A screen is an assay or biological assay that provides a tool that can be used to test for or establish the presence and level of a target activity in a specific sample. Bioassays in a screening program should be rapid, simple to conduct, relevant, capable of being automated, cost effective, and of the potential to deliver high throughput [15, 16, 69]. Appropriate technology should be used to permit low limits of detection. This last point is important because the concentration of desirable compounds is unknown in each sample and so it behooves one to strive for the lowest possible limit of detection. Screens should also be specific for the molecular or cellular therapeutic target of choice. Appropriate additional discriminatory tests outside of the focus of the chosen target activity, such as cytotoxicity measurement for cell-based assays, or isotype specificity tests for molecular assays, are valuable in that they provide additional information relative to the overall value of a potential hit. Furthermore, data generated from all screens in which samples are tested should be compared so that selective hits can be identified at an early rather than a late stage. Such a combination of specific screens, data comparisons, and discriminatory assays makes possible the earliest selection of the best hits for continued work and success. The screens must work in the presence of the compounds to be tested and accordingly must be compatible with a given molecule's physico-chemical characteristics. Accordingly, natural product screens need to be operational in the presence of solvents and buffered against extremes of pH and ionic strength and should not be affected by the presence of color. Screens should always incorporate the proper use of controls, both positive and negative. Screens should be bidirectional with regard to their output and have a defined and easily interpretable endpoint. Screens need to be capable of delivering quantifiable data. Screens can be designed in such a fashion as to monitor only a single biological activity. However, there is value to the approach of coupling biological screens to evaluate multiple general biological activities in addition to the target activity [38]. This is because a compound that does not provide a target hit may well generate value for itself by demonstrating some other unrelated and unexpected activity. Alternatively, compounds that do provide hits may well expand their value by demonstrating other unexpected types of biological activity. Additional considerations of the construction of screens, screening programs, and screen design are beyond the scope of this chapter, and the reader is instead referred to an excellent reference on the subject by Gad [54]. What will be addressed here are some of the common features of screens and screening specific to research on the development of natural products as drugs.

Before one can begin to screen, it is important to know for what one wants to screen. Drug discovery begins with basic ideas, ideas relevant to therapeutic targets and sources of compounds [40, 66, 164]. Therapeutic targets can arise from genomics, the molecular cloning of receptors and signaling molecules, a detailed understanding of physiology, biochemistry, and pathology, research into folklore or ethnomedicine, and knowledge of the traditional uses of natural products. Sources of compounds can present from existing chemical libraries, historical compound collections, natural product libraries, combinatorial libraries, rational chemical synthesis, general or targeted literature searches based upon existing knowledge or leads, and antisense oligonucleotides. The simultaneous consideration of all or at least several of these areas is essential to the original design of a screening program.

A totally random approach in the selection of a source can be coupled with mass screening, but such a path is typically not successful. But to be fair, the random approach is more likely to generate compound novelty [41]. Generally, approaches utilizing literature searching, existing chemical libraries and folklore, ethnomedicine, or traditional medicine are the most popular because of their cost effectiveness. It is worth noting that programs and selections based on or incorporating ethnomedicine or foklore are five times more effective in the ultimate generation of leads. The most effective approach is generally considered to be a mixing of as many components of the therapeutic arm with the compound source arm as opposed to the selection of any single aspect.

It is worth noting at this point that a new approach is being used in the field of natural products that leads to the generation of "unnatural natural product compounds" [40]. This approach is termed *biochemical combinatorial chemistry*. In this technique, appropriate secondary metabolic enzymes are isolated from a crude natural product mixture. These enzymes are then used to generate unnatural metabolites, which are isolated and then subjected to a bioassay procedure that couples a bioassay to an analytical procedure permitting structure elucidation.

What does one put into a screen? It is very tempting to consider the purification of natural products into their individual components before embarking on screening activities. However, this is a very economically challenging and financially unrealistic approach [65]. The classical approach is to design a screen or screening program that will permit the use of the assay or assays to provide guidance to successive steps in purification. An advantage of this is that the effectiveness and efficiency of purification can also be simultaneously evaluated at the same time as biological activity is being enhanced. Ultimately, after sufficient purification, a chemical structure can be determined for the active moiety. However, Grabley and Sattler [60] are of the opinion that the use of cost-effective physico-chemical and chemical screening procedures will facilitate biological screening because of an ability to provide purer extracts if not pure compounds for biological screening.

Some objectives should be kept in mind when preparing natural product extracts for their ultimate introduction into the screening process [70]. First, every attempt should be made to stop ongoing biological processes. Second, steps should be taken to provide chemical stability of the compounds in the extract. Third, efforts need to be made to minimize losses of material. Fourth, sample preparation costs need to be minimized. When trudging through the iterations required to purify mixtures, one should be sure to: (1) focus on the activity of interest, (2) focus on the compound(s) of interest, (3) eliminate nuisance materials such as cell parts, biopolymers, and other compounds not of interest, and (4) be sure to enrich the composition of compounds of interest.

What types of screens, assays, or bioassays should be used? Assays for activity can be performed at a variety of levels ranging from the molecular level to the whole organism. While it is true that the high-throughput screening of synthetic compounds generated by combinatorial methods may be best achieved at the molecular level, Duke et al. [40] are of the opinion that natural product screening should be performed at the highest level possible since more effort per compound has been invested in the discovery of each compound.

Historically, natural products have been subjected to what is termed bioassay-directed isolation. In this approach, a crude natural product mixture is subjected to fractionation and the individual fractions then bioassayed for specific biological activity. This process continues repetitively with comparison of individual fraction assay data to a bioassay database. When the data is shown to match a previously known profile, the process is terminated and the sample is discarded. If the data is shown to provide a new profile, the structure of the compound is determined. However, this approach can lead to the rediscovery of previously known compounds after significant effort has been expended.

Low-molecular-weight natural products from a variety of sources represent unique structural diversity. In order to more adequately and efficiently access this diversity, various new strategies improving targeting and direction have been developed [60]. Modern separation/chemical characterization approaches can eliminate much of this problem by identification of the compounds before they are subjected to bioassay. Indeed the coupling of such techniques to biological screens can improve the quality of the assay result and shorten research and development time frames. These new tandem approaches are termed fractionation-driven bioassays. While biological screening directly correlates to a predefined biological effect, physico-chemical and chemical screens do not. In this latter case, lead selection is based on either physico-chemical properties or chemical reactivities. In both cases, the first step is the chromatographic separation of compounds from the complex source mixture. In the second step, the physico-chemical properties or chemical reactivities of the separated compounds are analyzed. Both of these chemical-based strategies have proven to be of value as auxiliary or supplemental methods to biological approaches. Data generated from physico-chemical and/or chemical screening is very helpful in the de-replication or early identification and exclusion of known or otherwise unsuitable compounds that occur during high-throughput biological screening programs. Furthermore, the use of physico-chemical and chemical screening will aid in the establishment and building of natural-product-based compound libraries, which could then be used more broadly in testing programs.

There are two general types of new tandem assays. The first of these is referred to as the fractionation-driven bioassay. In this method, a crude natural product mixture is subjected to fractionation and the individual fractions then subjected to nuclear magnetic resonance (NMR) spectroscopy or mass spectrometry/mass spectrometry (MS/MS). The structures of the compounds in the individual fractions are identified; and, if they are known, their biological activity profiles are evaluated from an existing database. If the structures are unknown, then the compounds are subjected to bioassay. In an alternative approach, termed the isolate and assay approach, a crude natural product mixture is subjected to automated fractionation and purification. The individual fractions are then subjected to bioassay. Desirable biological activity serves as the trigger to subject the sample to NMR or MS/MS and ascertain the structure(s). If the material is a novel compound, the structure can be optimized. If the material is a previously known compound, it may well be discarded, depending on its biological activity or toxicological profile.

A broad range of screening technologies are currently available for use in screening for natural-product-based drugs [69]. For molecular targets, such procedures would include generalized solution-phase assays, immobilized substrate assays, scintillation proximity assays, and time-resolved fluorescence assays. For cell based targets, cell signaling, cell communication, cell receptor, and reported gene assays are available. Isolated subcellular systems are also available [156]. Bioassays can, if desired, incorporate lower level organisms, isolated vertebrate organs, or whole animals. In short, systems are limited only by the creativity and design of the screener.

Examples of screening programs can be readily found in the literature. Just to reference a few, Quinn [128, 129] has reported on his efforts on prospecting the biodiversity available in Queensland; Mehta and Pezzuto [112] have published on their program to identify cancer preventive agents from plants; El Sayed [41] has reported on his screening program for antiviral agents; Barrett [5] has written on his program to find novel antifungal agents; Yang and co-workers [167] has reported on his search for anti-HIV compounds from natural sources; and Bindseil and co-workers [9] have published on their experiences on screening with pure compounds.

The emergence of high-throughput screening (HTS) has permitted the rapid screening of extremely large collections of structurally diverse synthetic compounds against a variety of novel and diverse disease targets [19, 128, 129, 150]. However, despite initial hopes that HTS was the final solution for drug discovery, for reasons that will not be discussed here, HTS has not achieved that distinction. Nevertheless, HTS is still a powerful tool. HTS strategies focus on the ability to screen large libraries of compounds. However, the limiting factor in HTS is the ability to access large numbers of chemically diverse substances. Natural products are undoubtedly the greatest source of structural diversity. Accordingly, HTS of the unparalleled diversity that exists in natural product extracts offers the highest probability for discovery of novel lead compounds and should therefore be viewed as being complementary to compounds generated from combinatorial chemistry alone. The synergistic melding of HTS and natural products has started, and, as it progresses along its development path, exciting new breakthroughs will undoubtedly be presented.

As a compound generates interest through a variety of screening assays and progresses down the drug discovery path, certain questions need to be asked [5]. Is the chemical structure of the compound novel? Is the mechanism of action of the moiety novel or of utility? Is the biological activity of the compound useful? Is the potency of the compound reasonable? Is a proof of concept available? Is chemical modification of the structure possible? Is solubility a problem? Can the material be synthesized on a large scale? The ability to ask and answer these questions effectively early on in the process will be highly predictive of the ultimate success of a particular line of research. While there are no "correct" answers to these questions, as answers will be different depending on the therapeutic indication and other available therapeutic alternatives, they still need to be addressed to provide proper program focus.

Before ending this section, it is important to emphasize the necessity and importance of keeping detailed records as one initiates the screening process [70]. Maintenance of a secure physical inventory with a controlled environment and adequate records is another important detail. The use of bar code identifiers is very desirable for samples and relevant computer programs are readily available. It should be considered to be essential to establish a complete database, which should include the source, source location, isolation details, any relevant taxonomic information (kingdom, genus, and species), and any other relevant information (third-party suppliers of reagents, potential pathogenicity or toxicity, relevance to any international biodiversity treaties), preservation methods, age, position or location within a freezer, and the like. Other considerations in the design and management of a successful screening program could be limitation of access, storage of reference or backup samples, storage of a backup copy of inventory and database, establishment of tracking procedures, writing of standard operating practices (SOP) for sample handling, and temperature alarms.

1.7 ISOLATION AND PURIFICATION OF NATURAL PRODUCTS

Why do scientists working with natural products isolate and purify them? For one of two reasons: (1) to ascertain what the natural product is and (2) to carry out sufficient experimental work necessary to biologically characterize or profile the compound. It can be quite a sobering experience to look at a flask full of dark-colored, inhomogeneous sludge and liquid and realize that one is going to attempt to isolate just one particular type of molecule from all of the other materials that are present. To put this in perspective, typically the material sought after represents only about 0.0001 percent of the total biomass in the flask [15, 16]. Then, just to make things even more challenging, the desired molecules can also be bound with other materials and molecules present in the mixture, making the desired compound(s) even harder to purify. It is important to keep in mind that the isolation of natural products differs from that of the more prevalent biological macromolecules. This is because natural products are typically secondary metabolites and as such are smaller in size, chemically more diverse in structure, and present in smaller concentrations than the more homogeneous proteins, carbohydrates, lipids, nucleic acids, and the like.

This section will not present a condensed work on separation science and procedures as that is best left to any of the myriad analytical chemistry textbooks. What will be attempted here is to provide sufficient guidance on the isolation and purification of natural products so that proper focus can be assured in the design and implementation of a successful isolation and purification program.

Before initiating an isolation and purification, there are a number of basic questions that need to be asked and answered. First, what are you trying to isolate and purify? There are a number of different possible targets: (1) an unknown compound associated with a particular biological activity, (2) a previously known compound present in a specific organism, (3) a group of compounds within an organism that are all structurally related to each other, (4) all of the metabolites produced by one natural product source that are not produced by another closely related source, or (5) all of the molecules of a particular organism.

Second, why are you trying to isolate this material? While the asking of such a question might appear to be superficially inane, it is important to know why you want something so that you know how much of it you might need. Possible reasons for carrying out an extraction might be: (1) the generation and supply of larger amounts of an already known compound so that more extensive biological testing such as pharmacology and toxicology can be performed on the material, (2) the purification of a small amount of material for initial biological and chemical characterization to be performed, or (3) to purify sufficient material in order to conduct complete structural studies and further biological activity characterization.

Third, what type of purity is desired? If a natural product compound is to be used for biological testing, it is important to know not only the degree of purity of the material but also the nature of the impurities. It needs to be appreciated that the impurities themselves can contribute significantly to any biological activity observed in the screening program. If the material is to be used in more refined pharmacological or pharmacokinetic testing, then the material should generally be at least 99 percent pure. If, on the other hand, the material is to be used only for chemical characterization, the acceptable level of purity can range from 95 to 99 percent. Such a range of purity will generally be sufficient for the determination and assignation of a complex chemical structure via such techniques as NMR spectroscopy, infrared (IR) spectroscopy, and MS/MS spectrometry. It should be noted that if the compound under consideration is present in a high concentration in the starting material and a standard for that compound already exists, then structural confirmation can be achieved with less pure material and the associated purification scheme will be composed of fewer steps. Depending upon one's goals, varying degrees of purification may be acceptable. X-ray crystallographic studies will demand material of 99.9 percent purity, while detection of the presence or absence of a specific structural feature via analysis of the ultraviolet spectrum may tolerate purities down to a level of 50 percent. An important concept of purification is that the relationship between purity of compound achieved during natural product extraction and the amount of effort expended to achieve such a level of purity is almost exponential in nature. When starting with a crude, complex mixture, it is very easy to eliminate large components of unwanted material. However, as the purity begins to escalate, it can become infinitely more challenging to improve purity levels. For example, the effort required to go from 50 percent purity to 90 percent purity can pale in comparison to the effort required to go from 99.5 to 99.9 percent purity. In concordance with this, it is fair to state that the relationship between purity level and yield are also exponentially related. In a purification scheme no step delivers the desired material in 100 percent yield. Each extraction step results in the loss of material, and when working to attain very high levels of purity, losses can be extreme. While it may be necessary to take only very "centralized" cuts in a purification step, keep in mind that the "tail" cuts can themselves be later subjected to reprocessing.

Fourth, what type of fractionation should be used in the isolation and purification scheme? All separation processes involve the division of a mixture into a number of discrete fractions. This process is called fractionation. Such fractions can be physically separate such as the two phases of a liquid–liquid extraction or they may not be physically separate such as the continuous eluate from a chromatography column. The eluate from a chromatography column can then be artificially divided into fractions via the use of a fraction collector. The method of fractionation depends on the sample and the goals of the separation. Fractions are typically equal in size and can be large or small in volume. The collection of a large number of small fractions improves the probability that each fraction might contain pure compound. However, such an approach requires significant work in the analysis of each fraction. Additionally, this approach may spread the desired compound over so many fractions that if the target molecule(s) was present originally only in low levels, it may prove undetectable in any one of the fractions. Alternatively, if the separation process is cruder, employing the collection of only a few large volume fractions, a more rapid and facile tracking of the desired compound and its activity is possible.

Fifth, what is the nature of the compound? The answer to this question depends on how much is already known about the compound. General features that are useful at this stage of the project are acid/base properties (pK_a, pK_b), molecular charge, stability, and solubility (hydrophilicity/lipophilicity). If the target molecule is an unknown moiety, it is very likely that little of this information is known and all of it will have to be determined along the way. If one is isolating a known compound, much of this information will already be available. Finally, if the goal is to isolate a number of secondary metabolites rather than a single molecule, then the value of this step is less important, but an appreciation of the relative values of these parameters can still be useful with regard to understanding the characteristics of the mixture.

Sixth, where is the desired activity localized? Each potential source of a natural product source—whether it is plant, tree, moss, bacteria, vertebrate, invertebrate, insect, terrestrial, or marine based—has components or parts in which the desired activity or compound is present in greater concentration as opposed to other parts in which the compound is present in lesser amounts. To obviate any problems associated with dilution of the compound and its activity, the initial biomass should be selected on the basis of its content of the target biological activity.

Only with the thoughtful provision of answers to the above questions can one have a clear idea of what one is attempting to achieve and how to successfully secure the project goals. It should be obvious that there is no correct or incorrect protocol or standard operating procedure for the isolation and purification of natural products. Indeed, the final method or scheme itself is most likely to vary with the answers to the above questions as well as the natural product source and the specifics of the assays and biological assays that are to be used in the screening program. However, a consult of the literature is essential during the design and construction of an isolation and purification program for any natural product compound [149, 165]. While it is possible that extensive data may have already been published on the compound one is trying to isolate or compounds related to it, it is also entirely possible that nothing is known. Regardless, proper use of the natural products literature can facilitate the effort invested into the design and implementation of a specific isolation and purification program.

A variety of different techniques can be used for the isolation and purification of natural product compounds [15, 16, 82]. These techniques include, but are not limited to, solid-phase extraction [15, 16], high-performance liquid chromatography (HPLC) [15, 16], gradient high-performance liquid chromatography [15, 16], bioautography [165]; thin-layer chromatography (TLC) [15, 16], countercurrent chromatography [2, 49, 50], droplet countercurrent chromatography [74], vacuum column chromatography [165], desalting [149], liquid–liquid chromatography [75], paper chromatography [15, 16], ion exchange chromatography [15, 16], size exclusion chromatography [73], affinity chromatography [15, 16], acid–base switching technology [42], centrifugal partition chromatography [74], liquid–solid chromatography [15, 16], microwave-assisted extraction [80], pressurized solvent extraction [80], large-scale solvent extraction [42], and supercritical fluid extraction [94, 126]. While the theories along with the relative advantages and disadvantages behind each one of these procedures have not been discussed here, the listing will serve as a catalog of potential techniques available to the researcher. Specific details on any of these procedures can be obtained from any number of books on separation.

A debate still exists as to the timing of isolation and purification in the drug discovery process [65]. It is always tempting to isolate and purify before screening, but understandably this can present a challenge. Classical approaches have used a successful marriage between purification steps and bioassay activity assessment to isolate, identify, and fully characterize natural product compounds. It should be appreciated that with the advances in chromatographic and analytical techniques that have taken place over the last 15 years, the time required to proceed from an initial hit to an identified active compound should take no longer than for the resynthesis and purification of a potential active compound from a combinatorial library. Accordingly, the timing of isolation and purification in the natural product drug research and development timeline should not persist as such a point of contention as previously [9].

1.8 STRUCTURE IDENTIFICATION OF NATURAL PRODUCTS

The chemical structures of natural product compounds are tremendously diverse and can be very elegant in their nature [98, 115, 116, 170]. Such diversity can present a challenge to the analytical or medicinal chemist attempting to unravel the mystery of the chemical structure of an unknown material presented to him or her. However, modern technology has made structure identification simpler and faster. Today, scientists take for granted such techniques as MS, MS/MS, IR, Fourier transform infrared spectroscopy (FTIR), NMR, Fourier transform nuclear magnetic resonance spectroscopy (FTNMR), and others. It is beyond the scope of this chapter to discuss the theory and relative merits of each of these techniques, and the reader is urged to consult appropriate textbooks on analytical chemistry. However, it is worth mentioning that some particularly exciting developments in structure determination pertinent to the area of natural products have come from the field of computer-assisted structure elucidation (CASE). Several computer programs have become available for scientists to use and a number of publications reporting on the utility of this technological advancement have been forthcoming [154]. As has been previously described, the elucidation of the structure of a natural product begins with the collection of a crude material. This material is then subjected to a series of separation steps, usually involving chromatography, delivering in the end pure compound(s). Finally, a set of spectroscopic and spectrometric experiments are performed on the pure compound to delineate the structural characteristics. Such analysis may even reveal the two-dimensional or three-dimensional structure of the isolated chemical.

Time is money in the drug development business and in order to accelerate activities, the following steps of the structural elucidation process should be considered to be targets amenable to automation: (1) the choice of the smallest group of procedures that is most likely to reveal the unknown structure [NMR, FTNMR, MS, MS/MS, IR, FTIR, ultraviolet (UV), etc.], (2) the acquisition of data from the selected procedures, (3) the analysis of data from the selected procedures, and (4) the use of a computer program to construct the structure from collected spectroscopic and spectrometric data. Historically, structure identification has occurred after purification was complete. However, not infrequently in these cases, structural identification revealed that all of the previous laborious steps of purification had produced a compound that was already known or of an undesirable type. Now, the coupling of liquid chromatography (LC) or high-pressure liquid chromatography (HPLC) with such technologies as MS, MS/MS, and NMR has permitted the construction of devices that allow the injection of a crude sample, separation of the sample using automatically determined optimized conditions, and on the fly spectrometry or spectroscopy followed by CASE for each set of acquired spectra. The identification of unwanted compounds or de-replication should occur as early as possible in the natural product isolation and purification process to avoid the loss of time and funds. The development of coupled techniques has permitted the achievement of that goal. Indeed, coupled techniques such as LC/NMR/MS have now evolved and have potent application in the pharmaceutical field. As the sensitivity of instrumentation continues to improve, the value of these coupled techniques will increase even more.

A number of powerful aids for NMR spectroscopic interpretation have emerged and are now readily available from instrument manufacturers and include programs such as Auralia and AMIX [117]. However, any comprehensive CASE program will be based on a quality structure generator. To this point, only a few high-quality, pure structure generators have been developed. While over the years there have been a variety of programs that have been developed, they have been limited to those researched and developed by small, private groups. Now there are some highly capable programs that are commercially available, such as Assemble (Upstream Solutions GmbH, Zurich, Switzerland) and MOLGEN (http://www.molgen.de). A well-known deterministic CASE system that is often cited in the literature is CISOC-SES, which is now commercially available under the acronym NMR-SAMS (Spectrum Research, Madison, WI, USA). Another deterministic CASE program, COCON has been relatively recently introduced, and several examples demonstrating its value in structural elucidation have been reported [97, 154].

Deterministic algorithms have a limit to the size of the molecule with which they can work. To overcome this, a stochastic structure generator has been published for the computer-assisted structure elucidation of organic molecules [153]. The name of the program is SENECA. This program is written in Java and therefore is platform independent and allows for a simple plug-in mechanism for new spectroscopic data types. Theoretically, many different kinds of properties can be plugged into this system, as long as the property can be reliably calculated from a generated molecule.

Classical CASE systems can at least attempt to provide the two-dimensional structure of an unknown molecule. Now, with the greater exposure of, availability of, capabilities of, and demand for CASE programs, efforts are being made toward incorporation of the ability to confidently determine the three-dimensional structure of unknown molecules [68].

Advancements in chromatography, spectrometry, and spectroscopy together with breakthroughs in the coupling of these technologies are important steps in the production of a fully automated and integrated natural products structure determination instrument, which will provide significant advantage to the early, rapid, and facile identification of new natural-product-based drug opportunities.

1.9 SYNTHESIS OF NATURAL PRODUCTS

Once a natural product compound has been screened for biological activity, isolated, purified, its structure identified, and the pharmacological profile refined, the journey is not over. The molecule may turn out to be too complex in nature and too expensive to be synthesized. Indeed, when compared to a purely man-made synthetic alternative, many times the natural product compound is quickly eliminated from further consideration because of considerations of time and potential costs of synthetic production.

Any given natural product compound may possess unacceptable physicochemical, pharmacodynamic, pharmacokinetic, or bioavailability properties or demonstrate excessive toxicity and will therefore require optimization of its chemical structure. Optimization involves a dissection of the lead molecule and the synthetic addition, removal, replacement, or modification of substituent groups so as to enhance the utility and efficacy of the molecule. The synthesis of a complicated molecule is a very difficult task since every group and atom must be placed in a proper position and with the correct stereochemistry.

Such chemical structure modification or synthesis has been performed over decades by what might be termed more classical means. Indeed, the complete synthesis of natural products has been an area of interest for a long time, and the efforts to produce man-made natural products has provided significant challenge and learning opportunity over the years [21, 43, 53, 103, 115, 116, 125, 133, 151]. Because of the widespread chemical diversity that can be found in natural products, an ever-expanding collection of fascinating natural product compounds will continue to be presented to chemists for synthesis [162]. If one compares the chemical diversity of man-made synthetic products with the chemical diversity of natural product compounds, it quickly becomes apparent that there are significant qualitative differences between synthetics and natural products [114]. Natural product compounds contain more alcoholic and ether groups, while pure synthetic compounds possess more aromatics, amines, and amides. If one looks at group combinations, there are higher percentages of alcohol/ether, alcohol/ester, arene/alcohol, arene, alcohol, or ether functionalities in natural product compounds when compared to the synthetic compounds. However, pure man-made synthetic compounds are found to have combinations such as arene/amine, amine/amide, or amine/arene/amide in higher frequencies than natural product type of compounds. Finally, natural product compounds are found to more commonly possess bridgehead atoms and contain a greater number of rotatable bonds per molecule, chiral centers per molecule, and rings per molecule than pure man-made synthetic compounds. The importance of these differences is that they reveal and emphasize the complementarity of natural product compounds as a group with man-made synthetic compounds. Despite all of the knowledge and achievements that have been gained and the advances that have been made, classical synthetic organic chemistry will not alone unlock and open the potential of natural products to the pharmaceutical marketplace. Instead, the future lies in the synergistic union of classical organic chemistry with microbiology, biochemistry, combinatorial chemistry, and other fields to provide new synthetic strategies to generate natural-product-based drugs.

The history and specific techniques of combinatorial chemistry are beyond the scope of this chapter, and the reader is referred to appropriate textbooks for discussions on that topic. While it should be recognized that combinatorial chemistry is a perfect match for high-throughput screening because of its ability to produce large numbers of compounds in a short period of time. The promise of combinatorial chemistry to deliver more drug candidates within a shorter period of time has remained unfulfilled. What has been lacking in combinatorial chemistry is the skeletal structural novelty that natural products can provide [1, 12, 13, 59, 63, 111, 119, 120, 155, 158, 161, 163].

Over time, organic synthetic chemists have become interested in enzymes and their potential role in natural product synthesis [85, 109, 136]. Enzymes have great power as catalysts for regiospecific and stereocontrolled synthesis. These biological catalysts are very capable at room temperature of converting inexpensive substrates into value-added products at a significantly high throughput. However, barriers remain to the more expanded use of enzymes in organic synthetic chemistry. The most important of such obstacles includes the inability of enzymes to work with unnatural substrates.

Combinatorial biosynthesis utilizes enzymes from various natural product source biosynthetic pathways to create novel chemical structures [10, 27, 135].

The engineering of polyketide synthases has thus far been the central point of this activity and led to the production of several erythromycin analogs. The end result of such research activity will be the development of more rational and faster methods of production of new compounds for the development of therapeutic agents from natural products.

The research on and screening of natural products today is focusing on many different therapeutic indications. Fermentation broths and plant extracts have done well in delivering leads and genomics and molecular biology have done well in delivering targets. Regardless of the type of compounds involved, improved efficiencies in the design and synthesis of natural-product-based agonists and antagonists will be key to a realization of the full potential of natural products as drugs [1, 139, 140].

1.10 DEVELOPMENT OF NATURAL PRODUCTS

Regulatory Guidelines and Nonclinical Development

The classical model of drug development is composed of three phases: discovery, development, and marketing [108]. Discovery is the first of these phases and is composed of two essential components, drug discovery and drug design. Development is the second of these phases and is composed of two large components, preclinical studies and clinical studies [30, 56–58, 95, 108, 146]. The development phase, although lengthy, expensive, and time consuming is meant in its design and conduct to cull out undesirable compounds and ensure the safety and efficacy of compounds that successfully run the developmental gauntlet and decrease the hazard and risk of exposure for humans. We have touched on the drug discovery and design in earlier sections of this chapter. In the course of these phases, therapeutic targets have been selected and screens were designed to identify drug candidates. These same candidates were then isolated, purified, identified, and then structurally optimized in consideration of the molecule's biological activity profile. This process of optimization attempts to minimize toxicity and maximize therapeutic value, efficacy, and pharmacokinetic characteristics. Finally, research was performed to elucidate a mechanism of action for the potential drug and animal models identified to establish efficacy of the material. Pharmacology studies, a component of the discovery phase, confirm the efficacy of a potential drug. The availability of analytical methods is essential to establish the exposure of the test system to the drug candidate and the presence of the drug in various tissues and fluids. The absorption, distribution, metabolism, and excretion and elimination of the compound also need to be determined. Human in vitro P-450 studies are necessary to be performed to fully understand the hepatic metabolic pathways. These studies may also herald the potential for serious drug interactions that may have to be addressed during the development cycle.

Safety pharmacology studies are part of the development phase and the preclinical development program but can be performed either as part of the pharmacologic profiling of a drug candidate or as a prelude to the toxicology studies. These types of studies are a preliminary hazard assessment of the test article in key organ systems such as the central nervous system, cardiovascular system, gastrointestinal system, renal system, and pulmonary system and essential in the planning of human clinical trials. For an excellent and detailed discussion of safety pharmacology, the reader is referred to Gad [54].

Preclinical studies begin some of the more rigorous testing that a potential candidate must successfully endure and survive. It needs to be recognized that a decision to proceed with preclinical studies represents a major commitment of time, resources, effort, and money. The major objectives of any preclinical development program should be: (1) development of a Good Manufacturing Practices (GMP) synthetic method to produce the test article, (2) synthesis of a supply of test article that is adequate to permit the performance of all preclinical work and the initiation of clinical studies, (3) the creation of a usable and tolerable preclinical formulation(s) and efficacious clinical formulation(s), (4) complete pharmacologic and pharmacokinetic profiling of the test article, (5) performance of proper toxicology studies to support an Investigational New Drug (IND) application, and (6) construction of a complete and detailed informational platform to permit the recommendation of an initial human dose in phase I studies.

Nonclinical or preclinical studies are the same thing and are composed of studies on drug processes, pharmacology, and toxicology. Drug processes involve the scaling up of synthetic procedures or the bulk preparation of the drug [32, 146]. This means the development of a synthetic method that permits the synthesis of compound at a reasonable cost, in a reasonable amount of time in such a fashion that regulatory and GMP guidelines are satisfied. A major obstacle in the performance of preclinical studies is often the ability to merely "have drug." Drug supply is often overlooked in the drug development program and can cause costly delays. Related to this is the existence of some sort of formulation for the drug candidate that will permit effective dosing in preclinical studies. The lack of proper planning on a preclinical formulation (not the final clinical formulation) is oft overlooked, and this oversight can cause significant problems of delay or confounding toxicity. Also during the preclinical phase of the drug development paradigm, various analytical techniques must be developed and validated in a variety of species and biological milieu, analytical standards defined and made, compound stability profiles generated, and research into the proper human formulation initiated and conducted. Sometimes even radiolabeled drugs will be required to be made for some studies, a difficult, costly, and time-consuming project in and of itself.

Toxicology studies are performed to assess a drug candidate's potential safety. These studies are so critical that there are established procedures called Good Laboratory Practices (GLP), which specify how studies are to be done.

Toxicology studies have a profound quality assurance component, which facilitate inspection and permit validation of the data by a regulatory agency or reviewer. There are numerous different types of toxicology studies and the details of them are beyond the scope of this discussion, but the most comprehensive, detailed, and descriptive source is by Gad [56]. These studies typically include mutagenicity, genotoxicity, and cytotoxicity studies; acute studies; subchronic studies; chronic studies; reproductive and developmental toxicity studies; carcinogenicity studies; immunotoxicity assessments; various worker safety studies; and certain specialized toxicity studies, the conduct of which is determined by compound-specific issues or dose route. Toxicology studies should in their core design determine the target organ(s) of any potential toxicity, characterize the shape of the dose–response curve, ascertain the reversibility of any potential lesion, and facilitate the selection of a human dose in phase I clinical trials. An excellent example of preclinical and clinical development programs for a natural-product-based drug has been reported by Rowinsky et al. on the antineoplastic agent Taxol [141].

When a drug candidate completes the preclinical studies, an IND is filed and clinical trials are initiated. Clinical trials are composed of phases I to IV. Phase I studies represent the first exposure of a drug to humans. These studies examine the effects of single and multiple increasing doses in small numbers of normal and/or patient volunteers. Tolerance to the compound is also evaluated. Basic pharmacokinetic studies are performed in conjunction with the studies to aid in the determination of later dosage regimens and to fully characterize the routes of metabolism, excretion, and elimination and assess the presence and amounts of active and inactive or toxic metabolites. Typically only a small number of volunteers are involved. Phase II studies represent the first time that a drug candidate is tested in humans for efficacy. However, safety and tolerance are still monitored in these studies concurrently. In phase II studies, different dose–range finding studies are performed to optimize the dose of the drug, to maximize its efficacy, and to minimize any compound-associated intolerance. These studies typically involve up to several hundred patients and generally provide the first indications of the potential benefit of the drug as compared to its risk of exposure. The last clinical studies that need to be performed before submitting a complete information package for regulatory approval are phase III studies [New Drug Application (NDA) for filing in the United States]. In these studies, the drug candidate is utilized at the optimum dose, in the target patient population, and in exactly in the same fashion that the drug will be used if eventually approved and marketed. These studies will verify efficacy and safety and detect adverse reactions and contraindications. These are very large studies composed of hundreds to thousands of volunteer patients, depending upon the therapeutic indication. Phase IV studies are postmarketing surveillance studies, which are conducted after a drug has been approved for sale. These studies typically involve adverse reaction reporting, surveys and general sampling, and testing evaluations.

The design of a proper and detailed preclinical program is a challenging exercise, and the successful navigation of the program with its on-time completion should not be taken for granted. Effective project management with a timely resolution to all obstacles that will present themselves along the way can save significant amounts of money. Accordingly, consideration should always be given to the retention of a professional and suitably experienced preclinical project manager. The expenses incurred with the design and implementation of a preclinical development program is more than offset by the avoidance of cost and/or time overruns.

As stated earlier in this chapter, natural products can be interpreted by different people to mean different things. For purposes of this discussion, the term *natural products* refers to drugs of natural product origin and not to botanicals, herbals, or traditional medicine products. The development of botanicals, herbals, dietary supplements, and over-the-counter products is not relevant to this chapter and is covered elsewhere [152]. There are no regulations or guidelines that are specific to the development of therapeutic agents developed from natural products. Indeed, the development of a natural-product-based therapeutic candidate is more than adequately covered in existing governmental guidelines. These guidelines are available from a variety of governmental sources [*www.fda.gov* (U.S. FDA), *www.eudra.org* (European Union), *www.mhw.go.jp/english* (Japan), *www.ifpma.org* (ICH)] or references on preclinical development [51, 55–58, 105, 106]. The most comprehensive and detailed of these references is by Gad [56]. However, it should be noted that according to a draft document entitled "Botanical Drug Products Docket Number: OOD-1932" issued by the Office of Nutritional Products, Labeling and Dietary Supplements (HFS-800) in 2003, if a natural product has been sold in a crude or semipurified form as an over-the-counter drug and a sponsor wishes to file an IND on a moiety purified from such a preparation, then reduced documentation may be acceptable to the FDA.

Learning from the Mistakes of the Past in the Development of Natural Products

The very first sample of bark from the Pacific yew (*Taxus brevifolia*) was originally collected in 1962 by the U.S. Department of Agriculture as part of a plant screening program established by the Cancer Chemotherapy National Service Center of the National Cancer Institute (NCI) [32]. It was not until 1964 that a positive response was found for the extract in the KB cytotoxicity assay. Taxol was then identified as the active component of the mixture in 1969. Not unlike many other compounds being tested at that time, paclitaxel originally demonstrated moderate in vivo activity against P388 and L1210 murine leukemia models and was not considered to be a promising candidate for further development. However, strong activity demonstrated by Taxol against the B16 melanoma line introduced by the NCI in 1975 caused a reevaluation of the material. Indeed as a direct result of this additional research, Taxol was recommended as a candidate for preclinical development in 1977. Further work demonstrated strong activity against human tumor xenograft systems and stimulated hope of efficacious performance in the clinic. Taxol's mechanism of action was elucidated in 1979, formulation work completed, and toxicology studies started in 1980 [32, 100]. With the completion of nonclinical studies, approval was given for entry into phase I clinical trials in 1983 [28, 141]. The early clinical trials raised serious issues of toxicity. Indeed, further development of Taxol was almost discontinued. The problems, however, were determined to be related to the poor solubility of Taxol in aqueous systems and the necessity for a high dose, when compared to other antineoplastic agents of the time. The development of a suitable formulation required the use of Cremophor EL, a polyethoxylated castor oil derivative, which created a whole new set of challenges to be addressed [36]. Eventually, a safe clinical regimen was established reducing the incidence of allergic reactions, and Taxol proceeded into phase II clinical trials. However, at this point in development, faith and interest in the success of Taxol began to wane for a variety of reasons, and the priority of Taxol's development was again lowered. Indeed, the support for Taxol decreased to such a level that the production of bulk quantities of Taxol needed to proceed through clinical trials was severely cut back. This position forced the creation of a critical problem when Taxol demonstrated significant antitumor activity in phase II clinical trials in patients afflicted with ovarian cancer. The preliminary reports from clinical trials indicated an approximate 30 percent response rate, with some patients actually achieving remission [110]. Additional clinical trials supported the initial clinical response rate, and accordingly there was a strong demand for more Taxol. Concurrently, it was also discovered that Taxol demonstrated very favorable responses in patients afflicted with metastatic breast cancer [71]. Add to this the findings that favorable clinical responses to the administration of Taxol were also being found in patients afflicted with lung cancer, malignant melanoma, lymphomas, and cancers of the head and neck, and it is easy to appreciate the magnitude of the gap created between the supply of Taxol and the demand for Taxol [32]. Unfortunately, all of this favorable clinical performance was met with not just a poor supply of Taxol but an extremely disappointing inability to produce Taxol on a large scale. The crisis in the supply of paclitaxel was eventually resolved as a result of collaborative efforts between many groups [28, 32]. Such a crash collaborative effort, while admirable, came about as a result of poor planning.

Today, Taxol is viewed as a very important antineoplastic agent on the front line of cancer chemotherapy. Yet, despite its success, it must be sadly admitted that on several occasions during paclitaxel's development, Taxol almost did not make it to market. In an attempt to learn from the mistakes of the past, what lessons can be learned from the Taxol story? These are addressed in the next section when we talk of the future of natural products in drug discovery and development.

1.11 FUTURE OF NATURAL PRODUCTS

Current libraries providing fodder for the search for new therapeutic agents include, for the most part, but are not limited to, synthetic chemical libraries of purified natural products from a variety of different sources [27]. When combinatorial chemistry made its entry onto the drug discovery stage, many people considered this technique to be the ultimate solution for the discovery of potential new therapeutic agents. Accordingly, the perceived value of natural products in the drug discovery process began to pale. While combinatorial chemistry has demonstrated that it is certainly of value in the process of lead optimization, nature itself presents the most diverse and complete source of leads. A relatively recent approach that unites the strengths of combinatorial chemistry and natural product identification is a process referred to as combinatorial biosynthesis [10, 27, 135]. In this technique, the pathway leading to the production of a natural product is identified and the genetic basis of it elucidated. Genetic modifications of the organism are then made causing the production of different biologically active products. These products can then be evaluated for target therapeutic potency in various screens. While this approach has been utilized by some biotechnology companies, widespread acceptance in the pharmaceutical industry has been lacking to this point [27]. However, as acceptance develops, so will the identification of many new potential drug candidates.

To fully capitalize on the extensive biodiversity available to us in natural products, high-throughput screening processes need to be improved upon so that they can provide a higher degree of quantifiable rather than quantal (active versus not active) results, permitting a more adequate description of whether or not any given compound may be considered to be potentially active.

It is also important to realize that one should not rely on just one either in vitro or in vivo screening assay in a discovery program. Obviously, there has to be an economic component to screening programs with some degree of cost containment, but a realistic balance can be achieved, which provides a realistic cost-to-benefit ratio. The discovery and development of paclitaxel is a good example of this concern [32].

The early conduct of studies leading to the elucidation of the mechanism of action of a compound is important. The delineation of a novel mechanism of action can provide considerable impetus to the continued development of a chemical moiety. Irrespective of novelty, the establishment and correlation of a chemical structure with a mechanism of action can permit the categorization of a compound with other chemicals, potential therapeutic agents, and drugs and more rapidly facilitate a compound's potential for therapeutic use.

The efforts to identify potential leads for the treatment of a constellation of maladies and diseases are increasing at a hectic pace and the growth of numbers of actual compound candidates is markedly outpacing the availability of screening procedures. In order to accommodate the requisite testing associated with the generation of chemical leads, advances will need to be made in the methods of and approaches to the synthesis of natural products. Critical demands will be placed upon synthetic and medicinal chemists to produce pure chiral products in significant amounts. These products will undoubtedly be the result of complex synthetic pathways, which will incorporate catalysts and biocatalysts, soluble and immobilized elements, and biologically generated as well as man-made moieties. These multistep processes will need to be functional on both a low bench scale as well as a commercial scale. The development of paclitaxel suffered a severe developmental program setback because of an inability to produce adequate supplies at an appropriate time. In association with the development of effective synthetic pathways leading to the production of natural products, attention must also be paid to the required formulation of such products permitting human use, as was well demonstrated with Taxol.

Genetic research will continue to elucidate an ever-increasing number of potential pathophysiologic targets, which can be mated with hopeful novel therapeutic agents. This coupled with the fact that little of nature's natural products archives have really been adequately researched portends great opportunity. How big is this opportunity? Admittedly, there are no sound estimates for the actual total numbers of species of plants, shrubs, trees, insects, fungi, smuts, rusts, marine organisms, and the like that exist on land and in the waters in this world [24]. But, with regard to microorganisms, less than 0.1 percent of the total microorganisms present in the soil to date have been evaluated, simply because they cannot be cultured [27, 138]. Only about 70,000 of 2 to 5 million different fungi are considered to have been identified [27]. Similarly, only about 800,000 of approximately 20 million different insects have been identified.

Structure–activity studies on leads generated from natural products sources combined with computerized graphic model building will become increasingly more prevalent. Such activity in turn will result in the discovery of molecules with optimal activity, improved bioavailability, fewer side effects, and very desirable therapeutic indices [156].

However, abuse of our natural resources will certainly limit our ability to learn from nature [27, 78, 132]. Lack of proper conservation, the expanding pall of pollution, and the wanton destruction of forests, especially the rain forests, will, for example, diminish the opportunity to gain valuable knowledge from a staggering amount of indigenous biodiversity and obviate the possibility to generate a valuable therapeutic agent. Indeed, it is estimated that approximately one-eighth of all plant species are currently near the point of extinction [78]. Alternatively, the highly focused and single-minded research into natural products can in and of itself lead to the depletion of sources of natural products and even the eventual extinction of species [22].

In the hunt for natural products another obstacle that will have to be more commonly addressed is the issue of intellectual property rights [3, 11, 27, 32,

107, 160]. As various biotechnology and pharmaceutical companies attempt to expand their libraries with compounds generated from various biodiversity sources, many countries are beginning to impose significant demands on the use of such raw materials and are limiting access without the expression of adequate consideration to the "landowners" [124]. Of course, ultimately there is the potential for significant economic reward for involved countries, but the various forms of life, which historically have been considered to be free for the taking, are no longer free. The potential conflict between host countries and prospecting organizations will undoubtedly be resolved with the establishment of contingent compensation plans making use of rights and royalties coupled with the attainment of specific milestones. The building of the requisite trust to establish meaningful, lasting, and functional business relationships will require time, patience, education, and respect for the different cultures involved [140]. History will color all negotiations and ethnic and cultural differences will serve to make the process painfully slow.

The search for and the development of natural products will result in the creation of a variety of different types of alliances between industry, government, individuals, universities, and hopefully even foster a spirit of international cooperation [25, 26, 32, 86]. Over the last few years, those involved in the pharmaceutical industry have already seen the increasing numbers of partnerships, alliances, agreements, and other types of relationships forged between large pharmaceutical companies and smaller biotechnology companies, all for the purpose of seizing opportunities of new technology. The collaborations that currently exist in the field of chemotherapy can set an example for other therapeutic areas as well as drug treatment in general. Another pertinent example has been the formation of the International Cooperative Biodiversity Group (ICBG) Program [32]. The ICBG is a collection of academic, industrial, U.S. governmental organizations, and developing countries. This program is also jointly sponsored by the National Science Foundation (NSF), parts of the National Institutes of Health (NIH), the National Cancer Institute (NCI), the Fogarty International Center, the National Institute of Allergy and Infectious Diseases (NIAID), the National Heart, Lung, and Blood Institute (NHLBI), and the National Institute of Mental Health (NIMH). The purpose of this program is to facilitate research and drug discovery from natural sources. This program is of extreme value and promotes the identification, establishment of an inventory, conservation of natural resources, and economic development of financially challenged countries.

Thought should be given to the more expanded establishment of repository programs. The NCI has established a Natural Products Repository (NPR), which is an extremely valuable resource in the search for and discovery of new drugs. The opportunities are there, they just need to be found.

In conclusion, drug discovery can be significantly improved through the use of the knowledge to be gained from research into natural products. However, despite the powerful resource that natural products present to us, the knowledge that is locked therein can only be fully realized with proper management of the resources, the parallel development of ancillary technologies, and the fostering of open and shared communication.

REFERENCES

1. Abel, U., Koch, C., Speitling, M., Hansske, F. G. (2002). Modern methods to produce natural-product libraries. *Curr. Opin. Chem. Biol.*, *6*(4), 453–458.
2. Alvi, K. A. (2000). A strategy for rapid identification of novel therapeutic leads from natural products. In S. Cutler and H. Cutler (Eds.), *Biologically Active Natural Products*. CRC, Boca Raton, FL, pp. 185–195.
3. Artuso, A. (1997). Natural product research and the emerging market for biochemical resources. *J. Res. Pharmaceut. Econ.*, *8*(2), 3–23.
4. Banerji, A. (2000). Resurgence of natural product research—A Phoenix act, *PINSA-A. Proc. Ind. Natl. Sci. Acad., Part A: Phys. Sci.*, *66*(3/4), 383–392.
5. Barrett, D. (2002). From natural products to clinically useful antifungals. *Biochim. Biophys. Acta, 1587*, 224–233.
6. Barron, R. L., Vanscoy, G. J. (1993). Natural products and the athlete: Facts and folklore. *Ann. Pharmacother.*, *27*(5), 607–615.
7. Bauer, L. A. (1997). Individualization of drug therapy: Clinical pharmacokinetics and pharmacodynamics, basic concepts of pathophysiology and pharmacotherapy. In J. T. DiPiro, G. R. Matzke, L. M. Posey, R. L. Talbert, B. G. Wells, and G. C. Yee, (Eds.) (L. M. Posey, Sect. Ed.), *Pharmacotherapy, A Pathophysiologic Approach*, 3rd ed. Appleton & Lange, Stamford, CT, pp. 29–48.
8. Bhattaram, V. A., Graefe, U., Kohlert, C., Veit, M., Derendorf, H. (2002). Pharmacokinetics and bioavailability of herbal medicinal products. *Phytomedicine*, *9*(Suppl. 3), 1–33.
9. Bindseil, K. U., Jakupovic, J., Wolf, D., Lavayre, J., Leboul, J., van der Pyl, D. (2001). Pure compound libraries; a new perspective for natural product based drug discovery. *Drug Discov. Today*, *6*(16), 840–847.
10. Borman, S. (1998). Biosynthesis combinatorially. *Chem. Eng. News*, September 14, pp. 29–30.
11. Boyd, M. R. (1996). The position of intellectual property rights in drug discovery and development from natural products. *J. Ethnopharmacol.*, *51*(1–3), 17–25; discussion, 25–27.
12. Breinbauer, R., Vetter, I. R., Waldmann, H. (2002). From protein domains to drug candidates—Natural products as guiding principles in the design and synthesis of compound libraries. *Ang. Chem. Int. Ed.*, *41*(16), 2878–2890.
13. Breinbeuer, R., Manger, M., Scheck, M., Waldmann, H. (2002). Natural product guided compound library development. *Curr. Med. Chem.*, *9*, 2129–2145.
14. Burja, A. M., Banaigs, B., Abou-Mansour, E., Burgess, J. G., Wright, P. C. (2001). Marine cyanobacteria—A prolific source of natural products. *Tetrahedron, 57*, 9347–9377.
15. Cannell, R. J. P. (1998). Follow-up of natural product isolation. *Methods Biotechnol.*, *4*(Natural Products Isolation), 425–463.

16. Cannell, R. J. P. (1998). How to approach the isolation of a natural product. In R. J. P. Cannell (Ed.), *Methods in Biotechnology*, Vol. 4: *Natural Products Isolation*. Humana, Totowa, NJ, pp. 1–51.
17. Chan, K. (1995). Progress in traditional Chinese medicine. *Trends Pharmacol. Sci.*, *16*(6), 182–187.
18. Chung, I., Kim, Y., Ahn, J., Lee, H., Chen, G., Manji, H. K., Potter, W. Z., Pickar, D. (1995). Pharmacologic profile of natural products used to treat psychotic illnesses. *Psychopharmacol. Bull.*, *31*(1), 139–145.
19. Claeson, P., Bohlin, L. (1997). Some aspects of bioassay methods in natural-product research aimed at drug lead discovery. *Trends Biotechnol.*, *15/17*, 245–248.
20. Clark, A. M. (1996). Natural products as a resource for new drugs. *Pharmaceut. Res.*, *13/8*, 1133–1141.
21. Compostella, F., Lay, L. (2001). Critical surveys covering the year 2000: Total syntheses of natural products. In A. Corbella (Ed.), *Seminars in Organic Synthesis, Summer School, 26th*, Gargano, Italy, June 18–22, 2001. Societa Chimica Italiana, Rome, Italy, pp. 453–581.
22. Concannon, J. A., DeMeo, T. E. (1997). Goldenseal: Facing a hidden crisis. *Endangered Species Bull.*, *22*(6), 10–12.
23. Connolly, J. D. (1997). Natural products from around the world. *Revista Latinoamer. Quim.*, *25*(2), 77–85.
24. Cordell, G. A. (1998). *Nat. Prod. Updates*, *6*(1), 2.
25. Cordell, G. A. (1995). Natural products as medicinal and biological agents: Potentiating the resources of the rain forest. In P. F. Seidel, O. R. Gottlieb, and M. A. Colho Kaplan (Eds.), *Chemistry of the Amazon*. American Chemical Society Symposium Series, No. 588, Washington, DC, pp. 8–18.
26. Cordell, G. A. (1993). Pharmacognosy—New roots for an old science. In F. Z. Atta-ur-Rahman and Basha (Eds.), *Studies in Natural Products Chemistry*, Vol. 13: *Bioactive Natural Products (Part A)*. Elsevier Science, Amsterdam, pp. 629–700.
27. Cordell, G. A. (2002). Recent developments in the study of biologically active natural products. *ACGC Chem. Commun.*, *14*, 31–63.
28. Cragg, G. M., Schepartz S. A., Suffness M., Grever M. R. (1993). The taxol supply crisis. New NCI policies for handling the large-scale production of novel natural product anticancer and anti-HIV agents. *J. Nat. Prod.*, *56*, 1657–1668.
29. Cragg, G. M., Newman, D. J. (2001). Natural product drug discovery in the next millennium. *Pharmaceut. Biol.*, *39*(Suppl.), 8–17.
30. Cragg, G. M., Newman, D. J. (2001). Natural products drug discovery and development at the United Sates National Cancer Institute. In L. Yuan (Ed.), *Drug Discovery and Traditional Chinese Medicine: Science, Regulation, and Globalization*, [International Conference on Traditional Chinese Medicine: Science, Regulation and Globalization], 1st, College Park, MD, August 30–September 2, 2000 (2001) (meeting Date 2000). Kluwer Academic, Hingham, MA, pp. 19–32.
31. Cragg, G. M., Newman, D. J., Snader, K. M. (1997). Natural products in drug discovery and development. *J. Nat. Prod.*, *60*(1), 52–60.
32. Cragg, G. M. (1998). Paclitaxel (Taxol®): A success story with valuable lessons for natural product drug discovery and development. *Med. Res. Rev.*, *5*, 315–331.

33. Crews, D., Willingham, E., Skipper, J. K. (2000). Endocrine disruptors: Present issues, future directions. *Q. Rev. Biol.*, *75*(3), 243–260.
34. Da Rocha, A. B., Lopes, R. M., Schwartsmann, G. (2001). Natural products in anticancer therapy. *Curr. Opin. Pharmacol.*, *1*(4), 364–369.
35. Dahanukar, S. A., Kulkarni, R. A., Rege, N. N. (2000). Pharmacology of medicinal plants and natural products. *Ind. J. Pharmacol.*, *32*, S81–S118.
36. Davignon, J. P., Cradock J. C. (1987). In S. K. Carter and K. Hellman (Eds.), *Principles of Chemotherapy*. McGraw-Hill, New York, pp. 212–250.
37. Demain, A. L., Elander, R. P. (1999). The beta-lactam antibiotics: Past, present, and future. *Antonie Van Leeuwenhoek*, *75*(1/2), 5–19.
38. Demain, A. L. (1998). Microbial natural products: Alive and well in 1998. *Nat. Biotechnol.*, *16*(1), 3–4.
39. DerMarderosian, A., Beutler, J. A. (Eds.), *The Review of Natural Products*, 3rd ed. Facts and Comparisons, St. Louis, MO, pp. 824–826.
40. Duke, S. O., Dayan, F. E., Romagni, J. G., Rimando, A. M. (2000). Natural products as sources of herbicides: Current status and future trends. *Weed Rese.*, *40*(1), 99–111.
41. El Sayed, K. A. (2000). Natural products as antiviral agents. *Stud. Nat. Prod. Chem.*, *24* [Bioactive Natural Products (Part E)], 473–572.
42. Ellis, N. (2001). Extracting APIs from natural products. *Speciality Chem. Mag.*, *21*(6), 10–12.
43. Fallis, A. G. (1999). 1998 Alfred Bader Award Lecture. Tangents and targets: The synthetic highway from natural products to medicine. *Can. J. Chem.*, *77*(2), 159–177.
44. Farnsworth, N. R., Akerele, O., Bingel, A. S., Soejarto, D. D., Guo, Z. (1985). Medicinal plants in therapy. *Bull WHO*, *63*, 965–981.
45. Faulkner, D. J. (1998). Marine natural products. *Nat. Prod. Repts.*, *15*(2), 113–158.
46. Faulkner, D. J. (2000). Highlights of marine natural products chemistry (1972–1999). *Nat. Prod. Repts.*, *17*(1), 1–6.
47. Faulkner, D. J. (2000). Marine natural products. *Nat. Prod. Repts.*, *17*(1), 7–55.
48. Faulkner, D. J. (2002). Marine natural products. *Nat. Prod. Repts.*, *19*(1), 1–48.
49. Foucault, A. P., Chevolot, L. (1998). Counter-current chromatography: Instrumentation, solvent selection and some recent applications to natural product purification. *J. Chromatogr. A*, *808*(1–2), 3–22.
50. Foucault, A. P. (1997). Recent advances in purification of natural products by countercurrent chromatography. In P. Schreier (Ed.), *Natural Products Analysis: Chromatography, Spectroscopy, Biological Testing*, Symposium, Wuerzburg, Germany, September 1997. Vieweg, Wiesbaden, Germany, pp. 13–25.
51. Fuchs, R. L., Thomas, J. A. (Eds.) (2002). *Biotechnology and Safety Assessment*, 3rd ed. Academic Press, San Diego, CA.
52. Fugh-Berman, A., Cott, J. (1999). Dietary supplements and natural products as psychotherapeutic agents. *Psychosomatic Med.*, *61*(5), 712–728.
53. Furstner, A. (1999). Venturing into catalysis based natural product synthesis. *Synlett*, *10*, 1523–1533.

54. Gad, S. C. (2003). Principles of screening and study design. In *Safety Pharmacology in Pharmaceutical Development and Approval*. CRC, Boca Raton, FL, pp. 23–42.
55. Gad, S. C., Chengelis, C. P. (1998). *Acute Toxicology Testing*, 2nd ed. Academic, San Diego, CA.
56. Gad, S. C. (2002). *Drug Safety Evaluation*. Wiley-Interscience, New York.
57. Gad, S. C. (2001). *Regulatory Toxicology*, 2nd ed. Taylor & Francis.
58. Gad, S. C. (1995). *Safety Assessment for Pharmaceuticals*. Van Nostrand Reinhold.
59. Ganesan, A. (2001). Integrating natural product synthesis and combinatorial chemistry. *Pure Appl. Chem.*, 73(7), 1033–1039.
60. Grabley, S., Sattler, I. (2003). Natural products for lead identification: Nature is a valuable resource for providing tools. In A. Hillisch and R. Hingenfeld (Eds.), *Modern Methods of Drug Discovery*. Birkhäuser Verlag, Switzerland, pp. 87–107.
61. Grabley, S., Thiericke, R., Sattler, I. (2000). Tools for drug discovery: Natural product-based libraries. *Ernst Schering Res. Found. Workshop*, 32, 217–252.
62. Haefner, B. (2003). Drugs from the deep: Marine natural products as drug candidates. *Drug Discov. Today*, 8(12), 536.
63. Hall, D. G., Manku, S., Wang, F. (2000). Solution- and solid-phase strategies for the design, synthesis, and screening of libraries based on natural product templates: A comprehensive survey. *J. Combinatorial Chem.*, 3(2), 125–150.
64. Harvey, A. (2000). Strategies for discovering drugs from previously unexplored natural products. *Drug Discov. Today*, 5(7), 294–300.
65. Harvey, A. (2001). The continuing value of natural products to drug discovery. *GIT Lab. J.*, 5(6), 284–285.
66. Harvey, A. L. (1999). Medicines from nature: Are natural products still relevant to drug discovery? *Trends Pharmacol. Sci.*, 20(5), 196–198.
67. Havsteen, B. (1983). Flavonoids, a class of natural products of high pharmacological potency. *Biochem. Pharmacol.*, 32(7), 1141–1148.
68. Hemmer, M. C., Gasteiger, J. (2000). Prediction of three-dimensional molecular structures using information from infrared spectra. *Anal. Chim. Acta*, 420, 145–154.
69. Hill, D. C. (1998). Advanced screening technology and informatics for natural products drug discovery. In A. L. Harvey (Ed.), *Advances in Drug Discovery Techniques*. Wiley, Chichester, pp. 25–37.
70. Hilton, M. D. (2003). Natural products: Discovery and screening. In V. A. Vinci and S. R. Parekh (Eds.), *Handbook of Industrial Cell Cultures*. Humana, Totowa, NJ, pp. 107–136.
71. Holmes, F. A., Walters, R. S., Thierault, R. L., Forman, A. D., Newton, L. K., Raber, M. N., Buzdar, A. U., Frye, D. K., Hortobagyi, G. N. (1991). Phase II trial of taxol, an active drug in the treatment of metastatic breast cancer. *J. Natl. Cancer Inst.*, 83, 1797–1805.
72. Holt, G. A., Chandra, A. (2002). Herbs in the modern healthcare environment—An overview of uses, legalities, and the role of the healthcare professional. *Clin. Res. Regulatory Affairs (USA)*, 19, 83–107.

73. Horton, T. (1972). Large-scale gel filtration for purification of natural products. *Am. Lab.*, *4*(5), 83–84, 86–88, 90–91.
74. Hostettmann, K., Hamburger, M., Hostettmann, M., Marston, A. (1991). New developments in the separation of natural products. *Recent Adv. Phytochem.*, *25* (Modern Phytochemical Methods), 1–32.
75. Hostettmann, K., Marston, A. (1990). Liquid-liquid partition chromatography in natural product isolation. *Anal. Chim. Acta*, *236*(1), 63–76.
76. Jarvis, B. B. (2000). The role of natural products in evolution. *Recent Adv. Phytochem.*, *34* (Evolution of Metabolic Pathways), 1–24.
77. Jia, W., Liu, V. J. K., Tang, L. (2002). Chinese herbal drugs for diabetes. *Frontiers Biotechnol. Pharmaceut.*, *3*, 337–358.
78. Johnston, B. A. (1998). Major diversity loss: 1 in 8 plants in global study threatened. *HerbalGram*, *43*, 54.
79. Jung, M., Lee, S., Kim, H., Kim, H. (2000). Recent studies on natural products as anti-HIV agents. *Curr. Med. Chem.*, *7*(6), 649–661.
80. Kaufmann, B., Christen, P. (2002). Recent extraction techniques for natural products: Microwave-assisted extraction and pressurized solvent extraction. *Phytochem. Anal.*, *13*(2), 105–113.
81. Kaul, P. N., Joshi, B. S. (2001). Alternative medicine: Herbal drugs and their critical appraisal—Part II. *Prog. Drug Res.*, *57*, 1–75.
82. Keim, C., Ladisch, M. R. (2000). Bioseparation of natural products. *Prog. Biotechnol.*, *16* (Bioseparation Engineering), 15–20.
83. Kelecom, A. (1999). Chemistry of marine natural products: Yesterday, today and tomorrow. *Anais da Academia Brasileira de Ciencias*, *71*(2), 249–263.
84. Kerr, R. G., Kerr, S. S. (1999). Marine natural products as therapeutic agents. *Expert Opin. Therapeutic Patents*, *9*(9), 1207–1222.
85. Khosla, C. (2000). Natural product biosynthesis: A new interface between enzymology and medicine. *J. Org. Chem.*, *65*(24), 8127–8133.
86. Koch, C., Neumann T., Thiericke R., Grabley S. (1999). A central natural product pool—New approach in drug discovery strategies. In S. Grabley (Ed.), *Drug Discovery from Nature*. Hans-Knoll-Institut fur Naturstoff-Forschung Jena, Germany, pp. 51–55.
87. Koeller, J., Yee., G. C. (1997). Molecular biology and biotechnology drugs, basic concepts of pathophysiology and pharmacotherapy. In J. T. DiPiro, G. R. Matzke, L. M. Posey, R. L. Talbert, B. G. Wells, and G. C. Yee (Eds.), (L. M. Posey, Section Editor), *Pharmacotherapy, A Pathophysiologic Approach*, 3rd ed. Appleton & Lange, Stamford, CT, pp. 117–128.
88. Koh, H., Woo, S. (2000). Chinese proprietary medicine in Singapore: Regulatory control of toxic heavy metals and undeclared drugs. *Drug Safety*, *23*(5), 351–362.
89. Konig, G. M., Wright, A. D. (1996). Marine natural products research: Current directions and future potential. *Planta Med.*, *62*(3), 193–211.
90. Kroll, D. J. (2001). Concerns and needs for research in herbal supplement pharmacotherapy and safety. *J. Herbal Pharmacother.*, *1*(2), 3–23.
91. Lake, J. (2000). Natural product-derived treatments of neuropsychiatric disorders: Review of progress and recommendations. *Stud. Nat. Prod. Chem.*, *24*, No. Bioactive Natural Products (Part E), 1093–1137.

92. Lake, J. (2000). Psychotropic medications from natural products: A review of promising research and recommendations. *Alternative Ther. Health Med.*, *6*(3), 36, 39–45, 47–52.
93. Lang, Q., Wai, C. M. (2001). Supercritical fluid extraction in herbal and natural product studies—A practical review. *Talanta*, *53*(4), 771–782.
94. Lang, F., Keller, K., Ihrig, M., Oudtshoorn-Eckard, J., Moller, H., Srinivasan, S., Yu, H. (2001). Biopharmaceutical characterisation of herbal medicinal products. *Pharmazeutische Industrie*, *63*(10), 1005–1010.
95. Lee, K. H. (1999). Anticancer drug design based on plant-derived natural products. *J. Biomed. Sci.*, *6*(4), 236–250.
96. Liberra, K., Lindequist, U. (1995). Marine fungi—A prolific resource of biologically active natural products? *Pharmazie*, *50*(9), 583–588.
97. Lindel, T., Junker, J., Köck, M. (1997). COCON: From NMR correlation data to molecular constitutions. *J. Mol. Model*, *3*, 364–368.
98. Mabry, T. J. (2001). Selected topics from forty years of natural products research: Betalains to flavonoids, antiviral proteins, and neurotoxic nonprotein amino acids. *J. Nat. Products*, *64*(12), 1596–1604.
99. Majetich, G. Recent trends in the use of natural products and their derivatives as potential pharmaceutical agents. In S. J. Cutler and H. G. Cutler (Eds.), *Biologically Active Natural Products*. CRC, Boca Raton, FL, pp. 61–71.
100. Manfredi, J. J., Horwitz, S. B. (1984). Taxol: An antimitotic agent with a new mechanism of action. *Pharmacol. Ther.*, *25*, 83.
101. Mann, J. (2001). Natural products as immunosuppressive agents. *Nat. Prod. Repts.*, *18*(4), 417–430.
102. Mann, J. (2002). Natural products in cancer chemotherapy: Past, present and future. *Nat. Rev. Cancer*, *2*(2), 143–148.
103. Marko, I. (2001). Natural product synthesis: The art of total synthesis. *Science*, *294*(5548), 1842–1843.
104. Marriott, B. M. (2001). The role of dietary supplements in health. An overview in the United States. *Adv. Exper. Med. Biol.*, *492* (Nutrition and Cancer Prevention), 203–217.
105. Mathieu, M. (1997). *Biologics Development: A Regulatory Overview*, 2nd ed. PAREXEL International Corporation.
106. Mathieu, M., Evans, A. G. (1997). *New Drug Development A Regulatory Overview*, Fourth Edition, PAREXEL International Corporation.
107. Mays, T. D., Mazan, K. D. (1996). Legal issues in sharing the benefits of biodiversity prospecting, *J. Ethnopharmacol.*, *51*(1–3), 93–102; discussion 102-9.
108. McChesney, J. D. (2000). Commercialization of plant-derived natural products as pharmaceuticals: a view from the trenches. In S. J. Cutler and H. G. Cutler (Eds.), *Biologically Active Natural Products*. CRC, Boca Raton, Fla., pp. 253–264.
109. McCoy, M. (1999). BIOCATALYSIS GROWS FOR DRUG SYNTHESIS. *Chem. Eng. News*, January 4, pp. 10–13.
110. McGuire, W. P., Rowinsky, E. K., Rosenheim, N. B., Grumbine, F. C., Ettinger, D. S., Armstong, D. K., Donehower, R. C. (1989). Taxol: a unique antineoplastic agent

with significant activity in advanced ovarian epithelial neoplasms. *Annals of Internal Medicine*, *111*, 273–279.

111. Mehta, G., Singh, V. (2002). Hybrid systems through natural product leads: an approach towards new molecular entities. *Chem. Soc. Rev.*, *31*(6), 324–334.
112. Mehta, R. G., Pezzuto, J. M. (2002). Discovery of cancer preventive agents from natural products: from plants to prevention. *Curr. Oncol. Rep.*, *4*(6), 478–486.
113. Miles, D. H., Nguyen, C. L., Miles, D. H. (1998). Utilization of natural products for treatment of blood diseases. *Curr. Med. Chem.*, *5*(6), 421–440.
114. Müller, H., Brackhagen, O., Brunne, R., Henkel, T., Reichel, F. (2000). Natural products in drug discovery. *Ernst Schering Res. Found. Workshop*, *32*, 205–216.
115. Nakanishi, K. (1999). An historical perspective of natural products chemistry. *Comprehensive Nat. Prod. Chem.*, *8*, xxi–xxxviii.
116. Nakanishi, K. (1999). An historical perspective of natural products chemistry. In S. Ushio (Ed.), *Comprehensive Natural Products Chemistry*, Vol. 1. Elsevier Science B.V., Amsterdam, pp. 23–40.
117. Neidig, K. P., Geyer, M., Gorler, A., Antz, C., Saffrich, R., Beneicke, W., Kalbitzer, H. R. (1995). Aurelia, a program for computer-aided analysis of multidimensional NMR spectra. *J. Biomol. NMR*, *6*, 255–270.
118. Newman, D. J., Cragg, G. M., Snader, K. M., (2000). The influence of natural products upon drug discovery. *Nat. Prod. Repts.*, *17*, 215–234.
119. Nicolaou, K. C., Pfefferkorn, J. A. (2001). Solid phase synthesis of complex natural products and libraries thereof. *Biopolymers*, *60*(3), 171–193.
120. Nielsen, J. (2002). Combinatorial synthesis of natural products. *Curr. Opin. Chem. Biol.*, *6*(3), 297–305.
121. Nies, A. S. (2001). Principles of therapeutics. In J. G. Hardman and L. E. Limbird (Eds.), A. G. Gilman (Consulting Editor), *Goodman & Gilman's The Pharmacological Basis of Therapeutics*, McGraw-Hill, New York, pp. 45–66.
122. Nisbet, L. J., Moore, M. L. (1997). Will natural products remain an important source of drug research for the future? *Curr. Opin. Biotechnol.*, *8*(6), 708–712.
123. O'Keefe, B. R. (2001). Biologically active proteins from natural product extracts. *J. Nat. Prod.*, *64*, 1373–1381.
124. O'Neill, M. J., Lewis, J. A. (1993). The renaissance of plant research in the pharmaceutical industry. In A. D. Kinghorn and M. F. Balandrin (Eds.), *Human Medicinal Agents from Plants*, ACS Symposium No. 534. American Chemical Society, Washington, DC, pp. 48–55.
125. Patchett, A. (2002). 2002 Alfred Burger Award Address in Medicinal Chemistry. Natural products and design: Interrelated approaches in drug discovery. *J. Med. Chem.*, *45*(26), 5609–5616.
126. Puri, R. K. (1998). Supercritical fluid extraction and its applications in natural products. *Ind. J. Nat. Prod.*, *14*(2), 3–23.
127. Qin, G., Xu, R. (1998). Recent advances on bioactive natural products from Chinese medicinal plants. *Med. Res. Rev.*, *18*(6), 375–382.
128. Quinn, R. J. (1999). High-throughput screening in natural product drug discovery in Australia utilising Australia's biodiversity. *Drug Devel. Res.*, *46*(3/4), 250–254.

129. Quinn, R. J. (1999). QPRI's system for screening of natural products. *Proc. Phytochem. Soc. Eur.*, *43* (Bioassay Methods in Natural Product Research and Drug Development), 151–157.
130. Ram, V. J., Kumari, S., (2001). Natural products of plant origin as anticancer agents. *Drug News and Perspectives*, *14*(8), 465–482.
131. Rang, H. P., Dale, M. M., Ritter, J. M., Gardner, P. *Pharmacology.* Churchill Livingstone, New York, NY, 1995.
132. Raven, P. H. (1988). Our diminishing tropical forests. In E. O. Wilson (Ed.), *BioDiversity*. National Academy Press, Washington, DC, pp. 119–122.
133. Righi, P. (1999). Critical surveys covering the year 1998: Total syntheses of natural products. In A. Corbella (Ed.), *Seminars in Organic Synthesis, Summer School, 24th*, Gargano, Italy, June 14–18, 1999. Societa Chimica Italiana, Rome, Italy, pp. 401–518.
134. Robinson, D. H., Narducci, W. A., Ueda, C. T. (1997). Drug delivery and administration, basic concepts of pathophysiology and pharmacotherapy. In J. T. DiPiro, G. R. Matzke, L. M. Posey, R. L. Talbert, B. G. Wells, and G. C. Yee (Eds.) (Posey, L.M., Section Editor), *Pharmacotherapy, A Pathophysiologic Approach*, 3rd ed. Appleton & Lange, pp. 49–76.
135. Rodriguez, E., McDaniel, R. (2001). Combinatorial biosynthesis of antimicrobials and other natural products. *Curr. Opin. Microbiol.*, *4*(5), 526–534.
136. Roessner, C. A., Scott, A. I. (1996). Genetically engineered synthesis of natural products: From alkaloids to corrins. *Annu. Rev. Microbiol.*, *50*, 467–490.
137. Ross, E. M., Kenakin, T. P. (2001). Pharmacodynamics: Mechanisms of drug action and the relationship between drug concentration and effect, in J. G. Hardman, and L. E. Limbird (Eds.) (A. G. Gilman, Consulting Editor), *Goodman & Gilman's The Pharmacological Basis of Therapeutics*, 10th ed. McGraw-Hill, New York, pp. 31–44.
138. Rouhi, A. M. (1999). Chemistry from unknown microbes. *Chem. Eng. News*, *77*, January 4, pp. 21–22.
139. Rouhi, A. M. (1996). Natural products' scope expanding. IUPAC symposium showcases breadth of approaches chemists use to understand complex molecules from natural sources. *Chem. Eng. News*, *74*(43), 34–43.
140. Rouhi, A. M. (1997). Seeking drugs in natural products. *Chem. Eng. News*, *75*(14), 14–20.
141. Rowinsky, E. K., Cazenave L. A., Donehower R. C. (1990). Taxol: A novel investigational antimicrotubule agent. *J. Natl. Cancer Inst.*, *82*, 1247–1259.
142. Sandsborg, W. N. A., Rolfsen, N. (1999). Natural products in drug discovery and development. *Proc. Phytochem. Soc. Eur. Bioassay Methods Nat. Prod. Res. Drug Devel.*, *43*, 143–149.
143. Sausville, E., Johnson, J., Alley, M., Zaharevitz, D., Senderowicz, A. M. (2000). Inhibition of CDKs as a therapeutic modality. *Ann. N.Y. Acad. Sci.*, *910* (Colorectal Cancer), 207–222.
144. Scheuer, P. J. (1995). Marine natural products research: A look into the dive bag. *J. Nat. Prod. (Lloydia)*, *58*(3), 335–343.
145. Schoental, R. (1965). Toxicology of natural products. *Food Cosmetics Toxicol.*, *3*(4), 609–620.

146. Schuster, B. G. (2001). Demonstrating the validity of natural products as anti-infective drugs. *J. Alternative Complementary Med.*, *7* (Suppl. 1), S73–82.
147. Schwartsmann, G., Ratain, M. J., Cragg, G. M., Wong, J. E., Saijo, N., Parkinson, D. R., Fujiwara, Y., Pazdur, R., Newman, D. J., Dagher, R., DiLeone, L. (2002). Anticancer drug discovery and development throughout the world. *J. Clin. Oncol.* (Sept. 15), *20* (Suppl. 18), 47S–59S.
148. Shapiro, K., Gong, W. C. (2002). Natural products used for diabetes. *J. Am. Pharmaceut. Assoc.*, *42*, 217–226.
149. Shimizu, Y. (1998). Purification of water-soluble natural products. *Methods Biotechnol.*, *4* (Natural Products Isolation), 329–341.
150. Silva, C. J., Brian, P., Peterson, T. (2001). Screening of combinatorial biology libraries for natural products discovery. *Drugs Pharmaceut. Sci.*, *114* (Handbook of Drug Screening), 357–382.
151. Sorensen, E. J. (1999). Selected, recent developments in the chemical synthesis of biologically-active natural products. *Curr. Opin. Drug Discov. Devel.*, *2*(6), 606–630.
152. Spainhour, C. B. (2001). OTC drugs and nutraceuticals. In S. C. Gad (Ed.), *Regulatory Toxicology*, 2nd ed. Taylor & Francis, London, pp. 192–204.
153. Steinbeck, C. (2001). A platform-independent, distributed, and parallel system for computer-assisted structure elucidation in organic chemistry. *J. Chem. Inf. Comput. Sci.*, *41*, 1500–1507.
154. Steinbeck, C. (2001). The automation of natural product structure elucidation. *Curr. Opin. Drug Discov. Devel.*, *4*(3), 338–342.
155. Tsoi, C. J., Khosla, C. (1995). Combinatorial biosynthesis of "unnatural" natural products: The polyketide example. *Chem. Biol.*, *2*(6), 355–362.
156. Vlietinck, A. J., Apers, S. (2001). Biological screening methods in the search for pharmacologically active natural products. In C. Tringali (Ed.), *Bioactive Compounds from Natural Sources*. Taylor & Francis, London, pp. 1–29.
157. Volkman, J. K. (1999). Australasian research on marine natural products: Chemistry, bioactivity and ecology. *Marine Freshwater Res.*, *50*(8), 761–779.
158. Von Dohren, H. (1995). Peptides. *Biotechnology*, *28*, 129–171.
159. Waller, D. G., Renwick, A. G., Hillier, K. (2001). *Medical Pharmacology and Therapeutics*. W. B. Saunders, Harcourt Publishers Limited, Edinburgh.
160. Waterman, P. G. (1998). Natural products for drug discovery: An overview. In A. L. Harvey (Ed.), *Advances in Drug Discovery Techniques*. J Wiley, Chichester, pp. 13–23.
161. Wessjohann, L. A. (2000). Synthesis of natural-product-based compound libraries. *Curr. Opin. Chem. Biol.*, *4*(3), 303–309.
162. Whitehead, R. (1999). Natural product chemistry. *Annu. Repts. Progr. Chem., Sect. B: Organic Chem.*, *95*, 183–205.
163. Wilson, L. J. (2000). Recent advances in solid-phase synthesis of natural products, In K. Burgess (Ed.), *Solid-Phase Organic Synthesis*, J Wiley, New York, pp. 247–267.
164. Woodruff, H. B. (1999). Natural products from microorganisms. An Odyssey revisited. *Actinomycetologica*, *13*(2), 58–67.

165. Wright, A. (1998). Isolation of marine natural products. *Methods Biotechnol.*, *4* (Natural Products Isolation), 365–408.
166. Wrigley, S. K., Chicarelli-Robinson, M. (1997). Natural products research and Pharmaceuticals in the 1990's. *Annu. Repts. Med. Chem.*, *32*, 285–294.
167. Yang, S. S., Cragg, G. M., Newman, D. J., Bader, J. P. (2001). Natural product-based anti-HIV drug discovery and development facilitated by the NCI Development Therapeutics Program. *J. Nat. Prod.*, *64*(2), 265–277.
168. Yao, X., Hu, K., Peng, J., Qiao, S., Qiu, F., Dong, A., Wang, N., Cui, C., You, S., Shao, G., Chen, Y., Xu, S. (1998). Current status and prospects of research on natural medicinal products in China. *Int. Congr. Ser.*, *1157* (Towards Natural Medicine Research in the 21st Century), 445–455.
169. Yu, D., Chen, Y., Liang, X. (2000). Structural chemistry and biological activities of natural products from Chinese herbal medicines—part II. *Res. Commun. Mol. Pathol. Pharmacol.*, *108*(5–6), 393–436.
170. Yu, D., Chen, Y., Liang, X. (2001). Structural chemistry and biological activities of natural products from Chinese herbal medicines. *Emerging Drugs*, *1* (Molecular Aspects of Asian Medicines), 215–321.

2

CANCER CELL PROTEOMICS USING MOLECULAR APTAMERS

WEIHONG TAN, ZEHUI CAO, DIHUA SHANGGUAN,
YING LI, ZHIWEN TANG, PRABODHIKA MALLIKRATCHY, AND
HUI CHEN
University of Florida
Gainesville, Florida

2.1	INTRODUCTION	73
2.2	MOLECULAR DIAGNOSIS OF CANCERS	74
2.3	APTAMERS FOR TUMOR MARKER DETECTION	75
2.4	FLUORESCENCE ENERGY TRANSFER FOR APTAMER-BASED PDGF ANALYSIS	76
2.5	APTAMERS AS FLUORESCENCE ANISOTRPY PROBES FOR PROTEIN ANALYSIS	80
2.6	WHOLE-CELL-BASED APTAMER SELECTION FOR MULTIPLE TUMOR MARKERS	81
2.7	CONCLUSION	83
	References	83

2.1 INTRODUCTION

Tremendous progress has been made in understanding human cancers in recent decades. It is now well accepted that cancer is uncontrolled cell growth resulting from multiple mutated genes. These mutations could come from

---

Drug Discovery Handbook, by Shayne Cox Gad
Copyright © 2005 by John Wiley & Sons, Inc.

either an imperfect deoxyribonucleic acid (DNA) copying process in cells or damages to DNA by environmental mutagens. Despite our increasing understanding of cancers, finding a cure for cancers remains a great challenge. One of the important aspects of the cancer treatment is the accurate diagnosis of cancers. As late-stage cancers can only be treated with a very low success rate, early detection of benign tumors is critically important for patient survival.

With recent advances, especially the combination of positron emission tomography (PET) and computed tomography (CT), cancer-imaging techniques are becoming increasingly important in tumor detection [1]. In PET, radioactively linked glucose is injected into the patient. Since tumor cells are more active in growth and division than normal cells, they need more resources such as glucose from blood. Thus the accumulation of radioisotope-labeled glucose in tumor cells can be used to locate the tumor. Combined with CT, PET provides clear images that show the location, size, and shape of the tumor. This technique has been useful not only in tumor detection but also in determining the stage of the cancer, monitoring the progression of the tumor, and guiding cancer treatment. However, one problem with cancer imaging is that it is associated with expensive scanning equipment. The test itself is also expensive. More economical and readily available cancer diagnoses, such as those based on molecular recognition, may be more applicable in many situations.

2.2 MOLECULAR DIAGNOSIS OF CANCERS

The molecular diagnosis of cancers relies on biomarker molecules that are produced in higher than normal levels either directly by tumor cells or by the response of the human body to the presence of cancers. Detection of the biomarkers in a patient's body fluids can serve as the first step in cancer diagnosis and provide critical information to doctors as to whether or not a biopsy is needed. Tumor markers can be proteins or hormones. Some classic tumor markers include α-fetoprotein (AFP), carcinoembryonic antigen (CEA), and prostate-specific antigen (PSA). They are usually not very specific to a particular cancer as the level of one tumor marker can be elevated by more than one type of cancers. Another problem is that the presence of cancer does not necessarily cause a detectable level of tumor markers, especially in the early stages. Extra caution is thus needed in some cases to avoid false negatives.

With a better understanding of the genetic basis of cancers, some biomolecules may emerge as a new class of tumor markers with greater sensitivity and selectivity. The mutated genes in cancer cells usually lead to the expression of new proteins not found in normal cells, the overexpression of certain proteins that promote cell growth and division, such as growth factors and related proteins, or the underexpression of proteins that inhibit tumor cell proliferation, such as the p53 protein. One example of a protein capable of indicating cancers is the human epidermal growth factor receptor 2 (HER2), which is usually found on cell membranes. HER2 has been shown to be overexpressed in about 25 percent of all breast cancer patients [2]. Therefore, tests for the HER2 protein in tissue samples are recommended for breast cancer diagnosis. Overexpression of many other growth factors, including the insulin-like growth factor-I (IGF-I), the epidermal growth factor (EGF), and the platelet-derived growth factor (PDGF), are also related to tumor progression [3–5]. Sensitive detection of these proteins may serve as a good indication of both the presence and stage of cancers.

2.3 APTAMERS FOR TUMOR MARKER DETECTION

Over the last decades, antibodies have been the preferred recognition agents for protein detection. Their very high affinity and selectivity toward the target molecules ensure sensitive detection of specific proteins even in relatively complex samples. Antibodies for many important proteins have been developed, and some of them are routinely used for the clinical diagnosis of various diseases.

Despite these great successes in molecular diagnosis, antibodies possess some limitations in certain applications. First, the production of antibodies relies on an animal host and is a rather long process. This presents significant challenges for using antibodies in proteomic research, where thousands of proteins that are not well characterized need to be analyzed simultaneously. Second, as a special type of protein, antibodies are also sensitive to features of their surrounding environments, such as pH, temperature, ionic strength, and so forth. Labeling antibodies with fluorescent dyes or the immobilization of antibodies on a solid surface may alter their tertiary structures and thus their binding properties to targets. Furthermore, the long-term storage of antibodies may pose problems in maintaining their activities, which adds to the already high cost of using antibodies for protein probes.

Designed to mimic some of the antibodies' functions, a new class of molecules called aptamers may greatly challenge the role of antibodies in molecular diagnosis in the near future. Aptamers are nucleic acids that have high affinity and selectivity for their target molecules. By using the exponential enrichment (SELEX) process [6, 7] for the systematic evolution of ligands, oligonucleotide sequences can be isolated to recognize virtually any class of molecules [8]. The SELEX process begins with a library of synthesized oligonucleotides usually containing 10^{14} to 10^{15} random sequences. This library is then incubated with the target molecule of interest under certain conditions. The sequences that interact with and bind to the target molecules are isolated and amplified by PCR for the next round of incubation. This process is repeated until sequences that bind to the target with highest affinity and selectivity are determined. Compared to antibodies, aptamers have similar high affinity and selectivity for proteins [9, 10]. Due to their structural stability, aptamers can withstand harsher experimental conditions than antibodies and can be stored, modified, and immobilized without causing much degradation. It is also much easier and considerably cheaper to make aptamers using common DNA synthesizers.

Aptamers can be easily and site-specifically engineered with fluorescent labels for real-time protein recognition. By forming intramolecular signal transduction mechanisms, the labeled aptamers can report targets in homogenous solutions, eliminating the need to separate bound complex molecules from unbound molecules, which is often required in antibody-based protein assays. This translates into fast and convenient protein detection that is essential for large-scale high-throughput proteomic research. As an example, the real-time detection of PDGF in various approaches is demonstrated.

2.4 FLUORESCENCE ENERGY TRANSFER FOR APTAMER-BASED PDGF ANALYSIS

A member of a growth factor family, PDGF is involved in regulating cell growth and division. It is composed of a combination of primarily two types of subunits, A and B. Thus PDGF usually exists in three isoforms, PDGF-AA, AB, and BB. Overexpression of PDGF, especially the BB form, has been found in malignant tumors. A DNA aptamer that specifically binds to the PDGF-B chain was previously reported [11]. In physiological conditions, the stable conformation of the aptamer contains two regions of base pairing but is relatively unconstrained. Upon binding with PDGF, the aptamer forms a closely packed structure that includes a three-way helix junction with a conserved single-stranded loop at the junction point that is essential for binding (Fig. 2.1). Based on this fact, the sensitive detection of PDGF can be achieved using fluores-

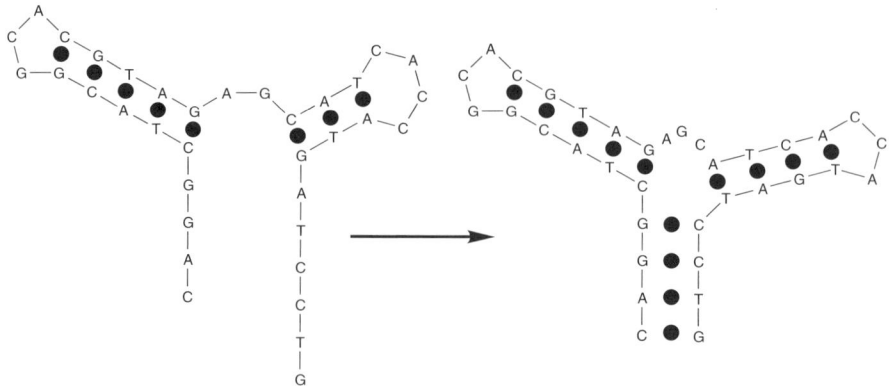

Figure 2.1 Conformational change of PDGF aptamer. Binding of PDGF causes the aptamer to change from a relatively unconstrained structure to a tightly packed one containing a three-way helical junction with a conserved single-stranded loop.

FLUORESCENCE ENERGY TRANSFER FOR APTAMER-BASED PDGF ANALYSIS 77

cence resonance energy transfer (FRET). The energy transfer between "donor" and "acceptor" molecules with overlapping emission and absorption spectra is highly dependent on the distance between these two molecules. Significant FRET usually happens with a distance less than 10 nm. This forms the basis for a FRET-based PDGF assay using aptamer. When the donor and acceptor are labeled at both ends of the PDGF aptamer (Fig. 2.2), binding of the aptamer to PDGF will reduce the distance between the two labels to the level at which the energy transfer becomes significant, thus revealing the presence of PDGF. Two types of energy transfer can be utilized. One of them uses a nonfluorescent molecule as the acceptor, which is usually called the quencher. Therefore, no emission of the acceptor will be observed and the only indication of FRET is the quenching of the donor fluorescence. In the other method of conducting FRET, the fluorescent acceptor produces an emission upon energy transfer.

The PDGF aptamer has been turned into a probe specific for PDGF by labeling the 5′ end with fluorescein and the 3′ end with a DABCYL quencher [12]. Due to its resemblance to molecular beacons used for nucleic acid detection, this probe is given the name molecular beacon aptamer (MBA). The addition of PDGF-BB as low as sub-nanomolar causes a detectable reduction of the fluorescence signal in real time as the fluorophore and quencher are

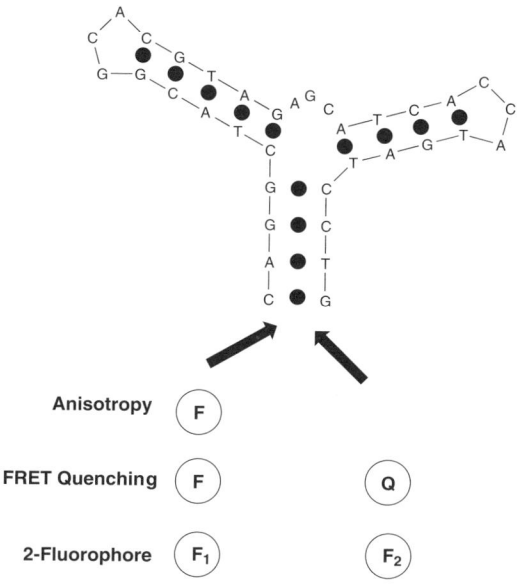

Figure 2.2 Engineering a PDGF probe from the PDGF aptamer. By strategic placement of single or double fluorophores (F) and a quencher (Q), the aptamer can be transformed into fluorescence probes based on fluorescence anisotropy, FRET quenching, and 2-fluorophore FRET. Positioning the labels at either end of the aptamer sequence minimizes interference with aptamer–protein binding.

brought together upon binding. This FRET-based assay leads to PDGF detection with not only high sensitivity but also good selectivity. A variety of extracellular proteins and unrelated growth factors are mixed with the MBA and show no noticeable fluorescence change. On the other hand, the other two variants of PDGF, PDGF-AA and PDGF-AB, can also cause a decrease of fluorescence in the MBA, but to a lesser extent due to their lower binding affinity to the aptamer. This difference in binding affinity is applied to construct an assay to selectively analyze variants of PDGF. A sample solution with each of the three isoforms of PDGF undergoes a serial dilution with a physiological buffer. The same amount of MBA is then added to the aliquots. Fluorescence is read in each aliquot by a microtiter plate reader and a dose–response curve is created. The three isoforms have clearly different dose-response curves, with the slopes of the curves decreasing from PDGF-BB to AB and then AA (Fig. 2.3). The slope change can be attributed to differences in binding affinity to the MBA as more protein with lower affinity is needed to obtain the same MBA quenching level. The variants of the same protein can be differentiated using the MBA-based assay.

The same approach can be used to monitor PDGF-BB in real biological samples such as cell media. Conditioned media from different cell lines, including human breast carcinoma cells HTB-26 and normal murine BALB/3T3 fibroblast cells, are diluted multiple times and incubated with a fixed amount of the MBA in a 96-well microtiter plate to obtain dose–response curves. Similarly, serial dilution of a PDGF-BB in a buffer solution was employed to build another dose–response curve. Comparing these three sets of results, one can see that the curves for pure PDGF-BB and HTB-26 cells have very similar slopes and maximum extents of quenching, indicating something in the cell

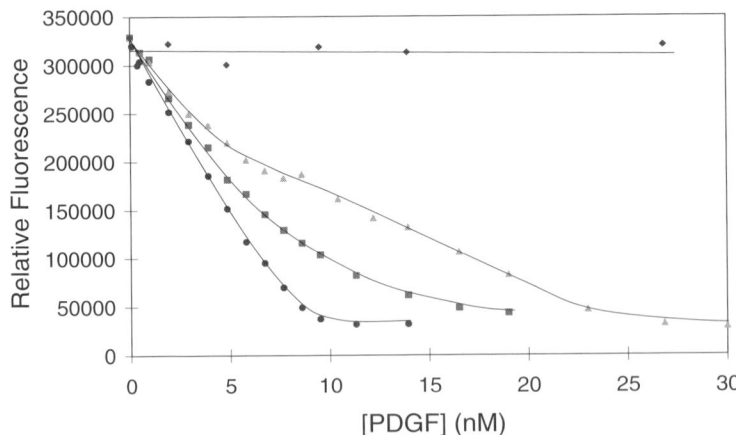

Figure 2.3 Dose-response curves of PDGF variants. Fluorescence signals of MBA for serial dilutions of PDGF-AA (triangle), PDGF-AB (square), PDGF-BB (circle), and denatured PDGF-BB (diamond). The concentration of the MBA was 10 nM.

media had almost the same affinity to the aptamer as pure PDGF-BB (Fig. 2.4). In contrast, the curve for BALB/3T3 cells has a flatter slope and less of a quenching effect. Since the slope of a dose–response curve is independent of the actual concentration of the target, but only related to the affinity between target and probe, it is likely that the BALB/3T3 cells do not produce PDGF-BB actively.

A two-fluorophore-based FRET assay should also be suitable for PDGF detection. Previously, this approach has been used for the detection of human α-thrombin with thrombin aptamer [13]. The aptamer behaves similarly to PDGF aptamer. When bound to the target protein, it takes a G-quadruplex conformation, yielding a more compact structure than in free solution. Two fluorophores with overlapping spectra are labeled on the aptamer. Binding of the aptamer to α-thrombin brings the two dyes into close proximity, thus triggering the energy transfer between the two molecules. Given that the donor emission declines and the acceptor emission increases as a result of the FRET, the ratio of the donor and acceptor emissions is measured instead of the fluorescence of a single dye. A large difference between the two emissions is expected before and after protein binding, leading to more sensitive detection.

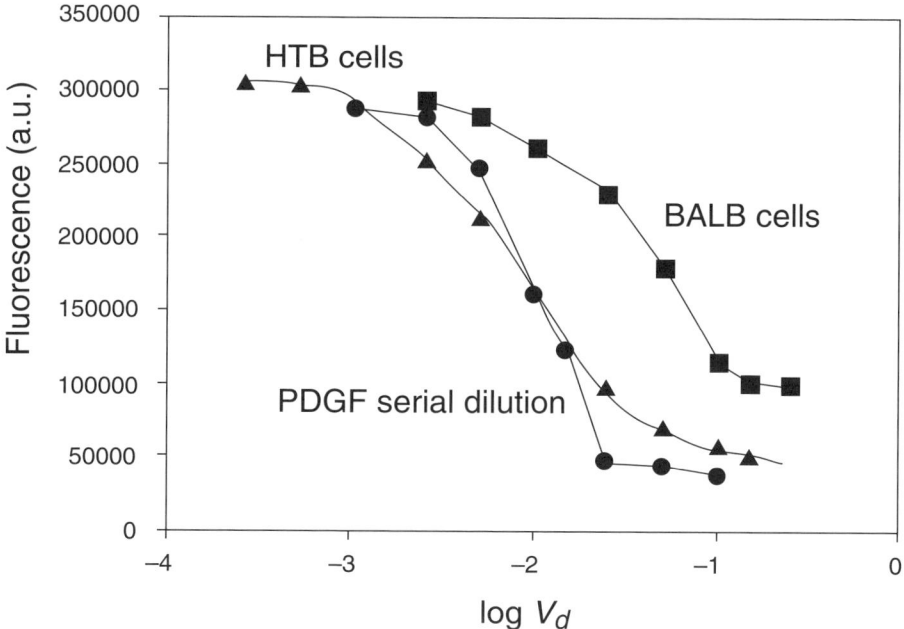

Figure 2.4 Cell sample dose–response curves. Fluorescence signals of mixtures of 10 nM of MBA with serial dilutions of protein preparations of conditioned cell media from HTB cells (triangle) and BALB cells (square). Serial dilutions of a 500-nM solution of PDGF-BB (circle) were used as a control. V_d is the dilution factor.

The two-fluorophore assay has a threefold lower detection limit than the fluorophore–quencher assay for α-thrombin. A similar method can be employed to improve the sensitivity of the aptamer-based PDGF assay.

Other reports using different strategies to incorporate FRET into aptamer-based assays have also successfully detected various targets in real time [13–15]. Another alternative approach links ribozyme with aptamer [16]. Binding between the aptamer and the target activates the ribozyme, and a nucleic acid with fluorophore and quencher labels is cleaved, resulting in fluorescence increase.

2.5 APTAMERS AS FLUORESCENCE ANISOTROPY PROBES FOR PROTEIN ANALYSIS

Fluorescence anisotropy (FA) has been widely used for studying biomolecular interactions. A relatively small fluorescent molecule, when bound to a relatively large molecule, will demonstrate an increased anisotropy. The principles behind FA can be explained as follows. When excited by a polarized light, those fluorescent molecules with absorption transition moments oriented along the electric vector of the incident light are preferentially excited. The rotational diffusion of the molecules on the excited state leads to depolarized emission. The extent of the depolarization can be described in terms of anisotropy (r). Anisotropy is usually measured using the parallel emission $I_\parallel$ and perpendicular emission $I_\perp$:

$$r = (I_\parallel - I_\perp)/(I_\parallel + 2 \bullet I_\perp)$$

Since FA is related to a molecule's ability to rotate in the excited state, the size as well as the molecular weight of the molecule plays an important role in determining this molecule's anisotropy.

The primary advantage of using aptamers as anisotropy probes is their relatively small sizes. Compared to monoclonal antibodies, they are in most cases much smaller than the target proteins. This makes aptamer-based anisotropy probes ideal for protein detection. With the aptamer conveniently labeled with one fluorophore, the anisotropy probe will report binding with a target protein via the increase in anisotropy of the fluorophore. Unlike the FRET-based assays, where the conformational change of the aptamer is essential for target detection, the aptamer anisotropy assay is not as heavily dependent on the structure of the aptamer probe. Therefore, it may be highly useful in applications where understanding of the aptamer structure is limited.

An anisotropy probe has been made by labeling the PDGF aptamer with one fluorescein dye [17]. Adding PDGF-BB to the aptamer results in an increased anisotropy of the aptamer due to the increased overall molecular weight upon complex formation. A detection limit of 2 nM PDGF-BB can be achieved. The selectivity of the assay is as good as that of the FRET-based assay. This can be attributed to the fact that various other proteins do not affect the anisotropy of the aptamer.

Aptamer-based assays using either FRET or fluorescence anisotropy as a signal transduction mechanism have proven to be ideal for sensitive and selective protein analysis. Their ability to detect targets in real time and homogeneous solutions holds great potential for easy, cheap, and fast applications in clinical diagnosis. As the in vitro selection of aptamers has evolved into an automated process [18], more and more aptamers for proteins, including important cancer biomarkers, are expected to be available in the near future.

2.6 WHOLE-CELL-BASED APTAMER SELECTION FOR MULTIPLE TUMOR MARKERS

Recent advances in proteomic research, especially the emergence of new analytical techniques, have begun to show great impact on cancer diagnosis. Based on the simultaneous detection of multiple protein markers, these new approaches have already demonstrated excellent sensitivity and specificity in cancer diagnosis that are beyond the capability of traditional methods. Evolved from matrix-assisted laser desorption ionization (MALDI) mass spectrometry, a new technique called surface-enhanced laser/desorption ionization (SELDI) is capable of identifying proteins with high throughput. With the help of SELDI, biomarkers for specific cancers have been discovered and successfully applied to real-world cancer diagnosis [19, 20]. The idea of this new method is to analyze the protein content of serum from patients with known cancer by SELDI. The acquired mass spectra are then compared to those obtained from healthy people. An iterative searching algorithm is used to identify the proteins in cancer patient serum that can best differentiate cancer patients from healthy people. Searching the serum from unknown patients for the newly found protein markers can reveal whether the patients have this specific cancer. Cancer patients, including those with early-stage cancer, are correctly diagnosed in most cases.

Despite the initial success, problems associated with the new strategy have begun to come to people's attention [21]. One of the issues is that two different laboratories using the same approach to study the same type of cancer have identified very different sets of protein markers, yet they both have success in cancer diagnosis using their biomarkers. This may indicate that the discovery of biomarkers using SELDI may be highly sensitive to sample preparation and experimental conditions. Therefore, great care needs to be taken in order to achieve comparable results across different laboratories. Furthermore, expensive instruments in this new method may prevent the application of the biomarkers in small clinics.

It is clear from these new advances that multiple protein markers produced by cancers, once identified, can lead to more sensitive and specific cancer diagnosis than using a single tumor marker. Therefore, it is of great importance to develop panels of probes for simultaneous biomarker detection in real time. One promising strategy for this purpose is to conduct aptamer selection targeted at a complex mixture of samples. In order to be applicable to cancer diagnosis, the use of cancer-cell-related complex protein mixtures as the target for selection is preferred.

The selection of high-affinity ligands for complex targets has been demonstrated both in theory [23] and in experiments for red blood cell membranes [22]. Since then, aptamers have been isolated for various complex systems, mostly cell membranes [24–27]. Since proteins not found or in low abundance on normal cell membranes may be present on tumor cell membranes, those proteins can serve as biomarkers for specific cancers. Whole-cell selection may yield a panel of aptamers that have high affinity and selectivity to the important tumor markers.

The procedural details of whole-cell selection vary between reports, but generally they incorporate the incubation of a random DNA or ribonucleic acid (RNA) library with target cells followed by the controlled elution of the sequences bound to the target cells. These sequences were subsequently amplified by polymerase chain reaction (PCR) and then reincubated with target cells iteratively. After several iterations (6 to 25), the pool of selected oligonucleotides was cloned and sequenced. A comparison of the sequence homology allowed the grouping of the aptamers into families that recognized distinct targets. For the discovery of tumor cell markers, since most abundant species on tumor cell membranes may not be related to cancer, it is advisable to do a counter selection using closely related normal cells. After incubation of the DNA or RNA library with the normal cells, only the unbound nucleic acids will proceed to be incubated with the tumor cells (Fig. 2.5). In this way, DNA or RNA molecules that bind to proteins that are present on both normal and tumor cells are not selected.

The selected aptamers can be engineered into protein-binding probes based on either FRET or fluorescence anisotropy. These probes can then be made into a kit for clinical diagnosis. Tissue samples from patients are incubated with the aptamer panel and the fluorescence signals can be read by something

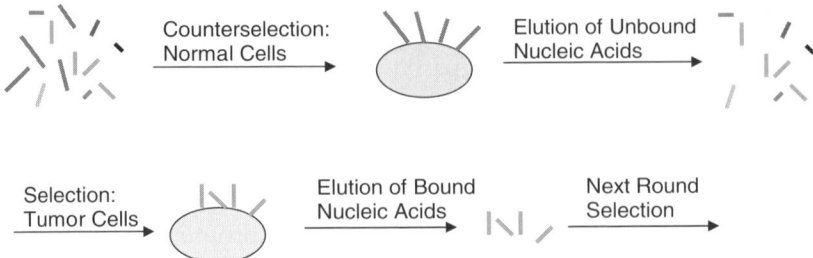

Figure 2.5 Schematic of cell-based aptamer selection and counter selection.

as simple as a microtiter plate reader. Positive results from most of the aptamers will be strong evidence for the presence of tumor cells that have been particularly targeted. As a result, cancer patients, including those with early-stage cancer, may be diagnosed with high confidence.

Another way to incorporate the selected aptamer panel into cancer diagnosis is to attach the aptamers with either radioactive or fluorescent labels and inject them into the patient's body. With relatively small molecular weights, aptamers should provide fast tumor penetration and blood clearance, which makes them excellent for high-contrast imaging [28]. A whole-body scanning will reveal the localization of the aptamers inside the body. Coexistence of the aptamers in a particular part of the body should give a strong indication of a tumor in that area. Because multiple tumor binding probes are utilized, the aptamer tests should be superior to traditional antibody tests in terms of specificity and reliability.

2.7 CONCLUSION

Aptamers have the potential to be excellent probes for the recognition of single protein targets as well as for a panel of proteins. Combined with fluorescence-based signal transduction mechanisms, the detection of proteins using aptamers can be both effective and convenient with high sensitivity and selectivity. To take advantage of the capability of aptamers in molecular recognition for cancer diagnosis, it is desirable to employ a set of aptamers to simultaneously analyze valuable tumor markers specific to a particular type of cancer. The whole-cell-based aptamer selection may be a promising new method for the rapid identification of nucleic acid ligands with high affinity for biomarkers on targeted tumor cells. It is reasonable to believe that the evolution of this new approach will eventually have a major impact on clinical cancer diagnosis.

Acknowledgement We thank our colleagues at the University of Florida for their help with this chapter. This work was supported by NIH GM66137 and NSF NIRT EF-0304569.

REFERENCES

1. Townsend, D. W., Carney, J. P., Yap, J. T., Hall, N. C. (2004). PET/CT today and tomorrow. *J. Nucl. Med.*, 45 (Suppl.) 1, 4S–14S.
2. Slamon, D. J., Godolphin, W., Jones, L. A., et al. (1998). Studies of the Her-2/Neu Proto-oncogene in human-breast and ovarian-cancer. *Science*, 244(4905), 707–712.
3. Chan, J. M., Stampfer, M. J., Giovannucci, E., et al. (1998). Plasma insulin-like growth factor-I and prostate cancer risk: A prospective study. *Science*, 279(5350), 563–566.

4. Osborne, C. K., Hamilton, B., Titus, G., Livingston, R. B. (1980). Epidermal growth factor stimulation of human breast cancer cells in culture. *Cancer Res.*, *40*(7), 2361–2366.
5. Ariad, S., Seymour, L., Bezwoda, W. R. (1991). Platelet-derived growth factor (PDGF) in plasma of breast cancer patients: Correlation with stage and rate of progression. *Breast Cancer Res. Treat.*, *20*(1), 11–17.
6. Tuerk, C., Gold, L. (1990). Systematic evolution of ligands by exponential enrichment: RNA ligands to bacteriophage T4 DNA polymerase. *Science*, *249*(4968), 505–510.
7. Ellington, A. D., Szostak, J. W. (1990). In vitro selection of RNA molecules that bind specific ligands. *Nature*, *346*(6287), 818–822.
8. Jayasena, S. D. (1999). Aptamers: An emerging class of molecules that rival antibodies in diagnostics. *Clin. Chem.*, *45*(9), 1628–1650.
9. Gold, L. (1995). Oligonucleotides as research, diagnostic, and therapeutic agents. *J. Biol. Chem.*, *270*(23), 1,3581–1,3584.
10. Xu, W., Ellington, A. D. (1996). Anti-peptide aptamers recognize amino acid sequence and bind a protein epitope. *Proc. Natl. Acad. Sci. USA*, *93*(15), 7475–7480.
11. Green, L. S., Jellinek, D., Jenison, R., et al. (1996). Inhibitory DNA ligands to platelet-derived growth factor B-chain. *Biochemistry*, *35*(45), 14413–14424.
12. Fang, X., Sen, A., Vicens, M., Tan, W. (2003). Synthetic DNA aptamers to detect protein molecular variants in a high-throughput fluorescence quenching assay. *ChemBioChem.*, *4*(9), 829–834.
13. Li, J. J., Fang, X., Tan, W. (2002). Molecular aptamer beacons for real-time protein recognition. *Biochem. Biophys. Res. Commun.*, *292*(1), 31–40.
14. Hamaguchi, N., Ellington, A., Stanton, M. (2001). Aptamer beacons for the direct detection of proteins. *Anal. Biochem.*, *294*(2), 126–131.
15. Nutiu, R., Li, Y. (2003). Structure-switching signaling aptamers. *J. Am. Chem. Soc.*, *125*(16), 4771–4778.
16. Hartig, J. S., Najafi-Shoushtari, S. H., Grune, I., et al. (2002). Protein-dependent ribozymes report molecular interactions in real time. *Nat. Biotechnol.*, *20*(7), 717–722.
17. Fang, X., Cao, Z., Beck, T., Tan, W. (2001). Molecular aptamer for real-time oncoprotein platelet-derived growth factor monitoring by fluorescence anisotropy. *Anal. Chem.*, *73*(23), 5752–5757.
18. Cox, J. C., Ellington, A. D. (2001). Automated selection of anti-protein aptamers. *Bioorg. Med. Chem.*, *9*(10), 2525–2531.
19. Petricoin, E. F., Ardekani, A. M., Hitt, B. A. et al. (2002). Use of proteomic patterns in serum to identify ovarian cancer. *Lancet*, *359*(9306), 572–577.
20. Petricoin, E. F., Ornstein, D. K., Paweletz, C. P. et al. (2002). Serum proteomic patterns for detection of prostate cancer. *J. Nat. Cancer Inst.*, *94*(20), 1576–1578.
21. Diamandis, E. P. (2003). Point—Proteomic patterns in biological fluids: Do they represent the future of cancer diagnostics? *Clin. Chem.*, *49*(8), 1272–1275.
22. Vant-Hull, B., Payano-Baez, A., Davis, R. H., Gold, L. (1998). The mathematics of SELEX against complex targets. *J. Mol. Biol.*, *278*(3), 579–597.

REFERENCES

23. Morris, K. N., Jensen, K. B., Julin, C. M., Weil, M., Gold, L. (1998). High affinity ligands from in vitro selection: Complex targets. *Proc. Natl. Acad. Sci. USA*, *95*(6), 2902–2907.
24. Blank, M., Weinschenk, T., Priemer, M., Schluesener, H. (2001). Systematic evolution of a DNA aptamer binding to rat brain tumor microvessels. Selective targeting of endothelial regulatory protein pigpen. *J. Biol. Chem.*, *276*(19), 1,6464–1,6468.
25. Daniels, D. A., Chen, H., Hicke, B. J., Swiderek, K. M., Gold, L. (2003). A tenascin-C aptamer identified by tumor cell SELEX: Systematic evolution of ligands by exponential enrichment. *Proc. Natl. Acad. Sci. USA*, *100*(26), 1,5416–1,5421.
26. Hicke, B. J., Marion, C., Chang, Y. F., et al. (2001). Tenascin-C aptamers are generated using tumor cells and purified protein. *J. Biol. Chem.*, *276*(52), 4,8644–4,8654.
27. Wang, C., Zhang, M., Yang, G., et al. (2003). Single-stranded DNA aptamers that bind differentiated but not parental cells: Subtractive systematic evolution of ligands by exponential enrichment. *J. Biotechnol.*, *102*(1), 15–22.
28. Cerchia, L., Hamm, J., Libri, D., Tavitian, B., de F, V. (2002). Nucleic acid aptamers in cancer medicine. *FEBS Lett.*, *528*(1–3), 12–16.

3

MOLECULAR SIMILARITY METHODS AND QSAR MODELS AS TOOLS FOR VIRTUAL SCREENING

JÜRGEN BAJORATH
B-IT, International Center for Information Technology
Rheinische Friedrich–Wilhelms–University Bonn
Bonn, Germany

3.1	INTRODUCTION	87
	Virtual Versus High-Throughput Screening	89
	Compound Sources	89
	Ligand- Versus Target-Based Virtual Screening	90
3.2	SELECTED SCIENTIFIC CONCEPTS FOR VIRTUAL SCREENING	91
	Molecular descriptors and chemical spaces	91
	Molecular similarity	92
	Quantitative Structure–Activity Relationship Analysis	95
3.3	GENERAL LIMITATIONS OF VIRTUAL SCREENING ANALYSIS	96
	Similar Property Principle Revisited	96
	Quantifying Molecular Similarity	97
	Potency and Selectivity Issues	98
3.4	SPECIFIC METHODS	99
	Two-Dimensional Substructure Searching	99
	Topological and Shape Representations	99
	Three-Dimensional Pharmacophores	100
	Molecular Fingerprints	102
	Two-Dimensional Versus Three-Dimensional Methods	104
	Clustering Techniques	106
	Partitioning Methods	108
	QSAR Models in Virtual Screening	112
	Drug Likeness	114

Drug Discovery Handbook, by Shayne Cox Gad
Copyright © 2005 by John Wiley & Sons, Inc.

3.5	VIRTUAL SCREENING PERFORMANCE	114
3.6	CONCLUSIONS AND OUTLOOK	116
	References	116

3.1 INTRODUCTION

In pharmaceutical research, computational (in silico, virtual) screening of compound databases has become an almost independent discipline [1, 2], although it is frequently discussed in the context of chemoinformatics [3, 4]. Its basic idea is to computationally screen two- or three-dimensional (2D or 3D) representations of compounds in large databases for molecules having desired biological activity. Similar to biological high-throughput screening (HTS), virtual screening (VS) is basically employed as a hit identification tool [1, 5] and cannot be expected to produce "magic bullets" [1] through some form of "computational magic." As will be discussed in this chapter, the algorithmic basis for many current VS methods is rather stringent, and compound selection criteria are derived in a scientifically rigorous manner.

The increasing attraction of VS approaches in recent years is in part due to the fact that significant computational power has become available cheaply and that many large-scale calculations can be readily carried out on desktop computing environments. Spurred on by such technological advances, a wealth of methods has been and continues to be developed for VS applications. Another reason for the attractiveness of VS methodologies is due to the fact that compound source databases have been dramatically growing in size over the past years. It is no longer uncommon for pharmaceutical compound inventories to contain a million compounds or more, and—based on our own experience—it is easily possible in today's chemistry market to collect 4 to 5 million compounds from vendor catalogs or other sources. The availability of so many compounds puts substantial constraints on hit identification efforts and more or less requires adding computational components to compound selection and screening repertoires. It is, therefore, not too surprising that there appears to be steadily increasing interest in VS approaches in the pharmaceutical industry, even outside "computer-heavy" environments. This chapter covers some major types of contemporary VS methods, in particular, those that are based on the analysis of molecular similarity and quantitative structure–activity relationships (QSAR). Prior to the introduction of specific concepts and methods, some more general aspects of VS and its relation to biological screening are discussed.

INTRODUCTION

Virtual Versus High-Throughput Screening

Despite many technological advances that continue to be made in the HTS arena toward miniaturization and increasingly higher throughput rates [6], HTS campaigns represent a costly affair for the pharmaceutical industry and often yield fairly limited success [7]. It is true that the costs per HTS experiment have been substantially reduced over the years in many cases, for example, simple enzyme assays that are nowadays often referred to as "penny screens." However, for more complex enzyme-linked immunosorbent assay (ELISA) or cell-based assays, reagent costs alone can easily amount to 50 cents or a dollar per compound well. Thus, a single-dose HTS screen of a compound deck containing 500,000 to a million molecules requires rather significant resources and does not even take into account required secondary assays or other follow-up studies. Thus, it is evident that "screening all compounds on all targets" is generally cost prohibitive. Therefore, preselection of compound sets by VS having a reasonable probability to act against a given target becomes attractive. Moreover, for time-consuming and expensive assays, for example, those depending on long-term culture of specific cell lines, computational analysis might initially replace HTS altogether, and small sets of compounds, perhaps tens to hundreds, are selected for testing in several rounds in order to obtain hit information with a minimum number of experiments.

In this context, it is often misunderstood that VS and HTS are in essence not competing but rather complementary disciplines. Of course, VS might sometimes replace HTS campaigns and vice versa, and virtual and high-throughput screeners may often develop preferences for rather different strategies in drug discovery (e.g., rationalization of compound selection versus increasingly higher throughput). However, this does not take away from the principal—and scientifically attractive—complementarity of VS and HTS, which is emphasized, for example, by the recent introduction of focused or sequential screening schemes [5, 8], as further discussed later on. Thus, as is the case for many research activities in drug discovery, a meaningful integration of experimental and computational screening approaches might often substantially increase the probability of ultimate success [9].

Compound Sources

Another aspect of VS that is often not well understood and might require some clarification is the source of compounds that are evaluated in silico. Are these compounds always virtual (and thus perhaps of limited synthetic feasibility)? Certainly, compounds are stored on the computer in a virtual format but—more often than not—these compounds are "real." Essentially, three sources of compounds can be distinguished for use in VS calculations. Compound inventories in the pharmaceutical industry are nowadays mostly cataloged and administered in computational compound management systems that provide more or less immediate access for VS. Thus, in these cases, compounds that have been synthesized and are physically stored and available for testing are a source for both HTS and VS. Virtual access to screening decks also provides an excellent basis for the integration of in silico and biological screening [5], at least if advanced compound handling and plating systems (e.g., for cherry picking of compounds) are available. Compound libraries assembled from public domain sources or vendor catalogs are another common source for VS. Although synthetic routes have been established for these compounds and they should, in principle, be available for acquisition, experience shows that, on average, often only about 50 percent of such public domain compounds can be readily acquired. Thus, for candidate compounds selected from such libraries, resynthesis is often required (making these compound sources "pseudoreal"). Large virtual combinatorial libraries currently represent the only truly theoretical compound source for VS [10]. Here combinatorial libraries are enumerated in silico or stored as reagents or molecular fragments for VS access. The underlying idea is to computationally select sets of compounds from vast chemical spaces as candidates for synthesis, dependent on the particular search problem [10]. In some cases, such virtual source libraries can be very large, literally containing billions of possible combinatorial compounds. The design and storage of libraries of this size would hardly be possible without almost unlimited hardware resources that have become available in recent years.

Ligand- Versus Target-Based Virtual Screening

Virtual screening techniques can principally be divided into target- and ligand-based approaches. The former critically depend on the availability of accurate three-dimensional structures of therapeutic targets, mostly proteins, and are often referred to as docking (of putative ligands into binding sites) [11, 12]. Structure-based VS attempts to identify active compounds de novo by focusing only on target structure information. By contrast, ligand-based VS depends on the availability of known active molecules, for example, enzyme inhibitors, receptor agonists, or antagonists, that are then used as templates for VS. Here the primary goal is the identification of molecules that exhibit biological activity similar to the templates but belong to different structural classes [1, 5]. This is often necessary to provide a starting point for a new lead optimization project focusing on molecules that are synthetically more accessible than the templates or have lower toxicity. In addition, such "lead hopping" exercises are often also carried out in order to circumvent patent positions of competitors in drug discovery. In practice, target- and ligand-based VS are not applied in a mutually exclusive manner. Whenever target structures and active compounds are available, these methods are used in concert in order to increase the probability of identifying novel hits. Also, the availability of experimental structures of target–ligand complexes makes it possible to closely combine these efforts, for example, by studying candidate compounds in the structural context of template molecules bound to receptor sites.

3.2 SELECTED SCIENTIFIC CONCEPTS FOR VIRTUAL SCREENING

In this section, major scientific concepts are introduced that provide a basis for the development of VS tools. In addition, some general limitations and caveats of VS analysis are also discussed in this section.

Molecular Descriptors and Chemical Spaces

Essentially all ligand-based VS calculations depend on the use of computationally implemented descriptors of molecular structure, physico-chemcial properties, or pharmacophore features. This is the case because the comparison of molecular characteristics beyond simple structural criteria requires the calculation of property values based on physical or mathematical models. Accordingly, thousands of different molecular descriptors have been reported over the years that often greatly vary in their degree of complexity and sophistication [13, 14]. Table 3.1 lists a few classes of molecular descriptors that are widely used for compound classification, QSAR, and VS applications. Descriptors are frequently divided into 1D, 2D, or 3D descriptors, dependent on the dimensionality of the molecular representation from which they are calculated [14]. According to this scheme, 1D descriptors are calculated from molecular composition formulas and include bulk properties such as molecular weight. Two-dimensional descriptors include, for example, molecular connectivity indices, graphs, or charge descriptors. Finally, 3D descriptors

TABLE 3.1 Desciptor Categories[a]

Descriptor Category	Examples
Physical properties	Molecular weight log $P(o/w)$
Atom and bond counts	Number of nitrogen atoms
	Number of aromatic atoms
	Number of rotatable bonds
Pharmacophore features	Number of hydrogen bond acceptors
	Total van der Waal surface areas of basic atoms
Charge descriptors	Total positive partial charge
	Dipole moment from partial charges
Connectivity and shape descriptors	Kier and Hall molecular shape indices
Surface area and volume	Solvent-accessible surface area
	Van der Waals volume

[a] Some type of widely used molecular descriptors are shown and representative examples are provided. It should be noted that categories often overlap and that many descriptors can be assigned to more than one class.

capture properties or structural features that are molecular conformation-dependent such as dipole moment, solvent-accessible surface, or specific pharmacophore arrangements. Furthermore, contributions of different types of descriptors can be combined into rather complex formulations, for example, descriptors combining molecular topology and electronic atom properties [15], descriptors that map features such as hydrophobic or polar character onto molecular surface representations [16], or composite descriptors that are calculated as linear combinations of many single descriptors and their specific contributions [17].

Molecular descriptors are selected for the definition of chemical spaces into which molecules are projected based on their calculated descriptor values and where qualitative or quantitative comparisons are carried out. Following this idea, n selected descriptors form the axes of an n-dimensional chemical reference space, and each test molecule is assigned a vector that determines its position in this space, as illustrated in Figure 3.1. Although underlying scientific concepts and details of many VS methods differ, molecular descriptors are fundamentally important for their application, and it is often critical to identify sets of descriptors that are suitable or preferred for a specific VS application. Automated descriptor selection has been facilitated by the application of machine learning techniques such as genetic algorithms [17, 18] or neural nets [19] or by identifying descriptors having highest information content in a compound collection under study [20].

Molecular Similarity

The introduction of the concept of molecular similarity has greatly influenced research in chemoinformatics and VS [21, 22]. The computational assessment

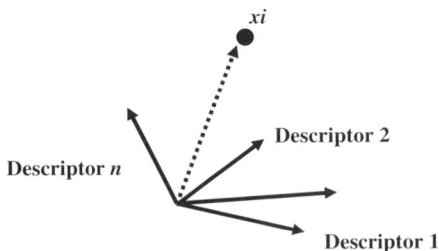

Molecule i: $xi = (xi1, xi2,\ldots,xin)$

For example: $xi = (a\_don(i), b\_ar(i),\ldots,sf\_hyd(i))$

Figure 3.1 Vector representation of a molecule in n-dimensional descriptor space. In this example, the position of test molecule i is determined by its values of n molecular descriptors used to define the chemical space (e.g., a_don, number of donor atoms; b_ar, number of aromatic bonds; sf_hyd, hydrophobic surface area).

of molecular similarity goes well beyond structural criteria or molecular topology and takes into account diverse molecular properties and, importantly, biological activity. The *similar property principle* states that similar molecules should have similar activity [21]. Despite its generality and exceptions (see below), it captures the essence of molecular similarity analysis and its relevance for ligand-based VS: If one can apply computationally derived models to identify molecular properties that are responsible for specific activity, beyond what one can "see," then it should also be possible, by application of such models, to identify molecules having similar activity. Figure 3.2 provides a few examples.

In descriptor spaces, molecular similarity can be assessed in qualitative terms (e.g., which molecules map to the same section in space?) or quantitative terms (what is the exact distance between these molecules?). Similarly,

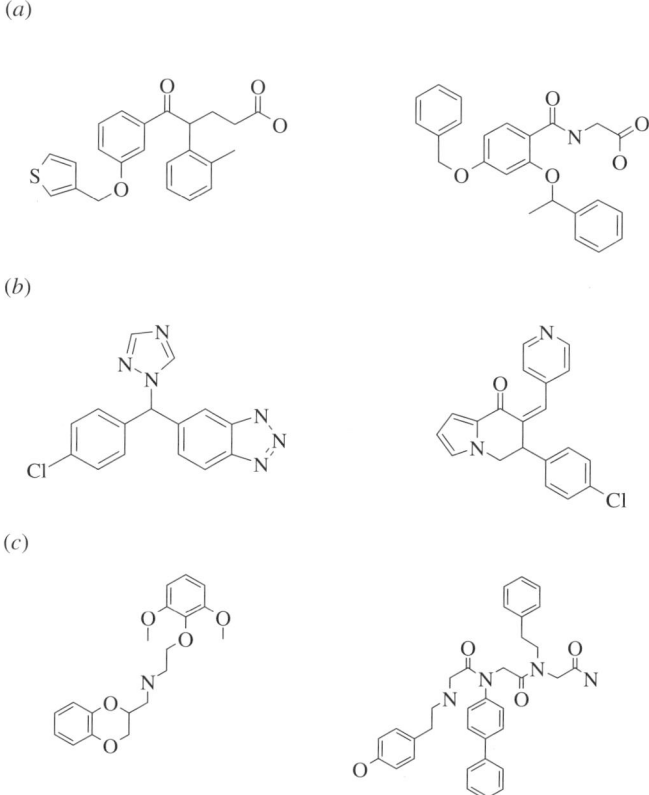

Figure 3.2 Remote similarity relationships that were successfully recognized using in-house developed similarity methods: (*a*) endothelin A antagonists, (*b*) aromatase inhibitors, (*c*) α1-adrenoreceptor ligands).

abstract molecular representations can be generated by transforming descriptor settings into bit strings of varying length, and the resulting bit patterns of these so-called fingerprints of molecules (described below in more detail) can be quantitatively compared to assess their degree of similarity. Regardless of the algorithms used, the quantitative evaluation of molecular similarity depends on mathematical measures [22, 23] to calculate the similarity of fingerprint-type representations or distances in chemical reference spaces. Some of the most popular similarity metrics are reported in Table 3.2.

It should also be noted that the concept of molecular diversity is related to but distinct from molecular similarity, although methods and metrics used for similarity and diversity analysis in part overlap. Major focal points of diversity analysis have included the design of chemically diverse (combinatorial) libraries [24, 25] and the selection of optimally diverse or representative subsets from large compound sources [25, 26]. In diversity analysis, it is typically attempted to evenly populate chemical reference spaces with compounds and design molecules that fulfill these requirements and are thus sufficiently dissimilar from each other. Creating chemical diversity in library design often involves the estimation of diversity distributions from the properties of either reagents or products of chemical reactions [25]. Recently, the design of chemically diverse libraries for screening has become less popular than it used to be in the mid- to late 1990s, and there is a clear trend to generate smaller compound libraries that are focused on therapeutic target classes or specific activities, which then often involves similarity comparison of candidate compounds and known active molecules [27].

TABLE 3.2 Similarity Coefficients and Distance Functions[a]

Tanimoto coefficient	$Tc = n_{ij}/(n_i + n_j - n_{ij})$
Dice coefficient	$Dc = n_{ij}/(n_i + n_j)/2$
Cosine coefficient	$Cc = n_{ij}/(n_i n_j)^{1/2}$
Hamming distance	$HD_{ij} = (n_i + n_j - 2n_{ij})$
Euclidean distance	$ED_{ij} = (n_i + n_j - 2n_{ij})^{1/2}$
Average distance	$D = \dfrac{\sum_{i=1}^{n}\sum_{j=1}^{n} D_{ij}}{n(n-1)}$

[a] Shown are similarity metrics and distance functions that are often used for bit string comparisons. In these formulations, n_i and n_j are the number of bits set on for molecules i and j, respectively, and n_{ij} is the number of bits in common to both molecules. D_{ij} is the distance between molecules i and j, D the average distance, and n the total number of molecules.

Quantitative Structure–Activity Relationship Analysis

In QSAR, the relationship between biological activity of a compound and its molecular properties is explored, and these properties are by necessity expressed or modeled using molecular descriptors. Standard QSAR analysis attempts to express biological activity as a linear combination of different descriptors, thus postulating the presence of a linear relationship between activity and relevant molecular properties [28]. Coefficients in these equations assign weights to descriptors according to the calculated importance of their contributions for predicting activity. Typically, a series of analogs with known activity is used as a learning set for model building. Resulting QSAR models are evaluated using cross-validation techniques and well-performing models are used to predict structural modifications that might increase potency. Thus, different from molecular similarity evaluation, standard QSAR analysis is usually limited to congeneric compound sets. In 2D-QSAR, model building is based on 2D representation of molecules and 2D descriptors. In 3D-QSAR, bioactive conformations of test compounds are predicted and conformation-dependent steric and electrostatic fields of these compounds are generated and probed on grid representations with pseudoatoms for favorable interaction sites [29]. Of course, such 3D descriptors can also be used in combination with 1D or 2D descriptors to derive QSAR models. The quality of 3D-QSAR models critically depends, first and foremost, on the correct prediction of bioactive conformations and also on correct alignment or superposition of test compounds (when comparing their fields). In order to address the uncertainties of correctly predicting bioactive conformations, the 4D-QSAR formalism was introduced [30, 31]. Here, rather than deriving models based on a single putative conformation, a conformational ensemble of test molecules is calculated for 3D-QSAR, which in essence adds conformational space as a fourth dimension to the modeling process. The 4D-QSAR approach does not alleviate the alignment problem, and superpositions of test molecules are usually generated and optimized in an automated fashion [32]. Higher dimensional QSAR models, with additional degrees of freedom, can be derived by, for example, implicit or explicit inclusion of binding site elements in the modeling process [33].

Like in molecular similarity analysis, an important question in QSAR is how to find the "best" descriptors for solving a specific structure–activity relationship problem. Experience and intuition are often used in combination with trial-and-error calculations, but QSAR modeling can also be coupled to machine learning techniques such as genetic algorithms or neural nets in order to identify preferred descriptor combinations. Furthermore, it is often required to reduce the number of descriptors and identify those that make the most important contributions to the predictions. Initial 3D-QSAR models typically involve a large number of descriptors and reducing their numbers increases the interpretability of the models and decreases their complexity, which often renders them less susceptible to calculation errors due to overfitting of data or overlearning (as these effects are often caused by redundancy of descriptor contributions). The elimination of input variables corresponds to a reduction in dimensionality of descriptor reference spaces for establishing QSAR equations. For linear structure–activity relationships, partial least-squares regression is often used, and other descriptor space dimension reduction techniques, as described later on, can also be applied. Neural network methods are increasingly employed in 3D-QSAR, not only for descriptor selection but also for dimension reduction and model simplification. For example, Kohonen self-organizing neural networks or maps, an unsupervised learning technique, project high-dimensional (descriptor) space representations onto 2D maps [34]. In addition to descriptor reduction through self-organizing maps, another attractive feature of neural network methods in 3D-QSAR is that these methods are designed to model nonlinear relationships [35], which presents a departure from the classical QSAR theme. The major drawback of applying neural network methods is their "black box" character, that is, the inability to extract actual models from neural network simulations and interpret them in physical or chemical terms.

3.3 GENERAL LIMITATIONS OF VIRTUAL SCREENING ANALYSIS

At this stage of the discussion, one should also comment on principal limitations of ligand-based VS that more or less apply regardless of the methods used. General limitations of QSAR analysis (e.g., correct prediction of bioactive conformations, alignment problem, overfitting of data and overtraining, linear versus nonlinear structure–activity relationships) also affect QSAR models that are adapted for VS. Therefore, limitations different from QSAR-specific ones will be discussed here, in particular, those that relate to molecular similarity analysis.

Similar Property Principle Revisited

The postulate that similar molecules should have similar activity is intuitive, although it is, of course, also known that structurally similar molecules do not necessarily need to display equivalent biological activity [36]. The identification of remote similarity relationships (i.e., exploring structural and property differences that are tolerated within a specific activity class) is a major focal point of molecular similarity analysis in pharmaceutical research. In sharp contrast to the basic ideas underlying this approach is a fact well known by medicinal chemists, namely that a single chemical change can render a molecule within a series of active analogs almost or completely inactive. Figure 3.3 shows an example to further illustrate this point. The possibility of these alternative scenarios, dependent on the type of the search problem under investi-

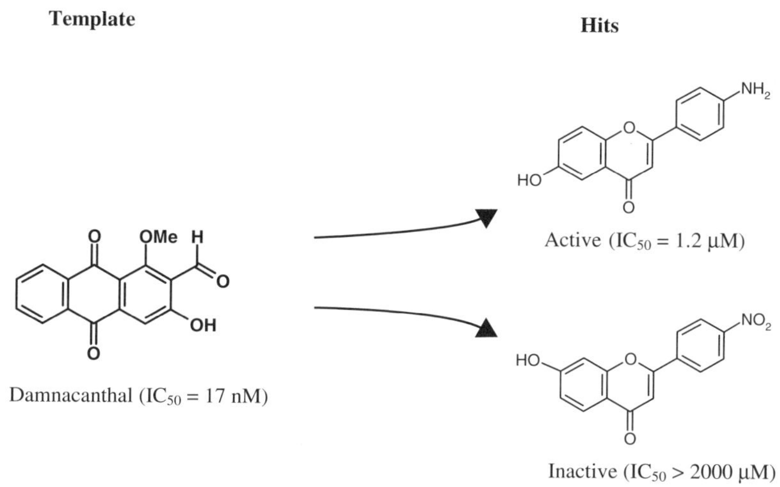

GENERAL LIMITATIONS OF VIRTUAL SCREENING ANALYSIS

Template   **Hits**

Damnacanthal (IC_{50} = 17 nM)

Active (IC_{50} = 1.2 μM)

Inactive (IC_{50} > 2000 μM)

Figure 3.3 Similarity paradox—a caveat for virtual screening. Results of a similarity search are shown using damnacanthal (a potent inhibitor of tyrosine kinase $p56^{lck}$) as a template. The search identified a number of molecules sharing the same core structure. However, minor modifications of this scaffold (typical for analogs) render compounds either active or inactive. Two examples are shown. Virtual screening calculations should identify these analogs as similar to damnacanthal. However, if by chance only the inactive analog would have been selected, VS analysis would have apparently failed, although a relevant scaffold was indeed identified. Adapted from [1].

gation, has substantial implication for VS analysis. For example, VS calculations are expected to recognize molecules as similar that share the same core structure. However, if only inactive analogs were selected for testing, VS analysis would have been considered a failure, although a relevant structural class was successfully identified. This conundrum has been termed the *similarity paradox* [1] and also led to the computational study of *latent hits* [37], that is, inactive compounds that can be transformed into active ones by very simple chemical modifications. For VS, this illustrates a potential danger associated with selection of only one or a few representative molecules from a series of compounds recognized as being similar to search templates. Compound selection strategies that generally address these difficulties are yet to be developed, and decisions are usually made on a case-by-case basis, dependent on the specifics of the search problem and the obtained results.

Quantifying Molecular Similarity

As mentioned above, one of the hallmarks of computational molecular similarity analysis is the ability to quantify similarity relationships, for example, by using metrics shown in Table 3.2. Similarity is expressed as a numerical value and relationships above a given threshold value are considered similar.

However, this convenient way of measuring similarity has some caveats. As an example, the Tanimoto coefficient (Tc), also shown in Table 3.2, is perhaps the most popular metric for quantitative comparison of bit string-type descriptors or fingerprints. This coefficient relates the number of bits that are set on in two-bit string representations to each other and quantifies fingerprint overlap. If the strings do not share any bits that are set on, the Tc value is 0 (no similarity) and if the bit settings are identical, Tc is 1 (maximum similarity). The calculation of Tc values for the assessment of molecular similarity has some inherent statistical limitations [38, 39]. More importantly, however, it is a priori not clear which threshold values correspond to "true" similarity. The comparison of studies using different fingerprint-type descriptors and molecular data sets illustrates this problematic issue. Patterson et al. evaluated different types of descriptors in the study of molecular similarity and established the concept of *neighborhood behavior* [40]. A major conclusion was that 85 percent of test compounds that displayed a Tc value of 0.85 in a fingerprint comparison to an active molecule were themselves active [40]. Similar findings were subsequently reported by others [41], and a Tc threshold value of 0.85 has in many instances become a guideline for the detection of biologically relevant similarity. However, in an analysis of various screening data sets, Martin et al. found that at a fingerprint similarity level of 0.85 only about 30 percent of detected compounds were active [42]. These findings are well in accord with observations that different similarity search methods frequently produce different results, dependent on the characteristics of activity classes that are studied [43]. Taken together, these findings demonstrate that results of similarity search calculations (and other VS methods) are generally influenced by the search tools applied, the data sets used, and the compound classes studied. Consequently, it is difficult to define generally applicable similarity threshold values or rules that unambiguously relate calculated and biological similarities to each other, which presents a potential caveat for VS analysis. Therefore, in some cases, Tc threshold values have been optimized for specific fingerprints and compound classes [4].

Potency and Selectivity Issues

Another limitation of VS has to do with the fact that one of its major goals is the detection of remote similarity relationships, that is, finding different structural classes having similar activity. For these calculations, optimized molecules are usually used as starting points that are typically taken from the patent literature or publications and have potency in the nanomolar range and high selectivity. It is often also attempted to find novel chemotypes as potential backup compounds for clinical candidates. Thus, template molecules are typically the result of a thorough lead optimization process. If VS calculations successfully identify a novel chemotype displaying similar activity, these molecules depart from optimized structural motifs and are therefore usually much less active. Simply put, successful VS analysis is expected to produce hits, not leads, which provide the starting point for a new medicinal chemistry program and optimization effort [1]. It is fairly typical for VS to produce hits with potency in the low micromolar range and moderate selectivity when highly potent templates are used [1]. This is sometimes not well understood and can lead to unrealistic expectations from VS technology.

3.4 SPECIFIC METHODS

In the following, different VS methods are discussed that have varying degrees of complexity. The first group of methods focuses on similarity searching, which involves quantitative assessment of molecular similarity. This is followed by clustering and partitioning techniques that can either be qualitative or quantitative in nature. Finally, various QSAR models for VS are discussed.

Two-Dimensional Substructure Searching

Perhaps the simplest form of VS, as we understand it today, is the search for matching substructures in compound databases. Early algorithms for substructural analysis did not utilize molecular descriptors but were based on graph representations where atoms are represented as nodes and bonds as edges [44]. In substructure analysis, comparison of molecular graphs aims at detecting subgraph isomorphism by comparing and matching node positions and edges [44]. This is a computationally expensive procedure and not amenable to screening of very large compound databases. Therefore, substructure search methods used as screening tools today mostly utilize dictionaries of molecular fragment-type descriptors or keys for comparisons and detect the presence or absence of such keys in database molecules [45]. For example, the publicly available 166 MACCS keys from MDL [46] are a popular set of structural fragment-type descriptors. Substructure searching can be directly linked to the analysis of biological activity, for example, by detecting recurrent fragments in active molecules and calculating the statistical significance of this frequency of occurrence relative to their distribution in other database molecules. This type of statistical substructure and property analysis has been implemented in the LeadScope program, which utilizes a dictionary containing more than 20,000 fragments [47]. For VS applications, a major limitation of substructure analysis is the difficulty of finding novel chemotypes, simply because only the presence or absence of predefined core structures can be detected. Thus, scaffold transitions can hardly be accomplished. By contrast, substructure methods are very effective when searching for analogs of active compounds or structurally related molecules.

Topological and Shape Representations

A number of specialized complex molecular descriptors have been designed that can be used as queries for similarity searching. For example, electrotopological indices combine molecular topology and electronic properties by capturing the valence state of each atom in a molecule and its electronic structure, taking into account perturbations caused by neighboring atoms [15, 48]. Each atom in a molecule is assigned a numerical E-state value and atom type. The atom types represented in a molecule can be used as a combined query and searched against atom types calculated for database molecules. Molecular comparisons are then carried out by calculating the Euclidian distance between atom type indices in template and test molecules [48]. By contrast, topomer shape descriptors are derived by fragmenting molecules in a defined way, assigning a rule-based conformation to each fragment, and calculating a molecular field for each fragment [49]. Thus, these descriptors capture whole-molecule shape information as a sum of fragment contributions. Molecular shape comparisons are carried out by calculating field differences for all possible fragment-by-fragment comparisons in two molecules. This fragment-based shape-matching algorithm has been shown to be applicable to similarity searching and used to screen large virtual libraries for active molecules [49].

Three-Dimensional Pharmacophores

Pharmacophores are generally defined as spatial arrangements of atoms or groups in molecules that are responsible for specific interactions with enzymes or receptors and the resulting biological activity. Similar to 2D substructures, pharmacophores can in some cases be derived from 2D molecular representations, but usually pharmacophore models are three dimensional and used for 3D database searching. Pharmacophore analysis has a long history in pharmaceutical research [50]. Early pharmacophore methods captured bond lengths and angles between specified atoms [51] or encoded shape information from atom sphere models [52]. Currently, pharmacophores are mostly represented as molecular (pseudo-) graphs where nodes correspond to pharmacophore points (atoms, groups) and edges to interpoint distances. Different features can be mapped to each point (e.g., hydrogen bond acceptors or aromatic groups) and interpoint distances can be binned in order to generate distance ranges. This structure makes it possible to differentiate similar pharmacophore arrangements from each other, rather than obtaining only yes/no answers when trying to match query and database pharmacophores. Conventionally used 3-point pharmacophores have a few years ago been extended to 4-point pharmacophore models [53]. Figure 3.4 shows a

Figure 3.4 Multiple-point pharmacophore models. The figure illustrates how pharmacophore models are derived from a known or predicted active conformation of a molecule. Atoms, groups, or features are represented as a graph from which three- and four-point pharmacophores are extracted. A four-point pharmacophore consist of four points and six distances (thereby defining a spatial arrangement). In a pharmacophore search, each point could represent several potential features and each distance might be divided into different ranges.

SPECIFIC METHODS

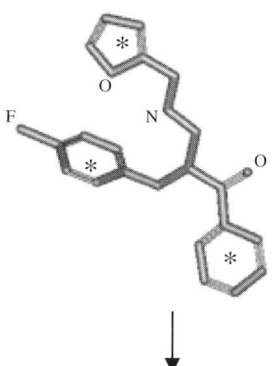

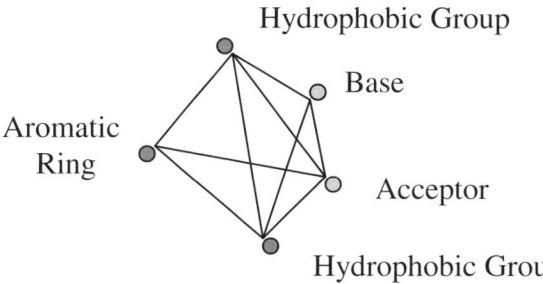

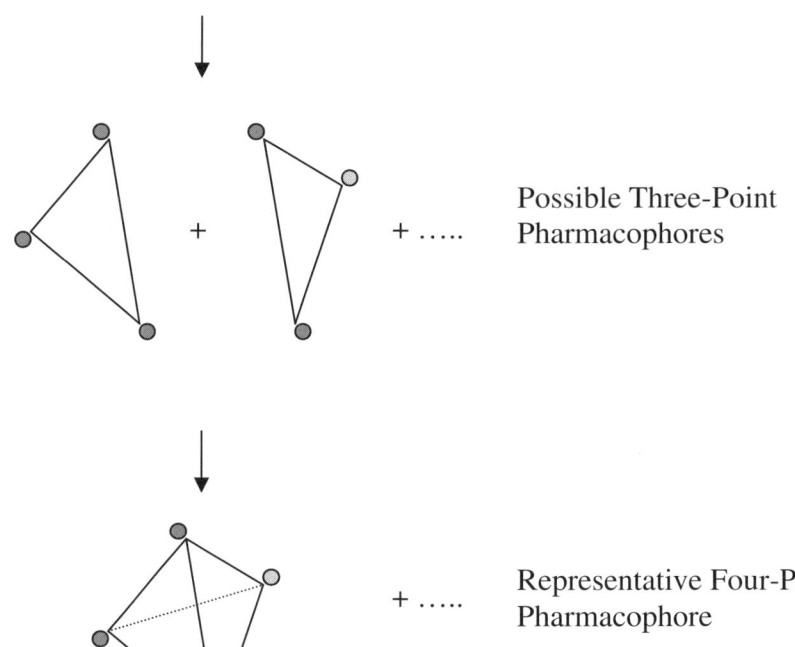

Possible Three-Point Pharmacophores

Representative Four-Point Pharmacophore

comparison of such 3- and 4-point pharmacophores. The introduction of 4-point pharmacophores has much increased the resolution of pharmacophore searching. For example, given a case where 6 features can be mapped to each point and each distance is divided into 10 ranges, a 3-point model could capture 33,000 potential pharmacophore arrangements whereas a 4-point pharmacophore would be able to monitor about 13 million.

Given the dramatic increase in computational performance in recent years, 3D pharmacophore searching can be readily applied to VS. Database searching for single pharmacophores has two major caveats. First, pharmacophore models typically represent a (nonconfirmed) hypothesis, as the pharmacophores are derived from the analysis of analogs, QSAR modeling, or inspection of crystal structures. Even if X-ray structures of active molecules in complex with their receptors are available, it is often not at all certain that attractive interactions that can be "seen" are indeed determinants of biological activity. Second, similar to the situation in conventional QSAR, it must generally be assumed that computed 3D structures of database molecules are representative of their bioactive conformations, which is probably the largest source of errors in pharmacophore searching. These difficulties have been addressed by the design of pharmacophore fingerprints, as discussed in the following.

Molecular Fingerprints

Bit string representations of various molecular descriptors are among the most popular methods for similarity searching. As already mentioned, such fingerprints are used in combination with various metrics (the Tanimoto coefficient being the most popular one) to quantify the degree of similarity between template and test molecules. Therefore, fingerprints need to be precalculated for database compounds and then compared to the corresponding fingerprints of query molecules. For VS analysis, fingerprints are attractive tools because bit string comparisons are very fast. The time-limiting step is usually the calculation of fingerprints of database compounds, which is required only once. It is important to note that the design and complexity of fingerprint representations often significantly differs, as further discussed below.

Two-Dimensional Fingerprints In simple fingerprints, each bit position accounts for the presence (if the bit is set on) or absence of a specific feature in a molecule. MACCS keys present an instructive example for this type of fingerprint [46]. If used in a bit string format (rather than as single substructure search queries), MACCS keys consist of a string of 166 bits, each of which accounts for the presence or absence of a specific structural key. BCI fingerprints (often consisting of 1024 bits) are also generated from different families of structural fragments [54], conceptually similar to MACCS keys. Any fragment dictionary-based approach has the intrinsic limitation that only cataloged fragments can be detected in test molecules, which may preclude the recognition of unique structural motifs. However, this limitation can sometimes be overcome by increasingly generalizing molecular fragment types, thus incorporating some degree of fuzziness in the analysis [54].

So-called minifingerprints are other short bit string representations, consisting of about 60 to 200 bits, that combine structural keys and molecular property descriptors specifically selected for their ability to correctly classify compounds according to their biological activity [55]. In these designs, value ranges of property descriptors (e.g., the number of hydrogen bond acceptors in a molecule) are incrementally encoded (e.g., no acceptors, no bit is set on; one or two acceptors, one bit is set on; three or four descriptors, two bits are set on, etc.). A special type of minifingerprint involves a binary transformation of molecular property descriptors based on statistical medians of descriptor value distributions in large compound databases [56]. This means that a bit position reserved for a particular descriptor is set to one if its value is larger than or equal to the median and set to zero if it is smaller. This novel coding scheme has made it possible to encode many property descriptors in a short bit string format. Although each descriptor contribution is monitored at a low level of resolution, simultaneous contributions of many different types of descriptors make these fingerprints very effective in similarity searching [56].

Daylight fingerprints represent a different class of 2D search tools [57]. Here all possible connectivity pathways in a molecule through up to seven bonds are recorded and mapped to a bit string format. Different from the simple designs discussed above (where each position is associated with a specific descriptor or value range), a hashing algorithm is applied to map connectivity pathways to overlapping bit segments. Daylight fingerprints are variable in length, dependent on hashing density, and popular forms consist of 1024 or 2048 bit positions.

Three-Dimensional Pharmacophore Fingerprints In pharmacophore fingerprints, each bit position typically monitors the presence or absence of one possible arrangement of a predefined three- or four-point pharmacophore in a test molecule, and these fingerprints therefore often consist of millions of bits [58–60]. Accordingly, they represent VS tools that are quite different from relatively simple 2D fingerprints. Possible pharmacophore arrangements are calculated based on fast systematic conformational search of test compounds. Pharmacophore fingerprinting elegantly circumvents some limitations of single pharmacophore searching. These fingerprints do not rely on accurate prediction of bioactive conformations and monitor possible spatial arrangements of functional groups at a high level of resolution. The underlying hypothesis is that increasing overlap of possible pharmacophore patterns in template and database compounds correlates with an increasing probability that these molecules share similar activity. Although pharmacophore fingerprinting is computationally demanding, it has become a much applied technique, given the large computational resources that are widely available by now.

Fingerprint Profiling and Scaling Methods have been introduced to tune fingerprint search calculations toward recognition of specific compound classes. When fingerprints are calculated and averaged for compounds belonging to diverse activity classes, compound class-specific bit patterns can often be observed [61]. Figure 3.5 shows examples of such fingerprint profiles. Dependent on the particular compound class, some bit positions may always, or almost always, be set on, thus defining consensus bit settings for this class. The application of scaling factors to such consensus bits emphasizes fingerprint profile differences during similarity search calculations and has been shown to lead to a significant increase in search performance [61], as reported in Table 3.3. Profiling and scaling can be applied to any 2D or 3D fingerprint representation where each bit monitors a specific feature but is much more difficult to apply to hashed or folded bit strings.

Two-Dimensional Versus Three-Dimensional Methods

At this stage of the discussion, it might be appropriate to comment on the relative performance of 2D and 3D descriptors and methods, a question that continues to be much debated in the literature. In principle, 3D methodologies should outperform 2D approaches, simply because molecules are active in three dimensions. However, difficulties involved in accurately calculating some conformation-dependent 3D properties and predicting bioactive compound conformations often reduce the accuracy of 3D methods to an extent that they are outperformed by simpler (and less error-prone) 2D approaches [4]. Moreover, 2D methods might also have some intrinsic advantages. For example, structural key-type descriptors implicitly encode much chemical information that might otherwise be difficult to explicitly calculate, which may—at least in part—explain their good performance in many applications [62]. Different studies have provided support for the superiority of either 3D [40, 59] or 2D [62, 63] methods. In some cases, successful recognition of remote similarity

Figure 3.5 Fingerprint profiles. Shown are representative profiles for two different biological activity classes calculated with a model fingerprint consisting of 61 bit positions (horizontal axis). The vertical axis records the frequency of bit occurrence within an activity class (i.e., 1 means the bit is always set on and 0 means it is never set on). Bits that are mostly or always set on define consensus positions for an activity class. The profiles of different activity classes show some notable differences that can be exploited in similarity searching by applying scale factors to class-specific consensus positions. Scaling increases the probability of identifying similar molecules.

SPECIFIC METHODS

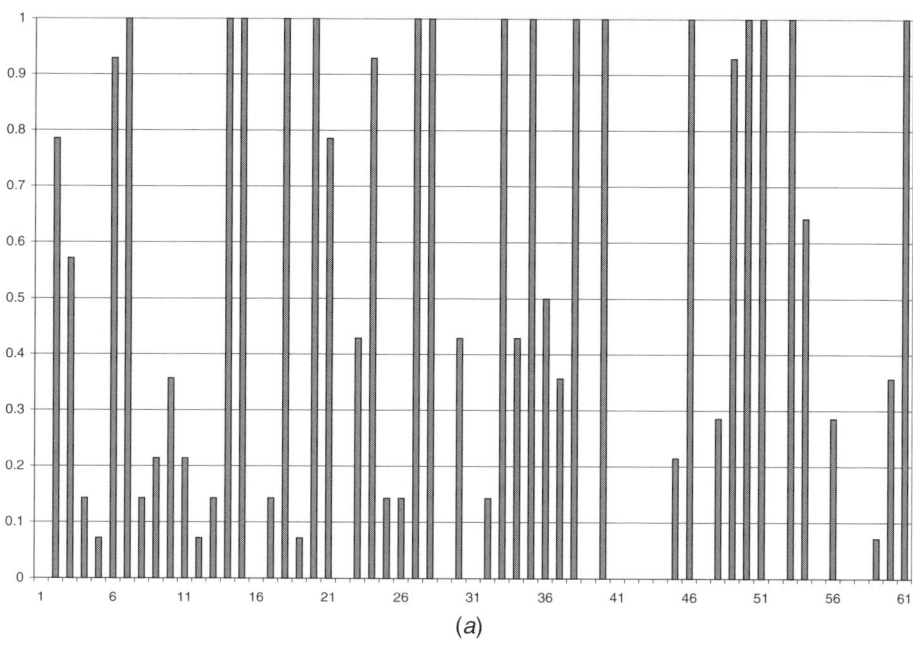

(a)

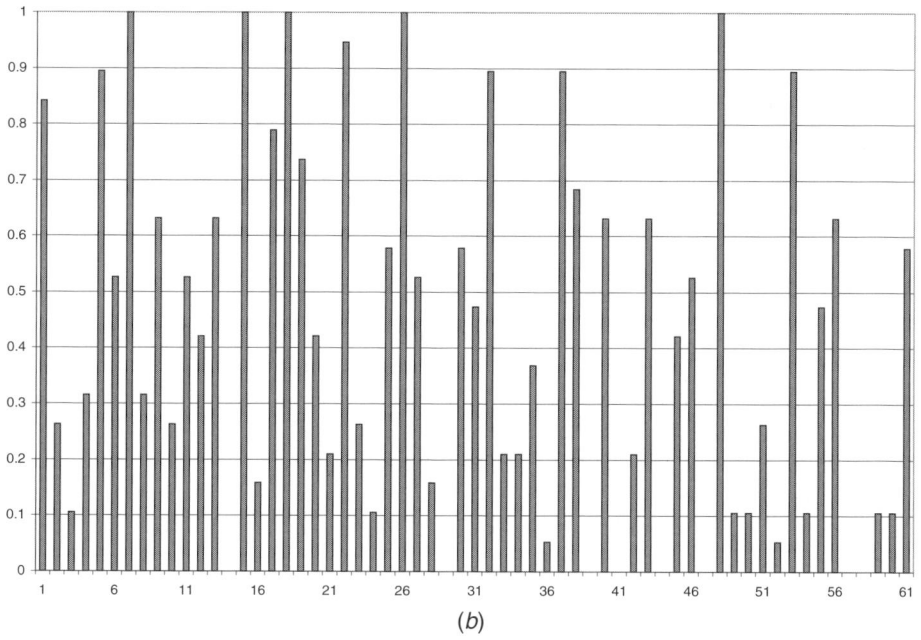

(b)

TABLE 3.3 Fingerprint Search Calculations and Scaling[a]

Fingerprint	Scaling	Tc	Percent Right	Percent Wrong
MACCS	Yes	0.94	47.9	0.030
	No	0.78	35.3	0.015
SE-MFP	Yes	0.96	46.5	0.012
	No	0.81	38.3	0.016
MP-MFP	Yes	0.90	58.7	0.018
	No	0.74	43.1	0.018

[a] The table summarizes the results of systematic similarity search calculations in a database containing 175 compounds belonging to 23 diverse activity classes and 5000 randomly selected synthetic compounds. Three fingerprints were tested, MACCS keys and two in-house designed fingerprints, SE-MFP and MP-MFP (described in [55] and [56], respectively). Overall best Tc similarity threshold values are also reported. Under scaling conditions, a scaling factor of 10 was uniformly applied to consensus positions with a bit frequency of at least 95 percent in the fingerprint profile of each activity class (see also Fig. 3.5). Here Percent Right reports the percentage of correctly identified active compounds averaged over all activity classes and Percent Wrong the average percentage of false positives (including both active and/or background compounds). As can be seen, fingerprint scaling leads to a significant increase in search performance, with MP-MFP producing best results.
Data were taken from [56].

relationships was thought to strictly depend on the application of 3D approaches [60], but these findings could nevertheless be well reproduced using much simpler 2D methods [64]. Simply put, it is probably not possible to claim that either 2D or 3D methods are generally superior since much depends on the specifics of each test case and application [4]. What can be said, however, is that 2D descriptors and methods perform surprisingly well in the analysis of molecular similarity and in many VS situations.

Clustering Techniques

Cluster analysis of compounds in chemical reference spaces has been applied for more than 20 years and continues to play an important role in pharmaceutical research [65]. The major goal of clustering is to group together sets of compound with similar structures and/or chemical features and distinguish them from others. Such compound classification techniques can assess molecular similarity in a qualitative or quantitative way. Fingerprint calculations can be based on single template molecules, whereas clustering or partitioning techniques become attractive whenever series of known active molecules are available. In a typical VS analysis, one adds these active template molecules to a compound database, finds descriptor solutions and clustering conditions that group active molecules together, and selects other compounds that occur in these clusters (and are thus considered similar) as candidates for testing. In the following, different clustering algorithms will be briefly discussed.

SPECIFIC METHODS

Hierarchical Versus Nonhierarchical Clustering Clustering methods can be divided into two major categories, hierarchical and nonhierarchical approaches [65]. Nonhierarchical clustering groups molecules together into a defined number of clusters, without considering intercluster relationships. Typically, nearest neighbors are calculated for each test molecule and two compounds are included in the same cluster if they share a prespecified minimum number of nearest neighbors. Hierarchical clustering, on the other hand, establishes relationships between initially generated clusters and is further divided into divisive or agglomerative approaches. In hierarchical-divisive clustering, the starting point is one large cluster that is subdivided based on distance calculations and similarity relationships (top-down approach). By contrast, in hierarchical-agglomerative clustering, one begins with singletons that are combined into increasingly larger clusters (bottom-up).

For both hierarchical and nonhierarchical approaches, the optimum clustering level and preferred input parameters are difficult to determine upfront and, therefore, cluster-level selection algorithms must usually be applied to determine conditions giving best results [66]. Such algorithms work, for example, by identifying clustering levels that balance the number of clusters with their average population or density. This can be accomplished by comparing intracluster distances between molecules in all clusters as well as distances between cluster centers. For chemistry applications, the most popular hierarchical and nonhierarchical methods have generally been Ward's (agglomerative) [67] and Jarvis–Patrick [68] clustering, respectively. For compound classification, Ward's clustering has been identified as a method of choice in a number of cases [62].

Multidomain Clustering For compound classification, hierarchical-agglomerative clustering with level selection has been implemented, in addition to self-organizing maps, in a decision-tree-like algorithm combining a variety of simulation techniques [69]. Here the final clustering step of compound sets, typically HTS data, is carried out using structural keys as descriptors, and for acceptable clusters the largest common substructure is determined. Clusters can be scored by calculating the average activity of the compounds they contain. If carried out in an iterative fashion, the algorithm makes it possible to identify small sets of compounds that have common substructures associated with biological activity. Moreover, molecules containing different substructures that modulate activity, so-called multidomain compounds, can appear in different clusters, thus revealing multiple structure–activity relationships [69]. Although this phylogenetic-like tree method was originally developed to process HTS data sets and extract structure–activity relationships from them, it can also be applied to VS by adding sets of known active compounds to databases and analyzing the content of clusters into which these molecules fall.

Partitioning Methods

In recent years, partitioning algorithms have become increasingly popular for compound classification, analysis of HTS results, and VS applications [70]. One of the major reasons for this popularity, among others, is the computational efficiency of many partitioning techniques. Like clustering, partitioning operates in chemical descriptor spaces. However, regardless of the algorithm, cluster analysis always involves the calculation and comparison of pairwise distances between compounds in reference spaces. Computationally, this is not problematic for small to medium-size compound data sets, but, when databases are large (containing millions of molecules), these pairwise distance calculations become more and more difficult and at some point prohibitive. By contrast, partitioning algorithms typically calculate subdivisions or sections of chemical space into which compounds fall, based on their descriptor coordinates or signature patterns. Thus, partitioning does not require pairwise distance comparisons and can therefore be applied to much larger compound databases than distance-based clustering.

Different classes of partitioning methods have been developed that are amenable to VS. Just as in cluster analysis, known active compounds are added to a source database that is then partitioned, and database compounds that occur in partitions with active molecules are considered as candidates for testing. Different from cluster analysis, partitioning does not produce inter-compound distances as a possible quantitative measure of molecular similarity (although such distances can be calculated retroactively). Thus, partitioning is usually more qualitative in nature.

Cell-Based Partitioning In recent years, cell-based partitioning has become one of the most popular analysis techniques in computational medicinal and combinatorial chemistry [70]. Here partitions or cells are generated to evenly segment chemical space by applying binning schemes to each descriptor axis. The dimensionality of the descriptor space and the number of bins per axis ultimately determine the number of cells, which is a major determinant of the outcome of compound classification calculations. Cell-based partitioning is generally carried out in low-dimensional descriptor spaces (e.g., six dimensions) for several reasons. In low-dimensional reference spaces with orthogonal axes, it is possible to limit space distortions due to descriptor correlation effects, avoid large numbers of unpopulated cells, and control cell population density. In addition, low-dimensional spaces can often be further reduced to three dimensions without loosing too much information, which makes it possible to visualize compound distributions, a rather important requirement for many chemical applications. Low-dimensional reference spaces can be constructed by using especially designed complex descriptors that combine different types of contributions (e.g., molecular connectivity and hydrogen bond potential) in an uncorrelated manner [71]. Alternatively, it might be required to reduce the dimensionality of originally defined high-dimensional descriptor spaces. Dimension reduction can be accomplished by calculating a small number of orthogonal descriptors axes from the original descriptors, for example, as linear combinations of these descriptors [17, 72], or by applying nonlinear functions that extract the most significant (descriptor-dependent) relationships between compounds and project compound distributions into lower-dimensional spaces [73]. Significant relationships or features would include, for example, detected similarity between compounds that share comparable activity.

Direct Space Methods Recent studies have shown that dimension reduction is not necessarily required for effective partitioning if original descriptor spaces are simplified. For example, by use of statistical medians of descriptor value distributions, as mentioned earlier, molecular property descriptors can be transformed into a binary format. Therefore, the median partitioning algorithm was developed that uses one property descriptor per partitioning step to iteratively divide a compound database into subpopulations above and below the median value of this descriptor [74]. According to this scheme, the choice of n descriptors and partitioning steps results in a total of 2^n unique partitions, each of which is characterized by a binary partition code. Thus, median partitioning operates in multidimensional, albeit simplified, descriptor spaces but does not involve dimension reduction. Essentially, every molecule is assigned a vector in multidimensional descriptor space with the length along each dimension being either zero or one (unity). As summarized in Table 3.4,

TABLE 3.4 Compound Classification by Cell-Based and Direct Space Partitioning[a]

Method	Dim	Bins	%P	P	nP	S	M	nM
Cell-based partitioning	6	7	55.2	57	175	82	19	60
Median partitioning	19	n/a	63.1	69	200	86	22	31

[a] For these calculations, 317 compounds belonging to 21 different activity classes were added to 2000 randomly selected synthetic molecules. This database was then partitioned using different methods. Column 2 (Dim) reports the dimensionality of the descriptor reference space and column 3 (Bins) the number of intervals on each axis for cell-based partitioning. In direct space partitioning, no binning scheme is applied. As a cell-based partitioning method, partitioning based on principal component analysis was used, as described in [72]. Principal component analysis is one of several dimension reduction methods. Column 4 (%P) reports percentage of active compounds that occur in so-called pure partitions. These partitions exclusively contain compounds sharing the same activity. In activity-based compound classification, this is the desired partitioning result. Column 5 (P) gives the total number of pure partitions, column 6 (nP) the total number of compounds in pure partitions, column 7 (S) the number of singletons, column 8 (M) the number of mixed partitions, and column 9 (nM) the total number of compounds in mixed partitions. Mixed partitions contain molecules having different activities or active and randomly selected molecules (classification errors). As can be seen, median partitioning in a 19-dimensional descriptor space compares favorably with cell-based partitioning in a 6-dimensional space when classifying these compounds according to their biological activity.
Data were taken from [74].

median partitioning can perform as well as or even better than cell-based partitioning in the classification of compounds according to diverse biological activities [74]. In its adaptation for VS of very large compound databases, median partitioning is carried out in a recursive manner [75]. With the aid of machine learning techniques, descriptor combinations are identified that successfully co-partition known active molecules added as "baits" to the database. At each recursion level, database compounds are pooled that occur in the same partitions as successfully co-partitioned baits, other database compounds are discarded, descriptor selection is reinitialized, and the process is repeated until only a small number of database compound remains (as potential candidates for testing). During the course of this analysis, these candidates have consistently been recognized as "similar" to bait molecules because they have always been found in active partitions.

Another recently developed direct space method, termed *dynamic mapping of consensus positions*, combines elements of binary descriptor transformation, partitioning, and bit string methods [76]. Starting from a large pool of binary-transformed descriptors, initially, the maximum number of descriptors is determined that have identical bit settings for all compounds belonging to the studied activity class. This defines a descriptor space where all compounds map exactly to the same (consensus) position. Then, in order to eliminate an increasing number of database compounds and ultimately retain only those that are most similar to the baits, the dimensionality (and resolution) of the descriptor space is increased in subsequent steps. This is accomplished by defining consensus positions that no longer require identical descriptor settings for all test compounds but permit one or more known actives to deviate in one or more descriptor settings from the others. This stepwise dimension extension eliminates most database compounds from activity class-specific consensus positions [76]. Like median partitioning, this approach makes use of simplified descriptor spaces, but it does no longer depend on machine learning techniques for descriptor selection. Moreover, it operates not only in multidimensional descriptor spaces but utilizes dimension extension as an approach to identify compounds that are similar to known actives. A unique feature of this mapping algorithm is that it generates compound class-specific descriptor reference spaces, which distinguishes it from many other classification methods. As a VS tool, dynamic mapping of consensus positions is very fast, permitting the rapid processing of hundreds of descriptors and millions of database compounds. Importantly, the method has achieved promising hit rates in large-scale VS analysis, as reported in Table 3.5.

Decision Tree Algorithms Another type of statistical partitioning method utilizes decision tree structures as the primary classification tool. The perhaps most popular decision tree method for the analysis of HTS data and VS is recursive partitioning [77]. Initially, learning sets with known active and inactive compounds are processed along a tree structure using two-state structural or topological descriptors. At every branch of the tree, descriptors are used to

SPECIFIC METHODS

TABLE 3.5 Virtual Screening by Recursive Median Partitioning and Dynamic Mapping of Consensus Positions[a]

Activity Class	DMC		RMP	
	HR (%)	RR (%)	HR (%)	RR (%)
I	26	9	12	18
I	50	18	3	11
III	74	29	3	5
IV	9	23	21	22
V	26	9	13	40

[a] Recursive median partitioning is abbreviated RMP and dynamic mapping of consensus positions is abbreviated DMC. Calculations were carried out on five different activity classes in a database consisting of ~1.3 million compounds. These are: I, benzodiazepines; II, cyclooxygenase-2 inhibitors; III, histamine H3 antagonists; IV, serotonin receptor ligands; V, tyrosine kinase inhibitors. In these VS calculations, between 21 and 61 active compounds were "hidden" in the database (active database compounds) as potential hits, and 10 bait molecules were used in each case. Hit rates (HR) were calculated by dividing the number of correctly identified active database compounds (hits) by the total number of selected database molecules, and recovery rates (RR) by dividing the number of hits by the total number of active database compounds. For both RMP and DMC calculations, hit and recovery rates were significant but DMC produced higher hit rates for four of the five activity classes that were studied.
Data were taken from [76].

successively divide the data set into compound subsets that have or do not have the tested features. If the partitioning analysis is successful, subsets that are significantly enriched with active compounds are obtained in terminal nodes that associate active molecules with specific descriptor settings and pathways. These activity-related descriptor combinations can then be used as predictive models to recursively partition large compound databases and search for novel active molecules [77]. Computationally, recursive partitioning is perhaps the most efficient partitioning method and millions of compounds and very large numbers of descriptors can be rapidly processed. The method has some known limitations. Different from multidomain clustering, as discussed above, active molecules can usually only occur in one terminal node. Furthermore, molecules with similar activity yet significant structural differences (remote similarity relationships) might be separated during the partitioning process and occur in different terminal nodes. In addition, the method is prone to high false-positive rates when predicting active compounds [78]. Therefore, several extensions of the recursive partitioning algorithm have recently been introduced that address some of these issues. For example, the use of multiple trees, so-called recursion forests, permits the application of consensus scoring schemes, which have been shown to produce significant hit rates in VS, while reducing false-positive rates [78]. In addition, tree structures have been developed for recursive partitioning that allow the simultaneous analysis and prediction of more than one property as a function of descriptor settings, for example, specificity in addition to activity [79]. This would make it possible, for example, to process a compound library that has been screened against multiple targets using a single tree structure and address selectivity questions.

QSAR Models in Virtual Screening

Despite its long-standing history in medicinal chemistry, QSAR techniques have only in recent years been adapted for VS. This might at least in part be due to the fact that QSAR analysis is predominantly applied during lead optimization, rather than hit identification. In essence, however, applying a QSAR model to screen a large compound library does not differ much from using it to analyze or predict a series of analogs. On the other hand, it is also clear that QSAR models are in principle better suited to search for analogs of known leads than for novel chemotypes, which might limit their application in VS. In the context of VS, QSAR overlaps with pharmacophore searching, as many pharmacophore models are based on QSAR, and the boundaries between these approaches are somewhat fluid. Recently, different types of QSAR models have been directly applied to VS.

Two-Dimensional QSAR Models   Due to the computational efficiency of 2D descriptor-based models, it is in principle straightforward to use 2D-QSAR models to virtually screen large compound databases. However, not many 2D-QSAR VS studies have been reported thus far. In a recent application, a number of 2D-QSAR models for cyclooxygenase-2 inhibitors were developed using different methodologies (including, e.g., partial least-squares linear regression as well as supervised and unsupervised neural networks) and applied in combination to develop a consensus prediction theme [80]. In this case, one searches for molecules that are consistently favored by several models. These QSAR models and consensus predictions were then applied in a preliminary screen of the database of the National Cancer Institute (~200,000 compounds) and several potentially interesting molecules were found [80].

Three-Dimensional QSAR Virtual Screens   Although the derivation of 3D-QSAR models is more complex and time consuming than 2D-QSAR modeling, there are more VS applications of 3D-QSAR, which is likely due to the fact that 3D-QSAR is more closely related to 3D pharmacophore searching. For example, in the discovery of novel VLA-4 antagonists, a 3D pharmacophore was constructed based on known crystallographic and structure–activity data and used to screen a combinatorial library. Compounds identified in this search were then subjected to 3D-QSAR analysis in order to better understand the different activity levels of identified hits [81]. Similarly, a pharmacophore search was combined with 3D-QSAR (comparative molecular field) analysis to identify hits and improve their potency [82]. In this case, the analysis went a step further and computed analogs were also studied by docking calculations, thus putting high emphasis on the exploitation of structural data for both ligands and receptor [82].

Four-Dimensional QSAR Although computationally more expensive, a multidimensional QSAR formalism was also adapted for in silico screening by including conformational ensembles (of glycogen phopsphorylase inhibitors) in 3D-QSAR analysis [83]. In this study, it was well recognized that screening (or design) of target-focused combinatorial libraries should be a prime application for QSAR-based VS since these libraries should already contain preferred structural motifs. Moreover, the analysis revealed that overfitting of 4D-QSAR models to training data—although usually known to represent a problem for QSAR predictions—might sometimes be beneficial for their application in VS simply because it favors the identification of library compounds that are likely to sample (and probe) molecular regions not well represented in training set molecules. Such findings might substantially aid in the design of second-generation focused libraries.

Probabilistic Methods QSAR-like approaches have also been developed for large-scale property predictions. Binary QSAR uses combinations of molecular descriptors to build relationships between compounds and their activity or other properties of interest [84]. The approach is based on Bayes's theorem, a mathematical formulation determining the probability that a result is due to a specific cause among multiple plausible causes. Binary QSAR makes use of learning sets to construct probability density functions that estimate properties of molecules from their descriptor values. Compared to standard QSAR approaches, this methodology is more qualitative in nature because it ultimately develops predictive models based on a binary formulation of activity that can be applied to large molecular test sets. Each test molecule is assigned a probability to be active and, based on a chosen threshold value, classified as either active or inactive. Similar to recursive partitioning, binary QSAR can be applied to efficiently process large HTS data sets and, using so-derived models, rapidly screen databases for molecules having a high probability to be active. An attractive feature of this probabilistic approach is that the derivation of predictive models does not depend on the availability of highly accurate data sets for learning. Thus, it can usually be applied to process raw HTS data. That binary QSAR calculations are not limited to activity predictions, but are more widely applicable, has been demonstrated, for example, by studies to systematically distinguish between naturally occurring and synthetic molecules [85].

Another conceptually similar but algorithmically distinct probabilistic machine learning technique that is experiencing increasing interest in the context of compound classification and VS is binary kernel discrimination [86]. Similar to binary QSAR, this methodology uses learning sets to calculate the probability of a molecule to be active (by applying kernel functions). However, the approach is currently more constrained in the choice of molecular descriptors and usually limited to two-state descriptors, similar to recursive partitioning.

Drug Likeness

Better understanding and predicting druglike or leadlike features of candidate compounds during the early phases of drug discovery has become a current focal point in computational medicinal chemistry and also influenced the development of VS methods. Neural network simulations have been extensively applied to build computational models that distinguish drugs from nondrugs [19]. However, as mentioned earlier, such models typically lack interpretability and, consequently, it has also been attempted to build simpler and more intuitive ones. These include decision tree models [87] and, in their simplest form, dictionaries of preferred and nonpreferred functional groups or rules that prioritize desired structural motifs and physico-chemical compound characteristics [88]. For VS, rule-based prediction schemes of druglikeness have become rather attractive and are widely used because they can be added as simple filter functions to screening calculations [5]. In addition, various computational models have been developed for ADME-related properties that can be carried out as virtual screens or added to VS methods. These include, for example, blood–brain barrier penetration [89] or oral availability, for which QSAR models reaching 70 percent prediction accuracy were developed based on a learning set consisting of known drugs [90].

3.5 VIRTUAL SCREENING PERFORMANCE

Having addressed general limitations of VS, one should also discuss how successful in silico screening has been thus far. Like any computational model, VS tools are generally rich in approximations and might frequently fail for technical reasons. Furthermore, as already mentioned earlier, different similarity search methods often produce different results [43], as the calculations are much influenced by specific features of studied compound classes. However, without doubt VS methods work and help to discover novel active molecules. Based on experience gathered over the past few years, novel hits with activity in the micromolar range are identified in about 75 percent of the VS projects carried out in our laboratory. In some cases, in-house developed VS methods, as described herein, were capable of identifying hits in situation where HTS had failed or was not feasible (e.g., because cell-based assays were too complex or time consuming to be formatted for HTS). In two other cases (structurally relatively simple) natural products with specific biological activity were chemically not accessible but could be replaced by *synthetic mimics* based on VS calculations, thus permitting the generation of analogs.

While obtaining desired VS results is far from being a routine procedure, a number of successful case studies have been reported in the literature for target structure- and ligand-based VS as well as for combinations of different 2D and 3D methodologies [5, 10]. Case studies have been discussed in this chapter in the context of specific methods. Some of these studies—but also others—are capable of illustrating the potential and current performance levels of VS in more general terms. For example, in a recent investigation, HTS and structure-based VS were carried out in parallel and successfully identified diverse inhibitors of tyrosine phosphatase-1B with activities in the low- to mid-micromolar range [91]. In this study, docking calculations ultimately produced significantly higher hit rates than random screening (by up to three orders of magnitude). In ligand-based VS, a combination of 3D pharmacophore search and 3D-QSAR analysis identified VLA-4 antagonists with activity in the sub-micromolar range [81], a potency level that is still an exception rather than the rule for many VS applications. In order to make use of all target- and ligand-related information that has become available, different types of VS methodologies are often combined, which further increases the probability of success. For example, when combining similarity searching, docking, and binary QSAR calculations, five new antagonists of the BCL-xL/Bak interaction were identified with potency in the low-micromolar range (and after testing fewer than 50 candidate molecules) [92]. When new hits were subsequently used to aid in the design of a target-focused library, seven additional active compounds were obtained [92]. Similarly, docking calculations were recently combined with 2D filtering methods and 3D pharmacophore searching in a study that discovered several new inhibitors of carbonic anhydrase [93].

As discussed in detail in this chapter, compound classification techniques (in particular, partitioning) have become increasingly attractive as similarity-based approaches for VS, due to their computational efficiency and success rates. For example, results of benchmark calculations reported for different partitioning methods illustrate this point. For a set of monoamine oxidase inhibitors, recursive partitioning produced a 15-fold improvement in hit rate over random selection [77]. Furthermore, in large-scale VS calculations, recursive median partitioning achieved hit rates between 3 and 21 percent for five different compound classes [75]. Even higher hit rates between 9 and 74 percent were obtained in a parallel study on these classes using dynamic mapping of consensus positions [76], a method conceptually related to median partitioning, as discussed (see Table 3.5). Although clustering and partitioning methods display significant promise as stand-alone VS tools, these techniques become particularly attractive when used in close combination with biological screening for iterative preselection of compound sets, as already indicated in the introductory sections. This process is often called sequential screening [8]. In these situations, VS has been shown in various independent investigations to increase hit rates by one to two orders of magnitude [5, 8], and this performance level can currently be considered as a standard in the field. For example, in a study targeting a number of cell-based anticancer assays, two rounds of sequential screening produced hit rates of up to 40% [5]. Moreover, an analysis of a number of different two-stage sequential screening investigation showed that testing of 10 to 20 percent of compounds in screening libraries was generally sufficient to identify 50 to 80 percent of potential hits [5]. These findings suggest that VS analysis often significantly reduces the number of compounds that need to be tested until novel molecules of interest are identified. It is important to note that achieving such efficiency levels does not require dramatic improvements in hit rates. For example, analysis of a large HTS data set (for an undisclosed target) has shown that up to fivefold increase in hit rates could be achieved by application of recursive partitioning [94] and that only ~20 percent of available database compounds needed to be screened in order to identify ~75 percent of potential hits [94]. These findings are well in accord with the numbers reported for two-stage sequential screens, as mentioned above, and have significant practical implications for organizing HTS campaigns.

Clearly, despite their documented promise, VS methods should not only be evaluated based on their performance as (stand-alone) hit identification tools but also based on their ability to complement experimental programs and increase their efficiency.

3.6 CONCLUSIONS AND OUTLOOK

Many molecular similarity methods have matured over the years and substantially contributed to making VS an important component of pharmaceutical research. In addition, novel methodologies continue to be developed at a fast pace to further increase the accuracy and computational efficiency of molecular similarity analysis and VS. Despite its predominant application in lead optimization, QSAR techniques have recently been adapted for VS. Moreover, QSAR-type probabilistic methods have been specifically designed for screening applications. Multidimensional QSAR modeling alleviates problems of accurately predicting bioactive conformations of ligands and will likely be more widely applied in the context of VS. It is also anticipated that ligand- and target-based design techniques will be more closely combined in the furture, as already exemplified by receptor-dependent QSAR modeling and methods for structure-based design of pharmacophores and 3D fingerprints. As far as experimental applications are concerned, currently available data suggest that the integration of VS and HTS is particularly promising, despite some practical constraints that still exist (such as limited availability of sufficiently flexible compound plating and management systems). However, there is little doubt that VS and HTS are highly complementary disciplines.

REFERENCES

1. Bajorath, J. (2002). Virtual screening: Methods, expectations, and reality. *Curr. Drug Discov., 2*(3), 24–28.

2. Walters, W. P., Stahl, M. T., Murcko, M. A. (1998). Virtual screening—An overview. *Drug Discov. Today, 3*, 160–178.
3. Bajorath, J. (2001). Rational drug discovery revisited: Interfacing experimental programs with bio- and chemo-informatics. *Drug Discov. Today, 6*, 989–995.
4. Bajorath, J. (2001). Selected concepts and investigations in compound classification, molecular descriptor analysis, and virtual screening. *J. Chem. Inf. Comput. Sci., 41*, 233–245.
5. Bajorath, J. (2002). Integration of virtual and high-throughput screening. *Nat. Rev. Drug Discov., 1*, 882–894.
6. Fox, S., Farr-Jones, S., Yund, M. A. (1999). High-throughput screening for drug discovery: Continually transitioning into new technologies. *J. Biomol. Screen., 4*, 183–186.
7. Lahana, R. (1999). How many leads from HTS? *Drug Discov. Today, 4*, 447–448.
8. Engels, M. F. M., Venkatarangan, P. (2001). Smart screening: Approaches to efficient HTS. *Curr. Opin. Drug. Discov. Devel., 4*, 275–283.
9. Owens, J. (2003). Jürgen Bajorath discusses integrating disciplines in the drug discovery industry. *Targets, 2*, 45–47.
10. Green, D. V. (2003). Virtual screening of virtual libraries. *Prog. Med. Chem., 41*, 61–97.
11. Brooijmans, N., Kuntz, I. D. (2003). Molecular recognition and docking algorithms. *Annu. Rev. Biophys. Biomol. Struct., 32*, 335–373.
12. Abagyan, R., Totrov, M. (2001). High-throughput docking for lead generation. *Curr. Opin. Chem. Biol., 5*, 375–382.
13. Livingstone, D. J. (2000). The characterization of chemical structures using molecular properties. A survey. *J. Chem. Inf. Comput. Sci., 40*, 195–209.
14. Xue, L., Bajorath, J. (2000). Molecular descriptors in chemoinformatics, computational combinatorial chemistry, and virtual screening. *Combin. Chem. High Throughput Screen., 3*, 363–372.
15. Hall, L. H., Kier, L. B. (2000). The E-state as the basis for molecular structure space definition and structure similarity. *J. Chem. Inf. Comput. Sci., 40*, 784–791.
16. Labute, P. (2000). A widely applicable set of descriptors. *J. Mol. Graph. Model., 18*, 464–477.
17. Xue, L., Bajorath, J. (2000). Molecular descriptors for effective classification of biologically active compounds based on principal component analysis identified by a genetic algorithm. *J. Chem. Inf. Comput. Sci., 40*, 801–809.
18. Gillet, V. J., Willett, P., Bradshaw, J. (1998). Identification of biological activity profiles using substructural analysis and genetic algorithms. *J. Chem. Inf. Comput. Sci., 38*, 165–179.
19. Sadowski, J. (2000). Optimization of chemical libraries by neural network methods. *Curr. Opin. Chem. Biol., 4*, 280–282.
20. Godden, J. W., Bajorath, J. (2003). An information-theoretic approach to descriptor selection for database profiling and QSAR modeling. *QSAR Comb. Sci., 22*, 487–497.
21. Johnson, M. A., Maggiora, G. M. (Eds) (1990). *Concepts and Applications of Molecular Similarity.* Wiley, New York.

22. Maggiora, G. M., Shanmugasundaram, V. (2004). Molecular similarity measures. In J. Bajorath (Ed.), *Chemoinformatics: Concepts, Methods, and Tools for Drug Discovery*. Humana, Totowa, NJ.
23. Willett, P., Barnard, J. M., Downs, G. M. (1998). Chemical similarity searching. *J. Chem. Inf. Comput. Sci., 38*, 983–996.
24. Martin, E. J., Blaney, J. M., Siani, M. A., Spellmeyer, D. C, Wong, A. K., Moos, W. H. (1995). Measuring diversity: Experimental design of combinatorial libraries for drug discovery. *J. Med. Chem., 38*, 1431–1436.
25. Martin, Y. C. (2001). Diverse viewpoints on computational aspects of molecular diversity. *J. Comb. Chem., 3*, 231–250.
26. Willett, P. (1999). Dissimilarity-based algorithms for selecting structurally diverse sets of compounds. *J. Comput. Biol., 6*, 447–457.
27. Xue, L., Godden, J. W., Stahura, F. L., Bajorath, J. (2001). A dual fingerprint-based metric for the design of focused combinatorial libraries and analogs. *J. Mol. Model., 7*, 125–131.
28. Free, S. M. Jr., Wilson, J. W. (1964). A mathematical contribution to structure-activity studies. *J. Med. Chem., 7*, 395–399.
29. Cramer, R. D. III, Patterson, D. E., Bunce, J. D. (1988). Comparative molecular field analysis (CoMFA). 1. Effect of shape on binding of steroids to carrier proteins. *J. Am. Chem. Soc., 110*, 5959–5967.
30. Hopfinger, A. J. (1980). A QSAR investigation of dihydrofolate reductase inhibition by Baker triazines based upon molecular shape analysis. *J. Am. Chem. Soc., 102*, 7196–7206.
31. Hopfinger, A. J., Wang, S., Tokarski, J. S., Jin, B., Albuquerque, M., Madhav, P. J., Duraiswami, C. (1997). Construction of 3D-QSAR models using the 4D-QSAR analysis formalism. *J. Am. Chem. Soc., 119*, 10509–10524.
32. Kearsley, S. K., Smith, G. M. (1990). An alternative method for the alignment of molecular structures: Maximizing electrostatic and steric overlap. *Tetrahedron Comput. Meth., 3*, 615–633.
33. Vedani, A., Dobler, M. (2002). 5D-QSAR: The key for simulating induced fit? *J. Med. Chem., 45*, 2139–2149.
34. Kohonen, T. (1989). *Self-Organization and Associative Memory*. Springer, Berlin.
35. Tetko, I. V., Kovalishyn, V. V., Livingstone, D. J. (2001). Volume learning algorithm artificial neural networks for 3D-QSAR studies. *J. Med. Chem., 44*, 2411–2420.
36. Kubinyi, H. (1998). Similarity and dissimilarity—A medicinal chemist's view. *Perspect. Drug Discov. Design, 11*, 225–252.
37. Mestres, J., Veeneman, G. H. (2003). Identification of "latent hits" in compound screening collections. *J. Med. Chem., 46*, 3441–3444.
38. Flower, D. R. (1998). On the properties of bit-string measures of chemical similarity. *J. Chem. Inf. Comput. Sci., 38*, 379–386.
39. Godden, J. W., Xue, L., Bajorath, J. (2000). Combinatorial preferences affect molecular similarity/diversity calculations using binary fingerprints and Tanimoto coefficients. *J. Chem. Inf. Comput. Sci., 40*, 163–166.
40. Patterson, D. E., Cramer, R. D. III, Ferguson, A. M., Clark, R. D., Weinberger, L. E. (1996). Neighborhood behavior: A useful concept for validation of "molecular diversity" descriptors. *J. Med. Chem., 49*, 3049–3059.

41. Matter, H. (1997). Selecting optimally diverse compounds from structure databases: A validation study of two-dimensional and three-dimensional descriptors. *J. Med. Chem., 40*, 1219–1229.
42. Martin, Y. C., Kofron, J. L., Traphagen, L. M. (2002). Do structurally similar molecules have similar biological activity? *J. Med. Chem., 45*, 4350–4358.
43. Sheridan, R. P., Kearsley, S. K. (2002). Why do we need so many chemical similarity search methods? *Drug Discov. Today, 7*, 903–911.
44. Barnard, J. M. (1993). Substructure searching methods. Old and new. *J. Chem. Inf. Comput. Sci., 33*, 532–538.
45. Merlot, C., Domine, D., Cleva, C., Church, D. J. (2003). Chemical substructures in drug discovery. *Drug Discov. Today, 8*, 594–602.
46. MACCS keys, MDL Information Systems, San Leandro, CA.
47. Roberts, G., Myatt, G. J., Johnson, W. P., Cross, K. P., Blower, P. (2000). LeadScope: Software for exploring large sets of screening data. *J. Chem. Inf. Comput. Sci., 40*, 1302–1314.
48. Kier, L. B., Hall, L. H. (2001). Database organization and searching with E-state indices. *SAR QSAR Environ. Res., 12*, 55–74.
49. Cramer, R. D. III, Poss, M. A., Hermsmeier, M. A., Caulfield, T. J., Kowala, M. C., Valentine, M. T. (1999). Prospective identification of biologically active structures by topomer similarity searching. *J. Med. Chem., 42*, 3919–3933.
50. Martin, Y. C. (1992). 3D database searching in drug design. *J. Med. Chem., 35*, 2145–2154.
51. Murray-Rust, P., Motherwell, S. (1978). Computer retrieval and analysis of molecular geometry. 1. General principles and methods. *Acta Chryst., 34B*, 2518–2526.
52. Sheridan, P., Venkataragharvan, R. (1987). Designing novel nicotinic agonists by searching a database of molecular shapes. *J. Comput.-Aided Mol. Design, 1*, 243–256.
53. Mason, J. S., Morize, I., Menard, P. R., Cheney, D. L., Hulme, C., Labaudiniere, R. F. (1999). New 4-point pharmacophore method for molecular similarity and diversity applications: Overview over the method and applications, including a novel approach to the design of combinatorial libraries containing privileged substructures. *J. Med. Chem., 42*, 3251–3264.
54. Barnard, J. M., Downs, G. M. (1997). Chemical fragment generation and clustering software. *J. Chem. Inf. Comput. Sci., 37*, 141–142.
55. Xue, L., Godden, J. W., Bajorath, J. (2003). Mini-fingerprints for virtual screening: Design principles and generation of novel prototypes based on information theory. *SAR QSAR Environ. Res., 14*, 27–40.
56. Xue, L., Godden, J. W., Stahura, F. L., Bajorath, J. (2003). Design and evaluation of a molecular fingerprint involving the transformation of property descriptor values into a binary classification scheme. *J. Chem. Inf. Comput. Sci., 43*, 1151–1157.
57. James, C. A., Weininger, D. *Daylight Theory Manual.* Daylight Chemical Information Systems, Irvine, CA.
58. McGregor, M. J., Muskal, S. M. (1999). Pharmacophore fingerprinting. 1. Application to QSAR and focused library design. *J. Chem. Inf. Comput. Sci., 39*, 569–574.

59. Mason, J. S., Cheney, D. L. (2000). Library design and virtual screening using multiple point pharmacophore fingerprints. *Pac. Symp. Biocomput., 5*, 576–587.
60. Bradley, E. K., Beroza, P., Penzotti, J. E., Grootenhuis, P. D. J., Spellmeyer, D. C., Miller, J. L. (2000). A rapid computational method for lead evolution: Description and application to α_1-adrenergic antagonists. *J. Med. Chem., 43*, 2770–2774.
61. Xue, L., Godden, J. W., Stahura, F. L., Bajorath, J. (2003). Profile scaling increases the similarity search performance of molecular fingerprints containing numerical descriptors and structural keys. *J. Chem. Inf. Comput. Sci., 43*, 1218–1225.
62. Brown, R. D., Martin, Y. C. (1996). Use of structure-activity data to compare structure-based clustering methods and descriptors for use in compound selection. *J. Chem. Inf. Comput. Sci., 36*, 572–584.
63. Matter, H., Pötter, T. (1999). Comparing 3D pharmacophore triplets and 2D fingerprints for selecting diverse compound subsets. *J. Chem. Inf. Comput. Sci., 39*, 1211–1225.
64. Xue, L., Stahura, F. L., Godden, J. W., Bajorath, J. (2001). Mini-fingerprints detect similar activity of receptor ligands previously recognized only by three-dimensional pharmacophore-based methods. *J. Chem. Inf. Comput. Sci., 41*, 394–401.
65. Willett, P. (1987). *Similarity and Clustering in Chemical Information Systems.* Research Studies Press, Letchworth.
66. Wild, D. J., Blankley, C. J. (2000). Comparison of 2D fingerprint types and hierarchy level selection methods for structural grouping using Ward's clustering. *J. Chem. Inf. Comput. Sci., 40*, 155–162.
67. Ward, J. H. (1963). Hierarchical grouping to optimize an objective function. *J. Am. Stat. Assoc., 58*, 236–244.
68. Jarvis, R. A., Patrick, E. A. (1973). Clustering using a similarity measure based on shared near neighbors. *IEEE Trans. Comput., C22*, 1025–1034.
69. Nicolaou, C. A., Tamura, S. Y., Kelley, B. P., Bassett, S. I., Nutt, R. F. (2002). Analysis of large screening data sets via adaptively grown phylogenetic-like trees. *J. Chem. Inf. Comput. Sci., 42*, 1069–1079.
70. Stahura, F. L., Bajorath, J. (2003). Partitioning methods for the identification of active molecules. *Curr. Med. Chem., 8*, 707–715.
71. Pearlman, R. S., Smith, K. M. (1998). Novel software tools for chemical diversity. *Perspect. Drug Discov. Design, 9*, 339–353.
72. Xue, L., Bajorath, J. (2002). Accurate partitioning of compounds belonging to diverse activity classes. *J. Chem. Inf. Comput. Sci., 42*, 757–764.
73. Agrafiotis, D. K., Lobanov, V. S. (2000). Nonlinear mapping networks. *J. Chem. Inf. Comput. Sci., 40*, 1356–1362.
74. Godden, J. W., Xue, L., Bajorath, J. (2002). Classification of biologically active compounds by median partitioning. *J. Chem. Inf. Comput. Sci., 42*, 1263–1269.
75. Godden, J. W., Furr, J. R., Bajorath, J. (2003). Recursive median partitioning for virtual screening of large databases. *J. Chem. Inf. Comput. Sci., 43*, 182–188.
76. Godden, J. W., Furr, J. R., Xue, L., Stahura, F. L., Bajorath, J. (2004). Molecular similarity analysis and virtual screening by mapping of consensus positions in binary-transformed descriptor spaces with variable dimensionality. *J. Chem. Inf. Comput. Sci., 44*, 21–29.

77. Rusinko, A. III, Farmen, M. W., Lambert, C. G., Brown, P. L., Young, S. S. (1999). Analysis of a large structure/biological activity data set using recursive partitioning. *J. Chem. Inf. Comput. Sci., 39*, 1017–1026.
78. van Rhee, A. M. (2003). Use of recursion forests in the sequential screening process: Consensus selection by multiple recursion trees. *J. Chem. Inf. Comput. Sci., 43*, 941–948.
79. Stockfisch, T. P. (2003). Partially unified multiple property recursive partitioning (PUMP-RP): A new method for predicting and understanding drug selectivity. *J. Chem. Inf. Comput. Sci., 43*, 1608–1613.
80. Baurin, N., Mozziconacci, J.-C., Arnoult, E., Chavatte, P., Marot, C., Morin-Allory, L. (In press). 2D QSAR consensus prediction for high-throughput virtual screening. An application to COX-2 inhibition modeling and screening of the NCI database. *J. Chem. Inf. Comput. Sci.*
81. Singh, J., van Vlijmen, H., Liao, Y., Lee, W. C., Cornebise, M., Harris, M., Shu, I. H., Gill, A., Cuervo, J. H., Abraham, W. M., Adams, S. P. (2002). Identification of potent and novel $\alpha 4\beta 1$ antagonists using in silico screening. *J. Med. Chem., 45*, 2988–2993.
82. Chen, H. F., Dong, X. C., Zen, B. S., Gao, K., Yuan, S. G., Panaye, A., Doucet, J. P., Fan, B. T. (2003). Virtual screening and rational drug design using structure generation system based on 3D-QSAR and docking. *SAR QSAR Environ. Res., 14*, 251–264.
83. Hopfinger, A. J., Reaka, A., Venkatarangan, P., Duca, J. S., Wang, S. (1999). Construction of a virtual high throughput screen by 4D-QSAR analysis: Application to a combinatorial library of glucose inhibitors of glycogen phosphorylase b. *J. Chem. Inf. Comput. Sci., 39*, 1151–1160.
84. Labute, P. (1999). Binary QSAR: A new method for the determination of quantitative structure activity relationships. *Pac. Symp. Biocomput., 4*, 444–455.
85. Stahura, F. L., Godden, J. W., Xue, L., Bajorath, J. (2000). Distinguishing between natural products and synthetic molecules by Shannon descriptor entropy analysis and binary QSAR calculations. *J. Chem. Inf. Comput. Sci., 40*, 1245–1252.
86. Harper, G., Bradshaw, J., Gittin, J. C., Green, D. V. S., Leach, A. R. (2001). Prediction of biological activity for high-throughput screening using binary kernel discrimination. *J. Chem. Inf. Comput. Sci., 41*, 1295–1300.
87. Wagener, M., van Geerestein, V. J. (2000). Potential drugs and non-drugs: Prediction and identification of important structural features. *J. Chem. Inf. Comput. Sci., 40*, 280–292.
88. Muegge, I. (2003). Selection criteria for drug-like compounds. *Med. Res. Rev., 23*, 302–321.
89. van de Waterbeemd, H., Camenisch, G., Folkers, G., Chretien, J. R., Raevsky, O. A. (1998). Estimation of blood-brain barrier crossing of drugs using molecular size and shape and H-bonding descriptors. *J. Drug Target., 6*, 151–165.
90. Yoshida, F., Topliss, J. G. (2000). QSAR model for drug human oral bioavailability. *J. Med. Chem., 43*, 2575–2585.
91. Doman, I. N., McGovern, S. L., Witherbee, B. J., Kasten, T. P., Kurumbail, R., Stallings, W. C., Connolly, D. T., Shoichet, B. K. (2002). Molecular docking and high-throughput screening for novel inhibitors of protein tyrosine phosphatase-1B. *J. Med. Chem., 45*, 2213–2221.

92. Stahura, F. L., Xue, L., Godden, J. W., Bajorath, J. (2002). Methods for compound selection focused on hits and application in drug discovery. *J. Mol. Graph. Model.*, *20*, 439–446.
93. Grüneberg, S., Stubbs, M. T., Klebe, G. (2002). Successful virtual screening for novel inhibitors of human carbonic anhydrase: Strategy and experimental confirmation. *J. Med. Chem.*, *45*, 3588–3602.
94. van Rhee, A. M., Stocker, J., Printzenhoff, D., Creech, C., Wagoner, P. K., Spear, K. L. (2001). Retrospective analysis of an experimental high-throughput screening data set by recursive partitioning. *J. Comb. Chem.*, *3*, 267–277.

4

SYSTEMS BIOLOGY: APPLICATIONS IN DRUG DISCOVERY

SEAN EKINS, ANDREJ BUGRIM, YURI NIKOLSKY, AND
TATIANA NIKOLSKAYA
GeneGo
St. Joseph, Michigan

4.1	DEFINITIONS OF SYSTEMS BIOLOGY	124
4.2	MODELING PATHWAYS, CELLS, WHOLE ORGANS, AND DISEASES	125
4.3	DATABASES AS A FOUNDATION FOR SYSTEMS BIOLOGY	127
4.4	HIGH-THROUGHPUT DATA	131
	DNA Sequence Variations	132
	Gene Expression Profiling	132
	Proteomics	135
	Metabolomics and Metabonomics	137
4.5	MINING HIGH-THROUGHPUT DATA	138
	Limitations of Clustering Approaches for Gene Expression Data	138
	Advanced Computational Methods for Interpretation of Microarray Data	139
4.6	AVAILABLE RESOURCES FOR BIOLOGICAL PATHWAY ANALYSIS	142
4.7	APPLICATIONS OF SYSTEMS BIOLOGY FOR DRUG DISCOVERY	146
	Glaucoma	146
	Breast Cancer	151
	Systems ADME/Tox	156
4.8	FUTURE DEVELOPMENT OF PATHWAY ANALYSIS SOFTWARE	159
	Reconstruction of Condition-Specific Molecular and Functional Networks	163
4.9	FUTURE OF SYSTEMS BIOLOGY APPLIED TO DRUG DISCOVERY	164
	References	165

Drug Discovery Handbook, by Shayne Cox Gad
Copyright © 2005 by John Wiley & Sons, Inc.

4.1 DEFINITIONS OF SYSTEMS BIOLOGY

The last several years have seen a paradigm shift for biology. Large-scale international collaborations using high-throughput biology methods have resulted in the sequencing and annotation of the human genome [1, 2], several other mammalian genomes, and hundreds of other genomes as well. The next addition to our growing knowledge base will occur over 2 to 4 years and result in the genomewide individual sets of polymorphisms. The ribonucleic acid (RNA) microarray-based expression technology has also evolved into an almost routine laboratory technique dominating proteomics, metabolomics [3], and metabonomics [4], which have also made substantial progress and will be described later.

Both government agencies and drug companies have invested heavily in these technologies, partly based on the expectation that they will individually or in combination lead to an increase in efficiency of drug discovery. However, despite the unprecedented technology progression and substantial advancement of human biology, genomics has so far failed to contribute to the drug discovery pipeline. In fact, the number of drugs approved by the Food and Drug Administration (FDA) decreased in recent years, with the average cost of bringing these new drugs to the market exceeding $1 billion including marketing costs and taking between 10 to 12 years [5]. The process of drug discovery is therefore lengthy and ultimately complex with several economic and technology factors contributing to this inefficiency. One problem is the poor utilization of already accumulated experimental data, mostly from preclinical research. Recently, robust statistical solutions for data analysis were introduced by bioinformatics companies, and these have been mainly used for data point clustering, visualization, and comparisons between experimental series. However, such statistics alone are not sufficient for meaningful data mining in the context of disease mechanisms. Most experts in industry and academia agree that only functional analysis of the data based on a fundamental understanding of the human biology can solve this bottleneck. The complexity of our own biology, therefore, requires a systemwide approach to genomic and other molecular data analysis. This brings us to the forefront of systems biology, which focuses on the structure and dynamics of complex life systems and is rapidly becoming a leading approach to the integration and mining of data in human biology.

Systems biology can be briefly defined as the integration of genetic, proteomic, transcriptomic, and metabonomic data using computational methods [4]. The tenet of systems biology has been described as identification of all the genes and proteins in an organism, a parts list, which is certainly insufficient to understand the whole. It is therefore apparent that understanding the assembly of these parts [6, 7] (Fig. 4.1) is truly the key to understanding life, form, and function. In what appears a sweeping rebuttal to reductionism, systems biology appears the new paradigm for interpretation of the biologically complex data sets we are now increasingly generating [7], with higher

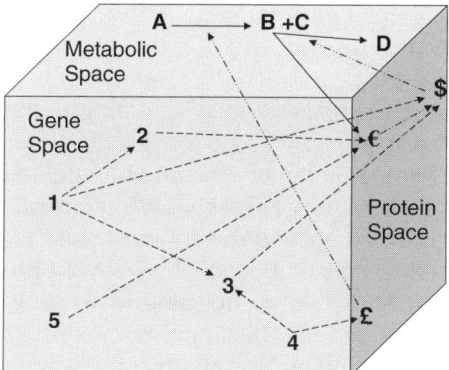

Figure 4.1 Schematic to illustrate the complex network of interconnections in the cell between gene space, protein space, and metabolic space. In gene space numbers represent distinct genes with arrows representing connections to other genes or transcription to proteins. In protein space the currency symbols represent unique proteins with arrows representing reactions catalyzed or protein–protein interactions. In metabolic space letters represent molecules that are substrates or metabolites with arrows representing reactions or effects on proteins. (This figure is available in full color at ftp://ftp.wiley.com/public/sci_tech_med/drug_discovery/.)

throughput technologies and the accumulated knowledge on human protein interactions defining certain conditions. Analysis of affected protein networks will allow identification of the sensitive pathways that have high utility as biomarkers for such conditions and new therapeutic targets. This new era for biology has in turn revolutionized the fields of expertise needed from contributors [8] such that engineers and computer scientists are just as likely as molecular biologists to be working on biological projects.

We will describe the various different tools and databases for this type of analysis, their applications, utility, and limitations. The integration of the methods of systems biology with traditional methods of computational biology and computational chemistry represents a powerful approach to combine experimental data, literature information, and predicted properties such that decisions can be made earlier in the development of a drug as to its potential for successful development.

4.2 MODELING PATHWAYS, CELLS, WHOLE ORGANS, AND DISEASES

Due to the use of high-throughput approaches the quality and quantity of available experimental biology observations is no longer a roadblock to understanding the whole cell. The challenge of requiring a data-driven approach for biology pressures computing science as it stands at present to develop advanced integrated tools that organize and structure the existing data for a complex system to make it more understandable. This is particularly important when we need to use this information for discovery of improved therapeutics and hypothesis testing in silico [9]. Computational models continue to be generated to explain complex cellular signaling pathways, metabolic cascades, and gene regulation [10–12]. These models ultimately resemble complex circuit diagrams, and analogies can be drawn with engineering with respect to feedback, robustness, and fragility [13, 14]. Laws identical to those derived for the electrical system may be obeyed during metabolic reactions [15, 16]. The simulation of these cell circuits with their high level of interconnections and organization is therefore possible in the same way that electrical circuits can be simulated prior to production. This approach may lead to the understanding and replication of experimental results through simulations [17].

Engineering terms are also being used with increasing frequency to describe transcriptional regulatory networks such as the role of hepatocyte nuclear factors (HNF) in control of pancreas and liver gene expression [18]. With the growing database of biological data surrounding metabolic pathways it becomes feasible to reverse engineer such circuit diagrams. One group used genetic programming (previously applied to the automated construction of electrical circuits) to build small metabolic pathways and genetic networks [19]. Such modeling approaches are likely to have a considerable impact on being able to recreate and compute the many simultaneous operations occurring within a cell. Much has been written regarding the fundamentals and difficulties of model construction and analysis [20], and the reader is referred to the studies in this section as well as those based on modeling microbes [21]. The result of such efforts can lead to a second example of modeling in systems biology, namely the "virtual cell" based on a heirachical system of molecules, reactions, and intracellular structure [22–24]. This type of approach enables the construction of a physiological model using biophysical knowledge and the application of physical laws. Once built, these types of models enable the effects of many variables to be studied, provide an understanding of functioning under stress, disease, gene knockout or knockin, and chemical modification [9]. This whole-cell approach has also been used to model the calcium wave in a neuronal cell [22]. The modeling and simulation of integrated biochemical and genetic processes for *Mycoplasma genitalium* has also been accomplished with E-CELL software [25]. All of these types of models represent a tool for iterative hypothesis testing, merging experimental data with computational/bioinformatics approaches. It has been suggested that in cells, molecules may interact with other cellular components in a manner dependent on their molecular properties and probalistic collisions with other reactive molecules [4]. Complex interactions also take place concurrently between endogenous metabolic pathways and the metabolism of xenobiotics requiring a systems biology approach [see later section on systems absorption, distribution, metabolism, excretion, and toxicology (ADME/Tox)]. For example, the need for in silico approaches to predict action on the cardiac cell is at present the most high profile due to the life-threatening consequences of cardiac arrhythmia [9]. However, drugs themselves seem to rarely bind to a single channel, receptor, or other protein, which makes predictions more difficult. Yet models of the whole heart have been generated to predict action potentials, the spread of activation and enable blood flow calculations [9, 26]. This is just one example of a complex cellular process that could be modeled more completely by understanding whole-cell and whole-organ behavior. It is likely that other whole-organ models will be created in a similar fashion until at some point they can all be integrated. The modeling of disease processes is also possible via *regulatory circuits* generated based on many input parameters that can be used to make predictions and validate therapeutic targets [27, 28]. These types of approaches might provide insights into overlapping molecular processes in disease such as oxidative stress as a trigger for Alzheimer's disease [29] and hypertension [30].

4.3 DATABASES AS A FOUNDATION FOR SYSTEMS BIOLOGY

Metabolism is certainly the most studied area of human biology with over a century of research leading to an understanding of metabolic networks as the major core effector in the living cell. Along with ion channel modulation, changing the concentration of metabolites (including, e.g., neurotransmitters and hormones) is a key method by which the organism acts in response to stimuli. Tissue-specific enzymatic reactions are the endpoints of complex vertical signaling events initiated by membrane receptors and amplified through signal transduction cascades and the ultimate activation of transcription factors. Overall, the interconnected metabolic pathways are the most fundamental network in the body. Gathering and organizing this type of knowledge on cellular pathways is a key first step in creating any system-level representation of a biological system. Since the 1970s, several teams in both industry and academia have systematically studied and organized databases on proteins, enzyme-encoding genes, and metabolic and cell-signaling pathways. Metabolic databases were generated for a wide diversity of organisms to centralize the growing quantities of biological information enabling access to scientists [31] for pathway reconstruction. The following list represents a summary of key databases of interest to those in the systems biology field, including many organism-specific and general-purpose databases for metabolism.

BIND Biomolecular Interaction Network Database (*http://www.bind.ca/*), stores full descriptions of molecular interactions, complexes, and pathways for virtually all components of molecular mechanisms and can be used to study networks of interactions, map pathways across taxonomic branches, or to generate information for kinetic simulations [32].

Biocarta A database of hundreds of metabolic and signaling pathways that integrates emerging proteomic information and summarizes important resources providing information for over 120,000 genes from multiple species (*www.biocarta.com*).

BioCyc Knowledge Library is the new name for EcoCyc and MetaCyc (*http://metacyc.org/*). It is a collection of pathway/genome databases. Each database in the BioCyc collection describes the genome and metabolic pathways of a single organism, with the exception of the MetaCyc database, which is a reference source on metabolic pathways from many organisms. The MetaCyc metabolic pathway database contains pathways from over 150 different organisms and describes metabolic pathways, reactions, enzymes, and substrate compounds gathered from a variety of literature sources.

BRENDA BRaunschweig ENzyme DAtabase (*http://www.brenda.uni-koeln.de/*) is a protein function database including metabolic and enzymatic information [33, 34].

CSNDB Cell Signaling Networks Database (*http://geo.nihs.go.jp/csndb/*) is a data and knowledge base for signaling pathways of human cells. It compiles the information on biological molecules, sequences, structures, functions, and biological reactions that transfer the cellular signals. Signaling pathways are compiled as binary relationships of biomolecules represented by graphs drawn automatically [35].

DIP Database of Interacting Proteins (*http://dip.doe-mbi.ucla.edu/*) catalogs experimentally determined interactions between proteins curated manually by expert curators and also automatically using computational approaches [36].

EcoCyc A database that describes the genome and the biochemical machinery of *Escherichia coli* K12 MG1655 (*http://ecocyc.org/*) and the functions of each of its molecular parts, to facilitate a system-level understanding.

EMP Enzymes and Metabolic Pathways (*http://emp.mcs.anl.gov/*) covers all aspects of enzymology and metabolism with about 30,000 records compiled from about 15,000 original journal publications [37]. EMP can also be used for metabolic reconstructions from sequenced genomes.

ERGO The ERGO system represents the development of a genome analysis strategy into a multidimensional environment, supporting both automatic and manual genomewide curation and integrating genomic information with biochemical data, literature, and high-throughput analysis into a comprehensive user-friendly network of metabolic and nonmetabolic pathways (*http://ergo.integratedgenomics.com/ERGO/*).

GenMAPP Gene MicroArray Pathway Profiler is a computer program designed for viewing and analyzing gene expression data on microarray pathway profiles (MAPPs) representing biological pathways and any other grouping of genes (*http://www.genmapp.org/*). GenMAPP facilitates the analysis of the large amounts of data produced in deoxyribonucleic acid

(DNA) microarray experiments and allows one to visualize gene expression data in a biological context [38].

GeNet Regulatory gene network provides information on the functional organization of regulatory gene networks acting at embryogenesis (http://www.csa.ru/Inst/gorb_dep/inbios/genet/genet.htm).

HPRD Human Protein Reference Database (http://www.hprd.org/) is a new resource for human-specific protein information that includes nearly 16,000 protein–protein interactions, 8000 proteins, posttranslational modifications, enzyme–substrate relations, and disease associations. Information is manually curated from the biomedical literature [39, 40].

HumanCyc A database that presents metabolic pathways as an organizing framework for the human genome. HumanCyc (http://humancyc.org/) provides the user with an extended dimension for functional analysis at the genomic level [41].

KEGG Kyoto Encyclopedia of Genes and Genomes (http://www.genome.ad.jp/kegg/) is possibly the best known database [42]. KEGG is a database of metabolic and regulatory data for systematic analysis of gene functions, including gene, pathway, and ligand databases enabling the comparison of two genome maps, manipulation of expression maps, graph comparison, and path computation.

MPW A collection of more than 3000 metabolic and functional diagrams representing functionality in 211 different organisms (http://bioresearch.ac.uk/whatsnew/detail/9020325.html) [43].

PKR A database of protein kinase information including tools for structural and computational analyses as well as links to related information maintained by others (http://www.sdsc.edu/kinases/) [44].

SENTRA A database of sensory signal transduction proteins in mostly microbes; includes information about sensory transduction proteins as well as annotated data from the SwissProt and EMBL databases [45].

SPAD The Signaling Pathway Database (http://www.grt.kyushu-u.ac.jp/eny-doc/) is used to provide an overview of signal transduction and is compiled in order to describe information on interactions between protein–protein, protein, and DNA as well as information on sequences of DNA and proteins. SPAD is divided to four categories based on extracellular signal molecules (growth factor, cytokine, and hormone) and stress, which initiate the intracellular signaling pathway.

STKE Signal Transduction Knowledge Environment (http://www.stke.org/) is an Internet-based resource for information on signal transduction.

SwissProt A curated protein sequence database with a high level of annotation including the description of the function of a protein, its domain structure, posttranslational modifications, and variants [46, 47].

TAR-GET A database of proteins of biotechnological and pharmaceutical interest (http://www-wit.mcs.anl.gov/target/). This database was created to support prioritization of targets for three-dimensional (3D) structural determination.

TRANSPATH Comprises data relating to molecules participating in signal transduction and the reactions they undergo, spanning a complex network of interconnected signaling components (*http://193.175.244.148/*) [48].

Unigene An experimental system for autopartitioning GenBank sequences into nonredundant gene-orientated clusters with each cluster representing a unique gene (*http://www.ncbi.nlm.nih.gov/entrez/query.fcgi?db=unigene*). At present there are thought to be approximately 106,219 human genes [49].

UMBBD The University of Minnesota biocatalysis/biodegradation database of xenobiotics (*http://umbbd.ahc.umn.edu/*) developed to provide information on microbial enzyme-catalyzed reactions that are important for biotechnology [50, 51].

WIT2 What is There? (*http://www-unix.mcs.anl.gov/compbio/*) is an integrated system for support of genetic sequence and comparative analysis of sequenced genomes and metabolic reconstructions from the sequence data. This database superceded PUMA—phylogeny of the unicellular organisms (*http:wit.mcs.anl.gov/WIT2/*), a database of metabolic pathways alignments [52].

Combined, these databases provide a high-quality, fairly comprehensive snapshot of human and other organism metabolism or signal transduction that most drug companies and large research centers have selectively licensed. Databases represent one component used for building and testing models of cellular pathways. For example, by integrating genomic and proteomic analyses of a systematically perturbed metabolic network, one group has been able to identify proteins regulated posttranscriptionally in the yeast–galactose–utilization pathway [53]. A similar approach has also been taken to build a genomic regulatory network for development in the sea urchin [54]. Being independent of each other, the databases currently available can be effectively networked using integration modules, such as SRS available from Lion Bioscience (*http://www.lionbioscience.com/solutions/e20472/e20475/index_eng.html*) and BioRS available from Biomax (*http://www.biomax.de/products/f_prod_BioRS.html*). However, neither of these systems can be used as a framework for data mining as they lack key features limiting their utility as follows: The amount of data on pathways, reactions, and interactions captured in these databases is insufficient to be used effectively for deciphering genomewide data sets. According to our own estimates, KEGG covers about 2500 products of human genes that are mostly metabolic enzymes. This represents approximately 15 percent of the human genes with known function. The generation of content for these databases has been slow with mainly voluntary contributions for certain protein families, which does not address the specific needs of drug discovery. These proteins need to be the most prominent drug targets such as G-protein coupled receptors (GPCRs) and other membrane receptors, kinases, and ion channels important for disease pathways. The collections of pathways in many of these databases (with the exception of ERGO, HumanCyc, HPRD, and Biocarta) are not organism specific, with the data having been collected from many species and different phylla (bacteria, fungi, plants, animals). The human-specific maps in databases like KEGG represent rather generic schemes that are the same for all organisms. The information on human enzymes may also be superimposed onto a network of generic pathways, which ultimately causes many errors and inconsistencies. However, only tissue, cell-type-specific, and expression time-adjusted maps are likely to be relevant, although these are not provided by any of these sources to date. Importantly, the architecture of these databases relies on molecular entities such as genes or proteins, as the elementary objects that are acceptable for isolated proteins and in agreement with the paradigm of *one gene—one protein* relationships as in bacterial systems. With human and other complex eukaryotic systems, this simplistic approach does not provide the functional links between objects, such as when all proteins are involved in the same pathway, or for the expression of all pathway related genes that encode these proteins in the context of the known human disease or disorder. Some of the available database resources provide the pathways in the form of predrawn metabolic maps with no database records for the interactions that constitute them. As a result, users are limited to retrieving one-dimensional information regarding an object, for example, a gene sequence, but not for function-related objects such as all relevant proteins in the pathway or all pathways involved in a disease. The maps themselves are generally static and cannot be supplemented by the users own proprietary data or pathways. With the exception of GenMAPP, few databases provide the tools for the incorporation and visualization of a user's high-throughput data on the pathways and maps. Finally, without links between clinical data and specific elements in the biochemical and cell-signaling networks, models of disease-related cellular pathways are not possible. There have been some novel attempts at using computational software to provide databases of the components of metabolic pathways. One example reproduced radiotracer experiments for substrate product relationships in *E. coli* by mapping the atoms derived from the molecular formulas and computing all pathways between any two given compounds [55]. At present this type of approach appears to have been rarely used.

4.4 HIGH-THROUGHPUT DATA

The dramatic increase in life science research output is largely due to the high-throughput technologies that can now be generated routinely. It is important to understand the limitations of this data as it plays a key role in systems biology modeling.

DNA Sequence Variations

Since it was first suggested that individual sequence variations can be used as markers for specific traits [56], detection and understanding of individual DNA sequence variability became mainstream in human biology and drug discovery. Specific sequence variations (polymorphisms) and their linear combinations in close vicinity on the chromosome (haplotypes) are known to be associated with predisposition to diseases and differential responses to drugs. Hence these can be used as molecular markers in genetic analysis on an unprecedented scale. A number of types of sequence-based markers were introduced over the last two decades for genetic analysis, including restriction fragment length polymorphisms (RFLPs), variable number tandem repeats (VNTRs), microsatellites, and finally single-nucleotide polymorphisms (SNPs) [57], which became widely used following completion of the draft human genome [2]. SNPs represent an inherited single-nucleotide variation in a given nucleotide in the genome (either in coding or noncoding regions), which is usually biallelic with a minor allele frequency of at least 1 percent [58]. In the human genome the saturation frequency of SNPs is predicted to reach 1 per 200 base pairs (bp), while the current average density of SNPs maps is about 1/1000 bp [59]. SNPs are more abundant in noncoding regions, represent a permanent feature of genomic DNA, are stable, inheritable, and easily detectible either by de novo DNA sequencing or by several methods for in vitro detection of nucleotide mismatches. SNPs can be detected by TaqMan fluorescence probes [60], mass spectrometry of short DNA fragments [61], pyrosequencing [62], Golden Gate assays [63], and resequencing by hybridization [59]. Independently of the detection method, SNP data fundamentally represents the percentage distribution of certain alleles in studied populations. About 6 million human SNPs have been detected of which approximately a quarter are publicly available via the SNP consortium (*http://snp.cshl.org/*), NCBI, and other sources and can be visualized using gene and chromosome browsers in software packages.

Gene Expression Profiling

Microarrays Complimentary DNA (cDNA) and oligonucleotide microarrays are the most widely used and robust tools for measuring gene expression on a large scale. Microarray technology is based on the fundamental property of complimentarity of genetic code in DNA and RNA [64, 65] and the classic technique of hybridization of nucleic acids, ubiquitous in biology for decades. Since the late 1980s, several technology breakthroughs have boosted the throughput several orders of magnitude and made it possible to evaluate transcription of a genome of any complexity in one experiment. In general, the total RNA is extracted from samples, amplified, labeled with fluorescent nucleotides, and hybridized to arrays with cDNA or oligonucleotide probes. After washing, the array is scanned by laser and visualized by specialized software. Oligonucleotide arrays measure the absolute level of each messenger RNA (mRNA), whereas cDNA arrays are typically hybridized with two samples labeled with red dye Cy5 and green dye Cy3 providing relative expression levels for each transcription unit [66]. Although based on the same hybridization reaction, these two array technologies are substantially different in terms of normalization and interpretation and are poorly comparable [67]. Microarrays are typically used for identification of a set of genes with changed level of expression in response to a certain condition(s). The statistical significance of differential expression is then monitored by various statistical tests (described in detail later). Microarray technology is widely applied throughput drug discovery for molecular disease subcategorization [68, 69], prediction of disease recurrence [70], treatment outcome [71], and potential drug toxicity [72–74]. Microarray gene expression profiles are the most reliable, high-throughput, and informative techniques available (excluding, perhaps, genomewide scan by sequencing). Compared with the hundreds to thousands of proteins and metabolites, oligonucleotide arrays may contain over a million distinct features (*http://www.affymetrix.com*). The microarray experiments themselves are run in a highly parallel manner and may contain hundreds to thousands of samples in large clinical studies. The major drawback of gene expression profiling is the distance of transcription from the assembly of active proteins and eventually phenotype (Fig. 4.2). In any cellular system, the information flow is defined by the main dogma of molecular biology: from DNA to mRNA to proteins. The process is more complicated and includes alternative splicing, RNA maturation, RNA editing, mRNA trafficking, protein trafficing, posttranslational modifications, protein folding, and assembly into active complexes. Eventually, active proteins perform certain cellular functions such as a metabolic transformation of malonyl into acetyl-CoA (Fig. 4.2), representing one reaction out of thousands of metabolic transformations in the human body regulated at multiple levels from the cell membrane receptors to core metabolism. Ultimately, the interpretation of gene expression profiles would benefit from combination with other available high-throughput data [75].

Although hundreds of large-scale microarray studies have been conducted, the vast majority of the data published is effectively unavailable for public use [76]. The best sources of the available data from microarrays are the Stanford Microarray Database, SMD (*http://genome-www5.stanford.edu/MicroArrat/SMD*), NCBI Gene Expression Omnibus (*http://www.ncbi.nlm.nih.gov/geo*), the database at the Whitehead Institute (*http://www-genome.wi.mit.edu/cgi-bin/cancer/datasets.cgi*), and the NIEHS (*http://dirniehs.nih.gov/microarray/datasets/home-pub.htm*) [66].

Serial Analysis of Gene Expression The serial analysis of gene expression (SAGE) method was developed for simultaneous detection and quantification of the mRNA content of eukaryotic cells [77]. It is based on the synthesis of cDNA from the total cellular mRNA pool, immobilization on streptavidin

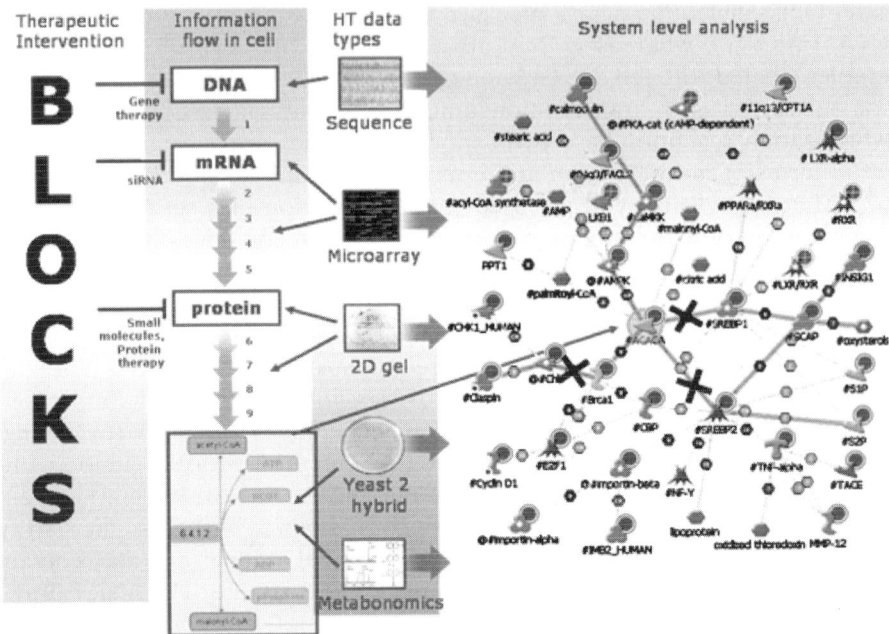

Figure 4.2 Utility of high-throughput data for providing information on the functional blocks and use for systems-level analysis. Nine levels of regulation of protein activity in a human cell can be summarized: (1) gene transcription, (2) mRNA processing and editing, (3) mRNA transport from nucleus, (4) mRNA stabilization, (5) protein translation, (6) protein transport, (7) folding and protein stabilization. (8) allosteric modulation, and (9) covalent modification. (This figure is available in full color at ftp://ftp.wiley.com/public/sci_tech_med/drug_discovery/.)

beads, and application of a combination of cloning with two restrictases and polymerase chain reaction (PCR) amplification to come up with hybrid DNA concateners consisting of 3′-end sequences of the expressed cDNAs (tags). The follow-up sequencing of multitag DNA molecules allows identification of the originally expressed genes by querying short tag sequences with sequence databases. The differential expression is determined by the analysis of quantities of gene products under different conditions. The method has two major advantages: It is quantitative and, unlike microarrays, does not require prior sequence information about the expressed gene sequences; however, the method allows recognition of mainly well-annotated genes. The short length of tag sequences (13 to 14 nucleotides) poorly suits sequence alignment and other bioinformatics techniques [78]. Also, the tags are not very specific, as in some cases different genes share the same tag [79]. Although not as widely used as microarrays (especially in pharmaceutical companies), SAGE analysis is inexpensive and can be used for the same applications. Some SAGE experiments are very elaborate and large scale, for instance, when drug sensitivity is tested on multiple lines of tumors and cell lines [80]. SAGE is often applied in combination with other technologies. Subsets of candidates of differentially expressed genes determined by SAGE can be placed on arrays and used for screening of large patient group [81]. A public SAGE data depository is available at NCBI (*http://www.genome.wi.mit.edu/cgi-bin/cancer/datasets.cgi*).

Proteomics

The term *proteome* was first coined to define the pool of proteins expressed by the genome [82]. As a field, proteomics deals with quantitative and qualitative multiparallel measurement of protein concentration and/or expression in whole-tissue samples [83]. Measuring protein expression directly is important because possessing a mature mRNA transcript is not sufficient for having a corresponding active protein, due to posttranslational modifications (PTMs), proteolysis, and other dynamic processes causing functional changes [84]. Expression proteomics studies typically involve identification of a set of proteins expressed in a certain tissue under differential conditions and quantification of protein abundance is achieved by a combination of two techniques. Two-dimensional gel electrophoresis (2-DE) allows separation of hundreds of soluble proteins via consecutive application of isoelectric focusing based on a protein's net charge. This is followed by separation based on size in sodium dodecyl sulfate (SDS) polyacrylamide gels [85, 86]. High-abundance proteins can be quantified from the gel using chromatography and fractioning. The method was of relatively limited use until the mid-1990s when advances in mass spectrometry (MS) and bioinformatics enabled the high-throughput identification and quantification of the individual proteins either from the spots on 2D gels or via a combination of chromatography and fractioning [87, 88]. The peptide mass fingerprint approach consists of automated or manual excision of the spots from the gels, protease treatment (using trypsin) of the protein mixture, followed by analysis with matrix-assisted laser desorption/ionization time-of-flight mass spectrometry (MALDI-TOF-MS) [89]. The peptide composition is then compared to species-specific peptide mass matching [90], and small proteins and those with unknown peptide composition are resolved using tandem MS allowing partial amino acid sequencing [91], followed by bioinformatics sequence alignment against protein sequence databases [87].

Compared to RNA expression analysis and SAGE, proteomics technology suffers from substantial problems including the absence of many important membrane-based receptors and channels that are insoluble proteins and missing from the data. High-abundance proteins such as albumins often obscure low-abundance proteins on 2-DE gels, and their effective separation requires complex prefractioning and other techniques [92, 93]. Proteomics is, however, quite widely used in research and clinical diagnosis of most complex diseases [94–96].

Most cellular processes are performed by assemblies of interacting proteins and proteomics is the only technology that allows experimental analysis of protein complexes and protein–protein interactions. The experimental arsenal of functional proteomics includes several ligand fishing techniques such as co-immunoprecipitation, sequential affinity tag purification [97], and biochemical interaction analysis–mass spectrometry [98]. Specific antibodies or tag-specific proteins are used as the bait for catching the interacting proteins as prey, followed by complex resolution by MS [84]. In two key studies, hundreds of yeast protein complexes were resolved by these methods, comprehensively covering the whole yeast proteome [99, 100].

Yeast Two Hybrid The most popular method for detection of protein–protein interactions is known as the yeast 2-hybrid system (Y2H [101]). Y2H is a simple genetic test based on modular structure of yeast transcriptional factors Gal 4 and Lex A, which are only active as dimers assembled on palindrome binding sites. Truncated for dimerization domains, the proteins can restore activity when fused to a pair of physically interacting proteins. The activity can be detected using color selection with the *LacZ* reporter or by complementation of auxotrophic mutants *leu2*, *ade2*, and *his3* [102]. The Y2H system is inexpensive and powerful for screening of protein libraries for interacting protein pairs, identification of unknown interacting genes, and detection of weak interactions that could be missed by proteomics methods [103]. This method is also a truly high-throughput method scalable for genomewide protein interactions screens. Some of the most important model organisms with sequenced genomes including yeast [104], fly *Drosophila melanogaster* [105], and worm, *Caenorhabditis elegans* [106] were mapped for interactions in this way. However, the system is known for its high (up to 50 percent) level of false-positive and false-negative interactions [107] due to interacting proteins that are overexpressed, subjected to folding in the foreign environment of yeast cells, and have to be trafficked into the nucleus prior to action. Filtering out false positives can be achieved by cross-referencing with other experimental data and superimposition of YH2 interactions onto networks of annotated pathways [108]. Specialty variations of Y2H have been developed, such as a reverse two-hybrid system for studying molecules disrupting the interactions and the split-ubiquitin system to analyze membrane proteins [102]. The raw Y2H data is therefore qualitative and represents simple tables of condition-specific interacting partners.

Protein Arrays Ever since the introduction of DNA chips in the early 1990s, creation of arrays with grids of active proteins became the holy grail of proteomics research. Unfortunately, the key DNA properties enabling the whole-genome arraying (active primary structure, PCR amplification, ease of labeling, stability, etc.) are not shared with proteins. Arraying of properly folded, nondenatured proteins and the development of detection techniques suitable for proteins of many different functions are nontrivial tasks not easily resolved yet [109]. The first proteins to be successfully used in array development were stable like antibodies [110]. As whole antibodies are large and inconvenient to work with, the technology was extended to antibody fragments [111], phage display-based expressed variable fragments [112], and oligonucleotides aptamers [113] as the source of content for arraying. MALDI targets investigated with affinity ligands, followed by affinity MS, are usually applied for capturing and detection of proteins of interest. Several techniques have also been developed for attaching proteins to surfaces such as the application of adenylate-treated glass, which reacts with primary amines [114], and tagging proteins with polyhistidine attached to nickel-containing slides [115]. To circumvent the differences in activity assays, the proteins are often arrayed according to functional target class (e.g., kinase arrays), although more recently multifunctional arrays have been developed.

Metabolomics and Metabonomics

Metabolomics and metabonomics are similar terms defining the profiling, measurement, and analysis of the components of the metabolome representing all low-molecular-weight molecules present in cells in a particular condition. Metabolomics can be defined as a discipline investigating metabolic regulation and fluxes in cells, and metabonomics is a larger scale, systemic determination of biochemical profiles and their regulation in biofluids and tissues [4]. Metabolites represent the endpoint of the organism's response to stimuli, the ultimate outcome of the complex information transfer network, signaling, and biochemical transformations such that metabolic profiling is therefore the most direct measure of physiology [116]. The composition of metabolites in samples can be very heterogeneous, and the measurement procedure often involves prefractioning, selective enrichment, and parallel measurement. The most robust methods in metabolomics are gas chromatography–mass spectrometry [(GC–MS), which resolves 500 to 1000 metabolites [117]], and liquid chromatography–mass spectrometry (LC–MS) [118]. These two methods, applied in combination, allow resolution of a thousand peaks for known organic small molecules and unidentified metabolites [119]. Unknown metabolites can be identified using a higher precision Fourier transformation mass spectrometry (FT-MS) [117]. The third approach is nuclear magnetic resonance (NMR) spectroscopy, applied mostly for detection of high-abundance known metabolites [120] and is particularly useful for metabolic fingerprinting used in clinical studies [121].

Interpretation of metabolic profiles of cells and tissues imposes arguably more problems than any other method described earlier, as metabolic fluxes and profiles cannot be explained by regulation of gene or protein expression alone [122]. The ultimate goal of quantification of all metabolites in a cellular system is currently impossible. This is due to lack of reproducible and robust automated analytical techniques, chemical heterogeneity of metabolites, lack of automated extraction techniques, and low resolution of existing hardware [123]. Moreover, taking into account complex relationships between endogenous and xenobiotic metabolism and the coexistence in the human body of up to 5000 species of metabolically active microorganisms, the full metabolic profile cannot be interpreted. This requires consideration of probabilistic concepts for determining symbiotic human metabolism [4]. Such complexity requires interpretation at the systems level, via integration of metabolomics with gene expression and proteomics data to unify metabolic, regulatory networks, and infer condition-specific metabolic networks [124], shifting our thinking of metabolic pathways as networks and subsystems [125].

4.5 MINING HIGH-THROUGHPUT DATA

The quantitative leap in experimental throughput required methods for comprehension of global gene expression experiments that have been useful in certain applications. Interactions could, however, be missed if analysis is restricted to expression patterns only. The statistical analysis of microarray data results in gene sets that are classifiers for certain conditions such as the disease type, but they do not explain biological mechanisms and cannot be directly linked to clinical observations or other phenotypes. Visualization-based approaches to high-throughput data lead into model-based approaches to provide broader understanding and insight. The state of the art at present are graphical approaches that link objects with vectors and other graphical annotations that can be queried [126].

Limitations of Clustering Approaches for Gene Expression Data

Three general statistical procedures are used for clustering of gene expression patterns on arrays including hierarchical clustering [127], *k*-value clustering [128], and self-organizing maps (SOM or Kohonen maps) [129]. The most common method is the hierarchical clustering procedure, which consists of a consecutive pairwise grouping of spots: The closest pairs are replaced by a single data point representing their average. Ultimately, this results in an expression pattern represented as a phylogenetic tree with a branch length corresponding to the degree of similarity between sets. In contrast, *K*-means clustering uses a predetermined K value that is the expected number of elements in the cluster. This approach, therefore, produces an unorganized compilation of clusters. A further approach is Bayesian clustering, which generates multiple recursive partitions into subgroups and assigns Bayesian probabilities to each group. This method requires a strong prior data distribution and is used in combination with other methods [130]. SOMs are better suited for exploratory data analysis as they impose a partial nonrigid, structure on clusters and facilitate visual interpretation [129, 131]. All these statistical methods have been applied to microarray-based disease subclassification by either unsupervised or supervised methods. In unsupervised clustering, the differential expression data is parsed into clusters of points statistically most related without prior knowledge of the phenotypic subcategories [132]. These different methods are therefore useful for various applications with each having attendant advantages and disadvantages. As regulation of mRNA expression constitutes only one level of biological control, these levels of complexity multiply in global genomewide experiments such that information can be lost or altered and clustering at the first level of expression can only be indirectly linked to eventual phenotype. This one-dimensional analysis is a limitation in the applicability of expression microarrays for studying complex diseases [6].

Advanced Computational Methods for Interpretation of Microarray Data

From Linear Pathways to Complex Biological Networks Traditionally, functional organization of a biological system was described in terms of pathways. Pathways were thought of as fairly small linear chains of biochemical reactions or signaling interactions that lead from a defined starting point (e.g., cell surface receptor) to a target effector (e.g., transcriptional factor). Such a description is in part due to the nature of biological research itself that until recently was inherently low throughput, with data scattered in tens of thousands of individual publications. In recent years, however, three major developments have taken place in this area in terms of pathway databases, natural language processing (NLP) algorithms for automatic extraction of pathway information online abstracts, and high-throughput techniques for determining potential protein–protein interactions (described above). Based on the data available from these technologies, it has become clear that organization of intra- and intercellular molecular processes is much more complex. Contrary to the previous model of fairly independent small pathways, it is now evident that known molecular processes can be linked into rather large highly interconnected networks. A recent review has described in considerable detail the theory behind network models, their architecture, and how they can be used to provide insight into the functional organization of the cell [125].

Further analysis of the existing databases on metabolic and signaling pathways shows that if taken together, known reactions and interactions would form a large cluster linked via molecular nodes shared among many processes [133]. If one tries to describe such clusters in terms of linear pathways, then billions of cascades can be generated. For example, an analysis of ~20,000 cell-signaling interactions culled from the literature (obtained from MetaCore, GeneGo Inc.) shows that such a set contains approximately 2 billion five-step pathways.

It is important to note that, taken together, information on molecular processes derived from different sources represents a universe of a putative biological functionality. Only a small fraction of it will be realized in a cell at any given time. Genes, proteins, and metabolites are tightly regulated on many different levels in molecular networks (Fig. 4.1). For example, only a fraction of genes are realized as protein products in a particular cell type. Additional constraints include spatial separation by subcellular localization, thermodynamic feasibility of a process in a specific environment, competition among processes for shared molecules and other physical, chemical, and biological limitations. Theoretically, a computational model of a full network can predict the dynamic behavior of a system. At present, however, dynamic models can only be built for a limited number of fairly well studied systems, and selection of which processes to model is limited by unavailable or inaccurate quantitative information for most of the tens of thousands of reactions and interactions. To add to the complexity, unique isoforms of the same protein expressed in different cells may vary dramatically in binding constants and other parameters. The lack of detailed quantitative knowledge is a major impediment for direct dynamical modeling of large molecular networks, hence requiring an alternative approach for modeling.

Elucidating Pathways and Interactions from the Literature There is also a growing interest in elucidating the pathways and protein–protein interactions from the literature. Differentially expressed genes can be clustered into functional groups based on their names or co-occurrence in MEDLINE texts [134, 135]. Different computational methods like hierachial clustering [136] and NLP (reviewed in [137]) have been applied for enhancing the biological relevance of clusters and literature coverage. However, only 30 percent of experimentally verified protein interactions were reported to correspond to pairs from MEDLINE abstracts [138] and false-positive interactions composed about 50 percent of NLP-extracted associations, partly due to poor synonym resolution [134]. A more recent publication used an NLP method, MedScan, to extract 2976 interactions between human proteins from MEDLINE abstracts with a precision of 91 percent for 361 randomly extracted protein interactions [139]. There are a number of databases covering different aspects of protein function that can be used for testing these methods, such as those described earlier. These are generally more biologically relevant sources of information on interacting proteins, particularly when extracted from full-text experimental literature (BIND, INTERACT [140], and DIP). However, these databases have relatively low coverage for human protein interactions. Subsequently, MedScan was able to extract novel (96 percent) information that was not in BIND and DIP [139]. The first attempts at consolidation of two approaches, that is, co-expression clustering and deduction of functional networks from literature and knowledge databases were performed for *E. coli* and yeast [141]. Overall the functional interpretation of expression data is in its infancy, and tools that have been developed have not been applied yet for the analysis of human and other higher eukaryotes. In many cases, there is a conserved gene expression pattern that is shared between species, and this corresponds to the same functional groups of orthologs. Similar genes are therefore regulated in the same manner across species, and these are usually involved in basic processes such as the cell cycle, metabolism, and secretion. New genes involved in these processes can therefore be identified via comparative analysis [142, 143] among the different species, and alterations in gene expression can be detected by microarray techniques [143].

Since many biological functions are carried out by discrete modules of physically interacting proteins [144], searching for the algorithms useful in identifying such interactive networks has become the key to microarray expression analysis and one of the methods for functional annotation for novel genes successfully demonstrated with yeast [145, 146]. Some algorithms can link the expressed genes with known transcriptional factors via the genomewide identification of DNA binding sites [147] or enable building of condition-specific probabilistic models [148]. The other published studies group the co-expressed genes based on the occurrence of common cis elements in the context of reference knowledge databases of transcriptional factors [149] of well-known metabolic pathways. For instance, MIPS (Munich Information Center for Protein Sequences) functional categories [150] may be used, which take into account pairwise sequence-based conservation between co-expressed proteins [151]. Microarray data is therefore often linked to curated knowledge databases on proteins, genes, and pathways such as SwissProt, KEGG, and others. Although robust and relatively comprehensive, the available analytical methods and databases are not sufficient for coping with the vast complexity of human biology and the elucidation of integrated cellular networks.

Making Sense of Biological Networks Systems biology is hence evolving to analyze all high-throughput experimental data in the context of the global physiology and disease pathology. This approach seeks to identify the networks of cellular pathways and corresponding physically interacting proteins [152]. The combination of all of these data types would be ideal to provide a truly multidimensional understanding of cell function. As direct dynamical modeling is currently limited to well-understood systems, a novel approach for making sense of large networks has been developed in recent years. This approach combines graph theory and statistical methods for finding network modules that represent potential active functional units. Two major directions of research exist in this area. The first represents attempts to understand the fundamental global organization of molecular networks. The second aims at finding potential functional motifs by studying local properties of a network and by combining protein or metabolic networks with data on gene expression for elucidation of condition-specific active modules. One important result achieved in this area is the realization that biological networks of different origins (e.g., metabolic, regulatory, protein interactions, networks for different organisms) share the same global architecture [105, 106, 125, 153]. Namely, these networks are scale free, meaning that distribution of node connectivity obeys the power law: $P(k) \sim k^{-\gamma}$, where k is the number of node links (also called node degree) and $P(k)$ is the fraction of nodes in the network with exactly k links. One consequence of such an architecture is the presence of so-called hub nodes, which are high-degree nodes connected to many low-degree nodes [125]. Such topological features give networks the property of robustness with removal of even a substantial fraction of nodes leaving the network connected [154]. This robustness may also have far-reaching implications for the selection of appropriate drug targets.

While understanding the global network architecture is an important first step, from the practical standpoint of drug discovery it is even more interesting to find disease-specific functional modules that may contain potential drug targets. This task is a much harder one as any complex network can be divided into subsets in many different ways, potentially generating billions of combinations. Different algorithms and criteria have been used in an attempt to automate parsing of such large networks into modules. One set of methods finds network modules based on network connectivity and various clustering algorithms. Examples include the Monte Carlo optimization method for finding tightly connected clusters of nodes [155], clustering of nodes based on the shortest path length distribution [156], and other graph clustering algorithms [157]. It was shown that many clusters that are found in such a way correspond to either known protein complexes or metabolic pathways [155]. Another way to approach finding functional modules is by analyzing network motifs [158]. Motifs are fairly simple subgraphs that share certain structural features, usually related to potential functionality. For example, feedback or feed-forward loops are typical motifs. In this approach a number of different motifs in the network are calculated and compared with a number of the same motifs in a randomly connected network. Those motifs in which the network is enriched may represent potential functional modules. In several studies performed to date motifs were found in regulatory networks of *E. coli* [159, 160] and in yeast protein interaction networks [161]. The next step in elucidation of functional modules is the analysis of high-throughput molecular data in the context of networks. By performing such analyses one may find mechanisms that are active under specific circumstances, such as disease or drug treatments (discussed in more detail later).

4.6 AVAILABLE RESOURCES FOR BIOLOGICAL PATHWAY ANALYSIS

The functional annotation of the filamentous fungus *Aspergillus nidulans* was to our knowledge one of the first attempts to reconstruct functionality of a complex eukaryotic organism based on the integration of Expressed Sequencing Tags (EST) data (from the Unigene EST collection) with a comprehensive literature-based collection of functional pathways metabolic pathways database (MPW) [43]. The study was based on a method of metabolic reconstruction, proposed for annotating prokaryotic genomes in the computational environment WIT [52, 162]. The method consisted of compiling metabolic blocks (pathways and subsystems) corresponding to the genetic component of an organism and connecting these blocks via intermediates into wire diagrams or metabolic reconstruction models. Over the past few years a new generation of computational biology software products has emerged for biological pathway construction and analysis such that the following are now commercially available.

Cell illustrator (*http://www.gene-networks.com*) is a tool for manual construction of pathways and simulating pathway mechanisms of action of both baseline and abnormal conditions.

DiMSim (*http://www.bio.cam.ac.uk/~mw263/bioinformatics.html*) is a methodology for simulating the flow of metabolites through networks of interacting metabolic reactions. This enables views of the reactions involved in metabolic processes in a graph format useful as the basis for simulations. The DiMSim approach is in marked contrast to other metabolic simulation systems that are based on differential equations, it is therefore able to model down to single molecules and cope with the special characteristics of biochemical systems [163].

Ingenuity pathways analysis (*http://www.ingenuity.com/*) enables the entry of a gene list from microarray and proteomics experiments that are then evaluated against a knowledge base to produce a gene network.

MetaCore (*www.genego.com*) relies on a unique, expert-curated database of pathways annotated with tissue specificity and subcellular localization and containing pathways for over 80 percent of known human genes. This data comes primarily from two sources: the original publications and expression data from the Unigene database. MetaCore also provides links between the pathways (or their elements) and disease states collected from the literature. Currently, there are more than 32,000 disease links classified into 6 major categories from the highest to lowest level verification, namely *cause, manifestations, hypothesis, animal, treatment*, and *no relation* and provides several different ways of exploring the data. The major components of MetaCore are described in more detail below.

Metabolic Component A comprehensive collection of metabolic pathways with interactive maps of the major functional blocks of human metabolism is organized in a three-tier architecture.

Cell-Signaling Component Cell-signaling and regulatory pathways important for diseases such as cancer (see later) are the major experimental content with approximately 20,000 signaling interactions assigned to over 6000 molecular objects. Some of these interactions are assigned to classes containing more than one molecule with signaling vertically integrated with other types of cellular processes.

Pathway Visualization Options Visualization options in MetaCore include over 300 fixed-curated maps for the major functional blocks of cell signaling and metabolism, color-coded for subcellular localization, and superimposed on the graphical images of the various human tissues (Fig. 4.3). In addition, there

144 SYSTEMS BIOLOGY: APPLICATIONS IN DRUG DISCOVERY

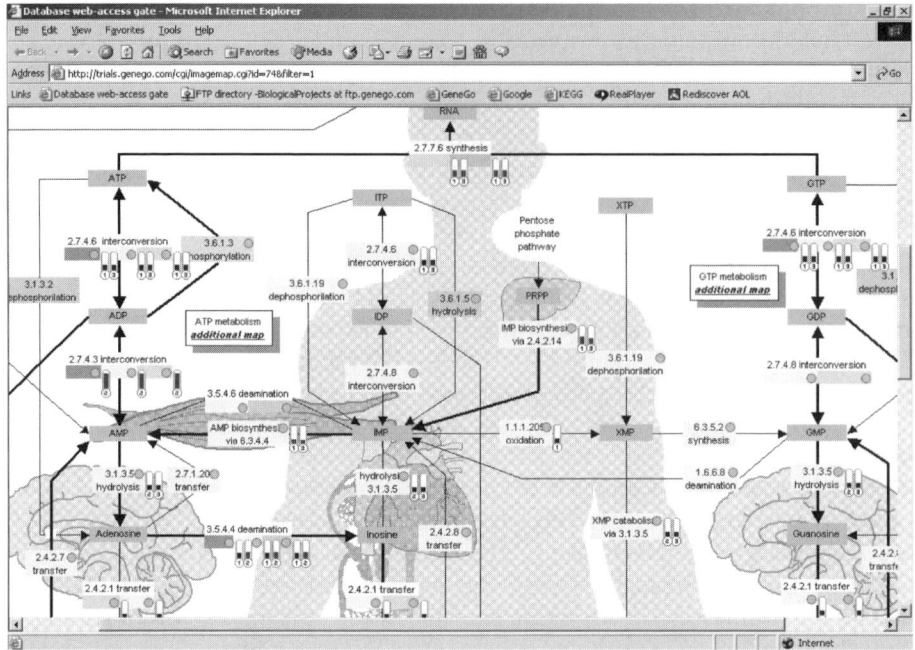

Figure 4.3 Interactive map for human purine metabolism from MetaCore. Metabolic pathways are shown as rectangular icons; the enzymes are tagged with EC numbers. The differentially expressed genes are uploaded directly from a microarray experiment and are marked next to the pathways as thermometer objects.

is a pull-down menu with a list of human EST libraries from the Unigene database. Computer-generated signaling and metabolic networks can also be produced that are highly interactive.

Data Visualization Options Both the curated maps and computer-generated networks can be used as a framework for visualizing gene expression (thermometer-like icons, Fig. 4.3) and other types of high-throughput data such as Affymetrix and Agilent microarrays, 2D proteomics data, SAGE, DNA sequence data, and molecular formulas for metabonomics. The visualization tools allow concurrent comparison of multiple experiments, with custom normalization and filtering features.

PathArt (*http://www.jubilantbiosys.com/pd.htm*) represents over 600 signaling and metabolic pathways, 5000 molecules and 7500 interactions incorporating data from peer-reviewed articles and patents enabling pathways and critical genes to be identified and compared across physiological diseases and organisms. The data covered is cell, tissue, and organism specific for human and mouse, with coverage of protein–protein interactions their specific domains, motifs, key binding residues, and experimental techniques.

Pathway Assist (*http://www.ariadnegenomics.com/products/pathway.html*) is a software tool for pathway creation, expansion, and visualization to allow the interpretation of gene regulation and protein–protein interactions using an NLP method (MedScan, described previously) [164] and can be used with the PathArt database. The software can utilize microarray data, allow data filtering and parsing from many sources, and currently describes more than 100,000 events of regulation, interaction, and modification between proteins, cell processes, and small molecules.

PathDB (*http://www.ncgr.org/pathdb/*) is both a data repository and a system for building and visualizing cellular networks targeted for the gene expression, proteomics, and metabolic profiling communities. Uses include finding all pathways and phenotypes associated with genes in a cluster or validating computational predicted associations with known biological data.

PubGene (*http://www.pubgene.org/*) can search through the millions of biology-related studies published for the names of genes and proteins and for sequence homologies between them. The articles are then searched for the occurrence of gene pairs. Based on this information, a map of all genes is made, based on the numerical information of gene pair frequencies. When combined with results of microarray analysis, the program can then estimate the importance of the various relationships between genes and proteins, utilizing the biological knowledge for human, mouse, rat, and yeast genes and proteins, from such databases as Medline, RefSeq, and PIR. A number of layers of metadata from various sources are included to allow for flexible information content display depending on the user's preference. PubGene uses in-house dictionaries of gene and protein names and gives the user synonyms resulting from a query [134].

VLX Biological Modeler (*http://www.teranode.com/applications/pathway modeling.php*) automatically translates pathway models into mathematical models based on biophysical and biochemical principles.

The Cognia Molecular System (*http://www.cognia.com/products_ cm_indepth.htm*) comes with a preloaded framework of information parsed from public data sources on over 100,000 proteins, genes, interactions, 250,000 compounds, and with links to over 35 external databases. This content is broad, covering many species and functions, roles, locations, and interactions that can be displayed as networks with NetworkBuilder. The visual output of the NetworkBuilder is filterable and customizable to extract the dynamically built output as well as component attributes within the database.

All of these products have unique underlying proprietary pathway databases. These are either human curated (GeneGo MetaCore, Jubilant PathArt) or are compiled from MEDLINE abstracts using automated text-mining tools (Ingenuity PathwayAssist). Alternatively, they may contain a combination of curated and automatically extracted data (Ariadne Pathway Analyst). Their respective performance obviously varies depending on the flexibility of the database schemas, the quantity and quality of the assembled data, and the algorithms implemented into their data-mining tools. To date there has been no comparison of the different approaches published, and it is likely that investigators in the industry may require access to multiple software systems to balance the overall strengths and weaknesses of individual methods.

4.7 APPLICATIONS OF SYSTEMS BIOLOGY FOR DRUG DISCOVERY

The following examples describe applications of gene network analysis software and systems reconstruction for particular topics relevant to drug discovery.

Glaucoma

Glaucoma is a progressive optic neuropathy characterized by structural damage to the optic nerve leading to blindness through loss of retinal ganglion cells. Risk factors in glaucoma include elevated intraocular pressure (IOP), age, race, family history, myopia, and diabetes. The growing body of evidence indicates there is a heightened risk of developing glaucoma among individuals with Alzheimer's disease [165–168]. The pathogenesis of the optic nerve neuropathy in glaucoma is a matter of debate. It is widely accepted that elevated IOP and a variety of factors may all contribute to primary insult of the optic nerve [169–171]. Cellular mechanisms involving vascular insufficiency [172, 173], vasospasm [174, 175], glutamate exitotoxicity [176], neurotoxic cytokine release [177, 178], abnormal metabolism, and autoimmune reaction have all been suggested [170]. There are numerous major factors contributing to the cumulative optic nerve insult and malnutrition in glaucoma (Fig. 4.4). Ultimately, glaucoma affects an estimated 70 million people worldwide, mostly 40 or older. This represents a fast growing market for pharmaceutical intervention because of the progressively aging population of the Western world. At present there are neither drugs on the market that address the disease mechanism or reliable molecular diagnostics for addressing susceptibility or severity. As is the case with many complex diseases, global gene expression profiling using microarrays is becoming an important powerful tool in glaucoma research.

Over the last 2 years, several groups studied differential gene expression in glaucomous human optic nerve head (ONH) astrocytes [179], monkey [180], and rat [181] disease models as well as response to corticoid hormones in different eye tissues [182, 183] and the response in normal ONH astrocytes to high pressure (a key risk factor associated with glaucoma [184]). In the pioneering work of Hernandes et al. [179] the expression of genes in cultured human astrocytes from glaucomous and normal ONH astrocytes were compared using standard Affymetrix U95Av2 arrays consisting of over 12,000 genes. Approximately 1700 genes were differentially regulated in diseased astrocytes relative to normal ONHs. In 150 of these genes a 5-fold or higher

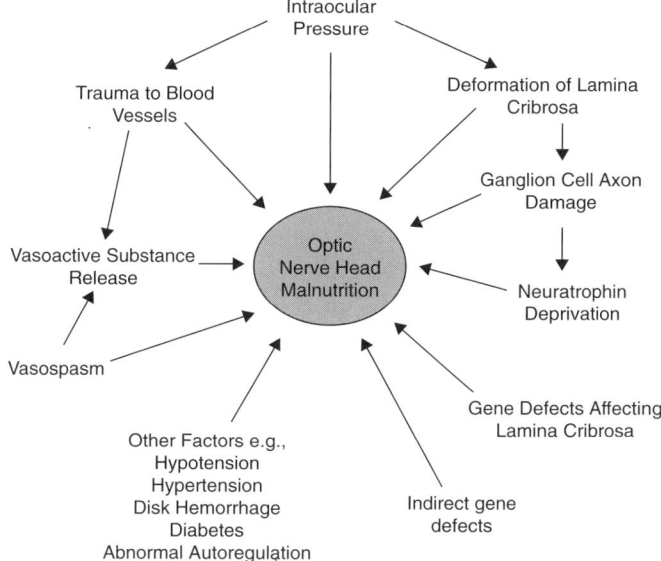

Figure 4.4 Some possible causes of initial insult to the optic nerve head in different glaucoma patients [171]. While increased intraocular pressure (IOP) is the most significant risk factor for glaucoma, RGCs cell death caused by optic nerve deformation may provide an explanation for a mechanical cause of glaucoma. While elevated IOP definitely plays a role in structural displacement of the ONH causing cytosceletal alteration, and loss of microtubules in RGC axons impedes retrograde axonal transport [258–260], it is conceivable that it also provides indirect insults via reduced blood flow and reactivation of microglia [178]. The role of vascular factors also is thought to be of significance to optic nerve and RGCs injury [172–175].

difference was noted, and these could be classified according to generic biological functions, including signal transduction, transcription, cell adhesion, proliferation, and metabolism. In the follow-up study [184], the same group interrogated the response of normal cultured astrocytes under increased hydrostatic pressure (HP) using the same molecular techniques and type of array. The expression of 596 genes was altered; 38 genes were upregulated and 24 downregulated over time with a threshold of 1.5 or higher. The 38 genes were also analyzed by hierarchical clustering analysis. The genes were of multiple cellular functions that were indicative of the systemic effect of high pressure on astrocytes. In two studies on animal models, glaucoma was induced in the retina of rat [181] and monkey [180] while differential expression was evaluated using total mRNA from whole retina as probes. The advantage of these models is that the genetic background is the same since one eye remains normal, and disease is induced in the other of the same animal. Sixty-two and 39 genes were differentially expressed in mild and severe glaucoma in monkey, respectively. In rat 81 genes were differentially expressed, although most were not the same genes. Finally, using dexamethasone the induction of gene expression was studied in cultured human trabecular meshwork (HTM) cells [183] and several other eye tissues [182]. Dexamethasone specifically induces expression of myocilin (TIGR/MYOC) gene in HTM cells and is linked to several types of glaucoma. When comparing these two studies, the results were neither consistent nor conclusive: 30 genes were upregulated over 2-fold in HTM-DEX cells in the Ishibashi study [183] while 249 genes were upregulated 4-fold or greater in the work of Lo et al. [182].

Microarray expression analysis in glaucoma has not as yet been used for subcategorizing different types of glaucoma, identifying prognostic gene lists for predicting disease outcome and response to treatment, or comparing human and animal studies. The approach of using enriched cell-type-specific primary cultures has been undertaken by Hernandez and co-authors [179], and this approach is likely the most promising since it uncovers the contribution of specific retina cell types in the development of neuropathy and allows the reconstruction of cell-specific alterations in molecular pathways implicated in the disease. We have compared the genes expressed in disease tissue (the endpoint of microarray analysis) from four recent studies in glaucoma and showed a lack of consistency between these different sample sets. Less than 5 percent of the genes were the same (on average) between any pair of experiments (Table 4.1). This lack of consistency in genes identified was somewhat surprising since the disease pathologies appear similar among monkey, rat, and human and because these species share about 99 percent of the same genes. Even more surprising is the inconsistency between expression in normal human astrocytes under high pressure (believed to be the key risk factor in glaucoma) and glaucomatous astrocytes. Since high pressure is believed to be a key risk factor for glaucoma or, at least, lead to the induction of glaucoma, a higher percentage of common genes were expected. These conclusions are based upon signature molecular profiles found when comparing between individuals for other diseases such as in breast cancer [130]. Although there are several potential reasons for such inconsistencies, the major problem is the application of analytical methods to tissues from different genetic backgrounds with different (and uncontrolled) environmental conditions.

We have performed a comprehensive computational gene network analysis of microarray expression in human glaucoma using these previously published data sets [179]. The raw expression data on about 1700 genes differentially regulated in glaucoma astrocytes from 4 glaucoma patients and 4 age-matched normal individuals was loaded in MetaCore. After data normalization and processing, multiple networks were then built using different algorithms and expression threshold ratios. One of many networks is presented on Figure 4.5. The threshold ratio of 2.5 and higher was set up for this network before we applied the stringent algorithm called *immediate interactions*. This algorithm allows the connection of only proteins (presented as coding genes), which are: (1) experimentally shown to physically interact pairwise and (2) being over- or underexpressed above the threshold ratio. The pathways are highlighted in Figure 4.5 as solid lines.

TABLE 4.1 Overlap Between Sets of Glaucoma-Relevant Differentially Expressed Genes from Recent Studies

	Hydrostatic (1.5-fold) [261]	Glaucoma (5-fold) [179]	Rat Retina (2-fold) [181]	Monkey Retina, Mild Glaucoma (1.5-fold) [180]	Monkey Retina, Severe Glaucoma (1.5-fold) [180]
Hydrostatic (1.5-fold) [261]		CDC42, FOS, VEGFC, Clusterin, HF1, C1q, C5, BMP4	VEGFC		
Glaucoma (5-fold) [179]	CDC42, FOS, VEGFC, Clusterin, HF1, C1q, C5, BMP4		C1q, c-Jun	C1q, c-Jun	C1q, c-Jun
Rat retina (2-fold) [181]	VEGFC	C1q, c-Jun		C1q, c-Jun, GFAP, ceruloplasmin	C1q, c-Jun, GFAP, ceruloplasmin
Monkey retina, mild glaucoma (1.5-fold) [180]		C1q, c-Jun	C1q, c-Jun, GFAP, ceruloplasmin		v-akt, chitinase 3-like 1, C4a, MHC class I DR alfa, FB, KIAA0805
Monkey retina, severe glaucoma (1.5-fold) [180]		C1q, c-Jun	C1q, c-Jun, GFAP, ceruloplasmin	v-akt, chitinase 3-like 1, C4a, MHC class I DR alfa, FB, KIAA0805	

Of particular interest are the grossly downregulated pathways of integrins, actin synthesis, actin-mediated signaling, TGF-beta1, NGF, collagen IV biosynthesis, and microtubule polymerization presented on the left hand of the network. Underexpression of some elements of these pathways is in good agreement with the observed phenotypic changes accompanying pathological reactivation and has been shown as a signature for glaucoma astrocytes [179, 185, 186]. Overexpressed pathways (on the right-hand side on the network) predominantly includes Nf-kB and AP-1/c-Fos/c-Jun-activated pathways. Both transcription factors Nf-kB and AP-1/c-Fos/c-Jun are strategically located in the watershed zone separating the up- and downregulated pathways. These genes are subsequently linked to the activation of downstream pathways via activation of transcription factors Rel, c-Myc, HMGY, and CREB, consistent with their known functions in promoting cell survival and amplification [187]. The network also indicates several activating interactions leading to potentially cytotoxic pathways and compounds such as interleukin-8 (IL-8), interleukin-6 (IL-6), amyloid precurson protein (APP)/β-amyloid, nitric oxide synthase (iNOS), and clusterin. The growing body of experimental data links the upregulation of these pathways with glaucoma and RGC death [188]. Separate smaller networks showing interactions within these individual nodes were generated using a smaller subset of genes (Figs. 4.6a–4.6c). Examples of the types of results and interpretation of the data are described next.

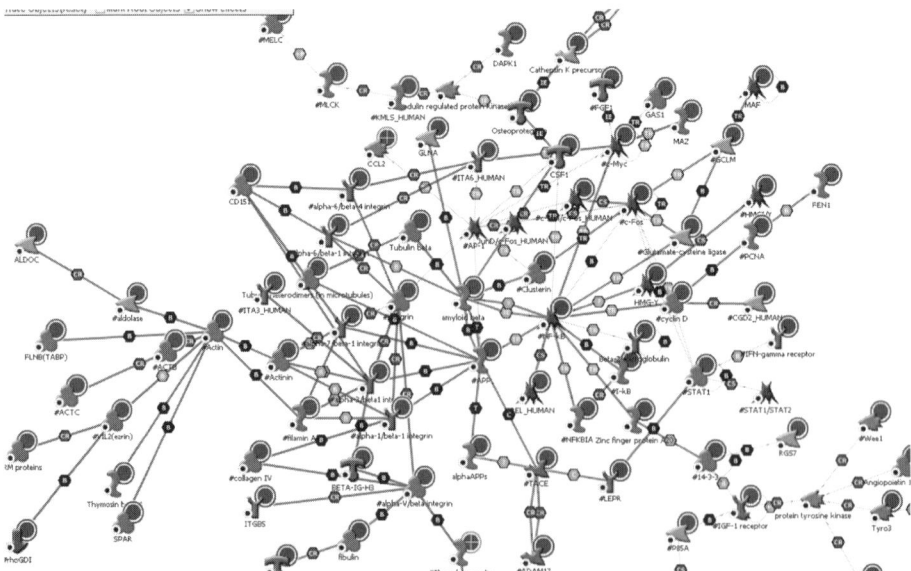

Figure 4.5 Functional network for differential gene expression in glaucoma. Pathways are marked in solid lines. Circles on the nodes mark the differentially expressed genes. The expression ratios and the page with gene/protein/interaction annotations are hyperlinked with nodes. (This figure is available in full color at ftp://ftp.wiley.com/public/sci_tech_med/drug_discovery/.)

The central node APP (Fig. 4.6a) links the upregulation of AP-1, NFkB, TGF-α2, protein kinase (PKC), and ADAM17/TACE pathways with the activation of amyloid precursor protein (APP) gene in glaucomatous ONHAs. The increased APP accumulation is likely to translate into the accumulation of neurotoxic β-amyloid peptide promoted by the c-Jun N-terminal kinase/mitogen activated protein kinase (JNK/MAPK) pathway, caspase3, reactive oxygen species (ROS) and β-amyloid itself. Others have reported similar results to these for reactive astrocytes in the central nervous system (CNS) [189]. APP/β-amyloid induces cell motility via binding to integrins and may become neurotoxic via astrocyte-mediated oxidative stress and mitochondrial dysfunction in neurons [190]. The interaction between two major nodes (Fig. 4.6b) also demonstrates the major mechanisms of iNOS induction in reactive astrocytes where it has been demonstrated to be neurotoxic in vitro and in vivo [177, 191]. The IL-1, NFkB, and tumor necrosis factor (TNF-α) pathways have been shown to play a role in iNOS overexpression in ONH [192]. In addition to a characterized role in complement regulation, clusterin has several debatable functions in the eye and CNS. Overexpression of clusterin is also neurotoxic via an astrocyte-mediated mechanism [193] and is highly upregulated in glaucomatous ONHAs possibly contributing to neurotoxicity either alone or in association with β-amyloid peptide (Fig. 4.6c).

The computational gene network analysis illustrates four proteins that play a key role in switching NFkB, transcriptional factor AP1, amyloid precursor protein APP, and misfolded β-amyloid peptide from downregulated pathways to upregulated pathways. Glaucoma may ultimately result from incorrect folding of APP, triggering a cascade of events to culminate in the loss of supportive and detoxification functions of ONH astroglia before finally leading to oxidative stress and apoptosis of retinal ganglion cells. This mechanism is similar to the cell death of neurons in Alzheimer's disease (AD) and was not previously shown experimentally for ONH astrocytes, as it cannot be deduced from a comparison of the lists of differentially regulated gene clusters. Possible parallels between neuron death in AD and death of ganglion neurons in glaucoma has been recently established in experimental studies [194, 195], as has the role of astrocytes as key mediators of β-amyloid-induced death [190]. The potential role of complement activation and clusterin accumulation have also been suggested from animal studies [193, 196] and is important to note that several of the key interactions established in the network analysis were not previously included in MetaCore.

Breast Cancer

Breast cancer is a complex disease or a group of underlying disease states that involves many tissues and cell subtypes [197]. Currently, the clinical subtyping of breast cancers is limited to indexing based on pathological staging and tumor grading [198, 199] and is carried out by standardized scoring algorithms [200]. These indexes are neither precise nor individualized, and their broad

Trace Objects (reset)
Mark Root Objects
Show effects
#TACE
TNF-alpha
#alpha-5/beta1 integrin
#alpha-1/beta-1 integrin
Tubulin heterodimers (in microtubules)
#NF-kB
#alpha-3/beta1 integrin
Fe65
#Tau
alphaAPPs
Fe65-like2
#AP-1
#TGF-beta 1
#APP
HSF1
#JNK(MAPKB-10)
amyloid beta
#gamma-Secretase complex
neurotrophin-3
NGF
BDNF
#VLDL (APOE)

(*a*)

NMDA receptor
#FOSL2
adrenomedullin
Nitric Oxide
#iNOS
#AP-1
#c-Fos
#Bax
#NF-kB

(*b*)

Trace Objects (reset
NGF
b-Myb
HSF1
#c-Fos
#AP-1
#Clusterin
#C5
amyloid beta
#SC5b-9 membrane attack complex

(*c*)

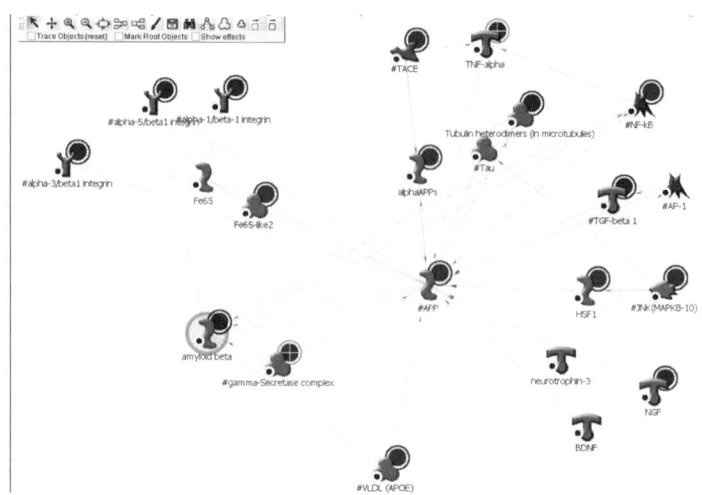

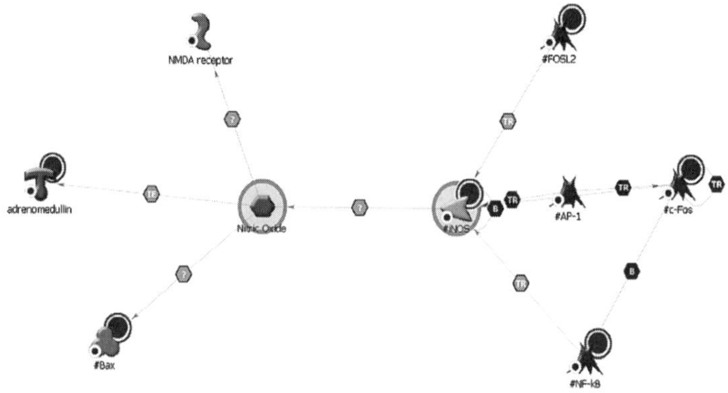

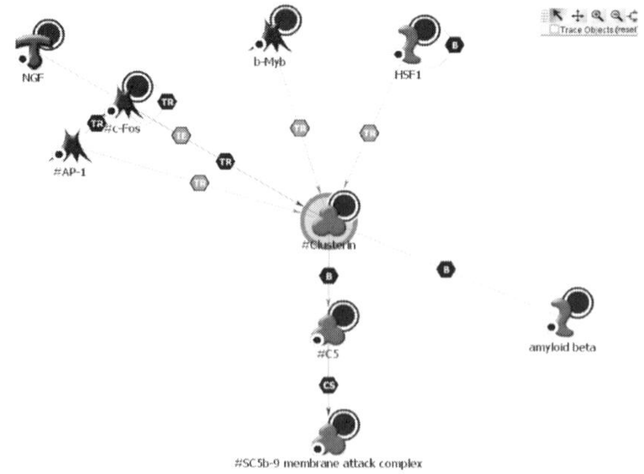

application results in overtreatment of many patients with chemo- and endocrine therapies [70]. Multiple phenotypical factors influence prognosis for breast cancer, including age, histological grade, tumor size, and lymph node and status of estrogen hormone receptor (ER) and *BRCA1* gene expression [201]. No single factor is a sufficient predictor of disease outcome. Global profiling of gene expression has quickly become the method of choice in breast cancer research for subcategorization of breast cancer with respect to ER state [69, 202–204], percentage of cells in the S phase of cell cycle [205], pathological staging and tumor grading [206], and a combination of gene expression/morphology subclasses [68, 207]. Overall, studies (Table 4.2) have shown microarray expression analysis can separate breast cancers into known and new categories with distinct pathological features (reviewed in [70]). ER-positive and ER-negative tumors can be distinguished by several dozen-strong classifier gene sets [69, 130]. ER-negative tumors can be further subdivided according to the long-term postoperational prognosis [204], and ER-positive tumors according to the patients' ethnic status [201]. More precise subdivision into five groups according to *ERBB2* oncogene overexpression, association with luminal subtypes, and ER overexpression was revealed in a multiyear study [68, 207, 208].

Microarray profiling has also proved useful for predicting development of metastases by detecting the presence of a signature set of differentially expressed genes in primary tumors [69, 130, 209] and for characterization of disease progression and tumor grading [206]. This is based on the hypothesis that cancer samples sharing the same patterns of expression clusters are expected to be similar in terms of pathology, drug sensitivity, and/or other phenotypes. Supervised analysis assumes prior separation of samples into predefined phenotypic classes. The analysis identifies those sets of genes for which expression is the most distinct between the classes and suffers from the flaws described previously [210]. This may explain the high rate of false positives in breast cancer prognosis based on array clustering [70], the exclusion of individual genes with high experimentally determined impact in breast cancer from a set of prognostic marker genes [69, 209], and the inconsistency between classifier gene sets [68]. Although surprisingly no significant difference between distinct pathological stages of the disease were detected in the earlier studies, there were altered expression signatures apparent between different tumor grades [206]. A subset of these genes correlated well with progression from DCIS (ductile carcinoma in situ) to IDC (invasive ductile carcinoma) stages. Microarray analysis has been used for predicting tumor sensitivity to chemotherapy treatment with docetaxel [211], adriamycin, and

Figure 4.6 Smaller networks for glaucoma microarray data. (*a*) The APP protein node and its immediate interaction space. (*b*) iNOS cross-activation network in glaucomatous astrocytes. (*c*) Potential role of clusterin in glaucoma pathology of astrocytes. The arrows indicate direction of protein interactions. (This figure is available in full color at ftp://ftp.wiley.com/public/sci_tech_med/drug_discovery/.)

TABLE 4.2 Major Gene Expression Studies and Data Sets in Breast Cancer

Study	Array	Samples/Cohort	Statistical Analysis	Reference
Discrimination of pathological stages	42,000 cDNAs	61 IDC, ILC tumors	Hierarchical clustering SAM, PAM	262
ER status, prognosis, and racial differences	8,064 cDNAs	36 primary tumors	Unsupervised PCA	201
Prediction of ER status, SPF (S phase of cell cycle)	6,728 cDNAs	48 primary tumors	Artificial neural networks	205
Postoperative prognosis	25,344 cDNAs: NCBI's Unigene	10 tumors: 5 dead, 5 survived in 5 years	Mann–Whitney test, permutation test	204
For ER(−) cancers	8,102 cDNAs[a]	115 malignant tumors	Unsupervised hierarch. clustering, PAM, Kaplan–Meir analysis	68
Subcategorization of tumors prognostic signatures	12,000 features U95Av2 (Affy)	89 primary tumors	Combined stat. tree model: k-means clustering, Bayesian classification Tree	130
Predictors of lymph node Status	12,000 cDNAs	36 tumors of different grades	Linear discriminant analysis	206
Signatures for pathological states, tumor grades	12,000 features U95Av2 (Aff)	24 tumors training set	Two-sample t-test, global permutation test, BRB	211
Prediction of therapeutic response to docetaxel	25,000 cDNAs	117 tumors from <55 years old patients	Unsupervised hierarchical clustering, 3-step supervised Classification	69
Subclassification, clinical prognosis (metastasis)	7,129 features HuGeneF1 (Affymetrix)	49 primary tumors	Binary regression model: SVD, Bayesian analysis	203
Clinical prognosis based on ER status				

[a] In this study, the authors also reanalyzed data from [69, 203].

cyclophosphamide [212], in spite of the very complex overall expression response to these drugs [211]. There are further limitations of these microarray studies as different groups have used unique gene sets on their arrays and no comparison studies or simple tag ID alignments have been performed to date. The selection of patient cohorts, sample size, experimental and analytical techniques all vary greatly between these studies. And as in the case of glaucoma the predictor genes suggested by each study are almost incomparable even when they predict similar outcomes [69, 130].

Not surprisingly, prediction of the disease course including metastasis and recurrence using prognostic signature gene sets is, however, prone to a high rate of false positives. The fundamental van't Veer study [69] had a false-positive rate of 27 percent in 5 years (nonmetastatic patients were incorrectly assigned to the metastatic group). Microarray efficiency in these studies is also questionable with several thousand genes on genomewide arrays displaying statistically significant differential expression versus healthy tissues, but only a few dozen of these genes are in the final sets of classifiers.

We have tested the analysis of computationally derived gene network signatures as a method of elucidating the relationship between gene expression changes induced by a compound and cellular endpoints. The approach is rooted in the idea of modular organization of large-scale networks of biological processes proposed by Lee Hartwell and colleagues [144] in which various types of cellular functionality are provided by relatively small, transient but tightly connected networks of molecules (5 to 25 nodes) engaged in performing specific functions. We analyzed the expression data from a study investigating the effects of 4-hydroxytamoxifen and estrogen treatment in MCF-7 breast cancer cell cultures [213]. After importing the data into MetaCore, filters were applied to retain only those genes that showed the strongest variance in expression (the top 2 percent, or approximately 40 genes) as a response to either treatment. The networks that link these strongly responsive genes to each other were recreated using the direct interactions algorithm allowing only genes from the subset to be present in the network. The result is several small networks of 2 to 15 genes and a handful of nonconnected nodes. The largest subnetworks for 4-hydroxytamoxifen (*a*) and estrogen (*b*) treatments are shown in Figure 4.7. These networks include mostly cell-cycle-related genes in agreement with the conclusion of hierarchical clustering showing that at the molecular level, both compounds act as agonists of cell proliferation, evoking very similar patterns of gene expression [213]. Unlike clustering, however, these computationally derived gene networks show the absence of cyclin D1 among 4-hydroxytamoxifen-induced genes, resulting in a significant difference in the network topology and overall fingerprint. For estrogen treatment there is a single large network that encompasses all cell-cycle-related genes induced by the treatment while the 4-hydroxytamoxifen network breaks into two smaller clusters [214]. These unique network signatures may be due to 4-hydroxytamoxifen acting as an antagonist of cell proliferation, while estrogen is an agonist [213] with network analysis identifying

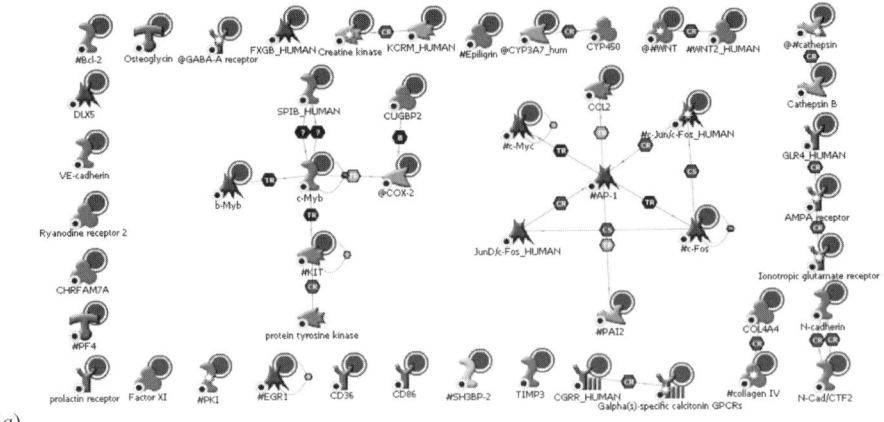

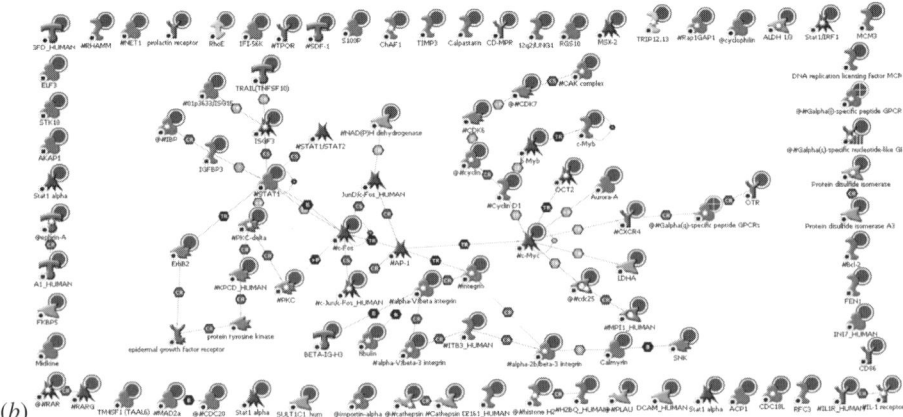

Figure 4.7 Clusters of highly responsive genes for the treatment of MC-7 breast cancer cells with (a) 4-hydroxytamoxifen and (b) estrogen after 24 hr. Both treatments induce similar sets of cell-cycle-related genes, but the topology of the networks is different, correlating with the unique impact of these treatments on cell proliferation. Genes toward the exterior possess fewer interactions than those located centrally. (This figure is available in full color at ftp://ftp.wiley.com/public/sci_tech_med/drug_discovery/.)

cyclin D1 as a responsible element well known as key for the G_1/S transition [215, 216].

Systems ADME/Tox

Understanding or computationally predicting the ADME/Tox properties of molecules earlier can help prevent late-stage clinical failure and has attracted considerable investment for research into developing these methods and models [217–219]. The majority of xenobiotics undergo metabolism via the cytochrome P450 (CYP) enzymes, which are capable of either inactivating or activating these molecules. To date specific CYP-substrate/inhibitor recognition interactions have been studied extensively and several quantitative structure–activity relationships (QSAR) and pharmacophore models have been built for a limited number of these enzymes in a reductionist manner [220–222]. It is clear the computational models need to be integrated with the technologies described earlier for systems biology [6, 217, 223] including metabolic databases and others containing ADME-associated proteins or pathways such as PharmaGKB [224], the human membrane transporter database [225], the ADME-AP database [226], and the nuclear receptor database [227]. There is considerable complexity that needs to be incorporated such as modeling of a single enzyme's catalysis of one molecule to multiple metabolites [228] that has been shown using commercially available cell-modeling software. This study showed that toxicity of a substrate is likely to be minimized when there are multiple metabolites, and this therefore may represent a biological advantage. To date the complex interconnections where one molecule may be metabolized by different enzymes with different kinetic constants and then the metabolite cascade may follow a unique pathway has not been modeled. In addition the complex kinetics [229] and general protein promiscuity [218] that have been observed in vitro for many of these drug metabolizing enzymes may be an important mechanism for enabling rapid clearance of toxins [230], and this in turn may be understood by using a systems biology approach [231].

Understanding the regulation of these enzymes is also important, with nuclear receptors identified as central to the many proteins involved in drug metabolism and transport. The pregnane X-receptor (PXR) is a transcriptional regulator of CYP3A, human MDR1, CYP2C8/9, CYP2B6, and other genes involved in the transport, metabolism, and biosynthesis of bile acids [232]. A second orphan nuclear receptor, the constitutive androstane receptor (CAR), has approximately 40 percent identity with PXR in the ligand binding domain. CAR accumulates in the nucleus, heterodimerizes with RXR, binds to the two Phenobarbital-responsive elements, and ultimately activates transcription of the CYP2B6 [233] and other genes in a manner similar to how PXR can activate CYP2B6 as well as CYP3A4 [232]. The complex interactions and cross-talk of these nuclear interactions urgently requires computational models that can predict binding to these proteins and enable visualization of connections between different genes and ligands [232, 234].

Global gene expression profiling has a role for evaluation of potential toxicity of drug candidates [235, 236]. Two companies have marketed content databases with patterns of gene expression on whole genome rat arrays in response to drugs and known toxic molecules. The Iconix Drug Matrix product (*http://www.iconixpharm.com/products/products_drugmatrix.html*) represents a collection of gene expression profiles (Amersham rat arrays with 8000 features) from liver of rats treated with about 450 drugs with known toxicological properties. GeneLogic used Affymetrix rat arrays in its ToxExpress product (*http://www.genelogic.com/solutions/toxexpress/*). Paradigm Genetics (now known as Icoria) is generating toxicogenomic data in collaboration with NIEHS, which would suggest a large microarray database will be created in the future that may be accessible. The EDGE^2 database, a public effort at the University of Wisconsin is striving to create a public database of mouse gene expression profiles after treatment with different toxic molecules [72]. In all of these efforts different clustering techniques are applied to analyze gene expression patterns and extract toxicity signatures.

Published studies using microarrays for toxicology include one group that treated rats with 52 hepatotoxins and used subtractive hybridization to create a final microarray enriched with liver genes, which was then benchmarked with 3 hepatotoxins. This study suggested hundreds of genes were affected by these xenobiotics, some of which had not been previously associated with toxicity of the compounds [237]. For example, phenobarbital treatment in mice affects the expression of 138 genes when assessed using the NIEHS ToxChip microarray, and approximately half of these genes are regulated by CAR [238]. These studies and databases of expression profiles will certainly be a valuable start for understanding the complexities of toxicity.

One product has currently been developed for systems ADME/Tox (MetaDrug, *www.genego.com*) that predicts the major metabolites from an input structure, scores them for the enzymes likely to be involved, and can use microarray data to visualize gene expression of the genes important for toxicology. A comprehensive set of over 3500 substrate–product reactions for 38 human cytochromes and other enzymes has been assembled. This data has been used in the assessment of product–substrate specificity for the CYPs [239], the classification of druglike molecules based on their K_m values for human CYPs [240] and for building models for predicting the rates (V_{max}) of N-dealkylation by CYP3A4 and CYP2D6 [241]. This complete combination of predictive and visualization tools with literature data is key to explain metabolism and toxicity for unknown compounds. Ultimately this type of combined approach will enable the prediction of idiosyncratic drug reactions that are often not detected until the drug has been released onto the market [242–244]. Our understanding of some of the factors resulting in drug-induced toxicity has expanded to focus on the molecular mechanisms involved. In particular, there has been considerable focus on hepatotoxicity mediated by drug–drug interactions seen as a major cause of the failure of clinical candidates. We can demonstrate the use of this approach using theophylline, a treatment for asthma. Cimetidine, diltiazem, ciprofloxacin, and enoxacin have all been shown to impair the elimination of theophyline, while phenytoin induces its metabolism. The interactions of all of these ligands with CYPs and other proteins can be visualized with MetaDrug (Fig. 4.8). Theophyline is shown with connections (as a substrate) to CYP1A1, CYP1A2, CYP1B1, CYP2C9, CYP2D6, CYP2E1, and CYP3A4 [244b, 244c]. The following would inhibit a range of these CYPs: cimetidine (CYP1A2, CYP2C9, CYP2D6, CYP2E1,

FUTURE DEVELOPMENT OF PATHWAY ANALYSIS SOFTWARE

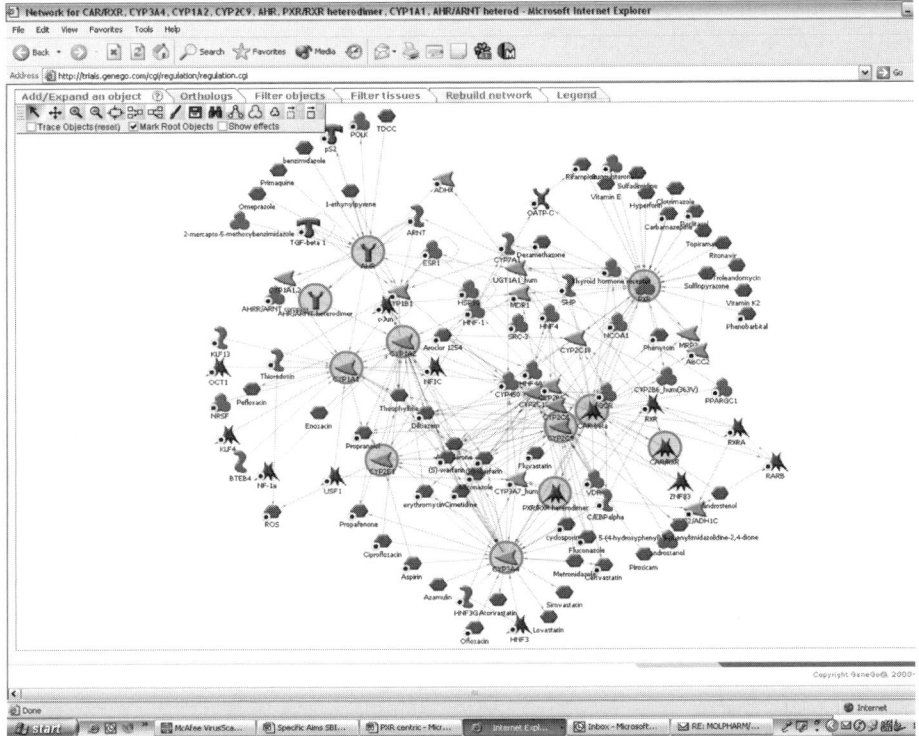

Figure 4.8 MetaDrug interaction map of key proteins and ligands linked to theophylline (circled) The network of cell signaling interactions built from the list of genes. This network has been automatically reconstructed, based on the interactions contained in the database. The genes from the original list are encircled. (This figure is available in full color at ftp://ftp.wiley.com/public/sci_tech_med/drug_discovery/.)

CYP3A4 [245]); diltiazem (CYP3A4 [246]); ciprofloxacin (CYP3A4 and CYP1A2 [247]), and enoxacin (CYP1A2 [248]). Phenytoin is a human PXR activator [249], which in turn will increase the levels of CYP3A4 and ultimately increase the clearance of theophylline by this route. Besides showing all these well-known interactions other potential regulatory and signaling pathways can be observed that may be important for further drug interactions with theophylline, which might not have been previously considered. Certainly there is potential for further integration of systems biology approaches into ADME/Tox [250].

4.8 FUTURE DEVELOPMENT OF PATHWAY ANALYSIS SOFTWARE

The merging of currently independent networks for signaling and metabolism and the enrichment of the mammalian network database with interactions shown for other organisms will be important for the future development of pathway analysis. All currently available genomic and proteomic databases are designed based on molecular entities, such as genes or proteins, as elementary objects [46, 251]. Recently, several public resources started linking their molecular data with the Gene Ontology (GO) "biological process" classification (*www.geneontology.org*) as knowledge of the biological roles of shared proteins is generally transferable across eukaryotes [252]. Such associations are useful in retrieving the subsets of molecular objects related to specific GO categories, but classification in itself does not provide connectivity among processes on the same level of hierarchy. Organizing relevant biological and chemical information around the core entities related to functional processes such as pathways, rather than molecular objects, may be advantageous and will require a novel data model (Fig. 4.9). Ultimately, a set of flexible many-to-many relationships will link functional and molecular information to more adequately capture the dual nature of biological systems. The physical objects (molecules, ions) and processes (reactions, binding, and transport) are linked as operational units by inputs and outputs exchanging matter and information.

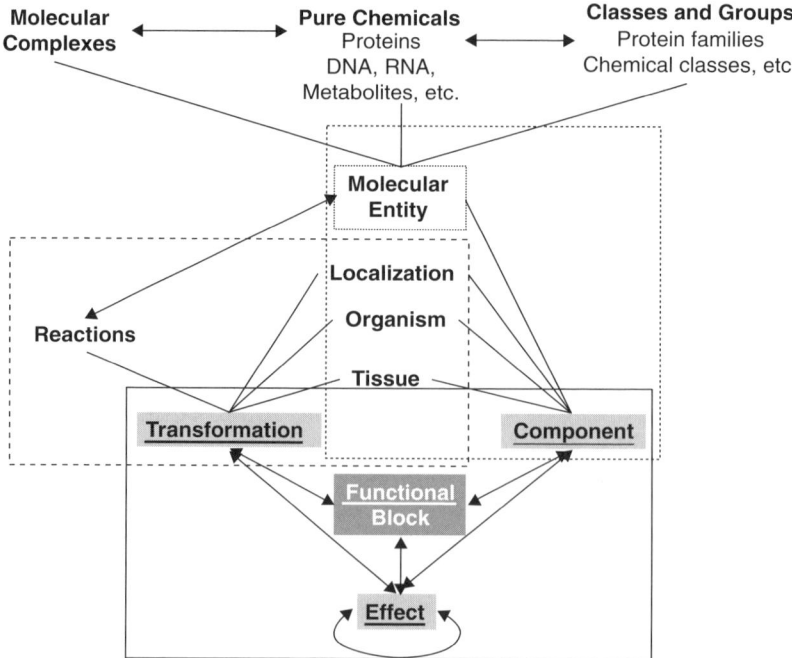

Figure 4.9 Proposed future database model. Borders enclose the database entities related to core functional entities: dotted border, entities related to the <u>component</u>; dashed line, entities related to the <u>transformation</u>; solid border, entities related to the <u>functional block</u>.

New algorithms will also be required for analyzing the networks of intracellular processes and the identification of self-consistent functional modules in them. In the living cell, the networks (particularly the signaling ones) are very dynamic and transient as they exist usually for a short time, in particular cellular compartments and under different conditions. Capturing the transient nature of cellular networking may be possible by defining the functional modules and the sets of related subnetworks operating in a concerted way and by developing an automated protocol for reconstruction of pathways and functional blocks involved in diseases and other conditions. This will produce functionally meaningful modules for cell signaling and metabolism. Recently, several systems biology teams developed methods for identification of tightly connected subnetworks as molecular complexes or functional modules [155, 156, 253, 254]. These studies were based on publicly available data on relatively simple yeast and bacterial protein interaction networks. The networks represented undirected interaction graphs with equal weights assigned to all interactions. Moreover, most of the data came from Y2H assays, strongly biased toward physical associations between proteins (such as complex formation), and known for its false-positive rate. The proteins that are related by other means (e.g., enzymes in a metabolic pathway or transient signal transduction interactions) could be missing from such data.

Future network analysis algorithms may also be based on the curated networks of cell signaling and metabolic processes, which contain additional information associated with the nodes and edges of the networks. The changes of expression of two molecules should correspond to the way they are connected in the network in terms of directionality and type of links between molecules. For example, in MetaCore this information describes three types of edges (activation, inhibition, and unspecified), 12 different mechanisms (e.g., phosphorylation, binding, and transcriptional regulation), and types of nodes (e.g., metabolic enzyme, second messenger, ion channel, and transcription factors). The previously developed algorithm [155] could then be used to cluster highly connected nodes found in a large network of protein interactions using an optimization approach. The sets of nodes are selected to maximize the function $Q(m,n) = 2m/[n(n-1)]$, where m is the number of interactions among n nodes. Thus, a cluster of proteins in which every member interacts with every other member will score the highest ($Q = 1$). The Monte Carlo optimization procedure starts with a randomly selected set of connected nodes and proceeds by moving selected nodes along the edges of the graph to maximize Q. Previously, this method was found to be effective in recognizing protein complexes and functional modules in the protein interaction network in yeast [155] in which the clusters were scored based on connectivity. Figure 4.10 shows graphs with the same underlying topology but a different directionality of the edges. In panel (a), one can find several directed paths containing 4 nodes. Altogether such paths connect 7 out of 9 nodes of the graph (marked by bold arrows). On panel (b) there are no directed paths of length 3. It is clear that graph (a) is more likely to represent a functional pathway than graph (b). The

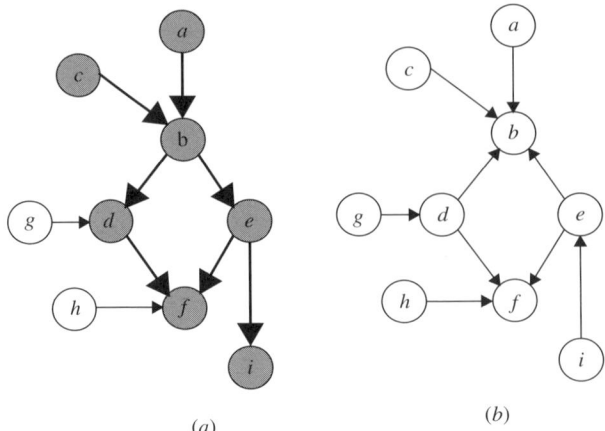

Figure 4.10 Schematic to demonstrate two graphs with identical topology but different directionality of edges. Graph (*a*) is more likely to represent a functional module as it contains longer directed paths than graph (*b*).

implementation of this method in a computational algorithm to find subgraphs containing long ordered paths (containing 4 or more nodes) will be valuable for gene network modeling in which the criterion for inclusion of a node will be whether it contributes to a longer directed path or allows new directed paths of sufficient length. For example, if a cluster of nodes *b*, *d*, *e*, *f* in Figure 4.10*a* were used as a seed, then the addition of node *a* would create a path of length 4. The consecutive inclusion of nodes *c* and *i* would not create longer paths but would create additional paths of length 4. An important way to assess the efficiency of the proposed algorithms is to evaluate the statistical significance of found modules with a random reassignment of the node connections leading to a different distribution of clusters [155]. Tight clusters are highly unlikely ($P < 10^{-5}$) when observing the distribution of clusters in a randomly connected network.

Comparing relatively small functional network modules of 5 to 25 nodes will be computationally tractable and several network comparison algorithms have recently been applied to the analysis of intracellular pathways. Ogata and co-workers [255] have described a heuristic graph comparison algorithm and its application for finding functionally related enzymes clusters (FREC) across 10 species of microorganisms [255]. An extension of this method has also been applied for finding the similarity between protein and gene expression networks [256]. More recently, an algorithm has been described that finds common interaction pathways by global alignment of protein interaction networks from the yeast *Saccharomyces cerevisiae* and a bacterium *Helicobacter pylori* [257]. New techniques that take advantage of annotated human interactions in the databases could be based on this algorithm. The construction of a multiparametric distance function between network modules from

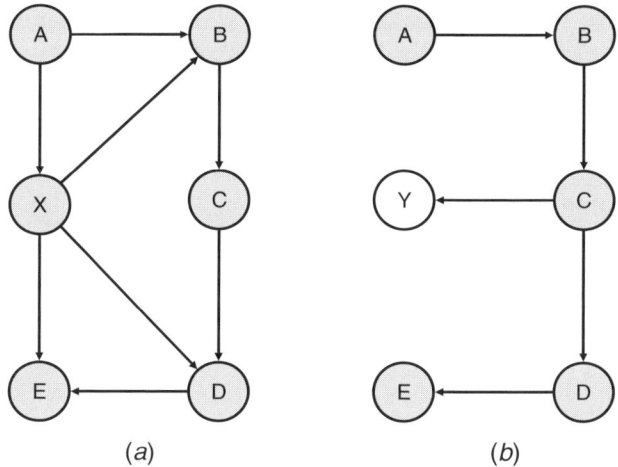

Figure 4.11 Schematic to illustrate how two network modules sharing 83 percent molecular composition can have significant topological differences. On network (*a*) the shortest distance between nodes A and E is 2, while on network (*b*) it is 4. The global connectivity of network (*a*) is also higher. (This figure is available in full color at ftp://ftp.wiley.com/public/sci_tech_med/drug_discovery/.)

different samples would be a first step. Then the computation of a matrix of distances among all network modules from different samples would be possible. Once the distances are computed, an appropriate clustering algorithm can be selected to find modules that correlate with the condition of interest. In many cases networks from different samples may have similar but nonidentical compositions but may or may not be substantially different in their topology. For example, the difference in just one node can change network topology as shown in Figure 4.11. It is important to distinguish between network modules that have similar molecular composition but different topologies. The distance between nodes could also be used such that A and E is 2 in Figure 4.11*a* and 4 in Figure 4.11*b*. Other network descriptors may include global connectivity or the connectivity of differentially expressed genes within the module, minimal path lengths connecting them (TopNet, 53), directionality of interactions, or P values for the distribution of highly differentiated genes. These approaches all represent unique measures of similarity/dissimilarity between network signatures and could be used to classify them.

Reconstruction of Condition-Specific Molecular and Functional Networks

The development of programs for the custom selection of functional modules based on the input sets of conditional gene expression data (both mRNA and protein expression) will be another important progression in computational gene network analysis software. Such input-related functional network modules will be relatively small and specific for the particular data set in question. In this case the genes or proteins of interest will be connected by the shortest, most direct and functionally most relevant interactions. This approach will rely on the topology of the underlying curated network encoding considerably more information and providing an understanding of affected genes up or downstream. Capturing the gene network will be more useful than lists of genes and will point to potential biomarkers. A final goal would be to use the integrated high-throughput data for the reconstruction of disease-specific molecular and functional networks (described earlier), analyze their topology and behavior, and ultimately rank potential drug targets according to their role in the mechanisms of diseases. This will require testing several different algorithms for reconstructing disease states and other conditions with the likely most promising direction coming from a data-intensive consensus approach. Some data are publicly available from studies with genes that demonstrate a different expression level between normal and disease states (*http://www.broad.mit.edu/resources.html; www.ebi.ac.uk*). This can be used to identify the gene clusters representing the affected functional pathways, and these could be grouped based on the connectivity of the underlying network. For each gene a set of nearest-neighbor genes on the network could be computed before identifying subsets of genes tightly connected either directly to each other or bridged by no more than one gene that is not a member of the original list. For every gene at least one of its nearest neighbors (or the central gene itself) is also a nearest neighbor of at least one other gene from the subset (Fig. 4.12).

4.9 FUTURE OF SYSTEMS BIOLOGY APPLIED TO DRUG DISCOVERY

We have seen in recent years a move toward generating high-throughput screening data. This has occurred earlier in drug discovery based around molecules of interest, particularly in terms of their physico-chemical properties and their biological activities toward both desirable and undesirable targets in isolation. This data is also being used to derive computational algorithms to predict these properties from molecular structure. Increasingly, higher content biological data is also being generated after cell or animal treatment and the levels of metabolites, genes, and proteins are determined. Combining the reductionist type of approach for a molecule (does it bind or not to a particular protein) with the effect on a whole system (what is the global effect on metabolism, gene expression, and transcription) will be important if we are to understand and predict efficacy and toxicity reliably. The applications of systems biology in drug discovery in the future are likely to use the vast amounts of qualitative and quantitative biological data collated in the various databases and network building algorithms to build predictive signatures for diseases and following treatment of cells or tissues with a myriad of molecules. Such a predictive approach will continue to be developed at the interface of

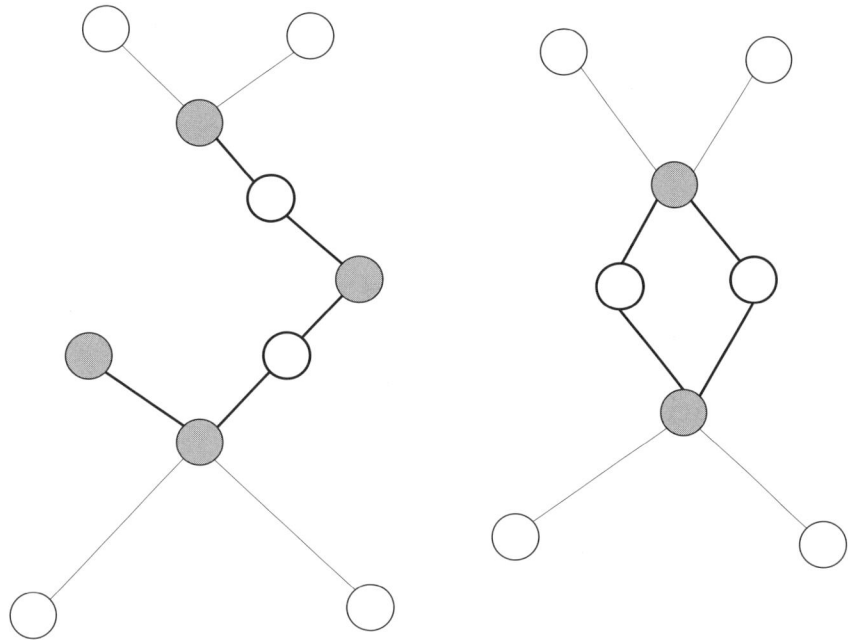

Figure 4.12 Schematic to show the identification of tight clusters from differentially expressed genes. Filled circles represent differentially expressed genes from the original set; open circles represent nearest neighbors; bold lines and circles represent genes and edges connecting genes into cluster.

cheminformatics and bioinformatics for application from target selection through clinical data analysis.

Acknowledgments We acknowledge our colleagues at GeneGo, collaborators at Chemical Diversity, Dr. Valery Shestopalov (University of Miami), and NIH grant 1-R43-GM069124–01 "In Silico Assessment of Drug Metabolism and Toxicity."

REFERENCES

1. Lander, E. S., Linton, L. M., Birren, B., Nusbaum, C., Zody, M. C., Baldwin, J., Devon, K., Dewar, K., Doyle, M., FitzHugh, W., Funke, R., Gage, D., Harris, K., Heaford, A., Howland, J., Kann, L., Lehoczky, J., LeVine, R., McEwan, P., McKernan, K., Meldrim, J., Mesirov, J. P., Miranda, C., Morris, W., Naylor, J., Raymond, C., Rosetti, M., Santos, R., Sheridan, A., Sougnez, C., Stange-Thomann, N., Stojanovic, N., Subramanian, A., Wyman, D., Rogers, J., Sulston, J., Ainscough, R., Beck, S., Bentley, D., Burton, J., Clee, C., Carter, N., Coulson, A., Deadman, R., Deloukas, P., Dunham, A., Dunham, I., Durbin, R., French, L., Grafham, D.,

Gregory, S., Hubbard, T., Humphray, S., Hunt, A., Jones, M., Lloyd, C., McMurray, A., Matthews, L., Mercer, S., Milne, S., Mullikin, J. C., Mungall, A., Plumb, R., Ross, M., Shownkeen, R., Sims, S., Waterston, R. H., Wilson, R. K., Hillier, L. W., McPherson, J. D., Marra, M. A., Mardis, E. R., Fulton, L. A., Chinwalla, A. T., Pepin, K. H., Gish, W. R., Chissoe, S. L., Wendl, M. C., Delehaunty, K. D., Miner, T. L., Delehaunty, A., Kramer, J. B., Cook, L. L., Fulton, R. S., Johnson, D. L., Minx, P. J., Clifton, S. W., Hawkins, T., Branscomb, E., Predki, P., Richardson, P., Wenning, S., Slezak, T., Doggett, N., Cheng, J. F., Olsen, A., Lucas, S., Elkin, C., Uberbacher, E., Frazier, M., et al. (2001). Initial sequencing and analysis of the human genome. *Nature, 409*(6822), 860–921.

2. Venter, J. C., Adams, M. D., Myers, E. W., Li, P. W., Mural, R. J., Sutton, G. G., Smith, H. O., Yandell, M., Evans, C. A., Holt, R. A., Gocayne, J. D., Amanatides, P., Ballew, R. M., Huson, D. H., Wortman, J. R., Zhang, Q., Kodira, C. D., Zheng, X. H., Chen, L., Skupski, M., Subramanian, G., Thomas, P. D., Zhang, J., Gabor Miklos, G. L., Nelson, C., Broder, S., Clark, A. G., Nadeau, J., McKusick, V. A., Zinder, N., Levine, A. J., Roberts, R. J., Simon, M., Slayman, C., Hunkapiller, M., Bolanos, R., Delcher, A., Dew, I., Fasulo, D., Flanigan, M., Florea, L., Halpern, A., Hannenhalli, S., Kravitz, S., Levy, S., Mobarry, C., Reinert, K., Remington, K., Abu-Threideh, J., Beasley, E., Biddick, K., Bonazzi, V., Brandon, R., Cargill, M., Chandramouliswaran, I., Charlab R.., Chaturvedi, K., Deng, Z., Di Francesco, V., Dunn, P., Eilbeck, K., Evangelista, C., Gabrielian, A. E., Gan, W., Ge, W., Gong, F., Gu, Z., Guan, P., Heiman, T. J., Higgins, M. E., Ji, R. R., Ke, Z., Ketchum, K. A., Lai, Z., Lei, Y., Li, Z., Li, J., Liang, Y., Lin, X., Lu, F., Merkulov, G. V., Milshina, N., Moore, H. M., Naik, A. K., Narayan, V. A., Neelam, B., Nusskern, D., Rusch, D. B., Salzberg, S., Shao, W., Shue, B., Sun, J., Wang, Z., Wang, A., Wang, X., Wang, J., Wei, M., Wides, R., Xiao, C., Yan, C., et al. (2001). The sequence of the human genome. *Science, 291*(5507), 1304–1351.

3. Fiehn, O. (2001). Combining genomics, metabolome analysis, and biochemical modelling to understand metabolic networks. *Compar. Funct. Genom., 2*, 155–168.

4. Nicholson, J. K., Wilson, I. D. (2003). Understanding global, systems biology: Metabonomics and the continuum of metabolism. *Natl. Rev. Drug Discov., 2*, 668–676.

5. Food and Drug Administration (FDA) (2004). *Innovation Stagnation: Challenge and Opportunity on the Critical Path to New Medicinal Products*, FDA, Washington, DC.

6. Kitano, H. (2002). Computational systems biology. *Nature, 420*, 206–210.

7. Kitano, H. (2002). Systems biology: A brief overview. *Science, 295*, 1662–1664.

8. Ideker, T. (2004). Systems biology 101—what you need to know. *Nat. Biotechnol., 22*, 473–475.

9. Noble, D., Colatsky, T. J. (2002). A return to rational drug discovery: Computer-based models of cells, organs and systems in drug target identification. *Emerging Ther. Targets, 4*, 39–49.

10. Baxter, D. A., Canavier, C. C., Clark, Jr. J. W., Byrne, J. H. (1999). Computational model for the serotonergic modulation of sensory neorons in Aplysia. *J. Neurophysiol., 82*, 2914–2935.

11. Bhalla, U. S. (2002). The chemical organization of signalling interactions. *Bioinformatics*, *18*, 855–863.
12. Kuroda, S., Schwieghofer, N., Kawato, M. (2001). Exploration of signal transduction pathways in cerebellar long-term depression by kinetic simulation. *J. Neurosci.*, *21*, 5693–5702.
13. Csete, M. E., Doyle, J. C. (2002). Reverse engineering of biological complexity. *Science*, *295*, 1664–1669.
14. McAdams, H. H., Arkin, A. P. (2000). Gene regulation: Towards a circuit engineering discipline. *Curr. Biol.*, *10*, R318–R320.
15. Yagil, G. (1976). Isotope exchange in biochemical networks application of an electric circuit technique to a biological reaction system. *J. Theor. Biol.*, *61*, 73–80.
16. Boiteux, A., Busse, H-G. (1989). Circuit analysis of the oscillatory state in glycolysis. *BioSystems*, *22*, 231–240.
17. McAdams, H. H., Shapiro, L. (1995). Circuit simulation of genetic networks. *Science*, *269*, 650–656.
18. Odum, D. T., Zizlsperger, N., Gordon, D. B., Bell, G. W., Rinaldi, N. J., Murray, H. L., Volkert, T. L., Schreiber, J., Rolfe, P. A., Gifford, D. K., Fraenkel, E., Bell, G. I., Young, R. A. (2004). Control of pancreas and liver gene expression by HNF transcription factors. *Science*, *303*, 1378–1381.
19. Koza, J. R., Keane, M. A., Streeter, M. J., Mydlowec, W., Yu, J., Lanza, G. (2003). *Genetic Programming IV: Routine Human-Competitive Machine Intelligence*. Kluwer Academic, Norwell, MA.
20. Wiechert, W. (2002). Modeling and simulation: Tools for metabolic engineering. *J. Biotechnol.*, *94*, 37–63.
21. Covert, M. W., Schilling, C. H., Famili, I., Edwards, J. S., Goryanin, I. I., Selkov, E., Palsson, B. O. (2001). Metabolic modeling of microbial strains in silico. *Trends Biochem. Sci.*, *26*, 179–186.
22. Schaff, J., Fink, C. C., Slepchenko, B., Carson, J. H., Loew, L. M. (1997). A general computational framework for modeling cellular structure and function. *Biophys. J.*, *73*, 1135–1146.
23. Schaff, J., Loew, L. M. The virtual cell. In R. B. Altman (Ed.), *Pacific Symposium on Biocomputing, Mauna Lani, Hawaii, 1999*, pp. 228–239.
24. Fink, C. C., Slepchenko, B., Moraru, I. I., Schaff, J., Watras, J., Loew, L. M. (1999). Morphical control of inositol-1,4,5-trisphosphate-dependent signals. *J. Cell Biol.* *147*, 929–935.
25. Tomita, M., Hashimoto, K., Takahashi, K., Shimizu, T. S., Matsuzaki, Y., Miyoshi, F., Saito, K., Tanida, S., Yugi, K., Ventner, J. C., Hutchison III, C. A. (1999). E-CELL: Software environment for whole-cell simulation. *Bioinformatics*, *15*, 72–84.
26. Noble, D. (2002). Modeling the heart—from genes to cells to the whole organ. *Science*, *295*, 1678–1682.
27. Musante, C. J., Lewis, A. K., Hall, K. (2002). Small- and large-scale biosimulation applied to drug discovery and development. *Drug Discov. Today*, *7*, S192–S196.
28. Bumol, T. F., Watanabe, A. M. (2001). Genetic information, genomic technologies, and the future of drug discovery. *JAMA*, *285*, 551–555.

29. Zhu, X., Raina, A. K., Lee, H. G., Casadesus, G., Smith, M. A., Perry, G. (2004). Oxidative stress signalling in Alzheimer's disease. *Brain Res.*, *1000*, 32–39.
30. Larkin, J. E., Frank, B. C., Gaspard, R. M., Duka, I., Gavras, H., Quakenbush, J. (2004). Cardiac transcriptional response to acute and chronic angiotensin II treatments. *Physiol. Genom*, *18*, 152–166.
31. Gerrard, J. A., Sparrow, A. D., Wells, J. A. (2001). Metabolic databases—what next? *Trends Biochem. Sci.*, *26*, 137–140.
32. Vasquez, A., Flammini, A., Maritan, A., Vespignani, A. (2003). Global protein function prediction from protein-protein interaction networks. *Nat. Biotechnol.*, *21*, 697–700.
33. Schomburg, I., Chang, A., Schomburg, D., (2002) BRENDA, enzyme data and metabolic information. *Nucleic Acids Res.*, *30*, 47–49.
34. Schomburg, I., Chang, A., Hofmann, O., Ebeling, C., Ehrentreich, F., Schomburg, D. (2002). BRENDA: A resource for the enzyme data and metabolic information. *Trends Biochem Sci.*, *27*, 54–56.
35. Takai-Igarashi, T., Kaminuma, T. (1999). A pathway finding system for the cell signaling networks database. *In Silico Biol.*, *1*, 129–146.
36. Xenarios, I., Rice, D. W., Salwinski, L., Baron, M. K., Markotte, E. M., Eisenberg, D. (2000). DIP: The database of interacting proteins. *Nucleic Acids Res.*, *28*, 289–291.
37. Selkov, E., Basmanova, S., Gaasterland, T., Goryanin, I., Gretchin, Y., Maltsev, N., Nenashev, V., Overbeek, R., Panyushkina, E., Pronevitch, L., Selkov, E. J., Yunis, I. (1996). The metabolic pathway collection from EMP: The enzymes and metabolic pathways database. *Nucleic Acids Res.*, *24*, 26–28.
38. Dahlquist, K. D., Salomonis, N., Vranizan, K., Lawlor, S. C., Conklin, B. R., (2002). GenMAPP, a new tool for viewing and analyzing microarray data on biological pathways. *Nat. Genet.*, *31*, 19–20.
39. Peri, S., Navarro, J. D., Kristiansen, T. Z., Amanchy, R., Surendranath, V., Muthusamy, B., Gandhi, T. K., Chandrika, K. N., Deshpande, N., Suresh, S., Rashmi, B. P., Shanker, K., Padma, N., Niranjan, V., Harsha, H. C., Talreja, N., Vrushabendra, B. M., Ramya, M. A., Yatish, A. J., Joy, M., Shivashankar, H. N., Kavitha, M. P., Menezes, M., Choudhury, D. R., Ghosh, N., Saravana, R., Chandran, S., Mohan, S., Jonnalagadda, C. K., Prasad, C. K., Kumar-Sinha, C., Deshpande, K. S., Pandey, A. (2004). Human protein reference database as a discovery resource for proteomics. *Nucleic Acids Res.*, *32*(Database issue), D497–501.
40. Peri, S., Navarro, J. D., Amanchy, R., Kristiansen, T. Z., Jonnalagadda, C. K., Surendranath, V., Niranjan, V., Muthusamy, B., Gandhi, T. K., Gronborg, M., Ibarrola, N., Deshpande, N., Shanker, K., Shivashankar, H. N., Rashmi, B. P., Ramya, M. A., Zhao, Z., Chandrika, K. N., Padma, N., Harsha, H. C., Yatish, A. J., Kavitha, M. P., Menezes, M., Choudhury, D. R., Suresh, S., Ghosh, N., Saravana, R., Chandran, S., Krishna, S., Joy, M., Anand, S. K., Madavan, V., Joseph, A., Wong, G. W., Schiemann, W. P., Constantinescu, S. N., Huang, L., Khosravi-Far, R., Steen, H., Tewari, M., Ghaffari, S., Blobe, G. C., Dang, C. V., Garcia, J. G., Pevsner, J., Jensen, O. N., Roepstorff, P., Deshpande, K. S., Chinnaiyan, A. M., Hamosh, A., Chakravarti, A., Pandey, A. (2003). Development of human protein reference

database as an initial platform for approaching systems biology in humans. *Genome Res.*, *13*(10), 2363–2371.

41. Karp, P. D., Riley, M., Saier, M., Paulsen, I. T., Collado-Vides, J., Paley, S. M., Pellegrini-Toole, A., Bonavides, C., Gama-Castro, S. (2002). The EcoCyc Database. *Nucleic Acids Res.*, *30*, 56–58.

42. Kanehisa, M. A., Goto, S., Kawashima, S., Nakaya, A. (2002). The KEGG databases at GenomeNet. *Nucleic Acids Res.*, *30*, 42–46.

43. Selkov, E. J., Grechkin, Y., Mikhailova, N., Selkov, E. (1998). MPW: The metabolic pathways database. *Nucleic Acids Res.*, *26*, 43–45.

44. Smith, C. M., Shindyalov, I. N., Veretnik, S., Gribskov, M., Taylor, S. S., Ten Eyck, L. F., Bourne, P. E. (1997). The protein kinase resource. *TIBS*, *22*, 444–446.

45. Maltsev, N., Marland, E., Yu, G. X., Bhatnagar, S., Lusk, R. (2002). Sentra, a database of signal transduction proteins. *Nucleic Acids Res.*, *30*, 349–350.

46. Bairoch, A., Apweiler, R. (2000). The SWISS-PROT protein sequence database and its supplement TrEMBL in 2000. *Nucleic Acids Res.*, *28*, 45–48.

47. Boeckmann, B., Bairoch, A., Apweiler, R., Blatter M-C., Estreicher, A., Gasteiger, E., Martin, M. J., Michoud, K., O'Donovan, C., Phan, I., Pilbout, S., Schneider, M. (2003). The SWISS-PROT protein knowledge base and its supplement TrEMBL in 2003. *Nucleic Acids Res.*, *31*, 365–370.

48. Schacherer, F., Choi, C., Gotze, U., Krull, M., Pistor, S., Wingender, E. (2001). The TRANSPATH signal transduction database: A knowledge base on signal transduction. *Bioinformatics*, *17*, 1053–1057.

49. Wheeler, D. L., Church, D. M., Federhen, S., Lash, A. E., Madden, T. L., Pontius, J. U., Schuler, G. D., Schriml, L. M., Sequeria, E., Tatusova, T. A., Wagner, L. (2003). Database resources of the national center for biotechnology. *Nucleic Acids Res.*, *31*, 28–33.

50. Ellis, L. B. M., Hou, B. K., Kang, W., Wackett, L. P. (2003). The University of Minnesota Biocatalysis/Biodegradation Database: Post-genomic data mining. *Nucleic Acids Res.*, *31*, 262–265.

51. Ellis, L. B. M., Speedie, S. M., Mcleish, R. (1998). Representing metabolic pathway information : An object orientated approach. *Bioinformatics*, *14*, 803–809.

52. Overbeek, R., Larsen, N., Pusch, G. D., D'Souza, M., Selkov, E. J., Kyrpides, N., Fonstein, M., Maltsev, N., Selkov, E. (2000). WIT: Integrated system for high-throughput genome sequence analysis and metabolic reconstruction. *Nucleic Acids Res.*, *28*, 123–125.

53. Ideker, T., Thorsson, V., Ranish, J. A., Christmas, R., Buhler, J., Eng, J. K., Bumgarner, R., Goodlett, D. R., Aebersold, R., Hood, L. (2001). Integrated genomic and proteomic analyses of a systematically perturbed network. *Science*, *292*, 929–934.

54. Davidson, E. H., Rast, J. P., Oliveri, P., Ransick, A., Calestani, C., Yuh C-H., Minokawa, T., Amore, G., Hinman, V., Arenas-Mena, C., Otim, O., Brown, C. T., Livi, C. B., Lee, P. Y., Revilla, R., Rust, A. G., Pan, Z. J., Schilstra, M. J., Clarke, P. J. C., Arnone, M. I., Rowen, L., Cameron, R. A., McClay, D. R., Hood, L.,

Bolouri, H. (2002). A genomic regulatory network for development. *Science*, *295*, 1669–1678.

55. Arita, M. (2003). In silico atomic tracing by substrate-product relationships in *Escherichia coli* intermediary metabolism. *Genome Res.*, *13*, 2455–2466.
56. Botstein, D., White, R. L., Skolnick, M., Davis, R. W. (1980). Construction of a genetic linkage map in man using restriction fragment length polymorphisms. *Am. J. Hum. Genet.*, *32*(3), 314–331.
57. Elahi, E., Kumm, J., Ronaghi, M. (2004). Global genetic analysis. *J. Biochem. Mol. Biol.*, *37*(1), 11–27.
58. Botstein, D., Risch, N. (2003). Discovering genotypes underlying human phenotypes: Past successes for mendelian disease, future approaches for complex disease. *Nat. Genet.*, *33* (Suppl.), 228–237.
59. Patil, N., Berno, A. J., Hinds, D. A., Barrett, W. A., Doshi, J. M., Hacker, C. R., Kautzer, C. R., Lee, D. H., Marjoribanks, C., McDonough, D. P., Nguyen, B. T., Norris, M. C., Sheehan, J. B., Shen, N., Stern, D., Stokowski, R. P., Thomas, D. J., Trulson, M. O., Vyas, K. R., Frazer, K. A., Fodor, S. P., Cox, D. R. (2001). Blocks of limited haplotype diversity revealed by high-resolution scanning of human chromosome 21. *Science*, *294*(5547), 1719–1723.
60. Marnellos, G. (2003). High-throughput SNP analysis for genetic association studies. *Curr. Opin. Drug Discov. Devel.*, *6*(3), 317–321.
61. Jurinke, C., van den Boom, D., Cantor, C. R., Koster, H. (2002). Automated genotyping using the DNA MassArray technology. *Methods Mol. Biol.*, *187*, 179–192.
62. Ronaghi, M., Pettersson, B., Uhlen, M., Nyren, P. (1998). PCR-introduced loop structure as primer in DNA sequencing. *Biotechniques*, *25*(5), 876–878, 880–882, 884.
63. Oliphant, A., Barker, D. L., Stuelpnagel, J. R., Chee, M. S. (2002). BeadArray technology: Enabling an accurate, cost-effective approach to high-throughput genotyping. *Biotechniques*, *56–58* (Suppl.), 60–61.
64. Watson, J. D., Crick, F. H. (1953). Genetical implications of the structure of deoxyribonucleic acid. *Nature*, *171*(4361), 964–967.
65. Watson, J. D., Crick, F. H. (1953). Molecular structure of nucleic acids; a structure for deoxyribose nucleic acid. *Nature*, *171*(4356), 737–738.
66. Butte, A. (2002). The use and analysis of microarray data. *Nat. Rev. Drug Discov.*, *1*(12), 951–960.
67. Kuo, W. P., Jenssen, T. K., Butte, A. J., Ohno-Machado, L., Kohane, I. S. (2002). Analysis of matched mRNA measurements from two different microarray technologies. *Bioinformatics*, *18*(3), 405–412.
68. Sørlie, T., Tibshirani, R., Parker, J., Hastie, T., Marron, J. S., Nobel, A., Deng, S., Johnsen, H., Pesich, R., Geisler, S., Demeter, S., Perov, C. M., Lonning, P. E., Brown, P. O., Borresen-Dale, A. L., Botstein, D. (2003). Repeated observation of breast tumor subtypes in independent gene expression data sets. *Proc. Natl. Acad. Sci. USA*, *100*, 8418–8423.
69. van 't Veer, L. J., Dai, H., van de Vijer, M. J., He, Y. D., van de Vijver, M. J., He, Y. D., Vant Veer, L. J., Dai, H., Hart, A. A., Voskoil, D. W., Schreiber, G. J., Peterse, J. L., Roberts, C., Marton, M. J., Parrish, M., Atsma, D., Nitteveen, A., Glas, A., Delahaye, L., van der Nelde, T., Bartelink, H., Rodenhuis, S., Rutgers, E. T.,

Freind, S. H., Bernards, R. (2002). Gene expression profiling predicts clinical outcome of breast cancer. *Nature*, *415*, 530–536.

70. Cleator, S., Ashworth, A. (2004). Molecular profiling of breast cancer: Clinical implications. *Br. J. Cancer. Res.*, *90*, 1120–1124.

71. Cardoso, F. (2003). Microarray technology and its effect on breast cancer (re)classification and prediction of outcome. *Breast Cancer Res.*, *5*(6), 303–304.

72. Thomas, R. S., Rank, D. R., Penn, S. G., Zastrow, G. M., Hayes, K. R., Pande, K., Glover, E., Silander, T., Craven, M. W., Reddy, J. K., Jovanovich, S. B., Bradfield, C. A. (2001). Identification of toxicologically predictive gene sets using cDNA microarrays. *Mol. Pharmacol.*, *60*, 1189–1194.

73. Waring, J. F., Jolly, R. A., Ciurlionis, R., Lum, P. Y., Praestgaard, J. T., Morfitt, D. C., Buratto, B., Roberts, C., Schadt, E., Ulrich, R. G. (2001). Clustering of hepatotoxins based on mechanism of toxicity using gene expression profiles. *Toxicol. Appl. Pharmacol.*, *175*, 28–42.

74. Gerhold, D., Lu, M., Xu, J., Austin, C., Caskey, C. T., Rushmore, T. (2001). Monitoring expression of genes involved in drug metabolism and toxicology using DNA microarrays. *Physiol. Genomics*, *5*, 161–170.

75. Coen, M., Ruepp, S. U., Lindon, J. C., Nicholson, J. K., Pognan, F., Lenz, E. M., Wilson, I. D. (2004). Integrated application of transcriptomics and metabonomics yields new insight into the toxicity due to paracetamol in the mouse. *J. Pharm. Biomed. Anal.*, *35*(1), 93–105.

76. Perou, C. M. (2001). Show me the data! *Nat. Genet.*, *29*(4), 373.

77. Velculescu, V. E., Zhang, L., Vogelstein, B., Kinzler, K. W. (1995). Serial analysis of gene expression. *Science*, *270*(5235), 484–487.

78. Yamamoto, M., Wakatsuki, T., Hada, A., Ryo, A. (2001). Use of serial analysis of gene expression (SAGE) technology. *J. Immunol. Methods*, *250*(1–2), 45–66.

79. Ryo, A., Kondoh, N., Wakatsuki, T., Hada, A., Yamamoto, N., Yamamoto, M. (2000). A modified serial analysis of gene expression that generates longer sequence tags by nonpalindromic cohesive linker ligation. *Anal. Biochem.*, *277*(1), 160–162.

80. Stein, W. D., Litman, T., Fojo, T., Bates, S. E. (2004). A Serial Analysis of Gene Expression (SAGE) database analysis of chemosensitivity: Comparing solid tumors with cell lines and comparing solid tumors from different tissue origins. *Cancer Res.*, *64*(8), 2805–2816.

81. Nacht, M., Ferguson, A. T., Zhang, W., Petroziello, J. M., Cook, B. P., Gao, Y. H., Maguire, S., Riley, D., Coppola, G., Landes, G. M., Madden, S. L., Sukumar, S. (1999). Combining serial analysis of gene expression and array technologies to identify genes differentially expressed in breast cancer. *Cancer Res.*, *59*(21), 5464–5670.

82. Wilkins, M. R., Sanchez, J. C., Gooley, A. A., Appel, R. D., Humphery-Smith, I., Hochstrasser, D. F., Williams, K. L. (1996). Progress with proteome projects: Why all proteins expressed by a genome should be identified and how to do it. *Biotechnol. Genet. Eng. Rev.*, *13*, 19–50.

83. Anderson, N. L., Anderson, N. G. (1998). Proteome and proteomics: New technologies, new concepts, and new words. *Electrophoresis*, *19*(11), 1853–1861.

84. Klein, J. B., Thongboonkerd, V. (2004). Overview of proteomics. *Contrib. Nephrol.*, *141*, 1–10.
85. O'Farrell, P. H. (1975). High resolution two-dimensional electrophoresis of proteins. *J. Biol. Chem.*, *250*(10), 4007–4721.
86. Klose, J. (1975). Protein mapping by combined isoelectric focusing and electrophoresis of mouse tissues. A novel approach to testing for induced point mutations in mammals. *Humangenetik*, *26*(3), 231–243.
87. Fountoulakis, M. (2000). Two-dimensional electrophoresis. In I. Wilson, C. Poole, and M. Cooke (Eds.), *Encyclopedia of Separation Science, II/Electrophoresis*, Academic, London, pp. 1356–1363.
88. Aebersold, R., Mann, M. (2003). Mass spectrometry-based proteomics. *Nature*, *422*(6928), 198–207.
89. Lahm, H. W., Langen, H. (2000). Mass spectrometry: A tool for the identification of proteins separated by gels. *Electrophoresis*, *21*, 2105–2114.
90. Henzel, W. J., Billeci, T. M., Stults, J. T., Wong, S. C., Grimley, C., Watanabe, C. (1993). Identifying proteins from two-dimensional gels by molecular mass searching of peptide fragments in protein sequence databases. *Proc. Natl. Acad. Sci. USA*, *90*(11), 5011–5015.
91. Mann, M., Wilm, M. (1994). Error-tolerant identification of peptides in sequence databases by peptide sequence tags. *Anal. Chem.*, *66*(24), 4390–4399.
92. Gorg, A., Obermaier, C., Boguth, G., Harder, A., Scheibe, B., Wildgruber, R., Weiss, W. (2000). The current state of two-dimensional electrophoresis with immobilized pH gradients. *Electrophoresis*, *21*(6), 1037–1053.
93. Zuo, X., Speicher, D. W. (2000). A method for global analysis of complex proteomes using sample prefractionation by solution isoelectrofocusing prior to two-dimensional electrophoresis. *Anal. Biochem.*, *284*(2), 266–278.
94. Hanash, S. (2003). Disease proteomics. *Nature*, *422*(6928), 226–232.
95. Zheng, P. P., Kros, J. M., Sillevis-Smitt, P. A., Luider, T. M. (2003). Proteomics in primary brain tumors. *Front Biosci.*, *8*, d451–d463.
96. Fountoulakis, M. (2004). Application of proteomics technologies in the investigation of the brain. *Mass. Spectrom. Rev.*, *23*(4), 231–258.
97. Rigaut, G., Shevchenko, A., Rutz, B., Wilm, M., Mann, M., Seraphin, B. (1999). A generic protein purification method for protein complex characterization and proteome exploration. *Nat. Biotechnol.*, *17*(10), 1030–1032.
98. Nelson, R. W., Nedelkov, D., Tubbs, K. A. (2000). Biosensor chip mass spectrometry: A chip-based proteomics approach. *Electrophoresis*, *21*(6), 1155–1163.
99. Gavin, A. C., Bosche, M., Krause, R., Grandi, P., Marzioch, M., Bauer, A., Schultz, J., Rick, J. M., Michon, A. M., Cruciat, C. M., Remor, M., Hofert, C., Schelder, M., Brajenovic, M., Ruffner, H., Merino, A., Klein, K., Hudak, M., Dickson, D., Rudi, T., Gnau, V., Bauch, A., Bastuck, S., Huhse, B., Leutwein, C., Heurtier, M. A., Copley, R. R., Edelmann, A., Querfurth, E., Rybin, V., Drewes, G., Raida, M., Bouwmeester, T., Bork, P., Seraphin, B., Kuster, B., Neubauer, G., Superti-Furga, G. (2002). Functional organization of the yeast proteome by systematic analysis of protein complexes. *Nature*, *415*(6868), 141–147.
100. Ho, Y., Gruhler, A., Heilbut, A., Bader, G. D., Moore, L., Adams, S. L., Millar, A., Taylor, P., Bennett, K., Boutilier, K., Yang, L., Wolting, C., Donaldson, I., Schan-

dorff, S., Shewnarane, J., Vo, M., Taggart, J., Goudreault, M., Muskat, B., Alfarano, C., Dewar, D., Lin, Z., Michalickova, K., Willems, A. R., Sassi, H., Nielsen, P. A., Rasmussen, K. J., Andersen, J. R., Johansen, L. E., Hansen, L. H., Jespersen, H., Podtelejnikov, A., Nielsen, E., Crawford, J., Poulsen, V., Sorensen, B. D., Matthiesen, J., Hendrickson, R. C., Gleeson, F., Pawson, T., Moran, M. F., Durocher, D., Mann, M., Hogue, C. W., Figeys, D., Tyers, M. (2002). Systematic identification of protein complexes in *Saccharomyces cerevisiae* by mass spectrometry. *Nature*, *415*(6868), 180–183.

101. Fields, S., Song, O. (1989). A novel genetic system to detect protein-protein interactions. *Nature*, *340*(6230), 245–246.

102. Cho, S., Park, S. G., Lee do, H., Park, B. C. (2004). Protein-protein interaction networks: From interactions to networks. *J. Biochem. Mol. Biol.*, *37*(1), 45–52.

103. Yang, M., Wu, Z., Fields, S. (1995). Protein-peptide interactions analyzed with yeast two-hybrid system. *Nucleic Acids Res.*, *23*, 1152–1156.

104. Ito, T., Chiba, T., Ozawa, R., Yoshida, M., Hattori, M., Sakaki, Y. (2001). A comprehensive two-hybrid analysis to explore the yeast protein interactome. *Proc. Natl. Acad. Sci. USA*, *98*(8), 4569–4574.

105. Giot, L., Bader, J. S., Brouwer, C., Chaudhuri, A., Kuang, B., Li, Y., Hao, Y. L., Ooi, C. E., Godwin, B., Vitols, E., Vijayadamodar, G., Pochart, P., Machineni, H., Welsh, M., Kong, Y., Zerhusen, B., Malcolm, R., Varrone, Z., Collis, A., Minto, M., Burgess, S., McDaniel, L., Stimpson, E., Spriggs, F., Williams, J., Neurath, K., Ioime, N., Agee, M., Voss, E., Furtak, K., Renzulli, R., Aanensen, N., Carrolla, S., Bickelhaupt, E., Lazovatsky, Y., DaSilva, A., Zhong, J., Stanyon, C. A., Finley, Jr., R. L., White, K. P., Braverman, M., Jarvie, T., Gold, S., Leach, M., Knight, J., Shimkets, R. A., McKenna, M. P., Chant, J., Rothberg, J. M. (2003). A protein interaction map of *Drosophila melanogaster*. *Science*, *302*(5651), 1727–1736.

106. Li, S., Armstrong, C. M., Bertin, N., Ge, H., Milstein, S., Boxem, M., Vidalain, P. O., Han, J. D., Chesneau, A., Hao, T., Goldberg, D. S., Li, N., Martinez, M., Rual, J. F., Lamesch, P., Xu, L., Tewari, M., Wong, S. L., Zhang, L. V., Berriz, G. F., Jacotot, L., Vaglio, P., Reboul, J., Hirozane-Kishikawa, T., Li, Q., Gabel, H. W., Elewa, A., Baumgartner, B., Rose, D. J., Yu, H., Bosak, S., Sequerra, R., Fraser, A., Mango, S. E., Saxton, W. M., Strome, S., Van Den Heuvel, S., Piano, F., Vandenhaute, J., Sardet, C., Gerstein, M., Doucette-Stamm, L., Gunsalus, K. C., Harper, J. W., Cusick, M. E., Roth, F. P., Hill, D. E., Vidal, M. (2004). A map of the interactome network of the metazoan *C. elegans*. *Science*, *303*(5657), 540–543.

107. Mrowka, R., Patzak, A., Herzel, H. (2001). Is there a bias in proteome research? *Genome Res.*, *11*(12), 1971–1973.

108. Jansen, R., Greenbaum, D., Gerstein, M. (2002). Relating whole-genome expression data with protein-protein interactions. *Genome Res.*, *12*(1), 37–46.

109. Lopez, M. F., Pluskal, M. G. (2003). Protein micro- and macroarrays: Digitizing the proteome. *J. Chromatogr.*, *787*, 19–27.

110. Haab, B. B. (2001). Advances in protein microarray technology for protein expression and interaction profiling. *Curr. Opin. Drug Discov. Devel.*, *4*(1), 116–123.

111. Borrebaeck, C. A., Ekstrom, S., Hager, A. C., Nilsson, J., Laurell, T., Marko-Varga, G. (2001). Protein chips based on recombinant antibody fragments: A highly

sensitive approach as detected by mass spectrometry. *Biotechniques*, *30*(5), 1126–1130, 1132.
112. Li, M. (2000). Applications of display technology in protein analysis. *Nat. Biotechnol.*, *18*, 1251–1256.
113. Jenkins, R. E., Pennington, S. R. (2001). Arrays for protein expression profiling: Towards a viable alternative to two-dimensional gel electrophoresis? *Proteomics*, *1*, 13–29.
114. Newman, J. R., Keating, A. E. (2003). Comprehensive identification of human bZIP interactions with coiled-coil arrays. *Science*, *300*(5628), 2097–2101.
115. Zhu, H., Bilgin, M., Bangham, R., Hall, D., Casamayor, A., Bertone, P., Lan, N., Jansen, R., Bidlingmaier, S., Houfek, T., Mitchell, T., Miller, P., Dean, R. A., Gerstein, M., Snyder, M. (2001). Global analysis of protein activities using proteome chips. *Science*, *293*(5537), 2101–2105.
116. Glassbrook, N., Beecher, C., Ryals, J. (2000). Metabolic profiling on the right path. *Nat. Biotechnol.*, *18*(11), 1142–1143.
117. Hall, R., Beale, M., Fiehn, O., Hardy, N., Sumner, L., Bino, R. (2002). Plant metabolomics: The missing link in functional genomics strategies. *Plant Cell*, *14*(7), 1437–1440.
118. Tolstikov, V. V., Fiehn, O. (2002). Analysis of highly polar compounds of plant origin: Combination of hydrophilic interaction chromatography and electrospray ion trap mass spectrometry. *Anal. Biochem.*, *301*(2), 298–307.
119. Stitt, M., Fernie, A. R. (2003). From measurements of metabolites to metabolomics: An "on the fly" perspective illustrated by recent studies of carbon-nitrogen interactions. *Curr. Opin. Biotechnol.*, *14*(2), 136–144.
120. Nicholson, J. K., Lindon, J. C., Holmes, E. (1999). "Metabonomics": Understanding the metabolic responses of living systems to pathophysiological stimuli via multivariate statistical analysis of biological NMR spectroscopic data. *Xenobiotica*, *29*(11), 1181–1189.
121. Lindon, J. C., Nicholson, J. K., Holmes, E., Everette, J. R. (2000). Metabonomics: Metabolic processing studied by NMR spectroscopy of biofluids. *Concepts Magn. Reson.*, *12*, 289–320.
122. ter Kuile, B. H., Westerhoff, H. V. (2001). Transcriptome meets metabolome: Hierarchical and metabolic regulation of the glycolytic pathway. *FEBS Lett.*, *500*(3), 169–171.
123. Goodacre, R., Vaidyanathan, S., Dunn, W. B., Harrigan, G. G., Kell, D. B. (2004). Metabolomics by numbers: Acquiring and understanding global metabolite data. *Trends Biotechnol.*, *22*(5), 245–252.
124. Weckwerth, W., Fiehn, O. (2002). Can we discover novel pathways using metabolomic analysis? *Curr. Opin. Biotechnol.*, *13*(2), 156–160.
125. Barabasi, A-L., Oltvai, Z. N. (2004). Network biology: Understanding the cell's functional organization. *Nature Rev. Genet.*, *5*, 101–113.
126. Gifford, D. K. (2001). Blazing pathways through genetic mountains. *Science*, *293*, 2049–2051.
127. Eisen, M. B., Spellman, P. T., Brown, P. O., Botstein, D. (1998). Cluster analysis and display of genome-wide expression patterns. *Proc. Natl. Acad. Sci. USA*, *95*, 14863–14868.

128. Tavazoie, S., Hughes, J. D., Campbell, M. J., Cho, R. J., Church, G. M. (1999). Systematic determination of genetic network architecture. *Nat. Genet.*, *22*, 281–286.
129. Tamayo, P., Slonim, D., Mesirov, J., Zhu, Q., Kitareewan, S., Dmitrovsky, E., Lander, E. S., Golub, T. R. (1999). Interpreting patterns of gene expression with self-organizing maps: Methods and application to hematopoietic differentiation. *Proc. Natl. Acad. Sci. USA*, *96*, 2907–2912.
130. Huang, E., Cheng, S. H., Dressman, H., Pittman, J., Tsou, M. H., Horng, C. F., Bild, A., Iversen, E. S., Liao, M., West, M., Nevins, J. R., Huan, A. T. (2003). Gene expression predictors of breast cancer outcomes. *Lancet*, *361*, 1590–1596.
131. Kohonen, T. (1997). *Self Organizing Maps*. Springer, Berlin.
132. Quackenbush, J. (2001). Computational analysis of microarray data. *Nat. Rev. Genet.*, *2*, 418–427.
133. Jeong, H., Tombor, B., Albert, R., Oltvai, Z. N., Barabasi, A. L. (2000). The large-scale organization of metabolic networks. *Nature*, *407*(6804), 651–654.
134. Jenssen, T. K., Laegreid, A., Komorowski, J., Hovog, E. (2001). A literature network of human genes for high-throughput analysis of gene expression. *Nat. Genet.*, *28*, 21–28.
135. Chaussabel, D., Sher, A. (2002). Mining microarray expression data by literature profiling. *Genome Biol.*, *3*, 0055.1–0055.16.
136. Raychaudhuri, S., Chang, J. T., Imam, F., Altman, R. B. (2003). The computational analysis of scientific literature to define and recognize gene expression clusters. *Nucleic Acids Res.*, *31*, 4553–4560.
137. Yandell, M. D., Majoros, W. H. (2002). Genomics and natural language processing. *Nat. Rev. Genet.*, *3*, 601–611.
138. Blaschke, K., Valencia, A. (2001). Can bibliographical pointers for known biological data be found automatically? Protein interactions as a case study. *Compar. Funct. Genomics*, *2*, 196–206.
139. Daraselia, N., Yuryev, A., Egorov, S., Novihkova, S., Nikitin, A., Mazo, I. (2003). Extracting human protein interactions from Medline using a full-sentence parser. *Bioinformatics*, *19*.
140. Elbeck, K., Brass, A., Paton, N., Hodgman, C. (1999). INTERACT: An object-oriented protein-protein interactions database. *Proc. Int. Conf. Intel. Syst. Mol. Biol.*, 87–94.
141. Herrgard, M. J., Covert, M. W., Palsson, B. O. (2003). Reconciling gene expression data with known genome-scale regulatory network structures. *Genome Res.*, *13*, 2423–2434.
142. Stuart, J. M., Segal, E., Koller, D., Kim, S. K. (2003). A gene-coexpression network for global discovery of conserved genetic modules. *Science*, *302*, 249–255.
143. Grigoryev, D. N., Ma, S-F., Irizarry, R. A., Ye, S. Q., Quakenbush, J., Garcia, J. G. N. (2004). Orthologous gene-expression profiling in multi-species models: Search for candidate genes. *Genome Biol.*, *5*, R34.1–R34.13.
144. Hartwell, L. H., Hopfield, J. J., Leibler, S., Murray, A. W. (1999). From molecular to modular cell biology. *Nature*, *402*, C47–C52.
145. Wyrick, J. J., Aparico, J. G., Chen, T., Barnett, J. D., Jennings, E. G., Young, R. A., Bell, S. P., Aparicio, O. M. (2001). Genome-wide distribution of ORC and MCM

proteins in *S. cerevisiae*: High-resolution mapping of replication origins. *Science*, *294*, 2357–2360.
146. Wu, L. F., Hughes, T. R., Davierwala, A. P., Robinson, M. D., Stoughton, R., Altschuler, S. J. (2002). Large-scale prediction of *Saccharomyces cerevisiae* gene function using overlapping transcriptional clusters. *Nat. Genet.*, *31*, 255–265.
147. Bar-Joseph, Z., Gerber, G., Simon, I., Gifford, D. K., Jaakola, T. S. (2003). Comparing the continuous representation of time-series expression profiles to identify differentially expressed genes. *Proc. Natl. Acad. Sci. USA*, *100*, 10146–10151.
148. Segal, E., Shapira, M., Regev, A., Pe'er, D., Botstein, D., Koller, D., Freidman, N. (2003). Module networks: Identifying regulatory modules and their condition-specific regulators from gene expression data. *Nat. Genet.*, *34*, 166–176.
149. Matys, V., Fricke, E., Geffers, R., Gossling, E., Haubrock, M., Hehl, R., et al. (2003). TRANSFAC®: Transcriptional regulation, from patterns to profiles. *NAR*, *1*, 374–378.
150. Ihmels, J., Friedlander, G., Bergmann, S., Sarig, O., Ziv, Y., Barkai, N. (2002). Revealing modular organization in the yeast transcriptional network. *Nat. Genet.*, *31*, 370–377.
151. van Noort, V., Snel, B., Huynen, M. A. (2003). Predicting gene function by conserved co-expression. *Trends Genet.*, *19*, 238–242.
152. Michnik, S. W. (2004). Proteomics in living cells. *Drug Discov. Today*, *9*, 262–267.
153. Wagner, A., Fell, D. A. (2001). The small world inside large metabolic networks. *Proc. R. Soc. Lond. B Biol. Sci.*, *268*(1478), 1803–1810.
154. Albert, R., Jeong, H., Barabasi, A. L. (2000). Error and attack tolerance of complex networks. *Nature*, *406*(6794), 378–382.
155. Spirin, V., Mirny, L. A. (2003). Protein complexes and functional modules in molecular networks. *Proc. Natl. Acad. Sci. USA*, *100*, 12123–12128.
156. Rives, A. W., Galitski, T. (2003). Modular organization of cellular networks. *Proc. Natl. Acad. Sci. USA*, *100*, 1128–1133.
157. Pereira-Leal, J. B., Enright, A. J., Ouzounis, C. A. (2004). Detection of functional modules from protein interaction networks. *Proteins*, *54*(1), 49–57.
158. Milo, R., Shen-Orr, S., Itzkovitz, S., Kashtan, N., Chklovskii, D., Alon, U. (2002). Network motifs: Simple building blocks of complex networks. *Science*, *298*, 824–827.
159. Shen-Orr, S. S., Milo, R., Mangan, S., Alon, U. (2002). Network motifs in the transcriptional regulation network of *Escherichia coli*. *Nat. Genet.*, *31*(1), 64–68.
160. Dobrin, R., Beg, Q. K., Barabasi, A. L., Oltvai, Z. N. (2004). Aggregation of topological motifs in the *Escherichia coli* transcriptional regulatory network. *BMC Bioinformatics*, *5*(1), 10.
161. Wuchty, S., Oltvai, Z. N., Barabasi, A. L. (2003). Evolutionary conservation of motif constituents in the yeast protein interaction network. *Nat. Genet.*, *35*(2), 176–179.
162. Selkov, E., Maltsev, N., Olsen, G. J., Overbeek, R., Whitman, W. B. (1997). A reconstruction of the metabolism of *Methanococcus jannaschii* from sequence data. *Gene*, *197*, GC11–26.

163. Xia, X-Q., Wise, M. J. (2003). DiMSim: A discrete-event simulator of metabolic networks. *J. Chem. Inf. Comput. Sci.*, *43*, 1011–1019.
164. Nikitin, A., Egorov, S., Daraselia, N., Mazo, I. (2003). Pathway studio—the analysis and navigation of molecular networks. *Bioinformatics*, *19*, 2155–2157.
165. Vickers, J. C. (1997). The cellular mechanism underlying neuronal degeneration in glaucoma: Parallels with Alzheimer's disease. *Aust. N. Z. J. Opthalmol.*, *25*, 105–109.
166. Janciauskiene, S., Krakau, T. (2001). Alzheimer's peptide: A possible link between glaucoma, exfoliation syndrome and Alzheimer's disease. *Acta Ophthalmol. Scand.*, *79*, 328–329.
167. Bayer, A. U., Ferrari, F. (2002). Severe progression of glaucomatous optic neuropathy in patients with Alzheimer's disease. *Eye*, *16*, 209–212.
168. Parisi, V. (2003). Correlation between morphological and functional retinal impairment in patients affected by ocular hypertension, glaucoma, demyelinating optic neuritis and Alzheimer's disease. *Semin. Opthalmol.*, *18*, 50–57.
169. Quigley, H. A. (1998). Selectivity in glaucoma injury. *Arch. Opthalmol.*, *116*, 396–398.
170. Wax, M. B., Tezel, G. (2002). Neurobiology of glaucomatous optic neuropathy: Diverse cellular events in neurodegeneration and neuroprotection. *Mol. Neurobiol.*, *26*, 45–55.
171. Osborne, N. N., Chidlow, G., Wood, J., Casson, R. (2003). Some current ideas on the pathogenesis and the role of neuroprotection in glaucomatous optic neuropathy. *Eur. J. Opthalmol.*, *13*, S19–S26.
172. Zhao, D. Y., Cioffi, G. A. (2000). Anterior optic nerve microvascular changes in human glaucomatous optic neuropathy. *Eye*, *14*, 445–449.
173. Anderson, D. R. (1999). Introductory comments on blood flow autoregulation in the optic nerve head and vascular risk factors in glaucoma. *Surv. Opthalmol.*, *43* (Suppl. 1), S5–S9.
174. Drance, S., Anderson, D. R., Schulzer, M. (2001). Collaborative Normal-Tension Glaucoma Study Group. Risk factors for progression of visual field abnormalities in normal-tension glaucoma. *Am. J. Ophthalmol.*, *131*, 699–708.
175. Gasser, P., Flammer, J. (1991). Blood-cell velocity in the nailfold capillaries of patients with normal-tension and high-tension glaucoma. *Am. J. Ophthalmol.*, *15*, 585–588.
176. Dreyer, E. B., Zurakowski, D., Schumer, R. A., Podos, S. M., Lipton, S. A. (1996). Elevated glutamate levels in the vitreous body of humans and monkeys with glaucoma. *Arch. Opthalmol.*, *114*, 299–305.
177. Yuan, L., Neufeld, A. H. (2000). Tumor necrosis factor-alpha: A potentially neurodestructive cytokine produced by glia in the human glaucomatous optic nerve head. *Glia*, *32*, 42–50.
178. Morgan, J., Caprioli, J., Koseki, Y. (1999). Nitric oxide mediates excitotoxic and anoxic damage in rat retinal ganglion cells cocultured with astroglia. *Arch. Opthalmol.*, *117*, 1524–1529.
179. Hernandez, M. R., Agapova, O. A., Yang, P., Salvador-Silva, M., Rocard, C. S., Aoi, S. (2002). Differential gene expression in astrocytes from human normal

and glaucomatous optic nerve head analyzed by cDNA microarray. *Glia, 38*, 45–64.

180. Miyahara, T., Kikuchi, T., Akimoto, M., Kurokawa, T., Shibuki, H., Yoshimura, N. (2003). Gene microarray analysis of experimental glaucomatous retina from cynomologous monkey. *Invest. Ophthalmol. Vis. Sci., 44*, 4347–4356.

181. Ahmed, F., Brown, K. M., Stephan, D. A., Morrison, J. C., Johnson, E. C., Tomarev, S. I. (2004). Microarray analysis of changes in mRNA levels in the rat retina after experimental elevation of intraocular pressure. *Invest. Ophthalmol. Vis. Sci., 45*, 1247–1258.

182. Lo, W. R., Rowlette, L. L., Caballero, M., Yang, P., Hernandez, M. R., Borras, T. (2003). Tissue differential microarray analysis of dexamethasone induction reveals potential mechanisms of steroid glaucoma. *Invest. Ophthalmol. Vis. Sci., 44*, 473–485.

183. Ishibashi, T., Nakazawa, M., Ono, H., Satoh, N., Gojobori, T., Fujiwara, S. (2003). Microarray analysis of embryonic retinoic acid target genes in the ascidian *Ciona intestinalis*. *Dev. Growth Differ., 45*, 249–259.

184. Yang, P., Agapova, O., Parker, A., Shannon, W., Pecen, P., Duncan, J., Salvador-Silva, M., Hernandez, M. R. (2004). DNA microarray analysis of gene expression in human optic nerve head astrocytes in response to hydrostatic pressure. *Physiol. Genomics, 27*, 157–169.

185. Hernandez, M. R. (2000). The optic nerve head in glaucoma: Role of astrocytes in tissue remodeling. *Prog. Retin. Eye Res., 19*, 297–321.

186. Swanson, R. A., Ying, W., Kauppinen, T. M. (2004). Astrocyte influences on ischemic neuronal death. *Curr. Mol. Med., 4*, 193–205.

187. Shaulian, E., Karin, M. (2002). AP-1 as a regulator of cell life and death. *Cell Biol., 4*, E131–E136.

188. Neufeld, A. H., Liu, B. (2003). Glaucomatous optic neuropathy: When glia misbehave. *Neuroscientist, 9*, 485–495.

189. Abraham, C. R. (2001). Reactive astrocytes and alpha1-antichymotrypsin in Alzheimer's disease. *Neurobiol. Aging, 22*, 931–936.

190. Abramov, A. Y., Canevari, L., Duchen, M. R. (2003). Changes in intracellular calcium and glutathione in astrocytes as the primary mechanism of amyloid neurotoxicity. *J. Neurosci., 23*, 5088–5095.

191. Liu, B., Neufeld, A. H. (2003). Activation of epidermal growth factor receptor signals induction of nitric oxide synthase-2 in human optic nerve head astrocytes in glaucomatous optic neuropathy. *Neurobiol. Dis., 13*, 109–123.

192. Yan, Z., Chen, Z., Chen, Z. (2000). Modulation of nitric oxide synthase isoenzymes in reperfused skeletal muscle. *Chin. J. Traumatol., 3*, 76–80.

193. Han, B. H., DeMattos, R. B., Dugan, L. L., Kim-Han, J. S., Brendza, R. P., Fryer, J. D., Kierson, M., Cirrito, J., Quick, K., Harmony, J. A., Aronow, B. J., Holtzman, D. M. (2001). Clusterin contributes to caspase-3-independent brain injury following neonatal hypoxia-ischemia. *Nat. Med., 7*, 338–343.

194. McKinnon, S. J. (2003). Glaucoma: Ocular Alzheimer's disease? *Front Biosci., 8*, S1140–S1156.

195. McKinnon, S. J., Lehman, D. M., Tahzib, N. G., Ransom, N. L., Reitsamer, H. A., Liston, P., LaCasse, E., Li, Q., Korneluk, R. G., Hauswirth, W. W. (2002). Bac-

uloviral IAP repeat-containing-4 protects optic nerve axons in a rat glaucoma model. *Mol. Ther.*, *5*, 780–787.

196. Singhrao, S. K., Neal, J. W., Rushmere, N. K., Morgan, B. P., Gasque, P. (2000). Spontaneous classical pathway activation and deficiency of membrane regulators render human neurons susceptible to complement lysis. *Am. J. Pathol.*, *157*, 905–918.

197. Jeffrey, S. S., Fero, M. J., Borresen-Dale, A. L., Botstein, D. (2002). Expression array technology in the diagnosis and treatment of breast cancer. *Mol. Interv.*, *2*, 101–109.

198. Allred, D. C., Mohsin, S. K., Fuqua, S. W. A. (2001). Histological and biological evolution of human premalignant breast disease. *Endocrinol. Relat. Cancer*, *8*, 47–61.

199. Dalton, L. W., Pinder, S. E., Elston, C. E., Ellis, I. O., Page, D. L., Dupont, W. D., Blamey, R. W. (2000). Histologic grading of breast cancer: Linkage of patient outcome with level of pathologist agreement. *Mod. Pathol.*, *13*, 730–735.

200. Eifel, P., Axelson, J. A., Costa, J., Crowley, J., et al. (2001). NIH Consensus Development Conference Statement: Adjuvant therapy for breast cancer. *J. Natl. Cancer Inst.*, *93*, 979–989.

201. Selaru, F. M., Yin, J., Olaru, A., Mori, Y., Xu, Y., Epstein, S. H., Sato, F., Deacu, E., Wang, S., Sterian, A., Fulton, A., Abraham, J. M., Shibata, D., Baquet, C., Stass, S. A., Meltzer, S. J. (2004). An unsupervised approach to identify molecular phenotypic components influencing breast cancer features. *Cancer Res.*, *64*, 1584–1588.

202. Gruvberger, S., Ringner, M., Chen, Y., Panavally, H., Saal, L. H., Borg, A., Ferno, M., Peterson, C., Meltzer, P. S. (2001). Estrogen receptor status in breast cancer is associated with remarkably distinct gene expression patterns. *Cancer Res.*, *61*, 5979–5984.

203. West, M., Blanchette, C., Dressman, H., Huang, E., Ishida, S., Spang, R., Zuzan, H., Olson, J. A., Jr., Marks, J. R., Nevins, J. R. (2001). Predicting the clinical status of human breast cancer by using gene expression profiles. *Proc. Natl. Acad. Sci. USA*, *98*, 11462–11467.

204. Nagahata, T., Onda, M., Emi, M., Nagai, H., Tsumagari, K., Fujimoto, T., Hirano, A., Sato, T., Nishikawa, K., Akiyama, F., Sakamoto, G., Kasumi, F., Miki, Y., Tanaka, T., Tsunoda, T. (2004). Expression profiling to predict postoperative prognosis for estrogen receptor-negative breast cancers by analysis of 25,344 genes on a cDNA microarray. *Cancer Sci.*, *95*, 218–225.

205. Gruvberger-Saal, S. K., Eden, P., Ringner, M., Badletorp, B., et al. (2004). Predicting continuous values of prognostic markers in breast cancer from microarray gene expression profiles. *Mol. Cancer Ther.*, *3*, 161–168.

206. Ma, X-J., Salunga, R., Tuggle, J. T., Gaudet, J., et al. (2003). Gene expression profiles of human breast cancer progression. *Proc. Natl. Acad. Sci. USA*, *100*, 5974–5979.

207. Sørlie, T., Perou, C. M., Tibshirani, R., Aas, T. (2001). Gene expression patterns of breast carcinomas distinguish tumor subclasses with clinical implications. *Proc. Natl. Acad. Sci. USA*, *98*, 10869–10874.

208. Perou, C. M., Sørlie, T., Eisen, M. B., et al. (2000). Molecular portraits of human breast cancers. *Nature, 406*, 747–752.
209. van de Vijer, M. J., He, Y. D., van 'T Veer, L. J., Dai, H., et al. (2002). Gene-expression signature as a predictor of survival in breast cancer. *N. Engl. J. Med., 347*, 1999–2009.
210. Schultze, A., Downwards, J. (2001). Navigating gene expression using microarrays—a technology review. *Nat. Cell Biol., 3*, E190–E195.
211. Chang, J. C., Wooten, E. C., Tsimelzon, A., Hilsenbeck, S. G., et al. (2003). Gene expression profiling for the prediction of therapeutic response to docetaxel in patients with breast cancer. *Lancet, 362*, 362–369.
212. Sotiriou, C., Powles, T. J., Dowsett, M., Jazaeri, A. A., et al. (2002). Gene expression profiles derived from fine needle aspiration correlate with response to systemic chemotherapy in breast cancer. *Breast Cancer Res., 4*, R5.
213. Hodges, L. C., Cook, J. D., Lobenhofer, E. K., Li, L., Bennett, L., Bushel, P. R., Aldaz, C. M., Afshari, C. A., Walker, C. L. (2003). Tamoxifen functions as a molecular agonist inducing cell cycle-associated genes in breast cancer cells. *Mol. Cancer Res., 1*, 300–311.
214. Nikolsky, Y., Ekins, S., Nikolskaya, T., Bugrim, A. (2005). A novel method for generation of signature networks as biomarkers from complex high throughput data. *Tox. Lett.* (in press).
215. Prall, O. W., Rogan, E. M., Musgrove, E. A., Watts, C. K., Sutherland, R. L. (1998). c-Myc or cyclin D1 mimics estrogen effects on cyclin E-Cdk2 activation and cell cycle reentry. *Mol. Cell Biol., 18*, 4499–4508.
216. Musgrove, E. A., Sarcevic, B., Sutherland, R. L. (1996). Inducible expression of cyclin D1 in T-47D human breast cancer cell is sufficient for Cdk2 activation and pRB hyperphosphorylation. *J. Cell Biochem., 60*, 363–378.
217. Ekins, S., Boulanger, B., Swaan, P. W., Hupcey, M., A. Z. (2002). Towards a new age of virtual ADME/TOX and multidimensional drug discovery. *J. Comput. Aided Mol. Des., 16*, 381–401.
218. Ekins, S. (2004). Predicting undesirable drug interactions with promiscuous proteins in silico. *Drug Discov. Today, 9*, 276–285.
219. Greene, N. (2002). Computer systems for the prediction of toxicity: An update. *Adv. Drug Del. Rev., 54*, 417–431.
220. Ekins, S., de Groot, M., Jones, J. P. (2001). Pharmacophore and three dimensional quantitative structure activity relationship methods for modeling cytochrome P450 active sites. *Drug Metab. Dispos., 29*, 936–944.
221. Jones, J. P., Mysinger, M., Korzekwa, K. R. (2002). Computational models for cytochrome P450: A predictive electronic model for aromatic oxidation and hydrogen abstraction. *Drug Metab. Dispos., 30*, 7–12.
222. Jones, J. P., Korzekwa, K. R. (1996). *Predicting the Rates and Regioselectivity of Reactions Mediated by the P450 Superfamily*, Vol. 272. Academic New York.
223. Bugrim, A., Nikolskaya, T., Nikolsky, Y. (2004). Early prediction of drug metabolism and toxicity: Systems biology approach and modeling. *Drug Discov. Today, 9*, 127–135.

224. Oliver, D. E., Rubin, D. L., Stuart, J. M., Hewett, M., Klein, T. E., Altman, R. B. (2002). Ontology development for a pharmacogenetics knowledge base. In *Pac Symp. Biocomp.*, pp. 88–99.
225. Yan, Q., Sadee, W. (2000). Human membrane transporter database: A web-accessible relational database for drug transport studies and pharmacogenomics. *AAPS Pharmsci.*, *2*, E20.
226. Sun, L. Z., Ji, Z. L., Chen, X., Wang, J. F., Chen, Y. Z. (2002). ADME-AP: A datbase of ADME associated proteins. *Bioinformatics*, *18*, 1699–1700.
227. Nakata, K., Yukawa, M., Komiyama, N., Nakano, T., Kaminuma, T. (2002). A nuclear receptor database that maps pathways to diseases. *Genome Infomatics*, *13*, 515–516.
228. Cook, D. L., Atkins, W. M. (1997). Enhanced detoxication due to distributive catalysis and toxic thresholds: A kinetic analysis. *Biochemistry*, *36*, 10802–10806.
229. Ekins, S., Stresser, D. M., Williams, J. A. (2003). In vitro and pharmacophore insights into CYP3A enzymes. *Trends Pharmacol. Sci.*, *24*, 191–196.
230. Ekins, S., Ring, B. J., Binkley, S. N., Hall, S. D., Wrighton, S. A. (1998). Autoactivation and activation of cytochrome P450s. *Int. J. Clin. Pharmacol. Therapeut.*, *36*, 642–651.
231. Atkins, W. M. (2004). Implications of the allosteric kinetics of cytochrome P450s. *Drug Discov. Today*, *9*, 478–484.
232. Ekins, S., Mirny, L., Schuetz, E. G. (2002). A ligand-based approach to understanding selectivity of nuclear hormone receptors PXR, CAR, FXR, LXRa and LXRb. *Pharm. Res.*, *19*, 1788–1800.
233. Sueyoshi, T., Kawamato, T., Zelko, I., Honkakoski, P., Negishi, M. (1999). The repressed nuclear receptor CAR responds to phenobarbital in activating the human CYP2B6 gene. *J. Biol. Chem.*, *274*, 6043–6046.
234. Mankowski, D. C., Ekins, S. (2003). Prediction of human drug metabolizing enzyme induction. *Curr. Drug Metab.*, *4*, 381–391.
235. Marton, M. J., DeRisi, J. L., Bennet, H. A., Iyer, V. R., Meyer, M. R., Roberts, C. J., Stoughton, R., Burchard, J., Slade, D., Dai, H., Bassett, D. E. J., Hartwell, L. H., Brown, P. O., Friend, S. H. (1998). Drug target validation and identification of secondary drug target effects using DNA microarrays. *Nat. Med.*, *4*, 1293–1301.
236. Gerhold, D., Jensen, R. V., Gullans, S. R. (2002). Better therapeutics through microarrays. *Nat. Genet.*, *32*, 547–552.
237. Waring, J. F., Cavet, G., Jolly, R. A., McDowell, J., Dai, H., Ciurlionis, R., Zhang, C., Stoughton, R., Lum, P. Y., Ferguson, A., Roberts, C. J. (2003). Development of a DNA Microarray for toxicology based on hepatoxin-regulated sequences. *Env. Health Perspect.*, *111*, 863–870.
238. Ueda, A., Hamadeh, H. K., Webb, H. K., Yamamoto, Y., Sueyoshi, T., Afshari, C. A., Lehmann, J. M., Negishi, M. (2002). Diverse roles of the nuclear orphan receptor CAR in regulating hepatic genes in response to phenobarbital. *Mol. Pharmacol.*, *61*, 1–6.
239. Korolev, D., Balakin, K. V., Nikolsky, Y., Kirillov, E., Ivanenkov, Y. A., Savchuk, N. P., Ivashchenko, A. A., Nikolskaya, T. (2003). Modeling of human cytochrome p450-mediated drug metabolism using unsupervised machine learning approach. *J. Med. Chem.*, *46*(17), 3631–3643.

240. Balakin, K. V., Ekins, S., Bugrim, A., Ivanenkov, Y. A., Korolev, D., Nikolsky, Y., Skorenko, S. A., Ivashchenko, A. A., Savchuk, N. P., Nikolskaya, T. (2004). Computational Kohonen self organizing maps for prediction of binding to human cytochrome P450 3A4. *Drug. Metab. Dispos*, *32*, 1183–1189.

241. Balakin, K. V., Ekins, S., Bugrim, A., Ivanenkov, Y. A., Korolev, D., Nikolsky, Y., Ivashchenko, A. A., Savchuk, N. P., Nikolskaya, T. (2004). Quantitative structure-metabolism relationship of the metabolic N-dealkylation rates. *Drug. Metab. Dispos*, *32*, 1111–1120.

242. Uetrecht, J. (2003). Screening for the potential of a drug candidate to cause idiosyncratic drug reactions. *Drug Discov. Today*, *8*, 832–837.

243. Williams, D. P., Park, B. K. (2003). Idiosyncratic toxicity: The role of toxicophores and bioactivation. *Drug Discov. Today*, *8*, 1044–1050.

244a. Li, A. P. (2002). A review of the common properties of drugs with idiosyncratic hepatotoxicity and the "multiple determinant hypothesis" for the manifestation of idiosyncratic drug toxicity. *Chem. Biol. Interact.*, *142*, 7–23.

244b. Ha, H. R., Chen, J., Freiburghaus, A. U., Follath, F. (1995). Metabolism of theophylline by cDNA-expressed human cytochromes P-450. *Br. J. Clin. Pharmacol.*, *39*, 321–326.

244c. Tjia, J. F., Colbert, J., Back, D. J. (1996). Theophylline metabolism in human liver microsomes: inhibition studies. *J. Pharmacol. Exp. Ther.*, *276*, 912–917.

245. Furuta, S., Kamada, E., Suzuki, T., Sugimoto, T., Kawabata, Y., Shinozaki, Y., Sano, H. (2001). Inhibition of drug metabolism in human liver microsomes by nizatidine, cimetidine and omeprazole. *Xenobiotica*, *31*(1), 1–10.

246. Sutton, D., Butler, A. M., Nadin, L., Murray, M. (1997). Role of CYP3A4 in human hepatic diltiazem N-demethylation: Inhibition of CYP3A4 activity by oxidized diltiazem metabolites. *J. Pharmacol. Exp. Ther.*, *282*(1), 294–300.

247. Loi, C. M., Parker, B. M., Cusack, B. J., Vestal, R. E. (1997). Aging and drug interactions. III. Individual and combined effects of cimetidine and cimetidine and ciprofloxacin on theophylline metabolism in healthy male and female nonsmokers. *J. Pharmacol. Exp. Ther.*, *280*, 627–637.

248. Sorgel. F., Kinzig, M. (1993). Pharmacokinetics of gyrase inhibitors, part 2: Renal and hepatic elimination pathways and drug interactions. *Am. J. Med.* *94*(3A), 56S–69S.

249. Luo, G., Cunningham, M., Kim, S., Burn, T., Lin, J., Sinz, M., Hamilton, G., Rizzo, C., Jolley, S., Gilbert, D., Downey, A., Mudra, D., Graham, R., Carroll, K., Xie, J., Madan, A., Parkinson, A., Christ, D., Selling, B., LeCluyse, E., Gan, L. S. (2002). CYP3A4 induction by drugs: Correlation between a pregnane X receptor reporter gene assay and CYP3A4 expression in human hepatocytes. *Drug Metab. Dispos.*, *30*(7), 795–804.

250. Ekins, S., Nikolsky, Y., Nikolskaya, T. (2005). Application of systems biology to absorption, distribution, metabolism, excretion and toxicity. *Trends Pharm. Sci.* (in press).

251. Benson, D. A., Karsch-Mizrachi, I., Lipman, D. J., Ostell, J., Rapp, B. A., Wheeler, D. L. (2002). GenBank. *Nucleic Acids Res.*, *30*, 17–20.

252. Ashburner, M., Ball, C. A., Blake, J. A., Botstein, D., Butler, H., Cherry, J. M., Davis, A. P., Dolinski, K., Dwight, S. S., Eppig, J. T., Harris, M. A., Hill, D. P., Issel-

Tarver, L., Kasarskis, A., Lewis, S., Matese, J. C., Richardson, J. E., Ringwald, M., Rubin, G. M., Sherlock, G. (2000). Gene Ontology: Tool for the unification of biology. *Nat. Genet.*, *25*, 25–29.

253. Snel, B., Bork, P., Huynen, M. A. (2002). The identification of functional modules from the genomic association of genes. *Proc. Natl. Acad. Sci. USA*, *99*, 5890–5895.

254. Wolf, D. M., Arkin, A. P. (2003). Motifs modules and games in bacteria. *Curr. Opin. Microbiol.*, *6*, 125–134.

255. Ogata, H., Fujibuchi, W., Goto, S., Kanehisa, M. A. (2000). Heuristic graph comparison algorithm and its application to detect functionally related enzyme clusters. *Nucleic Acids Res.*, *28*, 4021–4028.

256. Nakaya, A., Goto, S., Kanehisa, M. A. (2001). Extraction of correlated gene clusters by multiple graph comparison. *Genome Inform. Ser. Workshop Genome Inform.* *12*, 44–53.

257. Kelley, B. P., Sharan, R., Karp, R. M., Sittler, T., Root, D. E., Stockwell, B. R., Ideker, T. (2003). Conserved pathways within bacteria and yeast as revealed by global protein network alignment. *Proc. Natl. Acad. Sci. USA*, *100*, 11394–11399.

258. Jafari, S. S., Nielson, M., Graham, D. I., Maxwell, W. L. (1998). Axonal cytoskeletal changes after nondisruptive axonal injury. II. Intermediate sized axons. *J. Neurotrauma*, *15*, 955–966.

259. Jafari, S. S., Maxwell, W. L., Neilson, M., Graham, D. I. (1997). Axonal cytoskeletal changes after non-disruptive axonal injury. *J. Neurocytol.*, *26*, 207–221.

260. Quigley, H. A., Anderson, D. R. (1977). Distribution of axonal transport blockade by acute intraocular pressure elevation in the primate optic nerve head. *Ophthalmol. Vis. Sci.*, *16*, 640–644.

261. Salvador-Silva, M., Aoi, S., Parker, A., Yang, P., Pecen, P., Hernandez, M. R. (2004). Responses and signaling pathways in human optic nerve head astrocytes exposed to hydrostatic pressure in vitro. *Glia*, *45*, 364–377.

262. Zhao, H., Langerod, A., Ji, Y., Nowels, K. W., Nesland, J. M., Tibshirani, R., Bukholm, I. K., Karesen, R., Botstein, D., Borresen-Dale, A. L., Jeffrey, S. S. (2004). Different gene expression patterns in invasive lobular and ductal carcinomas of the breast. *Mol. Biol. Cell*, *15*, 2523–2536.

5

HIGH-THROUGHPUT FLOW CYTOMETRY

LARRY A. SKLAR, PETER C. SIMONS, ANNA WALLER,
SEAN M. BIGGS, SUSAN M. YOUNG, MARIUS OLAH,
CRISTIAN BOLOGA, TUDOR I. OPREA, ERIC R. PROSSNITZ,
AND BRUCE S. EDWARDS
University of New Mexico School of Medicine
University of New Mexico Health Sciences Center
Albuquerque, New Mexico

5.1	INTRODUCTION	186
5.2	OVERVIEW	187
	GPCR Signaling Pathways as Discovoery Targets	187
	Flow Cytometry Well Suited to Diversity Analysis	190
	Challenge of Throughput	192
	Challenge of Content	193
	Challenge of Informatics	194
	Screening	195
	Technolgy Associated with HyperCyt	196
5.3	PROTEIN EXPRESSION AND CHARACTERIZATION	196
	GPCR Constructs	196
	Soluble GPCRs Reconsitute with G Proteins	197
	FPR Assembles with Arrestin	197
	FPR Colocalization with Arrestin in Cells Parallels FPR Assemblies in Solution	198
	Arrestin Regulates Intracellular FPR Traffic	199
	FPR-$G_{i\alpha 2}$ Fusion Protein Functions Physiologically	200
	FPR Tail Assembly (Protein Domain Display)	201
	Proof of Principle for β2AR and Other GPCRs	202
5.4	SOLUBLE AND BEAD-BASED ASSEMBLIES	202
	FPR Assemblies Used to Characterize and Validate the Displays	204

Drug Discovery Handbook, by Shayne Cox Gad
Copyright © 2005 by John Wiley & Sons, Inc.

	β2AR Extends the Soluble GPCR on a Bead Concept	205
	Display of Cell Cycle Proteins Demonstrates Further Generality	206
5.5	HIGH-THROUGHPUT FLOW CYTOMETRY TECHNOLOGIES	207
	HyperCyt Exhibits Versatility and Compatibility with High Content	207
	HyperCyt Adds Microfluidic Mixing for Analysis of Soluble Compounds	207
	HyperCyt Cell Sorting Extends Microfluidic Mixing	208
	Sample Carryover in Microfluidic Mixing Systems	209
	Software Development	209
	FPR Ligand Screening Assay	210
5.6	VIRTUAL SCRENING	210
	Docking and Scoring	212
	Generation and Maintenance of Chemical Databases for VS	212
	Virtual Screening in the Context of High-Performance Computing	213
	FPR Case Study	216
5.7	SUMMARY	217
	References	219

5.1 INTRODUCTION

This chapter describes the creation of a discovery team in an academic environment. The team is specifically structured to perform four integrated activities: (1) expressing and characterizing membrane proteins of interest in appropriate quantities for analysis, (2) formatting the materials for analysis of cell physiology and molecular interactions in a novel manner, typically through solubilization of the components and display on particles, (3) performing screens of small-molecule activity as well as protein–protein interactions on a novel high-throughput flow cytometric instrument platform, and (4) integrating virtual screening and cheminformatics with activity screening to improve the overall efficiency of the discovery process. The results of the overall process are intended to provide insight into cell physiology, ligand–receptor and protein–protein interactions, the discovery of small molecules as probes for biological systems and, potentially, as leads in drug discovery. The process itself poses an interesting set of challenges and opportunities. The challenges arise in part from the need to produce material in appropriate quantity and with appropriate properties for screening. In addition, the screening needs to be performed in a mode that is targeted rather than brute force. The opportunities come not only from the discoveries that result from the research but are also derived from research into the nature of the discovery process itself and the insights that result from experiments into discovery. The purpose of this chapter is twofold: first, to provide a context for our particular approach to discovery research and, second, to provide specific examples of how the individual components of the process function in our laboratories.

5.2 OVERVIEW

Flow cytometry is proving to be a good match for G-protein-coupled receptor (GPCR) drug discovery, taking advantage of its versatility of assays, its ability to discriminate free and bound molecular complexes, and the incorporation of microfluidic principles to optimize sample delivery. A number of recent developments have suggested that flow cytometry would be a good match for chemical and biological diversity in general. With up to 30,000 human genes that encode proteins, the number of protein species may reach several hundred thousand including splice variants and posttranslational modification. In addition, there are upwards of 1 million natural polymorphisms in the human genome and other mutations that create protein variants. The analysis of protein–protein and protein–ligand interactions has broad-ranging implications for biomedical research.

The GPCRs are the most numerous molecular family in the human genome, representing 600 or more proteins that allow extracellular signals to induce cellular responses. Each GPCR could couple to as many as 1620 G-protein heterotrimers (27α, 5β, and 12γ subunits), relatively few of which are characterized. Signaling then proceeds through a host of effectors. The GPCR signals are terminated through steps involving phosphorylation by a family of kinases, interaction with arrestin family members, and processing inside the cell. While cell expression experiments (e.g., one GPCR and one G-protein heterotrimer) can typically resolve one set of interactions at a time, combinatorial approaches may resolve many interactions simultaneously. Because of the multiplicity of components, there has been little systematic cellular approach to GPCR diversity issues. Flat arrays and the flow cytometric analog, multiplex bead-based approaches, are proposed to meet this challenge [1]. Flow cytometric approaches will prove to be applicable both for understanding assemblies and mechanisms and, when combined with high-throughput (HyperCyt) and virtual screening, in small-molecule discovery. Moreover, these approaches have the potential to replace radioligands and to compete with the sensitivity and throughput of plasmon resonance. The introduction of soluble GPCR molecular assemblies in this arena has the potential of opening the door for other membrane molecules, and other signaling and response pathways. Here, we will focus on GPCR and related proteins (cell- and bead-based assemblies), a robust high-throughput (HT) flow cytometry platform (our HyperCyt), and the integration of virtual screening.

GPCR Signaling Pathways as Discovery Targets

The GPCRs represent the target of ~50 percent of the prescription drugs on the market. They are involved in virtually every physiological process in the human body, with ligands including light, odorants, amines, peptides, proteins, lipids, and nucleotides. GPCRs contain a common structural motif with seven transmembrane α helices. With the recent description of the three-dimensional crystal structure of rhodopsin in its inactive state, a greater, though still incomplete, understanding of the functions of this receptor family has been achieved [2, 3]. In addition to the activation of G proteins and separation of their α and βγ subunits (Fig. 5.1), GPCRs are regulated primarily by kinases, including second-messenger kinases and the family of G-protein-coupled receptor kinases (GRKs) [4]. Following GPCR phosphorylation by GRKs, arrestins associate with GPCRs [5]. The traditional role of arrestins has been to serve as desensitizing agents, preventing further association of GPCRs with G proteins. However, recent studies have demonstrated that arrestins can serve as adapters in the process of receptor internalization as well as scaffolds in the activation of numerous kinase pathways. Interactions between GPCRs and cellular proteins such as adaptins, rab GTPases, phosphatases, and ion channels have also been described [6]. The newly described regulators of G-protein signaling (RGS) proteins are downstream, controlling the kinetics and specificity of GPCR-mediated G-protein signals [7]. The localized tissue distributions and diverse structural elements of RGS proteins make them an intriguing

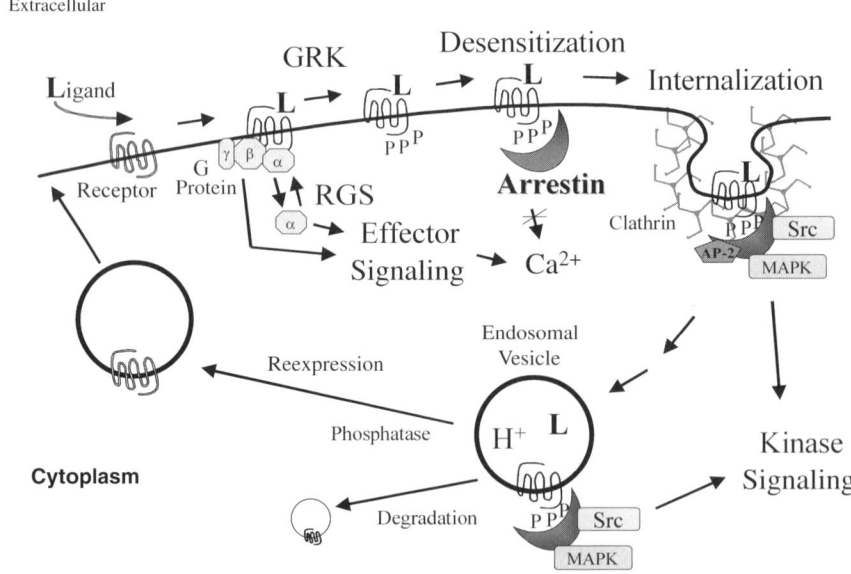

Figure 5.1 GPCR activation, termination, and processing. When a GPCR binds a ligand, the α subunit of the G-protein heterotrimer releases bound GDP and binds GTP, then separates from the βγ subunit. The GPCR gets phosphorylated by GRK, and the phosphorylated receptor binds arrestin. Both of these steps can lead to effector signaling. The GPCR is then internalized in an endosomal vesicle, from which it can be degraded or recycled back to the surface of the cell.

OVERVIEW 189

target for drug development [8]. Cytometric approaches that we employ for GPCR and related protein–protein interactions are listed in Table 5.1.

The GPCRs are divided into five major families based on sequence homologies. These include the class A (rhodopsin), class B (secretin-like), class C (metabotropic glutamate), class D (fungal pheromone), and class E [cyclic adenosine monophosphate (cAMP) receptors of dicyostelium]. The major class of physiological significance to humans is the class A group of GPCRs. In this class are the monoamine, peptide, hormone protein, olfactory, prostanoid, nucleotide-like, cannabinoid, platelet-activating factor, gonadotropin-releasing hormone, thyrotropin-releasing hormone, melatonin,

TABLE 5.1 Flow Cytometric Approaches to GPCR Physiology and Molecular Assembly

Physiology or Assembly	Approach
Expression of GPCR or signaling partner on cells [30–33, 98]	Analysis and sorting of transfected cells based on fluorescent ligand, antibody, or GFP fusion protein
GPCR physiology [30–33, 35, 36, 98, 99]	Analysis of cell response, GPCR internalization, GPCR desensitization, GPCR recycling
Solubilization of GPCR [30–33, 35–37, 98, 100]	Screen detergent for ability to produce receptor that binds to ligand beads or G-protein beads (see below)
Isolation and display of proteins [34–38]	Epitope-tagged proteins/epitope recognizing beads
Reconstitution of components for assembly analysis	Form complexes of GPCR and partners on particles
Assembly and disassembly kinetics [36, 37, 101–105]	Analyze kinetics with subsecond resolution with "rapid mix flow cytometer" (200-ms dead time)
GPCR phosphorylation status	Ability of GPCR to bind to arrestin beads
Ternary complex formation on G-protein beads [36, 37, 106]	Assembly of GPCR on G-protein beads with agonist binding (Figs. 5.8b, 5.8c, and 5.8d)
Display of GPCR ligand on ligand beads [37]	Competition of GPCR on ligand beads for agonist or antagonist binding (Fig. 5.8g)
Duplex assay of ligand beads and G-protein beads [37, 38]	Single step discrimination of agonist and antagonist
Display of protein domain on beads and multiplex [34]	Proteomic analysis of binding between domains of partners
G-protein multiplex [106]	Proteomic analysis of specificity of receptor binding to G proteins; screen for solubilization; screen for ligands; screen for phosphorylation

viral, lysosphingolipid, LTB4, and related orphan receptors. The largest subfamilies are the monoamine, peptide, and olfactory groups.

We have focused on the formyl peptide receptor (FPR), a member of the chemoattractant/chemokine subfamily of GPCRs involved in leukocyte trafficking and inflammatory responses to bacterial peptides [9–11]. FPR signals through G_i proteins and intracellular calcium. The ligands for FPR are short hydrophobic peptides, typically with an N-formyl group on the amino-terminal methionine, followed by two to five hydrophobic amino acids such as leucine, phenylalanine, norleucine, and tyrosine [12]. FPR, with a family of fluorescent ligands of varying affinity used to assess its presence and functional state [12], has been the prototype for the development of most of the assemblies and cell responses in Table 5.1. More recently we have examined the β2-adrenergic receptor (β2AR), the prototypic monoamine receptor. β2AR signals through G_s and adenylate cyclase. The contrasts between β2AR and FPR present notable opportunities. β2AR internalization depends on arrestin, whereas FPR internalization does not. It has been reported that β2AR specificity for G_s and G_i depend upon its phosphorylation status and ligand. β2AR has a constitutively active mutant that is predicted to assemble with G_s in the absence of ligand.

Currently, there are ~200 GPCRs with known ligands and ~150 orphan GPCRs [13]. There are also ~330 full-length human odorant-like GPCR genes known to date [14]. Since odorant ligands have been found for only a few of the potential odorant receptors, the odorant-like classification and terminology for a novel receptor are defined primarily by sequence homology [15]. Odorant-like receptor genes have been discovered in the olfactory epithelium, as might be expected, but they are also expressed in tissues lacking any known olfactory function, suggesting that some odorant-like GPCRs are not involved in the chemosensory process. Some GPCRs detect pheromones, and in humans, many pheromones are derived from steroid hormones. One olfactory-like GPCR has been shown to be expressed in the prostate (prostate-specific GPCR, or PSGR), and its expression is greatly increased in prostate tumors, suggesting a role in the process of carcinogenesis [16]. Another, GPR30, detects estrogen, previously thought only to activate nuclear receptors. GPR30 was cloned as an orphan receptor and subsequently shown to be activated by estrogen. A recent report identified a testosterone-activated GPCR [17], although the sequence has not yet been reported. As hundreds of cloned GPCRs remain orphans, there is great opportunity in the discovery of their ligands and drugs that modulate their activity.

Flow Cytometry Well Suited to Diversity Analysis

Biological diversity (numbers of proteins) and chemical diversity (numbers of compounds) present the analytical challenges of throughput, content, and informatics (Table 5.2). We anticipate that flow cytometry, which we envisioned as playing a role in drug discovery, can play a larger role in meeting

OVERVIEW 191

TABLE 5.2 Flow Cytometry for Diversity and Discovery Research

Technical Capability	Description
Sorting and analysis rates	Up to 50,000 events/s; 4.32 billion/day
Sample handling (HyperCyt) [19–24, 28, 41–43, 49, 103, 107–109]	Projected to 100,000 endpoint assays/day (100/min), small volume (1 µL); to 30 online mixes/min; to 10 sorted samples/min
Assays [18, 20, 21, 30–33, 98, 108]	Cell responses and populations, bead-based assemblies
Molecular assemblies [18, 36, 37, 86, 110–114]	Site numbers, reaction rates, K_d's, any two molecules
Sensitivity	Attomole to femtomole per sample
Number of parameters	Up to 12 parameters per assay
Content [35, 103]	Multiplex to 100 assays/sample; 10–20 plex assay/s (864,000/day)

the challenges of diversity by creating systems for measuring protein–protein interactions and cell responses with high throughput and high content. A flow cytometer is a multiparameter instrument that is normally used in a largely manual mode, working with individual samples. Literally thousands of different assays have been performed with a flow cytometer on cells or particles ranging from single parameter endpoint assays to time-dependent responses and complex assays involving multiple cell populations or simultaneous analysis of multiple responses. Because single particles are illuminated in a small volume and the signal is a pulse above a direct current background, the detection can resolve fluorescence on the particle from the volume around it without a wash step. The parameters include scatter signals and up to 10 or more fluorescence signals depending upon the number of lasers. Multiplexing is a technique that takes advantage of the multiparameter capability of flow cytometry. For example, an assay could use the green signal. The particles could be addressed using the red and orange signals to create a suspension array. Because flow cytometers are accessible to investigators in nearly all research environments, we believe that both the biological and instrumental platforms required for diversity analysis will be within their grasp.

The flow cytometer can be used as a generic platform for molecular assembly by attaching one molecular species to a particle using any number of tagging schemes (epitope tags, biotinylation, etc.). The particles become bright when a second species, labeled with a fluorophore, binds to the first. The labeling schemes include conjugation, fluorescent protein fusions, and the like. Tools are available for performing subsecond kinetic analysis. Our experience suggests that analysis can be extended to pairs of molecules of all types, including proteins, peptides, small molecules, deoxyribonucleic acid (DNA), ribonucleic acid (RNA), lipids, toxins, and so forth. The sensitivity is related to the number of particles, the fluorophores per particle, and the sample volume. It is practical to sample hundreds of particles in 1-μL volume with tens of thousands of molecules per particle. This translates into attomole quantities, or picogram amounts of a typical protein.

Challenge of Throughput

For the greatest impact, flow cytometry needs to deliver individual samples (microliter volumes) from multiwell plates at rates to 100,000 samples/day. Our HyperCyt system [18, 19] interfaces a flow cytometer and autosampler (Fig. 5.2). Commercial alternatives are ~30 times slower due to differences in the way samples are delivered and data is acquired. In HyperCyt, samples are separated from one another by air bubbles and delivered in a continuous stream. Likewise, the data are collected in a continuous stream, a plate representing a single data file, with the contents of each well separated by a gap introduced by the air bubble. The data are analyzed by proprietary software (FCSQuery) developed by Bruce Edwards. HyperCyt is a microsystem where microfluidic dimensions are at work in transporting samples to the flow cytometer.

We have addressed problems of mixing and carryover [20–24]. Mixing in HyperCyt requires breaking laminar flow and can be performed at sample delivery rates >1 sample/s. However, separation of samples by air bubbles in microfluidics leads to two types of carryover, of differing significance. Particle

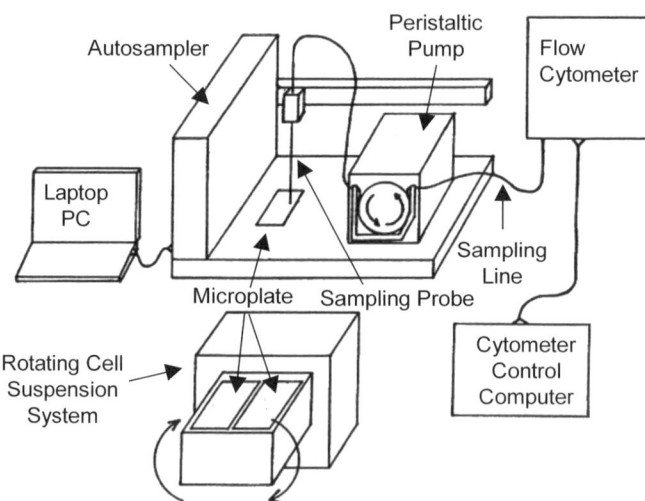

Figure 5.2 HyperCyt. A laptop PC controls both an autosampler and the cytometer control computer. Cells in a 96-well plate are kept in suspension with a rotating suspension system, which depends on surface tension to keep the fluid and cells in the wells while the plate is upside down. A peristaltic pump moves fluid and cells from the autosampler probe to the flow cytometer sample tube against the sheath flow pressure.

carryover from sample to sample is minimized by bubbles because particles tend to travel in the middle of the stream. Practically speaking, in an endpoint assay where the assay contents are premixed and developed in the well, a sample is contaminated with ~1 percent of the particles from the prior sample. For higher stringency, a wash plug can be easily added between samples at the cost of reducing throughput. Of greater significance, particularly for online assays, is fluid carryover. In this case, the assay (such as a cell response to a drug compound) develops online, and some of the compound from each sample is left behind as a thin film.

It is worth noting that flow cytometry is used in protein engineering or expression of combinatorial molecular libraries. Here, cells expressing a unique fluorescence phenotype can be analyzed and sorted at rates up to 50,000 per second (4.32 billion separate assays/24h). We have participated in a project with Novasite Inc. based on identifying protein variants according to cell responses. Here, each sample is mixed online with a compound and analyzed for response in a primary screen using HyperCyt. Sorting (cloning the variants) is performed as a secondary screen with HyperCyt for active compounds. Another drug discovery possibility is the "one-compound, one-bead" format [25]. In this scenario, appropriately labeled target proteins associate with a bead expressing a compound specific to the target. A modes-sized peptide array for nuclear receptors has already been reported [26]. As in the case of protein variants expressed on cells, the particles that become labeled could potentially be sorted at rates to 4.32 billion/day. However, particles ~100 μm diameter are required for the mass spectrometry step of identifying the compound on a particular bead. Sorting of one-bead, one-compound libraries has been achieved with the COPAS technology (www.unionbio.com) at rates to tens of thousands of particles/hour, adequate for screening libraries. This technology is a perfect match for our soluble GPCR-green fluorescent protein (GFP) fusion proteins, and could be used both in screening for active compounds and screening for detergents that keep receptors active.

Challenge of Content

Content in discovery research can be based on performing multiple measurements at a single time. High content usually requires multiparameter measurements at which flow cytometry excels. With digital signal processing, particle analysis rates in flow cytometry can easily reach 10,000 events/s with particle densities ~10^4 particles/μL, and sample delivery ~1 μL/s at 15 psi or less. High content has been achieved by multiplexing 100 distinct bead sets using two fluorescent signals from particle sets labeled with 1 of 10 different levels of brightness for each color (www.luminexcorp.com). Since the particle statistics could allow for 10- to 20-plex assays in each second of 5000 to 10,000 events, overall multiplex throughput for HyperCyt could approach 1 million assays/day. For drug discovery, a library of compounds could be tested against a family of receptors. Thus, a cell mixture could contain 10 sets of cells, each with a different GPCR, and each with a different address tag based on fluorescent intensity or color combinations. Each compound could be tested for specificity with respect to the members of the GPCR family. For clinical diagnostics, each well may contain a cell sample characterized by a complex assay such as immunophenotyping of blood leukocytes. With 10,000 cells stained with up to 10 fluorescent antibodies for distinct leukocyte subsets, one sample each second would allow dozens of leukocyte subsets to be identified.

Our team has recently considered one-step discrimination of GPCR agonists and antagonists. Using β2AR as the model receptor, we learned how to produce GPCR in soluble form as a GFP fusion protein that binds to G proteins displayed on beads (G-protein beads), and simultaneously to ligands displayed on distinct beads (ligand beads). The G-protein assembly forms in the presence of an agonist; the ligand assembly is inhibited by agonist or antagonist. A biplex of these two assays conducted simultaneously discriminates an agonist and antagonist in a single step. Moreover, a proteomic analysis results from displaying different G proteins as a multiplexed array, each GPCR evaluated for its ability to associate with a particular G protein. Content could be further enhanced by combining a cell response assay with a molecular assembly in the same sample well to evaluate physiology and assembly simultaneously.

Challenge of Informatics

To fully realize the potential of flow cytometry to meet the challenges of diversity, throughput, and content, informatic systems need to be integrated with the data collection efforts. Our group has provided persuasive evidence for acquiring a multiwell plate of samples as a single data file, where each cluster of events represents the contents of a single well [18–21, 27, 28]. This challenge is extended by multiplexing where each cluster of events represents 10 or more assays collected each second. In addition, the data need to be tracked back to the contents of the well, each well potentially containing a member of a chemical library. The data then need to be output in a format compatible with cheminformatic analysis of the assay result, the activity of the compound.

Since Beilstein and Chemical Abstracts Service have indexed ~15 to 20 million molecules to date, it is reasonable to assume that the medicinal chemistry space does not exceed 100 million compounds. This number is amenable to cheminformatics data handling, with computer capacity doubling every 2 years compared to the 5 to 10 percent annual growth of chemical space (note: this excludes the virtual chemical space). It is safe to assume, therefore, that tools and methodologies developed in cheminformatics will adequately handle large problems over the next 5 to 10 years. Chemical diversity, in a drug discovery sense, is one of the key issues in cheminformatics today. It is widely recognized that computational methods that handle molecular diversity—regardless of the chosen metric—need to operate under the restrictions imposed by pharmacokinetics (e.g., the desired properties of passive transcellular permeability, appropriate tissue distribution, aqueous solubility, metabolic stability, good excretion, and low toxicity profile, or ADMET), in addition to pharmacodynamics (efficacy, selectivity, potency). With numbers such as 10^{200} molecules accessible in the universe (attributed to D. Weininger) and 10^{16} chemicals accessible via combinatorial chemistry reactions (attributed to M. Geysen), it is not surprising that lead discovery, picking a needle in the haystack, is daunting.

Chemical vendors began to offer "diverse" chemical libraries in the early 1990s. By 2002, most of these companies had evolved high-quality, leadlike and druglike chemicals. Our in-house database, CADB08 (Commercially Available Data Base version 0.8), started with a collection of offerings from ChemNavigator (www.chemnavigator.com)—which covers over 100 vendors worldwide and offers over 7 million unique structures. Following duplicate removal, elimination of unwanted structures and property filtering, we obtained a database containing approximately 900,000 small, nonpeptide chemical structures. At the other end of the spectrum are efforts to mine the chemical space represented by natural-product-like molecules. This features a relatively high density of stereo centers, leading to structures of increased complexity. ChemBank, an ongoing project organized by the Initiative for Chemical Genomics (ICG) and being built by the Institute for Chemistry and Cell Biology (ICCB) (www.chembank.med.harvard.edu) has accumulated over 300,000 natural-product-like structures, based on the Diversity-oriented Organic Synthesis (DOS) approach [29]. Taken together, these two chemical collections represent a balanced, chemically diverse library that can be used for virtual screening.

Screening

Drug discovery in the last decade has undergone dramatic changes. Given the relatively low success rate of structure-based drug design in the early 1990s, the pharmaceutical industry reoriented itself to build the *high-throughput screening (HTS) machine*, whereby very large numbers of chemicals were screened for biological activities. Combinatorial chemistry, developed in part to justify the HTS machine, led to the development of large sets of chemicals for screening. At the same time, smaller subsets of libraries are dedicated to high-throughput (HT) nuclear magnetic resonance (NMR) screening where small molecules are tested for binding via NMR experiments, or to HT crystallography where small molecules are co-crystallized with proteins of interest. However, these two approaches are not applied, at an industrial scale, for GPCR targets, as there is only one crystal structure available (bovine rhodopsin). In the postgenomic era, brute force screening seems to be ramping down and is being replaced by targeted screens, on smaller, focused libraries.

These efforts now are likely to start with virtual screening, including docking ligands onto crystal structures. If crystal structures are not available, such as for most GPCR, ligands are docked into a homology model. An alternative is to perform pharmacophore searches, assuming that enough active molecules are known a priori to develop such pharmacophores. Selected members of a large, virtually screened library are then pooled into a sublibrary for physical screening. The output of the results can be used to select other compounds, guide further synthesis, and as a tool for predicting the ADMET characteristics of the active compounds. An iterative process of virtual and biological screening is within reach of teams that do not have the resources of biotech companies or pharmaceutical companies.

Technology Associated with HyperCyt

The potential for displaying virtually any molecule or cell response in a format compatible with HyperCyt (sampling rates >1/s with 10- to 20-plex assays) will make flow cytometry a powerful option for the real-time analysis of molecular interactions in drug discovery and diversity research. The strategy to achieve optimal outcome requires four complementary activities. Component 1 expresses and characterizes proteins relevant to signal transduction and termination. Component 2 displays functional proteins in a manner compatible with flow cytometric analysis. Component 3 implements robust HT screening. Component 4 implements virtual screening tools and selects compound libraries for HyperCyt. The resulting technological advances will allow us to define mechanisms of cell activation through GPCRs and to explore the diversity of molecular interactions in GPCR pathways.

5.3 PROTEIN EXPRESSION AND CHARACTERIZATION

There are two major goals of component 1 (protein expression and characterization). The first was to generate and characterize the tools needed to reconstitute signaling complexes on beads in component 2 (soluble and bead-based assemblies). To this end, we have engineered many signal transduction molecules appropriate for reconstitution studies. The second goal was to develop a battery of flow cytometric assays that could be used to evaluate GPCR pathways. Thus, the characterization of these expressed molecules in cells and in soluble systems prior to their display on beads led to significant biological advances.

GPCR Constructs

A significant fraction of our effort has been directed toward studying complexes between GPCR and G proteins and between GPCR and arrestins. We have generated wild-type, phosphorylation deficient mutants, and chimeric proteins including GPCR-GFP, FPR-G_i, FPR-G_i-GFP, and GPCR-G_i. We have made receptors with hexahistidine (his$_6$) or Myc tags at the amino terminus to facilitate purification and immobilization.

Soluble GPCRs Reconstitute with G Proteins

Our earliest studies focused on reconstituting the wild-type FPR with purified G proteins in solution as a precursor to reconstituting the complexes on beads. For these experiments we used a fluorometric assay to determine the dissociation of a fluorescent ligand from the solubilized FPR in detergent (Fig. 5.3a). The ligand dissociation rate differs between FPR coupled to a G protein and FPR alone. By determining the sensitivity of the dissociation of a fluorescent ligand to the presence of guanine nucleotide, which disrupts G-protein coupling, we demonstrated time- and concentration-dependent reconstitution of FPR with both endogenous and exogenous G proteins. Solubilized FPR with bound ligand and Gαi3 protein interacted with an affinity $\sim 10^{-6}$ M.

FPR Assembles with Arrestin

We extended our studies of FPR assemblies to those involving the protein arrestin. As arrestin predominantly interacts with phosphorylated FPR, we obtained FPR from cells that had been stimulated with ligand to achieve in vivo phosphorylation. We demonstrated a concentration- and time-dependent

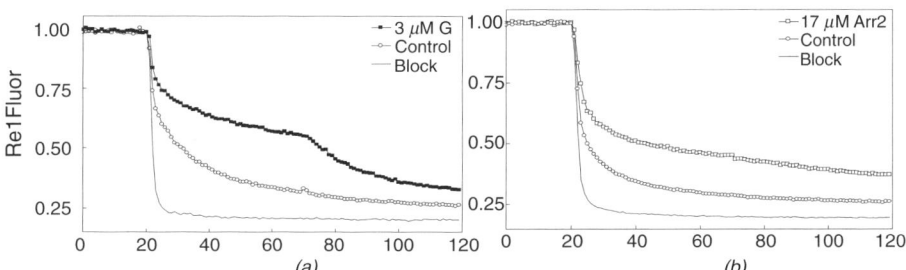

Figure 5.3 Fluorescent peptide assay of reconstitution of FPR with (a) G-protein and (b) arrestin. GTPγS (added at 70 s) distinguishes G-protein coupling, which is sensitive to the guanine nucleotide analog, from arrestin binding, which is insensitive. The data were obtained in a stirred fluorometer cuvet by addition of an antibody that quenches the fluorescence of unbound peptide, added at 20 s. When the receptor is blocked with excess nonfluorescent peptide (block), the peptide is rapidly quenched to about 0.25 of the initial fluorescence of the mixture. When there is only receptor and fluorescent peptide (control), the bound peptide remains fluorescent past 25 s, dissociates exponentially, and is immediately quenched. Both G protein and arrestin increase the affinity of the receptor for fluorescent ligand, resulting in higher binding and slower dissociation. Only the G protein is rapidly converted to an inactive state by the introduction of GTPγS at 70 s, and the fluorescent peptide dissociates quickly again [30].

reconstitution of liganded, phosphorylated FPR with exogenous arrestin-2 and -3 to form a high agonist affinity, nucleotide-insensitive complex with effective concentration EC_{50} values of 0.5 and 0.9 µM, respectively ([30], Fig. 5.3b). Moreover, we demonstrated that the addition of G proteins was unable to alter the ligand dissociation kinetics or induce a GTPγS-sensitive state of the phosphorylated FPR. The properties of phosphorylated FPR were reversible upon phosphatase treatment.

We further investigated the interactions between partially phosphorylated FPR and G proteins and arrestins. For these studies, we used FPR phosphorylation-deficient mutants (ΔA, ΔB, ΔC, ΔD). The ΔA and ΔB mutants contain only four of eight nonoverlapping potential phosphorylation sites, whereas the ΔC and ΔD each contain six of the eight potential phosphorylation sites. We demonstrated that phosphorylation of the wild-type FPR lowers its affinity for G protein, whereas mutant receptors lacking four potential phosphorylation sites (ΔA and ΔB) retain their ability to couple to G protein [31]. Phosphorylated mutant receptors lacking only two potential phosphorylation sites (ΔC and ΔD) did not couple to G protein but did bind arrestins.

FPR Colocalization with Arrestin in Cells Parallels FPR Assemblies in Solution

To corroborate the assemblies with the physiology of the proteins in vivo, we used confocal fluorescence microscopy to evaluate the interactions between FPR, as assessed by fluorescent ligand, and arrestins, as assessed with arrestin–GFP constructs (Fig. 5.4). We developed a novel series of red fluorescent ligands to allow simultaneous analysis of both proteins [32]. This approach demonstrated that, whereas stimulated wild-type FPR in whole cells colocalized with arrestin-2, the stimulated GPCR lacking four potential phosphorylation sites displayed no colocalization with arrestin-2 [31]. However, mutant receptors lacking only two potential phosphorylation sites are restored

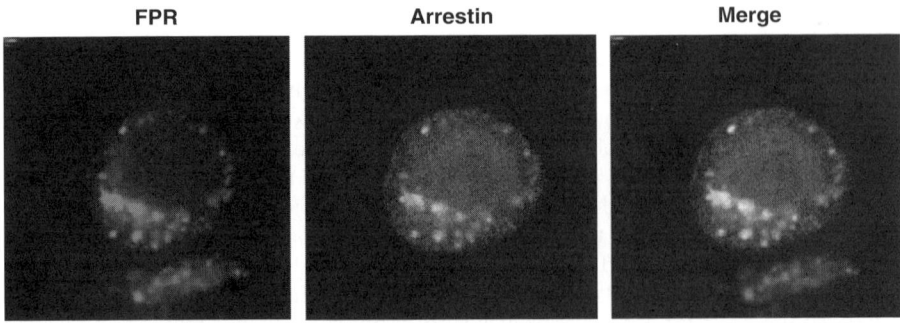

Figure 5.4 Colocalization of FPR (originally red ligand) with arrestin (originally green GFP) Following stimulation for 10 min with agonist. The brighter overlap demonstrates colocalization for a large fraction of the FPR [32].

in their ability to colocalize with arrestin-2 in vivo. Thus, we concluded that there is a submaximal level of FPR phosphorylation that simultaneously results in an inhibition of G-protein binding and an induction of arrestin binding. We proposed that phosphorylation alone may be sufficient to desensitize the FPR in vivo, raising the possibility that for certain GPCRs, desensitization may not be the primary function of arrestin.

We have subsequently evaluated mechanisms of arrestin activation and the role of the activation state of arrestin on the formation and properties of the FPR complex. Very little was known about the relationships between the sites of receptor phosphorylation, the resulting affinities of arrestin binding, and the ensuing mechanisms of receptor regulation for any given GPCR. To investigate these interactions, we used an active truncated mutant of arrestin (amino acids 1 to 382) and phosphorylation-deficient mutants of FPR described above. We demonstrated that the phosphorylation status of residues between amino acids 328 and 332 of the FPR is a key determinant that regulates the affinity of FPR for arrestins, whereas the phosphorylation status of residues between amino acids 334 and 339 regulates the affinity of FPR for agonist when arrestin is bound [32]. Again confocal fluorescence microscopy confirmed the receptor binding interactions in vivo. The results indicated that the agonist affinity state of FPR is principally regulated by phosphorylation at specific sites and is not simply a consequence of arrestin binding, contradicting current dogma. Furthermore, these studies were the first to demonstrate that GPCR agonist affinity and the affinity of arrestin binding to the phosphorylated receptor are regulated by distinct receptor phosphorylation sites.

Arrestin Regulates Intracellular FPR Traffic

Our most recent studies examined the role of arrestins in FPR trafficking in the cell (Fig. 5.5). Surprisingly, our previous results indicated that FPR internalization does not require arrestin. To further examine roles for arrestins in FPR function, we determined the effects of activating arrestin mutations on ternary complex formation (ligand, FPR, arrestin) and cellular trafficking. One such activating mutation, arrestin-2-3A protein, contains targeted mutations of three hydrophobic residues (I386A, V387A, F388A) in the carboxy-terminus that contribute to basal inactivity in the wild-type protein. Again FPR–arrestin interactions were confirmed in vivo by colocalization studies using confocal microscopy with arrestin–GFP chimeras and a fluorescent FPR agonist. To assess the effects of activated arrestin expression on receptor trafficking, we used a novel quantitative, in vivo flow cytometric assay. We demonstrated that activated, but not wild-type, arrestin expression inhibits FPR internalization [32]. We further demonstrated that arrestin-2-3A–GFP expression inhibits FPR recycling. In contrast to results with FPR, expression of activated arrestin had no effect on either the internalization or recycling rate of β2AR, which internalizes in an arrestin-dependent manner. These

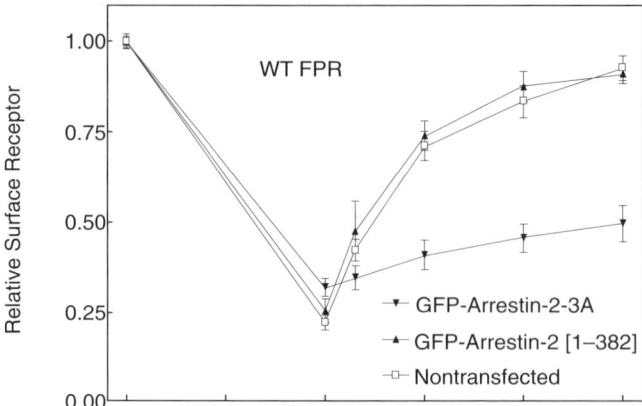

Figure 5.5 FPR recycling. Cells expressing FPR were stimulated with ligand to bring surface expression down to about 25 percent in all cases. The ligand was washed away, and the amount of receptor on the cell surface was measured over time by ligand binding. The rate of surface receptor reexpression was inhibited by an activated mutant of arrestin (arrestin-2A:I386A/V387A/F388A), suggesting that arrestin plays a critical role in receptor recycling in this system, rather than in receptor downregulation as in the visual system [32].

results indicate that arrestin is required for recycling of certain GPCRs, such as FPR, and that the activation state of arrestin regulates this process.

FPR-$G_{i\alpha2}$ Fusion Protein Functions Physiologically

As suggested above, phosphorylation of wild-type FPR might be sufficient to prevent coupling of GPCRs to G proteins in vivo. We investigated this phenomenon using a chimeric protein between FPR and the $G_{\alpha i2}$ subunit of heterotrimeric G proteins. The chimeric GPCR-G protein approach for component 2 could replace the requirement of providing an exogenous G_α subunit for each GPCR display. The functional capabilities of this chimeric protein were evaluated both in vivo, in stably transfected myeloid U937 cells, and in vitro, using our soluble reconstitution system [33]. The chimeric protein existed as a soluble complex containing βγ subunits. The chimeric protein mobilized intracellular calcium, desensitized normally in response to agonist, and was internalized (Fig. 5.6) and recycled at rates similar to those of the wild-type FPR. Confocal fluorescence microscopy revealed that internalized chimeric receptors colocalized with arrestin, as well as G protein. Soluble reconstitution experiments revealed that this chimeric receptor, even in its phosphorylated state, existed as a high ligand affinity G-protein complex, only partially prevented through the addition of arrestins. These results indicated that under conditions of high local G-protein concentrations, the phosphorylated form of the FPR is still capable of binding G proteins.

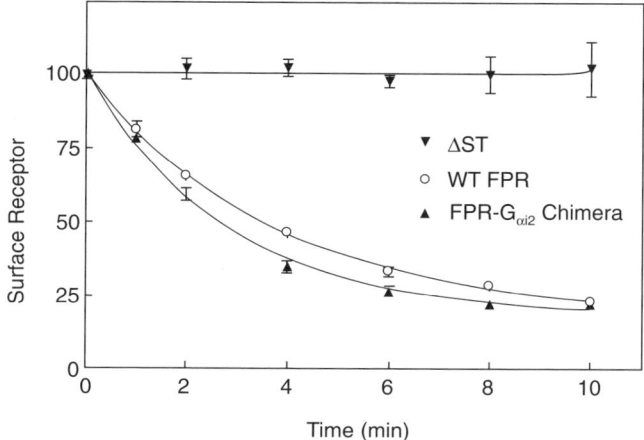

Figure 5.6 FPR internalization depends on phosphorylation. Internalization of wild-type (WT) FPR, a mutant lacking C-terminal phosphorylation sites, and a chimeric FPR-$G_{\alpha i2}$ protein were followed by ligand binding.

FPR Tail Assembly (Protein Domain Display)

Because GPCRs are integral membrane proteins, any assay of their function in solution requires the presence of detergent. As an alternative, we utilized peptides representing the unphosphorylated and phosphorylated FPR carboxyl terminus. These were chemically synthesized (e.g., biotin-MGQDFRE RLIHALPASLERALTEDS(P)TQT(P)SDT(P)TN(P)STLPSAEVELQAK-OH) to be homogeneous and bound to polystyrene beads via a biotin/streptavidin (SA) interaction. In experiments that excluded detergents, we chose fluorescein-conjugated arrestins to examine binding interactions between arrestins and the bead-bound FPR carboxyl terminus. Analyses were performed by flow cytometry (Fig. 5.7). Arrestin-2 and arrestin-3 bound to the 47mer FPR carboxyl-terminal peptide in a phosphorylation-dependent manner, with K_d values in the micromolar range [34]. Activated arrestin mutants that display phosphorylation-independent binding to intact receptors also bound the bead-bound FPR terminus in a phosphorylation-dependent manner, but with greater affinity than the full-length arrestins, with K_d values in the 5 to 50 nM range. The results suggested that the carboxyl terminus of arrestin is a critical determinant in regulating the binding affinity of arrestin for the phosphorylated domains of GPCRs, and furthermore that non-detergent-based approaches to investigate GPCR assemblies represent a viable alternative to approaches utilizing solubilized holo-GPCRs. To this end we have synthesized a number of carboxy-terminal peptides of varying length, both phosphorylated and not, from GPCRs including rhodopsin, β2AR, and vasopressin receptors. Preliminary data indicate that the minimal length for arrestin binding is between 16 and 21 amino acids. The number of phosphates and the amino acid sequence specificity remain to be characterized.

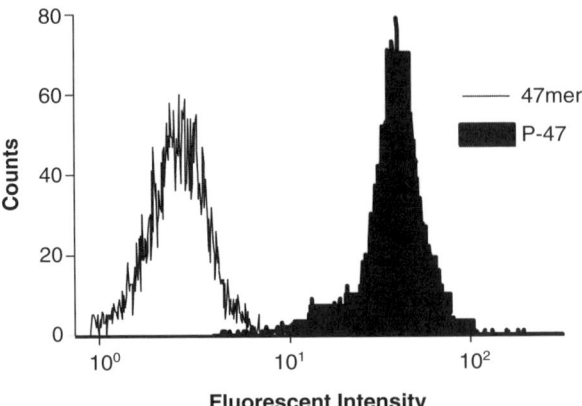

Figure 5.7 Protein domain display. Beads were coated with a synthetic peptide representing the 47 C-terminal amino acids of the FPR (phosphorylated or not), and fluoresceinated arrestin was allowed to bind to each of the beads. Beads with the phosphorylated peptide bound more arrestin, as expected from whole protein experiments.

Proof of Principle for β2AR and Other GPCRs

To extend studies carried out with FPR to other GPCRs, we have engineered a group of GPCR-GFP chimeric proteins. To date, the most important of these has been β2AR. β2AR is arguably the most studied and best understood of all GPCRs with well over 1000 listings in PubMed. Despite the extensive literature concerning β2AR, there are still many fundamental issues regarding its activation and function that remain poorly understood. In addition to reconstitution studies with FPR, similar studies with β2AR have provided significant insights into GPCR function as well as providing a novel screen for agonists and antagonists of this receptor. We have used FPR-GFP and β2AR-GFP chimeric proteins for bead-based assemblies and screening in component 2.

5.4 SOLUBLE AND BEAD-BASED ASSEMBLIES

Component 2 (soluble and bead-based assemblies) requires several elements. These include: (1) solubilizing and characterizing GPCRs produced by component 1, (2) evaluating and implementing particle chemistry for GPCR display, (3) validating physiology of expressed GPCRs by comparing cell and bead results, (4) analyzing quantitative aspects of each assembly (K_d, kinetics, etc), and (5) miniaturizing the protocols for each assembly to be compatible with screening by component 3. We have now succeeded in displaying solubilized GPCR complexes on beads in several ways (Fig. 5.8) and

SOLUBLE AND BEAD-BASED ASSEMBLIES

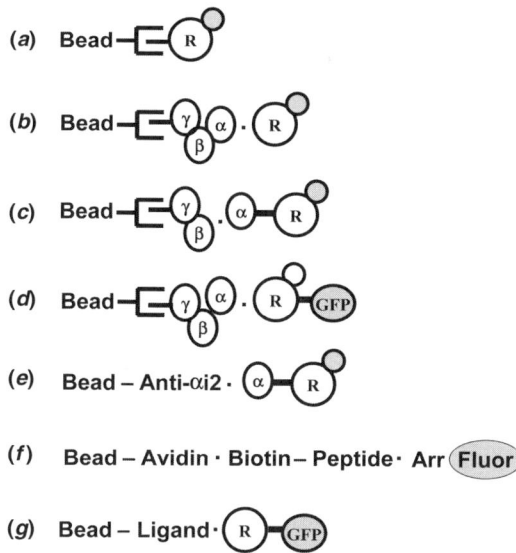

Figure 5.8 Representative bead assemblies. The fluorescent moiety in each case is shaded. Dots represent noncovalent interactions. These assemblies are described in the text.

TABLE 5.3 Principles for Protein Display on Particles

Bead surface	Hydrophilic preferred (dextran); streptavidin coated (not naked hydrophobic) latex works; we have not tried to coat functionalized latex.
Bead stability	Latex or dextran preferred; silica/glass may shatter.
Bead chemistry	Dextran compatible with chelated Ni, ligands, GST, fluorescent labels for multiplex.
Reagent display	Displaying the most expensive pure protein or ligand is preferred with nonspecific Ni chelate beads (cost efficient).
Reagent capture (GSH, biotin)	GSH and SA beads may be useful to capture proteins in a single step from cell lysates and display the protein for assembly and flow cytometry analysis.

made some observations regarding the principle of molecular display on particles (Table 5.3).

1. His6-tagged FPR were directly displayed on Ni-chelate beads and detected with fluorescent formyl peptides [35] (Fig. 5.8a).
2. G-protein βγ subunits were displayed using his6-tags on Ni-chelate beads or FLAG tags captured with biotinylated anti-FLAG antibody on SA beads. By adding G_α subunits, FPR and its fluorescent ligand (Fig. 5.8b, shaded ball) were captured [36, 37].

3. His6-tagged G-protein βγ subunits captured the FPR-$G_{\alpha i2}$ fusion protein (Fig. 5.8c), which was detected with fluorescent formyl peptide ligand [36].
4. His6-tagged G-protein βγ subunits, displayed with $G_{\alpha i}$ subunits, captured the GFP-FPR fusion protein in the presence of nonfluorescent ligand (Fig. 5.8d). This assembly has been used for other GFP-GPCR fusion proteins [36, 37].
5. Anti-$G_{\alpha i2}$ antibody was displayed on antibody-capture beads to bind the FPR-$G_{\alpha i2}$ fusion protein, and was detected by fluorescent formyl peptide ligand (Fig. 5.8e). This assembly was used to prove that the FPR and G_α subunits were in the fusion protein and serves as a representative alternative to Western blotting [33].
6. Biotinylated GPCR tail peptides were displayed on streptavidin beads for analysis of arrestin assembly and phosphorylation [34] (Fig. 5.7 and Fig. 5.8f).
7. The β2AR antagonist dihydroalprenolol (DHA) was displayed on dextran beads to capture β2AR-GFP [37] (Fig. 5.8g).
8. FPR and bacteriorhodopsin were reconstituted into lipid bilayers enclosing beads (not shown; unpublished observation).

Taken together, these displays represent an essentially complete proof of principle set for display of GPCRs with signal transduction and termination partners. We will discuss FPR and β2AR in more detail.

FPR Assemblies Used to Characterize and Validate the Displays

FPR, our first soluble GPCR, served as a prototype because it allowed direct comparison between displays obtained with fluorescent peptide ligand and fusion proteins [36] (Fig. 5.8a to 5.8d). The display was generalizable, demonstrated with two choices of epitope tags on the βγ, two types of beads, three FPR constructs, three G_α subunits, and two $G_{\beta\gamma}$ subunits. The results suggested that ~100,000 $G_{\beta\gamma}$ subunits/bead were displayed in appropriate orientation to bind FPR. We evaluated the time and concentration dependence of ternary complex formation (G protein, FPR, and ligand). As a result, affinities of ligand for FPR, ligand–FPR complex for G protein, and FPR-$G_{i\alpha2}$ fusion protein for $G_{\beta\gamma}$, were found to be consistent with comparable assemblies in detergent suspension.

The performance of the assembly was assessed in applications representing their potential to reveal ternary complex mechanisms [36]. First, we showed for a family of ligands that the affinity of FPR for its ligands varied in parallel with the assembly of FPR and G protein. Second, we showed that the beads could be used as a sensor to report the availability of FPR in solution, measuring competition between G protein on the bead and soluble G protein. Third, we showed kinetic discrimination between individual steps of ternary complex activation and disassembly. We showed that the disassembly of FPR from G protein in the presence of GTPγS and ligand occurs on the subsecond time frame. We also showed that the assembly was sensitive to the presence of RGS proteins (regulator of G protein signaling) in a manner consistent with their known ability to stimulate GTPase activity of the G proteins [36, 38].

β2AR Extends the Soluble GPCR on a Bead Concept

We solubilized the β2AR-GFP fusion protein and used it to approach a fundamental goal of drug discovery, the discrimination of agonist, antagonist, and a partial agonist, potentially in a single step [37]. For these assemblies we established two novel bead display systems and developed a multiplex approach for simultaneous analysis of the two assemblies, ligand beads (Fig. 5.8g) and G protein beads (Fig. 5.8d) using $G_{s\alpha}$ instead of $G_{i\alpha}$. We developed a ligand on a bead assembly by conjugating an antagonist (DHA) to a dextran bead. The assembly of β2AR-GFP was sensitive to the presence of agonist or antagonist (Fig. 5.9) and measured ligand affinity. The G-protein assembly (Fig. 5.10) was sensitive to the presence of agonists and resolved partial and full agonists. The behavior of the complexes on particles was essentially identical to the membrane-bound form of the receptor (Table 5.4) indicating that the physiological structure of β2AR is retained after solubilization. By modifying one set of beads to fluoresce red or using a different size [38] to resolve the two assays,

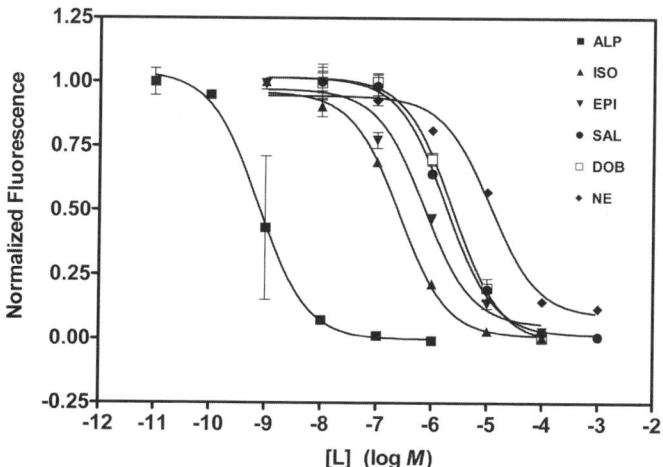

Figure 5.9 DHA bead competitive assay. Competition curves were generated as increasing ligand concentrations competed the β2AR-GFP off DHA beads. K_d values for nonfluorescent ligands were calculated from these data. The ligands used were alprenolol (ALP), isoproterenol (ISO), epinephrine (EPI), salbutamol (SAL), dobutamine (DOB), and norepinephrine (NE).

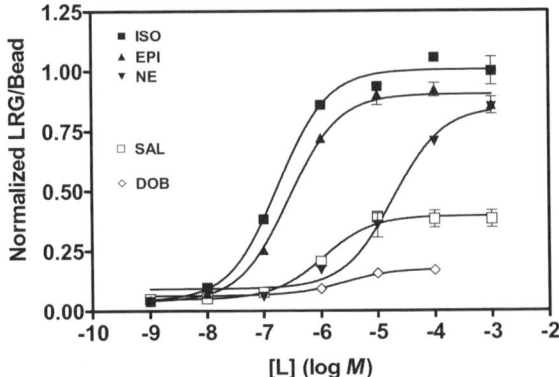

Figure 5.10 G-protein beads bind β2AR-GFP in the presence of β2AR agonists. Dose–response curves were generated as increasing agonist concentrations caused the β2AR-GFP to form ARG complexes on the G-protein beads. The ligands used were isoproterenol (ISO), epinephrine (EPI), norepinephrine (NE), salbutamol (SAL), and dobutamine (DOB).

TABLE 5.4 β2AR Assembly[a]

Ligand	LR K_d (nM)	LRG EC_{50} (nM)	LRG Efficacy
ALP	<1.3	NA	NA
ISO	220	180	1.0
EPI	680	280	0.9
NE	19,000	19,000	0.9
SAL	2,300	4,000	0.3
DOB	2,400	2,600	0.2

[a] Data obtained from three experiments as described for Figure 6.9, using DHA beads to obtain K_d values, and Figure 5.10, using G-protein beads to obtain EC_{50} values.

we could discriminate agonist and antagonist in a single step [37]. While a single-step primary screen would resolve agonists and antagonists, a dose–response secondary screen would resolve full and partial agonists. Molecules that block interactions between β2AR and G protein, uniquely, would not be active in the assembly of Fig. 5.9 but would compete with agonist in Fig. 5.10. Mathematical modeling of the ternary complex data with the soluble receptor systems is consistent with the idea that agonists promote higher affinity interaction of β2AR with G proteins than do partial agonists.

Display of Cell Cycle Proteins Demonstrates Further Generality

We collaborated with Dorota Skowyra (St. Louis University) on the system that controls cell cycle in yeast and has extensive homology with mammalian cell cycle control. There are three complexes: a substrate complex including

Sic, a ubiquitin ligase complex, and the proteasome complex that recognizes that ubiquitinated substrate, degrades it, and enables cell cycle progress. We have displayed the GST-tagged ubiquitin ligase through a GSH-bead, and evaluated binding of a GFP–Sic substrate complex [39]. Not only important in cell cycle, this display represents an opportunity to bridge the power of yeast genetics and HyperCyt.

5.5 HIGH-THROUGHPUT FLOW CYTOMETRY TECHNOLOGIES

We have developed two successive generations of HT flow cytometry sampling technology. The first, designated plug flow cytometry, used a reciprocating multiport flow injection valve to execute up to 10 endpoint assays per minute, 4 online mixing experiments per minute, and, in secondary screens, a 15-point concentration gradient of soluble compound in ~2 min [40–43]. The second-generation technology [19], designated HyperCyt, uses a peristaltic pump in combination with an autosampler to boost endpoint assay performance to rates in excess of 1 sample/s (Fig. 5.2). HyperCyt hardware and software have since been more fully developed and characterized for high content, microfluidic mixing, cell sorting, carryover, and screening.

HyperCyt Exhibits Versatility and Compatibility with High Content

We have validated cell-based endpoint assays for ligand binding, surface antigen expression, immunophenotyping, and cell adhesion. For example, at analysis rates of 1.5 s/sample a multiparameter (high content) lymphocyte immuno-phenotyping microassay in 96-well plates provided immunofluorescence results comparable to manual analysis (Fig. 5.11) [18]. High-content bead assays have been prototyped (not shown). In validation experiments with Terasaki microplates (sixty 10-μL wells per plate), all 60 cell samples were clearly resolved and reproducibly quantified over a 90-s analysis period [18]. The volume of the Terasaki wells is large enough that evaporative loss of fluid is not a problem but small enough that quantity-limited cells and reagents may be used at otherwise prohibitively high concentrations. Terasaki and 96-well "crystallization" microplates also enable a novel yet simple solution to the problem of cell settling. Uniform particle suspensions are maintained for an hour or more by periodically inverting the microplates in a "Ferris wheel" apparatus [18] (Fig. 5.2). Liquid surface tension prevents sample loss from inverted wells. This microvolume technology led to the potential application of HyperCyt in disease endpoint assays (signal transduction intermediates, cell adhesion) that can efficiently use limited cell numbers available from patient samples.

HyperCyt Adds Microfluidic Mixing for Analysis of Soluble Compounds

We have described two novel microfluidic mixing approaches for use in conjunction with HyperCyt. These approaches have been characterized with

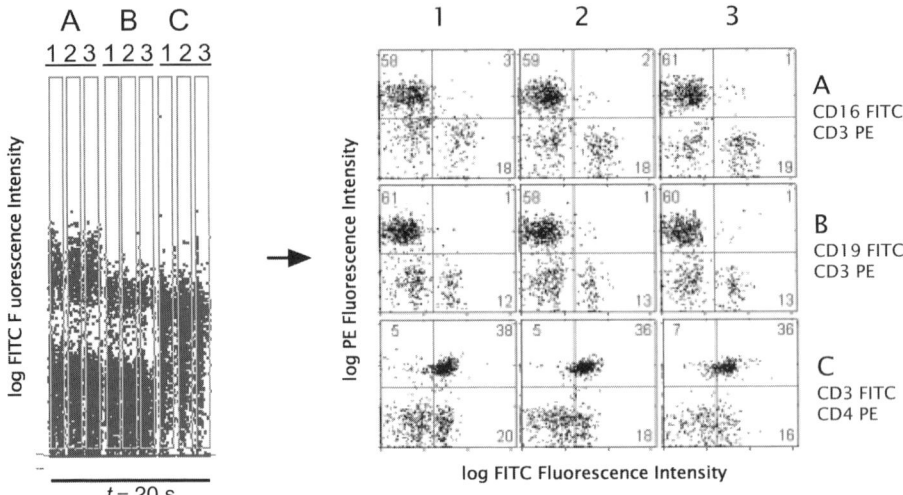

Figure 5.11 High-content, high-throughput immunophenotyping with HyperCyt. (*Left*) One parameter output with time. (*Right*) Replotted data, with high-content two-parameter output.

fluorescent cell calcium responses to sampled peptides as well as with a novel bead-based assay [biotin displacement of fluorescein isothiocyanate (FITC)-biotin on SA beads] that provides a fluorescence response analogous in time and kinetics to a cell calcium response [44]. The first approach used the pulsatile delivery of the HyperCyt peristaltic pump to generate mixing of particles with fluid compounds sampled from microplate wells [20]. The second approach uses a 100-μm magnetic "flea" contained within the HyperCyt delivery tubing and a magnetic stirrer [21]. Soluble compounds can be sampled from microplate wells and processed at rates to 10 samples/min [21].

HyperCyt Cell Sorting Extends Microfluidic Mixing

We have interfaced HyperCyt with a MoFlo flow cytometer and achieved online mixing combined with sorting of responding cells (Fig. 5.12). A relatively long sample aspiration time (10s) followed by 3 × 3s intersample washing enabled screening of FPR responses at a rate of 3 peptide concentrations/min with ~10,000 cells analyzed/concentration. When cells highly responsive to peptide were collected by sorting and expanded in culture for 12 to 30 days, the sort-purified cells showed enhanced sensitivity to low peptide concentrations and more sustained responses to all stimulatory peptide concentrations compared to the unsorted cell population from which they were derived [27]. A major technical advance changed the configuration of a flow nozzle to be compatible with the air bubbles between samples created by HyperCyt.

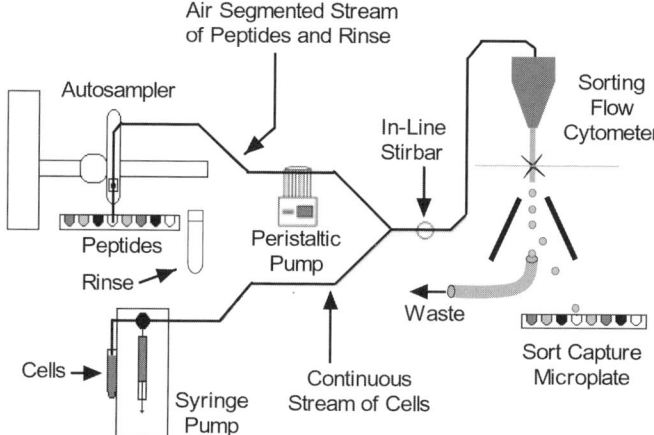

Figure 5.12 HyperCyt mixing and sorting to isolate ligand-responsive cells.

Sample Carryover in Microfluidic Mixing Systems

HyperCyt can deliver fluid samples of soluble compounds to mix with a continuous stream of cells at rates as fast as an autosampler can move from well to well. However, overall sample throughput is limited by unique aspects of intersample fluid carryover associated with the air-bubble segmented fluid flow of HyperCyt. Using a system in which samples are delivered to a spectrofluorometer, it has been possible to isolate and characterize factors important in such fluid carryover [23]. The most critical factor appears to be tubing internal surface area, while tubing junctions are of lesser importance (a contrast to particle carryover). Fluid carryover arises from a thin fluid film left on the inner tubing surface as the sample transits the length of the tube. The film acts as a transport mechanism to hold a small fraction of fluid from the sample as it passes by and deposits it into the subsequent sample.

Software Development

Software development has been central to progress in the automation and implementation of HyperCyt. FCSQuery, a Visual C^{++} program, was initially created to support the novel concept of time-resolved multisample data collection in which flow cytometry data from up to 96 samples are collected and stored in a single time-resolved data file. The program automatically detects time-resolved data clusters corresponding to individual samples, quantifies a range of statistical parameters for each sample and outputs the summary data to a Microsoft Excel spreadsheet. We are unaware of any commercial flow cytometry software with this capability.

A second Visual C^{++} program, CRFSoft, was created to enable real-time synchronization of the multiple components of the HyperCyt system. This is a multithreaded program by which a laptop computer controls the flow cytometer data acquisition computer, the peristaltic pump, and the autosampler while monitoring fluorescent "feedback" signals directly from the flow cytometer in real time [45]. In the initial implementation of this software, fluorescent beads sampled at the beginning and end of autosampling served as the synchronization signal to initiate and terminate data acquisition.

FPR Ligand Screening Assay

To validate HyperCyt for compound screening applications, FPR has been mutated to block internalization. The detection reagent is a fluorescein-conjugated peptide, fMLFK-FITC, that binds the FPR mutant with a K_d ~3 nM. The objective is to detect potential anti-inflammatory compounds analogous to the anti-inflammatory drug, cyclosporin H, that block binding of fMLFK-FITC to the FPR. The cells were distributed in a 96-well microplate, incubated with unlabeled blocking peptides, then added with 1.5 nM fMLFK-FITC to quantify free FPR. To determine the limits of assay sensitivity, two blocking peptides with low and intermediate FPR affinity were used between 10^{-7} and 10^{-4} M: 2Pep (fML, inhibition constant, K_i ~6000 nM) and 3Pep (fMLF, K_i ~170 nM). With HyperCyt sampling at 1 well/s, the resulting 96 time-resolved event clusters (Fig. 5.13a) were analyzed to determine the median fluorescence intensity (MFI) of FPR-bound fMLFK-FITC in each well (Fig. 5.13b). The MFI was 69 ± 10 fluorescence units (FU) in the absence of blocking peptide (Fig. 5.13b, solid line), while a high FPR affinity blocking peptide (4Pep, K_i ~3 nM) reduced it to 6 ± 1 FU (Fig. 5.13b and 5.13c). Blocking was considered significant if the MFI was more than 3 standard deviation (SD) below the mean of unblocked cells (Fig. 5.13b, dashed line) in all 8 replicate wells. This occurred with 10^{-4} M 2Pep and 10^{-4} to 10^{-6} M 3Pep. These results indicate that our screening assay will reliably detect compounds with micromolar K_i's. We have subsequently applied HyperCyt to screen the Prestwick Chemical Library (PCL) and found a single positive compound, sulfinpyrazone, which was previously identified as a member of a series of FPR antagonists (see below). This result confirms our approach with respect to the assay and the technology.

5.6 VIRTUAL SCREENING

Virtual screening [46, 47] is becoming, de facto, a complement to bioactivity screening [48]. We view the ability to perform virtual screening as relying on docking compounds and scoring binding interactions in the context of chemical databases and high-performance computing.

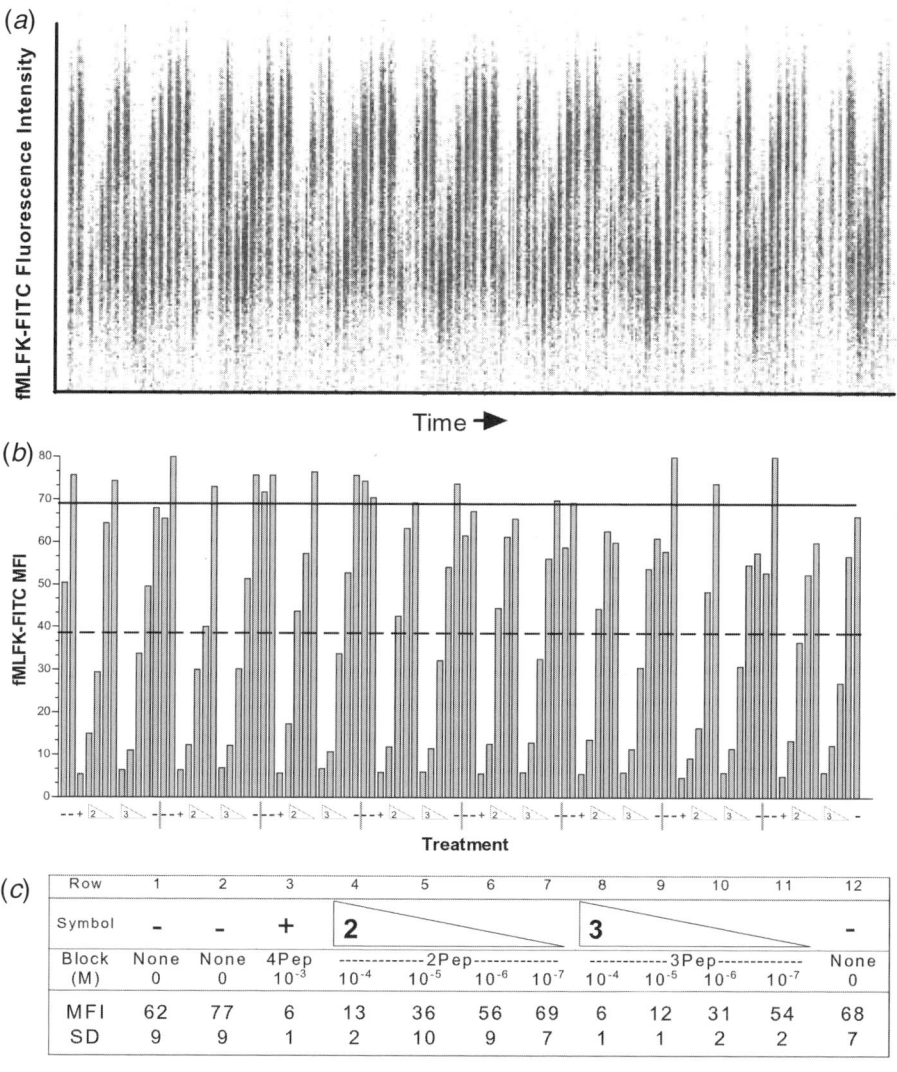

Figure 5.13 FPR ligand screening assay and performance. Two test peptides with low (2Pep or fML) and intermediate (3Pep or fMLF) FPR binding affinity were assessed for competitive binding activity in the FPR assay against 4Pep-FITC (fMLFK-FITC), using FPR-expressing cells. (*a*) Time-resolved clusters of cell fluorescence data, one cluster for each well, that were produced in conjunction with HyperCyte analysis. (*b*) Plot of the mean fluorescence intensity (MFI) of cells from each of the wells. The lines below the graph delineate each 12-well series. (*c*) Setup of the 12-well series: 2 unblocked controls, 1 blocked control, 4 concentrations of each test peptide, and one last unblocked control. This series was repeated 8 times on a 96-well plate.

Docking and Scoring

A steroidal scaffold that binds to β-tubulin, an anticancer target previously known to bind taxol and its derivatives [49], was identified by virtually screening the NCI (National Cancer Institute) open database. Given a three-dimensional (3D) structure of the receptor, virtual screening relies on docking and scoring to provide candidates for further analysis. DOCK [50], GOLD [51], FlexX [52], AutoDock [53], and FRED (www.eyesopen.com) are among the most widely used docking programs in the field. Most of these programs can dock libraries of single structures or even multiconformer libraries and are thus suitable for high-throughput searches.

Regardless of the choice of the docking software, the ability to score the ligands in an appropriate manner is limited. There are four major categories of scoring functions. First, knowledge-based methods [54–56] use Boltzman-weighted potentials of mean force, derived from statistical analyses of ligand–receptor interatomic contacts, based on available complexes in the Protein Data Bank (PDB) [57]. For example, SMoG [58–60], Muegge [61, 62], and DrugScore [63, 64] are implementations of this approach. Second, "Master equation" approaches [65] are semiquantitative estimates of the energetic contributions of various interaction types. The Williams approach [66, 67] was targeted at peptide–peptide interactions [68], while Rose devised a scoring function intended for protein targets [69, 70]. Third, regression-based methods [71, 72] take advantage of the available biological activity for training sets of ligand–receptor complexes extracted from the PDB. These include SCORE2 [72], VALIDATE [73], VALIDATEII [74], Pro_Score [75, 76], Jain [77], Horvath [46], and SCORE [78, 79]. Fourth, Poisson–Boltzman equation solvers address electrostatics and incorporate solvent effects, for example, ZAP [80]. Regression-based methods require biological input for parameterization, whereas the other three approaches do not. While perceived as a disadvantage, this type of scoring outperforms the others whenever biological results can capture target-specific information. None of the scoring schemes has been shown to be general, thus consensus scoring [81] has gained popularity. Recently, pharmacokinetic awareness [82] was also integrated into a novel scoring scheme [83] that allows for simultaneous optimization of binding affinity and pharmacokinetic properties.

Generation and Maintenance of Chemical Databases for VS

WOMBAT (WOrld of Molecular BioAcTivity) is a database that captures chemical structures and biological activities published in the medicinal chemistry literature. Its current release, WOMBAT 2003.2, contains 53,126 entries (47,872 unique structures), totaling 98,662 biological activities on 506 unique targets. Its activity list also includes 236 inactive compounds, 7982 activities "less than" and 159 activities "greater than" entries. WOMBAT 2003.2 contains 2148 different series from 2143 papers published in medicinal chemistry

TABLE 5.5 Target Class Distribution in WOMBAT 2003.2

Target Class	Compounds	Percentage
G-protein coupled receptors	19,839	37.3
Ion channels	6,090	11.5
Aspartyl proteases	2,656	5.00
Serine proteases	2,582	4.86
Kinases	2,111	3.97
Transporters	1,689	3.18
Cysteine proteases	504	0.95
Nuclear hormone receptors	448	0.84
Others	19,233	36.2

journals between 1975 and 2002 (over 90 percent coming from *Journal of Medicinal Chemistry*). The target class distribution of WOMBAT, a product from Sunset Molecular Discovery LLC (www.sunsetmolecular.com), is given in Table 5.5. A subset of 21,700 entries from WOMBAT, representing the first 40 percent of the database, was used to benchmark virtual screening tools based on FRED (Fast Rigid Exhaustive Docking), at the UNM High Performance Computing Center (HPC@UNM, www.hpc.unm.edu/).

CADB08 (Commercially Available Data Base version 0.8) contains 877,876 small, nonpeptide chemical structures. These structures, or their derivatives, are potentially available for purchase. CADB08 structures were sorted in the increasing order for rotatable bonds and molecular weight, then converted from SMILES [84] to 3D structures with Omicron [85], the OpenEye (OE) 3D-structure generator. The Omicron-generated structures were then submitted to OMEGA, the OE multiple-conformer generator, to sample conformational space for individual molecules, with an increment of 5 structures per rotor. We applied an energetic cutoff of 5 kcal to ensure that (mostly) room-temperature accessible conformers, allowing a computational error of 3 kcal, are represented for each molecule. Multiple conformers (occupying ~30 GB of disk space) are needed for docking with OE's FRED since this software is geared to rigid structures. The 1-conformer version of CADB08 was used to evaluate the performance of our virtual screening tools.

ChemBank has accumulated over 300,000 natural-product-like structures. CDLDB is the chemical collection from Chemical Diversity Labs, Inc., (CDL), from San Diego, CA (www.chemdiv.com). CDL has been marketing GPCR-focused libraries since 2001. The Prestwick Chemical Library (PCL) is an 880-compound library produced by Prestwick Chemical of Strasbourg, France. PCL contains out-of-patent drugs (85 percent) and known alkaloids (15 percent).

Virtual Screening in the Context of High-Performance Computing

CADB08 was virtually screened on Los Lobos, a nationally ranked Linux cluster with 256 dual-processor IBM Netfinity 4500 R (Intel Pentium-III 733

MHz) nodes, with Myrinet connectivity and 2 TB of storage space, from HPC@UNM. The OE software suite was ported to the MPI (message parsing interface) by Andrew Pineda (HPC@UNM), who wrote an MPI-based wrapper code in C for this task [86]. We note that virtual screening is considered an "embarassingly parallel" problem. The OE-MPI wrapper code manages output files from serial runs, summarizes the results (e.g., RMS values and scoring from docking runs), and provides a balanced workload across processors (Fig. 5.14). It handles errors (structures that cannot be docked are ignored without interrupting the virtual screening run), while making use of the high-speed communications network (Myrinet) on Los Lobos. OE-MPI is flexible, since we recently incorporated ROCS (Rapid Overlay of Chemical Structures [85]) into the VS protocol. ROCS is a Gaussian-shape volume overlap filter that limits the number of molecules, or conformers, to be docked by FRED into the binding site by first matching their 3D shape with the binding site. This can speed up the calculation by several orders of magnitude.

The OE-MPI code was benchmarked (Fig. 5.15) by Andy Pineda, Bryan Shiloff, and Tudor Oprea, against 21,700 molecules from WOMBAT, using the X-ray structures of the ligand-binding domain of the estradiol receptor α (complexed with the agonist estradiol) [87] (PDB: 1A52) and with the antagonist 4-hydroxy-tamoxifen [88] (PDB: 3ERT) using the protocol described in Fig. 5.16. SMILES input was sent to OMEGA for 1-conformer generation, then to FRED for docking and to FredA for scoring. This protocol recovered

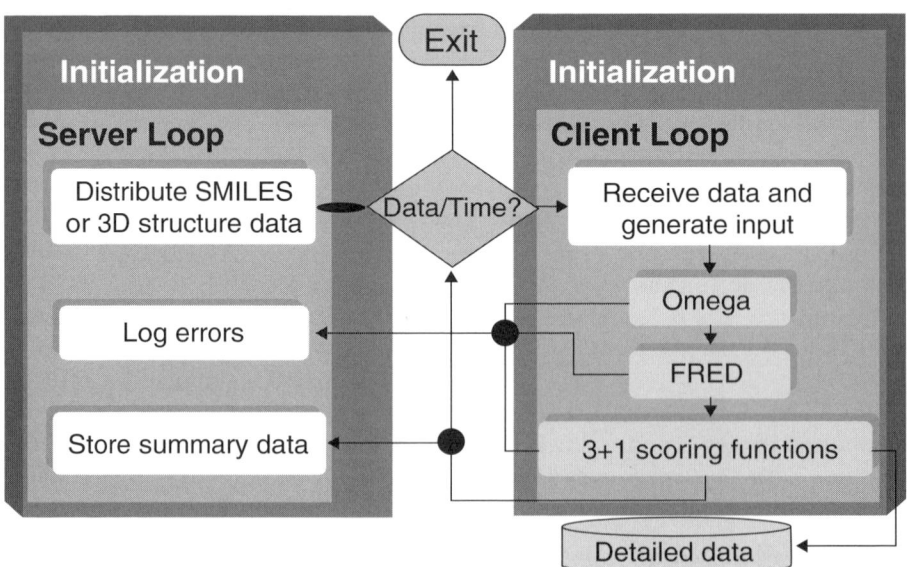

Figure 5.14 Client/server architecture for the OE-MPI code used to perform VS on Linux clusters.

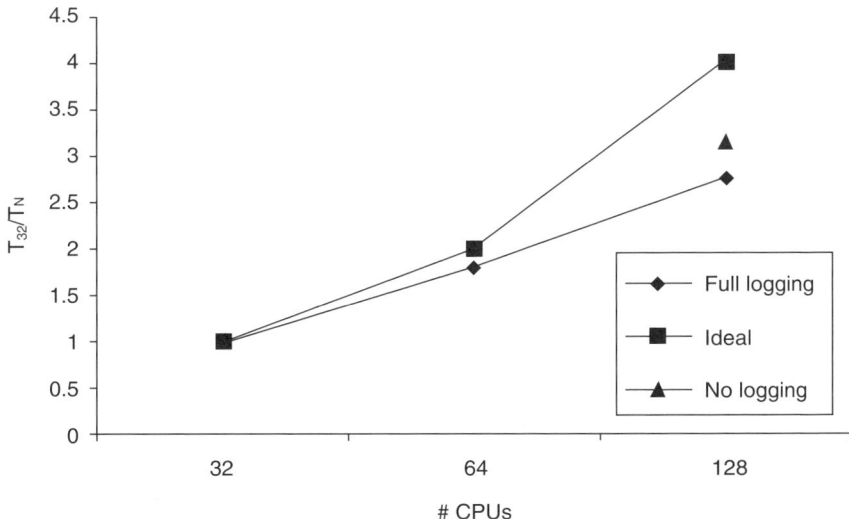

Figure 5.15 Evaluation of parallel speed increase for OE-MPI software. 21,000 SMILES from WOMBAT were used as a benchmark. The deviation from ideal is likely due to slow network I/O. Best performance is about 78 percent of ideal. 90+ percent should be attainable with I/O tuning. Best throughput was 67 SMILES per CPU/h (~207,000 molecules/day on 128 processors).

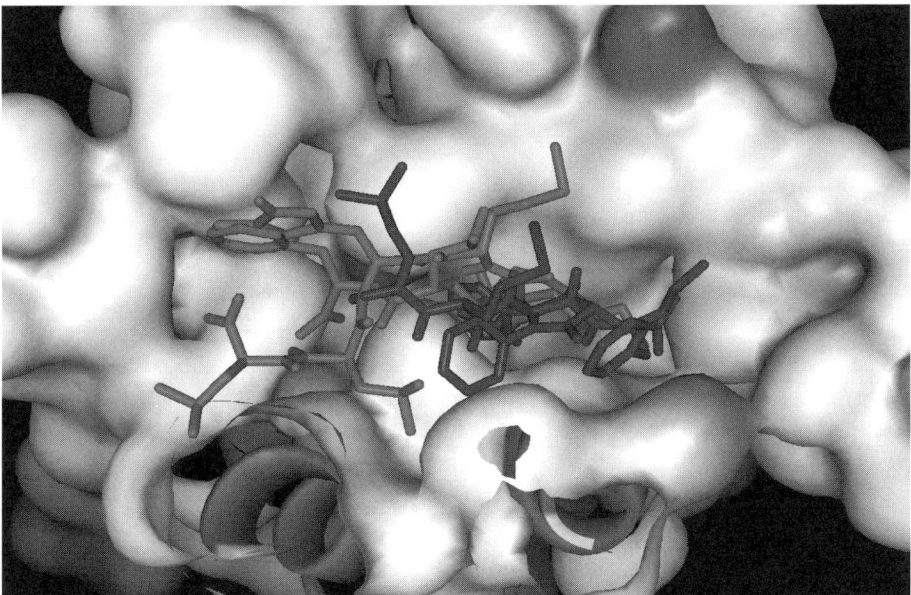

Figure 5.16 Binding modes of two formyl peptide agonists (light gray) and two formyl peptide antagonists (dark gray; cyclosporin H and phenylbutanone).

all five ER-α antagonists (nM) present in the 21,700 WOMBAT, in the top 10 FredA results, using the consensus scoring method. All 56 known ER-α actives from the 21,700 WOMBAT entries were recovered in the top 5000 molecules. This is not surprising since most of these control molecules are active only at the micromolar level. Random screening would require 21,700 experiments to recover the entire set. Screening <25 percent of the database to recover all actives is considered a good result, given the inaccuracies of the scoring functions. On the down side, FRED docked one conformer of Raloxifene (a potent ER-α antagonist) into the 1A52 structure (the agonist binding site), an artifact due to the high conformational flexibility of Raloxifene. However, Raloxifene was recovered among the top 3 hits of the 3ERT (tamoxifen, ER-α antagonist) docking.

We further evaluated the OE-MPI software on the 1-conformer version of CADB08, using the same protocol as in Fig. 5.14. The 3 scoring functions from FredA are: chemscore (regression), Gaussian shape score (steric), and screenscore (PMF); the +1 refers to the consensus score. OE-MPI generated 147,000 results per day on 198 processors (~30 molecules/CPU/h on <50 percent of the Los Lobos cluster). We were able to process 315,000 molecules in 2 consecutive runs. The difference between the benchmark results shown in Fig. 5.15 and the more realistic trials performed with CADB08 is due to increased chemical diversity and size. Given the combined size of CADB08 and ChemBank (1.2 million structures), and the simplifying assumption that each molecule has 50 conformers, we would require 408 CPU days on 198 processors per target at the rate described above. We anticipate that the introduction of the ROCS shape filter will reduce the VS effort by 50 percent or more.

FPR Case Study

Preliminary results indicate that it is possible to derive homology models for GPCRs and use these models to identify from a chemical database candidates for biological testing. In the absence of co-crystallized inhibitors bound to each target, we are incorporating additional information (e.g., mutation studies made possible by component 1) in our binding site definitions. We are using all available experimental results to further increase the probability that our models are correct and validate these models against amino acid mutation studies. Evidence from cross-linking and mutagenesis suggests that the ligand binding site is located between transmembrane (TM) helices, near the extracellular face of the membrane [89–93]. Sequence alignments of GPCRs have shown that they share a high degree of homology in the TM domains. The analysis by Baldwin on ~500 GPCR sequences identified 36 amino acids that were conserved in >70 percent of GPCRs [94, 95]. Based on these highly conserved residues, we derived several homology models of FPR TM helices and of most of the protein (1 to 337 domain) using as a template the 3D structure of bovine rhodopsin [96]. We refined these models using the experimental data from cross-linking and mutagenesis studies (not shown). The model that best agrees with cross-linking, mutagenesis, and ligand binding data, including only the 7 TM helices, was chosen for docking and pharmacophore modeling studies. Three agonists and four antagonists were docked in the FPR ligand binding site (Fig. 5.18) using AUTODOCK 3.05 [53]. The docking results suggested that antagonists have a different binding mode than agonists (Fig. 5.16), which allowed us to formulate a 3 point (one hydrogen bond, two hydrophobic) pharmacophore hypothesis specific for FPR antagonists.

Sulfinpyrazone, the PCL hit identified by HyperCyt, ranked eight in the virtual screening results (Table 5.6). Its measured affinity, 24 µM, confirms a 1991 report by Levesque et al. [97]. Our pharmacophore hypothesis, confirmed by sulfinpyrazone, formed the basis for selecting 4325 molecules from CDL collection (except for the GPCR-focused library) as an FPR-targeted subset that has, so far, yielded 68 primary hits from 9 distinct chemical series, and all but one are confirmed antagonists. Of these, 10 have measured K_i values below 10 µM. When comparing the FPR screening results of PCL, a library where no target-specific selection was exercised, with ~0.1 percent hit rate, and those of CDL-based FPR library, where a pharmacophore was used for the selection, with ~1.6 percent hit rate. The primary hits in the FPR library show consistent structure–activity relationships, which allow us to further prioritize the development of compounds with higher activity.

5.7 SUMMARY

We have developed a team for discovery research that has four highly integrated components. In the first component, expertise in cell and molecular biology and biochemistry is coupled to research into physiological pathways that regulate signal transduction and receptor processing. This expertise contributes to the expression of proteins, and the creation of cell- and particle-based assays for screening and protein–protein interactions. In the second component, expertise in biophysics, particle chemistry, fluorescence, flow cytometry, and mathematical modeling is coupled to research into ligand–receptor and protein–protein interactions. This expertise creates particle-based assays for multiplexed analysis and discovery of small molecule activities such as the discrimination of full and partial agonists. In the third component, expertise in flow cytometry hardware, software, and microfluidics allowed for the development of a novel instrument platform compatible with cell- and particle-based analysis. This expertise generated screening capabilities that have high-throughput, high-content, and real-time elements for small-molecule discovery. In the fourth component, expertise in computing and cheminformatics couple with the other three components to improve the efficiency of the small-molecule discovery process through the integration of virtual screening, chemical databases, and high-performance computing.

TABLE 5.6 Hits from Virtual Screening of the Prestwick Chemical Library for Binding to the FPR

Cpd ID	Structure	Molname	Cpd RANK	ID	Structure	Molname	RANK
278		Cinnarizine	1	312		Flunarizine	6
308		Pimozide	2	297		Benzydamine	7
292		Trazodone	3	290		Sulfinpyrazone	8
320		Fluphenazine	4	293		Glafenine	9
295		Pergolide	5	299		Mifepristone	10

REFERENCES

1. Nolan, J. P., Sklar, L. A. (2002). Suspension array technology: Evolution of the flat-array paradigm. *Trends Biotechnol.*, *20*, 9–12.
2. Luecke, H., Schobert, B., Lanyi, J. K., Spudich, E. N., Spudich, J. L. (2001). Crystal structure of sensory rhodopsin II at 2.4 angstroms: Insights into color tuning and transducer interaction. *Science*, *293*, 1499–1503.
3. Vaidehi, N., Floriano, W. B., Trabanino, R., Hall, S. E., Freddolino, P., Choi, E. J., Zamanakos, G., Goddard, W. A., 3rd (2002). Prediction of structure and function of G protein-coupled receptors. *Proc. Nat. Acad. Sci. USA*, *99*, 12622–12627.
4. Penn, R. B., Pronin, A. N., Benovic, J. L. (2000). Regulation of G protein-coupled receptor kinases. *Trends Cardiovasc. Med.*, *10*, 81–89.
5. Ferguson, S. S. (2001). Evolving concepts in G protein-coupled receptor endocytosis: The role in receptor desensitization and signaling. *Pharmacol. Rev.*, *53*, 1–24.
6. Pierce, K. L., Premont, R. T., Lefkowitz, R. J. (2002). Seven-transmembrane receptors. *Nat. Rev. Mol. Cell Biol.*, *3*, 639–650.
7. Ross, E. M., Wilkie, T. M. (2000). GTPase-activating proteins for heterotrimeric G proteins: Regulators of G protein signaling (RGS) and RGS-like proteins. *Ann. Rev. Biochem.*, *69*, 795–827.
8. Neubig, R. R. (2002). Regulators of G protein signaling (RGS proteins): Novel central nervous system drug targets. *J. Peptide Res.*, *60*, 312–316.
9. Sklar, L. A. (1987). Real-time spectroscopic analysis of ligand-receptor dynamics. *Annu. Rev. Biophys. Biophys. Chem.*, *16*, 479–506.
10. Sklar, L. A., Edwards, B. S., Graves, S. W., Nolan, J. P., Prossnitz, E. R. (2002). Flow cytometric analysis of ligand-receptor interactions and molecular assemblies. *Annu. Rev. Biophys. Biomol. Struct.*, *31*, 97–119.
11. Prossnitz, E. R., Sklar, L. A. (2004). Chemoattractant/complement receptors. In W. J. Lennarz and M. D. Lane (ed.), *Encyclopedia of Biological Chemistry*, vol. 1. pp. 425–429. Elsevier Science, San Diego.
12. Vilven, J. C., Domalewski, M., Prossnitz, E. R., Ye, R. D., Muthukumaraswamy, N., Harris, R. B., Freer, R. J., Sklar, L. A. (1998). Strategies for positioning fluorescent probes and crosslinkers on formyl peptide ligands. *J. Receptor Signal Transduct. Res.*, *18*, 187–221.
13. Vanti, W. B., Nguyen, T., Cheng, R., Lynch, K. R., George, S. R., O'Dowd, B. F. (2003). Novel human G-protein-coupled receptors. *Biochem. Biophys. Res. Commun.*, *305*, 67–71.
14. Mombaerts, P. (2001). The human repertoire of odorant receptor genes and pseudogenes. *Annu. Rev. Genom. Hum. Genet.*, *2*, 493–510.
15. Dryer, L., Berghard, A. (1999). Odorant receptors: A plethora of G-protein-coupled receptors. *Trends Pharmacol. Sci.*, *20*, 413–417.
16. Xia, C., Ma, W., Wang, F., Hua, S., Liu, M. (2001). Identification of a prostate-specific G-protein coupled receptor in prostate cancer. *Oncogene*, *20*, 5903–5907.
17. Estrada, M., Espinosa, A., Muller, M., Jaimovich, E. (2003). Testosterone stimulates intracellular calcium release and mitogen-activated protein kinases via a G protein-coupled receptor in skeletal muscle cells. *Endocrinology*, *144*, 3586–3597.

18. Ramirez, S., Aiken, C. T., Andrzejewski, B., Sklar, L. A., Edwards, B. S. (2003). High-throughput flow cytometry: Validation in microvolume bioassays. *Cytometry*, *53A*, 55–65.
19. Kuckuck, F. W., Edwards, B. S., Sklar, L. A. (2001). High throughput flow cytometry. *Cytometry*, *44*, 83–90.
20. Jackson, W. C., Kuckuck, F., Edwards, B. S., Mammoli, A., Gallegos, C. M., Lopez, G. P., Buranda, T., Sklar, L. A. (2002). Mixing small volumes for continuous high-throughput flow cytometry: Performance of a mixing Y and peristaltic sample delivery. *Cytometry*, *47*, 183–191.
21. Jackson, W. C., Bennett, T. A., Edwards, B. S., Prossnitz, E., Lopez, G. P., Sklar, L. A. (2002). Performance of in-line microfluidic mixers in laminar flow for high-throughput flow cytometry. *Biotechniques*, *33*, 220–226.
22. Truesdell, R. A., Vorobieff, P. V., Sklar, L. A., Mammoli, A. A. (2003). Mixing of a continuous flow of two fluids due to unsteady flow. *Phys. Rev. E*, *67*, 066304.
23. Bartsch, J. W., Waller, A., Mammoli, A. A., Buranda, T., Sklar, L. A., Tran, H. D., Edwards, B. S. (2004). An Investigation of liquid carry-over and sample residual for the Hypercyt® sample delivery system. *Anal. Chem.*, *76*, 3810–3817.
24. Truesdell, R., Bartsch, J., Buranda, T., Sklar, L. A., Mammoli, A. (In press). Direct measurement of mixing quality in pulsatile flow micromixer. *Fluids*.
25. Liu, R., Marik, J., Lam, K. S. (2002). A novel peptide-based encoding system for "one-bead one-compound" peptidomimetic and small molecule combinatorial libraries. *J. Am. Chem. Soc.*, *124*, 7678–7680.
26. Iannone, M. A., Consler, T. G., Pearce, K. H., Stimmel, J. B., Parks, D. J., Gray, J. G. (2001). Multiplexed molecular interactions of nuclear receptors using fluorescent microspheres. *Cytometry*, *44*, 326–337.
27. Young, S. M., Curry, M. S., Ransom, J. T., Ballesteros, J., Prossnitz, E. R., Sklar, L. A., Edwards, B. S. (2004). High throughput microfluidic mixing and multi-parametric sorting for bioactive compound screening. *J. Biomol. Screening*, *9*, 103–111.
28. Edwards, B. S., Andrzejewski, B., Sklar, L. A. (In preparation). Automation for Hypercyte®.
29. Schreiber, S. L. (2000). Target-oriented and diversity-oriented organic synthesis in drug discovery. *Science*, *287*, 1964–1969.
30. Key, T. A., Bennett, T. A., Foutz, T. D., Gurevich, V. V., Sklar, L. A., Prossnitz, E. R. (2001). Regulation of formyl peptide receptor agonist affinity by reconstitution with arrestins and heterotrimeric G proteins. *J. Biol. Chem.*, *276*, 49204–49212.
31. Bennett, T. A., Foutz, T. D., Gurevich, V. V., Sklar, L. A., Prossnitz, E. R. (2001). Partial phosphorylation of the N-formyl peptide receptor inhibits G protein association independent of arrestin binding. *J. Biol. Chem.*, *276*, 49195–49203.
32. Key, T. A., Foutz, T. D., Gurevich, V. V., Sklar, L. A., Prossnitz, E. R. (2003). N-formyl peptide receptor phosphorylation domains differentially regulate arrestin and agonist affinity. *J. Biol. Chem.*, *278*, 4041–4047.
33. Shi, M., Bennett, T. A., Cimino, D. F., Maestas, D. C., Foutz, T. D., Gurevich, V. V., Sklar, L. A., Prossnitz, E. R. (2003). Functional capabilities of an N-formyl peptide receptor-G(alpha)(i)(2) fusion protein: Assemblies with G proteins and arrestins. *Biochemistry*, *42*, 7283–7293.

34. Potter, R. M., Key, T. A., Gurevich, V. V., Sklar, L. A., Prossnitz, E. R. (2002). Arrestin variants display differential binding characteristics for the phosphorylated N-formyl peptide receptor carboxyl terminus. *J. Biol. Chem.*, *277*, 8970–8978.
35. Sklar, L. A., Vilven, J., Lynam, E., Neldon, D., Bennett, T. A., Prossnitz, E. (2000). Solubilization and display of G protein-coupled receptors on beads for real-time fluorescence and flow cytometric analysis. *Biotechniques*, *28*, 976–980, 982–985.
36. Simons, P. C., Shi, M., Foutz, T., Cimino, D. F., Lewis, J., Buranda, T., Lim, W. K., Neubig, R. R., McIntire, W. E., Garrison, J., Prossnitz, E., Sklar, L. A. (2003). Ligand-receptor-G-protein molecular assemblies on beads for mechanistic studies and screening by flow cytometry. *Mol. Pharmacol.*, *64*, 1227–1238.
37. Simons, P. C., Biggs, S. M., Waller, A., Foutz, T., Cimino, D. F., Guo, Q., Neubig, R. R., Tang, W. J., Prossnitz, E. R., Sklar, L. A. (2004). Real-time analysis of ternary complex on particles: Direct evidence for partial agonism at the agonist-receptor-G protein complex assembly step of signal transduction. *J. Biol. Chem.*, *279*, 13514–13521.
38. Waller, A., Simons, P., Prossnitz, E. R., Edwards, B. S., Sklar, L. A. (2003). High throughput screening of G-protein coupled receptors via flow cytometry. *Combinatorial Chem. High Throughput Screening*, *6*, 389–397.
39. Deffenbaugh, A. E., Scaglione, K. M., Zhang, L., Moore, J. M., Buranda, T., Sklar, L. A., Skowyra, D. (2003). Release of ubiquitin-charged Cdc34-S-Ub from the RING domain is essential for ubiquitination of the SCF(Cdc4)-bound substrate Sic1. *Cell*, *114*, 611–622.
40. Edwards, B. S., Kuckuck, F., Sklar, L. A. (1999). Plug flow cytometry: An automated coupling device for rapid sequential flow cytometric sample analysis. *Cytometry*, *37*, 156–159.
41. Edwards, B. S., Kuckuck, F. W., Prossnitz, E. R., Okun, A., Ransom, J. T., Sklar, L. A. (2001). Plug flow cytometry extends analytical capabilities in cell adhesion and receptor pharmacology. *Cytometry*, *43*, 211–216.
42. Edwards, B. S., Sklar, L. A. (2001). Plug flow cytometry. In J. P. Robinson (ed.), *Current Protocols in Cytometry*. Wiley, New York, pp. 1.17.11–11.17.10.
43. Edwards, B. S., Kuckuck, F. W., Prossnitz, E. R., Ransom, J. T., Sklar, L. A. (2001). HTPS flow cytometry: A novel platform for automated high throughput drug discovery and characterization. *J. Biomol. Screening*, *6*, 83–90.
44. Buranda, T., Lopez, G. P., Simons, P., Pastuszyn, A., Sklar, L. A. (2001). Detection of epitope-tagged proteins in flow cytometry: Fluorescence resonance energy transfer-based assays on beads with femtomole resolution. *Anal. Biochem.*, *298*, 151–162.
45. Edwards, B. S., Andrzejewski, B., Ramirez, S., Sklar, L. A. (2002). Multi-threaded integration of high throughput flow cytometry autosampling and analysis. *Cytometry, Suppl. 11*, 120.
46. Horvath, D. (1997). A virtual screening approach applied to the search for trypanothione reductase inhibitors. *J. Med. Chem.*, *40*, 2412–2423.
47. Walters, W. P., Stahl, M. T., Murcko, M. A. (1998). Virtual screening—an overview. *Drug Discov. Today*, *3*, 160–178.
48. Mestres, J. (2002). Virtual screening: A real screening complement to high-throughput screening. *Biochem. Soc. Trans.*, *30*, 797–799.

49. Wu, J. H., Batist, G., Zamir, L. O. (2001). Identification of a novel steroid derivative, NSC12983, as a paclitaxel-like tubulin assembly promoter by 3-D virtual screening. *Anticancer Drug Design*, *16*, 129–133.
50. Ewing, T. J., Makino, S., Skillman, A. G., Kuntz, I. D. (2001). DOCK 4.0: Search strategies for automated molecular docking of flexible molecule databases. *J. Comput. Aided Mol. Des.* *15*, 411–428.
51. Jones, G., Willett, P., Glen, R. C., Leach, A. R., Taylor, R. (1997). Development and validation of a genetic algorithm for flexible docking. *J. Mol. Biol.*, *267*, 727–748.
52. Kramer, B., Metz, G., Rarey, M., Lengauer, T. (1999). Ligand docking and screening with FlexX. *Med. Chem. Res.*, *9*, 463–478.
53. Osterberg, F., Morris, G. M., Sanner, M. F., Olson, A. J., Goodsell, D. S. (2002). Automated docking to multiple target structures: Incorporation of protein mobility and structural water heterogeneity in AutoDock. *Proteins*, *46*, 34–40.
54. Sippl, M. J. (1993). Boltzmann's principle, knowledge-based mean fields and protein folding. An approach to the computational determination of protein structures. *J. Comput. Aided Mol. Des.*, *7*, 473–501.
55. Sippl, M. J. (1995). Knowledge-based potentials for proteins. *Curr. Opin. Struct. Biol.*, *5*, 229–235.
56. Domingues, F. S., Koppensteiner, W. A., Jaritz, M., Prlic, A., Weichenberger, C., Wiederstein, M., Floeckner, H., Lackner, P., Sippl, M. J. (1999). Sustained performance of knowledge-based potentials in fold recognition. *Proteins, Suppl. 3*, 112–120.
57. Berman, H. M., Westbrook, J., Feng, Z., Gilliland, G., Bhat, T. N., Weissig, H., Shindyalov, I. N., Bourne, P. E. (2000). The Protein Data Bank. *Nucleic Acids Res.*, *28*, 235–242.
58. DeWitte, R. S., Shakhnovich, E. I. (1996). SMoG: De Novo design method based on simple, fast, and accurate free energy estimates. 1. Methodology and supporting evidence. *J. Am. Chem. Soc.*, *118*, 11733–11744.
59. DeWitte, R. S., Ishchenko, A. V., Shakhnovich, E. I. (1997). SMoG: De novo design method based on simple, fast, and accurate free energy estimates. 2. Case studies in molecular design. *J. Am. Chem. Soc.*, *119*, 4608–4617.
60. Ishchenko, A. V., Shakhnovich, E. I. (2002). SMall Molecule Growth 2001 (SMoG2001): An improved knowledge-based scoring function for protein-ligand interactions. *J. Med. Chem.*, *45*, 2770–2780.
61. Muegge, I., Martin, Y. C. (1999). A general and fast scoring function for protein-ligand interactions: A simplified potential approach. *J. Med. Chem.*, *42*, 791–804.
62. Muegge, I., Martin, Y. C., Hajduk, P. J., Fesik, S. W. (1999). Evaluation of PMF scoring in docking weak ligands to the FK506 binding protein. *J. Med. Chem.*, *42*, 2498–2503.
63. Gohlke, H., Hendlich, M., Klebe, G. (2000). Knowledge-based scoring function to predict protein-ligand interactions. *J. Mol. Biol.*, *295*, 337–356.
64. Sotriffer, C. A., Gohlke, H., Klebe, G. (2002). Docking into knowledge-based potential fields: A comparative evaluation of DrugScore. *J. Med. Chem.*, *45*, 1967–1970.

65. Ajay, Murcko, M. A. (1995). Computational methods to predict binding free energy in ligand-receptor complexes. *J. Med. Chem.*, *38*, 4953–4967.
66. Williams, D. H., Cox, J. P. L., Doig, A. J., Gardner, M., Gerhard, U., Kaye, P. T., Lal, A. R., Nicholls, I. A., Salter, C. J., Mitchell, R. C. (1991). Toward the semi-quantitative estimation of binding constants—guides for peptide binding in aqueous solution. *J. Am. Chem. Soc.*, *113*, 7020–7030.
67. Williams, D. H., Bardsley, B. (1999). Estimating binding constants—The hydrophobic effect and cooperativity. *Perspect. Drug Discov. Des.*, *17*, 43–59.
68. Mackay, J. P., Gerhard, U., Beauregard, D. A., Maplestone, R. A., Williams, D. H. (1994). Dissection of the contributions toward dimerization of glycopeptide antibiotics. *J. Am. Chem. Soc.*, *116*, 4573–4580.
69. Rose, P. W. (1997). Scoring methods in ligand design. In *Second UCSF Course in Computer-Aided Mol. Design*, San Francisco.
70. Marrone, T. J., Luty, B. A., Rose, P. W. (2000). Discovering high-affinity ligands from the computationally predicted structures and affinities of small molecules bound to a target: A virtual screening approach. *Perspect. Drug Discov. Des.*, *20*, 209–230.
71. Verkhivker, G., Appelt, K., Freer, S. T., Villafranca, J. E. (1995). Empirical free energy calculations of ligand-protein crystallographic complexes. I. Knowledge-based ligand-protein interaction potentials applied to the prediction of human immunodeficiency virus 1 protease binding affinity. *Protein Eng.* *8*, 677–691.
72. Bohm, H. J. (1998). Prediction of binding constants of protein ligands: A fast method for the prioritization of hits obtained from de novo design or 3D database search programs. *J. Comput. Aided Mol. Des.*, *12*, 309–323.
73. Head, R. D., Smythe, M. L., Oprea, T. I., Waller, C. L., Green, S. M., Marshall, G. R. (1996). VALIDATE: A new method for the receptor-based prediction of binding affinities of novel ligands. *J. Am. Chem. Soc.*, *118*, 3959–3969.
74. Marshall, G. R., Head, R., Ragno, R. (2000). Affinity prediction: The *sine qua non*. In E. D. Cera (Ed.), *Thermodynamics in Biology*. Oxford University Press, New York, pp. 87–111.
75. Eldridge, M. D., Murray, C. W., Auton, T. R., Paolini, G. V., Mee, R. P. (1997). Empirical scoring functions: I. The development of a fast empirical scoring function to estimate the binding affinity of ligands in receptor complexes. *J. Comput. Aided Mol. Des.*, *11*, 425–445.
76. Murray, C. W., Auton, T. R., Eldridge, M. D. (1998). Empirical scoring functions. II. The testing of an empirical scoring function for the prediction of ligand-receptor binding affinities and the use of Bayesian regression to improve the quality of the model. *J. Comput. Aided Mol. Des.*, *12*, 503–519.
77. Jain, A. N. (1996). Scoring noncovalent protein-ligand interactions: A continuous differentiable function tuned to compute binding affinities. *J. Comput. Aided Mol. Des.*, *10*, 427–440.
78. Wang, R. X., Liu, L., Lai, L. H., Tang, Y. Q. (1998). SCORE: A new empirical method for estimating the binding affinity of a protein-ligand complex. *J. Mol. Modeling*, *4*, 379–394.

79. Wang, R., Lai, L., Wang, S. (2002). Further development and validation of empirical scoring functions for structure-based binding affinity prediction. *J. Comput. Aided Mol. Des.*, *16*, 11–26.
80. Grant, J. A., Pickup, B. T., Nicholls, A. (2001). A smooth permittivity function for Poisson-Boltzmann solvation methods. *J. Computat. Chem.*, *22*, 608–640.
81. Wang, R., Wang, S. (2001). How does consensus scoring work for virtual library screening? An idealized computer experiment. *J. Chem. Inform. Comput. Sci.*, *41*, 1422–1426.
82. Oprea, T. I. (2002). Virtual screening in drug discovery: A viewpoint. *Molecules*, *7*, 51–62.
83. Zamora, I., Oprea, T., Cruciani, G., Pastor, M., Ungell, A. L. (2003). Surface descriptors for protein-ligand affinity prediction. *J. Med. Chem.*, *46*, 25–33.
84. Weininger, D. (1988). SMILES, a chemical language and information system. 1. Introduction to methodology and encoding rules. *J. Chem. Inform. Comput. Sci.*, *28*, 31–36.
85. Bostrom, J., Greenwood, J. R., Gottfries, J. (2003). Assessing the performance of OMEGA with respect to retrieving bioactive conformations. *J. Mol. Graphics Modelling*, *21*, 449–462.
86. Buranda, T., Sklar, L. A., Lopez, G., Simons, P., Huang, J., Perez-Luna, V. (2001). Fluorescence and FRET-based assays for biomolecules. Pending U.S. Patent and Registered Trademark.
87. Tanenbaum, D. M., Wang, Y., Williams, S. P., Sigler, P. B. (1998). Crystallographic comparison of the estrogen and progesterone receptor's ligand binding domains. *Proc. Nat. Acad. Sci. USA*, *95*, 5998–6003.
88. Shiau, A. K., Barstad, D., Loria, P. M., Cheng, L., Kushner, P. J., Agard, D. A., Greene, G. L. (1998). The structural basis of estrogen receptor/coactivator recognition and the antagonism of this interaction by tamoxifen. *Cell*, *95*, 927–937.
89. Quehenberger, O., Prossnitz, E. R., Cavanagh, S. L., Cochrane, C. G., Ye, R. D. (1993). Multiple domains of the N-formyl peptide receptor are required for high-affinity ligand binding. Construction and analysis of chimeric N-formyl peptide receptors. *J. Biol. Chem.*, *268*, 18167–18175.
90. Quehenberger, O., Pan, Z. K., Prossnitz, E. R., Cavanagh, S. L., Cochrane, C. G., Ye, R. D. (1997). Identification of an N-formyl peptide receptor ligand binding domain by a gain-of-function approach. *Biochem. Biophys. Res. Commun.*, *238*, 377–381.
91. Miettinen, H. M., Mills, J. S., Gripentrog, J. M., Dratz, E. A., Granger, B. L., Jesaitis, A. J. (1997). The ligand binding site of the formyl peptide receptor maps in the transmembrane region. *J. Immunol.*, *159*, 4045–4054.
92. Mills, J. S., Miettinen, H. M., Barnidge, D., Vlases, M. J., Wimer-Mackin, S., Dratz, E. A., Sunner, J., Jesaitis, A. J. (1998). Identification of a ligand binding site in the human neutrophil formyl peptide receptor using a site-specific fluorescent photoaffinity label and mass spectrometry. *J. Biol. Chem.*, *273*, 10428–10435.
93. Mills, J. S., Miettinen, H. M., Cummings, D., Jesaitis, A. J. (2000). Characterization of the binding site on the formyl peptide receptor using three receptor mutants and analogs of Met-Leu-Phe and Met-Met-Trp-Leu-Leu. *J. Biol. Chem.*, *275*, 39012–39017.

94. Baldwin, J. M. (1993). The probable arrangement of the helices in G protein-coupled receptors. *EMBO J.*, *12*, 1693–1703.
95. Baldwin, J. M., Schertler, G. F., Unger, V. M. (1997). An alpha-carbon template for the transmembrane helices in the rhodopsin family of G-protein-coupled receptors. *J. Mol. Biol.*, *272*, 144–164.
96. Palczewski, K., Kumasaka, T., Hori, T., Behnke, C. A., Motoshima, H., Fox, B. A., Le Trong, I., Teller, D. C., Okada, T., Stenkamp, R. E., Yamamoto, M., Miyano, M. (2000). Crystal structure of rhodopsin: A G protein-coupled receptor. *Science*, *289*, 739–745.
97. Levesque, L., Gaudreault, R. C., Marceau, F. (1991). The interaction of 3,5-pyrazolidinedione drugs with receptors for f-Met-Leu-Phe on human neutrophil leukocytes: A study of the structure-activity relationship. *Can. J. Physiol. Pharmacol.*, *69*, 419–425.
98. Bennett, T. A., Key, T. A., Gurevich, V. V., Neubig, R., Prossnitz, E. R., Sklar, L. A. (2001). Real-time analysis of G protein-coupled receptor reconstitution in a solubilized system. *J. Biol. Chem.*, *276*, 22453–22460.
99. Key, T. A., Vines, C. M., Wagener, B. M., Gurevich, V. V., Sklar, L. A., Prossnitz, E. R. (2000). Inhibition of chemoattractant N-formyl peptide receptor internalization by active arrestins. *Traffic.*, *2*, 87–99.
100. Biggs, S., Prossnitz, E., Sklar, L. (In press). Analysis in real-time of the assembly and disassembly of soluble G protein-coupled receptor-G protein complexes. Wiley, New York.
101. Seamer, L., Sklar, L. A. (2001). Time as a flow cytometric parameter. *Methods Cell. Biol.*, *63*, 169–183.
102. Seamer, L. C., Kuckuck, F., Sklar, L. A. (1999). Sheath fluid control to permit stable flow in rapid mix flow cytometry. *Cytometry*, *35*, 75–79.
103. Graves, S. W., Nolan, J. P., Sklar, L. A. (In press). Molecular assemblies, probes and proteomics in flow cytometry. In L. A. Sklar (Ed.), *Flow Cytometry in Biotechnology*. Oxford University Press, New York.
104. Wu, Y., Buranda, T., Lopez, G. P., Prossnitz, E. R., Sklar, L. A. (In preparation). Rapid kinetic instrumentation for GPCR assembly.
105. Graves, S., Nolan, J., Jett, J., Martin, J., Sklar, L. (2002). Nozzle design parameters and their effects on rapid sample delivery in flow cytometry. *Cytometry*, *47*, 127–137.
106. Simons, P. C., Vines, C. M., Key, T. A., Potter, R. M., Sklar, L. A., Prossnitz, E. R. (In press). Analysis of GPCR assemblies by flow cytometry. In L. A. Sklar (Ed.), *Flow Cytometry in Biotechnology*. Oxford University Press, New York.
107. Young, S. M., Curry, M. S., Ransom, J. T., Ballesteros, J. A., Prossnitz, E. R., Sklar, L. A., Edwards, B. S. (2004). High-throughput microfluidic mixing and multiparametric cell sorting for bioactive compound screening. *J. Biomol. Screening*, *9*, 103–111.
108. Sklar, L. A. (2003). High throughput screening for drug discovery by flow cytometry. In *Business Briefings: Future Drug Discovery*. World Markets Research Centre, London, p. 50.

109. Edwards, B. S., Sklar, L. A. (In press). Automation and high throughput. In L. A. Sklar (Ed.), *Flow Cytometry in Biotechnology*. Oxford University Press, New York.
110. Buranda, T., Huang, J. M., Perez-Luna, V. H., Schreyer, B., Sklar, L. A., Lopez, G. P. (2002). Biomolecular recognition on well-characterized beads packed in microfluidic channels. *Anal. Chem.*, 74, 1149–1156.
111. Piyasena, M., Buranda, T., Huang, J., Wu, Y., Sklar, L. A., and Lopez, G. P. (2004). Near simultaneous and real-time analysis of multiple analytes in affinity microcolumns using defined molecular assemblies on beads, *Anal. Chem.*, 76, 6266–6273.
112. Fu, Q., Rao, G. V. R., Ista, L. K., Xu, Sklar, L. A., Ward, T. L., Lopez, G. P. (In press). Synthesis of smart mesoporous materials. *MRS Proc.*
113. Buranda, T., Sklar, L. A., Lopez, G. P. (In press). Flow cytometry, beads and microchannels. In L. A. Sklar (Ed.), *Flow Cytometry in Biotechnology*. Oxford University Press, New York.
114. Nolan, J. P., Sklar, L. A. (1998). The emergence of flow cytometry for sensitive, real-time measurements of molecular interactions. *Nat. Biotechnol.*, 16, 633–638.

6

COMBINING NMR SPECTRAL INFORMATION WITH ASSOCIATED STRUCTURAL FEATURES TO FORM COMPUTATIONALLY NONINTENSIVE, RUGGED, AND OBJECTIVE MODELS OF BIOLOGICAL ACTIVITY

RICHARD D. BEGER, DAN A. BUZATU, AND JON G. WILKES
*National Center for Toxicological Research, Food and Drug Administration
Jefferson, Arkansas*

6.1	INTRODUCTION	228
	Nuclear Magnetic Resonance Spectral Information Content	229
	Biological Effect Modeling Based on NMR Spectra and Associated Structural Features	231
	Previous NMR Comparative Spectral Analysis Modeling Approaches	231
	Advantages of Modeling with Simulated Rather Than Experimental NMR Spectra	232
	Adding Structurally Assigned NMR Chemical Shift Information to 2D Templates	232
	Inspiration Based on Extension of Multidimensional NMR Techniques	233
6.2	METHODS AND EXAMPLES	236
	Predicted NMR Spectra	236
	Pattern Recognition Methods and Model Development	236
	Linear and Nonlinear Aspects	238
	Model Validation: LOO and LNO	238
	Procedure for Integrating NMR Spectral Information with Molecular Structure	239

Drug Discovery Handbook, by Shayne Cox Gad
Copyright © 2005 by John Wiley & Sons, Inc.

CoSCoSA Procedure for Incorporating Molecular Structure into NMR Spectra	245
Producing a ^{13}C and ^{15}N Heteronuclear Connectivity Matrix	271
6.3 FUTURE IMPROVEMENTS	278
Commercial Software Products	278
Improving the Algorithms for Binning Spectral Data	279
6.4 CONCLUSIONS	280
References	282

6.1 INTRODUCTION

Currently, quantitative structure–activity relationship (QSAR) and structure–activity relationship (SAR) models are used for drug discovery and to form ADME (absorption, distribution, metabolism, and excretion) models [1–4]. Three-dimensional (3D) QSAR models are typically based on physical fields obtained by superimposing each compound as a whole on a 3D grid; SAR modeling simplifies the process by breaking large compounds into secondary structural pieces but does not attempt to provide a physical basis for modeling the biological activity [1, 5–11]. The calculations used in building QSAR models depend on physical constants that vary in significance between structures of different types and may also exhibit nonlinear relationships for models based on a single structural type. Further, the selection of the most appropriate 3D structure for each molecule requires a number of assumptions. By trial and error, one can build a model for a finite number of molecules, adjusting conformational alignments to determine which orientations give the model the best relationship to known biological activities from a training set. However, the same process is much more prone to error when applied to structures for which the biological activity is unknown. The necessary conformational and other assumptions can give results that may be inaccurate, especially when applied for predictive purposes to structures of unknown activity.

This chapter covers an alternative strategy for model building known as quantitative spectrometric data–activity relationships (QSDARs). QSDAR models are based on the correlation between the triangular structure–spectra–activity relationship, whereas the aforementioned 3D QSAR models are based on the structure–activity relationship. 3D QSAR results show that receptor binding of a compound can be modeled successfully, if the model is carefully constructed and validated. These 3D QSAR models are based partly on electrostatics and partly on molecular geometry [5–9]. Mass spectra primarily reflect molecular geometry. Ultraviolet (UV), infrared (IR), and nuclear magnetic resonance (NMR) spectra reflect quantum mechanically constrained components that, like QSAR factors, depend on local electrostatics and geometry. UV, IR, and NMR spectra are all modulated by the principles of quantum mechanics. For the model-building purposes described here, an advantage of NMR is that spectral features (chemical shifts) can be assigned to the specific atom that produces them, whereas this is not generally possible for UV, IR, or mass spectra (MS).

Nuclear Magnetic Resonance Spectral Information Content

The ^{13}C NMR spectrum of a compound contains a pattern of frequencies that correspond directly to the quantum mechanical properties of a carbon nuclear magnetic dipole in a magnetic field. The pattern reflects the local electrostatic environment and electron orbital configuration of each carbon atom. Ab initio quantum mechanical calculations of ^{13}C chemical shift tensors in proteins reveal that they are dependent on the three-dimensional structural environment and electrostatic potential [12]. The NMR chemical shift tensor has two terms, a diamagnetic term and a paramagnetic term [13]. The diamagnetic term of an NMR chemical shift tensor is directly related to the electrostatic potential at its nucleus [13]. The paramagnetic term in the NMR chemical shift tensor is dependent upon the orbital configuration [13]. For ^{13}C NMR spectra the differences between the diamagnetic and paramagnetic term are very large so the spectral regions for different carbon orbital configurations are generally well-separated from each other.

Typically, ^{13}C NMR chemical shifts in the 0 to 100 ppm range are associated with carbon atoms that have sp^3 hybrid orbitals, with the more upfield shifts having a positive electrostatic potential (e.g., methyl groups) and the downfield shifts having a more negative electrostatic potential (e.g., ester bonds). ^{13}C NMR chemical shifts in the 100 to 220 ppm range are associated with carbon atoms that are sp^2 hybridized, with the more upfield shifts having a positive electrostatic potential (e.g., benzyl groups) and the downfield shifts having a more negative electrostatic potential (e.g., carbonyl groups). Superimposed on these basic relationships are effects arising from each molecule's carbon backbone as well as other carbon or heteroatom substituents. The effect of substituents on the ^{13}C NMR chemical shifts can be felt from as far as five bonds away or, depending on how the molecule has folded back upon itself, at even greater bond numbers directly through space. Thus, the final disposition of a molecule's NMR spectral features depends on structural and physical elements that, according to SAR theory, also determine the biological activity of the molecule. It follows that patterns can be discovered and a model can be built that correlate a finite training set of NMR spectra with the known biological characteristics of the compounds in the set. Once the patterns are validated, they can then be used for predicting the strength of the same biological activity for other compounds not used in the model-building process. This is the conceptual foundation of QSDAR.

It is possible to build spectrometric data–activity relationship (SDAR) and QSDAR models without associating the pattern of spectral features to the molecular geometry. Figure 6.1 shows the QSDAR modeling flowchart including binning, weighting, and validation steps. First, the structures of a set of compounds are generated and their ^{13}C NMR spectrum are predicted or found in a database. The set of compounds' ^{13}C NMR spectra are saved as a set of ordered pairs: chemical shift frequencies in parts per million and the area under the peak. The area under a specific chemical shift frequency is first normalized to an integer. A nondegenerate frequency is assigned an area of 100, a doubly degenerate frequency (two ^{13}C NMR chemical shifts at the same frequency) has an area of 200, and so forth. This initial normalization is done so that: (1) all the spectra have a similar signal-to-noise ratio and (2) line width variations due to differences in NMR instrumental field strengths, shimming, coupling to protons, temperature, pH, or solvent are eliminated. The bins define the number of chemical shift peaks within a ppm range. The bin width used in inputting the ^{13}C NMR spectra can be optimized by allowing the spectral window width to vary between 1.0 and 2.0 ppm and determining which width produces the best cross-validated models.

The QSDAR models can be defined using a selection of the most correlated individual bins. Alternatively, the models can use all the spectral bins but with the most significant ones more heavily weighted. A well-known method of producing optimal weighting is by calculating principal components (PCs)

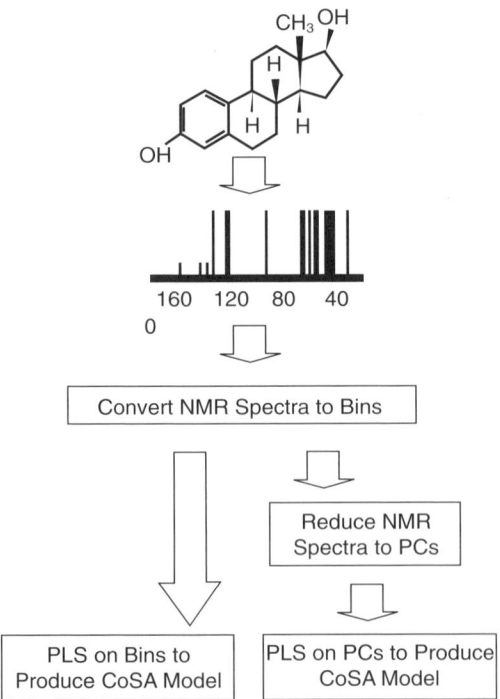

Figure 6.1 Procedural flowchart for QSDAR modeling.

of variation. Each PC contains a linear combination of the original bin intensities. The PCs are orthogonal axes or vectors ranked by rotation and translation in decreasing order of their capacity to distinguish differences among the training set spectra. Calculation of PCs produces a better quality model in cases where multiple spectral features interact to define the global SDAR.

Biological Effect Modeling Based on NMR Spectra and Associated Structural Features

The patterns in a one-dimensional SDAR or QSDAR model reflect the set of electrostatic and orbital relationships within a training set of molecules. However, the spectral features (chemical shifts) are not associated with particular atomic locations within the molecules. Since atoms in different parts of a large molecule may have similar chemical shifts and fill the same spectral bin, it follows that one-dimensional (1D) QSDAR may confound irrelevant with relevant spectral signals. An additional geometric dimension can be added to the spectral data, by indicating which chemical shifts are associated with two adjacent carbon atoms. This would produce a two-dimensional (2D) QSDAR model. Rather than a model in which bins at 220 and 50 ppm are associated with a particular biological effect, a 2D QSDAR model based on the same training set might suggest that the effect occurs only when these two chemical shifts are found on adjacent atoms. This improved basis for pattern definition is conceptually similar to 3D QSAR but with fundamental advantages. Unlike 3D QSAR, 2D QSDAR and analogous higher dimension models described later are developed without a hypothesized docking of each molecule onto a 3D grid, in relation to a known enzymatic binding site, or by comparison to the shape and size of a particular reference molecule. In 2D QSDAR, the second dimension simply correlates intramolecular spatial relations, two atoms at a time. This can be done without defining a standard molecular type with respect to which all other molecules are regarded as variants. Further, the energetically optimal conformation of each molecule interacting with the enzyme or other protein receptor mediating the biological effect does not need to be known. The subjective assumptions involved in 3D QSAR are unnecessary for 2D QSDAR. This provides a philosophically objective basis for modeling. Finally, it avoids the problematic assumption that the energy minimum conformation best represents each molecule's capacity to cause the biological effect. Below we will describe in more detail the modeling quality produced by the basic 1D QSDARs. Improvements obtained by using higher dimensions are then explained and discussed.

Previous NMR Comparative Spectral Analysis Modeling Approaches

We have used partial least-squares discriminant analysis (PLS-DA) to form a relationship between biological activity and experimental 1D ^{13}C NMR spectral data. This modeling technique is called comparative spectral analysis

(CoSA). We have developed an internally cross-validated and externally tested SDAR model of 108 diverse compounds binding to the estrogen receptor [14]. SDAR models segregated the 108 compounds into binding classifications based on relative strength with 20 strong, 15 medium, and 73 weak. Based only on ^{13}C NMR spectroscopic data, the statistical pattern recognition program with 22 Fisher-weighted PCs used 89.8 percent of the total variance and had a cross-validation of 75.0 percent. The first principal component accounted for 32.5 percent of the cumulative variance.

Advantages of Modeling with Simulated Rather Than Experimental NMR Spectra

The success of these first SDAR models based on experimental 1D ^{13}C NMR spectra led to look into the possibility of using predicted ^{13}C NMR spectra. ^{13}C NMR chemical shifts have been used to predict and refine chemical structures [15, 16]. Conversely, the chemical structure of a compound has been used to predict its ^{1}H and ^{13}C NMR chemical shifts [17–19]. The ability to use predicted over experimental NMR spectra in a QSDAR model is not necessary, but it saves time, money, and can prevent possible toxic exposures. Predicted ^{13}C NMR data points allow for the spectra to be independent of the solvent used. ACD/Labs CNMR predictor software (ACD/Labs, Toronto, Canada, www.acdlabs.com) was used to predict 1D ^{13}C NMR spectra that produced robust SDAR models of polychlorinated dibenzodioxins (PCDD) and polychlorinated dibenzofurans (PCDF) binding to aryl hydrocarbon receptor (AhR) [20] and rates of dechlorination for chlorobenzenes, chlorophenols, and chloroanilines [21].

The early success of qualitative SDAR models led us believe that QSDAR was feasible. Multiple linear regression (MLR) statistical techniques were used to form a relationship between predicted 1D ^{13}C NMR data and biological activity data. We were able to produce reliable QSDAR models of 30 steroids binding to the corticosterone binding globulin [22]. Another group of models quantified the tendency of 50 steroids to bind to the aromatase enzyme [23]. Binding to the aryl hydrocarbon receptor (AhR) was similarly modeled quantitatively for 26 PCDFs, 14 PCDDs, 12 polychlorinated biphenyls (PCBs), or for all of these compounds together [24]. These MLR 1D comparative spectral analysis (CoSA) models using simulated ^{13}C NMR data yielded higher cross-validated correlations than were seen with comparative molecular field analysis (CoMFA) methods.

Adding Structurally Assigned NMR Chemical Shift Information to 2D Templates

One limitation with 1D CoSA models is that they lack direct 3D structural information. The most obvious way to combine structural and spectral information is to establish one particular molecule as a best or normal representative of those causing a particular biological effect. Each carbon atom in this compound's backbone is numbered and all other compounds to be modeled must use the same backbone numbering system. Then, each compound's pattern is defined by the ordered pair (carbon number, chemical shift) rather than by the previously described system of (chemical shift bin number, occupancy number). This means that the chemical shifts have been assigned to the carbon atoms that produced them. The pattern as defined is correlated with the biological activity of each molecule. The resulting model combines structural information with the assigned simulated ^{13}C NMR chemical shifts. We have named this 1D SDAR method comparative structurally assigned spectra analysis (CoSASA) to distinguish it from the substantially different, unassigned methods previously described as CoSA. One supposes that the ability to include spatial relationships in SDAR modeling should improve the quality of the results. In fact, when used on the same spectral training set, we observed inferior results by CoSASA compared to CoSA [22, 23]. This unexpected result challenged our understanding of the factors affecting model quality.

Inspiration Based on Extension of Multidimensional NMR Techniques

Another way to combine structural and spectral information is to express geometrical information in spectral space. In the same way that 2D, 3D, and four-dimensional (4D) NMR experiments use additional spectral dimensions to reduce spectral overlap [25], we conceptualized an analogous way of correlating spectral and structural dimensions into a single multidimensional matrix. We hypothesized that spectra defined in such a matrix would produce improved QSDAR models. We also realized that the matrix definition would not limit compounds to those sharing the same backbone template. Compounds in the same model could have quite dissimilar structures: They could differ in the number and connectivity of carbon atoms, as well as the number and identity of constituents or atoms. Below we describe the way to define a multidimensional QSDAR data structure matrix.

A molecule's 3D connectivity matrix can be built by displaying all the possible carbon-to-carbon connectivities with their assigned carbon NMR chemical shifts on two dimensions and the distance between the two atoms in a third, orthogonal dimension. Figure 6.2 shows all the atom-to-atom connectivities and the 3D connectivity matrix for 3β-estradiol. The X axis represents the chemical shifts of carbon i, and the Y axis represents those of carbon j. The Z axis represents the distance between carbon i and carbon j (r_{ij}). By representing molecules with this matrix, the subjective superposition of molecules on a template is avoided. Other parameterization needed to develop typical 3D QSAR models is either avoided or minimized. For flexible molecules, some atom-to-atom distances can vary, so representations of multiple conformations in the 3D connectivity matrix format can be accommodated and will be explained later.

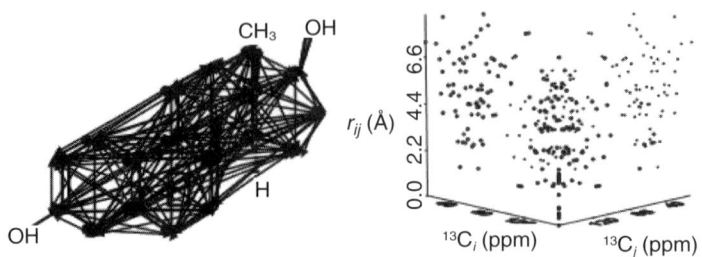

Figure 6.2 The 3D connectivity matrix of 3β-estradiol.

The connectivity matrix shown in Figure 6.2 is symmetrical about the x-y diagonal. That is, for every connection between atom i and atom j the identical relationship is represented across the diagonal at the connection between atom j and atom i. Along the X-Y diagonal at $r_{ij} = 0$ is the 1D ^{13}C NMR spectrum of 3β-estradiol. At $r_{ij} \approx 1.4$ Å are the nearest neighbor atom-to-atom connections and at $r_{ij} > 1.4$ Å are all the other distance related atom-to-atom connections.

The information in a 3D connectivity matrix determines a compound's structure, so the information in the matrix that is actually used for a model can be reduced to simplify and accelerate computations. One possible way to reduce the 3D matrix is to cut it into a set, arbitrarily, of four 2D spectral planes. The first 2D plane is the nearest neighbor through-bond connectivity plane. The three other 2D planes are constructed by binning or projecting the actual distances onto the nearest plane. This effectively compresses and greatly simplifies distance information along the Z axis. In addition to the bond-connectivity plane, there is a plane for somewhat greater atom-to-atom connections (2.0 Å $< r_{ij} < 3.6$ Å, perhaps two bond lengths but through space), another for medium range atom-to-atom connections (3.6 Å $< r_{ij} < 6.0$ Å), and a fourth for long-distance atom-to-atom connections ($r_{ij} > 6.0$ Å). Figure 6.3 shows the four 2D planes and all the atom-to-atom connections in each 2D plane for the reduced 3D connectivity matrix of 3β-estradiol. Similarities between the pattern of 2D spectral data associated with the biological activity of the training set compounds and the spectral data for the test compound are detected and used to determine whether the compound is predicted to exhibit the biological activity. We call this procedure comparative structural connectivity spectra analysis (CoSCoSA), to distinguish it from CoSA and CoSASA. We will now compare CoSCoSA modeling to actual or hypothetical NMR multidimensional analytical methods.

Standard NMR instrumental techniques include 2D ^{1}H-^{1}H correlation spectroscopy (COSY) [26] experiments in which connectivity relationships through three bonds are found for nearest neighbor protons with an off-diagonal cross peak. The ^{1}H-^{1}H COSY experiment is comparable to a 2D ^{13}C-^{13}C COSY experiment, though this experiment is not usually practical because

INTRODUCTION

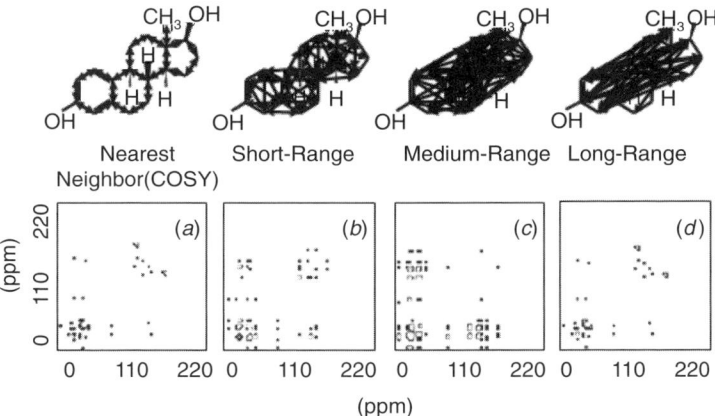

Figure 6.3 The 2D connectivity planes from the 3D connectivity matrix of 3β-estradiol.

of the low natural abundance of ^{13}C. The shortest distance layer of the connectivity matrix for ^{13}C 3D QSDAR methods described here corresponds in layout and information content seen in ^{13}C-^{13}C COSY experiments. Using the structurally assigned predicted spectra and adding the nearest neighbor information as cross peaks produces a simulated ^{13}C-^{13}C COSY layer in the matrix.

^{13}C-^{13}C COSY experiments have similarities to 2D ^{1}H-^{13}C heteronuclear single quantum correlation (HSQC) [27] and ^{1}H-^{13}C heteronuclear multiple quantum coherence (HMQC) [28] NMR experiments, both of which represent the connectivity of carbons and their attached protons. In practice, 2D ^{13}C-^{13}C COSY experiments are seldom run because small molecules are rarely fully ^{13}C labeled. Even if molecules are fully labeled with ^{13}C, through-bond connections are usually obtained directly from other NMR experiments like HCCH or indirectly from ^{1}H-^{1}H COSY with ^{13}C-^{1}H HMQC and HMBC [29].

In 3D QSDAR, 2D ^{13}C-^{13}C distance spectra that contain short-, medium-, and long-distance through-space connectivity spectral patterns are easily simulated using predicted spectra. Since the spectra are predicted based on the compound structures, the atom producing each chemical shift is known. It is then possible to estimate atom-to-atom distances (r_{ij}) for each pair of atoms and to associate the r_{ij} values with the corresponding spectral features. The shorter distance relationships in 3D QSDAR are analogous to 2D ^{1}H-^{1}H nuclear Overhauser effect spectroscopy (NOESY) NMR experiments where correlations through space are found for neighboring protons that are less than 5 Å apart. As with COSY, NOESY spectra are expressed as a matrix with off-diagonal cross peaks [30]. The volumes of the cross peaks in a NOESY experiment are dependent on the distance between the protons, the mixing time of the experiment, and the number of different nuclear Overhauser effect (NOE) spin diffusion pathways available for dipolar magnetization transfer. The volumes in a NOESY spectrum can be used to solve the three-dimensional structure of a protein by assigning short-range constraints to large volumes, medium-range constraints to medium volumes, and long-range constraints to small volumes [31]. ^{13}C-^{13}C NOESY experiments are for all practical purposes never executed, again because naturally occurring small compounds are not fully ^{13}C labeled. Since it is based on simulated spectra, the corresponding 3D QSDAR matrix is definable and its utility for modeling is practical.

6.2 METHODS AND EXAMPLES

Predicted NMR Spectra

Simulated ^{13}C NMR spectra were determined using the ACD/Labs CNMR predictor software (ACD, Toronto, Canada, www.acdlabs.com). In this and other NMR predictor programs [17–19], NMR spectra are calculated by a substructure similarity technique called hierarchically ordered spherical description of environment (HOSE) [32], which correlates similar substructures with similar NMR chemical shifts. Since some of the compounds were in the predictor software spectral database used for the HOSE calculation, their predicted spectra were equivalent to experimental NMR spectra. Since it used only simulated ^{13}C NMR spectra, QSDAR modeling was driven completely within the computer. No experimental data were acquired. The ^{13}C NMR chemical shift database used by ACD/Labs CNMR predictor contained 1.7 million chemical shifts from experimental NMR spectra of 140,000 compounds.

Pattern Recognition Methods and Model Development

A simple description of pattern recognition is the ability to discern a pattern or an analogous set of interrelationships within a set of data (i.e., to model a set of data). This basic concept underlies most scientific discoveries or theories. In a practical sense, pattern recognition is the process of creating a numerical model of a system or endpoint based on descriptors of that process or system. This turns out to be nothing more than mapping or approximating a function that describes the system. What is being modeled can vary greatly. It can be as simple as the amount of correction that needs to be applied to several mechanical parameters of a machine's arm to maintain a certain manufacturing tolerance, or a complex function such as the etiological interaction between several hundred biochemical parameters to produce a tumor.

Once it is established that a set of data contains some distinguishable pattern, a number of examples can be used to model that pattern. This is the model development component of the pattern recognition process. Examples from within data sets, that is, members of a data set containing unique descriptors, are used to establish models that describe the data sets as accurately as possible. This process is usually referred to as the training or learning stage, and it consists of training the pattern recognition method with repeated examples of input data (descriptors) and associated outcomes or endpoints. Training continues until the training set is modeled to the desired accuracy. A sufficient number of examples from a data set are required to enable the method (any pattern recognition method) to discover the pattern in the data and create an accurate model of that data. The number of examples that must be provided increases as the complexity of the pattern recognition problem modeled increases. In other words, the harder the question being asked is, the more examples that are needed to answer the question. This concept can be demonstrated with the SDAR and QSDAR techniques. For example, a simple SDAR model providing binary yes/no classification results for a biological or physiological endpoint (such as weak or strong receptor binding for a set of drug molecules) can be developed from a data set with as few as 10 examples. However, a QSDAR model of an analogous biological or physiological endpoint for a set of molecules, but producing an accurate quantitative result, usually requires at least 3 to 4 times that number of examples.

Pattern recognition model development can be broadly split into two categories, supervised methods and unsupervised methods. Supervised methods learn by experiencing multiple examples of both input stimuli and their associated outcomes. An example of this type of input–output relationship is the relation between the number of a patient's physiological parameters and his or her outcome after a medical procedure. In this example, the physiological data and the patient's outcome can be measured, and both types of data can be used to develop a model using a supervised method. Generalizations can be produced using this type of information and used to make predictions. There are many examples of supervised methods including principal components, multiple linear regression methods, and many types of artificial neural networks [33–35]. Unsupervised methods establish patterns in data differently compared to supervised methods. Input stimuli for a number of examples are provided to the method in a similar manner to the supervised method, but the outcome is not provided. Associations are established within the data between the examples based on the stimuli, and the result is a clustering of the data according to similarity. A simple example of this type of data structure is a population analysis of a community based on people's occupational parameters (education level, degree, type of job, income level, etc.). Self-organizing maps, decision trees, and k-means are some examples of frequently used unsupervised methods [33]. Since the SDAR and QSDAR methods are inherently supervised techniques, the remaining discussion will focus on these types of methods.

Linear and Nonlinear Aspects

Most pattern recognition methods fall into two main categories, linear and nonlinear methods. Linear methods use linear functions to map a relationship, and nonlinear methods use nonlinear functions to describe patterns. In performing pattern recognition and model development on a data set, it is ideal to use a method that is most aligned with the data structure (i.e., linear method to describe linear function, nonlinear method to describe nonlinear behavior in a data set). Although it is true that both types of methods can be used on both types of data structures, better generalization and predictive performance will be obtained from the nonlinear methods even when modeling data that has only slight nonlinear behavior.

Model Validation: LOO and LNO

The analysis of a developed model can be done using the leave-one-out (LOO) cross-validation procedure [1]. This procedure consists of excluding each example systematically from a training set, using the remaining examples to develop the model, and then using the model to predict the output property of the excluded example. For example, if a training set consists of a certain number of drug molecules and their respective binding affinities to a particular receptor, the LOO procedure would be used to methodically exclude one molecule and its binding affinity from the training set, and then use the developed model to predict the binding affinity of the excluded molecule. If the LOO procedure is performed n times, where n is the total number of examples in the data set, a cross-validation covariance coefficient can be calculated using the combined predicted results. The cross-validated r^2 (termed q^2) can be derived from

$$q^2 = 1 - \frac{\text{PRESS}}{\text{SD}}$$

where PRESS is the sum of the differences between the actual and predicted output data for each example during LOO cross-validation, and SD is the sum of the squared deviations between the measured and mean outputs of each example in the training set. We believe that q^2 is a more valid measure than r^2 for assessing the reliability of a mathematical model intended for predictive applications.

A modified version of the LOO cross-validation procedure is the leave-N-out (LNO) cross-validation procedure [36]. The LNO procedure is identical to the LOO procedure, but instead of removing only one example at a time from a data set for validation, N examples are removed from the training set for testing. The number of examples excluded (N) can be as many as desired, but typically N is not usually greater than 50 percent of the total data set. LNO is a more powerful procedure than LOO. It is a brute-force assessment of the

METHODS AND EXAMPLES 239

robustness and ruggedness of a model, especially when N represents a substantial percentage of the total data. It is a procedure that is not typically used, and, when used with good results, it is indicative of an exceptional predictive model.

Procedure for Integrating NMR Spectral Information with Molecular Structure

Figure 6.4 shows a flowchart of the procedures that were used to make comparative structurally assigned spectral analysis (CoSASA) models. CoSASA models are produced by using the assigned ^{13}C NMR chemical shifts at the 20 positions in the steroid backbone template as shown in Figure 6.5. CoSASA modeling is analogous to QSAR modeling in which the 3D grid used in QSAR corresponds to the CoSASA's structural template positions. Each CoSASA "bin" represents the chemical shift intensity of a particular carbon atom in the molecule. The structural information is contained in the carbon backbone numbering system and is readily available without expert interpretation of an NMR spectrum because each simulated ^{13}C NMR chemical shift is predicted based on an identifiable carbon situated in the molecular structure. For this mode of data representation, it is possible to develop two types of CoSASA QSDAR models, one based on the most correlated individual bins and another on a selection of the PCs calculated from the population levels of all the bins.

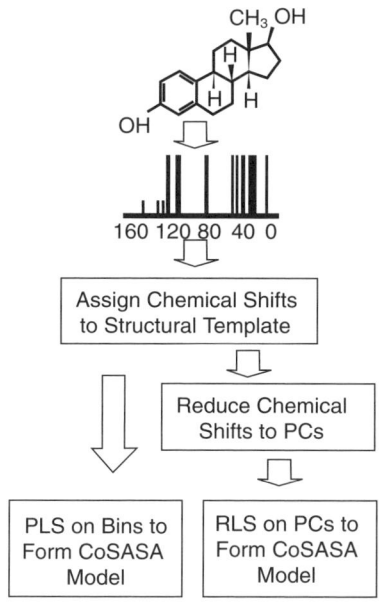

Figure 6.4 Procedural flowchart for CoSASA modeling.

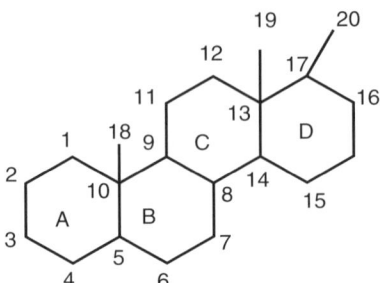

Figure 6.5 Numbering scheme of a steroid template for CoSASA modeling.

CoSASA Model of Corticosterone Binding Each of the 30 compounds specified in Table 6.1 and Figure 6.6 has known binding affinity to the corticosterone binding globulin. The assigned ^{13}C NMR spectra of the 30 steroids were placed in the steroid template positions shown in Figure 6.5 so MLR statistical analysis could be performed. The CoSASA model of corticosterone binding gave an explained variance (r^2) of 0.80 and a LOO cross-validated variance (q^2) of 0.73 [22]. The steroid backbone CoSASA model was based on the change in chemical shift frequency of atoms from positions 3, 14, and 20 on the steroid template. Positions 3 and 20 are near the steroid positions that previous QSAR models have shown are the active regions for corticosterone binding [37, 38]. The q^2 of 0.73 for the CoSASA model is slightly better than the QSAR q^2 of 0.68 but the CoSASA is much easier to build. We did not add more structurally assigned chemical shifts to the model because we wanted to limit the analysis to three variables, so that our CoSASA model would compare in this regard to the QSAR. This was a tactical choice. Without additional effort, more atoms could have been included and trends in q^2 as a function of number of bins used indicate that further improvements in predictive accuracy of the CoSASA model were possible.

CoSASA Model for the Aromatase Enzyme Each of the 50 compounds specified in Table 6.2 and Figure 6.7 have known binding affinity to the aromatase enzyme. The assigned ^{13}C NMR spectra of the 50 steroids were placed in the steroid template positions shown in Figure 6.5 so MLR statistical analysis could be performed. An aromatase CoSASA model was built by selecting the five most correlated assigned ^{13}C NMR chemical shifts [23]. The five ^{13}C NMR chemical shifts used in this model were the steroid template positions 3, 9, 6, 7, and 12. Individual ^{13}C NMR chemical shifts from atoms in the steroid template positions 3, 9, 6, and 7 all had correlations greater than 0.6 to the relative binding activity of the enzyme. The explained correlation to the binding activity data (r^2) of this model is 0.75 and the LOO cross-validated variance (q^2) is 0.66. An aromatase CoSASA model using the 5 most correlated principal components had an r^2 of 0.75 and the LOO q^2 of 0.67. In the 50-compound training set, position 3 showed the highest correlation to aromatase

TABLE 6.1 Structures of Corticosteroids Used in QSDAR Models of Corticosteroid Binding Globulin Data [22, 36, 37, 40, 51, 52]

Number	Structure[a]	R_1	R_2	R_3	R_4	R_5	R_6	R_7	R_8	R_9	R_{10}
1	SB	OH	H	H	H	OH	H				
2	SE	OH	OH	H	H			H	H	H	H
3	SC	=O	H	=O	H						H
4	SB	OH	H	H	H			H	H	H	H
5	SC	=O	OH	COCH$_2$OH	H			H	H	H	H
6	SC	=O	OH	COCH$_2$OH	OH			H	H	H	H
7	SC	=O	=O	COCH$_2$OH	OH						
8	SE	OH	=O		H						
9	SC	=O	H	COCH$_2$OH	H			H	H	H	H
10	SC	=O	H	COCH$_2$OH	OH			H	H	H	H
11	SB	=O	H	H	H	OH	H				
12	SD	OH	OH	H	OH						
13	SD	OH	OH	H	H						
14	SD	OH	=O	H	H	=O					
15	SB	H	OH	H	H						
16	SE	OH	COMe	H							
17	SE	OH	COMe	H							
18	SC	=O	H	COMe	H			H	H	H	H
19	SC	=O	H	COMe	OH			H	H	H	H
20	SC	=O	H	OH	H						
21	SF	=O	OH	COCH$_2$OH	OH						
22	SC	=O	OH	COCH$_2$OCOMe							
23	SC	=O	=O	COMe	H			H	H	H	H
24	SC	=O	H	COCH$_2$OH	H			OH	H	H	H
25[b]	SC	=O	H	OH	H			H	OH	H	H
26	SC	=O	H	COMe	OH			H	H	H	H
27	SC	=O	H	COMe	H			H	Me	H	H
28[b]	SC	=O	H	COMe	H			H	H	H	H
29	SC	=O	OH	COCH$_2$OH	OH			H	H	Me	H
30	SC	=O	OH	COCH$_2$OH	OH			H	H	Me	F

[a] Structures according to Figure 6.6.
[b] H (hydrogen) instead of Me at C_{10} steroid skeleton.

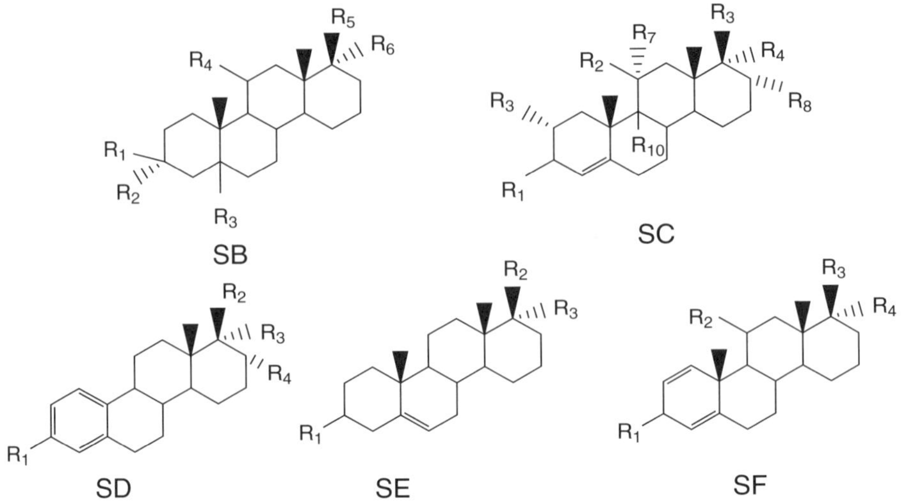

Figure 6.6 Corticosteroid binding steroids structures.

binding. The polar electronegative keto group at position 3 (compounds **20**, **23** to **25**, **34** to **50**) was easily detected and correlated to enzyme inhibition when the chemical shift of position 3 was between 198 and 200 ppm. In a previous QSAR model, a correlation was found between a polar electronegative region around positions 2 to 4 and inhibition of the enzyme [7]. The keto group at position 7 (compounds **1** to **8**, **14** to **16**) produced chemical shifts between 200 and 201 ppm and correlated to weak inhibitor activity. As before, a correlation between the chemical shift from the position 7 keto group and weak aromatase inhibition was found in the previous QSAR model [7, 23]. The weak activities of the androst-5-ene compounds were associated with chemical shifts between 117 and 127 ppm from position 6. Again, a similar correlation was found in QSAR models between all the androst-5-ene compounds and lower inhibitor activity to the enzyme [7]. In contrast, the strong correlation of the chemical shift from position 9 with aromatase inhibitor activity was unexpected based on the published QSAR model. The chemical shifts between 55 and 58 ppm in position 9 come from steroids with 6-akyl substitutions. The 6-alkyl substituted steroids are strong inhibitors of the enzyme. It is significant that a correlation around position 9 and inhibitor activity to the enzyme was not reported in the previous QSAR model, whereas the CoSASA QSDAR model based on chemical shifts from 5 atoms correctly showed a high correlation of position 9 to inhibitor activity [23]. The q^2 of 0.6 for the CoSASA model is slightly worse than the QSAR q^2 of 0.72 but, again, the spectral data model is substantially easier to build. As before, we did not include more structurally assigned chemical shifts in the model because—for purposes of comparison—we wanted to limit the CoSASA analysis to the same number of variables, 5, used in the corresponding QSAR. As with the coticosterone

METHODS AND EXAMPLES

TABLE 6.2 Structures of Steroids Used in QSDAR Models of Binding to Aromatase Enzyme [7, 23, 41]

Number	Binding Activity	Structure[a]	R_1	R_2	R_3	R_4	R_5
1	−2.92	A	CH$_2$OH	=O			
2	−3.54	A	CH$_2$OH	OH	H		
3	−3.00	A	CHO	=O			
4	−3.26	A	H	=O			
5	−2.62	A	Me	OH	H		
6	−3.06	B	CH$_2$OH	=O			
7	−2.14	B	CHO	=O			
8	−2.36	B	H	=O			
9	−1.89	D	CH$_2$OH	=O		H	
10	−2.88	D	CH$_2$OH	OH	H	H	
11	−2.03	D	CHO	=O		H	
12	−0.97	D	Me	=O		H	
13	−2.93	D	Me	=O		Br	
14	−1.28	A	Me	=O			
15	−1.23	B	Me	=O			
16	−2.61	B	Me	OH	H		
17	−2.36	D	Me	OH	H	H	
18	−0.65	F	=O				
19	−2.19	F	OH	H			
20	−1.03	H	H	H	H		
21	0.00	C	Me	=O		H	H
22	0.46	C	CH$_2$OH	=O		H	H
23	−0.84	H	CH$_2$OH	H	H		
24	0.15	H	Me	=O			
25	−0.13	E	=O		=O		CF$_2$
26	0.87	E	=O		H	H	CH$_2$
27	−0.51	E	OH	H	H	H	CH$_2$
28	−1.35	C	Me	OH	H	H	H
29	−0.67	C	CH$_2$OH	OH	H	H	H
30	−0.89	C	MeC(O)OCH$_2$	=O		H	H
31	−0.79	C	Me	=O		H	Br
32	−1.09	C	Me	=O		H	H
33	−1.08	C	CF$_3$	=O		H	H
34	0.56	I	Me				
35	0.87	J	Me				
36	1.56	I	C$_2$H$_5$				
37	0.94	J	C$_2$H$_5$				
38	0.94	I	C$_3$H$_7$				
39	0.78	J	C$_3$H$_7$				
40	0.65	I	C$_n$H$_9$				
41	0.53	J	C$_4$H$_9$				
42	0.21	I	CH(CH$_3$)$_2$				
43	0.04	J	CH(CH$_3$)$_2$				
44	−0.04	I	C$_6$H$_5$				
45	0.24	J	C$_6$H$_5$				
46	−0.24	I	CH$_2$C$_6$H$_5$				
47	0.61	J	CH$_2$C$_6$H$_5$				
48	0.91	I	CH=CH$_2$				
49	−0.32	I	C≡CH				
50	0.96	G					

[a] Structure column refer to Figure 5.7.

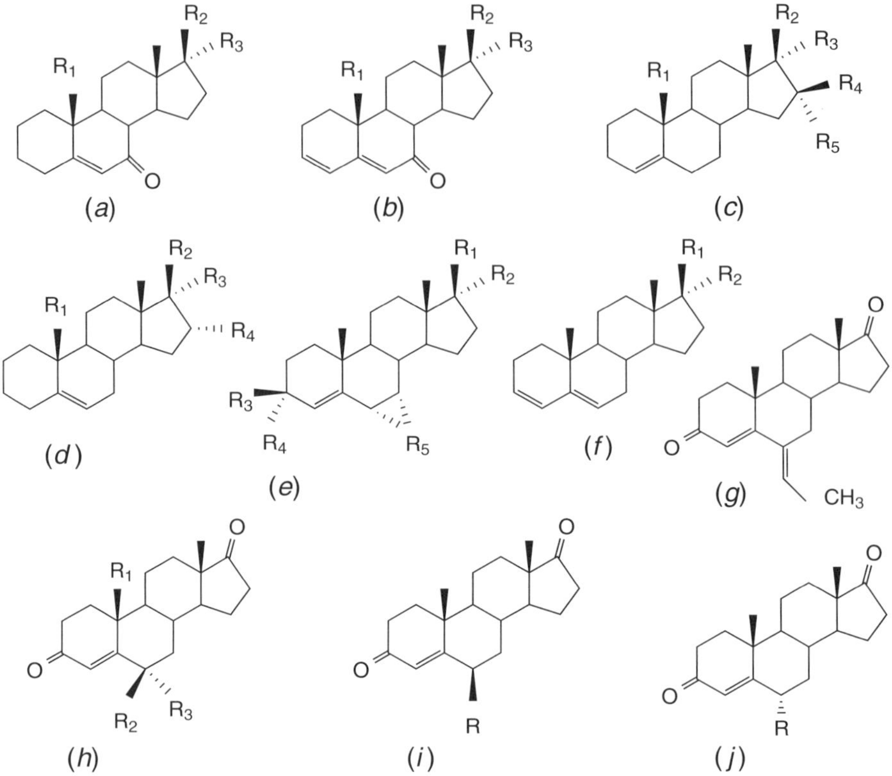

Figure 6.7 Aromatase enzyme steroid structures.

binding CoSASA, more atoms could have been added to improve model predictive accuracy.

CoSASA Model of Aryl Hydrocarbon Receptor Binding Figure 6.8a shows the backbone numbering system for 14 polychlorinated dibenzo-*p*-dioxins (PCDD), and Figure 6.8b shows the numbering system for 26 polychlorinated dibenzofurans (PCDF). The assigned ^{13}C NMR spectra of the 14 PCDD and 26 PCDF were placed in the template positions shown in Figures 6.8a and 6.8b so MLR statistical analysis could be performed. CoSASA models for PCDD and PCDF binding to the aryl hydrocarbon receptor were built from 12 assigned carbon chemical shifts as previously described [24, 38]. The 2D CoSASA model for the 26 PCDF compounds based on selection of 6 chemical shifts had an r^2 of 0.74 and a q^2 of 0.70. The q^2 of 0.70 for the CoSASA model is slightly worse than the QSAR q^2 of 0.72 but is much easier to build. The 2D CoSASA model for the 14 PCDD compounds based on selection of 5 chemical shifts had an r^2 of 0.81 and a q^2 of 0.53. The q^2 of 0.73 for the QSAR model is much better than the CoSASA q^2 of 0.53. The fact that the CoSASA

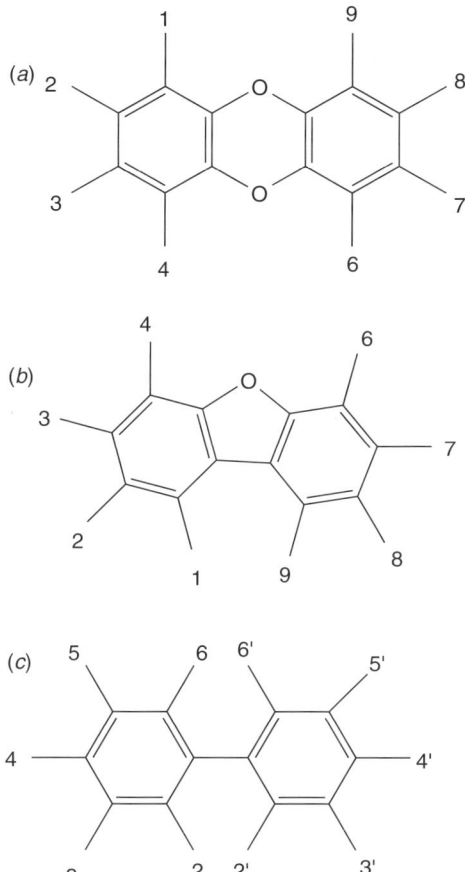

Figure 6.8 (*a*) Polychlorinated dibenzo-*p*-dioxins (PCDD). (*b*) Polychlorinated dibenzofurans (PCDF). (*c*) Poluchlorinated biphenyls (PCB).

model was far worse than the QSAR model may result from the large proportion of compounds in the training set that had 2 or 4 axes of symmetry. Symmetry information is lost when atoms included in the model are structurally described with respect to template location only.

CoSCoSA Procedure for Incorporating Molecular Structure into NMR Spectra

Figure 6.9 shows the flowchart on the comparative structural connectivity spectral analysis (CoSCoSA) modeling procedure [39–42]. The structures are used to predict the assigned 1D ^{13}C NMR spectra in this procedure. The resolution of the 2D spectra was typically reduced to 2.0 ppm in both dimensions

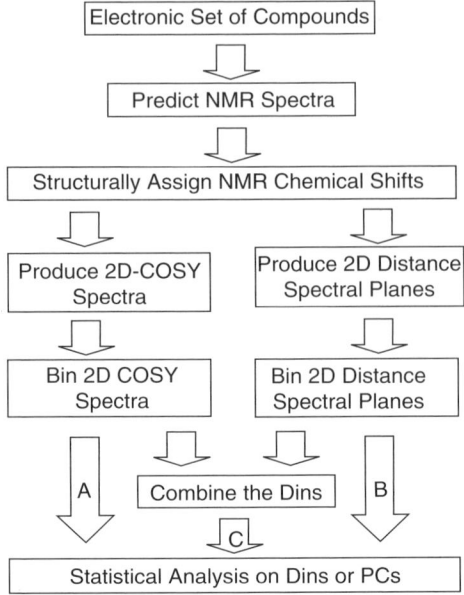

Figure 6.9 Procedural flowchart of CoSCoSA models.

to produce multiply populated bins for statistical analysis and to reduce the effects of uncertainties in the simulated spectra. The spectral widths were chosen because of convenience and prior success. The data in each 2D ^{13}C-^{13}C NMR spectral plane were saved as two-dimensional bins within a particular spectral range and normalized. A 2D bin with a single carbon-to-carbon connectivity was assigned an area of 100, and a bin with two carbon-to-carbon connections in a bin was assigned an area of 200, and so forth. This way of representing bin populations assured that all the carbon-to-carbon connections would have a similar signal-to-noise ratio.

Structural assignments from the predicted ^{13}C NMR spectra were used to produce predicted 2D ^{13}C-^{13}C COSY and theoretical 2D ^{13}C-^{13}C distance spectra. The arrows in Figure 6.10a show the through-bond neighboring carbon-to-carbon connections of a 3β-estradiol molecular backbone without any side chains. These through-bond carbon-to-carbon connections were used to simulate a 2D ^{13}C-^{13}C COSY spectrum of the steroid compounds. The arrows in Figures 6.3a to 6.3d show the short, medium, and long through space carbon-to-carbon connections. The atom-to-atom connections are broken into four 2D planes. The bin widths can be set by the model developer. In the case shown in Figure 6.3, the first plane represented nearest neighbor pairs, the second plane included short-distance atom-to-atom connections ($2.0\,\text{Å} < r_{ij} < 3.6\,\text{Å}$); the third plane, medium-distance atom-to-atom connections ($3.6\,\text{Å} < r_{ij} < 6.0\,\text{Å}$); and the fourth plane, all long-distance atom-to-atom connections ($r_{ij} > 6.0\,\text{Å}$).

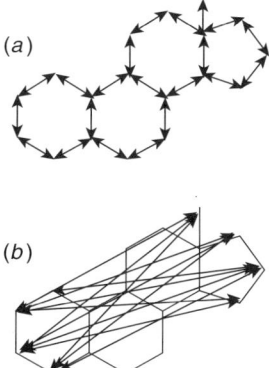

Figure 6.10 (*a*) Arrows represent the carbon-to-carbon through nearest-neighbor bond connections for the backbone of a steroid used in the predicted 2D ^{13}C-^{13}C COSY spectra. (*b*) Arrows represent the carbon-to-carbon long-range space connections used in the theoretical 2D ^{13}C-^{13}C distance spectra.

2D CoSCoSA Model of 130 Compounds Binding to the Estrogen Receptor

Table 6.3 shows the log of the relative binding activity (RBA) data for 130 structurally diverse compounds used to train 2D CoSCoSA models [38]. The data set was produced at the National Center for Toxicological Research (NCTR) using a competitive estrogen receptor (ER) binding assay with radiolabeled estradiol ([^{3}H]E$_2$) in rat uterine cytosol obtained from ovariectomized uteri of Sprague–Dawley rats [8, 43, 44]. This data set spanned 7 orders in magnitude ranging from a log(RBA) value of −4 for a compound with weak binding affinity to the estrogen receptor, to a log(RBA) of 2 for a compound with strong binding affinity to the estrogen receptor. For a particular molecule, the RBA to the estrogen receptor is defined as 100 times the ratio of the molar concentrations of 17β-estradiol and the competing compound required to decrease the amount of receptor-bound 17β-estradiol by 50 percent. Thus 17β-estradiol had an RBA of 100 and a $\log_{10}$(RBA) of 2.0.

For each of the 130 compounds, the ^{13}C 2D ^{13}C-^{13}C COSY NMR experiment was simulated using the ACD/Labs CNMR predictor software (ACD, Toronto, Canada, www.acdlabs.com). Figures 6.3*a* and 6.10*a*, shows the carbon-to-carbon nearest neighbor connections needed to produce the 2D ^{13}C-^{13}C COSY NMR spectral data to develop a model for 130 diverse compounds whose RBAs to the estrogen receptor are shown in Table 6.3 [39]. The simulated 2D ^{13}C-^{13}C COSY NMR spectra were formed by using the NMR spectral assignments obtained from predicted carbon chemical shifts to identify nearest neighboring carbon atoms and establish carbon-to-carbon through-bond connectivity spectral patterns of each compound. The 2D ^{13}C-^{13}C COSY spectra for all 130 compounds were binned into 2.0 ppm by 2.0 ppm square bins.

All statistical analysis was performed by Statistica software (Statsoft, Tulsa, OK). Because this was a very diverse data set for each CoSCoSA model, we

TABLE 6.3 130 Compounds and Training Set Predictions of Estrogen Receptor Binding [35, 39]

Compound Name	Compound Number	Exp. log(RBA)	16 Bin CoSCoSA log(RBA)	15 Bin + $L_{<7.5Å}$ CoSCoSA log(RBA)
Diethylstillbesterol	1	2.60	1.44	1.43
meso-Hexestrol	2	2.48	2.86	2.70
Ethinyl estradiol	3	2.28	1.44	1.43
4-Hydroxyestradiol	4	2.24	2.45	2.39
4-Hydroxytamoxifen	5	2.24	0.58	0.58
17β-Estradiol	6	2.00	2.36	1.85
α-Zearalenol	7	1.63	0.51	0.51
ICI182780	8	1.57	1.08	1.12
Dienestrol	9	1.57	1.44	1.43
α-Zearalanol	10	1.48	0.51	0.51
2-Hydroxyestradiol	11	1.47	1.26	1.32
Diethylstilbestrol monomethyl ether	12	1.31	1.44	1.43
3,3′-Dihydroxyhestrol	13	1.19	0.75	0.60
Droloxifene	14	1.18	1.63	1.63
Dimethylstibestrol	15	1.16	0.04	0.15
ICI164384	16	1.16	1.08	1.12
Moxestrol	17	1.14	1.44	1.43
17-Deoxyestradiol	18	1.14	0.62	0.79
2,6-Dimethylhexestrol	19	1.11	0.64	0.62
Estriol	20	0.99	0.62	0.79
Monomethyl ether hexestrol	21	0.97	0.51	0.93
Estrone	22	0.86	0.62	0.79
p-meso-Phenol	23	0.6	1.35	1.26
17α-Estradiol	24	0.49	1.17	0.79
Dihydroxymethoxychlorolefin	25	0.42	−0.10	−0.10
Mestranol	26	0.35	1.44	1.43
Zearalanone	27	0.32	0.51	0.51
Tamoxifen citrate	28	0.21	0.58	0.58
Toremifene citrate	29	0.14	0.58	0.58
α,α-Dimethylbethyl allenolic acid	30	−0.02	−0.04	−0.02
Coumestrol	31	−0.05	0.43	0.05
4-Ethyl-7-OH-(p-meoxyphenol) dihydro-1-benzopyran-2-one	32	−0.05	0.11	0.15
Nafoxidine	33	−0.14	0.58	0.58
Clomiphene citrate	34	−0.14	−0.47	−0.47
1,3,5-Estratrien-3,6α-17β-triol	35	−0.15	−0.60	−0.61
β-Zearalanol	36	−0.19	0.51	0.51
3-OH-estra-1,3,5-trien-16-one	37	−0.29	−0.08	0.25

TABLE 6.3 *Continued*

Compound Name	Compound Number	Exp. log (RBA)	16 Bin CoSCoSA log (RBA)	15 Bin + $L_{<7.5Å}$ CoSCoSA log (RBA)
3-Deoxyestradiol	38	−0.30	−1.47	−1.55
3,6,4′-Trihydroxyflavone	39	−0.35	−0.33	−0.35
Denistein	40	−0.36	−2.66	−2.61
4,4′-Dihroxystilbene	41	−0.55	−0.56	−0.51
Dihydroxymethoxychlor (HPTE)	42	−0.60	−1.47	−2.21
Monohydroxymethoxychlorolefin	43	−0.63	−0.10	−0.10
2,3,4,5-TetraCl-4′-biphenylol	44	−0.64	−1.61	−1.56
Norethynodrel	45	−0.67	−2.66	−2.61
2,2′,4,4′-Tetrahydroxybenzil	46	−0.68	−0.80	−0.81
β-Zearalenol	47	−0.69	0.51	0.51
4,6-Dihydroxyflavone	48	−0.82	−2.07	−1.95
Equol	49	−0.82	−0.66	0.05
Monohydroxymethoxychlor	50	−0.89	−2.07	−1.95
3β-Androstanediol	51	−0.92	−2.66	−2.61
Bisphenol B	52	−1.07	−2.66	−2.61
Phloretin	53	−1.16	−0.80	−0.81
Dietheylstilbestrol dimethyl ether	54	−1.25	−0.66	−0.68
2′,4,4′-Trihydroxychalcone	55	−1.26	−1.73	−1.71
2,5-Dichloro-4′-biphenylol	56	−1.44	−1.61	−1.56
4,4′-[1,2-ethanediyl)bisphenol	57	−1.44	−2.66	−2.61
17β-Estradiol-16β-OH-16-methyl-3-ether	58	−1.48	−1.34	−1.34
Aurin	59	−1.50	−0.56	−0.51
Nordihydroguariareticacid	60	−1.51	−2.66	−2.61
4-Nonylphenol	61	−1.53	−1.61	−1.56
Apigenin	62	−1.55	−2.07	−1.95
Kaempferol	63	−1.61	−2.66	−2.61
Daidzein	64	−1.65	−1.61	−1.56
3-Methylestriol	65	−1.65	−1.34	−1.34
4-Dodecylphenol	66	−1.73	−2.66	−2.61
2-Ethylhexyl-4-hydroxybenzoate	67	−1.74	−2.66	−2.61
4-*tert*-Octylphenol	68	−1.82	−2.66	−2.61
Phenolphthalein	69	−1.87	−1.47	−1.30
Kepone	70	−1.89	−2.66	−2.61
Heptyl-4-hydroxybenzoate	71	−2.09	−2.66	−2.61
Bisphenol A	72	−2.11	−2.66	−2.61
Naringenin	73	−2.13	−2.66	−2.61
4-Cl-4′-biphenylol	74	−2.18	−2.66	−2.61
3-Deoxyestrone	75	−2.2	−1.47	−1.55
4-Octylphenol	76	−2.31	−2.66	−2.61

TABLE 6.3 *Continued*

Compound Name	Compound Number	Exp. log (RBA)	16 Bin CoSCoSA log (RBA)	15 Bin + $L_{<7.5Å}$ CoSCoSA log (RBA)
Fisetin	77	−2.35	−2.14	−2.09
3′,4′,7-Trihydroxyisoflavone	78	−2.35	−2.66	−2.61
Biochanin A	79	−2.37	−2.66	−2.61
4-OH-chalcone	80	−2.43	−2.66	−2.61
4′-OH-chalcone	81	−2.43	−2.66	−2.61
2,2′-Methylenebis [4-chlorophenol)	82	−2.45	−2.07	−1.95
4,4′-Dihydroxybenzophenone	83	−2.46	−2.66	−2.61
Benzyl-4-hydroxybenzoate	84	−2.54	−2.66	−2.61
2,4-Dihyroxybenzophenone	85	−2.61	−2.66	−2.61
4′-Hydroxyflavanone	86	−2.65	−3.15	−2.96
3α-Androstanediol	87	−2.67	−2.66	−2.61
4-Phenethylphenol	88	−2.69	−2.66	−2.61
Prunetin	89	−2.74	−2.66	−2.61
Doisynoestrol	90	−2.74	−2.66	−2.61
Myricetin	91	−2.75	−2.66	−2.61
2-Cl-4-biphenylol	92	−2.77	−3.21	−2.61
Triphenylethylene	93	−2.78	−2.66	−2.61
3′-OH-flavanone	94	−2.78	−3.43	−3.31
Chalcone	95	−2.82	−2.66	−2.61
o,p′,-DDT	96	−2.85	−2.66	−2.61
4-Heptyloxyphenol	97	−2.88	−2.66	−2.61
Dihydrotestosterone	98	−2.89	−2.66	−2.61
Formononetin	99	−2.98	−2.66	−2.61
bis-[4-Hydroxyphenyl)methane	100	−3.02	−2.66	−2.61
p-Phenylphenol	101	−3.04	−2.66	−2.61
6-Hydroxyflavanone	102	−3.05	−2.14	−2.09
4,4′-Sulfonyldiphenol	103	−3.07	−1.47	−1.30
Butyl-4-hydroxybenzoate	104	−3.07	−2.66	−2.61
Diphenolic acid	105	−3.13	−2.66	−2.61
1,3-Diphenyltetramethyldisiloxane	106	−3.16	−2.66	−2.61
Propyl-4-hydroxybenzoate	107	−3.22	−2.66	−2.61
Ethyl-4-hydrobenzoate	108	−3.22	−2.66	−2.61
Phenol red	109	−3.25	−2.66	−2.61
3,3′,5,5′-TetraCl-4,4′-biphenyldiol	110	−3.25	−2.66	−2.61
4-tert-Amylphenol	111	−3.26	−2.66	−3.53
Baicalein	112	−3.35	−2.66	−2.61
Morin	113	−3.35	−2.66	−2.61
4-sec-Butylphenol	114	−3.37	−2.07	−1.95
4-Cl-3-methylphenol	115	−3.38	−2.66	−3.53
6-Hydroxyflavone	116	−3.41	−2.66	−2.61
4-Benzyloxyphenol	117	−3.44	−2.66	−2.61

METHODS AND EXAMPLES 251

TABLE 6.3 *Continued*

Compound Name	Compound Number	Exp. log (RBA)	16 Bin CoSCoSA log (RBA)	15 Bin + $L_{<7.5Å}$ CoSCoSA log (RBA)
3-Phenylphenol	118	−3.44	−2.14	−2.09
Methyl-4-hydrobenzoate	119	−3.44	−2.66	−3.53
2-*sec*-Butylphenol	120	−3.54	−3.20	−3.04
2,4′-Dichlorobiphenyl	121	−3.61	−2.66	−2.61
4-*tert*-Butylphenol	122	−3.61	−3.75	−3.53
2-Cl-4-methylphenol	123	−3.66	−2.66	−3.53
Phenolphthalin	124	−3.67	−2.66	−2.61
4-Cl-2-methylphenol	125	−3.67	−2.66	−3.53
7-Hydroxyflavanone	126	−3.73	−2.66	−2.61
3-Ethylphenol	127	−3.87	−2.90	−3.70
Rutin	128	−4.09	−2.66	−2.61
4-Ethylphenol	129	−4.17	−3.75	−3.53
4-Methylphenol	130	−4.50	−3.75	−3.53

used forward multiple linear regression (MLR) on a selected subset of spectral bins. After binning all 130 compounds, only 605 from the 7381 bins contained nonzero elements (called *hits*) in them. Of the 605 populated bins, only 337 contained more than one hit. From the remaining 337 multiply populated bins, an increasing number of the mostly highly correlated bins were selected by trial and error and used to construct MLR models until a model was obtained that had an r^2 greater than 0.8 and a F value greater than 30. We were able to identify 16 bins this way and produce a model that had an r^2 greater than 0.82, the F value was still increasing with the addition of a bin, and the p value of the bin added was significant ($p < 0.05$). We did not select any bins that had less than 2 hits. The reason for this is that a bin with one hit can inappropriately add to the r^2 of a model but cannot improve the LOO cross-validation (q_1^2). The use of a large number of very small, singly populated bins is the reason that Bursi and co-workers [45] had a high r^2 and very low q_1^2. To more rigorously test the validity of the CoSCoSA models, two L13O [10 percent of the data excluded) cross-validations were performed on each of the models. In these "leave-multiple-samples-out" experiments, the compounds omitted were varied and the results of the two corresponding experiments were averaged.

Figure 6.11*a* shows results of a CoSCoSA model based on the MLR analysis of 16 selected 2D bins from the ^{13}C-^{13}C COSY spectral data. The 16-bin COSY model for the 130 estrogenic compounds had an explained variance r^2 of 0.827, a LOO q_1^2 of 0.78, and an average L13O cross-validated variance (q_{13}^2) of 0.78 ± 0.01. The CoSCoSA model was based on COSY bins 28-12 (bin 1), 72-20 (bin 2), 54-28 (bin 3), 50-38 (bin 4), 64-56 (bin 5), 164-104 (bin

6), 152-108 (bin 7), 156-110 (bin 8), 126-112 (bin 9), 140-112 (bin 10), 142-112 (bin 11), 154-112 (bin 12), 154-114 (bin 13), 156-114 (bin 14), 128-116 (bin 15), and 126-120 (bin 16). All 2.0 ppm bins were written using the format a-b, where a and b are the ppm values corresponding to the two "connected" atoms. The MLR equation for the 16-bin CoSCoSA model is

$$\begin{aligned}\text{Predicted } \log_{10}(\text{RBA}) =\ & 0.00999 * \text{bin } 1 + 0.03173 * \text{bin } 2 + 0.0071 * \text{bin } 3 \\ & + 0.01196 * \text{bin } 4 + 0.02191 * \text{bin } 5 + 0.0093 * \text{bin } 6 \\ & + 0.2329 * \text{bin } 7 + 0.01324 * \text{bin } 8 + 0.00737 * \text{bin } 9 \\ & + 0.02558 * \text{bin } 10 + 0.0392 * \text{bin } 11 + 0.03094 * \text{bin } 12 \\ & - 0.00545 * \text{bin } 13 + 0.00526 * \text{bin } 14 + 0.00298 * \text{bin } 15 \\ & - 0.00768 * \text{bin } 16 \end{aligned}$$

All bins had more than three hits except for bins 152-108 and 140-112, each of which had only two hits. The correlation matrix for the 16 bins was calculated and only two sets of bins had a correlation between them greater than 0.5. The greatest average correlation between any bin with the other 17 bins was 0.04 and many of the average correlations were much lower than 0.04. The lack of a large correlation among bins suggests that the resulting patterns were based on essentially orthogonal data. The COSY bin 28-12 was most often associated with the CH_3 carbon connected to the CH_2 in the ethyl groups in diethylstilbestrol and hexestrol-like compounds. Twelve of the 14 compounds with a COSY hit in 28-12 had a $\log_{10}(\text{RBA})$ greater than –0.05. Compounds that populated a COSY bin at 154-112 were most often associated with the 3-carbon position connected to the 2-carbon position in the A-ring of 17β-estradiol like compounds. Nine of the 10 compounds with a COSY hit in bin 154-112 had a $\log_{10}(\text{RBA})$ greater than –0.05. Fourteen compounds had a hit in the COSY bin at 128-116. The 128-116 COSY bin was most often associated with the 2- to 3- and 5- to 6-carbon positions in a phenol ring. Twelve of the 14 compounds with a COSY hit in bin 128-116 had a $\log_{10}(\text{RBA})$ less than 0.60. The 24 compounds that had a hit or multiple hits in the COSY bin at 156-114 were most often associated with the hydroxylated carbon of a phenol ring connected to its two nearest neighboring carbons. Only 5 of the 24 compounds with a COSY hit in bin 156-114 had $\log_{10}(\text{RBA})$ less than –1.65. The 6 compounds that had a COSY bin at 64-56 were most often associated with the 2 carbons between the oxygen ester and the nitrodimethyl of tamoxifen-like compounds. Similar spectral-structural associations could be made for the other COSY bins effectively used in the CoSCoSA models.

Figure 6.11b shows results for the CoSCoSA model that was based on the MLR analysis of 15 selected 2D ^{13}C-^{13}C COSY bins plus the single distance variable. The distance variable, $L_{<7.5Å}$, was assigned a value of 1 when the maximum distance between nonhydrogen atoms in a compound was less than 7.5 Å (compact) and a value of zero for all other compounds. The $L_{<7.5Å}$ vari-

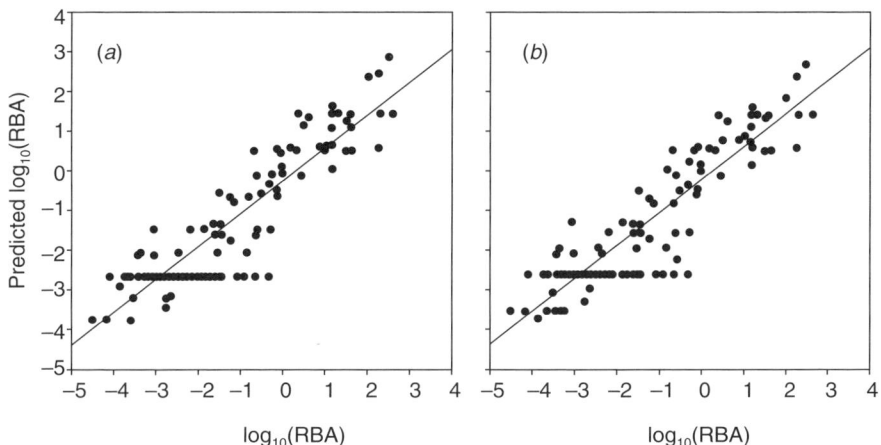

Figure 6.11 (a) 16 bin 2D-CoSCoSA model of estrogen binding. (b) 15 + L < 7.5 2D CoSCoSA model of estrogen binding.

able replaced the COSY bin at 154-114 (bin 13) in the previous 16-bin CoSCoSA model. The MLR equation for the 15-bin-with-$L_{<7.5Å}$ CoSCoSA model is

Predicted log(RBA) = 0.00969 * bin 1 + 0.03122 * bin 2 + 0.00637 * bin 3
+ 0.01066 * bin 4 + 0.02142 * bin 5 + 0.00902 * bin 6
+ 0.2263 * bin 7 + 0.01275 * bin 8 + 0.00732 * bin 9
+ 0.02507 * bin 10 + 0.03934 * bin 11 + 0.02666 * bin 12
+ 0.00526 * bin 14 + 0.00329 * bin 15 − 0.00701 * bin 16
− 0.91773 * $L_{<7.5Å}$

This 15-bin-with-$L_{<7.5Å}$ model had an r^2 of 0.83, a q_1^2 of 0.79, and an average q_{13}^2 of 0.78 ± 0.01. In this model, the $L_{<7.5Å}$ variable selected 9 compounds, all of which had a log(RBA) lower than −3.26. Smaller, compact molecules tended to bind weakly.

In Figures 6.11a and 6.11b, the line of compounds predicted to have a $\log_{10}$(RBA) of −2.60 is a set of compounds that did not have a hit in any of the 16 bins used to formulate the two CoSCoSA models. There were 56 compounds in the 16-bin CoSCoSA model that did not have any hit in one of the 16 bins. The 15-bin-with-$L_{<7.5Å}$ model had 52 compounds that did not have a hit in any of the 15 bins or $L_{<7.5Å}$. The removal of compounds with no hits from the CoSCoSA models did not change the r^2 or q^2 of the model by more than 2 percent. Three of the compounds in the 16-bin CoSCoSA model had residuals greater than 2 standard deviations (3β-androstanediol, genistein, and norethynodrel). In the 16-bin CoSCoSA model, only one other compound had a predicted residual greater than 2 standard deviations and it was 4-hydoxy-tamoxifen. The 15-bin-with-$L_{<7.5Å}$ model had 4 compounds with no hits that had residuals greater than 2 standard deviations. They are the same 3 compounds poorly modeled by the 16-bin CoSCoSA model plus 4,4′-sulfonyldiphenol. There were two compounds in the 15-bin-with-$L_{<7.5Å}$ CoSCoSA model with predicted residuals greater than 2 standard deviations, 4-hydoxy-tamoxifen and dihydroxymethoxychlor (HPTE). Almost all of the compounds with no hits in the 16 bins had experimental $\log_{10}$(RBA) lower than −1.0. The CoSCoSA models did not find a spectral relationship for these weakly binding compounds to the estrogen receptor. Most of the other bins in both CoSCoSA models were used to form a relationship between a spectral bin and binding to the estrogen receptor with a $\log_{10}$(RBA) stronger than −2.60.

To further test the ruggedness of CoSCoSA models of estrogen receptor binding, we predicted the $\log_{10}$(RBA)s of compounds from two published external data sets [46, 47]. The $\log_{10}$(RBA)s from those external data sets possessed a greater variability in binding activity. So, by a previously published method, a set of compounds that had their binding activity determined in all three labs [46, 47] was used to normalize the external data sets to the NCTR data [8, 48]. We then built the CoSCoSA models as before and used the resulting MLR equations to predict the $\log_{10}$(RBA) of the compounds in the test set. We used the published normalized $\log_{10}$(RBA) for 27 compounds from Waller and Kuiper data in our external testing of the CoSCoSA models [8]. However, many of the occupied bins for the new compounds from the external data set did not fall into the original 605 occupied bins. (The original set of bins comprised only 8.2 percent of the 2D COSY spectral plane.) We inferred that, in the different molecular context of the external data sets, NMR chemical shift information was expressed in adjacent but nonincluded bins. NMR chemical shifts exist along a continuum, and the process of binning them for this type of pattern recognition inherently compromises the pattern when it barely misses a selected bin.

To account for this source of confusion with the external data, we tried adding various fractions of "near-miss" signals into each compound's spectrum. With this in mind, we used the CoSCoSA model's MLR equation to predict the normalized log(RBA) of the compounds in the external test set. However, compounds from the external test set with bins that were one bin away (one of 8 bins surrounding a 2D bin) from the original 605 populated bins were modeled using none, one-quarter, and one-half of that bin's intensity in the nearest neighboring bin used in the original CoSCoSA model.

Table 6.4 shows the predictions for 27 compounds. The first 21 compounds show the predictions for Waller's data set [46] using both the 16-bin and 15-bin-plus-$L_{<7.5Å}$ model of estrogen binding. To make the predictions, we simulated the 2D spectra of the 21 compounds, again using ACD/Labs CNMR predictor software (ACD, Toronto, Canada, www.acdlabs.com). The simulated spectra of the test set were binned into the same 605 bins. However, many of the occupied bins for these compounds did not fall into the original 605 occu-

TABLE 6.4 External Test Set Predictions of Estrogen Receptor Binding[a]

Name	Normalized log (RBA)	16-CoSASA	15+$L_{<7.5Å}$-CoSCoSA	CoMFA
2-*tert*-Butylphenol	−4.55	−2.66	−3.53	−3.83
3-*tert*-Butylphenol	−4.82	−1.50 ± 0.64	−2.34 ± 0.66	−3.33
2,4,6,-triCl-4′-biphenylol	−0.16	−1.61	−1.56	−1.60
2-Cl-4,4′-biphenyldiol	−0.61	−1.61	−1.56	−1.49
2,6-Dichloro-4′-biphenylol	−1.11	−1.61	−1.56	−2.41
2,3,5,6,TetraCl-4,4′-biphenyldiol	−2.18	−1.61	−1.56	−0.82
2,2′,3,3′,6,6′-HexaCl-4-biphenylol	−2.74	−2.14	−2.08	−3.06
2,2′,3,4′,6,6′-HexaCl-4-biphenylol	−2.60	−1.61	−1.56	−2.48
2,2′,3,6,6′-PentaCl-4-biphenylol	−1.97	−1.61	−1.56	−3.07
2,2′5,5′-TetraCl-biphenyl	−2.67	−2.66	−2.61	−2.74
2,2′,4,4′,5,5′-HeaxCl-biphenyl	−2.83	−2.66	−2.61	−1.52
2,2′,4,4′,6,6′-HexaCl-biphenyl	−1.87	−2.66	−2.61	−1.83
2,2′,3,3′,5,5′-HeaxCl-6′-biphenylol	−2.69	−2.36	−2.29	−3.01
4′-Deoxyindenestrol	−1.37	2.89 ± 0.63	2.96 ± 0.67	−0.53
4′-Deoxyindenestrol	−0.23	2.89 ± 0.63	2.96 ± 0.67	0.111
5′-Deoxyindenestrol	−0.59	−0.61	−0.59	−1.00
5′-Deoxyindenestrol	0.35	−0.61	−0.59	−0.59
Indenestrol A (R)	1.08	3.95 ± 0.64	4.01 ± 0.67	0.29
Indenestrol A (S)	2.39	3.95 ± 0.64	4.01 ± 0.67	0.62
R 5020	−1.81	−2.45 ± 0.18	−2.48 ± 0.17	−0.70
Zearalenone	0.91	0.51	0.51	−0.12
5-Androstenediol	−0.49	−2.66	−2.61	−0.66
16a-Bromoestradiol	1.41	−0.11	0.05	0.33
16-Ketoestradiol	−0.38	−0.11	0.05	0.58
17-epi-Estriol	0.98	−0.11	0.05	−0.16
2-OH-estrone	−0.19	1.26	1.32	0.36
Raloxifene	1.34	0.17 ± 0.63	0.20 ± 0.66	−0.24

[a] The plus and minus sign reveals the variation seen when using none and one-half of a bin's intensity in a neighboring bins used to formulate the CoSCoSA model [35, 39].

pied bins (8.2 percent of the 2D COSY spectral plane). Therefore, if the simulated spectra did not fall into one of the original 605 populated bins, we put none, one-quarter, and one-half of the bin's intensity into the neighboring bin or bins used in the CoSCoSA model. We then built the CoSCoSA model as before and used its MLR equation to predict the $\log_{10}$(RBA) of the compounds in the test set. Only 6 of the 27 compounds from Waller and Kuiper's external data sets had binned COSY chemical shifts that were within one bin of those 16 bins used to formulate a CoSCoSA model. In Table 6.4, for these 6 compounds, we predicted the $\log_{10}$(RBA) using one-quarter intensity in a neighboring bin and plus or minus the deviation seen when predicting the $\log_{10}$(RBA) when using none and one-half intensity in the neighboring bin used for a CoSCoSA model. For Waller's test set and one quarter of a bin's intensity in neighboring bins, we achieved a q^2_{pred} of 0.50 for the 16-bin CoSCoSA model and a q^2_{pred} 0.57 for the 15-bin-plus-$L_{<7.5Å}$ CoSCoSA model. When using one half of a bin's intensity in a neighboring bin, we got a q^2_{pred} of 0.45 for the 16-bin CoSCoSA model and a q^2_{pred} 0.52 for the 15-bin-plus-$L_{<7.5Å}$ CoSCoSA model. Using none of a bin's intensity in a neighboring bin, we got a q^2_{pred} of 0.55 for the 16-bin CoSCoSA model and a q^2_{pred} of 0.62 for the 15-bin-plus-$L_{<7.5Å}$ CoSCoSA model. A comparative mean field analysis (CoMFA) model had a q^2_{pred} of 0.70 for Waller's normalized test set [8, 46]. When Indenstrol A (R), Indenestrol A (S), 4'deoxyindenestrol (R), and 4'-deoxyindenestrol (S) are removed from Waller's test set and none of a bin's intensity from neighboring bins, a q^2_{pred} of 0.59 for the 16-bin model and a q^2_{pred} 0.74 for the 15-bin-plus-$L_{<7.5Å}$ model are achieved. Further inspection of the predictions for Indenstrol A (R), Indenestrol A (S), 4'deoxyindenestrol (R), and 4'-deoxyindenestrol (S) revealed that one chemical shift prediction that was highly suspect (142 ppm). When we checked the structures used in the prediction of this chemical shift, all the compounds had chemical shifts from 134 to 139 ppm and not 142 ppm. On closer examination of the spectral prediction process, we found that the database structures used as the basis for prediction all had corresponding chemical shifts from 134 to 139 ppm, not 142 ppm. It appears that an error in the spectral prediction process may have contributed to poorer modeling results for these outliers.

3D-CoSCoSA Model of Corticosterone Binding A three-dimensional quantitative spectrometric data–activity relationship (3D QSDAR) model of corticosteroid binding globulin was developed by combining NMR spectral information with structural information to form a 3D connectivity matrix. The 3D connectivity matrix was built by displaying all possible carbon-to-carbon connections with their assigned carbon NMR chemical shifts and distances between the carbons. The matrix was simplified by compressing data into a 2-D ^{13}C-^{13}C COSY plane and selected theoretical 2D ^{13}C-^{13}C distance connectivity spectral slices to model binding for 30 steroids. Not all the atom-to-atom connectivity information is needed to develop a good CoSCoSA model. In the 2D CoSCoAS model of estrogen binding just presented, we used only the nearest neighbor COSY data. It is known for many steroids that the important binding sites are near the 3 and 17 positions of the molecule. Therefore, effective models can be built using only the nearest neighbor information and the long-range connections between areas surrounding positions 3 and 17 that are approximately 7.5 Å apart. To include information around positions 3 and 7, through-space carbon-to-carbon connections that were greater than 6.0 Å were included to produce a theoretical 2D ^{13}C-^{13}C distance connectivity spectrum that contains cross peaks, for atom i and atom j whenever the two carbons were greater than 6.0 Å apart. The 2D ^{13}C-^{13}C COSY and 2D ^{13}C-^{13}C distance connectivity spectra are symmetrical across the diagonal and for modeling purposes, only half of each individual spectrum contains all the necessary information. One-dimensional ^{13}C NMR spectra were not used in these CoSCoSA models because the 1D chemical shifts are highly correlated to the COSY and long-range cross peaks and do not provide additional information.

Structurally assigned ^{13}C NMR spectra of the 30 steroids in Figure 6.6 and Table 6.1 were used to produce the predicted 2D ^{13}C-^{13}C COSY and theoretical 2D ^{13}C-^{13}C distance spectra [41]. The arrows in Figure 6.3a show the through-bond neighboring carbon-to-carbon connections of a steroid backbone molecule without any side chains. These through-bond carbon-to-carbon connections were used to simulate 2D ^{13}C-^{13}C COSY spectra of the steroid compounds. The arrows in Figure 6.10b show all through-space carbon-to-carbon connections greater than 6.9 Å for β-estradiol. These through-space carbon-to-carbon connections and any other long-distance connections from side chains were used to produce the theoretical 2D ^{13}C-^{13}C distance connectivity spectra.

Four CoSCoSA models of binding to the corticosteroid binding globulin were built from the 2D COSY and 2D long-range distance spectra. One model used only the 2D COSY and the other, only the 2D long-range distance spectra. Also, COSY and distance spectra were combined in two different ways, either using the combined raw spectra (3D) or using the combined principal components (PCs) from the COSY and distance PCs calculated separately.

For the first case, 2D ^{13}C-^{13}C COSY spectra data were reduced to PCs. The PCs were used by forward multiple linear regression to produce the COSY model. Similarly, for the second approach, 2D ^{13}C-^{13}C distance connectivity data were reduced to PCs. These PCs were then used via forward multiple linear regression to produce a model of the long-distance connectivity data. The third type of model combined the raw 2D ^{13}C-^{13}C COSY and the 2D ^{13}C-^{13}C distance connectivity spectra into a single 3D data set before PC extraction [41]. These PCs were then used for forward multiple linear regression to produce the model for the three-dimensional representation of the 2D ^{13}C-^{13}C COSY and 2D ^{13}C-^{13}C distance spectra data. The fourth approach used the combined PCs from the individual 2D ^{13}C-^{13}C COSY and the 2D ^{13}C-^{13}C distance connectivity models to produce a combined through-bond and through-space CoSCoSA model.

The PCs for all CoSCoSA model were produced by evaluating the connectivity bins with forward multiple linear regression analysis and using only the most correlated PCs. The F test for many of the models continued to rise as more PCs were included in modeling. This process can lead to overfitting of the model. To avoid overfitting, we limited the number of PCs to either three or eight. Three reflected the number of independent components used in previous QSAR models for this endpoint. Eight reflected optimal results for this type of data.

Table 6.5 contains a comparison of the model performance parameters (n, r^2, q^2, and number of components) for corticosterone binding models based on QSAR [37], HE-state/E-state [52], E-state [52], SOM [53], combination QSAR/E-state [38], 2D CoSASA [22], 1D CoSA [22], the four multiple linear regression CoSCoSA models, and a CoSCoSA PD-ANN model validated two different ways (explained below). All four CoSCoSA models with 8 PCs have a strong correlation (r^2) and cross-validated variance (q^2) and are favorable when compared to the previous published models of binding to the corticosteroid binding globulin.

Figure 6.12a is a plot of the predicted binding versus experimental binding for the CoSCoSA 2.0-ppm resolution model based on ^{13}C-^{13}C COSY data. A model based on 8 PCs had an explained correlation (r^2) of 0.93 and a cross-validated variance (q^2) of 0.88, which indicates self-consistency and excellent predictive capability. Figure 6.12b is a plot of the predicted binding versus experimental binding for the CoSCoSA 2.0-ppm resolution model based on ^{13}C-^{13}C distance greater than 6.9 Å. Using 8 PCs, the r^2 of this model is 0.89 and the q^2 is 0.79. Figure 6.12c is a plot of the predicted binding versus experimental binding for the CoSCoSA 2.0-ppm resolution model based on the combined ^{13}C-^{13}C COSY and ^{13}C-^{13}C distance connectivity PCs. The r^2 of this model is 0.96 and the q^2 of this model is 0.92, which indicates excellent predictive capability. Finally Figure 6.12d is a plot of the predicted binding

TABLE 6.5 Corticosterone Model Performance Parameters n, r^2, q^2 and Number of Components

Model	n	r^2	q^2	Components
QSAR	31	0.72	0.68[a]	3 (PCs)
HE state/E-state	31	0.98[a]/0.96[b]	0.80[a]/0.76[b]	3[a] (PCs)/5[b] (PCs)
E-state	31	0.96[a]/0.96[b]	0.79[a]/0.67[b]	3[a](PCs)/4[b] (PCs)
SOM	31	0.85	—	3 (PCs)
QSAR/E-state	30	0.82	0.78	3 (atoms)
2D CoSASA	30	0.80	0.73	3 (atoms)
1D CoSA	30	0.80	0.78	3 (bins)
CoSCoSA COSY	30	0.84/0.93	0.74/0.88	3 (PCs)/8 (PCs)
CoSCoSA distance	30	0.66/0.89	0.38/0.79	3 (PCs)/8 (PCs)
CoSCoSA COSY + distance	30	0.84/0.96	0.74/0.92	3 (PCs)/8 (PCs)
3D[c] CoSCoSA	30	0.78/0.92	0.68/0.81	3 (PCs)/8 (PCs)
PD-ANN (3D[c]) CoSCoSA	30	0.96	0.78[d]	593:198:1
PD-ANN (3D[c]) CoSCoSA	30	0.96	0.73[e]	593:198:1

[a] 1.0-Å models.
[b] 2.0-Å models.
[c] 3D is the combination of COSY and distance data before PC extraction.
[d] A L3O q_3^2.
[e] A L10O q_{10}^2 [22, 36, 37, 40, 51, 52].

METHODS AND EXAMPLES

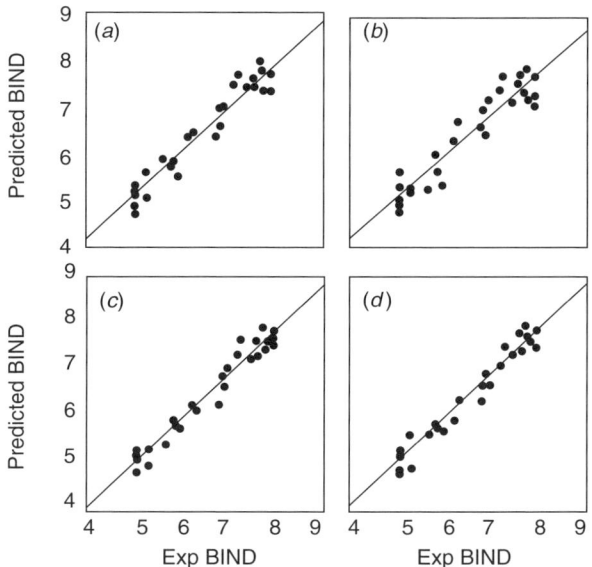

Figure 6.12 Plot of the predicted binding versus experimental binding for the corticosterone globulin. (a) The ^{13}C-^{13}C COSY data, (b) ^{13}C-^{13}C distance connectivity data greater than 6.9 Å, (c) the combined ^{13}C-^{13}C COSY and ^{13}C-^{13}C distance connectivity PCs, (d) the combined ^{13}C-^{13}C COSY and ^{13}C-^{13}C distance connectivity data before PCs are extracted from the combined data.

versus experimental binding for the CoSCoSA 2.0-ppm resolution model based on the combined ^{13}C-^{13}C COSY and ^{13}C-^{13}C distance connectivity spectral data before principal component extraction. The explained correlation (r^2) of this model is 0.92 and the cross-validated variance (q^2) of this model is 0.81.

Artificial Neural Network Model of 3D CoSCoSA Model of Corticosterone Binding Another model for predicting steroid binding activity to the corticosteroid binding globulin was developed using a parallel distributed-artificial neural network (PD-ANN). The PD-ANN is an algorithm that allows for simultaneous training and testing of multiple neural networks. The drawback to using an ANN is the time required to find an optimal network configuration. Traditionally this search has been performed in serial fashion, one configuration at a time, on one computer at a time. The PD-ANN takes advantage of an Internet-based parallel distribution scheme to perform the optimization task on multiple machines simultaneously. The result is a dramatic decrease in the time required for finding an optimal network configuration. For this study, the ANN consisted of an error back propagating, feed-forward Java code. The back-propagation algorithm in the PD-ANN was based on Rummelhart, Hinton, and Williams' generalized delta rule [49]. It was developed "in-house" and parallel distributed using JGravity, a Java-based parallel distribution platform [50].

More than 100 varying network configurations were evaluated over a couple of days using four high-speed Pentium personal computers. Two types of cross-validations were performed. In one case, each distributed neural network removed three examples and trained on the rest, rotating through the rest of the data on subsequent training sessions, and thus performing a comprehensive series of L3O cross-validations. In another case, 10 examples were left out, performing a L10O cross-validation.

The configuration consisting of 593 input, 198 hidden, and 1 output nodes provided the best validated results. This roughly corresponds to a hidden layer that is equal to $1/3n$, where n = number of input nodes. This is a common optimal input-hidden node relationship for three-layer, feed-forward neural networks [51]. The best results were obtained when the network was trained for 3400 cycles, with a learning rate of 0.1, and using a sigmoid transfer function. A sigmoid transfer function in the back-propagation algorithm required that the data (input and output) be scaled within a range between 0 and 1. The raw data for this model was actually scaled from 0.1 to 0.9.

All of the spectral bins based on the combination of ^{13}C-^{13}C COSY and ^{13}C-^{13}C distance spectra were used to develop the CoSCoSA PD-ANN model. The explained correlation (r^2) of this model is 0.96. Two cross-validated variance coefficients were calculated for this model. The leave-3-out (q_3^2) procedure yielded a value of 0.78, and the leave-10-out (q_{10}^2) procedure yielded a value of 0.73. These numbers illustrate a high degree of consistency and predictive capability for the PD-ANN model. All four CoSCoSA models based on 8 PCs have a q^2 greater than the 0.68 seen for the QSAR model. Three of the 4 CoSCoSA models based on 3 PCs have a q^2 greater than the 0.68 seen for the QSAR model. The only CoSCoSA model that did not have a q^2 greater than 0.68 was the ^{13}C-^{13}C distance connectivity model based on only 3 PCs. The HE-state and E-state models have a greater r^2 than all the QSDAR models, but these models are very computationally intensive with many distance formulas used for every point in the grid. Still, all the 2.0-ppm resolution CoSCoSA models based on 8 PCs have an r^2 greater than 0.89 and a q^2 greater than 0.79. All the CoSCoSA models with 8 PCs have a predictability that is much better or comparable to the predictability for QSAR, CoSA, CoSASA, HE-state/E-state, and E-state models. Our earlier CoSA QSDAR model started with only 256 spectral bins, a number subsequently reduced to 94 when all columns were removed that consistently had either zero population or only one nonzero entry. An analogous adjustment was made for these 3D models. The 2.0-ppm CoSCoSA models started with 6441 bins, a number then reduced to 271 for the ^{13}C-^{13}C COSY data and 322 for the ^{13}C-^{13}C distance connectivity data when all the columns with only zeroes were removed. The models used less than 5 percent of the available 2D connectivity spectral space.

A set of rules was used to combine all bins with only one hit into a populated nearest neighbor bin so that information content from their occupancy could be retained in the model. When several bins with multiple hits were equally close to the bin with one hit, the one-hit bin was added to the bin with the least number of hits. When all the bins with one hit were treated this way, the remaining bins for 2.0-ppm ^{13}C-^{13}C COSY modeling were reduced from 271 to 178. Similarly, the number of bins used for 2.0-ppm ^{13}C-^{13}C distance connectivity modeling was reduced from 322 to 194. Using the new data with treated one-hit bins, the resulting r^2 for the ^{13}C-^{13}C COSY model increased from 0.93 to 0.94 and q^2 increased from 0.88 to 0.89. The treated ^{13}C-^{13}C distance connectivity data produced a model where the r^2 increased from 0.89 to 0.91 and q^2 from 0.79 to 0.81. Using both the treated ^{13}C-^{13}C COSY and ^{13}C-^{13}C distance connectivity PCs for the type-four models, the r^2 decreased from 0.96 to 0.95 and q^2 from 0.92 to 0.90. In contrast, the adjusted spectra, when ^{13}C-^{13}C COSY and ^{13}C-^{13}C distance connectivity raw data were combined before the extraction of PCs, caused the r^2 to increase from 0.92 to 0.93 and q^2 increased from 0.81 to 0.84. The type-four (PC) approach to spectral combination gave superior results, but the effect of retaining information from the singly populated bins improved type-three (3D) results.

It is interesting to note that a strong predictive model was produced by the PD-ANN using the combined COSY and distance ^{13}C spectral data. There was no preprocessing (such as principal components or other types of regression analysis to extract features or bin population adjustments) performed on the data prior to its use. The model used the raw spectral information. The strength of the model is clearly demonstrated with the average q_{10}^2 of 0.73. The total data set consisted of only 30 compounds. Thus each time 10 compounds were left out, the data set use to build the model was depleted by 33 percent.

3D CoSCoSA Model of Binding to the Aromatase Enzyme The assigned ^{13}C NMR spectra of 50 steroids shown in Figure 6.7 and Table 6.2 were used to develop four types of CoSCoSA models of binding to the aromatase enzyme [42]. These four model types paralleled the four CoSCoSA models of the corticosteroid binding globulin: one model using nearest neighbor 2D COSY connectivities, a second model using 2D long-range distance connectivities only, a third model using COSY and long-distance spectra combined using the combined PCs derived by independent PC analysis of the two distance ranges, and the fourth model (3D) using the combined raw spectra as one might expect for a direct 3D representation of each molecule. Methods of statistical analyses also paralleled the earlier work. In this case, each of the four types of CoSCoSA model was built using the five most correlated PCs and, in a variant procedure, rebuilt for comparison using the number of most correlated PCs that produced a maximum result for the F value.

Table 6.6 contains a comparison of the model performance parameters (n, r^2, q^2, and number of components) used for QSAR [7], 1D CoSA [23], 2D CoSASA [23], and CoSCoSA models [42]. All four CoSCoSA models with PCs have a strong correlation (r^2) and cross-validated variance (q^2) and are favorable when compared to the previously published models of binding to

TABLE 6.6 Aromatase Model Performance Parameters n, r^2, and q^2 [7, 23, 41]

Model	Number of PCs	r^2	q^2
CoMFA	5	0.94	0.72
1D CoSA	5	0.78	0.71
1D CoSA	5 bins	0.82	0.77
2D CoSASA	5	0.75	0.67
2D CoSASA	5 atoms	0.74	0.66
2D CoSCoSA COSY	5	0.77	0.68
2D CoSCoSA COSY	9	0.89	0.89
2D 6-9 Å Distance	5	0.65	0.65
2D CoSCoSA 6-9 Å Distance	7	0.72	0.72
2D CoSCoSA COSY + Distance	5	0.77	0.68
2D CoSCoSA COSY + Distance	10	0.92	0.86
3D CoSCoSA	5	0.77	0.77
3D CoSCoSA	8	0.87	0.83

the aromatase enzyme. To further validate these results, the statistical process was tested by randomizing the binding activity data to see whether the CoSCosCA models could develop an invalid association based on meaningless spectral data–aromatase binding correlations. As hoped, excellent statistical correlation occured using actual binding data while the random data produced models having very low to zero correlations.

Figure 6.13a is a plot of the predicted binding versus experimental binding for the 2.0-ppm resolution CoSCoSA model based on ^{13}C-^{13}C COSY data and 9 PCs. The COSY model was the best model, and using 9 PCs it had an r^2 of 0.89 and a q^2 of 0.89, which indicates self-consistency and excellent predictive capability. Figure 6.13b is a plot of the predicted binding versus experimental binding for the 2.0-ppm resolution CoSCoSA model based on long-distance spectral data. The r^2 of this model was 0.72 and the q^2 was 0.72. Figure 6.13c is a plot of the predicted versus experimental binding for the 2.0 ppm resolution (PC) CoSCoSA model based on the combined ^{13}C-^{13}C COSY and ^{13}C-^{13}C distance connectivity PCs. The model used 10 PCs and had an r^2 of 0.92 and a q^2 of 0.86. Figure 6.13d is a plot of the predicted binding versus experimental binding for the 2.0-ppm resolution (3D) CoSCoSA model for ^{13}C-^{13}C COSY and ^{13}C-^{13}C distance connectivity spectral data when the two were combined before principal component extraction. The 3D model was based on 8 PCs and had an r^2 of 0.87 and a LOO cross-validated q^2 of 0.83 [42]. As was found for corticosteroid binding, the PC method of representing distance/chemical shift data associations produces better CoSCoSA models than the 3D approach.

All four CoSCoSA models of aromatase enzyme binding have a q^2 greater than the 0.72 seen for the QSAR model [7]. We believe it is valid to compare our models based on more than five PCs to QSAR models based on five components. The CoSCoSA models use spectral data in a "digital" or quantized

METHODS AND EXAMPLES 263

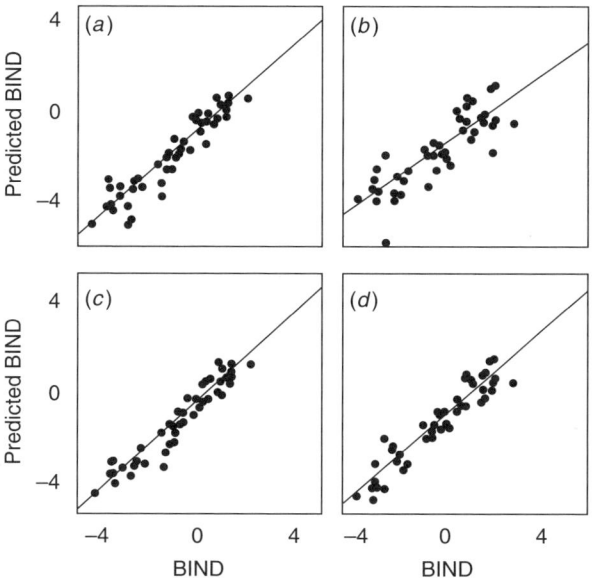

Figure 6.13 Plot of the predicted binding versus experimental binding for the aromatase enzyme. (a) The ^{13}C-^{13}C COSY data, (b) all the ^{13}C-^{13}C distance connectivity data greater than 6.9 Å, (c) the combined ^{13}C-^{13}C COSY and ^{13}C-^{13}C distance connectivity PCs, (d) the combined ^{13}C-^{13}C COSY and ^{13}C-^{13}C distance connectivity data before PCs are extracted from the combined data.

fashion, whereas QSAR models are based on "analog" or continuum format data. Our CoSCoSA models and QSAR models are based on similar physical phenomena. Both model types have access to electrostatic molecular features. The dynamic range of data used to build the modeling association differs significantly between these two approaches. However, under cross-validation it becomes clear that the CoSCoSA models have a better signal to noise (predictability) when more components are used. The use of an optimal rather than identical number of PCs for comparisons compensates for the data structure difference.

Another significant advantage of QSDAR compared to QSAR models is that the former require fewer assumptions to produce a model. During the calculation of electrostatic potentials in 3D QSAR modeling, assumptions are made for solvent effects, partial charges, and dielectrics. Further, only one structural conformation, typically the gas-phase minimum energy conformation, is assumed to represent each molecule to establish its tendencies with respect to the structure–activity relationship. Each of the QSAR assumptions and approximations are prone to produce significant error. Three-dimensional QSDAR modeling does not rely on those assumptions. ^{13}C NMR spectral data inherently reflect a Boltzmann's distribution of structural conformations to produce a "quantum mechanical energy" that represents the average structural local environment for every carbon atom in the molecule. This average quantum mechanical environment is obtained for a molecule in solution and not derived from gas-phase electrostatic interactions. These facts help explain the superior results of 3D QSDAR models in comparison to 3D QSAR.

Our simpler CoSA aromatase binding model started with 256 spectral bins, which were reduced to 87 when all the columns with only zeroes or with only one nonzero entry were removed. Our 2.0-ppm CoSCoSA model started with 6441 bins (3D overall, 2D in space). When all nonpopulated columns were removed, these were reduced in number to 280 for the ^{13}C-^{13}C COSY data and 397 for the ^{13}C-^{13}C distance connectivity data. These CoSCoSA models, for the particular training set and a 2.0-ppm resolution spectral bin width, used less than 6 percent of the available 2D connectivity spectral space.

The 2.0-ppm bins in QSDAR models tend to override errors from using predicted ^{13}C NMR data. Using bins considerably wider than typical instrumental resolution changes the information content of raw spectra and, by reducing the number of spectral bins, also reduces the computational intensity of modeling. Both results are desirable for modeling purposes. Instrument-dependent elements (shape, resolution) of the chemical shift peaks do not add artificial, irrelevant variability. At 2.0-ppm resolution, the CoSCoSA models retained enough spectral and structural information to represent the important activity associations. Also, wide spectral bins increase the number that are multiply populated, and so produce models of binding to the aromatase enzyme that have excellent predictive qualities under cross-validation.

3D CoSCoSA Models of Aryl Hydrocarbon Receptor Binding Previously reported log EC_{50} (median effective concentration) binding activity data [24, 54–60] were used to develop models of PCDD, PCDF, and PCB compounds. ^{13}C NMR spectra were simulated as before using ACD/Labs CNMR predictor software (ACD, Toronto, Canada, www.acdlabs.com) and used for QSDAR both CoSASA and CoSCoSA modeling. For these compound types, no chemical shift peaks were present outside the 107 to 159 ppm range. The use of predicted rather than experimentally measured NMR chemical shifts is not necessary to build the CoSCoSA and CoSASA models, but it saves time, money, and in this case prevents possible toxic exposures.

CoSCoSA and CoSASA models were produced by using the assigned ^{13}C NMR chemical shifts at the 12-carbon positions in the PCDF, PCDD, and PCB molecules, as shown in Figure 6.8. CoSASA models for PCDD and PCDF compounds were built directly from the 12 assigned carbon chemical shifts in each predicted spectrum, as previously described. No binning was involved for CoSASA models.

For 3D CoSCoSA, chemical shifts and atom assignments from the predicted spectra were used to form 2D through-bond COSY spectra and 2D through-space medium-range and long-range spectra. We reduced the resolution of all 2D spectra by populating 2.0-ppm bins. These dimensions were chosen, as before, to multiply populate bins for statistical analysis and to reduce

METHODS AND EXAMPLES 265

the effects of uncertainties from the use of simulated spectra. The 2.0-ppm bin spectral width was used successfully for AhR binding CoSA models [24].

CoSCoSA models were built from 2D bins in the plane of nearest neighbor through-bond (COSY) plane, a medium-range through-space distance plane, and a long-range through-space distance plane. The arrows in Figures 6.14a, 6.14b, and 6.14c show the through-bond nearest neighbor carbon-to-carbon connections of PCDF, PCDD, and PCB molecules. The arrows in Figures 6.14d and 6.14e show a selection of medium-range through-space connections, in this case from the 2 and 3 or 7 and 8 anchoring positions to the middle ring carbons in PCDF and PCDD molecules. Likewise for PCB molecules, in Figure 6.14f, we defined anchoring points as the 3, 4, and 5 positions and the 3′, 4′, and 5′ positions, so the "middle-ring" atoms consisted in this case of the two ring-connecting carbons only. The arrows in Figures 6.14g and 6.14h show the long-range through-space connections from the 2 and 3 or 7 and 8 anchoring positions to the opposite ring carbons in PCDF and PCDD molecules. Likewise for PCB molecules, in Figure 6.14i, the long-range connection from the 3, 4, and 5 positions and 3′, 4′, and 5′ positions to the carbons on the opposite ring. Since this long-range connectivity interaction overlapped the

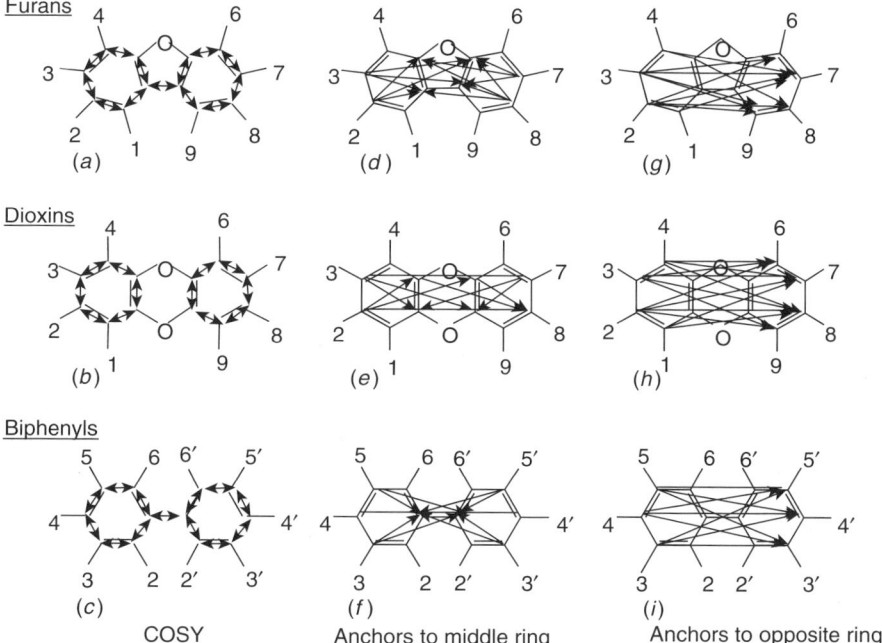

Figure 6.14 Arrows represent the 2D ^{13}C-^{13}C COSY spectra for (a) PCDFs, (b) PCDDs, and (c) PCBs. The arrows represent middle-range 2D ^{13}C-^{13}C distance spectra for (d) PCDFs, (e) PCDDs, and (f) PCBs. The arrows represent long-range 2D ^{13}C-^{13}C distance spectra for (g) PCDFs, (h) PCDDs, and (i) PCBs.

two anchoring points for each molecule, we choose only one anchor ring as the "origin" for which all long-range through-space connections originated. CoSCoSA models were built using the nearest neighbor 2D spectral plane only, the two anchoring structural elements through-distance 2D planes, and a combination of through-bond and through-space information. Since there were two anchors for every molecule, we might theoretically have separated the medium-range through-space distance connections from the outer ring anchors to middle of the compound 14 (Figs. 6.14d, 6.14e, and 6.14f) into two separate 2D planes. We did not do so because the training set was small and increasing the number of degrees of freedom with additional spectral dimensions would risk producing an overfitted model. In the 2D COSY, symmetry inherent in the nearest neighbor relationship produces a pair of arrows, only one of which is needed to represent each relationship. Therefore, only half the 2D COSY spectral connections were used. In contrast, the medium-range and long-range through-space distance spectra, defined using anchoring origins do not have the same symmetry and duplication. Therefore, the whole 2D spectral plane is needed and was used in model development.

Statistical analysis was performed by Statistica software (Statsoft, Tulsa, OK). For each CoSCoSA model, we used forward multiple linear regression (MLR) on the bins. We did not include in the models any bins with fewer than 2 hits. For the 26 PCDFs, forward MLR selection of 6 bins from COSY, through-space distance and combined COSY and through-space distance ranges were used to produce the corresponding CoSCoSA models. Forward MLR selection of 3 bins was used for COSY, through-space distance and combined CoSCoSA models of the 14 PCDDs and the 12 PCBs. The CoSCoSA model of all 52 compounds was built by forward MLR selection of 10 bins from the combined COSY and through-space distance-generated 2D spectral planes.

Figures 6.15a to 6.15c show plots of the predicted versus experimental binding for CoSCoSA models based on MLR of through-bond COSY spectral data for PCDFs, PCDDs, and PCBs, respectively. Figures 6.15d to 6.15f show graphs of the predicted versus experimental binding for CoSCoSA models based on MLR of through-space medium-range and long-range spectral data for PCDFs, PCDDs, and PCBs, respectively. Figures 6.15g to 6.15i show the predicted versus experimental binding for CoSCoSA models based on MLR of COSY combined with through-space medium-range and long-range spectral data for PCDFs, PCDDs, and PCBs, respectively. Tables 6.7 to 6.9 contain CoSCoSA parameters n, r^2, q^2, q_4^2, F, and 2D bins for the PCDF, PCDD, and PCB models, respectively [39].

Population of the COSY bin 153-113 was associated with 10 of the 11 PCDFs that had activities weaker than −7.0. Population of COSY bin 155-119 was associated with 12 of 16 compounds that had binding activities stronger than −7.0. Bins associated with strong or weak binding, once they are identified by COSY and other CoSCoSA methods, can be related to the substructure features that gave rise to the chemical shifts. The recognition of important

METHODS AND EXAMPLES

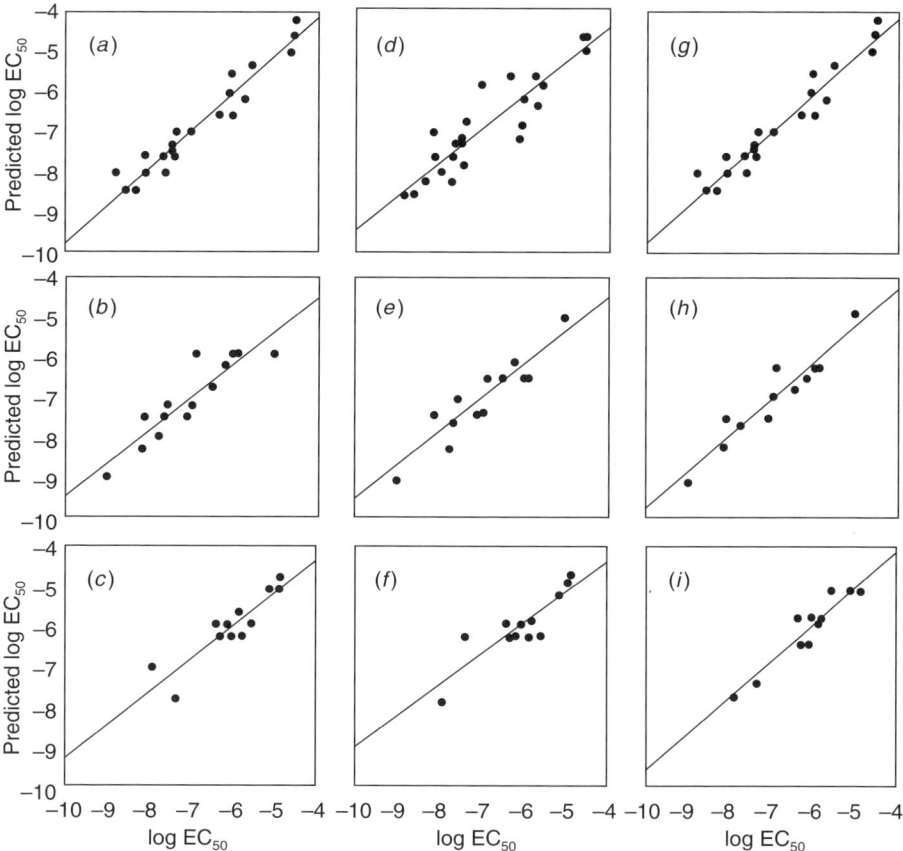

Figure 6.15 Graphs of the CoSCoSA models based on 2D ^{13}C-^{13}C COSY spectra for (*a*) PCDFs, (*b*) PCDDs, and (*c*) PCBs. The graphs of the CoSCoSA models based on 2D medium-range and long-range ^{13}C-^{13}C distance spectra for (*d*) PCDFs, (*e*) PCDDs, and (*f*) PCBs. The graphs of the CoSCoSA models based on 2D COSY and distance spectra for (*g*) PCDFs, (*h*) PCDDs, and (*i*) PCBs.

substructure features allows the inference of toxic features and other activity mechanisms. In this case, COSY bin 155-119 was populated when the ester oxygen of the furan was connected to the carbon with a chemical shift of 155 ppm. For PCDDs, the COSY bin at 143-123 was populated for all 8 compounds having binding activities stronger than −6.96. In this case, COSY bin 143-123 was populated when the ester oxygen of the dioxin group was connected to the carbon with a chemical shift of 143 ppm. For PCBs, the COSY bin at 137-125 was populated for both compounds having a binding activity stronger than −7.0. For PCDFs, the long-range bin 121-117 marked 4 of the 6 compounds that had activities weaker than −6.0. In the PCDD distance model, bin 127-143 was associated with all 8 compounds having binding activities stronger than −6.96.

The best COSY and through-space model for PCDFs turned out to be exactly the same as the COSY model for PCDF compounds. That is, for the PCDFs, the short-range connections all had better predicative strength than any of the longer distance associations. For PCDDs, the medium through-space bin 127-143 correctly identified all eight PCDD compounds with binding activities stronger than −6.96, as it had done in the less complicated spectral format. The COSY bin at 137-125 correctly identified both PCBs with a binding activity stronger than −7.0.

Figure 6.16 shows the predicted versus experimental binding for all 52 compounds, of all three structural types, using the combined COSY, medium- and long-range distance spectra. Statistical data for the combined CoSCoSA model are shown in Table 6.10, along with comparable data based on other methods. The r^2 for the PCDFs is 0.87, for PCDDs is 0.84, and for PCBs is 0.75. The CoSCoSA model is derived from ten 2D bins: 3 from COSY spectra, 4 from medium-range spectra, and 3 from long-range spectra. It is significant that each of these ten 2D bins represents peaks of only one compound type. One of the COSY and one of the medium-range bins have hits only for PCDDs. Another pair (a COSY and a medium-range bin) are occupied only for PCBs. The remaining 6 bins have hits only for PCDFs. From the combined COSY with through-space results, we conclude that the CoSCoSA models for PCDF, PCDD, and PCB compounds each contained relevant information sufficient to produce accurate models of binding affinity to the AhR. We also infer that building effective CoSCoSA models for diverse, multistructure-type data sets will require sufficient bins to represent the binding relationships for each of the structural types with a minimum of 2 bins each.

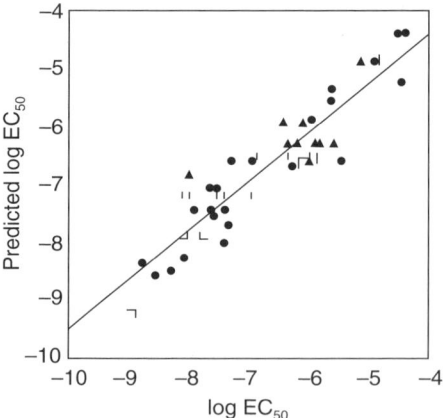

Figure 6.16 Plot of the predicted binding versus experimental binding for all 52 compounds using 2D COSY plus medium and long-range spectra. The PCDF compounds are represented by filled circles (●), PCDD compounds shown by open squares (□), and PCB compounds shown with filled triangles (▲).

Tables 6.7 to 6.9 address the question of how these CoSCoSA results compare to those from other QSDAR and QSAR modeling approaches. In Table 6.7 for PCDFs, two of the three CoSCoSA models had higher r^2 and q^2 values than the corresponding 1D CoSA model. All three CoSCoSA models for PCDFs had higher r^2 and q^2 than the 2D CoSASA model that associated spectral chemical shifts with structurally assigned locations. Our previous 1-ppm resolution CoSA model based on 5 bins for 26 PCDF compounds had an r^2 of 0.82 and q^2 of 0.72 [24]. A structural parameter model by Mekenyan and co-workers [61] to produce a 5-component model for 25 PCDF compounds (all 26 PCDF compounds except for 237-trichlorodibenzofuran) that had an r^2 of 0.85 and q^2 of 0.71. The best model for 39 dibenzofurans proposed by Turner used three infrared EVA components [62]. Since the CoSCoSA models were typically giving r^2 above 0.9 and q^2 above 0.8, we found a consistent and significant improvement, especially for the prediction accuracy, compared to previous art. In Table 6.8 for PCDDs, CoSCoSA results were superior to alternative methods in every case but one discussed in MeKenyan [61], which had approximately equivalent quality. In Table 6.9 for PCBs, the corresponding CoSCoSA models typically exceeded but at

TABLE 6.7 26 PCDF Compound Model Performance Parameters n, r^2, q^2, q_4^2, F, and 2D bins[a] **[24, 38, 60, 61]**

Model	n	r^2	q^2	q_4^2	F	2D Bins
Structural QSAR	5	0.85	0.72			
CoMFA	6 PC	0.85	0.72			
EVA	3 PC	0.96	0.73			
1D CoSA	5 bins	0.82	0.72		18.6	
2D CoSASA	6 atoms	0.74	0.70		9.1	
CoSCoSA COSY	6 bins	0.92	0.84	0.84	38.7	C 119-113, C 125-113, C153-113, C 127-119, C 155-119, C 127-125
CoSCoSA Mid + Long	6 bins	0.83	0.63	0.65	15.1	M 127-115, M 125-117, M 119-125, M 125-125, M 113-127, L 121-127
CoSCoSA COSY + Mid + Long	6 bins	0.92	0.84	0.84	38.7	C 119-113, C 125-113, C153-113, C 127-119, C 155-119, C 127-125

[a] C stands for COSY, M stands for medium-range, and L stands for long-range spectra.

TABLE 6.8 14 PCDD Compound Model Performance Parameters Bin n, r^2, q^2, q_2^2, F, and 2D bins[a] [24, 38, 60, 61]

Model	n	r^2	q^2	q_2^2	F	2D Bins
Structural QSAR	5	0.95	0.82			
COMFA	2 PC	0.88	0.73			
EVA	2 PC	0.85	0.48			
1D CoSA	3 bins	0.83	0.74		16.5	
2D CoSASA	5 atoms	0.81	0.53		6.7	
CoSCoSA COSY	3 bins	0.83	0.75	0.74	15.9	C 127-123, C 141-123, C 143-123
CoSCoSA Mid + Long	3 bins	0.83	0.75	0.71	16.3	M 123-141, M 125-141, M 127-141
CoSCoSA COSY + Mid + Long	3 bins	0.90	0.79	0.79	16.2	C 141-123, C 143-123, M 123-141

[a] C stands for COSY, M stands for medium-range, and L stands for long-range spectra.

TABLE 6.9 12 PCB Compound Model Performance Parameters n, r^2, q^2, q_2^2, F, and 2D Bins[a] [24, 38, 60, 61]

Model	n	r^2	q^2	q_2^2	F	2D Bins
COMFA	3 PC	0.87	0.49			
EVA	1 PC	0.72	0.16			
1D CoSA	3 bins	0.66	0.30		5.2	
CoSCoSA COSY	3 bins	0.82	0.58	0.58	12.2	C 135-125, C 127-127, C 133-131
CoSCoSA Mid + Long	3 bins	0.77	0.66	0.47	9.1	M 133-135, L 133-131, L 133-133
CoSCoSA COSY + Mid + Long	3 bins	0.91	0.80	0.80	26.3	C 137-125, M 133-137, M 131-139

[a] C stands for COSY, M stands for medium-range, and L stands for long-range spectra.

least equaled all previous methods, particularly for predictions from cross-validation experiments.

Table 6.10 shows the results for all 52 PCDF, PCDD, and PCB compounds. The combined COSY and through-space CoSCoSA models had a higher q^2 than the corresponding 1D CoSA model [39]. Additionally the CoSCoSA model was based on only ten 2D bins, whereas the CoSA model was based on fifteen 1D bins. The previous CoSA model for all 52 compounds had an r^2 of 0.87 and a q^2 of 0.67 [24], whereas the current CoSCoSA model had an r^2 of 0.85 and q^2 of 0.73 and q_4^2 of 0.52. When two outliers are removed from the cross-validation of the 52-compound CoSCoSA model, the q^2 and q_4^2 improve to 0.77. Both outliers occur when a compound has all zeros in every bin except

METHODS AND EXAMPLES

TABLE 6.10 All 52 PCDF, PCDD, and PCB Compound Model Performance Parameters n (parameters used), r^2, q^2, q_4^2, F, and 2D bins^a [24, 38, 60, 61]

Model	n	r^2	q^2	q_4^2	F	2D Bins
Structural QSAR	5	0.85	0.72			
CoMFA	6 PC	0.88	0.71			
EVA	3 PC	0.87	0.68			
1D CoSA	15 Bins	0.87	0.67		16.6	
CoSCoSA	10 Bins	0.85	0.73	0.52	24.0	C 121-117, C 141-123,
COSY +						C 137-125,
Mid + Long						M 119-117,
						M 129-119,
						M 133-135,
						M 123-141,
						L 119-123,
						L 119-125,
						L 121-127

^{a} C stands for COSY, M stands for medium-range, and L stands for long-range spectra.

for one column and that column has only two bin hits in it. When a column with only one remaining hit in it is used during the LOO or L4O cross-validation process, the linear regression β coefficient can change sign. One needs to be aware of this fact during CoSCoSa modeling. The CoSCoSA composite model for the 52 compounds represents a significant improvement over corresponding results from alternative modeling approaches [24, 61, 62].

Producing a ^{13}C and ^{15}N Heteronuclear Connectivity Matrix

Often, particularly for chemicals with potentially useful pharmaceutical value or toxicity, compound types to be modeled will contain atoms besides carbon, oxygen, and hydrogen. It that case, NMR spectra besides ^{13}C can be used. The most prominent atom that is both biologically important and for which accurate NMR prediction software is available is ^{15}N. Other NMR nuclei that could be used for SDAR or QSDAR modeling are ^{17}O, ^{19}F, and ^{31}P. Which of these might be useful would depend on the biological endpoint to model and the availability of an endpoint-characterized training set. Figure 6.17 shows a typical two-dimensional spectral connectivity layout for a ^{13}C-^{15}N heteronuclear CoSCoSA model. As before for multidimensional ^{13}C-^{13}C models, the symmetry-based data duplicates mean that only half of the spectra in the array are necessary to develop a model.

^{13}C-^{15}N Heteronuclear CoSCoSA Model of Antibiotic Cephalosporins

Figure 6.18 shows the structures of 17 cephalosporins, widely used antibiotics similar in structure and mode of action to penicillin. There are two nitrogen atoms in the backbone of cephalosporin molecules. These structures were used

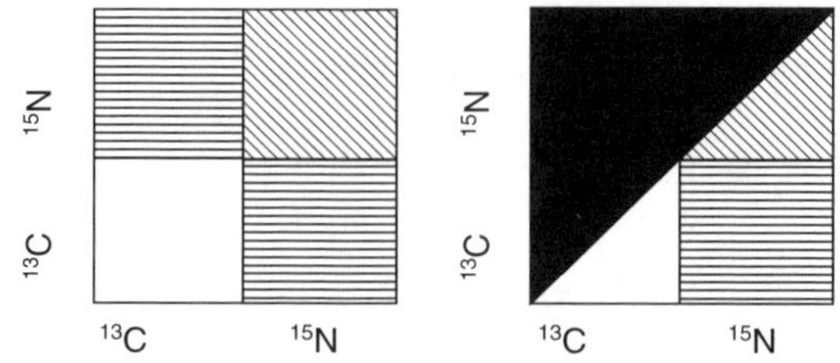

Figure 6.17 2D ^{13}C-^{15}N heteronuclear connectivity matrix.

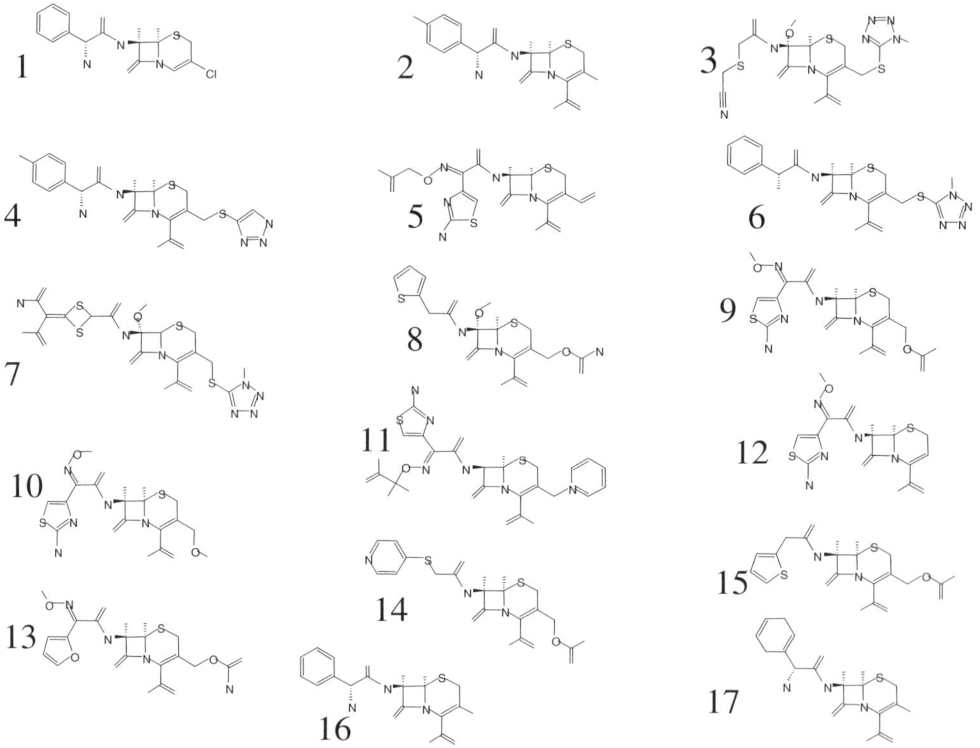

Figure 6.18 Celphasporin structures. Numbering of compounds corresponds to Table 6.11.

METHODS AND EXAMPLES   273

as an example to examine the potential of heteronuclear CoSCoSA methods to model the minimum inhibitory concentrations (MIC) of cephalosporin antibiotics, using only the through-bond (COSY-type) carbon-to-carbon and through-bond carbon-to-nitrogen 2D planes. In producing this 2D CoSCoSA model, for technical reasons we defined the endpoint as log(1/MIC) [63].

We used bin sizes of 3.0×3.0 ppm for carbon-to-carbon COSY connections and 10.0×3.0 ppm for the nitrogen-to-carbon direct connections. Nitrogen chemical shifts were predicted from software available on the ACD/Labs ILAB website (ACD, Toronto, Canada, *www.acdlabs.com*). Carbon shifts were predicted as before. In building the nitrogen-to-carbon connectivity matrix, 700.0 ppm was added to the predicted nitrogen chemical shifts, so that they fell in the range from 300 to 700 ppm. Thus, a single synthetic spectrum could be defined with carbon-to-carbon connectivity bins occupied from 0 to 240 ppm, and nitrogen-to-carbon from 300 to 700 ppm.

Forward regression MLR selected 4 bins from a total of 101 carbon-to-carbon and 48 carbon-to-nitrogen occupied bins. These were (−230 nitrogen, 156 carbon), (−280 nitrogen, 162 carbon), (−230 nitrogen, 168 carbon), and (135 carbon, 24 carbon). The CoSCoSA models were produced as described in previous examples, and the results are shown in Table 6.11 and Figure 6.19. Figure 6.19 shows the results of the cephalosporin CoSCoSA model with a correlation $r^2 = 0.92$, F value = 36.2, $P < .000005$, LOO of $q_1^2 = 0.88$, average L4O of

TABLE 6.11 Results of Cephalosporin CoSCoSA Modeling[a]

Compounds	Number in Figure 6.18	MIC	log(1/MIC)	Leave-1-Out log(1/MIC) Predicted	Leave-4-Out log(1/MIC) Predicted
Cefaclor	1	8	−0.90	−1.22	−1.23
Cefadroxil	2	8	−0.90	−0.90	−0.90
Cefmetazole	3	16	−1.20	−1.15	−1.15
Cefaperazone	4	16	−1.20	−1.15	−1.13
Cefixime	5	8	−0.90	−0.88	−1.20
Cefamandole	6	16	−1.20	−1.15	−1.15
Cefotetan	7	16	−1.20	−1.15	−1.13
Cefoxitin	8	16	−1.20	−1.15	−1.23
Cefotaxime	9	16	−1.20	−1.21	−1.23
Cefpodoxime	10	2	−0.30	−0.32	−0.32
Ceftazidime	11	16	−1.20	−1.21	−1.23
Ceftizoxime	12	8	−0.90	−0.85	−0.83
Cefuroxime	13	4	−0.60	−0.63	−0.42
Cephapirin	14	8	−0.90	−0.90	−0.90
Cephalothin	15	8	−0.90	−0.90	−0.90
Cephalexin	16	8	−0.90	−0.90	−0.90
Cephradine	17	8	−0.90	−0.90	−0.90

[a] MIC = minimal inhibitory concentration [62].

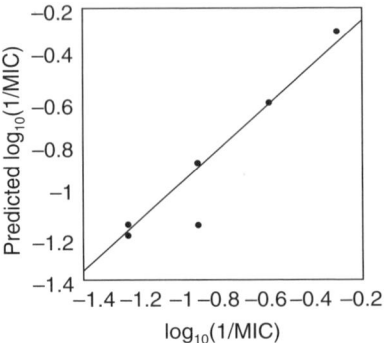

Figure 6.19 2D ^{13}C-^{15}N heteronuclear CoSCoSA model of celphasporin antibacterial activity.

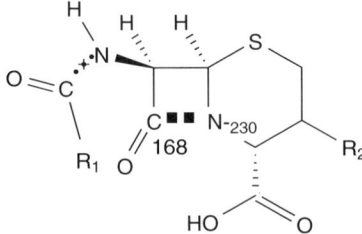

Figure 6.20 Structural explanation of bins selected in the 2D ^{13}C-^{15}N heteronuclear CoSCoSA model of celphasporin antibacterial activity. Dashed bond between carbon 168 and nitrogen (−230) corresponds to bond that undergoes acid hydrolysis and then binds to enzyme transpeptidase and irreversibly inactivates it. Dotted bond between carbon and nitrogen corresponds to bins reflecting carbon (156), nitrogen (−270) and carbon (162), nitrogen (−280) selected bins and may explain the backbone torsional angle dependence of dipeptides binding affinity to transpeptidase.

$q_4^2 = 0.79$, and standard deviation = 0.03. Figure 6.20 shows the structurally assigned interpretation of the chemical shifts used to formulate the CoSCoSA model of cephalosporins. The chemical shifts at (−230 nitrogen, 168 carbon) identifies the carbon-to-nitrogen bond where acid hydrolysis occurs and allows the celphalosporin to bind to bacterial enzyme transpeptidase, irreversibly inactivating the enzyme and stopping growth [64]. The chemical shifts at (−230 nitrogen, 156 carbon) and (−280 nitrogen, 162 carbon) identify another carbon-to-nitrogen bond in the middle of each celphalosporin compound. Since the enzyme transpeptidase binds to dipeptides, it is reasonable to assume that the other carbon-to-nitrogen bond [which looks like part of an amino acid backbone (carbonyl-to-amide-to-alpha carbon)] is involved via a hydrogen bond to the enzyme. The two different chemical shifts for the same carbon-to-nitrogen bond can represent the bond in two different configurations, two different electrostatic potentials, or more likely a combination of different configurations and associated electrostatic potentials. It has been shown that peptides interact with the penicillin binding transpeptidase proteins with binding energy dependent on the backbone torsion angles [65]. We mention this because most descriptions of cephalosporin activity mechanisms do not discuss the critical role of the second carbon-to-nitrogen bond. The CoSCoSA model including heteronuclear NMR peaks predicted that both carbon-to-nitrogen bonds contribute significantly to the antibiotic strength of cephalosporin. We tried modeling cephalosporin activity using ^{13}C COSY data alone, but the models were very poor with respect to all statistical measures of goodness of fit and leave-one-out cross-validation (data not shown). In this case, both ^{15}N and ^{13}C chemical shifts were very important in producing a reliable model of biological activity.

Using Molecular Dynamics of Compounds to Produce a 4D Connectivity Matrix A 4D-connectivity matrix can be defined as the sum of an arbitrary number, say 100, 3D connectivity matrices. In the simplified version of this 4D connectivity matrix concept, chemical shifts of atom *i* and atom *j* are assumed not to change as the molecule bends or twists with time, but the distance between atoms *i* and atom *j* would differ as a function of the molecular conformation. For any two atoms, molecular dynamics programs can estimate interatomic distance, a value that may change over some range and for which the percentage of time that the distance is within a certain bin will vary, depending on molecular connections, degrees of freedom, and the like. This concept applied to CoCoSA modeling would treat the distance between atoms as a potential variable rather than as a constant. A score of 100 in a 4D connectivity matrix will represent unvarying distances between two atoms as seen in bonds and also between more distant atoms if the molecules are very rigid. For two atoms in flexible molecules there will be a distribution of distance hits along the *z* axis varying from 1 to some maximum (most probable conformation). For all the possible atom pairs, distance distributions will be Gaussian, or skewed-Gaussian, functions when there is a single maximum distance. When there is more than one maximum, more complex distribution will be seen. In the simplified regime of CoSCoSA distance modeling, the occupancy pattern for a particular molecule and a particular through-space distance bin would be 100 if the full range of possible distances lay within that 2D bin. If 30 percent of the time, the molecular shape were such that the atomic distances lay outside that bin, the bin occupancy would be represented with 70, and the remaining 2D bin(s) would share the remaining 30 points. We shall now discuss why this way of representing molecular conformation characteristics confers a significant advantage when building a CoSCoSA model, particularly for cases in which the training set includes very diverse compound types.

Binding of a ligand to a receptor occurs because of a lowering of Gibbs free energy for the combined ligand–receptor system. As in physical and chemical systems, lowered Gibbs free energy of biological systems depends on two factors: favorable changes in enthalpy and/or entropy changes. Traditional 3D QSAR approaches are biased to reflect enthalpy changes, which ignore entropy contributions. Classical 3D QSAR models are based on electrostatic and stereospecific patterns in space as they correlate with biological activity. These correlations work because they reflect electrostatic differences and the corresponding changes of enthalpy that occur during binding. These QSAR models neglect entropy changes. Entropy calculations are routinely neglected because they are often misunderstood and because it is not easy to conceive a way in which such phenomena can be reflected in models that are built directly using molecular 3D structural conformations. Modelers typically use only the minimum energy conformer for each compound during model development, and this ignores the changes in molecular conformation that are statistically possible and that certainly occur during the compound's interaction with its enzymatic substrate when binding occurs.

Using the 4D Connectivity Matrix to Estimate Configurational Entropy

We have used molecular dynamics of 130 structurally diverse compounds from the previous 2D CoSCoSA model of estrogen binding to estimate the configurational entropy of each compound. A 4D connectivity matrix allows for multiple conformations of a molecule to be displayed. The multiple conformations can be put into an "entropy-like" equation to estimate the effect of configurational entropy [66–68]:

$$S \alpha 1/N \sum_{i<j} p_{ij} \ln p_{ij}$$

where p_{ij}, a probability that must lie between zero and one, is calculated as the percentage of time the distance between atoms i and j lie within the r_{ij} spatial distance bin, with this percentage then divided by 100 to express the occupancy within the probability range required by the equation.

Representation of entropy effects can and should be simplified, and this is possible by at least two methods. One method is to calculate the average r_{ij} distance for each molecule and express the entropy surrogate using each p_{ij} value as its number of standard deviations from the average. This type of data pretreatment is called autoscaling. The second method, analogous to the one used to simplify 3D connectivity matrices, is to break interatomic distance space into a number of 2D planes. For the 130 compound set, we used four 2D planes with distances of $r_{ij} < 2.5$ Å, 2.5 Å $< r_{ij} < 5.0$ Å, 5.0 Å $< r_{ij} < 7.5$ Å, and 7.5 Å $< r_{ij}$. As mentioned above, p_{ij} is defined as the percentage of time the distance between atoms i and j lies within the constraints of the specified 2D plane. A rigid molecule (or through-bond neighbors) will have distances between atoms that fluctuate very little, and p_{ij} for the occupied bin in those cases will equal to $100/100 = 1.0$, with all other ij bins unoccupied. The corresponding configurational entropy will be 0. A flexible molecule will have a distribution of p_{ij} across several 2D planes in the 4D matrix. The definitions above, applied to the flexible molecule, would be expected to yield a higher configuration entropy.

Molecular dynamics simulations were run using Hyperchem 7.0 (Hypercube Inc., Gainsville, FL) for 130 compounds. First, Polak-Ribiere conjugate gradient energy minimization was executed for 1000 iterations or until a minimized structure was obtained. The temperature was slowly raised to 310 K over 12 ps. (310 K was chosen to estimate an average body temperature, the presumed condition in which conformational variation is relevant to receptor binding.) Then, a 100-ps molecular dynamics trajectory was executed at 310 K and a structure was saved every 1 ps. The 100 structures from the trajectory were used to calculate p_{ij} for every nonhydrogen-to-nonhydrogen atomic pair.

Figure 6.21 shows the entropy of the unbound free compound versus estrogen receptor log(RBA) for the 130 compounds. The plot clearly shows that the more configurational entropy a compound has, the stronger it binds to the estrogen receptor. This relationship appears to violate the principles of physics, at least with respect to the expectations described two paragraphs above. We understand this counterintuitive relationship to derive from several interacting phenomena. First, the compounds compared here were initially selected for a QSAR model, presumably because they could be effectively modeled by electrostatics (enthalpy). Other compounds for which structure–activity data were available were not included in this QSAR test set. Some earlier projects from the same group had included almost 200 estrogens [43]. We, in turn, selected the same test set to facilitate comparisons of our results to the published QSAR. When one realizes the possible nonrandom nature of the test set compounds, the plot starts to make sense. If, based on enthalpy, a molecule has less capability to bind, it cannot have a large entropy loss upon binding. Likewise, a molecule with substantial enthalpic contribution to receptor binding can maintain binding capacity even with an unfavorable entropy

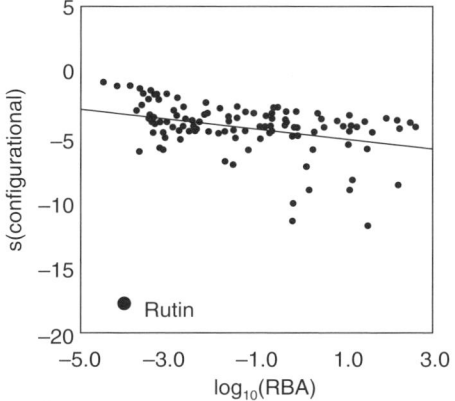

Figure 6.21 log(RBA) versus configurational entropy for 130 compounds that bind to the estrogen receptor. Rutin is shown with a larger circle.

term. We believe our apparently contraphysics entropy trend is a second-order consequence of the first-order relative relationships inherent in selection of this set compounds.

In Figure 6.21, rutin appears as a rather dramatic outlier compared to all other compounds modeled by our 4D QSDAR method. The result might be an inexplicable artifact. All modeling methods have weaknesses and perform better for some compounds than for others. There is a possible explanation for its behavior in the 4D QSDAR model. Experimentally, rutin has very low receptor binding. The rutin molecule is much larger than the others in this 130-member comparison set. Because of its size, it has large conformational entropy in its unbound state and this may be the reason it does not bind strongly to the estrogen receptor. The 3D QSAR would have us believe that the reason rutin does not bind strongly to the estrogen receptor is a result of strong steric hindrance and weak electrostatics. The 2D QSDAR model of estrogen receptor binding for rutin was fairly weak.

When it is bound, rutin's total entropy is probably not decreased very much because the majority of the molecule lies outside the binding site and is still conformationally free. The model attempts to estimate the entropic effect of binding as a function of the free state entropy for each test set molecule. For rutin, the 4D QSDAR model gives a poorer prediction than QSAR, which ignores entropy altogether. For a complete estimation of configurational entropy changes upon binding, one would need to compare the molecular dynamics trajectories of the compounds and the particular enzymatic substrate in both the free and bound states along with solution effects, rather than use only the range of free state conformations of the test set molecules.

The results described here illustrate for the first time how entropic phenomena can be realistically incorporated into QSDAR modeling. We intuit that larger data sets (from which outlier molecules have not been excluded to compensate for methodological limitations) will benefit from representation using entropic parameters. We suppose that, for a training set inclusive of the full range of structural variants, spectral pattern definition using only small molecule structural connectivity and conformational variability can provide significant access to entropy term contributions and improve the quality and utility of biological models. We are also beginning to examine other strategies that would make the 4D QSDAR spectral representation illustrated above less sensitive to the type of error demonstrated for rutin.

6.3 FUTURE IMPROVEMENTS

Commercial Software Products

ACD/Labs (ACD, Toronto, Canada, *www.acdlabs.com*) is the supplier of the most complete suite of spectroscopy tools allowing both experimental data processing for all general analytical data as well as the prediction of multinuclear 1D and 2D NMR data (H1, C13, N15, F19, P31). ACD/Labs released software modules in 2005 that enable both experimental and predicted 1D and 2D spectra to be passed to an add-on module for SDAR and QSDAR modeling. The company also intends to include pattern recognition methods that will correlate the spectra to biological and toxicological activity. Building 1D and 2D SDARs manually is not a complicated process, and the same is true in principle for 3D and 4D variants. However, for technical reasons, the automation of interatomic distance spectral binning, especially when based on molecular dynamic models, will greatly facilitate an otherwise tedious and impractical manual process. It is likely that this capability will be incorporated into a later version of the software.

Improving the Algorithms for Binning Spectral Data

CoSCoSA modeling is a powerful modeling technique because it uniquely combines quantum mechanical information from the chemical shifts with internal molecular distances. Optimizing the bin size through smart binning techniques, the distance cutoffs, multiple conformations of a structure, multiple spectra, and the number of distance connectivity spectra used in a CoSCoSA model may be needed or helpful for modeling biological, physical, chemical, ADME, or toxicological endpoints.

Potential Applications We have demonstrated various QSDAR models capable of predicting the strength of chemical interactions with biological systems. This kind of prediction is useful for estimating pure chemical toxicity or even risk assessment of environmental contaminants that are complicated mixtures. It is also useful for pharmaceutical lead compound identification and for qualifying new drugs with respect to side effects or unintended toxicity. In that regard, we believe the Food and Drug Administration (FDA) and other drug regulatory authorities could use QSDAR for anticipating problems, to specify the types of toxicity testing required for a new drug application. We have also shown the predictive value of QSDAR methods when the endpoint is essentially chemical (decomposition) but the process is mediated through the enzymatic capabilities of bacteria. This suggests possible QSDAR applicability in relation to bioremediation efforts.

The type of QSDAR method useful for a particular application depends on the amount of data available for the model training set, the variety of compound types to be modeled, the elemental composition of the compounds in the training set, and the predictive accuracy required. One-dimensional and 2D models are easier to set up and less computationally intensive to build and validate than 3D and 4D models. It is possible that some applications, such as drug lead discovery, may benefit from using the simpler methods early in a process and the more elaborate ones after the number of candidates has been reduced and a backbone scaffold has been selected.

The best computational approaches appear to depend less on the particular application and more on the overall size and character of the data set to be examined. When all members of the training set are structurally similar, 3D

QSDAR models can be built from PCs, as was shown for steroids binding to corticosterone binding globulin or aromatase enzyme or from bins, as was shown in cephalosporin antibacterial activity. When the training set molecules are not structurally similar, 3D QSDAR models are best built from selecting bins in the 2D planes of the 3D connectivity matrix by PLS forward regression, as was demonstrated for the 130 compounds interacting with the estrogen receptor. In the latter case, selected bins represented two chemical shifts (energies) separated by a specified distance. The chemical shifts typically signified the atom type in a specific environment. A selected bin in a CoSCoSA model represents two atom types at a specified distance, which are shown by the model to be positively or negatively associated with the biological activity being modeled. The MLR-selected bins can then be used to build a model of biological activity and to understand the mechanism underlying the biological activity. The selected bins can be used as structure-spectral anchors in models for lead identification or lead optimization of new drugs. This approach was shown for toxicity assessment application using PCDD, PCDF, and PCBs binding to the ArH receptor, but there is no reason the same concept cannot be applied to drug discovery.

The CoSCoSA modeling system can be applied to receptor binding systems for which the structure–activity relationship is unknown, a common situation faced by new drug discovery or lead optimization programs in the pharmaceutical industry. Producing QSAR models without detailed structural information is unreliable and is often based on intuition. QSDAR offers, in contrast, a more objective way of building models. Ironically, the use of spectral data–activity relationships allows an unbiased examination of specific structure–activity relationships, which can suggest a more meaningful way to discover antibiotics or other pharmaceuticals. This potential was demonstrated in the heteroatom NMR spectral model of cephalosporin activity.

The inclusion of molecular dynamics entropy information in QSDAR modeling promises access to information not available in QSAR approaches. The example shown illustrates a potential limitation of the technique when the molecular complexity of the training set compounds to be predicted is similar or diverse. If it is not reasonable to build a training set that spans a diverse range, QSAR methods blind to the deleterious effects of entropy may provide superior or complimentary models. One way to design compounds that have stronger binding to a receptor is to lower the configurational entropy in the unbound state while retaining the positive electrostatic contributions and minimal steric hindrances needed for strong binding affinity.

6.4 CONCLUSIONS

Structure and chemical shift information from the 3D connectivity matrix was used to produce very accurate models of biological binding activity. The 3D connectivity matrix uniquely combines quantum mechanical information from

CONCLUSIONS

the chemical shifts with nearest neighbor and a compound's internal atom-to-atom distance information. The combined information from COSY and long-range distance connectivity information from the 3D connectivity matrix was used to produce CoSCoSA models that were as accurate or more accurate and reliable than QSAR or E-state models based on separate calculations for electrostatics and steric interactions. The quality of results obtained from CoSCoSA models based on simulated NMR data should improve as improvement occurs in the spectral simulation software and prediction errors are reduced. Three-dimensional QSDAR modeling is a complementary technique that is unique and has widespread application to many informatics challenges. It is not intended to replace 3D QSAR and SAR modeling but to provide computational model developers a sensitive, alternate perspective on biological and toxicological properties. For example, in the case of cephalosporins binding to the transpeptidase enzyme, the CoSCoSA model can give multiple conformation information about binding mechanisms that could not be obtained from 3D QSAR or SAR models. The added advantage of this approach is that it can produce these kinds of results very rapidly.

CoSCoSA modeling is based on the unbiased distance-related distribution of a molecule's NMR spectral features, which directly corresponds to structural elements that, according to structure–activity relationship theory, determine the biological activity of the molecule. The diamagnetic term of ^{13}C NMR spectra is directly related to the electrostatic potential and has been shown by QSAR models and physical theory to be directly related to biological activity. The paramagnetic term of ^{13}C NMR spectra is related to the type of orbitals around a carbon nucleus. The orbitals around a nucleus have been shown to be directly related to bond energy, local molecular dynamics, local configurational entropy, and biological activity.

Three-dimensional QSDAR models are in a way a combination of conventional SAR and QSAR modeling methods. Instead of using a 3D grid to calculate electrostatic energies, the energies in the form of chemical shifts are put onto the compound itself in all the possible two-atom connections. The diamagnetic term of the chemical shift is directly dependent on the electrostatic energy of a nucleus; the paramagnetic term of the chemical is directly dependent on the type of orbital. So in this sense, 3D QSDAR models split up the electrostatic energies into electron orbital type. The distribution of distances between two atoms is related to molecular shape and configuration entropy. Three-dimensional QSDAR modeling is the only modeling method that can evaluate enthalpy and entropy effects at the same time.

Acknowledgments We thank Kathleen J. Holm, an NCTR summer student who aided in the development of the CoSCoSA models for estrogen receptor binding, cephalosporin antibacterial activity, and estimates of the configurational entropy.

REFERENCES

1. Cramer, R. D., Bunce, J. D., Patterson, D. E. (1988). Cross-validation, bootstrapping, and partial least squares compared with multiple regression in conventional QSAR studies. *Quant. Struct. Act. Relat.*, 7, 18–25.
2. Cramer, R. D., Paterson, D. E., Bunce, J. D. (1988). Comparative molecular field analysis (CoMFA). 1. Effect of shape on binding of steroids to carrier proteins. *J. Am. Chem. Soc.*, 110, 5959–5967.
3. Klopman, G. (1984). Artificial intelligence approach to structure-activity studies. Computer automated structure evaluation of biological activity of organic molecules. *J. Am. Chem. Soc.*, 106, 7315–7321.
4. Klopman, G. (1992). MULTICASE1. A hierarchial computer automated structure evaluation program. *Quant. Struct. Act. Rel.*, 11, 176–184.
5. Hansch, C., Leo, A. (1995). *Exploring QSAR—Fundamentals and Applications in Chemistry and Biology.* American Chemical Society, Washington DC.
6. Tong, W., Perkins, R., Xing, L., Welsh, W. J., Sheehan, D. M. (1997). QSAR models for binding of estrogenic compounds to estrogen receptor α and β subtypes. *Endocrinology*, 138, 4022–4025.
7. Oprea, T. I., Garcia, A. E. (1996). Three-dimensional quantitative structure-activity relationships of steroid aromatase inhibitors. *J. Comput. Aid. Mol. Des.*, 10, 186–200.
8. Shi, L. M., Fang, H., Tong, W., Wu, J., Perkins, R., Blair, R. M., Branham, W. S., Dial, S. L., Moland, C. L., D. M. Sheehan, D. M. (2001). QSAR models using a large diverse set of estrogens. *J. Chem. Inf. Comput. Sci.*, 41, 186–195.
9. Katritzky, A. R., Mu, L., Labanov, V. S., Karelson, M. (1996). Correlation of boiling points with molecular structure. 1. A training set of 298 diverse organics and a test set of 9 simple inorganics. *J. Phys. Chem.*, 100, 10400–10407.
10. Fujita, T., Iwasa, J., Hansch, C. (1964). A new substituent constant, π, derived from partition coefficient. *J. Am. Chem. Soc.*, 86, 5175–5180.
11. Branbury, S. P. (1995). Quantitative structure-activity relationship and ecological risk assessment: An overview of predictive aquatic toxicology research. *Toxicology*, 25, 67–89.
12. De Dios, A. C., Pearson, J. G., Oldfield, E. (1993). Secondary and tertiary structural effects on protein NMR chemical shifts: An *ab initio* approach. *Science*, 260, 1491–1496.
13. Emsley, J. W., Feeney, J., Sutcliffe, L. H. (1965). *High Resolution Nuclear Magnetic Resonance*, Vol. I. Pergamon, Oxford, pp. 1–287.
14. Beger, R. D., Freeman, J., Lay Jr., J., Wilkes, J., Miller, D. W. (2000). ^{13}C NMR and EI mass spectrometric data-activity relationship (SDAR) model of estrogen receptor binding. *Toxicol. Appl. Pharmacol.*, 169, 17–25.
15. Beger, R. D., Bolton, P. H. (1997). Protein ϕ and ψ dihedral restraints determined from multidimensional hypersurface correlations of backbone chemical shifts and their use in the determination of protein tertiary structures. *J. Biomol. NMR*, 10, 129–142.
16. Wishart, D. S., Sykes B. D. (1994). Chemical shifts as a tool for structure determination. *Methods Enzymol.*, 239, 363–392.

17. Lefebvre B., Williams, A. (2003). ^{13}C NMR chemical shift prediction: A comparison of methods and a case study analysis of Paclitaxel (TAXOL®). Poster presentation, 44th ENC.
18. Meiler, J., Meusinger, R., Will, M. (2000). Fast determination of 13 C NMR chemical shifts using artificial neural networks. *J. Chem. Inf. Comp. Sci.*, *40*, 1169–1176.
19. Kvasnicka, V. (1991). An application of neural networks in chemistry. Prediction of ^{13}C NMR chemical shifts. *J. Math. Chem.*, *6*, 63–76.
20. Shade, L., Beger, R. D., Wilkes. J. G. (2003). New computerized method for modeling binding affinities to the aryl hydrocarbon receptor using ^{13}C NMR spectra. *Environ. Toxicol. Chem.*, *22*, 501–509.
21. Beger, R. D., Freeman, J., Lay Jr., J. O., Wilkes, J. G., Miller, D. W. (2000). Producing ^{13}C NMR, infrared absorption and EI mass spectrometric data models of the monodechlorination of chlorobenzenes, chlorophenols, and chloroanilines. *J. Chem. Inf. Comput. Sci.*, *40*, 1449–1455.
22. Beger, R. D., Wilkes, J. G. (2001). Developing ^{13}C NMR quantitative spectrometric data-activity relationship (QSDAR) models of steroid binding to the corticosteroid binding globulin. *J. Comput. Aid. Mol. Des.*, *15*, 659–669.
23. Beger, R. D., Wilkes, J. G. (2001). ^{13}C NMR quantitative spectrometric data-activity relationship (QSDAR) models to the aromatase enzyme. *J. Chem. Inf. Comput. Sci.*, *41*, 1360–1366.
24. Beger, R. D., Wilkes, J. G. (2001). Models of polychlorinated dibenzodioxins, dibenzofurans, and biphenyls binding affinity to the aryl hydrocarbon receptor developed using ^{13}C NMR data. *J. Chem. Inf. Comput. Sci.*, *41*, 1322–1329.
25. Bax, A., Grzesiek, S. (1993). Methodological advances in protein NMR. *Acc. Chem. Res.*, *26*, 131–138.
26. Aue, W. P., Bartholdi, E., Ernst, R. R. (1976). Two-dimensional spectroscopy. Application to nuclear magnetic resonance. *J. Chem. Phys.*, *64*, 2229–2246.
27. Bodenhausen, G., Ruben, D. J. (1980). Natural abundance nitrogen-15 NMR by enhanced heteronuclear spectroscopy. *Chem. Phys. Lett.*, *69*, 185–189.
28. Bax, A., Griffey, R. H., Hawkins, B. L. (1983). Sensitivity-enhanced correlation of 15N and 1H chemical shifts in natural-abundance samples via multiple quantum coherence. *J. Am. Chem. Soc.*, *105*, 7188–7190.
29. Bax, A., Summers, M. F. (1986). ^{1}H and ^{13}C Assignments from sensitivity-enhanced detection of heteronuclear multiple-bond connectivity by 2D multiple quantum NMR. *J. Am. Chem. Soc.*, *108*, 2093–2094.
30. Kumar, A., Ernst, R. R., Wuthrich, K. (1980). A two-dimensional nuclear Overhauser enhancement (2D NOE) experiment for the elucidation of complete proton-proton cross-relaxation networks in biological macromolecules. *Biochem. Biophys. Res. Comms.*, *95*, 1–6.
31. Clore, G. M., Gronenborn, A. M., Nilges, M., Ryan C. A. (1987). Three-dimensional structure of potato carboxypeptidase inhibitor in solution. A study using nuclear magnetic resonance, distance geometry, and restrained molecular dynamics. *Biochemistry*, *26*, 8012–8023.
32. Bremser, W. (1978). HOSE—A novel substructure code. *Anal. Chim. Acta.*, *103*, 355–365.

33. Livingstone, D. (1995). *Data Analysis for Chemists*. Oxford University Press, New York.
34. Johnson, M., Maggiora, G. M. (1990). *Concepts and Applications of Molecular Similarity*. Wiley, New York.
35. Bishop, C. M. (1996). *Neural Networks for Pattern Recognition*. Oxford University Press, New York.
36. Shi, L. M., Fang, H., Tong, W., Wu, J., Perkins, R., M. Blair, R., Branham, W. S., Dial, S. L., Moland, C. L., Sheehan, D. M. (2001). QSAR models using a large diverse set of estrogens, *J. Chem. Inf. Comput. Sci.*, 41, 186–195.
37. Good, A. C., So, S. S., Richards, W. G. (1993). Structure-activity relationships from molecular similarity matrices. *J. Med. Chem.*, 36, 433–438.
38. De Gregorio, C., Kier, L. B., Hall, L. H. (1988). QSAR modeling with electrotopological state indices: Corticosteroids. *J. Comput.-Aided Mol. Design*, 2, 557–561.
39. Beger, R. D., Buzatu, D., Wilkes, J. G. (2002). Combining NMR spectral and structural data to form models of polychlorinated dibenzodioxins, dibenzofurans, and biphenyls binding to the AhR. *J. Comput.-Aided Mol. Design*, 16, 727–740.
40. Beger, R. D., Holm, K., Buzatu, D., Wilkes, J. G. (2003). Using simulated 2D ^{13}C-^{13}C NMR spectral data to model a diverse set of estrogens. *Internet Electronic J. Mol. Design*, 2, 435–453.
41. Beger, R. D., Buzatu, D., Wilkes, J. G., Lay, Jr., J. O. (2002). Developing comparative structural connectivity spectra analysis (CoSCoSA) models of steroid binding to the corticosteroid binding globulin. *J. Chem. Inf. Comput. Sci.*, 42, 1123–1131.
42. Beger, R. D., Wilkes, J. G. (2002). Comparative structural connectivity spectra analysis (CoSCoSA) models of steroids binding to the aromatase enzyme. *J. Mol. Recognition*, 15, 154–162.
43. Blair, R. M., Fang, H., Branham, W. S., Hass, B. S., Dial, S. L., Moland, C. L., Tong, W., Shi, L., Perkins, R., Sheehan, D. M. (2000). The estrogen receptor relative binding affinities of 188 natural and xenochemicals: Structural diversity of ligands. *Toxicol. Sci.*, 54, 138–153.
44. Branham, W. S., Dial, S. L., Moland, C. L., Hass, B. S., Blair, R. M., Fang, H., Shi, L., Tong, W., Perkins, R. G., Sheehan, D. M. (2002). Phytoestrogens and mycoestrogens bind to the rat uterine estrogen receptor, *J. Nutr.*, 132, 658–664.
45. Bursi, R., Dao, T., van Wilk, T., de Gooyer, M., Kellenbach, E., Verwer, P. (1999). Comparative spectra analysis (CoSA): Spectra as three-dimensional molecular descriptors for the prediction of biological activities. *J. Chem. Inf. Comput. Sci.*, 39, 861–867.
46. Waller, C. L., Opera, T. I., Chae, K., Park, H. K., Korach, K. S., Laws, S. C., Wiese, T. E., Kelce, W. R., Gray Jr., L. E. (1996). Ligand-based identification of environmental estrogens, *Chem. Res. Toxicol.*, 9, 1240–1248.
47. Kuiper, G. G. J. M., Carlsson, B., Grandien, K., Enmark, E., Haggblad, J., Nilsson, S., Gustafsson, J.-A. (1997). Comparison of the ligand binding specificity and transcript tissue distribution of estrogen receptors α and β. *Endocrinology*, 138, 863–870.
48. Fang, H., Tong, W., Perkins, R., Soto, A., Prechtl, N., Sheehan, D. M. (1998). Quantitative comparison of in vitro assays for estrogenic assays. *Environ. Health Prospect.*, 139, 723–729.

49. Rumelhart, D. E., McClelland, T. L. (1986). *Parallel Distributed Processing*. Brandford Books/MIT Press, Cambridge, MA.
50. *Jgravity*, version 1.0 beta-1 build-10. Titan Systems, LinCom Division.
51. Devillers, J. (1996). *Neural Networks in QSAR and Drug Design*. Academic, New York.
52. Kellogg, G. E., Kier, L. B., Gaillard, P., Hall, L. H. (1996). E-state fields: Applications to 3D QSAR. *J. Comput.-Aided Mol. Design*, *10*, 513–520.
53. Polanski, J. (1997). The receptor-like neural network for modeling corticosteroid and testosterone binding globulins. *J. Chem. Inf. Comput. Sci.*, *37*, 553–561.
54. Safe, S. (1990). Polychlorinated biphenyls (PCBs), dibenzo-*p*-dioxins (PCDDs), dibenzofurans (PCDFs), and related compounds: Environmental and mechanistic considerations which support the development of toxic equivalency factors (TEFs). *Crit. Rev. Toxicol.*, *21*, 50–88.
55. Bhandiera, S., Sawyer, T., Romkes, M., Zmudzka, B., Safe, L., Mason, G., Keys, B., Safe, S. (1984). Polychlorinated dibenzofurans (PCDFs): Effects of structure on binding to the 2,3,7,8-TDDD cytosolic receptor protein, AHH induction and toxicity. *Toxicology*, *32*, 131–144.
56. Bandiera, S., Safe, S., Okey, A. B. (1982). Binding of polychlorinated biphenyls classified as either phenobarbitone-, 3-methylcholanthrene- or mixed type inducers to cytosolic Ah receptor. *Chem. Biol. Interact.*, *39*, 259–277.
57. Poland, A., Knutson, J. C. (1982). 2,3,7,8-Tetrachlorodibenzo-*p*-dioxin and related halogenated aromatic hydrocarbons: Examination of the mechanism of toxicity. *Annu. Rev. Pharmacol. Toxicol.*, *22*, 517–554.
58. Poland, A., Glover, E., Kende, A. S. (1976). Stereospecific, high affinity binding of 2,3,7,8-tetrachlorodibenzo-*p*-dioxin by hepatic cytosol: Evidence that the binding species is the receptor for induction of aryl hydrocarbon hydroxylase. *J. Biol. Chem.*, *251*, 493–494.
59. Safe, S. (1984). Polychlorinated biphenyls (PCBs) and polybrominated biphenyls (PBBs): Biochemistry, toxicology and mechanism of action. *Crit. Rev. Toxicol.*, *13*, 319–395.
60. Safe, S. H. (1986). Comparative toxicology and mechanism of action of polychlorinated dibenzo-*p*-dioxins and dibenzofurans. *Annu. Rev. Pharmacol. Toxicol.*, *26*, 371–399.
61. Mekemyan, O. G., Veith, G. D., Call, D. J., Ankley, G. T. (1996). A QSAR evaluation of Ah receptor binding of halogenated aromatic xenobiotics. *Environ. Health Perspect.*, *104*, 1302–1310.
62. Turner, D. B., Willett, P., Ferguson, A. M., Heritage, T. (1997). Evaluation of novel infrared range vibration-based descriptor (EVA) for QSAR studies. 1. General application. *J. Comput.-Aided Design*, *11*, 409–422.
63. *Physicians' Desk Reference*, 54th ed. (2000). Medical Economics Company, Montvale, NJ.
64. Joklik, W. K., Willett, H. P., Amos, D. B., Wilfert, C. M. (1988). *Zinsser Microbiolgy*, 19th ed., pp. 128–160.
65. Grail, B. M., Payne, J. W. (2002). Conformational analysis of bacterial cell wall peptides indicates how particular conformations have influenced the evolution of

penicillin-binding proteins, b-lactam antibiotics and antibiotic resistance mechanisms. *J. Mol. Recognit.*, *15*, 113–125.
66. Compadre, R. L. L., Pearlstein, R. A., Hopfinger, A. J., Seydel, J. K. (1987). A quantitative structure-activity relationship analysis of some 4-aminophenyl sulfone antibacterial agents using linear free energy and molecular modeling methods. *J. Med. Chem.*, *30*, 900–906.
67. Pickett, S. D., Sternberg, M. J. E. (1993). Empirical scale of side-chain conformational entropy in protein folding. *J. Mol. Biol.*, *231*, 825–839.
68. Karplus, M., Kushick, J. N. (1981). Method for estimating the configurational entropy of macromolecules. *Macromolecules*, *14*, 325–332.

7

USING MICROSOFT EXCEL® AS A LABORATORY DATA MANAGEMENT TOOL

A. ERIK RUBIN, MARK F. RUSSO, AND WILLIAM NEIL
Bristol-Myers Squibb
New Brunswick, New Jersey

7.1	INTRODUCTION	287
7.2	EXCEL AS A LABORATORY DATA MANAGEMENT TOOL	288
	Deciding If Excel Is the Right Choice for Your Application	288
7.3	TAPPING INTO THE HIDDEN POWER OF EXCEL: VISUAL BASIC FOR APPLICATIONS	290
	Recording a Macro	290
	Fundamentals of VBA and Excel	291
7.4	READING, WRITING, AND PRESENTING DATA	304
	Clipboard	304
	Opening and Saving Data Files	306
	Reading and Writing Data Files Using VBA	307
	Importing Data from External Databases	319
	Automating the Generation of Reports	327
7.5	SUMMARY AND CONCLUSION	335

7.1 INTRODUCTION

Few and far between are laboratory scientists who have not faced a data management challenge of some kind in their careers. The chief reason for this is

Drug Discovery Handbook, by Shayne Cox Gad
Copyright © 2005 by John Wiley & Sons, Inc.

that data are, in fact, the primary deliverables from laboratory experiments. Data management troubles can range from simple tasks such as devising a system to store printouts of high-performance liquid chromatography (HPLC) chromatograms and link them to entries in one's notebook to complicated scenarios such as creating a data storage and visualization application for a high-throughput workflow that operates 24 h a day. With the advent of high-throughput screening (HTS) and high-throughput experimentation (HTE) strategies and techniques and their prevalence in modern laboratories, data collection and storage obstacles are more common than ever. In the midst of this chaotic setting, Excel can be a tremendous asset to someone faced with a data management predicament. The key to success, however, is knowing when Excel is the appropriate tool for the job.

This chapter will explore Excel's utility as a data management tool. We will look at criteria for selecting Excel as a data management solution, examine Visual Basic for Applications (Excel's built-in programming language), discuss common techniques for manipulating data in Excel, and consider strategies for automatic reporting of data.

7.2 EXCEL AS A LABORATORY DATA MANAGEMENT TOOL

Deciding If Excel Is the Right Choice for Your Application

Consider the following two data management challenges. Can you determine which one is appropriate for Excel?

Scenario 1 A chemical engineer wishes to measure the rate of filtration of a slurry of product in a reaction solvent during an isolation step. She accomplishes this by directing the filtrate to a collection vessel that sits atop a balance. She plots the mass of the collection vessel (which she reads off the balance display and records in her notebook) as a function of time. The engineer would like to automate this process by having a computer collect the mass from the balance every minute or so, store the data in a table, and plot the results on a graph. She would like to be able to e-mail the results to the other team members.

Scenario 2 A pharmaceutical company desires a compound management system to facilitate storage and retrieval of its hundreds of thousands of compounds. The data management software should enable tracking of the location of each compound, store information about each compound, relate the usage of each sample back to the original sample entry, and be accessible by hundreds of users through the company's intranet.

You probably guessed correctly. Scenario 1 is indeed the challenge that is appropriate for Excel. How did you choose? What tipped you off? Let us look more closely at the criteria for choosing Excel as a data management solution.

First and foremost, *Excel is not a database management system*. Consequently, rule number one is: If your application requires a database management system, Excel is not appropriate. A database program, such as Access or Oracle, is a better choice in that case. In fact, Excel and Access share many things in common. Both allow powerful queries for sorting and filtering, both run sophisticated calculations on stored data, and both store data in tabular format. In addition, both allow the generation of reports in multiple formats, both have a built-in programming language for customizations, and both can connect to external data sources such as files, databases, and instruments. So, how does one decide if the application requires a database such as Access or a spreadsheet such as Excel? There are a few simple guidelines that will help.

A database management system such as Access is appropriate when:

- Your data must be described in multiple tables that are related to each other. That is, you require a relational database.
- You must store a very large amount of data (thousands of entries).
- Your data include large text fields or binary data such as images.
- Many users will be interacting with and updating the data.

Excel is appropriate when:

- You can describe your data in a single table (you do not require a relational database).
- You are primarily performing calculations on, statistical comparisons of, or plotting your data (i.e., your data are primarily numerical).
- Your data set is of reasonable size (no more than 65,536 rows in theory, no more than a few thousand in practice).
- Only a few users will be interacting with and updating the data.

In addition to the technical guidelines above, there are two additional important criteria to consider. First, most scientists have Excel installed on their computers. Second, most scientists know how to use Excel already. Excel is so common in the scientific laboratory, it has been said that if chemists built planes, they'd probably be flown using Excel. User familiarity with the application environment and user interface is of tremendous benefit to an application developer and should be taken advantage of whenever possible.

Revisiting the two hypothetical data management challenges above, it is now clear that scenario 2 is completely inappropriate for Excel. However, closer inspection of scenario 1 reveals a few issues that are not addressed by simply *selecting* Excel as the data management tool. Recall that, in scenario 1, the engineer required that Excel collect data from the balance at predefined time intervals. How will Excel know when and how to get the data from the balance? The better part of this chapter is dedicated to learning how to get data into and out of Excel. The key is knowing how to tap into the hidden power of Excel. That is, learning how to exploit Excel's built-in programming language, Visual Basic for Applications.

7.3 TAPPING INTO THE HIDDEN POWER OF EXCEL: VISUAL BASIC FOR APPLICATIONS

As a Microsoft Excel user, you have undoubtedly come across Visual Basic for Applications (VBA) at some point even though you may not have realized it. Have you ever recorded a macro in Excel or used a macro recorded by someone else? Have you ever installed "software" from an instrument vendor that modifies Excel by adding a new menu item to the menu bar in order to expose some specific functionality? In fact, it is the built-in VBA functionality in Excel that enables both of these scenarios. In this section, we'll learn how.

Recording a Macro

Getting started with VBA and Excel is actually very easy: you merely create a macro. A macro is simply a routine or sequence of events (mouse clicks, key strokes, etc.) that is *recorded* and then, when the macro is run, played back exactly. Macros are very useful for lengthy tasks in Excel that are needed frequently. Recording a macro is a cinch. To begin, launch Excel and be sure that a blank workbook is open. On the Tools menu, point to Macros, and then click on Record New Macro. If you like, give your macro a name by entering it in the Macro name box. Click OK. You will see a new toolbar pop up that has a VCR-style stop button on it. You will need to click this button when you want to stop recording your macro. Type some numbers into the first five or six cells in column A on the active worksheet. Highlight those cells and then chart them using the Excel Chart Wizard (use the default wizard settings for this example). Click the Stop button to stop recording the macro. Your worksheet should look something like Figure 7.1. To test the macro, clear the worksheet (including the charts) and, on the Tools menu, point to Macro, then click on Macros. Select your macro from the list box and click Run. Your worksheet should look exactly as it did before clearing it (Fig. 7.1).

Under the hood, macros are nothing more than VBA subroutines that were written automatically by the Excel macro recorder as you performed the actions. In fact, the Excel macro recorder is the single most powerful instrument available to the Excel VBA programmer for learning how to write code to perform certain tasks. If you do not know how to do something, record a macro and look at the generated code. It is that simple. To view the code generated by the macro recorder, you must open the Excel Visual Basic Editor (VBE). To do this, on the Tools menu, point to Macro, and click on Visual Basic Editor. The VBE looks like, and in many ways is, a completely separate program embedded within Microsoft Excel. With the VBE open, you will see a window titled Project in the upper left-hand corner. This is the Project Explorer. In this window, you will see a tree structure with one or more root entries each called VBAProject (Fig. 7.2). In parentheses, after "VBAProject," the name of the workbook will be displayed (e.g., "Book2" in this

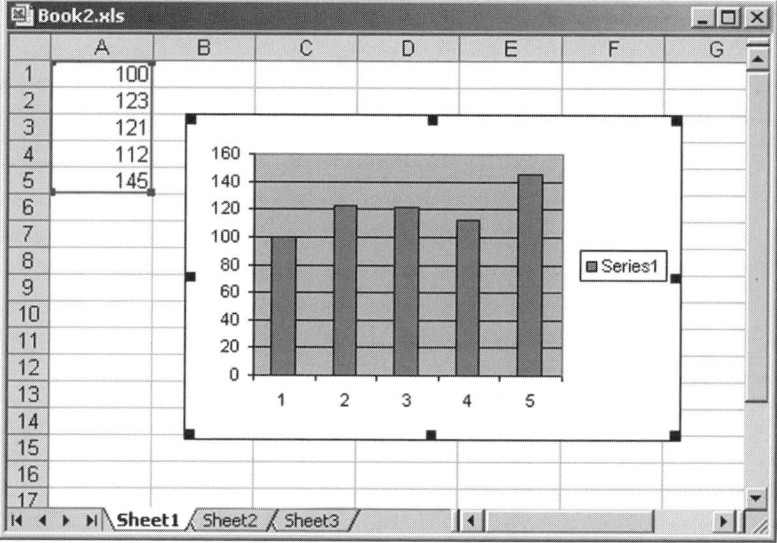

Figure 7.1 Macro output.

example). Expand the root node that corresponds to the workbook that you are working with, then expand the node with the name Modules. Double-click on the node called Module1. This will display the module that contains the code for your macro. Your screen should look something like Figure 7.2. The code that is displayed is raw VBA code. If you have never seen VBA code before, congratulations! You just wrote your first VBA code block without typing a single line. Now that we have seen VBA code, in the next section we will delve into the meaning of this code and learn how one can use VBA to manipulate Excel.

Fundamentals of VBA and Excel

To really become an expert VBA programmer in Excel, it takes a lot of patience and practice and it helps to have a project or a problem that you are trying to solve. There are many excellent texts written on the subject and new books are popping up continuously. The reader is encouraged to consult these references as his or her education progresses. Although a complete treatment of the interaction between VBA and Excel is neither feasible nor appropriate in this chapter, the authors believe that the key to understanding the power of VBA is appreciating four simple concepts that are fundamental to programming with VBA in Excel. These four concepts are:

- Understanding the structure of a VBA application (where the code is located)

292 USING MICROSOFT EXCEL AS A LABORATORY DATA MANAGEMENT TOOL

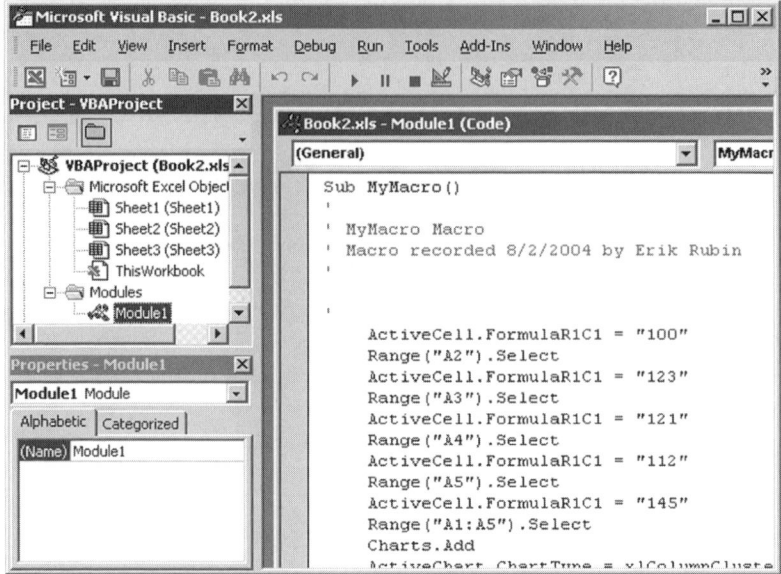

Figure 7.2 Visual basic editor (VBE) with macro code showing.

- Basic VBA language syntax (how to write the code)
- Event-driven code execution (when the code will run)
- Interacting with Excel objects (how to manipulate Excel through code)

In this section we will introduce each of these concepts to familiarize the reader. Later, we will address the application of these concepts to simple data management challenges.

Visual Basic Editor: Elements of a VBA Program As you probably have gleaned from the introduction to Excel macros, Excel VBA code is written in the Visual Basic Editor (VBE) and the code is stored in code modules. VBA code modules are associated with an Excel workbook and are actually embedded into the binary ".XLS" file (i.e., they are not stored as separate files). There are four types of VBA code modules in which you can write code. These are (1) standard modules, (2) user forms, (3) class modules, and (4) Microsoft Excel Object modules. Let us look at each of these types of code modules.

1. We have already seen standard modules. The Excel macro recorder added a standard module called Module1 to your VBA project when you recorded a new macro in the example above. To add a standard module to the VBA project yourself, switch to the VBE (ALT+F11) and activate the VBA project to which you want to add a module (to activate the project, click on your VBAProject in the project explorer). On the Insert menu, click on Module. A new module will be added to the Modules

node in the project explorer. A standard module is a common place to store procedures and functions for your VBA program. You can organize your code by using several modules and naming each module appropriately.
2. UserForms allow the VBA programmer to create custom dialog boxes that aid in interfacing with the end user. To add a UserForm to your project, click on UserForm on the Insert menu while in the VBE. A custom UserForm is a great tool when you need to collect information from a user in order to move to the next step in a program. A user form actually consists of two parts: a form designer and a code module. The form designer is where the programmer lays out the look and feel of the form and decides where the controls will be located. The code module is where the programmer writes code to respond to end-user actions on the form (e.g., when the user selects an item from a drop-down or clicks a button). Designing custom user forms is very simple, although we will not discuss them further here.
3. Class modules allow the programmer to create custom objects. A treatment of object-oriented programming is out of the scope of this chapter.
4. Excel Object modules, which are actually class modules, are associated with Excel workbooks and worksheets. There is one module called ThisWorkbook that is associated with each workbook, and each worksheet in the workbook has an associated sheet object module. You cannot add and remove sheet object modules in the traditional way. They are added and removed along with the actual Excel worksheets. You cannot delete the ThisWorkbook module ever. These modules can hold code in a manner similar to a standard module. These modules are especially useful, however, because they allow the programmer to respond to (write code for) specific events associated with the underlying object. For example, you could run initialization code whenever a workbook is opened, or run cleanup code when a workbook is closed. You could also prevent the deletion of a worksheet. We will look more closely at this topic when we address event-driven programming later in this chapter.

VBA Language Primer The Visual Basic for Applications language is like all programming languages at the root. Interim data are stored in variables, code stored in code modules are organized into callable procedures and functions, and the flow of a sequence of code is controlled by standard execution control constructs such as loops and conditional statements. In this part, we will introduce the VBA syntax associated with these core programming elements.

As mentioned, interim data are stored in variables. In VBA, variables are declared using the `Dim` keyword and can hold several different intrinsic data types. For example, to declare a variable of type integer, one would use a code statement similar to the one in Listing 7.1:

```
            Dim intMyInteger As Integer
```
Listing 7.1 Variable declaration in VBA.

This statement can occur anywhere in a code module and essentially defines a placeholder for your data called `intMyInteger`. As we will see, variables can be used within procedures and functions to manipulate the data the variables contain. There are several intrinsic data types available to the VBA programmer; Table 7.1 lists the most common and useful of the lot. When a variable data type listed in Table 7.1 is found after the `As` keyword in a variable declaration, it defines the type of data that is stored in the variable. For example, the variable `intMyInteger` above could not store the text value "Visual Basic for Applications", which is of type String.

When code is executed in VBA, it is actually executed by calling a procedure. The code is contained within the definition of the procedure. For example, in the macro recording example above, the macro code is contained within a procedure called "MyMacro" (Fig. 7.2). When the macro is run, the Excel macro execution engine actually searches for and runs the procedure that has the same name as the macro requested. VBA offers a number of ways to define procedures, and we will introduce the two most common: subroutines and functions.

A subroutine is a list of code statements (a code block) that is contained within `Sub` and `End Sub` statements. A subroutine does not return a value to the caller after its execution. In Listing 7.2, subroutine `MyFirstSubroutine` declares a variable of type String and then sets the value of that variable to "I love hockey". If you were to call this subroutine, it would execute the two statements and then exit uneventfully:

```
      Sub MyFirstSubroutine()
          Dim strMyString As String
          strMyString = "I love hockey"
      End Sub
```
Listing 7.2 A subroutine.

TABLE 7.1 Common Intrinsic VBA Data Types

Variable Data Type	Data Type Description
Boolean	True or False
Integer	An integer in the range –32,768 to 32,767
Long	An integer in the range –2,147,483,648 to 2,147,483,647
Double	A floating-point number roughly in the range –1.79769313486232E308 to 1.79769313486232E308
String	A sequence of characters
Variant	A data type that can hold any intrinsic data type

A function, on the other hand, is a code block that is enclosed by `Function` and `End Function` statements. Unlike subroutines, a function returns a value to the caller after it executes. For example, in Listing 7.3 the function `MyFirstFunction` simply returns the String value "I love hockey too". Note that the function is defined as a data type (String, in this case). This data type is the type associated with the returned value. If no type is indicated, the default type is Variant.

```
Function MyFirstFunction() As String
    MyFirstFunction = "I love hockey too"
End Function
```
Listing 7.3 A function.

In addition, one procedure can call another procedure. So, in Listing 7.4, when `MySecondSubroutine` is executed, `strMyString` will be set to the return value of `MyFirstFunction`, which is, if you recall from Listing 7.3, "I love hockey too".

```
Sub MySecondSubroutine ()
    Dim strMyString As String
    strMyString = MyFirstFunction()
End Sub
```
Listing 7.4 Example of a procedure calling another procedure.

As we learned, one procedure can call another and that procedure can call yet an other procedure, and so on, and so on. But who or what calls the first procedure in the sequence? The answer to that question is, in fact, one of the most important concepts associated with learning how to program VBA in Excel. We will answer that question in the next part of this section, event-driven programming. But before we move on, there are a few more basic VBA syntax topics we must cover.

Subroutines can take arguments. Arguments are variables that are defined right in the declaration of the subroutine and act as a way to pass information to a subroutine when it is called. For example, in Listing 7.5, function `TimesTwo` defines an argument, called `MyNumber`, of type integer. This function takes the number held in `MyNumber` and doubles it. The function then returns the doubled value to the caller.

```
Function TimesTwo (MyNumber As Integer) As Integer
    TimesTwo = MyNumber * 2
End Function
```
Listing 7.5 Passing arguments to procedures.

VBA offers standard methods of code execution control as well. These are code constructs that allow looping (or iteration) and condition evaluation (or selection). VBA offers two main looping constructs: For-Next and Do-Loop. Listing 7.6 demonstrates a For-Next loop that increments a variable named I from a value of 1 to 5. During each iteration, the value of I is added to another variable, J. When the code block is finished executing, J holds the value 15, which is the result of 1 + 2 + 3 + 4 + 5.

```
Dim I As Integer
Dim J As Integer
For I = 1 to 5
    J = J + I
Next I
```

Listing 7.6 A For-Next Loop example.

A Do-Loop, on the other hand, executes continuously while or until a condition is met. For example, the condition can be checked each time before a code block is executed as in Listing 7.7.

```
Dim I As Integer
Do While I < 10
    I = I + 1
Loop
```

Listing 7.7 A Do While-Loop example.

If the condition is verified at the end of each block execution, then the Until keyword is placed after the Loop keyword as in Listing 7.8. In fact, While and Until can be used interchangeably in both examples. It is more readable, however, to use the convention Do While-Loop and Do-Loop Until.

```
Dim I As Integer
Do
    I = I + 1
Loop Until I < 10
```

Listing 7.8 A Do-Loop Until example.

In addition to iteration execution control, VBA defines an important conditional construct: If-Then-Else. The If-Then-Else construct may also include an arbitrary number of conditions by adding Elseif statements. For example, consider Listing 7.9.

```
Dim MyNumber As Integer
MyNumber = 4
If MyNumber = 1 Then
    MsgBox "MyNumber is 1"
Elseif MyNumber = 5 Then
    MsgBox "MyNumber is 5"
Else
    MsgBox "MyNumber is neither 1 nor 5"
End If
```

Listing 7.9 An If-Then-Else Block.

The code in Listing 7.9 first declares an integer variable and sets its value to 4. The If-Then-Else block that follows opens with a statement that is interpreted as "If the number held in the variable MyNumber is equal to 1 then execute the code right underneath me (but before the next condition)." If the initial condition is not met (i.e., the number is not 1), the next condition is evaluated. In this case, the next condition reads "If the number held in the variable MyNumber is equal to 5 then execute the code right underneath me." If that condition is not met either, the next condition, Else, catches all remaining conditions. The If-Then-Else construct ends with the End If statement.

Event-Driven Programming Let us return, for a moment, to the VBA Primer part above, where we presented the "chicken and egg" scenario for code execution. We introduced the fact that one procedure can call another and that that procedure can in turn call another procedure. The question we left hanging was "Who calls the first procedure?" In VBA, this question is the same as asking "How do I run my program?" The answer is "by responding to events." We have seen an example of event-driven programming already: running our macro. To run the macro, we know to use the Excel macro player from the Tools menu. Another way to interpret this is that the macro code was run as a response to the event associated with the user selecting to play the macro from the Tools menu. Most of your code in a VBA program will be run in response to events. Some examples of events that you might code for are a user clicking a button, a user selecting a custom menu item, or a user closing or opening a workbook. In short, when and how your code will run will depend on one main thing: the events for which you coded responses.

Event-driven programming is best illustrated with an example. Let us say you wanted to modify Excel's behavior to include the following features for your workbook: When a new worksheet is added to the workbook, (a) the date and time that workbook was opened is printed in cell A1 of the new sheet, and (b) the date and time that the worksheet was created is printed in cell A2 of the new sheet. Solving the problem in this example involves two steps:

1. When the workbook opens, store the current date and time.
2. When a new worksheet is added, print the stored time from (1) in cell A1 of the new sheet and print the current date and time in cell A2.

Coding for these two steps involves responding to the event associated with a workbook opening and the event associated with a new worksheet being added to a workbook. Without programmatic events telling the programmer when these actions occur, it would be impossible to predict when a new sheet was added to the workbook, for example. Let us look at a VBA solution to the problem stated above. To begin, open Excel and switch to the VBE. While in the VBE, activate the VBAProject associated with your workbook by clicking on it in the Project Explorer. From the Insert menu in the VBE, click Module. In the newly created standard module, type the code in Listing 7.10.

```
Dim WorkbookOpenTime As Variant

Function GetWorkbookOpenTime() As Variant
    GetWorkbookOpenTime = WorkbookOpenTime
End Function

Sub SetWorkbookOpenTime()
    WorkbookOpenTime = Now
End Sub
```

Listing 7.10 Part 1 of the code for event-driven programming example.

The code in Listing 10 first declares a variable called `WorkbookOpenTime`. This variable will store the date and time when the workbook was opened. Next, a function called `GetWorkbookOpenTime` is defined, which will return the value of the variable `WorkbookOpenTime`. Finally, a subroutine, `SetWorkbookOpenTime`, is defined that sets `WorkbookOpenTime` to the current date and time. You have probably noticed the use of keyword `Now` in the code above. `Now` is a special function, built into VBA, that returns the current date and time.

To write code for the workbook events that we need for this example, open the `ThisWorkbook` code module by double clicking on it in the Project Explorer. At the top of the code window, you will see two drop-down boxes. From the left drop-down box, choose Workbook. With Workbook selected in the left box, the right drop-down box will list all of the available events that you can code for in Excel. Select Open from this menu. VBA automatically adds code to the module for you. The code is the outline (or stub code) for the event procedure that will be called when the workbook is opened. An event procedure is no different from the subroutines we discussed previously. There is one exception, however: The name of an event procedure is very specific. The name must be in the format `Object_Event`. In the present case, the event procedure name is `Workbook_Open`. In most cases, as in the present case, Excel writes the shell of the event procedure for you automatically. To add the stub code for the other event that we need to respond to, choose Workbook again from the left drop-down box, then click on NewSheet from the right drop-down box in the `ThisWorkbook` code module. This will add the event procedure shell for the NewSheet event. To respond to an event, all the programmer must do is write code between the `Sub` and `End Sub` statements in the event procedure. For this example, complete the automatically generated event procedure shells as in Listing 7.11.

```
Private Sub Workbook_Open()
    SetWorkbookOpenTime
End Sub

Private Sub Workbook_NewSheet(ByVal Sh As Object)
    Sh.Range("A1").Value = "Workbook Opened: " & _
        GetWorkbookOpenTime
    Sh.Range("A2").Value = "Sheet Created: " & Now
End Sub
```

Listing 7.11 Event procedure code.

That is all that is necessary to code a solution to our problem. In Listing 7.11, the first event procedure fires when the workbook is opened. The subroutine `SetWorkbookOpenTime` is called, which, if you recall from Listing 7.10, sets the variable `WorkbookOpenTime` to the current time. The second event procedure fires when any new worksheet is added to the workbook. Note that the event procedure code for the NewSheet event passes an argument, `Sh`, of type Object. This is a reference to an Excel object that represents the new worksheet that was added. This reference is used to place the workbook opened date in cell A1 and the worksheet creation date in cell A2. In the next part, we will explore Excel objects and their use in more detail.

Excel Objects Understanding Excel objects is as essential to programming in Excel as the steering wheel is to driving your car. Fully appreciating Excel objects, however, is not a simple task. The final part of this section is dedicated to presenting Excel objects. In order to properly introduce Excel objects we must first understand the concept of software objects.

Objects are software entities (blocks of code) that contain certain related functionality. The term used to describe the containment of this functionality is *encapsulation*; objects are said to *encapsulate* related functionality. In practice, software objects make their encapsulated functionality available to the programmer through procedures and variables, just like the procedures and variables that we introduced earlier in this chapter. With objects, however, variables are not usually directly accessible and must be read and set through properties that the object makes available. By convention, the procedures that an object makes available are called methods, and not subroutines or functions. In technical lingo, an object is said to *expose* its methods and properties, and the specific methods and properties that an object exposes is referred to as the object's *interface*. Syntactically, an object's interface is addressed through the so-called *dot notation*. For example, let us take an arbitrary object called MyInstrument and assume it exposes one property called Name (that returns a string containing the instrument name) and one method called Start (that starts the instrument). In VBA code, the method would be called using the syntax MyInstrument.Start and the property would be read using syntax similar to SomeStringVariable = MyInstrument.Name. In dot notation, the object's name and the property or method name are separated by a period.

Oftentimes, software objects correspond to entities that are readily identifiable. This is the case with many of the objects in Excel. For example, two of the main object types that programmers frequently interact with are the Workbook and Worksheet objects. These objects encapsulate all of the functionality associated with Excel Workbooks and Worksheets. The Workbook and Worksheet objects' interfaces allow the programmer, through VBA, to interact with, manipulate, and modify the Worksheets in an Excel workbook and the Workbook in an Excel file. Excel objects represent a bridge between the VBA core language and the Excel application. For this reason, interacting with Excel objects is the crux of programming VBA in Excel. There are hundreds of Excel objects and they are arranged in a hierarchical representation called the Excel *object model*. The object browser offers an excellent way to inspect the Excel object model and all of the properties and methods that Excel objects expose. To open the object browser, while in the VBE click on Object Browser in the View menu. With the object browser open, select Excel from the top drop-down menu (this will probably initially say ⟨All Libraries⟩ by default). The left pane of the browser lists all of the available Excel objects. The right pane lists all of the available properties and methods exposed by the object selected in the left pane.

Let us explore Excel objects more deeply by investigating the Worksheet object more closely. Every reader of this chapter is likely very familiar with the Excel worksheet concept. An Excel workbook is made up of a collection of Excel worksheets and the worksheets are where data, tables, and formulas are stored. What types of properties and methods can you imagine for the Worksheet object? The Worksheet object has a Name property, which changes the text displayed on the worksheet's tab. It has a Cells property that represents the cells on the worksheet. We will look at the Cells property when we address the Excel Range object later in this section. In fact, the Excel Worksheet object exposes 53 properties to the VBA programmer. Would it surprise you to learn that, included in its 27 methods, the Worksheet object exposes a Delete method that removes the worksheet from the workbook and a CheckSpelling method that verifies the spelling of words located in the worksheet's cells? In actuality, most of the Worksheet object's interface should be familiar to the everyday Excel user because it facilitates everyday worksheet tasks programmatically.

Let us return to the last example of the previous section on event-driven programming. In Listing 7.11, we wrote code to respond to the Workbook event, NewSheet, that fires when a new worksheet is added to the Excel workbook. In the event procedure, we made use of the Worksheet object, Sh, which is passed to the programmer as an argument when the event procedure is called by Excel. Sh, in this case is a variable that references a Worksheet object. The specific worksheet object that it references is the one that was just added to the workbook. We used the worksheet's Range property to place data in the newly added worksheet's cells. This code illustrates another feature of objects: An object's property can be an object too. In the code, we access the Worksheet object's Range property. This property returns an object of type Range (an object we will discus shortly), and it, in turn, exposes a property called Value that gives us access to the contents of the specific range or cell. Syntactically, this is the origin of the "dot-dot" notation that we see in the event procedure code in Listing 7.11.

When it comes to data management, there is no Excel object more important than the Range object. This is, of course, because the whole business of data management using Excel is getting data into and out of cells, and an Excel Range object is the object model representation of a cell in a worksheet. In fact, a Range object can be any collection of cells in a worksheet. It can be contiguous or disjoint. It can represent an entire row or an entire column. A range can even represent a random selection of cells on a worksheet. This may seem complicated at first. It may seem as though, unlike the Workbook and Worksheet objects, the Range object is not associated with any readily identifiable Excel component. At closer inspection, however, one sees that a Range object can be anything that you, as an Excel user, can select on a worksheet. For example, if you click on cell A1, you have just selected a range that comprises one cell. If you hold down the Ctrl key and click on cell B6, you now have selected a range that comprises two noncontiguous cells. In you click on the "H" column header, you selected a range that comprises an entire column. So, in fact, a Range object is identifiable after all.

Using a Range object associated with a worksheet is quite simple. A Worksheet exposes two key properties that return Range objects when they are called. These are the Cells and the Range properties. The key difference between these two properties is the format of the arguments that they take. The Cells property accepts two integers that refer to the row and the column of the cell that you would like to reference. For example, to obtain a reference to Cell B1, you could use the VBA code in the Listing 7.12.

```
Dim MyRange As Range
Set MyRange = ActiveSheet.Cells(1,2)
```

Listing 7.12 Using the Cells property of the Worksheet object.

Listing 7.12 highlights one of the most frustrating elements of learning how to program in VBA as a beginner. As a reader, you have just read through thousands of words introducing you to VBA syntax and Excel objects, but in these two lines of code in Listing 7.12, there are three things that have not been introduced anywhere in this chapter. The reality is that mastering VBA and Excel takes time, patience, and experience, and no single chapter can do the subject justice. Specifically, the new material presented here are the use of a Range type in a variable declaration, the use of the `Set` keyword, and the use of `ActiveSheet`. Just like the declaration of a variable of a type intrinsic to VBA (Table 7.1), a variable can be declared as any of the Excel objects in the object model. One can look at Excel objects as data types *intrinsic* to Excel. For example, to declare a variable of type Worksheet, one need only type `Dim wsht As Worksheet`. In the example above, `MyRange` is a variable of type Range. There is one difference between variables that are Excel object types and variables that are intrinsic data types. The difference is the use of the `Set` keyword during assignment. For example consider the code in Listing 7.13:

```
Dim J as Integer
J = 1000
```
Listing 7.13 Standard assignment to a VBA variable.

The above code is syntactically correct and will run with no problem. However, the code in Listing 14 would result in a error:

```
Dim R as Range
R = ActiveSheet.Cells(1,2)
```
Listing 7.14 Incorrect assignment of an object to a variable.

The difference is that assignment to object variables requires the `Set` keyword. So, the correct syntax for the above code is shown in Listing 7.15:

```
Dim R as Range
Set R = ActiveSheet.Cells(1,2)
```
Listing 7.15 Using the `Set` keyword to assign an object to a variable.

Finally, `ActiveSheet` is a globally available reference to a Worksheet object that represents the active worksheet. The active worksheet is the worksheet in the foreground of the active workbook. That is, the worksheet currently being "worked on" by the Excel user.

Getting back to the Range object, we see that the code `ActiveSheet.Cells(1,2)` returns a Range that represents the cell in row 1, column 2 of the active worksheet. That is, the code returns the Range object corresponding to cell B1. The Range property of the Worksheet object operates a little differently. In its most common syntax, the Range property accepts a string with A1-style notation as the argument. For Excel users, this is a familiar syntax because it is the same as the syntax used to reference cells in Excel formulas. For example, the following code would be used to reference the six cells in the Excel range A1:B3:

```
Set R = ActiveSheet.Range("A1:B3")
```
Listing 7.16 Using the Range property of the Worksheet object.

Like all objects, the Range object exposes properties and methods. The most useful Range properties to the beginner, however, are the Value and Formula properties. The Value property allows the programmer to read and write the contents of a cell represented by a Range. For example, the code in Listing 7.17 sets the contents of the cell A3 to "I just set this programmatically".

```
ActiveSheet.Range("A3").Value = _
        "I just set this programmatically"
```
Listing 7.17 Putting data into an Excel cell.

Similarly, the code in Listing 7.18 gets the contents of cell C4 and stores the result in a variable called MyVar:

```
Dim MyVar as Variant
MyVar = ActiveSheet.Cells(4,3).Value
```
Listing 7.18 Getting data out of an Excel cell.

The Formula property allows the programmer to set and get a cells formula through code. For example, the code in Listing 7.19 places the number 10 in cell A3 and then assigns the Excel formula "=A3/2" to cell A4. The result is displayed as the number 10 in cell A3 and the number 5 in cell A4.

```
ActiveSheet.Range("A3") = 10
ActiveSheet.Cells(4,1).Formula = "=A3/2"
```
Listing 7.19 Using the Formula property of the Range object.

In this section, we introduced the hidden power of Excel: Visual Basic for Applications. We began with macro recording, traveled through basic syntax, looked at event-driven programming, and culminated with Excel objects, where we learned how to get data into and out of Excel cells. The heart of data management programming in Excel is getting data into and out of Excel cells, and the remainder of this chapter will explore how we can build on what was introduced in this section to address more intricate data management challenges.

7.4 READING, WRITING, AND PRESENTING DATA

Excel has many powerful tools for managing and manipulating data. But, before these tools can be applied, your data must be moved into the cells of an Excel worksheet. This task is relatively straightforward for small, conveniently formatted data sets. Unfortunately, it can be very difficult for larger data sets that are stored in a format not recognized by Excel. After completing the desired data manipulation, frequently it is necessary to insert the data into a worksheet, display it graphically, or move the data out of Excel, into a file, another program, or a reporting tool. In this section we will describe how to move data into and out of Excel, and how to present it.

Clipboard

The simplest way to move data into and out of Excel is through the clipboard, using the copy and paste functions. When using your favorite program, if you highlight text, an image, or some other data, and then type Ctrl+C or select Copy from the Edit menu, the data you have highlighted are copied to the Windows clipboard. The format of the data on the clipboard will vary. Indeed, many times multiple formats are stored on the clipboard in order to maximize the number of ways in which the data can be pasted. For example, when you highlight and copy data from the cells of an Excel worksheet, the data are copied to the clipboard in a surprising number of formats. You can see this by copying data from cells in Excel, then selecting Paste Special from Microsoft Word's Edit menu. Multiple formats are listed in the Paste Special dialog that is displayed, including various formatted and unformatted textual options, and even the option to paste the data as an image. The more you understand about the formats used by various programs to write data to the clipboard, the better you will be able to transfer your data using the clipboard.

For exchanging data with Excel, the most useful clipboard format option is the basic Unformatted Text format. Understanding this will help you appreciate the result of pasting data copied from an Excel worksheet into a plain text editor, such as Notepad, or how to format data in a text editor to ensure that it is copied and pasted properly into Excel. The format used by Excel for its Unformatted Text paste option is simple: Data items in the cells of a row are separated by tab characters, and rows are terminated with two characters in sequence—a carriage return followed by a linefeed. You can see this by copying data from cells in Excel and pasting into a Microsoft Word document using the Paste Special option of the Edit menu. Once pasted, your data will likely be arranged in a nice tabular format. To see the special tab and line terminating symbols in Word, select the paragraph symbol (¶) from Word's Standard toolbar. Tab characters will be displayed as right-pointing arrows (→) and the carriage-return+linefeed sequence of two characters will be displayed as the paragraph symbol (¶). To prove that the data are in a natural format for Excel, select the text in Word and copy it to the clipboard. Return to Excel, highlight an empty cell and paste it. This time there is no need to use Paste Special. Your data will be separated automatically into the cells of your worksheet.

Other options are available for when the data are not in the preferred tab-delimited Excel format. Let us assume you have a microplate reader that saves measured data to a text file as eight rows, with each row containing 12 comma-delimited numbers. For example, consider the file in Figure 7.3, which has been opened in the Notepad text editor. If you select and copy the contents of the file, and paste them into Excel, you'll end up with something that looks like Figure 7.4. Excel will not automatically split the numerical values among the cells of the worksheet. Instead, each row in the file will be pasted into the cells of the first column of the worksheet. To split the data into adjacent cells, select the data in the first column and choose the Text to Columns option from the Data menu in Excel. From the wizard that is displayed after this option is selected, choose Delimited from the first step and click the Next button. On the second step of the wizard check Command and Space from Delimiters, and the "Treat consecutive delimiters as one" option (see Fig. 7.5). The Comma option splits each row at the comma characters. The Space and "Treat

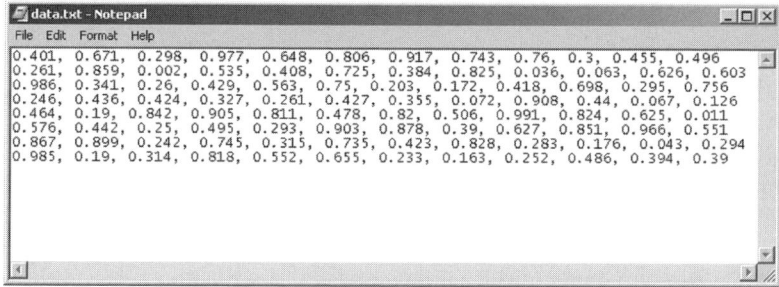

Figure 7.3 Typical data file generated by a laboratory instrument.

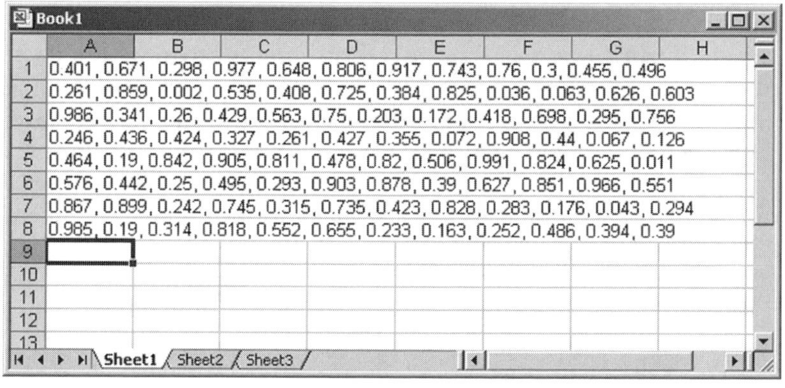

Figure 7.4 Comma-delimited data is pasted into the first cell of each row.

306 USING MICROSOFT EXCEL AS A LABORATORY DATA MANAGEMENT TOOL

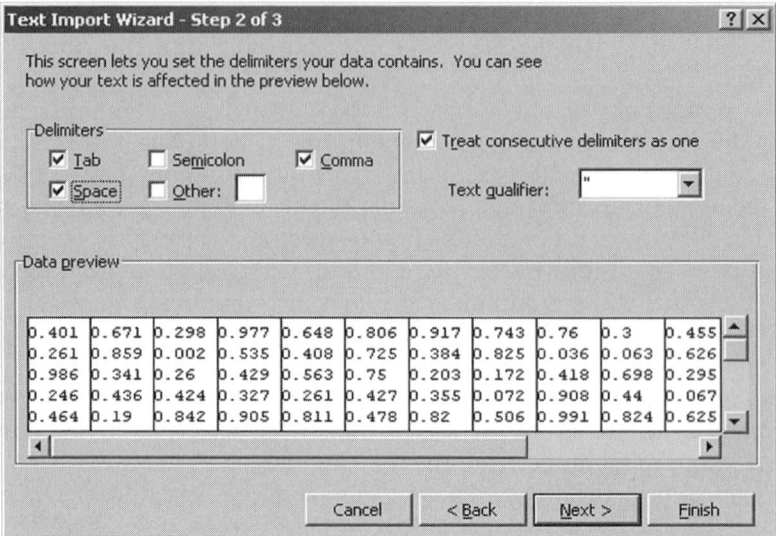

Figure 7.5 Second step of the Text to Columns wizard allows column dividers to be selected.

consecutive delimiters as one" options effectively eliminate the extraneous space that follows each comma in each row. Clicking the Finish button will cause the data to be split into the adjacent cells of a worksheet. If you paste data into the worksheet once again before Excel is closed, the same "Text to Columns..." options will be applied. If the copied data are formatted similarly, each line will be split into adjacent cells automatically without the need to select the "Text to Columns..." menu option. Once Excel is closed, the options are reset to their initial values and will need to be selected again the next time Excel is opened.

Opening and Saving Data Files

Text Import Wizard Some laboratory instruments have the ability to save measurements as native Excel files. It is much more common for an instrument to store measured data in plain text files. A data file may contain much more than a conveniently formatted table of numbers, but often, at some location in the file, the data are stored in a simple tabular format. Rather than open the file in a text editor and copy the data to the clipboard, Excel makes it possible to open the file directly into a worksheet. When attempting to open a text file, Excel recognizes that the file is plain text and displays the Text Import Wizard. Figure 7.6 is the dialog that will be displayed when you attempt to open the previous data.txt (Fig. 7.3) file directly in Excel using the Open menu option of the File menu. Note that this wizard is almost identical to the Text to Columns wizard. The Text Import Wizard gives you a couple of extra

READING, WRITING, AND PRESENTING DATA 307

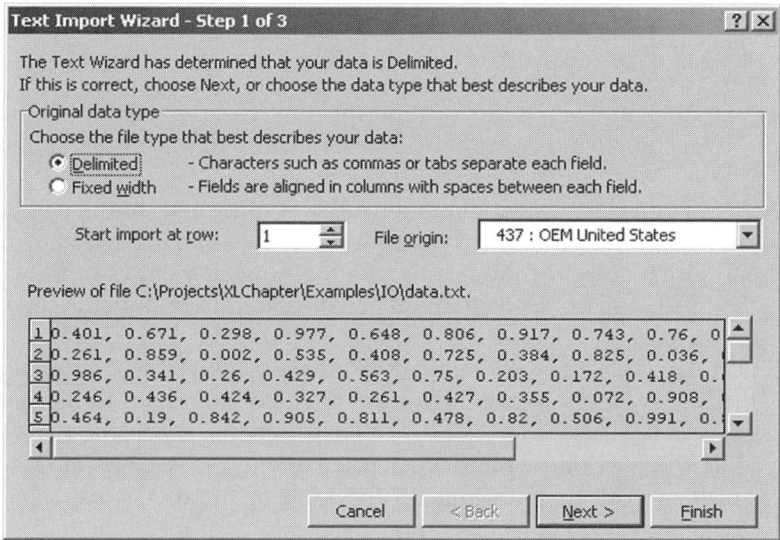

Figure 7.6 First step of the Text Import Wizard.

options, including the ability to skip rows in the file, which can be handy when you would like to skip over file header information. The Text Import Wizard process for splitting data into the cells of your worksheet is identical to the Text to Columns Wizard. This is a convenient option when data are stored in a file in a simple tabular format.

Saving Text and CSV Files Excel also provides an easy way to save data in a worksheet to plain text files rather than in Excel's native binary file format. Choosing the Save As option from the File menu will display the "Save As" dialog. In addition to specifying the name and location of the file to save, the file type can be selected. If the "Text (tab delimited)" option is selected, the data in the worksheet will be saved to a plain text file, where the data in each row is stored as a row in the text file, with data items separated by tab characters. The "CSV (comma delimited)" option in the "Save As" dialog will save the data in a similar format, with the only difference being that the data items are separated by commas.

Reading and Writing Data Files Using VBA

When plain text instrument files are written in a format that is not amenable to copy-and-paste or the Import Wizard, we must do the heavy lifting associated with opening the file, reading its contents, and transferring the data into the cells of a worksheet. In this section we will cover some of the more common techniques for opening, closing, reading, and writing data files with VBA and Excel.

Selecting Files to Open or Save The first step to accessing a file is to identify the file's name. In rare cases the file name will be fixed and can be hard-coded into a program. More commonly the user must be prompted to select or enter the file name when the VBA script runs. Excel provides two built-in functions to accomplish this; both are accessed through the Excel Application object.

The GetSaveAsFileName method of the Application object will display a standard Windows file dialog box and allow the user to navigate the file system and either select or enter a file name into which data can be written. GetSaveAsFileName takes a number of optional arguments, including InitialFileName, FileFilter, and Title. The InitialFileName argument is used to set the file name that will be displayed when the dialog is first opened. It is handy to set this to the most commonly used file name in order to save a few keystrokes. The Title parameter is a string that will be displayed as the caption of the dialog. It is common to set this argument to give the user of your script a hint on the kind of file that is expected. The FileFilter argument is a string that specifies one or more file name filters to be used by the dialog when selecting which files to be displayed. The function returns the selected file name, or if the user clicks the Cancel button, it returns False.

The subroutine in Listing 7.20 illustrates the use of GetSaveAsFilename. Note that we are using the line continuation character (_) in order to split a long command over several lines. In this example the vrnFileName variable is declared as type Variant. We use a Variant because the GetSaveAsFilename method can return multiple data types: When the user makes a selection, a String is returned representing the complete file path, otherwise the Boolean value False is returned when the dialog is cancelled.

```
Sub WriteAFile()

    Dim vrnFileName As Variant

    ' Ask the user for the file name
    vrnFileName = Application.GetSaveAsFilename( _
        InitialFileName:="data.csv", _
        FileFilter:= _
        "CSV Files (*.csv),*.csv, All Files (*.*),*.*", _
        Title:="Select Data File")
    ' Exit if the dialog is cancelled
    If vrnFileName = False Then Exit Sub

    ' Open and write the file here
End Sub
```

Listing 7.20 A subroutine illustrating the GetSaveAsFilename method of Application.

The `GetOpenFilename` method of the Application object displays a dialog that allows the user to enter or select a file that is intended to be opened. The method is called in a manner very similar to `GetSaveAsFilename`. One difference between the two methods is that `GetOpenFilename` does not accept the `InitialFileName` argument.

File Access Types In VBA, there are three ways to access data in a file. These are called sequential, random, and binary access. The main difference among these file access types is the way that data are read and written. Random and binary file access types usually are used to manipulate data stored in files in a form other than plain text. When a file is opened for random access, data must be stored in the file as fixed-size records; these records must be read and written as complete units. Binary file access permits individual bytes to be read and written. A common feature shared by these two file access types is that the data stored in the file can be accessed and manipulated in random order.

When a file is opened for sequential access, the data in the file can be read or written in a sequential manner only, that is, from the beginning of the file to the end. Unlike the other file access types, it is not possible to move randomly about a file opened for sequential access. Even though the sequential access type is much more limiting, it is by far the more common of the three access types because the sequential access type offers more convenient options for reading and writing plain text file data. We will consider only the sequential file access type in this section.

Opening and Closing Sequential Access Files Before data in a file can be accessed, the file must be opened. This is accomplished with the `Open` statement in VBA. The `Open` statement takes a number of arguments, including the name of the file to open, the file mode, and a unique file number.

In VBA unique numbers (integers) are used for nearly all of its native file-related commands. These numbers can take on any value between 1 and 511 but cannot be reused for a second file until the first file is closed. File numbers are assigned when a file is opened with the `Open` statement. It is possible to specify unused file numbers for the files that you open, but it is far easier to use a built-in VBA function called `FreeFile` to select available file numbers automatically. `FreeFile` is guaranteed to return a number that is not currently being used for another file.

The sequential file access type comes in three flavors, or modes: Input, Output, and Append. When opening a file with the Input mode keyword, you only will be able to read data from the file. Opening a file with the Output mode keyword will erase the file and allow you to start writing from the beginning. The Append mode keyword will open the file for writing but not erase its contents. Instead, writing will begin immediately after the end of the file. In other words, you will be appending to the file when you write.

When one is finished reading or writing a file, it must be closed. Closing a file is simply a matter of executing the `Close` statement, optionally followed by the pound sign (#) and the file number.

Listing 7.21 is a complete VBA subroutine called `ReadDataFromAFile`. It illustrates all the code necessary for selecting and opening a file for input, and closing it when finished. The `intFileNum` variable is declared as an integer and is assigned a new file number using the `FreeFile` statement. The `vrnFileName` variable is declared to hold the response from the `GetOpenFilename` function. The response will be either the complete file path containing the data to be read, or False if the dialog was cancelled.

The `Open` statement takes a number of arguments. The file name is given first as the `vrnFileName` variable, the `For Input` portion of the statement indicates that the file will be read, and the file number is assigned using the `As #intFileNum` portion of the statement. The file is closed at the end of the subroutine with the `Close` statement. The specific file to be closed is indicated by specifying the `#intFileNum` argument. Leaving out the argument will cause all open files to be closed.

```
Sub ReadDataFromAFile()
    Dim intFileNum As Integer  ' To hold the file number
    Dim vrnFileName As Variant ' To hold the file name

    intFileNum = FreeFile ' Get a file number

    ' Ask the user for the file name
    vrnFileName = Application.GetOpenFilename( _
        FileFilter:= _
        "CSV Files (*.csv),*.csv, All Files (*.*),*.*", _
        Title:="Read Data File")

    ' Exit if the dialog is cancelled
    If vrnFileName = False Then Exit Sub

    ' Open the file for Input
    Open vrnFileName For Input As #intFileNum

    ' Read data here

    ' Close the file before exiting
    Close #intFileNum
End Sub
```

Listing 7.21 Sample code for selecting a file name, opening the file, and closing it.

This subroutine can be tested by entering it into any code window of Excel's Visual Basic Environment. To run the code, move the cursor to a location within the subroutine and click the Run toolbar button (the blue right-pointing arrow on the standard toolbar of the VBE), or press the F5 key. The code will prompt for a file name, open the selected file, close the file, and exit in an uneventful manner. If you enter a file that does not exist, the dialog will not close. VBA will not create a file automatically when opening for Input. Conversely, when you open a file using the Output or Append modes, VBA will automatically create the file if it does not exist already.

In this section we have discussed how to open and close a file using the various sequential file access modes. In the next section we will discuss how to read and write these files.

Reading and Writing Sequential Access Files The most common way to write data to a file is to use the `Print #` statement. The first argument required by `Print #` is the number of the file used in the `Open` statement. Following the file number argument is a sequence of data items to be printed to the file. The number of data items is not fixed; essentially any number of items can be written with `Print #`.

Listing 7.22 includes a subroutine named `LogMessage`, which is a simple but powerful utility for keeping a log file. This subroutine can be very handy when there is a need to monitor the processing of a large amount of data or the progress of a complex program. Arguments passed to the `LogMessage` subroutine include the message to be written to the log file, followed by the path that identifies the location of the file. Because an `Open For Append` statement will create a file automatically if it does not exist, creating and maintaining a log file is as simple as calling the `LogMessage` subroutine.

In the body of `LogMessage`, the log file is opened, the message is written, and the file is closed. Because the file is opened For Append, the message will be added to the end of the file without erasing previously written data contained in the file. Note that the actual string written to the file is more than just the message that was passed to the subroutine. Concatenated to the front of the message is the current date and time followed by a colon.

The second subroutine in Listing 7.22 called `TestLog` simply shows how to call `LogMessage`. To run a test, enter both subroutines into any code module in the VBE, move your cursor to a point within the `TestLog` function and hit the F5 key. Check for the log.txt file at the specified location on your hard drive.

```
Sub LogMessage(strMessage As String, _
               strLogFile As String)

    Dim intFileNum As Integer

    intFileNum = FreeFile
    Open strLogFile For Append As #intFileNum
    Print #intFileNum, Now & ": " & strMessage
    Close #intFileNum

End Sub

Sub TestLog()
    LogMessage "Computation complete", "C:\log.txt"
End Sub
```
Listing 7.22 A simple subroutine for logging to a file, and a subroutine to test it.

Another statement that can be used to write data to a file is the `Write #` statement. Arguments expected by `Write #` are identical to `Print #`. The main difference between these two statements is that `Write #` will always write data to a file in a format that guarantees the data can be read in again by VBA. For example, all strings written using `Write #` are surrounded with double quotes ("), and dates are surrounded with hash symbols (#). Sometimes, this behavior is exactly what we want. Other times, when the file is to be read by humans, or another program that does not understand these special symbols, using `Write #` will only introduce unnecessary complexity.

```
Sub TestWriteStatement()
    Dim vrnFileName As Variant      ' File name
    Dim intFileNum As Integer       ' File number
    Dim strWell As String           ' Name of the well
    Dim blnRgntAdded As Boolean     ' True if some
                                    ' reagent was added
    Dim dblMeasurement As Double    ' Measurement taken
    Dim datTime As Date             ' Date-time of
                                    ' measurement

    strWell = "A01"  ' Assign test data

    blnRgntAdded = True
    dblMeasurement = 1.234
    datTime = Now
```
Listing 7.23 Testing the `Write #` statement.

```
    ' Get file name
    vrnFileName = Application.GetSaveAsFilename( _
        InitialFileName:="test.txt", _
        FileFilter:= _
        "Text Files (*.txt),*.txt,All Files (*.*),*.*", _
        Title:="Test Write Statement")

    ' Exit if cancelled
    If vrnFileName = False Then Exit Sub

    intFileNum = FreeFile    ' Set file number

    ' Write data to the file
    Open vrnFileName For Output As #intFileNum
    Write #intFileNum, strWell, blnRgntAdded, _
        dblMeasurement, datTime

    Close #intFileNum    ' Close the file
End Sub
```

Listing 7.23 *Continued*

After running the `TestWriteStatement` subroutine in Listing 7.23, the contents of the newly written file will look similar to the following:

"A01",#TRUE#,1.234,#2004-10-24 08:12:50#

The complement to `Write #` is the `Input #` statement. In addition to the file number, the `Input #` statement takes a list of variables that will be assigned to the values that are read from the file. The data in the file being read must be in the proper format—the same format used by the `Write #` statement.

Enter the code in Listing 7.24 into a code module in the VBE. Before running it, make sure your Immediate Window is displayed by selecting the View | Immediate Window menu option in the VBE. This is necessary in order to see the output that will be printed by the `Debug.Print` statement, which is included in the listing. When ready, run your subroutine as usual. Select the same file that was written by `TestWriteStatement` subroutine. The data written to that file will be read and printed in the Immediate Window.

```
Sub TestInputStatement()
    Dim vrnFileName As Variant      ' File name
    Dim intFileNum As Integer       ' File number
    Dim strWell As String           ' Name of the well
    Dim blnRgntAdded As Boolean     ' True if some
                                    ' reagent was added
```

Listing 7.24 Testing the `Input #` statement.

```
    Dim dblMeasurement As Double   ' Measurement taken
    Dim datTime As Date            ' Date-time of
                                   ' measurement

    ' Get file name
    vrnFileName = Application.GetOpenFilename( _
        FileFilter:= _
        "Text Files (*.txt),*.txt, All Files (*.*),*.*", _
        Title:="Test Input Statement")

    ' Exit if cancelled
    If vrnFileName = False Then Exit Sub

    ' Set file number and open
    intFileNum = FreeFile
    Open vrnFileName For Input As #intFileNum

    ' Read data, close file and print
    Input #intFileNum, strWell, blnRgntAdded, _
        dblMeasurement, datTime
    Close #intFileNum
    Debug.Print strWell, blnRgntAdded, _
        dblMeasurement, datTime
End Sub
```

Listing 7.24 *Continued*

The code in Listing 7.23 and Listing 7.24 write and read a single line of four data items. In practice, it is more likely that you will write many lines of data using a loop construct. For example, you may need to write or read 96 lines, one for each well of a microtiter plate.

When data are written to a file in a format that is not easily read using the `Input #` statement, a little more work is required. The entire line of data must be read as a single string first, and then the individual data items extracted from the string. We will discuss string manipulation in the next section. For now, let us look at how an entire line of data can be read using the `Line Input #` statement. Listing 7.25 illustrates how to use `Line Input #`. It is similar to `Input #`, with only a single string argument to hold the data that is read.

If you have a file named C:\test2.txt that contains the following line of data:

```
1.23; 3.45; 5.67; 7.89
```

running the code in Listing 7.25 will print that line to the Immediate Window exactly as it appears in the file. The data values, which are separated by semicolons, are not assigned to individual variables, as was the case with `Input #`

READING, WRITING, AND PRESENTING DATA 315

in Listing 24. Since the `Input #` statement expects data values to be separated by commas, using it on this file would result in a runtime error. Instead, the entire string of characters on the first line is read and assigned to the `strLine` string variable:

```
Sub TestLineInputStatement()
    Dim strFileName As String   ' File name
    Dim intFileNum As Integer   ' File number
    Dim strLine As String       ' Entire line of
                                ' data

    strFileName = "C:\test2.txt" ' Set file name

    ' Set file number and open file
    intFileNum = FreeFile
    Open strFileName For Input As #intFileNum

    'Read the entire line and close the file
    Line Input #intFileNum, strLine
    Close #intFileNum

    ' Print in the Immediate window
    Debug.Print strLine

End Sub
```

Listing 7.25 Reading an entire line from a file using the `Line Input #` statement.

Manipulating and Parsing Data from Strings   To extract data from strings that are read from a file, the characters that represent the data items must be dissected from the string and possibly converted to the proper VBA data types, for example, integers or double precision numbers. This dissection process is called *parsing*. VBA includes many functions for manipulating strings, which can be used to parse the desired data from a string. Table 7.2 lists several VBA functions that are useful for parsing strings. A short description is included with each function. For more detailed information, refer to the VBA Help documentation, which can be displayed from the Help menu in the VBE.

An Example   Let us consider a realistic example. Assume we have a file, the first few lines of which appear as in Figure 7.7. This is a sample data file written by an instrument that takes a single measurement from each well of a microtiter plate. Unfortunately, due to the way in which the instrument operates, the file starts with an unknown number of blank lines. The actual plate data section is initiated with a line that begins with the string "Plate:", and is followed by a bar code that has been assigned to the plate. The well

TABLE 7.2 VBA Functions Useful for Parsing Strings

Function	Description
InStr	Search one string looking for a specified substring contained within it. If found, return the position of the substring.
InStrRev	Similar to InStr, only search for the substring starting from the end of the string and proceeding back to the beginning.
Replace	Search a string, replacing all occurrences of one substring with another substring. Return the substituted string.
Split	Split a string on a specified delimiting substring, returning an array of delimited strings.
Join	Join an array of strings and return a single string. Optionally insert a specified delimiter between each string element in the joined string.
Trim	Remove all spaces from the beginning and end of a string and return the result.
Left	Return a substring composed of the specified number of characters at beginning of a string.
Right	Return a substring composed of the specified number of characters at end of a string.
Mid	Return a substring composed of a specified number of characters starting from a specified position within a string.
IsNumeric	Returns True if the argument (possibly a string) can be converted to a number.
CInt	Convert the argument (possibly a string) to an integer.
CLong	Convert the argument (possibly a string) to a long integer.
CDbl	Convert the argument (possibly a string) to a double precision number.
CDate	Convert the argument (possibly a string) to a date.

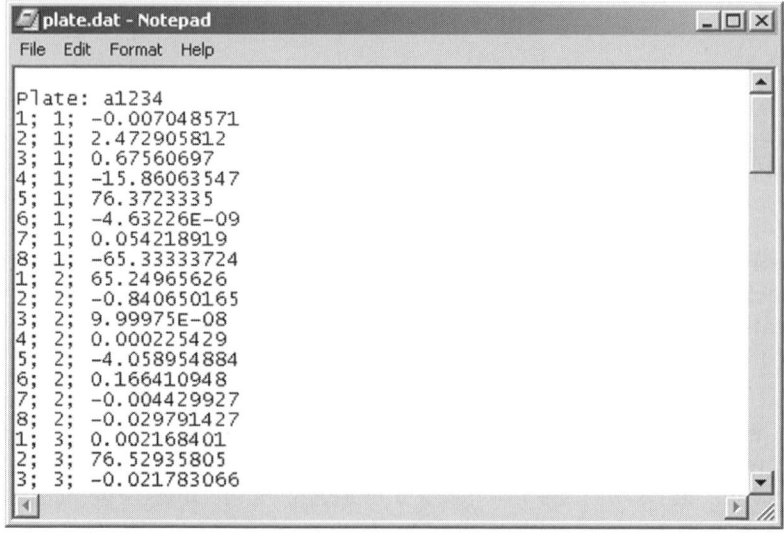

Figure 7.7 Sample Plate Data File.

READING, WRITING, AND PRESENTING DATA

measurements are formatted as three columns. The first column corresponds to the well row, the second column to the well column, and the last column to the measured well data. The challenge is to read the file and parse the data into the cells of an Excel spreadsheet. The data should be displayed in the spreadsheet in an 8 by 12 block of cells rather than in columns. The layout should correspond to that of a microtiter plate in a landscape orientation. In the following example we tackle the problem using the following three steps:

1. Finding the data section in the file
2. Reading and parsing the data from the file
3. Converting the strings to numbers.

In the next section we'll discuss how to transfer the data to the cells of a worksheet, and to format the worksheet to enhance the data presentation.

Listing 7.26 provides a complete subroutine for parsing, converting, and displaying all data in the Immediate Window. Subroutine comments help identify exactly what is being done. Note the use of several functions listed in Table 7.2.

After variables are declared, a file is selected and opened. First, all blank lines are read, trimmed of white space, tested, and discarded. The first non-blank line is checked for the proper format, and the bar code is removed using the `Mid` and `Trim` functions. Immediately following, 96 lines of data are read and split on the semicolon character into the `arrLine` string array using the `Split` function. The row and column designation are converted to Integers using the `CInt` conversion function. The measurements are converted to Doubles using the `CDbl` function and stored in the proper location of the `arrData` array. The subroutine completes with a couple of nested loops that print the data in the array to the Immediate Window.

```
Sub ParsePlateData1()
    ' Parse data into an array
    Dim vrnFileName As Variant   ' File name
    Dim intFileNum As Integer    ' File number
    Dim strLine As String        ' Line read
                                 ' from the file
    Dim strBarcode As String     ' Parsed bar code
    Dim arrLine() As String      ' Array to hold result
                                 ' of Split command
    Dim intRow As Integer        ' Current row
    Dim intCol As Integer        ' Current column
    ' Array to hold parsed plate data
    Dim arrData(1 To 8, 1 To 12) As Double
```
Listing 7.26 Parse plate data from a file.

```
' Loop counters
Dim I As Integer, J As Integer

' Select the file
vrnFileName = Application.GetOpenFilename( _
    FileFilter:= _
    "Text Files (*.dat),*.dat, All Files (*.*),*.*", _
    Title:="Select Data File")
' Exit if cancelled
If vrnFileName = False Then Exit Sub
' Get file number and open
intFileNum = FreeFile
Open vrnFileName For Input As #intFileNum

Line Input #intFileNum, strLine  ' Read line
strLine = Trim(strLine)  ' Trim whitespace

' Discard any blank lines
Do While strLine = ""
    Line Input #intFileNum, strLine
    strLine = Trim(strLine)
Loop

' Parse barcode
If Left(strLine, 6) = "Plate:" Then
    strBarcode = Trim(Mid(strLine, 7))
Else
    Debug.Print "Invalid file format"
    Exit Sub
End If

' Parse data
For I = 1 To 96  ' Loop over all Lines
    ' Read line
    Line Input #intFileNum, strLine
    ' Split on semicolon
    arrLine = Split(strLine, ";")
    ' Convert row ...
    intRow = CInt(arrLine(0))
    ' Convert column ...
    intCol = CInt(arrLine(1))
```

Listing 7.26 *Continued*

```
            ' Convert well data ...
            arrData(intRow, intCol) = _
                CDbl(arrLine(2))
    Next I

    Close #intFileNum  ' Close file

    For I = 1 To 8  ' Print data
        For J = 1 To 11
            Debug.Print arrData(I, J);
        Next J
        Debug.Print arrData(I, J)
    Next I

End Sub
```

Listing 7.26 *Continued*

Importing Data from External Databases

Another way to move data into Excel is to import it from an external database. A common practice of an organization is to store large amounts of diverse data in centralized databases. These data can form the basis of calculations that you may need to perform. Fortunately, Excel makes it easy to import data into a Worksheet from a centralized database.

Understanding Databases Databases come in many flavors and speak many languages. Perhaps the most common database type is the relational database. Microsoft Access, SQL Server, Oracle, and other databases are examples. A relational database is a collection of tables, where each table consists of a two-dimensional array of data organized as rows (records) and columns (fields). Tables 7.3 and 7.4 present data that make up two sample tables in a hypothetical relational database. The sample tables are named Compound_Data and Inventory_Data, respectively. Compound_Data holds data that describe compounds in a compound store. Inventory_Data holds data that describe the current inventory of each compound.

The two sample tables have a common field called Compound_ID. The data in this field are used to relate the records in the two tables, hence the name *relational* database. It is possible to query the tables in this sample database in order to select all inventory information from compounds with a molecular weight greater than 500. This requires an inspection of data in both tables, which is accomplished by *joining* the two tables on the common Compound_ID field values.

Another way to get a sense for the important elements of a relational database is to compare them to similar elements in Excel. Table 7.5 presents such

TABLE 7.3 Sample Table Named Compound_Data

Compound_ID	Name	MW
100001	Compound1	504.07
100002	Compound2	576.81
100003	Compound3	331.49
100004	Compound4	371.09
100005	Compound5	454.52
100006	Compound6	488.36
100007	Compound7	405.04
100008	Compound8	568.20

TABLE 7.4 Sample Table Named Inventory_Data

Compound_ID	Barcode	Tare_Weight	Current_Weight
100001	9999901	8.01	27.74
100001	9999902	8.57	14.90
100001	9999903	8.27	15.84
100002	9999904	8.11	28.02
100003	9999905	8.03	27.25
100004	9999906	8.82	11.05
100005	9999907	8.74	18.37
100006	9999908	8.74	17.48
100006	9999909	8.91	14.42
100007	9999910	8.03	21.05

TABLE 7.5 Similar Excel and Relational Database Elements

Excel	Relational Database
Workbook	Database files
Worksheet	Table
Column	Field
Row	Record

a comparison. Even though we can make this comparison, it is important to remember that Excel is not relational in nature; it is not possible to join data in the columns of Worksheets using standard Excel capabilities.

Accessing Databases using Excel and ODBC The multitude of languages spoken by the various database engines can make importing data a complex endeavor. Fortunately, there is a common language through which many database engines can be accessed. This common language is called Open DataBase Connectivity (ODBC). An ODBC driver is software that accepts a command in the common ODBC language and translates it into a database-specific form. Data obtained from a database-specific command is repackaged and returned in a common format. Almost all major relational database engines provide an ODBC driver. In fact, there are even ODBC drivers that expose data in properly formatted nonrelational databases, such as Excel Workbooks and flat files.

Early in this chapter, we stated that Excel is not a database management system. We used this statement as an argument against using Excel in circumstance that call for a database management system. Nevertheless, for illustration purposes we will show how to set up Excel as a simple database and access it through an ODBC connection. Due to the common language of ODBC, the method for importing data is identical for more powerful relational database engines.

Creating an Excel database is simple. As an example, open up a new Excel Workbook and rename one Worksheet by double clicking on its tab and entering Compound_Data. Enter data into the cells of this Worksheet, as shown in Figure 7.8. Do not forget to enter the field names in the first row. Rename another Worksheet to Inventory_Data and enter the data shown in Figure 7.9. Save the Workbook with the file name InvDatabase.xls.

The only remaining step necessary to turn the data in this Workbook into an ODBC-accessible database is to name the ranges in each Worksheet that contain table data. Highlight the range of cells containing data to be treated as a single table and enter the table name in the Name box, which is just to the left of the formula bar above the Worksheet area (See the circled element in Fig. 7.10). Once the name is entered, make sure to hit the Enter key, otherwise the name will not be assigned to the range. Set the name of the data range on the first sheet to Compound_Data, and the data range on the second sheet to Inventory_Data. Resave and close the modified Workbook.

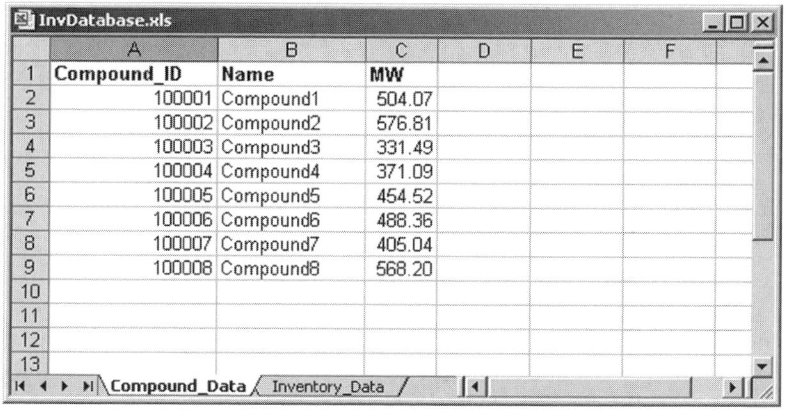

	A	B	C
1	Compound_ID	Name	MW
2	100001	Compound1	504.07
3	100002	Compound2	576.81
4	100003	Compound3	331.49
5	100004	Compound4	371.09
6	100005	Compound5	454.52
7	100006	Compound6	488.36
8	100007	Compound7	405.04
9	100008	Compound8	568.20

Figure 7.8 Compound_Data worksheet.

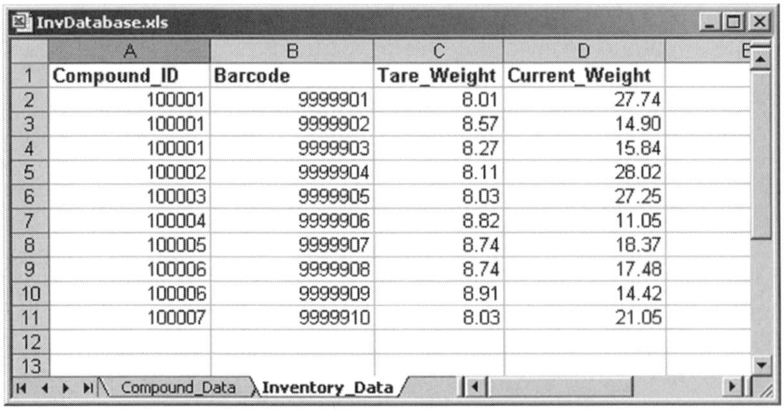

Figure 7.9 Inventory Data worksheet.

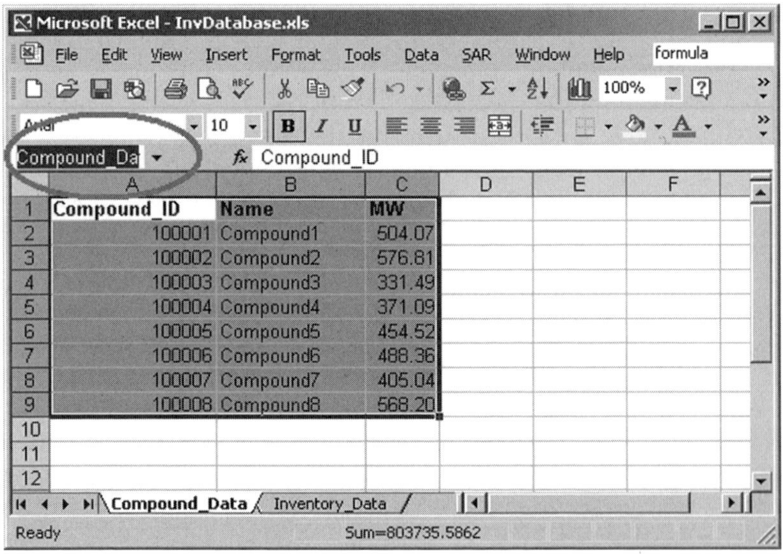

Figure 7.10 Creating an ODBC-accessible table in Excel by naming a range of cells.

Importing Data into Excel Data can be imported right into an Excel Worksheet from an external database using menu options. To see this in action, open a new Workbook and select the Data | Import External Data | Import Data ... menu option. You will see the Select Data Source dialog, similar to that shown in Figure 7.11. The dialog will show all the available data sources, including *Data Source Names* (DSN). A DSN contains all the information necessary for Excel to make an ODBC connection with an external data source.

READING, WRITING, AND PRESENTING DATA

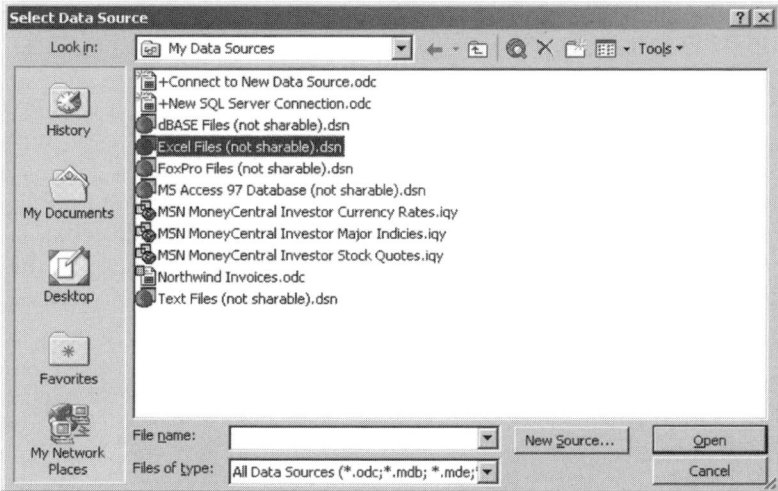

Figure 7.11 Select Data Source Dialog.

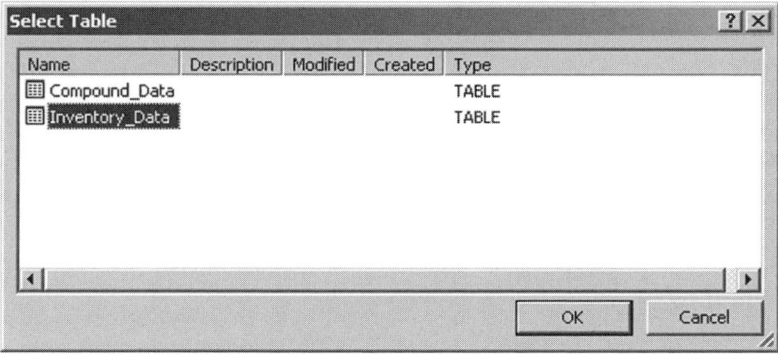

Figure 7.12 Tables in the InvDatabase.xls Excel database.

Excel is distributed with several preconfigured DSNs, including one for Excel itself.

Select the Excel Files DSN and click the Open button. Navigate to and select the InvDatabase.xls Excel file, and click OK. You should see a dialog that shows all tables in the Excel database (Fig. 7.12). Select one of the tables and click OK to view the Import Data dialog. If you click OK again, all the data from this table will be imported into your new Workbook. Instead, if you click the Edit Query... button from the Import Data dialog, Excel will display the Query Wizard from a helper application called Microsoft Query. The Microsoft Excel installation package includes Microsoft Query, but it is not installed by default. You may need to install this application from your Excel installation CDs before continuing.

324 USING MICROSOFT EXCEL AS A LABORATORY DATA MANAGEMENT TOOL

Follow the Query Wizard prompts. Select different fields from the available tables and filter the records to match your specified criteria. In Figure 7.13 we are restricting records from the Inventory_Data table to those with a Compound_ID field value of 100001. After proceeding through all the remaining Query Wizard prompts, the filtered records can be returned to your Excel Worksheet, as shown in Figure 7.14.

In fact, using the Query Wizard, it is possible to join data in multiple tables of the external Excel database, even though Excel is not relational. It is the Excel ODBC driver that imparts the relational capabilities on top of the Excel database—truly a nice added benefit.

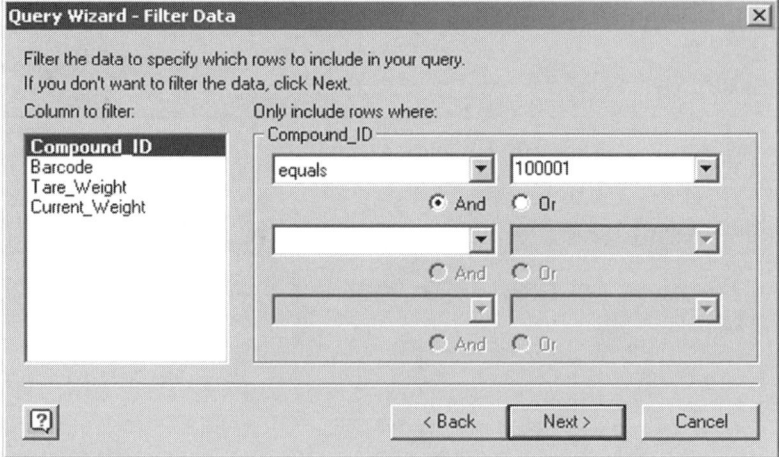

Figure 7.13 Filtering records of the Inventory_Data table to those with a Compound_ID of 100001.

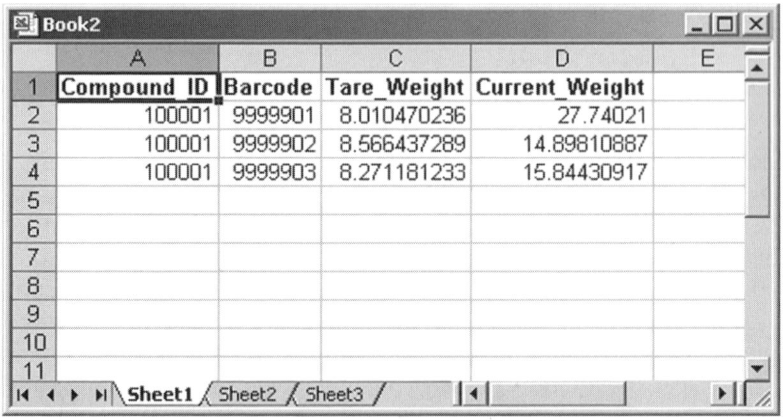

Figure 7.14 Records returned from the external database import.

Importing Data from an External Database with VBA We can make use of the Macro Recorder to learn how to automate the external data import process using VBA. By repeating the process of the previous section with the Macro Recorder activated, you will generate a subroutine that looks like the one in Listing 7.27. This listing has been edited slightly to make it more readable.

Listing 7.27 shows that three important strings were generated: a connection string, a path to a DSN, and a Structured Query Language (SQL) query. The connection string includes the pertinent information to locate the database. In this case, it is our InvDatabase.xls file stored in the root of the C drive. The DSN string is a path to the appropriate DSN file. In this case it points to the Excel File DSN installed with Excel.

The last string is the SQL query that tells the ODBC driver how to remove the data from the database. SQL is a standard language for manipulating data in a relational database. There are four basic statements used to manipulate data in a database using SQL.

SELECT—retrieves data
UPDATE—modifies existing data
INSERT—adds data
DELETE—removes data

The particular SQL query generated in this example is a SELECT statement. The basic syntax for a SELECT statement is as follows:

SELECT [Field Name(s)] FROM [Table Name(s)] WHERE [Criteria]

This requests that the data in [Field Name(s)] be returned from the tables in [Table Name(s)], subject to the conditions in [Criteria]. It is beyond the scope of this chapter to describe SQL in detail. A search of the Internet will reveal many good sources of information about SQL. Modifications of the SQL query in Listing 7.27 will allow you to customize the data import process, and rerun it as necessary.

```
Sub Macro1()
    Dim strConn As String
    Dim strSQL As String
    Dim strDSN As String

    ' Connection string
    strConn = _
        "ODBC;DSN=Excel Files;DBQ=C:\InvDatabase.xls;"
```

Listing 7.27 An automatically generated VBA Macro to select data from an external database.

```
    strConn = strConn & _
        "DefaultDir=C:\;DriverId=790;"
    strConn = strConn & _
        "MaxBufferSize=2048;PageTimeout=5;"

    ' DSN File
    strDSN = _
        "C:\Program Files\Common Files\ODBC\Data Sources\"
    strDSN = strDSN & "Excel Files (not sharable).dsn"

    ' SQL Query
    strSQL = "SELECT Inventory_Data.Compound_ID, "
    strSQL = strSQL & "Inventory_Data.Barcode, "
    strSQL = strSQL & "Inventory_Data.Tare_Weight, "
    strSQL = strSQL & "Inventory_Data.Current_Weight "
    strSQL = strSQL & "FROM `C:\InvDatabase`.Inventory_Data "
    strSQL = strSQL & "Inventory_Data "
    strSQL = strSQL & _
        "WHERE (Inventory_Data.Compound_ID=100001)"

    With ActiveSheet.QueryTables.Add( _
            Connection:=strConn, Destination:=Range("A1"))
        .CommandText = Array(strSQL)
        .Name = "Excel Files (not sharable)"
        .FieldNames = True
        .RowNumbers = False
        .FillAdjacentFormulas = False
        .PreserveFormatting = True
        .RefreshOnFileOpen = False
        .BackgroundQuery = True
        .RefreshStyle = xlInsertDeleteCells
        .SavePassword = True
        .SaveData = True
        .AdjustColumnWidth = True
        .RefreshPeriod = 0
        .PreserveColumnInfo = True
        .SourceConnectionFile = strDSN
        .Refresh BackgroundQuery:=False
    End With
End Sub
```

Listing 7.27 *Continued*

Setting up an Oracle ODBC connection As mentioned, ODBC can be used to access other databases, such a Microsoft Access or Oracle. It is only a matter of having the appropriate drivers installed and setting up the proper DSN. The following shows how to setup an ODBC connection for Oracle.

From the Windows Task Bar, select Start | Settings | Control Panel | Administrative Tools | Data Sources (ODBC). Click the Add . . . button and select an ODBC driver. In this example we will use the driver named Microsoft ODBC for Oracle (Fig. 7.15). Click Finish and complete the installation.

After this step you have the option to create an associated DSN. In the Create New Data Source dialog, enter a name, a description, your user name, and the Oracle server information (Fig. 7.16). If you are unsure of the proper parameters, try contacting your database administrator.

From this point you can access data in the Oracle database in a manner identical to the way in which data was accessed in the external Excel database. Make sure to use the Macro Recorder to regenerate the VBA in order to capture the proper connection string and DSN.

Automating the Generation of Reports

Reports are generated in Excel by inserting data into the cells of a Worksheet and formatting the cells to create an informative and pleasing presentation. In a previous section we saw how to read and write the value of a cell using the Range object's `Value` property. The next step is to take a closer look at

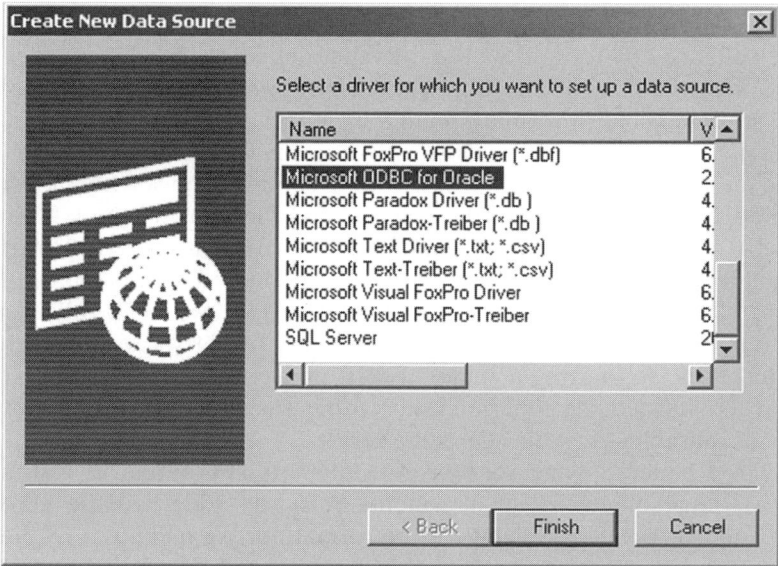

Figure 7.15 Creating a new ODBC Data Source for Oracle.

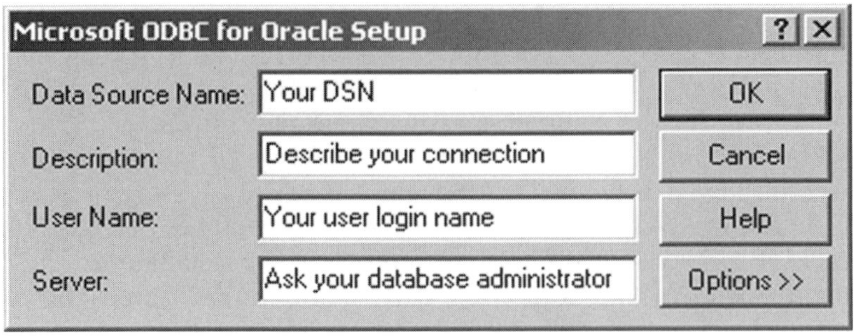

Figure 7.16 Configuring the Oracle DSN.

how to change a cell's appearance, and how to use that to convey important information about the data being reported.

Formatting Worksheets Everything about the way a Worksheet is formatted can be changed programmatically using VBA. This includes the way cell values are formatted, their alignment, and the font used, as well as cell borders and colors. This topic is much too large to cover in any detail here. Instead, we will investigate a few useful topics, and recommend that the interested reader make use of the Macro Recorder as a way to discovery how to programmatically change the appearance of a Worksheet.

When dealing with data that are associated with a microtiter plate or rack of tubes, it helps to display those data in a format that mimics the layout of the plate of tubes. Such displays are natural in Excel since the rectangular array of containers maps nicely onto a rectangular array of cells. The display can be enhanced by formatting the cells that delineate the plate or rack data, and marking the rows and columns with appropriate labels.

We are interested in formatting an 8 by 12 range of cells to hold an array of data associated with the wells of a 96-well microtiter plate. The goal is to outline the range with a thick border, to label columns above with the numbers "1" through "12" centered on the column, and the rows on the left with the letters "A" through "H," right justified. We want to encapsulate the formatting commands in a separate VBA subroutine so that it is possible to format any 8 by 12 Range as a plate. By taking this approach, we will be able to reuse the formatting function on any Range of cells. To get started, a macro was recorded while formatting a range of cells in accordance with our specifications. When finished, the generated subroutine was edited to make it more efficient and reusable—Listing 7.28 is the result.

Unlike a generated macro, first note that the subroutine in Listing 7.28 takes a Range object as an argument. This was added to the generated macro and represents the Range to be formatted. We make use of the With statement in the subroutine since all commands will be applied to the same Range.

READING, WRITING, AND PRESENTING DATA

The first formatting step is on the borders. Notice that a Range has a subordinate collection of Border objects appropriately called Borders. Each Border object itself has `LineStyle` and `Weight` properties. Each of the four Border objects in the `Borders` collection is accessed using a predefined constant, then their `LineStyle` and `Weight` properties are set appropriately. Note that we did not have to look up these objects, properties, or constants, since the Macro Recorder generated the code for us with all the appropriate keywords.

Following border formatting, we set the format of the numbers displayed within the Range. The Range object has a `NumberFormat` property, which takes a string that defines the format. Again, the Macro Recorder identified the property to set and its proper value. We only added a little editing.

Setting up column and row labels took a bit more work. The Macro Recorder gave us the `HorizontalAlignment` property and constants, but we added the loops and cell access using the `Cells` method of the Range object. One thing to note about the indexes of the `Cells` method is the ability to reach cells that fall outside the Range. The `FormatPlateDisplay` subroutine takes a Range argument that defines the 8 by 12 array of cells. To generate column and row labels we need to format cells in the row above the Range as well as cells in the column to the left of the Range. The column label format loop uses the loop index to set column values in the `Cells` method, but the row index is set to "0." The `Cells` method provides access to the cells in the parent Range starting at index "1." An index of "0" can be used to access cells just prior to the parent Range. In the column label format loop the row index of "0" provides access to the row above the parent Range that defines the plate display. Similarly, the row label format loop uses a column index of "0," which provides access to the column to the left of the parent Range. The final trick to note is the use of the `Chr$` function in the row label format loop. `Chr$` takes an integer and returns the corresponding ASCII encoded character. A capital "A" has an ASCII value of 65, with subsequent capital letters following in numeric order. We use this fact to generate the letters "A" through "H" by adding 64 to the row number, and passing this to the `Chr$` function.

To test the `FormatPlateDisplay` function we called it with an argument of `Sheet1.Range("B2:M9")`. The result is displayed in Figure 7.17.

```
Sub FormatPlateDisplay(Plate As Range)
    ' Format an 8x12 Range as a plate display
    Dim I As Integer

    With Plate

        ' Draw a thick border around the plate Range
        .Borders(xlEdgeLeft).LineStyle = xlContinuous
```

Listing 7.28 Subroutine to format a Range as a plate display.

```
            .Borders(xlEdgeLeft).Weight = xlThick
            .Borders(xlEdgeTop).LineStyle = xlContinuous
            .Borders(xlEdgeTop).Weight = xlThick
            .Borders(xlEdgeBottom).LineStyle = xlContinuous
            .Borders(xlEdgeBottom).Weight = xlThick
            .Borders(xlEdgeRight).LineStyle = xlContinuous
            .Borders(xlEdgeRight).Weight = xlThick

            ' Set number format for all cells in Range
            .NumberFormat = "0.00"

            ' Set column labels
            For I = 1 To 12
                .Cells(0, I).Value = I
                .Cells(0, I).HorizontalAlignment = xlCenter
            Next I

            ' Set row labels
            For I = 1 To 8
                .Cells(I, 0).Value = Chr$(I + 64)
                .Cells(I, 0).HorizontalAlignment = xlRight
            Next I

        End With

    End Sub
```

Listing 7.28 *Continued*

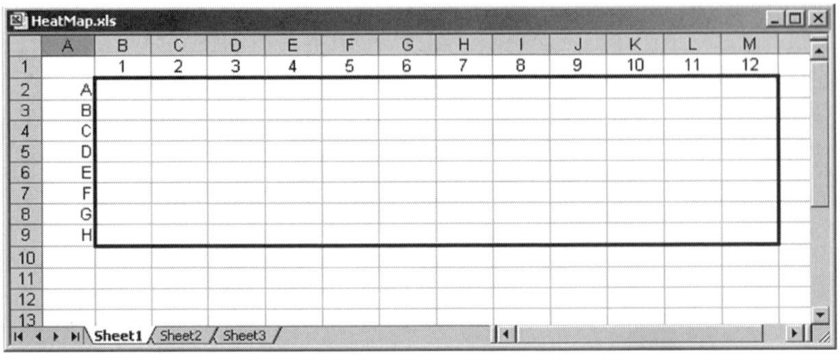

Figure 7.17 Result from calling the FormatPlateDisplay subroutine on Range B2:M9.

Generating a Heat Map Display One simple but powerful way to display data is through the use of a "heat map." In a heat map display, data values are rendered graphically using varying shades of two colors. The lower half of a numerical range is represented using shades that vary from a "cool" color to a neutral one, and the upper half of the range from the neutral color to "hot" color. In a sense, the colors indicate a level of heat associated with each data value. We will generate heat maps that use blue as the cool color, white as the neutral color, and red as the hot color.

To generate a heat map in Excel, the problem that we must solve is one of changing the background color of a cell based on the cell's value. It is possible do this in a rudimentary manner through the use of a feature in Excel called Conditional Formatting. Conditional Formats can be set for each cell through Excel's standard menu options. A maximum of three formats can be specified along with the conditions that must be satisfied in order for the associated format to be applied to the cell. Unfortunately, this is insufficient for generating a heat map because many more color shades and conditions can be required. VBA must be used to achieve the enhanced functionality necessary for generating a heat map.

Once again, by using the Macro Recorder to change a cell's background color, we learn that a Range object contains another object called `Interior`, which itself has a property called `ColorIndex`. Excel has a set of colors that can be used on various objects throughout its display. These colors are identified with an integer called a `ColorIndex`. Setting the `ColorIndex` property of the Range's `Interior` object will cause the background color of a cell to change to the one identified by the index value. With this in mind, we selected color indexes for five shades of blue, five shades of red, and white.

The HeatMap subroutine in Listing 7.29 is a complete procedure for inserting data values into the cells of a worksheet and overlying that with a heat map display. The subroutine takes two arguments: a Range object that represents the 8 by 12 block of cells on which to create the heat map and an 8 by 12 array of data values.

The first step of the `HeatMap` subroutine is to build an array of the color indexes identified for the heat map display. These color indexes are stored in an array called `arrClrIdx`. The color index to be used for a cell is determined by translating a cell's data value into an array index of the `arrClrIdx` array. The next two steps of the `HeatMap` subroutine are to set the data range that maps into the color gradient, and to call the `FormatPlateDisplay` subroutine given in Listing 7.28 in order to initialize the plate display. Following all initialization, a doubly nested loop over all data in the provided data array inserts the raw data value into the associated cell and then maps the data value into an index in the color index array, and uses that index to set the background color of the cell using the ColorIndex property of the Cell's Interior object.

In Listing 7.26 we parsed data from a file and loaded it into an array. The loaded data were printed to the Immediate Window. We can now modify this

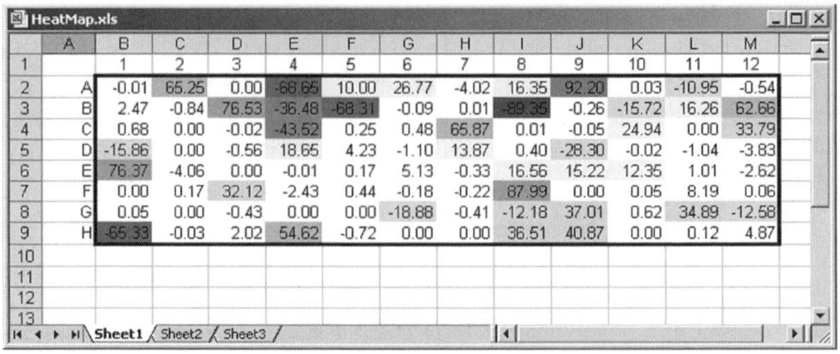

HeatMap.xls

	A	B	C	D	E	F	G	H	I	J	K	L	M
1		1	2	3	4	5	6	7	8	9	10	11	12
2	A	-0.01	65.25	0.00	-68.65	10.00	26.77	-4.02	16.35	92.20	0.03	-10.95	-0.54
3	B	2.47	-0.84	76.53	-36.48	-68.31	-0.09	0.01	-89.35	-0.26	-15.72	16.26	62.66
4	C	0.68	0.00	-0.02	-43.52	0.25	0.48	65.87	0.01	-0.05	24.94	0.00	33.79
5	D	-15.86	0.00	-0.56	18.65	4.23	-1.10	13.87	0.40	-28.30	-0.02	-1.04	-3.83
6	E	76.37	-4.06	0.00	-0.01	0.17	5.13	-0.33	16.56	15.22	12.35	1.01	-2.62
7	F	0.00	0.17	32.12	-2.43	0.44	-0.18	-0.22	87.99	0.00	0.05	8.19	0.06
8	G	0.05	0.00	-0.43	0.00	0.00	-18.88	-0.41	-12.18	37.01	0.62	34.89	-12.58
9	H	-65.33	-0.03	2.02	54.62	-0.72	0.00	0.00	36.51	40.87	0.00	0.12	4.87
10													
11													
12													
13													

Sheet1 Sheet2 Sheet3

Figure 7.18 Heat Map display generated from Listing 30.

example to display the data as a Heat Map. Listing 7.30 modifies Listing 7.26 by removing the print statements and calling the `HeatMap` subroutine with appropriate arguments. Figure 7.18 illustrates the result of running the code in Listing 7.30 and selecting an appropriate data file.

```
Sub HeatMap(Plate As Range, arrData() As Double)
    Dim I As Integer ' Loop counter
    Dim J As Integer ' Loop counter
    Dim dblMax As Double ' Maximum value in range
    Dim dblMin As Double ' Minimum value in range
    Dim intVal As Integer ' Color index
    Dim arrClrIdx(10) As Integer ' Color index array

    ' Set color indexes that define color gradient
    arrClrIdx(0) = 11 ' Blue (coolest)
    arrClrIdx(1) = 32
    arrClrIdx(2) = 41
    arrClrIdx(3) = 33
    arrClrIdx(4) = 34
    arrClrIdx(5) = 2  ' White
    arrClrIdx(6) = 27
    arrClrIdx(7) = 44
    arrClrIdx(8) = 45
    arrClrIdx(9) = 46
    arrClrIdx(10) = 3 ' Red (hottest)

    ' Set data range to map into color gradient
    dblMin = -100
    dblMax = 100
    ' Call the sheet formatting routine
```

Listing 7.29 Heat Map generation subroutine.

```vb
        FormatPlateDisplay Plate
        ' Insert data into cells, and color
        ' cells based on data Values
        For I = 1 To 8
            For J = 1 To 12
                ' Display data value in cell
                Plate.Cells(I, J).Value = _
                    arrData(I, J)
                ' Map value into color index array
                intVal = CInt( _
                    10# * (arrData(I, J) - dblMin) / _
                    (dblMax - dblMin))
                ' Set cell background color
                Plate.Cells(I, J).Interior.ColorIndex = _
                    arrClrIdx(intVal)
            Next J
        Next I
End Sub
```

Listing 7.29 *Continued*


```vb
Sub ParsePlateData()
    ' Parse data into an array
    ' and display as a heat map
    Dim vrnFileName As Variant   ' File name
    Dim intFileNum As Integer    ' File number
    ' Line read from the file
    Dim strLine As String
    Dim strBarcode As String     ' Parsed bar code
    ' Result of Split Command
    Dim arrLine() As String
    Dim intRow As Integer        ' Current row
    Dim intCol As Integer        ' Current column
    ' Parsed plate data
    Dim arrData(1 To 8, 1 To 12) _
        As Double
    ' Loop counters
    Dim I As Integer, J As Integer
    
    ' Select the file
    vrnFileName = Application.GetOpenFilename( _
        FileFilter:= _
```

Listing 7.30 Generate a Heat Map display using data loaded from a file.

```vba
                "Text Files (*.dat),*.dat, " & _
                " All Files (*.*),*.*", _
                Title:="Select Data File")
    ' Exit if cancelled
    If vrnFileName = False Then Exit Sub

    intFileNum = FreeFile ' Get file number and open
    Open vrnFileName For Input As #intFileNum
    Line Input #intFileNum, strLine ' Read line

    strLine = Trim(strLine) ' Trim whitespace
    Do While strLine = "" ' Discard any blank Lines
        Line Input #intFileNum, strLine
        strLine = Trim(strLine)
    Loop
    ' Parse bar code
    If Left(strLine, 6) = "Plate:" Then
        strBarcode = Trim(Mid(strLine, 7))
    Else
        Debug.Print "Invalid file format"
        Exit Sub
    End If
    ' Parse data
    For I = 1 To 96 ' Loop over all Lines
        Line Input #intFileNum, strLine ' Read line
        ' Split on semicolon
        arrLine = Split(strLine, ";")
        ' Convert row ...
        intRow = CInt(arrLine(0))
        ' Convert column ...
        intCol = CInt(arrLine(1))
        ' Convert well data ...
        arrData(intRow, intCol) = _
            CDbl(arrLine(2))
    Next I
    Close #intFileNum ' Close file

    ' Generate the HeatMap
    HeatMap Sheet1.Range("B2:M9"), arrData
End Sub
```

Listing 7.30 *Continued*

7.5 SUMMARY AND CONCLUSION

In this chapter we only have touched upon the power and utility of the pairing between Excel and the VBA programming language. Even though a complete discussion of the features of Excel and VBA would span several volumes and many thousands of pages, it should be clear that Excel is a vastly underutilized tool for managing data generated in a life sciences laboratory. There is an underappreciation for the wide diversity of data management tasks that can be handled by Excel and VBA.

It should also be clear that Excel is not a panacea; there are many laboratory data management situations for which Excel is not suited. As VBA and the functionality behind Excel are mastered, there is a tendency to attempt to use it to solve problems that require more sophisticated tools, such as a full-blown multiuser enterprise data management system, a document workflow system, or a collection of highly refined data mining utilities. Excel can be an excellent prototyping tool for these situations, but the skilled laboratory data manager will recognize when the time has come to benefit from more powerful tools.

For several years now Microsoft has been putting their efforts into the creation and enhancement of a new software development technology called .Net. The .Net Framework allows different programming languages and libraries to work together seamlessly. Microsoft Excel programming has been brought under the .Net Framework through a technology called Visual Studio for Applications (VSA). VSA will allow Excel scripting to be performed from one of the many .Net languages. Nevertheless, Microsoft has recognized that VBA is a powerful tool and that there are millions of lines of VBA code that have been written for Excel, and are in daily use. Given this reality, Microsoft has stated that it will continue to support VBA in Excel and other applications into the future.

As a powerful, full featured, and fully integrated programming environment for Excel, VBA remains a preferred choice for building Excel-based applications for managing laboratory data.

8

AGE OF REGULATION

SANDY WEINBERG
Fast Trak BioDefense, GE Healthcare
Atlantic, Georgia

GERALD J. WHARTENBY
Regulatory Consultant for GE Healthcare
Philadelphia, Pennsylvania

8.1	INTRODUCTION	337
8.2	ORIGINS OF THE FDA	339
8.3	21 CFR PART 11	343
8.4	RISK ASSESSMENT	344
8.5	RISK ASSESSMENT AND REGULATION	345
8.6	REGULATION IN DRUG DISCOVERY: THE FUTURE	349

8.1 INTRODUCTION

In examining regulations, often it is easy to become inundated with the complexity of existing rules and regulations. Like being in a maze, we travel through the twist and turns of current Food and Drug Administration (FDA) regulations often leading us in directions where a clear path is difficult to find. The number, import, and intricacy of the U.S. FDA regulations, requirements, guidelines, and draft documents are often so overwhelming as to make it difficult to see the underlying ideology that directs so many FDA measures.

As we travel this maze of regulations, there are three underlying precepts that together provide a foundation for and an understanding of all the myriad

Drug Discovery Handbook, by Shayne Cox Gad
Copyright © 2005 by John Wiley & Sons, Inc.

of FDA procedures: (1) proximal causality, (2) risk assessment, and (3) self-regulation. Having a clear understanding of these three precepts can help one understand future FDA actions.

Proximal causality is probably the most basic of FDA regulatory models. Let us consider a consumer who buys a product clearly labeled as containing dehydrated corn. The consumer has an expectation that the box contains pure corn kernels that are free from contaminants such as mice or rat droppings and also free from insect parts. Public health and consumer confidence are maintained as long as the label is accurate and the product is in fact free of contaminants. Consumers also need the assurance that this final product was processed properly from the beginning.

To ensure that this really does occur, regulators are authorized to investigate the packing plant itself to be certain that it is free of infestation. If the packing plant processes raw materials to eliminate those contaminants, regulators need to focus their attention on the cleanliness and operation of the processing procedure. It would also be necessary to inspect and regulate any and all equipment used in the process as well as the records of operation and maintenance of that equipment. And in the computerized world we live in, the computer systems too are subject to the eye of regulation.

Regulatory attention focuses over time on the earlier, inner steps of groundwork and progression, always moving backward to more proximal causalities of safety threats and associated problems. In part, that movement is a function of satisfaction with the more surface levels of control; in part, it is a function of deeper understanding of the ultimate causes of problems. Arguably an underlying causality of the trend derives from the reality that once investigators have achieved a satisfactory level of control for a surface issue they have the time and resources to dig deeper. It is clear that the trend to move backward to proximal causality is a response to consumer demand—standards of cleanliness are continually raised in response to public understanding. The same trend applies to the performance of drugs and their side effects, the reliability and therapeutic value of medical devices (heart pacemakers, dialysis machines), the purity of biologics (vaccines, blood products, etc.), and the quality and safety of all the other FDA regulated products (such as X-ray machines and microwave ovens) and their processes.

As we move through the regulatory maze, there is an ever-mounting cost to ensure the safety and quality of these various products. But does the cost of such regulations ever become a limiting factor?

Human blood provides an interesting example. While most blood transfusion donors are unpaid, there is a cost associated with the collection. As the screening cost rises, the resulting expense rises proportionately. To put it simply, if it cost $1.00 to collect a unit of blood, and only one in five units pass the screening process, the cost of a usable unit of blood now becomes $5.00. If the public health and safety is significantly increased, then the dollar factor is well worth it. But, if the screening provides only minimal value, do we not then need to look at a second method of regulation—risk assessment. It is clear that over the history of the FDA no regulated product is absolutely "risk free." So we must look at "benefits vs. risks."

Risk assessment is the second regulatory trend today. Simply put—it is the balancing of increased safety against the limited effect of the cost of implementing that safety factor. Having its origin in the medical device area, the concept of risk analysis considers the probability and potential severity of an adverse patient reaction, the overall performance of the system, and the process of manufacturing. Regulatory scrutiny is then assigned, and supporting evidence collected, in proportion to the degree of risk inherent in a given situation. Remembering that no regulated product is risk free and so consideration is given to a product with an eye toward risk vs. benefits especially in products used in life-threatening situations.

Self-regulation in the pharma industries is the ability to police oneself. The FDA has taken the consistent position that the industry is self-regulated and that it is the agency's job to oversee and police this self-regulation.

All companies and products falling under the wide umbrella of the FDA are subject to internal quality assurance requirements. Even if a product performed flawlessly, without any side effects, purity defects, or adverse reactions, the manufacturer would be justly criticized if it did not have in place a quality assurance program. It is the goal of the FDA and its industry partners to continually improve and oversee the minimization of potential problems.

Working in conjunction with the proximity causality and risk assessment trends, this emphasis on self-regulation results in a continuing shift of attention backwards in the laboratory and manufacturing processes. Users of laboratory information management systems (LIMS) now routinely investigate their software suppliers since the internal quality of an automated laboratory is dependent increasingly on the accuracy of the software that manages and interprets experimental results. In much the same way manufacturers extend their quality assurance to reach vendors of raw materials, of manufacturers equipment, and software control systems. The self-regulation, tempered by a risk assessment to put priority on areas of greatest potential danger, forces a proximal causality shift.

It is clear that we live in a world of regulation. Promoting public health and protecting consumers have placed us on the road to ensuring that effective and safe products reach the market place. All in all we are concerned with the quality of life expected by each member of our society. But how did all this come about? Let us take a step back and look into the history of the FDA.

8.2 ORIGINS OF THE FDA

It seems that the quality of life has been a major concern of people from the beginning of time. What they ate, how it was prepared, how they lived, concern over medical care and obviously the medications and drugs they took, and, of course, the medical devices that were recommended and used have been scrutinized over the history of humankind.

It has become a trend over this history of humans to look for help from authorities to ensure not only the safety of these devices and drugs but also the purity of products consumed and the effectiveness of the products as well.

How a product is prepared and how it is packaged and labeled are questions frequently asked over time. Is the product safe? All this has eventually laid the groundwork for future regulation laws. The cry to reform was loud and clear.

As far back as 1202, King John of England proclaimed the first English food law, the Assize of Bread, which prohibited adulteration of bread with such ingredients as ground peas or beans. By 1820, eleven physicians met in Washington, D.C., and established the U.S. Pharmacopeia, the first compendium of standard drugs for the United States. And it was President Lincoln who appointed a chemist, Charles M. Wetherill, to serve in the new Department of Agriculture, the beginning of the Bureau of Chemistry and the forerunner of the Food and Drug Administration. Since President Lincoln's time, there have been 100 major developments in the history of the FDA.

From the 1902 Biologics Control Act (passed to ensure purity and safety of serums, vaccines, and similar products used to prevent and treat humans) to the Food and Drugs Act (1906) passed to deal with misbranded and adulterated foods, drinks, and drugs to the 1938 Federal Food, Drug and Cosmetic Act (requiring new drugs to be shown safe before marketing) down to today where concern is questioned over temporary tattoos, each development has continued to lay the foundation of regulatory laws.

Today we question the members of Congress over the importation of drugs that may not have passed all the checks and balances in place here and ask ourselves how effective they are and is the risk worth taking based on the price of the drugs. The questions continue but the fundamental mission of the FDA is well defined.

The FDA's mission is clear:

1. To promote the public health by promptly and efficiently reviewing clinical research and taking appropriate action on the marketing of regulated products in a timely manner.
2. With respect to such products, protect the public health by ensuring that foods are safe, wholesome, sanitary, and properly labeled; human and veterinary drugs are safe and effective; there is reasonable assurance of the safety and effectiveness of devices intended for human use; cosmetics are safe and properly labeled; and public health and safety are protected from electronic product radiation.
3. Participate through appropriate processes with representatives of other countries to reduce the burden of regulation, harmonize regulatory requirements, and achieve appropriate reciprocal arrangements.

ORIGINS OF THE FDA 341

It is quite clear that the history of the FDA is rich, varied, and interconnected with the major political reformist trends of the last century. Buried in the detail of significant dates, regulatory revisions, and personalities are five aspects of the regulatory structure:

1. Accuracy of labeling
2. Growth
3. Oppositional
4. Speed the approval
5. Risk assessment

What is in a product, how those ingredients affect humans, what a product may be called, and how effective it is have laid a foundation of labeling accuracy at the very heart of the FDA's labeling concerns. The FDA has been most concerned with making certain that consumers as well as health-care professionals are provided with accurate information on consumer products.

In an article—"The Rise and Fall of Federal Food Standards in the United States: The Case of the Peanut Butter and Jelly Sandwich," Society for the Social History of Medicine, Spring Conference, 1999, Aberdeen, Scotland—Suzanne White Junod writes: "In 1906, Congress passed the U.S. Pure Food and Drugs Act. It was one of the first consumer protection acts passed in the United States." Dr. Junod further stated:

> For an act propelled into law through a focus on Food, the law was surprisingly and seriously flawed in its food provision. The first flaw was apparent even before the law was passed. Offshoots of the food industry (principally manufactures of rectified or blended whiskey and chemical preservatives including formaldehyde, borax, copper salts, salicylated, saccharin, and sodium benzoates) defeated a provision which would have allowed the Government to set standards for food products.

She continued:

> The second flaw was not quite so obvious until 1920s brought a heyday in inventive advertising, and the Great Depression of the 1930s created a market for cheap, inferior products. And this is where jelly first enters my sandwich standards analogy.

Concerned over strict food standards—these food industries inserted a so-called "Distinctive Name Proviso" into law. Dr. Junod stated in the article that this distinctive name proviso permitted the marketing of foods that would have otherwise been illegal under the 1906 act. She further stated in the case of jam and jelly, they would have been considered adulterated or misbranded under the law since they had so little fruit. Beautiful food dye hues, artificial pectin, and grass seeds, accompanied by expensive yet tasteful packaging, and promotion through clever advertising, all created a new kind of fabricated product. This product was given a fanciful and distinctive yet meaningless name—BRED-SPREAD. This product typifies the kind of inferior product that began to gain a foothold in the U.S. marketplace beginning in the 1920s.

Consumers were faced with a problem in labeling appearing on market products. Could one depend on the label to tell one the quality of food and how much of a certain product was really in the can?

But it was the FDA in 1933 that started the process of rewriting the 1906 Pure Food and Drugs Act in its efforts to rescue a potentially harmful industry.

The FDA writes: "Consumers rely on product labels to know what the product is and how it is to be used. The agency regulates what's on these labels to ensure that they are truthful and that they provide useable information that helps consumers make healthy, safe decisions when using the product" (FDA presentation entitled "Protecting Consumer", Protecting Public Health—PPT Slide 13 of 34).

From fake medical devices to almost jelly to patent medicines that claimed to heal just about everything, the FDA has continually launched a truth-in-labeling campaign to protect consumers from false claims to not only the purity of the product but to the effectiveness and safety of the product. Accuracy of labeling emerged as a significant FDA historical trend.

The second aspect of the regulation structure is "growth." Whether or not there has been a general growth of "big government" or of the quality of regulations is a matter of ongoing political debate. It is obvious that the growth of the number and variety of products and companies subject to FDA regulations is a matter of record. Through the history of the FDA we see an agency that constantly needs to respond to the challenges presented to the consumer by highly complex industries often producing highly complex products. The development of biotechnology, the discovery of new classes of pharmaceuticals, and the invention of new medical devices have all swollen the U.S. Pharmacopoeia in and lists of potential treatments.

The results of increasing demand and all but static supply (of regulations and regulatory budgets) has resulted in increasing long review times, decreasing common field investigation and inspection, and a need to find ways of rationing regulators energies without compromising public safety and health, a complex problem in itself.

But yet another aspect that gives rise to a regulatory pattern is the relationship of the product industry to the FDA. Is it an oppositional relationship by choice or history? Does a history of false claims from the early 1920s of certain product manufacturers to cure just about anything and the need to protect the consumer give rise to an oppositional relationship?

Have a sequence of events led to an oppositional atmosphere or has the development of the FDA over its history simply led it to the status of being the "watchdog?"

It is a fact that the FDA is the consumer *watchdog*. It stands in opposition to industry excesses, cut corners, and of course false claims.

The need of consumers for quick access to new products and, of course, scandals about unsafe or ineffective treatments has laid the groundwork for yet another aspect of the regulatory structure—speed the approval.

The problem of consumer demand for rapid release of promising and safe cures and treatments without regard to the science of testing and review represents another historical trend in the FDA. Whenever a new disease receives significant public attention, the FDA feels the pressure to accelerate the approval process. In recent years the tendency has been seen in human immunodeficiency virus/acquired immunodeficiency syndrome (HIV/AIDS) treatments, in terrorist-response vaccines and treatments, in Alzheimer's treatments, and in severe acute respiratory syndrome (SARS) diagnostics. The call to speed up the process is heard. But it is obvious to the industry that the FDA represents only a minor portion of the lengthy approval process that certainly can take up to 10 years.

We would rush to judgment to find fault only with the FDA. It is the range of quality and safety studies that take time, and while the FDA does have responsibilities to review and analyze those studies, blaming the FDA for the time required for approval and acceptance is nonproductive.

These four aspects have produced the final trend that is only recently receiving widespread attention. The first four trends force a rationing of regulatory effort and argue for focusing those limited efforts on areas of greatest quality concern to bring safe and effective products to market as rapidly as possible. The current course to use "risk assessment" as the rationing determinant is the result of those pressures.

The potential ultimate effect of adverse events, measuring both probability and severity, is used in risk assessment as a rationing factor.

8.3 21 CFR PART 11

Part 11 of section 21 of the Code of Federal Regulations (CFR) is an evolutionary result of the historical trends affecting the Food and Drug Administration. Part 11 deals with electronic records, with the systems that produce those records in automated laboratories, clinical testing environments, and manufacturing facilities, and with the electronic approvals and signatures of those records. The regulation includes specifications for the validation and testing of systems, for archiving and retrieval of records, and for training and standard operating support of user personnel.

In the proximal causality trend, Part 11 represents the regulation of a tool used in support of Good Manufacturing Practices (GMP), Good Laboratory Practices (GLP), and Good Clinical Practices (GCP) guidelines. In the risk assessment trend, Part 11 is clearly subject to a rationing of regulatory energy and evidence in accordance with the probability and severity of adverse events. Conventionally in the self-regulation trend, Part 11 requires that an organization establish standards for quality and prove conformity to those standards. In the development of the regulatory world we live in, Part 11 conforms with the initial focus on labeling (here, the accurate description of system requirements), with the growth of regulation (in response to the growth of automation), with the independence from and opposition to industry, imposing standards of quality and testing on the computer industry, with attempts at speeding the time to market (spearheaded by automating the review process), and with the use of risk assessment as the rationing factor, previously discussed in the blending of historical and regulatory trends.

In the remainder of this chapter you will find the six steps of conducting a risk assessment, a check list that outlines the broad scope of 21 CFR Part 11 and a drug discovery regulation checklist.

8.4 RISK ASSESSMENT

Risk assessment is the other side of benefits analysis. It examines the probability and severity of "negative" outcomes of a process, applicable to drug discovery, utilization of a pacemaker, manufacture of a drug, or any other process. However, risk assessment has a unique role: it can help determine the amount of compliance evidence required for acceptance. How closely the FDA will examine and how much time, energy, and expense should be invested in proving compliance will be determined by a risk assessment of the drug discovery process (or any other process).

To conduct a risk assessment, integrate these six steps into the process of planning and implementing a compliance strategy:

1. Utilizing the written requirement documentation for a specified system, device, computerized equipment, component, or application, identify the desired performance under each of the defined requirement application.
2. Utilizing historical experiences with the specified system, device, computerized equipment, component, or application, and/or utilizing historical experiences related to predicate systems, devices, computerized equipment, components, or applications, and/or utilizing industry standards related to the specified system, device, computerized equipment, component, or application, determine the alternate undesired performances for each defined requirement application.
3. Determine the probability of each occurrence performance, for both the desired performance and the undesired performances for each defined requirement. Probability of occurrence can be calculated utilizing historical experiences with the specified system, device, computerized equipment, component, or application, and/or utilizing historical experiences related to predicate systems, devices, computerized equipment, components, or applications, and/or utilizing industry standards related to the specified system, device, computerized equipment, component, or application.

4. Analyze each of the undesired performances to characterize the severity of that performance in terms of risk to human life, health, and/or safety.
   a. "High severity" is generally defined as loss of life, substantial loss of quality of life, and/or substantial disabling effect.
   b. "Medium severity" is generally defined as compromise of quality of life and/or some disabling effect.
   c. "Low severity" is generally defined as little or no effect on quality of life or on normal life activities.
5. For each undesired result calculate the risk according to the following grid:

SEVERITY	PROBABILITY	RISK
High severity	High probability	High risk
High severity	Medium probability	High risk
High severity	Low probability	Medium risk
Medium severity	High probability	Medium risk
Medium severity	Medium probability	Medium risk
Medium severity	Low probability	Low risk
Low severity	High probability	Medium risk
Low severity	Medium probability	Low risk
Low severity	Low probability	Low risk

6. Apply the table results to the validation protocol or policy to determine the appropriate level of testing and validation.

8.5 RISK ASSESSMENT AND REGULATION

Risk assessment provides a measure of degree of regulation in the process of regulating the drug discovery and development. Although regulation is applied to all aspects of the development process, two caveats mitigate that reality.

First, regulation of drug discovery and development is applied only retrospectively. That is, the end product and process are subject to FDA requirements ONLY IF the decision is made to move the discovered substance forward in the developmental process. Lines and paths that are abandoned or halted are not subject to regulatory review. That restriction may be of little practical value since all paths are considered potentially successful or they would not have been initiated. Most organizations, therefore, opt to integrate regulatory controls in anticipation of possible success.

The second caveat has a greater strategic and tactical value. The practical focus of the FDA is upon the tools utilized in discovery. While an educator may try to assess reading skills to determine a student's readiness to learn, regulatory attention focuses on the tools to assess the process. The student's reading skills and the discovery tools are both elements in the process and indicators for potential problems.

Since the drug discovery and development process is largely based upon the use of complex systems-automated laboratory equipment, molecular modeling software, statistical analysis systems, and so forth, those tools tend to be computers. The very nature of computers causes the most regulatory concern.

Computer systems are complex, based upon internalized and sometimes obscured decision rules. Computers are flexible, allowing multiple applications with appropriate changes in internal directions (programs). They can overwrite the changes made to those decision rules, directions, and even databases, often leaving little or no markers indicating changes were made.

These characteristics, coupled with the industry's increased reliance on the use of computer systems, have led to a focusing of a significant portion of regulatory energy on the automated tools of the field. In the drug discovery arena where reliance on computers is particularly high, the major regulatory focus, mitigated by risk assessment, has been on the systems that collect, analyze, manipulate, and report. The result of that focus is the newest of the FDA regulations, 21 CFR Part 11, Electronic Signatures and Electronic Archives.

Originally intended to define rules for accepting electronic signatures in lieu of human (paper) signatures on documents, Part 11 has been expanded to define virtually all aspects of computer control and of the regulation of computer systems. The requirement includes standards for documentation of the validation (testing and managerial control) of systems; of the training and operating procedure assistance to be provided to users; of security, change control, and disaster recovery assurances, etc. Here is a checklist that outlines the broad scope of 21 CFR Part 11:

- System is used in support of drug-related research, laboratory analysis, clinical research, manufacturing, production, and/or tracking.
- Record and/or signature system has been subjected to an appropriate and thorough system validation audit.
- Audit:
  - Conducted within 24 months
  - Conducted by independent or outside expert
  - Included review of:
    - Testing documentation
    - Development documentation
    - Standard Operating Procedure (SOP) documentation
  - Included inspection of operating environment
  - Trail review
    - Archive
- Change control

- Archive
- Disaster recovery
- Use
- Training
- System validation documentation collected
  - Evidence of requirements
  - Design approvals
  - Testing
  - Implementation
- Validation
  - Protocol
  - Team credentials
- Development documentation
  - Requirements/design document
  - Trace matrix
- Standard operating procedures
  - Use
  - Training
  - Change control
  - Archive
  - Disaster recovery
  - Audit trail review
- Testing
  - Boarder cases
  - Norm cases
  - Code review
- System Inventory
  - Hardware
  - Software
- Records are retained:
  - For appropriate length of time. Generally 10 years or two generations over treatment duration
  - In machine readable form
  - In human readable form
  - In heat proof, fire proof, flood-protected environment
- Records
  - Are appropriately labeled
  - Can be restored in reasonable (generally 72 h) time
- Procedures are in place to restrict access to data and records to appropriately authorized persons.

- Operation checks of system have been designed to assure appropriate functioning of hardware and software
- Audit trails
  - Preferably electronic and protected
  - Alternately manual and carefully monitored
  - Built into system to detect and identify data changed, including
  - Tracking of time
  - Date of change
  - Change agent
  - Reason for authorized change
- Password/password
- Password/key
- Password/biological
- Electronic signatures utilized only in system with:
  - Dual-level unique identifier authorizations
  - Internal procedures to assure that approved documents have not been modified (without authorization) from specified date and time
- Time system
  - Zulu time
  - Greenwich mean time (GMT)
  - Location-affixed time
  - Single time zone
  - Date system
  - U.S. (dd/mm/yy)
  - International (mm/dd/yy)
- Methodology has been implemented to assure the validity of input data
  - Might include
    - Dual confirmation of input
    - Use of check digits
    - Internal norm confirmations
    - Other techniques
- Systems users and administrators have:
  - Received appropriate regulatory and functional training
  - Ready and constant access to appropriately comprehensive, clear, applicable, timely, and management-approved SOPs
- All aspects of the electronic records and electronic signature systems in place have been designed to provide a level of security and control equal to or exceeding the equivalent controls inherent to manual (paper) systems.

The Part 11 requirement provides guidance for control of the system tools used in the drug development process. It serves as an outline of the FDA scrutiny of drug discovery and development. Evidence of compliance represents an insurance policy providing confidence that, should a drug move along the pipeline to a New Drug Application (NDA), the Food and Drug Administration will accept supporting data from the early stages of the development process.

8.6 REGULATION IN DRUG DISCOVERY: THE FUTURE

The regulation of drug discovery is an evolutionary process, balancing the public need for *access* to effective treatments and cures against the need to assure the *safety* of those emerging drugs. The U.S. Food and Drug Administration and equivalent agencies worldwide continually cope with that delicate balance, comparing the Hippocratic dictum "First, do no harm" with the demands of disease and injury victims and their families who plead "try something!"

The balance is currently maintained through three underlying principles of drug discovery regulation. First, the trend to *proximal causality* continually shifts the regulatory focus to the primal events in the pharmaceutical chain. In the future that focus will fall increasingly on the reliability of molecular models, of preclinical testing, and other early-stage procedures and computerized equipment used in the first stages of the drug discovery process.

Second, emerging awareness of the need to avoid the unnecessary and ineffective cost of excess regulation of products and procedures that represent little or no real safety threat has led to the growing utilization of a *risk assessment* in determining the appropriate depth of regulatory focus. The trend is expanding and generalizing: Risk assessment will stand as the most significant regulatory innovation of the decade. In many cases in which early drug discovery activities are so far removed from final production of drug product as to represent little real danger, the risk assessment will balance the trend to proximal causality to product a rationally limited regulatory involvement.

Finally, emerging as dominate in the self-canceling balance of proximal causality and risk, will be the continuing trend toward *self-regulation*. Expect the FDA to serve as a supra quality assurance unit periodically and remotely overseeing the drug discovery process to monitor the self-regulation of that process of drug discovery organizations themselves. The primary responsibility to establish appropriate and risk-assessed standards, to conform to those standards, to document that conformance, and to audit the process will lie with the pharma industries.

In such an evolving world the drug discovery process can be expected to appropriately balance public safety against development, leading ultimately to public assess to new drugs. It is an optimistic and realistic model.

Drug Discovery Regulation Checklist
- *Risk assessment*
  - Determination of nondesired alternate outcomes and results of the drug discovery operation
  - Calculation of the approximate probability of those occurrences, utilizing historical logs and/or results of predicate device operations
  - Determination of the potential severity (in terms of direct threat to human health and safety) of those occurrences
  - Resulting categorization of the operation as low, high, or medium risk
    - If risk is determined to be low, other steps can be:
      - Mitigated (limited testing in validation)
      - Eliminated (the vendor audit)
      - Reduced (the PQ)
- **Validation of the operation** including:
  - **System validation** of all automated components, including:
    - Documentation of system requirements and design
    - Development of a trace matrix
    - Development and exercise of appropriate test scripts
    - Review of the code itself
    - Analysis of the SOPs to assure:
      - Appropriate use
      - Archive
      - Disaster recovery
      - Change control
      - Training
  - **Process validation**, focusing on possible contamination of
    - Media, equipment, and final product, including but not limited to:
      - Bacterial contamination (both Gram positive and Gram negative)
      - Viral contamination
      - Material contamination
- **Part 11 audit** emphasizes:
  - An audit trail to track all data changes
  - If electronic signatures are in use:
    - Dual confirmation of identity
    - Locking of document after signature
    - Unambiguous time/date stamp
- **Archive** (electronic and human readable) of all files:
  - Control of data accuracy
  - Appropriate training

- **Audit of system vendor** either by:
  - Independent "expert witness" certifying compliance with appropriate regulations or
  - The end user organization
    - Key elements:
    - The criteria utilized
    - The credibility of the auditor
- **Installation qualification (IQ)**, using preestablished standards to assure appropriate initial implementation of the system or systems. The IQ may be conducted by the vendor, by the end-user organization, or by a combination of both.
- **Initial and periodic calibration**, in accordance with a metrology plan appropriate to the specific system or systems in use. Some systems are self-calibrating; a few others do not require recalibration after initial installation
- **Operational Qualification (OQ) and Performance Qualification (PQ)**, sometimes performed separately and (in some circumstances and systems) combined. The OQ assures the system is ready for use; the PQ assures that it is appropriately in use.
  - **Problem report**: A system for reporting (and reviewing) any encountered malfunctions, necessary changes, or other problems.

SIMULTANEOUS SCREENING OF MULTIPLE CELL LINES USING THE CELLCARD SYSTEM

OREN E. BESKE
Vitra Bioscience, Inc.
Mountain View, California

9.1	INTRODUCTION	354
9.2	CELLCARD TECHNOLOGY	356
9.3	GENERAL CONSIDERATIONS FOR MULTIPLEXING CELL-BASED ASSAYS	357
9.4	GENERAL PROTOCOL FOR USING THE CELLCARD SYSTEM	359
	Experiment Design and Plate Layout	359
	CellCard Preparation	360
	Tissue Culture	360
	CellCard Mixing and Dispensing into 96-Well Format	360
	Compound Addition and Assay Protocol Considerations	360
	CellCard Reading	361
	Image Analysis and Data Visualization	361
9.5	THEORETICAL EXAMPLE: APPLICATION OF CELLULAR ARRAYS TO THE Identification and Profiling of Cell Type Selective Antimitotic Compounds	362
9.6	SPECIFIC CASE STUDIES USING THE CELLCARD SYSTEM	363
9.7	MULTIPLEXED DOSE CURVE ANALYSIS USING THE TUNEL ASSAY	363
9.8	MULTIPLEXED ANTIPROLIFERATION SCREEN; IDENTIFYING CELL TYPE SELECTIVE Compounds	367
9.9	SUMMARY OF KEY FEATURES OF CELLCARD SYSTEM	370
	References	371

Drug Discovery Handbook, by Shayne Cox Gad
Copyright © 2005 by John Wiley & Sons, Inc.

9.1 INTRODUCTION

Historically, biochemical assays have vastly outnumbered cell-based assays in the screening environment of the drug discovery process. The popularity of Biochemical assays has been driven by their robustness, reproducibility, and inexpensive nature. As a result, biochemical assays are used in many steps of the drug discovery process including primary screening, lead identification, lead optimization, and the determination of chemical structure–activity relationships (SAR). Although robust and relatively easy to implement, many biochemical assays do not address important biological issues, that is, the complex cell biology surrounding most targets. The desire for more biologically relevant data from primary high-throughput screening (HTS) as well as downstream processes (i.e., lead identification and optimization) has resulted in a dramatic increase in the development of technologies specifically designed for cell-based assays and the use of cell-based assays in the drug discovery process (*http://www.hitechbiz.com*).

Traditionally, cell-based assays were slow, lacked reproducibility, and were prohibitively expensive. However, as our understanding of cell biology increases and new technologies emerge, cell-based screens are becoming faster, cheaper, more reproducible, and increasingly powerful. This is evident in the vast number of new biosensors, assays, and detection systems being developed for cell-based HTS [1–5]. In addition, the use of genetic tags, such as fluorescent proteins, have enabled scientist to develop unique assays that require no additional staining and provide a means to acquire kinetic as well as localization data (*http://www.bioimage.com*). Novel dyes (e.g., DiBAC, Fluo-3) have brought complicated cell biological processes to simple bench-top readers [6–9]. These bench-top readers encompass flexible new detection technologies that enable scientists to robustly acquire data from cell-based assays performed on a single microscope slide or in the context of a fully automated HTS facility.

Concomitant with these advances in cell biology and assay development there has been tremendous strides forward in the field of cellular imaging technologies. Recent years have seen multiple imagers brought to market (*http://www.cellomics.com, http://www.amershambiosciences.com, http://www.atto.com, http://www.moleculardevices.com, http://www.evotectechnologies.com*) with the ability to scan microtiter plates gathering multiple parameters from each well. These imaging technologies can be separated into two general categories: microimagers and macroimagers.

Macroimagers have the ability to image an entire microtiter plate with a single exposure. Since this approach acquires an image of the entire plate, macroimagers often have the ability to work with 96-, 384-, and 1536-well microtiter plates without having to significantly reconfigure the instrument. One example of this is the fluorecscence imaging plate reader, or FLIPR [7, 10, 11]. The instrument was developed to acquire quantitative kinetic measurements of calcium flux. It images the entire plate with a single exposure

INTRODUCTION

and therefore has a very fast read time, allowing for high-throughput operations. However, since these macroimagers acquire very low magnification images, they obtain average measurements of a cell population in much the same way photomultiplier tube (PMT) based readers do. As a result, it is not possible to identify heterogeneity within the cellular population, nor is it possible to gather subcellular information.

Microimaging technologies (*http://www.cellomics.com*, *http://www.evotechtechnologies.com*, *http://www.moleculardevices.com*) employ higher magnification approaches resulting in the ability to gain intracellular resolution. Typically, acquisition protocols can be highly customized ranging from 2× to 20× magnifications, 1 to 10 images per well, and 1 to 4 fluorescent labels. Microimagers have slower plate read times and provide fundamentally different information than macroimagers. While macroimagers enable rapid imaging of an entire plate, they are restricted to simple population-based intensity measurements from the well. Microimagers, by virtue of their ability to gain intracellular spatial resolution, allow scientist to analyze complex intracellular events such as nuclear translocation, nuclear morphology, and endosome formation [12–14]. The quantification of these cell biological phenotypes is gained through the use of sophisticated image analysis algorithms. These algorithms act upon the images and translate them into numbers that accurately represent the biology of interest. For example, using a single nuclear dye, these algorithms can determine nuclear size, density, condensation, fragmentation, and the like. In addition, the cellular samples can be stained with multiple fluorochromes in order to analyze multiple biological parameters from each well. In doing such, one can analyze multiple aspects of a complex biological process within a single well. This approach, often referred to as high content screening (HCS), has gained increasing popularity in recent years as it enables more information to be gathered from each microtiter well [15, 16].

The advent of DNA (deoxyribonucleic acid) arrays enabled scientists to gather more than one data point from a single experiment or microtiter well. This approach, widely used in the analysis of gene expression, has expanded to the analysis of the genome, proteins, and cells [17–20]. This parallel analysis of multiple analytes (often referred to as multiplexing) provides the researcher with the magnitude of an effect (the signal) as well as its selectivity (the comparison of the signal across analytes). By using multiplexing technologies, the data points come from a single reaction and therefore can be confidently compared to one another. That is, pipetting and multiple other sources of error cannot be the cause of any selectivity identified. Finally, multiplexing is a means to miniaturization. That is, since multiple data points are obtained from a single reaction, the amount of sample used per data point is greatly reduced, making it a very attractive approach when reagents are expensive or there is a rare sample to be analyzed, that is, patient primary cells. Multiplexing technologies decrease the cost as well as the reagent consumption for each data point.

Vitra Bioscience, Inc. (*http://www.vitrabio.com*) has developed a multiplexing technology that enables cell lines to be robustly multiplexed. This platform technology is called the CellCard system. The CellCard technology merges the ability to gather multiparametric intracellular data afforded by microimaging with the ability to simultaneously analyze a compounds potency and cell-type selectivity. This is achieved because the CellCard technology enables multiple cell lines to be assayed (multiplexed) in a single microtiter well [21, 22]. (add Beske 2004) This chapter describes the CellCard technology and how it can be applied to the determination of cell-type selectivity of bioactive molecules.

9.2 CELLCARD TECHNOLOGY

The CellCard technology is centered on an encoded microcarrier approach to multiplexing. By associating each cell type to a uniquely encoded microcarrier (CellCard), these cells can be uniquely identified and their response measured. The CellCards (Fig. 9.1a) are flat, rectangular, and contain a colored barcode. The encoding strategy employs two colored bands (small arrow) on either side of a clear data readout section (large arrow). This physical separation of the data from the encoding strategy ensures that there is no cross-talk or limitation of one imposed on the other. Therefore, the encoding bands do not limit the number of fluorophores that can be used to measure cellular parameters. The size and aspect ratio of the CellCards allow for approximately 80 to 100 of them to be automatically deposited in a single well of a 96-well microtiter plate (Fig. 9.1b). To perform a multiplex experiment using Cell-

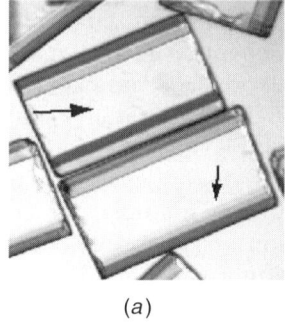
(a)

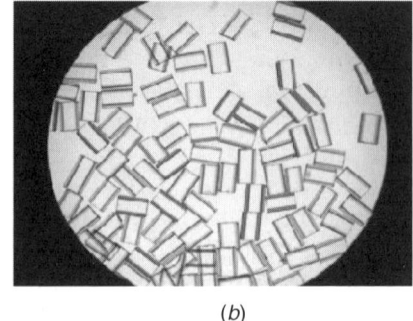
(b)

Figure 9.1 (a) The CellCards are flat and rectangular in aspect ratio. They are functionally separated into two parts: A colored barcode (small arrow) and an optically clear data readout section (large arrow). (b) The current implementation of the CellCard system is in 96-well microtiter plate format. This image shows multiple CellCard codes dispensed into a single well. There are approximately 80 CellCards in this well.

Cards is relatively simple (Scheme 9.1) and will be discussed in more detail later in the chapter. Briefly, adherent cells are grown on the surface of the CellCards with each cell type being uniquely associated with a code. That is, cell type 1 is grown on code 1 while cell type 2 is grown on code 2. After all of the cell types have adhered to the CellCards, they are mixed and then dispensed into a 96-well microtiter plate (Fig. 9.1b). In doing so, an aliquot of the encoded cell mixture is placed in each of the wells, generating a multiplexed assays in each well. Compounds are then added to the wells, incubated for the appropriate amount of time, and the cells stained for the biological parameters of interest. Finally, the CellCard array is decoded, and the cellular data captured and analyzed via the CellCard reader and associated software.

9.3 GENERAL CONSIDERATIONS FOR MULTIPLEXING CELL-BASED ASSAYS

Since the separate cell types to be analyzed are placed in a single well, there are some unique considerations that must be taken into account when deciding which cell types and assays are appropriate to be adapted to CellCards in a multiplexed format.

The living cells will be sharing a common extracellular environment during the compound incubation time. Therefore, one factor to take into consideration when choosing cell lines is the media conditions in which the incubation will take place. For example, if the cell lines to be assayed are all stable transfectants (i.e., overexpressing targets of interest) selected in Chinese hamster ovary (CHO) cells with a neomycin resistance marker, this media concern will be trivial since they are already being grown using common media conditions. On the other hand, if the cell lines are tumor cell lines derived from different tissues of origin (as presented later in this chapter) the situation is slightly different. If the cell lines all grow in a common media (i.e., RPMI +10 percent serum), there is no media concern. However, if a subset of the cell lines have been growing in different media, there may be a concern that common media-growth conditions are not being used. There are multiple possible solutions to remedy this issue. One solution is to adapt the cell lines all to a common media. This can be accomplished by passaging them for approximately 2 weeks (the time required is cell type and media type dependent) in the new media. Then, once adapted, the cell line should be recharacterized with respect to the relevant biology (i.e., doubling time or signaling ability) for the assay. If the biology is within a reasonable range for the screen, this cell line can now easily be multiplexed within the common media of the other cell types. This approach may not be appropriate for some cell types depending on the assay. Therefore, one may want to continue growing the cell lines in their independent media and only combine them into a common media for the duration of the compound incubation. Although this gets around the necessity to adapt

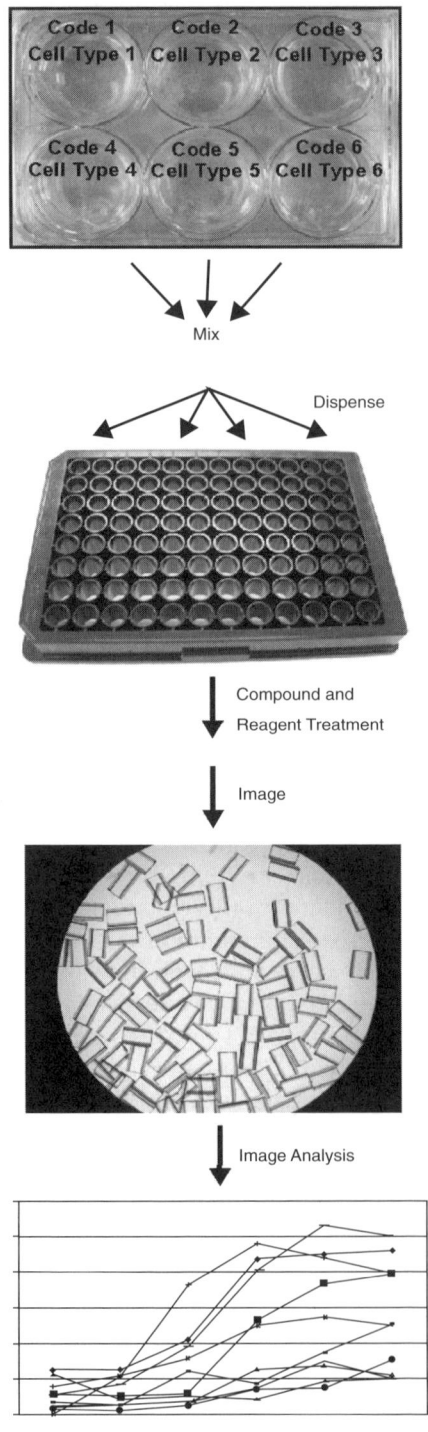

the cell line to a new growing condition, it could bring about other concerns when the cell line is exposed to the common media for the compound incubation time. Whatever approach is taken to address this issue, the new assay conditions and biological consequences must be characterized and be within an adequate range before performing the experiment. If this is not done, it may be difficult to interpret the data, since the new assay conditions may introduce undesired or uncharacterized effects on the biological parameters being measured.

The assay of choice and compound incubation conditions should also be appropriately chosen in order to gain the most robust biological information. With all of the cell types in a single well it is advantageous if their biological responses, as determined by the parameters being assayed, occur with similar kinetics. For example, take the case where cellular proliferation is being measured and one is screening for a compound that acts at a specific step in the cell cycle. Since the screen is designed to find inhibitors of a specific cell cycle event, the cell lines should have similar doubling times. In that way, the compound incubation time should be chosen that ensures that each of the cell types have transited through the cell cycle in order to allow the compound to exert its effect. For example, assume the compound incubation time is set at 48 h and the multiplexed experiment contains a cell line with a 24-h doubling time (fast growing) and another with a 72-h doubling time (slow growing). In this scenario some cells from the slow-growing cell line may not have experienced the appropriate molecular events of the cell cycle for the compound to exert its effects, while the faster growing cell line will have, on average, transited that critical step twice. In such a case, any lack of antiproliferative activity seen against the slow-growing line would be difficult to interpret. It could be due to the lack of cell cycle events or a lack of potency against that cell line.

As with the implementation of any new technology or assay, the appropriate assay development should be done in order to ensure that the assay conditions provide the optimal dynamic range and reproducibility.

9.4 GENERAL PROTOCOL FOR USING THE CELLCARD SYSTEM

Experiment Design and Plate Layout

A core software component of the CellCard system is the Experiment Manager application. This application provides a simple, easy-to-use interface

---

Scheme 9.1 The CellCard protocol starts with growing different cell types on unique codes in 6-well format. After mixing the cell cards and the attached cells they are dispensed into 96-well format. The wells are scanned after compound addition and assay reagents are added. Image analysis algorithms decode the cellular array and extract the biologically relevant data from the images.

to design and manage all aspects of the multiplexed experiment to be performed. It stores information (i.e., fluorochromes, analysis parameters, number of plates, etc.) about the assay being run in order to appropriately configure the CellCard dispenser and reader. In addition, other experimental details, for example, this software allows the user to log cellular and compound information (i.e., lot number, concentration), that are useful for downstream data analysis can also be managed with this software tool.

CellCard Preparation

In preparation for a CellCard experiment, the CellCards are first dispensed into 6-well plates. Before adding the CellCards, disposable ladles are placed at the bottom of the 6-well plate to aid in their removal prior to mixing them and their transfer into a microtiter plate (see below). Then, the appropriate amount, determined by the number of wells to be assayed, of CellCards are added to the ladles and dispersed into a monolayer.

Tissue Culture

The adherent cell lines to be multiplexed are maintained using standard tissue culture protocols in the media of choice. The cells should be split and accurately counted prior to plating them on the CellCards. Typically, 1 to 5×10^5 cells are seeded into the wells containing the CellCards. A range of seeding densities should be tested to ensure the proper confluence of cells for the particular assay that is to be performed. The cells should be allowed to adhere (under standard incubation conditions) for an appropriate amount of time to ensure robust adhesion to the particles. This is cell type specific and can range from 5 to 18 h. We recommend allowing the cells to adhere during an overnight incubation.

CellCard Mixing and Dispensing into 96-Well Format

After the cells have adhered to the CellCards, they are transferred into a 15-mL conical tube for mixing. A single 360° inversion is sufficient to mix the CellCards. The conical tube and the appropriate number of 96-well plates are then loaded onto the robotic CellCard dispenser. A detailed protocol for dispensing the CellCard mixture into 96-well plate format is provided in the CellCard users manual. After the CellCards have been dispensed into 96-well format, the plates are placed into cell culture incubation conditions for recovery prior to compound addition.

Compound Addition and Assay Protocol Considerations

Remove the 96-well plates containing the CellCards and attached cells from the incubator. Then, using a wand aspirator or a plate washer, remove media from the wells to ensure that the same residual volume is present in all wells. We recommend leaving 50 μL of residual media in each well. Add the appropriate amount of compound to each well and mix. We typically add 50 μL of a 2× solution of the compound. After the compounds have been added to each of the wells, return the plates to the cell culture incubator for the appropriate incubation time. This incubation time is assay dependent and can range from 30 min for a nuclear translocation assay to multiple days for a proliferation assay.

In general, only minor alterations in standard assay protocols need to be made when using the CellCard system. For example, incubation times with and concentrations of staining reagents remain the same with respect to an assay run on plates without CellCards. We recommend that all washes be done as gently as possible to limit CellCard agitation and possible sloughing off of cells from the particles. In addition, we recommend that the wells are not aspirated dry. We typically leave ~20 μL in each well at the end of a wash to minimize the potential removal of CellCards from the wells.

CellCard Reading

Before scaning the plates with the CellCard reader, the appropriate plate description file (generated in the experiment designer application) must be loaded. This will configure the reader to address the appropriate wells and acquire images of the appropriate assay-specific fluorochromes.

After the microtiter plate has been placed in the reader (either manually or via a robot), the image acquisition parameters must be set. The parameters for acquiring the bright-field images will be automatically set by the reader without user intervention. This will ensure that the CellCard codes will posses the proper hue's in the final image and robust code recognition. The default parameters for fluorescent image acquisition will be read in from the plate description file. However, these typically need to be adjusted by the user to ensure optimal image quality. A detailed protocol for this is provided in the user's manual. The acquisition parameters are set in such a way as to ensure that the images can be efficiently used by the analysis algorithms and accurately represents the biological readouts that are to be measured. If there are multiple plates in the experiment, these acquisition settings are set as the default for all subsequent plates to be imaged during the same session. Once the acquisition parameters are set, the plate is scanned by the imager and the images saved to the appropriate directories designated in the design of the experiment.

Image Analysis and Data Visualization

To extract the cell type specific data, the images acquired by the CellCard reader must be processed by a series of image analysis algorithms. The software will read in the assay-specific default image analysis parameters and proceed with the analysis. It is recommended that to run the analysis on at least two wells, that is, positive and negative controls, before analyzing the entire experiment. The CellCard system manual provides a detailed guide for how to accurately set the proper image analysis parameters.

The data tables generated by the CellCard analysis software are saved as a tab-delimited flat files. This data file can then be opened and visualized in the Vitra Bioscience data visualization software or other standard data visualization applications (e.g., Spotfire). In addition, the flat file structure is a standard one allowing the data to be parsed and uploaded into an enterprise database for further analysis and storage.

9.5 THEORETICAL EXAMPLE: APPLICATION OF CELLULAR ARRAYS TO THE IDENTIFICATION AND PROFILING OF CELL TYPE SELECTIVE ANTIMITOTIC COMPOUNDS

The development of antitumor drugs necessitates the determination of a compounds' potency as well as cell-type selectivity. The ability for 10 cell lines to be robustly assayed using the CellCard system enables both of these key pieces of information to be gathered simultaneously from a single microtiter well.

In addition, the multiplexing of such experiments results in a significant decrease in the time to gather data. Most proliferation assays require 48- to 72-h incubation times. Therefore, each experimental cycle from seeding cells to gathering data lasts roughly a week. Suppose there are 10 cell lines that are to be used to profile compound activity, and there are resources to run 30 plates a week. If these compounds are to be analyzed in a triplicate dose curve format (at 4 compounds and 1 cell type per plate) this means that the allocated resources can only accommodate12 compounds per week. However, if the 10 cell lines were multiplexed, the throughput would increase 10-fold. This would result in an overall throughput of 120 compounds per week. This would shorten the time required to gather the data on 120 compounds by a full month!

The choice of cell lines included in this type of profiling will determine the biological specificity being assayed. For example, one possible goal might be to determine the ability of a compound to inhibit proliferation across a series of cell lines from different tissues. The utility of this approach has been shown by the National Cancer Institute (NCI) in its compound screening efforts (*http://dtp.nci.nih.gov/index.html*). To this end multiple cell lines representing different tissues of origin would be grown on the cell cards, mixed, and assayed in each well. For example, by including two cell lines from each of the following tissues—colon, breast, lung, ovary, and prostate—one could be profiling the compounds ability to inhibit proliferation across the various tissue-derived cell lines.

Alternatively, the ultimate goal might be to determine the pathway or target selective effects of a compound. The validation of this approach was shown in the identification of a target selective compound that selectively inhibited k-Ras transformed cells [23]. This could be expanded to include other targets as well. For example, one could generate isogenic cell lines that differ in their genotypes at a single specific locus. One cell line might harbor a p53 mutation known to be associated with cancers while another may be a p53 knockout. In the same way, other cell lines could be generated to readout other pathways. When these are screened along with their parental cell type (the one used initially to generate the isogenic strains) compounds would be assayed for their ability to elicit a target or pathway selective response.

The choice of cell types to be used in the multiplex experiment is important to be certain. One must carefully decide on the exact biological question and the nature of the selectivity that is to be determined. In addition, it is critical that the kinetics of the biological response to be measured is similar for all cell types. That is, if the assay is designed to measure proliferation, it is advantageous to match doubling times of the cell types to be multiplexed. Once chosen, the combination of cells and the assay chosen to implement using the CellCard system will provide the user with the potency and selectivity information desired.

9.6 SPECIFIC CASE STUDIES USING THE CELLCARD SYSTEM

To demonstrate the feasibility of how to use this novel technology to profile compounds, we will present two case studies. In the first, the TUNEL assay was used to analyze the proapoptotic and antiproliferative effects of staurosporine across multiple cell lines. By comparing the dose-dependent effects on these two parameters, interesting cell type specific responses were identified. In the second example, a small set of compounds were analyzed in a single-dose format to simulate a screening scenario. This experiment was designed to identify antiproliferative compounds with cell type selectivity. In addition to measuring proliferation, this multiparametric assay also measures BrdU incorporation, a measure of cell cycle activity. By using these two parameters, the screen was able to identify cell type selective antiproliferative compounds using a single well.

9.7 MULTIPLEXED DOSE CURVE ANALYSIS USING THE TUNEL ASSAY

In this experiment six cell lines were simultaneously assayed and analyzed for their response to the potent proapoptotic compound staurosporine. The assay used was the TUNEL (Terminal Uracil Nick-End Label) assay, which enzymatically labels double-stranded DNA breaks, a hallmark of late-stage apoptosis. Analysis of the cellular staining from this procedure results in an apoptotic index parameter. This measurement has been normalized to a no-drug control and represents a fold increase over background. In addition to this measure, a DNA intercalating dye is used to label all nuclei and therefore serves not only to normalize the TUNEL data to cellular number but also to measure cell number as a measure of growth. For the growth parameter the data was normalized using the NCI paradigm (for an in-depth discussion please visit the NCI website at *http://dtp.nci.nih.gov/docs/compare/compare_methodology.html#specon*). Briefly, this approach results in a relative growth index such that a value of one indicates a cytostatic environment. A relative growth of greater than one represents proliferation, and numbers less than one indicate cell loss or cytotoxicity.

Figure 9.2 demonstrates how these two parameters, apoptotic index and relative growth, can be robustly derived from the same set of images. The cell line shown in Figure 9.2 (HT29s) demonstrates how this compound induces growth arrest and a concomitant induction of apoptosis.

However, this only represents one of the six cell lines that were present in each of the wells. Figure 9.3a shows the apopotic index and relative growth data for each of the cell lines. Immediately evident is the cell type specific differences that can be seen in these graphs. The magnitude of the responses, the slopes of the lines, the concentrations that yield maximal response all vary from cell type to cell type. A subset of these cell lines are shown in Figure 9.3b.

These graphs (Fig. 9.3b) show the response of A549, ADR-RES, and MCF7 cells. Note that despite the fact that the ED_{50s} for the apoptotic response are nearly identical for each of the cell types, the magnitude of the response is significantly diminished for MCF7 cells. A simple analysis of ED_{50} values would not have revealed this difference. Interestingly, even though the A549 and ADR-RES cells show superimposable responses when graphed for their apoptotic response, their relative growth profiles are different. The A549 cell proliferation curve is left shifted from the ADR-RES cells at the lower concentrations. This demonstrates that A549 proliferation is more sensitive to staurosporine than ADR-RES. The comparison of these multiple parameters in different cell lines enabled us to identify these significant cell type specific differences. The biological mechanisms underlying these differences are unknown. However, it demonstrates the power and utility of the CellCard system for comparing multiple parameters with the unequivocal knowledge that each data point is gathered from identical experimental conditions.

Figure 9.2 Multiparametric analysis of the TUNEL assay. (*a*) This graph demonstrates the apoptotic response of HT29 cells to staurosporine. The data has been normalized to represent a fold increase in apoptotic signal over background. (*b*) Analysis of the nuclear stain provides a growth index. The growth index has been normalized to a time zero control (NCI paradigm) such that no relative growth, a cytostatic situation, equals one. This relative growth plot displays the antiproliferative activity of staurosporine toward HT29 cells.

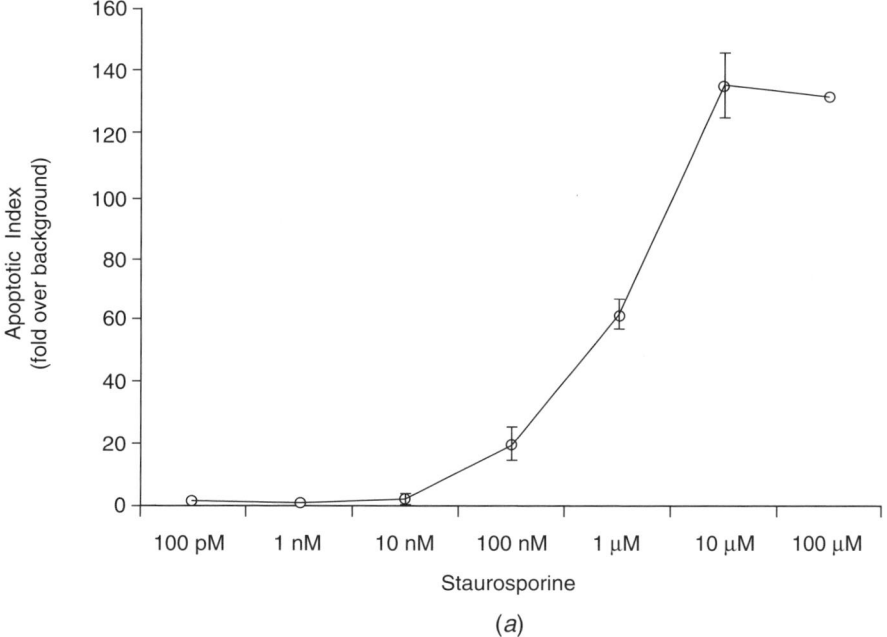

(a)

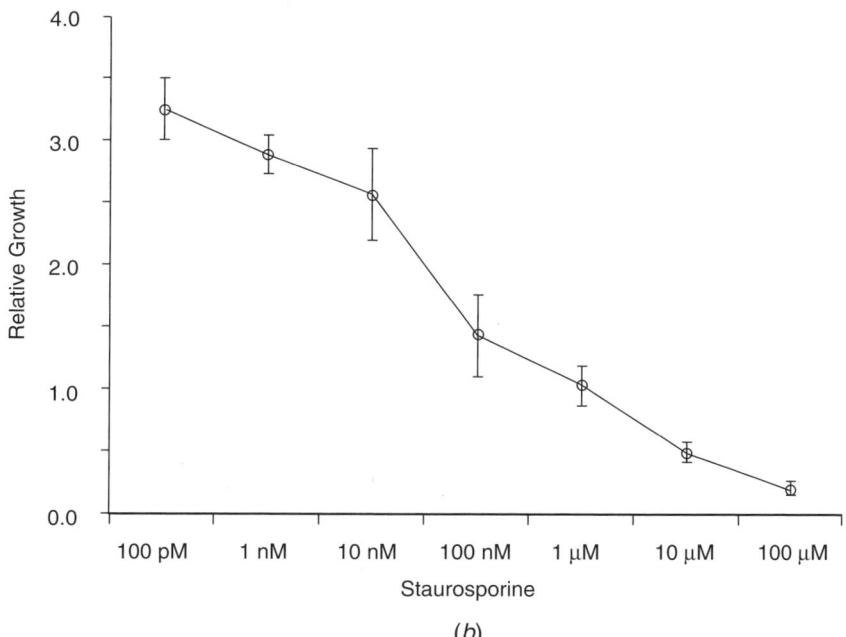

(b)

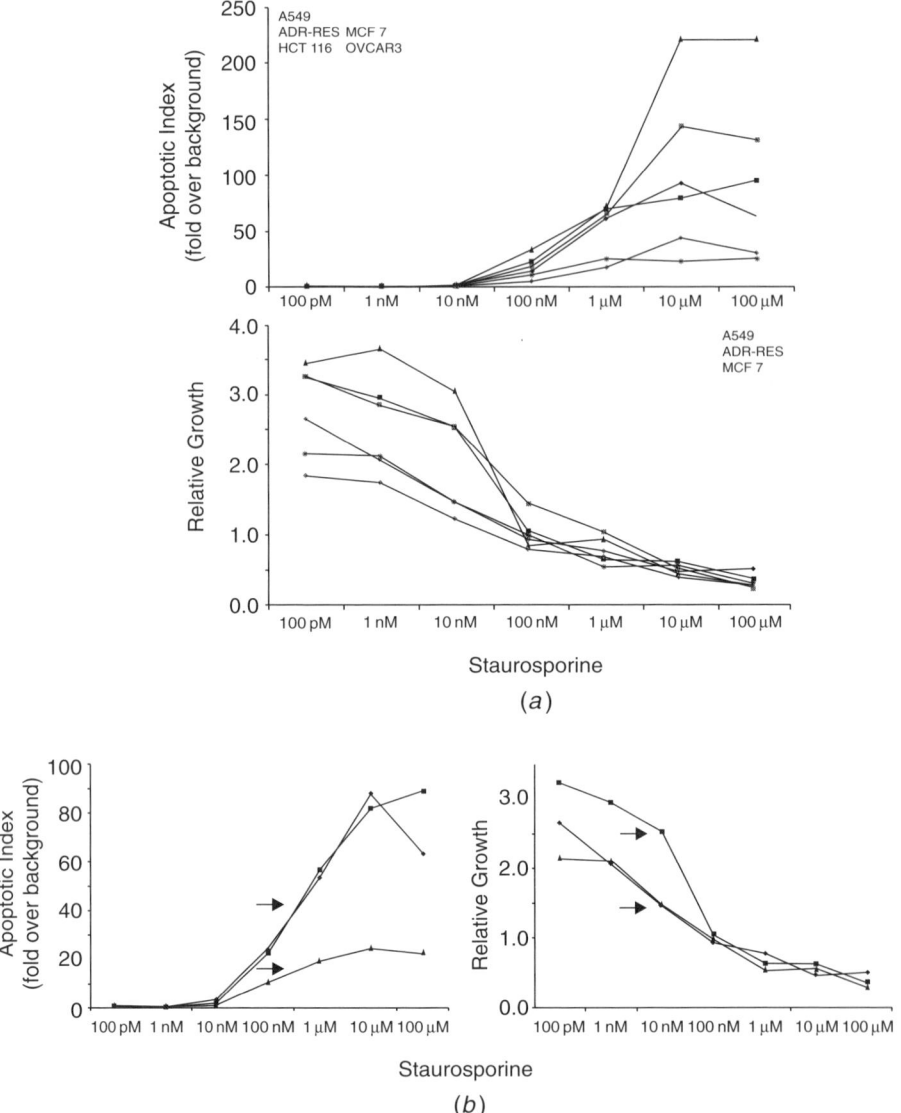

Figure 9.3 Multiplexed dose curve analysis of staurosporine using the TUNEL assay. (a) Six cell lines (A549—blue, ADR-RES—pink, HT29—aqua, MCF7—maroon, and OVCAR3—brown) were simultaneously assayed for their apoptotic and proliferative responses to increasing concentrations of staurosporine. (b) Three of the cell lines presented in (a) are shown for a more direct comparison of the parameters. Comparison of the parameters shows that even though the ED_{50} values for the apoptotic responses are identical for these cell lines (red arrows) their growth responses differ (blue and black arrows). Notably, all the data from each dose presented was generated from a single well. Therefore, all of this data in this figure was generated from seven wells.

9.8 MULTIPLEXED ANTIPROLIFERATION SCREEN; IDENTIFYING CELL TYPE SELECTIVE COMPOUNDS

A subset of the NCI's structural diversity set of compounds was screened for antiproliferative activity. The 10 cell lines used in this experiment came from the ATCC and represent tumors from varying tissues of origin (Table 9.1). Since these cell lines have a cell cycle time of approximately 20 to 30h, a compound incubation time of 48h was chosen. Experiments with camptothecin and taxol, known antiproliferative compounds, were performed to ensure that the incubation time and other assay parameters were appropriate (data not shown).

Each plate had a layout in which columns 2 to 11 of the 96-well plate were used for the compounds to be screened. Columns 1 and 12 contained positive and negative control conditions. The details of this layout are presented in Figure 9.4. The maroon squares represent negative control wells that were treated with 1 percent DMSO (dimethyl sulfoxide, the compound solvent). The green squares represent the positive control wells that were treated with 1µM camptothecin. The other wells in columns 1 and 12 represent other controls not relevant to this discussion. The test compounds were screened at a concentration of 1µM in 1 percent DMSO. After a 48-h incubation the cells were labeled with BrdU and stained according to protocol provided in the CellCard system manual. The plates were then imaged and analyzed using the CellCard system.

This assay provides the user with two key biological parameters for analysis; the first is proliferation. As part of the protocol, the nuclei are stained, and therefore this fluorochrome can be used to assess cell number and therefore the amount of cellular proliferation. In addition, the assay measures BrdU incorporation, a general measure of DNA polymerase activity. This activity is usually associated with the DNA synthesis phase (S-pahse) of the cell cycle,

TABLE 9.1 The Cell Types and Tissues of Origin Used in a Screen for Cell Type Selective Anti-Proliferative Compounds

Cell Type	Tissue of Origin
A549	Lung
Adr-Res	Unknown
HCT116	Colon
HT29	Colon
M14	Skin
MCF7	Breast
OVCAR3	Ovary
OVCAR5	Ovary
SKMEL2	Skin
SKMEL28	Skin

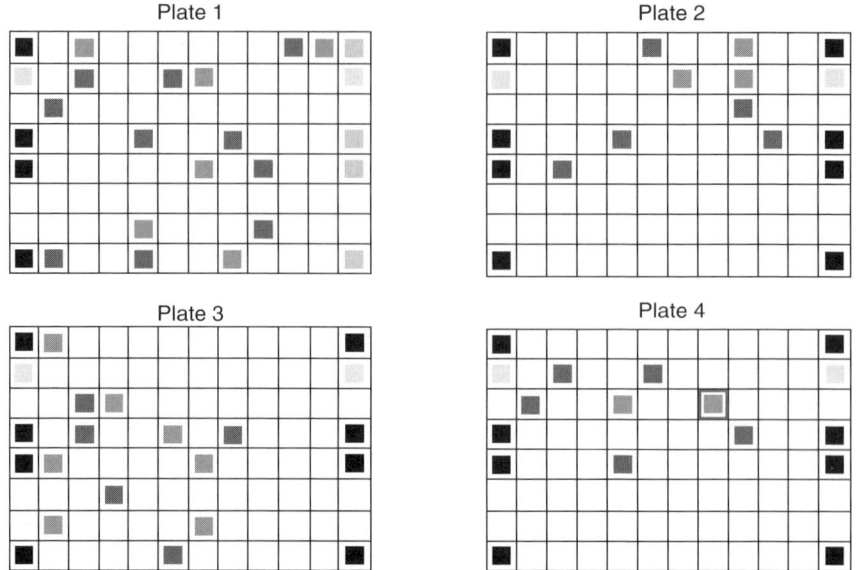

Figure 9.4 Summary of the plate layout and results from a compound screen designed to identify cell type selective antiproliferative compounds. The maroon squares represent negative control wells while the green squares are wells treated with 1 µM camptothecin, a known antiproliferative compound. The library compounds were dispensed in columns 2 to 11 at a final concentration on 1 µM. The blue squares represent nonselective antiproliferatives, whereas the red squares represent wells with cell type selective antiproliferative activity. The data from well C8 on plate 4 is shown in Figure 9.5.

and therefore quantification of BrdU incorporation is returned as an S-phase index. Since each cell type has different absolute values for these measures, the data was normalized to the no-drug (negative) controls. The normalization routine results in a value that is the percentage of the no-drug control. Therefore, these controls represent 100 percent activity. An example of these control values for plate 4 are shown on the left side of Figure 9.5.

In Figure 9.4 the red and blue squares show the hits that were identified on these plates. A hit was defined as any compound that inhibited proliferation of at least one cell type by at least 50 percent. The blue squares were nonselective compounds. That is, they inhibited the proliferation of all cell types by 50 percent or more. On the other hand, the red squares represent those compounds whose antiproliferative activity showed some cell type selectivity. In order to be selective the compound was required to have at least a twofold differential effect between at least two cell types. In addition, the cell type least effected by the compound had to show less than 50 percent inhibition (greater than 50 percent of control proliferation). As is evident from this figure, using the CellCard system compounds were simultaneously analyzed for their

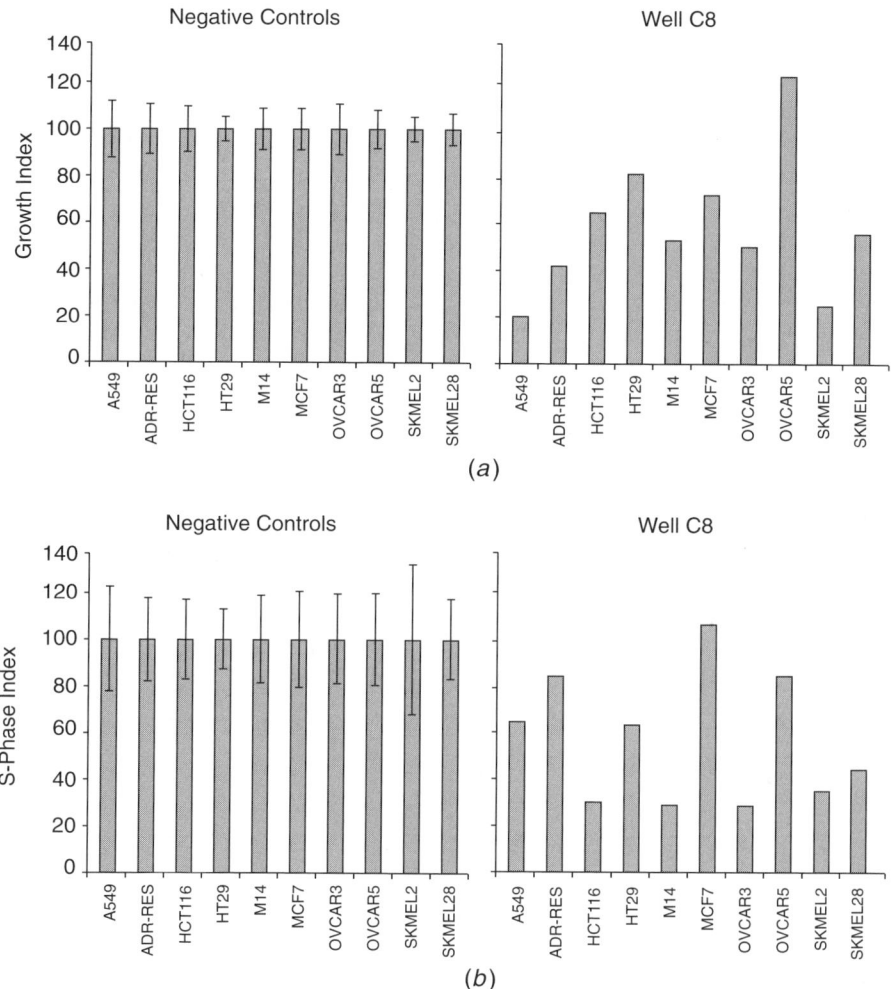

Figure 9.5 Single-well compound profiling. (*a*) Graphs of the growth index parameter. The graph on the left represents the average and standard deviations of the negative control wells on plate 4. The graph on the right shows the cell type sepcific profile of activity derived from well C8 on plate 4. (*b*) Graphs of S-phase index. As above, the graph on the left represents that data from the negative controls on plate 4. Similarly, the graph on the right shows the cell cycle (S-phase index) data generated from well C8 on plate 4.

potency and selectivity. This resulted in the identification of cell type selective antiproliferative compounds. Most importantly, the negative control wells showed very reproducible results on these plates.

The data from a single well, C8, in plate 4 is shown in Figure 9.5. The graph in Figure 9.5*a* demonstrates the variability of the antiproliferative response across the cell lines in this well and demonstrates the selectivity of the compound. To verify these results, the BrdU incorporation data was also analyzed. Keep in mind, both parameters are obtained from a single well and, therefore, the data from one parameter is directly comparable to the other. Further analysis shows that the BrdU data closely correlates with the proliferation data. For example, the OVCAR5 cell line displays no inhibition of proliferation or decrease in S-phase index when exposed to this compound. On the other hand, both the proliferation and S-phase index of the SKMEL2 cell line were significantly inhibited by this compound. Therefore, the use of the S-phase index as a confirmatory parameter served to provide a second data point on which to confirm the hits. When applied to a larger scale screening environment, this type of confirmatory analysis from the same cellular population should serve to significantly lower false-positive rates and provide a means to identify potential false negatives.

This screen yielded 20 unique, biologically relevant data points from each microtiter well. The direct comparison of these data points enables the identification of unique cell type specific difference and stimulates the design of unique screens. For example, hit compounds can be defined based on their profile across multiple parameters as well as cell types without calling into question the well-to-well, plate-to-plate, and/or day-to-day variability of the data. The data across cell types is generated within a single well in identical reaction conditions with a concomitant savings in reagents and time. Furthermore, depending on the cell lines chosen for the screen, biological mechanism of action could be screened for as well.

9.9 SUMMARY OF KEY FEATURES OF CELLCARD SYSTEM

The CellCard system developed by Vitra Bioscience provides a novel platform for cell-based assays. At the core of the platform are encoded particles referred to as CellCards. By growing individual cell types on CellCards with unique codes, the platform enables multiple cell types to be assayed simultaneously in 96-well format. This results in a miniaturization of the assays by decreasing the overall number of cells, reagents, and time required for each data point. This is especially valuable when rare reagents (i.e., primary cells, natural product compounds) are at stake. In this case, they can be used more efficiently and more data gathered from a single reagent lot. The case studies presented here demonstrate how comparing multiple parametric measurements across multiple cell types is a powerful way to profile compounds and identify cell type specific differences. Taken together, the data can be confidently compared knowing that some sources of variability (i.e., day-to-day) cannot be responsible for cell type specific differences identified.

Moving forward, the CellCard technology will continue to drive novel screen designs and compound profiling strategies.

REFERENCES

1. Gonzalez, J. E., Negulescu, P. A. (1998). Intracellular detection assays for high-throughput screening. *Curr. Opin. Biotechnol.*, *9*(6), 624–631.
2. Hertzberg, R. P., Pope, A. J. (2000). High-throughput screening: New technology for the 21st century. *Curr. Opin. Chem. Biol.*, *4*(4), 445–451.
3. Zaccolo, M., De Giorgi, F., et al. (2000). A genetically encoded, fluorescent indicator for cyclic AMP in living cells. *Nat. Cell. Biol.*, *2*(1), 25–29.
4. Durick, K., Negulescu, P. (2001). Cellular biosensors for drug discovery. *Biosens. Bioelectron.*, *16*(7–8), 587–592.
5. Numann, R., Negulescu, P. A. (2001). High-throughput screening strategies for cardiac ion channels. *Trends Cardiovasc. Med.*, *11*(2), 54–59.
6. Tretyn, A., Kado, R. T., et al. (1997). Loading and localization of Fluo-3 and Fluo-3/AM calcium indicators in sinapis alba root tissue. *Folia Histochem. Cytobiol.*, *35*(1), 41–51.
7. Smart, D., Jerman, J. C., et al. (2001). Characterisation using FLIPR of human vanilloid VR1 receptor pharmacology. *Eur. J. Pharmacol.*, *417*(1–2), 51–58.
8. Whiteaker, K. L., Gopalakrishnan, S. M., et al. (2001). Validation of FLIPR membrane potential dye for high throughput screening of potassium channel modulators. *J. Biomol. Screen.*, *6*(5), 305–312.
9. Baxter, D. F., Kirk, M., et al. (2002). A novel membrane potential-sensitive fluorescent dye improves cell-based assays for ion channels. *J. Biomol. Screen.*, *7*(1), 79–85.
10. Sullivan, E., Tucker, E. M., et al. (1999). Measurement of [Ca2+] using the fluorometric imaging plate reader (FLIPR). *Methods Mol. Biol.*, *114*, 125–133.
11. Sharif, N. A., Kelly, C. R., et al. (2003). Bimatoprost (Lumigan((R))) is an agonist at the cloned human ocular FP prostaglandin receptor: Real-time FLIPR-based intracellular Ca(2+) mobilization studies. *Prostaglandins Leukot. Essent. Fatty Acids*, *68*(1), 27–33.
12. Ding, G. J., Fischer, P. A., et al. (1998). Characterization and quantitation of NF-kappaB nuclear translocation induced by interleukin-1 and tumor necrosis factor-alpha. Development and use of a high capacity fluorescence cytometric system. *J. Biol. Chem.*, *273*(44), 28897–28905.
13. Conway, B. R., Minor, L. K., et al. (1999). Quantification of G-protein coupled receptor internatilization using G-protein coupled receptor-green fluorescent protein conjugates with the ArrayScantrade mark high-content screening system. *J. Biomol. Screen.*, *4*(2), 75–86.
14. Conway, B. R., Demarest, K. T. (2002). The use of biosensors to study GPCR function: Applications for high-content screening. *Receptors Channels*, *8*(5–6), 331–341.
15. Kapur, R. (2002). Fluorescence imaging and engineered biosensors: Functional and activity-based sensing using high content screening. *Ann. NY Acad. Sci.*, *961*, 196–197.
16. Dove, A. (2003). Screening for content—The evolution of high throughput. *Nat. Biotechnol.*, *21*(8), 859–864.

17. DeRisi, J., Penland, L., et al. (1996). Use of a cDNA microarray to analyse gene expression patterns in human cancer. *Nat. Genet.*, *14*(4), 457–460.
18. Ziauddin, J., Sabatini, D. M. (2001). Microarrays of cells expressing defined cDNAs. *Nature*, *411*(6833), 107–110.
19. Eickhoff, H., Konthur, Z., et al. (2002). Protein array technology: The tool to bridge genomics and proteomics. *Adv. Biochem. Eng. Biotechnol.*, *77*, 103–112.
20. Wu, R. Z., Bailey, S. N., et al. (2002). Cell-biological applications of transfected-cell microarrays. *Trends Cell. Biol.*, *12*(10), 485–488.
21. Beske, O. A. G. S. (2002). High-throughput cell analysis using multiplexed array technologies. *Drug Discov. Today*, *7*(18), S131–S135.
22. Beske (2004).
23. Torrance, C. J., Agrawal, V., et al. (2001). Use of isogenic human cancer cells for high-throughput screening and drug discovery. *Nat. Biotechnol.*, *19*(10), 940–945.

10

PROTEIN X-RAY CRYSTALLOGRAPHY IN DRUG DISCOVERY

PETER NOLLERT, MICHAEL D. FEESE, BART L. STAKER, AND HIDONG KIM
deCODE biostructures
Bainbride Island, Washington

10.1	INTRODUCTION	374
10.2	CRYSTALLIZATION	376
	Background	377
	Formats	380
	Membrane Proteins	381
	Crystallization Factors	382
	Observation and Documentation	384
	Crystallization Formulation Screens	384
	Cryopreservation for Cryocrystallography	385
	Crystal-Based Drug Discovery	386
	Derivatization for Anomalous Diffraction Experiments	388
	Crystallization Data Sources	388
	Protein Crystallization Demonstration Experiment	389
10.3	X-RAY DIFFRACTION EXPERIMENT	389
	Diffraction Theory	389
	X-ray Diffraction by Macromolecule Crystals	390
	Fourier Synthesis in X-ray Crystallography	392
	Bragg's Law and the Angular Dependence of X-ray Diffraction	396
	Electron Clouds and Thermal Motion	398
	X-ray Diffraction Data Collection in Practice	399
	Home Laboratory Diffraction Data Collection	402
	Synchrotron Diffraction Data Collection	403
10.4	X-RAY CRYSTAL STRUCTURE DETERMINATION	405
	Phase Problem	405

Drug Discovery Handbook, by Shayne Cox Gad
Copyright © 2005 by John Wiley & Sons, Inc.

	Structure Factor	405
	Heavy Atom Replacement Methods	406
	Multiple Isomorphous Replacement	407
	Anomalous Dispersion Methods	412
	Molecular Replacement	416
10.5	GENERATION AND ANALYSIS OF STRUCTURAL MODELS	420
	Aspects of Crystallographic Models of Macromolecules	420
	Building the Initial Model	422
	Refinement and Analysis of Structural Models	425
	Analysis and Preparation of Structural Models	432
	Crystallography–Drug Discovery Interface	433
10.6	EXAMPLES FOR THE USE OF X-RAY CRYSTALLOGRAPHY IN DRUG DISCOVERY	434
	Lead Optimization—Structure-Based Drug Design	436
	Antistructures	441
	Protein Therapeutics	442
	In silico Screening Based on Crystallographic Structural Models	442
	Crystallographic Screening	443
	Crystallographic Fragment Screening	444
	Site-Directed Leads via Fragment Tethering	445
	Structural Genomics	446
10.7	LIMITATIONS AND CHALLENGES OF X-RAY CRYSTALLOGRAPHY IN THE DRUG DISCOVERY PROCESS	447
	References	450

10.1 INTRODUCTION

The fundamental goal of applying protein X-ray crystallography to drug discovery is to increase its speed, quality, and its rate of success while reducing the cost. In conjunction with molecular biology, protein X-ray crystallography forms a seamless interface between target and lead discovery. Nucleic acid sequences from genetic studies provide the entry point for protein expression, the first step in crystallographic projects. The product of the crystallographic endeavor, the *structure*, is the accurate description of protein and ligand atom positions in three-dimensional space. Starting in the 1980s, protein X-ray crystallography has impacted the drug discovery process, primarily in the stage of lead optimization. Recently, however, by means of innovation and integration, protein crystallography has been transformed into an enabling technology now covering several stages of the drug discovery process (Fig. 10.1). Besides lead optimization, protein X-ray crystallography has affected drug discovery in (a) the identification of new drug targets, (b) the understanding of molecular target mechanisms, and (c) the discovery of new lead compounds.

The integration of protein X-ray crystallography into the early discovery process by joining it with chemical synthesis and assaying has become a very

INTRODUCTION

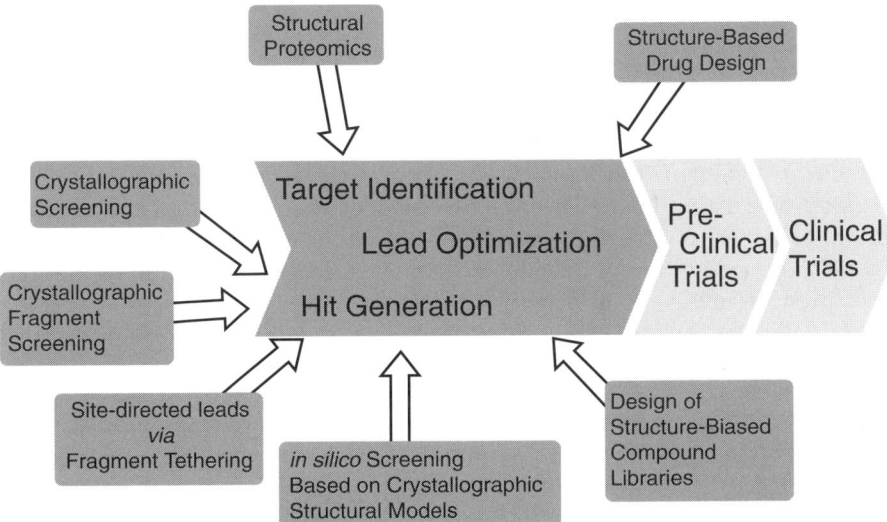

Figure 10.1 Impact of protein X-ray crystallography on and use of structural information for drug discovery purposes.

effective tool for lead optimization. Chemical synthesis can be iteratively directed toward more promising compounds and away from undesirable compounds based on insight into the interaction of ligands with the target protein. Thus, structure-based lead optimization has positively affected potency, selectivity profile, or pharmacokinetic properties of drug candidates. The application of protein X-ray crystallography impacts lead discovery by methods such as (a) X-ray crystallographic screening, (b) fragment screening, and (c) site-directed fragment tethering. Furthermore, structural information is being used to design and test target structure-based compound libraries (e.g., compound libraries specific for kinases) and to screen ligand compounds in silico with the goal of enriching useful compounds in virtual libraries prior to high-throughput assay-based screening.

X-ray crystallographic screening [1] combines the steps of lead identification, structural assessment, and subsequent lead optimization by structure determination of crystals that are soaked in compound cocktails. Ligands with highest affinities are selected by the crystal, and they are identified crystallographically. The utility of this strategy is exemplified in a section below.

Fragment screening is a variation of this theme and serves as a tool for assembling new lead compounds from small druglike fragments. Initially, libraries of leadlike fragments are formulated and co-crystallized or soaked in crystals. The resulting structures serve as a basis for combining novel chemical leads. The advantage over conventional assay-based screening is that very low affinity fragments with novel structures can be found.

Site-directed lead discovery by fragment tethering [2, 3] adds an additional layer of complexity to fragment screening, allowing the discovery of low-molecular-weight ligands. The strategy is based on the covalent modification of a target protein at a particular site on its surface via thiol-chemistry and the mass-spectrometric detection of weakly binding ligand precursors. Crystallography provides the tool to observe the binding mode and to direct the chemical synthesis of fused analogs, the latter having potentiated affinity.

Structure-based drug design consists of iterative processes aimed at the optimization of existing leads. The process includes the choice and the determination of a target structure, choosing a method for lead discovery and, crucially, the evaluation of drug leads [4]. A schematic description of the iterative process involved in structure-based drug design is shown in a later section (Fig. 10.22) [4]. The success of structure-based lead optimization is measured by improvements of affinity, specificity, or ADME properties.

Structure-based compound libraries can be obtained by so-called in silico screening. Crystallographic models are used to enrich compound libraries prior to conventional high-throughput screening [5]. Ligand structures are "docked" in silico into the binding pockets of protein structures, and the resulting modeled complexes often identify the correct ligand binding mode. More importantly, sets of accordingly treated compounds can be ranked by applying scoring functions that estimate ligand affinity, allowing the enrichment of actual chemical compound libraries with potentially good binders.

Structural genomics efforts seek to determine protein structures on the genome scale. The availability of many X-ray crystallographic structures is expected to deeply impact the discovery of new drugs and targets. Anticipated benefits are, for example, the discovery of new targets by assignment of functions to orphan targets, the improved quality of homology models for difficult targets, and the use of *antitargets* to decrease ligand cross-reactivity.

This chapter seeks to familiarize scientists with X-ray protein crystallographic techniques and their exemplified application for the purpose of drug discovery. Outlined are the individual steps that are required for determining X-ray crystallographic structures of proteins and protein–ligand complexes. First, the generation of crystals and their use in the X-ray diffraction experiment is described. Then state-of-the-art methods for crystal structure determination, model generation, and their refinement are reviewed, followed by an account of various ways of analyzing the resulting structural models. Several prominent examples are presented where X-ray crystallography has aided drug discovery. Finally, the limitations of the method are discussed.

10.2 CRYSTALLIZATION

Crystallography is the science of crystals. The word *crystallography*, a composite, derives from the Greek words *crustallos* or κρυσττaλος and *graphos* or γραφoς for *writing*. The word *crustallos* meaning "frozen" or "clear ice," captures several macroscopic crystal properties of these fascinating materials. Besides being transparent and solid, crystals are often facetted and, most importantly, have a high degree of internal order. They consist of regular three-dimensional lattices of molecules (see later section about the fundamentals of X-ray diffraction) and protein molecules with or without their bound drug compounds. Protein crystals consist of ordered protein and usually a sizeable fraction of disordered solvent, water. How are such crystals obtained?

Although there are no fixed rules as to how to grow protein crystals, the advent and combination of molecular biology, affinity tag purification tools, and laboratory liquid dispensing automation has helped to define generalized crystal growth procedures. Every protein, however, is different and a certain method and condition that works well for one particular protein is usually not applicable to the crystallization of other proteins. Furthermore, it is not possible to predict crystallization conditions on the basis of sequence. Therefore, the de novo crystallization of a particular protein poses a significant challenge.

While crystallographers have used systematic approaches to search for crystallization conditions, in most cases, preformulated screening kits are used (Table 10.1). Once a crystallization hit has been identified in a crystallization scouting trial, these initial conditions are often refined in order to grow crystals that are suitable for X-ray diffraction experiments. On the other hand, many protein crystallization conditions have been reported in the scientific literature and in databases (see below) and are therefore much simpler to carry out. Finally, comprehensive and practical advice on crystallization experiments can be found in Bergfors [6], McPherson [7, 8], and in Ducruix and Giegé [9].

Background

Protein crystallization occurs in three stages, nucleation, growth, and cessation of growth [10, 11]. Proteins crystallize from supersaturated solutions where the concentration exceeds their equilibrium solubility. The state of supersaturation depends on many factors such as the concentration and nature of the protein in question but also on salts and other components of the solution. It is this state of supersaturation that needs to be created in a crystallization experiment—the so-called crystallization setup—for crystal nuclei to form and crystals to grow (Fig. 10.2). The underlying principle is to alter the properties of the solvent (water), disrupt the interaction of water molecules with protein molecules, and increase the attractive interaction among protein molecules. This is generally done with (a) salts, (b) organic solvents, or (c) polymer compounds. When salts dissolve in water, their ions become hydrated and capture water molecules, the latter of which become unavailable to interaction with proteins. Therefore, the addition of salts such as ammonium sulfate may be used to favor protein–protein interactions, to form crystallization nuclei, and support crystal growth, thus *salting out* the protein.

Intriguingly, the opposite procedure, *salting in*, the removal of ions from a protein solution may also be used for the crystallization of proteins. Ions balance protein surface charges, and, once they are removed, protein molecules can balance their electrostatics by interacting with each other [12].

TABLE 10.1 Selection of Vendors of Crystallography-Related Products

Category	Items	Vendor / Company
Crystallization tools	Crystal screening formulations, various crystallography supplies	Hampton Research
	Crystal screening formulations, crystallization plates	Emerald BioSystems
	Crystal screening formulations,	Jena Bioscience
	Microfluidic crystallization devices	Fluidigm
	Crystallization plates	Nextal
	Crystallization plates	Corning
	Crystallization plates	Greiner
Crystallization trial robotics	CrystalMiner (database application for crystallization trials)	Emerald BioSystems
	CrysTel (integrated incubation and imaging system)	
	CrystalMonitor (automated crystallization imaging workstation)	
	MatrixMaker (automated solution formulation robot for the formulation of crystallization screening kits)	
	RoboFill (automated crystallization)	RoboDesign
	Odyssey (crystallization incubation and imaging system)	
	RoboMicroscope II (automated crystallization imaging workstation)	
	CrystalScore (automated crystallization imaging workstation)	Diversified Scientific
	925 PC Workstation (crystallization setup automation)	Gilson
	Crystal Farm (integrated incubation and imaging system)	Discovery Partners
	Rock Maker (software application for crystallization trials)	International
	Crystallization automation	Douglas instruments
X-ray diffraction instrumentation	X-ray diffraction instrumentation, detectors, optics and detectors, robotic sample changing system	Bruker
	X-ray diffraction instrumentation, detectors, optics, and detectors, robotic sample changing system	Mar Research
	X-ray diffraction instrumentation, detectors, optics, and detectors, robotic sample changing system	Rigaku MSC
	X-ray optics	Osmic

CRYSTALLIZATION

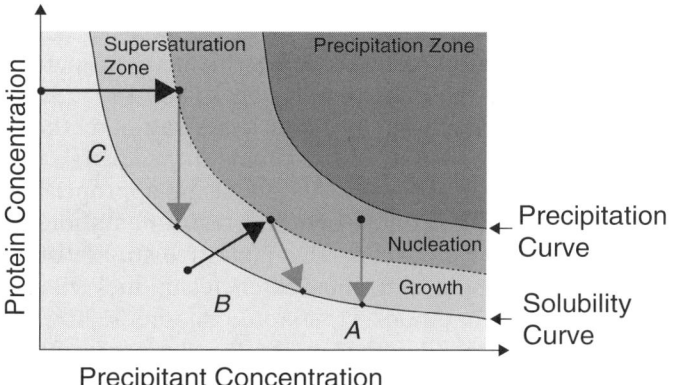

Figure 10.2 Schematic phase diagram of protein crystallization in the salting-out regime. Crystal nucleation is a critical phenomenon that may occur only in a certain area of the supersaturation zone and crystals may grow under conditions of supersaturation once nuclei have formed. (*A*) Pathway of a batch-type crystallization experiment. By mixing protein with the precipitant solution, the protein becomes supersaturated. Crystal nuclei form and crystals grow until the protein concentration in solution is saturated. (*B*) Pathway of a vapor diffusion-type crystallization experiment. The slow concentration process—via vapor diffusion—that follows mixing of the protein with the precipitant solution causes the protein to become supersaturated. The vapor diffusion process causes a concomitant increase in precipitant concentration thus extending the crystal growth process. (*C*) Pathway of a dialysis-type crystallization experiment. As the precipitant diffuses into the protein chamber, a state of protein supersaturation is reached. Once nuclei have formed, protein crystals may grow as long as the protein concentration remains supersaturated.

Organic solvents also bind water molecules, but in addition they lower the dielectric constant in the crystallization solution, thus enhancing electrostatic interactions between protein molecules. Finally, crowding agents such as hydrophilic polymers compete with protein molecules for hydration. This arises directly from water binding and indirectly via excluded-volume effects.

The mechanism of crystal nucleation is poorly understood. Once a protein solution is supersaturated and nuclei are present, crystals may grow on preformed crystal faces by deposition into lattice positions via diffusion and convection [13]. The crystalline state of matter is generally considered the thermodynamically most stable state, and the crystallization process can therefore be understood in terms of free energy minimization. Indeed, a lowering of the free energy by 12 to 25 kJ/mol has been measured for protein crystallization processes [14]. The last process in crystallization, the cessation of growth, may occur for a number of reasons such as limited protein supply or the poisoning of crystal growth surfaces by contaminations and crystalline defects.

Formats

A multitude of crystallization methods and formats are available, with batch and vapor diffusion being the most popular ones (Fig. 10.3). Since the quantity of protein sample is often limited, microcrystallization methods have been devised for batch, vapor diffusion, and other, more exotic, crystallization methods. They can all be carried out manually using appropriate plasticware (Table 10.1) or by appropriate liquid dispensing instrumentation such as multi-channel pipettors or dispensing robots. Comparative studies of protein crystallization by vapor diffusion and microbatch techniques yielded a similar effectiveness of the two methods [15]. Some methods, however, tend to succeed for certain targets while other methods are more useful for different protein targets.

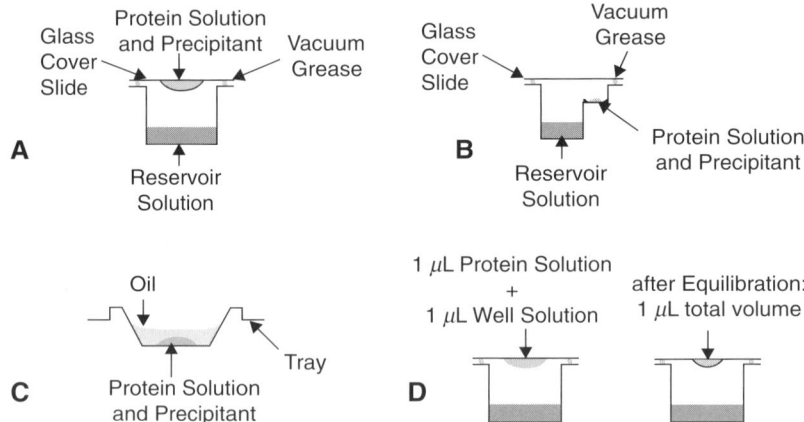

Figure 10.3 Schematic depiction of typical formats for vapor diffusion and batch crystallization experiments. (*a*) Hanging drop format of a vapor diffusion crystallization experiment. A small drop containing the protein and precipitant solution is applied onto a glass slide. The glass slide is attached to the crystallization chamber, the bottom of which is filled with the so-called reservoir solution. The concentrations of volatile components in the drop and reservoir solutions equilibrate over time through the vapor phase. Crystals form in the hanging drop. (*b*) Sitting drop format of a vapor diffusion crystallization experiment. A small drop containing the protein and precipitant solution are placed into a well. The crystallization chamber may be sealed with a glass cover slide or with transparent tape. Well and reservoir communicate via the vapor phase. Crystals form in the sitting drop. (*c*) "Under oil" format of a batch crystallization experiment. A small drop containing protein and precipitant solution is placed into a crystallization well. A layer of oil prevents dehydration. Crystals form in the drop. (*d*) Setup and equilibration of a vapor diffusion crystallization experiment with the hanging drop format. A 2-μL hanging drop (consisting of 1-μL protein solution and 1-μL precipitant solution from the reservoir) equilibrates via the vapor phase. Water transport occurs from the hanging drop to the reservoir because the hydrostatic pressure in the hanging drop is initially lower than in the reservoir solution. At equilibrium the hanging drop has a volume of 1 μL, with protein and precipitant concentration similar to that of the initially separated components.

Batch-type crystallization setups are conceptually simple experiments. Here, a protein solution is combined with a precipitant solution and crystals form within this solution. In order to prevent dehydration, oils may be added to seal off the crystallization experiment [16, 17] (Fig. 10.3c). Popular crystallization trays for microbatch crystallizations are so-called Terazaki plates, which provide small wells that can be used to hold final trial volumes of less than 1 µL [15] and 5 to 10 µL oil (silicon oil, paraffin oil, or mixtures). Oddly, the state of supersaturation is reached by dilution of the protein solution with a precipitant solution (A in Fig. 10.2). Therefore starting protein concentrations are often required to be chosen higher for this crystallization method than for the other methods.

Vapor diffusion crystallization experiments can be carried out in sitting drop or in hanging drop format. While the hanging drop arrangement is less prone to crystals sticking to a solid surface, it is less cumbersome to set up sitting drop crystallizations. In a hanging drop crystallization experiment, the protein solution is combined with the precipitant solution on a glass cover slide, inverted and, attached to a well containing the reservoir solution (Fig. 10.3a). Prior to the attachment, the rim of the well is beaded with vacuum grease in order to prevent dehydration of the crystallization chamber. The sitting drop format is very similar. Specialized crystallization plates are used that provide a platform to hold the drop (Fig. 10.3b). The chamber may be sealed off with transparent tape or with vacuum grease and a glass cover slide.

In vapor diffusion experiments, the precipitating solution is usually filled into the reservoir. Once protein and precipitation solution are combined and the crystallization chamber is sealed, the crystallization drop and the reservoir solution communicate via the vapor phase. At a typical 1:1 mixing ratio, the concentration of precipitant in the crystallization drop is only half of that in the reservoir solution (Fig. 10.3d). During the course of equilibration, the concentration of the precipitant increases to that in the reservoir, and, concomitantly, the drop volume decreases by about half (B in Fig. 10.2). The kinetics of this equilibration process and the pathway to supersaturation depend on the type of precipitant, the crystal chamber and drop geometry, and temperature. Crystals may form during or after equilibration.

Protein crystallization by dialysis is somewhat less popular but provides a high degree of control in the crystallization process [18]. First, a protein solution is filled into a small chamber that is closed by a dialysis membrane. Then a precipitant solution is brought into contact with that membrane and an exchange of small molecules and ions can occur. Exchange of proteins and other macromolecules are excluded by the dialysis membrane. Thus, salt concentrations can be increased slowly without *overshooting* into the precipitation zone (C in Fig. 10.2).

Membrane Proteins

About half of all drug targets constitute membrane proteins [19], proteins that are inserted in cellular membranes where they support critical functions such as solute transport, energy conversion, and signal transduction. Unfortunately, not a single structure of a transmembrane protein of pharmaceutical relevance is available to date. This is a result of the inherent difficulties of working with membrane proteins in the stages of expression, purification, and crystallization. There is, however, no principle reason for such proteins to systematically fail crystallization attempts. Indeed, many structures of membrane proteins have been determined once concerted efforts were taken, including structures of homologous proteins of the most prominent class of transmembrane protein drug targets, GPCRs (G-protein-coupled receptors). Several methods have been devised for the crystallization of this class of proteins, including the use of protein–detergent complexes, bicelles, and lipidic cubic phases [20, 21]. The latter method employs a lipid matrix as a crystallization facilitator, and the practicalities are described in detail by Nollert et al. [22].

Crystallization Factors

What are the factors that affect the crystal nucleation and growth processes? Of course, the quality and purity of the protein sample is of great importance since impurities may interact and disturb crystal packing. Fundamentally, the protein sequence, its organization into domains, and oligomerization state are decisive factors. The purity of proteins is usually judged in a semiquantitative way by sodium dodecyl sulfate (SDS) polyacrylamide gel electrophoresis (PAGE) and Coomassie brilliant blue staining. Purity levels above 90 to 95 percent are considered acceptable. For some proteins, however, crystallization can be a purification method, and crystals of minor contaminants have been grown from mixtures of many proteins. Particle homogeneity, that is, the uniformity of the oligomeric state is also an important factor, usually characterized by dynamic light scattering or by size exclusion chromatography (SEC). The latter chromatography is often applied as a final so-called "polishing step" in protein purification. Prior to use, protein samples are often centrifuged or filtrated to remove particulate contaminants. In order to prevent degradation, protease inhibitors and antimicrobials such as sodium azide or potassium cyanide (0.02 percent final concentration) are sometimes added. The concentration of the protein and the precipitation agent used have a tremendous effect on crystallization (Fig. 10.2). It is advisable to start crystallization experiments at high protein concentrations (greater than 10 mg/mL) and lower these once precipitate is detected in crystallization setups (Fig. 10.4b). Conversely, the protein and precipitant concentration may need to be increased if the drop remains clear (Fig. 10.4a).

Recombinant proteins are often expressed with tag fusions to assure high expression yield, solubility, protection from proteolysis, improved folding, and simple purification via affinity chromatography [23]. Their presence can either aid or impede crystallization, depending on the nature of the fusion, linker, and host protein. Several crystal structures have been obtained with their uncleaved large fusion proteins (including maltose-binding protein, thioredoxin, and glutathione-S-transferase) intact, where short three to five amino

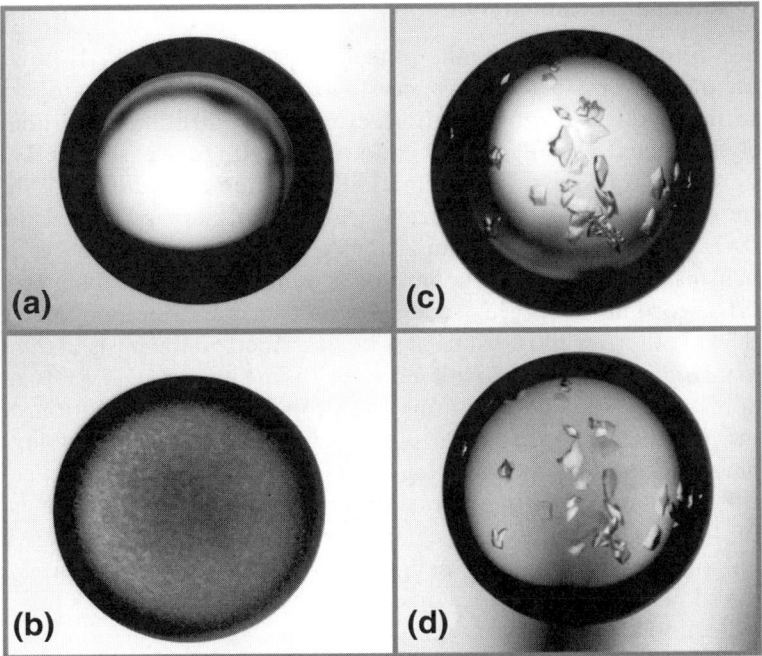

Figure 10.4 Images of hanging drop crystallization experiments. (*a*) Clear 1-μL drop at the outset of the crystallization experiment. (*b*) Precipitate. (*c*) Crystals of lysozyme inside a hanging drop. (*d*) Hanging drop with birefringent lysozyme crystals, imaged under cross-polarization setting. (This figure is available in full color at ftp://ftp.wiley.com/public/sci_tech_med/drug_discovery/.)

acid linkers were employed [24]. Alternatively, these tags may be removed with specific proteolytic enzymes that cleave appropriately engineered linker sites between the tag and the host protein. Once the protein is purified and concentrated, it may be stored via rapid freezing in liquid nitrogen. A cooling procedure proved beneficial that employs protein solution volumes below 50 μL and 0.2 mL ultrathin-walled polymerase chain reaction (PCR) tubes [25]. Fast thawing to room temperature was critical in order to prevent precipitation.

Besides precipitant type and concentration, further factors that affect crystallization include temperature, buffer type and concentration, the presence of additives, crystallization format, geometry, and other environmental parameters. It is not possible to screen all of these factors systematically. Several hundred crystallization experiments are therefore usually carried out varying the temperature (4 and 20°C) and formulations of precipitating agents, while all other parameters are kept constant. Subsequent fine screening may then be accomplished by systematically screening other factors. Rather exotic factors such as electric and magnetic fields have been identified to affect crystal quality. Their systematic use in the crystallography laboratory, however, is limited.

Observation and Documentation

Since it is often not known at the outset of crystallization trials how long it takes for protein crystals to grow, crystallization experiments are evaluated multiple times, days to months after setup. Crystallization experiments are routinely inspected visually with the aid of dissecting microscopes. These are operated under transmission illumination and often a pseudo-dark-field illumination option or crossed polarizers are used to increase crystal detection (Fig. 10.4). Microcrystals with dimensions of less than 5µm may be detected with such instruments. The outcome of crystallization experiments may be classified according to preset schemes, for example, "clear," "precipitate," "crystal," or "other," while crystals may be described on the basis of their shape (the crystal *habit*), size, and growth features. Taking images of crystals and crystallization setups helps documenting the progress of crystallization. Specialized software, such as Emerald Biosystem's CrystalMiner is available to assist in keeping track and in documenting crystallization trials. Its use in combination with automated imaging workstations such as CrystalMonitor is a very powerful tool for efficiently managing crystallization experiments. Although several algorithms are currently being developed by different groups to automate crystal detection, the optical inspection by crystallographers remains the main inspection route to date.

Intriguingly, the morphology of crystals need not correlate with their utility for X-ray diffraction purposes. Generally, large (>100µm) and faceted protein crystals yield higher resolution diffraction data than small (<50µm) crystals. Anecdotal evidence indicates though that small crystals may be of higher quality and some "ugly-looking" crystals diffract better than well-shaped ones. The gold standard in characterizing protein crystals is always a diffraction experiment.

Crystallization Formulation Screens

Crystallizing a given protein for the first time is analogous to searching for a needle in a haystack. Screening crystallization conditions is a multidimensional search because many factors affect crystallization (see below). Two approaches have been used to tackle this challenge: (a) systematic assessment of the protein's solubility and precipitation behavior and (b) screening with preformulated screening kits. The latter is very popular since it can yield crystals results without the need for individualized treatment for every target protein.

Crystallization screening kits are commercially available, many of them in 96-well format (Table 10.1). These contain sets of formulations, some of which have a proven track record to aid in the crystallization of proteins [26]. The formulations often consist of three components: salt, precipitating agent, and a pH setting buffer. Due to the numerous possible permutations, a so-called sparse matrix is selected consisting of a subset of all possible combinations [27]. Recently Majeed et al. [28] designed a *synergy screen* to overcome some of the limitations of these screens by combining multiple precipitation agents. If crystal hits are not obtained with the initial setup, follow-up experiments should be set up with lowered protein or precipitant concentrations where precipitate was observed and increased protein concentrations where clear drops were observed. In cases where no crystals are obtained from such trials, experimenters often switch to a different screening kit, or many screening kits are set up in parallel at the outset. Due to the stochastic nature of screening in general, it is impossible to know a priori how many crystallization experiments need to be set up in order to obtain a crystallization hit. Based on statistical analysis and a 1 to 2 percent likelihood for obtaining initial hits for typical proteins [29], it is estimated that screening 228 or 459 random crystallization conditions provides a 99 percent likelihood of observing at least one crystallization hit. The software program CRYSTOOL allows creating such random screens.

The systematic assessment of protein precipitation and solubility behavior is an iterative process. A particular precipitation agent is chosen and combined with the protein in batch-type crystallization experiments to map out a phase diagram similar to that shown in Figure 10.2. This is repeated for other precipitation agents if hits are not produced. Popular precipitation agents are ammonium sulfate, polyethylene glycols, Na/K phosphate, sodium chloride, and MPD [30].

Both crystallization approaches, systematic assessment or random screening, may provide either crystals suitable for X-ray diffraction experiments or poor hits that require refinement of crystallization conditions. In the latter case, fine screens varying one component concentration at a time are prepared in order to optimize crystal growth. In addition, crystal seeding experiments can be set up in which small or micro crystals serve as nucleation points for larger crystals useful for X-ray diffraction data collection [6].

Some of these different crystallization methodologies have been shown to work better for one protein than for others. Therefore, in de novo crystallization projects the type of crystallization method applied may be a useful parameter to test.

Cryopreservation for Cryocrystallography

In order to expose protein crystals to a beam of X-rays, individual crystals are either mounted within a glass capillary or they are "fished" with a filament loop and cooled to low temperatures, for example, that of liquid nitrogen [31, 32], cooled propane, ethane, CCl_3F, or $BrCF_3$. A gaseous nitrogen jet is the most popular method for cooling crystals and allows X-ray diffraction experiments to be carried out at temperatures around 100 K. This procedure reduces X-ray radiation damage and reduces protein motion, both increasing the diffraction data quality.

Water crystals diffract X-rays and produce characteristic powder diffraction rings, and, thus, their formation during the crystal cooling process should be avoided. This can be done by flash-cooling, which transforms liquid water into a glassy material. Water glass formation can be aided by the addition of so-called cryoprotectants such as glycerol, methyl-pentendiole, or trehalose. A popular strategy is to first prepare different cryoprotectant:precipitant solution mixing ratios, test their diffraction properties when frozen, and use the solution with the lowest cryoprotectant content that does not yield ice rings. Alternatively, ice formation on the protein crystal during the cooling process may be suppressed with high concentrations of certain salts. Lithium formate, chloride, and other highly soluble salts, so-called cryosalts, have been shown to suppress the formation of water ice upon cooling from ambient termperature to around 100 K [33]. Cyroprotectants may be avoided altogether by stripping surface-associated water from crystals by pulling them through oil, for example, perfluoropolyether, paratone-N, and paratone-N/mineral oil mixes [34].

Crystal-Based Drug Discovery

There are two ways to prepare crystals of target proteins with their bound small-molecule ligands. One is *co-crystallization* where the ligand–protein complex is crystallized, the other is *soaking* of pre-formed apo-crystals with a solution containing ligand molecules. While soaking may be very economical since a single crystallization setup can provide up to hundreds of usable crystals, not all ligands can be soaked in, with solubility issues posing a common problem. In co-crystallization experiments, at equilibrium two protein species are present, the apo-protein (no ligand bound) and the ligand-bound form. The relative fraction of these two forms is determined by the affinity constant K_a, the concentration of the ligand and the concentration of the protein. At an excess of ligand concentration for a simple single binding site model, the equilibrium binding is given by

$$[PL] = \frac{[P]_T K_a [L]}{1 + K_a [L]} \qquad (10.1)$$

with [PL] the concentration of the protein–ligand complex, $[P]_T$ the total protein concentration, and [L] the concentration of the free ligand. In many cases, it is difficult to estimate the proper concentrations in a crystallization setup. The goal is, of course, to populate most protein molecules with their ligand. As a rule of thumb, the concentration of a ligand to be used in a crystallization setup should exceed that of the protein by more than a factor of ten. A typical protein sample for crystallization is around 10 mg/mL protein. Depending on the molecular weight of the protein, the protein concentration will be around 250 µM. The ligand concentration should also exceed the dissociation constant of the ligand (which is the reciprocal association constant) by more than a factor of 10. Thus, if possible, a ligand concentration of more than 1 mM is often used.

Co-crystallization requires the formation of the ligand–protein complex prior to the crystallization process. This may be achieved simply by mixing of a protein solution with the ligand solution. In some cases where the off-rate of the ligand–protein complex is very low, a purification of the complex by size exclusion chromatography is feasible. It is evident that adding the small molecule, and possibly an organic solvent, usually DMSO (dimethylsulfoxide), and the formation of two protein species (apo-protein and ligand–protein complex) may alter the crystallization regime and thus the crystal quality or crystallization success. Therefore, co-crystallizations are frequently optimized for a particular ligand.

The ligand binding event often triggers subtle conformational transitions toward compacted protein structures, preventing the formation of those crystal contacts that are present in the respective apo-crystals. Alternatively, ligand binding may lead to the formation of crystals with a different packing arrangement and space group. This so-called ligand-depended crystal polymorphism [35] can be used as a tool for (a) binding or (b) the identification of initial crystallization conditions. The latter may be employed practically by screening a set of ligands as additives in co-crystallization experiments. As the formed ligand–protein complexes differ slightly in their conformations, different crystal contacts and packing arrangements are being sampled, thus increasing the chance of successful crystallization. Once a crystallization condition is identified, it may be used as a starting point for crystallizations with further ligands. This strategy is supported by the many cases where crystals of protein–ligand complexes have been obtained more readily than those of the respective apo-forms. Indeed, apo-forms of many ligand binding proteins have never been crystallized at all. It must be noted, however, that a particular crystallization condition that readily yields crystals of the apo-crystal form may not necessarily produce any crystals of ligand-bound forms. Most troubling, a given crystallization condition may not be compatible with binding of the ligand to the protein. Such a case may be envisaged where a protein crystallizes at low pH but does not bind its ligand at that pH.

Soaking compounds into pre-formed crystals is very popular because it short-cuts the possibly tedious process of identifying optimized co-crystallization conditions with a ligand. Soaking is performed by either (a) transferring the crystal into a solution consisting of precipitant solution and the ligand compound or by (b) addition of the ligand compound directly to the crystal-containing crystallization drop. It is difficult to estimate the preferred incubation time since the formation of the ligand–protein complex depends on the on-rate, diffusion of the compound within the crystal, and the component concentrations. In order to render hydrophobic drug leads soluble in aqueous solutions, organic solvents, detergents, or cyclodextrins have been employed.

The ligand compound or the solubilizing agent may have detrimental effects on the X-ray diffraction quality of crystals. In many cases visible cracks form upon soaking of ligands into apo-crystals. Evidently, substantial ligand-induced conformational transitions occur that may not be compatible with the original packing of the apo-crystal form. In such cases, co-crystallization experiments are warranted. Even though the formation of visible cracks renders crystals unusable for diffraction purposes, the effect may, for special cases, be used as an indicator for ligand binding [36].

Both soaking and co-crystallization may be employed for crystal-based drug discovery. The locations and binding modes of lead compounds and the specific interactions of the small-molecule atoms with those of the protein provide unique insight that can be used in several ways (see Section 10.6). Furthermore, cocktails of structurally diverse compounds or fractions of fragment libraries have been co-crystallized or soaked into apo-crystals and the ensuing crystal structures displayed high-affinity ligands [1], thus forming the basis of *crystallographic screening*.

Derivatization for Anomalous Diffraction Experiments

In order to obtain crystallographic phases crystals may be derivatized and used to collect anomalous X-ray diffraction data. In cases where proteins are expressed recombinantly, the protein may be labeled in vivo with selenium, substituting all sulfur-containing methionine residues by seleno-methionine. This method has become very popular due to the availability of tunable X-ray sources at synchrotrons. Chemical derivatization with heavy metals is a soaking or co-crystallization-based method. Cysteine, histidine, and methionine residues may react with heavy atoms such as mercury, gold, platinum, and iridium, whereas glutamate and aspartate can be complexed with lanthanides and actinides such as uranium and samarium [37]. Popular derivatizing reagents are K_2PtCl_4, $KAu(CN)_2$, $Hg(CH_3COO)_2$, $Pt(NH_3)_2Cl_2$, and $HgCl_2$ [38]. The binding of a particular heavy-metal compound to a protein may be screened prior to the crystallization and diffraction experiment by a simple native gel shift assay. Once a suitable reagent has been identified, the protocols for binding of heavy-metal compounds to proteins is similar to the binding of small-molecule compounds to form protein–drug complexes.

Crystallization Data Sources

More than 25,000 protein crystal structures have been deposited in the publicly accessible Protein Data Bank (PDB) (http://www.rcsb.org/pdb/). The crystallization conditions are available within the PDB files under section REMARK 280. As an example the crystallization conditions for hen egg lysozyme given in the structure report 3LZT are shown below:

```
REMARK 280 CRYSTALLIZATION CONDITIONS: BATCH METHOD USED. 1% PROTEIN
REMARK 280 SOLUTION IN 100MM SODIUM ACETATE PH 4.5-4.6. SODIUM
REMARK 280 NITRATE ADDED TO A CONCENTRATION OF 20MGS/ML. CRYSTALS
REMARK 280 GROWN AT ROOM TEMPERATURE.
```

Here, the batch crystallization method was used where a 10-mg/mL solution of hen egg white lysozyme in 100 mM sodium acetate buffer at pH 4.5 to 4.6 was combined with a precipitant solution containing a 20-mg/mL sodium nitrate solution. Scientific publications describing protein structures usually list crystallization conditions in their methods section. A specialized protein crystallization periodical does not exist. The International Union of Crystallography (IUCr), however, publishes a monthly journal *Acta Crystallographica D—Biological Crystallography* that hosts a section on crystallization papers, where crystallization methods, conditions, and technical advances in crystallization methodology are reported.

Finally, the crystallization database BMCD (Biological Macromolecule Crystallization Database), set up by Garry Gilliland, is a useful resource for finding crystallization conditions of previously crystallized proteins [30, 39]. Version 2.0 of the BMCD is available on the Web at http://wwwbmcd.nist.gov:8080/bmcd/bmcd.html.

Protein Crystallization Demonstration Experiment

Crystals of hen egg white lysozyme may be grown readily for demonstration purposes. A detailed procedure is given below for the crystallization of lysozyme according to the sitting drop vapor diffusion format.

1. Prepare a 100-mg/mL lysozyme solution by weighing out 100 mg of lyophilized chicken egg white lysozyme [Sigma (L7651)] into a tube and adding 1 mL of 50 mM sodium acetate at pH 4.5.
2. Prepare the precipitant solution; 30 percent (w/v) MPEG [mono-methyl polyethylene glycol 5000, Sigma (M7268)], 1 M sodium chloride, 50 mM sodium acetate at pH 4.5.
3. Fill 200 µL of the precipitant solution into the reservoir.
4. Pipet 2 µL from the reservoir solution into the crystallization well.
5. Add 2 µL of the lysozyme solution to the crystallization.
6. Crystals similar to those shown in Figure 10.4 appear within 1 h.

10.3 X-RAY DIFFRACTION EXPERIMENT

The information required to determine the three-dimensional crystal structure of a protein can be extracted from the X-ray diffraction data on a crystal of the target protein. In this section, basic diffraction theory and diffraction data collection practice will be discussed.

Diffraction Theory

Images of objects with dimensions ranging in size from a single cell to macroscopic scale are readily obtainable by visible light photography or microscopy.

Since the dimensions of the sample objects in light photography and microscopy are roughly the same order of magnitude, or larger, compared with the radiation wavelength being used, the light radiation is scattered by the object. The scattered radiation is then refocused by lens systems to give an image of the object.

To obtain images of individual molecules with dimensions on the order of 10^{-7}–10^{-9} m, radiation of wavelength much shorter than that of visible light, such as X-rays, is required. The atoms within an individual molecule will scatter X-rays, but there are no lens systems that can refocus the scattered X-rays to readily give an image of the target molecule. It is possible, however, to reconstruct a model of the target molecule by mathematical calculations on the X-rays scattered by the target molecule.

X-ray diffraction by protein crystals provides the data used in protein X-ray crystallographic structure determinations. Diffraction is the phenomenon whereby radiation is scattered by objects of dimensions comparable to the radiation wavelength. When the dimensions of the scattering objects and the incident radiation are comparable, the scattering objects act as new sources of the radiation and reemit the radiation in all directions (Fig. 10.5).

The radiation scattered by the diffracting objects are waves and, therefore, have both an amplitude and a phase component. Any given scattered wave will be to some degree in or out of phase with some other scattered wave. When two or more scattered waves are in phase with each other, there will be constructive interference, giving rise to diffraction maxima. When the scattered waves are out of phase with each other, there will be destructive interference, giving rise to extinctions of no intensity (Fig. 10.6). The resulting diffraction image will therefore have a distinct pattern, depending on the shape, multiplicity, and arrangement of the scattering objects.

Figure 10.7 shows examples of simple diffracting arrays. The samples in the first column could be holes, and the samples in the second column could be corresponding diffraction patterns from light passing through the array of holes. One thing to note is that the dimensions in the scattering array and the dimensions in the diffraction pattern are inversely proportional. In Figure 10.7, the widely spaced holes result in closely spaced diffraction maxima, while the closely spaced holes result in widely spaced diffraction maxima. These two-dimensional arrays of scattering objects result in two-dimensional diffraction patterns.

X-ray Diffraction by Macromolecule Crystals

The scattering of X-rays by protein crystals and their resulting diffraction patterns are completely analogous to the preceding examples of diffraction by slits and holes (Fig. 10.7). The spacings between atoms (bond lengths) in the protein crystal and the wavelengths of the X-rays are both on the order of 10^{-10} m (1 Å). Protein crystals will therefore scatter X-rays to give diffraction

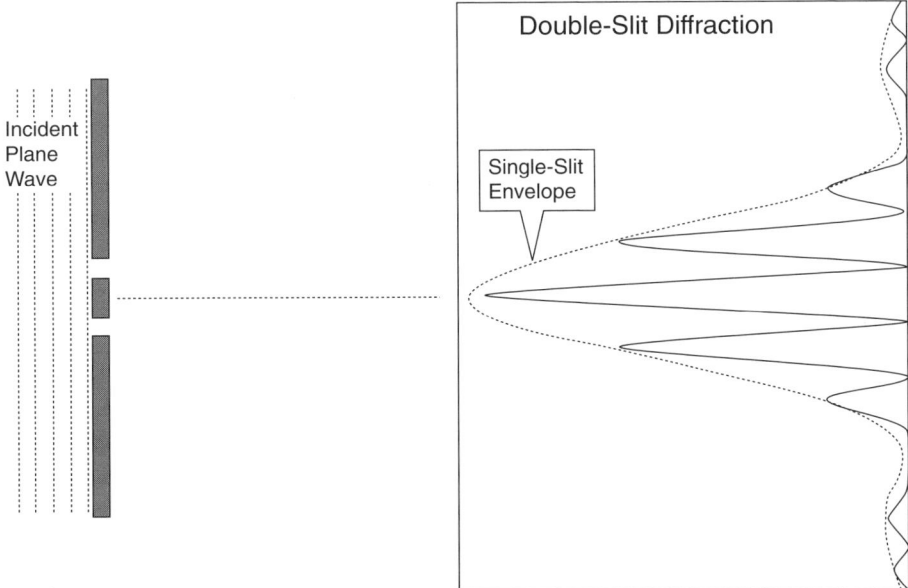

Figure 10.5 Diffraction of waves passing through two slits [40]. As radiation passes through a slit of width comparable to the radiation wavelength, the slit scatters the radiation in all directions. In this example, each of the two slits on the left act as emitters of the incident radiation. The radiation emitted by the two slits will interfere, resulting in the solid-line diffraction pattern on the right. (The relative heights of the peaks in the diffraction pattern correspond to intensity.) Interfering waves which do not have a large angular deviation from the incident radiation, are largely in phase with each other resulting in intensity maxima. As the angular deviation of the scattered waves from the incident radiation increases, the scattered waves are less in phase with each other, resulting in the general intensity falloff as one moves away from the center of the diffraction pattern. Shown in broken line is the diffraction pattern of a single slit.

images. In a protein X-ray crystallography experiment, a crystal of the target protein is placed in an intense X-ray beam of a particular wavelength λ. The protein crystal is a three-dimensional array of diffracting objects (the atoms and their electrons), which scatters the incident X-rays to produce a three-dimensional diffraction pattern. The three-dimensional diffraction pattern is typically recorded in two-dimensional sections by a detection plate. Figure 10.8 shows sample protein crystal diffraction images. Each of the dark spots in the diffraction images represents a diffracted X-ray whose intensities are maxima of constructive interference between scattered waves. In a typical protein diffraction data collection, tens or hundreds of thousands of such diffraction spots are collected from a single crystal.

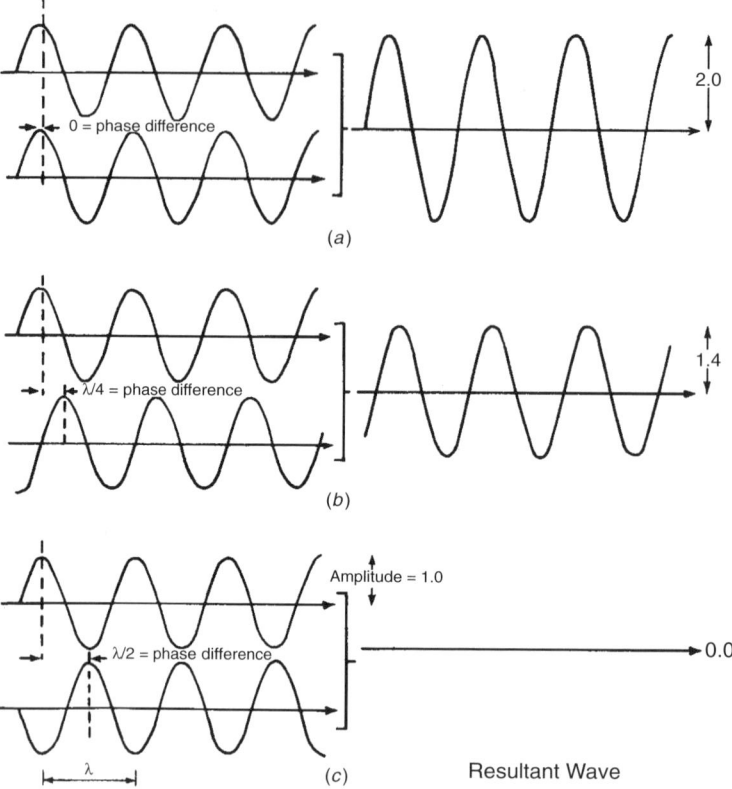

Figure 10.6 Constructive and destructive interference of waves [41]. The resultant wave produced by two or more interfering component waves is obtained by adding their amplitudes at corresponding points. (*a*) Total constructive interference. The two component waves are completely in phase. The maximum amplitude of the resultant wave is double that of each component wave. (*b*) Partial constructive interference. The two component waves are 90° out of phase. The resultant wave has an amplitude 1.4× the amplitude of each component wave. (*c*) Total destructive interference. The two component waves are 180° out of phase. The two component waves cancel each other out, and the resultant wave has zero amplitude.

In the analysis of the X-ray diffraction pattern, each of the spots (also referred to as reflections) is designated as F_{hkl}, wherein the set of three integers, hkl, are called Miller indices. The Miller indices give the position of the reflection relative to the orientation of the crystal during the data collection.

Fourier Synthesis in X-ray Crystallography

Since each of the spots in the diffraction pattern represents a wave (the diffracted X-rays), Fourier synthesis of these diffracted waves can be used to

X-RAY DIFFRACTION EXPERIMENT

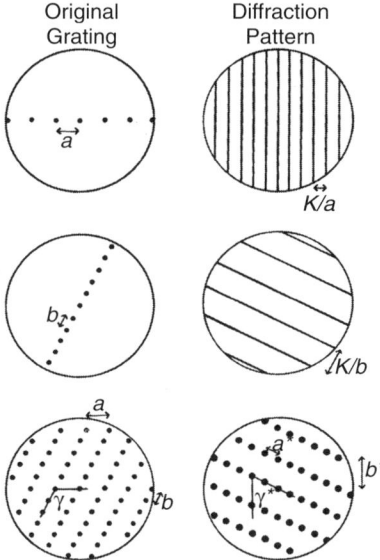

Figure 10.7 Schematic representations of diffracting arrays (left column) and their resultant diffraction patterns (right column). In the diffraction arrays, the black dots represent holes. In the diffracting arrays, the black lines and black dots represent intensity maxima of the diffracted radiation. Note that the sampling in the diffraction pattern is along the direction of the diffracting array. Comparing the top and middle examples shows that the sampling regions are farther apart in the diffraction pattern of the middle example, due to the closer spacing of the holes in the diffracting array. The bottom example shows a two-dimensional diffracting array. Its diffraction pattern shows discreet sampling in two directions, resulting in spots of intensity maxima in the diffraction pattern, as opposed to lines in the other two examples [41].

reconstruct the diffracting array, namely the electron density of the protein in the crystal. The atoms of the protein, known from its amino or nucleic acid sequence, are then built into the reconstructed electron density to give the three-dimensional structure of the target protein.

Fourier syntheses are mathematical calculations used to reconstruct any regularly repeating pattern, regardless of complexity, via the summation of relatively simple sine and cosine curves [Eq. (10.2)].

$$f(x) = a_0 + \sum_h (a_h \cos 2\pi hx + b_h \sin 2\pi hx) \quad h = 1, 2, 3, \ldots, n \quad (10.2)$$

As a very rudimentary example of a Fourier synthesis, consider the square wave in Figure 10.9. At first glance, it may appear impossible to reconstruct this function, especially its sharp corners, from smoothly curved sines and cosines. The summation of only four terms in the Fourier series, however, results in a function that is clearly beginning to take on the shape of the square

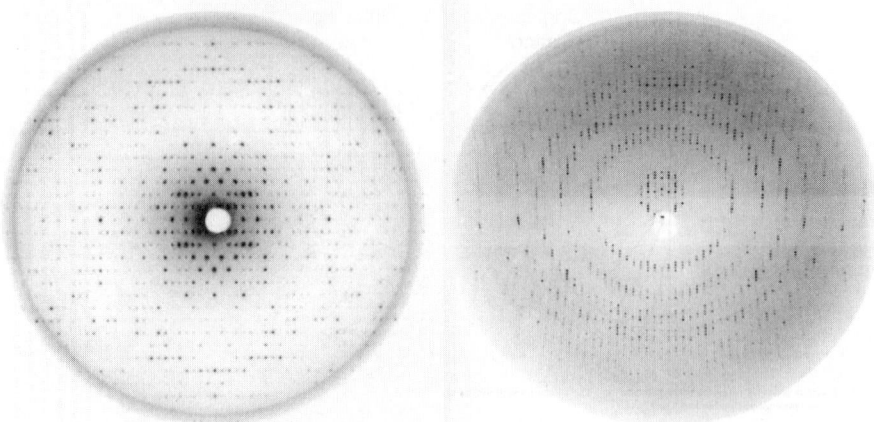

Figure 10.8 Representative diffraction patterns used in protein X-ray crystal structure determination. Each of the spots in these diffraction patterns is generated by diffracted X-rays. Note the general tendency for the intensities of the diffraction spots to decrease as one moves farther away from the center of the diffraction pattern.

wave. The addition of more Fourier terms will produce a resultant function that looks increasingly like the target square wave.

In protein X-ray crystallography, the "wave" that one tries to reconstruct is the electron density within the protein crystal. The electron density in the crystal is a regularly repeating three-dimensional function. The wave components [the sines and cosines in Eq. (10.2)] used to reconstruct the electron density of the target macromolecule are the individual spots of the X-ray diffraction pattern. Equation (10.3) gives the electron density of the target protein at a point (x, y, z):

$$\rho(x, y, z) = (1/V)\sum_{hkl}|F_{hkl}|\cos[2\pi(hx + ky + lz) - \alpha(hkl)] \qquad (10.3)$$

In Equation 10.3, the electron density at a three-dimensional coordinate position (x, y, z), $\rho(x, y, z)$, is expressed as the Fourier summation of all of the spots in the diffraction image. The diffraction spots, F_{hkl}, are waves, which have amplitude and phase components. Their amplitude components are expressed as $|F_{hkl}|$. Their phase components are the $\alpha(hkl)$ terms in Eq. 10.3, where $\alpha(hkl)$ is the angular phase of F_{hkl}. The $(1/V)$ factor is the reciprocal of the unit cell volume of the sample crystal. The Fourier summation of the electron density, Eq. 10.3, appears to lack the sine terms in the general Fourier summation, Eq. 10.2. Due to Friedel's Law (Section 10.4) generally holding true in protein diffraction, certain pairs of spots have the same intensity but opposite phase angles and the sine terms in Eq. 10.2 cancel out.

Only the amplitude components of the F_{hkl}'s are directly measurable by the X-ray detection hardware used in the diffraction experiment. The phase components are determined by mathematical procedures that are discussed in the next Section 10.4. The important thing to note is that in Eq. (10.3), as was the

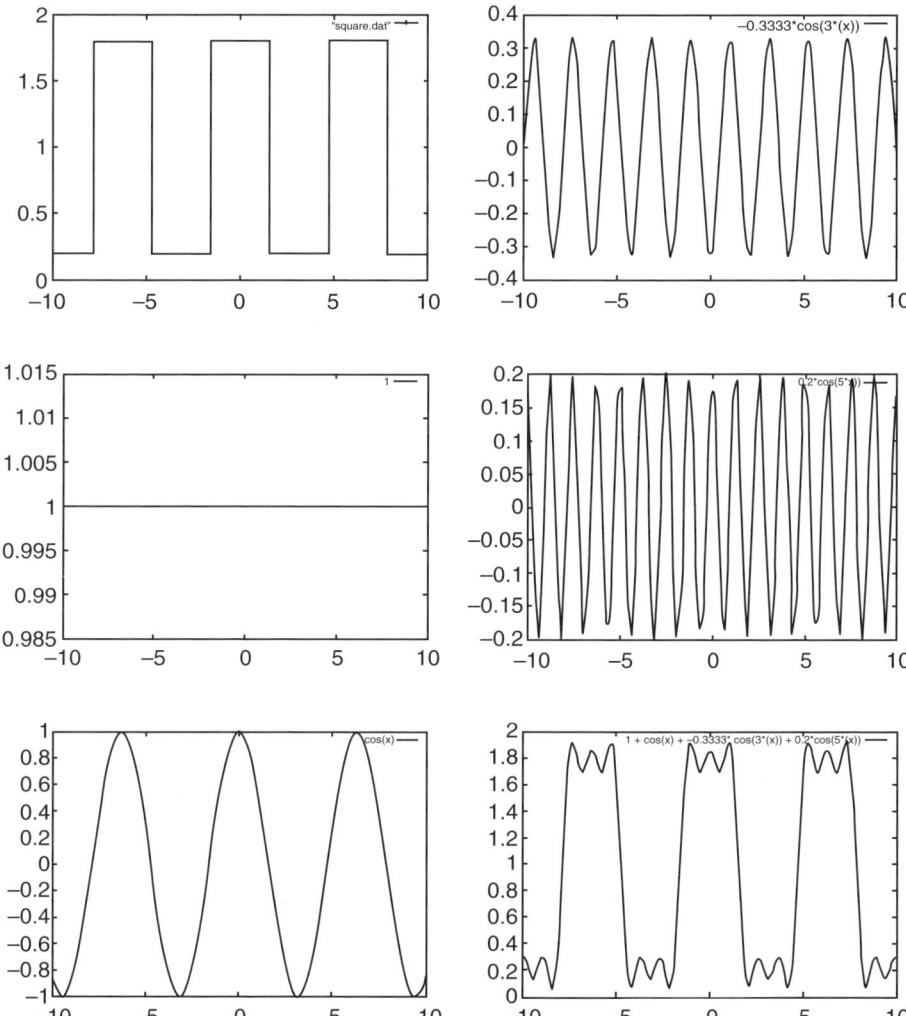

Figure 10.9 Reconstruction of a square-wave by Fourier synthesis. The target square-wave function is shown in the upper left. The summation of four cosine functions [the constant function is cos(0)] result in the function in the lower right, which is beginning to take on the step shape of the target square wave.

case for the example of the square wave (Fig. 10.9), the reconstructed electron density $\rho(x, y, z)$ becomes a more accurate representation of the true electron density as more terms F_{hkl} are added to the summation.

As seen in Figure 10.8, the X-ray diffraction pattern contains certain symmetry. The symmetry of the diffraction pattern is dictated by the internal symmetry of the protein crystal, more specifically, the symmetry of the unit cell of the protein crystal. The unit cell of a crystal can be described as a box with edges of length a, b, c and angles between these edges of α, β, γ. The unit cell of a crystal is the smallest portion of the crystal that can be used to generate the entire crystal by whole unit cell translations along the a, b, c cell edges. For protein crystals, the unit cell edges will be on the order of 50 to 400 Å. The unit cell is analogous to the scattering objects in Figure 10.7. A protein crystal with a large unit cell will produce a diffraction pattern with close spacings between the diffraction maxima, while a protein crystal with a small unit cell will produce a diffraction pattern with far spacings between the diffraction maxima. Because of this inverse relationship between unit cell size and diffraction pattern spacing, the diffraction pattern is called the reciprocal lattice of the actual crystal lattice.

Within each unit cell, there can be multiple copies of the protein molecule. These copies within a unit cell are related to all of the other copies in the unit cell by mathematical symmetry operations, such as rotations or translations. The smallest portion of the unit cell that can be used to generate the entire unit cell by the symmetry operations is called the asymmetric unit. The expression for the electron density [Eq. (10.3)] refers to an entire unit cell. In practice, since the contents of the unit cell can be generated by the mathematical symmetry operations, the X-ray crystallographic determination of protein structures generally refers to a single asymmetric unit of the crystal.

Bragg's Law and the Angular Dependence of X-ray Diffraction

The somewhat complicated phenomenon of diffraction by the three-dimensional array of a protein crystal can be conceptually simplified by treating diffraction as the reflection of the incident X-ray beam from planes within the crystal. These planes are planes of electrons in the crystal, with the electrons being concentrated at the atoms. The incident X-rays can be thought of as reflecting off of these planes, resulting in the spots in the diffraction pattern where the X-rays strike the detection plate. The diffraction spots are therefore termed *reflections*. The geometric formulation of diffraction as reflections from planes is shown in Figure 10.10.

In Figure 10.10, the incident X-rays of wavelength λ are represented by **1** and **2** and are reflected by planes P_1 and P_2 in the crystal with an interplanar separation d, resulting in the reflected X-rays **1'** and **2'**. The incident X-rays **1** and **2** make an angle θ with the planes P_1 and P_2. If the reflected X-rays **1'** and **2'** are to result in a beam of maximum intensity, the X-rays represented by **1** and **2**, and **1'** and **2'** must be in phase. In this geometric construction, for **1'** and **2'** to be in phase to produce an intensity maximum, the extra distance traveled by **2** and **2'** compared with the distance traveled by **1** and **1'** must be an integral number of wavelengths λ. Thus,

$$ACB = 2AC = n\lambda \tag{10.4}$$

where n is an integer. Since $AC/d = \sin\theta$, Eq. (10.4) can be rewritten as

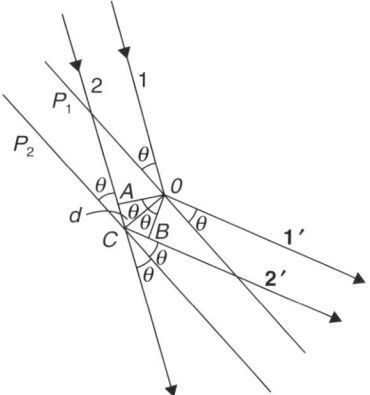

Figure 10.10 Geometric construction for the Bragg theory of diffraction [42]. Parallel planes of electrons in the crystal are represented by P_1 and P_2. The incident X-rays are represented by rays 1 and 2. Diffraction is treated as "reflection" of rays 1 and 2 off of the planes P_1 and P_2, resulting in the diffracted waves represented by rays **1'** and **2'**.

$$n\lambda = 2d \sin \theta \quad (10.5)$$

Equation (10.5) is Bragg's law, which was derived by Sir William H. Bragg and his son Sir William. L. Bragg, who were awarded the 1915 Nobel Prize in physics for their work in X-ray crystallography. The interplanar spacing d is the resolution of the particular diffracted wave. The resolution can be thought of as the resolvable distance between planes that "reflect" X-rays and lead to diffraction. It can be seen from Bragg's law that for constant n, diffracted X-rays with a greater angular divergence from the incident X-rays, those with high θ and farther away from the center of the diffraction pattern, result from reflections off of planes with small interplanar spacings d. Therefore, those reflections in the diffraction pattern with small interplanar spacings d and large θ are the high-resolution reflections.

The Bragg construction of diffraction also provides a description of the Miller indices hkl assigned to a particular reflection. The reflection F_{hkl} can be considered to be the resultant reflection off of the planes of constant spacing intersecting the unit cell edges a, b, and c, respectively h, k, and l times. A low-resolution reflection will have low integer values for hkl. For example, the reflection F_{100} is the reflection off the plane intersecting the a cell edge once, and never intersecting the b and c cell edges. This is the plane defined by the b and c cell edges. The spacing of this plane is the a unit cell edge. For a typical protein crystal, the resolution of the F_{100} reflection, d in Eq. (10.5), will be 50 to 400 Å. As the Miller indices hkl for a reflection increase, the spacings of the reflecting planes become smaller within the finite volume of the crystal's unit cell. Reflections F_{hkl} with higher integer values of hkl are therefore the higher resolution reflections in an X-ray diffraction data set.

Electron Clouds and Thermal Motion

Equation (10.5) gives only the positions of the diffracted X-rays in the diffraction pattern, and no indication of the intensity of the diffracted X-rays. The intensities of the reflections in the diffraction pattern are highly dependent on the content of the crystal and the angle θ of the diffracted X-ray. Because the electron clouds that scatter the X-rays have a finite volume, as θ increases, X-rays scattered by one part of the electron cloud are increasingly out of phase with X-rays scattered by another part of the electron cloud. Thus, higher resolution reflections with higher θ are naturally weaker in intensity than lower resolution reflections.

Thermal motion will also affect the intensities of scattered X-rays. The intensity fall-off shown in Figure 10.11a assumes that the carbon atom is stationary. The atoms in the crystal, however, will always be vibrating about their equilibrium points to some degree. This thermal motion effectively increases the volume of the electron cloud and further weakens the intensities of the higher resolution (high θ) reflections. Thermal motion is reflected in the exponential term $\exp[-B(\sin^2\theta)/\lambda^2]$, where B is a measure of the thermal motion of the atom. If $B = 0$, there is no thermal motion, and the exponential term goes to 1. In reality, $B > 0$. Figure 10.11b shows the effect of thermal motion on the intensities of scattered X-rays. When $B = 0$, the angular dependence of the diffracted X-ray intensities is the same as in Figure 10.12. When B increases, the intensities of the diffracted reflections are further weakened as θ increases.

Figure 10.8 shows the effects of the angular dependence of scattering on actual diffraction patterns for protein crystals. One can see that there is a

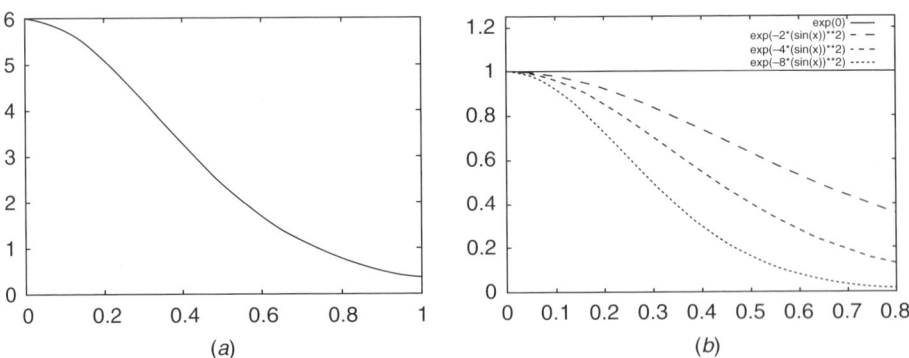

Figure 10.11 (a) Scattering factor of carbon as a function of angular deviation of the diffracted wave from the incident X-ray beam. The scattering factor f_C, plotted along the y axis, is a measure of the intensity of the diffracted wave at a particular angular deviation, $(\sin\theta)/\lambda$ (plotted along the x axis) from the undiffracted X-ray beam. (b) Effects of thermal motion on the angular dependence of X-ray scattering. The thermal motion factor $\exp[-B(\sin^2\theta)/\lambda^2]$ for values of B, ranging from 0 to 8. Atoms with higher B values have greater thermal motion. When $B = 0$, the function is constant and does not affect the natural intensity falloff shown in (a). As B increases, the intensity falloff shown in (a) becomes more rapid as $(\sin\theta)/\lambda$ increases.

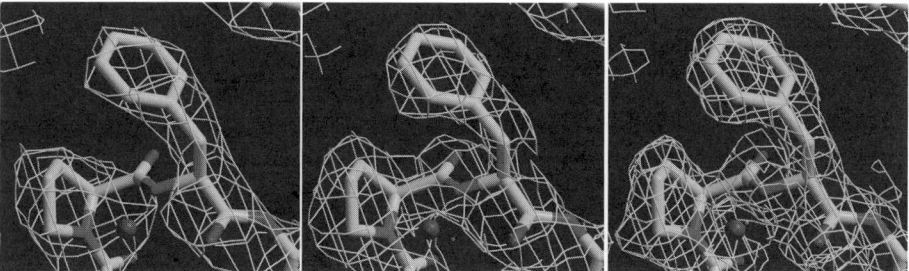

Figure 10.12 Electron density maps calculated for a protein at increasing resolutions: *(Left)* 3.5 Å, *(middle)* 2.8 Å, and *(right)* 2.1 Å. The electron density is shown in cyan chicken wire. Electron density is usually displayed as chicken wire contour lines around the protein model. The chicken wire contours correspond to ordered electron density in the crystal structure, and should therefore be superimposable on the model of the protein. (This figure is available in full color at ftp://ftp.wiley.com/public/sci_tech_med/drug_discovery/.)

general tendency for the intensities of the spots to weaken as their positions move farther from the center of the image. For any given protein crystal, the higher resolution reflections are more difficult to collect, if not unattainable, compared with the lower resolution reflections.

Similar to the example of the square wave reconstruction in Figure 10.9, adding more, higher resolution, reflection terms to the electron density Fourier transform [Eq. (10.1)], gives a more accurate representation of the electron density for the target protein. Figure 10.12 shows the effects of using higher resolution reflections in the calculation of electron density maps. At lower resolutions, the electron density gives the general shape of the target protein's electron density. At higher resolutions, the electron density reveals more details of the target protein structure.

X-ray Diffraction Data Collection in Practice

The collection of protein X-ray diffraction data requires three basic hardware components: X-ray source, goniometer, and X-ray detector. There are numerous configurations of these components and additional X-ray optical components that can be used. Figure 10.13 shows a schematic of an X-ray diffraction data collection system. An intense, monochromatic X-ray beam produced by an X-ray source strikes the sample protein crystal. The X-rays diffracted by the crystal are measured by an X-ray detector, on the opposite side of the crystal as the X-ray source. The protein crystal is mounted on a goniometer, which allows for very precise angular adjustments of the crystal position. During a data collection, the crystal will be rotated about one or more axes by the goniometer. Usually, the crystal is rotated about the ϕ axis, which is

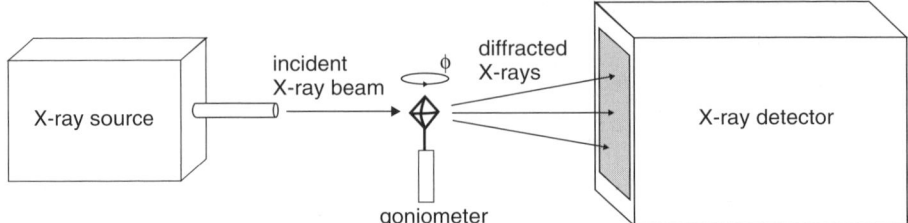

Figure 10.13 Schematic of the basic components of a protein X-ray diffraction data collection experiment. An X-ray source produces an intense X-ray beam. The sample protein crystal is mounted on a goniometer and placed in the X-ray beam. The goniometer allows precise angular positioning of the crystal. During data collection, the crystal is generally rotated about the φ axis. The diffracted X-rays are recorded by an X-ray detector.

normal to the incident X-ray beam. The X-ray detector can be positioned at an angular offset 2θ relative to the incident X-ray beam to collect higher resolution reflections [higher 2θ, Eq. (10.5)]. In Figure 10.13, the 2θ offset is 0°, that is, the incident X-ray beam, crystal, and X-ray detector are colinear.

Modern detectors are generally two-dimensional plates that record many diffraction spots simultaneously in a single X-ray exposure (Fig. 10.8). Data-processing software determines the crystal symmetry and unit cell parameters for the diffraction data, based on the pattern of the diffraction spots. The software then predicts positions on the detector where the reflections should fall and measures the intensities and positions of the reflections (Fig. 10.14). The measured intensities of the reflections are then converted to amplitudes, which are roughly the square root of the intensities.

The quality of the processed diffraction data can be gauged by certain statistical indicators. One of the most popular measures of data quality is R_{sym}. As discussed above, the protein X-ray diffraction pattern has certain symmetry. Reflections with Miller indices hkl are crystallographically constrained to have the same intensity as certain other reflections with different Miller indices due to the symmetry of the crystal. Therefore, while many total reflections, sometimes several hundreds of thousands, may be collected in an X-ray diffraction data set, the set of unique reflections is far less. A 2- to 20-fold excess, or redundancy, of reflections is usually collected in a protein X-ray diffraction data set. In theory, the reflection hkl and all of its symmetry related mates should have the same intensity. In practice, however, there will be differences in the measured intensities of these reflections. R_{sym} is the measure of how similar the intensities of these reflections are. The expression for R_{sym} is

$$R_{sym} = \sum\nolimits_{hkl} [|I - \langle I \rangle|/I] \qquad (10.6)$$

In Eq. (10.6), I is the intensity of a reflection with Miller indices hkl. $\langle I \rangle$ is the average intensity of that reflection and all of its symmetry-related reflections. In an ideal data set where all of the symmetry-related reflections have

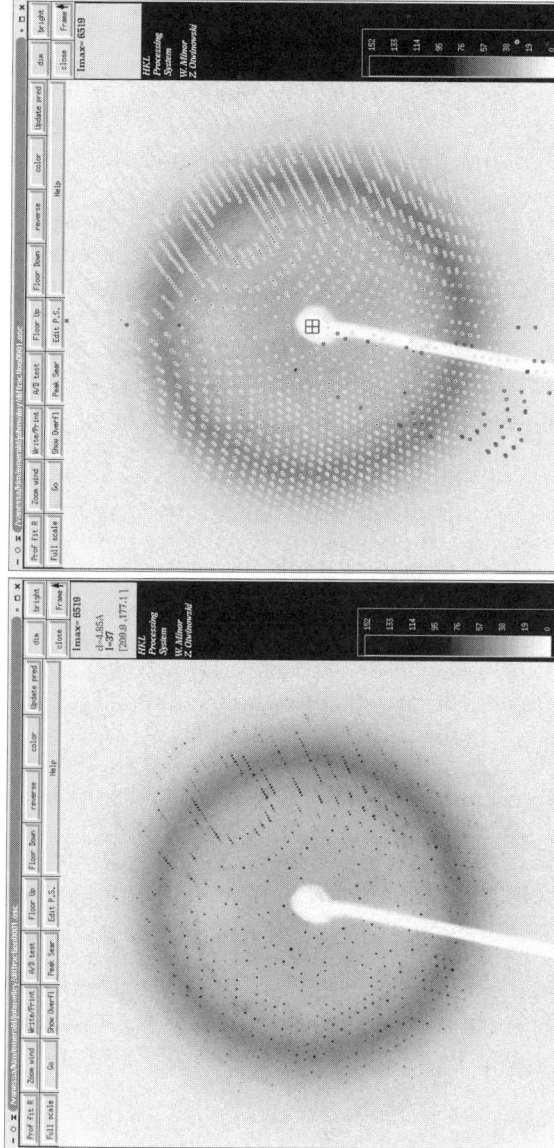

Figure 10.14 An example of diffraction data collection and processing. *(Left)* A protein crystal diffraction image. *(Right)* The reflections of the protein diffraction image with prediction circles calculated by the data-processing software. The prediction circles should coincide with the reflections. Images from the diffraction data-processing software HKL2000 [43].

identical measured intensities, the numerator in Eq. (10.6) is zero, and R_{sym} is zero. In practice, typical protein X-ray diffraction data sets have R_{sym}'s of 0.05 to 0.15 over the entire resolution range of the data, with R_{sym} increasing for higher resolution reflections. As previously mentioned, higher resolution reflections generally have weaker intensity and are more difficult to measure accurately. In general, a lower R_{sym} is considered an indication of a better quality data set.

One of the problems with using R_{sym} as an indicator of data quality is that R_{sym} is highly dependent on the redundancy of the data (the number of times a reflection is recorded). As more measurements are made for a reflection and its symmetry-related mates, R_{sym} increases [44], indicating deterioration of the data set. This is statistically counterintuitive since the quality of the data must improve with increased sampling (higher redundancy). To address this statistical flaw in R_{sym}, the indicator R_{mrgd}, which accounts for data redundancy, has recently been proposed [44]. Still, R_{sym} is more quoted as an indicator of data quality for protein X-ray diffraction data.

Another indicator of data quality that is statistically more robust than R_{sym} is I/σ, the signal-to-noise ratio of the measured intensity for a reflection. One of the most important aspects of I/σ is that it is more reliable than R_{sym} in defining the resolution limit of the data. The resolution limit is a critical descriptor of a protein diffraction data set. The resolution limit indicates how much detail should be expected in the electron density maps (Fig. 10.12) and also guides model refinement strategies (discussed below in Section 10.5). The resolution limit is sometimes taken as the resolution at which R_{sym} exceeds a certain threshold value, chosen by the experimenter. Typical threshold values for R_{sym} in setting the resolution limits of protein X-ray diffraction data range from 0.40 to 0.50. The experimenter can, however, intentionally or unintentionally manipulate R_{sym} to lower (better) values by such practices as reducing the number of measured reflections or rejecting reflections with weak intensity. In contrast, I/σ sets the resolution limit since by definition, it is the signal to noise ratio of the measurements and is less prone to experimenter manipulation [45]. An I/σ of 2 is often taken as the resolution limit of a protein X-ray diffraction data set. In most cases, though, both R_{sym} and I/σ are quoted for diffraction data in the primary protein crystallography literature.

Home Laboratory Diffraction Data Collection

The basic configuration shown in Figure 10.13 can be installed in any home laboratory and is now common throughout protein crystallography groups. There are several vendors offering home laboratory X-ray diffraction hardware. The purchase price of a complete home laboratory X-ray diffraction system is around US$500,000.

Most home laboratory systems use a copper rotating anode as the X-ray source. In such an X-ray source, an intense electron beam produced by heating a metal filament, the cathode, strikes a rotating cylinder of copper, the anode.

The electron beam excites the copper atoms in the anode, which emit X-ray radiation. The electronic energy spacing of copper is such that the emitted X-rays have $\lambda = 1.54$ Å (Cu K_α radiation). According to Bragg's law [Eq. (10.5)], the highest resolution diffraction data attainable with Cu K_α radiation is around 0.8 Å [42]. This maximum resolution is very rarely attained with most protein crystals, primarily due to crystalline imperfections. As shown in Figure 10.12, however, much more modest resolution limits are sufficient to confidently determine protein crystal structures.

The X-ray detector in a typical home laboratory system is an imaging plate coated with a phosphorescent material. The diffracted X-rays impinging on the imaging plate excite the phosphors on the plate. The plate is then scanned with a laser, and the phosphors release the stored energy from the X-rays as visible light [46]. The released light is then analyzed for the positions and intensities of the diffracted X-rays in the original image. Typically, a crystal is exposed to the incident X-ray beam for 5 to 30 min for one diffraction image. The readout of the image plate takes about 2 to 5 min. On some home laboratory systems, there are two or more imaging plates so that one plate can be exposed while the other plate is read, resulting in no latency for plate readout. The goniometer then reorients the crystal to the next angular position, and the next image is collected. For protein crystals, 90 to 180 images are typically collected in a complete data set.

Not shown in Figure 10.13 since it is not strictly required for diffraction is a cryostat. Protein crystals are very delicate and are subject to rapid radiation damage from the incident X-ray beam, most likely from the formation of free radicals in the crystal. Radiation damage continually degrades the diffraction resolution of a crystal. To prevent radiation damage, the sample crystal is mounted on the goniometer so that the crystal is in a constant jet of nitrogen gas near liquid nitrogen temperature. While such cooling of the crystal does not eliminate radiation damage, it usually retards radiation damage enough to allow collection of complete diffraction data from a single sample crystal. The use of cryostats in diffraction data collection is now practically universal in protein crystallography.

Synchrotron Diffraction Data Collection

The convenience of a home laboratory X-ray diffraction system is often mitigated by the relatively weak intensity of the incident X-ray beam. While the diffraction resolution limit of a Cu anode X-ray diffraction system is about 0.8 Å, protein crystals are usually very weakly diffracting, and diffraction resolution to 3.0 Å or better is a reasonable expectation with home laboratory X-ray sources.

To obtain higher resolution X-ray diffraction from protein crystals, data are often collected at synchrotron beamlines. Synchrotrons are large circular particle accelerators in which charged particles travel at relativistic speeds in an orbital path. Charged particles subjected to an acceleration emit radiation.

This radiation is emitted in the direction of travel of the charged particle if it had not been subjected to the acceleration.

In a synchrotron, electrons travel in a circular orbit. A centripetal acceleration is therefore placed on the electrons, which results in radiation being emitted tangential to the circular path of the electrons. The emitted radiation is then directed into experimental stations built as spurs off of the main synchrotron ring. Since the emitted radiation spans the electromagnetic spectrum, many different types of experimental stations can be built at a single synchrotron, each station selecting different wavelengths from the emitted radiation, such as ultraviolet or X-rays.

Synchrotron beamlines are capable of several orders of magnitude greater X-ray flux than home laboratory X-ray diffraction systems. Since the intensity of the incident X-ray beam is much greater, high-resolution reflections, which are barely measurable on home laboratory systems, can have useful intensities when collected at a synchrotron. The same protein crystal can therefore give much higher resolution diffraction data at a synchrotron.

In addition to high intensity, the wavelength of the incident X-ray beam at a synchrotron can be tuned. On a typical home laboratory Cu anode X-ray source, the wavelength of the incident radiation is fixed at 1.54Å. At a synchrotron, through the use of various optical elements, the wavelength of the incident X-ray beam entering an experimental station can be tuned over a range of several angstroms. The application of such wavelength tuning is critical in anomalous dispersion diffraction data collection, which will be discussed in the next section.

Overall, the components of a synchrotron X-ray diffraction system are as shown in Figure 10.13. The much greater intensity of the synchrotron incident X-ray beam, however, results in changes in the practice of data collection compared with the home laboratory X-ray system. Data collection is much faster at a synchrotron. A data set taking 2 to 3 days to collect on a home laboratory system can be collected in an hour at a synchrotron. Individual exposures requiring 5 to 30 min on a home laboratory system generally take 5 to 60 s at a synchrotron.

The rapid data collection possible at synchrotrons allows for different hardware compared with a home laboratory system. A major difference is in the type of X-ray detector. Most synchrotron beamlines are equipped with charge-coupled device (CCD) detectors. CCDs are light-sensitive integrated circuit chips. In a CCD X-ray detector, the diffracted X-rays impinge on a large phosphorescent face plate. The X-rays are converted to visible light by the phosphors. The visible light is then directed through an optical taper to demagnify the light down to the CCD chip [47]. One of the great advantages of using CCD detectors instead of imaging plates on a synchrotron beamline is the faster readout times of the CCD detectors, which are on the order of a few seconds instead of minutes, as with an imaging plate. With well-diffracting protein crystals, it is possible to collect a complete diffraction data set at a synchrotron in less than half an hour [48].

Another hardware becoming more common at synchrotron beamlines is the robotic sample mounter. These devices allow multiple crystals, on the order of 100, to be loaded into a magazine, which is generally filled with liquid nitrogen to keep the crystals frozen. The crystals are then automatically mounted and retrieved from the goniometer without the need for human intervention [49, 50].

Synchrotrons are massive public works projects, with physical scale on the order of kilometers, built by governmental agencies and costing around $1 billion. Some examples are the Advanced Photon Source at Argonne National Laboratory in Argonne, Illinois, and the SPring-8 synchrotron in Hyogo, Japan. While synchrotrons were originally built for high-energy particle physics experiments, their use in protein crystallography is growing rapidly. There are currently on the order of 50 to 100 synchrotron beamlines worldwide dedicated to protein crystallography, with new ones being commissioned [51].

10.4 X-RAY CRYSTAL STRUCTURE DETERMINATION

The previous section discussed general diffraction theory and diffraction data collection. This chapter will give an overview of determining the protein crystal structure from the diffraction data.

Phase Problem

As described in the previous Section 10.3, the electron density of a protein, and therefore its three-dimensional structure, can be reconstructed by the summation of sine and cosine terms. Such a summation is called a Fourier synthesis. The sine and cosine terms are the collected diffraction data, the reflections in the diffraction images (Fig. 10.8). These reflections are the X-rays diffracted by the crystal. Since these X-rays are waves, they have both amplitude and phase components.

The amplitude components of the diffraction data can be measured directly by the X-ray detector, which measures the intensities of the diffraction spots. The phase components, however, cannot be measured directly. This inability to directly measure the phase in diffraction data is referred to as the *phase problem*. In this chapter, various methods for determining phases for protein X-ray diffraction data will be discussed.

Structure Factor

In a diffraction image, each of the reflections identified by the Miller indices *hkl* are described mathematically by a structure factor F_{hkl}. The structure factor is the resultant vector of all of the waves diffracted in the direction of the spot *hkl* by all of the atoms in the crystal. The structure factor corresponding to a spot *hkl* is therefore the vector sum of the structure factors of each atom in the crystal. The structure factor of an individual atom in the crystal can be expressed as:

$$f_j = |f_j|e^{i\delta j} \qquad (10.7)$$

where $|f_j|$ is the magnitude of the structure factor, which can be measured directly, and $e^{i\delta j}$ is the phase, which cannot be measured directly. (i is the imaginary number square root of -1.) δj is the phase angle of the atom.

The phase of a structure factor F_{hkl} is the phase difference of the resultant diffraction of all of the atoms in the crystal unit cell with respect to the origin of the unit cell. In radians, the phase of each atom, δ, can be expressed as [42]:

$$\delta = 2\pi(hx + ky + lz) \qquad (10.8)$$

In this expression h, k, and l are the Miller indices, and x, y, and z describe the position of the atom in fractional coordinates of the unit cell. Since the structure factor F_{hkl} for a reflection with Miller indices hkl is the sum of all of the individual atom structure factors, the structure factor F_{hkl} can be expressed as:

$$F_{hkl} = \sum |f_j|e^{i\delta j} \qquad (10.9)$$

or

$$F_{hkl} = |F_{hkl}|e^{2\pi i(hx+ky+lz)} \qquad (10.10)$$

where $|F_{hkl}|$ is the directly measurable magnitude term, and $e^{2\pi i(hx + ky + lz)}$ is the phase term that cannot be directly measured. The structure factors are then summed as in Eq. (10.3) to calculate the electron density.

Although the phase of a reflection F_{hkl} cannot be directly measured by the X-ray detector, Eq. (10.8) shows that the phase is defined by the geometric coordinate terms hx, ky, and lz. Thus, by determining the positions of atoms within the crystal, the phases can be calculated. The remainder of this chapter will discuss methods for determining atomic positions to calculate phases for the diffraction data, and ultimately determine the three-dimensional crystal structures of proteins.

Heavy Atom Replacement Methods

Heavy atom replacement methods are used to determine structures of proteins for which no partial structures or reasonable approximations of the structures are available. These methods are suitable for determining the structures of proteins with new or unknown folds.

The determination of phases in crystallography is essentially the determination of one or more atomic positions within the crystal. While a direct visualization of specimen positions, like in visible or electron microscopy, is not possible in X-ray diffraction, atomic positions can be extracted from the phaseless diffraction data.

Multiple Isomorphous Replacement

As stated previously, the Fourier transform for the electron density using X-ray diffraction data [Eq. (10.3)] requires the phases of the diffracted X-rays, which in turn require positional information about the atoms in the crystal. Another Fourier transform calculated from the X-ray diffraction data, the Patterson function, does not require any phase or positional information. The Patterson function is expressed as:

$$P_u = (1/a)\sum_h |F|^2 \cos(2\pi hu) \qquad (10.11)$$

The Patterson function is mapped onto a coordinate system proportional to the direct unit cell dimensions of the crystal with the xyz coordinates of the direct unit cell converted to uvw in the Patterson map. In Eq. (10.11) for the Patterson function, the value of the Patterson function, in this case along the u axis, is a function of the h Miller index for a particular reflection and the square of the amplitude, $|F|^2$, of the reflection. Since the amplitudes of the reflections are experimentally measured, the Patterson function can be calculated directly from the diffraction data. Whereas in the electron density Fourier transform [Eq. (10.3)] the maxima correspond to atomic positions, the maxima in the Patterson function correspond to interatomic vectors. Furthermore, the intensity of the Patterson peaks is weighted by the atomic masses of the two atoms at the ends of the interatomic vectors.

In a protein consisting of nearly all carbon, nitrogen, and oxygen atoms (hydrogen atoms are generally not visible in protein X-ray diffraction), the Patterson function of interatomic vectors is practically uninterpretable. This is because all of the atoms are of similar mass so all of the peaks in the Patterson map are of similar value. If, however, there were a heavy atom in the crystal, such as mercury or lead, which had a much higher atomic mass than the carbons, nitrogens, and oxygens, the interatomic vectors between these heavy atoms would have a much higher value and would clearly stand out in the Patterson map.

In order to calculate interpretable Patterson functions, protein crystals can be soaked in micromolar to millimolar concentrations of heavy atom solutions to incorporate the heavy atoms into the protein crystals [37]. Some of the most common heavy atoms used for protein crystallography phasing are gold, lead, mercury, and platinum. These metals generally bind to the protein at sulfur-containing residues. As implied by the name *isomorphous replacement*, in soaking protein crystals with heavy atoms, it is important that the soaking procedure does not result in a significant change in the unit cell dimensions of the crystal. A crystal whose unit cell changes significantly upon heavy atom soaking will not be useful in the structure determination. In practice, soaking relatively fragile protein crystals in heavy atom solutions often results in damage to the crystal. The damage can range from simple reduction of diffraction resolution, which crystals can still be useful, to complete extinguishing of diffraction. It is not uncommon to screen tens or hundreds of different heavy atoms and soaking conditions to get a single heavy-atom-soaked protein crystal that diffracts and maintains the unit cell dimensions of the unsoaked protein crystal.

Due to the symmetry of the crystals, there are generally multiple copies of the protein molecule in the unit cell of the crystal, with one or more copies of the protein in the asymmetric unit of the crystal's unit cell. The heavy atoms will bind to each of the multiple protein atoms in the unit cell. The heavy atoms will therefore occupy all of the asymmetric units in the crystal. The mathematical symmetry operations relating each asymmetric unit to another result in the maxima in the Patterson map occurring at certain sections of the uvw Patterson space. Determining the atom positions from Patterson maps can be done manually by relatively simple algebraic equations. There are now, however, numerous software suites that automatically determine heavy atom positions from Patterson maps, and the modern protein crystallographer rarely, if ever, determines heavy atom positions manually.

X-ray diffraction data collected from a native protein crystal lacking heavy atoms (data set P) and diffraction data from a crystal of the same protein containing heavy atoms (data set PH) will differ in the measured intensities of their reflections solely by the contribution of the heavy atoms in data set PH. These differences in intensities can be used to calculate an isomorphous difference Patterson map, which will show peaks corresponding to the interatomic vectors between the heavy atoms (Fig. 10.15).

The isomorphous difference Patterson function calculated from data sets P and PH is expressed as:

$$P_u = (1/a)\sum_h |F_P - F_{PH}|^2 \cos(2\pi hu) \tag{10.12}$$

Compare this expression for the isomorphous difference Patterson with that of the general Patterson function [Eq. (10.11)]. In the isomorphous difference Patterson, the amplitude term is the difference between the amplitudes in the native protein and heavy-atom-derivatized protein data sets. The amplitude differences due to the presence of heavy atoms in data set PH will show up as strong peaks, allowing interpretation of the isomorphous difference Patterson maps.

The coordinates of the heavy atoms calculated from the positions of the Patterson peaks essentially determine the heavy atom substructure of data set PH. The amplitudes of the reflections from the heavy-atom-only substructure of data set PH are easily calculated by the appropriate summation of heavy atom scattering factors, which are listed in standard tables. The phases of the reflections from the heavy-atom-only substructure are calculated from Eq. (10.8). Once the heavy atom substructure has been determined, phases can be determined for the protein diffraction data.

X-RAY CRYSTAL STRUCTURE DETERMINATION

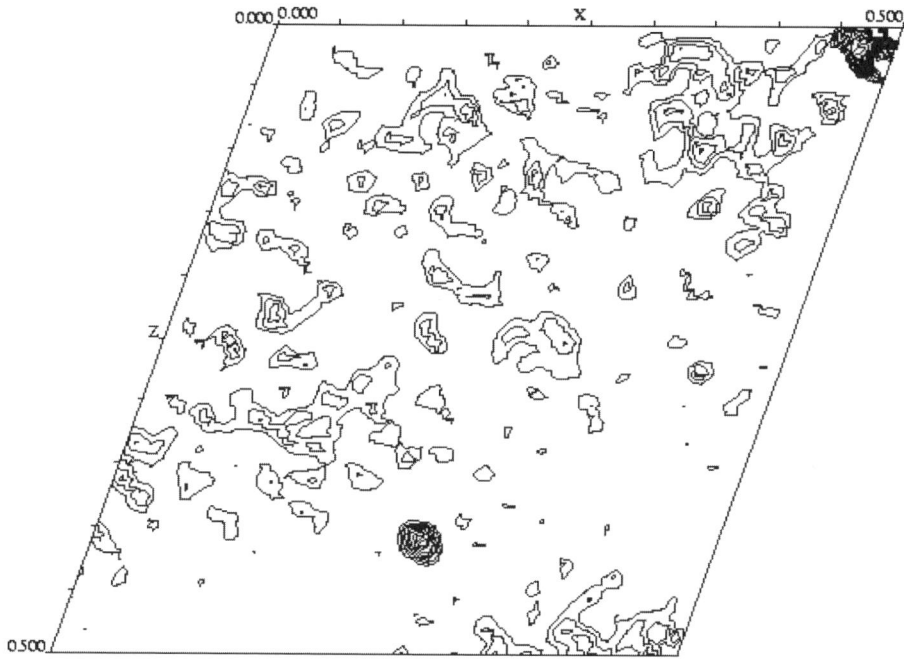

Figure 10.15 An isomorphous difference Patterson map calculated according to Eq. (10.12). The peak at approximately $X = 0.25$, $Z = 0.40$ corresponds to the interatomic vector between the heavy atoms in the heavy-atom-derivatized diffraction data set.

The general concept of phasing X-ray diffraction data with heavy atom replacement can be illustrated with Argand diagrams. Since each reflection F_{hkl} has an amplitude and phase component, the reflections can be represented as vectors in an Argand diagram (Fig. 10.16). The amplitude of the vector is determined from the measured intensity of the reflection F_{hkl}. The phase of F_{hkl} cannot be measured, so the phase angle in the Argand diagram is indeterminate from 0–2π radians.

The data sets P and PH can now be plotted on Argand diagrams to determine phases for the data set P. The only experimentally measurable difference between the diffraction data sets P and PH are the intensities of the reflections, which are due to the presence of the heavy atoms in data set PH. For a given reflection F_{hkl}, the superposition of the Argand diagrams for data set P and PH, with data set PH offset by the amplitude and phase of the heavy-atom-only substructure calculated for data set PH, shows the intersection of the P and PH phase circles at two points (Fig. 10.16). These two points of intersection are possible phases for the reflection F_{hkl}. Given only these two possible points, a good choice for the phase is the mean of the two points. Choosing the mean from just two points, however, can lead to a very large phase error. To break the phase ambiguity, diffraction data can be used from a second

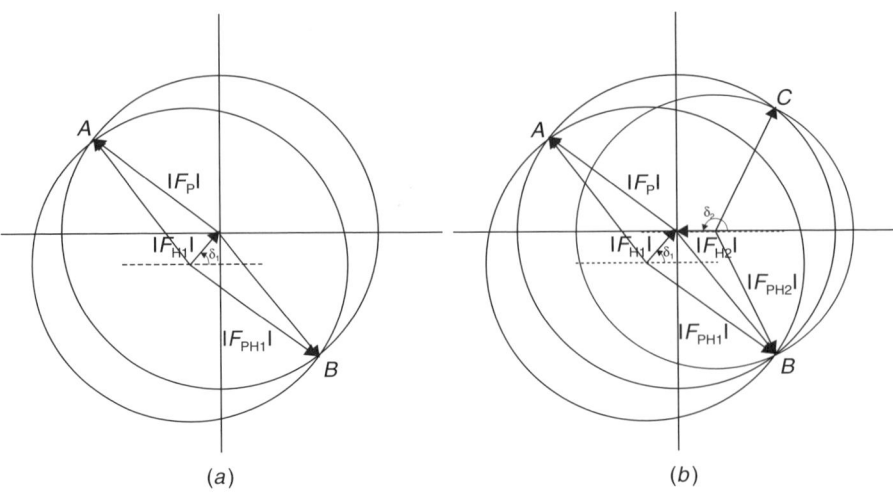

Figure 10.16 Argand diagrams illustrating phase determination. (a) The measured reflection F_P with Miller indices hkl from the protein crystal diffraction data is plotted as a vector with magnitude $|F_P|$. No phase information is present in the diffraction data, so the phase of the reflection is indeterminate from 0 to 2π radians as indicated by the light circle. The same reflection with Miller indices hkl from the heavy-atom-derivatized diffraction data is plotted as a vector with magnitude $|F_{PH1}|$, shown as the heavy circle. The reflection F_{PH1} is plotted offset from the reflection F_P by the magnitude, $|F_{H1}|$, and phase angle, δ_1, of the heavy-atom-only substructure, as determined from isomorphous Patterson maps. The points A and B where the light and heavy circles intersect are two possible phase angles for the protein reflection F_P. This A or B ambiguity is a graphical representation of the phase problem. (b) Resolution of the phase problem using a second heavy atom derivative. The reflections F_P and F_{PH1} are plotted as in (a). In addition, the reflection with the same Miller indices hkl from a second heavy-atom-derivatized diffraction data set is plotted as a vector with magnitude $|F_{PH2}|$. The reflection F_{PH2} is plotted offset from the reflection F_P by the magnitude, $|F_{H2}|$, and phase angle, δ_2, of the heavy-atom-only substructure in the second heavy-atom-derivative data set, as determined from isomorphous Patterson maps. The heavy circle from the second heavy atom data set also intersects the light circle of the protein-only diffraction data set at two points, B and C. One of these points, B, is common between the two possible phases indicated by the first heavy atom derivative and is the correct phase angle for the reflection F_P.

heavy-atom-derivatized crystal, data set PH2. After determining the amplitudes and phases of the heavy-atom-only substructure of data set PH2 by Patterson methods as described for data set PH, the reflection F_{hkl} from data set PH2 can be superimposed on the Argand diagrams for data sets P and PH (Fig. 10.16). With two, or more, heavy atom data sets, the correct phase is the point at which all of the phase circles intersect. Because of the requirement for more than one heavy atom derivative diffraction data set, this method is called multiple isomorphous replacement (MIR). Once the phases are deter-

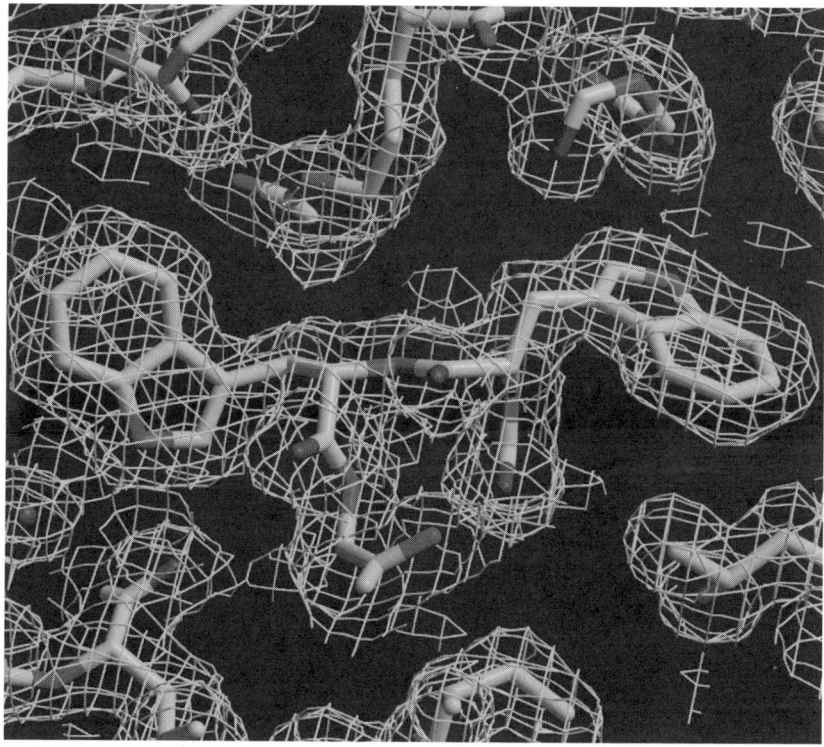

Figure 10.17 A $2F_o$–F_c electron density map calculated according to Eq. (10.13). (This figure is available in full color at ftp://ftp.wiley.com/public/sci_tech_med/drug_discovery/.)

mined for the measured reflections, incorporating these phases into the Fourier summation for the electron density [Eq. (10.3)] will give an electron density map, into which the model of the protein can be built.

After determination of the phases and building in the protein model, electron density maps are generally calculated by Fourier transforms using coefficients $2F_o - F_c$. The explicit expression for the electron density map calculation is [52]:

$$\rho(x,y,z) = (1/V)\sum_{hkl}(2|F_o|-|F_c|)\cos[2\pi(hx+ky+lz)-\alpha(hkl)] \quad (10.13)$$

In Eq. (10.13), $|F_o|$ and $|F_c|$ are, respectively, the amplitudes of the reflections from the experimental diffraction data and the structure model built into the electron density, and α is the calculated phase for the reflection with Miller indices hkl. This type of electron density map shows the electron density of the calculated model, and the difference electron density of the target protein structure and the calculated model [52]. An example of an electron density map calculated according to Eq. (10.13) is shown in Figure 10.17. The electron density is usually represented in chicken-wire contours into which the protein model can be built.

In practice, successful interpretation and application of MIR data to determine a protein crystal structure is not as straightforward as may appear in Figure 10.17. In MIR, it is assumed that the measured intensity differences between the native protein data set P and heavy atom data sets PHn are due solely to the presence of the heavy atoms in PHn. These differences in intensity are on the order of a few percent, so great accuracy and care is required in the measurement and processing of the diffraction data. Also, the unit cell parameters between the P and PHn crystals must not deviate significantly. It is not uncommon for a heavy atom soak of a protein crystal to distort the crystal and change the unit cell parameters. Such heavy-atom-derivatized crystals are then of little value in the structure determination. Because of its relatively stringent experimental requirements, MIR is becoming less and less popular compared with other methods, described below. Still, the general concepts of phase determination by MIR carry through to the other methods.

Anomalous Dispersion Methods

Friedel's Law As mentioned previously, X-ray diffraction can be treated as reflections of X-rays according to Bragg's law [Eq. (10.5)]. In protein X-ray crystallography, the X-rays can be thought to reflect off of planes of electrons (or atoms) in the crystal. Friedel's law [Eq. (10.14)] states that the intensities of the reflection F_{hkl} and the reflection $F_{\bar{h}\bar{k}\bar{l}}$ are equal ($\bar{h} = -h$):

$$I_{\bar{h}\bar{k}\bar{l}} = I_{\bar{h}\bar{k}\bar{l}} \tag{10.14}$$

That is, the intensities of the reflections with opposite Miller indices are equal. Applied to the reflection treatment of diffraction, Friedel's law simply states that the reflection intensity off of the front side of a plane and the back side of the same plane are equal. This assumption is valid when collecting diffraction data from native protein crystals that contain no heavy atoms, and even heavy-atom-derivatized crystals used in MIR under certain wavelengths of incident radiation.

Anomalous Dispersion Anomalous dispersion is the phenomenon whereby certain atoms under certain incident radiation wavelengths absorb some of the incident radiation, instead of simply scattering the radiation, as in normal diffraction. Typical anomalous scatterers used in protein X-ray crystallography are heavy atoms such as Hg, Au, and Pb, many of which are also used in MIR. Anomalous dispersion results in the breaking of Friedel's law. As a result, within a single diffraction data set, there are differences between the measured intensities of reflections that would otherwise be the same in the absence of anomalous scattering. These intensity differences can be used as in MIR to determine substructures of the anomalous scatterers in the crystal, which are then used to phase the diffraction data and determine the structures of the protein.

The main experimental requirement for anomalous dispersion X-ray diffraction data collection is similar to that of MIR, namely a protein crystal derivatized with an anomalous scatterer. Such derivatization can be done by soaking the protein crystal in a solution of an anomalous scatterer. In addition to the classic heavy metals used in MIR, halides can also give excellent anomalous diffraction data for protein structure determination [53].

The derivatization of crystals with an anomalous scatterer can also be done at the molecular biology level. Selenium is an excellent anomalous scatterer at wavelengths suitable for protein crystallography. Proteins can be expressed in bacteria and insect cells in a medium containing the amino acid Se-Met instead of Met, where the sulfur in Met is replaced with selenium. Crystals of the Se-Met protein then automatically contain an anomalous scatterer. Instead of soaking exogenous heavy atoms into a protein crystal and risking damage to the crystal, incorporation of an anomalous scatterer directly into the amino acid sequence of a protein reduces the crystal sample preparation steps and may preserve the diffraction power of the crystal.

For a given anomalous scatterer, the strength of the anomalous dispersion diffraction signal is dependent on the wavelength of the incident radiation. Therefore, another experimental requirement for anomalous X-ray diffraction data collection, which is not a requirement for MIR, is an incident radiation source tuned to the wavelength at which the anomalous dispersion signal is measurable for the anomalous scatterer in the crystal. Before the widespread use of synchrotron radiation, anomalous dispersion diffraction data were difficult to collect. Most protein diffraction data were collected on laboratory X-ray sources with fixed wavelength, usually Cu K_α, with $\lambda = 1.54$ Å. This wavelength was not optimized to measure the largest anomalous dispersion signal for the heavy atoms typically used as anomalous scatterers. Modern synchrotron beamlines now allow very precise tuning of the incident radiation to maximize the anomalous signal, making anomalous diffraction data collection routine.

Structure Determination by Anomalous Dispersion As discussed in the previous section on MIR, the unambiguous determination of phases from diffraction data generally requires the measurement of three different diffraction data sets, from the native protein crystal and at least two different heavy-atom-derivatized crystals. If one heavy-atom-derivatized crystal is available, and the heavy atom is also an anomalous scatterer, the anomalous signal from the heavy atom can be used to break the phase ambiguity associated with a single isomorphous data set (Fig. 10.16).

In a diffraction data set lacking any anomalous scatterers, there are numerous reflections, Bijvoet mates, which have the same measured intensity due to Friedels law [Eq. (10.14)]. If anomalous dispersion is present, the Bijvoet pair of reflections F_{hkl} and $F_{\bar{h}\bar{k}\bar{l}}$ does not have the same intensity. These differences in the measured intensities result in different intensity measurements for the reflections F_{hkl} and $F_{\bar{h}\bar{k}\bar{l}}$, whereas in the nonanomalous dispersion isomorphous replacement case these two reflections would have the same measured intensity. A single anomalous scatterer-derivatized crystal can thus give two contributions to the Argand diagram, for the reflections F_{hkl} and $F_{\bar{h}\bar{k}\bar{l}}$.

As in MIR, in anomalous dispersion methods, the positions of the anomalous scatterers are calculated from Patterson maps. The expression for an anomalous Patterson is

$$P_u = (1/a)\sum_h |F_+ - F_-|^2 \cos(2\pi h u) \qquad (10.15)$$

Compare this expression with that for the isomorphous difference Patterson [Eq. (10.12)]. In the anomalous Patterson expression, $|F_+ - F_-|^2$ is the square of the difference in amplitude between the reflection F_+ and its Bijvoet mate F_-, the amplitude difference resulting from anomalous dispersion. The positions of the anomalous scatterers used in phasing the diffraction data are determined from anomalous Patterson maps (Fig. 10.18). As in the isomorphous difference Patterson (Fig. 10.15), the peaks in the anomalous Patterson correspond to the interatomic vectors between the anomalous scatterers in the anomalous diffraction data set. Analysis of the anomalous Patterson peaks can determine the anomalous scatterer-only substructure in the anomalous diffraction data set. In modern protein crystallography, direct or ab initio methods, of which the mathematics will not be discussed here, are often used to determine positions from Patterson maps.

Combined with the data from the native crystal, the phase ambiguity can be broken using just a single anomalous scatterer derivative (Fig. 10.16). The two different measured reflections F_+ and F_- in the anomalous diffraction data

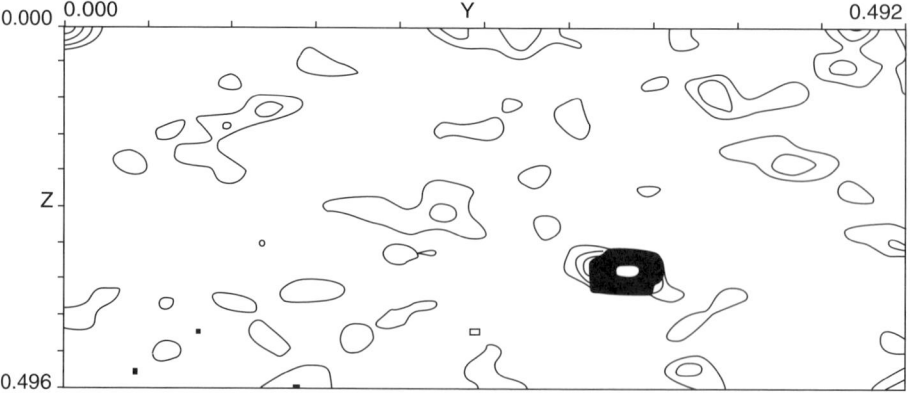

Figure 10.18 An anomalous Patterson map calculated according to Eq. (10.15). The major peak occurring at approximately $Y = 0.33$, $Z = 0.33$ corresponds to the interatomic vector between the anomalous scatterer atoms in the anomalous scatterer-derivatized diffraction data set.

set correspond to the reflections F_{PH1} and F_{PH2} in Figure 10.16. This method of using a single anomalous scatterer-derivatized crystal for phasing is called single isomorphous replacement with anomalous scattering (SIRAS).

Since the differences in measured intensities due to anomalous dispersion are small, on the order of 3 to 5 percent of the total measured intensity [54], anomalous dispersion data must be measured and processed with great accuracy and care. Diffraction data can be collected from additional derivative crystals prepared with different anomalous scatterers. The anomalous dispersion diffraction data from these additional derivative crystals can then be combined with the native protein and first-derivative data to give additional phasing information. Such use of multiple anomalous scatterer-derivatized crystals is called multiple isomorphous replacement with anomalous scattering (MIRAS).

Protein diffraction data can also be phased by purely anomalous dispersion data, without the use of diffraction data from an underivatized protein crystal. One of the earliest widespread uses of purely anomalous dispersion diffraction data in protein X-ray crystallography was multiple-wavelength anomalous dispersion (MAD). In MAD experiments, multiple data sets are collected from a single crystal derivatized with an anomalous scatterer, with each data set collected at different wavelengths. These wavelengths are at or near the wavelength that gives the maximum anomalous dispersion signal for the anomalous scatterer. At each wavelength, there will be more or less anomalous signal from the anomalous scatterer, which will result in changes in measured intensity for a particular reflection between the data sets.

The MAD diffraction experiments are entirely analogous to MIR diffraction experiments. Whereas in MIR diffraction experiments the wavelength remains fixed and multiple different heavy atom derivatives are used to collect data sets with differences in measured intensities, MAD diffraction experiments maintain the anomalous scatterer and use multiple different wavelengths to collect data sets with differences in measured intensities. MAD experiments can be considered as in situ isomorphous replacements where physics, rather than chemistry, is used to produce the change in scattering intensity at the site [55]. Once the positions of the anomalous scatterers are located, usually by Patterson methods, the phases for the reflections are determined as in the MIR case.

In addition to MAD, there is also single-wavelength anomalous dispersion (SAD) for phasing protein diffraction data. In SAD, X-ray diffraction data are collected from a single crystal derivatized with an anomalous scatterer at a single wavelength. The intensities of the reflections F_{hkl} and $F_{\overline{hkl}}$ are measured separately, essentially giving two data sets from a single crystal. As shown in Figure 10.16a, the Argand diagram analysis of only two diffraction data sets leads to a phase ambiguity. For a SAD diffraction data set, the two different amplitudes for the Bijvoet pair of reflections F_+ and F_- correspond to the reflections F_P and F_{PH1} in Figure 10.16a. A native diffraction data set collected from an isomorphous crystal lacking the anomalous scatterer can be used to break the phase ambiguity, which is the SIRAS technique described above. If, however, no additional isomorphous data sets are available, the correct phases can still be derived from purely SAD data.

Modern SAD protein structure determinations use probabilistic methods to determine initial phases and their reliability [53]. Once a set of phases are determined for a diffraction data set, an electron density map can be calculated [Eq. (10.13)]. The electron density maps can then be modified according to reasonable assumptions. A common electron density modification technique is solvent flattening. In the crystal, the interstitial regions between the protein molecules are occupied by solvent, which is usually disordered. This disordered solvent region should be relatively featureless, compared with the protein. By iteratively smoothing the electron density in the solvent region, the electron density features in the protein region will become enhanced until they become interpretable. This method of iterative electron density modification to determine protein structures from pure SAD data is called iterative single-wavelength anomalous scattering (ISAS) [56]. In general, the steps involved in determining protein structures from pure SAD data are determining the positions of the anomalous scatterers, determination of initial phases, and electron density modification until the electron density maps become interpretable.

For the crystal structure determination of proteins with unknown folds, anomalous dispersion techniques such as MAD and SAD are now much more popular than pure isomorphous replacement techniques such as MIR. One of the major advantages of anomalous dispersion over isomorphous replacement techniques is that anomalous dispersion data sets can all be collected from a single crystal. This avoids the problem of nonisomorphism, which can make MIR data collection and interpretation difficult. In MIR, multiple crystals derivatized with multiple different heavy atoms are used for data collection. Each heavy atom and its soaking conditions can have different effects on the crystal, sometimes causing significant distortions of the crystal unit cell. MIR data collected from crystals whose unit cells are significantly different from that of the native protein crystal (nonisomorphous data sets) are of little value for phasing. With MAD and SAD experiments, using cryo-data collection techniques, a single crystal derivatized with an anomalous scatterer can be used to collect all of the diffraction data sets. All of the data sets are therefore isomorphous since all of the data were collected from a single crystal.

Molecular Replacement

The isomorphous replacement and anomalous dispersion methods are required for determination of protein structures of unknown folds. These methods assume no prior knowledge of the target protein structure and involve the initial determination of a heavy atom or anomalous scatterer substructure. These substructures are then used to calculate phases for the protein diffraction data. If, however, there is reasonably accurate prior knowledge of the protein structure, molecular replacement (MR) methods can be used to determine the crystal structure of the target protein.

In MR structure determinations, the experimental diffraction data are probed with a search model that is assumed to be a good structural homolog of the target protein. The best search model would be the structure of the target protein itself. Structures of proteins with high sequence homology are also good search models in MR. In general, search models with more than 50 percent sequence identity to the target protein should give a correct structure of the target protein by MR [54]. Species isomorphs of a target protein, which generally share about 75 percent or greater sequence identity, often work for MR structure determinations.

As with isomorphous replacement and anomalous dispersion methods, the correct atomic positions must be determined in order to calculate phases for the diffraction data. In isomorphous replacement and anomalous dispersion methods, novel substructures are used as the atomic positions for initial phasing. In MR, the structure of the isolated target protein is already known or assumed to be very similar to that of a known protein structure. The MR calculations place the known structure, the search model, in the correct crystal lattice according to the experimental diffraction data. In a sense, isomorphous replacement and anomalous dispersion methods determine "protein" structures in that these methods assume no prior knowledge of the target protein structure. MR methods, in contrast, determine "crystal" structures in that the entire structure of the target protein is already known, and the calculations are merely (but not always so simply) fitting the known structure to the crystal lattice of the diffraction data.

Although there are many different algorithms and software packages for MR calculations, the basic computations are illustrated in Figure 10.19. The experimental diffraction data contains information about the structure and crystal lattice arrangement of the target protein. A suitable search model is first rotated so that its orientation is the same as the target protein in the experimental crystal with respect to a certain coordinate system. The rotation is usually expressed as a set of three Eulerian angles α, β, and γ. In this rotation search, the Patterson vectors for the various orientations of the search model are compared with those of the target protein in the diffraction data. As described previously, the Patterson vectors are the interatomic vectors in the protein in the crystal. One can imagine two types of interatomic vectors, intramolecular and intermolecular, the intramolecular vectors generally being shorter than the intermolecular vectors. The intramolecular vectors occur entirely within a single protein molecule, and their orientations define the orientation of the protein. In the rotational search in MR, the shorter intramolecular Patterson vectors in the search model are compared with those of the target protein in the diffraction data. The search model is then rotated until the intramolecular vectors of the search model and target protein have high correlation.

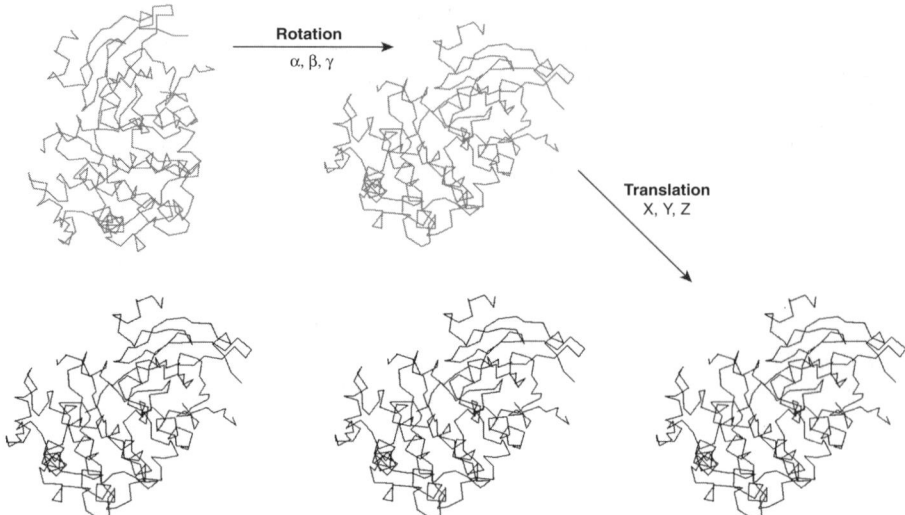

Figure 10.19 Schematic of an MR structure determination. (left) The target protein structure is drawn in black lines. The search model protein structure in an arbitrary orientation and position with respect to the target protein is drawn in red lines. (middle) The search model is rotated so that its orientation corresponds to that of the target protein structure as indicated by the intramolecular Patterson vectors in the experimental diffraction data. (right) The search model is translated so that it is correctly placed in the unit cell of the experimental diffraction data. (This figure is available in full color at ftp://ftp.wiley.com/public/sci_tech_med/drug_discovery/.)

After the rotational search, the correct placement of the oriented search model in the target protein crystal unit cell is determined by a translational search. In the translational search, the oriented search model is translated within the unit cell of the crystal until the crystal packing of the search model matches that of the experimental diffraction data. There are various methods for the translational search. In one method, the longer intermolecular Patterson vectors are compared with those of the experimental data at each translation of the search model. The highest correlation between the sets of intermolecular Patterson vectors is generally the correct solution for the crystal structure of the target protein.

Compared with isomorphous replacement and anomalous dispersion methods, crystal structure determination by MR has much less experimental requirements. No special derivatization of the sample crystal with heavy atoms or anomalous scatterers is required. Only a single data set from the crystal of the target protein is required. Also, no particular radiation wavelength is required for data collection. The ability to use diffraction data collected at any wavelength makes structure determination by MR readily accessible since all of the necessary diffraction data can be collected on a standard laboratory X-ray source (assuming adequate diffraction power of the crystals).

At this point, one may wonder why go through the effort of determining the "crystal" structure of the target protein by MR if the "protein" structure

X-RAY CRYSTAL STRUCTURE DETERMINATION

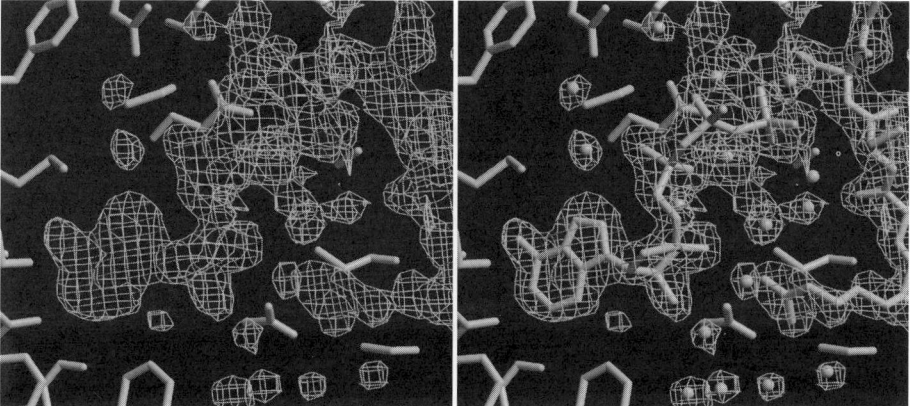

Figure 10.20 F_o–F_c electron density map calculated from an MR structure determination. *(Left)* The structure of a protein–ligand complex has been determined by MR using the protein-only structure as the search model. The F_o–F_c electron density map calculated from the protein-only MR solution reveals electron density (blue chicken wire contours) which does not correspond to the search model structure (yellow bonds). *(Right)* The structure of the bound ligand is built into the difference electron density in the F_o–F_c electron density map. Additional protein residues not present in the search model and bound waters are also built into the difference electron density. (This figure is available in full color at ftp://ftp.wiley.com/public/sci_tech_med/drug_discovery/.)

is already known or assumed to be that of the search model. MR crystal structure determination is required because the search model may not be complete. Although the overall structure of the target protein may already be known, the search model will lack certain structural features of the experimental data, such as amino acid mutations, insertions, and deletions, and the binding modes of ligands bound to the target protein. These additional features are not incorporated into the search model during MR calculations. The electron densities of these additional structures can only be calculated after determining the correct crystal structure of the target protein, and using phases determined from the correct placement of the protein in the crystal.

Once the search model has been correctly rotated and translated with respect to the diffraction data, difference electron density maps, using Fourier coefficients $F_o - F_c$, reveal structures that were not part of the search model. The explicit expression for this type of electron density map is

$$\rho(x, y, z) = (1/V)\sum_{hkl}(|F_o| - |F_c|)\cos[2\pi(hx + ky + lz) - \alpha(hkl)] \quad (10.16)$$

Compare this expression with that of the $2F_o - F_c$ electron density map [Eq. (10.13)]. The $F_o - F_c$ will show electron density in regions where the search model and the target protein structures differ and no electron density where the two structures are the same. Figure 10.20 illustrates the MR structure determination of a protein–compound complex. The experimental diffraction data were collected from a crystal of the protein–compound complex. A protein-only search model was used in the MR structure determination. The $F_o - F_c$ electron density map reveals the electron density of bound ligand, which was not present in the search model. Regions of the map where the search model and the target complex structure are the same are relatively featureless.

MR structure determination is now the crystallographic computation engine for X-ray structure-based drug design. Protein–compound complex structures are determined by MR from diffraction data collected from crystals of protein–compound complexes, either co-crystallized from protein–compound complex samples or having the compound soaked into protein crystals. Using MR structure determination, difference electron density maps can be calculated within minutes of collecting and processing the diffraction data to reveal the binding mode of the compound. Often, the protein–compound complex crystals are isomorphous with previous protein–compound complexes and the apo-protein crystals. In such cases, the MR solution is a trivial zero rotation and zero translation transformation of the search model (the target protein structure), and the model can proceed directly to structure refinement, which will be discussed in following sections.

All of the protein diffraction data phasing methods described above are now handled by automated software packages. These software include CNX (CNS), CCP4, PHASES, SOLVE, SHARP, SHELX, and EPMR, among others. Due to the great advances in the automation of crystallographic computing and computing power in general, the phase problem, which was once the central task of protein structure determination, is now less of a concern in protein X-ray crystallography.

10.5 GENERATION AND ANALYSIS OF STRUCTURAL MODELS

Aspects of Crystallographic Models of Macromolecules

The end result of crystallographic structure determination is the generation of an accurate three-dimensional model of the contents of the asymmetric unit of the crystal. In the simplest terms this model consists of the *spatial (xyz) coordinates* and elemental type for each atom. In most cases a great deal more information is included explicitly in the model or implied by convention or context. Protein Data Bank (PDB) file format coordinate records for each atom typically contain *xyz* coordinates, thermal motion parameters (*B factor* or temperature factor), and relative *occupancy* (0.0 to 1.0). B factors represent the thermal vibration of each atom (see previous section). Individual atom thermal motion can be modeled either as isotropic or anisotropic motion. In the isotropic case the atom is modeled as if it vibrates with equal amplitude in three orthogonal directions, producing a spherical time-averaged distribution of electron density. Isotropic thermal motion can be represented by a single B-factor parameter that is proportional to the mean-square amplitude of vibration. In the anisotropic case the atom is modeled with vibrations of differing amplitude in three orthogonal directions, producing an ellipsoidal time-averaged distribution of electron density. Complete description of anisotropic thermal motion requires six parameters (U_{11}, U_{22}, U_{33}, U_{12}, U_{13}, U_{23}) that are expressed as the mean-square amplitude of vibration in angstroms. These parameters serve to define the orientation of the axes of the ellipsoid and the mean-square amplitude of vibration along each axis. Atomic resolution data (1.0 Å resolution or better) is required to properly refine anisotropic B factors.

The last parameter, occupancy, determines the fraction of unit cells within the crystal that contain the given atom or molecule. This parameter is typically applied to ligands, ordered water molecules, and other ions and small molecules that are often observed bound to specific sites in proteins and other macromolecules. Because protein–protein contacts form the lattice interactions that make up the crystal, the atoms of the protein region of the crystal are in general constrained to have occupancies of 1.0. Exceptions to this are instances in which multiple conformations for individual residues or regions of the peptide chain can be discerned in the electron density maps. PDB format ATOM records for models with isotropic B factors contain at least 10 entries (atom number, atom name, residue type, chain name, residue number, X, Y, Z, B, Occ), but the PDB format allows a number of additional entries such as segment identifiers (equivalent to chain name) used by the XPLOR/CNS suite [57] and atom types and charges used by Refmac [58]:

```
ATOM   22  OH   TYR A  5   -42.861  -17.317  103.842  1.00  75.00   O
ATOM   23  N    ILE A  6   -35.203  -14.922  106.691  1.00  32.68   N
ATOM   24  CA   ILE A  6   -34.025  -14.142  106.955  1.00  31.42   C
ATOM   25  C    ILE A  6   -33.743  -13.217  105.782  1.00  30.29   C
ATOM   26  O    ILE A  6   -33.841  -13.620  104.644  1.00  51.93   O
ATOM   27  CB   ILE A  6   -32.833  -14.967  107.379  1.00  39.55   C
ATOM   28  CG1  ILE A  6   -32.291  -14.302  108.607  1.00  42.56   C
ATOM   29  CG2  ILE A  6   -31.732  -14.979  106.344  1.00  32.26   C
ATOM   30  CD1  ILE A  6   -31.813  -15.334  109.585  1.00  75.00   C
```

PDB format atom definitions for models with anisotropic B factors require two records. The first is a standard ATOM record with an equivalent B factor calculated from an average of U_{11}, U_{22}, and U_{33}. The second ANISOU card lists U_{11}, U_{22}, U_{33}, U_{12}, U_{13}, and U_{23} sequentially, but scaled by a factor of 10^4:

```
ATOM    107  N   GLY 13   12.681  37.302  -25.211  1.000  15.56         N
ANISOU  107  N   GLY 13   2406    1892    1614      198    519    -328  N
ATOM    108  CA  GLY 13   11.982  37.996  -26.241  1.000  16.92         C
ANISOU  108  CA  GLY 13   2748    2004    1679      -21    155    -419  C
ATOM    109  C   GLY 13   11.678  39.447  -26.008  1.000  15.73         C
```

```
ANISOU  109  C  GLY  13  2555   1955   1468     87    357    -109  C
ATOM    110  O  GLY  13  11.444 40.201 -26.971 1.000  20.93         O
ANISOU  110  O  GLY  13  3837   2505   1611    164   -121     189  O
ATOM    111  N  ASN  14  11.608 39.863 -24.755 1.000  13.68         N
ANISOU  111  N  ASN  14  2059   1674   1462     27    244     -96  N
```

Bonding and connectivity between atoms and residues is not in general defined explicitly in PDB format, but in most cases it is implicit in the relationship between atom names, residue names, residue numbers, and chain names. However, it is possible to explicitly define bonding and some non-bonding interactions with CONECT (general), LINK (general), SSBOND (disulfide bond), HYDBOND (hydrogen bond), and SLTBRG (salt bridge) records.

A complete description of the PDB format and all allowed records can be found at the Research Consortium for Structural Biology Web page (*www.rcsb.org*). In addition to the individual atom properties and connectivity, other parameters introduced during refinement such as noncrystallographic symmetry restraints, bulk solvent corrections, and TLS groups are important components of the final model. These aspects of model refinement will be discussed in more detail under "Refinement and Analysis of Structural Models."

Building the Initial Model

Building a de novo model of a macromolecule into initial electron density maps is often an arduous task. There are a number of software packages in common use (Quanta, XtalView, O, TurboFrodo, Coot) with more or less sophisticated tools to aid in interactive model building. The efficiency of the process depends on the quality of phases and the resolution of the map, the sophistication of the model-building software employed, and the crystallographer's familiarity with the molecule of interest as well as many general features of protein and nucleic acid structure.

Starting with only the primary amino acid sequence, a great deal can be learned about a given protein using freely available online database and analysis resources. The homology search engine BLAST (Basic Local Alignment Search Tool [59]), available from the National Center for Biotechnology Information (NCBI, *www.ncbi.nlm.nih.gov*), allows rapid identification of proteins with homologous regions of sequence, some of which might have known structures in the RCSB Protein Data Bank that can serve as templates for model building. Multiple sequence alignments of a homologous family of proteins will often reveal highly conserved residues that are crucial for structure or activity. JPRED [60] from the Barton group at the University of Dundee provides multiple sequence alignment with consensus secondary structure prediction. NCBI and the ExPASy (Expert Protein Analysis System) proteomics server of the Swiss Institute of Bioinformatics both provide excellent suites of tools for primary structure analysis. These tools allow identification of known amino acid motifs, such as substrate or cofactor binding motifs (e.g., the Rossman fold, P-loop motif, iron-sulfur clusters) and protein domains with known homologous structures that might be represented in the sequence of interest.

The SCOP ([61]; *scop.mrc-lmb.cam.ac.uk/scop/*) and CATH ([62]; *http://www.biochem.ucl.ac.uk/bsm/cath/*) online databases provide classification of protein structure with examples of each represented fold. An excellent resource for learning the basics of protein structure from peptide conformations to common folds is *Introduction to Protein Structure* by Carl Branden and John Tooze (Garland Science, UK). Knowledge of simple common aspects of tertiary structure such as the common helix-helix knobs-in-holes packing angles, the left-handed twist of β sheets and the right-handed twist of individual β strands, and the ways in which loop regions most commonly connect secondary structures will assist greatly in interpretation of blurry regions of electron density and recognition of secondary structure elements at low resolution. Another valuable reference is the classic study by P.N. Lewis, F.A. Momany, and H.A. Scheraga [63]. This study details the peptide conformations of the tight turns and chain reversals often encountered at the surfaces of proteins.

The main objectives of initial model building are to find the correct trace for the peptide chain and to make the correct sequence assignment for each residue. Before beginning actual model building, it is advisable to examine a number of electron density maps calculated to different resolutions and with different weighing [figure of merit or $\sigma(A)$] and density modification schemes (solvent flipping or flattening, histogram matching, or skeletonization). Less accurate phases for the highest resolution data might create noise that makes interpretation of the maps difficult. Initial maps calculated at lower resolution might reveal the trace of the peptide chain more clearly. The percentage of solvent content used in solvent flattening calculations can have a dramatic impact on the quality of the map, and the best estimate of the real solvent content will not necessarily produce the most interpretable electron density maps. Several values above and below the best estimate of the true value should be tried.

At this early stage density skeletonization, or "bones," calculations are a tremendous aid in determining the quality of electron density maps. The overall connectedness of the electron density can be gauged and secondary structure elements can often be recognized. Skeletonization calculations produce traces through the highest electron density points in the map. In well-phased electron density maps the bones will clearly reveal the trace of the peptide chain. In a poorly phased map the bones will be fragmented. Some arbitrary adjustment of the parameters used in the skeletonization calculations is often required to obtain the best results. If the map has been phased experimentally by isomorphous replacement or anomalous dispersion methods, it can be helpful to overlay the positions of the phasing sites onto the electron density. Heavy atoms are likely to bind to only a few reactive side chains [37], and if phases are derived from seleno-methionine anomalous dispersion data, the phasing positions will instantly reveal the location of all methionine residues in the structure. This information can dramatically speed up the initial chain tracing and sequence assignment. In addition, for heavy atom derivatives the maps should be examined for severe rippling in the vicinity of the phasing sites that indicates improper refinement of occupancy or B factor for the heavy atom. Adjusting these parameters until the rippling is minimized can make valuable improvements in the phases and thus in the overall quality of the maps.

Once suitable maps are obtained and optimal skeletonizations have been produced, chain tracing can begin. The most basic method of chain tracing is "baton" building in which a C_α-only trace is built interactively by sequentially adding about 3.5-Å-long baton segments. This is relatively straightforward with a map that is good enough to approximately locate the C_α positions by following the bones and observing the side-chain branch points. Some modern model-building software packages (e.g., QUANTA) have automated chain-tracing features that can build a C_α trace rapidly or even identify and automatically build secondary structure elements. It is also possible to use real-space rotation and translation searches to place models of secondary structures or even entire domains into initial electron density maps (FFFear, ACORN). These techniques can save a great deal of time by fitting large sections of standard secondary structure or approximately correct tertiary structure that can then be corrected interactively or by refinement software. Some modern model-building software packages are equipped with libraries of loop, turn, and random coil segments taken from known protein structures that can be screened rapidly for a good fit to difficult regions of electron density. As a general rule, segments that cannot be built with confidence should be left out of the initial model. Phase improvement by refinement of correctly built structure will improve the overall map and gradually allow interpretation of the difficult regions.

Once at least a partial chain trace is in hand, sequence assignment can begin. As mentioned above, phasing positions of heavy or anomalous atoms can occasionally be used as starting points for sequence assignment. Identification of the active site or bound ligands will aid in locating residues known to be involved in catalysis or binding. Aromatic residues such as tryptophan, tyrosine, and phenylalanine are often easily recognized because they are large and are most often located in the interior of proteins where their conformational entropy is low and the quality of the electron density is good. Because solvent flipping and flattening masks are not perfect, these procedures sometimes degrade the quality of electron density for surface loops and side chains. Disulfide bonds can often be recognized by strong connecting electron density. In fact, the electron density of disulfide bonds is often strong enough that it confuses early chain-tracing attempts, particularly automated procedures. If data has been collected using a long wavelength, such as Cu K_α, an anomalous difference map might reveal the locations of many sulfur atoms from methionine and cystine residues.

After sequence assignment a full-atom model of the peptide chain must be constructed and fit to the electron density. It is best to use a model-building software package that does automated full-atom model construction starting from a sequence-assigned C_α trace. The quality of auto-built models is often low, requiring manual adjustment to fit the electron density via additional interactive model building. Real space fitting, refinement, and regularization protocols allow this to be done rapidly. These algorithms can use either the electron density map contours (grid methods) or gradients calculated from these contours in least-squares algorithms to provide the best fitting position and conformation of residues or segments of peptide chain. Peptide conformations are fit either by iteratively adjusting or searching atom positions and torsions, by searching rotamer and conformation libraries, or by Monte Carlo methods. Grid-based methods have the smallest radius of convergence and are most appropriate for fitting single residues or side chains. Gradient methods have a larger radius of converge (several angstroms) and are more generally applicable, particularly in combination with Monte Carlo methods, for fitting longer segments of peptide chain. Because the radius of convergence of standard reciprocal space refinement techniques is small, it is important to obtain a reasonably good fit of atom positions in the initial model building.

Refinement and Analysis of Structural Models

Refinement is an iterative process that modifies the parameters of the model in a rational way to agree with the structure factor amplitudes derived from the observed diffraction intensities. This requires an error function that gauges the agreement between the structure factors predicted by the model and those derived from the observed data as well as an algorithm that alters the parameters of the model in such a way as to improve this agreement while maintaining a model that is both chemically and physically reasonable. Upon each iteration the model and the calculated phases improve, yielding better electron density maps that will allow the model to be extended into previously uninterpretable regions. These new electron density maps reveal more clearly the electron density for other well-ordered molecules such as waters of solvation, ions, and ligands. Unless extremely accurate experimental phases are obtained (e.g., from high-resolution MAD data), refinement is essential for the complete interpretation of the data.

Refinement Methods The error function used in protein refinement is typically either of the "least squares" or "maximum likelihood" type. Only the most cursory description of the mathematics of refinement can be presented here.

The least-squares method uses the sum-squared difference between the observed structure factors and those calculated from the model:

$$Q = \sum_{hkl} w_{hkl}(|F_o(hkl)| - k|F_c(hkl)|)^2 \qquad (10.17)$$

where $w(hkl)$ is a weight, often resolution dependent in macromolecular refinement, and k is a scaling constant that allows meaningful comparison between $|F_o|$ and $|F_c|$. The summation is over all m $|F_o|$, which constitutes a set of m equations. Each corresponding $|F_c|$ is a function of the n parameters of the model, which constitute n unknown parameters. The sum-square error function can then be represented as a system of equations in matrix form. For a given set of starting model parameters, the system of equations can be solved by iterative methods using a truncated Taylor expansion to represent $|F_c|$ as a function of the n parameters of the model and simplified block-diagonal least-squares matrices in which all terms that are not highly correlated (such as B factor and occupancy, if occupancy is being refined) are set to zero. Because of these approximations, multiple iterations, each corresponding to small shifts in the values of each model parameter, are required to reach convergence.

Maximum-likelihood methods are statistical and derive from Bayes's theorem. Bayes's theorem is expressed by the equation

$$P(A) * P(B|A) = P(B) * P(A|B) \qquad (10.18)$$

In words, Eq. (10.18) states: (the probability of A) times (the probability of B assuming that A is true) is equal to (the probability of B) times (the probability of A assuming that B is true). The corresponding expression for crystallographic refinement is

$$P(F_o) * P(\text{model}||F_o|) = P(\text{model}) * P(|F_o||\text{model}) \qquad (10.19)$$

It is more convenient to introduce F_c as a function of the model parameters and write

$$P(F_o) * P(|F_c|||F_o|) = P(|F_c|) * P(|F_o|||F_c|) \qquad (10.20)$$

However, the *prior probability* $P(|F_c|)$ and the *normalizing factor* $P(F_o)$ are constants and can be omitted without altering the essence of the relationship. The remaining terms are the *posterior probability* $P(|F_c|||F_o|)$ and the *likelihood* $P(|F_o|||F_c|)$. In the absence of the constants $P(|F_c|)$ and $P(F_o)$ this last term cannot be properly designated as a probability and is recast as the *likelihood* $L(|F_o|||F_c|)$ leaving the expression

$$P(|F_c|||F_o|) \propto L(|F_o|||F_c|) \qquad (10.21)$$

The total likelihood is expressed as the joint probability for all F_o:

$$L_{\text{total}} = \prod_{hkl} P(|F_{hkl,o}|||F_{hkl,c}|) \qquad (10.22)$$

but in practice the maximum-likelihood error function is often the negative log of the likelihood

$$-\log L_{\text{total}} = -\sum_{hkl} \log[P(|F_{hkl,o}|\,|\,|F_{hkl,c}|)] \qquad (10.23)$$

The precise functions used to calculate the likelihood are quite complex and will not be presented here.

Restraints and Parameterization in Refinement As described above, the relationship between observed diffraction data and the parameters of the model could be represented as a system of m equations (F_o) and n unknowns (parameters of the model that determine F_c). If the ratio of $m/n \geq 1$, the system is said to be determined or overdetermined and will have either one or infinitely many solutions. If the ratio of m/n is <1, the system is underdetermined and will have infinitely many solutions. In principle, this system of equations should have a solution corresponding to the exact structure within the crystal. In practice, because of experimental errors in the measurement of F_o, resolution limits of the data due to defects in the crystals, and limitations of our theoretical models, it is rarely possible to obtain a solution that defines the model precisely. In cases where the data are of atomic resolution (about 1.0 Å or better), it is possible to refine individual atoms according to their xyz coordinates, occupancy, and B factors (isotropic or anisotropic) and obtain an acceptable solution. The system of equations is determined or overdetermined with respect to these parameters.

When the data are not of atomic resolution, the system of equations is underdetermined with respect to the parameters of the model. An acceptable solution cannot be obtained unless additional restraints (or constraints) are added to the model in order to increase the effective number of observations that determine the model. Most of these restraints bear upon molecular geometry. Prior knowledge of the structure and chemistry of organic molecules from many sources (not the least of which are atomic resolution structures of organic compounds including peptides, amino acids, and nucleotides) allows the design of libraries of ideal values and standard deviations for the van der Waals radii, bond lengths, angles, dihedral angles, and planarity (of peptide bonds, aromatic rings, carboxylates, and amides) of the molecular structures of proteins and other macromolecules. These ideal values and standard deviations are applied as restraints that are designed to maintain a chemically and physically reasonable model during refinement. These restraints take the form of new terms added to the refinement error functions.

Generating proper library definitions and restraints for novel ligands is not a trivial problem, and an accurate definition is a prerequisite for correct interpretation of the data. It requires complete knowledge of the chemical structure, bonding, hybridization, protonation, ionization, and chirality. However, most common representations of molecules are highly simplified abstractions that do not properly describe the true nature of the chemical species involved.

In some cases only careful experiments or a high-resolution crystal structure of the compound will resolve uncertainties. In the absence of such data some guesses must be made. The standard two-dimensional chemical structure representation of the anticancer drug topotecan (Hycamptin) is given in Figure 10.21 as an example. It is not necessarily obvious from the two-dimensional representation of the structure that N3 of the tertiary amine substituent is sp^3 hybridized and could be protonated depending upon the pH, C10 is sp^3 hybridized and is not part of the conjugated ring system, or that the pyridone ring (D) is aromatic. This information was determined partly from standard chemical knowledge and partly from the crystal structure of a related compound, camptothecin iodoacetate [64].

Hydrogen atoms are usually excluded as part of the refined model unless the resolution is exceptionally high. This is due to their small contribution to X-ray diffraction (only one electron) and their large contribution to the total number of parameters in the model. Some refinement packages allow the model to be refined with riding hydrogens (hydrogens placed in standard positions with standard bond lengths and angles) added to the model. The hydrogens are not refined but do have an effect on refinement, somewhat in F_c, but mainly in the van der Waals terms, which can lead to improved distributions of torsion angles.

Further restraints can be added to take advantage of correlated vibrational properties. Simple B-factor restraints account for the physical reality that the vibrational amplitudes of covalently bonded atoms can be correlated. A recent advance in model parameterization has been the introduction of TLS groups in refinement [65]. TLS groups are substructures of the model (sometimes entire domains) that have correlated vibrational properties. These correlated vibrations are described by three tensors that designate the translational (T), librational (oscillation) (L), and screw (S) components of the motion. TLS groups account for an entire continuum of conformations of part of the model in a parsimonious manner with respect to the number of parameters required.

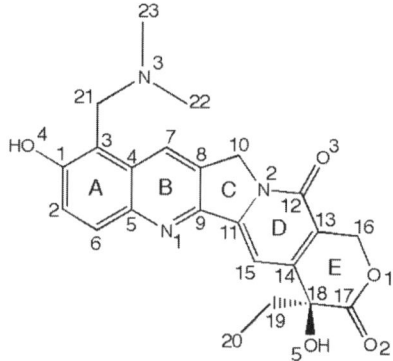

Figure 10.21 Standard two-dimensional chemical structure representation of topotecan.

GENERATION AND ANALYSIS OF STRUCTURAL MODELS

The greatest difficulty here is the definition of the TLS groups. Unless the data are of good enough quality and high enough resolution to properly refine anisotropic B factors, it is difficult to determine which regions of the structure have correlated vibrations (and in this case TLS groups will not be necessary). In general educated estimates and trial-and-error must guide the choice of TLS groups. An example of TLS group definition with the refined T, L, and S tensors is shown below.

```
REMARK   7 TLS DETAILS
REMARK   7 NUMBER OF TLS GROUPS :    2
REMARK   7
REMARK   7 TLS GROUP :     1
REMARK   7   NUMBER OF COMPONENTS GROUP :   1
REMARK   7   COMPONENTS       C SSSEQI    TO C SSSEQI
REMARK   7   RESIDUE RANGE  :   A   865      A  1110
REMARK   7   ORIGIN FOR THE GROUP (A):  45.6660  11.3419  41.2873
REMARK   7   T TENSOR
REMARK   7     T11:    0.0157 T22:     0.0641
REMARK   7     T33:    0.0294 T12:     0.0162
REMARK   7     T13:    0.0031 T23:     0.0101
REMARK   7   L TENSOR
REMARK   7     L11:    0.8473 L22:     1.0405
REMARK   7     L33:    1.1744 L12:     0.3695
REMARK   7     L13:   -0.0069 L23:     0.3502
REMARK   7   S TENSOR
REMARK   7     S11:   -0.0245 S12:   -0.0044 S13:     0.0036
REMARK   7     S21:   -0.0046 S22:    0.0421 S23:    -0.0484
REMARK   7     S31:   -0.1460 S32:    0.0620 S33:    -0.0176
```

If more than one chemically identical molecule is present in the asymmetric unit, improper or noncrystallographic symmetry (NCS) restraints (or constraints) can be applied that force the conformations of the molecules to be similar (or identical). If constraints are used, the molecules are forced to be identical and the effective number of the constrained parameters of the model is reduced by a factor of n for n-fold NCS. If restraints are used, the reduction in the effective number of parameters is less clear. Noncrystallographic symmetry is only useful in cases were the data are of poor quality or rather low resolution (about 3.0 Å or worse). In general the packing interactions of the NCS-related molecules will be different, and real differences in conformation of both the main chain and side chains can be expected and should not be ignored unless the data is of insufficient quality to observe them.

The parameters of the model discussed so far bear only upon the structure of the ordered phase of the crystal, but the disordered phase (solvent region) of the crystal, which can be 30 to 80 percent of the unit cell volume, also contributes to the intensity of F_o, particularly at low resolution. A suitable model for this contribution is required and can make significant improvement in the difference electron density of unbuilt regions of the model. The solvent X-ray scattering in essence introduces a contrast that significantly decreases the observed diffraction intensities relative to what would be observed if the disordered region of the crystal were vacuum. The time-average electron density in the solvent region of the crystal is constant and can be represented by the parameter K_{sol}, the ratio between the average solvent electron density and the average protein electron density. A second parameter B_{sol} is effectively a thermal factor that is used in a resolution-dependent exponential scaling term: Because different regions of the solvent do not scatter in phase, the intensity of solvent scattering is very weak at high resolution but quite strong at low resolution. The most common bulk solvent scaling function used is based upon Babinet's principle [66], which states that the Fourier transform of a mask and its complement have structure factors of the same amplitude but phases of opposite sign. For this method only the two parameters K_{sol} and B_{sol} are introduced to the model and only the amplitudes of the structure factors are changed. This function is used by most current refinement packages. A second method is the mask method in which the phases for the solvent scattering are calculated from a grid of dummy atoms filling the solvent region of the model. For this method K_{sol} and B_{sol} are also determined, but both the amplitudes and the phases of the structure factors are altered. This method is used by the XPLOR/CNS [57] refinement package.

Weighting and Judging Progress of Refinement The model restraint terms and the X-ray diffraction terms of the refinement error function both act upon the fundamental parameters of the model (xyz, B, occupancy). An overall weighting factor balances the contributions of these opposing aspects of the error function. Each restraint term is also given a weight that can be adjusted to give the desired final root mean square (rms) deviations from the ideal (library) values. For example, in least-squares refinement the error function with geometric restraints can be represented as:

$$Q = W_{X-ray}\left[\sum_{hkl} w_{hkl}(|F_o(hkl)| - k|F_c(hkl)|)^2\right]$$
$$+ \sum_{distance} w_{dist}(D_{ideal} - D_{model})^2$$
$$+ \sum_{angle} w_{angle}(\theta_{ideal} - \theta_{model})^2$$
$$+ \sum_{torsion} w_{torsion}(\chi_{ideal} - \chi_{model})^2 \quad (10.24)$$

and so on for all restraint terms

Proper choice of weights is essential for obtaining the best fit to the observed data while maintaining a chemically and physically reasonable model. The agreement between the model and the observed data is judged by the residual or R_{factor} defined as:

$$R = \sum ||F_o| - k|F_c|| / \sum |F_o| \quad (10.25)$$

The R_{factor} gives a normalized sum-error between the observed and calculated structure factors and is often expressed as a percentage. As the model comes to give a better representation of the observed data, $k|F_c|$ will approach the value of $|F_o|$ and the numerator in Eq. (10.25) will decrease. Thus the overall R_{factor} will decrease as refinement progresses. The R_{factor} is the most commonly sited metric of correctness of fit for refinement, but many others including correlation coefficients, figures of merit, $\sigma(A)$, and some variants of the R_{factor} are also often calculated and output by refinement software. Both least-squares and maximum-likelihood methods allow estimations of the standard uncertainties of the xyz coordinates, occupancies, and B factors. Lower standard uncertainties should correlate with improvements in R_{factor} and the other metrics. All refinement programs will output a list of rms deviations from ideality for all restraint terms. These data are a crucial guide for achieving a properly refined model and for judging the overall X-ray vs. geometry weight. In addition to the overall statistics for restraints, refinement programs will also list specific bonds, angles, torsions, and the like that deviate significantly (greater than three standard deviations at least) from the ideal values. These outliers often indicate regions of the model that require further interactive adjustment before refinement will converge.

In addition to judging the progress of refinement toward the best-fit model for the data, the refinement results must be used to gauge the appropriateness of the parameterization of the model. If the observation/parameter ratio is low, it is possible to overfit the model and produce a refined structure that appears to fit the data much better than it actually does. This is analogous to fitting a two-exponential decay with a three-exponential equation: The fit might very well be better, but the model parameters obtained will not represent the true nature of the data. An obvious case of this problem would be refinement with anisotropic B factors against data to no better than about 2.5 Å. The extra parameters will allow an apparent better fit to the data, but they will not represent true individual atom anisotropic thermal motion because it is not properly represented in the data.

In most cases the choice of model parameterization requires more subtle judgment, especially for restraints and weights. The free R_{factor} (R_{free}) was introduced as a means of validating the choice of model parameterization. R_{free} is a standard R_{factor}, but it is calculated against a small subset of the data (5 to 10 percent) that is set aside and not used in model refinement. The agreement between these data and the model, therefore, provides an unbiased gauge of the appropriateness of the model parameterization. If the model parameters are appropriate, the R_{factor} and R_{free} will both decrease as refinement progresses and the agreement between the model and the working data is improved. If the data are being overfit the R_{free} remains static or increases even though the R_{factor} continues to decrease. The standard R_{free} is a limited application of the notion of the jackknife test in which data are binned randomly and the model fit many times. Each time the model is fit with a different bin omitted and the result tested for agreement against the omitted data. This prevents the interpretation of the refinement being skewed by an unfortunate initial choice of test data for R_{free} calculations.

Analysis and Preparation of Structural Models

No refined crystallographic model is without errors and even the best refined model is unlikely to be an ideal starting place for computational work without some further modification. A number of protein structure validation software packages are available (ProCheck, WhatIf) to assist in locating and annotating errors. A first step is to make certain that the model represents the expected sequence. It is not uncommon to encounter residues in the model (commonly surface-exposed residues such as lysine, arginine, glutamate, and glutamine) that are represented as alanine because the side chains could not be distinguished in the electron density maps. In other cases, the residue might be represented with the proper name but have missing atoms. In high-resolution structures some residues might be modeled in several alternative conformations. For most computational modeling only one of the conformations can be chosen. Last but not least, inverted chiral centers must be identified and corrected. Some of this information could be included in the header of a standard PDB file. In addition, a PDB file typically represents the contents of the asymmetric unit of the crystal. In some cases the biologically relevant molecule could be only part of the asymmetric unit or it could span several asymmetric units. If the biologically relevant molecule does span more than one asymmetric unit, then the appropriate matrices for generating the complete molecule are most often included in the header of the PDB file.

Primary literature reporting protein X-ray crystal structures usually include a table, often called Table 1, summarizing the X-ray diffraction data. Results for the structural model refinement are also sometimes included in this table. A typical Table 1 is shown on page 433.

Geometric refinement statistics are a good indication of the overall quality of the model, but they can also mask specific instances of significant deviation from reasonable values. Geometric outliers (bonds, angles, dihedrals, and planes that deviate significantly from ideal values) are the first indication of problem areas of the structure. Any restraint that differs by more than 10 σ from that of the ideal value should be examined. A second crucial indicator of appropriate geometry is the Ramachandran plot [67]. The Ramachandram plot is a plot of the φ (C—N—C_α—C) and ψ (N—C_α—C—N) torsion angles of the peptide chain for each residue in the model. Each type of regular secondary structure has characteristic peptide chain geometry that corresponds to a particular region of the Ramachandran plot. In addition, steric hindrance restricts the peptide torsions to certain regions of the Ramachandran plot. When peptide torsions are observed outside of these regions, it is a clear indication of a problem in the model. One of the most common problems is a peptide flip in which the carbonyl function of the peptide chain should be pointed roughly in the opposite direction from its orientation in the model.

X-ray source	Cu rotating anode
X-ray detector	R-AXIS IV
Wavelength (Å)	1.54
Detector distance (mm)	250
$\Delta\phi$ per image (°)	1
Space group	$P2_12_12_1$
Unit cell	
a (Å)	62.7
b (Å)	64.2
c (Å)	107.7
Total reflections	231,651
Scaling resolution (Å)	50–2.87
Unique reflections	10,463
Overall completeness (%)	99.9
Last shell completeness (%)	99.9
Overall R_{sym}	0.050
Last shell R_{sym}	0.512
Overall I/σ	22.9
Last shell I/σ	2.4

Correcting the error and briefly refining or regularizing the model should bring the torsions back into the allowed regions. It should be noted that one rare but real feature of protein structure called the γ turn [68] can place residues into an apparenlty strongly disallowed region of the Ramachandran plot (φ = ~75°, ψ = ~−60°). Cis-peptides, almost always prolines, also have a distinctive signature in a Ramachandran plot. Prolines are particularly restricted due to the covalent bond between N and C_δ, while glycine, due to its lack of a side chain, has almost no restrictions on its peptide torsions.

In preparation for computational work particular attention should be paid to the protonation state and hydrogen bonding interactions in the model. The precise side-chain rotamers of histidine, glutamine, asparagine, and arginine residues define their hydrogen possibilities. However, the correct rotamers are not always clear in the electron density maps, and often there is not a great deal of attention paid to the appropriateness of possible hydrogen bonding interactions during refinement. In many cases the correct arrangement is not clear or several possible arrangements seem likely, and a best guess must be made.

Crystallography–Drug Discovery Interface

Protein crystal structures provide essential information for docking algorithms, ligand-growing algorithms, and for generation of structure-based pharmacophore models. Analysis of protein structure models could also reveal novel or unexpected binding sites (or even accessory pockets to expected binding sites) that might not be exploited by conventional SAR. This information could then be used to guide synthesis of new inhibitors that take full advantage of the potential for favorable interactions with the target protein. Many algorithms have been developed for active site and potential binding site identification based on electrostatic, hydrogen bonding, and hydrophobic interaction potentials. Other algorithms analyze surface topology for grooves and clefts that could accommodate small-molecule ligands of certain shapes.

X-ray crystallography is the only experimental technique that provides detailed knowledge of the structure of ligand binding sites in proteins. Knowledge of binding site structure opens the door to physical interpretation of structure–activity relationships among inhibitor compounds, and so it is an invaluable aid in drug discovery and design. The greatest value in crystallography comes from crystal structures of actual protein–ligand complexes. These crystal structures reveal the precise binding mode of compounds or even entire classes of compounds as well as revealing consequent conformational changes in the protein. In many cases the ligand-bound conformation of the protein is a much more useful target for computational drug design techniques than is the unbound conformation. Docking and ligand-growing algorithms, scoring functions, and molecular mechanics calculations are most often exquisitely sensitive to small changes in atomic coordinates of the binding site, as are real molecular interactions. Thus unbound conformations, homology models (unless derived from very highly homologous crystal structures), and other theoretical models of active sites are generally of less value in detailed computational modeling because they do not precisely represent the active site structure to which the ligands actually bind. Approximate models of binding sites can be of use in the generation of structure-based pharmacophore models, but the detailed structure of the actual ligand binding site is certainly more appropriate in most cases.

10.6 EXAMPLES FOR THE USE OF X-RAY CRYSTALLOGRAPHY IN DRUG DISCOVERY

Structure-based drug design is an integral tool in lead optimization of drug discovery programs in all major drug companies. In most cases the structure of the target protein with and without small molecules is investigated with X-ray crystallographic techniques, while nuclear magnetic resonance (NMR) techniques are frequently used to determine specific interactions of small molecules with their target molecules. The recent technical improvements seen in many areas of X-ray crystallography allow researchers to determine crystallographic structures of their target protein in complex with a new ligand within hours to days at moderate cost. This capability helps accelerating the work cycle of lead optimization, where new leads are synthesized and precise interactions on the atom level are being determined, to time scales of less than a single week. The work flow of structure-based drug design including target structure based in silico screening and lead optimization is lined out in Figure 10.22. In recent years fundamentally new applications such as crystallographic

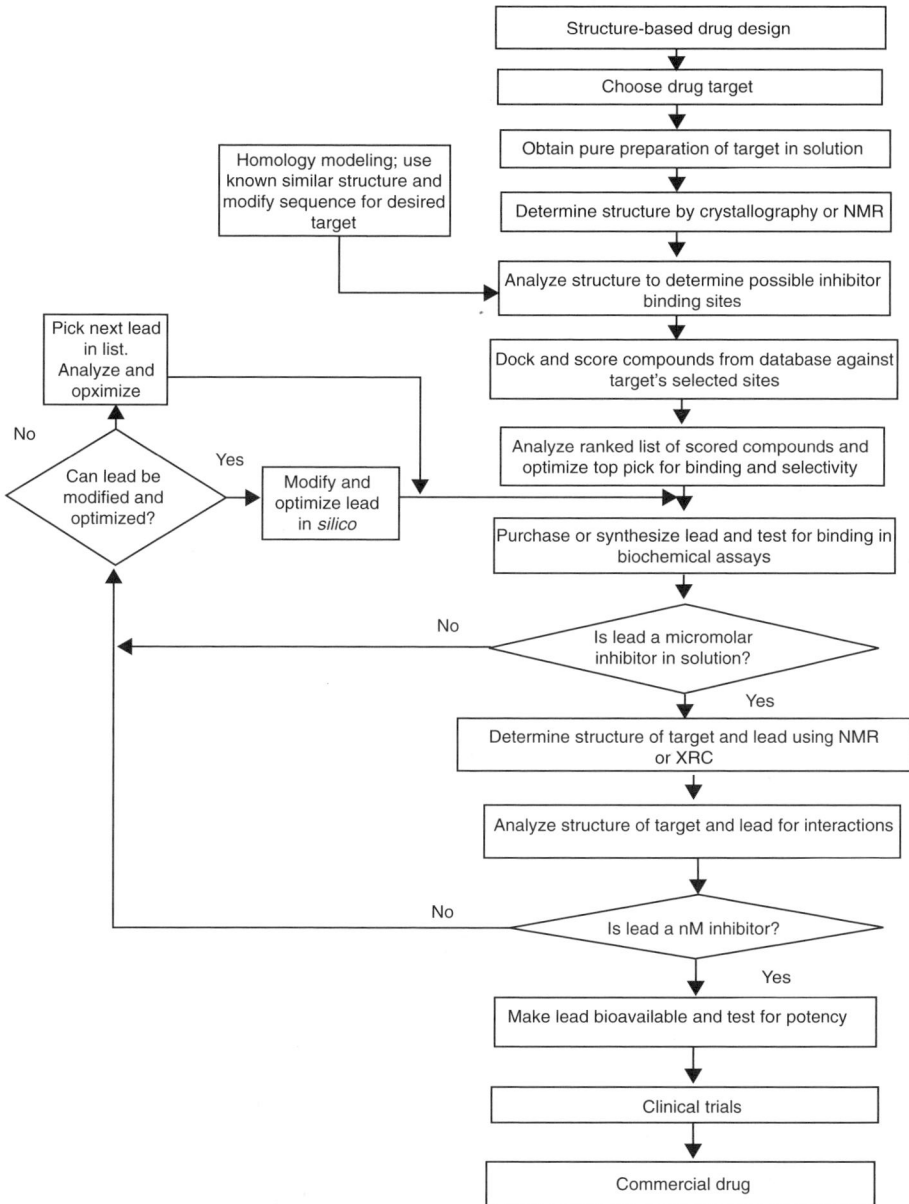

Figure 10.22 Schematic work flow of structure-based drug design. (Figure taken from Anderson [4].)

screening have been developed that apply X-ray crystallography as a versatile and instructive tool at different stages of the drug discovery process. Thus, X-ray crystallography can be applied, with different intentions, throughout the entire lead discovery and optimization process. The impact of structure-guided drug design on clinical agents is well documented [69, 70]. The following examples document the use of X-ray crystallography from its "classic" application in lead optimization to various specific applications in drug discovery (Fig. 10.1).

Lead Optimization—Structure-Based Drug Design

The process of incrementally changing a compound based on the analysis of its three-dimensional protein–drug complex is generally referred to as *structure-based drug design* (SBDD) [71]. The SBDD process is usually organized in a circular work flow (Fig. 10.22) and often driven by a collaboration of several groups within an organization that are proficient in several research disciplines. At least two groups are required, one responsible for chemical synthesis and the other for X-ray crystallographic analysis. Often additional researchers in biology, pharmacology, and modeling are involved. They supplement a drug discovery effort with testing of efficacy and toxicity of newly designed and synthesized compounds. The SBDD cycle often starts with a lead compound originating from a natural product source or one that has been identified by screening real or virtual compound libraries. A structure of the protein target in complex with the lead compound initiates the SBDD cycle. Chemical modifications are proposed based on the analysis of this structure, and new compounds are synthesized from such chemical hypotheses. The most important design factors are (a) a perfect complementary geometric fit of the ligand with that of the binding site, (b) low-energy conformations of ligand and protein, (c) complementary electrostatic potentials, (d) the formation of ionic or hydrogen bonds between functional groups, and (e) maximization of the hydrophobic contact area at the ligand–target interface. Since the binding of a small molecule to a protein surface is a complex phenomenon, specific predictions of the effect of certain changes of the ligand molecule are fraught with uncertainty. Generally, affinities increase with an increase of the hydrophobic interface. The free energy contribution of H bonds pays tribute to the balance of solvation energies and that of the formed or broken H bond. Even seemingly small alterations of single-ligand functionalities may have unexpected consequences. Therefore, newly synthesized compounds need to be tested for efficacy and toxicity, and new ligand–target complex structures need to be determined. A thorough analysis of these results stimulates a new round of SBDD. Generally, computer-aided design is lead by the need to optimize or maintain protein binding as well as enhance efficacy and decrease toxicity or other undesirable effects. Eventually new compounds are synthesized

EXAMPLES FOR THE USE OF X-RAY CRYSTALLOGRAPHY 437

and the SBDD cycle is repeated until an acceptable compound is discovered or the project is canceled.

Arguably the first marketed drug whose generation was significantly impacted by X-ray crystallography is the angiotensin-converting enzyme (ACE) inhibitor Captopril from Bristol Myers Squibb. Curiously though, structures of the actual target apo-ACE or those of ACE in complex with inhibitors were not available. Instead, in the 1970s investigators used the X-ray crystallographic structure of the closely related zinc protease bovine carboxypeptidase A in complex with benzylsuccinate to create a homology model of ACE. Using this homology model and the structure of the inhibitor complex investigators rationalized the synthesis of new drug compounds, eventually creating Captopril (Fig. 10.23), a highly selective compound with tight binding affinity and low side effects. Captopril was approved by the Food and Drug Administration (FDA) and marketed in 1981. Today, Captopril has undergone the usual lifecycle of approved drugs and is available as a generic.

The first successful drug to reach clinical use that was based on an SBDD development cycle is claimed by Merck Pharmaceuticals. Trusopt (generic dorzolamide) is an inhibitor of carbonic anhydrase II. The discovery of Trusopt was stimulated by X-ray crystallographic structures of carbonic anhydrase II in complex with acetazolamide and other sulfonamides. Rational design originating from the zolamide scaffold assisted in the discovery of inhibitor compounds with activities increased by three orders of magnitude (Fig. 10.24).

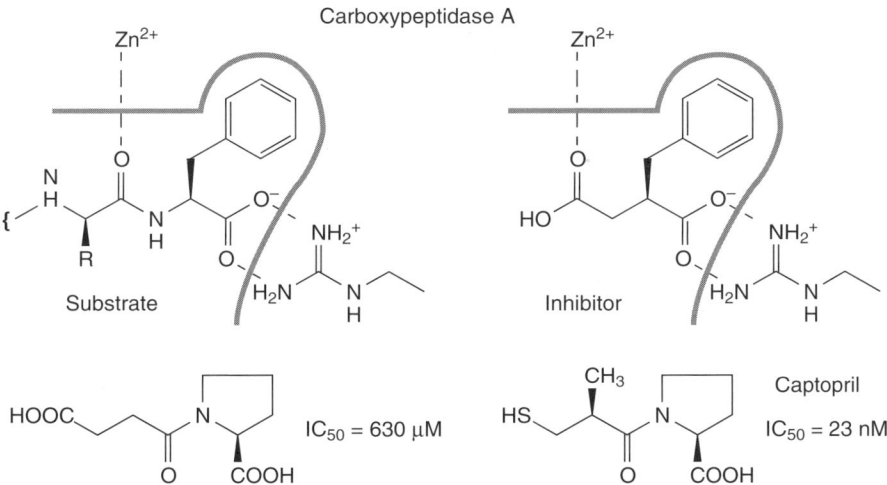

Figure 10.23 X-ray crystallographic data of carboxypeptidase A shows that a guanidinium group in an arginine residue and a zinc ion are crucial for complexing the natural substrate (top left) and the inhibitor compound benzylsuccinate (top right). The optimization of the first ACE lead N-succinoyl-prolin (bottom left) to Captopril (bottom right) was aided by rationalization of the binding interaction in the homologous enzyme. (Taken with permission from Kubinyi [72]).

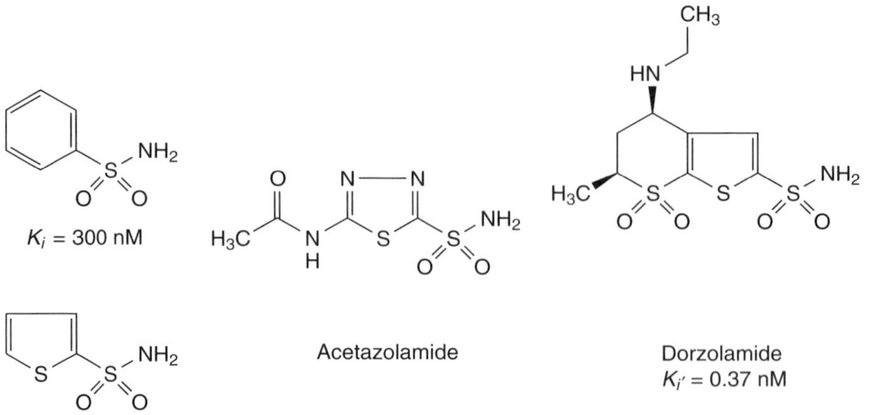

Figure 10.24 Carbonic anhydrase II inhibitors. Structures of carbonic anydrase in complex with moderately active compounds led to the discovery of the highly active compound dorzolamide.

HIV Protease Inhibitors The most profound impact that X-ray crystallography has had on current clinical practice and public health was the discovery of the human immune virus (HIV) protease inhibitors. HIV protease is a crucial enzyme important for replication of HIV. The HIV protease processes two of the three gene products of HIV into active enzymes and is required for viral reproduction. Inhibition of the enzyme can prevent HIV replication. Pharmaceutical companies had a strong background in the development of protease inhibitors, as many therapeutic targets are also proteases. The first X-ray crystallographic structure of HIV protease was published in 1989 and, within 8 years four separate compounds from four different companies, each developed using SBDD, were approved by the FDA for HIV treatment. The convergence of several unique circumstances aided this remarkable achievement: The expertise in pharmaceutical companies working on aspartyl proteases coupled with an intense public interest and generous financial support by the U.S. government provided the extraordinary context. Crucially, HIV protease is an enzyme amenable to facile co-crystallization with drug compounds, and the availability of appropriate structural information at a timely point in several drug discovery projects allowed a concentrated SBDD focus.

Figure 10.25 Pathway of the discovery of Mozenavir. Dupont's initial HIV protease inhibitors (compound 7) resulted from a computer-based three-dimensional search of known chemical entities using a pharmacophore hypothesis (top left). Scaffold entities 8, 9, and 10 were used as a base compound to add functional groups on the P1 and P2 pseudosymmetric compounds. The lower half of the panel shows the final compound DMP-450 (Mozenavir) with distances in Å to protein atoms of the two molecules of HIV protease in the crystal structure.

P1 ↔ P1′ 8.5–12.0Å

3.5–6.5Å

H-bond donor/acceptor

7 (DuPont Merck)

8

9

10

Gly-48

Ila-50 Ila-50′

Gly-48′

Asp-30

Asp-30′

Asp-25

11
(DuPont Merck)
K_i = 0.018 nM

Asp-25′

The discovery path of Mozenavir, an HIV protease inhibitor, nicely illustrates the SBDD process. Dupont began its HIV protease investigation using a pharmacophore hypothesis based on the X-ray crystallographic structure of the apo-enzyme HIV protease. This pharmacophore consisted of two lipophilic groups separated by 8.5 to 12 Å and coupled with one hydrogen bond donor/acceptors at a distance of 3.5 to 6.5 Å (Fig. 10.25). Extensive searches of the Cambridge Structural Database, composed of three-dimensional structures of small-molecule compounds, revealed a potential lead compound. The initial lead was pseudosymmetrical, with the two ends of the molecule being virtual mirror images of each other. This property was significant because it was known that HIV protease functions as an obligatory dimer. From the crystal structure it was known that the interface of the two proteins is composed of identical residues from each protease molecule, suggesting that inhibitor compounds may contain symmetrical moieties. Co-crystal structures with initial lead compounds revealed an important feature of the potential inhibitor binding site. One crucial observation was the realization that the methoxy groups of compound 7 (Fig. 10.25) could replace an ordered water molecule. Different scaffold compounds were then chosen to add methoxy- as well as other functional groups. Iterative SBDD cycles continued until several molecules were identified that had nanomolar affinities and favorable pharmacokinetic properties. Clinical development of the final compound, DMP450 (Mozenavir) was turned over to Triangle Pharmaceuticals and stopped in 2002 due to Mozenavir's side effects [70].

Protein Kinases Protein kinases are therapeutic targets for a variety of diseases. More than 500 kinases have been identified in the human genome, they are enzymes dependent on adenosine 5'-triphosphate (ATP) that phosphorylate other proteins. Most kinases act in cell-signaling pathways, phosphorylating other signaling proteins whose activity is then either turned on or turned off as a result of the attached negatively charged phosphate group. The catalytic domains of most kinases are structurally conserved; however, their mechanisms of regulation are distinctly different. The catalytic domains are composed of a bi-lobed structure consisting of a helical domain and a β-strand domain (Fig. 10.26). ATP is bound at the interface of the two domains, and this binding site has been deemed an attractive site for designing ATP competitive inhibitors that block ATP binding, thus preventing phosphorylation. Unfortunately, the ATP binding site is highly conserved in most kinases, thus rational design faces serious challenges in creating inhibitors selective for a specific kinase with little inhibition of other closely related kinases. SBDD of kinase inhibitors is popular because of the low experimental barriers. Conserved and unique residues can be identified in the inhibitor binding site and rational design attempts to maximize interactions with unique residues in the inhibitor binding site while minimizing interactions with conserved residues. The underlying hypothesis is that specificity can be created as interactions with the inhibitor are increased with residues specific for the kinase in question while cross-reactivity decreases for other kinases.

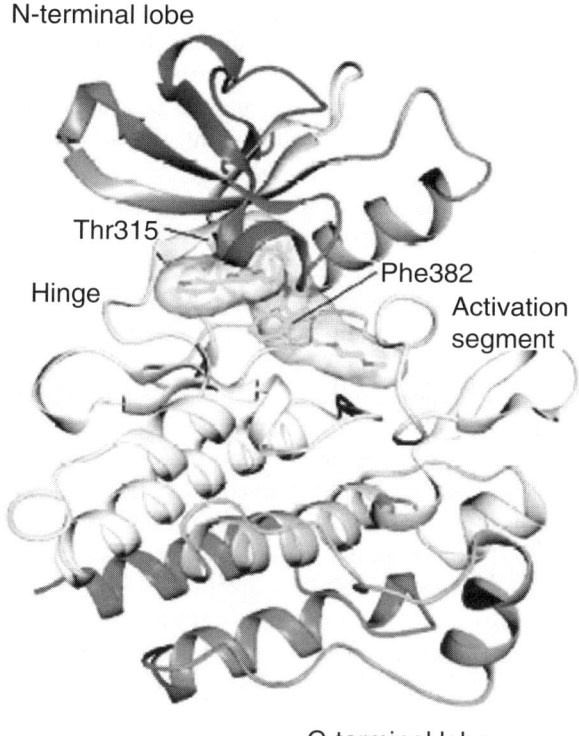

Figure 10.26 Overview of the structure of a kinase with an inhibitor bound. Kinases offer the possibility to design inhibitors based on stabilization of inactive conformations. (This figure is available in full color at ftp://ftp.wiley.com/public/sci_tech_med/drug_discovery/.)

Antistructures

As discussed, the main applications of X-ray crystallography in drug discovery and optimization projects are based on the analysis of the target and the interactions of targets with their ligands. The design efforts are aimed at strengthening the resulting complexes. The opposite approach, however, weakening the interaction of drug leads with some proteins may be used to one's advantage. The interacting proteins are not the actual targets but those that cause detrimental effects in drug efficacy. Such proteins are sometimes called *antitargets* and may be related enzymes with similar substrate binding pockets but with very different function, such as kinases or phosphatases.

Generally, weak binding of small-molecule drugs to serum albumin and detoxification proteins such as cytochrome P450 is a desired property. P450 enzymes are involved in the oxidative metabolization of most drugs and are often the source of drug-related side effects or their toxicity. Several P450 structures are available [73] and may be used for in silico docking studies and the published crystallization methods may be used to grow crystals for soaking or co-crystallization studies. The goal of such projects is to increase lead efficacy by defining lead modifications that weaken the interaction with antitarget proteins while maintaining the affinity to the target protein. It would be highly desirable to extend further this negative template design, that is, by including membrane-bound drug transporters, to modulate, based on molecular structural insight, the critical processes of absorption, distribution, metabolism, excretion, and toxicity. These possible applications clearly demonstrate that X-ray crystallography is headed toward effecting later stage discovery and early stage drug development projects.

Human serum albumins consist of three domains with six rather promiscuous ligand binding sites. Many drug leads bind to this protein causing a serious problem for lead discovery. In an instructive example, the feasibility of its "design away" approach was demonstrated for diflunisal, a nonsteroidal antiinflammatory cyclooxygenase inhibitor. Note that 99 percent of diflunisal in serum is unavailable to the target due to binding to human serum albumin. This requires high doses of up to 250 mg diflunisal to be administered, causing gastrointestinal irritation as a serious side effect. In a structure-based drug design effort diflunisal analogs were synthesized that were deemed to bind less efficiently to HSA-III (a human serum albumin subdomain). Several compounds were generated that exhibited more than 100-fold reduced binding to HSA-III (with only 10-fold reduction in affinity for full-length albumin). Significantly, several of these compounds maintained their activity against the actual target, cyclooxygenase-2 [74].

Protein Therapeutics

Protein therapeutics such as EPO and insulin are the hallmark of this fast-growing class of drugs. Hardly any new optimization program excludes the use of X-ray crystallography. Based on natural products, protein therapeutics are directly amenable to X-ray crystallographic investigation and subsequent redesign. Classes of protein or peptide therapeutics are monoclonal antibodies, cytokines, enzymes, and viral fusion inhibitors. In a landmark study Ewert et al. [75] used X-ray crystallographic structure-based antibody engineering to aid in the identification of residues that improve unsatisfactory antibody properties [75]. In a different case, an HIV entry inhibitor was designed based on CD4. The 27-amino-acid CD4 mimic interacted with gp120 and is bound to HIV particles with CD4-like affinity. This mini-CD4 is a prototype HIV-1 inhibitor and a potential component for vaccine formulations [76].

In silico Screening Based on Crystallographic Structural Models

X-ray crystallographic structures of proteins may be used to preselect a small number of compounds from compound libraries. Different types of computer-based algorithms such as docking [77] may be employed to predict the formation of ligand–protein complexes with single compounds or with entire compound libraries ("virtual screening" or "in silico screening"). This computational approach is particularly advantageous for targets where target structural information is readily available at no cost (i.e., in the public domain). A side-by-side comparison of assay-based high-throughput screening (HTS) with such virtual screening on the same target protein has been described by Doman et al. [5]. Numerous structures of PTP1B (tyrosine phosphatase 1 B), a diabetes type II drug target, are available from the Protein Data Bank. Molecular docking of ca. 235,000 compounds into the closed, ligand-removed structure of PTP1B yielded 365 high-scoring candidates that were tested for inhibition in enzyme-based assays. Of these candidates 34.8 percent inhibited PTP1B with IC_{50} values below 100 µM, representing an enrichment of 1700-fold (as compared to random screening). Conventional high-throughput screening on the other hand, yielded only 85 hits with IC_{50} values below 100µM out of 400,000 tested molecules (corresponding to a 0.021 percent hit rate). Interestingly, the hit lists were rather different and the hits generated by molecular docking appeared more druglike than the HTS hits, suggesting that the two screening techniques may be used in a complementary way. Indeed, the integration of virtual and high-throughput screening is judged to be a promising approach in modern lead discovery projects [78].

Crystallographic Screening

X-ray crystallographic screening is a modern combination of lead identification, immediate X-ray crystallographic structural evaluation, and subsequent lead optimization on crystals that are soaked with ligand mixtures. Ingeniously, the binding capability of some proteins in crystals is employed to "fish" and present tight binders. Those ligands with the highest affinity are identified and selected for further rounds of optimiziation [79]. Compared to conventional high-throughput screening, crystallographic screening (a) yields hits with activities in the mM to 30 µM range, (b) yields hits with evidently defined binding interactions, and (c) involves only minimal hit-to-lead synthetic chemistry efforts because follow-up libraries can be focused using detailed information from the crystal structure. Nienaber et al. [1] demonstrated that crystallographic screening can be performed in a rapid, efficient, and high-throughput fashion. They established the utility of the iterative process by discovering 8-aminopyrimidyl-2-aminoquinoline, a new class of anticancer urokinase inhibitors (Fig. 10.27). Initially, 61 compounds were divided into 9 separate mixtures with 6 to 8 compounds each. Care was taken to distribute into a particular cocktail those compounds that had the greatest degree of structural diversity in order to facilitate subsequent ligand identification based on the shape of F_o-F_c electron density maps. Nine urokinase crystals were soaked with individual cocktails, X-ray diffraction data was generated, and the resulting electron density maps were inspected. They showed the shape and orientation of compounds (Fig. 10.27 a). In one case two binders were present. Here, the removal of the prominent compound and resoaking allowed to identify the electron density of the second ligand. In the subsequent lead optimization step previous structure–activity relation ship (SAR) data was

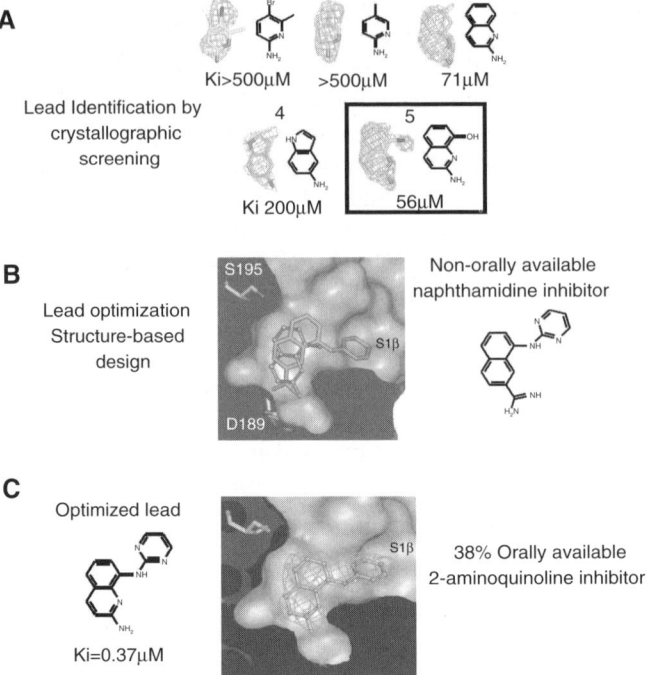

Figure 10.27 Urokinase lead identification via crystallographic screening and optimization [1]. (a) Initial F_o-F_c electron density maps for ligands that were identified from compound cocktail-soaked urokinase crystals. (b) Crystal structures of 8-aminopyrimidyl-2-naphtamidine (orange) and a 2-aminoquinoline lead (blue). (c) Structure and $2F_o-F_c$ electron density map for the optimized lead compound 8-aminopyrimidyl-2-aminoquinoline. (This figure is available in full color at ftp://ftp.wiley.com/public/sci_tech_med/drug_discovery/.)

included and lead to the development of 8-aminopyrimidyl-2-aminoquinoline, a ligand with a ca. 100-fold increased inhibitor potency (K_i = 0.37 μM) and a 38 percent oral availability, as determined by in vivo pharmacokinetic tests. This type of process is capable of identifying weaker binding ligands (1 mM) and is applicable where apo-crystals are available and tolerate soaking. Crystallographic screening may also be used to facilitate the validation of new targets, the development of assays and assist in assigning biochemical function to orphan targets.

Crystallographic Fragment Screening

A variation of this theme is crystallographic fragment screening. Here crystals are soaked with cocktails that contain small druglike fragments rather than complete leadlike compounds. Once several fragments are identified crystal-

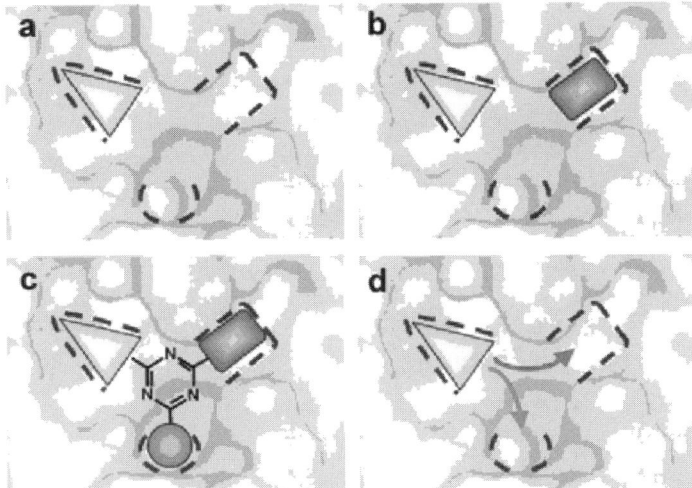

Figure 10.28 Schematic crystallographic fragment screening. Once fragments are identified (*a, b*) they can be joined (*c*) resulting in a leadlike compound or fragments may be developed along the lines of conventional structure-based drug design. (Figure taken from Blundell et al., 2002 [80].) (This figure is available in full color at ftp://ftp.wiley.com/public/sci_tech_med/drug_discovery/.)

lographically, they can be developed into new lead compounds (Fig. 10.28). Curiously, low-affinity small fragments that bind adjacent binding pockets can be joined and result in a larger molecule with increased affinity. Typically fragment libraries consisting of only a few hundred to a thousand compounds are screened. Crystallographic fragment screening is a new and promising technology employed by several biotechnology companies; however, specific examples for drug discovery have not yet been published in the scientific literature.

Rees et al.[81] discuss 25 examples for the successful application of the fragment-based lead discovery approach, some of them aided by crystallographic screening. They also formulate a "Rule of three" in which the average fragment is characterized as (a) having a mass of less than 300 Da, (b) having less than or equal to 3 hydrogen bond donors, (c) having less than or equal to 3 hydrogen bond acceptors, and (d) having a $c \log P$ of 3. In addition, the number of rotatable bonds was on average less than or equal to 3 and the polar surface area was about 60Å^2.

Site-Directed Leads via Fragment Tethering

An additional layer of complexity is added by generating site-directed leads via fragment tethering ([2]; Fig. 10.29). In a first step target proteins are covalently modified at a particular site on the surface. Mass-spectrometric detection allows the identification of weakly binding ligand precursors. In a second

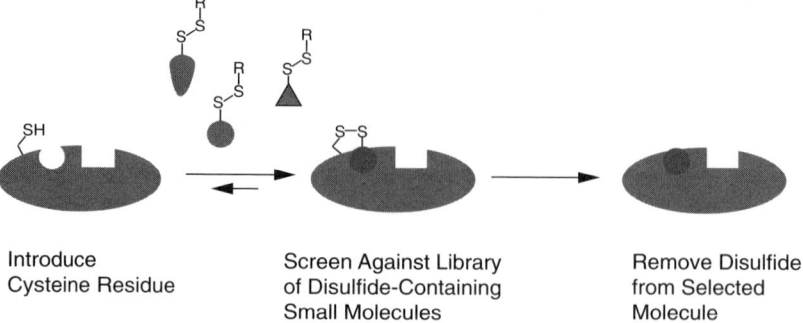

Figure 10.29 Schematic illustration of the fragment tethering approach.

step X-ray crystallography provides the tool to observe the interaction of the precursor with the protein and helps directing the chemical synthesis of fused analogs with potentiated affinity. This strategy was used to generate a potent inhibitor for the anticancer target thymidylate synthase [3]. Although thymidilate synthase contains an active site cystein, this reactive group can be introduced by surface mutatenesis (native amino acid to Cys). A library of 1200 disulfide-containing compounds was screened in pools of up to 100 compounds. Several disulfide adducts were detected by mass spectrometry. One of the selected compounds was investigated further: The inhibition constant K_i of the tether-free analog N-tosyl-D-proline was 1.1 mM. Subsequent crystallographic structure determination of the tymidilate synthase adduct and lead optimization improved the affinity over 3000-fold. This approach has been refined to extended tethereing where the first identified tethered compound serves as an anchor for the next fragment [82]. The tethering approach enables the nucleation of drug design efforts at specific sites on protein surfaces.

Structural Genomics

Current genome-wide *structural genomics* programs aim to determine representative structures for all proteins. The goal is to infer biological function from similar structures of known function [83]. Once representative structures are obtained, structural homology models of all members of each protein family can be built using these templates. This information is useful in selecting and classifying drug targets because the crystallographic structure holds information regarding protein function. The assignment of function by comparison on DNA (deoxyribonucleic acid) level fails for some proteins that have different sequence but similar fold. Furthermore, structures of proteins with bound natural ligands may expose their mechanism of action and allow researchers to act on this insight. In a proof-of-principle experiment

Zarembinski et al. [84] demonstrated that the function of a gene can indeed be assigned by X-ray crystallographic structure determination. They reported the 1.7-Å resolution structure of MJ0577, a protein form the hyperthermophile *Methanococcus jannaschii*. Unexpectedly, a bound ATP molecule was identified from the electron density map. Therefore, the protein was deduced to work as an ATPase or an ATP-mediated molecular switch. The latter function was confirmed by subsequent biochemical experiments. Similarly, a thymidilate synthase complementing protein was discovered from a structural genomics project by structure-based functional analysis, leading to its classification as an antibacterial drug target [85].

At first sight, this approach may seem to be a digression, but it addresses a grave problem in protein X-ray crystallography on new targets: There is no guarantee for obtaining a structural result. Current success rates for generating structures of novel protein targets are below 50 percent. However, once the scope of targets is increased, for example, including homologous proteins from various organisms, the probability of successful generation of X-ray crystallographic structures increases dramatically.

The number of potential drug targets has increased from about 500 in the mid-1990s to several thousand possible targets these days. It is the goal of structural genomics programs to provide structures of these new drug targets, even those of not yet identified targets or those proteins that may be used in the future for expedient homology model-building purposes once homologous proteins have been identified as useful drug targets. The formation of the Structural Genomics Consortium and its financial support by Glaxo-Smith-Kline pays tribute to the value of this approach. That consortium's goal is to determine 350 X-ray crystallographic structures of proteins directly related to human health, including proteins associated with cancer, neurological, and infectious diseases, within 3 years, starting from 2003.

The abundance of protein structural information has fueled large-scale bioinformatics approaches, aiding the drug discovery process at many stages. One particular application could be the improvement of the final drug product profile by systematic biasing drug lead optimization away from all known structures by treating their entirety as *antitargets*.

10.7 LIMITATIONS AND CHALLENGES OF X-RAY CRYSTALLOGRAPHY IN THE DRUG DISCOVERY PROCESS

Despite the outstanding track record of X-ray crystallography as an excellent tool for drug discovery purpose, its use has several limitations. The most serious ones result from (a) problems arising from protein flexibility, (b) shortcomings in computational methods to use structural information for the proper quantification of the energetics of protein–ligand interactions, and (c) difficulties in applying the crystallographic method to challenging targets such as membrane proteins.

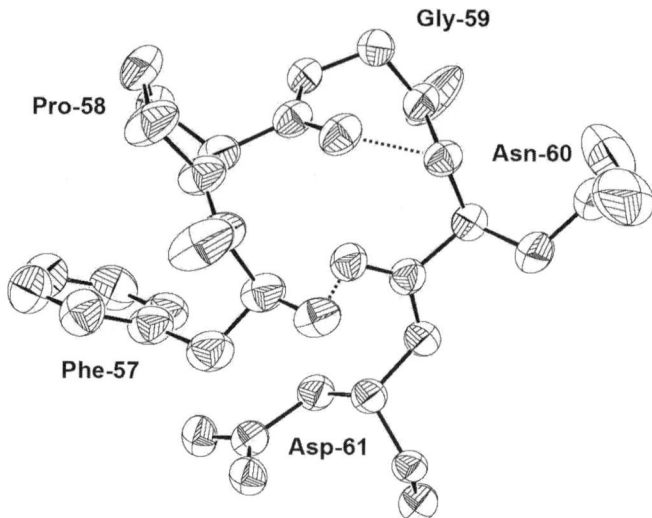

Figure 10.30 Elipsoidal representation of atom positions in a loop region in a serine protease depicting, for example, the high flexibility of the amide group of Asn-60 with respect to backbone atoms. Anisotropic displacement parameters can be calculated from atomic resolution data giving an impression of the amplitude and the direction of the mean displacement. (Figure taken from Schneider [86].)

Proteins are not as rigid as their visual representations in models suggest. Both proteins and small molecules are indeed flexible entities with thermal fluctuations, including bond vibrations in the picosecond range to slower nanosecond range side-chain rotations to very slow conformational changes that take place in time frames of micro to milliseconds. The amplitudes of the corresponding positional fluctuations are significantly large to affect ligand binding. Conversely, X-ray crystallographic structure describes protein structures as static, based on an apparent ensemble average at 100 K. The positional variations of atoms due to flexibility and crystal packing effects in standard resolution structures (1.5-to 3-Å resolution) is usually described by a single parameter, the B factor, the latter of which being a weak predictor of structural flexibility. Only few ultra-high-resolution structures of small proteins have allowed resolving alternative side-chain occupancies and anisotropic B-factor refinement [86] (Fig. 10.30).

The current standard, using a single model to describe instrinsically heterogeneous proteins, is an oversimplification that may lead to inaccurate models (Fig. 10.31). This shortcoming may be resolved by generating and refining an ensemble for alternative structures [87]. For interleukin 1β such models differed by as much as 1 Å between models.

There are numerous examples that demonstrate that the protein structure of a ligand bound form differs significantly from that of the ligand-free form or to that bound to a different ligand [88]. Apparently the ligand selects and

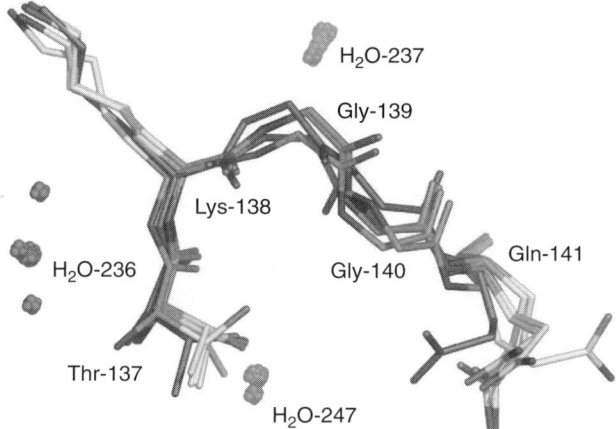

Figure 10.31 Structural heterogeneity in human interleukin1β. The ensemble of models displays considerable backbone variability, disordered side chains, and multiple locations of water molecules. The models were obtained from the same set of 2.3 Å resolution X-ray diffraction data, and refined to similar levels. (Image taken from DePristo et al. [87].) (This figure is available in full color at ftp://ftp.wiley.com/public/sci_tech_med/drug_discovery/.)

stabilizes a particular conformational state of the pool of many low-energy states that proteins can exist in at thermal equilibrium. In the process of SBDD this causes a high degree of unpredictability, making the method less useful. However, some proteins are less flexible, and their conformation hardly changes when ligands bind. These are the targets that are particularly susceptible to conventional SBDD efforts. There are only a few solutions to this fundamental predicament [89], notably the computationally intensive approach to treat the protein as a flexible entity and the tethering discovery approach. Understanding this limitation, however, may serve as the best antidote against the overuse of this tool.

Sanders et al. [90] point out a serious shortcoming of crystallographic screening. They described the discovery of competitive inhibitors for dihydroneopterin aldolase via crystallographic screening and demonstrated that several compounds with IC_{50} around 1μM were negative in crystal soaking experiments. Apparently the conformational shift associated with the binding of these missed compounds did not allow association to the protein in the preformed crystal.

The deficiencies of current computational methods to properly quantify the interactions of proteins with ligands is one of the consequences of molecular flexibility. But even more fundamentally, our current understanding of the energetics of ligand–protein interaction and hence their proper quantification by scoring functions is limited [91]. A weak point remains, for instance, in the description of entropic terms for binding interactions, although progress is being made and an energetic penalty of 10 to 30 kcal/mol is estimated for protein reorganization due to binding [92].

Finally, due to major technical bottlenecks X-ray crystallography cannot be applied to all drug discovery programs. The reason for this is the high uncertainty involved in obtaining any useful results with a given allocation of resources. The uncertainty is caused by the many risk-fraught steps involved in the crystallographic endeavor. Modern high-throughput crystallographic technologies seek to reduce this risk and have in some cases provided de novo structural information within less than 2 months at reasonable cost. This is not a given, though, since the outcome of a particular X-ray crystallographic project is unpredictable. This weakness reduces the applicability of X-ray crystallography to about half of all discovery projects in major pharmaceutical companies.

At the time of this writing, almost 30,000 protein structures were available from the PDB, an impressive record of the methodology's performance and significance. However, many structures for the presently ca. 500 proteins targeted by current drugs are not available because they have not been crystallized and their crystallographic structures have not been determined. This is particularly bothersome for membrane proteins, the latter of which represent half of the proteins that are targeted by today's marketed drugs [19]. Some 45 percent of molecular targets of known drugs are G-protein-coupled receptors (GPCR), proteins that have been notoriously difficult to deal with in the stages of overexpression, purification, and crystallization. Indeed, to date only one structure of such a protein has been determined, that of bovine rhodopsin, and several structures of similar bacterial homologs with seven transmembrane helices. It is anticipated that committed efforts in crystallographic structure determination projects can surmount the experimental barriers and yield structural information of this highly important protein class. The current state of SBDD on such valuable GPCR targets is reminiscent of the situation in the early 1970s when homology models based on carboxypeptidase A were successfully applied in the "design" of ACE inhibitors. Indeed, the first "virtually" discovered compound targeting a predicted structure of the serotonin receptor (a GPCR) entered phase 1 trials in 2004 and, Abbott's drug Atrasentan, targeting the endothelin-A receptor was optimized with the help of a homology model of the endothelin-A receptor (also a GPCR; [93]).

Acknowledgments We are greatly indebted to Dr. Wendy Sanderson and Dr. Ehmke Pohl for thoroughly reading and revising parts of this chapter.

REFERENCES

1. Nienaber, V. L., Richardson, P. L., Klighofer, V., Bouska, J. J., Giranda, V. L., Greer, J. (2000). Discovering novel ligands for macromolecules using X-ray crystallographic screening. *Nat. Biotechnol., 18*, 1105–1108.
2. Erlanson, D., Wells, J., Braisted, A. (2004). Tethering: Fragment-based drug discovery. *Annu. Rev. Biophys. Biomol. Struct., 33*, 199–223.

REFERENCES

3. Erlanson, D. A., Braisted, A. C., Raphael, D. R., Randal, M., Stroud, R. M., Gordon, E. M., Wells, J. A. (2000). Site-directed ligand discovery. *PNAS*, *19*, 9367–9372.
4. Anderson, A. C. (2003). The process of structure-based drug discovery. *Chem. Biol.*, *10*, 787–797.
5. Doman, T. N., McGovern, S. L., Witherbee, B. J., Kasten, T. P., Kurumbail, R., Stallings, W. C., Conolly, D. T., Shoichet, B. K. (2002). Molecular docking and high-throughput screening for novel inhibitors of protein tyrosine phosphatase-1B. *J. Med. Chem.*, *45*, 2213–2221.
6. Bergfors, T. M. (1999). *Protein Crystallization, Techniques, Strategies and Tips; a Lab Manual*. International University Line, La Jolla, CA.
7. McPherson, A. (1989). *Preparation and Analysis of Protein Crystals*. Krieger Publishing.
8. McPherson, A. (1998). *Crystallization of Biological Macromolecules*. Cold Spring Harbor Laboratory Press, Cold Spring Harbor, NY.
9. Ducruix, A., Giegé, G. (1992). *Crystallization of Nucleic Acids and Proteins, a Practical Approach*. IRL Press.
10. Weber, P. C. (1991). Physical principles of protein crystallizaiton. *Adv. Protein Chem.*, *41*, 1–36.
11. Durbin, S. D., Feher, G. (1996). Protein crystallization. *Annu. Rev. Phys. Chem.*, *47*, 171–204.
12. McPherson, A. (1990). Current approaches to macromolecular crystallization. *Eur. J. Biochem.*, *189*, 1–23.
13. Chernov, A. A. (2003). Protein crystals and their growth. *J. Struct. Biol.*, *142*, 3–21.
14. Drenth, J., Haas, C. (1992). Protein crystals and their stability. *J. Cryst. Growth*, *122*, 107–109.
15. Chayen, N. E. (1998). Comparative studies of protein crystallizaiton by vapour-diffusion and microbatch technique. *Acta Crystallogr.*, *D54*, 8–15.
16. Chayen, N. E., Shaw Stewart, P. D., Blow, D. M. (1992). Microbatch crystallization under oil—A new technique allowing many small-volume crystallization trials. *J. Cryst. Growth*, *122*, 176–180.
17. Chayen N. E. (1997). The role of oil in macromolecular crystallization. *Structure*, *5*, 1269–1274.
18. Zeppezauer, M. (1971). In W. B. Jakoby (Ed.), M*ethods in Enzymology*, Vol 22. Academic, New York and London.
19. Drews, J. (2000). Drug discovery: A historical perspective. *Science*, *287*, 1960–1964.
20. Iwata, S., (Ed.) (2003). *Methods and Results in Crystallization of Membrane Proteins*. International University Line, Biotechnology Series.
21. Hunte, C. C., von Jagow, G., Schagger, H (2003). *Membrane Protein Purification and Crystallization: A Practical Guide*. Academic, San Diego.
22. Nollert, P., Navarro, J., Landau, E. M. (2001). Crystallization of membrane proteins *in cubo*. *Methods Enzymol.*, *343*, 183–199.
23. Terpe, K. (2003). Overview of tag protein fusions: From molecular and biochemical fundamentals to commercial systems. *Appl. Microbiol. Biotechnol.*, *60*, 523–533.
24. Smyth, D. R., Mroziewicz, M. K., McGrath, W. J., Listwan, P., Kobe, B. (2003). Crystal structures of fusion proteins with large-affinity tags. *Protein Sci.*, *12*, 1313–1322.

25. Deng, J., Davies, D. R., Wisedchaisri, G., Wu, M., Hol, W. G. J., Mehlin, C. (2004). An improved protocol for rapid freezing of protein samples for long-term storage. *Acta Cryslallogr. D60*, 203–204.
26. Page, R., Grzechnik, S. K., Canaves, J., Spraggon, G., Kreusch, A., Kuhn, P., Stevens, R. C., Lesley, S. (2003). Shotgun crystallization strategy for structural genomics: An optimized two-tiered crystallization screen against the *Thermatoga maritima* proteome. *Acta Crystallogr. D, D59*, 1028–1037.
27. Jancarik, J., Kim, S.-H. (1991). Spare matrix sampling: A screening method for crystallization of proteins. *J. Appl. Crystallogr., 24*, 409–411.
28. Majeed, S., Ofek, G., Belachew, A., Huang, C., Zhou, T., Kwong, P. D. (2003). Enhancing protein crystallization through precipitant synergy. *Structure, 11*, 1–20.
29. Segelke, B. W. (2001). Efficiency analysis of sampling protocols used in protein crystallization screening. *J. Cryst. Growth, 232*, 553–562.
30. Gilliland, G. L., Tung, M., Ladner, J. (1996). The Biological Macromolecule Crystallization Database and NASA Protein Crystal Growth Archive. *J. Res. Natl. Inst. Stand. Technol., 101*, 309–320.
31. Garman, E. F., Schneider, T. R. (1997). Macromolecular cryocrystallography. *J. Appl. Crystallogr., 30*, 211–237.
32. Parkin, S., Hope, H. (1998). Macromolecular cryocrystallography: Cooling, mounting, storage and transportation of crystals. *J. Appl. Crystallogr., 31*, 945–953.
33. Rubinson, K. A., Ladner, J. E., Tordova, M., Gilliland, G. L. (2000). Cryosalts: Suppression of ice formation in macromolecular crystallography. *Acta Crystallogr., D56*, 996–1001.
34. Riboldi-Tunncliffe, A., Hilgenfeld, R. (1999). Cryocrystallography with oil—An old idea revived. *J. Appl. Crystallogr., 32*, 1003–1005.
35. Carter, C. W., Jr., Doublie, S., Coleman, D. E. (1994). Quantitative analysis of crystal growth. Tryptophanyl-tRNA synthetase crystal polymorphism and its relationship to catalysis. *J. Mol. Biol., 238*(3), 346–365.
36. Stewart, L., Clark, R., Behnke, C. (2002). High-thoughput crystallization and structure determination in drug discovery. *Drug Discov. Today, 7*(3), 187–196.
37. Petsko, G. A. (1985). Preparation of isomorphous heavy-atom derivatives, *Methods Enzymol., 114*, 147–156.
38. Rould, M. (1997). Screening for heavy-atom derivatives and obtaining accurate isomorphous differences. Methods Enzymol., 276, 112–147.
39. Gilliland, G. L. (1988). A biological macromolecule crystallization database: A basis for a crystallization strategy. *J. Cryst. Growth, 90*, 51–59.
40. HyperPhyscis web site: http://hyperphysics.phy-astr.gsu.edu/hbase/phyopt/mulslid.html#c4.
41. Glusker, J. P., Trueblood, K. N. (1972). *Crystal Structure Analysis: A Primer*. Oxford University Press, New York.
42. Stout, G. H., Jensen, L. H. (1989). *X-ray Structure Determination: A Practical Guide*. Wiley, New York, p. 202.
43. Otwinowski, Z., Minor, W. (1997). Processing of oscillation data collected in oscillation mode. *Methods Enzymol., 276*, 307–326.

44. Diedrichs, K., Karplus, K. A. (2002). Improved R-factors for diffraction in macromolecular crystallography. *Nat. Struct. Biol.*, *4*, 269–275.
45. Gewirth, D. (2003). *The HKL Manual.* pp. 89–90.
46. Fuji web site: http://home.fujifilm.com/products/science/ip/principle.html.
47. ADSC web site: http://www.adsc-xray.com/products.html.
48. Walsh, M. A., Dementieva, I., Evans, G., Sanishvili, R., Joachimiak, A. (1999). Taking MAD to the extreme: Ultrafast protein structure determination. *Acta Crystallogra.*, *D55*, 1168–1173.
49. Cohen, A. E., Ellis, P. J., Miller, M. D., Deacon, A. M., Phizackerley, R. P. (2002). An automated system to mount cryo-cooled protein crystals on a synchrotron beamline, using compact sample cassettes and a small-scale robot. *J. Appl. Crystallogr.*, *35*, 720–726.
50. Pohl, E., Ristau, U., Gehrmann, T., Jahn, D., Robrahn, B., Malthan, D., Dobler, H., Hermes, C. (2004). Automation of the EMBL Hamburg protein crystallography beamline BW7B. *J. Synchrotron Radiation*, *11*, 1–6.
51. ALS web site: http://www-als.lbl.gov/als/synchrotron_sources.html.
52. Drenth, J. (1994). *Principles of Protein X-ray Crystallography.* Springer-Verlag, New York, pp. 282–283.
53. Dauter, Z., Dauter, M., Dodson, E. (2002). Jolly SAD. *Acta Crystallogr.*, *D58*, 494–506.
54. McRee, D. (1993). *Practical Protein Crystallography.* Academic, San Diego, p. 145.
55. Hendrickson, W. A (1991). Determination of macromolecular structures from anomalous diffraction of synchrotron radiation. *Science*, *254*, 51–58.
56. Wang, B. C. (1985). Resolution of phase ambiguity in macromolecular crystallography. *Methods Enzymol.*, *115*, 90–112.
57. Brünger, A. T., Adams, P. D., Clore, G. M., DeLano, W. L., Gros, P., Grosse-Kunstleve, R. W., Jiang, J. S., Kuszewski, J., Nilges, M., Pannu, N. S., Read, R. J., Rice, L. M., Simonson, T., Warren, G. L. (1998). Crystallography and NMR System: A new software suite for macromolecular structure determination. *Acta Crystallogr.*, *D54*, 905–921.
58. Murshudov, G. N., Vagin, A. A., Dodson, E. J. (1997). Refinement of macromolecular structures by the maximum-likelihood method. *Acta Crystallogr*, *D53*, 240–255.
59. Altschul, S. F., Gish, W., Miller, W., Myers, E. W., Lipman, D. J. (1990). Basic local alignment search tool. *J. Mol. Biol.*, *215*, 403–410.
60. Cuff, J. A., Clamp, M. E., Siddiqui, A. S., Finlay, M, Barton, G. J. (1998). Jpred: A consensus secondary structure prediction server. *Bioinformatics*, *14*, 892–893.
61. Murzin, A. G., Brenner, S. E., Hubbard, T., Chothia, C. (1995). SCOP: A structural classification of proteins database for the investigation of sequences and structures. *J. Mol. Biol.*, *247*, 536–540.
62. Orengo, C. A., Michie, A. D., Jones, S., Jones, D. T., Swindells, M. B., Thornton, J. M. (1997). CATH: A hierarchic classification of protein domain structures. *Structure*, *5*(8), 1093–1108.
63. Lewis, P. N., Momany, F. A., Scheraga, H. A. (1973). Chain reversals in proteins. *Biochim. Biophys. Acta*, *303*, 211–229.

64. McPhail, A. T., Sim, G. A. (1968). The structure of camptothecin: X-ray analysis of camptothecin iodoacetate. *J. Chem. Soc. (B)*, *384*, 923–928.
65. Winn, M. D., Isupov, M. N., Murshudov, G. N. (2001). Use of TLS parameters to model anisotropic displacements in macromolecular refinement. *Acta Crystallogr.*, *D57*, 122–133.
66. Moews, P. C., Kretsinger, R. H. (1975). Refinement of the structure of carp muscle calcium-binding parvalbumin by model building and difference Fourier analysis. *J. Mol. Biol.*, *91*, 201–228.
67. Ramachandran, G. N., Ramakrishnan, C., Sasisekharan, V. (1963). Stereochemistry of polypeptide chain configurations. *J. Mol. Biol.*, *7*, 95–99.
68. Printz, M. P., Nemethy, G. (1972). The γ turn, a possible folded conformation of the peptide chain. Comparison with the β turn. *Macromolecules*, *5*, 755–758.
69. Hardy, L. W., Malikayil, A. (2003). The impact of structure-guided drug design on clinical agents. *Curr. Drug Discov.*, 15–20.
70. Klebe, G. (2000). Recent developments in structure-based drug design. *J. Mol. Med.*, *78*, 269–281.
71. Veerapandian, P. (Ed.) (1997). *Structure-Based Drug Design.* Marcel Dekker, New York.
72. Kubinyi, H. (1999). *J. Receptor Signal Transduction Res.*, *19*, 15–39.
73. Williams, P. A., Cosme, J., Vinkovic, D. M., Ward, A., Angove, H. C., Day, P. J., Vonrhein, C., Tickle, I. J., Jhoi, H. (2004). Crystal structures of human cytochrome P450 3A4 bound to metyrapone and progesterone. *Science*, *305*, 683–686.
74. Mao, H., Hajduk, P. J., Craig, R., Bell, R., Bore, T., Fesik, S. W. (2001). Rational design of diflunisal analogues with reduced affinity for human serum albumin. *J. Am. Chem. Soc.*, *123*, 10429–10435.
75. Ewert, S., Honegger, A., Plückthun, A. (2003). Structure-based improvement of the biophysical properties of immunoglobulin VH domains with a generalizable approach. *Biochemistry*, *42*, 1517–1528.
76. Stricher, M. L. F., Misse, D., Sironi, F., Pugniere, M., Barthe, I., Magne, X., Roumestand, C., Menez, A., Lusso, P. (2003). Rational design of a CD4 mimic that inhibits HIV-1 entry neutralization epitopes. *Nat. Biotechnol.*, *21*, 71–76.
77. Taylor, R. D., Jewsbury, P. J., Essex, J. W. (2002). A review of protein-small molecule docking methods. *J. Computer-Aided Mol. Des.*, *16*, 151–166.
78. Bajorath, J. (2002). Integration of virtual and high-throughput screening. *Nat. Rev. Drug Discov.*, *1*, 882–894.
79. Carr, R., Jothi, H., et al. (2002). Structure-based screening of low affinity compounds. *Drug Discov. Today*, *7*, 522–527.
80. Blundell, T. L., Jhoti, H., Abell, C. (2002). High-throughput crystallography for lead discovery in drug design. *Nature Reviews*, *1*, 45–54.
81. Rees, D. C., Congreve, M., Murray, C. W., Carr, R. (2004). Fragment-based lead discovery. *Nat. Rev.*, *3*, 660–672.
82. Erlanson, D. A., Lam, J. W., Wiesmann, C., Luong, T. N., Simmons, R. L., DeLano, W. L., Choong, I. C., Burdett, M. T., Flanagan, W. M., Lee, D., Gordon, E. M., O'Brien, T. (2003). In situ assembly of enzyme inhibitors using extended tethering. *Nat. Biotechnol.*, *21*(3), 308–314.

83. Zhang, Kim (2003). Overview of structural genomics: From structure to function. *Curr. Opin. Chem. Biol.*, 7, 28.
84. Zarembinski, T. I., Hung, L.-W., Mueller-Dieckmann, H.-J., Kim, K.-K., Yokota, H., Kim, R., Kim, S.-H. (1998). Structure-based assignment of the biochemical function of a hypothetical protein: A test case of structural genomics. *PNAS*, 95(26), 15189–15193.
85. Mathews, I. I., Deacon, A. M., Canaves, J. M., McMullan, D., et al. (2003). Functional analysis of substrate and cofactor complex structures of a thymidylate synthase complementing protein. *Structure*, 11, 677–690.
86. Schneider, T. R. (1996). What can we learn from anisotropic temperature factors? In E. Dodson, M. Moore, A. Ralph, and S. Bailey (Eds.), *Proceedings of the CCP4 Study Weekend*. SERC Daresbury Laboratory, Daresbury, UK, pp. 133–144.
87. De Pristo, M. A., de Bakker, P. I. W., Blundell, T. L. (2004). Heterogeneity and inaccuracy in protein structures solved by X-Ray crystallograph. *Structure*, 12, 831–838.
88. Davis, A., Teague, S. (1999). Hydrogen bonding, hydrophobic interactions, and failure of the rigid receptor hypothesis. *Angew. Chem. Int. Ed. Engl,.* 38, 736–749.
89. Teague, S. J. (2003). Implications of protein flexibility for drug discovery. *Nat. Rev. Drug Discov.*, 2, 527–541.
90. Sanders, W. J., Nienaber, V. L., Lerner, C. G., McCall, J. O., Merrick, S. M., Swanson, S. J., Harlan, J. E., Stoll, V. S., Stamper, G. F., Betz, S. F., Condroski, K. R., Meadows, R. P., Severin, J. M., Walter, K. A., Magdalinos, P., Jakob, C. G., Wagner, R., Beutel, B. A. (2004). Discovery of potent inhibitors of dihydroneopterin aldolase using CrystaLEAD high-throughput X-ray crystallographic screening and structure-directed lead optimization. *J. Med. Chem.*, 47, 1709–1718.
91. Reddy, M. R., Erion, M. D. (Eds.) (2004). *Free Energy Calculations in Rational Drug Design*. Kluwer Academic.
92. Verkhiveker, G. M., Bouzida, D., Gehlhaar, D. K., Rejto, P. A., Freer, S. T., Rose, P. W. (2002). Complexity and simplicity of ligand-macromolecule interactions: The energy landscape perspective. *Curr. Opin. Struct. Biol.*, 12, 197–203.
93. Thiel, K. A. (2004). Structure-aided drug design's next generation. *Nat. Biotechnol.*, 22(5), 513–519.

11

BIOLOGICAL AND CHEMISTRY ASSAYS AVAILABLE DURING DRUG DISCOVERY AND DEVELOPABILITY ASSESSMENT

DUANE B. LAKINGS
Drug Safety Evaluation Consulting, Inc.
Elgin, Texas

11.1	INTRODUCTION	457
11.2	DISCOVERY ASSAYS	460
	In Vitro Pharmacology Assays	460
	Analytical Chemistry Assays	462
	In Vivo Pharmacology Assays	464
11.3	DEVELOPABILITY ASSAYS	466
	Additional Pharmacology	468
	Formulation Development Assays	469
	Drug Delivery Assays	471
	Bioanalytical Chemistry Assays	474
	Pharmacokinetic Assays	476
	Drug Metabolism Assays	477
	Toxicology Assays	478
11.4	CONCLUSIONS	481

11.1 INTRODUCTION

The discovery and development of a novel therapeutic agent, whether a small organic molecule (commonly referred to as novel chemical entity, or NCE) or

Drug Discovery Handbook, by Shayne Cox Gad
Copyright © 2005 by John Wiley & Sons, Inc.

a macromolecule (such as a natural protein, modified protein, polypeptide, or oligonucleotide) requires scientific expertise from a number of different disciplines and an enormous amount of time (at least 5 to 6 years if everything goes smoothly but more realistically 10 to 12 years) and money (hundreds of millions of dollars). While humans may be the ultimate test species to ascertain the safety, efficacy, and pharmacokinetics of a potential new therapeutic agent, research studies conducted during the discovery phase in in vitro systems and in animal models can determine whether a discovery lead or group of leads:

1. Has the desired pharmacological profile (i.e., biological activity) for mediating a human disease process.
2. Has the necessary developability properties (e.g., sufficient aqueous solubility for delivery across membranes and metabolic stability or acceptable metabolism) to be a successful drug candidate.
3. Does not have a toxicity profile that could cause adverse experiences in humans at pharmacological doses.

At present, the drug discovery process consists of two distinct phases, like the two faces on a coin. On one side is the drug discovery *face* where researchers are charged with quickly finding new leads using protein, enzyme, or other targets that can be antagonized or agonized to possibly mediate a human disease or disorder. Some of these targets have been well characterized, but others have only recently been identified using genomic and proteomic research results. Compounds previously synthesized, combinatorial chemistry libraries, and/or newly prepared compounds (now commonly prepared using a rational drug design paradigm that models the active site of the target with the chemical structure to obtain a "better fit" and thus greater biological activity) are tested against the target using a high-throughput system (HTS) that can evaluate thousands of compounds in a short period of time. The compounds with the highest biological activity are designated *hits*. These hits are usually structurally modified and then retested in the HTS pharmacology screen to identify the pharmacophore (or the part of the chemical structure responsible for activity) and to determine if the addition or change of other chemical moieties increases or decreases the biological activity. This process is termed the structure–activity relationship (SAR) assessment. The modified hits with the highest biological activity in the in vitro pharmacology screen are designated *leads*. In years past, the lead with the highest activity in the screen was selected as the preclinical drug candidate. However, most of these leads have limited druglike properties, primarily because the active site of the target is lipophilic, and thus the most active compounds are also highly lipophilic (a property that causes limited aqueous solubility that in turn limits the ability of the compound to be delivered to the active site after administration to animal models or humans). Thus, most pharmaceutical and biotechnology firms now conduct some additional studies on these discovery leads.

On the other side of the coin is the lead optimization or developability assessment face where researchers conduct studies to determine if the new leads have the necessary druglike properties to reach the pharmacological site of action in sufficient concentration and for a sufficient duration to effectively interact with the in vivo target. These researchers use the expression *fail early, fail often* to define their charge of "killing", or removing from further consideration those discovery leads that do not have the desired characteristics or attributes for further development. This goal of fail early, fail often should be reached, whenever possible, before the initiation of clinical trial testing and preferably even before starting the definitive preclinical studies needed to support first-in-human clinical studies. Failing during expensive and time-consuming clinical trials should occur as infrequently as possible, especially for small biotechnology companies that have limited resources and where a clinical failure may result in having to close the doors.

Historically, for every 100 compounds screened for toxicity and biological activity in animal models, only 1 has the necessary pharmacology and safety profiles for evaluation in humans. Of those compounds tested in humans, only about 1 in 5 is successfully brought to the marketplace. This poor rate of success has been attributed to a number of factors, a primary one being that in vitro systems and animal models are not truly predictive of biological activity and/or safety in humans. The problem, however, may be that insufficient time and resources are put into first discovering the lead and then characterizing the lead or group of leads in pharmacology, drug metabolism, and toxicology models to first select the "optimal" compound (i.e., the lead with the desired druglike attributes needed for successful development) and then to critically evaluate the results to ensure that a compound with developability problems does not enter into preclinical and clinical development until those demerits are resolved. Instead of a rush from the first sign of biological activity in an in vitro pharmacology test to clinical trail evaluation, careful design, conduct, and interpretation of additional pharmacology, pharmacokinetic, and toxicology evaluations will detect "loser" compounds much earlier. This "weeding out of losers" will allow precious time and resources to be devoted to identifying discovery leads with a better chance than 1 in 5 of successfully completing clinical studies and becoming a marketed therapeutic agent. Being able to identify the 499 losers of the 500 discovery leads that have a potentially desirable biological activity as early as possible in the development process will save substantial time and resources.

Thus, one side of the drug discovery coin is to find as many novel leads as possible, and the other side is to identify the lead with the best chance of success and to "kill" as many of the other leads as possible. On both sides of the coin, researchers rely on various biological and chemical assays to evaluate the pharmacological activity and druglike properties of discovery leads.

This chapter summarizes some of these in vitro and in vivo assays available to and employed by researchers during the drug discovery and developability assessment phases of the discovery and development processes. The information presented should help researchers identify what assay systems are available and how these assay systems might be used to assist in first finding discovery hits and then in designing experiments to evaluate and optimize discovery leads as they progress through the drug discovery process and before they enter the preclinical research studies required to support regulatory agency submissions for first-in-human clinical trials.

11.2 DISCOVERY ASSAYS

The drug discovery process has undergone enormous changes during the past few years. After years of first synthesizing individual compounds, then purifying and obtaining physical and chemical characterization of the new chemicals, and finally testing the NCEs in in vitro and in vivo pharmacology models of a particular human disease, the pharmaceutical and biotechnology industry has embraced rationally designed combinatorial chemistry as a way to generate large numbers of organic compounds for evaluation. Other sources of compounds for evaluation include previously synthesized compounds, with some pharmaceutical companies having more than 100,000 unique chemical structures in their repositories. Using computer models of the active site of a target, *virtual* compounds with the appropriate structure to fit into and bind with the active site can be identified. Once these chemical structures have been determined, organic chemists synthesize series of compounds around these *structures*. These compounds are evaluated first in in vitro and then in in vivo pharmacological assays.

In Vitro Pharmacology Assays

Existing compounds or rationally designed, combinatorial chemistry-generated libraries are screened for biological activity, usually in an in vitro system in which a known biochemical process, which is thought to mimic a human disease or disorder, is agonized or antagonized. The increased number of compounds or mixtures of compounds to be tested has required pharmacologists to devise novel techniques or assays to screen rapidly for biological activity. Because the amount of material available from combinatorial chemistry syntheses is small (i.e., usually microgram quantities), miniaturization of the biological assay system is a necessary first requirement. The large number of compounds makes automation of the assay the second requirement. The use of first 96-well, then 384-well, and now larger titer plates and robotics to add the appropriate small amounts of combinatorial chemistry libraries and reagents, including the material that generates the signal for a positive result, provides such a system for rapid assessment of biological activity for a large number of compounds. Those assay wells that elicit a positive or negative response, depending on the biological test being conducted, are identified and if necessary, the compound(s) of interest is isolated and identified.

These in vitro pharmacology assays are commonly disease-target specific and are designed using one of two primary approaches. The first approach uses only the target, either isolated from cells or tissues that express the target or by incorporating the genetic code for the target into a cell system and, after expression, isolating the target. The target is then combined with known factors needed for activity and a detection system. These known factors might be an energy source, such as adenosine triphosphate (ATP), or a coenzyme, such as nicotinamide-adenine dinucleotide (NAD). Detection systems are usually target specific and are frequently a known substrate for the target that is or is not affected in the presence of an "active" discovery compound. The amount of change in the substrate or generation of the target–substrate product is a measure of the activity of the discovery compound being tested. The target may be bound to the titer plate or free in solution. Using robotics, aliquots of solutions containing the discovery compounds to be evaluated are added. Usually three or four concentrations, but sometimes more, of each discovery compound are tested. To describe the activity–concentration curves needed for evaluating and comparing the biological activities of a number of discovery leads, at least four levels of each compound need to be tested over a fairly wide concentration range. Since many of these discovery compounds have only limited aqueous solubility, a common practice is to prepare high-concentration solutions in dimethyl sulfoxide (DMSO) or another organic solvent, such as ethanol, and then dilute this solution with an aqueous buffer, such as phosphate buffered saline (PBS), to obtain the desired concentration range for testing. Even with this practice, some discovery compounds have insufficient solubility for effective testing against the target. The advantages of this approach are that the assay system is relatively simple, stable, and uniform across titer plates, and thus a large number of discovery compounds can be evaluated and compared in a short period of time. Disadvantages include that the discovery compounds do not have to be delivered (i.e., cross one or more membranes) to reach the target and that biotransformation products (i.e., metabolites) of the discovery compounds are not tested for activity.

The second approach used for in vitro pharmacology activity screening is to have the target expressed in a cell system. The "best" cell system would be the one where the target is expressed in the body and where the discovery compound will need to be delivered to be pharmacologically effective. This requires knowledge of where and how the target interacts with other proteins and systems to control or mediate a biological process, which when not functioning properly leads to a human disease or disorder. For "old" targets (i.e., targets that have been used for years for discovering new drugs), this functional knowledge has been mostly defined. However, for "novel" targets (i.e., targets that are being discovered through genomic and proteomic efforts), this knowledge is still being generated, and many of these novel targets may not be "drugable," which means that interaction with the target will not effectively medicate a disease process or that interaction with the target will cause undesirable effects in tissues or organs that express the target but are not involved in the disease process. Another approach used for a cell-based in vitro pharmacology assay is to incorporate the genetic code for the target, and possibly a detection system, into a cell, such as a bacteria or mammalian cell. The cell then expresses the target and the ability of discovery leads to interact with the target can be determined. While this cell-based approach has the advantage of requiring the discovery compounds to be delivered across at least one membrane, the cell line expressing the target may not be representative of the disease process to be mediated or contain the other proteins or enzymes involved with the target. Even so, an in vitro pharmacology assay in a cell system is considered more "robust" than an assay where the target is "free" in the assay system. Another advantage of a cell-based system, as will be discussed later in this chapter, is that information on the potential toxicity of a discovery compound can also be obtained by evaluating sufficiently high concentrations that may produce a toxic event, such as cell death.

Whichever approach is used, the in vitro pharmacology assay needs to be carefully characterized and, when possible, validated. Characterization includes determining what cofactors and other reagents are required for the target to be "active" and where in the body (i.e., which tissues and organs) the target is expressed. Validation means that agonizing or antagonizing the target in an in vivo animal model of the disease or, better yet, in humans produces the desired effect of mediating a disease process or disorder. Most of the newly defined targets have not yet been fully characterized or validated. However, many academic and pharmaceutical company laboratories are busy with these endeavors and will hopefully soon have a number of the novel targets sufficiently characterized so that discovery compounds that interact with these targets can be further evaluated in animal models of human diseases or disorders or in humans. In any case, those discovery compounds with the desired biological activity against a target in an in vitro pharmacology assay are designated hits and are evaluated further.

Analytical Chemistry Assays

Once the initial in vitro pharmacology assay results are available, the hits with the desired biological activity will be studied more intensively, usually in an in vivo animal model (discussed below) of the disease indication. Before these additional pharmacology assays can be conducted, a sufficient quantity of each discovery hit to be tested further will need to be synthesized, purified, identified, and characterized. The techniques employed for the synthesis and purification of the discovery hits are outside the scope of this chapter on assays; however, the identification and characterization of the discovery hits requires the use of analytical chemistry assays. For identification, the assay techniques most commonly employed are mass spectrometry (MS) and nuclear magnetic resonance (NMR) spectroscopy. MS can usually provide molecular weight (MW) information on the hits, and the MS fragmentation pattern can be used to determine key structural aspects of the molecules. The MW of the hits confirms the elemental composition (i.e., the number of carbons, hydrogens, nitrogens, oxygens, and other elements in the chemical structure) of the compounds that were synthesized. NMR spectra provide information on how the various moieties within the chemical structure interact, and, along with the MS fragmentation pattern, they are used to show that the compounds prepared are those with the in vitro biological activity. A common problem at this stage of the discovery process is that the hits being evaluated have poor purity, sometimes 50 percent or less, and thus the biological activity observed may not be only from the hits but also from pharmacologically active impurities or degradation products. Thus, each of the hits to be considered further should be characterized for purity, which requires an analytical chemistry assay that can separate a hit from impurities and/or degradants and that can detect not only the hit but also any impurity and/or degradant. Analytical chemistry techniques employed for determining the purity of hits are chromatographic assays such as thin-layer chromatography (TLC), high-performance liquid chromatography (HPLC), and for macromolecule leads, size exclusion chromatography and sodium dodecyl sulfate polyacrylamide gel electrophoresis (SDS–PAGE). The most common detection system for each of these assays is ultraviolet (UV) detection at a relatively low wavelength (i.e., 214 nm, since most organic compounds absorb light at this wavelength). Given the chemical structure of an analyte (i.e., a discovery hit), an experienced analytical chemist can usually select the appropriate assay method and the initial method parameter settings. Parameters that will need to be considered include the chromatographic column type (reversed–phase, normal phase, ion exchange, size exclusion) and size, the mobile phase and its composition (organic phase, buffer, and the percentage of each), the elution technique (isocratic or gradient), and the detection system (UV, refractive index, MS). The discovery lead is dissolved in the mobile phase and is injected. The chromatogram should contain one major peak (i.e., the discovery hit) and possibly a number of minor peaks, which may be impurities from the synthetic procedure. If more than one major peak is detected, further evaluations are immediately necessary since the compound of interest may be "contaminated" with another compound or may have partially degraded. Which of these major components has the desired biological activity will need to be assessed before additional testing is conducted.

If a discovery hit has one or more chiral centers, a stereospecific analytical chemistry method with the ability to separate all of the enantiomers will be necessary. Chiral chromatography or derivatization to form diastereoisomers (compounds with two asymmetric centers) are techniques commonly used. If a single enantiomer is being considered for further development, conversion to the other enantiomer is considered instability with the formed enantiomer being a degradation product. If possible, the enantiomers should be separated (using analytical chemistry techniques) or synthesized and tested in the in vitro pharmacological assay as soon as possible. If one enantiomer has more biological active than the other, further studies should be conducted on the "more active" enantiomer. If the enantiomers have similar biological activity, development of a racemic mixture is a possibility. However, regulatory authorities recommend that the pharmacology, pharmacokinetic, and toxicology profiles of the enantiomers be separately determined and the results be used to justify the development of a racemic mixture over a single enantiomer. Thus, evaluations on the enantiomers are needed as early as possible and should start with the in vitro pharmacology assay.

In Vivo Pharmacology Assays

After a sufficient quantity of the active compounds "discovered" during the in vitro pharmacology screening process have been prepared and characterized, additional studies are conducted to more fully evaluate the biological activity of these hits. An important requirement for this additional testing is having or developing an animal model, or in vivo pharmacological assay, that correlates to, or mimics, a disease or disorder in humans. Developing these animal models can be time consuming and expensive and is often complicated by the fact that the model may not simulate the disease as manifested in humans. Many important human diseases, including psychoses, depression, Alzheimer's disease, AIDS (acquired immunodeficient syndrome), and many cancers, do not yet have predictive animal models. In addition, most targets being discovered using genomic and proteomic efforts do not yet have an in vivo pharmacological model.

A necessary first step in the development of an in vivo pharmacological assay is to determine where in the body [i.e., the organ(s) or tissue(s) or particular cell types within an organ or tissue] the target is expressed and how interaction with the target will mediate the disease process. If a target is also expressed in organs or tissues that are not associated with the disease or disorder, interactions with the target in these organs/tissues may produce undesirable effects. Thus, targets that are widespread in the body may not be as "drugable" as a target that is expressed only in organ/tissue of interest [i.e., a brain neuron for a central nervous system (CNS) disorder]. Another important aspect is whether the target is within a cell and, if so, where within the cell; has the active site on the cell surface; or is excreted by the cell and thus is in extracelluar space. Targets that are intercellular may be more difficult to reach and may be involved in more than one biochemical pathway or cascade. Inhibiting or activating such a target may result in the desired effect but may also adversely affect other processes and thus lead to undesirable responses. Targets that are on cell surfaces are usually readily assessable and frequently are involved in a single process, such as providing a signal for controlling an intercellular reaction or assisting in the transport of an ion or biochemical through the cell wall. Targets that are in extracelluar space are also readily assessable but may have a wider distribution throughout the body. Information or where and why a target is expressed is critical for the successful definition of an animal model of a human disease or disorder.

After the animal model has been characterized, various dose levels of the hit(s), identified from screens of combinatorial chemistry libraries or other sources, are administered to the test species and a dose–response curve is generated. The most commonly used endpoint is the dose that provides a 50 percent effective dose response (the ED_{50}). Structural analogs of the hit(s) are frequently synthesized and tested in the same model to generate a family of curves with varying biological potencies or ED_{50}'s. These structural analogs may contain the addition or deletion of polar or nonpolar moieties that enhance or decrease the biological activity of the hit. Structural changes that enhance the biological activity of a hit are continued until a compound with the desired pharmacological profile or biological potency is obtained. This hit is now designated a lead. Frequently, a number of leads, each with a unique chemical structure, are identified.

At times, companies select the lead with the greatest biological activity in the in vivo pharmacological assay for further development. However, at this point in the drug discovery process, many unanswered questions still exist. Some of these concerns are:

1. Does the animal model used to identify the leads reflect, in most or all aspects, the disease in humans?
2. Is the delivery or extent of exposure of all the leads similar so that the generated dose–response curves accurately predict the compound with the greatest in vivo potency?
3. Are the route and frequency of administration and the formulation used to dose the pharmacological animal model similar to those proposed for preclinical animal safety studies and for human safety and efficacy evaluations?
4. Do the leads have similar pharmacokinetic and drug metabolism profiles so that the durations of exposure to the pharmacologically active compounds are similar?
5. Do any of the leads produce unacceptable toxicity in organs or physiological systems not involved in the desired pharmacology of the leads?

Before the more formal and definitive preclinical development begins, attempts should be made to answer as many of these questions as possible. The following section discusses some of the assays utilized in drug developability experiments that can be conducted relatively quickly and cheaply to ascertain whether a lead has the necessary biopharmaceutical, or druglike, characteristics needed for further development.

11.3 DEVELOPABILITY ASSAYS

Completed drug discovery research studies indicate that a compound (now designated a discovery lead) has sufficient in vitro and in vivo biological activity to mediate a disease process and thus has potential to become a human therapeutic agent. Is this lead now ready to be transferred from the discovery area to the preclinical development group? Should additional, nondefinitive research experiments be conducted to more fully characterize the properties of the lead? If more studies are considered necessary, what experiments should be conducted? If additional studies are to be performed, what types of biological and chemical assays are available to support these efforts?

Frequently, more than one discovery lead has the desired biological activity in the in vivo pharmacological assay. When that is the case, which of these leads should be selected for entry into the more formal preclinical drug development process? What criteria, other than biological potency, should be used for selecting this "optimal" lead from a group of leads? Can a series of research experiments (i.e., lead optimization studies) be identified for determining which of the leads have the druglike attributes considered necessary for successful development and which of the leads have undesirable properties or demerits that would be suggestive of potential development problems?

This section describes some of the biological and chemical research assays that could, and in most cases should, be used to evaluate the potential of a discovery lead to become a developmental candidate or to select the optimal lead from a group of discovery leads. Figure 11.1 shows where these developability experiments fit into the drug discovery and development process. These nondefinitive developability studies may also uncover problems that have to be resolved before the definitive preclinical development studies, required to support a regulatory agency submission for first-in-human clinical trials, are

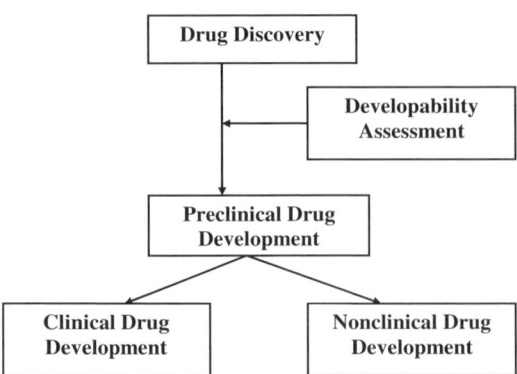

Figure 11.1 Where developability assessment experiments fit into the drug discovery and development process.

started and before the clinical protocols to evaluate the human safety, efficacy, and pharmacokinetics of a drug candidate are designed.

Before the preclinical drug development program is begun and because each discovery program is company and compound specific, a number of questions should be asked and answered so that the developability assessment research studies (both types and designs and whether or not appropriate assays are available) may be planned effectively and the timelines for their completion can be determined. These questions include, but are not limited to:

1. What is (are) the human disease indication(s) or disorder(s) to be studied in the clinic?
2. What are the proposed route and frequency of dosing for human clinical trails?
3. What is the estimated pharmacologic active substance (i.e., the discovery lead or an active metabolite) concentration in physiological fluids and how long should that concentration be maintained so that the desired biological response can be obtained?
4. What, if any, are the biological markers to monitor toxicity or therapeutic effectiveness in to-be-conducted preclinical, nonclinical, and clinical studies?

Additional questions need to be considered, including whether a lead can be synthesized and purified in sufficient quantity to support a development program, how to document and validate these processes, how to characterize and document the identity of and impurities present in the drug substance and the proposed drug product, and how to determine the stability of the drug substance and the proposed drug product. While each of these processes requires assays, these questions are outside the scope of this discussion in this chapter.

At least eight, and possibly more, scientific disciplines are involved in the developability characterization of a discovery lead or group of leads, and each of these disciplines utilize different assays, some biological and other chemical. As shown in Figure 11.2, these disciplines are in vivo pharmacology, analytical chemistry, solubility and stability, drug delivery, bioanalytical chemistry, animal pharmacokinetics, drug metabolism, and toxicology. The following sections provide some summary information on the assays for each of these scientific disciplines. The discussion is mostly for a single discovery lead; if more than one lead is being evaluated to select the lead with the most druglike properties, which depends on a variety of factors such as disease indication, route and frequency of dosing, and so forth, for further development, similar experiments could be conducted on each lead. Normally, these developability assessment experiments are conducted in an iterative fashion with the attributes considered most important evaluated first. Generally, in vitro evaluations,

Figure 11.2 Scientific disciplines involved in developability assessment.

which are less costly (both in terms of resources and time), are conducted before in vivo studies. If a lead or none of the leads in a group of leads has the desired profile for a given attribute, additional discovery leads should be identified and tested, like the SAR approach discussed earlier. Once a lead or a number of leads from a group has an acceptable profile for each attribute being evaluated, that lead or group of leads moves into the next stage of developability assessment experiments. This lead-optimization iterative process should be continued until a lead with the desired properties has been identified. At times, more than one lead will have the desired, or at least an acceptable, developability profile. When that pleasant occurrence happens, a primary lead can be designed with the other leads serving as backup candidates.

Additional Pharmacology

Preliminary pharmacology evaluations using in vitro or animal model assays developed and characterized during the discovery process will have shown that a lead interacts with a biological process suggestive of human therapeutic benefit. Depending on the design and extent of these early studies, additional pharmacology studies or assays may be needed to further characterize the dose, or physiologic fluid concentration, response curve using the proposed clinical route and frequency of administration. If possible, these pharmacology studies should be conducted in both males and females to show that the biological response is not gender specific. If the indication is a disease of age, older or younger animals, as appropriate, should be evaluated. The ED_{50} dose should be determined and that value divided into the no-observable-adverse-effect-level (NOAEL) dose in the same animal species, described later in the section on toxicology, estimates a therapeutic ratio (TR) or index (TI). If the TR is one or less, a lead will most likely elicit adverse effects in addition to the beneficial response. Unless the lead is for the treatment of a life-threatening disease, such as AIDS, some cancers, or certain CNS indications, a low TR is a warning sign that the lead may not have the necessary properties for further development. A TR of 5, preferably 10 or higher, indicates that a lead will most likely produce a pharmacological response before causing dose-limiting toxicities.

If possible, these developability pharmacology studies should be conducted with dosing to steady state unless the frequency of dosing to be used in clinical trials is as a single-dose or a few dose therapeutic. The number of doses required to reach steady state depends on the lead's pharmacokinetic profile (discussed later in the section on preliminary pharmacokinetics) in the same animal model. These multiple-dose studies provide information on the frequency of dosing necessary to maximize the biological response. This is particularly important for compounds that inhibit an enzymatic system or are effective only during certain phases of the cell cycle.

Another important aspect that should be considered when conducting these pharmacology evaluations is the identification and characterization of potential biological markers that may be predictive of therapeutic benefit. For some targets, markers may already be known, and, if so, assays for these markers should be adapted and incorporated into the studies. However, for many targets, predictive biological markers are not yet available. If the interaction of the lead with a target inhibits or enhances the formation of a particular biochemical, the concentration of that biochemical may be indicative of therapeutic response. Developing and utilizing assays for this marker first in animal models and then in human clinical studies may define a surrogate marker for tracking disease response or progress.

These pharmacology evaluations assist in the selection of dose levels and dosing regimens for preliminary and definitive toxicology studies and for initial human pharmacology or phase 1 safety and tolerance clinical trials. If the effective pharmacologic dose is unknown, underdosing and achieving no therapeutic response or overdosing and being unable to define an NOAEL dose are undesirable possibilities. In such cases, the development of a potential beneficial therapeutic agent could be inappropriately discontinued.

Formulation Development Assays

Nonclinical formulation definition of a lead is not usually studied in detail during the transition from discovery to preclinical development. The experiments necessary to define an acceptable formulation depend on the proposed clinical route of administration and usually require substantial quantities (i.e., milligram or gram amounts) of the lead. For a compound to be administered by intravenous injection or infusion, the formulation needs to be compatible with blood so that the compound does not precipitate when administered and has minimal local toxicity. Leads that are highly lipophilic or have limited aqueous solubility are the most likely compounds to have these types of problems. A low extent of, or a high variability in, absorption can cause problems for leads administered by other routes, such as oral, subcutaneous, or dermal. For compounds that are poorly absorbed, the amount reaching the site of action may be insufficient to elicit or to maintain a desired pharmacologic response. If the absorption is variable and the TR is low, a toxic response may be observed in some, but not all, animals or later in some humans. Experiments conducted by the author have shown that when the extent of absorption is 50 percent or more and the variability of absorption is less than 50 percent of the amount absorbed, which gives a 25 to 75 percent range of absorption, a lead has an acceptable bioavailability for further development. For leads with an extent of absorption less than 25 percent of the administered dose or a variability of absorption of more than 100 percent of the amount absorbed, other formulations with absorption enhancers or solubilizers might be evaluated to improve the drug delivery profile. If improvement in the drug delivery profile is not possible, the chances that a lead with low, variable absorption will become a therapeutic product are greatly reduced. The candidacy of such a compound should be carefully considered.

The analytical chemistry assay defined for characterizing a discovery hit should be further developed for the quantification of a lead in nonclinical formulations and should predict whether the compound degrades from the time of preparation to the time of dosing. An assay with this ability is called a stability-indicating assay. The physical and chemical properties of the lead usually suggest a technique (heat, light, pH) that can degrade the compound. Samples stored under nondegrading and degrading conditions are assayed by the stability-indicating assay. If the lead is not stable, formulation excipients possibly can be added to prevent the degradation or the lead can be maintained under conditions that provide sufficient stability for testing in animal models. However, when a lead has limited stability, either alone or in nonclinical formulations, the development of a clinical formulation with a sufficient shelf life for marketing is problematic. Again the candidacy of such a compound should be carefully considered.

For proteins and other large molecules, degradation may include changes in the secondary or tertiary structure, provided that rearrangement back to the original, biologically active structure does not occur. Stability-indicating assays for macromolecules should assess structural changes that cause reductions in biological activity. However, a protein may have a number of amino acids removed from one or both ends of the molecule and still retain some, and possibly all, biological activity relative to the intact protein. This chemically modified peptide may have different drug delivery and pharmacokinetic profiles or be more toxic than the parent protein. Structural modifications may not be apparent if biological activity alone is used to determine the amount of the macromolecule in a formulation. Thus, experiments to demonstrate that an assay method is stability indicating need to be carefully designed and conducted. For macromolecules, a specific chemical assay, such as HPLC or enzyme-linked immunosorbent assay (ELISA), and a biological potency assay, such as the in vitro pharmacological assay, may be necessary to determine the concentration and stability of a compound in a formulation.

The stability-indicating assay should be used to determine the amount of the lead in nonclinical formulations used for dosing animals in in vivo pharmacology, preliminary pharmacokinetic, drug metabolism, and toxicology studies. For single-dose studies, samples from each formulation at each dose level can be collected before dosing and after the completion of dosing. For multiple-dose studies, samples from each formulation used can be collected before the first dose and after the last dose or at selected other times. Results from these analyses ensure that the formulations contained the desired amount of the lead, that the concentrations did not change during the dosing period, and that the test species received the appropriate dose levels.

Another application of a stability-indicating assay is to evaluate the solubility profile of the discovery lead under the various conditions that the lead may experience during dosing and delivery to the site of pharmacological action. As mentioned earlier, a lead needs to have some aqueous solubility in order the effectively cross membranes. A lead to be orally administered will be subjected to high acidic conditions in the stomach and then to the more neutral conditions of the upper gastrointestinal (GI) tract before being absorbed into systemic circulation. Thus, stability of the lead is necessary under both acidic and neutral conditions. For most other routes of delivery, a lead will be subjected to neutral (i.e., physiological) to slightly basic conditions. The analytical chemistry assay developed earlier is usually capable of evaluating the solubility of a lead over the range of pHs. With slight modification, the robotic system used for the in vitro pharmacology assay can be adapted for the generation of samples for solubility assessments. Using the concentration of the lead that provides the desired biological activity as a starting point, the desired solubility range can be estimated. For a lead with high potency (i.e., pg/mL or ng/mL), aqueous solubility at the milligram/milliliter level may not be necessary. A high-concentration solution, frequently in an organic solvent such as DMSO, of the lead is prepared and small aliquots, usually microliter quantities, are added to aqueous solutions that contain a buffer, commonly phosphate buffer saline (PBS), and are adjusted in pH to give the desired pH range. After appropriate mixing, the samples are centrifuged or filtered to remove any precipitate (i.e., unsolubilized lead), and aliquots of the supernatant are assayed using the quantitative analytical chemistry method. The concentrations of the lead in the samples are used to prepare a solubility profile for that lead. Discovery leads that have poor aqueous solubility under conditions expected to be encountered during dosing and delivery to the site of pharmacological activity usually make poor drug candidates.

Drug Delivery Assays

Without an acceptable nonclinical formulation, the extent and variability of delivery may make interpretation of results from developability studies meaningless and prevent the continued development of a potentially useful therapeutic agent. The best formulation is of little or no use if the lead is not effectively delivered to the site of biologic action. One of the primary reasons discovery leads are not successful drug candidates is limited or insufficient transport across various membranes from the site of dosing to the site of activity. Only compounds administered intravenously to mediate a disease indication expressed in the cardiovascular system do not have to cross at least one membrane in order to reach the site of action. Thus, assessments of a lead's ability to cross membranes should be conducted as early as possible. Many pharmaceutical companies use delivery potential as a key indicator for whether or not a discovery lead should move into preclinical development.

Since most membranes are lipophilic in nature, a lead has to have some lipophilic characteristics in order to effectively diffuse into and across the membrane. However, in order to reach the membrane, the lead has to be dissolved in the surrounding media, which is aqueous. Thus, the lead also needs some hydrophilic properties in order to have sufficient aqueous solubility to be transported to the membrane. An estimate of a lead's ability to have both the lipophilic and hydrophilic characteristics necessary for effective delivery, primarily from the GI tract, can be determined from the chemical structure of the compound and using what is commonly called Lipinski's rules of five, which are four rules with cut-off numbers that are 5 or multiples of 5. These rules are:

1. A molecular weight of less than 500 daltons.
2. A $\log P$ (octanol–water coefficient) of less than 5.
3. Hydrogen bond donors (sum of hydroxyl and amine groups) less than 5.
4. Hydrogen bond receptors (sum of nitrogen and oxygen atoms) less than 10.

While these rules may be somewhat predictive of a lead's ability to cross membranes, not all compounds having the desired characteristics are orally absorbed or effectively transported across membranes. Thus, laboratory experiments, which need characterized assays, are required to determine if a lead will effectively be delivered to the site of biologic action.

A number of in vitro models or assays are available to evaluate the delivery potential of a lead. For a lead to be administered orally, the Caco-2 model is most commonly employed. Other in vitro systems are available to evaluate transport across other membrane types, including the blood-brain barrier (BBB) for CNS disorders, lung for pulmonary delivery, skin for topical delivery or to treat dermal disorders, and eye for ocular delivery. Each of these assay systems is summarized below. The analytical chemistry assay discussed earlier is usually capable of quantifying the lead in samples generated by these in vitro delivery models.

For determining the potential intestinal absorption and secretion of discovery leads, the most commonly used in vitro assay system employs Caco-2 cells, which were originally derived from a human colorectal carcinoma. When cultured on semipermeable membranes, Caco-2 cells differentiate into a highly functionalized epithelial barrier with remarkable morphological and biochemical similarity to the small-intestinal epithelium. The disappearance of a discovery lead in the reservoir above the confluent cell layer and the permeable membrane and the appearance of the lead and/or degradants or metabolites on the other side are measured using an analytical chemistry technique. Permeability coefficients obtained from Caco-2 cell transport studies have been shown to correlate with human intestinal absorption. Caco-2 assessments are commonly conducted in 24-well assay systems and can be configured for high-throughput systems.

A 96-well technique using a lipid membrane has been shown to have excellent correlation with other in vitro absorption systems (e.g., Caco-2 cells) and with in vivo results. When the assessment of the delivery potential of a large number of discovery leads, or even discovery hits, is needed, this membrane assay system may have application.

Whether attempting to assess the ability of discovery lead(s) for a CNS indication to reach the site of action or evaluating the potential toxicity of a lead(s) that should not be extensively delivered to the brain, assessment of transport across the BBB is a necessary first assessment. Primary microvascular cells from bovine brain tissue, which express the appropriate tight junctions and have the P-glycoprotein transport characteristics of the BBB, are commonly used for studying BBB permeability. The assay system and analytical chemistry techniques used for Caco-2 studies of GI permeability are employed.

Pulmonary delivery is a route of administration that has a number of advantages over oral administration. A discovery lead(s) administered to the lung can reach systemic circulation without having to cross the GI mucosa and without having to get past the liver, where many compounds are highly metabolized (i.e., have high first-pass metabolism). To evaluate the potential of a discovery lead for delivery through the lung, an in vitro assay system is available. Cultured, normal, human-derived tracheal/bronchial epithelial cells form a pseudostratified, highly differentiated model that closely resembles the epithelial tissue of the respiratory tract. Histological cross sections of both the in vitro tissue and a normal human bronchiole show a pseudostratified mucociliary phenotype. Numerous microvilli and cilia on the apical surface are present and confirm the presence of tight junctions. Cultures grown on cell culture inserts at the air–liquid interface provide gas-phase exposure of volatile materials and allow the measurement of transepithelial permeability for inhaled drug delivery studies on discovery leads.

Another potential route of delivery is dermal, and an in vitro assay system is available for evaluating the ability of a discovery lead(s) to penetrate through the skin and thus to reach systemic circulation. Normal human epidermal keratinocytes cultured on a permeable membrane produce a stratified, highly differentiated, organotypic tissue model of the human epidermis. The cultured tissue consists of metabolically and mitotically active cells organized into basal, spinous, and granular layers along a multilayered stratum corneum. The skin construct has an air-liquid interface and exhibits in vivo-like morphological and growth characteristics that allows discovery leads to be directly applied to the surface of the tissue. This in vitro system approximates the barrier of normal human skin. The topical mode of application of the lead mimics the route of human exposure for some disease indications. Of course, treatment of some skin disorders does not require delivery of a therapeutic agent across the skin and into the body. When that is the case, substantial delivery into systemic circulation may be considered a developability demerit since only local exposure is desired and necessary to effectively treat the disorder. The in vitro skin assay may then be used to predict which discovery leads have less delivery across the skin and thus be predictable of lower exposure to potential organs or tissues of toxicity.

If the human disease or disorder to be evaluated is expressed in the eye, the ability of a discovery lead(s) to be delivered to the site of biological activity can be evaluated using an in vitro assay that mimic the eye. A three-dimensional in vitro tissue construct models the human corneal epithelium. Normal, human-derived epidermal keratinocytes cultured on a permeable membrane form a stratified, squamous multilayered epithelium similar to that of the cornea. This in vitro construct has an air-liquid interface and exhibits in vivo-like morphological and growth characteristics allowing discovery leads to be directly applied to the surface of the tissue and to approximate in vivo conditions. Analysis of a discovery lead(s) concentration on the liquid side of the system provides a measure of delivery into the eye.

Bioanalytical Chemistry Assays

If not already available (which is common for newly discovered leads that may mediate a novel disease target), a bioanalytical chemistry method or assay needs to be defined and characterized for the quantification of the lead(s) in physiological fluids. This assay can then support experiments in some of the other scientific disciplines involved in assessing the developability of the lead and, after appropriate validation, the preclinical, nonclinical, and clinical development of a selected drug candidate. For preliminary studies, a bioanalytical chemistry assay should be characterized to demonstrate the range of reliable results, the lower and upper limits of quantification, specificity, accuracy, and precision. In addition, evaluations to determine the matrix to be used (blood, plasma, serum) should be conducted, and, once selected, the stability of the lead in that matrix type should be determined.

The first step in developing and characterizing a bioanalytical chemistry assay is to select the analytical technique. For an analyte with a molecular weight of less than 1000, instrumental methods such as HPLC coupled to an MS or liquid chromatography/mass spectrometry (LC/MS) or to a tandem or LC/MS/MS (a very sensitive and specific technique, and the most common method employed today by the pharmaceutical industry), or HPLC or gas chromatography (GC) with a variety of detectors, including UV, fluorescence, flame ionization (FID), and electron capture (EC), may be used. A macromolecular lead may require an ELISA or radioimmunoassay (RIA) method, both of which require the generation of antibodies for capture and detection of the macromolecule. Generation of these antibodies may not be possible in the timeline required for supporting developability assessment experiments. If the in vitro pharmacology assay has been sufficiently characterized (as discussed below), this assay may have application as a bioanalytical chemistry assay. Samples of the lead prepared in assay diluent and in a physiological fluid and over a large concentration range should be analyzed to show that the technique produces an appropriate signal to detect the analyte and to determine the potential interference caused by the matrix, which is termed assay specificity. The ability to quantify a lead in a physiological fluid may depend on the matrix. For example, serum is a poor choice when the analyte interacts with clotting factors and thus may be coprecipitated when blood is allowed to clot to produce serum. The matrix that gives the best recovery and has the least interference when the compound is added to whole blood should be selected.

The specificity of an assay evaluates the potential interferences from physiological matrix components in specimens from the different animal species to be used in pharmacology and toxicology experiments. Samples from each species to be studied are analyzed neat (with no added analyte or lead) and fortified with known amounts of the lead, and the results are calculated using a standard curve prepared in assay diluent. The response obtained from the neat samples indicate the level of interference from each matrix, and the calculated amounts in the fortified samples show the difference in absolute recovery from the matrix compared with the analyte in buffer. If the absolute recovery is low, that is, less than 50 percent, and/or highly variable, that is, greater than 25 percent, the assay may not have the desired characteristics to quantify the lead in collected physiological fluid specimens. Additional development should be expended on such a method so that the assay will provide reliable results that can be used to evaluate the pharmacokinetic profile of the lead.

Acceptable results from the above experiments suggest whether a bioanalytical chemistry method should be able to quantify a lead in a physiological matrix. The range and reliability of quantification of the assay are assessed through the preparation and analysis of standard curves, prepared in either diluted or undiluted matrix, and multiple samples fortified at two or more concentrations and commonly referred to as quality control (QC) samples. The standard curve responses that can be described by a mathematical equation (linear, quadratic, sigmoidal) define the range of reliable results. The lower limit of quantification (LLQ) is the lowest signal that can be accurately measured above background and should not be confused with the limit of detection (LOD), which is the lowest level that can be detected. The upper limit of quantification (ULQ) is the highest signal that can be defined by the response curve. The results from the analysis of QC samples, usually three or more at each concentration, provide information on precision, defined as the ability to obtain similar calculated concentrations from samples containing the same amounts of analyte, and accuracy, which is the ability to predict the actual concentration of the analyte in a sample.

The ability to measure a lead in a physiological fluid is not useful if the analyte is unstable during collection, processing, storage, or sample preparation. Thus, a nondefinitive stability study should be conducted to ensure that the compound does not degrade in blood during processing to obtain plasma or serum, during the time (hours, days, and weeks) and under the conditions (room temperature, refrigerated, frozen at −20 or −80°C) that collected specimens may be stored until analyzed, and during sample preparation, which includes thawing of frozen samples. The design of stability experiments is usually analyte and program specific. The results ensure that measured concentrations in unknown specimens reflect the amount of lead present at the time of collection.

Successful completion of the above experiments will characterize the assay method for use in evaluating a lead in animal models. If multiple leads are being evaluated, an assay capable of quantifying multiple leads using the same assay conditions should be the goal. If a lead is selected for further development, the method will need to be validated for each matrix and for each species before being used to support definitive toxicology, drug metabolism, and pharmacokinetic studies.

Pharmacokinetic Assays

The first animal pharmacokinetic (PK) study confirms that the bioanalytical chemistry assay is useful in characterizing the absorption and disposition profiles of a lead. The animal species for this study is usually the same as used in in vivo pharmacology evaluations, most likely a rodent. A study design or assay for a lead that has pharmacologic activity when administered orally to rats may consist of dosing at least two rats with intravenous (IV) bolus injections at a dose level between 25 and 50 percent of the pharmacologically active dose and dosing at least two rats orally (PO) at the pharmacologically active dose. Serial blood samples, collected from each rat and processed to obtain the desired physiological fluid, are analyzed by the developed and characterized bioanalytical chemistry assay. The number of samples and the collection times are program specific and are usually determined by the "desired" PK profile for the lead, which in turn is defined by the desired interaction of the lead with the target. Commonly, 10 samples are collected from each test species. After IV dosing, the collection times might be 0 (predose), 10, 20, 30, 45, 60, 90, 120, 240, and 480 min postdose. For the PO doses, the collection time series could be 0 (predose), 15, 30, 45, 60, 90, 120, 180, 240, and 480 min postdose. The goal of the specimen collection time series should be to have a sufficient number of samples available for defining important PK parameters, such as distribution and disposition rate constants and their associated half-lives, concentration maximum (C_{max}), and time (T_{max}) to C_{max} after non-IV dosing, and area under the concentration–time curve (AUC). The physiological fluid concentration of the lead versus time profiles after IV dosing provide preliminary information on the distribution and disposition kinetics of the lead. These IV results certify that the assay method is useful for quantifying the lead in specimens obtained from animals, predict the concentration ranges that can be expected in animal specimens at pharmacological and toxicological doses, and assist in determining the sampling times to be used in more definitive animal PK experiments. The plasma concentration versus time profiles after PO dosing provide preliminary information on the absorption kinetics and the absolute bioavailability of the lead. The design of additional animal PK studies depends on the results of the preliminary animal PK study, the theoretic kinetic profile needed to produce the desired pharmacology response, and the results from preliminary toxicology experiments (discussed below).

For most drug development programs, toxicology studies in two or more species are necessary. In this case, preliminary animal PK studies should be conducted in each species projected for use in animal safety studies. If differences in delivery or disposition exist between species and result in an enhanced or decreased toxicology profile, the PK profiles may explain, in part, the different toxicology profiles. If possible, physiological fluid specimens should be obtained from animals in the preliminary toxicology studies to determine the extent and uniformity of exposure, which is termed toxicokinetics (TK) over the evaluated dose range. Normally, three or four specimens from each animal are sufficient for TK evaluations, but this level of sampling may not be possible for all studies. A single specimen at one collection time can be obtained from one or two animals in a dose group, and the other animals in that dose group can be sampled at other times. Analyses of these specimens provide data on the extent of exposure but not on the uniformity of exposure within a dose group. For multiple-dose studies, specimens are usually obtained after the first dose and after the last, or next to last, dose. The results provide information on possible changes in exposure and on the accumulation potential of the lead and can be used to design multiple-dose animal PK and other development studies. If the change in disposition or accumulation is substantial, modification of the dosing regimen may be necessary to obtain the desired concentration profile after dosing to steady state.

Drug Metabolism Assays

The number and design of drug metabolism (DM) studies needed to characterize the fate of a lead in the body depend on the results from preliminary animal PK and toxicology studies. Commonly, the results from these in vivo experiments are not available during discovery, and developability assessments and in vitro DM assays are utilized to determine the metabolic stability and/or the extent of metabolism of a lead and to compare the extent of metabolism among various species, including humans. These in vitro experiments can be conducted in a variety of assay systems, including CYP450 isozymes (the enzymes responsible for most oxidative metabolism of drugs), microsomes, hepatocytes, or liver slices. Since hepatocytes contain both phase 1 (oxidative, hydrolysis, and reduction) and phase 2 (conjugation) metabolism systems and can be relatively easily obtained from pharmacology and toxicology animal species and from human donors, many researchers select this model for the first assessment of metabolism. If the results from hepatocytes show extensive metabolism, additional in vitro experiments are usually conducted first in microsomes to ascertain if oxidative metabolism is present and then in isolated CYP450 isozymes to determine which enzyme, or enzymes, is responsible. Extensive metabolism is not necessarily a death knell for a discovery lead. If rapid clearance from the body is a desired developability attribute for effectively treating a disease indication, rapid metabolism to inactive metabolites may be advantageous. However, for most disease indications, extensive metabolism may prevent delivery of a pharmacologically active substance to the site of biological action in sufficient concentration to produce the desired response. Thus, a lead that is extensively metabolized may not be a successful drug candidate.

Another reason for conducting in vitro metabolism studies early is to determine if species differences are present. Evaluating metabolism in the pharmacological and proposed toxicological animal species and in humans assists in selecting the species that are similar, at least in metabolism, to humans for definitive toxicology studies. If an animal species has limited metabolism while humans may have extensive metabolism, pharmacological and/or toxicological metabolites may be responsible for some, or all, of the biological activity or adverse effects in humans, and these responses would not be observed in the animal model. Conversely, if an animal species has extensive or different metabolism compared to humans, the pharmacological and/or toxicological profiles in that species would probably not be predictive of efficacy and/or safety in humans.

If desired, which is sometimes the case when metabolism is extensive, the metabolites generated from in vitro systems can be isolated and identified. After preparation of sufficient quantities for additional testing, these metabolites can be evaluated for pharmacological and/or toxicological potential. This author, like many drug discovery and development researchers, has found metabolites with equal or greater biological activity when compared to the parent compound. At times, these pharmacologically activity metabolites have more druglike attributes than the parent and can be developed either as a replacement of the parent or as a second-generation drug candidate.

Toxicology Assays

Toxicology studies or assays are conducted to define the safety profile of a compound and include definition of the NOAEL, maximum tolerated dose or MTD, potential organs of toxicity, and potential biochemical markers to detect and track toxic events. Most discovery leads that do not become therapeutic products have unacceptable toxicity in animals and/or humans. Before the definitive toxicology studies needed to support submission for first-in-human testing are initiated, a number of in vitro and animal experiments can be conducted to characterize the potential toxicity of a discovery lead. These early toxicology evaluations are usually conducted in the same species as used in pharmacology evaluations. As mentioned earlier, the lowest dose that has no toxicity, or an acceptable level of toxicity, is compared with the dose that gives the desired pharmacological response in the same animal species to obtain a TR or TI for that species.

A toxicology program to obtain toxicological characterization of a discovery lead should be accomplished through close interaction with the efforts of other scientists conducting other developability experiments. Before drug safety studies are conducted, a sufficient quantity of the lead should be available and characterized, using appropriately characterized analytical chemistry assays, so that testing is conducted with a known compound. If the lead requires formulation before dosing, the formulation should be the same for each study. If a change in the synthesis, purification, or formulation is necessary to improve the biopharmaceutical properties of the lead or the drug delivery profile, then some of the early toxicology studies should be repeated with the new formulation to determine whether the safety profile has been altered. These early safety studies do not need to be, but in many cases are, conducted in compliance with Good Laboratory Practice (GLP) regulations. However, these experiments should be designed and conducted as close as possible to the processes used for definitive GLP toxicology studies. Then, the results will be scientifically defensible and useful in predicting the toxicity expected from the GLP studies. Examples of the early toxicology studies or safety assays needed to characterize a lead include the following.

In Vitro Toxicology When a number of discovery leads has been identified and needs to be further evaluated to select the optimal lead for further evaluations, the potential for toxic effects may be determined using in vitro assays, such as cell-based systems or microarrays. By incubating various concentrations of the leads with cells, such as the pharmacological target cell or hepatocytes, and measuring an adverse effect, such as cell death or change in cell function, the compounds can be stratified as to toxicological potential. Since nephrotoxicity is second only to hepatotoxicity as a major problem associated with adverse experiences during clinical trials, both hepatocytes and primary human kidney cells are commonly used in in vitro toxicology evaluations. Similarly, microarrays that have systems considered predictive of toxic events can be used to determine which leads "turn on" or "turn off" these systems after a lead has been incubated with cells or administered to an animal model and appropriate samples collected. Defining the appropriate samples and when to collect them can be difficult. If the wrong samples are collected or the right samples are collected at the wrong times, the results may not be predictive of the toxicological profile of the discovery lead(s). While most, if not all, toxicologists think these in vitro systems cannot be used to predict toxicology in animal models or humans, the results may be useful in evaluating a group of discovery leads to determine which lead may have a more acceptable profile compared to the others.

An alternate approach for an in vitro toxicological assessment is to use a biological marker that is considered predictive of toxic events. One possibility for such a marker is glutathione S-transferases (GSTs), which are important phase 2 drug-metabolizing enzymes. GSTs are found in high concentrations in several organs of key toxicological importance, such as the kidney and liver, and GST expression is induced in the presence of many toxic substances. GSTs are rapidly and readily released in the event of toxicological insult, thus making GSTs very sensitive biomarkers of organ damage.

Discovery leads are incubated with hepatocytes from various species. After incubation and centrifugation to remove the cells, the supernatants are assayed for GST content using already available and characterized assays. If appropriately designed, in vitro metabolism studies conducted with hepatocytes could be the source of these samples. Since hepatocytes contain both phase 1 and phase 2 metabolism systems, this in vitro toxicological assessment provides information not only about the safety profile of a discovery lead but also on that of its metabolites. The concentration range evaluated should include a level near the projected pharmacologically active level, a level at 5 to 10 times the projected pharmacologically active level, and a level near the maximum solubility of the lead (determined during solubility assessments).

In Vivo Toxicology To evaluate the qualitative and quantitative single-dose toxicity of a drug candidate, a single dose at a number of dose levels is administered by the proposed clinical route and the animals are observed for 14 days after dosing. The acute study is not an LD_{50} (dose that is lethal to 50 percent of test species) study, which is not needed for overall risk assessment according to an ICH guideline. This ICH guideline suggests that the drug candidate dose levels include at least one that produces pharmacological activity and one that causes overt evidence of major or life-threatening toxicity and that a vehicle control group is included. The acute toxicity study should evaluate both the IV route and the intended clinical route of administration, unless the clinical route is IV. The studies are usually conducted in two relevant mammalian species, one of which is a nonrodent, and unless scientifically unjustifiable (i.e., a gender-specific human disease or disorder), should use equal numbers of males and females for each species evaluated. The test species is observed for 14 days after dosing and, as with all toxicology studies, all signs of toxicity with time of onset, duration of symptoms, and reversibility are recorded. Also, the time to first observations of lethality is recorded. Gross necropsies are performed on all animals sacrificed moribund, found dead, or terminated after 14 days of observation and the results are presented by group. An evaluation of results should include all observations made and a discussion of the toxicological findings and their implications to humans, taking into account the pharmacology of the lead, the proposed human therapeutic use and dose, and experience with related drugs. The highest no-toxic-effect dose and the highest nonlethal dose are noted. A similar design, without the 14-day observation period, can be employed to evaluate the potential toxicology profile of a lead or a group of leads.

Organs or tissues that are the targets of toxicity may be identified during a full histology workup of animals in each dose group and from the results obtained from the analyses of clinical chemistry samples. However, during developability assessment, a full histology assessment may not be necessary and a selected set of organs or tissues (i.e., those considered most likely to be organs of toxicity such as the liver, kidney, lung, heart, brain) may be evaluated. If possible, investigations into the biochemical mechanism of identified toxicity should be initiated. Results from these experiments can provide insight into potential toxicities for a lead or a class of leads in definitive toxicology studies, identify biological markers that predict a toxic event, and suggest conditions in human patients where administration of a drug candidate is contraindicated. If results from the early toxicology studies show that a lead has an unacceptable level of toxicity, the development candidacy of such a compound should be carefully considered.

If desired, the safety of a lead can be further assessed by conducting safety pharmacology and/or genotoxicity studies. These studies, which are to be completed prior to the initiation of human clinical testing, are more commonly conducted after selection of a drug candidate. However, if some discovery leads are considered "equal" after other developability assessments have been completed, the results from safety pharmacology or genotoxicity studies may be able to identify the "optimal" lead or determine that some leads do not have the desired profile and should not be selected as the drug candidate.

11.4 CONCLUSIONS

During drug discovery and developability assessment, many biological and chemical assays are needed for evaluating the biological potency of compounds and for characterizing the attributes and demerits of discovery leads. Generally, two different assay set types, like the two faces of a coin, are necessary. On one side, the discovery side, assays need to be defined and/or developed to evaluate thousands of compounds quickly and to determine which, if any, of these compounds have the ability to antagonize or agonize a target that is thought to play an important role in a human disease process or disorder. Pharmacology, both in vitro and in vivo, and analytical chemistry assays used to support this activity need to be designed with this goal in mind.

On the other side or face of the coin are developability assessment assays that are used to evaluate whether or not a discovery lead, or one of a group of leads, has the desired attributes considered necessary for successful development and does not have a major demerit that may be predictive of development problems.

In this author's opinion, many discovery leads are transferred to the preclinical development process with insufficient characterization to assess their development potential. This lack of knowledge usually results in poorly designed experiments that are not data productive and that, in many cases, have to be repeated when the drug candidate shows unexpected toxicity, low and variable delivery, instability or solubility problems, or unacceptable pharmacokinetic and drug metabolism profiles. In all too many cases, the recognition of these problem areas results in the termination of development for a potentially useful therapeutic agent. At best, the problems encountered cause a delay, at times substantial, in the development of a candidate while additional studies are conducted to elucidate the causes of the problems and minimize their effect. Then, the definitive development experiments have to be repeated.

The developability experiments, with their associated assays, can more fully characterize a discovery lead before the compound enters the definitive preclinical, then the nonclinical and clinical, drug development processes. The experimental designs and assays could also be used, with minor modifications, to evaluate a group of discovery leads with acceptable biological potency and thus to select the lead with the best characteristics, that is, the most druglike attributes and the least demerits, for further development.

If these developability experiments are completed as part of the transition from discovery to development, discovery leads that do not have the characteristics necessary to become therapeutic agents can be identified early and prevented (i.e., the fail early, fail often approach) from entering the development process. Analogs of a lead with unacceptable characteristics can be evaluated to find a development candidate that has more optimal properties. In addition, the results from the developability studies will allow the preclinical development studies to be designed and conducted in a timely, cost-efficient manner and thus most likely allow the candidate to have an earlier entry into the clinic.

12

STRATEGIES AND METHODS IN MONITORING AND TARGETING PROTEIN–PROTEIN INTERACTIONS

ARIANNA LOREGIAN AND GIORGIO PALÙ
University of Padova
Padova, Italy

12.1	INTRODUCTION	483
	Importance of Studying Protein–Protein Interactions	483
12.2	METHODS FOR THE STUDY OF PROTEIN–PROTEIN INTERACTIONS	485
	Two-Hybrid System	485
	Peptide Display Technologies	509
	Protein Microassays	517
	Protein Mass Spectrometry	528
	Computational Methods of Analysis of Protein–Protein Interactions	533
12.3	DISCOVERY OF INHIBITORS OF PROTEIN–PROTEIN INTERACTIONS	536
	Two-Hybrid System Variants	536
	Other in Vivo Genetic Selection Systems	540
	Inhibitors from Phage Display	541
	Microarrays and Small Molecules	541
	References	542

12.1 INTRODUCTION

Importance of Studying Protein–Protein Interactions

Protein–protein interactions are intrinsic to virtually every cellular process ranging from cell cycle control, DNA (deoxyribonucleic acid) replication,

---

Drug Discovery Handbook, by Shayne Cox Gad
Copyright © 2005 by John Wiley & Sons, Inc.

transcription, splicing, and translation, to intermediary metabolism, secretion, formation of cellular macrostructures and enzymatic complexes. Apart from a variety of stable protein–protein interactions, there are a plethora of transient protein–protein interactions that control and regulate a large number of cellular processes. All modifications of proteins involve such transient protein–protein interactions. Indeed, glycosyl transferases, acyl transferases, kinases, phosphatases, and proteases interact only transiently with their protein substrates. Such protein-modifying enzymes encompass a large number of fundamental processes such as signal transduction, cell growth, and metabolic pathways.

Since protein–protein interactions play a role in nearly all events that take place in a cell, information on the function of an unknown protein can be obtained by investigating its interaction with other proteins whose functions are already known. Thus, if the function of one protein is known, then the function of its binding partner is likely to be related. This concept has been called "guilt by association" and allows the researcher to employ a relatively small number of functionally characterized proteins and to quickly assign functions to their uncharacterized binding partners.

Moreover, alteration of protein–protein interactions is known to contribute to many diseases. As an example, tumor-forming viruses cause uncontrolled proliferation of the host cell by dissociating important protein–protein interactions between regulatory proteins of the cell cycle [1]. Hence, the manipulation of protein–protein interactions that contribute to disease is a potential therapeutic strategy. The contact surfaces of the protein complexes have unique structure and properties, and they are more conservative in comparison with the active site of enzymes. So they represent prospective targets for a new generation of drugs. During the last decade, numerous investigations were undertaken to find or design small molecules that block protein dimerization or heterologous protein–protein interactions [2].

To date, a variety of genetic and biochemical methods exist for studying protein–protein interactions and identifying inhibitors of such interactions. This chapter describes recent developments in proteomic research. In more detail, the first part of this chapter focuses on technologies recently developed in protein interaction investigation, that is, yeast two-hybrid screens, phage display, protein microarray technology, two-dimensional electrophoresis coupled to mass spectroscopy, and so forth. Different strategies are compared; problems that are encountered in studying protein–protein interactions, solutions to these problems, and advantages and limitations of various methods and techniques are also discussed.

The second part presents recent approaches to identify and characterize new inhibitory molecules that act by disrupting biologically relevant protein–protein interactions. In particular, screening strategies that employ variants of the technologies reviewed in the first part of this chapter are discussed.

12.2 METHODS FOR THE STUDY OF PROTEIN–PROTEIN INTERACTIONS

Two-Hybrid System

Principle of Yeast Two-Hybrid System The yeast two-hybrid system was originally developed by Fields and Song as a genetic assay to detect protein–protein interactions in a cellular setting [3]. It takes advantage of the finding that many eukaryotic transcription activators have at least two distinct functional domains, one that directs binding to a promoter DNA sequence and one that activates transcription [4, 5]. This fact was demonstrated by exchanging DNA binding domains and activation domains from one transcription factor to another while retaining its function. For example, it was shown that the activation domain of yeast GAL4 protein could be fused to the DNA-binding domain of *Escherichia coli* LexA to create a functional hybrid transcription activator in yeast [6].

In the classical yeast two-hybrid approach, a *bait* is constructed by fusing a protein X to the DNA binding domain (DBD) derived from a transcription factor and a *prey* is created by fusing a protein Y to the activation domain (AD) of a transcription factor. The bait and prey fusions are coexpressed in yeast, where the interaction of X and Y leads to the reconstitution of a functional transcription factor and induces the expression of a reporter gene(s) integrated in the region downstream of the DBD binding sites (Fig. 12.1). Commonly, auxotrophic markers (e.g., *HIS3* and *LEU2*) that can be selected for are used as reporter genes in combination with the *lacZ* gene encoding bacterial β-galactosidase. *LEU2* and *HIS3* allow selection of positive interactions by monitoring growth on selective plates lacking leucine or histidine, respectively, whereas *lacZ* expression can be easily measured using a colorimetric assay. Thus, the yeast two-hybrid system is devised to identify genes encoding proteins that physically interact with a given protein in vivo. The technique enables not only the characterization of known interaction couples but also the identification of new interacting partners and even embodies the technological means to manipulate protein–protein interactions.

The yeast two-hybrid system presents a number of major advantages over alternative biochemical and genetic approaches. First of all, it relies on an assay performed in vivo (in the yeast host cell), and thus it is not affected by the artificial conditions of in vitro assays. Second, it allows very high numbers of coding sequences to be assayed in a relatively simple experiment. In fact, an appealing feature of yeast two-hybrid system is the minimal requirements to start a screening. Only the complementary DNA (cDNA), full-length or even partial, of the gene of interest is required, in contrast to sometimes high quantities of purified proteins or good-quality antibodies needed in classical biochemical approaches. In addition, one of the major advantages of this system is that the identification of an interacting protein implies that at the same time the corresponding gene is cloned.

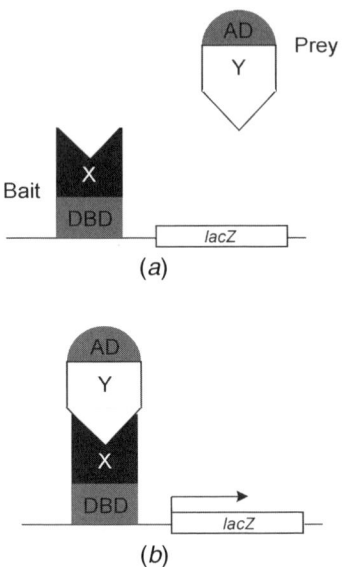

Figure 12.1 Principle of the yeast two-hybrid system. (*a*) Two chimeras, one (Bait) containing the DNA binding domain (DBD) fused to a protein X, and one (Prey) containing an activation domain (AD) fused to a protein Y, are cotransfected into an appropriate host strain. (*b*) If X and Y interact, the DBD and AD are brought into proximity and can activate transcription of a reporter gene(s) (here *lacZ*).

Weak and transient interactions, often the most interesting, can be more easily detected in two-hybrid assays since the genetic reporter gene strategy results in a significant amplification. By doing semiquantitative experiments, one can even interpret affinities from two-hybrid experiments. An exhaustive study of the sensitivity of the two-hybrid system with regard to the affinity of the protein–protein interaction has demonstrated that the strength of interaction, as predicted by the two-hybrid approach, usually correlates with that determined in vitro, allowing discrimination of low-, intermediate-, and high-affinity interactions [7]. The GAL4-based two-hybrid system can detect protein–protein interactions with a dissociation constant (K_d) of 10^{-6} M [8].

Since the emergence of the two-hybrid approach in 1989, a number of improvements have been developed that have increased its applicability. The two-hybrid system was initially predicted to be limited to the analysis of cytoplasmic proteins, as extracellular proteins are often *N*-glycosylated and contain disulfide bonds, both of which do not occur in the yeast nucleus [9]. However, several successes were reported with transmembrane receptors. Appropriate extracellular receptor–ligand interactions were demonstrated for the growth hormone, prolactin and growth-hormone-releasing receptors [10, 11]. Proteins of length as small as 8 to 10 residues [12] and as large as 778 residues [13] have been investigated in a two-hybrid study. Hydrophobic domains may affect expression but may be less problematic when expressed as a small percentage of the overall protein and when contained within the protein itself.

It should be noted, however, that the two-hybrid system cannot be used for all protein–protein interaction studies. For several experimental reasons, some proteins are not suited for this approach. Limitations and drawbacks of the two-hybrid system are listed in the next paragraphs.

First of all, a potential risk regards the use of artificially made fusion proteins. The fusion might change the actual conformation of the bait and/or prey and consequently alter its functionality. This misfolding might result in a limited activity or in the inaccessibility of binding sites. The best way to check the correct conformation of the protein of interest is to test its interaction with a known binding partner. This will only work if both fusion proteins are folded properly. Another way to circumvent the problem could be the reciprocal transfer of proteins, that is, switching the protein from the DBD fusion to the AD fusion and vice versa.

Since the readout of the two-hybrid system makes use of a transcriptional event, one of the most crucial initial experiments is to test whether the bait protein is able to activate transcription on its own. This can be a problem with transcriptional activators that naturally contain domains that activate the reporter genes when fused to the DBD. In addition, many proteins other than transcription factors have been shown to activate transcription when artificially fused to DBD [14]. Both classes of proteins are referred as *self-activators*. When dealing with self-activators, several approaches can be undertaken. First, the expression level of the DBD-X fusion can be decreased by using centromeric vectors and/or weaker promoters. Second, when using *HIS3* as a reporter gene, the concentration of 3-aminotriazole (3AT) can be increased to elevate the growth threshold of the host strain. The drug 3AT acts as a competitive inhibitor of the *HIS3*-encoded enzyme [8]. Under these conditions, it is expected that, even though the DBD-X fusion activates *HIS3* transcription to some extent, the DBD-X/AD-Y interaction leads to more *HIS3* expression to overcome the growth threshold imposed by 3AT [15]. However, these two strategies are not always successful, especially for strong self-activators. In this case, a "switched" two-hybrid system can be used [16], wherein the bait protein is fused to the activation domain (AD-X) and a DBD-Y library is screened.

One of the most ambiguous drawbacks is that the two-hybrid system makes use of yeast, *Saccharomyces cerevisiae*, as a host. The use of yeast can be seen as an advantage since yeast is closer to higher eukaryotics than those systems based on bacterial hosts or in vitro experiments. However, some interactions depend upon posttranslational modifications that do not, or inappropriately, occur in yeast. Such modifications are frequent and include phosphorylation, some types of glycosylation, and the formation of disulfide bridges. Some new versions of the two-hybrid system, however, try to overcome this problem by coexpressing the enzyme responsible for the posttranslational modification. Several plasmids have been designed to allow conditional expression of such a "third" partner. In addition to performing posttranslational modifications of one or both of the two interacting proteins, this third protein can directly contribute to the formation of a trimeric complex [17]. For instance, it has been shown that a tyrosine phosphorylation-dependent interaction could be detected when the corresponding kinase was coexpressed in yeast cells [18]. These different variations that involve third partners as native proteins, in the absence of any fused domains, are referred to as *tri-brid systems*.

Another limitation of the two-hybrid system is that both the DBD-X and the AD-Y fusions need to be translocated into the yeast nucleus. Thus, some two-hybrid vectors encode DBD and AD fused to a nuclear localization signal (NLS) to target the fusion proteins to the nucleus. However, proper localization can represent a major difficulty, especially when dealing with membrane-anchored proteins, extracellular proteins, or proteins that contain strong targeting signals. For this reason, some variations of the system have been designed for those proteins that are not active in the yeast nucleus, such as the *ubiquitin-based split-protein sensor* (USPS) system [19] and the *Sos recruitment system* (SRS) [20].

In addition, some proteins might become toxic upon expression in yeast. A number of proteins, such as homeobox gene products or cyclins are indeed toxic when expressed into the yeast nucleus. The use of an inducible promoter might circumvent the problem. Other proteins might proteolyze essential yeast proteins or the DBD and AD fusions.

It should also be kept in mind that it is impossible to exclude the possibility that a third protein Z is bridging the two interacting partners, and therefore a two-hybrid screen does not necessarily identify direct binary interactions. For example, some DBD-X fusions (e.g., DBD-lamin) are capable of allowing activation by the AD-Y fusions independent of a direct contact between X and Y. In addition, indirect interactions have been reported where an endogenous yeast protein acts as a bridge [21]. Nevertheless, the detection of such indirect protein–protein interactions can still be considered as an indication of a potential functional link between two proteins.

Finally, one of the major concerns for the yeast two-hybrid system are the so-called false positives, which actually include two different categories, namely technical and biological ones. The technical false positive is an apparent two-hybrid interaction that does not rely upon the assembly of two-hybrid proteins. Since the screen is based on the transcriptional activation of reporter genes, any event leading to an increase in the rate of transcription might be misinterpreted as a positive DBD-X/AD-Y interaction. The biological false positive means a bona fide two-hybrid interaction with no physiological relevance. Due to the so-called time/space constraints, it is potentially possible that two proteins, although able to interact, are never in close proximity to each other within the cell. The two proteins could be expressed in different cell types, or, even when present in the same cell, they could be localized in different intracellular compartments. Moreover, interacting proteins can be expressed at different times during embryogenesis or during homeostasis (e.g., at different points in the cell cycle). So once two interacting partners are selected, the biological significance of this interaction has to be investigated.

Most recently, modified two-hybrid strategies have been designed to increase the specificity in order to limit the above problems of false positives. Several strategies have been proposed. First, multiple reporter genes were designed for which the corresponding promoters are unrelated, containing different initiation sites and TATA boxes. Since many false positives of the two-hybrid method are promoter context dependent, the specificity of the assay is increased by testing the phenotypes conferred by different reporter genes in the same cell [22, 23]. Second, the expression level of the two fusion proteins was reduced by using truncated versions of the alcohol dehydrogenase 1 (ADH1) promoter and centromeric vectors, which are maintained at low copies in the yeast [24]. Third, mating strategies were introduced that enable parallel screening with many different baits [25, 26].

The quality of a two-hybrid screen can also be evaluated in terms of the number of expected interactions that were not detected. These are often referred to as *false negatives*. In many cases, the reasons for the lack of recovery of expected interactions are unknown, but several possibilities can be speculated. For example, the instability and/or the improper folding of a DBD-X or AD-Y fusion protein could affect its transcriptional activity. Alternatively, particular fusion proteins might be toxic and affect the viability of the host cells. In addition, the gene encoding an interacting protein may simply not be represented in the library due to a low complexity of the library or a bias in the representation of that clone.

Despite these and other limitations, the power of the yeast two-hybrid system is so enormous that it has quickly become the most frequently used assay to detect novel protein–protein interactions. A recent publication estimates that more than 50 percent of all interactions described in the literature have been detected using the yeast two-hybrid system [27].

Reagents Several comparable versions of yeast two-hybrid system are available. All systems share common elements. All use (1) a plasmid that directs the synthesis of a *bait*, that is, a known protein that is brought to DNA by being fused to a DBD, (2) a plasmid that directs the synthesis of a *prey(s)*, that is, a protein(s) fused to an AD and eventually other useful moieties, and finally (3) one or more reporter genes (*reporters*) with upstream binding sites for the bait. The versions differ in their specifics, and these details can be relevant to their successful use. The following paragraphs provide an overview of the most common yeast two-hybrid system reagents.

Target and Bait Vectors When performing a two-hybrid screen, the first decision to be made is the choice of the most appropriate vector system. A large number of different DBD- and AD-containing vectors are available. A list of the most commonly used plasmids in the yeast two-hybrid system together with their peculiar features is reported in Table 12.1. In addition, Parent et al. [28] have compiled an extensive list of yeast cloning vectors.

The most extensively used DBD and AD vectors are GAL4-based, probably because they were the first commercially available. The GAL4-based DBD vectors contain a portion of the yeast GAL4 protein [8, 29, 30]. This portion, encoded by residues 1 to 147, is sufficient to bind tightly to appropriate DNA binding sites, directs dimerization, and localizes fused proteins to the nucleus; it also contains a domain that weakly activates transcription from mammalian cell extracts in vitro, and it is thus conceivable that this domain may increase transcription resulting from weakly interacting proteins. The GAL4-based AD vectors usually contain the GAL4 activation domain II (aa 768–881).

Alternative systems make use of the DBD of the bacterial LexA protein and the AD of VP16 or the so-called B42AD. In LexA-based two-hybrid systems, the DBD is provided by the entire *E. coli* LexA protein [31, 32]. LexA carriers a dimerization domain at its C-terminus [33] and normally functions as a repressor of SOS genes in *E. coli* by binding LexA operator sequences that are part of the promoter [34]. When used in the yeast two-hybrid system, the LexA protein does not act as a repressor because the LexA operators are integrated upstream of the minimal promoter and coding region of the reporter gene. In yeast, LexA and most LexA derivatives enter the nucleus but are not necessarily nuclear localized. Both systems, GAL4 and LexA, have advantages and drawbacks that make the choice quite difficult. Since the LexA- and the GAL4-based two-hybrid systems have different properties, it is not unreasonable to imagine that some interactions might be detected differently in the two systems. Trying both will increase the chance of success.

Activation domain vectors can differ not only in the activation domain they carry but also in whether they contain other useful moieties such as epitopes tags and/or nuclear localization sequences [8, 29, 31, 32]. Some AD vectors, for example, pACT2, encode a hemaglutinin (HA) epitope tag (YPYDVPDYA in single-letter code), fused in frame to the activation domain [35]. This allows the protein to be detected with commercially available anti-HA antibodies in Western blot analysis. Some ADs are stronger than others. Although strong ADs should allow detection of weaker interactions, their expression can also be toxic to the cell due to poorly understood effects, either by subtraction of cofactors needed for transcription of other genes or by toxic effects that manifest when strong ADs are brought to DNA. Thus, it is possible that strong ADs may prevent detection of some interactions. AD fusion proteins also differ in whether they are expressed conditionally [31] or constitutively [8, 29]. Conditional expression allows the transcription phenotypes obtained in selections for interactors to be ascribed to the synthesis of the fusion protein, thus reducing the number of false-positive cells that grow because their reporters are aberrantly transcribed.

TABLE 12.1 Overview of Most Commonly Used Two-Hybrid Vectors

Plasmid Name	Selection Marker In Yeast	Selection Marker In E. coli	Functional Domain	Promoter	Comments
LexA-based Plasmids					
pBTM116	*TRP1*	Ap	LexA	ADH1 (truncated)	Basic plasmid used for cloning bait
pEG202	*HIS3*	Ap	LexA	ADH1 (full length)	Basic plasmid used for cloning bait
pJK202	*HIS3*	Ap	LexA + NLS	ADH1 (full length)	Same as pEG202 but incorporates an NLS between LexA and polylinker; used to ensure nuclear translocation of bait
pNLexA	*HIS3*	Ap	LexA		ADH promoter expresses polylinker followed by LexA; used with baits where N-terminal residues must remain unblocked
pGilda	*HIS3*	Ap	LexA	GAL1 (full length), inducible promoter, centromeric vector	Used for baits whose continuous presence is toxic
pEE2021	*HIS3*	Ap	LexA		An integrating form of pEG202 that can be targeted in *HIS3* following digestion with *KpnI*; used when lower expression levels of baits are required
pMW101	*HIS3*	Cm	LexA	ADH1 (full length)	Like pEG202, but with different antibiotic resistance marker
pMW103	*HIS3*	Km	LexA	ADH1 (full length)	Like pEG202, but with different antibiotic resistance marker

TABLE 12.1 *Continued*

Plasmid Name	Selection Marker In Yeast	Selection Marker In *E. coli*	Functional Domain	Promoter	Comments
pHybLex/Zeo	Zeocin	Zeocin	LexA	ADH1 (truncated)	Bait cloning vector with Zeocin marker; compatible with all two-hybrid systems
GAL4 DBD-based Plasmids					
pMA424	HIS3	Ap	GAL4 DBD	Original vector, 12 kb	
pGBT9	TRP1	Ap	GAL4 DBD	ADH1 (truncated)	Basic plasmid used for cloning bait
pAS1	TRP1	Ap	GAL4 DBD + HA	ADH1 (full length), CYH2	Incorporates a HA epitope for detection of baits
pAS2	TRP1	Ap	GAL4 DBD + HA	ADH1 (full length), CYH2	Incorporates a HA epitope for detection of baits
pAS2-1	TRP1	Ap	GAL4 DBD	ADH1 (full length), CYH2	
GAL4 AD-based Plasmids					
pGAD2F	LEU2	Ap	GAL4 AD	Original vector, 13 kb	
pGAD424	LEU2	Ap	GAL4 AD	ADH1 (truncated)	Basic plasmid used for cloning prey
pGAD10	LEU2	Ap	GAL4 AD	ADH1 (truncated)	Basic plasmid used for cloning prey
pJG4-5	TRP1	Ap	GAL4 AD + NLS + HA	GAL1	GAL1 promoter expresses GAL4 AD fused to NLS and HA; used to express cDNA libraries

Plasmid	Marker	Antibiotic	Fusion	Promoter	Description
pJG4-5I	TRP1	Ap	GAL4 AD + NLS + HA	GAL1	An integrating form of pJG4-5 that can be targeted in TRP1 following digestion with Bsu36I; used when lower expression levels of preys are required
pMW102	TRP1	Km	GAL4 AD + NLS + HA	GAL1	Like pJG4-5, but with different antibiotic resistance marker
pMW104	TRP1	Cm	GAL4 AD + NLS + HA	GAL1	Like pJG4-5, but with different antibiotic resistance marker
pACT1	LEU2	Ap	GAL4 AD	ADH1 (truncated), medium expression	
pACT2	LEU2	Ap	GAL4 AD + HA		Incorporates a HA epitope for detection of preys
Others					
pB42AD	TRP1	Ap	B42 + SV40 NLS + HA	GAL1 (full length), inducible promoter	GAL1 promoter expresses B42AD, SV40 NLS, HA epitope, polylinker
pYESTrp	TRP1	Ap	V5 epitope + SV40 NLS + B42	GAL1 (full length), inducible promoter	GAL1 promoter expresses SV40 NLS, B42AD, V5 epitope, polylinker; contains f1 ori and T7 promoter/flanking site; used to express cDNA libraries
pSD-10	URA3	Ap	VP16AD		

A key feature to be considered when choosing a vector for a successful screening is the promoter that regulates the expression level of the hybrid proteins. The 1500-bp full-length ADH1 promoter, which normally drives the expression of the metabolic enzyme alcohol dehydrogenase 1, allows high-level expression of sequences under its control. Expression from this promoter is maximal during logarithmic growth of the yeast cells and becomes repressed in late log phase by ethanol accumulation in the medium. Instead of this full-length promoter, some two-hybrid vectors, for example, pGAD424 and pGBT9, contain a truncated 410-bp ADH1 promoter. Expression from this promoter leads to low or very low levels of fusion protein expression [36]. The choice of the expression vector might be also influenced by the nature of the target. If the target is expected to interfere with the endogenous yeast metabolism, a lower expression might be beneficial. However, if the expression of the fusion protein needs to be assayed (e.g., by Western blot analysis), a higher expression level is more convenient.

The expression level depends not only on the promoter strength but also on the copy number of the plasmid. In most commonly used two-hybrid plasmids, the origin of replication is the 2μ origin. The 2μ plasmids are maintained stable and at high copy numbers (50 to 100 copies per cell) in yeast and function solely for their own replication [37]. It should be noted that in the context of a reverse two-hybrid system (see Reverse Two-Hybrid System), the expression levels of the fusion proteins should be maintained as low as possible since "background" interactions, which are more likely to occur at high protein expression levels, will kill the yeast cell. Therefore, the vectors used in the reverse two-hybrid systems make use of low-copy, centromeric vectors. Also when using toxic proteins, the use of centromeric plasmids in forward two-hybrid screens can be most helpful.

Finally, other characteristics that might influence the choice of a vector are of a more practical nature. In all vector systems care must be taken to maintain the proper reading frame when creating the fusion proteins. Thus, an important parameter to be considered when choosing a vector is the compatibility of the multiple cloning site. Another consideration concerns the marker gene(s). The most commonly used markers in yeast are genes encoding amino acid biosynthetic enzymes. Most two-hybrid yeast strains have a lesion in either *TRP1*, *LEU2*, *HIS3*, *ADE2*, and/or *URA3*, which enables selection for cells that were transformed with plasmids that carry the corresponding gene by growth in the absence of the appropriate amino acid(s). Alternatively, in a recently introduced vector (pHybLex/Zeo) zeocin is used as a selection marker.

Reporter Genes and Host Yeast Strains Reporter genes differ in the phenotypes they confer. Positive selection markers, such as *lacZ*, are widely used and provide colorimetric readout after an enzymatic assay. In the alternative, prototrophic markers (*LEU2* or *HIS3*) can be used and enable positive selection on media deficient for the specific amino acid encoded by the reporter gene, wherein a two-hybrid interaction results in cell growth. Reporter genes resulting in a sensitivity phenotype, such as G418 [38], *CYH2* [39], or *CAN1* [40], allow counter selection, wherein a productive protein–protein interaction abrogates cell growth (see Reverse Two-Hybrid System). In this case, yeast growth selection can be utilized to select mutations, proteins, peptides, or small molecules that dissociate target protein–protein interactions.

Many two-hybrid systems commonly use dual reporter systems to confirm the identification of a protein–protein interaction and/or putative novel binding partners isolated via a screen. For example, after an initial genetic selection based on growth assays, a secondary screen is performed with a second and different reporter gene such as *lacZ* to allow higher specificity [8, 31, 32]. This strategy allows one to address the question of the large numbers of transformants that need to be screened to properly survey the complexity of cDNA libraries.

Reporters also differ in the number and affinity of upstream binding sites for the bait and in the position of these sites relative to the transcription start point [31]. In the GAL4-based two-hybrid system, either the intact GAL1 upstream activating sequence (UAS), which contains four GAL4 binding sites, or an artificially constructed UAS consisting of three copies of the 17-mer consensus binding sequence is used. To avoid interference by endogenous GAL4 and GAL80 proteins, the yeast host strains used in the GAL4-based two-hybrid system must carry deletions both of the *GAL4* and *GAL80* genes. Due to the absence of these two proteins, the yeast cells grow more slowly as compared to the wild-type yeast. The use of bacterial LexA overcomes this problem. In LexA-based two-hybrid systems, expression of the reporter gene is under the control of six to eight copies of the LexA operator sequence and a minimal promoter.

Reporter genes can be integrated into the genome or reside on a plasmid. The disadvantage of having another plasmid containing the reporter gene, and thus the need for an additional auxotrophic marker, is compensated by several advantages. One of the major advantages of having the *lacZ* reporter gene on a high-copy-number plasmid is that weak signals are greatly amplified, which makes it possible to assay β-galactosidase activity directly on the selection plate by including X-gal in the medium. This avoids tedious replica and/or filter lift assays.

Finally, reporter genes differ in the number of molecules of the reporter gene product necessary to score the phenotype. These differences affect the strength of the protein–protein interactions that the system can detect. A list of the most commonly used yeast strains and reporter genes is given in Table 12.2.

Library Choice The source of DNA for the library is a key parameter for a successful screen. It is always a good idea to start with a library prepared from a tissue in which the target protein is known to be biologically relevant. A list of laboratory-made or commercially available cDNA or genomic libraries

TABLE 12.2 Overview of Most Commonly Used Yeast Strains and Their Reporter Genes

Strain	Reporter Gene(s)	Reporter Gene Regulated by	Origin of UAS
GAL4-Based			
Y187	lacZ	GAL4	GAL1 (= 4 × $UAS_{G\ 17\text{-mer}}$)
SFY526	lacZ	GAL4	GAL1 (= 4 × $UAS_{G\ 17\text{-mer}}$)
H7Fc	lacZ,	GAL4	3 × $UAS_{G\ 17\text{-mer}}$
	HIS3	GAL4	GAL1 (= 4 × $UAS_{G\ 17\text{-mer}}$)
YRG-2	lacZ,	GAL4	3 × $UAS_{G\ 17\text{-mer}}$
	HIS3	GAL4	GAL1 (= 4 × $UAS_{G\ 17\text{-mer}}$)
CG-1945	lacZ,	GAL4	3 × $UAS_{G\ 17\text{-mer}}$
	HIS3	GAL4	GAL1 (= 4 × $UAS_{G\ 17\text{-mer}}$)
Y190	lacZ,	GAL4	GAL1 (= 4 × $UAS_{G\ 17\text{-mer}}$)
	HIS3	GAL4	GAL1 (= 4 × $UAS_{G\ 17\text{-mer}}$)
LexA-Based			
EGY48	LEU2	6 × LexA op	
EGY191	LEU2	2 × LexA op	
L40	HIS3,	4 × LexA op	
	lacZ	8 × LexA op	

cloned in AD vectors is reported in Table 12.3. In most screens, cDNAs were derived from random- or oligo(dT)-primed RNA (ribonucleic acid). It should be kept in mind that, in contrast to genomic libraries, the relative representation of each cDNA closely reflects the endogenous expression level of the corresponding gene. Thus, interesting interacting proteins might be underrepresented, if their RNA is expressed at low levels in the cell. One solution is the use of normalized libraries. The process of normalizing cDNA libraries consists of reducing the representation of highly expressed cDNAs [41]. In addition, the choice of a random-primed versus an oligo(dT)-primed cDNA library can significantly influence the results of the screen. Discrete protein domains are more likely to be screened with random-primed libraries, while clones encoding nearly full-length proteins are enriched when oligo(dT)-primed libraries are used. In contrast, the complexity of genomic libraries is directly correlated to the size of the genome and to the number of independent clones that compose the library. For organisms that possess compact genomes, with small intergenic sequences and few introns, screening a genomic library instead of a cDNA library could be advantageous.

When screening libraries, a good representation is crucial. To screen a mammalian cDNA library until saturation, more than 5 to 10×10^6 yeast transformants need to be screened. In classical two-hybrid library preparations only one out of six fused cDNAs is in the correct frame, increasing the total number

TABLE 12.3 Overview of Laboratory-Made or Commercially Available cDNA or Genomic Libraries Cloned in AD Vectors

Host	Source	Vector (AD)	Supplier
Human Adult	293 (kidney) cells, cDNA	pACT2 (GAL4 AD)	Clontech
	HaCat (keratinocyte) cells, cDNA	pACT2 (GAL4 AD)	Clontech
	HeLa (cervical carcinoma) cells, exponentially growing, cDNA	pJG4-5 (GAL4 AD)	R. Brent
	HeLa cells, cDNA	pPC86 (GAL4 AD)	Invitrogen
	HeLa cells, cDNA	pYESTrp (B42)	Invitrogen
	HeLa cells, cDNA	pGADT7-Rec (GAL4 AD)	Clontech
	WI-38 cells (lung fibroblast), serum starved, cDNA	pJG4-5 (GAL4 AD)	R. Brent
	Jurkat cells (T-cell leukemia), exponentially growing, cDNA	pJG4-5 (GAL4 AD)	R. Brent
	Aorta, cDNA	pACT2 (GAL4 AD)	Clontech
	Bladder, cDNA	pYESTrp2 (B42)	Invitrogen
	Bone marrow, cDNA	pACT2 (GAL4 AD)	Clontech
	Brain, cDNA	pEXP-AD502 (GAL4 AD)	Invitrogen
	Brain, cDNA	pYESTrp (B42)	Invitrogen
	Brain, cDNA	pACT2 (GAL4 AD)	Clontech
	Breast, cDNA	pYESTrp2 (B42)	Invitrogen
	Breast tumor, cDNA	pYESTrp2 (B42)	Invitrogen
	Chondrocyte, cDNA	pACT2 (GAL4 AD)	Clontech
	Colon tumor, cDNA	pYESTrp2 (B42)	Invitrogen
	Erythroleukemia, cDNA	pACT2 (GAL4 AD)	Clontech
	Heart, cDNA	pPC86 (GAL4 AD)	Invitrogen
	Heart, cDNA	pEXP-AD502 (GAL4 AD)	Invitrogen
	Heart, cDNA	pACT2 (GAL4 AD)	Clontech
	Keratinocyte foreskin, cDNA	pGAD10 (GAL4 AD)	Clontech
	Kidney, cDNA	pYESTrp2 (B42)	Invitrogen

TABLE 12.3 *Continued*

Host	Source	Vector (AD)	Supplier
	Kidney, cDNA	pACT2 (GAL4 AD)	Clontech
	Leukocyte, cDNA	pACT2 (GAL4 AD)	Clontech
	Liver, cDNA	pJG4-5 (GAL4 AD)	J. Pugh
	Liver, cDNA	pPC86 (GAL4 AD)	Invitrogen
	Liver, cDNA	pYESTrp (B42)	Invitrogen
	Liver, cDNA	pACT2 (GAL4 AD)	Clontech
	Lung, cDNA	pPC86 (GAL4 AD)	Invitrogen
	Lung, cDNA	pYESTrp2 (B42)	Invitrogen
	Lung tumor, cDNA	pYESTrp2 (B42)	Invitrogen
	Lymph node, cDNA	pACT2 (GAL4 AD)	Clontech
	Lymphocyte, EBV-transformed peripheral blood lymphocytes, B-cell population, Ig+, cDNA	pACT2 (GAL4 AD)	Clontech
	Mammary gland, cDNA	pACT2 (GAL4 AD)	Clontech
	Ovary, cDNA	pYESTrp2 (B42)	Invitrogen
	Ovary, cDNA	pACT2 (GAL4 AD)	Clontech
	Pancreas, cDNA	pACT2 (GAL4 AD)	Clontech
	Placenta, cDNA	pYESTrp2 (B42)	Invitrogen
	Placenta, cDNA	pGADT7-Rec (GAL4 AD)	Clontech
	Prostate, cDNA	pPC86 (GAL4 AD)	Invitrogen
	Prostate, cDNA	pYESTrp2 (B42)	Invitrogen
	Prostate, cDNA	pACT2 (GAL4 AD)	Clontech
	Prostate, adenocarcinoma, cDNA	pPC86 (GAL4 AD)	Invitrogen
	Prostate, leiomyosarcoma, cDNA	pPC86 (GAL4 AD)	Invitrogen
	Skeletal muscle, cDNA	pPC86 (GAL4 AD)	Invitrogen
	Skeletal muscle, cDNA	pACT2 (GAL4 AD)	Clontech
	Spleen, cDNA	pPC86 (GAL4 AD)	Invitrogen
	Spleen, cDNA	pYESTrp2 (B42)	Invitrogen

Human Fetus	Spleen, cDNA	pACT2 (GAL4 AD)	Clontech
	Testes, cDNA	pYESTrp (B42)	Invitrogen
	Testes, cDNA	pACT2 (GAL4 AD)	Clontech
	Thymus, cDNA	pACT2 (GAL4 AD)	Clontech
	Brain, 22 weeks, cDNA	pJG4-5 (GAL4 AD)	R. Brent
	Brain, cDNA	pPC86 (GAL4 AD)	Invitrogen
	Brain, cDNA	pEXP-AD502 (GAL4 AD)	Invitrogen
	Brain, cDNA	pACT2 (GAL4 AD)	Clontech
	Kidney, cDNA	pACT2 (GAL4 AD)	Clontech
	Liver, cDNA	pYESTrp (B42)	Invitrogen
	Liver, cDNA	pACT2 (GAL4 AD)	Clontech
Mouse	CD4$^+$ T cells, cDNA	pJG4-5 (GAL4 AD)	V. Prasad
	Activated lymph nodes, cDNA	pPC86 (GAL4 AD)	Invitrogen
	Brain, cDNA	pPC86 (GAL4 AD)	Invitrogen
	Brain, cDNA	pACT2 (GAL4 AD)	Clontech
	7-day embryo, cDNA	pACT2 (GAL4 AD)	Clontech
	8.5-day embryo, cDNA	pPC86 (GAL4 AD)	Invitrogen
	10.5-day embryo, cDNA	pPC86 (GAL4 AD)	Invitrogen
	11-day Embryo, cDNA	pACT2 (GAL4 AD)	Clontech
	17-day Embryos, cDNA	pACT2 (GAL4 AD)	Clontech
	Embryonic fibroblasts, cDNA	pACT2 (GAL4 AD)	Clontech
	Kidney, cDNA	pPC86 (GAL4 AD)	Invitrogen
	Liver, cDNA	pACT2 (GAL4 AD)	Clontech
	Liver, cDNA	pACT2 (GAL4 AD)	Clontech
	Lymphoma, cDNA	pACT2 (GAL4 AD)	Clontech
	Testis, cDNA	pACT2 (GAL4 AD)	Clontech
Hamster	CHO (Chinese hamster ovary) cells, exponentially growing, cDNA	pJG4-5 (GAL4 AD)	V. Prasad

TABLE 12.3 *Continued*

Host	Source	Vector (AD)	Supplier
Rat	Brain, cDNA	pACT2 (GAL4 AD)	Clontech
	Brain, cDNA	pPC86 (GAL4 AD)	Invitrogen
	Liver, cDNA	pACT2 (GAL4 AD)	Clontech
	Liver, cDNA	pPC86 (GAL4 AD)	Invitrogen
	Lung, cDNA	pACT2 (GAL4 AD)	Clontech
Fish	Zebrafish, 1-month *Brachidanio rerio*, cDNA	pGAD10 (GAL4 AD)	Clontech
Xenopus laevis	Unfertilized oocyte, ages 1–3 years, injected with gonadotropin	pACT2 (GAL4 AD)	Clontech
Drosophila melanogaster	Adult whole, Canton S strain, cDNA	pACT2 (GAL4 AD)	Clontech
	Disc, cDNA	pJG4-5 (GAL4 AD)	R. Brent
	0- to 12-h embryos, cDNA	pJG4-5 (GAL4 AD)	R. Brent
	24-h embryos, cDNA	pACT2 (GAL4 AD)	Clontech
	Ovary, cDNA	pJG4-5 (GAL4 AD)	R. Brent
Caenorhabditis elegans	cDNA	pPC86 (GAL4 AD)	Invitrogen
Saccharomyces cerevisiae	Strain S228C, genomic DNA	pJG4-5 (GAL4 AD)	R. Brent
Schizosaccharomyces pombe	Strain SP972, exponentially growing, cDNA	pGAD GH (GAL4 AD)	Clontech

of independent clones to be screened to over a million. Making directional libraries of a relevant tissue or cell type could be a solution.

Screening of libraries selects for optimized interactions. Many isolates may not represent full-length cDNA. Indeed, it has been shown that protein subdomains can interact better than full-length proteins, probably reflecting domain function during folding of the protein. The best way to circumvent this problem is probably to clone only full-length cDNAs in the correct open reading frame. Although extremely time consuming, this approach was taken to establish the complete yeast protein linkage map (see Whole Genome Approaches Using the Two-Hybrid System).

For the library plasmid there is a major difference in the promoter used between the LexA and GAL4 system. While in the most commonly used plasmids of the GAL4 system, fusion proteins are weakly and constitutively expressed, they are cloned behind a stronger but inducible promoter in the LexA system. In one of the last versions of library plasmid used in the GAL4 system, pACT2, a truncated weakly expressing version of ADH1 is used.

Important is also the relative strength of the AD, related to its ability to initiate transcription. Both VP16 and the AD of GAL4 are strong activators making the system more sensitive. This might be needed for the detection of weak interactions but inevitably results in higher backgrounds. Therefore, the use of B42AD, which has intermediate transactivation activity, might be beneficial in conducting a screening.

New Developments Since the original description of the yeast two-hybrid system 16 years ago, a number of "variations on a theme" based on the initial idea have been described. In one set of variations, the classical configuration of the two-hybrid fusion proteins was modified to expand the range of possible protein–protein interactions that could be analyzed. For example, systems were developed to detect interactions that involve transcriptional activators or repressors, and membrane or extracellular proteins. In another set of variations, systems were developed to detect trimeric interactions or macromolecular interactions where a third protein activates or modifies one of the fused proteins and facilitates the protein–protein interaction (*three-hybrid systems*). Finally, the original concept was turned upside down and *reverse two-hybrid systems* were developed to identify mutations, peptides, or small molecules that dissociate protein–protein interactions. The next paragraphs describe the first two types of two-hybrid system variants; the reverse two-hybrid systems are discussed in the second part of the chapter. An additional review on progress and variations in two-hybrid and three-hybrid technologies is available [42].

Improvements of Classical Yeast Two-Hybrid System Refinements of vectors for use in the yeast two-hybrid system have been reviewed by Roder et al. [43]. A new host strain has been recently developed that is extremely sensitive to weak interactions and eliminates nearly all false positives using simple plate assays. This yeast strain contains three reporter genes: *HIS3, ADE2,* and *LacZ,* each under control of a different promoter (GAL1, GAL2, and GAL7, respectively) that responds to the same activator, GAL4 [23]. In another system, the target protein is expressed as a fusion with the DBD domain of the human estrogen receptor (ER) in a yeast strain containing an integrated *URA3* reporter gene driven by one or three ER response elements (EREs). In contrast to the *HIS3-* and *LEU2-*based systems, it can be assayed quantitatively by determining the activity of the reporter gene product orotidine-5'-monophosphate decarboxylase in cell-free extracts, and also allows negative selection by using 5-fluoroorotic acid [44].

Two-Bait Systems   The last few years have seen the advent of *two-bait systems* in which different baits are bound to DNA upstream of different reporters. These systems have been used to identify proteins that interact with different domains of a protein (Snf1) [45], different alleles of a protein (Ras) [46], and to identify mutant proteins that differentially bind to two known binding partners of a wild-type protein (Ste5) [47]. Combining data from these systems with data from classical two-hybrid systems likely can help dissect multimeric protein complexes [46]. In fact, two-bait systems, like other systems in which a third protein is expressed, can detect interactions that depend on modification or bridging by a third protein [48].

Sos Recruitment System   A recent variation of the yeast two-hybrid system, termed *Sos recruitment system* (SRS), is based on the Ras pathway in yeast and has the advantage of bypassing the reconstitution of a transcription factor [20]. The SRS is based on the observation that the mammalian guanosine 5'-diphosphate–guanosine 5'-triphosphate (GDP–GTP) exchange factor (GEF) hSos can only activate Ras when hSos is localized to the plasma membrane. In mammalian cells, this occurs by recruitment to the cytoplasmatic tail of activated growth factor receptors. The yeast *S. cerevisiae* requires a functional Ras signaling pathway for cell viability. A yeast strain containing a point mutation in the yeast GEF, *cdc25-2,* shows temperature-sensitive growth, but expression of hSos artificially targeted to the membrane by myristoylation or farnesylation can restore growth at nonpermissive temperature. Thus, the SRS uses the requirement of hSos to be recruited to the membrane to rescue growth as a method to detect protein–protein interactions. The target protein is fused to the GEF domain of hSos (GEF-X), and the partner protein or cDNA to be screened is fused to the Src membrane-anchoring (MA) domain (MA-Y). When these fusion proteins are coexpressed in a *cdc25-2* yeast strain, the cells grow at the nonpermissive temperature only if the two hybrid proteins interact; in this case, the GEF-X/MA-Y interaction allows recruitment of hSos to the membrane and rescue of the yeast *cdc25-2* mutation (Fig. 12.2).

The SRS may be the method of choice in studying interactions that involve transcriptional activators or repressors because it is not based on a transcriptional readout. Initially, the SRS was tested using the interacting AP-1 factors

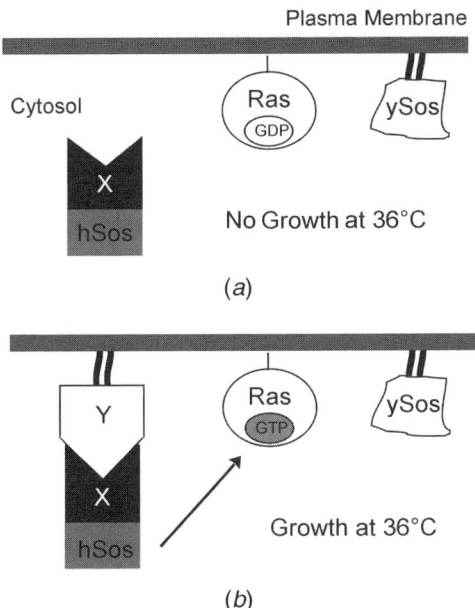

Figure 12.2 Principle of the SRS system. (*a*) In yeast cells carrying the *cdc25-2* mutation the yeast Ras guanyl exchange factor *cdc25* (ySos) is inactivated at nonpermissive temperature (36°C), leading to growth arrest. A bait X that is fused to the human Ras guanyl exchange factor Sos (hSos) cannot overcome the growth defect because the hSos-X fusion protein is located in the cytoplasm. (*b*) If an interacting protein Y is localized to the plasma membrane by means of a myristoylation signal, the interaction between X and Y recruits hSos-x to the membrane where it stimulates GDP exchange on Ras, thereby circumventing the *cdc25-2* mutation. Consequently, the yeast cells will grow at 36°C.

c-Jun and c-Fos, which are difficult to study in a conventional yeast two-hybrid system since both factors are strong self-activators [20, 49]. This system has been then successfully applied for identification of BRCA1 interacting partners [50]. In addition, certain proteins may function more physiologically when expressed in the cytoplasm rather than in the nucleus. For example, the SRS could be more suitable for examining interactions between proteins that require modification by cytoplasmic or membrane-associated enzymes.

However, a potential limitation of the SRS-based screening procedure is the isolation of false positives encoding Ras. In fact, mammalian Ras can bypass the requirement for a functional Ras GEF and represents a predictable false positive in this system. A possible solution to the problem is based on the observation that overexpression of GTPase activating protein (GAP) can suppress the bypass of the Cdc25 function by Ras [51]. However, it remains to be determined whether this improvement indeed significantly decreases the number of false positives.

The USPS System The *ubiquitin-based split-protein sensor* (USPS) system, originally developed by Johnsson and Varshavsky [19], relies on the properties of the ubiquitin protein. The ubiquitin protein consists of 76 amino acids and is highly conserved between all eukaryotes [52]. Its primary function is to act as a "tag" for degradation by being attached to proteins through ubiquitin conjugating enzymes. Once a protein becomes tagged with several ubiquitin moieties, it is transported to the 26S proteasome, where ubiquitin-specific proteases (UBPs) cut the peptide bond at a double glycine motif in the junction between the attached ubiquitins and the target protein. The released ubiquitin moieties are then recycled back to the cytoplasm, whereas the target protein is degraded by the 26S proteasome [53].

The USPS system takes advantage of the highly specific cleavage mediated by the UBPs. The expression of a fusion protein consisting of ubiquitin and a C-terminally attached reporter in yeast results in fast and complete cleavage by UBPs within minutes [19]. The folded structure of ubiquitin is crucial to the recognition and subsequent cleavage events by UBPs: expression of an N-terminal ubiquitin moiety carrying a point mutation in a hydrophobic core residue (NubG) together with the C-terminal ubiquitin moiety (Cub) in the same yeast cell does not result in cleavage by UBPs anymore, presumably because the partially unfolded NubG moiety does not recognize and bind to the Cub moiety. As the UBPs do not recognize Cub alone, no cleavage of the attached reporter takes place. In order to exploit the ubiquitin moieties as reporters for detecting protein–protein interactions, two binding partners, X and Y, are fused to NubG and Cub, respectively. Upon interaction of X and Y, the NubG and Cub moieties are brought into close proximity, resulting in a partial refolding of NubG, followed by reassociation of NubG and Cub into what the authors termed *split-ubiquitin*. Split-ubiquitin represents a good substrate for UBPs and, therefore, the attached reporter is cleaved off (Fig. 12.3). In the original description of the system, the reporter protein used was human dihydrofolate reductase (DHFR) expressed as a fusion to Cub, and cleavage of such a hybrid was used as an indication for X/Y interaction.

As the formation of split-ubiquitin and the subsequent cleavage by UBPs do not depend on any particular intracellular localization of the proteins, the split-ubiquitin system is suitable for the study of membrane proteins. This system has been successfully used to detect interaction among a variety of proteins, including sucrose transporters [54], yeast oligosaccharyl transferases [55], transmembrane proteins normally present in the endoplasmic reticulum [56, 57], and proteins involved in viral replication in *Arabidopsis* [58]. This two-hybrid variant was also proven capable of reconstituting the well-characterized homodimerization of the yeast Gcn4p leucine zipper domain. However, this strategy has not yet been adapted for the identification of novel interacting proteins.

Protein Three-Hybrid Systems The protein three-hybrid systems represent an additional adaptation of the yeast two-hybrid system that allows investi-

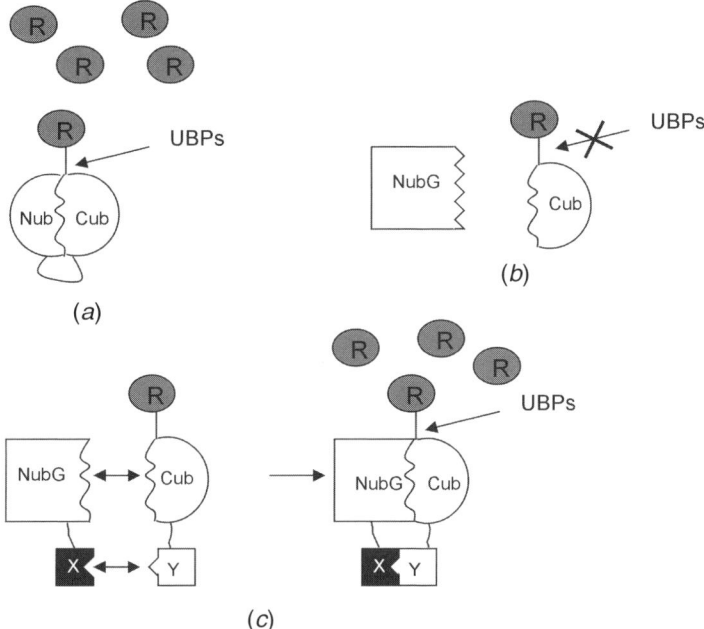

Figure 12.3 The ubiquitin-based split-protein system (USPS). (*a*) Native ubiquitin is recognized by UBPs, which cleave at a double-glycine motif located at the carboxy terminus of ubiquitin. A reporter protein (R) that is fused to the C-terminus of ubiquitin is released upon cleavage. (*b*) Cleavage at an extended loop separates ubiquitin into a C-terminal (Cub) and an N-terminal (Nub) domain. The introduction of a point mutation into Nub (NubG) causes its partial misfolding and therefore abolishes the binding of Nub to Cub. As UBPs do not recognize the isolated Cub domain, the reporter R is not cleaved from the C-terminus of Cub. (*c*) A protein X is fused to NubG and its interacting partner Y is fused to Cub. The interaction of X and Y brings NubG and Cub together, allowing the partial refolding of NubG and leading to their reassociation into split-ubiquitin. Like native ubiquitin, split-ubiquitin is recognized by UBPs, leading to the cleavage and release of the reporter R.

gation of ternary protein complexes. In fact, stable interaction of proteins X and Y may rely on the presence of a third protein Z. Protein Z either mediates the interaction or induces a conformational change in one of the proteins (e.g., X) so that it promotes interaction with another protein Y. One of these systems was employed by Zhang and Lauter [48] to investigate intracellular associations of proteins during initiation of the epidermal growth factor receptor (EGFR) signal transduction cascade. Within the yeast cells, the EGFR cytoplasmic domain was expressed as a DBD fusion protein and Sos as an AD fusion protein. To identify new interacting proteins, a mouse brain library fused to the SV40 T-antigen NLS was coexpressed from a third vector in conjunction with the EGFR–DBD and the Sos–AD hybrid proteins, leading to the identification of the mouse Grb2 protein [48]. Recently, another variant of this system was proposed by Tirode et al. [17]. This version uses both LexA and GAL4 plasmids and is based on the addition of a Met25 promoter for the third partner. This third partner is cloned on the same plasmid that already encodes the GAL4 and LexA hybrid proteins. By doing this, a selection marker is saved. Transcription of the third protein is repressed upon addition of methionine in the culture medium and provides an elegant way to perform negative controls [17].

An alternative use of a third protein coexpressed with the two-hybrid fusion proteins is exploiting the third protein's ability to modify or activate one of the fused proteins and facilitate the protein–protein interaction. As an example, the kinase three-hybrid system termed *tri-brid system* can detect protein–protein interactions that depend on posttranslational modifications. In fact, one of the limitations of the classical version of the yeast two-hybrid system is the lack of some crucial posttranslational modifications (such as tyrosine phosphorylation) in *S. cerevisiae* [59]. This was solved by introduction of a third component, that is, a cytosolic tyrosine kinase, which then phosphorylated substrates in the yeast cell. In principle, it should be possible to incorporate any posttranslational or allosteric regulation that is desired. However, care must be taken that the posttranslational modification is properly performed in yeast [60].

Finally, a yeast multiprotein system that allows simple and rapid detection of interactions among up to four proteins was recently developed by Loregian et al. [61]. This system, which further expands the three-hybrid system to include a fourth protein, could be most useful to dissect multiprotein complexes.

Mammalian Two-Hybrid   Although most two-hybrid systems use yeast, some mammalian variants do exist. In one, the gene of interest is expressed as a fusion with the GAL4 DBD. The second gene is expressed as a fusion with VP16 AD. A third vector, pG5CAT, contains the chloramphenicol acetyltransferase (*CAT*) reporter gene downstream of a GAL4 responsive UAS. These three vectors are cotransfected into a mammalian cell line. Interaction between proteins X and Y is assayed by measuring *CAT* gene expression by any standard method [62]. In another version, interaction of proteins fused to VP16 with a GAL4-derived bait drives expression of hygromycin B phosphotransferase, *CAT*, or CD4 cell surface antigen [63].

New Applications of the Yeast Two-Hybrid System   During the past decade, together with the evolution of the technique, new applications have emerged. Apart from the molecular dissection of known interactions and the identification of new interacting partners, several new problems can now be addressed by means of the two-hybrid technology.

Interaction Suppression   Interaction suppression makes use of the two-hybrid technique to evaluate the biological relevance of an interaction between two proteins. First, mutations that affect interaction of a target protein (A) to its partner (B) are identified. Once a mutant version of A is found, the influence of this mutation on the phenotype can be studied. However, if one finds mutants of protein A that do not bind protein B, one cannot exclude that also the interaction of protein A with an unknown protein C is abolished and accounts for the change in phenotype. Therefore, in a second step one has to search for binding suppressors, that is, a mutant version of B that is able to interact with the mutated version of A and restores the altered phenotype. This approach was taken to study the Ras/Raf pathway [64–66].

Protease Trap This system relies on the nuclear localization of the hybrid protein [67]. A functional transcription factor is fused to a domain that prevents it from translocating into the nucleus. If a protease site is cloned in between the transcription factor and this domain, it is possibile to search for a protease that cuts off the antinuclear localization domain enabling nuclear localization of the transcription factor and finally resulting in reporter gene expression. This approach could also be applied to screen for target sequences for a known protease by cloning random sequences in between the two functional domains.

Whole Genome Approaches Using the Yeast Two-Hybrid System One of the most spectacular and ambitious applications of the two-hybrid system consists of establishing so-called protein linkage maps (PLMs). These maps consist of all possible protein interactions that occur during the entire life span of a cell. PLMs might provide information on new functions of known proteins by discovery of unexpected interactions, identify cross-connections between cellular pathways, result in the identification of novel drug targets, and gain more insight in the overall cell system. Since it is a genetic system, the yeast two-hybrid system is well suited to high-throughput applications such as the identification of interactions that take place between all proteins expressed in a given cell or organism.

Currently, two approaches are being used to generate PLMs. In the so-called *matrix approach*, or *array approach*, a set of open reading frames (ORFs) is amplified using the polymerase chain reaction (PCR), cloned as both bait and prey (i.e., as a fusion to a DBD and as a fusion to an AD), and then expressed into isogenic yeast strains of opposite mating type. The yeast strain expressing a DBD fusion is then mated with an array of clones each expressing a different AD fusion [25]. Since yeast mating is a very efficient process, numerous combinations of DBD-X/AD-Y can be assayed simultaneously. Practically, this task is carried out by robots that transfer aliquots from a lawn of cells expressing one DBD fusion to arrays of cells each expressing a different AD fusion. This procedure is repeated for each strain expressing a DBD fusion, until all DBD fusions have been mated with the entire AD array. Positive interactions are selected through the ability of diploid yeast colonies containing an interacting fusion pair to grow on selective media. In order to eliminate the false positives arising from such an approach, the experiments are performed in duplicate, and only interactors found in both experiments are considered to be true positives. An advantage of the matrix approach is that it rapidly becomes clear which locations produce false positives, confirming that the system is working properly; if a particular AD hybrid in the array binds all DBD fusions, it most likely represents a false positive and should thus be discarded. This approach is nevertheless restricted by several limitations. First, only known or predicted proteins can be tested, and this is an obvious key problem in genome-wide projects. Second, the matrix approach enables the use of a single set of growth conditions, which precludes the possibility of using different selective conditions for each specific interaction. For example, the 3AT concentration cannot be adjusted to account for weak self-activation of some baits. Third, the use of full-length proteins for both the DBD-X and the AD-Y fusions might prevent the identification of several interactions due to various intrinsic problems such as folding, degradation, and toxicity.

A second approach in genome-wide yeast two-hybrid analysis is the so-called *library screening approach* in which DBD fusions are screened against complex libraries containing AD fusions of full-length ORFs or ORF fragments. In contrast to the matrix approach, this method does not separate the different AD fusions on an array. Instead, the library is divided into pools, and each yeast strain expressing a DBD fusion is mated with a library pool. Then, diploid cells containing an interacting protein pair are selected. The library screening approach is more sensitive than the matrix approach since it uses not only full-length ORFs but also random ORFs fragments. Often, a protein–protein interaction can be detected using fragments of the proteins of interest but not the full-length proteins, as they do not fold properly when expressed in yeast or are degraded [68]. The use of fragments often circumventes these problems. On the other hand, library screens are much more expensive and time consuming than matrix screens since they require the analysis of larger numbers of clones. In addition, the library plasmids encoding AD hybrids have to be isolated and sequenced from all selected clones in order to identify the interacting partners.

In the past few years, several systematic two-hybrid projects have been undertaken to analyze protein–protein interactions at a global level [68–72]. These are summarized in Table 12.4, together with relevant information about the type of cell or organism that was investigated, the method that was used to perform the screen, and the number of protein–protein interactions that were identified. The first PMLs were reported in 1994 for *Drosophila melanogaster* [73], followed by similar screenings for T7 bacteriophage [74] and a subset of yeast proteins involved in messenger RNA (mRNA) splicing [26]. Likewise, Walhout et al. [68] and Boulton et al. [71] mapped protein–protein interaction networks involved in *Caenorhabditis elegans* vulval development and DNA-damaging response components, respectively. Finally, Rain et al. [72] have recently built a large-scale protein–protein interaction map of the human gastric bacterial pathogen *Helicobacter pylori*, whose genome encodes 1590 putative ORFs [75]. A total of 261 baits were used to detect 1524 interactions, resulting in a protein interaction map covering much of the proteome.

The most comprehensive large-scale screening approaches reported to date focus on the yeast *S. cerevisiae*. Two groups studied all 6131 annotated yeast ORFs using both matrix and library screening approaches [69, 70, 76]. Initially, Ito et al. [76] performed a large-scale matrix screen using 159 ORFs that were cloned as DBD and AD fusions. They analyzed 430 matings (representing 10 percent of all possible combinations between DBD and AD fusions) and identified 175 interactions, of which 163 had not previously been reported. They recently completed their systematic approach, identifying a total of 841 interactions [70]. The second group performed both matrix and library screenings [69]. Using the matrix method, 192 ORFs were created as DBD fusions and then mated with the 6131 yeast ORFs fused to the AD, resulting in 281 protein–protein interactions. For the library screen, a library was made by pooling all 5345 AD-fused ORFs. These were then mated separately to the same 5345 ORFs fused to the DBD, yielding a total of 692 protein–protein interactions.

A surprising observation was made when comparing the results of Ito et al. [76] and Uetz et al. [69]. Despite the fact that both groups used the same yeast ORFs in their experiments, only 20 percent of all interactions detected in the two screens overlapped [70]. The reasons for this small overlap are difficult to explain, but most likely the differences are due to the different experimental settings. For example, the different DBD and AD plasmids used by the two groups may have affected the folding and expression level of the proteins, the use of PCR to amplify the yeast ORFs may have introduced mutations that abolish interactions, and, most importantly, the stringency of selection may have been different, eliminating interactions seen by one group from the other group's results. In summary, the large-scale screenings carried out so far indicate that even when using exhaustive library screens that potentially cover all interactions in a genome, it is still difficult to estimate what percentage of protein–protein interactions that occur in a cell or organism under investigation are actually identified in such screens. The small overlap between data sets of Ito et al. [76] and Uetz et al. [69] suggests that even within the subset of protein–protein interactions that can be detected using the yeast two-hybrid system, such screenings are still far from being exhaustive.

Peptide Display Technologies

Surface display of peptide and proteins is another approach to detect and characterize protein–protein interactions. The advantage of peptide display technology over other research tools and its great potential have been demonstrated by successful applications in diverse fields of biomedical research, including drug discovery [77–79].

TABLE 12.4 Overview of Genome-Wide Protein–Protein Interaction Studies

Organism	Predicted ORFs	Sample	Screen Method	Number of Interactions	References
T7 bacteriophage	55	Proteome	Matrix screen + Library screen	25	1
Vaccinia virus	266	Proteome	Matrix screen	37	2
Hepatitis C virus	10	Proteome	Matrix screen + Library screen	15	3
H. pylori	1,590	Subset of proteome	Library screen	1,524	4
S. cerevisiae	6,131	Nuclear proteins involved in pre-mRNA splicing pathway	Library screen	170	5
		RNA polymerase III subunits	Library screen	20	6
		SM-like proteins	Library screen	229	7
		Proteome	Library screen	841	8
		Proteome	Matrix screen	281	9
		Subset of proteome	Library screen	692	
C. elegans	19,293	Proteins involved in vulval development	Matrix screen	11	10
			Library screen	148	
		Proteins involved in DNA damage response	Matrix screen	26	11
			Library screen	165	
		26S proteasome subunits	Matrix screen	17	12
			Library screen	138	
D. melanogaster	13,600	Cell-cycle regulatory proteins	Matrix Screen	19	13

1. Bartel, P. L., Roecklein, J. A., SenGupta, D., Fields, S. (1996). A protein linkage map of *Escherichia coli* bacteriophage T7. *Nat. Genet, 12,* 72–77.
2. McCraith, S., Holtzman, T., Moss, B., Fields, S. (2000). Genome-wide analysis of vaccinia virus protein-protein interactions. *Proc. Nat. Acad. Sci. USA, 97,* 4879–4884.
3. Flajolet, M., Rotondo, G., Daviet, L., Bergametti, F., Inchauspe, G., Tiollais, P., Transy, C., Legrain, P. (2000). A genomic approach of the hepatitis C virus generates a protein interaction map. *Gene, 242,* 369–379.
4. Rain, J. C., Selig, L., De Reuse, H., Battaglia, V., Reverdy, C., Simon, S., Lenzen, G., Petel, F., Wojcik, J., Schachter, V., Chemama, Y., Labigne, A., Legrain, P. (2001). The protein-protein interaction map of *Helicobacter pylori. Nature, 409,* 211–215.
5. Fromont-Racine, M., Rain, J. C., Legrain, P. (1997). Toward a functional analysis of the yeast genome through exhaustive two-hybrid screens. *Nat. Genet., 16,* 277–282.
6. Flores, A., Briand, J. F., Gadal, O., Andrau, J. C., Rubbi, L., Van Mullem, V., Boschiero, C., Goussot, M., Marck, C., Carles, C., Thuriaux, P., Sentenac, A., Werner, M. (1999). A protein-protein interaction map of yeast RNA polymerase III. *Proc. Nat. Acad. Sci. USA, 96,* 7815–7820.
7. Fromont-Racine, M., Mayes, A. E., Brunet-Simon, A., Rain, J. C., Colley, A., Dix, I., Decourty, L., Joly, N., Ricard, F., Beggs, J. D., Legrain, P. (2000). Genome-wide protein interaction screens reveal functional networks involving Sm-like proteins. *Yeast, 17,* 95–110.
8. Ito, T., Chiba, T., Ozawa, R., Yoshida, M., Hattori, M., Sakaki, Y. (2001). A comprehensive two-hybrid analysis to explore the yeast protein interactome. *Proc. Nat. Acad. Sci. USA, 98,* 4569–4574.
9. Uetz, P., Giot, L., Cagney, G., Mansfield, T. A., Judson, R. S., Knight, J. R., Lockshon, D., Narayan, V., Srinivasan, M., Pochart, P., Qureshi-Emili, A., Li, Y., Godwin, B., Conover, D., Kalbfleisch, T., Vijayadamodar, G., Yang, M., Johnston, M., Fields, S., Rothberg, J. M. (2000). A comprehensive analysis of protein-protein interactions in *Saccharomyces cerevisiae. Nature, 403,* 623–627.
10. Walhout, A. J., Sordella, R., Lu, X., Hartley, J. L., Temple, G. F., Brasch, M. A., Thierry-Mieg, N., Vidal, M. (2000). Protein interaction mapping in *C. elegans* using proteins involved in vulval development. *Science, 287,* 116–122.
11. Boulton, S. J., Gartner, A., Reboul, J., Vaglio, P., Dyson, N., Hill, D. E., Vidal, M. (2002). Combined functional genomic maps of the *C. elegans* DNA damage response. *Science, 295,* 127–131.
12. Davy, A., Bello, P., Thierry-Mieg, N., Vaglio, P., Hitti, J., Doucette-Stamm, L., Thierry-Mieg, D., Reboul, J., Boulton, S., Walhout, A.J., Coux, O., Vidal, M. (2001). A protein-protein interaction map of the *Caenorhabditis elegans* 26S proteasome. *EMBO Reps., 2,* 821–828.
13. Finley, R. L. Jr., Brent, R. (1994). Interaction mating reveals binary and ternary connections between *Drosophila* cell cycle regulators. *Proc. Nat. Acad. Sci. USA, 91,* 12980–12984.

Phage Display Phage display is the most widely adopted molecular display technique. Smith first demonstrated in 1985 [80] that an *E. coli* phage can express a fusion protein bearing a foreign peptide on its surface and suggested that libraries of recombinant phages could be constructed and screened to identify proteins that bind to a specific partner. In general, phage display usually refers to an in vitro selection technique in which a peptide or a protein is genetically fused to a coat protein of a filamentous phage, such as M13, fd, and f1, resulting in display of the fused protein to the surface of the virion, while the DNA encoding the fusion resides within the phage.

Protein sequences are usually fused to the amino terminus of either the major coat protein, pVIII (about 3000 copies per phage), or a tail protein, pIII (5 copies per phage). Thus, a foreign DNA sequence inserted in these genes results in multiple copies of the fusion protein displayed by the phage. This is termed *polyvalent display*. In general, polyvalent display is limited to small peptides because larger inserts interfere with the function of the coat proteins, and the phage becomes poorly infective. In polyvalent display, the peptides are either displayed in an unconstrained form or in a constrained form by introducing flanking cysteine residues. The latter peptides are much less flexible, and peptides selected from constrained libraries have quite often higher affinities than those selected from unconstrained libraries. The minor coat protein pVI has been also used to display foreign peptide sequences fused at its C-terminus, a system that can be more suitable when a free C-terminal end is necessary for proper activity [81].

Next, a phage library is enriched for members that bind an immobilized target by a procedure called *panning* (Fig. 12.4) in which the phage library is incubated with the protein of interest bound to a plastic dish or well. After washing to remove nonspecifically bound phages, specifically bound phages are eluted either by competition with free target or by more stringent washes. The library is then amplified in *E. coli*, and the selection is repeated to enrich for phages possessing peptides specifically binding to the target protein. Multiple rounds of selection make this approach very efficient. Finally, the DNA is extracted from individual viral plaques and sequenced to determine the sequence of the peptides that bind to the target protein.

A major advance in phage display came with the development of the *monovalent systems* in which the coat protein fusion is expressed from a phagemid, and a helper phage supplies a large excess of the wild-type coat protein [82, 83]. Therefore, the phages are functional because the recombinant protein forms only a small amount of the total coat protein. The vast majority (>99 percent) of the population of phage virions displays either one or no copies of the fusion protein on their surface. Such phages can accommodate up to 100 kDa of heterologous protein without any significant decrease of phage infectivity. In addition, monovalent phage display avoids potential avidity effects observed with polyvalent display, in which the phage can attach to the absorbent at multiple points.

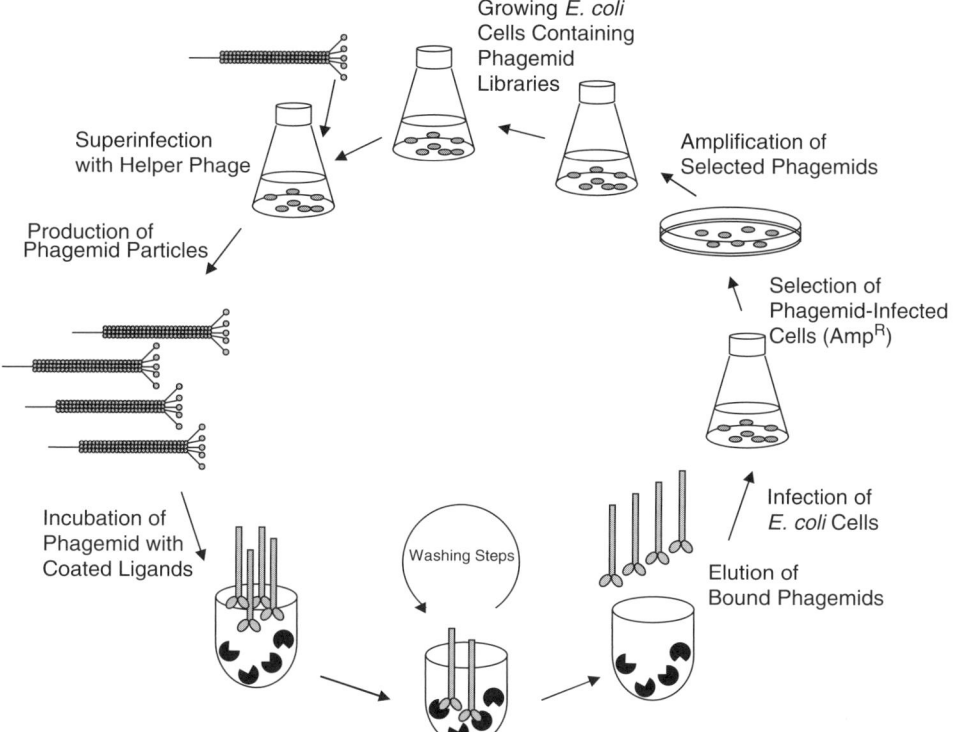

Figure 12.4 Scheme of the panning cycle. Phage particles displaying fusion proteins of peptide variants are produced by infecting phagemid libraries, i.e., *E. coli* cells carrying the phagemid as a plasmid DNA, with helper phage (*upper left*). These particles are incubated with a target of interest that has been immobilized on a plate or bead (*lower left*). Particles with low affinity for the target are washed away (*bottom*). The remaining particles are eluted by disrupting the binding interactions between the phage and target, e.g., with acid, and then used to infect fresh *E. coli* cells; clones are selected for antibiotic resistance (*right half*). The enriched clones are now ready to be packaged. The process is then repeated, resulting in stepwise enrichment of the phage pool in favor of the tightest binding sequences. After multiple rounds of selection/amplification, individual clones are characterized by DNA sequencing.

Consequently, phages were used for the display not only of short peptides but also of functional proteins such as antibody fragments, hormones, enzymes, and enzyme inhibitors, as well as for the selection of specific phage on the basis of functional interactions (antibody–antigen, hormone–hormone receptor, enzyme–enzyme inhibitor, etc.). Antibodies have been displayed on phages in the form of scFv or Fab' fragments using either the phage or the phagemid system. In the latter case, either the VH-CH1 or VL-CL chain is fused to pIII while the other chain is expressed without pIII fusion. Thus, phage display has been applied successfully not only to epitope mapping but also to identification of critical amino acids responsible for protein–protein interactions and to isolate leads for the discovery of new therapeutics (see the second part of this chapter).

Phage display presents several advantages for the study of protein–protein interactions. It is a rapid procedure and has been shown to be widely applicable. There are, however, limitations to the technique. In early examples of phage display, the length of the peptides displayed on bacteriophage virions was usually restricted to 6 to 15 amino acids because larger polypeptides (>10 residues for pVIII display) compromised phage infectivity and thus could not be efficiently displayed. The development of phagemid display systems solved this problem by fusing the peptides to an additional coat protein encoded by a phagemid vector and subsequently infecting with a helper phage [82]. Other intrinsic restrictions of display systems based on filamentous phage include their inability to display protein sequences toxic to bacterial cells and the fact that the fusion protein must be secreted across the bacterial membrane. An alternative system to alleviate the toxicity problem is represented by the use of bacteriophage λ, in which the synthesis of any toxic protein can be repressed during the lysogenic state and induced for a very short time just before cell lysis [84]. Furthermore, protein secretion is not required for phage λ display because this phage assembles intracellularly. In addition, λ-based systems offer the advantage of efficient construction and maintenance of very large libraries. Two λ-phage display systems have recently been described. One approach entails fusion to the V gene, which encodes a tail fiber, at the end corresponding to the C-terminus [85]. Alternatively, peptide sequences can be fused to the amino terminus of the D protein on the phage capsid [84]. Other phage vectors wherein displayed protein sequences are not secreted are the bacteriophage T4, where polypeptides can be exposed by fusion with the C-terminus of fibritin, a protein of the bacteriophage neck [86], and the bacteriophage T7 (see Reagents).

Other disadvantages of the phage display include the use of a bacterial host, which may preclude the correct folding or modification of some proteins, and the fact that all phage-encoded proteins are fusion proteins, which may limit the activity or the accessibility for binding of certain proteins.

Finally, one should keep in mind that, since the phage display libraries consist of fully randomized peptides displayed on phage, a binding partner identified in a particular panning experiment will not necessarily correspond to a "natural" ligand for the target. The panning procedure iteratively selects for those peptides that best bind the target under the panning conditions in vitro, without regard to the biological role of the target in vivo. For certain targets, such as antibodies with linear epitopes, the selected sequence will in all likelihood correspond to that region of the antigen recognized by the antibody. For targets that bind to large surfaces of a protein, or discontinuous regions of the primary sequence, the selected sequences are less likely to resemble the "natural" ligand. As a result, caution should be taken if one is planning on using the DNA corresponding to the selected sequences as probes when trying to clone any natural ligand proteins. In contrast, cDNA expression libraries are limited to natural proteins and, as a result, are much more likely to yield the native ligand for the target protein.

Other Peptide Display Systems   In the past few years, there have been numerous developments in the peptide display technology to make it applicable to a variety of protein–protein and protein–peptide interactions.

Bacterial Display Systems   A significant restriction of phage display is the unpredictable expression bias against some eukaryotic protein sequences expressed in *E. coli*, because incorporation of any protein fusion into phage particles depends upon the ability of *E. coli* to express that protein sequence in soluble form. A number of alternatives to phage display have been developed for the detection of protein–protein interactions in bacteria. One approach to physically link a protein to its gene is by fusion with the *lac* repressor [87]. Other methods of isolating peptides by attaching the encoded protein to its gene include affinity selection of nascent proteins translated on polysomes [88] and expression on the surface of bacteria [89]. Proteins or peptides have been displayed on the surface of *E. coli,* fused either to outer membrane proteins [90] or flagella [91]. However, like phage display, *E. coli* display is subjected to potential library bias due to expression in a prokaryotic host.

Eukaryotic Display Systems   In contrast to *E. coli*, yeast cells possess eukaryotic posttranslational processing enzymes and secretory machinery. Thus, a display system using *S. cerevisiae* as the host organism should overcome the library biases due to expression in a prokaryotic host. The yeast surface display system uses the α-agglutin mating adhesion receptor, which consists of two domains, Aga1 and Aga2 [92], to display recombinant protein on the surface of *S. cerevisiae*. Peptide sequences are fused to the carboxyl terminus of Aga2, and the Aga2 fusion is disulfide bonded to Aga1, which is in turn covalently linked to the yeast cell wall by phosphatidyl inositol glycan linkages. Therefore, each product of a DNA library is tethered to the surface of the yeast cell wall in a manner that makes it accessible to macromolecular recognition without steric hindrance from cell wall components. This could be of significant advantage when displaying peptides as candidate molecules for the disruption of protein–protein interactions (see the second part of this chapter).

A mammalian peptide display system also exists, wherein proteins of interest are targeted and anchored to the cell surface by cloning their genes in frame with an N-terminal secretion signal and the C-terminal transmembrane anchoring domain of platelet-derived growth factor receptor (see Reagents). In addition, a variety of eukaryotic viruses have been also proposed for the display of foreign peptides. Examples of these are the tobacco mosaic virus, in which *Plasmodium malariae* epitopes were displayed both on the surface of a loop region and at the C-terminus of the coat protein [93], and human rhinovirus 14, on whose surface a library of variants of the V3 loop of human immunodeficiency virus type 1 (HIV-1) gp120 has been displayed [94]. The eukaryotic display of peptide sequences has been successfully achieved also using the baculovirus *Autographa californica*, through expression of foreign peptide sequences as coat protein fusions [95].

In Vitro Display Technologies Some new display technologies have recently emerged that operate entirely in vitro and use PCR rather than cells to amplify genetic material [79]. For example, one of these systems relies on the unique properties of a replication initiator protein, P2A, encoded by the A gene of *E. coli* bacteriophage P2. This protein covalently attaches to the same molecule from which it has been expressed. By constructing a peptide library from random sequences linked to the A gene, the expressed P2A-peptide fusion moieties are directly tagged with the peptide coding sequence. These in vitro display technologies, by circumventing the need to introduce DNA into a host cell, could allow the construction of libraries that are several orders of magnitude larger (potentially up to 10^{15} molecules) than those obtainable when using whole cells or viruses for peptide display. Moreover, the linkage between the peptide sequence and its encoding nucleic acid leads to several advantages, including a greater control over binding conditions to the target molecule and the ease with which PCR-based mutagenesis and recombination can be performed.

The choice of the most appropriate expression system for the display of peptide libraries ultimately depends on the protein–protein interaction of interest. There would be no advantage, for example, in using a eukaryotic display system for the display of short sequence libraries when phage systems have proven so effective because of their simplicity, minimal cost, and ease of manipulation. On the other hand, eukaryotic display systems could be required for longer peptide sequences whose functional display needs proper posttranslational modifications.

Reagents In the last few years there has been significant progress in developing vectors for peptide display. In early versions of phage display, the foreign DNA was inserted into the phage genome between the N-terminal domains of gene III needed for infection and the C-terminal domain required for morphogenesis. In subsequent versions, the fusion site was moved to the N-terminus, which improved infectivity of the phage. Based on these findings, libraries displaying peptides of various length instead of gene fragments were created by various groups and successfully applied to the identification of protein ligands and epitope mapping.

Another important reason for improving the vectors for display of cDNA libraries was that fusion to either pIII or pVIII requires maintenance of the correct reading frame at both the junction with the signal sequence and with the C-terminus of the phage coat protein. Therefore, full-length cDNAs cannot be used as they contain stop codons. To overcome this problem, vectors have been developed that only require a single fusion at the end of the cDNA encoding the amino terminus [81, 96].

One of the currently most used versions is the Ph.D. display system, which enables the display of custom peptide libraries on the surface of bacteriophage M13 as coat protein fusions, creating a physical linkage between each displayed peptide and its encoding DNA sequence. This system utilizes the display vector M13KE, which is an M13 derivative with cloning sites engineered for N-terminal pIII fusion, resulting in a valency of 5 displayed peptides per phage. The use of a phage vector, rather then a phagemid, simplifies the intermediate amplification steps since neither antibiotic selection nor helper phage superinfection are required. However, displayed proteins longer than 20 to 30 residues have a deleterious effect on phage infectivity.

Large display libraries can be easily made using the T7Select vector. Peptides up to 50 residues can be displayed in large copy numbers (415 per phage); in the alternative, proteins up to 1200 amino acids can be displayed in low copy numbers (0.1 to 1 per phage) or midcopy numbers (10 per phage). Displayed peptides and proteins do not need to be exported through the periplasm and the cell membrane, as phage assembly takes place inside the *E. coli* cell and mature phages are realesed by cell lysis. Moreover, T7 grows rapidly, which saves time during cloning/screening procedures, and is extremely stable, expanding the variety of agents that can be used in panning.

The FliTrx Random Display Library system employs the same basic route of expression-fusion protein display as does phage display, but without the phage. The vehicle of choice is the FliTrx vector, which positions a diverse library of peptides in a flagellin (*fliC*)-thioredoxine (*trxA*) fusion protein (FliTrx). The recombinant protein is exported and assembled into partially functional flagella and displayed on the bacterial cell surface in a conformationally constrained form due to insertion of the peptide into the active-site loop of the thioredoxine protein. Peptides inserted in this loop have both their N- and C-terminus tethered by the rigid and stable tertiary fold of the thioredoxine molecule. With this system, phage infection and isolation steps are eliminated.

Finally, the pDisplay is a mammalian expression vector that is designed to target recombinant proteins to the surface of mammalian cells. Displayed proteins can be analyzed for their ability to interact with known or putative ligands. Proteins are targeted and anchored to the cell surface by cloning the gene of interest in frame with pDisplay's unique N-terminal cell surface targeting signal and the C-terminal transmembrane anchoring domain of platelet-derived growth factor receptor.

Protein Microarrays

Another approach for the high-throughput characterization of protein–protein interactions that has been recently developed exploits the protein microarray technology [97]. Here, proteins are expressed, purified, and attached to the surface of a microarray in a way that preserves their folding and their ability to specifically bind other proteins. Initially, protein chips were utilized to identify yeast genes encoding specific biochemical activities [98]. Recently, Zhu et al. [99] applied this technology to the study of protein–protein interactions.

Advances in microfabrication [100] and surface chemistry methods [101] have all allowed the development of several protein microarray formats, including the commercially available ProteinChips, functional protein arrays, antibody arrays, peptide arrays, interaction arrays, and surface plasmon resonance (SPR)-based arrays.

Apart from protein–protein interactions, protein microarrays can also be used to study other binding activities, such as protein–nucleic acid, protein–small molecule, and protein–drug interactions. Although these are all in vitro assays, the advantage is that the experimental conditions can be well controlled. For example, different cofactors or inhibitors can be included in the assays to vary the stringency of the binding reaction. Another advantage is that these assays are not biased toward abundant proteins. In addition, with adequate detection methods, protein microarrays can be used to identify the protein substrates of various enzymes such as proteases, methyl transferases, protein kinases, and phosphatases [102]. Finally, protein microarrays can be used to identify in vivo posttranslational modifications by probing for specific modifications, such as phosphorylation or glycosylation, using antibodies or lectins, respectively. Protein microarrays also have the potential advantage of allowing analysis of the kinetics of protein–protein interactions via real-time detection methods.

In summary, protein microarrays can be used to globally analyze the activities of proteins including their interaction with proteins, nucleic acids, lipids, carbohydrates, and small molecules. Because of its versatile and miniaturized nature, microarray technology is expected to enormously flourish in the near future. Different types of commonly used protein microarrays are listed in Table 12.5 and described in the following sections.

ProteinChips In the ProteinChip technology, proteins are exposed to chips with different surface features in parallel so that sets of proteins with common properties (hydrophilic, hydrophobic, charged, etc.) will attach to a particular type of surface. After washing to remove unbound proteins, proteins of interest can be enriched on the chip surface by selective washing, and/or protease digestion [103]. ProteinChips require very small sample amounts and can be used directly with biological fluids to provide information on post-translational modifications, protein structure, and other properties [103, 104]. Currently, ProteinChips have been used for a wide variety of applications including the identification of collagen binding proteins in *Lactobacillus* species [105] and the quantitation of prostate-specific membrane antigen levels [106].

Functional Protein Microarrays   Another approach for the exhaustive characterization of the biochemical activities of proteins of interest is using functional protein microarrays. In this technique, sets of proteins or even an entire proteome is overexpressed, purified, distributed in an array format, and then assayed for specific activities. These arrays can be used not only to screen for biochemical activities of interest but also to analyze posttranslational modifications and to detect binding to proteins, antibodies, small molecules, and drugs. The latter use has potentially powerful applications in drug discovery.

There are several types of functional protein arrays (reviewed in [102]): nanowells, which are miniature wells; thick absorbent surfaces, such as hydrogels; and solid surface supports, such as glass slides. Nanowell arrays typically contain wells 1 mm or less in diameter. The wells can be made of plastic material, such as polydimethylsiloxane (PDMS), or alternatively can be etched in glass. The nanowells have the advantges of compartmentalizing samples and of reducing evaporation. Using the nanowell arrays, Zhu et al. [98] characterized the kinase–substrate specificity of almost all (119 of 122) yeast kinases using 17 different substrates. The substrates were first covalently attached to the surface of the nanowells, and then the protein kinases were incubated with the substrates and [^{33}P]ATP (adenosine 5′-triphosphate). After washing away the kinases and unincorporated ATP, the nanowell chips were analyzed for phosphorylated substrates. Not only known kinase–substrate interactions were detected but also many novel activities were identified.

Glass microscope slides are also widely used for functional protein microarrays, as they are compatible with many commercial scanners. MacBeath and Schreiber [107] used this format to detect antibody–antigen interactions, protein kinase activities, and protein binding to small molecules. The major limitation in functional protein microarrays has been the preparation of proteins to analyze. This requires high-throughput expression and purification methods that can yield a large number of functionally active proteins. This problem was recently overcome (see High-Throughput Protein Production).

Antibody Microarrays   In antibody microarrays, antibodies are usually spotted onto a glass slide; then the arrays are incubated with a cell lysate or serum to allow the antigens present in the sample to bind to their cognate antibodies. The bound antigen is detected either by using radioactively labeled or fluorescently tagged proteins or by using a secondary antibody against each antigen of interest.

Low-density antibody arrays have been created that quantitate the levels of several proteins in serum [108, 109] and blood [110]. Recently, Sreekumar and colleagues [111] spotted 146 different antibodies on high-density glass arrays to characterize the differential expression of a number of antigens in LoVo colon carcinoma cells. They found that radiation treatment of the cells upregulated the levels of many interesting proteins, including tumor necrosis

TABLE 12.5 Overview of Current Protein Microarrays

Support	Binding	Advantages	Disadvantages	References
Ni-NTA-coated	Affinity binding	Low background; protein binding in uniform orientation; specific, strong, and high-density protein binding	Proteins have to be Hisx6 tagged	1
Avidin-coated	Affinity binding	Low background; protein binding in uniform orientation; specific, strong, and high-density protein binding	Proteins have to be biotinylated	2
DNA/RNA-coated	Hybridization	Low background; protein binding in uniform orientation; specific, strong, and high-density protein binding	Difficult production of labeled proteins	3
Nitrocellulose	Adsorption	High capacity of protein binding, no requirement of protein modification	High background, nonspecific adsorption, low-density protein binding	4, 5
Polylysine-coated	Adsorption	No requirement of protein modification	Nonspecific binding	6
Polyvinylidene fluoride	Adsorption	High capacity of protein binding, no requirement of protein modification	Random orientation of surface-bound proteins	7, 8
3D gel pad and agarose thin film	Diffusion	No requirement of protein modification, high capacity of protein binding	Difficult to make, not commercially available	9, 10
Epoxy-activated	Cross-linking	Strong and high-density protein binding, high-resolution detection methods	Protein binding in random orientation	11
Aldehyde-activated	Cross-linking	Strong and high-density protein binding, high-resolution detection	Protein binding in random orientation	1, 12

Gold-coated silicon	Cross-linking	Low background; specific, strong, and high-density protein binding; well suited for SPR and mass spectroscopy detection	Protein binding in random orientation; difficult to make, not commercially available	13, 14
Polydimethylsiloxane nanowells	Cross-linking	Specific, strong, and high-density protein binding, useful for biochemical assays	Protein binding in random orientation	11

1. Zhu, H., Bilgin, M., Bangham, R., Hall, D., Casamayor, A., Bertone, P., Lan, N., Jansen, R., Bidlingmaier, S., Houfek, T., Mitchell, T., Miller, P., Dean, R. A., Gerstein, M., Snyder, M. (2001). Global analysis of protein activities using proteome chips. *Science*, 293, 2101–2105.

2. Rowe, C. A., Tender, L. M., Feldstein, M. J., Golden, J. P., Scruggs, S. B., MacCraith, B. D., Cras, J. J., Ligler, F. S. (1999). Array biosensor for simultaneous identification of bacterial, viral, and protein analytes. *Anal. Chem.*, 71, 3846–3852.

3. Winssinger, N., Ficarro, S., Schultz, P. G., Harris, J. L. (2002). Profiling protein function with small molecule microarrays. *Proc. Nat. Acad. Sci. USA*, 99, 11139–11144.

4. Joos, T. O., Schrenk, M., Hopfl, P., Kroger, K., Chowdhury, U., Stoll, D., Schorner, D., Durr, M., Herick, K., Rupp, S., Sohn, K., Hammerle, H. (2000). A microarray enzyme-linked immunosorbent assay for autoimmune diagnostics. *Electrophoresis*, 21, 2641–2650.

5. Ge, H. (2000). UPA, a universal protein array system for quantitative detection of protein-protein, protein-DNA, protein-RNA and protein-ligand interactions. *Nucleic Acids Res.*, 28, e3.

6. Haab, B. B., Dunham, M. J., Brown, P. O. (2001). Protein microarrays for highly parallel detection and quantitation of specific proteins and antibodies in complex solutions. *Genome Biol.*, 2, RESEARCH0004.

7. Cahill, D. J. (2001). Protein and antibody arrays and their medical applications. *J. Immunol. Methods*, 250, 81–91.

8. Walter, G., Bussow, K., Cahill, D., Lueking, A., Lehrach, H. (2000). Protein arrays for gene expression and molecular interaction screening. *Curr. Opin. Microbiol.*, 3, 298–302.

9. Guschin, D., Yershov, G., Zaslavsky, A., Gemmell, A., Shick, V., Proudnikov, D., Arenkov, P., Mirzabekov, A. (1997). Manual manufacturing of oligonucleotide, DNA, and protein microchips. *Anal. Biochem.*, 250, 203–211.

10. Afanassiev, V., Hanemann, V., Wolfl, S. (2000). Preparation of DNA and protein micro arrays on glass slides coated with an agarose film. *Nucleic Acids Res.*, 28, E66.

11. Zhu, H., Klemic, J. F., Chang, S., Bertone, P., Casamayor, A., Klemic, K. G., Smith, D., Gerstein, M., Reed, M. A., Snyder, M. (2000). Analysis of yeast protein kinases using protein chips. *Nat. Genet.*, 26, 283–289.

12. MacBeath, G., Schreiber, S.L. (2000). Printing proteins as microarrays for high-throughput function determination. *Science*, 289, 1760–1763.

13. Houseman, B. T., Huh, J. H., Kron, S. J., Mrksich, M. (2002). Peptide chips for the quantitative evaluation of protein kinase activity. *Nat. Biotechnol.*, 20, 270–274.

14. Bieri, C., Ernst, O. P., Heyse, S., Hofmann, K. P., Vogel, H. (1999). Micropatterned immobilization of a G protein-coupled receptor and direct detection of G protein activation. *Nat. Biotechnol.*, 17, 1105–1108.

factor-related ligand, DNA fragmentation factors 40 and 45, and p53, and downregulated the levels of other proteins.

The major problem with antibody arrays is antibody specificity. Haab et al. [112] evaluated the reactivity of 115 antibodies with their cognate antigens. Protein microarrays containing either immobilized antibody or immobilized antigen were probed with antigens or antibodies, respectively. Only 30 percent of the antibody–antigen pairs showed specific binding, indicating that most antibodies are not suitable for quantitative detection. Nevertheless, for those antibodies that are specific, quantitation of antigen levels in a complex mixture could be obtained. In fact, in a later study they demonstrated that antibody microarrays could be used to obtain serum profiles [113].

SPR-Based Arrays Surface plasmon resonance (SPR) has been shown to be a versatile detection tool to analyze the kinetics of protein–protein interactions over a wide range of affinities, molecular weights, and binding rates [114–116]. Although the commercially available SPR chips are not yet high-throughput, Myszka and Rich [117] recently described a sensor surface with 64 individual immobilization sites in a single flow cell. Alternatively, Sapsford and colleagues [118] developed an antibody array biosensor and studied the kinetics of antigen–antibody interactions. Surface plasmon resonance–biomolecular interaction analysis (SPR–BIA) is another chip-based method that has been used for kinetic and thermodynamic studies of protein–protein interactions. Finally, in an adaptation of biosensor technology, soluble recombinant insulinlike growth factor receptor domains have been isolated on a chip to analyze protein–protein interactions [119]. This allowed the characterization of the binding of protein domains to other proteins and also provided information about the function of nonbinding domains.

Peptide Microarrays A modified version of protein microarrays is peptide microarrays, which can be used as potential ligands and as substrates for enzymatic activites when probing with proteins or other molecules. In one recent study, Houseman and colleagues [120] immobilized 9-mer peptide kinase substrates on a gold-coated glass surface to form a high-density peptide microarray and analyzed the phosphorylation of the peptides using fluorescence, SPR, and phosphoimaging. They could also quantitatively evaluate the effect of three kinase inhibitors. This study demonstrated the usefulness of coupling peptide chips with various detection methods to quantitatively analyze the dynamics of enzyme–substrate interactions, which could have potential applications in drug discovery.

Interaction Arrays In the field of protein–protein interaction assays, dot-blot filter arrays have been used to screen for specific interactions of immobilized proteins with other proteins. For example, Ge [121] has described a protein array for quantitative detection of protein binding to a variety of proteins, nucleic acids, and small molecules. A set of 48 proteins were spotted onto nitrocellulose filters and a ^{32}P-labeled glutathione *S*-transferase (GST) fusion of the human protein p52, known to be a transcriptional activator, was incubated with the array. He detected binding to nucleolin, which has been reported to have several functions, including ribosomal RNA processing and regulation of cell doubling time in cancer cells. In addition, interactions of DNA, RNA, or low-molecular-weight ligands with the immobilized molecules were shown. Such arrays could be miniaturized and therefore have the potential to be created in a microarray format.

Recent work by Zhu et al. [99] demonstrated the extraordinary power of array-based technology for proteomic studies. They cloned ~94 percent (>5800 of 6131) of the yeast ORFs in a yeast expression vector that expresses the proteins as N-terminally GST-Hisx6 double-tagged fusions and developed a high-throughput protein purification method using the GST tags. The purified proteins were then attached to Ni-NTA-coated glass slides using the Hisx6 tags. These microarrays could be utilized to characterize protein–protein interactions on a genome-wide scale. Using calmodulin as a model protein to probe the arrays, many known interactions could be confirmed and a set of novel binding proteins was detected. Analysis of the sequences of these proteins revealed the presence of a binding motif and therefore demonstrated the specificity of the binding interaction. Moreover, experiments aimed at detecting protein–lipid interactions demonstrated that this technology also enables the identification of proteins able to bind low-molecular-weight molecules. This suggests that an entire proteome can be immobilized onto a glass surface to directly screen for interactions with proteins and small molecules and therefore opens the possibility to examine an entire proteome directly for protein–drug interactions.

Reagents and Methods

High-Throughput Protein Production A major limiting factor influencing the development of protein arrays is the availability of adequate protein reagents. Proteins are more challenging to prepare for the microarray format than DNA. For successful implementation on chips, proteins need to be chemically stable to the derivatization and retain activity. Protein functionality is often dependent on the state of proteins, such as subcellular localization, posttranslational modifications, and binding to other proteins. Nevertheless, in recent years many research groups and companies have contributed enormous effort in developing high-throughput protein purification methods. The commonly used methods for preparing protein reagents include cloning as fusions to protein tags such as Hisx6 or GST, expression in a suitable host, and high-throughput purification [122]. High-throughput methods have been used to purify proteins from various host cells, including *E. coli*, yeast, insects, and human cell lines, allowing the preparation of microarrays containing over 100 proteins and even an entire proteome [99, 102, 122–128].

Alternatively, proteins can be produced using cell-free expression systems. For example, Keefe and Szostak [129] developed an mRNA display system in which each protein is in vitro translated and covalently linked through its C-terminus to the 3' end of its coding mRNA. More interestingly, He and Taussig [130] have reported a method for in vitro translation of Hisx6-tagged proteins directly onto Ni-NTA-coated surfaces via a protein in situ array (PISA), enabling single-chain antibodies to be arrayed for high-throughput immunoassays. This method has the advantage of synthesizing in situ proteins that may not be amenable to deposition from solution or to cell-based expression (e.g., insoluble proteins).

Antibodies used for protein microarrays can derive from polyclonal sera or can be hybridoma-derived mAbs, recombinant antibodies, or antibodies selected from phage display libraries.

Finally, while most approaches entail the attachment of purified proteins to solid supports, some approaches adopt on-chip synthesis of probes. These include the high-density peptide chemical synthesis approach. The so-called SPOT technology of immobilized peptides is being developed for a number of surfaces including chips and membranes to study several activities, including protein–protein interactions [131]. Membranes are made functional by chemical derivatization followed by spacer/linker attachment. Peptides are then created by standard Fmoc peptide synthesis. Such a synthetic strategy enables the use of peptides containing unnatural residues.

Supports Due to the susceptibility of protein activity to derivatization and immobilization chemistries, considerable effort has been directed toward optimizing the solid surface of protein arrays. It is important that the support retains proteins in an active state at high densities, can be printed in such a fashion that the proteins remain in a moisturized environment, and is compatible with most commercial arrayers and scanners. Soft substrates such as polyvinylidene fluoride (PVDF), polystyrene, and nitrocellulose membranes, which have been used to attach proteins in traditional biochemical assays (e.g., immunoblot and phage display), are often not suitable to protein microarrays [121, 122, 132]. These surfaces often do not allow high protein density, as the spotted material may spread on the surface, and/or they may not allow optimal signal-to-noise ratios [123, 127, 128, 133]. These problems have been overcome using other materials that have low fluorescence background and thus are compatible with most assays. The nature of the surface chemistry varies and includes porous acrylamide gel, membranes, plain glass, polymeric microwells, nanowells, and three-dimensional (3D) surface structures (Table 12.5).

Polyacrylamide gel and agarose microarrays, patterned on a glass surface, have been created by Guschin et al. [134] and Afanassiev et al. [135], respectively. Because both acrylamide and agarose tend to form hydrophilic and highly porous matrices, proteins can readily diffuse and therefore need to be immobilized by cross-linking to reactive ligands present in the matrix. Analytes are then added to these 3D arrays to perform the assays [134]. Because of the formation of 3D matrixes on the glass surface, the binding capacity is higher than that on a 2D surface; moreover, the moisturized environment prevents protein denaturation and thereby helps keep proteins in their active state. In addition to the sophisticated processes required to fabricate such 3D matrixes, the major disadvantage of the 3D arrays is that it is more difficult to change buffers and recover trapped molecules from these microarrays [102]. In contrast to 3D arrays, Zhu et al. [98] created an open nanowell structure on a PDMS surface supported by standard glass slides. The nanowells significantly reduce evaporation and minimize cross-contamination and background. Thanks to the open nanowell structure, different reagents and buffers can be sequentially added, which is essential for some biochemical assays. In addition, bound molecules can be easily recovered from the nanowells. Moreover, by covering the nanowells with gold, mass spectrometry and SPR analyses can be performed. The major disadvantage of this technology is that specialized equipment is required to load the nanowells at high density. Many researchers now directly array proteins and antibodies/antigens onto plain glass slides [98, 99, 107, 112, 136, 137]. To keep proteins in a wet environment during the arraying process, glycerol (30 to 40 percent) is added to the sample buffer and the spotting is performed in a moisturized environment [99, 107].

An additional consideration concerns the optimization of the chemistry used to attach the proteins to the support, as certain proteins are more prone to nonspecific adsorption and loss of functionality upon exposure to solid phases. To immobilize proteins to a solid substrate, the surface of the substrate has to be modified to achieve the maximum binding capacity. A convenient method is to coat the glass surface with a thin nitrocellulose membrane or poly-lysine such that proteins can be passively adsorbed to the modified surface through nonspecific interactions [121, 136, 138]. The attached proteins lay on the surface in random orientation and can be washed off under stringent washing conditions. Moreover, the background level is usually higher because of nonspecific attachment.

To achieve more specific and stronger protein binding, several groups have created reactive surfaces on glass that can covalently cross-link to proteins [98, 99, 107]. Generally, a self-assembled monolayer (SAM) is formed using a bifunctional silane cross-linker, which has one functional group that reacts with the hydroxyl groups on glass surface and another free one that can either directly react with primary amine groups of proteins (i.e., aldehyde or epoxy groups) or can be further chemically modified to reach maximum binding specificity [107, 139]. Gold-coated glass surface is another variation [120, 140]. To form a SAM on a gold surface, bifunctional thio-alkylene is usually utilized, which has a SH group that reacts with gold and another free one that reacts with target molecules. The advantage of using gold-coated surface is that mass spectrometry and SPR can eventually be used as detection methods to identify the bound molecules or to monitor the dynamics of the reactions, respectively [120, 140, 141]. This approach provides the opportunity to study dynamics of biochemical reactions in a high-throughput fashion and thus has great potential in drug discovery and biomedical research [141].

In the above covalent cross-linking approaches, because the reactive groups are also present in the side chains of proteins, it is conceivable that the proteins bind to the surface in a random fashion, which may alter the native conformation of proteins, reduce the activity of proteins, or make them inaccessible to interacting partners. Perhaps the best means of protein attachment is via highly specific affinity interactions [98, 133]. Proteins fused with a high-affinity tag at their N- or C-terminus are linked to the surface of the chip through this tag, and, thus, all of the attached proteins should orient uniformly away from the surface [99]. Using this method, immobilized proteins are more likely to remain in their native conformation, while the analytes have easier access to the active sites of proteins. This approach was first successfully demonstrated in attaching 5800 fusion proteins containing a Hisx6 tag onto a Ni-NTA-coated glass slide [99] (see, above, Interaction Arrays). Other affinity methods such as glutathione/GST and phosphonate/serine esterase cutinase ligand/protein tags have been also proposed [142].

Protein Delivery Systems (Arrayers) Arrays are created by spotting proteins onto the solid surface in an organized high-density format followed by immobilization to create a stable probe. Although a 96-format dot-blot instrument has been used to create low-density protein arrays on filters [121, 123, 138], high-density protein microarrays (>30,000 spots per slide) can be obtained using robotic printing tools ("arrayers"). Considerable expertise in this field has been achieved from work with DNA microarrays using photolithography, ink jet and contact printing, liquid dispensing, and piezoelectrics [143]. All of these methodologies have advantages and disadvantages related to the accuracy of dispensing and the compatibility with high-throughput arraying of proteins.

The contact printing arrayers deposit subnanoliter sample volumes directly to the surface using tiny pins with or without capillary slots. Because these contact printing robots cannot align their pins to the prefabricated structures and need to touch the surface, noncontact robotic printers, which use ink-jet technology, were used to deliver nanoliter to picoliter protein droplets to nanowells [98] and polyacrylamide gel packets [144]. Recently, electrospray deposition technology was applied to deliver dry proteins to a dextran-grafted surface [145]. This technology further decreased the spot size from ~150 to ~30 µm.

Other successful reports of protein arraying include a system for automated high-density protein delivery onto PVDF filters from liquid bacterial cultures [127]. Use of a thin (250 µm) transfer stamp enabled high spotting density (4,800 spots per slide) and detection of 10 pg of protein.

Finally, a group from Harvard University [107] has recently reported a method for generating high-density (16,000 spots/cm^2) protein arrays onto chemically derivatized (aldehyde-containing silane) glass slides in low (nanoliter) volume. The proteins were deposited onto the array in the presence of glycerol to avoid loss of functionality due to dehydration. The approach is suggested where it is intended to study protein–protein interactions or to carry out protein substrate identification.

Detection Methods The detection devices used in protein microarrays include charge-coupled device (CCD) cameras or laser scanners with confocal detection optics. Fluorescence detection methods are generally the preferred detection method because they are safe, simple, and extremely sensitive and can have very high resolution [128, 133]. Typically, a chip is either directly probed with a fluorescent molecule (e.g., fluorescently labeled protein) or in two steps by first using a tagged probe (e.g., biotin), which can then be detected using a fluorescently labeled affinity reagent (e.g., streptavidin). Some researchers made use of fluorescent cyanine dyes (e.g., Cy5) on antibody arrays, detecting human myeloma proteins [146]. Cyanine dyes were also used for a recent study of protein–protein interactions [147]. Another fluorescent labeling method is rolling circle amplification (RCA), which is extremely sensitive [148]. However, other detection methods have been proposed. For example, ELISA (enzyme-linked immunosorbent assay) was first used to detect proteins for both glass arrays [134] and filter arrays [124, 136]. Zhu et al. [98] and Ge [121] have used radioisotope labeling to study kinase–substrate interactions in nanowells, and protein–protein, protein–DNA, and protein–drug interactions on filter arrays, respectively.

Because labeling molecules can sometimes affect protein activity and are restricted to the available detection channels, nonlabeling methods have advantages as a direct detection approach for protein microarrays. As an example, the surface-enhanced laser desorption and ionization (SELDI) technology adapted mass spectrometry as a readout system to analyze differential protein expression on spot arrays [149, 150]. Protein extracts are incubated on different spots of the same adsorptive surface chemistry (e.g., cation–anion exchange material, hydrophobic surface, etc.). After washing away unbound proteins, the whole variety of nonspecifically captured target proteins can be analyzed using SELDI mass spectroscopy.

Label-free SPR-based detection systems can also be used [114, 151]. Intrinsic Bioprobes have linked a SPR-based biosensor to MALDI-TOF (matrix-assisted laser desorption ionization–time of flight) mass spectrometry, designed as a chip-based proteomic tool for providing information about the function and structure of proteins [101]. This allows the real-time analysis of protein interactions with a sensitivity in the low femtomolar range.

Time-resolved fluorimetry (TRF) offers the advantage of greater sensitivity due to a lower background signal [152]. This is achieved by the use of pulsed excitation and time-gated detection. An adaptation of an immunoassay recently used antibodies to detect protein on a solid surface. The antibodies were covalently attached to a unique oligonucleotide primer. This novel approach permitted a rolling circle amplification reaction for increased sensitivity by PCR amplification of the associated DNA tag [137, 153, 154].

Atomic force microscopy (AFM) is a powerful detection technology capable of detecting protein interactions at the level of a single molecule [155]. Protiveris is developing protein biosensors, based on chemomechanical actuation of silicon microcantilevers in a principle similar to AFM. This unique technology should provide very high sensitivity without the need for derivatization. In a recent example applied to a biological system, microcantilever-based arrays have been used to detect prostate-specific antigens [156].

Protein Mass Spectrometry

For organisms with known genomes, recent advances in protein mass spectrometry (MS) have vastly facilitated the dissection of interactions in protein complexes [157]. Proteins and tryptic peptides from these complexes can be analyzed by MALDI–TOF [157], sequences deducted from their mass, and the sequences compared with a database of predicted proteins encoded by the organism's genome. If mass alone cannot predict the exact sequence, fragmentation methods (nanoelectrospray tandem mass spectrometry) can be used to produce stretches of up to 16 residues of sequence from femtomolar amounts of protein fragments [158].

So far, such methods have been used to identify caspase 8 as an interactor with immunoprecipitated CD95 [159] and to characterize the protein complement of a number of multiprotein complexes including the yeast spindle pole body [160], the yeast spliceosomal U1 small nuclear ribonucleoprotein (snRNP) [161], the yeast anaphase promoting complex [162], the spliceosome [163], and *trans*-Golgi network-derived transport vesicles [164].

Clearly, mass spectrometric methods have tremendous potential. As its sensitivity and ease of use increase, MS will come to complement biological methods for detecting and analyzing protein interactions and may eventually supplant them. However, other than its considerable cost and the high level of technical sophistication required, the most significant limitation of MS for analysis of protein interactions is that the protein complexes first need to be isolated by physical methods such as electrophoresis [158]. Many proteins will not be detected because, for example, they are too small, large, scarce, alkaline, or acidic to be analyzed by two-dimensional gel analysis, or because the interaction is too weak and transient to allow affinity purification.

Reagents and Methods
Separation and Isolation of Proteins by 2D PAGE An established technique for protein separation is two-dimensional polyacrylamide gel electrophoresis (2D PAGE), which separates proteins based on two properties: their size and their isoelectric point. Two-dimensional PAGE is capable of profiling many thousands of proteins on a single matrix with exquisite resolution, separating isoforms differing by a single posttranslational modification. Posttranslational modifications detectable by 2D PAGE include deamidation [165], phosphorylation [166], and glycosylation [167, 168]. Thus, 2D PAGE is an open assay

approach, which allows the separation and detection of proteins from a wide variety of sources without the need for any prior knowledge of function.

In recent years, there have been several technical advances that have led to improvement in the quality of 2D PAGE separations. Most notably, these include the introduction of immobilized pH gradients (IPGs), which provide better reproducibility and increased resolution [169–172]. Originally, the pH gradient for isoelectric focusing in 2D PAGE was created by carrier ampholytes. These are mixtures of a few hundred different homologs of amphoteric buffers, synthesized together in one reaction flask. The mixtures contain buffers with isoelectric points evenly distributed over a wide pH spectrum (i.e., from pH 3 to pH 10), and they have high buffering power at their isoelectric points. When an electric field is applied, they begin to migrate according to their charges toward the anode or the cathode, and automatically form stable pH gradients. The pH gradient works well when protein separation is performed in native conditions. However, in order to obtain a high-resolution 2D protein separation, electrophoresis is usually performed under denaturing conditions, which destabilizes the pH gradient and increases the running time.

Today, with the development of IPGs [173], the pH gradients in these gels are prepared by copolymerizing acrylamide monomers with acrylamide derivatives containing carboxylic and tertiary amino groups. Because the buffering groups that form the gradient are fixed, the pH gradient cannot drift and is not influenced by the sample composition. The use of IPGs has allowed many methodical innovations for 2D PAGE. Premanufactured gel strips and instruments are available as commercial products. The immobilized strips are of various lengths and cover various pH ranges, including broad pH range (pH 3 to 10), narrower ranges (pH 4 to 7 and pH 6 to 11), and even a range of a single pH unit increment. The method can be applied to different experimental purposes, and the separation can be improved by using larger gel format and narrow pH range immobilized strips.

The ability of 2D PAGE to resolve thousands of proteins in one gel makes it an unbeatable technology, and it is expected to remain in use for another decade. However, although a powerful technique, there are limitations to 2D PAGE-based approaches. First, it is rather labor intensive and requires a high level of skill. Moreover, 2D PAGE has proven resistant to automation [174, 175] and image analysis remains a bottleneck. Second, without prior enrichment the approach can lack the sensitivity of other assay formats, such as ELISA [176, 177]. Thus, scarce proteins either cannot or can hardly be detected on 2D gels. Although this problem can be circumvented by increasing the sample amount, there may be a risk of overloading the system and reducing the resolution [177]. Finally, proteins with extreme pH (below 3 or above 10) are not separated [169], but focused as vertical lines on both sides. In addition, many large or hydrophobic proteins do not enter the gel during the first dimension and are therefore refractory to separation and quantitation by 2D PAGE [178, 179].

Methods of Protein Detection and Image Analysis   There are many ways to detect proteins in 2D gels. The most common methods are Coomassie blue staining [180, 181] and silver staining [182, 183]. Other methods, including [^{35}S]-Met or ^{14}C radiolabeling [184, 185], zinc imidazole [186, 187], colloidal gold [188, 189], India ink [97, 190], ponceau S, and amido black [181], can also be used in different applications to achieve better sensitivity in particular cases. However, there are some drawbacks. For example, glycoproteins are not stained by Coomassie blue [180], and many organic dyes are unsuitable for protein detection on PVDF membranes if samples are to be used for MALDI-TOF.

After 2D electrophoresis and protein visualization by staining, images of gels are digitized for computer analysis by an image scanner or fluorescent scanner and are then subjected to analysis by special image analysis software (either PDQUEST from BioRad or ImageMaster from Amersham Pharmacia Biotech). The 2D patterns are very complex, and special software tools are required to find differentially expressed proteins in a series of gels, such as posttranslational modified proteins or up- and downregulated proteins. Image spots on the gels are initially detected, manually edited, and then matched.

LC-MS-Based Approaches   An alternative to 2D PAGE for the separation of proteins is liquid chromatography (LC). Reverse-phase high-pressure liquid chromatography (RP-HPLC) has the advantage of using a liquid phase compatible with direct injection into a mass spectrometer (LC-MS) for identification. Unfortunately, proteins are not generically amenable to high-quality separation by RP-HPLC. This problem has been solved by digesting the proteins to peptides with a protease, usually trypsin. The resulting peptides behave in a more predictable manner than the intact proteins and can be readily separated by RP-HPLC. This approach has meant that analysis of protein mixtures by multidimensional LC in conjunction with MS is becoming feasible [191]. To date, the LC-MS approach has been successfully used to characterize protein complexes [191, 192] and to specifically look at protein–protein interactions [193].

Protein Identification Using Mass Spectrometry   After resolving the protein mixtures, the next step is protein identification. Proteins can usually be identified by peptide mass fingerprinting (PMF) [187, 194–196] and database searching. In PMF, the peptide masses of unknown proteins are compared to the predicted masses of peptides from the theoretical digestion of proteins in a database. First, the protein spots are excised and digested with a protease (e.g., trypsin or chymotrypsin). The digest is then applied onto a sample plate and coated with matrix. The matrix is typically a small energy-absorbing molecule such as α-cyano-4-hydroxycinnamic acid or 2,5-dihydroxybenzoic acid. The analyte is spotted, together with the matrix, on the sample plate and allowed to evaporate, resulting in the formation of crystals. The plate is then put into the MALDI-TOF mass spectrometer, and the laser is automatically targeted to specific places on the plate, and peptide mass spectra are then obtained. If the protein is digested with trypsin, the trypsin autolytic fragment peaks can serve as internal standards for mass calibration. Several software packages are available to perform database matching such as Protein Prospector [197], ProFound [198], and MASCOT [199] (see Table 12.6). The more numbers of peptides match to a protein in database, the more likely the protein identification is. The advantage of using PMF is that the protein identification process is fast and user-friendly. But the success of the method can be compromised by several factors: (1) mass redundancy of peptides with the same masses but different amino acid compositions can cause ambiguity in protein identification; (2) PMF cannot identify posttranslationally modified peptides since such information is not present in the database; (3) PMF cannot analyze samples containing a mixture of proteins since they generate mixtures of peptides after the digestion; and (4) insufficient peptides are obtained in the PMF to submit to the database search, that is, there is insufficient data to identify the unknown protein.

When such problems occur, it is necessary to use postsource decay (PSD) to confirm the result. Because PSD can deduce the amino acid sequence of peptides from normal, posttranslationally modified, or novel proteins of interest, this can significantly increase the accuracy of the protein identification and sometimes leads to the discovery of new proteins. With peptide sequencing, the amino acid sequence of unknown peptides can be obtained and then used to search the database to identify the protein from which it was derived. Unlike PMF, PSD can be used for gels containing more than one protein. This advantage greatly improves the protein identification process since protein bands from one-dimensional (1D) gel can be identified whether homogenous or not. The PMF data can be supplemented with partial amino acid sequence information along the result of database search, so an unsuccessful search of the protein database with the PMF data may be reversed with an additional partial sequence. The drawback of PSD is that it is not user-friendly since the process is not easily automated. As a result, MS analysis and database searching takes considerable time and must be performed by an experienced operator.

Affinity Tagging and Mass Spectroscopy Affinity purification has long been used to study interaction of several proteins in a multiprotein complex. Several methods have been used for isolating protein complexes, of which the most common is the use of GST fusions [200] where the protein of interest whose binding partners need to be determined is expressed as a fusion protein with a cleavable GST tag and immobilized to a solid support. The immobilized protein is incubated with a cell extract containing the target proteins so that multiprotein complexes can be formed. The protein complexes are then cleaved off from the GST tag, purified, and the bound proteins identified by 2D PAGE and MS [200].

Because no a priori knowledge of the interacting proteins is needed, this approach can be used to identify novel interactions between known proteins

TABLE 12.6 Publicly Available Database Matching Tools for Protein Identification from Mass Spectrometry Data

Name	URL
Mascot	http://www.matrixscience.com
Mowse	http://www.hgmp.mrc.ac.uk/Bioinformatics/Webapp/mowse
MS-Tag	http://www.prospector.ucsf.edu
Pepsea	http://www.pepsea.protana.com
ProFound	http://www.proteometrics.com
Protein Prospector	http://www.prospector.ucsf.edu/
SEQUEST	http://www.fields.scripps.edu/sequest
Sonar	http://www.proteometrics.com

in a cellular pathway or new interacting partners. The purification of a protein complex not only helps to identify its components but also provides information on the relationship between apparently unrelated cellular pathways [200]. A crucial requirement for the characterization of protein complexes is high affinity between the tagged protein and the other complex elements to ensure that components of the protein complex are not lost during the purification process [201].

Using this approach, the human spliceosome multiprotein complex has been isolated using biotinylated RNA as a bait and analyzed with a 2D gel to identify 19 novel factors [163]. Other protein complexes that have been characterized using this approach include proteins found in the yeast nuclear-pore complex and the GroEL chaperonin system [202].

The high-throughput potential of this approach has been demonstrated by two recent reports on the systematic characterization of protein complexes in yeast [203, 204]. These two high-throughput studies differed in their experimental design. In one study, Ho et al. [204] constructed 725 inducible FLAG epitope–tagged fusions, which were overexpressed and purified together with their associated proteins. This study identified more than 3000 protein–protein interactions involving 1578 yeast proteins. In another study, Gavin et al. [203] integrated a tandem affinity purification (TAP) tag cassette by homologous recombination at the 3' end of each ORF and used haploid growing cells. The advantages of this strategy are that in the resulting yeast strains there is no competition for the "untagged" protein and that expression is under the control of the endogenous promoters. After TAP and sodium dodecyl sulfate (SDS)–PAGE, peptide mass fingerprinting by MALDI-MS was used to identify the proteins. Starting with a set of more than 500 chromosomal tagged genes, Gavin et al. [203] were able to purify and subsequently characterize 232 protein complexes encompassing 1440 yeast proteins. The differences in the results of these two studies likely reflect the fact that one study used endogenous protein levels, whereas the other used overexpressed proteins. The overexpressed proteins are likely to interact with more proteins, but they may also associate with proteins that they do not normally bind and thereby yield false positives.

Compared with the yeast two-hybrid approach and protein microarrays, affinity purification coupled to MS detection offers some major advantages but also has some drawbacks. On the whole, this approach is more physiological because protein–protein interactions are analyzed in vivo, rather than reconstituted ex vivo or in vitro. The approach is not restricted to one particular cell type or organism. Importantly, it also reveals which proteins and isoforms belong to a given cellular proteome. Whereas in the yeast two-hybrid system both binding partners are expressed as fusion proteins, in the affinity purification approach only one component of the complex is expressed as a fusion protein, reducing the probability of steric interference. On the other hand, a clear advantage of the two-hybrid system and of protein microarrays is that these strategies are economical and more easily amenable to automation for high-throughput approaches. They also are better suited for the detection of transient or unstable interactions. It is important to note that the yeast two-hybrid system and protein microarrays yield information about binary physical interactions, which are not necessarily mapped by affinity purification approaches and vice versa. Therefore, all methods are highly complementary and should be applied in combination to produce a comprehensive protein–protein interaction map of a given cell.

Computational Methods of Analysis of Protein–Protein Interactions

Computational methods play an important role at all stages of the characterization of protein–protein interactions. They are used to predict putative interactions, to validate the results of high-throughput interaction screens, and to analyze the protein networks inferred from interaction databases.

Publicly accessible databases of protein–protein interactions are most useful for the analysis of protein interaction data. Several databases that are currently available (Table 12.7) provide access to both experimental data and the results of different computational methods of inference. For example, molecular interaction (MINT) is a database designed to store data on functional interactions between proteins. Beyond cataloging binary complexes, MINT was conceived to store other types of functional interactions, including enzymatic modifications of one of the interacting partners. Presently, MINT stores 4568 interactions, 782 of which are indirect or genetic interactions [205]. The database of interacting proteins (DIP) documents experimentally determined protein–protein interactions. At the moment, it contains 10,432 interactions, 90 percent of which are from high-throughput experiments in microorganisms [27]. Entries describing interactions among proteins from mammalian proteomes are approximately 750. BIND (Biomolecular Interaction Network Database) has a somewhat larger scope and provides information about bimolecular interactions, complexes, and pathways [206]. BIND contains 5939 interactions, approximately 300 of which describe interactions among

TABLE 12.7 Databases of Protein–Protein Interactions

Database	URL	Comments	References
BIND	http://www.bind.ca	Interaction database that archives biomolecular interaction, complex and pathway information	1
DIP	http://dip.doe-mbi.ucla.edu	Collections of experimentally determined protein–protein interactions	2
InterDOM	http://InterDom.lit.org.sg	This database focuses on providing supporting evidence for the detected protein interactions based on putative protein domain interactions	3
LiveDIP	http://dip.doc-mbi.ucla.edu/ldip.html	Extension of DIP providing access to information on functional states of protein complexes	4
MINT	http://cbm.bio.uniroma2.it/mint		5
MIPS	http://mips.gsf.de	*S. cerevisiae* specific; also provides information on genetic interactions	6
PREDICTOME	http://predictome.bu.edu	Compilation of functional link predictions with experimental, genome-scale yeast two-hybrid data	7

STRING	http://www.bork.embl-heidelberg.de/STRING	Compilations of functional link predictions based on gene proximity, common evolutionary history (phylogenetic profiles), and domain fusion events	8
The GRID	http://biodata.mshri.on.ca/grid/servlet/Index	Compilation of BIND, MIPS, and several genome-scale datasets; *S. cerevisiae* specific	9

1. Bader, G. D., Donaldson, I., Wolting, C., Ouellette, B. F., Pawson, T., Hogue, C. W. (2001). BIND—The Biomolecular Interaction Network Database. *Nucleic Acids Res., 29*, 242–245.
2. Xenarios, I., Salwinski, L., Duan, X. J., Higney, P., Kim, S. M., Eisenberg, D. (2002). DIP, the Database of Interacting Proteins: A research tool for studying cellular networks of protein interactions. *Nucleic Acids Res., 30*, 303–305.
3. Ng, S. K., Zhang, Z., Tan, S. H., Lin, K. (2003). InterDom: A database of putative interacting protein domains for validating predicted protein interactions and complexes. *Nucleic Acids Res., 31*, 251–254.
4. Duan, X. J., Xenarios, I., Eisenberg, D. (2002). Describing biological protein interactions in terms of protein states and state transitions: The LiveDIP database. *Mol. Cell. Proteomics, 1*, 104–116.
5. Zanzoni, A., Montecchi-Palazzi, L., Quondam, M., Ausiello, G., Helmer-Citterich, M., Cesareni, G. (2002). MINT: A Molecular INTeraction database. *FEBS Lett., 513*, 135–140.
6. Mewes, H. W., Frishman, D., Guldener, U., Mannhaupt, G., Mayer, K., Mokrejs, M., Morgenstern, B., Munsterkotter, M., Rudd, S., Weil, B. (2002). MIPS: A database for genomes and protein sequences. *Nucleic Acids Res., 30*, 31–34.
7. Mellor, J. C., Yanai, I., Clodfelter, K. H., Mintseris, J., DeLisi, C. (2002). Predictome: A database of putative functional links between proteins. *Nucleic Acids Res., 30*, 306–309.
8. von Mering, C., Huynen, M., Jaeggi, D., Schmidt, S., Bork, P., Snel, B. (2003). STRING: A database of predicted functional associations between proteins. *Nucleic Acids Res, 31*, 258–261.
9. Breitkreutz, B. J., Stark, C., Tyers, M. (2003). The GRID: The General Repository for Interaction Datasets. *Genome Biol., 4*, R23.

mammalian proteins. Even if these databases are still largely incomplete, they have provided bioinformaticians and biologists in general with a unique and easily searchable depository of interaction information.

12.3 DISCOVERY OF INHIBITORS OF PROTEIN–PROTEIN INTERACTIONS

Protein–protein interfaces are considered to be new prospective drug targets. Indeed, given the ubiquitous nature of protein–protein interactions, and the knowledge that inappropriate protein–protein binding can lead to disease, it is not surprising that protein–protein interactions have attracted the attention of scientists in the pharmaceutical industry and elsewhere who are interested in producing inhibitors for use as biochemical tools or therapeutic agents. There are ample examples in the literature of the use of antibodies, dominant negative proteins, or peptides to inhibit particular protein–protein assemblies. In contrast, there are only a few examples of small "druglike" molecules that disrupt protein–protein interactions and can act inside cells, but the number of successes is rapidly growing [207, 208].

A variety of strategies can be adopted to identify and characterize compounds that disrupt protein–protein interactions. Here, only recent advances in screening strategies that employ variants of the technologies reviewed in the first part of this chapter are discussed.

Two-Hybrid System Variants

Variants of the classical yeast two-hybrid system known as *reverse two-hybrid* [1, 209] and *split-hybrid* [210] systems have been recently developed for identifying peptides and/or small molecules that dissociate protein–protein interactions. These extensions of the yeast two-hybrid system from a basic tool of exploratory research to a tool for drug discovery take advantage of several key aspects of yeast. *S. cerevisiae* provides a robust organism that can easily be manipulated by molecular techniques, the complete genome information is available, reagents are inexpensive, and assays are easily configured to automation for high-throughput screen design. On the other hand, a primary constraint of these permutations of the two-hybrid system may be the permeability of yeast to inhibitor molecules. Therefore, the applicability of these systems seems to be limited to small molecules and, in any case, preliminary studies should be performed to evaluate the permeability properties of each molecule. If standard yeast strains prove relatively impermeable to the desired molecule, this barrier may be circumvented by utilizing yeast permeability mutants such as *PDR5*, *erg6*, or *SNQ2* [211], although the viability of such strains may present additional complications. Alternatively, inhibitory peptides dissociating a target protein–protein interaction could be identified by a *reverse three-hybrid system,* wherein yeast cells expressing the two interacting fusion proteins are co-transformed with a third vector expressing the peptide library in a nonfused form or fused to the SV40 T-antigen NLS, by analogy to three-hybrid systems that we and others recently proposed to detect ternary protein–protein interactions [48, 61].

Reverse Two-Hybrid Systems The reverse two-hybrid system is an "upside-down" version of the classical yeast two-hybrid system in which the interaction between target proteins causes the activation of a toxic or lethal marker gene. All reverse two-hybrid systems make use of yeast strains in which productive protein–protein interactions increase the expression of a counterselectable marker that is toxic under particular conditions. Under these conditions, dissociation of an interaction provides a selective advantage, thereby facilitating detection. A few growing yeast colonies in which hybrids fail to interact can be identified among millions of nongrowing colonies expressing interacting proteins.

Several reverse two-hybrid systems have been proposed. The first reverse two-hybrid system uses a yeast strain that is resistant to cycloheximide due to the presence of a mutant *CYH2* gene [39]. This strain also contains the wild-type *CYH2* allele under the transcriptional control of the GAL1 promoter. The *CYH2* gene encodes the L29 ribosomal protein and confers cycloheximide sensitivity [212] and is dominant over the *cyh2* allele, which produces cycloheximide resistance. Using this reporter gene, a productive protein–protein interaction of GAL4-based fusion proteins drives expression of the *CYH2* gene, and the yeast strain becomes sensitive to cycloheximide, resulting in abrogation of cell growth when cells are plated on selective media containing cycloheximide. Disruption of the protein–protein interaction results in a failure of *CYH2* gene transcription, and cells retain the cycloheximide-resistant phenotype; thus, the strain is able to grow on media containing cycloheximide.

Another reverse two-hybrid system makes use of the counterselectable marker *URA3* (Fig. 12.5). The *URA3* gene encodes an enzyme, orotidine-5′-phosphate decarboxylase [213], involved in uracil biosynthesis, that confers uracil prototrophy allowing cell growth on uracil-deficient media. The same enzyme can also catalyze the conversion of a nontoxic analog, 5-fluoroorotic acid (5-FOA), into a toxic product, 5-fluorouracil [214]. This conveniently allows both positive and negative growth selection, on medium lacking uracil or on medium containing 5-FOA, respectively. Hence, mutations that prevent an interaction can be selected from large libraries or randomly generated alleles. Similarly, molecules that dissociate or prevent an interaction could be selected from large libraries of peptides or compounds [209, 215].

In the past few years, the reverse two-hybrid system has been applied successfully to the selection of dissociative molecules. In the first description of this approach, Vidal et al. [209] demonstrated that peptides corresponding to portions of adenovirus E1A protein were able to disrupt the pRB/E2F1 interaction. In another study, the interaction between the activin receptor R1 and

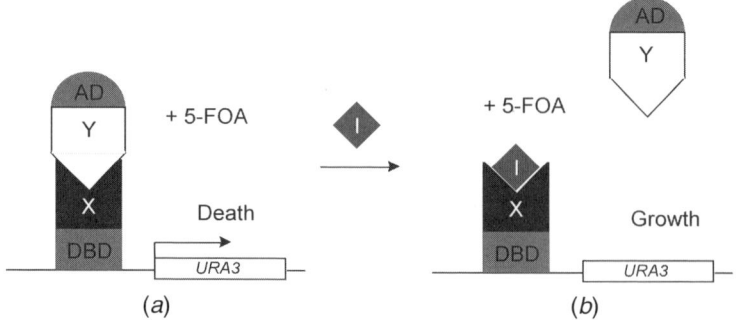

Figure 12.5 Reverse two-hybrid system designed to identify small-molecule inhibitors of protein–protein interactions. (*a*) Interaction of protein X, fused to the DNA binding domain (DBD), with protein Y, fused to the activating domain (AD) of a transcription factor, recruits the AD proximal to a promoter containing DBD binding sites upstream of an *URA3* reporter gene. This reconstitutes a functional transcriptional activator that in turn induces the synthesis of Ura3p, leading to yeast death in medium containing the pro-toxin 5-fluoroorotic acid (5-FOA). (*b*) If a small molecule (I, inhibitor) disrupts the intracellular protein–protein interaction, proximity of the AD to the promoter is removed, and transcription of the *URA3* gene is abolished, allowing cell survival in the presence of 5-FOA.

the immunophilin protein FKBP12 was shown to be dissociated by nanomolar concentrations of the small molecule FK506 [216]. Recently, a novel compound, WAY141520, which inhibits the interaction between β3 and α1B subunits of N-type calcium channels, has been selected from a collection of ~150,000 molecules [217]. This study demonstrated that small-molecule inhibitors can indeed be identified from large libraries using the reverse two-hybrid system.

Split-Hybrid System An alternative approach to inhibition studies of protein–protein interactions, which has been termed the *split-hybrid system*, uses the *E. coli* tetracycline repressor (TetR) and operator as a coupled two-step sequential reporter gene system [210]. In this system, expression of TetR is induced following protein–protein interaction. TetR consequently binds to a *tetR* operator-regulated *HIS3* gene, repressing its transcription (Fig. 12.6). In these conditions, cell growth is inhibited on media deficient in histidine. Dissociation of the target protein–protein interaction therefore prevents *tetR* transcription, which in turn restores *HIS3* synthesis and allows cells to grow in the absence of histidine. The system can be modulated by addition of tetracycline, which relieves the repression in a dose-dependent manner, to selection media. The system's utility was demonstrated by testing interactions of cyclic adenosine 5′-monophosphate (cAMP) response element binding protein (CREB) and its coactivator protein CREB binding protein (CBP). The interaction of CREB and CBP depends on phosphorylation of a critical

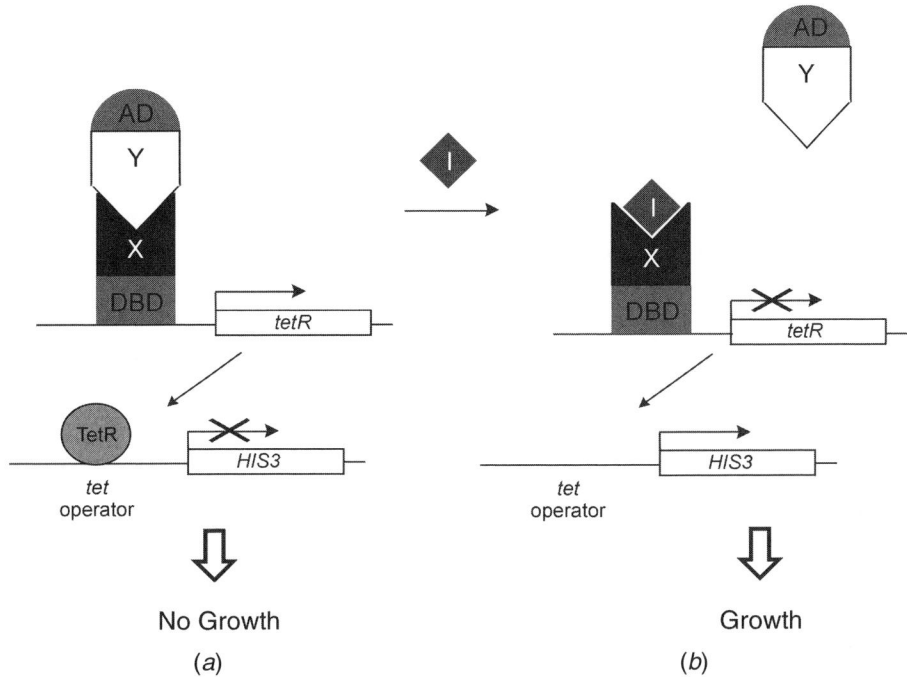

Figure 12.6 General schematic of the yeast split-hybrid system. (*a*) The interaction between one protein (X) fused to a DNA binding domain (DBD) and another (Y) fused to a transcriptional activation domain (AD) promotes the expression of tetracycline repressor (TetR), which binds to *tet* operator upstream of the *HIS3* reporter gene, preventing the growth of yeast cells in histidine-deficient media. (*b*) The disruption of the X–Y interaction by an inhibitor molecule (I) prevents expression of TetR. This, in turn, allows expression of *HIS3* reporter gene and hence growth of yeast cells in histidine-deficient media.

serine residue in CREB. Coexpression of wild-type CREB and CBP fusion proteins results in the activation of the *tetR* gene and production of TetR protein. TetR protein binds to the *tet* operator of the sequential reporter gene and effectively prevents *HIS3* gene expression. No cell growth was observed on histidine-deficient media. Prevention of the CREB/CBP interaction via mutation of the critical serine, or other disruptive mutations, resulted in histidine prototrophy, since the *tetR* gene product is not produced and hence the *HIS3* gene is not repressed [210]. In addition to mutagenesis studies, the split-hybrid system could allow the screening of compounds that dissociate relevant protein–protein interactions.

Disruption of Protein–Protein Interactions by Peptide Aptamers In a third approach, Geyer et al. [218] used a modified yeast two-hybrid system to identify peptide aptamers that potentially disrupted protein–protein interactions.

Peptide aptamers are proteins that contain a conformationally constrained peptide region of variable sequence displayed from a scaffold, for example, E. coli thioredoxin. Geyer et al. expressed aptamers at high levels in cells engineered to express the bait and prey proteins at low concentrations and identified disrupted interactions by decrease of the positive signal. Using this system, they selected peptide aptamers that overcame the cell cycle arrest in S. cerevisiae [218].

Other in Vivo Genetic Selection Systems

Approaches conceptually similar to the above yeast two-hybrid system variants, but based on prokaryotic cells, do exist. One of these systems relies on the properties of the repressor protein cI of bacteriophage λ [219]. The cI repressor protein binds to its operator as a homodimer. Each monomer has two distinct domains, a C-terminal dimerization domain and an N-terminal DNA binding domain. In the system proposed by Park and Raines [219], a protein of interest (X) is fused to the N-terminal DNA binding domain of cI (NcI) to create a hybrid repressor (NcI-X). The dimerization of protein X reconstitutes a functional repressor that is able to bind DNA and to repress the transcription of *tet* and *lacZ* reporter genes (Fig. 12.7). The reporter genes are under control of the λP_R promoter in a reporter plasmid that also drives the expression of NcI-X. Thus, E. coli cells transformed with the reporter plasmid show a LacZ$^-$Tet^S phenotype. After cotransformation with a plasmid expressing a peptide library, bacteria expressing dissociative peptides show a LacZ$^+$Tet^R phenotype, growing on media containing tetracycline and forming blue colonies. The power of this approach has been recently demonstrated by selecting nine-residue peptides from a combinatorial library that dissociate the HIV-1 protease dimer [219]. Even though Park and Raines [219] used this genetic selection method for targeting a homodimer, they suggest that the

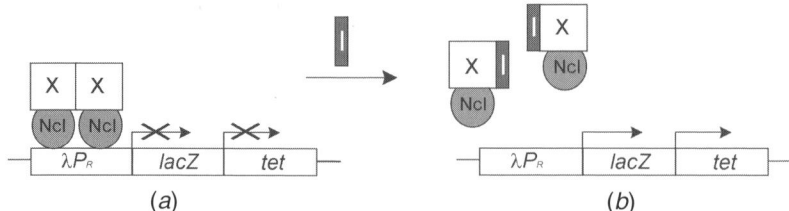

Figure 12.7 Genetic selection for inhibitors of a homodimeric protein–protein interaction. (*a*) The dimer of a hybrid λ repressor, NcI-X, composed of the N-terminal DNA binding domain of λ repressor protein (NcI) and a protein (X) able to dimerize, turns off *lacZ* and *tet* reporter genes by binding the λP_R promoter. (*b*) The expression of *lacZ* and *tet* reporter genes is turned on when an inhibitor (I) dissociates the NcI-X homodimer. In these conditions, cells expressing the *tet* and *lacZ* reporter genes survive and form blue colonies in the presence of tetracycline and X-Gal.

same approach could be applicable to identify molecules that disrupt a heterodimeric protein complex, as the assembly of effective heterodimeric hybrid repressor proteins has been reported [220]. Moreover, although Park and Raines panned a peptide library, this method may also be used to select small-molecule inhibitors.

Inhibitors from Phage Display

Phage display is another technology that can produce specific binding agents for any target molecule and is especially well suited for identifying peptides that bind specific protein domains and dissociate protein complexes. Display of peptides on the surface of bacteriophage virions has become an important means of generating peptide libraries. The very large size of either random libraries or pool of individual variants of a single sequence that can be generated means that complex mixtures can be screened. In fact, display of a repertoire of different peptide sequences on phages can enable the simultaneous analysis of >10^9 molecules in a single assay. The problems intrinsic to analysis of small amounts of a selected chemical are thus overcome by the simple and economical process of viral replication. Moreover, high-throughput technology for automated phage display selection has been also developed using picking-spotting robots [221].

Thus, phage display has been applied successfully not only to identify critical amino acids responsible for protein–protein interactions but also to isolate leads for the discovery of new therapeutics [78]. Several antagonist of protein–protein interactions have been discovered using phage libraries. These include peptides blocking IL-1α binding to type I IL-1 receptor [222], insulin-like growth factor 1 binding to its regulatory binding protein IGFBP-1 [223], the angiogenesis factor VEGF binding to its cell-surface receptor KDR [224, 225], IgG Fc binding to staphylococcal protein A [226], and HIV-1 gp120 binding to CD4 cell receptor [227]. All of these peptides are between 10 and 20 residues in length, still a bit large to transform easily into small-molecule drug candidates. They are also quite hydrophobic and tend to cover a large surface on the target protein [225, 226]. In any case, display technology can provide a complementary tool with combinatorial chemistry, by identifying lead peptides that can be the basis for the design of small molecules that disrupt specific protein–protein interactions.

Microarrays and Small Molecules

Recently, several groups have proposed methods for identifying small-molecule inhibitors using a microarray format. MacBeath and Schreiber [107] immobilized small organic compounds from a combinatorial chemistry library to create a high-density small-molecule microarray. Single resin beads from combinatorial synthesis were placed in 96-well plates, and the organic molecules were chemically released from the beads. The organic molecules were diluted, spotted, and covalently attached on derivatized glass slides. These microarrays were then incubated with fluorescently labeled target proteins to identify new ligands. Similarly, Winssinger et al. [228] constructed a library of small molecules by tethering them to a peptide nucleic acid (PNA) tag. These PNA tags provide the basis for the structure of the corresponding small molecules and immobilize them at specific sites on the chip. The study also showed that the immobilized small molecules could withstand stringent washing conditions. As a test case, these arrays were used to identify a small-molecule inhibitor of a caspase. As a third example, Kuruvilla et al. [229] constructed a small-molecule microarray containing a collection of 3780 structurally complex 1,3-dioxane compounds to dissect the function of a yeast protein, Ure2, which is a central regulator of the nitrogen metabolic pathway. The library of small molecules was synthesized with a technology platform based on one bead–one stock solution and parsed out to form the small-molecule microarray [230, 231]. The microarray was then probed with fluorescently labeled Ure2. One compound (uretupamine) was identified as specifically inhibiting Ure2 function in a subsequent reporter assay. An analysis of gene expression profiles determined that uretupamine inhibits a particular function of Ure2 without affecting other functions of the protein [232].

The microarray format used in these approaches enables high-throughput screening with minimum amounts of small molecules. This strategy is therefore expected not only to be useful to select ligands of every protein of interest in an entire proteome but also to allow parallel high-throughput screening of small molecules that can disrupt relevant protein–protein interactions, which may greatly facilitate the development of pharmaceutical agents.

REFERENCES

1. Vidal, M., Endoh, H. (1999). Prospects for drug screening using the reverse two-hybrid system. *Trends Biotechnol.*, *17*, 374–381.
2. Loregian, A., Marsden, H. S., Palù, G. (2002). Protein–protein interactions as targets for antiviral chemotherapy. *Rev. Med. Virol.* *12*, 239–262.
3. Fields, S., Song, O. (1989). A novel genetic system to detect protein–protein interactions. *Nature, 340*, 245–246.
4. Hope, I. A., Struhl, K. (1986). Functional dissection of a eukaryotic transcriptional activator protein, GCN4 of yeast. *Cell, 46*, 885–894.
5. Keegan, L., Gill, G., Ptashne, M. (1986). Separation of DNA binding from the transcription-activating function of a eukaryotic regulatory protein. *Science, 231*, 699–704.
6. Brent, R., Ptashne, M. (1985). A eukaryotic transcriptional activator bearing the DNA specificity of a prokaryotic repressor. *Cell, 43*, 729–736.
7. Estojak, J., Brent, R., Golemis, E. A. (1995). Correlation of two-hybrid affinity data with in vitro measurements. *Mol. Cell. Biol., 15*, 5820–5829.

8. Durfee, T., Becherer, K., Chen, P. L., Yeh, S. H., Yang, Y., Kilburn, A. E., Lee, W. H., Elledge, S. J. (1993). The retinoblastoma protein associates with the protein phosphatase type 1 catalytic subunit. *Genes Devel., 7*, 555–569.
9. Fields, S., Sternglanz, R. (1994). The two-hybrid system: An assay for protein–protein interactions. *Trends Genet., 10*, 286–292.
10. Young, K. H., Ozenberger, B. A. (1995). Investigation of ligand binding to members of the cytokine receptor family within a microbial system. *Ann. N.Y. Acad. Sci., 766*, 279–281.
11. Kajkowski, E. M., Price, L. A., Pausch, M. H., Young, K. H., Ozenberger, B. A. (1997). Investigation of growth hormone releasing hormone receptor structure and activity using yeast expression technologies. *J. Receptor Signal Transduc. Res., 17*, 293–303.
12. Heery, D. M., Kalkhoven, E., Hoare, S., Parker, M. G. (1997). A signature motif in transcriptional co-activators mediates binding to nuclear receptors. *Nature, 387*, 733–736.
13. Loregian, A., Bortolozzo, K., Boso, S., Sapino, B., Betti, M., Biasolo, M. A., Caputo, A., Palù, G. (2003). The Sp1 transcription factor does not directly interact with the HIV-1 Tat protein. *J. Cell. Physiol., 196*, 251–257.
14. Ma, J., Ptashne, M. (1987). A new class of yeast transcriptional activators. *Cell, 51*, 113–119.
15. Yasugi, T., Benson, J. D., Sakai, H., Vidal, M., Howley, P. M. (1997). Mapping and characterization of the interaction domains of human papillomavirus type 16 E1 and E2 proteins. *J. Virol., 71*, 891–899.
16. Du, W., Vidal, M., Xie, J. E., Dyson, N. (1996). RBF, a novel RB-related gene that regulates E2F activity and interacts with cyclin E in Drosophila. *Genes Devel., 10*, 1206–1218.
17. Tirode, F., Malaguti, C., Romero, F., Attar, R., Camonis, J., Egly, J. M. (1997). A conditionally expressed third partner stabilizes or prevents the formation of a transcriptional activator in a three-hybrid system. *J. Biol. Chem., 272*, 22995–22999.
18. Osborne, M. A., Dalton, S., Kochan, J. P. (1995). The yeast tribrid system—genetic detection of trans-phosphorylated ITAM-SH2-interactions. *Bio/Technology, 13*, 1474–1478.
19. Johnsson, N., Varshavsky, A. (1994). Split ubiquitin as a sensor of protein interactions in vivo. *Proc. Natl. Acad. Sci. USA, 91*, 10340–10344.
20. Aronheim, A., Zandi, E., Hennemann, H., Elledge, S. J., Karin, M. (1997). Isolation of an AP-1 repressor by a novel method for detecting protein–protein interactions. *Mol. Cell. Biol., 17*, 3094–3102.
21. Neville, M., Stutz, F., Lee, L., Davis, L. I., Rosbash, M. (1997). The importin-beta family member Crm1p bridges the interaction between Rev and the nuclear pore complex during nuclear export. *Curr. Biol., 7*, 767–775.
22. Bartel, P., Chien, C. T., Sternglanz, R., Fields, S. (1993). Elimination of false positives that arise in using the two-hybrid system. *Biotechniques, 14*, 920–924.
23. James, P., Halladay, J., Craig, E. A. (1996). Genomic libraries and a host strain designed for highly efficient two-hybrid selection in yeast. *Genetics, 144*, 1425–1436.

24. Chevray, P. M., Nathans, D. (1992). Protein interaction cloning in yeast: Identification of mammalian proteins that react with the leucine zipper of Jun. *Proc. Natl. Acad. Sci. USA, 89*, 5789–5793.
25. Bendixen, C., Gangloff, S., Rothstein, R. (1994). A yeast mating-selection scheme for detection of protein–protein interactions. *Nucleic Acids Res., 22*, 1778–1779.
26. Fromont-Racine, M., Rain, J. C., Legrain, P. (1997). Toward a functional analysis of the yeast genome through exhaustive two-hybrid screens. *Nat. Genet., 16*, 277–282.
27. Xenarios, I., Fernandez, E., Salwinski, L., Duan, X. J., Thompson, M. J., Marcotte, E. M., Eisenberg, D. (2001). DIP: The Database of Interacting Proteins: 2001 update. *Nucleic Acids Res., 29*, 239–241.
28. Parent, S. A., Fenimore, C. M., Bostian, K. A. (1985). Vector systems for the expression, analysis and cloning of DNA sequences in *S. cerevisiae*. *Yeast, 1*, 83–138.
29. Chien, C. T., Bartel, P. L., Sternglanz, R., Fields, S. (1991). The two-hybrid system: A method to identify and clone genes for proteins that interact with a protein of interest. *Proc. Natl. Acad. Sci. USA, 88*, 9578–9582.
30. Harper, J. W., Adami, G. R., Wei, N., Keyomarsi, K., Elledge, S. J. (1993). The p21 Cdk-interacting protein Cip1 is a potent inhibitor of G1 cyclin-dependent kinases. *Cell, 75*, 805–816.
31. Gyuris, J., Golemis, E., Chertkov, H., Brent, R. (1993). Cdi1, a human G1 and S phase protein phosphatase that associates with Cdk2. *Cell, 75*, 791–803.
32. Vojtek, A. B., Hollenberg, S. M., Cooper, J. A. (1993). Mammalian Ras interacts directly with the serine/threonine kinase Raf. *Cell, 74*, 205–214.
33. Brent, R. (1982). Regulation and autoregulation by lexA protein. *Biochimie, 64*, 565–569.
34. Ebina, Y., Takahara, Y., Kishi, F., Nakazawa, A., Brent, R. (1983). LexA protein is a repressor of the colicin E1 gene. *J. Biol. Chem., 258*, 13258–13261.
35. Green, N., Alexander, H., Olson, A., Alexander, S., Shinnick, T. M., Sutcliffe, J. G., Lerner, R. A. (1982). Immunogenic structure of the influenza virus hemagglutinin. *Cell, 28*, 477–487.
36. Legrain, P., Dokhelar, M. C., Transy, C. (1994). Detection of protein–protein interactions using different vectors in the two-hybrid system. *Nucleic Acids Res., 22*, 3241–3242.
37. Fradet, Y., Tardif, M., Parent-Vaugeois, C. (1985). [The use of multiparameter flow cytometry in the detection and evaluation of human bladder tumors]. *Union Medicale du Canada, 114*, 778–780.
38. Hadfield, C., Jordan, B. E., Mount, R. C., Pretorius, G. H., Burak, E. (1990). G418-resistance as a dominant marker and reporter for gene expression in *Saccharomyces cerevisiae*. *Curr. Genet., 18*, 303–313.
39. Leanna, C. A., Hannink, M. (1996). The reverse two-hybrid system: A genetic scheme for selection against specific protein/protein interactions. *Nucleic Acids Res., 24*, 3341–3347.
40. Ahmad, M., Bussey, H. (1986). Yeast arginine permease: Nucleotide sequence of the *CAN1* gene. *Curr. Genet., 10*, 587–592.

41. Soares, M. B., Bonaldo, M. F., Jelene, P., Su, L., Lawton, L., Efstratiadis, A. (1994). Construction and characterization of a normalized cDNA library. *Proc. Natl. Acad. Sci. USA, 91*, 9228–9232.
42. Drees, B. L. (1999). Progress and variations in two-hybrid and three-hybrid technologies. *Curr. Opin. Chem. Biol., 3*, 64–70.
43. Roder, K. H., Wolf, S. S., Schweizer, M. (1996). Refinement of vectors for use in the yeast two-hybrid system. *Anal. Biochem., 241*, 260–262.
44. Le Douarin, B., Pierrat, B., vom Baur, E., Chambon, P., Losson, R. (1995). A new version of the two-hybrid assay for detection of protein–protein interactions. *Nucleic Acids Res., 23*, 876–878.
45. Jiang, R., Carlson, M. (1996). Glucose regulates protein interactions within the yeast SNF1 protein kinase complex. *Genes Devel., 10*, 3105–3115.
46. Xu, C. W., Mendelsohn, A. R., Brent, R. (1997). Cells that register logical relationships among proteins. *Proc. Natl. Acad. Sci. USA, 94*, 12473–12478.
47. Inouye, C., Dhillon, N., Durfee, T., Zambryski, P. C., Thorner, J. (1997). Mutational analysis of STE5 in the yeast *Saccharomyces cerevisiae*: Application of a differential interaction trap assay for examining protein–protein interactions. *Genetics, 147*, 479–492.
48. Zhang, J., Lautar, S. (1996). A yeast three-hybrid method to clone ternary protein complex components. *Anal. Biochem., 242*, 68–72.
49. Struhl, K. (1988). The JUN oncoprotein, a vertebrate transcription factor, activates transcription in yeast. *Nature, 332*, 649–650.
50. Yu, X., Wu, L. C., Bowcock, A. M., Aronheim, A., Baer, R. (1998). The C-terminal (BRCT) domains of BRCA1 interact in vivo with CtIP, a protein implicated in the CtBP pathway of transcriptional repression. *J. Biol. Chem., 273*, 25388–25392.
51. Aronheim, A. (1997). Improved efficiency sos recruitment system: Expression of the mammalian GAP reduces isolation of Ras GTPase false positives. *Nucleic Acids Res., 25*, 3373–3374.
52. Goldstein, G., Scheid, M., Hammerling, U., Schlesinger, D. H., Niall, H. D., Boyse, E. A. (1975). Isolation of a polypeptide that has lymphocyte-differentiating properties and is probably represented universally in living cells. *Proc. Natl. Acad. Sci. USA, 72*, 11–15.
53. Hershko, A., Ciechanover, A. (1992). The ubiquitin system for protein degradation. *Annu. Rev. Biochem., 61*, 761–807.
54. Reinders, A., Schulze, W., Kuhn, C., Barker, L., Schulz, A., Ward, J. M., Frommer, W. B. (2002). Protein–protein interactions between sucrose transporters of different affinities colocalized in the same enucleate sieve element. *Plant Cell, 14*, 1567–1577.
55. Stagljar, I., Korostensky, C., Johnsson, N., te Heesen, S. (1998). A genetic system based on split-ubiquitin for the analysis of interactions between membrane proteins *in vivo*. *Proc. Natl. Acad. Sci. USA, 95*, 5187–5192.
56. Wang, B., Nguyen, M., Breckenridge, D. G., Stojanovic, M., Clemons, P. A., Kuppig, S., Shore, G. C. (2003). Uncleaved BAP31 in association with A4 protein at the endoplasmic reticulum is an inhibitor of Fas-initiated release of cytochrome c from mitochondria. *J. Biol. Chem., 278*, 14461–14468.

57. Wittke, S., Dunnwald, M., Albertsen, M., Johnsson, N. (2002). Recognition of a subset of signal sequences by Ssh1p, a Sec61p-related protein in the membrane of endoplasmic reticulum of yeast *Saccharomyces cerevisiae*. *Mol. Biol. Cell, 13*, 2223–2232.
58. Tsujimoto, Y., Numaga, T., Ohshima, K., Yano, M. A., Ohsawa, R., Goto, D. B., Naito, S., Ishikawa, M. (2003). *Arabidopsis* TOBAMOVIRUS MULTIPLICATION (TOM) 2 locus encodes a transmembrane protein that interacts with TOM1. *EMBO J., 22*, 335–343.
59. Lim, M. Y., Dailey, D., Martin, G. S., Thorner, J. (1993). Yeast MCK1 protein kinase autophosphorylates at tyrosine and serine but phosphorylates exogenous substrates at serine and threonine. *J. Biol. Chem., 268*, 21155–21164.
60. Dixon, E. P., Johnstone, E. M., Liu, X., Little, S. P. (1997). An inverse mammalian two-hybrid system for beta secretase and other proteases. *Anal. Biochem., 249*, 239–241.
61. Loregian, A., Bortolozzo, K., Boso, S., Caputo, A., Palù, G. (2003). Interaction of Sp1 transcription factor with HIV-1 Tat protein: Looking for cellular partners. *FEBS Lett., 543*, 61–65.
62. Buchert, M., Schneider, S., Adams, M. T., Hefti, H. P., Moelling, K., Hovens, C. M. (1997). Useful vectors for the two-hybrid system in mammalian cells. *Biotechniques, 23*, 396–398, 400, 402.
63. Luo, Y., Batalao, A., Zhou, H., Zhu, L. (1997). Mammalian two-hybrid system: A complementary approach to the yeast two-hybrid system. *Biotechniques, 22*, 350–352.
64. Van Aelst, L., White, M. A., Wigler, M. H. (1994). Ras partners. *Cold Spring Harbor Symp. Quantitative Biol., 59*, 181–186.
65. White, M. A., Nicolette, C., Minden, A., Polverino, A., Van Aelst, L., Karin, M., Wigler, M. H. (1995). Multiple Ras functions can contribute to mammalian cell transformation. *Cell, 80*, 533–541.
66. Khosravi-Far, R., White, M. A., Westwick, J. K., Solski, P. A., Chrzanowska-Wodnicka, M., Van Aelst, L., Wigler, M. H., Der, C. J. (1996). Oncogenic Ras activation of Raf/mitogen-activated protein kinase-independent pathways is sufficient to cause tumorigenic transformation. *Mol. Cell. Biol., 16*, 3923–3933.
67. van Criekinge, W., Cornelis, S., Van De Craen, M., Vandenabeele, P., Fiers, W., Beyaert, R. (1999). GAL4 is a substrate for caspases: Implications for two-hybrid screening and other GAL4-based assays. *Mol. Cell. Biol. Res. Communs., 1*, 158–161.
68. Walhout, A. J., Sordella, R., Lu, X., Hartley, J. L., Temple, G. F., Brasch, M. A., Thierry-Mieg, N., Vidal, M. (2000). Protein interaction mapping in *C. elegans* using proteins involved in vulval development. *Science, 287*, 116–122.
69. Uetz, P., Giot, L., Cagney, G., Mansfield, T. A., Judson, R. S., Knight, J. R., Lockshon, D., Narayan, V., Srinivasan, M., Pochart, P., Qureshi-Emili, A., Li, Y., Godwin, B., Conover, D., Kalbfleisch, T., Vijayadamodar, G., Yang, M., Johnston, M., Fields, S., Rothberg, J. M. (2000). A comprehensive analysis of protein–protein interactions in *Saccharomyces cerevisiae*. *Nature, 403*, 623–627.
70. Ito, T., Chiba, T., Ozawa, R., Yoshida, M., Hattori, M., Sakaki, Y. (2001). A comprehensive two-hybrid analysis to explore the yeast protein interactome. *Proc. Natl. Acad. Sci. USA, 98*, 4569–4574.

71. Boulton, S. J., Gartner, A., Reboul, J., Vaglio, P., Dyson, N., Hill, D. E., Vidal, M. (2002). Combined functional genomic maps of the *C. elegans* DNA damage response. *Science, 295*, 127–131.
72. Rain, J. C., Selig, L., De Reuse, H., Battaglia, V., Reverdy, C., Simon, S., Lenzen, G., Petel, F., Wojcik, J., Schachter, V., Chemama, Y., Labigne, A., Legrain, P. (2001). The protein–protein interaction map of *Helicobacter pylori*. *Nature, 409*, 211–215.
73. Finley, R. L., Jr., Brent, R. (1994). Interaction mating reveals binary and ternary connections between *Drosophila* cell cycle regulators. *Proc. Natl. Acad. Sci. USA, 91*, 12980–12984.
74. Bartel, P. L., Roecklein, J. A., SenGupta, D., Fields, S. (1996). A protein linkage map of *Escherichia coli* bacteriophage T7. *Nat. Genet., 12*, 72–77.
75. Tomb, J. F., White, O., Kerlavage, A. R., Clayton, R. A., Sutton, G. G., Fleischmann, R. D., Ketchum, K. A., Klenk, H. P., Gill, S., Dougherty, B. A., Nelson, K., Quackenbush, J., Zhou, L., Kirkness, E. F., Peterson, S., Loftus, B., Richardson, D., Dodson, R., Khalak, H. G., Glodek, A., McKenney, K., Fitzegerald, L. M., Lee, N., Adams, M. D., Venter, J. C., et al. (1997). The complete genome sequence of the gastric pathogen *Helicobacter pylori*. *Nature, 388*, 539–547.
76. Ito, T., Tashiro, K., Muta, S., Ozawa, R., Chiba, T., Nishizawa, M., Yamamoto, K., Kuhara, S., Sakaki, Y. (2000). Toward a protein–protein interaction map of the budding yeast: A comprehensive system to examine two-hybrid interactions in all possible combinations between the yeast proteins. *Proc. Natl. Acad. Sci. USA, 97*, 1143–1147.
77. Johnsson, K., Ge, L. (1999). Phage display of combinatorial peptide and protein libraries and their applications in biology and chemistry. *Curr. Top. Microbiol. Immunol., 243*, 87–105.
78. Lowman, H. B. (1997). Bacteriophage display and discovery of peptide leads for drug development. *Annu. Rev. Biophys. Biomol. Struct., 26*, 401–424.
79. FitzGerald, K. (2000). *In vitro* display technologies—new tools for drug discovery. *Drug Discov. Today, 5*, 253–258.
80. Smith, G. P. (1985). Filamentous fusion phage: Novel expression vectors that display cloned antigens on the virion surface. *Science, 228*, 1315–1317.
81. Jespers, L. S., Messens, J. H., De Keyser, A., Eeckhout, D., Van den Brande, I., Gansemans, Y. G., Lauwereys, M. J., Vlasuk, G. P., Stanssens, P. E. (1995). Surface expression and ligand-based selection of cDNAs fused to filamentous phage gene VI. *Bio/Technology, 13*, 378–382.
82. Bass, S., Greene, R., Wells, J. A. (1990). Hormone phage: An enrichment method for variant proteins with altered binding properties. *Proteins, 8*, 309–314.
83. Lowman, H. B., Bass, S. H., Simpson, N., Wells, J. A. (1991). Selecting high-affinity binding proteins by monovalent phage display. *Biochemistry, 30*, 10832–10838.
84. Sternberg, N., Hoess, R. H. (1995). Display of peptides and proteins on the surface of bacteriophage lambda. *Proc. Natl. Acad. Sci. USA, 92*, 1609–1613.
85. Maruyama, I. N., Maruyama, H. I., Brenner, S. (1994). Lambda foo: A lambda phage vector for the expression of foreign proteins. *Proc. Natl. Acad. Sci. USA, 91*, 8273–8277.
86. Efimov, V. P., Nepluev, I. V., Mesyanzhinov, V. V. (1995). Bacteriophage T4 as a surface display vector. *Virus Genes, 10*, 173–177.

87. Cull, M. G., Miller, J. F., Schatz, P. J. (1992). Screening for receptor ligands using large libraries of peptides linked to the C terminus of the lac repressor. *Proc. Natl. Acad. Sci. USA, 89*, 1865–1869.
88. Mattheakis, L. C., Bhatt, R. R., Dower, W. J. (1994). An *in vitro* polysome display system for identifying ligands from very large peptide libraries. *Proc. Natl. Acad. Sci. USA, 91*, 9022–9026.
89. Francisco, J. A., Campbell, R., Iverson, B. L., Georgiou, G. (1993). Production and fluorescence-activated cell sorting of *Escherichia coli* expressing a functional antibody fragment on the external surface. *Proc. Natl. Acad. Sci. USA, 90*, 10444–10448.
90. Lang, H. (2000). Outer membrane proteins as surface display systems. *Int. J. Med. Microbiol., 290*, 579–585.
91. Westerlund-Wikstrom, B. (2000). Peptide display on bacterial flagella: Principles and applications. *Int. J. Med. Microbiol., 290*, 223–230.
92. Boder, E. T., Wittrup, K. D. (1997). Yeast surface display for screening combinatorial polypeptide libraries. *Nat. Biotechnol., 15*, 553–557.
93. Turpen, T. H., Reinl, S. J., Charoenvit, Y., Hoffman, S. L., Fallarme, V., Grill, L. K. (1995). Malarial epitopes expressed on the surface of recombinant tobacco mosaic virus. *Bio/Technology, 13*, 53–57.
94. Resnick, D. A., Smith, A. D., Gesiler, S. C., Zhang, A., Arnold, E., Arnold, G. F. (1995). Chimeras from a human rhinovirus 14-human immunodeficiency virus type 1 (HIV-1) V3 loop seroprevalence library induce neutralizing responses against HIV-1. *J. Virol., 69*, 2406–2411.
95. Grabherr, R., Ernst, W. (2001). The baculovirus expression system as a tool for generating diversity by viral surface display. *Combinatorial Chem. High Throughput Screening, 4*, 185–192.
96. Crameri, R., Suter, M. (1993). Display of biologically active proteins on the surface of filamentous phages: A cDNA cloning system for selection of functional gene products linked to the genetic information responsible for their production. *Gene, 137*, 69–75.
97. Li, K. W., Geraerts, W. P., van Elk, R., Joosse, J. (1989). Quantification of proteins in the subnanogram and nanogram range: Comparison of the AuroDye, FerriDye, and India ink staining methods. *Anal. Biochem., 182*, 44–47.
98. Zhu, H., Klemic, J. F., Chang, S., Bertone, P., Casamayor, A., Klemic, K. G., Smith, D., Gerstein, M., Reed, M. A., Snyder, M. (2000). Analysis of yeast protein kinases using protein chips. *Nat. Genet., 26*, 283–289.
99. Zhu, H., Bilgin, M., Bangham, R., Hall, D., Casamayor, A., Bertone, P., Lan, N., Jansen, R., Bidlingmaier, S., Houfek, T., Mitchell, T., Miller, P., Dean, R. A., Gerstein, M., Snyder, M. (2001). Global analysis of protein activities using proteome chips. *Science, 293*, 2101–2105.
100. Folch, A., Toner, M. (2000). Microengineering of cellular interactions. *Annu. Rev. Biomed. Eng., 2*, 227–256.
101. Nelson, R. W., Nedelkov, D., Tubbs, K. A. (2000). Biosensor chip mass spectrometry: A chip-based proteomics approach. *Electrophoresis, 21*, 1155–1163.
102. Zhu, H., Snyder, M. (2001). Protein arrays and microarrays. *Curr. Opin. Chem. Biol., 5*, 40–45.

103. Fung, E. T., Thulasiraman, V., Weinberger, S. R., Dalmasso, E. A. (2001). Protein biochips for differential profiling. *Curr. Opin. Biotechnol., 12*, 65–69.
104. Merchant, M., Weinberger, S. R. (2000). Recent advancements in surface-enhanced laser desorption/ionization-time of flight-mass spectrometry. *Electrophoresis, 21*, 1164–1177.
105. Howard, J. C., Heinemann, C., Thatcher, B. J., Martin, B., Gan, B. S., Reid, G. (2000). Identification of collagen-binding proteins in *Lactobacillus* spp. with surface-enhanced laser desorption/ionization-time of flight ProteinChip technology. *Appl. Environ. Microbiol., 66*, 4396–4400.
106. Xiao, Z., Jiang, X., Beckett, M. L., Wright, G. L., Jr. (2000). Generation of a baculovirus recombinant prostate-specific membrane antigen and its use in the development of a novel protein biochip quantitative immunoassay. *Protein Expression and Purification, 19*, 12–21.
107. MacBeath, G., Schreiber, S. L. (2000). Printing proteins as microarrays for high-throughput function determination. *Science, 289*, 1760–1763.
108. Moody, M. D., Van Arsdell, S. W., Murphy, K. P., Orencole, S. F., Burns, C. (2001). Array-based ELISAs for high-throughput analysis of human cytokines. *Biotechniques, 31*, 186–190, 192–184.
109. Huang, R. P., Huang, R., Fan, Y., Lin, Y. (2001). Simultaneous detection of multiple cytokines from conditioned media and patient's sera by an antibody-based protein array system. *Anal. Biochem., 294*, 55–62.
110. Wiese, R., Belosludtsev, Y., Powdrill, T., Thompson, P., Hogan, M. (2001). Simultaneous multianalyte ELISA performed on a microarray platform. *Clin. Chem., 47*, 1451–1457.
111. Sreekumar, A., Nyati, M. K., Varambally, S., Barrette, T. R., Ghosh, D., Lawrence, T. S., Chinnaiyan, A. M. (2001). Profiling of cancer cells using protein microarrays: Discovery of novel radiation-regulated proteins. *Cancer Res., 61*, 7585–7593.
112. Haab, B. B., Dunham, M. J., Brown, P. O. (2001). Protein microarrays for highly parallel detection and quantitation of specific proteins and antibodies in complex solutions. *Genome Biol., 2*, RESEARCH0004.
113. Miller, J. C., Butler, E. B., Teh, B. S., Haab, B. B. (2001). The application of protein microarrays to serum diagnostics: Prostate cancer as a test case. *Disease Markers, 17*, 225–234.
114. McDonnell, J. M. (2001). Surface plasmon resonance: Towards an understanding of the mechanisms of biological molecular recognition. *Curr. Opin. Chem. Biol., 5*, 572–577.
115. Nieba, L., Nieba-Axmann, S. E., Persson, A., Hamalainen, M., Edebratt, F., Hansson, A., Lidholm, J., Magnusson, K., Karlsson, A. F., Pluckthun, A. (1997). BIACORE analysis of histidine-tagged proteins using a chelating NTA sensor chip. *Anal. Biochem., 252*, 217–228.
116. Salamon, Z., Brown, M. F., Tollin, G. (1999). Plasmon resonance spectroscopy: Probing molecular interactions within membranes. *Trends Biochem. Sci., 24*, 213–219.
117. Myszka, D. G., Rich, R. L. (2000). Implementing surface plasmon resonance biosensors in drug discovery. *Pharm. Sci. Technol. Today, 3*, 310–317.

118. Sapsford, K. E., Liron, Z., Shubin, Y. S., Ligler, F. S. (2001). Kinetics of antigen binding to arrays of antibodies in different sized spots. *Anal. Chem.*, 73, 5518–5524.
119. Linnell, J., Groeger, G., Hassan, A. B. (2001). Real time kinetics of insulin-like growth factor II (IGF-II) interaction with the IGF-II/mannose 6-phosphate receptor: The effects of domain 13 and pH. *J. Biol. Chem.*, 276, 23986–23991.
120. Houseman, B. T., Huh, J. H., Kron, S. J., Mrksich, M. (2002). Peptide chips for the quantitative evaluation of protein kinase activity. *Nat. Biotechnol.*, 20, 270–274.
121. Ge, H. (2000). UPA, a universal protein array system for quantitative detection of protein–protein, protein-DNA, protein-RNA and protein-ligand interactions. *Nucleic Acids Res.*, 28, e3.
122. Cahill, D. J. (2001). Protein and antibody arrays and their medical applications. *J. Immunol. Methods*, 250, 81–91.
123. Stoll, D., Templin, M. F., Schrenk, M., Traub, P. C., Vohringer, C. F., Joos, T. O. (2002). Protein microarray technology. *Frontiers Biosci.*, 7, c13–32.
124. Bussow, K., Cahill, D., Nietfeld, W., Bancroft, D., Scherzinger, E., Lehrach, H., Walter, G. (1998). A method for global protein expression and antibody screening on high-density filters of an arrayed cDNA library. *Nucleic Acids Res.*, 26, 5007–5008.
125. Braun, P., Hu, Y., Shen, B., Halleck, A., Koundinya, M., Harlow, E., LaBaer, J. (2002). Proteome-scale purification of human proteins from bacteria. *Proc. Natl. Acad. Sci. USA*, 99, 2654–2659.
126. Albala, J. S., Franke, K., McConnell, I. R., Pak, K. L., Folta, P. A., Rubinfeld, B., Davies, A. H., Lennon, G. G., Clark, R. (2000). From genes to proteins: High-throughput expression and purification of the human proteome. *J. Cell. Biochem.*, 80, 187–191.
127. Lueking, A., Horn, M., Eickhoff, H., Bussow, K., Lehrach, H., Walter, G. (1999). Protein microarrays for gene expression and antibody screening. *Anal. Biochem.*, 270, 103–111.
128. Haab, B. B. (2001). Advances in protein microarray technology for protein expression and interaction profiling. *Curr. Opin. Drug Discov. Devel.*, 4, 116–123.
129. Keefe, A. D., Szostak, J. W. (2001). Functional proteins from a random-sequence library. *Nature*, 410, 715–718.
130. He, M., Taussig, M. J. (2001). Single step generation of protein arrays from DNA by cell-free expression and *in situ* immobilisation (PISA method). *Nucleic Acids Res.*, 29, E73–73.
131. Reineke, U., Volkmer-Engert, R., Schneider-Mergener, J. (2001). Applications of peptide arrays prepared by the SPOT-technology. *Curr. Opin. Biotechnol.*, 12, 59–64.
132. Walter, G., Bussow, K., Cahill, D., Lueking, A., Lehrach, H. (2000). Protein arrays for gene expression and molecular interaction screening. *Curr. Opin. Microbiol.*, 3, 298–302.
133. Templin, M. F., Stoll, D., Schrenk, M., Traub, P. C., Vohringer, C. F., Joos, T. O. (2002). Protein microarray technology. *Trends Biotechnol.*, 20, 160–166.
134. Guschin, D., Yershov, G., Zaslavsky, A., Gemmell, A., Shick, V., Proudnikov, D., Arenkov, P., Mirzabekov, A. (1997). Manual manufacturing of oligonucleotide, DNA, and protein microchips. *Anal. Biochem.*, 250, 203–211.

135. Afanassiev, V., Hanemann, V., Wolfl, S. (2000). Preparation of DNA and protein micro arrays on glass slides coated with an agarose film. *Nucleic Acids Res., 28*, E66.

136. Joos, T. O., Schrenk, M., Hopfl, P., Kroger, K., Chowdhury, U., Stoll, D., Schorner, D., Durr, M., Herick, K., Rupp, S., Sohn, K., Hammerle, H. (2000). A microarray enzyme-linked immunosorbent assay for autoimmune diagnostics. *Electrophoresis, 21*, 2641–2650.

137. Schweitzer, B., Kingsmore, S. F. (2002). Measuring proteins on microarrays. *Curr. Opin. Biotechnol., 13*, 14–19.

138. de Wildt, R. M., Mundy, C. R., Gorick, B. D., Tomlinson, I. M. (2000). Antibody arrays for high-throughput screening of antibody-antigen interactions. *Nat. Biotechnol., 18*, 989–994.

139. Heyse, S., Vogel, H., Sanger, M., Sigrist, H. (1995). Covalent attachment of functionalized lipid bilayers to planar waveguides for measuring protein binding to biomimetic membranes. *Protein Sci., 4*, 2532–2544.

140. Bieri, C., Ernst, O. P., Heyse, S., Hofmann, K. P., Vogel, H. (1999). Micropatterned immobilization of a G protein-coupled receptor and direct detection of G protein activation. *Nat. Biotechnol., 17*, 1105–1108.

141. Rich, R. L., Day, Y. S., Morton, T. A., Myszka, D. G. (2001). High-resolution and high-throughput protocols for measuring drug/human serum albumin interactions using BIACORE. *Anal. Biochem., 296*, 197–207.

142. Hodneland, C. D., Lee, Y. S., Min, D. H., Mrksich, M. (2002). Selective immobilization of proteins to self-assembled monolayers presenting active site-directed capture ligands. *Proc. Natl. Acad. Sci. USA, 99*, 5048–5052.

143. Lin, S. C., Tseng, F. G., Huang, H. M., Huang, C. Y., Chieng, C. C. (2001). Microsized 2D protein arrays immobilized by micro-stamps and micro-wells for disease diagnosis and drug screening. *Fres. J. Anal. Chem., 371*, 202–208.

144. Arenkov, P., Kukhtin, A., Gemmell, A., Voloshchuk, S., Chupeeva, V., Mirzabekov, A. (2000). Protein microchips: Use for immunoassay and enzymatic reactions. *Anal. Biochem., 278*, 123–131.

145. Avseenko, N. V., Morozova, T. Y., Ataullakhanov, F. I., Morozov, V. N. (2002). Immunoassay with multicomponent protein microarrays fabricated by electrospray deposition. *Anal. Chem., 74*, 927–933.

146. Silzel, J. W., Cercek, B., Dodson, C., Tsay, T., Obremski, R. J. (1998). Mass-sensing, multianalyte microarray immunoassay with imaging detection. *Clin. Chem., 44*, 2036–2043.

147. Marx, J. (2000). Medicine. DNA arrays reveal cancer in its many forms. *Science, 289*, 1670–1672.

148. Lizardi, P. M., Huang, X., Zhu, Z., Bray-Ward, P., Thomas, D. C., Ward, D. C. (1998). Mutation detection and single-molecule counting using isothermal rolling-circle amplification. *Nat. Genet., 19*, 225–232.

149. Weinberger, S. R., Morris, T. S., Pawlak, M. (2000). Recent trends in protein biochip technology. *Pharmacogenomics, 1*, 395–416.

150. Davies, H., Lomas, L., Austen, B. (1999). Profiling of amyloid beta peptide variants using SELDI Protein Chip arrays. *Biotechniques, 27*, 1258–1261.

151. Mullett, W. M., Lai, E. P., Yeung, J. M. (2000). Surface plasmon resonance-based immunoassays. *Methods, 22*, 77–91.
152. Luo, L. Y., Diamandis, E. P. (2000). Preliminary examination of time-resolved fluorometry for protein array applications. *Luminescence, 15*, 409–413.
153. Schweitzer, B., Roberts, S., Grimwade, B., Shao, W., Wang, M., Fu, Q., Shu, Q., Laroche, I., Zhou, Z., Tchernev, V. T., Christiansen, J., Velleca, M., Kingsmore, S. F. (2002). Multiplexed protein profiling on microarrays by rolling-circle amplification. *Nat. Biotechnol., 20*, 359–365.
154. Schweitzer, B., Wiltshire, S., Lambert, J., O'Malley, S., Kukanskis, K., Zhu, Z., Kingsmore, S. F., Lizardi, P. M., Ward, D. C. (2000). Inaugural article: Immunoassays with rolling circle DNA amplification: A versatile platform for ultrasensitive antigen detection. *Proc. Natl. Acad. Sci. USA, 97*, 10113–10119.
155. Stolz, M., Stoffler, D., Aebi, U., Goldsbury, C. (2000). Monitoring biomolecular interactions by time-lapse atomic force microscopy. *J. Struct. Biol., 131*, 171–180.
156. Wu, G., Datar, R. H., Hansen, K. M., Thundat, T., Cote, R. J., Majumdar, A. (2001). Bioassay of prostate-specific antigen (PSA) using microcantilevers. *Nat. Biotechnol., 19*, 856–860.
157. Andersen, J. S., Svensson, B., Roepstorff, P. (1996). Electrospray ionization and matrix assisted laser desorption/ionization mass spectrometry: Powerful analytical tools in recombinant protein chemistry. *Nat. Biotechnol., 14*, 449–457.
158. Jensen, O. N., Wilm, M., Shevchenko, A., Mann, M. (1999). Sample preparation methods for mass spectrometric peptide mapping directly from 2-DE gels. *Methods Mol. Biol., 112*, 513–530.
159. Muzio, M., Chinnaiyan, A. M., Kischkel, F. C., O'Rourke, K., Shevchenko, A., Ni, J., Scaffidi, C., Bretz, J. D., Zhang, M., Gentz, R., Mann, M., Krammer, P. H., Peter, M. E., Dixit, V. M. (1996). FLICE, a novel FADD-homologous ICE/CED-3-like protease, is recruited to the CD95 (Fas/APO-1) death-inducing signaling complex. *Cell, 85*, 817–827.
160. Wigge, P. A., Jensen, O. N., Holmes, S., Soues, S., Mann, M., Kilmartin, J. V. (1998). Analysis of the *Saccharomyces* spindle pole by matrix-assisted laser desorption/ionization (MALDI) mass spectrometry. *J. Cell Biol., 141*, 967–977.
161. Neubauer, G., Gottschalk, A., Fabrizio, P., Seraphin, B., Luhrmann, R., Mann, M. (1997). Identification of the proteins of the yeast U1 small nuclear ribonucleoprotein complex by mass spectrometry. *Proc. Natl. Acad. Sci. USA, 94*, 385–390.
162. Zachariae, W., Shevchenko, A., Andrews, P. D., Ciosk, R., Galova, M., Stark, M. J., Mann, M., Nasmyth, K. (1998). Mass spectrometric analysis of the anaphase-promoting complex from yeast: Identification of a subunit related to cullins. *Science, 279*, 1216–1219.
163. Neubauer, G., King, A., Rappsilber, J., Calvio, C., Watson, M., Ajuh, P., Sleeman, J., Lamond, A., Mann, M. (1998). Mass spectrometry and EST-database searching allows characterization of the multi-protein spliceosome complex. *Nat. Genet., 20*, 46–50.
164. Shevchenko, A., Keller, P., Scheiffele, P., Mann, M., Simons, K. (1997). Identification of components of trans-Golgi network-derived transport vesicles and detergent-insoluble complexes by nanoelectrospray tandem mass spectrometry. *Electrophoresis, 18*, 2591–2600.

165. Sarioglu, H., Lottspeich, F., Walk, T., Jung, G., Eckerskorn, C. (2000). Deamidation as a widespread phenomenon in two-dimensional polyacrylamide gel electrophoresis of human blood plasma proteins. *Electrophoresis, 21*, 2209–2218.
166. Kaufmann, H., Bailey, J. E., Fussenegger, M. (2001). Use of antibodies for detection of phosphorylated proteins separated by two-dimensional gel electrophoresis. *Proteomics, 1*, 194–199.
167. Taniguchi, N., Ekuni, A., Ko, J. H., Miyoshi, E., Ikeda, Y., Ihara, Y., Nishikawa, A., Honke, K., Takahashi, M. (2001). A glycomic approach to the identification and characterization of glycoprotein function in cells transfected with glycosyltransferase genes. *Proteomics, 1*, 239–247.
168. Fivaz, M., Vilbois, F., Pasquali, C., van der Goot, F. G. (2000). Analysis of glycosyl phosphatidylinositol-anchored proteins by two-dimensional gel electrophoresis. *Electrophoresis, 21*, 3351–3356.
169. Gorg, A., Obermaier, C., Boguth, G., Harder, A., Scheibe, B., Wildgruber, R., Weiss, W. (2000). The current state of two-dimensional electrophoresis with immobilized pH gradients. *Electrophoresis, 21*, 1037–1053.
170. Gorg, A., Obermaier, C., Boguth, G., Weiss, W. (1999). Recent developments in two-dimensional gel electrophoresis with immobilized pH gradients: Wide pH gradients up to pH 12, longer separation distances and simplified procedures. *Electrophoresis, 20*, 712–717.
171. Hoving, S., Voshol, H., van Oostrum, J. (2000). Towards high performance two-dimensional gel electrophoresis using ultrazoom gels. *Electrophoresis, 21*, 2617–2621.
172. Wildgruber, R., Harder, A., Obermaier, C., Boguth, G., Weiss, W., Fey, S. J., Larsen, P. M., Gorg, A. (2000). Towards higher resolution: Two-dimensional electrophoresis of *Saccharomyces cerevisiae* proteins using overlapping narrow immobilized pH gradients. *Electrophoresis, 21*, 2610–2616.
173. Bjellqvist, B., Ek, K., Righetti, P. G., Gianazza, E., Gorg, A., Westermeier, R., Postel, W. (1982). Isoelectric focusing in immobilized pH gradients: Principle, methodology and some applications. *J. Biochem. Biophys. Methods, 6*, 317–339.
174. Lopez, M. F. (2000). Better approaches to finding the needle in a haystack: Optimizing proteome analysis through automation. *Electrophoresis, 21*, 1082–1093.
175. Quadroni, M., James, P. (1999). Proteomics and automation. *Electrophoresis, 20*, 664–677.
176. Gygi, S. P., Corthals, G. L., Zhang, Y., Rochon, Y., Aebersold, R. (2000). Evaluation of two-dimensional gel electrophoresis-based proteome analysis technology. *Proc. Natl. Acad. Sci. USA, 97*, 9390–9395.
177. Corthals, G. L., Wasinger, V. C., Hochstrasser, D. F., Sanchez, J. C. (2000). The dynamic range of protein expression: A challenge for proteomic research. *Electrophoresis, 21*, 1104–1115.
178. Santoni, V., Kieffer, S., Desclaux, D., Masson, F., Rabilloud, T. (2000). Membrane proteomics: Use of additive main effects with multiplicative interaction model to classify plasma membrane proteins according to their solubility and electrophoretic properties. *Electrophoresis, 21*, 3329–3344.

179. Molloy, M. P., Phadke, N. D., Maddock, J. R., Andrews, P. C. (2001). Two-dimensional electrophoresis and peptide mass fingerprinting of bacterial outer membrane proteins. *Electrophoresis, 22*, 1686–1696.
180. Goldberg, H. A., Domenicucci, C., Pringle, G. A., Sodek, J. (1988). Mineral-binding proteoglycans of fetal porcine calvarial bone. *J. Biol. Chem., 263*, 12092–12101.
181. Sanchez, J. C., Ravier, F., Pasquali, C., Frutiger, S., Paquet, N., Bjellqvist, B., Hochstrasser, D. F., Hughes, G. J. (1992). Improving the detection of proteins after transfer to polyvinylidene difluoride membranes. *Electrophoresis, 13*, 715–717.
182. Rabilloud, T. (1992). A comparison between low background silver diammine and silver nitrate protein stains. *Electrophoresis, 13*, 429–439.
183. Hochstrasser, D. F., Merril, C. R. (1988). "Catalysts" for polyacrylamide gel polymerization and detection of proteins by silver staining. *Appl. Theor. Electrophor., 1*, 35–40.
184. Garrels, J. I., Franza, B. R., Jr. (1989). The REF52 protein database. Methods of database construction and analysis using the QUEST system and characterizations of protein patterns from proliferating and quiescent REF52 cells. *J. Biol. Chem., 264*, 5283–5298.
185. Latham, K. E., Garrels, J. I., Solter, D. (1993). Two-dimensional gel analysis of protein synthesis. *Methods Enzymol., 225*, 473–489.
186. Ortiz, M. L., Calero, M., Fernandez Patron, C., Patron, C. F., Castellanos, L., Mendez, E. (1992). Imidazole-SDS-Zn reverse staining of proteins in gels containing or not SDS and microsequence of individual unmodified electroblotted proteins. *FEBS Lett., 296*, 300–304.
187. James, P., Quadroni, M., Carafoli, E., Gonnet, G. (1993). Protein identification by mass profile fingerprinting. *Biochem. Biophys. Res. Commun., 195*, 58–64.
188. Yamaguchi, K., Asakawa, H. (1988). Preparation of colloidal gold for staining proteins electrotransferred onto nitrocellulose membranes. *Anal. Biochem., 172*, 104–107.
189. Eckerskorn, C., Strupat, K., Karas, M., Hillenkamp, F., Lottspeich, F. (1992). Mass spectrometric analysis of blotted proteins after gel electrophoretic separation by matrix-assisted laser desorption/ionization. *Electrophoresis, 13*, 664–665.
190. Hughes, J. H., Mack, K., Hamparian, V. V. (1988). India ink staining of proteins on nylon and hydrophobic membranes. *Anal. Biochem., 173*, 18–25.
191. Link, A. J., Eng, J., Schieltz, D. M., Carmack, E., Mize, G. J., Morris, D. R., Garvik, B. M., Yates, J. R., 3rd. (1999). Direct analysis of protein complexes using mass spectrometry. *Nat. Biotechnol., 17*, 676–682.
192. Washburn, M. P., Wolters, D., Yates, J. R., 3rd. (2001). Large-scale analysis of the yeast proteome by multidimensional protein identification technology. *Nat. Biotechnol., 19*, 242–247.
193. Muller, D. R., Schindler, P., Towbin, H., Wirth, U., Voshol, H., Hoving, S., Steinmetz, M. O. (2001). Isotope-tagged cross-linking reagents. A new tool in mass spectrometric protein interaction analysis. *Anal. Chem., 73*, 1927–1934.
194. Jensen, O. N., Podtelejnikov, A. V., Mann, M. (1997). Identification of the components of simple protein mixtures by high-accuracy peptide mass mapping and database searching. *Anal. Chem., 69*, 4741–4750.

195. Mann, M., Hojrup, P., Roepstorff, P. (1993). Use of mass spectrometric molecular weight information to identify proteins in sequence databases. *Biol. Mass Spectrom., 22*, 338–345.

196. Yates, J. R., 3rd, Speicher, S., Griffin, P. R., Hunkapiller, T. (1993). Peptide mass maps: A highly informative approach to protein identification. *Anal. Biochem., 214*, 397–408.

197. Clauser, K. R., Baker, P., Burlingame, A. L. (1999). Role of accurate mass measurement (+/− 10 ppm) in protein identification strategies employing MS or MS/MS and database searching. *Anal. Chem., 71*, 2871–2882.

198. Zhang, W., Chait, B. T. (2000). ProFound: An expert system for protein identification using mass spectrometric peptide mapping information. *Anal. Chem., 72*, 2482–2489.

199. Perkins, D. N., Pappin, D. J., Creasy, D. M., Cottrell, J. S. (1999). Probability-based protein identification by searching sequence databases using mass spectrometry data. *Electrophoresis, 20*, 3551–3567.

200. Pandey, A., Mann, M. (2000). Proteomics to study genes and genomes. *Nature, 405*, 837–846.

201. Naaby-Hansen, S., Waterfield, M. D., Cramer, R. (2001). Proteomics-post-genomic cartography to understand gene function. *Trends Pharmacol. Sci., 22*, 376–384.

202. Blobel, G., Wozniak, R. W. (2000). Proteomics for the pore. *Nature, 403*, 835–836.

203. Gavin, A. C., Bosche, M., Krause, R., Grandi, P., Marzioch, M., Bauer, A., Schultz, J., Rick, J. M., Michon, A. M., Cruciat, C. M., Remor, M., Hofert, C., Schelder, M., Brajenovic, M., Ruffner, H., Merino, A., Klein, K., Hudak, M., Dickson, D., Rudi, T., Gnau, V., Bauch, A., Bastuck, S., Huhse, B., Leutwein, C., Heurtier, M. A., Copley, R. R., Edelmann, A., Querfurth, E., Rybin, V., Drewes, G., Raida, M., Bouwmeester, T., Bork, P., Seraphin, B., Kuster, B., Neubauer, G., Superti-Furga, G. (2002). Functional organization of the yeast proteome by systematic analysis of protein complexes. *Nature, 415*, 141–147.

204. Ho, Y., Gruhler, A., Heilbut, A., Bader, G. D., Moore, L., Adams, S. L., Millar, A., Taylor, P., Bennett, K., Boutilier, K., Yang, L., Wolting, C., Donaldson, I., Schandorff, S., Shewnarane, J., Vo, M., Taggart, J., Goudreault, M., Muskat, B., Alfarano, C., Dewar, D., Lin, Z., Michalickova, K., Willems, A. R., Sassi, H., Nielsen, P. A., Rasmussen, K. J., Andersen, J. R., Johansen, L. E., Hansen, L. H., Jespersen, H., Podtelejnikov, A., Nielsen, E., Crawford, J., Poulsen, V., Sorensen, B. D., Matthiesen, J., Hendrickson, R. C., Gleeson, F., Pawson, T., Moran, M. F., Durocher, D., Mann, M., Hogue, C. W., Figeys, D., Tyers, M. (2002). Systematic identification of protein complexes in *Saccharomyces cerevisiae* by mass spectrometry. *Nature, 415*, 180–183.

205. Zanzoni, A., Montecchi-Palazzi, L., Quondam, M., Ausiello, G., Helmer-Citterich, M., Cesareni, G. (2002). MINT: A Molecular INTeraction database. *FEBS Lett., 513*, 135–140.

206. Bader, G. D., Donaldson, I., Wolting, C., Ouellette, B. F., Pawson, T., Hogue, C. W. (2001). BIND—The Biomolecular Interaction Network Database. *Nucleic Acids Res., 29*, 242–245.

207. Cochran, A. G. (2000). Antagonists of protein–protein interactions. *Chem. Biol., 7*, R85–94.

208. Toogood, P. L. (2002). Inhibition of protein–protein association by small molecules: Approaches and progress. *J. Med. Chem., 45*, 1543–1558.
209. Vidal, M., Brachmann, R. K., Fattaey, A., Harlow, E., Boeke, J. D. (1996). Reverse two-hybrid and one-hybrid systems to detect dissociation of protein–protein and DNA-protein interactions. *Proc. Natl. Acad. Sci. USA, 93*, 10315–10320.
210. Shih, H. M., Goldman, P. S., DeMaggio, A. J., Hollenberg, S. M., Goodman, R. H., Hoekstra, M. F. (1996). A positive genetic selection for disrupting protein-protein interactions: Identification of CREB mutations that prevent association with the coactivator CBP. *Proc. Natl. Acad. Sci. USA, 93*, 13896–13901.
211. Shah, N., Klausner, R. D. (1993). Brefeldin A reversibly inhibits secretion in *Saccharomyces cerevisiae*. *J. Biol. Chem., 268*, 5345–5348.
212. Kaufer, N. F., Fried, H. M., Schwindinger, W. F., Jasin, M., Warner, J. R. (1983). Cycloheximide resistance in yeast: The gene and its protein. *Nucleic Acids Res., 11*, 3123–3135.
213. Rose, M., Botstein, D. (1983). Structure and function of the yeast URA3 gene. Differentially regulated expression of hybrid beta-galactosidase from overlapping coding sequences in yeast. *J. Mol. Biol., 170*, 883–904.
214. Boeke, J. D., LaCroute, F., Fink, G. R. (1984). A positive selection for mutants lacking orotidine-5′-phosphate decarboxylase activity in yeast: 5-Fluoro-orotic acid resistance. *Mol. Gen. Genet., 197*, 345–346.
215. Vidal, M., Braun, P., Chen, E., Boeke, J. D., Harlow, E. (1996). Genetic characterization of a mammalian protein–protein interaction domain by using a yeast reverse two-hybrid system. *Proc. Natl. Acad. Sci. USA, 93*, 10321–10326.
216. Huang, J., Schreiber, S. L. (1997). A yeast genetic system for selecting small molecule inhibitors of protein–protein interactions in nanodroplets. *Proc. Natl. Acad. Sci. USA, 94*, 13396–13401.
217. Young, K., Lin, S., Sun, L., Lee, E., Modi, M., Hellings, S., Husbands, M., Ozenberger, B., Franco, R. (1998). Identification of a calcium channel modulator using a high throughput yeast two-hybrid screen. *Nat. Biotechnol., 16*, 946–950.
218. Geyer, C. R., Colman-Lerner, A., Brent, R. (1999). "Mutagenesis" by peptide aptamers identifies genetic network members and pathway connections. *Proc. Natl. Acad. Sci. USA, 96*, 8567–8572.
219. Park, S. H., Raines, R. T. (2000). Genetic selection for dissociative inhibitors of designated protein–protein interactions. *Nat. Biotechnol., 18*, 847–851.
220. Hu, J. C. (1995). Repressor fusions as a tool to study protein–protein interactions. *Structure, 3*, 431–433.
221. Walter, G., Konthur, Z., Lehrach, H. (2001). High-throughput screening of surface displayed gene products. *Combinatorial Chem. High Throughput Screening, 4*, 193–205.
222. Yanofsky, S. D., Baldwin, D. N., Butler, J. H., Holden, F. R., Jacobs, J. W., Balasubramanian, P., Chinn, J. P., Cwirla, S. E., Peters-Bhatt, E., Whitehorn, E. A., Tate, E. H., Akeson, A., Bowlin, T. L., Dower, W. J., Barrett, R. W. (1996). High affinity type I interleukin 1 receptor antagonists discovered by screening recombinant peptide libraries. *Proc. Natl. Acad. Sci. USA, 93*, 7381–7386.
223. Lowman, H. B., Chen, Y. M., Skelton, N. J., Mortensen, D. L., Tomlinson, E. E., Sadick, M. D., Robinson, I. C., Clark, R. G. (1998). Molecular mimics of insulin-

like growth factor 1 (IGF-1) for inhibiting IGF-1: IGF-binding protein interactions. *Biochemistry, 37*, 8870–8878.
224. Fairbrother, W. J., Christinger, H. W., Cochran, A. G., Fuh, G., Keenan, C. J., Quan, C., Shriver, S. K., Tom, J. Y., Wells, J. A., Cunningham, B. C. (1998). Novel peptides selected to bind vascular endothelial growth factor target the receptor-binding site. *Biochemistry, 37*, 17754–17764.
225. Wiesmann, C., Christinger, H. W., Cochran, A. G., Cunningham, B. C., Fairbrother, W. J., Keenan, C. J., Meng, G., de Vos, A. M. (1998). Crystal structure of the complex between VEGF and a receptor-blocking peptide. *Biochemistry, 37*, 17765–17772.
226. DeLano, W. L., Ultsch, M. H., de Vos, A. M., Wells, J. A. (2000). Convergent solutions to binding at a protein–protein interface. *Science, 287*, 1279–1283.
227. Ferrer, M., Harrison, S. C. (1999). Peptide ligands to human immunodeficiency virus type 1 gp120 identified from phage display libraries. *J. Virol., 73*, 5795–5802.
228. Winssinger, N., Ficarro, S., Schultz, P. G., Harris, J. L. (2002). Profiling protein function with small molecule microarrays. *Proc. Natl. Acad. Sci. USA, 99*, 11139–11144.
229. Kuruvilla, F. G., Shamji, A. F., Sternson, S. M., Hergenrother, P. J., Schreiber, S. L. (2002). Dissecting glucose signalling with diversity-oriented synthesis and small-molecule microarrays. *Nature, 416*, 653–657.
230. Blackwell, H. E., Perez, L., Stavenger, R. A., Tallarico, J. A., Cope Eatough, E., Foley, M. A., Schreiber, S. L. (2001). A one-bead, one-stock solution approach to chemical genetics: Part 1. *Chem. Biol., 8*, 1167–1182.
231. Clemons, P. A., Koehler, A. N., Wagner, B. K., Sprigings, T. G., Spring, D. R., King, R. W., Schreiber, S. L., Foley, M. A. (2001). A one-bead, one-stock solution approach to chemical genetics: Part 2. *Chem. Biol., 8*, 1183–1195.
232. Chen, J. (2002). Protein modulators made to order. *Chem. Biol., 9*, 543–544.

13

HIGH-THROUGHPUT SCREENING: EVOLUTION OF TECHNOLOGY AND METHODS

MARTYN N. BANKS, ANGELA M. CACACE, JONATHAN O'CONNELL, AND JOHN G. HOUSTON
Bristol Myers Squibb Co.
Pharmaceutical Research Institute
Wallingford, Connecticut

13.1	INTRODUCTION	560
13.2	EVOLUTION AND INTEGRATION OF HT PLATFORMS INTO THE DISCOVERY PROCESS	563
	Evolution of Instrumentation and Screening Formats	563
	Miniaturization	565
	Evolution of Cellular HTS Assays	566
	Evolution of HTS Technologies into the Drug Discovery Process	567
	Quality Control in the HTS Process	569
13.3	HTS AUTOMATED SYSTEMS: TECHNOLOGY AND PROCESS	570
	Plate Design	570
	Compound Management	572
	Liquid Handling	577
	Accuracy and Precision	578
	Assay Detection	579
	Automation	581
	Data Analysis	582
13.4	ADVANCES IN HIGH-THROUGHPUT BIOASSAY TECHNOLOGY	586
	Overview of HTS Assay Formats	587
	Reagent Production	587
	Designing Bioassays for HTS	591

Drug Discovery Handbook, by Shayne Cox Gad
Copyright © 2005 by John Wiley & Sons, Inc.

Fluorescent Methods in HTS	594
Cell-Based Assay Design	595
Additional Advances in Cellular Assay Formats	597
References	598

13.1 INTRODUCTION

Over the last 20 years high-throughput screening (HTS) has become a successful, reliable component of the drug discovery process. The evolution of science and technology surrounding HTS approaches has progressed rapidly over this same time period: from its origins in low-throughput testing of natural product extracts for antimicrobial activity to today's heavily automated, industrialized drug discovery process that enables the screening of millions of compounds against a range of complex, biological targets. During this time, we have seen an array of unique technology solutions and waves of standardization come to the fore to create the process we now recognize and use.

So, what is the best description of HTS today? To a large extent, it can still be defined by the scale of its operation and its reliance on serendipity for a successful outcome. A process by which hundreds of thousands to millions of compounds are tested for activity against disease targets of interest with the goal of identifying truly active, progressible "hits" [1].

Creative insights and smart ideas have significantly improved our ability to identify a greater number and quality of hits, especially in the last 5 years, as each of the core components that constitute the lead discovery process have been systematically enhanced.

Compound collections, assay design techniques, HTS technology platforms, and data capture/analysis systems are the essential core components of the lead discovery process, and each will be discussed in detail in this chapter. Figure 13.1 describes a typical lead discovery process highlighting these key functions.

Compound collections are typically constructed from a variety of different sources such as in-house "legacy" compounds, external acquisitions, combinatorial libraries, and natural product samples. The science around improving the "diversity" and quality of the compounds in such a collection has been significantly enriched since the mid-1980s. At that time a paucity of knowledge and experience created compound collections of poor quality and dubious lineage. Today, compound collections are recognized as major assets within a companies discovery arsenal and are treated as such. Compounds are now typically stored in custom-designed, fully automated warehouses where compounds are formatted for screening, monitored for quality, and instantly retrievable for testing.

INTRODUCTION

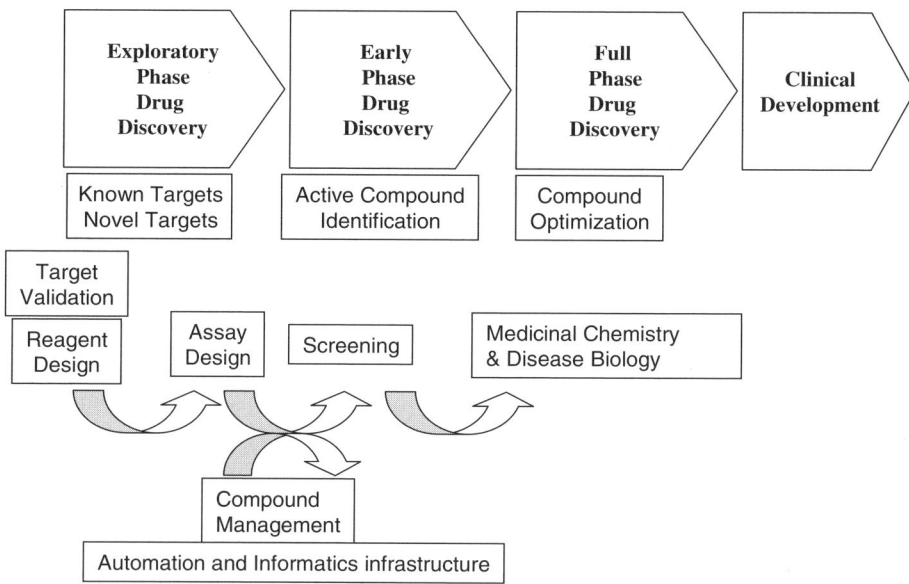

Figure 13.1 Overview of the drug discovery process.

Judicious pruning and regrowth of the screening deck can greatly enhance the quality and diversity of the collection and significantly improve the success rates of the HTS campaigns (Bristol Myers Squibb in-house experience).

Design of a physiologically relevant and pharmacologically accurate assay is also a crucial part of the HTS process and can materially affect the successful outcome of a screening campaign. Converting a target of interest into a bioassay suitable for scale-up to HTS can be a rate-limiting step in the lead discovery process. The complex matrix of parameters that have to be taken into account in the design of an assay can include choice of assay mode (agonist or antagonist), availability of standard protocols, reagent availability, and scalability. More specific design elements, once a basic format and protocol have been chosen, will include optimization of key assay parameters such as pH, temperature, time or cell number, type of media, and the like. Today an array of assay design choices are available to the HTS specialist that are tried and trusted in the HTS environment.

Just as the choices for assay design have improved over the years, so has the range of HTS technology platforms used to run HTS campaigns. This area of HTS support can trace its roots to the days of ad hoc user groups sharing knowledge about automated solutions for labor-intensive tasks such as compound dispensing or reagent transfers. Nowadays a full range of automated equipment is available to the HTS specialist from the relatively simple, such as modular workstations that can carry out plate-to-plate transfers, or reagent dispensing, or have detection capability, to the more complex, such as the multitasking, turnkey robotic arm, and track systems.

Some companies (e.g., GlaxoSmithKline) have taken the automation route to possibly its ultimate endpoint with the construction of drug discovery factories where the process of drug discovery is completely housed in a custom-built, fully automated building. For a basic description of the types of automation approaches available to the typical HTS lab, one can read Bojanic et al. [2].

The final major component of the lead discovery process is data management operations. For capturing the millions of individual data points emanating from a screening run to their visual display, detailed analysis, and interpretation, a whole host of in-house and commercially available software packages have now been developed.

High-throughput screening as a discovery process has benefited greatly from significant advances in automation, microtiter plate design, and bioassay techniques, as well as taking advantage of custom-designed compound and data management systems. It has evolved into a fully integrated, seamless process that the majority of pharmaceutical and biotechnology companies rely on to generate lead compounds to enable their drug discovery programs.

The HTS story brings together a plethora of different technologies, skill sets, and insights from a range of industries. This chapter will attempt to review some of the main features of the HTS platform that are essential for the effective functioning of the drug discovery process as well as illuminating some of the evolutionary milestones in HTS techniques over the last 20 years.

The twin demands of increasing the capacity to test compounds while maintaining, or even reducing, costs, has encouraged screening groups and the HTS support industry to come up with ever-changing screening formats utilizing state-of-the-art automation and miniaturization techniques. However, the development of an array of high-throughput screening formats starting from simple bacterial lawn assays, through spot and wash techniques, to homogeneous assay formats, each brought with them a set of technology problems that needed to be solved. Even today HTS scientists still need to consider a range of process issues such as how to get compounds to the assay, which type of assay to use, how to store compounds in readily available sets, guaranteeing the purity and stability of stored compounds, finding rapid and reliable ways to dispense compounds and reagents, development of sensitive detection devices to measure assay products, effective data capture and analysis tools to deal with the volume of data being generated. Each of these problems and questions will be touched on in this chapter.

This chapter will review the variety of screening plate formats that have been developed from the 96-well plate to the high-density 3456 plates and the implications of their usage on the supporting technologies and assay designs. Automation support from compound and reagent handling as well as compound storage architecture and reagent production will be discussed. The range of bioassay techniques now available for use by HTS scientists will be summarized.

13.2 EVOLUTION AND INTEGRATION OF HT PLATFORMS INTO THE DISCOVERY PROCESS

The real impetus for HTS as we know it today came in the late 1970s. Even though the first prototypical microtiter plate was described in 1954 [3], the ubiquitous 96-well plate became popular with the increasing use of enzyme-linked immunosorbent assays (ELISA) during the late 1970s [4, 5]. The strengthening of three technology streams—reaction vessels, methods to reproducibly deliver small volumes of reagents, and methods of measuring the endpoint of an assay (absorbance, fluorescence, and luminescence)—has driven the evolution of HTS over the last 20 years.

Three other major developments have also been drivers for HTS evolution. The development of precise nonseparation bioassays, the ability to organize and distribute large numbers of chemicals [6], and the analysis of large volumes of data were critical for building the HTS infrastructure. The integration of these technologies from chemistry, biology, engineering, and information technology are the critical elements in performing a cost-effective and timely high-throughput screen [2]. We will describe the developments that led to the process of HTS including new instrumentation, homogeneous bioassays, assay miniaturization, cell-based assays, and strategies to deal with the large number of compounds identified in HTS (Fig. 13.2).

Evolution of Instrumentation and Screening Formats

This evolution of instrumentation and screening formats is best described by example of the evolving methodologies used to identify compounds that affect

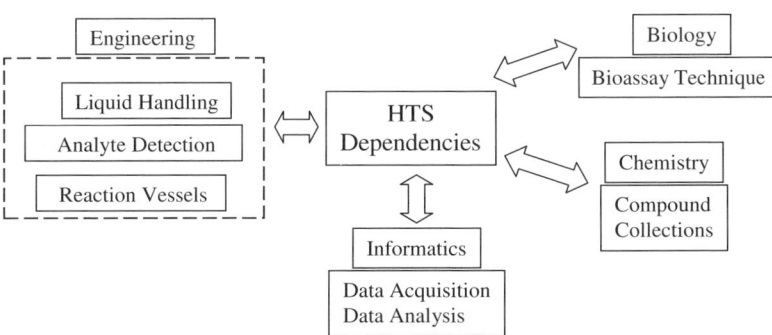

Figure 13.2 HTS evolution has been dependent on the development of a whole range of different disciplines.

the activity of tyrosine kinases. Tyrosine kinases are critical for effective second-messenger signaling in many different cell types, and aberrant activity has been demonstrated in a variety of oncogenic states. More recently, anticancer drugs that modulate tyrosine kinase activity have shown efficacy in the clinic [7].

One of the first screening methods for identifying compounds that affect tyrosine kinase activity involves incubating enzyme, radiolabeled [^{32}P]–adenosine triphospate (ATP), and substrate in a tube. The phosphorylated substrate was then spotted onto P81 phosphocellulose paper. After washing to remove residual [^{32}P]-ATP, the amount of phosphorylated substrate was quantified in a γ counter [8]. This method was very tedious and low throughput. Improvements in throughput were made by running the assay in 96-well microtiter plates, spotting reaction products onto a large sheet of P81 paper using a 12-channel pipette, and then washing the entire sheet to remove radiolabeled ATP. The dried paper was then exposed to photographic film, and spots that produce lower levels of exposure were indicative of tyrosine kinase inhibition. To determine the degree of inhibition, individual zones corresponding to each well of the microtiter plate were cut from the P81 paper and placed separately into a scintillation counter for Cherenkov counting. Alternatively, the film could be semiquantitatively assessed based on the intensity of the developed spot or zone. This approach was further refined by the introduction of the phosphoimagers and gas flow imagers [9] in the mid- to late 1980s. These innovations provided a means of quantitation that eliminated the need to cut up individual squares.

The next important step in the evolution of screening formats was the use of immunoassays to quantitate levels of phosphotyrosine [10]. In this screening method, a synthetic tyrosine kinase substrate, for example, PolyGlu-Tyr, was adsorbed onto the surface of a microtiter well. Unbound material was removed using a 96-well plate washer, and enzyme, ATP, and putative inhibitors were incubated in the substrate-coated well. After a brief incubation, the contents of the well were removed and the phosphorylated substrate was detected by a phosphotyrosine-specific antibody using stardard ELISA techniques. This new format offered significant advantages; it was nonisotopic, it increased parallel operations, and used the plate readers and washers that had been developed by engineers to support the growing ELISA market. However, the method also had significant disadvantages; the assay still involved multiple steps and washing the microtiter plates was tedious. Today, when it is commonplace to screen 1 million compounds, this format would involve 10,400 plates and hundreds of liters of wash fluids. The reader should be aware that there were significant scientific issues with this methodology, specifically the nonphysiological nature of the substrate and its solid-phase disposition in the assay.

Importantly during this period, new technologies were emerging that allowed the development of a filtration assay. Harvesters that took the contents of 96-well microtiter plates through a filter mat containing 96 zones in a

6 × 16 array were developed based on a concept designed by Warner and Potter [11, 12]. The filter mat was automatically washed as part of the harvesting action and was then ready for quantitation. A range of substrates could now be used, including protein substrates that could be acid precipitated and the radiolabeled product captured on a glass fiber mat. In addition, new scintillation counters were developed that could read the incorporated radioactivity of the entire filter. The new counters used 6 photomultipliers that significantly reduced the time required to measure the whole filter mat.

A significant improvement in high-throughput screening came with the introduction of the homogeneous assay, where the reaction could be measured without a separation stage. This was critical to HTS evolution because it allowed higher precision, a more rapid throughput, and led the way for further miniaturization. One early problem with homogeneous assays was the susceptibility to give false-positive results, for example, as in scintillation proximity assays (SPA) [13]. In a scintillation proximity assay, beads containing scintillant are coated with substrate and incubated with the tyrosine kinase, radiolabeled [^{33}P]-ATP and putative inhibitors. Compounds that interfere with tyrosine kinase activity decrease the amount of radiolabeled [^{33}P] in proximity to the bead containing the scintillant and subsequently decrease the light output. At that time, the SPA beads used emit light at 420 nm; however, a significant number of colored compounds will absorb light in this part of the spectrum. By reducing the light output, this data could be misinterpreted as inhibition of tyrosine kinase activity. Concomitant with the emergence of SPA was the development of the 96-well scintillation counter (Topcount, Packard Instruments, USA, and Microbeta, Wallac Oy, Finland), a necessary prerequisite for HTS because this new instrument accepted the 96-well plate directly and had the ability to quench correct for colored compounds to a certain extent.

Miniaturization

Until the early 1990s the 96-well plate was the standard format for HTS, however, the development of new assay technologies and pressure from HTS laboratories facilitated the design of new microplate formats. The size of the wells was decreased to yield 384 wells that had the same footprint as a 96-well plate. This reduced the volume of assays by a factor of 4. Instrumentation companies rapidly adapted to this change with liquid-handling devices that had better spatial resolution and were capable of delivering smaller volumes. Detection devices were also improved to give higher sensitivity and higher throughput.

The late 1990s brought further miniaturization and changes in plate design with higher density plates such as 864-well plates from Affymax, 9600-well plates from Dupont Pharmaceuticals, and 3456-well plates from Aurora Biosciences. The use of plates with higher well densities required significant adaptation of liquid handling and detection technologies. These changes in plate design posed significant problems not seen in the evolution of 96-well to 384-well platforms. For that transition, plate readers for absorbance and fluorescence measurements were dependent on photomultiplier tubes that required significant read times. A 96-well scintillation counter with a mask could be adapted to allow each well to be read in a 384-well assay, however, the read time increased by a factor of 4, which compromised throughput. Measurement of 1536-well plate assays was improbable using photomultiplier technology. The implementation of charged coupled devices (CCD) that through a telecentric lens could image the whole plate in one measurement allowed measurement of assays using higher density plates [14]. This imaging innovation also had other advantages. For example, the same instrument could be used for luminescence, fluorescence, and fluorescent polarization readouts from any microtiter plate format. In addition, the CCD chip has a light detection sensitivity that is more red shifted than photomultiplier tubes and therefore could better accommodate a wide range of fluorescent dyes. New SPA beads were developed to work with CDD imagers that shifted the wavelength of the emitted light to 620 nm; this in turn reduced the false-positive interference from colored compounds. Today, we have automated assay platforms that can routinely run low-volume assays (5 to 10 µL) with high precision and process the evaluation of millions of compounds in 2 to 3 weeks. Advancements in bioassay technology, instrumentation, control software, and efficient processes have now provided the drug hunter with very powerful tools [2].

Evolution of Cellular HTS Assays

Cell-based assays for HTS first emerged in the mid-1980s. These assays were known as *black box screens* where growth factors, cytokines, or hormones produced a change in growth of cells, but it was not clear how these signaling events culminated in a cellular response. The early measurements of cellular proliferation used [^{3}H]-thymidine incorporation as an indicator of deoxyribonucleic acid (DNA) synthesis [15–17]. It was not until the late 1980s that redox dyes (3-[4,5-dimethylthiazol-2-yl]-2,5 diphenyltetrazolium bromide), which measure the oxidative potential of the mitochondria as an indicator of cell proliferation, became available for the HTS scientist. This enabled the development of homogeneous proliferation assays that were amenable to automated systems and could be measured using simple plate readers [18, 19]. Advances in molecular biology were critical to the engineering of mammalian cells that could express targets, cofactors, and reporters for cell-based HTS [20]. The earliest reporter assays, described in 1982, included chloramphenicol acetyltransferase and β-galactosidase, and these were rapidly followed by luciferase, aequorin, and green fluorescent proteins (from *A. victoria*), and secreted alkaline phosphatase [21–26]. Reporter assays enabled the study of complex cellular signaling pathways that could be modulated pharmacologically. For luciferase-based assays, luminometers with integrated pipetting are required to measure flash luminescence [27]. The advance of "glow" luminescence enabled conventional microplate scintillation counter technology to capture and measure light from these assays [28]. Reporters such as β-lactamase and renilla luciferase, which measure real-time changes in gene expression in living cells, have also been developed [29–31]. In addition, there are synthetic versions of click beetle luciferase genes for dual-color measurements in multiplexed reporter assays allowing multiple signaling pathways to be interrogated as one time [32]. Another major advance in cell-based screening came with the ability to directly measure changes in the levels of second messengers (calcium, cyclic adenosine monophosphate (cAMP), inositol-3-phosphate (IP3), and mitogen activated protein (MAP) kinase) within the cell in an automated screening environment [33–36]. Advances in automated fluorescence microscopy enabled more complex measurements within cells such as phenotypic changes, intracellular protein translocation, and intracellular posttranslational modifications [37, 38]. Until recently with the development of instruments like Patch Xpress, patch clamp for single-cell electrophysiological measurements eluded automation [39]. Advances such as those mentioned above have rapidly increased the number of cell-based screens and also allow the screener to run assays against targets in the appropriate cellular context.

Evolution of HTS Technologies into the Drug Discovery Process

The developments discussed here describe an HTS technology base capable of assaying a wide range of in vitro and cellular targets, which has resulted in the identification of many active compounds than can be progressed into medicinal chemistry programs. To prioritize which compounds progress in the drug discovery process, medicinal chemists consider several factors, including potential liabilities, compounds selectivity, drug–drug interaction problems, drug metabolism, and absorption or permeability issues.

Figure 13.3 shows a typical flow of activity through a traditional drug discovery program. Primary assays are performed to give feedback on a medicinal chemistry hypothesis; secondary assays are then performed to provide more detailed data. This then is followed by selectivity information and then liability information. At each step, it is typical to reduce the number of compounds being progressed. This serial cycle then repeats as the chemistry hypothesis evolves. Compression of this cycle improves the efficiency of drug discovery. One way to achieve this increase in efficiency is to use parallel approaches, such that data on primary bioassay, selectivity, and potential liability can be assessed together by the medicinal chemist (Fig. 13.4). This requires many assays to be run simultaneously but on fewer compounds than a typical HTS.

One major difference between the HTS operation and the parallel lead optimization process is the way that compounds are moved through the process. For HTS a large collection of compounds is assembled, usually divided into a series of master compound plates, all of the same format, and all

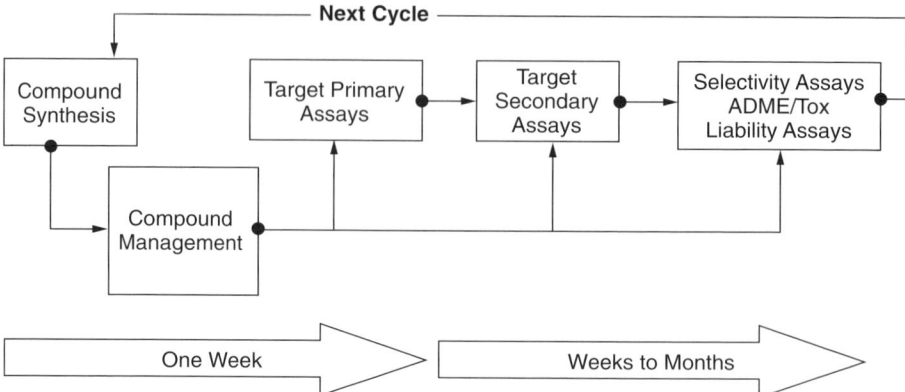

Figure 13.3 Serial workflow diagram that is typical of many drug discovery programs.

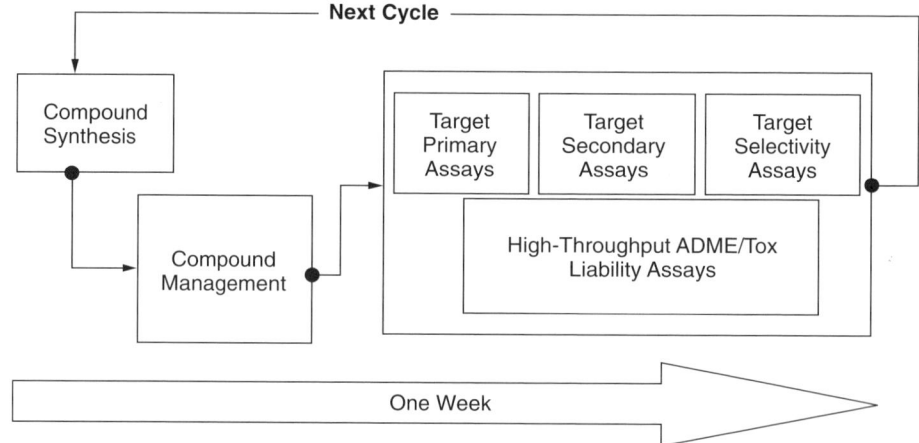

Figure 13.4 Parallel workflow diagram demonstrates the advantage of the process infrastructure that has been developed for high-throughput screening.

screened for bioactivity against a target. For lead optimization a small number of compounds arrive from a series of different programs. These compounds then have to be routed to a range of different assay suites that may exist in different assay format, for example, 96-well, 384-well, or higher. For a given drug discovery program, the type of bioassay changes as it progresses toward late-stage drug discovery and new assays complement or replace older ones. Also, critical compounds may need to be checked against a broad range of assays, rather than just the program-specific assays. This process is summarized in Figure 13.5.

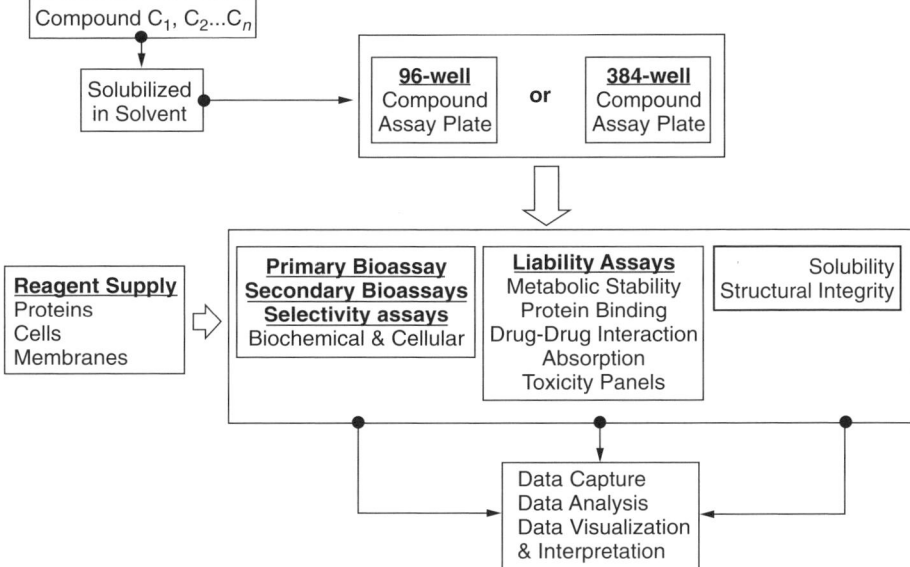

Figure 13.5 Detailed flowchart of the compound distribution, reagent supply, assay, and data analysis that needs to be synchronized to achieve an efficient parallel lead optimization process.

Quality Control in the HTS Process

The quality of the data from HTS assays is dependent upon the supply of robust reagents, and this is an integral part of the process. Some reagents such as proteins and membranes can be stockpiled and quality controlled off-line. For cell-based assays, cells may need to be provided at the time of the assay. New automated cell culture machines greatly facilitate this process (e.g., SelecT from the Automation Partnership, UK).

Assay accuracy and precision is critical to the success of HTS strategies and is constantly monitored through the use of standard compounds and blank plates. Hardware quality control requires preventative maintenance and daily calibration. By integrating quality control ranges in the robotic control software, the assay performance can automatically monitor and alert scientists to any issues in prosecuting the assay. By successfully completing the quality control limits set on the assay, the data then moves directly to databases that intelligently process the raw data. An example can be found in drug concentration response analysis where automated curve-fitting routines are embedded into data analysis software and performed automatically. Here curves are fit upon comparison to a range of pharmacological rules. These rules determine whether there is a reasonable curve that fits the rules applied to minimal or maximal activity or slope. If there is a reasonable pharmacological fit of the data, it then streams into a data-reporting information technology (IT) tool.

The scientist is alerted when data points fail to fit using the IT tool, and the responsible individual can make the necessary judgment call to repeat the experiment.

For selectivity or liability one might want to test the compound at a concentration significantly higher than that used in the primary bioassay. Therefore, it is important to understand the solubility of the compound under the assay conditions. From our own experience, a compound's liability can be overlooked due to poor solubility in the liability or selectivity assay, and these false negatives can misdirect a drug discovery program. Inclusion of structural integrity analysis of the compound being tested into the screening process is also an important quality control parameter. Structural integrity information confirms the identity of the compound tested in the bioassay and also verifies that the compound has not degraded during the process. In the past 20 years, we have witnessed a dramatic change in the way that bioassays are performed from both a scientific and technology perspective, and this has been driven by multidisciplinary teams involved in engineering, biology, and chemistry.

13.3 HTS AUTOMATED SYSTEMS: TECHNOLOGY AND PROCESS

The last 10 years have seen a significant increase in the size of pharmaceutical compound collections from hundreds of thousands to millions. With the advent of HTS in the early 1990s, a typical lead discovery organization would have been expected to screen between 10 and 20,000 compounds through 15 targets a year. Today, many companies can screen 1 to 5 million compounds through 20 to 100 targets per year with each campaign lasting only a matter of weeks [40].

The areas that will be discussed here with relevance to screening large numbers of compounds are: advances in plate design, compound management, liquid handling, assay detection, automation for HTS, and data analysis.

Plate Design

One of the most significant advances in HTS technology over the last 15 years has been the ability to reduce assay volume and increase microtiter plate well density. In the mid-1990s, the predominant assay format was the 96-well plate [41]. By the late 1990s to the early 2000s, assay formats moved to much higher well-density plates (Table 13.1) [41–44]. Advances in plate design required a combination of enhanced liquid handling technology to deliver smaller volumes and enhanced reader technology to measure assay output from >96-well plates. The evolution into the 384-well plate was dependent upon the ability of liquid handling equipment and assay detectors to manage both the 96- and 384-well plate. For example, instrumentation with 96 tips can pipette in 4 quadrants to a 384-well plate and likewise for the transition from 384-well

TABLE 13.1 Assay Formats Including Plate Density and Reaction Volumes

Plate Format	Well Layout	Max Volume (µL)	Typical Assay Volume (µL)
96	12 × 8	250	100
384	24 × 16	80	50
1536	48 × 32	10	5
3456	72 × 48	2.2	2

TABLE 13.2 HTS Assay Plate Types

Clear	Absorbance assays	Clear light path for light to pass through
White solid	Luminescence	Reflective surface to maximize light output
White, clear bottom	Luminescence	Reflective but allows bottom read
Black solid	Fluorescence	Minimizes background
Black, clear bottom	Fluorescence	Low background but allows bottom read

plate to 1536-well plate (4 × 384). The ultra low volume 3456-well plate (36 × 96) was produced by Aurora, when the 96-well plate was still the predominant assay format for HTS and was considered a more revolutionary plate development process. The principles of assay design remain the same regardless of plate format. In reality, the success of an HTS assay depends on the ability to dispense liquids in the volume ranges indicated in Table 13.1. Other critical factors to consider when adopting a higher density plate format for HTS include length of detection in terms of instrument read times and changes in surface area to volume ratios. In the case of fluorescence assay detection, the read times vary, and this is dependent on the choice of instrument, for example, a single device based on a photomultiplier tube (PMT) could take up to 15 min to read an entire 1536-well plate on a well-by-well basis. This means that well number 1536 will have been incubated for 15 min longer than well number 1 unless the assay is stopped and has a stable signal. Finally, the plastic binding surface is larger in ratio to total volume in higher density formats. While this has not been reported to be an issue, in our hands, this surface-to-volume ratio will have an effect if the assay reagents are particularly sticky or hydrophobic. Hydrophilic coated plates can help minimize this effect. The number and variety of different plate types that are available is quite significant within each of these formats. Table 13.2 describes the range of plate types that are currently available for HTS.

There are also a number of plate coatings that are available for a multitude of bioassay applications. There are too many to discuss here, but they include antibody coatings to bind proteins, scintillant coatings for specific radiochemical applications, poly-D-Lysine, and collagen for cellular adhesion and the previously mentioned low binding plates.

Three different plastics are used for microtiter plates throughout the HTS process. The first is polypropylene, which is preferred for compound storage because it is solvent tolerant and has low binding properties for compounds. The second is polystyrene, which is the preferred assay plate due to rigidity for automation, flexibility to be colored (as described in Table 13.2), and use for a variety of assay technologies. The third is cyclo-olephin copolymer (COC), which is a newer plate type with potential for compound storage since it has lower binding properties for compounds and has excellent rigidity as well as optical clarity.

Compound Management

As discussed above, the compound collection of most pharmaceutical companies has grown significantly and collections are now in the 1 to 5 million compound range [45]. This presents challenges for both storage and distribution. A number of key considerations are discussed below.

Composition Historically, the pharmaceutical compound collection has typically included legacy compounds that a company has made for discovery programs over many years. The main issue with these legacy compounds is that this type of collection lacks diversity since it was made to target specific proteins. Examples of this from our own experience include work that generated many β-lactams for antimicrobial work and a large number of steroids in the late 1980s. Still, today a large number of compounds within screening decks are made up of β-lactams and steroids for exactly this reason. Consequently, there has been a focus within pharmaceutical companies to increase the number of compounds within screening decks to increase chemical diversity and remove compounds that are unlikely to progress to lead compounds.

In HTS, it is important to screen a chemically diverse set of compounds to find novel leads. However, measurement of diversity has been, and still is, a very contentious issue. In order to have an optimal screening deck that covers all permutations of chemical diversity, a compound collection would have to be on the order of 10^{60} compounds [46]. Clearly, it is not possible to screen such a large number of compounds, so efforts are made to sample this diversity and to populate the collection with compounds that have potential for progression into the drug discovery process. This is no trivial task, and yet it is critical to the success of HTS because any HTS campaign can only be as successful as the compounds it tests. The success of an HTS campaign is measured by the number of novel chemotypes that progress into full medicinal chemistry programs, where the definition of a chemotype is a compound or set of compounds that have a unique structure or chemical composition.

There are many companies that build libraries of compounds and sell them to pharmaceutical companies to populate their compound collections. This is a popular approach since the cost of synthesizing compounds is high from a personel perspective. However, it can also mean that the same library is being tested by many companies, often against the same targets. In this case, the chance of finding a compound that is different from a competitor is low, and so this shifts the emphasis to the speed with which a company can discover and progress a compound first. Additionally, the use of combinatorial libraries can mean that chemical diversity within a library is poor since they tend to be built on common scaffolds. However, if these libraries contain active compounds based on similar chemical scaffolds, then some level of structure–activity relationships (SAR) can be derived very rapidly from the HTS.

Another approach to organizing a compound collection is to group compounds into categories of potential activity, that is, kinase or G-protein-coupled receptor (GPCR) activity. This enables a smaller subset of compounds to be tested against targets that are either very early in the discovery process or considered too challenging (i.e., cost prohibitive, limited reagents, etc.) to screen the entire compound collection [47, 48].

Wet or Dry Compound Storage Wet or dry refers to storing compounds as solids or dissolved in solvent. The generic standard for supplying compound to a bioassay is in dimethyl sulfoxide (DMSO). This is the standard for solubilizing the majority of compounds in a screening collection, but there are a number of significant issues relating to storage of compounds in a universal solvent such as DMSO [49]. First, a number of compounds are not soluble in DMSO. Thus, the predicted concentration of compound will never be attained in the bioassay. Second, DMSO is very hydroscopic, so, when exposed to the atmosphere, it adsorbs moisture and will continue to do so until it reaches equilibrium at about 70 percent DMSO, 30 percent H_2O. Now the compound is diluted, and solubility under these aqueous conditions may be adversely effected, both of which alter the effective compound concentration in the bioassay. Third, DMSO can dissolve certain plate materials and these may have an adverse effect on the bioassay. Finally, DMSO is reactive with some compounds, especially in the presence of oxygen, thereby effecting the integrity of compounds in storage.

There have been many different solutions to address the above issues that vary in complexity with the ultimate goal of maintaining a completely dry environment for the storage and distribution of compounds. Recent literature [49] suggests that it is best to store compounds in DMSO in an inert, dry atmosphere at room temperature, thus avoiding freeze–thaw cycles. The investment to achieve such storage conditions may be significant, especially when the size of the compound collection is considered. For example, Figure 13.6 shows a photograph of the Haystack liquid store at Bristol-Myers Squibb that is two stories high. Not only must the environment where the compounds are stored be controlled, but the environment where compounds are dissolved, transferred to and from the store, replicated for screening, and stored for screening must also be controlled. Given the cost of full environmental control of

Figure 13.6 Photo of Haystack (The Technology Partnership, Cambridge, UK). The haystack compound storage system at Bristol-Myers Squibb showing the "wet" tube store.

their compound collections, some companies have taken the route of making up compounds in either 90 percent or 70 percent DMSO from the outset and then using less extensive environmental control. While less expensive, this compromise will reduce the aqueous solubility of certain compounds.

Long-term storage of compounds is optimal when compounds are stored as dry samples and preferably at low temperature. Thus, most companies have a dry-compound store, often in tubes as well as a wet store. At periodic intervals, samples from the dry store are taken to replenish the wet store in DMSO.

Compound Storage in Plate Formats Compound storage in plate formats should reflect the most typical HTS usage plate format. Storing compounds in 384-well format is most efficient if screens are run in a 384-well format. When preparing a compound plate for a bioassay, the replication cycle for compound transfer is from one plate to a daughter plate (or many identical daughter plates), rather than four 96-well plates to one 384-well plate. However, this may not be the most efficient process when individual compound samples are required since an entire plate would need to be removed just to pick an individual compound. Consequently, multiple formats are used to store compounds.

For example, the Haystack system at Bristol-Myers Squibb in Connecticut (Fig. 13.6) has three components. The first is a dry store where compounds are stored as powders or films. The second is a tube store that stores compounds in 100 percent DMSO in sealed microtubes. These microtubes are made from the dry tubes by adding volatile solvent to the dry tubes, transferring the required volume out, and then drying the sample within the original tube back down for storage. These microtubes are used to generate master plates for HTS and for the selection of individual compounds. The master plates for HTS are then stored in the third area of Haystack where they can be retrieved to make assay plates for HTS.

Replication for HTS As discussed above, storage temperature and environmental conditions will have a significant effect on the transfer of compounds into screening plates, otherwise known as the compound replication process, so only generalities are discussed here. The key points to consider are that freeze–thaw cycles should be minimized as should DMSO exposure to the atmosphere. Figure 13.7 represents a typical compound replication process.

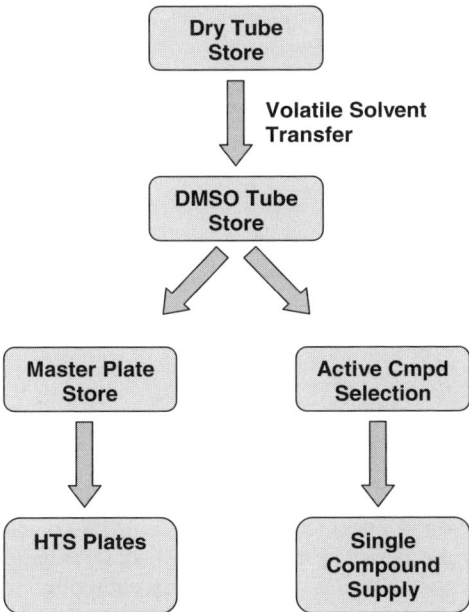

Figure 13.7 Compound storage. Flowchart representing the typical process for compound storage and retrieval using the Haystack automated store at Bristol Myers Squibb Co.

Dry compound from a tube store is used to make a number of master plates in DMSO. Periodically, a master plate is used to make a number of screening plates ready for bioassays.

The important factors to consider are the tolerance of the assay to DMSO and the concentration of compound. A number of biochemical assays can tolerate higher DMSO concentrations in the 2 to 5 percent range that enables plates to be used directly, but cellular assays rarely tolerate higher than 1 percent DMSO and inevitably require an intermediate dilution unless nanoliter dispensing is available (see next section). Typical compound concentrations for screening are in the 10 µM range and typical compound storage concentrations are in the 2 mM range.

Nanoliter Compound Dispensing As discussed above, some assays have a low tolerance for DMSO. This means that as the volume of the bioassay decreases when moving from a 384-well to 1536-well plate, the volume of DMSO to be added needs to decrease proportionally. For example, if a typical 1536-well assay volume is 5 µL, to attain a DMSO concentration of 1 percent without using an intermediate aqueous dilution, a compound addition of 50 nL is required [50, 51].

The value of avoiding an intermediate aqueous dilution during screening is illustrated in Figure 13.8. Here, IC_{50} determinations for 1000 compounds were performed in a standard enzyme assay via two routes. The first using an inter-

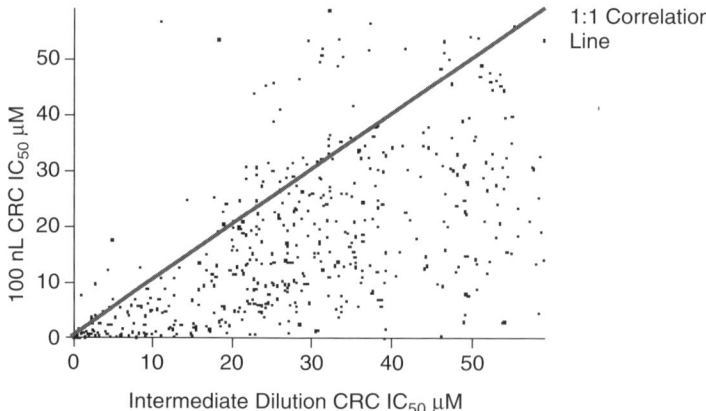

Figure 13.8 Correlation of intermediate vs. 100% DMSO compound transfers where the graph represents the correlation of 1000 IC_{50} determinations performed for an enzyme-based target via two routes. (1) Making up compound in 100% DMSO, creating a 3-fold, 10-point serial dilution in 100% DMSO. Performing a 10-fold dilution in water and transferring 1 µL to the assay plate. Represented on the x axis. (2) Making up the compound in 100% DMSO, creating a 3-fold, 10-point serial dilution in 100% DMSO. Transferring 100 µL direct to the assay plate. Represented on the y axis.

mediate aqueous dilution and the second using a 100-nL dispense of compound in 100 percent DMSO directly to the assay plate. As can be seen, the majority of compounds appear more potent via the 100-nL method. The enzyme used in this assay had a reasonably hydrophobic binding pocket, and a number of the active compounds tested were hydrophobic in nature. The difference in potency occurs because of the decrease in solubility of hydrophobic compounds in aqueous solution. Thus, if the compound is at a high concentration in 100 percent DMSO, it is more likely to drop out of solution when it is diluted to a low concentration in aqueous buffer.

It is only in recent years that the technology has been made available to dispense compounds in the low nanoliter range [50]. A significant advance came in 2003 with the onset of acoustic dispensing. This technology uses ultrasound to eject nanoliter droplets from the surface of liquid in a plate that contains compounds and fires the droplets into an inverted destination plate above. This liquid handling technique is very precise and accurate, and, because this dispense method is noncontact, no washing pipette tips are involved.

Closed-Loop Screening Closed-loop screening is a term used to describe the process where an assay is run continuously from initial screening to IC_{50} determination. The initial screen assays a large collection of compounds at a single compound concentration, and then active compounds are confirmed on that same system from the same source compound plate. Critical to this process is the ability to carry and store a large number of compound plates on the screening platform. Thus, when the assay is running, HTS assay plates can be prepared from these on-line compound plates. The clear advantage in this approach is that reconfirmed positive compounds are tested continuously from the original plates stored on the assay system, and the dependency on a compound management department to select these compounds from a large store is circumvented. This benefit has become more important as the size of compound collections has increased, for example, if a compound collection is 1 million compounds and of these 1 percent are active, then 10,000 compounds will need to be supplied to a confirmatory assay.

Liquid Handling

The key challenge for either the bench scientist or the automated screening platform is reliably dispensing all components of the assay with sufficient accuracy and precision. Working in the larger plate volumes such as 100 µL for the 96-well plates means that individual assay components need to be dispensed in the volume range of 10 to 50 µL. As the well density of the plate increases, the volume of each well decreases, and accurate liquid dispensing becomes a more significant challenge. Table 13.3 summarizes the different types of liquid handling devices that are available, along with their working volumes. However, before getting into detail here, it is important to consider

TABLE 13.3 Summary of Liquid Handling Types[a]

Type	Plates	Volume Range (µL)
8- to 16-tip variable span	96, 384 (some 1536)	0.5–100
8- to 16-tip fixed span	96, 384 (some 1536)	0.5–100
96 head, tips	96, 384	2–100
384-head, tips	384, 1536	0.5–50
Peristaltic pump	96, 384	5–200
Solenoid valve	96, 384 1536, 3456	0.2–10
Pin tools	96, 384 1536, 3456	0.005–0.1
Acoustic	96, 384 1536, 3456	0.002–0.2

[a] This table represents typical averages only.

the parameters that define the liquid handling performance: accuracy and precision.

Accuracy and Precision

Liquid handling needs to be both precise and accurate. The term *accuracy* simply refers to the question: "Does this instrument dispense the absolute volume that has been requested?" Thus if 10 µL were to be dispensed and the average of all dispenses was 10 µL, then instruments would have excellent accuracy. The variation in dispense volume is more a question of precision. Consequently, accuracy is often represented as percent change from nominal. For example, an accuracy of 10 percent on a 10 µL dispense means that all dispenses will be between 9 and 11 µL. For the majority of instrumentation used in HTS, it is expected that they will have an accuracy of <10 percent of nominal.

Precision is a measure of how variable a dispense is and takes into account the standard deviation of all dispenses. Precision is typically expressed in terms of the coefficient of variation expressed as a percentage (%CV). It is standard practice to aim for liquid handler performances in the with %CV values of <5 percent.

It is imperative to continuously monitor the performance of an instrument to prevent waste of valuable reagents and data. Both precision and accuracy are essential to the understanding of instrument performance. For example (Table 13.4), analyzing the accuracy alone would indicate that either instrument 1 or 3 performed the best, since in each case the average dispense is 10 µL. If one measures precision alone, it would indicate that instrument 2 performed best since it has the lowest %CV, indicating that all dispenses are very similar. Close interpretation of the data, however, suggests that instrument 1 performed with variable dispense. Instruments 2 and 3 are more difficult to differentiate as 2 gives precise data but over dispenses, yet 3 exhibits more variation but is more accurate. It is also important to test the instrument with

TABLE 13.4 Dispense Volumes for Specific Instrument[a]

	Instrument 1	Instrument 2	Instrument 3
Dispense volumes (µL)	8.1	10.3	9.1
	9.5	10.2	11
	9.1	10.2	9.5
	1.6	10.2	9.4
	12	9.9	10.5
	8.8	10.1	10
	10.2	10	10.7
	11.2	9.9	11
	12.6	10.3	9.2
	13.1	10.1	10.4
	9.9	10.4	10.2
	14	10.4	9.2
Average dispense	10.0	10.2	10.0
% CV	32.2	1.7	7.1
Number outside 10%	7.0	0.0	2.0

[a] Shows the actual volume dispensed when a 10-µL volume was requested over 12 replicates. This data is used to analyze precision and accuracy.

a range of buffers and reagents. The precision and accuracy will vary significantly with the reagents that are used. For example, glycerol may stabilize proteins but it increases viscosity. Bovine serum albumin (BSA) may also stabilize the assay reagents but it is liable to foam as do detergents. Additionally, it is very important to examine if the liquid handler creates a pattern following dispensing that could indicate that a pipette tip is clogged or not functioning. For example, if an instrument has 384 tips, each well is dispensed independently into a 384-well plate, but in a 1536-well plate each tip dispenses to 4 adjacent wells so it is important to look for differences in blocks of 4. If an instrument has 8 tips that dispense in a linear pattern across a plate, then differences between rows or dispense patterns within a row may occur. For example, the Labsystems MultiDrop dispenses in the pattern indicated in Figure 13.9. Panel 1 demonstrates an obvious dispense pattern where the columns show a low initial dispense. Panel 2 shows a typical dispense with no patterns and a good level of precision. Finally, the liquid handling instruments are validated as part of transitioning a bioassay from the laboratory bench to an automated system. This final test ensures that the instrument will function with precision and accuracy during routine use with a variety of assay reagents including liquids of varying viscosities, beads, and cells.

Assay Detection

In HTS there are a number of assay technologies that use a range of detection methods. This list includes absorbance, fluorescence, fluorescence reso-

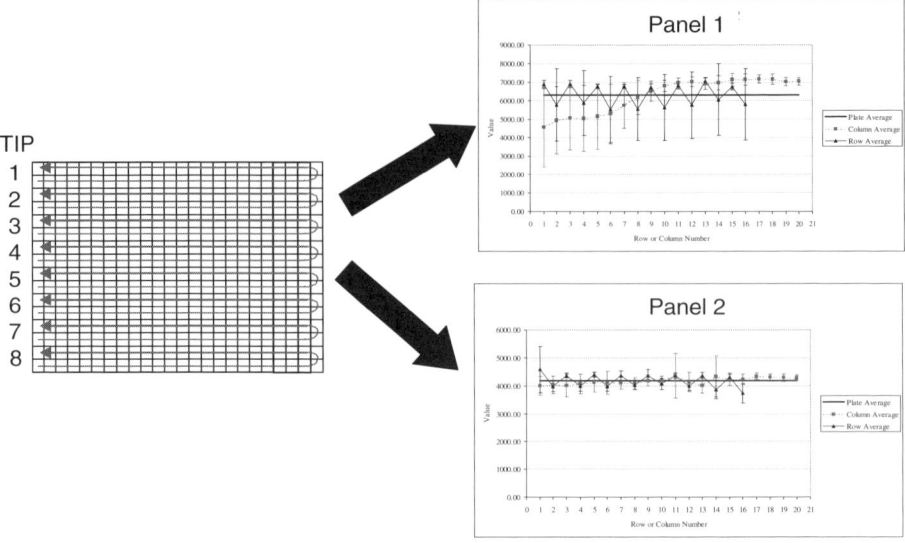

Figure 13.9 Dispense patterns from an 8-tip device in a 384-well plate shows the dispense pattern on the left, starting in row B and dispensing across the plates and back down row A. Panel 1 shows a significant pattern when row and column averages are plotted. Panel B demonstrates a typical "good" dispense from the same instrument.

nance energy transfer (FRET), homogeneous time resolved fluorescence (HTRF), flash luminescence, glow luminescence, SPA, and fluorescence life time (see Section 13.4). There have been significant advances in the field of assay detection over the last 5 years [2]. This has primarily been as a result of the development of multimodality imaging technology. As has been mentioned previously in this chapter, the majority of detection methodologies have traditionally been based around photomultiplier tube (PMT) technology. This has the advantage of being very sensitive but realistically means that each PMT can measure one well at a time. This was not an issue when the majority of HTS assays were based in the 96-well plate, but as the well density of plates has increased, the limits of PMT technology have been exceeded. For example, if an SPA assay were being run in a 384-well plate, it takes about 1 min to acquire enough signal from each well via a PMT to get acceptable assay data. Certain scintillation counters have as many as 12 PMT tubes, so they can measure up to 12 wells at once, but this still translates to about 40 min to read a single 384-well plate. For other technologies such as fluorescence, FRET, and HTRF, the throughput limits of PMT-based readers were not reached until assays were executed in 1536-well plates.

Imaging technology uses highly cooled CCD chips and telecentric lenses to measure signal from entire plates at one time [14]. The highly specialized cameras are cooled to temperatures as low as −103°C to reduce electronic noise in the image and increase sensitivity. Combining this with excitation methods to enable illumination of the entire plate, read times have been reduced to as little as 2 s for most fluorescent technologies and 3 min for SPA. This was achieved independently of plate format.

As the complexity of reader technology has increased, so has the cost. Consequently, the trend is toward multimodality instruments that enable a number of technologies to be read on a single instrument. However, there are some technologies that require dedicated instrumentation due to their specific needs. Examples of these include flash luminescence, which requires integrated liquid handling to dispense reagents at the exact same time as the image of the plate is being taken to capture the burst of light from a very rapid response. There are also ion-channel-related assays that measure the flux of ions within cells, and these assays have dedicated systems with online pipetting such as the FLIPR (Molecular Devices, California).

Automation

As the number of plates assayed increased, the desire to fully automate the process also increased. By the mid-1990s, the typical HTS lab had a significant automation infrastructure to run screens. This automation consists of a robot arm to move the plate between various liquid handlers, incubators, and detectors usually no different to the ones that the HTS scientist would use at the bench. To tie all this together is a piece of software called a scheduler that controls both the robot arm and the individual devices attached to the system. Automating the HTS process does offer some clear advantages in efficiency and productivity [42, 52, 53]. First, the scheduler enables each plate to be treated identically to the one that went before it so that incubation times are identical from plate to plate. By contrast, manual screening tends to batch process plates, which means that plate 1 often differs in incubation times from the last plate in a batch. Second, the robot is able to run 24 h a day and 7 days a week if desired. Finally, the lab scientist is freed up to process data, design other assays, and perform auxiliary tasks while the robot is running the assay.

While there have been few revolutionary advances in HTS robotics over the last 15 years, they are significantly more robust and mechanically precise than earlier robotic systems. The major advances have typically come from the new liquid handling or detection instrumentation that is integrated on these automated systems. Advances in these peripheral systems have greatly increased the rate at which plates are processed through the individual steps. Now the rate limiting step is the speed of the robot arm. One solution is that robotic developers have linked robot arms together to increase efficiency. An example of this is the Dimension 4 system (CRS Thermo, Canada), which has a belt to move plates up and down to the system and then separate robot arms for each peripheral instrument integrated onto the automated system.

Scheduling Software Scheduling software can be described as dynamic, static, or perpetual. The definition of each varies but is generally considered as follows. In the case of a static scheduler, the software predetermines every move that the robot is to make for each step of the assay and the number of plates required. This is based on preprogrammed timing for each individual move of the robotic arm. The benefit is that the robotic operator knows exactly what the robot will do, when each plate will be processed, and what the overall timing or variation for the process will be. The downside is that it is not possible to change once the schedule has started, it is inflexible to pauses or errors that may occur during a screening process on the automated system and no conditional steps can be programmed. In the case of the dynamic scheduler, the software calculates each move in real time and no moves are predetermined. The advantage is that priorities can be assigned to certain tasks, conditional steps can be programmed, for example, error recovery is more flexible. The downside is that it is more complex to program with a strong reliance on a well-determined cycle time for each plate to prevent the occurrence of deadlocks where the robot will grind to a halt because it physically cannot complete a move because equipment is occupied. Lastly, the perpetual scheduler calculates each step in real time but also adapts and generates each move as the process continues. Examples of such applications are crystal growth and cell culture where incubation times are not known and the next step is not initiated until a specific criteria has been reached indicating the robot should perform a specific function.

Reagent Management Reagent stability during continuous HTS is a critical factor in generating accurate and precise data over time. It is critical to empirically determine the stability of all the reagents to find optimal conditions, typically overnight or for about 16 h. The standard approach to extend the life of a reagent on a system is to cool it to 4°C. This is easily achievable by the use of vessels chilled with water jackets, peltier devices, or refrigerators.

There are a number of additional factors that have to be considered in addition to temperature when automating the dispensation of biological reagents. These include nonspecific binding to tubing or the vessel, temperature variations over time, stirring, and shear force to name a few. The binding of proteins to the surface of vessels, tubing, and tips is a significant issue, but these can be minimized with the use of detergents, carrier proteins, and the like. From an automation perspective, if the dispenser carries reagents through tubing and that reagent sits in the tubing for the period of a cycle time or 5 min. This creates a number of issues: (1) reagent binding to the tubing, (2) temperature variation between reagent that sat in the tube and reagent that is drawn from the storage vessel, and (3) the potential for reagent settling in the tubing. These issues are usually best solved by using empty and prime cycles to empty the tubing back to the storage vessel between dispenses.

Data Analysis

Given the number of compounds screened in today's HTS labs, it is obvious that a huge amount of data is generated. The days of manually processing data were left behind a number of years ago. HTS labs now have fully integrated data systems that automatically extract data from robotic systems, process it, apply quality control (QC) filters, upload to databases, and report the analyzed data. Approaches to this vary significantly between companies, some preferring to buy commercial software and others preferring to write their own custom software; however, the principles remain relatively similar.

Data Capture and QC Once a plate is read on either a robotic instrument or a stand-alone detector, the data should be immediately fed to a database where it can be analyzed. This prevents the user from having to manually move around vast numbers of data files with the potential for introducing error. Analysis of the data may simply involve normalizing the data to generate a percent inhibition relative to controls or it may involve a more complex calculation such as a rate determination, relative to previous reads of the same plate.

Once the plate has been analyzed, it must then pass predetermined QC parameters. Typically, this involves the statistic Z' (Fig. 13.10, Table 13.5) [54].

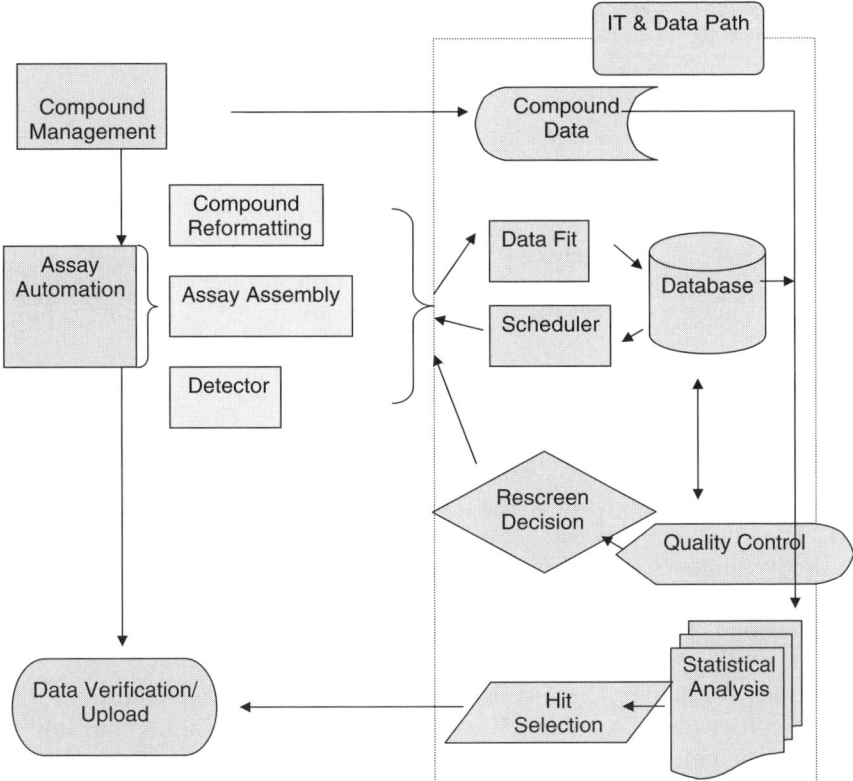

Figure 13.10 Informatics architecture that support a high-throughput screening operation. [Adapted from *DDT*, Vol, 6, No. 12 (2001).]

TABLE 13.5 Statistical Parameters

Statistic	Formula
The *mean* is a measure of the central value of a set of data.	$\bar{x} = \frac{1}{b}\sum_{i=1}^{b} x_i$
The *standard deviation* measures the variation present in the data.	$s = \sqrt{\frac{1}{b-1}\sum_{i=1}^{b}(x_i - \bar{x})^2}$
The *CV* expresses the standard deviation of the data as a percentage of the mean.	$CV = \frac{s}{\bar{x}} \times 100$
The *signal window* measures the distance between the distributions of the totals and the blanks.	$\dfrac{(\bar{x}_3 - 3*s_3) - (\bar{x}_1 + 3*s_1)}{\sqrt{\dfrac{s_1^2 + s_3^2}{2}}}$
The *Z* statistic measures the signal window as a fraction of the distance between the means of the distribution	$1 - \dfrac{3*s_3 + 3*s_1}{\bar{x}_3 - \bar{x}_1}$
The *signal-to-noise* (S/N) ratio can be considered the "signal strength" of the screen divided by the average variability of the screen.	$S/N = \dfrac{(\bar{x}_3 - \bar{x}_1)}{\sqrt{s_3^2 + s_1^2}}$
The *signal-to-background* ratio (S/B)	$S/B = \dfrac{\bar{x}_3}{\bar{x}_1}$

Software will allow the user to determine appropriate QC parameters, such as $Z' > 0.5$, and then determine a process path for that plate depending on the result. The data automatically progresses to corporate databases if the QC criteria are met. Data that fails to meet these criteria are flagged, and, if the failure continues to repeat for the next plates, automated screening is halted.

IC$_{50}$ analysis   In today's HTS environment, the number of IC$_{50}$ curves generated is in the thousands. This can take days to process manually, so many companies have developed software that automatically processes and analyzes IC$_{50}$ data. In our experience, with software written in-house at Bristol Myers Squibb (BMS), we have reduced a day's data analysis time to about 15 min of work. The impact of this has been significant since it has enabled many more IC$_{50}$ determinations to be performed.

Hit Selection   The output of an HTS is a set of compounds that show a broad range of activity. A key step when performing an HTS is determining which compounds are active. These compounds are commonly referred to as a hits. For example, when screening an assay to look for compounds that inhibit a particular enzyme, a wide range of inhibition values will be obtained. This is highlighted in Figure 13.11. In order to identify which compounds are active, a cut-off threshold must be selected. There are a number of critical factors that need to be considered when selecting a threshold.

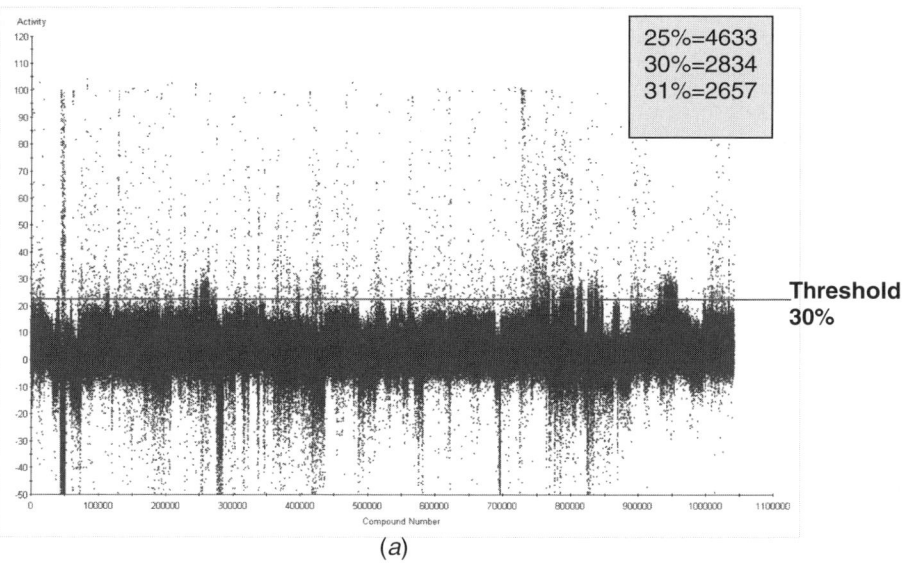

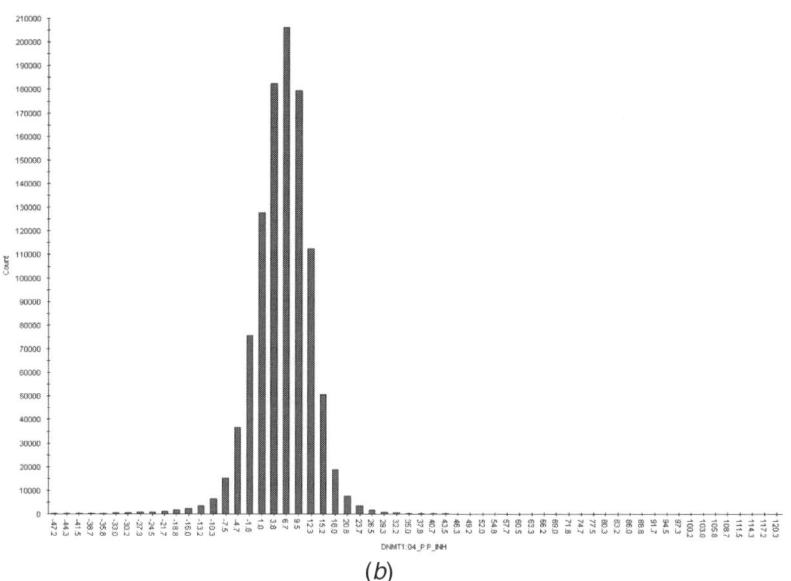

Figure 13.11 (*a*) Scatter plot and (*b*) histogram showing activity for over 1 million compounds in a single assay. The graph in (*a*) shows the activity of over 1 million compounds tested in a methyl transferase biochemical assay. Each compound was tested in single point at 10 μM. Illustrated on the graph is the threshold used to select hits. Any compound falling above this threshold was called a primary hit. The inset box gives the number of compounds that fall above a range of threshold values. (*b*) Histogram illustrating the same data.

First, the quality of the data must meet predetermined criteria as described previously. Second, the range of activity is usually a distribution, with the median close to 0 percent inhibition (Fig. 13.11). The spread of this distribution helps to determine where the threshold should be set, which is typically outside of three standard deviations from the mean. Thus, an active compound will have a discrete activity that separates it from the activity of the vast majority of compounds tested. In the example given in Figure 13.11, the majority of compounds have an activity of less than 30 percent inhibition, and, therefore, this was designated the cut-off threshold. Finally, from a pragmatic perspective, the number of hits generated needs to be within the range that the compound management department can accommodate in terms of picking compounds for further evaluation.

13.4 ADVANCES IN HIGH-THROUGHPUT BIOASSAY TECHNOLOGY

The field of high-throughput screening has evolved in large part due to the need to reduce the cost (money and people) of screening large sets of compounds (1 million or larger) against molecular targets of therapeutic interest. This movement away from biological assays that are performed in test tubes to assays that are performed in multiwell dishes drives the necessity to examine assay biotechnologies that are robust, homogenous, and amenable to integration on automated systems. The true advances came with technologies that moved away from complex multiple-step assays with tube transfers to homogenous assays where steps could now be eliminated but the target biology remains intact. This movement to higher density homogeneous formats has occurred concurrently with advances in detector technology, liquid handling, robotic integration tools, and informatics infrastructure.

The scale and complexity of the screening approach taken can largely depend upon the resources and technological infrastructure available to the organization. In companies with limited budgets, multiple screening formats and approaches may be constrained. However, in those companies with large R&D budgets multiple approaches to screening a well-validated target can be implemented. A bioassay for a HTS campaign is generally decided upon by a team of people including disease biologists, medicinal chemists, and HTS professionals. This team defines the goals for the screen. These goals include understanding the pharmacological outcome (i.e., agonist or antagonist) to be achieved, knowledge surrounding the target type and how to assay for its activity in a biologically relevant system, and finally bioassay connectivity to an appropriate preclinical model of disease. The HTS bioassay chosen must be robust and reproducible from day to day and demonstrate low intra- and interplate variability. There is also the choice of whether to approach the target from a biochemical or cell-based assay. For a complete discussion on the debate between biochemical or cell-based assay approaches to drug discovery,

ADVANCES IN HIGH-THROUGHPUT BIOASSAY TECHNOLOGY

see references 55 to 58. The method selected to meet the screening objectives will most likely be a method that is homogeneous and does not require any complex filtration or multiple removal steps.

Overview of HTS Assay Formats

In general the choice of high-throughput assay format is guided by the goals of the screen and also practical availability of reagents. Some of the pragmatic issues that enter into how to screen a target relate to (1) amount of protein available or cellular supply, (2) substrate requirements, (3) pharmacological relevance of the cellular system, (4) capacity to follow-up the target, and (5) cost to screen a given compound collection. Pharmaceutical targets can be screened using either cell-based or biochemical assays. Cell-based assays include assays that capture binding events, measurement of changes in cellular metabolism, and physiological changes in intracellular second messengers, regulation of intracellular enzymes, and modulation of intracellular signal components. While cell-based assays provide context for pharmacological targets in their native environment, the outcome of these assays can be less straightforward then biochemical assays. Some examples of cell-based assays include the following: calcium flux, voltage-sensitive dyes, reporter assays, whole-cell binding, cAMP detection, physiological changes, protein translocation, and enzymatic activation. Table 13.6 summarizes cell-based assays and detection methods.

Biochemical assays are generally preferred for intracellular targets including cytoplasmic enzymes. A well-optimized biochemical assay will, in general, demonstrate less data scatter than a cell-based approach. In addition, the presence of a single target in the well makes the follow-up process less complex. The compounds identified from a biochemical screen do not have to be cell permeable nor do they need to demonstrate significant potency. Therefore, they may actually be more chemically diverse than those identified from a cellular screen. Biochemical assays can be separated into two categories: separation based (product is measured after isolation from starting material) and homogeneous (no separation required). Enzymatic assays can be run as kinetic or endpoint assays, and basic mechanistic considerations will be discussed below. Examples of biochemical assays include kinase, protease, metabolic enzymes, and enzymes that modify nucleotides. Table 13.7 summarizes biochemical approaches to HT bioassays for these targets along with detection methods.

Reagent Production

Over the last several years, the demand for cell-based assays has increased dramatically. This has been driven by the requirement to support hit identification, lead evaluation in preclinical model systems, and assays to identify compound liabilities. Multiple approaches can be taken to develop cell-based reagents for drug discovery and include transiently transfected cell populations and stable cell line development [59]. Robust transient transfec-

TABLE 13.6 Summary of Cell-Based Assays and Detection Methods for HTS[a]

Assay Type	Application	Detection Method
Proliferation assays	Identify cytotoxic compounds	Alamar blue (FLINT) MTT/MTS (Absorption) TUNEL (TRF)
Reporter assays	Signal transduction pathways and direct NHR assays	Luciferase (luminescence) β-lactamase (HTRF)
Second-messenger assays	cAMP, IP3, MAP kinase, etc.	TRF, FP, α screen, enzyme complementation, etc.
Ion flux assays	Intracellular changes in calcium or other ions Measurements of changes in membrane	Fluo-4 (FLINT), Aequorin (flash luminescence), membrane potential dyes (FLINT)
High content screening assays	Intracellular events, physiological or morphologic changes, protein translocation or posttranslational modification, multiparametric cytotoxicity, RNA/DNA content, receptor	Automated fluorescence microscopy (varying quality depending on optical configuration, excitation and emission detectors)

[a]This is not a comprehensive list.

TABLE 13.7 Summary of Biochemical Assays and Detection Methods for HTS[a]

Assay Type	Application
Time-resolved fluorescence	Kinases, second messengers, proteases, helicases, protein–protein interactions, etc.
Fluorescence polarization	Kinases, phosphatases, receptor–ligand binding, protein–protein interactions, etc.
Enzyme complementation	cAMP, IP3, kinases, other enzymes, protein–protein interactions, etc.
Luminescence	Enzymes that utilize ATP, etc.
Scintillation proximity assays	Receptor–ligand binding, enymatic assays, protein–protein interactions

[a]This is not a comprehensive list.

CellmateTM

Flaskmaster

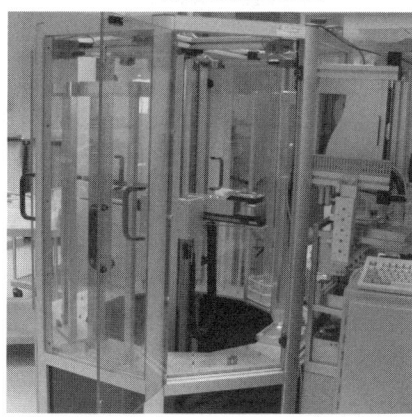

Figure 13.12 Automated infrastructure for bulk production of cellular reagents for HTS. The Cellmate (The Technology Partnership, Cambridge, UK). The Cellmate can reform bulk production of up to 500+ flasks, 1–2 cell lines/day and cell harvest by scraping. Unattended operation of the Cellmate is made possible with the addition of the Flaskmaster (BMS) to feed flasks into and out of the instrument.

tion technologies including lipid-based, viral, and electroporation methods have been developed and are used to support in vitro screening assays. Automated solutions have been applied to scale-up these reagents (Cellmate/Flaskmaster; Fig. 13.12) [60]. During screening, quality control associated with these transfection methods can be monitored through the use of the fluorescent proteins with spectral properties that do not interfere with the corresponding assay formats. Stable cell line development can occur in parallel with the generation of these transient transfected cell populations. Flow cytometry has become an integral component in the process of stable cell line development and is used to monitor, sort, and assess the pharmacology of cell-based targets relative to their corresponding expression levels (epitope tagging, IRES GFP constructs, fluorescent physiological indicators, etc.). Cell line selection for screening requires the ability to process multiple cell clones in parallel using a variety of assay formats in parallel.

The logistics of managing the "just-in-time" need for mammalian cells with active screening efforts requires careful planning and coordination. For example, at BMS, requests for cell line development, maintenance, transfer, mycoplasma testing, FACS analysis, and cryogenic preservation can be made using web-based requesting tools similar to the tools used for compound requesting. A cell-tracking database allows screening groups to track resources, quality control, and document issues that develop pertaining to particular cell lines. Finally, integration of the SelecT fully automated cell culture system allows for the maintenance, quality control, and supply of cells for screening purposes. This system is equipped with scheduling software that

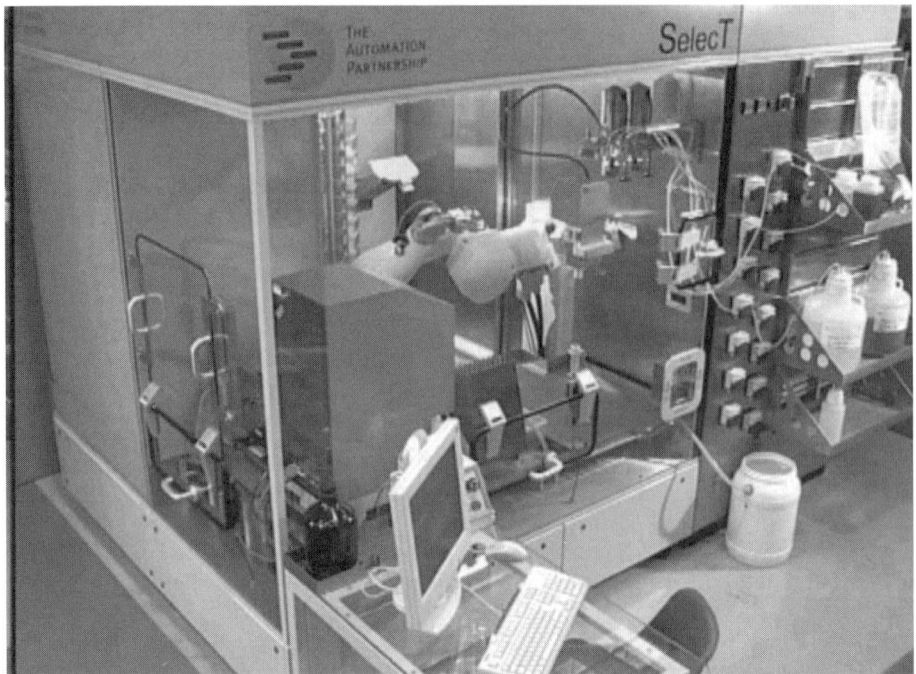

Figure 13.13 The SelecT (The Technology Partnership, Cambridge, U.K.). A fully automated cell culture system performs routine maintenance of up to 50 cell lines per day. It manages growth, incubation, and routine passaging of multiple cell lines with unattended operation. The SelecT performs cell counting, viability measurements, and cell plating into microtiter plates.

allows multiple cell lines to be processed as necessary for screening (see Fig. 13.13). In addition, the system has added consistency to cell-based assays, reduced cell waste, and allowed staff to have continuous access to cells. Many groups responsible for the production of reagents have been moving to bulk production and cryogenic preservation of cell-based reagents for screening; thus, decoupling the just-in-time need for living cells for HTS assays. In this way, one quality-controlled batch of cells can be used throughout the screening campaign. These approaches improve data quality and consistency in HTS.

Biochemical reagents including purified proteins, membrane vesicles, and the like, can be produced in a variety of expression systems including *Escherichia coli*, pichia, insect cells, and mammalian cells. The advent of recombination-based cloning systems allows for the rapid generation of multiple expression constructs in parallel. The capacity for purification of protein from multiple expression systems and parallel purification schemes is very labor intensive and requires quality control analyses to ensure that proteins are purified to near homogeneity, have not formed aggregates. The scale up process is not linear with respect to protein production when compared to the original growth conditions in a test tube, therefore, both growth conditions and purification processes need to be optimized further for bulk reagent production.

Designing Bioassays for HTS

When approaching an HTS campaign to identify small-molecule modulators of targets for drug discovery, one must consider the pharmacological outcome as well as the quality of the data produced from the assay. The importance of data quality relates to statistical validation in HTS and this was previously discussed in Section 13.3. Briefly, a Z' of >0.5 [54] indicates that there is acceptable statistical variation of the data surrounding the maximal signal and the background or minimal signal in the assay, therefore, the assay has the potential to reliably identify active compounds from inactive compounds. The balance is choosing an assay that exhibits sufficient signal, is mechanistically sound, can be run rapidly, and is cost effective.

Mechanistic Considerations for Biochemical Assays Important considerations for the design of enzymatic assays for HTS include the rate at which an enzyme catalyzes conversion of substrate to product, the enzyme concentration, and substrate concentration. For a complete review of mechanistic enzymology, see Ref. 61. Kinetic assays determine the slope/rate of the enzymatic reaction, require multiple measurements over time, exhibit excellent correlation over time for known actives, and are less prone to compound interference. If one performs an enzymatic reaction as an endpoint assay, inhibitors may act to slow the enzyme rate but will be missed in an endpoint HTS because the reaction product is measured at a time point after the enzymatic reaction has run to completion. Another important experimental variable for enzymatic HTS assay design is the enzyme concentration that has a linear effect on reaction velocity. If one chooses an enzyme concentration that is too high, it may minimize the apparent inhibitor effect ($IC_{50} = 1/2$ [Enz]). Competitive enzyme inhibitors compete with substrate for binding to the free enzyme. In this case, assays run at high substrate concentration, inhibit the formation of the enzyme/inhibitor complex, and are less likely to identify competitive inhibitors. Therefore, it is recommended that screening assays be run at substrate concentrations below the K_m. A second class of inhibitor binds exclusively to the enzyme/substrate complex, known as uncompetitive inhibitors, and these are more likely to be detected at high substrate concentrations above K_m. A third type is a mixed inhibitor, and these bind both the free enzyme and the enzyme/substrate complex. These are less sensitive to substrate concentration. A good understanding of how a particular inhibitor modality may act under the physiological conditions that are appropriate for the target pharmacology provides a significant advantage in designing the appropriate biochemical assay for the target. When the modality of inhibition is unknown, it is always best practice to run the assay where K_m equals substrate concentration. The use of natural substrates is recommended for HTS assays as this more closely approximates the in vivo situation. The use of full-length enzymes as opposed to activated catalytic domains is strongly recommended as there may be differences in catalysis by the full-length enzyme in vivo. Order of reagent addition may also be an important consideration during an HTS assay. This will depend on the stability of reagents and the mechanism of the reaction. Addition of a substrate to an enzyme may help stabilize the enzyme or stabilizing agents can be added (BSA, detergents, glycerol, etc.). Running an enzymatic assay under equilibrium conditions is important, and so inhibitors that are "slow binders" may require a longer preincubation time. It is generally recommended that compounds be incubated for a period of between 5 and 15 min.

Ligand–Receptor Binding Interactions Although radiometric assays are beginning to decrease with the growing use of fluorescence assay technologies, radiometric assays still comprise between 20 and 50 percent of all screens performed [62]. In the future, label-free methods to measure biophysical interactions may become standard, and these methods include optical and dielectric methods and acoustic biosensors. Indeed innovative cellular assay technologies have been developed based on radio-frequency spectrometry and bioimpedance measurements [63]. Receptor-mediated signal transduction events can be linked to bioimpedance changes by performing dielectric spectroscopy of cells across a spectrum of frequencies (1 KHz to 110 MHz) [63]. The advantage of this technology is that a single assay technique can be used to monitor G_i, G_s, and G_q without the need of fluorescent labels, tagged proteins, or promiscuous G proteins.

The use of radiolabeled ligands in filtration assays has historically been used to measure high-affinity binding interactions for cell surface receptors and has been replaced by SPA [64–66]. SPA is the current binding assay format of choice for the GPCR (7 transmembrane receptors) HTS assay. The need for separation has been eliminated using a homogeneous technique that relies on the excitation of a scintillant incorporated into microbeads or plates upon binding of the radiolabeled ligand to a receptor immobilized on the surface of a lectin-coated surface of the solid support. Radioactive ligand in close proximity to the receptor results in the emission of light. Free radioligand in solution is too distant from the scintillant, and the β particle energy is dissipated into the aqueous environment. Major advances in this technology over the past 5 years include the development of beads that emit light at 615 nm in contrast to the traditional SPA bead, which emits light around 420 nm. This longer wavelength is well suited for detection by the CCD cameras in modern imagers and results ultimately in fewer problems with color quench by test compounds. These systems provide the added advantage of sensitivity.

When designing an SPA HTS assay, there are many assay conditions that need to be optimized. A multiparameter experiment is performed to determine an acceptable signal window (see Section 13.3). In such a matrixed experiment, the buffer conditions, amount of receptor, bead concentration, and ligand concentrations are co-varied to guide the designer to choose conditions for the HTS assay such that it is statistically robust (see discussion of Z' in Section 13.3). The K_d is then determined following optimization of bead settling/equilibration time, and specific assay conditions have been determined. The K_d is defined as the concentration of ligand required at which half the receptor molecules are bound at equilibrium. This variable can then be approximated by titrating a fixed receptor population with increasing concentrations of hot ligand and measuring the fraction of occupied receptor. The K_d can only be determined accurately under equilibrium conditions. Generally, if an HTS assay is run at equilibrium, there will be agreement between the HTS binding assay and the functional cell-based assays. An example of correlation of the K_d of a ligand binding to a receptor using standard filtration methods versus the apparent K_d determined empirically in a SPA is shown in Figure 13.14. The reversibility of radiolabeled ligand binding and discrimination of specific binding from nonspecific binding or background signal can then be determined by mixing a fixed concentration of receptor with primary ligand to establish the receptor:ligand concentration. The K_a for the unlabeled compound should equal the K_d unless the label has had an effect on the binding, and this would be demonstrated in a typical competitive displacement curve. The plateau at high competitive ligand concentrations generally represents the increase in nonspecific binding signal in the assay. One may choose to further evaluate a nonspecific signal by using other competitive ligands of diverse mechanism and structure. But, generally, in screening one is looking for compounds that displace the radioligand. The signal in a radioligand SPA is determined in part by the concentration of ligand used in the assay. Assays run at up to 10 times the ligand concentration tend to demonstrate maximal signal intensity. The problems associated with HTS assays that utilize such high ligand concentrations include high nonspecific background signal and the

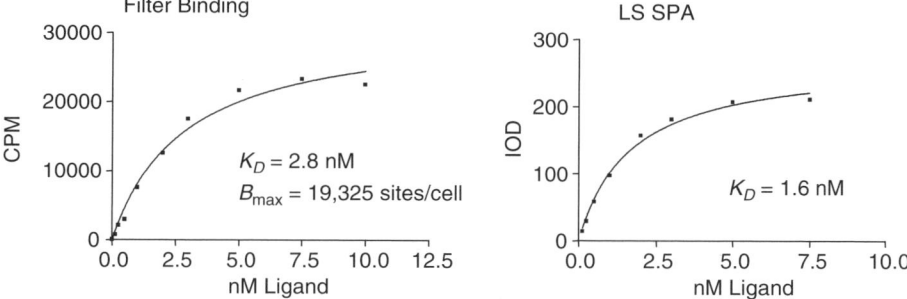

Figure 13.14 Shows ligand saturation binding analysis of the same GPCR using either standard filter binding or Leadseeker SPA. In this analysis the K_D values are within a twofold of one another and in agreement with internal historical data. This also allows the HT screener to report apparent K_i values for competitive compounds identified during HTS.

limited ability to detect compounds with lower affinity for the receptor. Obviously, the goal would be to run the assay at or close to the K_d such that the assay is sensitive. Since the goal is to identify compounds that exert their activity under normal physiological conditions, it is important to compare K_d values of ligands in the HTS assay compared to a more physiological assay. For a review of receptor–ligand interaction see Kanakin et al. [67]. Additional mechanistic considerations need to be made for functional assays for GPCRs. A complete discussion of orthosteric versus allosteric modulation of GPCRs as well as an inverse, partial, and full agonism and mechanisms of reversible or insurmountable antagonism are beyond the scope of this chapter [68].

Fluorescent Methods in HTS

Many of the methods used to identify small-molecule modulators of drug targets involve fluorescence. Being that fluorescence-based detection bioassay technologies tend to be homogeneous and sensitive, they are very common choices for assaying a variety of target types. Fluorescent methods include fluorescence intensity (FLINT), fluorescence polarization (FP), fluorescence resonance energy transfer (FRET), fluorescent proteins (GFP and RFP), time-resolved fluorescence (TRF), subcellular imaging (SCI), fluorescence lifetime (FL), and fluorescent methods associated with detecting single molecules like fluorescence correlation spectroscopy (FCS) and one- and two-dimensional fluorescence intensity distribution analysis (FIDA) [69, 70]. FP is the quantitation of single or complex biomolecule rotation and the polarization of the emitted fluorescence. Generally, the fluorescent indicator is linked to an antigen or a ligand, which then is bound by an antibody or receptor thereby causing a large mass change. Polarization values represent the rotation rates or the related mathematical parameter, anisotropy. This widely used method is volume-independent and can be used in high-density, low-volume HTS formats [71].

Moving away from radiation-dependent energy transfer along with improvements in dyes and detection technologies has reinvigorated an interest in FRET-based assays where fluorophore pairs with overlapping emission and excitation spectrum exchange energy as donor and acceptor fluors. This assay format is suitable for interaction of proteins and cell-based assay formats; for example, antibody—antigen interactions, protease substrate that separates the donor and acceptor fluors. The FRET principle can be applied with long lifetime lanthanide-based fluorescent dyes (Eu^{3+} cryptate) as a homogeneous time-resolved fluorescence, or HTRF, that enable discrimination between fluorophore and signals form interfering assay substituents like compounds [72]. Proprietary HTRF reagents have been designed to assay receptor activation, proteases, kinases, and protein–protein interactions under commercial names such as Trace from CisBio and LANCE from Perkin-Elmer [69].

Fluorescence lifetime assays use time-domain measurements whereby fluorophores are excited by short, high-frequency laser pulses, and the lifetime of a fluorophore is determined. Detectors of FL have been developed though technically challenging (Tecan, Molecular Devices, and Evotec). The full benefit of this technology remains to be seen as the magnitude of changes in FL are difficult to predict and are empirically determined.

Fluorescent methods for HTS have recently achieved single-molecule detection that are very sensitive, information rich, and assayed in reduced reaction volumes using confocal optics where de-convoluted one- and two-dimensional photon count distributions provide data at the molecular level. Readout parameters include polarization, inhibition, molecular brightness, and concentration of multiple assay components [69]. For HTS the most powerful application of FCS is FIDA, enabling direct measure of ligand binding states applied to receptor–ligand interactions [73].

Fluorescent Methods in Cell-Based Bioassays Cell-based functional assays using fluorescent probes initially were developed using organic small molecules. These are highly fluorescent and have been used to monitor changes in real-time gene expression in reporter assays, measuring local changes in ion concentrations, membrane potential, pH, and other physiological changes [55, 74, 75]. Caged fluorescent calcium-sensitive dyes such as Fluo-4 are loaded as membrane-permeable esters that are then cleaved by intracellular esterases and exhibit altered quantum yield upon calcium binding. An example of Fluo-4 measurement of calcium flux in response to a cognate ligand binding to a G_q-coupled 7 TM receptor is shown in Figure 13.15. These changes in intracellular calcium can be measured using a FLIPR (for more detail see the following section). Other cell-based and FRET-based dyes have been developed to detect changes in membrane potential as well as reporter assays that measure changes in real-time gene expression of the reporter gene β-lactamase [76–79].

Attempts to resolve intracellular events that occur using fluorescent-labeled ribonucleic acid (RNA), DNA, or proteins have allowed tracking of movement or changes in intracellular localization triggered by signaling cascades or physiologic changes in organelles. Now proteins can be fused to GFP or RFP to take advantage of their fluorescent properties to examine spatial and temporal resolution using fluorescence microscopy, and this subcellular imaging has become known as high content screening (HCS) [80, 81].

Cell-Based Assay Design

When designing a cell-based assay multiple considerations need to be assessed at the outset. The first consideration is evaluation of the cell line to be used prior to expressing the target: Can enough cells be produced to support the screening effort, do the cells senesce over time, and so forth? Other important variables for cell line selection at the outset of assay design include the growth characteristics of the given cell type, whether the cells are prone to changes in phenotype or morphology, and will these changes have the potential to impact

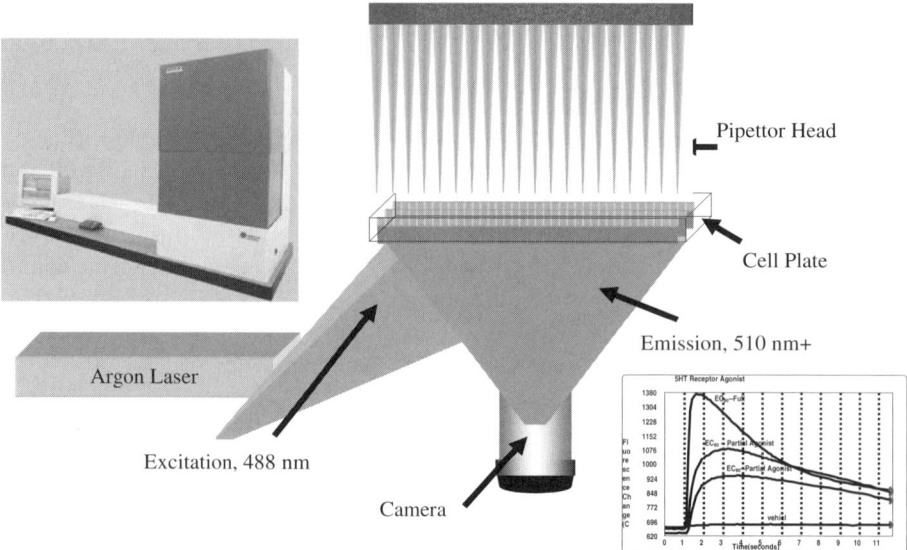

Figure 13.15 Depicts a FLIPR (Molecular Devices) fluorometric imaging system that is useful for quantitating kinetic changes in intracellular fluorescence. The inset FLIPR trace shows a range of G protein coupled receptor agonist responses; full and partial efficacy as compared to vehicle control.

the outcome of the assay? Characterization of target expression (either endogenous or exogenous) as well as any critical cofactors that may need to be expressed in the cell line to ensure the assay will perform (i.e., Gα subunits, reporter constructs, enzyme subunits, etc.) over time is also critical. For example, one key question could be: Is expression consistent from passage to passage? Once the cell line has been characterized for routine maintenance and expression, it is then expanded for assay design.

The cell culture optimization within the microtiter plates is an important part of assay design for cell-based screening. Cell number, assay media including serum, and other additives are monitored and tested along with the assay components. As a key example, the critical variables for designing a functional 7TM-R assay will be discussed. There are multiple parameters such as cellular dissociation conditions, cellular dispensing, cell density in the well, and media components in the well that are critical to optimal cell plating for a FLIPR calcium flux assay. In this assay, we measure changes in intracellular calcium upon ligand binding to a G_q-coupled 7TM-R. After plating, the next day the calcium indicator, Fluo-4 AM, is added to the cells in a HEPES-buffered saline solution. The composition of the buffer for certain cell-based assays is critical to the overall performance, that is, ionic concentration and composition can be critical for performing calcium flux assays or measuring membrane potential. For a review of membrane permeability and membrane potential see [82]. If the media needs to be removed and aspirated prior to addition of the fluorescent indicator, aspiration tip heights need to be adjusted on the liquid handling device to ensure that the monolayer of cells is intact. The concentration and choice of fluorescent calcium indicator needs to assessed in the assay. Finally, the agonist and antagonist pharmacology can be assessed in the assay following optimization of liquid handling on the FLIPR including pipet height, addition volumes. For a schematic representation of the FLIPR and inset agonist kinetic analysis see Figure 13.15.

Additional Advances in Cellular Assay Formats

Ion channels have long eluded pharmaceutical research because of the difficulties developing functional assays that adequately measure changes in membrane potential in response to ligands or currents. Fluorescent dyes have been developed that can be used to measure changes in membrane potential. When combined with fluorescent plate readers that contain on-line pipettors, rapid changes in fluorescence in response to ion channel openers and blockers can be measured [83, 84]. Although these surrogate assays for ion channels allow for high-throughput hit identification, they do not replace single-cell patch recordings. Many companies have developed automated methods to perform electrophysiological measurements on cells, and these include Robocyte, OpusXpress, IonWorks, and Patch Express [39, 85]. These advances will allow higher throughput ion channel drug discovery and liability measurements to become a reality and enable therapeutics to be developed against this difficult target class.

Other cellular assay formats use reporter-based assays that measure the effects of compounds that modulate signal transduction cascades. Reporters such as β-lactamase, luciferase, and nitroreductase are commonly useful in functional cell-based screening for a variety of target types including receptors, enzymes, protein–protein, and nuclear hormone receptors. Development of click beetle luciferase reporters now enable multiple pathways to be interrogated within a single cell [32].

Enzyme fragment complementation is another flexible cell-based assay format that is based on high-affinity complementation of a small fragment of β-galactosidase (Pro-label) to an inactive deletion mutant of the enzyme (enzyme acceptor) and generates a signal via hydrolysis of a substrate [87]. This assay format has enabled direct measurement of a wide variety of second messengers and signal transducers within cells including cAMP, IP3, MAP kinase, and others. This method can also be useful for measuring protein–protein interactions within cells [87].

With advances in automated fluorescence microscopy came the evolution of high content screening (HCS). HCS has made possible complex measurements in cellular morphology and physiology, changes in intracellular protein translocation and posttranslational modifications, measurements of receptor internalization, and multiparameter cytotoxicity measurements. For a detailed review of HCS applications and instrumentation see reference 38.

The evolution of HTS bioassay technologies has greatly enhanced the quality of data produced from screening large compound collections, amount of information added to the compounds following screening, and ultimately improved decision making by medicinal chemistry to facilitate the drug discovery process.

REFERENCES

1. Houston, J. G., Banks, M. N. (2003). High throughput screening for lead discovery. In D. J. Abraham (Ed.), *Burger's Medicinal Chemistry and Drug Discovery*, Vol. 2. Wiley, New York, pp. 37–69.
2. Bojanic, D., Keighley, W. W., Russell, M. J., Wood, T. P. (1997). Factors for the successful integration of assays, equipment, robotics, and software. In J. P. Devlin (Ed.), High Throughput Screening: the Discovery of Bioactive Substances. Marcel Dekker, New York, pp. 493–508.
3. Takatsy, G. (1955). The use of spiral loops in serological and virological methods. *Acta Microbiol. Hung.*, *3*, 191–202.
4. Engvall, E., Perlmann, P. (1971). Enzyme lined immunoassay (ELISA) quantitative assay of immunoglobulin G. *Immunochemistry*, *8*, 871–874.
5. Voller A., Bidwell, D. E. (1985). Enzyme immunoassays. In W. P. Collins (Ed.), *Alternative Immunoassays*. Wiley, New York, pp. 77–86.
6. Wood, T., Keghley, W. (2004). Automation sample supply for high throughput screening. *Eur. Pharm. Rev.*, *9*, 68–73.
7. Jarmaat, M. L., Giaccone, G. (2003). Small molecule epidermal growth factor receptor tyrosine kinase inhibitors. *Oncologist*, *8*, 576–586.
8. Foulkes, J. G., Chow, M., Gorka, C., Frackelton, A. R., Baltimimore, D. (1985). Purification and characterization of a protein tyrosine kinase encoded by the Abelson murine leukemia virus. *J. Biol. Chem.*, *260*, 8070–8077.
9. Filthut, H. (1989). A new detector for radiochromatography and radio-labelled multisample distribution. *J. Planas Chrom. Modern TLC*, *3*, 198–202.
10. Schraag, B., Staal, G. E. J., Adriaansen-Slot, S. S., Salden, M., Rijkesen, G. (1993). Standardization of an enzyme-linked immunosorbent assay for the determination of tyrosine kinase activity. *Anal. Biochem.*, *211*, 233–239.
11. Potter, C. G., Warner, G., Yrjonen, T., Soine, E. (1986). A liquid scintillation counter specifically designed for samples deposited on a flat matrix. *Phys. Med. Biol.*, *31*, 361–369.
12. Warner, G. T., Potter, C. G., Yrjonen, T., Soinl, E. (1985). A new design for a liquid scintillation counter for micro samples using flat bed geometry. *Int. J. Appl. Radiat. Isot.*, *36*, 819–821.
13. Cook, N. D. (1996). Scintillation proximity assay a versatile high throughput screening technology. *Drug Discov. Today*, *1*, 287–294.
14. Ramm, P. (1999). Imaging systems in assay screening. *Drug Discov. Today*, *4*, 401–410.
15. Jakob, W., Heder, G., Halle, W. (1981). DNA synthesis in cultures of calf aortic endothelial cells treated with corpus luteum extract: Its inhibition by ^{3}H-thymidine. *Exp. Pathol.*, *20*, 41–50.

16. Hoy, C. A., Lewis, E. D., Schimke, R. T. (1990). Perturbation of DNA replication and cell cycle progression by commonly used [^{3}H] thymidine labeling protocols. *Mol. Cell. Biol.*, *10*, 1584–1592.
17. Hu, V. W., Black, G. E., Torres-Duarte, A., Abramson, F. P. (2002). ^{3}H-thymidine is a defective tool with which to measure rates of DNA synthesis. *FASEB J.*, *16*, 1456–1457.
18. Borenfreund, E., Babich, H., Martin-Alguacil, N. (1988). Comparisons of two in vitro cytotoxicity assays—The natural red (NR) and tetrazolium MTT tests. *Toxicol. in Vitro*, *2*, 1–6.
19. Mosmann, T. (1983). Rapid colorimetric assay for cellular growth and survival: Application to proliferation and cytotoxicity assays. *J. Immunol. Methods*, *65*, 55–63.
20. Maxwell, I. H., Maxwell, F. (1988). Electroporation of mammalian cells with firefly luciferase expression differ markedly among cell types. *DNA*, *7*, 557.
21. Gorman, C., Moffat, L. F., Howard, B. H. (1982). Recombinant genomes which express chloramphenicol acetyltransferase in mammalian cells. *Mol. Cell. Biol.*, *2*, 1044.
22. An, G., Kidaka, K., Siminovitch, K. (1982). Expression of bacterial β-galactosidase in animal cells. *Mol. Cell. Biol.*, *2*, 1628.
23. Bronstein, I., Fortin, J., Stanley, E., Stewart, G. S. A. B., Kricka, L. J. (1994). Chemiluminescent and bioluminescent reporter gene assays. *Anal. Biochem.*, *219*, 169.
24. Pazzagali, M., Devine, J. H., Peterson, D. O., Baldwin, T. O. (1992). Use of bacterial and firefly luciferases as reporter genes in DEAE-dextran-mediated transfection of mammalian cells. *Anal. Biochem.*, *204*, 315.
25. Fisher, M., Harbron, S., Rabin, B. R. (1995). An amplified chemiluminescent assay for the detection of alkaline phosphatase. *Anal. Biochem.*, *227*, 73.
26. Chalfie, M., Tu, Y., Euskirchen, G., Ward, W. W., Prasher, D. C. (1994). Green fluorescent protein as a marker for gene expression. *Science*, *263*, 802.
27. Kolb, A., Roelant, C., Scheirer, W. (1994). Reagents and instruments for luminescence assays in drug discovery. In A. K. Campbell, L. J. Kricka, and P. E. Stanley (Eds.), *Bioluminiescence and Chemiluminescence*. Wiley, Chichester, p. 357.
28. Stanley, P. E. (1989). A review of bioluminescent ATP techniques in rapid microbiology. *J. Biolum. ChemiLumin.*, *4*, 375–380.
29. Yang, Y., Thomason, D. B. (1993). An easily synthesized photolyzeable luciferase substrate for *in vivo* luciferase activity measurement. *BioTechniques*, *15*, 848.
30. Tanahashi, H., Ito, T., Inoye, S., Tsuji, F. I., Sakaki, Y. (1990). Photoprotein aequorin: Use as a reporter enzyme in studying gene expression in mammalian cells. *Gene*, *96*, 249.
31. Zlokarnik, G. (2000). Fusions to beta-lactamase as a reporter for gene expression in live mammalian cells. *Methods Enzymol.*, *326*, 221–244.
32. Wood, K. V., Lam, A. Y., Seliger, H. H., McElroy, W. D. (1989). Complementary DNA coding click beetle luciferases can elicit bioluminescence of different colors. *Science*, *233*, 700.
33. Gabriel, D., Verner, M., Pfrifer, M. J., Daser, B., Tenzillon, L., Bouhelal, R. (2003). HTS technologies for direct cyclic AMP measurement. *Assay Drug Rev.*, *1*(2), 291–302.

34. Berridge, M. J., Downes, C. P., Hanley, M. R. (1982). Lithium amplifies agonist dependent phosphotidylinositol in brain and salivary glands. *Biochem. J.*, *206*, 587–595.
35. Wilson, D. B., Connell, T. M., Bross, T. E., Majerus, P. W., Sterman, W. R. Tyter, A. N. Rubin, L. J., Brown, J. E. (1985). Isolation and characterization of the inositol cyclic phosphate products of polyphosphoinositide cleavage by phospholipase C. Physiological effects in permeabilized platelets and Limulus photoreceptor cells. *J. Biol. Chem.*, *260*, 13496–13501.
36. Gutkind, J. S. (1998). Cell growth control by G-protein-coupled receptors from signal transduction to signal integration. *Oncogene*, *17*, 1331–1342.
37. Beske, O. E., Goldband, S. (2002). HT cell analysis using multiplexed assay technologies. *DDT*, *7*(18, Suppl.), S131–S135.
38. Zamanova, L., Schenk, A., Valler, M. J., Nienhaus, U., Heilker, R. (2003). Confocal optics microscopy for biochemical and cellular HTS. *Drug Discov. Today*, 8(23), 1085–1093.
39. *The PatchXpress Users Newsletter,* Vol. II (March 2004). Axon Instrument, Mol. Devices., CA, USA.
40. Houston, J. G., Banks, M. (1997). The chemical-biological interface: Developments in automated and miniaturized screening technology. *Curr. Opin. Biotechnol.*, *8*, 734–740.
41. Burbaum, J. J. (2000). The evolution of miniaturized well plates. *J. Biomol. Screen.*, *5*, 5–8.
42. Banks, M. (2000). Approaches to high throughput toxicity screening. In C. K. Goldfard and W. Pucell (Eds.), *Automation and Technology for HTS in Drug Development.* Yaler and Francis, New York, pp. 9–29.
43. Wolcke, J., Ullman, D. (2001). Miniaturized HTS technologies. *Drug Discov. Today*, *6*, 637–646.
44. Mere, L., Bennert, T., Coassin, P., England, P., Hamman, B., Rink, T., Zimmerman, S., Negulescu, P. (1999). Miniaturized FRET assays and microfluidics: Key components for ultra-high-throughput screening. *Drug Discov. Today*, *4*, 363–369.
45. Spencer, P. A. The challenges of managing a compound collection. *Eur. Pharmaceut. Rev.*, *19*, 51–75.
46. Bohacek, R., McMartin, C., Guida, W. (1996). The art and practice of structure-based drug design: A molecular modeling perspective. *Med. Results Rev.*, *16*, 3–50.
47. Miller, M. A. (2002). Chemical database techniques in drug discovery. *Nat. Rev. Drug Discov.*, *1*, 220–227.
48. Mestres, J. (2002). Virtual screening: A real screening complement to high throughput screening. *Biochem. Soc. Trans.*, *30*, 797–799.
49. Cheng, X., Hochlowski, J., Tang, H., Hepp, D., Beckner, C., Kantor, S., Smitt, R. (2003). Studies on repository compound stability in DMSO under various conditions. *J. Biomol. Screen.*, *8*, 292–304.
50. Comley, J. (2002). Continued miniaturization of assay technologies drives market for nanolitre dispensing. *Drug Discov. World*, *5*(3), 43–53.
51. Comley, J. (2002). Nanolitre dispensing: On the point of delivery. *Drug Discovery World*, *3*(3), 43–46.

52. Oldenburg, K. R. (1999). Automation basics: Robotics vs workstations. *J. Biomol. Screen.*, *4*, 53–54.
53. Wildey, M. J., Homon, C. A., Hutchins, B. (1999). Allegro™: Moving the bar upwards. *J. Biomol. Screen.*, *4*, 115–118.
54. Zhang, J. H., Chung, T. D. Y., Oldenburg, K. R. (1999). A simple statistical parameter for use in evaluation and validation of high throughput screening assays. *J. Biomol. Screen.*, *4*, 67–73.
55. Johnston, P. A. (2002). Cellular platforms for HTS: Three case studies. *Drug Discov. Today*, *7*(6), 353–363.
56. Pagliaro, L., Praestegaard, M. (2001). Transfected cell lines as tools for high throughput screening: A call for standards. *J. Biomol. Screen.*, *6*(3), 133–136.
57. Moore, K., Rees, S. (2001). Cell-based versus isolated target screening: How lucky do you feel? *J. Biomol. Screen.*, *6*(2), 69–74.
58. Gonzalez, J. E., Negulescu, P. A. (1998). Intracellular detection assays for high-throughput screening. *Curr. Opin. Biotechnol.*, *9*, 624–631.
59. Pagliaro, L., Praestegaard, M. (2001). Transfected cell lines as tools for high throughput screening. *Call for Standards*, *6*(3), 133–135.
60. Bernard, C., Conners, D., Barber, L., Jayachandra, S., Bullen, A., Cacace, A. (2004). Adjunct automation to the cellmate cell culture robot., *J.A.L.A.*, *9*(4), 209–217.
61. Copeland, R. A. (2003). Mechanistic considerations in HTS. *Anal. Biochem.*, *320*, 1–12.
62. Hertzberg, R. P., Pope, A. J. (2000). High-throughput screening: new technology for the 21st century. *Curr. Opin. Chem. Biol.*, *4*, 445–451.
63. Ciambrone, G. J., Liu, V. F., Lin, D. C., McGuinness, R. P., Leung, G. K., Pitchford, S. (2004). Cellular dielectric spectroscopy: A powerful new approach to label-free cellular analysis. *J. Biomol. Screen.*, *9*(6), 467–480.
64. Major, J. (1999). What is the future of HTS? *J. Screen.*, *4*, 119.
65. Bosworth, N., Tower, P. (1989). Scintillation proximity assay. Nature, *341*, 167–168.
66. Sittampalam, G., Kahl, S., Janzen, Q. (1997). HTS: Advances in assay technologies. *Curr. Opin. Chem. Biol.*, *1*, 384–391.
67. Kenakin, T. (2004). Principles: receptor theory in pharmacology. *Trends Pharmacol. Sci.*, *25*(4), 186–192.
68. Kenakin, T. (2004). G Protein coupled receptors and allosteric machines. *Receptors and Channels.*, *10*(2), 51–60.
69. Gribbon, P., Sewing, A. (2003). Fluorescent readouts in HTS: No gain without pain? *Drug Discov. Today*, *8*(22), 1035–1043.
70. Eggeling, C., Brand, L., Ullmann, D., Jager, S. (2003). Highly sensitive fluorescence detection technology currently available for HTS. *Drug Discov. Today*, *8*(14).
71. Turconi, S., Shea, K., Ashman, S. (2001). Real experiences of uHTS: A prototypic 1536 well fluorescence anisotropy-based uHTS screen and application of well-level quality control procedures. *J. Biomol. Screen.*, *6*, 275–290.
72. Bazin, H., Trinquet, E., Mathis, G. (2002). Time resolved amplification of cryptate emission: A versatile technology to trace biomolecular interactions. *J. Biotechnol.*, *82*, 233–250.

73. Scheel, A. A., Funsch, B., Busch, M., Gradl, G., Pschoff, J., Lohse, M. J. (2001). Receptor-ligand interactions studied with homogeneous fluorescence-based assays suitable for miniaturized screening. *J. Biomol. Screen.*, *6*, 11–18.
74. Kunapoli, P., Ransom, R., Murphy, K. L., Pettibone, D., Kerby, J., Grimwood, S., Zuck, P., Hodder, P., Lacson, R., Hoffman, I., Inglese, J., Strulovici, B. (2003). Development of an intact cell reporter gene beta-lactamase assay for G protein-coupled receptors for high throughput screening. *Anal. Biochem.*, *314*, 16–29.
75. Chambers, C., Smith, F., Williams, C., Marlos, S., Liu, Z. H., Hayer, P., Ciaramella, G., Keighley, W., Gribbon, P., Sewing, A. (2003). Measuring intracellular calcium fluxes in high throutput mode. *Comb. Chem. High Throughput Screen.*, *6*, 355–362.
76. Baxter, D. F., Kirk, M., Garcia, A. F., Raimond, A., Holmquist, M. H., Flint, K. K., Bojanic, D., Distefano, P. S., Curtis, R. (2002). A novel membrane potential-sensitive fluorescent dye improves cell-based assays for ion channels. *J. Biomol. Screen.*, *7*, 79–85.
77. Farinas, J., Chow, A. W., Nada, H. G. (2001). A microfluidic device for measuring cellular membrane potential. *Anal. Biochem.*, *295*, 138–142.
78. Gonzalez, J. E., Maher, M. P. (2002). Cellular fluorescent indicators and voltage/ion probe reader (VIPR™): Tools for ion channel and receptor drug discovery. *Receptors Channels*, *8*, 283–295.
79. Zlokarnick, G., Negulescu, P.A., Knapp, T., Mere, L., Burres, N., Feng, L., Witney, M., Roemer, K., Tsien, R.Y. (1998). Quantitation of transcriptional and clonal selection of single living cells with beta lactamase as reporter. *Science*, *279*, 84–88.
80. Chiesa, A., Rapizzi, E., Tosella, V., Pinton, P., de Virglio, M., Fogarty, K. E., Rizzuto, R. (2001). Recombinant aequorin and green fluorescent protein as valuable tools in the study of cell signaling. *Biochem. J.*, *355*, 1–12.
81. Lippincott-Schwartz, J., Patterson, G. H. (2003). Development and use of fluorescent protein markers in living cells. *Science*, *300*, 87–90.
82. Falconer, M., Smith, F., Surah-Narwal, S., Congrave, G., Liu, Z., Hayter, P., Ciarmella, G., Keighley, W., Haddock, P., Waldron, G., Sewing, A. (2002). High-throughput screening for ion channel modulators. *J. Biomol. Screen.*, *7*, 460–465.
83 Cooper, M. A. (2003). Biosensor profiling of molecular interactions in pharmacology. *Curr. Opin. Pharmacol.*, *3*, 557–562.
84. Tang, W., Kang, J., Wu, X., Ranme, D., Shen, H., Li, Z., Dunnington, D., Garyantes, T. (2001). Development and evaluation of high throughput functional assay methods for hERG potassium channel. *J. Biomol. Screen.*, *6*, 325–331.
85. Schroader, K., Neagle, B., Trezise, D. J., Worley, J. (2003). IonWorks HT: A new high throughput electrophysiology measurement platform. *J. Biomol. Screen.*, *8*, 50–64.
86. Scheirer, W. (1997). Reporter gene assay applications. In J. P. Devlin (Ed.), *HTS: The Discovery of Bioactive Substances*. Marcel Dekker, New York, pp. 401–412.
87. Eglen, R. M., Pribilla, I. (2004). High-throughput screening using label-free technologies. *J. Biomol Screen.*, *9*(6), 465–466.

14

METAL-ENHANCED FLUORESCENCE: APPLICATION TO HIGH-THROUGHPUT SCREENING AND DRUG DISCOVERY

KADIR ASLAN,[1] IGNACY GRYCZYNSKI,[2] JOANNA MALICKA,[2] JOSEPH R. LAKOWICZ,[2] AND CHRIS D. GEDDES[1,2]

[1]*Institute of Fluorescence, University of Maryland Biotechnology Institute Baltimore, Maryland*
[2]*Center for Fluorescence Spectroscopy, University of Maryland School of Medicine Baltimore, Maryland*

14.1	METAL–FLUOROPHORE INTERACTIONS	604
14.2	METAL SUBSTRATES FOR METAL-ENHANCED FLUORESCENCE	609
	Silver Island Films for MEF	610
	Silver Colloid Films for MEF	610
	Anisotopic Silver Nanostructures for MEF	615
	Laser-Deposited Silver for MEF	618
	Electrochemically Deposited Silver for MEF	619
	Electroplating of Silver for MEF	622
	Roughened Silver Electrodes for MEF	622
	Silver Fractal-Like Structures on Glass Substrates for MEF	625
14.3	METAL-ENHANCED FLUORESCENCE SENSING: APPLICATION TO SENSING IN PLATE WELLS	630
	Enhanced DNA/RNA Detection Using MEF (Hybridization Assays)	630
	Enhanced DNA Labels	637
	Over-labeled Proteins as Ultrabright Probes	639
	Enhanced and Selective Excitation Using Multiphoton Excitation and Metallic Nanoparticles	643
	Enhanced Energy Transfer on Silver Surfaces	645

Drug Discovery Handbook, by Shayne Cox Gad
Copyright © 2005 by John Wiley & Sons, Inc.

Ratiometric Sensing on Surfaces	649
MEF Solution-Based Assays	653
Directional Emission	654
14.4 ROLE OF MEF IN HTS AND DRUG DISCOVERY	660
14.5 CLOSING REMARKS	662
References	664

14.1 METAL-FLUOROPHORE INTERACTIONS

Fluorescence detection is the basis of most assays in high-throughput screening and drug discovery today. Although fluorescence is a highly sensitive technique, there is always a need for reduced detection limits and/or small copy-number detection. Fluorophore detectability is usually limited by autofluorescence of the samples and/or the photostability of the fluorophores. In an effort to overcome these problems, we have been investigating the use of metallic nanostructures to favorably modify the spectral properties of fluorophores [1–4]. We refer to the use of fluorophore–metal interactions as radiative decay engineering (RDE) or metal-enhanced fluorescence (MEF).

In the present review, we describe the effects of different silver nanostructures, which were prepared by various methods in our laboratories, on the emission intensity and photostability of fluorophores that are frequently used in many biological assays. The silver nanostructures consist of subwavelength-size nanoparticles of silver deposited on inert substrates. We show that proximity to silver nanostructures results in a preferential increase in intensity of low-quantum-yield fluorophores and that the lifetimes decrease as the intensities increase. We subsequently discuss the use of MEF for its applications in high-throughput screening and drug discovery.

It is well known that the presence of a nearby metal surface (m) can favorably increase the radiative decay rate by addition of a new rate Γ_m (Fig. 14.1a) [1, 2]. The metallic surface can cause Förster-like quenching with a rate (k_m), can concentrate the incident field (E_m), and can, importantly for sensing applications, increase the radiative decay rate (Γ_m). These new phenomena typically occur at different distances from the metal surface as depicted in Figure 14.2. As the value of Γ_m increases, that is, the spontaneous rate at which a fluorophore emits photons, the quantum yield increases while the lifetime decreases. In order to illustrate this point, we calculated the lifetime and quantum yield for fluorophores with an assumed natural lifetime $\tau_N = 10\,\text{ns}$, $\Gamma = 10^8\,\text{s}^{-1}$ and various values for the nonradiative decay rates and quantum yields. The values of k_{nr} varied from 0 to $9.9 \times 10^7\,\text{s}^{-1}$, resulting in quantum yields from 1.0 to 0.01. Suppose now the metal results in increasing values of Γ_m. Since Γ_m is a rate process returning the fluorophore to the ground state, the lifetime decreases as Γ_m becomes comparable and larger than Γ (Fig.

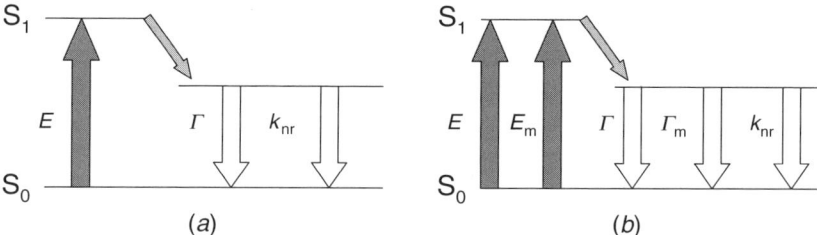

Figure 14.1 Classical Jablonski diagram for (*a*) the free space condition and (*b*) the modified form in the presence of metallic particles, islands, colloids, or silver nanostructures. E excitation, E_m, metal-enhanced excitation rate; Γ_m, radiative rate in the presence of metal. For our studies, we do not consider the effects of metals on k_{nr}.

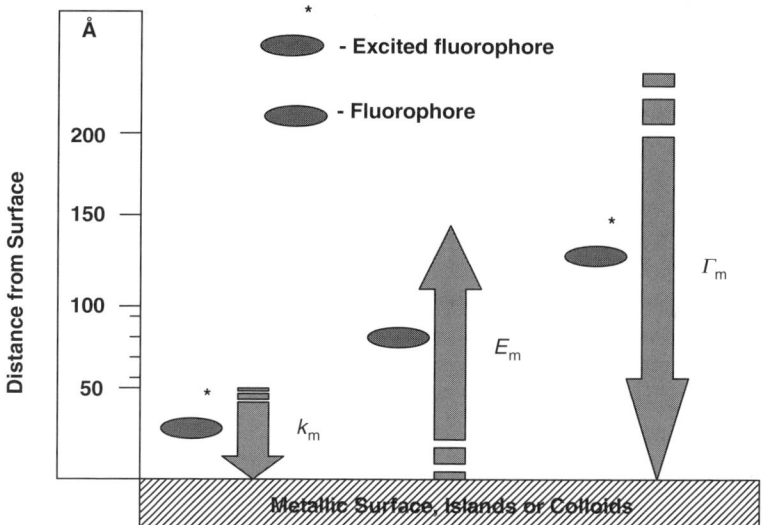

Figure 14.2 Predicted distance dependencies for a metallic surface on the transitions of a fluorophore. Γ_m modifications typically occur at distances larger than 50 Å from the metal surface.

14.3*a*). In contrast, as Γ_m / Γ increases, Q_m increases but no change is observed for fluorophores where $Q_0 = 1$ (Fig. 14.3*b*).

As a result of these calculations, we have predicted that the metallic surfaces can create new unique fluorophores with increased quantum yields and shorter lifetimes. Figure 14.4 illustrates that the presence of a metal surface within close proximity of a fluorophore with low quantum yield ($Q_0 = 0.01$) increases its quantum yield ~10-fold, resulting in brighter emission, while reducing its lifetime 10-fold, resulting in an enhanced photostability of the

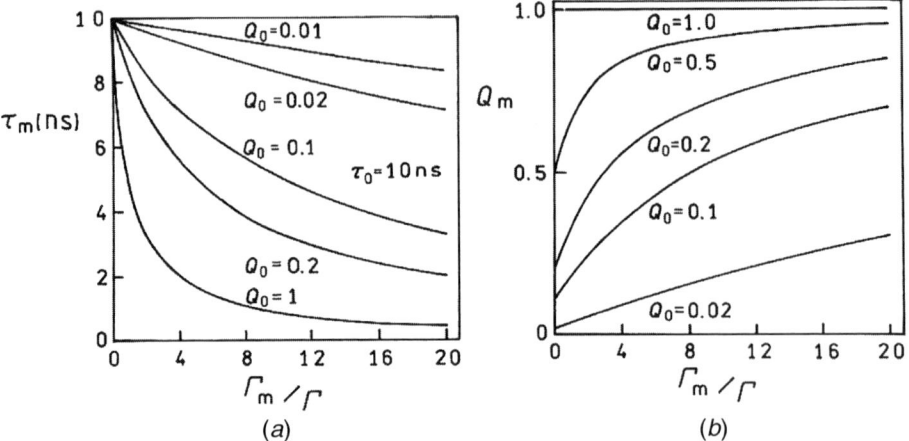

Figure 14.3 Effect of an increase in radiative decay rate (Γ_m / Γ) on the lifetime and quantum yield of a fluorophore.

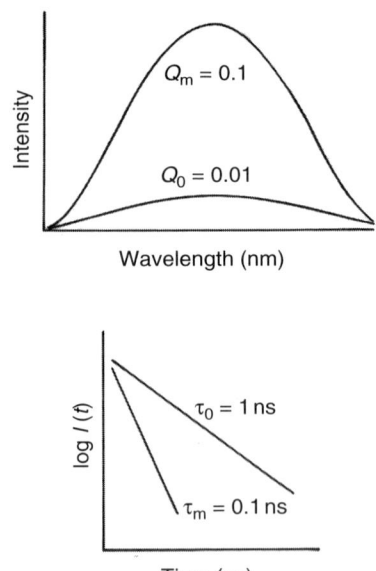

Figure 14.4 Metallic surfaces can create unique fluorophores with high quantum yields and short lifetimes. These are likely to find multifarious applications in metal-enhanced fluorescence sensing.

fluorophore due to spending less time in an excited state, that is, less time for oxidation and/or other processes.

We speculated that the properties of the fluorophores with different quantum yields would be affected differently by the presence of nearby metal due to the difference in radiative rate modifications, with a view to understanding metal–fluorophore interactions for sensing applications. We tested this hypothesis by investigating the changes in emission properties of two fluorophores with similar absorption/emission spectra but with different quantum yield (Rhodamine B, quantum yields, $Q_0 = 0.48$ and Rose Bengal, $Q_0 = 0.02$). Figure 14.5 summarizes our observations with these fluorophores. The emission from Rhodamine B on silver island films (SiFs) was ≈20 percent higher as compared to the emission from the glass side. On the other hand, emission from Rose Bengal on SiFs was ≈5-fold higher than those on the glass side. The emission spectra in Figure 14.5 do not solely demonstrate an increase in the radiative decay rate. However, the demonstration of such an increase can be done by additional intensity decay (lifetime) measurements. Figure 14.6

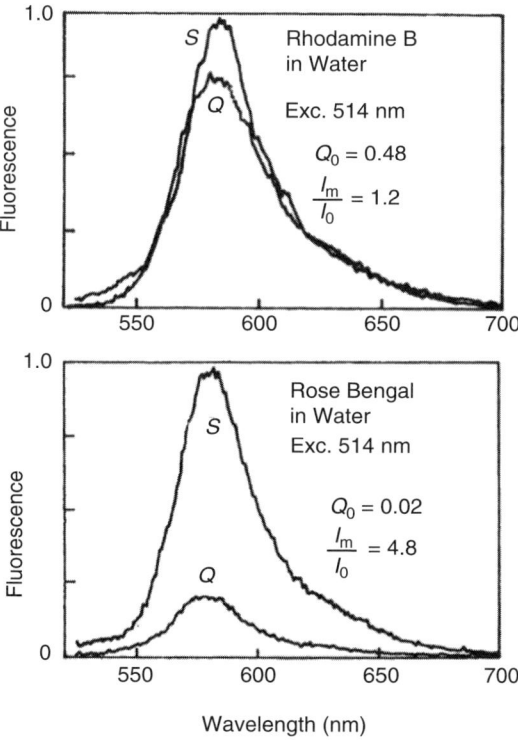

Figure 14.5 Effect of silver island films (SiFs) on the emission spectra of (*a*) Rhodamine B and (*b*) Rose Bengal. These probes were chosen to demonstrate metal-enhanced fluorescence as they both have similar spectral characteristics but different quantum yields. Adapted from Ref. 2.

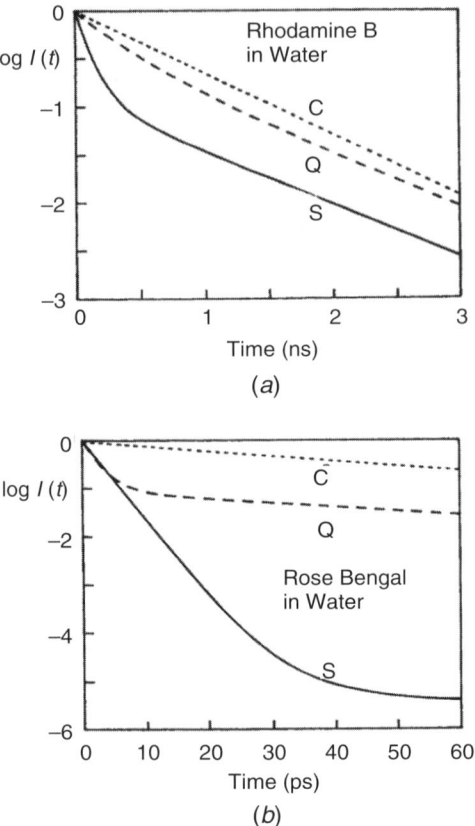

Figure 14.6 Intensity decays for (*a*) Rhodamine B and (*b*) Rose Bengal in cuvette (C), between quartz slides (Q), and between silvered slides (S). Adapted from Ref. 2.

shows that the intensity decay is more rapid near SiFs. We interpret this decrease in lifetime as due to the interactions of the fluorophore with the metal surface. Hence an increase in intensity accompanied by a reduction in lifetime can only be explored by a radiative rate modification; compare Figures 14.1 and 14.2 and the following equations:

$$Q_m = \frac{\Gamma + \Gamma_m}{\Gamma + \Gamma_m + k_{nr}}$$

$$\tau_m = \frac{1}{\Gamma + \Gamma_m + k_{nr}}$$

The results from Rhodamine B and Rose Bengal (Figs. 14.5 and 14.6) were consistent with our expectations that the presence of metal increases the emis-

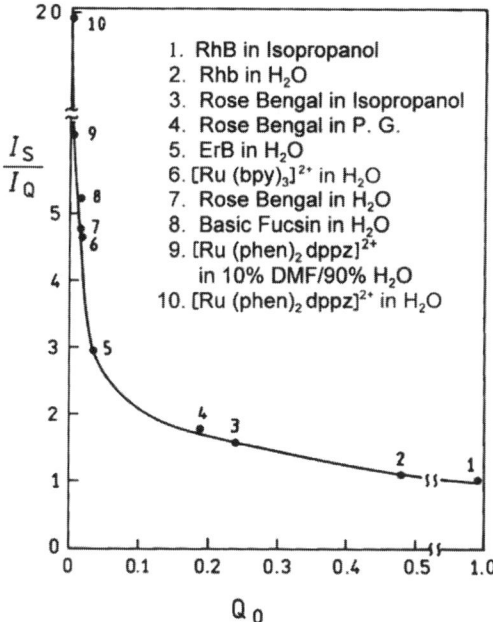

Figure 14.7 Effect of silver island films on the quantum yields of both high and low quantum yield fluorophores. I_S, intensity between silvered plates, I_Q, intensity between quartz plates (control sample). Adapted from Ref. 2.

sion intensity (quantum yield) and decreases the lifetime of the fluorophores. However, one could be concerned with possible artifacts due to dye binding to the surfaces or other unknown effects. For this reason, we examined a number of additional fluorophores between uncoated quartz plates and between silver island films. In all cases, the emission was more intense for the solution between the silver islands. For example, [Ru(bpy)$_3$] and [Ru(phen)$_2$dppz] have quantum yields near 0.02 and 0.001, respectively. A larger enhancement was found for [Ru(phen)$_2$dppz] than for [Ru(bpy)$_3$] (Fig. 14.7). The enhancements for 10 different fluorophore solutions are shown in Figure 14.7. In all cases, a lower bulk-phase quantum yield results in larger enhancements for samples between silver island films. We rationalized that the simple enhancement factor is $\approx 1/Q_0$, where Q_0 is the quantum yield (intensity) in the absence of metal.

14.2 METAL SUBSTRATES FOR METAL-ENHANCED FLUORESCENCE

The effects of metals in close proximity to fluorophores have been predicted theoretically [5–7] and are analogous to the increases in Raman signals observed for surface-enhanced Raman scattering (SERS), *except* that the effects on fluorescence are due to through-space interactions, which do not require molecular contact between the fluorophores and the metal. For use in medical and biotechnology applications, such as diagnostic or microfluidic devices, it would be useful to obtain metal-enhanced fluorescence (MEF) at desired locations in the measurement device, that is, *MEF on demand*. We subsequently investigated various methods for the preparation of silver nanostructures, and used indocyanine green (ICG) and fluorescein-labeled proteins and oligonucleotides to test our new potential sensing platforms, noting the widespread use of ICG and fluorescein in medical applications due their Food and Drug Administration (FDA) approval for use in humans.

Silver Island Films for MEF

In recent years we have been reporting our observations on the favorable effects of silver nanoparticles deposited randomly on glass substrates (silver island films, SiFs) for increasing the intensities and photostability of fluorophores, particularly those with low quantum yields [2–4, 8]. A typical absorption spectrum of SiFs, which display absorption maximum near 430 nm, is given in Figure 14.8a. Figure 14.8b also shows the schematic for a fluorophore between two SiFs (note that the configuration of slides were used for the experiments throughout this work, unless otherwise indicated). We have investigated the effects of SiFs on the properties of ICG, when bound to human serum albumin (HSA). The emission spectra of ICG–HSA bound to quartz and SiFs are shown in Figure 14.8b. The intensity of ICG is increased approximately 10-fold on the SiFs as compared with quartz. The spectral shape is similar both on quartz and SiFs. We found the same amount of increase in the emission of ICG whether the surfaces were coated with HSA, which already contained bound ICG, or if the surfaces were first coated with HSA followed by exposure to a dilute solution of ICG. From ongoing studies of albumin-coated surfaces, we estimated that the same amount of HSA binds to each surface, with the difference in binding being less than a factor of 2. Hence the observed increase in the intensity on SiFs is not due to the increased ICG–HSA binding but rather due to a change in the quantum yield and/or rate of excitation of ICG near metallic silver nanoparticles [8].

Silver Colloid Films for MEF

In contrast to silver island films, the preparation of colloidal suspensions of silver yields homogeneously sized spherical silver particles. An advantage of a colloidal suspension is the possibility of injection for medical imaging (silver and gold colloids are widely used in medicine), such as in retinal angiography, which could widely benefit from an enhanced ICG and fluorescein quantum yield and photostability. Thus, we also investigated the effects of spherical silver particles on the emission of ICG. The sample geometry was similar to

METAL SUBSTRATES FOR METAL-ENHANCED FLUORESCENCE

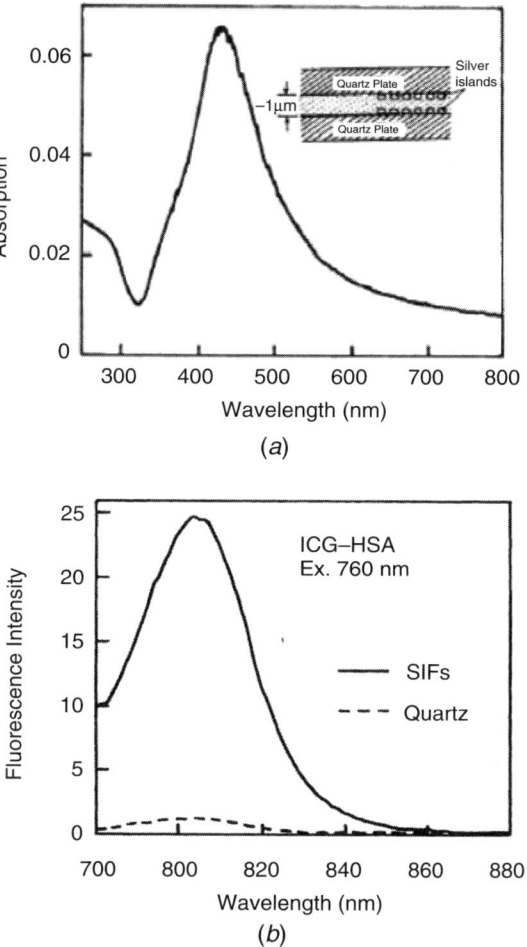

Figure 14.8 Typical absorption (*a*) spectrum silver island films (SiFs). Schematic for a fluorophore solution between two SiFs (inset). (*b*) Fluorescence spectra of ICG–HSA on SiFs and on quartz. Adapted from Ref. 8.

that of SiFs, except the spherical silver particles were prepared separately and were immobilized to 1-amino-3-propylethoxy silane (APS)-coated glass slides by immersing the glass in a solution of the particles (Fig. 14.9*a*). Figure 14.9 shows an absorption spectrum, typical of our colloid-coated APS glass slides. The absorption peak centered near 430 nm is typical of colloidal silver particles with subwavelength dimensions. An atomic force microscopy (AFM) image of silver-colloid-coated glass slides shows that the size of the silvercolloids was smaller than 50 nm with partly aggregated sections (Fig. 14.9*b*). The surfaces were incubated with ICG–HSA to obtain a monolayer surface coating. The emission spectra showed a ≈ 30-fold larger intensity on the

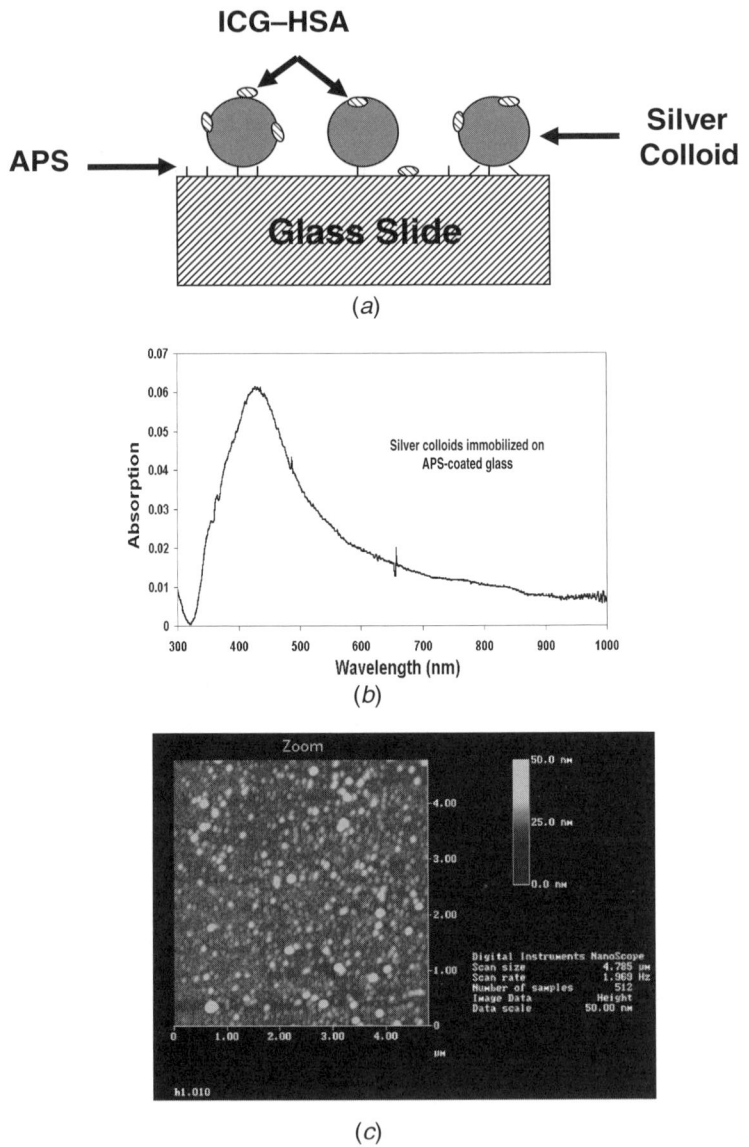

Figure 14.9 (*a*) Glass surface geometry. (*b*) APS is used to functionalize the surface of the glass with amine groups, which readily bind silver colloids, absorption spectrum of silver colloids on APS-coated glass, and (*c*) AFM image of a silver colloid coated glass. APS; 1-amino-3-propylethoxy silane. Adapted from Ref. 9.

surfaces coated with silver colloids (Fig. 14.10) [9]. The fluorescence intensities typically increased with the concentration of APS used to treat the cleaned surfaces (Fig. 14.11a), which also appeared to correlate with the optical density of silver colloids at 430 nm (not shown). We also measured the lifetimes of ICG on both surfaces and observed a significant reduction of the lifetimes on the silver colloids (Fig. 14.11b), providing additional evidence that the increase

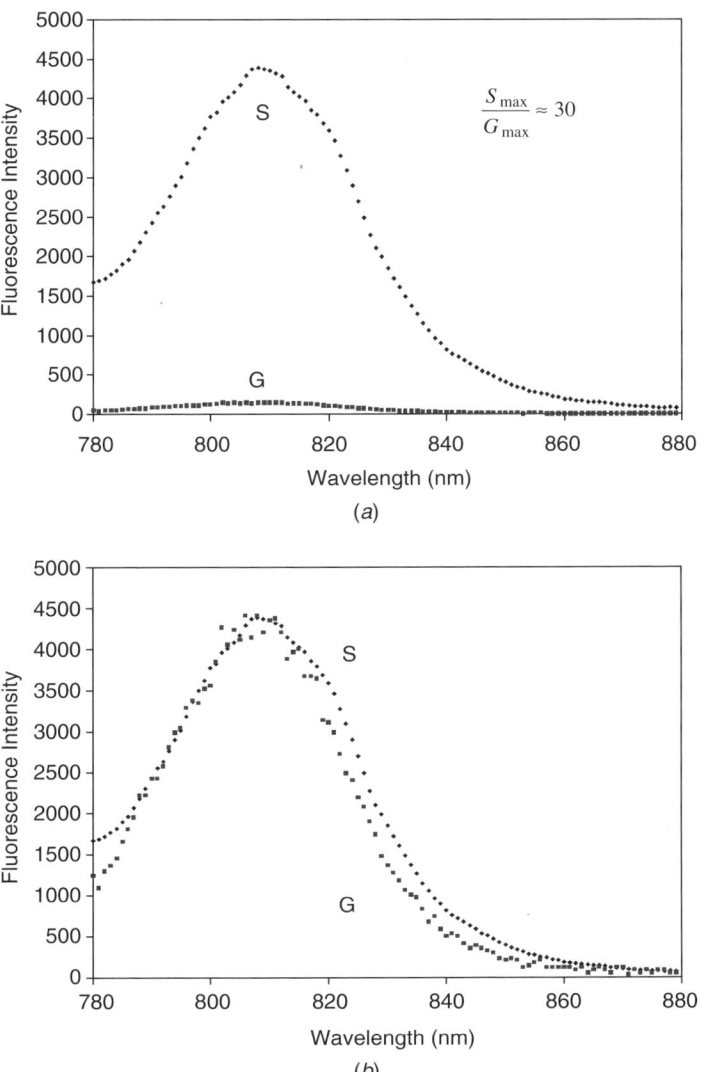

Figure 14.10 (a) Fluorescence intensity of ICG–HSA coated glass, G, and above silver colloids, S, Ex = 760 nm. (b) Fluorescence intensities normalized to the intensity on silver colloids. Adapted from Ref. 9.

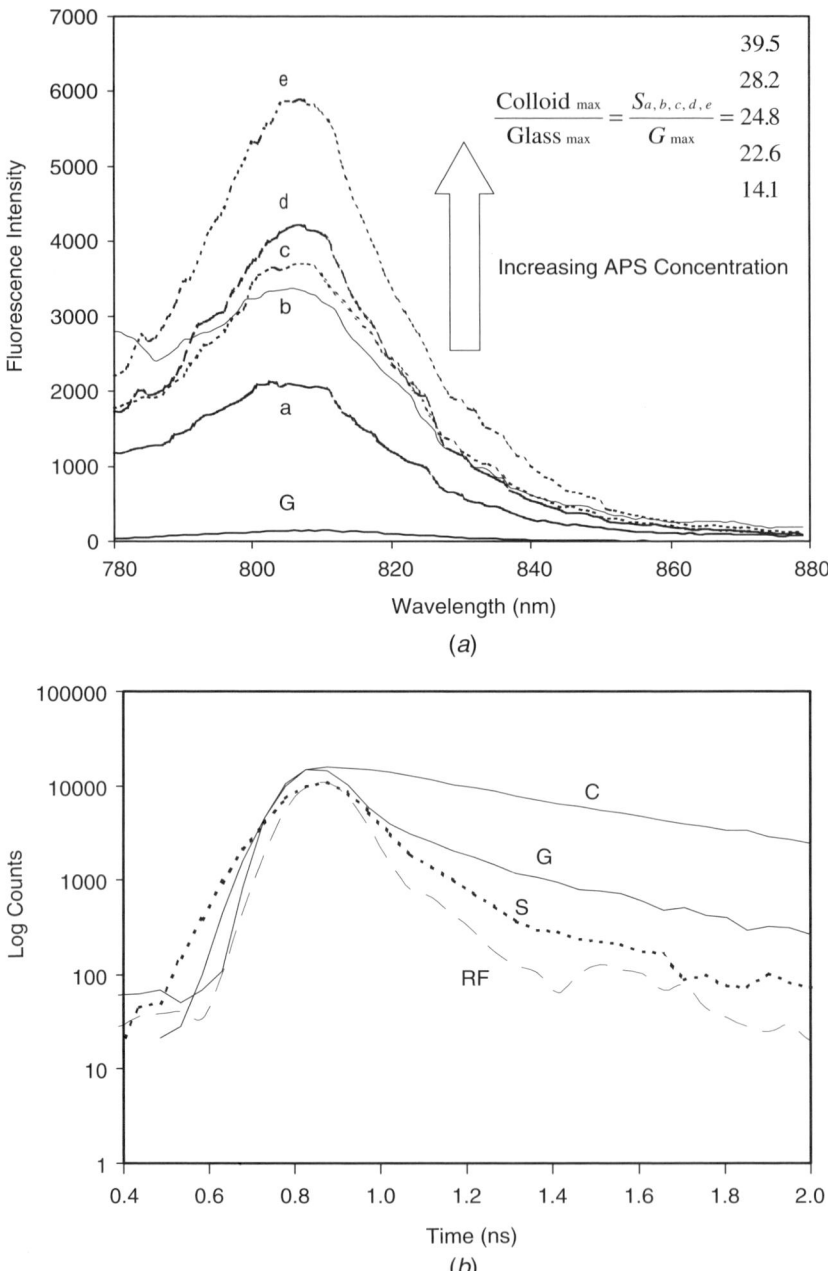

in intensity is in fact due to a modification of the radiative decay rate, Γ_m, by the close proximity to silver colloids [9].

Anisotropic Silver Nanostructures for MEF

The optical properties of nonspherical particles have been attracting many researchers since the early twentieth century. Figure 14.12a shows typical absorption spectra of silver colloids, which can be calculated for the small particle limit ($r << \lambda$) from the complex dielectric constant of the metal [10, 11]. The absorption spectra are dependent on the shape of the particles as well as the dielectric constant of the surrounding medium, with elongated spheroids displaying longer absorption wavelengths.

Several groups have considered the effects of metallic spheroids on the spectral properties of nearby fluorophores [12–16]. A typical model is shown in Figure 14.12b for a prolate spheroid with an aspect ratio of a/b. The particle is assumed to be a metallic ellipsoid with a fluorophore positioned near the particle. The fluorophore is located outside the particle at a distance r from the center of the spheroid and a distance d from the surface. The fluorophore is located on the major axis and can be oriented parallel or perpendicular to the metallic surface. The presence of a metallic particle can have dramatic effects on the radiative decay rate of a nearby fluorophore. Figure 14.12c shows the radiative rates expected for a fluorophore at various distances from the surface of a silver particle and for different orientations of the fluorophore transition moment. The most remarkable effect is for a fluorophore perpendicular to the surface of a spheroid with a/b = 1.75. In this case the radiative rate can be enhanced by a factor of 1000-fold or greater. The effect is much smaller for a sphere (a/b = 1.0) and much smaller for a more elongated spheroid (a/b = 3.0) when the optical transition is not in resonance.

We have developed a methodology for depositing silver nanorods with controlled loadings onto glass substrates. The absorption spectra of silver nanorods deposited on glass substrates are shown in Figure 14.13a. Silver nanorods display two distinct surface plasmon peaks; transverse and longitudinal, which typically appear at ≈420 and ≈650 nm, respectively. In our experiments, the longitudinal surface plasmon peak shifted and increased in absorbance as more nanorods are deposited on the surface of the substrates.

◄──────────

Figure 14.11 Fluorescence intensity of ICG–HSA on silver colloids as a function of increased APS used. Cleaned glass slides were initially soaked in: (a) 0.1, (b) 0.25, (c) 0.5, (d) 1.0, and (e) 1.25 percent (v/v) APS solution for 4 h, washed and soaked in colloid solution for 4 days. (a) G—Fluorescence intensity of ICG–HSA deposited on 0.5 percent APS covered glass. ICG coverage for the glass controls, i.e., for the different percent APS-coated glass slides, was approximately constant. (b) Complex intensity decays of ICG–HSA in a cuvette (buffer), C, on glass slides, G, and silver colloids, S. RF—Instrumental response function. Adapted from Ref. 9.

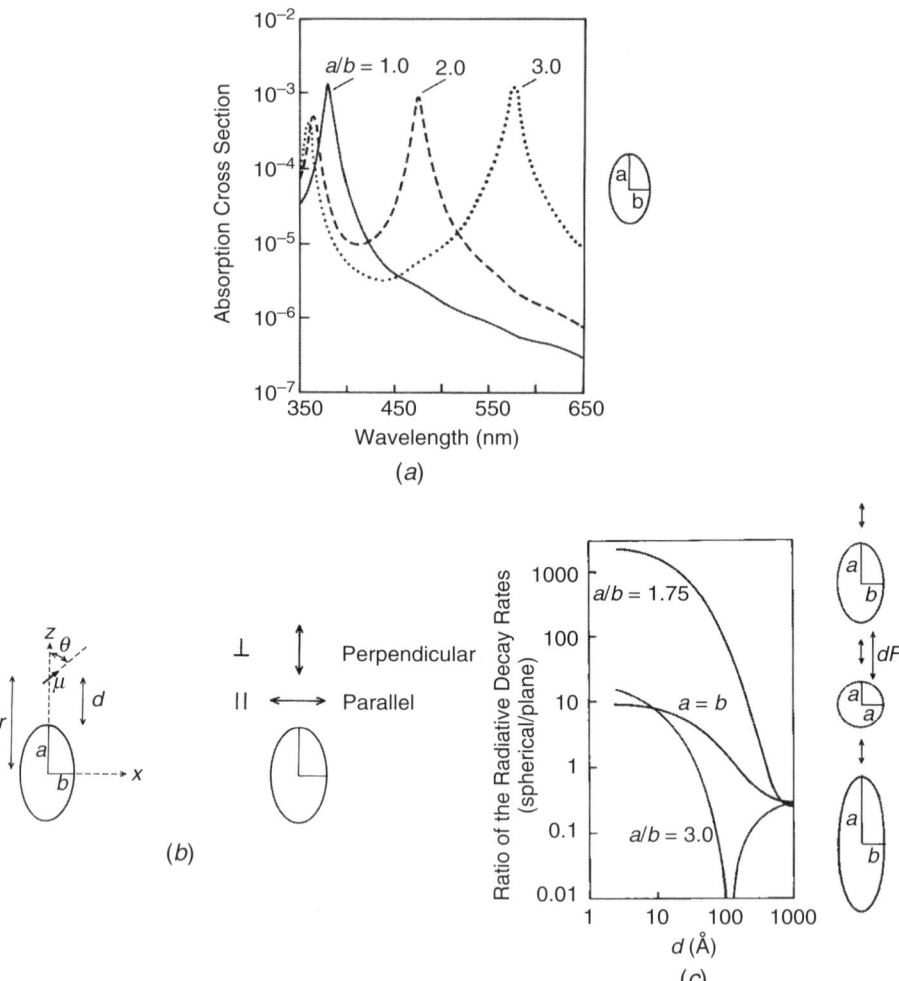

Figure 14.12 (*a*)Calculated absorption cross section of silver nanostructures. (*b, c*) Effect of a metal spheroid on the radiative decay rate. Adapted from Ref. 1.

In parallel to these measurements, we have observed an increase in the size of the nanorods (by AFM, data not shown). In order to compare the extent of enhancement of fluorescence with respect to the extent of loading of silver nanorods deposited on the surface, we have arbitrarily chosen the value of absorbance at 650 nm as a means of loading of the nanorods on the surface. This is because the 650-nm band is solely attributed to the longitudinal absorbance of the nanorods.

Figure 14.13(*b*) shows the fluorescence emission intensity of ICG–HSA measured from both glass and on silver nanorods, and the enhancement factor

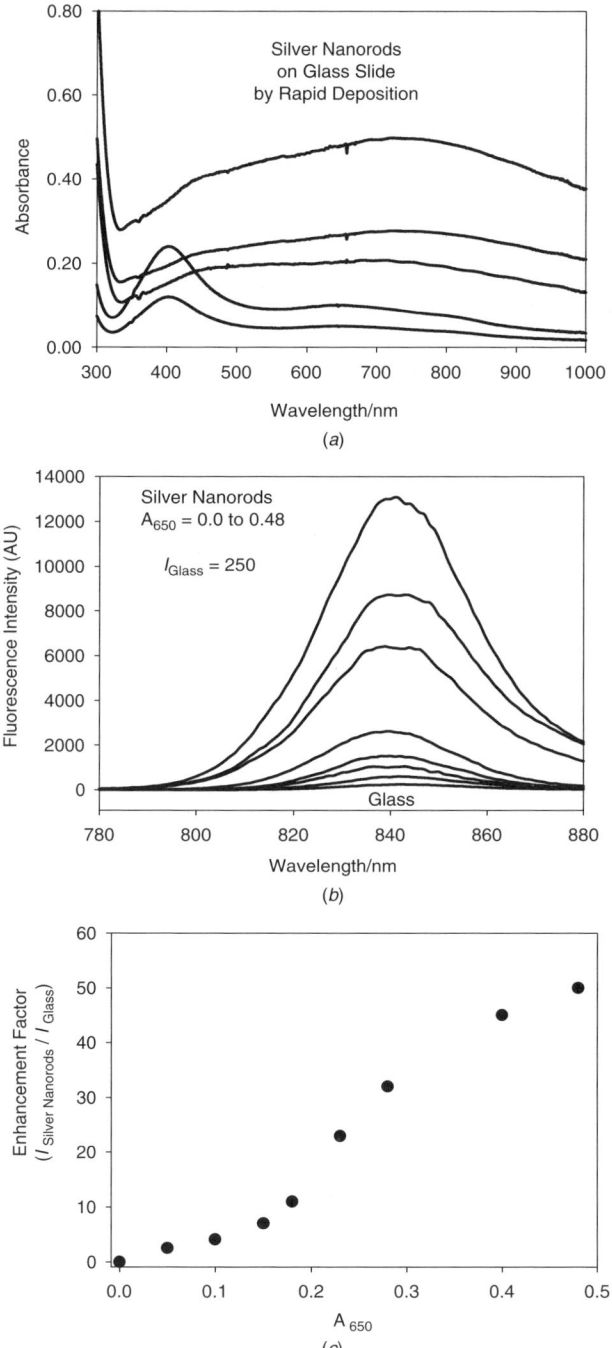

Figure 14.13 (*a*)Absorption spectra of silver nanorods, (*b*) emission spectra of ICG–HSA on silver nanorods, and (*c*) enhancement factor versus the absorbance.

versus the loading density of silver nanorods. We have obtained up to 50-fold enhancement in emission of ICG on silver nanorods when compared to the emission on glass (Fig. 14.13c). We also measured the lifetime of ICG on both glass and silver nanorods and observed a significant reduction in the lifetime of ICG on silver nanorods, providing the evidence that an increased emission is due to radiative rate modifications, Γ_m.

Laser-Deposited Silver for MEF

It would be useful to obtain MEF at desired locations in a measurement device for use in medical and biotechnology applications, such as diagnostic or microfluidic devices, or even plate wells. While a variety of methods could be used, we reasoned that the light-directed deposition of silver would be widely applicable. In a typical preparation of light-induced deposition of silver on glass slides; a 180-μL solution of silver colloids (optical density = 0.1) was syringed between the glass microscope slide and the plastic cover slip, which created a microsample chamber 0.5 mm thick (Fig. 14.14). The sample chamber was irradiated by a HeCd laser beam, with a power of ~8 mW, which was collimated but defocused using a 10× microscope objective [numerical aperture (NA)] 0.40, to provide illumination over a 0.5-mm-diameter spot [17]. This resulted in deposition of silver on the surface of the glass substrate.

We examined the emission spectrum of ICG–HSA when bound to illuminated or nonilluminated regions of the APS-treated slides. For APS-treated slides (Fig. 14.15) the intensity of ICG was increased about sevenfold in the

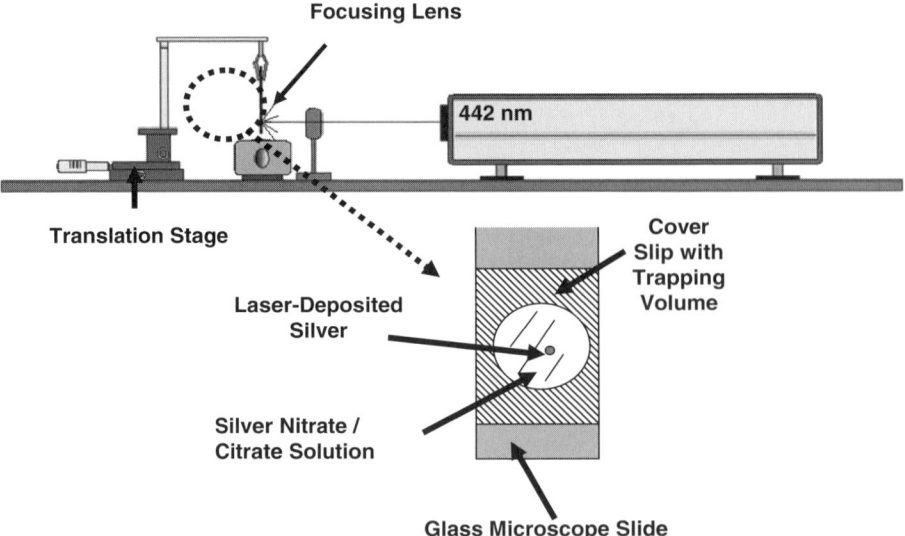

Figure 14.14 Experimental setup for light-induced deposition of silver on APS-coated glass microscope slides. Adapted from Ref. 17.

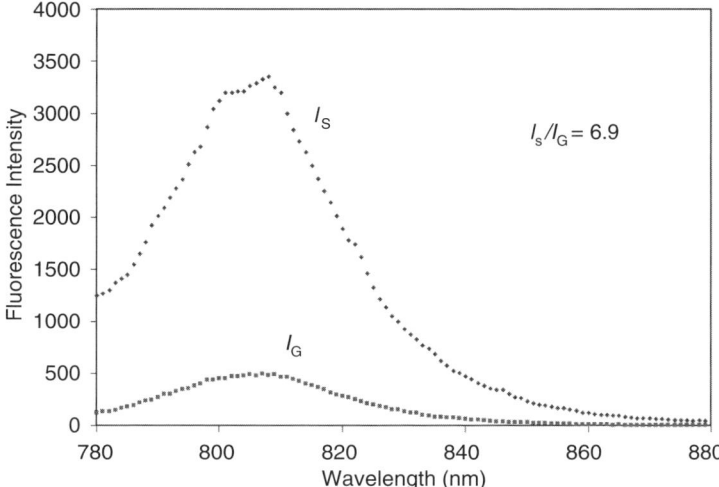

Figure 14.15 Fluorescence spectra of ICG–HSA on glass and on light-deposited silver. I_S, intensity on silver, I_G, intensity on glass. Adapted from Ref. 17.

regions with laser-deposited silver. We examined the photostability of ICG–HSA when bound to glass or laser-deposited silver. We reasoned that ICG molecules with shortened lifetimes should also be more photostable because there is less time for photochemical processes to occur. The intensity of ICG–HSA was recorded with continuous illumination at 760 nm. When excited with the same incident power, the fluorescence intensities, when considered on the same intensity scale, decreased somewhat more rapidly on the silver (Fig. 14.16a). Alternatively, one can consider the photostability of ICG when the incident intensity is adjusted to result in the same initial signal intensities on silver and glass. In this case (Fig. 14.16b) photobleaching is slower on the silver surfaces. The fact that the photobleaching is not accelerated for ICG on silver indicates that the increased intensities on silver *are likely not due* to an increased rate of excitation.

Electrochemically Deposited Silver for MEF

One of our methods for silver deposition onto glass/quartz substrates was to pass a controlled current between two silver electrodes ($9 \times 35 \times 0.1$ mm, separated by 10 mm) in pure water under illumination with laser (Fig. 14.17a) [18]. For the production of silver colloids, a simple constant-current generator circuit (60 μA) was used. After 30 min of current flow, a clear glass microscope slide was positioned within the cuvette (no chemical glass surface modifications) and was illuminated (HeCd, 442 nm). We observed silver deposition on the glass microscope slide, the amount depending on the illumination time. Simultaneous electrolysis and 442-nm laser illumination resulted in the

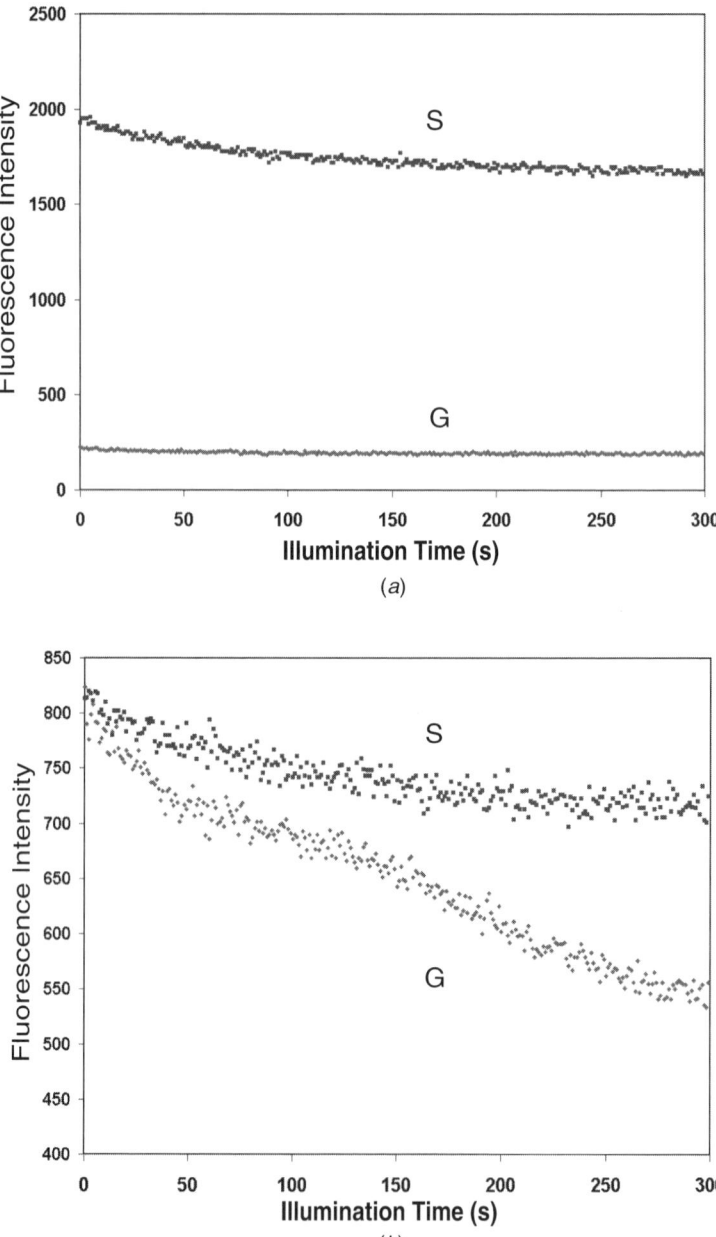

Figure 14.16 Photostability of ICG–HSA on (G) glass, and (S) laser-deposited silver, measured with the same excitation power at (*a*) 760 nm, and (*b*) with the laser power at 760 nm adjusted for the same initial fluorescence intensity. Laser-deposited samples were made by focusing 442-nm laser light onto APS-coated glass slides immersed in a $AgNO_3$ citrate solution for 15 min. The OD of the sample was ~0.3. Adapted from Ref. 17.

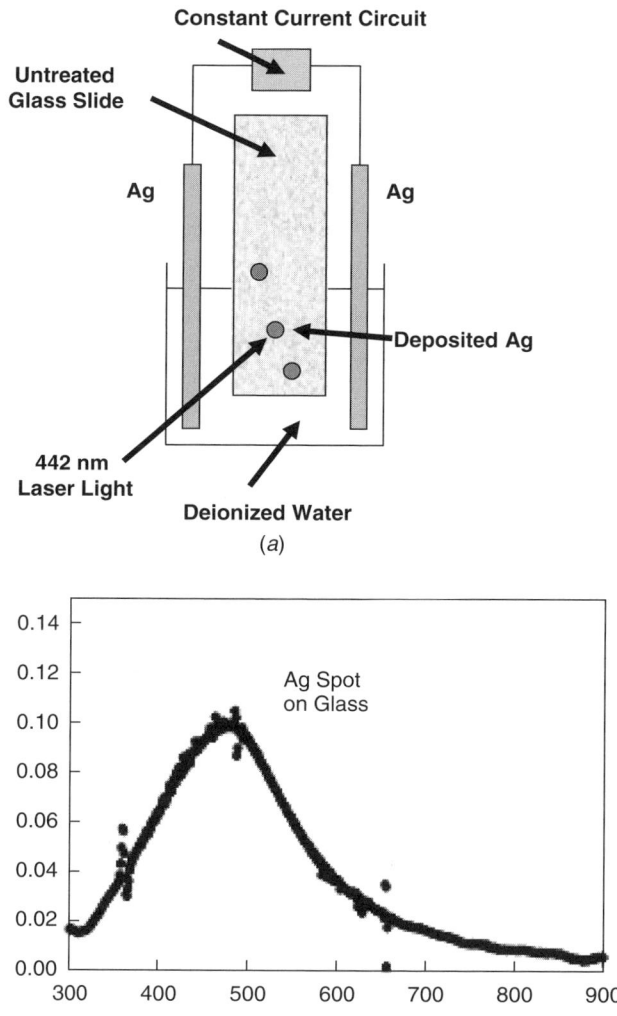

Figure 14.17 Light-deposited silver produced electrochemically. (*a*) Constant current circuit and, (*b*) absorbance spectrum of silver spot on glass. Adapted from Ref. 18.

deposition of metallic silver in the targeted illuminated region of the glass substrate. An absorption spectrum of the deposited silver is shown in Figure 14.17*b*. A single absorption band is present on glass indicating that the silver particles are somewhat spherical. We examined ICG–HSA when coated on glass (G) or silver particles (S). The emission intensity was increased about 18-fold on the silver particles (Fig. 14.18). We found a dramatic increase

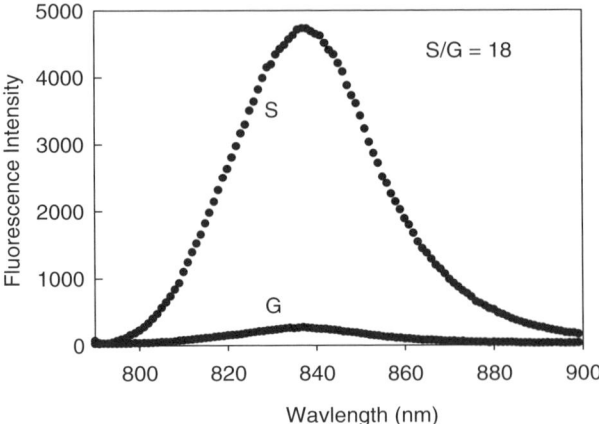

Figure 14.18 Fluorescence spectra of ICG–HSA on glass (G) and on light-deposited silver (S). Adapted from Ref. 18.

in the photostability near the silver particles (Fig. 14.19). This very encouraging result indicates that a much higher signal can be obtained from each fluorophore prior to photodestruction, and that more photons can be obtained per fluorophore before the ICG on the silver eventually degrades [18].

Electroplating of Silver for MEF

In this method, the silver cathode electrode (as shown in the scheme in Fig. 14.17) was replaced with an ITO-coated glass electrode (Fig. 14.20a). The current was again 60 μA. After a short period of time, silver readily deposited on the ITO surface (no laser illumination), the extent of which was again dependent on the exposure time. Two maxima were found on the ITO absorption spectrum, which eventually formed one large broad band. This suggests that the particles are elongated and display both transverse and longitudinal resonances (not shown). Enhanced fluorescence emission from ICG–HSA was also found for silver particles on ITO (Fig. 14.20b), and there also appeared to be a small blue shift on silvered ITO [18].

Roughened Silver Electrodes for MEF

In a typical preparation, silver electrodes are placed 10 mm apart in deionized water (Fig. 14.21) [19]. A constant current of 60 μA is supplied across the two electrodes for 10 min by a constant-current generator. Figure 14.22 shows the time-dependent growth of the silver nanostructures on the silver cathode. In comparison, the anode was relatively unperturbed. Three electrodes were coated with ICG–HSA and studied; the roughened cathode, the anode, and an unroughened electrode. Essentially no emission was seen from ICG–HSA on

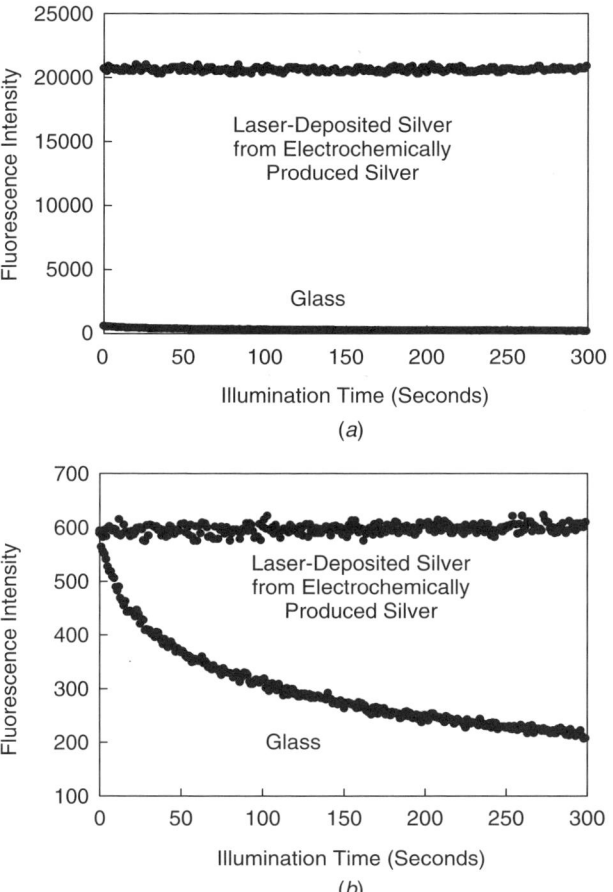

Figure 14.19 Photostability of ICG–HSA on glass and laser-deposited silver produced via electrolysis (a) measured using the same excitation power at 760 nm and (b) with power *adjusted* to give the same initial fluorescence intensities. In all the measurements, vertically polarized excitation was used, while the fluorescence emission was observed at the magic angle, that is, 54.7°. Adapted from Ref. 18.

an unroughened, bright silver surface (Ag in Fig. 14.23a, the control). However, a dramatically larger signal was observed on the roughened cathode, and a somewhat smaller signal was observed on the anode. In all our experiments we typically found that the roughened silver cathode was ~20 to 100-fold more fluorescent than the unroughened control Ag electrode. In comparison the Ag anode was 5 to 50 times more fluorescent than the Ag control. When we increased the time for roughening to over 1 h, the intensities of both electrodes after coating with ICG–HSA were essentially the same, but still 50-fold more fluorescent than the unroughened Ag control. The emission spectra on the two electrodes probably had the same emission maximum,

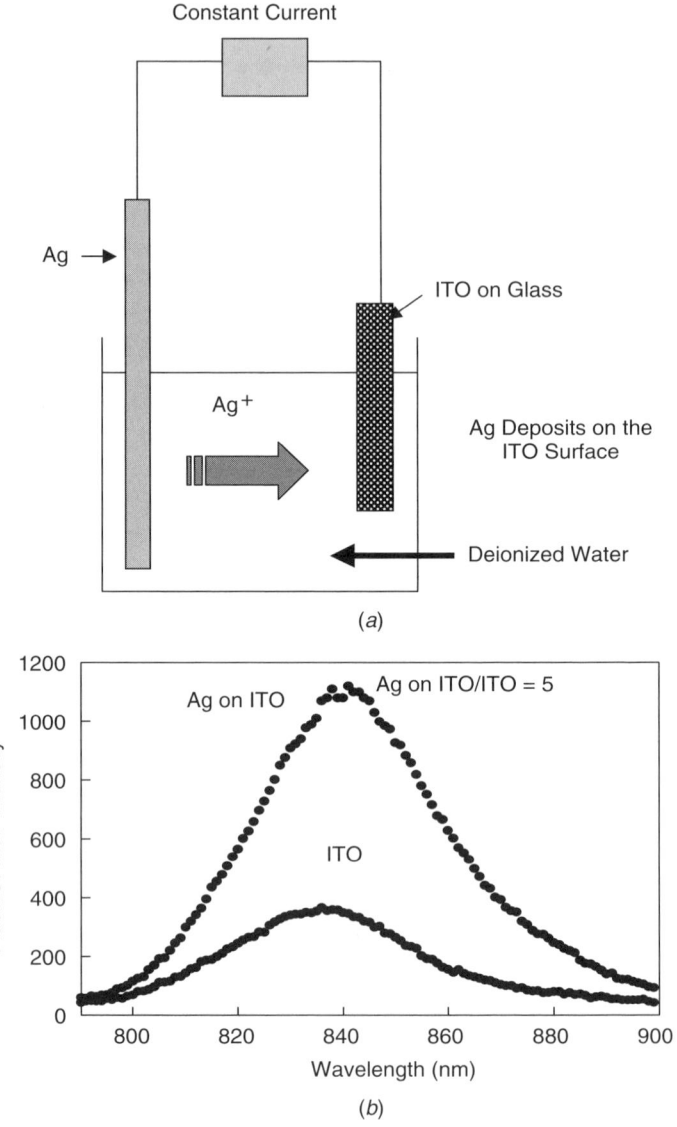

Figure 14.20 (*a*) Electroplating of silver on substrates and (*b*), fluorescence emission spectra of ICG–HSA on ITO and silver deposited on ITO. Adapted from Ref. 18.

where the slight shift seen in Figure 14.23*a* is thought to be due to the filters used to reject the scattered light. It should be noted that the amount of material coated on both surfaces was approximately the same, and the effect was not due to an increased surface area and therefore increased protein coverage on the roughened surface [19].

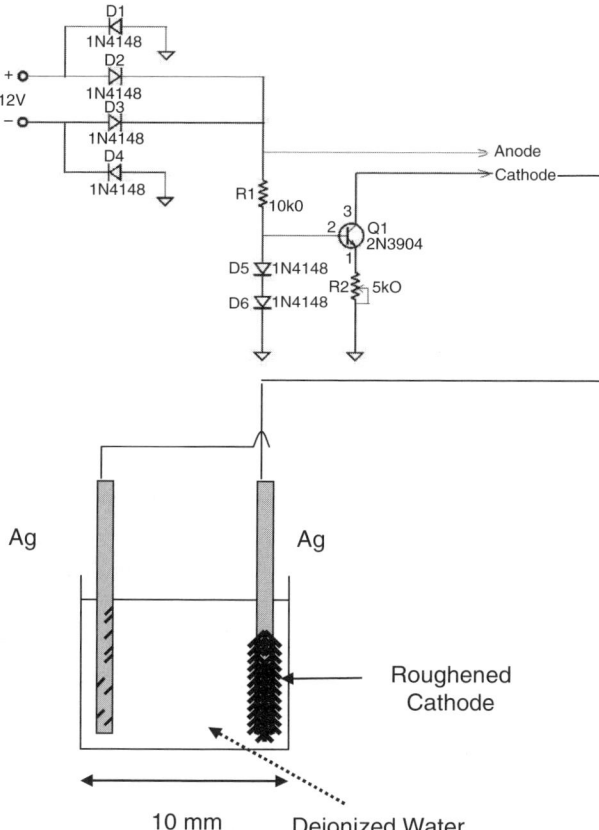

Figure 14.21 Experimental setup for the production of roughened silver electrodes. Adapted from Ref. 18.

To place our findings in context with the enhanced Raman signals observed with chloride-dipped electrodes, we examined the Ag cathode after dipping in 10^{-4} M NaCl for 1 h. In contrast to SERS electrodes, we observed a decrease in fluorescence intensity for ICG–HSA chloride-dipped electrodes (Fig. 14.23b), confirming our hypothesis of MEF being a through-space phenomenon as compared to SERS, which is known to be due to a contact interaction.

Silver Fractal-Like Structures on Glass Substrates for MEF

Using a similar method to produce roughened silver electrodes and electroplated silver, fractal-like silver structures were generated on glass using two silver electrodes held between two glass microscope slides (Fig. 14.24a) [20]. A direct current of 10 μA was passed between the electrodes for about 10 min, during which the voltage started near 5 V and decreased to 2 V. During the

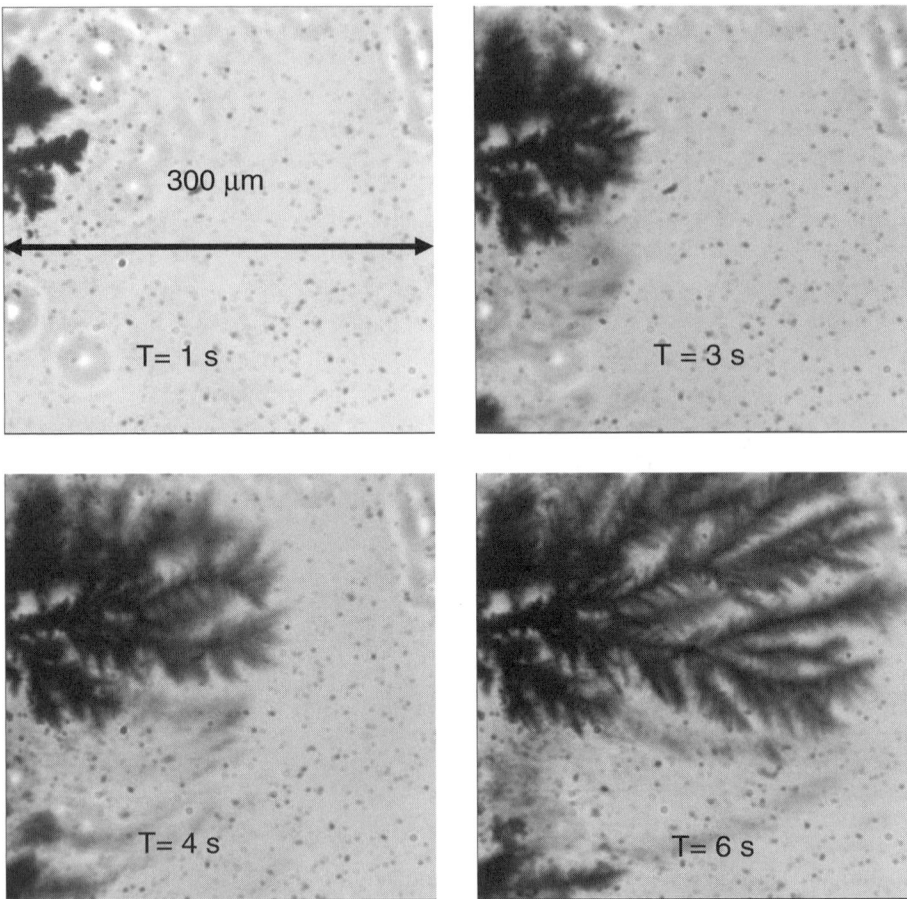

Figure 14.22 Fractal-like silver growth on the silver cathode as a function of time, visualized using transmitted light. This structure was characteristic of the whole electrode. Adapted from Ref. 19.

current flow, fractal silver structures grew on the cathode and then on the glass near the cathode (Fig. 14.24), thus producing silver nanostructures on glass, as compared to those grown on silver electrodes as described in the previous section. Similarly to the silver electrodes, the structures grew rapidly but appeared to twist as they grew. Following passage of the current, the silver structures on glass were coated with FITC–HSA overnight.

For the FITC–HSA-coated fractal silver surfaces on glass, we were able to measure a fluorescence image very similar to that of the fractal silver surface alone (bright-field image) using the same apparatus (Fig. 14.25). Interestingly, regions of high and low fluorescence intensity were observed (Fig. 14.25*b*).

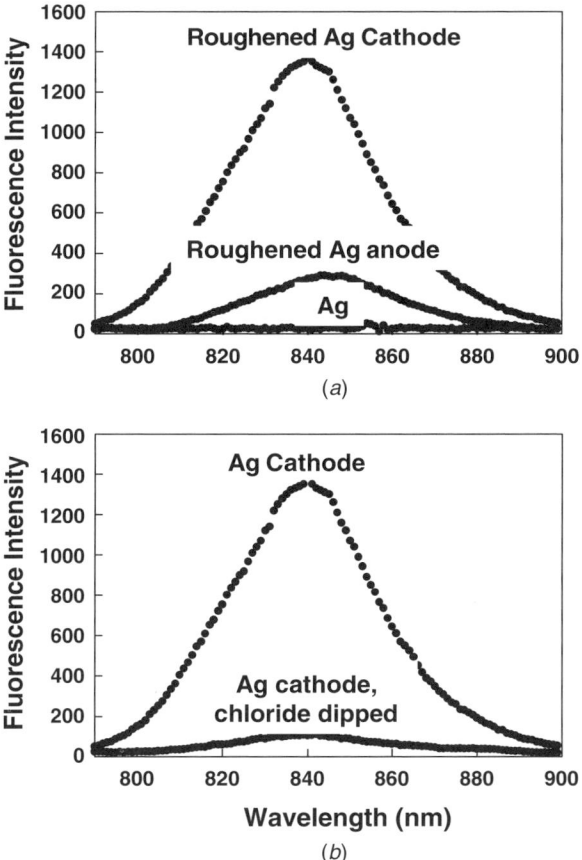

Figure 14.23 Fluorescence emission spectra of ICG–HSA on (*a*) roughened silver electrodes and (*b*) silver cathode. Adapted from Ref. 19.

This result is roughly consistent with recent SERS data, which showed the presence of intense signals that appeared to be located between clusters of particles [21, 22]. Figure 14.26 represents the intensity of FITC–HSA on silvered glass. The emission intensities range from 100-fold (position 6) to 600-fold (position 1) greater than the signal from FITC–HSA on unsilvered glass, that is, the control sample. We recognized that some of the increase in intensity of FITC–HSA could be due to binding of more FITC–HSA to silver structures with large surface areas. We note that the fluorescein is not quenched on the surface, probably because the size of an HSA molecule positions the fluorescein about 40 Å from the surface, which is near the distance for maximal radiative rate and therefore fluorescence enhancement; compare Figure 14.2.

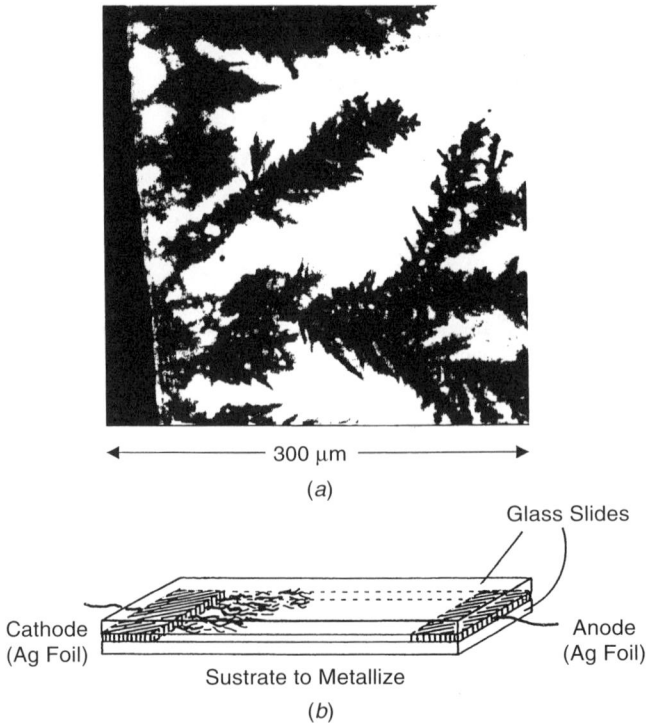

Figure 14.24 (*a*) Bright-field image of fractal-like silver surfaces on glass. (*b*) Configuration for creation of fractal-like silver surfaces on glass. Adapted from Ref 20.

We also studied the photostability of FITC on the fractal silver surface and for comparison on silver island films and uncoated quartz. Although the relative photobleaching is higher on fractal silver, the increased rate of photobleaching is less than the increase in intensity (Fig. 14.27). From the areas under these curves we estimate ≈16-fold and ≈160-fold more photons can be detected from the FITC–HSA on SiFs or fractal silver, respectively, relative to quartz, before photobleaching [20].

As we have shown, metallic silver can be deposited on a variety of surfaces by a variety of methods. The deposited silver is still useful for metal-enhanced fluorescence, and indeed some deposits seem better for overall fluorescence intensity enhancements, for example, silver fractals. Given the versatility of these deposition techniques, we can readily envisage their widespread use in metal-enhanced fluorescence sensing applications and indeed in drug discovery. In the following sections we show some actual enhanced fluorescence sensing schemes that are a direct result of metal-enhanced fluorescence and our knowledge gained from studying different silvered surfaces.

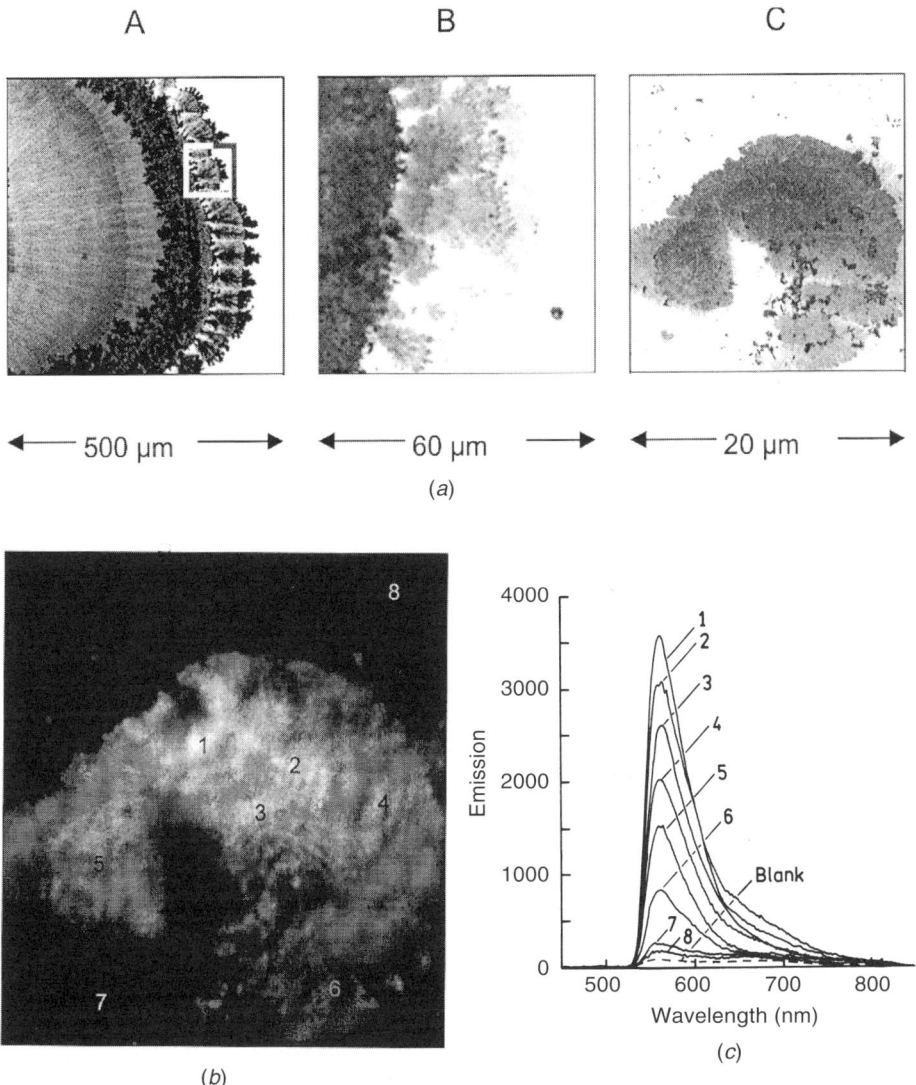

Figure 14.25 (*a*) Silver nanostructures deposited on glass during electroplating (A). Panels B and C are consecutive magnification of the marked area on panel A; bright-field image. (*b*) Fluorescence image of FITC–HSA deposited on the silver structure above and (*c*) the emission spectra of the numbered areas shown on the right. Adapted from Ref. 20.

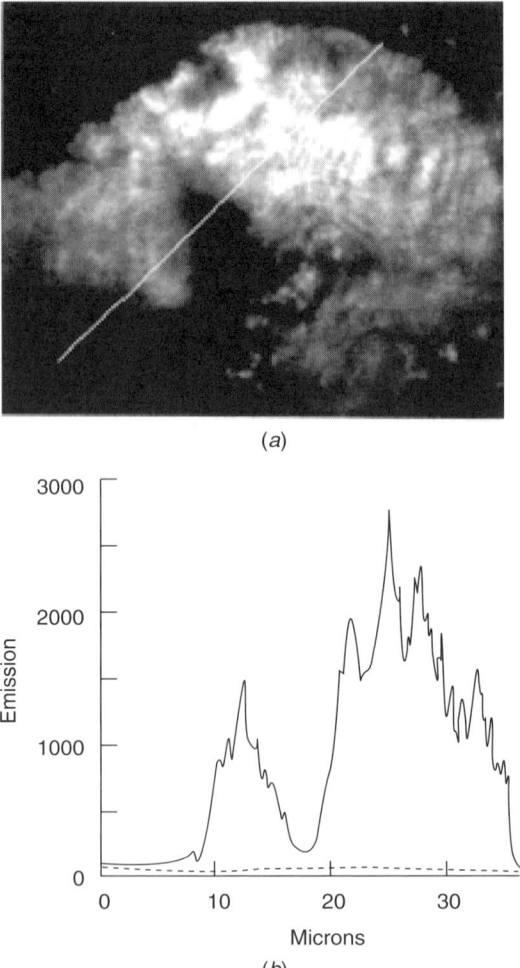

Figure 14.26 (*a*) Diagonal scan of the emission intensity of FITC–HSA. (*b*) Dashed line is the intensity observed across a line of equivalent length across unsilvered glass. Adapted from Ref. 20.

14.3 METAL-ENHANCED FLUORESCENCE SENSING: APPLICATION TO SENSING IN PLATE WELLS

Enhanced DNA/RNA Detection Using MEF (Hybridization Assays)

Detection of deoxyribonucleic acid (DNA) hybridization is the basis of a wide range of biotechnology and diagnostic applications, such as gene chips [23, 24], polymerase chain reaction (PCR) [25, 26], and for fluorescence in situ hybridization [27]. In all these applications increased sensitivity is desirable,

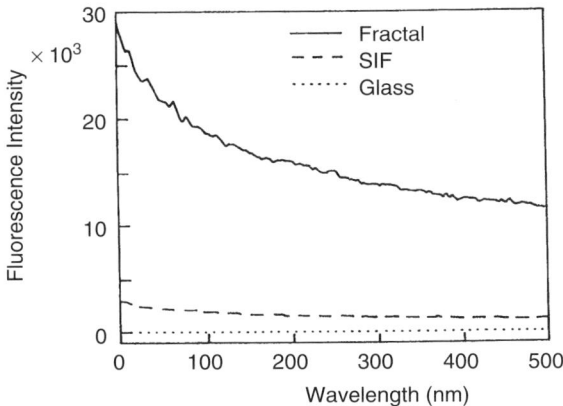

Figure 14.27 Photostability of FITC–HSA on silver fractals, SiFs, and glass. Adapted from Ref. 20.

particularly for detection of a small number of copies of biohazard agents. Also, it would be valuable to have a general approach to detect the changes in fluorescence intensity upon hybridization. In general, the detectability of a fluorophore is determined by two factors, the extent of *background emission* from the sample, and the *photostability* of the fluorophore. A highly photostable fluorophore can undergo about 10^6 excitation–relaxation cycles prior to photobleaching, yielding about 10^3 to 10^4 measured photons per fluorophore [28, 29]. Background emission from the samples can easily overwhelm weak emission signals.

In a recent report, we described an approach that should provide a readily measurable change in fluorescence intensity in DNA hybridization formats [30]. Our approach increases the intensity relative to the background and increases the number of detected photons per fluorophore molecule by a factor of 10-fold or more. Figure 14.28a shows the sequence and structure of the oligomers used in these experiments. The thiolated oligonucleotide ss-DNA-SH was used as the capture sequence, which bound spontaneously to the silver particles. The sample containing the silver-bound DNA was positioned in a fluorometer (geometry in Fig. 14.29) followed by addition of ss-Fl-DNA, which is an amount approximately equal to the amount of silver-bound capture DNA.

The fluorescence intensity began to increase immediately upon mixing and leveled off after about 20 min (Fig. 14.29a). We believe this increase in intensity is due to localization of ss-Fl-DNA near the silver particles by hybridization with the capture oligomer (Fig. 14.28c). In control experiments, we hybridized ss-DNA-SH with ss-Fl-DNA prior to deposition on silver particles. We found a similar 12-fold increase in intensity upon immobilization on silver as compared to an equivalent amount of ds-Fl-DNA-SH in solution.

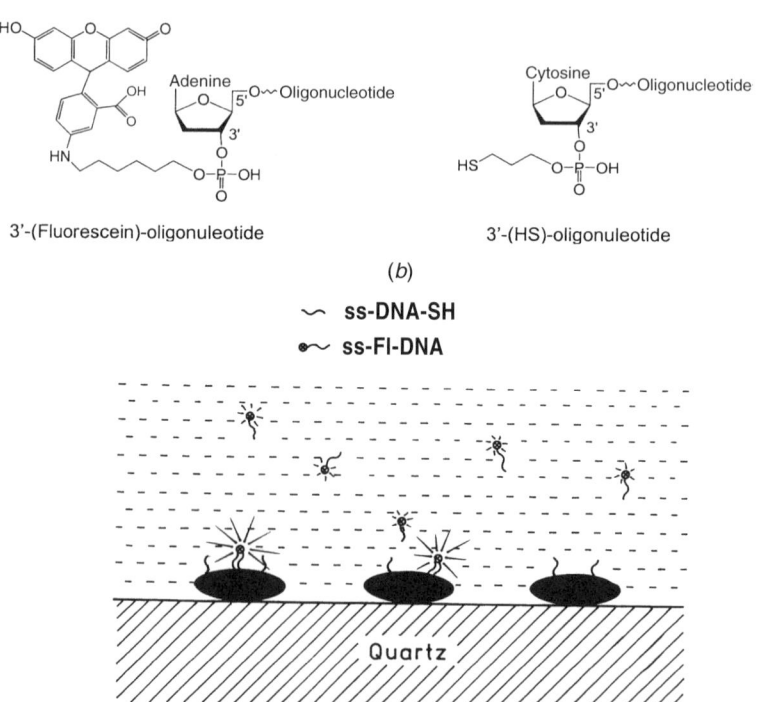

Figure 14.28 Structures of DNA oligomers. The lower panel shows a schematic of the oligomers bound to silver particles. Adapted from Ref. 30.

We examined the emission spectra of ss-Fl-DNA before and after hybridization to form ds-Fl-DNA-SH (Fig. 14.30). The fluorescence intensity was found to be 12-fold higher for the bound form. This dramatic increase can be seen visually by the photographs in Figure 14.30b. There was no detectable shift in the emission spectra. The intensity increase was reversed by melting the DNA at 80°C (Fig. 14.31) and increased once again upon cooling and presumed rehybridization. The intensity did not recover completely upon

METAL-ENHANCED FLUORESCENCE SENSING 633

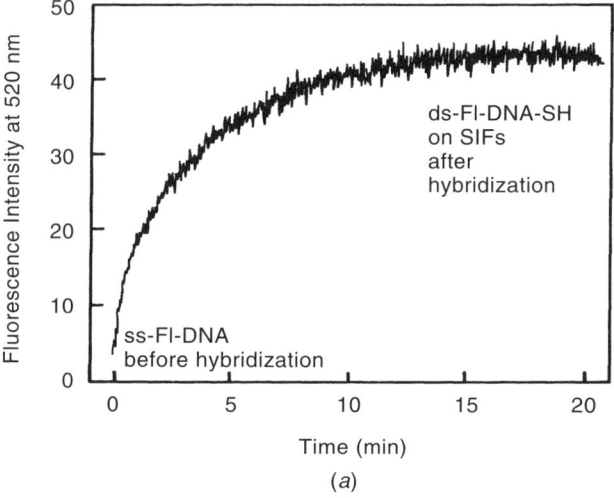

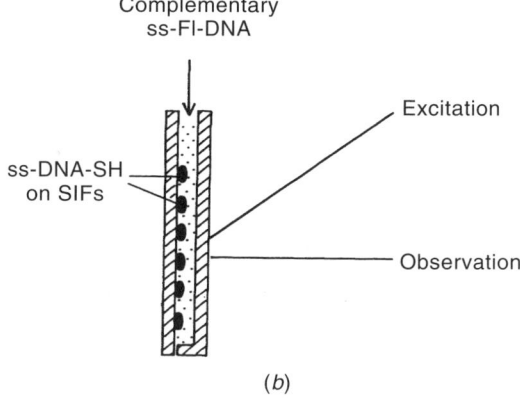

Figure 14.29 (a) Time-dependent hybridization of ss-Fl-DNA to ss-DNA-SH. (b) Lower panel shows the sample configuration. Adapted from Ref. 30.

slow cooling, which may be due to loss of capture DNA from the silver surfaces.

Figure 14.32 shows the frequency-domain intensity decays of the single-stranded fluorescein-labeled oligomer in solution and the double-stranded oligomer when bound to silver particles. The lifetime is dramatically shortened for the silver-bound oligomer, which strongly supports our conclusion that the intensity increase is due to localization of the fluorophore near the silver surfaces. In control experiments, we found that emissions of fluorescein in the single- and double-stranded oligos were similar to within 10 percent. The double-stranded form displayed an approximate 10 percent smaller intensity.

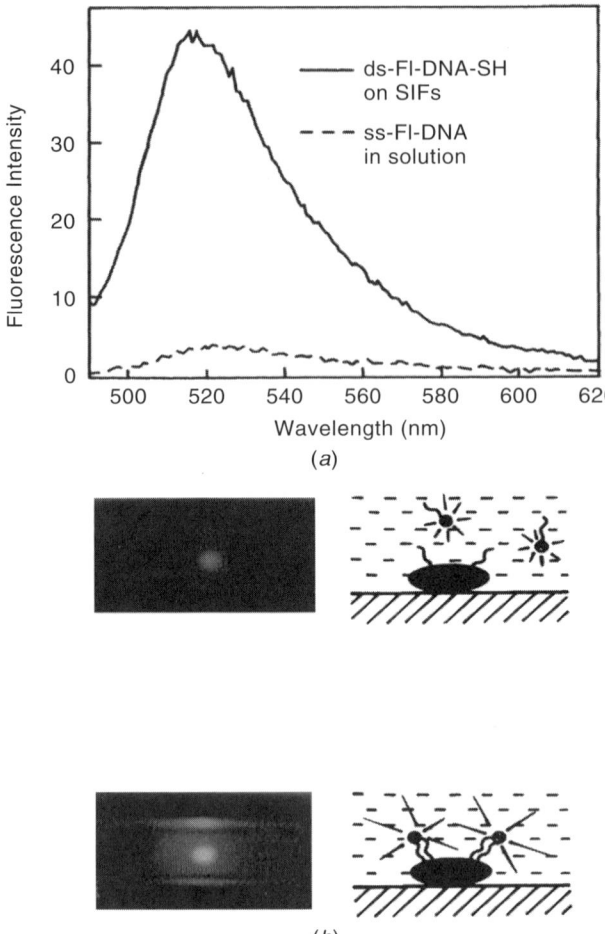

Figure 14.30 (*a*) Emission spectra of ss-Fl-DNA in solution (dashed line) and bound (solid line) to silver particles. Roughly the same number of molecules of ss-Fl-DNA and ds-Fl-DNA-SH was in the illuminated area. (*b*) The lower panel shows photographs of ss-Fl-DNA in solution (top) and ds-Fl-DNA-SH on SiFs (bottom). Adapted from Ref. 30.

Hence the differences in intensity and lifetime between the solution and silver-bound forms are not due to effects of hybridization on the fluorescein probe. It is interesting to note that there is no detectable 4-ns component for the sample with silver particles, indicating that all the emission is due to the silver-bound DNA.

The differences in lifetime of fluorescein between the solution and silver-bound form suggested an alternative approach to measuring hybridization. The emission phase angle and modulation measurements could be useful in

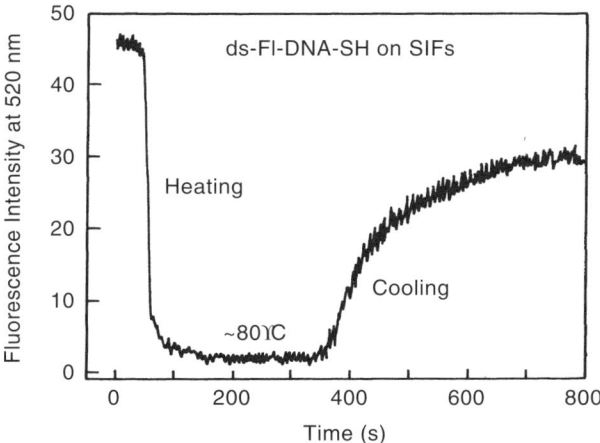

Figure 14.31 Thermal melting of ds-Fl-DNA-SH on a SiFs near 80°C and rehybridization of Fl-DNA-SH on SiFs. Adapted from Ref. 30.

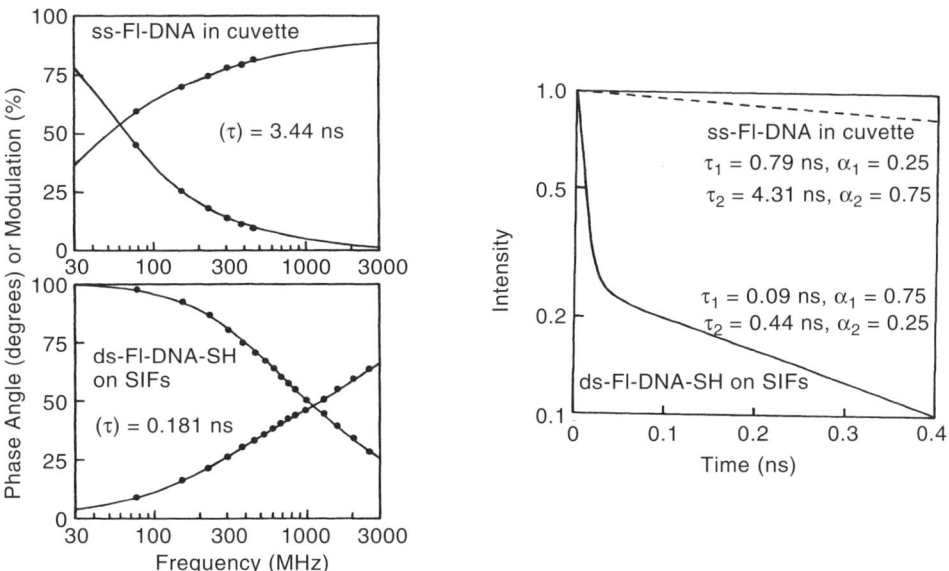

Figure 14.32 Intensity decays of ss-Fl-DNA in solution and ds-Fl-DNA-SH on SiFs measured in (*a*) the frequency-domain and (*b*) reconstructed in the time domain. Adapted from Ref. 30.

detecting DNA hybridization; since these values depend on the fluorescence lifetime, they were expected to change upon hybridization. These measurements revealed a rapid decrease in phase angle and increase in modulation following addition of ss-Fl-DNA to the silver-bound capture oligomer (Fig. 14.33). These changes are due to the decrease in lifetime upon binding to the capture oligomers on the silver particles. The changes in phase and modulation (Fig. 14.33) occur somewhat more rapidly than the change in intensity (Fig. 13.29). This difference occurs because the phase and modulation are intensity-weighted parameters. It is important to note that phase angle and modulation measurements are mostly independent of total intensity, to within

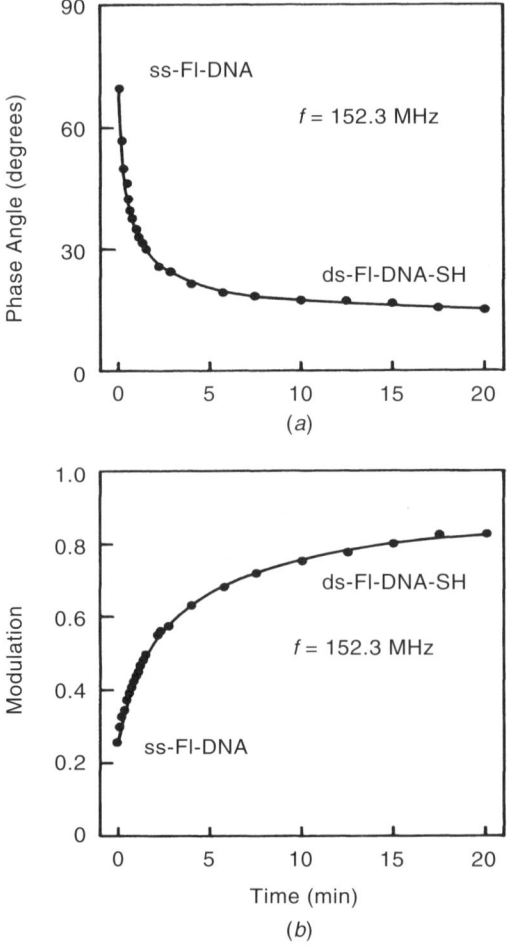

Figure 14.33 (*a*) Phase angle and (*b*) modulation of ss-Fl-DNA upon hybridization to silver-bound ss-DNA-SH. The light modulation frequency and detection frequency were 152.3 MHz. Adapted from Ref. 30.

the limitations of the instrumentation and the extent of background fluorescence in the sample [31]. This is important because the extent of hybridization can be measured using optical components such as optical fibers where the intensity may vary, or in microwell plates, where the well-to-well intensity may vary due to the plate or adsorbing species in the sample.

We examined the intensity of Fl-DNA-SH on the silver surfaces with continuous illumination (Fig. 14.34). As a comparison we examined a similar biotinylated oligomer bound to a silvered quartz slide using biotinylated bovine serum albumin and avidin. We found that fluorescein in our samples deposited on silver particles photobleaches more slowly than when deposited on a protein monolayer. It should be noted that with the protein monolayer the Fl-DNA was uniformly deposited on the entire silvered slide, that is, also between silver particles. The lifetimes measured for the sample on the protein monolayer contained a long, ≈4-ns component. This component indicates that there is some fluorescein bound to the glass surface between the silver particles.

Enhanced DNA Labels

We also investigated the effects of SiFs on the emission spectral properties of the fluorophore-labeled DNA. Emission spectra of Cy3-DNA and Cy5-DNA on glass and on SiFs are shown in Figure 14.35. The emission intensity is increased two- to threefold between SiFs as compared to between the quartz slides for Cy3-DNA and Cy5-DNA, respectively. The slightly larger increase in emission intensity for Cy5-DNA compared to Cy3-DNA is consistent with

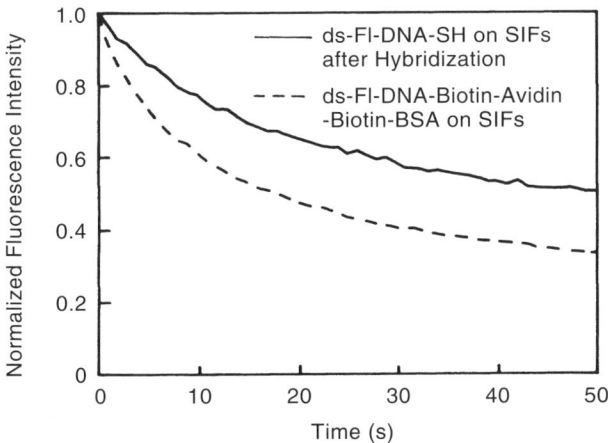

Figure 14.34 Photostability of ds-Fl-DNA-SH bound to SiFs (solid line). For comparison we show the photostability of a similar biotinylated oligomer bound to a protein monolayer of BSA–biotin–avidin uniformly deposited on SiFs (dashed line). In this case some of Fl-DNA–biotin are distant from and not affected by the silver particles. Adapted from Ref. 30.

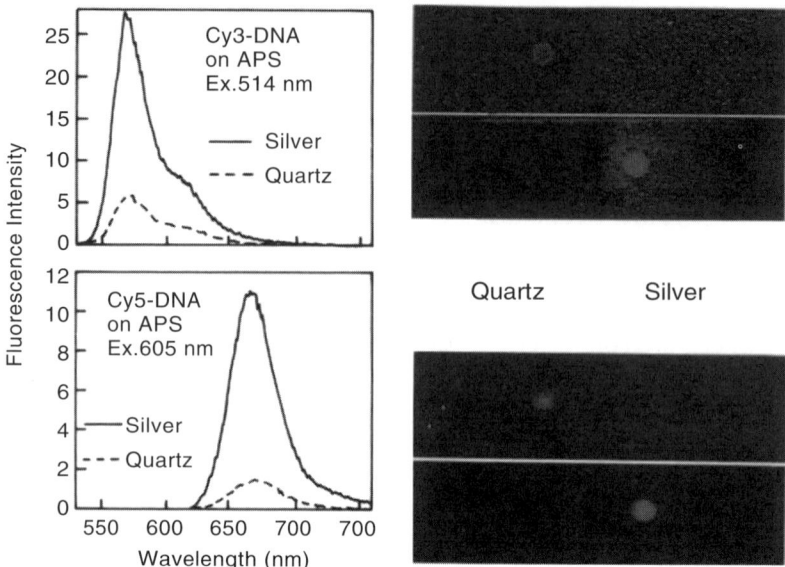

Figure 14.35 Emission spectra of Cy3-DNA (*top left*) and Cy5-DNA (*bottom left*) between quartz plates with and without SiFs. Photographs of the corresponding fluorophores. The emission from Cy3 on quartz (left side of picture top right) is very weak, as compared to silver side (right side of the picture). The emission from Cy5 on quartz (left side of the picture bottom right) is very weak, as compared to silver side (right side of the picture). Clearly, the presence of silver results in dramatic increases in observed fluorescence intensities. Adapted from Ref. 32.

the results where larger enhancements are observed with low quantum yield fluorophores. Figure 14.35 also shows the photographs of the labeled oligomers on quartz and on SiFs. The emission from the labeled-DNA on quartz is almost invisible but is brightly visible on the SiFs. This difference in intensity is due to an increase in the photonic mode density near the fluorophore, which in turn results in an increase in the radiative decay rate and quantum yield of the fluorophores. We note that the photographs are taken through emission filters, and the increase in emission intensity is not due to an increased excitation scatter from the silvered plates [32].

Photostability of the labeled DNA was studied by measuring the emission intensity during continuous illumination at a laser power of 20 mW (Fig. 14.36). The intensity initially dropped rapidly, but became more constant at longer illumination times. Although not a quantitative result, examination of these plots visually suggests slower photobleaching at longer times in the presence of silver particles compared with quartz slides.

Another widely used fluorophore in DNA detection is 1,19-[1,3-propanediylbis [(dimethylimino)-3,1-propanediyl]]bis[4-[(3-methyl-2(3H)-

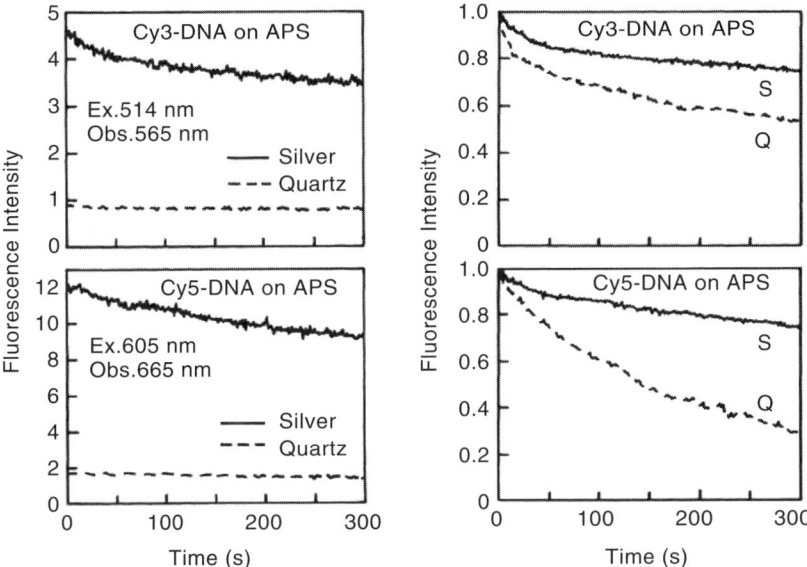

Figure 14.36 Photostability of Cy3-DNA and Cy5-DNA between quartz plates with (S) and without (Q) silver island films. The laser power was 20 mW. Adapted from Ref. 32.

benzoxazolylidene) methyl]]-quinolinum tetraiodide (YOYO-1) [33–35]. This fluorophore contains multiple positive charges, binds strongly to DNA, and displays almost no fluorescence in water. As a result, it is useful for the observation of DNA in electrophoretic gels. We measured the emission spectra of YOYO-1–DNA when bound to quartz or silver via the protein layers (Fig. 14.37). The emission intensity is 15-fold higher on the SiFs than on quartz [36]. This dramatic intensity increase can be seen in the upper panels, which are photographs of equal amounts of YOYO-1–DNA spotted on a slide.

Over-labeled Proteins as Ultrabright Probes

Fluorescein is one of the most widely used probes in immunoassays or immunostaining of biological specimens with specific antibodies. However, fluorescein is prone to self-quenching, which is due to Forster resonance energy transfer between nearby fluorescein molecules (homo-transfer) [37]. As a result, the intensity of labeled protein does not increase with higher increased extents of labeling, but actually decreases. Figure 14.38 shows the spectral properties of FITC–HSA with molar labeling ratios (L) ranging from 1-to-1 ($L = 1$) to 1-to-9 ($L = 9$) (the samples had the same optical density at 490 nm). The relative intensity decreased progressively with increased labeling. The insert in Figure 14.38 shows the intensities normalized to the same amount of protein, so that the relative fluorescein concentration increases ninefold along

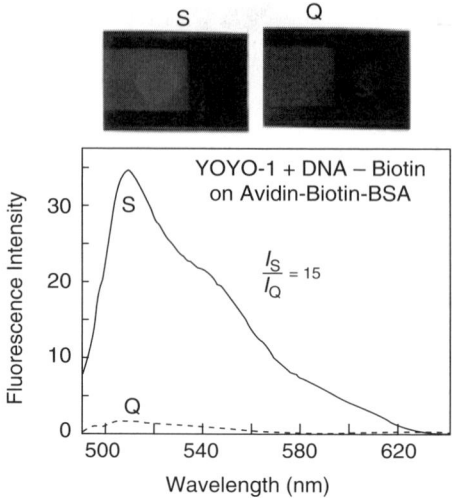

Figure 14.37 Emission spectra of YOYO-1-labeled DNA bound to the quartz (Q) and silver (S) surfaces. The upper panels show a real-color photograph of labeled DNA spotted on the silver (left) and quartz (right) surfaces. Adapted from Ref. 36.

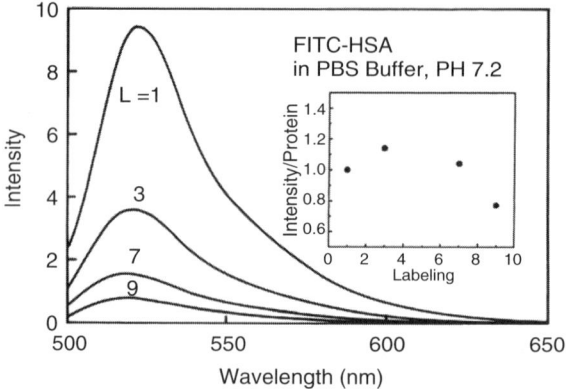

Figure 14.38 Dependence of emission intensity FITC–HSA on the degree of labeling. Adapted from Ref. 38.

the x axis. It is important to note that the intensity per labeled protein molecule does not increase and in fact decreases, as the labeling ratio is increased from 1 to 9.

However, we found that the self-quenching could be largely eliminated by the close proximity to SiFs [38], as can be seen from the emission spectra for labeling ratios of 1 and 7 (Fig. 14.39), and from the dependence of the intensity on the extent of labeling (Fig. 14.40). We speculate that

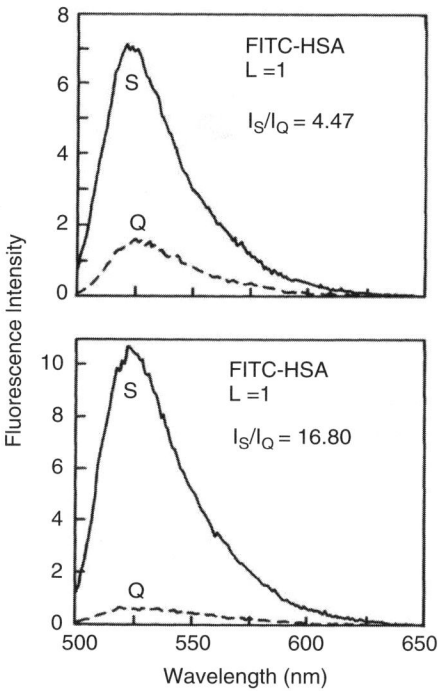

Figure 14.39 Emission spectra of FITC–HSA with different degrees of labeling on quartz (Q) and on SiFs (S). Silver island films *release* the quenched fluorescence of overlabeled proteins. Adapted from Ref. 38.

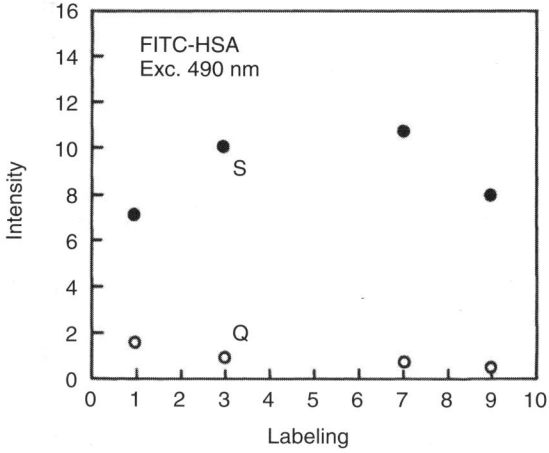

Figure 14.40 Emission intensity of FITC–HSA at 520 nm vs. different degrees of labeling on quartz and on SiFs (S). Adapted from Ref. 38.

the *decrease* in self-quenching is due to an increase in the rate of radiative decay, Γ_m.

It is informative to consider how silver-enhanced fluorescence, particularly of a heavily labeled sample, can be used for improved assays and sensing. Figure 14.41 shows emission spectra of a quartz plate coated with FITC–HSA to which we adjusted the concentration of Rhodamine B (0.25 µM) to result in an approximate 1.5-fold larger Rhodamine B intensity. In this example, one can consider the Rhodamine B to be simple autofluorescence or any other interference signal. When the same conditions are used for FITC–HSA on silver with $L = 1$, the fluorescein emission is now 2- to 3-fold higher than that of Rhodamine B (Fig. 14.41b). When using the heavily labeled sample ($L = 7$), the fluorescein emission becomes more dominant (Fig. 14.41c). This suggests that the use of overlabeled proteins increases signal intensity and suppresses the unwanted background autofluorescence.

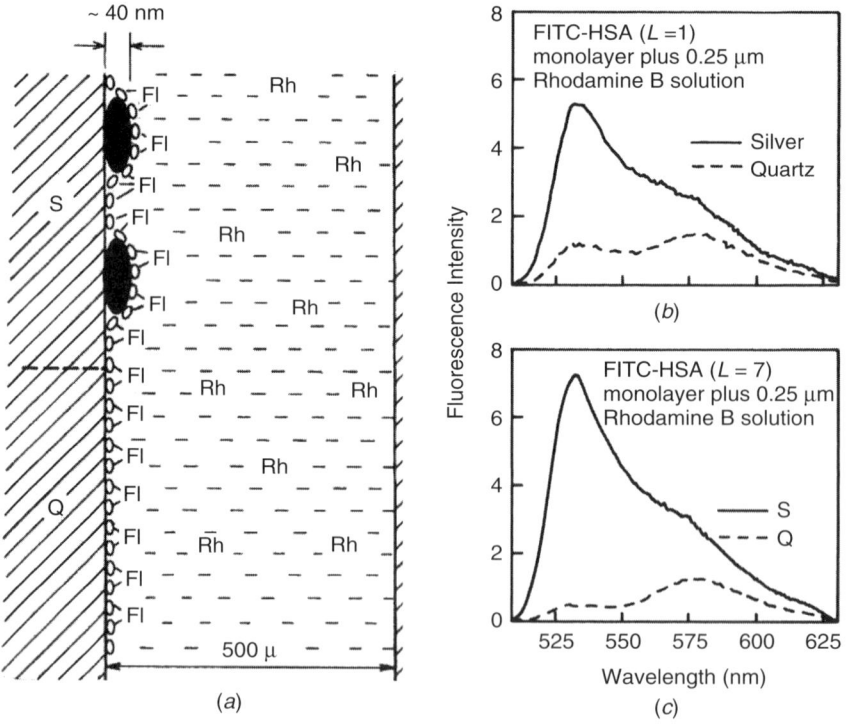

Figure 14.41 Emission spectra of a monolayer of (b) FITC–HSA $L = 1$ and (c) $L = 7$ containing 0.25 mM Rhodamine B between the quartz plates (Q) or one SIF (S). (a) Schematic of the sample with bound fluorescein and free Rhodamine B. Adapted from Ref. 38.

Enhanced and Selective Excitation Using Multiphoton Excitation and Metallic Nanoparticles

For multiphoton excitation, most of the excitation occurs at the focal point of the excitation, where the local intensity is the highest. We reasoned that this property might provide us with the opportunity to obtain enhanced fluorescence from fluorophores on the localized regions of our silver island films.

Recently, we have reported our findings on the *enhanced* and *localized* multiphoton excitation of Rhodamine B (RhB) fluorescence near metallic silver islands [39]. Figure 14.42 shows the emission spectra of 10^{-4} M RhB between silver island films with 2-photon excitation at 852 nm. The increase in fluorescence emission intensity for RhB molecules adjacent to metallic silver islands

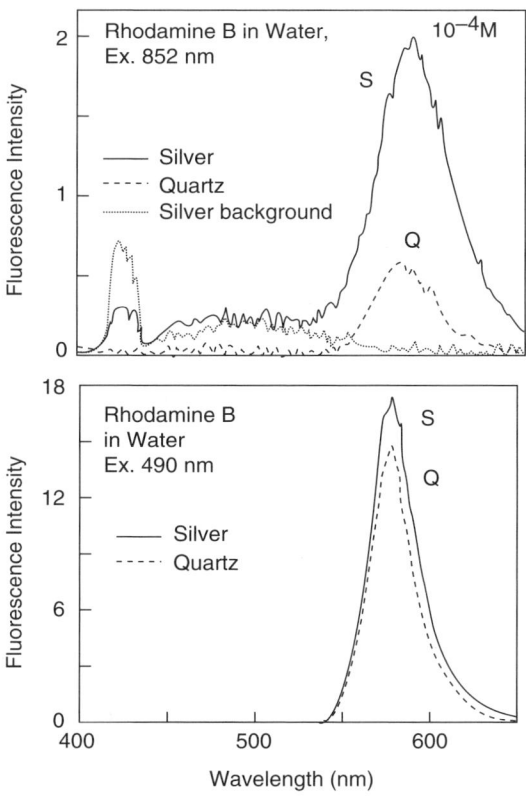

Figure 14.42 (*a*) Emission spectra of 10^{-4} M RhB between silver island films (S) with 2-photon excitation at 852 nm (2) from a Tsunami mode-locked Ti:sapphire laser, 80 MHz repetition rate, 90 fs pulse and about 0.5 W average power. Also shown are the emission spectra observed from uncoated quartz slides (Q), and silver islands alone without RhB. (*b*) RhB between silver islands, S(-), or quartz plates, Q(. . .), with 1-photon excitation at 490 nm. Adapted from Ref. 39.

is accompanied by a reduction in lifetime, compared to that observed using 1-photon excitation.

Given the high quantum yield of RhB ($Q_0 = 0.48$), these results can be explained by the metallic particles significantly increasing the E_m around the RhB molecules. Moreover, given the sample geometry and the absence of any notable increase in emission intensity using 1-photon excitation, as well the fact that the 1-photon mean lifetime remained essentially unchanged both in the presence and absence of silver, suggests that enhanced 2-photon excitation is *localized* to regions in close proximity to the silver islands. That is, a much more dramatic enhancement is possible for multiphoton excitation.

For a 2-photon absorption process the rate of excitation is proportional to the square of the incident intensity. This suggests that 2-photon excitation could be enhanced by a factor of 3.8×10^8. Such an enhancement in the excitation rate is thought to provide selective excitation of fluorophores *near to* metal islands or colloids, even if the solution contains a considerable concentration of other fluorophores that could undergo 2-photon excitation at the same wavelength but are more distant from the metals surface (Fig. 14.43). This interpretation is borne out by the fact that given the overwhelming excess of high quantum yield RhB in this sample geometry (96 percent of solution is too distant for *any* fluorophore–metal effect), the fluorescence lifetime is still shorter than that typically observed for bulk solution RhB, in the absence of metal. Also, for our samples we found that for 1-photon excitation, the photostability of RhB was not affected by the presence or absence of silver islands (Fig. 14.44b). However, for 2-photon excitation, an increased photostability was observed for RhB in the presence of silver islands (Fig. 14.44a). These results are consistent with shorter lifetime observed for RhB between silver islands and with our interpretation that multiphoton excitation occurs preferentially near to the silver islands.

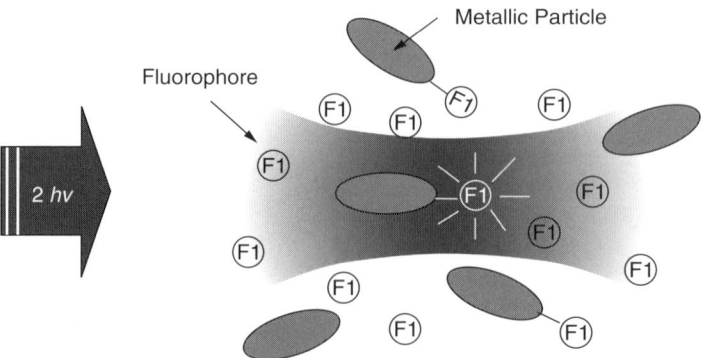

Figure 14.43 Pictorial representation of the preferential multiphoton excitation of fluorophores in close proximity to metal, in the presence of free fluorophore, Fl. Adapted from Ref. 39.

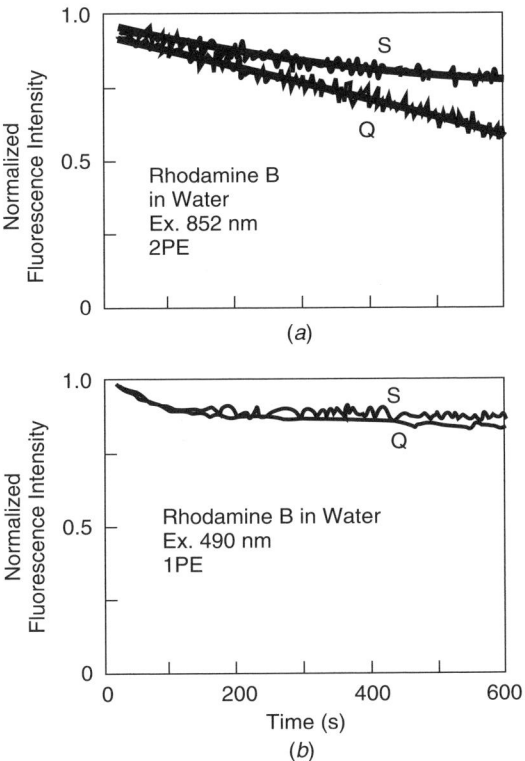

Figure 14.44 (a) Photostability of RhB between quartz slides (Q), and silver island films (S), with 2-photon excitation at 852 nm, and (b) with 1-photon excitation at 490 nm. The 490-nm excitation was from an argon ion laser attenuated to about 10 mW. Fluorescence was observed at 580 nm. Adapted from Ref. 39.

Enhanced Energy Transfer on Silver Surfaces

In a recent report, we have described the effects of metallic particles on fluorescence resonance energy transfer (RET) between donors and acceptors covalently bound to DNA [40]. Theoretical studies have predicted increased rates of energy transfer over distances as large as 700 Å near silver particles of appropriate size and shape [41, 42]. To test this prediction, we prepared a double helical DNA oligomer, 23 base pairs long, with a donor and acceptor placed at opposite sides about 75 Å apart. Because the Forster distance (R_0) value is near 50 Å, little energy transfer is expected under free-space conditions. We used steady-state and time-resolved fluorescence to determine the effects of silver island films on RET between the widely spaced donor–acceptor pairs. We examined the donor-labeled and acceptor-labeled oligomers separately, in the absence of RET, to determine the effects of the silver particles. Figure 14.45a shows the emission spectra of 7-amino-3-((((propyl)amino)car-

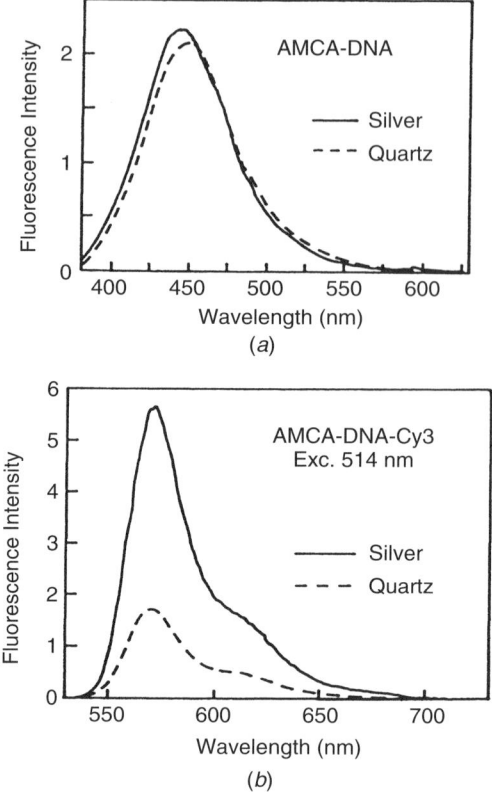

Figure 14.45 Emission spectra of (a) the donor AMCA-DNA between quartz plates (– – –) and between silver island films (———). (b) Emission spectra of the directly excited acceptor in AMCA-DNA-Cy3 between quartz plates (– – –) and between silver island films (———). Adapted from Ref. 40.

bonyl)methyl)-4-methyl coumarin-6-sulfonic acid (AMCA)-DNA between uncoated quartz plates and between silver island films. Figure 14.45b also shows the emission spectra of the donor–acceptor pair AMCA-DNA-Cy3 with no excitation of the donor and direct excitation of the acceptor at 514 nm. The intensity of the donor-alone AMCA-DNA was essentially unchanged by the silver particles. In contrast the intensity of DNA-Cy3 was increased several-fold. The different effects are consistent with the effects expected for high and low quantum yield fluorophores. It is not possible to increase a quantum yield above unity, and the larger increases in quantum yield (observed fluorescence intensity) are obtained for lower quantum yield fluorophores.

We subsequently examined the emission spectra of the donor- and acceptor-labeled DNA between quartz plates and silver island films (Fig. 14.46). An increase in the acceptor emission is seen in the donor-normalized spectra (Fig. 14.46c). It is difficult to judge the extent to which the increased acceptor emis-

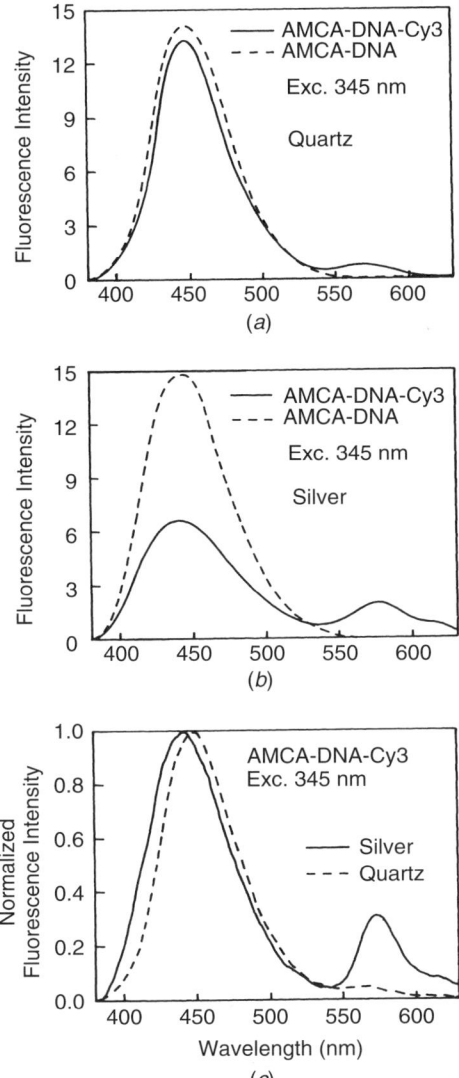

Figure 14.46 Emission spectra of AMCA-DNA donor and AMCA-DNA-Cy3 the donor–acceptor pair on (*a*) quartz and (*b*) silver. (*c*) Shows normalized emission spectra of AMCA-DNACy3 donor–acceptor between quartz (- -) and between silver island films (—). Adapted from Ref. 40.

sion is due to increased energy transfer or due to the increase intensity of Cy3 shown in Figure 14.45. Nonetheless, it was clear that much of the Cy3 emission from AMCA-DNA-Cy3 between the silver islands is due to RET.

Our analysis of the frequency-domain intensity decays of the AMCA-DNA-Cy3 suggested that R_0 contributed to the 30 percent of the donor emission (not shown). The 30 percent fraction of the strongly affected donor–acceptor pairs is in disagreement with the estimated 4 percent volume of the sample, which is close to these silver particles. To clarify this discrepancy we repeated our experiments with different sample geometry, using one SiFs instead of two SiFs. That is, one side of the sandwich was a SiFs and the other an unsilvered quartz plate. This was accomplished by coating different slides with one-third or two-thirds of the area with silver, so that there were regions of the sample between two quartz plates, between one quartz and one SiFs, or between two silver island films (Fig. 14.47a). When there was only one SiFs, we found *no* increase in energy transfer (Fig. 14.47) and *no* decrease in the donor lifetime.

It is difficult to anticipate the future uses of long-range resonance energy transfer because all present RET assays have been designed to position the donors and acceptors within the upper range of Forster distances near 50 Å. One possibility for metal-enhanced RET is detection of target sequences with larger numbers of base pairs (Fig. 14.48). Shorter D–A distances can be detected between quartz and larger distances between two silvered plates. One can also imagine the use of induced long-range RET for analysis of chromosomes with fluorescence in situ hybridization (FISH). As currently performed, the emission spectra of the FISH samples reflects the location of specific

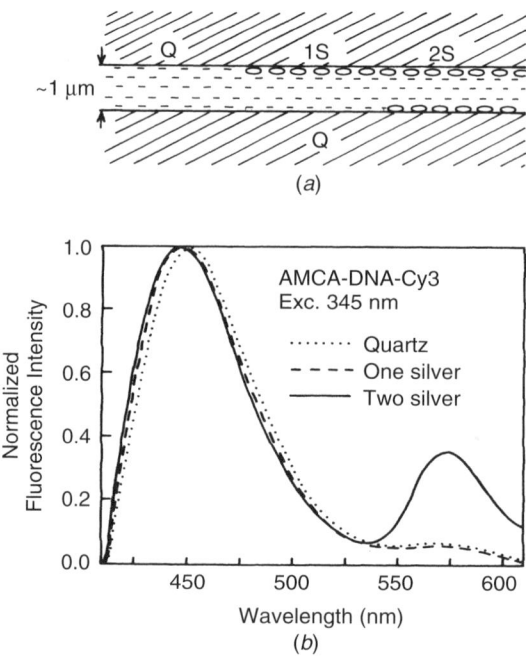

Figure 14.47 Normalized emission spectra of AMCA-DNA-Cy3 recorded on quartz (Q), one silvered slide (1 S), and two silvered slides (2 S). Adapted from Ref. 40.

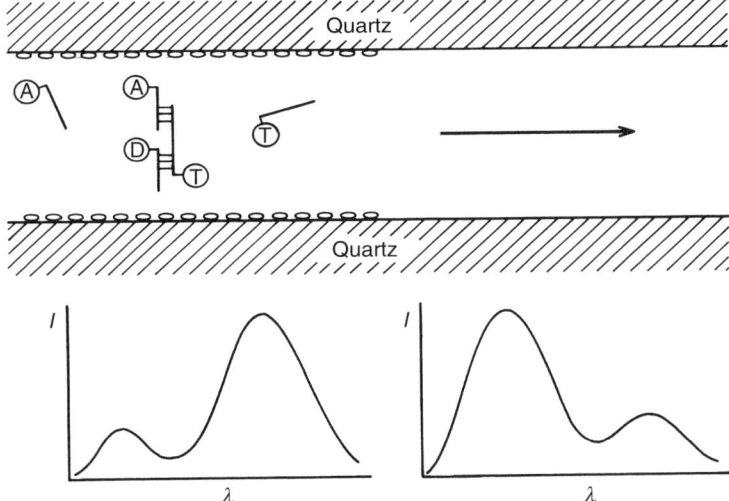

Figure 14.48 Detection of DNA sequences using long-range RET on a silver particle surface. D, A, and T are the donor, acceptor, and target nucleotides, respectively.

sequences, and RET does not normally occur between the labeled oligonucleotides used in these hybridizations. This situation, however, may change for labeled chromosomes; especially if placing a solution in a microcavity-type system results in RET over hundreds of angstroms. Further studies by our laboratories are under way.

Ratiometric Sensing on Surfaces

We have also developed new methodologies for attaching well-known pH probes (Dextran-SNAFL-2, Molecular Probes) onto silver island films. We hypothesized that SiFs could increase the sensitivity of these probes based on our observations with other fluorophores. We have developed two different methods to immobilize pH probes on to glass substrates (for control samples) and silver island films.

Immobilization of dextran-SNAFL-2 onto glass substrates is done by the following procedure: Precleaned glass substrates were soaked in a 2 percent v/v solution of 3-(glycidoxypropyl) trimethoxysilane (GOPTES) in ethanol for 2 h. Then, the GOPTES-coated glass substrates were removed from the solution and rinsed several times with ethanol. In order to remove unbound deposited materials, the substrates were additionally placed in ethanol in the ultrasonic bath for 30 sec. This procedure was repeated at least four more times. After drying the substrates in a stream of dry nitrogen, they were placed in a basic solution of dextran/SNAFL-2 conjugate and incubated for 20 h. Then, the glass substrates were removed from dextran solution and rinsed with deionized water, and dried in a stream of dry nitrogen (Fig. 14.49).

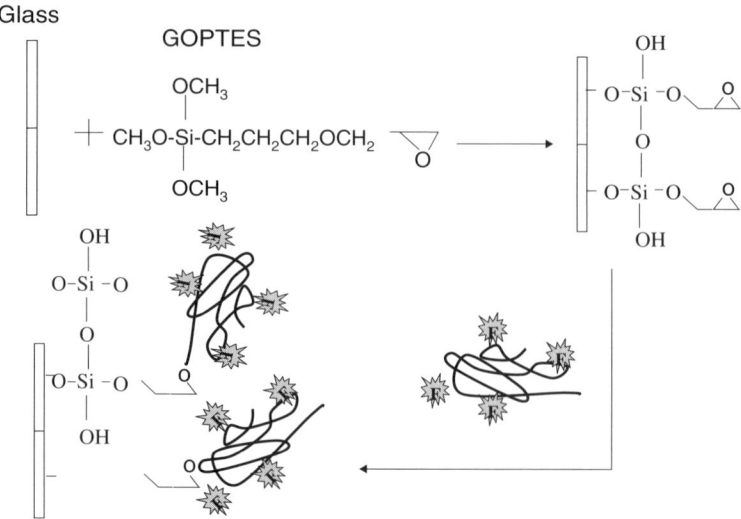

Figure 14.49 Experimental procedure for immobilization of pH and Ca probes onto glass slides.

Immobilization of dextran-SNAFL-2 onto SiFs is done by the following procedure as depicted in Figure 14.50. SiFs were prepared according to the procedure as reported in the literature. Then, the SiFs were soaked in a 2 percent v/v solution of 3-(aminopropyl) triethoxysilane (APS) in ethanol for 2h. The APS-coated SiFs were removed from the solution and rinsed several times with ethanol to remove the unbound APS. APS-coated SiFs were subsequently soaked in a 2 percent v/v solution of GOPTES in ethanol for 2h. The unbound GOPTES were removed by rinsing the substrates with ethanol several times. After drying the substrates in a stream of dry nitrogen, they were placed in a basic solution of dextran/SNAFL-2 conjugate and incubated for 20h. The glass substrates were then removed from the dextran solution and rinsed with deionized water and then dried in a stream of dry nitrogen.

SNAFLs (seminaftofluoresceins) and SNARFs (seminaftorhodafluors) are classes of pH probes with wavelength ratiometric pH sensing capabilities. The acid form of the probe displays a spectrum with an emission peak at 552 nm and base form at 635 nm (Fig. 14.51). Large spectral shifts between both forms of the probe readily allows for convenient pH sensing using the intensity ratio at two emission wavelengths.

We have thus measured fluorescence emission intensity of dextran-SNAFL-2, which is immobilized onto glass and SiFs, and we determined the enhancement factor for both the acid form (pH 6.8) and the base form (pH 8.9). The acid form of SNAFL-2 displayed a larger enhancement (average of 43.4) than that of the base form (average of 31.1). One explanation of this observation is that the acid form of SNAFL-2 (which is protonated in excited state) inter-

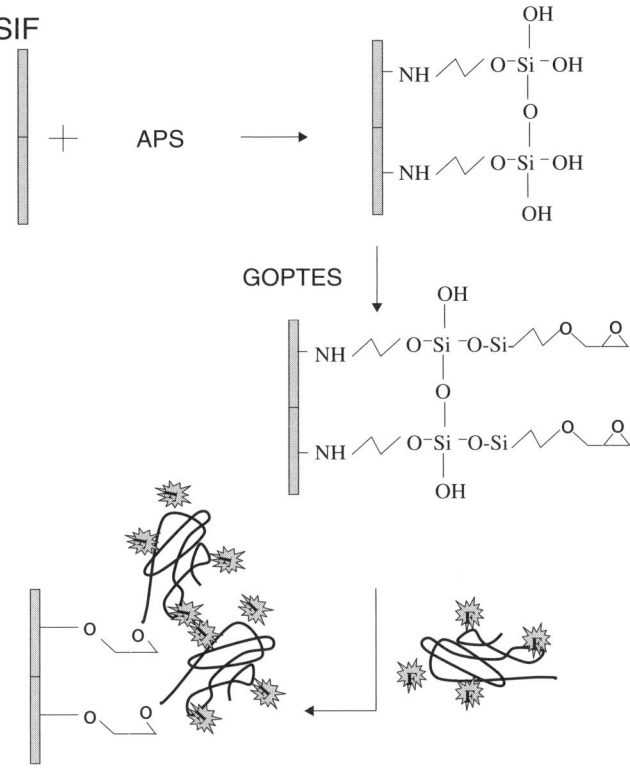

Figure 14.50 Experimental procedure for immobilization of pH and Ca probes onto SiFs.

acts more strongly with the silver nanoparticles, as its energy is closer to a nanoparticle resonance (absorbance peak for SiFs is about 440 nm), than that of base form.

We have determined that pH sensitivity of SNAFL-2 is retained, if not better, on SiFs (Fig. 14.52), which is a similar ratiometric response obtained from both when SNAFL-2 is in solution or immobilized onto SiFs. Figure 14.53 shows the photostability of SNAFL-2 immobilized onto glass and SiFs. When SNAFL-2 is immobilized onto the glass surface, it photobleaches quickly, but photobleaches substantially slower on SiFs. High-intensity enhancement and improved photostability are desired features of probes for chemical sensing and cellular imaging, allowing the use of reduced probe concentrations (less toxicity, less problem with probe loading), or the use of a lower excitation intensity (less damage to cells). Additionally, the short data acquisition times provides for less exposure damage to cells. Another advantage of a higher sample intensity is the ability to use low-power excitation sources such as LEDs (light-emitting diodes) or laser diodes, substantially reducing any instrumentation costs.

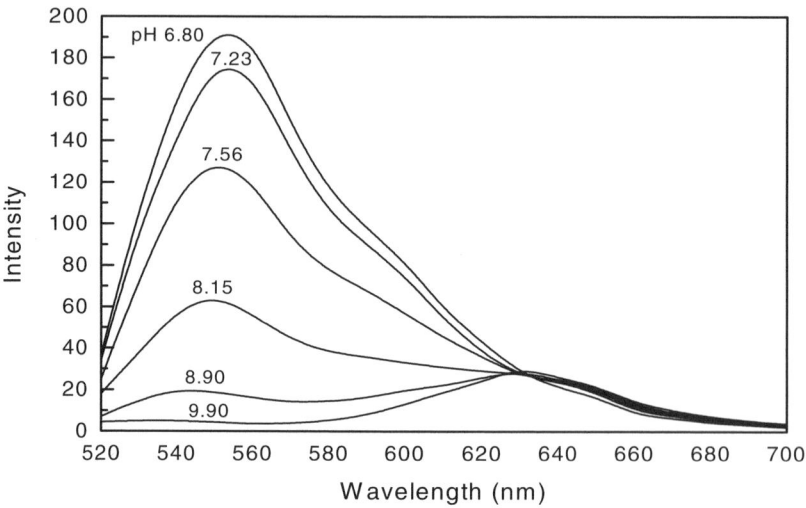

Figure 14.51 Emission spectra of SNAFL 2-Dextran in various pH solutions (PBS buffer) with 130 mM KCl. The excitation wavelength was 488 nm and the temperature was 23°C.

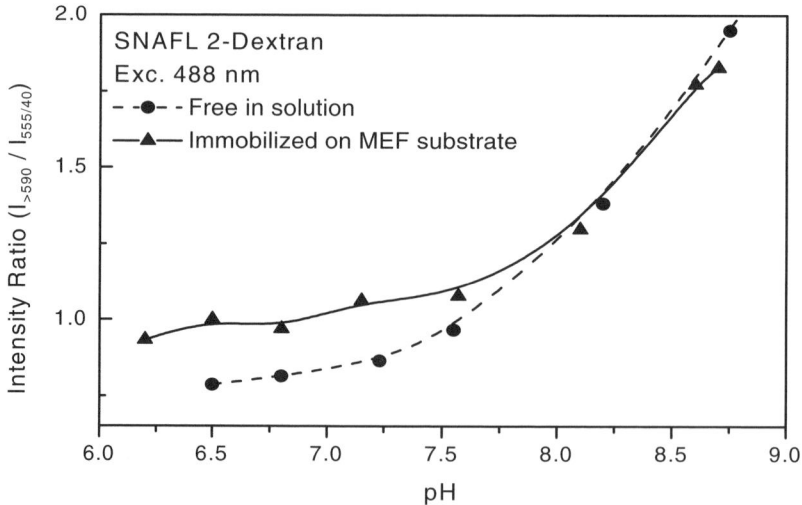

Figure 14.52 pH sensitivity of SNAFL-2 is retained in the presence of silver nanoparticles. The intensity on MEF substrate is about 30 to 40 times brighter than intensity of probe immobilized on bare glass slide, demonstrating enhanced ratiometric sensitivity.

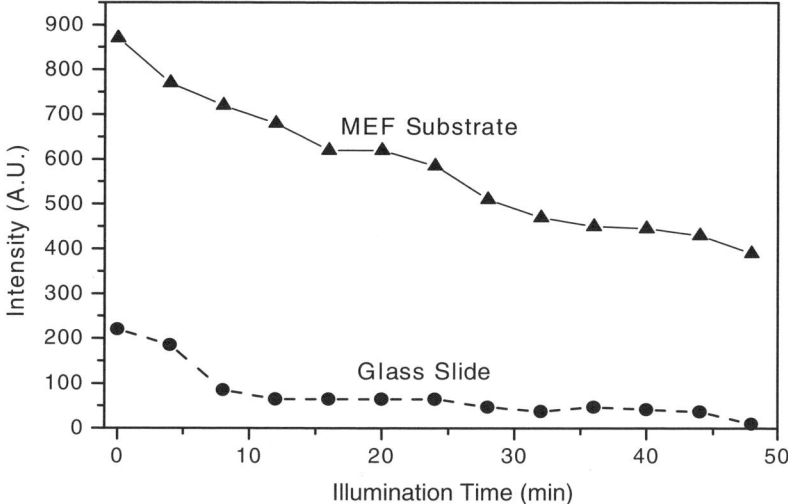

Figure 14.53 Photostability of SNAFL 2-Dextran (pH 6.3) immobilized on the MEF substrate and glass slide. The illumination conditions were the same for each slide: Green LED (535 nm, 0.5 mW), microscope objective 20× and CCD camera was used as the intensity detector.

MEF Solution-Based Assays

In order to illustrate the usefulness of MEF in solution-based biosensing applications, including high-throughput screening (HTS) plate wells, we have prepared silica-coated silver spheres in suspension. In this regard, silver spheres were prepared by using the following procedure: 2 mL of 1.16 mM trisodium citrate solution was added dropwise to a heated (95°C) aqueous solution of 0.65 mM of $AgNO_3$ while stirring. The mixture was kept heated for 10 min, and then it was cooled to room temperature. This procedure yields silver spheres with sizes in the range of 30 to 80 nm. The surface of the silver spheres were modified with APS in ethanol. The APS-coated silver spheres were resuspended in predetermined amount of water and NH_4OH. Then, a solution of tetraethylorthosilicate (TEOS) was rapidly injected, the reaction continuing for a period of time to control thickness of the SiO_2 coating layer. The amount of TEOS was calculated based on the total area of silver spheres and the desired shell thickness, assuming complete conversion of TEOS to silica (Fig. 14.54). Figure 14.54*b* shows a typical transmission electron microscopy (TEM) image of the silica-coated silver spheres.

In order to demonstrate that the silica-coated silver spheres can result in metal-enhanced fluorescence, we have utilized the well-known biotin–streptavidin interactions. These interactions occur in relatively fast reaction times (20 min) and one of the strongest biological interactions found in nature (dissociation constant, $K_D = 10^{-15}$ M). For this purpose, we have further modi-

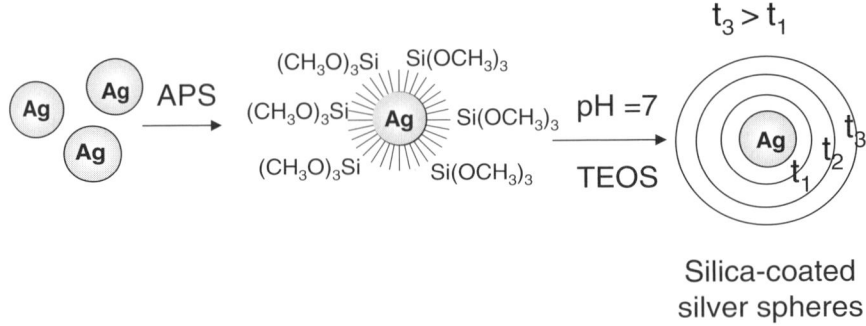

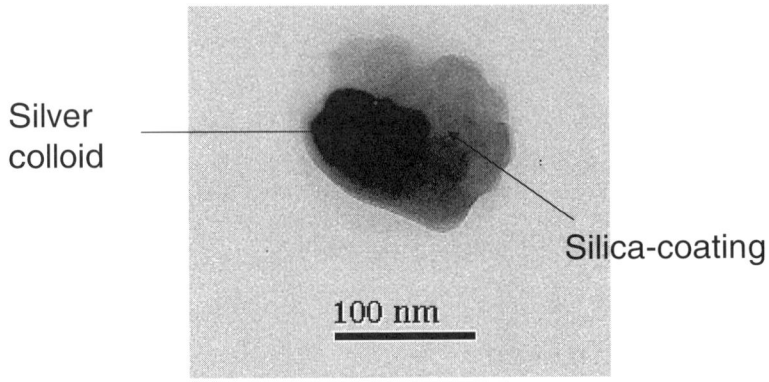

Figure 14.54 (*a*) Experimental procedure for coating of silver spheres with silica, and (*b*) a representative TEM image of silica-coated silver spheres.

fied the surface of the silica-coated silver spheres with bovine serum albumin (BSA)–biotin. Then, we have aggregated the biotinylated-silver spheres with Cy3-labeled streptavidin (Fig. 14.55). Figure 14.56 shows the fluorescence emission intensities recorded from the aggregated biotinylated-silver spheres (no label) and from the nonaggregated biotinylated-silver spheres (control samples, C1 and C2), respectively. Aggregation of the labeled biotinylated-silver spheres resulted in approximately 3- to 5-fold higher fluorescence intensity than the nonaggregate system or the aggregated system with no label. Interestingly, the Cy3 lifetime was also reduced in accordance with a radiative rate modification.

Directional Emission

We have recently reported a new method for fluorescence detection that promises to increase sensitivity by up to 1000-fold [43]. Our new method

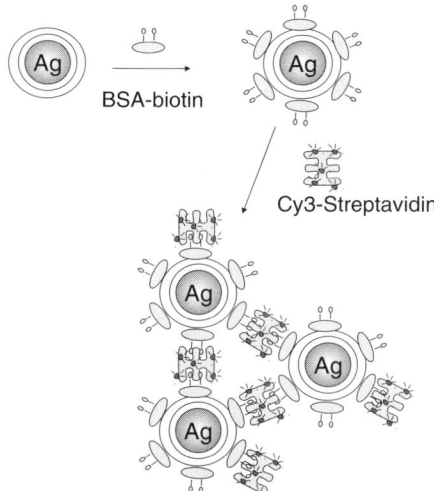

Figure 14.55 Experimental procedure for coating of silver spheres with biotinylated-BSA and aggregation of silica-coated silver spheres with Cy3-labeled streptavidin.

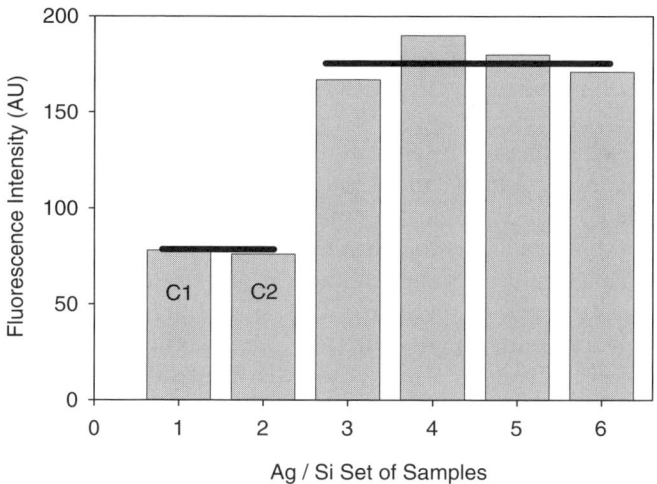

Figure 14.56 Fluorescence emission intensity recorded from the aggregated silica-coated silver spheres (samples 3–6, four separate samples) and control samples (samples 1–2).

depends on the coupling of excited fluorophores with the surface plasmons present in thin metal films, typically silver, gold, or aluminum. This new phenomenon, surface plasmon-coupled emission (SPCE) occurs for fluorophores 20 to 250 nm from the metal surface, where the emission occurs over a narrow angular distribution, converting normally isotropic emission into easily collected directional emission. The interaction is *independent* on the mode of excitation, that is, it does not require evanescent wave or surface plasmon excitation. With typical optical components, the collection efficiency is 1 percent or less, however, our new approach promises to couple up to 50 percent of the emission from unoriented and appropriately distanced fluorophores. We believe our new findings offer both unique and exciting possibilities for high-sensitivity fluorescence detection, as distal fluorescence (autofluorescence) from fluorophores or species from the metal surface only weakly couples. In addition SPCE is highly polarized, and autofluorescence can be further discriminated against by collecting only the polarized component or the light emanating at the appropriate angle. Further, different emission wavelengths couple at different angles allowing spectral discrimination without additional dispersive optics.

In our preliminary studies of SPCE, a continuous 50-nm-thick silver film on a glass substrate was spin coated with varying thicknesses of sulforhodamine 101 (S101) doped PVA [43]. Figure 14.57a shows the configuration that was used for SPCE studies. This combined sample was positioned on a precise rotary stage, which allows excitation and observation at any desired angle relative to the vertical axis along the cylinder. Two modes of excitation were considered in our studies (Fig. 14.57b).

We have found that the mode of excitation does not matter for SPCE [43–45]. That is, an excited fluorophore should couple with the surface plasmons whether it is excited by the surface plasmon evanescent field or directly using the reverse Kretschmann (RK) configuration. Figure 14.58 shows the dependence of the emission intensity on observation angle with RK excitation, noting that the incident field cannot induce surface plasmons [1]. On the prism side (back side B) of the sample, the emission is very sharply distributed at ±47° or ±50° from the normal axis for both the 15- (top) and 30-nm (bottom) films, respectively. This small difference in angle is due to the thickness of the PVA film. The intensity observed on the front (F) side of the sample was much lower and not sharply distributed at any particular angle, as illustrated in Figure 14.58. Higher emission intensity is observed on the front side of the thicker sample (bottom), which is consistent with lower efficiency coupling into the surface plasmon for fluorophores more distal from the metal surface.

We additionally examined the polarization of SPCE and found that the emission is polarized in the plane of incidence (*p*-polarized) irrespective of the polarization of the excitation. This polarization of the directional emission is an interesting find and proves that it is due to coupling of the surface plasmons, and that the polarization of the SPCE is independent of the polarization of the normal incidence excitation. This suggests that the emission dipoles parallel to the plane of incidence couple into the surface plasmon, and dipoles perpendicular to the plane of incidence do not result in SPCE, or at least display only very weak coupling [43, 45]. Figure 14.59 shows the emission for S101 in PVA observed with a hemispherical prism and RK excitation, where the hemispherical prism allows for the SPCE cone to be observed.

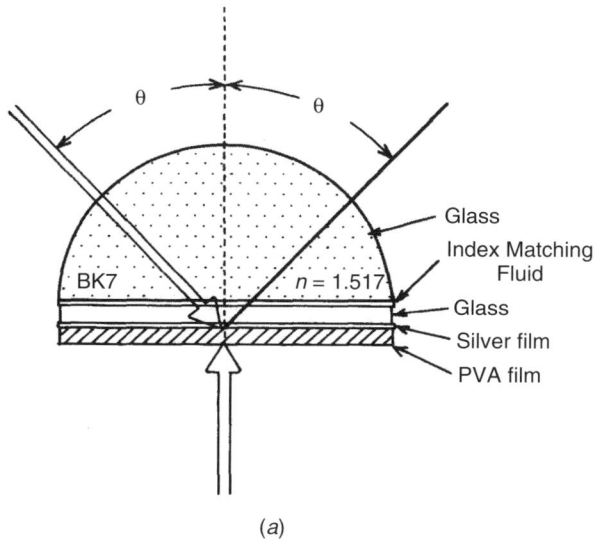

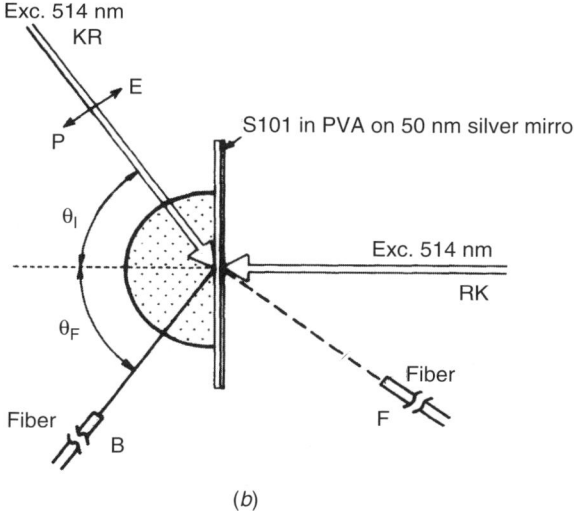

Figure 14.57 (*a*) Configuration of hemicylinder and spin-coated PVA-fluorophore slide, and (*b*) experimental geometry for the measurement of free space emission (F) and SPCE (B) with the Kretschmann (KR) and reverse Kretschmann (RK) configuration. Adapted from Ref. 43.

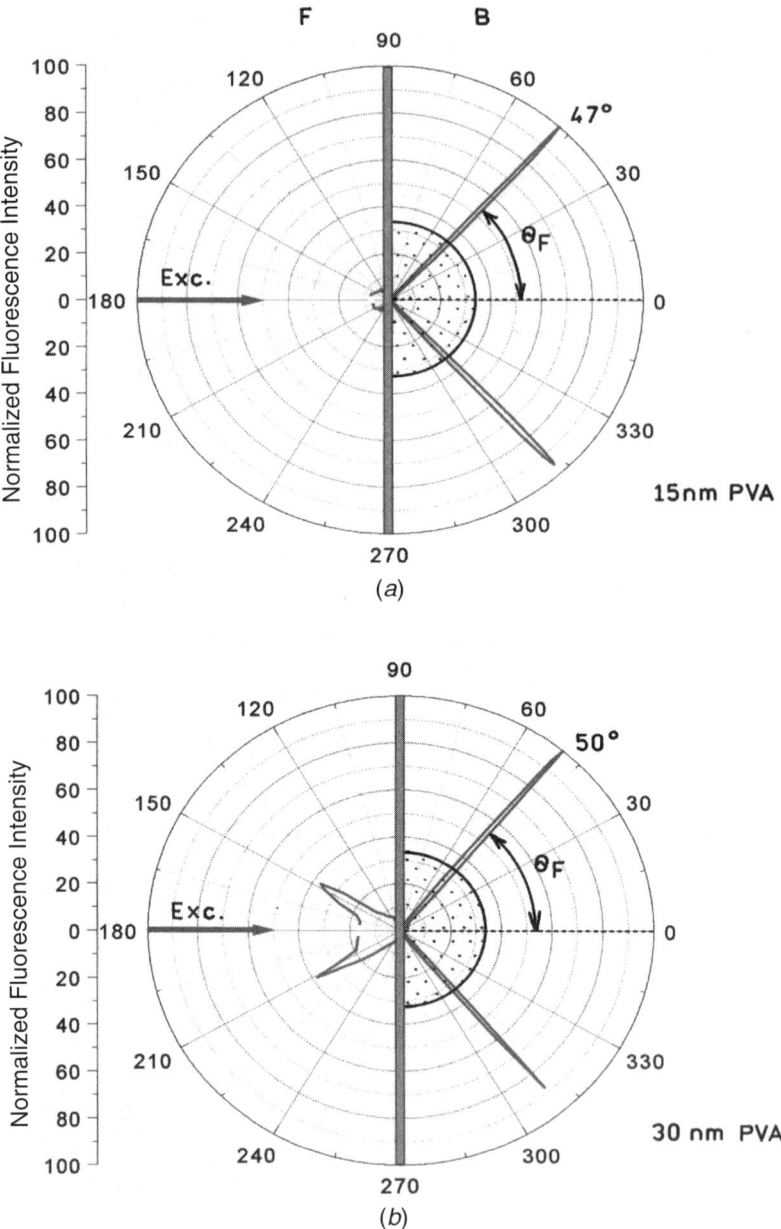

Figure 14.58 Angular dependence of S101 emission excited using the reverse Kretschmann configuration. The PVA thickness was approximately (*a*) 15 nm and (*b*) 30 nm. Adapted from Ref. 43.

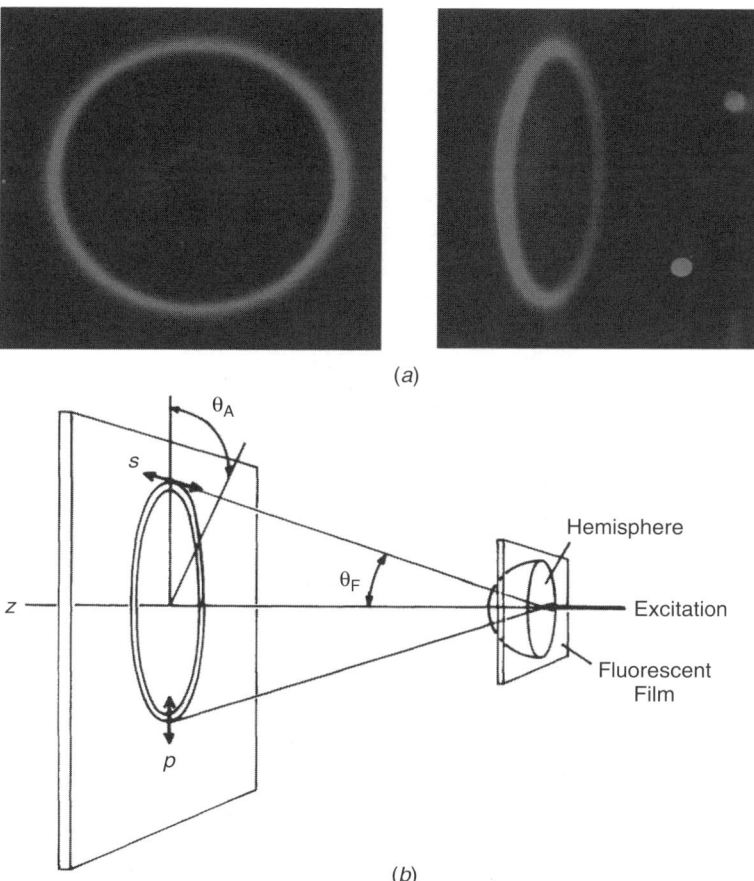

Figure 14.59 (*a*) Emission for S101 in PVA observed with a hemispherical prism and RK excitation and (*b*) the cone of emission with a hemispherical prism. The emission was incident on tracing paper and photographed through a LWP 550 filter. Adapted from Ref. 43.

We also have found that SPCE strongly depends on wavelength. This suggests that fluorophores with different emission maxima will display SPCE at different angles. Subsequently, Figure 14.60 shows the SPCE from a mixture of fluorophores using RK excitation and a hemispherical prism, with 532-nm excitation. We recorded the emission spectra at different observation angles where the spectra are clearly distinct at each angle (not shown), with the shorter wavelengths occurring for larger angles [45]. This remarkable find suggests the potential for multifluorophore surface assays featuring the intrinsic spectral resolution of SPCE.

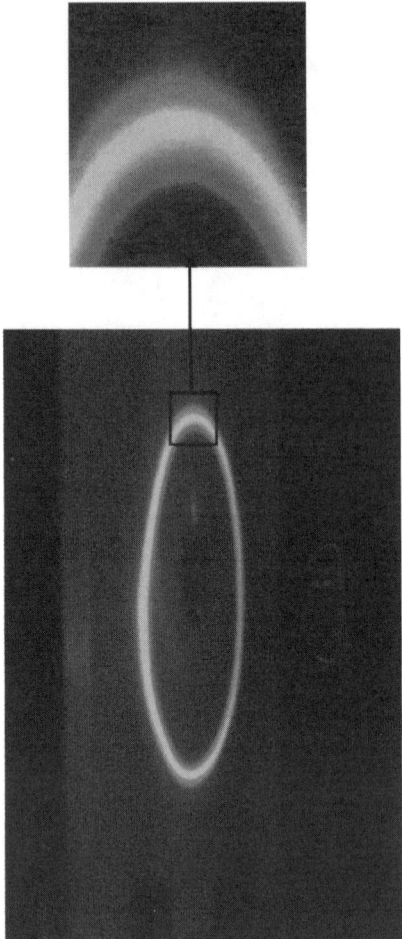

Figure 14.60 Photograph of SPCE from a mixture of S101, Rhodamine 123, and Pyridine 2, using RK excitation and a hemispherical prism. Initially the fluorophore concentrations were optimized to give similar free space fluorescence intensities. Adapted from Ref. 43.

14.4 ROLE OF MEF IN HTS AND DRUG DISCOVERY

Metal-enhanced fluorescence appears to be most suitable to the fluorescence assays used in drug discovery and DNA analysis. As we have shown, silver can be readily deposited on glass or polymer substrates by a variety of methods. Silver colloids are easily prepared and can be attached to surfaces functionalized with amine or sulfhydryl groups. One can imagine the bottom of multiwell plates or DNA arrays being coated with silver particles. A variety of new assay formats are possible. Assays could be based on the lightning rod effect, that is, E_m modifications. The biochemical affinity interactions could bring the fluorophore close to the metal surface, for localized excitation, eliminating the washing steps. Another approach could be to use low quantum yield fluorophores, and the increased quantum yield of fluorophores brought in close proximity with the metal (Fig. 14.61). These effects might be coupled with another remarkable property of metal–fluorophore interactions. If the metal is close to a semitransparent metallic surface, the emission can couple into the metal and become directional rather than isotropic (Fig. 14.62), offering the possibility of a new high-sensitivity detection system for HTS wells.

Metal-enhanced fluorescence is not limited to single metallic particles. Theory predicted that the lightening rod effect is much stronger between two metallic spheres than for isolated spheres. If the biochemical affinity reaction brings particles into closer proximity, then excitation fields may be substantially increased in the spaces between particles (Fig. 14.63). Multiphoton excitation is also known to be increased near metallic surfaces and may be even more efficient between metal particles. Also, proximity to the particles can result in long-range energy transfer (Fig. 14.64), which would allow selective detection of macromolecular complexes. Scattering from corrugated surfaces

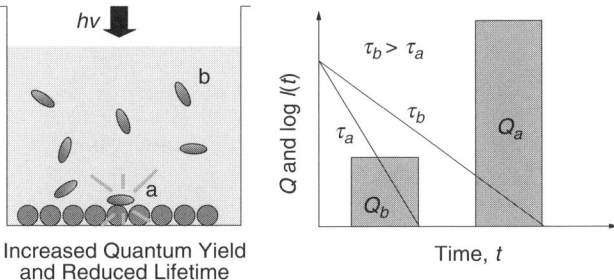

Figure 14.61 Potential uses of metal-enhanced fluorescence in drug discovery based on local increases in quantum yield. Adapted from Ref. 46.

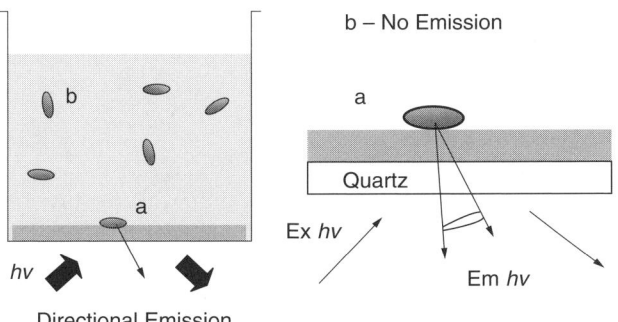

Figure 14.62 Potential uses of metal-enhanced fluorescence in drug discovery based on directional emission. Adapted from Ref. 46.

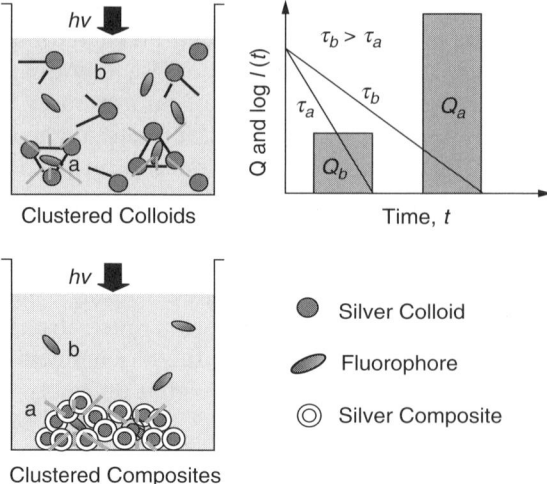

Figure 14.63 Potential uses of metal-enhanced fluorescence based on colloid clustering. Adapted from Ref. 46.

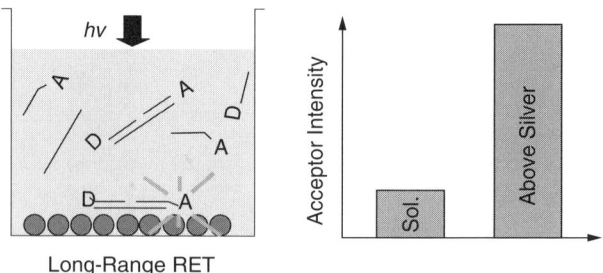

Figure 14.64 Potential uses of metal-enhanced fluorescence based or long range RET. Adapted from Ref. 46.

could also be used to increase the detection efficiency of fluorophores (Fig. 14.65).

14.5 CLOSING REMARKS

In this invited review chapter, we have summarized the favorable effects of fluorophores in close proximity to metallic nanostructures. These favorable effects include increased fluorescence intensities (increased quantum yields), increased probe photostability (reduced fluorescence lifetimes), and increased rates of excitation and energy transfer.

Metallic silver can be deposited by a variety of methods, depending on the sensing application, on demand if required, and using fairly simple apparatus

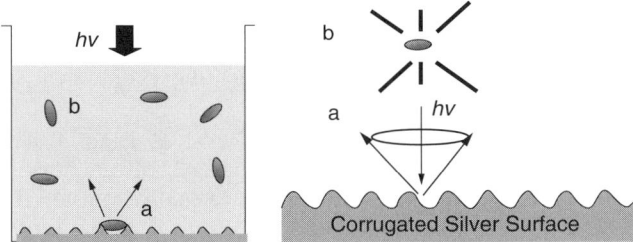

Figure 14.65 Potential uses of metal-enhanced fluorescence in drug discovery based on corrugated surfaces. Adapted from Ref. 46.

and biologically inert chemicals. The metal–fluorophore effects we have observed to date offer unique perspectives in fluorescence sensing, providing for improved background suppression, increased detection limits, and localized excitation near to silver nanostructures. While our results to date are for the most part preliminary, we believe MEF will find many applications in sensing, especially in high-throughput screening and in drug discovery.

Acknowledgments Work was supported by the NIH, National Center for Research Resources, RR-01889. Financial support to J. R. Lakowicz, C. D. Geddes, and I. Gryczynski from University of Maryland Biotechnology Institute (UMBI) is also acknowledged.

ACRONYMS AND SYMBOLS

Γ	radiative decay rate in the absence of metal
Γ_m	radiative decay rate in the presence of metal
APS	1-amino-3-propylethoxy silane
E	excitation rate
E_m	metal-enhanced excitation rate
HAS	human serum albumin
ICG	indocyanine green
KR	Kretschmann
k_m	quenching rate in the presence of metal
k_{nr}	nonradiative rates
MEF	metal-enhanced fluorescence
Q_0	quantum yield
r	particle diameter
RET	fluorescence resonance energy transfer
RK	reverse Kretschmann
SERS	surface-enhanced raman scattering
SiFs	silver island films
SPCE	surface plasmon-coupled emission

REFERENCES

1. Lakowicz, J. R. (2001). Radiative decay engineering: Biophysical and biomedical applications. *Anal. Biochem.*, 298, 1–24.
2. Lakowicz, J. R., Shen, Y., D'Auria, S., Malicka, J., Fang, J., Gryczynski, Z., Gryczynski, I. (2002). Radiative decay engineering 2. Effects of silver island films on fluorescence intensity, lifetimes, and resonance energy transfer. *Anal. Biochem.*, 301, 261–277.
3. Lakowicz, J. R., Shen, Y., Gryczynski, Z., D'Auria, S., Gryczynski, I. (2001). Intrinsic fluorescence from DNA can be enhanced by metallic particles. *Biochem. Biophys. Res. Commun.*, 286, 875–879.
4. Gryczynski, I., Malicka, J., Shen, Y., Gryczynski, Z., Lakowicz, J. R. (2002). Multiphoton excitation of fluorescence near metallic particles: Enhanced and localized excitation. *J. Phys. Chem. B*, 106, 2191–2195.
5. Gersten, J., Nitzan, A. (1981). Spectroscopic properties of molecules interacting with small dielectric particles. *J. Chem. Phys.*, 75, 1139–1152.
6. Weitz, D. A., Garoff, S., Gersten, J. I., Nitzan, A. (1983). The enhancement of Raman scattering, resonance Raman scattering and fluorescence from molecules absorbed on a rough silver surface. *J. Chem. Phys.*, 78(9), 5324–5338.
7. Kummerlen, J., Leitner, A., Brunner, H., Aussenegg, F. R., Wokaun, A. (1983). *Mol. Phys.*, 80, 1031.
8. Malicka, J., Gryczynski, I., Geddes, C. D., Lakowicz, J. R. (2003). Metal-enhanced emission from indocyanine green: A new approach to in vivo imaging. *J. Biomed. Opt.*, 8(3), 472–478.
9. Geddes, C. D., Cao, H., Gryczynski, I., Gryczynski, Z., Fang, J., Lakowicz, J. R. (2003). Metal-enhanced fluorescence due to silver colloids on a planar surface: Potential applications of indocyanine green to in vivo imaging. *J. Phys. Chem. A*, 107, 3443–3449.
10. Link, S., El-Sayed, M. A. (2000). Shape and size dependence of radiative, nonradiative and photothermal properties of gold nanocrystals. *Int. Rev. Phys. Chem.*, 19, 409–453.
11. Kreibig, U., Vollmer, M., Toennies, J. P. (1995). *Optical Properties of Metal Clusters*. Springer-Verlag, Berlin.
12. Gersten, J., Nitzan, A. (1981). Spectroscopic properties of molecules interacting with small dielectric particles. *J. Chem. Phys.*, 75, 1139–1152.
13. Chew, H. (1987). Transition rates of atoms near spherical surfaces. *J. Chem. Phys.*, 87, 1355–1360.
14. Philpott, M. R. (1975). Effect of surface plasmons on transitions in molecules. *J. Chem. Phys.*, 62(5), 1812–1817.
15. Chance, R. R., Prock, A., Silbey, R. (1978). Molecular fluorescence and energy transfer near interfaces. *Adv. Chem. Phys.*, 37, 1–65.
16. Weitz, D. A., Garoff, S., Gersten, J. I., Nitzan, A. (1983). The enhancement of Raman scattering, resonance Raman scattering and fluorescence from molecules absorbed on a rough silver surface. *J. Chem. Phys.*, 78(9), 5324–5338.
17. Geddes, C. D., Parfenov, A., Lakowicz, J. R. (2003). Photodeposition of silver can result in metal-enhanced fluorescence. *Appl. Spectrosc.*, 57(5), 526–531.

18. Geddes, C. D., Parfenov, A., Roll, D., Fang, J., Lakowicz, J. R. (2003). Electrochemical and laser deposition of silver for use in metal-enhanced fluorescence. *Langmuir, 19*(15), 6236–6241.
19. Geddes, C. D., Parfenov, A., Roll, D., Gryczynski, I., Malicka, J., Lakowicz, J. R. (2003). Silver fractal-like structures for metal-enhanced fluorescence: Enhanced fluorescence intensities and increased probe photostabilities. *J. Fluorescence, 13*(3), 267–276.
20. Parfenov, A., Gryczynski, I., Malicka, J., Geddes, C. D., Lakowicz, J. R. (2003). Enhanced fluorescence from fluorophores on fractal silver surfaces. *J. Phys. Chem. B, 107*(34), 8829–8833.
21. Michaels, A. M., Jiang, J., Brus, L. (2000). Ag nanocrystal junctions as the site for surface-enhanced Raman scattering of single rhodamine 6G molecules. *J. Phys. Chem. B, 104*, 11965–11971.
22. Michaels, A. M., Nirmal, M., Brus, L. E. (1999). Surface enhanced Raman spectroscopy of individual rhodamine 6G molecules on large Ag nanocrystals. *J. Am. Chem. Soc., 121*, 9932–9939.
23. Brown, P. O., Botstein, D. (1999). Exploring the new world of the genome with DNA microarrays. *Nat. Genet. Suppl., 21*, 33–37.
24. Schena, M., Heller, R. A., Theriault, T. P., Konrad, K., Lachenmeier, E., Davis, R. W. (1998). Microarrays: Biotechnology's discovery platform for functional genomics. *TIBTECH, 16*, 301–306.
25. Komurian-Pradel, F., Paranhos-Bacala, G., Sodoyer, M., Chevallier, P., Mandrand, B., Lotteau, V., Andre, P. (2001). Quantitation of HCV RNA using real-time PCR and fluorimetry. *J. Virol. Methods, 95*, 111–119.
26. Walker, N. J. (2002). A technique whose time has come. *Science, 296*, 557–559.
27. Difilippantonio, M. J., Ried, T. (2003). Technicolor genome analysis. In J. R. Lakowicz (Ed.), *Topics in Fluorescence Spectroscopy*, Vol. 7: *DNA Technology*. Kluwer Academic/Plenum, New York, pp. 291–316.
28. Soper, S. A., Nutter, H. L., Keller, R. A., Davis, L. M., Shera, E. B. (1993). The photophysical constants of several fluorescent dyes pertaining to ultrasensitive fluorescence spectroscopy. *Photochem. Photobiol., 57*, 972–977.
29. Amrbose, W. P., Goodwin, P. M., Jett, J. H., VanOrden, A., Werner, J. H., Keller, R. A. (1999). Single molecule fluorescence spectroscopy at ambient temperature. *Chem. Rev., 99*, 2929–2956.
30. Malicka, J., Gryczynski, I., Lakowicz, J. R. (2003). DNA hybridization assays using metal-enhanced fluorescence. *Biochem. Biophys. Res. Commun., 306*, 213–218.
31. Szmacinski, H., Lakowicz, J. R. (1994). Lifetime-based sensing. In J. R. Lakowicz (Ed.), *Topics in Fluorescence Spectroscopy*. Plenum, New York, pp. 295–334.
32. Lakowicz, J. R., Malicka, J., Gryczynski, I. (2003). Silver particles enhance emission of fluorescent DNA oligomers. *BioTechniques, 34*, 62–68.
33. Rye, H., Yue, S., Wemmer, D., Quesada, M. A., Haugland, R. P., Mathies, R. A., Glazer, A. N. (1992). Stable fluorescent complexes of double-stranded DNA with bis-intercalating asymmetric cyanine dyes: Properties of applications. *Nucleic Acids Res., 20*, 2803–2812.

34. Benson, S. C., Mathies, R. A., Glazer, A. N. (1993). Heterodimeric DNA binding dyes designed for energy transfer: Stability and applications of the DNA complexes. *Nucleic Acids Res.*, *21*, 5720–5726.
35. Benson, S. C., Zeng, Z., Glazer, A. N. (1995). Fluorescence energy transfer cyanine heterodimers with high affinity for double-stranded DNA. *Anal. Biochem.*, *231*, 247–255.
36. Lakowicz, J. R., Malicka, J., Gryczynski, I. (2003). Increased intensities of YOYO-1-labeled DNA oligomers near silver particles. *Photochem. Photobiol.*, *77*(6), 604–608.
37. Forster, Th. (1948). Intermolecular energy migration and fluorescence. *Ann. Phys.*, *2*, 55–75.
38. Lakowicz, J. R., Malicka, J., D'Auria, S., Gryczynski, I. (2003). Release of the self-quenching of fluorescence near silver metallic surfaces. *Anal. Biochem.*, *320*, 13–20.
39. Lakowicz, J. R., Gryczynski, I., Malicka, J., Gryczynski, Z., Geddes, C. D. (2002). Enhanced and localized multiphoton excited fluorescence near metallic silver islands: Metallic islands can increase probe photostability. *J. Fluorescence*, *12*, 299–302.
40. Malicka, J., Gryczynski, I., Fang, J., Kusba, J., Lakowicz, J. R. (2003). Increased resonance energy transfer between fluorophores bound to DNA in proximity to metallic silver particles. *Anal. Biochem.*, *315*, 160–169.
41. Hua, X. M., Gersten, J. I., Nitzan, A. (1985). Theory of energy transfer between molecules near solid state particles. *J. Chem. Phys.*, *83*, 3650–3659.
42. Gersten, J. I., Nitzan, A. (1984). Accelerated energy transfer between molecules near a solid particle. *Chem. Phys. Lett.*, *104*(1), 31–37.
43. Gryczynski, I., Malicka, J., Gryczynski, Z., Lakowicz, J. R. (2004). Radiative decay engineering 4. Experimental studies of surface plasmon-coupled directional emission. *Anal. Biochem.*, *324*, 170–182.
44. Liebermann, T., Knoll, W. (2000). Surface plasmon field-enhanced fluorescence spectroscopy. *Colloid Surf.*, *171*, 115–130.
45. Lakowicz, J. R. (2004). Radiative decay engineering 3: Surface plasmon-coupled directional emission. *Anal. Biochem.*, *324*, 153–169.
46. Geddes, C. D., Gryczynski, I., Malicka, J., Gryczynski, Z., Lakowicz, J. R. (2003). Metal-enhanced fluorescence: Potential applicatiosn in HTS. *Comb. Chem. High Throughput Screen.*, *6*, 109–117.

15

METHODS FOR THE DESIGN AND ANALYSIS OF REPLICATE-EXPERIMENT STUDIES TO ESTABLISH ASSAY REPRODUCIBILITY AND THE EQUIVALENCE OF TWO POTENCY ASSAYS

BRIAN J. EASTWOOD, AMY K. CHESTERFIELD, MARY C. WOLFF, AND CHRISTIAN C. FELDER

Eli Lilly and Company
Indianapolis, Indiana

15.1	INTRODUCTION	668
15.2	BIOMOLECULAR SCREENING ASSAY CONCENTRATION–RESPONSE MODELS AND OUTCOMES	669
	Affinity Models and Parameter Estimates	669
	Muscarinic M_1 Receptor $GTP\gamma^{35}S$ Assay	670
15.3	ESTIMATING POTENCY PRECISION	671
	Replicate-Experiment Protocol	671
	Statistical Model for Replicate-Experiment Protocol	672
	Replicate-Experiment Outcomes	672
	Model Parameter and Outcome Estimates	673
	Acceptance Criteria	674
	Example Data	674
	Methods for Interlaboratory Comparisons	676

Drug Discovery Handbook, by Shayne Cox Gad
Copyright © 2005 by John Wiley & Sons, Inc.

15.4 STATISTICAL PROPERTIES OF ESTIMATES AND ACCEPTANCE CRITERIA	678
Data-Generating Model	678
Properties of Estimated Outcomes	679
Properties of Acceptance Criteria	683
References	687

15.1 INTRODUCTION

In pharmaceutical drug discovery molecules are evaluated in biochemical assays to ascertain the potency of the molecule acting on the proposed biological target. Initially molecules are typically run at a single concentration in a high-throughput screen. Then molecules exhibiting substantial activity are run again in a concentration–response assay to estimate the potency of the molecule for the biological target. From this leads are identified, and then new chemical entities are developed to improve the potency against the biological target and further enhance other properties such as pharmacokinetics and margin of safety. These new chemical entities are often assessed directly in the concentration–response assay to develop and refine the structure–activity relationship.

Statistical parameters estimated in validation studies to establish the reliability of single-concentration assays, such as the Signal Window and Z'-Factor, are well established [1–3]. Eastwood et al. [4] showed that these statistics do not ensure that potency estimates from concentration–response assays will be sufficiently reliable for structure–activity relationship use, and proposed the minimum significant ratio (MSR) as a statistical parameter to estimate the precision of potency estimates from concentration–response assays. They also proposed a standard experimental protocol to evaluate the precision of estimates derived from concentration response assays. In this chapter the statistical models describing the data generated from these protocols are developed and the statistical properties of the outcomes described in Eastwood et al. [4] are examined. This chapter is organized as follows: Section 15.2 contains a description of a model commonly used to describe biochemical concentration–response assays in pharmaceutical drug discovery. Concentration–response data from a human M_1 muscarinic [^{35}S]guanosine-5'o-(3-thio)triphosphate (GTPγ^{35}S) assay is described and fit by this model. Section 15.3 contains a description of the replicate-experiment protocols proposed in Eastwood et al. [4], a statistical model for data derived from these experiments, and acceptance criteria for concentration–response assays. The methods are applied to the GTPγ^{35}S assay to estimate the precision of the potency estimates. Methods are proposed to extend the replicate-experiment protocol to cover transferring an assay to a new laboratory. Section 15.4 contains statistical distribution theory describing the estimators and outcomes and simulation studies further describing the properties of the parameters.

15.2 BIOMOLECULAR SCREENING ASSAY CONCENTRATION–RESPONSE MODELS AND OUTCOMES

Affinity Models and Parameter Estimates

See Kenakin [5] for a general discussion of affinity models. For a concentration–response assay of a stimulatory, agonist, or activation nature a mechanistic model commonly used to fit the data is the four-parameter logistic (4PL) model:

$$z(c) = b + \frac{t-b}{1+(\text{EC}_{50}/c)^s} \tag{15.1}$$

where c is the concentration of the compound, z is the assay response, b and s are the bottom and slope (nuisance) parameters, EC_{50} is the concentration giving a half-maximal response, and t is the top or relative efficacy parameter. The EC_{50} is the affinity, or potency, estimate for such data. Usually the data have been normalized to separately measured responses representing no stimulus and maximal stimulus of a reference agonist. There are slight differences in the models for receptor binding and inhibitory assays. In receptor binding assays $t = 100$, $b = 0$, and $s = -1$ are theoretical values when the data have been normalized in this manner, and $s = -1$ holds for responses that have not been normalized as well. For inhibitory assays normalized in the manner described above, $t = 100$, and for some types of competitive inhibition, $b = 0$. For both receptor binding and inhibition assays the potency (half-maximal response parameter) is denoted as the IC_{50}.

There are many different forms of the 4PL equation and a commonly used form in discovery research, notably by the software package Graphpad/Prism [6], is

$$z(c) = b + \frac{t-b}{1+10\wedge[(\log \text{EC}_{50} - \log(c))s]} \tag{15.2}$$

where $\log(c)$ is the base 10 logarithm of the concentration and $\wedge$ means exponent.

The EC_{50} parameter estimate is right-skewed, whereas the $\log \text{EC}_{50}$ parameter estimate is nearly symmetric [7]. Both EC_{50} and $\log \text{EC}_{50}$ estimates are approximately normally distributed [8], but since the $\log \text{EC}_{50}$ estimate is less skewed, it will be better approximated by a normal distribution in small sample sizes [9].

Muscarinic M_1 Receptor $GTP\gamma^{35}S$ Assay

A replicate-experiment protocol was carried out for a human M_1 muscarinic receptor $GTP\gamma^{35}S$ binding assay. Details on the assay protocol are as follows: $GTP\gamma^{35}S$ binding was determined using the antibody capture scintillation proximity method [10]. Briefly, 100 µL homogenized membrane preparations from CHO cells over expressing human M_1 muscarinic receptors (20 to 4 fmol/well; Perkin-Elmer Life Sciences, Boston, MA) were incubated for 30 min with 50 µL compound (11-point half-log serial dilutions starting at 100 µM concentration) in a 96-well format. Fifty microliters of $GTP\gamma^{35}S$ (500 pM final concentration; Perkin-Elmer Life Sciences, Boston, MA) were added per well and incubated for 30 min. Following incubation, membranes were solubilized by addition of 20 µL 3 percent Nonidet P40 (0.27 percent final concentration; Roche, Indianapolis, IN) per well and incubated for 30 min. Twenty microliters rabbit polyclonal antibodies to $G\alpha_{q/11}$ (1/400 final concentration; Covance, Denver, PA) were added per well and incubated for 60 min. Anti-rabbit SPA PVT antibody binding beads (1.25 mg/well final concentration; Amersham Life Sciences, Piscataway, NJ) were suspended in 20-mL assay buffer and 50 µL were added per well and incubated for 3 h. Assay plates were then centrifuged for 10 min at $1000g$, and radioactivity was counted for 1 min per well on a Wallac Trilux MicroBeta plate counter (Perkin-Elmer Life Sciences, Boston, MA). All incubations were performed at room temperature in a 20 mM N-(2-hydroxyethyl)piperazine-N′-(2-ethanesulfonic acid) (HEPES), 100 mM NaCl, 5 mM $MgCl_2$, pH 7.4 assay buffer.

All data were analyzed using GraphPad/Prism [11] to determine concentration–response curves by fitting data using a 4PL model. Responses for test compounds were normalized to 100 µM oxotremorine-M.

Figure 15.1 contains concentration–response data and least-squares fits of the 4PL equation for acetylcholine (Fig. 15.1a) and carbachol (Fig. 15.1b). The $\log EC_{50}$ estimates can be transformed to linear estimates by the delta method [12]. For this model the formulas are

$$\widehat{EC_{50}} = 10\wedge(\widehat{\log EC_{50}}) \quad \text{and} \quad SE(\widehat{EC_{50}}) = SE(\widehat{\log EC_{50}})\widehat{EC_{50}}\ln(10) \quad (15.3)$$

for the standard error (SE). Using these formulas gives EC_{50} estimates and standard errors for acetylcholine and carbachol of 276.7 (40.1) nM and 2871 (398) nM. Note that the standard error of carbachol is nearly 10-fold higher than the standard error of acetylcholine, which is consistent with the lognormal distribution assumption of EC_{50} parameter estimates.

Based on the point estimates acetylcholine is 10.4-fold more potent than carbachol. However, the statistical variation of the assay is currently unknown, and hence it is necessary to determine the assay variation to know if this ratio of potencies is within normal variation limits in the assay or if this represents a real potency difference between the two molecules. The standard errors of the parameter estimates can be used, but, as discussed in Eastwood et al. [4], these are unreliable as the data-generating process does not satisfy the statis-

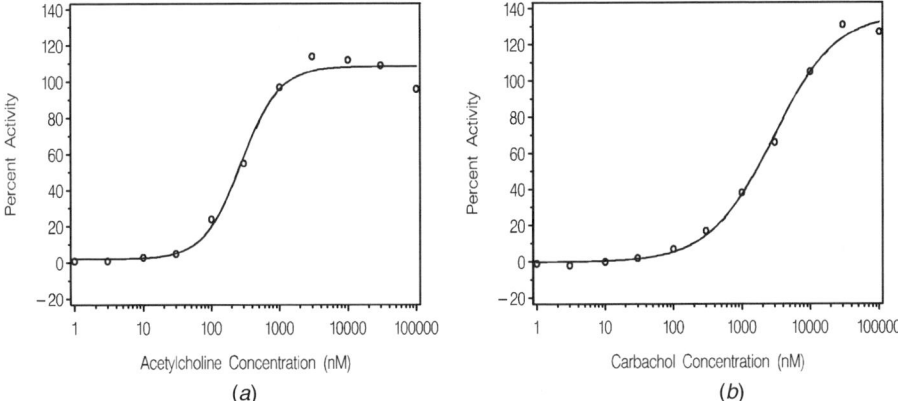

Figure 15.1 Concentration–response curves of (*a*) acetylcholine and (*b*) carbachol in the GTPγ^{35}S hM1 receptor assay. Four-parameter logistic estimates (standard errors) for acetylcholine are $b = 2.1$ (3.39), $t = 108.4$ (3.31), $s = 0.96$ (0.11), and log $EC_{50} = 2.44$ (0.063), and for carbachol are $b = -0.2$ (2.10), $t = 137.3$ (5.17), $s = 0.94$ (0.10), and log $EC_{50} = 0.46$ (0.060).

tical assumptions underlying the least-squares model, and estimates are often derived on a small number of data points (in this case 11 points are in the concentration–response curves). Eastwood et al. [4] proposed using a replicate-experiment protocol to estimate assay precision. In Sections 15.3 and 15.4 we develop the statistical model for the replicate-experiment protocol and examine the properties of the parameter estimates.

15.3 ESTIMATING POTENCY PRECISION

Replicate-Experiment Protocol

The precision and accuracy of the biological parameter estimates can not be assured, or even estimated, from validation methods such as Signal Window or Z'-factor employed to validate the single concentration assays [4]. Methods to formally examine the precision have not been standardized in drug discovery. Eastwood et al. [4] proposed a set of experiments based on well-known clinical methods described in Bland and Altman [13]. The protocol, briefly summarized from [4], is to evaluate a set of molecules in two independent runs of the assay and examine the agreement in the resulting potency estimates. Note that we assume experimental iterations require changes across days, and therefore throughout this chapter we use the terms *run* and *day* synonymously. More important than the time lapse between runs is the assumption of independence. By that we mean two complete iterations of the entire assay process including fresh preparation of buffers and other experimental materials. This protocol then is a refinement of an interobserver agreement study where the "observers" are two runs from the same study.

Methods for examining agreement between two measurements are numerous. See Carrasco and Jover [14], for a summary and classification of these methods. The methods used here are derived from those described by Bland and Altman [13]. Using the classification scheme in Carrasco and Jover [14], the techniques described here are a disaggregate approach as they evaluate each component of the measurement model.

Statistical Model for Replicate-Experiment Protocol

We first describe a statistical model for the replicate-experiment protocol of two runs of each compound within a single laboratory. Let X_t and Y_t be the log potencies obtained for run 1 and run 2, respectively, of compound t, for $t = 1, \ldots, n$. The statistical model we use is to assume $X_t = \mu_t + e_t$ and $Y_t = \mu_t + f_t$, where μ_t is the true log potency, and e_t and f_t are the measurement errors for run 1 and run 2, respectively. We assume further that $\mu_t \sim (\mu, \sigma_C^2)$, $e_t \sim N(\mu_X, \sigma^2)$, and $f_t \sim N(\mu_Y, \sigma^2)$. Note that no assumption is made about the distribution of the compound's true values other than that each compound has a possibly different potency and that the potency distribution does have a finite mean and variance.

We do assume that the measurement errors are independent, identically and normally distributed, independent across days, and that the measurement standard deviation does not vary by day. The normal distribution assumption is reasonable here as the log potency is a parameter from a nonlinear regression model and hence has an approximate normal distribution [8]. The standard deviation, σ, will typically not be equal to the standard error reported by the nonlinear regression fitting program. See [4] for a discussion of this point.

Each day may have a different mean. An alternative to the latter assumption is to assume that there is a run-day variance component, and that μ_X and μ_Y are realizations from this distribution. In the simulation work to evaluate sample size and acceptance criterion properties, we make this assumption and examine the effect of interday variation on the properties of the estimators. However, for analysis this assumption is not necessary, and instead we consider the day effects to be fixed, or alternatively, conditional on the day realizations.

Replicate-Experiment Outcomes

Eastwood et al. [4] identify three outcomes for the replicate-experiment protocol: the minimum significant ratio (MSR), the mean ratio (MR), and limits of agreement (LsA). Their derivations are as follows.

On any particular day, let LP_s and LP_t be the log potencies for two molecules s and t. Then under the same assumptions stated for the replicate-experiment protocol, the difference in log potencies is distributed as $LP_s - LP_t \sim N(\mu_s - \mu_t, 2\sigma^2)$. Therefore to test the null hypothesis $H_0: \mu_s = \mu_t$, the test statistic $T = LP_s - LP_t$ has a $N(0, 2\sigma^2)$ distribution under the null hypothesis. Thus if $Z_{1-\alpha/2}$ is the $(1 - \alpha/2)$th percentile from a normal distribution, one would reject the null hypothesis if $|T| > Z_{1-\alpha/2}\sqrt{2}\sigma$, or alternatively if the ratio of the lower potency (i.e., higher numerical value) to the higher potency estimate exceeded $10\wedge(Z_{1-\alpha/2}\sqrt{2}\sigma)$. Therefore, $10\wedge(Z_{1-\alpha/2}\sqrt{2}\sigma)$ is the minimum significant ratio for level α, or MSR_α. Taking α to be 5 percent, and rounding to one significant digit gives the formula $MSR = 10\wedge(2\sqrt{2}\sigma)$. Throughout this chapter, unless denoted otherwise, MSR is determined using $Z_{1-\alpha/2} = 2$.

The mean ratio (MR) is the average ratio across the compound set between the two days. To see this, note that

$$MR = (10\wedge\mu_X)/(10\wedge\mu_Y) = 10\wedge(\mu_X - \mu_Y)$$

and that

$$\left(\frac{1}{n}\right)\sum E[X_t - Y_t] = \mu_X - \mu_Y$$

The limits of agreement (LsA) are as defined by Bland and Altman [13]. In terms of population parameters on the log scale, they are can be written as

$$LsA_\alpha = 10\wedge[(\mu_X - \mu_Y) \pm Z_{1-\alpha/2}\sigma_d]$$

where $\sigma_d = \sqrt{2}\sigma$ is the standard deviation of the paired difference. Translating to the linear scale the results are

$$LsA_\alpha \triangleq (LsA_{L,\alpha}, LsA_{U,\alpha}) = (MR/MSR_\alpha, MR \times MSR_\alpha)$$

A key consideration in the selection of these outcomes is that all are independent of $\{\mu_t: t = 1, \ldots, n\}$, that is, they do not depend on the actual potency of compounds chosen. In practice, it would be appropriate to choose a set with a large range in potencies from a generalizability perspective, but theoretically the outcomes are independent of the potency distribution.

Model Parameter and Outcome Estimates

Let (x_t, y_t), $t = 1, \ldots, n$ be the observed log potencies for the n compounds, and let d_t and m_t be difference and mean log potency across the two runs for the n compounds. Bland and Altman [13] recommend plotting m_t versus d_t. However, we prefer to plot $r_t = 10\wedge d_t$ versus $gm_t = 10\wedge m_t$ with each plotted on a logarithmic scale as potency ratios and values are more familiar to our biologists and chemists than log potencies. Each run's average error, μ_x and μ_y, is not individually estimable, but their difference, $\mu_x - \mu_y$, is estimated by $\bar{x} - \bar{y}$.

The average potency ratio, $\bar{r} = \bar{x}/\bar{y} = MR$, is plotted as a horizontal line on the ratio-geometric mean scatter plot.

The data are naturally paired, and so under the null hypothesis $H_0: \mu_X = \mu_Y$ $\sqrt{n/2}\log(MR)/\sigma$ has a t distribution with $n-1$ degrees of freedom. Let s_d be the sample standard deviation of d_t the difference in log potencies. Then $\hat{\sigma} = s_d/\sqrt{2}$ is the usual unbiased estimate of σ. If $t_{0.975,n-1}$ is the 97.5th percentile from a distribution with $n-1$ degrees of freedom and if $M = 10\wedge(t_{0.975,n-1}\, s_d/\sqrt{n})$, then a 95 percent confidence interval for the MR is given by $(MR_L, MR_U) = (\widehat{MR}/M, \widehat{MR} \times M)$. The null hypothesis is rejected if the confidence interval does not contain the value 1.0.

The mean ratio limits do not describe a region that contains the data but instead a region that contains the true mean ratio MR. As Bland and Altman [13] show, this interval is not relevant for describing individual agreement, and they recommend plotting ± 2 standard deviations as the limits of agreement, which corresponds approximately to $\alpha = 0.05$. Thus the estimated MSR is $\widehat{MSR} = 10\wedge(2s_d)$, and the sample limits of agreement are $\widehat{LsA} \triangleq (\widehat{LsA_L}, \widehat{LsA_U}) = (\widehat{MR}/\widehat{MSR}, \widehat{MR} \times \widehat{MSR})$.

Note that $\hat{\mu}_t = (x_t + y_t)/2$, $t = 1, \ldots, n$ and $\hat{\mu} = (\hat{\mu}_1 + \ldots \hat{\mu}_n)/n$ are the compound estimates and overall compound mean. However, these parameters are not of primary interest in the replicate-experiment protocol. In fact, unless the sample of molecules tested is a random sample from the population, $\hat{\mu}$ will be a biased estimator of μ.

Acceptance Criteria

See Eastwood et al. [4] for a discussion of MSR requirements from pharmacological validation and medicinal chemistry structure–activity relationship perspectives.

The MSR can be directly estimated from the replicate-experiment protocol. Sufficient evidence pertaining to the across-run variation can only be obtained over several runs of the assay, which is beyond the scope of most preproduction evaluation efforts and not discussed here. In the replicate-experiment protocol, across-run variation impacts the magnitude of the MR, and therefore of the LsA, and so preliminary decisions can be made based on either the MR or LsA outcomes. Simulation evidence presented in Section 15.4 suggests that a sample size of 20 to 30 compounds works reasonably well for criteria designed to reject assays if the assay variation, taking into account across-run variation, exceeds a threefold potency ratio.

Example Data

A replicate-experiment protocol was conducted using 24 compounds for the muscarinic human M_1 GTPγ^{35}S assay. Each compound was run on two separate days. Tables 15.1 and 15.2 contain the complete 4PL model parameter

TABLE 15.1 Four-Parameter Logistic Regression Parameter Estimates for Run 1 of GTPγ³⁵S M1 Agonist Assay

Compound	b Est (SE)	t Est (SE)	s Est (SE)	log EC$_{50}$ Est (SE)
C1	−5.3 (4.29)	110.6 (3.79)	1.19 (0.23)	2.30 (0.077)
C2	2.1 (3.39)	108.4 (3.31)	1.57 (0.32)	2.44 (0.063)
C3	−4.7 (1.24)	80.0 (3.13)	0.96 (0.11)	3.43 (0.074)
C4	−0.2 (2.10)	137.3 (5.17)	0.94 (0.10)	3.46 (0.060)
C5	0.3 (1.11)	91.2 (2.34)	0.87 (0.07)	3.28 (0.045)
C6	2.5 (1.83)	38.2 (0.82)	1.18 (0.21)	1.53 (0.107)
C7	−3.3 (2.68)	74.0 (3.24)	0.75 (0.12)	2.68 (0.095)
C8	−4.9 (2.26)	94.1 (10.98)	0.58 (0.10)	3.38 (0.188)
C9	−0.6 (2.99)	106.8 (2.56)	1.15 (0.16)	2.27 (0.058)
C10	−2.6 (1.04)	88.2 (2.09)	1.05 (0.09)	3.41 (0.035)
C11	−2.9 (0.53)	58.6 (0.86)	1.16 (0.07)	3.14 (0.024)
C12	0.7 (1.41)	114.0 (2.41)	1.03 (0.08)	3.15 (0.038)
C13	1.3 (3.23)	62.9 (2.02)	1.30 (0.30)	1.81 (0.090)
C14	3.0 (1.01)	84.1 (1.93)	1.38 (0.14)	2.42 (0.036)
C15	1.3 (1.27)	98.9 (2.34)	1.09 (0.10)	2.26 (0.040)
C16	4.5 (3.88)	51.7 (4.61)	1.18 (0.56)	1.74 (0.203)
C17	−7.0 (4.21)	93.1 (3.67)	0.84 (0.15)	2.34 (0.096)
C18	−1.6 (1.11)	25.5 (0.79)	0.99 (0.17)	2.07 (0.082)
C19	−1.5 (0.78)	32.3 (1.06)	1.51 (0.25)	3.01 (0.054)
C20	0.3 (1.46)	22.1 (2.02)	1.28 (0.56)	2.98 (0.172)
C21	0.9 (1.61)	37.3 (1.83)	0.95 (0.22)	2.64 (0.113)
C22	1.0 (1.44)	29.5 (1.24)	1.05 (0.24)	2.13 (0.104)
C23	0.1 (2.42)	58.9 (1.92)	1.19 (0.24)	2.16 (0.081)
C24	−1.1 (0.92)	30.1 (1.27)	0.92 (0.15)	2.87 (0.084)

estimates for run 1 and run 2, respectively, calculated using Eq. (15.2). Table 15.3 contains the EC$_{50}$ parameter estimates and standard errors for both runs, calculated using the delta method of Eq. (15.3). Acetylcholine is compound C2 and carbachol is compound C4. Figure 15.2 is the plot of the potency ratios versus the geometric mean potency. There does not appear to be any systematic trend between agreement and geometric mean potency, and all ratios appear inside the LsA estimates, which are 0.68 to 1.55. The MSR estimate is 1.51. There appears to be little evidence of interrun variation as the MR estimate is 1.02 with 95 percent confidence interval ±10 percent ($M = 1.1$).

To return to the issue at the end of Section 15.2, the potency ratio of carbachol to acetylcholine was 10.4 in run 1, 9.51 in run 2, and both are well above the MSR estimate of 1.51. This analysis confirms that in this assay there is a real potency difference between acetylcholine and carbachol.

TABLE 15.2 Four-Parameter Logistic Regression Parameter Estimates for Run 2 of GTPγ^{35}S M1 Agonist Assay

Compound	b Est (SE)	t Est (SE)	s Est (SE)	log EC$_{50}$ Est (SE)
C1	−2.7 (2.33)	102.5 (2.01)	1.15 (0.13)	2.28 (0.046)
C2	1.7 (3.35)	98.5 (3.38)	1.43 (0.30)	2.49 (0.073)
C3	1.9 (0.91)	78.8 (2.05)	0.97 (0.08)	3.46 (0.040)
C4	4.0 (0.53)	116.7 (1.29)	0.97 (0.03)	3.47 (0.018)
C5	4.5 (0.91)	77.8 (1.51)	1.06 (0.09)	3.13 (0.037)
C6	8.7 (0.90)	35.2 (0.48)	1.51 (0.23)	1.48 (0.048)
C7	4.8 (1.63)	67.7 (1.71)	1.17 (0.17)	2.55 (0.060)
C8	5.1 (2.53)	84.9 (6.86)	0.68 (0.14)	3.39 (0.152)
C9	−2.5 (2.60)	104.4 (2.10)	1.25 (0.15)	2.34 (0.046)
C10	0.7 (0.95)	92.5 (1.84)	1.17 (0.09)	3.34 (0.032)
C11	0.4 (1.60)	66.4 (2.78)	1.17 (0.20)	3.23 (0.071)
C12	3.3 (1.86)	111.1 (2.46)	1.15 (0.12)	3.05 (0.042)
C13	3.9 (1.59)	71.6 (0.95)	0.99 (0.09)	1.88 (0.043)
C14	5.2 (1.56)	90.8 (2.19)	1.74 (0.23)	2.43 (0.036)
C15	2.1 (1.32)	99.2 (1.76)	1.66 (0.16)	2.37 (0.027)
C16	1.5 (1.78)	58.2 (1.14)	1.17 (0.14)	1.55 (0.049)
C17	−3.1 (1.68)	103.2 (1.38)	1.06 (0.08)	2.23 (0.033)
C18	−1.2 (1.80)	35.3 (1.41)	0.85 (0.17)	2.22 (0.107)
C19	0.9 (0.87)	43.5 (1.30)	1.31 (0.19)	3.10 (0.053)
C20	4.0 (1.82)	30.2 (2.39)	2.11 (1.45)	3.00 (0.133)
C21	1.7 (0.98)	48.0 (1.09)	1.22 (0.15)	2.63 (0.050)
C22	2.9 (1.45)	35.6 (1.13)	1.24 (0.27)	2.13 (0.086)
C23	2.6 (1.84)	70.4 (1.67)	1.01 (0.12)	2.17 (0.057)
C24	−0.2 (3.00)	44.6 (3.92)	0.58 (0.16)	2.72 (0.191)

Methods for Interlaboratory Comparisons

Often the follow-up work is conducted in a different laboratory than the screening laboratory. In those instances it is necessary to validate that the protocols in the two labs yield similar results. To examine this property, the replicate-experiment protocol is modified so that the screening laboratory run evaluates the compound set a single time, and the confirmation laboratory conducts the replicate-experiment protocol as outlined above. In addition to the comparisons of the two runs in the confirmation laboratory, the screening laboratory's run will be compared to the first run of the confirmation laboratory.

This is a "minimalist" form of an interlaboratory comparison study and is based on the assumption that the screening laboratory's assay has previously been evaluated and in production, whereas the confirmation laboratory is just initiating the evaluation process. In situations that are different from this scenario, more elaborate evaluation protocols may be possible and/or

TABLE 15.3 EC$_{50}$ Estimates for Runs 1 and 2 of GTPγ^{35}S M1 Agonist Assay

Compound	Run 1 EC$_{50}$	(SE)	Run 2 EC$_{50}$	(SE)
C1	201.4	(35.8)	191.4	(20.4)
C2	276.7	(40.1)	308.3	(51.6)
C3	2697.7	(459.4)	2904.0	(268.3)
C4	2870.8	(397.9)	2930.9	(122.2)
C5	1923.1	(198.4)	1345.9	(114.8)
C6	33.8	(8.4)	30.4	(3.3)
C7	481.9	(105.1)	354.0	(48.9)
C8	2404.4	(1040.3)	2426.6	(849.9)
C9	186.2	(24.9)	216.8	(23.0)
C10	2546.8	(207.5)	2208.0	(163.9)
C11	1383.6	(78.0)	1690.4	(276.4)
C12	1412.5	(124.1)	1127.2	(110.0)
C13	64.4	(13.4)	76.2	(7.5)
C14	263.6	(21.7)	269.2	(22.3)
C15	182.0	(17.0)	232.8	(14.5)
C16	54.3	(25.4)	35.5	(4.0)
C17	218.3	(48.2)	171.4	(13.1)
C18	116.4	(21.9)	164.8	(40.5)
C19	1025.7	(127.6)	1253.1	(153.6)
C20	948.4	(374.5)	1006.9	(308.4)
C21	434.5	(113.2)	427.6	(49.1)
C22	134.6	(32.3)	136.1	(26.9)
C23	144.5	(27.1)	146.6	(19.2)
C24	737.9	(142.7)	528.4	(232.7)

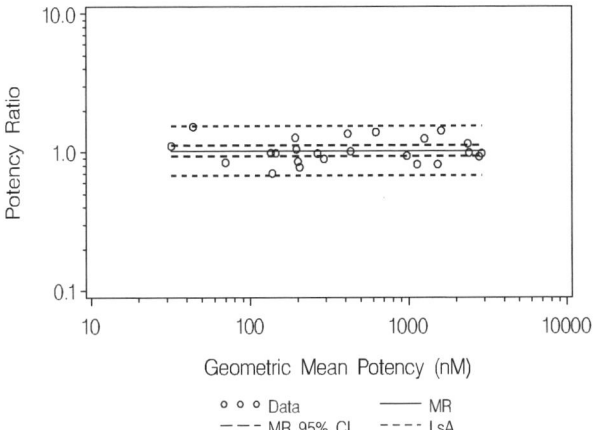

Figure 15.2 Potency ratio versus geometric mean potency results from the replicate-experiment study for the GTPγ^{35}S hM1 agonist assay. $\widehat{MSR} = 1.51$, $\widehat{MR} = 1.02$ [95% confidence interval = 0.94–1.12, $p = 0.57$], and $\widehat{LsA} = (0.68, 1.55)$.

appropriate. For example, if the monitoring history of the production laboratory is very extensive, then an extension to these methods has been proposed to compare assay variations across laboratories [15].

If we change the notation so that X_t and Y_t are the log potency estimates from laboratories 1 and 2 for compound t, $t = 1, \ldots, n$, then a model to describe this situation would be $X_t = \mu_t + l_1 + e_t$ and $Y_t = \mu_t + l_2 + f_t$. Assumptions about μ_t, e_t, and f_t are as before. The parameters l_1 and l_2 are fixed-effect parameters representing the difference in average log potency as measured by laboratories 1 and 2 versus "true" log potency of the molecule. An assumption we are making is that the laboratory bias is constant across all molecules.

Note that "interlaboratory" comparison methods may also be used to compare two different protocols within the same laboratory. These intralab comparisons could include personnel differences, equipment differences, cell line changes, or assay differences based on a cloned cell line versus a native tissue assay.

Under this model the minimum significant ratio, $MRS_\alpha = 10\wedge(Z_{1-\alpha/2}\sqrt{2}\sigma)$ is defined as before. However, the mean ratio is now $MR = 10\wedge(l_1 - l_2 + \mu_X - \mu_Y)$. Note that none of the four parameters, l_1, l_2, μ_X, and μ_Y, is individually estimable, neither is l_1-l_2 nor $\mu_X-\mu_Y$. The limits of agreement are $LsA_\alpha = (MR/MSR_\alpha, MR\times MSR_\alpha)$ as before.

The outcomes are estimated as per the replicate-experiment protocol.

15.4 STATISTICAL PROPERTIES OF ESTIMATES AND ACCEPTANCE CRITERIA

Data-Generating Model

The model of the data-generating process is a slight generalization of that developed in Section 15.3. We now assume that on any run r the log potency of molecule t is given by $Z_{rt} = \mu_t + \rho_r + \varepsilon_{rt}$, for $r = 1, 2$ and $t = 1, \ldots, n$, where μ_t is the true log potency of compound t, ρ_r is a random run (day) shift of all molecules, and ε_{rt} is the random measurement error. We assume $\varepsilon_{rt} \sim N(0,\sigma^2)$, $\rho_r \sim N(0,\sigma_R^2)$, and that ρ_r and ε_{rt} are statistically independent. The model of Section 15.3 is just a special case of this model, where we let μ_X and μ_Y be two specific runs and consider them fixed throughout the replicate experiment. Then $e_t = \mu_X + \varepsilon_{1t}$ and $f_t = \mu_Y + \varepsilon_{2t}$.

We make this generalization because run effects are not controllable and molecules are usually compared across runs instead of within runs. A statistically significant difference between potency estimates of two molecules then holds if the ratio of the two potencies (weaker to stronger) exceeds the minimum significant ratio that takes into account both within- and across-run variation. Using the same arguments as in Section 15.3, the critical value is $OvrMSR_\alpha = 10\wedge[Z_{1-\alpha/2}\sqrt{2(\sigma_R^2 + \sigma^2)}]$, and as with the MSR, $Z_{1-\alpha/2} = 2$ unless explicitly defined otherwise. In this section we will examine the statistical properties of $\widehat{MSR}$, $\widehat{MR}$, and $\widehat{LsA}$ for various values of OvrMSR, MSR, and n. Note that because $\sigma_R^2 \geq 0$ OvrMSR $\geq$ MSR, which imposes limits on the simulation boundaries.

Properties of Estimated Outcomes

In this section we look at the statistical properties of $\widehat{MSR}$, $\widehat{MR}$, and $\widehat{LsA}$ under the assumption that the log potency estimates are normally distributed. We will examine their properties through both probability distribution theory and Monte Carlo simulation. We will examine cases for MSR varying from 1 to 4, and for OvrMSR varying from 1 to 8, subject to the constraint that OvrMSR $\geq$ MSR (which must hold by definition). Since none of the estimates depends upon the actual potencies of the molecules, the log potencies are left unspecified. The sample size is varied from 10 to 40 compounds. All simulation data sets are generated and estimators calculated within a data step using Version 8.2 of the SAS system [16]. Plots are produced by Proc Plot Version 8 of SAS/Graph [17]. Simulations are conducted using 5000 Monte Carlo repetitions.

Under the data-generating model of the previous section, if we condition on μ_X and μ_Y, then $T = \sqrt{n}(\overline{X} - \overline{Y})/s_d$ has a noncentral $t_{n-1,\delta}$ distribution with noncentrality parameter $\delta = \sqrt{n}(\mu_X - \mu_Y)/(2\sigma)$. As mentioned in Section 15.3, under the null hypothesis, $H_0: \mu_X = \mu_Y$, T has a central t distribution and can be used to test that null hypothesis. However, under the data-generating model from the beginning of Section 15.4, this hypothesis is false unless $\sigma_R^2 = 0$, and therefore this hypothesis is of little interest. In fact, testing it may lead to counterintuitive conclusions, as the noncentrality parameter δ is a decreasing function of the assay measurement error σ, hence of the MSR. Consequently, the more precise an assay is within each run, the more likely one will conclude there is a statistically significant difference between two runs of the assay.

The phrase "conditioning on μ_X and μ_Y" means that we consider the two runs as fixed and test a hypothesis about those two particular runs. But, if we consider the two days as just representative runs of the assay production process, then the two run day means are themselves random variables and not fixed values. Under this "unconditional" analysis, the statistic T does not have a t distribution and does not have any well-characterized probability distribution. Therefore to examine this distribution we used Monte Carlo simulation.

Figure 15.3 shows the effect of OvrMSR, MSR, and n on the distribution of $\widehat{MR}$. Figure 15.3a shows box plots of the distribution of MR for different combinations of OvrMSR and MSR, using a sample size of 25 compounds. For all combinations the median $\widehat{MR}$ is 1.0, but as OvrMSR increases the variability and skewness of $\widehat{MR}$ increase. However, for a fixed

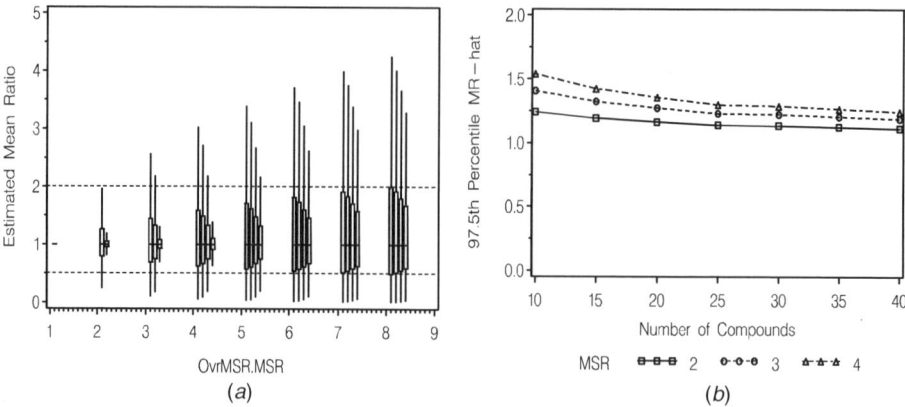

Figure 15.3 (*a*) Box plot of distribution of $\widehat{MR}$ versus OvrMSR.MSR (i.e., whole number indicates the OvrMSR, decimal indicates the MSR) using 25 compounds. (*b*) 97.5th percentile of $\widehat{MR}$ versus number of compounds by MSR when OvrMSR = MSR.

value of OvrMSR the variability of $\widehat{MR}$ decreases as MSR increases, and the distribution becomes more symmetric. These changes occur because $SE(\bar{X} - \bar{Y}) = \sqrt{2}(\sigma_R^2 + \sigma^2/n)$, and hence holding OvrMSR fixed while increasing MSR effectively lowers σ_R^2 and increases σ^2. Figure 15.3*b* shows the insensitivity of the distribution of $\widehat{MR}$ to the sample size. When $\sigma_R^2 = 0$ (i.e., OvrMSR = MSR), sample size has its greatest effect, specifically a slight decrease in the variability of $\widehat{MR}$ as the sample size increases. The effect is even smaller at positive values of across-run variation.

Figure 15.4 shows the effect of OvrMSR, MSR, and *n* on the precision of $\widehat{MR}$. Since a 95 percent confidence interval for MR is obtained by multiplying and dividing $\widehat{MR}$ by *M*, where *M* is as defined in Section 15.3, we examine precision by examining the distribution of *M*. Figure 15.4*a* shows the distribution of *M* versus both OvrMSR and MSR for a fixed sample size of 25. Note that the confidence interval precision does not depend upon OvrMSR, but the precision does depend upon the actual value of MSR. The lack of a relationship between *M* and OvrMSR is predicted from the distributional theory, as is the increasing relationship between *M* and MSR. Figure 15.4*b* shows the effect of sample size on confidence interval precision. With less than a 5 percent error rate, if MSR ≤ 4, then a sample size of 25 compounds is sufficient to ensure that $\widehat{MR}$ is within 50 percent of the true value. Note that the 97.5th percentile of *M* is an increasing function of MSR, not because the median value is increasing but instead because the variability and skewness of the distribution are increasing as MSR increases, as shown in Figure 15.4*a*.

Figure 15.5 shows the effect of OvrMSR, MSR, and *n* on the power to reject the null hypothesis $H_0: \mu_X = \mu_Y$. Figure 15.5*a* shows for a fixed sample size of 25 compounds that power increases as OvrMSR increases provided the MSR

STATISTICAL PROPERTIES OF ESTIMATES AND ACCEPTANCE CRITERIA 681

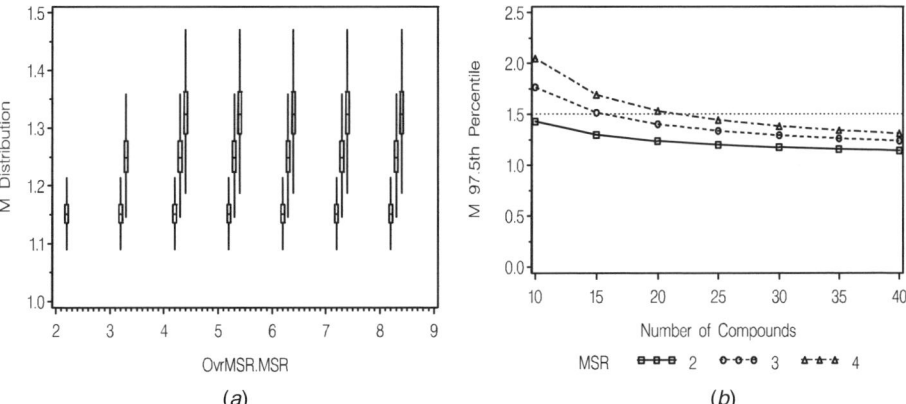

Figure 15.4 (*a*) Box plot of distribution of 95% confidence interval multiplier, M, for MR versus OvrMSR.MSR (i.e., whole number indicates the OvrMSR, decimal indicates the MSR) using 25 compounds. (*b*) 97.5th percentile of M versus number of compounds by MSR when OvrMSR = MSR.

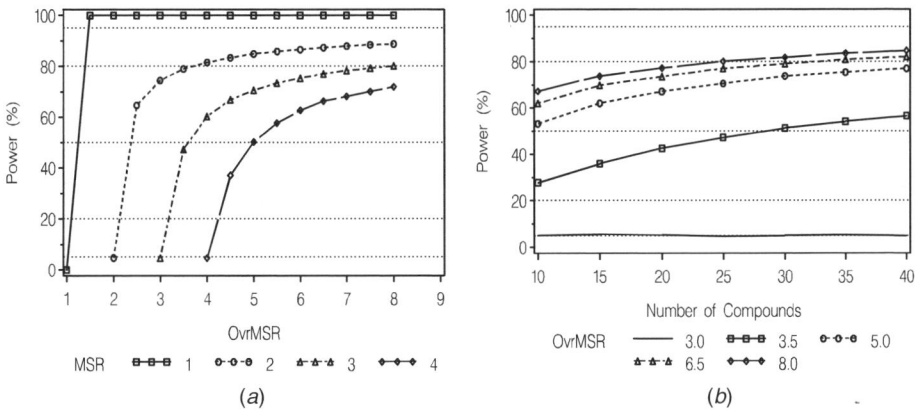

Figure 15.5 (*a*) Power to reject null hypothesis H_0: $\mu_X = \mu_Y$ versus OvrMSR by MSR for a sample of size 25 compounds. (*b*) Power versus number of compounds by OvrMSR when MSR = 3.

is fixed. However, it also shows that for a fixed OvrMSR power increases as MSR decreases. Hence the lower the measurement error, the more likely one will conclude there is across-run variation, that is, the better the assay is within-run, the more likely it will detect across-run variation. Figure 15.5*b* shows the power versus sample size relationship for various values of OvrMSR, holding MSR fixed at a value of 3. A sample size of 25 will have between 50 and 80 percent power to detect across-run variation, whereas a sample of size 10 will not exceed 70 percent power at any likely value of OvrMSR. Note, however, that across-run variation is usually present, and that because of the decreasing function of power versus MSR (holding the other values fixed) a straight hypothesis test is easily misinterpreted and not recommended.

The distribution of MSR parallels that of the precision of $\widehat{MR}$. Figure 15.6 shows the effect of OvrMSR, $\widehat{MSR}$, and n on the distribution of $\widehat{MSR}$. Figure 15.6a illustrates that the distribution of MSR is independent of OvrMSR for a sample size of 25 compounds. However, both the average value and variance of $\widehat{MSR}$ are increasing functions of the true MSR. Note that the distribution of MSR is approximately symmetric for all MSR ≤ 4 at this sample size. Figure 15.6b shows the value of the 97.5th percentile of $\widehat{MSR}$ versus the sample size for MSR = 2, 3, and 4. The results show that for 25 compounds $\widehat{MSR}$ will be within 50 percent of its true value, whereas a sample size of 20 or less will fail to achieve this objective.

Finally, we examine properties of the limits of agreement. These limits are intended to provide an interval that contains the upper and lower bounds on potency ratios if two estimates are obtained on the same compound. Figure 15.7 examines the coverage properties of LsA as a function of OvrMSR, MSR, and n. Figure 15.7a illustrates the probability that the LsA interval will not contain all the points, which varies from 25 percent at $n = 10$ compounds to over 90 percent at $n = 40$ compounds. Note that except for the degenerate case of no assay variation (MSR = 1 or OvrMSR = 1), the coverage probability is independent of the assay variation, both within and across run. Figure 15.7b shows the probability that more than 5 percent of the sample will lie outside the limits of agreement (i.e., more than 1 compound for $n < 30$ and more than 2 compounds for n between 30 and 40). As in Figure 15.7a, except for degenerate cases, this probability depends only upon the sample size. At $n = 10$ com-

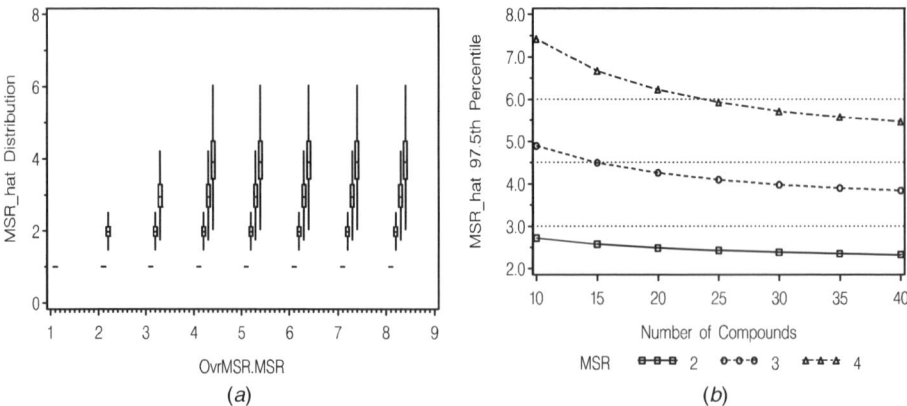

Figure 15.6 (a) Box plot of distribution of $\widehat{MSR}$ versus OvrMSR.MSR (i.e., whole number indicates the OvrMSR, decimal indicates the MSR) using 25 compounds. (b) 97.5th percentile of $\widehat{MSR}$ versus number of compounds by MSR when OvrMSR = 8 (note the relationship holds for any value of OvrMSR > MSR).

STATISTICAL PROPERTIES OF ESTIMATES AND ACCEPTANCE CRITERIA 683

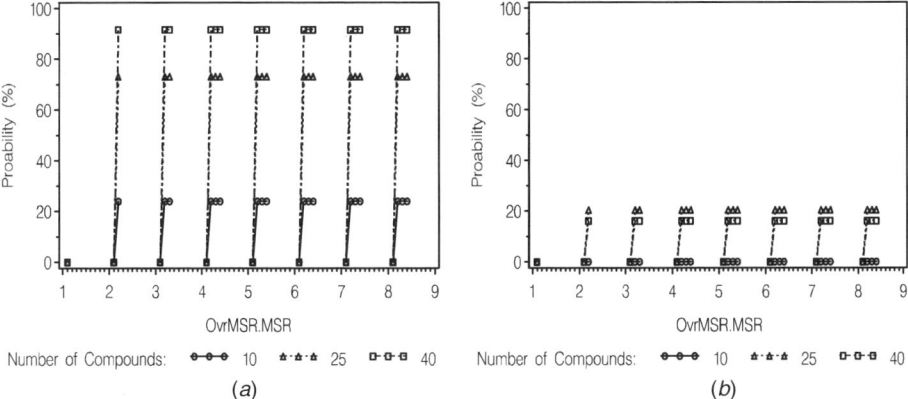

Figure 15.7 (*a*) Probability (%) that $\widehat{\text{LsA}}$ does not contain all tested compounds versus OvrMSR.MSR (i.e., whole number indicates the OvrMSR, decimal indicates the MSR) by 10, 25, and 40 compounds. (*b*) Probability (%) that more than 5% of sample (1 compound for $n < 30$ and 2 compounds for $30 \leq n \leq 40$) lies outside $\widehat{\text{LsA}}$ versus OvrMSR.MSR by 10, 25, and 40 compounds.

pounds there is a negligible probability that increases to about 20 percent at $n = 40$ compounds. This latter probability is higher than the nominal rate that results from basing the interval on an estimated value of the assay variation that is not included in the nominal error rate.

Figure 15.8 examines the distribution of the more extreme value of the limits of agreement. This is defined as the value "furthest" away from 1.0, which is equal to $\widehat{\text{LsA}}_{\max} = \max\{1/\widehat{\text{LsA}}_L, \widehat{\text{LsA}}_U\}$. For a sample size of 25 compounds Figure 15.8*a* shows box plots of the distribution of $\widehat{\text{LsA}}_{\max}$. Note the median, variation about the median, and skewness are all increasing functions of both OvrMSR and MSR. For MSR = 3, Figure 15.8*b* shows the relationship of both OvrMSR and *n*. $\text{LsA}_{\max}$ is an increasing function of the former, and relatively independent of the latter.

Properties of Acceptance Criteria

In Section 15.3 three outcomes were defined from the replicate-experiment protocol: the MSR, MR, and LsA. Each of these outcomes is positively correlated with one or both components of assay variation, and in the preceding section we examined the sampling distributions of each. In this section we look at the statistical properties of the following three rejection rules or tests:

$$\widehat{\text{MSR}} > C \qquad (15.4)$$

$$\widehat{\text{MR}} > C \qquad (15.5)$$

$$\widehat{\text{LsA}} \not\subset [1/C, C] \qquad (15.6)$$

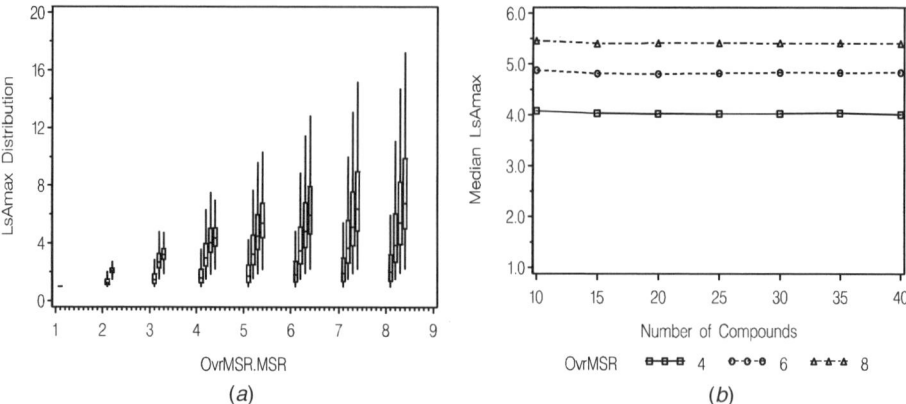

Figure 15.8 (a) Box plot of distribution of $\widehat{\text{LsA}}_{\max}$ versus OvrMSR.MSR (i.e., whole number indicates the $\widehat{\text{OvrMSR}}$, decimal indicates the $\widehat{\text{MSR}}$) using 25 compounds. (b) 97.5th percentile of $\widehat{\text{LsA}}_{\max}$ versus number of compounds by OvrMSR when MSR = 3.

where the critical value, C, is greater than one in each case (and is usually different in the three tests). Note that in the case of Eq. (15.6), an equivalent statement to that above is to reject the assay if $\widehat{\text{LsA}}_{\max} > C$. Throughout this section we estimate the probability of rejecting an assay, defined as the power of the test. The setting of the critical value has to be, of course, related to the biological and chemical requirements of the assay. Throughout this section we assume that the objectives are to reject assays if OvrMSR ≥ 3.

We showed in the previous section that $\widehat{\text{MSR}}$ is estimated to within approximately 50 percent, positively correlated to within-run assay variation, and independent of across-run assay variation. Figure 15.9 shows the effect of OvrMSR, MSR, C, and n on power. Figure 15.9a shows that for $n = 25$ compounds and $C = 3$, power is independent of OvrMSR (except for the degenerate case of OvrMSR = 0). For MSR = 2, power is less than 1 percent, is just less than 50 percent when MSR = 3, and is over 90 percent for MSR ≥ 4. Figure 15.9b shows the effect of sample size, critical value, and MSR on power when OvrMSR > MSR. The results indicate setting $C = 2.5$ will result in high rejection rates of potentially good assays (i.e., when MSR = 2), especially when combined with a low sample size ($n = 10$). Conversely, setting $C = 3.5$ results in less than 80 percent power to reject the assay when MSR = 4 at all sample sizes. Small sample sizes ($n < 15$) only have adequate power for low critical values, and a sample size of 20 to 30 is required to have high power to reject assays whose MSR = 4. The conclusion from this analysis is that Eq. (15.4) alone will only suffice in those cases where across-run assay variation is negligible, but in cases where the within-assay variation exceeds 4.0, this criterion will have high probability of rejecting the assay.

We showed in Figure 15.4b that $\widehat{\text{MR}}$ seldom exceeds 1.5 in cases of no across-run assay variation. So in Figure 15.10 we examine the power about the

STATISTICAL PROPERTIES OF ESTIMATES AND ACCEPTANCE CRITERIA

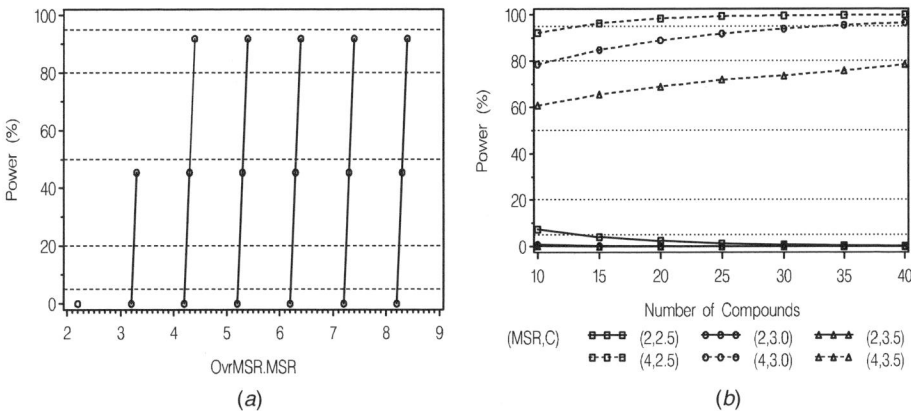

Figure 15.9 (*a*) Power (%) that $\widehat{\text{MSR}} > 3$ versus OvrMSR.MSR (i.e., whole number indicates the OvrMSR, decimal indicates the MSR) for a sample of size 25 compounds. (*b*) Power versus number of compounds by MSR and *C* when OvrMSR > MSR.

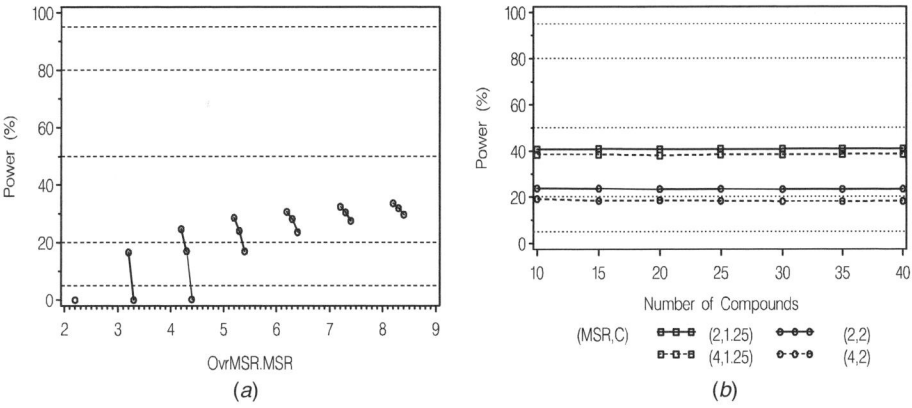

Figure 15.10 (*a*) Power (%) that $\widehat{\text{MR}} > 1.5$ versus OvrMSR.MSR (i.e., whole number indicates the OvrMSR, decimal indicates the MSR) for a sample of size 25 compounds. (*b*) Power versus number of compounds by MSR and *C* when OvrMSR = 8.

value $C = 1.5$. Figure 15.10*a* shows the effect of OvrMSR and MSR on power for $n = 25$ compounds and $C = 1.5$. The power is a very flat function of OvrMSR, and power remains below 40 percent for all values examined. Moreover, for a fixed OvrMSR, power decreases as MSR increases. Figure 15.10*b* shows the effect of *n*, MSR, and *C* on power for the extreme case of OvrMSR = 8. Note that power is invariant to sample size with this test, power decreases as MSR increases as noted previously, and that even a critical value of $C = 1.25$ only achieves about 40 percent power. However, the combination of MSR = 2 and OvrMSR = 3 will fail this test about more than 30 percent of the time, and hence setting a low threshold is not recommended. The conclusion from this analysis is that test 2 [Eq. (15.5)] has inadequate power.

From Figure 15.8a we showed that for OvrMSR = 3 median $\widehat{LsA}_{max}$ values were between 2 and 3, depending upon the value of MSR, and from Figure 15.8b we showed that median $\widehat{LsA}_{max}$ values exceed 4 in the set MSR = 3 and OvrMSR > 4. Thus, in Figure 15.11 we examine the power of test 3 [Eq. (15.6)] around a critical value of $C = 3$. Figure 15.11a shows the effect of OvrMSR and MSR on power for $n = 25$ compounds. For OvrMSR ≥ 4 over 80 percent power is obtained in all cases except where MSR = 2. Note that in all such cases the power in test 3 [Eq. (15.6)] is significantly enhanced over the power in test 1 [Eq. (15.4)]. However, note that for OvrMSR = 3 power may exceed 50 percent in those cases where the across-run component of variation is negligible. Figure 15.11b shows the effect of n, C, and MSR on power when OvrMSR = 8. At MSR = 4 high power (>90 percent) is maintained for all sample sizes and choices of C. At MSR = 2 adequate power (>80 percent) is obtained only through setting a low threshold ($C = 2.5$). The conclusion from this analysis is that test 3 [Eq. (15.6)] has excellent power to detect high OvrMSR except in those cases of low MSR.

The overall recommendation is that rejection rules be based on a combination of tests 1 and 3, [Eqs. (15.4) and (15.6)] and that test 2 [Eq. (15.5)] should be avoided. Note that if the same critical value is used in both tests 1 and 3 [Eqs. (15.4) and (15.6)], then rejection in test 1 [Eq. (15.4)] implies rejection in test 3 [Eq. (15.6)]. Therefore, to avoid redundancy, the critical value for test 3 [Eq. (15.6)] should be set higher than that of test 1 [Eq. (15.4)], but the optimal settings cannot be determined without reference to the chemistry and

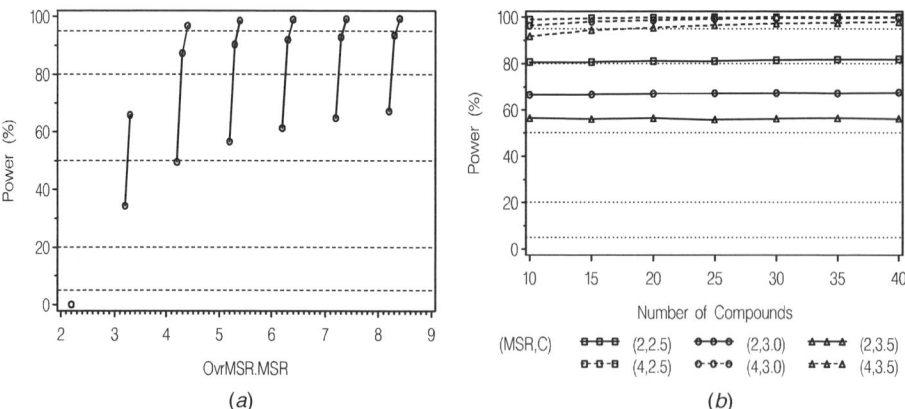

Figure 15.11 (a) Power (%) that $\widehat{LsA}_{max} > 3$ versus OvrMSR.MSR (i.e., whole number indicates the OvrMSR, decimal indicates the MSR) for a sample of size 25 compounds. (b) Power versus number of compounds by MSR and C when OvrMSR = 8.

biology requirements. Consequently, final settings are beyond the scope of this chapter.

Acknowledgment The authors wish to acknowledge numerous discussions with Viswanath Devanarayan, Mark Farmen, Barry Sawyer, and Philip Iversen that have helped refine the replicate-experiment protocol since it was conceived.

REFERENCES

1. Sittampalam, G. S., Iversen, P. W., Boadt, J. A., Kahl, S. D., Bright, S., Zock, J. M., Janzen, W. P., Lister, M. D. (1997). Design of signal windows in high throughput screening assays for drug discovery. *J. Biomol. Screen., 2*, 159–169.
2. Zhang, J.-H., Chung, T. D. Y., Oldenburg, K. R. (1999). A simple statistical parameter for use in evaluation and validation of high throughput screening assays. *J. Biomol. Screen., 4*, 67–73.
3. Taylor, P. B., Stewart, F. P., Dunnington, D. J., Quinn, S.T., Schulz, C. K., Vaidya, K. S., Kurali, E., Tonia, R. L., Xiong, W. C., Sherrill, T. P., Snider, J. S., Terpstra, N. D., Hertzberg, R. P. (2000). Automated assay optimization with integrated statistics and smart robotics. *J. Biomol. Screen., 5*, 213–225.
4. Eastwood, B. J., Farmen, M. W., Iversen, P. W, Craft, T. J., Smallwood, J. K., Garbison, K. E., Delapp, N. W., Smith, G. F. (Submitted for publication). The minimum significant ratio: A statistical parameter to characterize the reproducibility of potency estimates from concentration-response assays. *J. Biomol. Screen.*
5. Kenakin, T. (1997). *Pharmacologic Analysis of Drug-Receptor Interaction,* 3rd ed. Lippencott-Raven, Philadelphia, pp. 242–289.
6. Molulsky, H., Christopoulos, A. (2003). *Fitting Models to Biological Data Using Linear and Nonlinear Regression: A Practical Guide to Curvefitting.* Graphpad Software, San Diego, CA, pp. 259–260.
7. Ratkowsky, D. A., Reedy, T. J. (1986). Choosing near-linear parameters in the four-parameter logistic model for radioligand and related assays. *Biometrics, 42*, 575–582.
8. Gallant, A. R. (1987). *Nonlinear Statistical Models.* Wiley, New York, pp. 16–18.
9. Serfling, R. J. (1980). *Approximation Theorems in Mathematical Statistics.* Wiley, New York, p. 33.
10. DeLapp, N. W., McKinzie, J. H., Sawyer, B. D., Vandergriff, A., Falcone, J., McClure, D., Felder, C. C. (1999). Determination of [^{35}S] guanosine-5'-O-(3-thio) triphosphate binding mediated by cholinergic muscarinic receptors in membranes from Chinese hamster ovary cells and rat striatum using an anti-G protein scintillation proximity assay. *J. Pharmacol. Exper. Therapeut., 289*(2), 946–955.
11. Motulsky, H. (2003). *Graphpad/Prism® User's Guide, Version 4.* Graphpad Software, San Diego, CA.
12. Bishop, Y. M. M., Feinberg, S. E., Holland, P. W. (1975). *Discrete Multivariate Analysis: Theory and Practice.* MIT Press, Cambridge, MA, pp. 486–487.

13. Bland, J. M., Altman, D. G. (1986). Statistical methods for assessing agreement between two methods of clinical measurement. *Lancet, 1*, 307–310.
14. Carrasco, J. L., Jover, L. (2003). Estimating the generalized concordance correlation coefficient through variance components. *Biometrics, 59*, 849–858.
15. Eastwood, B. J. (2001). Comparing two measurement devices: Review and extensions to estimate new device variability. *Computing Science and Statistics, 33*, 462–471.
16. SAS Institute (1999). *SAS ® Language Reference: Dictionary, Version 8*. SAS Institute, Cary, NC, pp. 226–626.
17. SAS Institute (1999). *SAS/GRAPH® Software: Reference, Version 8*. SAS Institute, Cary, NC, pp. 801–859.

16

COUPLED LUMINESCENT METHODS IN DRUG DISCOVERY: 3-MIN ASSAYS FOR CYTOTOXICITY AND PHOSPHATASE ACTIVITY

MICHAEL J. COREY
Chromos
Seattle, Washington

ROBERT J. KINDERS

16.1	INTRODUCTION	690
	Today's Assay Requirements	690
	Fluorescence versus Luminance	691
	Scintillation Proximity Assays	691
	Coupled Enzyme Assays	692
	Coupled Luminescent Assays as Alternatives to Fluorescent Assays	693
	Chemiluminescent Assays	693
	Coupled Bioluminescent Assay Methods	694
	Measurement of Cytotoxicity by Release Assays	700
	Measurement of Phosphatase Activity	701
16.2	THE DEATHTRAK 3-MIN CYTOTOXICITY ASSAY	701
	Introduction to DeathTRAK	701
	Modes of Use of DeathTRAK in Drug Discovery	702
	Summary of Cytotoxicity/Proliferation Methods	711
16.3	PHOSTRAK: A VERY RAPID, HOMOGENEOUS GENERAL PHOSPHATASE ASSAY	712
	Measurement of Phosphatase Activities	712
	Introduction to PhosTRAK	713

*The methods described herein are covered by US Patent#6811990 and are licensed to Cell Technology, Inc., Mountain View, CA (www.celltechnology.com).

Drug Discovery Handbook, by Shayne Cox Gad
Copyright © 2005 by John Wiley & Sons, Inc.

Applicability of PhosTRAK Method to Various Types of Phosphatases	713
Representative PhosTRAK Results with Diverse Phosphatases	715
Limitations Due to Complexity and Thermodynamic Considerations	716
PhosTRAK Summary	717
16.4 COUPLED LUMINESCENT ASSAYS IN PRACTICE: SPECIFICATIONS AND LIMITATIONS	718
Assay Specifications	718
Backward Reactions	721
16.5 SUMMARY AND CONCLUSIONS	722
16.6 FORMULATIONS	722
Reaction Master Mixes (Cocktails) and Dilution Buffers	722
PhosTRAK Formulations	723
Lytic Formulations	725
References	726

16.1 INTRODUCTION

Today's Assay Requirements

The prototypical high-throughput screening (HTS) operation in today's drug discovery environment is directed to identification of modulators of enzyme activities or receptor binding. There was a time when the inhibition constant of a single transition-state analog was determined by a graduate student nervously keeping atmospheric CO_2 out of his vessel as he titrated evolved acid with excruciating care, and *receptor binding assays* referred to a postdoc grinding up a mouse organ to mix with her precious radiolabeled ligand. There was such a time—but today's practitioner of lead identification in a drug discovery enterprise feels little nostalgia for it. Current demands on discovery groups require that 10^5 to 10^6 or more molecules be tested in a single experiment, quickly, accurately, and at acceptable expense. The various revolutions in biotechnology of the past 30 years have largely given way to evolutionary changes, favoring faster, smaller, operationally simpler, and more information-intensive assay methods. Developments in instrumentation that have contributed to this process are outside of the current subject matter, but biochemical manipulations may also be applied, ideally leading to ever more quantifiable photons from fewer molecular events in a shorter time frame. This process is most readily apparent in the shift from chromogenic and other traditional methods to fluorescent, luminescent, and scintillation proximity assays, a trend that will accelerate as assay volumes enter the submicroliter range.

Coupled luminescent assays are only one class of an expanding array of sensitive methods. At present, coupled luminescent methods are underutilized, largely due to the lack of widespread understanding of their value or even of their existence. What follows is a brief treatment of several alternative means of generating many photons from small samples in a short time

INTRODUCTION

interval. To understand why coupled luminescent methods are the most efficient and sensitive available in addressing certain types of problems, while they are marginal or unsuitable for others, some insight into the nature, theory, and practice of alternative techniques is needed.

Fluorescence versus Luminance

From the scientist's viewpoint, nature has bestowed powerful gifts on the phenomenon of fluorescence, including the associated techniques of fluorescence quenching, environmentally sensitive spectral shifts that are subject to synthetic modification, fluorescence polarization (FP), and fluorescence resonance energy transfer (FRET). Some of these technologies are already widely disseminated, such as the use of FP in drug immunoassays (as in the Abbott TDx). It is safe to predict that for rapid, specific, and exquisitely sensitive measurements of intermolecular association events, fluorescent technologies will dominate the field for at least several decades. There are also many clever fluorescent methods of measuring enzyme activities, but enzyme catalysis is intrinsically different from binding processes in that enzyme-catalyzed reactions are almost always observed indirectly (i.e., by their products). To put it another way, in studying enzyme activity and inhibition, one frequently deals with the concentrations of particular molecules, rather than the presence or absence of molecular complexes. This important difference represents an opportunity for luminescent assay methods to play a role, since a variety of means of coupling the concentrations of specific molecules to light emission are known, and more are sure to be developed. Possible reasons for preferring a luminescent assay are: (1) The sensitivity can be even more exquisite than that of fluorimetry [1, 2]; (2) there is no need to develop or attach a fluorophore; (3) in many cases the readout is almost immediate, which is not the case with many fluorescent binding assays; (4) no expensive lamp is needed; (5) coupled luminescent assays can be made operationally simpler; and (6) for some reactions luminance may be the only available approach to a rapid, homogeneous format.

Scintillation Proximity Assays

A thorough discussion of the pros and cons of scintillation proximity assays (SPAs) in drug development HTS is beyond the scope of this chapter, but certain aspects of SPA that may cause one to choose or reject this option in favor of fluorescent or luminescent assays can be briefly described. Assays employing radioisotopes have generally been operationally complex, even apart from the exposure and disposal problems, because a labor-intensive separation of "bound" from "unbound" isotopically labeled molecules has been necessary. The SPA concept avoids this difficulty and promotes radioactive assays to the realm of homogeneous assays suitable for HTS, by providing a means of distinguishing bound from unbound in situ. This is accomplished by including a solid-phase matrix (usually a bead) containing both the target (unlabeled) molecule and a chemical scintillant. Binding of the labeled compound brings the radioisotope into proximity with the scintillant, leading to a luminance signal upon disintegration of a bound, but not an unbound, atom of the label. Because many ligands can be appropriately labeled and many targets can be successfully immobilized, this assay method is quite general. One finds in reviewing the literature that a high proportion of SPA articles report development of new applications. Thus SPA represents an approach that is quite likely to provide a successful, homogeneous method, even if one knows relatively little about the system and possesses only a conventional scintillation counter (or luminometer). However, the disadvantages of SPA compared to a fluorescent or luminescent method are important. Although technically homogeneous, in that no physical separations are required, successful application of SPA often requires multiple addition and incubation steps, some of which may be prolonged and involve shaking or other manipulations. For example, an SPA for inorganic phosphate has a very good limit of detection of ~10^{-7} M phosphate but requires 2 h and multiple steps and is a single-point assay [3]. An HTS method of identifying protein tyrosine phosphatase 1B inhibitors makes clever use of a catalytically inactive active-site mutant to turn the problem into a binding assay, but the method employs seven reagent additions and three shaking incubations, illustrating both the flexibility and the operational difficulties attending SPA [4]. The entire Raf/MEK/ERK signal transduction pathway has been reconstituted in vitro and studied by a homogeneous method employing SPA, greatly improving the chances of identifying modulators of the pathway; however, the method takes 8 h, including a 6-h settling step [5]. Finally, there is some question as to whether SPAs provide the same quality of information in terms of sensitivity and dose–response curves as competing fluorescent methods [6]. In short, SPAs are an excellent way of turning apparently intractable screening problems into homogeneous assays; but they are unlikely to be the method of choice if one-step or two-step fluorescent or luminescent alternatives are available.

Coupled Enzyme Assays

Coupled enzyme assays have a distinguished history. The "standard" clinical and research assays for many enzymes are coupled assays, often with spectrophotometric readouts. In a case that may be taken as typical of the thought processes involved in development of such an assay, researchers investigating antibiotic function were faced with the problem of detecting D-alanine and found a chromogenic answer [7, 8]. While this solution is effective in an academic research setting, it does not provide the throughput needed by drug discovery groups and is limited by the extinction coefficient of the molecule being measured. Another group subsequently developed an assay in which horseradish peroxidase (HRP) was generated in a coupled assay and yielded a fluorescent signal [9]. However, luminescent HRP substrates are also available

INTRODUCTION

with greater sensitivity [10], and this may eventually prove to be a further stage in the evolution of D-alanine assays.

Coupled Luminescent Assays as Alternatives to Fluorescent Assays

The issue of coupling of enzyme activities is in a sense independent of the fluorescence/luminance choice; however, in practice, fluorescent and coupled luminescent readouts often represent competing alternatives. This is because, although there are many fortuitous exceptions, most enzyme-catalyzed reactions and intermolecular associations do not yield conveniently robust and distinguishable fluorescence changes or luminance signals on their own. Thus the problem of detecting the molecular event often reduces itself to the question of how to cause the product(s) of the event to release photons (or inhibit their release). Introduction of a fluorophore that is perturbed by the event is one method; coupling of the products to light generation is another. Advocates of luminance technology could present theoretical reasons in support of a claim for greater sensitivity, including the following: (1) Luminance detection requires no lamp, so that the sample chamber can be darker, lowering the background; (2) unlike fluorescent methods, luminance detection requires no monochromator or bandpass filter to block the exciting wavelengths, so that every photon contributes to the signal; (3) luminescence is inherently isotropic, allowing the detector to cover as great a spherical arc as the engineering permits; this is not true of fluorescence, which requires angular separation of the detector from the incident light. The theoretical advantages in sensitivity appear to be borne out in practice, at least in some cases (for examples and analyses, see [1, 2]). Fluorescence has inherent counterbalancing advantages, including the possibility of pumping virtually any desired level of excitation energy into the system, and the more refined approach of using modern synthetic chemistry to tune the fluorophore(s) to precise specifications. It is hoped that the methods and approaches presented herein will help the reader to imagine how both possibilities might contribute to his or her particular goals.

Chemiluminescent Assays

In modern parlance *bioluminescence* refers to light production by an enzymatic system (such as a luciferase) or other light-emitting protein (such as aequorin [11]), while *chemiluminescence* is photon emission by a (usually) small molecule in the absence of a catalyst. For a number of enzymes it is possible to design substrates that, upon catalysis, yield a product that can emit a photon. Thus the primary reaction generates light, and no coupling enzyme is required. This method has led to significant improvements in detection capabilities [12]. However, the limitations of the method in the HTS environment are often significant: (1) The enzyme substrate is unnatural, and inhibition effects are therefore hard to interpret; (2) in a related point, the K_m of a chemiluminescent substrate is frequently much greater (i.e., the binding is much weaker) than that of the natural substrate, allowing competition by weak competitive inhibitors that are not useful as lead compounds; (3) luminescent substrates often have special chemical, solvent, temperature, or pH requirements necessitating extra steps in the assay process; (4) the requirements mentioned under (3) may make the reaction incompatible with the biochemical system under study; and (5) chemiluminescent substrates are not available for many reactions.

Coupled Bioluminescent Assay Methods

The two major types of luciferase used in coupled reactions are bacterial luciferases, whose light emission is coupled to oxidation of reduced nicotinamide adenine dinucleotide (NADH), and beetle luciferases, which use the free energy of hydrolysis of adenosine triphosphate (ATP) to generate light. Systems employing aequorin as the luminescent protein can also be engineered to yield signals in response to the activities of certain enzymes involved in Ca^{2+} mobilization and signaling [13], including the very important class of G-protein-coupled receptors (GPCRs). These systems are of considerable interest in drug discovery. In general, there is little overlap between the Ca^{2+} mobilization processes that are subject to these methods and the phosphate-dependent reaction series described herein, although a coupled luminescent assay for the guanosine triphophatase (GTPase) activity of GPCRs is not inconceivable.

Coupled Methods Employing Bacterial Luciferases Bacterial luciferases [14] have yielded a number of useful coupled assay methods, including means of quantifying a range of small molecules with excellent sensitivity (ethanol [15], folate [16], glucose [17], lactic acid [18], bile acids [19]). Some reactions, including the reactions involving glyceraldehyde-3-phosphate dehydrogenase discussed below, allow a choice of either bacterial or beetle luciferases as coupling enzymes. The author's decision to concentrate on firefly luciferase has been based primarily on practical considerations, including the poor aqueous solubility of the long-chain aldehyde substrates of the bacterial enzyme, which may require a separate addition and mixing step; the tendency of the bacterial signal to decay more rapidly; the existence of stabilized and engineered systems employing the firefly enzyme; and the fact that the firefly enzyme has been more fully characterized. Moreover, methods of ATP detection are more likely to be generally applicable in drug discovery since the enzymes of critical interest that are involved in phosphate metabolism outnumber those that generate NAD(P)H.

Coupled Assays with Firefly Luciferase Many in vivo reactions create ATP, and many others that are coupled to hydrolysis of ATP in vivo can be made to run backwards in vitro. In principle, the reaction rate of any enzyme involved in these pathways can be monitored by coupling production (or consumption) of ATP to light generation by firefly luciferase. In practice, many of the potential coupled systems are unworkable in vitro, but a number of enzymatic reactions or reaction series have been combined with luciferase, some of which are discussed below. As with the chemiluminescent assays, some of the methods are intended for detection of small molecules, and others are adaptable to detection of either small molecules or enzymatic activities, depending on what is omitted, that is, what is the limiting reagent. See [20] for examples of very early uses of coupled methods with firefly luciferase for detection of small molecules.

Role of Coupled ATP/Luciferase Reactions in Drug Discovery Thousands of enzymes are necessary components of metabolism, but a large proportion of the enzymes of phosphate metabolism are specifically involved in information flow. As a result, many of them not only play essential roles in life-sustaining pathways but are also vital elements of reaction series that are subject to *information failure*. Many phosphorylation and dephosphorylation reactions are critical mediators of major human health problems such as cancer, diabetes, autoimmunity, and inflammation. These features make these enzymes highly attractive targets for drug development, based on the philosophy that subtle perturbations of the body's signaling systems, rather than gross, "brute-force" interventions in metabolic flows, are the best route to highly specific and effective drugs. Of the enzymes that are currently the major targets of HTS, only proteases are *not* directly involved in phosphate metabolism. The other major groups, including kinases, phosphatases, deoxyribonucleic acid (DNA) polymerases and many nucleases, phosphodiesterases, and the whole generic category of "ATPases," can be made to change the concentration of ATP in a reaction vessel and can thus yield a luminance signal via luciferase. Still other enzymes can be coupled to phosphate metabolism by the inclusion of "bridging" enzymes. If we restrict ourselves to the desirable trait of yielding a *positive* signal (i.e., an increase in luminance when the enzyme is active), it might appear that we have lost most kinases and polymerases for thermodynamic reasons, but this is not the case (see below). In short, over half of the assays carried out in today's drug discovery endeavors (on a frequency basis) are adaptable to ATP/luciferase-coupled systems.

Advantages of Coupled ATP/Luciferase Systems Many of the advantages of these systems can be defined in terms of what is *not* required. Typically a coupled luminescent assay employing the firefly luciferase can be developed as a one-step assay in which a reagent cocktail is added and measurement begins immediately (although multistep assays may be preferred for particular applications). This represents a significant savings in operational complexity, leading to lower labor, equipment, and materials costs and fewer failure modes. Because the assays are generally very fast, instrument utilization is minimized; in effect, more work can be done with fewer instruments and less space. This effect will be increasingly important as phased-array luminometers become more common. These instruments can read entire plates in 2 s or less, but this speed is largely wasted if the biochemistry of the assay requires prolonged incubations or multiple steps. Coupled luminescent assays are frequently the only means of taking full advantage of the capabilities of these instruments. Generation of ATP by the reaction system also provides a continuous readout. This yields the advantages of (1) flexibility in data reduction, (2) the possibility of prolonged incubation to achieve the desired sensitivity, and (3) the opportunity for a second reading if the initial reading is somehow missed. The components of these coupled assays are often less expensive than the fluorophores, labeled peptides, and antibodies used in other assay methods; even using 96-well plates, it is quite realistic to expect reagent costs of $0.01 to $0.05 per well or less, and this cost should scale down with effective miniaturization. The homogeneous format also eliminates the need for washing and separation steps and reduces generation of hazardous waste. Finally, recent developments in stabilization of firefly luciferase and other methods of enhancing and prolonging luminescent signals will continue to add value to these methods.

Adaptation of Coupled Luminescent Assays to High-Throughput Screening
Because of their operational simplicity, it can be relatively simple to adapt a coupled luminescent assay to use in HTS, and the sensitivity usually allows reduction to a denser plate format. However, there are also special requirements that must be met. In using firefly luciferase, some consideration must be given to its stability. If available, one can employ a stabilized luciferase; otherwise, reductants are beneficial, and it may be necessary to keep the luciferase cold and protect it from light until use, depending on the wait time. Another issue that does not arise as frequently in other kinds of assays is the purity of certain assay components. Coupled assays that yield ATP require an ATP precursor, such as adenosine diphosphate (ADP), adenosine monophosphate (AMP), or cyclic AMP (cAMP), as a substrate. If the substrate is contaminated with ATP, even in small amounts, there will be a significant background signal. ATP can generally be separated from these other components by the use of any of several methods [22–24]. Similarly, commercial enzyme preparations are often contaminated with other enzymes. If one of the contaminating enzymes is the substance under test (as in the DeathTRAK assay; see below), then the user must purify the enzyme, inactivate the troublesome contaminant in a way that will not cause subsequent inhibition during the screening process, or accept a dynamic background signal.

Fortunately, these background sources are usually very constant, and as a result they generally have surprisingly small effects on the quality of the results, including the Z values. However, to achieve this predictability, the user must be sensitive to such issues as the order of reagent addition. For example, if ADP is present in the reagent cocktail, and one of the coupling enzymes is slightly contaminated with one or more of the enzymes capable of transferring phosphate moieties among nucleotides, then mixing of the ADP with the

INTRODUCTION   697

coupling enzymes should occur only very shortly before assay initiation; otherwise a high background must be expected. Finally, scientists involved in HTS are used to the idea that many fluorophores and other reagents must be protected from light, but a researcher developing or employing a coupled luminescent assay should try to develop a new way of thinking about light. In these assay methods, light is not merely a readout or a damaging form of radiation; it is also potentially a *reagent*. As such, it can drive a "backward" coupled reaction if care is not taken to avoid this possibility, and the result can be an undesirable *lag phase* in the reaction that can lower the quality of the data and degrade throughput (see Section 16.4).

Limitations and Considerations   There are other considerations that may influence the choice of a coupled luminescent assay. Primary among these is the appropriate pairing of a screening strategy with the objectives of a study. As a simple example, suppose one is screening for a novel antibiotic for use against a specific bacterium, using midlog cultures. If one is willing to accept any effect that reduces the bacterial count, including bacteriostasis, wall or membrane lysis, metabolic poisons, and genotoxic effects, then it is hard to improve on a simple turbidometric growth assay (although the sensitivity of a coupled luminescent enzyme-release assay may be an advantage in working with early-phase cultures). However, if the researcher is interested in mechanism, or wishes to distinguish killers from bacteriostatic agents, or wishes to measure lysis and proliferation separately, then a more sophisticated approach may be required. An example from mammalian systems may also be warranted. Suppose a worker wishes to screen a completely random library for druglike characteristics against an enzyme target. The primary assay should be simple from a biochemical point of view since this will help to alleviate the problem of systemically generated false hits. This may be a good case for an SPA approach since these systems are often less sensitive to interferents. Moreover, since one has little a priori information about a given "hit," the secondary screening process should be rigorous and information intensive. Now suppose that one is screening with a *targeted* library, that is, a library of entities with at least one common structural element that is known to be selective for the desired enzyme. Biochemical simplicity is now less important as an assay parameter since a much lower proportion of false hits is to be expected. Analytical sensitivity and precision are now driving attributes of the primary screening method, while the secondary screening method should probe selectivity, including both false-hit modes of the primary screen and biochemical or biological characteristics that are independent of the readout of the primary. A very rapid, ultrasensitive coupled luminescent assay may be the best choice for the primary method.

   The application of these principles to the use of coupled luminescent assays leads to two conclusions: (1) The more one knows about one's library, the more appropriate it may be to choose a rapid, ultrasensitive assay involving relatively complex biochemistry; and (2) if a coupled method is used with a random library, then the secondary screening method should be directed to the elements of the system that are most likely to yield false positives (such as downstream signal-generating enzymes).

Examples of Coupled Luminescent Assays Employing Firefly Luciferase See [25] for an excellent source of theory, methods, and practical information about firefly luciferase, including coupled assays. The number of assay types in coupled-luminescent category is still small compared to the potential breadth of the field, but expansion is accelerating for at least two reasons: (1) As mentioned above, the drive to smaller samples and faster instrumentation places an increasingly great premium on rapid, ultrasensitive methods; and (2) the availability of thermostable luciferases adds considerably to the flexibility of these methods. Examples provided include assays for kinases and DNA polymerases, as well as the glyceraldehyde-3-phosphate dehydrogenase release cytotoxicity assay and phosphate/phosphatase assay, which are the experimental foci of this chapter.

KINASES Kinases are the second most important molecular target in HTS, after GPCRs. Promega sells an HTS kinase assay kit (Kinase-Glo) of which the readout is luciferase detection of the consumption of ATP by the kinase. The kit represents a successful approach, but coupled systems of this type are inherently burdened by the "negative-signal" characteristic. In other words, the kinase yields a *negative* luminance readout, and inhibition of the kinase *increases* the readout. Although in some cases negative-signal systems prove to be the only or best alternative, the problems of such methods are not merely the counterintuitive nature of the scheme. In such a scheme, a significant amount of the ATP must be consumed to generate a signal; otherwise one is faced by the statistical pitfalls of analyzing a small difference between two large numbers. But if significant amounts of ATP are consumed, it is easy to lose the "initial-rate" assumption that is dear to enzymologists. In practice, this means that conclusions drawn about kinetic parameters (such as inhibition constants) may not be as reliable as if they were measured differently. However, the most important drawback of a negative-signal system is one that may be invisible to the user: How does one interpret an absence of light? Did the kinase do its job and consume the ATP, indicating that the tested compound is inactive, or was this the next Captopril or Viagra, but one that yielded no signal in the assay because it adventitiously blocked the luciferase system as well? This problem is usually avoided in positive-signal systems, which are more likely to generate false-positive hits than false negatives. The user must judge the likelihood and potential cost of each type of failure mode in deciding whether to use a negative-signal assay.

The general impression is that a kinase is an enzyme that transfers a phosphate group from ATP to a target substrate molecule. However, kinases, like other enzymes, are not exempt from the principle of microscopic reversibility, and, with no special considerations in play (such as off-gassing or further conversion of a product), kinase-catalyzed reactions are readily reversible in practice. This means that they can generate ATP, which means that their activity can be measured with firefly luciferase. Some scientists without specific enzymology training are uncomfortable with the idea of screening so-called reverse reactions; however, thermodynamic theory is absolutely clear on the point that if a molecule inhibits the forward reaction, it must inhibit the reverse reaction, and to the same extent. In any case, the *forward vs. reverse* distinction is itself not always clear-cut. (Is there a kinetic difference between an "alcohol dehydrogenase" and an "aldehyde reductase"?) Kinases were so named because the first described kinases activated proteins by phosphorylating them, thereby catabolizing ATP; but what of creatine kinase? This enzyme is very proficient at *synthesizing* ATP, and in fact the reaction is thermodynamically downhill in this direction under physiological conditions. A method of measuring creatine kinase activity by luciferase detection of evolved ATP was described some 25 years ago [26]. The original method is clever but not startling since the reaction is thermodynamically favorable. The idea was subsequently broadened in a more recent patent application to include reactions in which ATP synthesis is thermodynamically disfavored [27]. This category embraces nearly all kinases, although it does not mean that all kinases can be assayed in this way; however, the apparently unfavorable ΔG of the reactions successfully handled implies that the thermodynamic issue in itself is not a major barrier to development of "backward" kinase reaction schemes. Nevertheless, thermodynamics may still prove to be an important limitation on what can be achieved in other systems, especially those in which natural hydrolysis of ATP is not coupled to an energy-storing process that can be readily reversed in vitro. In any case, the contrast between the convenient, homogeneous one-step nature of this assay and the relatively slow, clumsy HTS methods in current use for kinase screening indicates that coupled luminescent methods have a promising future in this area.

PHOSPHOROLYSIS Scientists at Promega Corporation have developed a coupled luminescent assay (READIT) for the activity of nucleic acid polymerases, with extensions to detection of specific sequences, including single nucleotide polymorphisms [28–30]. This is a highly promising area in which coupled luminescent assays are likely to surpass competing methods in efficiency and accuracy. Given the central importance of nucleic acids in biology, the authors believe that the use of this technology is likely to expand greatly in the next few years. General awareness of coupled luminescent methods may be raised as a side effect.

PHOSPHODIESTERASES AND BEYOND Although "biologicals" are making impressive inroads in terms of investigational new drug (IND) submissions, much of drug discovery can still be regarded as a process of characterizing small molecules and their effects. Since this is not far removed from the process of *detecting* small molecules, synergies may be achieved and methods borrowed from seemingly distant fields for use in drug discovery. Given the wide range of enzymes involved in phosphate metabolism, a little imagination can readily yield a new idea or application that appears far removed from the initial objective. A single hypothetical example originating in the world of food science, where detection of caffeine is of interest, is sufficient to demonstrate the principle. Caffeine inhibits a cAMP-dependent phosphodiesterase (PDE); cAMP is a product of the nominal reaction of adenylate cyclase, an enzyme that ordinarily consumes ATP. However, it should be possible to employ mass-action principles by supplying cAMP and pyrophosphate to the adenylate cyclase in the presence of PDE. If caffeine is not present, the PDE will exhaust the cAMP and depress the luminescent signal. However, if the PDE is inhibited by caffeine, the cAMP will give rise to light production instead. (Here a negative-signal assay design is presented, despite the potential disadvantages cited above.) Caffeine itself is not of great current interest in drug discovery, but other PDE inhibitors are of major significance [31–33]. Many such schemes are possible.

Measurement of Cytotoxicity by Release Assays

Cytotoxicity has been measured by monitoring the release of various molecules from damaged cells for several decades. A modern example of such a method is the coupled fluorescent lactate dehydrogenase (LDH) release assay CytoTox-ONE from Promega. This method is operationally very convenient, has moderate sensitivity [limit of detection (LoD) ~200 nucleated mammalian cells], is moderately rapid (10 min), and is quite expensive, listing at >$25.00 per 96-well plate. Another approach is to transfect target cells with an enzyme that can then yield a fluorescent or luminescent signal upon cell lysis. Two enzymes that have been employed in this way are firefly luciferase itself and B-galactosidase [34]. The method obviously requires a separate transfection step, but this can be done in a bulk fashion and adds little to the cost or labor required. Unfortunately, neither luciferase nor B-galactosidase is well suited to the purpose; luciferase has a short half-life under test conditions, while B-galactosidase is not readily released into the assay medium by membrane rupture alone and requires additional steps for a successful assay. Release of naturally present alkaline phosphatase can be monitored by means of a luminescent substrate, but only a few types of cells release measurable quantities of this enzyme on lysis [35].

The important glycolytic enzyme glyceraldehyde-3-phosphate dehydrogenase (G3PDH) has also been used for decades to measure cytotoxicity by spectrophotometric means [36]. The original coupled luminescent assay method for G3PDH was published in 1997 [37]. This method was extremely sensitive, with an LoD of 0.03 nucleated mammalian cell or 1 rabbit erythrocyte. However, it was impractical for use in HTS, not only because three separate transfers were involved but also because it was not a homogeneous method:

the Reagent cocktail was not adapted for use with live cells, which had to be removed by centrifugation or filtering to avoid a false signal due to leakage from these cells. The G3PDH assay method known as DeathTRAK was developed to address these problems [38].

Measurement of Phosphatase Activity

Currently the most widespread methods of assessing phosphatase activity in drug discovery programs are the use of radioactive phosphorus tracers (especially with SPA) and antibody-based detection of phosphorylated or dephosphorylated target molecules. The coupled luminescent method proposed herein, PhosTRAK [38], is a very rapid and sensitive alternative. For the phosphatases tested to date, the assay is truly a one-step method: All reaction components, the phosphatase, and modulators are mixed and the readout is obtained as the phosphatase reaction proceeds (see Section 16.3).

16.2 DEATHTRAK 3-MIN CYTOTOXICITY ASSAY

Introduction to DeathTRAK

DeathTRAK is a homogeneous, one-step assay (see Fig. 16.1 for a depiction of the biochemical scheme of DeathTRAK). The limit of detection is well under one nucleated mammalian cell. (Means of achieving still greater sensitivity are discussed in Section 16.4.) As an essential glycolytic enzyme, G3PDH is abundantly present in all known cells, and the DeathTRAK method is therefore applicable to quantification of any cell type. The two major modes of use are in assessment of membrane rupture or damage and in measurement of total biomass. It should also be noted that like other release methods, DeathTRAK is poorly suited to detection of intracellular events, including fatal events, that do not affect membrane integrity. For example, release assays are not generally the first choice for detecting apoptosis, although they may well be useful in distinguishing apoptosis from necrosis if used in combination with a metabolic assay that is sensitive to intracellular events.

Three enzymatic reactions are depicted in the biochemical scheme of DeathTRAK presented in Figure 16.1. G3PDH is the test enzyme. In the normal flow of glycolysis, the free energy associated with the reducing capacity of the C–H bond of carbon-1 of glyceraldehyde-3-phosphate is captured in the formation of ATP from ADP and inorganic phosphate by the sequential reactions catalyzed by G3PDH and phosphoglycerokinase (PGK). The reaction series is exergonic, and the components are all readily available. Firefly luciferase and the small molecules needed for its activity are included in the reaction cocktail. The coupled reaction series produces light immediately upon mixing, assuming there is a source of G3PDH.

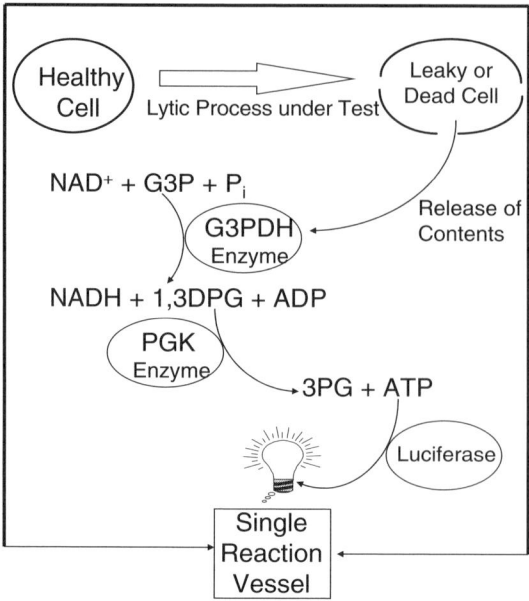

Figure 16.1 Biochemical scheme of DeathTRAK homogeneous cytotoxicity assay. NAD^+, nicotinamide adenine dinucleotide (oxidized form); G3P, glyceraldehyde-3-phosphate; P_i, PO_4^{3-}; G3PDH, glyceraldehyde-3-phosphate dehydrogenase; NADH, nicotinamide adenine dinucleotide (reduced form); 1,3DPG, 1,3-diphosphoglycerate; ADP, adenosine triphosphate; PGK, phosphoglycerokinase; 3PG, 3-phosphoglycerate; ATP, adenosine triphosphate.

Modes of Use of DeathTRAK in Drug Discovery

Applications of DeathTRAK to Measurement of Cytotoxic Processes Many death modes lead to membrane rupture, including natural killer (NK), lymphokine-activated killer (LAK), cytotoxic T lymphocytes (CTL), and complement attack, pore-formers, detergents, various other antibiotics, and osmolysis. Numerous molecules that modulate these processes are under consideration as drug candidates, and other aspects of the processes await investigation by drug discovery groups. Thus the potential applications of DeathTRAK in screening for drugs that modulate natural cytotoxicity are widespread. DeathTRAK and the earlier glyceraldehyde-3-phosphate dehydrogenase/phosphoglycerokinase/luciferase (GPL) assay were originally developed to measure complement activity [39]. Figure 16.2 shows the results of a complement assay using DeathTRAK, in which the effects of an anti-Factor I antibody were assessed. Figure 16.3 shows the results of an assay of cytotoxic T lymphocytes (CTL), selected and expanded by the proprietary Rapid Expansion Method (CellExSys, Inc.; [40] and references therein), using the same assay protocol. The DeathTRAK method appears to have major advantages

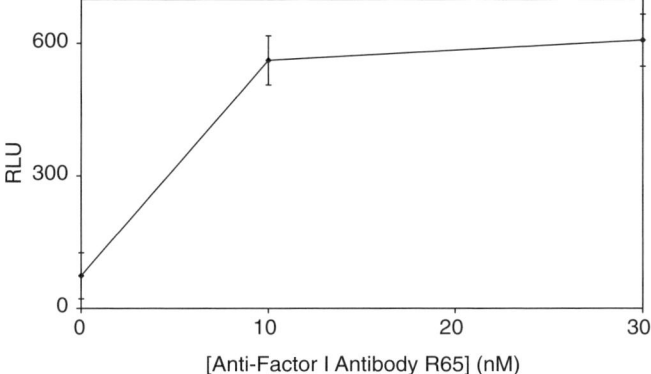

Figure 16.2 Enhancement of complement-mediated lysis of PC-3 prostate cancer cell line measured by DeathTRAK. Human complement serum (Sigma) was preincubated with the indicated concentrations of rat monoclonal antibody R65 (Alidex, Inc., Redmond, WA) against human complement Factor I, following by complement assays as previously described [39].

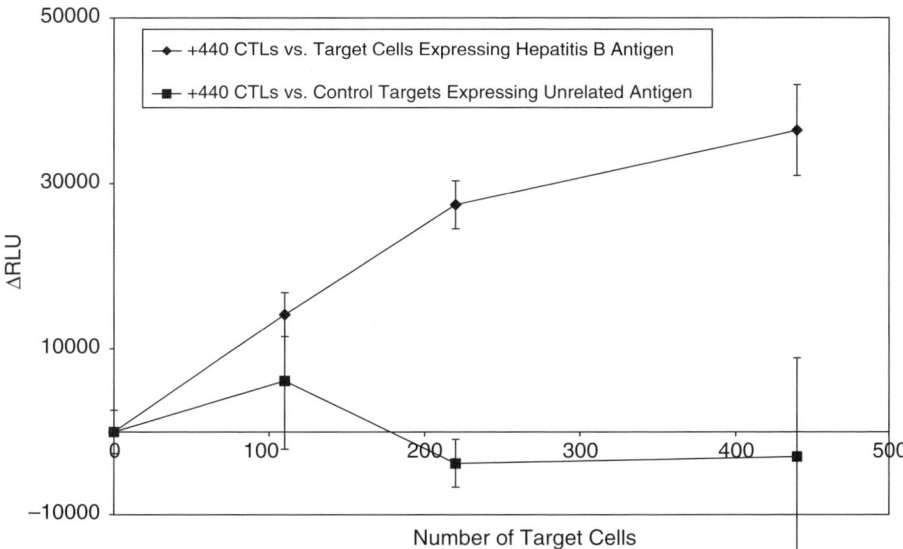

Figure 16.3 Measurement of cytotoxic T lymphocyte activity against autologous cells expressing specific hepatitis B antigen. CTLs were grown and isolated by the rapid expansion method, which is proprietary to CellExSys, Inc. and is described in [40]. Targets were EBV-immortalized lymphocytes from the same patient, infected with a vaccinia vector containing a recombinant gene for an HBV-specific antigen. Control targets were infected with an unrelated antigen. The infection process induces a certain degree of leakiness, which accounts for most of the scatter seen in the control reactions. Cells at 10× were incubated for 4 h at 37°C, whereupon 1/10 of the reaction volume (5 µL) was transferred to 45 µL of DeathTRAK cocktail (Section 16.6) and read on a TopCount scintillation counter. Reported data are the differences between the initial read and a subsequent read at 2.33 min.

over the standard ^{51}Cr-release assay in assessment of CTL activity, some of which were enumerated above.

Biomass Measurement (Competitive Comparison) Nearly all assays commonly referred to as *proliferation assays* actually measure either total biomass or integrated metabolism. Currently, the most important true biomass assays used in drug discovery are turbidometry and measurements of ATP release. As mentioned above, turbidometric methods are very good ways of quantifying bacteria and certain other microbes under controlled circumstances. However, they yield no mechanistic information and do not distinguish between healthy, morbid, and dead cells. This lack of specificity is more serious in dealing with mammalian cells, especially since many types are adherent or semiadherent, and cell debris may interfere with the turbidometric signal.

ATP Release Assays ATP release is a much more sensitive way of measuring numbers of live cells than turbidometry, but since ATP has a short half-life outside of living cells, these are actually *viability* assays. Using this method, it is possible to obtain separate signals for prokaryotes and eukaryotes by using so-called somatic cell lytic agents, which do not lyse prokaryotes, followed by either lytic detergents or solutions specific for the type of prokaryote under study (see below).

The advantages of ATP release assays are important [41]. The method is rapid, relatively sensitive, and linear over a very wide range. It also has the important advantage relative to more complex release assays that it has been well characterized in a number of systems. However, the disadvantages are also sufficient to motivate a researcher to choose a different assay under some circumstances. The rate of ATP release from various cell types is surprisingly variable, possibly because so much of the compound is trapped inside of various intracellular organelles; multiple membrane systems must therefore be ruptured before a consistent readout can be obtained. Thus the speed and convenience of the method are good but not always excellent. The sensitivity is also approximately 30- to 100-fold poorer than that available from coupled enzyme systems. However, the most important drawback is the inherent statistical flaw in using these assays for measurements of cytotoxicity, in that for sensitive detection of cell death, one must subtract two large numbers, each with its associated error, to obtain the signal strength. Finally, many of the commercial ATP release kits are expensive, a significant limitation in the HTS setting.

Metabolic Assays The other important type of *proliferation* assay is really a method of measuring metabolic rate. These assays include, for example, the MTT, XTT, WST-1, and Alamar Blue techniques. (The BrdU-incorporation assay [42, 43] as currently practiced does not qualify as a high-throughput method.) All of these methods rely on chemical reactions that take place in intact mitochondria. When mitochondrial transmembrane redox potential is lost, the reactions slow or stop. Thus the coupling between the signal and the viability of the cells under study is quite good. However, these assays are inherently flawed in several ways [44]. First, the signal obtained represents the *integral* of metabolism. If the researcher performs a 2.5-h experiment, and all the cells suddenly die at 2.3h, the signal will look almost the same as if they were still alive. (Cytotoxicity can easily exhibit this type of nonlinear time dependence if one is studying a highly cooperative system such as complement.) Moreover, some fairly broad categories of compounds can cause direct changes in the redox signal in the absence of cytotoxic effects [45, 46], and the test reagents are believed to interfere with metabolism in some cases. The assays are also ill-suited to measurements of cytotoxicity for the same statistical reasons associated with ATP release assays.

DeathTRAK in Biomass Measurement A rapid, ultrasensitive assay for total biomass may be an attractive alternative to a metabolism-based viability assay or cell-counting procedure. Speed, accuracy, and easy adaptability to automation are possible reasons. G3PDH can be released from cells and measured in a one-step procedure. Because the enzyme is both catalytically efficient and abundantly present, a signal can generally be obtained in 30s or less, although the researcher may decide on a 3-min assay to attain the specified sensitivity. In fact the assay can be continued until the desired sensitivity is achieved. Figure 16.4 shows the results of 3-min assays of dilutions of detergent-permeabilized cells. Data reduction can be accomplished by automated linear fitting of the time-dependent readouts; however, single-point measurements after a fixed length of time are also quite valid. This is illustrated in Figures 16.2 and 16.5, in which the linear-fit method, using all data collected over 20 min, can be compared with a single readout at 2.6min (see Fig. 16.2). The two graphs yield the same information, and this is generally the case except at very low signal strength.

DeathTRAK Cytotoxicity/Proliferation Dual Mode Unlike ATP, the G3PDH molecule is relatively stable in a cell culture supernatant and can be stabilized further. Because of this, it is possible, following a challenge such as a potential antibiotic or candidate cancer drug, to measure both cytotoxicity (as represented by total membrane lysis or damage) and live biomass in the same sample, simply by adding detergent and remeasuring. This is an alternative to viability/cytotoxicity methods reported in the past, which generally have the same drawbacks as the independent assays that are being combined (e.g., [3-(4,5-dimethylthiazol-2-yl)]-5-(3-carboxymethoxyphenyl)-2-(4-sulfophenyl)-2H-tetrazolium (MTS) reduction and LDH release [47]). The dual-mode DeathTRAK process requires 6min and is depicted schematically in Figure 16.6. Figure 16.7 shows an example of the results obtained with this dual-mode *cytotoxicity/biomass* or *cytotoxicity/proliferation* assay. In this case the assay method was used to measure the dose that is lethal to 50 percent of test subjects (LD_{50}) (cytotoxicity) of a detergent (Nonidet P-40), which was also used

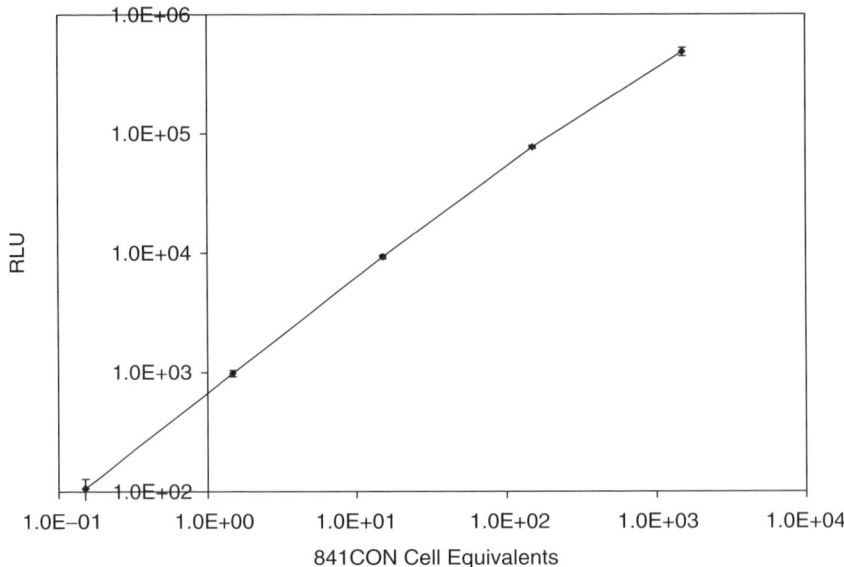

Figure 16.4 Linear response of DeathTRAK in cell counting. 841CON (a normal human colonic epithelium-derived cell line) cells were counted, adjusted to 3×10^6/mL, killed by addition of 0.2% Nonidet P-40, and diluted to yield the indicated cell equivalents in 5-μL aliquots, which were mixed (in triplicate) with 45 μL of DeathTRAK cocktail (Section 16.6). A two-point readout is shown, consisting of the difference between the reading at 2.7 min and the initial reading. The background rate has been subtracted from all points. Error bars (standard deviations) are shown but are small. The 0.15-cell point reading was approximately 50 percent higher than background and was statistically distinguishable from background by T test ($p = .0014$), with a Z factor of 0.13. The Z factor of the 1.5-cell point was 0.78.

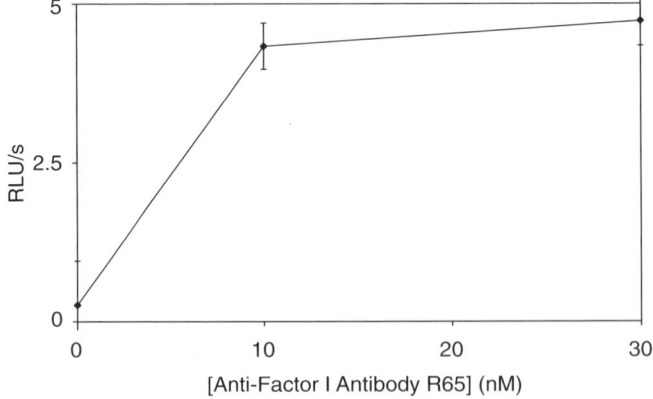

Figure 16.5 Antibody-mediated effects on complement lysis of PC-3 cells measured by one-step DeathTRAK assay (linear-fit analysis). The same experiment as in Figure B is depicted. Linear fits of data gathered up to 20 min after initiation are graphed.

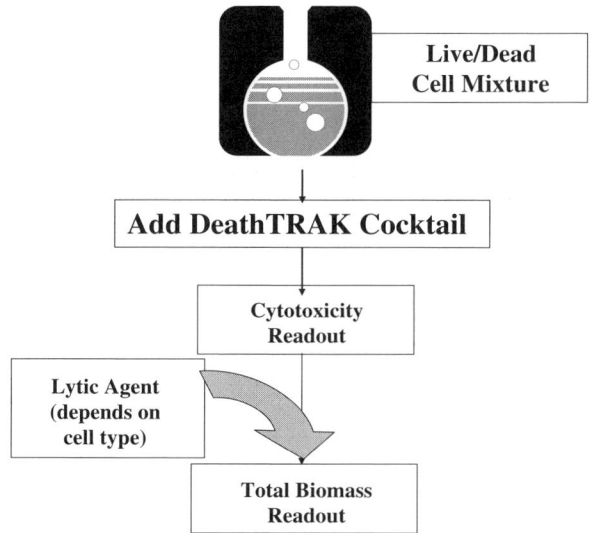

Biomass − Cytotoxicity = Intact Cell Count

Figure 16.6 Scheme of the DeathTRAK cytotoxicity/proliferation dual mode. The data obtained directly are the cytotoxicity measurement (i.e., enzyme released due to the process under test) and the biomass or total cell count (i.e., the sum of the cytotoxicity measurement and the viability readout at the end of the experiment). The viability readout is obtained by subtracting the first value from the second.

as the total lysis agent for the proliferation measurement. One might expect some evidence of interference between the two modes when the same substance is used as both test agent and assay reagent, but fortunately, although DeathTRAK is inhibited about 45 percent by 0.2% Nonidet P-40, this inhibition effect is nearly independent of detergent concentration above 0.2 percent, and very little interference was seen. Thus the "proliferation" signal is virtually constant, reflecting the constant total number of cells seeded into the wells. The calculated LD_{50} obtained from the cytotoxicity readout of the experiment is in agreement with the known critical Micelle concentration of Nonidet P-40. Equivalent results were obtained for the HL-60, PC-3, and T-24 cells lines, except that killing T-24 required a significantly higher concentration of Nonidet P-40 (~2.5-fold greater). In a library-screening paradigm, one can obtain both cytotoxicity (representing direct lysis or membrane damage) and proliferation data (represent static or growth inhibitory effects) from a single plate in the same fashion.

CYTOTOXICITY/PROLIFERATION MEASUREMENTS OF GRAM-NEGATIVE BACTERIA
Gram-negative prokaryotes may be studied in the same way. Figures 16.8 and 16.9 depict respective cytotoxicity and biomass measurements of *Escherichia coli* strain K1 (generously provided by Dr. Craig Rubens of Children's

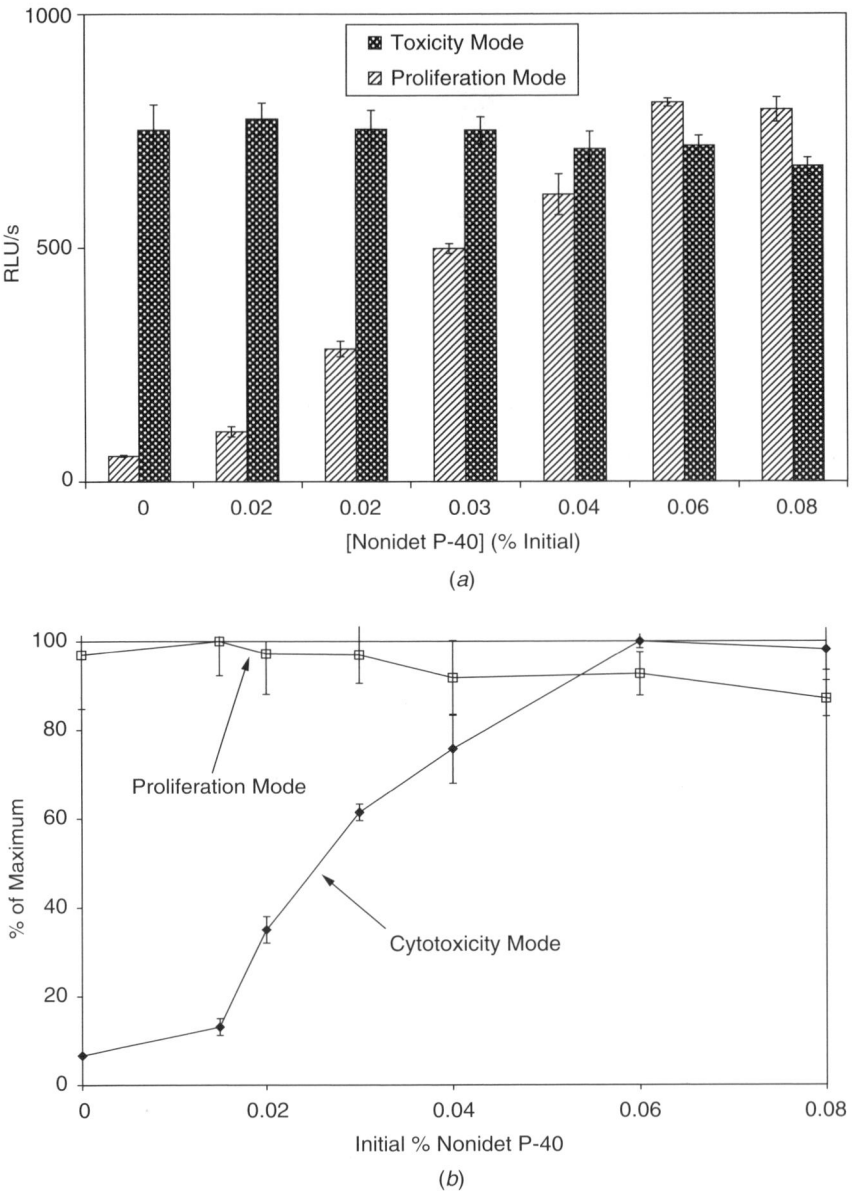

Figure 16.7 Cytotoxicity/proliferation mode with 841CON cells. (*a*) The detergent Nonidet P-40 acted as both the toxin and the total lysis reagent. Detergent was added at the indicated concentration to 1000 841CON cells that had grown overnight in 100 μL IMDM, whereupon standard DeathTRAK cocktail (100 μL) was added and luminance was read for 3 min; 10 μL of 0.42% Nonidet P-40 was then added to each well and the proliferation measurement was taken for 3 min. The data reported are the linear fits from 1 to 3 min of each data set. All time-linear correlation coefficients were >0.999. Note that the displayed errors (standard deviations) incorporate both assay scatter and errors in plating the cells. (*b*) A simple visual presentation, similar to a screening report, of the respective effects on toxicity and proliferation of the evaluated drug (Nonidet P-40) in the assay.

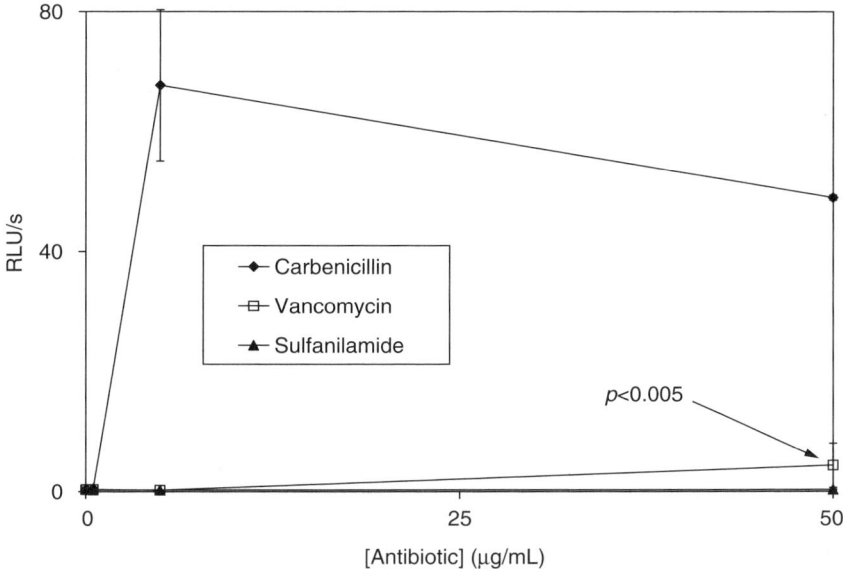

Figure 16.8 DeathTRAK cytotoxicity measurements of the effects of three antibiotics on *E. coli* strain K1. Bacteria were grown to midlog phase, washed into PBS, and resuspended to an A_{600} of 1.549 (~2.18×10^8/mL). These were further diluted to a nominal 2×10^5/mL, whereupon a 10 percent volume of 10 mM dithiothreitol/1 percent PICGUWS (a protease inhibitor; Sigma) was added to the cells; 55 µL of cells were distributed to each well and antibiotics or vehicle (PBS) were added to the indicated final concentrations (reactions in duplicate). After 3 h, 45 µL of the standard DeathTRAK cocktail was added and the luminance was read for 2.5 min. The linear fit is reported (correlations were 0.99 or greater except for one, 0.97).

Hospital and Regional Medical Center, Seattle, Washington) challenged by carbenicillin, vancomycin, and sulfanilamide antibiotics. An estimate of the LD_{50} of carbenicillin can be made from the data shown in Figure 16.9, but no mechanistic information is available; however, the cytotoxicity wing of the same assay shows that carbenicillin causes a loss of cell wall integrity as part of the killing process. Vancomycin has only a small effect on viability of this strain under these conditions but could be identified as a hit in the cytotoxicity wing, since the release data, though much lower than the carbenicillin data, are significantly different from zero (Fig. 16.8). As expected, sulfanilamide was not toxic to *E. coli* K1 under these conditions.

CYTOTOXICITY/PROLIFERATION MEASUREMENTS OF GRAM-POSITIVE BACTERIA Similar data are shown for a Gram-positive organism in Figures 16.10 and 16.11. The proliferation/viability panel shows that all three antibiotics were effective against this group-A streptococcus (also provided by Dr. Craig Rubens), but the LD_{50} are distinct and are easily estimated with considerable

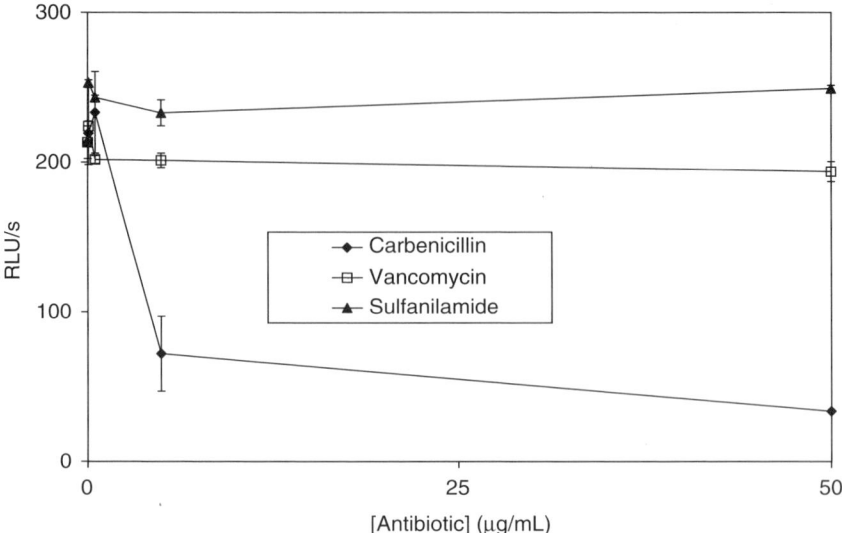

Figure 16.9 DeathTRAK proliferation/viability measurements of the effects of three antibiotics on *E. coli* K1: 10 µL of G⁻ lytic cocktail (Section 16.6) was added to the reactions shown in Figure 16.8 and the luminance was read for a further 2.5 min. Minimum linear correlation: 0.997.

precision from the data obtained in this 3-min assay. The toxicity results for the Gram-positives are more complex, and here it should be noted that Gram-positives are "leakier" than Gram-negatives in DeathTRAK assays. Thus, the toxicity data for Gram-positives amount to a complex average of killing effects, cytostatic effects, and leakage effects. It might be concluded that such a seriously convoluted readout is not useful, but when the data are combined with the viability observations, there may be mechanistic information available that is not apparent from either alone. One interpretation of the toxicity data is that gentamicin kills most of the cells (thereby reducing total synthesis of the G3PDH test enzyme and lowering the overall signal) and makes the rest somewhat leaky; carbenicillin makes them very leaky; and vancomycin is the most effective killer but causes no leakiness beyond the natural rate. Whether or not this sort of the analysis is worthwhile in a given system is a question that must be answered by the individual researcher, but in any case, an antibiotic screening strategy in which baseline readouts for cytotoxicity and viability are established during the development work, and significant anomalies in *either* type of readout are investigated further, is likely to miss fewer useful molecules than methods employing single readouts. Note that if both kinds of data are desired in a very rapid assay, it is necessary to lyse the cells very quickly between the two measurements. This can be accomplished for

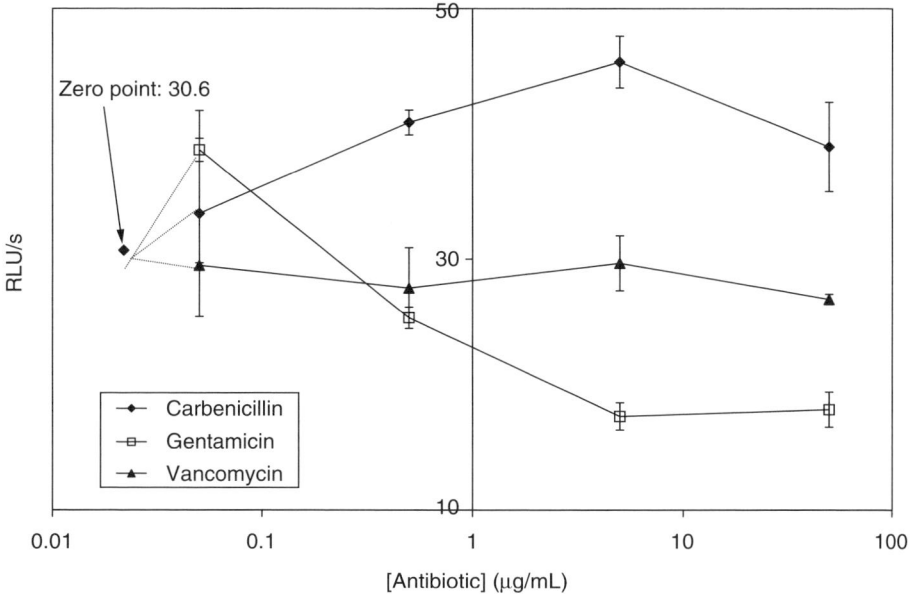

Figure 16.10 Cytotoxicity readout of group-A streptococcus challenged with three antibiotics, measured by DeathTRAK. Group-A Streptococcus (GAS) were grown overnight in THY medium, washed 2X with THY, and resuspended at approximately 4×10^6/mL in the same medium; 50-µL aliquots of the cells were transferred to microtiter wells and antibiotic or vehicle (PBS) was added in 5µL (duplicates). GAS were then incubated for 90 min at 37°C with 240 rpm shaking. The DeathTRAK cocktail (45 µL) was then added directly and luminance was read for 4.3 min; linear fits are reported. Toxicity readout is a sum of natural leakiness of this organism and leakiness due to membrane damage. Error bars are standard deviations.

both the Gram-negatives and Gram-positives studied in these experiments, but the suggested lytic formulations are different (see Section 16.6), and there is no certainty that the same formulations will be effective with other species or strains. Different bacteria, and perhaps even different strains of the same species, may be resistant to lysis by particular detergents and/or pore formers, and new lytic formulations may be needed when such organisms are employed as targets. This stands in contrast to the situation with mammalian cells in which all cell types tested were very rapidly lysed by 0.2% Nonidet P-40.

Summary of Cytotoxicity/Proliferation Methods

Apart from speed, the major parameters influencing the choice of a cytotoxicity assay for use in HTS in a drug discovery paradigm include sensitivity, ease

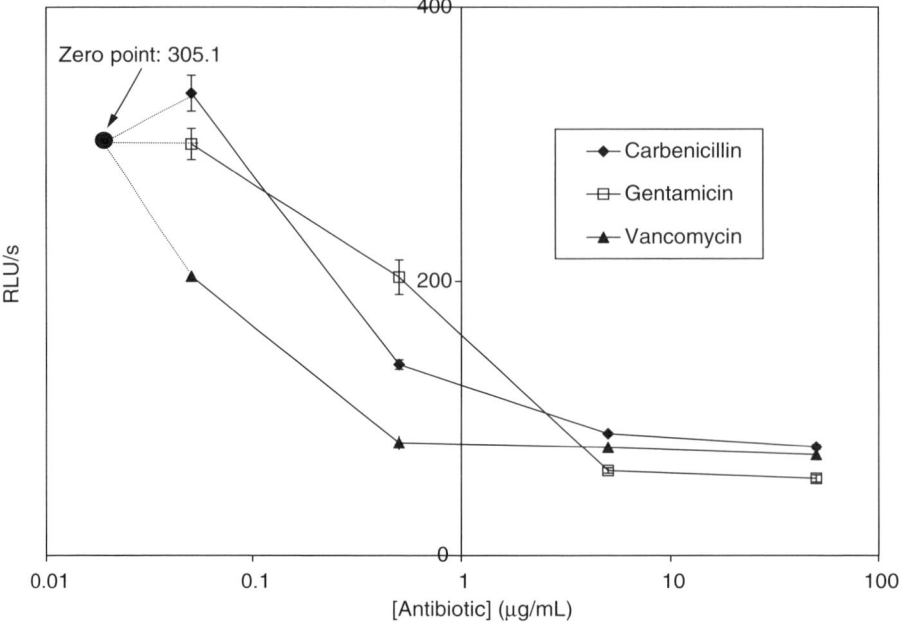

Figure 16.11 Proliferation/viability readout of group-A streptococcus challenged with three antibiotics, measured by DeathTRAK: 10 µL of G^+ lytic cocktail (Section 16.6) was added to the reactions shown in Figure 16.10 and the luminance was read for a further 2.3 min; the linear fit is reported. Error bars (standard deviations) are displayed but most are small.

of operation and adaptability to HTS, cost, information quality, and statistical characteristics. From many of these viewpoints, a coupled luminescent assay such as DeathTRAK has a number of clear advantages. However, in managing such a program, the extent of validation and historical acceptance of a procedure also comes into play, and these considerations must be weighed against the perceived benefits.

16.3 PHOSTRAK: A VERY RAPID, HOMOGENEOUS GENERAL PHOSPHATASE ASSAY

Measurement of Phosphatase Activities

The reaction catalyzed by phosphatases is not the simple reverse of the kinase reaction since no ATP is involved in the phosphatase reaction; nevertheless, in the biological sense, phosphatases may be considered as the dephosphorylating counterpart of kinases, and their importance in biological signal transduction is comparable, if not equal, to that of kinases. The search for modulators of phosphatase activities with druglike characteristics is an intensive and accelerating part of current drug discovery efforts [48, 49]. A substantial portion of the screening is carried out by methods involving specific antibody recognition of reactants or products [49, 50]. While there are various levels of sophistication associated with these methods, ranging all the way from immunoprecipitation [51] to advanced FRET and FP techniques (some of which employ novel high-affinity ligands instead of or in addition to antibodies [49]), these methods are generally slow because the association process is slow, frequently require separations and/or multiple addition, incubation, or mixing steps (although FP assays are often operationally simple), and are often expensive because of the antibodies and chemical conjugates involved. One alternative method that is not available with kinases (without additional coupling enzymes) is direct detection of liberated inorganic phosphate. This possibility is exploited in the malachite green assay [52]. This method has the advantage of using natural substrates, and the materials are inexpensive, but the method is a fairly slow, multistep process. Finally, a rapid chemiluminescent assay has been developed [53]; this assay yields speed and sensitivity, but the major disadvantage is that this technique employs a substrate that is so different from the natural substrates of most phosphatases that any inhibition data gathered would be highly suspect, simply because for most phosphatases the K_m of the unnatural substrate is so poor that the reaction is too easy to inhibit. Moreover, the method requires multiple steps. Thus there is still a need for a rapid, sensitive, homogeneous assay that can make use of natural or quasi-natural substrates.

Introduction to PhosTRAK

PhosTRAK is a coupled luminescent method of measuring the concentration of free phosphate, and, by extension, of phosphatase activity [38]. The biochemical scheme of PhosTRAK is related to that of DeathTRAK (compare Figs. 16.1 and 16.12), and similar speed and sensitivity are achieved; however, performance of PhosTRAK is distinct in a number of respects. Phosphate-free buffers must be used since free phosphate (hereafter, P_i) is the limiting reagent. A more significant difference is that DeathTRAK is a small set of related methods (depending on the cell type), whereas PhosTRAK is essentially a template on which a specific phosphatase assay can be developed.

Applicability of PhosTRAK Method to Various Types of Phosphatases

Phosphatases are a large class of enzymes with varying characteristics; some exhibit metal dependence, while others, especially the class known as acid phosphatases, have pH activity profiles that are significantly different from those of most enzymes. Protein phosphatases have been recognized as critical

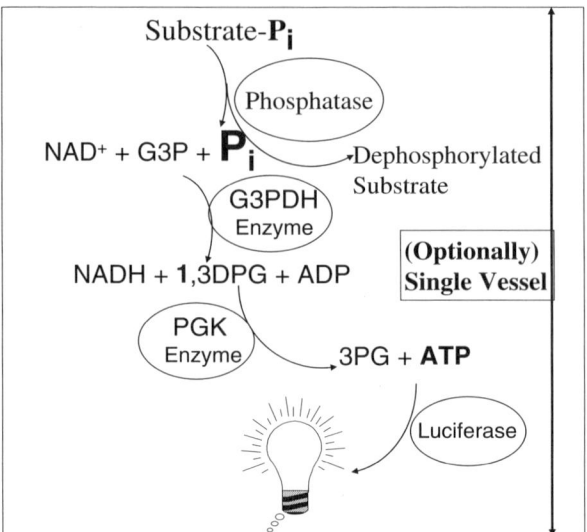

Figure 16.12 Biochemical scheme of PhosTRAK. NAD^+, nicotinamide adenine dinucleotide (oxidized form); G3P, glyceraldehyde-3-phosphate; P_i, PO_4^{3-}; G3PDH, glyceraldehyde-3-phosphate dehydrogenase; NADH, nicotinamide adenine dinucleotide (reduced form); 1,3DPG, 1,3-diphosphoglycerate; ADP, adenosine triphosphate; PGK, phosphoglycerokinase; 3PG, 3-phosphoglycerate; ATP, adenosine triphosphate.

mediators of signaling pathways for some time (see [54–58] for recent reviews of aspects of this field), but many important phosphatases act on small molecules, such as inositol and its derivatives [59, 60] or phosphatidic acids [61, 62]. It is evident from work described herein and by analogy with other assay methods that a fairly small range of substrates is sufficient to handle a large subset of the protein phosphatases [e.g., Thr-Arg-Asp-Ile-Tyr-(PO_3)-Glu-Thr-Asp-Tyr-(PO_3)-Tyr-(PO_3)-Arg-Lys from the insulin receptor [63] and/or the Fischer substrate Glu-Asn-Asp-Tyr-(PO_3)-Ile-Asn-Ala-Ser-Leu [64] should be useful with most protein tyrosine phosphatases], but other phosphatases require individual treatment, including custom-synthesized substrates in some cases. Experiments and common sense suggest that the great majority of phosphatase activities can be measured in one-step reactions using a PhosTRAK-based scheme, although many will require some initial development work. However, there are likely to be a number of phosphatases (acid and alkaline phosphatases are obvious examples) that require separate reaction conditions for the hydrolysis and detection steps, leading to two-step methods, and there may also be phosphatases for which coupled luminescent schemes are simply unworkable or inappropriate.

Representative PhosTRAK Results with Diverse Phosphatases

Figures 16.13 to 16.16 present some of the data that have been obtained to date with PhosTRAK. The *E. coli* λ phosphatase [65] was assayed with unmodified casein as a substrate (Fig. 16.13). Casein has enough natural phosphorylation to provide an excellent, low-cost substrate for many protein phosphatases. While casein is not the natural substrate of these phosphatases, it is arguably much superior to synthetic, nonproteinaceous substrates as an analog of natural protein substrates. The leukocyte antigen-related phosphatase [66] was assayed with the Fischer phosphopeptide substrate [64]. The sensitivity is excellent, but no saturation was seen in this concentration regime (Fig. 16.14). The T-Cell Protein Tyrosine Phosphatase [67, 68] (TCPTP, Sigma) was also successfully assayed in the same system (data not shown). Note that leukocyte antigen-related phosphatase (LARP) and TCPTP were assayed in identical buffers, while the λ phosphatase required only the addition of manganese to this buffer. Calcineurin, however, is a very unusual enzyme, with complex kinetics and cofactor requirements [69]. It was chosen for study primarily because it was thought to represent a special challenge, and, indeed, development of a single-step assay of calcineurin required a modest amount of additional work, and the 3-min goal in a very low-cost assay was not achieved, evidently because this was insufficient time for the autoactivation

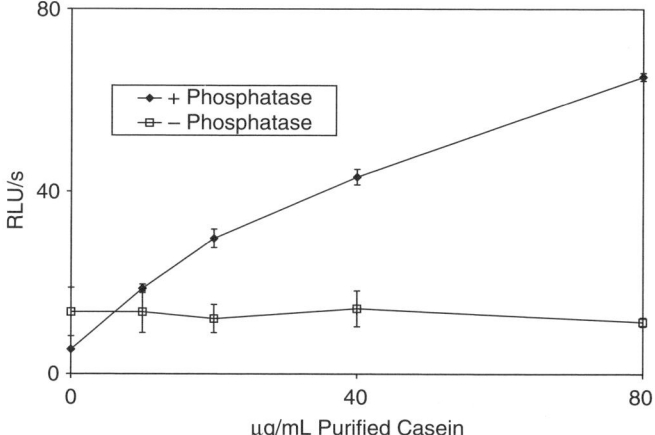

Figure 16.13 *E. coli* λ-phosphatase activity measured by PhosTRAK. λ-Phosphatase was assayed with purified but unmodified α-casein (Sigma) as substrate. Casein was dissolved in H_2O at 400 μg/mL and added in 1/10 of the final assay volume to yield the indicated concentrations. The PhosTRAK cocktail (Section 16.6) with or without λ phosphatase was made 2 mM in $MnCl_2$. Substrate was added in 1/10 volume and luminance was read. Data from 2 to 6 min were taken for analysis; linear fits are reported.

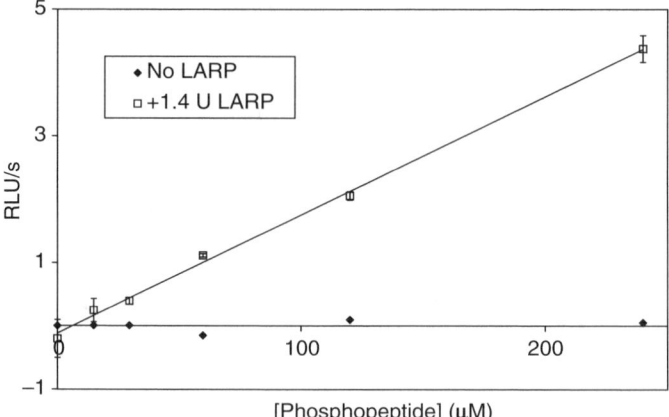

Figure 16.14 Substrate dependence of LAR phosphatase/Fischer substrate reaction measured by PhosTRAK. Measurements were taken with or without 1.4 units LARP (1 unit hydrolyzes 1.0 nmol of *p*-nitrophenylphosphate per minute at 30°C, pH 7.0). The substrate was dissolved at 3 mM in DMSO, and the DMSO concentration was equalized across all reactions at 4 percent; 45 µL of PhosTRAK cocktail with or without 31.1 units/mL LARP were aliquoted into each assay well, the substrate was added in 5 µL, and the luminance was read for 3 min. The background rate without enzyme and substrate was subtracted from the data.

process to act. However, a strong signal was evident at 12 min (Fig. 16.15), using an amount of calcineurin corresponding to approximately 6 percent of the quantity used in the CalBiochem assay based on malachite green. Assays performed in the presence of the calcineurin autoinhibitory peptide [70] required preincubation for the autoactivation and inhibition processes, but the enzyme activity was still measured in a 3-min endpoint reaction (Fig. 16.16).

Limitations Due to Complexity and Thermodynamic Considerations

Although it is possible in principle to couple virtually any reaction to ATP or NADH production by using a sufficiently large and complex system, the bounds of pragmatism enter the assay design process in several ways. If numerous enzymes are present, the difficulties of interpretation are multiplied. Some enzymes may have conflicting requirements for small molecules, pH, cofactors, ionic strength, or other environmental factors. Thermodynamic principles impose other significant restrictions on the possibilities. For example, numerous reactions and reaction series involving ATP produce free phosphate or pyrophosphate, and these reactions can be reversed to yield a direct assay for free phosphate; however, the G3PDH-PGK system is one of a limited number of simple reaction series in which ATP is produced from free phosphate

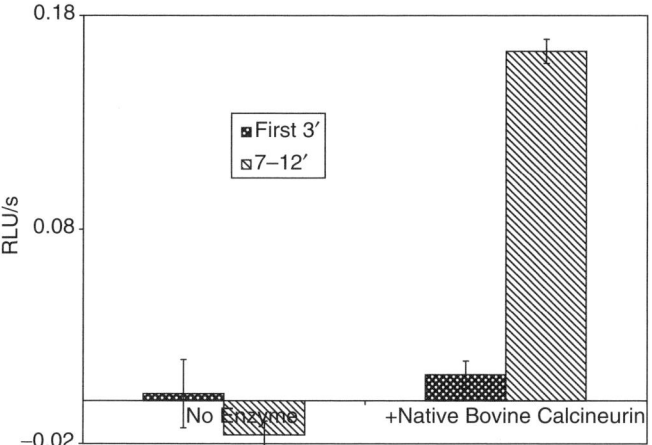

Figure 16.15 PhosTRAK measurement of inhibition of calcineurin by autoinhibitory peptide. Recombinant human calcineurin Aα and B coexpressed subunits (protein phosphatase 2B) were assayed in a buffer containing calmodulin, $MnCl_2$, and other components as recommended by the supplier (CalBiochem, San Diego, CA) with 150 µM RII peptide (DLDVPIPGRFDRRV(PO₃)SVAAE) in the presence of 0, 10, or 30 µM autoinhibitory peptide (AIP) in a 25-µL volume in triplicate. The enzyme with or without AIP was incubated without substrate for 15 min at room temperature. RII peptide was then added and the reaction was transferred to 30°C for 67 min; 5 µL of the reaction, containing 2.2 units of calcineurin (the unit is defined as the amount of enzyme that releases 1 pmol of P_i per minute from the RII peptide at 30°C, pH 7.4), was then transferred to a microtiter well with 45 µL of the standard PhosTRAK cocktail, except that the amounts of PGK and G3PDH enzymes were 50 and 66 percent greater, respectively. Luminance was read after 3.02 min. No enzyme background was subtracted.

exothermically, and this appears to be a key element of the ease and flexibility of PhosTRAK. Alternative systems involving substrate-level phosphorylation are also possible, and some pathways suggest the possibility of other coupled luminescent enzyme assays with applications in bacteriology and antibiotic screening [71].

PhosTRAK Summary

There remains considerable development and validation work to do for PhosTRAK. The availability of ultrapure reagents would likely improve the dynamic range by an order of magnitude at each end. Extensive inhibitor studies would clarify the importance of spurious inhibitory modes due to the presence of multiple enzymes and confirm or refute the hypothesis that targeted libraries are likely to perform best with this assay. The breadth of applications of coupled luminescent assays such as PhosTRAK will gradually

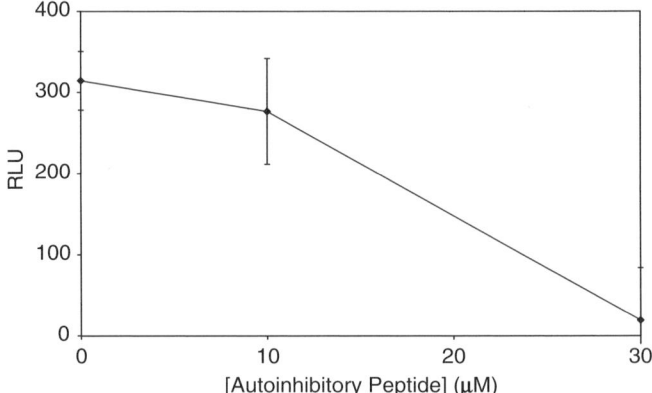

Figure 16.16 Autoactivation of native bovine calcineurin observed by one-step PhosTRAK assay. Forty units of enzyme were incubated in duplicate with a combination of the calcineurin reaction and PhosTRAK cocktail modified as described in the legend to Figure 16.15. Luminance was measured every 20 s for 12 min. Linear fits of the first 3 min and the interval from 7 to 12 min are shown.

become apparent as additional formulations permit one-step reactions with a broader range of enzymes. In still other coupled systems, the phosphatase reaction may merely be an intermediate coupling step downstream from a different enzyme of interest.

16.4 COUPLED LUMINESCENT ASSAYS IN PRACTICE: SPECIFICATIONS AND LIMITATIONS

Assay Specifications

Specification parameters of the coupled luminescent assays described herein include speed, sensitivity, operational simplicity, Z values, and cost.

Speed and Linear Response With the exception of the 12-min calcineurin assay described above, the one-step assays described herein can be accomplished in 3 min from initiation. If a single reading is to be taken, a G3PDH standard curve should be included in the experiment to assure that the readings are within the linear range of the response. If biochemical saturation is observed (i.e., a signal outside the linear range of a G3PDH standard curve), the data from 0 to 2 min may be usable, if they have been captured; alternatively, the samples can be diluted or the reaction cocktail adjusted appropriately (usually by adding ADP). Given that signal generation begins immediately in most cases, if a very rapid luminescence reader is available, it may be possible to develop an assay with a plate turnover under 30 s, although a sophisticated injection scheme would be required.

Sensitivity
DeathTRAK Sensitivity/Limit of Detection The DeathTRAK assay as described, using vendor-supplied materials without further purification, is capable of distinguishing the G3PDH released by 0.15 nucleated mammalian cell equivalent from the background signal. An acceptable Z value is achieved at about one nucleated mammalian cell.

PhosTRAK Sensitivity/Limit of Detection PhosTRAK as described, using vendor-supplied materials without further purification, can distinguish 3 pmol of P_i from the background signal.

Operational Characteristics Both DeathTRAK and PhosTRAK are fundamentally one-step assays, in which the reaction under study occurs simultaneously with the measurement, and only a single reagent addition is required. However, this does not imply that no other steps are needed under any circumstances. For example, for cytotoxicity measurements by DeathTRAK, the usual practice is to initiate the (potentially) cytotoxic process, wait for an effect, and then add the reaction cocktail and read. In the case of PhosTRAK, true one-step protocols appear to be possible for many or most phosphatases, but there are exceptions, as noted. A related issue is *how* to add the cocktail. If all the components of the coupled reaction are stored together while the test reactions wait for equipment, or for other plates to be read, undesirable side reactions can occur (as well as the backward reaction; see below). Usually it is possible to inject the critical enzymes separately from the other components or to include an essential factor in the test well instead of the reagent reservoir; if not, the reaction cocktail can be mixed just before the run. Another matter is whether to keep components cold. Obviously, this depends on their stability and how long they will have to be stored before use. Again, with a little thought given to protocol design, it may be possible to keep critical reagents cold in a small volume that will have little effect on the temperature in the well after injection. In any case, temperature equilibration is fairly rapid inside a typical luminometer, and as a result, temperature artifacts with significant consequences are relatively rare.

Z Values The Z (or Z′) value or factor [72] is often reported as a measure of assay "quality." Calculation of the Z factor [Eq. (16.1)] takes into account both the signal strength and the scatter of the positive and background signals, while properly ignoring the absolute background level, essentially conveying the ability of the assay as performed to distinguish signal variations from noise:

$$Z = 1 - \frac{3(\text{STDEV}_{\text{positive}} + \text{STDEV}_{\text{negative}})}{\text{AVG(positive control)} - \text{AVG(negative control)}} \quad (16.1)$$

where STDEV is the standard deviation, AVG is the arithmetic mean, and "positive control" refers to the maximum signal obtained without inhibition.

Of interest is the fact that the absolute level of the background signal does not enter into the calculation of the Z factor; only the variabilities and the difference between positive and negative signals matter. The implication of this for coupled luminescent systems is that limited contamination of assay components by reagents under test (such as ATP in the DeathTRAK system, and either ATP or P_i in the PhosTRAK system) has little or no effect on the Z factor since the contamination introduces a significant but *constant* background signal. In a sense this corresponds to reality in that what is important is the statistical distinction between signal and background, rather than the absolute level of either. However, this can be carried too far: Some sources of noise scale with the background signal strength, and a very high background level can cause biochemical and/or detector saturation or even unwanted side reactions.

The Z factor is useful in that it informs the researcher as to whether a desired degree of change in a phenomenon will be detectable at a high confidence level. However, the Z factor has inherent limitations and is sometimes misapplied. The Z factor describes the performance of an assay in a given formulation, but, since error modes typically observed in HTS often do not scale with signal strength, it is often possible to "buy" a high Z factor by simply using more of the target agent; this, however, requires the use of more of the test compounds as well to make use of the increased amount of target. Moreover, the value of a high Z factor is frequently overestimated: Strong changes in signal are usually the most important, and a Z factor of 0.5 or greater is generally quite adequate for detection of these interactions. Rather than the magnitude of the Z factor, the *cost* of achieving an acceptable Z factor is usually a better basis for decision making. Conversely, when reagents are inexpensive, higher Z factors may permit identification of less pronounced effects that may aid in deconvoluting structure–function relationships or provide lead compounds subject to improvements. The assays described herein have been optimized for sensitivity, rapidity, and low cost; however, the very nature of one-step assays permits fewer opportunities for the introduction of measurement errors, and the Z values of these and other operationally simple assays often reflect this simplicity. Representative Z values obtained were as follows: Figure 16.5, 0.77; Figures 16.7a and 16.7b, 0.87; Figure 16.8, 0.97; Figure 16.9, 0.82. Figure 16.10 exhibits Z values of <0.5 because the leakiness of the Gram-positive organisms leads to scatter in the control; however, in Figure 16.11, the biomass data, an average Z value of 0.77 ± 0.16, is obtained over all the runs at 0.5 to 50μg/mL of the antibiotics. The Z value of the λ phosphatase assay (Fig. 16.13) is 0.90. In the LARP assay (Fig. 16.14), Z values of 0.60 ± 0.21 were obtained at 15 to 120μM substrate. Despite the low signal strength, the Z value of the one-step calcineurin assay depicted in Figure 16.15 is 0.80, reflecting the excellent separation between signal and background. Typical Z values of DeathTRAK and PhosTRAK assays using HTS conditions are 0.6 to 0.9.

Cost Even at full retail prices, the cost of these assays per 45-μL aliquot is less than $0.05. Most of this is due to the luciferase, which can be cloned and expressed, or purified, or obtained in bulk.

Backward Reactions

In reaction series in which time-dependent accumulation of ATP is an essential component of the signal, care must be taken to avoid artifactual fluctuations in the ATP concentration due to backward reactions of luciferase. These reactions can occur if all three products of the canonical luciferase reaction are present, that is, AMP, pyrophosphate, and light. The kind of problem this can create is depicted in Figure 16.17. Here the components of the Death-TRAK reaction series have been mixed and intentionally incubated in the presence of room light for several minutes before initiation of the test reaction. Because there is a small amount of G3PDH contaminating the PGK preparation, the test reaction can proceed slowly even in the absence of added G3PDH, yielding products of the luciferase reaction. A pseudo-steady-state develops in which the concentration of ATP is maintained at a slightly elevated level by the action of the backwards luciferase reaction on these products, together with available photons from ambient light. When the reaction is initiated by adding the sample, the microtiter plate is simultaneously transferred to the dark chamber of the luminometer. This alters the pseudo-steady-state concentration of ATP since high levels of light are no longer available and the backwards reaction cannot occur. The concentration of ATP slowly drops until a new "dark pseudo-steady-state" is achieved, at which point the response to the sample can be measured accurately. As Figure 16.17 illustrates, the delay can be as long as 10 min. One half of the plate was covered with an opaque plate seal during the preincubation; this reduced but did not eliminate the lag phase. The solution is simple in theory and usually, in practice: Once the full reagent cocktail has been mixed, it should be protected from light.

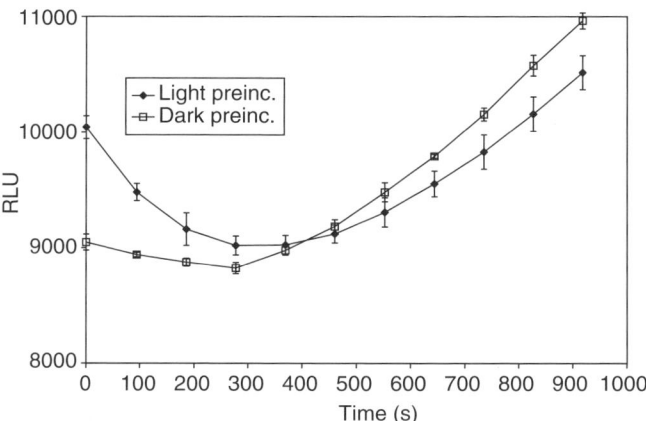

Figure 16.17 Demonstration of light-dependent lag phase of DeathTRAK. Standard DeathTRAK assays using detergent-lysed Raji cell (in triplicate) were preincubated under a fluorescent lamp for 10 min and immediately transferred to luminometer chamber for reading. Covering half of the plate with an opaque plate seal reduced but did not eliminate the lag phase.

Another approach is to supply the elements of the cocktail in two separate injections, or provide a portion of it with the sample, so that the test reaction cannot occur before initiation. The time-linear correlation coefficients reported herein, many greater than 0.999, were obtained either by protecting the cocktails from light or by adding the PGK component immediately before assay initiation.

16.5 SUMMARY AND CONCLUSIONS

Coupled luminescent technology has wide potential. Two novel assay methods based on this technology are described in detail herein, but the range of potential applications is great and has been little explored to date. As these examples make clear, the practitioner of coupled luminescent assays can typically expect very rapid and sensitive readouts with minimal sample and reagent usage. However, there are idiosyncrasies and limitations inherent to the method, many of which are associated with the presence of multiple catalytic activities; only the use of proper controls and secondary screens can assure correct interpretation of the results. It is hoped that this review will aid the drug discovery researcher in deciding whether and how coupled luminescent techniques can advance the goals of his or her program.

16.6 FORMULATIONS

Reaction Master Mixes (Cocktails) and Dilution Buffers

DeathTRAK Standard Cocktail   The standard cocktail is used for measurement of both eukaryotes and prokaryotes. The cocktail may be made as specified, using aliquots of the 4XGP cocktail, which may be stored at −70°C for up to 2 years. Alternatively, the nonlabile cocktail (below) may be stored for up to 1 year.

IMDM mammalian growth medium	31.2%
Phosphate-buffered saline (PBS)	16.9%
ATP Assay Diluent (Sigma)	38.1%
4XGP cocktail (see below)	8.5%
ATP Assay Cocktail (Sigma)	4.2%
Phosphoglycerokinase*	0.11 units/mL (diluted with PGK diluent, below)
Dithiothreitol	0.0004 M final
ADP	3 µM final*

*In working with some systems, the ADP and phosphoglycerokinase concentrations may be adjusted upward to improve sensitivity, especially if the static background level and rate of increase (respectively) are not important concerns. If extreme sensitivity and a very wide dynamic range are needed, it may be fruitful to make or obtain an ADP preparation that is substantially free of ATP [22–24] and eliminate or inactivate the G3PDH contamination in the PGK preparation. Unit definition for PGK: "One unit will convert 1.0 µmol of 1,3-diphosphoglycerate to 3-phosphoglycerate per min at pH 6.9 at 25°C."

4XGP Cocktail
80% 5X PGK diluent (below)
0.004 M nicotinamide adenine dinucleotide (oxidized form)
0.012 M glyceraldehyde-3-phosphate
0.004 M dithiothreitol
QS to volume with H_2O

5X PGK Diluent
0.5 M triethanolamine
0.25 M NaH_2PO_4
0.005 M EDTA (ethylenediaminetetraacetic acid)
0.5 mg/mL bovine serum albumin (fraction V)

Nonlabile Cocktail In the alternative composition, which is easier to reconstitute but has a shorter shelf life, all the reagents except the ATP assay mix, phosphoglycerokinase, and ADP are mixed together and stored at −70°C in suitable aliquots. This method has the advantage that the ADP solution can be made fresh to adjust to the ATP contamination level of a particular lot, or to take advantage of purified ADP if available.

1X G3PDH Diluent
To 200 parts 5X PGK diluent add:
  1 part 1 M dithiothreitol
  799 parts H_2O

PhosTRAK Formulations

PhosTRAK Standard Cocktail
0.05 M Tris-HCl (final) (pH 7.4)
38% ATP assay diluent
4.2% ATP assay mix
PGK = 4.5×10^{-4} units/mL
G3PDH = 4.9×10^{-4} units/mL
8.1% Phosphate-free 4XGP cocktail (below)

QS to volume with H_2O
(Add shortly before assay in minimal volume)
6 µM ADP
3 µM G3P

Unit definition for G3PDH: "One unit will reduce 1.0 µmol of 3-phosphoglycerate to D-glyceraldehyde-3-phosphate per minute in a coupled system with 3-phosphoglyceric phosphokinase at pH 7.6 at 25°C."

Phosphate-Free GP Cocktail
83.5% 5× phosphate-free PGK diluent
0.004 M dithiothreitol
0.0042 M NAD^+
QS to volume with H_2O

Phosphate-Free 5X PGK Diluent
0.5 M triethanolamine
0.25 M Tris pH 7.4
0.005 M EDTA
0.5 mg/mL bovine serum albumin (fraction V)
Titrate to pH 7.3 with HCl; QS to volume with H_2O

Phosphate-Free Nonlabile Cocktail (NLC) An NLC can also be made up for the PhosTRAK reaction, as follows. However, if long-term storage is intended, it should be kept in mind that G3P is more stable at mildly acidic pH. Hydrolyzed G3P will, of course, contribute to the background rate of the PhosTRAK signal. G3P can be omitted from the NLC and added just prior to the assay, or the pH and buffer strength of the NLC can be adjusted to suit the storage requirements.

0.05 M Tris-HCl (final) (pH 7.4)
38% ATP assay diluent
8.1% Phosphate-free 4XGP cocktail (below)
6 µM ADP
3 µM G3P
0.001 M dithiothreitol
QS to volume with H_2O
PGK, G3PDH, ATP assay mix, and (optionally) G3P are added shortly before the run.

Lytic Formulations

Although these formulations were found to be effective with the cell types and strains specified, it is clear that some strains are resistant to certain detergents and/or antibiotics (such as polymyxin B). In the case of mammalian cells, no cell type tested has proven to be resistant to Nonidet P-40, but the possibility that such a cell type exists still remains. Thus the lytic formulation under consideration must be carefully tested with the cells to be used for screening.

Mammalian Cells and Gram-Positive Prokaryotes For mammalian cells and Gram-positive prokaryotes, add the lytic agent in 10 percent of the culture volume to achieve the following final concentrations (before addition of DeathTRAK cocktail):

0.2% Nonidet P-40 or Igepal
0.001 M dithiothreitol

In the modes of use tested and envisioned, the assay cocktail is added immediately after the lytic agent; thus no pH buffering is required in the lytic agent. However, if a contemplated application involves prolonged incubation in the presence of the lytic agent, it may be desirable to buffer the lytic agent to prevent inactivation of G3PDH. The use of protease inhibitors may also be indicated in such cases.

Gram-Negative Prokaryotes For Gram-negatives, the lytic agent is made up in PBS and added in 10 percent of the culture volume to yield final concentrations (prior to addition of the DeathTRAK cocktail) as follows:

3000 units/mL polymyxin B
25 mg/mL chicken egg-white lysozyme

ABBREVIATIONS

3PG	3-phosphoglycerate
4XGP	4X glyceraldehyde-3-phosphate dehydrogenase/ phosphoglycerokinase cocktail
ADP	adenosine diphosphate
AMP	adenosine monophosphate
ATP	adenosine triphosphate
BrdU	bromodeoxyuridine
cAMP	cyclic adenosine monophosphate
CTL	cytotoxic T lymphocyte
DTT	dithiothreitol

FP	fluorescence polarization
FRET	fluorescence resonance energy transfer
G3P	glyceraldehyde-3-phosphate
G3PDH	glyceraldehyde-3-phosphate dehydrogenase
GPCR	G-protein coupled receptor
GPL	glyceraldehyde-3-phosphate dehydrogenase/ phosphoglycerokinase/luciferase
HTS	high-throughput screening
IND	investigational new drug
K_m	Michaelis constant
LAK	lymphokine-activated killer
LARP	leukocyte antigen-related phosphatase
LDH	lactate dehydrogenase
MTT	3-[4,5-dimethylthiazol-2-yl]-2,5-diphenyltetrazolium bromide
NAD⁺	nicotinamide adenine dinucleotide (oxidized form)
NADH	nicotinamide adenine dinucleotide (reduced form)
NADPH	nicotinamide adenine dinucleotide phosphate (reduced form)
NK	natural killer
NLC	nonlabile cocktail
PBS	phosphate-buffered saline
PDE	phosphodiesterase
PGK	phosphoglycerokinase
P_i	inorganic phosphate = PO_4^{3-}
RLU	relative luminance units
SPA	scintillation proximity assay
TCPTP	T-cell protein tyrosine phosphatase
WST-1	2-(4-iodophenyl)-3-(4-nitrophenyl)-5-(2,4-disulfophenyl)-2H-tetrazolium
XTT	2,3-Bis(2-methoxy-4-nitro-5-sulfophenyl)-2H-tetrazolium-5-carboxanilide

REFERENCES

1. Berthold, F. (1990). In K. Van Dyke and R. Van Dyke (Eds.), *Luminescence Immunoassay and Molecular Applications*. CRC Press, Boca Raton, FL, pp. 11–25.
2. Khalil, O. S. (2002). In J. E. Butler (Ed.), *Immunochemistry of Solid Phase Immunoassay*. CRC Press, Boca Raton, FL, pp. 67–83.
3. Jeffery, J. A., Sharom, J. R., Fazekas, M., Rudd, P., Welchner, E., Thauvette, L., White, P. W. (2002). An ATPase assay using scintillation proximity beads for high-throughput screening or kinetic analysis. *Anal. Biochem.*, 304, 55–62.
4. Skorey, K. I., Kennedy, B. P., Friesen, R. W., Ramachandran, C. (2001). Development of a robust scintillation proximity assay for protein tyrosine phosphatase 1B using the catalytically inactive (C215S) mutant. *Anal. Biochem.*, 291, 269–278.

5. McDonald, O. B., Chen, W. J., Ellis, B., Hoffman, C., Overton, L., Rink, M., Smith, A., Marshall, C. J., Wood, E. R. (1999). A scintillation proximity assay for the Raf/MEK/ERK kinase cascade: High-throughput screening and identification of selective enzyme inhibitors. *Anal. Biochem.*, 268, 318–329.
6. Sills, M. A., Weiss, D., Pham, Q., Schweitzer, R., Wu, X., Wu, J. J. (2002). Comparison of assay technologies for a tyrosine kinase assay generates different results in high throughput screening. *J. Biomol. Screen.*, 7, 191–214.
7. Schindler, P. W., Konig, W., Chatterjee, S., Ganguli, B. N. (1986). Improved screening for beta-lactam antibiotics. A sensitive, high-throughput assay using DD-carboxypeptidase and a novel chromophore-labeled substrate. *J. Antibiotics (Tokyo)*, 39, 53–57.
8. Voss, K., Galensa, R. (2000). Determination of L- and D-amino acids in foodstuffs by coupling of high-performance liquid chromatography with enzyme reactors. *Amino Acids*, 18, 339–352.
9. Gutheil, W. G., Stefanova, M. E., Nicholas, R. A. (2000). Fluorescent coupled enzyme assays for D-alanine: Application to penicillin-binding protein and vancomycin activity assays. *Anal. Biochem.*, 287, 196–202.
10. Sudhaharan, T., Reddy, A. R. (1999). A bifunctional luminogenic substrate for two luminescent enzymes: Firefly luciferase and horseradish peroxidase. *Anal. Biochem.*, 271, 159–167.
11. Ross, W. N. (1989). Changes in intracellular calcium during neuron activity. *Annu. Revi. Physiol.*, 51, 491–506.
12. Li, F., Zhang, C., Guo, X., Feng, W. (2003). Chemiluminescence detection in HPLC and CE for pharmaceutical and biomedical analysis. *Biomed. Chromatogr.*, 17, 96–105.
13. Dupriez, V. J., Maes, K., Le Poul, E., Burgeon, E., Detheux, M. (2002). Aequorin-based functional assays for G-protein-coupled receptors, ion channels, and tyrosine kinase receptors. *Receptors and Channels*, 8, 319–330.
14. Hastings, J. W., Gibson, Q. H. (1963). Intermediates in the bioluminescent oxidation of reduced flavin mononucleotide. *J. Biol. Chem.*, 238, 2537–2554.
15. Anderstam, B., Gutierrez, A., Lundin, A., Alvestrand, A. (1998). A luminometric assay for determination of ethanol in microdialysates. *Scand. J. Clin. Lab. Invest.*, 58, 89–96.
16. Huang, W., Feltus, A., Witkowski, A., Daunert, S. (1996). Homogeneous bioluminescence competitive binding assay for folate based on a coupled glucose-6-phosphate dehydrogenase—Bacterial luciferase enzyme system. *Anal. Chem.*, 68, 1646–1650.
17. Wieland, E., Wilder-Smith, E., Kather, H. (1986). Automatic bioluminescent glucose determination using commercially available reagent kits coupled to the bacterial NAD(P)H-linked luciferase system. *J. Clin. Chem. Clin. Biochem.*, 24, 399–403.
18. Robrish, S. A., Curtis, M. A., Sharer, S. A., Bowen, W. H. (1984). The analysis of picomole amounts of L(+)- and D(−)-lactic acid in samples of dental plaque using bacterial luciferase. *Anal. Biochem.*, 136, 503–508.
19. Schoelmerich, J., Hinkley, J. E., Macdonald, I. A., Hofmann, A. F., DeLuca, M. (1983). A bioluminescent assay for 12-alpha-hydroxy bile acids using immobilized enzymes. *Anal. Biochem.*, 133, 244–250.

20. Nygaard, S. F., Rorth, M. (1969). An enzymatic assay of 2,3-diphosphoglycerate in blood. *Scand. J. Clin. Lab. Invest.*, *24*, 399–403.

21. Loos, J. A., Prins, H. K. (1971). A mechanized system for the determination of ATP + ADP, 2,3-diphosphoglycerate, glucose 1,6-diphosphate and lactate in small amounts of blood cells. *Circulation*, *43*, 1141–1146.

22. de Korte, D., Haverkort, W. A., van Gennip, A. H., Roos, D. (1985). Nucleotide profiles of normal human blood cells determined by high-performance liquid chromatography. *Anal. Biochem.*, *147*, 197–209.

23. Crescentini, G., Stocchi, V. (1984). Fast reversed-phase high-performance liquid chromatographic determination of nucleotides in red blood cells. *J. Chromatogr.*, *290*, 393–399.

24. Rochette-Egly, C., Kempf, J., Egly, J. M. (1979). A new chromatographic method using immobilized acriflavin for measuring cyclic AMP in cells prelabeled with radioactive adenine. *J. Cyclic Nucleotide Res.*, *5*, 397–406.

25. Lundin, A. (2000). Use of firefly luciferase in ATP-related assays of biomass, enzymes, and metabolites. *Methods Enzymol.*, *305*, 346–401.

26. Lundin, A., Gerhardt, W., Lindberg, K., Lovgren, T., Nordlander, R., Nyquist, O., Styrelius, I. (1979). Determination of creatine kinase B subunit activity by continuous monitoring of ATP—A new bioluminescent technique applied to clinical chemistry. *Clin. Biochem.*, *12*, 214–215.

27. Welch, A. R. (2000). High throughput assay method for enzymes which metabolically hydrolyze nucleoside triphosphates and an assay system therefore. Patent No. WO 00/18950, assigned to Diagnon Corporation.

28. Learish, R. D., Shultz, J., Ho, S., Bulleit, R. F. (2002). Small-scale telomere repeat sequence content assay using pyrophosphorolysis coupled with ATP detection. *Biotechniques*, *33*, 1349–1353.

29. Rhodes, R. B., Lewis, K., Shultz, J., Huber, S., Voelkerding, K. V., Leonard, D. G., Tsongalis, G. J., Kephart, D. D. (2001). Analysis of the factor V Leiden mutation using the READIT™ assay. *Mol. Diagnostics*, *6*, 55–61.

30. Tsongalis, G. J., Rainey, B. J., Hodges, K. A. (2001). READIT: A novel technology used in the interrogation of nucleic acid sequences for single-nucleotide polymorphisms. *Exper. Mol. Pathol.*, *71*, 222–225.

31. Rocco, P. R., Momesso, D. P., Figueira, R. C., Ferreira, H. C., Cadete, R. A., Legora-Machado, A., Koatz, V. L., Lima, L. M., Barreiro, E. J., Zin, W. A. (2003). Therapeutic potential of a new phosphodiesterase inhibitor in acute lung injury. *Eur. Respir. J.*, *22*, 20–27.

32. Reffelmann, T., Kloner, R. A. (2003). Therapeutic potential of phosphodiesterase 5 inhibition for cardiovascular disease. *Circulation*, *108*, 239–244.

33. Boolell, M., Allen, M. J., Ballard, S. A., Gepi-Attee, S., Muirhead, G. J., Naylor, A. M., Osterloh, I. H., Gingell, C. (1996). Sildenafil: An orally active type 5 cyclic GMP-specific phosphodiesterase inhibitor for the treatment of penile erectile dysfunction. *Int. J. Impotence Res.*, *8*, 47–52.

34. Schafer, H., Schafer, A., Kiderlen, A. F., Masihi, K. N., Burger, R. (1997). A highly sensitive cytotoxicity assay based on the release of reporter enzymes, from stably transfected cell lines. *J. Immunol. Methods*, *204*, 89–98.

35. Kasatori, N., Urayama, T., Mori, T., Ishikawa, F. (1994). Cytotoxicity test based on luminescent assay of alkaline phosphatase released from target cells [in Japanese]. *Rinsho Byori*, *42*, 1050–1054.

36. Rich, G. T., Dawson, A. P., Pryor, J. S. (1984). Glyceraldehyde-3-phosphate dehydrogenase release from erythrocytes during haemolysis. No evidence for substantial binding of the enzyme to the membrane in the intact cell. *Biochem. J.*, *221*, 197–202.

37. Corey, M. J., Kinders, R. J., Brown, L. G., Vessella, R. L. (1997) A very sensitive coupled luminescent assay for cytotoxicity and complement-mediated lysis. *J. Immunol. Methods*, *207*, 43–51.

38. Corey, M. J., Kinders, R. J. (2002). Methods and compositions for coupled luminescent assays. U. S. patent application No. 10/071,350, assigned to M. J. Corey.

39. Corey, M. J., Kinders, R. J., Poduje, C. M., Bruce, C. L., Rowley, H., Brown, L. G., Hass, G. M., Vessella, R. L. (2000). Mechanistic studies of the effects of anti-factor H antibodies on complement-mediated lysis. *J. Biol. Chem.*, *275*, 12917–12925.

40. Hardwick, A., McMillen, D., Martinez, J., Austin, A., Posey, A., Ave-Teel, C., Maples, P., Schneider, S. (2003). Clinical-scale production of antigen-specific T cells directed against hepatitis B virus. *Bioproc. J.*, *2*, 27–31.

41. Slater, K. (2001). Cytotoxicity tests for high-throughput drug discovery. *Curr. Opin. Biotechnol.*, *12*, 70–74.

42. Crissman, H. A., Steinkamp, J. A. (1987). A new method for rapid and sensitive detection of bromodeoxyuridine in DNA-replicating cells. *Exper. Cell Res.*, *173*, 256–261.

43. Maybaum, J., Kott, M. G., Johnson, N. J., Ensminger, W. D., Stetson, P. L. (1987). Analysis of bromodeoxyuridine incorporation into DNA: Comparison of gas chromatographic/mass spectrometric, CsCl gradient sedimentation, and specific radioactivity methods. *Anal. Biochem.*, *161*, 164–171.

44. Niu, Q., Zhao, C., Jing, Z. (2001). An evaluation of the colorimetric assays based on enzymatic reactions used in the measurement of human natural cytotoxicity. *J. Immunol. Methods*, *251*, 11–19.

45. Bruggisser, R., von Daeniken, K., Jundt, G., Schaffner, W., Tullberg-Reinert, H. (2002). Interference of plant extracts, phytoestrogens and antioxidants with the MTT tetrazolium assay. *Planta Med.*, *68*, 445–448.

46. Natarajan, M., Mohan, S., Martinez, B. R., Meltz, M. L., Herman, T. S. (2000). Antioxidant compounds interfere with the 3-[4,5-dimethylthiazol-2-yl]-2,5-diphenyltetrazolium bromide cytotoxicity assay. *Cancer Detect. Prevent.*, *24*, 405–414.

47. Wong, J. K., Kennedy, P. R., Belcher, S. M. (2001). Simplified serum- and steroid-free culture conditions for high-throughput viability analysis of primary cultures of cerebellar granule neurons. *J. Neurosci. Methods*, *110*, 45–55.

48. Doman, T. N., McGovern, S. L., Witherbee, B. J., Kasten, T. P., Kurumbail, R., Stallings, W. C., Connolly, D. T., Shoichet, B. K. (2002). Molecular docking and high-throughput screening for novel inhibitors of protein tyrosine phosphatase-1B. *J. Med. Chem.*, *45*, 2213–2221.

49. Parker, G. J., Law, T. L., Lenoch, F. J., Bolger, R. E. (2000). Development of high throughput screening assays using fluorescence polarization: Nuclear receptor-ligand-binding and kinase/phosphatase assays. *J. Biomol. Screen.*, 5, 77–88.
50. Mendoza, L. G., McQuary, P., Mongan, A., Gangadharan, R., Brignac, S., Eggers, M. (1999). High-throughput microarray-based enzyme-linked immunosorbent assay (ELISA). *Biotechniques*, 27, 778–780, 782–786, 788.
51. Agazie, Y. M., Hayman, M. J. (2003). Development of an efficient "substrate-trapping" mutant of Src homology phosphotyrosine phosphatase 2 and identification of the epidermal growth factor receptor, Gab1, and three other proteins as target substrates. *J. Biol. Chem.*, 278, 13952–13958.
52. Baykov, A. A., Evtushenko, O. A., Avaeva, S. M. (1988). A malachite green procedure for orthophosphate determination and its use in alkaline phosphatase-based enzyme immunoassay. *Anal. Biochem.*, 171, 266–270.
53. Bronstein, I. Y. (2000). Method of detecting a substance using enzymatically-induced decomposition of dioxetanes. U.S. Patent No. RE36,536, assigned to Tropix, Inc.
54. Charbonneau, H., Tonks, N. K. (1992). 1002 Protein phosphatases? *Annu. Rev. Cell Biol.*, 8, 463–493.
55. Tonks, N. K. (2003). PTP1B: From the sidelines to the front lines! *FEBS Lett.*, 546, 140–148.
56. Neel, B. G., Gu, H., Pao, L. (2003). The 'Shp'ing news: SH2 domain-containing tyrosine phosphatases in cell signaling. *Trends Biochem. Sci.*, 28, 284–293.
57. Van Hoof, C., Goris, J. (2003). Phosphatases in apoptosis: To be or not to be, PP2A is in the heart of the question. *Biochim. Biophys. Acta*, 1640, 97–104.
58. Irie-Sasaki, J., Sasaki, T., Penninger, J. M. (2003). CD45 regulated signaling pathways. *Curr. Top. Med. Chem.*, 3, 783–796.
59. Atack, J. R. (1996). Inositol monophosphatase, the putative therapeutic target for lithium. *Brain Res. Rev.*, 22, 183–190.
60. Jiang, H., Harris, M. B., Rothman, P. (2000). IL-4/IL-13 signaling beyond JAK/STAT. *J. Allergy Clin. Immunol.*, 105, 1063–1070.
61. Sciorra, V. A., Morris, A. J. (2002). Roles for lipid phosphate phosphatases in regulation of cellular signaling. *Biochim. Biophys. Acta*, 1582, 45–51.
62. Kanoh, H., Sakane, F., Imai, S., Wada, I. (1993). Diacylglycerol kinase and phosphatidic acid phosphatase—Enzymes metabolizing lipid second messengers. *Cell Signaling*, 5, 495–503.
63. Lee, J. P., Cho, H., Bannwarth, W., Kitas, E. A., Walsh, C. T. (1992). NMR analysis of regioselectivity in dephosphorylation of a triphosphotyrosyl dodecapeptide autophosphorylation site of the insulin receptor by a catalytic fragment of LAR phosphotyrosine phosphatase. *Protein Sci.*, 1, 1353–1362.
64. Daum, G., Solca, F., Diltz, C. D., Zhao, Z., Cool, D. E., Fischer, E. H. (1993). A general peptide substrate for protein tyrosine phosphatases. *Anal. Biochem.*, 211, 50–54.
65. Zhuo, S., Clemens, J. C., Hakes, D. J., Barford, D., Dixon, J. E. (1993). Expression, purification, crystallization, and biochemical characterization of a recombinant protein phosphatase. *J. Biol. Chem.*, 268, 17754–17761.

REFERENCES

66. Streuli, M., Krueger, N. X., Tsai, A. Y., Saito, H. (1988). A family of receptor-linked protein tyrosine phosphatases in humans and *Drosophila*. *Proc. Natl. Acad. Sci. USA*, *86*, 8698–8702.
67. Ibarra-Sanchez, M. J., Simoncic, P. D., Nestel, F. R., Duplay, P., Lapp, W. S., Tremblay, M. L. (2000). The T-cell protein tyrosine phosphatase. *Semin. Immunol.*, *12*, 379–386.
68. Hao, L., Tiganis, T., Tonks, N. K., Charbonneau, H. (1997). The noncatalytic C-terminal segment of the T cell protein tyrosine phosphatase regulates activity via an intramolecular mechanism. *J. Biol. Chem.*, *272*, 29322–29329.
69. Martin, B., Pallen, C. J., Wang, J. H., Graves, D. J. (1985). Use of fluorinated tyrosine phosphates to probe the substrate specificity of the low molecular weight phosphatase activity of calcineurin. *J. Biol. Chem.*, *260*, 14932–14937.
70. Perrino, B. A. (1999). Regulation of calcineurin phosphatase activity by its autoinhibitory domain. *Arch. Biochem. Biophys.*, *372*, 159–165.
71. Kapatral, V., Bina, X., Chakrabarty, A. M. (2000). Succinyl coenzyme A synthetase of *Pseudomonas aeruginosa* with a broad specificity for nucleoside triphosphate (NTP) synthesis modulates specificity for NTP synthesis by the 12-kilodalton form. *J. Bacteriol.*, *182*, 1333–1339.
72. Zhang, J. H., Chung, T. D., Oldenburg, K. R. (1999). A simple statistical parameter for use in evaluation and validation of high throughput screening assays. *J. Biomol. Screen.*, *4*, 67–73.

17

DESIGN AND PHARMACEUTICAL APPLICATIONS OF PRODRUGS

Tomi Järvinen
University of Kuopio
Kuopio, Finland

Jarkko Rautio
University of Kuopio
Kuopio, Finland

Mar Masson
University of Iceland
Reykjavik, Iceland

Thorsteinn Loftsson
University of Iceland
Reykjavik, Iceland

17.1	INTRODUCTION	734
17.2	DEFINITION OF THE CONCEPTS	735
	Prodrug Concept	735
	Double-Prodrug Concept	735
	Soft Drug Concept	737
	Chemical Delivery System Concept	737
17.3	REQUIREMENTS FOR CLINICALLY USEFUL PRODRUGS	738
17.4	BIOREVERSIBLE PRODRUG STRUCTURES FOR THE MOST COMMON FUNCTIONAL GROUPS	740
	Prodrugs for Carboxyl Groups	741
	Prodrugs for Phosphate and Phosphonate Groups	744
	Prodrugs for Hydroxyl Groups	747
	Prodrugs for Amines and Amides	750

Drug Discovery Handbook, by Shayne Cox Gad
Copyright © 2005 by John Wiley & Sons, Inc.

17.5	PHARMACEUTICAL APPLICATIONS OF THE	
	PRODRUG STRATEGY	752
	Improved Drug Absorption	752
	Improved Aqueous Solubility	766
	Prolonged Duration of Action	769
	Improved Drug Targeting	770
	Improved Formulation and Delivery of Peptides	779
	Reduced Side Effects	783
	Macromolecular Prodrugs	784
	References	785

17.1 INTRODUCTION

The process of drug discovery has been changed considerably during the last decade. Modern drug discovery technologies such as computer-aided drug design, combinatorial chemistry, and high-throughput pharmacological screens have provided large numbers of highly potent and selective drug candidates. However, optimal physico-chemical and biopharmaceutical requirements for the formulation and delivery of these drugs are not selected by these procedures, and thus the new drug candidates often have poor "druglike" properties and a high risk of failure in subsequent preclinical or clinical development [1]. Thus, both drug formulation and delivery must be carefully considered throughout the drug design process.

There are various approaches that can be applied to overcome poor properties of drug formulation and delivery. Some of these problems can be overcome by dosage form design. Another approach is to develop various drug analogs, but the risk in this approach is the irreversible modification of bioactive center(s) and subsequent loss of therapeutic effect. An important strategy approach to improve poor drug formulation and delivery properties is prodrug technology. Historically, the term *prodrug* was first introduced about 40 years ago to describe compounds that undergo biotransformation prior to exhibiting pharmacological effects [2]. In the 1970s and 1980s, the prodrug approach was primarily applied to improve physico-chemical, biopharmaceutical, and/or pharmacokinetic properties of drugs that already were on the market. Currently, prodrug design and research should be more involved in the lead optimization step during the drug discovery process (Fig. 17.1).

The prodrug approach is a valuable tool to further optimize potent structures and to solve potential formulation or delivery problems. Currently, about 5 percent of all worldwide approved drugs can be classified as prodrugs, and this number is still increasing [3]. Despite these high numbers, understanding the full potential of the prodrug approach in modern drug discovery and development has only just begun, and many prodrug innovations remain to be discovered.

DEFINITION OF THE CONCEPTS

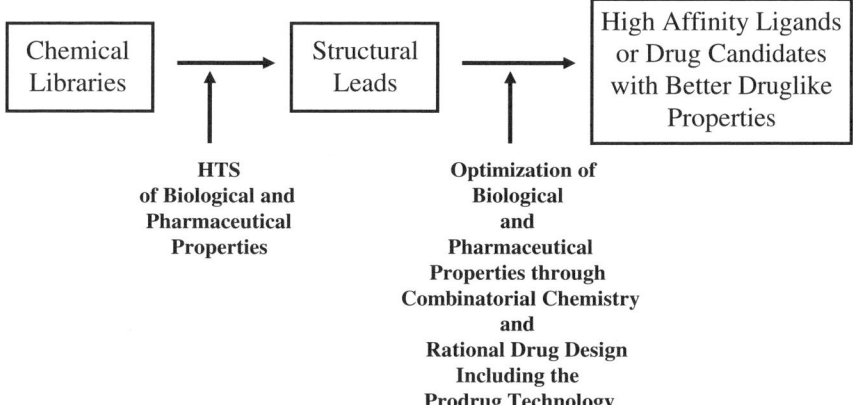

Figure 17.1 Modern drug discovery paradigm (modified from lectures by Prof. Valentino Stella and Prof. Ronald Borchardt).

17.2 DEFINITION OF THE CONCEPTS

Prodrug Concept

Prodrugs are pharmacologically inactive derivatives of drug molecules that require a chemical or enzymatic transformation in order to release the active parent drug in vivo, prior to exerting a pharmacological effect (Fig. 17.2) [4]. Prodrugs have also been called reversible or bioreversible derivatives, latentiated drugs, or biolabile drug–carrier conjugates, but the term *prodrug* is now the standardized term. In most cases, prodrugs are simple chemical derivatives that are one or two chemical or enzymatic steps away from the parent drug. In certain cases, a prodrug may consist of two pharmacologically active drugs that are coupled together in a single molecule, so that each drug acts as a promoiety for the other drug. Such derivatives are called co-drugs [5].

The major aim in designing prodrugs is to overcome limitations of a parent drug that would otherwise hinder its clinical use. Drug analogs are also used to overcome such limitations (Fig. 17.2), but facile modifications of the functional group(s) or the bioactive center(s) may result in unacceptable changes in the desired pharmacological profile, compared to the parent drug, which is not a problem in the case of prodrugs.

Double-Prodrug Concept

Various drawbacks of prodrug derivatives can be overcome by preparing a prodrug from another prodrug, the result being a double prodrug (pro-

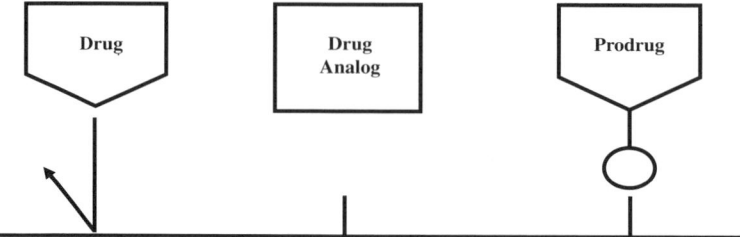
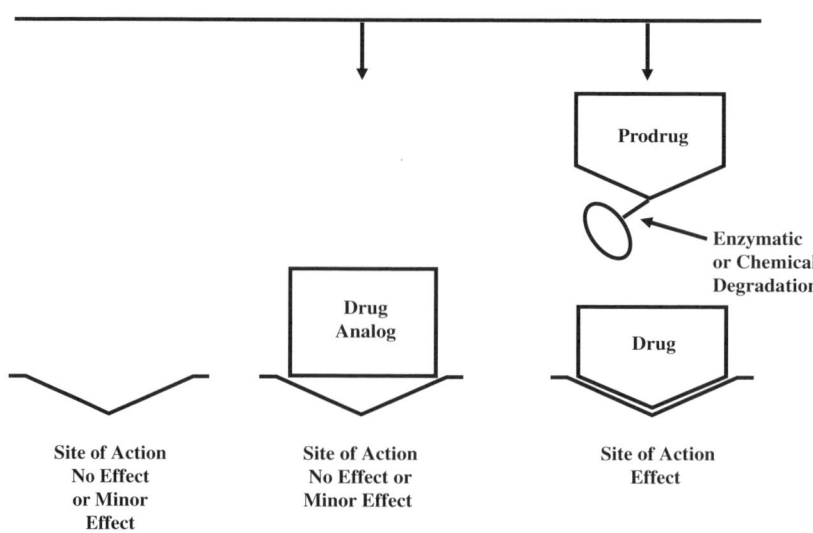

Figure 17.2 Basic principle of the prodrug concept.

prodrug). For example, stability problems of a prodrug, especially in solutions, can sometimes be solved by preparing a double prodrug that requires enzymatic hydrolysis before spontaneous chemical release of the parent drug (Fig. 17.3). The double-prodrug concept has been applied, for example, in development of fosphenytoin [6], prodrugs of 9-[(R)-2-(phosphonomethoxy)propyl]adenine (tenofovir) [7], and pilocarpine prodrugs [8]. The term *double prodrug* should not be confused with the term *bi-functional*

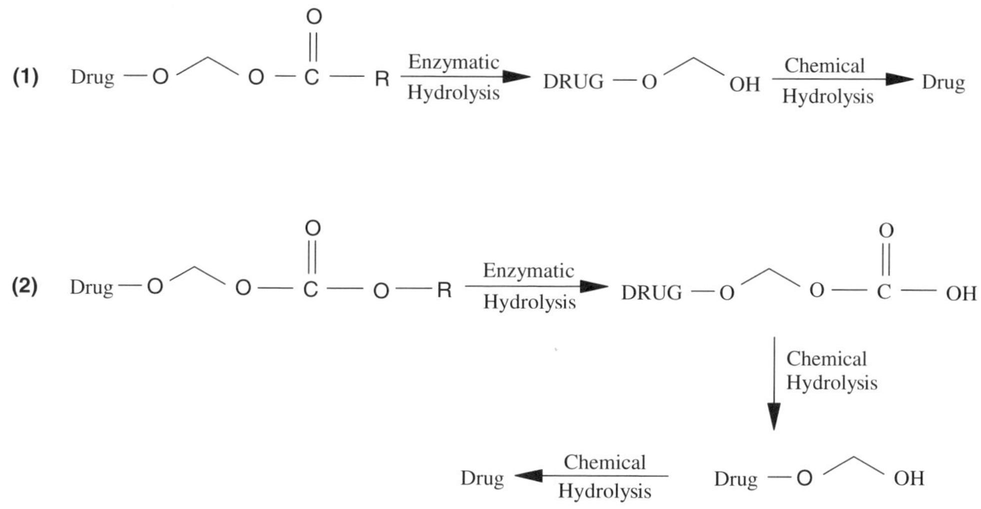

DEFINITION OF THE CONCEPTS 737

(1) Drug—O—⁀—O—C(=O)—R $\xrightarrow{\text{Enzymatic Hydrolysis}}$ DRUG—O—⁀—OH $\xrightarrow{\text{Chemical Hydrolysis}}$ Drug

(2) Drug—O—⁀—O—C(=O)—O—R $\xrightarrow{\text{Enzymatic Hydrolysis}}$ DRUG—O—⁀—O—C(=O)—OH

$\downarrow$ Chemical Hydrolysis

Drug—O—⁀—OH

Drug $\xleftarrow{\text{Chemical Hydrolysis}}$ Drug—O—⁀—OH

Figure 17.3 General mechanism for the bioconversion of (1) acyloxymethyl and (2) alkoxycarbonyloxymethyl double prodrug structures.

prodrug, which indicates a prodrug that has been modified at two functional groups on the parent molecule.

Soft Drug Concept

Soft drugs, unlike prodrugs, are active drugs that are designed to undergo a predicable and controllable deactivation in vivo after achieving their therapeutic role. Soft drugs are active agents that have a metabolically weak structure built into the molecule. After exerting their therapeutic effect at the desired site, soft drugs are rapidly metabolized to form inactive metabolites, which in turn are rapidly metabolized to prevent unwanted pharmacological activity or toxicity. The concept of soft drugs was introduced in the 1970s, and since then various classes of soft drugs have been developed such as soft analgetics, soft antimicrobials, soft β-blockers, and soft corticosteroids [9–11]. Soft drugs are most commonly designed for topical (e.g., ophthalmic, nasal, dermal, and pulmonary) applications. Some of the commercially available soft drugs are, for example, Brevibloc (soft β-blocker esmonol), Ultiva (soft opioid analgetic remifentanyl), and Alrex/Lotemax (soft corticosteroid loteprednol etabonate). If the prodrug approach is applied to a soft drug, the resulting product is known as a *pro-soft drug* [12].

Chemical Delivery System Concept

A chemical delivery system (CDS) is an inactive drug derivative that undergoes several predictable enzymatic transformations via inactive intermediates and finally delivers the active drug to the site of action. In an ideal case, the active drug molecule is released from its CDS only in the target organ or tissue, and thus CDS remains inactive elsewhere in the body and is susceptible to rapid elimination [10]. The main focus of CDS drug design has been to target a variety of drugs to the central nervous system, and applications for ophthalmic use already exist. In some applications it can be argued whether or not the appropriate term is prodrug or CDS. Nowadays the term *retrometabolic drug design* is often used to include both soft drug design and chemical delivery systems.

17.3 REQUIREMENTS FOR CLINICALLY USEFUL PRODRUGS

The major goal in designing a prodrug is to overcome various physico-chemical, biopharmaceutical, and/or pharmacokinetic problems that may be associated with a parent drug molecule, which otherwise would be of limited clinical use. The prodrug approach should also be considered when other dosage forms from the active compound are considered. For example, the prodrug approach is extremely useful in developing orally active compounds from an intravenous (i.v.) active compound. The common barriers to drug formulation and delivery most often addressed by the prodrug approach are:

1. Low aqueous solubility
   Prevents the development of aqueous-based formulations (e.g., solutions, drops, infusions).
   Leads to both rate-limited dissolution and variable oral bioavailability.
2. Low lipid solubility
   Results in low membrane permeation and low oral bioavailability.
   Hinders delivery across barriers such as cornea and skin (e.g., topical drug delivery).
   Limits the design of lipid-based formulations.
3. Short duration of action (due to rapid elimination from the body or strong first-pass metabolism)
   Necessitates frequent administration of a drug that often leads to poor patient compliance.
4. Lack of site specificity (e.g., poor brain, tumor, or colon targeting)
   May lead to undesirable systemic effects.
5. Poor taste or odor
   May lead to poor patient compliance.
6. Side effects
   May lead to safety concerns.
   May also lead to poor patient compliance.

7. Economic barriers

   For example; patent and product lifetime may be prolonged by prodrugs with improved formulation and/or drug delivery properties.

A successful prodrug design requires that the reasons for developing prodrugs are clearly defined and understood. The following points must be evaluated in order to confirm that the prodrug concept may be clinically useful:

1. Identification of the suitable functional groups on the parent molecule.
2. Identification of the precise drug formulation or delivery problem.
3. Identification of the physico-chemical/pharmaceutical properties required in order to overcome the specific problem.
4. Selection of a bioreversible prodrug structure that has the proper physico-chemical/pharmaceutical properties to overcome the problem.
5. The selected prodrug-structure can be economically synthesized.

The requirements for a clinically useful prodrug are dependent on dosage form and administration route. In general, prodrugs should fulfill at least most of the following criteria:

1. Synthesis and purification methods for prodrug should be feasible.
2. Adequate chemical stability

   Adequate shelf life in the final formulation.

   Prodrug will not degrade before reaching the site of action.
3. Adequate aqueous solubility

   Allowing dissolution and aqueous-based formulation development.
4. Optimal lipophilicity, to confirm efficient permeation across biological membranes

   $\log P$ value of 2 is considered to be optimal for gastrointestinal (GI) absorption, for example.
4. The prodrug must release the parent drug in vivo at a reasonable rate

   before, during, or after absorption or at the specific site of action.
5. The prodrug should be nontoxic and tasteless, and the promoiety released after bioactivation must also be nontoxic.

Both solubility and permeability are considered fundamental parameters for oral drug absorption. A biopharmaceutical classification scheme (BSC) has been proposed to categorize drugs into four groups, based on their aqueous solubility and permeability properties (Fig. 17.4) [13]. Drug molecules that belong to class I are the most promising, from the bioavailability and drug delivery point of view. If drug molecules in classes II to IV have suitable functional groups for prodrug derivatization, the prodrug approach can be applied in order to change molecules from classes II to IV to class I (Fig. 17.4).

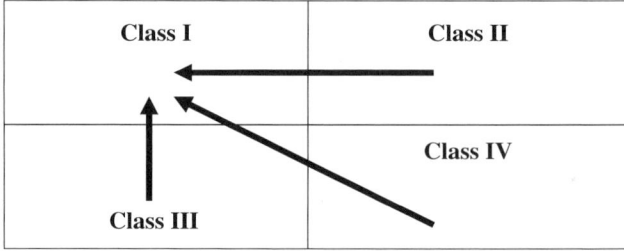

Figure 17.4 Biopharmaceutic drug classes and the effect of prodrugs on biopharmaceutics classification.

17.4 BIOREVERSIBLE PRODRUG STRUCTURES FOR THE MOST COMMON FUNCTIONAL GROUPS

Prodrugs are commonly formed by linking a promoiety to a nucleophilic carboxyl, phosphate, hydroxyl, or amino group on the parent drug molecule via a labile linkage. The labile linkage is usually designed to be stable in a pharmaceutical formulation and be readily cleaved by enzymes in vivo to release the parent drug after administration. This labile linkage can also be susceptible to nonenzymatic chemical cleavage. The release of the parent drug can then be triggered by a pH shift from a relatively acidic (or basic) formulation to a physiological pH of 7.4 in vivo. However, the utility of the latter approach is somewhat limited by the difficulty of achieving sufficient stability in the pharmaceutical formulation [14].

Key enzymes that are most commonly targeted for the cleavage of a labile prodrug linkage include various esterases and alkaline phosphatases, which are ubiquitous in mucosal membranes, blood, liver, and other organs [15]. A variety of other enzymes [16], such as aminoacylases, cystein conjugate β-lysase, γ-glutamyltransferases, dipeptidases, amino- and carboxypeotidases, oxoprolinase, β-glucuronidase [17], and azoreductase [18] can also be targeted for the bioconversion of a prodrug. However, this type of enzymatic activity is not ubiquitous in all tissues, and prodrugs converted by these enzymes are, therefore, more suitable for organ-targeted drug delivery.

Many older drugs, as well as some new ones, are prodrugs that do not contain promoieties as defined above. For example, nabumetone undergoes a rapid liver oxidase catalyzed biotansformation to the bioactive form (Fig. 17.5) [19]. Similarly, various antiviral and anticancer nucleosidic drugs are transformed to their corresponding bioactive monophosphates and triphosphates in vivo in a kinase-mediated step. In such cases, these in vivo metabolic transformations are sometimes difficult to predict from in vitro investigations, and the metabolic processing of these drugs can vary considerably between and within animal species. The discussion in this chapter will, therefore, be limited to the former type of prodrugs, where a well-defined promoiety is linked to a suitable nucleophilic group on a parent drug molecule.

Prodrugs for Carboxyl Groups

In general, the pK_a of a drug carboxyl group is in the range of 2.5 to 5.5. In the upper gastric fluids (pH 1.3), the carboxyl group will be neutral, but mainly ionized in the small intestine and other physiological fluids (pH 7.4). Ionization does not necessarily prevent drug absorption, especially in the case of oral dosing, because at least some part of the drug will be in the uncharged form. However, masking the carboxyl group can reduce some side effects, such as local irritation. Passive absorption can also be enhanced by forming neutral lipophilic prodrugs.

The most common prodrugs are those requiring a hydrolytic bioconversion that is mediated by an enzymatic reaction. Therefore, prodrugs derived from carboxyl groups are commonly obtained by formation of esters [20], which are substrates for various esterases found throughout the blood, intestinal mucosa, liver, and other organs [21]. Moreover, it is possible to develop ester prodrugs that have higly variable aqueous solubility or lipophilicity characteristics by carefully selecting the prodrug moiety. Examples of promoieties for the carboxyl group are shown in Fig. 17.6.

Alkyl and Aryl Ester Prodrugs Prodrugs of angiotensin-converting enzyme (ACE) inhibitors are some of the most successful prodrugs in clinical use today

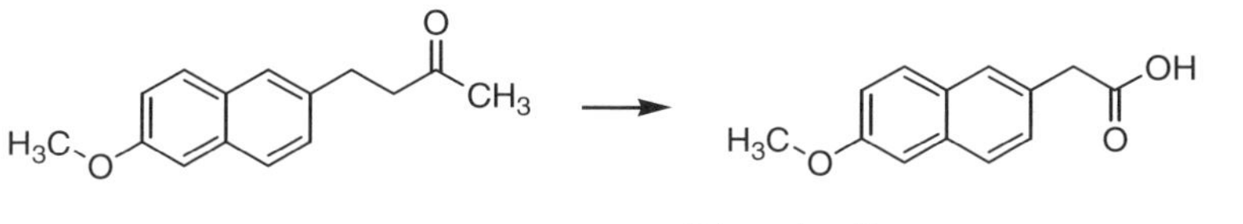

Nabumetone

Bioactive Form:
6-methoxy-2-naphthylacetic acid (6-MNA)

Figure 17.5 Bioconversion of nabumetone to the bioactive form.

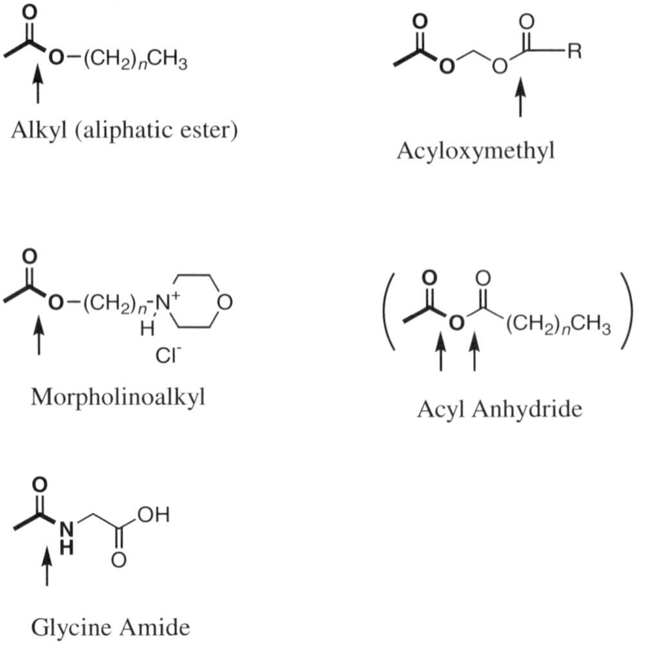

Figure 17.6 Examples of promoieties for carboxyl groups. The arrows show the biolabile linkages.

[16]. Various ester derivatives have been investigated to improve the oral bioavailability of enalaprilate. While the proline carboxyl esters convert to diketopiperazine in an intramolecular cyclization reaction that is chemically unstable [22], esters on the N-carboxyalkyl group are chemically more stable and absorbed well after oral administration. A prototype of this class of prodrugs is enalapril, which is a monoethyl ester of the ACE inhibitor enalaprilate (Fig. 17.7). Other ACE inhibitors that have been introduced since enalapril, such as benzapril and ramipril, are also mono ethyl esters of diacids.

The main disadvantage of simple alkyl ester prodrugs is that they will tend to reduce aqueous solubility, relative to the parent drug. Since aqueous solubility and dissolution rates can be a limiting factor in drug absorption, low aqueous solubility can be addressed by introducing an amino group on the alkyl chain. A series of morphinoalkyl (ethyl, propyl, butyl) esters of indomethacin and naproxen [23] have been investigated in vitro and in vivo. These prodrugs were freely soluble in simulated gastric fluid and aqueous pH 7.4 phosphate buffer. The octanol/pH 7.4 phosphate buffer distribution coeffcient was increased by 40- to 1500-fold through prodrug formation, relative to the parent drug, whereas the octanol/simulated gastric fluid (pH 1.3) distribution coefficient decreased 3- to 20-fold. The esters were relatively stable

BIOREVERSIBLE PRODRUG STRUCTURES

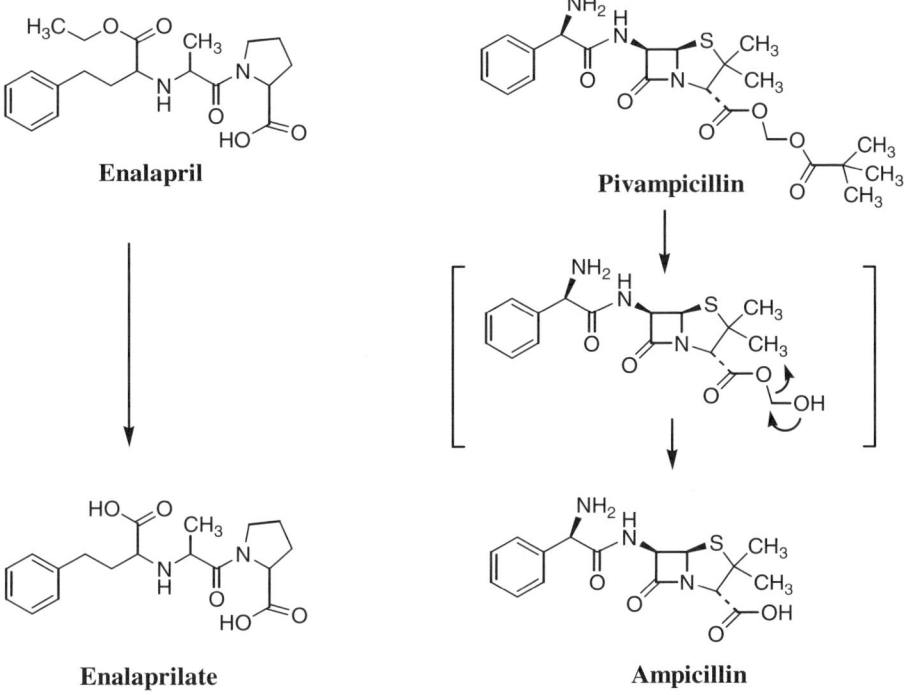

Figure 17.7 Examples of prodrugs with modification on the carboxyl groups and their bioconversion.

at pH 7.4 with a half-life of 10 h, but they were rapidly converted to the parent drug in rat plasma with half-life of 1.2 to 31 min. These prodrugs improved the bioavailability of the parent drug only slightly, but gastric mucosal injury was significantly reduced.

Acyloxyalkyl Prodrugs Bioconversion of simple alkyl esters may sometimes be inefficient in humans, due to steric hindrance around the carboxyl group on the drug molecule, which limits the therapeutic potential. α-Acyloxymethyl derivatives are double esters, from which the terminal ester moiety is first removed through an enzyme-catalyzed hydrolysis, with the formation of a highly unstable hydroxymethyl ester intermediate that subsequently releases both formaldehyde and the parent drug. In α-acyloxyalkyl esters that contain a longer alkyl chain (e.g., ethyl, propyl, butyl), hydrolysis may take place at the carbonyl of the parent drug, rather than the carbonyl of the promoiety [24].

As an example, various acyloxyalkyl esters of β-lactam antibiotics are well absorbed and efficiently hydrolyzed by esterases in vivo and in vitro, while prodrugs that are simple esters of the C-3 thiazolidine carboxyl group are ineffective in humans, due to limited hydrolysis [25]. The pivaloyloxymethyl derivative of ampicillin has been clinically introduced as pivampicillin (Fig. 17.7). More examples of ampicillin prodrugs are discussed in Section 17.5.

Amides In general, amides are more stable toward hydrolysis than esters, but various amides formed with amino acids are in vivo substrates for endopeptidase [21]. For example, glycine amide prodrugs of ketoprofen and naproxen have been evaluated for efficacy in mice and gastrointestinal irritation in rats [26]. Hydrolysis in serum is relatively slow, but these prodrugs exhibit similar anti-inflammatory activity as the parent drugs. Such prodrugs contain a free carboxylic acid group, and the physico-chemical properties are, therefore, not significantly altered with the prodrug modification. However, gastrointestinal irritation is still somewhat reduced with the glycine amide prodrugs.

Prodrugs for Phosphate and Phosphonate Groups

The phosphate and phosphonate groups are negatively charged zwitterions at nearly all physiological pH values, making them very polar and poorly absorbed from the GI tract. Moreover, drugs containing these groups also tend to have a low volume of distribution, and they are, therefore, subject to rapid and efficient renal clearance [27]. Prodrug strategies for phosphate and phosphonate functional groups usually mask the charge by chemical derivation to obtain neutral and better absorbed prodrugs (Fig. 17.8).

Prodrugs for Phosphate Groups Phosphate prodrugs are primarily prodrugs for 5'-*O*-monophosphates nucleoside analogs. The unmodified 5'-*O*-monophosphate esters have very limited oral bioavailability, due to poor absorption and rapid dephosphorylation in the gastrointestinal tract [27]. Lipophilic dialkyl ester prodrugs are often well absorbed, but their bioconversion is sometimes slow and inefficient. For example, simple dialkyl phosphate esters (5'-*O*-monophosphate) of the antiviral drug zidovudine (AZT) have no activity toward inhibiting the human immunodeficiency virus (HIV) replication [28] because of poor bioconversion to AZT.

The in vivo enzymatic conversion of phosphate esters is catalyzed by phosphodiesterases and phosphonodiesterases [27]. The phosphate ester linkage to the nucleside is liable to both chemical and enzymatic cleavage. Phosphate ester prodrugs tend, therefore, to be converted to the parent nucleoside rather than the monophosphate nucleotide. One way to circumvent this problem is to form homo- or hetero-dinucleoside phosphate derivatives of biologically active nucleoside analogs. Hydrolysis of the ester bond produces one molecule of nucleoside and one molecule of the nucleotide. These two products should then be synergistically active in the cell. The $5' \rightarrow 5'$ homo- and hetero-dinucleoside phosphates of the antiviral nucleoside analogs AZT, ddC, and ddI (Fig. 17.9), have been investigated for activity against HIV [27, 29]. However, the hydrolysis of these compounds was inefficient and, thus, these prodrugs exhibited lower activity than the parent drugs.

BIOREVERSIBLE PRODRUG STRUCTURES

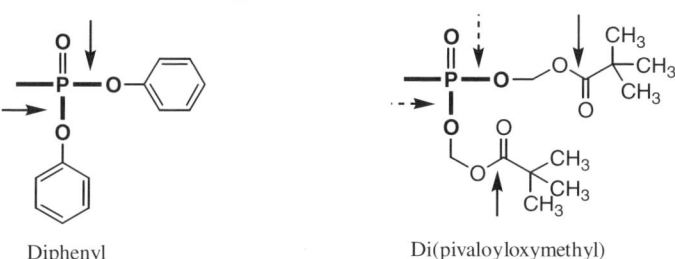

Figure 17.8 Examples of promoieties for phosphate and phosphonate groups. The arrows show the biolabile linkages.

An attempt to better control the bioconversion of the prodrug is an acyloxymethyl phosphate ester, from which the acyloxy group is cleaved by esterases, and the resulting hydroxymethyl ester derivative is rapidly eliminated to release formaldehyde and the phosphate group. For example, from di(pivaloyloxymethyl) phosphate esters the first pivaloyloxy group is efficiently removed, whereas the hydrolytic cleavage of the second acyloxymethyl ester group can be considerably slower, due to lower affinity of the ionized compound for the enzyme [27]. A mechanistic study for the enzymatic conversion of di(pivaloyloxymethyl) AZT monophosphate (AZTMP) has shown that the hydrolytic cleavage of the first pivaloyloxmethyl group is catalyzed by esterases, whereas the second group is mainly removed through hydrolysis of the phosphate ester bond by phosphodiesterases [30].

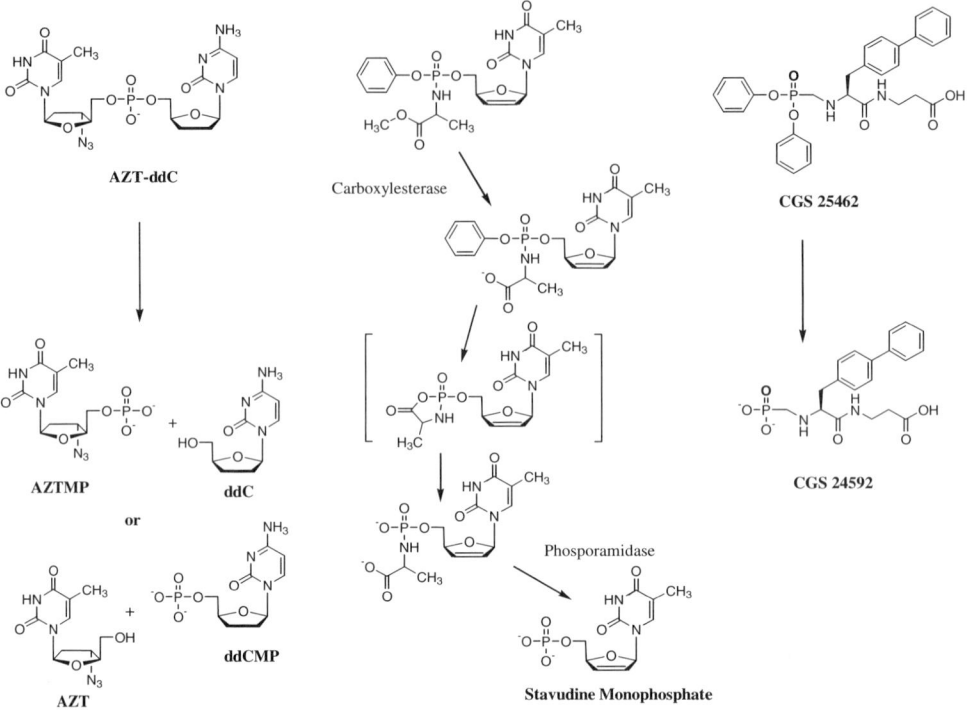

Figure 17.9 Examples of prodrugs for phosphates and phosphonates and their bioconversion.

Diphosphate 1,2-diacyl glycerol derivatives have been investigated as prodrugs for nucleoside monophosphates. In this case, the phospholipid prodrug is targeted to the cellular membrane, and monophosphate is released in a phosphatase-catalyzed hydrolysis reaction. For example, acyclovir diphosphate dimyrstoylglycerol can be effective against the HIV virus in thymidine kinase deficient cell lines, whereas acyclovir and acyclovir monophosphate are ineffective [31].

A study of the degradation pathway for stavudine-5'-(phenyl methoxy-L-alaninyl) phosphate (Fig. 17.9) showed that the methoxy ester is first cleaved in a esterase-catalyzed reaction, followed by a spontaneous chemical hydrolysis of the aryl phosphoester, which is intramolecularly catalyzed by the free carboxylic acid group. The amino acid moiety is then efficiently removed in a reaction catalyzed by phosphoroamidases [32]. This prodrug was equally effective against HIV in normal and kinase-deficient cell lines. Because the bioconversion is dependent on the chemical structure of the amino acid and the nucleoside moiety, the methoxy-L-valinyl derivatives are stable toward enzymatic hydrolysis and have very little efficacy against HIV.

Prodrugs for Phosphonate Groups The C-P bond linking a phosphonate group to the parent drug structure is not subject to enzymatic hydrolysis, and the selection of suitable promoieties is, therefore, less problematic than in the case of phosphate groups.

The most simple prodrug approach for increasing the lipophilicity of phosphonates is by esterification of the phosphonic OH groups. Several alkyl and aryl esters of phosphonates have been made, but they usually suffer from their high stability in biological media. In a series of dialkyl prodrugs, 9-[2-(phosphonomethoxy)ethoxy]adenine bioconversion to the parent drug depended on the chemical nature of the alkyl chain [33]. As the size of the alkyl chain increased, the prodrugs broke down more efficiently to the monoester, but the monoester failed to release the parent phosphonate. Similar results were obtained from a study of dialkyl prodrugs of 9-[2-(phosphonomethoxy)ethyl]adenine (PMEA) [34].

A series of dialkyl and diaryl prodrugs were studied in order to increase oral bioavailability of the neutral endopeptidase inhibitor CGS 24592 [35]. The diethyl and dicyclohexyl prodrugs were resistant to hydrolysis, whereas the diphenyl (Fig. 17.9) and di(acyloxyalkyl) derivatives were efficiently hydrolyzed in human plasma. Moreover, they significantly increased the bioavailability from <3 percent in rats.

Bis(acyloxymethyl) esters have been developed to overcome the hydrolytic resistance that has been observed with alkyl and aryl esters phosphonates. Investigations of prodrugs to increase the bioavailability of bisphosphonate antiosteoporosis drugs have shown that tri- and di(pivaloyloxymethyl) prodrugs of clodronic acid rapidly degrade in liver homogenates to release clodronic acid in quantitative amounts, whereas the benzoyloxypropyl esters were not subject to either enzymatic or chemical hydrolysis [36]. Monoamine derivatives were rapidly hydrolyzed at pH 7.4, but the reaction was not subject to enzymatic catalysis [37]. This suggests that the monoamine derivatives could be useful intermediates in the design of enzymatically labile prodrugs.

Prodrugs for Hydroxyl Groups

Prodrugs are commonly derived from existing hydroxyl groups on the parent drug molecule, usually by ester formation (Fig. 17.10). The esters are typically formed by a condensation reaction between the hydroxyl group and a carboxylic acid or by reaction with an acid halide. This condensation is commonly carried out by using a coupling agent, such as 1-ethyl-3-(3-dimethylamopropyl)carbodiimide and a catalyst such as 4-dimethylaminopyridine (DMAP). Both aliphatic and aromatic esters can be substrates for nonspecific esterases, which are widely distributed throughout various tissues and blood serum. Phosphate esters are substrates for phosphatases found in blood and as intestinal-brush border-bound enzymes.

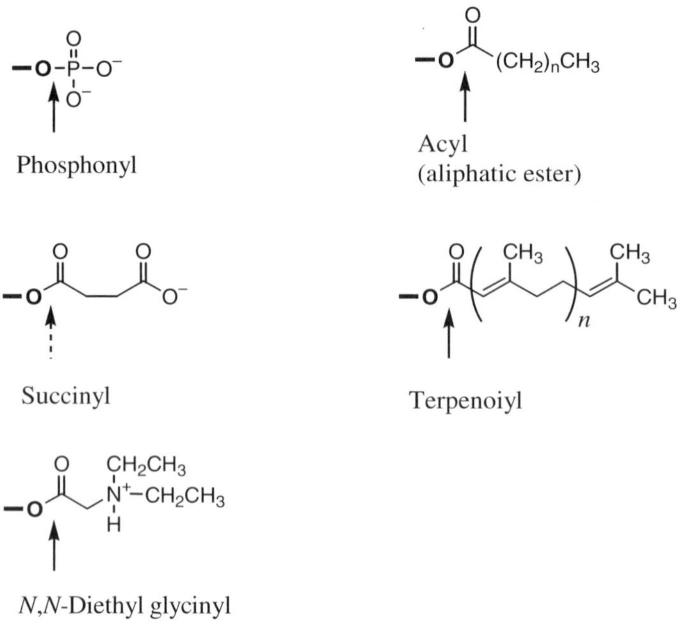

Figure 17.10 Examples of promoieties for hydroxyl groups. The arrows indicate the biolabile linkages.

Water-Soluble Prodrugs Water-soluble prodrugs are commonly obtained as phosphates, succinates, or amino acid esters of the hydroxyl group. For example, succinate esters and phosphate esters of corticosterioids have been used for intravenous administration. In general, the succinate esters have a limited chemical stability [20], due to intramolecular hydrolytic catalysis of the ester bonds by the neighboring carboxyl group [38]. These compounds are, therefore, marketed as a freeze-dried powder, which is dissolved just prior to administration. Due to anionic charges, the succinate esters are relatively poor substrates for carboxylesterases. Therefore, they are only slowly converted after intravenous administration and partially eliminated unchanged in urine [39, 40].

Phosphate esters are more stable toward chemical hydrolysis and can be formulated as aqueous solutions with an acceptable shelf life [20]. Phosphate esters of corticosteroids, such as methylpredinsolone (Fig. 17.11) [38] are rapidly and quantitatively converted to the parent drug in vivo.

Amino acid esters, which are substrates for endopeptidases in vivo [21], are also considered as suitable promoieties that improve aqueous solubility. The synthesis of these prodrugs involves either condensation with N-protected amino acids or reaction of an amine with an α-halogenacetate ester of the

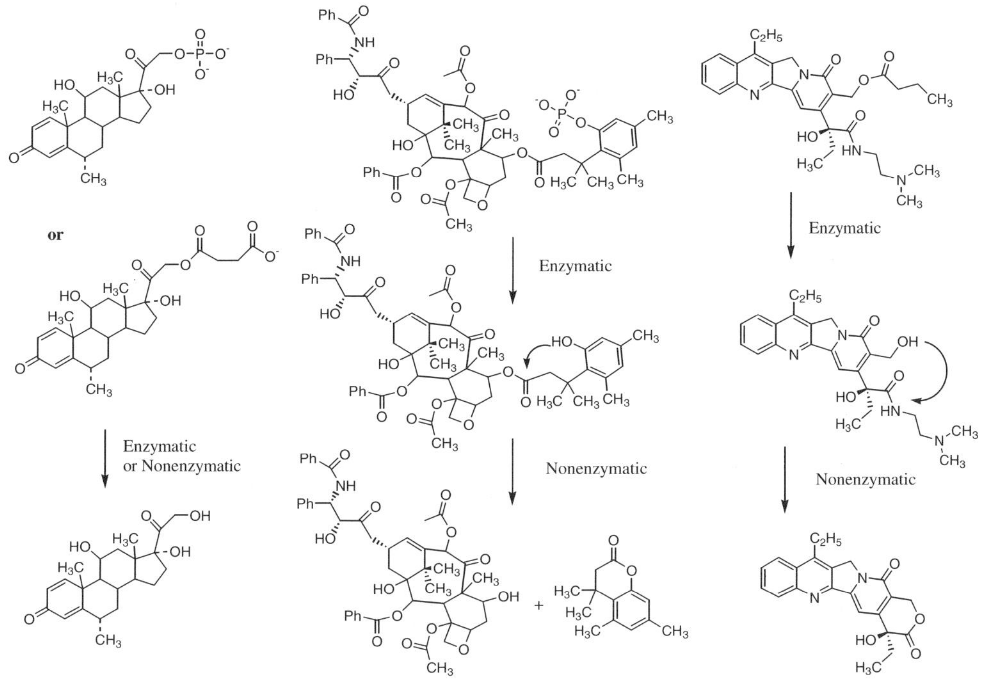

Methylprednisolone **Taxol** **7-Ethylcamptothecin**

Figure 17.11 Prodrugs derived from hydroxyl groups and their bioconversion.

parent drug. Amino acid derivatives such as *N*-propyl-, *N*,*N*-dimethyl, and diethtyl glycine esters, 4-morpholinoacetate, and methyl-1-pirperazinoacetate [41, 42] were freely water-soluble and their half-lives for bioconversion were less than 30 min. However, their chemical stability was limited. In general, α-amino acid esters tend to be unstable in aqueous solution due to an electron-withdrawing effect of the protonated amino group [20].

Hydrophilic corticosteroid prodrugs can also be obtained by forming a glycoside derivative from the primary hydroxyl group. Glycosides can be relatively stable in the small intestine but easily bioconverted by bacterial glycosidases in the colon. This type of prodrug, for example, dexamethasone-β-D-glucuronide [17], is therefore intended for colon-specific delivery.

Lipophilic Prodrugs Lipophilic produgs of drugs containing a hydroxyl group can be obtained by esterification with a carboxylic acid that has a non-polar side chain. The purpose of producing a more lipophilic prodrug is to enhance both absorption and transmembrane delivery of polar drugs. For example, various lipophilic linear aliphatic esters of metronidazole have been investigated for dermal delivery. Enzymatic conversion in human plasma and skin homogenates is faster with long-chain aliphatic esters, such as caproyl esters, than with shorter chains, especially acetyl and propionyl esters. In vitro investigations have shown that metronidazole acetate can penetrate through excised mouse skin and is partially metabolized to metronidazole [43].

Promoieties to Overcome Steric Hindrance Steric hindrance in a drug structure can affect the enzymatic lability of esters formed at the hydroxyl group. In such cases, the lack of enzymatic lability can be addressed by introducing a suitable spacer between the modified hydroxyl group and the ester function. The ester is then cleaved in an enzymatic reaction, followed by a rapid intramolecular chemical reaction of the spacer group to release the parent drug.

Taxol phosphates have, for example, been investigated as prodrugs for the poorly soluble anticancer drug taxol [44]. Simple phosphate esters were more soluble than the parent drug, but they were also stable toward alkaline phosphatase in plasma and consequently ineffective in vivo. The steric hindrance around the phosphate ester was reduced by introducing an oxyphenylpropionate spacer group (Fig. 17.11). These phosphate esters were readily cleaved in purified alkaline phosphatase preparations, and the remaining hydroxyphenylpropionate group was eliminated through an intramolecular lactonization reaction. However, these prodrugs were not converted in animal plasma, apparently due to tight protein binding, and consequently had no in vivo activity.

Prodrugs of Lactones Prodrugs of lactones can be obtained by forming an ester or an amide with the hydroxyl group in the open ring of a lactone. When incubated in plasma, they revert to the parent drug in a two-step process, with initial enzymatic degradation of the ester or amide on the hydroxyl group followed by spontaneous ring closure to the lactone. This approach has been used for water-soluble prodrugs of camptothecin derivatives (Fig. 17.11), where an amide is formed from the carboxyl group of the open lactam and an enzyme-labile acyl moiety is attached to the hydroxyl group [45, 46].

Prodrugs for Amines and Amides

Nucleophilic amino groups are present in many drugs, and they can be found as either primary, secondary, or tertiary aliphatic or aromatic amines. They can also be part of resonance-stabilized systems, such as heterocyclic aromatic groups, amides, imides, or guanidines. Depending on structure, the pK_a of a protonated basic amino group can vary from less than 2 to more than 11. Amino groups in resonance-stabilized systems can also act as weak acids. Acidic amino groups will generally be neutral at physiological pH values, and the charge of a moderately basic amino groups is not necessarily a barrier to drug absorption. However, amino groups are often subjected to modification, especially if readily esterifiable hydroxyl or carboxyl groups are not present in the drug structure [14]. Various promoieties have been investigated for amines, with particular consideration toward the nature of the amino group (Fig. 17.12).

N-*Acyl Prodrugs* The *N*-acylation of a basic amino group gives an amide, which has limited utility in prodrug design due to their relative in vivo stability [14]. However, certain amides can be substrates for specific enzymes in vivo. For example, the γ-glutamyl derivatives of dopamine are cleaved by γ-glutamyl transferases in the kidneys and have been considered for kidney-targeted drug delivery [14, 47]. Biolabile prodrugs can, however, sometimes be obtained by acylation of relatively acidic amides or imides. For example, the half-life of the N^3-acetyl derivative of 5-fluorouracil (Fig. 17.13) is 40 min at pH 7.4 (37°C) and only about 4 min in 80 percent human plasma [14].

N-*Mannich Bases* *N*-Mannich bases have been considered as chemically reversible prodrugs for NH-acidic compounds (e.g., amides, imides) and both aliphatic and aromatic amines. *N*-Mannich bases decompose in aqueous, alkaline, and slightly acidic solutions. However, they are relatively stable in acidic formulations, and the release of the parent drug is, therefore, triggered by a shift to higher pH values after administration. For example, the *N*-Mannich base formed between salicylamide and piperidine is relatively stable below pH 4, whereas the half-life for hydrolysis at pH 7.4 (37°C) is 14 min [48]. The *N*-Mannich base promoiety can also be utilized for water-soluble prodrugs of drugs with NH-acidic amino (amides) groups. For example, various *N*-Mannich bases of carbamazepine (Fig. 17.13) have been developed as water-soluble prodrugs for parenteral administration [49].

N-α-*Acyloxymethyl and N-Phosphoroxymethyl Derivatives* Probably the most successful and common approach to prodrugs from amines is to form *N*-α-acyloxymethyl or *N*-phosphoroxymethyl derivatives. The bioconversion of these prodrugs undergoes a two-step process; first an enzymatic cleavage of the ester bond, catalyzed by esterases or phosphatases, followed by a spontaneous chemical decomposition of the *N*-hydroxymethyl intermediate, which releases formaldehyde and the parent drug. Useful prodrugs cannot be obtained by *N*-α-acyloxymethylation of primary or secondary amines due to the extreme lability of such derivatives [14]. This approach is, therefore, limited to acidic imides or amides. For example, various such promoieties have been considered for the sparingly soluble and poorly absorbed anticonvulsant phenytoin. The water-soluble *N*-phosphoroxymethyl derivative of fosphenytoin is now marketed as Cerebyx and is further discussed in Section 17.5. The *N*-α-acyloxymethyl and *N*-phosphoryloxymethyl promoieties have also been used to obtain either water-soluble or lipophilic prodrugs of theophylline (Fig. 17.13) [50], allopurinol [51], 6-mercaptopurine [52], and 5-fluorouracil [53].

Promoieties for Basic Amino Groups:

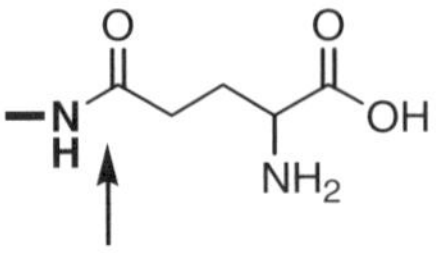

N-γ-Glutamyl

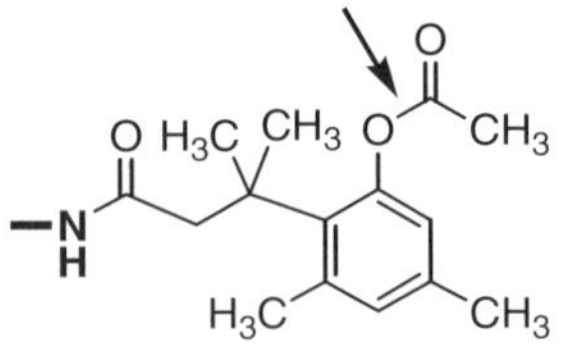

N-3(2'-Acetoxy-4',6'dimethylphenyl) 3,3-Propionyl

Promoieties for Acidic Amino Groups:

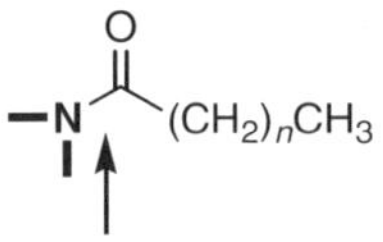

N-Acyl

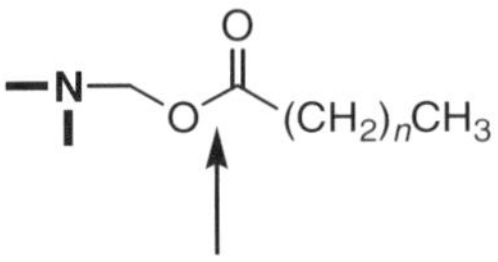

N-Acyloxymethyl

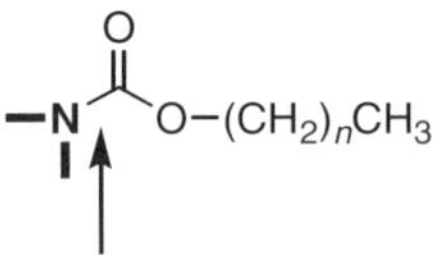

N-Acyloxycarbonyl

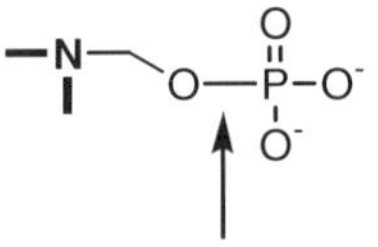

N-Phosphoroxymethyl

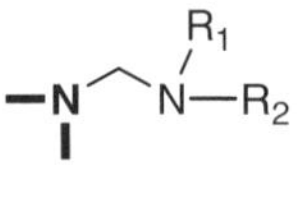

N-Mannich base

Figure 17.12 Examples of promoieties for amino groups. The arrows indicate the biolabile linkages.

17.5 PHARMACEUTICAL APPLICATIONS OF THE PRODRUG STRATEGY

Improved Drug Absorption

Oral Drug Delivery For orally administered drugs, the intestinal epithelia serves as physical barrier by forming tight cellular junctions that restrict paracellular drug delivery into the systemic circulation. Consequently, efficacious drugs must possess either optimal physico-chemical properties (e.g., lipophilic-

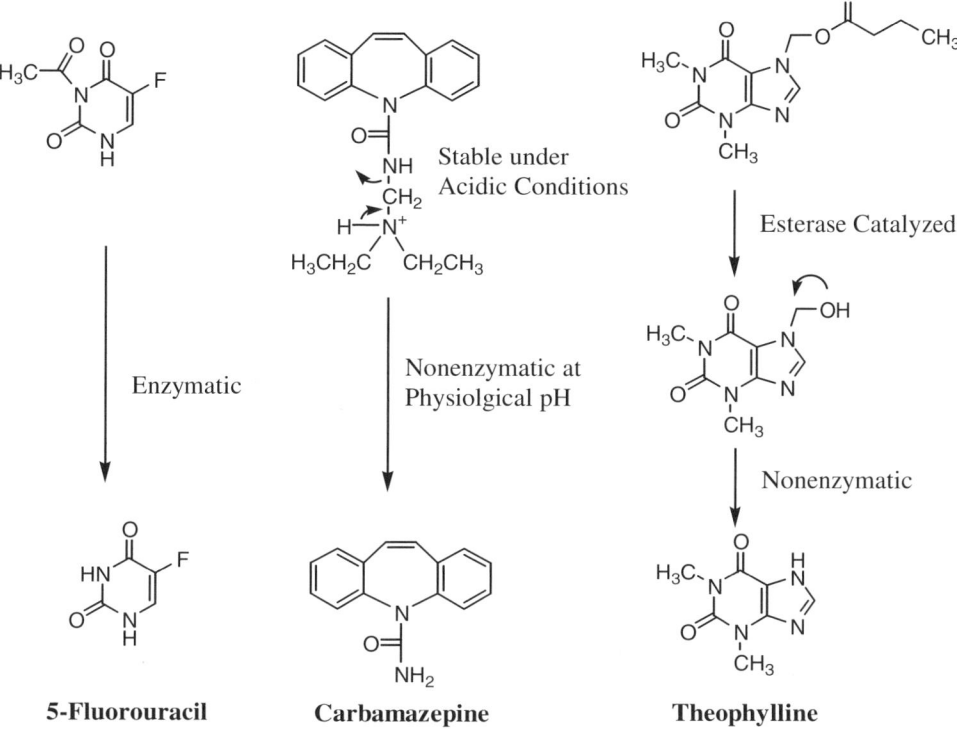

Figure 17.13 Prodrugs derived from amino and amide groups and their bioconversion.

ity, sufficient molecular weight, ionic charge, hydrogen-bonding potential, or structural conformation) in order to passively diffuse via the transcellular route, or the drug must have structural features that allow it to be taken up by one of the endogenous transport mechanisms present at the intestinal epithelium [54]. Other prerequisites for a drug's effective oral absorption are sufficient aqueous solubility at the absorption site and adequate chemical stability in the GI tract or sufficient stability during first-pass metabolism.

Lipophilic Prodrugs Since nonfacilitated and largely nonspecific passive transcellular transport is the most common absorption route in oral drug delivery, lipophilicity can be increased to achieve better diffusion across the lipoidal biological membranes, and thus better oral bioavailability. This approach has been one of the major research areas of prodrug research. Well-known and successful examples of oral prodrugs are various lipophilic ampicillin derivatives. Ampicillin is highly polar and in a zwitterionic form in the pH range of the gastrointestinal tract and has an oral bioavailability of only 30 to 40 percent. A transient increase in its lipophilicity by esterification resulted in clinically useful ester prodrugs of ampicillin (Fig. 17.14), namely pivampicillin (pivaloylmethyl ester), bacampicillin (ethoxycarbonylethyl ester), and talampicillin (phtalidyl ester), which are all well-absorbed and rapidly hydrolyzed to ampicillin either during or after absorption [25].

More recent examples, for example, tenofovir disoproxil, adefovir dipivoxil, ximelagatran, and BIBR 1048, show further proof of the concept. In particular, the acyclic nucleoside phosphonate 9-[2-(phosphonylmethoxy) propyl]adenine (PMPA, tenofovir) (Fig. 17.15) has potent and selective activity against HIV. However, high hydrophilicity of the phosphonic acid moiety has been postulated to account for its poor oral bioavailability (less than 0.1 percent in Caco-2 cell monolayer model). On the basis of its favorable in vitro chemical and enzymatic degradation properties, and octanol/water partition coefficient [log P (pH 6.5) = 1.3], bis(isopropyloxycarbonyloxymethyl)PMPA [bis(POC)-PMPA, tenofovir disoproxil] was selected for further development [55]. In clinical studies, tenofovir disproxil has been well tolerated, with an oral bioavailability in fasted patients of approximately 25 percent [56]. Similarly, the lipophilic bis[(pivaloyloxy)methyl] ester prodrug adefovir (PMEA), a close hydrophilic analog of tenofovir (Fig. 17.15), enhanced the oral bioavailability up to 30 to 40 percent, compared to only 12 percent when adefovir alone was administered in HIV-infected patients [57]. Recently, both tenofovir disoproxil and adefovir dipivoxil have been approved for the market. Furthermore, lipophilic prodrugs such as ximelagatran and BIBR 1048 (Fig. 17.16) may be the first examples of orally administered thrombin inhibitors [58].

Carrier-Mediated Transport This means of transport is particularly important where drugs or prodrugs are either polar or charged, and passive transcellular absorption is neglible. Therefore, the targeting of intestinal epithelial membrane transporters has become an attractive approach for improving the oral bioavailability of poorly absorbed drugs. By using a bioreversible prodrug derivatization strategy, a drug with low membrane permeability is converted

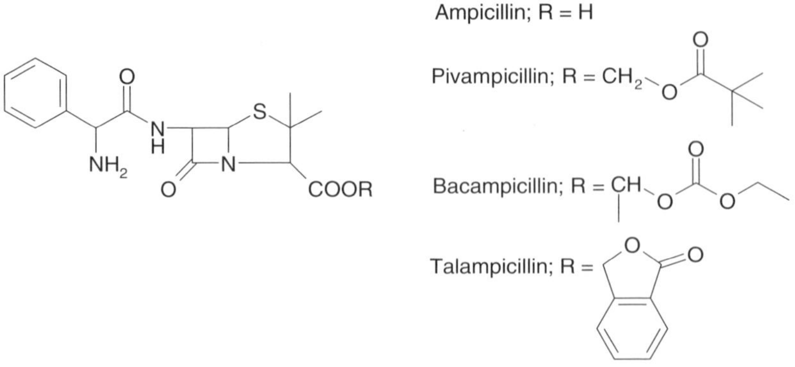

Figure 17.14 Prodrugs of ampicillin.

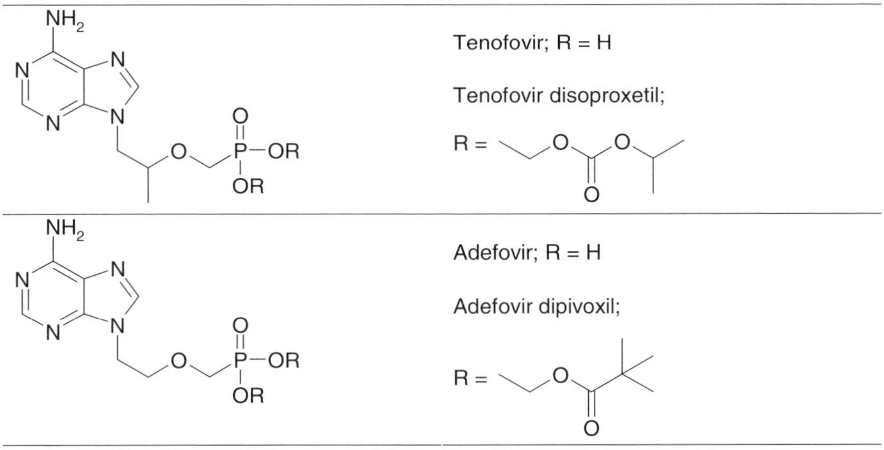

Figure 17.15 Prodrugs of tenofovir and adefovir.

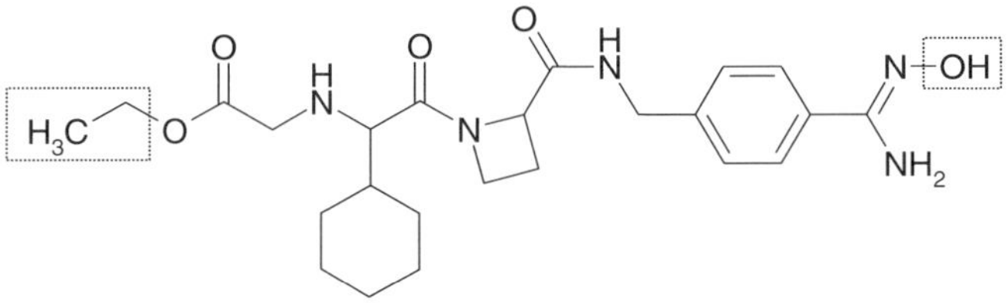

Ximelagatran (promoieties marked with the rectangle)

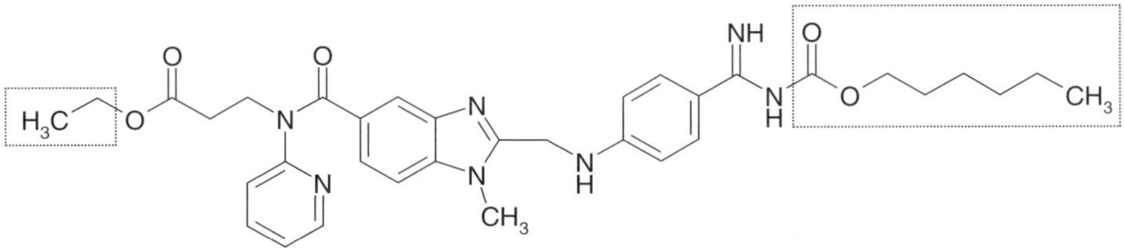

BIBR 1048 (promoieties marked with the rectangle)

Figure 17.16 Prodrugs of thrombin inhibitors.

to a prodrug that is a substrate for the endogenous transport system. The nutrient transport systems have mainly been investigated and utilized for the enhancement of intestinal absorption, and these primarily include peptide, amino acid, bile acid, and glucose transporters. Other systems seem to have narrow structural requirements that currently make them less relevant for this purpose. Among the membrane transporters, the peptide transport mechanism appears to be most promising as it is widely distributed throughout the small intestine and shows low substrate specificity, together with high transport capacity [54, 59, 60]. Therefore, in addition to endogenous peptides, various drugs can be recognized as substrates by peptide transporters, even drugs that do not include amino acid residues.

A good example of a prodrug that exploits intestinal transporters is valacyclovir, the L-valyl ester of acyclovir (Fig. 17.17). Acyclovir is a specific and selective inhibitor of viral herpes replication and is neither highly lipid nor aqueous soluble, with limited and variable oral bioavailability (15 to 21 percent) that decreases with increasing doses [61]. Conversely, valacyclovir has an oral bioavailability of three to five times higher than that of acyclovir itself. Moreover, valacyclovir is rapidly and extensively hydrolyzed to acyclovir after oral administration [62, 63]. Studies of its transport mechanisms have demonstrated that the improved oral bioavailability is due to its active transport by the human intestinal peptide transporter (PEPT1) [64]. Similarly, biolabile L-valyl ester prodrugs of AZT and ganciclovir afforded 3- to 10-fold better intestinal permeability than their parent drugs, and their membrane transport was mediated predominantly by the peptide transporter [65, 66], although they do not possess an amino acid residue in their structures.

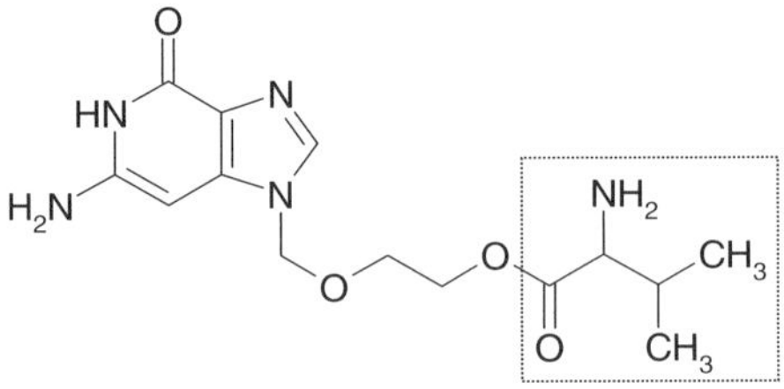

Figure 17.17 Valacyclovir (promoiety is marked by the rectangle).

Others For poorly water-soluble drugs, dissolution in the GI tract can be a rate-limiting step for intestinal absorption, leading to poor and variable oral bioavailability. For example, milproxifene phosphate has a higher aqueous solubility than its insoluble parent drug, milproxifene, which results in a nearly 10-fold increase in permeation across the Caco-2 cell monolayer, due to its better dissolution properties [67]. Increasing aqueous solubility by the prodrug approach is more thoroughly discussed below.

Due to its great instability in acidic environment, as well as its hydrophilic character, the semisynthetic β-lactam antibiotic carbenicillin has been only used parenterally in clinical practice. Its facilitated oral absorption was achieved by bioreversible esterification of the side-chain carboxyl group, which resulted not only in improved lipophilicity but also in greater acid stability [68]. Carbenecillin is released from the prodrug carindacillin with relative ease by enzymatic hydrolysis, following absorption.

Dermal Drug Delivery The skin, and in particular its most superficial layer, the stratum corneum, provides a unique and impressive resistance to molecular transport in either direction. Most existing drugs do not possess the requisite physico-chemical properties for efficient percutaneous absorption, and therapeutic levels of these agents cannot be achieved in either tissue or blood. It has been demonstrated in numerous studies that the ideal properties for a drug molecule to readily diffuse across the skin are both adequate hydrophilicity and lipophilicity (i.e., adequate water and lipid solubility) and an optimal partition coefficient [69]. Especially the partition coefficient seems to be crucially important in establishing a high initial drug concentration within the superficial skin layers. These optimal features can often be achieved by the prodrug approach, which is a very efficient means of manipulating drug–skin and drug–vehicle interactions. Prodrugs have been studied in both topical drug delivery (local treatment preferred) and transdermal drug delivery (systemic treatment preferred) [70, 71]. The principle of dermal and transdermal prodrug delivery is illustrated in Figure 17.18.

Topical Drug Delivery Naproxen is a nonsteroidal anti-inflammatory drug (NSAID) with both analgesic and antipyretic properties. It is suitable for the treatment of both topical and some soft-tissue disorders. Unfortunately, its topical administration in humans shows a bioavailability of only 1 to 2 percent, which may be attributed to a very polar and ionizable carboxylic acid group. Various naproxen prodrugs (Table 17.1) were studied in attempts to increase the topical penetration of naproxen by transiently masking the carboxylic acid by esterification [24, 73–75]. Acyloxyalkyl, aminoacyloxyalkyl, and aminoalkyl esters of naproxen readily hydrolyzed to naproxen in vitro and possessed highly variable aqueous solubilities and lipophilicities. The highly lipophilic acyloxyalkyl esters did not enhance the dermal permeation of naproxen across

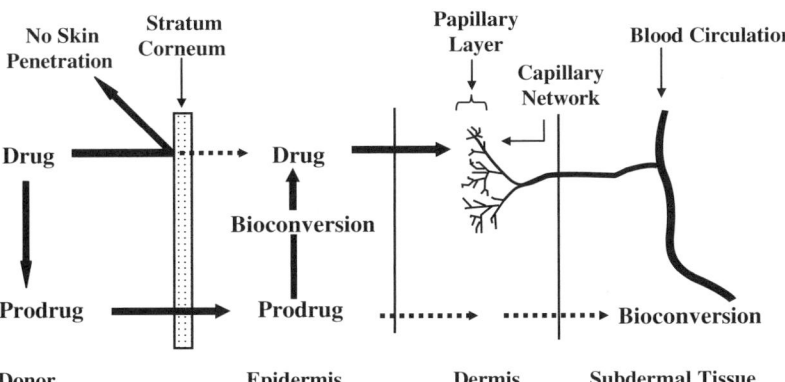

Figure 17.18 An illustration of the dermal and transdermal prodrug concept (modified from Ref. 72). In topical (i.e., dermal) drug delivery, bioconversion of the prodrug into the parent drug must take place by enzymes in the skin immediately after application, but when a systemic effect is desired (transdermal drug delivery), release of the parent drug can take place in the cutaneous circulation as well.

TABLE 17.1 Apparent Partition Coefficients, Aqueous Solubilities, and Steady-State Fluxes of Naproxen and Its Prodrugs

Structure: H₃C—O—[naphthalene]—CH(CH₃)—C(=O)—O—R

Compound	R	Log P_{app}[a] pH 7.4	Aqueous Solubility pH 7.4 (mM)	Flux[b] (nmol/cm²·h)
Naproxen	H	0.30 ± 0.03	101.9 ± 1.3	6.5 ± 0.6
Acetyloxy-ethyl ester	—(CH₂)₂—O—CO—CH₃	3.50 ± 0.03	0.060 ± 0.003	0.4 ± 0.1
Methyl-piperazinyl-acyloxyethyl ester	—(CH₂)₂—O—CO—CH₂—N⟨piperazinyl⟩NCH₃	1.16 ± 0.08	4.1 ± 0.4	24.6 ± 1.0[c]
Methyl-piperazinyl-ethyl ester	—(CH₂)₂—N⟨piperazinyl⟩NCH₃	2.29 ± 0.02	0.37 ± 0.03	58.5 ± 13.0[c]
Methyl-piperazinyl-butyl ester	—(CH₂)₄—N⟨piperazinyl⟩NCH₃	2.44 ± 0.09	1.13 ± 0.09	27.7 ± 1.8[c]

[a] P_{app} is the apparent partition coefficient between 1-octanol and phosphate buffer (0.16 M) at room temperature.
[b] The delivery of total naproxen species through excised human skin in vitro in isotonic phosphate buffer (0.05 M, pH 7.4) at 37°C.
[c] $p < .05$ compared to naproxen (ANOVA, Fisher's PLSD test).

the human skin in vitro, most probably due to their very low aqueous solubilities. In contrast, members in a series of aminoacyloxyalkyl esters (amino acids as the promoiety) and aminoalkyl esters had a combination of adequate aqueous solubility and lipophilicity, and gave significant in vitro improvements in the skin permeation of naproxen (Table 17.1). Moreover, the prodrugs hydrolyzed simultaneously during permeation, which is of great interest to move forward in the development of topical dosage forms.

It is also interesting note that at pH 7.4, where the highest flux values of both naproxen and prodrugs were achieved, significantly lower concentrations of prodrugs were used in the donor compartment than that of naproxen to achieve dermal transport. The permeability coefficients, which, unlike flux, are independent of donor concentration, were higher (up to 998-fold) for almost all prodrugs when compared to that of naproxen at pH 7.4. However, at pH 5.0 only two prodrugs exhibited higher permeability coefficients than naproxen, despite their flux values being lower than those for any other studied compound. Because the permeability of the studied compounds increased from aqueous solutions as the solubility of the compounds in these solutions decreased, the high permeability coefficient of these prodrugs can be explained by their very low aqueous solubility and decreased concentration in the applied vehicle, when compared to other studied compounds. Therefore, the flux, which measures the mass of material transported through the skin, may be a more relevant therapeutic parameter than the permeability coefficient.

Transdermal Drug Delivery Propranolol, an adrenergic β-blocker, is a widely used and clinically effective cardiovascular agent that is indicated in the treatment of angina pectoris, hypertension, and cardiac arrhythmia. However, low and variable bioavailability follows the oral admistration of propranolol, which may be attributed to extensive stereoselective hepatic first-pass metabolism. Because the transdermal route of administration allows for systemic drug absorption while eliminating hepatic first-pass effects, the relatively impermeable propranolol was made more lipophilic by the formation of diastereomeric ester prodrugs [76]. Subsequent in vitro diffusion studies across rat skin demonstrated a 3- to 12-fold higher flux value associated with propranolol than with the prodrugs. On the other hand, the permeability coefficient was higher for all prodrugs, due to their lower solubilities in the donor phase, compared to that of propranolol. Based on the hydrolysis data in conjunction with the flux values, it could be concluded that drugs having both optimum partitioning characteristics and resistance to enzymatic hydrolysis during permeation were able to afford the highest transdermal permeation.

In summary, the design of dermal prodrugs (topical or transdermal) not only includes the optimization of physico-chemical properties (lipid and aqueous solubility) but also considers the evaluation of enzymatic degradation properties within permeated membranes. When local treatment is preferred, the active parent drug should be released during permeation or immediately afterwards. When systemic treatment is preferred, the parent drug can be released into the systemic circulation as well.

Ophthalmic Drug Delivery In general, the site of action for ophthalmic drugs is located inside the eye. Thus, topical delivery of eyedrops into the lower cul-de-sac of the eye is the most common method of drug treatment in ocular disease. After topical administration, an aqueous eyedrop solution mixes with tear fluid to be dispersed over the eye surface. However, various precorneal factors (i.e., drainage of instilled solution, noncorneal absorption, induced lacrimation) limit ocular absorption by shortening the cornea contact time of applied drugs. These factors, and the corneal barrier itself, limit penetration of a topically administered ophthalmic drug. As a result, only a few percent of the applied dose is delivered into the intraocular tissues, the major part (50 to 99 percent) being absorbed into the systemic circulation (Fig. 17.19), where it can cause various unwanted side effects.

The cornea is generally considered to be a major pathway for the ocular permeation of topically applied drugs [77]. Compared to many other epithelial tissues (e.g., bronchial, intestinal, nasal, tracheal), the corneal epithelium is relatively impermeable but less than the stratum corneum of the skin [78]. For corneal permeation, lipophilicity of the drug seems to be the most important property, and both parabolic [79] and sigmoidal [80] curves have been

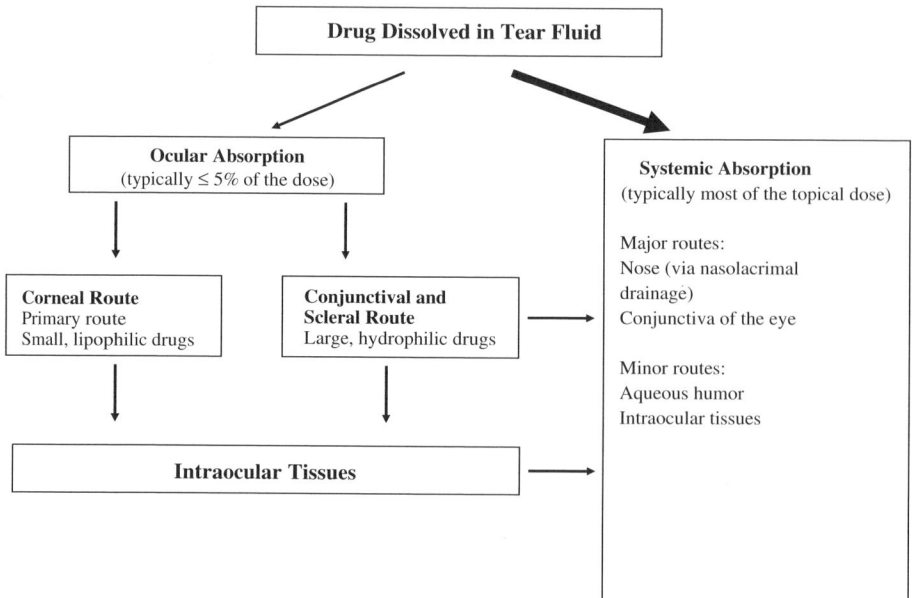

Figure 17.19 Main absorption routes of a topically applied ophthalmic drug.

used to describe the influence of drug lipophilicity on corneal drug permeation. The optimum apparent partition coefficient (P_{app}; octanol/pH 7.4 buffer) for corneal drug absorption is in the range of 100 to 1000 [81, 82], which is consistent with the lipophilic nature of the corneal epithelium [83]. Aqueous solubility is another drug property important for the efficacy of ophthalmic drug delivery. The surface of the eye is constantly being cleaned and moistened by the aqueous tear fluid. Thus, it is difficult for drug molecules to be absorbed by the corneal epithelium unless they are soluble in the tear film. In addition, the drug's water solubility must be sufficient to enable the formulation of aqueous eyedrops. The dilemma is that an ideal potential ophthalmic drug should be simultaneously both water soluble and lipid soluble, and only a few molecules can fulfill these severe criteria. Because of this fact, various pharmaceutical technologies, such as prodrugs and cyclodextrins, have been applied to improve the physico-chemical properties of ophthalmic drugs.

Ocular absorption of a drug can be substantially enhanced by increasing its lipophilicity (hydrophilic compounds) or aqueous solubility (poorly water-soluble compounds). Prodrug derivatization, as a method to achieve these objectives, is a potential and a very useful approach that was introduced to ophthalmology about 20 years ago when the ocular absorption of epinephrine was substantially improved by this approach.

Dipivalyl epinephrine (dipivefrine) is a dipivalic acid diester of epinephrine (Fig. 17.20) that releases epinephrine after corneal absorption [84, 85]. Dipivefrine penetrates the human cornea 17 times faster than epinephrine [86] due to the fact that dipivefrine is 600 times more lipophilic (at pH 7.2) than epinephrine [87]. Consequently, a smaller topical dose of dipivefrine achieves a similar therapeutic effect in the eye [88] and the smaller prodrug dose results in decreased systemic levels of epinephrine. Compared to the conventional 2 percent epinephrine hydrochloride eyedrop, a 0.1 percent dipivefrine eyedrop results in an only slightly less effective IOP lowering effect, while side effects are greatly reduced [89].

Prostaglandin analogs represent a new class of active ocular hypotensive agents. Commercially available latanoprost, travoprost, and unoprostone (Fig. 17.21) are lipophilic isopropyl ester prodrugs that rapidly hydrolyze in the cornea to the biologically active prostaglandin [90, 91, and references cited therein]. The present lipophilic prodrugs achieve efficient ocular absorption and make the topical administration possible with decreased systemic side effects. Various lipophilic prodrugs have been studied in order to enhance ophthalmic absorption of drugs such as pilocarpine [8, 92], timolol [93], and tilisolol [94].

Another possible prodrug strategy for enhancing drug absorption is a prodrug design based on the structures of membrane transporters. However, very little is known about the presence of various transporters on the corneal epithelium. For example, hydrophilic L-valyl ester of acyclovir is threefold more permeable across the intact rabbit cornea than the parent drug [95]. This enhanced permeability suggests the presence of a carrier-mediated transport

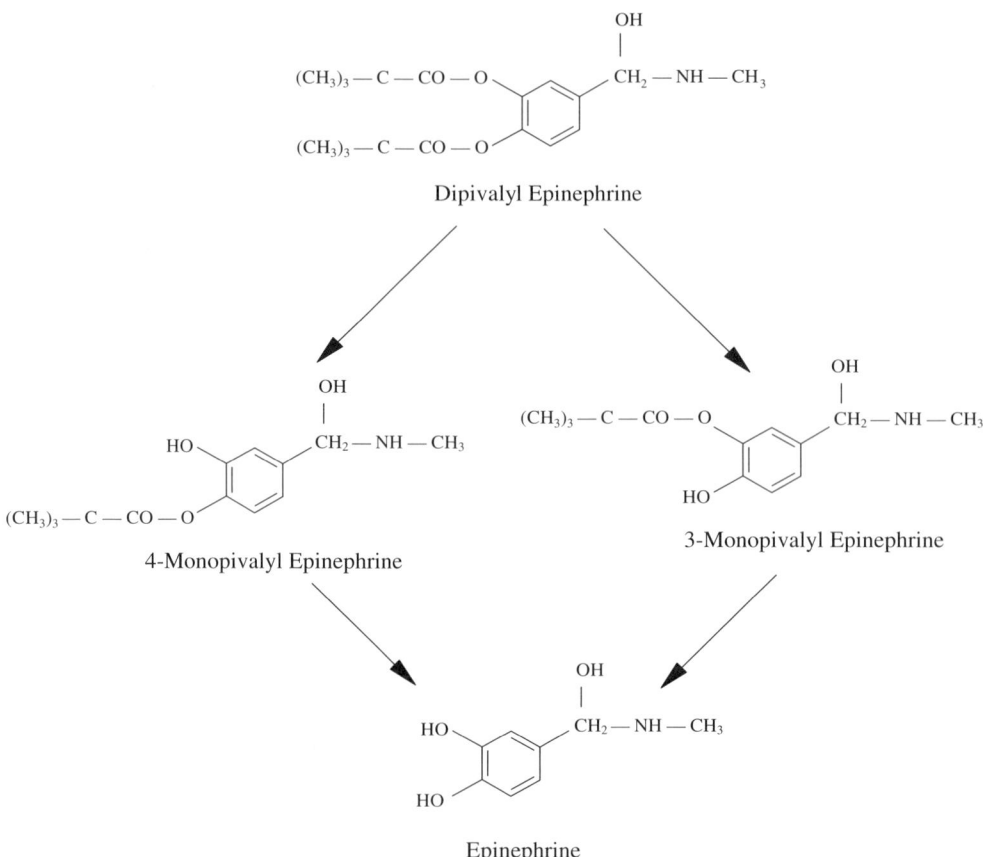

Figure 17.20 Bioconversion of dipivalyl epinephrine.

Nasal Drug Delivery Systemic drug delivery via the nasal cavity circumvents the gastrointestinal degradation and first-pass elimination associated with oral drug delivery. Moreover, the nasal route can be an attractive alternative to parenteral administration. In the healthy nose, epithelium is covered by a 2- to 4-mm layer of aqueous mucus, which is renewed approximately every 10 min. This aqueous mucus layer is one of three physical barriers for drug absorption by the nasal cavity. The other two barriers are the lipophilic epithelium lining the nasal cavity and an effective enzymatic pathway, both of which result in presystemic drug degradation. The enzymatic barrier in nasal administration is, however, not as efficient as in drug metabolism in the gastro-

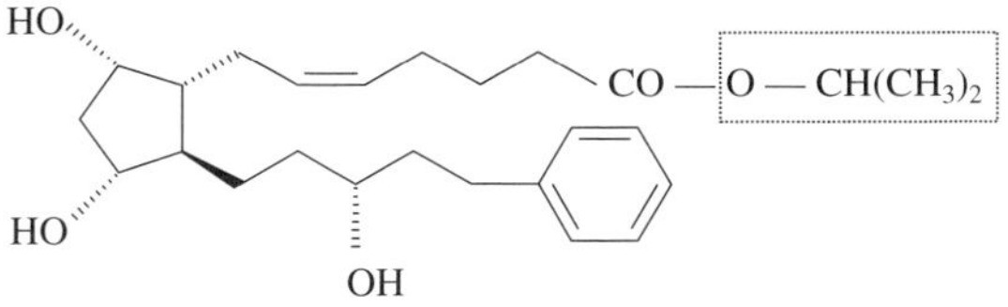

Latanoprost

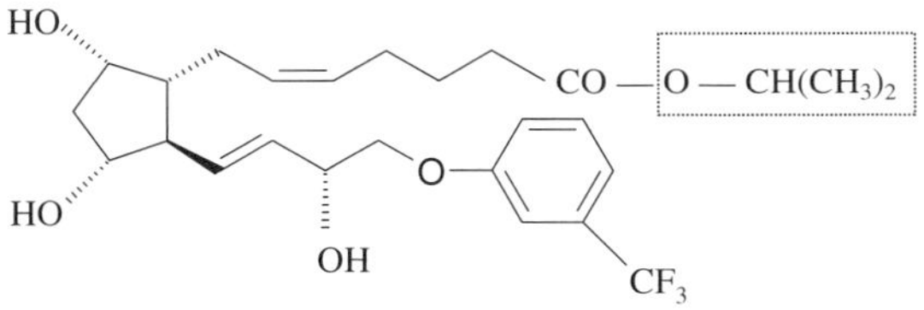

Travoprost

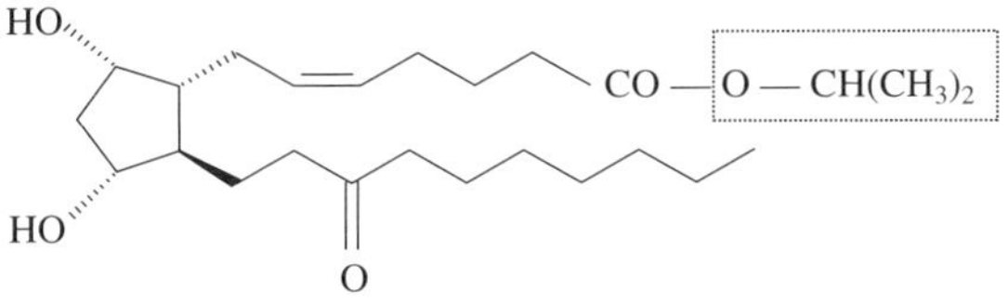

Unoprostone

Figure 17.21 Chemical structures of latanoprost, travoprost, and unoprostone (pro-moieties are marked by the rectangle).

intestinal tract. Drug absorption from the nasal cavity is predominately passive. Small hydrophilic drugs are absorbed via passive diffusion through aqueous channels, whereas lipophilic drugs are absorbed via the transcellular route. Drug molecules are rapidly absorbed from the nasal cavity, resulting in plasma profiles that are frequently comparable to those obtained after intravenous delivery. However, the absorption rates decrease with increasing molecular weight and fall off sharply when the weight exceeds 1000 daltons [96, 97].

Water-Soluble Prodrugs Enhancement of nasal drug delivery can be obtained by the formation of water-soluble derivatives of lipophilic drugs. Such derivatives will not only increase the amount of dissolved drug in the mucus, resulting in enhanced drug diffusion through the aqueous mucus barrier, but also allows the dose to be dissolved in a small volume (i.e., less than 200 μL) of an aqueous nasal vehicle. Testosterone, for example, undergoes extensive first-pass metabolism when given orally. The drug is completely absorbed in rats after nasal administration, but it is difficult to formulate a sufficient drug dosage (about 3 mg) for intranasal delivery in humans due to its low aqueous solubility (0.01 mg/mL). Through the formation of a water-soluble prodrug ester, testosterone 17β-N, N-dimethylglycinate hydrochloride, the solubility was increased to over 100 mg/mL. The resulting bioavailability of testosterone after nasal administration of the prodrug to rats was reported to be nearly 100 percent [98].

The nasal administration of benzodiazepines is another example. Benzodiazepines are generally well, but slowly, absorbed from the gastrointestinal tract, reaching maximum serum concentrations about 2 h after oral administration. Benzodiazepines are used to treat panic attacks and seizures and as sedatives before anesthesia and surgery, where they are administered intravenously for rapid onset of drug action. As in the case of testosterone, the nasal administration of benzodiazepines is hampered by their low aqueous solubility. It is possible to increase their solubility by opening the diazepine ring of the benzodiazepine molecule; that is, by the formation of a prodrug. For benzodiazepines, such as alprazolam, midazolam, and triazolam, this ring opening is completely reversible (Fig. 17.22). In human serum in vitro at 37°C, the half-life for the first-order rate constant for the ring-closing reaction was estimated to be less than 2 min for both alprazolam and midazolam [99]. An aqueous nasal formulation containing the midazolam prodrug had an absolute bioavailability of 64 to 73 percent [99, 100].

Lipophilic Prodrugs Lipophilic prodrugs are thought to be absorbed from the nasal cavity via the transcellular route and, thus, their absorption is strongly affected by their lipophilicity. Acyclovir is a hydrophilic antiviral agent that does not readily permeate biological membranes [101]. The nasal absorption of acyclovir was enhanced significantly by forming lipophilic prodrug esters of the parent drug, but the results indicate that enzymatic hydrolysis of the esters prior to absorption (i.e., by the enzymatic barrier) reduced the amount of drug absorbed (Table 17.2) [102, 103]. Further studies

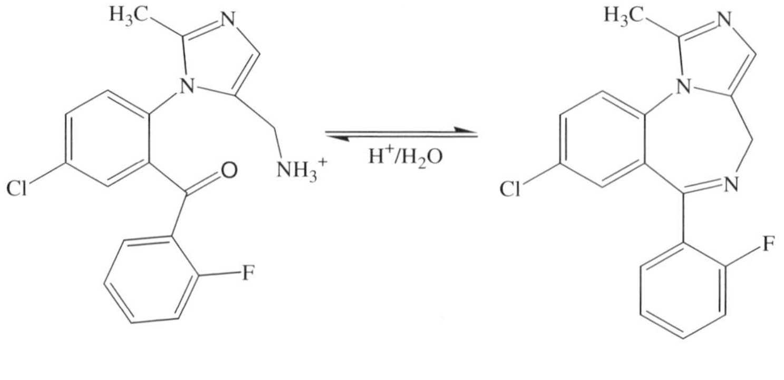

Figure 17.22 The prodrug–midazolam equilibrium.

TABLE 17.2 Structure of Acyclovir and Acyclovir Ester Prodrugs, Their Aqueous Solubilities, Partition Coefficients, Half-lives in Rat Plasma, and Percentages of Absorption in a Rat Nasal Perfusion Model [102, 103]

R =	Solubility[a] (mM)	PC[b]	$t_{1/2}$ (min)	Percent Uptake in 90 min
H	11.2	0.06	—	0.0
—C(=O)—(CH$_2$)$_2$CH$_3$	4.6	0.83	25.6	9.7
—C(=O)—C(CH$_3$)$_3$	1.5	2.01	105.0	14.8
—C(=O)—(CH$_2$)$_3$CH$_3$	1.6	2.35	16.2	17.2
—C(=O)—(CH$_2$)$_4$CH$_3$	0.7	8.58	3.6	30.1

[a] Solubility in pH 7.4 isotonic phosphate buffer at 37°C.
[b] Partition coefficient between 1-octanol and pH 7.4 isotonic phosphate buffer at 37°C.

in rat nasal mucosal homogenate showed that the rate and extent of prodrug degradation assumes a direct relationship with the acyl side-chain length. Lengthening of the linear acyl chain from three to eight carbons resulted in more than 600-fold increase in first-order degradation rate constants. Esters with branched acyl chains were more resistant to nasal carboxylesterase-mediated cleavage, and, thus, the branched esters not only increased the lipophilicity of acyclovir but also minimized mucosal hydrolysis, resulting in better drug absorption from the nasal cavity.

Enhanced Brain Delivery   The anti-Parkinson drug levodopa is a zwitterion and, thus, both hydrophilic and water insoluble. Both lipophilicity and water solubility of the drug was increased through the formation of alkyl prodrug esters. Nasal administration of the prodrugs resulted in rapid and complete absorption into the systemic circulation and improved dopamine delivery to the central nervous system (CNS) when compared to intravenous administration of the prodrug. These results indicate that the prodrugs were delivered from the nasal cavity to the cerebral spinal fluid via the olfactory epithelium [104, 105].

Transport-Mediated Absorption   Some amino acids have been shown to be absorbed via active transport from the nasal cavity and can be used as transport moieties for the active absorption of drugs [106]. For example, studies have indicated that an L-aspartate β-ester prodrug of acyclovir improved the nasal absorption of acyclovir by an active transport system [107].

Prodrugs of Peptides   Both peptides and protein drugs are susceptible to rapid enzymatic degradation. For example, the aminopeptidase-catalyzed hydrolysis of Leu-enkephalin and Met-enkephalin is responsible for the rapid deactivation of these peptides at various mucosal sites. Formation of 4-imidazolidinone derivatives of enkephalins almost totally prevented aminopeptidase cleavage of the drugs. These derivatives were readily converted to the parent peptides by spontaneous hydrolysis [108]. In has been suggested that such prodrugs can enhance the bioavailability of peptides after nasal administration [109].

Improved Aqueous Solubility

The efficient aqueous solubility of a drug molecule is a prerequisite for the preparation of aqueous-based solutions for drug delivery. Moreover, aqueous solubility is also one of the key physico-chemical properties in the oral drug absorption process. Two prodrug approaches have been introduced to enhance the aqueous solubility of a poorly soluble drug: (1) decreasing the parent drug's melting point by prodrug derivatization and/or (2) introducing an ionizable/polar promoiety to the parent drug, with the latter being the more common approach [15]. The most commonly utilized ionizable and, therefore, solubilizing prodrug structures are presented in Table 17.3.

TABLE 17.3 Examples of Water-Soluble Prodrug Structures for Poorly Water-Soluble Drugs

Prodrug Derivative	Molecular Structure
Phosphate esters	Drug—O—PO_3^{-2}
α-Amino acid esters and amides (from alcohols X = O; from amines X = NH; R = alkyl/aryl group)	Drug—X—C(=O)—CH(R)—NH_3^+
Amino acid amides (from carboxylic acids; R = alkyl/aryl group)	Drug—C(=O)—N(H)—CH(R)—COO^-
Hemiesters (typically n = 2–4)	Drug—O—C(=O)—$(CH)_n$—COO^-
Aminoalkyl esters (R = alkane group; R_1 and R_2 = H and/or alkane group, or heterocycle)	Drug—O—C(=O)—R—$\overset{+}{N}HR_1R_2$
Trialkylammonium (quaternary ammonium alkyl) esters (R = alkane group; R_1, R_2, and R_3 = alkyl groups)	Drug—O—C(=O)—R—$\overset{+}{N}R_1R_2R_3$
Glucose derivatives (as an example glucuronide)	Drug—O—spacer—O—(glucuronide ring with COO^-, HO, OH, OH)
Sulfate esters	Drug—O—SO_3^-

Phosphates A number of phosphoric acid esters have been developed as potential water-soluble prodrugs. Many are marketed as injectable dosage forms while only a few are used in oral dosage forms [67]. The phosphate promoiety can be introduced either directly or via various spacer groups to the parent drug molecule that contains a hydroxyl or various amino functional groups, where it is enzymatically cleaved with relative ease by endogenous alkaline phosphatases.

Etoposide Phosphate A representative example of a phosphate prodrug is etoposide phosphate, which is currently in clinical use. Due to its poor aqueous solubility and chemical instability, etoposide requires a complex IV formulation and must be diluted to a concentration of less than 0.4 mg/mL to avoid precipitation. Monophosphorylation of the phenolic group of etoposide increases the aqueous solubility up to 20 mg/mL and, furthermore, improves its chemical stability. After IV administration to patients, rapid and quantitative bioconversion of etoposide phosphate to etoposide was observed. The efficacy of the phosphate prodrug also proved to be bioequivalent to etoposide, with analogous therapeutic effects [110].

Fosphenytoin   Phenytoin is a sparingly water-soluble (20 to 25 µg/mL) weakly acidic (pK_a = 8.3) drug. In comparison, fosphenytoin [(phenytoin-3-yl)methyl phosphate] is a phosphonooxymethyl prodrug of phenytoin that has high aqueous solubility (140 mg/mL), which also provides good chemical stability within neutral and slightly basic pH ranges. The bioconversion process involves a phosphatase-catalyzed dephosphorylation, followed by spontaneous decomposition of the hydroxymethyl intermediate, eventually releasing the parent drug and formaldehyde (Fig. 17.23). In epileptic patients, both intra-

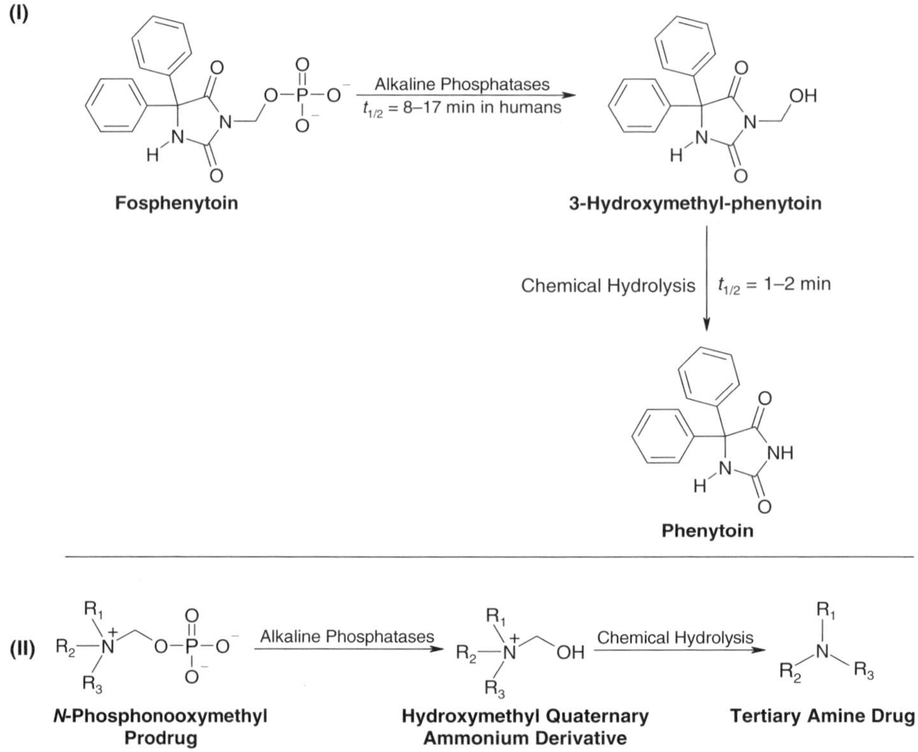

Figure 17.23   Bioconversion of (I) fosphenytoin and (II) *N*-phosphonooxymethyl prodrugs.

muscular and intravenous administrations of fosphenytoin led to its rapid and quantitatively bioconversion to phenytoin with half-lives ranging from 8 to 17 min. Moreover, the blood levels of phenytoin following intramuscular administration of fosphenytoin were far superior to those achieved from phenytoin sodium alone. The pharmacokinetics, pharmacodynamics, and safety properties of fosphenytoin have been thoroughly reviewed by Stella [6]. The same prodrug strategy has been extended to various drugs containing a tertiary amino group, which were then derivatized to N-phosphonooxymethyl prodrugs [111–113]. Again, highly water-soluble prodrugs were substrates for alkaline phosphatase that cleaved the phosphoric acid monoester bond, followed by spontaneous in vivo chemical hydrolysis to the tertiary amine and formaldehyde (Fig. 17.23).

Other Prodrug Structures A number of hemiester prodrugs (mainly hemisuccinate esters) have been studied as water-soluble prodrugs [114–116]. These derivatives usually improve aqueous solubility of the parent drug, but the limited amount of published information indicates that hemiester prodrugs generally possess limited chemical stability in aqueous solutions and give a slow and incomplete hydrolysis to the parent drug in human plasma. In contrast, their bioconversion appears relatively fast in the presence of hepatic hydrolases.

Amino acid prodrugs have been evaluated as water-soluble derivatives of various alcohols [42, 117] and amines [118, 119], as the corresponding ester or amide prodrugs. In addition to good aqueous solubility, the use of amino acids as a promoiety provides rapid and quantitative bioconversion by ubiquitous esterases and/or peptidases. While amino acid ester prodrugs show usually relatively rapid chemical hydrolysis, which precludes the development of solution formulations, amino acid amide prodrugs, in general, exhibit greater chemical stability while still maintaining enzymatic susceptibility.

Prolonged Duration of Action

In general, prodrugs have been designed to prolong the duration of a particular drug action by two means: (1) inactive prodrug is delivered into the body at a sustained rate, where it rapidly releases the parent drug, or (2) after administration the prodrug releases the parent drug over a prolonged and relatively well-defined period of time. Primary goals for the prolongation of drug action can be an improvement in patient compliance by eliminating the need for frequent dosing, and by reducing side effects through a steady drug concentration.

The slow release for very lipophilic prodrug derivatives of several steroids (e.g., testosterone, nandrolone) and neuroleptics (e.g., fluphenazine, flupenthixol, haloperidol) from the site of intramuscular injection results in a prolonged duration of action. Once released into the tissue fluid or blood, prodrugs are rapidly bioconverted in most cases with no attenuation of their

770 DESIGN AND PHARMACEUTICAL APPLICATIONS OF PRODRUGS

therapeutic action. In the case of the neuroleptic fluphenazine (Fig. 17.24), the onset of action generally appears between 24 to 72 h after injection, and then continues for 1 to 8 weeks with an average duration of 3 to 4 weeks [120].

The prodrug approach has also been evaluated to obtain sustained drug action as a means of prolonging release of the parent drug. Longer duration of drug action has been achieved, for example, with a prodrug of insulin [121], which prolongs a normoglucemic profile in diabetic rats, due to a slow bioconversion of insulin from (9-fluorenylmethoxycarbonyl-SO_3H)$_3$-derivatized insulin, and with a bisdimethylcarbamate prodrug (bambuterol) of the bronchodilator terbutaline [122, 123], which after oral administration is predominantly bioconverted to terbutaline outside the lungs via a cascade of hydrolysis and oxidation reactions (Fig. 17.25). Moreover, prodrugs of various antiglaucoma drugs such as pilocarpine, timolol, and epinephrine have a slow bioconversion to their parent drugs, achieving a longer duration of action that is, however, closely related to improved ocular permeation. A prodrug of epinephrine was described above.

Improved Drug Targeting

Targeted or site-specific drug delivery is one of the ultimate goals in controlled drug delivery. In prodrug design, targeting can be achieved by site-directed drug delivery or site-specific drug bioactivation (Table 17.4). With site-directed drug delivery, such as localized drug delivery (e.g., dermal and ocular drug delivery), the bioreversible prodrug is selectively or primarily transported, as an intact prodrug, to the site of drug action. On the other hand, with site-specific bioactivation, the prodrug can be widely distributed all over the body but undergoes bioactivation and exerts activity only at the desired site. The most common prodrug applications in targeted drug delivery are described as follows. In many prodrug applications, either site-directed drug delivery or site-specific bioactivation has been combined in order to achieve successful targeted drug delivery.

Tumor Targeting The physiological barriers of solid tumors, which are aggregated masses of cells caused by changes or mutation in genes that control

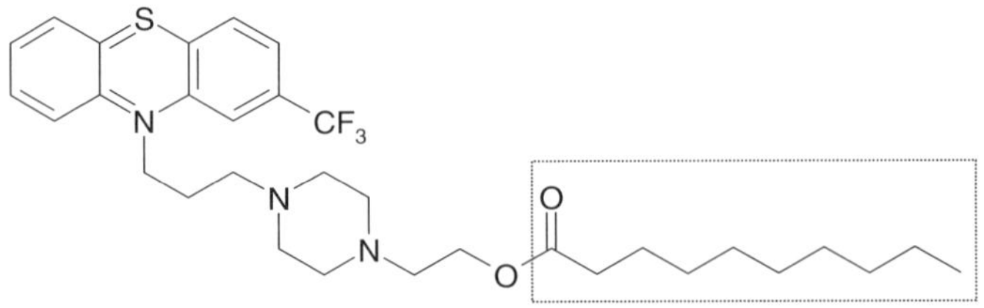

Figure 17.24 Fluphenazine decanoate (lipophilic promoiety marked by the rectangle).

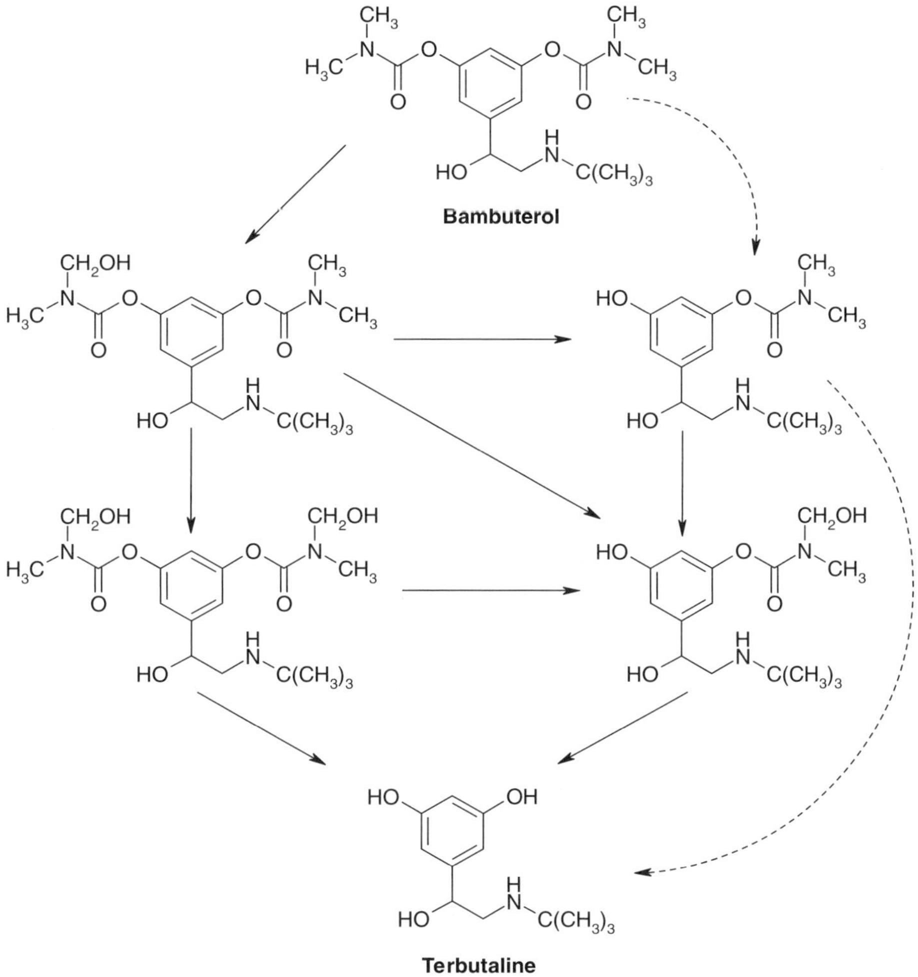

PHARMACEUTICAL APPLICATIONS OF THE PRODRUG STRATEGY 771

Figure 17.25 Bioconversion of bambuterol to terbutaline via hydrolytic (broken arrows) and oxidative (solid arrows) pathways.

healthy cell division, considerably restrict cancer chemotherapy. Compared with the normal, ordered vasculature of healthy tissues, blood vessels in tumors are highly abnormal and have distended capillaries with leaky walls and sluggish flow, leading to inconsistent drug delivery. Furthermore, the increased intratumoral interstitial pressure limits the penetration of drugs into tumors. Highly irregular blood flow results in localized hypoxia, which subsequently leads to tumor resistance to certain anticancer agents and to radiotherapy as well. However, increased bioreductive enzyme expression is an adaptive strategy that solid tumors use to detoxify anticancer drugs [124–126], and the hypoxic environment further contributes to the increased reductase activity of

TABLE 17.4 Prodrug Approaches to Improve Drug Targeting

Site-Directed Drug Delivery	Site-Specific Bioactivation
Prodrug is selectively or primarily transported to the site of drug action: • Localized drug delivery (e.g., dermal and ocular drug delivery) • Prodrug improves drug delivery to the specific tissue (e.g., improved CNS drug delivery)	Prodrug is widely distributed throughout the body: • Bioactivation takes place only at the desired site of drug action

tumors. Due to the high proliferation rates of tumor cells, in addition to bioreductive activity, the levels of other enzymes such as acid phosphatase, uridine phosphorylase, β-glucuronidase, and γ-glutamyl transpeptidase are often elevated in these cells and have been exploited in targeted prodrug–tumor delivery [124]. Recent prodrug approaches have been actively applied to achieve very precise and direct effects at the tumor site of action, with minimal effects to the rest of the body. Therefore, an effective prodrug must be selectively cleaved to the active drug by specific activating enzymes at the site of action. For tumor drug delivery, the first studied prodrug-activating enzymes were endogenous to specific tumor types (e.g., sarcoma 180, NK lymphoma, Walker carcinoma 256, Novikoff hepatoma), such as β-glucuronidase and aryl sulfatase, and a NAD(P)H dehydrogenase isozyme that sensitized tumors to aniline mustard and 5-aziridinyl-2,4-dinitrobenzamide, respectively. To expand the range of tumors susceptible to this mode of therapy, prodrug-activating enzymes were initially targeted to tumor cells by using antibodies and more recently genes. These approaches are referred to as ADEPT (antibody-directed enzyme prodrug therapy) and GDEPT (gene-directed enzyme prodrug therapy), VDEPT (virus-directed enzyme prodrug therapy), or GPAT (gene–prodrug activation therapy), respectively. The latter approach is also known as "suicide gene therapy." Some specific examples of hypoxia-selective prodrugs, and both ADEPT and GDEPT technologies in targeted tumor delivery are described in the following section.

Hypoxia-Selective Prodrugs Despite highly irregular blood flow to the solid tumor, with the development of oxygen-depleted areas (hypoxia) that result in poor drug delivery, hypoxic cells also present a major therapeutic opportunity for tumor targeting via bioreductive prodrugs. The substantial selectivity of hypoxia-selective drugs depends upon the presence of overexpressed reductase enzymes in tumor cells that reduce bioreductive prodrugs to active cytotoxic radicals under hypoxia. In normal cells, under aerobic conditions, the reduced radical is oxidized to the nontoxic prodrug with the concomitant production of a superoxide radical (Fig. 17.26). There are three main classes of hypoxia-selective prodrugs that are in clinical use or are being developed for

PHARMACEUTICAL APPLICATIONS OF THE PRODRUG STRATEGY

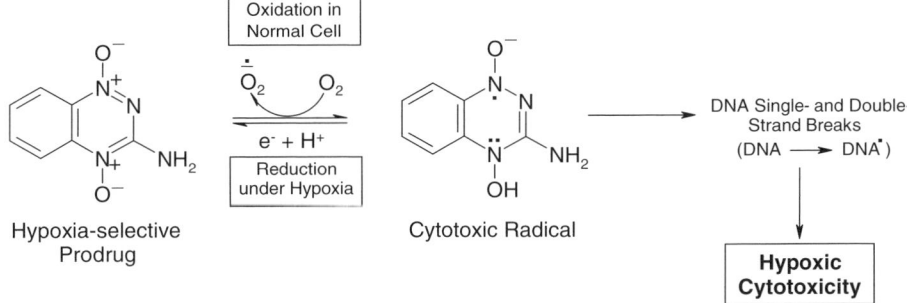

Figure 17.26 Schematic presentation of bioreduction for a hypoxia-selective prodrug (tirapazamine, TPZ) to its cytotoxic free radical, causing preferential toxicity to hypoxic cells.

clinical use: quinone-based compounds, nitroimidazoles, and *N*-oxides [124–127].

Among the quinones of clinical interest that act as hypoxia-selective prodrugs are mitomycin C, its *N*-methyl derivative porfiromycin, and aziridinylquinone EO-9. All are structurally similar and are converted upon reductive metabolism to semiquinones and hydroquinones, which probably alkylate nucleophiles [e.g., DNA (deoxyribonucleic acid)] through covalent binding. Although EO-9 demonstrated excellent antitumor activity in hypoxic tumor cells and in a solid tumor animal model, it failed in clinical trials due to its short half-life (rapid elimination) and poor distribution. Moreover, mitomycin C has demonstrated only modest selectivity toward hypoxic cells and its clinical utility is also limited because of severe side effects. Despite promising preclinical data, and an acceptable toxicity profile, porfiromycin is considered to be inferior to mitomycin C.

The most extensively studied compounds in a second class of hypoxia-selective prodrugs are 2-nitroimidazoles, of which RSU1069 has been shown to be highly efficient both in vitro and in vivo. RSU1069 containing an aziridinyl alkylating group in the side chain (dual function) showed increased hypoxia selectivity compared to its nonalkylating nitroimidazole analogs and was bioconverted into a bifunctional alkylating drug by reduction. In further drug development, its bromoethyl prodrug RB6145 was shown to form RSU1069 rapidly under physiological conditions and have less systemic toxicity and improved pharmacokinetic properties compared to RSU1069. Clinical trials of RB6145 were, however, aborted due to irreversible cytotoxicity toward retinal cells.

The *N*-oxide, tirapazamide (TPZ), is the first drug introduced into clinical use as a selective hypoxia-activated prodrug. TPZ undergoes 1-electron bioreduction to monodeoxygenated toxic products (Fig. 17.26) in the presence of cytochrome P450, NADPH : cytochrome P450 oxidoreductase, xanthine oxidase, and aldehyde dehydrogenase. Under hypoxia, the formed oxidizing radical leads to DNA single- and double-strand breaks, which kills the tumor cell.

Antibody-Directed Enzyme Prodrug Therapy (ADEDT) In the ADEPT design (Fig. 17.27), prodrug-activating enzymes are targeted toward human tumor xenografts by conjugating them to tumor-selective monoclonal antibodies. An antitumor antibody is covalently conjugated to an enzyme not normally present in the extracellular fluid or on cell membranes (e.g., bacterial or viral enzymes), and these conjugates are then localized in the tumor via systemic administration. After allowing for the conjugate to clear from the blood, a prodrug is administered. The activation of a prodrug to form a chemotherapeutic drug then takes place by reacting with the pretargeted enzyme, which was delivered onto the tumor surface. ADEPT has progressed into several phase I clinical trials (e.g., CMDA prodrug with a murine monoclonal antibody linked to the bacterial enzyme carboxypeptidase) and has shown evidence of tumor response. Although giving high tumor to blood ratios of antibody–enzyme conjugate, the targeting systems require the administration of a clearing antibody in addition to the antibody–enzyme conjugate. Moreover, the paucity of tumor-specific antigens has limited the applicability of ADEPT. This limitation has in turn led to the development of GDEPT, in which vectors are used to deliver suicide genes that encode specific prodrug-activating enzymes for tumor cells. The ADEPT approach has been reviewed recently [128, 129].

Gene-Directed Enzyme Prodrug Therapy (GDEPT) In GDEPT (Fig. 17.27), an inactive prodrug can be activated to release a cytotoxic drug by an enzyme that has been delivered via a gene to the tumor for expression. GPAT is a

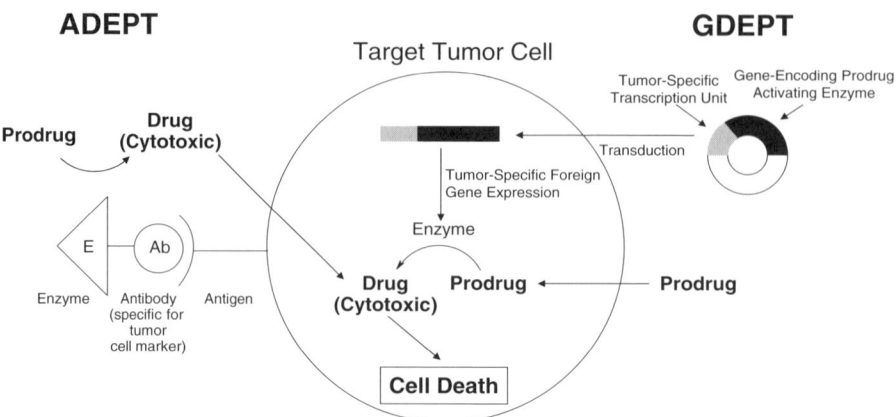

Figure 17.27 Outline of antibody-directed enzyme prodrug therapy (ADEPT) and gene-directed enzyme prodrug therapy (GDEPT).

variation of GDEPT, which uses known transcriptional differences between normal and tumor cells to achieve the selective expression of a prodrug-metabolizing enzyme [128]. The enzymes used are from nonmammalian origin, such as viral tyrosine kinase, bacterial cytosine deaminase, carboxypeptidase G2, purine nucleoside phosphorylase, and nitroreductase or enzymes of human origin that are absent or are expressed at only low concentrations in tumor cells, such as deoxycytidine kinase and cytochrome P450. A number of GDEPT approaches have been described in the past decade [130, 131]. A prototypical example of this approach is centered on inserting a thymine kinase gene into the herpes simplex virus (HSV-TK), followed by combination treatment with the prodrug ganciclovir (GCV). Another widely studied example is the bacterial gene cytosine deaminase (CD), which sensitizes tumor cells to the antifungal drug, 5-fluorocytosine (5-FC). As a result 5-FC is bioconverted to an anticancer drug, 5-fluorouracil (5-FU). Both of these approaches have been tested in clinical trials; HSV-TK-GCV against, for example, mesothelioma, ovarian cancer, glioblastoma, breast cancer, melanoma, multiple myeloma, and astrocytoma, and CD-5-FC has been used against colon and breast carcinomas. In addition to these classical examples, many other enzyme–prodrug combinations have been explored for GDEPT [132, 133]. Despite promising in vitro and in vivo results, the antitumor effect in clinical trials has not proved promising, primarily due to difficulty in achieving selective gene delivery to a sufficient number of tumor cells.

Brain Targeting According to prevailing consensus only small, nonionic, and lipid-soluble molecules can diffuse across the blood–brain barrier (BBB) from the systemic circulation, whereas larger, more water-soluble, ionic molecules do not readily cross the BBB. There are exceptions, however, and some large, polar molecules can be transported into the brain via endogenous transporters. Therefore, the strategies used to improve drug delivery and targeting to the brain utilize passive drug uptake for lipophilic prodrugs and also more sophisticated approaches, such as chemical drug delivery system and carrier-mediated transport.

Lipophilic Prodrugs A series of aliphatic and aromatic lipophilic ester prodrugs of the anticancer agent chlorambucil have been developed to increase efficacy in the treatment of brain tumors [134, 135]. While short-chain aliphatic ester (e.g., methyl and propyl) and the aromatic esters (e.g., phenylmethyl and phenylethyl) were hydrolyzed too rapidly in rat and human plasma in vitro, long-chain aliphatic esters (hexyl and octyl) had more favorable bioconversion rates. However, as lipophilic prodrugs, they were highly bound to proteins and did not readily enter the brain. The sterically more hindered chlorambucil-tertiary butyl ester was subsequently developed in an attempt to maintain the desired slower bioconversion rate while having reduced lipophilicity and protein binding. Significant amounts of chlorambucil-tertiary butyl ester did enter the rat brain, with a 35-fold greater increase compared to an equimolar dose of chlorambucil [135]. However, it exhibited poor bioconversion to chlorambucil in brain tissue and despite an enhanced brain delivery ratio, chlorambucil-tertiary butyl ester did not demonstrate anticancer activity superior to the equimolar administration of chlorambucil. Thus, increased lipophilicity over the parent drug alone does not ensure improved drug efficacy. While enhanced lipophilicity may improve permeation across the BBB, it also tends to increase both uptake into other tissues, causing an increased tissue burden and interaction with plasma proteins. Moreover, both bioconversion selectivity (serum vs. brain) and rate of bioconversion in the targeted tissue should be taken into account when designing bioreversible prodrugs for brain targeting.

Chemical Drug Delivery System (CDS) The principle of CDS, in addition to providing access to the brain by increasing the lipophilicity of a drug, exploits specific properties of the BBB to lock drugs in the brain on arrival and prevent them from recrossing the BBB. Linking an active drug molecule to a bioremovable lipophilic targetor moiety, 1,4-dihydro-*N*-methylnicotinic acid (dihydrotrigonelline), for example, results in a derivative that readily distributes throughout the body and brain after administration due to its lipophilic character. Once inside the brain parenchyma, the lipophilic dihydrotrigonelline is enzymatically oxidized to the ionic quaternary salt via the ubiquitous $NAD(P)H \leftrightarrow NAD(P)^+$ coenzyme system (Fig. 17.28). The acquisition of charge has a dual effect: accelerating the rate of systemic elimination of this derivative and capturing the ionic drug-targetor inside the brain. Subsequently, slow release of the drug from the targetor can result in a sustained and brain-specific release of free active drug, and the targetor is readily removed from the brain by active processes. The CDS has been explored with a wide variety of hydroxy- and amino-containing drugs [136, 137], and considerably increased brain targeting has been achieved, for example, zidovudine (AZT), ganciclovir, benzylbenicillin, and estradiol.

Among all CDSs, estradiol-CDS is in the most advanced stage of investigation [137]. Estradiol is a lipophilic drug with an octanol/water partition coefficient ($\log P$) of 3.3, and derivatization with 1,4-dihydrotrigonelline as a targetor further increases lipophilicity ($\log P = 4.5$), thus enabling better transport across the BBB. Oxidation of the targetor moiety leads to a more ionic and less lipophilic form, with a $\log P$ value of only −0.14. A study in rats found that the concentration of estradiol in rat brain was elevated four to five times longer after estradiol-CDS administration than after estradiol treatment. Moreover, clinical evaluations suggested a potent central effect with only slight elevations in systemic estrogen levels.

Carrier-Mediated Transport Carrier-mediated drug delivery takes advantage of facilitative endogenous transport systems that are present in brain endothelial cells. A number of carrier transport systems in the BBB are responsible for brain uptake of nutrients and their analogs, such as glucose, amino acids,

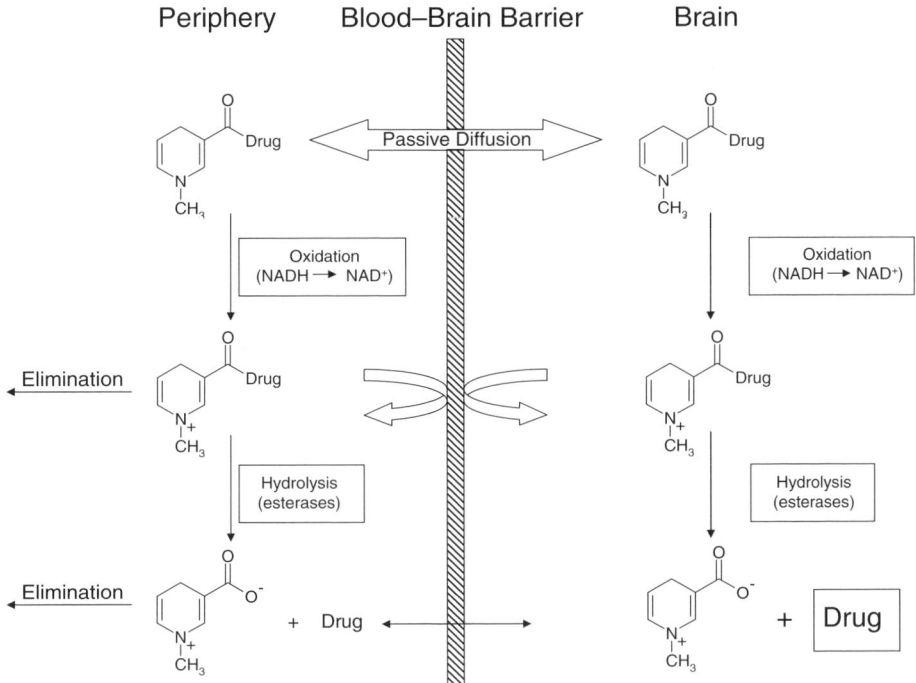

Figure 17.28 Principle of the chemical drug delivery system (CDS).

choline, vitamins, low-density lipoprotein (LDL), and nucleosides in the systemic circulation [138, 139]. Of the nutrient transport systems present at the BBB, the carriers for glucose and large neutral amino acids (LAT) have been estimated to have a sufficiently high transport capacity that hold promise for significant drug delivery to the brain. From these two systems, the glucose carrier is very restrictive in its requirements, and only those molecules closely resembling D-glucose (e.g., D-mannose, D-galactose) are transported. By contrast, the large neutral amino acids transport (LAT) system has the highest maximum transport rate and is less specific for its substrates and, thus, is capable of transporting numerous endogenous and exogenous amino acids of great structural variety. A classic example of the LAT-mediated transport is the anti-Parkinson's agent dopamine. Dopamine itself is poorly transported across the BBB, while its precursor molecule L-dopa is transported into the brain by the LAT system. Once in brain, L-dopa is decarboxylated to dopamine. Other amino acid mimetic drugs in clinical use that enter the brain predominantly via LAT-mediated transport are α-methyldopa (a centrally acting antihypertensive) and gabapentin (an antiepileptic).

In an analogous way to LAT-mediated transport, the glucose transporter was employed to facilitate the penetration of glucose prodrugs of chlorambucil into the brain [140]. All glucose-conjugated prodrugs were able to inhibit the transport of [^{14}C]glucose by the glucose transporter (GLUT1) isoform in a concentration-dependent manner and displayed an inhibition curve similar to that of the inhibition by glucose itself. Especially the 6-*O*-glucose ester derivative of chlorambucil showed activity at a 160-fold lower concentration than chlorambucil itself. Similarly, glucose and galactose derivatives of the anticolvulsant agent nipecotic acid [141], and the potential neuroprotective agent 7-chlorokynurenic acid [142], have been explored as a possible means of improving brain delivery of the active parent drug by providing suitable substrates for active membrane transporters. Moreover, ascorbic acid derivatives of nipecotic acid were shown to competitively inhibit the ascorbate transporter SVCT2, which is expressed in the choroid plexus [143]. In vivo studies with rats showed increased significance in latency for the appearance of pentylentetrazol-induced tonic convulsions than nipecotic acid, demonstrating better absorption compared to the nonconjugated parent drug.

Colon Targeting To achieve successful colon targeting, a drug needs to be protected from absorption and degradation in the upper gastrointestinal tract and then released into the proximal colon. The best known prodrugs that target the colon are azo-bond derivatives of 5-aminosalicylic acid (5-ASA). These prodrugs are formed by attaching a hydrophilic promoiety to 5-ASA and are absorbed from the upper gastrointestinal tract to a very limited extent due to their polar nature, but they are converted to their constituent entities by azoreductases produced by anaerobic colonic bacteria. Carrier molecules that have been used include, for example, sulfapyridine in sulfasalazine, 4-amino benzoyl-β-alanine in balsalazine, and *p*-aminohippurate (4-amino benzoyl glycine) in ipsalazine (Fig. 17.29) [144, 145]. The first prodrug introduced for this purpose, sulfasalazine, causes side effects due to the formation of sulfapyridine, from the degradation of the prodrug. Balsalazine has been found to be therapeutically more effective in clinical trials with fewer side effects. However, in spite of promising pharmacokinetic results, ipsalazine has not developed further. Olsalazine, a dimer of two 5-ASA molecules linked by an azo-bond, has shown promising results in clinical trials (Fig. 17.29).

Similarly, hydrophilic glycoside and glucuronide prodrugs of drugs such as dexamethasone, prednisolone, budesonide, naloxone, and nalmefene, which release the active parent drug by bacterial glycosidases and glucuronidases, have been studied for their colon-targeting potential [144, 145]. In vivo studies in rats revealed that nearly 60 percent of an oral 21-β-D-glucosidic prodrug of dexamethasone reached the colon in the form of the free drug. In the case of prednisolone-β-D-glucoside, 15 percent of the given dose reached the colon. When unmodified steroids were administered orally, they were absorbed from the small intestine and less that 1 percent of dose reached the colon.

Kidney Targeting Renal-specific drug targeting may be an attractive approach under the following conditions: when drugs reaching the kidney cause unwanted extrarenal effects, or the intrarenal transport of a drug is not optimal in relation to the target cell within the kidney, or when a drug is

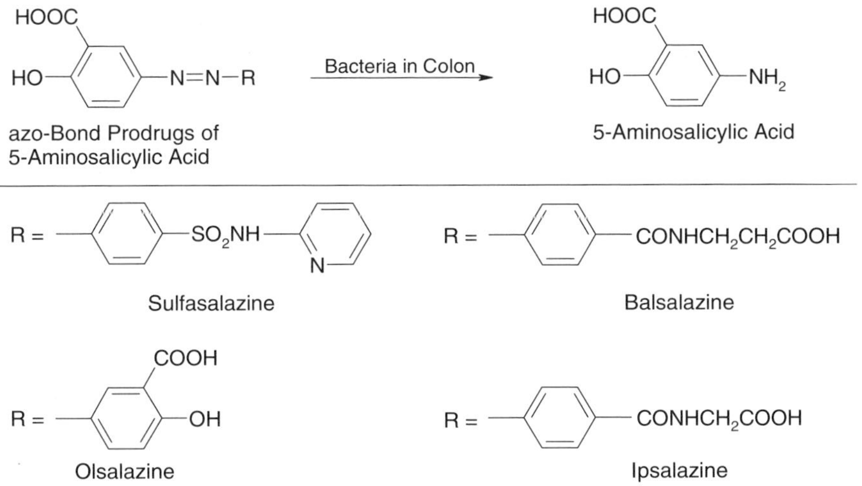

Figure 17.29 Azo-bound prodrugs of 5-aminosalicylic acid (5-ASA) and their bioconversion.

inactivated before reaching the site of action in the kidneys, or when pathological conditions such as abnormalities in glomerular filtration, tubular secretion, or when the occurrence of proteinuria can affect the normal renal distribution of a drug.

Most research on targeted drug delivery to the kidney has focused on the development of amino acid prodrugs [146]. The strategy is based upon the relative high concentration of bioconverting enzymes in the proximal tubular cell, either cytosolic enzymes, such as L-amino acid decarboxylase, β-lyase, and N-acetyl transferase, or enzymes at the brush border of the proximal tubule, such as γ-glutamyl transpeptidase. These prodrugs are designed to be activated by one or two enzymes in order to release the active parent drug. For example, dopamine showed a high renal specificity following the intraperitoneal administration in mice of the double prodrug, γ-glutamyl-L-dopa [147]. This double prodrug first requires γ-glutamyl transpeptidase-catalyzed cleavage of the γ-glutamyl linkage, and consequently the resulting L-dopa is decarboxylated to dopamine by L-amino acid decarboxylase. Targeted delivery to the kidney has also been achieved with N-acetyl-γ-glutamyl prodrugs of sulfamethoxazole [148]. On the other hand, L-γ-glutamyl-sulfamethoxazole, which requires only γ-glutamyl transpeptidase for its activation, did not show renal selectivity [149]. Up to now, results of the prodrug approach in renal targeting are still rather disappointing and without a clear therapeutic impact.

Improved Formulation and Delivery of Peptides

Bioactive peptides have become an important class of drugs with the recent advantages in biotechnology, peptide synthesis, and modern screening technologies. Although numerous bioactive peptides and proteins have been identified, they are usually limited to parenteral use, due to unfavorable biopharmaceutical properties. The hydrolysis of peptides is catalyzed by a variety of digestive and mucosal peptidases and proteases with broad substrate specificities. This enzymatic activity is a significant barrier for the absorption of peptides, and they are often extensively degraded before and after entering the systemic circulation. Therapeutic peptides tend to be relatively hydrophilic, which limits their penetration through the cellular membrane. Thus, they are not absorbed via the transcellular route, but primarily via a less efficient paracellular route or by cellular endocytosis. Efflux system activity also limits the absorption of peptides and peptidomimetics. The oral bioavailability of peptides is, therefore, typically less than 1 to 2 percent [150]. Systematic clearance is also rapid and the in vivo half-life is generally less than 30 min [150, 151].

Limited stability of peptides can often be addressed by designing peptidomimetic drugs with modified peptide backbones [152]. However, the clinical development peptidomimetics is still hampered by inadequate bioavailability, due to limited absorption and low aqueous solubility.

The usual objectives for producing a prodrug from a peptide is to improve both enzymatic stability and membrane permeation properties. Prodrugs for improved solubility have also been investigated. The C-terminal carboxyl and N-terminal amino groups are commonly selected for modification, but the amino acid side chains and the peptide backbone are also targeted. Many innovative prodrugs have been reported, but the development of clinically successful peptide prodrugs still remains a challenge. These prodrugs must sufficiently resist enzymatic degradation and be relatively lipophilic to allow absorption, but at the same time they must be sufficiently labile to release the parent peptide prior to systematic clearance.

Prodrugs to Improve Aqueous Solubility Prodrugs that increase the aqueous solubility of the immunosuppressive hydrophobic cyclic undecapeptide cyclosporin A have been developed, based on the *N,O*-acyl migration reaction [153] (Fig. 17.30). The *O*-acyl-peptide isomers can be formed under mild acidic conditions, but they will revert back to the normal *N*-acyl form under physiological conditions. The *O*-acyl isocyclosporin A is quantitatively converted back to cyclosporin A under simulated physiological conditions, but the half-life for this reaction is more than 4 h at pH 7.4 (37°C) [154], which is considered to be too slow as a useful peptide prodrug [153].

The aqueous solubilities of the *O*-acyl prodrugs of the HIV protease inhibitors KNI-720, KNI-272, and KNI-727 (Fig. 17.30) are more than 4000-fold higher than that of the parent drugs [155, 156]. These prodrugs are rapidly ($t_{1/2} < 1$ min) and quantitatively converted back to the parent drugs under simulated physiological conditions (pH 7.4, 37°C).

Prodrugs for Improved Enzymatic Stability Endopeptidases are important barriers against the absorption of peptides such as ACTH, oxytocin, various

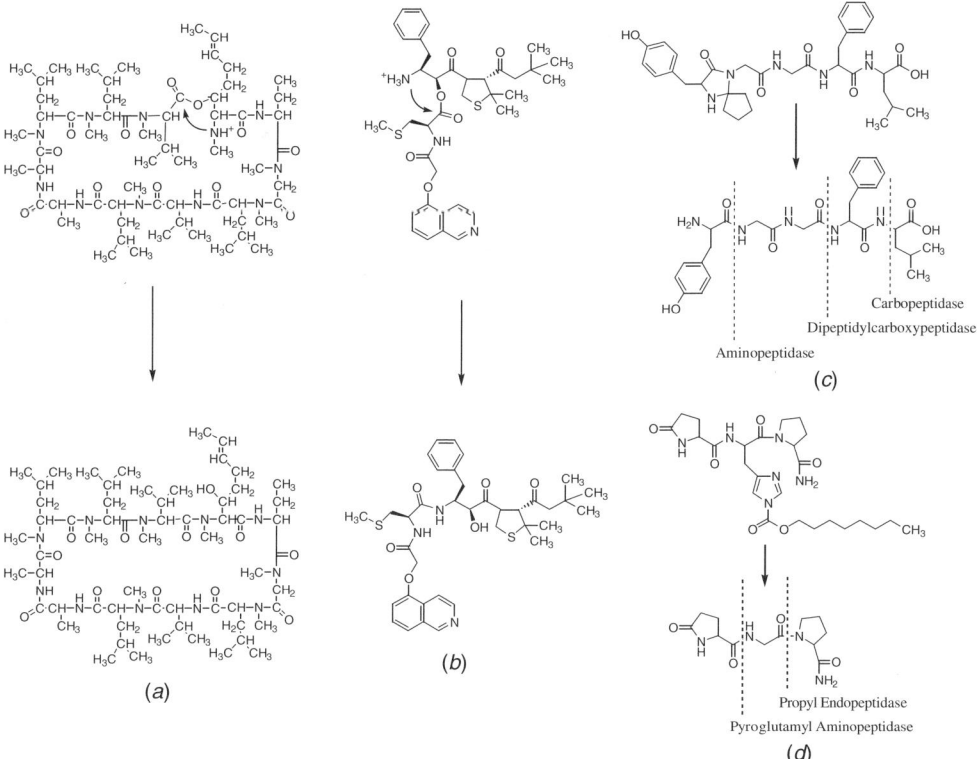

Figure 17.30 Prodrugs of (*a*) cyclosporin A and (*b*) KNI-727 based on the *O*,*N*-acyl migration reaction, the (*c*) 4-imidazolidone prodrug of Leu-enkephaline and (*d*) *N*-octyloxycabonyl prodrug of TRH.

rennin inhibitors, and enkephalins [157]. Prodrugs derived from both C-terminal and N-terminal residues can protect a peptide from the degradation catalyzed by endopeptidases. For example, derivation of the C-terminal carboxylic group of the pseudopeptide rennin inhibitor SR42128 with a highly lipophilic 1,3-dipalmytoyl acetyl glycerol promoiety prevents α-chymotrypsin-catalyzed degradation [158].

An α-aminoamide moiety appears in a large number of peptides. These peptides can be condensed with aldehydes or ketones to form 4-imidazolidone prodrugs that will protect the parent peptide from an aminopeptidase-catalyzed degradation. These prodrugs are spontaneously hydrolyzed in aqueous solutions. The rate of hydrolysis depends on the structure of the parent peptide and the carbonyl component [157]. For example, the 4-imidazolidone derivatives of Leu-enkephalin that are formed with either acetaldehyde, acetone, or cyclopentanone (Fig. 17.30) decompose at pH 7.4 (37°C) with half-lives of 30, 10.9, and 3.1 h, respectively [157].

Side-chain modification has also been tried to protect against enzyme-catalyzed degradation. This approach has, for example, been used for prodrugs of the CNS-active thyrotropin-releasing hormone (TRH). Although this peptide has been suggested for the treatment of various conditions, its clinical utility is limited by a short plasma half-life (6 to 8 min), due to rapid enzymatic degradation, as well as limited transport across the blood–brain barrier [159].

Modifications on the imidazole group of the histidine residue of TRH to form N-alkoxycarbonyl prodrugs greatly increased stability toward TRH-specific pyroglutamyl aminopeptidases found in serum, and the half-life in plasma was also improved [157, 159]. Although the lipophilicity increased significantly with the octyloxycarbonyl prodrug (Fig. 17.30), this did not result in any improvement in penetration through rabbit and rat intestinal tissue or Caco-2 cell monolayers. This lack of penetration enhancement was explained as a result of enzymatic cleavage by carboxypeptidase and nonspecific esterase.

Prodrugs can also be obtained by forming an ester with hydroxyl groups on amino acid moieties. The O-pivaloyl ester of the hydroxyl group on tyrosine of the antidiuretic peptide desmopressin is an example of such a prodrug. This prodrug is more stable toward α-chymotrypsin-catalyzed degradation than the parent peptide [160].

Prodrug for Increased Membrane Permeation Bioreversible cyclization of the peptide backbone is now one of the most promising approaches for peptide prodrugs with improved enzymatic stability and membrane permeability. The cyclization is achieved by linking the N-terminal amino group and the C-terminal carboxyl groups together through an enzyme-labile promoiety. Cyclic-acyloxyalkyl-based, phenylpropionic acid, or coumarine-based peptide prodrugs have been investigated (Fig. 17.31). These promoieties are cleaved in a slow esterase-catalyzed step that is followed by a rapid chemical reaction that releases the parent peptide [151]. Lacking terminal residues, these prodrugs are not substrates for endopeptidase. Lipophilicity is also increased, due to the lipophilic nature of the promoieties by masking the charge of terminal residues, due to the fact that the cyclic prodrugs exist in unique solution structures, some of which contain a high degree of intramolecular hydrogen bonding [161]. However, the effect of cyclization on permeation varies significantly with the structure of the peptide and promoiety. Lipophilic cyclic peptides are absorbed via the transcellular route, but the net flux is often limited by substrate properties for the polarized efflux systems (e. g., MRP and P-glycoprotein) [151]. For example, the cyclic phenylpropionic acid prodrugs of the opioid peptides Leu-enkephalin and DADLE exhibited 1680- and 77-fold increases in flux through Caco-2 cell monolayers, respectively, compared to those of the parent peptides. In contrast, very little permeation was observed with the cyclic acyloxyalkyl prodrugs of these peptides. This difference was mainly explained by the fact that only the acyloxyalkyl prodrugs were substrates for the efflux systems in Caco-2 cells [162, 163].

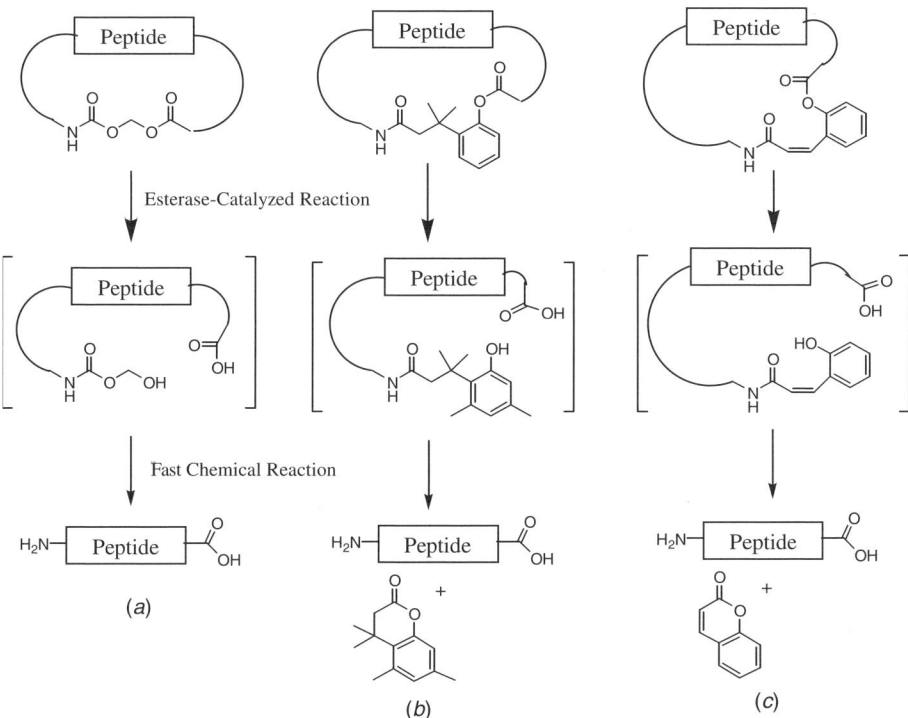

Figure 17.31 The bioconversion of (*a*) cyclic acyloxyalkyl, (*b*) phenylpropionic acid, and (*c*) coumarine-based peptide prodrugs.

Reduced Side Effects

Prodrugs have been utilized to minimize both local and systemic side effects. In general, improved drug targeting results in decreased side effects. Prolonged release of an active drug from a prodrug derivative also decreases side effects due to decreased plasma or specific tissue peak concentrations. In addition, the utilization of topical drug administration with prodrugs, instead of systemic administration, may decrease systemic side effects, or water-soluble prodrugs may allow IV administration without pain or irritation at an injection site. The pain or irritation at an injection site may be due to irritable formulation (low or high pH, co-solvents) or probable precipitation of the drug molecule at the injection site.

Fosphenytoin is an excellent example of how the prodrug approach can be used to develop an IV dosage form that decreases both side effects and local irritation [6]. Sodium phenytoin for IV and IM (intramuscular) administration was previously formulated as a very alkaline solution (pH near 12) that contained 10 percent ethanol and 40 percent propylene glycol. This formulation must be given slowly and minimally diluted to avoid precipitation and serious side effects. At least, the formulation only results in local tissue irritation.

Fosphenytoin, a water-soluble and bioreversible prodrug of phenytoin, increased the aqueous solubility of phenytoin by about 4000-fold at physiological pH and allows preparation of an aqueous and well-tolerated IV formulation.

Many topically applied ophthalmic drugs are absorbed into the systemic circulation, giving rise to possible systemic side effects. The general prodrug strategy is to improve the corneal absorption of a drug molecule by increasing lipophilicity, thus reducing the installed dose, systemic absorption, and side effects. For example, lipophilic prodrugs of timolol [164] and epinephrine [89] have been developed to decrease systemic side effects via improved corneal absorption.

The GI irritation resulting from NSAIDs is thought to be due to two different mechanisms: a direct contact of acidic drug on the GI mucosal and an inhibition of prostaglandin biosynthesis. Recent studies have confirmed that this GI side effect from NSAIDs are mainly a result of inhibiting COX-1, which regulates prostaglandin synthesis for normal cell activity. Thus, alternative formulations, administration routes, and prodrugs should not be able to eliminate all these side effects. However, direct contact of the NSAIDs has been shown to produce GI lesions, and, thus, the temporary masking of the carboxylic moiety has been proposed as a possible method to reduce or even abolish GI side effects [165, 166]. Thus, most oral prodrugs of NSAIDs have been developed to overcome GI-tract irritation, such as ibuprofen [167], indomethacin [168], ketoprofen [169], and naproxen [170].

Macromolecular Prodrugs

A variety of biological and synthetic macromolecules have been conjugated either directly or with various spacer groups to drug molecules by means of a biodegradable bond to make either macromolecular prodrugs or polymeric

TABLE 17.5 Macromolecular Promoieties for Prodrug Design

Macromolecular Promoiety	Reference
Proteins	
albumin	172
antibodies	129, 172
Polysaccharides	
dextrans	173
Polyamino acids	
polyaspartamide	174, 175
polyglutamic acid	176
Miscellaneous	
N-(2-hydroxypropyl)methacrylamide (HPMA)	177
Styrene-maleic acid anhydride copolymer (SMA)	178, 179
Polyethylene glycol (PEG)	180–182

prodrugs. The macromolecular prodrugs offer several advantages over other drug delivery methods such as: (1) improved drug targeting, (2) improved aqueous solubility, (3) extended duration of action, and (4) improved chemical stability. An increased understanding of the biological factors most relevant to macromolecular prodrugs has enabled the design of such therapeutics that are sufficiently interesting to justify clinical testing, and several macromolecular prodrugs have already entered clinical trials [171] as anticancer agents. Low-molecular-weight prodrugs are the primary subject of the present chapter and, therefore, a more detailed description of some of the most commonly utilized macromolecules as promoieties can be found in the references listed in Table 17.5.

REFERENCES

1. Venkatesh, S., Lipper, R. A. (2000). Role of the development scientist in the compound lead selection and optimization. *J. Pharm. Sci.*, 89, 145–154.
2. Alber, A. (1958). Chemical aspects of selective toxicity. *Nature*, 16, 421–423.
3. Krise, J. P., Stella, V. J. (2003). Prodrugs: A year of renewed interest in an old concept. *AAPS Newsmag.*, 16, 25.
4. Stella, V. J., Charman, W. N. A., Naringrekar, V. H. (1985). Prodrugs. Do they have advantages in clinical practice? *Drugs*, 29, 455–473.
5. Leppänen, J., Huuskonen, J., Nevalainen, T., Gynther, J., Taipale, H., Järvinen, T. (2002). Design and synthesis of a novel L-dopa-entacapone codrugs. *J. Med. Chem.*, 45, 1379–1382.
6. Stella, V. J. (1996). A case for prodrugs: Fosphenytoin. *Adv. Drug Delivery Rev.*, 19, 311–330.
7. Shaw, J. P., Sueoko, C. M., Oliyai, R., Lee, W. A., Arimilli, M. N., Kim, C. U., Cundy, K. C. (1997). Metabolism and pharmacokinetics of novel oral prodrugs of 9-[(R)-2-(phosphonomethoxy)propyl]adenine (PMPA) in dogs. *Pharm. Res.*, 14, 1824–1829.
8. Bundgaard, H., Falch, E., Larsen, C., Mosher, G. L., Mikkelson, T. J. (1986). Pilocarpine prodrugs II: Synthesis, stability, bioconversion, and physicochemical properties of sequentially labile pilocarpine acid diesters. *J. Pharm. Sci.*, 75, 775–783.
9. Thorsteinsson, T., Loftsson, T., Masson, M. (2003). Soft antibacterial agents. *Curr. Med. Chem.*, 10, 1129–1136.
10. Bodor, N., Buchwald, P. (1997). Drug targeting via retrometabolic approaches. *Pharmacol. Ther.*, 76, 1–27.
11. Bodor, N., Buchwald, P. (2000). Soft drug design: General principals and recent applications. *Med. Res. Rev.*, 20, 58–101.
12. Bodor, N. (1985). Prodrugs versus soft drugs. In H. Bundgaard (Ed.), *Design of Prodrugs*. Elsevier Science, Amsterdam, pp. 333–354.
13. Amidon, G. L., Lennernas, H., Shah, V. P., Crison, J. R. (1995). A theoretical basis for a bipharmaceutic classification: The correlation of in vitro drug product dissolution and in vivo bioavailability. *Pharm. Res.*, 12, 413–420.

14. Bundgaard, H. (1985). Formation of prodrugs of amines, amides, ureides, and imides. *Methods Enzymol.*, *112*, 347–359.
15. Fleisher, D., Bong, R., Stewart, B. H. (1996). Improved oral drug delivery: Solubility limitations overcome by the use of prodrugs. *Adv. Drug Delivery Rev.*, *19*, 115–130.
16. Taylor, M. D. (1996). Improved passive oral drug delivery via prodrugs. *Adv. Drug Delivery Rev.*, *19*, 131–148.
17. Haeberlin, B., Rubas, W., Nolen, H. W., Friend, D. R. (1993). In-vitro evaluation of dexamethasone-beta-D-glucuronide for colon-specific drug-delivery. *Pharm. Res.*, *10*, 1553–1562.
18. Chourasia, M. K., Jain, S. K. (2003). Pharmaceutical approaches to colon targeted drug delivery systems. *J. Pharm. Pharm. Sci.*, *6*, 33–66.
19. Cabri, W., Di Fabio, R. (2000). *From Bench to Market*. Oxford University Press, Oxford, p. 148.
20. Prokai, L., Prokai-Tatarai, K. (1999). In P. K. Gupta and G. A. Brazeau (Eds.), *Prodrugs. Injectable Drug Development. Techniques to Reduce Pain and Irritation*. Interpharm, Denver, CO, pp. 267–306.
21. Sinkula, A. A., Yalkowsky, S. H. (1975). Rationale for design of biologically reversible drug derivatives—prodrugs. *J. Pharm. Sci.*, *64*, 181–210.
22. Menard, J., Patchett, A. A. (2001). Angiotensin-converting enzyme inhibitors. *Adv. Protein Chem.*, *56*, 13–75.
23. Tammara, V. K., Narurkar, M. M., Crider, A. M., Khan, M. A. (1993). Synthesis and evaluation of morpholinoalkyl ester prodrugs of indomethacin and naproxen. *Pharm. Res.*, *10*, 1191–1199.
24. Rautio, J., Taipale, H., Gynther, J., Vepsäläinen, J., Nevalainen, T., Järvinen, T. (1998). In vitro evaluation of acyloxyalkyl esters as dermal prodrugs of ketoprofen and naproxen. *J. Pharm. Sci.*, *87*, 1622–1628.
25. Ferres, H. (1983). Pro-drugs of β-lactam antibiotics. *Drugs of Today*, *19*, 499–538.
26. Shanbhag, V. R., Crider, A. M., Gokhale, R., Harpalani, A., Dick, R. M. (1992). Ester and amide prodrugs of ibuprofen and naproxen—Synthesis, antiinflammatory activity, and gastrointestinal toxicity. *J. Pharm. Sci.*, *81*, 149–154.
27. Krise, J. P., Stella, V. J. (1996). Prodrugs of phosphates, phosphonates, and phosphinates. *Adv. Drug Delivery Rev.*, *19*, 287–310.
28. McGuigan, C., Turner, S., Nicholls, S. R., Oconnor, T. J., Kinchington, D. (1994). Haloalkyl phosphate derivatives of act as inhibitors of HIV—Studies in the phosphate region. *Antiviral Chem. Chemother.*, *5*, 162–168.
29. Schott, H., Ludwig, P. S., Immelmann, A., Schwendener, R. A. (1999). Synthesis and *in vitro* anti-HIV activities of amphiphilic heterodinucleoside phosphate derivatives containing the 2′,3′-dideoxynucleosides ddC, AZT and ddI. *Eur. J. Med. Chem.*, *34*, 343–352.
30. Pompon, A., Lefebvre, I., Imbach, J. L., Kahn, S., Farquhar, D. (1994). Decomposition pathways of the mono-(pivaloyl-oxymethyl) and bis(pivaloyl-oxymethyl) esters of azidothymidine 5′-monophosphate in cell extract and in tissue-culture medium—An application of the online Isrp-cleaning HPLC technique. *Antiviral Chem. Chemother.*, *5*, 91–98.

31. Hostetler, K. Y., Parker, S., Sridhar, C. N., Martin, M. J., Li, J. L., Stuhmiller, L. M., Vanwijk, G. M. T., Vandenbosch, H., Gardner, M. F., Aldern, K. A., Richman, D. D. (1993). Acyclovir diphosphate dimyristoylglycerol—A phospholipid prodrug with activity against acyclovir-resistant herpes-simplex virus. *Proc. Nat. Acad. Sci. USA*, 90, 11,835–11,839.

32. Saboulard, D., Naesens, L., Cahard, D., Salgado, A., Pathirana, R., Velazquez, S., McGuigan, C., De Clercq, E., Balzarini, J. (1999). Characterization of the activation pathway of phosphoramidate triester prodrugs of stavudine and zidovudine. *Mol. Pharmacol.*, 56, 693–704.

33. Serafinowska, H. T., Ashton, R. J., Bailey, S., Harnden, M. R., Jackson, S. M., Sutton, D. (1995). Synthesis and in vivo evaluation of prodrugs of 9-[2-(phosphonomethoxy)ethoxy]adenine. *J. Med. Chem.*, 38, 1372–1379.

34. Starrett, J. E. Jr, Tortolani, D. R., Russell, J., Hitchcock, M. J., Whiterock, V., Martin, J. C., Mansuri, M. M. (1994). Synthesis, oral bioavailability determination, and in vitro evaluation of prodrugs of the antiviral agent 9-[2-(phosphonomethoxy)ethyl]adenine (PMEA). *J. Med. Chem.*, 37, 1857–1864.

35. Delombaert, S., Erion, M. D., Tan, J., Blanchard, L., Elchehabi, L., Ghai, R. D., Sakane, Y., Berry, C., Trapani, A. J. (1994). N-Phosphonomethyl dipeptides and their phosphonate prodrugs, a new-generation of neutral endopeptidase (Nep, Ec-3.4.24.11) inhibitors. *J. Med. Chem.*, 37, 498–511.

36. Niemi, R., Vepsäläinen, J., Taipale, H., Järvinen, T. (1999). Bisphosphonate prodrugs: Synthesis and in vitro evaluation of novel acyloxyalkyl esters of clodronic acid. *J. Med. Chem.*, 42, 5053–5058.

37. Niemi, R., Pennanen, H., Vepsäläinen, J., Taipale, H., Järvinen, T. (1998). Bisphosphonate prodrugs: Synthesis and in vitro evaluation of novel partial amides of clodronic acid. *Int. J. Pharm.*, 174, 111–115.

38. Larsen, C., Johansen, M. (1987). Macromolecular prodrugs. 4. Kinetics of hydrolysis of metronidazole monosuccinate dextran ester conjugates in aqueous-solution and in plasma—Sequential release of metronidazole from the conjugates at physiological pH. *Int. J. Pharm.*, 35, 39–45.

39. Mollmann, H., Rohdewald, P., Barth, J., Mollmann, C., Verho, M., Derendorf, H. (1988). Comparative pharmacokinetics of methylprednisolone phosphate and hemisuccinate in high-doses. *Pharm. Res.*, 5, 509–513.

40. Kong, A. N., Jusko, W. J. (1991). Disposition of methylprednisolone and its sodium succinate prodrug in vivo and in perfused liver of rats—Nonlinear and sequential 1st-pass elimination. *J. Pharm. Sci.*, 80, 409–415.

41. Bundgaard, H., Larsen, C., Thorbek, P. (1984). Prodrugs as drug delivery systems. 26. Preparation and enzymatic-hydrolysis of various water-soluble amino-acid esters of metronidazole. *Int. J. Pharm.*, 18, 67–77.

42. Mahfouz, N. M., Hassan, M. A. (2001). Synthesis, chemical and enzymatic hydrolysis, and bioavailability evaluation in rabbits of metronidazole amino acid ester prodrugs with enhanced water solubility. *J. Pharm. Pharm.*, 53, 841–848.

43. Másson, M., Thorsteinsson, T., Sigurdsson, T. H., Loftsson, T. (2000). Lipophilic metronidazole derivatives and their absoprtion through hairless mouse skin. *Die Pharmazie*, 55, 369–371.

44. Ueda, Y., Matiskella, J. D., Mikkilineni, A. B., Farina, V., Knipe, J. O., Rose, W. C., Casazza, A. M., Vyas, D. M. (1995). Novel, water-soluble phosphate derivatives of 2′-ethoxy carbonylpaclitaxel as potential prodrugs of paclitaxel—Synthesis and antitumor evaluation. *Bioorganic Med. Chem. Lett.*, 5, 247–252.

45. Sawada, S., Yaegashi, T., Furuta, T., Yokokura, T., Miyasaka, T. (1993). Chemical modification of an antitumor alkaloid, 20(S)-camptothecin–E-lactone ring-modified water-soluble derivatives of 7-ethylcamptothecin. *Chem. Pharm. Bull.*, 41, 310–313.

46. Yaegashi, T., Sawada, S., Nagata, H., Furuta, T., Yokokura, T., Miyasaka, T. (1994). Synthesis and antitumor-activity of 20(S)-camptothecin derivatives—A-ring-substituted 7-ethylcamptothecins and their E-ring-modified water-soluble derivatives. *Chem. Pharm. Bull.*, 42, 2518–2525.

47. Bodor, N., Sloan, K. B., Higuchi, T., Sasahara, K. (1977). Improved delivery through biological-membranes. 4. Prodrugs of L-dopa. *J. Med. Chem.*, 20, 1435–1445.

48. Johansen, M., Bundgaard, H. (1980). Pro-drugs as drug delivery systems. 13. Kinetics of decomposition of N-Mannich bases of salicylamide and assessment of their suitability as possible pro-drugs for amines. *Int. J. Pharm.*, 7, 119–127.

49. Bundgaard, H., Johansen, M., Stella, V., Cortese, M. (1982). Pro-drugs as drug delivery systems. 21. Preparation, physicochemical properties and bioavailability of a novel water-soluble pro-drug type for carbamazepine. *Int. J. Pharm.*, 10, 181–192.

50. Redden, R., Douglas, J. E., Burke, M. J., Horrobin, D. F. (1998). In vitro hydrolysis of polyunsaturated fatty acid N-acyloxymethyl derivatives of theophylline. *Int. J. Pharm.*, 165, 87.

51. Bundgaard, H., Falch, E., Jensen, E. (1989). A novel solution-stable, water-soluble prodrug type for drugs containing a hydroxyl or an NH-acidic group. *J. Med. Chem.*, 32, 2503–2507.

52. Sloan, K. B., Hashida, M., Alexander, J., Bodor, N., Higuchi, T. (1983). Prodrugs of 6-thiopurines—Enhanced delivery through the skin. *J. Pharm. Sci.*, 72, 372–378.

53. Mollgaard, B., Hoelgaard, A., Bundgaard, H. (1982). Pro-drugs as drug delivery systems. 23. Improved dermal delivery of 5-fluorouracil through human-skin via N-acyloxymethyl pro-drug derivatives. *Int. J. Pharm.*, 12, 153–162.

54. Yang, C., Tirucherai, G. S., Mitra, A. K. (2001). Prodrug based optimal drug delivery via membrane transporter/receptor. *Expert Opin. Biol. Ther.*, 1, 159–175.

55. Shaw, J. P., Sueoka, C. M., Oliyai, R., Lee, W. A., Arimilli, M. N., Kim, C. U., Cundy, K. C. (1997). Metabolism and pharmacokinetics of novel oral prodrugs of 9-[(R)-2-(phosphonomethoxy)propyl]adenine (PMPA) in dogs. *Pharm. Res.*, 14, 1824–1829.

56. Gallant, J. E., Deresinski, S. (2003). Tenofovir disoproxil fumarate. *Clin. Infect. Dis.*, 37, 944–950.

57. Cundy, K. C., Barditch-Crovo, P., Walker, R. E., Collier, A. C., Ebeling, D., Toole, J., Jaffe, H. S. (1995). Clinical pharmacokinetics of adefovir in human immunodeficiency virus type 1-infected patients. *Antimicrob. Agents Chemother.*, 39, 2401–2405.

REFERENCES

58. Gustafsson, D. (2003). Oral direct thrombin inhibitors in clinical development. *J. Internal Med.*, *254*, 322–334.
59. Han, H. K., Amidon, G. L. (2000). Targeted prodrug design to optimize drug delivery. *AAPS PharmSci*, *2*, E6.
60. Steffansen, B., Nielsen, C. U., Brodin, B., Eriksson, A. H., Andersen, R., Frokjaer, S. (2004). Intestinal solute carriers: An overview of trends and strategies for improving oral drug absorption. *Eur. J. Pharm. Sci.*, *21*, 3–16.
61. de Miranda, P., Good, S. S. (1992). Species differences in the metabolism and disposition of antiviral nucleoside analogues: 1. Acyclovir. *Antiviral Chem. Chemother.*, *3*, 1–8.
62. Weller, S., Blum, M. R., Doucette, M., Burnette, T., Cederberg, D. M., de Miranda, P., Smiley, M. L. (1993). Pharmacokinetics of the acyclovir pro-drug valaciclovir after escalating single- and multiple-dose administration to normal volunteers. *Clin. Pharm. Ther.*, *54*, 595–605.
63. Beutner, K. R., Friedman, D. J., Forszpaniak, C., Andersen, P. L., Wood, M. J. (1995). Valaciclovir compared with acyclovir for improved therapy for herpes zoster in immunocompetent adults. *Antimicrob. Agents Chemother.*, *39*, 1546–1553.
64. Guo, A., Hu, P., Balimane, P. V., Leibach, F. H., Sinko, P. J. (1999). Interactions of a nonpeptidic drug, valacyclovir, with the human intestinal peptide transporter (hPEPT1) expressed in a mammalian cell line. *J. Pharmcol. Exper. Ther.*, *289*, 448–454.
65. Han, H., de Vrueh, R. L., Rhie, J. K., Covitz, K. M., Smith, P. L., Lee, C. P., Oh, D. M., Sadee, W., Amidon, G. L. (1998). 5′-Amino acid esters of antiviral nucleosides, acyclovir, and AZT are absorbed by the intestinal PEPT1 peptide transporter. *Pharm. Res.*, *15*, 1154–1159.
66. Sugawara, M., Huang, W., Fei, Y. J., Leibach, F. H., Ganapathy, V., Ganapathy, M. E. (2000). Transport of valganciclovir, a ganciclovir prodrug, via peptide transporters PEPT1 and PEPT2. *J. Pharm. Sci.*, *89*, 781–789.
67. Heimbach, T., Oh, D. M., Li, L. Y., Forsberg, M., Savolainen, J., Leppanen, J., Matsunaga, Y., Flynn, G., Fleisher, D. (2003). Absorption rate limit considerations for oral phosphate prodrugs. *Pharm. Res.*, *20*, 848–856.
68. Li, Y. H., Ito, K., Tsuda, Y., Kohda, R., Yamada, H., Itoh, T. (1999). Mechanism of intestinal absorption of an orally active beta-lactam prodrug: Uptake and transport of carindacillin in Caco-2 cells. *J. Pharmcol. Exper. Ther.*, *290*, 958–964.
69. Sloan, K. B., Wasdo, S. (2003). Designing for topical delivery: Prodrugs can make the difference. *Med. Res. Rev.*, *23*, 763–793.
70. Sloan, K. B. (1989). Prodrugs for dermal delivery. *Adv. Drug Delivery Rev.*, *3*, 67–101.
71. Sloan, K. B. (1992). Functional group considerations in the development of prodrugs approaches to solving topical delivery problems. In K. B. Sloan (Ed.), *Prodrugs; Topical and Ocular Drug Delivery*. Marcel Dekker, New York, pp. 17–116.
72. Chien, Y. W. (1983). Logics of transdermal controlled drug administration. *Drug Devel. Ind. Pharm.*, *9*, 497–520.

73. Rautio, J., Nevalainen, T., Taipale, H., Vepsäläinen, J., Gynther, J., Pedersen, T., Järvinen, T. (1999). Synthesis and in vitro evaluation of aminoacyloxyalkyl esters of 2-(6-methoxy-2-naphthyl)propionic acid as novel naproxen prodrugs for dermal drug delivery. *Pharm. Res.*, *16*, 1172–1178.
74. Rautio, J., Nevalainen, T., Taipale, H., Vepsäläinen, J., Gynther, J., Laine, K., Järvinen, T. (2000). Synthesis and in vitro evaluation of novel morpholinyl- and methylpiperazinylacyloxyalkyl prodrugs of 2-(6-methoxy-2-naphthyl)propionic acid (naproxen) for topical drug delivery. *J. Med. Chem.*, *43*, 1489–1494.
75. Rautio, J., Nevalainen, T., Taipale, H., Vepsäläinen, J., Gynther, J., Laine, K., Järvinen, T. (2000). Piperazinylalkyl prodrugs of naproxen improve in vitro skin permeation. *Eur. J. Pharm. Sci.*, *11*, 157–163.
76. Udata, C., Tirucherai, G., Mitra, A. K. (1999). Synthesis, stereoselective enzymatic hydrolysis, and skin permeation of diastereomeric propranolol ester prodrugs. *J. Pharm. Sci.*, *88*, 544–550.
77. Doane, M. G., Jensen, A. D., Dohlman, C. H. (1978). Penetration routes of topically applied eye medications. *Am. J. Ophthalmol.*, *85*, 383–386.
78. Rojanasakul, Y., Wang, L-Y., Bhat, M., Glover, D. D., Malanga, C. J., Ma, K. H. (1992). The transport barrier of epithelia: A comparative study on membrane permeability and charge selectivity in the rabbit. *Pharm. Res.*, *9*, 1029–1034.
79. Chien, D-S., Sasaki, H., Bundgaard, H., Buur, A., Lee, V. H. L. (1991). Role of enzymatic lability in the corneal and conjunctival penetration on timolol ester prodrugs in the pigmented rabbit. *Pharm. Res.*, *8*, 728–733.
80. Wang, W., Sasaki, H., Chien, D.-S., Lee, V. H., L. (1991). Lipophilicity influence on conjunctival drug penetration in the pigmented rabbit: A comparison with corneal penetration. *Curr. Eye Res.*, *10*, 571–579.
81. Schoenwald, R. D., Huang, H.-S. (1983). Corneal penetration behavior of beta-blocking agents I: Physicochemical factors. *J. Pharm. Sci.*, *72*, 1266–1272.
82. Schoenwald, R. D., Ward, R. L. (1978). Relationship between steroid permeability across excised rabbit cornea and octanol-water partition coefficients. *J. Pharm. Sci.*, *67*, 787–789.
83. Sasaki, H., Yamamura, M., Mukai, T., Nishida, K., Nakamura, J., Nakashima, M., Ichikawa, M. (1999). Enhancement of ocular drug penetration. *Crit. Rev. Ther. Drug Carrier Syst.*, *16*, 85–146.
84. Hussain, A., Truelove, J. E. (1976). Prodrug approaches to enhancement of physicochemical properties of drugs IV: Novel epinephrine prodrug. *J. Pharm. Sci.*, *65*, 1510–1512.
85. Anderson, J. A. (1980). Systemic absorption of topical ocularly applied epinephrine and dipivefrin. *Arch. Ophthalmol.*, *98*, 350–353.
86. Mandell, A. I., Stentz, F. (1978). Dipivalyl epinephrine: A new pro-drug in the treatment of glaucoma. *Ophthalmology*, *85*, 268–275.
87. Wei, C., Anderson, J. A., Leopold, I. (1978). Ocular absorption and metabolism of topically applied epinephrine and a dipivalyl ester of epinephrine. *Investigative Ophthalmol. Vis. Sci.*, *17*, 315–321.
88. Kaback, M. B., Podos, S. M., Harbin, Jr. T. S., Mandell, A., Becker, B. (1976). The effects of dipivalyl epinephrine on the eye. *Am. J. Ophthalmol.*, *81*, 768–772.

89. Kohn, A. N., Moss, A. P., Hargett, N. A., Ritch, R., Smith, H., Podos, S. M. (1979). Clinical comparison of dipivalyl epinephrine and epinephrine in the treatment of glaucoma. *Am. J. Ophthalmol.*, 87, 196–201.

90. Susanna, R., Chew, P., Kitazawa, Y. (2002). Current status of prostaglandin therapy; Latanoprost and unoprostone. *Surv. Ophthalmol.*, 47, S97–S104.

91. Netlans, P. A., Landry, T., Sullivan, E. K., Andrew, R., Silver, L., Weiner, A., Mallick, S., Dickerson, J., Bergamini, W. V. W., Robertson, S. M., Davis, A. D. (2001). Travoprost compared with latanoprost and timolol in patients with open-angle glaucoma or ocular hypertension. *Am. J. Ophthalmol.*, 132, 472–484.

92. Suhonen, P., Järvinen, T., Lehmussaari, K., Reunamäki, T., Urtti A. (1996). Rate control of ocular pilocarpine delivery with bispilocarpic acid diesters. *Int. J. Pharm.*, 127, 85–94.

93. Chang, S.-C., Bundgaard, H., Buur, A., Lee, V. H. L. (1988). Low dose O-butyryl timolol improves the therapeutic index of timolol in the pigmented rabbit. *Investigative Ophthalmol. Vis. Sci.*, 29, 626–629.

94. Sasaki, H., Igarashi, Y., Nishida, K., Nakamura, J. (1993). Ocular delivery of the β-blocker, tilisolol, through the prodrug approach. *Int. J. Pharm.*, 93, 49–60.

95. Anand, B. S., Mitra, A. K. (2002). Mechanism of corneal permeation of L-valyl ester of acyclovir: Targeting the oligopeptide transporter on the rabbit cornea. *Pharm. Res.*, 19, 1194–1202.

96. Washington, N., Washington, C., Wilson, C. G. (2001). Nasal drug delivery. In *Physiological Pharmaceutics. Barriers to Drug Absorption*, 2nd ed., Taylor & Francis, London, pp. 199–220.

97. Pezron, I., Tirucherai, G. S., Duvvuri, S., Mitra, A. K. (2002). Prodrug strategies in nasal drug delivery. *Expert Opin. Ther. Patents*, 12, 331–340.

98. Hussain, A. A., Al-Bayatti, A. A., Dakkuri, A., Okochi, K., Hussain, M. A. (2002). Testosterone 17β-N,N-dimethylglycinate hydrochloride: A prodrug with a potential for nasal delivery of testosterone. *J. Pharm. Sci.*, 91, 785–789.

99. Loftsson, T., Gudmundsdottir, H., Sigurjonsdottir, J. F., Sigurdsson, H. H., Sigfusson, S. D., Masson, M., Stefansson, E. (2001). Cyclodextrin solubilization of benzodiazepines: Formulation of midazolam nasal spray. *Int. J. Pharm.*, 212, 29–40.

100. Gudmundsdottir, H., Sigurjonsdottir, J. F., Masson, M., Fjalldal, O., Stefansson, E., Loftsson, T. (2001). Intranasal administration of midazolam in a cyclodextrin based formulation: Bioavailability and clinical evaluation in humans. *Pharmazie*, 56, 963–966.

101. Cox, D. S., Raje, S., Gao, H. L., Salama, N. N., Eddington, N. D. (2002). Enhanced permeability of molecular weight markers and poorly bioavailable compounds across Caco-2 cell monolayers using the absorption enhancer, zonula occludens toxin. *Pharm. Res.*, 19, 1680–1688.

102. Shao, Z., Hoffman, A. J., Mitra, A. K. (1994). Biodegradation characteristics of acyclovir 2′-esters by respiratory carboxylesterases: Implications in prodrug design for intranasal and pulmonary drug delivery. *Int. J. Pharm.*, 112, 181–190.

103. Shao, Z., Park, G.-B., Krishnamoorthy, R., Mitra, A. K. (1994). The physicochemical properties, plasma enzymatic hydrolysis, and nasal absorption of acyclovir and its 2′-ester prodrugs. *Pharm. Res.*, 11, 237–242.

104. Hussain, A. A. (1998). Intranasal drug delivery. *Adv. Drug Delivery Rev.*, 29, 39–49.
105. Kao, H. D., Traboulsi, A., Itoh, S., Dittert, L., Hussain, A. (2000). Enhancement of systemic and CNS specific delivery of L-dopa by the nasal administration of its water soluble prodrugs. *Pharm. Res.*, 17, 978–984.
106. Yang, C., Mitra, A. K. (2001). Nasal absorption of tyrosine-linked model compounds. *J. Pharm. Sci.*, 90, 340–347.
107. Yang, C., Gao, H., Mitra, A. K. (2001). Chemical stability, enzymatic hydrolysis, and nasal uptake of amino acid ester prodrugs of acyclovir. *J. Pharm. Sci.*, 90, 617–624.
108. Rasmussen, G. J., Bundgaard, H. (1991). Prodrugs of peptides. 15. 4-Imidazolidinone prodrug derivatives of enkephalins to prevent aminopeptidase-catalyzed metabolism in plasma and absorptive mucosae. *Int. J. Pharm.*, 76, 113–122.
109. Tirucherai, G. S., Pezron, I., Mitra, A. K. (2002). Novel approaches to nasal delivery of peptides and proteins. *S. T.P. Pharma Sci.*, 12, 3–12.
110. Witterland, A. H. I., Koks, C. H. W., Beijnen, J. H. (1996). Etoposide phosphate, the water soluble prodrug of etoposide. *Pharm. World Sci.*, 18, 163–170.
111. Krise, J. P., Narisawa, S., Stella, V. J. (1999). A novel prodrug approach for tertiary amines. 2. Physicochemical and *in vitro* enzymatic evaluation of selected *N*-phosphonooxymethyl prodrugs. *J. Pharm. Sci.*, 88, 922–927.
112. Krise, J. P., Charman, W. N., Charman, S. A., Stella, V. J. (1999). A novel prodrug approach for tertiary amines. 3. *In vivo* evaluation of two *N*-phosphonooxymethyl prodrugs in rats and dogs. *J. Pharm. Sci.*, 88, 928–932.
113. Krise, J. P., Zygmunt, J., Georg, G. I., Stella, V. J. (1999). Novel prodrug approach for tertiary amines: Synthesis and preliminary evaluation of *N*-phosphonooxymethyl prodrugs. *J. Med. Chem.*, 42, 3094–3100.
114. Larsen, C., Kurtzhals, P., Johansen, M. (1988). Kinetics of regeneration of metronidazole from hemiesters of maleic acid, succinic acid and glutaric acid in aqueous buffer, human plasma and pig liver homogenate. *Int. J. Pharm.*, 41, 121–129.
115. Robinson, R. P., Reiter, L. A., Barth, W. E., Campeta, A. M., Cooper, K., Cronin, B. J., Destito, R., Donahue, K. M., Falkner, F. C., Fiese, E. F., Johnson, D. L., Kuperman, A. V., Liston, T. E., Malloy, D., Martin, J. J., Mitchell, D. Y., Rusek, F. W., Shamblin, S. L., Wright, C. F. (1996). Discovery of the hemifumarate and (alpha-L-alanyloxy)methyl ether as prodrugs of an antirheumatic oxindole: Prodrugs for the enolic OH group. *J. Med. Chem.*, 39, 10–18.
116. Hendler, S. S., Sanchez, R. A., Zielinski, J. (2001). Water soluble pro-drugs of propofol. U. S. Patent 6,254,853 B1.
117. Hudkins, R. L., Iqbal, M., Park, C. H., Goldstein, J., Herman, J. L., Shek, E., Murakata, C., Mallamo, J. P. (1998). Prodrug esters of the indolocarbazole CEP-751 (KT-6587). *Bioorganic Med. Chem. Lett.*, 8, 1873–1876.
118. Bradshaw, T. D., Chua, M. S., Browne, H. L., Trapani, V., Sausville, E. A., Stevens, M. F. (2002). *In vitro* evaluation of amino acid prodrugs of novel antitumour 2-(4-amino-3-methylphenyl)benzothiazoles. *Br. J. Cancer*, 86, 1348–1354.
119. Nam, N. H., Kim, Y., You, Y. J., Hong, D. H., Kim, H. M., Ahn, B. Z. (2003). Water soluble prodrugs of the antitumor agent 3-(3-amino-4-methoxy)phenyl]-2-(3,4,5-trimethoxyphenyl)cyclopent-2-ene-1-one. *Bioorganic Med. Chem.*, 11, 1021–1029.

120. Marder, S. R., Aravagiri, M., Wirshing, W. C., Wirshing, D. A., Lebell, M., Mintz, J. (2002). Fluphenazine plasma level monitoring for patients receiving fluphenazine decanoate. *Schizophr. Res.*, *53*, 25–30.
121. Shechter, Y., Tsubery, H., Fridkin, M. (2003). Suspensions of pro-drug insulin greatly prolong normoglycemic patterns in diabetic rats. *Biochem. Biophys. Res. Commun.*, *307*, 315–321.
122. Svensson, L. A., Tunek, A. (1988). The design and bioactivation of presystemically stable prodrugs. *Drug Metabolism Rev.*, *19*, 165–194.
123. Tunek, A., Levin, E., Svensson, L. A. (1988). Hydrolysis of 3*H*-bambuterol, a carbamate prodrug of terbutaline, in blood from humans and laboratory animals *in vitro*. *Biochem. Pharmcol.*, *37*, 3867–3876.
124. Sinhababu, A. K., Thakker, D. R. (1996). Prodrugs of anticancer agents. *Adv. Drug Delivery Rev.*, *19*, 241–273.
125. Brown, J. M., Giaccia, A. J. (1998). The unique physiology of solid tumors: Opportunities (and problems) for cancer therapy. *Cancer Res.*, *58*, 408–416.
126. Naylor, M. A., Thomson, P. (2001). Recent advances in bioreductive drug targeting. *Mini Rev. Med. Chem.*, *1*, 17–29.
127. Wardman, P., Dennis, M. F., Everett, S. A., Patel, K. B., Stratford, M. R., Tracy, M. (1995). Radicals from one-electron reduction of nitro compounds, aromatic *N*-oxides and quinones: The kinetic basis for hypoxia-selective, bioreductive drugs. *Biochem. Soc. Symp.*, *61*, 171–194.
128. Xu, G., McLeod, H. L. (2001). Strategies for enzyme/prodrug cancer therapy. *Clin. Cancer Res.*, *7*, 3314–3324.
129. Jung, M. (2001). Antibody directed enzyme prodrug therapy (ADEPT) and related approaches for anticancer therapy. *Mini Rev. Med. Chem.*, *1*, 399–407.
130. Aghi, M., Hochberg, F., Breakefield, X. O. (2000). Prodrug activation enzymes in cancer gene therapy. *J. Gene Med.*, *2*, 148–164.
131. Chen, L., Waxman, D. J. (2002). Cytochrome P450 gene-directed enzyme prodrug therapy (GDEPT) for cancer. *Curr. Pharm. Des.*, *8*, 1405–1416.
132. Niculescu-Duvaz, I., Spooner, R., Marais, R., Springer, C. J. (1998). Gene-directed enzyme prodrug therapy. *Bioconjugate Chem.*, *9*, 4–22.
133. Springer, C. J., Niculescu-Duvaz, I. (2000). Approaches to gene-directed enzyme prodrug therapy (GDEPT). *Adv. Exper. Med. Biol.*, *465*, 403–409.
134. Genka, S., Deutsch, J., Shetty, U. H., Stahle, P. L., John, V., Lieberburg, I. M., Ali-Osman, F., Rapoport, S. I, Greig, N. H. (1993). Development of lipophilic anticancer agents for the treatment of brain tumors by the esterification of water-soluble chlorambucil. *Clin. Exper. Metastasis*, *11*, 131–140.
135. Greig, N. H., Genka, S., Daly, E. M., Sweeney, D. J., Rapoport, S. I. (1990). Physicochemical and pharmacokinetic parameters of seven lipophilic chlorambucil esters designed for brain penetration. *Cancer Chemother. Pharmcol.*, *25*, 311–319.
136. Bodor, N., Buchwald, P. (1999). Recent advances in the brain targeting of neuropharmaceuticals by chemical delivery systems. *Adv. Drug Delivery Rev.*, *36*, 229–254.
137. Bodor, N., Buchwald, P. (2002). Barriers to remember: Brain-targeting chemical delivery systems and Alzheimer's disease. *Drug Discov. Today*, *7*, 766–774.

138. Tsuji, A., Tamai, I. I. (1999). Carrier-mediated or specialized transport of drugs across the blood-brain barrier. *Adv. Drug Delivery Rev.*, 5, 277–290.

139. Tamai, I., Tsuji, A.(2000). Transporter-mediated permeation of drugs across the blood-brain barrier. *J. Pharm. Sci.*, 89, 1371–1388.

140. Halmos, T., Santarromana, M., Antonakis, K., Scherman, D. (1996). Synthesis of glucose-chlorambucil derivatives and their recognition by the human GLUT1 glucose transporter. *Eur. J. Pharmcol.*, 318, 477–484.

141. Bonina, F. P., Arenare, L., Palagiano, F., Saija, A., Nava, F., Trombetta, D., de Caprariis, P. (1999). Synthesis, stability, and pharmacological evaluation of nipecotic acid prodrugs. *J. Pharm. Sci.*, 88, 561–567.

142. Battaglia, G., La Russa, M., Bruno, V., Arenare, L., Ippolito, R., Copani, A., Bonina, F., Nicoletti, F. (2000). Systemically administered D-glucose conjugates of 7-chlorokynurenic acid are centrally available and exert anticonvulsant activity in rodents. *Brain Res.*, 860, 149–156.

143. Manfredini, S., Pavan, B., Vertuani, S., Scaglianti, M., Compagnone, D., Biondi, C., Scatturin, A., Tanganelli, S., Ferraro, L., Prasad, P., Dalpiaz, A. (2002). Design, synthesis and activity of ascorbic acid prodrugs of nipecotic, kynurenic and diclophenamic acids, liable to increase neurotropic activity. *J. Med. Chem.*, 45, 559–562.

144. Kearney, A. S. (1996). Prodrugs and targeted drug delivery. *Adv. Drug Delivery Rev.*, 19, 225–239.

145. Chourasia, M. K., Jain, S. K. (2003). Pharmaceutical approaches to colon targeted drug delivery systems. *J. Pharm. Pharm. Sci.*, 6, 33–66.

146. Haas, M., Moolenaar, F., Meijer, D. K., de Zeeuw, D. (2002). Specific drug delivery to the kidney. *Cardiovasc. Drugs Ther.*, 16, 489–496.

147. Wilk, S., Mizoguchi, H., Orlowski, M. (1978). γ-Glutamyl dopa: A kidney-specific dopamine precursor. *J. Pharmcol. Exper. Ther.*, 206, 227–232.

148. Orlowski, M., Mizoguchi, H., Wilk, S. (1980). *N*-acyl-gamma-glutamyl derivatives of sulfamethoxazole as models of kidney-selective prodrugs. *J. Pharmcol. Exper. Ther.*, 212, 167–172.

149. Lee, M. R. (1990). Five years' experience with gamma-L-glutamyl-L-dopa: A relatively renally specific dopaminergic prodrug in man. *J. Autonomic Pharmcol.*, 10, 103–108.

150. Gangwar, S., Pauletti, G. M., Wang, B. H., Siahaan, T. J., Stella, V. J., Borchardt, R. T. (1997). Prodrug strategies to enhance the intestinal absorption of peptides. *Drug Discov. Today*, 2, 148–155.

151. Borchardt, R. T. (1999). Optimizing oral absorption of peptides using prodrug strategies. *J. Controlled Release*, 62, 231–238.

152. Ripka, A. S., Rich, D. H. (1998). Peptidomimetic design. *Curr. Opin. Chem. Biol.*, 2, 441–452.

153. Oliyai, R. (1996). Prodrugs of peptides and peptidomimetics for improved formulation and delivery. *Adv. Drug Delivery Rev.*, 19, 275–286.

154. Oliyai, R., Stella, V. J. (1992). Kinetics and mechanism of isomerization of cyclosporine-A. *Pharm. Res.*, 9, 617–622.

155. Hamada, Y., Matsumoto, H., Kimura, T., Hayashi, Y., Kiso, Y. (2003). Effect of the acyl groups on O → N acyl migration in the water-soluble prodrugs of HIV-1 protease inhibitor. *Bioorganic Med. Chem. Lett.*, 13, 2727–2730.

156. Hamada, Y., Ohtake, J., Sohma, Y., Kimura, T., Hayashi, Y., Kiso, Y. (2002). New water-soluble prodrugs of HIV protease inhibitors based on O → N intramolecular acyl migration. *Bioorganic Med. Chem.*, *10*, 4155–4167.

157. Bundgaard, H. (1992). The utility of the prodrug approach to improve peptide absorption. *J. Controlled Release*, *21*, 63–72.

158. Delie, F., Couvreur, P., Nisato, D., Michel, J. B., Puisieux, F., Letourneux, Y. (1994). Synthesis and *in-vitro* study of a diglyceride prodrug of a peptide. *Pharm. Res.*, *11*, 1082–1087.

159. Friis, G. J. (2000). Peptide and protein derivatives. In S. Frokjaer and L. Hovgaard (Eds.), *Pharmaceutical. Formulation Development of Peptides and Proteins*. Taylor & Francis, London, pp. 206–219.

160. Kahns, A. H., Buur, A., Bundgaard, H. (1993). Prodrugs of peptides. 18. Synthesis and evaluation of various esters of desmopressin (Ddavp). *Pharm. Res.*, *10*, 68–74.

161. Gudmundsson, O. S., Jois, S. D. S., Vander Velde, D. G., Siahaan, T. J., Wang, B., Borchardt, R. T. (1999). The effect of conformation on the membrane permeation of coumarinic acid-, and phenylpropionic acid-based cyclic prodrugs of opioid peptides. *J. Peptide Res.*, *53*, 383–392.

162. Bak, A., Gudmundsson, O. S., Friis, G. J., Siahaan, T. J., Borchardt, R. T. (1999). Acyloxyalkoxy-based cyclic prodrugs of opioid peptides: Evaluation of the chemical and enzymatic stability as well as their transport properties across Caco-2 cell monolayers. *Pharm. Res.*, *16*, 24–29.

163. Gudmundsson, O. S., Nimkar, K., Gangwar, S., Siahaan, T., Borchardt, R. T. (1999). Phenylpropionic acid-based cyclic prodrugs of opioid peptides that exhibit metabolic stability to peptidases and excellent cellular permeation. *Pharm. Res.*, *16*, 16–23.

164. Chang, S.-C., Bundgaard, H., Buur, A., Lee, V. H. L. (1987). Improved corneal penetration of timolol by prodrugs as a means to reduce systemic drug load. *Investigative Ophthalmol. Vis. Sci.*, *29*, 487–491.

165. Mahmud, T., Rafi, S. S., Scott, D. L., Wriilesworth, J. M., Bjarnason, I. (1996). Nonsteroidal antiiflammatory drugs and uncoupling of mitochondrial oxidative phosphorylation. *Arthritis Rheum.*, *39*, 1998–2003.

166. Ogiso, T., Iwaki, M., Tanino, T., Nagai, T., Ueda, Y., Muraoka, O., Tanabe, G. (1996). Pharmacokinetics of indomethacin ester prodrugs: Gastrointestinal and hepatic toxicity and the hydrolytic capacity of various tissues in rats. *Biol. Pharm. Bull.*, *19*, 1178–1183.

167. Samara, E., Avnir, D., Ladkani, D., Bialer, M. (1995). Pharmacokinetic analysis of diethylcarbonate prodrugs of ibuprofen and naproxen. *Biopharm. Drug Disposition*, *16*, 201–210.

168. Bonina, F., Trombetta, D., Borzi, A., De Pasquale, A., Saija, A. (1997). 1-Ethylazacycloalkan-2-one indomethacin esters as new oral prodrugs: Chemical stability, enzymatic hydrolysis, anti-inflammatory activity and gastrointestinal toxicity. *Int. J. Pharm.*, *156*, 245–250.

169. Jaksic, P., Mlinaric-Majerski, K., Zorc, B., Dumic, M. (1996). Macromolecular prodrugs. VI. Kinetic study of poly[α,β-(N-2-hydroxyethyl-DL-aspartamide]-ketoprofen hydrolysis. *Int. J. Pharm.*, *135*, 177–182.

170. Mahfouz, N. M., Omar, F. A., Aboul-Fadl, T. (1999). Cyclic amide derivatives as potential prodrugs II: N-hydroxymethylsuccinimide-/istatin esters of some

NSAIDs as prodrugs with an improved therapeutic index. *Eur. J. Med. Chem., 34*, 551–562.

171. Duncan, R., Gac-Breton, S., Keane, R., Musila, R., Sat, Y. N., Satchi, R., Searle, F. (2001). Polymer-drug conjugates, PDEPT and PELT: Basic principles for design and transfer from the laboratory to clinic. *J. Controlled Release, 74*, 135–146.
172. Bickel, U., Yoshikawa, T., Pardridge, W. M. (2001). Delivery of peptides and proteins through the blood-brain barrier. *Adv. Drug Delivery Rev., 46*, 247–279.
173. Mehvar, R. (2000). Dextrans for targeted and sustained delivery of therapeutic and imaging agents. *J. Controlled Release, 69*, 1–25.
174. Lovrek, M., Zorc, B., Boneschans, B., Butula, I. (2000). Macromolecular prodrugs. VIII. Synthesis of polymer-gemfibrozil conjugates. *Int. J. Pharm., 200*, 59–66.
175. van der Merwe, T., Boneschans, B., Zorc, B., Breytenbach, J., Zovko, M. (2002). Macromolecular prodrugs: X. Kinetics of fenoprofen release from PHEA-fenoprofen conjugate. *Int. J. Pharm., 241*, 223–230.
176. Li, C. (2002). Poly(L-glutamic acid)—Anticancer drug conjugates. *Adv. Drug Delivery Rev., 54*, 695–713.
177. Kopecek, J., Kopeckova, P., Minko, T., Lu, Z. (2000). HPMA copolymer-anticancer drug conjugates: Design, activity, and mechanism of action. *Eur. J. Pharm. Biopharm., 50*, 61–81.
178. Maeda, H., Sawa, T., Konno, T. (2001). Mechanism of tumor-targeted delivery of macromolecular drugs, including the EPR effect in solid tumor and clinical overview of the prototype polymeric drug SMANCS. *J. Controlled Release, 74*, 47–61.
179. Maeda, H. (2001). SMANCS and polymer-conjugated macromolecular drugs: Advantages in cancer chemotherapy. *Adv. Drug Delivery Rev., 46*, 169–185.
180. Greenwald, R. B., Conover, C. D., Choe, Y. H. (2000). Poly(ethylene glycol) conjugated drugs and prodrugs: A comprehensive review. *Crit. Rev. Ther. Drug Carrier Sys., 17*, 101–161.
181. Greenwald, R. B. (2001). PEG drugs: An overview. *J. Controlled Release, 74*, 159–171.
182. Greenwald, R. B., Choe, Y. H., McGuire, J., Conover, C. D. (2003). Effective drug delivery by PEGylated drug conjugates. *Adv. Drug Delivery Rev., 55*, 217–250.

18

GABA AND GLUTAMATE RECEPTOR LIGANDS AND THEIR THERAPEUTIC POTENTIAL IN CNS DISORDERS

ULF MADSEN, HANS BRÄUNER-OSBORNE,
JEREMY R. GREENWOOD, TOMMY N. JOHANSEN,
POVL KROGSGAARD-LARSEN, TOMMY LILJEFORS,
MOGENS NIELSEN, AND BENTE FRØLUND
Danish University of Pharmaceutical Sciences
Copenhagen, Denmark

18.1	THERAPEUTIC PROSPECTS FOR GABA AND GLUTAMATE RECEPTOR LIGANDS: GENERAL ASPECTS	798
	GABA and Glutamate Receptor Ligands as Potential Drugs in Neurological Diseases	799
	Central Nervous System Ischemia	800
	Alzheimer's Disease	800
	Other Diseases and Disease Conditions	801
18.2	GABA: INHIBITORY NEUROTRANSMITTER	802
	Classification of the GABA System and Therapeutic Targets	802
	GABA Biosynthesis and Metabolism	803
	GABA Reuptake	804
	The $GABA_A$ Receptor Complex	807
	$GABA_B$ Receptor Ligands	823
	$GABA_C$ Receptor Agonists and Antagonists	828
18.3	GLUTAMATE: EXCITATORY NEUROTRANSMITTER	829
	Classification of Glutamic Acid Receptors	829
	NMDA Receptor Ligands	830
	AMPA Receptor Ligands	837

Drug Discovery Handbook, by Shayne Cox Gad
Copyright © 2005 by John Wiley & Sons, Inc.

KA Receptor Agonists and Antagonists	843
Structure of Ionotropic Glutamate Receptors	846
Metabotropic Glutamate Receptor Ligands	851
Structure of Metabotropic Glutamate Receptors	863
References	868

18.1 THERAPEUTIC PROSPECTS FOR GABA AND GLUTAMATE RECEPTOR LIGANDS: GENERAL ASPECTS

The complex functions of the mammalian central nervous system (CNS) are primarily determined by two superior systems: (1) excitation by the major excitatory amino acid transmitter, (S)-glutamic acid (Glu), which depolarizes neurons via a large number of receptor subtypes, and (2) inhibition by 4-aminobutyric acid (GABA), which hyperpolarizes neurons via multiple receptor subtypes. Although the roles of amino acid neurotransmitters in the CNS are far from being fully mapped out, accumulating evidence from a variety of neurochemical and electrophysiological studies seems to indicate that a majority, perhaps all, central neurons are under excitatory and inhibitory control by Glu and GABA [1–4].

It is interesting to note that the brain and spinal cord functions are essentially determined by two simple amino acids. However, the transmission processes mediated by Glu and GABA are very complex and apparently highly regulated. The functions and interactions of the amino acid neurotransmitter systems are further complicated by the fact that Glu is the biosynthetic precursor of GABA. Although GABA predominantly hyperpolarizes neurons, depolarizing actions of GABA are also observed, primarily at the early stages of development of the brain. In addition, there is evidence to suggest that a third mechanism may play a fundamental role in the function of the brain, namely disinhibition. The indirect neuronal excitation implies synaptic contact between two inhibitory neurons. This indirect excitatory mechanism in the CNS has never been unequivocally proven or disproven, but a number of apparently paradoxical observations have been explained on the basis of disinhibition.

Both Glu and GABA operate through ionotropic as well as metabotropic receptors. Whereas the former classes of receptors, iGluRs and $GABA_{A,C}$, mediate fast neurotransmission, the metabotropic receptors, mGluRs and $GABA_B$, are involved in slow neurotransmission processes. All of these subclasses of receptors are heterogeneous and may be involved in a variety of CNS disorders and disease conditions. Thus, all subtypes of Glu and GABA receptors are potential therapeutic targets [5–7].

In light of the ubiquitous presence of these amino acid receptors in the CNS, it is not straightforward to design Glu and GABA receptor ligands as drugs with selective effects on diseases involving degenerating or derailed amino acid neurotransmission in restricted brain areas. Glu and GABA receptor subtypes are, however, unevenly distributed in the CNS, and this fact opens up the prospect of developing receptor ligands as drugs with a regioselective effect on particular brain disorders [1–7].

There are numerous examples of therapeutic agents acting as agonists or antagonists at metabotropic (G-protein-coupled) receptors in other CNS neurotransmitter systems, suggesting that from a drug development point of view the metabotropic $GABA_B$ receptors and the mGluRs would be particularly attractive. On the other hand, the ionotropic $GABA_{A,C}$ receptors and the iGluRs, which mediate fast neurotransmission, represent complex challenges as drug targets. Thus, full antagonists or agonists at ionotropic Glu or GABA receptors are less likely to be useful therapeutic agents. Whereas the former type of ligands may cause severe CNS disturbances via blockade of fundamental neurotransmission processes, Glu agonists are likely to exert neurotoxic effects in the human CNS, as observed in numerous animal models. Full $GABA_A$ agonists, which show effective receptor desensitization in model studies, may also desensitize receptors in human brain tissue after systemic administration and thus provoke undesirable "functional antagonism." These issues make partial agonists targeting ionotropic Glu or GABA receptors more attractive as therapeutic agents. Medicinal chemists are faced not only with the challenges of designing partial agonists with optimal levels of efficacy for each disease with malfunctioning ionotropic amino acid receptor mechanisms, but also of developing principles for the design of such ligands.

Modulation of ionotropic amino acid receptors by compounds interacting noncompetitively with the receptors via binding to physiological modulatory sites or nonphysiological binding sites at receptor protein interfaces are likely to play increasingly important roles as therapeutic agents. A notable example of this category of drugs is the benzodiazepines, which interact with binding sites of no apparent physiological function, but located at protein interfaces in heteropentameric $GABA_A$ receptor complexes.

GABA and Glutamate Receptor Ligands as Potential Drugs in Neurological Diseases

It is well known that Glu can act as a neurotoxin, especially when energy supply is compromised. This has given rise to the proposal that injury to neurons in many neurological disorders may be caused, at least in part, by overstimulation of Glu receptors (excitotoxicity) [8]. At least to some extent, this may involve reversal of Glu transporters. Such neurological conditions range from acute insults such as stroke, trauma, and epilepsy, to chronic neurodegenerative states such as Huntington's disease, Parkinson's disease, AIDS (acquired immunodeficiency syndrome) dementia, and amyotrophic lateral sclerosis [1, 8].

Central Nervous System Ischemia

Stroke was the first clinical indication considered for Glu receptor antagonists because the evidence for an etiological role of excitotoxic mechanisms is the most convincing. A very large number of studies have been performed with iGluR antagonists in experimentally induced ischemic stroke in laboratory animals, and, in most cases, these model experiments have revealed neuroprotective effects elicited by different subclasses of competitive and noncompetitive iGluR antagonists.

Interpretation of these results and extrapolation of the observations to the human clinic are, however, complicated by the different effects of iGluR antagonists in models of focal and global ischemia, and the quite pronounced influence on the results of the experimental condition. A number of iGluR antagonists entered early clinical trials for stroke, and, with different patterns and degrees of severity, these compounds have produced severe and frequently unacceptable side effects such as psychotomimetic effects, respiratory depression, and cardiovascular dysregulation.

These predominantly negative clinical results seem to emphasize that blockade of iGluRs by full antagonists has limited therapeutic potential. In principle, low-efficacy iGluR agonists that thus predominantly antagonize the receptors may have some therapeutic potential. Alternatively, functional iGluR partial agonism, established by coadministration of an iGluR antagonist and an iGluR agonist at a fixed ratio [9] may find therapeutical use. Although difficult to establish as a practical approach, functional partial agonism does, in principle, present a flexible approach to therapies based on the concept of partial agonism.

Little is known about the prospects for mGluR ligands, antagonists or perhaps agonists, in ischemia, but there generally is a growing interest in this heterogeneous group of receptors as therapeutic targets.

Alzheimer's Disease

Alzheimer's disease (AD) is a neurodegenerative disorder characterized by cell loss, pathological changes in neuronal transmission, and the formation of senile plaques and neurofibrillary tangles. The cholinergic transmitter system is particularly vulnerable to degeneration, and loss of cholinergic pathways in the CNS may largely explain the cognitive impairments and memory deficiencies that are prominent symptoms of AD patients [1].

In the progression of the neurodegenerative processes in the brains of AD patients, not only cholinergic neurons but also other transmitter systems, including Glu and GABA neurons, are increasingly affected. It has been proposed that the β-amyloid content of the senile plaques renders neighboring neurons vulnerable to excitotoxicity, which is likely to contribute to the multifactorial disease mechanism in AD.

Whereas neither full Glu receptor agonists nor antagonists appear to have obvious therapeutic prospects in AD, partial Glu agonists or functional partial

Glu agonists with appropriately balanced agonist/antagonist profiles may, at least in principle, hold interest as neuroprotective agents or drugs for symptomatic treatment of AD patients.

Since Glu neurotransmitter systems, like cholinergic neuronal pathways, play a key role in learning and memory functions, various GluR ligands have been studied in animal models of these CNS functions. There has recently been a particular pharmacological interest in compounds showing positive noncompetitive modulatory effects at iGluRs. Such compounds have been shown to improve learning and memory in model systems, but, obviously, potentiation of iGluR functions may provoke excitotoxicity, emphasizing that therapeutic targeting of iGluRs to AD is a difficult balancing trick.

The results of studies on different $GABA_A$ receptor ligands in animal models relevant to learning and memory seem to support GABAergic therapeutic approaches in AD. Whereas administration of $GABA_A$ agonists impairs learning and memory in animals via inhibition of cholinergic pathways, memory enhancement is observed after administration of a $GABA_A$ antagonist. Similarly, compounds showing negative noncompetitive modulatory effects at $GABA_A$ receptors enhance performance in learning and memory tasks in animal models.

The lack of positive and particularly negative effects of high-efficacy partial $GABA_A$ agonists in AD patients is of interest and may support the view that low-efficacy partial $GABA_A$ agonists hold clinical potential in AD.

Other Diseases and Disease Conditions

As the result of extensive neurobiological, molecular pharmacological, animal behavioral, and clinical studies, a steadily increasing number of potential therapeutic targets among the Glu and GABA neurotransmitter systems are being identified. Compounds potentiating $GABA_A$ receptor functions, including benzodiazepines, and agents that increase GABA levels via inhibition of the GABA-metabolizing enzyme, GABA-transaminase (GABA-T), or GABA reuptake systems all show antiepileptic effects. Different competitive and noncompetitive Glu receptor antagonists are currently undergoing clinical trials in epilepsy [1, 2].

Growing evidence suggests involvement of Glu transmitter systems in schizophrenia [1, 10, 11]. It is possible that schizophrenia is a reflection of a pattern of imbalances in the CNS and not simply a unidirectional change in neurotransmission. Several lines of research suggest that Glu neurotransmitter mechanisms contribute to the pathophysiology of schizophrenia. Based on extensive animal behavioral studies, drugs that target mGluRs may be considered more attractive than iGluR ligands therapeutically, as well as regards side effect. Undoubtedly animal behavioral and clinical studies will steadily grow in this area.

The number of diseases and disease conditions potentially susceptible to Glu and GABA receptor ligand therapeutic interventions is increasing. Different GluR ligands have been subjected to intensive studies in various models of pain, and a high-efficacy partial GABA$_A$ agonist has been shown in the human clinic to possess potent nonopioid analgesic effects [2]. This latter category of compounds also is of major clinical interest as agents capable of reestablishing normal sleep architecture in patients suffering from sleep disorder [2, 12].

18.2 GABA: INHIBITORY NEUROTRANSMITTER

Classification of the GABA System and Therapeutic Targets

The GABA system is the major inhibitory neurotransmitter system in the CNS, and GABAergic mechanisms are directly involved in a variety of physiological and behavioral processes and in many neurological illnesses. The GABA neurotransmitter system includes a number of synaptic processes (Fig. 18.1), such as the various enzymes and mechanisms involved in synthesis, release, and metabolism of GABA, the transmembrane GABA transporters, and the GABA receptors (GABA$_A$, GABA$_B$, and GABA$_C$). Each of these entities presents a potential drug target, but so far only the GABA receptors, the GABA transporters, and the GABA-metabolizing enzyme GABA-T have proven therapeutically useful.

Although many drugs interacting with the GABA neurotransmitter system have been clinically helpful in the treatment of anxiety, epilepsy, sleep

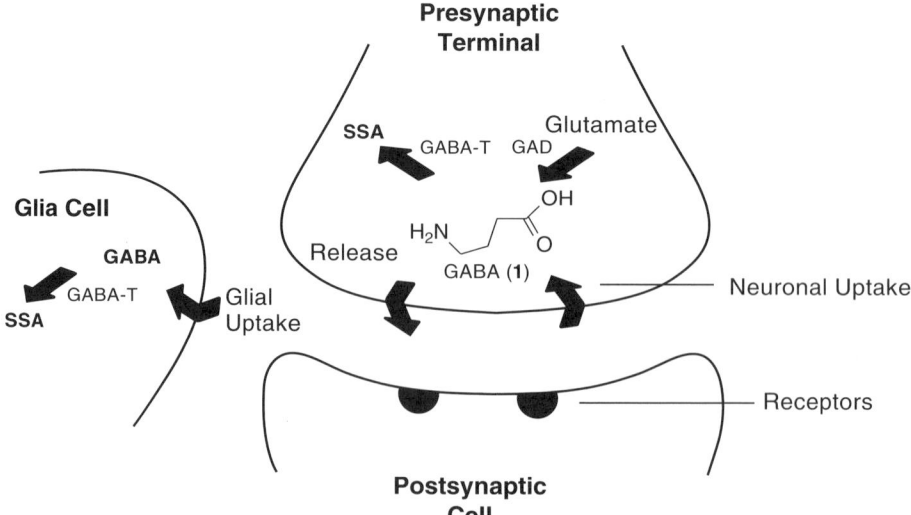

Figure 18.1 Schematic illustration of the biochemical pathways, transport mechanisms, and receptors at a GABA-operated synapse. GAD: Glutamate decarboxylase; GABA-T: GABA transferase; SSA: Succinic acid semialdehyde.

disorders, alcohol withdrawal, and in the induction and maintenance of anesthesia, there are still problems that need to be solved. Therefore, major efforts are underway to improve the therapeutic spectrum of the presently available drugs, reduce their side effects, and discover new therapeutic tools and targets.

GABA Biosynthesis and Metabolism

The main pathway of GABA synthesis is the decarboxylation of Glu, catalyzed by Glu decarboxylase (GAD), which employs pyridoxal-5′-phosphate (PLP) as a cofactor. GABA (**1**) is released into synaptic clefts by depolarization of presynaptic neurons and is rapidly removed from the synapses by reuptake into both glia and presynaptic nerve terminals, where it is initially catabolized by transamination into succinic semialdehyde. The major pathway of its degradation is via transamination with α-ketoglutarate, catalyzed by the PLP-dependent enzyme GABA-T. This enzymatic degradation takes place within GABAergic neurons, as well as in surrounding glia cells. It has been shown that inhibition of the activity of GABA-T results in an increase in available GABA, which makes it a therapeutic target in, for example, neurological disorders associated with abnormally low levels of GABA [13–15].

Inhibitors of GABA Metabolism Based on the knowledge of the mechanism of inactivation of GABA-T, a number of inhibitors of this enzyme have been developed [16–20]. These compounds are typically analogs of GABA, containing appropriate, notably unsaturated, substituents at the γ-carbon adjacent to the γ-amino group. Via a deprotonation at the γ-carbon, the substituents are converted by GABA-T into electrophiles, which react with nucleophilic groups at or near the active site of the enzyme and thereby inactivate the enzyme irreversibly [19]. The receptor protein nucleophile that participates in this reaction is known to be Lys329, which links PLP to the active site [21].

Although GABA-T, like other PLP-dependent enzymes, does not show strict stereospecificity with respect to inactivation by mechanism-based inactivators, such inhibitors do react with the enzyme in a stereoselective manner. Thus the *S* forms of the GABA-T inhibitors Vigabatrin (**2**) and its fluoromethyl derivative (**3**) are more active as GABA-T inactivators than the respective *R*-isomers [22] (Fig. 18.2). These observations are in good agreement with results obtained from modeling studies of the inhibitors using the reported crystal structure of the enzyme [21]. The α-carboxylate of the inhibitor–PLP complex interacts with Arg192, which positions the *S*-proton of the γ-carbon appropriately for abstraction by Lys329.

The therapeutic potential of the GABA metabolic system is illustrated by Vigabatrin (**2**), which is utilized in the treatment of epilepsy [23], and is in clinical trials for treatment of drug addiction [24–26]. Unfortunately, the high risk of irreversible visual field defects, frequently associated with Vigabatrin

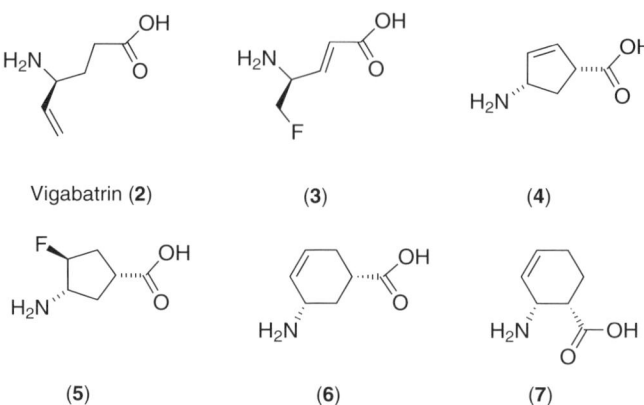

Figure 18.2 Structures of some GABA-T inhibitors, compound **4** being inactive.

treatment, continues to cause concern [27, 28]. However, a recent study on rats has revealed that low-dose Vigabatrin maintains therapeutic efficacy for the treatment of cocaine addiction with reduced risk for emergence of visual defects [29].

In order to optimize the inactivator–enzyme interaction, conformationally rigid analogs of Vigabatrin and 4-amino-5-halopentanoic acid have been developed [20, 30]. Unlike the cyclopentene analog (**4**) of Vigabatrin, which does not inactivate GABA-T, the saturated 5-fluoro analog (**5**) exhibits inactivation properties similar to the corresponding open-chain analogs.

The apperance of the crystal structure of GABA-T [21] initiated the development of a series of cyclohexene analogs of Vigabatrin that possess slightly increased flexibility about the double bond [16, 17]. The analogs **6** and **7** irreversibly inhibit GABA-T, whereas the corresponding *trans*-isomers are competitive reversible inhibitors of the enzyme. Once bound, the proton of the *trans*-isomer, which needs to be removed for the reaction to take place, is not oriented toward the active site (Lys 329). Thus, no irreversible reaction takes place [17]. Interestingly, similar effects on drug addiction have been observed for competitive reversible GABA-T inhibitors [25, 31].

GABA Reuptake

The reuptake of GABA into surrounding neurons and glial cells is affected by selective GABA transporters that belong to the superfamily of sodium- and chloride-dependent neurotransmitter transporters. Four different transporters have been identified displaying regionally distinct distributions within the CNS [32]. The four subtypes have been cloned from different species by different research groups leading to different acronyms, bringing the numbering of the transporters into confusion [32]. Whereas the homologous transporters cloned from rat, mouse, and human have been numbered GAT-1, GAT1, and GAT-1, respectively, the mouse GAT3 and GAT4 correspond to the rat and human GAT-2 and GAT-3. The betaine/GABA transporter from humans (BGT-1) corresponds to the mouse GAT2. Based on studies of the regional distribution of GABA transporters, GAT1 seems to be the most abundant in the CNS, and it appears to be predominantly localized in neurons. GAT3 and GAT4 seem to be preferentially expressed in glial cells, while GAT2 is equally expressed in neurons and in glial cells [32–38].

Inhibition of GABA uptake has been shown to enhance the synaptic efficacy of GABA and to increase the extracellular concentration of GABA in the brain [39, 40]. The anticonvulsant effects seen in different seizure models following the administration of GABA uptake inhibitors substantiates the functional importance of these observations. In addition to their anticonvulsant activities, in vivo and in vitro studies indicate that GABAergic agents, including reuptake inhibitors, possess neuroprotective [41, 42], cognition-enhancing [43], and pain-relieving properties [44].

It has been established that the transport systems show different substrate specificities, which makes selective pharmacological manipulation of the GABA transport system possible [45]. For the purpose of stimulating GABA neurotransmission, the most effective strategy for such therapeutic intervention seems to be a selective blockade of glial GABA uptake, in order to elevate the level of GABA taken up by the neuronal carrier.

Inhibitors of GABA Reuptake The classical GABA transport inhibitors nipecotic acid (**8**) and guvacine (**9**) are effective inhibitors of both neuronal and glial GABA uptake [46–49] (Fig. 18.3). Introducing small substituents at the amino group of nipecotic acid (**8**) or guvacine (**9**) results in compounds with reduced affinity for GABA transport carriers [50], whereas large lipophilic derivatives such as *N*-(4,4-diphenyl-3-butenyl)nipecotic acid (**12**) are much more potent than the parent compounds [51]. By contrast with nipecotic acid (**8**) and guvacine (**9**), such lipophilic compounds are able to cross the blood–brain barrier (BBB) and have been shown to be potent anticonvulsants in animal models. This development has led to the compound Tiagabine (**13**), now marketed as a therapeutic adjunct for the treatment of epilepsy [23]. Due to nonoptimal pharmacokinetic profile of Tiagabine (**13**) [52] a large number of lipophilic analogs of nipecotic acid (**8**) have been synthesized, that include an electronegative moiety in the carbon chain (e.g., compounds **14** and **15**) [53–57]. Although some of these analogs have turned out to be more potent than Tiagabine (**13**), a new therapeutic candidate apparently has not emerged from these studies.

Although effective as an anticonvulsant, Tiagabine (**13**) does not display glial selectivity and has been shown to be GAT-1 selective [58]. By the developing *exo*-THPO (**10**) and analogs thereof, a certain degree of selectivity for the glial transport system has been achieved [59]. *N*-Me-*exo*-THPO (**11**)

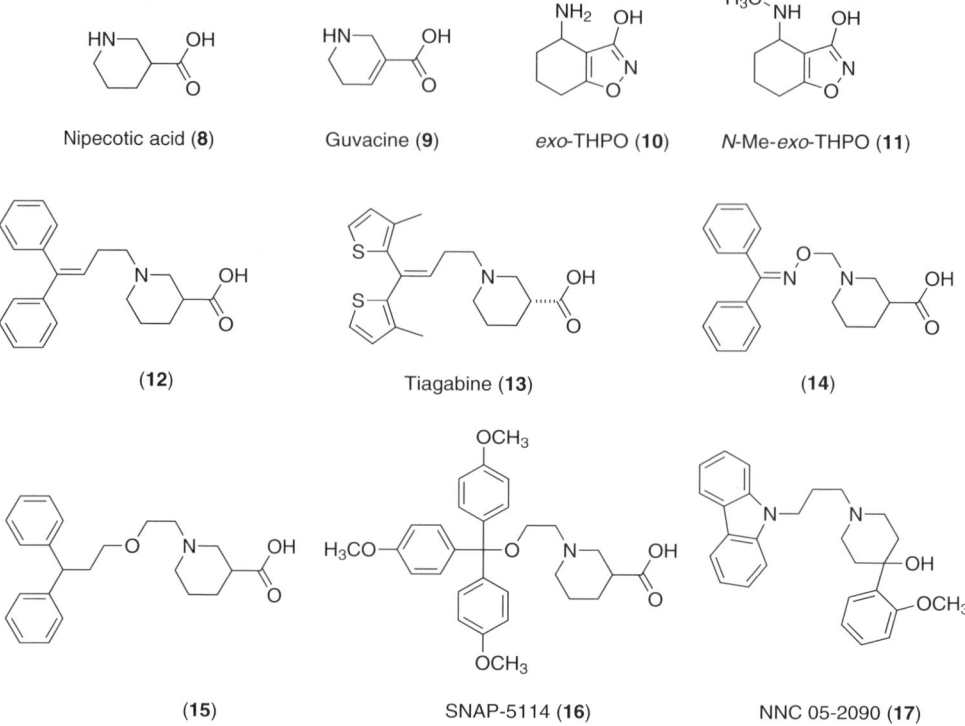

Figure 18.3 Structures of some GABA uptake inhibitors.

exhibits more than 10-fold higher inhibitory potency for glial transport than for neuronal transport of GABA [59]. Although Tiagabine (**13**) is approximately 1000-fold more potent than *exo*-THPO (**10**) and its *N*-alkylated analogs as an inhibitor of GABA uptake in cultured neuronal cells and in GAT1 expressing cells, the anticonvulsant potencies of *exo*-THPO (**10**) and its analogs have been shown to be only 3 to 7-fold less than that of Tiagabine (**13**) [58]. Thus, the anticonvulsant efficacies of *exo*-THPO analogs seem to correlate better with their ability to inhibit glial rather than neuronal GABA transport [58]. Although glia selective, both *exo*-THPO analogs and liphophilic analogs of nipecotic acid and guvacine interact selectively with GAT1. Very few inhibitors of GABA transport have been shown to display selectivity for other GABA transporters (GAT2, -3, or -4). SNAP-5114 (**16**) [60, 61] has been reported to inhibit GAT2-4, and a group of compounds exemplified by NNC 05-2090 (**17**) seem to inhibit GAT2 and GAT4 [62]. Further development of subtype-selective compounds would be useful for studying the pharmacological and physiological importance of the GAT subtypes and may help understand the different anticonvulsant profiles observed for different inhibitors [63].

The GABA$_A$ Receptor Complex

The GABA$_A$ receptor is a member of the superfamily of ligand-gated ion channels (Fig. 18.4). The structure and function of this group of receptors displays a high degree of complexity, which has been the subject of a number of excellent reviews over the last decade [5, 64–70]. The GABA$_A$ receptor is an oligomeric protein that forms a complex containing a considerable number of separate but allosterically interacting binding sites (Fig. 18.4b). This is reflected in the structural diversity of compounds acting at GABA$_A$ receptors, including important drugs such as benzodiazepines, barbiturates, and neurosteroids. The heterogenity of the GABA$_A$ receptor in the brain is pronounced. The exact number of different receptor subtypes is not known, but so far a total of 19 different subunits belonging to several subunit classes have been cloned (α1-6, β1-3, γ1-3, δ, ϵ, θ, π, and ρ1-3). GABA$_A$ receptors are built up as pentameric assemblies of different families of receptor subunits, as illustrated in Figure 18.4a. It is believed that the most abundant receptor subtype consists of two α_1, two β_2, and one γ_2 subunit in a well-defined structural arrangement forming a pore/channel in the cell membrane that is permeable to chloride ions.

Site-directed mutagenesis studies have shown that the binding site(s) for GABA, GABA$_A$ agonists, and competitive antagonists are located at the interface between α and β subunits, whereas the binding site(s) for benzodiazepines are located at the interface between α and γ subunits.

The GABA$_A$ receptors display different pharmacology and regional distribution depending on their subunit combination suggesting that GABA$_A$ receptor ligands with subunit-specific effects may provide drugs with improved clinical utility.

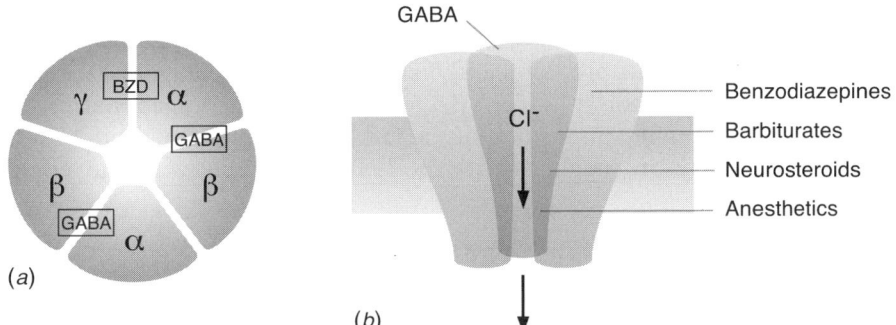

Figure 18.4 (a) Schematic model of the pentameric structure of the GABA$_A$ receptor complex indicating the position of the binding sites for GABA and benzodiazepines (BZD). (b) Schematic illustration of the GABA$_A$ receptor complex indicating the chloride ion channel and additional binding sites.

The GABA$_A$ Receptor Agonists and Antagonists In spite of the very distinct and specific structural requirements for GABA$_A$ receptor agonists, conformational restriction of different parts of the GABA (**1**) molecule and bioisosteric replacement of the carboxyl group have led to a series of specific GABA$_A$ agonists. Some of these compounds have played a key role in the pharmacological characterization of GABA$_A$ receptors.

Muscimol (**18**) (Fig. 18.5), a constituent of the mushroom *Amanita muscaria*, has been highly useful in the investigation of the GABA$_A$ receptors [48]. However, the fact that muscimol (**18**) is nonselective, being both a GABA$_A$ receptor agonist [71, 72] and a substrate for GABA-T [73], and moreover a neurotoxin, prevents its therapeutic application. Because the 3-isoxazolol moiety has proved to be efficient as a bioisosteric replacement for the carboxyl group of GABA, muscimol (**18**) has been extensively used as a lead structure in the development of GABA$_A$ receptor ligands.

The conformationallly restricted bicylic analog THIP (**19**), a specific GABA$_A$ agonist [74], has been shown to be devoid of the neurotoxic properties of muscimol (**18**) and metabolically stable, in contrast to muscimol (**18**). THIP (**19**) shows nonopioid analgesic effects [75] and is in clinical trials as a novel hypnotic agent. Unlike GABA, muscimol (**18**) and THIP (**19**) are capable of penetrating the BBB after systemic administration, in spite of their zwitterionic structures [76].

Based on the structure of THIP (**19**), a series of monoheterocyclic GABA$_A$ agonists has been developed, including isonipecotic acid (**20**) and isoguvacine (**21**) [74, 77]. THIP (**19**) and the equipotent analog isoguvacine (**21**) are now used as standard GABA$_A$ agonists.

Bicuculline (**22**) [78] (Fig. 18.6) and bicuculline methochloride (BMC) [79] are classical GABA$_A$ antagonists. SR 95531(**23**) belongs to a group of arylaminopyridazine analogs that show potent and selective competitive GABA$_A$ receptor antagonist effect [80, 81]. SR 95531(**23**) is now used as a standard GABA$_A$ antagonist.

The nonannulated THIP analogs, 4-PIOL (**24**) [82], the corresponding sulfur analog thio-4-PIOL (**25**) and **26** [83] represent a group of low-efficacy GABA$_A$ agonists that predominantly show GABA$_A$ antagonist profiles [83–85]. As reported for other GABA$_A$ antagonists, the functional consequences of the subunit composition of the GABA$_A$ receptors on thio-4-PIOL (**25**) pharmacology are negligible [86, 87]. This contrasts with the effects

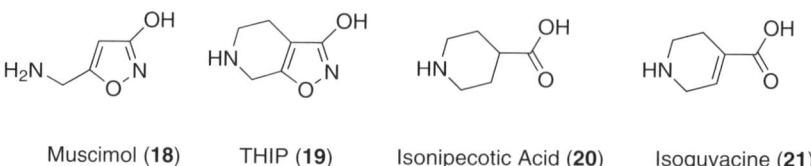

Muscimol (**18**)   THIP (**19**)   Isonipecotic Acid (**20**)   Isoguvacine (**21**)

Figure 18.5 Structures of some GABA$_A$ receptor agonists.

GABA: INHIBITORY NEUROTRANSMITTER

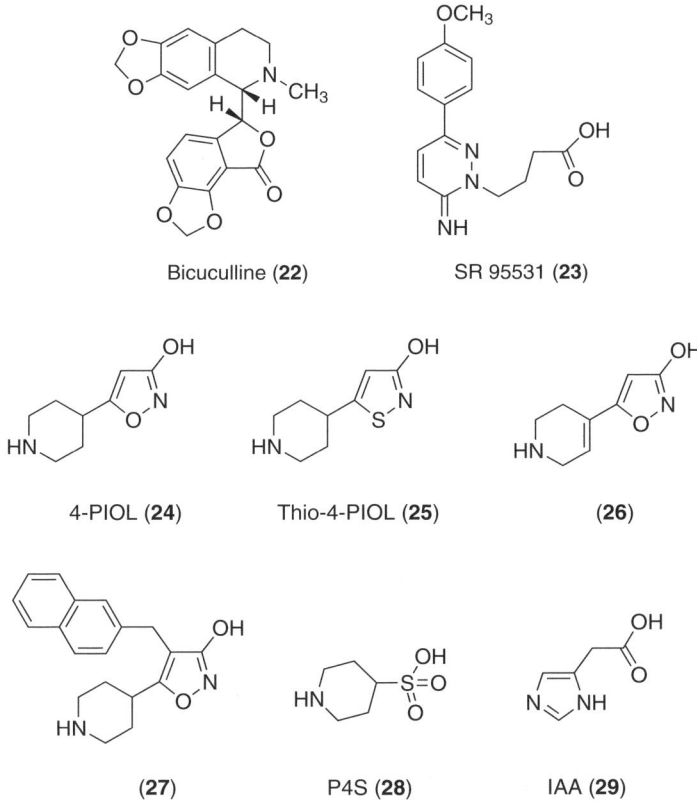

Figure 18.6 Structures of some GABA$_A$ receptor antagonists (**22**, **23**, and **27**) and some partial GABA$_A$ receptor agonists (**24–26**, **28**, and **29**).

observed on the pharmacology of GABA$_A$ agonists [86]. Importantly, unlike full and highly efficacious partial GABA$_A$ agonists, repeated administration of 4-PIOL (**24**) has been reported to not cause significant desensitization of GABA$_A$ receptors [85, 88].

Based on the structure of 4-PIOL (**24**), a series of potent and selective competitive GABA$_A$ antagonists has been developed, as exemplified by compound **27**, with potencies in the low nanomolar range (see also the following section) [89]. Very little information has been obtained regarding direct acting full GABA$_A$ receptor agonist or antagonists in clinical studies. Therapeutic use of GABA$_A$ receptor agonists and antagonists that can interact with all GABA$_A$ receptor subtypes in the CNS may be associated with side effects, such as tolerance and convulsions. In this regard, partial agonists with various levels of efficacy, displaying either predominantly agonist or antagonist profiles, could be of therapeutic relevance, as exemplified by THIP.

THIP (**19**) and the heterocylic GABA isosteres piperidine-4-sulfonic acid (P4S, **28**) and imidazole-4-acetic acid (IAA, **29**) show the characteristics of partial agonists [90]. Binding studies have shown only relatively low selectivity of $GABA_A$ agonists for $GABA_A$ receptors containing different α or β subunits [87]. By contrast, electrophysiological investigations using heteromeric ($α_xβ_3γ_2$) human $GABA_A$ receptors expressed in *Xenopus* oocytes have revealed that the relative efficacy as well as the potency displayed by these compounds is highly dependent on the receptor subunit combination [86]. The patterns of $GABA_A$ receptor subunit dependence of the maximal responses of THIP (**19**) and IAA (**29**) are qualitatively the same as that seen for P4S (**28**) (Fig. 18.7). Agonist potencies are frequently highest at receptors containing $α_5$ or especially $α_6$ subunits and generally lowest at $α_4$-containing receptors. Coexpression of different α subunits in combination with β and γ subunits has been shown to result in unpredictable changes in both efficacy and potency of $GABA_A$ receptor agonists [88, 90, 91].

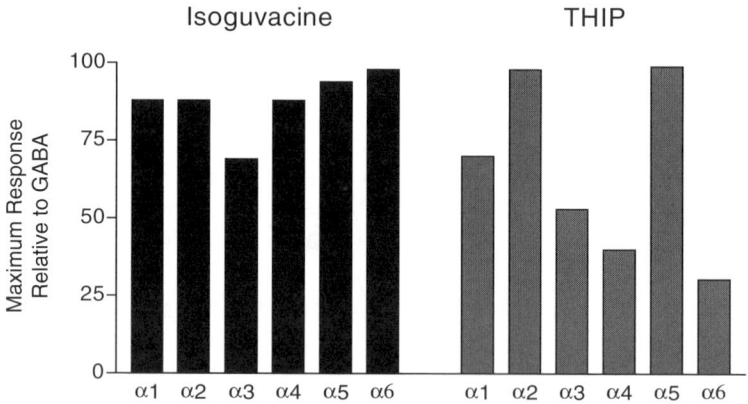

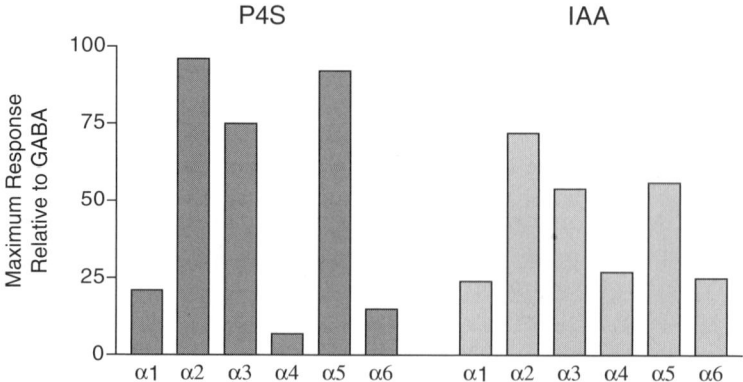

Figure 18.7 Subunit-dependent efficacy of isoguvacine (**21**), THIP (**19**), P4S (**28**), and IAA (**29**) on $α_xβ_3γ_{2S}$ containing $GABA_A$ receptors expressed in oocytes. Maximum response at every receptor combination was determined using 3 mM GABA [86].

Another example of the consequences of the variation of subunits is the effect of substituting of a δ subunit for the γ_2 subunit in a receptor complex containing $\alpha_4\beta_3$ subunits. THIP (**19**) behaves as a partial agonist at $\alpha_4\beta_3\gamma_2$-containing receptors, while the compound behaves as a superagonist at $\alpha_4\beta_3\delta$-containing receptors, with a maximum response 60 percent larger than that of GABA, suggesting that GABA acts as a partial agonist at the $\alpha_4\beta_3\delta$ receptor subtype [92, 93]. Interestingly, δ-containing receptors show a 50-fold higher affinity for GABA than γ-containing receptors [94]. The α_4 and the δ subunits are co-localized in the thalamus and hippocampus [95, 96], and some of these receptors are presumed to be extrasynaptic [97]. This restricted expression and putative extrasynaptic location makes them potentially interesting as drug targets.

It has been shown that the hypnotic effect of THIP (**19**) is elicited by a mechanism different from those of benzodiazepines and ethanol [98]. Most likely, this effect is mediated by benzodiazepine-insensitive α_4-containing receptors. These receptor subtypes are located in regions distinctly different from the most abundant α_1-containing receptors that are supposed to be responsible for the effects of currently used hypnotics [99]. These observations have led to the suggestion that THIP (**19**) has a major effect on α_4- and δ-containing extrasynaptic receptors [98].

Despite the fact that the amino acid residues contributing to the binding site of GABA appear to be conserved within the α and β subunits [69], the functional consequences of binding are highly dependent on the subunit composition of individual $GABA_A$ receptors. Moreover, the pharmacological response seems to be determined not only by α and β subunits but by the interaction of all subunits present in the receptor complex. The determinants of these changes in pharmacological profiles that results in functional selectivity are still not clarified, making the development of compounds with a predefined receptor subtype-selectivity profile difficult.

Receptor and Pharmacophore Models of the GABA Site of the $GABA_A$ Receptor Complex   There is currently no experimentally determined three-dimensional structure of the $GABA_A$ receptor complex. Thus, the study of the structure–activity relationship (SAR) and the design of novel $GABA_A$ ligands cannot be based on experimental structural information about the receptor protein. The $GABA_A$ receptors belong to the same superfamily as the nicotinic acetylcholine receptors, and a soluble homopentameric acetylcholine binding protein (AChBP) from the snail *Lymnaea stagnalis* has been isolated [100]. This protein is weakly homologous to the extracellular ligand-binding domain of the nicotinic acetylcholine receptor. The protein has been successfully crystallized and its three-dimensional structure has been solved by X-ray crystallography [101]. The AChBP structure has been used for homology building of the extracellular ligand-binding domain of $GABA_A$ receptors [102–104]. The receptor models obtained may be used as tools for the design of mutagenesis experiments. However, it should be noted that the AChBP protein only shares approximately 18 percent sequence identity with $GABA_A$ receptor subunits, making sequence alignment and homology modeling questionable. Thus, given the low level of sequence identity, it is not straightforward to use these models for the prediction of ligand affinity or SAR studies.

In the absence of an experimentally determined three-dimensional (3D) structure for the $GABA_A$ receptor, rationalization of SARs and design of novel ligands may be attempted by the use of pharmacophore models. Such a model is developed on the basis of analysis of the molecular properties of known ligands and their receptor binding data [105, 106]. A well-developed pharmacophore model may be employed to design new ligands and/or to search databases for new compounds that are compatible with the model [105, 106].

Most attempts to develop pharmacophore models for ligands binding to the GABA site of $GABA_A$ receptors have used the assumption that the 3-isoxazolol rings in, for example, muscimol (**18**), THIP (**19**), and 4-PIOL (**24**) (Figs. 18.5 and 18.6) are located in very similar regions in the binding cavity and thus may be superimposed in a pharmacophore model. Furthermore, it has been assumed that the relative spatial locations of the nitrogen atom and the 3-isoxazolol ring in THIP (**19**) define the bioactive conformation of muscimol (**18**). Based on these assumptions, a number of molecular modeling studies have been carried out in order to develop pharmacophore models for $GABA_A$ receptor agonists and competitive antagonists [107, 108].

However, it has not been possible to rationalize crucial experimental data by using models in which the 3-isoxazolol rings of muscimol (**18**), THIP (**19**), and 4-PIOL (**24**) are superimposed. For instance, it has been shown that replacement of the 3-isoxazolol ring in these molecules by an isothiazolol ring (e.g., compound **25**) has significantly different effects on the receptor affinity [72, 83, 109]. This indicates that the 3-isoxazolol rings in the three classes of compounds do not occupy identical positions in the receptor cavity. Furthermore, high-level quantum chemical ab initio calculations indicate that the assumed THIP-like bioactive conformation of muscimol has a very high conformational energy [110], which is not compatible with its high affinity. In addition, the introduction of a methyl group in the 4-position of the 3-isoxazolol ring of muscimol (**18**) severely inhibits interaction with the $GABA_A$ receptor recognition site [111]. This effect probably reflects steric repulsion between the methyl group and the receptor, which makes a direct superimposition of the 3-isoxazolol rings of muscimol (**18**) and THIP (**19**) highly questionable.

The equipotent activities of 4-PIOL (**24**) and compound **26** [83] clearly demonstrates that the distance between the nitrogen atom and the 3-isoxazolol ring in the bioactive conformation of 4-PIOL (**24**) is significantly larger than it is in THIP (**19**) and muscimol (**18**). A pharmacophore model that overcomes these problems has been developed [89, 112, 113] by noting that an arginine residue is a suitable binding partner to the carboxylate group of the endogenous ligand GABA, as well as to the 3-isoxazolol anions of muscimol (**18**), THIP (**19**), and 4-PIOL (**24**). High-level quantum chemical ab initio calculations show that a bidentate interaction between the arginine side chain and the ligand is compatible with observed SARs [114]. The involvement of an arginine residue in the binding of muscimol (**18**) has been demonstrated by site-directed mutagenesis studies [115–117]. The flexibility of the arginine side chain makes it feasible to accommodate muscimol (**18**) and 4-PIOL (**24**) in the same binding pocket, by assuming that the side chain adopts different conformations when binding to the two ligands, as illustrated in Figure 18.8.

Figure 18.9 displays the deduced binding modes of muscimol (**18**), THIP (**19**), and 4-PIOL (**24**). It should be noted that the 3-isoxazolol rings of the three compounds do not overlap, which is in agreement with experimental evidence (see above). The black tetrahedrons in Figure 18.9 denote sterically "forbidden" regions. The pharmacophore model indicates that the 4-position in 4-PIOL (**24**) (marked by an arrow) does not correspond to the "forbidden" 4-position in muscimol (**18**) (see above). Thus, by contrast with muscimol

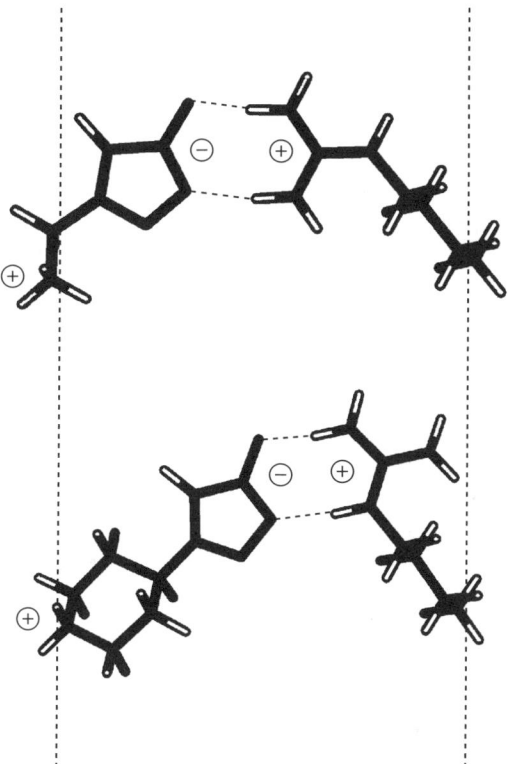

Figure 18.8 Proposed binding mode of muscimol (**18**) and 4-PIOL (**24**) to the GABA$_A$ receptor employing two different conformations of an arginine side chain. (This figure is available in full color at ftp://ftp.wiley.com/public/sci_tech_med/drug_discovery/.)

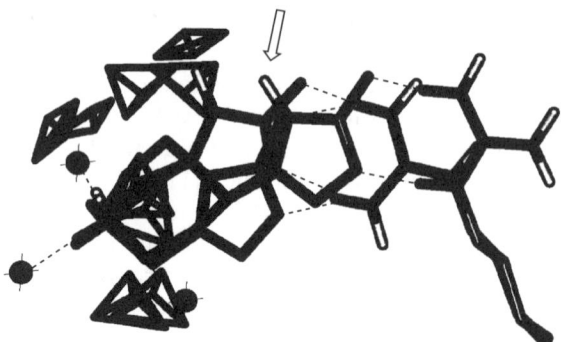

Figure 18.9 Pharmacophore model of GABA$_A$ receptor agonists showing the proposed binding modes of muscimol (**18**), 4-PIOL (**24**), and THIP (**19**). Dashed lines indicate hydrogen bond interactions. The cyan-colored tetrahedrons indicate positions of methyl groups in GABA$_A$ agonists that cause strong steric repulsion with the receptor. The arrow points at the 4-isoxazolyl position in 4-PIOL (**24**). The magenta-colored spheres indicate sites with which the ammonium group in muscimol (**18**) interacts via hydrogen bonds. Hydrogens on carbon have been omitted for clarity, except for hydrogens at the 4-isoxazolyl position of muscimol and 4-PIOL. (This figure is available in full color at ftp://ftp.wiley.com/public/sci_tech_med/drug_discovery/.)

derivatives, substitution at the 4-position of 4-PIOL (**24**) may be allowed according to the model. This prediction has resulted in the successful design of a number of 4-substituted 4-PIOL analogs with alkyl, phenylalkyl, diphenylalkyl, and naphthylalkyl substituents, demonstrating the presence of a receptor pocket of considerable dimensions in the vicinity of the 4-position of 4-PIOL (**24**) [89, 113].

Compound **27** displays the highest affinity of the 4-substituted 4-PIOLs with an increase in affinity by a factor of about 200 compared to 4-PIOL (**24**). Compound **27**, fitted to the pharmacophore model, is shown in Figure 18.10.

Ligands for the Benzodiazepine Site and Other Allosteric Sites of the GABA$_A$ Receptor Complex The GABA$_A$ receptor complex is the target of a large number of structurally diverse compounds, some of which are pharmacologically active and used clinically. Agents that increase chloride ion flux directly or indirectly through the GABA$_A$ receptor have sedative, anxiolytic, anaesthetic, hypnotic, and/or anticonvulsive action. Compounds that decrease chloride ion current may induce convulsions, anxiety, or increase cognitive functions. These modulatory effects are produced by compounds such as benzodiazepines, ethanol, general anesthetics, barbiturates, and neuroactive steroids via a wide range of distinct allosteric binding sites within the pentameric, proteinous receptor complex [5, 69, 118, 119].

The benzodiazepine binding site was defined in 1977 as the binding site for 1,4-benzodiazepines [e.g., diazepam (**30**) (Fig. 18.11)] in a rat membrane suspension. Since then a wide variety of structurally diverse nonbenzodiazepine

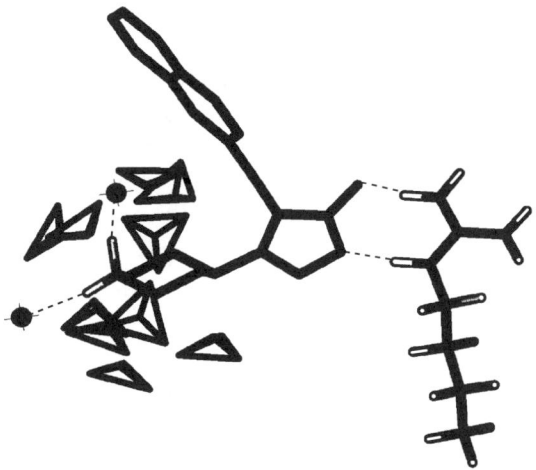

Figure 18.10 Compound **27** fitted to the GABA$_A$ pharmacophore model. Hydrogens on carbon have been omitted for clarity. (This figure is available in full color at ftp://ftp.wiley.com/public/sci_tech_med/drug_discovery/.)

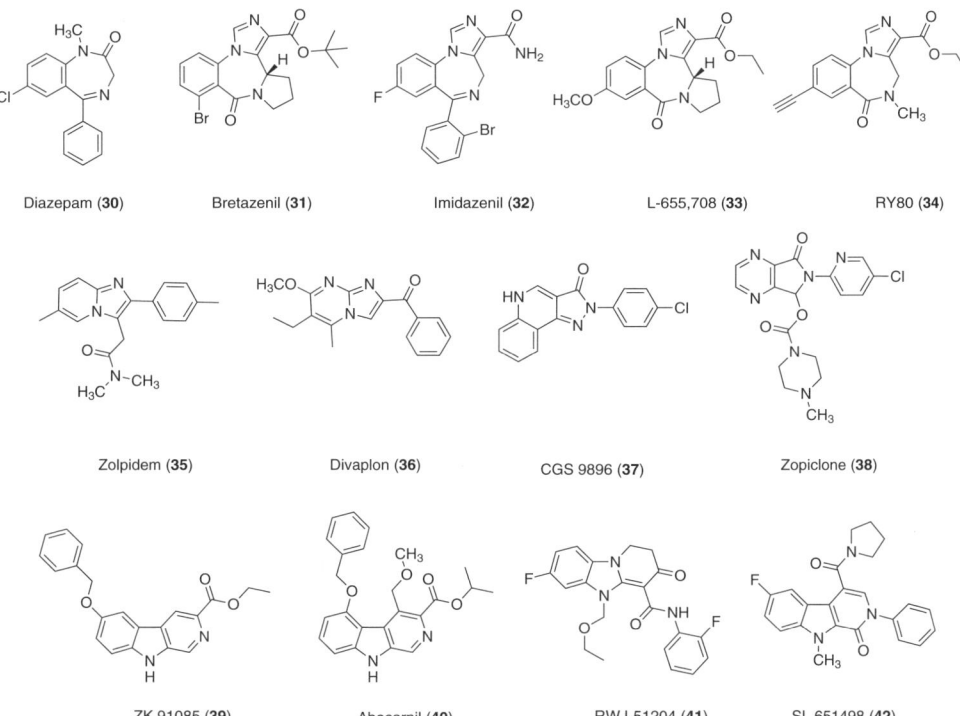

Figure 18.11 Structures of some ligands for the benzodiazepine binding site.

815

compounds, as illustrated in Figures 18.11 and 18.12, have been found to interact with this binding site. These classes of substances include: imidazopyridines [Zolpidem (**35**)], imidazopyrimidines [Divaplon (**36**)], pyrazoloquinolinones [CGS 9896 (**37**)], cyclopyrrolones [Zopiclone (**38**)], β-carbolines [Abecarnil (**40**)], pyrazolopyrimidine [Zaleplon (**44**)], and triazolopyridazines [L-838,417 (**45**)] [67, 120]. The benzodiazepine binding site lies within the GABA$_A$ receptor complex, through which the various ligands exert their effects.

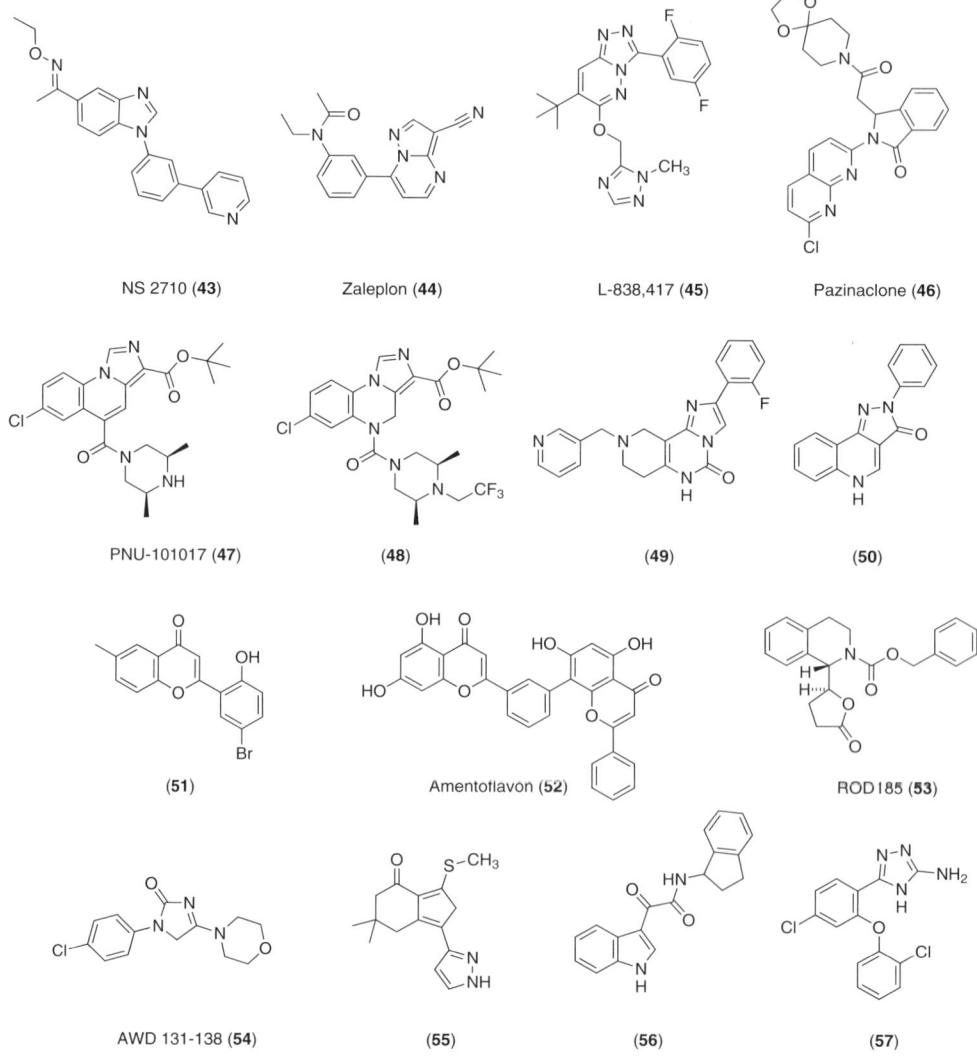

Figure 18.12 Structures of some ligands for the benzodiazepine binding site.

Knowledge of the multiplicity and complexity of the benzodiazepine binding sites has increased dramatically over time, from one site, to two sites, to an unknown number of different regionally distinct subtypes in the mammalian brain. The complexity has been further enhanced by the discovery of the agonist–antagonist–inverse agonist principle [121].

Studies of benzodiazepine sites at $GABA_A$ receptor complexes are performed using (a) radioactive labeled ligands in vitro or in vivo (see, e.g., [122–124]); (b) by fluorescence based methods, including recombinant $GABA_A$ receptor complexes, which have been made cation (Ca^{2+}) permeable via specific point mutations in subunits [92, 125, 126]; (c) by ^{36}Cl-flow measurements [127]; or (d) by electrophysiological techniques. Genetically modified mice (so-called knockin mice) have been used to differentiate the various pharmacological effects of agonists on the benzodiazepine binding site [99, 128]. The biological materials used for assaying benzodiazepine binding sites include: brain membrane preparations and brain slices; recombinant receptors [human or rat complementary deoxyribonucleic acid (cDNA) or messenger ribonucleic acid (mRNA)]; any combination of α_x ($x = 1, 2, 3, 5$), β_y ($y = 1, 2, 3$), and γ_2 expressed in *Xenopus* oocytes, HEK 293 cells, CHO cells, S_f9 cells, or by construction of stable $GABA_A$ receptor-expressing cell lines.

The clinical effects of nonselective benzodiazepines such as diazepam (**30**) include sedation, muscle relaxation, anxiolysis, and memory impairment. Furthermore, the drugs can induce tolerance and dependence. Using the above-mentioned tools for studying the $GABA_A$ receptor complex, the benzodiazepine pharmacology of different $GABA_A$ subtypes is emerging, and the observed clinical effects of, for instances, diazepam (**30**) are being partly dissected. The α subunit seems to be a determinant for the pharmacology displayed by benzodiazepines. The sedative effect is suggested to be mediated primarily by α_1-containing $GABA_A$ receptors, whereas α_2-containing receptors seem to be important for different anxiolytic effects [99, 128–130]. The very distinct distribution of α_5-containing $GABA_A$ receptors, being limited primarily to the hippocampus, which is known to be of importance in learning and memory, makes this group of receptors interesting as therapeutic targets. These observations have been used in the development of novel compounds with more specific actions.

In the following discussion selected ligands for the benzodiazepine site at the $GABA_A$ receptor complex are listed (see Figs. 18.11 and 18.12):

Bretazenil (**31**) and *Imidazenil* (**32**) are high-affinity partial agonists of the benzodiazepine binding sites (nonselective) [131]. *L-655,708* (**33**) is a partial inverse agonist. The substance shows higher affinity for α_5-containing receptors than for α_1-containing receptors [132]. *RY 80* (**34**) shows about 50 times higher affinity for α_5-containing receptors than for α_1-, α_4-, or α_6-containing receptors [133]. *ZK 91085* (**39**) is a β-carboline derivative with high affinity for the benzodiazepine binding sites. The substance stimulates $GABA_A$ receptor function both via the benzodiazepine site and the Loreclezole site (see below)

[134]. *RWJ-51204* (**41**) is described as a high-affinity partial agonist. The compound shows more anxiolytic activity and less sedation in mice and monkeys but not in rats [135]. *SL 651498* (**42**) shows stronger binding affinity for α_1- and α_2-containing receptors compared to α_5-containing receptors. The compound is a full agonist at α_2- and α_3-containing receptors, while it has a partial agonist profile at α_1- and α_5-containing receptors [136].

NS 2710 (ME 3127) (**43**) is described as possessing functional subtype selectivity and anxiolytic activity [137]. *Zaleplon* (**44**) is a nonbenzodiazepine hypnotic drug. It has a lower EC_{50} value in potentiating GABA effects at α_1-containing receptors compared to α_2-, α_3-, and α_5-containing receptors [138]. *L-838,417* (**45**) has anxiolytic action. The efficacy of the substance shows subtype selectivity: enhanced responses at α_2-, α_3-, and α_5-containing receptors, while no modulation of the GABA response is found at α_1-containing receptors [99]. The substance is devoid of the sedative and amnesic effects usually described for agonists of the benzodiazepine binding site. *Pazinaclone (DN 2327)* (**46**) is reported to be a partial agonist, although the substance shows the same adverse effects as full agonists [139]. *PNU-101017* (**47**) has high affinity for the benzodiazepine binding sites and shows partial agonist-like inhibition of induced convulsions in mice, as well as a weak anxiolytic effect. Besides the benzodiazepine binding site, the compound may act via a second low-affinity site at $GABA_A$ receptors producing a negative allosteric effect, whereas the trifluoroethyl analog **48**, produces a positive allosteric effect [140, 141]. *Derivatives of imidazopyridopyrimidine* (e.g., **49**) show high affinity and some selectivity toward $GABA_A$ receptor subtypes. One derivative is reported to be an α_2/α_3-selective partial agonist showing in vivo activity [142]. *Derivatives of pyrazoloquinolinones* (e.g., **50**) have been found in several studies to be very potent ligands for benzodiazepine sites [143]. Naturally occurring and synthetic *flavone derivatives* (**51**, **52**) are inhibitors of benzodiazepine binding in vitro [144, 145]. Amentoflavone (**52**), a biflavonoid, was identified as the first nonnitrogenous substance showing high affinity for benzodiazepine binding sites [146]. *ROD 185* (**53**) is reported to bind to benzodiazepine sites as well as to a novel allosteric modulatory site at the $GABA_A$ receptor complex [147]. *AWD 131–138* (**54**) is a low-affinity partial agonist. The compound has anticonvulsant and anxiolytic properties in rodents [148]. *Derivatives of benzothiophenone* (e.g., **55**) show higher affinity for α_5-containing receptors compared to α_1-, α_2-, and α_3-containing receptors. Some analogs have inverse agonist activity at α_5-containing receptors; enhancement of cognitive performance is seen in rat behavioral models [149]. *Derivatives of N-(indol-3-yl glyoxyl)amine* (e.g., **56**) bind with higher affinity to α_1-containing receptors than for α_2-, α_3-, and α_5-containing receptors [150, 151]. *Derivatives of phenyltriazole* (e.g., **57**) show anticonvulsant and sedative effects, although less potent than diazepam in animal studies [152].

Apart from the benzodiazepine binding site, a number of $GABA_A$ sites exist that recognize specific chemical patterns. In the following discussion, selected ligands are listed for some of these nonbenzodiazepine allosteric sites at the $GABA_A$ receptor complex (see Fig. 18.13):

Barbiturates [e.g., pentobarbital (**58**)], *neurosteroids*, and *neuroactive steroids* [e.g., α-THDOC (**59**) and alphaxolone (**60**)] are selective allosteric modulators of $GABA_A$ receptors [68, 153–155]. Although many actions of the barbiturates and the steroids are similar, they are believed to interact with distinct binding sites within the $GABA_A$ receptor complex. *Anesthetics* generally have low affinity for $GABA_A$ receptor complexes, but defined amino acid mutations in α and β subunits have been found to be critical for the action of general anesthetics [156]. As an example Enflurane (**61**) potentiates GABA currents in $\alpha_3\beta_3\gamma_2$ subunit containing receptors [157]. *Loreclezole* (**62**) is an anticonvulsive compound that potentiates $GABA_A$ receptor function, depending on the β subunit: β_2- or β_3-containing receptors have much higher affinity for Loreclezole (**62**) compared to β_1-containing receptors [158]. *α-EMTBL* (**63**) has no affinity for the benzodiazepine binding sites or the GABA binding site; it shows a profile similar to that of Loreclezole (**62**) [159]. *Etomidate* (**64**) is an anxiolytic compound that potentiates GABA-induced currents depending on the type of β subunit present in the receptor complex, as seen for Loreclezole (**62**) [160, 161]. *Etifoxin* (**65**) is an anxiolytic substance, which potentiates GABA-induced currents, depending on the type of β subunit present in the receptor complex as seen for Loreclezole (**62**) [162]. *Mefenamic acid* (**66**) has anticonvulsant and proconvulsant activity. The substance potentiates $GABA_A$ receptor function in vitro. It exhibits a profile comparable to Loreclezole (**62**) [163]. *MDL 26,479 (Suritozole)* (**67**) is reported to be active in animal models that are predictive of cognitive enhancement and antidepressant activity. The compound has been suggested to act at the $GABA_A$ receptor complex at a site different from the benzodiazepine binding site [164]. *Tracazolate* (**68**) possesses anxiolytic and anticonvulsant activity. The compound potentiates GABA-induced currents in recombinant receptors [165].

Propofol (**69**) activates $GABA_A$ receptors in the absence of GABA [166]. *SB-205384* (**70**) potentiates GABA-induced currents. The substance prolongs the half-life of the decay of current after GABA removal. Its effect is selective for α_3-containing receptors [167]. *Retigabine* (**71**) is an antiepileptic drug. The substance interacts allosterically with the $GABA_A$ receptor complex, possibly via a novel site coupled to the GABA binding site. The compound is also a potassium channel opener [168]. *Fluoxetine* (**72**) is reported to affect $GABA_A$ receptor function via a novel modulatory site. The substance increases the response to subthreshold concentrations of GABA but does not alter the maximum response. The α_5-containing receptors show less response compared to other α-containing receptors [169]. *Clozapine* (**73**) is an antipsychotic drug. The compound is reported to inhibit GABA-evoked chloride ion currents at $\alpha_1\beta_2\gamma_{2L}$ $GABA_A$ receptors expressed in *Xenopus* oocytes [170].

A wide range of different ligands are known to block the chloride channel [2]. These compounds could be defined as negative allosteric modulators of

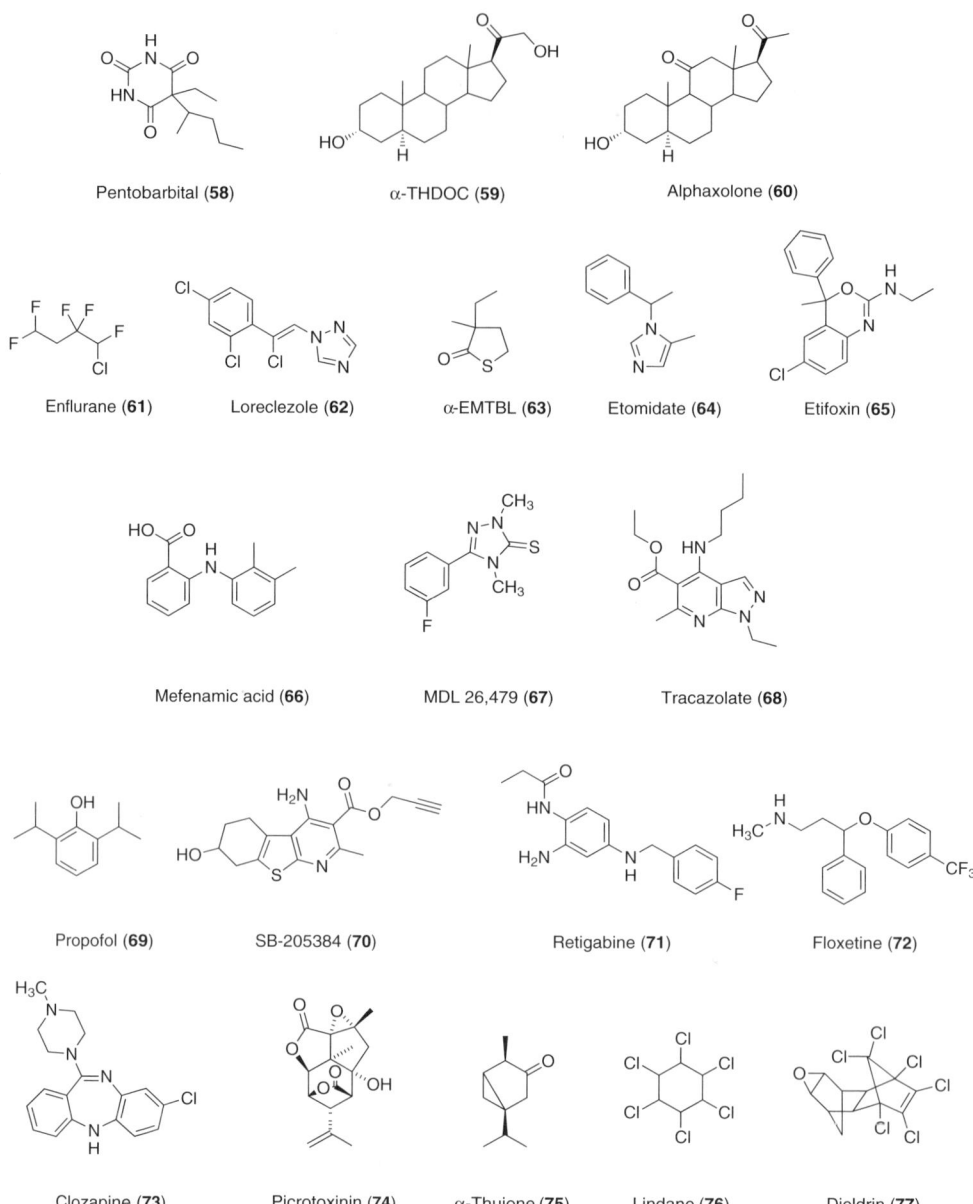

Figure 18.13 Structures of some ligands for nonbenzodiazepine allosteric binding sites at the GABA$_A$ receptor.

GABA$_A$ receptors (or noncompetitive GABA$_A$ receptor antagonists). *Picrotoxinin* (**74**) is a potent convulsant. α-*Thujone* (**75**), which is an ingredient of absinthe, has been found to interact with the GABA$_A$ receptor complex via the picrotoxin binding site [171]. *Lindane* (**76**) and *Dieldrin* (**77**) are insecticides known to inhibit GABA$_A$ receptor function [172].

Receptor and Pharmacophore Models of the Benzodiazepine Site of the GABA$_A$ Receptor Complex The X-ray crystallographic structure of a soluble homopentameric AChBP [101] described in the previous section has been employed via homology models in attempts to obtain structural information about the benzodiazepine binding site [102, 103]. The amino acid residues forming the benzodiazepine binding site have been predicted, and a mechanism for the mediation of positive cooperativity at the benzodiazepine site has been proposed [103]. As mentioned earlier the low level of sequence identity between AChBP and the corresponding parts of GABA$_A$ receptors necessitates extensive validation of models before any conclusions regarding their usefulness can be drawn.

A number of pharmacophore models of ligand binding at the benzodiazepine site of the GABA$_A$ receptor complex have been reported [144, 173–179]. The most comprehensive and well-validated pharmacophore model reported so far is that of Cook and co-workers [174, 175]. The model is based on 136 different ligands from 10 structurally different classes of compounds. In developing the model it was assumed that agonists, antagonists, and inverse agonists share the same binding pocket. The elements of the pharmacophore model are shown in Figure 18.14. H1 and A2 denote hydrogen bond donor and acceptor sites, respectively, whereas H2/A3 is a bifunctional hydrogen bond donor/acceptor site. L1, L2, L3, and L$_{Di}$ are lipophilic regions and S1, S2, and S3 are regions of steric repulsive ligand-receptor interactions (receptor-essential volumes). As shown in Figure 18.14, 1,4-benzodiazepines such as diazepam (**30**) are proposed to bind in a different binding mode than the more planar ligands such as the pyrazoloquinolone CGS 9896 (**37**). Diazepam and CGS 9896 both interact with H1, H2/A3, L1, and L2 sites, but the L3 site is only occupied by 1,4-benzodiazepines, and the L$_{Di}$ site only by compounds similar to CGS 9896.

This model has been widely used for the design of novel ligands [180, 181] and for SAR analyses [182–184]. The validity and usefulness of the model have been further demonstrated by an extensive study of flavone derivatives [144, 178]. It should be noted that flavones were not included in the original development of the model. As shown in Figure 18.15, these studies have resulted in an extension of the model built by Cook and co-workers [174, 175], by the identification of two new regions of steric ligand–receptor repulsions, S4 and S5, and the proposal of a large region (denoted "channel" in Fig. 18.15) in which a wide variety of substituents may be accommodated. This region is proposed to be partly lipophilic, with possibilities for hydrogen bonding. The extension of this region may be a channel in which larger substituents are only

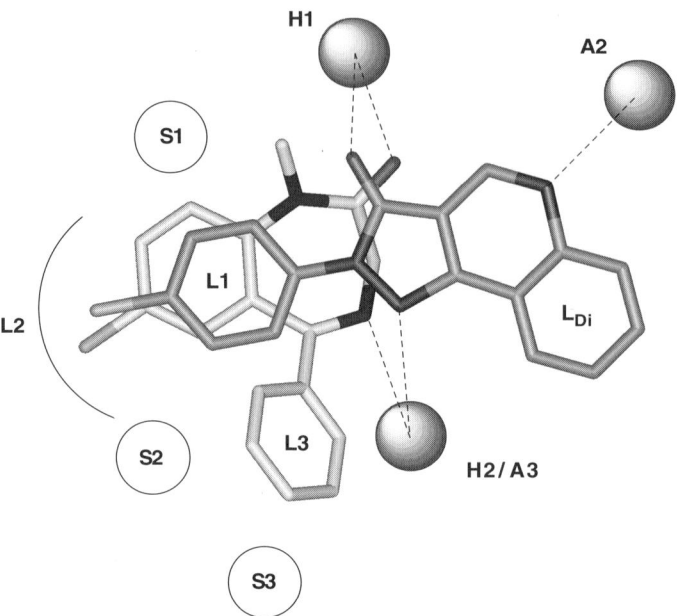

Figure 18.14 Pharmacophore model developed by Cook and co-workers displaying the proposed binding modes of diazepam (**30**) and CGS 9896 (**37**). Dashed lines indicate hydrogen bonds. Hydrogens on carbon are omitted for clarity. (This figure is available in full color at ftp://ftp.wiley.com/public/sci_tech_med/drug_discovery/.)

partly desolvated. Such a channel may exist at the interface between an α and a γ subunit in the $GABA_A$ receptor where the benzodiazepine binding site is most probably located.

By using the model, a number of high-affinity flavone derivatives have been designed. The highest affinity flavonoid (**51**) is shown fitted to the model in Figure 18.15 ($K_i = 0.9$ nM, [^{3}H]-Ro 15–1788 binding to rat cortical membrane preparations) [178]. Flavone itself displays an affinity of 4200 nM. Thus, the use of the pharmacophore model has resulted in an increase in the affinity by a factor of 4700 via the introduction of three properly positioned substituents.

The pharmacophore model displayed in Figure 18.15 has been converted into a search query and used for database searching with the Catalyst software suite (Accelrys, Inc.). A number of hits have been identified, and these may be used as interesting starting points for the design of novel ligands for benzodiazepine sites [185].

Current developments of pharmacophore models for benzodiazepine sites focus on models for specific receptor subtypes [133, 186, 187]. Most of this work has been performed in the context of the pharmacophore model developed by Cook and co-workers [174, 175], described above. As more data for subtype-selective compounds become available, such pharmacophore models will undoubtedly be further developed to a level where they can be effectively

GABA: INHIBITORY NEUROTRANSMITTER

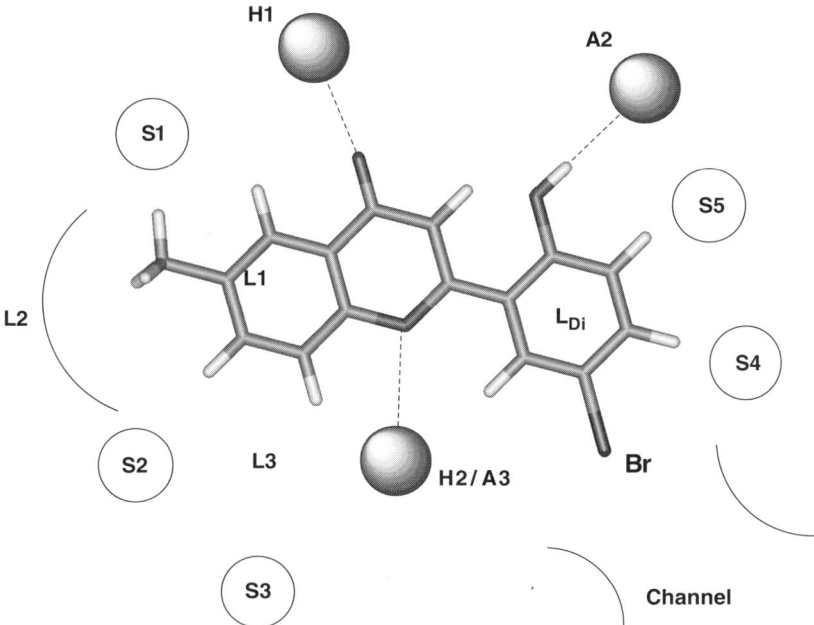

Figure 18.15 Proposed binding mode of 5′-bromo-2′-hydroxy-6-methylflavone (**51**) in the extended pharmacophore model. Dashed lines indicate hydrogen bonds. (This figure is available in full color at ftp://ftp.wiley.com/public/sci_tech_med/drug_discovery/.)

used for the design of novel subtype-selective ligands. It may also be within reach to employ pharmacophore models to elucidate the molecular properties involved in the actions of agonism and inverse agonism [174, 188, 189].

GABA$_B$ Receptor Ligands

The GABA$_B$ Receptor GABA$_B$ receptors were first identified in the early 1980s on the basis of pharmacological responses to the agonist baclofen (**78**) and insensitivity to the GABA$_A$ antagonist bicuculline (**22**) (Fig. 18.6) [190, 191], but resisted cloning until the late 1990s. GABA$_B$ receptors belong to the family of G-protein-coupled receptors (GPCR), which upon activation cause a decrease in calcium and an increase in potassium membrane conductance and inhibit cyclic adenosine 5′-monophosphate (cAMP) formation. The response is thus inhibitory and leads to, for example, hyperpolarization and decreased neurotransmitter release [192]. Specifically, the receptor belongs to the family C of GPCRs, with similarity to the mGluRs (see Metabotropic glutamate Receptor Ligands), which are characterized by a large extracellular amino-terminal domain (ATD) and a 7-transmembrane (7TM) domain (Fig. 18.16) [193].

Cloning revealed that the receptors consist of two subtypes, GABA$_{B(1)}$ and GABA$_{B(2)}$, which form a heterodimeric receptor through a coiled-coil inter-

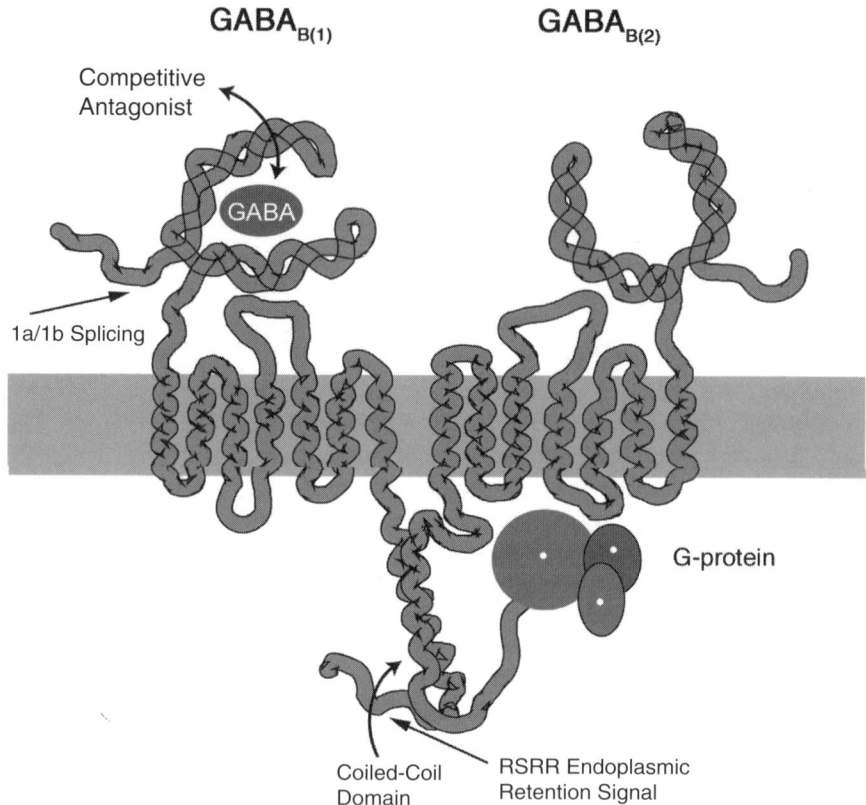

Figure 18.16 Schematic illustration of the $GABA_{B(1)}$ and $GABA_{B(2)}$ subunits forming the heterodimeric $GABA_B$ receptor. GABA (**1**) binds in the cleft of the amino-terminal domain of $GABA_{B(1)}$, which leads to G-protein activation through the $GABA_{B(2)}$ subunit. The localization of the coiled-coil domain, the RSRR endoplasmic retention signal and the $GABA_{B(1a)/(1b)}$ splice site has been noted. Adapted from [194]. (This figure is available in full color at ftp://ftp.wiley.com/public/sci_tech_med/drug_discovery/.)

action in the intracellular carboxy-terminus (Fig. 18.16) [195–198]. The roles of the individual subunits in the receptor complex are just beginning to emerge. $GABA_{B(1)}$ contains the orthosteric ligand-binding site [199, 200] and is retained in the endoplasmatic reticulum (ER) in the absence of $GABA_{B(2)}$ [195–198]. It has been shown that $GABA_{B(1)}$ features an RSRR motif adjacent to the α-helical coil, which is responsible for ER retention by an as yet unknown mechanism [201–203]. Formation of the coiled-coil interaction by heteromerization appears to mask the RSRR motif, allowing release of the $GABA_{B(1)}$ subunit from the ER, and surface trafficking of the heteromeric complex. However, $GABA_{B(2)}$ is not merely a chaperone that ensures proper trafficking of the receptor complex to the plasma membrane. Data from several studies indicates that this subunit is mainly responsible for coupling to the G-protein [204–208].

In spite of intense research, the GABA$_{B(1,2)}$ heterodimer remains the only functional GABA$_B$ receptor described so far. Since it has become virtually axiomatic that neurotransmitter receptor systems are composed of pharmacologically distinct subgroups, it was expected that GABA$_B$ receptors would form a family of related genes. Therefore, the discovery that GABA$_B$ receptors arise from the expression of only two genes comes as a surprise and significantly limits the therapeutic potential of GABA$_B$ selective ligands. This is despite the fact that pharmacological differences have been observed in vivo and in vitro, which suggested the existence of multiple receptor subtypes [192].

GABA$_B$ Receptor Agonists As already mentioned, the GABA$_B$ receptor is selectively activated by the GABA (**1**) analog baclofen (**78**), of which the *R*-form is the active enantiomer. Baclofen (**78**) was developed as a lipophilic derivative of GABA (**1**), in an attempt to enhance the BBB penetrability of the endogenous ligand. Since 1972, racemic baclofen (**78**) has been marketed for the treatment of spasticity due to its muscle-relaxant effects [209]. This action appears largely due to a reduction in neurotransmitter release onto motoneurons in the ventral horn of the spinal cord. GABA$_B$ receptor agonists also represent attractive drug targets in the pharmacotherapy of various neurological and psychiatric disorders, including neuropathic pain, anxiety, depression, absence epilepsy, and drug addiction. However, the (side) effects of agonists, principally sedation, tolerance, and muscle relaxation, limit their utility for the treatment of many of these diseases. Unfortunately, the lack of GABA$_B$ receptor subtypes makes the therapeutic prospects of agonists rather dim.

Although a number of GABA$_B$ receptor agonists have been synthesized, the number of selective and potent agonists for the GABA$_B$ receptor is limited, and (*R*)-baclofen still remains one of the more potent and selective agonists for the GABA$_B$ receptor. So far, the most successful strategy toward increasing potency has been bioisosteric replacement of the carboxyl group of GABA (**1**) with a phosphinic acid moiety. Thus, the phosphinic acid analog of GABA (**1**), CGP27492 (**79**), is the most potent GABA$_B$ agonist reported to date, being approximately 10-fold more potent than GABA (**1**) [210].

Competitive GABA$_B$ Receptor Antagonists The first GABA$_B$ receptor antagonists to be reported were phaclofen (**80**) and saclofen (**81**) (Fig. 18.17), the respective phosphonic acid and sulfonic acid bioisosteric analogs of baclofen (**78**) [211, 212]. 2-OH-Saclofen (**82**) is the most potent antagonist obtained using these scaffolds [213], however, and with potencies in the mid-micromolar range there was room for improvement. Furthermore, these compounds were unable to penetrate the BBB, which made in vivo studies difficult.

In an attempt to improve the pharmacology of the GABA$_B$ agonist CGP27492 (**79**), a new series of selective and potent GABA$_B$ antagonists,

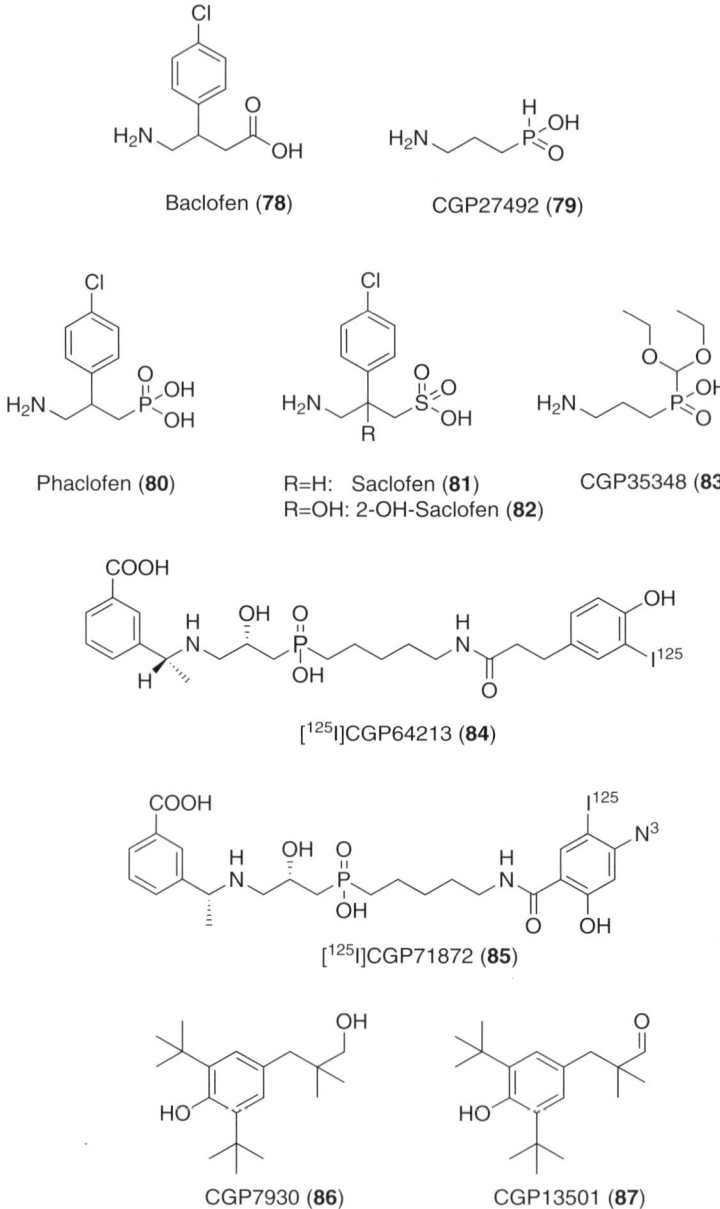

Figure 18.17 Chemical structures of ligands used to characterize and clone GABA$_B$ receptors. The recently identified positive allosteric modulators CGP7930 (**86**) and CGP13501 (**87**) are expected to broaden the spectrum of therapeutic applications for GABA$_B$ drugs.

capable of penetrating the BBB after systemic administration, were discovered [214]. It was found that substituents larger than methyl conferred antagonism, compounds such as CGP35348 (**83**) displaying low-micromolar potency. However, it was not possible to break the nanomolar barrier with this series of compounds, which was finally achieved by substituting the amino group of the γ-aminopropylphosphinic acids with branched benzyl substituents. In this series, many sub- to low-nanomolar compounds have been reported, including the radioligands [^{125}I]CGP64213 (**84**) and [^{125}I]CGP71872 (**85**), which were crucial tools for the functional cloning of the receptor [215].

The development of antagonists capable of penetrating the BBB has allowed in vivo analysis of the therapeutic potentials of such ligands. Accordingly, it has been suggested, based on animal studies, that GABA$_B$ antagonists could improve cognitive processes and be beneficial against epilepsy and depression. However, further studies are needed to validate these effects, and again severe side effects seem difficult to avoid due to the absence of receptor subtypes.

Positive Allosteric Modulators of GABA$_B$ Receptors Ca^{2+} allosterically modulates the potency of GABA (**1**) but not baclofen (**78**), causing a potentiation of GABA (**1**) responses [216–218]. The EC$_{50}$ for Ca^{2+} potentiation is 37 μM, and it is thus likely that the Ca^{2+} site on the GABA$_B$ receptors is saturated under normal physiological conditions [217]. However, it is possible that Ca^{2+} modulation plays a role in pathophysiological situations characterized by a significant drop in extracellular Ca^{2+}. A series of substituted phenols with positive allosteric modulator effect has been described [219]. These compounds, CGP7930 (**86**) and CGP13501 (**87**) (Fig. 18.17), are small organic molecules and thus attractive from a drug development perspective. Like Ca^{2+}, these compounds do not activate the receptor directly but increase the potency and affinity of classical GABA$_B$ agonists including GABA (**1**) [219]. The compounds do not displace binding of radiolabeled orthosteric ligands and therefore do not bind to the GABA binding site. So far the allosteric binding site has not been localized and characterized, but certain positive allosteric modulators at mGluR1 and calcium-sensing receptors have been shown to bind within the 7TM domain [220, 221]. This suggests that CGP7930 (**86**) and CGP13501 (**87**) also bind to the 7TM domain of GABA$_{B(1)}$ and/or GABA$_{B(2)}$, but other binding sites have not been ruled out [222]. Due to their dependence on GABA (**1**) for activity, the novel allosteric modulators are expected to show an improved side effect profile compared to GABA$_B$ agonists, which act indiscriminately on GABA$_B$ receptors. Unlike agonists, these drugs should not induce tolerance because they do not cause prolonged receptor activation leading to desensitization and internalization. If allosteric modulators exhibit the therapeutic potential of baclofen (**78**), while being devoid of major side effects and muscle relaxant properties, they will considerably broaden the spectrum of therapeutic applications of GABA$_B$ ligands.

GABA$_C$ Receptor Agonists and Antagonists

Shortly after the identification of the GABA$_B$ receptor, a third class of GABA receptors, the GABA$_C$ receptors, was discovered [223]. These receptors are not blocked by bicuculline (**22**) nor modulated by benzodiazepines or barbiturates, which typically affect GABA$_A$ receptors [223, 224]. By contrast with GABA$_B$ receptors, GABA$_C$ receptors are not activated by baclofen (**78**) or inhibited by phaclofen (**80**) [223].

GABA$_C$ receptors are the most incompletely characterized of the GABA receptors, but in the past decade major progress toward understanding the molecular nature of GABA$_C$ receptors has been made [225, 226]. To date three different subtypes of GABA$_C$ receptors have been identified in humans (ρ_{1-3}) [227–229], where the ρ_1 subunit is predominantly located in the retina, the ρ_2 subunit is found in most brain regions [230], and the ρ_3 subunit is found in the hippocampus [231]. The membrane topology is assumed to be very similar to that of GABA$_A$ receptors, but GABA$_C$ receptors are believed to consist of only ρ subunits (ρ_{1-3}). The three isoforms of the ρ subunits can assemble into homomeric or pseudoheteromeric chloride channels showing a pharmacology different from the GABA$_A$ receptors [68, 232].

CACA (**88**) has been the key compound in identifying the GABA$_C$ receptors [223, 233] (Fig. 18.18). This compound is a moderately potent partial GABA$_C$ agonist and is inactive at GABA$_A$ receptors. The *trans*-isomer TACA (**89**) shows no preference for neither of the two receptor classes [233]. CAMP (**90**) is a selective GABA$_C$ agonist that is inactive at GABA$_A$ receptors [234]. TPMPA (**91**) was the first selective antagonist for GABA$_C$ receptors to be developed and is at least 100 times more potent as an antagonist at GABA$_C$ receptors than at GABA$_A$ receptors [235, 236]. Recently, the isosterically

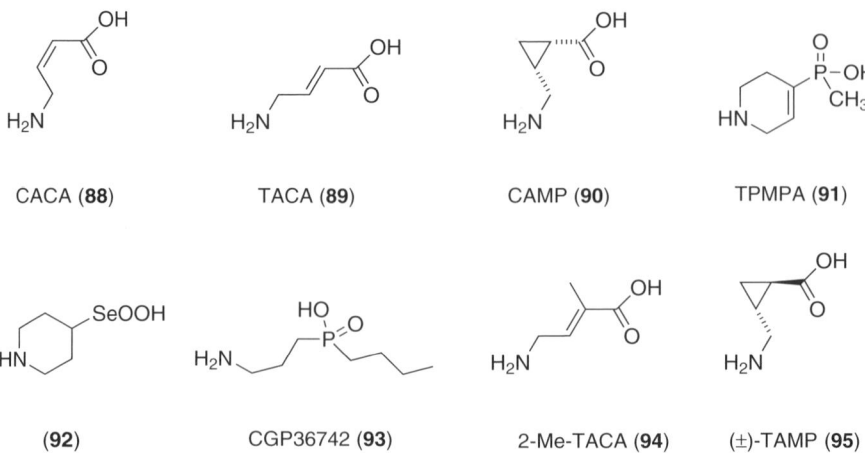

Figure 18.18 Structures of some ligands for the GABA$_C$ receptor.

derived seleninic acid analog of isonipecotic acid, compound **92**, has been reported to be more potent than TPMPA (**91**) at $GABA_C$ ρ_1 receptors and to display a higher selectivity than TPMPA (**86**) for $GABA_C$ receptors relative to $GABA_A$ receptors [237].

Interestingly, the $GABA_C$ receptors share several ligands with the $GABA_A$ and $GABA_B$ receptors. Most $GABA_A$ agonists seem to have some agonist/antagonist action at $GABA_C$ receptors, exemplified by THIP (**19**) and P4S (**28**), which are partial agonists at $GABA_A$ receptors and competitive antagonists at $GABA_C$ receptors [238]. Muscimol (**2**), isoguvacine (**21**), and IAA (**29**) act as partial $GABA_C$ receptor agonists [238, 239]. Several phosphinic acid analogs of GABA, such as CGP36742 (**93**), have been identified as $GABA_C$ antagonists [240]. This group of compounds was originally developed as $GABA_B$ receptor antagonists [210, 214].

Certain ligands have been shown to act selectively or preferentially at specific ρ subunits. As an example, TPMPA (**91**) is eight times more potent at ρ_1 than at ρ_2 $GABA_C$ receptors expressed in *Xenopus* oocytes [241] and half as potent at recombinant ρ_3 $GABA_C$ receptors than at ρ_1 [242]. 2-Methyl-TACA (**94**) is a competitive antagonist at human ρ_1 $GABA_C$ receptors, a partial agonist at human ρ_2 $GABA_C$ receptors, but without effect at rat ρ_3 $GABA_C$ receptors [241], whereas ($\pm$)-TAMP (**95**) is a potent antagonist at rat ρ_3 $GABA_C$ receptors and a partial agonist at ρ_1 and ρ_2 $GABA_C$ receptors [242–244].

Based on studies in the retina, where $GABA_C$ receptors predominate, it has been established that this class of receptors plays a unique functional role in retinal signal processing [245, 246]. To clarify the functional relevance of the different $GABA_C$ receptor subtypes, the emerging pharmacological differences between recombinant ρ_1, ρ_2, and ρ_3 are of significant importance.

18.3 GLUTAMATE: EXCITATORY NEUROTRANSMITTER

Classification of Glutamic Acid Receptors

Glu operates, as described in Section 18.1, via two main receptor classes, iGluRs and mGluRs, each consisting of three groups of receptor subtypes [1, 2]. The three heterogeneous groups of iGluRs are named after selective agonists, N-methyl-D-aspartic acid (NMDA), 2-amino-3-(3-hydroxy-5-methyl-4-isoxazolyl)propionic acid (AMPA), and kainic acid (KA) receptors. Native iGluRs assemble as homo- or heterotetramers, from subunits within each receptor type: NMDA receptors from NR1, NR2A-D, and NR3A,B; AMPA receptors from GluR1-4; and KA receptors from GluR5-7 and KA1,2 (Fig. 18.19) [247, 248]. At many central synapses, Glu activates a mixed population of NMDA and AMPA/KA receptors. Fast excitatory transmission is mainly mediated by AMPA/KA receptors, since the NMDA receptor ion channel is blocked by Mg^{2+} at normal resting potential (see next section).

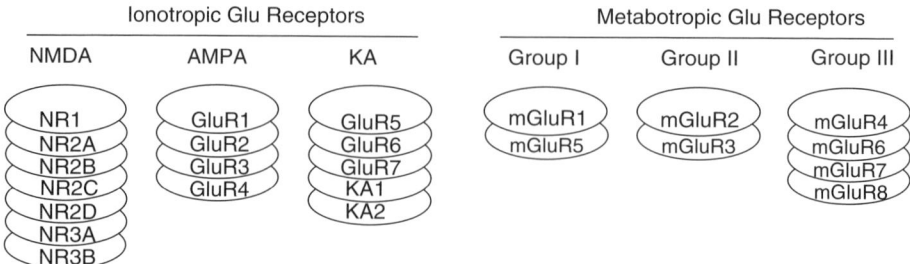

Figure 18.19 Schematic illustration of Glu receptor multiplicity. Each circle represents a receptor subunit or a receptor subtype for the ionotropic and metabotropic receptors.

The mGluRs lead to slower and longer lasting neuronal effects and modulate the activity of iGluRs. mGluRs are divided into groups I, II, and III based on pharmacology, signal transduction mechanisms, and sequence homology [2, 248]. Group I mGluRs stimulate phospholipase C and include the cloned subtypes mGluR1,5. Groups II and III mGluRs both inhibit cAMP formation and consist of the subtypes mGluR2,3 and mGluR4,6-8, respectively (Fig. 18.19). mGluRs assemble as dimeric structures and probably as homomers [249]. It is now generally agreed that both iGluRs and mGluRs play important roles in the healthy as well as the diseased CNS and that all subtypes of these receptors are potential targets for therapeutic intervention in a number of diseases (see Section 18.1).

NMDA Receptor Ligands

The NMDA receptors include a number of different binding sites and thus several potential targets for therapeutic attack. NMDA receptors are cation channels that flux Na^+, K^+, and Ca^{2+}, of which the latter is implicated in the neurotoxicity observed after excessive receptor stimulation. NMDA receptors are unique among ligand-gated ion channels in that they require two different agonists for activation, Glu (**96**) (Fig. 18.20) and glycine (**127**) (Fig. 18.23), and at the same time membrane depolarization in order to relieve a blockade by Mg^{2+} [247, 248]. Polyamines such as spermine and spermidine modulate the activity of NMDA receptors in a biphasic manner [2, 249]. At micromolar concentrations polyamines enhance the activity of the ion channel, whereas at higher, probably nonphysiological, concentrations the ion channels are blocked in a voltage-dependent manner. Furthermore, zinc, redox modulators, and NO have been shown to inhibit the NMDA receptors via allosteric sites [250].

In the normal brain, NMDA receptors are fundamental to development and function because of their involvement in synaptic plasticity and neuronal signaling processes, including mechanisms of learning and memory. Furthermore,

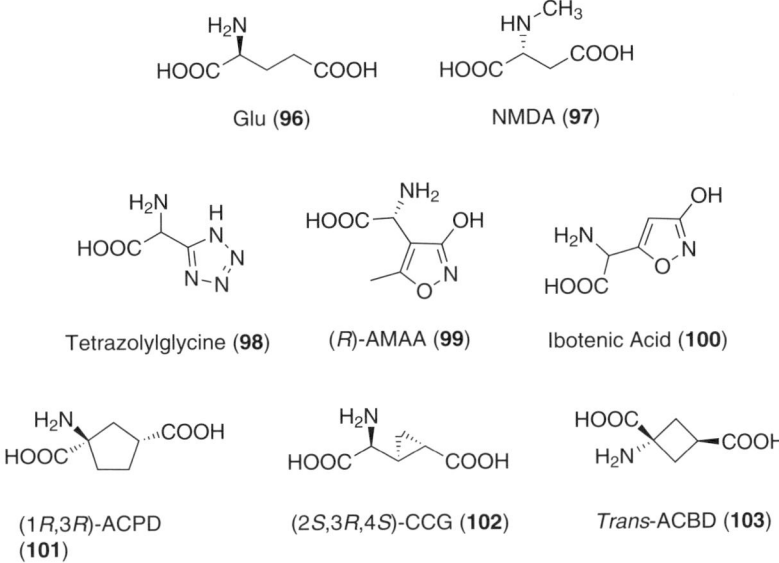

Figure 18.20 Structures of Glu (**96**) and some NMDA receptor agonists.

NMDA receptor-induced neurotoxicity is intimately involved in a number of neuronal disorders as previously described (see Section 18.1).

Functional NMDA receptors are heteromers, typically consisting of NR1 and NR2 subunits [251, 252] and probably possessing a tetrameric structure [247], although this has not been thoroughly investigated as in the case of AMPA receptors (see below). The Glu binding site is located on NR2 subunits and the glycine binding site on NR1 subunits [251]. A combination of NR1 and NR3 subunits expressed in *Xenopus* oocytes form receptors that are activated by glycine alone [251], although the existence or physiological significance of such receptors in the CNS is not known.

Recently, it has been shown that NMDA receptors have a direct protein–protein interaction with the G-protein-coupled dopamine receptor subtype D1 [253]. The study of NMDA receptor signaling mechanisms and the search for therapeutic agents are still areas of active research. In the following sections ligands for the major sites located at the NMDA receptor complex will be described.

NMDA Receptor Agonists NMDA receptors are unusual among Glu recognition sites in as much as the majority of the more potent ligands, both agonists and antagonists, posses an *R*-configuration about the α-amino acid center. NMDA (**97**) itself represents the only known agonist in which *N*-methylation does not lead to reduced affinity [254]. Other potent NMDA agonists have been developed, particularly by replacement of the distal acidic group and/or by conformational restriction of the three essential functional groups, namely the α-amino group, the α-carboxyl group, and the ω-acidic moiety. Among the potent NMDA agonists with a different distal acidic group are tetrazolylglycine [255] (**98**) and (*R*)-AMAA [256] (**99**), exemplifying two widely used carboxyl group bioisosteric groups, the tetrazole and the 3-isoxazolol, respectively. (*R*)-AMAA (**99**) and other Glu ligands have been developed using the naturally occurring neurotoxin ibotenic acid (**100**) as a lead [257]. Ibotenic acid (**100**) is, apart from being a potent NMDA agonist, a potent agonist of some mGluR subtypes and a somewhat weaker agonist at other Glu receptor types.

Conformational restriction of the Glu backbone leading to Glu analogs with a folded conformation has led to a number of potent and selective NMDA agonists [258, 259]. (1*R*,3*R*)-ACPD (**101**) [260], (2*S*,3*R*,4*S*)-CCG (**102**) [261], and *trans*-ACBD (**103**) [262] are examples of such carbocyclic acidic amino acids, exemplifying that the higher activity does not always reside with the *R* form.

Competitive NMDA Receptor Antagonists A large number of potent and selective competitive NMDA antagonists (Fig. 18.21) have been developed over the past two decades, and the availability of these compounds has greatly facilitated studies of the physiological and pathophysiological roles of NMDA receptors [2]. The distance between the two acidic groups in NMDA antagonists is typically one or three C–C bonds longer than in Glu. Many potent ligands have successfully been developed using ω-phosphonic acid analogs such as (*R*)-APV (**104**) as lead structures [263]. Combination of an ω-

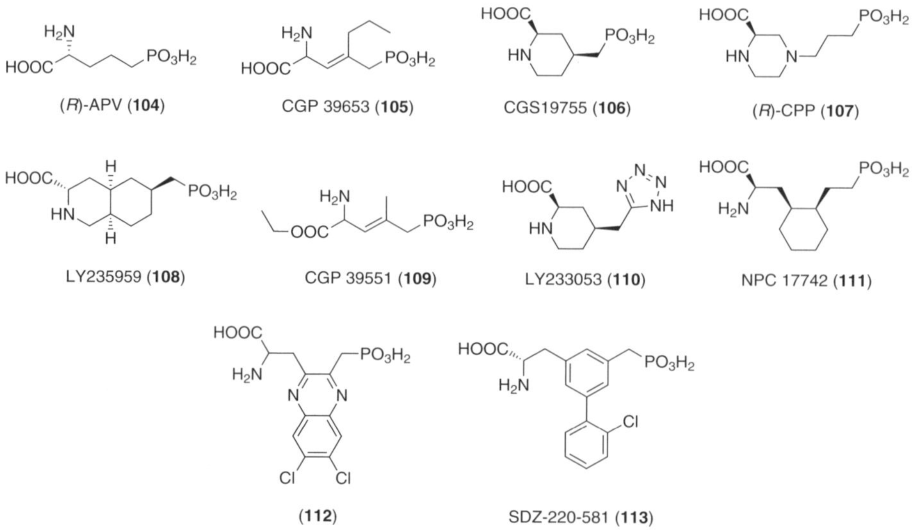

Figure 18.21 Structures of some competitive NMDA receptor antagonists.

phosphonate group, a long carbon backbone, and conformational restriction has led to different series of potent antagonists.

Conformational restriction has been achieved by use of double bonds (CGP 39653 (**105**) [264]) and ring systems (CGS19755 (**106**) [265] and (*R*)-CPP (**107**) [266]), or bicyclic structures (LY235959 (**108**) [267]). These antagonists have shown very effective neuroprotective properties in various in vitro models. However, many of these compounds suffer from poor BBB penetration. The esterified analog CGP 39551 (**109**) [264], which is a prodrug, shows good oral bioavailability and anticonvulsant activity. LY233053 (**110**) [268] represents another class of antagonist with a tetrazole ring as the terminal acidic group. Substitution of the tetrazole for a phosphono group has limited effect on in vitro activity and shows improved bioavailability.

The carbocyclic analog NPC 17742 (**111**) [269] is active after systemic administration. The quinoxalin-alanine analog **112** [270] shows high antagonist potency with concomitant affinity for the glycine site. Related to the latter two analogs is a series of phenylalanine-based antagonists, represented by SDZ-220-581 (**113**) [271], and the ability of **113** to penetrate the BBB may be explained by increased lipophilicity, though active transport has been suggested [272, 273].

Uncompetitive and Noncompetitive NMDA Receptor Antagonists The dissociative anesthetics PCP (**114**) and ketamine (**115**) block the NMDA receptor channel in a use-dependent manner [274] (Fig. 18.22). Thus, initial agonist activation of the channel is a prerequisite in order for such uncompetive antagonists to gain access to the binding site, which is situated within the ion channel. The antagonists eventually become trapped within the ion channel, and this may result in very slow kinetics. MK-801 [275] (**116**) has been developed as a very effective uncompetitive NMDA antagonist and has been extensively investigated to probe the therapeutic utility of such compounds, notably for the treatment of ischemic insults such as stroke [276]. MK-801 (**116**) and related high-affinity ligands have, however, shown severe side effects, including psychotomimetic effects, neuronal vacuolization, and impairment of learning and memory [277, 278]. Ligands with lower affinity, such as memantine (**117**) [279], dextromethorphan (**118**) [280], and remacemide (**119**) [281] have shown improved therapeutic indexes. Memantine (**117**) is being used for the treatment of AD and Parkinson's disease [279, 282] and may also have potential in the treatment of AIDS dementia [283]. The fast kinetics and low affinity of memantine (**117**) compared to MK-801 (**116**) may explain the absence of severe side effects. Remacemide (**119**) also shows a favorable side effect profile as compared to other uncompetitive antagonists and may have potential in the treatment of Huntington's chorea and Parkinson's disease [281, 284].

Two substituted aminoamides with low-affinity uncompetitive NMDA antagonist activities have been described, namely the ethynyl-substituted analog **120** [285] of the clinical antidepressant agent, milnacipran, and

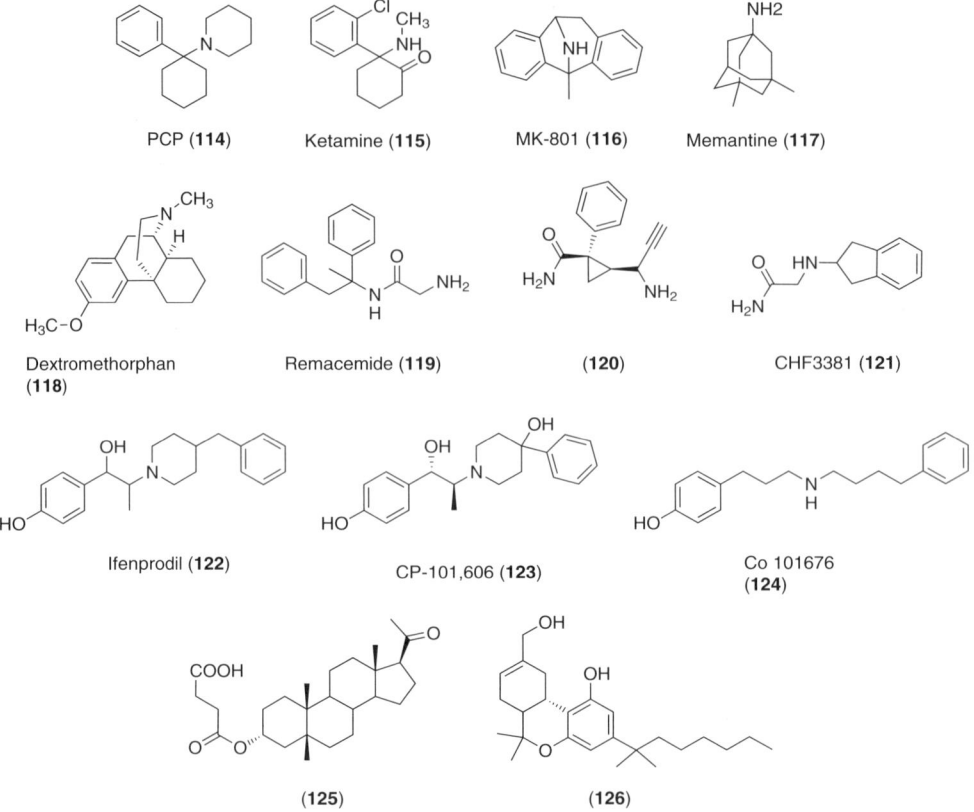

Figure 18.22 Structures of some uncompetitive (**114–121**) and noncompetitive (**122–126**) NMDA receptor antagonists.

CHF3381 [286] (**121**). Compound **120** also is a potent serotonin uptake inhibitor, and like the parent compound milnacipran, which shows no serious side effects and good BBB penetration, these analogs may represent a new class of antidepressants [285]. CHF3381 (**121**) has been studied in animal models and shows potential as a neuroprotective agent with antiepileptic and antinociceptive activity [286–288].

Several compounds with noncompetitive activity at NMDA receptors have been described, and these compounds most likely do not bind to the same site as the uncompetitive ligands. Ifenprodil (**122**) and CP-101,606 (**123**) represent an important series of noncompetitive NMDA receptor antagonists. These compounds are active in ischemia models and as anticonvulsants and antinociceptive agents [289, 290]. Early clinical trials with these analogs have been disappointing due to unwanted effects. The side effect profiles do, however, seem to be significantly improved as compared to, for example, competitive NMDA antagonists [291]. Co 101676 (**124**) represents a series of structurally simplified ifenprodil analogs [292]. The sterol hemisuccinate (**125**) has been shown to be a noncompetitive NMDA antagonist and to be protective against NMDA-induced seizures in mice [293]. The cannabinoid dexanabinol [294] (**126**) shows noncompetitive NMDA antagonism and significant neuroprotective effects.

Glycine Co-agonist Site The excitatory co-agonist site for glycine (**127**) (Fig. 18.23) at the NMDA receptor is named the glycine$_B$ receptor. This receptor site is different from the inhibitory glycine receptors found primarily in the spinal cord of the mammalian CNS, where glycine activates strychnine-sensitive ionotropic receptors named glycine$_A$ receptors [295]. Glycine$_B$

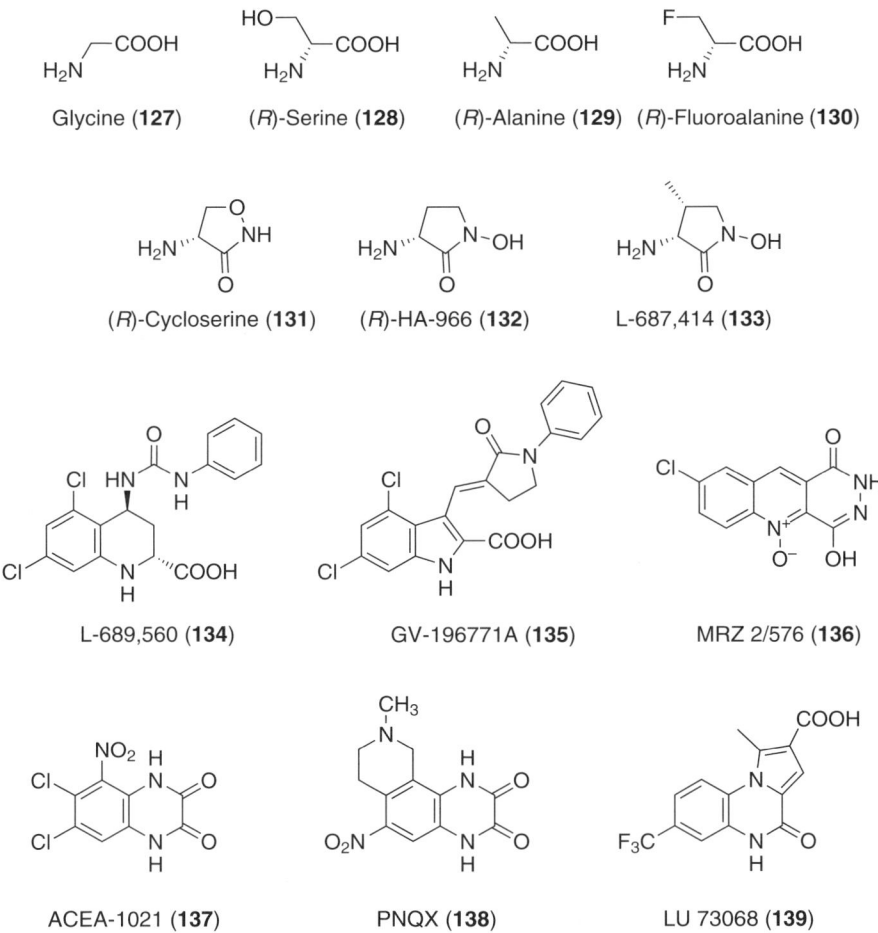

Figure 18.23 Structures of glycine (**127**) and some compounds showing agonist (**128–130**), partial agonist (**131–133**), or antagonist (**134–139**) activity at the glycine$_B$ receptor site at the NMDA receptor complex.

receptors seem to modulate the level of activity at NMDA receptors. A certain concentration level of glycine is always present in the synapse. Thus, Glu is activating the NMDA receptors, whereas the level of glycine can modulate this activity and possibly control receptor desensitization [296]. (R)-Serine (**128**) is a potential endogeneous agonist at glycine$_B$ receptors [297, 298], and other small α-amino acids such as (R)-alanine (**129**) and (S)-fluoroalanine (**130**) enantioselectively bind to and activate glycine$_B$ receptors [299].

Limited success of competitive NMDA receptor antagonists as therapeutic agents has focused attention on the glycine$_B$ site. (R)-Cycloserine (**131**) and (R)-HA-966 (**132**) are partial glycine agonists, capable of penetrating the BBB after systemic administration. These pyrrolidine analogs have become useful tools for studying the in vivo role of the glycine$_B$ site [300, 301]. (R)-Cycloserine (**131**) has shown promising effects in the treatment of schizophrenia and AD [251, 302]. The methylated analog L-687,414 (**133**) is more potent as an antagonist than the parent compound, (R)-HA-966 (**132**) [303]. L-687,414 (**133**) has neuroprotective properties in animal models, and it does not prevent hippocampal LTP, which is often seen after administration of full antagonists. This indicates that partial agonists may have therapeutic advantages as compared to full antagonists in terms of fewer side effects [304]. A number of glycine$_B$ antagonists have also been developed. L-689,560 (**134**) [305] displays high potency and is derived from the endogenous compound kynurenic acid, the first glycine$_B$ antagonist reported. Further development of such structures has led to GV196771A [306] (**135**). The potent pyridophtalazindione MRZ 2/576 (**136**) represents another class of glycine$_B$ receptor antagonist that exhibits neuroprotective and antinociceptive properties in animal models [307, 308]. Quinoxalinediones such as ACEA-1021 [309] (**137**) and PNQX [310] (**138**) show selective antagonist activity at the glycine$_B$ site (see below). Other quinoxalinediones as LU 73068 (**139**) have shown mixed glycine$_B$ site and AMPA/KA receptor affinities and produce potent anticonvulsive effects [311].

Subtype-Selective NMDA Ligands and Their Therapeutic Potential A number of competitive, uncompetitive, noncompetitive NMDA antagonists and glycine$_B$ site antagonists have shown promise in various animal models. However, when these compounds have been administered to humans, severe side effects have limited or prevented their therapeutic use. A better understanding of the heterogeneity of NMDA receptors may, however, provide new therapeutic possibilities. Subtype-selective agents could potentially lack such side effects.

Noncompetitive antagonists such as ifenprodil (**122**), CP-101,606 (**123**), and Co 101676 (**124**) have received much attention because of their subtype selectivity and their mode of action, specifically their activity-dependent blockade of NR2B-containing NMDA receptors [291, 312]. Ifenprodil (**122**) inhibits NR1/NR2B receptors at least 150-fold more than NR1/NR2C or NR1/NR2D receptors [313].

NMDA receptor blocking agents, such as (R)-CPP (**107**), ketamine (**115**), and dextromethorphan (**118**) have shown antinociceptive activity in chronic neurogenic pain models as well as in humans [280, 314, 315]. Unfortunately, the effects of the compounds are often seen at doses causing the typical side effects described earlier for NMDA receptor antagonists. The future development of analgesic agents may depend on the identification of subtype-selective agents, and much interest is focused on NR2B selective NMDA antagonists [291, 316]. NMDA blocking agents also show anxiolytic properties. However, it should be emphasized that even modest cognitive side effects are unlikely to be acceptable for this clinical application.

The affinity of glycine$_B$ receptor ligands is dependent on the presence of the NR2 subtype, although the glycine binding site is located on the NR1 subunit [317]. GV196771A (**135**) shows some selectivity for heteromers containing NR2A/2B subunits in combination with NR1, and has shown activity in animal models of neuropathic pain [306].

Certain uncompetitive antagonists such as MK-801 (**116**) (slower kinetics), dextromethorphan (**118**), and memantine (**117**) have shown some NR2C subtype selectivity, although there is some controversy about the selectivity of memantine (**117**) as an NMDA receptor ligand [279, 318].

AMPA Receptor Ligands

AMPA Receptor Agonists A large number of selective and potent AMPA receptor agonists have been developed by substitution of a heterocyclic bioisosteric group for the distal carboxylate group of Glu. For example, the heterocycles 1,2,4-oxadiazole-3,5-dione, 3-isoxazolol, and uracil, as represented by quisqualic acid (**140**), AMPA (**144**), and (S)-willardiine (**141**), respectively, have been incorporated into numerous AMPA receptor agonists (Fig. 18.24). The natural product quisqualic acid (**140**) was the first agonist in use for pharmacological characterization of AMPA receptors, but due to nonselective action it was later replaced by AMPA (**144**). Analogs of AMPA containing aliphatic, aromatic, or heteroaromatic substituents at the 5-position of the isoxazole ring have been synthesized and pharmacologically characterized [319–321]. Replacement of the 5-methyl group of AMPA by small alkyl groups, such as ethyl or trifluoromethyl, resulted in compounds with potencies in the same range as that of AMPA and that show similar intrinsic activity [322, 323]. Substitution of the methyl group of AMPA for hydrogen resulted in demethyl-AMPA (**145**), which has 300 times lower agonist potency in the rat cortical wedge assay system, but only 10 times lower affinity [321]. This apparent loss of agonist activity of demethyl-AMPA in the rat cortical wedge may be explained by an observed affinity for the cloned Glu transporters EAAT1 and EAAT2, though not for EAAT3 [324]. Thus, demethyl-AMPA may be a substrate for the former two transporters and may be removed from the synaptic cleft. Increasing the size of the 5-alkyl substituent of AMPA to three or more carbon atoms markedly reduces agonist potency [321, 325].

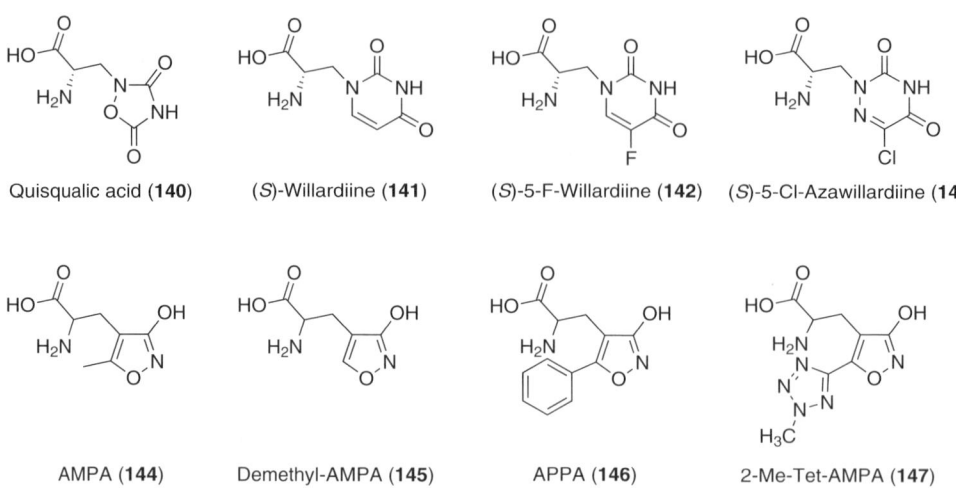

Figure 18.24 Structures of AMPA (**144**) and some AMPA receptor agonists.

Whereas all alkyl-substituted AMPA analogs synthesized so far have shown either full agonism or inactivity at AMPA receptors, the 5-phenyl-substituted analog of AMPA, APPA (**146**), shows the characteristics of a weak partial AMPA receptor agonist, with an efficacy of approximately 60 percent relative to that of AMPA [326]. Quite surprisingly, (*S*)-APPA turned out to be a full AMPA receptor agonist, while (*R*)-APPA proved to be a weak competitive AMPA receptor antagonist [327]. In order to elucidate the SAR of these 5-substituted AMPA-analogs further, a number of 5-membered heteroaromatic substituents have been introduced, including an unsubstituted 5-tetrazolyl (Tet-AMPA), 1-methyl-5-tetrazolyl (1-Me-Tet-AMPA), and 2-methyl-5-tetrazolyl (2-Me-Tet-AMPA) (**147**) [319]. Whereas Tet-AMPA and 1-Me-Tet-AMPA are devoid of agonist activity at AMPA receptors, 2-Me-Tet-AMPA (**147**) proved to be the most potent AMPA receptor agonist synthesized so far, at least 10 times more potent than AMPA [319]. This suggests that steric as well as electronic properties play important roles for the binding of ligands to the receptor.

Electrophysiological studies of the enantiomers of 2-Me-Tet-AMPA (**147**) have shown that the excitatory properties reside exclusively with the *S*-enantiomer [328]. Compared to AMPA (**144**), which does not distinguish between the individual cloned AMPA receptors [328, 329], (*S*)-2-Me-Tet-AMPA shows some selectivity for homomers of GluR3 and GluR4 over receptors constructed from GluR1 or GluR1/GluR2, in electrophysiological studies in *Xenopus* oocytes.

Uracil and 6-azauracil are both effective bioisosteres of the distal carboxylate group of Glu. The natural product (*S*)-willardiine (**141**), in which the pK_a of the uracil moiety is considerably higher than that of the distal carboxylate group of Glu [330], showed low, but measurable, affinity for both AMPA and

KA receptors and induces a rapidly desensitizing response in hippocampal neurons [331]. Introduction of electron-withdrawing substituents in the 5-position of the uracil moiety, increasing the acidity of the heterocycle markedly, improves the receptor affinity and has provided some of the most potent and subtype-selective AMPA and KA receptor agonists yet identified [332]. SAR studies on 5-halogen-substituted willardiines have shown that AMPA receptor affinity and potency in hippocampal neurons were dependent on the size and the electron-withdrawing effect of the substituent [331–335]. (S)-5-F-Willardiine (**142**) (Fig. 18.24) is the most potent agonist in this series, whereas willardiine analogs with sterically larger substituents apparently were not so well accommodated by AMPA receptors [332]. 5-Halogen-substituted 6-azawillardiines showed an SAR somewhat similar to that of the 5-substituted willardiines with (S)-5-Cl-6-azawillardiine (**143**) (Fig. 18.24) possessing highest AMPA receptor affinity [332]. Furthermore, (S)-5-F-willardiine (**142**) and (S)-willardiine (**141**) induce strongly desensitizing agonist responses at AMPA receptors, whereas willardiine analogs containing larger substituents, such as (S)-5-I-willardiine (**176**) (Fig. 18.29), produce much weaker desensitization comparable to that of KA [331]. At cloned AMPA receptors, the willardiine series of compounds show remarkable subtype selectivity, having 10 to 20 times higher affinity for GluR1 or GluR2 than GluR4 [332]. By contrast, the 6-azawillardiine analogs, like AMPA, did not distinguish significantly between the individual AMPA receptor subtypes [332].

The isoxazole-based Glu homolog (S)-Br-HIBO (**148**) also shows AMPA receptor subtype selectivity, preferring GluR1 over GluR3 in receptor binding and functional assays [336] (Fig. 18.25). The structural basis for this selectivity was first modeled and confirmed by mutation studies [337], followed by

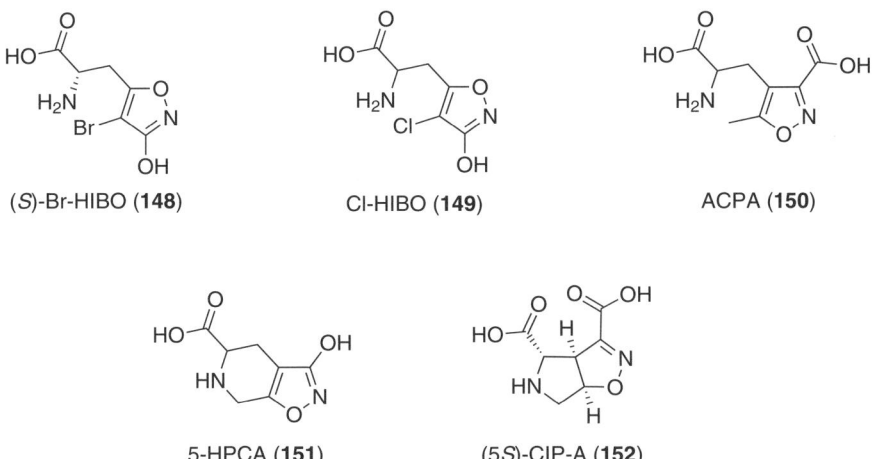

Figure 18.25 Structures of AMPA receptor selective Glu homologs (**148–150**) and AMPA receptor selective bicyclic Glu analogs (**150, 151**).

an X-ray crystallographic characterization [338]. As for the willardiines, the selectivity of (S)-Br-HIBO (**148**) is intimately related to preferential hydrogen bonding of the distal anion of (S)-Br-HIBO (**148**) through a water molecule to a nonconserved tyrosine residue of GluR2 [335, 337, 338]. A bromo-chloro replacement gave the racemic compound Cl-HIBO (**149**), which was recently reported to be a strongly desensitizing full agonist at GluR1, showing increased receptor selectivity ratios in electrophysiology experiments, ranging from 275 to 1600 for GluR1/2 over GluR3/4 [339]. Replacing the 3-isoxazolol group of AMPA by a 3-carboxyisoxazole unit gives the Glu homolog ACPA (**150**), which is a selective AMPA receptor agonist that is more potent than AMPA on cortical neurons as well as in *Xenopus* oocytes expressing cloned AMPA receptors [340, 341]. The potent excitatory AMPA receptor activity of ACPA (**150**) has been shown to reside with the S-enantiomer [342].

Although the 3-isoxazolol and the uracil ring systems have been the most widely used bioisosteres for the distal carboxyl group of Glu in the design of AMPA receptor agonists, other acidic heterocyclic units have proven useful as well. As mentioned in the beginning of this section, AMPA receptors were originally identified pharmacologically by using the naturally occurring Glu analog quisqualic acid (**140**) [343, 344] (Fig. 18.24). This 1,2,4-oxadiazole-3,5-dione analog of Glu was later shown to possess other Glu receptor activities, in particular potent mGluR agonist effects. As a consequence, quisqualic acid (**140**) has been replaced by the highly selective compounds AMPA (**144**) [345, 346] and (S)-F-willardiine (**142**) [347], as standard agonist at AMPA receptors. Conformational restriction of the skeleton of Glu has played an important role in the design of selective GluR ligands. However, only few structurally rigid AMPA receptor-selective Glu analogs have been reported. One such example is the cyclized analog of AMPA, 5-HPCA (**151**) [76] (Fig. 18.25), which recently has been resolved [348].

Interestingly, the pharmacological effects of 5-HPCA (**151**) reside exclusively with the *R*-enantiomer, in striking contrast to the usual stereoselectivity trend among AMPA receptor agonists. To examine the structural basis of the observed enantiopharmacology, (*R*)- and (*S*)-5-HPCA have been docked into the GluR2 ligand binding site. These studies showed that due to the rigid nature of the compound (*R*)-5-HPCA, but not (*S*)-5-HPCA, can present almost all of the required pharmacophore elements to the receptor [348]. Whereas (*R*)-5-HPCA selectively displaces [^{3}H]AMPA binding, the 3-carboxyisoxazolineproline, CIP-A, a bicyclic Glu analog, displays affinity for both AMPA and KA receptors [349]. This dual affinity resides exclusively with the 5S-isomer, (5S)-CIP-A (**152**) (Fig. 18.25) [350].

Competitive and Noncompetitive AMPA Receptor Antagonists Early pharmacological studies on AMPA and KA receptors were hampered by the lack of selective and potent antagonists. The discovery of the quinoxaline-2,3-diones CNQX (**153**) and DNQX (**154**) was a breakthrough since these compounds are quite potent antagonists, although nonselective [351] (Fig. 18.26). Subsequently, the more potent analog NBQX (**155**) was shown to be neuroprotective in cerebral ischemia and to have improved AMPA receptor selectivity compared to CNQX (**153**) [352]. However, NBQX (**155**) failed in clinical trials because of nephrotoxicity due to a limited aqueous solubility but nonetheless has become a valuable tool for research. DNQX (**154**) has played a key role in elucidating the binding mode of competitive antagonists, as it was the first antagonist co-crystallized with the GluR2 ligand-binding domain [353] (see Structure of Ionotropic Glutamate Receptors). Attempts to improve the aqueous solubility of such antagonists without losing activity at AMPA receptors, by introducing appropriate polar substituents onto the quinoxaline-2,3-dione ring system, have been highly successful and have resulted in very potent AMPA receptor antagonists, as exemplified by ZK200775 (**156**) [354] and compound **157** [355].

Another series of potent and selective competitive AMPA receptor antagonists based on the isantin oxime skeleton includes NS 1209 (**158**), which shows long-lasting neuroprotection in animal models of ischemia and an increased aqueous solubility compared to NBQX (**155**) [356] (Fig. 18.26). In recent years, several series of imidazo[1,2-*a*]indeno[1,2-*e*]pyrazin-based com-

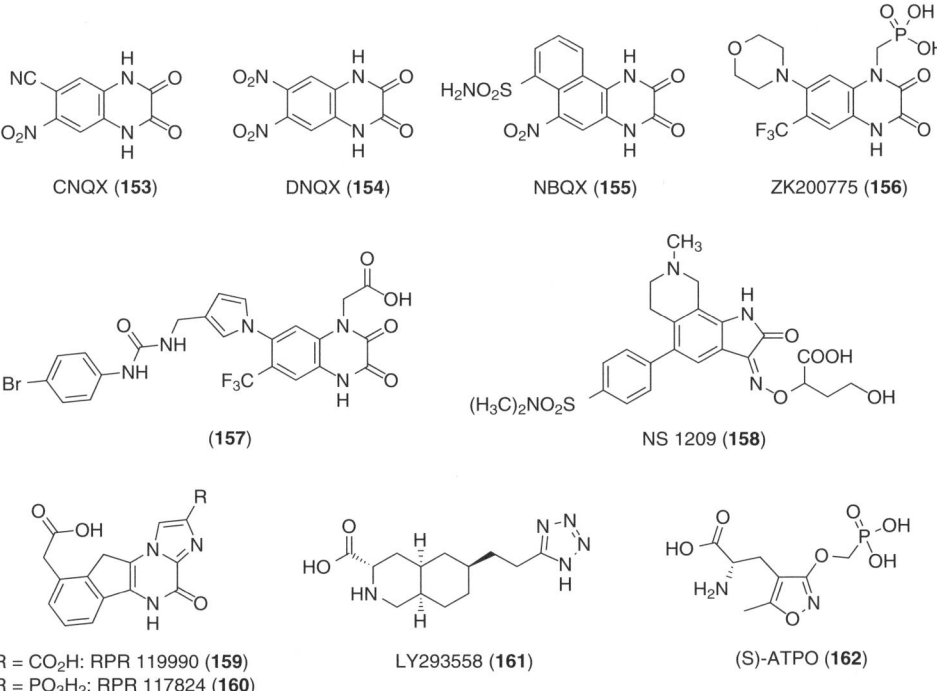

Figure 18.26 Structures of some competitive AMPA receptor antagonists.

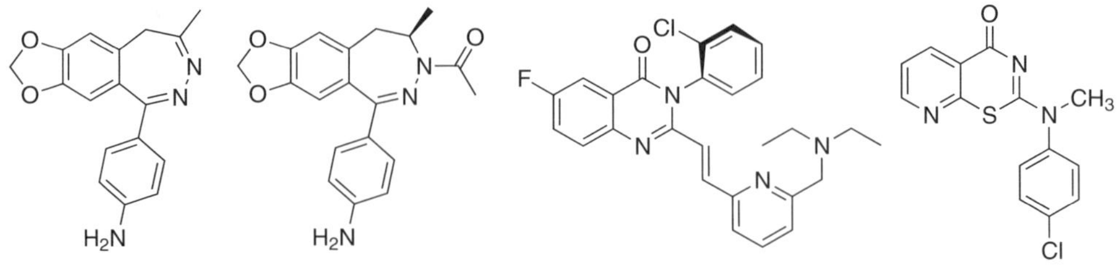

Figure 18.27 Structures of some noncompetitive AMPA receptor antagonists.

petitive antagonists have been identified [357–360]. The initial compounds displayed high, although nonselective, affinity for AMPA and NMDA receptors, but through extensive SAR studies, highly potent and selective compounds such as RPR 119990 (**159**) and RPR 117824 (**160**) have been identified. However, a drawback of this series of compounds is their lack of oral activity [361]. At least two classes of amino-acid-containing compounds, based on decahydroisoquinoline-3-carboxylic acid [362] and AMPA [363], have been found to be competitive AMPA receptor antagonists. LY293558 (**161**), a member of the former class, is systemically active, although it shows significant antagonist effects at KA receptors in addition to its potent AMPA receptor blocking effects [329, 364, 365]. The AMPA receptor antagonist (S)-ATPO (**162**), which was designed using AMPA as a lead structure, has a carbon backbone longer than that which normally confers AMPA receptor agonism [366].

The 2,3-benzodiazepines, such as GYKI 52466 (**163**) and Talampanel (**164**), represent a class of noncompetitive AMPA receptor antagonists that have enabled effective pharmacological separation of AMPA and KA receptor-mediated events [367] (Fig. 18.27). These compounds appear to bind to sites distinct from the agonist recognition site, and are thus negative allosteric modulators [368, 369]. Talampanel (**164**), currently under clinical development as a treatment for multiple sclerosis, epilepsy, and Parkinson's disease [367], may inhibit AMPA receptor function even in the presence of high levels of Glu [367, 370]. A number of structurally similar potent noncompetitive AMPA receptor antagonists with anticonvulsant activity, typified by CP-465,022 (**165**) and YM928 (**166**), have been identified [371, 372]. CP-465,022 (**165**), the (+)-atropisomer of CP-392,110, interacts stereoselectively with a binding site identical to that of the 2,3-benzodiazepines [371, 373].

Modulatory Agents at AMPA Receptors The agonist-induced desensitization of AMPA receptors can be markedly inhibited by a number of structurally dissimilar AMPA receptor potentiators known as AMPA-kines (Fig. 18.28), including aniracetam (**167**), cyclothiazide (**168**), and in particular CX-516 (**169**), which has been shown to improve memory function in aged rats [374].

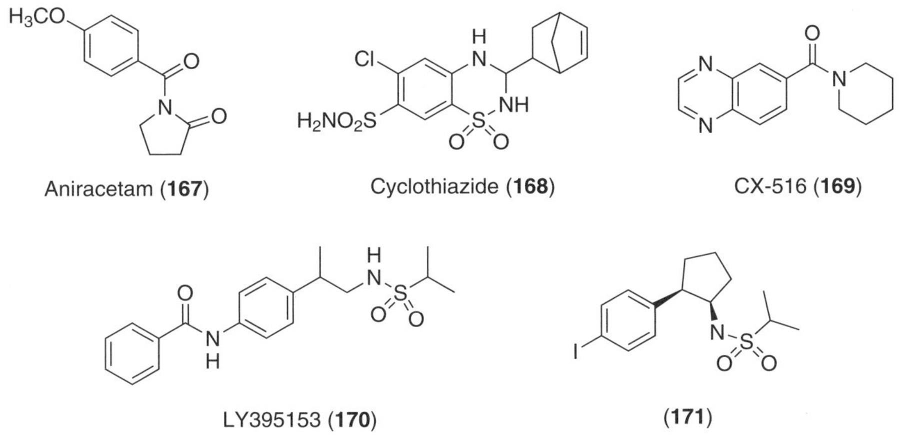

GLUTAMATE: EXCITATORY NEUROTRANSMITTER 843

Aniracetam (**167**)   Cyclothiazide (**168**)   CX-516 (**169**)

LY395153 (**170**)   (**171**)

Figure 18.28 Structures of some positive modulators of AMPA receptors.

These AMPA-kines positively modulate ion flux via stabilization of receptor subunit interface contacts and subsequent reduction in the degree of desensitization [375]. A series of more potent arylpropylsulfonamide-based AMPA-kines has been identified, including LY395153 (**170**) [376]. [^{3}H]LY395153 labels a modulatory site on AMPA receptors with nanomolar affinity but remains unclear whether all of the structurally diverse AMPA-kines share the same binding site [377]. Incorporation of the sulfonamidopropyl moiety of the arylpropylsulfonamide potentiators into a 2-substituted 1-sulfonamidocyclopentane ring has provided compound **171**, the most effective AMPA-kine yet described [378].

KA Receptor Agonists and Antagonists

The pharmacology and pathophysiology of KA receptors are far less well understood than for AMPA receptors. However, identification of selective agonists and competitive antagonists has developed the field of KA receptor research, especially over the last 5 years, and has provided insight into roles of these receptors in the CNS. For a number of years, KA (**172**) and domoic acid (**173**) have been used as standard KA receptor agonists despite their activities at AMPA receptors, characterized by nondesensitizing responses at these receptors (Fig. 18.29). More selective KA receptor agonist activities have been described for (*S*)-ATPA (**174**) [329, 379], (*S*)-thio-ATPA (**175**) [380], and (*S*)-5-I-willardiine (**176**) [332, 381]. These compounds exhibit some selectivity among low-affinity KA receptor subtypes, namely a preference for GluR5 over GluR6 [332, 379, 380]. Studies on isolated dorsal roots showed that ATPA and (*S*)-5-I-willardiine (**176**) are, respectively, 10- and 100-fold more potent than KA [381]. Whereas (*S*)-ATPA (**174**) and KA (**172**) induce strongly desensitizing responses in *Xenopus* oocytes expressing homomeric KA receptors of

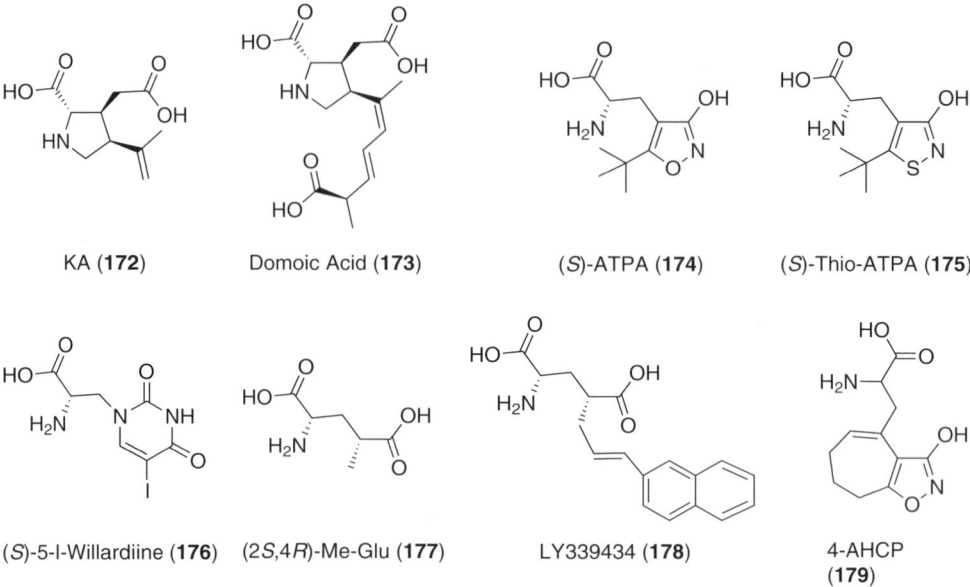

Figure 18.29 Structures of KA (**172**) and some KA receptor agonists.

the GluR5 subtype, (*S*)-thio-ATPA (**175**) produce only partially desensitizing responses [380]. (*S*)-ATPA (**174**), (*S*)-thio-ATPA (**175**), and (*S*)-5-I-willardiine (**176**) are structurally related to potent AMPA agonists discussed in earlier sections, illustrating that the structural characteristics required for activation of GluR1-4 and GluR5 receptors are very similar. However, the presence of the relatively bulky and lipophilic *tert*-butyl- or iodo-substituents of these compounds is apparently the major determinant of the observed receptor selectivity. The ligand-binding domain of GluR2 has recently been crystallized in complex with (*S*)-ATPA (**174**) [382] (see next section).

Among the four possible stereoisomers of the 4-methyl-substituted analog of Glu, only the 2*S*,4*R*-isomer (**177**) shows selectivity for KA receptors, with a binding affinity comparable to that of KA. Replacement of the 4-methyl group of (2*S*,4*R*)-Me-Glu (**177**), a ligand that shows limited selectivity between GluR5 and GluR6, by a range of bulky, unsaturated substitutents containing alkyl, aryl, or heteroaryl groups [383–386] has yielded a number of interesting GluR5 receptor-selective compounds including LY339434 (**178**). LY339434 shows approximately a 100-fold selectivity for GluR5 over GluR6 and no affinity for GluR1, 2, or 4 receptors [383].

The conformationally restricted 3-isoxazolol-containing Glu homolog, 4-AHCP (**179**), was initially demonstrated to be a very potent AMPA agonist at rat spinal neurons [387]. In more recent studies, the excitatory effect of 4-AHCP (**179**) has been shown to reside with the *S*-enantiomer, which is equipotent with (*S*)-ATPA (**174**) and 35 to 115 times more potent at GluR5 than at cloned AMPA receptors or at GluR6/KA2 [388].

Whereas a large number of selective competitive AMPA receptor antagonists have been identified, particularly over the last decade, only a few selective KA receptor antagonists have been reported. Most of these KA receptor-preferring antagonists are structurally related to AMPA receptor antagonists and show at least some affinity toward AMPA receptors further to that KA receptor affinity, reflecting that similar structural features are required for activation and blockade of AMPA and some KA receptors, in particular GluR5.

One of the first reported KA receptor-preferring antagonists was the isantin oxime, NS 102 (**180**), which shows some selectivity toward low-affinity [^{3}H]KA sites [389] as well as antagonist effect at homomeric GluR6 [390] (Fig. 18.30). However, low aqueous solubility has limited the use of NS 102 (**180**) as a pharmacological tool. A number of decahydroisoquinoline-based acidic amino acids, including LY382884 (**181**), has been characterized as competitive GluR5-selective antagonists that exhibit antinociceptive effects [391]. In the course of SAR studies using this decahydroisoquinoline skeleton, excellent receptor selectivity been observed for compound **182**, which shows approximately 1000-fold higher affinity for GluR5 than for GluR1–4, GluR6, or GluR7. In addition, a group of quinoxalinedione-derived antagonists structurally related to the AMPA receptor antagonist **157** (Fig. 18.26), and with high affinity for GluR5, have been reported [392]. One example is BSF 91594 (**183**), which shows more than 50-fold selectivity for GluR5 over other KA receptor subtypes or AMPA receptors.

Most recently, a series of arylureidobenzoic acids have been reported as the first compounds with noncompetitive antagonist activity at GluR5 [393]. The most potent ligands, exemplified by compound **184**, exhibit more than 50-fold selectivity for GluR5 over GluR6 or the AMPA receptor subtypes. This com-

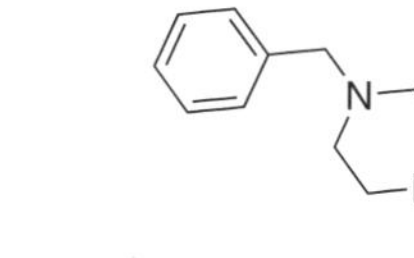

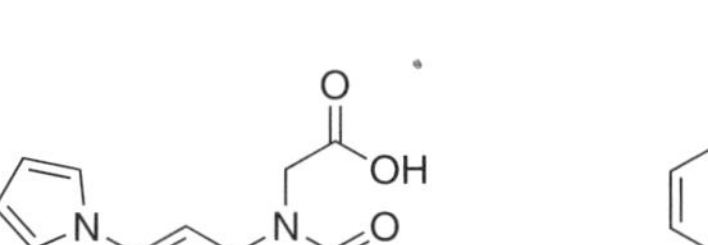

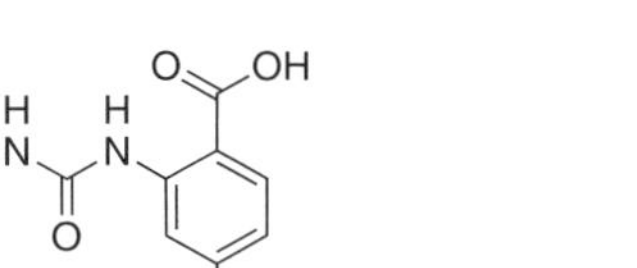

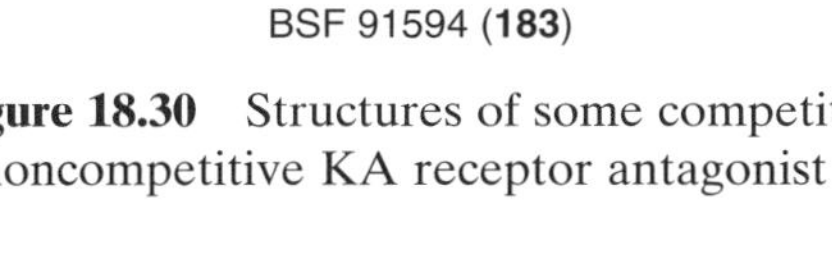

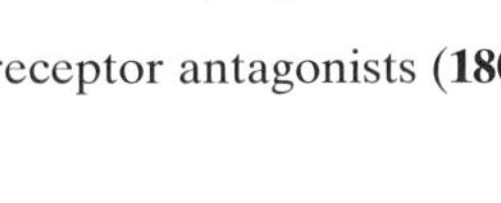

Figure 18.30 Structures of some competitive KA receptor antagonists (**180–183**) and a noncompetitive KA receptor antagonist (**184**).

pound may prove valuable to further pharmacological research and as a lead in the search of therapeutic glutamatergic agents.

Structure of Ionotropic Glutamate Receptors

iGluR subunits may be considered as modular constructions of three distinct domains (Fig. 18.31): (1) a transmembrane segment consisting of three transmembrane helices and a reentrant loop, distantly related to K^+ channels such as KcsA, forming the walls of the pore; (2) an extracellular ligand binding domain (S1S2) composed of two lobes connected by a hinge ("venus flytrap"), distantly related to glutamine binding protein (QBP); and (3) an N-terminal domain structurally related to periplasmic binding proteins such as leucine–isoleucine–valine binding protein (LIVBP), and which appears to play a role in oligomerization [394]. It is currently believed that iGluRs assemble as dimers-of-dimers, although the evidence is weaker for NMDA receptors than for non-NMDA receptors [251]. Numerous studies, including electro-

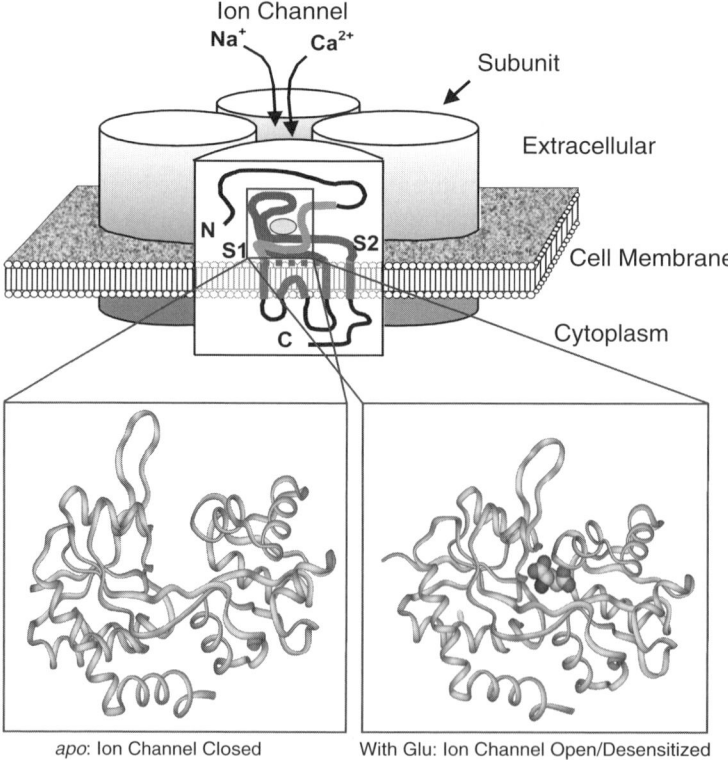

Figure 18.31 Illustration of the tetrameric topology of AMPA receptors (*top*), showing *apo* and agonized state of the ligand binding core (*below*). (This figure is available in full color at ftp://ftp.wiley.com/public/sci_tech_med/drug_discovery/.)

physiology [395, 396], electron microscopy [397], and X-ray crystallography [335] now support a tetrameric structure for the AMPA receptor that resembles some K$^+$ channels (Fig. 18.31).

Our understanding of the competitive binding site of iGluRs has been greatly enhanced by the development of a soluble construct of S1S2 that possesses many of the properties of the intact receptor [398]. Experimentation with linkers of various lengths connecting the two half-domains led to the crystallization of GluR2-S1S2, first with the partial agonist KA (**172**) [399], then with a wide variety of ligands as well as various point mutations. A bacterial homolog known as GluR0 has also been crystallized, and recently, a construct of the glycine-binding NMDA subunit, NR1. The currently available crystal structures are listed in Table 18.1.

As can be seen from Table 18.1, different competitive ligands (agonists, partial agonists, antagonists) induce different conformational changes in GluR2-S1S2. Parallel structural studies on NMDA receptors, long the subject of pharmacophore modeling [291], have now commenced. Structure-based design, docking, and homology modeling can be performed, using the structural repertoire presented by these crystallographic snapshots to take account of receptor flexibility [339, 388, 406, 407]. While the molecular dynamics of the GluR2 ligand binding core have been probed, as well as that of the related GlnBP, simulations have as yet not been able to reproduce the process of domain closure or opening, possibly because of the relatively long simulation times required [408, 409]. At GluR2, and presumably at other receptors in this class with highly conserved binding sites, Glu binds with a partially folded conformation [353]. The binding mode of Glu at Glu-sensitive NMDA receptors has not yet been demonstrated, although on the basis of pharmacophore modeling, it is at least as folded as at AMPA receptors. The amino acid moieties of agonists at GluR2 appear to be zwitterionized and are involved in ion-pair interactions—the ligand α-carboxylate with R485 in domain 1 and the ammonium with E705 in domain 2. By contrast, the terminal acidic moieties of Glu and AMPA agonists, while apparently ionized at the receptor, are not involved in ion pairing but rather accept a number of hydrogen bonds from receptor hydroxy and amide groups, as well as participating in a water-mediated hydrogen bond network. The receptor shows substantial flexibility in the distal region, with peptide side chains and water molecules rearranging to accomodate the ligand. For example, the 3-isoxazolol moiety of AMPA (**144**) does not bind in a similar position to Glu's ω-carboxylate, whereas that of 2-Me-Tet-AMPA (**147**) does (Fig. 18.32) [338, 353].

The structural details of the binding modes of two distinct classes of AMPA receptor antagonists are known: the unionized planar quinoxalinedione DNQX [353] and the tri-ionized AMPA analog (*S*)-ATPO (**162**) [338]. Neither of these ligands fully exploit the domain-open antagonized binding site, and neither come into close contact with nonconserved residues (Fig. 18.33). There is therefore substantial scope for the development of subtype-specific non-NMDA receptor antagonists.

TABLE 18.1 List of Available Crystal Structures of iGluR Binding Domains in Complex with Various Ligands

Receptor	Ligand	Ligand Class	Domain Closure	Reference
NR1 (NMDA)	Glycine (**127**)	Endogenous agonist		251
	(*R*)-Serine (**128**)	Endogenous agonist		251
	(*R*)-Cycloserine (**131**)	Agonist		251
	Dichlorokynurenic acid	Antagonist		251
GluR2 (AMPA)	Glu (**96**)	Endogenous agonist	20°	353
	AMPA (**144**)	Agonist	20°	353
	Demethyl-AMPA (**145**)	Agonist	20°	400
	2-Me-Tet-AMPA (**147**)	Agonist	21°	338
	(*S*)-ATPA (**174**)	Agonist (GluR5 selective)	21°	382
	Thio-ATPA (**175**)	Agonist (GluR5 selective)	18°	401
	Br-HIBO (**148**)	Agonist (GluR1,2 selective)	17°	338
	ACPA (**150**)	Agonist	20°	338
	Quisqualic acid (**140**)	Agonist	20°	402
	(*S*)-Willardiine (**141**)	Agonist (GluR1,2 selective)	17°	334, 335
	(*S*)-F-Willardiine (**142**)	Agonist (GluR1,2 selective)	16°	334, 335
	(*S*)-Br-Willardiine	Agonist	15°	334, 335
	(*S*)-I-Willardiine (**176**)	Agonist (GluR5 selective)	11°	334, 335
	KA (**172**)	Partial agonist	12°	399

Construct	Ligand	Notes	Angle	Ref
	DNQX (**154**)	Antagonist	5°	353
	(*S*)-ATPO (**162**)	Antagonist	2–5°	403
	Apo	(None)	0°	353
GluR2 (Y702F)	ACPA (**150**), (*S*)-Br-HIBO (**148**)			338
GluR2 (L650T)	AMPA (**144**), quisqualic acid (**140**), KA (**172**)			404
GluR2 (L650T, L483Y)	AMPA (**144**)			404
GluR2 (L483Y)	AMPA (**144**), DNQX (**154**)			375
GluR2 (N754D)	KA (**172**)			375
GluR2 (N754S)	Glu (**96**) + cyclothiazide (**196**)	Allosteric modulator (desensitization blocker)		375
GluR0	Glu (**96**), (*R*)-serine (**128**), *apo*			405

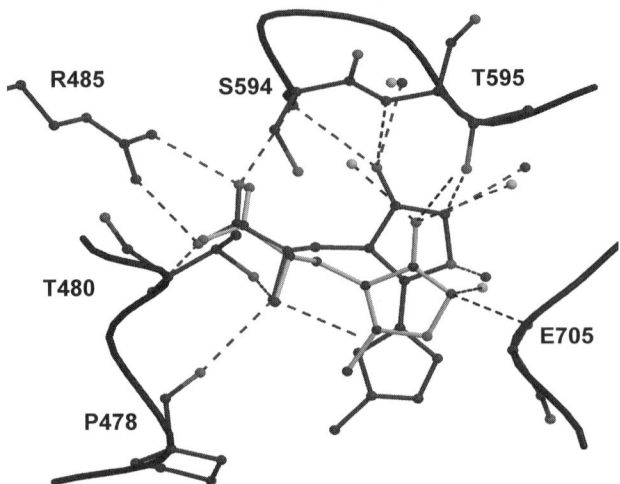

Figure 18.32 Full agonists AMPA (**144**) and 2-Me-Tet-AMPA (**147**) co-crystallized with GluR2-S1S2. Note the difference in the positioning of the isoxazole rings and the differences in accompanying water molecules and hydrogen bond network. (This figure is available in full color at ftp://ftp.wiley.com/public/sci_tech_med/drug_discovery/.)

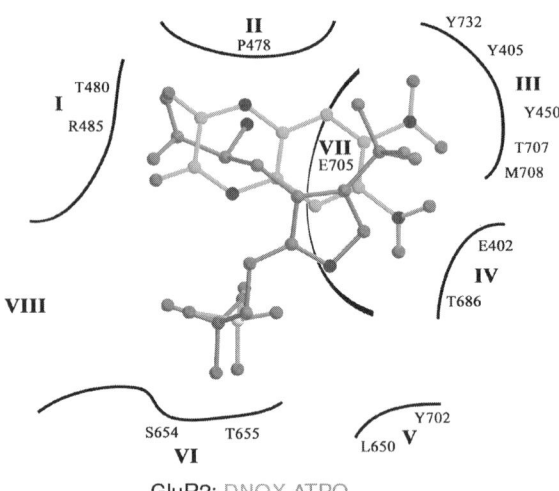

Figure 18.33 Competitive antagonists DNQX (**154**) and ATPO (**162**) co-crystallized with GluR2-S1S2, showing different binding modes of the two classes of antagonists. Regions III–VI show some variations between AMPA and KA receptors, while only region V varies among AMPA receptor subtypes. (This figure is available in full color at ftp://ftp.wiley.com/public/sci_tech_med/drug_discovery/.)

The high homology among closely related iGluR subunits (see Table 18.2), for example, GluR1-7, allows models to be built from the X-ray structures of GluR2 with confidence [337, 339, 388, 406, 410]. While the competitive binding sites of GluR1,2 are essentially identical, as are GluR3,4, the single nonconserved residue between these two subclasses of AMPA receptors, Y/F702, has been open to exploitation, even though it only interacts with ligands via a water molecule. Thus, a series of selective GluR1,2 ligands has been designed and synthesized, and the basis of selectivity is well understood [339, 388]. The deliniation between GluR5 [which weakly binds both AMPA (**144**) and KA (**172**)] and the AMPA-preferring GluR1-4 is sharper; L/V650 and M/S708 make direct steric contact with the ligand. GluR5-selective ligands such as (*S*)-4-AHCP (**179**) and (*S*)-thio-ATPA (**175**) exploit the altered shape of the binding cavity [388]. The development of specific agonists and antagonists for other subtypes is ongoing, a task made easier by such detailed structural information.

As with many ion channels, iGluRs exhibit desensitization, that is, that following rapid activation and opening of the channel, the receptor undergoes a conformational change on a millisecond timescale, characterized by low conductance even in the continued presence of agonist [247]. In general, full agonists cause rapid and almost complete desensitization, whereas partial agonists are weakly desensitizing. Much is now known about the transmission of free energy of binding of agonists and partial agonists to domain closure, channel opening, and subsequent desensitization, even though we lack atomic resolution structural detail of the full-length receptor [334]. Compounds modulating desensitization are of particular interest as therapeutic agents, and great inroads have been made toward unravelling the structural mechanism of desensitization [375]. While the S1S2 construct used in crystallization experiments has generally been of the *flop* splice variant of GluR2, some of the properties of the *flip* isoform have been investigated by introducing the key mutation N754S, known to confer sensitivity to the desensitization blocker cyclothiazide (**168**) [411].

This crystal structure (Fig. 18.34) reveals that cyclothiazide (**168**), and presumably other allosteric modulators of AMPA receptors, affect desensitization by binding to and stabilizing the interface between two subunit monomers [375]. It is likely that other AMPA receptor modulators, both positive and negative, act via this interface—a promising target for structure-based drug design.

Metabotropic Glutamate Receptor Ligands

The cloning of the mGluRs and the evidence that has subsequently emerged on their potential utility as drug targets in a variety of neurological disorders has encouraged medicinal chemists to design ligands targeted at the mGluRs. Over the first decade since mGluRs were cloned, our knowledge of the SAR of these receptors has increased dramatically and led to the discovery of new

TABLE 18.2 Binding Site Homology of Selected iGluRs[a]

	402	450	478	485	650	654	686	702	705	723
NR1	I C T	K F G	A P L T I	E R A	T V K	Q S **S**	A I Q	A F I	W **D** S A	G F G
GluR2	L **E** S	K Y G	A P L T I	V **R** E	T L **D**	S G **S**	T T A	A Y L	L E S **T**	G Y G
GluR4	M **E** S	K Y G	A P L T I	V **R** E	T L **D**	S G **S**	T T A	A F L	L E S **T**	G Y G
GluR5	L **E** E	K Y G	A P L T I	V **R** E	A V R	D G **S**	N S D	A L L	M E S **T**	G Y G
GluR6	L **E** E	K Y G	A P L T I	V **R** E	A V E	D G A	S N E	A F L	M E S **T**	G Y G
KA1	L **E** N	V Y G	A G L T I	E **R** E	T I H	G G **S**	S T E	A F L	L E S **T**	G Y G
δ2	L **E** E	K Y G	S A L T I	D **R** E	T V L	D S A	E S Q	A F V	W **D** A A	G Y G
Function	"Lock"	Pocket	—NH_3^+	—COO^-	Pocket	Distal anion	"Lock"	Pocket	—NH_3^+	Pocket

[a] Numbering according to GluR2. Residues in bold are known to interact directly with ligands or line the binding site.

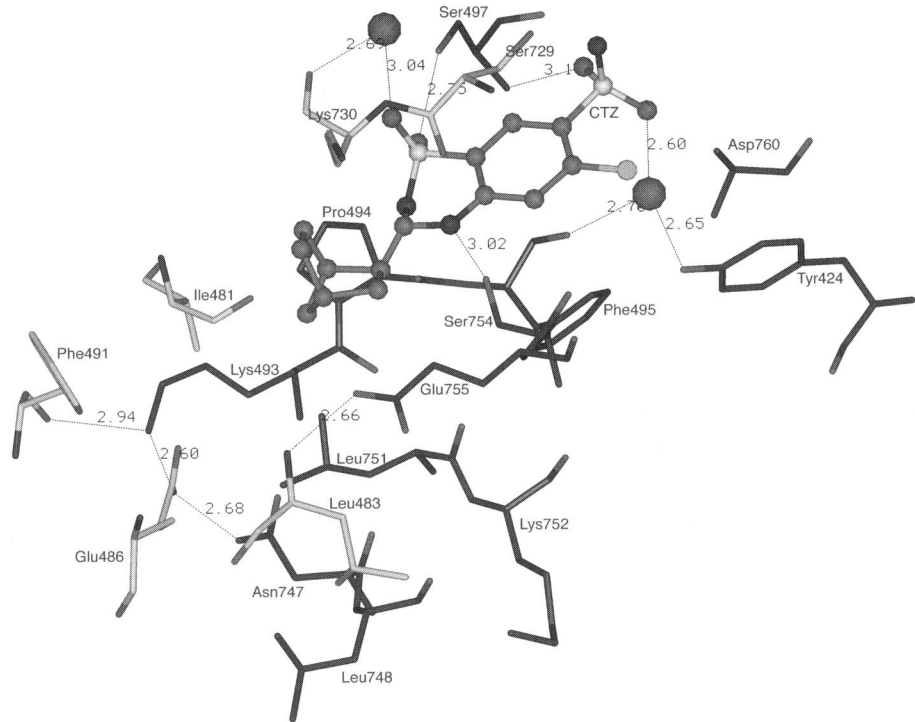

Figure 18.34 Cyclothiazide (**168**) bound to the (N754S)GluR2-S1S2 dimer interface. Interfacial residues colored black or cyan according to monomer of origin. (This figure is available in full color at ftp://ftp.wiley.com/public/sci_tech_med/drug_discovery/.)

potent and selective ligands [412]. The vast majority of these ligands are Glu analogs incorporating a glycine moiety and a distal acidic function, but positive and negative allosteric modulators with no structural similarity to Glu have also been discovered. In the following sections we will describe some of the principles of drug design that have been employed by medicinal chemists in order to design ligands with increased receptor selectivity and potency.

Metabotropic Glutamate Receptor Agonists

Conformationally Constraint Analogs The first agonist to show selectivity for mGluRs over to iGluRs was (1*S*,3*R*)-ACPD (**185**) [413]. When tested on the eight cloned mGluRs it acts rather nonselectively [414–420], but it has been used extensively as a template for design of new mGluR ligands, some of which are shown in Figure 18.35. Within these series of compounds it is interesting to note the profound changes in receptor selectivity, potency, and efficacy caused by relatively small structural changes. Introduction of a nitrogen atom in the C4 position of **185** gives (2*R*,4*R*)-APDC (**186**), which displays an increased potency for group II receptors compared to the parent compound

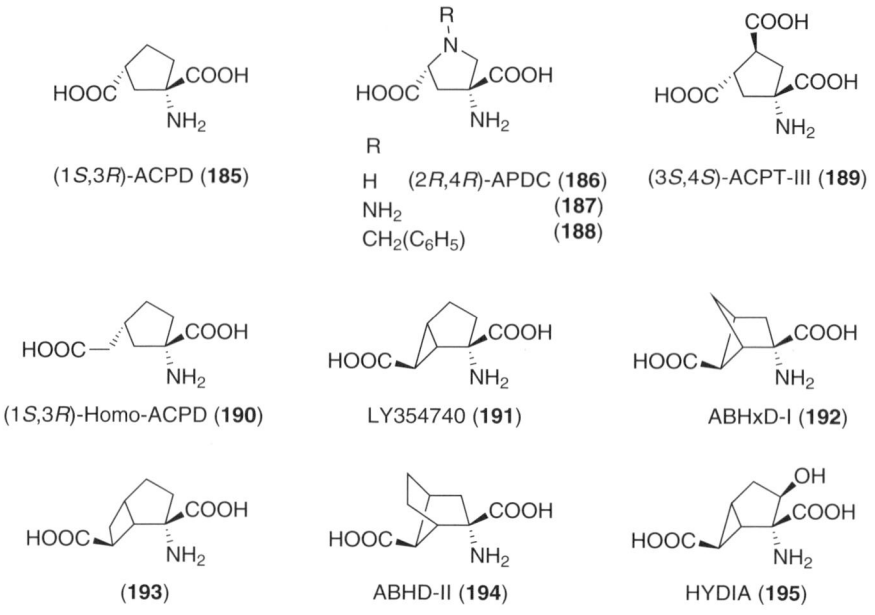

Figure 18.35 Structures of some mGluR agonists.

while losing affinity for group I and III receptors [421–423]. N1-substitutions of **186** have been reported to cause significantly reduced potency at group II receptors [423–425]. Interestingly, however, 1-amino-APDC (**187**) acts as a partial group II agonist [425], and 1-benzyl-APDC (**188**) shifts selectivity from mGluR2 and mGluR3 toward mGluR6 [423]. Similarly, addition of a carboxylic acid moiety at the corresponding C4-position of **185** to give (3*S*,4*S*)-ACPT-III (**189**) provides agonist potency at mGluR4 comparable to Glu, as well as weak antagonist effects at mGluR1 and mGluR2 [426].

In 1997 we reported that (1*S*,3*R*)-homo-ACPD (**190**), the homolog of **185**, possesses increased selectivity for mGluR2 as compared to mGluR1 and mGluR4, albeit with lower potency and efficacy than the parent compound [427]. At the same time, a dramatic 10,000-fold increase in potency at group II receptors was reported by further restraining the conformation of **190** by introducing a single bond between the side chain and the 5-position in the cyclopropane ring to give LY354740 (**191**) [421, 428, 429]. Compound **191** displays low-nanomolar agonist potency at mGluR2 and mGluR3, low-micromolar agonist potency at mGluR6 and mGluR8, while showing no activity at the remaining mGlu receptors.

The synthesis and pharmacology of ABHxD-I (**192**) has been reported [430]. Interestingly, this compound displays agonist potency comparable to Glu at mGluR1 to mGluR6; thus the compound is less selective than the parent compound **185**. The observation has been of key importance in developing early models of the mGluR binding site. Compound **192** is quite a rigid molecule, which adopts a conformation corresponding to the extended anti–anti conformation of Glu. The observation that the compound is a potent agonist on all three mGluR groups led the authors to suggest that Glu adopts the same extended conformation at all three receptor groups, and that group selectivity is thus not a consequence of different conformations but rather a consequence of other factors such as steric hindrance [430]. This hypothesis has later been confirmed by several more extensive pharmacophore models [431–433]. One notable example of these sharp steric requirements can be seen by comparing highly selective group II agonist **191** with the nonselective agonist **192**. Both compounds prefer conformations corresponding to the extended anti–anti conformation of Glu [423, 430]. However, overlaying the compounds with Glu (all in the anti–anti conformation) shows that **191** does occupy extra volume compared to **192** [430]. Accordingly, it has been suggested that group II receptors can accept this extra volume, whereas group I and III receptors cannot [430]. The unique pharmacological profile of **191** is further highlighted by other close analogs such as compound **193** and ABHD-II (**194**), which are inactive and weak group I/II agonists, respectively [434, 435].

The 2-(carboxycyclopropyl)glycines are also conformationally restricted analogs of Glu, albeit with more flexibility than the bicyclic compounds described above. In agreement with the pharmacophore models, it is also the isomers [in particular (2S,3R,4S)-CCG, **102**] (Fig. 18.20) with the extended conformation that are agonists at mGluRs [436]. Also in agreement with the findings from pharmacophore modeling showing that it is the fully extended conformation of Glu that recognizes mGluR1, mGluR2, and mGluR4, (2S,3R,4S)-CCG (**102**) has been shown to display potent agonist activity at members of all three mGluR groups, albeit with some preference for the group II receptors [436, 437]. The high potency and group II selectivity of **191** has inspired the synthesis of analogs, a few of which are shown in Figure 18.35. The 3-hydroxy analog of **191**, HYDIA (**195**), has been reported [438]. It is interesting to note that such a modest change in the parent compound converts the compound from an agonist into a potent group II antagonist.

Extension of Backbone Chain Length Based on the observation that (S)-2-aminoadipic acid (**196**) [439–442], the homolog of Glu, displays increased mGluR selectivity compared to Glu, we have further studied the effects of backbone extension (Fig. 18.36). (S)-2-Aminoadipic acid (**196**) selectively activates mGluR2 and mGluR6, whereas it has no effect on mGluR1, mGluR4, or mGluR5. Working along these lines led to the discovery that (S)-homo-AMPA (**197**) is a specific agonist at mGluR6 with no activity at mGluR1-5 or mGluR7 [440, 441]. Since mGluR4, mGluR6, and mGluR7 all belong to group III, it was surprising to find that (S)-2-aminoadipic acid (**196**) and (S)-homo-AMPA (**197**) only activate mGluR6 within this group. It is also interesting to note the dramatic pharmacological difference between AMPA (**144**) and its homolog **197** at AMPA receptors. As described previously, the former is a

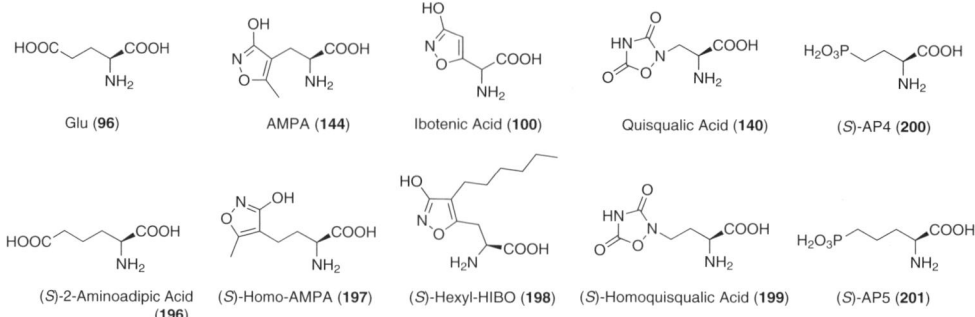

Figure 18.36 Structures of some Glu analogs showing effects at iGluRs and/or at mGluRs (*upper row*) and the corresponding homologs that interact preferentially with mGluRs (*lower row*).

highly selective agonist, whereas the latter is completely inactive at AMPA receptors [441]. In addition to their previously described effects as AMPA receptor agonists, ibotenic acid (**100**) and quisqualic acid (**140**) were shown in the mid-1980s to stimulate IP accumulation indicative of metabotropic effects [443]. The cloning of mGluRs confirmed these results by showing that both compounds activate group I receptors [417, 444, 445]. Whereas quisqualic acid (**140**) is more potent than Glu at group I mGluRs, ibotenic acid (**100**) is less potent.

Furthermore, the latter compound also shows activity at group II receptors [442]. The pharmacological effects of the homologs of ibotenic acid (**100**) and quisqualic acid (**140**) has been reported [442]. Interestingly, (S)-Br-HIBO (**148**) (Fig. 18.25) is a selective (albeit weak) antagonist at mGluR1 and mGluR5. Introducing more lipophilic/bulky side chains in the 4-position increases the activity of the compounds as group I antagonists. HIBO analogs with small side chains also activate AMPA receptors, but in contrast to the mGluR1,5 antagonism, this activity diminishes with increasing size of the side chain. This trend peaks at six carbons providing (S)-hexyl-HIBO (**198**), which is a fairly potent group I antagonist devoid of activity at AMPA receptors [446]. (S)-Homoquisqualic acid (**199**) also produces effects quite different from the parent compound **140**. On mGluR5 (S)-homoquisqualic acid (**199**) is also an agonist albeit with 500-fold lower activity than quisqualic acid (**140**), whereas the homolog displays surprising antagonism at mGluR1 [447]. Furthermore, (S)-homoquisqualic acid (**199**) is an agonist at mGluR2 in contrast to the parent compound [447].

(S)-AP4 (**200**) is also a classical mGluR agonist that in the early 1980s was shown to reduce evoked release of Glu via a presynaptic mechanism [448]. Cloning of group III receptors has shown that (S)-AP4 (**200**) is an agonist at these receptors, some 10-fold more potent than Glu, and that many of the effects of this ligand in the CNS can be ascribed to activation of group III receptors [416, 418, 419, 449]. Again, the homolog, (S)-AP5 (**201**), shows a rather different pharmacological profile from the parent compound. (S)-AP5 (**201**) also activates group III receptors, but with a markedly lower potency [447, 450]. However, it also antagonizes mGluR2, unlike the parent compound **200** [447, 451].

Taken together, the results from the homologs clearly demonstrate that the distance between the glycine moiety and the distal acidic group is of vital importance for the pharmacological activity. As pointed out, the selectivity profile of the homologs are often quite different from the parent compounds. Furthermore, in several cases, agonists have been converted into antagonists when the backbone chain length is increased. While the effects are not predictable at a glance, the results show that chain length is a parameter that is worth taking into consideration when analogs of newly identified mGluR "hits" are being designed.

Bioisosteres We have already described a number of compounds in which the distal carboxylic acid of Glu has been replaced by an acidic bioisostere (see Fig. 18.36). Compounds such as quisqualic acid (**140**) (bioisosteric group: 1,2,4-oxadiazol-3,5-dione), ibotenic acid (**100**), and (S)-homo-AMPA (**197**) (bioisosteric group: 3-isoxazolol) and (S)-AP4 (**200**) (bioisosteric group: phosphonic acid) show that all three groups of mGluRs are capable of accepting bioisosteres to replace the distal carboxyl group. The increased receptor selectivity shown by these compounds compared to Glu also demonstrates that the region of the receptor, in which the distal carboxylic acid binds, differs among the receptor subtypes. Thus, further research into new bioisosteres may be a fruitful path to new subtype-selective mGluR ligands. This argument is further substantiated by group I selective agonists employing phenol as the distal acidic group (see Fig. 18.37). For example, (S)-3,5-DHPG (**202**) is a selective agonist at mGluR1 and mGluR5 [452–454], equipotent with Glu on both group I subtypes.

Competitive Metabotropic Glutamate Receptor Antagonists

Phenylglycines One of the first potent mGluR antagonists to be reported was (S)-4CPG (**203**), which antagonizes (1S,3R)-ACPD-induced (**185**) IP formation in cerebral cortical slices [455]. Likewise, (S)-4CPG (**203**) has been shown to antagonize IP formation by mGluR1, and furthermore, the compound seems to be an agonist [453] or antagonist [454] at mGluR2 with no effect on mGluR4. (S)-4CPG (**203**) has been used extensively as a template for designing further potent and selective antagonists at mGluR1. Whereas it has been debated whether (S)-4CPG (**203**) is an agonist or antagonist at mGluR2 [453, 454], it is generally agreed that (S)-4C3HPG (**204**) is an mGluR1 antagonist/mGluR2 agonist and that the α-methylated analog, (S)-M4CPG (**207**), is an antagonist at both subtypes [453, 454]. It has been shown that the antagonist potency is increased by methylation at the 2-position of the phenyl ring. Thus (+)-4C2MPG (**205**) and 4C3H2MPG (**206**) are both approximately five-

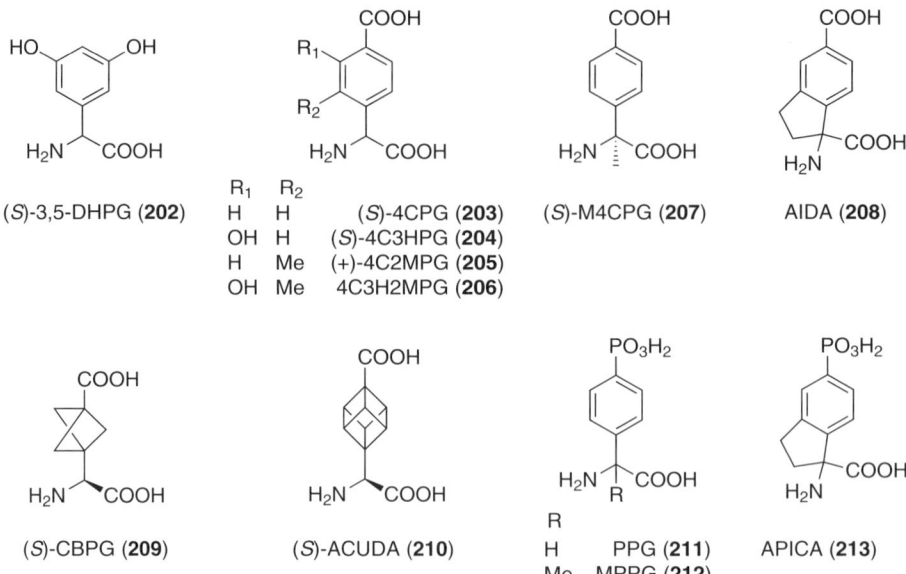

Figure 18.37 Structures of some mGluR ligands.

fold more potent than the nonmethylated parent compounds [456]. AIDA (**208**), a compound that can be viewed as a cyclized analog of either **207** or **205**, has also been shown to possess mGluR1 antagonist activity, albeit with lower potency than the noncyclic compounds [457, 458]. The 4-carboxyphenylglycines discussed above show mGluR1 antagonist potencies in the 5- to 200-μM range. Interestingly, the potencies are somewhat affected by the agonist used to determine the functional potency [459]. Furthermore, it is notable that the 4-carboxyphenylglycines show selectivity for the mGluR1 subtype with nil or weak activities at the closely related mGluR5 subtype [442, 458–460]. One exception to this rule is the α-thioxanthylmethyl analog LY393675 (**223**), which has been shown to be equipotent at mGluR1 and mGluR5, with low-micromolar potency [461]. Likewise, (*S*)-hexyl-HIBO (**198**) is equipotent as an antagonist at mGluR1 and mGluR5 [446].

Based on modeling studies, it has been suggested that the antagonist action of the 4-carboxyphenylglycines is not caused by the phenyl ring in this series of compounds, but rather by the coplanar arrangement of the glycine moiety and the distal acidic group caused by the flat nonflexible phenyl ring [462]. In order to test this hypothesis, (*S*)-CBPG (**209**) and (*S*)-ACUDA (**210**) were synthesized [463, 464]. Both of these compounds retain the coplanar arrangement of the glycine moiety and the distal acidic group, and in agreement with their hypothesis, both compounds proved to be moderately potent mGluR1 antagonists with nil or weak activity at mGluR2, mGluR4, and mGluR5 [463, 464].

The phosphonic acid bioisostere of **203**, (PPG, **211**), is a potent group III agonist [465]. Based on the results from the 4-carboxyphenylglycines, the phosphonic acid bioisostere of **207**, MPPG (**212**) [466], and that of **208**, APICA (**213**) [467] have been designed. By analogy with results showing that α-methylation of **203** converts the compound from being an agonist into an antagonist at mGluR2 [267], **212** is an antagonist at group III mGluRs [467–469]. Compound **212** is slightly more potent as an antagonist at mGluR2, which is rather surprising given that neither (S)-AP4 (**200**) nor PPG (**211**) have any appreciable effect on this receptor subtype [467–469]. APICA (**213**) shows a pharmacological profile quite similar to that of MPPG (**212**), being slightly less potent as an mGluR2 antagonist than the parent compound [467].

α-Substitutions. As eluded to above, α-methylation has been widely used to derive antagonists from agonists. Some of these analogs are shown in Figure 18.38. Maintaining the selectivity profiles as of their antagonistic parent compounds, MAP4 (**214**) and MCCGI (**215**) antagonize mGluR2 and mGluR4, respectively, albeit with significantly reduced antagonist potency compared to the parent agonist [469, 470]. As mentioned above, α-methylation of the potent and selective group III agonist PPG (**211**), results in the mixed group II/III antagonist MPPG (**212**) [467–469]. Another example of swhich in group selectivity and loss of potency obtained by α-methylation is (S)-MetQUIS (**216**), which is a moderately potent antagonist at mGluR2 and a very weak antagonist at mGluR1 and mGluR5 [471]. Although some important compounds have been obtained, when the results of α-methylation are viewed as a whole, this does not seem to be a generally viable strategy for designing antagonists.

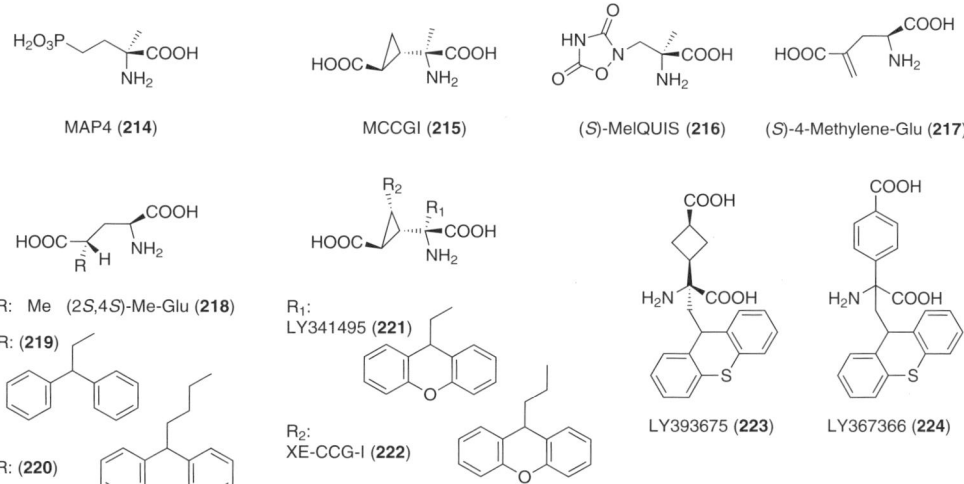

Figure 18.38 Structure of some mGluR ligands.

Substituting agonists with bulky, lipophilic side chains has been a much more successful approach to the design of potent antagonists. Two of the pioneering compounds in this class are 4-substituted analogs of Glu such as **219** and **220**. Both compounds are specific antagonists for mGluR2 and mGluR3 with potencies in the low-micromolar range [472, 473]. Interestingly, compounds with small substituents in the same position, such as (2S,4S)-Me-Glu (**218**) or (S)-4-methylene-Glu (**217**), are more potent agonists at mGluR2 than Glu, with some activity at mGluR1 but without appreciable activity at mGluR4 [474]. Thus by increasing the bulk and lipophilicity at the 4-position of these fly-swatter substituents, the selectivity for group II is retained, and even increased, but the compounds are converted from agonists to antagonists. More importantly, by contrast with α-methylation, the potency of the antagonists is retained compared to the parent agonist compounds.

These results have inspired the design of a number of new potent antagonists with fly-swatter side chains. MCCGI (**215**), the methylated analog of **102** (Fig. 18.25), was first substituted with fly-swatters in the 2-position. It was found that the most potent compound was obtained with a xanthylmethyl subsitituent conferring LY341495 (**221**), which displays antagonist activity at group II receptors in the low-nanomolar range [475, 476]. However, like the parent compound (2S,3R,4S)-CCG (**102**), LY341495 (**221**) also shows affinity for other subtypes, especially mGluR8 [477]. The synthesis and pharmacology of analogs with fly-swatter substituents at the 3′-position were recently reported. In this series, the compound with the xanthylethyl substituent, XE-CCG-I (**222**) was found to be the most potent antagonist, with potencies at mGluR2 and mGluR3 comparable to **221** [478]. Finally, the fly-swatter principle has been successfully applied to 3-carboxycyclobutylglycine, for example, LY393675 (**223**) [479] and to 4CPG (**203**), for example, LY367366 (**224**) [461]. Compounds **223** and **224**, respectively, show submicromolar and low-micromolar potency at both group I subtypes, which for the latter is quite remarkable given that most other phenylglycine analogs prefer the mGluR1 subtype [442, 458–461].

It can be concluded that in their antagonized state receptors from all three mGluR groups can accommodate quite large and lipophilic side chains in a variety of positions. Furthermore, compared with small α substituents such as methyl groups, which most often confer antagonists with reduced potency, the large fly-swatter substitutents in most cases confer antagonism and with increased potency. Thus the fly-swatter principle seems to be an attractive strategy for obtaining potent subtype-selective antagonists.

Allosteric Modulators of Metabotropic Glutamate Receptors

Noncompetitive Antagonists In 1996, CPCCOEt (**225**), a non-amino-acid compound with no structural similarity with Glu, was reported as a group I selective antagonist [480]. Since then, it has been shown that the compound is a selective mGluR1 noncompetitive antagonist with no mGluR5 activity, acting at the 7TM region rather than the agonist-binding ATD [481–483]. A number of other non-amino-acid mGluR antagonists have been discovered (see Fig. 18.39). BAY36-7620 (**226**) and EM-TBPC (**227**) are also mGluR1-specific antagonists acting at the 7TM domain with nanomolar potencies [484, 485]. The enol ether Ro 64-5229 (**228**) shows submicromolar noncompetitive antagonist potency at mGluR2 with no activity at mGluR1, mGluR3, mGluR4, mGluR5, NMDA, or AMPA receptors [486, 487].

SIB-1893 (**229**) and MPEP (**230**) have been reported [488, 489] to be selective, noncompetitive antagonists with nanomolar potencies at mGluR5, MPEP (**230**) being the most potent [488, 489]. Like CPCCOEt (**225**), MPEP (**230**) has been shown to act at the 7TM region rather than the agonist-binding ATD [490]. Recently, it was reported that SIB-1893 (**229**) and MPEP (**230**) also act as positive allosteric modulators of mGluR4, albeit with somewhat lower

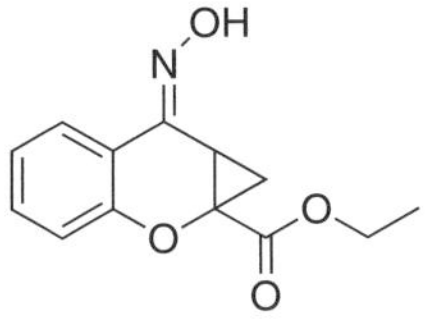

CPCCOEt (**225**)

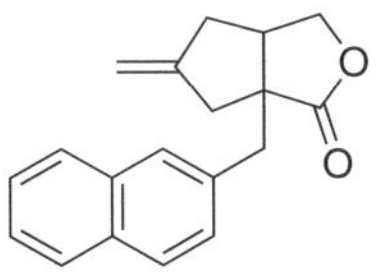

BAY36-7620 (**226**)

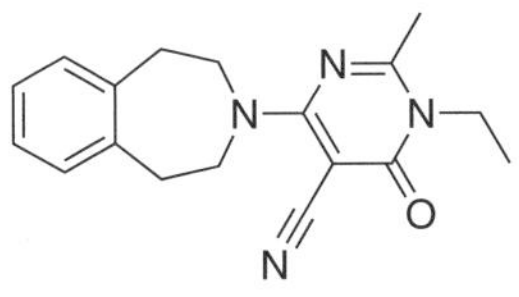

EM-TBPC (**227**)

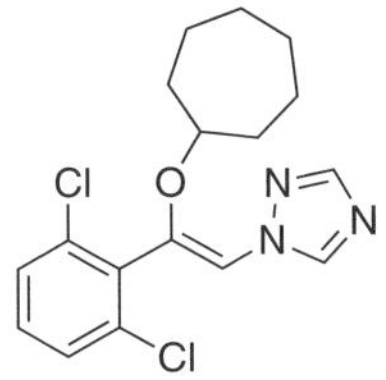

Ro 64-5229 (**228**)

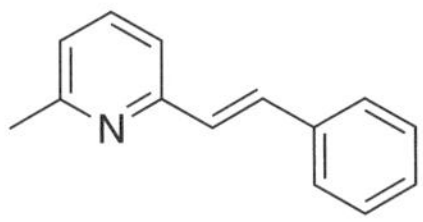

SIB-1893 (**229**)

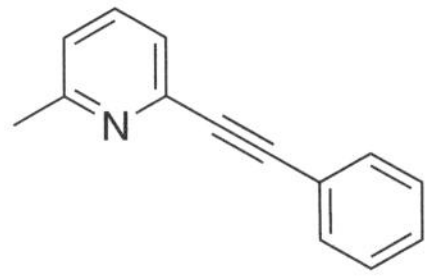

MPEP (**230**)

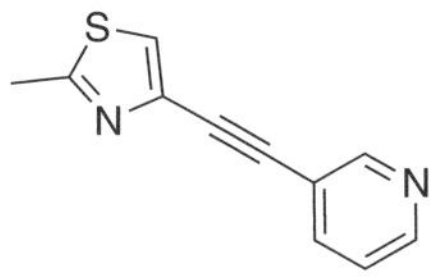

MTEP (**231**)

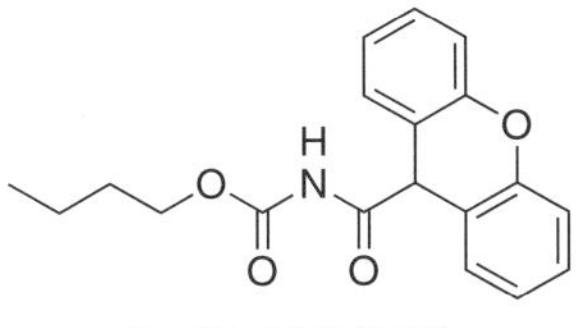

Ro 67-4853 (**232**)

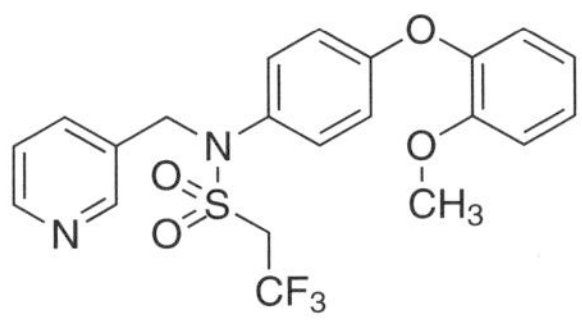

LY487379 (**233**)

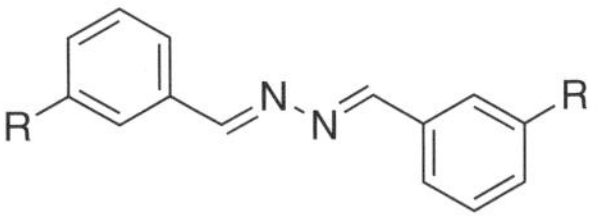

R	
F	DFB (**234**)
OCH_3	DMeOB (**235**)

Figure 18.39 Structures of some noncompetitive mGluR antagonists and positive allosteric modulators.

potency than the antagonism at mGluR5 [491]. Furthermore, MPEP (**230**) antagonizes NMDA receptors with low-micromolar potency [492], which has led to the design of the analog MTEP (**231**), which is slightly more potent than **230** as an antagonist at mGluR5 and shows no NMDA antagonist activity [492].

Positive Allosteric Modulators As already mentioned in the previous section, SIB-1893 (**229**) and MPEP (**230**) act as positive allosteric modulators at mGluR4. The allosteric effect is dependent upon Glu activation, and the compounds are thus unable to activate the mGluR4 receptor directly. Instead, the compounds enhance the response mediated by Glu, causing a left-ward shift of concentration–response curves and an increase in the maximum response [491]. Other compounds with similar pharmacological profiles at other mGluR subtypes have also been reported, including the mGluR1-specific positive allosteric modulator Ro 67-4853 (**232**) [220], the mGluR2-specific compound LY487379 (**233**) [493], and the mGluR5-specific compound DFB (**234**) [494]. Interestingly, DMeOB (**235**), which is a close analog of DFB, acts as a noncompetitive antagonist at mGluR5 [494], which together with the dual effects of SIB-1893 (**229**) and MPEP (**230**) at mGluR4 and mGluR5, illustrates the delicate balance of the activation mechanism of this class of G-protein-coupled receptors.

The allosteric modulators described in the preceding paragraphs have been discovered by high-throughput screening of compound libraries. It is interesting to note the diverse structures that have come out of these screens and that the compounds are highly selective even within the mGluR groups I, II, and III. As has been demonstrated for several of the compounds, most if not all of the noncompetitive antagonists act outside the agonist-binding pocket located in the ATD, which could very well be the explanation for the high degree of subtype selectivity. Given that the binding of the endogenous agonist Glu is of immense importance for proper receptor function, it is not surprising that this domain has been more highly conserved during evolution than the rest of the receptor. Likewise, it is not surprising that ligands acting outside the agonist-binding pocket show a higher degree of receptor selectivity than ligands that are structurally similar with Glu and act within the pocket. With this in mind, high-throughput screening is a useful step in the discovery of new subtype-specific ligands.

Metabotropic Glutamate Receptors as Putative Drug Targets As previously described, group I mGluRs stimulate phopholipase C and thus lead to cell excitation, whereas group II and III mGluRs inhibit adenylate cyclase, and thus cause cell inhibition. It has also been described that Glu is involved in many neurodegenerative, psychiatric, and cognitive disorders characterized by either hyper- or hypofunction of the glutamatergic system. Together, these observations have led to the pursuit of group I antagonists and group II/III agonists as potential therapeutics for disorders characterized by glutamatergic hyperfunction and likewise group I agonists and group II/III antagonists for disorders characterized by glutamatergic hyperfunction. Testing these ideas was initially hampered by a lack of selective, potent ligands, but the recent development of such compounds [e.g., LY354740 (**191**) and the allosteric modulators described in the previous section] has greatly enhanced our repertoire and confirmed that the mGluRs do indeed hold promise as novel therapeutic targets. It is beyond the scope of this chapter to describe these results in detail, and the reader is thus referred to several recent articles that review the preclinical results of mGluR ligands as drug candidates [495–498].

Structure of Metabotropic Glutamate Receptors

The current state of knowledge regarding the structural detail of mGluRs resembles that of iGluRs. Through recent breakthroughs in X-ray crystallography, we have obtained a series of snapshots of the extracellular binding domain of one of the subtypes at atomic resolution, mGluR1 [499, 500]. mGluRs belong to family C of the 7TM GPCRs, and at present, insight into the membrane-spanning region is limited to comparison with crystal structures such as that of rhodopsin, a family A GPCR [222]. There is strong evidence that mGluRs operate as homodimers, covalently bound via a disulfide bridge [501]. mGluRs are distinguished by the presence of a large extracellular N-terminal superdomain, which is homologous to a class of periplasmic binding proteins, such as L-arabinose binding protein [502]. Note the two large globular domains (I and II) connected by a hinge region and separated from the 7TM domain by a conserved cysteine-rich loop (Fig. 18.40).

While crystallization of the full-length receptor is unlikely in the near future, the first successful crystallization of an extracellular domain was reported in 2000, specifically mGluR1 in the presence and absence of the endogenous ligand Glu [398]. Of particular interest is the fact that according to these structures, the monomers can undergo a conformational change, or 70° "twist" with respect to one another. Furthermore, Glu induces domain closure of only one of the superdomains of the dimer, even though the ligand is bound to both ("closed–open", Fig. 18.41). Thus this crystal structure gives ligand design hints for both competitive agonists and antagonists and also suggests a role for the dimer interface. More recently [500], mGluR1 was co-crystallized with an antagonist, (*S*)-M4CPG (**207**), which confirms that competitive mGluR antagonists stabilize the "open–open" untwisted state. In addition, mGluR1 was co-crystallized in the presence of the positive modulator Gd^{3+} and Glu, revealing a "closed–closed" twisted state. Gd^{3+} appears to stabilize the twisted state via coordination to residues in the dimer interface, known previously to accommodate an engineered negatively modulatory Zn^{2+} site [503]. This leads to the following picture of mGluR activation (Fig. 18.41), where the free energy of ligand binding induces domain closure and twisting,

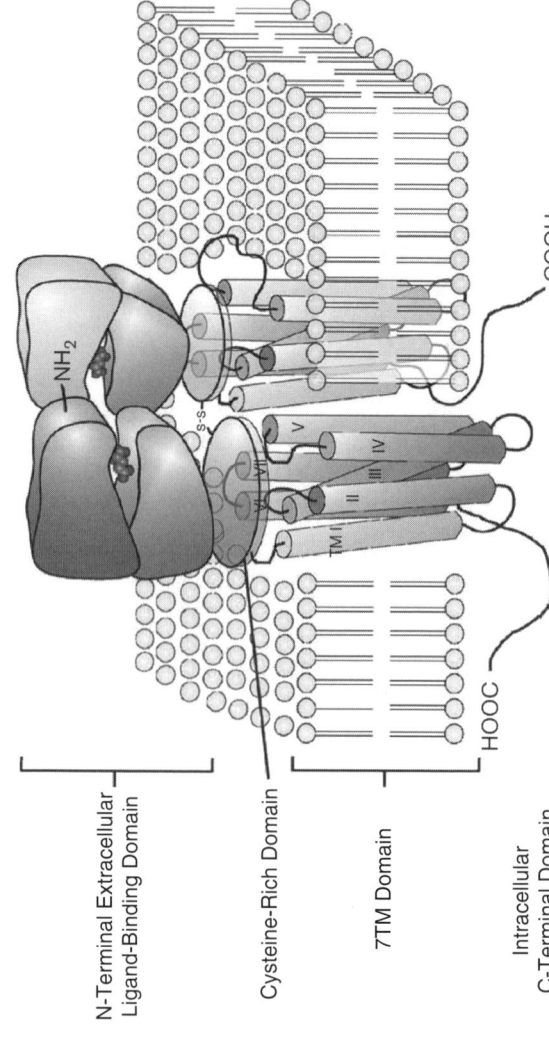

Figure 18.40 Illustration of the topology of the mGluR dimer. The structure of the N-terminal superdomain has been established by X-ray crystallography for mGluR1, whereas the 7TM region has been homology modeled based on the structure of rhodopsin (illustration from [495]). (This figure is available in full color at ftp://ftp.wiley.com/public/sci_tech_med/drug_discovery/.)

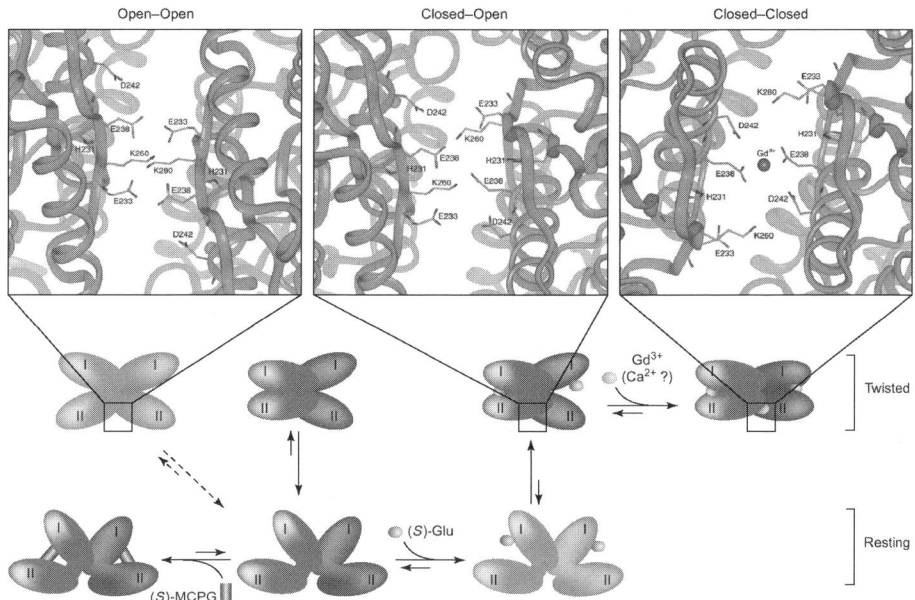

Figure 18.41 Illustration of the induced twisting and domain closure of the extracellular portion of mGluR1, presumably coupled to signal transduction via the 7TM domain. The changes in the dimer interface are highlighted above (illustration from [222]). (This figure is available in full color at ftp://ftp.wiley.com/public/sci_tech_med/drug_discovery/.)

which is then mechanically transferred via the 7TM domain to the G-protein by as yet unknown transformations.

Although the binding site sequence homology between mGluR1 and the other mGluRs is lower than that seen among iGluRs (Table 18.3), a number of attempts have been made to model other members of the family with a view to structure-based design and explanations of subtype selectivity. Such models have been particularly useful in conjunction with site-directed mutagenesis and subsequent pharmacological assays [504–507]. Whereas the distal but not the α-carboxylate of Glu forms an ion pair with the domain-closed iGluRs, the situation at mGluR is reversed: The distal carboxylate binds to a positively charged hydrophilic receptor region. As with iGluRs, binding is mediated by water molecules involved in hydrogen bond networks, which may be disrupted or displaced by other ligands.

While Glu binds to mGluRs in an extended conformation, there may be minor differences in the binding mode between subgroups, especially in the positioning and orientation of the distal carboxylate [508]. Note that while the residues that bind the α-amino acid are highly conserved, there is very little conservation about the distal carboxylate (Fig. 18.42). The binding site differences between the three groups of mGluRs may be exploited relatively easily for the design of subtype-specific ligands. Within members of a given group,

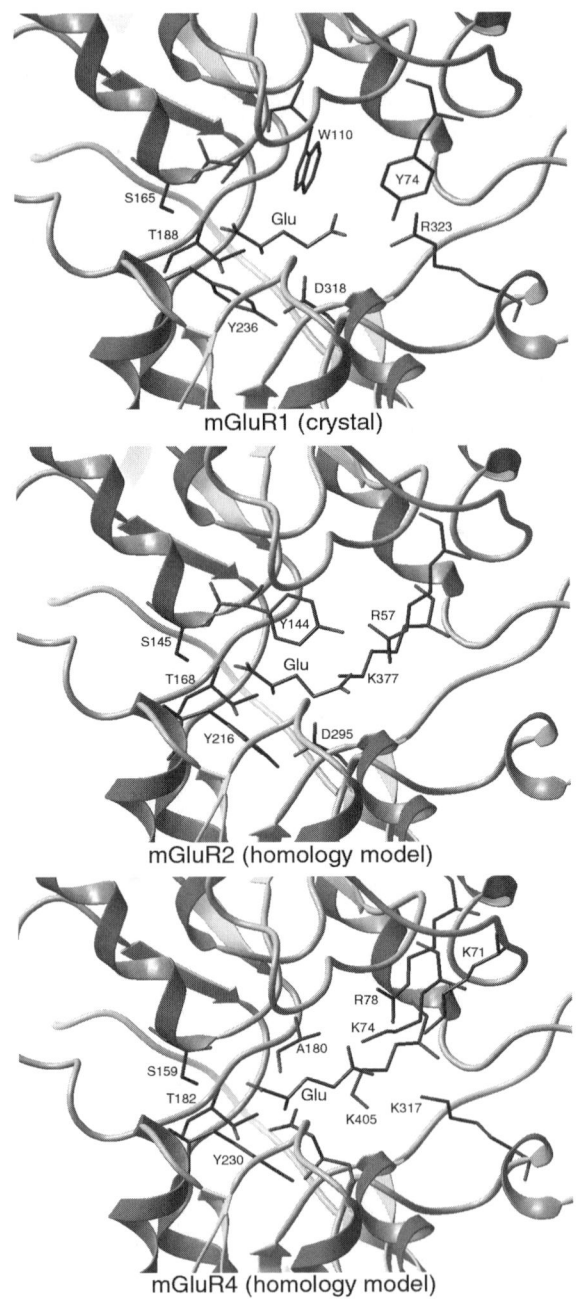

Figure 18.42 Domain-closed binding sites of mGluR1, mGluR2, and mGluR4 as representatives of groups I, II, and III mGluRs. mGluR2 and mGluR4 have been modeled from mGluR1 ([504, 508] and M. B. Hermit and J. R. Greenwood, unpublished data). (This figure is available in full color at ftp://ftp.wiley.com/public/sci_tech_med/drug_discovery/.)

TABLE 18.3 Binding Site Homology of mGluRs[a]

Residue	74	78	110	163	164	165	166	186	187	188	189	208	211	236	292	293	318	319	320	323	409
mGluR1	**Y**	**R**	W	G	S	S	S	**S**	A	**T**	S	**D**	Q	**Y**	**E**	**G**	**D**	G	W	**R**	**K**
mGluR5	Y	R	W	G	S	S	S	S	A	T	S	D	Q	Y	E	G	D	G	W	R	K
mGluR2	R	R	S	S	Y	S	D	A	S	T	S	D	Q	Y	R	S	D	G	W	L	K
mGluR3	R	R	S	S	Y	S	S	A	S	T	S	D	Q	Y	R	S	D	G	W	Q	K
mGluR4	K	R	S	S	G	S	S	A	S	T	A	D	Q	Y	N	E	D	S	W	K	K
mGluR6	Q	R	S	S	A	S	S	A	S	T	A	D	Q	Y	N	E	D	S	W	K	K
mGluR7	N	R	S	S	G	S	S	A	S	T	A	D	Q	Y	N	D	D	S	W	K	K
mGluR8	K	R	S	A	A	S	S	A	S	T	A	D	Q	Y	N	E	D	S	W	K	K

[a] Residues within 6 Å of Glu bound to mGluR1 listed. Residues in bold known to interact with Glu in mGluR1. Numbering according to mGluR1 (M. B. Hermit, unpublished data).

binding site conservation renders large selectivity differences difficult to obtain for competitive ligands.

REFERENCES

1. Bräuner-Osborne, H., Egebjerg, J., Nielsen, E. Ø., Madsen, U., Krogsgaard-Larsen, P. (2000). Ligands for glutamate receptors: Design and therapeutic prospects. *J. Med. Chem.*, *43*, 2609–2645.
2. Egebjerg, J., Schousboe, A., Krogsgaard-Larsen, P. (2002). *Glutamate and GABA Receptors and Transporters*. Taylor & Francis, London.
3. Barnard, E. A. (1992). Receptor classes and the transmitter-gated ion channels. *Trends. Biochem. Sci.*, *17*, 368–374.
4. Monaghan, D. T., Wenthold, R. J. (1997). *The Ionotropic Glutamate Receptors*. Humana Press, Totowa.
5. Johnston, G. A. (1996). GABAA receptor pharmacology. *Pharmacol. Ther.*, *69*, 173–198.
6. Krogsgaard-Larsen, P., Frølund, B., Jørgensen, F. S., Schousboe, A. (1994). $GABA_A$ receptor agonists, partial agonists and antagonists. Design and therapeutic prospects. *J. Med. Chem.*, *37*, 2489–2505.
7. Zefirova, O. N., Zefirov, N. S. (2000). Physiologically active compounds interacting with glutamate receptors. *Russ J. Org. Chem.*, *36*, 1231–1258.
8. Carlsson, M., Carlsson, A. (1990). Interactions between glutamatergic and monoaminergic systems within the basal ganglia—implications for schizophrenia and Parkinson's disease. *Trends Neurosci.*, *13*, 272–276.
9. Ebert, B., Madsen, U., Søby, K. K., Krogsgaard-Larsen, P. (1996). Functional partial agonism at ionotropic excitatory amino acid receptors. *Neurochem. Int.*, *29*, 309–316.
10. Dean, B., Hussain, T., Hayes, W., Scarr, E., Kitsoulis, S., Hill, C., Opeskin, K., Copolov, D. L. (1999). Changes in serotonin2A and GABA(A) receptors in schizophrenia: Studies on the human dorsolateral prefrontal cortex. *J. Neurochem.*, *72*, 1593–1599.
11. Ulas, J., Cotman, C. W. (1993). Excitatory amino acid receptors in schizophrenia. *Schizophrenia Bull.*, *19*, 105–117.
12. Lancel, M., Faulhaber, J. (1996). The GABAA agonist THIP (gaboxadol) increases non-REM sleep and enhances delta activity in the rat. *Neuroreport*, *7*, 2241–2245.
13. Sherif, F., Ahmed, S. S., Eriksson, L. (1995). Brain aminobutyrate aminotransferase activity in Alzheimer's disease. *Neurodegeneration*, *4*, 114–115.
14. Sherif, F. M., Ahmed, S. S. (1995). Basic aspects of GABA-transaminase in neuropsychiatric disorders. *Clin. Biochem.*, *28*, 145–154.
15. Jones-Davis, D. M., Macdonald, R. L. (2003). GABA(A) receptor function and pharmacology in epilepsy and status epilepticus. *Curr. Opin. Pharmacol.*, *3*, 12–18.
16. Choi, S., Silverman, R. B. (2002). Inactivation and inhibition of gamma-aminobutyric acid aminotransferase by conformationally restricted vigabatrin analogues. *J. Med. Chem.*, *45*, 4531–4539.

17. Choi, S., Storici, P., Schirmer, T., Silverman, R. B. (2002). Design of a conformationally restricted analogue of the antiepilepsy drug Vigabatrin that directs its mechanism of inactivation of gamma-aminobutyric acid aminotransferase. *J. Am. Chem. Soc.*, *124*, 1620–1624.

18. Johnson, T. R., Silverman, R. B. (1999). Syntheses of (*Z*)- and (*E*)-4-amino-2-(trifluoromethyl)-2-butenoic acid and their inactivation of gamma-aminobutyric acid aminotransferase. *Bioorg. Med. Chem.*, *7*, 1625–1636.

19. Nanavati, S. M., Silverman, R. B. (1989). Design of potential anticonvulsant agents: Mechanistic classification of GABA aminotransferase inactivators. *J. Med. Chem.*, *32*, 2413–2421.

20. Qiu, J., Silverman, R. B. (2000). A new class of conformationally rigid analogues of 4-amino-5-halopentanoic acids, potent inactivators of gamma-aminobutyric acid aminotransferase. *J. Med. Chem.*, *43*, 706–720.

21. Storici, P., Capitani, G., De Biase, D., Moser, M., John, R. A., Jansonius, J. N., Schirmer, T. (1999). Crystal structure of GABA-aminotransferase, a target for antiepileptic drug therapy. *Biochemistry*, *38*, 8628–8634.

22. Bouclier, M., Jung, M. J., Lippert, B. (1979). Stereochemistry of reactions catalysed by mammalian-brain L-glutamate 1-carboxy-lyase and 4-aminobutyrate: 2-Oxoglutarate aminotransferase. *Eur. J. Biochem.*, *98*, 363–368.

23. Bialer, M., Johannessen, S. I., Kupferberg, H. J., Levy, R. H., Loiseau, P., Perucca, E. (2002). Progress report on new antiepileptic drugs: A summary of the Sixth Eilat Conference (EILAT VI). *Epilepsy Res.*, *51*, 31–71.

24. Dewey, S. L., Morgan, A. E., Ashby, C. R., Jr., Horan, B., Kushner, S. A., Logan, J., Volkow, N. D., Fowler, J. S., Gardner, E. L., Brodie, J. D. (1998). A novel strategy for the treatment of cocaine addiction. *Synapse*, *30*, 119–129.

25. Gerasimov, M. R., Schiffer, W. K., Brodie, J. D., Lennon, I. C., Taylor, S. J., Dewey, S. L. (2000). Gamma-aminobutyric acid mimetic drugs differentially inhibit the dopaminergic response to cocaine. *Eur. J. Pharmacol.*, *395*, 129–135.

26. Stromberg, M. F., Mackler, S. A., Volpicelli, J. R., O'Brien, C. P., Dewey, S. L. (2001). The effect of gamma-vinyl-GABA on the consumption of concurrently available oral cocaine and ethanol in the rat. *Pharmacol. Biochem. Behav.*, *68*, 291–299.

27. Paul, S. R., Krauss, G. L., Miller, N. R., Medura, M. T., Miller, T. A., Johnson, M. A. (2001). Visual function is stable in patients who continue long-term vigabatrin therapy: Implications for clinical decision making. *Epilepsia*, *42*, 525–530.

28. Nousiainen, I., Mantyjarvi, M., Kalviainen, R. (2001). No reversion in vigabatrin-associated visual field defects. *Neurology*, *57*, 1916–1917.

29. Schiffer, W. K., Marsteller, D., Dewey, S. L. (2003). Sub-chronic low dose gamma-vinyl GABA (vigabatrin) inhibits cocaine-induced increases in nucleus accumbens dopamine. *Psychopharmacol. (Berl.)*, *168*, 339–343.

30. Qiu, J., Pingsterhaus, J. M., Silverman, R. B. (1999). Inhibition and substrate activity of conformationally rigid vigabatrin analogues with gamma-aminobutyric acid aminotransferase. *J. Med. Chem.*, *42*, 4725–4728.

31. Ashby, C. R., Jr., Paul, M., Gardner, E. L., Gerasimov, M. R., Dewey, S. L., Lennon, I. C., Taylor, S. J. (2002). Systemic administration of 1*R*,4*S*-4-amino-cyclopent-2-

ene-carboxylic acid, a reversible inhibitor of GABA transaminase, blocks expression of conditioned place preference to cocaine and nicotine in rats. *Synapse, 44*, 61–63.

32. Borden, L. A. (1996). GABA transporter heterogeneity: Pharmacology and cellular localization. *Neurochem. Int., 29*, 335–356.

33. Ribak, C. E., Tong, W. M., Brecha, N. C. (1996). Astrocytic processes compensate for the apparent lack of GABA transporters in the axon terminals of cerebellar Purkinje cells. *Anat. Embryol. (Berl.), 194*, 379–390.

34. Ribak, C. E., Tong, W. M., Brecha, N. C. (1996). GABA plasma membrane transporters, GAT-1 and GAT-3, display different distributions in the rat hippocampus. *J. Comp. Neurol., 367*, 595–606.

35. Johnson, J., Chen, T. K., Rickman, D. W., Evans, C., Brecha, N. C. (1996). Multiple gamma-aminobutyric acid plasma membrane transporters (GAT-1, GAT-2, GAT-3) in the rat retina. *J. Comp. Neurol., 375*, 212–224.

36. Yang, C. Y., Brecha, N. C., Tsao, E. (1997). Immunocytochemical localization of gamma-aminobutyric acid plasma membrane transporters in the tiger salamander retina. *J. Comp. Neurol., 389*, 117–126.

37. De Biasi, S., Vitellaro-Zuccarello, L., Brecha, N. C. (1998). Immunoreactivity for the GABA transporter-1 and GABA transporter-3 is restricted to astrocytes in the rat thalamus. A light and electron-microscopic immunolocalization. *Neuroscience, 83*, 815–828.

38. Minelli, A., DeBiasi, S., Brecha, N. C., Zuccarello, L. V., Conti, F. (1996). GAT-3, a high-affinity GABA plasma membrane transporter, is localized to astrocytic processes, and it is not confined to the vicinity of GABAergic synapses in the cerebral cortex. *J. Neurosci., 16*, 6255–6264.

39. Ebert, U., Krnjevic, K. (1990). Systemic CI-966, a new gamma-aminobutyric acid uptake blocker, enhances gamma-aminobutyric acid action in CA1 pyramidal layer in situ. *Can. J. Physiol. Pharmacol., 68*, 1194–1199.

40. Richards, D. A., Bowery, N. G. (1996). Comparative effects of the GABA uptake inhibitors, tiagabine and NNC-711, on extracellular GABA levels in the rat ventrolateral thalamus. *Neurochem. Res., 21*, 135–140.

41. Leker, R. R., Neufeld, M. Y. (2003). Anti-epileptic drugs as possible neuroprotectants in cerebral ischemia. *Brain Res. Rev., 42*, 187–203.

42. Rekling, J. C. (2003). Neuroprotective effects of anticonvulsants in rat hippocampal slice cultures exposed to oxygen/glucose deprivation. *Neurosci. Lett., 335*, 167–170.

43. O'Connell, A. W., Fox, G. B., Kjoller, C., Gallagher, H. C., Murphy, K. J., Kelly, J., Regan, C. M. (2001). Anti-ischemic and cognition-enhancing properties of NNC-711, a gamma-aminobutyric acid reuptake inhibitor. *Eur. J. Pharmacol., 424*, 37–44.

44. Backonja, M. M. (2002). Use of anticonvulsants for treatment of neuropathic pain. *Neurology, 59*, S14–S17.

45. Krogsgaard-Larsen, P., Frølund, B., Frydenvang, K. (2000). GABA uptake inhibitors. Design, molecular pharmacology and therapeutic aspects. *Curr. Pharm. Des., 6*, 1193–1209.

46. Johnston, G. A., Curtis, D. R., Beart, P. M., Game, C. J., McCulloch, R. M., Twitchin, B. (1975). Cis- and trans-4-aminocrotonic acid as GABA analogues of restricted conformation. *J. Neurochem.*, *24*, 157–160.
47. Johnston, G. A. R., Krogsgaard-Larsen, P., Stephanson, A. L., Twitchin, B. (1976). Inhibition of the uptake of GABA and related amino acids in rat brain slices by the optical isomers of nipecotic acid. *J. Neurochem.*, *26*, 1029–1032.
48. Krogsgaard-Larsen, P., Johnston, G. A. R. (1975). Inhibition of GABA uptake in rat brain slices by nipecotic acid, various isoxazoles and related compounds. *J. Neurochem.*, *25*, 797–802.
49. Schousboe, A., Thorbek, P., Hertz, L., Krogsgaard-Larsen, P. (1979). Effects of GABA analogues of restricted conformation on GABA transport in astrocytes and brain cortex slices and on GABA receptor binding. *J. Neurochem.*, *33*, 181–189.
50. Krogsgaard-Larsen, P., Falch, E., Larsson, O. M., Schousboe, A. (1987). GABA uptake inhibitors: Relevance to antiepileptic drug research. *Epilepsy Res.*, *1*, 77–93.
51. Ali, F. E., Bondinell, W. E., Dandridge, P. A., Frazee, J. S., Garvey, E., Girard, G. R., Kaiser, C., Ku, T. W., Lafferty, J. J., Moonsammy, G. I., Oh, H. J., Rush, J. A., Setler, P. E., Stringer, O. D., Venslavsky, J. W., Volpe, B. W., Yunger, L. M., Zirkle, C. L. (1985). Orally active and potent inhibitors of "gamma"-aminobutyric acid uptake. *J. Med. Chem.*, *28*, 653–660.
52. Genton, P., Guerrini, R., Perucca, E. (2001). Tiagabine in clinical practice. *Epilepsia*, *42* (Suppl. 3), 42–45.
53. Andersen, K. E., Lau, J., Lundt, B. F., Petersen, H., Huusfeldt, P. O., Suzdak, P. D., Swedberg, M. D. (2001). Synthesis of novel GABA uptake inhibitors. Part 6: Preparation and evaluation of N-Omega asymmetrically substituted nipecotic acid derivatives. *Bioorg. Med. Chem.*, *9*, 2773–2785.
54. Andersen, K. E., Sorensen, J. L., Huusfeldt, P. O., Knutsen, L. J., Lau, J., Lundt, B. F., Petersen, H., Suzdak, P. D., Swedberg, M. D. (1999). Synthesis of novel GABA uptake inhibitors. 4. Bioisosteric transformation and successive optimization of known GABA uptake inhibitors leading to a series of potent anticonvulsant drug candidates. *J. Med. Chem.*, *42*, 4281–4291.
55. Andersen, K. E., Sorensen, J. L., Lau, J., Lundt, B. F., Petersen, H., Huusfeldt, P. O., Suzdak, P. D., Swedberg, M. D. (2001). Synthesis of novel gamma-aminobutyric acid (GABA) uptake inhibitors. 5. (1) Preparation and structure-activity studies of tricyclic analogues of known GABA uptake inhibitors. *J. Med. Chem.*, *44*, 2152–2163.
56. Knutsen, L. J., Andersen, K. E., Lau, J., Lundt, B. F., Henry, R. F., Morton, H. E., Naerum, L., Petersen, H., Stephensen, H., Suzdak, P. D., Swedberg, M. D., Thomsen, C., Sorensen, P. O. (1999). Synthesis of novel GABA uptake inhibitors. 3. Diaryloxime and diarylvinyl ether derivatives of nipecotic acid and guvacine as anticonvulsant agents. *J. Med. Chem.*, *42*, 3447–3462.
57. Falch, E., Krogsgaard-Larsen, P. (1989). GABA uptake inhibitors containing mono- and diarylmethoxyalkyl N-substituents. *Drug. Des. Delivery.*, *4*, 205–215.
58. White, H. S., Sarup, A., Bolvig, T., Kristensen, A. S., Petersen, G., Nelson, N., Pickering, D. S., Larsson, O. M., Frølund, B., Krogsgaard-Larsen, P., Schousboe, A. (2002). Correlation between anticonvulsant activity and inhibitory action on glial

gamma-aminobutyric acid uptake of the highly selective mouse gamma-aminobutyric acid transporter 1 inhibitor 3-hydroxy-4-amino-4,5,6,7-tetrahydro-1,2-benzisoxazole and its N-alkylated analogs. *J. Pharmacol. Exp. Ther.*, *302*, 636–644.

59. Falch, E., Perregaard, J., Frølund, B., Søkilde, B., Buur, A., Hansen, L. M., Frydenvang, K., Brehm, L., Bolvig, T., Larsson, O. M., Sanchez, C., White, H. S., Schousboe, A., Krogsgaard-Larsen, P. (1999). Selective inhibitors of glial GABA uptake: Synthesis, absolute stereochemistry and pharmacology of the enantiomers of 3-hydroxy-4-amino-4,5,6,7-tetrahydro-1,2-benzisoxazole (exo-THPO) and analogues. *J. Med. Chem.*, *42*, 5402–5414.

60. Borden, L. A., Dhar, T. G., Smith, K. E., Branchek, T. A., Gluchowski, C., Weinshank, R. L. (1994). Cloning of the human homologue of the GABA transporter GAT-3 and identification of a novel inhibitor with selectivity for this site. *Receptors Channels*, *2*, 207–213.

61. Dhar, T. G., Borden, L. A., Tyagarajan, S., Smith, K. E., Branchek, T. A., Weinshank, R. L., Gluchowski, C. (1994). Design, synthesis and evaluation of substituted triarylnipecotic acid derivatives as GABA uptake inhibitors: Identification of a ligand with moderate affinity and selectivity for the cloned human GABA transporter GAT-3. *J. Med. Chem.*, *37*, 2334–2342.

62. Thomsen, C., Sørensen, P. O., Egebjerg, J. (1997). 1-(3-(9H-carbazol-9-yl)-1-propyl)-4-(2-methoxyphenyl)-4-piperidinol, a novel subtype selective inhibitor of the mouse type II GABA-transporter. *Br. J. Pharmacol.*, *120*, 983–985.

63. Dalby, N. O. (2000). GABA-level increasing and anticonvulsant effects of three different GABA uptake inhibitors. *Neuropharmacology*, *39*, 2399–2407.

64. Macdonald, R. L., Olsen, R. W. (1994). GABAA receptor channels. *Annu. Rev. Neurosci.*, *17*, 569–602.

65. Sieghart, W. (1995). Structure and pharmacology of "gamma"-aminobutyric acidA receptor subtypes. *Pharmacol. Rev.*, *47*, 181–234.

66. Doble, A., Martin, I. L. (1996). *The GABA$_A$/Benzodiazepine Receptor as a Target for Psychoactive Drugs*. Springer-Verlag, Heidelberg.

67. Barnard, E. A., Skolnick, P., Olsen, R. W., Mohler, H., Sieghart, W., Biggio, G., Braestrup, C., Bateson, A. N., Langer, S. Z. (1998). International Union of Pharmacology. XV. Subtypes of gamma-aminobutyric acid$_A$ receptors: classification on the basis of subunit structure and receptor function. *Pharmacol. Rev.*, *50*, 291–313.

68. Chebib, M., Johnston, G. A. R. (2000). GABA-Activated ligand gated ion channels: Medicinal chemistry and molecular biology. *J. Med. Chem.*, *43*, 1427–1447.

69. Korpi, E. R., Grunder, G., Luddens, H. (2002). Drug interactions at GABA(A) receptors. *Prog. Neurobiol.*, *67*, 113–159.

70. Frølund, B., Ebert, B., Kristiansen, U., Liljefors, T., Krogsgaard-Larsen, P. (2002). GABA$_A$ Receptor ligands and their therapeutic potentials. *Curr. Med. Chem.*, *2*, 817–832.

71. Krogsgaard-Larsen, P., Brehm, L., Schaumburg, K. (1981). Muscimol, a psychoactive constituent of *Amanita muscaria*, as a medicinal chemical model structure. *Acta. Chem. Scand. Ser B*, *B35*, 311–324.

REFERENCES

72. Krogsgaard-Larsen, P., Hjeds, H., Curtis, D. R., Lodge, D., Johnston, G. A. R. (1979). Dihydromuscimol, thiomuscimol and related heterocyclic compounds as GABA analogues. *J. Neurochem.*, *32*, 1717–1724.
73. Fowler, L. J., Lovell, D. H., John, R. A. (1983). Reaction of muscimol with 4-aminobutyrate aminotransferase. *J. Neurochem.*, *41*, 1751–1754.
74. Krogsgaard-Larsen, P., Johnston, G. A. R., Lodge, D., Curtis, D. R. (1977). A new class of GABA agonist. *Nature*, *268*, 53–55.
75. Kendall, D. A., Browner, M., Enna, S. J. (1982). Comparison of the antinociceptive effect of gamma-aminobutyric acid (GABA) agonists: Evidence for a cholinergic involvement. *J. Pharmacol. Exp. Ther.*, *220*, 482–487.
76. Krogsgaard-Larsen, P., Brehm, L., Johansen, J. S., Vinzents, P., Lauridsen, J., Curtis, D. R. (1985). Synthesis and structure-activity studies on excitatory amino acids structurally related to ibotenic acid. *J. Med. Chem.*, *28*, 673–679.
77. Krogsgaard-Larsen, P., Roldskov-Christiansen, T. (1979). GABA agonists. Synthesis and structure-activity studies on analogs of isoguvacine and THIP. *Eur. J. Med. Chem. Chim. Ther.*, *14*, 157–164.
78. Curtis, D. R., Duggan, A. W., Felix, D., Johnston, G. A. R. (1970). GABA, bicuculline and central inhibition. *Nature*, *226*, 1222–1224.
79. Johnston, G. A. R., Beart, P. M., Curtis, D. R., Game, C. J. A., McCulloch, R. M., MacLachlan, R. M. (1972). Bicuculline methochloride as a GABA antagonist. *Nature (New Biol.)*, *240*, 219–220.
80. Wermuth, C. G., Biziére, K. (1986). Pyridazinyl-GABA derivatives: A new class of synthetic $GABA_A$ antagonists. *Trends Pharmacol. Sci.*, *7*, 421–424.
81. Wermuth, C. G., Bourguignon, J.-J., Schlewer, G., Gies, J.-P., Schoenfelder, A., Melikian, A., Bouchet, M.-J., Chantreux, D., Molimard, J.-C., Heaulme, M., Chambon, J.-P., Biziere, K. (1987). Synthesis and structure-activity relationships of a series of aminopyridazine derivatives of γ-aminobutyric acid acting as selective GABA-A antagonists. *J. Med. Chem.*, *30*, 239–249.
82. Byberg, J. R., Labouta, I. M., Falch, E., Hjeds, H., Krogsgaard-Larsen, P., Curtis, D. R., Gynther, B. D. (1987). Synthesis and biological activity of a GABA-A agonist which has no effect on benzodiazepine binding and structurally related glycine antagonists. *Drug. Des. Delivery.*, *1*, 261–274.
83. Frølund, B., Kristiansen, U., Brehm, L., Hansen, A. B., Krogsgaard-Larsen, P., Falch, E. (1995). Partial $GABA_A$ receptor agonists. Synthesis and in vitro pharmacology of a series of nonannulated analogs of 4,5,6,7-tetrahydroisoxazolo[4,5-c]pyridin-3-ol (THIP). *J. Med. Chem.*, *38*, 3287–3296.
84. Mortensen, M., Frølund, B., Jørgensen, A. T., Liljefors, T., Krogsgaard-Larsen, P., Ebert, B. (2002). Activity of novel 4-PIOL analogues at human $\alpha_1\beta_2\gamma_{2\sigma}$ $GABA_A$ receptors—correlation with hydrophobicity. *Eur. J. Pharmacol.*, *451*, 125–132.
85. Kristiansen, U., Lambert, J. D. C., Falch, E., Krogsgaard-Larsen, P. (1991). Electrophysiological studies of the $GABA_A$ receptor ligand, 4-PIOL, on cultured hippocampal neurones. *Br. J. Pharmacol.*, *104*, 85–90.
86. Ebert, B., Mortensen, M., Thompson, S. A., Kehler, J., Wafford, K. A., Krogsgaard-Larsen, P. (2001). Bioisosteric determinants for subtype selectivity of ligands for heteromeric $GABA_A$ receptors. *Bioorg. Med. Chem. Lett.*, *11*, 1573–1577.

87. Ebert, B., Thompson, S. A., Suonatsou, K., McKernan, R., Krogsgaard-Larsen, P., Wafford, K. A. (1997). Differences in agonist/antagonist binding affinity and receptor transduction using recombinant human gamma-aminobutyric acid type A receptors. *Mol. Pharmacol.*, *52*, 1150–1156.
88. Hansen, S. L., Ebert, B., Fjalland, B., Kristiansen, U. (2001). Effects of $GABA_A$ receptor partial agonists in primary cultures of cerebellar granule neurons and cerebral cortical neurons reflect different receptor subunit compositions. *Br. J. Pharmacol.*, *133*, 539–549.
89. Frølund, B., Jørgensen, A. T., Tagmose, L., Stensbøl, T. B., Vestergaard, H. T., Engblom, C., Kristiansen, U., Sanchez, C., Krogsgaard-Larsen, P., Liljefors, T. (2002). A novel class of potent 4-arylalkyl substituted 3-isoxazolol $GABA_A$ antagonists: synthesis, pharmacology and molecular modeling. *J. Med. Chem.*, *45*, 2454–2468.
90. Ebert, B., Wafford, K. A., Whiting, P. J., Krogsgaard-Larsen, P., Kemp, J. A. (1994). Molecular pharmacology of γ-aminobutyric acid type A receptor agonists and partial agonists in oocytes injected with different α, β and γ receptor subunit combinations. *Mol. Pharmacol.*, *46*, 957–963.
91. Verdoorn, T. A. (1994). Formation of heteromeric gamma-aminobutyric acid type A receptors containing two different alpha subunits. *Mol. Pharmacol.*, *45*, 475–480.
92. Adkins, C. E., Pillai, G. V., Kerby, J., Bonnert, T. P., Haldon, C., McKernan, R. M., Gonzalez, J. E., Oades, K., Whiting, P. J., Simpson, P. B. (2001). Alpha4beta3delta GABA(A) receptors characterized by fluorescence resonance energy transfer-derived measurements of membrane potential. *J. Biol. Chem.*, *276*, 38934–38939.
93. Brown, N., Kerby, J., Bonnert, T. P., Whiting, P. J., Wafford, K. A. (2002). Pharmacological characterization of a novel cell line expressing human alpha(4)beta(3)delta GABA(A) receptors. *Br. J. Pharmacol.*, *136*, 965–974.
94. Saxena, N. C., Macdonald, R. L. (1996). Properties of putative cerebellar gamma-aminobutyric acid A receptor isoforms. *Mol. Pharmacol.*, *49*, 567–579.
95. Pirker, S., Schwarzer, C., Wieselthaler, A., Sieghart, W., Sperk, G. (2000). GABA(A) receptors: immunocytochemical distribution of 13 subunits in the adult rat brain. *Neuroscience*, *101*, 815–850.
96. Sur, C., Farrar, S. J., Kerby, J., Whiting, P. J., Atack, J. R., McKernan, R. M. (1999). Preferential coassembly of α4 and δ subunits of the gamma-aminobutyric acid$_A$ receptor in rat thalamus. *Mol. Pharmacol.*, *56*, 110–115.
97. Nusser, Z., Mody, I. (2002). Selective modulation of tonic and phasic inhibitions in dentate gyrus granule cells. *J. Neurophysiol.*, *87*, 2624–2628.
98. Storustovu, S., Ebert, B. (2003). Gaboxadol: In vitro interaction studies with benzodiazepines and ethanol suggest functional selectivity. *Eur. J. Pharmacol.*, *467*, 49–56.
99. McKernan, R. M., Rosahl, T. W., Reynolds, D. S., Sur, C., Wafford, K. A., Atack, J. R., Farrar, S., Myers, J., Cook, G., Ferris, P., Garrett, L., Bristow, L., Marshall, G., Macaulay, A., Brown, N., Howell, O., Moore, K. W., Carling, R. W., Street, L. J., Castro, J. L., Ragan, C. I., Dawson, G. R., Whiting, P. J. (2000). Sedative but not anxiolytic properties of benzodiazepines are mediated by the $GABA_A$ receptor α1 subtype. *Nat. Neurosci.*, *3*, 587–592.

100. Smit, A. B., Syed, N. I., Schaap, D., van Minnen, J., Klumperman, J., Kits, K. S., Lodder, H., van der Schors, R. C., van Elk, R., Sorgedrager, B., Brejc, K., Sixma, T. K., Geraerts, W. P. (2001). A glia-derived acetylcholine-binding protein that modulates synaptic transmission. *Nature*, *411*, 261–268.

101. Brejc, K., van Dijk, W. J., Klaassen, R. V., Schuurmans, M., van Der Oost, J., Smit, A. B., Sixma, T. K. (2001). Crystal structure of an ACh-binding protein reveals the ligand-binding domain of nicotinic receptors. *Nature*, *411*, 269–276.

102. Cromer, B. A., Morton, C. J., Parker, M. W. (2002). Anxiety over GABA(A) receptor structure relieved by AChBP. *Trends Biochem. Sci.*, *27*, 280–287.

103. Trudell, J. R. (2002). Unique assignment of inter-subunit association in GABA$_A$ $\alpha_1 \beta_3 \gamma_2$ receptors by molecular modelling. *Biochim. Biophys. Acta*, *1565*, 91–96.

104. Ernst, M., Brauchart, D., Boresch, S., Sieghart, W. (2003). comparative modeling of GABA(A) receptors: Limits, insights, future developments. *Neuroscience*, *119*, 933–943.

105. Liljefors, T., Pettersson, I. (2002). Computer-aided development and use of three-dimensional pharmacophore models. In P. Krogsgaard-Larsen, T. Liljefors, and U. Madsen (Eds.), *Textbook of Drug Design and Discovery*. Taylor and Francis, pp. 86–116, London.

106. Güner, O. F. (2000). *Pharmacophore Perception, Development, and Use in Drug Design.* International University Line, La Jolla.

107. Galvez-Ruano, E., Aprison, M. J., Robertson, D. H., Lipkowitz, K. B. (1995). Identifying agonistic and antagonistic mechanisms operative at the GABA receptor. *J. Neurosci. Res.*, *42*, 666–673.

108. Buur, J. R. B., Hjeds, H., Krogsgaard-Larsen, P., Jørgensen, F. S. (1993). Conformational analysis and molecular modelling of a partial GABAA agonist and a glycine antagonist related to the GABAA agonist, THIP. *Drug Des. Discov.*, *10*, 213–229.

109. Krogsgaard-Larsen, P., Mikkelsen, H., Jacobsen, P., Falch, E., Curtis, D. R., Peet, M. J., Leah, J. D. (1983). 4,5,6,7-Tetrahydroisothiazolo[5,4-c]pyridin-3-ol and related analogues of THIP. Synthesis and biological activity. *J. Med. Chem.*, *26*, 895–900.

110. Brehm, L., Frydenvang, K., Hansen, L. M., Norrby, P.-O., Krogsgaard-Larsen, P., Liljefors, T. (1997). Structural features of muscimol, a potent GABA$_A$ receptor agonist. Crystal structure and quantum chemical ab initio calculations. *Struct. Chem.*, *8*, 443–451.

111. Krogsgaard-Larsen, P., Johnston, G. A. R. (1978). Structure-activity studies on the inhibition of GABA binding to rat brain membranes by muscimol and related compounds. *J. Neurochem.*, *30*, 1377–1382.

112. Tagmose, L. (2000). A pharmacophore model for GABA$_A$ receptor agonists. Ph.D. Thesis, Royal Danish School of Pharmacy, Copenhagen.

113. Frølund, B., Tagmose, L., Liljefors, T., Stensbøl, T. B., Engblom, C., Kristiansen, U., Krogsgaard-Larsen, P. (2000). A novel class of potent 3-isoxazolol GABA$_A$ antagonists: Design, synthesis, and pharmacology. *J. Med. Chem.*, *43*, 4930–4933.

114. Tagmose, L., Hansen, L. M., Norrby, P.-O., Liljefors, T. (2000). Differences in agonist binding pattern for the GABA$_A$ and the AMPA receptors illustrated by high-level ab initio calculations. In K. Gundertofte and F. S. Jørgensen (Eds.),

Molecular Modelling and Prediction of Bioactivity. Kluwer Academic/Plenum Publishers, pp. 365–366, New York.

115. Westh-Hansen, S. E., Witt, M. R., Dekermendjian, K., Liljefors, T., Rasmussen, P. B., Nielsen, M. (1999). Arginine residue 120 of the human GABA$_A$ receptor α_1 subunit is essential for GABA binding and chloride ion current gating. *NeuroReport*, *10*, 2417–2421.

116. Boileau, A. J., Evers, A. R., Davis, A. F., Czajkowski, C. (1999). Mapping the agonist binding site of the GABAA receptor: Evidence for a beta-strand. *J. Neurosci.*, *19*, 4847–4854.

117. Hartvig, L., Lükensmejer, B., Liljefors, T., Dekermendjian, K. (2000). Two conserved arginines in the extracellular N-terminal domain of the GABA(A) receptor alpha(5) subunit are crucial for receptor function. *J. Neurochem.*, *75*, 1746–1753.

118. Nutt, D. J., Malizia, A. L. (2001). New insights into the role of the GABA(A)-benzodiazepine receptor in psychiatric disorder. *Br. J. Psychiatry*, *179*, 390–396.

119. Mohler, H., Fritschy, J. M., Rudolph, U. (2002). A new benzodiazepine pharmacology. *J. Pharmacol. Exp. Ther.*, *300*, 2–8.

120. Atack, J. R. (2003). Anxioselective compounds acting at the GABAA receptor benzodiazepine binding site. *Curr. Drug. Target CNS Neurol. Disord.*, *2*, 213–232.

121. Braestrup, C., Schmiechen, R., Neef, G., Nielsen, M., Petersen, E. N. (1982). Interaction of convulsive ligands with benzodiazepine receptors. *Science*, *216*, 1241–1243.

122. Arbilla, S., Allen, J., Wick, A., Langer, S. Z. (1986). High affinity [^{3}H]zolpidem binding in the rat brain: An imidazopyridine with agonist properties at central benzodiazepine receptors. *Eur. J. Pharmacol.*, *130*, 257–263.

123. Katsifis, A., Mardon, K., Mattner, F., Loc'h, C., McPhee, M. E., Dikic, B., Kassiou, M., Ridley, D. D. (2003). Pharmacological evaluation of (S)-8-[(123)I]iodobretazenil: A radioligand for in vivo studies of central benzodiazepine receptors. *Nucl. Med. Biol.*, *30*, 191–198.

124. Quirk, K., Blurton, P., Fletcher, S., Leeson, P., Tang, F., Mellilo, D., Ragan, C. I., McKernan, R. M. (1996). [^{3}H]L-655,708, a novel ligand selective for the benzodiazepine site of GABAA receptors which contain the alpha 5 subunit. *Neuropharmacology*, *35*, 1331–1335.

125. Farrar, S. J., Whiting, P. J., Bonnert, T. P., McKernan, R. M. (1999). Stoichiometry of a ligand-gated ion channel determined by fluorescence energy transfer. *J. Biol. Chem.*, *274*, 10100–10104.

126. Jensen, M. L., Timmermann, D. B., Johansen, T. H., Schousboe, A., Varming, T., Ahring, P. K. (2002). The beta subunit determines the ion selectivity of the GABAA receptor. *J. Biol. Chem.*, *277*, 41438–41447.

127. Smith, A. J., Alder, L., Silk, J., Adkins, C., Fletcher, A. E., Scales, T., Kerby, J., Marshall, G., Wafford, K. A., McKernan, R. M., Atack, J. R. (2001). Effect of alpha subunit on allosteric modulation of ion channel function in stably expressed human recombinant gamma-aminobutyric acid(A) receptors determined using (36)Cl ion flux. *Mol. Pharmacol.*, *59*, 1108–1118.

128. Rudolph, U., Crestani, F., Benke, D., Brunig, I., Benson, J. A., Fritschy, J. M., Martin, J. R., Bluethmann, H., Mohler, H. (1999). Benzodiazepine actions

mediated by specific gamma-aminobutyric acid$_A$ receptor subtypes. *Nature*, *401*, 796–800.

129. Crestani, F., Martin, J. R., Mohler, H., Rudolph, U. (2000). Mechanism of action of the hypnotic zolpidem in vivo. *Br. J. Pharmacol.*, *131*, 1251–1254.
130. Low, K., Crestani, F., Keist, R., Benke, D., Brunig, I., Benson, J. A., Fritschy, J. M., Rulicke, T., Bluethmann, H., Mohler, H., Rudolph, U. (2000). Molecular and neuronal substrate for the selective attenuation of anxiety. *Science*, *290*, 131–134.
131. Costa, E., Auta, J., Grayson, D. R., Matsumoto, K., Pappas, G. D., Zhang, X., Guidotti, A. (2002). GABAA receptors and benzodiazepines: A role for dendritic resident subunit mRNAs. *Neuropharmacology*, *43*, 925–937.
132. Navarro, J. F., Buron, E., Martin-Lopez, M. (2002). Anxiogenic-like activity of L-655,708, a selective ligand for the benzodiazepine site of GABA(A) receptors which contain the alpha-5 subunit, in the elevated plus-maze test. *Prog. Neuropsychopharmacol. Biol. Psychiatry*, *26*, 1389–1392.
133. Huang, Q., He, X., Ma, C., Liu, R., Yu, S., Dayer, C. A., Wenger, G. R., McKernan, R., Cook, J. M. (2000). Pharmacophore/receptor models for GABA(A)/BzR subtypes (α1β3γ2, α5β3γ2, and α6β3γ2) via a comprehensive ligand-mapping approach. *J. Med. Chem.*, *43*, 71–95.
134. Thomet, U., Baur, R., Scholze, P., Sieghart, W., Sigel, E. (1999). Dual mode of stimulation by the beta-carboline ZK 91085 of recombinant GABA(A) receptor currents: Molecular determinants affecting its action. *Br. J. Pharmacol.*, *127*, 1231–1239.
135. Dubinsky, B., Vaidya, A. H., Rosenthal, D. I., Hochman, C., Crooke, J. J., DeLuca, S., DeVine, A., Cheo-Isaacs, C. T., Carter, A. R., Jordan, A. D., Reitz, A. B., Shank, R. P. (2002). 5-Ethoxymethyl-7-fluoro-3-oxo-1,2,3,5-Tetrahydrobenzo[4,5]imidazo[1,2a]pyridine-4-N-(2-fluorophenyl)carboxamide (RWJ-51204), a new nonbenzodiazepine anxiolytic. *J. Pharmacol. Exp. Ther.*, *303*, 777–790.
136. Griebel, G., Perrault, G., Simiand, J., Cohen, C., Granger, P., Depoortere, H., Francon, D., Avenet, P., Schoemaker, H., Evanno, Y., Sevrin, M., George, P., Scatton, B. (2003). SL651498, a GABAA receptor agonist with subtype-selective efficacy, as a potential treatment for generalized anxiety disorder and muscle spasms. *CNS Drug. Rev.*, *9*, 3–20.
137. Hood, S. D., Argyropoulos, S. V., Nutt, D. J. (2000). Agents in development for anxiety disorders. *CNS Drugs*, *6*, 421–431.
138. Sanna, E., Busonero, F., Talani, G., Carta, M., Massa, F., Peis, M., Maciocco, E., Biggio, G. (2002). Comparison of the effects of zaleplon, zolpidem, and triazolam at various GABA(A) receptor subtypes. *Eur. J. Pharmacol.*, *451*, 103–110.
139. Inui, Y., Yamamoto, M., Awasaki, Y., Nishida, N. (1996). A nonbenzodiazepine partial agonist, S-(+)-DN-2327, has minimal physical dependence-producing liability, but shows cross-dependence on barbital in rats. *Prog. Neuropsychopharmacol. Biol. Psychiatry*, *20*, 1197–1211.
140. Tang, A. H., Franklin, S. R., Carter, D. B., Sethy, V. H., Needham, L. M., Jacobsen, E. J., Von Voigtlander, P. F. (1997). Anxiolytic-like effects of PNU-101017, a partial agonist at the benzodiazepine receptor. *Psychopharmacol. (Berl.)*, *131*, 255–263.
141. Im, H. K., Im, W. B., Pregenzer, J. F., Stratman, N. C., VonVoigtlander, P. F., Jacobsen, E. J. (1998). Two imidazoquinoxaline ligands for the benzodiazepine site

sharing a second low affinity site on rat GABA(A) receptors but with the opposite functionality. *Br. J. Pharmacol.*, *123*, 1490–1494.

142. Albaugh, P. A., Marshall, L., Gregory, J., White, G., Hutchison, A., Ross, P. C., Gallagher, D. W., Tallman, J. F., Crago, M., Cassella, J. V. (2002). Synthesis and biological evaluation of 7,8,9,10-tetrahydroimidazo[1,2-*c*]pyrido[3,4-e]pyrimdin-5 (6*H*)-ones as functionally selective ligands of the benzodiazepine receptor site on the GABA(A) receptor. *J. Med. Chem.*, *45*, 5043–5051.

143. Savini, L., Chiasserini, L., Pellerano, C., Biggio, G., Maciocco, E., Serra, M., Cinone, N., Carrieri, A., Altomare, C., Carotti, A. (2001). High affinity central benzodiazepine receptor ligands. Part 2: Quantitative structure-activity relationships and comparative molecular field analysis of pyrazolo[4,3-*c*]quinolin-3-ones. *Bioorg. Med. Chem.*, *9*, 431–444.

144. Dekermendjian, K., Kahnberg, P., Witt, M.-R., Sterner, O., Nielsen, M., Liljefors, T. (1999). Structure-activity relationships and molecular modelling analysis of flavonoids binding to the benzodiazepine site of the rat brain GABAA receptor complex. *J. Med. Chem.*, *42*, 4343–4350.

145. Marder, M., Paladini, A. C. (2002). GABA(A)-receptor ligands of flavonoid structure. *Curr. Top. Med. Chem.*, *2*, 853–867.

146. Nielsen, M., Frokjaer, S., Braestrup, C. (1988). High affinity of the naturally-occurring biflavonoid, amentoflavon, to brain benzodiazepine receptors in vitro. *Biochem. Pharmacol.*, *37*, 3285–3287.

147. Sigel, E., Baur, R., Furtmueller, R., Razet, R., Dodd, R. H., Sieghart, W. (2001). Differential cross talk of ROD compounds with the benzodiazepine binding site. *Mol. Pharmacol.*, *59*, 1470–1477.

148. Yasar, S., Bergman, J., Munzar, P., Redhi, G., Tober, C., Knebel, N., Zschiesche, M., Paronis, C. (2003). Evaluation of the novel antiepileptic drug, AWD 131–138, for benzodiazepine-like discriminative stimulus and reinforcing effects in squirrel monkeys. *Eur. J. Pharmacol.*, *465*, 257–265.

149. Chambers, M. S., Atack, J. R., Broughton, H. B., Collinson, N., Cook, S., Dawson, G. R., Hobbs, S. C., Marshall, G., Maubach, K. A., Pillai, G. V., Reeve, A. J., MacLeod, A. M. (2003). Identification of a novel, selective GABA(A) alpha5 receptor inverse agonist which enhances cognition. *J. Med. Chem.*, *46*, 2227–2240.

150. Collins, I., Davey, W. B., Rowley, M., Quirk, K., Bromidge, F. A., McKernan, R. M., Thompson, S. A., Wafford, K. A. (2000). *N*-(indol-3-ylglyoxylyl)piperidines: High affinity agonists of human GABA-A receptors containing the alpha1 subunit. *Bioorg. Med. Chem. Lett.*, *10*, 1381–1384.

151. Primofiore, G., Settimo, F. D., Taliani, S., Marini, A. M., Novellino, E., Greco, G., Lavecchia, A., Besnard, F., Trincavelli, L., Costa, B., Martini, C. (2001). Novel *N*-(arylalkyl)indol-3-ylglyoxylylamides targeted as ligands of the benzodiazepine receptor: Synthesis, biological evaluation, and molecular modeling analysis of the structure-activity relationships. *J. Med. Chem.*, *44*, 2286–2297.

152. Akbarzadeh, T., Tabatabai, S. A., Khoshnoud, M. J., Shafaghi, B., Shafiee, A. (2003). Design and synthesis of 4*H*-3-(2-phenoxy)phenyl-1,2,4-triazole derivatives as benzodiazepine receptor agonists. *Bioorg. Med. Chem.*, *11*, 769–773.

153. Covey, D. F., Evers, A. S., Mennerick, S., Zorumski, C. F., Purdy, R. H. (2001). Recent developments in structure-activity relationships for steroid modulators of GABA(A) receptors. *Brain Res. Brain Res. Rev.*, *37*, 91–97.

154. Lambert, J. J., Belelli, D., Harney, S. C., Peters, J. A., Frenguelli, B. G. (2001). Modulation of native and recombinant GABA(A) receptors by endogenous and synthetic neuroactive steroids. *Brain Res. Rev.*, 37, 68–80.

155. Lambert, J. J., Belelli, D., Hill-Venning, C., Peters, J. A. (1995). Neurosteroids and GABAA receptor function. *Trends Pharmacol. Sci.*, 16, 295–303.

156. Krasowski, M. D., Harrison, N. L. (1999). General anaesthetic actions on ligand-gated ion channels. *Cell. Mol. Life Sci.*, 55, 1278–1303.

157. Siegwart, R., Jurd, R., Rudolph, U. (2002). Molecular determinants for the action of general anesthetics at recombinant alpha(2)beta(3)gamma(2)gamma-aminobutyric acid(A) receptors. *J. Neurochem.*, 80, 140–148.

158. Wingrove, P. B., Wafford, K. A., Bain, C., Whiting, P. J. (1994). The modulatory action of loreclezole at the gamma-aminobutyric acid type A receptor is determined by a single amino acid in the beta 2 and beta 3 subunit. *Proc. Natl. Acad. Sci. USA*, 91, 4569–4573.

159. El Hadri, A., Abouabdellah, A., Thomet, U., Baur, R., Furtmuller, R., Sigel, E., Sieghart, W., Dodd, R. H. (2002). N-Substituted 4-amino-3,3-dipropyl-2(3H)-furanones: New positive allosteric modulators of the GABA(A) receptor sharing electrophysiological properties with the anticonvulsant loreclezole. *J. Med. Chem.*, 45, 2824–2831.

160. Belelli, D., Lambert, J. J., Peters, J. A., Wafford, K., Whiting, P. J. (1997). The interaction of the general anesthetic etomidate with the gamma-aminobutyric acid type A receptor is influenced by a single amino acid. *Proc. Natl. Acad. Sci. USA*, 94, 11031–11036.

161. Belelli, D., Muntoni, A. L., Merrywest, S. D., Gentet, L. J., Casula, A., Callachan, H., Madau, P., Gemmell, D. K., Hamilton, N. M., Lambert, J. J., Sillar, K. T., Peters, J. A. (2003). The in vitro and in vivo enantioselectivity of etomidate implicates the GABAA receptor in general anaesthesia. *Neuropharmacology*, 45, 57–71.

162. Hamon, A., Morel, A., Hue, B., Verleye, M., Gillardin, J. M. (2003). The modulatory effects of the anxiolytic etifoxine on GABA(A) receptors are mediated by the beta subunit. *Neuropharmacology*, 45, 293–303.

163. Halliwell, R. F., Thomas, P., Patten, D., James, C. H., Martinez-Torres, A., Miledi, R., Smart, T. G. (1999). Subunit-selective modulation of GABAA receptors by the non-steroidal anti-inflammatory agent, mefenamic acid. *Eur. J. Neurosci.*, 11, 2897–2905.

164. Miller, J. A., Dudley, M. W., Kehne, J. H., Sorensen, S. M., Kane, J. M. (1992). MDL 26,479: A potential cognition enhancer with benzodiazepine inverse agonist-like properties. *Br. J. Pharmacol.*, 107, 78–86.

165. Thompson, S. A., Wingrove, P. B., Connelly, L., Whiting, P. J., Wafford, K. A. (2002). Tracazolate reveals a novel type of allosteric interaction with recombinant gamma-aminobutyric acid(A) receptors. *Mol. Pharmacol.*, 61, 861–869.

166. Krasowski, M. D., Hong, X., Hopfinger, A. J., Harrison, N. L. (2002). 4D-QSAR analysis of a set of propofol analogues: Mapping binding sites for an anesthetic phenol on the GABA(A) receptor. *J. Med. Chem.*, 45, 3210–3221.

167. Meadows, H. J., Kumar, C. S., Pritchett, D. B., Blackburn, T. P., Benham, C. D. (1998). SB-205384: A GABA(A) receptor modulator with novel mechanism of action that shows subunit selectivity. *Br. J. Pharmacol.*, 123, 1253–1259.

168. van Rijn, C. M., Willems-van Bree, E. (2003). Synergy between retigabine and GABA in modulating the convulsant site of the GABAA receptor complex. *Eur. J. Pharmacol.*, *464*, 95–100.

169. Robinson, R. T., Drafts, B. C., Fisher, J. L. (2003). Fluoxetine increases GABA(A) receptor activity through a novel modulatory site. *J. Pharmacol. Exp. Ther.*, *304*, 978–984.

170. Asproni, B., Pau, A., Bitti, M., Melosu, M., Cerri, R., Dazzi, L., Seu, E., Maciocco, E., Sanna, E., Busonero, F., Talani, G., Pusceddu, L., Altomare, C., Trapani, G., Biggio, G. (2002). Synthesis and pharmacological evaluation of 1-[(1,2-diphenyl-1*H*-4-imidazolyl)methyl]-4-phenylpiperazines with clozapine-like mixed activities at dopamine D(2), serotonin, and GABA(A) receptors. *J. Med. Chem.*, *45*, 4655–4668.

171. Hold, K. M., Sirisoma, N. S., Ikeda, T., Narahashi, T., Casida, J. E. (2000). Alpha-thujone (the active component of absinthe): Gamma-aminobutyric acid type A receptor modulation and metabolic detoxification. *Proc. Natl. Acad. Sci. USA*, *97*, 3826–3831.

172. Casida, J. E. (1993). Insecticide action at the GABA-gated chloride channel: Recognition, progress, and prospects. *Arch. Insect. Biochem. Physiol.*, *22*, 13–23.

173. Borea, P. A., Gilli, G., Bertolasi, V., Ferretti, V. (1987). Stereochemical features controlling binding and intrinsic activity properties of benzodiazepine-receptor ligands. *Mol. Pharmacol.*, *31*, 334–344.

174. Zhang, W., Koehler, K. F., Zhang, P., Cook, J. M. (1995). Development of a comprehensive pharmacophore model for the benzodiazepine receptor. *Drug. Des. Discov.*, *12*, 193–248.

175. He, X., Zhang, C., Cook, J. M. (2001). Model of the BzR binding site: Correlation of the data from site-directed mutagenesis and the pharmacophore/receptor model. *Med. Chem. Res.*, *10*, 269–308.

176. Hong, X., Hopfinger, A. J. (2003). 3D-pharmacophores of flavonoid binding at the benzodiazepine GABA(A) receptor site using 4D-QSAR analysis. *J. Chem. Inf. Comput. Sci.*, *43*, 324–336.

177. Huang, X., Liu, T., Gu, J., Luo, X., Ji, R., Cao, Y., Xue, H., Wong, J. T., Wong, B. L., Pei, G., Jiang, H., Chen, K. (2001). 3D-QSAR model of flavonoids binding at benzodiazepine site in GABAA receptors. *J. Med. Chem.*, *44*, 1883–1891.

178. Kahnberg, P., Lager, E., Rosenberg, C., Schougaard, J., Camet, L., Sterner, O., Ostergaard Nielsen, E., Nielsen, M., Liljefors, T. (2002). Refinement and evaluation of a pharmacophore model for flavone derivatives binding to the benzodiazepine site of the GABA(A) receptor. *J. Med. Chem.*, *45*, 4188–4201.

179. Marder, M., Estiu, G., Blanch, L. B., Viola, H., Wasowski, C., Medina, J. H., Paladini, A. C. (2001). Molecular modeling and QSAR analysis of the interaction of flavone derivatives with the benzodiazepine binding site of the GABA(A) receptor complex. *Bioorg. Med. Chem.*, *9*, 323–335.

180. Diaz-Arauzo, H., Evoniuk, G. E., Skolnick, P., Cook, J. M. (1991). The agonist pharmacophore of the benzodiazepine receptor. Synthesis of a selective anticonvulsant/anxiolytic. *J. Med. Chem.*, *34*, 1754–1756.

181. Diaz-Arauzo, H., Koehler, K. F., Hagen, T. J., Cook, J. M. (1991). Synthetic and computer assisted analysis of the pharmacophore for agonists at benzodiazepine receptors. *Life Sci., 49*, 207–216.

182. Cox, E. D., Diaz-Arauzo, H., Huang, Q., Reddy, M. S., Ma, C., Harris, B., McKernan, R., Skolnick, P., Cook, J. M. (1998). Synthesis and evaluation of analogues of the partial agonist 6-(propyloxy)-4-(methoxymethyl)-beta-carboline-3-carboxylic acid ethyl ester (6-PBC) and the full agonist 6-(benzyloxy)-4-(methoxymethyl)-beta-carboline-3-carboxylic acid ethyl ester (Zk 93423) at wild type and recombinant GABAA receptors. *J. Med. Chem., 41*, 2537–2552.

183. Da Settimo, A., Primofiore, G., Da Settimo, F., Marini, A. M., Novellino, E., Greco, G., Gesi, M., Martini, C., Giannaccini, G., Lucacchini, A. (1998). N'-Phenylindol-3-ylglyoxylohydrazide derivatives: Synthesis, structure-activity relationships, molecular modeling studies, and pharmacological action on brain benzodiazepine receptors. *J. Med. Chem., 41*, 3821–3830.

184. Da Settimo, A., Primofiore, G., Da Settimo, F., Marini, A. M., Novellino, E., Greco, G., Martini, C., Giannaccini, G., Lucacchini, A. (1996). Synthesis, structure-activity relationships, and molecular modeling studies of N-(indol-3-ylglyoxylyl)benzylamine derivatives acting at the benzodiazepine receptor. *J. Med. Chem., 39*, 5083–5091.

185. Kahnberg, P., Howard, M. H., Liljefors, T., Nielsen, M., Nielsen, E.ø., Sterner, O., Pettersson, I. (2004). The use of a pharmacophore model for identification of novel ligands for the benzodiazepine binding site of the GABA$_A$ receptor *J. Mol. Graph. Model., 23*, 253–261.

186. He, X., Huang, Q., Ma, C., Yu, S., McKernan, R., Cook, J. M. (2000). Pharmacophore/receptor models for GABA(A)/BzR α2β3γ2, α3β3γ2 and α4β3γ2 recombinant subtypes. Included volume analysis and comparison to α1β3γ2, α5β3γ2, and α6β3γ2 subtypes. *Drug. Des. Discov., 17*, 131–171.

187. Filizola, M., Harris, D. L., Loew, G. H. (2000). Development of a 3D pharmacophore for nonspecific ligand recognition of α1, α2, α3, α5, and α6 containing GABA(A)/benzodiazepine receptors. *Bioorg. Med. Chem., 8*, 1799–1807.

188. Allen, M. S., LaLoggia, A. J., Dorn, L. J., Martin, M. J., Costantino, G., Hagen, T. J., Koehler, K. F., Skolnick, P., Cook, J. M. (1992). Predictive binding of beta-carboline inverse agonists and antagonists via the CoMFA/GOLPE approach. *J. Med. Chem., 35*, 4001–4010.

189. Allen, M. S., Tan, Y. C., Trudell, M. L., Narayanan, K., Schindler, L. R., Martin, M. J., Schultz, C., Hagen, T. J., Koehler, K. F., Codding, P. W., et al. (1990). Synthetic and computer-assisted analyses of the pharmacophore for the benzodiazepine receptor inverse agonist site. *J. Med. Chem., 33*, 2343–2357.

190. Bowery, N. G., Hill, D. R., Hudson, A. L., Doble, A., Middlemiss, D. N., Shaw, J., Turnbull, M. (1980). (−)-Baclofen decreases neurotransmitter release in the mammalian CNS by an action at a novel GABA receptor. *Nature, 283*, 92–94.

191. Hill, D. R., Bowery, N. G. (1981). ^{3}H-Baclofen and ^{3}H-GABA bind to bicuculline-insensitive GABA$_B$ sites in rat brain. *Nature, 290*, 149–152.

192. Bowery, N. G., Bettler, B., Froestl, W., Gallagher, J. P., Marshall, F., Raiteri, M., Bonner, T. I., Enna, S. J. (2002). International Union of Pharmacology. XXXIII.

Mammalian gamma-aminobutyric acid(B) receptors: Structure and function. *Pharmacol. Rev.*, *54*, 247–264.

193. Kaupmann, K., Huggel, K., Heid, J., Flor, P. J., Bischoff, S., Mickel, S. J., McMaster, G., Angst, C., Bittiger, H., Froestl, W., Bettler, B. (1997). Expression cloning of GABA(B) receptors uncovers similarity to metabotropic glutamate receptors. *Nature*, *386*, 239–246.

194. Marshall, F. H., Jones, K. A., Kaupmann, K., Bettler, B. (1999). GABAB receptors—the first 7TM heterodimers. *Trends Pharmacol. Sci.*, *20*, 396–399.

195. Kaupmann, K., Malitschek, B., Schuler, V., Heid, J., Froestl, W., Beck, P., Mosbacher, J., Bischoff, S., Kulik, A., Shigemoto, R., Karschin, A., Bettler, B. (1998). GABA(B)-receptor subtypes assemble into functional heteromeric complexes. *Nature*, *396*, 683–687.

196. Jones, K. A., Borowsky, B., Tamm, J. A., Craig, D. A., Durkin, M. M., Dai, M., Yao, W. J., Johnson, M., Gunwaldsen, C., Huang, L. Y., Tang, C., Shen, Q., Salon, J. A., Morse, K., Laz, T., Smith, K. E., Nagarathnam, D., Noble, S. A., Branchek, T. A., Gerald, C. (1998). GABA(B) receptors function as a heteromeric assembly of the subunits GABA(B)R1 and GABA(B)R2. *Nature*, *396*, 674–679.

197. White, J. H., Wise, A., Main, M. J., Green, A., Fraser, N. J., Disney, G. H., Barnes, A. A., Emson, P., Foord, S. M., Marshall, F. H. (1998). Heterodimerization is required for the formation of a functional GABA(B) receptor. *Nature*, *396*, 679–682.

198. Kuner, R., Kohr, G., Grunewald, S., Eisenhardt, G., Bach, A., Kornau, H. C. (1999). Role of heteromer formation in GABAB receptor function. *Science*, *283*, 74–77.

199. Malitschek, B., Schweizer, C., Keir, M., Heid, J., Froestl, W., Mosbacher, J., Kuhn, R., Henley, J., Joly, C., Pin, J. P., Kaupmann, K., Bettler, B. (1999). The N-terminal domain of gamma-aminobutyric acid(B) receptors is sufficient to specify agonist and antagonist binding. *Mol. Pharmacol.*, *56*, 448–454.

200. Kniazeff, J., Galvez, T., Labesse, G., Pin, J. P. (2002). No ligand binding in the GB2 subunit of the GABA(B) receptor is required for activation and allosteric interaction between the subunits. *J. Neurosci.*, *22*, 7352–7361.

201. Margeta-Mitrovic, M., Jan, Y. N., Jan, L. Y. (2000). A trafficking checkpoint controls GABA(B) receptor heterodimerization. *Neuron*, *27*, 97–106.

202. Pagano, A., Rovelli, G., Mosbacher, J., Lohmann, T., Duthey, B., Stauffer, D., Ristig, D., Schuler, V., Meigel, I., Lampert, C., Stein, T., Prezeau, L., Blahos, J., Pin, J., Froestl, W., Kuhn, R., Heid, J., Kaupmann, K., Bettler, B. (2001). C-terminal interaction is essential for surface trafficking but not for heteromeric assembly of GABA(b) receptors. *J. Neurosci.*, *21*, 1189–1202.

203. Calver, A. R., Robbins, M. J., Cosio, C., Rice, S. Q., Babbs, A. J., Hirst, W. D., Boyfield, I., Wood, M. D., Russell, R. B., Price, G. W., Couve, A., Moss, S. J., Pangalos, M. N. (2001). The C-terminal domains of the GABA(b) receptor subunits mediate intracellular trafficking but are not required for receptor signaling. *J. Neurosci.*, *21*, 1203–1210.

204. Galvez, T., Duthey, B., Kniazeff, J., Blahos, J., Rovelli, G., Bettler, B., Prezeau, L., Pin, J. P. (2001). Allosteric interactions between GB1 and GB2 subunits are required for optimal GABA(B) receptor function. *EMBO J.*, *20*, 2152–2159.

205. Margeta-Mitrovic, M., Jan, Y. N., Jan, L. Y. (2001). Function of GB1 and GB2 subunits in G protein coupling of GABA(B) receptors. *Proc. Natl. Acad. Sci. USA*, 98, 14649–14654.

206. Havlickova, M., Prezeau, L., Duthey, B., Bettler, B., Pin, J. P., Blahos, J. (2002). The intracellular loops of the GB2 subunit are crucial for G-protein coupling of the heteromeric gamma-aminobutyrate B receptor. *Mol. Pharmacol.*, 62, 343–350.

207. Duthey, B., Caudron, S., Perroy, J., Bettler, B., Fagni, L., Pin, J. P., Prezeau, L. (2002). A single subunit (GB2) is required for G-protein activation by the heterodimeric GABA(B) receptor. *J. Biol. Chem.*, 277, 3236–3241.

208. Robbins, M. J., Calver, A. R., Filippov, A. K., Hirst, W. D., Russell, R. B., Wood, M. D., Nasir, S., Couve, A., Brown, D. A., Moss, S. J., Pangalos, M. N. (2001). GABA(B2) is essential for G-protein coupling of the GABA(B) receptor heterodimer. *J. Neurosci.*, 21, 8043–8052.

209. Campbell, S. K., Almeida, G. L., Penn, R. D., Corcos, D. M. (1995). The effects of intrathecally administered baclofen on function in patients with spasticity. *Phys. Ther.*, 75, 352–362.

210. Froestl, W., Mickel, S. J., Hall, R. G., von Sprecher, G., Strub, D., Baumann, P. A., Brugger, F., Gentsch, C., Jaekel, J., Olpe, H. R., Rihs, G., Vassout, A., Waldmeier, P. C., Bittiger, H. (1995). Phosphinic acid analogues of GABA. 1. New potent and selective GABAB agonists. *J. Med. Chem.*, 38, 3297–3312.

211. Kerr, D. I., Ong, J., Prager, R. H., Gynther, B. D., Curtis, D. R. (1987). Phaclofen: A peripheral and central baclofen antagonist. *Brain Res.*, 405, 150–154.

212. Kerr, D. I., Ong, J., Johnston, G. A., Abbenante, J., Prager, R. H. (1989). Antagonism at GABAB receptors by saclofen and related sulphonic analogues of baclofen and GABA. *Neurosci. Lett.*, 107, 239–244.

213. Kerr, D. I., Ong, J., Johnston, G. A., Abbenante, J., Prager, R. H. (1988). 2-Hydroxysaclofen: An improved antagonist at central and peripheral GABAB receptors. *Neurosci. Lett.*, 92, 92–96.

214. Froestl, W., Mickel, S. J., von Sprecher, G., Diel, P. J., Hall, R. G., Maier, L., Strub, D., Melillo, V., Baumann, P. A., Bernasconi, R., Gentsch, C., Hauser, K., Jaekel, J., Karlsson, G., Klebs, K., Maltre, L., Mondadori, C., Olpe, H. R., Walmeier, P. C., Bittiger, H. (1995). Phosphinic acid analogues of GABA. 2. Selective, orally active GABAB antagonists. *J. Med. Chem.*, 38, 3313–3331.

215. Froestl, W., Bettler, B., Bittiger, H., Heid, J., Kaupmann, K., Mickel, S. J., Strub, D. (2003). Ligands for expression cloning and isolation of GABA(B) receptors. *Farmaco*, 58, 173–183.

216. Jensen, A. A., Madsen, B. E., Krogsgaard-Larsen, P., Bräuner-Osborne, H. (2001). Pharmacological characterization of homobaclofen on wild type and mutant GABA(B)1b receptors coexpressed with the GABA(B)2 receptor. *Eur. J. Pharmacol.*, 417, 177–180.

217. Galvez, T., Urwyler, S., Prezeau, L., Mosbacher, J., Joly, C., Malitschek, B., Heid, J., Brabet, I., Froestl, W., Bettler, B., Kaupmann, K., Pin, J. P. (2000). Ca(2+) requirement for high-affinity gamma-aminobutyric acid (GABA) binding at GABA(B) receptors: Involvement of serine 269 of the GABA(B)R1 subunit. *Mol. Pharmacol.*, 57, 419–426.

218. Pin, J. P., Parmentier, M. L., Prezeau, L. (2001). Positive allosteric modulators for gamma-aminobutyric acid(B) receptors open new routes for the development of drugs targeting family 3 G-protein-coupled receptors. *Mol. Pharmacol.*, *60*, 881–884.
219. Urwyler, S., Mosbacher, J., Lingenhoehl, K., Heid, J., Hofstetter, K., Froestl, W., Bettler, B., Kaupmann, K. (2001). Positive allosteric modulation of native and recombinant gamma-aminobutyric acid(B) receptors by 2,6-di-*tert*-butyl-4-(3-hydroxy-2,2-dimethyl-propyl)-phenol (CGP7930) and its aldehyde analog CGP13501. *Mol. Pharmacol.*, *60*, 963–971.
220. Knoflach, F., Mutel, V., Jolidon, S., Kew, J. N., Malherbe, P., Vieira, E., Wichmann, J., Kemp, J. A. (2001). Positive allosteric modulators of metabotropic glutamate 1 receptor: Characterization, mechanism of action, and binding site. *Proc. Natl. Acad. Sci. USA*, *98*, 13402–13407.
221. Hauache, O. M., Hu, J., Ray, K., Xie, R., Jacobson, K. A., Spiegel, A. M. (2000). Effects of a calcimimetic compound and naturally activating mutations on the human Ca2+ receptor and on Ca2+ receptor/metabotropic glutamate chimeric receptors. *Endocrinology*, *141*, 4156–4163.
222. Jensen, A. A., Greenwood, J. R., Bräuner-Osborne, H. (2002). The dance of the clams: Twists and turns in the family C GPCR homodimer. *Trends Pharmacol. Sci.*, *23*, 491–493.
223. Drew, C. A., Johnston, G. A., Weatherby, R. P. (1984). Bicuculline-insensitive GABA receptors: Studies on the binding of (−)-baclofen to rat cerebellar membranes. *Neurosci. Lett.*, *52*, 317–321.
224. Shimada, S., Cutting, G., Uhl, G. R. (1992). Gamma-aminobutyric acid A or C receptor? Gamma-aminobutyric acid rho 1 receptor RNA induces bicuculline-, barbiturate-, and benzodiazepine-insensitive gamma-aminobutyric acid responses in *Xenopus* oocytes. *Mol. Pharmacol.*, *41*, 683–687.
225. Enz, R. (2001). GABA(C) receptors: A molecular view. *Biol. Chem.*, *382*, 1111–1122.
226. Zhang, D., Pan, Z. H., Awobuluyi, M., Lipton, S. A. (2001). Structure and function of GABA(C) receptors: A comparison of native versus recombinant receptors. *Trends Pharmacol. Sci.*, *22*, 121–132.
227. Cutting, G. R., Curristin, S., Zoghbi, H., O'Hara, B., Seldin, M. F., Uhl, G. R. (1992). Identification of a putative gamma-aminobutyric acid (GABA) receptor subunit rho2 cDNA and colocalization of the genes encoding rho2 (GABRR2) and rho1 (GABRR1) to human chromosome 6q14–q21 and mouse chromosome 4. *Genomics*, *12*, 801–806.
228. Cutting, G. R., Lu, L., O'Hara, B. F., Kasch, L. M., Montrose-Rafizadeh, C., Donovan, D. M., Shimada, S., Antonarakis, S. E., Guggino, W. B., Uhl, G. R., et al. (1991). Cloning of the gamma-aminobutyric acid (GABA) rho 1 cDNA: A GABA receptor subunit highly expressed in the retina. *Proc. Natl. Acad. Sci. USA*, *88*, 2673–2677.
229. Bailey, M. E., Albrecht, B. E., Johnson, K. J., Darlison, M. G. (1999). Genetic linkage and radiation hybrid mapping of the three human GABA(C) receptor rho subunit genes: GABRR1, GABRR2 and GABRR3. *Biochim. Biophys. Acta*, *1447*, 307–312.

230. Enz, R., Cutting, G. R. (1999). GABAC receptor rho subunits are heterogeneously expressed in the human CNS and form homo- and heterooligomers with distinct physical properties. *Eur. J. Neurosci.*, *11*, 41–50.

231. Boue-Grabot, E., Roudbaraki, M., Bascles, L., Tramu, G., Bloch, B., Garret, M. (1998). Expression of GABA receptor rho subunits in rat brain. *J. Neurochem.*, *70*, 899–907.

232. Johnston, G. A. (2002). Medicinal chemistry and molecular pharmacology of GABA(C) receptors. *Curr. Top. Med. Chem.*, *2*, 903–913.

233. Johnston, G. A. R., Krogsgaard-Larsen, P., Stephanson, A. (1975). Betel nut constituents as inhibitors of gamma-aminobutyric acid uptake. *Nature (London)*, *258*, 627–628.

234. Duke, R. K., Chebib, M., Balcar, V. J., Allan, R. D., Mewett, K. N., Johnston, G. A. (2000). (+)- and (−)-*cis*-2-Aminomethylcyclopropanecarboxylic acids show opposite pharmacology at recombinant rho(1) and rho(2) GABA(C) receptors. *J. Neurochem.*, *75*, 2602–2610.

235. Murata, Y., Woodward, R. M., Miledi, R., Overman, L. E. (1996). The first selective antagonist for a GABAC receptor. *Bioorg. Med. Chem. Lett.*, *6*, 2073–2076.

236. Ragozzino, D., Woodward, R. M., Murata, Y., Eusebi, F., Overman, L. E., Miledi, R. (1996). Design and in vitro pharmacology of a selective gamma-aminobutyric acidC receptor antagonist. *Mol. Pharmacol.*, *50*, 1024–1030.

237. Krehan, D., Frølund, B., Krogsgaard-Larsen, P., Kehler, J., Johnston, G. A., Chebib, M. (2003). Phosphinic, phosphonic and seleninic acid bioisosteres of isonipecotic acid as novel and selective GABA(C) receptor antagonists. *Neurochem. Int.*, *42*, 561–565.

238. Woodward, R. M., Polenzani, L., Miledi, R. (1993). Characterization of bicuculline/baclofen-insensitive (rho-like) gamma-aminobutyric acid receptors expressed in *Xenopus* oocytes. II. Pharmacology of gamma-aminobutyric acidA and gamma-aminobutyric acidB receptor agonists and antagonists. *Mol. Pharmacol.*, *43*, 609–625.

239. Qian, H., Dowling, J. E. (1993). Novel GABA responses from rod-driven retinal horizontal cells. *Nature*, *361*, 162–164.

240. Chebib, M., Vandenberg, R. J., Froestl, W., Johnston, G. A. (1997). Unsaturated phosphinic analogues of gamma-aminobutyric acid as GABA(C) receptor antagonists. *Eur. J. Pharmacol.*, *329*, 223–229.

241. Chebib, M., Mewett, K. N., Johnston, G. A. (1998). GABA(C) receptor antagonists differentiate between human ρ1 and ρ2 receptors expressed in *Xenopus* oocytes. *Eur. J. Pharmacol.*, *357*, 227–234.

242. Vien, J., Duke, R. K., Mewett, K. N., Johnston, G. A., Shingai, R., Chebib, M. (2002). *trans*-4-Amino-2-methylbut-2-enoic acid (2-MeTACA) and (+/−)-*trans*-2-aminomethylcyclopropanecarboxylic acid ((+/−)-TAMP) can differentiate rat ρ3 from human ρ1 and ρ2 recombinant GABA(C) receptors. *Br. J. Pharmacol.*, *135*, 883–890.

243. Kusama, T., Spivak, C. E., Whiting, P., Dawson, V. L., Schaeffer, J. C., Uhl, G. R. (1993). Pharmacology of GABA ρ 1 and GABA α/β receptors expressed in *Xenopus* oocytes and COS cells. *Br. J. Pharmacol.*, *109*, 200–206.

244. Kusama, T., Wang, T. L., Guggino, W. B., Cutting, G. R., Uhl, G. R. (1993). GABA ρ 2 receptor pharmacological profile: GABA recognition site similarities to ρ 1. *Eur. J. Pharmacol.*, 245, 83–84.

245. McGillem, G. S., Rotolo, T. C., Dacheux, R. F. (2000). GABA responses of rod bipolar cells in rabbit retinal slices. *Vis. Neurosci.*, 17, 381–389.

246. Matsui, K., Hasegawa, J., Tachibana, M. (2001). Modulation of excitatory synaptic transmission by GABA(C) receptor-mediated feedback in the mouse inner retina. *J. Neurophysiol.*, 86, 2285–2298.

247. Dingledine, R., Borges, K., Bowie, D., Traynelis, S. F. (1999). The glutamate receptor ion channels. *Pharmacol. Rev.*, 51, 7–61.

248. Hollmann, M., Heinemann, S. (1994). Cloned glutamate receptors. *Annu. Rev. Neurosci.*, 17, 31–108.

249. Ransom, R. W., Stec, N. L. (1988). Cooperative modulation of [^{3}H]MK-801 binding to the *N*-methyl-d-aspartate receptor ion channel complex by l-glutamate, glycine and polyamines. *J. Neurochem.*, 51, 830–836.

250. Lynch, D. R., Guttmann, R. P. (2002). Excitotoxicity: Perspectives based on *N*-methyl-d-aspartate receptor subtypes. *J. Pharmacol. Exp. Ther.*, 300, 717–723.

251. Furukawa, H., Gouaux, E. (2003). Mechanisms of activation, inhibition and specificity: Crystal structures of the NMDA receptor NR1 ligand-binding core. *EMBO J.*, 22, 2873–2885.

252. Tikhonova, I. G., Baskin, II, Palyulin, V. A., Zefirov, N. S., Bachurin, S. O. (2002). Structural basis for understanding structure-activity relationships for the glutamate binding site of the NMDA receptor. *J. Med. Chem.*, 45, 3836–3843.

253. Lee, F. J., Xue, S., Pei, L., Vukusic, B., Chery, N., Wang, Y., Wang, Y. T., Niznik, H. B., Yu, X. M., Liu, F. (2002). Dual regulation of NMDA receptor functions by direct protein-protein interactions with the dopamine D1 receptor. *Cell*, 111, 219–230.

254. Collingridge, G. L., Watkins, J. C. (1994). *The NMDA Receptor*. Oxford University Press, Oxford.

255. Lunn, W. H. W., Schoepp, D. D., Calligaro, D. O., Vasileff, R. T., Heinz, L. J., Salhoff, C. R., O'Malley, P. J. (1992). dl-Tetrazol-5-ylglycine, a highly potent NMDA agonist: Its synthesis and NMDA receptor efficacy. *J. Med. Chem.*, 35, 4608–4612.

256. Madsen, U., Frydenvang, K., Ebert, B., Johansen, T. N., Brehm, L., Krogsgaard-Larsen, P. (1996). NMDA receptor agonists. Resolution, absolute stereochemistry and pharmacology of the enantiomers of 2-amino-2-(3-hydroxy-5-methyl-4-isoxazolyl)acetic acid (AMAA). *J. Med. Chem.*, 39, 183–190.

257. Krogsgaard-Larsen, P., Ebert, B., Lund, T. M., Bräuner-Osborne, H., Sløk, F. A., Johansen, T. N., Brehm, L., Madsen, U. (1996). Design of excitatory amino acid receptor agonists, partial agonists and antagonists: Ibotenic acid as a key lead structure. *Eur. J. Med. Chem.*, 31, 515–537.

258. Kyle, D. J., Patch, R. J., Karbon, E. W., Ferkany, J. W. (1992). NMDA receptors: Heterogeneity and agonism. In P. Krogsgaard-Larsen and J. J. Hansen (Eds.), *Excitatory Amino Acid Receptors: Design of Agonists and Antagonists.* Ellis Horwood, pp. 121–161, Chichester.

259. Jane, D. E., Olverman, H. J., Watkins, J. C. (1994). Agonists and competitive antagonists: Structure-activity and molecular modelling studies. In G. L.

Collingridge, J. C. Watkins (Eds.), *The NMDA Receptor.* Oxford University Press, pp. 31–104, Oxford.

260. Curry, K., Peet, M. J., Magnuson, D. S. K., McLennan, H. (1988). Synthesis, resolution, and absolute configuration of the isomers of the neuronal excitant 1-amino-1,3-cyclopentanedicarboxylic acid. *J. Med. Chem., 31,* 864–867.

261. Shinozaki, H., Ishida, M., Shimamoto, K., Ohfune, Y. (1989). A conformationally restricted analogue of l-glutamate, the (2S, 3R,4S) isomer of l-α-(carboxycyclopropyl)glycine, activates the NMDA-type receptor more markedly than NMDA in the isolated rat spinal cord. *Brain Res., 480,* 355–359.

262. Allan, R. D., Hanrahan, J. R., Hambley, T. W., Johnston, G. A. R., Mewett, K. N., Mitrovic, A. D. (1990). Synthesis and activity of a potent *N*-methyl-d-aspartic acid agonist, *trans*-1-aminocyclobutane-1,3-dicarboxylic acid, and related phosphonic and carboxylic acids. *J. Med. Chem., 33,* 2905–2915.

263. Ornstein, P. L., Klimkowski, V. J. (1992). Competitive NMDA receptor antagonists. In P. Krogsgaard-Larsen and J. J. Hansen (Eds.), *Excitatory Amino Acid Receptors: Design of Agonists and Antagonists.* Ellis Horwood, pp. 183–200, Chichester.

264. Fagg, G. E., Olpe, H.-R., Pozza, M. F., Baud, J., Steinmann, M., Schmutz, M., Portet, C., Baumann, P., Thedinga, K., Bittiger, H., Allgeier, H., Heckendorn, R., Angst, C., Brundish, D., Dingwall, J. G. (1990). CGP 37849 and CGP 39551: Novel and potent *N*-methyl-d-aspartate receptor antagonists with oral activity. *Br. J. Pharmacol., 99,* 791–797.

265. Lehmann, J., Hutchison, A. J., McPherson, S. E., Mondadori, C., Schmutz, M., Sinton, C. M., Tsai, C., Murphy, D. E., Steel, D. J., Williams, M. (1988). CGS 19755, a selective and competitive *N*-methyl-d-aspartate-type excitatory amino acid receptor antagonist. *J. Pharmacol. Exp. Ther., 246,* 65–75.

266. Lehmann, J., Schneider, J., McPherson, S., Murphy, D. E., Bernard, P., Tsai, C., Bennett, D. A., Pastor, G., Steele, D. J., Boehm, C., Cheney, D. L., Liebermann, J. M., Williams, M., Wood, P. L. (1987). CPP, a selective *N*-methyl-d-aspartate (NMDA)-type receptor antagonist: Characterization *in vitro* and *in vivo. J. Pharmacol. Exp. Ther., 240,* 737–746.

267. Hutchison, A. J., Williams, M., Angst, C., de Jesus, R., Blanchard, L., Jackson, R. H., Wilusz, E. J., Murphy, D. E., Bernard, P. S., Schneider, J., Campbell, T., Guida, W., Sills, M. A. (1989). 4-(Phosphonoalkyl)- and 4-(phosphonoalkenyl)-2-piperidinecarboxylic acids: Synthesis, activity at *N*-methyl-d-aspartic acid receptors, and anticonvulsant activity. *J. Med. Chem., 32,* 2171–2178.

268. Ornstein, P. L., Schoepp, D. D., Arnold, M. B., Leander, D., Lodge, D., Paschal, J. W., Elzey, T. (1991). 4-(Tetrazolylalkyl)piperidine-2-carboxylic acids. Potent and selective *N*-methyl-d-aspartic acid receptor antagonists with a short duration of action. *J. Med. Chem., 34,* 90–97.

269. Ferkany, J. W., Hamilton, G. S., Patch, R. J., Huang, Z., Borosky, S. A., Bednar, D. L., Jones, B. E., Zubrowski, R., Willetts, J., Karbon, E. W. (1993). Pharmacological profile of NPC 17742 [2R,4R,5S-(2-amino-4,5-(1, 2-cyclohexyl)-7-phosphonoheptanoic acid)], a potent, selective and competitive *N*-methyl-d-aspartate receptor antagonist. *J. Pharmacol. Exp. Ther., 264,* 256–264.

270. Baudy, R. B., Greenblatt, L. P., Jirkovsky, I. L., Conklin, M., Russo, R. J., Bramlett, D. R., Emrey, T. A., Simmonds, J. T., Kowal, D. M., Stein, R. P., Tasse, R. P. (1993). Potent quinoxaline-spaced phosphono alpha-amino acids of the AP-6

type as competitive NMDA antagonists: Synthesis and biological evaluation. *J. Med. Chem.*, 36, 331–342.

271. Urwyler, S., Laurie, D., Lowe, D. A., Meier, C. L., Muller, W. (1996). Biphenyl-derivatives of 2-amino-7-phosphonoheptanoic acid, a novel class of potent competitive N-methyl-d-aspartate receptor antagonist—I. Pharmacological characterization in vitro. *Neuropharmacology*, 35, 643–654.

272. Urwyler, S., Campbell, E., Fricker, G., Jenner, P., Lemaire, M., McAllister, K. H., Müller, W. (1996). Biphenyl-derivatives of 2-amino-7-phosphonoheptanoic acid, a novel class of potent competitive N-methyl-d-aspartate receptor antagonist—II. Pharmacological characterization in vivo. *Neuropharmacology*, 35, 655–669.

273. Li, J. H., Bigge, C. F., Williamson, R. M., Borosky, S. A., Vartanian, M. G., Ortwine, D. F. (1995). Potent, orally active, competitive N-methyl-d-aspartate (NMDA) receptor antagonists are substrates for a neutral amino acid uptake system in Chinese hamster ovary cells. *J. Med. Chem.*, 38, 1955–1965.

274. Anis, N. A., Berry, S. C., Burton, N. R., Lodge, D. (1983). The dissociative anaesthetics, ketamine and phencyclidine, selectively reduce excitation of central mammalian neurones by N-methyl-aspartate. *Br. J. Pharmacol.*, 79, 565–575.

275. Wong, E. H. F., Kemp, J. A., Priestly, T., Knight, A. R., Woodruff, G. N., Iversen, L. L. (1986). The anticonvulsant MK-801 is a potent N-methyl-d-aspartate antagonist. *Proc. Natl. Acad. Sci. USA*, 83, 7104–7108.

276. Buchan, A. M. (1990). Do NMDA antagonists protect against cerebral ischemia: Are clinical trials warranted? *Cerebrovasc. Brain Metab. Rev.*, 2, 1–26.

277. Herrling, P. L. (1997). *Excitatory Amino Acids: Clinical Results with Antagonists*. Academic Press, London.

278. Iversen, L. L., Kemp, J. A. (1994). Non-competitive NMDA antagonists as drugs. In G. L. Collingridge and J. C. Watkins (Eds.), *The NMDA Receptor*. Oxford University Press, Oxford.

279. Parsons, C. G., Danysz, W., Quack, G. (1999). Memantine is a clinically well tolerated N-methyl-d-aspartate (NMDA) receptor antagonist—a review of preclinical data. *Neuropharmacology*, 38, 735–767.

280. Nelson, K. A., Park, K. M., Robinovitz, E., Tsigos, C., Max, M. B. (1997). High-dose oral dextromethorphan versus placebo in painful diabetic neuropathy and postherpetic neuralgia. *Neurology*, 48, 1212–1218.

281. Greenamyre, J. T., Eller, R. V., Zhang, Z., Ovadia, A., Kurlan, R., Gash, D. M. (1994). Antiparkinsonian effects of remacemide hydrochloride, a glutamate antagonist, in rodent and primate models of Parkinson's disease. *Ann. Neurol.*, 35, 655–661.

282. Rabey, J. M., Nissipeanu, P., Korczyn, A. D. (1992). Efficacy of memantine, an NMDA receptor antagonist, in the treatment of Parkinson's disease. *J. Neural. Transm. Park. Dis. Dement. Sect.*, 4, 277–282.

283. Lipton, S. A. (1994). HIV-related neuronal injury. Potential therapeutic intervention with calcium channel antagonists, NMDA antagonists. *Mol. Neurobiol.*, 8, 181–196.

284. Kieburtz, K., Feigin, A., McDermott, M., Como, P., Abwender, D., Zimmerman, C., Hickey, C., Orme, C., Claude, K., Sotack, J., Greenamyre, J. T., Dunn, C.,

Shoulson, I. (1996). A controlled trial of remacemide hydrochloride in Huntington's disease. *Mov. Disord.*, *11*, 273–277.

285. Shuto, S., Ono, S., Imoto, H., Yoshii, K., Matsuda, A. (1998). Synthesis and biological activity of conformationally restricted analogues of milnacipran: (1*S*, 2*R*)-1-phenyl-2-[(*R*)-1-amino-2-propynyl]-*N*,*N*-diethylcyclopropanecarboxamide is a novel class of NMDA receptor channel blocker. *J. Med. Chem.*, *41*, 3507–3514.

286. Villetti, G., Bregola, G., Bassani, F., Bergamaschi, M., Rondelli, I., Pietra, C., Simonato, M. (2001). Preclinical evaluation of CHF3381 as a novel antiepileptic agent. *Neuropharmacology*, *40*, 866–878.

287. Villetti, G., Bergamaschi, M., Bassani, F., Bolzoni, P. T., Maiorino, M., Pietra, C., Rondelli, I., Chamiot-Clerc, P., Simonato, M., Barbieri, M. (2003). Antinociceptive activity of the *N*-methyl-d-aspartate receptor antagonist *N*-(2-indanyl)-glycinamide hydrochloride (CHF3381) in experimental models of inflammatory and neuropathic pain. *J. Pharmacol. Exp. Ther.*, *306*, 804–814.

288. Tarral, A., Dostert, P., Guillevic, Y., Fabbri, L., Rondelli, I., Mariotti, F., Imbimbo, B. P. (2003). Safety, pharmacokinetics, and pharmacodynamics of CHF 3381, a novel *N*-methyl-d-aspartate antagonist, after single oral doses in healthy subjects. *J. Clin. Pharmacol.*, *43*, 901–911.

289. Gotti, B., Duverger, D., Bertin, J., Carter, C., Dupont, R., Frost, J., Gaudilliere, B., MacKenzie, E. T., Rousseau, J., Scatton, B., Wick, A. (1988). Ifenprodil and SL 82.0715 as cerebral anti-ischemic agents. I. Evidence for efficacy in models of focal cerebral ischemia. *J. Pharmacol. Exp. Ther.*, *247*, 1211–1221.

290. Araki, T., Kogure, K., Nishioka, K. (1990). Comparative neuroprotective effects of pentobarbital, vinpocetine, flunarizine and ifenprodil on ischemic neuronal damage in the gerbil hippocampus. *Res. Exp. Med. (Berl.)*, *190*, 19–23.

291. Chenard, B. D., Menniti, F. S. (1999). Antagonists selective for NMDA receptors containing the NR2B subunit. *Curr. Pharm. Des.*, *5*, 381–404.

292. Tamiz, A. P., Whittemore, E. R., Zhou, Z. L., Huang, J. C., Drewe, J. A., Chen, J. C., Cai, S. X., Weber, E., Woodward, R. M., Keana, J. F. (1998). Structure-activity relationships for a series of bis(phenylalkyl)amines: Potent subtype-selective inhibitors of *N*-methyl-d-aspartate receptors. *J. Med. Chem.*, *41*, 3499–3506.

293. Weaver, C. E., Marek, P., Park-Chung, M., Tam, S. W., Farb, D. H. (1997). Neuroprotective activity of a new class of steroidal inhibitors of the *N*-methyl-D-aspartate receptor. *Proc. Natl. Acad. Sci. USA*, *94*, 10450–10454.

294. Feigenbaum, J. J., Bergmann, F., Richmond, S. A., Mechoulam, R., Nadler, V., Kloog, Y., Sokolovsky, M. (1989). Nonpsychotropic cannabinoid acts as a functional *N*-methyl-D-aspartate receptor blocker. *Proc. Natl. Acad. Sci. USA*, *86*, 9584–9587.

295. Stephenson, F. A., Turner, A. J. (1998). *Amino Acid Neurotransmission*. Portland Press, London.

296. McBain, C. J., Mayer, M. L. (1994). *N*-methyl-D-aspartic acid receptor structure and function. *Physiol. Rev.*, *74*, 723–760.

297. Wolosker, H., Blackshaw, S., Snyder, S. H. (1999). Serine racemase: A glial enzyme synthesizing D-serine to regulate glutamate-*N*-methyl-D-aspartate neurotransmission. *Proc. Natl. Acad. Sci. USA*, *96*, 13409–13414.

298. Mothet, J. P., Parent, A. T., Wolosker, H., Brady, R. O., Jr., Linden, D. J., Ferris, C. D., Rogawski, M. A., Snyder, S. H. (2000). D-Serine is an endogenous ligand for the glycine site of the N-methyl-D-aspartate receptor. *Proc. Natl. Acad. Sci. USA*, 97, 4926–4931.
299. McBain, C. J., Kleckner, N. W., Wyrick, S., Dingledine, R. (1989). Structural requirements for activation of the glycine coagonist site of N-methyl-D-aspartate receptors expressed in *Xenopus* oocytes. *Mol. Pharmacol.*, 36, 556–565.
300. Kemp, J. A., Leeson, P. D. (1993). The glycine site of the NMDA receptor—five years on. *Trends Pharmacol. Sci.*, 14, 20–25.
301. Leeson, P. D., Iversen, L. L. (1994). The glycine site on the NMDA receptor: Structure-activity relationships and therapeutic potential. *J. Med. Chem.*, 37, 4053–4067.
302. Schwartz, B. L., Hashtroudi, S., Herting, R. L., Schwartz, P., Deutsch, S. I. (1996). D-Cycloserine enhances implicit memory in Alzheimer patients. *Neurology*, 46, 420–424.
303. Leeson, P. D., Williams, B. J., Baker, R., Ladduwahetty, T., Moore, K. W., Rowley, M. (1990). Effects of five-membered ring conformation on bioreceptor recognition: Identification of 3R-amino-1-hydroxy-4R-methylpyrrolidin-2-one (L-687,414) as a potent glycine/N-methyl-D-aspartate receptor antagonist. *J. Chem. Soc. Chem. Commun.*, 1578–1580.
304. Priestley, T., Marshall, G. R., Hill, R. G., Kemp, J. A. (1998). L-687,414, a low efficacy NMDA receptor glycine site partial agonist in vitro, does not prevent hippocampal LTP in vivo at plasma levels known to be neuroprotective. *Br. J. Pharmacol.*, 124, 1767–1773.
305. Leeson, P. D., Carling, R. W., Moore, K. W., Moseley, A. M., Smith, J. D., Stevenson, G., Chan, T., Baker, R., Foster, A. C., Grimwood, S., Kemp, J. A., Marshall, G. R., Hoogsteen, K. (1992). 4-Amino-2-carboxytetrahydroquinolines. Structure-activity relationships for antagonism at the glycine site of the NMDA receptor. *J. Med. Chem.*, 35, 1954–1968.
306. Carignani, C., Ugolini, A., Pinnola, V., Belardetti, F., Trist, D. G., Corsi, M. (1998). NMDA receptor subunit characterization of the glycine site antagonist GV196771A and its action on the spinal cord wind-up. *Naunyn Schmiedebergs Arch. Pharmacol.*, 358, P1119.
307. Wenk, G. L., Baker, L. M., Stoehr, J. D., Hauss-Wegrzyniak, B., Danysz, W. (1998). Neuroprotection by novel antagonists at the NMDA receptor channel and glycineB sites. *Eur. J. Pharmacol.*, 347, 183–187.
308. Williams, M., Kowaluk, E. A., Arneric, S. P. (1999). Emerging molecular approaches to pain therapy. *J. Med. Chem.*, 42, 1481–1500.
309. Cai, S. X., Kher, S. M., Zhou, Z. L., Ilyin, V., Espitia, S. A., Tran, M., Hawkinson, J. E., Woodward, R. M., Weber, E., Keana, J. F. (1997). Structure-activity relationships of alkyl- and alkoxy-substituted 1,4-dihydroquinoxaline-2,3-diones: Potent and systemically active antagonists for the glycine site of the NMDA receptor. *J. Med. Chem.*, 40, 730–738.
310. Nikam, S. S., Cordon, J. J., Ortwine, D. F., Heimbach, T. H., Blackburn, A. C., Vartanian, M. G., Nelson, C. B., Schwarz, R. D., Boxer, P. A., Rafferty, M. F. (1999). Design and synthesis of novel quinoxaline-2,3-dione AMPA/GlyN receptor antagonists: Amino acid derivatives. *J. Med. Chem.*, 42, 2266–2271.

311. Potschka, H., Loscher, W., Wlaz, P., Behl, B., Hofmann, H. P., Treiber, H. J., Szabo, L. (1998). LU 73068, a new non-NMDA and glycine/NMDA receptor antagonist: Pharmacological characterization and comparison with NBQX and L-701,324 in the kindling model of epilepsy. *Br. J. Pharmacol.*, *125*, 1258–1266.

312. Mott, D. D., Doherty, J. J., Zhang, S., Washburn, M. S., Fendley, M. J., Lyuboslavsky, P., Traynelis, S. F., Dingledine, R. (1998). Phenylethanolamines inhibit NMDA receptors by enhancing proton inhibition. *Nat. Neurosci.*, *1*, 659–667.

313. Williams, K. (1993). Ifenprodil discriminates subtypes of the N-methyl-D-aspartate receptor: Selectivity and mechanisms at recombinant heteromeric receptors. *Mol. Pharmacol.*, *44*, 851–859.

314. Kristensen, J. D., Svensson, B., Gordh, T., Jr. (1992). The NMDA-receptor antagonist CPP abolishes neurogenic "wind-up pain" after intrathecal administration in humans. *Pain*, *51*, 249–253.

315. Eide, P. K., Jorum, E., Stubhaug, A., Bremnes, J., Breivik, H. (1994). Relief of postherpetic neuralgia with the N-methyl-D-aspartic acid receptor antagonist ketamine: A double-blind, cross-over comparison with morphine and placebo. *Pain*, *58*, 347–354.

316. Marino, M., Valenti, O., Conn, P. J. (2003). Glutamate receptors and Parkinson's disease: Opportunities for intervention. *Drugs Aging*, *20*, 377–397.

317. Buller, A. L., Larson, H. C., Schneider, B. E., Beaton, J. A., Morrisett, R. A., Monaghan, D. T. (1994). The molecular basis of NMDA receptor subtypes: Native receptor diversity is predicted by subunit composition. *J. Neurosci.*, *14*, 5471–5484.

318. Monaghan, D. T., Larsen, H. (1997). NR1 and NR2 subunit contributions to N-methyl-D-aspartate receptor channel blocker pharmacology. *J. Pharmacol. Exp. Ther.*, *280*, 614–620.

319. Bang-Andersen, B., Lenz, S. M., Skjærbæk, N., Søby, K. K., Hansen, H. O., Ebert, B., Bøgesø, K. P., Krogsgaard-Larsen, P. (1997). Heteroaryl analogues of AMPA. Synthesis and quantitative structure-activity relationships. *J. Med. Chem.*, *40*, 2831–2842.

320. Falch, E., Brehm, L., Mikkelsen, I., Johansen, T. N., Skjærbæk, N., Nielsen, B., Stensbøl, T. B., Ebert, B., Krogsgaard-Larsen, P. (1998). Heteroaryl analogues of AMPA. 2. Synthesis, absolute stereochemistry, photochemistry, and structure-activity relationships. *J. Med. Chem.*, *41*, 2513–2523.

321. Sløk, F. A., Ebert, B., Lang, Y., Krogsgaard-Larsen, P., Lenz, S. M., Madsen, U. (1997). Excitatory amino-acid receptor agonists. Synthesis and pharmacology of analogues of 2-amino-3-(3-hydroxy-5-methylisoxazol-4-yl)propionic acid. *Eur. J. Med. Chem.*, *32*, 329–338.

322. Madsen, U., Frølund, B., Lund, T. M., Ebert, B., Krogsgaard-Larsen, P. (1993). Design, synthesis and pharmacology of model compounds for indirect elucidation of the topography of AMPA receptor sites. *Eur. J. Med. Chem.*, *28*, 791–800.

323. Madsen, U., Ebert, B., Krogsgaard-Larsen, P., Wong, E. H. F. (1992). Synthesis and pharmacology of (RS)-2-amino-3-(3-hydroxy-5-trifluoromethyl-4-isoxazolyl)propionic acid, a potent AMPA receptor agonist. *Eur. J. Med. Chem.*, *27*, 479–484.

324. Stensbøl, T. B., Uhlmann, P., Morel, S., Eriksen, B. L., Felding, J., Kromann, H., Hermit, M. B., Greenwood, J. R., Bräuner-Osborne, H., Madsen, U., Junager, F., Krogsgaard-Larsen, P., Begtrup, M., Vedsø, P. (2002). Novel 1-hydroxyazole

bioisosteres of glutamic acid. Synthesis, protolytic properties, and pharmacology. *J. Med. Chem.*, *45*, 19–31.

325. Lauridsen, J., Honoré, T., Krogsgaard-Larsen, P. (1985). Ibotenic acid analogues. Synthesis, molecular flexibility, and *in vitro* activity of agonists and antagonists at central glutamic acid receptors. *J. Med. Chem.*, *28*, 668–672.

326. Christensen, I. T., Reinhardt, A., Nielsen, B., Ebert, B., Madsen, U., Nielsen, E. Ø., Brehm, L., Krogsgaard-Larsen, P. (1989). Excitatory amino acid agonists and partial agonists. *Drug Des. Del.*, *5*, 57–71.

327. Ebert, B., Lenz, S. M., Brehm, L., Bregnedal, P., Hansen, J. J., Frederiksen, K., Bøgesø, K. P., Krogsgaard-Larsen, P. (1994). Resolution, absolute stereochemistry, and pharmacology of the *S*-(+)- and *R*-(−)-isomers of the apparent partial AMPA receptor agonist (*R*,*S*)-2-amino-3-(3-hydroxy-5-phenylisoxazol-4-yl)propionic acid [(*R*,*S*)-APPA]. *J. Med. Chem.*, *37*, 878–884.

328. Vogensen, S. B., Jensen, H. S., Stensbøl, T. B., Frydenvang, K., Bang-Andersen, B., Johansen, T. N., Egebjerg, J., Krogsgaard-Larsen, P. (2000). Resolution, configurational assignment, and enantiopharmacology of 2-amino-3-[3-hydroxy-5-(2-methyl-2*H*-tetrazol-5-yl)isoxazol-4-yl]propionic acid, a potent GluR3- and GluR4-preferring AMPA receptor agonist. *Chirality*, *12*, 705–713.

329. Clarke, V. R. J., Ballyk, B. A., Hoo, K. H., Mandelzys, A., Pellizzari, A., Bath, C. P., Thomas, J., Sharpe, E. F., Davies, C. H., Ornstein, P. L., Schoepp, D. D., Kamboj, R. K., Collingridge, G. L., Lodge, D., Bleakman, D. (1997). A hippocampal GluR5 kainate receptor regulating inhibitory synaptic transmission. *Nature*, *389*, 599–602.

330. Hill, R. A., Wallace, L. J., Miller, D. D., Weinstein, D. M., Shams, G., Tai, H., Layer, R. T., Willins, D., Uretsky, N. J., Danthi, S. N. (1997). Structure-activity studies for alpha-amino-3-hydroxy-5-methyl-4-isoxazolepropanoic acid receptors: Acidic hydroxyphenylalanines. *J. Med. Chem.*, *40*, 3182–3191.

331. Patneau, D. K., Mayer, M. L., Jane, D. E., Watkins, J. C. (1992). Activation and desensitization of AMPA/kainate receptors by novel derivatives of willardiine. *J. Neurosci.*, *12*, 595–606.

332. Jane, D. E., Hoo, K., Kamboj, R., Deverill, M., Bleakman, D., Mandelzys, A. (1997). Synthesis of willardiine and 6-azawillardiine analogs: Pharmacological characterization on cloned homomeric human AMPA and kainate receptor subtypes. *J. Med. Chem.*, *40*, 3645–3650.

333. Hawkins, L. M., Beaver, K. M., Jane, D. E., Taylor, P. M., Sunter, D. C., Roberts, P. J. (1995). Binding of the new radioligand (*S*)-[^{3}H]AMPA to rat brain synaptic membranes: Effects of a series of structural analogues of the non-NMDA receptor agonist willardiine. *Neuropharmacology*, *34*, 405–410.

334. Jin, R., Gouaux, E. (2003). Probing the function, conformational plasticity, and dimer-dimer contacts of the GluR2 ligand-binding core: Studies of 5-substituted willardiines and GluR2 S1S2 in the crystal. *Biochemistry*, *42*, 5201–5213.

335. Jin, R., Banke, T. G., Mayer, M. L., Traynelis, S. F., Gouaux, E. (2003). Structural basis for partial agonist action at ionotropic glutamate receptors. *Nat. Neurosci.*, *6*, 803–810.

336. Coquelle, T., Christensen, J. K., Banke, T. G., Madsen, U., Schousboe, A., Pickering, D. S. (2000). Agonist discrimination between AMPA receptor subtypes. *Neuroreport*, *11*, 2643–2648.

337. Banke, T. G., Greenwood, J. R., Christensen, J. K., Liljefors, T., Traynelis, S. F., Schousboe, A., Pickering, D. S. (2001). Identification of amino acid residues in GluR1 responsible for ligand binding and desensitization. *J. Neurosci.*, *21*, 3052–3062.
338. Hogner, A., Kastrup, J. S., Jin, R., Liljefors, T., Mayer, M. L., Egebjerg, J., Larsen, I. K., Gouaux, E. (2002). Structural basis for AMPA receptor activation and ligand selectivity: Crystal structures of five agonist complexes with the GluR2 ligand-binding core. *J. Mol. Biol.*, *322*, 93–109.
339. Bjerrum, E. J., Kristensen, A. S., Pickering, D. S., Greenwood, J. R., Nielsen, B., Liljefors, T., Schousboe, A., Bräuner-Osborne, H., Madsen, U. (2003). Design, synthesis, and pharmacology of a highly subtype-selective GluR1/2 agonist, (*RS*)-2-amino-3-(4-chloro-3-hydroxy-5-isoxazolyl)propionic acid (Cl–HIBO). *J. Med. Chem.*, *46*, 2246–2249.
340. Madsen, U., Wong, E. H. F. (1992). Heterocyclic excitatory amino acids. Synthesis and biological activity of novel analogues of AMPA. *J. Med. Chem.*, *35*, 107–111.
341. Wahl, P., Madsen, U., Banke, T., Krogsgaard-Larsen, P., Schousboe, A. (1996). Different characteristics of AMPA receptor agonists acting at AMPA receptors expressed in *Xenopus* oocytes. *Eur. J. Pharmacol.*, *308*, 211–218.
342. Johansen, T. N., Stensbøl, T. B., Nielsen, B., Vogensen, S. B., Frydenvang, K., Sløk, F. A., Bräuner-Osborne, H., Madsen, U., Krogsgaard-Larsen, P. (2001). Resolution, configurational assignment and enantiopharmacology at glutamate receptors of 2-amino-3-(3-carboxy-5-methyl-4-isoxazolyl)propionic acid (ACPA) and demethyl-ACPA. *Chirality*, *13*, 523–532.
343. Watkins, J. C., Evans, R. H. (1981). Excitatory amino acid transmitters. *Annu. Rev. Pharmacol. Toxicol.*, *21*, 165–204.
344. Monaghan, D. T., Bridges, R. J., Cotman, C. W. (1989). The excitatory amino acid receptors. *Annu. Rev. Pharmacol. Toxicol.*, *29*, 365–402.
345. Krogsgaard-Larsen, P., Honore, T., Hansen, J. J., Curtis, D. R., Lodge, D. (1980). New class of glutamate agonist structurally related to ibotenic acid. *Nature*, *284*, 64–66.
346. Watkins, J. C., Krogsgaard-Larsen, P., Honoré, T. (1990). Structure-activity relationships in the development of excitatory amino acid receptor agonists and competitive antagonists. *Trends Pharmacol. Sci.*, *11*, 25–33.
347. Bleakman, D., Lodge, D. (1998). Neuropharmacology of AMPA and kainate receptors. *Neuropharmacology*, *37*, 1187–1204.
348. Vogensen, S. B., Greenwood, J. R., Varming, A. R., Brehm, L., Pickering, D. S., Nielsen, B., Liljefors, T., Clausen, R. P., Johansen, T. N., Krogsgaard-Larsen, P. (2004). A stereochemical anomaly: The cyclised (*R*)-AMPA analogue (*R*)-3-hydroxy-4,5,6,7-tetrahydroisoxazolo[5,4-c]pyridine-5-carboxylic acid [(*R*)-5-HPCA] resembles (*S*)-AMPA at glutamate receptors. *Org. Biomol. Chem.*, *2*, 206–213.
349. Conti, P., De Amici, M., De Sarro, G., Stensbøl, T. B., Bräuner-Osborne, H., Madsen, U., De Micheli, C. (1998). Synthesis and pharmacology of a new AMPA-kainate receptor agonist with potent convulsant activity. *J. Med. Chem.*, *41*, 3759–3762.
350. Conti, P., De Amici, M., De Sarro, G., Rizzo, M., Stensbøl, T. B., Bräuner-Osborne, H., Madsen, U., Toma, L., De Micheli, C. (1999). Synthesis and enantiopharmacology of new AMPA-kainate receptor agonists. *J. Med. Chem.*, *42*, 4099–4107.

351. Honoré, T., Davies, S. N., Drejer, J., Fletcher, E. J., Jacobsen, P., Lodge, D., Nielsen, F. E. (1988). Quinoxalinediones: Potent competitive non-NMDA glutamate receptor antagonists. *Science*, *241*, 701–703.

352. Sheardown, M. J., Nielsen, E. Ø., Hansen, A. J., Jacobsen, P., Honoré, T. (1990). 2,3-Dihydroxy-6-nitro-7-sulfamoyl-benzo(*F*)quinoxaline. A neuroprotectant for cerebral ischemia. *Science*, *247*, 571–574.

353. Armstrong, N., Gouaux, E. (2000). Mechanisms for activation and antagonism of an AMPA-sensitive glutamate receptor: Crystal structures of the GluR2 ligand binding core. *Neuron*, *28*, 165–181.

354. Turski, L., Huth, A., Sheardown, M., McDonald, F., Neuhaus, R., Schneider, H. H., Dirnagl, U., Wiegand, F., Jacobsen, P., Ottow, E. (1998). ZK200775: A phosphonate quinoxalinedione AMPA antagonist for neuroprotection in stroke and trauma. *Proc. Natl. Acad. Sci. USA*, *95*, 10960–10965.

355. Lubisch, W., Behl, B., Hofmann, H. P. (1997). Pyrrolylquinoxalinediones: The importance of pyrrolic substitution on AMPA receptor binding. *Bioorg. Med. Chem. Lett.*, *7*, 1101–1106.

356. Nielsen, E. Ø., Varming, T., Mathiesen, C., Jensen, L. H., Møller, A., Gouliaev, A. H., Watjen, F., Drejer, J. (1999). SPD 502: A water-soluble and in vivo long-lasting AMPA antagonist with neuroprotective activity. *J. Pharmacol. Exp. Ther.*, *289*, 1492–1501.

357. Mignani, S., Bohme, G. A., Birraux, G., Boireau, A., Jimonet, P., Damour, D., Genevois-Borella, A., Debono, M. W., Pratt, J., Vuilhorgne, M., Wahl, F., Stutzmann, J. M. (2002). 9-Carboxymethyl-5*H*,10*H*-imidazo[1,2-*a*]indeno[1,2-*e*]pyrazin-4-one-2-carbocylic acid (RPR117824): Selective anticonvulsive and neuroprotective AMPA antagonist. *Bioorg. Med. Chem.*, *10*, 1627–1637.

358. Stutzmann, J. M., Bohme, G. A., Boireau, A., Damour, D., Debono, M. W., Genevois-Borella, A., Jimonet, P., Pratt, J., Randle, J. C., Ribeill, Y., Vuilhorgne, M., Mignani, S. (2001). Synthesis of anticonvulsive AMPA antagonists: 4-Oxo-10-substituted-imidazo[1,2-*e*]pyrazin-2-carboxylic acid derivatives. *Bioorg. Med. Chem. Lett.*, *11*, 1205–1210.

359. Jimonet, P., Bohme, G. A., Bouquerel, J., Boireau, A., Damour, D., Debono, M. W., Genevois-Borella, A., Hardy, J. C., Hubert, P., Manfre, F., Nemecek, P., Pratt, J., Randle, J. C., Ribeill, Y., Stutzmann, J. M., Vuilhorgne, M., Mignani, S. (2001). Bioisosteres of 9-carboxymethyl-4-oxo-imidazo[1,2-*a*]indeno-[1,2-*e*]pyrazin-2-carboxylic acid derivatives. Progress towards selective, potent in vivo AMPA antagonists with longer durations of action. *Bioorg. Med. Chem. Lett.*, *11*, 127–132.

360. Stutzmann, J. M., Bohme, G. A., Boireau, A., Damour, D., Debono, M. W., Genevois-Borella, A., Imperato, A., Jimonet, P., Pratt, J., Randle, J. C., Ribeill, Y., Vuilhorgne, M., Mignani, S. (2000). 4,10-Dihydro-4-oxo-4*H*-imidazo[1,2-*a*]indeno[1,2-*e*]pyrazin-2-carboxylic acid derivatives: Highly potent and selective AMPA receptors antagonists with in vivo activity. *Bioorg. Med. Chem. Lett.*, *10*, 1133–1137.

361. Canton, T., Bohme, G. A., Boireau, A., Bordier, F., Mignani, S., Jimonet, P., Jahn, G., Alavijeh, M., Stygall, J., Roberts, S., Brealey, C., Vuilhorgne, M., Debono, M. W., Le Guern, S., Laville, M., Briet, D., Roux, M., Stutzmann, J. M., Pratt, J. (2001). RPR 119990, a novel alpha-amino-3-hydroxy-5-methyl-4-

isoxazolepropionic acid antagonist: Synthesis, pharmacological properties, and activity in an animal model of amyotrophic lateral sclerosis. *J. Pharmacol. Exp. Ther.*, 299, 314–322.

362. Ornstein, P. L., Arnold, M. B., Allen, N. K., Bleisch, T., Borromeo, P. S., Lugar, C. W., Leander, J. D., Lodge, D., Schoepp, D. D. (1996). Structure-activity studies of 6-(tetrazolylalkyl)-substituted decahydroisoquinoline-3-carboxylic acid AMPA receptor antagonists. 1. Effects of stereochemistry, chain length, and chain substitution. *J. Med. Chem.*, 39, 2219–2231.

363. Madsen, U., Bang-Andersen, B., Brehm, L., Christensen, I. T., Ebert, B., Kristoffersen, I. T., Lang, Y., Krogsgaard-Larsen, P. (1996). Synthesis and pharmacology of highly selective carboxy and phosphono isoxazole amino acid AMPA receptor antagonists. *J. Med. Chem.*, 39, 1682–1691.

364. Ornstein, P. L., Arnold, M. B., Augenstein, N. K., Lodge, D., Leander, J. D., Schoepp, D. D. (1993). (3SR,4aRS,6RS,8aRS)-6-[2-(1H-tetrazol-5-yl)ethyl] decahydroisoquinoline-3-carboxylic acid: A structurally novel, systemically active, competitive AMPA receptor antagonist. *J. Med. Chem.*, 36, 2046–2048.

365. Bleakman, R., Schoepp, D. D., Ballyk, B., Bufton, H., Sharpe, E. F., Thomas, K., Ornstein, P. L., Kamboj, R. K. (1996). Pharmacological discrimination of GluR5 and GluR6 kainate receptor subtypes by (3S,4aR,6R,8aR)-6-[2-(1(2)H-tetrazole-5-yl)ethyl]decahydroisdoquinoline-3-carboxylic acid. *Mol. Pharmacol.*, 49, 581–585.

366. Møller, E. H., Egebjerg, J., Brehm, L., Stensbøl, T. B., Johansen, T. N., Madsen, U., Krogsgaard-Larsen, P. (1999). Resolution, absolute stereochemistry, and enantiopharmacology of the GluR1–4 and GluR5 antagonist 2-amino-3-[5-*tert*-butyl-3-(phosphonomethoxy)-4-isoxazolyl]propionic acid. *Chirality*, 11, 752–759.

367. Sólyom, S., Tarnawa, I. (2002). Non-competitive AMPA antagonists of 2,3-benzodiazepine type. *Curr. Pharm. Des.*, 8, 913–939.

368. Donevan, S. D., Rogawski, M. A. (1993). GYKI 52466, a 2,3-benzodiazepine, is a highly selective, noncompetitive antagonist of AMPA/kainate receptor responses. *Neuron*, 10, 51–59.

369. Zorumski, C. F., Yamada, K. A., Price, M. T., Olney, J. W. (1993). A benzodiazepine recognition site is associated with the non-NMDA glutamate receptor. *Neuron*, 10, 61–67.

370. Anderson, B. A., Harn, N. K., Hansen, M. M., Harkness, A. R., Lodge, D., Leander, J. D. (1999). Synthesis and anticonvulsant activity of 3-aryl-5H-2,3-benzodiazepine AMPA antagonists. *Bioorg. Med. Chem. Lett.*, 9, 1953–1956.

371. Welch, W. M., Ewing, F. E., Huang, J., Menniti, F. S., Pagnozzi, M. J., Kelly, K., Seymour, P. A., Guanowsky, V., Guhan, S., Guinn, M. R., Critchett, D., Lazzaro, J., Ganong, A. H., DeVries, K. M., Staigers, T. L., Chenard, B. L. (2001). Atropisomeric quinazolin-4-one derivatives are potent noncompetitive alpha-amino-3-hydroxy-5-methyl-4-isoxazolepropionic acid (AMPA) receptor antagonists. *Bioorg. Med. Chem. Lett.*, 11, 177–181.

372. Ohno, K., Tsutsumi, R., Matsumoto, N., Yamashita, H., Amada, Y., Shishikura, J., Yatsugi, H. I., Okada, M., Sakamoto, S., Yamaguchi, T. (2003). Functional characterization of YM928, a novel moncompetitive alpha-amino-3-hydroxy-5-methyl-4-isoxazolepropionic acid (AMPA) receptor antagonist. *J. Pharmacol. Exp. Ther.*, 306, 66–72.

373. Menniti, F. S., Chenard, B. L., Collins, M. B., Ducat, M. F., Elliott, M. L., Ewing, F. E., Huang, J. I., Kelly, K. A., Lazzaro, J. T., Pagnozzi, M. J., Weeks, J. L., Welch, W. M., White, W. F. (2000). Characterization of the binding site for a novel class of noncompetitive alpha-amino-3-hydroxy-5-methyl-4-isoxazolepropionic acid receptor antagonists. *Mol. Pharmacol.*, *58*, 1310–1317.

374. Granger, R., Deadwyler, S., Davis, M., Moskovitz, B., Kessler, M., Rogers, G., Lynch, G. (1996). Facilitation of glutamate receptors reverses an age-associated memory impairment in rats. *Synapse*, *22*, 332–337.

375. Sun, Y., Olson, R., Horning, M., Armstrong, N., Mayer, M., Gouaux, E. (2002). Mechanism of glutamate receptor desensitization. *Nature*, *417*, 245–253.

376. Zarrinmayeh, H., Bleakman, D., Gates, M. R., Yu, H., Zimmerman, D. M., Ornstein, P. L., McKennon, T., Arnold, M. B., Wheeler, W. J., Skolnick, P. (2001). [^{3}H]N-2-(4-(N-benzamido)phenyl)propyl-2-propanesulfonamide: A novel AMPA receptor potentiator and radioligand. *J. Med. Chem.*, *44*, 302–304.

377. Linden, A. M., Yu, H., Zarrinmayeh, H., Wheeler, W. J., Skolnick, P. (2001). Binding of an AMPA receptor potentiator ([^{3}H]LY395153) to native and recombinant AMPA receptors. *Neuropharmacology*, *40*, 1010–1018.

378. Shepherd, T. A., Aikins, J. A., Bleakman, D., Cantrell, B. E., Rearick, J. P., Simon, R. L., Smith, E. C., Stephenson, G. A., Zimmerman, D. M., Mandelzys, A., Jarvie, K. R., Ho, K., Deverill, M., Kamboj, R. K. (2002). Design and synthesis of a novel series of 1,2-disubstituted cyclopentanes as small, potent potentiators of 2-amino-3-(3-hydroxy-5- methyl-isoxazol-4-yl)propanoic acid (AMPA) receptors. *J. Med. Chem.*, *45*, 2101–2111.

379. Stensbøl, T. B., Borre, L., Johansen, T. N., Egebjerg, J., Madsen, U., Ebert, B., Krogsgaard-Larsen, P. (1999). Resolution, absolute stereochemistry and molecular pharmacology of the enantiomers of ATPA. *Eur. J. Pharmacol.*, *380*, 153–162.

380. Stensbøl, T. B., Jensen, H. S., Nielsen, B., Johansen, T. N., Egebjerg, J., Frydenvang, K., Krogsgaard-Larsen, P. (2001). Stereochemistry and molecular pharmacology of (S)-thio-ATPA, a new potent and selective GluR5 agonist. *Eur. J. Pharmacol.*, *411*, 245–253.

381. Thomas, N. K., Hawkins, L. M., Miller, J. C., Troop, H. M., Roberts, P. J., Jane, D. E. (1998). Pharmacological differentiation of kainate receptors on neonatal rat spinal motoneurones and dorsal roots. *Neuropharmacology*, *37*, 1223–1237.

382. Lunn, M. L., Hogner, A., Stensbøl, T. B., Gouaux, E., Egebjerg, J., Kastrup, J. S. (2003). Three-dimensional structure of the ligand-binding core of GluR2 in complex with the agonist (S)-ATPA: Implications for receptor subunit selectivity. *J. Med. Chem.*, *46*, 872–875.

383. Small, B., Thomas, J., Kemp, M., Hoo, K., Ballyk, B., Deverill, M., Ogden, A. M., Rubio, A., Pedregal, C., Bleakman, D. (1998). LY339434, a GluR5 kainate receptor agonist. *Neuropharmacology*, *37*, 1261–1267.

384. Pedregal, C., Collado, I., Escribano, A., Ezquerra, J., Dominguez, C., Mateo, A. I., Rubio, A., Baker, S. R., Goldsworthy, J., Kamboj, R. K., Ballyk, B. A., Hoo, K., Bleakman, D. (2000). 4-Alkyl- and 4-cinnamylglutamic acid analogues are potent GluR5 kainate receptor agonists. *J. Med. Chem.*, *43*, 1958–1968.

385. Bunch, L., Johansen, T. H., Bräuner-Osborne, H., Stensbøl, T. B., Johansen, T. N., Krogsgaard-Larsen, P., Madsen, U. (2001). Synthesis and receptor binding affinity of new selective GluR5 ligands. *Bioorg. Med. Chem.*, *9*, 875–879.

386. Valgeirsson, J., Christensen, J. K., Kristensen, A. S., Pickering, D. S., Nielsen, B., Fischer, C. H., Bräuner-Osborne, H., Nielsen, E. ø., Krogsgaard-Larsen, P., Madsen, U. (2003). Synthesis and in vitro pharmacology at AMPA and kainate preferring glutamate receptors of 4-heteroarylmethylidene glutamate analogues. *Bioorg. Med. Chem.*, *11*, 4341–4349.
387. Krogsgaard-Larsen, P., Nielsen, E. Ø., Curtis, D. R. (1984). Ibotenic acid analogues. Synthesis and biological *in vitro* activity of conformationally restricted agonists at excitatory amino acid receptors. *J. Med. Chem.*, *27*, 585–591.
388. Brehm, L., Greenwood, J. R., Hansen, K. B., Nielsen, B., Egebjerg, J., Stensbøl, T. B., Bräuner-Osborne, H., Sløk, F. A., Kronborg, T. T., Krogsgaard-Larsen, P. (2003). (S)-2-Amino-3-(3-hydroxy-7,8-dihydro-6H-cyclohepta[d]isoxazol-4-yl)propionic acid, a potent and selective agonist at the GluR5 subtype of ionotropic glutamate receptors. Synthesis, modeling, and molecular pharmacology. *J. Med. Chem.*, *46*, 1350–1358.
389. Johansen, T. H., Drejer, J., Wätjen, F., Nielsen, E. ø. (1993). A novel non-NMDA receptor antagonist shows selective displacement of low-affinity [^{3}H]kainate binding. *Eur. J. Pharmacol. Mol. Pharmacol. Sect.* *246*, 195–204.
390. Verdoorn, T. A., Johansen, T. H., Drejer, J., Nielsen, E. ø. (1994). Selective block of recombinant GluR6 receptors by NS-102, a novel non-NMDA receptor antagonist. *Eur. J. Pharmacol.*, *269*, 43–49.
391. O'Neill, M. J., Bond, A., Ornstein, P. L., Ward, M. A., Hicks, C. A., Hoo, K., Bleakman, D., Lodge, D. (1998). Decahydroisoquinolines: Novel competitive AMPA/kainate antagonists with neuroprotective effects in global cerebral ischaemia. *Neuropharmacology*, *37*, 1211–1222.
392. Lubisch, W., Behl, B., Henn, C., Hofmann, H. P., Reeb, J., Regner, F., Vierling, M. (2002). Pyrrolylquinoxalinediones carrying a piperazine residue represent highly potent and selective ligands to the homomeric kainate receptor GluR5. *Bioorg. Med. Chem. Lett.*, *12*, 2113–2116.
393. Valgeirsson, J., Nielsen, E. Ø., Peters, D., Varming, T., Mathiesen, C., Kristensen, A. S., Madsen, U. (2003). 2-Arylureidobenzoic acids: Selective noncompetitive antagonists for the homomeric kainate receptor subtype GluR5. *J. Med. Chem.*, *46*, 5834–5843.
394. Madden, D. R. (2002). The structure and function of glutamate receptor ion channels. *Nat. Rev. Neurosci.*, *3*, 91–101.
395. Rosenmund, C., Stern-Bach, Y., Stevens, C. F. (1998). The tetrameric structure of a glutamate receptor channel. *Science*, *280*, 1596–1599.
396. Ayalon, G., Stern-Bach, Y. (2001). Functional assembly of AMPA and kainate receptors is mediated by several discrete protein–protein interactions. *Neuron*, *31*, 103–113.
397. Safferling, M., Tichelaar, W., Kummerle, G., Jouppila, A., Kuusinen, A., Keinanen, K., Madden, D. R. (2001). First images of a glutamate receptor ion channel: Oligomeric state and molecular dimensions of GluRB homomers. *Biochemistry*, *40*, 13948–13953.
398. Kuusinen, A., Arvola, M., Keinanen, K. (1995). Molecular dissection of the agonist binding site of an AMPA receptor. *EMBO J.*, *14*, 6327–6332.
399. Armstrong, N., Sun, Y., Chen, G. Q., Gouaux, E. (1998). Structure of a glutamate-receptor ligand-binding core in complex with kainate. *Nature*, *395*, 913–917.

400. Kasper, C., Lunn, M. L., Liljefors, T., Gouaux, E., Egebjerg, J., Kastrup, J. S. (2002). GluR2 ligand-binding core complexes: Importance of the isoxazolol moiety and 5-substituent for the binding mode of AMPA-type agonists. *FEBS Lett.*, *531*, 173–178.
401. Lunn, M. L. (2001). From gene to crystal structure—a study of the ligand binding domain of AMPA and kainate receptors. Ph.D. Thesis, Royal Danish School of Pharmacy.
402. Jin, R., Horning, M., Mayer, M. L., Gouaux, E. (2002). Mechanism of activation and selectivity in a ligand-gated ion channel: Structural and functional studies of GluR2 and quisqualate. *Biochemistry*, *41*, 15635–15643.
403. Hogner, A., Greenwood, J. R., Liljefors, T., Lunn, M. L., Egebjerg, J., Larsen, I. K., Gouaux, E., Kastrup, J. S. (2003). Competitive antagonism of AMPA receptors by ligands of different classes: Crystal structure of ATPO bound to the GluR2 ligand-binding core, in comparison with DNQX. *J. Med. Chem.*, *46*, 214–221.
404. Armstrong, N., Mayer, M., Gouaux, E. (2003). Tuning activation of the AMPA-sensitive GluR2 ion channel by genetic adjustment of agonist-induced conformational changes. *Proc. Natl. Acad. Sci. USA*, *100*, 5736–5741.
405. Chen, G. Q., Cui, C., Mayer, M. L., Gouaux, E. (1999). Functional characterization of a potassium-selective prokaryotic glutamate receptor. *Nature*, *402*, 817–821.
406. Campiani, G., Morelli, E., Nacci, V., Fattorusso, C., Ramunno, A., Novellino, E., Greenwood, J., Liljefors, T., Griffiths, R., Sinclair, C., Reavy, H., Kristensen, A. S., Pickering, D. S., Schousboe, A., Cagnotto, A., Fumagalli, E., Mennini, T. (2001). Characterization of the 1H-cyclopentapyrimidine-2,4(1H,3H)-dione derivative (S)-CPW399 as a novel, potent, and subtype-selective AMPA receptor full agonist with partial desensitization properties. *J. Med. Chem.*, *44*, 4501–4504.
407. Johansen, T. N., Greenwood, J. R., Frydenvang, K., Madsen, U., Krogsgaard-Larsen, P. (2003). Stereostructure-activity studies on agonists at the AMPA and kainate subtypes of ionotropic glutamate receptors. *Chirality*, *15*, 167–179.
408. Arinaminpathy, Y., Sansom, M. S., Biggin, P. C. (2002). Molecular dynamics simulations of the ligand-binding domain of the ionotropic glutamate receptor GluR2. *Biophys. J.*, *82*, 676–683.
409. Pang, A., Arinaminpathy, Y., Sansom, M. S., Biggin, P. C. (2003). Interdomain dynamics and ligand binding: Molecular dynamics simulations of glutamine binding protein. *FEBS Lett.*, *550*, 168–174.
410. Pentikäinen, O. T., Settimo, L., Keinanen, K., Johnson, M. S. (2003). Selective agonist binding of (S)-amino-3-(3-hydroxy-5-methyl-4-isoxazolyl)propionic acid (AMPA) and (2S-(2a,3b,4b))-2-carboxy-4-(1-methylethenyl)3-pyrrolidine-acetic acid (kainate) receptors: A molecular modeling study. *Biochem. Pharmacol.*, *66*, 2413–2425.
411. Partin, K. M., Bowie, D., Mayer, M. L. (1995). Structural determinants of allosteric regulation in alternatively spliced AMPA receptors. *Neuron*, *14*, 833–843.
412. Augelli-Szafran, C. E., Schwarz, R. D. (2003). Metabotropic glutamate receptors: Agonists, antagonists and allosteric modulators. *Annu. Rep. Med. Chem.*, *38*, 21–30.
413. Irving, A. J., Schofield, J. G., Watkins, J. C., Sunter, D. C., Collingridge, G. L. (1990). 1S,3R-ACPD stimulates and L-AP3 blocks Ca^{2+} mobilization in rat cerebellar neurons. *Eur. J. Pharmacol.*, *186*, 363–365.

414. Aramori, I., Nakanishi, S. (1992). Signal transduction and pharmacological characteristics of a metabotropic glutamate receptor, mGluR1, in transfected CHO cells. *Neuron*, 8, 757–765.

415. Tanabe, Y., Masu, M., Ishii, T., Shigemoto, R., Nakanishi, S. (1992). A family of metabotropic glutamate receptors. *Neuron*, 8, 169–179.

416. Tanabe, Y., Nomura, A., Masu, M., Shigemoto, R., Mizuno, N., Nakanishi, S. (1993). Signal transduction, pharmacological properties, and expression patterns of two rat metabotropic glutamate receptors, mGluR3 and mGluR4. *J. Neurosci.*, 13, 1372–1378.

417. Abe, T., Sugihara, H., Nawa, H., Shigemoto, R., Mizuno, N., Nakanishi, S. (1992). Molecular characterization of a novel metabotropic glutamate receptor mGluR5 coupled to inositol phosphate/Ca^{2+} signal transduction. *J. Biol Chem.*, 267, 13361–13368.

418. Nakajima, Y., Iwakabe, H., Akazawa, C., Nawa, H., Shigemoto, R., Mizuno, N., Nakanishi, S. (1993). Molecular characterization of a novel retinal metabotropic glutamate receptor mGluR6 with a high agonist selectivity for L-2-amino-4-phosphonobutyrate. *J. Biol Chem.*, 268, 11868–11873.

419. Okamoto, N., Hori, S., Akazawa, C., Hayashi, Y., Shigemoto, R., Mizuno, N., Nakanishi, S. (1994). Molecular characterization of a new metabotropic glutamate receptor mGluR7 coupled to inhibitory cyclic AMP signal transduction. *J. Biol. Chem.*, 269, 1231–1236.

420. Saugstad, J. A., Kinzie, J. M., Shinohara, M. M., Segerson, T. P., Westbrook, G. L. (1997). Cloning and expression of rat metabotropic glutamate receptor 8 reveals a distinct pharmacological profile. *Mol. Pharmacol.*, 51, 119–125.

421. Monn, J. A., Valli, M. J., Massey, S. M., Hansen, M. M., Kress, T. J., Wepsiec, J. P., Harkness, A. R., Grutsch, J. L., Jr., Wright, R. A., Johnson, B. G., Andis, S. L., Kingston, A., Tomlinson, R., Lewis, R., Griffey, K. R., Tizzano, J. P., Schoepp, D. D. (1999). Synthesis, pharmacological characterization, and molecular modeling of heterobicyclic amino acids related to (+)-2-aminobicyclo[3.1.0]hexane-2,6-dicarboxylic acid (LY354740): Identification of two new potent, selective, and systemically active agonists for group II metabotropic glutamate receptors. *J. Med. Chem.*, 42, 1027–1040.

422. Schoepp, D. D., Salhoff, C. R., Wright, R. A., Johnson, B. G., Burnett, J. P., Mayne, N. G., Belagaje, R., Wu, S., Monn, J. A. (1996). The novel metabotropic glutamate receptor agonist 2R,4R-APDC potentiates stimulation of phosphoinositide hydrolysis in the rat hippocampus by 3,5-dihydroxyphenylglycine: Evidence for a synergistic interaction between group 1 and group 2 receptors. *Neuropharmacology*, 35, 1661–1672.

423. Tückmantel, W., Kozikowski, A. P., Wang, S., Pshenichkin, S., Wroblewski, J. T. (1997). Synthesis, molecular modeling, and biology of the 1-benzyl derivative of APDC—an apparent mGluR6 selective ligand. *Bioorg. Med. Chem. Lett.*, 7, 601–606.

424. Valli, M. J., Schoepp, D. D., Wright, R. A., Johnson, B. G., Kingston, A. E., Tomlinson, R., Monn, J. A. (1998). Synthesis and metabotropic glutamate receptor antagonist activity of N1-substituted analogs of 2R,4R-4-aminopyrrolidine-2,4-dicarboxylic acid. *Bioorg. Med. Chem. Lett.*, 8, 1985–1990.

425. Kozikowski, A. P., Araldi, G. L., Tückmantel, W., Pshenichkin, S., Surina, E., Wroblewski, J. T. (1999). 1-Amino-APDC, a partial agonist of group II metabotropic glutamate receptors with neuroprotective properties. *Bioorg. Med. Chem. Lett.*, *9*, 1721–1726.

426. Acher, F. C., Tellier, F. J., Azerad, R., Brabet, I. N., Fagni, L., Pin, J. P. (1997). Synthesis and pharmacological characterization of aminocyclopentanetricarboxylic acids: New tools to discriminate between metabotropic glutamate receptor subtypes. *J. Med. Chem.*, *40*, 3119–3129.

427. Bräuner-Osborne, H., Madsen, U., Mikiciuk-Olasik, E., Curry, K. (1997). New analogues of ACPD with selective activity for group II metabotropic glutamate receptors. *Eur. J. Pharmacol.*, *332*, 327–331.

428. Monn, J. A., Valli, M. J., Massey, S. M., Wright, R. A., Salhoff, C. R., Johnson, B. G., Howe, T., Alt, C. A., Rhodes, G. A., Robey, R. L., Griffey, K. R., Tizzano, J. P., Kallman, M. J., Helton, D. R., Schoepp, D. D. (1997). Design, synthesis and pharmacological characterization of (+)-2-aminobicyclo[3.1.0]hexane-2,6-dicarboxylic acid (LY354740): A potent, selective and orally active group 2 metabotropic glutamate receptor agonist possessing anticonvulsant and anxiolytic properties. *J. Med. Chem.*, *40*, 528–537.

429. Schoepp, D. D., Johnson, B. G., Wright, R. A., Salhoff, C. R., Mayne, N. G., Wu, S., Cockerham, S. L., Burnett, J. P., Belegaje, R., Bleakman, D., Monn, J. A. (1997). LY354740 is a potent and highly selective group II metabotropic glutamate receptor agonist in cells expressing human glutamate receptors. *Neuropharmacology*, *36*, 1–11.

430. Kozikowski, A. P., Steensma, D., Araldi, G. L., Tückmantel, W., Wang, S., Pshenichkin, S., Surina, E., Wroblewski, J. T. (1998). Synthesis and biology of the conformationally restricted ACPD analogue, 2-aminobicyclo[2.1.1]hexane-2,5-dicarboxylic acid-I, a potent mGluR agonist. *J. Med. Chem.*, *41*, 1641–1650.

431. Jullian, N., Brabet, I., Pin, J. P., Acher, F. C. (1999). Agonist selectivity of mGluR1 and mGluR2 metabotropic receptors: A different environment but similar recognition of an extended glutamate conformation. *J. Med. Chem.*, *42*, 1546–1555.

432. Costantino, G., Macchiarulo, A., Pellicciari, R. (1999). Pharmacophore models of group I and group II metabotropic glutamate receptor agonists. Analysis of conformational, steric, and topological parameters affecting potency and selectivity. *J. Med. Chem.*, *42*, 2816–2827.

433. Bessis, A.-S., Jullian, N., Coudert, E., Pin, J.-P., Acher, F. (1999). Extended glutamate activates metabotropic glutamate receptor types 1, 2 and 4: Selective features at mGluR4 binding site. *Neuropharmacology*, *38*, 1543–1551.

434. Kozikowski, A. P., Araldi, G. L., Flippen-Anderson, J., George, C., Pshenichkin, S., Surina, E., Wroblewski, J. (1998). Synthesis and metabotropic glutamate receptor activity of a 2-aminobicyclo[3.2.0]heptane-2,5-dicarboxylic acid, a molecule possessing an extended glutamate conformation. *Bioorg. Med. Chem. Lett.*, *8*, 925–930.

435. Tellier, F., Acher, F., Brabet, I., Pin, J. P., Azerad, R. (1998). Aminobicyclo[2.2.1]heptane dicarboxylic acids (ABHD), rigid analogs of ACPD and glutamic acid: Synthesis and pharmacological activity on metabotropic receptors mGluR1 and mGluR2. *Bioorg. Med. Chem.*, *6*, 195–208.

436. Hayashi, Y., Tanabe, Y., Aramori, I., Masu, M., Shimamoto, K., Ohfune, Y., Nakanishi, S. (1992). Agonist analysis of 2-(carboxycyclopropyl)glycine isomers for cloned metabotropic glutamate receptor subtypes expressed in Chinese hamster ovary cells. *Br. J. Pharmacol.*, *107*, 539–543.

437. Brabet, I., Parmentier, M. L., De Colle, C., Bockaert, J., Acher, F., Pin, J. P. (1998). Comparative effect of L-CCG-I, DCG-IV and gamma-carboxy-L-glutamate on all cloned metabotropic glutamate receptor subtypes. *Neuropharmacology*, *37*, 1043–1051.

438. Adam, G., Hennig, M., Kolczewski, S., Ohresser, S., Wichmann, J., Woltering, T., Mutel, V. (1999). Synthesis and receptor pharmacology of HYDIA; a new group II mGlu receptor antagonist. *Neuropharmacology*, *38*, A1.

439. Thomsen, C., Hansen, L., Suzdak, P. D. (1994). L-Glutamate uptake inhibitors may stimulate phosphoinositide hydrolysis in baby hamster kidney cells expressing mGluR1a via heteroexchange with L-glutamate without direct activation of mGluR1a. *J. Neurochem.*, *63*, 2038–2047.

440. Bräuner-Osborne, H., Sløk, F. A., Skjærbæk, N., Ebert, B., Sekiyama, N., Nakanishi, S., Krogsgaard-Larsen, P. (1996). A new highly selective metabotropic excitatory amino acid agonist: 2-amino-4-(3-hydroxy-5-methylisoxazol-4-yl)butyric acid. *J. Med. Chem.*, *39*, 3188–3194.

441. Ahmadian, H., Nielsen, B., Bräuner-Osborne, H., Johansen, T. N., Stensbøl, T. B., Sløk, F. A., Sekiyama, N., Nakanishi, S., Krogsgaard-Larsen, P., Madsen, U. (1997). (*S*)-Homo-AMPA, a specific agonist at the mGlu6 subtype of metabotropic glutamic acid receptors. *J. Med. Chem.*, *40*, 3700–3705.

442. Bräuner-Osborne, H., Nielsen, B., Krogsgaard-Larsen, P. (1998). Molecular pharmacology of homologues of ibotenic acid at cloned metabotropic glutamic acid receptors. *Eur. J. Pharmacol.*, *350*, 311–316.

443. Nicoletti, F., Meek, J. L., Iadarola, M. J., Chuang, D. M., Roth, B. L., Costa, E. (1986). Coupling of inositol phospholipid metabolism with excitatory amino acid recognition sites in rat hippocampus. *J. Neurochem.*, *46*, 40–46.

444. Masu, M., Tanabe, Y., Tsuchida, K., Shigemoto, R., Nakanishi, S. (1991). Sequence and expression of a metabotropic glutamate receptor. *Nature*, *349*, 760–765.

445. Houamed, K. M., Kuijper, J. L., Gilbert, T. L., Haldeman, B. A., O'Hara, P. J., Mulvihill, E. R., Almers, W., Hagen, F. S. (1991). Cloning, expression, and gene structure of a G protein-coupled glutamate receptor from rat brain. *Science*, *252*, 1318–1321.

446. Madsen, U., Bräuner-Osborne, H., Frydenvang, K., Hvene, L., Johansen, T. N., Nielsen, B., Sánchez, C., Stensbøl, T. B., Bischoff, F., Krogsgaard-Larsen, P. (2001). Synthesis and pharmacology of 3-isoxazolol amino acids as selective antagonists at group I metabotropic glutamate receptors. *J. Med. Chem.*, *44*, 1051–1059.

447. Bräuner-Osborne, H., Krogsgaard-Larsen, P. (1998). Pharmacology of (*S*)-homoquisqualic acid and (*S*)-2-amino-5-phosphonopentanoic acid [(*S*)-AP5] at cloned metabotropic glutamate receptors. *Br. J. Pharmacol.*, *123*, 269–274.

448. Koerner, J. F., Cotman, C. W. (1981). Micromolar L-2-amino-4-phosphonobutyric acid selectively inhibits perforant path synapses from lateral entorhinal cortex. *Brain Res.*, *216*, 192–198.

449. Duvoisin, R. M., Zhang, C. X., Ramonell, K. (1995). A novel metabotropic glutamate receptor expressed in the retina and olfactory bulb. *J. Neurosci.*, *15*, 3075–3083.
450. Johansen, P. A., Chase, L. A., Sinor, A. D., Koerner, J. F., Johnson, R. L., Robinson, M. B. (1995). Type 4a metabotropic glutamate receptor: Identification of new potent agonists and differentiation from the L-(+)-2-amino-4-phosphonobutanoic acid-sensitive receptor in the lateral perforant pathway in rats. *Mol. Pharmacol.*, *48*, 140–149.
451. Cartmell, J., Adam, G., Chaboz, S., Henningsen, R., Kemp, J. A., Klingelschmidt, A., Metzler, V., Monsma, F., Schaffhauser, H., Wichmann, J., Mutel, V. (1998). Characterization of [^{3}H]-(2S,2'R,3'R)-2-(2',3'-dicarboxy-cyclopropyl)glycine ([^{3}H]-DCG-IV) binding to metabotropic mGlu$_2$ receptor-transfected cell membranes. *Br. J. Pharmacol.*, *123*, 497–504.
452. Ito, I., Kohda, A., Tanabe, S., Hirose, E., Hayashi, M., Mitsunaga, S., Sugiyama, H. (1992). 3,5-Dihydroxyphenyl-glycine: A potent agonist of metabotropic glutamate receptors. *Neuroreport*, *3*, 1013–1016.
453. Hayashi, Y., Sekiyama, N., Nakanishi, S., Jane, D. E., Sunter, D. C., Birse, E. F., Udvarhelyi, P. M., Watkins, J. C. (1994). Analysis of agonist and antagonist activities of phenylglycine derivatives for different cloned metabotropic glutamate receptor subtypes. *J. Neurosci.*, *14*, 3370–3377.
454. Thomsen, C., Boel, E., Suzdak, P. D. (1994). Actions of phenylglycine analogs at subtypes of the metabotropic glutamate receptor family. *Eur. J. Pharmacol.*, *267*, 77–84.
455. Eaton, S. A., Jane, D. E., Jones, P. L., Porter, R. H., Pook, P. C., Sunter, D. C., Udvarhelyi, P. M., Roberts, P. J., Salt, T. E., Watkins, J. C. (1993). Competitive antagonism at metabotropic glutamate receptors by (S)-4-carboxyphenylglycine and (RS)-alpha-methyl-4-carboxyphenylglycine. *Eur. J. Pharmacol.*, *244*, 195–197.
456. Clark, B. P., Baker, S. R., Goldsworth, J., Harris, J. R., Kingston, A. E. (1997). (+)-2-Methyl-4-carboxyphenylglycine (LY367385) selectively antagonises metabotropic glutamate mGluR1 receptors. *Bioorg. Med. Chem. Lett.*, *7*, 2777–2780.
457. Pellicciari, R., Luneia, R., Costantino, G., Marinozzi, M., Natalini, B., Jakobsen, P., Kanstrup, A., Lombardi, G., Moroni, F., Thomsen, C. (1995). 1-Aminoindan-1,5-dicarboxylic acid: A novel antagonist at phospholipase C-linked metabotropic glutamate receptors. *J. Med. Chem.*, *38*, 3717–3719.
458. Moroni, F., Lombardi, G., Thomsen, C., Leonardi, P., Attucci, S., Peruginelli, F., Torregrossa, S. A., Pellegrini-Giampietro, D. E., Luneia, R., Pellicciari, R. (1997). Pharmacological characterization of 1-aminoindan-1,5-dicarboxylic acid, a potent mGluR1 antagonist. *J. Pharmacol. Exp. Ther.*, *281*, 721–729.
459. Brabet, I., Mary, S., Bockaert, J., Pin, J.-P. (1995). Phenylglycine derivatives discriminate between mGluR1- and mGluR5-mediated responses. *Neuropharmacology*, *34*, 895–903.
460. Kingston, A. E., Burnett, J. P., Mayne, N. G., Lodge, D. (1995). Pharmacological analysis of 4-carboxyphenylglycine derivatives: Comparison of effects on mGluR1α and mGluR5a subtypes. *Neuropharmacology*, *34*, 887–894.
461. Clark, B. P., Harris, J. R., Kingston, A. E., McManus, D. (1998). Alpha-substituted phenylglycines as group I metabotropic glutamate receptor antagonists. *XVth Eur. Int. Symp. Med. Chem.*, 183.

462. Costantino, G., Pellicciari, R. (1996). Homology modeling of metabotropic glutamate receptors. (mGluRs) structural motifs affecting binding modes and pharmacological profile of mGluR1 agonists and competitive antagonists. *J. Med. Chem.*, *39*, 3998–4006.

463. Pellicciari, R., Raimondo, M., Marinozzi, M., Natalini, B., Costantino, G., Thomsen, C. (1996). (S)-(+)-2-(3′-Carboxybicyclo[1.1.1]pentyl)glycine, a structurally new group i metabotropic glutamate receptor antagonist. *J. Med. Chem.*, *39*, 2874–2876.

464. Pellicciari, R., Costantino, G., Giovagnoni, E., Mattoli, L., Brabet, I., Pin, J. P. (1998). Synthesis and preliminary evaluation of (S)-2-(4′-carboxycubyl)glycine, a new selective mGluR1 antagonist. *Bioorg. Med. Chem. Lett.*, *8*, 1569–1574.

465. Gasparini, F., Bruno, V., Battaglia, G., Lukic, S., Leonhardt, T., Inderbitzin, W., Laurie, D., Sommer, B., Varney, M. A., Hess, S. D., Johnson, E. C., Kuhn, R., Urwyler, S., Sauer, D., Portet, C., Schmutz, M., Nicoletti, F., Flor, P. J. (1999). (RS)-4-Phosphonophenylglycine, a potent and selective group III metabotropic glutamate receptor agonist, is anticonvulsive and neuroprotective in vivo. *J. Pharmacol. Exp. Ther.*, *289*, 1678–1687.

466. Jane, D. E., Pittaway, K., Sunter, D. C., Thomas, N. K., Watkins, J. C. (1995). New phenylglycine derivatives with potent and selective antagonist activity at presynaptic glutamate receptors in neonatal rat spinal cord. *Neuropharmacology*, *34*, 851–856.

467. Ma, D., Tian, H., Sun, H., Kozikowski, A. P., Pshenichkin, S., Wroblewski, J. T. (1997). Synthesis and biological activity of cyclic analogues of MPPG and MCPG as metabotropic glutamate receptor antagonists. *Bioorg. Med. Chem. Lett.*, *7*, 1195–1198.

468. Thomsen, C., Bruno, V., Nicoletti, F., Marinozzi, M., Pellicciari, R. (1996). (2S,1′S,2′S,3′R)-2-(2′-Carboxy-3′-phenylcyclopropyl)glycine, a potent and selective antagonist of type 2 metabotropic glutamate receptors. *Mol. Pharmacol.*, *50*, 6–9.

469. Gomeza, J., Mary, S., Brabet, I., Parmentier, M. L., Restituito, S., Bockaert, J., Pin, J.-P. (1996). Coupling of metabotropic glutamate receptors 2 and 4 to Gα15, Gα16, and chimeric G αq/i proteins: Characterization of new antagonists. *Mol. Pharmacol.*, *50*, 923–930.

470. Johansen, P. A., Robinson, M. B. (1995). Identification of 2-amino-2-methyl-4-phosphonobutanoic acid as an antagonist at the mGlu(4a) receptor. *Eur. J. Pharmacol.*, *290*, R1–R3.

471. Kozikowski, A. P., Steensma, D., Varasi, M., Pshenichkin, S., Surina, E., Wroblewski, J. T. (1998). α-Substituted quisqualic acid analogs: New metabotropic glutamate receptor group II selective antagonists. *Bioorg. Med. Chem. Lett.*, *8*, 447–452.

472. Wermuth, C. G., Mann, A., Schoenfelder, A., Wright, R. A., Johnson, B. G., Burnett, J. P., Mayne, N. G., Schoepp, D. D. (1996). (2S,4S)-2-Amino-4-(4,4-diphenylbut-1-yl)-pentane-1,5-dioic acid: A potent and selective antagonist for metabotropic glutamate receptors negatively linked to adenylate cyclase. *J. Med. Chem.*, *39*, 814–816.

473. Escribano, A., Ezquerra, J., Pedregal, C., Rubio, A., Yruretagoyena, B., Baker, S. R., Wright, R. A., Johnson, B. G., Schoepp, D. D. (1998). (2S,4S)-amino-4-(2,2-

diphenylethyl)pentanedioic acid selective group 2 metabotropic glutamate receptor antagonist. *Bioorg. Med. Chem. Lett.*, *8*, 765–770.

474. Bräuner-Osborne, H., Nielsen, B., Stensbøl, T. B., Johansen, T. N., Skjærbæk, N., Krogsgaard-Larsen, P. (1997). Molecular pharmacology of 4-substituted glutamic acid analogues at ionotropic and metabotropic excitatory amino acid receptors. *Eur. J. Pharmacol.*, *335*, R1–R3.

475. Ornstein, P. L., Bleisch, T. J., Arnold, M. B., Wright, R. A., Johnson, B. G., Schoepp, D. D. (1998). 2-Substituted (2SR)-2-amino-2-((1SR,2SR)-2-carboxycycloprop-1-yl)glycines as potent and selective antagonists of group II metabotropic glutamate receptors. 1. Effects of alkyl, arylalkyl, and diarylalkyl substitution. *J. Med. Chem.*, *41*, 346–357.

476. Kingston, A. E., Ornstein, P. L., Wright, R. A., Johnson, B. G., Mayne, N. G., Burnett, J. P., Belagaje, R., Wu, S., Schoepp, D. D. (1998). LY341495 is a nanomolar potent and selective antagonist of group II metabotropic glutamate receptors. *Neuropharmacology*, *37*, 1–12.

477. Wu, S., Wright, R. A., Rockey, P. K., Burgett, S. G., Arnold, J. S., Rosteck, P. R., Jr., Johnson, B. G., Schoepp, D. D., Belagaje, R. M. (1998). Group III human metabotropic glutamate receptors 4, 7 and 8: Molecular cloning, functional expression, and comparison of pharmacological properties in RGT cells. *Brain. Res. Mol. Brain. Res.*, *53*, 88–97.

478. Pellicciari, R., Costantino, G., Marinozzi, M., Macchiarulo, A., Amori, L., Josef Flor, P., Gasparini, F., Kuhn, R., Urwyler, S. (2001). Design, synthesis and preliminary evaluation of novel 3′-substituted carboxycyclopropylglycines as antagonists at group 2 metabotropic glutamate receptors. *Bioorg. Med. Chem. Lett.*, *11*, 3179–3182.

479. Chen, Y., Bacon, G., Sher, E., Clark, B. P., Kallman, M. J., Wright, R. A., Johnson, B. G., Schoepp, D. D., Kingston, A. E. (2000). Evaluation of the activity of a novel metabotropic glutamate receptor antagonist (+/−)-2-amino-2-(3-*cis* and *trans*-carboxycyclobutyl-3-(9-thioxanthyl)propionic acid) in the in vitro neonatal spinal cord and in an in vivo pain model. *Neuroscience*, *95*, 787–793.

480. Annoura, H., Fukunaga, A., Uesugi, M., Tatsuoka, T., Horikawa, Y. (1996). A novel class of antagonists for metabotropic glutamate receptors, 7-(hydroxyimino)cyclopropan[*b*]chromen-1*a*-carboxylates. *Bioorg. Med. Chem. Lett.*, *6*, 763–766.

481. Hermans, E., Nahorski, S. R., Challiss, R. A. (1998). Reversible and noncompetitive antagonist profile of CPCCOEt at the human type 1α metabotropic glutamate receptor. *Neuropharmacology*, *37*, 1645–1647.

482. Litschig, S., Gasparini, F., Rüegg, D., Stoehr, N., Flor, P. J., Vranesic, I., Prézeau, L., Pin, J.-P., Thomsen, C., Kuhn, R. (1999). CPCCOEt, a noncompetitive metabotropic glutamate receptor 1 antagonist, inhibits receptor signaling without affecting glutamate binding. *Mol. Pharmacol.*, *55*, 453–461.

483. Bräuner-Osborne, H., Jensen, A. A., Krogsgaard-Larsen, P. (1999). Interaction of CPCCOEt with a chimeric mGlu$_{1b}$ and calcium sensing receptor. *Neuroreport*, *10*, 3923–3925.

484. Carroll, F. Y., Stolle, A., Beart, P. M., Voerste, A., Brabet, I., Mauler, F., Joly, C., Antonicek, H., Bockaert, J., Müller, T., Pin, J. P., Prézeau, L. (2001). BAY36-7620: A potent non-competitive mGlu$_1$ receptor antagonist with inverse agonist activity. *Mol. Pharmacol.*, *59*, 965–973.

485. Malherbe, P., Kratochwil, N., Knoflach, F., Zenner, M. T., Kew, J. N., Kratzeisen, C., Maerki, H. P., Adam, G., Mutel, V. (2003). Mutational analysis and molecular modeling of the allosteric binding site of a novel, selective, noncompetitive antagonist of the metabotropic glutamate 1 receptor. *J. Biol Chem.*, 278, 8340–8347.
486. Kolczewski, S., Adam, G., Stadler, H., Mutel, V., Wichmann, J., Woltering, T. (1999). Synthesis of heterocyclic enol ethers and their use as group 2 metabotropic glutamate receptor antagonists. *Bioorg. Med. Chem. Lett.*, 9, 2173–2176.
487. Adam, G., Kolczewski, S., Ohresser, S., Wichmann, J., Woltering, T., Mutel, V. (1999). Synthesis, structure-activity relationship and receptor pharmacology of new non competitive, non-amino acid, mGlu2 receptor selective antagonists. *Neuropharmacology*, 10, A1.
488. Varney, M. A., Cosford, N. D., Jachec, C., Rao, S. P., Sacaan, A., Lin, F. F., Bleicher, L., Santori, E. M., Flor, P. J., Allgeier, H., Gasparini, F., Kuhn, R., Hess, S. D., Velielebi, G., Johnson, E. C. (1999). SIB-1757 and SIB-1893: Selective, noncompetitive antagonists of metabotropic glutamate receptor type 5. *J. Pharmacol. Exp. Ther.*, 290, 170–181.
489. Gasparini, F., Lingenhöhl, K., Stoehr, N., Flor, P. J., Heinrich, M., Vranesic, I., Biollaz, M., Allgeier, H., Heckendorn, R., Urwyler, S., Varney, M. A., Johnson, E. C., Hess, S. D., Rao, S. P., Sacaan, A. I., Santori, E. M., Veliçelebi, G., Kuhn, R. (1999). 2-Methyl-6-(phenylethynyl)-pyridine (MPEP), a potent, selective and systemically active mGlu$_5$ receptor antagonist. *Neuropharmacology*, 38, 1493–1503.
490. Pagano, A., Ruegg, D., Litschig, S., Stoehr, N., Stierlin, C., Heinrich, M., Floersheim, P., Prezeau, L., Carroll, F., Pin, J. P., Cambria, A., Vranesic, I., Flor, P. J., Gasparini, F., Kuhn, R. (2000). The non-competitive antagonists 2-methyl-6-(phenylethynyl)pyridine and 7-hydroxyiminocyclopropan[*b*]chromen-1*a*-carboxylic acid ethyl ester interact with overlapping binding pockets in the transmembrane region of group I metabotropic glutamate receptors. *J. Biol Chem.*, 275, 33750–33758.
491. Mathiesen, J. M., Svendsen, N., Bräuner-Osborne, H., Thomsen, C., Ramirez, M. T. (2003). Positive allosteric modulation of the human metabotropic glutamate receptor 4 (hmGluR4) by SIB-1893 and MPEP. *Br. J. Pharmacol.*, 138, 1026–1030.
492. Cosford, N. D., Tehrani, L., Roppe, J., Schweiger, E., Smith, N. D., Anderson, J., Bristow, L., Brodkin, J., Jiang, X., McDonald, I., Rao, S., Washburn, M., Varney, M. A. (2003). 3-[(2-Methyl-1,3-thiazol-4-yl)ethynyl]-pyridine: A potent and highly selective metabotropic glutamate subtype 5 receptor antagonist with anxiolytic activity. *J. Med. Chem.*, 46, 204–206.
493. Schaffhauser, H., Rowe, B. A., Morales, S., Chavez-Noriega, L. E., Yin, R., Jachec, C., Rao, S. P., Bain, G., Pinkerton, A. B., Vernier, J. M., Bristow, L. J., Varney, M. A., Daggett, L. P. (2003). Pharmacological characterization and identification of amino acids involved in the positive modulation of metabotropic glutamate receptor subtype 2. *Mol. Pharmacol.*, 64, 798–810.
494. O'Brien, J. A., Lemaire, W., Chen, T. B., Chang, R. S., Jacobson, M. A., Ha, S. N., Lindsley, C. W., Schaffhauser, H. J., Sur, C., Pettibone, D. J., Conn, P. J., Williams, D. L., Jr. (2003). A family of highly selective allosteric modulators

of the metabotropic glutamate receptor subtype 5. *Mol. Pharmacol.*, *64*, 731–740.

495. Spooren, W., Ballard, T., Gasparini, F., Amalric, M., Mutel, V., Schreiber, R. (2003). Insight into the function of Group I and Group II metabotropic glutamate (mGlu) receptors: Behavioural characterization and implications for the treatment of CNS disorders. *Behav. Pharmacol.*, *14*, 257–277.

496. Gasparini, F., Kuhn, R., Pin, J. P. (2002). Allosteric modulators of group I metabotropic glutamate receptors: Novel subtype-selective ligands and therapeutic perspectives. *Curr. Opin. Pharmacol.*, *2*, 43–49.

497. Moldrich, R. X., Chapman, A. G., De Sarro, G., Meldrum, B. S. (2003). Glutamate metabotropic receptors as targets for drug therapy in epilepsy. *Eur. J. Pharmacol.*, *476*, 3–16.

498. Flor, P. J., Battaglia, G., Nicoletti, F., Gasparini, F., Bruno, V. (2002). Neuroprotective activity of metabotropic glutamate receptor ligands. *Adv. Exp. Med. Biol.*, *513*, 197–223.

499. Kunishima, N., Shimada, Y., Tsuji, Y., Sato, T., Yamamoto, M., Kumasaka, T., Nakanishi, S., Jingami, H., Morikawa, K. (2000). Structural basis of glutamate recognition by a dimeric metabotropic glutamate receptor. *Nature*, *407*, 971–977.

500. Tsuchiya, D., Kunishima, N., Kamiya, N., Jingami, H., Morikawa, K. (2002). Structural views of the ligand-binding cores of a metabotropic glutamate receptor complexed with an antagonist and both glutamate and Gd3+. *Proc. Natl. Acad. Sci. USA*, *99*, 2660–2665.

501. Romano, C., Yang, W. L., O'Malley, K. L. (1996). Metabotropic glutamate receptor 5 is a disulfide-linked dimer. *J. Biol Chem.*, *271*, 28612-28616.

502. Costantino, G., Macchiarulo, A., Pellicciari, R. (2001). Homology model of the closed, functionally active, form of the amino terminal domain of mGlur1. *Bioorg. Med. Chem.*, *9*, 847–852.

503. Jensen, A. A., Sheppard, P. O., Jensen, L. B., O'Hara, P. J., Bräuner-Osborne, H. (2001). Construction of a high affinity zinc binding site in the metabotropic glutamate receptor mGluR1: Noncompetitive antagonism originating from the amino-terminal domain of a family C G-protein-coupled receptor. *J. Biol Chem.*, *276*, 10110–10118.

504. Clausen, R. P., Bräuner-Osborne, H., Greenwood, J. R., Hermit, M. B., Stensbøl, T. B., Nielsen, B., Krogsgaard-Larsen, P. (2002). Selective agonists at group II metabotropic glutamate receptors: Synthesis, stereochemistry, and molecular pharmacology of (*S*)- and (*R*)-2-amino-4-(4-hydroxy[1,2,5]thiadiazol-3-yl)butyric acid. *J. Med. Chem.*, *45*, 4240–4245.

505. Bessis, A. S., Rondard, P., Gaven, F., Brabet, I., Triballeau, N., Prezeau, L., Acher, F., Pin, J. P. (2002). Closure of the Venus flytrap module of mGlu8 receptor and the activation process: Insights from mutations converting antagonists into agonists. *Proc. Natl. Acad. Sci. USA*, *99*, 11097–11102.

506. Sato, T., Shimada, Y., Nagasawa, N., Nakanishi, S., Jingami, H. (2003). Amino acid mutagenesis of the ligand binding site and the dimer interface of the metabotropic glutamate receptor 1. Identification of crucial residues for setting the activated state. *J. Biol Chem.*, *278*, 4314–4321.

507. Yao, Y., Pattabiraman, N., Michne, W. F., Huang, X. P., Hampson, D. R. (2003). Molecular modeling and mutagenesis of the ligand-binding pocket of the mGlu3 subtype of metabotropic glutamate receptor. *J. Neurochem.*, *86*, 947–957.
508. Hermit, M. B., Greenwood, J. R., Nielsen, B., Bunch, L., Jørgensen, C. G., Vestergaard, H. T., Stensbøl, T. B., Sanchez, C., Krogsgaard-Larsen, P., Madsen, U., Bräuner-Osborne, H. (2004). Ibotenic acid and thioibotenic acid: A remarkable difference in activity at group III metabotropic glutamate receptors. *Eur. J. Pharmacol.*, *486*, 241–250.

19

CARDIAC SARCOLEMMAL ATP-SENSITIVE POTASSIUM CHANNEL ANTAGONISTS: NOVEL ISCHEMIA-SELECTIVE ANTIARRHYTHMIC AGENTS

GEORGE E. BILLMAN
The Ohio State University
Columbus, Ohio

19.1	INTRODUCTION	909
19.2	EFFECTS OF MYOCARDIAL ISCHEMIA ON EXTRACELLULAR POTASSIUM	911
19.3	EFFECT OF EXTRACELLULAR POTASSIUM ON VENTRICULAR ARRHYTHMIAS	913
19.4	EFFECT OF ATP-SENSITIVE POTASSIUM CHANNEL ANTAGONISTS ON VENTRICULAR ARRHYTHMIAS	914
	Nonselective ATP-Sensitive Potassium Channel Antagonists	915
	Selective ATP-Sensitive Potassium Channel Antagonists	917
	Proarrhythmic Effects of ATP-Sensitive Potassium Channel Agonists	923
19.5	SUMMARY	924
	References	925

19.1 INTRODUCTION

Sudden cardiac death (defined as unexpected death from cardiac causes that occurs within 1 h of the onset of symptoms [1]) remains the leading cause of

Drug Discovery Handbook, by Shayne Cox Gad
Copyright © 2005 by John Wiley & Sons, Inc.

death in industrialized countries, accounting for between 300,000 and 500,000 deaths each year in the United States [2–5]. Holter monitoring studies reveal that these sudden deaths most frequently (up to 93 percent) resulted from ventricular tachyarrhythmias [6–8]. Despite the enormity of this problem, the development of safe and effective antiarrhythmic agents remains elusive. In fact, several initially promising antiarrhythmic drugs have actually been shown to increase, rather than to decrease, the risk for arrhythmic death in patients recovering from myocardial infarction. For example, the now infamous Cardiac Arrhythmia Suppression Trial (CAST study [9]) demonstrated that, although class I antiarrhythmic drugs (i.e., drugs that block sodium channels) effectively suppressed premature ventricular contractions, some of these compounds (flecainide and encainide) increased the risk for arrhythmic cardiac death. In a similar manner, many class III antiarrhythmic drugs (drugs that prolong refractory period, most likely via modulation of potassium channels) have been shown to prolong QT interval, to promote the life-threatening tachyarrhythmia torsade de pointes (i.e., polymorphic ventricular tachycardia in which the QRS waves seem to "twist" around the baseline), and to increase cardiac mortality in some patient populations [10, 11]. Unfortunately, only a few drugs have been clinically proven to reduce cardiac mortality in high-risk patients, such as patients recovering from myocardial infarction. To date, only β-adrenergic receptor antagonists and amiodarone, a class III antiarrhythmic drug that also blocks β-adrenergic receptors, have been shown to reduce sudden cardiac death [4, 12–15]. Even the best currently available therapies have not been completely successful in the suppression of malignant arrhythmias, and they also frequently exhibit untoward side effects. For example, mortality following myocardial infarction remains high among patients with substantial ventricular dysfunction, even with β-adrenergic receptor antagonist therapy. The 1-year mortality is 10 percent or higher, with sudden death accounting for approximately one-third of the deaths in these high-risk patients [16]. In a similar manner, the long-term use of amiodarone is limited due to adverse side effects, including pulmonary fibrous and thyroid toxicity [124]. Clearly, the "ideal" antiarrhythmic drug has yet to be discovered.

The discouraging clinical experience with most antiarrhythmic drugs indicates that these agents may attack the wrong therapeutic target. Although these drugs were effective against some types of arrhythmias, they clearly failed to prevent those arrhythmias that culminate in death. In order to develop agents that will reduce sudden death, one must first identify those factors that trigger the malignant arrhythmias. As previously noted, ventricular tachyarrhythmias are responsible for sudden cardiac death. Only a minority of these patients had a known history of heart disease prior to the collapse, yet up to 90 percent of all sudden death patients were subsequently shown to have underlying coronary artery disease [2]. Thus, it is likely that myocardial ischemia plays a crucial role in the induction of the lethal arrhythmias in these patients. If this hypothesis is correct, then it is crucial to identify the factor or factors that render the ischemic myocardium vulnerable to arrhythmia formation. Once identified, it should then be possible to develop therapeutic interventions that correct these ischemically induced changes in cardiac electrical stability and, thereby, prevent sudden death. The most effective antiarrhythmic agents would be those that target the ischemic myocardium with little or no action on the normal (i.e., nonischemic) cardiac tissue. As a consequence of this selectivity, one would expect that these drugs should also have a low propensity for proarrhythmic events.

It is well established that myocardial ischemia provokes abnormalities in the biochemical homeostasis of individual cardiac cells. These intracellular changes culminate in the disruption of cellular electrophysiologic properties, and as a result life-threatening alterations in cardiac rhythm, such as ventricular fibrillation, frequently occur. A number of different chemical substances have been proposed as possible causative factors in the genesis of ventricular fibrillation during myocardial ischemia, including catecholamines, amphiphilic products of lipid metabolism, various peptides, cytosolic calcium accumulation, and increases in extracellular potassium [17–19]. An accumulating body of evidence suggests that the activation of specific potassium channels, the adenosine triphosphate (ATP)-sensitive potassium channels, leads to potassium efflux from the ischemic cells, thereby inducing the electrophysiologic changes that are ultimately responsible for the formation of lethal cardiac arrhythmias [20–22]. As such, drugs that selectively inhibit these channels should prove to be particularly effective in the prevention of sudden cardiac death. Therefore, this chapter will first discuss the relationship between ischemically induced alterations in extracellular potassium and arrhythmia formation and then evaluate the antiarrhythmic potential of ATP-sensitive potassium channel antagonists, drugs that may act selectively on the ischemic myocardium.

19.2 EFFECTS OF MYOCARDIAL ISCHEMIA ON EXTRACELLULAR POTASSIUM

Myocardial ischemia provokes both rapid increases in extracellular potassium and reductions in action potential duration. Harris and co-workers [23, 24] were the first to show that extracellular potassium rises dramatically after coronary artery ligation, correlating with the onset of ventricular arrhythmias. They further demonstrated that intracoronary injections of potassium chloride provoked electrocardiographic (ECG) changes and triggered ventricular arrhythmias similar to those induced by myocardial ischemia [23, 24]. They proposed that changes in extracellular potassium represented a major factor in the development of malignant arrhythmias during ischemia. Subsequent studies that used ion-selective electrodes to measure potassium activity directly have confirmed these earlier observations [25–27]. Extracellular potassium increased within the first 15 s and reached a plateau within 5 to 10 min after the interruption of coronary perfusion [25–28]. Furthermore, regional differences or inhomogeneities of potassium accumulation were recorded, accompanied by corresponding differences in ventricular electrical activity [25, 26, 28]. This extracellular potassium accumulation resulted primarily from increases in potassium efflux rather than from decreased potassium influx due to inhibition of the Na^+/K^+-ATPase [30–32].

A growing body of evidence demonstrates that the ischemically induced potassium accumulation and the corresponding reductions in action potential duration result, primarily, from the opening of ATP-sensitive potassium channels [33]. Using the patch clamp technique, Trube and Hescheler [34] were the first to record single ATP-sensitive potassium channel activity. Noma [35] and Trube and Hescheler [34] further demonstrated that reductions in cellular ATP induced by cyanide exposure evoked an outward potassium current. They proposed that the activation of an ATP-sensitive potassium channel might be responsible for the reductions in action potential duration induced by hypoxia. Several studies further implicated the activation of this current in the changes in cardiac action potential and extracellular potassium accumulation during myocardial ischemia [36–46]. The ATP-sensitive potassium channel inhibitor, glibenclamide, for example, has been shown either to attenuate or to abolish reductions in action potential duration in hypoxic myocytes [37, 38], isolated cardiac tissue [41–43, 45, 46], and regionally or globally ischemic hearts [36, 40, 47]. This sulfonylurea drug also reduced extracellular potassium accumulation induced by ischemia [39, 43–45]. Conversely, ATP-sensitive potassium channel agonists (pinacidil, cromakalim) exacerbated ischemically induced reductions in action potential duration, as well as promoted extracellular potassium accumulation [36–38, 40–43, 47–50].

However, it must be emphasized that the ATP-sensitive potassium channel is activated only at low ATP concentrations with half-maximum suppression of channel opening at 20 to 100µM [34, 35, 51, 52], yet intracellular concentrations are normally much higher (5 to 10mM). Furthermore, cytosolic ATP levels remain in the millimolar range for the first 10 min of hypoxia, well after potassium accumulation begins [32, 53] Therefore, the role that this channel plays in the response to myocardial ischemia can be questioned [33]. A number of studies provide an explanation for this apparent paradox. Cardiac tissue contains a very high density of ATP-sensitive potassium channels. As a consequence, only a small increase in the open-state probability (<1% of maximum) was required to shorten action potential duration during ischemia [54–56]. Indeed, Faivre and Findlay [54] found, in guinea pig myocytes, that the opening of only 30 channels (less than 1 percent of the population) provoked a 50 percent reduction in action potential duration. They concluded that "physiologically relevant activity of the K_{ATP} channel in cardiac membrane is confined to a very small percentage of the possible cell K_{ATP} current, and thus intracellular ATP would not have to fall very far before the opening of K_{ATP} channels would influence cardiac excitability" [54]. In a similar manner, Deutsch et al. [57] correlated potassium current during hypoxia in an intact rabbit papillary muscle. They showed that during hypoxia, a significant shortening of action potential duration (blocked by glibenclamide) occurred when tissue ATP levels fell by approximately 25 percent. They also concluded that only modest changes in cellular ATP were required to induce major changes in cardiac electrical properties.

The activation of the ATP-sensitive potassium channel may also contribute significantly to the S-T segment changes associated with myocardial ischemia. In anesthetized open-chest dogs [58] the intracoronary injection of the ATP-sensitive potassium channel opener, pinacidil, elicited elevations in the S-T segment very similar to those induced by myocardial ischemia. Abnormal electrocardiographic (T wave) changes suggestive of alterations of repolarization have also been reported in up to 30 percent of patients taking pinacidil as an antihypertensive medication [59]. Conversely, glibenclamide attenuated the S-T segment elevations induced by the occlusion of the left anterior descending coronary artery [60]. Similar results were obtained in conscious dogs. Billman et al. [61] demonstrated that either glibenclamide or the cardioselective ATP-sensitive potassium channel antagonist HMR 1098 attenuated ischemically induced S-T segment changes. They further reported that these drugs prevented ischemically induced increases in the descending portion of the T wave (an index of the transmural dispersion of repolarization [62]). Finally, myocardial ischemia failed to alter the ECG of mice in which the *Kir 6.2* gene (the gene responsible for the pore forming subunit of the cardiac ATP-sensitive potassium channel [63]) had been removed [64]. The large changes in the ECG that were induced by ligation of the left coronary artery in the wild-type control mice could be suppressed by prior treatment with the cardioselective ATP-sensitive potassium channel antagonist HMR 1098 [64]. It therefore seems likely that the activation of the ATP-sensitive potassium channel contributes significantly to alterations in cardiac electrical stability induced by myocardial ischemia, leading to the formation of malignant arrhythmias.

19.3 EFFECT OF EXTRACELLULAR POTASSIUM ON VENTRICULAR RHYTHM

Cardiac arrhythmias can result from either abnormalities in impulse generation or impulse conduction [65]. The extracellular accumulation of potassium induced by myocardial ischemia can provoke these perturbations in cardiac electrical activity. Elevations in extracellular potassium promote the depolarization of the tissue surrounding the ischemic regions, as noted above. The flow of this injury current (electrotonic current flow between ischemic and normal cells) has been implicated as a potential cause for the initiation of premature ventricular beats. Under normal conditions ventricular cells do not display a spontaneous rhythm. However, an automatic rhythm can be produced when the cells are partially depolarized [65]. Coronel et al. [29], in fact, demonstrated an increased excitability of normal tissue near the border of the ischemia, a region of the heart in which extracellular potassium concentrations were also increased.

Changes in action potential duration induced by alterations in potassium efflux can also provoke abnormalities of impulse conduction. As noted above, increased potassium efflux from ischemic tissue triggers a reduction in action potential duration. A major factor contributing to ventricular fibrillation, particularly during myocardial ischemia, is a dispersion or inhomogeneity of the refractory period that is related to regional differences in action potential duration [65]. This allows for the fragmentation of impulse conduction during ensuing beats. As previously noted, the activation of the ATP-sensitive potassium channel produced large reductions in action potential duration [36–43, 45–47], which are inhibited by glibenclamide [36–38, 40–42, 47] but exacerbated by ATP-sensitive potassium channel agonists [43, 48, 66–68]. A differential ATP sensitivity of the ATP-dependent potassium channel has also been reported between endocardial and epicardial cells during ischemia such that smaller reductions in ATP were necessary to activate potassium channels located in epicardial tissue [69]. As a result, an inhomogeneity in extracellular potassium and refractory period, as well as a gradient in action potential duration, was recorded between the epicardial and endocardial tissue [69–71]. Nonuniformities in refractory period could set the stage for the formation of irregular reentrant pathways and ventricular arrhythmias. In a similar manner, the ATP-sensitive potassium channel antagonist glibenclamide was shown to attenuate ischemically induced reductions in the refractory period [61, 72]. In contrast, activation of the ATP-sensitive potassium current with pinacidil elicited a marked dispersion of repolarization and refractory period between the epicardium and endocardium, leading to the development of extrasystoles [73]. These effects could be abolished by glibenclamide [73]. More recently, Coromilas et al. [74] demonstrated that the pinacidil restored excitability in the tissue adjacent to the ischemic region (i.e., the epicardial border zone). This reactivation of formerly inexcitable tissue led to the formation of reentrant circuits and the induction of ventricular tachycardia. Thus, the activation of the ATP-sensitive potassium channel could contribute significantly to the induction of malignant arrhythmias by changing impulse generation (depolarization-induced changes in automaticity), conduction (refractory period dispersion), or a combination of both.

19.4 EFFECT OF ATP-SENSITIVE POTASSIUM CHANNEL ANTAGONISTS ON VENTRICULAR ARRHYTHMIAS

As noted above, it is now generally accepted that the activation of the ATP-sensitive potassium channel during myocardial ischemia provokes a potassium efflux and reductions in action potential duration that lead to dispersion of repolarization. Since heterogeneity of repolarization plays a crucial role in the induction of ventricular fibrillation, drugs that prevent ATP-sensitive potassium channel activation should be particularly effective in the suppression of malignant arrhythmias induced by ischemia.

Nonselective ATP-Sensitive Potassium Channel Antagonists

The sulfonylurea drug glibenclamide (Fig. 19.1) prevented hypoxia-induced reductions in action potential duration in single cells, isolated hearts, and intact animals [36, 37, 40–43, 45–47, 75–79]. This drug also attenuated the ischemically induced changes in the S-T segment in intact anesthetized or unanesthetized animals [60, 61]. Therefore, one would predict that glibenclamide should also prevent arrhythmias that arise as a consequence of these ischemically induced changes in cardiac electrical properties.

Glibenclamide prevented arrhythmias induced by ischemia in a variety of experimental models (Table 19.1). For example, this drug abolished ventricular arrhythmias induced by ischemia in isolated hearts [40, 80–83]. Furthermore, both glibenclamide and glimepride reduced blood glucose and decreased the incidence of irreversible ventricular fibrillation induced by reperfusion (after a 6-min coronary occlusion) in anesthetized rats [84, 140]. Glibenclamide also reduced the incidence of life-threatening arrhythmias induced by coronary artery ligation in the conscious rat [85] and anesthetized rabbit [86], as well as improved survival in anesthetized rats during ischemia and reperfusion [87]. This drug also inhibited arrhythmias associated with the intracellular calcium overload induced by ouabain [140, 144]. Billman et al. [61, 88] further demonstrated that glibenclamide prevented ventricular fibrillation induced by the combination of acute ischemia during submaximal exercise in animals previously shown to be susceptible to sudden death. It should be noted, however, that in these animals glibenclamide significantly reduced the exercise, and reactive hyperemia induced increases in coronary blood flow, as well as depressed ventricular function (large reductions in left ventricular dP/dt maximum) [61, 88]. In the isolated working rabbit heart, glibenclamide provoked an immediate decrease in coronary blood flow reducing forward flow to zero [68]. Thus, glibenclamide has potent vasoconstrictor effects due to the inhibiton of ATP-sensitive potassium channels located on the coronary vascular smooth muscle.

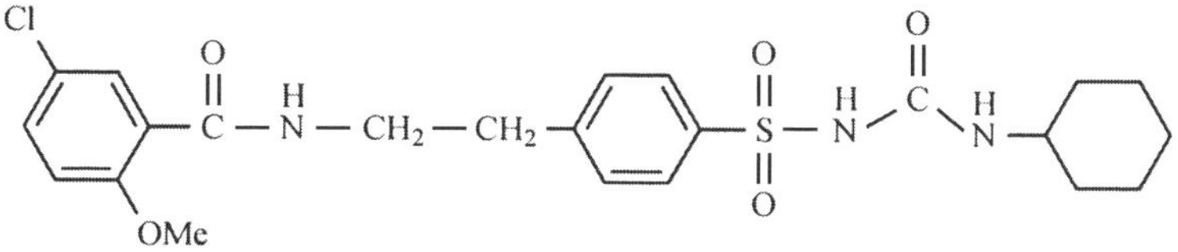

Glibenclamide

Figure 19.1 Chemical structure of glibenclamide.

TABLE 19.1 Effect of Nonselective ATP-Sensitive Potassium Channel Antagonist on Ischemic Cardiac Arrhythmias—Preclinical Studies[a]

Model	Result	Reference
Rabbit heart	↓ VF	Pogasta et al. [81]
Rat heart	↓ VF	Wolleben et al. [83]
Rat heart	↓ VF	Kantor et al. [40]
Dog and rabbit hearts	No Effect	Smallwood et al. [47]
Guinea pig heart	↓ VF Duration	Gwilt et al. [80]
Anesthetized rat	↓ VF	Zhang et al. [140]
Rat heart	↓ VF Duration	Bril et al. [141]
Guinea pig tissue	↓ Re-entrant PVCs	Pasnani and Ferrier [142]
Conscious dog	↓ VF	Billman et al. [88]
Rat heart	↓ VF	Tosaki et al. [82]
Rat heart	No Effect	Rees and Curtis [143]
Conscious rat	↓ VF	Lepran et al. [85]
Anesthetized rat	↓ VF	Baczko et al. [87]
Anesthetized rabbit	↓ VF	Barrett and Walker [86]
Conscious dog	↓ VF	Billman et al. [61]
Anesthetized rat	↓ VF	El Reyani et al. [84]

[a] VF, ventricular fibrillation; PVCs, premature ventricular contractions (activations).

There are a few clinical studies that illustrate the antiarrhythmic potential of ATP-sensitive potassium channels (Table 19.2, 89–92]. Cacciapuoti et al. [89] found that glibenclamide significantly reduced the frequency and severity of ventricular arrhythmias recorded during transient ischemia in non-insulin-dependent diabetic patients with coronary artery disease. This drug, however, did not affect nonischemic arrhythmias nor did it change the number or the length of the ischemic episodes. Glibenclamide significantly reduced the incidence of ventricular fibrillation in non-insulin-dependent patients with acute myocardial infarction [92]. Recently, the effects of sulfonylurea drugs on the incidence of ventricular arrhtyhmias (24-h Holter monitoring) were evaluated in nondiabetic and diabetic patients with decompensated congestive heart failure [93]. These authors found that diabetic patients receiving sulfonylurea agents (glibenclamide or glipizide) exhibited a significantly lower incidence of both repetitive ventricular beats and runs of nonsustained ventricular tachycardia as compared to either nondiabetic patients or diabetic patients who did not receive sulfonlyurea drugs. It is interesting to note that glibenclamide also reduced the S-T segment elevation and prevented ventricular fibrillation induced by chest impact (a baseball delivered at 30 mph to the chest of anesthetized swine [94]). The authors concluded that "selective K_{ATP} channel activation may be a pivotal mechanism in sudden death resulting from low-energy chest wall trauma" (i.e., Commotio Cordis) "in young people during sporting activities" [94].

TABLE 19.2 Effect of Nonselective ATP-Sensitive Potassium Channel Antagonists on Cardiac Arrhythmias—Clinical Studies[a]

Patient Population	Result	Reference
Non-insulin-dependent diabetics treated with glibenclamide ($n = 19$)	Significant reduction in the frequency of PVCs and nonsustained VT	Cacciapuoti et al. [89]
Non-insulin-dependent diabetics ($n = 232$, 106 treated with glibenclamide, 126 with other hypoglycemic drugs) and nondiabetic patients all with acute myocardial infarction	Incidence of VF or sustained VT reduced in the glibencalimide-treated patients. Cardiovascular mortality highest in nonglibenclamide-treated diabetic patients	Lomuscio et al. [92] Lomuscio and Fiorentini [91]
Retrospective analysis of acute myocardial infarction patients ($n = 5,715,745$ diabetic patients)	VF rates similar in nondiabetic (11.0%) and diabetic patients taking glibenclamide (11.8%) but much higher in diabetic patients taking gliclazide (18%) or insulin (22.8%)	Davis et al. [90]
Diabetic patients with decompensated heart failure ($n = 207$)	Significant reduction in incidence of repetitive ventricular beats and nonsustained VT in patients treated with either gilbenclamide or glipizide	Aronson et al. [93]

[a] VF, ventricular fibrillation; VT, ventricular tachycardia; PVCs, premature ventricular contractions (activations).

It is important to emphasis that glibenclamide is not selective for cardiac tissue. As noted above, this drug can profoundly reduce coronary blood flow via actions on vascular smooth muscle and can also promote insulin release, thereby provoking hypoglycemia [61]. These noncardiac actions would limit the antiarrhythmic potential of glibenclamide in the clinic. Cardioselective compounds should have fewer side effects and would therefore provide a better therapeutic option than the nonselective ATP-sensitive potassium channel antagonist glibenclamide.

Selective ATP-Sensitive Potassium Channel Antagonists

Several different ATP-sensitive potassium channel subtypes have been identified (Table 19.3). The ATP-sensitive potassium channel consists of a pore-forming subunit coupled to a sulfonylurea receptor [95–99, 139]. The functional channel forms as a hetero-octomer composed of a tetramer of the

TABLE 19.3 Possible ATP-Sensitive Potassium Channel Subtypes: Combinations of the Pore Forming Units (Kir 6.1 or Kir 6.2) and the Sulfonylurea Receptors (SUR1, SUR2A, and SUR2B)[a]

Channel	Tissue	Effect of Activation	Effect of Inhibition	Reference
Kir6.1/SUR1	Mitochondria (?)	Ischemic preconditioning	Prevents ischemic preconditioning	Liu et al. [100]
Kir6.1/SUR2B	Vascular smooth muscle and coronary endothelial cells	Vasodilation and increase blood low	Prevent vasodilation, promote vasoconstriciton	Suszuki et al. [63] Schnizler et al. [145]
Kir6.2/SUR1	Pancreatic β-cells	Insulin secretion		Aguilar-Bryan et al. [146]
Kir6.2/SUR1	Neurons	Reduce excitability or neurotransmitter release	Increase excitability or neurotransmitter release	Liss and Roeper [147]
Kir6.2/SUR2A	Cardiac muscle	Decrease action potential duration	Prevent decrease in action potential duration	Inagaki et al [97]
Kir6.2/SUR2A	Skeletal muscle	Prevent glucose uptake—fatigue	Glucose uptake	Chutkow et al. [148] Gong et al. [149]
Kir6.2/SUR2B	Nonvascular smooth muscle	Muscle relaxation (e.g., urinary bladder tone)	Prevent muscle relaxation	Gopalakrishnan et al. [150]

[a] Note that neither specific actions nor a target tissue for Kir6.1/SUR2A have yet been identified.

pore and four sulfonyl receptor subunits. At present, two different pore-forming subunits have been identified, both of which produce an inward rectifier potassium current (Kir 6.1 and Kir 6.2) [64, 100]. Three different sulfonylurea receptor subtypes have been isolated; SUR1 (on pancreatic islet cells), SUR2A (on cardiac tissue), and SUR2B (on vascular smooth muscle) [96, 99, 101, 102]. Thus, six different potassium channel pore and sulfonylurea receptor combinations are possible (Table 19.3). Suzuki et al. [63] and Manning-Fox et al. [98] recently demonstrated that Kir 6.2 and Kir 6.1 were required for cardiac and vascular smooth muscle ATP-sensitive potassium channel activity, respectively. They concluded that Kir 6.2/SUR2A most likely forms the cardiac cell membrane ATP-sensitive potassium channel, while Kir 6.1/SUR2B is located on vascular smooth muscle. In a similar manner, Lui et al. [100] reported that the actitivy to mitochondrial ATP-sensitive potassium channels was most closely mimicked by the Kir 6.1/SUR1 subtype, an conclusion that has not yet been confirmed. Given this apparent tissue specificity and the limited number of possible channel–receptor pairings, it should be possible to develop compounds that selectively inhibit (or activate) a particular ATP-sensitive potassium channel subtype. In particular, a drug that selectively blocks the Kir 6.2/SUR2A subtype should prevent ischemically induced changes in cardiac electrical properties (e.g., reductions in action potential duration) and thereby prevent arrhythmias without the untoward side effects noted for the nonselective ATP-sensitive channel antagonist glibenclamide.

HMR 1883, a Cardioselective ATP-Potassium Channel Antagonists
The sulfonylthiourea drug (1-[5-[2–5-chloro-*o*-ansamide)ethyl]2-methoxyphenyl]sulfonyl]-3-methylthiourea) HMR 1883 (Fig. 19.2) and its sodium salt HMR 1098 were recently developed to block the cardiac ATP-sensitive potassium channel [103]. HMR 1883 inhibited the sarcolemmal cardiac ATP-sensitive potassium channel activated by the channel opener rilmakalin at a much lower concentration (guinea pig myocytes IC_{50} = 0.6 – 2.2 µM) than was required to promote insulin release (9- to 50-fold higher concentration was required to block rat pancreatic insulinoma, RIN m5F, cells) [104]. In a similar manner, HMR 1098 reversed the action potential shortening induced by rilmakalin in human cardiomycytes (IC_{50} = 0.42 ± 0.008 µM) and proved to be more potent in acidic condtions (IC_{50} = 0.24 ± 0.009 µM) as would occur during myocardial ischemia [105]. Indeed, this drug attenuated hypoxia-induced shortening of cardiac action potential in rat and guinea pig tissue [104]. In contrast to glibenclamide, HMR 1883 did not alter hypoxia-induced increases in coronary blood flow in Langendorff perfused guinea pig hearts [104, 106]. HMR 1098 also inhibited Rb+ efflux through Kir 6.2/SUR2A channels expressed in HEK293 cells (IC_{50} = 181 nM), demonstrating that this system may be used to screen for compounds with a high affinity for the cardiac sarcolemmal channel [106]. Significantly, HMR 1883 did not alter flavoprotein autofluorescence, an index of mitochondrial redox state [107]. In

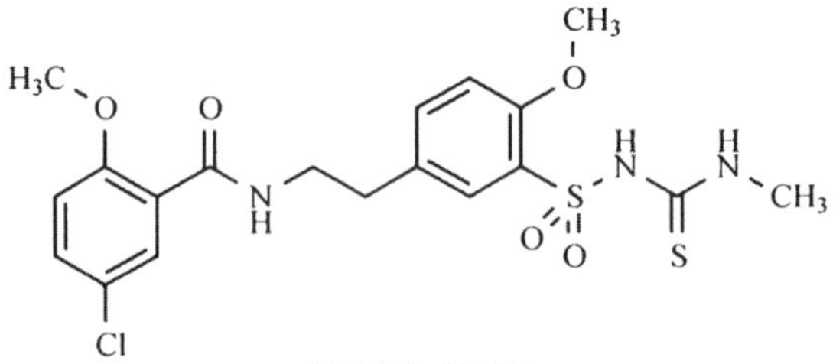

Figure 19.2 Chemical structure of HMR 1883.

contrast, 5-hydroxydecanoic acid, a selective blocker of mitochondrial channels, completely inhibited the flavoprotein fluorescence [108]. In agreement with these in vitro findings, HMR 1883 did not prevent the cardioprotective effects induced by ischemic preconditioning in either rat [109] or rabbit [110]. Accumulating evidence suggests that the activation of mitochondrial ATP-sensitive potassium channels plays a crucial role in the mechanical protection that results from ischemic preconditioning [111–113]. Recently, Liu et al. [100] demonstrated the mitochondrial ATP-sensitive potassium channel most closely resembles the Kir 6.1/SUR1 subtype. Thus, HMR 1883 can inhibit cardiac membrane ATP-sensitive potassium channels with miminal effects on mitochondrial channels.

HMR 1883 also prevented ischemically induced changes in the S-T segment in anesthetized mice [64], anesthetized swine [114], or conscious dogs [115]. A similar response was also noted for glibenclamide but not for 5-hydroxydecanoic acid [116]. In the conscious dogs with healed myocardial infarctions both HMR 1883 and glibenclamide prevented ischemically induced reductions in effective refractory period [61]. Furthermore, HMR 1883 significantly reduced monophasic action potential shortening induced by coronary artery occlusion in anesthetized pigs [117] and prevented an ischemically induced dispersion of ventricular repolarization in isolated Langendorff perfused rabbit hearts [118].

As would be predicted based upon these electrophysiological actions, HMR 1883/1098 prevented ischemically induced arrhtyhmias in a number of models (Table 19.4). For example, HMR 1883 reduced cardiac mortality in anesthetized pigs [113, 119] and prevented ventricular fibrillation induced by myocardial ischemia and reperfusion in rats [120]. HMR 1098 also significantly reduced arrhythmias induced by programmed electrical stimulation in dogs with healed myocardial infarctions without altering blood glucose levels [121]. In a similar manner, both glibenclamide and HMR 1883 [61], but not 5-hydroxydecanoic acid [116], significantly reduced the incidence of ventricular fibrillation induced by myocardial ischemia in conscious dogs with healed anterior wall myocardial infarctions. HMR 1098, but not 5-hydoxydecanoic acid, also prevented ventricular fibrillation induced by the ATP-sensitive

TABLE 19.4 Effect of Cardioselective ATP-Sensitive Potassium Channel Antagonists on Ischemic Cardiac Arrhythmias[a]

Model	Result	Reference
Anesthetized pig	HMR 1883, ↓ VF	Bohn et al. [119]
Conscious dog	HMR 1883, ↓ VF	Billman et al. [61]
Anesthetized pig	HMR 1883, ↓ VF	Wirth et al. [114]
Anesthetized rat	HMR 1883, ↓ VF	Wirth et al. [120]
Anesthetized dog	HMR 1098, ↓ arrhythmias induced by programmed electrical stimulation dogs with healed myocardial infarctions	Zhu et al. [121]
Rabbit heart	HMR 1098 ↓ VF by programmed electrical stimulation	Behrens et al. [118]
Rabbit heart	HMR 1098, ↓ VF induced by pinadicil	Fischbach et al. [122]
Conscious dog	HMR 1402, ↓ VF	Billman et al. [123]

[a] VF, ventricular fibrillation.

potassium channel agonist pinacidil in isolated rabbit hearts subjected to hypoxic perfusion [122]. These data strongly suggest that the selective opening of sarcolemmal ATP-sensitive potassium channels during ischemia provokes ventricular fibrillation while activation of the mitochodrial channels does not. Since, as noted above, ischemic preconditioning may result at least in part from the activation of mitochondrial channels, it may be possible to develop drugs that selectively activate these channels. However, nonspecific ATP-sensitive potassium channels should be avoided due the enhanced risk for malignant arrhythmias that would result from the activation of the sarcolemmal channels.

In contrast to glibenclamide, HMR 1883 did not alter plasma insulin or blood glucose levels in these animals [61]. Furthermore, glibenclamide but not HMR 1883 significantly reduced exercise-induced increase in mean cornary blood flow and provoked large reductions in left ventricular dP/dt maximum (both at rest and during exercise) [61]. Thus, HMR 1883 appears to act selectively on the cardiac cell membrane ATP-sensitive potassium channel and thereby prevents ischemically induced arrhythmias with minimal effects on either pancreatic or smooth muscle channels.

HMR 1402, a Cardioselective ATP-Sensitive Potassium Antagonists As noted above, at least six different K_{ATP} channels are possible. Recent evidence suggests that the Kir 6.2/SUR2A combination is restricted to cardiac muscle [63, 98]. Thus, substances that preferentially inhibit this channel should display selectivity for cardiac tissue. Recently, a second cardioselective ATP-sensitive

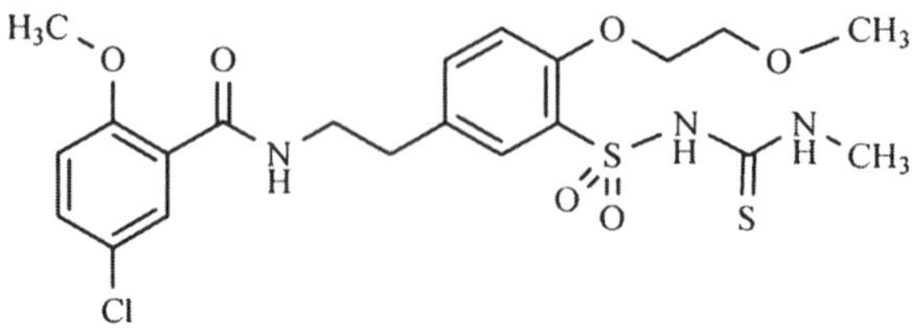

Figure 19.3 Chemical structure of HMR 1402.

potassium channel antagonist, HMR 1402 (Fig. 19.3), 1-[[5-[2(5-chloro-*o*-anisamido)ethyl]-β-methoxyethoxyphenyl]sulfonyl]-3-methylthioura has been developed [99]. This compound is structural similar to HMR 1883/1098. As was noted for HMR 1883/1098, HMR 1402 had no significant effects on action potential duration (APD_{90}), the resting potential, the amplitude of the phase 1 of the action potential, or on the upstroke velocity [123]. These data suggest that HMR 1402 did not affect potassium channels (i.e., I_{K1}, I_K, and I_{Ks}) or sodium channel at rest. This conclusion was directly confirmed by patch-clamp experiments performed in either rat or guinea pig ventricular myocytes [123]. In contrast, HMR 1402 potently blocked the rilmakalim-activated K_{ATP} channels in both guinea pig papillary muscles and rat ventricular myocytes [123]. At an external pH of 6.0, this inhibition was approximately 6.1 times more potent than that reported for HMR 1883 (IC_{50} for HMR 1402: 98 nM [123] versus IC_{50} for HMR 1883: 0.6 μM [104]). Similarly, hypoxia consistently elicited a marked reduction in APD_{90} that was potently antagonized by HMR 1402 [106, 123]. This inhibition of the hypoxia-induced shortening of the action potential duration was more potent for HMR 1402 than has been previously reported for HMR 1883 [104]. For example, 0.5 μM of HMR 1883 had no significant effect [104], while 0.3 μM of HMR 1402 produced a significant inhibition of reductions in APD_{90} induced by hypoxia [123]. Thus, one may conclude that HMR 1402 is more potent in blocking rilmakalim-activated and hypoxia-activated K_{ATP} channels than HMR 1883.

In contrast, HMR 1402 exhibited slightly more potent inhibition of vascular smooth and pancreatic β-cells than had been previously reported for HMR 1883/1098. For example, HMR 1402, in contrast to HMR 1883/1098, reduced hypoxia-induced increases in coronary flow at low concentrations [104, 123]. However, the same dose of glibenclamide (10 μM) provoked much larger reductions in coronary flow under normoxic conditions as well as hypoxic conditions [104]. As such, in isolated guinea pig hearts, HMR 1402 was more potent in inhibiting hypoxia-induced vasodilation than HMR 1883 but was still much less potent than glibenclamide. In a similar manner, HMR 1402 only partially inhibited the effect of diazoxide on the pancreatic β-cell membrane potential ($IC_{50} = 3.9$ μM [123]). This inhibition was somewhat more potent than that of HMR 1883 (IC_{50} approximately 20 μM) [98, 104] but considerably less than that of glibenclamide (IC_{50} = 9.3 nM) [104]. HMR 1402, like HMR 1883/1098, did not alter either plasma insulin or blood glucose levels in conscious dogs [123]. This was in marked contrast to the pronounced hypoglycemia and the increase in plasma insulin provoked by glibenclamide [61, 123] These in vitro and in vivo data strongly suggest that HMR 1402 acts preferentially on cardiac K_{ATP} channels but with slightly less selectively than HMR 1883/1098.

As one might predict, HMR 1402 was found to reduce the incidence of ventricular fibrillation induced by myocardial ischemia, protecting seven of eight animals tested [123]. This protection was very similar to that noted for both HMR 1883 and glibenclamide [61, 88]. However, it is important to emphasize that, in contrast to the actions of either HMR 1402 or HMR 1883, glibenclamide significantly reduced both exercise and reactive hyperemia-induced increases in coronary blood flow, as well as depressed ventricular function (large reductions in left ventricular dP/dt maximum) in animals [61, 88]. Therefore, nonselective K_{ATP} channel antagonist may protect against ischemic arrhythmias but not without potentially significant adverse side effects.

Proarrhythmic Effects of ATP-Sensitive Potassium Channel Agonists

In contrast to ATP-sensitive potassium channel antagonists, channel agonists induced arrhythmias during ischemia in both isolated hearts and intact animals [83, 125, 127]. The ATP-sensitive potassium channel agonist pinacidil facilitated ventricular fibrillation during myocardial ischemia in isolated rat [83], guinea pig [127] or rabbit [125] hearts with reduced potassium levels. Di Diego and Antzelevitch [73] found that the activation of the ATP-sensitive potassium channel with pinacidil caused marked inhomogeneities of refractory period, which provoked extrasystoles. These extrasystoles were blocked by glibenclamide in strips of isolated canine myocardium. They concluded that this dispersion of repolarization and refractory period formed a substrate for reentry. Indeed, the ATP-sensitive potassium channel opener chromakalim reduced effective refractory period and increased vulnerability for reentrant arrhythmias [67]. Furthermore, this drug also increased interventricular dispersion of refractory period and induced ventricular fibrillation in 5 of 12 isolated rabbit hearts under normoxic conditions [68]. Chi et al. [126] further demonstrated that pinacidil increased the frequency of ventricular fibrillation during myocardial ischemia in unanesthetized dogs. This drug had no effect on arrhythmias induced by electrical stimulation in normally perfused tissue [126]. More recently, Coromilas et al. [74] demonstrated that the pinacidil restored excitability in the tissue adjacent to the ischemic region (i.e., the epicardial border zone). This reactivation of formerly inexcitable tissue led to the formation of reentrant circuits and the induction of ventricular tachycardia.

Although proarrhythmic events have not been associated with ATP-sensitive potassium channel agonists in the clinical setting [128, 129], these drugs can alter ECG parameters that have been linked to arrhythmia formation. Indeed, abnormal T-wave changes suggestive of alteration in cardiac repolarization were reported in 30 percent of the hypertensive patients taking pinacidil [59]. In the presence of ischemia, these cardiac electrophysiological actions of nonselective ATP-sensitive potassium channel agonists could become more pronounced, increasing the propensity for life-threatening arrhythmias. However, nicorandil reduced the number of episodes of nonsustained ventricular tachycardia in patients with unstable angina [130]. This drug also reduced the episodes of transient ischemia. Thus, it is likely that the antiarrhythmic effects of this drug probably resulted secondarily from reductions in ischemia rather than a direct action on the ventricular myocardium. Remme and Wilde [131] concluded that since clinical studies of ATP-sensitive potassium channel agonists have not reported any major adverse effects on cardiac rhythm, the proarrhythmic potential of these agents may be "overestimated."

In contrast to these nonselective agonists, mitochondrial antagonists may lack proarrhythmic effects. For example, the novel mitochondrial agonist, BMS-191095 [132], reduced myocardial infarction size and preserved mechanical function during ischemia/reperfusion without adversely affecting cardiac electrophysiological parameters or the induction of arrhythmias [133].

When considered together, these reports indicate that the activation of sarcolemmal ATP-sensitive potassium channels promote arrhythmias, particularly in the setting of acute myocardial ischemia. In contrast, the selective activation of mitochondrial ATP-sensitive potassium channels contribute to ischemic preconditioning and thereby reduce the mechanical dysfunction induced by ischemia without affecting cardiac electrical stability. Therefore, the "ideal" modulator of ATP-sensitive potassium channels would both inhibit sarcolemmal channels and increase the activation (open) of mitochondrial channels in cardiac tissue.

19.5 SUMMARY

The activation of cardiac cell membrane ATP-sensitive potassium channels during myocardial ischemia promotes potassium efflux, reductions in action potential duration, and inhomogeneities in repolarization, thereby creating a substrate for reentrant arrhythmias. Drugs that block this channel should be particularly effective antiarrhythmic agents. Indeed, it is interesting to note that many currently available antiarrhythmic drugs, including verapamil, mibefradil, quinidine, lidocaine, and amiodarone, have also been reported to inhibit ATP-sensitive potassium channels at therapeutic concentrations [134–137]. Of particular note, amiodarone selectively inhibited sarcolemmal ATP-sensitive potassium channels without affecting mitochondrial channels [138]. Therefore, the inhibition of the ATP-sensitive potassium channel may be required for antiarrhythmic actions during myocardial ischemia. HMR

1883 (or its sodium salt HMR 1098) or HMR 1402 selectively blocks cardiac sarcolemmal ATP-sensitive potassium channels. As such, these drugs attenuate ischemically induced changes in cardiac electrical properties, thereby preventing malignant arrhythmias without the untoward effects of other drugs. Since, as noted above, the ATP-sensitive potassium channel only becomes active as ATP levels fall, these drugs have the added advantage that they would have effects *only* on ischemic tissue with little or no effect noted on normal tissue. Thus, selective antagonists of the cardiac cell surface ATP-sensitive potassium channel may represent the first truly ischemia-selective antiarrhythmic medications and should be free of the proarrhythmic actions that have plagued many of the currently available antiarrhythmic drugs.

REFERENCES

1. Torp-Pedersen, C., Kober, L., Elming, H., Burchart, H. (1997). Classification of sudden and arrhythmic death. *PACE, 20*, 2545–2552.
2. Abildstrom, S. Z., Kobler, L., Torp-Pedersen, C. (1999). Epidemiology of arrhythmic and sudden death in the chronic phase of ischemic heart disease. *Cardiac Electrophysiol. Rev., 3*, 177–179.
3. Gillum, R. (1989). Sudden coronary death in the United States 1980–1985. *Circulation, 79*, 756–765.
4. Zipes, D. P., Wellens, H. J. (1998). Sudden cardiac death. *Circulation, 98*, 2334–2351.
5. Zheng, Z-J., Croft, J. B., Giles, W. H., Mensah, G. A. (2001). Sudden cardiac death in the United States, 1989 to 1998. *Circulation, 104*, 2158–2163.
6. Bayes de Luna, A., Coumel, P., LeClercq, J. F. (1989). Ambulatory sudden cardiac death: Mechanisms of production of fatal arrhythmia on the basis of data from 157 cases. *Am. Heart. J., 117*, 151–159.
7. Hinkle, L. E. J., Thaler, H. T. (1982). Clinical classification of cardiac deaths. *Circulation, 65*, 457–464.
8. Greene, H. L. (1990). Sudden arrhythmic cardiac death: Mechanisms, resuscitation and classification: The Seattle perspective. *Am. J. Cardiol., 65*, 4B–12B.
9. Echt, D. S., Liebson, P. R., Mitchell, L. B., Peters, R. W., Obiasmanno, D., Barker, A. H., Arensberg, D., Baker, A., Freedman, L., Greene, H. L., Hunter, M. L., Richardson, D. W. (1991). Mortality and morbidity in patients receiving encainide, flecainide, or placebo. *N. Engl. J. Med., 324*, 782–788.
10. Sager, P. T. (1999). New advances in class III antiarrhythmic drug therapy. *Curr. Opin. Cardiol., 14*, 15–23.
11. Waldo, A. L., Camm, A. J., de Ruyter, H., Friedman, P. L., MacNeil, D. J., Pauls, J. F., Pitt, B., Pratt, C. M., Schwartz, P. J., Veltri, E. P., for the SWORD Investigators. (1996). Effect of *d*-sotalol on mortality in patients with left ventricular dysfunction after recent and remote myocardial infarction. *Lancet, 348*, 7–12.
12. Amiodarone Trials Meta-Analysis Investigators. (1997). Effects of prophylactic amiodarone on mortality after acute myocardial infarction and in congestive heart

failure: Meta-analysis of individual data from 6500 patients in randomized trials. *Lancet, 350,* 1417–1424.
13. Held, P., Yusuf, S. (1989). Early intravenous beta-blockade in acute myocardial infarction. *Cardiology, 76,* 132–143.
14. Held, P. H., Yusuf, S. (1993). Effects of beta-blockers and Ca^{2+} channel blockers in acute myocardial infarction. *Eur. Heart. J., 14*(Suppl. F), 18–25.
15. Kendall, M. J., Lynch, K. P., Hjalmarson, A., Kjekshus, J. (1995). β-Blockers and sudden cardiac death. *Ann. Intern. Med., 123,* 353–367.
16. Buxton, A. E., Lee, K. L., Fisher, J. D., Josephson, M. E., Prystowsky, E. N., Hafley, G. (1999). A randomized study of the prevention of sudden death in patients with coronary artery disease. *N. Engl. J. Med., 341,* 1882–1890.
17. Opie, L. H., Nathan, D., Lubbe, W. F. (1979). Biochemical aspects of arrhythmogenesis. *Am. J. Cardiol., 43,* 131–148.
18. Curtis, M. J., Pugsley, M. K., Walker, M. J. A. (1993). Endogenous chemical mediators of ventricular arrythmias in ischaemic heart disease. *Cardiovasc. Res., 27,* 703–719.
19. Billman, G. E. (1991). The antiarrhythmic and antifibrillatory effects of calcium antagonists. *J. Cardiovasc. Pharmacol., 18*(Suppl. 10), S107–S117.
20. Billman, G. E. (1994). The role of the ATP-sensitive K^+ channel in K^+ accumulation and cardiac arrhythmias during myocardial ischemia. *Cardiovasc. Res., 28,* 762–769.
21. Wilde, A. A. M. (1993). Role of ATP-sensitive K^+ channel current in ischemic arrhythmias. *Cardiovasc. Drugs Ther., 7,* 521–526.
22. Wilde, A. M. M., Aksnes, G. (1995). Myocardial potassium loss and cell depolarization in ischaemia and hypoxia. *Cardiovasc. Res., 29,* 1–15.
23. Harris, A. S. (1966). Potassium and experimental coronary occlusion. *Am. Heart. J., 71,* 797–802.
24. Harris, A. S., Bisteni, A., Russell, R. A., Brigham, J. C., Firestone, J. E. (1954). Excitory factors in ventricular tachycardia resulting from myocardial ischemia: Potassium a major excitant. *Science, 119,* 200–203.
25. Coronel, R., Fiolet, J.W., Wilms-Schopman, F. J., Schaapherder, A. F., Johnson, T. A., Gettes, L. S., Janse, M. J. (1988). Distribution of extracellular potassium and its relation to electrophysiologic changes during acute myocardial ischemia in the isolated perfused porcine heart. *Circulation, 77,* 1125–1138.
26. Hill, J. L., Gettes, L. S. (1980). Effect of acute coronary artery occlusion on local myocardial extracellular K^+ activity in swine. *Circulation, 61,* 768–778.
27. Kleber, A. G. (1984). Extracellular potassium accumulation in acute myocardial ischemia. *J. Mol. Cell. Cardiol., 16,* 389–394.
28. Johnson, T. A., Engle, C. L., Boyd, L. M., Koch, G. G., Gwinn, M., Gettes, L. S. (1991). Magnitude and time course of extracellular potassium inhomogeneities during acute ischemia in pigs. Effect of verapamil. *Circulation, 83,* 622–634.
29. Coronel, R., Wilms-Schopman, F. J., Opthof, T., van Capelle, F. J., Janse, M. J. (1991). Injury current and gradients of diastolic stimulation threshold, TQ potential, and extracellular potassium concentration during acute regional ischemia in the isolated perfused pig heart. *Circ. Res., 68,* 1241–1249.

30. Kleber, A. G. (1983). Resting membrane potential, extracellular potassium activity, and intracellular sodium activity during acute global ischemia in isolated perfused guinea pig hearts. *Circ. Res.*, *52*, 442–450.

31. Rau, E. E., Shine, K. I., Langer, G. A. (1977). Potassium exchange and mechanical performance in anoxic mammalian myocardium. *Am. J. Physiol. Heart Circ. Physiol.*, *232*, H85–H94.

32. Shine, K. I., Douglas, A. M., Ricchiuti, N. (1977). ^{42}K exchange during myocardial ischemia. *Am. J. Physiol. Heart Circ. Physiol.*, *232*, H564–H570.

33. Wilde, A. M. M. (1998). ATP and the role of I_{KATP} during acute myocardial ischaemia: Controversies revive. *Cardiovasc. Res.*, *35*, 181–183.

34. Trube, G., Hescheler, J. (1984). Inward-rectifying channels in isolated patches of the heart cell membrane: ATP-dependence and comparison with cell-attached patches. *Pfugers Arch.*, *401*, 178–184.

35. Noma, A. (1983). ATP-regulated K^+ channels in cardiac muscle. *Nature*, *305*, 147–148.

36. Bekheit, S. S., Restivo, M., Boutjdir, M., Henkin, R., Gooyandeh, K., Assadi, M., Khatib, S., Gough, W. B., El-Sherif, N. (1990). Effects of glyburide on ischemia-induced changes in extracellular potassium and local myocardial activation: A potential new approach to the management of ischemia-induced malignant ventricular arrhythmias. *Am. Heart. J.*, *119*, 1025–1033.

37. Benndorf, K., Friedrich, M., Hirche, H. (1991). Anoxia opens ATP regulated K channels in isolated heart cells of the guinea pig. *Pflugers Arch.*, *419*, 108–110.

38. Fosset, M., De Weille, J. R., Green, R. D., Schmid-Antomarchi, H., Lazdunski, M. (1988). Antidiabetic sulfonylureas control action potential properties in heart cells via high affinity receptors that are linked to ATP-dependent K^+ channels. *J. Biol. Chem.*, *263*, 7933–7936.

39. Hicks, M. N., Cobbe, S. M. (1991). Effect of glibenclamide on extracellular potassium accumulation and the electrophysiological changes during myocardial ischaemia in the arterially perfused interventricular septum of rabbit. *Cardiovasc. Res.*, *25*, 407–413.

40. Kantor, P. F., Coetzee, W. A., Carmeliet, E. E., Dennis, S. C., Opie, L. H. (1990). Reduction of ischemic K^+ loss and arrhythmias in rat hearts: Effect of glibenclamide, a sulfonylurea. *Circ. Res.*, *66*, 478–485.

41. Nakaya, H., Takeda, Y., Tohse, N., Kanno, M. (1991). Effects of ATP-sensitive K^+ channel blockers on the action potential shortening in hypoxic and ischaemic myocardium. *Br. J. Pharmacol.*, *103*, 1019–1026.

42. Ruiz-Petrich, E., Leblanc, N., deLorenzi, F., Allard, Y., Schanne, O. F. (1992). Effects of K^+ channel blockers on the action potential of hypoxic rabbit myocardium. *Br. J. Pharmacol.*, *106*, 924–930.

43. Vanheel, B., De Hemptinne, A. (1992). Influence of KATP channel modulation on net potassium efflux from ischaemic mammalian cardiac tissue. *Cardiovasc. Res.*, *26*, 1030–1039.

44. Venkatesh, N., Lamp, S. T., Weiss, J. N. (1991). Sulfonylureas, ATP-sensitive K^+ channels, and cellular K^+ loss during hypoxia, ischemia, and metabolic inhibition in mammalian ventricle. *Circ. Res.*, *69*, 623–637.

45. Weiss, J. N., Venkatesh, N., Lamp, S. T. (1992). ATP-sensitive K$^+$ channels and cellular K$^+$ loss in hypoxic and ischaemic mammalian ventricle. *J. Physiol. (Lond.)*, *447*, 649–673.

46. Wilde, A. A., Escande, D., Schumacher, C. A., Thuringer, D., Mestre, M., Fiolet, J. W., Janse, M. J. (1990). Potassium accumulation in the globally ischemic mammalian heart. A role for the ATP-sensitive potassium channel. *Circ. Res.*, *67*, 835–843.

47. Smallwood, J. K., Ertel, P. J., Steinberg, M. I. (1990). Modification by glibenclamide of the electrophysiological consequences of myocardial ischaemia in dogs and rabbits. *Naunyn Schmiedebergs Arch. Pharmacol.*, *342*, 214–220.

48. Edwards, G., Weston, A. H. (1993). The pharmacology of ATP-sensitive potassium channels. *Annu. Rev. Pharmacol. Toxicol.*, *33*, 597–637.

49. Mitani, A., Kinoshita, K., Fukamachi, K., Sakamoto, M., Kurisu, K., Tsuruhara, Y., Fukumura, F., Nakashima, A., Tokunaga, K. (1991). Effects of glibenclamide and nicorandil on cardiac function during ischemia and reperfusion in isolated perfused rat hearts. *Am. J. Physiol. Heart Circ. Physiol.*, *261*, H1864–H1871.

50. Venkatesh, N., Stuart, J. S., Lamp, S. T., Alexander, L. D., Weiss, J. N. (1992). Activation of ATP-sensitive K$^+$ channels by cromakalim: Effects on cellular K$^+$ loss and cardiac function in ischemic and reperfused mammalian ventricle. *Circ. Res.*, *71*, 1324–1333.

51. Findlay, I. (1988). ATP4- and ATP.Mg inhibit the ATP-sensitive K+ channel of rat ventricular myocytes. *Pflugers Arch.*, *412*, 37–41.

52. Nichols, C. G., Lederer, W. J. (1990). The regulation of ATP-sensitive K$^+$ channel activity in intact and permeabilized rat ventricular myocytes. *J. Physiol. (Lond.)*, *423*, 91–110.

53. Rovetto, M. J., Whitmer, J. T., Neely, J. R. (1973). Comparison of the effects of anoxia and whole heart ischemia on carbohydrate utilization in isolated working rat hearts. *Circ. Res.*, *32*, 699–711.

54. Faivre, J. F., Findlay, I. (1990). Action potential duration and activation of ATP-sensitive potassium current in isolated guinea-pig ventricular myocytes. *Biochim. Biophys. Acta*, *1029*, 167–172.

55. Nichols, C. G., Lederer, W. J. (1991). Adenosine triphosphate-sensitive potassium channels in the cardiovascular system. *Am. J. Physiol. Heart Circ. Physiol.*, *261*, H1675–H1686.

56. Nichols, C. G., Ripoll, C., Lederer, W. J. (1991). ATP-sensitive potassium channel modulation of the guinea pig ventricular action potential and contraction. *Circ. Res.*, *68*, 280–287.

57. Deutsch, N., Klitzner, T. S., Lamp, S. T., Weiss, J. N. (1991). Activation of cardiac ATP-sensitive K$^+$ current during hypoxia: Correlation with tissue ATP levels. *Am. J. Physiol. Heart Circ. Physiol.*, *261*, H671–H676.

58. Kubota, I., Yamaki, M., Shibata, T., Ikeno, E., Hosoya, Y., Tomoike, H. (1993). Role of ATP-sensitive K$^+$ channel on ECG ST segment elevation during a bout of myocardial ischemia. A study on epicardial mapping in dogs. *Circulation*, *88*, 1845–1851.

59. Goldberg, M. R. (1988). Clinical pharmacology of pinacidil, a prototype for drugs that affect potassium channels. *J. Cardiovasc. Pharmacol.*, *12*(Suppl. 2), S41–S47.

60. Kondo, T., Kubota, I.,Tachibana, H., Yamaki, M., Tomoike, H. (1996). Glibenclamide attenuates peaked T wave in early phase of myocardial ischemia. *Cardiovasc. Res.*, *31*, 683–687.

61. Billman, G. E., Englert, H. C., Schoelkens, B. A. (1998). HMR 1883, a novel cardioselective inhibitor of the ATP-sensitive potassium channel; Part II: Effects on susceptibility to ventricular fibrillation induced by myocardial ischemia in conscious dogs. *J. Pharmacol. Exp. Ther.*, *286*, 1465–1473.

62. Yan, G. X., Antzelevitch, C. (1998). Cellular basis for the normal T wave and the electrocardiographic manifestations of the long-QT syndrome. *Circulation*, *98*, 1928–1936.

63. Suzuki, M., Li, R. A., Miki, T., Uemura, H., Sakamoto, N., Ohmoto-Sekine, Y., Tamagawa, M., Ogura, T., Seino, S., Marban, E., Nakaya, H. (2001). Functional roles of cardiac and vascular ATP-sensitive potassium channels clarified by Kir6.2-knockout mice. *Circ. Res.*, *88*, 570–577.

64. Li, R. A., Leppo, M., Miki, T., Seino, S., Marban, E. (2000). Molecular basis of electrocardiographic ST-segment elevation. *Circ. Res.*, *87*, 837–839.

65. Wit, A. L., Janse, M. J. (1989). Electrophysiological mechanisms of ventricular arrhythmias resulting from myocardial ischemia and infarction. *Physiol. Rev.*, *69*, 1049–1169.

66. Krause, E., Englert, H., Gögelein, H. (1995). Adenosine triphosphate-dependent K$^+$ currents activated by metabolic inhibition in rat ventricular myocytes differ from those elicited by the channel opener rilmakalim. *Pflugers Arch.*, *429*, 625–635.

67. Uchida, T., Yashima, M., Gotoh, M., Qu, Z., Garfinkel, A., Weiss, J. N., Fishbein, M. C., Mandel, W. J., Chen, P. S., Karagueuzian, H. S. (1999). Mechanism of acceleration of functional reentry in the ventricle: Effects of ATP-sensitive potassium channel opener. *Circulation*, *99*, 704–712.

68. Wolk, R., Cobbe, S. M., Kane, K. A., Hicks, M. N. (1999). Relevance of inter- and intraventricular electrical dispersion to arrhythmogenesis in normal and ischaemic rabbit myocardium: A study with cromakalim, 5-hydroxydecanoate and glibenclamide. *J. Cardiovasc. Pharmacol.*, *33*, 323–334.

69. Furukawa, T., Kimura, S., Furukawa, N., Bassett, A. L., Myerburg, R. J. (1992). Potassium rectifier currents differ in myocytes of endocardial and epicardial origin. *Circ. Res.*, *70*, 91–103.

70. Gilmour, R. F. Jr, Zipes, D. P. (1980). Different electrophysiological responses of canine endocardium and epicardium to combined hyperkalemia, hypoxia, and acidosis. *Circ. Res.*, *46*, 814–825.

71. Kimura, S., Bassett, A. L., Kohya, T., Kozlovskis, P. L., Myerburg, R. J. (1986). Simultaneous recording of action potentials from endocardium and epicardium during ischemia in the isolated cat ventricle: Relation of temporal electrophysiologic heterogeneities to arrhythmias. *Circulation*, *74*, 401–409.

72. Tweedie, D., Henderson, C., Kane, K. (1993). Glibenclamide, but not class III drugs, prevents ischaemic shortening of the refractory period in guinea-pig hearts. *Eur. J. Pharmacol.*, *240*, 251–257.

73. Di Diego, J. M., Antzelevitch, C. (1993). Pinacidil-induced electrical heterogeneity and extrasystolic activity in canine ventricular tissues. Does activation of

ATP-regulated potassium current promote phase 2 reentry? *Circulation*, 88, 1177–1189.
74. Coromilas, J., Costeas, C., Deruyter, B., Dillon, S. M., Peters, N. S., Wit, A. L. (2002). Effects of pinacidil on electrophysiological properties of epicardial border zone of healing canine infarcts: Possible effects of K_{ATP} channel activation. *Circulation*, 105, 2309–2317.
75. Bellemin-Baurreau, J., Poizot, A., Hicks, P. E., Rochette, L., Armstrong, J. M. (1994). Effects of ATP-dependent K^+ channel modulators on an ischemia-reperfusion rabbit isolated heart model with programmed electrical stimulation. *Eur. J. Pharmacol.*, 256, 115–124.
76. Dhein, S., Pejman, P., Krusemann, K. (2000). Effects of the $I_{K.ATP}$ blockers glibenclamide and HMR1883 on cardiac electrophysiology during ischemia and reperfusion. *Eur. J. Pharmacol.*, 398, 273–284.
77. Hamada, K., Yamazaki, J., Nagao, T. (1998). Shortening of monophasic action potential duration during hyperkalemia and myocardial ischemia in anesthetized dogs. *Jpn. J. Pharmacol.*, 76, 149–154.
78. Koumi, S. I., Martin, R. L., Sato, R. (1997). Alterations in ATP-sensitive potassium channel sensitivity to ATP in failing human hearts. *Am. J. Physiol. Heart Circ. Physiol.*, 272, H1656–H1665.
79. MacKenzie, I., Saville, V. L., Waterfall, J. F. (1993). Differential class III and glibenclamide effects on action potential duration in guinea-pig papillary muscle during normoxia and hypoxia/ischaemia. *Br. J. Pharmacol.*, 110, 531–538.
80. Gwilt, M., Henderson, C. G., Orme, J., Rourke, J. D. (1992). Effects of drugs on ventricular fibrillation and ischaemic K^+ loss in a model of ischaemia in perfused guinea-pig hearts in vitro. *Eur. J. Pharmacol.*, 220, 231–236.
81. Pogatsa, G., Koltai, M. Z., Balkanyi, I., Devai, I., Kiss, V., Koszeghy, A. (1988). The effect of various hypoglycaemic sulphonylureas on the cardiotoxicity of glycosides and arrhythmogenic activity due to myocardial ischaemia. *Acta Physiol. Hung.*, 71, 243–250.
82. Tosaki, A., Engelman, D. T., Engelman, R. M., Das, D. K. (1995). Diabetes and ATP-sensitive potassium channel openers and blockers in isolated ischemic/reperfused hearts. *J. Pharmacol. Exp. Ther.*, 275, 1115–1123.
83. Wolleben, C. D., Sanguinetti, M. C., Siegl, P. K. (1989). Influence of ATP-sensitive potassium channel modulators on ischemia-induced fibrillation in isolated rat hearts. *J. Mol. Cell. Cardiol.*, 21, 783–788.
84. El Reyani, N. E., Bozdogan, O., Baczko, I., Lepran, I., Papp, J. G. (1999). Comparison of the efficacy of glibenclamide and glimepiride in reperfusion-induced arrhythmias in rats. *Eur. J. Pharmacol.*, 365, 187–192.
85. Lepran, I., Baczko, I., Varro, A., Papp, J. G. (1996). ATP-sensitive potassium channel modulators: Both pinacidil and glibenclamide produce antiarrhythmic activity during acute myocardial infarction in conscious rats. *J. Pharmacol. Exp. Ther.*, 277, 1215–1220.
86. Barrett, T. D., Walker, M. J. (1998). Glibenclamide does not prevent action potential shortening induced by ischemia in anesthetized rabbits but reduces ischemia-induced arrhythmias. *J. Mol. Cell. Cardiol.*, 30, 999–1008.
87. Baczko, I., Lepran, I., Papp, J. G. (1997). KATP channel modulators increase survival rate during coronary occlusion-reperfusion in anaesthetized rats. *Eur. J. Pharmacol.*, 324, 77–83.

88. Billman, G. E., Avendano, C. E., Halliwill, J. R., Burroughs, J. M. (1993). The effects of the ATP-dependent potassium channel antagonist glyburide on coronary blood flow and susceptibility to ventricular fibrillation. *J. Cardiovasc. Pharmacol.*, *21*, 197–204.
89. Cacciapuoti, F., Spiezia, R., Bianchi, M., Lama, D., D'Avino, M., Varricchio, M. (1991). Effectiveness of glibenclamide on myocardial ischemic ventricular arrhythmias in non-insulin dependent diabetes mellitus. *Am. J. Cardiol.*, *67*, 843–847.
90. Davis, T. M., Parsons, R. W., Broadhurst, R. J., Hobbs, M. S., Jamrozik, K. (1998). Arrhythmias and mortality after myocardial infarction in diabetic patients. Relationship to diabetes treatment. *Diabetes Care*, *21*, 637–640.
91. Lomuscio, A., Fiorentini, C. (1996). Influence of oral antidiabetic treatment on electrocardiac alterations induced by myocardial infarction. *Diabetes Res. Clin. Pract.*, *31*(Suppl), S21–S26.
92. Lomuscio, A., Vergani, D., Marano, L., Castagnone, M., Fiorentini, C. (1994). Effects of glibenclamide on ventricular fibrillation in non-insulin-dependent diabetics with acute myocardial infarction. *Coron. Artery Dis.*, *5*, 767–771.
93. Aronson, D., Mittleman, M. A., Burger, A. J. (2003). Effects of sulfonylurea agents and adenosine triphosphate dependent potassium channel antagonists on ventricular arrhythmias in patients with decompensated heart failure. *PACE*, *26*, 1254–1261.
94. Link, M. S., Wang, P. J., VanderBrink, B. A., Avelar, E., Pandian, N. G., Maron, B. J., Estes, N. A. (1999). Selective activation of the K^+_{ATP} channel is a mechanism by which sudden death is produced by low-energy chest-wall impact (Commotio cordis). *Circulation*, *100*, 413–418.
95. Gögelein, H. (2001). Inhibition of cardiac ATP-dependent potassium channels by sulfonylurea drugs. *Curr. Opin. Invest. Drugs.*, *2*, 71–80.
96. Gögelein, H., Hartung, J., Englert, H. C. (1999). Molecular basis, pharmacology and physiological role of cardiac K_{ATP} channels. *Cell Physiol. Biochem.*, *9*, 227–241.
97. Inagaki, N., Gonoi, T., Clement, J. P., Namba, N., Inazawa, J., Gonzalez, G., Aguilar-Bryan, L., Seino, S., Bryan, J. (1995). Reconstitution of I_{KATP}: An inward rectifier subunit plus the sulfonylurea receptor. *Science*, *270*, 1166–1170.
98. Manning-Fox, J. E. M., Kanji, H. D., French, R. J., Light, P. E. (2002). Cardioselectivity of the sulphonylurea HMR 1098: Studies on native and recombinant cardiac and pancreatic K-ATP channels. *Br. J. Pharmacol.*, *135*, 480–488.
99. Englert, H. C., Heitsch, H., Gerlach, U., Knieps, S. (2003). Blockers of the ATP-sensitive potassium channel SURA/Kir6.2: A new approach to prevent sudden cardiac death. *Curr. Med. Chem.*, *1*, 253–271.
100. Liu, Y., Ren, G., O'Rourke, B., Marban, E., Seharaseyon, J. (2001). Pharmacological comparison of native mitochondrial K_{ATP} channels with molecularly defined surface K_{ATP} channels. *Mol. Pharmacol.*, *59*, 225–230.
101. Gribble, F. M., Tucker, S. J., Seino, S., Ashcroft, F. M. (1998). Tissue specificity of sulfonylureas: Studies on cloned cardiac and beta-cell K_{ATP} channels. *Diabetes*, *47*, 1412–1418.
102. Reimann, F., Proks, P., Ashcroft, F. M. (2001). Effects of mitiglinide (S 21403) on Kir6.2/SUR1, Kir6.2/SUR2A and Kir6.2/SUR2B types of ATP-sensitive potassium channel. *Br. J. Pharmacol.*, *132*, 1542–1548.

103. Englert, H. C., Gerlach, U., Goegelein, H., Hartung, J., Heitsch, H., Mania, D., Scheidler, S. (2001). Cardioselective K (ATP) channel blockers derived from a new series of m-anisamidoethylbenzenesulfonylthioureas. *J. Med. Chem.*, *44*, 1085–1098.

104. Gögelein, H., Hartung, J., Englert, H. C., Scholkens, B. A. (1998). HMR 1883, a novel cardioselective inhibitor of the ATP-sensitive potassium channel. Part I: Effects on cardiomyocytes, coronary flow and pancreatic beta-cells. *J. Pharmacol. Exp. Ther.*, *286*, 1453–1464.

105. Kaab, S., Zwermann, L., Barth, A., Hinterseer, M, Englert, H. C., Gögelein, H. R., Nabauer, M. (2003). Selective block of sarcolemmal IKATP in human cardiomyocytes using HMR 1098. *Cardiovasc. Drug. Ther.*, *17*, 435–441.

106. Weyermann, A., Vollert, H., Busch, A. E., Bleich, M., Gögelein, H. (2004). Inhibitors of ATP-sensitive potassium channels in guinea pig isolated ischemic hearts. *Naunyn Schmiedebergs Arch. Pharmacol.*, *369*, 374–381.

107. Sato, T. (1999). Signaling in late preconditioning: Involvement of mitochondrial K_{ATP} channels. *Circ. Res.*, *85*, 1113–1114.

108. Liu, Y., Sato, T., O'Rourke, B., Marban, E. (1998). Mitochondrial ATP-dependent potassium channels: Novel effectors of cardioprotection? *Circulation*, *97*, 2463–2469.

109. Fryer, R. M., Eells, J. T., Hsu, A. K., Henry, M. M., Gross, G. J. (2000). Ischemic preconditioning in rats: Role of mitochondrial K_{ATP} channel in preservation of mitochondrial function. *Am. J. Physiol. Heart. Circ. Physiol.*, *278*, H305–H312.

110. Jung, O., Englert, H. C., Jung, W., Gögelein, H., Scholkens, B. A., Busch, A. E., Linz, W. (2000). The K_{ATP} channel blocker HMR 1883 does not abolish the benefit of ischemic preconditioning on myocardial infarct mass in anesthetized rabbits. *Naunyn Schmiedebergs Arch. Pharmacol.*, *361*, 445–451.

111. Gross, G. J., Fryer, R. M. (1999). Sarcolemmal versus mitochondrial ATP-sensitive K^+ channels and myocardial preconditioning. *Circ. Res.*, *84*, 973–979.

112. O'Rourke, B. (2000). Myocardial K_{ATP} channels in preconditioning. *Circ. Res.*, *87*, 845–855.

113. Grover, G. J., Garlid, K. D. (2000). ATP-sensitive potassium channels: A review of their cardioprotective pharmacology. *J. Mol. Cell. Cardiol.*, *32*, 677–695.

114. Wirth, K. J., Rosenstein, B., Uhde, J., Englert, H. C., Busch, A. E., Scholkens, B. A. (1999). ATP-sensitive potassium channel blocker HMR 1883 reduces mortality and ischemia-associated electrocardiographic changes in pigs with coronary occlusion. *J. Pharmacol. Exp. Ther.*, *291*, 474–481.

115. Billman, G. E., Houle, M. S., Englert, H. C., Goegelein, H. (1999). Ischemically-induced changes in the T-wave and susceptibility to sudden death: Evidence that activation of the ATP-sensitive potassium channel may contribute to ventricular fibrillation. *Circulation*, *100*(Suppl. I), 1–52. 1999.

116. Billman, G. E., Englert, H. C., Goegelein, H., Busch, A. (2001). Selective sarcolemmal, but not selective mitochondrial, ATP-sensitive potassium channel antagonists prevent ventricular fibrillation induced by ischaemia. *Eur. Heart. J.*, *22*(Abstract Suppl.), 246.

117. Wirth, K. J., Uhde, J., Rosenstein, B., Englert, H. C., Gögelein, H., Scholkens, B. A., Busch, A. E. (2000). K_{ATP} channel blocker HMR 1883 reduces monophasic

action potential shortening during coronary ischemia in anesthetised pigs. *Naunyn Schmiedebergs Arch. Pharmacol.*, *361*, 155–160.

118. Behrens, S., Zabel, M., Janssen, A., Barbierato, M., Schultheiss, H. P. (2001). Influence of a new ATP-sensitive potassium-channel antagonist (HMR 1098) on ventricular fibrillation inducibility during myocardial ischemia. *Eur. Heart. J.*, *22*(Abstract Suppl.), 546.

119. Bohn, H., Englert, H. C., Schoelkens, B. A. (1998). The K_{ATP} channel blocker HMR 1883 attenuates the effects of ischemia on MAP duration and improves survival during LAD occlusion in anaesthetized pig. *Br. J. Pharmacol.*, *124*, 23P.

120. Wirth, K. J., Klaus, E., Englert, H. G., Scholkens, B. A., Linz, W. (1999). HMR 1883, a cardioselective K (ATP) channel blocker, inhibits ischaemia- and reperfusion-induced ventricular fibrillation in rats. *Naunyn Schmiedebergs Arch. Pharmacol.*, *360*, 295–300.

121. Zhu, B. M., Miyamoto, S., Nagawa, Y., Wajima, T., Hashimoto, K. (2003). Effect of sarcolemmal K-ATP blocker HMR 1098 on arrhythmias induced by programmed electrical stimulation in canine old myocardial infarction model: Comparison with glibenclamide. *J. Pharmacol. Sci.*, *93*, 106–113.

122. Fischbach, P. S., White, A., Barrett, T. D., Lucchesi, B. R. (2004). Risk of ventricular proarrhythmia with the selective opening of the myocardial sarcolemmal versus mitochondrial ATP-gated potassium channel. *J. Pharmacol. Exp. Ther.*, *309*, 554–559.

123. Billman, G. E., Houle, M. S., Englert, H. C., Gögelein, H. (2004). Effects of a novel cardioselective ATP-sensitive potassium channel antagonist 1-[[5-[2(5-chloro-*o*-anisamido)ethyl]-β-methoxyethoxyphenyl]sulfonyl]-3-methylthioura, sodium salt (HMR 1402), on susceptibility to ventricular fibrillation induced by myocardial ischemia: In vitro and in vivo studies. *J. Pharmacol. Exp. Ther.*, *309*, 182–192.

124. Nattel, S. (2000). Class III drugs: Amiodarone, bertylium, ibutilide, and sotalol. In D. P. Zipes, J. Jalife (Eds), *Cardiac Electrophysiology from Cell to Bedside*, 3rd ed. W. B. Saunders, Philadelphia, pp. 921–932.

125. Chi, L., Black, S. C., Kuo, P. I., Fagbemi, S. O., Lucchesi, B. R. (1993). Actions of pinacidil at a reduced potassium concentration—A direct cardiac effect possibly involving the ATP-sensitive potassium channel. *J. Cardiovasc. Pharmacol.*, *21*, 179–190.

126. Chi, L., Uprichard, A. C. G., Lucchesi, B. R. (1990). Profibrillatory actions of pinacidil in a conscious canine model of sudden coronary death. *J. Cardiovasc. Pharmacol.*, *15*, 452–464.

127. D'Alonzo, A. J., Zhu, J. L., Darbenzio, R. B., Dorso, C. R., Grover, G. J. (1998). Proarrhythmic effects of pinacidil are partially mediated through enhancement of catecholamine release in isloated perfused guinea-pig hearts. *J. Mol. Cell. Cardiol.*, *30*, 415–423.

128. Friedel, H. A., Brogden, R. N. (1990). Pinacidil—A review of its pharmacodynamic and pharmacokinetic properties and therapeutic potential in the treatment of hypertension. *Drugs*, *39*, 929–967.

129. Krumenacker, M., Roland, E. (1992). Clinical profile of nicorandil—An overview of its hemodynamic properties and therapeutic efficacy. *J. Cardiovasc. Pharmacol.*, *20*(Suppl. 3), S93–S102.

130. Patel, D. J., Purcell, H. J., Fox, K. M. (1999). Cardioprotection by opening of the K_{ATP} channel in unstable angina. *Eur. Heart J.*, *20*, 51–57.
131. Remme, C. A., Wilde, A. M. M. (2000). K_{ATP} channel opener, myocardial ischemia and arrhythmias—Should the electrophysiologist worry? *Cardiovasc. Drug Ther.*, *14*, 17–22.
132. Grover, G. J., D'Alonzo, A. J., Garlid, K. D., Bajgar, R., Lodge, N. J., Sleph, P. G., Darbenzio, R. B., Hess, T. A., Smith, M. A., Paucke, P., Atwal, K. S. (2001). Pharmacologic characterization of BMS-191095, a mitochondrial K_{ATP} opener with no peripheral vasodilator or cardiac action potential shortening activity. *J. Phamacol. Exp. Ther.*, *297*, 1184–1192.
133. Grover, G. J., D'Alonzo, A. J., Darbenzio, R. B., Parhaqm, C. S., Hess, T. A., Bathala, M. S. (2002). In vivo characterization of the mitochondrial selective K_{ATP} opener (3*R*)-*trans*-4-(4-chlorophenyl)-*N*-(1*H*-imidazol-2-ylmethyl)dimethyl-2H-1-benzopyran-6-carbonitril monohydrochloride (BMS-191095): Cardioprotective, hemodynamic, and electrophysiological effects. *J. Pharmacol. Exp. Ther.*, *303*, 132–140.
134. Colatsky, T. J., Follmer, C. H., Starmer, C. F. (1990). Channel specificity in antiarrhythmic drug action. Mechanism of potassium channel block and its role in suppressing and aggravating cardiac arrhythmias. *Circulation*, *82*, 2235–2242.
135. Haworth, R. A., Goknur, A. B., Berkoff, H. A. (1989). Inhibition of ATP-sensitive potassium channels of adult rat heart cells by antiarrhythmic drugs. *Circ. Res.*, *65*, 1157–1160.
136. Holmes, D. S., Sun, Z. Q., Porter, L. M., Bernstein, N. E., Chinitz, L. A., Artman, M., Coetzee, W. A. (2000). Amiodarone inhibits cardiac ATP-sensitive potassium channels. *J. Cardiovasc. Electrophysiol.*, *11*, 1152–1158.
137. Olschewski, A., Brau, M. E., Olschewski, H., Hempelmann, G., Vogel, W. (1996). ATP-dependent potassium channel in rat cardiomyocytes is blocked by lidocaine. Possible impact on the antiarrhythmic action of lidocaine. *Circulation*, *93*, 656–659.
138. Sato, T., Takizawa, T., Saito, T., Kobayashi, S., Hara Nakaya, H. (2003). Amiodarone inhibits sarcolemmal but not mitchondrial K-ATP channels in guinea pig ventricular cells. *J. Pharmacol. Exp. Ther.*, *307*, 955–960.
139. Gribble, F. M., Reimann, F. (2003). Sulphonylurea action revisited: The postcloning era. *Diabetologia*, *46*, 875–891.
140. Zhang, H. L., Li, Y. S., Fu, S. X., Yang, X. P. (1991). Effects of glibenclamide and tolbutamine on ischemia-induced and ouabain-induced arrhythmias and membrane potentials of ventricular myocardium from rat and guinea-pig. *Acta Pharmacol. Sinica*, *12*, 398–402.
141. Bril, A., Laville, M-P, Gout, B. (1992). Effects of glibenclamide on ventricular arrhythmias and cardiac function in ischaemia and reperfusion in isolated rat heart. *Cardiovasc. Res.*, *26*, 1069–1076.
142. Pasnani, J. S., Ferrier, G. R. (1992). Differential effects of glyburide on premature beats and ventricular tachycardia in an isloated tissue model of ischemia and reperfusion. *J. Pharmacol. Exp. Ther.*, *262*, 1076–1084.
143. Rees, S. A., Curtis, M. J. (1995). Pharmacological analysis in rat of the role of the ATP-sensitive potassium channels as a potential target for antifibrillatory

intervention in acute myocardial-ischemia. *J. Cardiovasc. Pharmacol.*, 26, 280–288.

144. Yazar, A. Polat, G., Levant, A., Kaygusuz, A., Camdeviren, H., Buyukafsar, K. (2002). Effects of glibenclamide, metaformin and insulin on the incidence and latency of death by ouabian-induced arrhythmias. *Pharmacol. Res.*, 45, 183–187.

145. Schnitzler, M. M. Y., Derst, C., Daut, J., Preisig-Muller, R. (2000). ATP-sensitive potassium channels in capillaries isloated from guinea-pig heart. *J. Physiol. (Lond.)*, 525, 307–317.

146. Aguilar-Bryan, L., Nichols, C. G., Wechsler, S. W., Clement, J. P., Boyd, A. E., Gonzales, G., Herrera-Sosa, H., Nguy, K., Bryan, J., Nelson D. (1995). Cloning of the beta cell high affinity sulfonylurea receptor: A regulator of insulin secretion. *Science*, 268, 423–426.

147. Liss, B., Roeper, J. (2001). Molecular physiology of neuronal K-ATP channels. *Mol. Membr. Biol.*, 18, 117–127.

148. Chutkow, W. A., Samuel, V., Hansen, P. A., Pu, J., Valdivia, C. R., Makielshi, J. C., Burant, C. F. (2001). Disruption of Sur2-containg K_{ATP} channels enhances insulin-stimulated glucose uptake in skeletal muscle. *Proc. Natl. Acad. Sci. (USA)*, 98, 11760–11764.

149. Gong, B., Miki, T., Seino, S., Renaud, J. M. (2000). A K_{ATP} channel deficiency affects resting tension, not contractile force, during fatigue in skeletal muscle. *Am. J. Physiol.*, 279, C1351–1358.

150. Gopalakrishnan, M., Whiteaker, K. L., Molinari, E. J., Davis-Taber, R., Scott, V. E. S., Shieh, C. C., Buckner, S. A., Milicic, I., Cain, J. C., Postl, S., Sullivan, J. P., Brioni, J. D. (1999). Characterization of the ATP-sensitive potassium channels (K-ATP) expressed in guinea pig bladder smooth muscle cells. *J. Pharmacol. Exp. Ther.*, 289, 551–558.

20

FACTORS INFLUENCING THE EFFICACY OF MEDIATOR-SPECIFIC ANTI-INFLAMMATORY, GLUCOCORTICOID, AND ANTICOAGULANT THERAPIES FOR SEPSIS

PETER C. MINNECI,[1,2] KATHERINE J. DEANS,[1,2] MICHAEL HALEY[1]
XIZHONG CUI,[1] CHARLES NATANSON,[1] AND PETER Q. EICHACKER,[1]
[1]*National Institutes of Health, Bethesda, Maryland*
[2]*Massachusetts General Hospital, Boston, Massachusetts*

20.1	INTRODUCTION	938
20.2	MEDIATOR-SPECIFIC ANTI-INFLAMMATORY THERAPIES IN SEPSIS	939
	Role of Inflammatory Mediators in the Pathophysiology of Sepsis	939
	Mediator-Specific Anti-inflammatory Agent Trials	939
	Factors Influencing the Efficacy of Anti-inflammatory Agents in Sepsis	940
20.3	GLUCOCORTICOIDS IN SEPSIS	943
	Glucocorticoid Trials	943
	Factors Influencing the Effects of Glucocorticoids	943
	Comparison of Glucocorticoid Versus Mediator-Specific Anti-inflammatory Agents	945
20.4	ANTICOAGULANTS IN SEPSIS	946
	Role of the Coagulation Pathways in Sepsis	946
	Clinical Experience with the Anticoagulant Agents	946
	Factors Influencing the Efficacy of Anticoagulant Agents	946
20.5	CONCLUSIONS	948
	References	950

Drug Discovery Handbook, by Shayne Cox Gad
Copyright © 2005 by John Wiley & Sons, Inc.

20.1 INTRODUCTION

Sepsis is a major cause of mortality for hospitalized patients worldwide with a mortality rate estimated to be between 30 and 60 percent [1, 2]. During sepsis, cardiovascular dysfunction is produced by bacterial infection with the release of bacterial toxins and the production of harmful host mediators. Septic shock is characterized by a hyperdynamic state including increased cardiac output, decreased systemic vascular resistance, and compromised tissue perfusion secondary to hypotension from loss of vascular integrity and maldistribution of blood flow within the microcirculation. Treatment goals have conventionally included maintenance of blood pressure and identification and eradication of the source of infection. An additional goal of treatment, which is still in evolution, is to interrupt maladaptive responses by the host that contribute to the pathogenesis of this lethal syndrome.

A large amount of evidence now supports the concept that septic shock relates in part to excessive activation of host inflammation and inappropriate activation of coagulation [3–7]. During severe infection, local host defenses, are overcome with the release of microbes and microbial toxins into the intravascular space with rapid activation of systemic host defenses, including the complement, coagulation, and kallikrein-kinin systems and several different types of inflammatory cells. This activation leads to the release of other proinflammatory mediators, including cytokines and oxygen radicals, with further amplification of the immune response. Platelet activation and the formation of microvascular thrombi lead to localized tissue ischemia and other inflammatory stimuli. The host produces endogenous anti-inflammatory and anticoagulant responses to counter this inflammatory activation. However, in some patients, these counterregulatory responses are insufficient, and inflammation and coagulation continue with vasodilatation and tissue injury progressing to shock, organ damage, and death. Interruption of these pathophysiologic events causing septic shock has been the focus of much drug research and development over the past 30 years.

Early attempts to limit excessive inflammation during sepsis with high-dose glucocorticoids were unsuccessful. Because this failure was attributed in part to the nonspecific anti-inflammatory effects of steroids, agents were developed to selectively inhibit specific host inflammatory mediators believed to be the most harmful during sepsis. More recently, based on growing evidence that sepsis may be associated with a state of relative adrenal insufficiency, doses of glucocorticoids much lower than those originally tested have been studied. Finally, to counter activation of the coagulation system during sepsis, anticoagulant agents have been developed and are now under study. To date only low-dose steroids and activated protein C, an anticoagulant, have shown encouraging results in subgroups of patients. However, growing experience in this field now demonstrates that there are important factors capable of influencing the efficacy of each of these different therapies. Accounting for these factors may greatly improve this therapeutic approach. This chapter will outline the rationale for targeting inflammatory and coagulant pathways during sepsis, review the clinical experience with mediator-specific anti-inflammatory agents, steroids, and anticoagulants, and highlight factors that growing evidence suggests may influence their efficacy.

20.2 MEDIATOR-SPECIFIC ANTI-INFLAMMATORY THERAPIES IN SEPSIS

Role of Inflammatory Mediators in the Pathophysiology of Sepsis

Evidence of several types supports the role of individual inflammatory mediators in the pathogenesis of septic shock. Many of the manifestations of sepsis can be produced in normal animals and humans by administering purified host inflammatory mediators. Intravenous infusion of either recombinant tumor necrosis factor (TNF)-α or interleukin (IL)-1 to normal animals or humans produces septic physiology characterized by fever, tachycardia, vasodilatation, hypotension, myocardial dysfunction, hemoconcentration, and metabolic acidosis [3, 8–13]. Furthermore, when administered in high doses, these agents are lethal in animal models [14]. The role of the host inflammatory response in the development of septic shock is further supported by the demonstration of increased circulating levels of inflammatory mediators in both animal models of sepsis and septic patients. During meningococcemia, a very lethal form of sepsis, high levels of TNF or other inflammatory mediators were associated with death [15]. Finally, the involvement of the inflammatory response in the development of septic shock is supported by evidence from animal models of sepsis in which treatment with selective anti-inflammatory mediators prevents the physiologic changes associated with sepsis and improves survival [16–31].

Mediator-Specific Anti-inflammatory Agent Trials

Several different types of mediator-specific anti-inflammatory agents have now been tested in preclinical and clinical trials. Each of these agents was developed to inhibit the effects of a specific proinflammatory mediator thought to play a role in the development of septic shock [3, 5, 32–34]. Agents that have been tested include IL-1 receptor antagonists, antibradykinin agents, anti-TNF antibodies, soluble TNF receptors, platelet activating factor receptor antagonists, platelet activating factor acetylhydrolases, and antiprostaglandin agents. Despite being highly beneficial in preclinical trials, none of these agents demonstrated a significant benefit in their individual clinical trials [35–61] (Table 20.1). However, a small but significant treatment effect of the mediator-specific anti-inflammatory agents was demonstrated in a recent meta-analysis that combined all of these clinical trials ($n = 27$) as a group [62] (Fig. 20.1). The treatment mortality rate was 35.5 percent compared to a control mortality rate of 37.5 percent. The odds ratio of survival with the mediator-specific anti-inflammatory agents was 1.09 (95 percent confidence

TABLE 20.1 Clinical Trials of Mediator-Specific Anti-inflammatory Agents[a]

Class of Agent	Mechanism of Action	Number of Clinical Trials	Odds Ratio of Survival [62, 111] (95% Confidence Interval)
Anti-TNF mAb	Antibody-specific inactivation of circulating TNF	9	1.10 (0.97–1.25)
Soluble TNF receptors	Receptor-specific inactivation of circulating TNF	3	0.95 (0.78–1.16)
IL-1 receptor antagonists	Receptor-specific inactivation of IL-1	3	1.18 (0.97–1.44)
Prostaglandin antagonists	Cyclooxygenase inhibition of prostaglandin production	3	1.22 (0.78–1.58)
Bradykinin antagonists	Inhibition of the kallikrein/kinin cascade	2	0.91 (0.65–1.27)
PAF antagonists	Receptor-specific inactivation or enzyme degradation of PAF	7	1.10 (0.93–1.30)

[a] Legend: TNF-mAb, tumor necrosis factor monoclonal antibody; PAF, platelet activating factor; IL-1: Interleukin-1.

interval (CI) 1.01 to 1.18, $p = .03$). Hence, these agents resulted in an absolute risk reduction of 2 percent and a relative risk reduction of 7 percent. The improvement in the odds ratio of survival in this meta-analysis supports the involvement of the systemic inflammatory response in the development of the sepsis syndrome. However, it can be seen in this meta-analysis that the larger individual clinical trials (>250 patients) more accurately captured the treatment effect of these agents compared to the smaller individual clinical trials (<250 patients) (Fig. 20.1). Based on this analysis, an individual trial of any one of these mediator-specific agents would require approximately 6000 patients to demonstrate a statistically significant treatment effect.

Factors Influencing the Efficacy of Anti-inflammatory Agents in Sepsis

Clinical trials of the mediator-specific anti-inflammatory agents in septic patients did not demonstrate the significant treatment effects that were shown

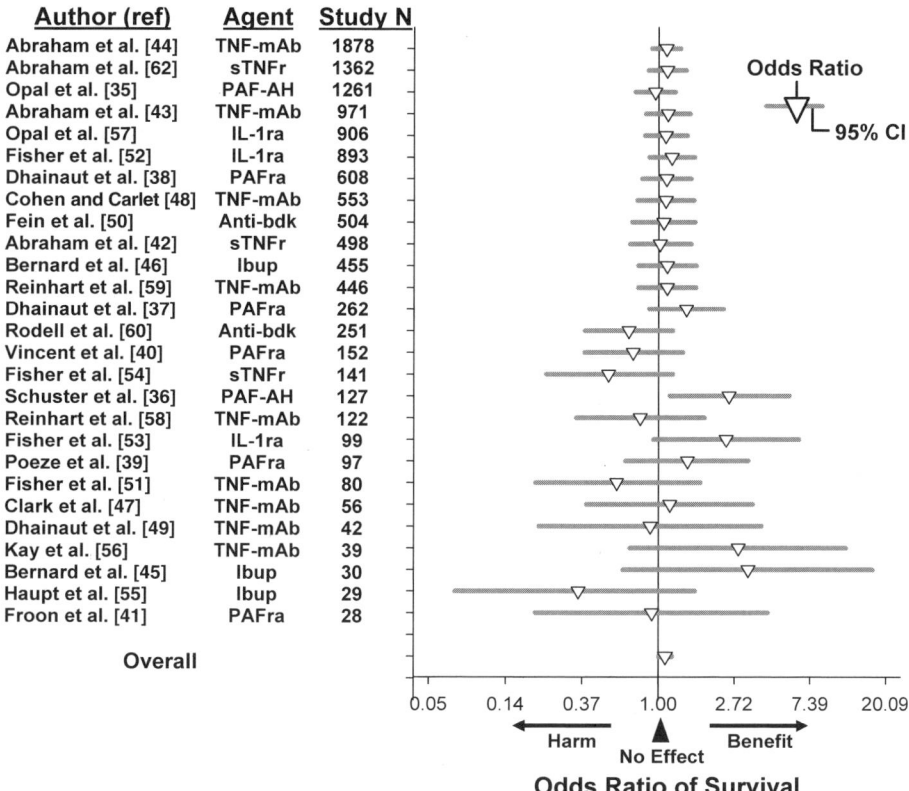

Figure 20.1 Odds ratio of survival for clinical trials of mediator-specific anti-inflammatory agents: The odds ratio of survival for each clinical trial of an anti-inflammatory agent is shown as the inverted triangle with the solid line representing its 95% confidence interval [35–61]. The trials are arranged by the number of patients in each trial. The treatment effect becomes more consistent as the size of the trial increases. Overall, there is a significant improvement in survival with treatment ($p = .03$). Legend: TNF-mAb, tumor necrosis factor monoclonal antibody; sTNFr, soluble tumor necrosis factor receptor; IL-1ra, interleukin-1 receptor antagonist; PAFra, platelet activating factor receptor antagonist; PAF-AH, platelet activating factor acetylhydrolase; Ibup, ibuprofen; Anti-bdk, anti bradykinin agent.

in the preclinical trials that tested them. Potential reasons discussed to explain these disparate results have included differences in the type or route of infection. It has also been suggested that the underlying risk of death associated with sepsis may influence the treatment effect of anti-inflammatory agents [63].

A related series of studies recently assessed the influence of several of these factors on the efficacy of the anti-inflammatory agents in sepsis [64]. These studies included an analysis of past published preclinical and clinical trials as well as prospective experiments. Using metaregression analysis, the investigators examined the effect of these agents in relationship to changes in the type, severity, or site of infection. Of these factors, severity of infection and its associated risk of death, as measured by control group mortality rate, was found to profoundly influence the treatment effect of several different mediator-specific anti-inflammatory agents. Although very beneficial when the risk of death was great, these agents became less beneficial and were potentially harmful as this risk decreased (Fig. 20.2). This relationship is important because published preclinical trials of mediator-specific anti-inflammatory agents were performed with a very high risk of death (median control mortality rate 88 percent) as compared to the clinical trials that were performed at a significantly lower risk of death (median control mortality rate 41 percent) (Fig. 20.2). Hence, the preclinical trials of these agents were carried out at a risk of death level where these agents are very beneficial compared to the clinical trials, which had a level of risk at which these agents would have little benefit.

Furthermore, predicted levels of efficacy of these agents in clinical trials based on the demonstrated relationship between severity of illness and treatment effect in the preclinical trials would have been similar to the actual effects that were reported in the clinical trials of these agents. This suggests that the small treatment effects of these agents detected in their clinical trials could have been predicted if the relationship between treatment effect and

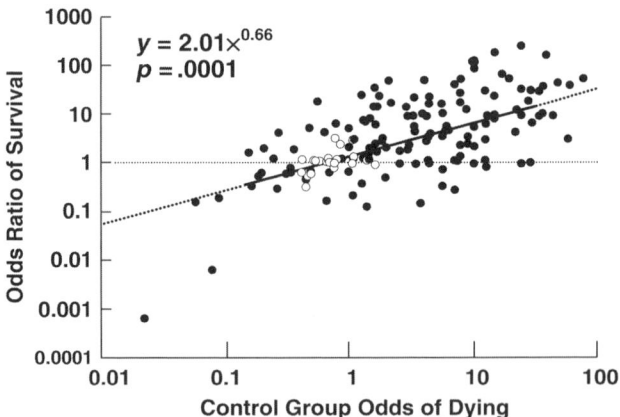

Figure 20.2 Relationship between treatment effect of mediator-specific anti-inflammatory agents and severity of illness. A weighted regression line representing the relationship between control odds of dying and the odds ratio of survival with treatment for published and prospective animal studies (closed circles) [64] and in the clinical trials (open circles) of mediator-specific anti-inflammatory agents [35–61]. There is a significant relationship ($p = .0001$) between the treatment effect of these agents and the control odds of dying. These agents are beneficial at a high risk of death and have no effect or are harmful at a low risk.

risk of death had been taken into account. To be beneficial clinically, these agents may have to be directed to those patients with a high risk of death.

20.3 GLUCOCORTICOIDS IN SEPSIS

Glucocorticoid Trials

Glucocorticoids were the first type of anti-inflammatory agent tested in the treatment of sepsis. Steroids have been shown to have anti-inflammatory properties and to improve vascular reactivity [65]. As with other anti-inflammatory agents, administration of glucocorticoids in animal models of sepsis led to improved outcome [28, 66–71]. However, early clinical trials using short courses of very high doses of glucocorticoids to block excessive inflammation demonstrated varying efficacy [72–79]. Three meta-analyses of the prospective randomized trials of high-dose glucocorticoids were performed [5, 80, 81]. Two of these analyses demonstrated no beneficial effects of steroid therapy[80, 81], and the third meta-analysis demonstrated harmful effects of steroid therapy after exclusion of a single outlying study (odds ratio of survival of 0.70, 95 percent confidence interval 0.556 to 0.91, $p = .008$) [5] (Fig. 20.3). The harmful effects of steroids in these trials may have been due to an increased incidence and severity of secondary infections caused by the immunosuppressive effects of high-dose steroid therapy [74, 78, 79].

More recent studies have demonstrated that relative adrenal suppression and adrenal hyporesponsiveness occur in septic patients and correlate with increased mortality [82–84]. The results of these studies led to renewed interest in the potential benefits of steroid therapy in the treatment of septic shock. Five recent studies were performed using lower glucocorticoid doses, termed either *stress* or *physiologic* doses, in septic patients who required persistent vasopressor support [85–89]. Each of these trials demonstrated beneficial effects of steroids on survival and/or shock reversal. A meta-analysis combining the results of these trials demonstrated a consistent and significant beneficial effect of steroids on survival (odds ratio, 95 percent confidence interval, p value) (1.52, 1.03 to 2.27, $p = .036$) and shock reversal (4.79, 2.07 to 11.11, $p = .001$). [90] (Fig. 20.3). Furthermore, analysis of the three trials that used an ACTH stimulation test to classify patients as either "responders" (normal adrenal function) or "nonresponders" (relative adrenal insufficiency), revealed similar beneficial effects of steroids on survival and shock reversal in both groups [90].

Factors Influencing the Effects of Glucocorticoids

Analysis of all of the randomized controlled clinical sepsis trials of steroids demonstrated that their effects were dose dependent with high doses being harmful and low doses being beneficial ($p = .004$) [90]. In addition, compared

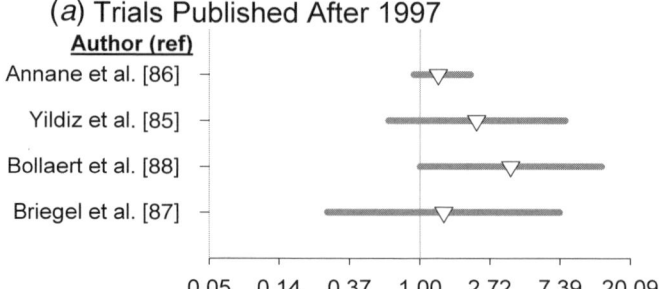

(a) Trials Published After 1997

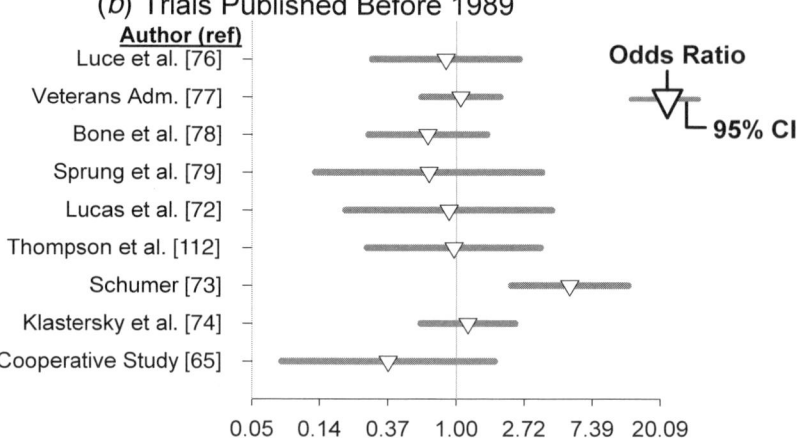

(b) Trials Published Before 1989

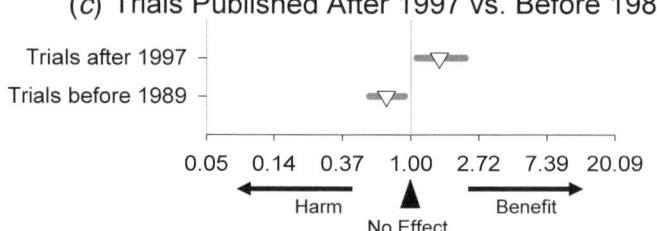

(c) Trials Published After 1997 vs. Before 1989

Odds Ratio of Survival

to the earlier trials of high-dose steroid therapy, the more recent trials of low-dose steroids were performed in sicker patients with higher control mortality rates (mean, p value) (57 vs. 34 percent, $p = .06$) and who were more likely to be on vasopressors (100 vs. 65 percent, $p = .03$). Furthermore treatment in the more recent trials was begun later in the course of the illness (median, p value) (23 vs. <2 h, $p = .02$) and continued for a longer period of time (6 vs. 1 day, $p = .004$) [90]. Therefore, in contrast to high-dose steroids, which decreased survival rates during sepsis, physiologic dose steroids reversed shock and increased survival in patients with established vasopressor-dependent septic shock. However, other factors that differed between earlier and later trials may have also altered the effects of these agents.

Comparison of Glucocorticoid Versus Mediator-Specific Anti-inflammatory Agents

Several reasons may explain why physiologic dose steroids appeared more beneficial than mediator-specific anti-inflammatory agents in the treatment of sepsis. First, steroids may not only limit inflammatory injury but also reverse a state of relative adrenal insufficiency [88]. Second, although a significant linear relationship between treatment effect and control mortality was not demonstrated among the clinical steroid trials, it is worth noting that the average control mortality rate in the beneficial low-dose glucocorticoid trials was >55 percent. This control mortality rate with steroids was higher than in clinical trials testing mediator-specific anti-inflammatory agents, possibly because patients were only enrolled if they had persistent vasopressor requirement in the former but not the latter. The need for vasopressors in patients receiving steroids may have been a marker for an increased risk of death. As a result, the beneficial glucocorticoid trials were performed in patients with risks of death consistent with those where mediator-specific agents would also have been beneficial [64]. Therefore, the differing effects of these agents may

◀──────────────────────────────────

Figure 20.3 Odds ratio of survival for clinical trials of glucocorticoids. The odds ratio of survival (open triangles) and 95% confidence intervals (horizontal lines) with glucocorticoid therapy in sepsis trials. In the four recently published sepsis trials of (*a*) low-dose steroids, the effects of steroids were similar, and (*c*) when combined in a meta-analysis, there was a significant improvement in the odds ratio of survival ($p = .036$) [85–88]. The effects of steroids on the odds ratio of survival in the 9 sepsis trials of high-dose steroids published before 1989 were variable; however, with exclusion of one trial [73], which was a significant outlier, the effects of steroids in the other (*b*) eight trials on the odds ratio of survival were consistent [65, 72, 74, 76–79, 112]. When these eight trials were combined in a meta-analysis, there was a significant decrease in the odds ratio of survival with steroid therapy during sepsis ($p = .008$). Thus, (*c*) the effects of steroids on survival rates in the recent low-dose sepsis trials are significantly different than those in the older trials of high-dose steroids.

be explained by differences in the severity of illness of the patients studied and their underlying risks of death.

20.4 ANTI-COAGULANTS IN SEPSIS

Role of the Coagulation Pathways in Sepsis

Activation of the coagulation system during sepsis is supported by the frequent occurrence of disseminated intravascular coagulation and the presence of intravascular thrombi at autopsy in septic patients [6, 7]. During sepsis, increased levels of activated coagulation factors, tissue factor and tissue factor pathway inhibitor, and decreased fibrinogen levels have been documented [6]. The intrinsic and extrinsic coagulation pathways are activated by lipopolysaccharides in the bacterial cell wall, which can stimulate tissue factor production. The subsequent formation of thrombin can further activate both procoagulant and anticoagulant pathways. In addition, coagulation system activation potentiates the host inflammatory response through endothelial cell activation with increased cytokine production and leukocyte adhesion [6, 91–93]. These coagulation abnormalities correlate with the development of organ dysfunction and are associated with increased mortality [94, 95]. Based on these interactions and associations, anticoagulants, including antithrombin-III, tissue factor pathway inhibitor, and activated protein C, have been developed and studied as treatments for severe sepsis and septic shock.

Clinical Experience with the Anticoagulant Agents

Based on successful preclinical trials, there have been 12 randomized controlled clinical trials of anticoagulant agents in septic patients, 8 of antithrombin-III (AT-III), 2 of tissue factor pathway inhibitor (TFPI), and 2 of activated protein C (APC) [96–107]. The overall odds ratio of survival with treatment in the 2 APC trials was 1.31 (95 percent CI 1.07 to 1.61, $p = .009$), in the 2 TFPI trials was 1.01 (95 percent CI 0.84 to 1.23, $p = .86$), and in the 8 AT-III trials was 1.03 (95 percent CI 0.88 to 1.20, $p = .72$). The treatment effects in all of these trials agents were consistent, and there was an overall trend toward a small improvement in survival with anticoagulant therapies (odds ratio of survival 1.09, 95 percent CI 0.98 to 1.21, $p = .098$) (Fig. 20.4). Furthermore, each of these agents was associated with significant increases in risk of bleeding [94, 96, 106].

Factors Influencing the Efficacy of Anti-Coagulant Agents

As with both mediator-specific and glucocorticoid anti-inflammatory agents, severity of infection and its associated risk of death may also alter the effects of anticoagulants. Of all the anti-inflammatory and anticoagulant agents now

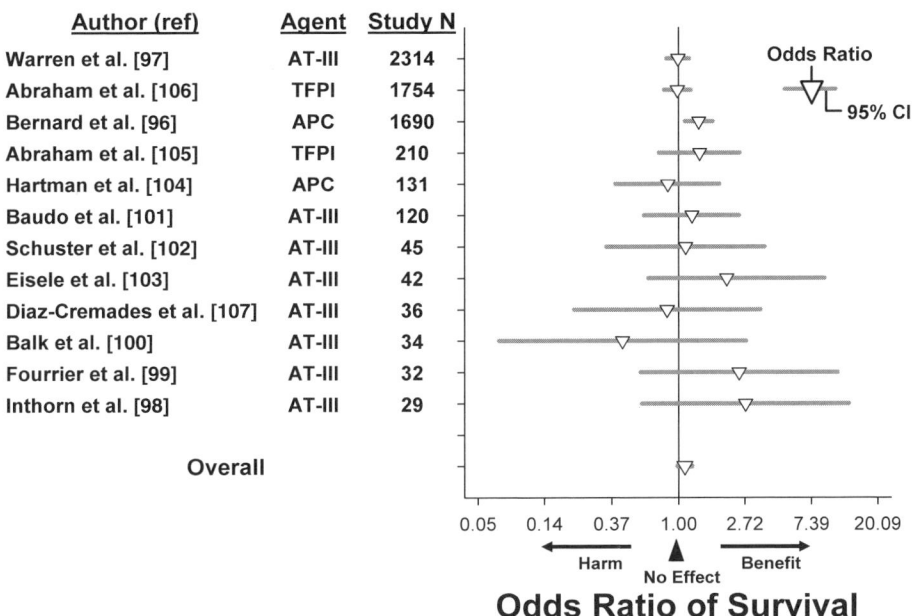

Figure 20.4 Odds ratio of survival for clinical trials of the anticoagulant agents. The odds ratio of survival for each clinical trial of an anticoagulant is shown as the inverted triangle with the solid line representing its 95 percent confidence interval [96–107]. Overall, there is a trend toward improvement in survival with treatment ($p = .10$). APC, activated protein C; AT-III, antithrombin III; TFPI, tissue factor pathway inhibitor.

tested in sepsis, only APC was shown to have a significant beneficial effect on survival in a single phase III trial [96]. In a trial of 1690 patients, APC, which has both anti-inflammatory and antithrombotic properties, significantly ($p = .005$) reduced mortality rates from 30.8 percent in controls to 24.7 percent in treated patients [96]. In this trial, however, APC demonstrated a relationship between treatment effect and risk of death that was similar to the one previously noted in preclinical and clinical trials testing mediator-specific anti-inflammatory agents [64]. It was most beneficial in patients with a high risk of death and less beneficial and on the side of harm in patients with a low risk of death (Table 20.2). Because of this relationship, the use of APC was limited by the Food and Drug Administration (FDA) to patients with severe sepsis (sepsis associated with acute organ dysfunction) and a high risk of death. Furthermore, the FDA requested that the manufacturer of APC perform phase IV trials and clarify the agent's effects in patients with a low risk of death. That trial has reportedly been terminated after enrollment of only 2000 of the 11,000 anticipated subjects for futility.

The significant relationship between treatment effect and severity of illness in the phase III trial of APC suggests that this factor could alter the efficacy

TABLE 20.2 Results of Phase III Trial of Activated Protein C [96] Stratified by Severity of Illness[a]

APACHE II Score	Control Mortality (%)	Odds Ratio of Survival
3–19	15	0.77
20–24	26	1.19
25–29	36	1.82
30–52	49	1.56

[a] Legend: APACHE, acute physiology, age, and chronic health evaluation.

of the other anticoagulant agents. In fact an earlier meta-analysis of 10 of these prospective trials and 1 retrospective suggested that such a relationship might exist [94]. However, with addition of the most recent phase III trial testing TFPI and one other prospective trial and the exclusion of the retrospective trial, this relationship was no longer significant. Furthermore, analysis of subgroups within the phase III trials testing TFPI and antithrombin III failed to show a relationship between risk of death and the efficacy of these. Additional analysis of the large phase III trials testing APC, AT-III, and TFPI suggests that another factor, concurrent treatment with heparin, may have confounded the results of these trials and, in the case of the latter two, potentially masked the influence of severity of sepsis [96, 97, 106]. In each of these trials, the experimental study agents were less beneficial in patients who received concurrent heparin therapy than in those that did not [108] (Fig. 20.5). This relationship was consistent across all three trials but was strongest in the APC trial. Although there are several potential reasons for these results, one is that potential beneficial anticoagulant effects of heparin may have negated those of the study drugs [108, 109]. Consistent with this possibility, the patients in the placebo groups of these trials that received heparin therapy had a better outcome than patients in the placebo groups not receiving heparin (Fig. 20.6). Furthermore, in a phase III trial examining the effect of prophylactic low-molecular-weight heparin in critically ill patients, many of whom were likely septic, heparin therapy was reported to significantly improve clinical outcome [110]. Therefore, heparin, which could have its own beneficial effects on survival in critically ill patients, may have decreased the ability of the anticoagulant agents to improve survival.

20.5 CONCLUSIONS

Newer treatment agents for sepsis directed at the host's response during infection have had differing survival effects that appear dependent, at least in part, on several different factors. Mediator-specific anti-inflammatory agents have been shown overall to have a small beneficial effect on survival. However, the

CONCLUSIONS

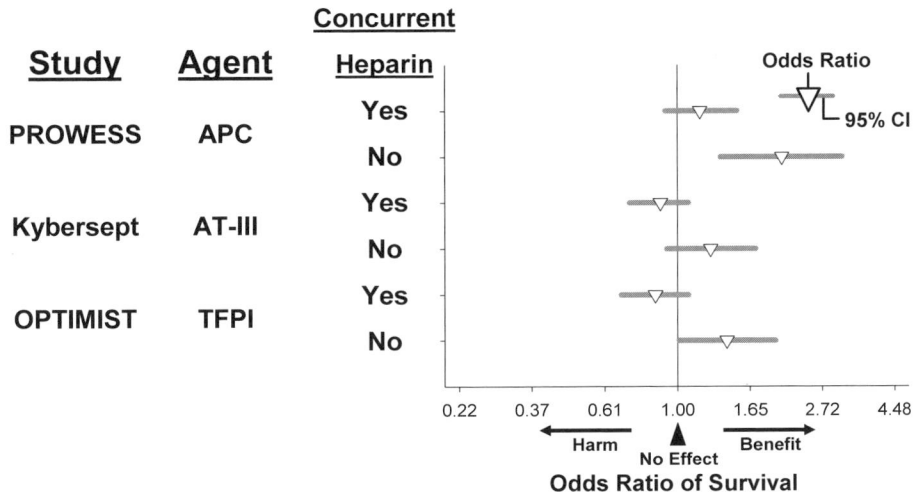

Figure 20.5 Effect of concurrent heparin therapy on the odds ratio of survival in the anticoagulant trials. The odds ratio of survival (triangle) and 95 percent confidence intervals (horizontal lines) for the large phase III trials of the three different anticoagulant agents (APC, AT-III, TFPI) in patients who either were or were not receiving concurrent heparin [96, 97, 106]. The treatment effect of each study agent was less beneficial or harmful in patients receiving heparin compared to patients not receiving heparin.

treatment effects of these agents are likely related to the patients' risk of death. In patients at a higher risk of death, these agents may be more beneficial, and in those at lower risk they appear ineffective or potentially harmful. The disparate treatment effects of these agents in clinical and preclinical trials can be explained by this relationship.

The effects of glucocorticoids in sepsis are dose dependent with high doses having a harmful effect and low doses being beneficial. Analysis of the glucocorticoid trials suggests that control mortality rates were higher in the beneficial trials and lower in the harmful ones. While these are consistent with the relationship between severity of illness and treatment effect noted for mediator-specific anti-inflammatory, a significant relationship was not found in the steroid trials. This may be secondary to an overwhelming influence of drug dose on treatment effect.

As a class anticoagulant agents did not have a beneficial effect on survival and did not demonstrate a relationship between risk of death and treatment effect. However, activated protein C did demonstrate a beneficial effect on survival with a relationship between treatment effect and severity of illness that was very similar to the mediator-specific anti-inflammatory agents. Furthermore, the treatment effect of the anticoagulant agents may have been confounded by concurrent heparin therapy. This potential interaction will require further study.

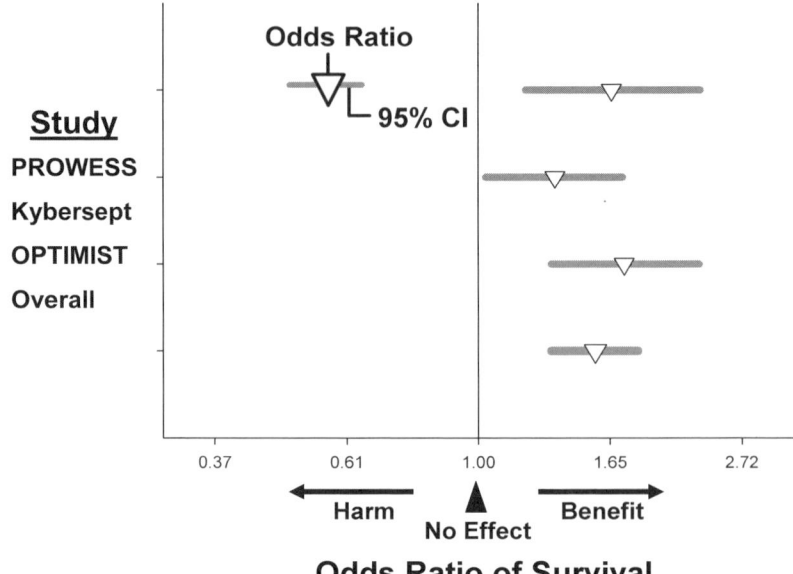

Figure 20.6 Efficacy of heparin in the placebo groups of the anticoagulant trials. The odds ratio of survival (triangle) with heparin treatment and 95 percent confidence intervals (horizontal lines) in each of the placebo groups of the phase III trials of APC (PROWESS trial), AT-III (Kybersept trial), and TFPI (OPTIMIST trial) [96, 97, 106]. Heparin treatment was beneficial in all placebo groups and resulted in an overall improvement in the odds ratio of survival ($p < .0001$).

Our evolving clinical experience with novel therapies for sepsis, including mediator-specific anti-inflammatory agents, glucocorticoids, and anticoagulant agents, suggests that several factors may alter their effects. Based on this knowledge, these agents may only be beneficial in particular subgroups of patients and therefore require further testing in targeted populations. Development of future agents for sepsis should include careful preclinical and clinical testing that account for factors such as the severity of illness, the dose-dependent effects of the therapies under study, and the influence of concurrent therapies with competing effects.

REFERENCES

1. Center for Disease Control: Increase in national hospital discharge survey rates for septicemia—United States, 1979–1987. *MMWR*, 1990, *39*, 31–34.
2. Angus, D. C., Linde-Zwirble, W. T., Lidicker, J., Clermont, G., Carcillo, J., Pinsky, M. R. (2001). Epidemiology of severe sepsis in the United States: Analysis of incidence, outcome, and associated costs of care. *Crit. Care Med.*, *29*, 1303–1310.

REFERENCES

3. Natanson, C., Hoffman, W. D., Suffredini, A. F., Eichacker, P. Q., Danner, R. L. (1994). Selected treatment strategies for septic shock based on proposed mechanisms of pathogenesis. *Ann. Intern. Med.*, *120*, 771–783.
4. Quezado, Z. M., Banks, S. M., Natanson, C. (1995). New strategies for combatting sepsis: The magic bullets missed the mark . . . but the search continues. *Trends Biotechnol.*, *13*, 56–63.
5. Zeni, F., Freeman, B., Natanson, C. (1997). Anti-inflammatory therapies to treat sepsis and septic shock: A reassessment. *Crit. Care Med.*, *25*, 1095–1100.
6. Camerota, A. J., Creasey, A. A., Patla, V., Larkin, V. A., Fink, M. P. (1998). Delayed treatment with recombinant human tissue factor pathway inhibitor improves survival in rabbits with gram-negative peritonitis. *J. Infect. Dis.*, *177*, 668–676.
7. Park, C. T., Creasey, A. A., Wright, S. D. (1997). Tissue factor pathway inhibitor blocks cellular effects of endotoxin by binding to endotoxin and interfering with transfer to CD14. *Blood*, *89*, 4268–4274.
8. Suffredini, A. F., Fromm, R. E., Parker, M. M., Brenner, M., Kovacs, J. A., Wesley, R. A., Parrillo, J. E. (1989). The cardiovascular response of normal humans to the administration of endotoxin. *N. Engl. J. Med.*, *321*, 280–287.
9. Alexander, R. B., Ponniah, S., Hasday, J. and Hebel, J. R. (1998). Elevated levels of proinflammatory cytokines in the semen of patients with chronic prostatitis/chronic pelvic pain syndrome. *Urology*, *52*, 744–749.
10. Natanson, C., Eichenholz, P. W., Danner, R. L., Eichacker, P. Q., Hoffman, W. D., Kuo, G. C., Banks, S. M., MacVittie, T. J., Parrillo, J. E. (1989). Endotoxin and tumor necrosis factor challenges in dogs simulate the cardiovascular profile of human septic shock. *J. Exp. Med.*, *169*, 823–832.
11. Waage, A., Espevik, T. (1988). Interleukin 1 potentiates the lethal effect of tumor necrosis factor alpha/cachectin in mice. *J. Exp. Med.*, *167*, 1987–1992.
12. Tracey, K. J., Beutler, B., Lowry, S. F., Merryweather, J., Wolpe, S., Milsark, I. W., Hariri, R. J., Fahey, T. J., 3rd, Zentella, A., Albert, J. D., et al. (1986). Shock and tissue injury induced by recombinant human cachectin. *Science*, *234*, 470–474.
13. Okusawa, S., Gelfand, J. A., Ikejima, T., Connolly, R. J., Dinarello, C. A. (1988). Interleukin 1 induces a shock-like state in rabbits. Synergism with tumor necrosis factor and the effect of cyclooxygenase inhibition. *J. Clin. Invest.*, *81*, 1162–1172.
14. Eichacker, P. Q., Hoffman, W. D., Farese, A., Banks, S. M., Kuo, G. C., MacVittie, T. J., Natanson, C. (1991). TNF but not IL-1 in dogs causes lethal lung injury and multiple organ dysfunction similar to human sepsis. *J. Appl. Physiol.*, *71*, 1979–1989.
15. Waage, A., Halstensen, A., Espevik, T. (1997). Association between tumor necrosis factor in serum and fatal outcome in patients with meningococcal disease. *Lancet*, *1*, 355–357.
16. Ohlsson, K., Bjork, P., Bergenfeldt, M., Hageman, R., Thompson, R. C. (1990). Interleukin-1 receptor antagonist reduces mortality from endotoxin shock. *Nature*, *348*, 550–552.
17. Wakabayashi, G., Gelfand, J. A., Burke, J. F., Thompson, R. C., Dinarello, C. A. (1991). A specific receptor antagonist for interleukin 1 prevents *Escherichia coli*-induced shock in rabbits. *Faseb J.*, *5*, 338–343.

18. Fischer, E., Marano, M. A., Van Zee, K. J., Rock, C. S., Hawes, A. S., Thompson, W. A., DeForge, L., Kenney, J. S., Remick, D. G., Bloedow, D. C., et al. (1992). Interleukin-1 receptor blockade improves survival and hemodynamic performance in *Escherichia coli* septic shock, but fails to alter host responses to sublethal endotoxemia. *J. Clin. Invest.*, 89, 1551–1557.
19. Beutler, B., Milsark, I. W., Cerami, A. C. (1985). Passive immunization against cachectin/tumor necrosis factor protects mice from lethal effect of endotoxin. *Science*, 229, 869–871.
20. Suitters, A. J., Foulkes, R., Opal, S. M., Palardy, J. E., Emtage, J. S., Rolfe, M., Stephens, S., Morgan, A., Holt, A. R., Chaplin, L. C., et al. (1994). Differential effect of isotype on efficacy of anti-tumor necrosis factor alpha chimeric antibodies in experimental septic shock. *J. Exp. Med.*, 179, 849–856.
21. Bagby, G. J., Plessala, K. J., Wilson, L. A., Thompson, J. J., Nelson, S. (1991). Divergent efficacy of antibody to tumor necrosis factor-alpha in intravascular and peritonitis models of sepsis. *J. Infect. Dis.*, 163, 83–88.
22. Mathison, J. C., Wolfson, E., Ulevitch, R. J. (1988). Participation of tumor necrosis factor in the mediation of gram negative bacterial lipopolysaccharide-induced injury in rabbits. *J. Clin. Invest.*, 81, 1925–1937.
23. Fiedler, V. B., Loof, I., Sander, E., Voehringer, V., Galanos, C., Fournel, M. A. (1992). Monoclonal antibody to tumor necrosis factor–alpha prevents lethal endotoxin sepsis in adult rhesus monkeys. *J. Lab. Clin. Med.*, 120, 574–588.
24. Emerson, T. E., Jr., Lindsey, D. C., Jesmok, G. J., Duerr, M. L., Fournel, M. A. (1992). Efficacy of monoclonal antibody against tumor necrosis factor alpha in an endotoxemic baboon model. *Circ. Shock*, 38, 75–84.
25. Eskandari, M. K., Bolgos, G., Miller, C., Nguyen, D. T., DeForge, L. E., Remick, D. G. (1992). Anti-tumor necrosis factor antibody therapy fails to prevent lethality after cecal ligation and puncture or endotoxemia. *J. Immunol.*, 148, 2724–2730.
26. Silva, A. T., Bayston, K. F., Cohen, J. (1990). Prophylactic and therapeutic effects of a monoclonal antibody to tumor necrosis factor-alpha in experimental gram-negative shock. *J. Infect. Dis.*, 162, 421–427.
27. Jesmok, G., Lindsey, C., Duerr, M., Fournel, M., Emerson, T., Jr. (1992). Efficacy of monoclonal antibody against human recombinant tumor necrosis factor in *E. coli*-challenged swine. *Am. J. Pathol.*, 141, 1197–1207.
28. Hinshaw, L. B., Archer, L. T., Beller-Todd, B. K., Coalson, J. J., Flournoy, D. J., Passey, R., Benjamin, B., White, G. L. (1980). Survival of primates in LD100 septic shock following steroid/antibiotic therapy. *J. Surg. Res.*, 28, 151–170.
29. Hinshaw, L. B., Emerson, T. E., Jr., Taylor, F. B., Jr., Chang, A. C., Duerr, M., Peer, G. T., Flournoy, D. J., White, G. L., Kosanke, S. D., Murray, C. K., et al. (1992). Lethal *Staphylococcus aureus*-induced shock in primates: Prevention of death with anti-TNF antibody. *J. Trauma*, 33, 568–573.
30. Tracey, K. J., Fong, Y., Hesse, D. G., Manogue, K. R., Lee, A. T., Kuo, G. C., Lowry, S. F., Cerami, A. (1987). Anti-cachectin/TNF monoclonal antibodies prevent septic shock during lethal bacteraemia. *Nature*, 330, 662–664.
31. Opal, S. M., Cross, A. S., Kelly, N. M., Sadoff, J. C., Bodmer, M. W., Palardy, J. E., Victor, G. H. (1990). Efficacy of a monoclonal antibody directed against tumor necrosis factor in protecting neutropenic rats from lethal infection with *Pseudomonas aeruginosa*. *J. Infect. Dis.*, 161, 1148–1152.

32. Natanson, C., Danner, R. L., Reilly, J. M., Doerfler, M. L., Hoffman, W. D., Akin, G. L., Hosseini, J. M., Banks, S. M., Elin, R. J., MacVittie, T. J., et al. (1990). Antibiotics versus cardiovascular support in a canine model of human septic shock. *Am. J. Physiol.*, *259*, H1440–H1447.
33. Wheeler, A. P., Bernard, G. R. (1999). Treating patients with severe sepsis. *N. Engl. J. Med.*, *340*, 207–214.
34. Parrillo, J. E. (1993). Pathogenetic mechanisms of septic shock. *N. Engl. J. Med.*, *328*, 1471–1477.
35. Opal, S., Laterre, P. F., Abraham, E., Francois, B., Wittebole, X., Lowry, S., Dhainaut, J. F., Warren, B., Dugernier, T., Lopez, A., Sanchez, M., Demeyer, I., Jauregui, L., Lorente, J. A., McGee, W., Reinhart, K., Kljucar, S., Souza, S., Pribble, J. (2004). Recombinant human platelet-activating factor acetylhydrolase for treatment of severe sepsis: Results of a phase III, multicenter, randomized, double-blind, placebo-controlled, clinical trial. *Crit. Care Med.*, *32*, 332–341.
36. Schuster, D. P., Metzler, M., Opal, S., Lowry, S., Balk, R., Abraham, E., Levy, H., Slotman, G., Coyne, E., Souza, S., Pribble, J. (2003). Recombinant platelet-activating factor acetylhydrolase to prevent acute respiratory distress syndrome and mortality in severe sepsis: Phase IIb, multicenter, randomized, placebo-controlled, clinical trial. *Crit. Care Med.*, *31*, 1612–1619.
37. Dhainaut, J. F., Tenaillon, A., Le Tulzo, Y., Schlemmer, B., Solet, J. P., Wolff, M., Holzapfel, L., Zeni, F., Dreyfuss, D., Mira, J. P., et al. (1994). Platelet-activating factor receptor antagonist BN 52021 in the treatment of severe sepsis: A randomized, double-blind, placebo-controlled, multicenter clinical trial. BN 52021 Sepsis Study Group. *Crit. Care Med.*, *22*, 1720–1728.
38. Dhainaut, J. F., Tenaillon, A., Hemmer, M., Damas, P., Le Tulzo, Y., Radermacher, P., Schaller, M. D., Sollet, J. P., Wolff, M., Holzapfel, L., Zeni, F., Vedrinne, J. M., de Vathaire, F., Gourlay, M. L., Guinot, P., Mira, J. P. (1998). Confirmatory platelet-activating factor receptor antagonist trial in patients with severe gram-negative bacterial sepsis: A phase III, randomized, double-blind, placebo-controlled, multicenter trial. BN 52021 Sepsis Investigator Group. *Crit. Care Med.*, *26*, 1963–1971.
39. Poeze, M., Froon, A. H., Ramsay, G., Buurman, W. A., Greve, J. W. (2000). Decreased organ failure in patients with severe SIRS and septic shock treated with the platelet-activating factor antagonist TCV-309: A prospective, multicenter, double-blind, randomized phase II trial. TCV-309 Septic Shock Study Group. *Shock*, *14*, 421–428.
40. Vincent, J. L., Spapen, H., Bakker, J., Webster, N. R., Curtis, L. (2000). Phase II multicenter clinical study of the platelet-activating factor receptor antagonist BB-882 in the treatment of sepsis. *Crit. Care Med.*, *28*, 638–642.
41. Froon, A. M., Greve, J. W., Buurman, W. A., van der Linden, C. J., Langemeijer, H. J., Ulrich, C., Bourgeois, M. (1996). Treatment with the platelet-activating factor antagonist TCV-309 in patients with severe systemic inflammatory response syndrome: A prospective, multi-center, double-blind, randomized phase II trial. *Shock*, *5*, 313–319.
42. Abraham, E., Glauser, M. P., Butler, T., Garbino, J., Gelmont, D., Laterre, P. F., Kudsk, K., Bruining, H. A., Otto, C., Tobin, E., Zwingelstein, C., Lesslauer, W., Leighton, A. (1997). p55 Tumor necrosis factor receptor fusion protein in

the treatment of patients with severe sepsis and septic shock. A randomized controlled multicenter trial. Ro 45-2081 Study Group. *JAMA, 277,* 1531–1538.

43. Abraham, E., Wunderink, R., Silverman, H., Perl, T. M., Nasraway, S., Levy, H., Bone, R., Wenzel, R. P., Balk, R., Allred, R., et al. (1995). Efficacy and safety of monoclonal antibody to human tumor necrosis factor alpha in patients with sepsis syndrome. A randomized, controlled, double-blind, multicenter clinical trial. TNF-alpha MAb Sepsis Study Group. *JAMA, 273,* 934–941.

44. Abraham, E., Anzueto, A., Gutierrez, G., Tessler, S., San Pedro, G., Wunderink, R., Dal Nogare, A., Nasraway, S., Berman, S., Cooney, R., Levy, H., Baughman, R., Rumbak, M., Light, R. B., Poole, L., Allred, R., Constant, J., Pennington, J., Porter, S. (1998). Double-blind randomised controlled trial of monoclonal antibody to human tumour necrosis factor in treatment of septic shock. NORASEPT II Study Group. *Lancet, 351,* 929–933.

45. Bernard, G. R., Reines, H. D., Halushka, P. V., Higgins, S. B., Metz, C. A., Swindell, B. B., Wright, P. E., Watts, F. L., Vrbanac, J. J. (1991). Prostacyclin and thromboxane A2 formation is increased in human sepsis syndrome. Effects of cyclooxygenase inhibition. *Am. Rev. Respir. Dis., 144,* 1095–1101.

46. Bernard, G. R., Wheeler, A. P., Russell, J. A., Schein, R., Summer, W. R., Steinberg, K. P., Fulkerson, W. J., Wright, P. E., Christman, B. W., Dupont, W. D., Higgins, S. B., Swindell, B. B. (1997). The effects of ibuprofen on the physiology and survival of patients with sepsis. The Ibuprofen in Sepsis Study Group. *N. Engl. J. Med., 336,* 912–918.

47. Clark, M. A., Plank, L. D., Connolly, A. B., Streat, S. J., Hill, A. A., Gupta, R., Monk, D. N., Shenkin, A., Hill, G. L. (1998). Effect of a chimeric antibody to tumor necrosis factor-alpha on cytokine and physiologic responses in patients with severe sepsis—a randomized, clinical trial. *Crit. Care Med., 26,* 1650–1659.

48. Cohen, J., Carlet, J. (1996). INTERSEPT: An international, multicenter, placebo-controlled trial of monoclonal antibody to human tumor necrosis factor-alpha in patients with sepsis. International Sepsis Trial Study Group. *Crit. Care Med., 24,* 1431–1440.

49. Dhainaut, J. F., Vincent, J. L., Richard, C., Lejeune, P., Martin, C., Fierobe, L., Stephens, S., Ney, U. M., Sopwith, M. (1995). CDP571, a humanized antibody to human tumor necrosis factor-alpha: Safety, pharmacokinetics, immune response, and influence of the antibody on cytokine concentrations in patients with septic shock. CPD571 Sepsis Study Group. *Crit. Care Med., 23,* 1461–1469.

50. Fein, A. M., Bernard, G. R., Criner, G. J., Fletcher, E. C., Good, J. T., Jr., Knaus, W. A., Levy, H., Matuschak, G. M., Shanies, H. M., Taylor, R. W., Rodell, T. C. (1997). Treatment of severe systemic inflammatory response syndrome and sepsis with a novel bradykinin antagonist, deltibant (CP-0127). Results of a randomized, double-blind, placebo-controlled trial. CP-0127 SIRS and Sepsis Study Group. *JAMA, 277,* 482–487.

51. Fisher, C. J., Jr., Opal, S. M., Dhainaut, J. F., Stephens, S., Zimmerman, J. L., Nightingale, P., Harris, S. J., Schein, R. M., Panacek, E. A., Vincent, J. L., et al. (1993). Influence of an anti-tumor necrosis factor monoclonal antibody on cytokine levels in patients with sepsis. The CB0006 Sepsis Syndrome Study Group. *Crit. Care Med., 21,* 318–327.

52. Fisher, C. J., Jr., Dhainaut, J. F., Opal, S. M., Pribble, J. P., Balk, R. A., Slotman, G. J., Iberti, T. J., Rackow, E. C., Shapiro, M. J., Greenman, R. L., et al. (1994). Recombinant human interleukin 1 receptor antagonist in the treatment of patients with sepsis syndrome. Results from a randomized, double-blind, placebo-controlled trial. Phase III rhIL-1ra Sepsis Syndrome Study Group. *JAMA*, *271*, 1836–1843.

53. Fisher, C. J., Jr., Slotman, G. J., Opal, S. M., Pribble, J. P., Bone, R. C., Emmanuel, G., Ng, D., Bloedow, D. C., Catalano, M. A. (1994). Initial evaluation of human recombinant interleukin-1 receptor antagonist in the treatment of sepsis syndrome: A randomized, open-label, placebo-controlled multicenter trial. The IL-1RA Sepsis Syndrome Study Group. *Crit. Care Med.*, *22*, 12–21.

54. Fisher, C. J., Jr., Agosti, J. M., Opal, S. M., Lowry, S. F., Balk, R. A., Sadoff, J. C., Abraham, E., Schein, R. M., Benjamin, E. (1996). Treatment of septic shock with the tumor necrosis factor receptor: Fc fusion protein. The Soluble TNF Receptor Sepsis Study Group. *N. Engl. J. Med.*, *334*, 1697–1702.

55. Haupt, M. T., Jastremski, M. S., Clemmer, T. P., Metz, C. A., Goris, G. B. (1991). Effect of ibuprofen in patients with severe sepsis: A randomized, double-blind, multicenter study. The Ibuprofen Study Group. *Crit. Care Med.*, *19*, 1339–1347.

56. Kay, C. (1996). Paper presented at Cambridge Health Institute's Designing better drugs and clinical trials for sepsis/SIRS: Reducing mortality to patients and suppliers. Washington, DC.

57. Opal, S. M., Fisher, C. J., Jr., Dhainaut, J. F., Vincent, J. L., Brase, R., Lowry, S. F., Sadoff, J. C., Slotman, G. J., Levy, H., Balk, R. A., Shelly, M. P., Pribble, J. P., LaBrecque, J. F., Lookabaugh, J., Donovan, H., Dubin, H., Baughman, R., Norman, J., DeMaria, E., Matzel, K., Abraham, E., Seneff, M. (1997). Confirmatory interleukin-1 receptor antagonist trial in severe sepsis: A phase III, randomized, double-blind, placebo-controlled, multicenter trial. The Interleukin-1 Receptor Antagonist Sepsis Investigator Group. *Crit. Care Med.*, *25*, 1115–1124.

58. Reinhart, K., Wiegand-Lohnert, C., Grimminger, F., Kaul, M., Withington, S., Treacher, D., Eckart, J., Willatts, S., Bouza, C., Krausch, D., Stockenhuber, F., Eiselstein, J., Daum, L., Kempeni, J. (1996). Assessment of the safety and efficacy of the monoclonal anti-tumor necrosis factor antibody-fragment, MAK 195F, in patients with sepsis and septic shock: A multicenter, randomized, placebo-controlled, dose-ranging study. *Crit. Care Med.*, *24*, 733–742.

59. Reinhart, K., Menges, T., Gardlund, B., Harm Zwaveling, J., Smithes, M., Vincent, J. L., Tellado, J. M., Salgado-Remigio, A., Zimlichman, R., Withington, S., Tschaikowsky, K., Brase, R., Damas, P., Kupper, H., Kempeni, J., Eiselstein, J., Kaul, M. (2001). Randomized, placebo-controlled trial of the anti-tumor necrosis factor antibody fragment afelimomab in hyperinflammatory response during severe sepsis: The RAMSES Study. *Crit. Care Med.*, *29*, 765–769.

60. Rodell, T. C., Scharschmidt, L., Knaus, W. A. (1995). CP-0127 SIRS and Sepsis Study Group: Results of a multi-center randomized, placebo-controlled trial of CP-0127, a novel bradykinin antagonist, in patients with SIRS and sepsis. *Shock*, *3*, 60.

61. Abraham, E., Laterre, P. F., Garbino, J., Pingleton, S., Butler, T., Dugernier, T., Margolis, B., Kudsk, K., Zimmerli, W., Anderson, P., Reynaert, M., Lew, D., Lesslauer, W., Passe, S., Cooper, P., Burdeska, A., Modi, M., Leighton, A., Salgo, M., Van der Auwera, P. (2001). Lenercept (p55 tumor necrosis factor receptor fusion

protein) in severe sepsis and early septic shock: A randomized, double-blind, placebo-controlled, multicenter phase III trial with 1,342 patients. *Crit. Care Med.*, 29, 503–510.

62. Minneci, P. C., Deans, K. J., Banks, S. M., Eichacker, P. Q., Natanson, C. (2004). Should we continue to target the platelet-activating factor pathway in septic patients? *Crit. Care Med.*, 32, 585–588.

63. Knaus, W. A., Harrell, F. E., Jr., LaBrecque, J. F., Wagner, D. P., Pribble, J. P., Draper, E. A., Fisher, C. J., Jr., Soll, L. (1996). Use of predicted risk of mortality to evaluate the efficacy of anticytokine therapy in sepsis. The rhIL-1ra Phase III Sepsis Syndrome Study Group. *Crit. Care Med.*, 24, 46–56.

64. Eichacker, P. Q., Parent, C., Kalil, A., Esposito, C., Cui, X., Banks, S. M., Gerstenberger, E. P., Fitz, Y., Danner, R. L., Natanson, C. (2002). Risk and Efficacy of Antiinflammatory Agents: Retrospective and confirmatory studies of sepsis. *Am. J. Respir. Crit. Care Med.*

65. Cooperative Study Group (1963). The effectiveness of hydrocortisone in the management of severe infection. *JAMA*, 183, 462–465.

66. Fabian, T. C., Patterson, R. (1982). Steroid therapy in septic shock. Survival studies in a laboratory model. *Am. Surg.*, 48, 614–617.

67. Hinshaw, L. B., Beller-Todd, B. K., Archer, L. T., Benjamin, B., Flournoy, D. J., Passey, R., Wilson, M. F. (1981). Effectiveness of steroid/antibiotic treatment in primates administered LD100 *Escherichia coli*. *Ann. Surg.*, 194, 51–56.

68. Hinshaw, L. B., Archer, L. T., Beller-Todd, B. K., Benjamin, B., Flournoy, D. J., Passey, R. (1981). Survival of primates in lethal septic shock following delayed treatment with steroid. *Circ. Shock*, 8, 291–300.

69. Hinshaw, L. B., Beller, B. K., Archer, L. T., Flournoy, D. J., White, G. L., Phillips, R. W. (1979). Recovery from lethal *Escherichia coli* shock in dogs. *Surg. Gynecol. Obstet.*, 149, 545–553.

70. White, G. L., Archer, L. T., Beller, B. K., Hinshaw, L. B. (1978). Increased survival with methylprednisolone treatment in canine endotoxin shock. *J. Surg. Res.*, 25, 357–364.

71. Beller, B. K., Archer, L. T., Passey, R. B., Flournoy, D. J., Hinshaw, L. B. (1983). Effectiveness of modified steroid-antibiotic therapies for lethal sepsis in the dog. *Arch. Surg.*, 118, 1293–1299.

72. Lucas, C. E., Ledgerwood, A. M. (1984). The cardiopulmonary response to massive doses of steroids in patients with septic shock. *Arch. Surg.*, 119, 537–541.

73. Schumer, W. (1976). Steroids in the treatment of clinical septic shock. *Ann. Surg.*, 184, 333–341.

74. Klastersky, J., Cappel, R., Debusscher, L. (1971). Effectiveness of betamethasone in management of severe infections. A double-blind study. *N. Engl. J. Med.*, 284, 1248–1250.

75. Bennett, I. L., Finland, M., Hamborger, M., Kass, E. H., Lepper, M., Waisbren, B. A. (1963). The effectiveness of hydrocortisone in the management of severe infection. *JAMA*, 183, 462–465.

76. Luce, J. M., Montgomery, A. B., Marks, J. D., Turner, J., Metz, C. A., Murray, J. F. (1988). Ineffectiveness of high-dose methylprednisolone in preventing parenchymal lung injury and improving mortality in patients with septic shock. *Am. Rev. Respir. Dis.*, 138, 62–68.

77. Veterans Administration Cooperative Study Group (1987). Effect of high-dose glucocorticoid therapy on mortality in patients with clinical signs of systemic sepsis. The Veterans Administration Systemic Sepsis Cooperative Study Group. *N. Engl. J. Med.*, *317*, 659–665.
78. Bone, R. C., Fisher, C. J., Jr., Clemmer, T. P., Slotman, G. J., Metz, C. A., Balk, R. A. (1987). A controlled clinical trial of high-dose methylprednisolone in the treatment of severe sepsis and septic shock. *N. Engl. J. Med.*, *317*, 653–658.
79. Sprung, C. L., Caralis, P. V., Marcial, E. H., Pierce, M., Gelbard, M. A., Long, W. M., Duncan, R. C., Tendler, M. D., Karpf, M. (1984). The effects of high-dose corticosteroids in patients with septic shock. A prospective, controlled study. *N. Engl. J. Med.*, *311*, 1137–1143.
80. Cronin, L., Cook, D. J., Carlet, J., Heyland, D. K., King, D., Lansang, M. A., Fisher, C. J., Jr. (1995). Corticosteroid treatment for sepsis: A critical appraisal and meta-analysis of the literature. *Crit. Care Med.*, *23*, 1430–1439.
81. Lefering, R., Neugebauer, E. A. (1995). Steroid controversy in sepsis and septic shock: A meta-analysis. *Crit. Care Med.*, *23*, 1294–1303.
82. Rothwell, P. M., Udwadia, Z. F., Lawler, P. G. (1991). Cortisol response to corticotropin and survival in septic shock. *Lancet*, *337*, 582–583.
83. Moran, J. L., Chapman, M. J., O'Fathartaigh, M. S., Peisach, A. R., Pannall, P. R., Leppard, P. (1994). Hypocortisolaemia and adrenocortical responsiveness at onset of septic shock. *Intens. Care Med.*, *20*, 489–495.
84. Soni, A., Pepper, G. M., Wyrwinski, P. M., Ramirez, N. E., Simon, R., Pina, T., Gruenspan, H., Vaca, C. E. (1995). Adrenal insufficiency occurring during septic shock: Incidence, outcome, and relationship to peripheral cytokine levels. *Am. J. Med.*, *98*, 266–271.
85. Yildiz, O., Doganay, M., Aygen, B., Guven, M., Keleutimur, F., Tutuu, A. (2002). Physiological-dose steroid therapy in sepsis [ISRCTN36253388]. *Crit. Care*, *6*, 251–259.
86. Annane, D., Sebille, V., Charpentier, C., Bollaert, P. E., Francois, B., Korach, J., Capellier, G., Cohen, Y., Azoulay, E., Troche, G., Chaumet-Riffaut, P., Bellisant, E. (2002). Effect of treatment with low doses of hydrocortisone and fludrocortisone on mortality in patients with septic shock. *JAMA*, *288*, 862–871.
87. Briegel, J., Forst, H., Haller, M., Schelling, G., Kilger, E., Kuprat, G., Hemmer, B., Hummel, T., Lenhart, A., Heyduck, M., Stoll, C., Peter, K. (1999). Stress doses of hydrocortisone reverse hyperdynamic septic shock: A prospective, randomized, double-blind, single-center study. *Crit. Care Med.*, *27*, 723–732.
88. Bollaert, P. E., Charpentier, C., Levy, B., Debouverie, M., Audibert, G., Larcan, A. (1998). Reversal of late septic shock with supraphysiologic doses of hydrocortisone. *Crit. Care Med.*, *26*, 645–650.
89. Chawla, K., Kupfer, Y., Goldman, I., Tessler, S. (1999). Hydrocortisone reverses refractory shock. *Crit. Care Med.*, *27*, A33 (abstract).
90. Minneci, P. C., Deans, K. J., Banks, S. M., Eichacker, P. Q., Natanson, C. (2004). Dose dependent effects of steroids on survival rates and shock during sepsis: A meta-analysis. *Ann. Intern. Med.*
91. Schaub, R. G., Simmons, C. A., Koets, M. H., Romano, P. J., 2nd, Stewart, G. J. (1984). Early events in the formation of a venous thrombus following local trauma and stasis. *Lab. Invest.*, *51*, 218–224.

92. Coughlin, S. R. (1994). Thrombin receptor function and cardiovascular disease. *Trends Cardiovasc. Med.*, *4*, 77–83.

93. Lorant, D. E., Patel, K. D., McIntyre, T. M., McEver, R. P., Prescott, S. M., Zimmerman, G. A. (1991). Coexpression of GMP-140 and PAF by endothelium stimulated by histamine or thrombin: A juxtacrine system for adhesion and activation of neutrophils. *J. Cell. Biol.*, *115*, 223–234.

94. Freeman, B. D., Zehnbauer, B. A., Buchman, T. G. (2003). A meta-analysis of controlled trials of anti-coagulant therapies in patients with sepsis. *Shock*, *20*, 5–9.

95. Kidokoro, A., Iba, T., Fukunaga, M., Yagi, Y. (1996). Alterations in coagulation and fibrinolysis during sepsis. *Shock*, *5*, 223–228.

96. Bernard, G. R., Vincent, J. L., Laterre, P. F., LaRosa, S. P., Dhainaut, J. F., Lopez-Rodriguez, A., Steingrub, J. S., Garber, G. E., Helterbrand, J. D., Ely, E. W., Fisher, C. J., Jr. (2001). Efficacy and safety of recombinant human activated protein C for severe sepsis. *N. Engl. J. Med.*, *344*, 699–709.

97. Warren, B. L., Eid, A., Singer, P., Pillay, S. S., Carl, P., Novak, I., Chalupa, P., Atherstone, A., Penzes, I., Kubler, A., Knaub, S., Keinecke, H. O., Heinrichs, H., Schindel, F., Juers, M., Bone, R. C., Opal, S. M. (2001). Caring for the critically ill patient. High-dose antithrombin III in severe sepsis: A randomized controlled trial. *JAMA*, *286*, 1869–1878.

98. Inthorn, D., Hoffmann, J. N., Hartl, W. H., Muhlbayer, D., Jochum, M. (1997). Antithrombin III supplementation in severe sepsis: Beneficial effects on organ dysfunction. *Shock*, *8*, 328–334.

99. Fourrier, F., Chopin, C., Huart, J. J., Runge, I., Caron, C., Goudemand, J. (1993). Double-blind, placebo-controlled trial of antithrombin III concentrates in septic shock with disseminated intravascular coagulation. *Chest*, *104*, 882–888.

100. Balk, R. A., Bedrosian, C. L. M., Baughman, R., Eisele, B., Keinecke, H. O., Bone, R. C. (1995). Prospective, double-blind, placebo-controlled, trial of ATIII substitution in sepsis. *Intens. Care Med.*, *21*, S17.

101. Baudo, F., Caimi, T. M., de Cataldo, F., Ravizza, A., Arlati, S., Casella, G., Carugo, D., Palareti, G., Legnani, C., Ridolfi, L., Rossi, R., D'Angelo, A., Crippa, L., Giudici, D., Gallioli, G., Wolfler, A., Calori, G. (1998). Antithrombin III (ATIII) replacement therapy in patients with sepsis and/or postsurgical complications: A controlled double-blind, randomized, multicenter study. *Intens. Care Med.*, *24*, 336–342.

102. Schuster, H. P., Eisele, B., Keinecke, H. O., Heinrichs, H., Mescheder, A., Knaub, S. (1997). S-AT III Study: Antithrombin III in patients with sepsis. *Intens. Care Med.*, *23*(Suppl. 1), S76.

103. Eisele, B., Lamy, M., Thijs, L. G., Keinecke, H. O., Schuster, H. P., Matthias, F. R., Fourrier, F., Heinrichs, H., Delvos, U. (1998). Antithrombin III in patients with severe sepsis. A randomized, placebo-controlled, double-blind multicenter trial plus a meta-analysis on all randomized, placebo-controlled, double-blind trials with antithrombin III in severe sepsis. *Intens. Care Med.*, *24*, 663–672.

104. Hartment, D. L., Bernard, G. R., Rosenfeld, B. A., Helterbrand, J. D., Yan, S. B., Fisher, C. J., Jr. (1998). Recombinant human activated protein C (rhAPC) improves coagulation abnormailites associated with severe sepsis. *Intens. Care Med.*, *24*(Suppl. 1), S77.

105. Abraham, E., Reinhart, K., Svoboda, P., Seibert, A., Olthoff, D., Dal Nogare, A., Postier, R., Hempelmann, G., Butler, T., Martin, E., Zwingelstein, C., Percell, S., Shu, V., Leighton, A., Creasey, A. A. (2001). Assessment of the safety of recombinant tissue factor pathway inhibitor in patients with severe sepsis: A multicenter, randomized, placebo-controlled, single-blind, dose escalation study. *Crit. Care Med., 29*, 2081–2089.

106. Abraham, E., Reinhart, K., Opal, S., Demeyer, I., Doig, C., Rodriguez, A. L., Beale, R., Svoboda, P., Laterre, P. F., Simon, S., Light, B., Spapen, H., Stone, J., Seibert, A., Peckelsen, C., De Deyne, C., Postier, R., Pettila, V., Artigas, A., Percell, S. R., Shu, V., Zwingelstein, C., Tobias, J., Poole, L., Stolzenbach, J. C., Creasey, A. A. (2003). Efficacy and safety of tifacogin (recombinant tissue factor pathway inhibitor) in severe sepsis: A randomized controlled trial. *JAMA, 290*, 238–247.

107. Diaz-Cremades, J. M., Lorenzo, R., Sanchez, M., Moreno, M. J., Alsar, M. J., Bosch, J. M., Fajardo, L., Gonzalez, D., Guerrero, D. (1994). Use of antithrombin III in critical patients. *Intens. Care Med., 20*, 577–580.

108. Haley, M., Cui, X., Minneci, P. C., Deans, K. J., Natanson, C., Eichacker, P. Q. (2004). Recombinant human activated protein C in sepsis: Assessing its clinical use. *Am. J. Med. Sci., 324*, 215–219.

109. Davidson, B. L., Geerts, W. H., Lensing, A. W. (2002). Low-dose heparin for severe sepsis. *N. Engl. J. Med., 347*, 1036–1037.

110. Goldhaber, S. Z. (2003). PREVENT (PRospective EValuation of Dalteparin Efficacy in Immobilized PatieNts Trial). Presented at the Society of Critical Care Medicine Critical Care Congress in San Antonio, TX.

111. Sevransky, J., Natanson, C. (1999). Published clinical trials in sepsis: An update. *Sepsis, 2*, 11–19.

112. Thompson, W. L., Gurley, H. T., Lutz, B. A., Jackson, D. L., Kyols, L. K., Morris, I. A. (1976). Inefficacy of glucocorticoids in shock (double-blind study). *Clin. Res., 24*, 258A (abstract).

21

COMBINATORIAL CHEMISTRY IN THE DRUG DISCOVERY PROCESS

NATHAN T. ROSS, BRIAN R. MCNAUGHTON,
AND BENJAMIN L. MILLER
University of Rochester
Rochester, New York

21.1	INTRODUCTION TO COMBINATORIAL CHEMISTRY	962
	Introduction	962
	Origins of Combinatorial Chemistry	963
	Solid-Phase Synthesis of Biopolymers	965
21.2	TYPES OF LIBRARY SYNTHESIS	971
	Early Examples of Parallel Synthesis	971
	Split-Pool Synthesis	973
	On-Bead Screening	974
21.3	WHO WON? IDENTIFYING HITS FROM SCREENS OF SPLIT-POOL LIBRARIES	974
	Recursive Deconvolution	976
	Tagging Methods	977
	Binary Encoding	977
	Infrared Coded Resins	979
	Radio-Frequency Tagging and the Irori Corporation	980
21.4	OTHER COMBINATORIAL SYNTHETIC TECHNIQUES	981
	Photolithographic Synthetic Methods	981
	High-Throughput Methods	983
21.5	TAG-FREE METHODOLOGY	985
	Dynamic Combinatorial Chemistry and Allied Methods	985
	Click Chemistry and in Situ Click Chemistry	992
	Thiol-Based in Situ Assembly of Macromolecular Ligands	997
	Structure–Activity Relationship by Nuclear Magnetic Resonance	998

Drug Discovery Handbook, by Shayne Cox Gad
Copyright © 2005 by John Wiley & Sons, Inc.

21.6	LIBRARY DESIGN ISSUES	1001
21.7	CONCLUSIONS	1004
	References	1004

21.1 INTRODUCTION TO COMBINATORIAL CHEMISTRY

Introduction

The global pharmaceutical industry stands as a testament to the power of fundamental research in chemistry and biology and the impact such efforts can have on human health and human history. Many of the medicines we now take for granted—from penicillin (**1**) and quinine (**2**) to more modern drugs such as Taxol (**3**) and vancomycin (**4**) are the products of drug discovery efforts rooted in "traditional" medicinal chemistry. As discussed in detail throughout this text, such efforts arose initially based on a desire to understand and improve the physiological activity of natural products: compounds isolated from natural sources such as fungi, plants, and marine sponges. A highly simplified (or perhaps simplistic) view of traditional medicinal chemistry is that it is a linear endeavor: A compound is synthesized, or a natural product modified, based on an initial hypothesis of activity. After carrying out a bioassay of the new compound, the hypothesis is modified, another new compound is synthesized, and so on until a structure with the desired properties (receptor affinity, binding selectivity, and pharmacokinetic properties) is finally obtained. In many cases, this can be an exceptionally long process: The road from initial isolation to approval of the Food and Drug Administration (FDA) for Taxol was roughly 30 years [1]! Without exception, it is a remarkably expensive and uncertain process. As has been noted by others: "For every 10,000 or more compounds synthesized each year in an average pharmaceutical company, less than one makes it to the market. Any method that allows the pharmaceutical chemist to increase the likelihood of synthesizing an active analogue or to increase his ability to find, or even design, novel leads is of enormous commercial interest" [2].

INTRODUCTION TO COMBINATORIAL CHEMISTRY 963

3   4

Although this statement was written in an account describing research on computer-aided drug design, it can easily be applied to combinatorial chemistry. Broken down into segments, the drug discovery process typically involves target identification, target validation, lead identification, and lead optimization [3]. In a little over a decade, combinatorial chemistry has evolved as an attempt to significantly shorten the lead identification and lead optimization portions of drug development. Although most precisely defined as the exhaustive recombination of sets of reagents (following the mathematical definition of combinatorial), combinatorial chemistry has become synonymous with any process that allows for the synthesis of arbitrarily large sets of different compounds, known as libraries, simultaneously.

Origins of Combinatorial Chemistry

Combinatorial chemistry largely arose out of two paradigm-altering ideas: solid-supported synthesis and molecular biology. Molecular biology has been critical both because of the significantly greater range of targets it has made available and because of the recognition that much of nature is in essence a "combinatorial chemistry experiment." In addition to obvious parallels at the genetic level, combinatorial strategies can even be seen in naturally occurring small-molecule "libraries" of biologically active compounds. For example, the pupa of the squash beetle *Epilachna borealis* secretes a combinatorial library of macrocyclic polyamines as a defensive strategy (Scheme 21.1) [4]. Solid-supported synthesis, developed initially as a tool primarily for biochemists and molecular biologists, has become ubiquitous in the field. Because of this, we will describe it in some detail. Several other parallel developments, including high-throughput screening, automation (robotics), and computer-aided molecular design, have also significantly impacted the development of

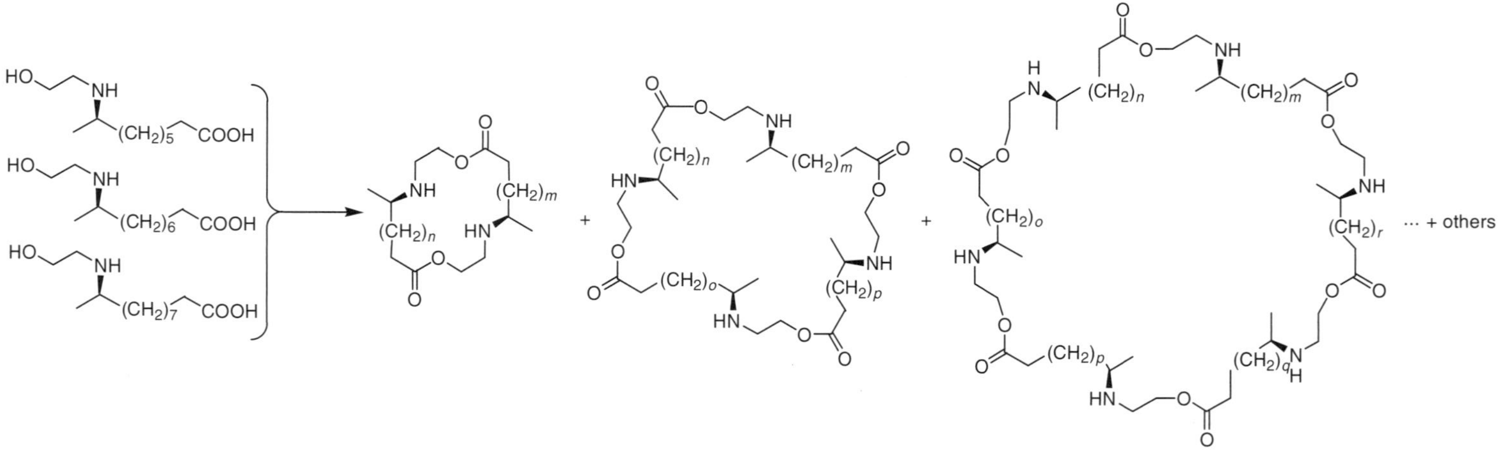

Scheme 21.1 Beetle-synthesized combinatorial library of defensive compounds.

combinatorial chemistry. We should note, however, that this review is by no means intended to be comprehensive. The combinatorial chemistry literature has grown explosively over the past few years, and we have undoubtedly neglected many important contributions. We refer those readers interested in more detailed coverage of the field to the many monographs now available [5], as well as to the primary literature.

Solid-Phase Synthesis of Biopolymers

Solid-phase synthesis emerged from pioneering efforts by Merrifield and co-workers in the 1960s targeted at simplifying the synthesis of longer peptides (four or more amino acids at that time) [6]. Merrifield's work was driven by two primary assumptions. First, attachment of a peptide to a solid support during synthesis was anticipated to significantly improve reaction yields, since excess reagents could be used to drive reactions to completion. Second, unreacted materials remaining in solution at the end of each synthetic step could be subsequently filtered away, simplifying isolation of the final product. In execution, an amino acid with a protected amino terminus was first covalently attached to functionalized polystyrene resin via its carboxy terminus (Scheme 21.2). Next, the amino terminus was deprotected and reacted with the carboxy terminus of a second amino acid, forming a peptide bond. After the last amino acid was attached, treatment of the resin with a strong acid cleaved the peptide from the polymer matrix. Merrifield showed that this process was highly reproducible, and when the quantities of reactants were optimized, very high reaction yields could be obtained.

Merrifield found that making this process effective requires first selecting a resin made from a functionalized, inert chemical polymer. Although several materials were initially tested, polystyrene cross-linked with divinylbenzene proved to be the most advantageous. This was due to its insolubility in organic and aqueous solvents, its ability to be functionalized via standard aromatic chemistry, and its porous nature (5). Polystyrene resin beads are also able to swell to different degrees in different organic solvents, allowing for some control of reagent penetration into the bead depending on the specific solvent employed [7].

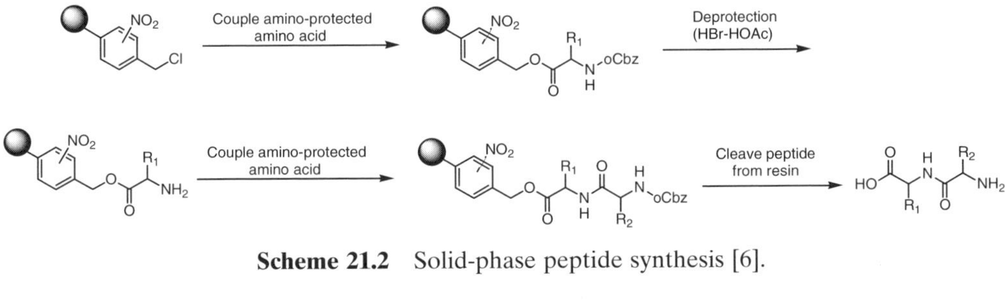

Scheme 21.2 Solid-phase peptide synthesis [6].

R = [structure showing chloromethylated phenyl group]

5

The resin most closely associated with Merrifield's work is chloromethyl polystyrene-co-divinylbenzene. Providing an initial proof of concept, Merrifield and co-workers synthesized the tetrapeptide Leu-Ala-Gly-Val, using a scheme analogous to that shown in Scheme 21.2. Following cleavage of the peptide from resin, analysis proved that it was in fact the desired product. Moreover, the purity of the crude material obtained following resin cleavage was high enough to indicate that couplings were very nearly quantitative. These initial results validated Merrifield's hypotheses and set the stage for more complex syntheses.

As effective and efficient as Merrifield's new solid-phase synthesis was, it was not completely without problems. The conditions required to cleave the carbobenzoxy (Cbz) protecting group, HBr-HOAc, were so acidic that some of the peptide was simultaneously cleaved from the resin beads. Initial attempts to circumvent this problem involved modification of the chloromethyl resin by nitration or bromination to increase its acid stability. When this still failed to completely prevent cleavage of material from the resin during deprotection, subsequent syntheses substituted the significantly more acid-labile *t*-butoxycarbonyl (*t*-BOC) group for Cbz. Use of the *t*-BOC protecting group allowed the Merrifield group to synthesize a nonapeptide, the plasma kinin bradykinin (**6**), in just eight days and in an overall yield of 68 percent [8]. This synthesis represented an important advance in the field for several reasons. First, it incorporated several modifications to the initial protocols, which both streamlined the synthetic process and made it more compatible with the synthesis of longer peptides. Second, the bradykinin synthesis demonstrated that the solid-phase methodology was compatible with reactive side chains (in this case, serine and arginine). However, what is perhaps most significant about the synthesis of bradykinin was that it was the first small molecule synthesized on solid-phase resin using automation [9]. The ability to automate solid-phase synthesis was one of the first goals that Merrifield set for his laboratory at the outset of their work, and it had been accomplished here.

L-ARG-L-PRO-L-PRO-GLY-L-PHE-L-SER-L-PRO-L-PHE-L-ARG

Bradykinin, (**6**)

Another landmark synthesis from the Merrifield laboratory was that of valinomycin (**7**), a cyclic dodecadepsipeptide derived from D/L valine, D-α-hydroxyisovaleric acid, and L-lactic acid. Representing a significant leap in complexity over the tetrapeptide structure described just 5 years previously, the synthesis of valinomycin featured solution-phase production of four building blocks via esterification reactions. These were subsequently incorporated into the solid-phase synthesis as in Merrifield's previous efforts.

Valinomycin (**7**)

It only took a few years after Merrifield's first publication describing solid-phase synthesis of peptides for others to apply the new methodology to their fields of study. In 1965, Letsinger and Mahadevan (who had also described an alternative strategy for solid-phase peptide synthesis shortly after Merrifield) published a report describing the solid-phase synthesis of deoxycytidylyl-

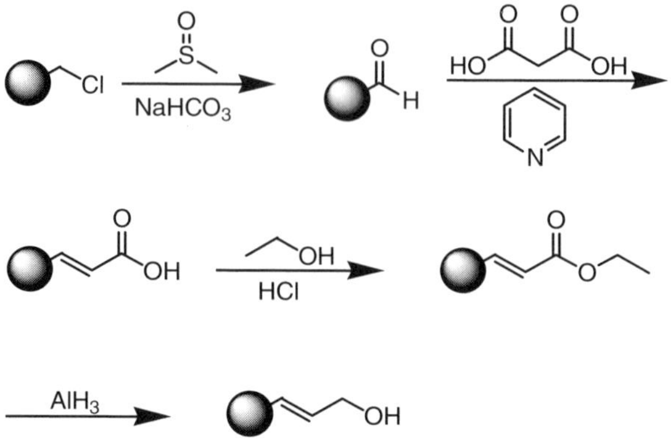

Scheme 21.3

(3′-5′)-thymidine on polystyrene resin [10]. Much like peptide solid-phase synthesis, purification during synthesis was much easier than traditional solution-phase methods. However, unlike Merrifield's peptide synthesis, oligonucleotide synthesis was not very rapid due to more than one step in the synthesis requiring over 2 days of reaction time. Subsequent advances by other authors (vide infra) have enabled solid-phase oligonucleotide synthesis to become the highly optimized, highly mechanized, and inexpensive process it is today.

Several years later, in 1971, Frechet and Schuerch developed a protocol to modify Merrifield's chloromethylated polystyrene resin (Scheme 21.3), making the first solid-phase synthesis of an oligosaccharide possible [11]. Oligosaccharides present a significantly greater synthetic challenge than oligonucleotide or polypeptide synthesis, in part due to the elaborate protecting schemes that are required to direct reactions to the appropriate hydroxyl group. First, the sugar monomers must have a "stable" protecting group on all of the hydroxyl groups that will not be participating in forming the new glycosidic bond. Second, the hydroxyl that is going to be participating in the new glycosidic bond must have a comparatively labile protecting group, which can be easily cleaved without cleaving the stable protecting group or cleaving the saccharide from the resin itself. This last reason is why Frechet and Schuerch [11] had to modify Merrifield's chloromethylated resin to make it stable to conditions that would remove both the "labile" and "stable" protecting groups. Scheme 21.4 outlines the general procedure for the synthesis of oligosaccharides on solid phase. Solid- and solution-phase oligosaccharide synthesis has continued to be an area of significant interest. Notable advances include work from the Danishefsky [12], Wong [13], and Seeberger [14] groups. Seeberger in particular has been successful in adapting standard peptide synthesis instrumentation to the production of complex oligosaccharides.

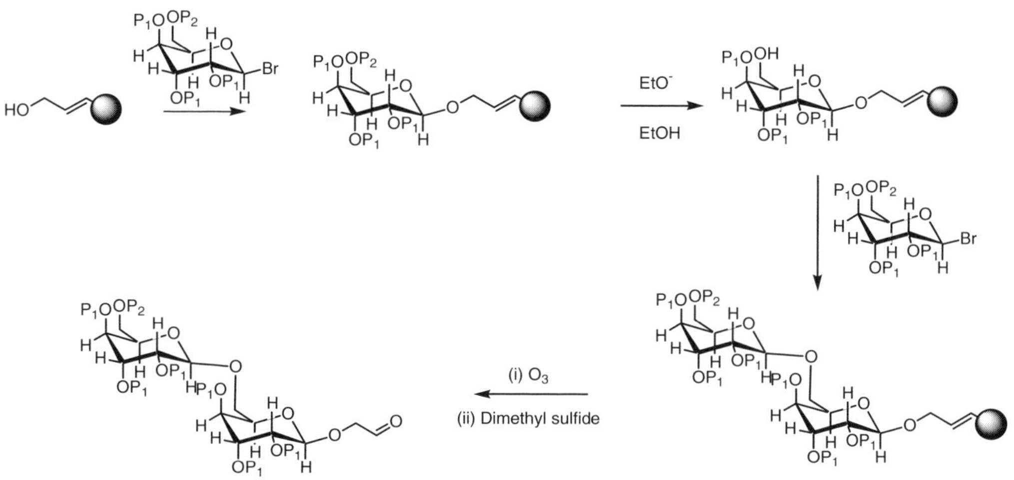

Scheme 21.4 Solid-phase disaccharide synthesis [11].

On the shoulders of the pioneering work just described, researchers have used solid-phase chemistry to advance research in fields spanning organic synthesis and chemical biology. A particularly notable extension of Merrifield's peptide synthesis was reported by Kent and co-workers in their construction of backbone engineered human immunodeficiency virus (HIV) protease on solid support [15]. Protein engineering via site-directed mutagenesis is a standard, widely used technique. However, such modifications are necessarily limited to the incorporation of genetically coded (natural) amino acids. Cellular methods for the incorporation of nonnatural amino acids into proteins exist but are generally highly labor intensive [16]. Conversely, solid-phase synthesis provides a means to incorporate nonnatural amino acids, β-amino acids, peptoids (oligomers of N-alkyl glycine), and fully nonpeptide segments into the growing peptide chain.

In the Kent group's approach, HIV-1 protease was prepared through the ligation of synthetic peptide segments, corresponding to the two halves of the HIV-1 protease monomer [15]. The resulting active HIV-1 protease was fused through a thioester replacement at the natural peptide bond Gly^{51}–Gly^{52}, a region sensitive to mutational changes (Scheme 21.5). Furthermore, this chemospecific ligation allowed fusion of the two peptides without the use of side-chain protective groups.

Along similar lines, the development and advancement of solid supported oligonucleotide synthesis has provided researchers a vehicle to synthesize and perturb desired deoxyribonucleic acid (DNA) sequences in a facile manner. In their landmark report, Khorana and co-workers cited the use of solid support nucleotide synthesis to prepare a 77 base-pair gene, corresponding to yeast alanine transfer ribonucleic acid (tRNA) [17]. The synthetic scheme

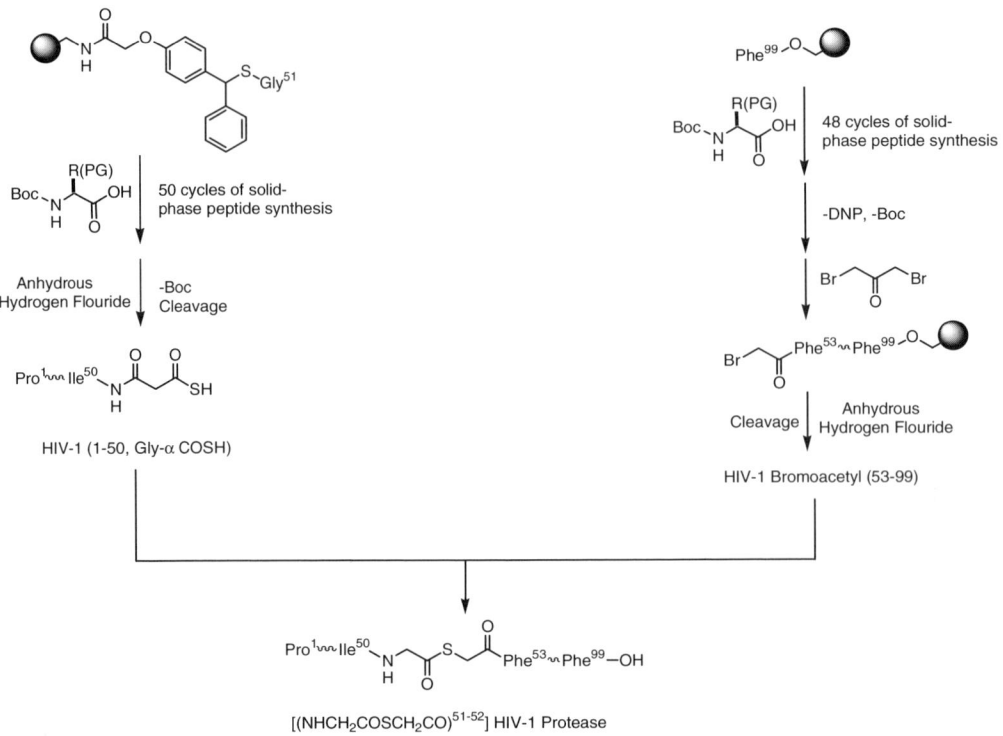

Scheme 21.5 Solid-phase synthesis of backbone-engineered HIV-1 protease.

called for solid support chemical synthesis of 15 oligonucleotide segments, containing 3′ and 5′ hydroxyl groups terminating each end, respectively. Phosphorylation of the 5′ hydroxyl group was then performed with the use of T_4 polynucleotide kinase, followed by annealing and ligase-induced head-to-tail joining of the appropriate segments, as a function of bihelical complexation. Khorana's group followed this work with the synthesis of a biologically functional *Escherichia coli* tyrosine tRNA suppressor gene, consisting of 202 base pairs [18]. Since these reports, numerous other groups have continued research pertaining to the synthesis of biologically functional genes including human insulin A and B chains [19] and the human leudocyte α_1-interferon gene, containing 514 base pairs [20].

In parallel with the development of solid-phase methods for biopolymer synthesis, a small but dedicated cadre of researchers pursued the solid-phase synthesis of more traditionally "organic" molecules. Because of space limitations, we will not discuss these developments in detail. Reviews of this work are recommended, however, since they set the stage for the explosion in solid-phase organic synthetic methods that occurred later [21, 22].

21.2 TYPES OF LIBRARY SYNTHESIS

Early Examples of Parallel Synthesis

Before the advent of combinatorial chemistry, there existed no means to rapidly generate a wide variety of small molecules with the potential to bind molecular targets. The bulk of druglike molecules were natural products or synthetic analogs, which typically required laborious efforts to synthesize or derive from undifferentiated mixtures. In the late 1970s, the advances in solid-phase synthesis led to the idea that one might be able to make large numbers of compounds simultaneously. If one could also then speed up the screening process, this might provide a completely new paradigm for compound production and analysis. One of the first examples of this new approach came in the early 1980s, when Richard Houghten used his "tea bag" method to make over 500 peptides in a matter of days [23]. Houghton's mesh pouch, which resembled a tea bag (Fig. 21.1), was dipped in solutions of amino acid and coupling reagents. The pouches were transferred from one solution to the next until the desired peptides were made. At roughly the same time, Ronald Frank and his colleagues were working on developing a method to combinatorially synthesize oligonucleotides on cellulose disks, by a technique that they called the filter disk method [24]. After initial functionalization of the cellulose, nucleotides were sequentially added by dipping the disks in reaction vials con-

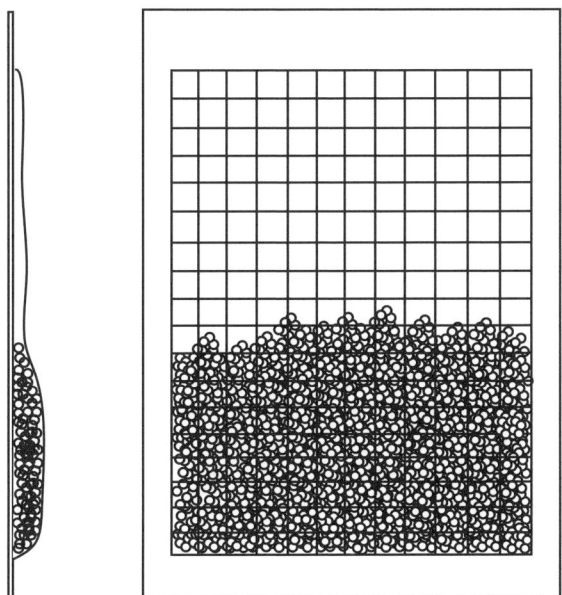

Figure 21.1 Houghten "tea bag".

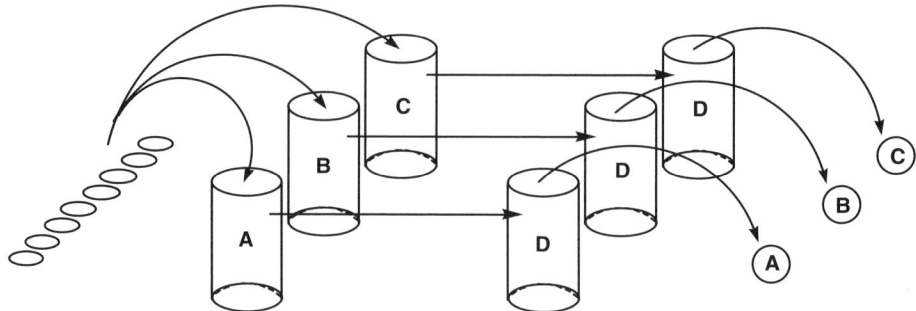

Figure 21.2 Filter disk method. Individual cellulose disks are separately treated with reagents (A, B, C) in reaction vessels and may undergo subsequent reaction under identical conditions (D) or undergo further diversification.

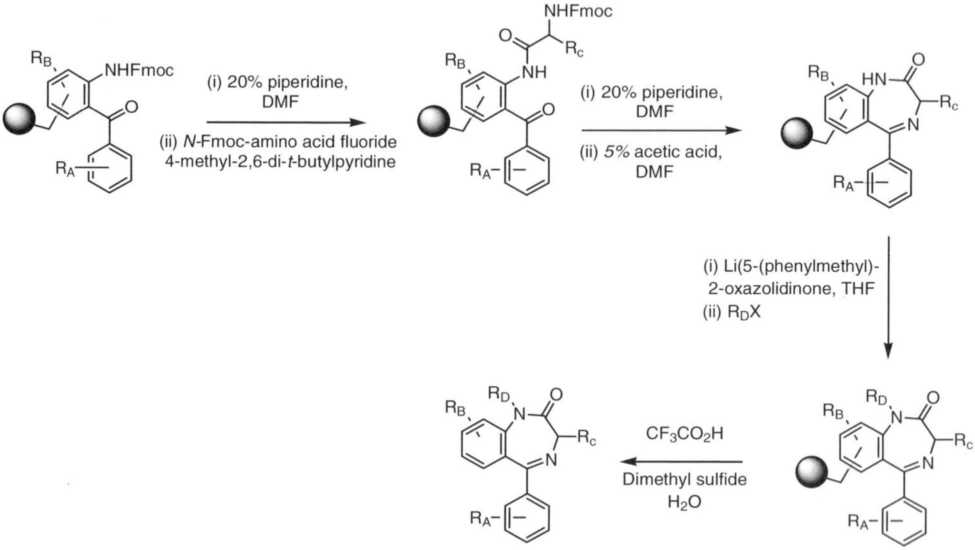

Scheme 21.6 1,4-Benzodiazepine synthesis [25].

taining the desired nucleotide (Fig. 21.2). Both of these methods fall into the category of parallel synthesis since mixing of reaction products was not possible. The solid support is also critical here: Although highly parallel solution-phase reactions have emerged as an important area of combinatorial synthesis, most early efforts (and a significant fraction of the current art) relied on such solid-supported synthesis. Parallel synthesis of "druglike" small molecules was then a logical extension of these efforts. One relatively early report, by Bunin and Ellman, provided an indication of the potential power of this methodology via the solid-phase synthesis of a series of 1,4-benzodiazepines (Scheme 21.6) [25].

Split-Pool Synthesis

As one might imagine, the complexity and logistical overhead associated with parallel synthesis increases significantly as the size of the target library increases. For the production of very large (up to several million compounds) libraries, the split-pool approach (also known as "one bead, one compound") is frequently used. Split-pool synthesis, developed in parallel by the Furka and Lam laboratories, relies (as its name implies) on the ability to iteratively *split* resin into separate reaction vessels, perform a desired synthetic step, then *mix* the beads together again following completion of the reaction [26]. Split-pool is a "true combinatorial" technique in the mathematical sense of the word, in that exhaustive recombination of components takes place. As shown schematically in Figure 21.3 for the production of a simple 27-compound library via three synthetic steps, one starts with a solid support, usually polystyrene or polystyrene-graft-polyethylene glycol (PEG) resin beads, and splits them into a predetermined number of reaction vessels. The number of vessels corresponds to the number of different ways that the library will be diversified at the first synthetic step. After the first reaction, in which each vessel is treated with a different reagent, the pools are brought back together. The beads are then mixed as efficiently as possible, and split into equivalent groups of beads (or reaction vessels) again. The number of pools again corresponds to the number of different couplings to be done at that synthetic step. Each of these pools gets a unique residue coupled to it, and they are mixed back together into one large pool. This process is repeated until the synthesis of the library is complete. Each "split" step in the synthesis is often described as a *diversification step*.

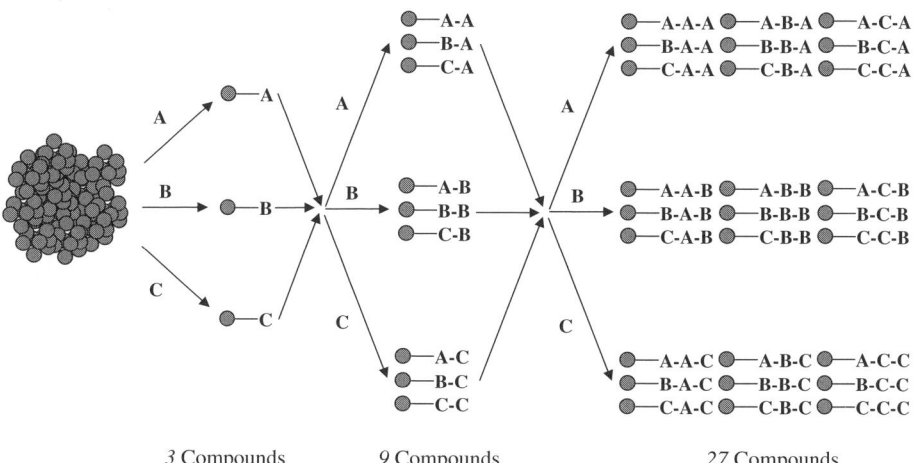

Figure 21.3 Split-pool synthesis of a 27-compound library on resin beads, via iterative reaction with reagents A, B, and C.

Split-pool synthesis is a powerful and efficient tool for creating arbitrarily large libraries quickly. Two important features of this synthetic strategy deserve mention. First, an attractive aspect of the split-pool approach is that each individual resin bead carries many copies of a single compound, and not a mixture of different compounds (assuming all reactions go to completion with no side products). This allows for on-bead screening to be used to determine if a ligand will bind to a given target, and circumvents the need for analyzing and purifying thousands (or millions) of compounds before screening. Second, split-pool methodology, with rare exceptions (see description of radio frequency tagging below), does not provide final compounds in a format that allows for purification prior to screening. Furthermore, following a split-pool scheme, each bead usually does not carry a readily discernible history with it, requiring a secondary method for identification of the compound it carries.

On-Bead Screening

Once one has synthesized a library on solid phase using either multiple parallel or split-mix methods, the next challenge is to identify the compound (or compounds) in the library that have the desired receptor-binding or enzyme-inhibiting activity. For libraries derived from multiple parallel synthesis, this typically involves cleavage of the compound from the bead and screening using one of several high-throughput techniques. For split-pool libraries, cleavage of individual compounds from millions of beads is generally not practical (although there have been some efforts in that direction; vide infra). In these cases, on-bead screening is generally employed. Shown schematically in Figure 21.4, on-bead screening involves treatment of the full library with a solution of a labeled target receptor. After an incubation period followed by rinsing away excess receptor, visualization according to the particular label used (e.g., fluorescence for a rhodamine-tagged receptor, or treatment with streptavidin-alkaline phosphatase for a biotin-tagged receptor) allows for those beads binding the target to be readily visualized. Of course, control experiments (analogous to the "blocking" step of a Western blot) are essential to provide confidence in the "hit".

21.3 WHO WON? IDENTIFYING HITS FROM SCREENS OF SPLIT-POOL LIBRARIES

Once a hit is identified from a split-pool library, how is the structure of the compound determined? As we have already described, individual beads generally do not carry their synthetic history with them. For oligonucleotide and polypeptide libraries, this is not a problem. Oligonucleotides can be identified by subjecting individual resin beads to polymerase chain reaction (PCR)

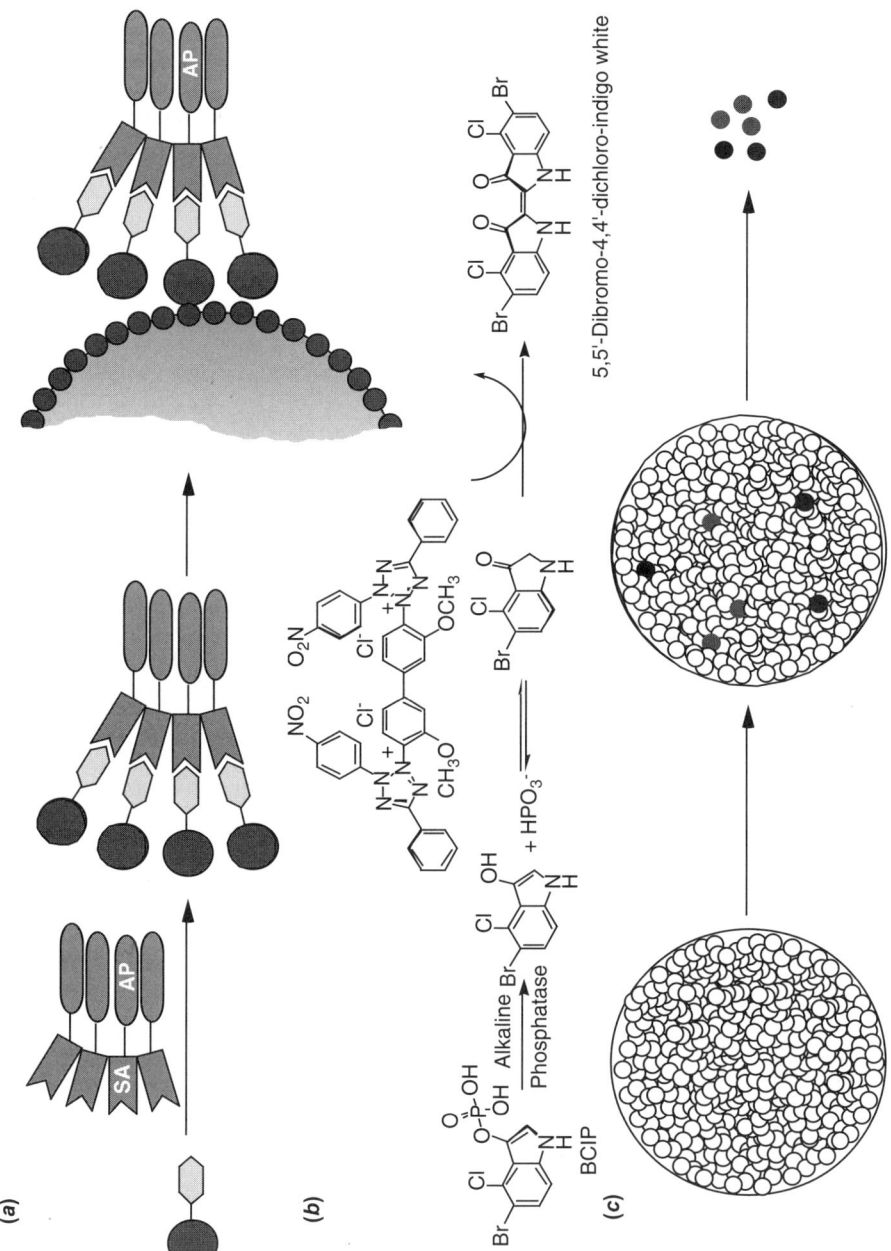

Figure 21.4 On-bead screening. (*a*) Treatment of a biotin-tagged protein with tetrameric streptavidin–alkaline phosphatase, followed by interaction with a resin bead. (*b*) Visualization reaction catalyzed by alkaline phosphatase, leading to precipitation of a purple dye on the target bead. (*c*) Staining of hits following deposition of alkaline phosphatase reaction product.

amplification, and often polypeptides can be identified by standard sequencing methods (i.e., Edman degradation) carried out on single beads [27] or via ladder synthesis coupled with mass spectrometry [28]. In rare instances, direct identification of "active" library members from single beads by mass spectrometry is possible, particularly if some reagents incorporate specific isotopes designed to enhance their identification [29, 30]. However, most often it is necessary to "deconvolute" large libraries via an iterative procedure or incorporate a readily identifiable surrogate "tag" for each reagent into every step of the synthesis.

Recursive Deconvolution

One of the earliest methods for determining the identity of "active" compounds from mixtures of synthetic library members, termed recursive deconvolution, was first described by Sydney Brenner and co-workers in November 1994 [31]. Recursive deconvolution actually starts during library synthesis and goes hand in hand with split-pool synthesis. After doing the first split and coupling of the split-pool synthesis, an aliquot of each batch of resin is removed and set aside for later deconvolution. The nonaliquotted beads are carried on through the next diversification step of the split-pool synthesis. After the second coupling, aliquots from each batch are again taken aside and saved. This process is continued throughout the synthesis, up until the last varied position (also known as a diversification step). At the last varied position the resin is split into individual batches for coupling just as before. However, after this coupling has taken place, the resin in not pooled back together. The rest of the synthesis is carried out in individual batches. On-bead screening is conducted on the individual batches of resin corresponding to the residue at the last varied position, and the batch yielding the most hits is determined to have the best residue at the last varied position. That "best" residue is then coupled to the resin that was aliquotted out after the coupling of the residue just prior to the last varied residue. The synthesis of the molecules is completed in separate batches again. However, this time the batches correspond to the individual residues at the second to last varied position. Screening is again done on-bead, and the batch that yields the most hits is determined to be the one with the best residue at the second to last varied residue. This process continues until all of the varied positions are deconvoluted.

Although recursive deconvolution requires a significant investment of time and material, it still possesses several advantages over other identification methods. For instance, no chemical entities are coupled to the solid-phase resin other than those that make up the library members, eliminating the possibility for chemical tags to produce false positives or false negatives in screening. Also, recursive deconvolution does not require library analysis by high-performance liquid chromatography (HPLC), ultraviolet (UV), infrared (IR), or mass spectrometry.

Tagging Methods

If one does not wish to use a deconvolution strategy, an alternative is to add an additional, readily detectable reagent at each step of the synthesis that tags each reagent used [32]. Initial examples of this strategy included tagging peptide libraries with oligonucleotides [33, 34]. After selection of a bead as "active," PCR amplification of the oligonucleotide on the bead provided the code corresponding to the peptide. However, for a variety of reasons (including reagent cost and compatibility), other tagging methods were sought.

Binary Encoding

One of the most innovative tagging protocols was first described by Still and co-workers in 1993 [35, 36]. In their initial report, a series of photolabile, gas chromatograph detectable, chemical tags was employed to deduce amino acid sequences in a 117,649-member peptide-based library. Still's method operates on a binary basis where, for example, three tags can be used to specifically designate one of seven peptide building blocks. For instance, a binary library readout of 101 would indicate tag 1 was present at position 1, tag 2 was not present at position 2, and tag 3 was present at position 3. Three binary positions, therefore, can account for seven unique library members (001, 010, 011, 100, 101, 110, 111). In their initial report, Still and co-workers used 18 tags (3 × 6) to identify 117,649 individual library members [35, 36]. To do this, at each step in the synthesis a series of three unique tags was coupled to less than 1 percent of the total attachment sites on each bead of resin. The residue encoded by those three tags was then coupled to the additional 99+ percent of the resin's functionalized surface. This was repeated until the synthesis was complete, at which point the library was screened on bead. Once a colorimetric readout had been observed, the bead of interest was removed from the rest of the beads. It was then subjected to ultraviolet light, which cleaved the tags from the resin. The tags, each of which has a unique retention time on a gas chromatograph, were then analyzed, and the sequence of the interacting peptide deduced [36].

Figure 21.5 shows an example of encoding 31 reagents at a diversification step with 6 tags. The chromatographic output of "011010" in this case would indicate that reagent 24 had been incorporated into the compound undergoing analysis on the selected bead. Importantly, this methodology is applicable to libraries of essentially *any* type of compound, as long as the chemistry employed does not modify the tags themselves. Subsequent publication of a set of tags able to undergo carbenoid insertion into the resin bead itself obviated the need for resin beads (and subsequent synthetic steps) to carry amino groups [37].

Several other tagging strategies have been described, including cleavable HPLC-separable tags [38], tags that provide specific IR signatures [39], and even simple tagging schemes based on one-dimensional ordering of beads on

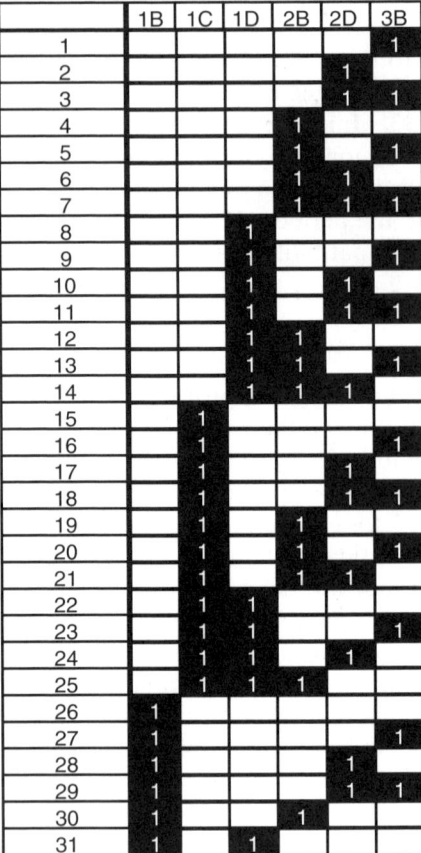

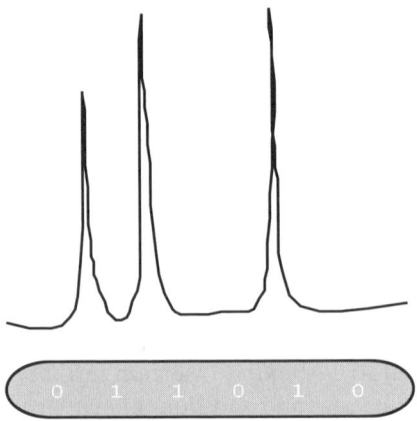

Figure 21.5 Binary encoding.

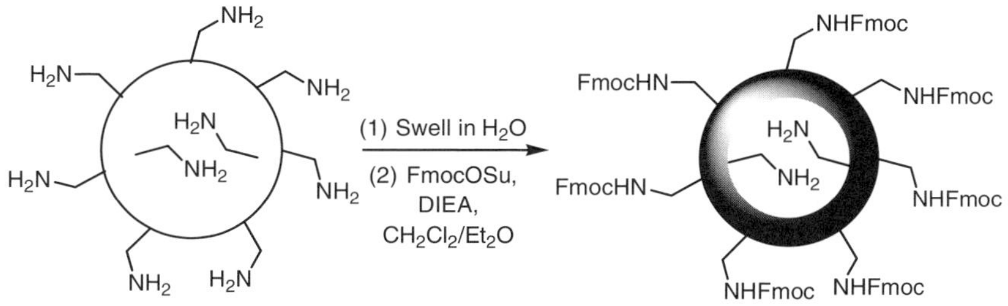

Scheme 21.7 Topological segregation of reaction functionality.

a string [40, 41]. A particularly interesting recent variation of tagging has been reported by the Lam group [42]. In this "biphasic" approach (Scheme 21.7), amino terminated PEG–polystyrene graft polymer beads are first swelled in water, then treated with a solution of 9-fluorenylmethoyloxycarbonyl-N-hydroxysuccinimide (FmocOSu) and base in ether/dichloromethane. Because the ether/dichloromethane mixture is unable to penetrate into the water-saturated interior of the bead, Fmoc protection of only the outer layer of beads occurs. *Topological segregation* of reactive groups on the bead subsequently allows for selective tagging of the interior of the bead, while library members are displayed on the external surface. In principle, this strategy has the advantage of eliminating interference (and potential "false positives") by the tag in on-bead screening.

Infrared Coded Resins

An exceptionally innovative strategy for library encoding is the use of the resin bead itself as a means of increasing deconvolution efficiency [43, 44]. Fenniri and co-workers [43, 44] have developed a method to slightly alter the copolymer composition of the Merrifield resin to provide unique bands in the IR and Raman spectrum. These chemically inert resin modifiers do not affect the reactivity of the resin, nor do they have an appreciable affect on resin swelling properties. Also, no library members need to be altered to incorporate tags for deconvolution, as in other earlier methods. All of the resin's surface is therefore available for library synthesis, while providing a spectroscopic readout of which type of resin bead on which the library has been built. This readout, which Fenniri has termed a *barcode* (the similarity to the barcodes on consumer goods is immediately apparent from Fig. 21.6), can be used to effectively "tag" the first position in a library synthesis and has been speculated to be able to be adapted for complete library deconvolution [43]. Considerable effort has been invested by the Fenniri group in demonstrating that the barcode resins have similar reactivity profiles to standard commercial resins, and that the presence of the synthetic ligand on the bead does not interfere with reading the barcode [44].

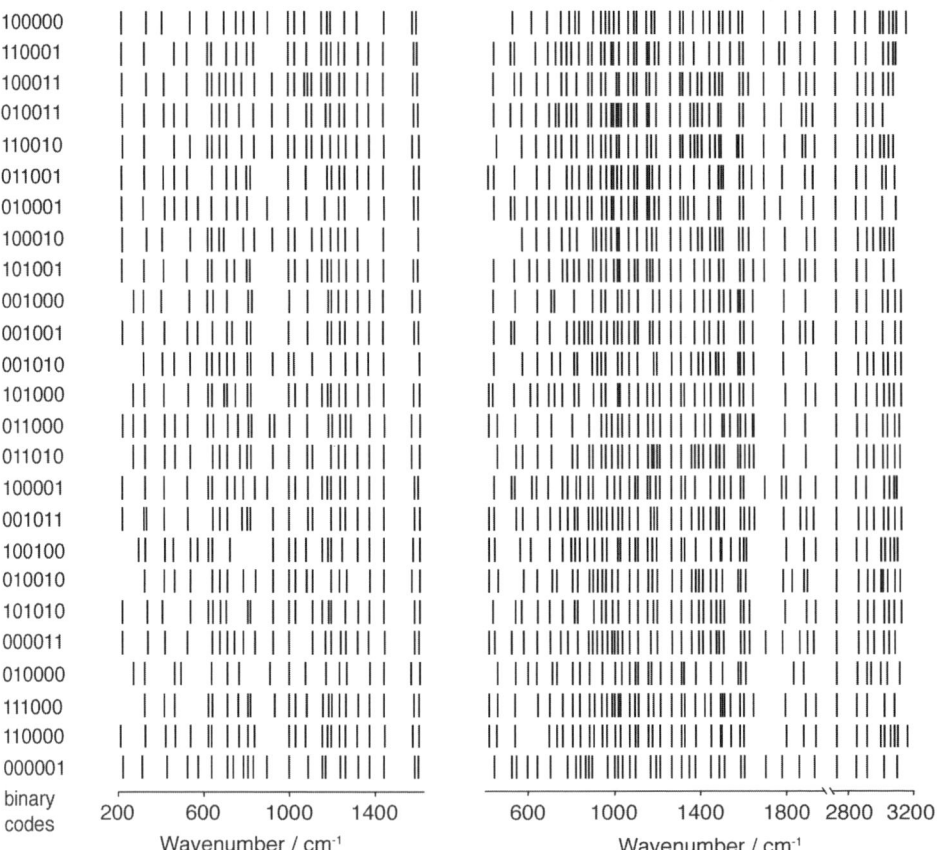

Figure 21.6 IR absorption bands used to generate a resin "barcode" [43].

Radio-Frequency Tagging and the Irori Corporation

In 1995, the Armstrong and Nicolaou groups simultaneously published an elegant solution to the problem of combinatorial library tagging [45, 46]. Much like laboratory animals and consumer goods are now tracked, both groups employed radio-frequency (RF) encoding to track the synthesis history of groups of beads. Effectively, a solution-permeable container (typically an inert material such as Teflon) of beads containing an RF chip (Fig. 21.7) is prepared. At each step in the synthesis, the IR code of each container is recorded, allowing for its entire synthetic history to be recovered following library screening. Importantly, the process can be automated, and using highly miniaturized systems even applied to the synthesis of moderately large split-pool libraries.

Further miniaturization (and greater library diversity) has been demonstrated by the Irori Corporation, using optical scanning of resin containers labeled with a black and white indexing grid. As a demonstration of this tech-

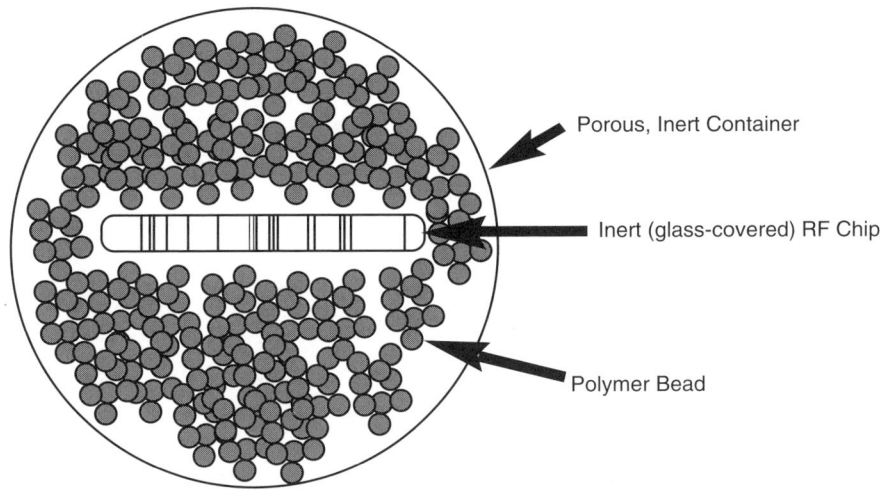

Figure 21.7 Radio-frequency encoding.

nology, designated the "NanoKan" system, Nicolaou and co-workers synthesized a 10,000-member library of benzopyrans in NanoKans using split-pool synthesis (Scheme 21.8) [47]. Given that NanoKans are large relative to the size of individual resin beads, the synthesis of very large (>10,000-compound) libraries is probably not practical using this system. However, the highly automated nature of the encoding process, ability to produce relatively large (milligram) quantities of compounds, and rapid decoding attributes make this an attractive solution for those laboratories able to afford the material and equipment costs.

21.4 OTHER COMBINATORIAL SYNTHETIC TECHNIQUES

Photolithographic Synthetic Methods

Like its evolution out of early efforts in the solid-phase synthesis of biopolymers, combinatorial chemistry has likewise coevolved with high-throughput screening. As previously mentioned, combinatorial libraries are often screened as single compounds arrayed in 96-, 384-, or 1024-well plates, with even higher densities on the horizon. The increasing miniaturization is in part driven by the relatively small amount of compound available from a single resin bead, but also out of a desire to limit the amount of the target biomolecule required for any single assay. One area of research that has revolutionized genetic analysis, and that is beginning to make its way into combinatorial chemistry, is the idea of the microarray. In terms of the density of different spatially addressable compounds that can be produced per unit area, the champion methodology here is undoubtedly the photolithographic process first

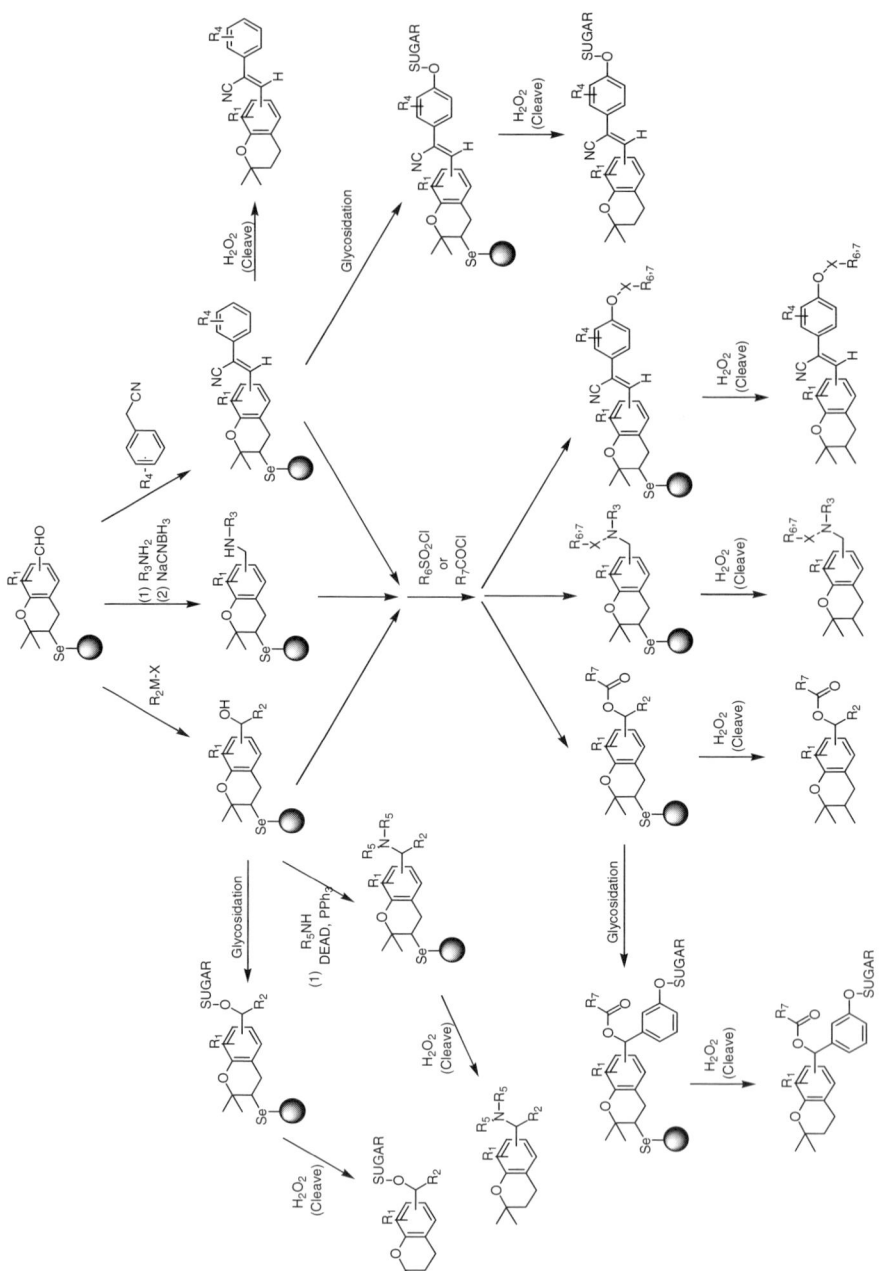

Scheme 21.8 Nicolaou benzopyran library.

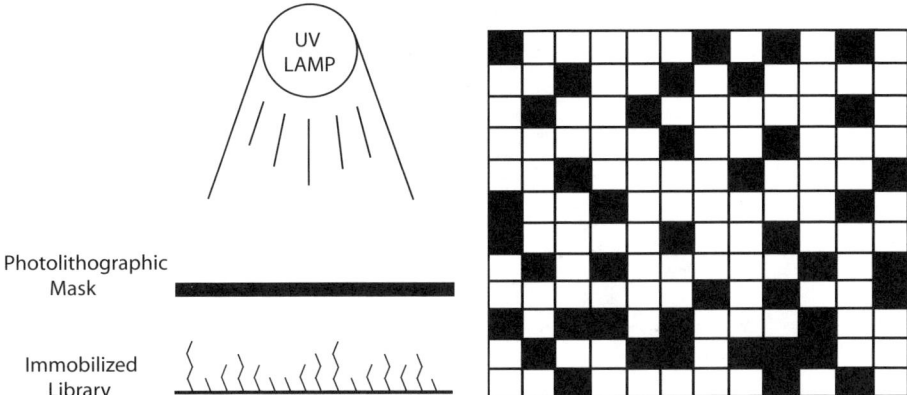

Figure 21.8 Light-directed synthesis (*left*) via a photolithographic mask (*right*).

described in 1991 by Fodor and co-workers [48]. Taking a page from the microelectronics industry, photolithographic synthesis of microarrays relies on the availability of photodeprotectable reagents: protected monomers that will become an array of oligomers (bio- or otherwise), or, potentially, protected reagents that can form an array of small molecules. Beginning with a substrate (glass or a thin gold film) derivatized with a photocleavable group, light is shined through a photolithographic mask to selectively deprotect particular regions of the substrate (Fig. 21.8). Following UV deprotection of an area of the chip, a library building block is coupled to all of the available, deprotected, sites. Changing the photolithographic mask subsequently allows for other areas of the chip to undergo deprotection. The process is repeated until all of the desired library members are generated. On chip screening of labeled targets can then easily be performed, and no deconvolution of binders is necessary because the patterns of the photolithographic masks used to create the immobilized library are recorded during synthesis. Initially developed primarily as a method for peptide synthesis, photolithographic synthesis of oligonucleotides as pursued by the genomics company Affymetrix has become a major industry. In principle, such a method could also be applied to the synthesis of druglike molecules; however, to our knowledge, this has yet to be demonstrated successfully. A recent report by Kodakek and co-workers describing the photolithographic synthesis of arrays of peptoids further illustrates the potential power of this method [49].

High-Throughput Methods

An alternative to the photolithographic technique is the pin-based spotting method developed by Khodursky and co-workers [50]. Although arrays produced using this method are not capable of yielding the spot density of Affymetrix chips, they can be produced using equipment that is significantly

Figure 21.9 Commercial microarrayer. The print head, capable of holding 384 pins, is just below the center of the photograph.

less complex and expensive. Still at the early stages of being applied to small-molecule screening, spotted arrays have already shown considerable promise in the rapid screening of libraries. The arrayer itself is effectively a three-axis robot equipped with a print head bearing precisely machined, steel quill pens (or "pins") (Fig. 21.9). Computer control of the arrayer allows for precise deposition of nanoliter-sized droplets on a glass slide, producing an array of potentially many thousands of spots.

One early demonstration of the utility of spotted small-molecule arrays was reported by Schreiber and Coworkers [51, 52]. Following synthesis of a split-pool library on *macrobeads* (large resin beads capable of carrying greater amounts of each individual library member), resin beads were individually deposited into 384-well plates. In-plate cleavage of library members was carried out, followed by evaporative removal of the solvent. The residue remaining was taken up in dimethyl sulfoxide (DMSO), and spotted onto glass slides using an arrayer analogous to that described by Yanofsky [50]. Subsequent screening using a fluorescence-tagged protein provided ready identification of those compounds on the array with the desired binding ability.

The use of macrobeads in combination with microtiter plate cleavage allows for the generation of stock solutions of the library members. Therefore, libraries can be screened many times, against many different ligands. However, as with all other split-pool strategies, some form of tagging is necessary to provide a positive identification of library hits. Use of small-molecule microarrays is clearly a technique that is on the rise in academia. Recent reports have included a description of small-molecule arrays targeting profiling caspase activation [53], small-molecule ligands for human IgG [54], and arrays of synthetic oligosaccharides [55].

21.5 TAG-FREE METHODOLOGY

Dynamic Combinatorial Chemistry and Allied Methods

In addition to the methods described above, combinatorial chemistry has spawned several research efforts designed to mimic nature's methods of deriving compounds with a desired property. In some cases, the goal is to mimic evolution: selecting and amplifying high-affinity binders from dynamic pools of chemical structures. In other cases the goal is to mimic biosynthesis: templating the construction of a molecule from a library of starting materials allowed to interact with a binding pocket. For some of these new techniques, receptor-templated synthesis *and* evolution are coupled.

Evolution is a central tenet of modern biology. The phenotype of an organism is, of course, a reflection of the constant process of mutation, selection, and amplification operating on a vast scale at the molecular (genetic) level (Fig. 21.10). In the mid-1990s, several research groups, including ours, began reporting proof-of-principle studies of a conceptually new direction in combinatorial chemistry. This field of dynamic combinatorial chemistry (DCC) seeks to mimic what nature so elegantly achieves. In a dynamic combinatorial library, molecular building blocks react in a reversible fashion, affording a continuous interconversion of all possible library members.

Dynamic combinatorial libraries (DCLs) are most often constructed using covalently linked functional groups formed under reversible conditions, such as esters, imines, hydrazones, borates, disulfides, and others [56]. Libraries constructed through hydrogen bonding networks, ionic bond formation, or coordination to a metal center have also been reported [57]. In the dynamic approach, an equilibrating system containing various numbers of reactive

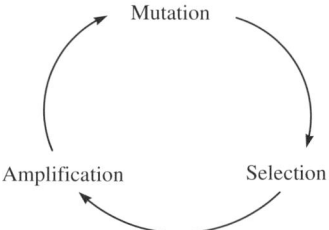

Figure 21.10 Darwinian evolution.

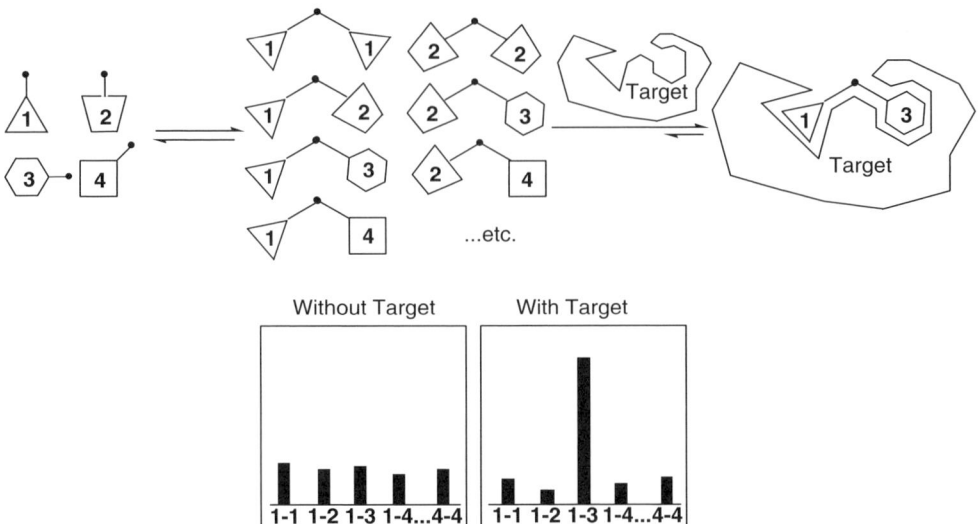

Figure 21.11 Target-directed selection and amplification of chosen library members from a dynamic equilibrium.

building blocks provides a broad pool of interconverting compounds. For example, a mixture containing (c) aldehydes and (d) amines theoretically generates ($c \times d$) imines. When exposed to a target receptor, library members undergo a selection based on binding affinity. Because formation of each library member from its constituent components is reversible (i.e., allowing "mutation" to occur), LeChatelier's principle requires that the equilibrium shift in favor of those compounds "sequestered" by binding to the target. In principle, this means that the highest affinity binder will be subject to the greatest amplification. Selected member(s) can then be monitored by enhancement of mole fraction via HPLC and/or mass spectrometry (MS) or by other suitable methods (Fig. 21.11).

Several research groups have published proof-of-concept illustrations for the selection and amplification of individual DCL species based on target affinity [58]. Eliseev and co-workers describe the selection and amplification of Z/Z di-carboxylate ligand upon addition of a guanidinium group, from a photochemically induced Z/Z, E/E, and E/Z equilibrium [59]. Orientation of the carboxylate moieties directly influences the ability to bind the target. As shown in Scheme 21.9, only the Z/Z compound exhibits a conformation that affords hydrogen bonding between both carboxylates to the arginine center. After several cycles of photochemical equilibrarion and affinity-chromatographic removal of the Z/Z isomer, HPLC analysis revealed a dramatic and selective enhancement in the mole fraction of the Z/Z isomer [59].

A fundamental concern when planning a DCL is ensuring the production of a library exhibiting unbiased representation of all library members. With

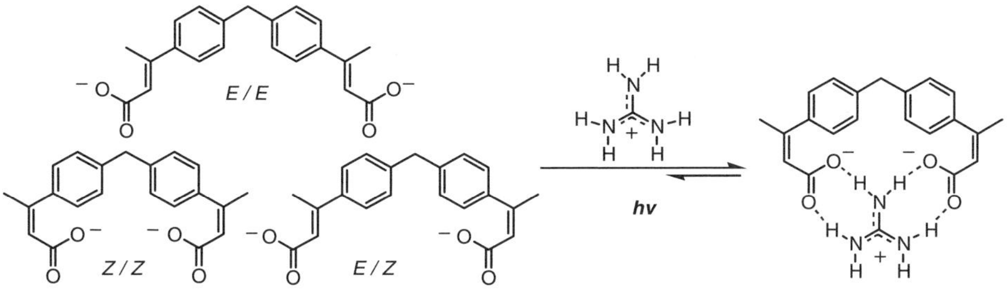

Scheme 21.9 Eliseev's selection and amplification of a DCL member based on guanidinium affinity.

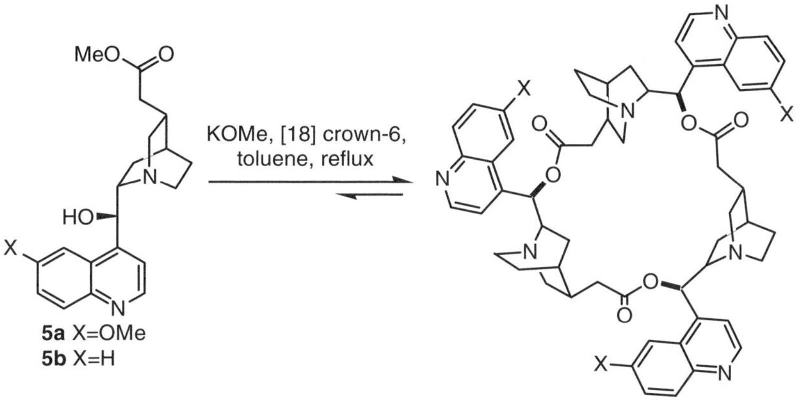

Scheme 21.10 Quantitative formation of the cyclic trimer of cinchonidine derived hydroxyl ester **5**.

library selection under thermodynamic control, intrinsic properties of starting materials and library products can affect the observed selection as well as binding to the target. Significant differences in reactivity between individual library components, or self-selection within those components, can result in a DCL containing a limited number of members constituting the majority of the molecular pool, while other theoretical library members are underrepresented or entirely absent.

In several studies, this self-selection property has been explored explicitly. Such an example has been published by Rowan et al. [60] in which a cinchonidine-derived hydroxyl ester (**5**) was reacted with potassium methoxide under thermodynamic conditions, which theoretically could generate numerous cyclic and polymeric compounds. Instead, only the cyclic trimer was obtained in quantitative yield (Scheme 21.10) [60]. From the perspective of lead discovery, this research demonstrates the importance of assuring a diverse representation of library members when planning a DCL. In general, researchers have had success in constructing DCLs using a variety of reactions that undergo rapid reversion, including transesterification.

In situ target-induced selection and amplification makes DCC an ideal technique for lead discovery of compounds that bind to biological macromolecules. Because of the in situ amplification technique, reversible reactions chosen must be amenable to physiological conditions, or a bi-phasic or compartmental approach must be used. For example, researchers have shown that thiols undergo facile disulfide exchange under mild conditions in aqueous media [61]; however, when using proteins as a target, component exchange with cysteine residues is a concern. In general, numerous other reversible reactions including imination, hydrazone formation, and olefin metathesis have been employed to prepare DCLs in aqueous media [56]. When using reactions not amenable to a physiological environment, researchers have cited the use of compartmentalization through a semipermeable membrane through which library members formed in an organic phase can pass, and subsequently encounter the chosen target in an aqueous phase [62].

Because of its potential for building up binding structures in a modular fashion, as well as the intended parallel with genetic evolution, DCC would seem to be an ideal method for the identification of new compounds targeting nucleic acids. We have described the selection and amplification of equilibrating metal coordinated salicylaldimine and salicylamide complexes in the development of DNA [63] and RNA [64] ligands, respectively. In part, these efforts were an expansion of several other groups' reports that *nonlabile* organotransition metal complexes could bind DNA and RNA with interesting results [65].

In our initial report, we described the first utilization of immobilized DNA to drive the selection and amplification of ligand(s) from a dynamic library of zinc(II) complexes. Six salicylaldimine "monomers" (**7** to **12**) were reacted with excess zinc(II) chloride, providing a library of 36 unique bis(salicylaldiminato)zinc complexes (**6**) (Fig. 21.12). Given the ability of zinc(II) to bind a broad range of functionality with variable coordination numbers and geometries, it is likely the total library diversity was much greater. Two independent libraries, one consisting of compounds **7** to **12** and the other containing compounds **7** to **12** and excess zinc(II) chloride were incubated in affinity columns constructed of ds(T·A) cellulose. Following a 2-h incubation of the libraries on resin, solutions were eluted and lyophilized. Analysis of the mixtures was performed by hydrolysis of the complexes with trifluoroacetic acid, followed by derivatization of the amines with excess 2-naphthoyl chloride (to allow for UV detection) and separation/structure elucidation by reverse-phase HPLC. Comparing relative abundance of monomers present after elution provided information about which components bound dsDNA, that is, less monomer present after elution represents binding of that monomer to the immobilized dsDNA.

After multiple analyses and control experiments, monomer **8** proved to be most strongly retained by the dsDNA affinity column in the presence of

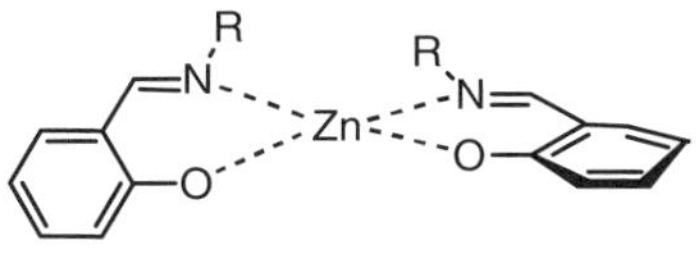

Figure 21.12 Salicylaldimine monomers (**7** to **12**) and hypothetical zinc(II)-mediated bis(salicylaldiminato) complex.

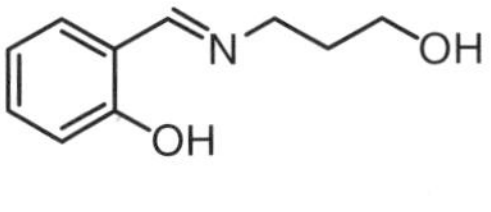

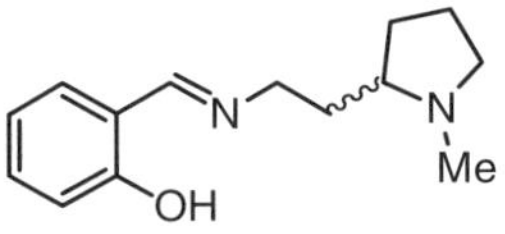

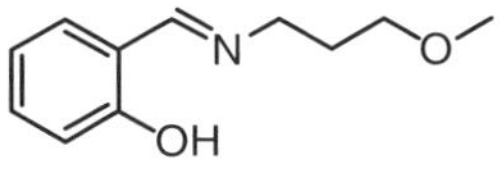

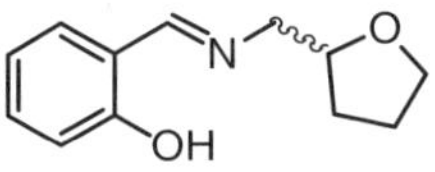

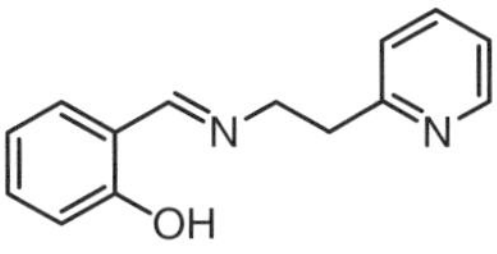

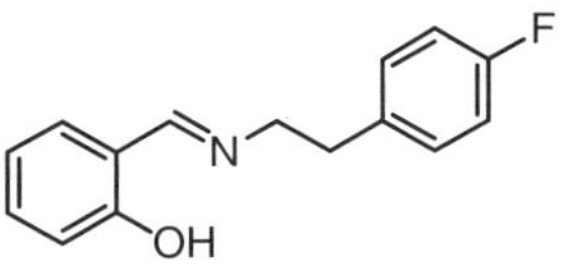

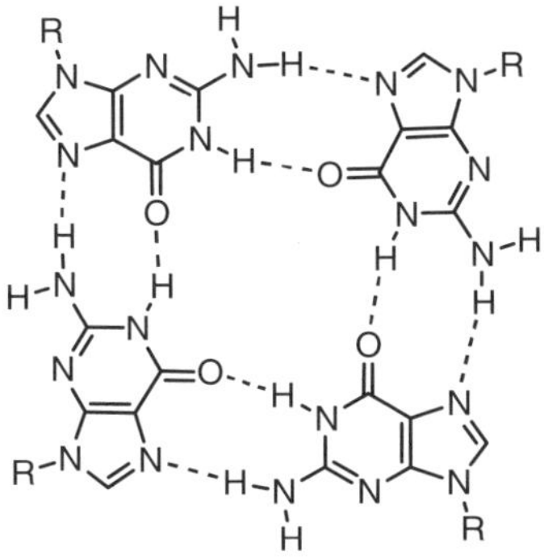

Figure 21.13 Hydrogen bonding network in the guanine tetrol motif.

zinc(II). Subsequent UV titration experiments provided a dissociation constant of 1.1 µM, significantly stronger than other possible bis(salicylaldiminato)zinc complexes tested.

Balasubramanian et al. recently reported using a dynamic combinatorial strategy in an effort to discover ligands capable of selectively binding guanine quadruplex DNA [66]. The tertiary structure of guanine quadruplex structure has shown interesting activity involving the regulation of telomere length and exhibits various motifs that differentiate it from B-DNA. For example, the terminal guanine tetrad provides potential for hydrophobic or π-π ligand association (Fig. 21.13). In previous work, this group demonstrated that the tetrapeptide sequence F-R-H-R exhibits selective binding to the quadruplex motif [67].

Based on this study, Balasubramanian and co-workers [66] subsequently used disulfide exchange to construct a dynamic library using a thiol-terminated F-R-H-R tetrapeptide **P** and a thiol-containing acridone derivative **A**, proposed to associate with the hydrophobic guanidine tetrol. Equilibrium was established in the presence of glutathione to mediate disulfide exchange between (**A**) and (**P**), as well as diversify the library. Upon equilibration, a biotinylated sequence 5'-biotin(GTTAGG)$_5$ was incubated in the dynamic library. After target-induced reequilibration, disulfide exchange was halted by acidification of the medium, followed by removal of the ligand–target complex through immobilization on streptavidin-coated beads. Subsequent heat-induced denaturation allowed for the release and characterization of selected ligands (Scheme 21.11). From the nine possible library members, introduction of the guanine quadruplex increased the molar concentration of the hetero complex **A–P** and the peptide dimer **P–P** 400 and 500 percent, respectively. This dynamic approach yielded the first report of a short peptide sequence that displayed selective binding to guanine quadruplex DNA.

Lehn and co-workers [68] have demonstrated that a DCL approach can be used to probe the binding site of the plant lectin Conconavalin A (Con A). Lehn et al. employed acylhydrazone formation through the intermolecular condensation of various hydrazines and aldehydes, reversible in a pH range of 3 to 7. Six monosaccharide benzaldehydes and nine core component hydrazines were used, offering one, two, or three hydrazine reactive sites, thereby producing both monovalent and multivalent DCL members (Scheme 21.12).

Complete equilibration of the 15 library components afforded a dynamic library containing 474 members. Initial screening was performed using an enzyme-linked lectin assay (ELLA) [69]. In the ELLA assay, microtiter plates were coated with yeast mannan (a known substrate of Con A). Peroxidase-

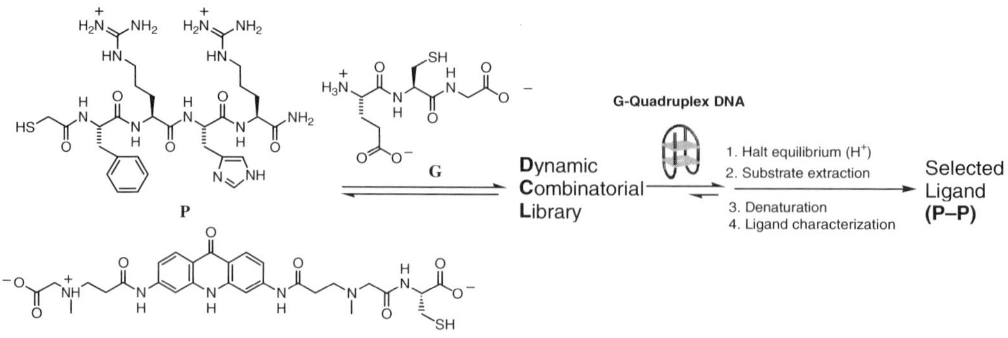

Scheme 21.11 Balasubramanian's dynamic approach toward G-quadruplex DNA ligands.

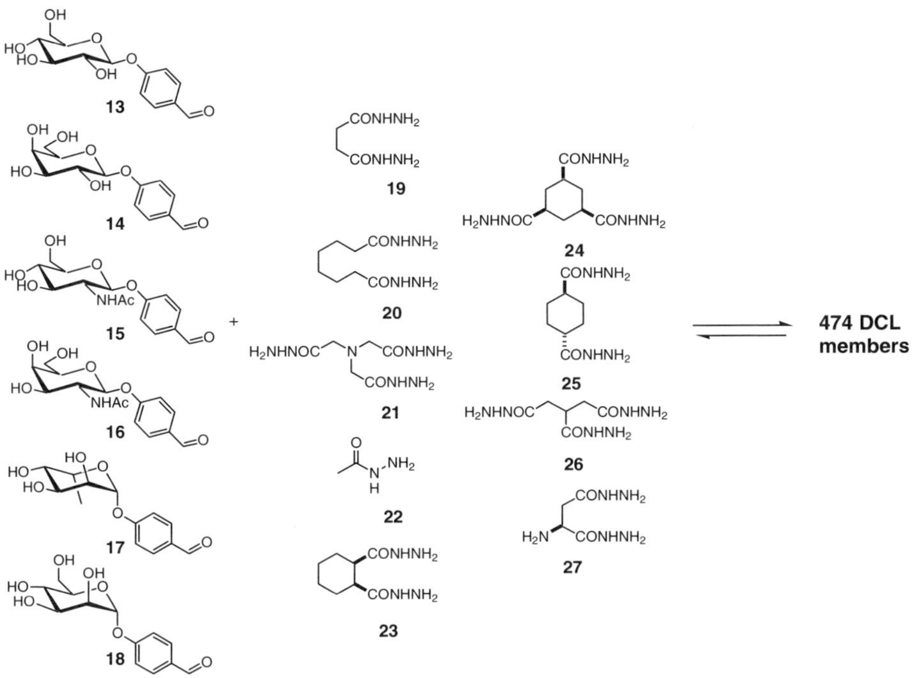

Scheme 21.12 Dynamic approach toward Concavalin A ligand [68].

labeled Con A was then introduced in the presence of the library. Binding of Con A by library members was then monitored as a function inhibition activity of the enzyme. Initial results indicated more than one inhibitor of enzyme activity, that is, various members displayed similar binding coefficients. Using a dynamic deconvolution method, 15 sublibraries were made, each omitting one component, thereby excluding any library members consisting of that component [70]. As a result, sublibraries that showed a decrease in inhibitory effect illustrate the importance of that missing component, and therefore any library members containing them. After full analysis, Lehn and co-workers [68] found that the mannose carbohydrate **18** was the most active saccharide, and the two trivalent hydrazines **21** and **26** were the most active linkers. From libraries containing only those constituents, the trimannose compound (**18**)$_3$-**21** was found to be the best inhibitor. Comparison to methyl-α-D-mannoside yielded an estimated IC$_{50}$ value of 22 μm, a 36-fold increase over the monosaccharide.

As evidenced by the above example, larger DCLs can produce various library members that have similar receptor binding affinity, making it difficult to quantify amplification of a single library member. Kazlauskus et al. have reported the use of irreversible destruction of unbound or weakly bound ligands, thereby making analysis of selected ligands in a large molecular pool an easier task [71]. In their study, numerous aryl sulfonamides were prepared

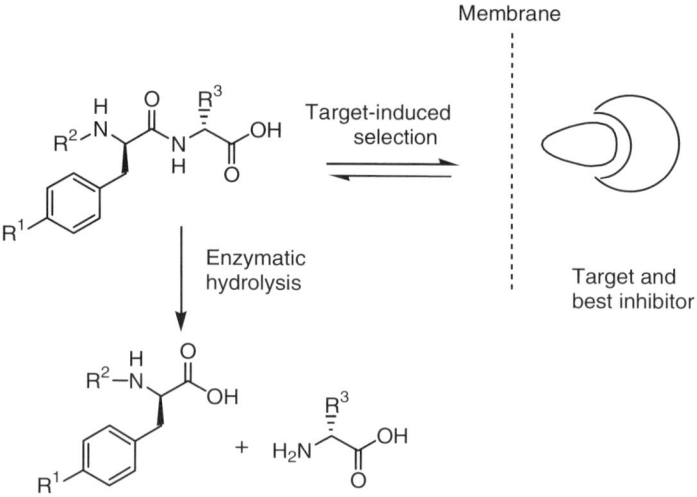

Scheme 21.13 Kazlauskus' destruction/amplification approach.

and screened against carbonic anhydrase. After subjection to the target, the protease *pronase* was added, which enzymatically hydrolysized unbound or weakly bound ligands, while maintaining the integrity of ligands with high target affinity, allowing for subsequent analysis (Scheme 21.13).

Recent reports by Severin [72] and by Sanders and co-workers [73] place the idea of target-mediated selection and amplification of high-affinity binders from dynamic combinatorial libraries on firmer mathematical footing. In essence, both researchers note that the library member amplified to the greatest extent is not necessarily that with the highest affinity to the target, if the DCL experiment is not designed properly. This potential difficulty decreases proportionately as the size of the DCL increases.

While dynamic combinatorial chemistry has yet to produce a major success (in terms of a clinical lead compound) in the same sense that other techniques have, several groups have provided proof of concept experiments indicating that substrate-induced selection can afford detectable amplification of "active" molecules within a library. Recent reports by an industrial group led by Eliseev targeting neuraminidase have demonstrated that DCC does have the potential to be more than just an academic endeavor [74]. We anticipate that the dynamic combinatorial approach will prove to be a novel resource for lead discovery and will continue to provide fundamental insight on aspects of molecular recognition.

Click Chemistry and in Situ Click Chemistry

As we have already described, the solid-phase approach to combinatorial synthesis has survived and prospered because of two distinct features. First, solid-phase synthesis allows for the use of excess reagents, driving otherwise moderate reactions toward completion. Second, solid-phase reactions allow for product purification through filtration, bypassing more complex chromatographic procedures. However, despite these attributes, solid-phase chemistry can suffer severe limitations as a tool for the pharmaceutical industry. For example, driving reactivity with the use of excess reagents in the mobile phase becomes wasteful with respect to both reagents and solvent. Second, the use of a traceless linker [75] is often required when preparing drug candidates on solid support, complicating synthetic design. Third, the ability to gain diversity relies on consistent reactivity in the presence of various functionality. Finally, very large scale solid-phase synthesis is prohibitively cumbersome and expensive. Dynamic combinatorial chemistry is one response to these issues; other researchers have recognized that solution-phase multiparallel synthesis using near perfect reactions represents an attractive alternative. One version of this idea has been explored in detail by Sharpless and co-workers, who use the term *click chemistry* to describe the design of libraries relying exclusively on reactions that are exceptionally high yielding and functionality tolerant [76]. *Click reactions* result from a high thermodynamic driving force ($\Delta G \leq -20\,\text{kcal/mol}$) and thereby afford selective reactivity in the presence of various moieties.

In a typical click chemistry reaction sequence, building blocks are pieced together using well-studied reactions on readily available starting materials. For example, olefins can be extensively decorated through the use of cycloadditions such as 1,3-dipolar and the Diels–Alder reaction, nucleophilic opening of strained rings, dihydroxylation, and aziridination, while carbonyl moieties can be modified via hydrazone formation or other "irreversible" nucleophilic addition. While an argument can be made that click chemistry is limited by its definition (the use of only a few perfect reactions), it has been shown that great diversity can be achieved when using only a few reaction types [77]. As an example, the generation of *spring-loaded* cyclic electrophiles such as epoxides or aziridines, and subsequent nucleophilic ring-opening reactions from olefinic starting material offers a diverse library of molecules (Scheme 21.14). Other possibilities, including olefin heterocyclization and subsequent nucleophilic ring-opening reactions, are dependable, stereospecific, often highly regioselective, can be performed either in water or in the absence of solvent, and provide the product in nearly quantitative yield.

Because of these attractive aspects, click chemistry is increasingly being used for drug discovery and biomedical research applications, in both academic and industrial laboratories [78]. For example, Fokin, Sharpless, and Wong have reported the use of click chemistry in the formation of an inhibitor of α-1,3-fucosyltransferase (Fuc-T), an enzyme that catalyzes the final glycosylation in the biosynthesis of many important saccharides [79]. The key step in this transformation is a manganese(II)-catalyzed nucleophilic displacement of L-fucose from guanosine diphosphate β-L-fucose (GDP-fucose) from a glycoconjugate acceptor (Fig. 21.14).

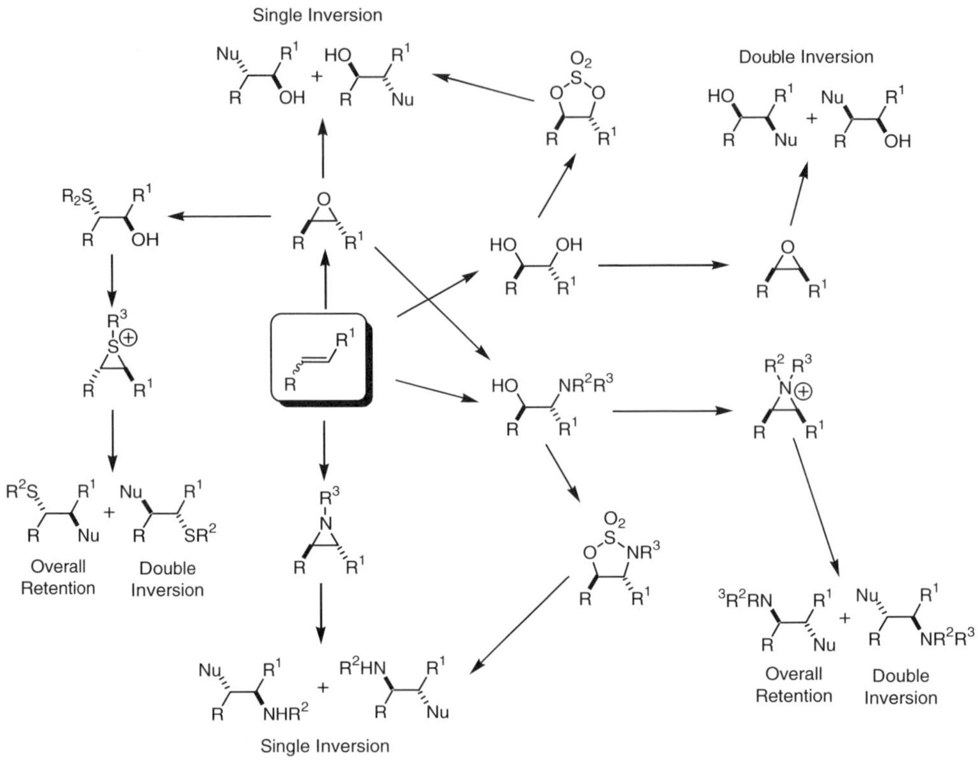

Scheme 21.14 Generation of diverse components through the formation of cyclic electrophiles from olefins and their oxidation products.

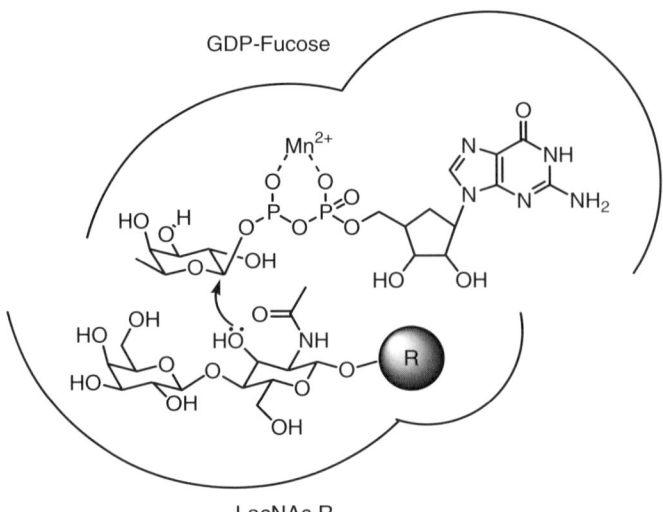

Figure 21.14 Fucosyltransferase-catalyzed transfer of β-lucose to acceptor glycoconjugate.

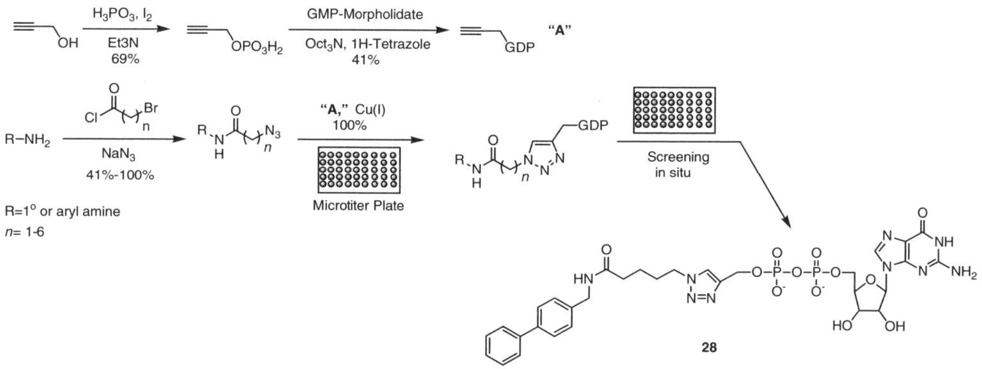

Scheme 21.15 Triazole synthesis in microtiter plate and subsequent in situ fucosyltransferase inhibitor detection.

The majority of the binding affinity for Fuc-T lies in the GDP moiety; however, it has been demonstrated that interaction of aromatic or aliphatic functionality in the neighboring hydrophobic pocket can increase affinity 70-fold. Therefore, a library of aromatic and aliphatic GDP-triazole candidates were prepared and screened for inhibition. The triazole formation is accomplished using Huisgen's Cu(I)-catalyzed 1,3-dipolar cycloaddition between an alkyne and azide [80]. This reaction was chosen because it proceeds in water, produces near quantitative yield, and is effective in the presence of vast functionality, granting a "click" seal of approval. Thus, 85 azide compounds were synthesized utilizing various hydrophobic groups and tether lengths of 2 to 6 carbons (Scheme 21.15).

After completion of triazole formation, compounds were screened for inhibitory activity in situ with the use of a pyruvate kinase/lactate dehydrogenase-coupled enzyme assay. The library was screened against human α-1,3-fucosyltransferase VI (Fuc-T VI), where compound **28** was found to show the highest binding affinity. Steady-state kinetic studies show compound **28** is a competitive inhibitor against GDP-fucose, with a $K_{i(\text{comp})} = 62$ nM, making this compound the first nanomolar inhibitor of Fuc-Ts. Compound **28** was then tested for inhibitory activity against other glycosytransferases and nucleotide binding enzymes. Fokin, Sharpless, and Wong [79] concluded that compound **28** is highly selective toward Fuc-T VI, showing weak inhibition of α-Fuc-T III and α-Fuc-T VI, and no inhibition of galactosyltransferases α-1,3-GalT and β-1,4-GalT. Weak inhibition was observed against guanylate kinase, and no inhibition was detected against pyruvate kinase. The above scenario is a prime example of the use of a click reaction to create moderately sized libraries in near quantitative yield, from which screens can be conducted without the need for purification of compounds.

The initial discovery by Mock and co-workers of a dramatic enhancement in the reaction rate of the Huisgen 1,3-dipolar cycloaddition of azides and alkynes, a relatively slow reaction under "normal" conditions, displayed for the

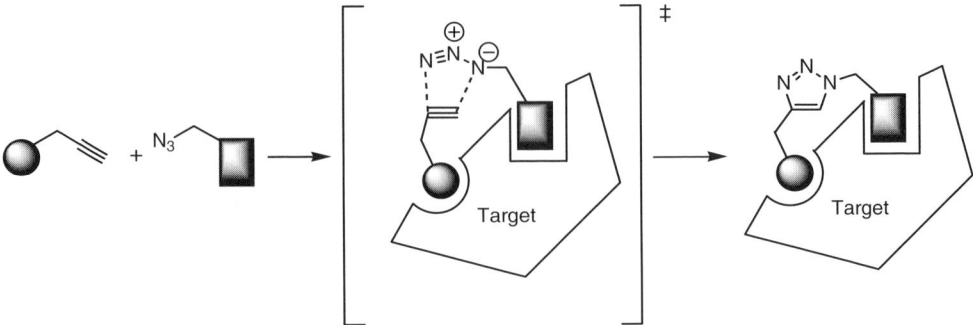

Scheme 21.16 In situ target-catalyzed ligand formation using Cu(I)-catalyzed triazole formation.

first time the use of biomacromolecules as reaction vessels [81]. Rate acceleration is presumably brought about through the binding of reaction components to a biomacromolecule, thereby providing a unique microenvironment (Scheme 21.16). This finding has prompted researchers to investigate an in situ approach to ligand synthesis. Click reactions, benign to physiological conditions, afford the ability to create small-molecule ligands in the presence of their biological substrate, without the potential for side reactions with biomolecular functionality such as amino acid side chains.

Sharpless and co-workers [82] have reported the use of triazole formation, a favored click reaction, as a method for irreversible bond formation in the presence of a desired target. In their approach, Sharpless et al. targeted acetylcholine esterase (AChE), an essential component in neurotransmitter function in the central and peripheral nervous systems. The enzyme displays two neighboring binding sites, located within a narrow gorge approximately 20 Å in depth [83]. Both binding sites have known ligands exhibiting tacrine and phenanthridinium motifs [84]. Using these known ligand motifs of differing linker lengths, decorated with alkynyl and azide functionality, triazole formation was allowed to proceed in the presence of AChE. Because the rate of triazole formation at room temperature in the absence of enzyme is negligible, any detected bi-dentate triazole formed is done so solely from enzyme-assisted rate acceleration. After sufficient incubation in the presence of *Electrophorus* AChE, the mixture was analyzed via desorption–ionization mass spectrometry (DIOS) [85]. Of the 98 potential bivalent ligands, including 34 regioisomeric pairs of *syn* and *anti* tacrine/phenanthridinium triazoles, and 15 regioisomeric pairs of tacrine/tacrine triazoles, an observable amount of only one bidentate ligand was found, TZ2-PA6 (Scheme 21.17).

Preparation of the selected triazole in the absence of *Electrophorus* AChE provided authentic material in high yield in approximateley 1:1 *syn*:*anti* ratio. Regioisomers were separated by HPLC and individually analysized. HPLC comparison shows that the enzyme exclusively produces the *syn* triazole

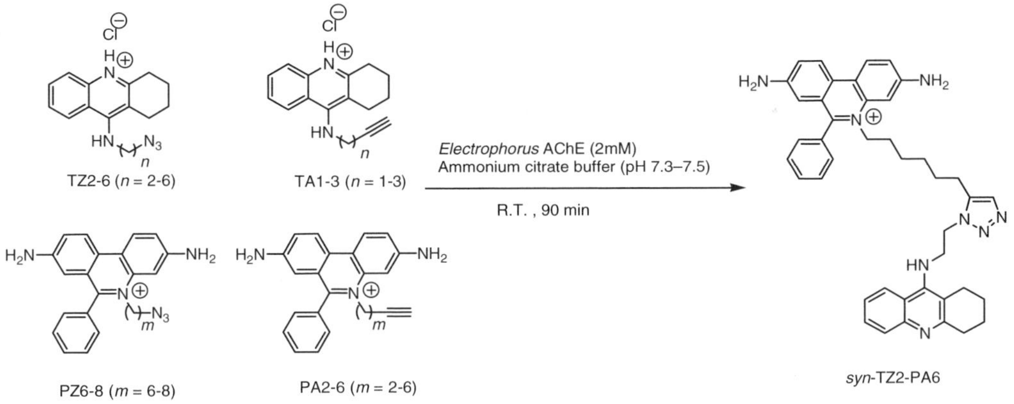

Scheme 21.17 Azide (Z), acetylene (A), tacrine (T), and phenanthridinium (P). Substrate-induced selection of AChE.

isomer. It was later found that the *syn* isomer has a K_d of 77 to 410 fM, 140 times stronger than the *anti* isomer.

Click chemistry continues to provide researchers facile ways to create diverse libraries from which biological systems can be perturbed and information regarding molecular recognition can be gained. Conceptually, click chemistry also provides motivation for the discovery of new high-yielding, selective, and functionality-tolerant synthetic methodology.

Thiol-Based in Situ Assembly of Macromolecular Ligands

The above sections outlined studies in which biological macromolecules were used as reaction vessels, inducing rate acceleration in the formation of high-affinity ligands, or amplification of high-affinity ligands from dynamic libraries. In a related effort, Sunesis Pharmaceuticals has recently described a "tethering" technique that uses a native or engineered cysteine thiol present in close proximity to a macromolecule's active site(s) to covalently retain a relatively low-affinity ligand, from which a multidentate high-affity ligand can be synthesized in situ [86]. After subsequent thiol deprotection, a variety of disulfides are incubated in situ in the presence of a suitable reducing agent such as 2-mercaptoethanol. Like the dymanic combinatorial model, disulfides containing chemical architecture compatible with neighboring binding sites preferentially undergo disulfide exchange with a ligand's covalently attached thiol, thereby capturing a multidentate lead compound that can be elucidated through mass spectrometry (Fig. 21.15).

In their initial report, Sunesis researchers focused on the enzyme thymidylate synthase, (TM), an essential component in the biosynthetic pathway of dTMP and dUMP, an interesting target for antifungal and anticancer studies [87]. *E. coli* TM contains a cysteine residing in the active site, as well as four

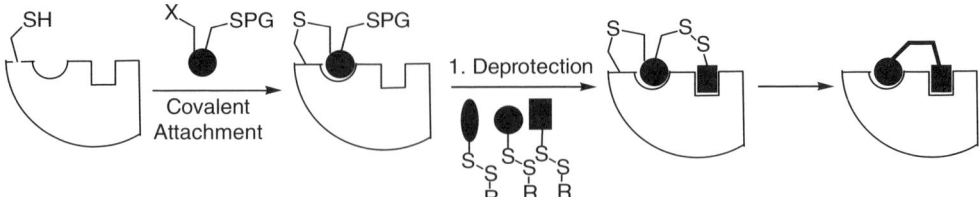

Figure 21.15 In situ assembly of high-affinity ligands through covalent capture.

cysteine residues buried within the protein. Prior research suggested the active-site cysteine residue (C146) to be most susceptible to modification. This was confirmed by incubating both wild-type and mutant TM (C146S) in the disulfide cystamine. It was observed that the wild-type enzyme selectively formed disulfide likage at the active site cysteine [87]. Therefore, disulfide libraries were constructed using solid support methodology to couple mono-BOC-cystamine to readily available carboxylic acids, sulfonyl chlorides, isocyanates, isothiocyanates, as well as oxime, and N-tosyl-D-proline-based disulfide libraries. Upon cleavage and sufficient purification, disulfide library members were dissolved in DMSO to a final concentration of 50 or 100 nM. Typical experimental protocol then calls for the addition of 1 μL DMSO solution, containing 8 to 15 disulfide library members to be incubated in 49 μL of approximately 15 μM protein suspended in 25 mM potassium phosphate/1 mM 2-mercaptoethanol buffer (pH 7.5). After sufficient incubation (30 min) the solution is analyzed via HPLC/MS, therefore providing structure elucidation for attached ligands (with respect to their atomic mass of the ligand). Upon screening 1200 library members, HPLC/MS analysis showed a preferential capture of N-tosyl-D-proline analogs over other library members. Quantitative structure–activity relationship (SAR) studies show a fair amount of flexibility is afforded to the tosyl phenyl ring, as t-Butyl can readily replace methyl moiety without great loss of affinity. Proline cyclic architecture appears to be important in retaining affinity, replacing this moiety with phenylalanine, phenylglycine, or pyrrole results in loss of affinity. Lastly, the phenylsulfonamide scaffold appears essential, as methyl proline is observed as a high-affinity ligand (Fig. 21.16). Recently, Sunesis has used the tether-based approach to discovered lead compounds for various protein active sites and small-molecule inhibitors of protein–protein interaction such as cytokine interleukin-2 and the IL-2α receptor [88].

Structure–Activity Relationship by Nuclear Magnetic Resonance

Structural analysis of proteins by nuclear magnetic resonance (NMR) spectroscopy continues to be an extensive area of research [89]. More recently, NMR has proven to be a valuable tool for ligand discovery as well.

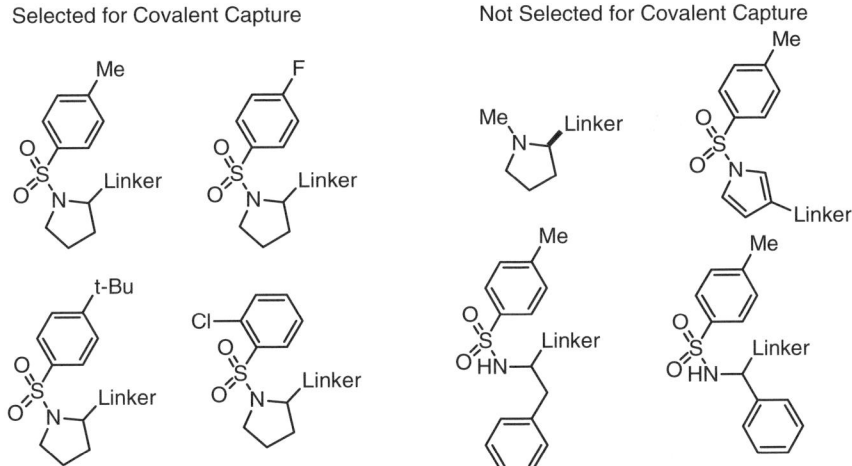

Figure 21.16 Covalent capture of ligands by tethering and SAR analysis.

*S*tructure–*a*ctivity *r*elationships by NMR, or simply SAR by NMR, developed by Fesik and co-workers, makes use of the well-known ability to detect the binding of ligands to various sites of a protein through ^{15}N or ^{1}H amide chemical shift changes in two-dimensional ^{15}N-heteronuclear single quantum correlation (^{15}N-HSQC) spectra [90].

Using ^{15}N-labeled protein, spectra can be compiled in a short period of time, allowing numerous ligands to be screened quickly. Once small-molecule ligands are discovered, affinity to the target binding site is optimized through the synthesis and screening of analogous compounds. Relatively low-affinity small-molecule ligand(s) for neighboring binding sites are then found and optimized through identical techniques. Proximal small molecules can be linked synthetically, producing a multidentate ligand with enhanced affinity (Fig. 21.17).

In their initial study, Fesik and co-workers used the SAR by NMR approach to derive a high-affinity ligand for the well-studied FK506 binding protein (FKBP) [90]. FKBP was initially screened against a library from which many compounds were found to bind with dissociation constants in the millimolar and micromolar range. A trimethoxyphenyl pipecolinic acid derivative, **29**, showed the highest affinity with a K_d of 2.0 μM. The binding site of this small molecule was found to be similar to that of similar moiety within the FK506 structure, and therefore an effort was made to find small-molecule receptors for nearby binding sites. Compounds were then screened in the presence of ^{15}N-FKBP, saturated in **29**. Those studies provided a small molecule that bound to FKBP with an affinity of $K_d = 0.8$ mM, as determined from the changes in ^{15}N-HSQC spectrum. Optimization of this low-affinity ligand

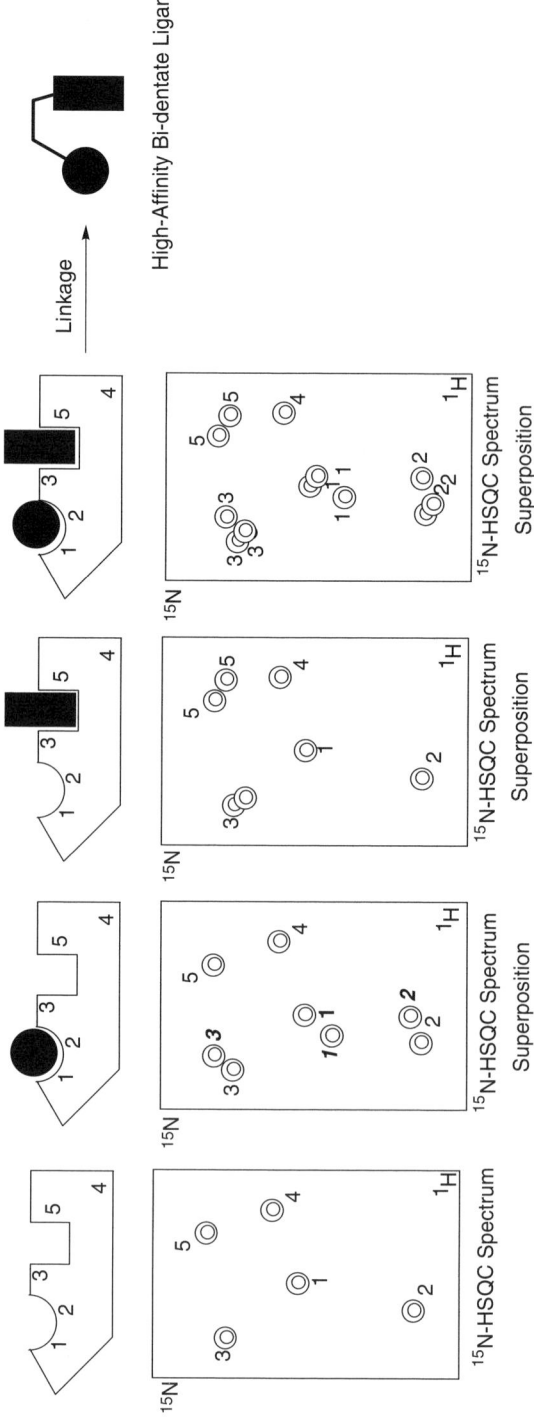

Figure 21.17 ^{15}N-HSQC analysis and development of a multicomponent ligand.

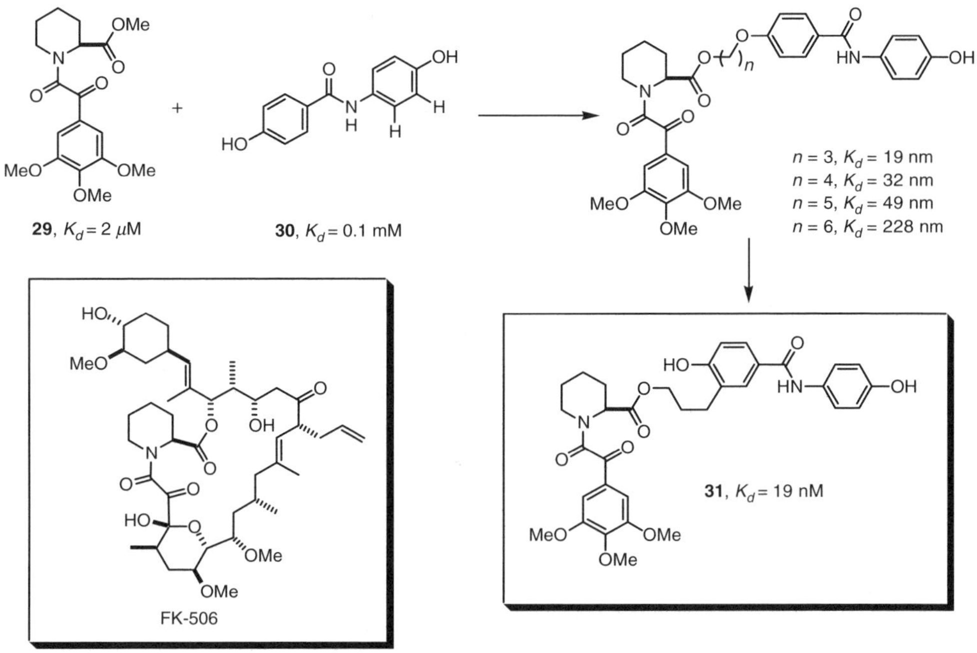

LIBRARY DESIGN ISSUES

Scheme 21.18 Development of high-affinity ligand for FK506 binding protein (FKBP).

provided compound **30**, which bound with a $K_d = 0.1$ mM. Subsequent analysis showed molecules **29** and **30** bound FKBP in sufficient proximity to allow for a chemical linkage to be made. Again, SAR by NMR experiments were performed using ^{15}N-HSQC spectroscopy, connecting the two low-molecular-weight pieces with linkers of various lengths. One molecule (**31**) incorporating a linker of three carbons provided a ligand with an affinity of $K_d = 19$ nM (Scheme 21.18).

This novel approach toward high-affinity ligand discovery has recently been used by numerous research groups both in academia and industry to develop inhibitors for biological substrates such as stromelysin [91], the single-chain hepatitis C virus NS3 protease/NA4A cofactor [92], and human papillomavirus E2 protein [93]. As high-throughput NMR systems (including, e.g., HPLC-NMR) capable of very fast two-dimensional spectral acquisition become more widely available, we can anticipate that this technique will find broad usage.

21.6 LIBRARY DESIGN ISSUES

As described above, early combinatorial chemistry efforts grew initially out of solid-phase synthesis studies on biopolymers. Consequently, much of the early history of combinatorial chemistry is focused on biopolymers and related compounds that could be formed using reactions closely analogous to those used in biopolymer synthesis. Although these approaches allowed the synthesis of large *numbers* of compounds, the actual *structural diversity* of such libraries was somewhat limited. Ellman's demonstration of the combinatorial synthesis of benzodiazepines (Scheme 21.6), along with several contemporary efforts, both demonstrated that combinatorial chemistry could be applied to the production of "druglike" molecules and led to the concept of the "decorated scaffold" as a means of providing conformationally rigid combinatorial libraries. While such libraries provided access to a conceptually new type of library structure, structural diversity within the library itself was still somewhat limited. In part because of these limitations, it was recognized that molecular weight and hydrophobicity of library compounds were straying beyond what might be considered optimal to lead to successful drugs. This observation was codified by Lipinski and co-workers into the "rule of 5," stating that a potential reason for the relatively low hit-to-lead success rate of many early combinatorial chemistry and high-throughput screening efforts was that libraries strayed beyond the structural norms (molecular weight less than 500, fewer than 5 hydrogen-bond donating groups, fewer than 10 hydrogen-bond accepting groups, log P over 5) for the vast majority of commercial drugs [94]. Of course, one can also view the rule of 5 as a challenge for developing new compounds that might successfully break it, but the primary lesson was that it pays to think carefully about library design, structure space, and the overall goal (drug development, in this case) of the combinatorial/high-throughput screening effort.

In recent years, two parallel developments (one computational and one conceptual) have set the stage for improving library design. On the computational side [95], researchers in academia and industry have devoted considerable effort to the development of tools capable of describing *structural diversity* and *structural similarity* in mathematical terms [96, 97]. Most chemists are capable of describing the similarities and differences among small groups of compounds, but rigorous analysis of large libraries is a significantly more difficult task. For example, describing a 1000-compound library in terms of 4 simple descriptors (molecular weight, number of H-bond acceptors, number of H-bond donors, number of chiral centers) yields a four-dimensional graph with 1000 points! For complex libraries, more precise descriptors such as BCUT values [98, 99] or four-point pharmacophore fingerprints [100, 101] can be used along side computerized reagent selection methods in the planning stages of a library. In combination with computer-aided drug design methodologies, an area of significant interest is the "virtual" screening of libraries generated in silico.

In parallel with advances in computational methodology, the conceptual advance of *diversity-oriented synthesis* is also beginning to influence the manner in which combinatorial libraries are designed and executed [102]. Diversity-oriented synthesis (and the related concept of *function-oriented syn-*

LIBRARY DESIGN ISSUES

thesis [103]) is in many respects a response to target-oriented synthesis. Target-oriented synthesis is perhaps the most familiar mode of thinking for organic chemists: Given a particular natural product or med-chem lead structure, how does one bring simple commercially available starting materials and the vast body of reaction knowledge available to the development of an efficient method for its construction? Target-oriented synthesis relies on retrosynthetic analysis (thinking about chemical synthesis in reverse). In contrast, diversity-oriented synthesis focuses on the development of pathways for the construction of collections of conformationally restricted molecules with complex, diversified structures, and dense functionality. Diversity-oriented synthesis is forward thinking in that reactions are chosen not based on their ability to bring the chemist closer to a single-target structure but rather contribute diversity (via an ability to accommodate multiple input reagents) and complexity (via an ability to form multiple bonds).

An example of a diversity-oriented synthesis developed by the Schreiber group is shown in Scheme 21.19 [104]. Ugi four-component condensation is used to set up an intramolecular Diels–Alder cycloaddition. Subsequent "bidirectional" amide alkylation with allyl bromide, followed by a tandem

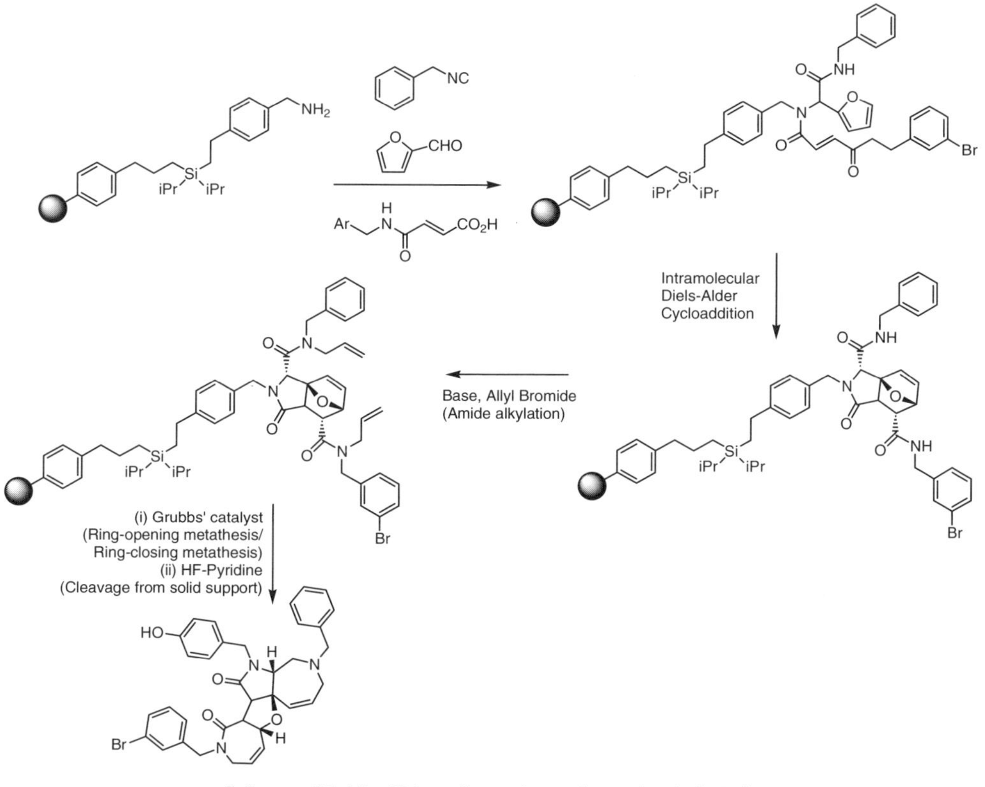

Scheme 21.19 Diversity-oriented synthesis [104].

ring-opening metathesis/ring-closing metathesis reaction, provides the final product. Cleavage from the resin bead is then accomplished by treatment with HF-pyridine. Further diversification is possible via elaboration of the bromoarene or phenolic groups.

21.7 CONCLUSIONS

In a relatively short period of time, combinatorial chemistry has become a widely used tool for synthetic chemists looking to streamline lead identification and for those needing novel structures to drive a plethora of studies in chemistry, biology, materials science, and beyond. Combinatorial chemistry has proven to be a valuable method for streamlining the development of new catalysts [105, 106] and new types of arrayed sensors [107]. We can anticipate that combinatorial techniques will be an important part of every chemist's repertoire for the foreseeable future.

REFERENCES

1. Rowinsky, E. K., Cazenave, L. A., Donehower, R. C. (1990). Taxon: A novel investigational antimicrotubule agent. *J Ntl. Cancer Inst.*, 82, 1247–1259.
2. Sheridan, R. P., Venkataraghavan, R. (1987). New methods in computer-aided drug design. *Acc. Chem. Res.*, 20, 322–329.
3. Crafford, C. (2003). The drug discovery gridiron. *Modern Drug Discov.*, 25–26.
4. Schroder, F. C., Farmer, J. J., Attygalle, A. B., Smedley, S. R., Eisner, T., Meinwald, J. (1998). Combinatorial chemistry in insects: A library of defensive macrocyclic polyamines. *Science*, 281, 428–431.
5. Examples include: (a) Nicolaou, K. C., Hanko, R., Hartwig, W. (Eds.) (2002). *Handbook of Combinatorial Chemistry.* Wiley-VCH. (b) Gordon, E. M., Kerwin, J. F., Jr. (Ed.) (1998). *Combinatorial Chemistry and Molecular Diversity in Drug Discovery.* Wiley-Liss, New York. (c) Czarnik, A. W., DeWitt, S. H. (1997). *A Practical Guide to Combinatorial Chemistry*. American Chemical Society, Washington, DC.
6. Merrifield, R. B. (1963). Solid phase peptide synthesis. I. The synthesis of a tetrapeptide. *J. Am. Chem. Soc.*, 85, 2149–2154.
7. Hodge, P. (1997). Polymer-supported organic reactions: What takes place in the beads? *Chem. Soc. Rev.*, 26, 417–424.
8. Merrifield, R. B. (1964). Solid phase peptide synthesis. III. An improved synthesis of bradykinin. *Biochemistry*, 3, 1385–1390.
9. Merrifield, R. B. (1965). Automated synthesis of peptides. *Science*, 150, 178–185.
10. Letsinger, R. L., Mahadevan, V. (1965). Oligonucleotide synthesis on a polymer support. *J. Am. Chem. Soc.*, 87, 3526–3527.
11. Frechet, J. M., Schuerch, C. (1971). Solid-phase synthesis of oligosaccharides. I. Preparation of the solid support. Poly[*p*-(1-propen-3-ol-1-yl)styrene]. *J. Am. Chem. Soc.*, 93, 492–496.

12. Dudkin, V. Y., Orlova, M., Geng, X., Mandal, M., Olson, W. C., Danishefsky, S. J. (2004). Toward fully synthetic carbohydrate-based HIV antigen design: On the critical role of bivalency. *J. Am. Chem. Soc.*, *126*, 9560–9562.
13. Ritter, T. K., Mong, K.-K. T., Liu, H., Nakatani, T., Wong, C.-H. (2003). A programmable one-pot oligosaccharide synthesis for diversifying the sugar domains of natural products: A case study of vancomycin. *Angewante Chem. Int. Ed.*, *42*, 4657–4660.
14. Love, K. R., Seeberger, P. H. (2004). Automated solid-phase synthesis of protected tumor-associated antigen and blood group determinant oligosaccharides. *Angewante Chem. Int. Ed.*, *43*, 602–605.
15. Schnolzer, M., Kent, S. (1992). Constructing proteins by dovetailing unprotected synthetic peptides: Backbone-engineered HIV protease. *Science*, *256*, 221–225.
16. Zhang, Z., Alfonta, L., Tian, F., Bursulaya, B., Uryu, S., King, D. S., Schultz, P. G. (2004). Selective incorporation of 5-hydroxytryptophan into proteins in mammalian cells. *Proc. Natl. Acad. Sci. USA*, *101*, 8882–8887.
17. Holley, R., Apgar, J., Everett, G., Madison, J., Marquette, M., Merrill, S., Penswick, J., Zamir, A. (1965). Structure of a ribonucleic acid. *Science*, *147*, 1462–1465.
18. Belagaje, R., Brown, E., Gait, M., Khoana, H., Norris, K. (1979). Total synthesis of a tyrosine suppressor transfer RNA gene. XIII. Synthesis of deoxyribopolynucleotide segments corresponding to the nucleotide sequence −1 to −29 in the promoter region. *J. Biol. Chem.*, 5754–5763.
19. Goeddel, D., Kleid, D., Bolivar, F. (1979). Expression in *Escherichia coli* of chemically synthesized genes for human insulin. *Proc. Natl. Acad. Sci. USA*, *76*, 106–110.
20. Edge, M., Greene, A., Heathcliffe, G., Meacock, P., Schuch, W., Scalon, D., Atkinson, T., Newton, C., Markham, A. (1981). Synthesis of human leudocyte α1-interon gene. *Nature*, *292*, 756–760.
21. Hodge, P. (1997). Polymer-supported organic reactions: What takes place in the beads? *Chem. Soc. Rev.*, *26*, 417–424.
22. Frechet, J. M. J. (1981). Synthesis and applications of organic polymers as supports and protecting groups. *Tetrahedron*, *37*, 663–683.
23. Houghten, R. A. (1985). General method for the rapid solid-phase synthesis of large numbers of peptides: Specificity of antigen-antibody interaction at the level of individual amino acids. *Proc. Natl. Acad. Sci.*, *82*, 5131–5135.
24. Frank, R., Heikens, W., Heisterberg-Moutsis, G., Blocker, H. (1983). A new general approach for the simultaneous chemical synthesis of large numbers of oligonucleotides: Segmental solid supports. *Nucleic Acids Res.*, *11*, 4365–4377.
25. Bunin, B. A., Ellman, J. A. (1992). A general and expedient method for the solid-phase synthesis of 1,4-benzodiazepine derivatives. *J. Am. Chem. Soc.*, *114*, 10997–10998.
26. Lam, K. S., Salmon, S. E., Hersh, E. M., Hruby, V. J., Kazmierski, W. M., Knapp, R. J. (1991). A new type of synthetic peptide library for identifying ligand-binding activity. *Nature*, *354*, 82–84.
27. Yan, B. (2002). Single bead analysis in combinatorial chemistry. *Curr. Opin. Chem. Biol.*, *6*, 328–332.
28. Chait, B. T., Wang, R., Beavis, R. C., Kent, S. B. H. (1993). Protein ladder sequencing. *Science*, *262*, 89–92.

29. Lane, S. J., Pipe, A. (2000). Single bead and hard tag decoding using accurate isotopic difference target analysis-encoded combinatorial libraries. *Rapid Commun. Mass Spectrom.*, *14*, 782–793.
30. Geysen, H. M., Wagner, C. D., Bodnar, W. M., Markworth, C. J., Parke, G. J., Schoenen, F. J., Wagner, D. S., Kinder, D. S. (1996). Isotope or mass encoding of combinatorial libraries. *Chem. Biol.*, *3*, 679–688.
31. Erb, E., Janda, K. D., Brenner, S. (1994). Recursive deconvolution of combinatorial chemical libraries. *Proc. Natl. Acad. Sci.*, *91*, 11422–11426.
32. Brenner, S., Lerner, R. A. (1992). Encoded combinatorial chemistry. *Proc. Natl. Acad. Sci. USA*, *89*, 5381–5383.
33. Needels, M. C., Jones, D. G., Tate, E. H., Heinkel, G. L., Kochersperger, L. M., Dower, W. J., Barrett, R. W., Gallop, M. A. (1993). Generation and screening of an oligonucleotide-encoded synthetic peptide library. *Proc. Natl. Acad. Sci. USA*, *90*, 10700–10704.
34. Debaene, F., Mejias, L., Harris, J. L., Winssinger, N. (2004). Synthesis of a PNA-encoded cysteine protease inhibitor library. *Tetrahedron*, *60*, 8677–8690.
35. Ohlmeyer, M. H. J., Swanson, R. N., Dillard, L. W., Reader, J. C., Asouline, G., Kobayashi, R., Wigler, M., Still, W. C. (1993). Complex synthetic chemical libraries indexed with molecular tags. *Proc. Natl. Acad. Sci.*, *90*, 10922–10926.
36. Nestler, H. P., Bartlett, P. A., Still, W. C. (1994). A general method for molecular tagging of encoded combinatorial chemistry libraries. *J. Organic Chem.*, *59*, 4723–4724.
37. Nestler, H. P., Bartlett, P. A., Still, W. C. (1994). A general method for molecular tagging of encoded combinatorial chemistry libraries. *J. Organic Chem.*, *59*, 4723–4724.
38. Ni, Z.-J., Maclean, D., Holmes, C. P., Murphy, M. M., Ruhland, B., Jacobs, J. W., Gordon, E. M., Gallop, M. A. (1996). Versatile approach to encoding combinatorial organic syntheses using chemically robust secondary amine tags. *J. Med. Chem.*, *39*, 1601–1608.
39. Rahman, S. S., Busby, D. J., Lee, D. C. (1998). Infrared and Raman spectra of a single resin bead for analysis of solid-phase reactions and use in encoding combinatorial libraries. *J. Organic Chem.*, *63*, 6196–6199.
40. Krchnak, V., Padera, V. (2003). The encore technique: A novel approach to directed split-and-pool combinatorial synthesis. *Methods Enzymol.*, *369*, 112–124.
41. Shukla, R., Sasaki, Y., Krchnak, V., Smith, B. D. (2004). Identification of synthetic phosphatidylserine translocases from a combinatorial library prepared by directed split-and-pool synthesis. *J. Comb. Chem.*, *6*, 703–709.
42. (1) Wang, X., Zhang, J., Song, A., Lebrilla, C. B., Lam, K. S. (2004). Encoding method for OBOC small molecule libraries using a biphasic approach for ladder-synthesis of coding tags. *J. Am. Chem. Soc.*, *126*, 5740–5749. (2) Liu, R., Marik, J., Lam, K. S. (2002). A novel peptide-based encoding system for "one-bead one-compound" peptidomimetic and small molecule combinatorial libraries. *J. Am. Chem. Soc.*, *124*, 7678–7680.
43. Fenniri, H., Ding, L., Ribbe, A. E., Zyrianov, Y. (2001). Barcoded resins: A new concept for polymer-supported combinatorial library self-deconvolution. *J. Am. Chem. Soc.*, *123*, 8151–8152.

44. Fenniri, H., Chun, S., Ding, L., Zyrianov, Y., Hallenga, K. (2003). Preparation, physical properties, on-bead binding assay and spectroscopic reliability of 25 barcoded polystyrene-poly(ethylene glycol) graft copolymers. *J. Am. Chem. Soc.*, *125*, 10546–10560.

45. Moran, E. J., Sarshar, S., Cargill, J. F., Shahbaz, M. M., Lio, A., Mjalli, A. M. M., Armstrong, R. W. (1995). Radio frequency tag encoded combinatorial library method for the discovery of tripeptide-substituted cinnamic acid inhibitors of the protein tyrosine phosphatase PTP1B. *J. Am. Chem. Soc.*, *117*, 10787–10788.

46. Nicolaou, K. C., Xiao, X.-Y., Parandoosh, Z., Senyei, A., Nova, M. P. (1995). Radiofrequency encoded combinatorial chemistry. *Angewante Chem. Int. Ed.*, *34*, 2289–2291.

47. Nicolaou, K. C., Pfefferkorn, J. A., Mitchell, H. J., Roecker, A. J., Barluenga, S., Cao, G.-Q., Affleck, R. L., Lillig, J. E. (2000). Natural product-like combinatorial libraries based on privileged structures. 2. Construction of a 10,000-membered benzopyran library by directed split-and-pool chemistry using Nanokans and optical encoding. *J. Am. Chem. Soc.*, *122*, 9954–9967.

48. Fodor, S. P. A., Read, J. L., Pirrung, M. C., Stryer, L., Liu, A. T., Solas, D. (1991). Light-directed, spatially addressable parallel chemical synthesis. *Science*, *251*, 767.

49. Li, S., Bowerman, D., Marthandan, N., Klyza, S., Luebke, K. J., Garner, H. R., Kodadek, T. (2004). Photolithographic synthesis of peptoids. *J. Am. Chem. Soc.*, *126*, 4088–4089.

50. Khodursky, A. B., Peter, B. J., Cozzzarelli, N. R., Botstein, D., Brown, P. O., Yanofsky, C. (2000). DNA microarray analysis of gene expression in response to physiological and genetic changes that affect tryptophan metabolism in *Escherichia coli*. *Proc. Natl. Acad. Sci. USA*, *97*, 12170–12175.

51. Blackwell, H. E., Perez, L., Stavenger, R. A., Tallarico, J. A., Eatough, E. C., Foley, M. A., Schreiber, S. L. (2001). A one-bead, one-stock solution approach to chemical genetics: Part 1. *Chem. Biol.*, *8*, 1167–1182.

52. Clemons, P. A., Koehler, A. N., Wagner, B. K., Sprigings, T. G., Spring, D. R., King, R. W., Schreiber, S. L., Foley, M. A. (2001). A one-bead, one-stock solution approach to chemical genetics: Part 2. *Chem. Biol.*, *8*, 1183–1195.

53. Winssinger, N., Ficarro, S., Schultz, P. G., Harris, J. L. (2002). Profiling protein function with small molecule microarrays. *Proc. Natl. Acad. Sci. USA*, *99*, 11139–11144.

54. Uttamchandani, M., Walsh, D. P., Khersonsky, S. M., Huang, X., Yao, S. Q., Chang, Y.-T. (In press). Microarrays of tagged combinatorial triazine libraries in the discovery of small-molecule ligands of human IgG. *J. Comb. Chem*.

55. Adams, E. W., Ratner, D. M., Bokesch, H. R., McMahon, J. B., O'Keefe, B. R., Seeberger, P. H. (2004). Oligosaccharide and glycoprotein microarrays as tools in HIV glycobiology: Glycan-dependent gp120/protein interactions. *Chem. Biol.*, *11*, 875–881.

56. For representative examples see: (1) Miller, B., Karan, C. (2001). RNA-selective coordination complexes identified via dynamic combinatorial chemistry. *J. Am. Chem. Soc.*, *123*, 7455–7456. (2) Hochgurtel, M., Biesinger, R., Kroth, H., Piecha, D., Hofmann, M., Krause, S., Schaaf, O., Nicolau, C., Eliseev, A. V. (2003). Ketones as builing blocks for dynamic combinatorial libraries: Highly active neuraminidase inhibitors generated via selection pressure of the biological target. *J.*

Med. Chem., *46*, 356–358. (3) Otto, S., Furlan, R., Sanders, J. (2002). Selection and amplification of hosts from dynamic combinatorial libraries of macrocyclic disulfides. *Science*, *297*, 590–593.

57. Huc, I., Krishe, M. J., Funeriu, D. P., Lehn, J. M. (1999). Dynamic combinatorial chemistry: Target H-bonding directed assembly of receptors based on bipyridine-metal complexes. *Eur. J. Inorganic Chem.*, 1415–1420.
58. Berl, V., Huc, I., Lehn, J., DeCaian, A., Fisher, J. (1999). Induced fit selection of a barbiturate receptor from a dynamic structural and conformational/configurational library. *Eur. J. Organic Chem.*, *11*, 3089–3094.
59. Eliseev, A., Nelen, M. (1997). Use of molecular recognition to drive chemical evolution: Controlling the composition of an equilibrating mixture of simple arginine receptors. *J. Am. Chem. Soc.*, *119*, 1147–1148.
60. Rowan, S., Hamilton, D., Brady, P., Sanders, J. (1997). Automated recognition, sorting, and covalent self-assembly by predisposed building blocks in a mixture. *J. Am. Chem. Soc.*, *119*, 2578–2579.
61. Whitesides, G., Lees, W. (1993). Equilibrium constants for thiol-disulfide interchange reactions: A coherent, corrected set. *J. Organic Chem.*, *58*, 642–647.
62. Miller, B., Klekota, B. (2003). Biphasic reaction vessel and method of use. 6,599,754, July 29.
63. Klekota, B., Hammond, M., Miller, B. (1997). Generation of novel DNA-binding compounds by selection and amplification from self-assembled libraries. *Tetrahedr. Lett.*, *38*, 8639–8642.
64. Karan, C., Miller, B. (2001). RNA-selective coordination complexes identified via dynamic combinatorial chemistry. *J. Am. Chem. Soc.*, *123*, 7455–7456.
65. For an early example, see: Pyle, A. M., Long, E. C., Barton, J. K. (1989). Shape-selective targeting of DNA by phenanthrenequinone diimine rhodium-III photocleaving agents. *J. Am. Chem. Soc.*, *111*, 4520–4522.
66. Balasubramanian, S., Whitney, A., Ladame, S. (2004). Templated ligand assembly by using G-quadruplex DNA and dynamic covalent chemistry. *Angewante Chem. Int. Ed.*, *43*, 1143–1146.
67. Schouten, J. A., Ladame, S., Mason, S. J., Cooper, M. A., Balasubramanian, S. (2003). G-quadruplex-specific peptide-hemicyanine ligands by partial combinatorial selection. *J. Am. Chem. Soc.*, *125*, 5594–5595.
68. Lehn, J., Ramström, O., Lohmann, S., Bunyapaiboonsri, T. (2004). Dynamic combinatorial carbohydrate libraries: Probing the binding site of concanavalin A lectin. *J. Eur. Chem.*, *10*, 1711–1715.
69. Page, D., Roy, R. (1997). Synthesis and biological properties of mannosylated starbust poly(amidoamine) dendrimers. *Bioconjugate Chem.*, *8*, 714–723.
70. Lehn, J., Bunyapaiboonsri, T., Ramström, O., Lohmann, S., Peng, L., Goeldner, M. (2001). Dynamic deconvolution of a pre-equilibrated dynamic combinatorial library of acetylcholinesterase inhibitors. *ChemBioChem*, *2*, 438–444.
71. Kazlauskus, R., Cheeseman, J., Corbett, A., Shu, R., Croteau, J., Gleason, J. (2002). Amplification of screening sensitivity through selective destruction: Theory and screening of a library of carbanic anhydrase inhibitors. *J. Am. Chem. Soc.*, *124*, 5692–5701.
72. Severin, K. (2004). The advantage of being virtual—Target-induced adaptation and selection in dynamic combinatorial libraries. *Eur. J. Chem.*, *10*, 2565–2580.

73. Corbett, P. T., Otto, S., Sanders, J. K. M. (2004). Correlation between host–guest binding and host amplification in simulated dynamic combinatorial libraries. *Eur. J. Chem.*, *10*, 3139–3143.
74. Hochguertel, M., Kroth, H., Piecha, D., Hofmann, M. W., Nicolau, C., Krause, S., Schaaf, O., Sonnenmoser, G., Eliseev, A. V. (2002). Target-induced formation of neuraminidase inhibitors from in vitro virtual combinatorial libraries. *Proc. Natl. Acad. Sci. USA*, *99*, 3382–3387.
75. James, I. W. (1999). Linkers for solid phase organic synthesis. *Tetrahedron*, *55*, 4855–4956.
76. Kolb, H., Finn, M., Sharpless, K. (2001). Click chemistry. *Angewante Chem. Int. Ed.*, *40*, 2004–2021.
77. Taylor, S., Taylor, A., Screiber, S. (2004). Synthetic strategy toward skeletal diversity via solid-supported, otherwise unstable reactive intermediates. *Angewante Chem. Int. Ed.*, *43*, 1681–1685.
78. Kolb, H., Sharpless, B. (2003). The growing impact of click chemistry on drug discovery. *Drug Discov. Today*, *8*, 1128–1137.
79. Lee, L., Mitchell, M., Huang, S., Fokin, V., Sharpless, B., Wong, C. (2003). A potent and highly selective inhibitor of human α-1,3-fucosytransferase via click chemistry. *J. Am. Chem. Soc.*, *125*, 9588–9589.
80. Huisgen, R. (1984). *1,3-Dipolar Cycloaddition Chemistry*. Wiley, New York, pp. 1–176.
81. Mock, W., Irra, T., Wepseic, J., Manimaran, T. (1983). Cycloaddition induced by cucurbituril. A case of Pauling principle catalysis. *J. Organic Chem.*, *48*, 3619–3620.
82. Lewis, W., Green, L., Grynszpan, F., Radic, Z., Carlier, P., Talyor, P., Finn, M., Sharpless, B. (2002). Click chemistry in situ: Acetylcholinesterase as a reaction vessel for the selective assembly of a femtomolar inhibitor from an array of building blocks. *Angewante Chem. Int. Ed.*, *41*, 1053–1057.
83. Sussman, J., Harel, M., Frowlow, F., Oefner, C., Goldman, A., Toker, L., Silman, I. (1991). Atomic structure of acetycholinesterase from *Torpedo californica*: A prototype acetylcholine-binding protein. *Science*, *253*, 872–879.
84. Radic, Z., Taylor, P. (2001). Interaction kinetics of reversible inhibitors and substrates with acetylcholinesterase and its fasciculin 2 complex. *J. Biol. Chem.*, *276*, 4622–4633.
85. Wei, J., Burkiak, J., Siuzdak, G. (1999). Desorption–ionization mass spectrometry on porous silicon. *Nature*, *401*, 243–246.
86. Erlanson, D., Lam, J., Wiesmann, C., Luong, T., Simmons, R., DeLano, W., Choong, I., Burdett, M., Flanagan, M., Lee, D., Gordon, E., O'Brien, T. (2003). In situ assembly of enzyme inhibitors using extended tethering. *Nat. Biotechnol.*, *21*, 308–314.
87. Erlanson, D., Braisted, A., Raphael, D., Rangal, M., Stroud, R., Gordon, E., Wells, J. (2000). Site-directed ligand discovery. *Proc. Natl. Acad. Sci. USA*, *97*, 9367–9372.
88. Erlanson, D., McDowell, R., O'Brien, T. (2004). Fragment based drug discovery. *J. Med. Chem.*, *47*, 3464–3482.
89. Homans, S. (2004). NMR spectroscopy tools for structure-aided drug design. *Angewante Chem. Int. Ed.*, *43*, 290–300.
90. Shuker, S., Hajdek, P., Meadows, R., Fesik, S. (1996). Discovering high-affinity ligands for proteins: SAR by NMR. *Science*, *274*, 1531–1534.

91. Shuker, S., Hajdek, P., Meadows, R., Nettesheim, D., Olejniczak, E., Sheppard, G., Steinman, D., Carrera, G., Marcotte, P., Severin, J., Walter, K., Smith, H., Gubbins, E., Simmer, R., Holzman, T., Morgan, D., Davidsen, S., Summers, J., Fesik, S. (1997). Discovery of potent nonpeptide inhibitors of stromelysin using SAR by NMR. *J. Am. Chem. Soc.*, *119*, 5818–5827.

92. Wyss, D., Arasappan, A., Senior, M., Wang, Y., Beyer, B., Njoroge, G., McCoy, M. (2004). Non-peptidic small molecule inhibitors of the single-chain hepatitis C virus NS3 protease/NS4A cofactor complex discovered by structure-based NMR screening. *J. Med. Chem.*, *47*, 2486–2498.

93. Shuker, S., Hajdek, P., Dinges, J., Miknis, G., Merlock, M., Middleton, T., Kempf, D., Egan, D., Walter, K., Robins, T., Holtzman, T., Fesik, S. (1997). NMR based discovery of lead inhibitors that block DNA binding of the human papillomavirus E2 protein. *J. Med. Chem.*, *40*, 3144–3150.

94. Lipinski, C. A., Lombardo, F., Dominy, B. W., Feeney, P. J. (1997). Experimental and computational approaches to estimate solubility and permeability in drug discovery and development settings. *Adv. Drug Delivery Rev.*, *23*, 3–25.

95. Evensen, E., Eksterowicz, J. E., Stanton, R. V., Oshiro, C., Grootenhuis, P. D. J., Bradley, E. K. (2003). Comparing performance of computational tools for combinatorial library design. *J. Med. Chem.*, *46*, 5125–5128.

96. Ghose, A. K., Viswanadhan, V. N. (Ed.) (2001). *Combinatorial Library Design and Evaluation*. Marcel Dekker, New York.

97. Martin, Y. C. (2001). Diverse viewpoints on computational aspects of molecular diversity. *J. Comb. Chem.*, *3*, 231–250.

98. Pirard, B., Pickett, S. D. (2000). Classification of kinase inhibitors using BCUT descriptors. *J. Chem. Inf. Comput. Sci.*, *40*, 1431–1440.

99. Gao, H. (2001). Application of BCUT metrics and genetic algorithm in binary QSAR analysis. *J. Chem. Inf. Comput. Sci.*, *41*, 402–407.

100. Lyne, P. D., Kenny, P. W., Cosgrove, D. A., Deng, C., Zabludoff, S., Ashwell, S., Wendoloski, J. J. (2004). Identification of compounds with nanomolar binding affinity for checkpoint kinase-1 using knowledge-based virtual screening. *J. Med. Chem.*, *47*, 1962–1968.

101. Manallack, D. T., Pitt, W. R., Gancia, E., Montana, J. G., Livingstone, D. J., Ford, M. G., Whitley, D. C. (2002). Selecting screening candidates for kinase and G protein-coupled receptor targets using neural networks. *J. Chem. Inf. Comput. Sci.*, *42*, 1256–1262.

102. Schreiber, S. L. (2000). Target-oriented and diversity-oriented organic synthesis in drug discovery. *Science*, *287*, 1964–1969.

103. Wender, P. A., Baryza, J. L., Brenner, S. E., Clarke, M. O., Craske, M. L., Horan, J. C., Meyer, T. (2004). Function oriented synthesis: The design, synthesis, PKC binding and translocation activity of a new bryostatin analog. *Curr. Drug Discov. Techn.*, *1*, 1–11.

104. Lee, D., Sello, J. K., Schreiber, S. L. (2000). Pairwise use of complexity-generating reactions in diversity-oriented organic synthesis. *Organic Lett.*, *2*, 709–712.

105. Francis, M. B., Jacobsen, E. N. (1999). Discovery of novel catalysts for alkene epoxidation from metal-binding combinatorial libraries. *Angewante Chem. Int. Ed.*, *38*, 937–941.

106. Garbacia, S., Touzani, R., Lavastre, O. (2004). Image analysis as a quantitative screening test in combinatorial catalysis: Discovery of an unexpected ruthenium-based catalyst for the Sonogashira reaction. *J. Comb. Chem.*, *6*, 297–300.
107. McCleskey, S. C., Griffin, M. J., Schneider, S. E., McDevitt, J. T., Anslyn, E. V. (2003). Differential receptors create patterns diagnostic for ATP and GTP. *J. Am. Chem. Soc.*, *125*, 1114–1115.

22

HERBAL MEDICINES AND ANIMAL MODELS OF GASTROINTESTINAL DISEASES

C.H. CHO
The University of Hong Kong, Hong Kong, China

J.K.S. KO
School of Chinese Medicine, The Hong Kong Baptist University, Hong Kong, China

22.1	EXPERIMENTAL MODELS OF GASTRIC LESIONS AND ULCERATION	1014
	Acute Gastric Mucosal Injury	1014
	Evaluation of Acute Gastric Lesions	1015
	Chronic Gastric Ulcer	1015
	Acute and Chronic Duodenal Ulcers	1016
22.2	EXPERIMENTAL MODELS OF INFLAMMATORY BOWEL DISEASE (IBD)	1016
	Trinitrobenzene Sulfonic Acid/Ethanol	1017
	Dintrobezene Sulfonic Acid/Ethanol	1017
	Dextran Sulfate Sodium	1017
	Assessment of Colitis Severity	1018
22.3	ANIMAL MODELS OF GASTROINTESTINAL CANCER	1019
	Gastric Cancer Models	1019
	Colorectal Cancer Models	1020
22.4	HERBAL MEDICINES AS POTENTIAL THERAPEUTIC AGENTS FOR ULCERS AND MALIGNANT DISEASE IN THE GASTROINTESTINAL TRACT	1021
	General	1021
	Polysaccharides in General	1022
	Polysaccharides from Mushroom and Seaweed	1023

Drug Discovery Handbook, by Shayne Cox Gad
Copyright © 2005 by John Wiley & Sons, Inc.

Polysaccharides from Other Plants 1024
Phenolic Compounds and Terpenoids from Medicinal Plants 1024
References 1028

22.1 EXPERIMENTAL MODELS OF GASTRIC LESIONS AND ULCERATION

Animal models of gastric and duodenal ulcers are the key in studying the pathogenesis of gastrointestinal (GI) diseases and testing antiulcer agents. They provide essential tools to screen herbal medicines for their antiulcer and anticancer effects. These models ideally should mimic human diseases, be easy to produce, and reproducible to allow continuous replication in various laboratories. Tables 22.1 and 22.2 summarize the methodologies for these animal models.

Acute Gastric Mucosal Injury

These animal models are chemically or psychologically induced and cause rapidly developing diffuse surface epithelial injury in the gastric mucosa. This is evident by multiple hemorrhagic erosions. The lesions are usually induced by intragastric administration of chemicals (e.g., concentrated ethanol, HCl, or nonsteroidal anti-inflammatory drugs) or by psychological stress (e.g., cold restraint) in fasted rats. Among these models, ethanol and nonsteroidal anti-inflammatory drugs are probably the most widely used for testing

TABLE 22.1 Models of Mucosal Damage (Acute and Chronic) in the Stomach[a]

Damaging Agents	Dose and Route of Administration	Animals	Type of Damage	
			Acute	Chronic
Ethanol	50–100%, 1 mL, IG	Rat	Erosion	—
HCl	0.5–0.5 N, 1 mL, IG	Rat	Erosion	—
Aspirin (acidified with 0.2 N HCl)	50–100 mg/kg, IG	Rat, mouse	Erosion to ulcer	—
Indomethacin	10–50 mg/kg, IG or SC	Rat	Erosion to ulcer	—
Stress	Restraint + cold (4°C) for 2 h	Rat	Erosion to ulcer	—
Acetic acid	Local or topical application of acetic acid (dosage in text)	Rat	—	Ulcer

[a] All animals should be fasted at least 18 h before induction of gastric damage. IG = intragastric administration; SC = subcutaneous injection.

EXPERIMENTAL MODELS OF GASTRIC LESIONS AND ULCERATION

TABLE 22.2 Models of Duodenal Ulcers[a]

Damaging Agents	Dose and Route of Administration	Species
Secretogogues (histamine, carbachol, pentagastrin)	Carbachol 0.5 µg/kg, SC for 24–48 hr + Histamine 50 µg/kg/min, SC for 24–48 h	Rat
	Carbachol 0.5 µg/kg/min, SC for 24–48 h + Pentagastrin 2 µg/kg/min, SC for 24–48 h	Rat
	Histamine 90 µg/kg, SC × 8, 30 min apart	Guinea pig
Cysteamine HCl	280–300 mg/kg, IG × 3 (e.g., 9 a.m., 12 noon, 3 pm) or 400 mg/kg, IG × 2	Rat or mouse

[a] IG = intragastric administration; SC = subcutaneous injection.

antiulcer agents in animals. Most acute erosions in the stomach induced by these methods heal rapidly in a few days without scar formation. These lesions do not reulcerate spontaneously, an enigmatic feature of ulcer disease in humans.

Evaluation of Acute Gastric Lesions

Lesions can be evaluated semiquantitively in the gastric mucosa by using a grade (e.g., scale 0 to 3) or quantitatively by planimetry of the hemorrhagic areas [1]. Since the lesions are usually uniform in width and depth, some investigators measure the length of hemorrhagic erosions. Indirect assessment of mucosal lesions is also possible by measuring the amount of blood loss into the gastric lumen either as ^{51}Cr-labeled red blood cells [2] or as the concentration of hemoglobin [3]. Biochemical measurement of released enzyme has also been used as the endpoint of evaluation [4].

Chronic Gastric Ulcer

The acetic acid method, although artificial, represents one of the few reproducible models of chronic gastric ulcer. There are several methods of applying the acetic acid to the stomach, the most popular involving injection of a small volume (0.05 mL) of 20 to 30 percent (v/v) acetic acid into the submucosal layer through the serosa of the anterior wall of the glandular stomach in fasted rats. The site of injection is readily recognized by the localized swelling and blanching of the serosa at the injection site. Some modifications consist of topical application of 0.06 mL of 100 percent acetic acid into a mold over the serosa and allowing it to remain for 60 s [5], or luminal application of 60 percent (v/v) of acetic acid (0.12 mL) with the help of a syringe and a ring clamped between the anterior and posterior walls of the stomach for 45 s and withdrawn through the same syringe [6]. All these methods produce an ulcer of unified size after 24 h and that can be maintained for at least 14 days. The ulcer can be easily quantified by measuring the diameter or the area of the ulcer crater. With the acetic acid model, histological evaluation is essential to assess the degree of necrosis, inflammation, epithelial regeneration, and proliferation of fibrous connective tissue, as well as angiogenesis at the ulcer margin and base.

Acute and Chronic Duodenal Ulcers

The duodenal ulcer models fall into two groups: those produced by secretogogues that induce multiple, irregular ulcers and those caused by cysteamine-like agents that induce solitary or kissing ulcers (Table 22.2). Hydrochloric acid seems to be essential for duodenal ulceration. Experimental duodenal ulcers can be induced in rats and guinea pigs by giving secretogogues such as histamine, cholinergic drugs, and pentagastrin [7, 8], which are similar to Zollinger–Ellison syndrome or other human hypersecretory states. The severity of the ulcer may be evaluated semiquantitatively with a scale of 0 to 3, where 0 = normal mucosa, 1 = erosion, 2 = deep ulcer involving the muscularis propria, or 3 = penetrating ulcer (into liver or pancreas). The intensity of the ulcer process and the degree of inflammation and healing should be evaluated by light microscopy.

The duodenal ulcers induced in rats by cysteamine and its derivatives are very similar to those seen in humans, on the basis of functional and morphologic criteria [9]. Ulcers can be induced by this method in both rats and mice. It should be emphasized that animals should not be fasted because food restriction increases the toxicity of the ulcer-inducing agent. The ulcers develop mainly on the antimesenteric or anterior wall of the duodenum. They frequently perforate or penetrate into the liver or pancreas. These ulcers are sharply demarcated craters with a necrotic center. Measurement of the largest diameter of the ulcer can be used to calculate the ulcer area. The same 0 to 3 scale as for ulcers induced by secretogogues provides a semiquantitative evaluation of ulcer severity.

22.2 EXPERIMENTAL MODELS OF INFLAMMATORY BOWEL DISEASE (IBD)

Ample evidence suggests that the mechanisms underlying initiation, progression, and chronicity of IBD are related to the immune system, genetic susceptibility, microbiologic aspects, and the environment. In order to address all these issues, an ideal experimental model of IBD is needed for studying the early events, dissection of the interactions among different inflammatory components, and identification of immunological processes and genes involved. However, to date, no such single and ideal IBD model exists that exactly reflects the complexity of these multifactorial disorders. Nevertheless, there are many experimental models of IBD that work differentially to simulate its various forms, such as ulcerative colitis (UC) and Crohn's disease (CD) in humans [10]. Despite inherent limitation in their similarity to human forms of IBD, studies of experimental colitis models have advanced our understanding of its pathogenesis [10, 11]. They also provide opportunities to evaluate potential therapeutic agents for this disease in the colon [12]. Current IBD models include chemical- and hapten-induced colitis, which are used in the current experimental settings to assess the efficacy of pharmacological agents in animals.

Trinitrobenzene Sulfonic Acid/Ethanol

2,4,6-Trinitrobenzene sulfonic acid (TNBS)/ethanol-induced colitis is the most commonly used animal model in various laboratories [11] and was initially established by Morris and co-workers in 1989 [13]. Intrarectal administration of 30 mg TNBS in ethanol (50 percent, v/v) vehicle is injected slowly through a polythene catheter via the rectum into the lumen of the colon, 8 cm proximal to the anus. This results in acute inflammation with ulcers that evolves into chronic inflammation of the distal colon in rats [13] and mice [14]. This approach is based on the hypotheses that in CD, an increase in mucosal permeability results in entry to the lamina propria of a luminal antigen that is not adequately cleared by the mucosal immune system. TNBS is a hapten that, when coupled to a substance of high molecular weight (such as tissue proteins), elicits immunological responses. Ethanol is known as a "barrier breaker" that causes widespread acute mucosal damage when instilled into the distal colon of rats at a concentration of at least 30 percent. The combined administration of TNBS and ethanol results in a severe, transmural, granulomatous inflammation of the distal colon. Once induced, healing is continuous after the initial injury and there are no relapses. The duration of inflammation lasts for at least 8 weeks [10]. It has been well characterized and its macroscopic and microscopic features share many of the histopathological and clinical features of human CD [11, 15, 16].

Dinitrobenzene Sulfonic Acid/Ethanol

2,4-Dinitrobenzene sulfonic acid (DNBS), a hapten molecule, was developed to produce inflammation closely resembling human UC [16]. Its action is based on the hypersensitivity of intestinal mucosa toward antigen that also occurs in IBD development. Again in this model, 50 percent ethanol is used to remove the protective mucus coating together with 30 mg DNBS given as an enema inserted intrarectally into an anesthetized rat about 8 cm proximal to the anus to induce colitis. In the DNBS model, numerous colonic lesion and other characteristics of UC can be found. Although the inflammatory responses in the DNBS model are similar to those induced by other haptens, such as TNBS [13], granulomatous lesions are not found.

Dextran Sulfate Sodium

Dextran sulfate sodium (DSS), a sulfated polysaccharide of molecular size of 54,000, can be used to induce acute and chronic colitis in rats and mice. This model is highly reproducible and resembles human UC [17]. To induce acute colitis, animals are given 3 to 10 percent DSS in drinking water for 7 to 9 days. By the end of the treatment period, animals show signs of diarrhea and gross rectal bleeding. Colon length is reduced and its weight is increased indicating that an inflammatory type of edema occurs. In addition, histological tissue damage includes multiple mucosal erosion and crypt abscesses. For chronic inflammatory studies, animals are treated with 5 percent DSS in drinking water for 7 days followed by a rest period of 7 to 10 days. This cycle of treatment is repeated several times to produce a relapse/remission pattern of inflammation in the colon [18]. In this repeated process of inflammatory injury, cell turnover is rapid and massive, and genetic damage may be caused by random mutation. Faulty genes may appear and as a result dysplasia appears in the mucosa, a clear precursor to malignancy. Low-grade and high-grade dysplasia is accompanied by adenoma followed by carcinoma in the colon [19]. This is a typical inflammation–adenoma–carcinoma model in animals, which is useful for screening both anti-inflammatory and anticancer agents in animals.

Assessment of Colitis Severity

Morphological Assessment To measure mucosal damage in TNBS and DNBS models, the distal colon is removed, opened, and rinsed thoroughly in ice-cold normal saline. The lesion area is recorded and then measured using a grid. In the assessment of the inflammatory response, 8 cm of the distal colon or the whole colon is weighed. The ratio of colon to body weight is used to assess the degree of colon edema. The whole colon length is also measured to indicate the severity of colonic inflammation [19, 20].

Histopathological Assessment In the DSS colitis model, a longitudinal section of colonic tissue is obtained and fixed in 10 percent formalin. The tissue is processed and embedded in a paraffin block, which is then sectioned at 5 µm and stained with the hematoxylin-eosin to assess damage. The severity of damage is graded according to the system described by Okayasu and co-workers [17], in which 0 = normal; 1 = focal inflammatory cell infiltration including polymorphonuclear leukocytes; 2 = inflammatory cell infiltration, gland dropout, and crypt abscess; and 3 = mucosal ulceration.

In the inflammation produced by TNBS and DNBS, tissue processing and staining are the same as for the DSS model. Microscopic damage is measured as follows: 0 = colon with normal architecture; 1 = damage limited to surface epithelium; 2 = focal ulceration limited to the mucosa; 3 = focal and transmural ulceration and inflammation; 4 = extensive transmural lesion bordered by normal mucosa; and 5 = extensive transmural lesion involving the entire section [21].

Biochemical Assessment Colonic tissues are collected and immediately frozen in liquid nitrogen and stored at −70°C until determination for myeloperoxidase (MPO) activity according to the method described by Suzuki and coworkers [22]. The supernatant of a homogenate is mixed with hydrogen peroxide and 3,3′,5,5′-tetramethylbenzidine prepared in sodium phosphate buffer (pH 5.4). The final mixture is measured at 450 nm using a spectrophotometer with horseradish as standard. The value of MPO activity is represented as U/g tissue. This enzyme is also taken as an index of neutrophil infiltration.

22.3 ANIMAL MODELS OF GASTROINTESTINAL CANCER

Gastric Cancer Models

Human Gastric Cancer Cell Implantation Model   Direct injection of human gastric cancer cells into the gastric wall is a recently adopted method designed to induce tumor growth in situ in athymic nude mice (Table 22.3). In this method, a population of 10^6 gastric cancer cells (either KKLS, MKN-28, or TMK-1) are being injected into the stomach wall in anesthetized animals. Sizable tumors can be notified after a few weeks to a month depending on the carcinogenic potential of the human gastric cancer cell lines [23–25].

Chemical Induction Model   Chemical carcinogens may exert their effect directly on contact at the point of application or may require biochemical activation by the host before becoming carcinogenic. Many of these activation reactions are controlled by both endogenous or genetic factors and exogenous environmental elements. Some carcinogens cause cancer in animal models after a single dose, whereas other chemicals require multiple exposures. The incidence and latent period before overt cancer is found may be a function of all these variables. Carcinogens are distinct from other drugs and chemicals

TABLE 22.3   Models of Gastric Cancer

Type of Model	Cell Types or Carcinogens	Animals	Method	Duration of Induction
Cancer cell implantation model	Human gastric cancer cells (KKLS, MKN-28, TMK-1)	Athymic nude mice	Direct injection of cells (10^6) into the gastric wall	Few weeks
Chemical induction model	N-methyl-N-nitrisourea (MNU)	Balb/c mice	240 ppm in drinking water	50 weeks
	N-methyl-N′-nitro-N-nitrosoguanidine (MNNG)	Wistar rats	25 µg/ml in drinking water	52 weeks

in that they may not cause acute toxic effects and their adverse action may require a long latent period. In animals, certain carcinogens given in large quantity may require a relatively short period of time. Some examples of these are illustrated in Table 22.3 and are discussed below.

Balb/c mice are given 240 ppm N-methyl-N-nitrosourea (MNU) in drinking water daily during alternative weeks for a total of 10 weeks and killed at the end of the 50th week. More than 70 percent survive and about 80 percent have adenocarcinoma of both differentiated and undifferentiated types [26].

Male Wistar rats are treated with 25 µg/mL N-methyl-N'-nitro-N-nitrosoguanidine (MNNG) in tap water for 52 weeks. About 60 percent of the animals develop differentiated adenocarcinomas in the glandular stomach at the antral site [27]. Higher concentrations of MNNG produce higher incidences of cancer [28].

Colorectal Cancer Models

Inflammation-Associated Model Experimental studies indicate that dextran sulfate sodium in drinking water either at a concentration of 5 percent (w/v) for 63 days or 180 days or combined with exposure to cigarette smoke generate dysplasia and/or adenocarcinoma in about 30 to 80 percent of the mice [17, 19, 29, 30]. This ulcerative colitis-associated colonic adenoma formation in mice provides an animal model closer to humans in whom with inflammation-associated neoplasia is one of the major pathways in the development of colorectal cancer. This animal model is also useful to study the relationship between colorectal cancer and cigarette smoking, which is one of the major risk factors for colorectal cancer in humans [19, 30].

Genetically Modified Model Mice with mutation of the *APC* gene or defects in the *TCTβ* and *p53* genes, have been developed to generate high percentages of animals with multiple intestinal adenomas or nodular masses in the cecum to the proximal colon by 4 months, respectively [31, 32].

Chemical Induction Models Cancers of the small bowel and colon can be induced by several carcinogens depending on route of administration (Table 22.4). The most effective carcinogens appear to be the natural product cycasin and the synthetic analogs 1,2-dimethylhydrazine (DMH) and azoxymethane (AOM) [33]. In rats, AOM can be given by subcutaneous (SC) injection once weekly for 2 weeks at a dose of 15 mg/kg body weight. Rats are killed 8, 23, and 38 weeks after the last dose of AOM to assess tumor growth and invasiveness. Most of the colon tumors obtained at week 38 are invasive or non-invasive adenocarcinomas [34].

The DMH requires metabolic activation, and intestinal microflora may play a role in the metabolism of this colon carcinogen. Both mice and rats are commonly used studying colon cancer induced by DMH. In mice, 30 mg/kg DMH is injected SC once weekly for 6 weeks [35]. However, in rats, a smaller dose (25 mg/kg) is administered once a week for 16 weeks [36]. They are killed

TABLE 22.4 Models for Carcinogenesis in Large Bowel[a]

Chemical	Animal	Route	Organ Involved
Cycasin	Rat	Oral	Large bowel
1,2-Dimethylhydrazine	Rat, mouse, hamster	Oral, SC, IP, IR	Small and large bowel
Azoxymethane	Rat, mouse	Oral, SC, IR	Small and large intestine
N-methyl-N'-nitro-N-nitroguanidine	Rat	IR	Large intestine

[a]Routes of administration: oral (OR), subcutaneous (SC), intraperitoneal (IP), and intrarectal (IR).

5 to 12 weeks after the last dose of the carcinogen for histopathological study. The DMH-induced colon tumors are very close to human colon cancer with regard to morphology, pattern of growth, and clinical manifestations.

The direct alkylating agent, MNNG, which does not require metabolic activation, is a potent topical carcinogen. Intrarectal administration of MNNG at a dose of 1 to 3 mg/rat/week for 20 weeks induces colon tumors in 100 percent of rats [37]. This is the most reliable model for the topical and selective production of tumors in the distal colon and rectum. The use of rats, especially the Fischer (F344) strain, seems to be appropriate because rat colons have light and electron microscopic morphologies as well as histochemical properties that are quite similar to those of humans.

22.4 HERBAL MEDICINES AS POTENTIAL THERAPEUTIC AGENTS FOR ULCERS AND MALIGNANT DISEASES IN THE GASTROINTESTINAL TRACT

The experimental ulcer models in animals discussed above provide a platform for studying the mechanisms of action and for screening potential antiulcer and anticancer agents derived from different herbal medicines. In fact, a number of herbal medicines are known to protect against a variety of GI disorders including those in the upper and lower GI tract described above. These medicines contain various chemical structures, have different properties, and are derived from numerous sources. In this chapter we focus on three major types of chemical compounds from various kinds of herbs: polysaccharides, terpenoids, and polyphenols. Their implications for different GI disorders and their mechanisms of action are discussed.

General

The three biological processes that play key roles in maintaining the integrity and balancing the growth of tissues are cell proliferation, apoptosis, and angiogenesis. The degrees of ulcer healing and tumor growth are greatly influenced by the levels of these biological activities. We focus on how the different herbal compounds can modulate these processes to promote ulcer healing and inhibit tumor growth in the GI tract. Many herbal medicines contain a large amount of polysaccharides and thousands of phenolic and terpenoid compounds. In this chapter, perspectives on the application of polysaccharides, terpenoids, and phenolic compounds from herbal extracts of different medicinal plants that have demonstrated antiulcer and anticarcinogenic properties in the GI tract are discussed. In addition, some novel herbs that possess anti-inflammatory and/or immunomodulatory actions and have antitumor potential are introduced.

Polysaccharides in General

Polysaccharides from various herbal medicines have been well studied for their immunomodulatory action and antitumor effect. These actions and effects may be due to stimulation of the phagocytic activity and anti-inflammatory activity. They also increase hematopoiesis in the bone marrow and promote the proliferation of several types of hematopoietic precursor cells. These findings suggest that polysaccharides of plant origins have significant pharmacological actions on somatic cells with high proliferative capacity. Recently, increasing evidence implicates that plant polysaccharides have protective and repairing activities in the GI tract similar to those from the cells in this system; they also share similar biological properties.

There are various sources of polysaccharides with different compositions. Table 22.5 summarizes the sources of polysaccharides and their pharmacological actions on the GI tract. These polysaccharides have different molecular structures and are mostly sulfated molecules with uronic acid as part of the

TABLE 22.5 Sources of Polysaccharides and Their Pharmacological Actions on the Gastrointestinal Tract and Liver

Sources	Pharmacological Actions	References
Angelica sinensis	Antiulcer and antitumor in the stomach and liver	55, 59
	Promote gastric cell migration and proliferation	56, 57 58
	Antihepatic damage	59
Asparagus racemosus	Antioxidant in hepatic cells	60
Aloe vera	Anti-inflammatory bowel disease	62
Bupleurum falcatum	Antiulcer in the stomach	52, 53
Basidiomycetes mushroom	Antitumor in the stomach and colon	47, 48
Panax ginseng	Antiulcer in the stomach	54
Ulva lactuca (green seaweed)	Inhibits colon cancer cell proliferation	40
Cladosiphon fucoidan	Gastric protection and *H. pylori* detachment from mucin	51, 52

residue. The sulfated polysaccharides, which contribute to the active components in extracts, produce a variety of pharmacological actions [38–40]. Desulfation abolished their inhibitory action on the proliferation of arterial smooth muscle cells [41]. Some of the actions of these compounds are partially dependent on the size of the polysaccharide molecules and the length of the polysaccharides determining the effectiveness of drugs in different biological systems [42, 43]. Experimental evidence shows that the β-1,3 or β-1,6 linked polysaccharides are responsible for the different pharmacological actions on vascular and tumor cells, while their analog, the α-1,4 linkage can be digested by the saccharidases to monomers in the GI tract [44, 45]. Indeed, the β-glucan receptors were first identified on human monocytes as phagocytic receptors that initiate phagocytosis [46].

Polysaccharides from Mushroom and Seaweed

The orally bioactive glucans and proteoglycans isolated from mushrooms are currently the most promising class of immunoceuticals. They are capable of augmenting the key pathways of host immunity. Higher *Basidiomycetes* mushrooms have been used in folk medicine throughout the world since ancient times. It has been known for many years that certain mushrooms of higher *Basidiomycetes* origin are effective against cancers of the stomach and esophagus [47, 48]. However, the active components responsible for these activities are largely unknown. Several antitumor polysaccharides such as hetero-β-glucans and their protein complexes have been isolated from medicinal mushrooms. Using standard methods of fractionation and purification of polysaccharides, Chihara isolated a water-soluble antitumor polysaccharide from the fruiting bodies of *Lentinus edodes* in 1978 [49]. This was found to be a 1,3-β-D-glucan with 1,6-β-D-glucopyranoside branches. Clinical data suggest that these lentinan polysaccharides can prolong the life span of patients with advanced and recurrent stomach and colorectal cancer, while displaying little toxicity [44]. Lentinan's antitumor activity is stronger than that of polysaccharides from other fungi or higher plants.

Two proteoglycans from *Coriolus versicolor*—the polysaccharides-K (PSK) and polysaccharides-peptide (PSP) have been demonstrated to extend survival by 5 years or beyond in patients with cancers of the stomach, colon, and esophagus. PSP significantly improves quality of life, provides substantial pain relief, and enhances immune status in a majority of patients with cancers of the stomach, esophagus, and other organs. The uses and the mechanisms of action of these mushroom glucans and proteoglycans for GI cancer were reviewed by Kidd in 2000 [50]. The major action is to enhance anticancer immunity by increasing the numbers of tissue and blood-borne phagocytic cells. Both PSK and PSP may function as antigenic stimuli to enhance cytotoxic killer activity directed at the tumor.

Seaweed provides another source of polysaccharides. The green seaweed *Ulvans lactuc* contains sulfated polysaccharides that can inhibit a colonic epithelial cancer cell line but have no effect on normal colonocytes. It has been reported that sulfated polysaccharides from *Cladosiphon fucoidan* confer gastric protection and are associated with *Helicobacter pylori* detachment from gastric mucin [51, 52]. These data suggested that they may be useful for patients with *H. pylori*–positive gastric ulcers by reducing the risk of gastric cancer.

Polysaccharides from Other Plants

The acidic polysaccharide fraction from the roots of *Bulpeurum falcatum*, irrespective of the route of administration, inhibits the formation of gastric lesions induced by necrotizing agents such as ethanol and HCl-ethanol [53]. The protection is due to antisecretory activity in the stomach, stimulation of the defensive mechanism through mucus secretion, and a radical scavenging effect in the gastric mucosa. However, such protection was independent of prostaglandin [53, 54]. Similarly, acidic polysaccharides from the leaves of Panax ginseng also inhibit the formation of gastric lesions through the same mechanisms [55].

Recent studies showed that a crude extract from the roots of *Angelica sinensis* mainly consisting of polysaccharides, prevents neutrophil-dependent ulcers, such as indomethacin- and ethanol-induced damage in rat stomachs [56]. It also enhances ulcer healing in the gastric mucosa, perhaps through its anti-inflammatory action. In normal rat gastric epithelial cells, the same type of extract stimulates cell proliferation and migration [56–58], the two major types of biological processes responsible for ulcer healing. Furthermore, it was demonstrated that the action is mediated in part through stimulation of the expression of epidermal growth factor and *c*-myc at the ulcer site followed by activation of ornithine decarboxylase activity [57, 58]. The protection extends to the liver damage provoked by paracetamol and carbon tetrachloride. This effect is partially mediated through the inhibition of nitric oxide synthase in hepatocytes [59]. Polysaccharides from the same herb also inhibit the invasion and metastasis of hepatocellular carcinoma cells in vitro [60]. In addition to the polysaccharides from *A. sinensis*, other natural polysaccharides from plants have protective effects on the liver. For example, a purified aqueous fraction from *Asparagus racemosus* shows potent antioxidant properties in mitochondrial membranes isolated from rat liver cells [61] and an aqueous extract from the seeds of *Celosia argentea*, containing an acidic polysaccharide, protects against chemically and immunologically induced hepatitis [62]. Recently, an aloe vera gel, a mucilaginous aqueous extract, was shown to have therapeutic potential in inflammatory bowel disease in humans [63].

Phenolic Compounds and Terpenoids from Medicinal Plants

Curcumin and Phenolic Compounds Numerous phenolic substances present in fruits or medicinal plants have been found to possess potential cancer chemopreventive activities [64]. One of these is curcumin, a yellow pigment isolated from the rhizome of *Curcuma longa* L., Zingiberaceae, which is extensively used for imparting color and flavor to food. Although it has been used as a chemopreventive agent in Asia for years, investigations of its mode of action remain limited. Turmeric, the powdered form of *C. longa*, is a traditional medicinal remedy that exerts an antioxidant effect with antimutagenic activity [65]. One of the likely reasons for its action is the phenolic nature of its active ingredient curcumin (diferuloylmethane) and the other curcuminoid derivatives in the plant extract, such as feruloyl-*p*-hydroxy-cinamoylmethane and *bis*-(*p*-hydroxy-cinamoyl) methane [66]. Among these, curcumin has been extensively studied as a chemopreventive agent against cancers in various models, including its cytotoxicity to human HCT-15 and HT-29 colon cancer cells [67]. Curcumin is known to have immunomodulatory functions. Studies revealed upregulation of intestinal mucosal $CD4^+$ T cells and B cells by curcumin, which could be essential for its tumor-inhibitory activity [68].

Chemopreventive activity of synthetic curcumin had been observed in colon carcinogenesis models [69]. In addition, it had demonstrated anticarcinogenic effects on papillomas and squamous cell carcinomas of the forestomach, as well as on adenomas and adenocarcinomas of the duodenum in mice [70]. Inhibition of cell proliferation and induction of apoptosis via the Fas signaling pathway had been suggested to be the events responsible for such effects [71]. Given its specific mode of action, some suggested a correlation with the modulation of arachidonic acid metabolism [72, 73], while others had proposed the involvement of nuclear transcriptional factors like AP-1 (activator protein-1) and nuclear factor-kappaB (NF-κB) [74]. Recent data suggested that IκBα, the inhibitory unit of NF-κB, could be cleaved by caspase-3, followed by subsequent NF-κB activation and eventually promote tumor cell death [75]. Other than the transcriptional deactivation pathway, curcumin had also demonstrated an inhibitory action on intracellular signals from protein kinase C [76]. All these phenomena imply that inhibition of colon carcinogenesis by curcumin involves modulation at different cellular and molecular levels. As a matter of fact, the steps in deregulating signaling pathways and/or directly modulating gene transcription could theoretically be targeted as potential sites for chemopreventive intervention. *Curcuma* extract can be administered safely to colorectal cancer patients at doses of up to 2.2 g daily, which is equivalent to about 180 mg of curcumin [77]. With a clear understanding of the precise mechanistic pathways of the chemopreventive actions exerted by the curcuminoid compounds, *Curcuma* extract and curcumin will certainly play a permissive role in the development of target-oriented agents against colon carcinogenesis and the pathogenesis of other GI cancers.

Licorice and Its Constituents Terpenoids and Polyphenols During the search for new chemopreventive agents, some plant triterpenoids have been found to show anticarcinogenic properties. Licorice root (*radix Glycyrrhizae glabra*) is one of the oldest and most frequently used botanicals in Chinese medicine. In the United States and many Asian countries, licorice products are mostly used as flavoring and sweetening agents in the food and tobacco industries. Cytotoxic and radical scavenging activities of *G. radix* extract have been reported [78]. Essential constituents of licorice root include triterpenoids, such as glycyrrhizin and its aglycone glycyrrhizic acid, along with various polyphenols and polysaccharides [79]. Among these, glycyrrhizic acid (making up 4 to 20 percent of the root) consists of a glucuronic acid moiety attached to a steroid-like triterpenoid glycyrrhetinic acid, which could be responsible for the antioxidation and anti-inflammatory effects (being an inhibitor of both lipoxygenase and cyclooxygenase). The activation of glucuronidation suggests a possible role in the induction of xenobiotic detoxification in the liver. Besides its mild estrogenic effects, glycyrrhizin also encourages the production of hormones such as hydrocortisone and inhibits cyclooxygenase-2 (COX-2) activity, which may further explain its anti-inflammatory and anticarcinogenic properties.

Glycyrrhizae radix is the active ingredient in many Chinese/oriental herbal complex formulations. Experimental studies of these decoctions indicated significant potential to inhibit deoxyribonucleic acid (DNA)–synthesizing enzyme activity in colonic carcinomas [80] and to prevent formation of aberrant crypt foci in a colonic carcinogenesis model [81]. A phytopharmaceutical containing *Glycyrrhizae* extract (Revitonil tablets) had demonstrated strong immunostimulating potential [82]. Both licorice root and glycyrrhizin modulate immune activity by inducing interferon and interleukin production in mice [83, 84]. Glycyrrhizin has been reported to stimulate host resistance against solid tumors as well as to inhibit colon tumor cell activity [85]. However, glycyrrhizin is contraindicated in patients with hypertension, heart failure, renal diseases, and liver cirrhosis. Deglycyrrhinized licorice is currently under investigation for its safety and potential clinical use.

Other than glycyrrhizin, about 300 different polyphenols (making up 1 to 5 percent of the root) also possess antioxidative properties and have antitumor potential [86]. Many of them are flavonoids and isoflavonoids, which have weaker estrogenic activity than the triterpenoids. Glabridine, a major isoflavan in licorice root, possesses some estrogen-like activities and a strong antiproliferative action in human breast cancer cells [87]. Similar effects can also be produced by the chalcone isoliquiritigenin, despite its smaller quantity in the licorice root than other flavonoids and triterpenoids [88]. In fact, the tumor growth inhibition produced by isoliquiritigenin seems to be associated with its potential in inducing cell cycle arrest [89]. Besides growth inhibition, isoliquiritigenin also possesses a potent lipoxygenase (LOX) inhibitory effect, rather than COX inhibition [90]. Another possible antitumor action of isoliquiritigenin is its antitube formation effect, which contributes much to the antiangiogenic action of licorice root extract [91]. Alternatively, glabridine and the isoflavene glabrene had both demonstrated activity against *H. pylori*. This could contribute at least in part, to the gastroprotection of the whole extract of licorice root, which contains chemopreventive agents useful against peptic ulcer and gastric cancer [92]. Despite there not being much direct evidences that nonglycyrrhizin polyphenols from *Glycyrrhizae* extract can exert antitumor action on GI cancers, the fact that isoliquiritigenin can significantly suppress carcinogen-induced aberrant crypt focus formation in mouse colon [93] is at least an indication of their chemopreventive potential against intestinal cancers.

Other Herbs with Terpenoids *Tripterygium wilfordii* Hook F is a Chinese medicinal herb that belongs to the Celastraceae family. Historically, extracts of *T. wilfordii* were used in traditional Chinese medicine and later found to be effective in the treatment of inflammatory/autoimmune disorders such as rheumatoid arthritis [94]. The polysaccharide moiety from *Tripterygium* extract has exhibited profound immunosuppressive properties, inhibiting tumor necrosis factor-α (TNF-α) and cell adhesion molecule expression in human monocytes [95]. It is now known that some diterpenoid components in *T. wilfordii* such as triptolide exert their anti-inflammatory and immunosuppressant effects by inhibiting cytokine production by T lymphocytes [96, 97]. PG490-88 (14-succinyl triptolide sodium salt) is a semisynthetic compound derived from the diterpene triepoxide, or triptolide (PG490), that comes from the *T. wilfordii* extract. PG490-88 alone caused a dose-dependent inhibition of COLO 205 colorectal tumor growth as well as tumor regression. Moreover, PG490 blocked the induction of NF-κB by TNF-α and p53 transcriptional activity induced by chemotherapy, which implicated its synergistic action with other DNA-damaging chemotherapeutic drugs [98]. PG490-88 is now in phase I clinical trials for patients with solid tumors. A recent study had shown that PG490 inhibits metastasis of solid tumors from the stomach [99]. All these findings indeed illustrate the potential of *T. wilfordii* and its diterpenoid component triptolide in treating GI cancers.

The root barks of the Chinese tree *Pseudolarix kaempferi* (Tujingpi), also called Golden Larch Bark in the Western world, contains active medicinal compounds called pseudolaric acids. It is traditionally used in Chinese medicine to treat fungal infections. Pseudolaric acids A, B, C, and D are novel diterpenes extracted from *P. kaempferi*. Among these, pseudolaric acid B is the major antifungal constituent while it also demonstrates cytotoxic action against colorectal cancer cells [100]. In addition, other novel triterpene lactones isolated from the herb also possessed cytotoxicity against human colon cancer cells [101]. Recently, it was reported that pseudolaric acid B can act as a ligand to activate the nuclear peroxisome proliferator activated receptors (PPAR) α, γ, and δ [102]. Both gene and protein expressions of the nuclear hormone receptor PPARγ were upregulated in rodent colon tumors and in certain human colon cancer cells [103]. Since PPARγ is a ligand-modulated transcription factor, it may provide a novel target for chemopreventive strategies for colorectal cancer. The ability of PPARγ activation to concomitantly inhibit COX-2 expression and induce apoptosis in colon cancer cells illustrates its crucial role in anticarcinogenesis [104]. A similar consequence of PPARγ activation was also observed in human gastric cancer cells through induction

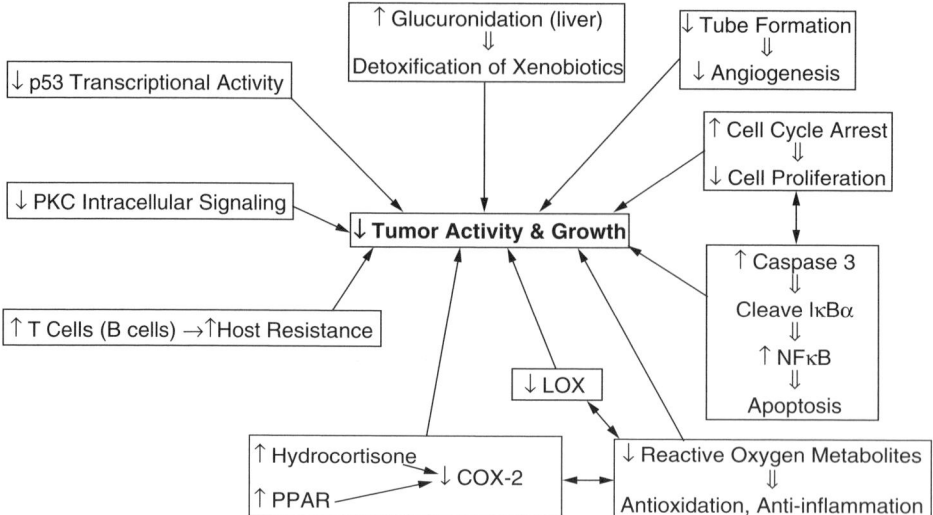

Figure 22.1 Proposed mechanistic actions of terpenes and phenolic compounds.

of apoptosis together with G1 cell cycle arrest [105]. In this respect, it appears that *P. kaempferi* and its terpenoid components should be a target for active investigation for the development of effective chemotherapeutic strategies in the twenty-first century. The proposed mechanistic actions for the different terpenoids and phenolic compounds reported in this chapter are summarized in Figure 22.1.

REFERENCES

1. Szabo, S., Trier, J. S., Brown, A., Schnoor, J., Homan, H. D., Bradford, J. C. (1985). A quantitative method for assessing the extent of experimental gastric erosions and ulcer. *J. Pharmacol. Methods*, 13, 59–66.
2. Domschke, S., Domschke, W. (1984). Gastroduodenal damage due to drugs, alcohol and smoking. *Clin. Gastroenterol.*, 13, 405–436.
3. Lichtenberger, L. M., Graziani, L. A., Dial, E. J., Butler, B. D., Hills, B. A. (1983). Role of surface active phospholipases in gastric cytoprotection. *Science*, 219, 1327–1329.
4. Whittle, B. J. R., Steel, G. (1985). Evaluation of rat gastric mucosa by a prostaglandin analogue using cellular enzyme marker and histological techniques. *Gastroenterology*, 88, 315–327.
5. Szabo, S., Cho, C. H. (1988). Animal models for studying the role of eicosanoids in peptic ulcer disease. In K. Hillier (Ed.), *Eicosanoids and the Gastrointestinal Tract*. MTP Press, Lancaster, UK, pp. 75–102.

6. Tsukimi, Y., Okabe, S. (1994). Validity of kissing gastric ulcers induced in rats for screening of antiulcer drugs. *J. Gastroenterol. Hepatol.*, *9*, S60–S65.
7. Robert, A., Stout, T. C., Dale, J. E. (1970). Production by secretogogues of duodenal ulcers in the rat. *Gastroenterology*, *59*, 95–102.
8. Cho, C. H., Pfeiffer, C. J. (1981). Gastrointestinal ulceration in the guinea pig in response to dimaprit, histamine and H_1- and H_2-blocking agents. *Digest. Dis. Sci.*, *26*, 306–311.
9. Szabo, S. (1978). Animal model of human disease. Duodenal ulcer disease. Animal model: Cysteamine-induced acute and chronic duodenal ulcer in the rat. *Am. J. Pathol.*, *93*, 273–276.
10. Elson, C. O., Sartor, R. B., Tennyson, G. S., Riddell, R. H. (1995). Experimental models of inflammatory bowel disease. *Gastroenterology*, *109*, 1344–1368.
11. Kim, H. S., Berstad, A. (1992). Experimental colitis in animal models. *Scand. J. Gastroenterol.*, *27*, 529–537.
12. Sartor, R. B. (1997). How relevant to human inflammatory bowel diseases are current animal models of intestinal inflammation? *Aliment. Pharmacol. Ther.*, *11*(Suppl. 3), 89–97.
13. Morris, G. P., Beck, P. L., Herridge, M. S., Depew, W. T., Szewczuk, M. R., Wallace, J. L. (1989). Hapten-induced model of chronic inflammation and ulceration in the rat colon. *Gastroenterology*, *96*, 795–804.
14. Chin, K. W., Barrett, K. E. (1994). Mast cells are not essential to inflammation in murine model of colitis. *Digest. Dis. Sci.*, *39*, 513–525.
15. Yamada, Y., Marshall, S., Specian, R. D., Grisham, M. B. (1992). A comparative analysis of two models of colitis in rats. *Gastroenterology*, *102*, 1524–1534.
16. Hawkins, J. V., Emmel, E. L., Feuer, J. J., Nedelman, M. A., Harvey, C. J., Kennedy, A. R., Lichtenstein, G. R., Billings, P. C. (1997). Protease activity in a hapten-induced model of ulcerative colitis in rats. *Digest. Dis. Sci.*, *42*, 1969–1980.
17. Okayasu, I., Hatakeyama, S., Yamada, M., Ohkusa, T., Inagaki, Y., Nakaya, R. (1990). A novel method in the induction of reliable experimental acute and chronic ulcerative colitis in mice. *Gastroenterology*, *98*, 694–702.
18. Copper, H. S., Murphy, S., Kido, K., Yoshitake, H., Flanigan, A. (2000). Dysplasia and cancer in the dextran sulfate sodium mouse colitis model. Relevance to colitis-associated neoplasia in the human: A study of histopathology, B-catenin and p53 expression and the role of inflammation. *Carcinogenesis*, *21*, 757–768.
19. Liu, E. S. L., Ye, Y. N., Shin, V. Y., Yuen, S. T., Leung, S. Y., Wong, B. C. Y., Cho, C. H. (2003). Cigarette smoke exposure increase ulcerative colitis-associated colonic adenoma formation in mice. *Carcinogenesis*, *24*, 1407–1413.
20. Guo, X., Wang, W. P., Cho, C. H. (1999). Involvement of neutrophils and free radicals in the potentiating effects of passive cigarette smoking on inflammatory bowel disease in rats. *Gastroenterology*, *117*, 884–892.
21. Wallace, J. L., Keenan, C. M. (1990). An orally active inhibitor of leukotriene synthesis accelerates healing in a rat model of colitis. *Am. J. Physiol.*, *258*, G527–G537.
22. Suzuki, K., Ota, H., Sasagawa, S., Sakatani, T., Fijikura, T. (1983). Assay method for myeloperoxidase in human polymorphonuclear leukocytes. *Anal. Biochem.*, *132*, 345–352.

23. Jung, Y. D., Mansfield, P. F., Akagi, M., Takeda, A., Liu, W., Bucana, C. D., Kicklin, D. J., Ellis, L. M. (2002). Effects of combination anti-vascular endothelial growth factor receptor and anti-epidermal growth factor receptor therapies on the growth of gastric cancer in a nude mouse model. *Eur. J. Cancer*, 38, 1133–1140.
24. Shin, V. Y., Wu, W. K. K., Liu, E. S. L., Ye, Y. N., So, W. H. L., Koo, M. W. L., Liu, E. S. L., Luo, J. C., Cho, C. H. (2004). Nicotine promotes gastric tumor growth and neovascularization by activating extracellular regulated signal kinase and cycylooxygenase-2. *Carcinogenesis*, 25, 2487–2495.
25. Takahashi, Y., Mai, M., Nishioka, K. (2000). α-Difluoromethylornithine induces apoptosis as well as anti-angiogenesis in the inhibition of tumor growth and metastasis in human gastric cancer model. *Int. J. Cancer*, 85, 243–247.
26. Yamachika, T., Nakanishi, H., Inada, K. I., Tsukamoto, T., Shimizu, N., Kobayashi, K., Fukushima, S., Tatematsu, M. (1998). N-methyl-N-nitrosourea concentration-dependent, total intake-dependent, induction of adenocarcinomas in the glandular stomach of Balb/c mice. *Jpn. J. Cancer Res.*, 89, 385–391.
27. Yano, H., Tatsuta, M., Iishi, H., Baba, M., Sakai, N., Uedo, N. (1999). Attenuation by d-limonene of sodium chloride-enhanced gastric carcinogenesis induced by N-methyl-N'-nitro-N-nitrosoguanidine in Wistar rats. *Int. J. Cancer*, 82, 665–668.
28. Uedo, N., Tatsuta, M., Bara, M., Hirasawa, R., Iishi, H., Yano, H., Sakai, N., Uehara, H., Nakaizumi, A. (1999). Inhibition by rat C-erbB-2/neu antisense oligonucleotide of gastric carcinogenesis induced by N-methyl-N'-nitrosoguanidine in Wistar rats. *Int. J. Cancer*, 83, 670–673.
29. Yamada, M., Ohkusa, T., Okayasu, I. (1992). Occurrence of dysplasia and adenocarcinoma after experimental chronic ulcerative colitis in hamsters induced by dextran sulphate sodium. *Gut*, 33, 1521–1527.
30. Ye, Y. N., Liu, E. S. L., Shin, V. Y., Wu, W. K. K., Cho, C. H. (2004). Contributory role of 5-lipoxygenase and its association with angiogenesis in the promotion of inflammation-associated colonic tumorigenesis by cigarette smoking. *Toxicology*, 203, 179–188.
31. Kado, S., Uchida, K., Funabashi, H., Iwata, S., Nagata, Y., Ando, M., Onoue, M. (2001). Intestinal microflora are necessary for development of spontaneous adenocarcinoma of the large intestine in T-cell receptor β chain and p53 double-knockout mice. *Cancer Res.*, 61, 2395–2398.
32. Rao, C. V., Cooma, I., Rodriguez, J. G. R., Simi, B., El-Bayoumy, K., Reddy, B. S. (2000). Chemoprevention of familial adenomatous polyposis development in the APC^{min} mouse model by 1,4 phenylene *bis* (methylene) selenocyanate. *Carcinogenesis*, 21, 617–621.
33. Bralow, S. P., Weisburger, J. H. (1976). Experimental carcinogenesis in the digestive organs. *Clin. Gastroenterol.*, 5, 527–542.
34. Rao, C. V., Hirose, Y., Indranie, C., Reddy, B. S. (2001). Modulation of experimental colon tumorigenesis by types and amounts of dietary fatty acids. *Cancer Res.*, 61, 1927–1933.
35. Symolon, H., Schmelz, E. M., Dillehay, D. L., Merrill Jr., A. H. (2004). Dietary soy sphingolipids suppress tumorigenesis and gene expression in 1,2-dimethyldrazine-treated CG1 mice and $Apc^{Min/+}$ mice[1,2]. *J. Nutr.*, 134, 1157–1161.
36. Nakaji, S., Ishiguro, S., Iwane, S., Ohta, M., Sugawara, K., Sakamoto, J., Fukuda, S. (2004). The prevention of colon carcinogenesis in rats by dietary cellulose is

greater than the promotive effect of dietary lard as assessed by repeated endoscopic observation. *J. Nutr.*, *134*, 935–939.

37. Narisawa, T., Magadia, N. E., Weisburger, J. H., Wynder, E. L. (1974). Promoting effect of bile acids on colon carcinogenesis after intrarectal instillation of N-methyl-N'-nitro-N-nitrosoguanidine in rats. *J. Natl. Cancer Inst.*, *53*, 1093–1097.

38. Benitz, W. E., Lessler, D. S., Coulson, J. D., Bernfield, M. (1986). Heparin inhibits proliferation of fetal vascular smooth muscle cells in the absence of platelet-derived growth factor. *J. Cell. Physiol.*, *127*, 1–7.

39. Klein-Soyer, C., Beretz, A., Cazenave, J. P., Wittendorp-Rechenmann, E., Vonesch, J. L., Rechenmann, R. V., Driot, F., Maffrand, J. P. (1989). Sulfated polysaccharides modulate effects of acidic and basic fibroblast growth factors on repair of injured confluent human vascular endothelium. *Arterioscler. Thromb. Vasc. Biol.*, *9*, 147–153.

40. Kaeffer, B., Benard, C., Lahaye, M., Blottiere, H. M., Cherbut, C. (1999). Biological properties of ulvan, a new source of green seaweed sulfated polysaccharides on cultured normal and cancerous colonic epithelial cells. *Planta Med.*, *65*, 527–531.

41. Vischer, P., Buddecke, E. (1991). Different action of heparin and fucoidan on arterial smooth muscle cell proliferation and thrombospondin and fibronectin metabolism. *Eur. J. Cell Biol.*, *56*, 407–414.

42. Sudhalter, J., Folkman, J., Svah, C. M., Bergendal, K., D'Amore, P. A. (1989). Importance of size, sulfation, and anticoagulant activity in the potentiation of acidic fibroblast growth factor by heparin. *J. Biol. Chem.*, *264*, 6892–6897.

43. Logeart, D., Prigent-Richard, S., Boisson-Vidal C., Chaubet, F., Durand, P., Jozefonvicz, J., Letourneur, D. (1997). Fucans, sulfated polysaccharides extracted from brown seaweeds, inhibit vascular smooth muscle cell proliferation. II. Degradation and molecular weight effect. *Eur. J. Cell Biol.*, *74*, 385–390.

44. Chihara, G. (1992). Recent progress in immunopharmacology and therapeutic effects of polysaccharides. *Devel. Biol. Stand.*, *77*, 191–197.

45. Miao, H. Q., Ishai-Michaeli, R., Peretz, T., Vlodavsky, I. (1995). Laminarin sulfate mimics the effects of heparin on smooth muscle cell proliferation and basic fibroblast growth factor-receptor binding and mitogenic activity. *J. Cell. Physiol.*, *164*, 482–490.

46. Czop, J. K., Kay, J. (1991). Isolation and characterization of β-glucan receptors on human mononuclear phagocytes. *J. Exper. Med.*, *173*, 1511–1520.

47. Yang, Q. Y., Jong, S. C. (1989). Medicinal mushroom in China. *Mushroom Sci.*, *12*, 631–643.

48. Wasser, S. P., Weis, A. L. (1999). Therapeutic effects of substances occurring in higher Basidiomycetes mushroom: A modern perspective. *Crit. Rev. Immunol.*, *19*, 65–96.

49. Chihara, G. (1978). Antitumor and immunological properties of polysaccharides from fungal origin. *Mushroom Sci.*, *9*, 797–814.

50. Kidd, P. M. (2000). The use of mushroom glucans and proteoglycans in cancer treatment. *Alt. Med. Rev.*, *5*, 4–27.

51. Shibata, H., Kimura-Takagi, I., Nagaoka, M., Hashimoto, S., Aiyyama, R., Ueyama, S., Yokokura, T. (2000). Properties of fucoidan from *Cladosiphon okamuranus tokida* in gastric mucosal protection. *Biofactors*, *11*, 235–245.

52. Shibata, H., Iimuro, M., Uchiya, N., Kawamori, T., Nagaoka, M., Ueyama, S., Hashimoto, S., Yokokura, T., Sugimura, T., Wakabay, K. (2003). Preventive effects of *Cladosiphon fucidan* against *Helicobacter pylori* infection in Mongolian gerbils. *Helicobacter*, 8, 59–65.

53. Sun, X. B., Matsumoto, T., Yamada, H. (1991). Effects of a polysaccharide fraction from the roots of *Bupleurum falcatum L.* on experimental gastric ulcer models in rats and mice. *J. Pharm. Pharmacol.*, 43, 699–704.

54. Yamada, H. (1995). Structure and pharmacological activity of pectic polysaccharides from the roots of *Bulpleurum falcatum L. Folia Pharmacol. Jpn.*, 106, 229–237.

55. Sun, X. B., Matsumoto, T., Yamada, H. (1992). Anti-ulcer activity and mode of action of the polysaccharide fraction from the leaves of Panax ginseng. *Planta Med.*, 58, 432–435.

56. Cho, C. H., Mei, Q. B., Shange, P., Lee, S. S., So, H. L., Guo, X., Li, Y. (2000). Study of the gastrointestinal protective effects of polysaccharides from *Angelica sinensis* in rats. *Planta Med.*, 66, 348–351.

57. Ye, Y. N., Koo, W. W. L., Li, Y., Matsui, H., Cho, C. H. (2001). *Angelica sinensis* modulates migration and proliferation of gastric epithelial cells. *Life Sci.*, 68, 961–968.

58. Ye, Y. N., Liu, E. S. L., Shin, V. Y. Cho, C. H. (2001). A mechanistic study of proliferation induced by *Angelica sinensis* in a normal gastric epithelial cell line. *Biochem. Pharmacol.*, 61, 1439–1448.

59. Ye, Y. N., Li, Y., So, H. L., Cho, C. H. (2001). Protective effects of polysaccharide-enriched fraction from *Angelica sinensis* on hepatic injury. *Life Sci.*, 69, 637–646.

60. Shang, P., Qian, A. R., Yang, T. H., Jia, M., Mei, Q. B., Cho, C. H., Zhao, W. M., Chen, Z. N. (2003). Experimental study of anti-tumor effects of polysaccharides from *Angelica sinensis*. *World J. Gastroenterol.*, 9, 1963–1967.

61. Jayashree, P. K., Krutin, K. B., Thomas, P. A. D. (2000). Antioxidant properties of *Asparagus racemosus* against damage induced by γ-radiation in rat liver mitochondria. *J. Ethnopharmacol.*, 71, 425–435.

62. Hase, K., Kadota, S., Basnet, P., Takahashi, T., Namba, T. (1996). Protective effect of celosian, an acidic polysaccharide, on chemically and immunologically induced liver injuries. *Biol. Pharm. Bull.*, 19, 567–572.

63. Langmead, L., Feakins, R. M., Goldthorpe, S., Holt, H., Tsironi, E., De Silva, A., Jewell, D. P., Rampton, D. S. (2004). Randomized, double-blind, placebo-controlled trial of oral aloe vera gel for active ulcerative colitis. *Aliment. Pharmacol. Ther.*, 19, 739–747.

64. Surh, Y-J. (1999). Molecular mechanisms of chemopreventive effects of selected dietary and medicinal phenolic substances. *Mutat. Res.*, 428, 305–327.

65. Conney, A. H., Lou, Y. R., Xie, J. G., Osawa, T., Newmark, H. L., Liu, Y., Chang, R. L., Huang, M. T. (1997). Some perspectives on dietary inhibition of carcinogenesis: Studies with curcumin and tea. *Proc. Soc. Exper. Biol. Med.*, 216, 234–245.

66. Anto, R. J., George, J., Babu, K. V. D., Rajasekharan, K. N., Kuttan, R. (1996). Antimutagenic and anticarcinogenic activity of natural and synthetic curcuminoids. *Mutat. Res.*, 370, 127–131.

67. Syr, W-R., Shen, C-C., Don, M-J., Ou, J-C., Lee, G-H., Sun, C-M. (1998). Cytotoxicity of curcuminoids and some novel compounds from *Curcuma zedoaria*. *J. Nat. Prod.*, *61*, 1531–1534.

68. Churchill, M., Chadburn, A., Bilinski, R. T., Bertagnolli, M. M. (2000). Inhibition of intestinal tumors by curcumin is associated with changes in the intestinal immune cell profile. *J. Surg. Res.*, *89*, 169–175.

69. Kawamori, T., Lubet, R., Steele, V. E., Kellof, G. J., Kaskey, R. B., Rao, C. V., Reddy, B. S. (1999). Chemopreventive effect of curcumin, a naturally occurring anti-inflammatory agent, during the promotion/progression stages of colon cancer. *Cancer Res.*, *59*, 597–601.

70. Huang, M-T., Lou, Y-R., Ma, W., Newmark, H. L., Reuhl, K. R., Conney, A. H. (1994). Inhibitory effects of dietary curcumin on forestomach, duodenum, and colon carcinogenesis in mice. *Cancer Res.*, *54*, 5841–5847.

71. Moragoda, L., Jaszewski, R., Majumdar, A. P. N. (2001). Curcumin induced modulation of cell cycle and apoptosis in gastric and colon cancer cells. *Anticancer Res.*, *21*, 873–878.

72. Plummer, S. M., Holloway, K. A., Manson, M. M., Munks, R. J. L., Kaptein, A., Farrow, S., Howells, L. (1999). Inhibition of cyclo-oxygenase 2 expression in colon cells by the chemopreventive agent curcumin involves inhibition of NF-κB activation via the NIK/IKK signaling complex. *Oncogene*, *18*, 6013–6020.

73. Rao, C. V., Rivenson, A., Simi, B., Reddy, B. S. (1995). Chemoprevention of colon carcinogenesis by dietary curcumin, a naturally occurring plant phenolic compound. *Cancer Res.*, *55*, 259–266.

74. Baeuerle, P. A., Henkek, T. (1994). Function and activation of NF-kappa B in the immune system. *Annu. Rev. Immunol.*, *12*, 141–179.

75. Reuther, J. Y., Baldwin, A. S. Jr. (1999). Apoptosis promotes a caspase-induced amino-terminal truncation of I-kappa-B-alpha that functions as a stable inhibitor of NF-kappa B. *J. Biol. Chem.*, *274*, 20664–20670.

76. Liu, J. Y., Lin, S. J., Lin, J. K. (1993). Inhibitory effects of curcumin on protein kinase C activity induced by 12-O-tetradecanoyl-phorbol-13-acetate in NIH 3T3 cells. *Carcinogenesis*, *14*, 857–861.

77. Sharma, R. A., McLelland, H. R., Hill, K. A., Ireson, C. R., Euden, S. A., Manson, M. M., Pirmohamed, M., Marnett, L. J., Gescher, A. J., Steward, W. P. (2001). Pharmacodynamic and pharmacokinetic study of oral curcuma extract in patients with colorectal cancer. *Clin. Cancer Res.*, *7*, 1894–1900.

78. Nemoto, Y., Satoh, K., Toriizuka, K., Hirai, Y., Tobe, T., Sakagami, H., Nakashima, H., Ida, Y. (2002). Cytotoxic and radical scavenging activity of blended herbal extracts. *In Vivo*, *16*, 327–332.

79. Wang, Z. Y., Nixon, D. W. (2001). Licorice and cancer. *Nutr. Cancer*, *39*, 1–11.

80. Sakamoto, S., Mori, T., Sawaki, K., Kawachi, Y., Kuwa, K., Kudo, H., Suzuki, S., Sugiura, Y., Kasahara, N., Nagasawa, H. (1993). Effects of kampo (Japanese herbal) medicine "sho-saiko-to" on DNA-synthesizing enzyme activity in 1,2-dimethylhydrazine-induced colonic carcinomas in rats. *Planta Med.*, *59*, 152–154.

81. Yoo, B. H., Lee, B. H., Kim, J. S., Kim, N. J., Kim, S. H., Ryu, K. W. (2001). Effects of Shikunshito-Kamiho on fecal enzymes and formation of aberrant crypt foci induced by 1,2-dimethylhydrazine. *Biol. Pharm. Bull.*, *24*, 638–642.

82. Wagner, H., Jurcic, K. (2002). Immunological studies of Revitonil, a phytopharmaceutical containing *Echinacea purpurea* and *Glycyrrhiza glabra* root extract. *Phytomedicine*, 9, 390–397.
83. Tomoda, M., Shimizu, N., Kanari, M., Gonda, R., Arai, S., Okuda, Y. (1990). Characterization of two polysaccharides having activity on the reticuloendothelial system from the root of *Glycyrrhiza uralensis*. *Chem. Pharm. Bull.*, 38, 1667–1671.
84. Dai, J. H., Iwatani, Y., Ishida, T., Terunuma, H., Kasai, H., Iwakula, Y., Fujiwara, H., Ito, M. (2001). Glycyrrhizin enhances interleukin-12 production in peritoneal macrophages. *Immunology*, 103, 235–243.
85. Chung, J. G., Chang, H. L., Lin, W. C., Wang, H. H., Yeh, C. C., Hung, C. F., Li, Y. C. (2000). Inhibition of N-acetyltransferase activity and DNA–2-aminofluorene adducts by glycyrrhizic acid in human colon tumour cells. *Food Chem. Toxicol.*, 38, 163–172.
86. Rafi, M. M., Vastano, B. C., Zhu, N., Ho, C. T., Ghai, G., Rosen, R. T., Gallo, M. A., DiPaola, R. S. (2002). Novel polyphenol molecule isolated from licorice root (*Glycyrrhiza glabra*) induces apoptosis, G2/M cell cycle arrest, and Bcl-2 phosphorylation in tumor cell lines. *J. Agric. Food Chem.*, 50, 677–684.
87. Tamir, S., Eizenberg, M., Somjen, D., Stern, N., Shelach, R., Kaye, A., Vaya, J. (2000). Estrogenic and antiproliferative properties of glabridin from licorice in human breast cancer cells. *Cancer Res.*, 60, 5704–5709.
88. Maggiolini, M., Statti, G., Vivacqua, A., Gabriele, S., Rago, V., Loizzo, M., Menichini, F., Amdo, S. (2002). Estrogenic and antiproliferative activities of isoliquiritigenin in MCF7 breast cancer cells. *J. Steroid Biochem. Mol. Biol.*, 82, 315–322.
89. Kanazawa, M., Satomi, Y., Mizutani, Y., Ukimura, O., Kawauchi, A., Sakai, T., Baba, M., Okuyama, T., Nishino, H., Miki, T. (2003). Isoliquiritigenin inhibits the growth of prostate cancer. *Eur. Urol.*, 43, 580–586.
90. Yamamoto, S., Aizu, E., Jiang, H., Nakadate, T., Kiyoto, I., Wang, J. C., Kato, R. (1991). The potent anti-tumor-promoting agent isoliquiritigenin. *Carcinogenesis*, 12, 317–321.
91. Kobayashi, S., Miyamoto, T., Kimura, I., Kimura, M. (1995). Inhibitory effect of isoliquiritin, a compound in licorice root, on angiogenesis in vivo and tube formation in vitro. *Biol. Pharm. Bull.*, 18, 1382–1386.
92. Fukai, T., Marumo, A., Kaitou, K., Kanda, T., Terada, S., Nomura, T. (2002). Anti-*Helicobacter pylori* flavonoids from licorice extract. *Life Sci.*, 71, 1449–1463.
93. Baba, M., Asano, R., Takigami, I., Takahashi, T., Ohmura, M., Okada, Y., Sugimoto, H., Arika, T., Nishino, H., Okuyama, T. (2002). Studies on cancer chemoprevention by traditional folk medicines XXV. Inhibitory effect of isoliquiritigenin on azoxymethane-induced murine colon aberrant crypt focus formation and carcinogenesis. *Biol. Pharm. Bull.*, 25, 247–250.
94. Tao, X., Lipsky, P. E. (2000). The Chinese anti-inflammatory and immunosuppressive herbal remedy *Tripterygium wilfordii* Hook F. *Rheum. Dis. Clin. N. Am.*, 26, 29–50.
95. Luk, J. M., Lai, W., Tam, P., Koo, M. W. L. (2000). Suppression of cytokine production and cell adhesion molecule expression in human monocytic cell line THP-1 by *Tripterygium wilfordii* polysaccharide moiety. *Life Sci.*, 67, 155–163.

96. Chen, B. J. (2001). Triptolide, a novel immunosuppressive and anti-inflammatory agent purified from a Chinese herb *Tripterygium wilfordii* Hook F. *Leukem. Lymph.*, 42, 253–265.
97. Qiu, D., Zhao, G., Aoki, Y., Shi, L., Uyei, A., Nazarian, S., Ng, J. C., Kao, P. N. (1999). Immunosuppressant PG490 (Triptolide) inhibits T-cell interleukin-2 expression at the level of purine-box/nuclear factor of activated T-cells and NF-κB transcriptional activation. *J. Biol. Chem.*, 274, 13443–13450.
98. Fidler, J. M., Li, K., Chung, C., Wei, K., Ross, J. A., Gao, M., Rosen, G. D. (2003). PG490-88, a derivative of triptolide, causes tumor regression and sensitizes tumors to chemotherapy. *Mol. Cancer Ther.*, 2, 855–862.
99. Jiang, X. H., Wong, B. C., Lin, M. C., Zhu, G. H., Kung, H. F., Jiang, S. H., Yang, D., Lam, S. K. (2001). Functional p53 is required for triptolide-induced apoptosis and AP-1 and nuclear factor-κB activation in gastric cancer cells. *Oncogene*, 20, 8009–8018.
100. Pang, D. J., Li, Z. L., Hu, C. Q., Chen, K., Chang, J. J., Lee, K. H. (1990). The cytotoxic principles of *Pseudolarix kaempferi*: Pseudolaric acid-A and -B and related derivatives. *Planta Med.*, 56, 383–385.
101. Chen, G. F., Li, Z. L., Pan, D. J., Tang, C. M., He, X., Xu, G. Y., Chen, K., Lee, K. H. (1993). The isolation and structural elucidation of four novel triterpene lactones, pseudolarolides A, B, C, and D, from *Pseudolarix kaempferi*. *J. Nat. Prod.*, 56, 1114–1122.
102. Jaradat, M. S., Noonan, D. J., Wu, B., Avery, M. A., Feller, D. R., Jardat, M. S. (2002). Pseudolaric acid analogs as a new class of peroxisome proliferators-activated receptor agonists. *Planta Med.*, 68, 667–671.
103. DuBois, R. N., Gupta, R., Brockman, J., Reddy, B. S., Krakow, S. L., Lazar, M. A. (1998). The nuclear eicosanoid receptor, PPARγ, is aberrantly expressed in colonic cancers. *Carcinogenesis*, 19, 49–53.
104. Yang, W-L., Frucht, H. (2001). Activation of the PPAR pathway induces apoptosis and COX-2 inhibition in HT-29 human colon cancer cells. *Carcinogenesis*, 22, 1379–1383.
105. Sato, H., Ishihara, S., Kawashima, K., Moriyama, N., Suetsugu, H., Kazumori, H., Okuyama, T., Rumi, M. A. K., Fukuda, R., Nagasue, N., Kinoshita, Y. (2000). Expression of peroxisome proliferator-activated receptor (PPAR)γ in gastric cancer and inhibitory effects of PPARγ agonists. *Br. J. Cancer*, 83, 1394–1400.

23

ENDOCRINE AND METABOLIC AGENTS

BRIAN L. FURMAN
*University of Strathclyde, Strathclyde Institute for Biomedical Sciences
Glasgow, Scotland*

23.1	HYPOTHALAMIC HORMONES, THEIR ANALOGS, AND ANTAGONISTS	1039
	Thyrotrophin-Releasing Hormone	1039
	Corticotrophin-Releasing Hormone	1040
	Growth-Hormone-Releasing-Hormone	1043
	Gonadotrophin-Releasing Hormone	1043
	Somatostatin	1045
23.2	ANTERIOR PITUITARY HORMONES	1047
	Growth Hormone (Somatotropin)	1047
	Follicle-Stimulating Hormone	1048
	Luteinizing Hormone	1049
23.3	NEUROHYPOPHYSEAL HORMONES	1049
	Vasopressin	1049
	Oxytocin	1052
23.4	THYROID HORMONES AND DRUGS MODULATING THYROID FUNCTION	1053
	Iodide	1055
	Thionamide Drugs	1055
	Potassium Perchlorate	1056
23.5	DRUGS AFFECTING BONE METABOLISM	1057
	Parathyroid Hormone	1057
	Calcitonin	1058
	Bisphosphonates	1058
	Mithramycin	1059
	Calcimetics	1059

Drug Discovery Handbook, by Shayne Cox Gad
Copyright © 2005 by John Wiley & Sons, Inc.

23.6	ADRENAL CORTICOSTEROIDS	1060
	Glucocorticoids	1060
	Inhibitors of Glucocorticoid Synthesis	1063
	Aldosterone and Other Mineralocorticoids	1064
	Mineralocorticoid Antagonists	1065
23.7	ANDROGENS	1066
	Androgen Receptors	1066
	Anabolic Steroids	1067
	Antiandrogens	1068
	5α-Reductase Inhibitors	1068
23.8	ESTROGENS	1069
	Estrogen Receptors	1069
	Estrogen Agonists	1070
	Antagonists at Estrogen Receptors	1070
	Selective Estrogen Receptor Modulators (SERMs)	1070
	Aromatase Inhibitors	1070
23.9	PROGESTINS	1071
	Agonists at Progesterone Receptors	1072
	Progesterone Receptors	1072
23.10	TREATMENT OF DIABETES MELLITUS	1073
	Insulin	1073
	Biguanides	1077
	Thiazolidinediones	1077
	Sulfonylureas	1079
	Meglitinide Analogs	1080
	α-Glucosidase Inhibitors	1081
	Glucagon-like Peptide-1 (GLP-1) and Glucose-Dependent Insulinotropic Polypeptide (GIP)	1082
	Exendin-4	1082
	Glucagon	1083
23.11	ANTIOBESITY AGENTS	1083
	Orlistat	1085
	References	1086

This chapter addresses the properties of endocrine hormones, their receptors, hormone analogs acting as agonists and antagonists, and the potential therapeutic applications. The general approach has been to classify the hormones according to their main, or classical, anatomical origin, although the disease classification is used in the case of hormones relevant to diabetes mellitus and obesity. The information is referenced using reviews or, where, appropriate, primary literature.

23.1 HYPOTHALAMIC HORMONES, THEIR ANALOGS AND ANTAGONISTS

Several peptide hormones are synthesized in the hypothalamus. These include oxytocin and vasopressin, which are synthesized in the paraventricular and supraoptic nuclei of the hypothalamus and stored in the posterior lobe of the pituitary gland, to which they are transported along axons in association with neurophysins. Oxytocin and vasopressin will be discussed under Neurohypophyseal Hormones (see later). The other hormones of the hypothalamus have a key role in the regulation of the secretory activity of the anterior pituitary. These are growth-hormone-releasing hormone (GHRH; somatorelin), thyrotropin-releasing hormone (TRH; protirelin), corticotrophin-releasing hormone (CRH), and gonadotropin-releasing hormone (GnRH; gonadorelin) and are secreted into the median eminence from where they are transported to the anterior lobe of the pituitary along long portal blood vessels. There is considerable evidence that TRH, GHRH, and CRH not only stimulate secretion of anterior pituitary hormones but also have trophic effects on their target pituitary cells [1]. The hypothalamic hormones are normally secreted in a pulsatile manner. This pulsatile nature of secretion is physiologically important, and alteration in the pulse frequency changes the secretory response of the anterior pituitary, this being best studied in relation to GnRH [2]. These peptides and/or their synthetic analogs are used in clinical practice either as diagnostic or therapeutic agents [3]. The actions of these hormones, however, extend beyond their effects on the secretion of the hormones of the anterior pituitary. Their potential therapeutic applications, and those of their analogs and antagonists, similarly may be unrelated to the originally identified physiological roles of the hormones.

Thyrotrophin-Releasing Hormone

Thyrotrophin-releasing hormone (TRH; protirelin; thyroliberin) is the major hypothalamic regulator of the secretion of thyrotrophin [4] and therefore ultimately of the thyroid hormones thyroxine and triiodothyronine. However, a large amount of TRH is found in other regions of the brain, where it may modulate other neurotransmitters, with which it may be colocalized, or even function as a neurotransmitter itself [5]. Many actions other than stimulation of thyrotrophin secretion have been described for TRH and attributed to centrally mediated effects. These include increased arousal/wakefulness, improvements in memory and learning, locomotor activation, antidepressant activity, anticonvulsant activity, antinociceptive activity, reduction in food and water intake, increased gastric acid secretion, increased gastrointestinal contractility, and complex cardiovascular actions [5]. TRH is also found in the periphery, especially in the gastrointestinal and genitourinary tracts [5]. It is inactivated by a selective TRH-degrading ectoenzyme (pyroglutamyl aminopeptidase II) found on the surface of neurons, which appears to provide the only mechanism for inactivating this peptide. The TRH breakdown product cyclo(His-Pro) (histidyl proline diketopiperazine) has both agonist and antagonist activity at the TRH receptor.

Thyrotropin-Releasing Hormone Receptors TRH acts through two G-protein-coupled receptors, TRH-R1 and TRH-R2, which show different distributions in the brain and periphral tissues but which show identical affinities for TRH and its analogs [6]. These receptors are coupled through $G\alpha_q/G\alpha_{11}$ to the phospholipase C-activated hydrolysis of phosphatidylinositol.

The widespread actions suggest that there may be many interesting potential therapeutic applications and the pharmacology of TRH is in its infancy. Orally active analogs (e.g., taltirelin; CG-3703) have been produced and shown to increase wakefulness in a canine model of narcolepsy and to be neuroprotective [7, 8].

Corticotrophin-Releasing Hormone

Corticotropin-releasing hormone, also known as corticotropin-releasing factor (CRF) is a hypothalamic hormone that acts on the anterior pituitary to stimulate the secretion of corticotropin, thereby playing a key role in regulating the synthetic/secretory activity of the adrenal cortex [9]. CRF is also widely distributed in the central nervous system and in the periphery, and its functions probably extend beyond its role in stimulating corticotropin secretion. For example, its release activates the sympathetic nervous system, and thus the hormone is pivotal to the entire stress response. In addition to regulating the stress response, CRF and related peptides have a wide variety of actions in the central nervous system and in the periphery, possibly regulating anxiety, mood, feeding, inflammation, gastric emptying, and blood pressure [10]. There is considerable evidence for a link between hyperactive CRF pathways and depression and anxiety. At least three, additional, naturally occurring, related mammalian peptides are now known. These are urocortin, stresscopin/urocortin III (SCP/UCNIII), and stresscopin-related peptide/urocortin II (SRP/UCNII). Urocortin is also distributed throughout the central nervous system and in the periphery, but its central distribution is more restricted than that of CRF, being expressed strongly in the Edinger–Westphal nucleus and the lateral superior olive, where CRF is not expressed [11]. Urocortin is also found in the hippocampus, basal ganglia, medial septum, paraventricular nucleus, and the lateral hypothalamus [12].

Corticotropin-Releasing Factor Receptors Two main CRF receptors have been identified, CRF_1 and CRF_2, for which the endogenous ligands show different selectivities. CRF has a much higher affinity (13 to 17-fold) for the CRF_1 receptor relative to the CRF_2 receptor [10]. There are at least two subtypes of the CRF_2 receptor, CRF_{2A} and CRF_{2B} [10]. Both CRF receptors are G-protein-coupled, seven-transmembrane receptors, signaling mainly through G_s to increase cyclic adenosine monophosphate (cAMP) but also through G_q, activating the phospholipase C pathway [10].

Actions of CRF and Related Peptides Numerous studies have characterized the role of CRF in coordinating the behavioral and cardiovascular responses to stress [12, 13]. Central, but not systemic, administration of CRF produces behavioral activation (e.g., increased locomotor activity) in a range of species. It suppresses food intake and enhances behavioral responses to stress [12]. Urocortin also elevates plasma corticotrophin when administered into the rat brain, and stress results in increased urocortin messenger ribonucleic acid (mRNA) expression in the hypothalamus and midbrain, suggesting a role for urocortin in regulating the stress response [12].

The cardiovascular effects of CRF depend on its route of administration. Intracerebroventricular administration in all species produces increases in blood pressure, heart rate, and cardiac output mediated by activation of the sympathetic nervous system [13]. On the other hand, a species-dependent hypotensive effect is observed on peripheral injection [13]. CRF, in high doses in the human, appears to have vasodilator effects [13], and this may contribute to hypotension, causing syncope. The cardiovascular actions of peripherally administered urocortin are more marked than those of CRF. In rats and mice, urocortin produces a long-lasting hypotensive effect, mediated through vasodilatation, and an increase in heart rate [13], although in the sheep, urocortin increases blood pressure due to a pronounced increase in cardiac output, resulting from a direct effect in increasing cardiac contractility. The peripheral cardiovascular actions of urocortin appear to be mediated by the CRF_2 receptor [13]. Urocortin is also cardioprotective against ischemia, an effect mediated through the CRF_2 receptor [11]. Like CRF, urocortin has been shown to have a number of central and peripheral actions. Thus in various species it suppresses appetite, delays gastric emptying, increases locomotor activity, and is anxiogenic [11, 13].

Although an anti-inflammatory effect of CRF would be anticipated through release of adrenal glucocorticoids, CRF also has direct effects in augmenting proinflammatory cytokine (TNFα, IL-1β) production from macrophages [14]. Urocortin also exerts direct proinflammatory effects and produced mast cell degranulation and an increase in vascular permeability in rat skin [15].

Effects on nonvascular smooth muscle may also be important. Incubation of human myometrial cells with CRF, but not urocortin II or urocortin III, for 8 to 16h significantly induced mRNA and protein expression of constitutive but not inducible nitric oxide synthase (NOS) isoforms. This action resulted in increased activity of soluble guanylyl cyclase. CRF also caused acute activation of the membrane-bound guanylyl cyclase. These effects appeared to be mediated via the CRF_1 receptor [16].

Therapeutic Applications The only current use of CRF is in the differential diagnosis of Cushing's syndrome/Cushing's disease and other adrenal disorders [17]. Patients with pituitary Cushing's disease show an exaggerated increase in plasma corticotrophin and cortisol concentrations in response to CRF injection, whereas those with Cushing's syndrome of adrenal origin, or with the ectopic adrenocorticotropic hormone (ACTH) syndrome, show no response. In normal humans, the disappearance curve of ovine CRF from plasma was biexponential, with a plasma half-life of ~6 to 12 min for the fast component and ~55 to 75 min for the slow component [18, 19]. The long half-life appears to be associated with a prolonged biological effect.

Corticotrophin-Releasing Factor Antagonists There is enormous interest in developing selective antagonists at the CRF_1 receptor because of their potential as anxiolytics and antidepressants.

Antalarmin is a selective antagonist at the CRF_1 receptor. It appears to be highly selective for the CRF_1 receptor and potently (K_i 1.3 to 1.9 nM) competes for binding with ^{125}I-labeled ovine CRF in tissues expressing predominantly CRF_1 receptors and does not displace CRF in the heart, which expresses predominantly CRF_2 receptors [20].

Chronic administration of antalarmin to rats produced effects consistent with antagonism of the hypothalamic-pituitary-adrenal (HPA) axis, that is, reductions in plasma concentrations of corticotropin and corticosterone and signs of chronic understimulation of the adrenal cortex (reduced width and vascularization of the zona fasciculata; increased number of apoptotic cells in the zona fasciculata) [22]. It also inhibited CRF-induced corticotropin release [20]. Antalarmin inhibited restraint-stress-induced gastric ulceration in rats [23] and ameliorated the behavioural changes produced in the chronic mild stress model of depression in mice [21].

Consistent with direct proinflammatory effects of endogenous CRF via the CRF_1 receptor, antalarmin significantly inhibited carageenin-induced subcutaneous inflammation in rats [20] and adjuvant arthritis in rats [24] and also prolonged survival in a mouse model of endotoxin shock [14]. It also antagonized the endogenous CRF-like peptide urocortin in producing an increased vascular permeability in rat skin [15].

R121919 is a nonpeptide, water-soluble pyrrolopyrimidine that binds with high affinity to human CRF_1 receptors [10]. It is an antagonist at CRF_1 receptors and was shown to be antidepressant in a clinical study [25].

CP-154,526 is a nonpeptide antagonist at CRF_1 receptors is closely related to antalarmin. It antagonizes CRH- and stress-induced neuroendocrine and behavioral effects and has also been shown to be antidepressant in a rat model [26].

Antagonists at the CRF_2 Receptors The peptide antisauvagine-30 is more than 300-fold selective for the CRF_2 receptor compared with the CRF_1 receptor. Currently, there are no nonpeptide antagonists at the CRF_2 receptor [27].

Growth-Hormone-Releasing Hormone

Growth-hormone-releasing hormone (GHRH; somatorelin) is a potent stimulator of the secretion of somatotropin. GHRH acts via a G-protein-coupled receptor to stimulate the synthesis and secretion of GH via activating adenylyl cyclase and increasing cAMP [28]. It also potently activates the MAP kinase pathway in pituitary somatotrophs. A synthetic analog, sermorelin, which has an amino acid composition identical to the N-terminal 29 amino acids of natural GHRH, is used in the diagnosis of abnormalities of GH secretion. A number of peptide antagonists of GHRH (e.g., MZ-5-156; JV-1-36) have been synthesised and have been shown not only to suppress GH secretion but also to inhibit the growth of a number of tumors in vivo and in vitro [29].

Gonadotrophin-Releasing Hormone

Gonadotrophin-releasing hormone (GnRH; gonadorelin) is secreted by the hypothalamus and regulates the secretion of the pituitary gonadotrophins follicle-stimulating hormone (FSH) and luteinizing hormone (LH) [30]. These hormones in turn regulate the secretory and gametogenic functions of the ovaries and testes. Administration of GnRH produces a rapid rise in the plasma concentrations of both FSH and LH. Several synthetic analogs (triptorelin, goserelin, histrelin, buserelin, deslorelin, leuprorelin, and nafarelin) have been developed. These are also agonists at the GnRH receptor, which is a G-protein-coupled ($G_{q/11}$) receptor but are more potent than the native hormone and are more resistant to proteolysis. For example, goserelin and buserelin are 20 times more potent than GnRH, while histrelin is 150 to 200 times more potent than the native hormone [3, 31]. Like GnRH they bind with high affinity to the GnRH receptor on anterior pituitary cells (e.g., [32]) and acutely stimulate the secretion of LH and FSH. However, prolonged, continuous exposure leads to inhibition of secretion by desensitization of the pituitary gonadotropes [33, 34]. There is evidence for direct effects on the ovaries [35] and on tumor cells [36, 37]. High concentrations (10^{-4} M) of triptorelin had a direct inhibitory effect on the growth of an actively proliferating androgen responsive-prostatic cell line, an effect that was inhibited by the GnRH-receptor antagonist cetrorelix, while low concentrations (10^{-7} M) promoted growth [37].

A second form of GnRH has been identified in the brain, peripheral nervous system and other tissues and is referred to as GnRH-II. There is a distinct receptor for this hormone, which is distributed in the brain similarly to GnRH-II, especially in areas associated with sexual arousal. It is also found in reproductive tissues. Its function is unknown. GnRH-II stimulates gonadotrophin secretion through the GnRH-I receptor [38].

Therapeutic Applications Although GnRH agonists acutely stimulate the secretion of gonadotropins and have been used to induce ovulation, their main therapeutic uses depend on the downregulation of the GnRH receptor function on the pituitary gonadotropes that occurs during continuous administration, with the resultant inhibition of gonadotropin secretion and, therefore, the inhibition of the follicular development in the female and inhibition of gonadal steroid hormone production in both sexes [39]. This is seen both in the human and in experimental animals. For example, a single injection of microencapsulated histrelin [10 to 300 µg/kg subcutaneous (SC)] induced a dose-dependent disruption of normal estrous cyclical activity in rats, resulting in persistent diestrous-like vaginal cytology. Desensitization was preceded by stimulation; LH secretion was maximal within 8 h and then gradually declined, remaining at diestrous levels from days 7 to 28 [40]. This desensitization forms the basis of their current and potential clinical applications in the treatment of endometriosis, uterine leiomyoma (fibroids), malignant neoplasms (especially prostatic carcinoma), central precocious puberty, and in assisted reproduction protocols. Deslorelin is used currently primarily in veterinary practice to induce ovulation in mares and cattle [41]. GnRH agonists have been proposed as potential contraceptive agents, although this use remains experimental. Because they are peptides, they cannot be given by mouth, but some (e.g., buserelin, nafarelin) are available as nasal sprays. Predictably, women may experience menopausal-like symptoms (hot flushes, increased sweating, vaginal dryness, dyspareunia, loss of libido) some weeks after starting treatment. Withdrawal bleeding may occur during the first few weeks of treatment and breakthrough bleeding may occur during prolonged treatment. Prolonged treatment for several months (as in the treatment of endometriosis) leads to significant loss of bone density [42]. Men being treated for prostatic carcinoma may experience dangerous, and sometimes painful, disease flare and bone pain at the initiation of treatment, as a consequence of the agonist action producing a transient increase in androgen secretion. This may be minimized by the initial use of an antiandrogen, such as flutamide or cyproterone [43].

Gonadotropin-Releasing Hormone Antagonists While the continuous administration of GnRH agonists is highly effective in downregulating of the GnRH receptor function on the pituitary gonadotropes, these agents have the major, although predictable, disadvantage of initially stimulating the receptors, this being responsible for the initial disease flare seen in men being treated for prostatic carcinoma. Thus, peptide antagonists of GnRH have been developed. Early attempts to produce effective antagonists resulted in compounds that produced marked histamine release, as a result of mast cell degranulation [44, 45]. The antagonists that have been developed include cetrorelix and ganirelix, which are both currently available for clinical use, as well as antide, degarelix, abarelix, and antarelix. Unlike the GnRH agonists these compounds are devoid of any initial stimulatory actions. Because these agents are peptides, they still require to be administered parenterally. Potent, selective nonpeptide antagonists are under development, some of which (CMPD1, NBI-42902, and TAK-013) are intended for oral administration [38].

Somatostatin

Somatostatin is a naturally occurring peptide found throughout the central and peripheral nervous systems, the D cells of the pancreatic islets and gastrointestinal tract, and in many other structures. It inhibits the secretion of growth hormone (GH; somatotropin), insulin, glucagon, and gut hormones and is generally inhibitory to gastrointestinal motility and exocrine secretion [46]. Somatostatin also has a pronounced antiproliferative effect and promotes cellular apoptosis. It exists as two active forms—somatostatin-14 and somatostatin-28. Its secretion is regulated by many humoral, neural, and nutrient factors [46].

Somatostatin Receptors The actions of somatostatin are mediated via high-affinity, membrane-bound, G-protein-coupled somatostatin receptors (SSTRs) of which five subtypes have been identified (SSTR1 to SSTR5). These subtypes have different tissue distribution and show somewhat different binding affinities (in the nanomolar range) for somatostatin-14 and somatostatin-28 but very marked differences in affinities for somatostatin analogs such as lanreotide and octreotide [46].

The intracellular signaling pathways mediating the effects of somatostatin is complex. All five receptors are negatively coupled to adenylyl cyclase, and the inhibitory actions of somatostatin on hormone secretion are mediated via inhibition of cyclic-AMP formation, activation of potassium channels, reduction in intracellular Ca^{2+}, and activation of the serine/threonine protein phosphatase calcineurin [46]. The antiproliferative effects of somatostatin may be mediated via activation of protein tyrosine phosphatases (SHP-2).

Somatostatin Analogs Two synthetic somatostatin analogs are currently available for clinical use, lanreotide and octreotide, while a third, vapreotide, is undergoing clinical evaluation. Both bind with high affinity to somatostatin receptors (SSTRs), showing high selectivity for SSTR2 and SSTR5 relative to other SSTRs [46]. In clonal (CHO)-K1 cells stably expressing hSSTRs1–5 the binding affinities of lanreotide, expressed as IC_{50} values, are 0.75 nM at SSTR2 and 5.2 nM at SSTR5 (IC_{50}), compared with 2330 nM at SSTR1, 107 nM at SSTR3, and 2100 nM at SSTR4. Its EC_{50} in inhibiting growth hormone secretion from primary human fetal pituitary cultures was 2.3 nM [47]. The affinities of octreotide for these receptors in the same system were similar.

Several nonpeptide agonists have been synthesized. Some of these have markedly different binding affinities for different somatostatin receptors. Thus L-797, 591 shows very high affinity for the SSTR1 receptor (IC_{50} 1.4 nM), with 120-fold selectivity for this receptor compared with, SSTR4, 1300-fold selectivity compared with SSTR2, 1600-fold selectivity compared with SSTR3, and 2570-fold selectivity compared with SSTR5 [46]. Others in the same series showed high selectivity for SSTR2 (L-779, 976), SSTR3 (L-796, 778), and SSTR4 (L-803, 087). L-817, 818 shows very high affinity for SSTR5, with 8-fold selectivity compared with SSTR1 and about 130 to 200-fold selectivity compared with SSTR2, SSTR3, and SSTR4. These agents are now being used to obtain information about the somatostatin receptor subtypes that mediate the many effects of this hormone.

Somatostatin Receptor Antagonists   Peptides have been developed that are antagonists at somatostatin receptors. CYN-154806 is a potent octapeptide antagonist at the SST2 receptor, with at least 100-fold lower potencies at other receptor subtypes [48]. Others peptides have been developed with high selectivity for SSTR3 [49].

SRA880 was the first selective nonpeptide antagonist at the SSTR1 receptor to be developed. It is a competitive antagonist at the SSTR1 receptor with much lower affinities for the other receptor subtypes [50]. Similarly, nonpeptide (BN81644 and BN81674) antagonists have been developed with high affinity for the SSTR3 receptor [51].

The further development of potent and selective agonists and antagonists at the somatostatin receptor will enable a full exploration of the role of somatostatin and in developing novel therapies. Although it is clearly of enormous value to have highly selective agonists, there is also an argument for developing agonists with a broad spectrum of receptor activity, as they may be useful in neuroendocrine tumors expressing multiple somatostatin receptors.

Therapeutic Applications   Lanreotide and octreotide are used to treat acromegaly and neuroendocrine tumors, especially symptoms associated with carcinoid tumors with features of carcinoid syndrome, VIPomas, glucagonomas, and the Zollinger-Ellison syndrome. They have also been used for many other conditions including the control of bleeding of esophageal varices, diarrhea due to acquired immunodeficiency syndrome (AIDS) or chemotherapy, and breast cancer. In patients with acromegaly, lanreotide provides consistent suppression of GH levels to <5 mU/L in approximately 50 to 65 percent of all cases [52]. Most neuroendocrine gastrointestinal tumors express somatostatin receptors, and these somatostatin analogs produce clinical improvement mediated via a direct inhibitory effect on hormone production from the tumors in 30 to 70 percent of patients. Induction of tumor cell apoptosis has been described when high doses of lanreotide (12 mg/day) are administered. Eventually, however, all patients show escape from the therapy with regard both to hormonal production and tumor growth [53]. Radiolabeled lanreotide and octreotide may be used for the diagnosis and treatment of neuroendocrine tumors [52]. Somatotostatin-receptor binding peptides have been developed containing a chelator to enable radiolabeling and an apoptosis-inducing peptide [54].

23.2 ANTERIOR PITUITARY HORMONES

The anterior pituitary synthesizes and secretes a number of hormones, many of which are directly involved in the regulation of other endocrine glands. These include thyrotrophin (thyroid-stimulating hormone; TSH), corticotrophin (adrenocorticotrophic hormone; ACTH), and the two gonadotrophins, follicle-stimulating hormone (FSH) and luteinizing homone (LH). The other major hormones of the anterior pituitary are growth hormone (GH; somatotropin) and prolactin. Anterior pituitary hormones, or their human recombinant analogs, are used where there is a deficiency in the particular anterior pituitary hormone (e.g., somatotropin deficiency) or where an enhanced activity of the endogenous hormone is required to produce a therapeutic action.

Growth Hormone (Somatotropin)

Growth hormone is secreted in a pulsatile manner from the anterior pituitary gland. Its secretion is regulated by a balance between the actions of the stimulatory GHRH and the inhibitory somatostatin. Moreover, a third hormone, ghrelin, which is abundantly expressed in the stomach and present in the hypothalamus is also a potent stimulator of GH secretion. About 45 percent of the hormone in the circulation is bound to a binding protein, which is similar to the extracellular domain of the growth hormone receptor and which markedly (10-fold) prolongs the half-life of the circulating hormone. Its effects are mediated by a specific GH receptor, one molecule of somatropin binding to two GH receptors, with the formation of a ligand-occupied receptor dimer [55]. Its actions are numerous and the GH receptor is expressed ubiquitously. Actions include stimulation of lipolysis, increased amino acid transport into muscle cells, chondrocyte clonal expansion, phosphorus retention by the kidney, increased calcium and phosphate absorption from the small intestine, increased synthesis of 1,25 dihydroxyvitamin D_3, and promotion of linear growth in bones. It has direct actions on adipocytes (increasing lipolysis) and on hepatocytes (increasing gluconeogenesis).

Many, but not all, of the actions of somatotropin are mediated by two factors, now known as insulin-like growth factor I (IGF-I, somatomedin C) and insulin-like growth factor II (IGF-II, somatomedin A) [56]. These factors are probably produced in all tissues, although the liver is probably the major source of circulating hormones. The IGFs are structurally related to proinsulin and act through similar receptor mechanisms, although having their own, distinct high-affinity receptors [56]. There are six high-affinity serum binding proteins that modulate the actions of the IGFs, either increasing or decreasing their effects [57]. The IGFs play a key role in preadolescent growth, produce rapid metabolic effects, and have long-term, autocrine growth-promoting actions [58]. IGF-I also regulates the secretion of somatotropin by a negative feedback effect.

Therapeutic Applications Somatropin (human recombinant growth hormone) is used before epiphyseal fusion for the treatment of short stature due to decreased or absent secretion of endogenous hormone or due to gonadal dysgenesis. The response is monitored by determination of the serum concentration of IGF-I. It is also used to treat wasting associated with AIDS and severe illness. Because of its anabolic actions, it is prone to abuse by athletes. Overdose will induce symptoms of acromegaly.

Pegvisomant—An Antagonist at the Growth Hormone Receptor   Pegvisomant is a novel recombinant, polypeptide GH analog with nine mutations, eight of which increase the affinity for the GH receptor. However, the mutation at gly(120) to Arg(120) prevents dimerization of the receptor and results in GH antagonism. The mutated protein is covalently bound to polyethylene glycol, reducing renal clearance and prolonging its biological half-life [59]. It suppresses IGF-I secretion in normal subjects [59] and, importantly, in patients with acromegaly [60]. Pegvisomant suppresses serum IGF-I, producing normalization and improvement in symptoms in around 97 percent of patients with acromegaly. In terms of IGF-I normalization, it is the most effective treatment to date [61].

Follicle-Stimulating Hormone

One of two gonadotrophins (the other being luteinizing hormone, or LH) secreted by the anterior pituitary, FSH is a glycoprotein, the α chain of which is identical to that of LH, the receptor selectivity being conferred by the β chain. Follitrophin is recombinant follicle-stimulating hormone. Two follitrophins are available (follitrophin α and follitrophin β), which differ slightly in their carbohydrate structures. Its actions are those of FSH, which stimulates the development, maturation, and secretory activity of the follicles in women and spermatogenesis in men.

Follicle-Stimulating Hormone Receptors   Follitrophin produces its effects by activating the FSH receptor, a G-protein-coupled receptor with a large extracellular domain. Activation of the receptor leads to increased cAMP formation [62]. Although currently there are no readily available FSH receptor antagonists, a selective antagonist, (7-[4-[Bis-(2-carbamoyl-ethyl)-amino]-6-chloro-(1,3,5)-triazin-2-ylamino)-4-hydroxy-3-(4-methoxy-phenylazo)-naphthalene]-2-sulfonic acid), at the human and rat FSH receptor has been developed that prevents FSH-induced cAMP accumulation and steroidogenesis in vitro, as well as inhibiting ovulation in rats [63].

Therapeutic Applications   The main use of FSH is in the treatment of female infertility, either where ovulation does not occur spontaneously, including women with polycystic ovarian disease, or to produce controlled ovarian hyperstimulation to induce the development of multiple follicles in medically assisted reproduction (e.g., in vitro fertilization and embryo transfer) [64]. An ovarian hyperstimulation syndrome may occur in ~8.6 percent of patients receiving recombinant FSH [65]. It may also be effective in male infertility in patients with severe isolated hypogonadotropic hypogonadism, in whom spermatogenesis could be achieved after 18 months of treatment with follitrophin, preceded by 6 months of treatment with hCG (LH) [66].

Luteinizing Hormone

Luteinizing hormone is the second of the two gonadotrophins secreted by the anterior pituitary. It is a glycoprotein, the α chain of which is identical to that of FSH, the receptor selectivity being conferred by the β chain. Human chorionic gonadotrophin (which mimics luteinizing hormone) has been widely used, but human recombinant LH is now available. Its main use is in combination with FSH in the treatment of female infertility, either where ovulation does not occur spontaneously, including women with polycystic ovarian disease, or to produce controlled ovarian hyperstimulation to induce the development of multiple follicles in medically assisted reproduction (e.g., in vitro fertilization and embryo transfer) [3].

The LH receptor is a G-protein-coupled receptor, stimulation of which activates adenylyl cyclase, increasing levels of cAMP, as well as activating phospholipase C [67]. In the latter respect the LH receptor differs from the FSH receptor. Chorionic gonadotrophin has high affinity for the LH receptor but is much less active at the FSH receptor. There do not appear to be any LH receptor antagonists.

23.3 NEUROHYPOHYSEAL HORMONES

The neurohypophysis releases oxytocin and vasopressin (arginine vasopressin in most mammals), which are cyclic nonapeptides synthesized in the paraventricular and supraoptic nuclei of the hypothalamus, from where they are transported along neurons for storage in, or release from, nerve terminals in the posterior lobe of the pituitary. They are transported in neurosecretory vesicles with their carrier proteins neurophysin I and neurophysin II. Both are powerful smooth-muscle contractants, but they have distinctive individual physiology and their actions are mediated via different receptors. Like other hypothalamic hormones, their biology extends far beyond their classical endocrine actions.

Vasopressin

The classical endocrine functions of vasopressin are the facilitation of water reabsorption by the kidney and the contraction of vascular smooth muscle cells, vasopressin being one of the most potent circulating pressor peptides

[68]. Vasopressin is now also recognized, along with CRF, as a physiological regulator of corticotropin secretion. It has numerous nonendocrine actions and acts as a neurotransmitter in the central nervous system [69]. In the latter context it can modulate social behavior in a variety of species [70] and influences memory [71]. Vasopressin increases the release of von Willebrand factor and Factor VIII from the vascular endothelium.

Vasopressin has been implicated in congestive heart failure and in depressive and anxiety disorders. Thus it has become increasingly important to understand the receptors mediating the effects of vasopressin and therefore to develop selective agonists and antagonists at the various receptors.

Receptors Mediating the Effects of Vasopressin Three main receptors have been identified, referred to as V_{1a}, V_{1b}, and V_2 receptors. The receptors show marked species variability, such that, for example, agonists that are highly selective for the human form of a particular subtype may be nonselective and much less potent at the rat form. Vasopressin receptors are G-protein-coupled receptors, but the different receptors are coupled through different G proteins. Thus the V_2 receptor is coupled through G_s and activates adenylyl cyclase, increasing the production of cAMP. On the other hand, the V_{1a} and V_{1b} receptors are coupled through $G_{q/11}$, and its activation results in increased hydrolysis of phosphatidylinositol [72].

The V_2 vasopressin receptor is present in the collecting duct of the kidney, where it mediates vasopressin-induced antidiuresis, maintaining normal plasma osmolality, blood volume, and blood pressure. Activation of this receptor results in the incorporation of the water channel aquaporin-2 into the apical membrane of collecting duct cells [73]. Contraction of vascular smooth muscle cells is mediated via the V_{1a} vasopressin receptor and stimulation of corticotropin secretion by the V_{1b} receptors [69]. Stimulation of the release of factor-VIII-related antigen and von Willebrand factor from vascular epithelium is mediated by V_2 receptors.

Agonists and Antagonists at Vasopressin Receptors d[D-3-Pal2]VP (Deamino[D-3-(pyridyl)Ala2,Arg8]VP) is a full agonist at the pituitary in stimulating the secretion of corticotrophin [74]. It is a V_{1b} vasopressin receptor agonist, a weak V_2 vasopressin receptor agonist, and a weak V_{1a} vasopressin receptor antagonist in the rat [75].

F-180, a vasopressin (VP) structural analog, shows high binding affinity ($K_i = -11\,nM$) for the human and bovine $V_{(1a)}$ receptor subtype and is selective for this receptor subtype, at which it is potent in stimulating the accumulation of inositol phosphate [76]. It shows more than 400-fold selectivity for the V_{1a} compared with V_{1b} receptors and even higher selectivity relative to the V_2 receptor. At the oxytocin receptor it behaves as an antagonist. However, it has only weak affinity for the rat V_{1a} receptor. F-180 is the first selective V_{1a} agonist described for human and bovine vasopressin receptors.

Chlorpropamide, the antidiabetic sulfonylurea, is known to augment vasopressin-mediated antidiuresis, and it has been used to treat cranial diabetes insipidus but is ineffective in patients with nephrogenic diabetes insipidus. It was shown to be a weak inverse agonist for the V_2 vasopressin renal receptor and to upregulate vasopressin receptors both in vivo and in vitro incubation [77].

SSR149415 was the first nonpeptide, orally active drug with nanomolar, selective affinity for mammalian V_{1b} receptors. It exhibits potent antagonist properties both in vitro and in vivo, antagonizing effects of vasopressin in increasing intracellular $[Ca^{2+}]$, promoting platelet aggregation, stimulating vascular smooth muscle cell proliferation, producing hypertension, and causing coronary vasospasm. It displays potent anxiolytic and antidepressant-like activities [78].

SR121463 is a highly selective, orally active V_2 receptor antagonist and possesses powerful oral aquaretic properties in various animal species and in humans. SR121463 is well-tolerated and dose dependently increases urine output and decreases urine osmolality. Other orally active V_2 receptor antagonists are OPC-31260, OPC-41061 (tolvaptan), VPA-985 (lixivaptan), SR121463, VP-343, and FR-161282. YM-087 (conivaptan), JTV-605, and CL-385004 are mixed V_{1a}/V_2 receptor antagonists [79].

Therapeutic Applications Vasopressin itself and its analog dDAVP are used for the treatment of central diabetes insipidus, which results from inadequate secretion of the antidiuretic hormone. Peptide analogs have variable bioavailability even via the intranasal route and orally active, nonpeptide agonist agents, such as OPC51803, are needed.

Another use of vasopressin analogs is in acute variceal hemorrhage, in which condition terlipressin (triglycyl lysine vasopressin), the synthetic analog of lysine vasopressin (the porcine vasopressin), is frequently used. DDAVP (desmopressin-1-deamino, 8-D-arginine vasopressin—dDAVP), a selective V_2 agonist, is used widely to treat von Willebrand disease, hemophilia A, and several platelet disorders. Its actions in these conditions may be mediated by stimulating the release of Factor-VIII-related antigen from vascular epithelium and Factor VIII coagulant activity from liver and other unidentified sites, in addition to stimulating the release of von Willebrand factor and tissue plasminogen activator [80, 81].

As indicated above, selective and orally active compounds are now available. These will provide new information about the pathophysiological role of vasopressin, as well as providing new therapies. V_2 receptor blockade could be of interest in several water-retaining diseases such as the syndrome of inappropriate antidiuretic hormone secretion (SIADH), liver cirrhosis, and congestive heart failure. Nonpeptide vasopressin V_2 receptor antagonists have shown beneficial effects in rat, canine, and porcine models of congestive heart failure and have reached phase II and phase IIIa stages of clinical development, in which they were shown to improve fluid status, osmotic balance, and hemodynamic status [82]. The anxiolytic and antidepressant properties of V_{1b} receptor antagonists suggest that this class of drugs has potential in the treatment of stress-related disorders, anxiety, and depression [78].

Oxytocin

The classical effects of oxytocin are stimulation of uterine contractility, particularly in the later stages of pregnancy and the ejection of milk from the lactating mammary gland in response to suckling. However, oxytocin receptors are very widely distributed in the brain and in peripheral tissues and numerous other actions have been described including anxiolytic effects, induction of maternal behavior, satiety, stimulation of insulin and glucagon secretion, natriuresis, impairment of memory, control of reproductive behavior, analgesia, and central control of the cardiovascular system. It has been implicated in penile erection [83] and seminal ejaculation [84]. Not all of these actions are seen in all species and their physiological importance remains to be established.

Receptors Mediating the Effects of Oxytocin Unlike vasopressin, there appears to be only one oxytocin receptor, which is a G-protein-coupled receptor coupled to $G_{q/11}$ and stimulating phospholipase C, resulting in the generation of inositol trisphosphate and diacylglycerol [85]. Oxytocin has also been shown to induce desensitization of the oxytocin receptor, which occurs in a time span that would be relevant to clinical use [85]. However, the overall response to oxytocin (OT) is influenced by the hormonal milieu, in particular the estrogen/progesterone ratio [86]. Thus estrogens stimulate OT receptor expression in uterus, pituitary, kidney, and the ventromedial nucleus of the hypothalamus. On the other hand, the very high concentrations of progesterone produced by the placenta during pregnancy may block oxytocin receptor signaling in the uterus through a nongenomic mechanism, and this may contribute to uterine quiescence [85].

Agonists and Antagonists at the Oxytocin Receptor [Thr^4,Gly^7]OT is a potent oxytocic agent that is a useful pharmacological tool [87]. It is very selective for the rat, but not the human, oxytocin receptor, having no vasopressor or antidiuretic activity. It has high affinity for hippocampal oxytocin receptors and a very low affinity for central vasopressin and peripheral V_{1a}, V_{1b}, and V_2 vasopressin receptors.

Atosiban is a synthetic analog of oxytocin and acts as a competitive antagonist at the oxytocin receptor. Blockade of this receptor prevents oxytocin from stimulating inositol triphosphate production and inhibits influx of calcium into myometrial cells. The oxytocin-driven release of prostaglandins is also prevented [88].

Therapeutic Applications of Oxytocin and Its Antagonists Oxytocin is principally used to induce or augment labor and in the treatment or prevention of postpartum hemorrhage. The antagonist atosiban is used in the management of preterm labor (labor occurring before 37 weeks of gestation), where it can be used to delay imminent preterm labor from 24 to 33 weeks of gestation, thereby improving neonatal survival. It may be as effective as β_2-agonists, such as ritodrine, but is associated with fewer cardiovascular side effects [89, 90]. Myometrial sensitivity to oxytocin increases with increasing gestational age, as the receptor is progressively upregulated and atosiban is much more effective in later pregnancy [91].

23.4 THYROID HORMONES AND DRUGS MODULATING THYROID FUNCTION

Thyroxine (T_4) and tri-iodothyronine (liothyronine, T_3) are synthesized in the follicular epithelial cells from tyrosine residues found in the thyroid protein thyroglobulin. Iodide transport is a key step in the biosynthesis of T_4 and T_3, and it is actively transported into thyroid follicle cells by a mechanism that selectively concentrates iodide against a large concentration gradient [92]. The transporter co-transports Na^+ and iodide and is referred to as the sodium/iodide symporter [92]. Tyrosine residues are iodinated to form mono- and di-iodotyrosines, which then undergo oxidative coupling to form the active hormones, which are then stored as part of the thyroglobulin molecule within the thyroid follicle. Thryoglobulin, containing the attached thyroid hormones, is secreted at the apical surface of the thyrocytes into the follicle lumen, where it is stored. Thyroid hormone synthesis and secretion are controlled by the anterior pituitary, which secretes thyrotropin under the influence of the hypothalamic hormone thyrotropin-releasing hormone (protirelin; TRH).

T_4 is the main secretory product of the thyroid follicle cells and is converted to the active hormone, T_3 in the tissues under the influence of microsomal deiodinase enzymes which remove the iodine from the "outer" ring (5'-deiodination). Removal of the iodine from the "inner," tyrosyl ring (5-deiodination) produces an inactive substance known as reverse tri-iodothyronine (rT_3) [94]. T_4 is highly (>99 percent) bound to plasma protein (mostly to thyroid-binding globulin and transthyretin), with only a small amount of free hormone (FT_4) being available for biological activity.

The liver is a major site of metabolism, although deiodination to T_3 probably occurs in most tissues [93]. In the presence of normal thyroid function, the plasma half-life of thyroxine is about 6 to 7 days, while that of T_3 is about 1 to 2 days.

Receptors Mediating the Effects of Thyroxine and Tri-iodothyronine The classical effects of T_4 in maintaining the metabolic activity of tissues by inducing the production of key enzymes are generally accepted to be mediated via specific nuclear thyroid hormone receptors, which bind the active T_3 and of which there are two main types, TRα and TRβ [94].

Around 30 genes have been described that are regulated by thyroid hormones, including sarcoplasmic reticulum calcium adenosinetriphosphatase (ATPase), α-myosin heavy chain, malic enzyme [95], and uncoupling proteins [96]. However, as seen with steroids (see later), there is evidence for rapid, nongenomic actions of thyroid hormones mediated via nonnuclear, cell surface, G-protein-coupled receptors. These actions can be produced by T_4 without conversion to T_3. Some of these actions seem to be mediated via the mitogen-activated protein kinase (MAP kinase) cascade and are concerned with modulating the actions of growth factors and certain cytokines [97]. A potent, nongenomic action in activating cardiac myocyte sodium channels was demonstrated [98], and this action was shared by T_4 and T_3, as well as some other iodinated tyrosines, but not by rT_3. There is also evidence for direct, nongenomic effects of thyroid hormones on mitochondrial function, via mitochondrial receptors [99].

Therapeutic Applications Thyroxine is used in all diseases associated with clinical hypothyroidism, where it will reverse the signs and symptoms of hypothyroidism. The dose is titrated for individual patients depending on clinical response and measurement of plasma-free thyroid hormone concentration. Although active orally, absorption is variable and it is given by injection in myxedema coma.

The adverse effects of thyroxine and tri-iodothyronine are identical to the symptoms of thyrotoxicosis (anginal pain, cardiac arrhythmias, palpitations, tachycardia, tachypnoea, diarrhea, vomiting, tremors, restlessness, excitability, insomnia, flushing, sweating, excessive weight loss, and headache).

Thyroid Hormone Analogs T_3 lowers serum cholesterol but is not used to treat hypercholesterolemia because of its effect in producing tachycardia and cardiac dysrrhythmias. Similarly it exerts a positive inotropic effect, but its ability to produce tachycardia in the same doses precludes its use in heart failure. Analogs of thyroid hormones have been synthesized in an attempt to exploit either the positive inotropic or cholesterol-lowering effects but without adverse effects on the heart. For example, CGS 23425 reduced low-density lipoprotein (LDL) cholesterol by 44 percent in hypercholesterolemic rats but had no effect on atrial contractility and force at doses up to 1000-fold greater [100]. On the other hand, the thyroid hormone analog DITPA (3,5-diiodothyropropionic acid) improved systolic as well as diastolic function through effects on cardiac muscle and is a promising potential new treatment for heart failure [101]. Recent work has provided a molecular basis for the differences between the pharmacological profiles of these two analogs. While DITPA had affinity for both thyroid receptors, CGS 23425 was shown to bind selectively to the β form. This difference was reflected in differences in gene regulation by the two compounds, with DITPA and T_3, but not CGS 23425, upregulating genes encoding cardiac contractile proteins, sarcoplasmic reticulum calcium

ATPase, and several proteins of mitochondrial oxidative phosphorylation [102].

Iodide

Iodide, which is a fundamental constituent of T_4 and T_3, has complex effects on the thyroid, where it can inhibit the synthesis and secretion of T_4 and T_3, as well as reducing the cellularity and vascularity of the gland in Graves' disease. It may also produce hyperthyroidism in certain circumstances (Jod–Basedow effect) [103].

In normal individuals excess iodide inhibits thyroid hormone synthesis (acute Wolff–Chaikoff effect), this being mediated probably by the formation of iodopeptide(s) that temporarily inhibit thyroid peroxidase (TPO) mRNA and protein synthesis and, therefore, thyroglobulin iodinations. After a few days, the organification of intrathyroidal iodide resumes and the normal synthesis of T_4 and T_3 returns ("escape" from the Wolff–Chaikoff effect). This occurs through downregulation of the sodium iodine symporter (NIS), thus decreasing the intrathyroidal inorganic iodine concentration and permitting the $TPO–H_2O_2$ system to resume normal activity. However, in the newborn and the fetus, and in euthyroid patients with autoimmune thyroiditis, this escape phenomenon is not achieved and hypothyroidism results [104, 105]. The ability of iodide to reduce thyroid cellularity and vascularity in patients with Graves' disease is not clearly understood, but iodide was shown to cause apoptosis in rat and porcine thyroid [106, 107].

Iodide is used in the treatment of Graves' disease to prepare for thyroid surgery. It induces euthyroidism and reduces the size and vascularity of the gland. It is administered for 8 to 10 days before surgery, after the patient is rendered euthyroid using an antithyroid drug such as methimazole [108]. Iodide also protects the thyroid against radiation-induced damage due to accidental exposure to radioactive iodine. In adults with normal thyroid function, ingestion of 100 mg of iodide just before exposure to radioactive iodine blocks at least 95 percent of the thyroid uptake of radioactivity [109].

Thionamide Drugs

The thionamides carbimazole, methimazole, methylthiouracil, and propylthiouracil inhibit the thyroid peroxidases (TPO) involved in catalyzing the iodination of tyrosine residues in thyroglobulin and the oxidative coupling of iodinated tyrosines. This inhibition is competitively antagonized by iodide at low, but not higher, drug concentrations [110]. It has also been suggested that carbimazole and methimazole reduces the autoimmunity underlying Graves' disease, since successful treatment is associated with suppression of thyroid receptor autoantibodies [111]. Carbimazole is rapidly converted to the active compound methimazole [112]. Propylthiouracil additionally inhibits the D1 isoform of the deiodinase enzyme (iodothyronine 5'-deiodinase) that catalyzes the conversion of thyroxine (T_4) to the active tri-iodothyronine (T_3) in most peripheral tissues. This may contribute to the more rapid onset of its therapeutic effect compared with that of methimazole. It also inhibits the 5-deiodination of T_3, which is an inactivating step; theoretically this could counter the reduced formation of T_3 [113], but there does not appear to be any clinical evidence for this. Apart from the effects on deiodinase, the actions and adverse effects of propythiouracil are broadly similar to those of carbimazole and methimazole. Methimazole or propylthiouracil are useful for the production of experimental hypothyroidism in the rat. For example, 0.03 percent methimazole in the drinking water for 4 weeks reduced serum free T4 concentrations by 92 percent [114].

Therapeutic Applications Carbimazole is used in Europe and the United Kingdom, whereas methimazole is used in the United States for the treatment of Graves' disease and other forms of hyperthyroidism. Propylthiouracil is also used. They may be used alone or, in the treatment of Graves' disease, a "block and replace" regime may be used, in which large doses of antithyroid drugs are given continuously, along with thyroxine to prevent iatrogenic hypothyroidism. The rationale for this is partly the belief that large doses of antithyroid drugs suppress the underlying autoimmunity. However, there is no evidence that such regimens are more effective than others [115].

These drugs are generally well tolerated, the major adverse effects appearing to have an immune origin. The most dangerous adverse effect is agranulocytosis, which occurs in 0.1 to 0.5 percent of patients [116]. Agranulocytosis has been suggested to occur almost exclusively during the first 10 weeks of treatment and is probably related to the drug dose [117]. Recombinant human granulocyte colony-stimulating factor (rhG-CSF) may shorten the recovery time in patients with methimazole-induced agranulocytosis. Hepatotoxicity is another rare, but acknowledged, adverse effect [118], with propylthiouracil being more likely than other antithyroid drugs to produce liver injury, which may be asymptomatic [119].

Potassium Perchlorate

Potassium perchlorate competitively inhibits the active iodide transport mechanism, the sodium/iodide symporter, in the thyroid, which has the capacity to selectively concentrate iodide against a large concentration gradient [92]. Iodide transport is a key step in the biosynthesis of the thyroid hormones thyroxine (T_4) and tri-iodothyronine (liothyronine, T_3). Although not now used for the treatment of Graves' disease because it may cause aplastic anemia, it is used to treat thyrotoxicosis associated with the antiarrhythmic agent amiodorone, which is an iodine-rich compound with complex effects on thyroid function. Amiodarone-induced thyrotoxicosis occurs in 2 to 12 percent of patients receiving this drug and is due to iodine-induced excessive thyroid hormone synthesis in patients with nodular goiter or latent Graves' disease or to a thyroid-destructive process caused by amiodarone or iodine [92].

23.5 DRUGS AFFECTING BONE METABOLISM

The prevalence of osteoporosis is high affecting, for example, around 17 percent of postmenopausal women in the United States and up to 30 percent of women older than 65. Given the increasing number of aged people, this presents a major problem because of the attendant morbidity and mortality due to fractures of the hip, vertebrae, and distal forearm. Hip fracture is associated with 20 percent mortality and 50 percent permanent loss in function. Thus there is great interest in developing safe and effective therapies to prevent osteoporosis. Agents affecting bone metabolism comprise a heterogeneous group of drugs. Their main applications are in the treatment and prevention of osteoporosis associated with the menopause or with the use of glucocorticoids [120, 121], Paget's disease [122], and in the treatment of hypercalcemia of malignancy [123].

Parathyroid Hormone

Parathyroid hormone (PTH), the main secretory product of the parathyroid gland, plays a key role in calcium homeostasis and in the regulation of bone turnover. It is secreted in response to a reduction in extracellular calcium and produces increased resorption of calcium from bone, increased renal reabsorption of calcium and increased renal excretion of phosphate. It also stimulates the formation of calcitriol (1,25-dihydroxycholecalciferol; 1,25 dihydroxy-D_3) in the kidney. While the effects of continuous, high levels of parathyroid hormone lead to a net bone loss, there is strong evidence for an anabolic effect of parathyroid hormone on bone. Its secretion in response to a reduction of calcium is controlled by a calcium sensing receptor on the cell membrane of the parathyroid chief cells. This is a G-protein-coupled receptor, activation of which leads to increases in phosphatidylinositol turnover and sustained increases in intracellular calcium and an inhibition of adenylyl cyclase and of PTH secretion [124].

The effects of PTH on bone are exerted through alterations in the formation and activity of the osteclasts and osteoblasts. The regulation of the formation of osteoclasts, the bone resorbing cells, is achieved through a balance between the production and action on osteoclast-precursor cells of two substances osteoprotegerin and RANKL secreted by osteoblasts. RANKL, a member of the TNF (Tumor Nevosis Factor) superfamily, normally binds to its receptor RANK on the osteoclast-precursor cells to promote their development into osteoclasts.

Osteoprotegerin (OPG) is the natural inhibitor of RANKL, binding to RANKL and preventing its interaction with RANK, thereby inhibiting osteoclastogenesis [125]. The osteoclastogenic activity of PTH occurs primarily by suppression of OPG gene expression in early osteoblasts and elevation of RANKL gene expression in mature osteoblasts [126].

Parathyroid hormone exerts its effects through binding to a class 2 G-protein-coupled receptor that coupled through G_s activates adenylyl cyclase, increasing cAMP and activating PKA (protein kinase A) [127]. It is also coupled through G_q to phospholipase C, resulting in increased phosphatidyl inositol breakdown, increasing the formation of inositol trisphosphate and diacylglycerol [128]. This receptor also binds PTH-related protein (PTHrP). There are now known to be at least two PTH receptors, PTH-1R and PTH-2R. PTH-2R has only 51 percent homology with PTH-1R but recognizes PTH. However, it does not bind to PTHrP. The natural ligand for this receptor is thought to be TIP39 (tubero-infundibular peptide of 39 aa).

Anabolic Actions of PTH on Bone The anabolic actions of PTH when administered intermittently can be attributed to stimulation of osteoblast differentiation and reduction in osteoblast apotosis [129]. During continuous administration, the effects on osteoclast formation predominates, leading to a net loss of bone. The anabolic effect of PTH is shared by PTHrP and by N-terminal PTH fragments, including recombinant human PTH(1–34), now known as teriparatide [130].

Calcitonin

The thyroid also secretes calcitonin (CT) from the parafollicular or C cells, which are found among the thyroid follicles in mammals. Calcitonin lowers serum calcium concentrations and the secretion of calcitonin is regulated by the extracellular [Ca^{2+}]. Calcitonin is extremely potent (femtomolar concentrations) in inhibiting the resorptive function of mature osteoclasts [131] and is used in Paget's disease of bone, postmenopausal osteoporosis, hypercalcemia, and bone pain due to malignant neoplasms.

Calcitonin Receptors The CT receptor is a G-protein-coupled receptor, which signals through multiple heterotrimeric G proteins, G_s, G_i, and G_q, leading to the activation of adenylyl cyclase and phospholipase C. Adenylyl cyclase and phospholipase C appear to mediate different components of the responses to CT. For example, cAMP-dependent mechanisms lead to reduced osteoclast motility, while protein-kinase-C-dependent events may mediate CT-induced retraction of the cell. CT also activates the Erk1/2 pathway, which may play an important role in mediating the effects of CT on osteoclasts [131].

Bisphosphonates

This class of agents includes alendronate, etidronate, pamidronate, ibandronate, risedronate, clodronate, tiludronate, and zoledronate, which are variously used as the free acid or as the sodium salts as inhibitors of bone resorption. They are analogs of pyrophosphate, which has been long-known to inhibit calcium phosphate dissolution in vitro [132]. There are two classes of bisphosphonates; the aminobisphosphonates (e.g., pamindronate, ibendronate, risedronate, zoledronate) potently and selectively inhibit the mevalonate pathway by inhibition of farnesyl diphosphate synthase, inhibiting the conversion of geranylpyrophosphate to farnesylpyrophosphate, with a resultant depletion of geranylgeranylpyrophosphate and the inhibition of protein prenylation, including the prenylation of small GTPases. This results in osteoclast apoptosis and inhibition of bone resorption [132–134]. On the other hand, the non-aminobisphosphonates (clodronate, etidronate, tiludronate) do not inhibit farnesyl diphosphate synthase but become incorporated into the phosphate chain of ATP-containing molecules, rendering them nonhydrolysable and cytotoxic to osteoclasts [132, 133].

Therapeutic Applications Bisphosphonates are used widely to prevent/treat loss of bone due to glucocorticoid treatment, postmenopausal estrogen loss or to malignant disease, and in treating hypercalcemia of malignancy and Paget's disease of the bone. High doses of aminobisphosphonates are very effective in hypercalcemia of malignancy and in reducing the risk of skeletal morbidity and fractures in patients with breast cancer or multiple myeloma, with a low incidence of adverse effects [135]. There is interest in using bisphosphonates to target other drugs (e.g., antineoplastic agents, anti-inflammatory drugs) to bone because of their high affinity for this tissue [136]. Bisphosphonates are poorly absorbed from the gastrointestinal tract, with an oral bioavailability of around 1 to 2 percent, but accumulate in bone, from which their clearance is very slow.

Mithramycin

Mithramycin is an antineoplastic antibiotic that complexes with deoxyribonucleic acid (DNA) in the presence of divalent cations such as magnesium. It has also been used in Paget's disease. Mithramycin has a direct inhibitory effect on bone resorption and is an mRNA synthesis inhibitor [137]. It acts directly on bone, as evidenced by its concentration-dependent (0.1 to 100 nM) inhibition of osteoclastic bone resorption in the in vitro bone slice assay, with a maximum inhibition of ~66 percent [138]. Mithramycin is used to control the hypercalcemia and hypercalciuria of malignancy and to reduce bone turnover in Paget's disease. Because of its marked toxicity, it is generally used only when other measures have failed.

Calcimimetics

Calcimimetics comprise a new class of drug acting at the parathyroid cell calcium sensing receptor (CaSR) to increase its sensitivity to extracellular calcium, which will lead to a decrease in circulating PTH levels. This will result in a reduction in serum calcium concentration, which makes these drugs promising as treatments for hyperparathyroidism and parathyroid carcinoma. There are two classes of calcimimetics. Class I calcimimetics are agonists that do not require the presence of calcium and include several organic and inorganic polyvalent cations such as gadolinium, neomycin, and spermine. These agents lack selectivity for the CaSR and are not used therapeutically to lower serum calcium. On the other hand, class II calcimimetics are low-molecular-weight, orally active phenylalkylamine compounds, which are ineffective in the absence of calcium. R568 (*N*-(3-[2-chlorophenyl]propyl)-(R)-α-methyl-3-methoxybenzylamine) is the most extensively studied type II calcimimetic compound [124]. Others being investigated include cinacalcet (AMG073) and NPS 568.

23.6 ADRENAL CORTICOSTEROIDS

Glucocorticoids

The main naturally occurring glucocorticoid in the human is cortisol (hydrocortisone), whereas in rodents corticosterone fulfills that role. Glucocorticoids are secreted primarily by the zona fasciculata of the adrenal cortex. They have numerous actions including stimulating hepatic glycogen synthesis, promoting gluconeogenesis, impairing peripheral glucose utilization, and exerting immunomodulatory and anti-inflammatory effects. Many of their actions are exerted at the genome level, effects being observed within 30 min to 18 h. Additionally, glucocorticoids produce rapid effects observed within seconds or minutes; these actions are described as nongenomic, as their onset time is far too rapid to require altered protein synthesis.

Glucocorticoid Receptors Glucocorticoids are lipophilic and produce their classical effects by stimulating glucocorticoid receptors, which are members of the nuclear receptor superfamily that includes receptors for sex hormones such as androgens, estrogens and progestins, mineralocorticoids, thyroid hormones, and vitamin D. There are two types of corticosteroid receptors. Type I is the mineralocorticoid receptor, which also binds glucocorticoids and is expressed primarily in the kidney, colon, and salivary and sweat glands, as well as in the hippocampus [139]. The type II receptor is the glucocorticoid receptor (GR), which is expressed ubiquitously as two major splice variants GRα and GRβ. GRα mediates the classical effects of glucocorticoids and the function of GRβ is unclear, although there is evidence that it can act as a dominant negative inhibitor of GRα. Changes in the ratio of GRα:GRβ forms of the receptor could explain relative glucocorticoid resistance in inflammatory diseases [139]. Glucocorticoids enter cells and interact with cytoplasmic GRα, which are bound to heat shock proteins (hsp90). Release of these chaperone molecules and phosphorylation of GRα result in homodimerization of the receptor–ligand complex, which migrates to the cell nucleus. Domains within the N-terminal region of the receptor interact with either negative or positive glucocorticoid response elements (GREs) in promoter regions of particular genes with consequential initiation or repression of transcription. This mechanism results in the induction of a number of proteins that may contribute to the anti-inflammatory effect of glucocorticoids, including MAP kinase phosphatase-1 (MKP-1), glucocorticoid-induced leucine zipper (GILZ), and annexin I (Anx-A1, formerly known as lipocortin-1) (Fig. 23.1).

Another important mechanism of glucocorticoid action involves indirect inhibition of transcription by interaction of the ligand-activated GRα with

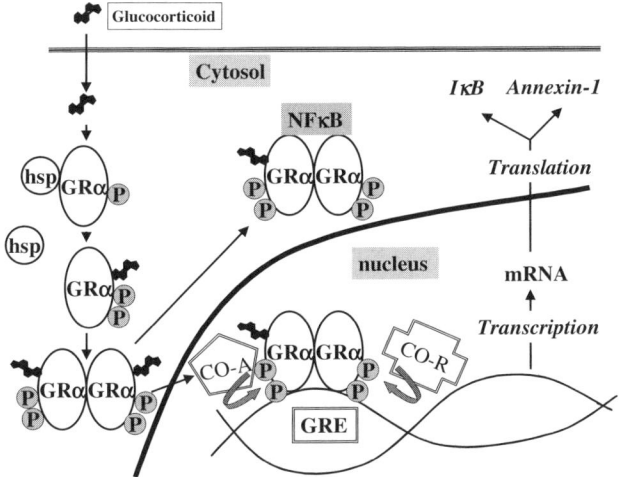

Figure 23.1 Schematic diagram showing some mechanisms of action of glucocorticoids [139, 140]. Glucocorticoids enter the cell and interact with cytoplasmic GRα, which are bound to heat shock proteins (hsp), the function of which with other molecules is to maintain the GR in a conformation that allows ligand binding to occur. Release of these molecules and hyperphosphorylation of GRα result in homodimerization of the receptor–ligand complex, which migrates to the cell nucleus. There is also evidence for a role for hsp in the trafficking of the ligand-bound GR to the nucleus. Domains within the N-terminal region of the receptor interact with either negative or positive glucocorticoid response elements (GREs) in promoter regions of particular genes with consequential initiation or repression of transcription. The ligand-activated receptor on binding to DNA may also recruit coactivator (CO-A) or co-repressor (CO-R) protein complexes that enhance the effect of the glucocorticoid receptor in activating or repressing transcription. Activation of transcription results in the induction of a number of proteins that may contribute to the anti-inflammatory effect of glucocorticoids, including annexin 1 and IκB, which is a protein normally bound to the transcription factor NF-κB and which prevents it from translocating to the nucleus. The ligand-activated GRα may also interact directly with transcription factors such as NF-κB and AP-1, preventing them from binding to DNA and effecting transcription.

transcription factors such as NF-κB and AP-1. This interaction occurs in the cytoplasm (Fig. 23.1). NF-κB plays a key role in inflammation by activating the transcription of pivotal inflammatory cytokines such as TNFα, IL-1β, and GM-CSF, and therefore antagonism of this transcription factor contributes importantly to the anti-inflammatory effect of glucocorticoids [139].

Mechanisms Underlying Nongenomic Activities of Glucocorticoids It is unclear if the rapid glucocorticoid actions, which do not directly affect the transcription or translation of inflammatory genes, involve membrane-bound glucocorticoid receptors, or are mediated via classical cytoplasmic receptors without their required nuclear translocation or interaction with transcription factors [140].

Therapeutic Applictions Glucocorticoids are used to replace endogenous cortisol in adrenal insufficiency, for suppression of systemic and local inflammatory and allergic disorders (rheumatoid arthritis, asthma, allergic rhinitis, contact and atopic dermatitis, allergic conjunctivitis, and ulcerative colitis), in the treatment of shock, congenital adrenal hyperplasia, cerebral edema, nausea and vomiting associated with chemotherapy, psoriasis, and in the palliative management of leukemias and lymphomas. Hydrocortisone was the first glucocorticoid to be used, and many compounds have been introduced that have different potencies and oral bioavailabilities. These include prednisolone, methylprednisolone, betamethasone, dexamethasone, deflazocort, and triamcinolone, which may be used systemically, and aclometasone, beclometasone, clobetasol, diflocortolone, fluocinonide, fluocortolone, halobetasol, halcinonide, and mometasone, which are used topically. Beclometasone, fluticasone, budesonide, and mometasone can also be used by inhalation in the treatment of asthma. Most of the adverse effects of glucocortocoids are predictable from the physiological actions of cortisol and are related to their metabolic, immunomodulatory, and anti-inflammatory effects. These include adrenal suppression as a result of negative feedback actions on the hypothalamus and anterior pituitary, stunted growth in children, elevation of blood glucose, muscle wasting, impaired wound healing, skin atrophy, bruising, increased susceptibility to infection, and osteoporosis with associated risk of vertebral and long-bone fractures. These adverse effects markedly limit the use of glucocorticoids. The adverse effects are miminized by using topical/local treatment instead of systemic treatment, although topical and inhaled steroids can be absorbed in sufficient quantities to cause systemic effects, including adrenal suppression.

Attempts to Dissociate the Anti-inflammatory and Adverse Effects of Glucocorticoids A new class of drugs, described as *dissociated glucocorticoids*, is under development. These compounds are characterized by a separation of the transactivating effects resulting from the activation of GR from those effects that repress transcription. Dissociated steroids potently inhibit

AP-1 or NF-κB, resulting in reduction of the formation of inflammatory mediators (e.g., IL-1β) but produce only weak transactivation via GREs. This should, for example, eliminate the metabolic effects of glucocorticoids. RU 24858 showed a promising profile of anti-inflammatory activity both in vitro (inhibition of lipopolysacchoride-induced IL-1β production) and in vivo (inhibition of cotton pellet granuloma; thymolysis), although it still possessed transactivational properties and exerted osteoporotic activites in vivo [140]. On the other hand, a nonsteroidal GR agonist, ZK216348, suppressed inflammation in rodent skin inflammation models, without elevating blood glucose [140]. This is a potentially very exciting line of research that may yield glucocorticoid-like anti-inflammatory agents with a much enhanced safety profile. However, it should be noted that ZK 216348 still suppressed the HPA axis.

As indicated above, glucocorticoids produce part of their inflammatory effect through increasing the formation of Anx-A1; and recombinant Anx-A1 or derived peptides can mimic many of the anti-inflammatory actions of glucocorticoids both in vitro and in animal models of acute and chronic inflammation. Their effects include inhibition of leukocyte adhesion and migration, induction of apoptosis in inflammatory cells, suppression of the production of inflammatory mediators by inhibition of PLA_2 (phospholipase A_2), COX-2 (cyclooxygenase-2), and iNOS (inducible nitric oxide synthase) and the induction of the anti-inflammatory cytokine IL-10 [140]. Development of analogs of these peptides may yield useful anti-inflammatory drugs.

Glucocorticoid Antagonists Glucocorticoid antagonists have been synthesized. RU486 (mifepristone) potently inhibits the glucocorticoid activitity of dexamethasone and has high affinity for the glucocorticoid and progesterone receptors, displays partial agonist activity at the glucocorticoid receptors in some tissues, and behaves as a competitive antagonist in others.

Inhibitors of Corticosteroid Synthesis

Corticosteroids (chiefly cortisol and aldosterone) are synthesized from cholesterol via pregnenolone, utilizing a number of cytochrome P450 enzymes [141]. Thus, P450 cholesterol side-chain cleavage enzyme (CYP11A) catalyses the conversion of cholesterol to pregnenolone, which is then converted to progesterone under the influence of 3β-hydroxysteroid dehydrogenase. Progesterone may be 17-hydroxylated to 17α-hydroxyprogesterone (using steroid 17α-hydroxylase, CYP17) giving rise to cortisol via 11-deoxycortisol, the conversion of which to cortisol is catalyzed by steroid 11β-hydroxylase (CYP11B1). Alternatively, progesterone may be converted to deoxycorticosterone under the influence of steroid 21-hydroxylase (CYP21), giving rise to aldosterone (via corticosterone and 18-hydroxycorticosterone) under the influence of aldosterone synthase (CYP11B2). Several drugs inhibit one or more of the enzymes involved in these pathways resulting in inhibition of corticosteroid synthesis.

Therapeutic Applications Inhibitors of corticosteroid synthesis are used in the treatment of Cushing's syndrome and of certain cancers. Metyrapone, which inhibits steroid 11β-hydroxylase, is also used in the differential diagnosis of corticotrophin-dependent Cushing's syndrome.

Aminoglutethimide is an inhibitor of steroid synthesis in the adrenal cortex, where it inhibits a number of cytochrome P450-mediated steps, including cholesterol side-chain cleavage enzyme. It also inhibits the cytochrome P450 aromatase enzyme responsible for the conversion of androgens to estrogens [142]—see later.

Mitotane, a metabolite of the insecticide DDT, inhibits adrenal steroid synthesis and is cytotoxic to adrenal cortical cells. It is used in the treatment of inoperable adrenal tumors and in some patients with Cushing's disease. The mechanism of action of mitotane remains unclear, but part of its action may be due to inhibition of cyclic-AMP generation in response to corticotrophin [143]. It may inhibit other steps in steroidogenesis, including inhibition of 11 β-hydroxylase [144]. It is also cytotoxic to adrenal cortical cells [145, 146]; at least in dog adrenocortical cells there is no clear separation of the concentration–response curves in terms of inhibition of steroid production and reduced cell viability, each being evident at 3μM and probably maximal at 100μM [146].

Trilostane is a selective, reversible, competitive inhibitor of 3β-hydroxysteroid dehydrogenase, a key enzyme involved in the synthesis of steroids. It inhibits the conversion of pregnenolone to progesterone, therefore inhibiting the synthesis of progesterone and therefore of the adrenal, gonadal, and placental steroids derived from progesterone, including cortisol, aldosterone, androgens, and estrogens. It does not inhibit the synthesis of cortisol from progesterone, 17-hydroxyprogesterone or 11-deoxycortisol in guinea pig isolated adrenal cortical cells [147, 148]. Trilostane may be used in the treatment of Cushing's syndrome [149], although responses may be variable. It has also found use in the treatment of breast cancer in postmenopausal women [150] who have experienced a relapse after initial treatment with estrogen receptor antagonists. Trilostane was shown to be an effective alternative to mifepristone for the midterm termination of pregnancy, when administered prior to misoprostil [151].

The antifungal agent *ketoconazole* is also a potent inhibitor of steroid biosynthesis, by inhibiting steroid 11β-hydroxylase and also, more potently, the conversion of 17α-hydroxyprogesterone to androstenedione under the influence of 17α-hydroxylase [152]. It has been used in Cushing's syndrome. *Etomidate*, an anesthetic agent, is a potent inhibitor of 11β-hydroxylase [153], which limits its use to induction, rather than maintenance of anesthesia and prevents its use in long-term sedation in traumatized patients [154].

Aldosterone and Other Mineralocorticoids

Aldosterone is the endogenous mineralocorticoid and is of fundamental importance in the maintenance of water and electrolyte balance, producing sodium and water retention and potassium excretion. It regulates sodium absorption in the distal convoluted tubule of the kidney and has analogous actions at other epithelial sites such as the colon [155]. Aldosterone is synthesized primarily in the zona glomerulosa of the adrenal cortex, which unlike other regions of this gland expresses aldosterone synthase (CYP11B2) [156]. There is now much evidence that aldosterone synthase is expressed in cardiac and vascular tissue and that aldosterone is synthesized in these tissues [157]. Aldosterone secretion is regulated primarily by the renin-angiotensin system through angiotensin II, which acts directly on the zona glomerulosa to stimulate hormone synthesis [156]. Potassium also stimulates aldosterone synthesis, through a direct action on glomerulosa cells. Although cortisol has high affinity for the mineralocorticoid receptor, it normally lacks significant mineralocorticoid activity because it is rapidly oxidized by the high affinity, type 2, 11β-hydroxysteroid dehydrogenase expressed in high levels in the kidney and colon [158]. Mutations in the gene for this enzyme (*HSD11B2* gene) result in the syndrome of apparent mineralocorticoid excess due to excessive activation of the mineralocorticoid receptor by cortisol. Other endogenously produced compounds with mineralocorticoid activity are 11-deoxycorticosterone, 18-hydroxydeoxycorticosterone, corticosterone, and 19-nordeoxycorticocosterone [141]. Apart from its fundamentally important physiological role, aldosterone is clearly important in pathophysiology, where it makes an important contribution in cardiac remodeling in heart failure [159] and in various forms of hypertension. It may also contribute to endothelial dysfunction, reduced fibrinolysis, and cardiac arrhythmias.

The only mineralocorticoid in general use is fludrocortisone, which is a synthetic, orally active 9-α fluorinated analog of cortisol and that at normal therapeutic doses has only mineralocorticoid activity. Its mineralocorticoid potency relative to that of cortisol has been attributed to its resistance to 11-β-oxidation [160] by type 2, 11β-hydroxysteroid dehydrogenase. Its main use is in replacement therapy in adrenal cortical insufficiency, and it is the drug of first choice in the treatment of orthostatic hypotension [161].

Receptors Mediating the Effects of Aldosterone Aldosterone acts at the mineralocorticoid receptor (MR) and its classical effects are exerted through modulation of the transcription rates of various genes, including those encoding subunits of the epithelial sodium channel (ENaC) and the Na$^+$K$^+$ATPase [162]. It is now widely accepted that aldosterone exerts rapid nongenomic actions [163], which may be mediated by the classical MR or by other, as yet ill-defined receptors. The role of membrane-bound aldosterone receptors is uncertain, but there is evidence for their existence and for their resistance to classical aldosterone MR antagonists [164].

Mineralocorticoid Antagonists

Mineralocorticoid antagonists prevent the action of the aldosterone at the MR. In view of the pathophysiological role of aldosterone, these drugs have a potentially major value in the treatment of heart failure. A major randomized trial (RALES—Randomized ALdactone Evaluation Study) showed that addition of the aldosterone antagonist spironolactone to standard therapy in patients with congestive heart failure reduced mortality by 30 percent [165, 166]. This effect was reproducible in a rat model of heart failure, in which administration of another aldosterone antagonist, canrenone (a metabolite of spironolactone), attenuated left ventricular remodelling and improved cardiac function [167]. Spironolactone produces adverse effects including gynecomastia, breast tenderness, menstrual irregularities, and impotence. Other aldosterone antagonists have been developed including mexrenone and eplerenone. These drugs have similar affinity to spironolactone for the mineralocorticoid receptor but had a much reduced affinity for androgen and progesterone receptors and have markedly reduced antiandrogenic and progestagenic effects compared to spironolactone in animals [168] and humans [169, 170].

23.7 ANDROGENS

The principal natural androgens are testosterone and dihydrotestosterone, which is produced from testosterone in tissues expressing 5α-reductase activity. Testosterone is secreted primarily by the testes and significant amounts, primarily as the precursor dehydroepiandrosterone, originate in the adrenal cortex, which is the main source of androgens in the female. Androgens play a key role in male reproduction, being responsible for the development and maintenance of the reproductive ducts (epididymis, vas deferens, ejaculatory ducts), glands (prostate, seminal vesicles, bulbourethral glands), and the external genitalia. They are also required for spermatogenesis and the development and maintenance of the male secondary sexual characteristics. Androgens play a key role in the masculinization of external genitalia (function of dihydrotestosterone) and the development of the epididymis, vas deferens, and seminal vesicles from the Wolffian duct (function of testosterone) in the male fetus. The therapeutic applications of androgens are currently quite limited and they are used primarily for androgen-replacement therapy in hypogonadal males.

Androgen Receptors

There appears to be a single androgen receptor, which is a member of the nuclear receptor superfamily, typically having an N-terminal domain implicated in activation of transcription, a DNA binding domain and a ligand binding domain. Like glucocorticoid receptors, binding of the ligand causes the cytoplasmic receptor to dissociate from heat shock proteins leading to dimerization and phosphorylation of the receptor and its translocation to the nucleus. Within the nucleus the receptor binds to the androgen response element (ARE). This results in recruitment of hormone- and cell-specific coactivators and activation of transcription [171].

The androgen receptor is widely distributed and, in addition to the male reproductive tract, it is expressed in many tissues including the female external genitalia, skeletal muscle, skin, sebaceous glands, hair follicles, and the central nervous system.

Anabolic Steroids

The anabolic action of the main male sex hormone testosterone is well known and this property is retained in synthetic derivatives such as nandrolone (19-nortestosterone). Synthetic, 17α-alkylated compounds (oxymethalone, stanozolol, oxandrolone, danazol) are orally active and, in animals, show selectivity for anabolic versus classical androgenic activity, although the evidence for such selectivity in the human is unclear. These agents are used for their anabolic properties in several disease states [172]. All can produce concomitant androgenic side effects and the 17α-alkylated compounds may produce hepatotoxicity. The anabolic actions explain their abuse by athletes who often consume much greater than therapeutic doses, in an attempt to increase muscle mass. Whether performance is actually increased by these agents is uncertain, although a randomized, double-blind, placebo-controlled trial showed that a weekly, large dose of testosterone enanthate increased muscle strength, especially when combined with exercise [173]. Large doses of these drugs have been shown to upregulate androgen receptors, and this may contribute to their anabolic effects [174].

Anabolic steroids act at the androgen receptor. As there is only one androgen receptor, differences in effects in different tissues remain unexplained. Reduced androgenicity of nandrolone is explained by the reduction in its activity by 5α-reductase. Thus it has reduced activity in tissues expressing high levels of this enzyme and full activity in tissues that express low levels of, or no, 5α-reductase [175, 176]. There is also evidence that they act as antiglucocorticoids by competing for binding at the glucocorticoid receptor [177] and thus may exert an anticatabolic effect.

There is great interest in separating the anabolic and androgenic effects of androgens. Although there is only one androgen receptor, the effect of the androgen–receptor combination may vary in a cell-context-specific way because of the different expression of coactivators in different tissues. Moreover, binding of different agents may result in particular receptor conformations that recruit different coactivators. Thus, by analogy with SERMs (see below), the concept of selective androgen receptor modulators (SARMs) has been proposed [178]. Some evidence to support this possibility is derived from the development of a nonsteroidal androgenic agonist agent showing increased efficacy and potency for anabolic effects (levator ani weight) relative to androgenic effects (weight of prostate/seminal vesicles) in castrated rats [179]. Although the tissue selectivity was not very marked, this may indicate the feasibility of this approach for separating androgenic and anabolic properties.

Antiandrogens

Compounds interfering with the actions of androgens may be steroidal (cyproterone acetate; mifepristone) or nonsteroidal [flutamide; hydroxyflutamide (the active metabolite of flutamide); nilutamide; bicalutamide]. While the nonsteroidal agents are selective for the androgen receptor, the steroidal compounds display partial agonist activity at this receptor and also antagonist activity at progesterone and glucocorticoid receptors [180]. Binding of antagonists to the androgen receptor prevents the receptor conformation change required to effect the activation of transcription [180].

Nilutamide, flutamide, and bicalutamide are used alone, or in combination with gonadorelin analogs, in the treatment of advanced prostate cancer. These drugs have also had limited evaluation in the management of benign prostatic hypertrophy and also to treat hirsutism and androgenic alopecia in women with polycystic ovarian disease [181]. When used in men, common, predictable adverse effects are gynecomastia, possibly accompanied by galactorrhea, decreased libido, and inhibition of spermatogenesis.

5α-Reductase Inhibitors

The conversion of testosterone to dihydrotestosterone, which has a much higher affinity for the androgen receptor, is an important activating mechanism. This conversion is catalyzed by the enzyme 5α-reductase, which is expressed in two isoforms, 5α-reductase type 1 and 5α-reductase type 2, which have only ~50 percent homology and which have different enzymic properties and tissue distribution [182]. The normal prostate, genital skin, seminal vesicles, dermal papilla, and developing fetal male external genitalia predominantly express type 2, whereas the type 1 isoform is the predominant form in established prostate cancer, the scalp, nongenital skin, and the sebaceous gland [182]. There are also marked species differences, which lead to marked differences in the potencies and selectivities of inhibitors, for example, when tested against human enzymes compared with rat enzymes. Thus there is enormous potential for developing type-selective inhibitors that will have applications in benign prostatic hyperplasia (BPH), prostatic cancer, androgenic alopecia, hirsutism, and acne. Finasteride is an azasteroid drug that is selective for the type 2 isoenzyme and that is marketed for the treatment of BPH, in which it causes a marked reduction in serum dihydrotestosterone concentrations, accompanied by a 15 to 25 percent reduction in gland size and reduction in urinary symptoms. Duasteride is a mixed type 1/type 2 inhibitor. Many nonsteroidal compounds, with high selectivity for one or other isoforms are under development [182].

23.8 ESTROGENS

The estrogens are secreted primarily by the gonads with smaller amounts originating in the adrenal cortex. The main physiological estrogen is estradiol-17β. Estrogens play key roles in the development and maintenance of the adult, female reproductive tract, and female secondary sexual characteristics but also have important effects in nonreproductive tissues, particularly bone, the cardiovascular system, and the brain. Estrogens are also implicated in breast and uterine tumor progression.

Estrogen Receptors

Like receptors for glucocorticoids, androgens, and progestins, the estrogen receptor (ER) belongs to the superfamily of nuclear receptors mediating their effects through alteration in transcription. The estrogen receptor, like the glucocorticoid receptor, is bound to heat shock proteins in the resting state, these being displaced upon hormone binding. The activated receptor dimerizes and interacts with specific DNA estrogen response elements. The conformation of the ligand-bound receptor depends on whether the ligand is an agonist or an antagonist. Binding of a full agonist displaces co-repressor proteins and recruits coactivator proteins that then result in gene transcription. On the other hand antagonist binding favors continued interaction with co-repressors and no transcription occurs. Other drugs, such as tamoxifen, a selective estrogen receptor modulator (SERM), promote a conformation that allows partial interaction with both co-repressors and coactivators. The level of different coactivators and co-repressors may be cell-type dependent, which may explain cell-context-dependent activities of different ligands at the estrogen receptor [183].

Two estrogen receptors are now known—ERα and ERβ. These are both nuclear receptors that, although having considerable homology at the DNA binding domains and therefore similar binding affinities for estrogen response elements, have relatively low overall sequence homology of ~47 percent [184, 185]. These receptors show different tissue distribution. For example, ERα is predominant in uterus, prostate stroma, ovarian thecal cells, and bone, whereas ERβ is expressed predominantly in colon, ovarian granulosa cells, bone marrow, and brain dorsal raphe [185].

Estrogens, like other steroids, can also exert rapid (within minutes), nongenomic effects, some of which are prevented by estrogen antagonists but others of which are not and appear to involve cAMP-mediated mechanisms [186]. For example, the rapid vasodilator effect of estradiol is prevented by the estrogen antagonist ICI 182, 780 and appears to be mediated by endothelial-derived nitric oxide produced under the influence of eNOS activated via a MAP kinase-dependent pathway through ERα [187].

Estrogen Agonists

The main natural estrogen is estradiol-17β but other natural estrogens include estriol and estrone. The main synthetic estrogens in use are ethinyl estradiol and mestranol, which is inactive until metabolized to ethinyl estradiol. Stilbestrol is a potent, nonsteroidal, synthetic estrogen agonist. Until recently, there were no agonists or antagonists selective for ERα compared with ERβ receptors or vice versa. However, propyl pyrazole triol [PPT; 4-propyl-1,3,5-Tris (4-hydroxyphenyl) pyrazole] appears to have selective agonist activity at Erα and diarylpropiol nitrile appears to have selective ERβ agonist activity [188]. These drugs should facilitate the characterization of responses mediated via the two different estrogen receptors.

Antagonists at Estrogen Receptors

Clomiphene and ICI 182,780 behave as pure antagonists at the estrogen receptor. An ERα-selective antagonist, methyl-piperidino-pyrazole, has also been described [188].

Selective Estrogen Receptor Modulators (SERMs)

Tamoxifen was originally considered as an estrogen antagonist and, accordingly, is widely used in the treatment of estrogen-dependent breast cancer. While clearly acting as an antagonist in the breast, tamoxifen functions as an agonist in bone, uterus, and the cardiovascular system [183]. Raloxifene, on the other hand, shows no agonist effects in the uterus but is an agonist in bone [183]. Because of this behavior as agonists in some tissues and antagonists in others, these drugs and others (e.g., idoxifene, droloxifene, torimifene, droloxifene, ospemifene, lasofoxifene, and arzoxiphene) are referred to as SERMs (selective estrogen receptor modulators).

Aromatase Inhibitors

Aromatase is responsible for the last stage in the synthesis of estrogens, either from testosterone to give estradiol or from the androgenic testosterone precursor androstanedione to give estrone, which can then be converted into estriol or estradiol. There is much interest in developing potent selective inhibitors of this enzyme, as these have application in the treatment of estrogen-dependent conditions such as breast cancer [189] and endometriosis [190] and in promoting ovulation by removing the estrogen-mediated negative feedback inhibition of gonadotrophin secretion [191]. The first agent to be used clinically was aminoglutethamide (see earlier) but is nonselective, inhibiting several steps in steroid biosynthesis and also inducing aromatase activity [192].

This induction may explain the failure of aminoglutethimide therapy in some cancer patients. Steroidal inhibitors, such as the androstanedione analogs exemstane and formestane, bind irreversibly to the catalytic site of the enzyme. They are administered parenterally and have androgenic activity. Nonsteroidal, imidazole compounds (anastrozole, letrozole) bind to the cytochrome P450 site of the enzyme. They can be administered orally and have no androgenic activity [189].

Therapeutic Applications of Agents Acting at Estrogen Receptors Agonists are used extensively in hormone replacement therapy following gonadal failure or to replace estrogen in postmenopausal women suffering from menopausal symptoms (hot flushes, vaginal dryness, postmenopausal osteoporosis). They may be used alone or together with a progestin, which is indicated in postmenopausal women with an intact uterus, in order to minimize the risk of endometrial cancer associated with estrogenic stimulation of the uterus. While women benefit hugely from the relief of menopausal symptoms afforded by estrogen replacement therapy and clearly have a reduced risk of osteoporosis (as evidenced by hip fractures) and colorectal cancer, there is controversy concerning whether or not long-term hormone replacement therapy in the menopause is associated with an increased risk of stroke and coronary heart disease, as suggested by a large, randomized trial [193]. Certainly there is no evidence to support the estrogen-mediated protection from coronary artery disease that would be predicted from the very low prevalence of myocardial infarction in premenopausal women.

The other major use is in pharmacological contraception, where they are administered along with a progestin, resulting in failure of ovulation as a result of suppression of the hypothalamic secretion of GnRH and inhibition of gonadotrophin secretion. The progestin component also prevents estrogen-mediated endometrial development and changes in cervical mucus.

The major use of antagonists is in treating cancer of the breast. The SERM tamoxifen is the main drug used for this purpose. SERMs such as raloxifene, which are devoid of uterine stimulant activity but retain agonist activity in bone, are used to treat postmenopausal osteoporosis. Additionally, estrogen antagonists have potential in the treatment of endometriosis.

23.9 PROGESTINS

The natural progestin progesterone is secreted from the corpus luteum during the luteal phase of the menstrual cycle, where it prevents further endometrial proliferation and stimulates endometrial gland branching and secretion. Progesterone plays a key role in mammary gland development and in maintaining pregnancy. In pregnancy, the placenta is the major source of progesterone after the first trimester.

Progesterone Receptors The progesterone receptor is a typical member of the superfamily of nuclear receptors. There are two isoforms of the progesterone receptor—PR-A and PR-B—and these show different tissue expression and function. For example, in the mouse uterus, where progesterone inhibits the proliferative effect of estrogen, it appears that PR-A mediates the inhibitory response, while PR-B may actually stimulate proliferation [194]. Both PR isoforms are expressed in the mouse mammary gland and although both may be required for mammary gland development, ablation of PR-A in knockout mice did not affect the ability of PR-B to mediate the effect of progesterone in producing ductal side branching and development of the alveoli, suggesting the predominant role of PR-B in this tissue [194]. As with glucocorticoids and estrogens, progestins can produce nongenomic effects. For example, progesterone rapidly (within 2 to 5 min) stimulates insulin secretion in isolated islets and perfused pancreas and reduces glucose uptake by adipocytes [195]. It also inhibits proliferation in human transformed cell lines from the uterine cervix, cells that lack steroid receptors and in which this effect of the hormone is not blocked by progesterone antagonists [196].

Agonists at Progesterone Receptors

Many compounds are available with agonist activity at progesterone receptors. Their pharmacology is complex because many, either directly or through their metabolites, bind to other steroid receptors resulting, for example, in estrogenic activity in some (norethisterone, lynestenol, norethindorel), androgenic (norethisterone, lynestenol, levonorgestrel, gestodene) and glucocorticoid (chlormadinone acetate, megestrol acetate, medroxyprogesterone acetate), and antimineralocorticoid (drospirenone, gestodene) activities [197]. It is therefore very difficult to make general statements about the pharmacology of these agents.

Progesterone Antagonists

The best known antagonist of progesterone is mifepristone (RU 486), which is used to terminate pregnancy and which may be used for postcoital contraception. Antagonists at the progesterone receptor also have contraceptive potential as a consequence of their ability to prevent endometrial maturation. By analogy with SERMs, there is interest in developing selective progestin receptor modulators.

Therapeutic Applications of Progestins Progestins are widely used in pharmacological contraception, where they are used either alone or in combination with the estrogens ethinyl estradiol or mestranol or along with estrogens in hormone replacement therapy. The controversy concerning whether or not long-term hormone replacement therapy in the menopause is associated with an increased risk of stroke and coronary heart disease, as suggested by the Women's Health Initiative trial [193], may partly relate to the use in that study of the progestin medroxyprogesterone acetate, which has weak androgenic activity.

23.10 TREATMENT OF DIABETES MELLITUS

The prevalence of diabetes mellitus, especially of Type 2 diabetes, is increasing, a world total of >300 million being predicted by 2030. This high prevalence combined with the associated increased mortality and morbidity, primarily as a result of macrovascular disease and microvascular long-term complications make diabetes mellitus a major health problem. Overwhelming evidence, especially from the DCCT study [198] for Type 1 diabetes and the UKPDS study [199] for Type 2 diabetes suggests that good metabolic control would markedly reduce mortality and morbidity.

While Type 1 diabetes mellitus is due to autoimmune destruction of the β cells and the consequent severe insulin deficiency, it is now broadly accepted that Type 2 diabetes results from both peripheral insulin insensitivity and impaired insulin secretion [200]. There is progressive islet β-cell failure in patients with Type 2 diabetes and a reduction in the β-cell mass [201]. Thus as well as developing drugs that improve target tissue responsiveness to insulin, the islet β cell remains an important target for the development of drugs for treating Type 2 diabetes. Drugs are required that both enhance insulin secretion in response to normal, physiological, meal-related nutrient stimuli and that prevent the progressive loss of islet β-cell mass. Augmentation of meal-related insulin secretion by drugs should be glucose dependent. This will ensure that increased insulin secretion occurs when required, the corollary being that insulin secretion would remain at a basal rate between meals, avoiding hyperinsulinemia and thus hypoglycemia, as seen with sulphonylureas, and the potential cardiovascular risks associated with chronic hyperinsulinemia and insulin resistance [202, 203]. Physiological, meal-related insulin secretion occurs rapidly in response to absorbed nutrients, primarily glucose. Glucose stimulates insulin secretion following its transport into and metabolism in the β cell, generating ATP, which closes K_{ATP} channels, resulting in depolarization and calcium influx (Fig. 23.2). The effect of glucose on insulin secretion can be amplified by activation of both adenylyl cyclase and phospholipase C catalyzed pathways [204]. The early rapid insulin secretory response to glucose is important in maintaining normal blood glucose, and it is widely accepted that loss of this early-phase insulin secretion is an important and early event in the development of Type 2 diabetes [205].

Insulin

Insulin is used to replace the deficient hormone in the treatment of diabetes, and currently there is no alternative therapy for Type 1 diabetes. Insulin is also used in the treatment of Type 2 diabetes when this cannot be adequately controlled by orally active antidiabetic drugs. The aim of treatment using insulin is to maintain euglycemia (a plasma glucose level of 4 to 7 mmol/L) without causing hypoglycemia. However, good control is difficult to achieve because of the difficulty of administering insulin in a way that mimics physiological

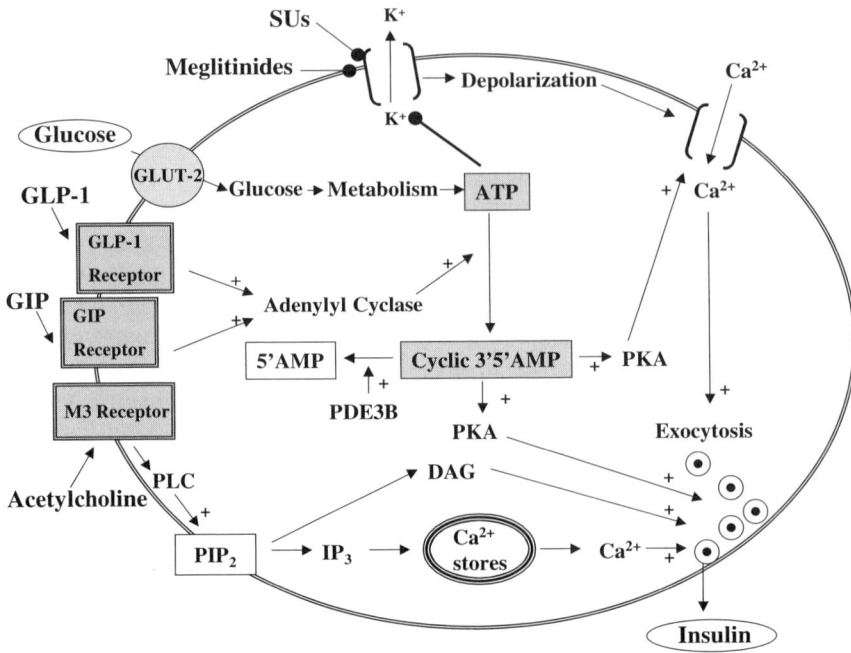

Figure 23.2 Schematic illustration of glucose-induced insulin secretion and its amplification mechanisms. Glucose is transported into the pancreatic islet β cell using the transporter GLUT-2. It is metabolized producing ATP. The resultant increase in the ATP:ADP ratio inhibits the plasma membrane K_{ATP} channel, blocking potassium efflux and resulting in depolarization of the β-cell membrane. Depolarization activates L-type voltage-sensitive calcium channels, allowing influx of Ca^{2+} ions, which trigger exocytosis. The sulfonylureas (SUs) and meglitinide analogs also block the K_{ATP} channel by binding to its SUR1 subunit. The effect of glucose is amplified by the second-messenger cyclic-3'5'AMP, inositol 1,4,5-trisphosphate (IP_3), and diacylglycerol (DAG). Cyclic-3'5'AMP is produced from ATP as a consequence of activation of adenylyl cyclase, primarily by incretin peptides glucagon-like peptide-1 (GLP-1) and glucose-dependent insulinotropic polypeptide (GIP) secreted by the intestines in response to nutrients. The effect of cyclic-3'5'AMP is terminated by the phosphodiesterase isoform PDE3B. Cyclic-3'5'AMP activates protein kinase A (PKA), which acts at a number of sites (releasing Ca^{2+} from intracellular stores, enhancing Ca^{2+} through L-type channels, and augmenting the effects of Ca^{2+} on exocytosis) to augment the effects of glucose and also produces effects independent of PKA. IP_3 and DAG are produced from the membrane lipid phosphatidyl inositol 4,5-bisphosphate (PIP_2) by the activation of phospholipase C (PLC). Vagally released acetylcholine acting through the muscarinic M_3 cholinoceptor is probably the major physiological activator of this mechanism. IP_3 activates the release of Ca^{2+} from intracellular stores and DAG activates protein kinase C (PKC), which amplifies the effect of Ca^{2+} in promoting exocytosis (a + associated with → denotes activation/stimulation; • denotes inhibition).

insulin secretion, with rapid peaks during and immediately following a meal and low, basal concentrations between meals.

Insulin preparations are now largely based upon human insulin prepared by enzymic modification of porcine insulin, by chemical combination of the A and B chains produced using genetically modified bacteria, or from a precursor produced by yeast modified by recombinant DNA technology.

Insulin analogs, with different onsets and durations of action, have been produced. In solution, soluble insulin forms hexamers, which must dissociate for the absorption of insulin from a subcutaneous injection site. If amino acid substitutions are made in the insulin B chain (B28 proline replace by lysine and B29 lysine by proline yielding lispro insulin, or B28 proline by aspartate yielding insulin aspart), the tendency to form hexamers is diminished and this increases bioavailability so that the analogs are absorbed much more rapidly [206]. On the other hand, elongation of the C-terminal end of the B chain of the insulin molecule with two arginine residues and substitution of A21 asparagine by glycine elevates the isoelectric point of insulin from pH 5.4 toward neutral. This analog, insulin glargine, crystallizes in the subcutaneous injection site, providing a very long delay in absorption. Long-acting insulin analogs injected once daily may be combined with the rapid onset, short-acting analogs injected with each meal, which should provide a much more physiological plasma insulin profile than could be achieved using conventional insulin preparations [207].

Mechanism of Action of Insulin The insulin receptor is a tetrameric protein belonging to the subfamily of receptor tyrosine kinases. Binding of insulin to the α subunit activates the tyrosine kinase activity of the β subunit, leading to receptor autophosphorylation and also to tyrosine phosphorylation of the insulin receptor substrate (IRS) adaptor proteins. The phosphorylation of tyrosine sites on IRS allow its interaction with SH2 (Src-homology-2) proteins, including Grb2, which leads to activation of MAP kinase and alterations in protein synthesis and the p85 regulatory subunit of PI-3-kinase (phosphatidylinositol 3-kinase). PI-3 kinase plays a pivotal role in insulin action in enhancing glucose uptake by mobilization of GLUT-4 glucose transporters in skeletal muscle and adipose tissue and, via PKD1 [PtdIns(3,4,5) P_3-dependent protein kinase], in activating PKB (protein kinase B), which itself inactivates GSK3 (glycogen synthase kinase 3). Inactivation of GSK3 leads to activation of glycogen synthase and of the initiation factor eIF2B, resulting, respectively, in increased glycogen synthesis and increased protein synthesis [208–210] (Fig. 23.3).

Understanding the insulin receptor-mediated signaling may allow the development of small, nonpeptide molecules that activate the signaling pathways downstream of the receptor. For example, this could be achieved by a molecule that mimicked PtdIns(3,4,5) P_3 in enabling PDK1 to activate PKB. Similarly, a drug that inhibited GSK3 would also mimic effects of insulin [208].

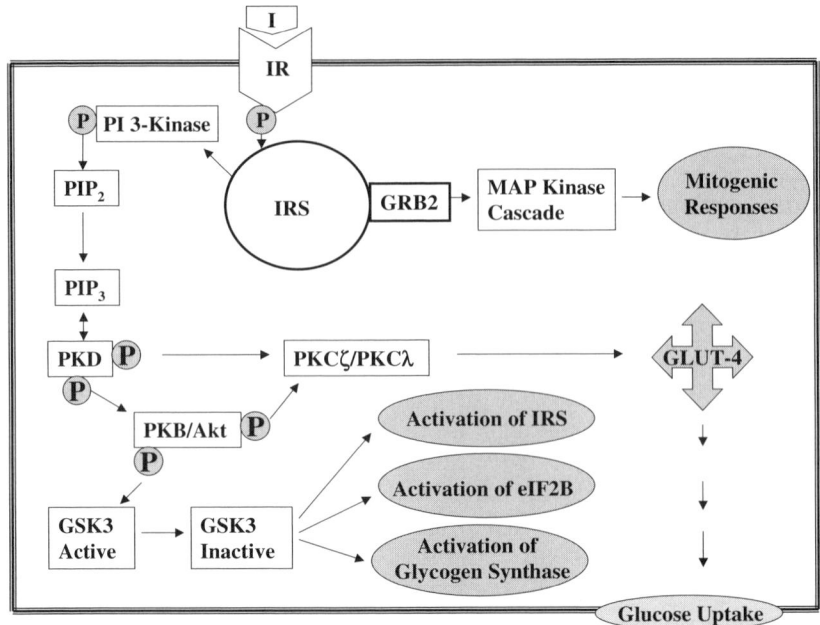

Figure 23.3 Mechanism of action of insulin [208–210]. Insulin (I) binds to the α subunits of the insulin receptor (IR), resulting in activation of the tyrosine kinase of the β subunit. Autotyrosine phosphorylation of the receptor results in its interaction with and phosphorylation of insulin receptor substrates (IRS). Phosphorylated sites on IRS interact with various proteins containing Src homology (SH) domains. These include phosphatidylinositol 3-kinase (PI 3-kinase) and growth factor receptor binding protein 2 (Grb2). Interaction with Grb2 results in activation of the MAP kinase cascade and hence mitogenic responses. Activation of PI 3-kinase leads to phosphorylation of the membrane lipid phosphatidyl inositol 4,5-bisphosphate (PIP_2) producing phosphatidyl inositol (3,4,5)-trisphosphate (PIP_3). PIP_3 interacts with protein kinase B (PKB/akt), recruiting it to the membrane and allowing its phosphorylation and activation by phosphatidyl inositol (3,4,5)-trisphosphate-dependent protein kinase (PDK). PKB phosphorylates a number of proteins, including glycogen synthase kinase 3 (GSK3), the activity of which normally inhibits the activity of several enzymes including glycogen synthase, eukaryotic initiation factor 2B (eIF2B), and IRS. Phosphorylation of GSK3 inactivates the enzyme increasing the activity of these enzymes. Thus synthesis of glycogen and protein are activated and insulin signaling through IRS is augmented. PDK (possibly through PKB) also phosphorylates atypical protein kinase C (PKCξ/λ), which activates the trafficking of the glucose transporter (GLUT-4) to the plasma membrane, thereby stimulating glucose uptake (P→ denotes phosphorylation).

Biguanides

The biguanides are derivatives of the compound biguanide (guanylguanidine) that exert a blood glucose-lowering effect in Type 2 diabetes mellitus. Metformin is now the only biguanide in clinical use [211, 212], following the withdrawal of phenformin and buformin in most countries in the late 1970s due to a high incidence of associated lactic acidosis. Metformin, which has a much lower risk of lactic acidosis, is used widely in the treatment of Type 2 diabetes. Biguanides do not increase plasma insulin concentrations and do not cause severe hypoglycemia; hence they are regarded as antihyperglycemic (rather than hypoglycemic) agents. The United Kingdom Prospective Diabetes Study (UKPDS), a large, randomized, controlled trial over >10 years in Type 2 diabetic patients, noted that patients who received metformin as initial oral antidiabetic drug therapy had lower morbidity and mortality from macrovascular disease [213]. Since glycemic control achieved with metformin was similar to that achieved with other antidiabetic agents (sulfonylureas and insulin) studied in the trial, it is possible that additional effects of metformin may be involved. These could include an improved lipid profile and increased fibrinolysis, which could be related to the fact that metformin does not elevate, and may reduce, serum concentrations of insulin.

Metformin increases insulin sensitivity in liver and muscle, suppressing hepatic gluconeogenesis and glycogenolysis and enhancing insulin-stimulated glucose uptake into skeletal muscle. It also exerts an insulin-independent suppression of fatty acid oxidation and a reduction in hypertriglyceridemia. These effects collectively reduce insulin resistance and glucotoxicity in Type 2 diabetes [214]. There is considerable evidence that AMP-activated protein kinase (AMPK) is an important target for metformin, which activates this enzyme in vitro and in vivo [215]. Activation of AMPK produces very similar effects to those of metformin, decreasing hepatic glucose production and gluconeogenesis, inhibiting lipolysis, lipogenesis, and lipogenic gene expression, and increasing skeletal muscle glucose uptake [216]. In addition to its widespread use in Type 2 diabetes mellitus, metformin is increasingly being used to treat the polycystic ovary syndrome (PCOS), where it leads to a decrease in serum insulin and androgen levels, as well as improved in ovulatory function [217]. It may also be useful in obesity and a randomized double-blind placebo-controlled trial in nondiabetic, obese, hyperinsulinemic adolescents on a low-calorie diet showed metformin to produce weight loss, a decrease in body fat, reduced hyperinsulinemia, and enhanced insulin sensitivity compared to placebo [218].

Thiazolidinediones

Thiazolidinediones (TZDs), also termed *glitazones*, are used as antidiabetic agents for the treatment of Type 2 diabetes mellitus. Several have been developed and rosiglitazone and pioglitazone are in widespread use, although troglitazone was withdrawn soon after marketing because of idiosyncratic hepatotoxicity. They appear to lower blood glucose by increasing the sensitivity to insulin, which can be demonstrated both in animals and humans. For example, in patients with Type 2 diabetes, troglitazone produced a 30 percent increase in insulin-stimulated glucose disposal [219].

Receptors for Thiazolidinediones Thiazolidinediones act as agonists of the nuclear peroxisome proliferator-activated receptor-γ (PPARγ) at which they bind with very high affinity. PPARγ are nuclear receptors belonging to a superfamily of transcription factors (PPARα, PPARδ, and PPARγ) and that regulate gene transcription in response to small lipophilic ligands, prostaglandin J2 being one natural agonist at this receptor. PPARγ are most strongly expressed in adipose tissue and weakly expressed in liver and skeletal muscle [219–222]. The receptor is found as a heterodimer with the retinoid receptor (RXR), which, in the resting state, binds a co-repressor complex that inhibits transcription. Binding of a thiazolidinedione results in dissociation of the co-repressor complex and recruitment of coactivators, leading to increased transcription.

The mechanisms underlying the increase insulin sensitivity in response to activation of PPARγ are complex and may involve both direct effects on muscle and indirect effects mediated by actions on adipose tissue. Thus activation of PPARγ in adipose tissue leads to increased release of adipocyte-derived hormones that increase insulin sensitivity (adiponectin) and a decrease in the release of those that decrease insulin sensitivity (resistin, TNFα, IL-6). Leptin production is reduced, although this cannot explain an increase in insulin sensitivity. Activation of PPARγ diminishes the release of free fatty acids from adipocytes, which itself will augment insulin sensitivity in muscle, due to the well-known effect of free fatty acids in inhibiting insulin-stimulated glucose uptake by skeletal muscle [219–222]. Paradoxically, mice in which PPARγ are partially deleted (PPARγ +/−) are more insulin sensitive than wild-type animals and have less adipose tissue and put on less weight on a high-fat diet. This raises speculation that PPARγ antagonists may be of value in treating obesity and Type 2 diabetes. There is therefore now interest in developing, by analogy with SERMs, selective PPARγ modulators (SPPARMs) that have mixed agonist/antagonist properties [222].

Therapeutic Applications The TZDs are used alone or in combination with sulfonylureas or meglitinide analogs as oral antidiabetic agents in the treatment of Type 2 diabetes mellitus [223]. They are ineffective in the absence of insulin and therefore are not used in Type 1 diabetes. The drugs promote weight gain, which is potentially important disadvantage when treating Type 2 diabetes. This is partly due to an increase in the number of adipocytes and is predictable from the known ability of PPARγ activation to promote adipocyte differentiation. Thus, as discussed above, the development of PPARγ antagonists and of SPPARMs is a valid objective and compounds of this type

(SR-202—PPARγ antagonist; MCC555—SPPARM), which are antidiabetic and which prevent weight gain in experimentally obese, hyperglycaemic animals are available.

Sulfonylureas

Sulfonylureas have been classified as so-called *first-generation* agents (tolbutamide, chlorpropamide, tolazamide, and acetohexamide) and *second-generation* agents (glibenclamide, glipizide, gliclazide, glimepiride, and gliquidone). The second-generation agents are significantly more potent, although there is no clear evidence of superiority of the second-generation drugs in terms of therapeutic effectiveness. The major differences among the sulfonylureas, apart from potency, are pharmacokinetic. Some agents are long-acting (chlorpropamide, glibenclamide) while others have much shorter durations of action (tolbutamide, gliquidone). Sulfonylureas are extensively bound to plasma proteins and are metabolized mainly in the liver to compounds that may be inactive in the case of some agents (e.g., gliquidone) or active in the case of others (e.g., acetohexamide, tolazamide). Tolbutamide, glibenclamide, glipizide, and glimepiride are metabolized by CYP2C9 and genetic polymorphisms of this enzyme may markedly influence their pharmacokinetics [224, 225].

A prolonged controversy resulted from a study suggesting that tolbutamide might produce an increase in cardiovascular mortality [226]. However, the UK Prospective Diabetes Study Group showed no increased cardiovascular mortality with sulfonylureas in a large, randomized, controlled trial over 10 years in which blood glucose was intensively controlled [213]. Indeed, sulfonylureas significantly reduced diabetic microvascular endpoints. It must be pointed out, however, that the UKPDS used different sulfonylureas from the one (tolbutamide) used in the UGDP [226] study.

Receptors for Sulfonylureas Sulfonylureas work primarily by stimulating insulin secretion following binding to a high-affinity sulfonylurea receptor subunit (SUR) of the β-cell potassium-sensitive ATP channel (K_{ATP} channel). Binding results in blocking K efflux through the $K_{IR}6.2$ channel, depolarization of the β cell, opening of voltage-sensitive Ca^{2+} channels, influx of Ca^{2+}, and insulin secretion [227]. There are different types of SUR (SUR1, SUR2A, SUR2B), the expression of which differs among different tissues. Thus the β-cell potassium-sensitive K_{ATP} channel contains SUR1, whereas skeletal and cardiac muscle contain SUR2A. Different sulfonylureas have different affinities for the various SURs. For example, tolbutamide and gliclazide are significantly more potent in blocking SUR1 (expressed with $K_{IR}6.2$ in *Xenopus* oocytes) relative to SUR2A, whereas glibenclamide and glimepiride are equipotent at both [228], although other findings [229] showed glibenclamide to bind with much higher affinity to SUR1, relative to SUR2. There has been much debate about the role of extrapancreatic effects of sulfonylureas in their metabolic actions. Extrapancreatic effects probably result, at least in part indirectly from the improved blood glucose control, because chronic hyperglycemia may itself produce insulin resistance [230]. Nevertheless, in vitro effects have been demonstrated on both skeletal muscle and adipose tissue [231]. Clearly, there are K_{ATP} channels in many peripheral tissues, including skeletal muscle [228]. Moreover, disruption of the SUR2A containing K_{ATP} channels in skeletal-muscle-enhanced glucose uptake [232]. Thus extrapancreatic effects in stimulating glucose uptake by skeletal muscle might relate to block of SUR2A-containing K_{ATP} channels. Another intriguing question is whether or not block of heart, vascular smooth muscle, or cardiac muscle K_{ATP} channels, which contain SUR2, might be manifest as some form of detrimental action on these systems.

Therapeutic Applications Sulfonylureas are widely used in the treatment of Type 2 diabetes. Because they work primarily through stimulating the secretion of insulin, they are of no use in the treatment of established Type 1 diabetes, in which there is extensive autoimmune destruction of the β cells. Patients who are older than 40 years and have had diabetes for less than 5 years are most likely to respond. After 10 years only about 50 percent of patients remain satisfactorily controlled by sulfonylureas. This is referred to as secondary failure and may be attributable to β-cell failure and/or dietary noncompliance [233]. There is evidence that improved control can be achieved by combining insulin with sulfonylureas [234].

Although their pharmacology is broadly similar, individual drugs apparently have unique properties. There is evidence that gliclazide has antiplatelet and antioxidant activity independent of its effects on glycemic control [235]. These effects have been attributed to the azabicyclo ring, which is unique to gliclazide among the sulfonylureas. It remains to be seen in long-term, large-scale, randomized, controlled trials if these effects confer additional benefits on patients with Type 2 diabetes, in terms of a reduction in diabetic complications and/or reduced mortality.

The main adverse effect of sulfonylureas is hypoglycemia. Pharmacokinetic considerations are important in determining which agents are more likely to produce hypoglycemia as an adverse effect. Chlorpropamide and glibenclamide are more likely to produce hypoglycemia than are shorter acting drugs such as tolbutamide, glipizide, gliclazide, and gliquidone, especially when renal insufficiency is present [236].

Meglitinide Analogs

It is widely accepted that loss of early-phase insulin secretion is an important and early event in the development of Type 2 diabetes [205]. Thus agents restoring or replacing this phase of insulin release have a strong theoretical basis for treating this condition. It was shown that a compound, later named meglitinide, that was closely related to the nonsulfonylurea moiety of glibenclamide augmented insulin secretion through blockade of K_{ATP} channel and was hypoglycemic [237]. Two meglitinide analogs, repaglinide and nateglinide, are currently used for treating patients with Type 2 diabetes mellitus, mitiglinide being in clinical development. For a variety of reasons, these agents are less likely than sulfonylureas to produce hypoglycemia as adverse effect doses [205, 238]. Nateglinide and repaglinide are rapidly absorbed from the gastrointestinal tract. They are also rapidly eliminated through extensive hepatic metabolism (by CYP2C9 and CYP3A4) [205, 239, 240]. Moreover, they appear to show glucose dependency in stimulating insulin secretion and may, at least for nateglinide, dissociate more quickly from the receptor and allow more rapid K_{ATP} channel reactivation [241]. Because of their short action, they are administered before each meal, allowing a more physiological pattern of insulin secretion, and reducing the problem of hypoglycemia caused by missed meals.

Mechanism of Action The mechanism of action of meglitinide analogs is similar to that of the sulfonylureas. They block the β-cell K_{ATP} channel by combining with its SUR1 (sulfonylurea receptor) subunit. However, there is evidence for differences among sulfonylureas, repaglinide and nateglinide, in their interaction with the receptor. Meglitinide binds with similar affinities to the SUR1, SUR2A, and SUR2B receptors (transiently expressed in COS-7 cells), which are subunits of different K_{ATP} channels [229]. Other studies in electrophysiological experiments showed that mitiglinide was more potent in blocking SUR1 relative to SUR2, although repaglinide was equipotent in blocking both [228]. The short action of nateglinide may be partly related to the rapid reactivation of the β-cell K_{ATP} channel following removal of the drug [241]. Site-directed mutagenesis studies indicate that nateglinide binds to a similar site to glibenclamide but a different site from repaglinide, another meglitinide analog [242]. Nevertheless, rat islets desensitized by prolonged exposure to sulfonylureas were still capable of secreting insulin in response to nateglinide [241].

α-Glucosidase Inhibitors

α-Glucosidases are intestinal brush border enzymes responsible for the hydrolysis of oligosaccharides such as sucrose to glucose; they have relative selectivity for sucrase. Inhibition of their hydrolysis prevents carbohydrate absorption as disaccharides are poorly absorbed. Three drugs, acarbose, voglibose, and miglitol, which are competitive inhibitors of α-glucosidase are currently in use. As they are reversible inhibitors, carbohydrate absorption is delayed, not prevented [243]. Inhibitors of these enzymes thus lower postprandial glucose concentrations and are adjuncts for the treatment of diabetes mellitus. Thus these agents can smooth the daily blood glucose fluctuations in patients with diabetes mellitus and reduce glycosylated hemoglobin concentrations. When used as an adjunct to insulin and diet, they also lower postprandial blood glucose in Type 1 diabetes and produce a modest reduction in glycosylated hemoglobin [244]. They are also effective in Type 2 diabetes when used alongside diet alone or when coadministered with sulfonylureas, metformin, or insulin [245]. A clinical trial suggested that acarbose might prevent Type 2 diabetes [246], although this finding has been challenged.

Glucagon-Like Peptide-1 (GLP-1) and Glucose-Dependent Insulinotropic Polypeptide (GIP)

The GLP-1, secreted by intestinal L cells, and GIP, secreted by intestinal K cells, are known as incretins and are secreted from the small intestine in response to the presence of glucose and other nutrients in the gut. These hormones augment glucose-induced insulin secretion in a glucose-dependent manner through activating adenylyl cyclase, leading to an increase in islet β-cell cyclic-AMP [247–249]. Their importance is clearly illustrated by the marked glucose intolerance and impairment of insulin secretion seen in mice lacking receptors for both hormones [250]. GLP-1 is also a potent inhibitor of glucagon secretion [251]. GLP-1 and GIP stimulate insulin gene transcription, pancreatic islet cell proliferation and β-cell replication, promote differentiation of pancreatic ductal cells into glucagon- and insulin-producing cells through a PDX-1 dependent pathway, and inhibit β-cell apoptosis [247–249]. They also suppress gastric emptying. GLP-1 was found to improve blood glucose in experimental animals and in patients with Type 2 diabetes, leading to the idea that GLP-1 and GIP might form the basis for novel treatments. The use of the native hormones is severely limited by their very short duration of action, as both are rapidly degraded by the enzyme dipeptidyl peptidase IV [DPPIV]. Indeed the N-terminally truncated products of this enzymic degradation [GLP-1(6–36) and GIP(3–42)] are antagonists of the respective hormones. Two approaches have been used to address this problem. First, attempts are being made to find analogs of the hormones that are resistant to DPPIV-mediated degradation. At least one stable analog (liraglutide-NN 2211) of GLP-1 is in clinical trial. Moreover, exenatide, a synthetic form of exendin-4, is undergoing clinical evaluation (see Exendin-4). Analogs of GIP stable to DPPIV are being studied in experimental animals, and several of these have been shown be insulinotropic and to be antihyperglycemic [249]. The second approach is to use inhibitors of DPPIV to prolong the effect of the endogenous incretins. Several orally active DPPIV inhibitors have been developed (e.g., NVPDPP728 and P32/98 NVP-LAF237) and have shown effectiveness in experimental and clinical diabetes [252, 253].

Exendin-4

Exendin-4 is a 39-amino-acid peptide isolated from the venom of the gila monster lizard *Heloderma suspectum* and is a member of the glucagon superfamily of peptide hormones [254]. At the N-terminus the replacement of His7 and Ala8 by His7 and Gly8 makes exendin-4 resistant to cleavage by the DPP-

IV, which probably explains its long duration of action compared with GLP-1 [255]. It is structurally related to GLP-1 and interacts with the GLP-1 receptor, elevating cyclic-AMP in the islet β cell and augmenting glucose-induced insulin secretion [256]. It also increases insulin sensitivity and is effective in lowering blood glucose in several animal models of diabetes [257] and in humans [258, 259]. Like GLP-1 it increases β-cell replication and neogenesis resulting in increased β-cell mass [260], reduces gastric emptying, and diminishes calorific intake [258].

Glucagon

Glucagon is a polypeptide hormone secreted by the A cells (α cells) of the pancreatic islets and is an important hormone in preventing a fall in the blood glucose concentration [261]. It is secreted in response to hypoglycemia, and there is strong evidence for a major autonomic component in this response [261]. By activating adenylyl cyclase, glucagon stimulates glycogenolysis and gluconeogenesis in hepatocytes [262] and lipolysis in adipose tissue [263]. Glucagon also relaxes smooth muscle through an increase in cAMP [264].

Glucagon Receptors Glucagon mediates its effects by the glucagon receptor, a member of the G-protein-coupled superfamily of seven-helical transmembrane receptors (GPCRs) [265]. It is coupled through G_s to adenylyl cyclase increasing cAMP, but its ability to increase intracellular Ca^{2+} in hepatocytes probably additionally involves activation of the PLC/inositol phosphate pathway [265].

Therapeutic Applications The main use of glucagon is in the treatment of acute insulin-induced hypoglycemia, where the patient has lost consciousness and cannot take oral glucose. It is ineffective in starvation or where liver glycogen stores are depleted. Because of the role of glucagon in the metabolic disturbances of diabetes mellitus [266], there may be a role for antagonists at the glucagon receptor in treating diabetes and both peptide and nonpeptide antagonists are being developed [267–270]. The nonpeptide agent Bay 27-9955 is orally active and was shown in the human to antagonize glucagon in elevating plasma glucose and increasing glucose production [268]. NNC 25-2504 is another, very potent (picomolar to low nanomolar binding affinity) nonpeptide, stable, orally active antagonist [271].

23.11 ANTIOBESITY AGENTS

Obesity is a major risk factor for Type 2 diabetes mellitus, dyslipidemia, myocardial infarction, stroke, and cancers of the breast and colon. A 5 to 10 percent reduction in body weight is associated with a clinically significant improvement in associated morbidity. However, it is a difficult condition to manage and sustained weight loss requires not only effective drugs but also concomitant adjustments to lifestyle and diet. Very few drugs are currently available and the anorexigenic drugs fenfluramine and dexfenfluramine, which have a central effect mediated through promoting neuronal release of serotonin and inhibiting its reuptake [272], were withdrawn in 1997 because of association with pulmonary hypertension and cardiac valve abnormalities [273]. The only appetite suppressant currently available is sibutramine, which blocks the neuronal uptake of serotonin and norepinephrine [274].

The concept of distinct feeding and satiety centers, previously regarded, respectively, as the lateral hypothalamic area and the ventromedial hypothalamus, has been superseded by the idea of distinct neuronal populations that express particular anorexigenic or orexigenic peptides and activation or inhibition of which switch feeding on or off according to the nutritional status [275]. Appetite, satiety, and feeding behavior are integrated by a number of neuropeptides in the hypothalamus. The orexigenic peptides include neuropeptide Y, expressed in neurons largely originating in the arcuate nucleus (NPY—probably acting through the Y1 and Y5 receptors), the orexins (primarily orexin-A), melanin-concentrating hormone (MCH), and agouti-related protein (Agrp—an antagonist of the melanocortin MC3 and MC4 receptors and expressed in NPY neurons). Lipid endocannabinoids may also play an important role in maintaining appetite. On the other hand, the hypothalamic anorectic peptides include α-melanocyte stimulating hormone (αMSH—acting through MC3 and MC4 receptors) and cocaine and amphetamine-related transcript (CART) [275, 276]. The hypothalamic neuronal circuits modifying feeding behavior are markedly influenced by a number of nutrient and hormonal signals from the periphery. Glucose is a satiety signal, and there are glucose-sensitive neurons in the hypothalamus, which are inhibited by NPY and activated by orexins. Gold-thioglucose-induced damage to these neurons results in hyperphagia and obesity [275]. Insulin acting as a satiety signal probably modulates these neurons. Cholecystokinin (CCK), pancreatic polypeptide, and PYY 3-36 and the preproglucagon-derived peptides glucagon-like peptide 1 (GLP-1) and oxyntomodulin, produced in the small intestine, are also anorectic [275, 276]. On the other hand, ghrelin from the stomach is orexigenic. In broad terms, the orexigenic peptides activate the NPY/Agouti-related peptide neuronal systems and inhibit the anorexogenic neurons, whereas the anorexigenic peptides do the opposite [275, 276] (Fig. 23.4). All of these peptides and their receptors are potential targets for the development of antiobesity agents. Many agents are under development, although none has yet reached the market.

The discovery that adipose tissue is an active endocrine tissue, secreting hormones (e.g., resistin) that influence the sensitivity to insulin as well as food intake (e.g., leptin, which acts as an anorexigenic signal, stimulating melanocortin neurons) and that this tissue is a major target for the thiazolidinedione-agonists at the PPARγ (peroxisome proliferator-γ) activated receptors [277] has revealed more potential targets. Mitochondrial uncoupling

ANTIOBESITY AGENTS

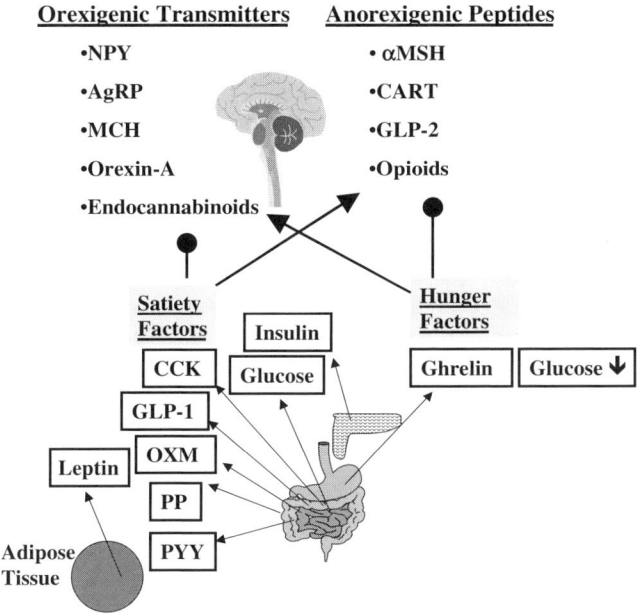

Figure 23.4 Regulation of appetite and feeding behavior [275–277]. Distinct neuronal populations express particular anorexigenic or orexigenic neurotransmitter peptides (or lipids). The activity of these neurons is influenced by signals from the periphery, especially the gut. Signals activating feeding increase the activity of orexigenic circuits and concomitantly inhibit anorexigenic neuronal activity, and vice versa for satiety signals. NPY, neuropeptide Y; AgRP, agouti-related protein; MCH, melanin-concentrating hormone; αMSH, α-melanocyte-stimulating hormone; CART, cocaine and amphetamine-regulated transcript; GLP-2, glucagons-like peptide-2; CCK, cholecystokinin; GLP-1, glucagons-like peptide-1; OXM, oxyntomodulin.

proteins in muscle and adipose tissue are also potential targets for new antiobesity agents [278] including agonists at β_3-adrenoceptors [279].

Orlistat

Orlistat is an irreversible and highly selective inhibitor of pancreatic and gastric lipase and therefore inhibits the hydrolysis of triglyceride (triacylglycerol) in the gut lumen. It binds covalently to the serine residue on the active site [280]. It shows around 1000-fold selectivity for lipase compared with amylase, trypsin, or chymotrypsin. As triacylglycerol hydrolysis is a prerequisite for intestinal fat absorption, lipase inhibition leads to reduced fat absorption and therefore to decreased availability of fat within the body. This is the basis for the widespread use of orlistat in the treatment of obesity [280, 281]. Orlistat is highly selective for lipase and has no significant effect in inhibiting amylase, trypsin, or chymotrypsin [280]. Its very poor gastrointestinal absorption (around 1 percent) results in very low plasma concentrations and therefore absence of systemic adverse effects [280].

Therapeutic Applications   Orlistat is used for the treatment of obesity in patients who have demonstrated their ability to lose weight as a result of dietary control and increased physical activity. Weight loss after 12 months averages 8 to 10 percent when used with an appropriate hypocaloric diet.

Plasma levels of fat-soluble vitamins are reduced [281], but the potential clinical consequences of this has not been evident. Studies in volunteers have shown a reduced plasma cholecystokinin response to meals, which may suggest an increased risk of gallstones, especially in obese patients who are already at greater risk [281]. Gastrointestinal effects (flatus, oily spotting, fecal urgency, steatorrhea) are quite common and are predictable for a drug that prevents fat absorption.

Acknowledgment   I am very grateful to Pat Owen for her editorial assistance.

REFERENCES

1. Levy, A. (2002). Physiological implications of pituitary trophic activity. *J. Endocrinol.*, *174*, 147–155.
2. Marshall, J. C., Eagleson, C. A., McCartney, C. R. (2001). Hypothalamic dysfunction. *Mol. Cell. Endocrinol.*, *183*, 29–32.
3. Parker, K. L., Schimmer, B. P. (2001). Pituitary hormones and their hypothalamic releasing factors. In J. G. Hardman, L. E. Limbird, A. Goodman Gilman (Eds), *Goodman & Gilman's the Pharmacological Basis of Therapeutics*, 10th ed. McGraw-Hill, New York, pp. 1541–1562.
4. Persani, L. (1998). Hypothalamic thyrotropin-releasing hormone and thyrotropin biological activity. *Thyroid*, *8*, 941–946.
5. Nillni, E. A., Sevarino, K. A. (1999). The biology of pro-thyrotropin-releasing hormone-derived peptides. *Endocr. Rev.*, *20*, 599–648.
6. Sun, Y., Lu, X., Gershengorn, M. C. (2003). Thyrotropin-releasing hormone receptors—Similarities and differences. *J. Mol. Endocrinol.*, *30*, 87–97.
7. Nishino, S., Arrigoni, J., Shelton, J., Kanbayashi, T., Dement, W. C., Mignot, E. (1997). Effects of thyrotropin-releasing hormone and its analogs on daytime sleepiness and cataplexy in canine narcolepsy. *J. Neurosci.*, *17*, 6401–6408.
8. Urayama, A., Yamada, S., Kimura, R., Zhang, J., Watanabe, Y. (2002). Neuroprotective effect and brain receptor binding of taltirelin, a novel thyrotropin-releasing hormone (TRH) analogue, in transient forebrain ischemia of C57BL/6J mice. *Life Sci.*, *72*, 601–607.
9. Vale, W., Spiess, J., Rivier, C., Rivier, J. (1981). Characterization of a 41-residue ovine hypothalamic peptide that stimulates secretion of corticotropin and beta-endorphin. *Science*, *213*, 1394–1397.

10. Dautzenberg, F. M., Hauger, R. L. (2002). The CRF peptide family and their receptors: Yet more partners discovered. *Trends Pharmacol. Sci.*, 23, 71–77.
11. Latchman, D. S. (2002). Urocortin. *Int. J. Biochem. Cell. Biol.*, 34, 907–910.
12. Smagin, G. N., Heinrichs, S. C., Dunn, A. J. (2001). The role of CRH in behavioral responses to stress. *Peptides*, 22, 713–724.
13. Parkes, D. G., Weisinger, R. S., May, C. N. (2001). Cardiovascular actions of CRH and urocortin: An update. *Peptides*, 22, 821–827.
14. Agelaki, S., Tsatsanis, C., Gravanis, A., Margioris, A. N. (2002). Corticotropin-releasing hormone augments pro-inflammatory cytokine production from macrophages in vitro and in lipopolysaccharide-induced endotoxin shock in mice. *Infect. Immun.*, 70, 6068–6074.
15. Singh, L. K., Boucher, W., Pang, X., Letourneau, R., Seretakis, D., Green, M., Theoharides, T. C. (1999). Potent mast cell degranulation and vascular permeability triggered by urocortin through activation of corticotropin-releasing hormone receptors. *J. Pharmacol. Exp. Ther.*, 288, 1349–1356.
16. Aggelidou, E., Hillhouse, E. W., Grammatopoulos, D. K. (2002). Up-regulation of nitric oxide synthase and modulation of the guanylate cyclase activity by corticotropin-releasing hormone but not urocortin II or urocortin III in cultured human pregnant myometrial cells. *Proc. Natl. Acad. Sci. USA*, 99, 3300–3305.
17. Newell-Price, J., Trainer, P., Besser, M., Grossman, A. (1998). The diagnosis and differential diagnosis of Cushing's syndrome and pseudo-Cushing's states. *Endocr. Rev.*, 19, 647–672.
18. Nicholson, W. E., DeCherney, G. S., Jackson, R. V., DeBold, C. R., Uderman, H., Alexander, A. N., Rivier, J., Vale, W., Orth, D. N. (1983). Plasma distribution, disappearance half-time, metabolic clearance rate, and degradation of synthetic ovine corticotropin-releasing factor in man. *J. Clin. Endocrinol. Metab.*, 57, 1263–1269.
19. Schulte, H. M., Chrousos, G. P., Booth, J. D., Oldfield, E. H., Gold, P. W., Cutler, G. B. Jr, Loriaux, D. L. (1984). Corticotropin-releasing factor: Pharmacokinetics in man. *J. Clin. Endocrinol. Metab.*, 58, 192–196.
20. Webster, E. L., Lewis, D. B., Torpy, D. J., Zachman, E. K., Rice, K. C., Chrousos, G. P. (1996). In vivo and in vitro characterization of antalarmin, a nonpeptide corticotropin-releasing hormone (CRH) receptor antagonist: Suppression of pituitary ACTH release and peripheral inflammation. *Endocrinology*, 137, 5747–5750.
21. Ducottet, C., Griebel, G., Belzung, C. (2003). Effects of the selective nonpeptide corticotropin-releasing factor receptor 1 antagonist antalarmin in the chronic mild stress model of depression in mice. *Prog. Neuropsychopharmacol. Biol. Psychiatry*, 27, 625–631.
22. Bornstein, S. R., Webster, E. L., Torpy, D. J., Richman, S. J., Mitsiades, N., Igel, M., Lewis, D. B., Rice, K. C., Joost, H. G., Tsokos, M., Chrousos, G. P. (1998). Chronic effects of a nonpeptide corticotropin-releasing hormone type I receptor antagonist on pituitary-adrenal function, body weight, and metabolic regulation. *Endocrinology*, 139, 1546–1555.
23. Gabry, K. E., Chrousos, G. P., Rice, K. C., Mostafa, R. M., Sternberg, E., Negrao, A. B., Webster, E. L., McCann, S. M., Gold, P. W. (2002). Marked suppression of

gastric ulcerogenesis and intestinal responses to stress by a novel class of drugs. *Mol. Psychiatry*, 7, 474–483.

24. Webster, E. L., Barrientos, R. M., Contoreggi, C., Isaac, M. G., Ligier, S., Gabry, K. E., Chrousos, G. P., McCarthy, E. F., Rice, K. C., Gol, P. W., Sternberg, E. M. (2002). Corticotropin releasing hormone (CRH) antagonist attenuates adjuvant induced arthritis: Role of CRH in peripheral inflammation. *J. Rheumatol.*, 29, 1252–1261.

25. Zobel, A. W., Nickel, T., Kunzel, H. E., Ackl, N., Sonntag, A., Ising, M., Holsboer, F. (2000). Effects of the high-affinity corticotropin-releasing hormone receptor 1 antagonist R121919 in major depression: The first 20 patients treated. *J. Psychiatr. Res.*, 34, 171–181.

26. Seymour, P. A., Schmidt, A. W., Schulz, D. W. (2003). The pharmacology of CP-154,526, a non-peptide antagonist of the CRH1 receptor: A review. *CNS Drug Rev.*, 9, 57–96.

27. Mackay, K. B., Stiefel, T. H., Ling, N., Foster, A. C. (2003). Effects of a selective agonist and antagonist of CRF2 receptors on cardiovascular function in the rat. *Eur. J. Pharmacol.*, 469, 111–115.

28. Mayo, K. E., Miller, T., DeAlmeida, V., Godfrey, P., Zheng, J., Cunha, S. R. (2000). Regulation of the pituitary somatotroph cell by GHRH and its receptor. *Recent Prog. Horm. Res.*, 55, 237–266.

29. Camanni, F., Ghigo, E., Arvat, E. (1998). Growth hormone-releasing peptides and their analogs. *Front. Neuroendocrinol.*, 19, 47–72.

30. Bremner, W. J., Huhtaniemi, I., Amory, J. K. (2001). Pituitary gonadotrophins and their disorders. In K. L. Becker (Ed.), *Principles and Practice of Endocrinology and Metabolism*, Vol. 3. Lippincott Williams & Wilkins, Philadelphia, pp. 170–177.

31. Barradell, L. B., McTavish, D. (1993). Histrelin. A review of its pharmacological properties and therapeutic role in central precocious puberty. *Drugs*, 45, 570–588.

32. Grosse, R., Schmid, A., Schoneberg, T., Herrlich, A., Muhn, P., Schultz, G., Gudermann, T. (2000). Gonadotropin-releasing hormone receptor initiates multiple signaling pathways by exclusively coupling to G(q/11) proteins. *J. Biol. Chem.*, 275, 9193–9200.

33. Ortmann, O., Diedrich, K. (1999). Pituitary and extrapituitary actions of gonadotrophin-releasing hormone and its analogues. *Hum. Reprod.*, 14(Suppl. 1), 194–206.

34. Surrey, E. S., Schoolcraft, W. B. (2000). Evaluating strategies for improving ovarian response of the poor responder undergoing assisted reproductive techniques. *Fertil. Steril.*, 73, 667–676.

35. Yano, T., Yano, N., Matsumi, H., Morita, Y., Tsutsumi, O., Schally, A. V., Taketani, Y. (1997). Effect of luteinizing hormone-releasing hormone analogs on the rat ovarian follicle development. *Horm. Res.*, 48(Suppl. 3), 35–41.

36. McArdle, C. A., Franklin, J., Green, L., Hislop, J. N. (2002). Signalling, cycling and desensitisation of gonadotrophin-releasing hormone receptors. *J. Endocrinol.*, 173, 1–11.

37. Ravenna, L., Salvatori, L., Morrone, S., Lubrano, C., Cardillo, M. R., Sciarra, F., Frati, L., Di Silverio, F., Petrangeli, E. (2000). Effects of triptorelin, a

gonadotropin-releasing hormone agonist, on the human prostatic cell lines PC3 and LNCaP. *J. Androl.*, *21*, 549–557.

38. Herbst, K. L. (2003). Gonadotropin-releasing hormone antagonists. *Curr. Opin. Pharmacol.*, *3*, 660–666.

39. Brogden, R. N., Buckley, M. M., Ward, A. (1990). Buserelin. A review of its pharmacodynamic and pharmacokinetic properties, and clinical profile. *Drugs*, *39*, 399–437.

40. Gunnet, J. W., Demarest, K. T., Hahn, D. W., Ericson, E., McGuire, J. L. (1991). Effects of a sustained release formulation of the gonadotrophin-releasing hormone agonist histrelin on serum concentrations of gonadotrophins and oestradiol, and ovarian LH/human chorionic gonadotrophin receptor content in the rat. *J. Endocrinol.*, *131*, 211–218.

41. Johnson, C. A., Thompson, D. L. Jr., Cartmill, J. A. (2002). Pituitary responsiveness to GnRH in mares following deslorelin acetate implantation to hasten ovulation. *J. Anim. Sci.*, *80*, 2681–2687.

42. Dawood, M. Y. (1994). Hormonal therapies for endometriosis: Implications for bone metabolism. *Acta. Obstet. Gynecol. Scand. Suppl.*, *159*, 22–34.

43. Cook, T., Sheridan, W. P. (2000). Development of GnRH antagonists for prostate cancer: New approaches to treatment. *Oncologist*, *5*, 162–168.

44. Lee, C. H., Van Antwerp, D., Hedley, L., Nestor, J. J. Jr., Vickery, B. H. (1989). Comparative studies on the hypotensive effect of LHRH antagonists in anesthetized rats. *Life Sci.*, *45*, 697–702.

45. Rabinovici, J., Rothman, P., Monroe, S. E., Nerenberg, C., Affe, R. B. (1992). Endocrine effects and pharmacokinetic characteristics of a potent new gonadotropin-releasing hormone antagonist (Ganirelix) with minimal histamine-releasing properties: Studies in postmenopausal women. *J. Clin. Endocrinol. Metab.*, *75*, 1220–1225.

46. Patel, Y. C. (1999). Somatostatin and its receptor family. *Front. Neuroendocrinol.*, *20*, 157–198.

47. Shimon, I., Yan, X., Taylor, J. E., Weiss, M. H., Culler, M. D., Melmed, S. (1997). Somatostatin receptor (SSTR) subtype-selective analogues differentially suppress in vitro growth hormone and prolactin in human pituitary adenomas. Novel potential therapy for functional pituitary tumors. *J. Clin. Invest.*, *100*, 2386–2392.

48. Feniuk, W., Jarvie, E., Luo, J., Humphrey, P. P. (2000). Selective somatostatin sst(2) receptor blockade with the novel cyclic octapeptide, CYN-154806. *Neuropharmacology*, *39*, 1443–1450.

49. Reubi, J. C., Schaer, J. C., Wenger, S., Hoeger, C., Erchegyi, J., Waser, B., Rivier, J. (2000). SST3-selective potent peptidic somatostatin receptor antagonists. *Proc. Natl. Acad. Sci. USA*, *97*, 13973–13978.

50. Hoyer, D., Nunn, C., Hannon, J., Schoeffter, P., Feuerbach, D., Schuepbach, E., Langenegger, D., Bouhelal, R., Hurth, K., Neumann, P., Troxler, T., Pfaeffli, P. (2004). SRA880, in vitro characterization of the first non-peptide somatostatin sst(1) receptor antagonist. *Neurosci. Lett.*, *361*, 132–135.

51. Poitout, L., Roubert, P., Contour-Galcera, M. O., Moinet, C., Lannoy, J., Pommier, J., Plas, P., Bigg, D., Thurieau, C. (2001). Identification of potent non-peptide somatostatin antagonists with sst(3) selectivity. *J. Med. Chem.*, *44*, 2990–3000.

52. Stewart, P. M., James, R. A. (1999). The future of somatostatin analogue therapy. *Baillieres Best Pract. Res. Clin. Endocrinol. Metab.*, *13*, 409–418.
53. Eriksson, B., Oberg, K. (1999). Summing up 15 years of somatostatin analog therapy in neuroendocrine tumors: Future outlook. *Ann. Oncol.*, *10*(Suppl. 2), S31–S38.
54. Bernard, B., Capello, A., van Hagen, M., Breeman, W., Srinivasan, A., Schmidt, M., Erion, J., van Gameren, A., Krenning, E., de Jong, M. (2004). Radiolabeled RGD-DTPA-Tyr3-octreotate for receptor-targeted radionuclide therapy. *Cancer Biother. Radiopharm.*, *19*, 173–180.
55. Herrington, J., Carter-Su, C. (2001). Signaling pathways activated by the growth hormone receptor. *Trends Endocrinol. Metab.*, *12*, 252–257.
56. LeRoith, D., Bondy, C., Yakar, S., Liu, J. L., Butler, A. (2001). The somatomedin hypothesis: 2001. *Endocr. Rev.*, *22*, 53–74.
57. Baxter, R. C. (2000). Insulin-like growth factor (IGF)-binding proteins: Interactions with IGFs and intrinsic bioactivities. *Am. J. Physiol. Endocrinol. Metab.*, *278*, E967–E976.
58. O'Dell, S. D., Day, I. N. (1998). Insulin-like growth factor II (IGF-II). *Int. J. Biochem. Cell. Biol.*, *30*, 767–771.
59. Thorner, M. O., Strasburger, C. J., Wu, Z., Straume, M., Bidlingmaier, M., Pezzoli, S. S., Zib, K., Scarlett, J. C., Bennett, W. F. (1999). Growth hormone (GH) receptor blockade with a PEG-modified GH (B2036-PEG) lowers serum insulin-like growth factor-1 but does not acutely stimulate serum GH. *J. Clin. Endocrinol. Metab.*, *84*, 2098–2103.
60. Van der Lely, A. J., Hutson, R. K., Trainer, P. J., Besser, G. M., Barkan, A. L., Katznelson, L., Klibanski, A., Herman-Bonert, V., Melmed, S., Vance, M. L., Freda, P. U., Stewart, P. M., Friend, K. E., Clemmons, D. R., Johannsson, G., Stavrou, S., Cook, D. M., Phillips, L. S., Strasburger, C. J., Hackett, S., Zib, K. A., Davis, R. J., Scarlett, J. C., Thorner, M. O. (2001). Long-term treatment of acromegaly with pegvisomant, a growth hormone receptor antagonist. *Lancet*, *358*, 1745–1749.
61. Parkinson, C., Trainer, P. J. (2001). The place of pegvisomant in the management of acromegaly. *Expert Opin. Investig. Drugs*, *10*, 1725–1735.
62. Simoni, M., Gromoll, J., Nieschlag, E. (1997). The follicle-stimulating hormone receptor: Biochemistry, molecular biology, physiology, and pathophysiology. *Endocr. Rev.*, *6*, 739–773.
63. Arey, B. J., Deecher, D. C., Shen, E. S., Stevis, P. E., Meade, E. H. Jr, Wrobel, J., Frail, D. E., Lopez, F. J. (2002). Identification and characterization of a selective, nonpeptide follicle-stimulating hormone receptor antagonist. *Endocrinology*, *143*, 3822–3829.
64. Daya, S. (2002). Updated meta-analysis of recombinant follicle-stimulating hormone (FSH) versus urinary FSH for ovarian stimulation in assisted reproduction. *Fertil. Steril.*, *77*, 711–714.
65. Lenton, E., Soltan, A., Hewitt, J., Thomson, A., Davies, W., Ashraf, N., Sharma, V., Jenner, L., Ledger, W., McVeigh, E. (2000). Induction of ovulation in women undergoing assisted reproductive techniques: Recombinant human FSH (follitropin alpha) versus highly purified urinary FSH (urofollitropin HP). *Hum. Reprod.*, *15*, 1021–1027.

66. Bouloux, P., Warne, D. W., Loumaye, E. (2002). Efficacy and safety of recombinant human follicle-stimulating hormone in men with isolated hypogonadotropic hypogonadism. *Fertil. Steril.*, 77, 270–273.
67. Ascoli, M., Fanelli, F., Segaloff, D. L. (2002). The lutropin/choriogonadotropin receptor, a 2002 perspective. *Endocr. Rev.*, 23, 141–174.
68. Abboud, F. M., Floras, J. S., Aylward, P. E., Guo, G. B., Gupta, B. N., Schmid, P. G. (1990). Role of vasopressin in cardiovascular and blood pressure regulation. *Blood Vessels*, 27, 106–115.
69. Barberis, C., Tribollet, E. (1996). Vasopressin and oxytocin receptors in the central nervous system. *Crit. Rev. Neurobiol.*, 10, 119–154.
70. Rose, J. D., Moore, F. L. (2002). Behavioral neuroendocrinology of vasotocin and vasopressin and the sensorimotor processing hypothesis. *Front. Neuroendocrinol.*, 23, 317–341.
71. Alescio-Lautier, B., Paban, V., Soumireu-Mourat, B. (2000). Neuromodulation of memory in the hippocampus by vasopressin. *Eur. J. Pharmacol.*, 405, 63–72.
72. Birnbaumer, M. (2000). Vasopressin receptors. *Trends Endocrinol. Metab.*, 11, 406–410.
73. Knepper, M. A. (1998). Long-term regulation of urinary concentrating capacity. *Am. J. Physiol.*, 275, F332–F333.
74. Schwartz, J., Derdowska, I., Sobocinska, M., Kupryszewski, G. (1991). A potent new synthetic analog of vasopressin with relative agonist specificity for the pituitary. *Endocrinology*, 129, 1107–1109.
75. Derick, S., Cheng, L. L., Voirol, J., Stoev, S., Giacomini, M., Wo, N. C., Szeto, H. H., Ben Mimoun, M., Andres, M., Gaillard, R. C., Guillon, G., Manning, M. (2002). [1-Deamino-4-cyclohexylalanine]arginine vasopressin: A potent and specific agonist for vasopressin V_{1b} receptors. *Endocrinology*, 143, 4655–4664.
76. Andres, M., Trueba, M., Guillon, G. (2002). Pharmacological characterization of F-180: A selective human V(1a) vasopressin receptor agonist of high affinity. *Br. J. Pharmacol.*, 135, 1828–1836.
77. Durr, J. A., Hensen, J., Ehnis, T., Blankenship, M. S. (2000). Chlorpropamide upregulates antidiuretic hormone receptors and unmasks constitutive receptor signaling. *Am. J. Physiol. Renal Physiol.*, 278, F799–F808.
78. Griebe, G., Simiand, J., Stemmelin, J., Gal, C. S., Steinberg, R. (2003). The vasopressin V1b receptor as a therapeutic target in stress-related disorders. *Curr. Drug Targets CNS Neurol. Disord.*, 2, 191–200.
79. Serradeil-Le Gal, C., Wagnon, J., Valette, G., Garcia, G., Pasca, M., Maffrand, J. P., Le Fur, G. (2002). Nonpeptide vasopressin receptor antagonists: Development of selective and orally active V1a, V2 and V1b receptor ligands. *Prog. Brain Res.*, 139, 197–210.
80. Mannucci, P. M. (2001). How I treat patients with von Willebrand disease. *Blood*, 97, 1915–1919.
81. Kaufmann, J. E., Vischer, U. M. (2003). Cellular mechanisms of the hemostatic effects of desmopressin (DDAVP). *J. Thromb. Haemost.*, 1, 682–689.
82. Thibonnier, M. (2003). Vasopressin receptor antagonists in heart failure. *Curr. Opin. Pharmacol.*, 3, 683–687.

83. Andersson, K. E. (2001). Oxytocin-containing neurons in the PVN project to spinal autonomic nuclei. *Pharmacol. Rev.*, 53, 417–450.
84. Filippi, S., Vignozzi, L., Vannelli, G. B., Ledda, F., Forti, G., Maggi, M. J. (2003). Role of oxytocin in the ejaculatory process. *Endocrinol. Invest.*, 26(3, Suppl.), 82–86.
85. Gimpl, G., Fahrenholz, F. (2001). The oxytocin receptor system: Structure, function, and regulation. *Physiol. Rev.*, 81, 629–683.
86. Zingg, H. H., Laporte, S. A. (2003). The oxytocin receptor. *Trends Endocrinol. Metab.*, 14, 222–227.
87. Lowbridge, J., Manning, M., Haldar, J, Sawyer, W. H. (1977). Synthesis and some pharmacological properties of [4-threonine, 7-glycine]oxytocin, [1-(L-2-hydroxy-3-mercaptopropanoic acid), 4-threonine, 7-glycine]oxytocin (hydroxy[Thr4, Gly7]oxytocin), and [7-glycine]oxytocin, peptides with high oxytocic-antidiuretic selectivity. *J. Med. Chem.*, 20, 120–123.
88. Lamont, R. F. (2003). The development and introduction of anti-oxytocic tocolytics. *Br. J. Obstet. Gynaecol.*, 110, 108–112.
89. World-wide Atosiban Versus Beta-agonists Study Group (2001). Effectiveness and safety of the oxytocin antagonist atosiban versus beta-adrenergic agents in the treatment of preterm labor. *Br. J. Obstet. Gynaecol.*, 108, 133–142.
90. Goodwin, T. M., Valenzuela, G. J., Silver, H., Creasy, G., and the atosiban study group (1996). Dose ranging study of the oxytocin antagonist atosiban in the treatment of preterm labor. *Obstet. Gynecol.*, 88, 331–336.
91. Fuchs, A. R., Fuchs, F., Husslein, P., Soloff, M. S. (1984). Oxytocin receptors in the human uterus during pregnancy and parturition. *Am. J. Obstet. Gynecol.*, 150, 734–741.
92. Wolff, J. (1998). Perchlorate and the thyroid gland. *Pharmacol. Rev.*, 50, 89–105.
93. Hernandez, A., St Germain, D. L. (1997). Selenodeiodinases and their role in thyroid hormone activation and inactivation. *Curr. Opin. Endocrinol. Diabetes*, 4, 333–340.
94. Brent, G. A. (1994). The molecular basis of thyroid hormone action. *N. Engl. J. Med.*, 331, 847–853.
95. Yen, P. M. (2001). Physiological and molecular basis of thyroid hormone action. *Physiol. Rev.*, 81, 1097–1142.
96. Barbe, P., Larrouy, D., Boulanger, C., Chevillotte, E., Viguerie, N., Thalamas, C., Trastoy, M. O., Roques, M., Vidal, H., Langin, D. (2001). Triiodothyronine-mediated up-regulation of UCP2 and UCP3 mRNA expression in human skeletal muscle without coordinated induction of mitochondrial respiratory chain genes. *FASEB J.*, 15, 13–15.
97. Davis, P. J., Tillmann, H. C., Davis, F. B., Wehling, M. (2002). Comparison of the mechanisms of nongenomic actions of thyroid hormone and steroid hormones. *J. Endocrinol. Invest.*, 25, 377–388.
98. Huang, C. J., Geller, H. M., Green, W. L., Craelius, W. (1999). Acute effects of thyroid hormone analogs on sodium currents in neonatal rat myocytes. *J. Mol. Cell. Cardiol.*, 31, 881–893.
99. Wrutniak-Cabello, C., Casas, F., Cabello, G. (2001). Thyroid hormone action in mitochondria. *J. Mol. Endocrinol.*, 26, 67–77.

100. Taylor, A. H., Stephan, Z. F., Steele, R. E., Wong, N. C. (1997). Beneficial effects of a novel thyromimetic on lipoprotein metabolism. *Mol. Pharmacol.*, *52*, 542–547.

101. Spooner, P. H., Morkin, E., Goldman, S. (1999). Thyroid hormones and thyroid hormone analogues in the treatment of heart failure. *Coron. Artery Dis.*, *10*, 395–399.

102. Adamson, C. R., Maitra, N., Bahl, J. J., Greer, K., Klewar, S., Hoying, J., Morkin, E. J. (2004). Regulation of gene expression in cardiomyocytes by thyroid hormone and thyroid hormone analogs, 3,5-diiodothyropropionic acid and CGS 23425. *Pharmacol. Exp. Ther.*, *311*, 164–171.

103. Stanbury, J. B., Ermans, A. E., Bourdoux, P., Todd, C., Oken, E., Tonglet, R., Vidor, G., Braverman, L. E., Medeiros-Neto, G. (1998). Iodine-induced hyperthyroidism: Occurrence and epidemiology. *Thyroid*, *8*, 83–100.

104. Eng, P. H., Cardona, G. R., Fang, S. L., Previti, M., Alex, S., Carrasco, N., Chin, W. W., Braverman, L. E. (1999). Escape from the acute Wolff-Chaikoff effect is associated with a decrease in thyroid sodium/iodide symporter messenger ribonucleic acid and protein. *Endocrinology*, *140*, 3404–3410.

105. Markou, K., Georgopoulos, N., Kyriazopoulou, V., Vagenakis, A. G. (2001). Iodine-induced hypothyroidism. *Thyroid*, *11*, 501–510.

106. Burikhanov, R. B., Matsuzaki, S. (2000). Excess iodine induces apoptosis in the thyroid of goitrogen-pretreated rats in vivo. *Thyroid*, *10*, 123–129.

107. Langer, R., Burzler, C., Bechtner, G., Gartner, R. (2003). Influence of iodide and iodolactones on thyroid apoptosis. Evidence that apoptosis induced by iodide is mediated by iodolactones in intact porcine thyroid follicles. *Exp. Clin. Endocrinol. Diabetes*, *111*, 325–329.

108. Kaplan, E. L. (2001). Surgery of the thyroid gland. In K. Becker (Ed.), *Principles and Practice of Endocrinology and Metabolism*, 3rd ed. Lippincott, Williams & Wilkins, Philadelphia, pp. 440–445.

109. Verger, P., Aurengo, A., Geoffroy, B., Le Guen, B. (2001). Iodine kinetics and effectiveness of stable iodine prophylaxis after intake of radioactive iodine: A review. *Thyroid*, *11*, 353–360.

110. Taurog, A. (1976). The mechanism of action of the thioureylene antithyroid drugs. *Endocrinology*, *98*, 1031–1046.

111. Peakman, M., Hussain, M., Cundy, T., Vergani, D. (1989). Increased activated T-lymphocytes and normal thyrotropin receptor antibody levels in Graves' disease in long-term remission. *J. Clin. Lab. Immunol.*, *30*, 1–5.

112. Nakashima, T., Taurog, A. (1979). Rapid conversion of carbimazole to methimazole in serum; evidence for an enzymatic mechanism. *Clin. Endocrinol.*, *10*, 637–648.

113. Heinen, E., Herrmann, J., Mosny, D., Moreno, F., Teschke, R., Kruskemper, H. L. (1981). Inhibition of peripheral deiodination of 3, 5, 3′-triiodothyronine: An adverse effect of propylthiouracil in the treatment of T3-thyrotoxicosis. *J. Endocrinol. Invest.*, *4*, 331–334.

114. Zhang, L., Parratt, J. R., Beastall, G. H., Pyne, N. J., Furman, B. L. (2002). Streptozotocin diabetes protects against arrhythmias in rat isolated hearts: Role of hypothyroidism. *Eur. J. Pharmacol.*, *435*, 269–276.

115. Lucas, A., Salinas, I., Rius, F., Pizarro, E., Granada, M. L., Foz, M., Sanmarti, A. (1997). Medical therapy of Graves' disease: Does thyroxine prevent recurrence of hyperthyroidism? *J. Clin. Endocrinol. Metab.*, *82*, 2410–2413.

116. Bartalena, L., Bogazzi, F., Martino, E. (1996). Adverse effects of thyroid hormone preparations and antithyroid drugs. *Drug Saf.*, *15*, 53–63.
117. Meyer-Gessner, M., Benker, G., Lederbogen, S., Olbricht, T., Reinwein, D. (1994). Antithyroid drug-induced agranulocytosis: Clinical experience with ten patients treated at one institution and review of the literature. *J. Endocrinol. Invest.*, *17*, 29–36.
118. Woeber, K. A. (2002). Methimazole-induced hepatotoxicity. *Endocr. Pract.*, *8*, 222–224.
119. Liaw, Y. F., Huang, M. J., Fan, K. D., Li, K. L., Wu, S. S., Chen, T. J. (1993). Hepatic injury during propylthiouracil therapy in patients with hyperthyroidism. A cohort study. *Ann. Intern. Med.*, *118*, 424–428.
120. Schneider, D. L. (2000). New therapies for the prevention and treatment of osteoporosis. *Curr. Opin. Endocrinol. Diabetes*, *7*, 310–319.
121. Canalis, E. (2000). Glucocorticoid-induced osteoporosis. *Curr. Opin. Endocrinol. Diabetes*, *7*, 320–324.
122. Siris, E. S. (1999). Goals of treatment for Paget's disease of bone. *J. Bone. Miner. Res.*, *14*, 49–52.
123. Body, J. J. (2000). Current and future directions in medical therapy: Hypercalcemia. *Cancer*, *88*(12, Suppl.), 3054–3058.
124. Cohen, A., Silverberg, S. J. (2002). Calcimimetics: Therapeutic potential in hyperparathyroidism. *Curr. Opin. Pharmacol.*, *2*, 734–739.
125. Zaidi, M., Blair, H. C., Moonga, B. S., Abe, E., Huang, C. L. (2003). Osteoclastogenesis, bone resorption, and osteoclast-based therapeutics. *J. Bone Miner. Res.*, *18*, 599–609.
126. Huang, J. C., Sakata, T., Pfleger, L. L., Bencsik, M., Halloran, B. P., Bikle, D. D., Nissenson, R. A. (2004). PTH differentially regulates expression of RANKL and OPG. *J. Bone Miner. Res.*, *19*, 235–244.
127. Gardella, T. J., Juppner, H. (2001). Molecular properties of the PTH/PTHrP receptor. *Trends Endocrinol. Metab.*, *12*, 210–217.
128. Mannstadt, M., Juppner, H., Gardella, T. J. (1999). Receptors for PTH and PTHrP: Their biological importance and functional properties. *Am. J. Physiol.*, *277*(5, Pt. 2), F665–F675.
129. Hock, J. M., Krishnan, V., Onyia, J. E., Bidwell, J. P., Milas, J., Stanislaus, D. (2001). Osteoblast apoptosis and bone turnover. *J. Bone Miner. Res.*, *16*, 975–984.
130. Fox, J. (2002). Developments in parathyroid hormone and related peptides as bone-formation agents. *Curr. Opin. Pharmacol.*, *2*, 338–344.
131. Zaidi, M., Inzerillo, A. M., Moonga, B. S., Bevis, P. J., Huang, C. L. (2002). Forty years of calcitonin—Where are we now? A tribute to the work of Iain Macintyre, FRS. *Bone*, *30*, 655–663.
132. Fleisch, H. (2002). Development of bisphosphonates. *Breast Cancer Res.*, *4*, 30–34.
133. Rogers, M. J., Gordon, S., Benford, H. L., Coxon, F. P., Luckman, S. P., Monkkonen, J., Frith, J. C. (2000). Cellular and molecular mechanisms of action of bisphosphonates. *Cancer*, *88*(12, Suppl.), 2961–2978.
134. Frith, J. C., Rogers, M. J. (2003). Antagonistic effects of different classes of bisphosphonates in osteoclasts and macrophages in vitro. *J. Bone Miner. Res.*, *18*, 204–212.

135. Ross, J. R., Saunders, Y., Edmonds, P. M., Patel, S., Wonderling, D., Normand, C., Broadley, K. (2004). A systematic review of the role of bisphosphonates in metastatic disease. *Health Technol. Assess.*, *8*, 1–176.

136. Uladag, H. (2002). Bisphosphonates as a foundation of drug delivery to bone. *Curr. Pharm. Des.*, *8*, 1929–1944.

137. Kiang, D. T., Loken, M. K., Kennedy, B. J. (1979). Mechanism of the hypocalcemic effect of mithramycin. *J. Clin. Endocrinol. Metab.*, *48*, 341–344.

138. Hall, T. J., Schaeublin, M., Chambers, T. J. (1993). The majority of osteoclasts require mRNA and protein synthesis for bone resorption in vitro. *Biochem. Biophys. Res. Commun.*, *195*, 1245–1253.

139. Umland, S. P., Schleimer, R. P., Johnston, S. L. (2002). Review of the molecular and cellular mechanisms of action of glucocorticoids for use in asthma. *Pulm. Pharmacol. Ther.*, *15*, 35–50.

140. Goulding, N. J. (2004). The molecular complexity of glucocorticoid actions in inflammation: A four ring circus. *Curr. Opin. Pharmacol.*, *4*, 629–636.

141. White, P. C. (2001). Synthesis and metabolism of corticosteroids. In K. L. Becker (Ed.), *Principles and Practice of Endocrinology and Metabolism*, 3rd ed. Lippincott, Williams & Wilkins, Philadelphia, pp. 704–714.

142. Santen, R. J., Misbin, R. I. (1981). Aminoglutethimide: Review of pharmacology and clinical use. *Pharmacotherapy*, *1*, 95–120.

143. Lacroix, M., Hontela, A. (2003). The organochlorine o,p'-DDD disrupts the adrenal steroidogenic signaling pathway in rainbow trout (*Oncorhynchus mykiss*). *Toxicol. Appl. Pharmacol.*, *190*, 197–205.

144. Touitou, Y., Bogdan, A., Auzeby, A., Dommergues, J. P. (1979). Glucocorticoid and mineralocorticoid pathways in two adrenocortical carcinomas: Comparison of the effects of o,p'-dichlorodiphenyldichloroethane, aminoglutethimide and 2-p-aminophenyl-2-phenylethylamine in vitro. *J. Endocrinol.*, *82*, 87–94.

145. Fang, V. S. (1979). Cytotoxic activity of 1-(o-chlorophenyl)-1-(p-chlorophenyl)-2,2-dichloroethane (mitotane) and its analogs on feminizing adrenal neoplastic cells in culture. *Cancer Res.*, *39*, 139–145.

146. Morishita, K., Okumura, H., Ito, N., Takahashi, N. (2001). Primary culture system of adrenocortical cells from dogs to evaluate direct effects of chemicals on steroidogenesis. *Toxicology*, *165*, 171–178.

147. Lambert, F., Corcelle-Cerf, F., Lammerant, J., Kolanowski, J. (1984). On the specificity of the inhibitory effect of trilostane and aminoglutethimide on adrenocortical steroidogenesis in guinea pig. *Mol. Cell. Endocrinol.*, *37*, 115–120.

148. Lambert, A., Frost, J., Mitchell, R., Robertson, W. R. (1986). On the assessment of the in vitro biopotency and site(s) of action of drugs affecting adrenal steroidogenesis. *Ann. Clin. Biochem.*, *23*, 225–229.

149. Semple, C. G., Beastall, G. H., Gray, C. E., Thomson, J. A. (1983). Trilostane in the management of Cushing's syndrome. *Acta Endocrinol. (Copenh.)*, *102*, 107–110.

150. Ingle, J. N., Krook, J. E., Schaid, D. J., Everson, L. K., Mailliard, J. A., Long, H. J., McCormack, G. W. (1990). Evaluation of trilostane plus hydrocortisone in women with metastatic breast cancer and prior hormonal therapy exposure. *Am. J. Clin. Oncol.*, *13*, 93–97.

151. le Roux, P. A., Tregoning, S. K., Zinn, P. M., van der Spuy, Z. M. (2002). Inhibition of progesterone secretion with trilostane for mid-trimester termination of pregnancy: Randomized controlled trials. *Hum. Reprod.*, *17*, 1483–1489.
152. Weber, M. M., Lang, J., Abedinpour, F., Zeilberger, K., Adelmann, B., Engelhardt, D. (1993). Different inhibitory effect of etomidate and ketoconazole on the human adrenal steroid biosynthesis. *Clin. Investig.*, *71*, 933–938.
153. Dorr, H. G., Kuhnle, U., Holthausen, H., Bidlingmaier, F., Knorr, D. (1984). Etomidate: A selective adrenocortical 11 beta-hydroxylase inhibitor. *Klin. Wochenschr.*, *62*, 1011–1013.
154. Preziosi, P., Vacca, M. (1998). Adrenocortical suppression and other endocrine effects of etomidate. *Life Sci.*, *42*, 477–489.
155. Fuller, P. J. (1997). Aldosterone's effects and mechanism of action. *Curr. Opin. Endocrinol. Diabetes*, *4*, 218–224.
156. Rainey, W. E., White, P. (1998). Functional adrenal zonation and regulation of aldosterone biosynthesis. *Curr. Opin. Endocrinol. Diabetes*, *5*, 175–182.
157. Slight, S. H., Joseph, J., Ganjam, V. K., Weber, K. T. (1999). Extra-adrenal mineralocorticoids and cardiovascular tissue. *J. Mol. Cell. Cardiol.*, *31*, 1175–1184.
158. Ferrari, P., Krozowski, Z. (2000). Role of the 11beta-hydroxysteroid dehydrogenase type 2 in blood pressure regulation. *Kidney Int.*, *57*, 1374–1381.
159. Delcayre, C., Swynghedauw, B. (2002). Molecular mechanisms of myocardial remodeling. The role of aldosterone. *J. Mol. Cell. Cardiol.*, *34*, 1577–1584.
160. Oelkers, W., Buchen, S., Diederich, S., Krain, J., Muhme, S., Schoneshofer, M. (1994). Impaired renal 11 beta-oxidation of 9 alpha-fluorocortisol: An explanation for its mineralocorticoid potency. *J. Clin. Endocrinol. Metab.*, *78*, 928–932.
161. Oldenburg, O., Kribben, A., Baumgart, D., Philipp, T., Erbel, R., Cohen, M. V. (2002). Treatment of orthostatic hypotension. *Curr. Opin. Pharmacol.*, *2*, 740–747.
162. Cole, T. J., Pierce, D. (2001). Mineralocorticoid target genes. *Curr. Opin. Endocrinol. Diabetes*, *8*, 118–123.
163. Funder, J. W. (2001). Non-genomic actions of aldosterone: Role in hypertension. *Curr. Opin. Nephrol. Hypertens.*, *10*, 227–230.
164. Christ, M., Wehling, M. (1999). Rapid actions of aldosterone: Lymphocytes, vascular smooth muscle and endothelial cells. *Steroids*, *64*, 35–41.
165. Pitt, B., Zannad, F., Remme, W. J., Cody, R., Castaigne, A., Perez, A., Palensky, J., Wittes, J. (1999). The effect of spironolactone on morbidity and mortality in patients with severe heart failure. *N. Engl. J. Med.*, *341*, 709–717.
166. Soberman, J. E., Weber, K. T. (2000). Spironolactone in congestive heart failure. *Curr. Hypertens. Rep.*, *2*, 451–456.
167. Cittadini, A., Monti, M. G., Isgaard, J., Casaburi, C., Stromer, H., Di Gianni, A., Serpico, R. Saldamarco, L., Vanasia, M., Sacca, L. (2003). Aldosterone receptor blockade improves left ventricular remodeling and increases ventricular fibrillation threshold in experimental heart failure. *Cardiovasc. Res.*, *58*, 555–564.
168. de Gasparo, M., Joss, U., Ramjoue, H. P., Whitebread, S. E., Haenni, H., Schenkel, L., Kraehenbuehl, C., Biollaz, M., Grob, J., Schmidlin, J. (1987). Three new epoxyspirolactone derivatives: Characterization in vivo and in vitro. *J. Pharmacol. Exp. Ther.*, *240*, 650–656.

169. McMahon, E. G. (2001). Recent studies with eplerenone, a novel, selective aldosterone receptor antagonist. *Curr. Opin. Pharmacol.*, *1*, 190–196.
170. Pitt, B., Remme, W., Zannad, F., Neaton, J., Martinez, F., Roniker, B., Bittman, R., Hurley, S., Kleiman, J., Gatlin, M. (2003). Eplerenone Post-Acute Myocardial Infarction Heart Failure Efficacy and Survival Study Investigators. Eplerenone, a selective aldosterone blocker, in patients with left ventricular dysfunction after myocardial infarction. *N. Engl. J. Med.*, *348*, 1309–1321.
171. Gobinet, J., Poujol, N., Sultan, Ch. (2002). Molecular action of androgens. *Mol. Cell. Endocrinol.*, *198*, 15–24.
172. Shahidi, N. T. (2001). A review of the chemistry, biological action, and clinical applications of anabolic-androgenic steroids. *Clin. Ther.*, *23*, 1355–1390.
173. Bhasin, S., Storer, T. W., Berman, N., Yarasheski, K. E., Clevenger, B., Phillips, J., Lee, W. P., Bunnell, T. J., Casaburi, R. (1997). Testosterone replacement increases fat-free mass and muscle size in hypogonadal men. *J. Clin. Endocrinol. Metab.*, *82*, 407–413.
174. Kuhn, C. M. (2002). Anabolic steroids. *Recent Prog. Horm. Res.*, *57*, 411–434.
175. Bergink, E. W., Janssen, P. S., Turpijn, E. W., van der Vies, J. (1985). Comparison of the receptor binding properties of nandrolone and testosterone under in vitro and in vivo conditions. *J. Steroid Biochem.*, *22*, 831–836.
176. Sundaram, K., Kumar, N., Monder, C., Bardin, C. W. (1995). Different patterns of metabolism determine the relative anabolic activity of 19-norandrogens. *J. Steroid Biochem. Mol. Biol.*, *53*, 253–257.
177. Zhao, J., Bauman, W. A., Huang, R., Caplan, A. J., Cardozo, C. (2004). Oxandrolone blocks glucocorticoid signaling in an androgen receptor-dependent manner. *Steroids*, *69*, 357–366.
178. Negro-Vilar, A. (1999). Selective androgen receptor modulators (SARMs): A novel approach to androgen therapy for the new millennium. *J. Clin. Endocrinol. Metab.*, *84*, 3459–3466.
179. Yin, D., Gao, W., Kearbey, J. D., Xu, H., Chung, K., He, Y., Marhefka, C. A., Veverka, K. A., Miller, D. D., Dalton, J. T. (2003). Pharmacodynamics of selective androgen receptor modulators. *J. Pharmacol. Exp. Ther.*, *304*, 1334–1340.
180. Berrevoets, C. A., Umar, A., Brinkmann, A. O. (2002). Antiandrogens: Selective androgen receptor modulators. *Mol. Cell. Endocrinol.*, *198*, 97–103.
181. Rittmaster, R. S. (1999). Antiandrogen treatment of polycystic ovary syndrome. *Endocrinol. Metab. Clin. North Am.*, *28*, 409–421.
182. Occhiato, E. G., Guarna, A., Danza, G., Serio, M. (2004). Selective non-steroidal inhibitors of 5 alpha-reductase type 1. *J. Steroid Biochem. Mol. Biol.*, *88*, 1–16.
183. McDonnell, D. P., Connor, C. E., Wijayaratne, A., Chang, C. Y., Norris, J. D. (2002). Definition of the molecular and cellular mechanisms underlying the tissue-selective agonist/antagonist activities of selective estrogen receptor modulators. *Recent Prog. Horm. Res.*, *57*, 295–316.
184. Ruff, M., Gangloff, M., Wurtz, J. M., Moras, D. (2000). Estrogen receptor transcription and transactivation: Structure-function relationship in DNA- and ligand-binding domains of estrogen receptors. *Breast Cancer Res.*, *2*, 353–359.
185. Gustafsson, J. A. (2003). What pharmacologists can learn from recent advances in estrogen signalling. *Trends Pharmacol. Sci.*, *24*, 479–485.

186. Segars, J. H., Drigger, P. H. (2002). Estrogen action and cytoplasmic signaling cascades. Part I: Membrane-associated signaling complexes. *Trends Endocrinol. Metab*, *13*, 349–354.
187. Chen, Z., Yuhanna, I. S., Galcheva-Gargova, Z., Karas, R. H., Mendelsohn, M. E., Shaul, P. W. (1999). Estrogen receptor alpha mediates the nongenomic activation of endothelial nitric oxide synthase by estrogen. *J. Clin. Invest.*, *103*, 401–406.
188. Harrington, W. R., Sheng, S., Barnett, D. H., Petz, L. N., Katzenellenbogen, J. A., Katzenellenbogen, B. S. (2003). Activities of estrogen receptor alpha- and beta-selective ligands at diverse estrogen responsive gene sites mediating transactivation or transrepression. *Mol. Cell. Endocrinol.*, *206*, 13–22.
189. Campos, S. M. (2004). Aromatase inhibitors for breast cancer in postmenopausal women. *Oncologist*, *9*, 126–136.
190. Zeitoun, K. M., Bulun, S. E. (1999). Aromatase: A key molecule in the pathophysiology of endometriosis and a therapeutic target. *Fertil. Steril.*, *72*, 961–969.
191. Mitwally, M. F., Casper, R. F. (2002). Aromatase inhibition for ovarian stimulation: Future avenues for infertility management. *Curr. Opin. Obstet. Gynecol.*, *14*, 255–263.
192. Kao, Y. C., Okubo, T., Sun, X. Z., Chen, S. (1999). Induction of aromatase expression by aminoglutethimide, an aromatase inhibitor that is used to treat breast cancer in postmenopausal women. *Anticancer Res.*, *19*, 2049–2056.
193. Anderson, G. L., Limacher, M., Assaf, A. R., Bassford, T., Beresford, S. A., Black, H., Bonds, D., Brunner, R., Brzyski, R., Caan, B., Chlebowski, R., Curb, D., Gass, M., Hays, J., Heiss, G., Hendrix, S., Howard, B. V., Hsia, J., Hubbell, A., Jackson, R., Johnson, K. C., Judd, H., Kotchen, J. M., Kuller, L., LaCroix, A. Z., Lane, D., Langer, R. D., Lasser, N., Lewis, C. E., Manson, J., Margolis, K., Ockene, J., O'Sullivan, M. J., Phillips, L., Prentice, R. L., Ritenbaugh, C., Robbins, J., Rossouw, J. E., Sarto, G., Stefanick, M. L., Van Horn, L., Wactawski-Wende, J., Wallace, R., Wassertheil-Smoller, S. (2004). Women's Health Initiative Steering Committee. Effects of conjugated equine estrogen in postmenopausal women with hysterectomy: The Women's Health Initiative randomized controlled trial. *JAMA*, *291*, 1701–1712.
194. Conneely, O. M., Mulac-Jericevic, B., DeMayo, F., Lydon, J. P., O'Malley, B. W. (2002). Reproductive functions of progesterone receptors. *Recent Prog. Horm. Res.*, *57*, 339–355.
195. Sutter-Dub, M. T. (2002). Rapid non-genomic and genomic responses to progestogens, estrogens, and glucocorticoids in the endocrine pancreatic B cell, the adipocyte and other cell types. *Steroids*, *67*, 77–93.
196. Sager, G., Orbo, A., Jaeger, R., Engstrom, C. (2003). Non-genomic effects of progestins—Inhibition of cell growth and increased intracellular levels of cyclic nucleotides. *J. Steroid Biochem. Mol. Biol.*, *84*, 1–8.
197. Schindler, A. E., Campagnoli, C., Druckmann, R., Huber, J., Pasqualini, J. R., Schweppe, K. W., Thijssen, J. H. (2003). Classification and pharmacology of progestins. *Maturitas*, *46*(Suppl. 1), S7–S16.
198. The Diabetes Control and Complications Trial Research Group (1993). The effect of intensive treatment of diabetes on the development and progression of long-term complications in insulin-dependent diabetes mellitus. *N. Engl. J. Med.*, *329*, 977–986.

199. Turner, R. C. (1998). The U.K. Prospective Diabetes Study. A review. *Diabetes Care*, *21*(Suppl. 3), C35–38.
200. Kahn, S. E. (2003). The relative contribution of insulin resistance annd beta cell dysfunction to the pathophysiology of Type 2 diabetes. *Diabetologia*, *46*, 3–19.
201. Porte, D. Jr., Kahn, S. E. (2001). Beta-cell dysfunction and failure in type 2 diabetes: Potential mechanisms. *Diabetes*, *50*(Suppl. 1), S160–163.
202. Juhan-Vague, I., Alessi, M. C. (1997). PAI-1, obesity, insulin resistance and risk of cardiovascular events. *Thromb. Haemost.*, *78*, 656–660.
203. Bastard, J. P., Pieroni, L., Hainque, B. (2000). Relationship between plasma plasminogen activator inhibitor 1 and insulin resistance. *Diabetes Metab. Res. Rev.*, *16*, 192–201.
204. Howell, S. L., Jones, P. M., Persaud, S. J. (1994). Regulation of insulin secretion: The role of second messengers. *Diabetologia*, *37*(Suppl. 2), S30–35.
205. Dornhorst, A. (2001). Insulinotropic meglitinide analogues. *Lancet*, *358*, 1709–1716.
206. Vajo, Z., Duckworth, W. C. (2000). Genetically engineered insulin analogs: Diabetes in the new millenium. *Pharmacol. Rev.*, *52*, 1–9.
207. Owens, D. R. (2002). Opimising glycaemic control—The role of long-acting insulin analogues in basal insulin therapy. *Br. J. Diabetes Cardiovasc. Dis.*, *2*, 403–407.
208. Alessi, D. R. (2001). Discovery of PDK1, one of the missing links in insulin signal transduction. Colworth Medal Lecture. *Biochem Soc. Trans.*, *29*(Pt. 2), 1–14.
209. Saltiel, A. R., Kahn, C. R. (2001). Insulin signalling and the regulation of glucose and lipid metabolism. *Nature*, *414*, 799–806.
210. Khan, A. H., Pessin, J. E. (2002). Insulin regulation of glucose uptake: A complex interplay of intracellular signalling pathways. *Diabetologia*, *45*, 1475–1483.
211. Schafer, G. (1983). Biguanides: A review of history, pharmacodynamics and therapy. *Diabetes Metab. Rev.*, *9*, 148–163.
212. Bailey, C. J. (1992). Biguanides and NIDDM. *Diabetes Care*, *15*, 755–772.
213. UKPDS. (1998). Intensive blood glucose control with sulfonylureas or insulin compared with conventional treatment and risk of complications in patients with type 2 diabetes (UKPDS 33). UK Prospective Diabetes Study Group. *Lancet*, *352*, 837–853.
214. Radziuk, J., Bailey, C. J., Wiernsperger, N. F., Yudkin, J. S. (2003). Metformin and its liver targets in the treatment of type 2 diabetes. *Curr. Drug Targets Immune Endocr. Metab. Disord.*, *3*, 151–169.
215. Zhou, G., Myers, R., Li, Y., Chen, Y., Shen, X., Fenyk-Melody, J., Wu, M., Ventre, J., Doebber, T., Fujii, N., Musi, N., Hirshman, M. F., Goodyear, L. J., Moller, D. E. (2001). Role of AMP-activated protein kinase in mechanism of metformin action. *J. Clin. Invest.*, *108*, 1167–1174.
216. Hardie, D. G. (2003). Minireview: The AMP-activated protein kinase cascade: The key sensor of cellular energy status. *Endocrinology*, *144*, 5179–5183.
217. Seli, E., Duleba, A. J. (2004). Treatment of PCOS with metformin and other insulin-sensitizing agents. *Curr. Diabetes Rep.*, *4*, 69–75.

218. Kay, J. P., Alemzadeh, R., Langley, G., D'Angelo, L., Smith, P., Holshouser, S. (2001). Beneficial effects of metformin in normoglycemic morbidly obese adolescents. *Metabolism, 50*, 1457–1461.
219. Olefsky, J. M., Saltiel, A. R. (2000). PPAR gamma and the treatment of insulin resistance. *Trends Endocrinol. Metab., 11*, 362–368.
220. Desvergne, B., Wahli, W. (1999). Peroxisome proliferator-activated receptors: Nuclear control of metabolism. *Endocr. Rev., 20*, 649–688.
221. Auwerx, J. (1999). PPAR-γ, the ultimate thrifty gene. *Diabetologia, 42*, 1033–1049.
222. Rangwala, S. M., Lazar, M. A. (2004). Peroxisome proliferator-activated receptor gamma in diabetes and metabolism. *Trends Pharmacol. Sci., 25*, 331–336.
223. Krentz, A. J., Bailey, C. J., Melander, A. (2000). Thiazolidinediones for type 2 diabetes. New agents reduce insulin resistance but need long term clinical trials. *BMJ, 321*, 252–253.
224. Niemi, M., Backman, J. T., Neuvonen, M., Neuvonen, P. J., Kivsito, K. T. (2001). Effects of rifampicin on the pharmacokinetics and pharmacodynamics of glyburide and glipizide. *Clin. Pharmacol. Ther., 69*, 400–406.
225. Niemi, M., Cascorbi, I., Timm, R., Kroemer, H. K., Neuvonen, P. J., Kivisto, K. T. (2002). Glyburide and glimepiride pharmacokinetics in subjects with different CYP2C9 genotypes. *Clin. Pharmacol. Ther., 72*, 326–332.
226. Meinert, C. L., Knatterud, G. L., Prout, T. E., Klimt, C. R. (1970). A study of the effects of hypoglycemic agents on vascular complications in patients with adult-onset diabetes. II. Mortality results. *Diabetes, 19*(Suppl.), 789–830.
227. Aguilar-Bryan, L., Bryan, J. (1999). Molecular biology of adenosine triphosphate-sensitive potassium channels. *Endocr. Rev., 20*, 101–135.
228. Gribble, F. M. Reimann, F. (2003). Sulphonylurea action revisited: The post-cloning era. *Diabetologia, 46*, 875–891.
229. Meyer, M., Chudziak, F., Schwanstecher, C., Schwanstecher, M., Panten, U. (1999). Structural requirements of sulphonylureas and analogues for interaction with sulphonylurea receptor subtypes. *Br. J. Pharmacol., 128*, 27–34.
230. Kurowski, T. G., Lin, Y., Luo, Z., Tsichlis, P. N., Buse, M. G., Heydrick, S. J., Ruderman, N. B. (1999). Hyperglycemia inhibits insulin activation of Akt/protein kinase B but not phosphatidylinositol 3-kinase in rat skeletal muscle. *Diabetes, 48*, 658–663.
231. Muller, G., Geisen, K. (1996). Characterization of the molecular mode of action of the sulfonylurea, glimepiride, at adipocytes. *Horm. Metab. Res., 28*, 469–487.
232. Chutkow, W. A., Samuel, V., Hansen, P. A., Pu, J., Valdivia, C. R., Makielski, J. C., Burant, C. F. (2001). Disruption of Sur2-containing K(ATP) channels enhances insulin-stimulated glucose uptake in skeletal muscle. *Proc. Natl. Acad. Sci. USA, 98*, 11760–11764.
233. Goldfine, A. B., Maratos-Flier, E. (2001). Oral agents for the treatment of type 2 diabetes. In K. L. Becker (Ed.), *Principles and Practice of Endocrinology and Metabolism*, 3rd ed. Lippincott, Williams & Wilkins, Philadelphia pp. 1344–1348.
234. Riddle, M. C., Schneider, J. (1998). Beginning insulin treatment of obese patients with evening 70/30 insulin plus glimepiride versus insulin alone. Glimepiride Combination Group. *Diabetes Care, 21*, 1052–1057.

235. Jennings, P. E. (2000). Vascular benefits of gliclazide beyond glycemic control. *Metabolism*, *49*(10, Suppl. 2), 17–20.
236. Harrower, A. D. (1996). Pharmacokinetics of oral antihyperglycaemic agents in patients with renal insufficiency. *Clin. Pharmacokinet.*, *31*, 111–119.
237. Henquin, J. C. (1990). Established, unsuspected and novel pharmacological insulin secretagogues. In C. J. and Bailey, P. R. Flatt (Eds), *New Antidiabetic Drugs*, 1st ed. Smith Gordon and Company, London, pp. 93–106.
238. Hanefeld, M., Bouter, K. P., Dickinson, S., Guitard, C. (2000). Rapid and short-acting mealtime insulin secretion with nateglinide controls both prandial and mean glycemia. *Diabetes Care*, *23*, 202–207.
239. Weaver, M. L., Orwig, B. A., Rodriguez, L. C., Graham, E. D., Chin, J. A., Shapiro, M. J., McLeod, J. F., Mangold, J. B. (2001). Pharmacokinetics and metabolism of nateglinide in humans. *Drug Metab. Dispos.*, *29*(4, Pt. 1), 415–421.
240. Hatorp, V. (2002). Clinical pharmacokinetics and pharmacodynamics of repaglinide. *Clin. Pharmacokinet.*, *41*, 471–483.
241. Hu, S., Wang, S., Fanelli, B., Bell, P. A., Dunning, B. E., Geisse, S., Schmitz, R., Boettcher, B. R. (2000). Pancreatic β-cell K_{ATP} channel activity and membrane-binding studies with nateglinide: A comparison with sulfonylureas and repaglinide. *J. Pharmacol. Exp. Ther.*, *293*, 444–452.
242. Hansen, A. M., Christensen, I. T., Hansen, J. B., Carr, R. D., Ashcroft, F. M., Wahl, P. (2002). Differential interactions of nateglinide and repaglinide on the human beta-cell sulphonylurea receptor 1. *Diabetes*, *51*, 2789–2795.
243. Salvatore, T., Giugliano, D. (1996). Pharmacokinetic-pharmacodynamic relationships of acarbose. *Clin. Pharmacokinet.*, *30*, 94–106.
244. Hollander, P., Pi-Sunyer, X., Coniff, R. F. (1997). Acarbose in the treatment of type 1 diabetes. *Diabetes Care*, *20*, 248–253.
245. Holman, R. R., Cull, C. A., Turner, R. C. (1999). A randomized double blind trial of acarbose in type 2 diabetes shows improved glycemic control over 3 years (U.K. Prospective Diabtes Study 44). *Diabetes Care*, *22*, 960–964.
246. Chiasson, J. L., Josse, R. G., Gomis, R., Hanefeld, M., Karasik, A., Laakso, M. (2002). STOP-NIDDM Trail Research Group. Acarbose for prevention of type 2 diabetes mellitus: The STOP-NIDDM randomised trial. *Lancet*, *359*, 2072–2077.
247. Doyle, M. E., Egan, J. M. (2001). Glucagon-like peptide-1. *Recent Prog. Horm. Res.*, *56*, 377–399.
248. Drucker, D. J. (2001). Minireview: The glucagon-like peptides. *Endocrinology*, *142*, 521–527.
249. Gault, V. A., Flatt, P. R., O'Harte, F. P. (2003). Glucose-dependent insulinotropic polypeptide analogues and their therapeutic potential for the treatment of obesity-diabetes. *Biochem. Biophys. Res. Commun.*, *308*, 207–213.
250. Preitner, F., Ibberson, M., Franklin, I., Binnert, C., Pende, M., Gjinovci, A., Hansotia, T., Drucker, D. J., Wollheim, C., Burcelin, R., Thorens, B. (2004). Gluco-incretins control insulin secretion at multiple levels as revealed in mice lacking GLP-1 and GIP receptors. *J. Clin. Invest.*, *113*(4), 635–645.
251. Creutzfeldt, W. O., Kleine, N., Willms, B., Orskov, C., Holst, J. J., Nauck, M. A. (1996). Glucagonostatic actions and reduction of fasting hyperglycemia by exoge-

nous glucagon-like peptide I(7-36) amide in type I diabetic patients. *Diabetes Care*, *19*, 580–586.

252. Ahren, B. (2003). Gut peptides and type 2 diabetes mellitus treatment. *Curr. Diabetes Rep.*, *3*, 365–372.

253. Wiedeman, P. E., Trevillyan, J. M. (2003). Dipeptidyl peptidase IV inhibitors for the treatment of impaired glucose tolerance and type 2 diabetes. *Curr. Opin. Investig. Drugs*, *4*, 412–420.

254. Eng, J. (1992). Exendin peptides. *Mt. Sinai J. Med.*, *59*, 147–149.

255. Drucker, D. J. (1998). Glucagon-like peptides. *Diabetes*, *47*, 159–169.

256. Goke, R., Fehmann, H. C., Linn, T., Schmidt, H., Krause, M., Eng, J., Goke, B. (1993). Exendin-4 is a high potency agonist and truncated exendin-(9-39)-amide an antagonist at the glucagon-like peptide 1-(7-36)-amide receptor of insulin secreting beta-cells. *J. Biol. Chem.*, *268*, 19650–19655.

257. Young, A. A., Gedulin, B. R., Bhavsar, S., Bodkin, N., Jodka, C., Hansen, B., Denaro, M. (1999). Glucose-lowering and insulin-sensitizing actions of exendin-4: Studies in obese diabetic (ob/ob, db/db) mice, diabetic fatty Zucker rats, and diabetic rhesus monkeys (*Macaca mulatta*). *Diabetes*, *48*, 1026–1034.

258. Edwards, C. M., Stanley, S. A., Davis, R., Brynes, A. E., Frost, G. S., Seal, L. J., Ghatei, M. A., Bloom, S. R. (2001). Exendin-4 reduces fasting and postprandial glucose and decreases energy intake in healthy volunteers. *Am. J. Physiol. Endocrinol. Metab.*, *281*, E155–E161.

259. Egan, J. M., Clocquet, A. R., Elahi, D. (2002). The insulinotropic effect of acute exendin-4 administered to humans: Comparison of nondiabetic state to type 2 diabetes. *J. Clin. Endocrinol. Metab.*, *87*, 1282–1290.

260. Xu, G., Stoffers, D. A., Habener, J. F., Bonner-Weir, S. (1999). Exendin-4 stimulates both beta-cell replication and neogenesis, resulting in increased beta-cell mass and improved glucose tolerance in diabetic rats. *Diabetes*, *48*, 2270–2276.

261. Taborsky, G. J. Jr, Ahren, B., Havel, P. J. (1998). Autonomic mediation of glucagon secretion during hypoglycemia: Implications for impaired alpha-cell responses in type 1 diabetes. *Diabetes*, *47*, 995–1005.

262. Exton, J. H. (1987). Mechanisms of hormonal regulation of hepatic glucose metabolism. *Diabetes Metab. Rev.*, *3*, 163–183.

263. Schade, D. S., Woodside, W., Eaton, R. P. (1979). The role of glucagon in the regulation of plasma lipids. *Metabolism*, *28*, 874–886.

264. Gu, Z., Jensen, R. T., Maton, P. N. (1992). A primary role for protein kinase A in smooth muscle relaxation induced by adrenergic agonists and neuropeptides. *Am. J. Physiol.*, *263*(3, Pt. 1), G360–364.

265. Mayo, K. E., Miller, L. J., Bataille, D., Dalle, S., Goke, B., Thorens, B., Drucker, D. J. (2003). International Union of Pharmacology. XXXV. The glucagon receptor family. *Pharmacol. Rev.*, *55*, 167–194.

266. Unger, R. H. (1978). Role of glucagon in the pathogenesis of diabetes: The status of the controversy. *Metabolism*, *27*, 1691–1709.

267. Madsen, P., Knudsen, L. B., Wiberg, F. C., Carr, R. D. (1998). Discovery and structure-activity relationship of the first non-peptide competitive human glucagon receptor antagonists. *J. Med. Chem.*, *41*, 5150–5157.

268. Petersen, K. F., Sullivan, J. T. (2001). Effects of a novel glucagon receptor antagonist (Bay 27-9955) on glucagon-stimulated glucose production in humans. *Diabetologia*, *44*, 2018–2024.

269. Ladouceur, G. H., Cook, J. H., Hertzog, D. L., Jones, J. H., Hundertmark, T., Korpusik, M., Lease, T. G., Livingston, J. N., MacDougall, M. L., Osterhout, M. H., Phelan, K., Romero, R. H., Schoen, W. R., Shao, C., Smith, R. A. (2002). Integration of optimized substituent patterns to produce highly potent 4-aryl-pyridine glucagon receptor antagonists. *Bioorg. Med. Chem. Lett.*, *12*, 3421–3424.

270. Ying, J., Ahn, J. M., Jacobsen, N. E., Brown, M. F., Hruby, V. J. (2003). NMR solution structure of the glucagon antagonist [desHis1, desPhe6, Glu9]glucagon amide in the presence of perdeuterated dodecylphosphocholine micelles. *Biochemistry*, *42*, 2825–2835.

271. Madsen, P., Ling, A., Plewe, M., Sams, C. K., Knudsen, L. B., Sidelmann, U. G., Ynddal, L., Brand, C. L., Andersen, B., Murphy, D., Teng, M., Truesdale, L., Kiel, D., May, J., Kuki, A., Shi, S., Johnson, M. D., Teston, K. A., Feng, J., Lakis, J., Anderes, K., Gregor, V., Lau, J. (2002). Optimization of alkylidene hydrazide based human glucagon receptor antagonists. Discovery of the highly potent and orally available 3-cyano-4-hydroxybenzoic acid [1-(2,3,5,6-tetramethylbenzyl)-1H-indol-4-ylmethylene]hydrazide. *J. Med. Chem.*, *45*, 5755–5775.

272. Rothman, R. B., Baumann, M. H. (2002). Serotonin releasing agents. Neurochemical, therapeutic and adverse effects. *Pharmacol. Biochem. Behav.*, *71*, 825–836.

273. Gross, S. B., Lepor, N. E. (2000). Anorexigen-related cardiopulmonary toxicity. *Rev. Cardiovasc. Med.*, *1*, 80–89.

274. Finer, N. (2002). Sibutramine: Its mode of action and efficacy. *Int. J. Obes. Relat. Metab. Disord*, *26*(Suppl. 4), S29–S33.

275. Williams, G., Bing, C., Cai, X. J., Harrold, J. A., King, P. J., Liu, X. H. (2001). The hypothalamus and the control of energy homeostasis: Different circuits, different purposes. *Physiol. Behav.*, *74*(4–5), 683–701.

276. Druce, M. R., Small, C. J., Bloom, S. R. (2004). Minireview: Gut peptides regulating satiety. *Endocrinology*, *145*, 2660–2665.

277. Steppan, C. M., Lazar, M. A. (2002). Resistin and obesity-associated insulin resistance. *Trends Endocrinol. Metab.*, *13*, 18–23.

278. Crowley, V., Vidal-Puig, A. J. (2001). Mitochondrial uncoupling proteins (UCPs) and obesity. *Nutr. Metab. Cardiovasc. Dis.*, *11*, 70–75.

279. Weyer, C., Gautier, J. F., Danforth, E. Jr. (1999). Development of beta 3-adrenoceptor agonists for the treatment of obesity and diabetes—An update. *Diabetes Metab.*, *25*(1), 11–21.

280. Ballinger, A., Peikin, S. R. (2002). Orlistat: Its current status as an anti-obesity drug. *Eur. J. Pharmacol.*, *440*(2–3), 109–117.

281. Clapham, J. C., Arch, J. R., Tadayyon, M. (2001). Anti-obesity drugs: A critical review of current therapies and future opportunities. *Pharmacol. Ther.*, *89*, 81–121.

24

RESPIRATORY VIRUSES

PAUL D. OLIVO
Apath, LLC,
St. Louis, Missouri

24.1	INTRODUCTION	1106
24.2	SEASONAL PATTERNS	1106
24.3	PUBLIC HEALTH AND ECONOMIC IMPACT	1107
24.4	CONTROL OF VIRAL RESPIRATORY INFECTIONS	1107
24.5	SPECIFIC VIRUS PATHOGENS	1108
	Picornaviridae	1108
	Coronaviridae	1109
	Paramyxoviridae	1109
	Orthomyxoviridae	1111
	Bunyaviridae	1112
	Adenoviridae	1113
	Herpesviridae	1113
24.6	MAJOR CLINICAL SYNDROMES	1114
	Common Cold	1114
	Pharyngitis	1116
	Laryngitis	1116
	Acute Laryngotracheobronchitis (Croup)	1116
	Tracheitis and Tracheobronchitis	1116
	Bronchiolitis	1117
	Pneumonia	1117
24.7	DIAGNOSIS OF VIRAL RESPIRATORY INFECTION	1118
24.8	EXISTING ANTIVIRAL THERAPY	1120
24.9	CHALLENGES OF ANTIVIRAL THERAPY FOR RESPIRATORY VIRUS INFECTIONS	1121
24.10	STRATEGIES FOR DEVELOPMENT OF ANTIVIRAL THERAPY	1122

Drug Discovery Handbook, by Shayne Cox Gad
Copyright © 2005 by John Wiley & Sons, Inc.

24.11	DEVELOPMENT OF ANTIVIRALS AGAINST RESPIRATORY VIRUSES	1124
24.12	ANIMAL MODELS	1126
24.13	SUMMARY	1127
	References	1128

24.1 INTRODUCTION

The respiratory system is very susceptible to microbial invasion. Despite the numerous defense mechanisms that the body uses to counteract microbial invasion, respiratory tract infections are among the most common human maladies. Viruses cause disease by one of several mechanisms. They cause cell death, organ malfunction, or elicit an inflammatory reaction that can account for many of the symptoms and pathology of infection. Pneumonia, or infection within lung tissue, is one of the leading causes of mortality in the world, and viruses contribute either directly or indirectly to a significant fraction of cases of pneumonia.

Viruses are submicroscopic entities of between 20 and 300nm in diameter. They consist of a nucleic acid genome, which can be either RNA (ribonucleic acid) or DNA (deoxyribonucleic acid), and a protein coat called a capsid. Some viruses also are enveloped in a lipid membrane. Viruses lack metabolic organelles and thus have no ability to generate energy and can only replicate within living cells. Viruses, therefore, exist on the border between inanimate chemistry and life forms.

Based on their overall structure and genetic organization, animal viruses are classified into dozens of different families. Classification is not based on disease because there is no relationship between the type of virus and the diseases they cause. Several hundred antigenically distinct viruses from at least seven virus families are associated with sporadic or epidemic respiratory infections in children and adults [1].

24.2 SEASONAL PATTERNS

Many viral respiratory diseases are associated with a seasonal variation in incidence. Influenza virus and respiratory syncytial virus (RSV) infections occur predominantly in winter months with an overlap of peak periods of incidence [2]. Rhinoviruses are isolated throughout the year but their peak incidence is spring and fall [3]. Adenoviruses and herpesviruses in general do not show seasonal variation with the exception of varicella-zoster virus (chicken pox), which is more common in late winter and early spring.

24.3 PUBLIC HEALTH AND ECONOMIC IMPACT

Respiratory viral infections have a major impact on health. The rate of respiratory viral infections has been calculated based on surveys to be over 85 illnesses per 100 persons per year in the United States, making acute viral respiratory infections among the most common illnesses experienced by people within the United States and probably worldwide [4]. Between a fourth and a half of these illnesses lead people to seek medical attention and cause a significant amount of lost work and school days [5]. The overall burden of disease related to respiratory virus infections, including mortality, hospitalizations, medical costs (asthma, etc.), and societal costs (i.e., lost productivity, etc.) is immense. The resulting negative economic impact of viral respiratory disease in the United States is huge.

Mortality due to acute respiratory infection in otherwise healthy children and adults is rare except for influenza during epidemics. However, acute respiratory infections (viral and bacterial) are the leading cause of morbidity and mortality in children, with several million deaths worldwide and 100,000 deaths per year in the United States. In the 2003/2004 season there were 121 influenza-related deaths in children [6]. In developing countries acute respiratory infections are a major cause of childhood mortality. Measles, for example, which has been successfully controlled by vaccination in developed countries, remains among the major cause of mortality worldwide [7].

24.4 CONTROL OF VIRAL RESPIRATORY INFECTIONS

There are four fundamental modalities that are used to control respiratory infections within a population.

1. Public health measures
2. Vaccination
3. Antiviral therapy
4. Personal hygiene and etiquette

No one of these modalities alone can be counted upon to ensure that society can minimize the negative impact of these diseases on the population. However, particular modalities can predominate in importance for certain viruses. For example, during the 2002/2003 outbreak of severe acute respiratory syndrome (SARS), public health measures, and specifically rapid identification and isolation of infected patients, were critical to the successful efforts to control the outbreak [8]. For influenza, a combination of public health measures and vaccination are the most important control vehicles. Antiviral therapy has not played a major role in the overall management of respiratory infections. Hopefully this will change in the future. Personal hygiene is very important, and the lack of appropriate hygiene enhances person-to-person transmission of many respiratory viruses. It is particularly important for individuals with respiratory virus infections such as influenza to minimize transmission to others by avoiding school and work during the time that they are coughing and sneezing. If this is not possible, a face mask should be worn and close contact should be avoided or minimized.

24.5 SPECIFIC VIRUS PATHOGENS

Viruses of many types (enveloped/nonenveloped; DNA/RNA, etc.) cause respiratory tract infections, and there is no relationship between type of disease and the nature of the viral structure or genome. In up to 30 percent of lower respiratory infections no pathogen is identified. Although this is in large part due to diagnostic difficulties, this statistic suggests that many new viruses remain to be discovered. In fact, in the past several decades a new human virus has been discovered approximately every 2 years. Viruses from seven families cause most respiratory tract infections. A brief overview of each family and several of the specific viruses that cause respiratory infections is provided. For a more detailed description of these viruses and their replication cycles, the reader is directed to other chapters in this text and various review articles and books [9, 10].

Picornaviridae

Picornaviridae are nonenveloped, positive-sense RNA viruses that include several viruses that cause respiratory infections including rhinoviruses and enteroviruses. Rhinoviruses are the largest genus of the Picornaviridae family and the single most common cause of the common cold. Rhinoviruses have also been increasingly recognized as a cause of hospitalizations related to acute respiratory tract infections in the United States. Since their discovery in 1956, over 100 different serotypes have been identified. This antigenic variability contributes to the high number of rhinovirus infections that individuals experience during their lifetime.

The rhinovirus virion structure has been extensively studied, and the knowledge gained from this has been put to practical use in the development of a potent rhinovirus inhibitor (see below). The rhinovirus genome consists of a single strand of RNA of 7.2 kilobases that codes for a single polyprotein that is proteolytically cleaved by 2 viral proteases during and following translation into 11 proteins. This proteolysis is a key target of strategies to develop antirhinovirus drugs. Nevertheless, despite the prospect for new drugs, treatment of rhinovirus infections will be difficult. Studies with human volunteers have shown that the incubation period of rhinovirus infection of the upper respiratory tract is as short as 12 h [11, 12]. This makes the window of therapeutic opportunity quite short.

Coronaviridae

The Coronaviridae family is a large family of enveloped, positive-sense RNA viruses with a distinct crown-shaped (Latin, *corona* = crown) virion that is created by projections of the virus spike protein. At about 29 kilobases, the RNA genome of members of this family is the largest genome among the animal RNA viruses. The family includes two genera that contain many human and animal viruses. The genus *Coronavirus*, which consists of three antigenically distinct groups, contains all the known human pathogens in this family. There are four human coronaviruses that have been well characterized: human coronavirus 229E (HCoV-229E), HCoV-OC43, SARS-associated coronavirus (SARS-CoV), and the recently described coronavirus, HCoV-NL63. HCoV-229E and HCoV-OC43 are the prototype strains of group 1 and group 2, respectively, and, along with numerous other strains, are a major cause of upper respiratory tract disease. HCoV-NL63 was recently isolated from a 7-month-old child suffering from bronchiolitis and conjunctivitis [13]. The complete genome sequence of HCoV-NL63 indicates that this virus is not a recombinant but rather a new group 1 coronavirus. Screening of clinical specimens from individuals suffering from respiratory illness identified seven additional HCoV-NL63-infected individuals, indicating that the virus is widely spread within the human population [13].

Severe acute respiratory syndrome is a viral respiratory illness caused by SARS-associated coronavirus (SARS-CoV) [14]. It was first described following cases in 2003 in the Guandong Province of southern China and gained international attention following an outbreak that spread to numerous countries via a businessman who traveled to a Hong Kong hotel. The disease has an insidious onset and a long incubation period (4 to 14 days). Symptoms progress slowly but steadily in the first few weeks. Fever, cough, and severe shortness of breath are the major symptoms and diarrhea occurs in up to a quarter of cases. There are no definitive laboratory findings but lymphopenia (low lymphocyte count in blood) and increased lactate dehydrogenase (LDH) are frequent. The virus can be grown from numerous samples but most routinely from respiratory secretions and stool. Of course, it is not advisable for clinical laboratories to attempt to culture the virus unless they are equipped with special biosafety containment facilities. Although SARS-CoV is highly contagious during the disease, fortunately it is not contagious prior to symptoms or following recovery. This greatly facilitates control of spread of the disease and makes isolation of cases an effective public health measure. As is true of most severe viral respiratory diseases, treatment is supportive. Therapy with corticosteroids with or without ribavirin has been used without apparent efficacy [8].

Paramyxoviridae

The Paramyxovidae family includes a large number of enveloped, negative-sense RNA viruses, many of which cause respiratory tract infections. These include measles virus, respiratory syncytial virus (RSV), parainfluenza virus (PIV), and human metapneumovirus (HMPV). Measles virus causes a highly contagious childhood exanthem (rash syndrome) that has become rare in the United States since universal vaccination was instituted in the 1960s. Malnourished children are susceptible to serious measles pneumonia, which remains a major cause of childhood morbidity and mortality in the developing world. The focus of measles control is on maximizing the prevalence of vaccination. Measles is in fact an eradicable disease like smallpox and polio.

By the age of 2 virtually every child has been infected with RSV [15]. It is the leading cause of serious lower respiratory tract infection in infants and children [2]. Infants with lung or cardiac diseases are at increased risk for RSV infection, and bone marrow transplant recipients are susceptible to life-threatening interstitial pneumonia due to RSV [16, 17]. More recently, RSV has been recognized as an important cause of lower respiratory tract infection in the elderly [18, 19]. Infection early in life is associated with an increased likelihood of reactive airway disease later, possibly related to the predilection for RSV to induce the mechanisms involved in the development of asthma and allergy in children [20–22].

Despite significant effort, therapy of severe RSV infections is suboptimal and an effective vaccine remains elusive [15]. Ribavirin has reasonably antiviral activity against RSV and is used clinically. However, its use has declined because of side effects and difficulty of administration. There are significant ongoing efforts to develop an effective RSV vaccine and more effective antiviral drugs. Because each has its drawbacks and relative merits, it is likely that both vaccines and antivirals will have a place in the control of RSV-related disease.

Palivizumab (neutralizing monoclonal antibody) (Synagis) and respiratory syncytial virus immune globulin intravenous (RSV-IGIV) are used prophylactically to prevent severe lower respiratory tract infections caused by RSV in high-risk infants such as very young children with chronic lung disease and preterm infants [23, 24]. Recently, RSV immunoprophylaxis has been also recommended for infants and children with congenital heart disease [25]. Such immunoprophylaxis has been shown to be effective in that monthly administration of palivizumab during the RSV season results in a 45 to 55 percent decrease in the rate of hospitalization attributable to RSV. Other measures have been proposed to reduce the risk of RSV infection in high-risk infants such as avoidance of child-care centers during the RSV season [25].

Parainfluenza viruses types 1, 2, 3, and 4 (PIV-1, PIV-2, PIV-3, and PIV-4) were isolated about 50 years ago from children with lower respiratory tract infections. PIV-3 is the most prevalent, especially in very young children, but most children are infected with all three viruses by 5 years of age. Whereas RSV is overall more likely to be associated with severe lower respiratory tract infection (LRT) than PIV, the overlap of the clinical spectrum between the two viruses is large. Reinfection of children and adults with RSV and PIV are common because immunity is short-lived. Prior infection, however, may reduce the severity of subsequent infections. There is no specific antiviral, vaccine, or immunoprophylaxis for PIV.

In 2001 Osterhaus and colleagues isolated a new virus from patients with acute respiratory tract infections [26]. The virus, named human metapneumovirus (HMPV), is a paramyxovirus virus related to avian pneumoviruses. Since that time investigators throughout the world have isolated HMPV from many patients with acute respiratory tract infection. Various studies have shown that HMPV is only occasionally isolated from asymptomatic children but accounts for a significant percentage of upper respiratory tract infections. Recently, it was shown to be a significant cause of lower respiratory tract infections during the first year of life [27]. Overall, HMPV causes a spectrum of disease very similar to RSV [27].

Orthomyxoviridae

Influenza viruses are members of the large Orthomyxoviridae family of viruses. Orthomyxoviruses are enveloped, negative-sense RNA viruses with genomes that are composed of multiple RNA molecules called segments. Influenza A and B viruses cause a devastating amount of human morbidity and mortality throughout the world. Influenza A viruses are classified based on the type of hemagglutinin (H1 to H15) and neuraminidase (N1 to N9) they contain. Combinations of these two surface glycoproteins define each subtype. Outbreaks of human disease have been limited to combinations of H1, H2, or H3, together with N1 or N2. However, all subtypes have been found in viruses isolated from aquatic birds, which are the natural reservoir for influenza viruses. Hemagglutinin binds to sialic acid-containing receptors on respiratory epithelial cells. Neutralizing antibody to influenza is directed to hemagglutinin and thus it is the most critical immunogen in vaccines.

Influenza viruses, like all RNA viruses, have a penchant for rapid mutation. New strains emerge each year as a result of mutations in surface glycoproteins, a process referred to as antigenic drift. A major change in the hemagglutinin and neuraminidase combination is referred to as antigenic shift. This occurs via a reassortment of the viral genome segments when two influenza viruses infect the same host and genomic segments are exchanged. Pandemic outbreaks occur when reassortment involves human and animal influenza viruses that have not recently circulated in the population. Transmission of avian strains containing H5, H7, or H9 directly to humans followed by adaptation to human-to-human transmission may be another source of pandemic strains [27a]. Worldwide pandemics such as occurred in 1918, 1958, and 1968 have received much attention due to the large number of deaths that occurred during those years. However the number of deaths that have occurred due to influenza viruses during interpandemic years is still substantial. Each winter 10 to 20 percent of the U.S. population is infected with influenza, and there are tens of thousands of deaths and over 100,000 hospitalizations [29].

Influenza virus in healthy persons is usually a tracheobronchitis. Pneumonia and other complications are rare. Serious complications from influenza virus infection occur more frequently in pregnant women, the very young, the very old, and among persons with chronic diseases or immunodeficiency. Pneumonia, both primary viral and secondary bacterial, can lead to respiratory failure that is the most common cause of death related to influenza [30]. Myocarditis, pericarditis, and rhabdomyolysis are rare complications of influenza virus infection. The mechanism of these pathological complications is not clear but may be due to cyokines rather than a direct result of invasion of heart or muscle tissue [31].

Two types of influenza vaccines have been approved for use. The formaldehyde-inactivated intramuscular vaccine has been in use for several decades and the live, attenuated intranasal vaccine was approved in 2003. Both vaccines are trivalent in that they consist of two strains of influenza A virus and one strain of influenza B virus. The strains used in the vaccines are based on the recommendation of a Food and Drug Administration (FDA) Advisory Panel that meets each February, and its decision is based on worldwide surveillance data.

Two neuraminidase inhibitors were recently approved as antiviral therapy for influenza (see below). They are clearly effective at reducing symptoms especially if taken early in the course of the disease. They are very expensive, and their impact on management of influenza will depend on more rapid and point-of-care diagnostic methods to avoid overuse. Recently, it was shown that postexposure prophylaxis of household contacts of those with influenza reduces the secondary spread of influenza in families when the initial household case is treated [32].

Bunyaviridae

The Bunyaviridae is a large family of enveloped, negative-sense, segmented RNA viruses that contains five genera. Only one of these genera, *Hantavirus*, contains viruses that cause respiratory disease in humans. A number of hantaviruses are known to cause a recently recognized severe respiratory syndrome called Hantavirus cardiopulmonary syndrome or Hantavirus pulmonary syndrome (HPS) [33, 34]. Hantaviruses are distinct from other members of the Bunyaviridae family in that they are not transmitted by arthropod vectors but are rather transmitted directly from aerosolized excreta of rodents, which are chronically and asymptomatically infected. HPS was first recognized after an outbreak in 1993 in the Four Corners region at the intersection of New Mexico, Arizona, Colorado, and Utah. A cluster of fatal cases of pneumonia in young adults was eventually linked to a newly discovered *Hantavirus* called Sin Nombre virus (SNV), which was found to be enzootic in the deer mouse (*Peromyscus maniculatis*). Since then, retrospective studies have identified various hantaviruses as the cause of previously unrecognized outbreaks and a number of subsequent outbreaks have been described throughout the Western Hemisphere. Fortunately person-to-person transmission from HPS patients does not occur, but in a large outbreak in Argentina caused by the Andes hantavirus, there was evidence of person-to-person transmission [35]. In the United Sates there have been over 300 cases of HPS with a case fatality rate of 30 to 40. Cases occur throughout the year with most cases in spring and summer. The illness usually begins with a fever, myalgias, and cough. After several days to a week these symptoms progress to shortness of breath, a rapid onset of pulmonary edema with hypoxia, and, in some cases, shock related to poor cardiac output and arrythmias. Death can occur quickly after pulmonary symptoms develop, especially if mechanical ventilation is not instituted rapidly. Ribavirin, which has antiviral activity in cell culture and appears to demonstrate efficacy in animal models, unfortunately did not look promising in an uncontrolled clinical trial [36].

Adenoviridae

Adenoviruses are ubiquitous nonenveloped DNA viruses that cause nonserious infections in immunocompetent individuals. Transplantation recipients and other immunocompromised patients, however, are susceptible to serious adenovirus respiratory tract infections including pneumonia. Treatment options are limited to ribavirin or cidofovir, each of which is associated with significant toxicity.

Herpesviridae

The Herpesviridae family consists of many large enveloped DNA viruses that include eight distinct types of viruses that cause human disease. Despite similarity in structure and genetic organization, the herpesviruses have a wide spectrum of biological behavior and cause a wide spectrum of disease. The name (Latin, *herpes* = creep) is an anachronism related to the skin eruptions that occur with infection by several members of the family. Herpesviruses are not among the most common causes of viral respiratory tract infections. Many herpesviruses, however, can infect the respiratory tract, and in particular clinical settings several herpesviruses can cause serious pneumonia [37].

Varicella-zoster virus (VZV) causes both chicken pox (a primary infection in children) and shingles (reactivation from a latent infection) that occurs primarily after the fourth or fifth decade of life. Adults who escaped infection with VZV during childhood are at risk for a serious VZV infection as an adult. Primary infection with VZV in adults is frequently associated with a severe interstitial pneumonia that can be life-threatening. The live attenuated VZV vaccine may reduce the number of susceptible adults, but it is unclear whether the vaccine provides the life-long immunity that is associated with a natural infection. VZV, although less susceptible to acyclovir than herpes simplex 1 and 2 viruses, is inhibited by acyclovir and its derivatives.

Another serious pneumonia caused by a herpesvirus is the often fatal human cytomegalovirus (HCMV) interstitial pneumonia. HCMV generally causes an asymptomatic infection of immunocompetent individuals but is a major problem among immunocompromised patients such as those with acquired immunodeficiency syndrome (AIDS), patients undergoing cancer chemotherapy, and, most especially, organ and bone marrow transplant patients. Improved therapy for AIDS, more selective cancer therapy, and better agents for controlling graft rejection have all reduced the severity and prevalence of HCMV-associated infections, but it remains a significant clinical problem. Several different drugs are available to treat HCMV infections, but resistance to these has been an ongoing issue.

24.6 MAJOR CLINICAL SYNDROMES

Infection of the respiratory tract can be conveniently divided into upper and lower respiratory tract infections, or URI and LRI, respectively. URI and LRI can be further characterized by certain clinical syndromes that are associated with infection of particular anatomical regions of the respiratory tract. Below is a brief overview of these syndromes. Table 24.1 includes a list of most of the viruses associated with each of these syndromes. It must be borne in mind, however, that respiratory infections are often a continuum in which multiple anatomical sites are usually involved in the infection.

Common Cold

The term *cold* refers to an acute, self-limited syndrome of the upper respiratory tract especially the nasal passages, paranasal sinuses, and the pharynx. Colds are one of the most common illnesses and are caused by a variety of RNA viruses. Adults average 2 to 3 colds per year and children 6 to 8 colds per year [38]. Symptoms include rhinitis (runny nose, nasal stuffiness, sneezing, etc.) and pharyngitis (sore throat) often associated with chills and fever. Other less frequent symptoms include cough, hoarseness, headache, fatigue, and malaise. Findings on physical exam include nasal discharge and pharyngeal erythema but are generally nonspecific. Colds are generally self-limited, lasting approximately 9 to 10 days. Rhinoviruses cause 40 to 50 percent of colds, and the remaining are caused by a variety of RNA viruses including coronaviruses, enteroviruses, PIV, RSV, and influenza A and B viruses.

Transmission of common cold viruses occurs by direct inoculation of virus into the upper respiratory tract and infection of ciliated nasal epithelial cells [39]. Infiltration of polymorphonuclear leukocytes then follows probably as a result of cytokine production. The number of infected cells is not high, and there is little cellular damage that is directly due to viral replication. Most of the symptoms are attributable to the inflammatory mediators triggered by the virus. Interestingly, in contrast to allergic rhinitis, histamine does not appear

TABLE 24.1 Association of Specific Viruses with Respiratory Syndromes

Viruses	Common Cold	Pharyngitis	Tracheobronchitis	Croup[m]	Bronchiolitis[n]	Pneumonia[p]
Influenza[a]	Infrequent[j]	Common	Common	Common	Common	Common
RSV[b]	Infrequent	Infrequent	Common	Major	Common[o]	Common
PIV1-3[c]	Infrequent	Common	Common	Common	Major	Infrequent
HMPV[d]	Infrequent	Common	Common	Common	Common	Infrequent
Measles			Infrequent	Infrequent		Infrequent[q]
Rhinovirus	Major					Infrequent
Enterovirus	Common	Common			Infrequent	Infrequent
Coronavirus	Major	Infrequent			Infrequent	Infrequent
HIV[e]	Infrequent[k]					
Adenovirus		Infrequent	Infrequent	Infrequent	Infrequent	Infrequent
VZV[f]	Infrequent[l]					Infrequent
CMV[g]						Infrequent[r]
HSV[h]	Infrequent				Infrequent	Infrequent
EBV[i]	Infrequent[l]					Infrequent

[a] Influenza viruses A and B. The relative importance varies from year to year.
[b] Respiratory syncytial virus.
[c] Parainfluenza viruses types 1, 2, and 3. For purpose of this table differences among the three types are not considered.
[d] Human metapneumovirus.
[e] Human immunodeficiency virus.
[f] Varicella-zoster virus.
[g] Cytomegalovirus.
[h] Herpes simplex virus.
[i] Epstein-Barr virus.
[j] Infrequent as an overall cause, but may be a common cause in certain clinical circumstances; see specific footnotes.
[k] As a component of primary HIV infection.
[l] As a component of infectious mononucleosis.
[m] Acute laryngotracheobronchitis of children.
[n] In children.
[o] RSV is a major cause of croup in infants.
[p] In immunocompetent patients. For immunocompromised patients see specific footnotes.
[q] Measles is a common cause of pneumonia in malnourished patients in the developing world.
[r] CMV is a common cause of pneumonia in organ transplant recipients.

to play a significant role in the pathophysiology of colds [40, 41]. Complications of colds include secondary bacterial infections of the paranasal sinuses and middle ear and exacerbations of asthma and emphysema. Viruses that cause the common cold are also implicated in otitis media (inner ear inflammation) especially in children [42, 43].

Pharyngitis

Inflammation of the pharynx, or pharyngitis, is commonly referred to as a sore throat. Viruses cause about one third of cases of acute pharygitis, bacteria cause another one third, and the remaining one third are of unknown etiology. Viral pharygitis primarily occurs as part of a larger symptom complex such as the common cold, influenza, infectious mononucleosis, and primary infection with human immunodeficiency virus (HIV) [44].

Laryngitis

Acute laryngitis, or inflammation of the vocal cords, is commonly seen in primary care settings. It generally presents as hoarseness and a painful cough. As with pharyngitis, it often accompanies other URI symptoms such as cold symptoms. All the major respiratory viruses have been implicated in the etiology of acute laryngiits.

Acute Laryngotracheobronchitis (Croup)

Croup is a viral infection of the respiratory tract in young children that results in inflammation of the subglottic region. It leads to dyspnea (shortness of breath), a so-called seal's bark cough, and inspiratory sounds referred to as stridulous notes or simply as stridor. This characteristic sound is pathognomonic to the caregiver familiar with this syndrome. The origin of the term *croup* is not certain, but it is thought to have been derived from an Anglo-Saxon term that meant to cry out in a shrill voice. Although the term *croup* has been used for a number of respiratory infections associated with compromised airflow such as diphtheria, today it is used exclusively for viral laryngotracheobronchitis. Croup is relatively common in young children, with most cases occurring between 3 months and 3 years of age. PIV, most commonly PIV-3, and RSV are the most commonly implicated pathogens.

Tracheitis and Tracheobronchitis

Tracheitis rarely occurs as an isolated infection and is usually associated with URI and/or bronchitis. It is usually associated with severe pain in the chest due to upper airway damage and inflammation. Infection of the trachea and large bronchi leads to impaired mucociliary clearance mechanisms that, in turn, predisposes to bacterial secondary infection [45].

Bronchiolitis

Bronchiolitis is an acute lower respiratory tract viral infection of young children that is characterized by acute onset of wheezing associated with cough and respiratory distress. The vast majority of cases are caused by RSV or PIV. Other respiratory viruses account for the remainder of cases, although a small percentage is associated with the bacterium *Mycoplasma pneumonia*. Viral replication in the bronchial epithelial cells precipitates an inflammatory reaction with infiltration of neutrophils, basophils, and eosinophils and release of inflammatory mediators. Inflammatory changes such as edema and cellular necrosis with sloughing lead to obstruction of small airways, which accounts for the wheezing and hyperinflation of the lungs. Wheezing occurs commonly, but the inflammatory process precipitated in virus-induced bronchiolitis is not the same as occurs in allergic reaction or asthma. Distinguishing viral bronchiolitis from the first episode of asthma or from the many other causes of wheezing requires clinical acumen. Therapy is primarily supportive with oxygen, and, if necessary, mechanical ventilation for hospitalized infants. Bronchodilators and corticosteroids are used frequently, but their efficacy has not been proven. Ribavirin administered by aerosol is reserved for those at high risk for severe infection. Most normal children recover fully from bronchiolitis, but there is an increased risk of subsequent lower respiratory tract disease and reactive airway disease especially in those with bronchiolitis early in life.

Pneumonia

Pneumonia is an infection of the lung parenchyma that leads to inflammatory changes that compromise alveolar gas exchange. Obviously, pneumonia can be the most serious of the respiratory tract infections caused by viruses because it can lead to hypoxia that, if persistent, may result in widespread organ dysfunction. There is a wide spectrum of clinical presentations of pneumonia depending on the immune status of the patient and the specific viral pathogen. Serious viral pneumonia in healthy older children and adults is relatively uncommon. Virtually all the respiratory viruses have been associated with pneumonia in healthy patients, but the morbidity is generally low. The SARS outbreak was particularly unusual and worrisome because many of the fatalities occurred in healthy young adults. Another exception is the hantavirus pulmonary syndrome, which has a high mortality rate in young adults. Fortunately, the number of cases to date has been small. Influenza virus causes pneumonia especially during significant pandemics. One of the most serious consequences of influenza is secondary bacterial pneumonia. Any healthy adult who did not have chicken pox as a child, or who was not vaccinated, is at risk for varicella pneumonia, which can be serious.

The very young, the elderly, those with chronic cardiac disease or pulmonary disease, and immunocompromised patients are at higher risk for certain types of viral pneumonia that can be associated with significant morbidity and mortality. RSV bronchopneumonia can be very serious in preterm infants. In highly immunocompromised patients, such as bone marrow transplant recipients, RSV pneumonia is associated with a high rate of mortality. Human cytomegalovirus is another frequent cause of life-threatening pneumonia in organ transplant patients. Virtually the entire spectrum of human viral pathogens has been described in association with pneumonia in severely immunocompromised patients [46, 47].

The pathogenesis of viral pneumonia is complex and has differential features depending on the host and virus. Pneumonia may occur in a situation where the virus enters the body via the respiratory tract and remains localized, it may extend from the lungs into a systemic viral infection, or it may have entered the lungs as a consequence of a systemic infection. Obviously the pathogenesis will vary with each of these scenarios. In general, viral pneumonia involves a combination of viral-mediated cell death of respiratory bronchiolar epithelial cells, hypersecretion of mucus by mucus-secreting cells, and inflammatory changes such as infiltration of neutrophils and monocytes, edema, and focal hemorrhage. Alveoli can show marked changes including capillary edema, hemorrhage, and formation of hyaline membranes, which explains the poor gas exchange that develops. Viral pneumonia is a diffuse process that generally involves the entire lung and on radiographs appears as a diffuse reticular pattern. This is in contrast to bacterial pneumonia, which is often limited to individual lobes of the lung. The relative prominence of the various pathological changes varies with the virus, the host, and the severity of the pneumonia. In HPS, for example, the most dramatic process is severe alveolar capillary leak that leads to significant fluid buildup in the air sacs. This in turn causes a serious respiratory distress-like syndrome characterized by dramatic hypoxia.

24.7 DIAGNOSIS OF VIRAL RESPIRATORY INFECTION

The clinical presentation of any viral infection of the respiratory tract does not enable a specific diagnosis. There are, in general, no pathognomonic signs and symptoms that distinguish viral respiratory tract infections from bacterial infections or from other causes. The pattern of signs, symptoms, the radiographic (X-ray) pattern, and routine laboratory results can promote a high index of suspicion but do not lead to a specific etiology. Upper respiratory tract infections are characterized by rhinorrhea (runny nose), sneezing, cough, and sore throat with or without fever. Lower respiratory tract infections are most often characterized by a nonproductive (i.e., dry) cough and shortness of breath, associated with generalized symptoms of fever, malaise, and myalgias. The physical findings include particular auscultatory (stethoscope) sounds such as rales, crackles, and wheezing. Cyanosis (blue-colored skin of the digits, lips, etc.) may be observed when hypoxemia is present. For certain viruses such as measles, there may be a characteristic rash.

The history and the clinical setting are critical factors in the diagnostic evaluation. Where the patient lives, whether there have been other cases, whether there has been known contact with an individual already diagnosed, whether the infection occurs during the season when certain viral infections are prevalent, for example, all have a bearing on the likelihood of a particular diagnosis. In addition, the history combined with the clinical presentation will help direct the diagnostic effort.

Diagnostic tests for viral diseases fall in to four categories: serology, antigen detection, viral culture, and nucleic acid testing. Serology involves testing the patient for antibodies to a specific agent. It generally involves measuring acute (during symptoms) and convalescent (after resolution) antibody titers. A rise in titer (usually fourfold) is consistent with recent infection but obviously only provides a retrospective diagnosis. Recently, measurement of specific IgM antibodies has allowed the diagnosis to be made in the acute setting. IgM tests are not available for all pathogens and are of variable sensitivity and specificity. Antigen detection has become the modality of choice for rapid diagnosis of acute viral infections. The most common method used involves obtaining cells from an appropriate site such as a scraping of the posterior pharynx or a bronchoalveolar lavage specimen. Cell-associated specific viral antigens are then detected by direct immunofluoresence, which involves the use of a fluorescently labeled monoclonal antibody directed against a viral protein. Such direct fluorescent antibody (DFA) tests are available for RSV, PIV, adenovirus, and influenza and have high specificity (>90 percent) and reasonably good sensitivity (70 to 90 percent). Viral culture is the most definitive modality for making the diagnosis, but when performed by traditional methods can take 3 to 10 days. Newer methods of culture, especially when combined with DFA, can reduce the time to detect positive cultures to 1 to 3 days. Nucleic acid detection (NAD) allows for the detection of even trace amounts of viral genome in patient samples. The most commonly used methods involve amplification of the viral nucleic acid by polymerase chain reaction (PCR) either directly for DNA viruses or following reverse transcription for RNA viruses. Detection of the amplified nucleic acid can be performed by a variety of methods, but fluorescent-based real-time detection methods are supplanting older methods. The cost and complexity of these tests, however, has limited their routine use.

A critical issue for any diagnostic evaluation is having a clear idea of what tests are appropriate to order. Use of any of the specific tests requires a certain index of suspicion because ordering a panel of tests is prohibitively expensive as a routine mode of operating. Even viral culture is better performed when certain agents are considered in advance because the type of cell lines and the conditions of culturing can affect the detection of particular viruses. The ultimate goal of viral diagnostic testing is the use of multiplexing, that is, the ability to detect many agents from the same specimen using one test. Further supporting this approach is the fact that up to 15 percent of respiratory tract infections are caused by more than one agent (mixed infection). Effective antiviral therapy requires effective diagnostic modalities. Rapid, point-of-care, inexpensive, multiplexed, sensitive, and specific diagnostic tests will enable antiviral therapy to be more effective, especially for acute viral diseases such as most viral respiratory infections.

24.8 EXISTING ANTIVIRAL THERAPY

Significant progress has been made in the development of antiviral drugs. Just 30 years ago, there were no FDA-approved antiviral drugs. Today, there are over 40 such drugs [48]. Still, the applicability of these drugs continues to be limited. More than half the FDA-approved antiviral drugs, for example, are for use in treating infections caused by human immunodeficiency virus (HIV), and most of the remaining drugs are for use in treating herpesvirus infections. The first oral antiviral was a drug for herpes, acyclovir, developed by the laboratory of Gertrude B. Elion at Burroughs-Wellcome in the 1960s. Acyclovir is a nucleoside analog that fits into the binding pocket of herpes DNA polymerase and terminates the enzyme's growing nucleic acid chain. High-dose acyclovir is efficacious for VZV pneumonia. Another nucleoside analog, ganciclovir, as well as a nonnucleoside DNA polymerase inhibitor, foscarnet, are available to control HCMV pneumonia in immunocompromised patients. Their distinct toxicity and resistance profiles make them a useful pair during the management of these difficult pneumonias.

There are only a handful of small-molecule antiviral drugs available for treatment of the RNA viruses that cause the majority of respiratory tract infections. Consequently, for many RNA viruses, there are only limited, if any, therapeutic options available. Two drugs, amantadine (Symmetrel) and its derivative rimantadine (Flumadine), have been available for many years for prophylactic use in patients at high risk for influenza virus such as residents of nursing homes [49, 50]. These drugs interfere with proton transport of the viral transmembrane protein, M2, but are only effective against influenza A virus.

Recently, two drugs, oseltamivir (Tamiflu) and zanamivir (Relenza), that are inhibitors of the influenza neuraminidase enzyme were approved for treatment of influenza A and B [28, 51]. As expected, their efficacy depends on initiation of therapy as soon as possible. They have been shown to be effective when used prophylactically during an outbreak [32]. This so-called post-exposure prophylaxis may be the most effective use of these drugs. Neither the M2 or neuraminidase inhibitors have been studied in cases of severe pneumonia.

Ribavirin is the only available antiviral drug for treatment of other respiratory viruses [52–56]. It has been used in a small-particle aerosol formulation to treat severe bronchiolitis caused by RSV. The efficacy of ribavirin in RSV bronchopneumonia is controversial, and its use has been limited to treatment of high-risk infants. The increased use of prophylactic monoclonal antibody has contributed to the reduced use of ribavirin. There is limited data on the efficacy of ribavirin in other viral respiratory diseases. It may be efficacious for PIV and measles infections, but it has not shown to be of value in HPS [36]. There remains the potential for combining ribavirin with other antivirals. For example, ribavirin has limited antiviral effect on hepatitis C virus (HCV) but definitely adds to the therapeutic potency of interferon-α. Whether this approach can be successfully applied to respiratory viral infections remains to be seen. Combination of biological agents such as mAbs and small-molecule antiviral drugs is also an intriguing possibility. The concept of using neutralizing monoclonal anti-RSV antibodies to decrease the long-term morbidity associated with RSV infection in children by reducing RSV replication has also been proposed [57].

24.9 CHALLENGES OF ANTIVIRAL THERAPY FOR RESPIRATORY VIRUS INFECTIONS

There are many difficulties associated with the development of antiviral therapies. The manner in which viruses replicate and their close association with cellular processes during their life cycle makes finding agents with a selective effect on the virus difficult. In addition, there are often multiple strains, serotypes, and sequence variants of any given virus (quasi-species). This inherent variation is related to the high frequency of mutation of viruses, which leads to the rapid selection of viruses that are drug resistant.

Another difficulty is the uncertain correlation between the level of viral replication and the extent of clinical disease. Generally, it is safe to say that viral replication is necessary for most viruses to cause disease, but the amount and duration of viral replication may not correlate temporally with the degree of pathology. Therefore, a therapeutic intervention that focuses on inhibition of viral replication may be inadequate to ameliorate either symptoms or pathology unless it is instituted either before infection or within a short period after infection. As an example, despite the fact that influenza is characterized by significant virus replication in respiratory epithelial cells, there is now growing evidence that the cytokines produced early at the site of infection mediate many of the clinical and pathological manifestations [58]. These cytokines include interferon-α (IFN-α), tumor necrosis factor-α (TNF-α), interleukin-1 (IL-1), interleukin-6 (IL-6), interleukin-8 (IL-8), and various chemokines. Studies of respiratory secretions of experimentally infected humans have shown that peak cytokine levels directly correlated with virus replication and disease. However, cytokines are clearly not the only factor contributing to disease, and they are also likely to be essential for resolution of the infection [58].

Both immunologic and nonimmunologic factors have been implicated in the pathogenesis of RSV-induced disease. RSV bronchiolitis has been proposed to result from production of Th2-type cytokines, although this clearly cannot be the total explanation. Interferon-γ may contribute to RSV-induced wheezing, possibly through induction of leukotriene release. Pulmonary surfactant has recently been thought to be an important factor in RSV-mediated severe bronchiolitis. Airway damage and hyperreactivity induced during RSV infection can alter long-term lung function and may lead to chronic asthma. While the mechanisms underlying hyperreactivity of the airways are unclear, in a mouse model airway hyperreactivity response was associated with inappropriate production of cytokines, particularly IL-13 [59]. Chemokines have been also thought to be an important contributor to the symptoms that arise during RSV infection, but their role has not been fully elucidated. Mice treated with anti-RANTES (CCL5) antibody demonstrated significant decreases in airway hyperreactivity, suggesting that RANTES (CCL5) is an important mediator of the pathophysiological responses seen in RSV infection [60].

Inflammation and mucus overproduction are partially responsible for RSV-induced disease in infants. In other studies using a murine model, it was shown treatment with anti-CXCR2 antibody resulted in decreased airway hyperreactivity and decreased mucus in the bronchoalveolar lavage fluid. Studies with CXCR2 (−/−) *knockout* mice also demonstrated the important of CXC chemokines in RSV-induced mucus production and airway hyperreactivity [61]. There is also evidence that the nuclear transcription factor NF-κB controls the expression of many genes involved in the inflammatory and immunomodulatory processes that occur during RSV infection. As a possible central activator of inflammatory and innate immune responses to RSV, NF-κB may be a potential target [62]. Additional therapeutic strategies will be promoted by better understanding of the pathogenesis of RSV-induced disease and other viral respiratory tract infections [63].

24.10 STRATEGIES FOR DEVELOPMENT OF ANTIVIRAL THERAPY

There are a variety of approaches that can be applied to the development of antiviral therapies. Cytokines, cytokine inhibitors, monoclonal antibodies directed against the virus, and RNA interference (RNAi)-based agents are some of the modalities under development [64]. Small molecules that inhibit viral replication generally receive the most attention. Such inhibitors are usually focused on viral targets such as viral enzymes (e.g., RNA polymerase, protease, helicase, etc.) or other viral proteins (e.g., fusion proteins). Small molecules that target viral RNA are a promising possibility but have received limited attention [65].

The replication of viruses can be described in terms of a series of key processes that occur sequentially starting with viral attachment to the cell and ending with the exit of new progeny virus from the cell. Each of these events involves the interaction of essential viral proteins with cellular machinery, and each specific activity involved in this process is a potential target of an inhibitory molecule. The best targets are those that are the most genetically and biochemically tractable. Virally encoded proteins that have an easily measurable activity fit this criterion the best. Reverse genetics can be used to prove that the particular viral protein is required for viral replication, which validates it as a target for an effective antiviral compound. This relatively straightforward validation process contrasts sharply with other areas of drug discovery in which target validation is often one of the most challenging tasks.

The replication of all viruses depends on various viral enzymes and cofactors encoded by the viral genome. Many of these enzymes are therefore excellent targets for antiviral compounds. For RNA viruses these enzymes include RNA-dependent RNA polymerases and their cofactors that replicate the viral genome and synthesize messenger RNA (mRNA), RNA helicases that unwind duplex regions of RNA during replication, and proteases that are involved in the proteolytic processing of viral proteins to their active form. For DNA viruses DNA polymerases and DNA helicases are key targets. Inhibitors that target the virus–cellular interaction at the cellular membrane also represent an effective strategy. Such inhibitors can target either the viral attachment protein or the cellular receptor.

In addition to viral proteins, cellular proteins that are critical for the replication of particular viruses may represent valuable targets for antiviral development. Cellular enzymes that have been shown to be potential targets of antiviral compounds include inosine monophosphate dehydrogenase (IMPDH), S-adenosylhomocysteine (SAH) hydrogenase, cytidine triphosphate (CTP) synthetase, and α-glucosidase [66, 67]. It is generally thought that inhibitors of host enzymes would be associated with more toxicity than inhibitors of viral targets. This assumption is not necessarily true for several reasons. First, much drug-related toxicity is unrelated to the mechanism of therapeutic action. Second, the vast majority of medicinals target cellular enzymes and most are safe even when taken chronically. Third, many antiviral drugs that target viral enzymes are associated with significant toxicity (e.g., HIV protease inhibitors). Finally, cellular enzymes are likely to exist that are essential for viral replication, but their activity may not be essential for health of the cell because of redundancy. Such enzymes would, of course, be excellent targets for antiviral drugs.

Screening programs to identify compounds with antiviral activity involve two general methods: *targeted* screening and *unbiased* screening. In the targeted approach, a particular biochemical target is chosen, and then inhibitors of the chosen target are screened. The chosen target is typically an enzyme or a receptor that is known or thought to be essential to viral replication. In the alternative approach, unbiased screening, inhibitors of viral replication are sought without a priori concern for the target. Unbiased screening necessarily involves the use of cell culture for virus replication. This is due, in part, to the fact that cellular targets are likely targets of many antiviral agents. Although cell-based screening has been used successfully throughout the drug discovery field, it has historically been problematic when screening for antiviral compounds because it requires inoculation of an infectious virus onto the cells, which complicates the process of screening large libraries of compounds.

Partial viral replication systems have been developed to circumvent the problems associated with cell-based cultures using whole viral systems. In partial viral systems, viral genomes lacking one or more genetic elements essential for complete replication are used to accomplish viral genomic replication without producing the infectious virus. This is particularly important for viral agents such as SARS-CoV, which is classified as biohazard level 3 (BL-3) and for BL-4 agents such as Ebola virus. A screening process that utilizes these incomplete viral genomes can identify inhibitors of any biochemical pathway involved in viral genome replication, transcription, and translation. This allows for screening with respect to multiple possible targets. These targets do not have to be known, thus making the screening process unbiased. In addition, the targets are prevalidated, given that inhibition of RNA replication is directly relevant to the viral disease process. Screening with partial viral replication systems additionally is advantageous because complex viral replication pathways can be easily monitored by measuring levels of viral RNA or expression of a reporter gene (e.g., luciferase, β-galactosidase, secreted alkaline phosphatase, green fluorescent protein, etc.) present in the replicon or defective genome. The major drawback of unbiased, cell-based screening, whether it involves infectious virus or partial genomes, is that the target of an identified antiviral molecule is not immediately known. Efforts to identify the target and ultimately determine the mechanism must be undertaken to enable the process of optimizing the lead compound.

24.11 DEVELOPMENT OF ANTIVIRALS AGAINST RESPIRATORY VIRUSES

The approval of two inhibitors of the influenza virus neuraminidase (NA), oseltamivir (Tamiflu) and zanamivir (Relenza), has opened the door for the development of small-molecule inhibitors of other respiratory viruses. Unfortunately, the door remains only partially open and progress has been slow for a variety of reasons. Another potent NA inhibitor being developed by BioCryst (Birmingham, AL) failed to demonstrate significant efficacy in phase III trials. Many other NA inhibitors have been described that show potent in vitro activity against influenza virus. Whether they will be developed is unclear. In addition to neuraminidase inhibitors, compounds are under development that target other influenza virus replication processes. Interaction of the influenza virus with the cell membrane is an important process during replication of most enveloped viruses. Ongoing work has identified a number of compounds that inhibit interaction of the viral envelop with the cellular membrane. A polyoxometalate (PM523) is a potent inhibitor of influenza, respiratory syncytial virus, and measles virus and acts by inhibiting fusion between the viral envelope and the cellular membrane. Resistance to PM523 mapped to sequences that encoded amino acids in hemagglutinin (HA) that are in the interface of the trimers of HA. Another potential target of anti-influenza virus is the mRNA capping process. Cap formation of influenza mRNA utilizes the 5'-mGpppXm of host mRNA. Inhibitors of cap formation have been described that act on the PB2 viral protein, which is a component of the viral RNA polymerase complex. A metabolic product of ribavirn, 1,2,4-triazole carboxamide (T-CONH2) is inhibitory for influenza replication in vitro and in vivo [68]. Endonuclease inhibitors also have been described.

Pleconaril [Picovir (ViroPharma, Exton, PA)], which binds to the capsid of many picornaviruses including rhinoviruses, is an inhibitor of virus uncoating [69, 70]. Clinical trials showed efficacy against the common cold, but unfortunately it was not approved by the FDA, which sought additional studies on drug interactions [71]. Cell culture-based transdominant genetic techniques were used to discover a peptide inhibitor of human rhinovirus that acts on intercellular adhesion molecule 1 (ICAM-1) [72, 73]. Inhibitors of other rhinovirus targets such as the 3C (e.g., AG7088) and 2A proteases and the RNA polymerase are also under development [74–77].

The increasing recognition of the significant disease burden caused by RSV has stimulated activity in anti-RSV discovery. At one time as many as 20 anti-RSV R&D programs were ongoing in the pharmaceutical/biotechnology industry. Unfortunately, this activity has not borne fruit and has not been sustained. Drug discovery efforts against RSV have primarily focused on fusion inhibitors that target the RSV F protein. One such fusion inhibitor (V-14637; ViroPharma) showed potent activity in preclinical studies and underwent early clinical trials in an inhalation formulation, but its development was suspended by the company. Other fusion inhibitors have been described that have potent anti-RSV activity in cell culture and in animal models. Unfortunately, clinical development is not being pursued. Recently, representatives from Arrow Therapeutics (London, GB) have described an inhibitor (A60444) that apparently targets the nucleocapsid (N) protein. Its biochemical mechanism of action has not been described. It is undergoing phase I clinical trials.

In addition to small molecules other approaches to antiviral therapy are being studied. These include antisense oligonucleotides and RNA interference [64, 78–80]. Other investigators have searched for RSV adhesion receptors on mammalian cells as possible antiviral targets. Soluble human lectins, complex polysaccharides, and a low-molecular-weight selectin antagonist, TBC1269, were used to characterize and isolate the RSV receptor on a human epithelial cell line (Hep2 cells). Selectin antagonists, fucoidan and TBC 1269, inhibit RSV infection both in vitro and in a mouse model of infection. Annexin II has been proposed as a potential RSV receptor on Hep2 cells. Recombinant annexin II binds to RSV G protein, heparin, and plasminogen, and the binding is inhibited by a selectin antagonist, TBC1269. These workers have suggested that inhibitors of annexin II could have potential in treating RSV infection [81].

There has been significant effort to develop an effective therapy for SARS-CoV. In a cynomolgus macaque model of SARS, pegylated interferon showed some promise as a potential treatment [82]. Calpain and β-D-N^4-hydroxycytodine have been shown to be inhibitors of SARS-CoV, but their clinical potential is unclear [83]. Rational design based on X-ray crystal structure has been used to design drugs for SARS-CoV 3CL protease.

24.12 ANIMAL MODELS

Animal models are an important component of antiviral development because they provide an indication of in vivo efficacy prior to initiating human clinical trials. However, there are few animal models for human respiratory viruses and most of those that exist are suboptimal. Mouse models have many advantages for testing antivirals. The available genomic information, transgenic strains, immunodeficient strains, immunologic reagents, and the like make murine models of infection highly tractable. Unfortunately, mice are not fully permissive for many human respiratory viruses, and infection of mice with respiratory viruses does not always mimic the relevant human disease [84].

Significant effort has been devoted to the development of animal models of RSV infection using mice, cotton rats, guinea pigs, ferrets, hamsters, calves, and nonhuman primates [85, 86]. These models have been used to test vaccines and to determine the mechanism of vaccine-enhanced disease seen with a formalin-inactivated RSV vaccine [87]. They have also provided a means for testing efficacy and safety of prophylactic and therapeutic immunotherapy [57]. The value of these animal models for testing small-molecule therapeutic drugs, however, is less clear.

The cotton rat (*Sigmodon hispidus*) has been promoted as a useful model for several respiratory viruses including influenza, parainfluenza virus, and RSV [88–90]. After inoculation of the nares with RSV, animals develop pulmonary pathology. Unfortunately, the cotton rat is only "semipermissive" for RSV in that the recoverable virus never exceeds the innoculum. Nonhuman primates such as African green monkeys and Rhesus macaques have been used for studies of vaccines, monoclonal antibodies, and antivirals. The high cost and increasing lack of availability are major drawbacks to the use of primate models. Animal models must be used with the realization that RSV disease in humans has many pathophysiological features. The clinical manifestations depend upon age, genetic makeup, immunologic status, and underlying disease, and there is no single animal model that duplicates all the various forms of RSV disease [86].

A small-animal model is important for the development of new therapies for rhinoviruses. However, existing mouse models of infection are difficult to work with, and until recently mouse cell lines were thought to be generally nonpermissive for HRV replication in vitro. Recently, a virus of the minor-receptor group, HRV1B, was shown to be able to infect and replicate in a mouse respiratory epithelial cell line (LA-4) more efficiently than in a mouse fibroblast cell line (L) [91]. In contrast the major-receptor group virus, HRV16, requires human intercellular adhesion molecule-1 (ICAM-1) for cell entry and therefore cannot infect LA-4 cells. However, transfection of in-vitro-transcribed HRV16 RNA resulted in the replication of viral RNA and production of infectious virus. Furthermore, these workers showed that expression of a chimeric ICAM-1 molecule, comprising mouse ICAM-1 with extracellular domains 1 and 2 replaced by the equivalent human domains, rendered the otherwise nonpermissive mouse respiratory epithelial cell line susceptible to entry and efficient replication of HRV16. These observations suggest that the development of mouse models of respiratory tract infection by major as well as minor group HRV are possible [91].

A significant effort has been applied to the development of a SARS-CoV animal model. A cynomolgus macaque model of SARS-CoV has been used by one group to test antiviral agents [82]. Murine models have also been studied, but their relevance to human infection is uncertain.

24.13 SUMMARY

There are significant challenges associated with the development of antiviral therapy for respiratory virus infections. These challenges fall into four broad categories: (1) virological, (2) pathophysiological, (3) clinical, and (4) economical. Viruses are obligate intracellular pathogens, and their intimate association with cells during their life cycles makes them very difficult to inhibit compared with extracellular or free-living pathogens such as most bacteria. Another problem is the high rate of mutation of viruses, especially RNA viruses. Thus they have a high predilection for developing resistance to antiviral agents.

Another problematic issue is that the majority of the symptoms associated with viral respiratory tract infections are triggered by the virus rather than directly mediated by viral replication. After initial infection, respiratory viruses replicate rapidly, and the body responds with a strong innate and acquired immune response that triggers symptoms that can last weeks. Therefore, in order to have an impact on disease symptoms and pathology, therapy, even with an extremely potent inhibitor of viral replication, needs to be initiated within 24 to 48h of onset of symptoms. This is obviously a very narrow treatment window. From the clinical standpoint this issue is problematic as well. Symptoms are not seen during the incubation period, and there is often a delay in clinical suspicion.

This situation results ultimately in a delay in definitive diagnosis. Consideration of therapy of viral respiratory tract infections, therefore, is not likely to be initiated in a timely manner. Finally, for a variety of reasons antiviral therapies are generally expensive, and, in the existing medical economic structure, an expensive antiviral therapy is not likely to be used presumptively prior to when a definitive diagnosis has been made. All of these issues make it difficult to start therapy early at the time it is likely to be the most effective.

In summary, there has been significant progress made in the development of antiviral drugs during the past two decades. Most of the progress, however, has been made in the treatment of chronic viral diseases such as HIV-associated AIDS. Treatment of acute viral disease, including most respiratory viral infections, is fraught with many difficulties. Nevertheless, despite these difficulties, we are poised to enter a new era in antiviral therapy. Respiratory viruses cause an immense amount of human morbidity and mortality, and antiviral therapy is an important weapon to control these agents. The development of antiviral drugs to treat respiratory virus infections will be one of the great challenges of the twenty-first century.

REFERENCES

1. Mackie, P. L. (2003). The classification of viruses infecting the respiratory tract. *Paediatr. Respir. Rev.*, *4*, 84–90.
2. Centers for Disease Control and Prevention (CDC) (1999). Update: respiratory syncytial virus activity–United States, 1997–98 season. *JAMA*, *281*, 127.
3. Arruda, E., Pitkaranta, A., Witek, T. J., Jr., Doyle, C. A., Hayden, F. G. (1997). Frequency and natural history of rhinovirus infections in adults during autumn. *J. Clin. Microbiol.*, *35*, 2864–2868.
4. Department of Health and Human Services (DHHS) (1992). *Current Estimates from the National Health Interview Survey, 1992*. DHHS, Hyattsville, MD, pp. 94–1517.
5. Monto, A. S., Sullivan, K. M. (1993). Acute respiratory illness in the community. Frequency of illness and the agents involved. *Epidemiol. Infect.*, *110*, 145–160.
6. Gerberding, J. L., Morgan, J. G., Shepard, J. A., Kradin, R. L. (2004). Case records of the Massachusetts General Hospital. Weekly clinicopathological exercises. Case 9-2004. An 18-year-old man with respiratory symptoms and shock. *N. Engl. J. Med.*, *350*, 1236–1247.
7. Clements, C. J., Strassburg, M., Cutts, F. T., Torel, C. (1992). The epidemiology of measles. *World Health Stat. Q.*, *45*, 285–291.
8. Wenzel, R. P., Edmond, M. B. (2003). Managing SARS amidst uncertainty. *N. Engl. J. Med.*, *348*, 1947–1948.
9. Richman, R. D., Whitely, R. J., Hayden, F. G. (Eds.) (2002). *Clinical Virology*. ASM Press, Washington, DC.
10. Flint, S. J., Enquist, L. W., Racaniello, V. R., Skalka, A. M. (2003). *Principles of Virology*. ASM Press, Washington, DC.
11. Harris, J. M. 2nd, Gwaltney, J. M., Jr. (1996). Incubation periods of experimental rhinovirus infection and illness. *Clin. Infect. Dis.*, *23*, 1287–1290.
12. Gwaltney, J. M., Jr., Hendley, J. O., Patrie, J. T. (2003). Symptom severity patterns in experimental common colds and their usefulness in timing onset of illness in natural colds. *Clin. Infect. Dis.*, *36*, 714–723.

13. Van Der Hoek, L., Pyrc, K., Jebbink, M. F., Vermeulen-Oost, W., Berkhout, R. J., Wolthers, K. C., Wertheim-Van Dillen, P. M., Kaandorp, J., Spaargaren, J., Berkhout, B. (2004). Identification of a new human coronavirus. *Nat. Med.*, *10*, 368–373.
14. Vijayanand, P., Wilkins, E., Woodhead, M. (2004). Severe acute respiratory syndrome (SARS): A review. *Clin. Med.*, *4*, 152–160.
15. Hall, C. B. (1999). Respiratory syncytial virus: A continuing culprit and conundrum. *J. Pediatr.*, *135*, 2–7.
16. Hertz, M. I., Englund, J. A., Snover, D., Bitterman, P. B., McGlave, P. B. (1989). Respiratory syncytial virus-induced acute lung injury in adult patients with bone marrow transplants: A clinical approach and review of the literature. *Medicine (Baltimore)*, *68*, 269–281.
17. Krinzman, S., Basgoz, N., Kradin, R., Shepard, J. A., Flieder, D. B., Wright, C. D., Wain, J. C., Ginns, L. C. (1998). Respiratory syncytial virus-associated infections in adult recipients of solid organ transplants. *J. Heart Lung Transplant*, *17*, 202–210.
18. Falsey, A. R. (1998). Respiratory syncytial virus infection in older persons. *Vaccine*, *16*, 1775–1778.
19. Han, L. L., Alexander, J. P., Anderson, L. J. (1999). Respiratory syncytial virus pneumonia among the elderly: An assessment of disease burden. *J. Infect. Dis.*, *179*, 25–30.
20. Sigurs, N., Bjarnason, R., Sigurbergsson, F., Kjellman, B. (2000). Respiratory syncytial virus bronchiolitis in infancy is an important risk factor for asthma and allergy at age 7. *Am. J. Respir. Crit. Care Med.*, *161*, 1501–1507.
21. Kneyber, M. C. J., Steyerberg, E. W., de Groot, R., Moll, H. A. (2000). Long-term effects of respiratory syncytial virus (RSV) bronchiolitis in infants and young children: A quantitative review. *Acta Paediatr.*, *89*, 654–660.
22. Papadopoulos, N. G., Psarras, S., Manoussakis, E., Saxoni-Papageorgiou, P. (2003). The role of respiratory viruses in the origin and exacerbations of asthma. *Curr. Opin. Allergy Clin. Immunol.*, *3*, 39–44.
23. Atkins, J. T., Karimi, P., Morris, B. H., McDavid, G., Shim, S. (2000). Prophylaxis for respiratory syncytial virus with respiratory syncytial virus-immunoglobulin intravenous among preterm infants of thirty-two weeks gestation and less: Reduction in incidence, severity of illness and cost. *Pediatr. Infect. Dis. J.*, *19*, 138–143.
24. Carter, B. S. (2000). Palivizumab (synagis): Counting "costs" and values. *Pediatrics*, *106*, 1168–1169.
25. Meissner, H. C., Long, S. S. (2003). Revised indications for the use of palivizumab and respiratory syncytial virus immune globulin intravenous for the prevention of respiratory syncytial virus infections. *Pediatrics*, *112*, 1447–1452.
26. van den Hoogen, B. G., de Jong, J. C., Groen, J., Kuiken, T., de Groot, R., Fouchier, R. A., Osterhaus, A. D. (2001). A newly discovered human pneumovirus isolated from young children with respiratory tract disease. *Nat. Med.*, *7*, 719–724.
27. Williams, J. V., Harris, P. A., Tollefson, S. J., Halburnt-Rush, L. L., Pingsterhaus, J. M., Edwards, K. M., Wright, P. F., Crowe, J. E., Jr. (2004). Human metapneumovirus and lower respiratory tract disease in otherwise healthy infants and children. *N. Engl. J. Med.*, *350*, 443–450.

27a. Nicholson, K. G., Wood, J. M., Zambon, M. (2003). Influenza. *Lancet, 362,* 1733–1745.

28. Nicholson, K. G., Aoki, F. Y., Osterhaus, A. D., Trottier, S., Carewicz, O., Mercier, C. H., Rode, A., Kinnersley, N., Ward, P. (2000). Efficacy and safety of oseltamivir in treatment of acute influenza: A randomised controlled trial. Neuraminidase Inhibitor Flu Treatment Investigator Group. *Lancet, 355,* 1845–1850.

29. Thompson, W. W., Shay, D. K., Weintraub, E., Brammer, L., Cox, N., Anderson, L. J., Fukuda, K. (2003). Mortality associated with influenza and respiratory syncytial virus in the United States. *JAMA, 289,* 179–186.

30. Purcell, K., Fergie, J. (2004). Concurrent serious bacterial infections in 912 infants and children hospitalized for treatment of respiratory syncytial virus lower respiratory tract infection. *Pediatr. Infect. Dis. J., 23,* 267–269.

31. Hayden, F. G., Fritz, R., Lobo, M. C., Alvord, W., Strober, W., Straus, S. E. (1998). Local and systemic cytokine responses during experimental human influenza A virus infection. Relation to symptom formation and host defense. *J. Clin. Invest., 101,* 643–649.

32. Hayden, F. G., Belshe, R., Villanueva, C., Lanno, R., Hughes, C., Small, I., Dutkowski, R., Ward, P., Carr, J. (2004). Management of influenza in households: A prospective, randomized comparison of oseltamivir treatment with or without postexposure prophylaxis. *J. Infect. Dis., 189,* 440–449.

33. Prochoda, K., Mostow, S. R., Greenberg, K. (1993). Hantavirus-associated acute respiratory failure. *N. Engl. J. Med., 329,* 1744.

34. Fabbri, M., Maslow, M. J. (2001). Hantavirus pulmonary syndrome in the United States. *Curr. Infect. Dis. Rep., 3,* 258–265.

35. Khan, A. S., Young, J. C. (2001). Hantavirus pulmonary syndrome: At the crossroads. *Curr. Opin. Infect. Dis., 14,* 205–209.

36. Chapman, L. E., Mertz, G. J., Peters, C. J., Jolson, H. M., Khan, A. S., Ksiazek, T. G., Koster, F. T., Baum, K. F., Rollin, P. E., Pavia, A. T., Holman, R. C., Christenson, J. C., Rubin, P. J., Behrman, R. E., Bell, L. J., Simpson, G. L., Sadek, R. F. (1999). Intravenous ribavirin for hantavirus pulmonary syndrome: Safety and tolerance during 1 year of open-label experience. Ribavirin Study Group. *Antivir. Ther., 4,* 211–219.

37. Greenberg, S. B. (1994). Respiratory herpesvirus infections. An overview. *Chest, 106,* 1S–2S.

38. Gwaltney, J. M., Jr., Hendley, J. O., Simon, G., Jordan, W. S., Jr. (1966). Rhinovirus infections in an industrial population. I. The occurrence of illness. *N. Engl. J. Med., 275,* 1261–1268.

39. Arruda, E., Boyle, T. R., Winther, B., Pevear, D. C., Gwaltney, J. M., Jr., Hayden, F. G. (1995). Localization of human rhinovirus replication in the upper respiratory tract by in situ hybridization. *J. Infect. Dis., 171,* 1329–1333.

40. Naclerio, R. M., Proud, D., Kagey-Sobotka, A., Lichtenstein, L. M., Hendley, J. O., Gwaltney, J. M., Jr. (1988). Is histamine responsible for the symptoms of rhinovirus colds? A look at the inflammatory mediators following infection. *Pediatr. Infect. Dis. J., 7,* 218–222.

41. Naclerio, R. M., Proud, D., Lichtenstein, L. M., Kagey-Sobotka, A., Hendley, J. O., Sorrentino, J., Gwaltney, J. M. (1988). Kinins are generated during experimental rhinovirus colds. *J. Infect. Dis.*, *157*, 133–142.

42. Pitkaranta, A., Virolainen, A., Jero, J., Arruda, E., Hayden, F. G. (1998). Detection of rhinovirus, respiratory syncytial virus, and coronavirus infections in acute otitis media by reverse transcriptase polymerase chain reaction. *Pediatrics*, *102*, 291–295.

43. Heikkinen, T., Chonmaitree, T. (2003). Importance of respiratory viruses in acute otitis media. *Clin. Microbiol. Rev.*, *16*, 230–241.

44. Valle, S. L. (1987). Febrile pharyngitis as the primary sign of HIV infection in a cluster of cases linked by sexual contact. *Scand. J. Infect. Dis.*, *19*, 13–17.

45. Nugent, K. M., Pesanti, E. L. (1983). Tracheal function during influenza infections. *Infect. Immun.*, *42*, 1102–1108.

46. Ison, M. G., Hayden, F. G. (2002). Viral infections in immunocompromised patients: What's new with respiratory viruses? *Curr. Opin. Infect. Dis.*, *15*, 355–367.

47. Garbino, J., Gerbase, M. W., Wunderli, W., Kolarova, L., Nicod, L. P., Rochat, T., Kaiser, L. (2004). Respiratory viruses and severe lower respiratory tract complications in hospitalized patients. *Chest*, *125*, 1033–1039.

48. De Clercq, E. (2001). Antiviral drugs: Current state of the art. *J. Clin. Virol.*, *22*, 73–89.

49. Nicholson, K. G. (1996). Use of antivirals in influenza in the elderly: Prophylaxis and therapy. *Gerontology*, *42*, 280–289.

50. Nicholson, K. G., Wiselka, M. J. (1991). Amantadine for influenza A. *BMJ*, *302*, 425–426.

51. Hayden, F. G., Osterhaus, A. D., Treanor, J. J., Fleming, D. M., Aoki, F. Y., Nicholson, K. G., Bohnen, A. M., Hirst, H. M., Keene, O., Wightman, K. (1997). Efficacy and safety of the neuraminidase inhibitor zanamivir in the treatment of influenzavirus infections. GG167 Influenza Study Group. *N. Engl. J. Med.*, *337*, 874–880.

52. Randolph, A. G., Wang, E. E. (1996). Ribavirin for respiratory syncytial virus lower respiratory tract infection. A systematic overview. *Arch. Pediatr. Adolesc. Med.*, *150*, 942–947.

53. Randolph, A. G., Wang, E. E. (2000). Ribavirin for respiratory syncytial virus infection of the lower respiratory tract. *Cochrane Database Syst. Rev.*, CD000181.

54. Sidwell, R. W., Robins, R. K., Hillyard, I. W. (1979). Ribavirin: An antiviral agent. *Pharmacol. Ther.*, *6*, 123–146.

55. Snell, N. J. (2001). Ribavirin—Current status of a broad spectrum antiviral agent. *Expert Opin. Pharmacother.*, *2*, 1317–1324.

56. Tam, R. C., Lau, J. Y., Hong, Z. (2001). Mechanisms of action of ribavirin in antiviral therapies. *Antivir. Chem. Chemother.*, *12*, 261–272.

57. Mejias, A., Chavez-Bueno, S., Rios, A. M., Saavedra-Lozano, J., Fonseca Aten, M., Hatfield, J., Kapur, P., Gomez, A. M., Jafri, H. S., Ramilo, O. (2004). Anti-respiratory syncytial virus (RSV) neutralizing antibody decreases lung inflammation, airway obstruction, and airway hyperresponsiveness in a murine RSV model. *Antimicrob. Agents Chemother.*, *48*, 1811–1822.

58. Van Reeth, K. (2000). Cytokines in the pathogenesis of influenza. *Vet. Microbiol.*, 74, 109–116.
59. Tekkanat, K. K., Maassab, H. F., Cho, D. S., Lai, J. J., John, A., Berlin, A., Kaplan, M. H., Lukacs, N. W. (2001). IL-13-induced airway hyperreactivity during respiratory syncytial virus infection is STAT6 dependent. *J. Immunol.*, 166, 3542–3548.
60. Tekkanat, K. K., Maassab, H., Miller, A., Berlin, A. A., Kunkel, S. L., Lukacs, N. W. (2002). RANTES (CCL5) production during primary respiratory syncytial virus infection exacerbates airway disease. *Eur. J. Immunol.*, 32, 3276–3284.
61. Miller, A. L., Strieter, R. M., Gruber, A. D., Ho, S. B., Lukacs, N. W. (2003). CXCR2 regulates respiratory syncytial virus-induced airway hyperreactivity and mucus overproduction. *J. Immunol.*, 170, 3348–3356.
62. Haeberle, H. A., Takizawa, R., Casola, A., Brasier, A. R., Dieterich, H. J., Van Rooijen, N., Gatalica, Z., Garofalo, R. P. (2002). Respiratory syncytial virus-induced activation of nuclear factor-kappaB in the lung involves alveolar macrophages and toll-like receptor 4-dependent pathways. *J. Infect. Dis.*, 186, 1199–1206.
63. van Schaik, S. M., Welliver, R. C., Kimpen, J. L. (2000). Novel pathways in the pathogenesis of respiratory syncytial virus disease. *Pediatr. Pulmonol.*, 30, 131–138.
64. Ge, Q., McManus, M. T., Nguyen, T., Shen, C. H., Sharp, P. A., Eisen, H. N., Chen, J. (2003). RNA interference of influenza virus production by directly targeting mRNA for degradation and indirectly inhibiting all viral RNA transcription. *Proc. Natl. Acad. Sci. USA*, 100, 2718–2723.
65. Hwang, S., Tamilarasu, N., Kibler, K., Cao, H., Ali, A., Ping, Y. H., Jeang, K. T., Rana, T. M. (2003). Discovery of a small molecule Tat-trans-activation-responsive RNA antagonist that potently inhibits human immunodeficiency virus-1 replication. *J. Biol. Chem.*, 278, 39092–39103.
66. De Clercq, E. (2002). Strategies in the design of antiviral drugs. *Nat. Rev., Drug Discov.*, 1, 13–25.
67. Leyssen, P., Charlier, N., Paeshuyse, J., De Clercq, E., Neyts, J. (2003). Prospects for antiviral therapy. *Adv. Virus Res.*, 61, 511–553.
68. Shigeta, S. (2001). Targets of anti-influenza chemotherapy other than neuraminidase and proton pump. *Antivir. Chem. Chemother.*, 12 (Suppl. 1), 179–188.
69. Rogers, J. M., Diana, G. D., McKinlay, M. A. (1999). Pleconaril. A broad spectrum antipicornaviral agent. *Adv. Exp. Med. Biol.*, 458, 69–76.
70. Romero, J. R. (2001). Pleconaril: A novel antipicornaviral drug. *Expert Opin. Investig. Drugs*, 10, 369–379.
71. Hayden, F. G., Herrington, D. T., Coats, T. L., Kim, K., Cooper, E. C., Villano, S. A., Liu, S., Hudson, S., Pevear, D. C., Collett, M., McKinlay, M. (2003). Efficacy and safety of oral pleconaril for treatment of colds due to picornaviruses in adults: Results of 2 double-blind, randomized, placebo-controlled trials. *Clin. Infect. Dis.*, 36, 1523–1532.
72. Poritz, M. A., Malmstrom, S., Schmitt, A., Kim, M. K., Zharkikh, L., Kamb, A., Teng, D. H. (2003). Isolation of a peptide inhibitor of human rhinovirus. *Virology*, 313, 170–183.

73. Turner, R. B., Wecker, M. T., Pohl, G., Witek, T. J., McNally, E., St. George, R., Winther, B., Hayden, F. G. (1999). Efficacy of tremacamra, a soluble intercellular adhesion molecule 1, for experimental rhinovirus infection: A randomized clinical trial. *JAMA*, *281*, 1797–1804.

74. Reich, S. H., Johnson, T., Wallace, M. B., Kephart, S. E., Fuhrman, S. A., Worland, S. T., Matthews, D. A., Hendrickson, T. F., Chan, F., Meador, J., 3rd, Ferre, R. A., Brown, E. L., DeLisle, D. M., Patick, A. K., Binford, S. L., Ford, C. E. (2000). Substituted benzamide inhibitors of human rhinovirus 3C protease: Structure-based design, synthesis, and biological evaluation. *J. Med. Chem.*, *43*, 1670–1683.

75. Patick, A. K., Binford, S. L., Brothers, M. A., Jackson, R. L., Ford, C. E., Diem, M. D., Maldonado, F., Dragovich, P. S., Zhou, R., Prins, T. J., Fuhrman, S. A., Meador, J. W., Zalman, L. S., Matthews, D. A., Worland, S. T. (1999). In vitro antiviral activity of AG7088, a potent inhibitor of human rhinovirus 3C protease. *Antimicrob. Agents Chemother.*, *43*, 2444–2450.

76. Murray, M. A., Janc, J. W., Venkatraman, S., Babe, L. M. (2001). Peptidyl diazomethyl ketones inhibit the human rhinovirus 3C protease: Effect on virus yield by partial block of P3 polyprotein processing. *Antivir. Chem. Chemother.*, *12*, 273–281.

77. Wang, Q. M., Johnson, R. B., Jungheim, L. N., Cohen, J. D., Villarreal, E. C. (1998). Dual inhibition of human rhinovirus 2A and 3C proteases by homophthalimides. *Antimicrob. Agents Chemother.*, *42*, 916–920.

78. Player, M. R., Barnard, D. L., Torrence, P. F. (1998). Potent inhibition of respiratory syncytial virus replication using a 2-5A-antisense chimera targeted to signals within the virus genomic RNA. *Proc. Natl. Acad. Sci. USA*, *95*, 8874–8879.

79. Torrence, P. F., Powell, L. D. (2002). The quest for an efficacious antiviral for respiratory syncytial virus. *Antivir. Chem. Chemother.*, *13*, 325–344.

80. Joost Haasnoot, P. C., Cupac, D., Berkhout, B. (2003). Inhibition of virus replication by RNA interference. *J. Biomed. Sci.*, *10*, 607–616.

81. Malhotra, R., Ward, M., Bright, H., Priest, R., Foster, M. R., Hurle, M., Blair, E., Bird, M. (2003). Isolation and characterisation of potential respiratory syncytial virus receptor(s) on epithelial cells. *Microbes Infect.*, *5*, 123–133.

82. Haagmans, B. L., Kuiken, T., Martina, B. E., Fouchier, R. A., Rimmelzwaan, G. F., Van Amerongen, G., Van Riel, D., De Jong, T., Itamura, S., Chan, K. H., Tashiro, M., Osterhaus, A. D. (2004). Pegylated interferon-alpha protects type 1 pneumocytes against SARS coronavirus infection in macaques. *Natl. Med.*, *10*, 290–293.

83. Barnard, D. L., Hubbard, V. D., Burton, J., Smee, D. F., Morrey, J. D., Otto, M. J., Sidwell, R. W. (2004). Inhibition of severe acute respiratory syndrome-associated coronavirus (SARSCoV) by calpain inhibitors and beta-D-N4-hydroxycytidine. *Antivir. Chem. Chemother.*, *15*, 15–22.

84. Cook, P. M., Eglin, R. P., Easton, A. J. (1998). Pathogenesis of pneumovirus infections in mice: Detection of pneumonia virus of mice and human respiratory syncytial virus mRNA in lungs of infected mice by in situ hybridization. *J. Gen. Virol.*, *79*, 2411–2417.

85. Graham, B. S., Perkins, M. D., Wright, P. F., Karzon, D. T. (1988). Primary respiratory syncytial virus infection in mice. *J. Med. Virol.*, *26*, 153–162.

86. Byrd, L. G., Prince, G. A. (1997). Animal models of respiratory syncytial virus infection. *Clin. Infect. Dis.*, *25*, 1363–1368.
87. Peebles, R. S., Jr., Sheller, J. R., Collins, R. D., Jarzecka, K., Mitchell, D. B., Graham, B. S. (2000). Respiratory syncytial virus (RSV)-induced airway hyperresponsiveness in allergically sensitized mice is inhibited by live RSV and exacerbated by formalin-inactivated RSV. *J. Infect. Dis.*, *182*, 671–677.
88. Ottolini, M. G., Porter, D. D., Hemming, V. G., Hensen, S. A., Sami, I. R., Prince, G. A. (1996). Semi-permissive replication and functional aspects of the immune response in a cotton rat model of human parainfluenza virus type 3 infection. *J. Gen. Virol.*, *77*, 1739–1743.
89. Blanco, J. C., Pletneva, L., Boukhvalova, M., Richardson, J. Y., Harris, K. A., Prince, G. A. (2004). The cotton rat: An underutilized animal model for human infectious diseases can now be exploited using specific reagents to cytokines, chemokines, and interferons. *J. Interferon Cytokine Res.*, *24*, 21–28.
90. Ottolini, M., Blanco, J., Porter, D., Peterson, L., Curtis, S., Prince, G. (2003). Combination anti-inflammatory and antiviral therapy of influenza in a cotton rat model. *Pediatr. Pulmonol.*, *36*, 290–294.
91. Tuthill, T. J., Papadopoulos, N. G., Jourdan, P., Challinor, L. J., Sharp, N. A., Plumpton, C., Shah, K., Barnard, S., Dash, L., Burnet, J., Killington, R. A., Rowlands, D. J., Clarke, N. J., Blair, E. D., Johnston, S. L. (2003). Mouse respiratory epithelial cells support efficient replication of human rhinovirus. *J. Gen. Virol.*, *84*, 2829–2836.

25

STRATEGIES IN THE DESIGN OF ANTIVIRAL DRUGS

Erik De Clercq and Piet Herdewijn
Rega Institute for Medical Research, Katholieke Universiteit Leuven
Leuven, Belgium

25.1	INTRODUCTION	1135
25.2	VIRUS ADSORPTION INIBITORS	1140
25.3	VIRUS–CELL FUSION INHIBITORS	1141
25.4	INHIBITORS OF VIRAL DNA OR RNA SYNTHESIS	1147
	Viral DNA Polymerase Inhibitors	1147
	Reverse Transcriptase Inhibitors	1155
	Acyclic Nucleoside Phosphonates	1158
	Inhibitors of Processes Associated with Viral RNA Synthesis	1161
25.5	VIRAL PROTEASE INHIBITORS	1163
25.6	VIRAL NEURAMINIDASE INHIBITORS	1169
25.7	IMP DEHYDROGENASE INHIBITORS	1171
25.8	*S*-ADENOSYLHOMOCYSTEINE HYDROLASE INHIBITORS	1173
25.9	CONCLUSION	1177
	References	1177

25.1 INTRODUCTION

Effective vaccines have led, or may lead, to the eradication of important viral pathogens such as smallpox, polio, measles, mumps, and rubella. But other viral diseases, particularly human immunodeficiency virus (HIV) and hepatitis C

Drug Discovery Handbook, by Shayne Cox Gad
Copyright © 2005 by John Wiley & Sons, Inc.

virus (HCV), have so far proven intractable to the vaccine approach. The need for effective antiviral drugs is further emphasized by the lack of vaccines for most respiratory tract virus infections [adeno, rhino, parainfluenza, and respiratory syncytial virus (RSV)], the widely occurring human papilloma viruses (HPV), herpesviruses [herpes simplex virus types 1 and 2 (HSV-1, -2), varicella-zoster virus (VZV), Epstein–Barr virus (EBV), cytomegalovirus (CMV), and human herpesviruses types 6, 7, and 8 (HHV-6, -7, -8)], and the vast array of hemorrhagic fever viruses. Although vaccines have been developed for hepatitis B virus (HBV) and influenza (type A and B), they have not been able to eliminate the need for effective chemotherapeutic agents.

Many new antiviral drugs have been licensed in recent years (Table 25.1), most of which are used for the treatment of HIV infections. Of the current armamentarium [1] of over 30 drugs, 19 are anti-HIV, 5 are anti-CMV, 5 are anti-HSV and anti-VZV, 1 is anti-RSV, 3 are anti-hepatitis B, 2 are anti-hepatitis C, and 2 are anti-influenza. But there is considerable room for improvement, as these compounds are not invariably efficacious or well tolerated. The emergence of drug–virus resistance and drug-related side effects are among the mainstay reasons for further refinement of antiviral drug design and development.

In antiviral drug design and targeting, one could, in principle, aim at either viral proteins (enzymes) or cellular proteins (enzymes). The first approach is likely to yield more specific, less toxic compounds with a narrow spectrum of antiviral activity and a higher likelihood of drug resistance development, whereas the second approach may afford antiviral compounds with a broader activity spectrum, less chance of resistance development but higher likelihood for toxicity. Both routes should be worth exploring, the preferred route being dictated by both the nature of the virus and the targets that the virus or its host cell have to offer.

As exemplified for HIV (Fig. 25.1), the viral life cycle encompasses a number of critical steps starting from the attachment of the virus to the cell and finishing by the release of the progeny virions from the cell. While the "normal" cytolytic viruses replicate their genome and express their genes autonomously, that is, independently from the host cell metabolism, the replicative cycle of retroviruses gets closely associated with the host cell, in

---

Figure 25.1 The viral life cycle, as exemplified by the human immunodeficiency virus. Viral life cycles have several specific steps, many of which are targets for antiviral drugs. After virus adsorption, enveloped viruses enter the cell by virus–cell fusion. For HIV, which is a retrovirus with an RNA genome, replication of the genome occurs after reverse transcription and integration into the host–cell chromosome. For DNA viruses such as herpesviruses, the genome is replicated by a viral DNA polymerase. After transcription to RNA, translation and proteolytic processing of the precursor polypeptide, viral proteins assemble at the cell membrane, where they bud to release new virions.

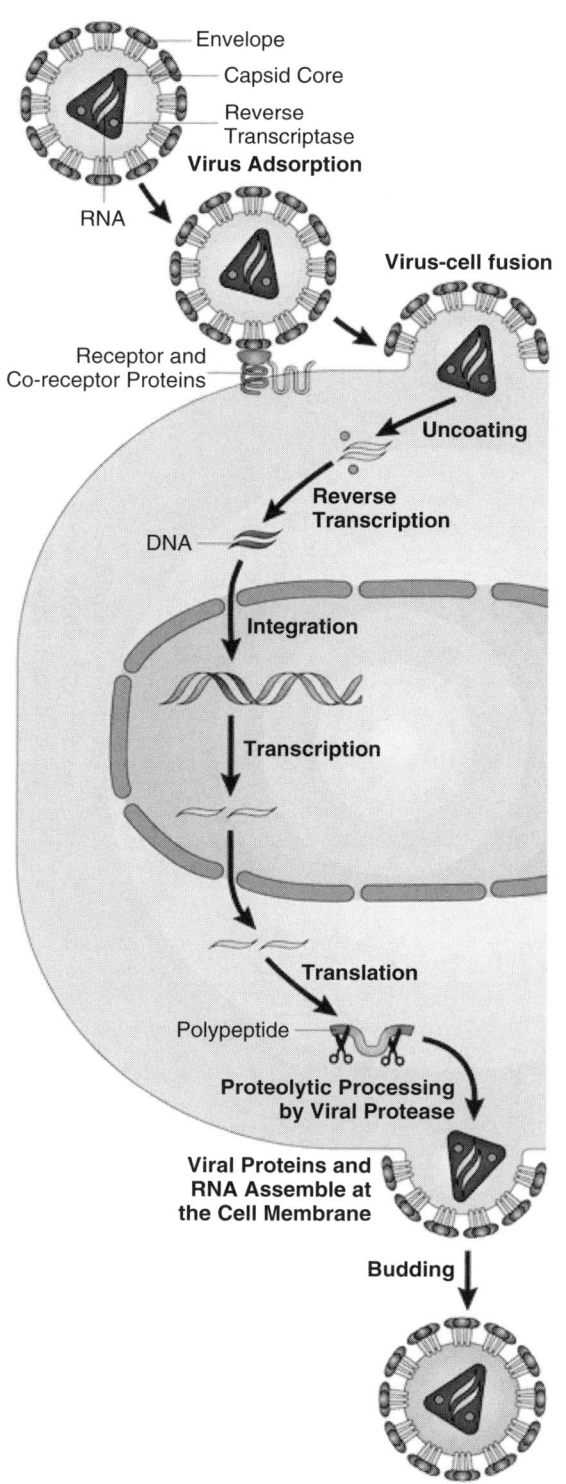

TABLE 25.1 Approaches to Antiviral Drug Design That Have Led to Licensed Drugs

Approach	Target Virus(es)	Compounds Approved	Selected Compounds in (pre)Clinical Development for the Indicated Target Virus
Virus adsorption inhibitors	HIV, HSV, CMV, RSV, and other enveloped viruses		Polysulfates, polysulfonates, polycarboxylates, polyoxometalates, chicoric acid, zintevir, cosalane derivatives, cyanovirin, negatively charged albumins
Virus–cell fusion inhibitors	HIV, RSV, and other paramyxoviruses	Enfuvirtide	HIV: AMD3100 and SCH-C derivatives
Viral DNA polymerase inhibitors	Herpesviruses (HSV-1, -2, VZV, CMV, EBV, HHV-6, -7, -8)	Acyclovir, valaciclovir, ganciclovir, valganciclovir, penciclovir, famciclovir, brivudin,[a] foscarnet, fomivirsen	Bicyclic furopyrimidine nucleoside and nonnucleoside analogs, A5021, cyclohexenylguanine Helicase-primase inhibitors: BILS 179BS and BAY 57-1293
Reverse transcriptase inhibitors	HIV	NRTIs (nucleoside reverse transcriptase inhibitors): zidovudine, didanosine, zalcitabine, stavudine, lamivudine,[b] abacavir NNRTIs (nonnucleoside reverse transcriptase inhibitors): nevirapine, delavirdine, efavirenz	Emtricitabine, amdoxovir, BCH-10652, racivir, reverset, elvucitabine, alovudine Capravirine, dapivirine, UC781, DPC083, SJ-3366, (+)-calanolide A

Acyclic nucleoside phosphonates (NtRTIs (nucleotide reverse transcriptase inhibitors)	DNA viruses (polyoma-, papilloma-, herpes-, adeno-, and poxviruses), HIV, HBV	CMV: cidofovir HBV: adefovir dipivoxil HIV: tenofovir disoproxil fumarate Disoproxil fumarate
Inhibitors of processes associated with viral RNA synthesis	HIV HCV	HIV: integrase inhibitors HCV: NS5B polymerase inhibitors HCV: NS3 helicase inhibitors
Viral protease inhibitors	HIV, herpesviruses, rhinoviruses, HCV	HIV: saquinavir, ritonavir, indinavir, nelfinavir, amprenavir, lopinavir HIV: atazanavir, mozenavir, tipranavir Human rhinovirus: AG7088 HCV: NS3 protease inhibitors RWJ-270201, A-192558, A-315675
Viral neuraminidase inhibitors	Influenza A and B virus	Zanamivir, oseltamivir[c]
IMP dehydrogenase inhibitors	HCV, RSV	Ribavirin[d] Mycophenolic acid, EICAR, VX-497
S-adenosylhomocysteine hydrolase inhibitors	(−)RNA hemorrhagic fever viruses (e.g, Ebola)	3-Deazaneplanocin A

[a] Brivudin is approved in certain European countries (i.e., Germany).
[b] Lamivudine is also approved for the treatment of HBV.
[c] In addition to zanamivir and oseltamivir, amantadine and rimantadine have been approved as anti-influenza drugs, but these compounds are targeted at the viral uncoating process, not the viral neuraminidase.
[d] For the treatment of HCV infections ribavirin is used in combination with interferon-α.

that, subsequent to the reverse transcription (ribonucleic acid → deoxyribonucleic acid (RNA → DNA)] step, the resulting proviral DNA becomes integrated into the cellular genome and then follows the "classical" transcription and translation processes. Here, we focus primarily on approaches targeted at specific processes in viral infection (Fig. 25.1), including virus adsorption, virus–cell fusion, viral DNA or RNA synthesis (and cellular components associated therewith), and viral enzymes (protease, neuraminidase). Two cellular enzymatic processes, namely IMP dehydrogenase and S-adenosylhomocysteine hydrolase, could also be envisaged as targets for certain classes of antiviral agents, and will therefore be addressed as well.

25.2 VIRUS ADSORPTION INHIBITORS

Numerous polyanionic compounds [e.g., polysulfates such as polyvinylalcohol sulfate, polysulfonates such as polyvinylsulfonate (Fig. 25.2), polycarboxylates, polynucleotides such as zintevir, polyoxometalates, negatively charged albumins] have been shown to inhibit HIV replication by preventing virus attachment (adsorption) to the cell surface. All these negatively charged polymers may be expected to interact with the positively charged amino acids in the V3 loop (which is rich in R and K residues) of the HIV glycoprotein gp120. In doing so [2], the polyanions shield the V3 loop and thus hamper the binding of the HIV virions with heparin sulfate, the primary binding site at the cell surface before a more specific binding can occur with the CD4 receptor on $CD4^+$ cells.

Heparin sulfate is widely expressed on animal cells and is involved in virus–cell binding for a broad array of enveloped viruses, including HSV [3] and dengue virus [4]. So, polysulfates, polysulfonates, and other polyanionic substances that interfere with the target cell binding of these enveloped viruses may be effective in the treatment and prophylaxis of such infections. In the management of HIV infections, polyanionic substances may have a major role as vaginal microbicides, as, when applied in an appropriate formulation, they may successfully prevent sexual transmission of HIV infection. Moreover, these polyanions are not only active against HIV but also HSV and other sexually transmitted disease (STD) pathogens such as *Neisseria gonorrhoeae* and *Chlamydia trachomatis* [5].

Although polyanions may have multiple sites of interaction (i.e., virus adsorption, reverse transcriptase, integrase), virus attachment to the cells would be the preferred target from a therapeutic viewpoint, as it is the first opportunity to curtail the viral life cycle, and the polyanions do not need to enter the cells (which would be problematic). The interaction of polyanionic substances at this level can also be considered specific, since repeated passage of HIV in the presence of polyanions can lead to resistance mediated by mutations in the envelope glycoprotein gp120, particularly in the V3 loop (K269E,

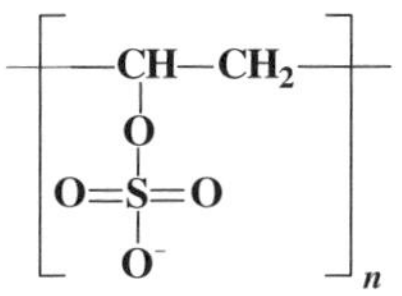

PVAS

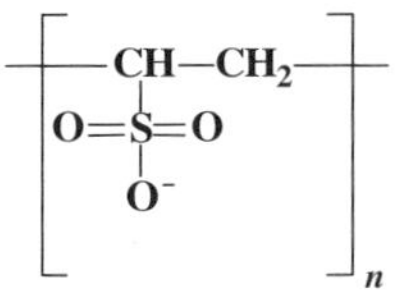

PVS

Viral Adsorption Inhibitors

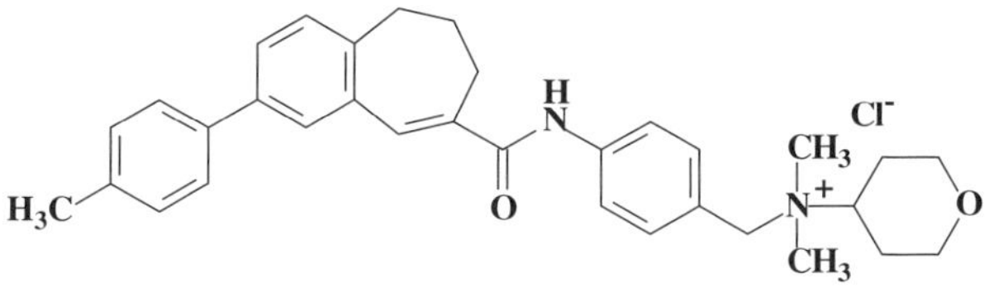

TAK779
CCR5 antagonist

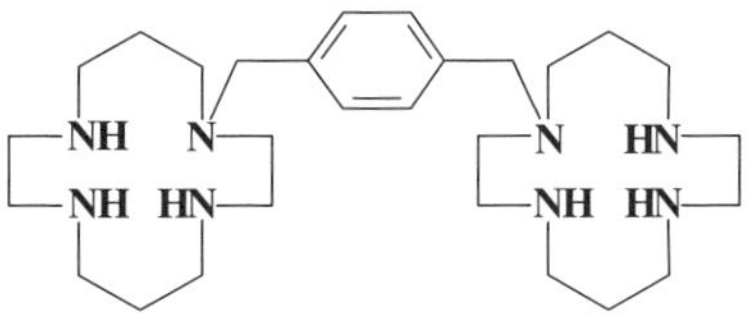

AMD3100
CXCR4 antagonist

Virus–Cell Fusion Inhibitors

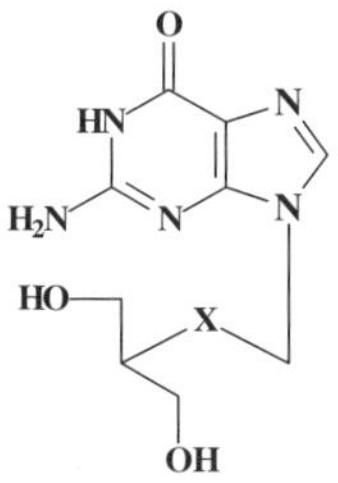

Acyclic guanosine analogue
(X = O, ganciclovir)
(X = CH_2, penciclovir)
Viral DNA Polymerase Inhibitors

Figure 25.2 Basic (skeletal) pharmacophores of the classes of antiviral agents described in this review. (IMP, inosine monophosphate; NRTI, nucleoside reverse transcriptase inhibitor; NNRTI, nonnucleoside reverse transcriptase inhibitor; PVS, polyvinylsulfonate; PVAS, polyvinylalcohol sulfate; SAH, *S*-adenosylhomocysteine.)

Q278H, N293D), as originally shown for dextran sulfate [6] and subsequently for zintevir [7] and negatively charged albumins [8].

25.3 VIRUS–CELL FUSION INHIBITORS

Enveloped viruses, as a rule, enter their host cells by fusion between the viral envelope and cellular plasma membrane (Fig. 25.1). This fusion process is basically similar for different enveloped virus families (i.e., retro-, paramyxo-, her-

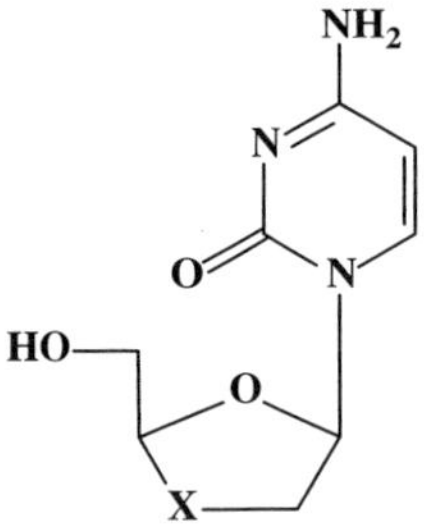

NRTI
(X = CHN₃, AZT)

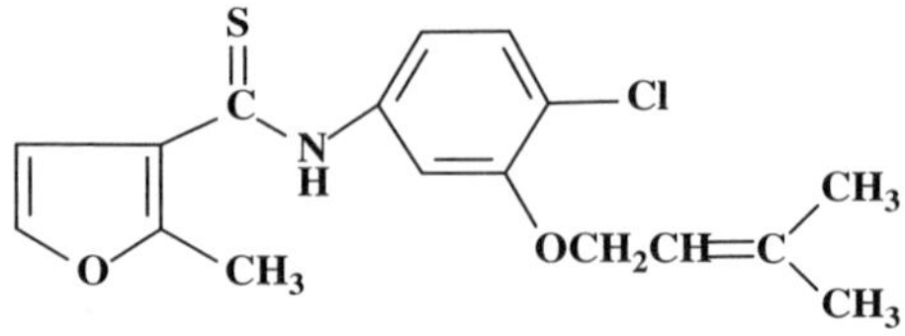

NNRTI (UC781)

Reverse Transcriptase Inhibitors

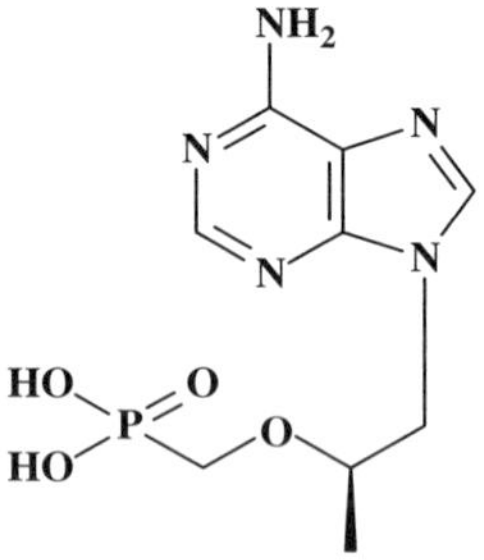

Tenofovir
Acyclic Nucleoside Phosphonate

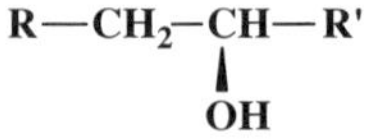

Peptidomimetic

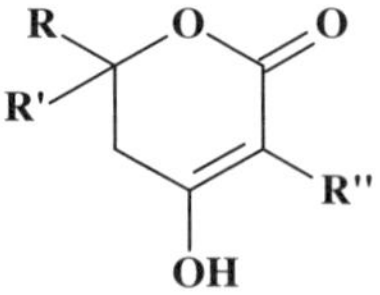

Nonpeptidomimetic

Protease Inhibitors

Figure 25.2 *Continued*

pesviruses), but for HIV it is preceded by the interaction of the viral glycoprotein gp120 with its co-receptor (CXCR4 for T-tropic or X4 HIV strains, CCR5 for M-tropic or R5 strains). CXCR4 and CCR5 normally act as the receptors for the CXC chemokine SDF-1 (stromal-cell derived factor), CC chemokines RANTES (regulated upon activation, normal T-cell expressed and secreted), and MIP-1α and -1β (macrophage inflammatory proteins), respectively. The fact that CXCR4 and CCR5, quite incidentally, are used by

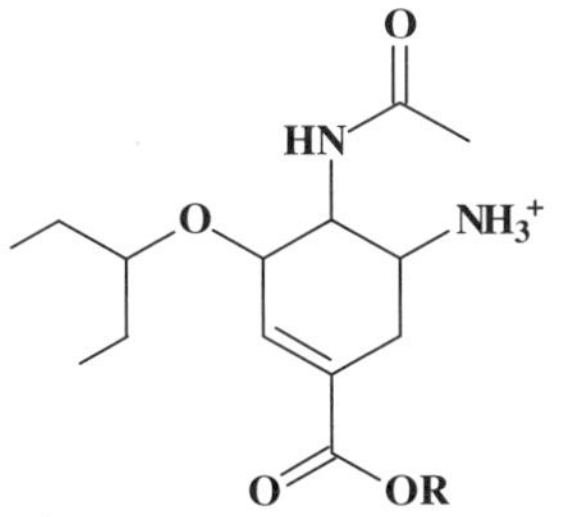

Oseltamivir (R = CH$_2$CH$_3$)
Neuraminidase Inhibitor

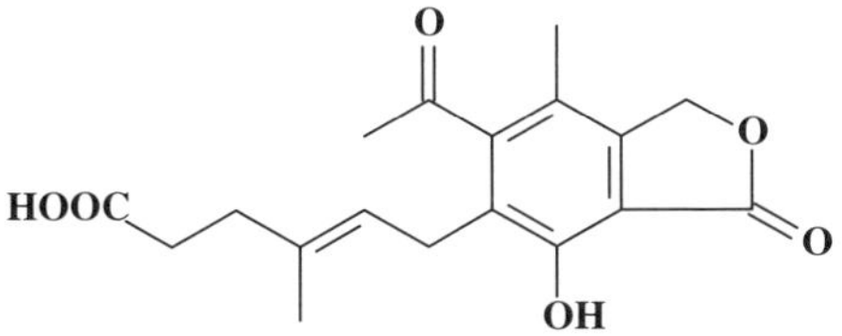

Mycophenolic Acid

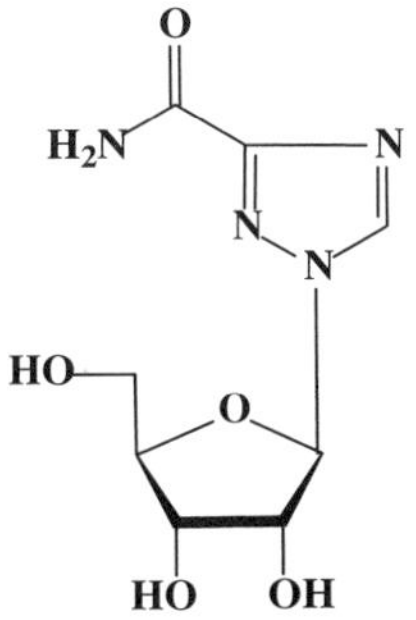

Ribavirin

IMP Dehydrogenase Inhibitors

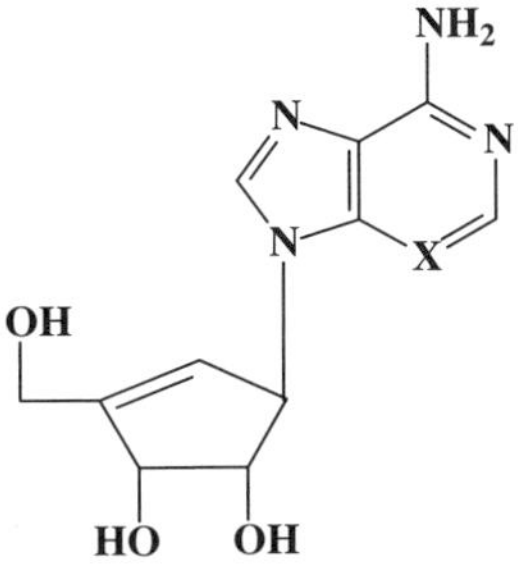

X = N or CH
SAH Hydrolase Inhibitors

Figure 25.2 *Continued*

HIV as co-receptors to enter the cells has prompted the search for CXCR4 and CCR5 antagonists, which, through blockade of the corresponding co-receptor, may be expected to block HIV entry into the cells.

This has now been demonstrated with a number of compounds, the most prominent among the CXCR4 antagonists being the bicyclam AMD3100 (Fig. 25.2) [9, 10] and the best documented among the CCR5 antagonists being

TAK-779 (Fig. 25.2) [11, 12]. An extensive SAR study of phenylenebis(methylene) linked bis-tetraazamacrocycles that inhibit HIV replication has been described. Depending on the substitution of the phenylenebis(methylene) linker, submicromolar anti-HIV activity was exhibited by analogs bearing macrocycles of 12 to 14 ring members. Identical macrocyclic rings are not required for activity. However, substitution of an acyclic polyamine equivalent for one of the cyclam rings resulted in a substantial reduction of anti-HIV potency, clearly establishing the importance of the constrained macrocyclic structure. The parasubstituted phenylene gave a more active compound than the metasubstituted phenylene. The antiviral activity of the bicyclam analogs appeared to be relativly insensitive to the presence of electron-withdrawing or electron-donating substituents on the phenyl ring. The activity of pyridine-linked bicyclams is dependent on the way the heteroaromatic linker is connecting the cyclam rings. For example, 2,6- and 3,5-pyridine linked bicyclams (**3** and **4**) are more potent than the 2,5- and 2,4-substituted pyridine-linked compounds (**5** and **6**). The position of the four amino groups of the tetraazamacrocyclic ring is important. The Py[iso-14]ane N_4 compound (**7**) is less active than the parent compound, and introduction of a pyridine group results in a further reduction in biological activity (**8**). However, single pyridine insertion in the concept of an inversed macrocyclic framework (**9**) exhibited anti-HIV-1 potency surpassing that of the saturated, aliphatic counterparts (Fig. 25.3) [13–15].

The site of interaction of TAK-779 with the transmembrane helices of CCR5 has been mapped (Fig. 25.4) [12], and, likewise, crucial amino acid residues involved in the binding of AMD3100 to its receptor CXCR4 have been identified [16]. AMD3100 acts on the CXCR4 receptor through binding to Asp^{171} in TM-IV and Asp^{262} in TM-VI, with each of its cyclam moieties. This is demonstrated in the molecular model of the main ligand binding pocket of the CXCR4 receptor wherein AMD3100 was manually docked in a manner that respects the above interactions. AMD3100 could prevent the receptor from changing into an active conformation (Fig. 25.5).

Recently, a new CCR5 antagonist, SCH-C (SCH 351125), was announced as an orally bioavailable inhibitor of M-tropic R5 strains, capable of suppressing R5 HIV-1 infection both in vitro and in vivo (SCID-hu Thy/Liv mice) [21]. The clinical potential of the CXCR4 and CCR5 antagonists in the management of HIV infections remains to be proven. To ensure maximal coverage (both X4 and R5 strains), dual CXCR4/CCR5 antagonists should be developed or single CCR5 and CXCR4 antagonists should be combined.

The interaction of gp120 with its co-receptor (CCR5 or CXCR4) triggers a series of conformational changes in the gp120/gp41 complex, ultimately leading to the formation of a "trimer-of-hairpins" structure in gp41, which is a bundle of six α-helices: three α-helices formed by the COOH-terminal regions packed in an antiparallel manner with three α-helices formed by the NH_2-terminal regions. The fusion peptide region, located at the extreme NH_2-terminus, will insert into the cellular membrane, while the COOH-terminal

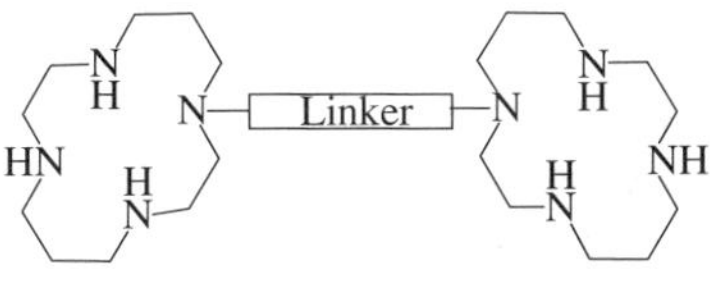

VIRUS–CELL FUSION INHIBITORS

	Linker	EC$_{50}$ (µM)	CC$_{50}$ (µM)
1	 1,4-xylyl	0.0042	> 421
2	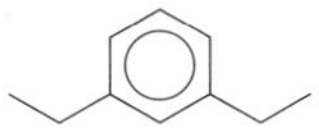 1,3-xylyl	0.0337	> 421
3	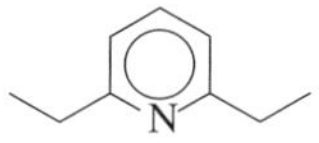 2,6-lutidinyl	0.025	> 405
4	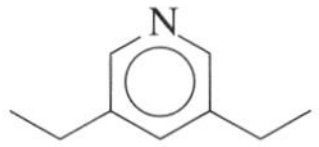 3,5-lutidinyl	0.032	> 395
5	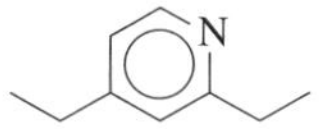 2,4-lutidinyl (N at 3)	16.367	17
6	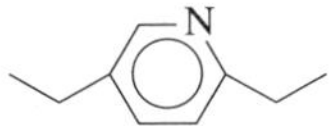 2,6-lutidinyl (N at 2)	0.908	18
7	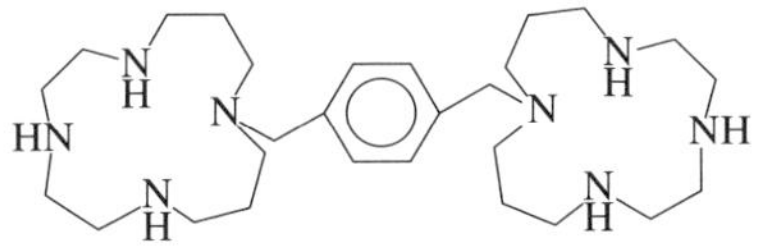 (fused bis-macrocycle)	0.034	> 421
8	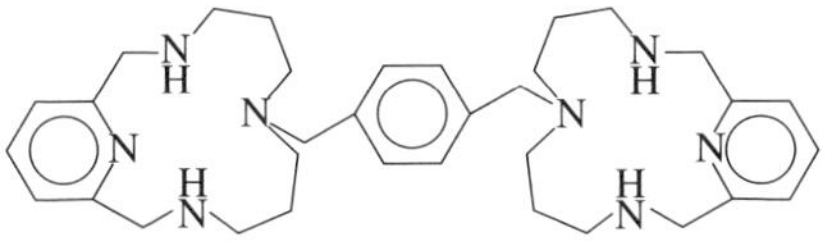 (pyridyl-fused bis-macrocycle)	0.534	199
9	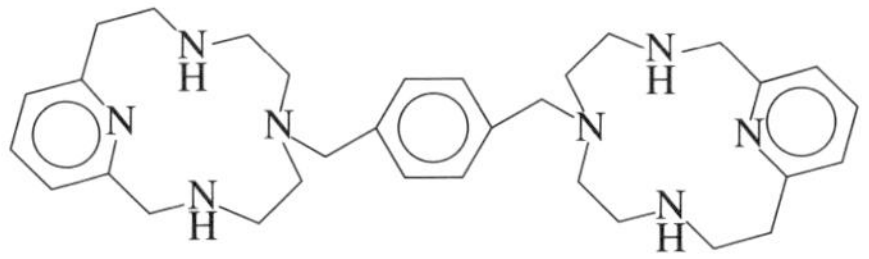 (pyridyl-fused bis-macrocycle)	0.008	194

Figure 25.3 Anti-HIV-1 (III$_B$)data for a series of bis-azamacrocycles showing the importance of the linker and the constitution of the cyclam moiety.

region remains anchored in the viral envelope: in this sense the trimer-of-hairpins motif brings the two membranes together, and agents that interfere with the formation of the gp41 trimer-of-hairpins structure may be expected to inhibit the fusion process [18].

Several constructs have been designed to interfere with the gp41-mediated fusion process: the so-called 5-helix that binds the COOH-terminal region of

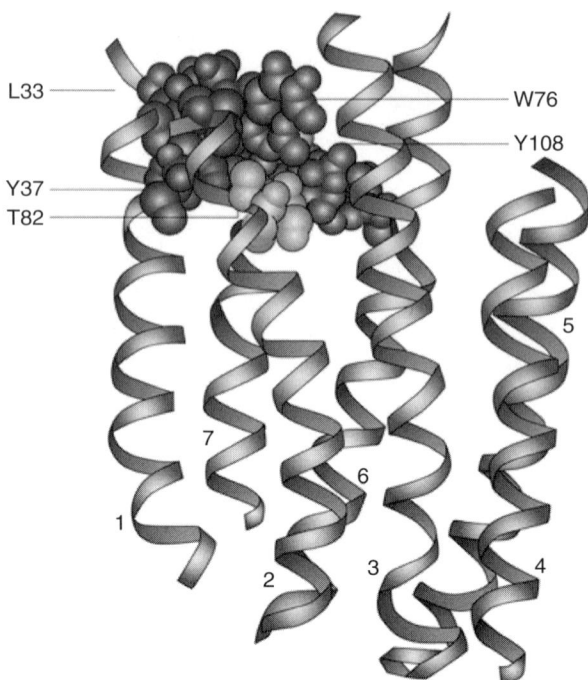

Figure 25.4 Interactions of antiviral drugs with their protein targets. A structural model of CCR5 complexed with TAK-779 (Fig. 25.2) viewed from within the plane of the membrane [12]. The indicated cluster of amino acids in the TAK779 binding site includes several aromatic residues (Y37, W76, Y108) that might form favorable interactions with the aromatic rings of TAK779.

gp41 [22], D-peptide inhibitors that dock into the pocket formed by the α-helices of gp41 [23], and T-20 (enfuvirtide, previously called DP-178, a synthetic 36-amino-acid peptide corresponding to residues 127 to 162 of the ectodomain of gp41). Enfuvirtide disrupts the conformational changes associated with membrane fusion. T-20 has proved effective in reducing the plasma HIV levels in humans, thus providing the proof-of-concept that viral entry can be successfully blocked in vivo [24]. Disadvantages are that enfuvirtide needs a twice-daily subcutaneous injection for delivery and that its manufacturing process is expensive and complicated. Worldwide clinical studies have indicated that enfuvirtide, in combination with three to five other antiretroviral drugs, is able to effect an incremental decrease in viral load [25, 26].

Insight into the HIV fusion process should help in designing fusion inhibitors for other viruses as well, as trimer-of-hairpins motifs may also be predicted [16] for other virus families, including paramyxoviridae (parainfluenza, measles, respiratory syncytial virus). In fact, peptides (like T-20) for each of these paramyxoviruses have been shown to block viral fusion [27].

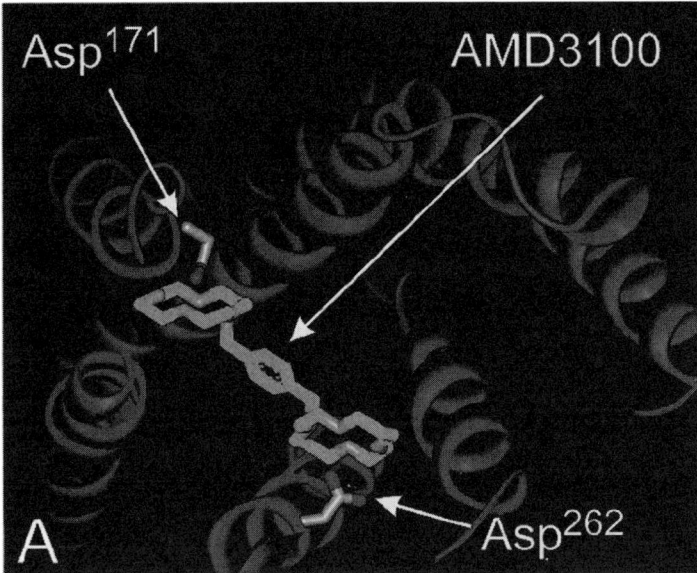

Figure 25.5 Molecular model of the main ligand binding pocket of the CXCR4 receptor with AMD3100. The receptor model is built over the rhodospin model of Palczewski et al. [17]. The conformation of AMD3100 is based on structural requirements of high antiviral effect of AMD3100 [18, 19] and the crystallographic X-ray structure of 6,6′-spiro-bis(1,4,8,11-tetraazacyclotetradecane)-dinickel(II)tetraperchlorate [20] obtained from the Cambridge Structural Database.

Also, a cobalt-chelating complex (CTC-96) that has been shown to inhibit infection by herpesviruses (i.e., HSV) through blocking fusion [28] may be expected to possess an extended antiviral activity spectrum, given the premise that enveloped viruses belonging to different virus families share an analogous process of membrane fusion.

25.4 INHIBITORS OF VIRAL DNA OR RNA SYNTHESIS

Viral DNA Polymerase Inhibitors

Being DNA viruses, the herpesviruses heavily rely for their replication on their own genome-encoded DNA polymerase. In contrast with the retroviruses (Fig. 25.1), herpesviruses do not have a reverse transcription step in their replicative cycle, which means that their genome can be replicated by the viral DNA polymerase after the latter has been expressed in the virus-infected cell. All the antiviral agents that are currently available for the treatment of herpesvirus (i.e., HSV-1, HSV-2, VZV, CMV) infections are nucleoside analogs: They belong to either the class of the acyclic guanosine analogs (i.e., acyclovir, penciclovir, ganciclovir, and their oral prodrug forms valaciclovir, famciclovir, and valganciclovir, respectively) or thymidine analogs (i.e., brivudin) (Fig. 25.2 and Table 25.1). All these compounds are targeted at the viral DNA polymerase, but before they can interact with viral DNA synthesis, they need to be phosphorylated intracellularly to the triphosphate form. The first (and, for brivudin, also the second) phosphorylation step is ensured by the HSV- or VZV-encoded thymidine kinase, or CMV-encoded protein kinase and is thus confined to the virus-infected cells, which explains the specific antiviral action of the established antiherpetic compounds. Subsequent phosphorylations are performed by cellular kinases. In their triphosphate form, the nucleoside analogs then interact with the viral DNA polymerase as either a competitive inhibitor or an alternate substrate with respect to the natural substrate (i.e., dGTP for the guanosine analogs, dTTP for the thymidine analogs). If the acyclic guanosine analogs act as alternate substrates, their incorporation prevents further chain elongation (Fig. 25.6a).

An important issue with respect to antiherpesvirus (HSV-1, HSV-2, VZV) agents is their intracellular phosphorylation. Thymidine kinase (TK) is a crucial enzyme in the salvage pathway of thymidine 5′-triphosphate, which is a precursor of the thymidine incorporation into DNA. The substrate specificity of human TK is essentially limited to thymidine. The viral enzyme has a broader substrate specificity than the human enzyme and also guanine nucleoside analogs are accepted as substrates. The viral TK is important for the pathogenicity of the virus in humans, and HSV TK is a potential target for drugs that might prevent reactivation of HSV when used prophylactically.

Modified nucleosides have a much higher affinity for HSV-1 TK than for cellular TK, so that phosphorylation may occur preferentially in virus-infected cells. As mentioned before, the antiherpes nucleoside analogs can be classified in guanosine analogs and thymidine analogs. Acyclovir is the prototype of the guanine series. Acyclovir has high selectivity and *quasi*-negligible side effects, but it has rather poor efficacy. Although the triphosphate of acyclovir is an excellent inhibitor of DNA synthesis because of its strong chain-terminating effect, acyclovir itself is a poor substrate for TK. Analogs providing good binding (low K_m/K_{cat}) to both viral TK and viral DNA polymerase, while retaining low toxicity and high specificity for the viral TK over the cellular TK, would lead to significantly improved chemotherapy.

The prototype of a pyrimidine nucleoside with antiherpes activity is 5-(*E*)-(2-bromovinyl)-2′-deoxyuridine (BVDU, brivudin). BVDU is phosphorylated preferentially by the HSV-1 encoded dThd/dCyd kinase to the 5′-monophosphate and the 5′-diphosphate; the latter is further phosphorylated to the 5′-triphosphate by cellular kinase(s). BVDU can be incorporated into the DNA. Thus, a common characteristic for all antiherpesvirus (HSV, VZV) nucleosides is that they have to be phosphorylated by the virus-specific kinase in infected cells. The modes of binding of the guanine and of the pyrimidine nucleosides to the viral TK are different, which has been demonstrated by co-crystallization experiments [29]. The guanine moiety of acyclovir is bound

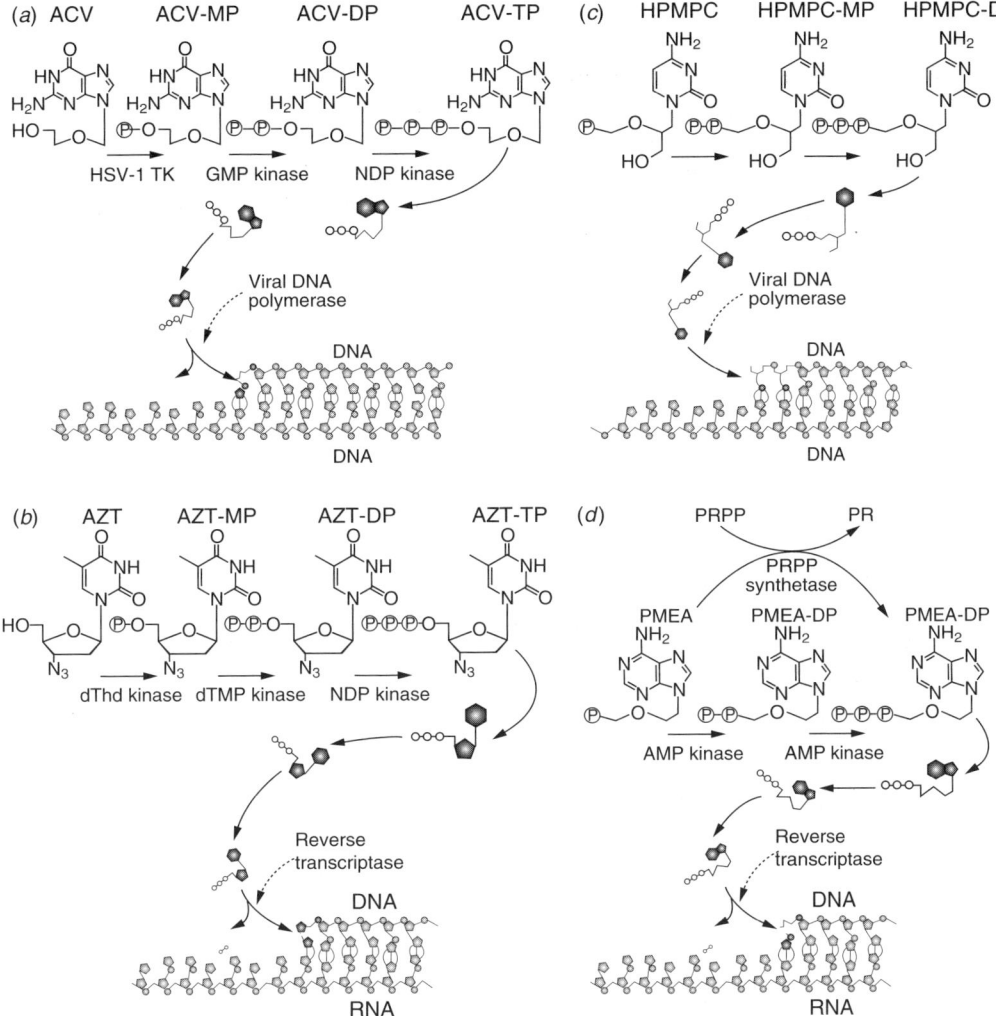

Figure 25.6 Examples of antiviral nucleoside analogs acting by a chain termination mechanism. (*a*) Acyclovir (ACV) targets viral DNA polymerases such as the herpesvirus DNA polymerase. Before it can interact with viral DNA synthesis, it needs to be phosphorylated intracellularly, in three steps, to the triphosphate form. The first phosphorylation step is ensured by the HSV-encoded thymidine kinase (TK), and is thus confined to the virus-infected cells. (*b*) Azidothymidine (AZT) targets the HIV reverse transcriptase and also needs to be phosphorylated, in three steps, to the triphospate form before it can interfere with reverse transcription. (*c*) Cidofovir (HPMPC), an acyclic nucleotide analog, which can be viewed as an acyclic nucleoside analog extended by a phosphonate moiety, targets viral DNA polymerases, and is active against DNA viruses whether or not they encode for a specific viral thymidine kinase. In contrast with acyclovir and azidothymidine, cidofovir requires only two phosphorylations to be converted to the active (triphosphate) form. (*d*) Adefovir (PMEA)—also an acyclic nucleoside phosphonate—is active against retroviruses and hepadnaviruses; like cidofovir, adefovir needs only two phosphorylations and can thus bypass the nucleoside kinase reaction that limits the activity of dideoxynucleoside analogs such as AZT.

with hydrogen bond pairing being made with Gln-125 via the 1-NH and 6-carbonyl group; also Gln-125 is close enough for a possible further hydrogen bond with the 2-amino group. The 6-carbonyl group of acyclovir is close enough to the guanidinium group of Arg-176 to form a hydrogen bond. The hydroxyl group of acyclovir, lying close to the position occupied by the 5'-OH of dTh and interacts with the side chains of Arg-163 and Glu-83 and a nearby water (Fig. 25.7).

BVDU shows a mode of binding to HSV-TK, which is similar to that of deoxythymidine, as highlighted by hydrogen bonding and nonpolar interactions. The deoxyribose makes hydrogen bond interactions via its 3'-OH with Tyr-101 and Glu-225, via its 5'-OH with Glu-83, and via its 5'-OH and a water molecule with Arg-163. The base makes van der Waals contacts with Met-128 and Ile-100 on one side, and with Tyr-172 on the other, and makes pairwise hydrogen bond interactions via its 4-carbonyl and 3-NH groups with the amide of Gln-125.

The 2-carbonyl group of BVDU is hydrogen bonded to two water molecules, which, in turn, interact with the guanidium group of Arg-176, the hydroxyl group of Tyr-101, and the side-chain carbonyl of Gln-125. The bulky 5-substituent of BVDU occupies the deep space available in the neighborhood of residues Trp-88, Tyr-132, Arg-163, and Ala-167. This accommodation is at the expense of a relocation of the side chain of Tyr-132, shifted away from the ligand to make room for the substituent group (Fig. 25.8).

Thus, the larger guanine group of acyclovir occupies an environment defined by the same residues as the pyrimidine ring of BVDU, and the volume occupied by guanine is approximately coplanar with that occupied by uracil. However, the guanine is located much closer to Tyr-101 and Arg-176, and direct hydrogen bonding is observed in the case of acyclovir. Hydrogen bonding with Gln-125 involves a conformational shift of the side chain, as well as 180° rotation of the amide.

However, the determination of the way a molecule binds in the active site of an enzyme is only the first step in the drug discovery process. Analysis of the structural factors influencing the affinity of the compounds for the enzyme should be followed by calculations of binding energies in function of conformation, which may lead to a quantitative explanation of activity in function of binding strength. This model may then be used for further drug discovery. These studies have been performed with a series of 5-substituted 2'-deoxyuridine substrates of the HSV-1 TK [30]. More than 80 percent of the total interaction energy of thymidine comes from the interaction with only six residues, namely the four hydrogen bond partners Glu-83, Glu-225, Gln-125, and Tyr-101 and residues Tyr-172 and Met-128, which sandwich the nucleoside base by means of stacking interactions. Substitution of the 5-methyl group of thymidine by a bulkier, unsaturated group such as a thiophene has no significant effect on the major binding interactions involving the base and sugar moieties. The 5-substituent is positioned in a small pocket and the accessible portion of this pocket is mainly formed by Tyr-132 (a small displacement of

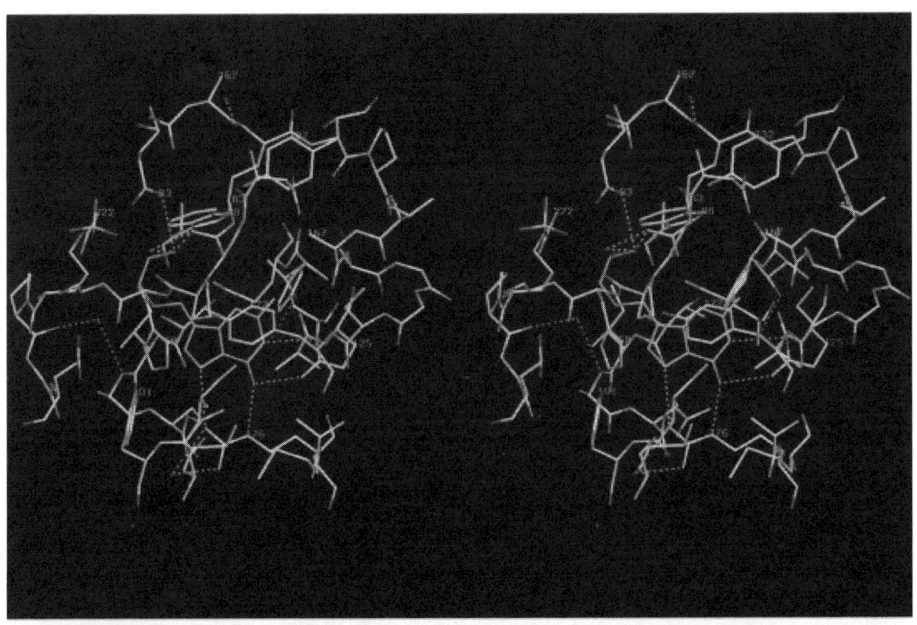

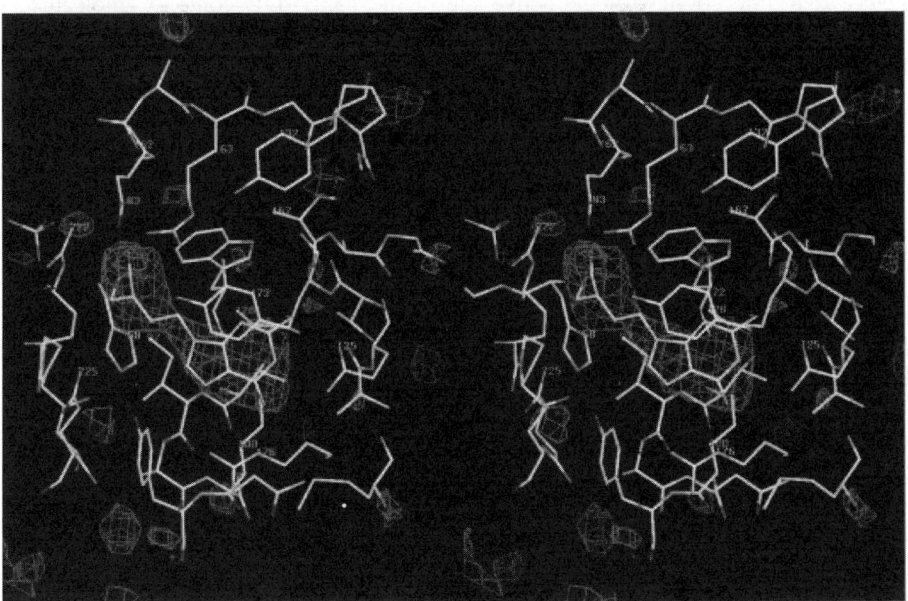

Figure 25.7 *Top*: Stereo view of the binding of aciclovir to TK (molecule I), showing active-site residues and intermolecular hydrogen bonding. *Bottom*: Stereo view of the binding of penciclovir to TK (molecule I) superimposed on difference Fourier map contoured at 3σ density.

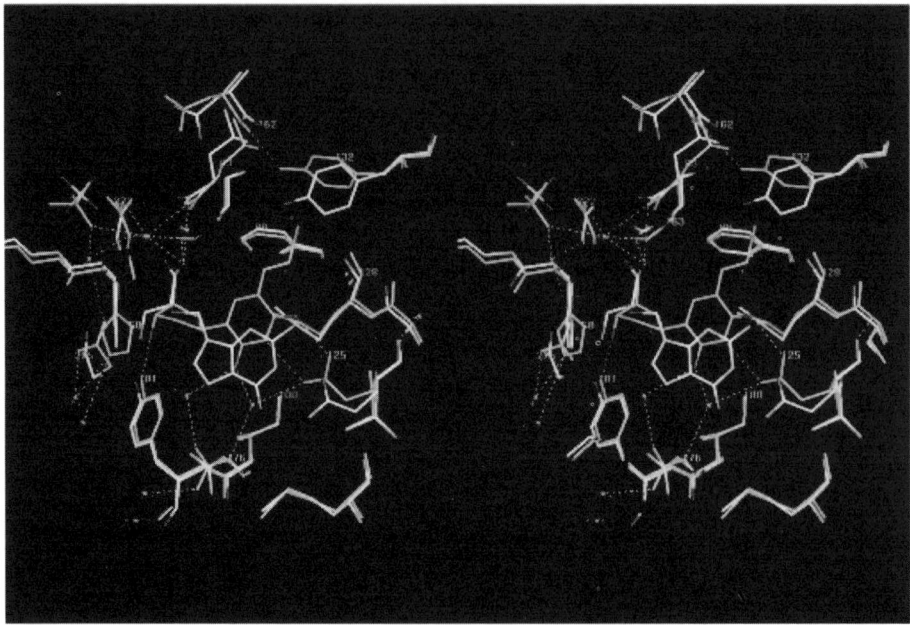

Figure 25.8 Stereo view of the active site of the TK:BVDU complex (molecule I) overlaid on the active site of the TK/ganciclovir complex after alignment of enzyme molecules. Intermolecular hydrogen bonding is shown for the TK/BVDU complex.

Tyr-132 is needed to enlarge this pocket to accommodate the bulky substituent). The observed affinity differences of 18 different 5-substituted 2'-deoxyuridines was rationalized based on a binding energy–affinity relationship study. Interaction energies between substrate and enzyme, under the form of nonbonded van der Waals and electrostatic forces, were calculated for each complex and for each experimental setup to provide a measure of the tightness of binding. In order to account for possible differences in hydrophobicity between the compounds, free energies of solvation for each of the 18 substrates were also calculated. The differences in solvation-free energy between the 18 compounds can be largely attributed to the different substitution patterns of position 5. The best correlation was found for the case where the dielectric constant is treated as a distance-dependent function and where only the residues positioned within 3 Å of the substrate were allowed to relax. The least-squares equation for this particular case can be given by

$$\ln(\mathrm{IC}_{50}) = 0.177\ \mathrm{IE} + \Delta G_{\mathrm{sol}} RT + 51.9$$

in which IE stands for interaction energy in kJ/mol, ΔG_{sol} is the solvation free energy of the substrate in kJ/mol, RT is expressed in kJ/mol, and IC_{50} is expressed in µM. A plot of this equation is given in Fig. 25.9. It is clear that the affinity for this series of 5-substituted 2′-deoxy uridine substrates can be improved either by increasing the side chain's interaction energy or by increasing the side chain's hydrophobicity (Fig. 25.10 and Fig. 25.11).

Is there room for improvement? As the acyclic guanosine analogs are not readily taken up orally, their oral bioavailability has been improved by formulating them as prodrug forms (valaciclovir, famciclovir, and valganciclovir). The success obtained with acyclovir, valaciclovir, and famciclovir in the treatment of HSV and VZV infections has impeded further progress in this area. Yet, brivudin, which is considerably more potent than acyclovir and penciclovir as an anti-VZV agent, represents an important alternative for the treatment of VZV infections. Although brivudin is active in the nanomolar concentration range against VZV replication, its potency can still be superseded by bicyclic furopyrimidine nucleoside analogs bearing a long alkyl or alkylaryl side chain attached to the furane ring [35, 36]. The mechanism of action of these exquisitely potent and selective anti-VZV agents remains to be elucidated, although there is no doubt that their specificity for VZV is governed by the virus-encoded thymidine kinase.

Within the class of the guanosine analogs, several new congeners have been described, namely A-5021 [37, 38] and the D- and L-enantiomers of cyclohexenylguanine [39]. As far as it has been determined, these compounds seem to have an activity spectrum and mode of action similar to that of acyclovir, but further studies seem warranted to verify whether these new guanosine analogs may be endowed with an extended spectrum of activity [i.e., against HHV-6, -7, and -8, which are not (particularly) sensitive to acyclovir or brivudin] or increased in vivo efficacy, or improved pharmacokinetics, or other properties that make them worth pursuing.

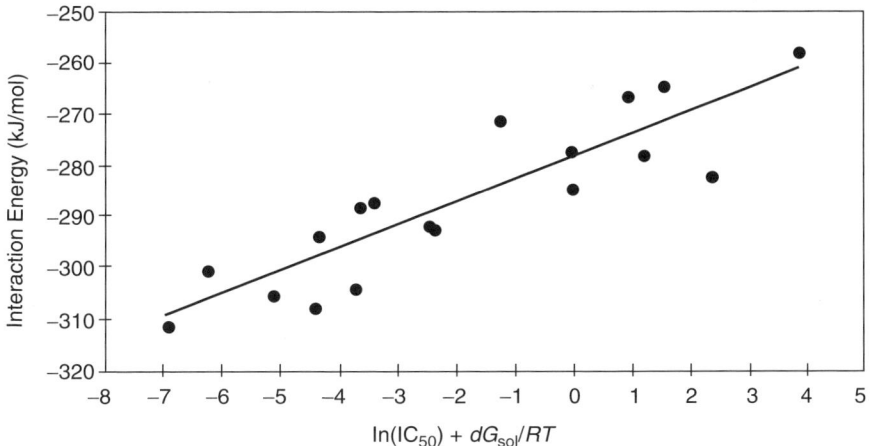

Figure 25.9 Plot of $\ln(IC_{50}) + \Delta G_{sol}/RT$ vs. interaction energy.

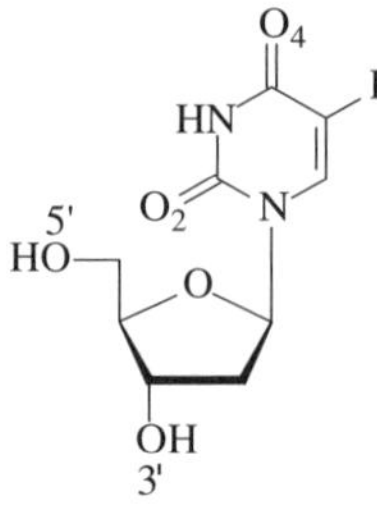

	-R	IC$_{50}$ (µM)		-R	IC$_{50}$ (µM)
T	—Me	1.0	9	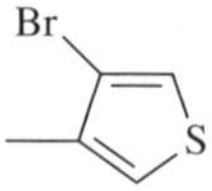 3-Br-thienyl	143
1	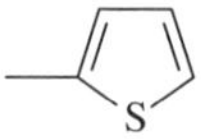 thienyl	2.4	10	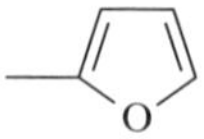 furyl	2.9
2	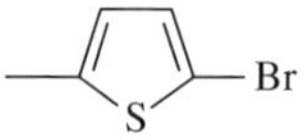 5-Br-thienyl	3.5	11	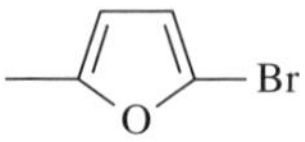 5-Br-furyl	51
3	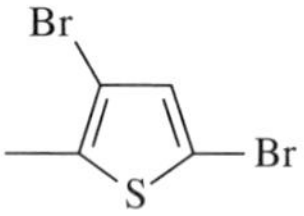 3,5-diBr-thienyl	62	12	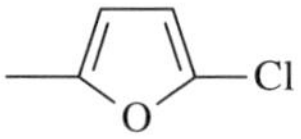 5-Cl-furyl	28
4	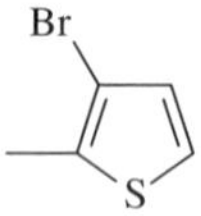 3-Br-thienyl	61	13	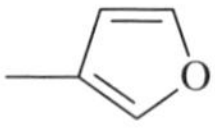 furyl (3-)	1.5
5	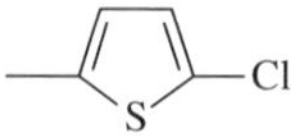 5-Cl-thienyl	2.3	14	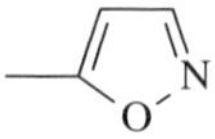 isoxazolyl	36
6	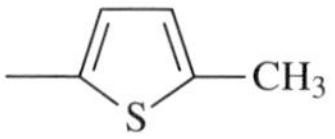 5-CH$_3$-thienyl	2.3	15	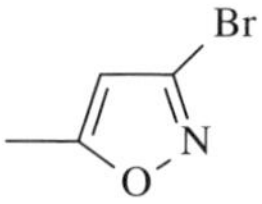 Br-isoxazolyl	34
7	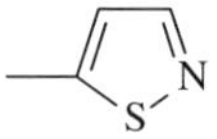 isothiazolyl	3.1	16	 phenyl	102
8	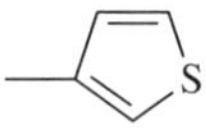 3-thienyl	4.0	17	 —CH=CH—Br	0.3

Figure 25.10 The compounds with their numbering label and experimental affinity for HSV-1 thymidine kinase. Affinity data were obtained from the literature [31–33] and are the concentrations required to inhibit thymidine phosphorylation by 50 percent against radiolabeled thymidine used at a concentration of 1 µM. All affinities were measured under the same conditions, and highly similar K_m values (0.25 µM) were found in all cases [34].

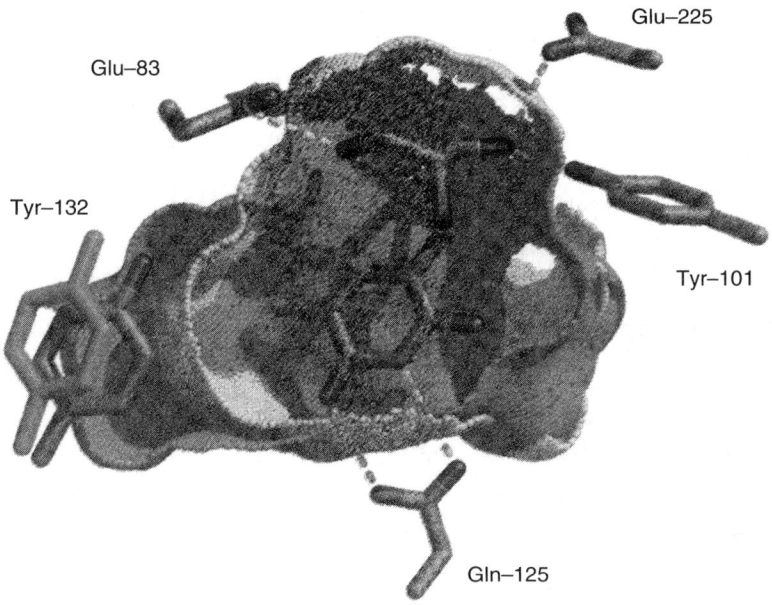

Figure 25.11 Only a minor small position shift of Tyr-132 is needed to enlarge the pocket of the 5-methyl group of thymidine so that it becomes large enough to occupy furane or thiophene groups. The bottom part of the solvent-accessible surface of the pocket is as found in the crystal structure; the new pocket after a small position shift of Tyr-132 is also shown.

In addition to the viral DNA polymerase, the HSV DNA helicase–RNA primase, an enzyme composed of the HSV gene products UL5, UL8, and UL52 that in terms of the viral DNA synthesis process precedes the DNA polymerase activity, has been identified as an attractive target for the design of new inhibitors of HSV-1 and HSV-2 infection: BILS 179BS and BAY 57-1293 are two examples of such helicase [40, 41]. These compounds exhibited remarkable in vivo efficacy in animal models of HSV-1 and HSV-2 infection, combined with excellent oral bioavailability. The antiviral activity of BAY 57–1293 was quoted as superior to all compounds currently used to treat HSV infections [42]. These data obviously validate the use of helicase–primase inhibitors for the treatment of HSV infections but leave us with the question of whether this strategy would also work with herpesviruses other than HSV or with DNA viruses at large.

Reverse Transcriptase Inhibitors

As is evident from Figure 25.1, the reverse transcriptase plays an essential role in the replicative cycle of retroviruses such as HIV, as it ensures the formation of proviral DNA that will then be integrated into the host cell genome and passed onto all the progeny cells resulting from the once infected parent cell. The substrate (dNTP) binding site of HIV reverse transcriptase (RT) has proven to be an attractive target for nucleosidic HIV inhibitors: No less than six nucleoside analogs—zidovudine (azidothymidine, AZT), didanosine (dideoxyinosine), zalcitabine (dideoxycytidine), stavudine (didehydro-dideoxythymidine, d4T), lamivudine (3′-thiadideoxycytidine), and abacavir—have been licensed as anti-HIV drugs (Fig. 25.2, Table 25.1), and several others such as emtricitabine [43], amdoxovir [44], BCH-10652 [(±)-2′-deoxy-3′-oxa-4′-thiocytidine, dOTC) [45], 5-fluoro-substituted dOTC [46], 5-fluoro-substituted d4C (reverset) [47], its L-counterpart (elvucitabine, ACH-126443) [48], and alovudine (MIV-310, the 3′-fluoro counterpart of AZT) [49] are in (advanced) clinical development. All these dideoxynucleoside analogs act according to a similar "recipe": As exemplified for AZT (Fig. 25.6b), they must be phosphorylated intracellularly consecutively by three cellular kinases, a nucleoside kinase, a nucleoside 5′-monophosphate kinase, and a nucleoside 5′-diphosphate kinase, to the corresponding 5′-triphosphate derivative, before the latter can interact, as a chain terminator, with the reverse transcription (RNA → DNA) reaction. One of the mechanisms by which resistance to AZT may arise is through removal of the chain-terminating residue, a kind of repair reaction involving pyrophosphorolysis, that may be regarded as the opposite of the reverse transcriptase reaction. Not all chain terminators are readily removed, for example, the acyclic nucleoside phosphonate derivative tenofovir (PMPA; Fig. 25.2) is not (see below), and, in this sense, PMPA should be less prone to resistance development than the regular nucleoside analogs.

Lamivudine is different from the other anti-HIV nucleosides because it has the L-configuration. Natural nucleosides and most of their antiviral analogs possess the D-configuration. The biological activity of a nucleoside is dependent on the nature of the base moiety: Thus the aglycon moiety at position C1 is considered as absolutely necessary. As the nucleosides have to be phosphorylated to become active, the presence of a hydroxylmethyl group at position C4 is likewise a prerequisite for biological activity. Thus, the configuration of the C1 and C4 atoms of a nucleoside is of crucial importance. The difference between a D-nucleoside and an L-nucleoside in the dideoxy series is in fact not very pronounced. When the base moiety and the hydroxymethyl group are considered as reference points, the ring oxygen function is situated on the backside in D-nucleosides and on the front side in L-nucleosides, opposite to the C2–C3 bond (Fig. 25.12) [50]. D- and L-dideoxynucleosides overlap very well. Therefore, it is not surprising that antiviral activity has been found in the series of L-dideoxyribonucleoside (e.g., L-5-dideoxyribocytosine, L-fluorodideoxycytosine, L-oxathiolane series). These are mainly cytosine nucleosides, which may be a reflection of the kinases involved not being very selective. Indeed, similar antiviral activity of D- and L-dideoxynucleosides is not a general rule, as demonstrated by carbovir, where the antiviral activity is more dependent on the stereochemistry of the carbocyclic ring. Abacavir can be distinguished from the other anti-HIV nucleosides because it is a carbocyclic

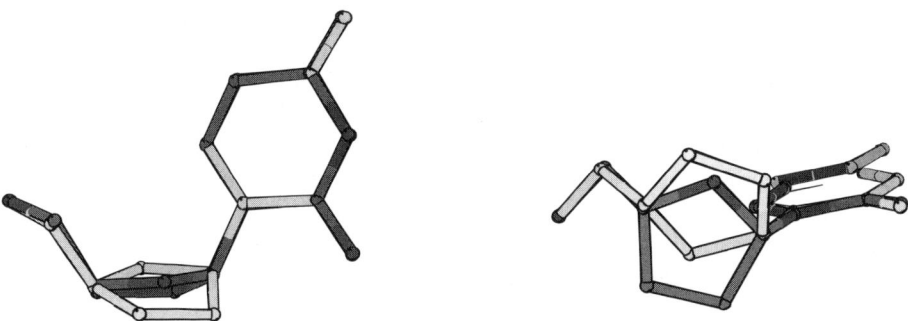

Figure 25.12 Molecular overlap of D- and L-dideoxynucleosides showing the subtle difference in sugar orientation between both optical antipodes.

nucleoside analog. It can be considered as a prodrug of the phosphates of (−)carbovir, and a correlation has been demonstrated between carbovir triphosphate formation and anti-HIV activity. Abacavir, however, does not show the same toxicity as carbovir. It allows the nucleoside carbovir, which caused problems in animals, to be bypassed. The pharmacokinetic, distribution, and toxicity profile of abacavir is distinct from and improved over that of (−)carbovir. Structurally, abacavir belongs to the carbocyclic nucleosides, which means that the ring oxygen atom is replaced by a methylene group. As a result, the anomeric center is removed, as is the stereoelectronic influence of the ring oxygen atom. The structure–activity relationship (SAR) of carbocyclic nucleosides is therefore different from that of normal nucleosides.

The first phosphorylation step that converts the 2′,3′-dideoxynucleoside analogs to their 5′-monophosphate can be considered as the bottleneck in the overall metabolic pathway leading to the formation of the active 5′-triphosphate metabolites. If certain dideoxynucleoside analogs (e.g., 2′,3′-dideoxyuridine) are not active against HIV under conditions where others are, this stems from their poor, or lack of, phosphorylation at the nucleoside kinase level. Therefore, attempts have been made at constructing prodrugs of 2′,3′-dideoxynucleoside 5′-monophosphate that deliver the 5′-monophosphate derivatives on cellular uptake, which can then be converted to the corresponding 5′-di- and 5′-triphosphate derivatives. This approach thus bypasses the initial kinase dependency and has been concretized by the design of the phosphoramidate [51,52] and cyclosaligenyl [53,54] prodrugs of d4T 5′-monophosphate. Both prodrugs were found to efficiently deliver within the cells d4T 5′-monophosphate, which, after conversion to its 5′-triphosphate, afforded anti-HIV activity under conditions where the nucleoside (due to inefficient phosphorylation) did not. It remains to be established whether this nucleoside kinase bypass strategy also yields increased antiviral efficacy in vivo.

All the aforementioned 2′,3′-dideoxynucleoside analogs, in their 5′-triphosphate form, act as competitive substrates/inhibitors with respect to the natural substrates (dNTPs) at the catalytic site of HIV RT, and, as HBV uses a similar

RT in its life cycle, all these compounds may be expected to inhibit HBV replication as well. This premise has been borne out particularly for lamivudine, which is currently licensed for the treatment of chronic HBV infections. Such an extended activity spectrum cannot be anticipated for a second class of RT inhibitors, referred to as NNRTIs (for nonnucleoside reverse transcriptase inhibitors), which interact with an allosteric, nonsubstrate binding ("pocket") site on the RT of HIV-1. This binding pocket is located at about 10 Å distance from the substrate-binding site (Fig. 25.13) [55]. This pocket does not exist in unliganded RT and does not occur in RTs other than HIV-1 RT, or, if it does, only the HIV-1 RT pocket offers the required allowances for interactions with the NNRTIs: that is, stacking interactions with the aromatic amino acids Y181, Y188, W229, and Y318; electrostatic interactions with K101, K103, and E138; van der Waals interactions with L100, V106, Y181, G190, W229, L234, and Y318; and hydrogen bonding with the main-chain peptide bonds [56]. A model for the interaction of a representative NNRTI, the thiocarboxanilide UC781 (Fig. 25.2), with HIV RT is shown in Fig. 25.14 [57].

The NNRTIs are notorious for rapidly eliciting virus–drug resistance resulting from mutations at amino acid residues that surround the NNRTI-binding site. In the clinic, the most prominent mutations engendering resistance to NNRTIs are the K103N and Y181C mutations. While, at present, only three NNRTIs (i.e., nevirapine, delavirdine, and efavirenz) have been formally licensed, and several others, that is, UC781 [57], capravirine (S-1153, AG 1549) [58], dapivirine (TMC 125) [59], DPC083 [60], SJ-3366 [61], and (+)-calanolide [62], are in clinical development, it is obvious that in the future design of new NNRTIs, not only potency and safety but also resilience to drug resistance mutations should come into play [63]. It is noteworthy that some amino acids, such as W229 and Y318 that surround the NNRTI binding site, do not seem apt to mutate, or if they do, they lead to a "suicidal" loss of RT activity [64]. Such immutable amino acids may be prime targets for the rational design of new NNRTIs.

Acyclic Nucleoside Phosphonates

The acyclic nucleoside phosphonates can be conceived as acyclic nucleoside analogs extended by a phosphonate moiety. The phosphonate group is equi-

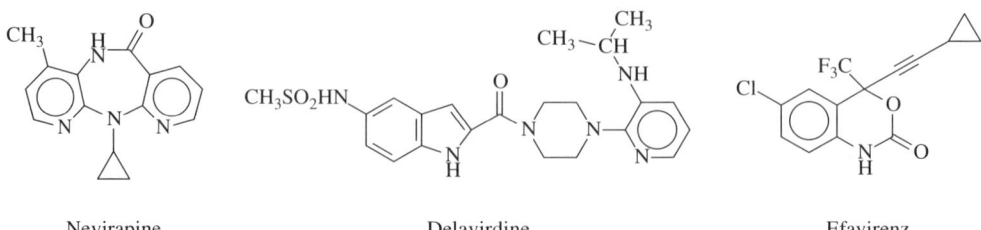

Nevirapine Delavirdine Efavirenz

Figure 25.13 Structure of nonnucleoside reverse transcriptase inhibitors.

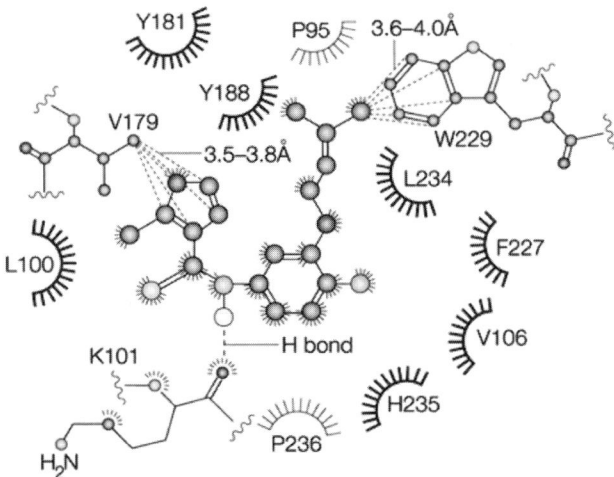

Figure 25.14 Interaction of HIV-1 RT with UC781. Features stabilizing the complex between the human immunodeficiency virus 1 (HIV-1) reverse transcriptase (RT) and the nonnucleoside reverse transcriptase inhibitor UC781 (Fig. 25.2). The hydrogen bond with K101, and the two methyl group–aromatic ring interactions are shown explicitly. Other main hydrophobic contacts are shown with bold lines; minor ones are shown with faint lines [53]. Standard CPK coloring is used.

valent to a phosphate group, but unlike phosphate, phosphonate can no longer be cleaved through esterases that would normally convert nucleoside monophosphates back to their nucleoside form. Consequently, acyclic nucleoside phosphonates may be expected to show a broadened antiviral activity spectrum as compared to that of the acyclic nucleoside analogs (acyclovir, . . .) and dideoxynucleoside analogs (zidovudine, . . .). On the one hand, they should be active against those DNA viruses that do not encode for a specific viral thymidine kinase (TK) or protein kinase (PK) or have become resistant to the nucleoside analogs through TK or PK deficiency. On the other hand, they should also be able to bypass the nucleoside kinase reaction that limits the activity of the dideoxynucleoside analogs against retroviruses (HIV) and hepadnaviruses (HBV).

These premises were fulfilled on both scores, albeit by different types of acyclic nucleoside phosphonates: cidofovir (HPMPC), as to the broad-spectrum activity against DNA viruses; and adefovir (PMEA) and tenofovir (PMPA, Fig. 25.2), as to the activity against retro- and hepadnaviruses.

Although their eventual activity spectrum is different, both types of compounds share a common strategy in their mode of action: They both need two (instead of three) phosphorylation steps to be converted to their active (diphosphorylated) metabolites, which then act as chain terminators in the DNA polymerase reaction [HPMPC (Fig. 25.6c)] or reverse transcriptase reaction [PMEA, PMPA (Fig. 25.6d)]. For HPMPC to shut down viral DNA synthesis, the incorporation of two consecutive HPMPC units is required [65], whereas for PMEA one such incorporation suffices [66]. In both cases, the acyclic nucleotides remain stably incorporated, presumably because the phosphonate group prevents repair enzymes excising these nucleotides.

The "era" of the acyclic nucleoside phosphonates started with the description of the broad-spectrum anti-DNA virus activity of the adenine derivative HPMPA [67]. These nucleotide analogs are isopolar and sterically similar to the natural nucleotides. It is rather surprising that none of the furanose nucleosides with a 5′-phosphonate moiety exhibit appreciable antiviral activity, which might be partly due to inefficient cellular uptake, partly due to diminished affinity for the metabolizing or target enzymes. The acyclic nucleotide analogs, however, demonstrate antiviral activity, although the structural window with respect to biological activity is rather narrow. Potent antiviral activity has been found mainly in the HPMP (hydroxypropoxymethylphosphonate) series and PME (phosphonomethoxyethyl) series. The hydroxyl group in the HPMP series may be replaced by a fluorine or hydrogen atom. The PME pyrimidine derivatives do not show inhibitory effect on viral multiplication. Phosphonomethylether nucleosides have unusual pharmacokinetic properties. They are very slow to pass biological membranes presumably due to the negatively charged phosphonate moiety. This charge also entails the low oral bioavailability of the phosphonates. Nucleosides phosphonates are not taken up into cells by the nucleoside transport mechanism. There limitations can successfully be addressed by utilizing a prodrug strategy, that is, pivaloyloxymethyl, isopropyloxycarbonyloxymethyl, or hexadexyloxypropyl esters. When chemically or enzymatically deacylated, the hydroxyalkyl phosphonate quickly decomposes to release the phosphonic acid and the corresponding aldehyde or ketone. Based on stability, solubility, and improved oral bioavailability, the bis(POC) derivative of tenofovir and the bis(POM) derivative of adefovir were selected for development and have now been approved for the treatment of HIV and HBV infections, respectively.

The cytosine compound, cidofovir (HPMPC), which appeared to be less harmful to the host in preliminary toxicity experiments, was developed as an antiviral drug [68] and approved for clinical use in the treatment of CMV retinitis in AIDS (acquired immunodeficiency syndrome) patients. Cidofovir also holds great potential for the treatment of several other DNA virus infections, namely TK-deficient HSV and VZV infections that are resistant to acyclovir (or brivudin), and herpesvirus infections at large (EBV, HHV-6, HHV-7, HHV-8), human papillomavirus (HPV) infections [i.e., pharyngeal, esophageal, and laryngeal papillomatosis, plantar and genital warts (condylomata acuminata), and cervical intraepithelial neoplasia], polyomavirus infections (i.e., progressive multifocal leukoencephalopathy), adenovirus infections (i.e., epidemic keratoconjunctivitis), poxvirus infections [i.e., smallpox, monkeypox, cowpox, orf, molluscum contagiosum (for the role of cidofovir in the treatment of poxvirus infections, see also Refs. 69–71)]. Adefovir and tenofovir, the two other protagonists from the acyclic nucleoside phosphonate group, have in the mean time progressed, both in their oral prodrug forms, adefovir dipivoxil and tenofovir disoproxil (with fumarate added as salt), to final licensing for the treatment of HBV and HIV infections, respectively.

In contrast to all other antiviral drugs, acyclic nucleoside phosphonates have a particularly long intracellular half-life (one to several days), thus allowing infrequent dosing (only once daily for adefovir and tenofovir, or once weekly or every other week for cidofovir), and do not easily lead to resistance, even after prolonged treatment (for more than one or two years). No drug metabolic interactions are known for the acyclic nucleoside phosphonates, which means that they can readily be added onto any drug (combination) regimen, as has been particularly shown for tenofovir in the treatment of HIV infections.

Inhibitors of Processes Associated with Viral RNA Synthesis

Gene expression (i.e., transcription to viral RNA) of the genome of retroviruses such as HIV is not possible without integration of the proviral DNA into the host chromosome (Fig. 25.1). Thus the enzyme involved—integrase—has been considered as an attractive target for chemotherapeutic intervention. Numerous integrase inhibitors have been described [72, 73]; of those none, however, with sufficient specificity to be further pursued as an integrase-targeted drug. The problem with integrase inhibitors is that, while they could be effective in an enzyme-based assay, their anti-HIV activity in cell culture may be masked by cytotoxicity. And even if selective anti-HIV activity in cell culture is noted, caution should be exercised in unconditionally attributing this activity to inhibition of the integration process, as the compounds concerned may well owe their anti-HIV activity to an action targeted elsewhere. This has proven to be the case for the anionic compounds zintevir [7] and L-chicoric acid [74], two integrase inhibitors that owe their anti-HIV activity primarily to an interaction with the viral envelope gp120, and thus fall in the category of the polyanionic inhibitors of virus adsorption (see Section 25.2). Up till now, the only compounds that qualify as genuine integrase inhibitors are diketo acid derivatives (i.e., L-731,988 and L-708,906) [75–77] and pyranodipyrimidine (PDP) derivatives (i.e., V-165) [78]. These compounds were found to inhibit HIV-1 replication in cell culture, on the one hand, and to inhibit the strand transfer function of the integrase (the other catalytic function of the enzyme being endonucleolytic cleavage of the terminal dinucleotide GT from the 3′-end), on the other hand. These two events could be causally linked as mutations in the HIV-1 integrase conferred resistance to the inhibitory effects of the compounds on both strand transfer and HIV-1 infectivity [75]. Clinical trials with the diketo acid derivatives (S-1360 and L-870180) have recently been initiated.

At the transcription level, HIV gene expression may be inhibited by compounds that interact with cellular factors that bind to the long terminal repeat

(LTR) promoter and that are needed for basal-level transcription, such as NF-κB inhibitors [79], but greater specificity may be expected from those compounds that specifically inhibit the transactivation of the HIV LTR promoter by the viral trans-acting transactivator (Tat) protein. The Tat protein interacts specifically with a responsive element, called TAR, located at the beginning of the viral messenger RNA (mRNA) transcribed from the LTR promotor, thereby enhancing ("transactivating") the transcription process. Several compounds have been described as inhibitors of the transcription process, for example, fluoroquinolines [80] and bistriazoloacridone derivatives such as temacrazine [81]. The latter was found to block HIV-1 RNA transcription starting from the HIV LTR promoter without interfering with the transcription of any cellular genes. The peptide analog CGP64222, which is structurally reminiscent of the amino acid 48–56 sequence RKKRRQRRR of the Tat protein, was originally designed to act as a Tat antagonist [82]. However, although CGP64222 is able to interact with the Tat-driven transcription process, its anti-HIV activity in cell culture is mediated primarily by an interaction with CXCR4, the co-receptor for X4 HIV strains [83].

Viral RNA transcription could also be affected by targeting cyclin-dependent kinases (cdks), which are required for the replication of many viruses (including HIV). Indeed, flavopiridol, a typical inhibitor of cdks (particularly cdk9, which is involved in the Tat-driven transcription process), has proven to be effective in blocking HIV infectivity [84].

Capping and methylation of HIV pre-mRNAs are coupled to the elongation by polymerase II. Binding of the capping enzyme and cap methyltransferase to polymerase II depends on phosphorylation of its carboxyl-terminal domain. It has been recently demonstrated that the co-transcriptional capping of HIV mRNA is stimulated by the Tat protein, consequently to its binding to the capping enzyme [85]. These findings implicate capping as an elongation checkpoint critical to HIV gene expression and thus corroborate earlier observations that S-adenosylmethionine-dependent methylations play an important role in the Tat-dependent transactivation of transcription from the LTR promoter [86]. They also offer an explanation for the inhibitory effects observed with S-adenosylhomocysteine hydrolase inhibitors (see Section 25.8) on Tat-dependent transactivation and HIV replication [86].

One of the virus infections in the greatest need of antiviral therapy is HCV, and, here, two specific enzymatic functions associated with viral RNA synthesis could be envisaged as targets for the design of new antiviral agents, namely the nonstructural protein 3 (NS3)-associated helicase and the nonstructural protein 5B (NS5B) RNA-dependent RNA polymerase. Crystal structures of both enzymes are available [87–89], and both enzymatic activities have been characterized in sufficient detail [90, 91] to facilitate the development of effective HCV chemotherapeutics. For the helicase, there is no precedent, but for the RNA polymerase there is, and the experience gathered from the studies with the HIV RT inhibitors may be of paramount importance when targeting the HCV RNA polymerase, especially if, as it appears [91], this enzyme shows similar kinetics to the HIV RT. Meanwhile, the first indications of compounds targeted at the NS5B polymerase (i.e., the benzimidazole derivative JTK-003) have appeared [89].

25.5 VIRAL PROTEASE INHIBITORS

Viral proteases play a critical role in the life cycle of many different viruses, including retroviruses such as HIV, herpesviruses, picornaviruses (rhino), and flaviviruses such as HCV, and hence viral proteases have been favored as targets for antiviral agents [92]. Their role is essentially based upon the cleavage of a newly expressed precursor polyprotein into smaller, mature viral proteins, termed functional (if endowed with enzymatic activity) or structural (if being part of the virion structure). For example, in HIV replication, HIV protease cleaves the gag and gag-pol precursor proteins to the structural proteins (p17, p24, p9, p7) and functional proteins (protease, reverse transcriptase/RNase H, and integrase). HIV protease inhibitors have been tailored to the peptidic linkages (e.g., F-P, F-L, and F-T) in the gag and gag-pol precursor proteins that are cleaved by the protease, and have been extensively modeled in the active site of the enzyme. All protease inhibitors that are currently licensed for the treatment of HIV infection, namely saquinavir, ritonavir, indinavir, nelfinavir, amprenavir, and lopinavir, share the same structural determinant (Fig. 25.2), that is, an hydroxyethylene core (instead of the normal peptidic linkage) that makes them nonscissile, "peptidomimetic," substrate analogs for the HIV protease. The hydroxyethylene hallmark is also present in the azadipeptide atazanavir (BMS-232632) [93], expected to be licensed soon for the treatment of HIV infection.

The HIV protease belongs to the aspartyl enzymes and has a dimeric structure (Fig. 25.15) [94]. The monomeric units, both identical 99-residue polypeptides, each contribute symmetrically to the catalytic site, and the enzyme is a C_2-symmetric dimer. The protease is unique in its specificity, being able to hydrolyze almost any peptide bond, given an optimal distribution of amino acids in P_4 through $P_{4'}$. The hallmark of the aspartyl proteases is the presence of two catalytic β-carboxyl groups from the side chains of two aspartyl residues brought into close proximity by the protein fold. In the case of the dimeric HIV-1 protease, each monomer chain donates a single Asp residue at position 25 in the polypeptide chain (Asp_{25} and $Asp_{25'}$). One of the Asp is more acidic (pK_a 3.4 to 3.7) than the other Asp (pK_a 5.5 to 6.5). The HIV protease participates in general acid–general base catalysis where one of the catalytic Asp is protonated and the other is not. The nucleophile in catalysis is, most probably, an activated water molecule. The mechanism is given is Scheme 25.1. The amide hydrate is considered as the transition state. This mechanism is relevant for inhibitor design, that is, making HIV-1 protease inhibitors as transition-state mimics (nonhydrolysable isosters with tetrahedral geometry).

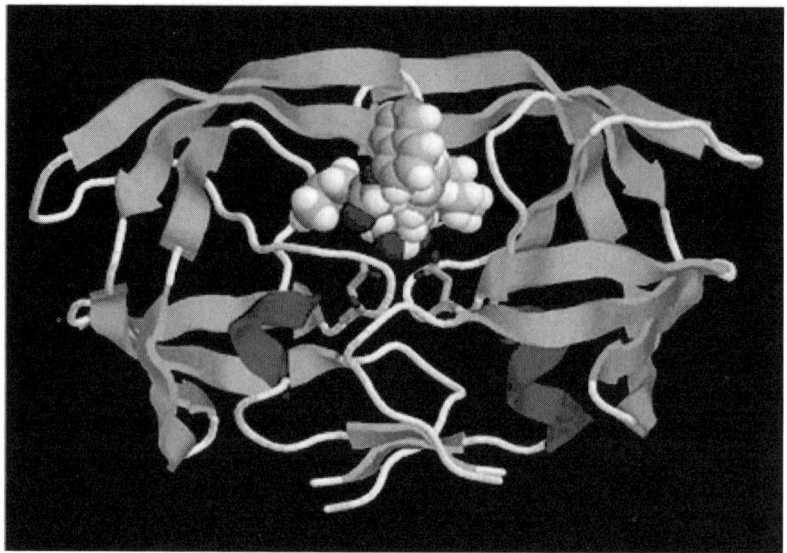

Figure 25.15 Interaction of HIV protease with KNI272. Ribbon diagram of human immunodefiency virus (HIV) protease complexed with the peptidomimetic protease inhibitor KNI272; derived from the crystal structure [94]. The inhibitor is shown as a space-filling model, and the two active-site aspartic acids are shown as sticks; both have standard CPK coloring.

In the enzyme, the catalytic triads, Asp_{25}-Thr_{26}-Gly_{27}, and $Asp_{25'}$-$Thr_{26'}$-$Gly_{27'}$ lie on the floor of the cavity with the catalytic Asp_{25} and $Asp_{25'}$ in near coplanar orientation. The ceiling of the catalytic site is composed of a pair of six-amino-acid segments (-Ile_{47}-Gly-Gly-Ile-Gly-Gly_{52} . . .) and of the corresponding region 47'–52', in the other monomer, which forms the flap. The flaps are flexible structures and they play an important role in substrate and inhibitor binding. Upon binding the inhibitors, the HIV-1 protease undergoes a conformational change, where the flaps move from an open conformation to a conformation where the ligand is embraced. A water molecule with approximate tetrahedral coordination bridges the P_2 and $P_{1'}$ carbonyl groups of the inhibitor to the amide groups of Ile_{50} and $Ile_{50'}$ of the flaps via hydrogen bonds. The hydroxyethylene group of the transition-state inhibitor is normally aligned with the catalytic carboxylates of Asp_{25} and $Asp_{25'}$.

The core structure for tight binding of an inhibitor consists of four side chains corresponding to the substrate residues P_2–$P_{2'}$. In the first protease inhibitor that has reached the market, saquinavir, it was observed that the hydroxyethylamine transition-state mimetic was readily adapted to the $P_{1'}$ proline containing substrate sequence and that the activity of the inhibitor could be markedly improved in potency when the amino acid proline at the $P_{1'}$ site was replaced by the (S,S,S)-decahydro-isoquinoline-3-carbonyl group.

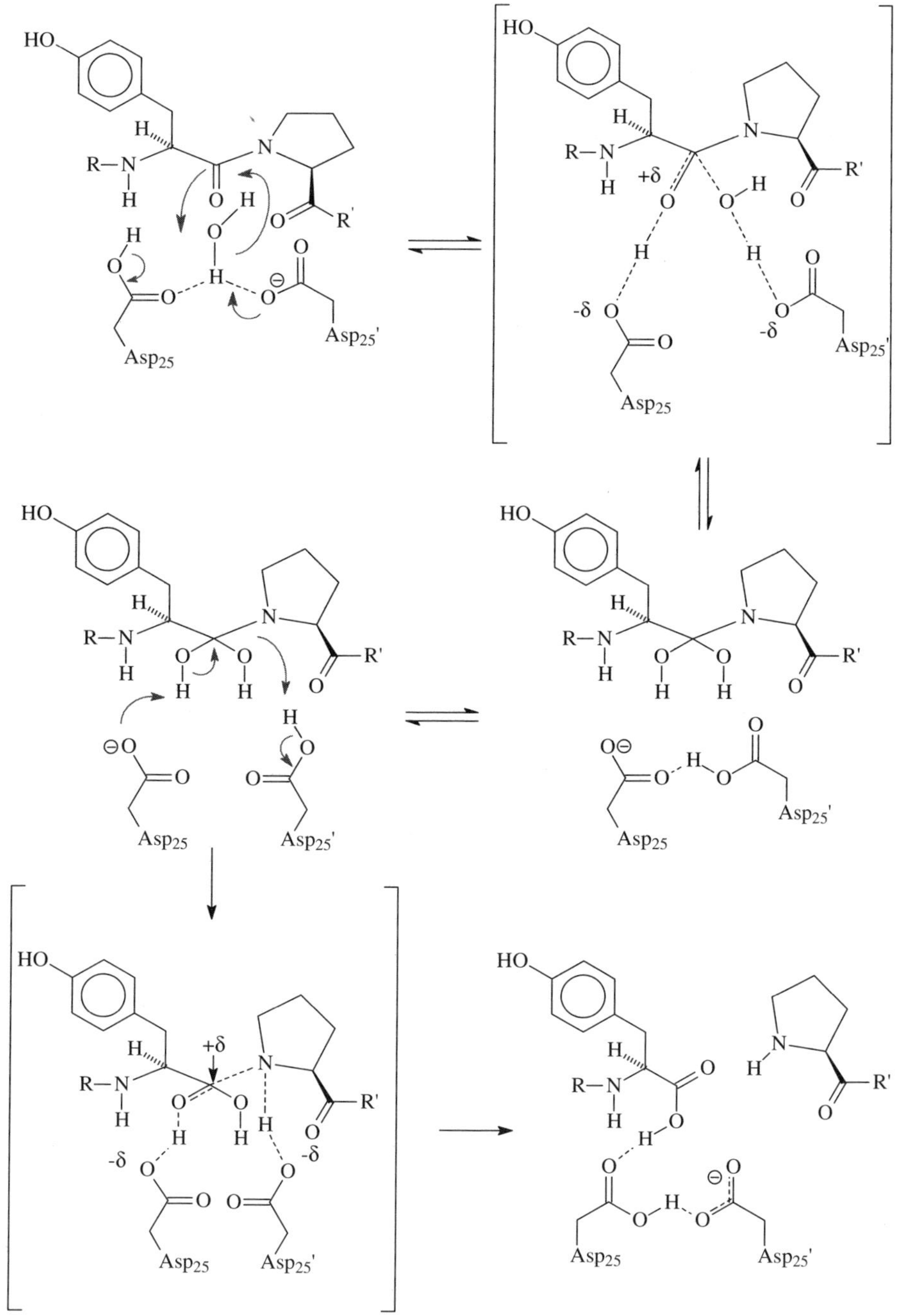

Scheme 25.1 Proposed chemical mechanism of the peptidolytic reaction of HIV-1 protease based on kinetic and structural data. Proposed hydrogen bonds are indicated by dashed lines and are based in part on molecular modeling of the enzyme-intermediate complex [96].

An interesting example of de novo drug design is the cyclic urea mozenavir (DMP-450) as a nonpeptidic inhibitor of HIV-1 protease [95], where the structural water linking the flap residues Ile_{50} and $Ile_{50'}$ is displaced (Fig. 25.16). This is an energetically favorable process. The carbonyl group replacing the structural water molecule is part of a cyclic urea because this function is an excellent hydrogen bond acceptor. The diol moiety is hydrogen binding to Asp_{25} and $Asp_{25'}$. The other substituent ensures high complementarity with the enzyme subsites. However, the virus was able to mount considerable resistance against this compound, which was not further developed, also because of poor oral bioavailability in humans and highly variable blood levels. Mozenavir is, perhaps, one of the nicest examples of rational drug design, attesting that the most "rational" approaches may fail at the end because of (unpredictable) lack of clinical relevance.

The HIV protease inhibitors can be designed using a stepwise approach of lead optimization. This is demonstrated here by the discovery of saquinavir. The discovery of saquinavir started with the identification of a protected dipeptide as a weak enzyme inhibitor. An increase in potency was obtained by introduction of a P2 asparagine residue (Fig. 25.17). The replacement of

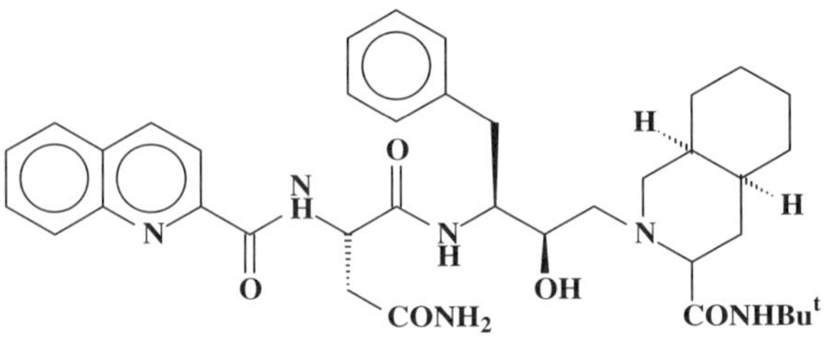

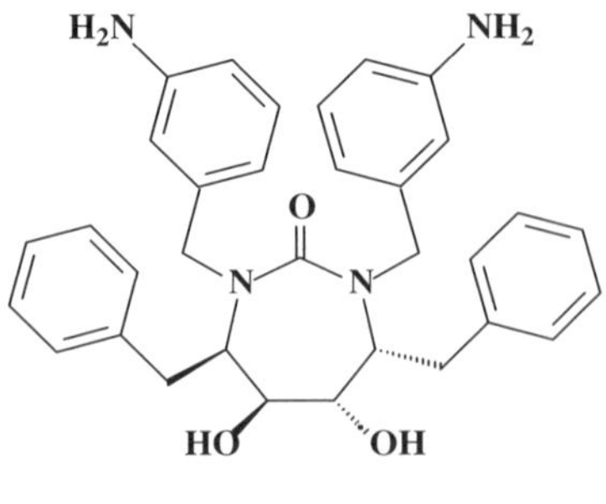

Figure 25.16 Structure of saquinavir and DPM-323. Saquinavir has been developed based on a systematic optimization of a core structure. DPM-323 is the result of a rational design approach, but the compound never reached the market because of failure in late-stage development.

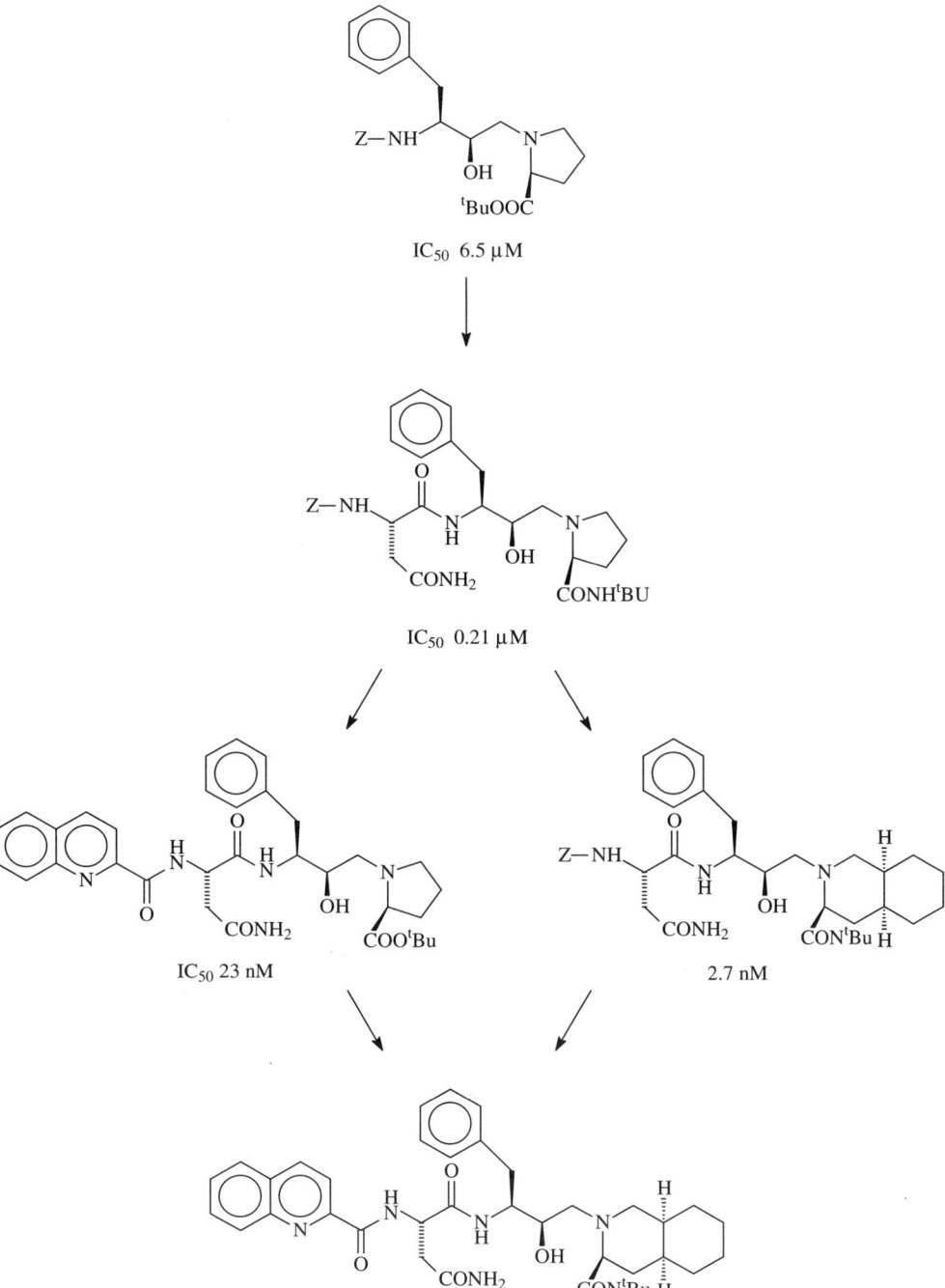

Figure 25.17 Lead optimization of a dipeptide protease inhibitor leading to the discovery of saquinavir.

the Z group (in the C-terminal 'Butylester analog) by a 2-quinolinyl group gave a compound that inhibited HIV-1 at nanomolar concentrations (~20nM). A systematic investigation of proline replacement led to the discovery of the decahydroisoquinoline substituent as optimal C-terminal (IC$_{50}$: 2.7nM). An increase in potency was obtained by combining both substituents, leading to the synthesis of saquinavir, which inhibits both HIV-1 and HIV-2 with IC$_{50}$ values <1 nM. The compound has a low but sufficient oral bioavailability.

The HIV protease inhibitors (PIs) have proven to be valuable therapeutics in drug combination schedules [HAART (highly active antiretroviral therapy)] with NRTIs and NNRTIs, in the treatment of HIV infections. Yet, they are met by compounding factors such as difficulties in drug adherence, drug–drug interactions, overlapping resistance patterns, and long-term side effects such as lipodystrophy and cardiovascular and metabolic disturbances (i.e., diabetes). This has prompted the search for new, nonpeptidic inhibitors of HIV protease, with (instead of the peptidomimetic hydroxyethylene core) cyclic urea, 4-hydroxycoumarin, L-mannaric acid, or 4-hydroxy-5,6-dihydro-2-pyrone (Fig. 25.2) [as in tipranavir (PNU-140690)] as the central scaffold [97, 98]. Such compounds should show little, if any, cross-resistance with the peptidomimetic inhibitors [99]; their in vivo efficacy, pharmacokinetics, and short- and long-term safety profiles remain to be established.

Whether the PI approach would be as successful for tackling herpes-, picorna-, and flaviviruses, as it turned out to be for HIV, remains to be seen. While the HIV protease is an "aspartate" protease, herpesvirus proteases belong to the "serine" proteases with SHH as the catalytic triad [100]. Several nonpeptidic inhibitors of herpesviral protease (i.e., CMV protease, which is also referred to as assemblin because of its role in the CMV assembly process) have been described, for example, thieno[2,3-d]oxazinones [101], aryl hydroxylamine derivatives [102], monobactams [103], pyrrolidine-5,5-*trans*-lactams [104], and 1,4-dihydroxynaphthalene and naphthoquinones [105]. While a useful exercise at targeting the herpesviral protease, all these efforts should be viewed as a prelude to further investigations on the in vitro and in vivo inhibitory effects of these compounds on virus replication.

Further advanced is the structure-assisted design of mechanism-based, irreversible, inhibitors of human rhinovirus 3C protease (a "cysteine" protease that is involved in the proteolytic cleavage of the viral precursor polyprotein to both capsid and functional proteins required for RNA replication). This work has yielded a wealth of compounds with potent activity against multiple rhinovirus serotypes [106–111]. Of the human rhinovirus 3C protease inhibitors, AG7088, which was shown to inhibit rhinovirus replication even when added late in the virus life cycle [112], and ruprintrivir [113] have proceeded to clinical trials for the prevention and treatment of the common cold.

The HCV protease is a "serine" protease that is encoded by the nonstructural NS3 protease domain and is responsible for the proteolytic cleavage of the nonstructural NS3, NS4A, NS4B, NS5A, and NS5B from the viral polyprotein (the NS4A protein then binds to the NS3 protein and enhances its proteolytic activity). The HCV NS3-4A protease is remarkably similar to the pestiviral NS3-4A protease [as found in BVDV (bovine viral diarrhea virus)] [114] and has been intensively pursued as a target for the design of inhibitors. Again, as for the herpesviral serine protease, several inhibitors, both peptide-based [115, 116], nonpeptidic [117], and macrocyclic [89], of the HCV NS3-4A protease have been identified, but as there is no cell culture assay available for HCV, their activity against HCV infectivity could not be assessed. Given the similarities of the HCV and BVDV NS3-4A proteases, it would seem justifiable to evaluate putative HCV protease inhibitors for their activity against BVDV replication, for which cell culture assay systems have been established. In addition, HCV protease inhibitors, as well as HCV helicase and polymerase could and should also be evaluated in the subgenomic replicon system [118].

25.6 VIRAL NEURAMINIDASE INHIBITORS

Influenza virus (both A and B) has adopted a unique replication strategy in using one of its surface glycoproteins, hemagglutinin, to bind to the target cell receptor [containing a terminal sialic acid (*N*-acetylneuraminic acid, NANA)], and the other surface glycoprotein, neuraminidase, to leave the cell after the viral replicative cycle has been completed. The viral neuraminidase is thus needed for the elution of the newly formed virus particles from the cells. In addition, the viral neuraminidase may also promote viral movement through the respiratory tract mucus, thus enhancing viral infectivity. Therefore, the influenza viral neuraminidase has been envisaged as a suitable target for the design of specific inhibitors. Neuraminidase (sialidase) is a tetrameric glycoprotein with a subunit molecular weight of 60 kDa. The enzyme is a glycohydrolase cleaving α-ketosidically linked terminal *N*-acetyl-D-neuraminic acid residues on the mucosal sialoglycoproteins (Fig. 25.18).

An interesting observation is that 18 of the residues in the active site of influenza A and B sialidase sequences that have been characterized, are conserved. Fifteen of the conserved residues are charged. Eight of the strain-invariant active-site residues are positioned to make direct contact with *N*-acetyl-D-neuraminic acid bound into the catalytic site, with a further tier of 10 residues that appears to establish a structural scaffold for the catalytic site. A stabilized sialosyl cation transition-state intermediate is involved when influenza virus sialidase catalyzes the release of *N*-acetyl-D-neuraminic acid from α-sialosides. *N*-acetyl-D-neuraminic acid (as an anomeric mixture) inhibits influenza A virus neuraminidase with a K_i of 10^{-3} M. More potent inhibitors are those based on the unsaturated transition-state analogs with a glycal structure (2-deoxy-2,3-didehydro-*N*-acetylneuraminic acid) (Fig. 25.19).

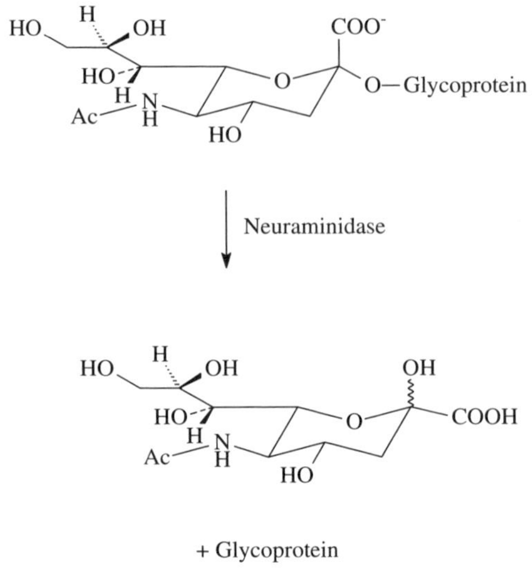

Figure 25.18 The glycohydrolase activity of neuraminidase.

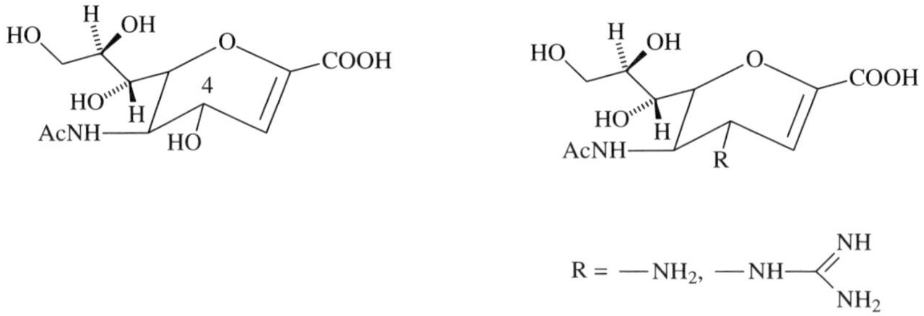

Figure 25.19 Unsaturated transition-state analogs that inhibit influenza A virus neuraminidase.

One of the most important interactions between *N*-acetyl-D-neuraminic acid and its unsaturated congener with the enzyme is that between the carboxylate of both of these sialic acids and a triargininyl cluster (Arg_{118}, Arg_{292}, and Arg_{371}). This interaction is important both enzymatically—on binding, the sugar ring is flattened out prior to the cleavage of the glycosidic bond—and structurally, the electrostatic interaction between the negatively charged carboxylate and the positively charged amine groups of the arginines being a major contributor to the binding of substrates and substrate-based inhibitors. Based on modeling experiments, a series of 4-substituted-4-deoxy-*N*-acetyl-D-neuraminic acid derivatives were synthesized. The 4-amino analog has a K_i of 4×10^{-8} M against influenza virus neuraminidase, which represents an increase in binding affinity when compared with 2-deoxy-2,3-didehydro-*N*-

acetylneuraminic acid (K_i of 4×10^{-6} M). The replacement of the amino group with the larger more basic guanidino functionality led to an even more potent inhibitor (K_i of 3×10^{-11} M). The guanidino substituent is situated within hydrogen bond distance with two conserved acid residues. Zanamivir selectively inhibits influenza virus sialidase, while other sialidases from both bacterial and mammalian sources are not inhibited. The second approach toward new sialidase inhibitors has been concentrated on the synthesis of carbocyclic mimetics and the introduction of a lipophilic substituent in place of the glycerol side chain, which led to very potent orally active compounds (i.e., oseltamivir) for the treatment of influenza virus infections.

Computer-assisted drug design, led to the identification of zanamivir (GG167) as a specific and potent inhibitor of the enzyme and of the in vitro and in vivo replication of both influenza A and B virus [119]. Zanamivir was tailored to interact with the conserved amino acid residues within the active site of influenza A and B virus neuraminidase, and its inhibitory effect on the enzyme has proved to be predictive of the susceptibility of clinical isolates to the drug [120]. Meanwhile, zanamivir was shown to be efficacious and safe in the treatment (by inhalation) of influenza virus infections [121], and the drug has been licensed for clinical use.

Zanamivir has to be given by inhalation because of its poor oral bioavailability. In attempts to identify potentially orally bioavailable inhibitors, a series of carbocyclic transition-state-based analogs were developed in which the polar glycerol and guanidino groups (as present in zanamivir) were replaced by a lipophilic (3-pentyloxy) side chain and amino group, respectively, to give GS 4071 [122]. X-ray crystallographic studies showed that these groups of GS 4071 could be accommodated within the active site of the neuraminidase (Fig. 25.20). As aimed for, GS 4071, when administered as the ethyl ester prodrug (GS 4104), turned out to be orally bioavailable and was found to be effective in protecting mice and ferrets against influenza infection [123]. Subsequently, GS 4104 (oseltamivir) was found to be effective and safe in the oral treatment and prevention of influenza virus infections [124, 125] and has been licensed for clinical use.

Zanamivir and oseltamivir have paved the way for the development of similarly, structure-based, designed neuraminidase inhibitors, such as the cyclopentane derivative RWJ-270201 (with a comparable, or even better, efficacy profile in the murine influenza model) [126–128] and the pyrrolidine derivatives A-192558 and A-315675 [129–131]. The clinical potential of RWJ-270201, A-192558, and/or A-315675 in the prevention and/or treatment of influenza virus infections in humans still needs to be established.

25.7 IMP DEHYDROGENASE INHIBITORS

IMP dehydrogenase is a key enzyme in the de novo biosynthesis of purine mononucleotides: It is responsible for the NAD-dependent oxidation of IMP

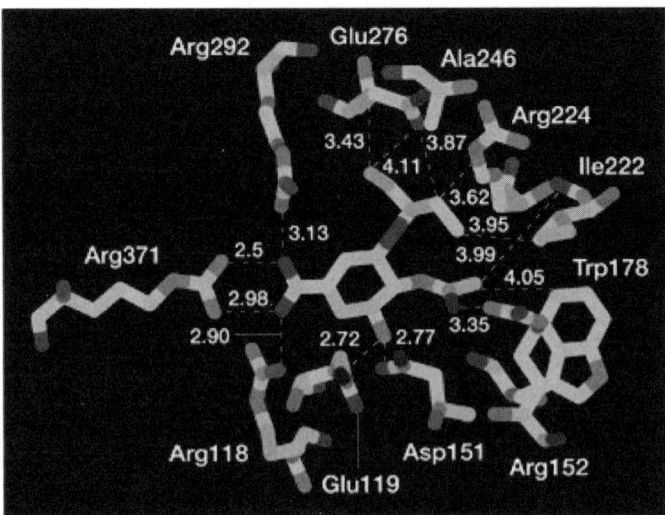

Figure 25.20 Interaction of influenza neuraminidase with oseltamivir. The picture shows binding interactions of the neuraminidase inhibitor oseltamivir (Fig. 25.2) with influenza neuraminidase, derived from the crystal structure.

(inosine 5′-monophosphate) to XMP (xanthosine 5′-monophosphate), which is then further converted to GMP, GDP, and GTP, and from GDP, via dGDP, also to dGTP. Inhibitors of IMP dehydrogenase may be expected to affect both RNA and DNA synthesis, via reduction of the intracellular pools of GTP and dGTP, respectively. Although IMP dehydrogenase is a cellular target, inhibitors targeted at this enzyme may be expected to preferentially inhibit viral RNA and/or DNA synthesis, as there is an increased need for such syntheses in virus-infected cells.

IMP dehydrogenase can be targeted by two types of inhibitors: competitive or uncompetitive with respect to the normal substrate, IMP. To the first category belongs ribavirin, which has been officially approved for clinical use as an aerosol for the treatment of RSV infections and in combination with interferon-α for the treatment of HCV infections.

To the second category belongs mycophenolic acid (Fig. 25.2), an immunosuppressing agent, which has been approved as its morpholinoethyl ester prodrug, for the prevention of acute allograft rejection following kidney transplantation. The X-ray crystal structure of IMP dehydrogenase, complexed with mycophenolic acid at the active site, has been reported at high resolution (2.6 Å) [132]. New congeners of both ribavirin (Fig. 25.2) (i.e., EICAR) [133] and mycophenolic acid (i.e., VX-497) [134] have an activity spectrum as broad as ribavirin, but considerably greater potency. This activity spectrum encompasses both DNA and RNA viruses, and among the latter, picorna-, toga-, flavi-, bunya-, arena-, reo-, rhabdo-, and, particularly, ortho- and paramyxoviruses.

Mycophenolic acid has marked activity against yellow fever virus and, in addition, markedly potentiates the inhibitory effects of acyclic guanosine analogs (acyclovir, penciclovir, ganciclovir) against HSV, VZV, and CMV infections, which could be of great clinical utility in organ transplant recipients subject to these infections [135]. Also, mycophenolic acid potentiates the activity of guanine-derived dideoxynucleoside analogs, such as abacavir, against HIV [136], which may be further exploited as a new combination strategy in the treatment of HIV infections.

While IMP dehydrogenase inhibitors should, in their own right, be further explored for their potential in the treatment of various (+)RNA and (−)RNA virus infections, including hemorrhagic fever virus infections, current interest is mainly focused on their use in combination with (pegylated) interferon-α in the treatment of HCV infections. This stems from the successful responses that have been seen following treatment of chronic hepatitis C with ribavirin in combination with interferon-α (in patients who did not respond to interferon alone) [137].

Recently, ribavirin was shown to act as an RNA virus mutagen, forcing RNA viruses into a lethal accumulation of errors, dubbed "error catastrophe" [138, 139]. The antiviral activity of ribavirin may then result from the lethal mutagenic effect following incorporation of ribavirin into the viral genome, and, obviously, this lethal mutagenesis may be enhanced by the inhibitory effect of ribavirin (in its 5′-monophosphate form) on IMP dehydrogenase and the consequent decrease in cellular GTP pools (as mentioned above). The ability of ribavirin to force RNA viruses into error catastrophe has so far been shown only with poliovirus [138, 139], and it remains to be verified whether the theory also holds for other RNA viruses (i.e., HCV) and other ribavirin analogs (i.e., EICAR).

25.8 S-ADENOSYLHOMOCYSTEINE HYDROLASE INHIBITORS

S-adenosyl-L-homocysteine (AdoHcy, SAH) hydrolase is a key enzyme in methylation reactions depending on S-adenosylmethionine (SAM) as the methyl donor, including those methylation reactions that are required for the maturation of viral mRNA. In particular, (−)RNA viruses are critically dependent on these methylations for the stability and functioning of their mRNA. All capped, methylated structures consist of a N7-methyl guanosine residue linked at the 5′-hydroxyl group to the 5′-end of the mRNA strand by a triphosphate linkage. Most 5′-capped methylated structures also contain a methyl group on the 2′-hydroxy group of the penultimate nucleotide (Fig. 25.21). This 5-capped, methylated structure has been shown to protect mRNA form 5′-end nuclease digestion. Methylation of the capped structure increases the affinity for ribosome binding to the 5′-end of this mRNA during formation of the translational initiation complex. Uncapped or undermethylated viral mRNA is less effectively translated into viral proteins. Both

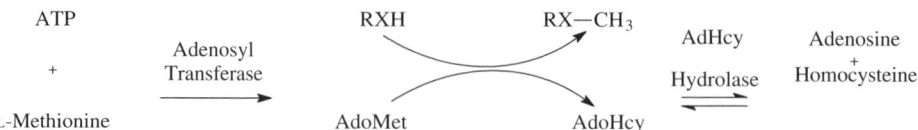

Figure 25.21 Structure of capped mRNA showing the methylated terminal guanine and the methylated 2'-hydroxyl groups.

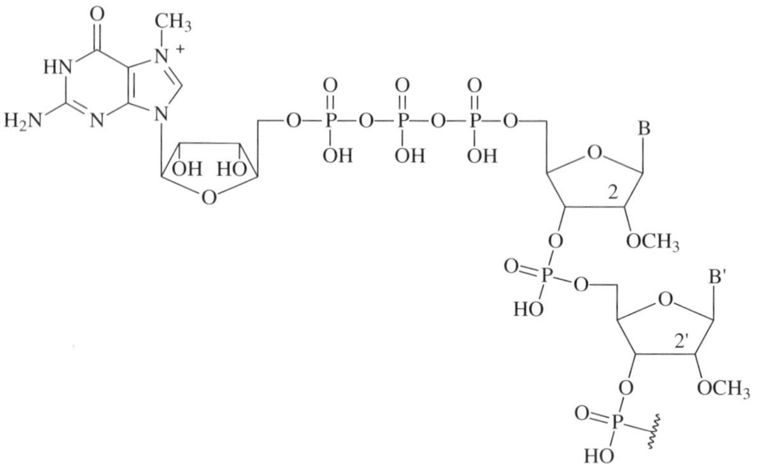

Figure 25.22 Adenosylmethione (synthesized from ATP and L-methionine) functions as methyldonor in a methyltransferase reaction and is converted to adenosyl homocysteine. AdoHcy hydrolase hydrolyses AdoHcy to adenine and homocysteine.

guanine-7 methyltransferase and nucleoside-2' methyltransferase coded by vaccinia virus require AdoMet as a methyl donor (Fig. 25.22). SAH is both a product and inhibitor of the methyltransferase reactions; however, SAH is rapidly hydrolyzed by SAH hydrolase into homocysteine and adenosine, and this prevents the accumulation of SAH that would otherwise lead to an inhibition of the SAM-dependent methylation reactions.

Mammalian AdoHcy hydrolase is a homotetramer containing one NAD^+ per subunit. The first step in the enzymatic reaction involves oxidation of the 3'-hydroxyl group of S-adenosyl-L-homocysteine to form 3'-keto-AdoHcy, resulting in the conversion of NAD^+ to NADH. The 3'-keto group increases the acidity of the C-4' proton, allowing for abstraction of this proton by a base in the active site of the enzyme. Subsequently, β-elimination of homocysteine results in the formation of the intermediate 3'-keto-4',5'-didehydro-5'-deoxyadenosine. Addition of water to the 5'-position of this intermediate affords 3'-keto-adenosine, which is then reduced by the enzyme-bound NADH, resulting in the formation of adenosine and regeneration of NAD^+.

Inhibitors of the SAH hydrolase may be expected to lead to an accumulation of SAH, and consequent inhibition of the methylation reactions. Again, as noted for IMP dehydrogenase, SAH hydrolase is a cellular target, but as virus replication increases the need for such methylations, SAH hydrolase inhibitors may confer selective antiviral activity that may well vary from one virus to another depending on their individual methylation needs.

Various adenosine analogs, that is, carbocyclic adenosine, carbocyclic 3-deazaadenosine, neplanocin A, 3-deazaneplanocin A, and their 5'-nor derivatives, have been described as potent inhibitors of SAH hydrolase [140]. Carbocyclic adenosine and replanocin A belong to the first generation of AdoHcy hydrolase inhibitors. Carbocyclic adenosine is a reversible, competitive inhibitor with a K_i of 5 nM. Neplanocin A, in contrast, is an irreversible, tightly binding inhibitor with a K_i of 8.4 nM. Neplanocin A inactivates the hydrolase by a "cofactor depletion" mechanism, converting the NAD^+ cofactor to its inactive form (NADH) with simultaneous oxidation of neplanocin A to 3'-keto-neplanocin A. This 3'-keto derivative is tightly bound to the NADH form of the enzyme (Fig. 25.23). Unfortunately, inhibitors of the first generation have a common problem of cellular toxicity, which precludes the clinical use of these compounds as antiviral agents. In attempts to design more specific inhibitors of AdoHcy hydrolase, two different approaches have been followed. One approach is the replacement of the adenine moiety with 3-deazaadenine, resulting in 3-deazaaristeromycin and 3-deazaneplanocin A. Another approach involves removing of the 4'-hydroxymethyl substituent, which would preclude 5'-phosphorylation by adenosine kinase and deamination by adenosine deaminase (Fig. 25.24).

These compounds are indeed not substrates for either adenosine deaminase and adenosine kinase, but they retain potent inhibitory activity against AdoHcy hydrolase. Inhibition of AdoHcy hydrolase by 3-deazaneplanocin and 3-deazaaristeromycin was reported to be reversible and competitive, with K_i values of 0.05 and 3 nM, respectively. In contrast, the 5'-nor derivatives have been shown to inactivate AdoHcy hydrolase irreversibly. These inhibitors serve as substrates for the enzyme's oxidative activity, resulting in reduction of the enzyme-bound NAD^+ to NADH. The reaction stops at this point since these inhibitors are not substrate for the "hydrolytic activity" of the enzyme. These second-generation AdoHcy hydrolase inhibitors exhibit

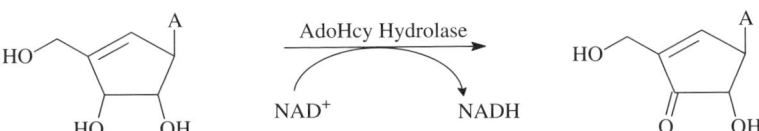

Figure 25.23 Enzymatic oxidation of neplanocin to its 3'-keto derivative with concommitant reduction of NAD^+ to NADH. The 3'-keto derivative of neplanocin is bond to the NADH coenzyme of SAH hydrolase.

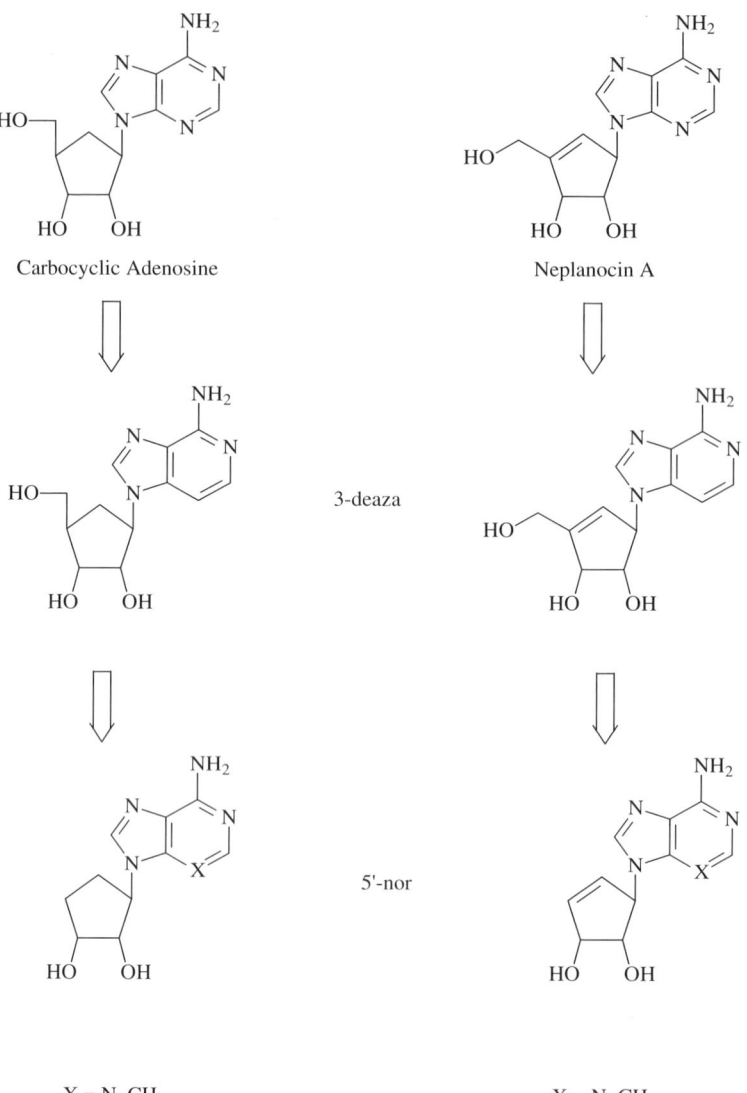

Figure 25.24 Approaches followed to discover new SAH hydrolase inhibitors, i.e., replacement of the N-3 atom of the adenine base by CH and removal of the 4′-hydroxy methyl group of the sugar moiety.

broad-spectrum antiviral activity, whereas their cytotoxicity is considerably lower than that of the parent compounds. A third type of mechanism-base inhibitors of AdoHcy hydrolase are defined as compounds that use the hydrolytic activity of the enzyme to convert a prodrug to a potent drug in the active site of the enzyme, from which (E)-5′,6′-didehydro-6′-deoxy-6′-halo-homoadenosines are representative examples [141].

All SAH hydrolase inhibitors possess a characteristic antiviral activity spectrum, encompassing, in particular, poxviruses, (±)RNA viruses (reo) and (−)RNA viruses (bunya-, arena-, rhabdo-, filo-, ortho-, and paramyxoviruses). This includes a number of hemorrhagic fever viruses such as Ebola hemorrhagic fever. In fact, a mouse model for Ebola hemorrhagic fever has been developed [142], and the SAH hydrolase inhibitors carbocyclic 3-deazaadenosine [143] and 3-deazaneplanocin A [144] were found to protect the animals against an otherwise lethal Ebola virus infection. SAH hydrolase inhibitors thus offer real potential for the treatment of hemorrhagic fever virus infections.

25.9 CONCLUSION

The strategies reviewed here for interfering with the key events in the viral replicative cycle have the potential to target virtually all important human viral pathogens. Several such strategies, namely those aimed at viral DNA synthesis, viral polyprotein cleavage, and viral release from the cells (through the viral neuraminidase) have already provided a number of effective and useful therapeutics for the treatment of herpesvirus (HSV-1, HSV-2, VZV, CMV), retrovirus (HIV), hepadna (HBV), and influenza virus infections. Further improvements along these lines, following the same approaches, may in the future, yield more efficacious and more selective antiviral compounds. This should by no means detract from other approaches, not addressed here, that may also be envisaged to target viral compounds or virus-associated events, such as agents that specifically bind to the picornaviral capsids (i.e., pleconaril), or the HIV nucleocapsid p7 [i.e., 2,2′-dithiobisbenzamides (DIBAs)], or the influenza virus A matrix [adamantanamine derivatives (i.e., amantadine)], or glycosylation inhibitors (i.e., deoxynojirimycin derivatives), or antisense oligonucleotides or ribozymes targeted at selected viral mRNAs, or gene therapy approaches, or immunotherapy, and so forth. Whatever approach or strategy is followed to combat virus infections, the highest profit is likely to be expected if two or more of these different strategies are combined, especially in the treatment of chronic virus infections such as HIV, HBV, and HCV.

Acknowledgments We thank Christiane Callebaut and Chantal Biernaux for their invaluable editorial assistance.

REFERENCES

1. De Clercq, E. (2001). Antiviral drugs: Current state of the art. *J. Clin. Virol.*, 22, 73–89.
2. Gallaher, W. R., Ball, J. M., Garry, R. F., Martin-Amedee, A. M., Montelaro, R. C. (1995). A general model for the surface glycoproteins of HIV and other retroviruses. *AIDS Res. Hum. Retrovir.*, 11, 191–202.

3. Shukla, D., Liu, J., Blaiklock, P., Shworak, N. W., Bai, X., Esko, J. D., Cohen, G. H., Eisenberg, R. J., Rosenberg, R. D., Spear, P. G. (1999). A novel role for 3-*O*-sulfated heparan sulfate in herpes simplex virus 1 entry. *Cell*, *99*, 13–22.

4. Chen, Y., Maguire, T., Hileman, R. E., Fromm, J. R., Esko, J. D., Linhardt, R. J., Marks, R. M. (1997). Dengue virus infectivity depends on envelope protein binding to target cell heparan sulfate. *Nat. Med.*, *3*, 866–871.

5. Herold, B. C., Bourne, N., Marcellino, D., Kirkpatrick, R., Strauss, D. M., Zaneveld, L. J., Waller, D. P., Anderson, R. A., Chany, C. J., Barham, B. J., Stanberry, L. R., Cooper, M. D. (2000). Poly(sodium 4-styrene sulfonate): An effective candidate topical antimicrobial for the prevention of sexually transmitted diseases. *J. Infect. Dis.*, *181*, 770–773.

6. Esté, J. A., Schols, D., De Vreese, K., Van Laethem, K., Vandamme, A. M., Desmyter, J., De Clercq, E. (1997). Development of resistance of human immunodeficiency virus type 1 to dextran sulfate associated with the emergence of specific mutations in the envelope gp120 glycoprotein. *Mol. Pharmacol.*, *52*, 98–104.

7. Esté, J. A., Cabrera, C., Schols, D., Cherepanov, P., Gutierrez, A., Witvrouw, M., Pannecouque, C., Debyser, Z., Rando, R. F., Clotet, B., Desmyter, J., De Clercq, E. (1998). Human immunodeficiency virus glycoprotein gp120 as the primary target for the antiviral action of AR177 (zintevir). *Mol. Pharmacol.*, *53*, 340–345.

8. Cabrera, C., Witvrouw, M., Gutierrez, A., Clotet, B., Kuipers, M. E., Swart, P. J., Meijer, D. K., Desmyter, J., De Clercq, E., Esté, J. A. (1999). Resistance of the human immunodeficiency virus to the inhibitory action of negatively charged albumins on virus binding to CD4. *AIDS Res. Hum. Retrovir.*, *15*, 1535–1543.

9. Schols, D., Struyf, S., Van Damme, J., Esté, J. A., Henson, G., De Clercq, E. (1997). Inhibition of T-tropic HIV strains by selective antagonization of the chemokine receptor CXCR4. *J. Exp. Med.*, *186*, 1383–1388.

10. De Clercq, E. (2000). Inhibition of HIV infection by bicyclams, highly potent and specific CXCR4 antagonists. *Mol. Pharmacol.*, *57*, 833–839.

11. Baba, M., Nishimura, O., Kanzaki, N., Okamoto, M., Sawada, H., Iizawa, Y., Shiraishi, M., Aramaki, Y., Okonogi, K., Ogawa, Y., Meguro, K., Fujino, M. (1999). A small-molecule, nonpeptide CCR5 antagonist with highly potent and selective anti-HIV-1 activity. *Proc. Natl. Acad. Sci. USA*, *96*, 5698–5703.

12. Dragic, T., Trkola, A., Thompson, D. A., Cormier, E. G., Kajumo, F. A., Maxwell, E., Lin, S. W., Ying, W., Smith, S. O., Sakmar, T. P., Moore, J. P. (2000). A binding pocket for a small molecule inhibitor of HIV-1 entry within the transmembrane helices of CCR5. *Proc. Natl. Acad. Sci. USA*, *97*, 5639–5644.

13. Bridger, G. J., Skerlj, R. T., Thornton, D., Padmanabhan, S., Martellucci, S. A., Henson, G. W., Abrams, M. J., Yamamoto, N., De Vreese, K., Pauwels, R., De Clercq, E. (1995). Synthesis and structure-activity relationships of phenylenebis(methylene)-linked bis-tetraazamacrocycles that inhibit HIV replication. Effects of macrocyclic ring size and substituents on the aromatic linker. *J. Med. Chem.*, *38*, 366–378.

14. Bridger, G. J., Skerlj, R. T., Padmanabhan, S., Martellucci, S. A., Henson, G. W., Abrams, M. J., Joao, H. C., Witvrouw, M., De Vreese, K., Pauwels, R., De Clercq, E. (1996). Synthesis and structure-activity relationships of phenylenebis(methylene)-linked bis-tetraazamacrocycles that inhibit human immunodeficiency virus

replication. 2. Effect of heteroaromatic linkers on the activity of bicyclams. *J. Med. Chem.*, *39*, 109–119.

15. Bridger, G. J., Skerlj, R. T., Padmanabhan, S., Martellucci, S. A., Henson, G. W., Struyf, S., Witvrouw, M., Schols, D., De Clercq, E. (1999). Synthesis and structure-activity relationships of phenylenebis(methylene)-linked bis-azamacrocycles that inhibit HIV-1 and HIV-2 replication by antagonism of the chemokine receptor CXCR4. *J. Med. Chem.*, *42*, 3971–3981.

16. Hatse, S., Princen, K., Gerlach, L. O., Bridger, G., Henson, G., De Clercq, E., Schwartz, T. W., Schols, D. (2001). Mutation of Asp^{171} and Asp^{262} of the chemokine receptor CXCR4 impairs its coreceptor function for human immunodeficiency virus-1 entry and abrogates the antagonistic activity of AMD3100. *Mol. Pharmacol.*, *60*, 164–173.

17. Palczewski, K., Kumasaka, T., Hori, T., Behnke, C. A., Motoshima, H., Fox, B. A., Le Trong, I., Teller, D. C., Okada, T., Stenkamp, R. E., Yamamoto, M., Miyano, M. (2000). Crystal structure of rhodopsin: A G protein-coupled receptor. *Science*, *289*, 739–745.

18. Bridger, G. J., Skerlj, R. T., Thornton, D., Padmanabhan, S., Martellucci, S. A., Henson, G. W., Abrams, M. J., Yamamoto, N., De Vreese, K., Pauwels, R., De Clercq, E. (1995). Synthesis and structure-activity relationships of phenylenebis(methylene)-linked bis-tetraazamacrocycles that inhibit HIV replication. Effects of macrocyclic ring size and substituents on the aromatic linker. *J. Med. Chem.*, *38*, 366–378.

19. Joao, H. C., De Vreese, K., Pauwels, R., De Clercq, E., Henson, G. W., Bridger, G. J. (1995). Quantitative structural activity relationship study of bis-tetraazacyclic compounds. A novel series of HIV-1 and HIV-2 inhibitors. *J. Med. Chem.*, *38*, 3865–3873.

20. McAuley, A., Subramanian, S., Whitcombe, T. W. (1987). Synthesis and X-ray crystal-structure of a C-spirbi-[cyclam nickel (II)] complex (cyclam = 1,4,8,11-tetraazacyclotetradecane). *J. Chem. Soc.*, *8*, 539–541.

21. Strizki, J. M., Xu, S., Wagner, N. E., Wojcik, L., Liu, J., Hou, Y., Endres, M., Palani, A., Shapiro, S., Clader, J. W., Greenlee, W. J., Tagat, J. R., McCombie, S., Cox, K., Fawzi, A. B., Chou, C. C., Pugliese-Sivo, C., Davies, L., Moreno, M. E., Ho, D. D., Trkola, A., Stoddart, C. A., Moore, J. P., Reyes, G. R., Baroudy, B. M. (2001). SCH-C (SCH 351125), an orally bioavailable, small molecule antagonist of the chemokine receptor CCR5, is a potent inhibitor of HIV-1 infection *in vitro* and *in vivo*. *Proc. Natl. Acad. Sci. USA*, *98*, 12718–12723.

22. Root, M. J., Kay, M. S., Kim, P. S. (2001). Protein design of an HIV-1 entry inhibitor. *Science*, *291*, 884–888.

23. Eckert, D. M., Malashkevich, V. N., Hong, L. H., Carr, P. A., Kim, P. S. (1999). Inhibiting HIV-1 entry: Discovery of D-peptide inhibitors that target the gp41 coiled-coil pocket. *Cell*, *99*, 103–115.

24. Kilby, J. M., Hopkins, S., Venetta, T. M., DiMassimo, B., Cloud, G. A., Lee, J. Y., Alldredge, L., Hunter, E., Lambert, D., Bolognesi, D., Matthews, T., Johnson, M. R., Nowak, M. A., Shaw, G. M., Saag, M. S. (1998). Potent suppression of HIV-1 replication in humans by T-20, a peptide inhibitor of gp41-mediated virus entry. *Nat. Med.*, *4*, 1302–1307.

25. Lalezari, J. P., Henry, K., O'Hearn, M., Montaner, J. S. G., Piliero, P. J., Trottier, B., Walmsley, S., Cohen, C., Kuritzkes, D. R., Eron, J. J. Jr., Chung, J., DeMasi, R.,

Donatacci, L., Drobnes, C., Delehanty, J., Salgo, M. (2003). Enfuvirtide, an HIV-1 fusion inhibitor, for drug-resistant HIV infection in North and South America. *N. Engl. J. Med.*, 348, 2175–2185.

26. Lazzarin, A., Clotet, B., Cooper, D., Reynes, J., Arastéh, K., Nelson, M., Katlama, C., Stellbrink, H.-J., Delfraissy, J.-F., Lange, J., Huson, L., DeMasi, R., Wat, C., Delehanty, J., Drobnes, C., Salgo, M. (2003). Efficacy of enfuvirtide in patients infected with drug-resistant HIV-1 in Europe and Australia. *N. Engl. J. Med.*, 348, 2186–2195.

27. Lambert, D. M., Barney, S., Lambert, A. L., Guthrie, K., Medinas, R., Davis, D. E., Bucy, T., Erickson, J., Merutka, G., Petteway, S. R. Jr. (1996). Peptides from conserved regions of paramyxovirus fusion (F) proteins are potent inhibitors of viral fusion. *Proc. Natl. Acad. Sci. USA*, 93, 2186–2191.

28. Schwartz, J. A., Lium, E. K., Silverstein, S. J. (2001). Herpes simplex virus type 1 entry is inhibited by the cobalt chelate complex CTC-96. *J. Virol.*, 75, 4117–4128.

29. Champness, J. N., Bennett, M. S., Wien, F., Visse, R., Summers, W. C., Herdewijn, P., De Clercq, E., Ostrowski, T., Jarvest, R. L., Sanderson, M. R. (1998). Exploring the active site of herpes simplex virus type-1 thymidine kinase by X-ray crystallography of complexes with aciclovir and other ligands. *Proteins: Structure, Functions and Genetics*, 32, 350–361.

30. De Winter, H., Herdewijn, P. (1997). Understanding the binding of 5-substituted 2′-deoxyuridine substrates to thymidine kinase of herpes simplex virus type-1. *J. Med. Chem.*, 39, 4727–4737.

31. Wigerinck, P., Pannecouque, C., Snoeck, R., Claes, P., De Clercq, E., Herdewijn, P. (1991). 5-(5-Bromothien-2-yl)-2′-deoxyuridine and 5-(5-chlorothien-2-yl)-2′-deoxyuridine are equipotent to (*E*)-5-(2-bromovinyl)-2′-deoxyuridine in the inhibition of herpes simplex virus type I replication. *J. Med. Chem.*, 34, 2383–2389.

32. Wigerinck, P., Snoeck, R., Claes, P., De Clercq, E., Herdewijn, P. (1991). Synthesis and antiviral activity of 5-heteroaryl-substituted 2′-deoxyuridines. *J. Med. Chem.*, 34, 1767–1772.

33. Wigerinck, P., Kerremans, L., Claes, P., Snoeck, R., Maudgal, P. C., De Clercq, E., Herdewijn, P. (1993). Synthesis and antiviral activity of 5-thien-2-yl-2′-deoxyuridine analogues. *J. Med. Chem.*, 36, 538–543.

34. Bohman, C., Balzarini, J., Wigerinck, P., Van Aerschot, A., Herdewijn, P., De Clercq, E. (1994). Mechanism of cytostatic action of novel 5-(thien-2-yl)- and 5-(furan-2-yl)-substituted pyrimidine nucleoside analogues against tumor cells transfected by the thymidine kinase gene of herpes simplex virus. *J. Biol. Chem.*, 269, 8036–8043.

35. McGuigan, C., Yarnold, C. J., Jones, G., Velazquez, S., Barucki, H., Brancale, A., Andrei, G., Snoeck, R., De Clercq, E., Balzarini, J. (1999). Potent and selective inhibition of varicella-zoster virus (VZV) by nucleoside analogues with an unusual bicyclic base. *J. Med. Chem.*, 42, 4479–4484.

36. McGuigan, C., Barucki, H., Carangio, A., Blewett, S., Andrei, G., Snoeck, R., De Clercq, E., Balzarini, J., Erichsen, J. T. (2000). Highly potent and selective inhibition of varicella-zoster virus by bicyclic furopyrimidine nucleosides bearing an aryl side chain. *J. Med. Chem.*, 43, 4993–4997.

37. Iwayama, S., Ono, N., Ohmura, Y., Suzuki, K., Aoki, M., Nakazawa, H., Oikawa, M., Kato, T., Okunishi, M., Nishiyama, Y., Yamanishi, K. (1998). Antiherpesvirus

activities of (1′S,2′R)-9-[[1′,2′-bis(hydroxymethyl)-cycloprop-1′-yl]methyl] (A-5021) in cell culture. *Antimicrob. Agents Chemother.*, 42, 1666–1670.

38. Ono, N., Iwayama, S., Suzuki, K., Sekiyama, T., Nakazawa, H., Tsuji, T., Okunishi, M., Daikoku, T., Nishiyama, Y. (1998). Mode of action of (1′S,2′R)-9-[[1′,2′-bis(hydroxymethyl)cycloprop-1′-yl]methyl]guanine (A-5021) against herpes simplex virus type 1 and type 2 and varicella-zoster virus. *Antimicrob. Agents Chemother.*, 42, 2095–2102.

39. Wang, J., Froeyen, M., Hendrix, C., Andrei, G., Snoeck, R., De Clercq, E., Herdewijn, P. (2000). The cyclohexene ring system as a furanose mimic: Synthesis and antiviral activity of both enantiomers of cyclohexenylguanine. *J. Med. Chem.*, 43, 736–745.

40. Crute, J. J., Grygon, C. A., Hargrave, K. D., Simoneau, B., Faucher, A. M., Bolger, G., Kibler, P., Liuzzi, M., Cordingley, M. G. (2002). Herpes simplex virus helicase-primase inhibitors are active in animal models of human disease. *Nature Med.*, 8, 386–391.

41. Kleymann, G., Fischer, R., Betz, U. A., Hendrix, M., Bender, W., Schneider, U., Handke, G., Eckenberg, P., Hewlett, G., Pevzner, V., Baumeister, J., Weber, O., Henninger, K., Keldenich, J., Jensen, A., Kolb, J., Bach, U., Popp, A., Maben, J., Frappa, I., Haebich, D., Lockhoff, O., Rubsamen-Waigmann, H. (2002). New helicase-primase inhibitors as drug candidates for the treatment of herpes simplex disease. *Nature Med.*, 8, 392–398.

42. Betz, U. A. K., Fischer, R., Kleymann, G., Hendrix, M., Rubsamen-Waigmann, H. (2002). Potent in vivo antiviral activity of the herpes simplex virus primase-helicase inhibitor BAY 57–1293. *Antimicrob. Agents Chemother.*, 46, 1766–1772.

43. Rousseau, F. S., Kahn, J. O., Thompson, M., Mildvan, D., Shepp, D., Sommadossi, J. P., Delehanty, J., Simpson, J. N., Wang, L. H., Quinn, J. B., Wakeford, C., van der Horst, C. (2001). Prototype trial design for rapid dose selection of antiretroviral drugs: An example using emtricitabine (Coviracil). *J. Antimicrob. Chemother.*, 48, 507–513.

44. Furman, P. A., Jeffrey, J., Kiefer, L. L., Feng, J. Y., Anderson, K. S., Borroto-Esoda, K., Hill, E., Copeland, W. C., Chu, C. K., Sommadossi, J. P., Liberman, I., Schinazi, R. F., Painter, G. R. (2001). Mechanism of action of 1-β-D-2,6-diaminopurine dioxolane, a prodrug of the human immunodeficiency virus type 1 inhibitor 1-β-D-dioxolane guanosine. *Antimicrob. Agents Chemother.*, 45, 158–165.

45. Stoddart, C. A., Moreno, M. E., Linquist-Stepps, V. D., Bare, C., Bogan, M. R., Gobbi, A., Buckheit, R. W. Jr., Bedard, J., Rando, R. F., McCune, J. M. (2000). Antiviral activity of 2′-deoxy-3′-oxa-4′-thiocytidine (BCH-10652) against lamivudine-resistant human immunodeficiency virus type 1 in SCID-hu Thy/Liv mice. *Antimicrob. Agents Chemother.*, 44, 783–786.

46. Otto, M. J., Arastèh, K., Schulbin, H., Beard, A., Cartee, L., Liotta, D. C., Schinazi, R. F., Murphy, R. L. (2002). Single and multiple dose pharmacokinetics and safety of the nucleoside Racivir® in male volunteers. HIV DART 2002. *Front. Drug Devel. for Antiretroviral Therapies*, Naples, Florida, USA, 15–19 December 2002. *Antiviral Res., Abst.*, no. 044.

47. Schinazi, R. F., Mellors, J., Bazmi, H., Diamond, S., Garber, S., Gallagher, K., Geleziunas, R., Klabe, R., Pierce, M., Rayner, M., Wu, J. T., Zhang, H., Hammond, J., Bacheler, J., Manion, D. J., Otto, M. J., Stuyver, L., Trainor, G., Liotta, D. C.,

Erickson-Viitanen, S. (2002). DPC 817: A cytidine nucleoside analog with activity against zidovudine- and lamivudine-resistant viral variants. *Antimicrob. Agents Chemother.*, 46, 1394–1401.

48. Dunkle, L. M., Gathe, J. C., Pedevillano, D. E., Oshana, S. C., Pottage, J. C., Rice, W. C. (2002). ACH-126,443 (β-L-Fd4C) shows potent anti-HIV activity in proof-of-princple study in treatment-experienced patients with M184V and other reverse transcriptase mutations. HIV DART 2002, Frontiers in Drug Development for Antiretroviral Therapies, Naples, Florida, USA, 15–19 December 2002. *Antiviral Res., Abst.*, no. 069.

49. Katlama, C., Ghosn, J., Tubiana, R., Wirden, M., Valantin, M. A., Harmenberg, J., Mardh, G., Westling, C., Calvez, V. (2002). MIV-310 reduces HIV viral load in patients failing multiple antiretroviral therapy: Results from a 4-week phase II study. HIV DART 2002, Naples, Florida, USA, 15–19 December 2002. Abst., no. 071.

50. Herdewijn, P. (1997). Structural requirements for antiviral activity in nucleosides. *Drug Discov. Today*, 2, 235–242.

51. Siddiqui, A. Q., McGuigan, C., Ballatore, C., Zuccotto, F., Gilbert, I. H., De Clercq, E., Balzarini, J. (1999). Design and synthesis of lipophilic phosphoramidate d4T-MP prodrugs expressing high potency against HIV in cell culture: Structural determinants for in vitro activity and QSAR. *J. Med. Chem.*, 42, 4122–4128.

52. Saboulard, D., Naesens, L., Cahard, D., Salgado, A., Pathirana, R., Velazquez, S., McGuigan, C., De Clercq, E., Balzarini, J. (1999). Characterization of the activation pathway of phosphoramidate triester prodrugs of stavudine and zidovudine. *Mol. Pharmacol.*, 56, 693–704.

53. Meier, C., Lorey, M., De Clercq, E., Balzarini, J. (1998). *cyclo*Sal-2′,3′-dideoxy-2′,3′-didehydrothymidine monophosphate (*cyclo*Sal-d4TMP): Synthesis and antiviral evaluation of a new d4TMP delivery system. *J. Med. Chem.*, 41, 1417–1427.

54. Balzarini, J., Aquaro, S., Knispel, T., Rampazzo, C., Bianchi, V., Perno, C. F., De Clercq, E., Meier, C. (2000). Cyclosaligenyl-2′,3′-didehydro-2′,3′-dideoxythymidine monophosphate: Efficient intracellular delivery of d4TMP. *Mol. Pharmacol.*, 58, 928–935.

55. De Clercq, E. (1998). The role of non-nucleoside reverse transcriptase inhibitors (NNRTIs) in the therapy of HIV-1 infection. *Antiviral Res.*, 38, 153–179.

56. Jonckheere, H., Anné, J., De Clercq, E. (2000). The HIV-1 reverse transcription (RT) process as target for RT inhibitors. *Med. Res. Rev.*, 20, 129–154.

57. Esnouf, R. M., Stuart, D. I., De Clercq, E., Schwartz, E., Balzarini, J. (1997). Models which explain the inhibition of reverse transcriptase by HIV-1-specific (thio)carboxanilide derivatives. *Biochem. Biophys. Res. Commun.*, 234, 458–464.

58. Fujiwara, T., Sato, A., el-Farrash, M., Miki, S., Abe, K., Isaka, Y., Kodama, M., Wu, Y., Chen, L. B., Harada, H., Sugimoto, H., Hatanaka, M., Hinuma, Y. (1998). S-1153 inhibits replication of known drug-resistant strains of human immunodeficiency virus type 1. *Antimicrob. Agents Chemother.*, 42, 1340–1345.

59. Andries, K., de Béthune, M.-P., Kukla, M. J., Azijn, H., Lewi, P. J., Janssen, P. A. J., Pauwels, R. (2000). R165335-TMC125, a novel non nucleoside reverse transcriptase inhibitor (NNRTI) with nanomolar activity against NNRTI resistant HIV

strains. Abstracts of the 40th Interscience Conference on Antimicrobial Agents and Chemotherapy, Toronto, Canada, 17–20 September 2000, no. 1840.

60. Corbett, J. W., Ko, S. S., Rodgers, J. D., Jeffrey, S., Bacheler, L. T., Klabe, R. M., Diamond, S., Lai, C. M., Rabel, S. R., Saye, J. A., Adams, S. P., Trainor, G. L., Anderson, P. S., Erickson-Viitanen, S. K. (1999). Expanded-spectrum nonnucleoside reverse transcriptase inhibitors inhibit clinically relevant mutant variants of human immunodeficiency virus type 1. *Antimicrob. Agents Chemother.*, 43, 2893–2897.

61. Buckheit, R. W. Jr., Watson, K., Fliakas-Boltz, V., Russell, J., Loftus, T. L., Osterling, M. C., Turpin, J. A., Pallansch, L. A., White, E. L., Lee, J. W., Lee, S. H., Oh, J. W., Kwon, H. S., Chung, S. G., Cho, E. H. (2001). SJ-3366, a unique and highly potent nonnucleoside reverse transcriptase inhibitor of human immunodeficiency virus type 1 (HIV-1) that also inhibits HIV-2. *Antimicrob. Agents Chemother.*, 45, 393–400.

62. Creagh, T., Ruckle, J. L., Tolbert, D. T., Giltner, J., Eiznhamer, D. A., Dutta, B., Flavin, M. T., Xu, Z. Q. (2001). Safety and pharmacokinetics of single doses of (+)-calanolide a, a novel, naturally occurring nonnucleoside reverse transcriptase inhibitor, in healthy, human immunodeficiency virus-negative human subjects. *Antimicrob. Agents Chemother.*, 45, 1379–1386.

63. Ren, J., Nichols, C., Bird, L. E., Fujiwara, T., Sugimoto, H., Stuart, D. I., Stammers, D. K. (2000). Binding of the second generation non-nucleoside inhibitor S-1153 to HIV-1 reverse transcriptase involves extensive main chain hydrogen bonding. *J. Biol. Chem.*, 275, 14316–14320.

64. Pelemans, H., Esnouf, R., De Clercq, E., Balzarini, J. (2000). Mutational analysis of Trp-229 of human immunodeficiency virus type 1 reverse transcriptase (RT) identifies this amino acid residue as a prime target for the rational design of new non-nucleoside RT inhibitors. *Mol. Pharmacol.*, 57, 954–960.

65. Xiong, X., Smith, J. L., Chen, M. S. (1997). Effect of incorporation of cidofovir into DNA by human cytomegalovirus DNA polymerase on DNA elongation. *Antimicrob. Agents Chemother.*, 41, 594–599.

66. Balzarini, J., Hao, Z., Herdewijn, P., Johns, D. G., De Clercq, E. (1991). Intracellular metabolism and mechanism of anti-retrovirus action of 9-(2-phosphonylmethoxyethyl)adenine, a potent anti-human immunodeficiency virus compound. *Proc. Natl. Acad. Sci. U.S.A.*, 88, 1499–1503.

67. De Clercq, E., Holý, A., Rosenberg, I., Sakuma, T., Balzarini, J., Maudgal, P. C. (1986). A novel selective broad-spectrum anti-DNA virus agent. *Nature*, 323, 464–467.

68. De Clercq, E. (1993). Therapeutic potential of HPMPC as an antiviral drug. *Rev. Med. Virol.*, 3, 85–96.

69. De Clercq, E. (2001). Vaccinia virus inhibitors as a paradigm for the chemotherapy of poxvirus infections. *Clin. Microbiol. Rev.*, 14, 382–397.

70. De Clercq, E. (2002). Cidofovir in the treatment of poxvirus infections. *Antiviral Res.*, 55, 1–13.

71. De Clercq, E. (2002). Cidofovir in the therapy and short-term prophylaxis of poxvirus infections. *Trends Pharmacol. Sci.*, 23, 456–458.

72. Pommier, Y., Pilon, A. A., Bajaj, K., Mazumder, A., Neamati, N. (1997). HIV-1 integrase as a target for antiviral drugs. *Antiviral Chem. Chemother.*, 8, 463–483.

73. Pommier, Y., Marchand, C., Neamati, N. (2002). Retroviral integrase inhibitors year 2000: Update and perspectives. *Antiviral Res.*, *47*, 139–148.
74. Pluymers, W., Neamati, N., Pannecouque, C., Fikkert, V., Marchand, C., Burke, T. R. Jr., Pommier, Y., Schols, D., De Clercq, E., Debyser, Z., Witvrouw, M. (2000). Viral entry as the primary target for the anti-HIV activity of chicoric acid and its tetra-acetyl esters. *Mol. Pharmacol.*, *58*, 641–648.
75. Hazuda, D. J., Felock, P., Witmer, M., Wolfe, A., Stillmock, K., Grobler, J. A., Espeseth, A., Gabryelski, L., Schleif, W., Blau, C., Miller, M. D. (2000). Inhibitors of strand transfer that prevent integration and inhibit HIV-1 replication in cells. *Science*, *287*, 646–650.
76. Wai, J. S., Egbertson, M. S., Payne, L. S., Fisher, T. E., Embrey, M. W., Tran, L. O., Melamed, J. Y., Langford, H. M., Guare, J. P. Jr., Zhuang, L., Grey, V. E., Vacca, J. P., Holloway, M. K., Naylor-Olsen, A. M., Hazuda, D. J., Felock, P. J., Wolfe, A. L., Stillmock, K. A., Schleif, W. A., Gabryelski, L. J., Young, S. D. (2000). 4-Aryl-2,4-dioxobutanoic acid inhibitors of HIV-1 integrase and viral replication in cells. *J. Med. Chem.*, *43*, 4923–4926.
77. Pluymers, W., Pais, G., Van Maele, B., Pannecouque, C., Fikkert, V., Burke, T. R. Jr., De Clercq, E., Witvrouw, M., Neamati, N., Debyser, Z. (2000). Inhibition of human immunodeficiency virus type 1 integration by diketo derivatives. *Antimicrob. Agents Chemother.*, *46*, 3292–3297.
78. Pannecouque, C., Pluymers, W., Van Maele, B., Tetz, V., Cherepanov, P., De Clercq, E., Witvrouw, M., Debyzer, Z. (2002). New class of HIV integrase inhibitors that block viral replication in cell culture. *Curr. Biol.*, *12*, 1169–1177.
79. Daelemans, D., Vandamme, A.-M., De Clercq, E. (1999). Human immunodeficiency virus gene regulation as a target for antiviral chemotherapy. *Antiviral Chem. Chemother.*, *10*, 1–14.
80. Baba, M., Okamoto, M., Kawamura, M., Makino, M., Higashida, T., Takashi, T., Kimura, Y., Ikeuchi, T., Tetsuka, T., Okamoto, T. (1998). Inhibition of human immunodeficiency virus type 1 replication and cytokine production by fluoroquinoline derivatives. *Mol. Pharmacol.*, *53*, 1097–1103.
81. Turpin, J. A., Buckheit, R. W. Jr., Derse, D., Hollingshead, M., Williamson, K., Palamone, C., Osterling, M. C., Hill, S. A., Graham, L., Schaeffer, C. A., Bu, M., Huang, M., Cholody, W. M., Michejda, C. J., Rice, W. G. (1998). Inhibition of acute-, latent-, and chronic-phase human immunodeficiency virus type 1 (HIV-1) replication by a bistriazoloacridone analog that selectively inhibits HIV-1 transcription. *Antimicrob. Agents Chemother.*, *42*, 487–494.
82. Hamy, F., Felder, E. R., Heizmann, G., Lazdins, J., Aboul-ela, F., Varani, G., Karn, J., Klimkait, T. (1997). An inhibitor of the Tat/TAR RNA interaction that effectively suppresses HIV-1 replication. *Proc. Natl. Acad. Sci. U.S.A.*, *94*, 3548–3553.
83. Daelemans, D., Schols, D., Witvrouw, M., Pannecouque, C., Hatse, S., Van Dooren, S., Hamy, F., Klimkait, T., De Clercq, E., Vandamme, A. M. (2000). A second target for the peptoid Tat/transactivation response element inhibitor CGP64222: Inhibition of human immunodeficiency virus replication by blocking CXC-chemokine receptor 4-mediated virus entry. *Mol. Pharmacol.*, *57*, 116–124.
84. Chao, S. H., Fujinaga, K., Marion, J. E., Taube, R., Sausville, E. A., Senderowicz, A. M., Peterlin, B. M., Price, D. H. (2000). Flavopiridol inhibits P-TEFb and blocks HIV-1 replication. *J. Biol. Chem.*, *275*, 28345–28348.

85. Chiu, Y. L., Ho, C. K., Saha, N., Schwer, B., Shuman, S., Rana, T. M. (2002). Tat stimulates cotranscriptional capping of HIV mRNA. *Mol. Cell*, *10*, 585–597.

86. Daelemans, D., Esté, J. A., Witvrouw, M., Pannecouque, C., Jonckheere, H., Aquaro, S., Perno, C. F., De Clercq, E., Vandamme, A.-M. (1997). S-Adenosylhomocysteine hydrolase inhibitors interfere with the replication of human immunodeficiency virus type 1 through inhibition of the LTR transactivation. *Mol. Pharmacol.*, *52*, 1157–1163.

87. Dymock, B. W., Jones, P. S., Wilson, F. X. (2000). Novel approaches to the treatment of hepatitis C virus infection. *Antiviral Chem. Chemother.*, *11*, 79–96.

88. Kim, J. L., Morgenstern, K. A., Griffith, J. P., Dwyer, M. D., Thomson, J. A., Murcko, M. A., Lin, C., Caron, P. R. (1998). Hepatitis C virus NS3 RNA helicase domain with a bound oligonucleotide: The crystal structure provides insights into the mode of unwinding. *Structure*, *6*, 89–100.

89. Tan, S. L., Pause, A., Shi, Y., Sonenberg, N. (2002). Hepatitis C therapeutics: Current status and emerging strategies. *Nature Rev. Drug Disc.*, *1*, 867–881.

90. Paolini, C., De Francesco, R., Gallinari, P. (2000). Enzymatic properties of hepatitis C virus NS3-associated helicase. *J. Gen. Virol.*, *81*, 1335–1345.

91. Carroll, S. S., Sardana, V., Yang, Z., Jacobs, A. R., Mizenko, C., Hall, D., Hill, L., Zugay-Murphy, J., Kuo, L. C. (2000). Only a small fraction of purified hepatitis C RNA-dependent RNA polymerase is catalytically competent: Implications for viral replication and in vitro assays. *Biochemistry*, *39*, 8243–8249.

92. Patick, A. K., Potts, K. E. (1998). Protease inhibitors as antiviral agents. *Clin. Microbiol. Rev.*, *11*, 614–627.

93. Robinson, B. S., Riccardi, K. A., Gong, Y. F., Guo, Q., Stock, D. A., Blair, W. S., Terry, B. J., Deminie, C. A., Djang, F., Colonno, R. J., Lin, P. F. (2000). BMS-232632, a highly potent human immunodeficiency virus protease inhibitor that can be used in combination with other available antiretroviral agents. *Antimicrob. Agents Chemother.*, *44*, 2093–2099.

94. Erickson, J. W., Burt, S. K. (1996). Structural mechanisms of HIV drug resistance. *Ann. Rev. Pharmacol. Toxicol.*, *36*, 545–571.

95. Hodge, C. N., Aldrich, P. E., Bacheler, L. T., Chang, C. H., Eyermann, C. J., Garber, S., Grubb, M., Jackson, D. A., Jadhav, P. K., Korant, B., Lam, P. Y., Maurin, M. B., Meek, J. L., Otto, M. J., Rayner, M. M., Reid, C., Sharpe, T. R., Shum, L.L., Winslow, D. L., Erickson-Viitanen, S. (1996). Improved cyclic urea inhibitors of the HIV-1 protease: Synthesis, potency, resistance profile, human pharmacokinetics and X-ray crystal structure of DMP 450. *Chem. Biol.*, *3*, 301–314.

96. Hyland, L. J., Tomaszek, T. A. Jr., Meek, T. D. (1991). Human immunodeficiency virus-1 protease. 2. Use of pH rate studies and solvent kinetic isotope effects to elucidate details of chemical mechanism. *Biochemistry*, *30*, 8454–8463.

97. Turner, S. R., Strohbach, J. W., Tommasi, R. A., Aristoff, P. A., Johnson, P. D., Skulnick, H. I., Dolak, L. A., Seest, E. P., Tomich, P. K., Bohanon, M. J., Horng, M. M., Lynn, J. C., Chong, K. T., Hinshaw, R. R., Watenpaugh, K. D., Janakiraman, M. N., Thaisrivongs, S. (1998). Tipranavir (PNU-140690): A potent, orally bioavailable nonpeptidic HIV protease inhibitor of the 5,6-dihydro-4-hydroxy-2-pyrone sulfonamide class. *J. Med. Chem.*, *41*, 3467–3476.

98. Hagen, S. E., Domagala, J., Gajda, C., Lovdahl, M., Tait, B. D., Wise, E., Holler, T., Hupe, D., Nouhan, C., Urumov, A., Zeikus, G., Zeikus, E., Lunney, E. A., Pavlovsky, A., Gracheck, S. J., Saunders, J., VanderRoest, S., Brodfuehrer, J. (2001). 4-Hydroxy-5,6-dihydropyrones as inhibitors of HIV protease: The effect of heterocyclic substituents at C-6 on antiviral potency and pharmacokinetic parameters. *J. Med. Chem.*, *44*, 2319–2332.

99. Larder, B. A., Hertogs, K., Bloor, S., van den Eynde, C. H., DeCian, W., Wang, Y., Freimuth, W. W., Tarpley, G. (2000). Tipranavir inhibits broadly protease inhibitor-resistant HIV-1 clinical samples. *AIDS*, *14*, 1943–1948.

100. Waxman, L., Darke, P. L. (2000). The herpesvirus proteases as targets for antiviral chemotherapy. *Antiviral Chem. Chemother.*, *11*, 1–22.

101. Jarvest, R. L., Pinto, I. L., Ashman, S. M., Dabrowski, C. E., Fernandez, A. V., Jennings, L. J., Lavery, P., Tew, D. G. (1999). Inhibition of herpes proteases and antiviral activity of 2-substituted thieno-[2,3-d]oxazinones. *Bioorg. Med. Chem. Lett.*, *9*, 443–448.

102. Smith, D. G., Gribble, A. D., Haigh, D., Ife, R. J., Lavery, P., Skett, P., Slingsby, B. P., Stacey, R., Ward, R. W., West, A. (1999). The inhibition of human cytomegalovirus (hCMV) protease by hydroxylamine derivatives. *Bioorg. Med. Chem. Lett.*, *9*, 3137–3142.

103. Ogilvie, W. W., Yoakim, C., Do, F., Hache, B., Lagace, L., Naud, J., O'Meara, J. A., Deziel, R. (1999). Synthesis and antiviral activity of monobactams inhibiting the human cytomegalovirus protease. *Bioorg. Med. Chem.*, *7*, 1521–1531.

104. Borthwick, A. D., Angier, S. J., Crame, A. J., Exall, A. M., Haley, T. M., Hart, G. J., Mason, A. M., Pennell, A. M., Weingarten, G. G. . (2000). Design and synthesis of pyrrolidine-5,5-*trans*-lactams (5-oxo-hexahydro-pyrrolo[3,2-*b*]pyrroles) as novel mechanism-based inhibitors of human cytomegalovirus protease. 1. The α-methyl-*trans*-lactam template. *J. Med. Chem.*, *43*, 4452–4464.

105. Matsumoto, M., Misawa, S., Chiba, N., Takaku, H., Hayashi, H. (2001). Selective nonpeptidic inhibitors of herpes simplex virus type 1 and human cytomegalovirus proteases. *Biol. Pharm. Bull.*, *24*, 236–241.

106. Dragovich, P. S., Webber, S. E., Babine, R. E., Fuhrman, S. A., Patick, A. K., Matthews, D. A., Lee, C. A., Reich, S. H., Prins, T. J., Marakovits, J. T., Littlefield, E. S., Zhou, R., Tikhe, J., Ford, C. E., Wallace, M. B., Meador J. W. 3rd, Ferre, R. A., Brown, E. L., Binford, S. L., Harr, J. E., DeLisle, D. M., Worland, S. T. (1998). Structure-based design, synthesis, and biological evaluation of irreversible human rhinovirus 3C protease inhibitors. 1. Michael acceptor structure-activity studies. *J. Med. Chem.*, *41*, 2806–2818.

107. Dragovich, P. S., Webber, S. E., Babine, R. E., Fuhrman, S. A., Patick, A. K., Matthews, D. A., Reich, S. H., Marakovits, J. T., Prins, T. J., Zhou, R., Tikhe, J., Littlefield, E. S., Bleckman, T. M., Wallace, M. B., Little, T. L., Ford, C. E., Meador, J. W. 3rd, Ferre, R. A., Brown, E. L., Binford, S. L., DeLisle, D. M., Worland, S. T. (1998). Structure-based design, synthesis, and biological evaluation of irreversible human rhinovirus 3C protease inhibitors. 2. Peptide structure-activity studies. *J. Med. Chem.*, *41*, 2819–2834.

108. Dragovich, P. S., Prins, T. J., Zhou, R., Fuhrman, S. A., Patick, A. K., Matthews, D. A., Ford, C. E., Meador, J. W. 3rd, Ferre, R. A., Worland, S. T. (1999). Structure-based design, synthesis, and biological evaluation of irreversible human rhinovirus

3C protease inhibitors. 3. Structure-activity studies of ketomethylene-containing peptidomimetics. *J. Med. Chem.*, 42, 1203–1212.

109. Dragovich, P. S., Prins, T. J., Zhou, R., Webber, S. E., Marakovits, J. T., Fuhrman, S. A., Patick, A. K., Matthews, D. A., Lee, C. A., Ford, C. E., Burke, B. J., Rejto, P. A., Hendickson, T. F., Tuntland, T., Brown, E. L., Meador J. W. 3rd, Ferre, R. A., Harr, J. E., Kosa, M. B., Worland, S. T. (1999). Structure-based design, synthesis, and biological evaluation of irreversible human rhinovirus 3C protease inhibitors. 4. Incorporation of P1 lactam moieties as L-glutamine replacements. *J. Med. Chem.*, 42, 1213–1224.

110. Dragovich, P. S., Prins, T. J., Zhou, R., Brown, E. L., Maldonado, F. C., Fuhrman, S. A., Zalman, L. S., Tuntland, T., Lee, C. A., Patick, A. K., Matthews, D. A., Hendrickson, T. F., Kosa, M. B., Liu, B., Batugo, M. R., Gleeson, J. P., Sakata, S. K., Chen, L., Guzman, M. C., Meador, J. W. 3rd, Ferre, R. A., Worland, S. T. (2002). Structure-based design, synthesis, and biological evaluation of irreversible rhinovirus 3C protease inhibitors. 6. Structure-activity studies of orally 2-pyridone-containing peptidomimetics. *J. Med. Chem.*, 45, 1607–1623.

111. Matthews, D. A., Dragovich, P. S., Webber, S. E., Fuhrman, S. A., Patick, A. K., Zalman, L. S., Hendrickson, T. F., Love, R. A., Prins, T. J., Marakovits, J. T., Zhou, R., Tikhe, J., Ford, C. E., Meador, J. W., Ferre, R. A., Brown, E. L., Binford, S. L., Brothers, M. A., DeLisle, D. M., Worland, S. T. (1999). Structure-assisted design of mechanism-based irreversible inhibitors of human rhinovirus 3C protease with potent antiviral activity against multiple rhinovirus serotypes. *Proc. Natl. Acad. Sci. U.S.A.*, 96, 11000–11007.

112. Patick, A. K., Binford, S. L., Brothers, M. A., Jackson, R. L., Ford, C. E., Diem, M. D., Maldonado, F., Dragovich, P. S., Zhou, R., Prins, T. J., Fuhrman, S. A., Meador, J. W., Zalman, L. S., Matthews, D. A., Worland, S. T. (1999). In vitro antiviral activity of AG7088, a potent inhibitor of human rhinovirus 3C protease. *Antimicrob. Agents Chemother.*, 43, 2444–2450.

113. Hsyu, P.-H., Pithavala, Y. K., Gersten, M., Penning, C. A., Kerr, B. M. (2002). Pharmacokinetics and safety of an antirhinoviral agent, ruprintrivir, in healthy volunteers. *Antimicrob. Agents Chemother.*, 46, 392–397.

114. Tautz, N., Kaiser, A., Thiel, H.-J. (2000). NS3 serine protease of bovine viral diarrhea virus: Characterization of active site residues, NS4A cofactor domain, and protease-cofactor interactions. *Virology*, 273, 351–363.

115. Llinàs-Brunet, M., Bailey, M., Fazal, G., Ghiro, E., Gorys, V., Goulet, S., Halmos, T., Maurice, R., Poirier, M., Poupart, M. A., Rancourt, J., Thibeault, D., Wernick, D., Lamarre, D. (2000). Highly potent and selective peptide-based inhibitors of the hepatitis C virus serine protease: Towards smaller inhibitors. *Bioorg. Med. Chem. Lett.*, 10, 2267–2270.

116. Bennett, J. M., Campbell, A. D., Campbell, A. J., Carr, M. G., Dunsdon, R. M., Greening, J. R., Hurst, D. N., Jennings, N. S., Jones, P. S., Jordan, S., Kay, P. B., O'Brien, M. A., Underwood, J., Raynham, T. M., Wilkinson, C. S., Wilkinson, T. C., Wilson, F. X. (2001). The identification of α-ketoamides as potent inhibitors of hepatitis C virus NS3-4A proteinase. *Bioorg. Med. Chem. Lett.*, 11, 355–357.

117. Sing, W. T., Lee, C. L., Yeo, S. L., Lim, S. P., Sim, M. M. (2001). Arylalkylidene rhodanine with bulky and hydrophobic functional group as selective HCV NS3 protease inhibitor. *Bioorg. Med. Chem. Lett.*, 11, 91–94.

118. Bartenschlager, R. (2002). Hepatitis C virus replicons, potential role for drug development. *Nature Rev. Drug Disc.*, *1*, 911–916.
119. von Itzstein, M., Wu, W. Y., Kok, G. B., Pegg, M. S., Dyason, J. C., Jin, B., Van Phan, T., Smythe, M. L., White, H. F., Oliver, S. W., Colman, P. M., Varghese, J. N., Ryan, D. M., Woods, J. M., Bethell, R. C., Hotham, V. J., Cameron, J. M., Penn, C. R. (1993). Rational design of potent sialidase-based inhibitors of influenza virus replication. *Nature*, *363*, 418–423.
120. Barnett, J. M., Cadman, A., Gor, D., Dempsey, M., Walters, M., Candlin, A., Tisdale, M., Morley, P. J., Owens, I. J., Fenton, R. J., Lewis, A. P., Claas, E. C., Rimmelzwaan, G. F., De Groot, R., Osterhaus, A. D. (2000). Zanamivir susceptibility monitoring and characterization of influenza virus clinical isolates obtained during phase II clinical efficacy studies. *Antimicrob. Agents Chemother.*, *44*, 78–87.
121. Hayden, F. G., Osterhaus, A. D., Treanor, J. J., Fleming, D. M., Aoki, F. Y., Nicholson, K. G., Bohnen, A. M., Hirst, H. M., Keene, O., Wightman, K. (1997). Efficacy and safety of the neuraminidase inhibitor zanamivir in the treatment of influenza virus infections. *N. Engl. J. Med.*, *337*, 874–880.
122. Kim, C. U., Lew, W., Williams, M. A., Liu, H., Zhang, L., Swaminathan, S., Bischofberger, N., Chen, M. S., Mendel, D. B., Tai, C. Y., Laver, W. G., Stevens, R. C. (1997). Influenza neuraminidase inhibitors possessing a novel hydrophobic interaction in the enzyme active site: design, synthesis, and structural analysis of carbocyclic sialic acid analogues with potent anti-influenza activity. *J. Am. Chem. Soc.*, *119*, 681–690.
123. Mendel, D. B., Tai, C. Y., Escarpe, P. A., Li, W., Sidwell, R. W., Huffman, J. H., Sweet, C., Jakeman, K. J., Merson, J., Lacy, S. A., Lew, W., Williams, M. A., Zhang, L., Chen, M. S., Bischofberger, N., Kim, C. U. (1998). Oral administration of a prodrug of the influenza virus neuraminidase inhibitor GS 4071 protects mice and ferrets against influenza infection. *Antimicrob. Agents Chemother.*, *42*, 640–646.
124. Hayden, F. G., Treanor, J. J., Fritz, R. S., Lobo, M., Betts, R. F., Miller, M., Kinnersley, N., Mills, R. G., Ward, P., Straus, S. E. (1999). Use of the oral neuraminidase inhibitor oseltamivir in experimental human influenza: randomized controlled trials for prevention and treatment. *JAMA*, *282*, 1240–1246.
125. Nicholson, K. G., Aoki, F. Y., Osterhaus, A. D., Trottier, S., Carewicz, O., Mercier, C. H., Rode, A., Kinnersley, N., Ward, P. (2000). Efficacy and safety of oseltamivir in treatment of acute influenza: A randomised controlled trial. *Lancet*, *355*, 1845–1850.
126. Smee, D. F., Huffman, J. H., Morrison, A. C., Barnard, D. L., Sidwell, R. W. (2001). Cyclopentane neuraminidase inhibitors with potent in vitro anti-influenza virus activities. *Antimicrob. Agents Chemother.*, *45*, 743–748.
127. Sidwell, R. W., Smee, D. F., Huffman, J. H., Barnard, D. L., Bailey, K. W., Morrey, J. D., Babu, Y. S. (2001). In vivo influenza virus-inhibitory effects of the cyclopentane neuraminidase inhibitor RWJ-270201. *Antimicrob. Agents Chemother.*, *45*, 749–757.
128. Bantia, S., Parker, C. D., Ananth, S. L., Horn, L. L., Andries, K., Chand, P., Kotian, P. L., Dehghani, A., El-Kattan, Y., Lin, T., Hutchinson, T. L., Montgomery, J. A., Kellog, D. L., Babu, Y. S. (2001). Comparison of the anti-influenza virus activity of RWJ-270201 with those of oseltamivir and zanamivir. *Antimicrob. Agents Chemother.*, *45*, 1162–1167.

129. Wang, G. T., Chen, Y., Gentles, R., Sowin, T., Kati, W., Muchmore, S., Giranda, V., Stewart, K., Sham, H., Kempf, D., Laver, W. G. (2001). Design, synthesis, and structural analysis of influenza neuraminidase inhibitors containing pyrrolidine cores. *J. Med. Chem.*, *44*, 1192–1201.

130. De Goey, D. A., Chen, H. J., Flosi, W. J., Grampovnik, D. J., Yeung, C. M., Klein, L. L., Kempf, D. J. (2002). Enantioselective synthesis of antiinfluenza compound A-315675. *J. Org. Chem.*, *67*, 5445–5453.

131. Hanessian, S., Bayrakdarian, M., Luo, X. (2002). Total synthesis of A-315675: A potent inhibitor of influenza neuraminidase. *J. Am. Chem. Soc.*, *124*, 4716–4721.

132. Sintchak, M. D., Fleming, M. A., Futer, O., Raybuck, S. A., Chambers, S. P., Caron, P. R., Murcko, M. A., Wilson, K. P. (1996). Structure and mechanism of inosine monophosphate dehydrogenase in complex with the immunosuppressant mycophenolic acid. *Cell*, *85*, 921–930.

133. De Clercq, E., Cools, M., Balzarini, J., Snoeck, R., Andrei, G., Hosoya, M., Shigeta, S., Ueda, T., Minakawa, N., Matsuda, A. (1991). Antiviral activities of 5-ethynyl-1-β-D-ribofuranosylimidazole-4-carboxamide and related compounds. *Antimicrob. Agents Chemother.*, *35*, 679–684.

134. Markland, W., McQuaid, T. J., Jain, J., Kwong, A. D. (2000). Broad-spectrum antiviral activity of the IMP dehydrogenase inhibitor VX-497: A comparison with ribavirin and demonstration of antiviral additivity with alpha interferon. *Antimicrob. Agents Chemother.*, *44*, 859–866.

135. Neyts, J., Andrei, G., De Clercq, E. (1998). The novel immunosuppressive agent mycophenolate mofetil markedly potentiates the antiherpesvirus activities of acyclovir, ganciclovir, and penciclovir in vitro and in vivo. *Antimicrob. Agents Chemother.*, *42*, 216–222.

136. Margolis, D., Heredia, A., Gaywee, J., Oldach, D., Drusano, G., Redfield, R. (1999). Abacavir and mycophenolic acid, an inhibitor of inosine monophosphate dehydrogenase, have profound and synergistic anti-HIV activity. *J. Acquir. Immune Defic. Syndrom.*, *21*, 362–370.

137. Saracco, G., Ciancio, A., Olivero, A., Smedile, A., Roffi, L., Croce, G., Colletta, C., Cariti, G., Andreoni, M., Biglino, A., Calleri, G., Maggi, G., Tappero, G. F, Orsi, P. G., Terreni, N., Macor, A., Di Napoli, A., Rinaldi, E., Ciccone, G., Rizzetto, M. (2001). A randomized 4-arm multicenter study of interferon alfa-2b plus ribavirin in the treatment of patients with chronic hepatitis C not responding to interferon alone. *Hepatology*, *34*, 133–138.

138. Crotty, S., Maag, D., Arnold, J. J., Zhong, W., Lau, J. Y., Hong, Z., Andino, R., Cameron, C. E. (2000). The broad-spectrum antiviral ribonucleoside ribavirin is an RNA virus mutagen. *Nature Med.*, *6*, 1375–1379.

139. Crotty, S., Cameron, C. E., Andino, R. (2001). RNA virus error catastrophe: Direct molecular test by using ribavirin. *Proc. Natl. Acad. Sci. U.S.A.*, *98*, 6895–6900.

140. De Clercq, E., Cools, M., Balzarini, J., Marquez, V. E., Borcherding, D. R., Borchardt, R. T., Drach, J. C., Kitaoka, S., Konno, T. (1989). Broad-spectrum antiviral activities of neplanocin A, 3-deazaneplanocin A, and their 5′-nor derivatives. *Antimicrob. Agents Chemother.*, *33*, 1291–1297.

141. Yuan, C. S., Wnuk, S. F., Liu, S., Robins, M. J., Borchardt, R. T. (1994). (E)-5′,6′-didehydro-6′-deoxy-6′-fluorohomoadenosine: A substrate that measures the

hydrolytic activity of S-adenosylhomocysteine hydrolase. *Biochemistry*, *33*, 12305–12311.

142. Bray, M., Davis, K., Geisbert, T., Schmaljohn, C., Huggins, J. (1999). A mouse model for evaluation of prophylaxis and therapy of Ebola hemorrhagic fever. *J. Infect. Dis.*, *179*(Suppl. 1), S248–S258.

143. Huggins, J., Zhang, Z.-X., Bray, M. (1999). Antiviral drug therapy of filovirus infections: S-adenosylhomocysteine hydrolase inhibitors inhibit Ebola virus in vitro and in a lethal mouse model. *J. Infect. Dis.*, *179*(Suppl. 1), S240–S247.

144. Bray, M., Driscoll, J., Huggins, J. W. (2000). Treatment of lethal Ebola virus infection in mice with a single dose of an S-adenosyl-L-homocysteine hydrolase inhibitor. *Antiviral Res.*, *45*, 135–147.

26

PROTEIN KINASE INHIBITORS IN DRUG DISCOVERY

KEYKAVOUS PARANG AND GONGQIN SUN
University of Rhode Island
Kingston Rhode Island

26.1	PHOSPHORYLATION IN SIGNAL TRANSDUCTION PATHWAYS	1193
26.2	CLASSIFICATION OF PROTEIN KINASES	1194
26.3	GENERAL MECHANISTIC FEATURES OF PROTEIN PHOSPHORYLATION BY PROTEIN KINASES	1195
26.4	GENERAL STRUCTURAL FEATURES OF PROTEIN KINASES	1196
26.5	ACTIVATION OF PROTEIN KINASES	1198
26.6	PROTEIN KINASES IN HUMAN DISEASES	1199
26.7	IMPORTANT PROTEIN KINASES AS THERAPEUTIC TARGETS	1199
	Protein Tyrosine Kinases	1199
	Serine/Threonine Kinases	1202
26.8	PROTEIN KINASES AS TARGETS FOR INHIBITOR DESIGN	1205
	General Overview of Protein Kinase Inhibitors (PKIs)	1205
	Alternative Strategies in Designing Protein Kinase Inhibitors	1214
26.9	SELECTED INHIBITORS OF PROTEIN KINASES ON MARKET OR IN CLINICAL TRIALS	1215
	Receptor Tyrosine Kinase Inhibitors	1216
	Nonreceptor Tyrosine Kinase Inhibitors	1221
	Serine/Threonine Kinase Inhibitors	1225
26.10	PROSPECTS AND FUTURE DIRECTIONS	1232
	References	1232

Drug Discovery Handbook, by Shayne Cox Gad
Copyright © 2005 by John Wiley & Sons, Inc.

ABBREVIATIONS AND GLOSSARY

ACS	acute coronary syndromes
AML	acute myelogenous leukemia
ALL	acute lymphoblastic leukemia
Bcr-Abl	Abelson tyrosine kinase fused to the breakpoint cluster region
bFGF	basic fibroblast growth factor
CDK	cyclin-dependent protein kinase
CDKI	CDK inhibitor
Csk	C-terminal Src kinase
DYRK	dual specificity, tyrosine-phosphorylated and regulated kinase
EGF	epidermal growth factor
EGFR	EGF receptor
Erb	estrogen receptor subtype b
ERK	extracellular signal-regulated protein kinase
FDA	U.S. Food and Drug Administration
FGF	fibroblast growth factor
FGFR	FGF receptor
Flk-1/KDR	fetal liver kinase-1/kinase insert domain containing receptor (VEGFR-2, VEGFR type 2)
Flt-1	fms-like tyrosine kinase-1
GSK3	glycogen synthase kinase 3
HER	human epidermal growth factor receptor
IC_{50}	50% inhibitory concentration
IR	insulin receptor
IRK	insulin receptor kinase
IRS	insulin receptor substrate
JAK	Janus kinases
JNK	c-jun N-terminal kinase
KDR	kinase insert domain containing receptor
Lck	lymphocyte kinase
MAP	mitogen-activated protein
MAPK	mitogen-activated protein kinase
MAPKK	MAP kinase kinase
MEK-1	MAPK/ERK kinase 1
MKK	MAP kinase kinase
mTOR	mammalian target of rapamycin
NGF	nerve growth factor
NRTK	nonreceptor tyrosine kinases
PDGF	platelet-derived growth factor
PDGFR	PDGF receptor
PKA	protein kinase A; cAMP-dependent protein kinase
PKC	protein kinase C

PKI	protein kinase inhibitor
PTK	protein tyrosine kinase
PTP	protein tyrosine phosphatase
RA	rheumatoid arthritis
Ras	rat sarcoma virus oncogene
Rb	retinoblastoma tumor suppressor gene
RTK	receptor tyrosine kinase
S	DNA synthesis stage
SAPK	stress-activated protein kinases
SH2	Src homology 2
STKs	serine/threonine kinases
Tek/Tie-2	tunica internal endothelial cell kinase type 2
Tie-1	tunica internal endothelial cell kinase type 1
TNF	tumor necrosis factor
VEGF	vascular endothelial growth factor

26.1 PHOSPHORYLATION IN SIGNAL TRANSDUCTION PATHWAYS

Approximately 2 percent of eukaryotic genes encode protein kinases (PKs) [1, 2]. Approximately 518 PKs are predicted in the human kinome based on the information from the human genome sequence [3]. PKs have essential roles in cell-signaling pathways by interacting with extracellular ligands such as growth factors and hormones and transmitting signals across the cell membrane to the cytoplasm and the nucleus. Signal transduction is mediated by PKs through catalyzing the transfer of the γ-phosphoryl group from adenosine triphosphate (ATP) to the hydroxyl group of defined tyrosine (Tyr), serine (Ser), and/or threonine (Thr) residues in various critical protein substrates (Fig. 26.1). These amino acids are usually located in the consensus sequences on the protein substrate. Phosphorylated protein substrates have distinct functional properties in many different cell-signaling pathways. Abnormal phosphorylation is often a cause or consequence of diseases. In addition to the signals arising from receptor–ligand interactions as seen with growth factors,

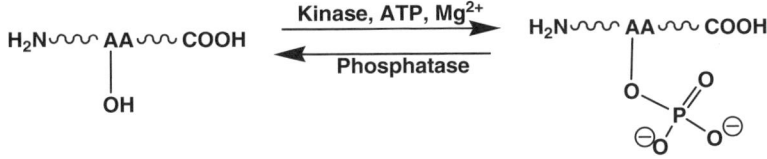

Figure 26.1 General scheme for phosphorylation and dephosphorylation reactions catalyzed by PKs and PPs, respectively. PKs phosphorylate the hydroxyl group of amino acids (AA-OH = tyrosine serine, or threonine) on the protein substrate in the presence of Mg^{2+} and ATP. PPs reverse the phosphorylation reaction by cleaving the O–P bond in phosphoproteins, to generate the free hydroxyl group in proteins.

environmental factors such as mechanical deformation by cell stretch or shear stress will activate PKs. On the other hand, protein phosphatases (PPs) remove the phosphate groups from distinct phosphoproteins in a reverse reaction to counterbalance the phosphorylation reaction.

26.2 CLASSIFICATION OF PROTEIN KINASES

Although 518 members of the human protein kinase superfamily can be classified into at least 7 distinct groups and many families and subfamilies [3], PKs are usually divided into two large families based on their substrate specificities: protein tyrosine kinases (PTKs) (e.g., Src tyrosine kinases, Bcr-Abl) and serine/threonine kinases (STKs) [e.g., mitogen-activated protein kinases (MAP kinases), protein kinase C (PKC), protein kinase A (PKA)] (Fig. 26.2). PTKs such as Src kinases and several of serine/threonine kinases such as MAP kinases play a crucial role in signal transduction pathways and regulating various cellular functions (e.g., proliferation, differentiation). Several serine/threonine kinases such as cyclin-dependent kinases (CDKs) are involved in cell cycle checkpoint control.

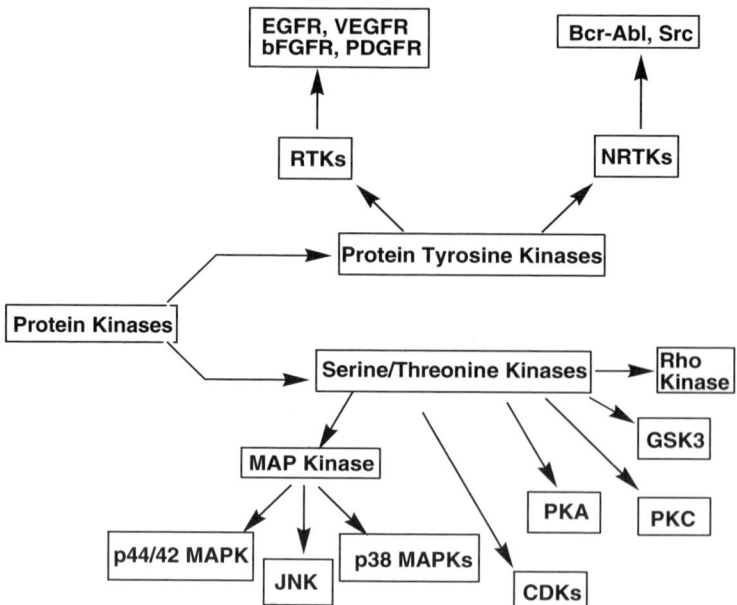

Figure 26.2 Classification of PKs involved in signaling pathways into PTKs and STKs. PTKs are subdivided into the receptor tyrosine kinases (RTKs) and nonreceptor tyrosine kinases (NRTKs). Important PKs known as drug targets for each class are shown here.

It is estimated that there are approximately 100 PTKs. PTKs are subdivided into two families: receptor tyrosine kinases (RTKs; e.g., growth factor receptor kinases) and nonreceptor tyrosine kinases (NRTKs; e.g., Bcr-Abl, Src) [4]. RTKs are transmembrane cell surface proteins that are activated by the binding of ligands to their extracellular domains. The binding leads to receptor oligomerization and phosphorylation/activation of the intracellular domains by inducing autophosphorylation. Activated and phosphorylated intracellular domains then activate downstream signaling proteins such as NRTKs in the cytoplasm [4]. RTKs include the receptors for ligands, such as insulin (insulin receptor kinase; IRK), and many growth factors, such as epidermal growth factor (EGF), basic fibrobalst growth factor (FGF), platelet-derived growth factor (PDGF), vascular endothelial growth factor (VEGF), and nerve growth factor (NGF). NRTKs include members of the Src family, Fak, Jak, Abl, and Zap70. NRTKs can be regulated and activated by different mechanisms such as RTKs and other cell surface receptors such as receptors of the immune system and G-protein-coupled receptors [4].

26.3 GENERAL MECHANISTIC FEATURES OF PROTEIN PHOSPHORYLATION BY PROTEIN KINASES

To design protein kinase inhibitors (PKIs), it is important to understand the mechanistic features of the particular target kinase. Although 518 PKs share a conserved catalytic domain in sequence and structure, the catalysis and phosphorylation reaction is regulated differently by these enzymes. There is still an incomplete understanding of the mechanisms used by PKs in the phosphorylation reaction. The phosphorylation reaction occurs in the active site of PKs by the formation of a ternary complex consisting of a protein substrate, MgATP, and a PK [5]. ATP and protein substrate bind to ATP and substrate binding sites, respectively, in the catalytic domain of kinases. The transfer of the phosphoryl group takes place directly from ATP to the protein substrate through dissociative or associative mechanisms in a three-step process: (1) the initial binding of substrate to a preformed PK–ATP complex, (2) the transfer of phosphoryl group to hydroxyl group, and (3) the release of phosphorylated protein from the PK–ADP (adenosine diphosphate) complex (Fig. 26.3). In a fully dissociative mechanism, the bond between γ-phosphorus and ADP (leaving group) is broken in the transition state before the formation of the bond between the nucleophile (tyrosine, threonine, or serine) and phosphorus. In an associative mechanism transition state, a significant amount of bond is formed between the nucleophile and phosphorus, before the complete bond breakage between the phosphorus and the leaving group [6–8]. The distinction between the associative and dissociative mechanisms may be critical in designing transition-state and mechanism-based inhibitors of PKs. Inhibitors can be designed that could potentially inhibit the binding of ATP (ATP

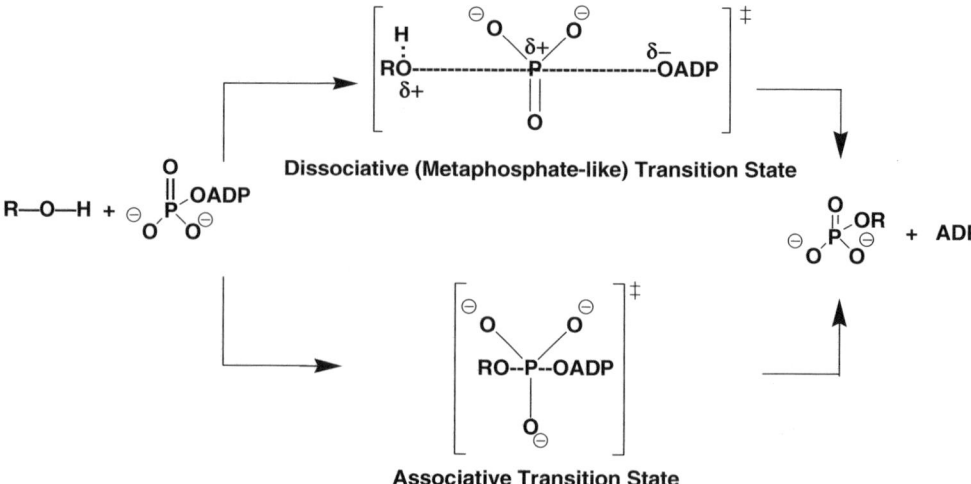

Figure 26.3 Dissociative versus associative transition states for phosphoryl transfer.

binding site inhibitors), protein substrate (substrate binding site inhibitors), or both (bisubstrate analog inhibitors).

26.4 GENERAL STRUCTURAL FEATURES OF PROTEIN KINASES

The three-dimensional structures of several individual PKs have been determined by X-ray crystallography. Protein Data Bank (PDB) comprises over 160 PKs, protein kinase mutants, and protein kinase:ligand complexes. This wealth of structural information of PKs and PK–ligand complexes can be used for structure-based rational design of novel and effective PK inhibitors. For example, it has been shown from these structures that the backbone carbonyl and amide groups of the protein backbone located in the hinge region of ATP binding site potentially form three hydrogen bonds with ATP binding site inhibitors through two hydrogen bond acceptors and one hydrogen bond donor [9]. The detailed information about protein kinase structures and the interaction mode with their inhibitors have been reviewed extensively elsewhere [4, 10–13] and is beyond the scope of this chapter. Some of the crystal structures of the kinase domain of several RTKs and NRTKS that are clinically important include IRK [14], insulin-like growth factor-1 receptor [15], c-Src tyrosine kinase [16, 17], FGFR (FGF receptor) [18, 19], VEGFR [20], lymphocyte kinase (Lck) [21, 22], hematopoetic cell kinase (Hck) [23, 24], Bruton's tyrosine kinase (BtK) [25], and Abl kinase [26].

All PKs have an ~260-amino-acid conserved kinase domain folding into two structurally dissimilar lobes, N-terminal and C-terminal domains that are joined by a linker peptide coil [27]. The kinase domain is responsible for catalytic activity and the phosphorylation reaction. The N-terminal domain of PKs or nucleotide binding site is generally ~90 amino acids, which recognizes and binds ATP utilized by all kinases. Many of the amino acid residues comprising the nucleotide binding pocket are highly conserved forming five β strands (β1 to β5) and at least one conserved helix (αC). The C-terminal lobe is predominantly α-helical and is not conserved among kinases varying in size, sequence, and topology containing the substrate binding groove, activation loop, and catalytic residues [28] (Fig. 26.4). The transfer of the phosphoryl group takes place directly between ATP and protein substrate in the kinase domain. In addition to kinase domain, RTKs comprise of multidomain membrane-spanning proteins.

Discussing more detailed structural features of all PKs is beyond the scope of this chapter. Cytoplasmic nonreceptor kinases contain several other domains in addition to the kinase domain. One of the well-studied examples is Src kinases that are potential drug targets and can serve as an example for the interesting structure and function that PTKs possess.

Src family kinases share common structural motifs and are made up of five components/domains: N-terminal or the fatty acid acylation domain, which targets the kinases to the plasma membrane, Src homology 3 (SH3), Src homology 2 (SH2), kinase domain (catalytic, including ATP and substrate binding sites), and CT noncatalytic regulatory domain (see Fig. 26.4) [17]. SH3 and SH2 are protein–protein interaction domains that have both adaptor and regulatory functions [17] by binding to their respective ligands [29]. SH3 domains are ~60 amino acids in length and bind proline-rich regions of several signaling proteins [16, 23, 30]. SH2 domains are modules of approximately 100

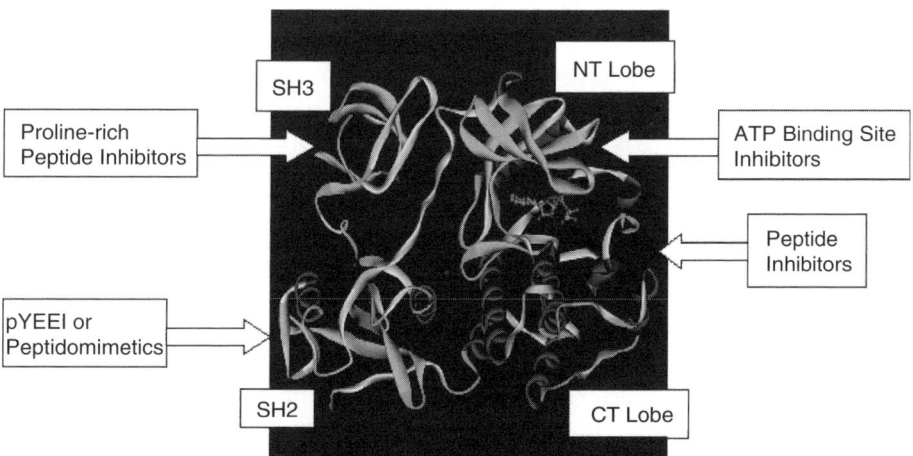

Figure 26.4 General structure of Src tyrosine kinases and potential sites for inhibitors intervention. AMP-PNP, a ligand in kinase domain, is represented by ball-and-stick vectors (PDB 2SRC) [17].

amino acids that have evolved to recognize and bind to phosphotyrosine sequences located on proteins [17, 31] in response to extracellular signals [31–33]. The phosphotyrosyl group binds in a deep phosphotyrosine binding pocket of SH2 domain where it is stabilized through a large number of hydrogen bonds and electrostatic interactions with positively charged amino acids. In addition to stabilizing the inactive conformation of the kinases, the SH2 domain mediates interactions of the Src family kinases with other cellular proteins such as FAK, p130^{cas}, p85, PI3K, and p68^{sam} [34–37] propagating downstream signaling by recruiting the SH2 domain-containing protein to its proper signaling complex and/or by altering its enzymatic activity. Inappropriate signaling through Src kinases has been linked to many pathological conditions, providing an impetus to understand the mechanism of SH2 domain recognition and to design SH2 domain binding inhibitors [29].

26.5 ACTIVATION OF PROTEIN KINASES

The activity of PKs is tightly regulated because of their central role in signal transduction [12, 14, 38]. Different kinases utilize various mechanisms to become activated such as transphosphorylation of specific residues within the activation loop in IRK, and the human Lck kinase [22, 39–43] leads to a dramatic conformational change in the structure of kinase. The conformational change exposes the substrate binding site for substrate binding by repositioning the key catalytic residues that are involved in binding to protein substrate and catalytic reaction. Another mechanism for regulation of kinase activity is phosphorylation sites N- and C-terminal to the kinase domain as seen in EphB2 receptor kinase domain [44]. Both of these mechanisms have been seen in Src family kinases [14, 16, 23, 45]. There are two major Src phosphorylation sites that have opposite effects on Src activity, Tyr416 and Tyr527. Tyr416, an autophosphorylation site, is located in a region called the activation loop of the kinase catalytic domain, which is ordered as an α-helix in inactive Src [17]. Phosphorylation of Tyr416 enhances Src activity. Tyr527 is located in the C-terminal tail of the molecule and, when phosphorylated by another PK called C-terminal Src kinase (Csk), binds to the SH2 domain of Src [17, 23]. Phosphorylation of Tyr527 by Csk leads to the kinase inactivation. This interaction locks the c-Src molecule in an inhibited closed conformation. Displacement of SH2 and/or SH3 domains, either by C-terminal tail dephosphorylation or by competitive binding of optimal SH2/SH3 domain ligands, allows the kinase domain to switch from a closed to an open conformation [17]. The kinase reactivation by different pTyr-containing peptides related to phosphotyrosine–glutamic acid–glutamic acid–isoleucine (pYEEI) known to bind the SH2 domain of Src has been demonstrated [46–49]. Src is known to be activated by SH2-mediated interaction with autophosphorylated PDGF receptor [50], and by dual SH3/SH2 interactions with sequences in the focal adhesion kinase FAK [51] and Sin [52]. In the inactive conformation of Src kinases, the

SH2 and kinase domains lie in opposite sides. A dramatic structural reorientation between the SH2 domain and the active site regions will occur upon the activation of the enzyme. In the active conformation, the SH2 domain is positioned closer to the active site region in a manner that promotes substrate binding in kinase domain [23, 53].

26.6 PROTEIN KINASES IN HUMAN DISEASES

Protein kinases are involved in most aspects of normal cellular signaling pathways and function and numerous biological events, such as apoptosis, differentiation, cell growth, and metabolism; therefore, aberrant kinase activity is believed to contribute to several diseases such as cancer [2, 54], diabetes [55, 56], cardiovascular [57], neurodegenerative [58] and inflammatory disorders [59, 60], and restenosis [61].

Protein kinase family has now become one of the most important targets for anticancer drug discovery since overexpression and enhanced PK activity have been directly associated with cell transformation, carcinogenesis, and many forms of human cancer [2, 54]. v-Src, a protein tyrosine kinase, was discovered more than 26 years ago as the first oncogene [62]. Since then it has been reported that over 80 percent of the oncogenes and proto-oncogenes involved in human cancers encode PTKs [2].

Activation or inhibition of a kinase in a specific disease state is not sufficient to consider it as putative therapeutic targets since dysregulation can be a consequence of the disease and may not be a contributing factor to disease pathology. Additional evidences such as genetic or physiological/cell biological data are needed to implicate a PK as a cause and not the effect of the disease and as an attractive target.

26.7 IMPORTANT PROTEIN KINASES AS THERAPEUTIC TARGETS

Herein, some of the PKs validated as targets for drug discovery are introduced. We will discuss a number of inhibitors developed against some of these kinases in Section 26.9.

Protein Tyrosine Kinases

Protein tyrosine kinases are enzymes that catalyze phosphorylation of tyrosine in many proteins by the transfer of the γ-phosphoryl group from ATP. While tyrosine phosphorylation represents a minor subset of protein phosphorylation in the cell in terms of level, it is involved in several cellular responses such as mitogenesis [63], differentiation [64], migration [65], and survival [66]. Two subfamilies of PTKs (RTKs and NRTKs) are discussed here. Typically, RTKs are linked to growth factor receptors and NRTKs are associated with the protein products of proto-oncogenes, such as *v*-src [67]. PTKs serve as molecular targets for many of the investigational new agents.

Receptor Tyrosine Kinases The RTKs play instrumental roles in cancer development and have been well studied and extensively explored in anticancer drug discovery in the last decade. RTKs include EGF and its receptor (EGFR), tunica internal endothelial cell kinase type 1 (Tie-1), tunica internal endothelial cell kinase type 2 (Tek/Tie-2), VEGF and its receptor (VEGFR-2 or Flk-1/KDR), bFGF and its receptor (FGFR), and PDGF and its receptor (PDGFR). EGFR, bFGFR, VEGFR, and PDGFR kinases are discussed here.

EGF Receptor Tyrosine Kinases The EGF stimulates the proliferation of a large variety of cells such as epithelial cells and fibroblasts [68] by binding to and activating the EGFR and initiating intracellular signaling [68–70]. The EGF receptor kinases are also known as the EGF family of type I tyrosine kinases or ErbB tyrosine kinases. All four members in this family of RTKs, HER-1 (c-ErbB1/EGFR), HER-2 (c-ErbB2/neu), HER-3 (c-ErbB3), and HER-4 (c-ErbB4) [71, 72], are closely homologous and share common structural features such as an extracellular ligand binding domain and a cytoplasmic tyrosine kinase domain.

The EGFRs are implicated in the development and progression of certain human tumors [73], including head and neck, colon, breast, lung, prostate, and ovarian cancers. All EGFR kinases are frequently overactive in cancer cells. HER-1 is significantly overexpressed in a large variety of epithelial cancers, such as lung and breast cancers [74]. HER-2 (Neu, ErbB2) was identified in 1985 and soon became an attractive target for anticancer drug design since it is overexpressed, most commonly by gene amplification, in a number of human malignancies, including gastric, breast, colon, ovarian, and non-small-cell lung carcinoma [74]. Several pharmaceutical companies are interested in the development of agents targeting EGFR kinases.

VEGF and bFGF Receptor Tyrosine Kinases Two other important growth factor receptor targets are VEGFR-2 or Flk-1/KDR [75, 76] and bFGF-R, which have been associated to tumor angiogenesis. VEGFR-2 and FGFR are expressed exclusively in endothelial cells [75]; thus, they are very attractive targets for design of angiogenesis inhibitors. Some studies indicated that immunoneutralization of FGF had little or no effect on tumor angiogenesis [77, 78]. In spite of this FGFR has been explored in several diseases such as cancer, diabetic retinopathy, atherosclerosis, and rheumatoid arthritis [19]. The role of VEGF as a key regulator of angiogenesis has been confirmed [79]. There are two VEGF receptor tyrosine kinases: VEGFR1 (Flt-1) and VEGFR-2. There is now a general consensus that VEGFR2 is the major key mediator of the mitogenesis and angiogenesis in endothelial cells and is activated by binding of VEGF, a mitogen-specific protein for vascular endothelial cells [75]. It is proposed that the antiangiogenic therapy through inhibition of the

VEGFR is a safe and well-tolerated approach in treatment of cancer patients and may selectively target the tumor-associated vessels and reduce tumor-induced edema.

PDGF Receptor Tyrosine Kinases The PDGFRs have been reported to be overexpressed in many cancers of mesenchymal of glial origin and, thus, can also be a valuable target for these types of malignancies [80, 81]. The role of Flt-3, a member of the PDGFR tyrosine kinase family, in a variety of cancers, especially acute myelogenous leukemia (AML), has been clearly demonstrated [80].

Nonreceptor Tyrosine Kinase Inhibitors The NRTKs include about one third of all PTKs and are found in cytoplasm signaling downstream of RTKs. Several important targets for anticancer drug design have been identified in the last 10 years by studying the NRTK subfamily [82] such as Bcr-Abl and Src family of kinases.

Bcr-Abl The Abelson PTK (c-Abl) is a nuclear protein tyrosine kinase with the unknown biological function. c-Abl is proposed to function in sensing the integrity of the genome and promoting programmed cell death. Bcr is a multifunctional cytosolic serine–threonine kinase that may play a role in regulating the activity of the Rho subfamily of small G proteins. The Philadelphia chromosome is formed by a reciprocal translocation between chromosome 9 and 22, resulting in the formation of a novel hybrid gene, which produces Bcr-Abl (the breakpoint cluster region fused to the Abelson leukemia virus), an oncoprotein responsible for Philadeliphia chromosome-positive chronic myelogenous leukemia (CML). This chromosome is found in more than 95 percent of patients with CML and about 15 to 20 percent of patients with adult acute lymphoblastic leukemia (ALL) [83] making Bcr-Abl an excellent target for the drug discovery activity against these types of cancer. Unlike c-Abl and Bcr, Bcr-Abl is both cytosolic and nuclear.

Src Kinases The Src family of PTKs, Src, Yes, Lck, Fyn, Lyn, Fgr, Hck, Blk, Frk, and Yrk are NRTKs [84, 85] found widely expressed in mammalian cells. Src tyrosine kinases, *v*-Src and *c*-Src, are prototypes of this well-established superfamily of PTKs. The Src family of PTKs has become key targets for both basic research and drug discovery over recent years. Src has significant structural homology to all nine other proteins in this superfamily. Enhanced Src tyrosine kinases activity has been directly linked to T-cell activation, mitogenesis, differentiation, cell transformation, and oncogenesis [54].

The Src exists as an intriguing therapeutic target for drug discovery with respect to cancer. Src has been associated with several different cancers including colon [86] and breast [87] cancers for which transformed phenotypes have been correlated with Src mutations and/or overexpression [54, 84, 88] of Src tyrosine kinase activities [89]. Src has also been implicated in the development of osteoporosis and inflammation-mediated bone loss [90–93]. Src plays multifunctional roles in osteoblast and osteoclast activities. In osteoblasts, Src recently has been shown to be a negative regulator of osteoblast functional activity and implicated as a mediator of sex steroid-induced antiapoptotic signaling in osteoblasts [94, 95]. In osteoclasts, Src is important for osteoclast motility, activation (ruffled border formation), survival, and adhesion by mediating various signal transduction pathways [96–99]. Src is involved in bone remodeling (resorption and formation) and bone-related diseases, such as osteoporosis, rheumatoid arthritis, periodontal disease, Paget's disease, hypercalcemia of malignancy, and metastasis of certain cancers to bone [85].

Lyn activity is elevated in patients with acute myelogenous leukemia. Lck is a positive activator in T-cell signaling, and its activation is involved in autoimmune diseases such as rheumatoid arthritis [100]. Thus, Src kinases are important targets for drug discovery in several diseases such as bone-related diseases, cancer, and autoimmune diseases [54, 88, 100–104].

Serine/Threonine Kinases

MAP Kinases (ERK, JNK, p38 MAPKs) Exposure of cells to external stresses and extracellular stimuli such as ischemia/reperfusion [105, 106] will activate several members of the mitogen-activated protein kinase (MAPK) family, followed by other downstream signals [107, 108]. MAP kinases are STKs that are activated by dual phosphorylation of threonine and tyrosine residues of the Thr-X-Tyr segment in a loop located adjacent to the active site [108–110]. Prototypic members of the family present in mammalian cells include (1) extracellular signal-regulated kinases (ERK1/2; also known as p44/42 MAPK), (2) c-jun N-terminal kinases [(JNKs; also called stress-activated protein kinases (SAPKs)], (3) ERK5/big MAP kinase 1 (BMK1), and (4) p38 MAPKs. Each of these subgroups plays distinct roles in cell growth and differentiation. Each MAP kinase is phosphorylated by specific upstream kinases. Phosphorylated and activated MAP kinases move to the nucleus and regulate the phosphorylation of various substrates such as transcription factors (Fig. 26.5), which are involved in regulating the expression of specific sets of genes and thus mediate a specific response to the stimulus [111–116].

One important pathway causing the activation of MAPKs is the Ras-dependent MAPK cascade that is activated by growth factors (Fig. 26.5). The uncontrolled activation of this pathway by the overexpression of particular growth factor receptor tyrosine kinases or their mutation to constitutively active forms is now known to cause cancer. Ras activation begins its downstream signaling through association with the Raf (a serine/threonine kinase). The GTP (guanidine triphosphate)-bound dependent interaction of Ras with Raf is required for Raf kinase activation. Activated Raf phosphorylates and activates downstream kinases such as mitogen-activated ERK-activating kinase (MEK) and MAPK, respectively [117].

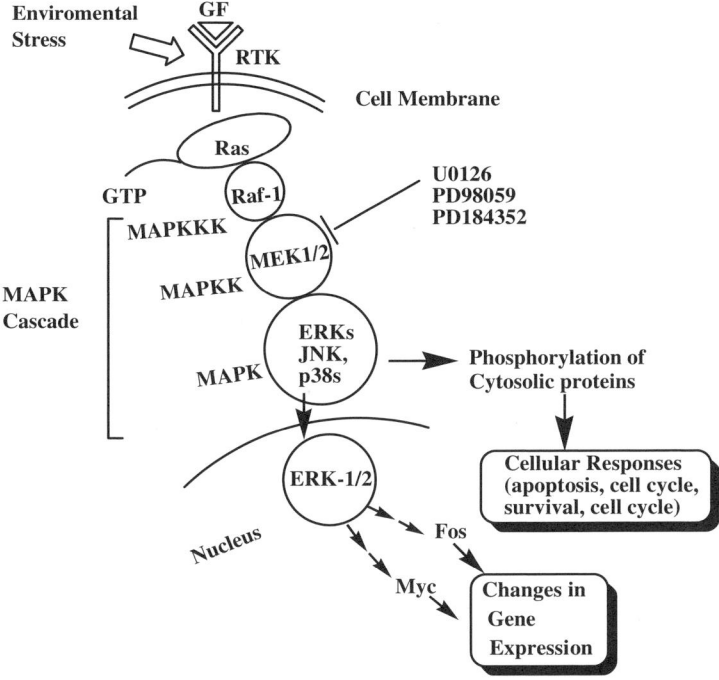

Figure 26.5 Classical Ras-MAPK pathway and potential sites as drug targets. The binding of growth factors to extracellular domains of RTKs activates tyrosine kinase activities in their intracellular domains, which, in turn, activate the Ras cascade. Activation of Ras leads to sequential phosphorylation and activation of a series of enzymes involved in MAPK cascade that finally transmit the stimulatory external signals received from cytoplasmic membrane into the nucleus. Growth-factor-induced activation of pathway often leads to cell cycle progression. All MAPKs are part of a three-tiered cascade whereby a MAPKKK (in this case Raf-1) activate the downstream kinases MAPKKs (in this case MEK1/2), which, in turn, activate MAPKs (in this case, ERKs). Several MEK inhibitors discussed in text are shown here.

Cyclin-Dependent Kinases (CDKs) Cell cycle is tightly regulated by members of the CDK family and cyclin regulatory subunits [118, 119]. CDKs are a family of STKs that are important in controlling entry into and progression through the cell cycle, and their activity is regulated at several levels [120] such as the binding of each CDK to a specific cyclin partner. As cells respond to the presence of mitogens, they enter the cell cycle, and the level of cyclin D1 increases leading to the activation of CDK4 and CDK6. These kinases phosphorylate the retinoblastoma tumor suppressor protein pRb, thereby abrogating its inhibition of members of the E2F family of transcription factors. The inhibition of the transcription factors triggers a program of gene expression that results in entry into the S phase of the cell cycle. The deregulation of the pRb pathway and CDK4 are important in cancer progression [121]. Tumors may express abnormally high levels of the cyclin D1 protein, contain multiple copies of the cyclin *D1* gene [122], or contain mutations or deletions of the *pRb* gene [123–125].

Protein Kinase C (PKC) and Protein Kinase A (PKA) The PKC superfamily is a member of STKs that are stimulated by Ca^{2+}, phospholipid, and diacylglycerol and play a critical role in intracellular signal transduction for cell proliferation and differentiation, gene expression, the control of cell differentiation and growth, and vascular permeability and proliferation. Thus, PKC enzymes are involved in cardiovascular disease, cancer, ischemia, inflammation, and central nervous system (CNS) disorders [126]. The mammalian PKC family consists of 12 different isosymes as α, βI, βII, γ, δ, ε, ζ, η, θ, ι, λ, and μ that are classified into 3 subfamilies of novel, conventional, and atypical on the basis of their cofactor requirements [127]. PKC isoforms have specific cellular localization in various tissues and cell types [128] and distinct and in some cases opposing roles in cell growth and differentiation [129, 130]. For example, the induction of β isoform in response to hyperglycemia in aortic, cardiac, retinal, and renal tissues is shown to be associated with diabetic manifestations, such as neuropathy, nephropathy, angiopathy, macular edema, retinopathy, and cardiomyopathy [131, 132]. Increased levels of PKC have been associated in several cell lines including lung [133], gastric carcinomas [134], and breast [135] with malignant transformation. However, no clear relationship has been observed in vivo. For example, with PKCβ expression being found to increase, remain the same, or decrease in colon tumors when compared with normal epithelium [136].

The PKA family represents another group of STKs and has at least two isoforms, type I and type II [137] and is known to mediate cyclic adenosine monophosphate (cAMP) effects and is involved in many endocrine and nonendocrine cell responses. The PKA holoenzyme is a tetramer consisting of a dimeric regulatory subunit (R2) and two monomeric catalytic subunits (C2). Both PKC- and PKA-mediated signaling cascades have crucial roles in glial cell proliferation and differentiation [138, 139].

Rho Kinase The ROCKs, or Rho kinases, are STKs that are involved in a variety of biochemical signal transductions in the cells. In addition to cytoskeleton rearrangement, cell migration, motility, proliferation, and neurite outgrowth, this enzyme has a crucial role in cardiovascular function such as smooth muscle contraction through calcium sensitization, an event that controls vascular vessel tone [140, 141]. Thus, ROCK inhibitors have been designed as therapeutic agents for cardiovascular diseases.

Phosphoinositide 3-kinase (PI3K) The PI3K has a protein kinase fold. Upon growth factor receptor stimulation, the PI3K family of enzymes is recruited and produces 30 phosphoinositide lipids that act as second messengers by binding to and activating diverse cellular target proteins [142]. For example, the production of the phosphoinositide polyphosphate PI(3,4,5)P3, or PIP3, by PI3K has critical roles in pathways governing apoptosis, differentiation, cell proliferation, and migration; hence, abnormal regulation of PIP3 levels and the levels of their lipid products are associated with a variety of cancer types.

It is believed that resistance to radiation treatment in a number of cancers is due to activation of the PI3K pathway, suggesting that inhibition of PI3K could be an important strategy to reduce resistance in cancer radiation treatment [143].

Glycogen Synthase Kinase 3 (GSK3) The GSK-3 is another cytosolic STK found in two closely related isoforms (α and β) with high homology (ca. 90 percent) at the catalytic domain [144] but distinct functional roles. Both isoforms are expressed ubiquitously in mammalian tissues. GSK3 is one of many signaling components downstream from the insulin receptor (IR). For example, GSK-3β plays a critical role in cancer (via angiogenesis, apoptosis, and tumorigenesis), CNS function (via the proteins tau and β-catenin), and glucose homeostasis [145]. Insulin-dependent glycogen synthesis is activated by the inhibition of GSK-3-dependent phosphorylation leading to lower plasma glucose. Thus, GSK-3β is a potential target for inhibitor design and treating type II diabetes [132, 146]. Additionally, due to critical roles of GSK-3 in cancer and CNS function, GSK inhibitors may have therapeutic potential for treating neurodegenerative diseases, bipolar disorder, stroke, and cancer [147]. Several potential inhibitors have been recently reviewed [148].

26.8 PROTEIN KINASES AS TARGETS FOR INHIBITOR DESIGN

Many reviews of protein kinase inhibitor discovery have appeared over the past 14 years [4, 11, 12, 38, 149, 150], and several central themes have emerged. This chapter will focus on inhibitors directed against critical molecular targets discussed above that are approved or are in clinical trials.

General Overview of Protein Kinase Inhibitors (PKIs)

A number of agents have been approved by FDA as PKIs. These inhibitors include small molecules such as STI-571 (Gleevec) [151], ZD-1839 (Iressa) [152], and Sirolimus (Rapamycin) [153] and monoclonal antibodies such as Herceptin (Trastuzumab) [154], Erbitux (Cetuximub) [155, 156], and Bevacizumab (Avastin) [79] (Tables 26.1 and 26.2).

Several other PKIs are in clinical trials that target different stages of signal transduction such as binding to RTKs, intracellular signaling, and cell cycle regulation [150, 157, 158]. There are several issues to be considered in designing PKIs. First, the inhibitors must exhibit selectivity against a family or a member of PKs. Second, the PKs in different systems should be monitored

TABLE 26.1 Selected Inhibitors of Protein Tyrosine Kinases in Market or Clinical Trials

Kinase Target	Agent	Trial (Disease)	Sponsor
ABL (c-Kit, PDGFR)	Gleevec (STI-571)	Approved (CML, GIST)	Novartis
EGFR	Iressa (ZD1839)	Approved (Lung cancer)	AstraZeneca
	Tarceva (OSI-774)	Phase III (Cancer)	OSI/Roche/Genentech
	Erbitux (Cetuximab)	Approved (Cancer)	Imclone
	ABX-EGF (mAb)	Phase II (Cancer)	Abgenix
	MDX-447 (mAb)	Phase I (Cancer)	Merck KgaA
	EMD 72000 (mAb)	Phase I (Cancer)	Merck KgaA
	Genistein	Phase II (Cancer)	NCI
	R3	Phase II (Cancer)	Center of Molecular Immunology
	EMD-72000 (EMD-6200)	Phase I/II	Merck
	RH3 (mAb)	Phase II (Cancer)	York Medical Bioscience Inc.
EGFR, ERB2R	CI1033/PD183805	Phase II (Cancer)	Pfizer
	EKB569	Phase I (Cancer)	Wyeth-Ayerst
	GW2016	Phase II (Cancer)	GlaxoSmithKline
	PKI166	Phase I (Cancer)	Novartis

Target	Compound	Status	Company
VEGFR (PDGFR, FGFR)	SU6668	Phase I (Cancer)	Pharmacia Corporation
PDGFR (Flt-3)	CT53518	Phase I (Cancer)	Millennium Pharmaceuticals
VEGFR	SU5416	Phase III (Cancer)	Pharmacia Corporation
	PTK787/ZK222584	Phase III (Cancer)	Novartis/Schering-Plough
	Anti-VEGFR2 mAb	Phase I (Cancer)	Imclone
	Bevacizumab (Avastin)	Approved (Cancer)	Genentech
	AE-941 (GW786034)	Phase II/III (Cancer)	AEterna
	CEP-7055	Phase I (Cancer)	Cephalon
	CEP-5214	Phase I (Cancer)	Cephalon
	VEGF-trap	Phase I/II (Cancer)	Regeneron
VEGFR (EGFR)	ZD6474	Phase II (Cancer)	AstraZeneca
VEGFR (PDGFR)	SU11248	Phase II (Cancer)	Sugen
VEGFR (RAF)	Bay 43-9006	Phase III (Cancer)	Bayer/Onyx
VEGFR (PDGFR, c-kit, Flt-3)	AG013676	Phase II (Cancer)	Pfizer
NGFR, Trk	CEP-2583	Phase II (Cancer)	Cephalon
HER-2/neu	17-AAG	Phase I (Cancer)	Kosan
	Trastuzumab (mAb)	Approved (Cancer)	Genentech
	2C4 (mAb)	Phase I/II (Cancer)	Genentech
	CP-724,714	Phase I (Cancer)	OSI Pharmaceuticals/Pfizer
	TAK-165	Phase I (Cancer)	Takeda
	MDX-210 (mAb)	Phase I (Cancer)	Novartis

TABLE 26.2 Selected Inhibitors of Protein Serine/Threonine Kinases in Market or Clinical Trials

Kinase Target	Agent	Trial (Disease)	Sponsor
PKC, c-Kit, PDGFR	PKC412	Phase II (Cancer, retinopathy)	Novartis
PKC	ISIS3521	Phase III (Cancer)	ISIS Pharmaceuticals
	CGP41251	Phase II (Cancer)	Novartis
	UCN-01	Phase I/II (Cancer)	Kyowa Hakko Kogyo
	Bryostatin-1	Phase II (Cancer)	Biotek
PKC-β	LY333531	Phase I (Cancer)	Eli Lilly
		Phase II/III (Diabetic neuropathy)	
	LY317615	Phase I/II (Cancer)	Eli-Lilly
CDKs	Flavopiridol	Phase II (Cancer)	Aventis
	E7070	Phase I (Cancer)	EISAI
	BMS-387032	Phase I (Cancer)	Bristol-Myers Squibb
	UCN-01	Phase II (Cancer)	Kyowa Hakko Kogyo
	CYC202	Phase I (Cancer)	Cyclacel
CDK2	CYC202	Phase I (Cancer)	Cyclacel
MKK1	PD184352	Phase I (Cancer)	Pfizer
MEK	PD184352	Phase II (Cancer)	Pfizer
	U-0126	Phase I (Cancer)	Promega
MLK	CEP-1347	Phase II (Neurodegeneration)	Cephalon

Target	Compound	Phase (Indication)	Company
RAF	BAY43-9006	Phase II (Cancer)	Onyx Pharmaceuticals/Bayer
	ISIS5132	Phase II (Cancer)	Isis pharmaceuticals
	L-779,450	Phase II (Cancer)	Merck
Ras	ISIS2503	Phase II (Cancer)	Isis pharmaceuticals
	SCH66336	Phase II (Cancer)	Schering-Plough
	BMS214662	Phase I (Cancer)	Bristol-Myers Squibb
	R115777	Phase II/III (Cancer)	Johnson and Johnson
mTOR	CCI779	Phase II (Cancer)	Wyeth-Ayerst
	RAD001	Phase I (Cancer)	Novartis
	Rapamycin	Phase II/II (Immunosuppressant) Approved (Immunosuppressant)	Wyeth-Ayerst
p38-MAPK	VX702	Phase II (Inflammation; ACS)	Vertex Pharmaceuticals
	BIRB-796	Phase III (Inflammation; RA; Crohn's; Psoriasis; Endotoxic shock)	Boehringer Ingelheim
	SCIO-323	Phase I (RA; Stroke; Diabetes)	Scios, Inc.
	SCIO-469	Phase II (RA; Crohn's)	Scios, Inc.
PDK1	UCN-01	Phase I/II (Cancer)	Kyowa Hakko Kogyo

since modulation of the same PK by an inhibitor may have beneficial effects and deleterious effects in two different systems. For example, inhibiting a PK required for triggering apoptosis could reduce ischemia-induced cell death in terminally differentiated cardiomyocytes but might also promote tumor formation in other organs or cell types [159]. Finally, even selective inhibition of a dysregulated PK in a particular disease state may be deleterious in a long-term use for other systems in which that same PK is not dysregulated but instead serves essential functions. For example, Herceptin (Trastuzumab, Genentech) is a monoclonal antibody that inhibits overexpressing cell surface HER2 tyrosine kinase receptor in patients with breast cancers. On the other hand, inhibition of HER2 by herceptin can cause severe cardiac dysfunction in some women receiving the therapy, suggesting a critical role for this receptor in cardiomyocyte survival [154, 159].

The comprehensive discussion of the structure–activity relationships of PKIs is beyond the scope of this chapter. For more details, readers can refer to some recent excellent reviews [13, 150, 157, 160–171]. Monovalent ligands targeting the ATP binding site, the substrate binding site, and the SH2 domain of clinically important PKs and some of novel approaches in designing PKIs such as allosteric binding site and bisubstrate analog inhibitors are discussed here.

ATP Binding Site Inhibitors The PKs can be inhibited by compounds competing with ATP for binding to ATP binding site on the enzyme molecule. In general, ATP binding site inhibitors contain a rigid, largely planar structure, an aromatic ring system to complement the ATP binding pocket, and a hydrogen bond donor–acceptor pair. These groups of compounds exhibit similar hydrogen bond attachment points at the interlobe linker (hinge region) through their chemical scaffolds [172]. These hydrogen bond interactions are complemented with an additional stabilizing network of hydrophobic interactions between the planes of their aromatic ring systems and hydrophobic amino acid side chains in the binding pocket [28].

Selectivity and potency have to be considered in designing ATP binding site inhibitors. In recent years, considerable progress has been made to increase specificity and potency of inhibitors toward PKs. This strategy has led to the development of two drugs, Iressa and Gleevec [60]. Several other compounds are in advanced clinical trials.

Selectivity These compounds have structural similarities with ATP and bind to the enzyme instead of ATP; therefore, the kinase cannot phosphorylate proteins and signaling halts. ATP binding site inhibitors of PKs were initially thought to be unsuitable drug targets. The high degree of structural conservation in this binding site within all PKs led to the initial belief that highly selective small-molecule PKIs targeting the ATP pocket would be difficult to generate [158, 173–177].

Since the kinase protein family contains approximately 518 members, the discovery of highly specific kinase inhibitors has been slow, but substantial progress has been made. Indeed, contrary to initial concerns, there is sufficient sequence and conformational diversity in ATP binding site [177] to identify selective ATP-competitive inhibitors. The screening of unbiased chemical libraries has led to identification of several relatively selective ATP competitors [178, 179]. Several different kinds of inhibitors directed toward the ATP binding site [160, 161, 180–182] are now in clinical development as anticancer agents (Tables 26.1 and 26.2), immunosuppressant (Rapamycin) [153], neurodegeneration [CEP-1347 (KT7515)] [183], and antidiabetic neuropathy (LY333531) [184, 185]. ATP binding site inhibitors have structure diversity and include purines, quinazolines, pyrazolopyrimidines, pyrrolopyrimidines, pyridopyrimidines, maleimides, oxindoles, phthalimides, isoquinoline, naphthyridinones, quinolines, and several other classes of compounds [160, 161, 180]. Some of these compounds have shown potent inhibitory action against particular kinases (K_i in nanomolar–picomolar range) [150, 160, 180–182] and can even distinguish between closely related tyrosine kinases (e.g., between Src kinases) [186, 187] and so are very promising drug leads. Although selective lead inhibitors competitive with ATP have been identified for specific PKs indicating that the initial belief was misguided, however, the process is still labor intensive mainly due to the high structural homology between protein kinase active site regions.

One approach to achieve selectivity is to change the substitution patterns in one category of the PKIs. This strategy has been used in the case for several benzopyranone-containing compounds (flavonoids) (e.g., flavopiridol, quercetin, genistein, apigenin, desmal, and PD98059) (Fig. 26.6). Flavonoids, in general, are competitive inhibitors with respect to ATP and lack selectivity for PTKs over STKs [188, 189]. By establishing the structure–activity relationships of flavonoids, relatively selective inhibitors have been identified [163, 190]. Flavopiridol is a CDK inhibitor (e.g., CDK1, CDK2, CDK4, and CDK7) in clinical development [191]. Genistein has been shown to inhibit tyrosine kinases such as the EGFR and pp60^{src}, but scarcely to inhibit the activity of serine and threonine kinases such as cAMP-dependent kinases [192, 193]. PD98059 (2′-amino-3′-methoxyflavone) is a potent, selective, and cell-permeable inhibitor of MAP kinase kinase (also known as MAPK/ERK kinase or MEK kinase) by selectively blocking the activity of MEK and/or inhibiting the phosphorylation and activation of MAP kinase [194–199].

In general, the goal is an IC$_{50}$ that is at least 100-fold lower for the target kinase over other enzymes. Sometimes it is necessary to develop not entirely selective inhibitors. For example, the antiangiogenic compound SU6668 elicits potent antiangiogenic and proapoptotic effects in vivo by inhibition of KDR, FGFR1, and PDGFRβ receptors [81]. In cancer, a combination of CDKs inhibitory and antiapoptotic (e.g., Akt) effects by a drug targeting Bcr-Abl

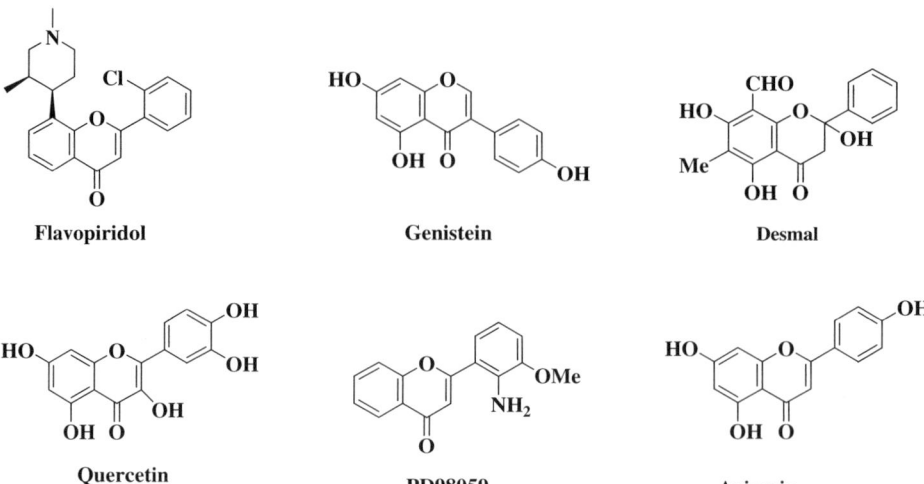

Figure 26.6 Chemical structures of some flavonoids having protein kinase inhibition activity.

could be beneficial for anticancer effects. However, a p38-MAPK inhibitor for treatment of inflammatory diseases cannot be tolerated to inhibit CDKs or Akt because of enhanced toxicity.

Potency The PKs and a plethora of many nonkinase enzymes utilize ATP as a common substrate with different K_m values. The ATP binding site inhibitors must have sufficient potency to compete with the high ATP concentration (2 to 10 mM) in vivo [158]. Enzymes with a low K_m for ATP could be inhibited if ATP binding site inhibitors do not bind to the target kinase with an extremely high affinity: several orders of magnitude higher than that of ATP. In vivo, the inhibitor will be present in concentrations typically in the mid to high nanomolar range, whereas the intracellular concentration of ATP is millimolar. As a result, the ATP-binding-site-directed inhibitors usually exhibit a significantly lower potency under in vivo conditions than they do in vitro.

Substrate Binding Site Inhibitors Inhibitors competing with the peptide or protein substrate for binding to substrate binding sites have also been described [180, 200–202]. These peptides are frequently in the form of uncharged phenylalanine analogs [13, 203] and comprise of peptides containing the substrate consensus motif to generate greater specificity. PKs exhibit more sequence diversity in substrate binding site and peptide and protein substrate recognition; therefore, peptide-based substrate binding site inhibitors with high specificity might be easier to discover, and they might prevent phosphorylation of some substrates but not others. These peptide-based inhibitors do not need to compete with physiologic ATP. Despite generating high selectivity toward PKs, substrate peptide-based inhibitors have relatively weak inhibitory performance (K_i in the high micromolar range) compared to the most potent ATP analogs [180]. There are issues such as bioavailability and susceptibility to proteolytic degradation that have to be considered when designing these inhibitors. Some closely related PKs exhibit overlapping specificities with respect to small peptides. Other than peptide-based inhibitors, no small organic molecules have been discovered to show a competition pattern at the protein–substrate binding site of PKs, possibly due to the fact that the substrate binding groove is relatively solvent exposed and shallow.

Allosteric Inhibitors Regions outside the ATP binding site such as the activation loop can be used and accessed for designing inhibitors. Binding of inhibitors to this region may cause significant conformational changes in their target kinases [28]. These compounds are likely to be more selective than compounds that simply compete directly with ATP alone. For example, BIRB 796, a slow-binding inhibitor of p38 MAP kinase, was designed based on structure optimization of several compounds and their mode of binding with P38 MAP kinase in the crystal structure. BIRB 796 binding to a region that is normally occupied by the phenylalanine of the Asp-Phe-Gly ("DFG") motif converts the enzyme to a catalytically inactive conformation by changing the conformation of the active site [204]. Gleevec, a recently approved drug against CML, interacts with the aromatic side chain of the conserved phenylalanine of the DFG motif and displaces the activation loop from its normal position, resulting in trapping the unactivated form of c-Abl [28] (PDB codes 1FPU [26], 1IEP [205]).

SH2 Domain Inhibitors SH2-containing targets in PTKs have been the subject of great interest by several researchers [206]. Ligands that antagonize the Src SH2 domain-dependent protein–protein interactions may provide possible therapeutic agents [29, 207–210]. SH2 domains generally recognize specific short peptide motifs containing a phosphorylated Tyr (pTyr) in the context of specific residues found within 3 to 5 amino acids C-terminal to the pTyr [31]. The Src SH2 domain has been the subject of intensive research. The SH2 inhibitors will be discussed when considering Src kinase inhibitors (Section 26.9) in greater details.

Bisubstrate Analog Inhibitors Bisubstrate analog inhibitors could be designed to enhance specificity and potency of PKIs. Bisubstrate analog inhibitors are designed to mimic two natural ligands that simultaneously associate with two essential binding sites of kinases. The combination of two substrates required by the target enzyme into a single molecule by an appropriate linker makes it less likely that both components will be recognized by other enzymes; therefore, enhanced enzyme specificity may be expected [211]. Additionally, if one of the substrates demonstrates specific binding to its binding pocket, a greater selectivity may arise [212, 213]. Using bisubstrate analog inhibitors for PKs allows converting a low-affinity peptide substrate to a selective and moderately potent inhibitor for these enzymes with affinities at a low micromolar–high nanomolar range.

In general, an optimal bisubstrate analog inhibitor for PKs should have the following properties. The inhibitor should (1) inhibit the binding of both natural substrates to their binding pockets particularly by competitive inhibition, (2) be designed based on the distance, geometry, and orientation of natural substrates and their essential chemical scaffolds during the transition state, (3) enhance specificity against one particular PK, and (4) show optimal physico-chemical properties, bioavailability, and low toxicity. During the last 16 years, many different approaches have been examined for the possibility of using bisubstrate analog inhibitors against PKs [214]. These strategies fall into four categories: (1) sulfonamides and sulfonylbenzoyl derivatives [186, 215–217], (2) carboxylic acid derivatives [218–220], (3) phosphodiester derivatives [221], and (4) dipeptidyl derivatives [53, 222, 223]. The first three categories target ATP binding site and substrate binding site of PKs. The last category of compounds was designed to target SH2 and SH3 domains or the SH2 domain and substrate binding site. None of these strategies have all the characteristics for an optimal bisubstrate analog inhibitor for PKs, but substantial progress has been made in this direction.

Most bisubstrate analogs have been designed to mimic the phosphate donor (ATP) and the acceptor components (Ser-, Thr-, or Tyr-containing peptides) based in part on mechanistic and structural features of a predicted transition state for PKs. We synthesized a potent and selective peptide–ATP bisubstrate inhibitor ($K_i = 370\,nM$) that binds to ATP binding site and substrate binding site of the IRK (Fig. 26.7) that behaved as a competitive inhibitor versus both ATP and peptide substrates based on the kinetic analysis. A specific distance (5.7 Å) between these two components was required to achieve potent inhibition [224]. The X-ray crystallographic studies indicated that both the nucleotide and peptide portions of the bisubstrate inhibitor were binding to ATP and protein substrate binding sites, respectively (Fig. 26.7). The bisubstrate analog inhibitor was only a modest inhibitor against Csk since the preferred substrate peptide sequence for Csk is quite different from IRK.

Alternative Strategies in Designing Protein Kinase Inhibitors

A number of alternative strategies have been developed for designing PKIs including substrate docking site, irreversible, and molecular chaperone inhibitors. In addition to substrate binding site, it has been shown that MAP kinases [225–227] and Csk [228] employ remote-docking-based mechanisms for substrate recognition. We anticipate that elucidation of the substrate recognition mechanisms will reveal novel approaches in developing protein substrate-competitive inhibitors.

A number of irreversible inhibitors of the erbB family of protein tyrosine kinases have been designed [229] based on the reaction of the nucleophilic Cys thiol in ATP binding site with Michael acceptor groups in the inhibitors

Figure 26.7 Designed bisubstrate analog inhibitor for IRK and the three-dimensional structure of the bivalent tyrosine kinase inhibitor (ball-and-sticks vectors) bound to the ATP binding site and substrate binding site of IRK catalytic site (PDB code IGAG) [224].

such as CI-1033 (PD183805) [230] and EKB-569 [231] that are now in clinical trials.

Molecular chaperones such as heat shock protein 90 (Hsp90) are involved in stabilizing the conformation of PKs [232]. A number of compounds such as geldanamycin, radicicol, and 17-AAG have been studied as Hsp90 inhibitors and indirect PKIs [232–234]. It remains to be seen if any of these strategies will translate to potential therapeutic agents.

26.9 SELECTED INHIBITORS OF PROTEIN KINASES ON MARKET OR IN CLINICAL TRIALS

Tables 26.1 and 26.2 show most of the PKIs currently in clinical trials or on the market for several diseases. There are two types of inhibitors, monoclonal antibodies (mAbs), which are directed at the extracellular domain of various RTKs, and small-molecule inhibitors. Most of PKIs are targeted against cancer, but several PKIs are designed for the treatment of a number of chronic conditions, including inflammatory and cardiovascular diseases. Both active and inactive protein kinase structures have been used in inhibitor binding studies. Detailed discussion of all these inhibitors is beyond the scope of this chapter. Herein several PKIs approved or in advanced phases of clinical studies are discussed based on validated target enzymes.

Receptor Tyrosine Kinase Inhibitors

Recently, considerable effort in cancer drug discovery has been directed at the growth factors and growth factor receptors. There are two approaches for designing inhibitors against RTKs, targeting the extracellular ligand binding and intracellular kinase domains. Several agents such as antibodies have been designed to disrupt phosphorylation sequence by competing for receptor binding. Alternatively agents have been designed to prevent tyrosine kinase autophosphorylation. Several of the compounds developed as RTK inhibitors target more than one closely related receptor from this family.

Targeting the Extracellular Domains of EGFR It has been found that compounds that interfere with EGF binding to its receptor have utility as antiprolifereative agents [235]. HER-2 receptor is a member of the EGFR family of RTKs that is involved in cell proliferation and differentiation in developing embryo and in adult tissues. It is known that cancerous breast cancer tissues overexpress the HER2 receptors on the cell surface, triggering the cell division and multiplication in tumor growth. Approximately 30 percent of all women with metastatic breast cancer overexpress the HER2 protein [236]. The HER2 extracellular domain has a constitutively open structure based on shown structural studies [237]. Genentech Inc. developed the humanized monoclonal antibody Trastuzumab directed against the extracellular domain of HER-2. The mAb received an approval of the FDA in 1998 under the name Herceptin that binds to HER2 receptor sites on the surface of breast cancer cells, blocking the receptor sites, and preventing the tumor growth by interrupting the growth signal to intracellular domains. The structure of HER2 receptor in complex with Herceptin Fab shows that Herceptin binds HER2 on the C-terminal portion of domain IV (Fig. 26.8). Herceptin proved to be effective in 15 percent of patients with HER2-overexpressing metatstatic breast cancer [154].

Chimeric antibody Erbitux (Cetuximub) developed by Imclone is another antibody that blocks the extracellular domain of ErbB1 and prevents signaling that triggers colorectal tumor growth. Erbitux was approved in February 2004 for the treatment of metastatic colorectal cancer [155, 156]. Other promising EGFR antibodies include Thermacin h-R3 (Cima-her) for the treatment of head and neck tumors in combination with radiotherapy.

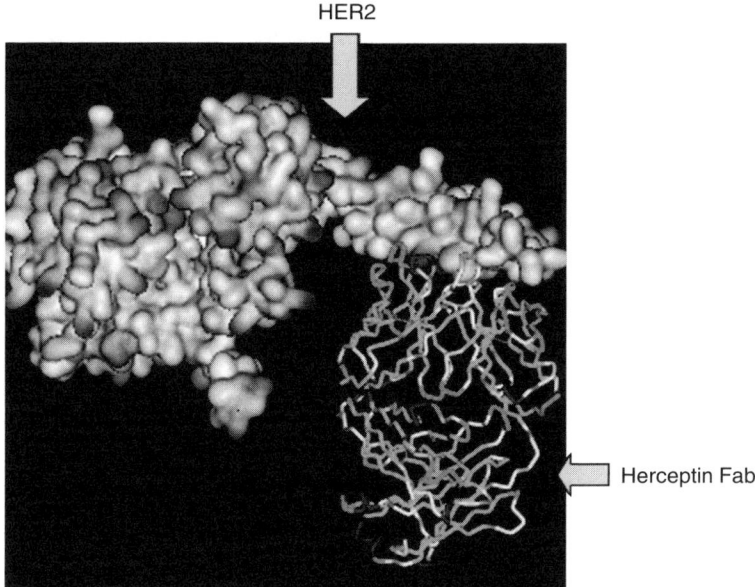

Figure 26.8 Herceptin binding site. The backbone of HER2 is shown in surface format contacting the Herceptin Fab (tube format). The HER2 extracellular domain has an open structure and is a noncatalytic domain (PDB code 1N8Z) [237].

Targeting the Intracellular Domains of EGFR Several potent EGFR inhibitors that bind to the intracellular kinase domain of EGFR are currently on the market or in clinical trials as anticancer agents (Table 26.1). Among these are ZD1839 (Iressa, gefitinib, Fig. 26.9), developed by AstraZeneca, and OSI-774 (Tarceva, Fig. 26.9), co-developed by OSI Pharmaceutical, Roche, and Genentech.

ZD1839 showed marked efficacy against several cancers, for example, prostate cancer and non-small-cell lung carcinoma [238], in human clinical trials and is approved by FDA for the treatment of metastatic non-small-cell lung cancer [239]. Clinical studies with this and other EGF receptor antagonists have not been uniformaly successful [240, 241]. For example, pulmonary toxicity was observed in several patients [242]. ZD1839 exhibited specificity for the growth inhibition of EGF-stimulated cells as compared to non-EGF-stimulated cells in cell cultures of KB human tumors tyrosine kinases with an IC_{50} of 0.08 to 0.09 µM and 3.64 µM, respectively. ZD1839 did not exhibit any significant inhibitory effect on other related kinases such as MAPK, PKC, and KDR [238].

OSI-774 (Tarceva, erlotinib), an orally active and reversible potent inhibitor of EGFR tyrosine kinase with an IC_{50} of 2 nM, has shown promising results in clinical trials in patients with non-small-cell lung carcinoma, head and neck, prostate, and cervical cancers (29 to 34 percent disease stabilization)

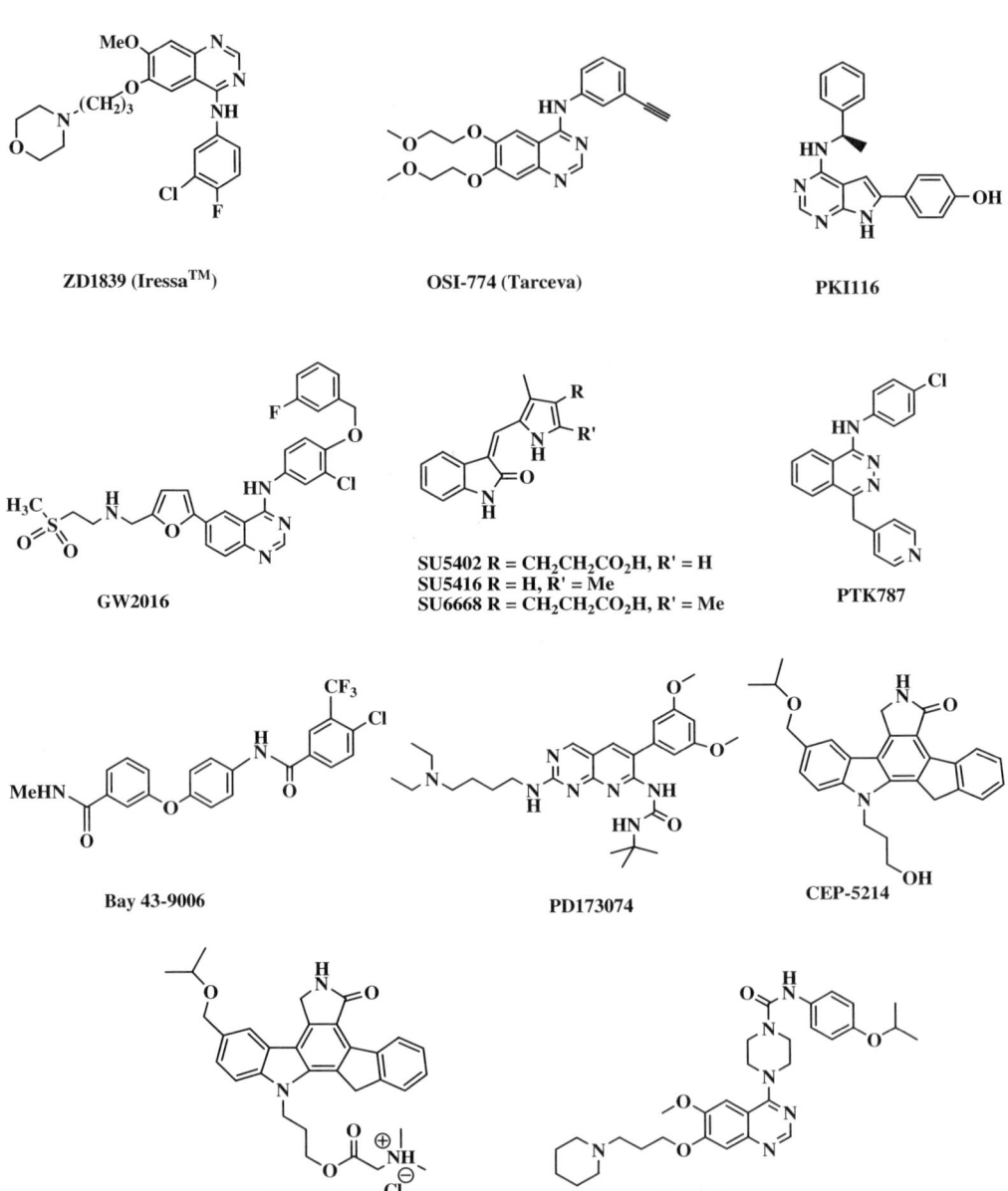

Figure 26.9 Chemical structures of RTKIs.

[243]. OSI-774 is currently in phase III clinical trials for several cancers such as pancreatic cancer and non-small-cell lung cancer. Structural studies have shown that OSI-774 binds to ATP binding site of [243] in the active form of EGFR [244].

A dual inhibitor of the HER-1 and HER-2 tyrosine kinases developed by Novartis, PKI116 is in phase I clinical trials against head and neck, breast, prostate, urinary, bladder, ovary, gliomas, and gastrointestinal tract cancers (Fig. 26.9). PKI116 had IC_{50} values in the nanomolar range for the inhibition of the intracellular domain of HER-1 and HER-2 tyrosine kinases [245]. GW2016 (Fig. 26.9) is a dual tyrosine kinase inhibitor of ErbB-2 and EGFR developed by GlaxoSmithKline for the treatment of solid tumors [246].

VEGFR and FGFR Inhibitors Several RTKIs have been designed to inhibit angiogenesis [247] as VEGFRIs that have application in anticancer drug development. In angiogenesis the new blood vessels are formed for solid tumors [248].

Bevacizumab (Avastin; rhuMab VEGF), an anti-VEGF humanized antibody developed by Genentech, has been approved by the FDA as an antiangiogenic anticancer drug and as a first-line therapy for metastatic colorectal cancer. Bevacizumab has a long terminal half-life of 17 to 21 days in humans [79].

The indolin-2-one family of antiangiogenic molecules, SU5402, SU5416, and SU6668 (Fig. 26.9), are potent inhibitors of VEGFR tyrosine kinase activity. Members of this family have a functionalized methylpyrrole ring attached to C3. The structure of the inhibitor SU5402 (Fig. 26.9) bound to FGFR1 tyrosine kinase domain (Fig. 26.10) [19] identifies the orientation of this ring in the complex. Specificity of SU5402 for FGFR1 derives in part from a hydrogen bond between the carboxyl function of SU5402 and the side chain of Asn^{568} at the end of the hinge (Fig. 26.10).

SU5416 (Fig. 26.9) was developed by Sugen Inc. (Pharmacia) as a highly potent antiangiogenic antitumor agent that advanced to phase III clinical trials against glioma, Kaposi's sarcoma, and colorectal cancers by selectively targeting VEGFR-2 [248]. SU5416 enters the endothelial cells and binds to intracellular domains of VEGFR. SU5416 failed to exhibit efficacy in combination with chemotherapy [249]. Clinical trials for SU5416 have been halted by Pharmacia recently, but it remains a valuable biological tool for study of signal transduction pathways involving VEGF and its receptor (VEGFR).

SU6668, a close analog of SU5416, is another clinical angiogenesis inhibitor developed by Sugen Inc. (Fig. 26.9) and is undergoing various phase I clinical trials against advanced solid tumors. SU6668 is less selective than SU5416 and inhibits two other tyrosine kinases including EGFR and PDGFR in addition to VEGFR2 [81, 250, 251].

Figure 26.10 Crystal structure of the tyrosine kinase domain of FGFR1 (ribbon) in complex with SU5402 inhibitor (ball-and-sticks vectors) (PDB code 1FGI) [19]. The hydrogen bonding is shown between the carboxyl function of SU5402 and the side chain of Asn^{568}.

PTK787/ZK222584 (Fig. 26.9) developed by Novartis/Schering has been introduced to target the VEGFR family [252–256] such as VEGFR1 and VEGFR2. PTK787 is effective in combination therapy against solid tumors but has limited success in monotherapy. PTK787 is in phase III clinical trials for colorectal cancer and is one of the most selective VEGFR kinase inhibitor described to date [257–259].

Other nonselective inhibitors targeting several RTKs such as VEGFRs, PDGFR, c-kit, and Flt-3 have been introduced [260]. These compounds are in several stages of clinical trials and have to be monitored for their toxicity profile because of their nonselective inhibitory effects. Bay 43–9006 (Fig. 26.9), an RAF kinase inhibitor and a nonselective inhibitor of RTKs including VEGFRs, is currently in phase III trials for metastatic renal-cell carcinoma [79]. PD173074 (Fig. 26.9) is another nonselective inhibitor of RTKs that is a potent inhibitor of the type-1 FGFR tyrosine kinase (FGFR-1) and VEGFR-2 tyrosine kinases [261].

CEP-7055, a water-soluble prodrug of CEP-5214 (Fig. 26.9), is currently in phase I clinical trials as an antiangiogenic antitumor agent. CEP-7055 was reported to be a potent inhibitor of all three types of VEGF receptor tyrosine kinases, VEGFR-1, -2, and -3 tyrosine kinases, with IC_{50} values in nanomolar range and inhibited tumor formation and growth up to 90 percent in a variety of human tumor models [262]. The prodrug CEP-7055 shows a similar pharmacodynamic of the drug CEP-5214 but has a better pharmacokinetic profile.

PDGFR Inhibitors SU6668 has been shown to be (Fig. 26.9) a nonselective PDGFR tyrosine kinase inhibitor since it also inhibits both VEGFR-2 and EGFR tyrosine kinases at similarly low concentrations [81, 251]. Recently, CT53518 (also known as MLN518) (Fig. 26.9) developed by Millenium Pharmaceuticals as a potent inhibitor of Flt-3, entered clinical trials against AML with impressive clinical results. CT53518 delays disease progression in a mouse model of Flt-3-associated AML and is proapoptotic in a human AML cell line [260].

Nonreceptor Tyrosine Kinase Inhibitors

Bcr-Abl Inhibitor STI-571 (Fig. 26.11), an ATP binding site inhibitor developed by Novartis, is very potent against tumorigenic fusion protein Bcr-Abl as described by Druker and co-workers [187, 263]. This compound was approved by the FDA in May 2001 for the treatment of CML under the name Gleevec (Glivec, imatinib mesylate) [181, 182], as the first small-molecule PTKI. Detailed pharmacological profile of STI-571 has been extensively reviewed [150, 157, 245, 264]. The drug induces 100 percent remission in the

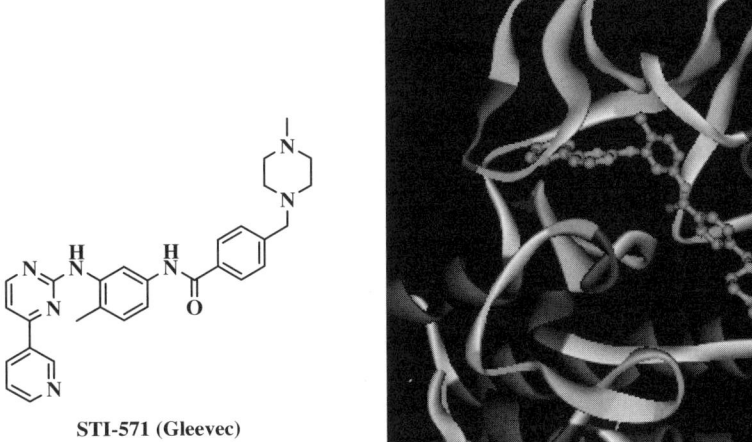

Figure 26.11 Crystal structure of Gleevec (ball-and-stick vectors) with the catalytic domain of cAbl (ribbon). Gleevec targets a unique inactive conformation (PDB code 1IEP) [205].

prefully malignant chronic phase of the disease, but in advanced phases such as aggressive blast stage of the disease tends to be relatively short-lived, due to the development of de novo resistance to the drug by *BCR-ABL* gene amplification, multiple-point mutations within the kinase domain of Abl, and reactivation of the tyrosine kinase activity of the Bcr-Abl oncoprotein [242, 265, 266]. Azam et al. [242] proposed that point mutations cause steric hinderance of drug occupancy and allosteric destabilization of the autoinhibited Abl conformation. Three other kinases have been reported to be inhibited by Gleevec: the stem cell factor receptor (c-Kit), Abl-homolog Arg, and PDGF receptor [267, 268]; therefore, its use for the treatment of gastrointestinal stromal tumors in which c-Kit is overexpressed by mutation has been explored [269–271].

An inactive distinct conformation of Bcr-Abl is targeted by STI-571, making it more selective than other PTKI inhibitors. The pyridine and the pyrimidine rings of the drug are surrounded by a hydrophobic pocket in the ATP binding site (Fig. 26.11). The nonaromatic scaffold of the molecule is between the activation domain and the C helix, locking the kinase in an inactive conformation [26, 205]. The aspartic acid of the DFG motif in the activation domain cannot coordinate the magnesium at the catalytic site due to flipping of aspartic acid and phenyl alanine in this motif in the inactive conformation. The gatekeeper residue in Abl is Thr^{315}, to which Gleevec hydrogen bonds. A bulkier nonpolar residue replaces the threonine in many other kinases such as Src kinases; hence access to the pocket by STI-571 is blocked in these kinases [272]. A number of other kinases use this extra space created by Thr^{315} that is not used by ATP for competing selectively with ATP [273–275].

PD173955 is a more potent inhibitor of Abl than Gleevec [205]; however, it can bind to both active (Fig. 26.12) and inactive conformations of Abl. Therefore, PD173955 is a less selective inhibitor than Gleevec [276].

Src Kinase Inhibitors Although no drug has been approved as Src kinase inhibitor yet, research efforts in this direction have advanced significantly in recent years utilizing structure-based design, combinatorial chemistry, and high-throughput screening in designing numerous potent Src inhibitors with varying chemical structures [166]. It is hoped that Src inhibitors will soon emerge as useful therapeutics for cancer therapy. As mentioned in Section 26.4, Src kinases are made up of five components/domains: N-terminal, Src homology 2 (SH2), Src homology 3 (SH3), kinase (catalytic, including NT and CT lobes), and C-terminal noncatalytic domain (Fig. 26.4). Currently, ATP binding site is an important target for the design of Src kinase inhibitors through an ATP-competitive inhibition mechanism. Substrate binding site and SH2, SH3 domains have also been employed as important targets for Src kinase inhibition [29, 277].

Src SH2 Domain Inhibitors Antagonizing the protein–protein interactions of SH2 domains with activated receptors using peptides and peptidomimiet-

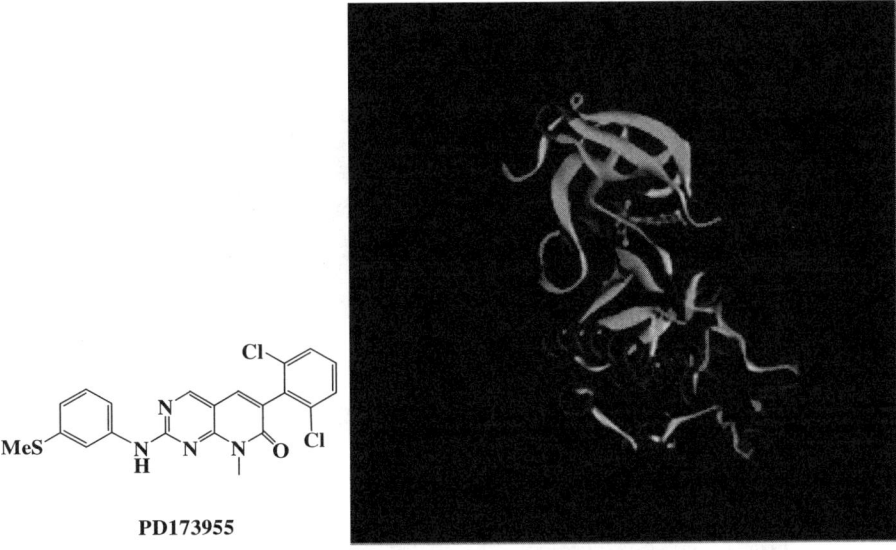

Figure 26.12 Crystal structure of PD173955 (ball-and-sticks vectors) targeting the active conformation of Abl (ribbon) (PDB code 1M52) [205].

ics, has been pursued to inhibit the signaling pathways. For example, tetrapeptide pYEEI, a peptide with high affinity (K_d, 100 nM) for the Src SH2 domain [31], has two major interactions, that is, (i) electrostatic interactions between the pTyr phosphate moiety of the peptide and Arg residues embedded within a deep positively charged phosphotyrosine binding pocket (P site) on the SH2 domain and (ii) hydrophobic interactions between the lipophilic Ile side chain of the consensus sequence peptide and the hydrophobic pocket (P + 3 site) of the SH2 domain [278]. The crystal structure of pYEEI in complex with the Src SH2 domain reveals a very characteristic binding mode that can be described as a two-pronged plug engaging a two-hole socket (Fig. 26.13) [279]. The EE motif lies across the flat surface of the protein and plays the role of a linker delivering the two residues pTyr and Ile to respective binding pockets [280–282]. Tyrosyl motif-based analogs directed against pTyr binding pocket of the SH2 domain usually mimic highly charged phosphate functional groups, since recognition of ligands by SH2 domains mimics the electrostatic interactions of two negative charges and the ester oxygen for phosphate in phosphotyrosyl residues with positively charged amino acids, αA and βB arginine residues, located in pTyr binding pocket [283, 284]. Additionally pTyr residues themselves are not suitable for inhibitor design, due to the enzymatic liability of the phosphate esters to protein tyrosine phosphatases and the poor bioavailability of the doubly ionized phosphate moieties. The design of SH2 inhibitors has focused on peptidomimetic

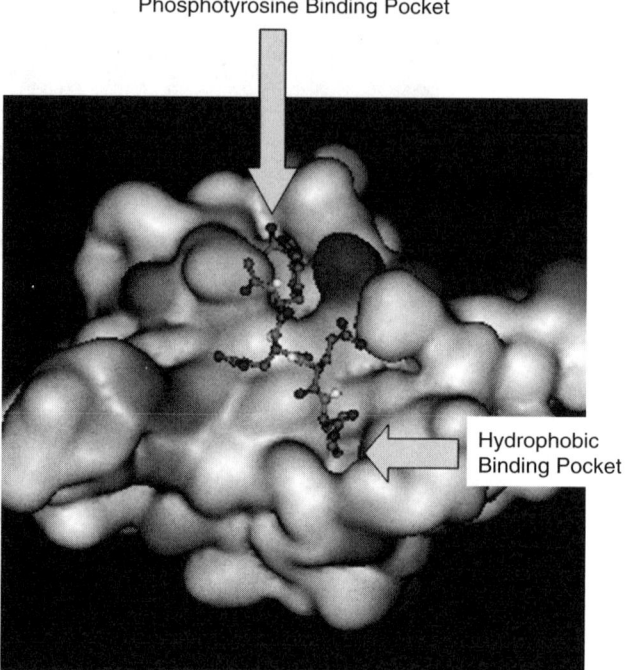

Figure 26.13 X-ray crystal structure and interaction sites of pYEEI and the Src SH2 domain (PDB code 1SHD) [279].

modifications of cognate peptide sequences [285]. SH2-domain-directed pTy mimetics include phosphonate-based pTyr mimetics such as phosphonomethylphenylalanine (Pmp) and its analogs [286–294], carboxylic-acid-based pTyr mimetics such as malonyltyrosine or phenylalanine analogs and their derivatives such as carboxymethyl phenylalanine [292, 295–298], and uncharged pTyr mimetics [299], and conformationally constrained peptides [300]. Several of these inhibitors could enhance the kinase catalytic activity by switching the closed inactive to the open active conformation [46–49] by disrupting the intramolecular interactions between the Tyr^{527}-phosphorylated C-terminal tail and the SH2 domain. As a result, side effects might appear by the Src SH2 domain inhibitors. Much more attention has to be paid to the outcome of the Src SH2 inhibition and possible kinase activation.

Src Protein Substrate Binding Site Inhibitors Development of peptide inhibitors based on the peptide substrate motifs has been reported [301, 302]. The identified inhibitors are highly selective for $p60^{c-src}$ over other closely related Src family PTKs such as Lck and Lyn. However, with few exceptions, most of the inhibitors identified by this approach are only moderately potent with IC_{50} values in the micromolar range [13]. The substrate site targeted moiety, CIYKYYF, is reported to be one of the most potent peptides developed against p60$^{c\text{-}src}$ PTK (IC$_{50}$ = 0.5 μM) [303]. However, no structural, kinetics, and mechanistic studies have been proposed to confirm whether this peptide binds to the substrate binding site.

Src ATP Binding Site Inhibitors There exists a plethora of templates available to design ATP binding site inhibitors as Src kinase inhibitors [166, 304]. Although there are no Src inhibitor drugs on the market yet, several companies (e.g., Wyeth/Ayerst, Novartis, Ariad, Pfizer, Sugen/Pharmacia) are actively engaged in this area of drug discovery to create therapeutic agents as Src inhibitors for both osteoporosis and cancer treatment. AP22408 was developed as a bone-targeted Src kinase inhibitor that exhibited high affinities for hydroxyapatite and potent in vitro inhibition of Src-dependent, osteoclasts-mediated bone resoption [102, 304–306].

Several nonselective experimental compounds such as natural products (herbimycin A, radicolol, geldanamycin) [307–309], pyrazole[3,4-*d*]pyrimidine (PP1), PP2, SKI606, PD173955, PD180970, and SU6656 (Fig. 26.14) have been used for targeting the Src family of kinases. PP1 and PP2 may not be an optimal choice as Src family kinase inhibitors since they do not discriminate between different members of the Src family. For example, PP1 and PP2 inhibit Lck with IC$_{50}$ values of 50 to 60 nM. On the other hand, PP1 and PP2 also inhibited Csk, SAPK2a/p38 with IC$_{50}$ values of 520 to 1430 nM) [310]. The compound SU6656 is reported to inhibit Lck with an IC$_{50}$ of 0.04 μM as one of the most potent Lck inhibitors [179, 311]. A number of these compounds also inhibit Bcr-Abl, leading to cell cycle arrest and apopotosis of various Bcr-Abl-positive cell lines tested [312, 313]. For example, PP1 inhibits c-Abl (IC$_{50}$ = 30 nM) [310]. No molecules that demonstrate high selectivity between different Src family kinases have been identified so far. The natural product damnacanthal (Fig. 26.14; IC$_{50}$ for Lck, 17 nM) was 20- and 7-fold more selective for Lck over Fyn and Src, respectively [314]. The 3-(*N*-phenyl)carbamoyl-2-iminochromene (Fig. 26.14) [315] inhibited c-Src (IC$_{50}$ =120 nM), was < 285-fold more selective over Lyn, 570-fold more selective over Fyn, and 18-fold more selective with respect to Lck [310].

In general many of the Src kinase inhibitors published so far lack sufficient specificity desirable for clinical application or even pharmacological tools. Only few inhibitors have been tested without provoking toxicity in animal models. Src kinases inhibition seems to be still attractive since their inhibition might influence several important characteristics of tumorigenesis [84].

Serine/Threonine Kinase Inhibitors

Serine/threonine kinases represent attractive targets for designing PKIs and therapeutic intervention. For example, the inhibition of various CDKs has been used in cancer treatment [316–318]. Fasudil (Fig. 26.15) is in the clinic and approved for treatment of cerebral vasospasm [319]. It has a significant

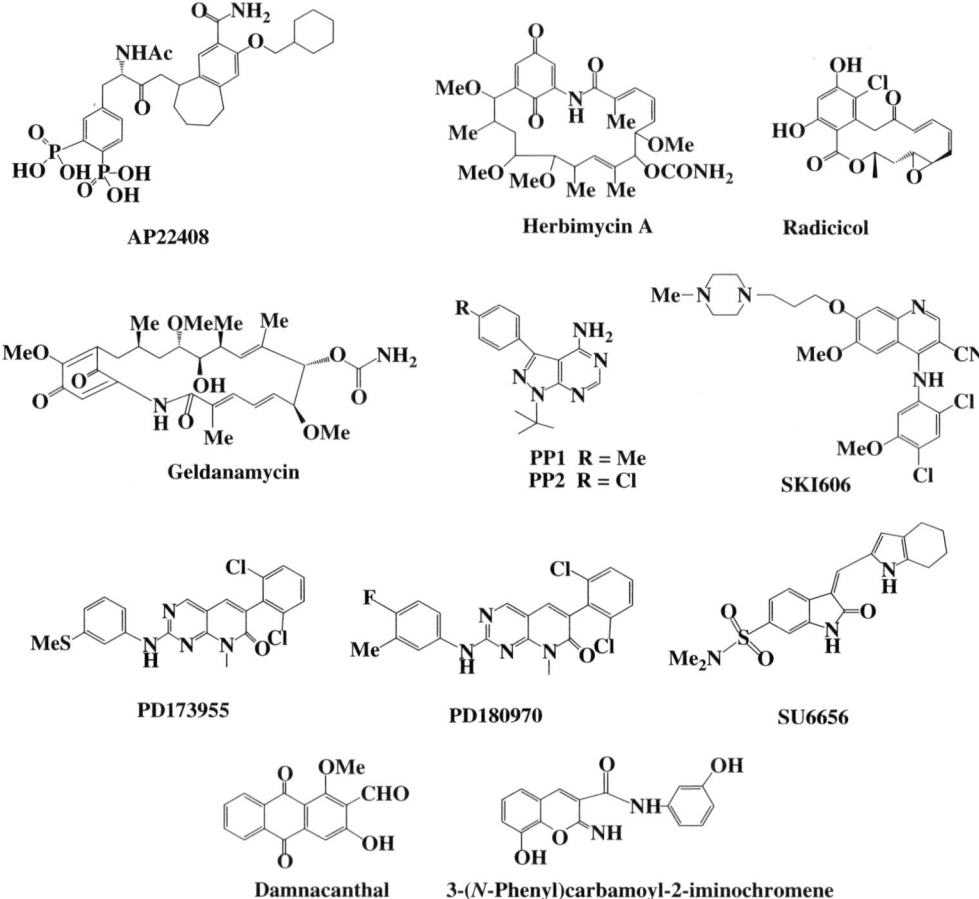

Figure 26.14 ATP binding site inhibitors as Src Kinase inhibitors.

vasodilatory effect attributed to its potent inhibition of Rho-kinase signaling to myosin light-chain kinase. Some of important serine/threonine kinase inhibitors are discussed here.

MAP Kinase Inhibitors Three families of stress-activated MAPKs are the p38-MAPKs, JNKs, and ERKs. We will describe important MAP kinase inhibitors in some detail.

P38 MAPK Inhibitors P38 MAP kinase such as p38α is activated in response to cellular stresses, growth factors, and cytokines. Activated p38α activates other kinases that phosphorylate heat shock proteins and transcription factors. Several proinflammatory and inflammatory cytokines, such as tumor necrosis factor-α (TNF-α) and interleukin 1β, are activated by increased activ-

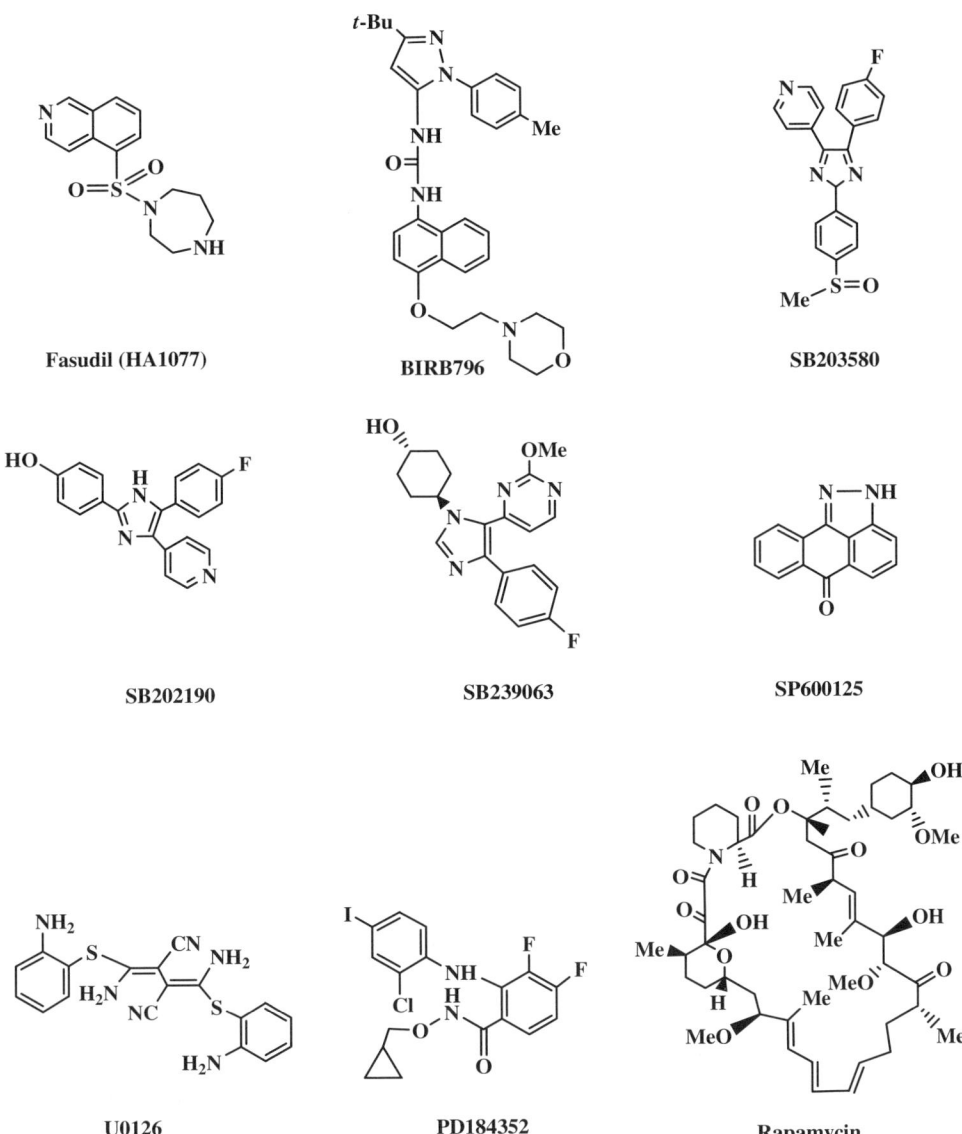

Figure 26.15 Examples of MAP kinase inhibitors.

ity of p38 MAPK. These cytokines have been implicated in a number of diseases associated with inflammation, suggesting that p38 selective inhibitors could be used as therapeutic agents for the treatment of a number of inflammatory diseases such as rheumatoid arthritis [59, 167], autoimmune diseases [320–330], cancer, diabetes, and acute myocardial infarctions [331–336]. Two members of the p38 MAPK family, p38α and p38β, are activated by ischemia.

Hence, the inhibition of p38-MAPKs might have therapeutic applications for treating acute coronary syndromes (ACS) [159]. Several companies have developed p38 inhibitors [337–340].

BIRB 796, a diaryl urea, [204, 341] (Fig. 26.15) is a selective and potent p38α inhibitor currently in phase IIb/III clinical trials for the treatment of rheumatoid arthritis and Crohn's disease. BIRB 796 binding to the Phe residue in the conserved DFG motif of activation loop induces a large conformational change in the enzyme. The enzyme with modified conformation cannot bind to ATP [204]. In p38α this residue is much smaller, allowing access for interaction by BIRB 796, but the entry to this pocket is blocked by bulky gatekeeper amino acid side chains such as Phe[80] in CDK2 and Asp or Glu residues in most MAPK family members.

As described above, MAPKs regulate the production of cytokines and chemokines and upregulation of adhesion molecules on endothelial cells [159, 342]; hence, preventing the release of inflammatory cytokines and chemokines is a potentially promising strategy for the treatment of ACS. VX-702 (structure not disclosed) is a p38MAP kinase inhibitor in phase II clinical trials for ACS [343]. p38 MAP kinase inhibition by VX-702 leads to degradation of the cytokine messenger ribonucleic acids (mRNAs), including those coding for interleukin (IL)-1/ß, TNF, IL-10, IL-6, interferon (IFN), MIP1/ß, and IL-8 [159].

SB203580 [59, 332, 344], SB202190 [178], and SB239063 [345] (Fig. 26.15) are potent and specific ATP-competitive inhibitors of p38 MAP Kinase. SB239063 [345], a selective p38 MAP kinase inhibitor, was reported to have beneficial effects in the intact rat model of ischemia-reperfusion (I/R) injury, produced a dramatic reduction in the myocardial inflammatory response by reducing upregulation of P-selectin and intercellular adhesion molecule, and reduced neutrophil accumulation within the ischemic zone [159].

JNK Inhibitors Upstream activators of the MAPKs such as JNKs can be targeted to inhibit MAPK signaling [346]. This approach might reduce toxicity MAPKs. JNKs are activated by at least 12 different MAPK kinase kinases (MAPKKKs) and 2 MAPK kinases (MAPKKs), hence inhibiting JNKs can block signaling of all of these MAPKKKs and MAPKKs. JNK was suggested to be an attractive target for the treatment of chronic inflammatory disease, apoptotic cell death, and cancer [347]. SP600125 (Fig. 26.15), an inhibitor of JNK, was reported to prevent the expression of several anti-inflammatory genes in cell-based assays [347–349] but is a relatively weak and nonselective inhibitor [179]. The development of more selective JNK inhibitors will be needed to determine whether designing of specific and potent inhibitors of JNK is possible.

ERK Inhibitors The protein kinases MAPK ERK kinase (MEK)1/2, which is involved in activating the ERK family of MAPKs (Fig. 26.5), and mTOR (mammalian target of rapamycin) have critical roles in cell cycle progression.

ERK inhibitors, such as U0126, PD184352, and rapamycin/sirolimus (Fig. 26.15), are in clinical trials for the treatment of a variety of cancers [350]. U0126 and PD184352 are noncompetitive ATP binding site inhibitors that maintain kinases in an inactive state by preventing their phosphorylation by upstream activating kinases such as Raf. Rapamycin also inhibits the protein kinase mTOR [351] and is currently used successfully as an immunosuppressant and an inhibitor of in-stent restenosis [153].

CDK Inhibitors A number of groups have identified CDK inhibitors. The inhibition of CDK2 should arrest cells in G1 phase and prevent them from entering the cell cycle. Therefore, inhibitors of CDK2 may have utility in the treatment of proliferative diseases such as cancer and psoriasis [352–364].

Flavopiridol (L86-8275, HMR1275) (Fig. 26.6), a synthetic flavonoid analog, is a CDK1, CDK2, and CDK4 inhibitor in clinical trials as a cancer therapy and is the first CDK inhibitor that is not competitive with ATP. Flavopiridol has shown activity in phase I and phase II trials with unknown mechanism [365]. It is not yet clear if the antiproliferative properties of flavopiridol are due to its ability to inhibit transcription through CDK9/cyclin T, its inhibition of cell cycle CDKs or through action on other targets.

Indirubin (Fig. 26.16) proved to be an inhibitor of CDK2 and CDK5/p25 [366]. Aminothiazole-based compounds are relatively selective CDK2 inhibitors compared with CDK1 and CDK4. The results of phase I clinical trials with BMS 387032 have recently been reported, and phase II trials are planned [367]. Olomoucine (Fig. 26.16) was one of the first CDK inhibitors to be developed [368]. Two of its derivatives, roscovitine [369] and purvalanol [370] (Fig. 26.16), are more potent ATP-competitive inhibitors of CDK1, CDK2, and CDK5 and have been used extensively to inhibit these PKs in cell-based assays. Roscovitine and purvalanol inhibit CDK2 activity with IC_{50} values of 0.25 and 0.1 μM, respectively. Roscovitine inhibited DYRK1A with an IC_{50} of 3.1 μM but had little effect on the other tested PKs. Purvalanol A is a 2.5-fold more potent inhibitor of CDK2 but is less selective than roscovitine and inhibits DYRK1A, ribosomal S6 kinase 1 (RSK1), ERK2, Lck, and Csk [179]. R-roscovitine (CYC202) is currently undergoing phase II clinical trials for advanced breast cancer and stage IIIB/IV non-small-cell lung cancer in combination with standard chemotherapy regimes [371].

PKC Inhibitors The important role of PKC in cancer renders it a potentially suitable target for anticancer therapy. Several PKC inhibitors, such as UCN-01 and LY-317615 (Fig. 26.16), are now in clinical trials as anticancer agents [372]. Ruboxistaurin (LY-333531) (Fig. 26.16) is a selective inhibitor of PKC-β that is in phase III clinical trials for the treatment of macular edema, neuropathy, and diabetic retinopathy [373].

The microbial alkaloid staurosporine (Fig. 26.16) is an antiproliferative agent and potent inhibitor of PKC. Staurosporine is a competitive inhibitor of PKC's conserved ATP binding site. Staurosporine has poor specificity for PKC

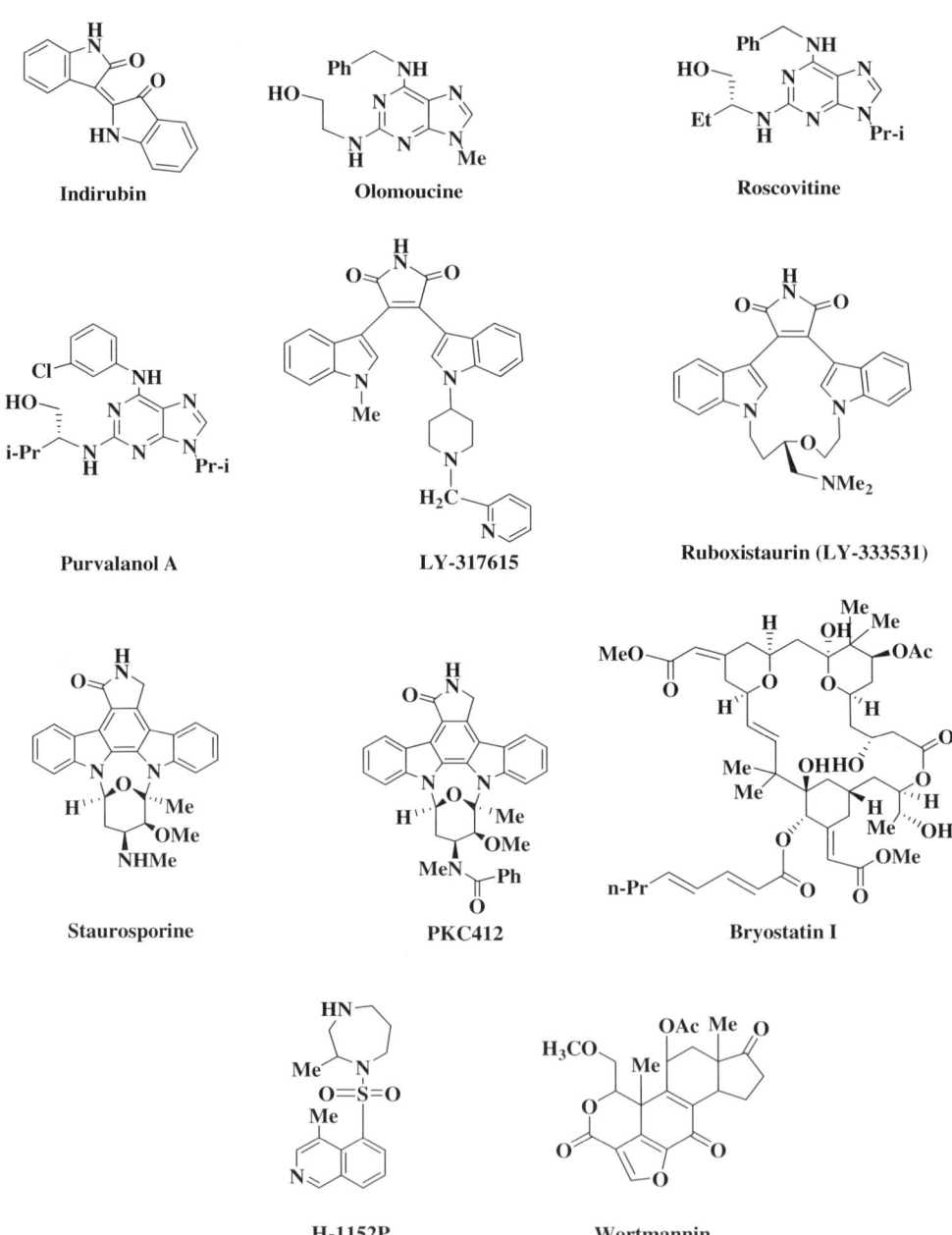

Figure 26.16 Examples of CDK, PKC, PI3K, and Rho kinase inhibitors.

and its isoforms and has inhibitory activity toward STKs and PTKs [374]. Staurosporine derivatives such as PKC412 (*N*-benzoyl-staurosporine) [375] and UCN-01 (7-hydroxystaurosporine) [376, 377] exhibit improved selectivity, with a potentially better therapeutic index in vivo and have been reported to enhance the effects of other cytotoxic agents.

The bryostatins have been shown to have promising antineoplastic and immunomodulatory activity in preclinical models. Bryostatin I (Fig. 26.16) is the prototype of this class of agents [378]. Bryostatins are potent activators of cPKC and nPKC subfamilies. However, in the presence of activating ligands, such as the tumor-promoting phorbol esters, bryostatins act as antagonists [379] possibly due to differential isoform activation [380] or nuclear translocation [381].

Antisense oligonucleotides have been developed to achieve PKC isoform-specific inhibition by inhibiting expression of target mRNA sequences. ISIS3521, an antisense phosphorothioate oligonucleotide to PKCα, has been investigated in a phase I trial [382].

Rho-kinase Inhibitors Inhibition of Rho-kinase helps blood vessels to relax and increases the blood supply to cardiac tissue. Fasudil (HA1077) (Fig. 26.15) progressed to human clinical trials in the early 1990s. It was approved in Japan in 1995 for the treatment of cerebral vasospasm. Fasudil had no marked side effects over a 2-week period when given by intravenous injection. At micromolar concentrations, HA1077 inhibits several PKs, such as the Rho-dependent protein kinase ROCK, but it is unclear whether its clinical efficacy results from inhibition of this or other PKs, or whether it is due to a nonkinase effect. ROCK can constrict blood vessels by inhibiting smooth-muscle myosin phosphatase. A recent structure determination [383] of PKA in complex with Fasudil and with a more potent Rho-kinase inhibitor H-1152P (Fig. 26.16), which differs by only two methyl groups, demonstrated characteristic binding within the ATP site. It is suggested that it is the combination of residues at the ligand binding site, which generates a uniquely shaped inhibitor binding pocket that confers selectivity.

PI3K Inhibitors The structure of Wortmanin (Fig. 26.16), a selective PI3K inhibitor, in complex with PI3K has been reported [384]. By inhibition of PI3K, Wortmanin has been shown to inhibit osteoclasts-induced bone resorption in vitro [385]. mTOR is a downstream target of the PI3K pathway that plays a role in control of cellular growth, size, and cell proliferation. Its direct inhibitor rapamycin (Fig. 26.15) (386) was the first agent to be identified used as an immunosuppressive agent. A related compound CCI779 is in clinical phase II for renal and glioblastoma cancer in which the level of phosphatidylinositol-3,4,5-triphosphate is elevated [387].

GSK3 Inhibitors A number of small-molecule inhibitors of GSK3 for the treatment of type 2 diabetes have been proposed, including indirubins, paullones, hymenialdisine, aminopyrimidines, and aminopyridines such as CT 99021, CT 20026, and maleimides. The discussion of these inhibitors is beyond the scope of this chapter, but they have been recently reviewed [148].

26.10 PROSPECTS AND FUTURE DIRECTIONS

Approval of several PKIs by FDA has led to the conclusion that, in the future, several diseases associated with dysregulated PKs can be treated. Herein we discussed some PKIs approved or in advanced clinical trials. PKs continue to be considered as molecular targets for cancer and other diseases, including inflammation, osteoporosis, atherosclerosis, restenosis, and stroke. PKs not only play roles in the development of diseases but also function in pathways that regulate the most basic of normal cellular processes; hence, toxicity remains a major concern. In spite of all this progress, specificity remains an area of concern for designing PKIs due to PKs structural similarities. Additionally, it is not possible to test the drugs against all kinases, and side effects are expected even after approval of drugs by FDA. Despite these obstacles, PKIs have created a lot of promises for a wide variety of diseases.

Acknowledgment We acknowledge the financial support from BRIN/INBRE Program of the National Center for Research Resources, NIH, Grant number 1 P20 RR16457.

REFERENCES

1. Hunter, T. (1987). A thousand and one protein kinases. *Cell*, *50*, 823–829.
2. Hunter, T. (2000). Signaling—2000 and beyond. *Cell*, *100*, 113–127.
3. Manning, G., Whyte, D. B., Martinez, R., Hunter, R., Sudarsanam, S. (2002). The protein kinase complement of the human genome. *Science*, *298*, 1912.
4. Hubbard, S. R., Till, J. H. (2000). Protein tyrosine kinase structure and function. *Annu. Rev. Biochem.*, *69*, 373–398.
5. Ho, M., Bramson, H. N., Hansen, D. E., Knowles, J. R. Kaiser, E. T. (1988). Stereochemical course of the phospho group transfer catalyzed by cAMP-dependent protein-kinase. *J. Am. Chem. Soc.*, *110*, 2680–2681.
6. Aqvist, J., Kolmodin, K., Florian, J., Warshel, A. (1999). Mechanistic alternatives in phosphate monoester hydrolysis: What conclusions can be drawn from available experimental data? *Chem. Biol.*, *6*, R71–R80.
7. Admiraal, S. J., Herschlag, D. (2000). The substrate-assisted general base catalysis model for phosphate monoester hydrolysis: Evaluation using reactivity comparisons. *J. Am. Chem. Soc.*, *122*, 2145–2148.
8. Mildvan, A. S. (1997). Mechanisms of signaling and related enzymes. *Proteins*, *29*, 401–416.
9. Uckun, F. M., Mao, C. (2004). Tyrosine kinases as new molecular targets in treatment of inflammatory disorders and leukemia. *Curr. Pharm. Des.*, *10*, 1083–1091.

10. Williams, D. H., Mitchell, T. (2002). Latest developments in crystallography and structure-based design of protein kinase inhibitors as drug candidates. *Curr. Opin. Pharmacol.*, 2, 1–7.
11. Scapin, G. (2002). Structural biology in drug design: Selective protein kinase inhibitors. *Drug Discov. Today*, 7, 601–611.
12. Hubbard, S. (2002). Protein tyrosine kinases: Autoregulation and small molecule inhibition. *Curr. Opin. Struct. Biol.*, 12, 735–741.
13. Al-Obeidi, F. A., Wu, J. J., Lam, K. S. (1998). Protein tyrosine kinases: Structure, substrate specificity, and drug discovery. *Biopolymers*, 47, 197–223.
14. Hubbard, S. R. (1999). Structural analysis of receptor tyrosine kinases. *Prog. Biophys. Mol. Biol.*, 71, 343–358.
15. Munshi, S., Kornienko, M., Hall, D. L., Reid, J. C., Waxman, L., Stirdivant, S. M., Darke, P. L., Kuo, L. C. (2002). Crystal structure of the Apo, unactivated insulin-like growth factor-1 receptor kinase. Implication for inhibitor specificity. *J. Biol. Chem.*, 277, 38797–38802.
16. Xu, W., Harrison, S. C., Eck, M. J. (1997). Three-dimensional structure of the tyrosine kinase c-Src. *Nature*, 385, 595–602.
17. Xu, W., Doshi, A., Lei, M., Eck, M. J., Harrison, S. C. (1999). Crystal structures of c-Src reveal features of its autoinhibitory mechanism. *Mol. Cell*, 3, 629–638.
18. Mohammadi, M., Schlessinger, J., Hubbard, S. R. (1996). Structure of the FGF receptor tyrosine kinase domain reveals a novel autoinhibitory mechanism. *Cell*, 86, 577–587.
19. Mohammadi, M., McMahon, G., Sun, L., Tang, C., Hirth, P., Yeh, B. K., Hubbard, S. R., Schlessinger, J. (1997). Structures of the tyrosine kinase domain of fibroblast growth factor receptor in complex with inhibitors. *Science*, 276, 955–960.
20. McTigue, M. A., Wickersham, J. A, Pinko, C., Showalter, R. E., Parast, C. V., Tempczyk-Russell, A., Gehring, M. R., Mroczkowski, B., Kan, C. C., Villafranca, J. E., Appelt, K. (1999). Crystal structure of the kinase domain of human vascular endothelial growth factor receptor 2: A key enzyme in angiogenesis. *Structure with Folding & Design*, 7, 319–330.
21. Zhu, X., Kim, J. L., Newcomb, J. R., Rose, P. E., Stover, D. R., Toledo, L. M., Zhao, H., Morgenstem, K. A. (1999). Structural analysis of the lymphocyte-specific kinase Lck in complex with non-selective and Src family selective kinase inhibitors. *Structure with Folding & Design*, 7, 651–661.
22. Yamaguchi, H., Hendrickson, W. A. (1996). Structural basis for activation of human lymphocyte kinase Lck upon tyrosine phosphorylation. *Nature*, 384, 484–489.
23. Sicheri, F., Moarefi, I., Kuriyan, J. (1997). Crystal structure of the Src family tyrosine kinase Hck. *Nature*, 385, 602–609.
24. Schindler, T., Sicheri, F., Pico, A., Gazit, A., Levitzki, A., Kuriyan, J. (1999). Crystal structure of Hck in complex with a Src family-selective tyrosine kinase inhibitor. *Mol. Cell*, 3, 639–648.
25. Mao, C., Zhou, M., Uckun, F. M. (2001). Crystal structure of Bruton's tyrosine kinase domain suggests a novel pathway for activation and provides insights into the molecular basis of X-linked agammaglobulinemia. *J. Biol. Chem.*, 276, 41435–41443.

26. Schindler, T., Bornmann, W., Pellicena, P., Miller, W. T., Clarkson, B., Kuriyan, J. (2000). Structural mechanism for STI-571 inhibition of abelson tyrosine kinase. *Science*, *289*, 1938–1942.

27. Taylor, S. S., Radzio-Andzelm, E. (1994). Three protein kinase structures define a common motif. *Structure*, *2*, 345–355.

28. Stout, T. J., Foster, P. G., Matthews, D. J. (2004). High-throughput structural biology in drug discovery: Protein kinases. *Curr. Pharm. Des.*, *10*, 1069–1082.

29. Sawyer, T. K. (1998). Src homology-2 domains: Structure, mechanisms, and drug discovery. *Biopolymers*, *47*, 243–261.

30. Sicheri, F., Kuriyan, J. (1997). Structures of Src-family tyrosine kinases. *Curr. Opin. Struct. Biol.*, *7*, 777–785.

31. Songyang, Z., Shoelson, S. E., Chaudhuri, M., Gish, G., Pawson, T., Haser, W. G., King, F., Roberts, T., Ratnofsky, S., Lechleider, R. J., Neel, B. G., Birge, R. B., Fajardo, J. E., Chou, M. M., Hanatusa, H., Schaffhausen, B., Cantley, L. C. (1993). SH2 domains recognize specific phosphopeptide sequences. *Cell*, *72*, 767–778.

32. Sadowski, I., Stone, J. C., Pawson, T. (1986). A noncatalytic domain conserved among cytoplasmic protein-tyrosine kinases modifies the kinase function and transforming activity of Fujinami sarcoma virus P130gag-fps. *Mol. Cell. Biol.*, *6*, 4396–4408.

33. Cantley, L. C., Auger, K. R., Carpenter, C., Duckworth, B., Graziana, A., Kapeller, R., Soltoff, S. (1991). Oncogenes and signal transduction. *Cell*, *64*, 281–302.

34. Fukui, Y., Hanafusa, H. (1991). Requirement of phosphatidylinositol-3 kinase modification for its association with p60src. *Mol. Cell. Biol.*, *11*, 1972–1979.

35. Schaller, M. D., Hilderbrand, J. D., Shannon J. D., Fox, J. W., Vines, R. R., Parsons, J. T. (1994). Autophosphorylation of the focal adhesion kinase, pp125FAK, directs SH2 dependent binding of pp60src. *Mol. Cell. Biol.*, *14*, 1680–1688.

36. Taylor, S. J., Shalloway, D. (1994). An RNA-binding protein associated with Src through its SH2 and SH3 domains in mitosis. *Nature*, *368*, 867–871.

37. Petch, L. A., Bockholt, S. M., Bouton, A., Parsons, J. T., Burridge, K. (1995). Adhesion-induced tyrosine phosphorylation of the p130 src substrate. *J. Cell. Sci.*, *108*, 1371–1379.

38. Johnson, L. N., Noble, M. E., Owen, D. J. (1996). Active and inactive protein kinases: Structural basis for regulation. *Cell*, *85*, 149–158.

39. Adams, J. A. (2003). Activation loop phosphorylation and catalysis in protein kinases: Is there functional evidence for the autoinhibitor model? *Biochemistry*, *42*, 601–607.

40. Toker, A. N., Newton, A. C. (2000). Cellular signaling: Pivoting around PDK1. *Cell*, *103*, 185–188.

41. Till, J. H., Ablooglu, A. J., Frankel, M., Bishop, S. M., Kohanski, R. A., Hubbard, S. R. (2001). Crystallographic and solution studies of an activation loop mutant of the insulin receptor tyrosine kinase: Insights into kinase mechanism. *J. Biol. Chem.*, *276*, 10049–10055.

42. Ablooglu, A. J., Frankel, M., Rusinova, E., Ross, J. B. A., Kohanski, R. A. (2001). Multiple activation loop conformations and their regulatory properties in the insulin receptor's kinase domain. *J. Biol. Chem.*, *276*, 46933–46940.

43. Orr, J. W., Newton, A. C. (1994). Requirement for negative charge on "activation loop" of protein kinase C. *J. Biol. Chem.*, *269*, 27715–27718.
44. Wybenga-Groot, L. E., Baskin, B., Ong, S. H., Tong, J., Pawson, T., Sicheri, F. (2001). Structural basis for autoinhibition of the Ephb2 receptor tyrosine kinase by the unphosphorylated juxtamembrane region. *Cell*, *106*, 745–757.
45. Moarefi, H., LaFevre-Bernt, M., Sicheri, F., Huse, M., Lee, C.-H., Kuriyan, J., Miller, W. T. (1997). Activation of the Src-family tyrosine kinase Hck by SH3 domain displacement. *Nature*, *385*, 650–653.
46. Mandine, E., Jean-Baptiste, V., Vayssiere, B., Gofflo, D., Benard, D., Sarubbi, E., Deprez, P., Baron, R., Superti-Furga, G., Lesuisse, D. (2002). High-affinity Src-SH2 ligands which do not activate Tyr(527)-phosphorylated Src in an experimental in vivo system. *Biochem. Biophys. Res. Commun.*, *298*, 185–192.
47. Liu, X., Brodeur, S. R., Gish, G., Songyang, Z., Cantley, L. C., Laudano, A. P., Pawson, T. (1993). Regulation of c-Src tyrosine kinase activity by the Src SH2 domain. *Oncogene*, *8*, 1119–1126.
48. Alonso, G., Koegl, M., Mazurenko, N., Courtneidge, S. A. (1995). Sequence requirements for binding of Src family tyrosine kinases to activated growth factors. *J. Biol. Chem.*, *270*, 9840–9848.
49. Boerner, R. J., Kassel, D. B., Barker, S. C., Ellis, B., DeLacy, P., Knight, W. B. (1996). Correlation of the phosphorylation states of pp60^{c-src} with tyrosine kinase activity: The intramolecular pY530-SH2 complex retains significant activity if Y419 is phosphorylated. *Biochemistry*, *35*, 9519–9525.
50. Erpel, T., Courtneidge, S. A. (1995). Src family protein tyrosine kinases and cellular signal transduction pathways. *Curr. Opin. Cell Biol.*, *7*, 176–182.
51. Thomas, J. W., Ellis, B., Boerner, R. J., Knight, W. B., Wite, G. C., II, Shaller, M. D. (1998). SH2- and SH3-mediated interactions between focal adhesion kinase and Src. *J. Biol. Chem.*, *273*, 577–583.
52. Alexandropoulos, K., Baltimore, D. (1996). Coordinate activation of c-Src by SH3- and SH2 binding sites on a novel, p130CAS-related protein, Sin. *Genes Devel.*, *10*, 1341–1355.
53. Profit, A. A., Lee, T. R., Niu, J., Lawrence, D. S. (2001). Molecular rulers: An assessment of distance and spatial relationships of Src tyrosine kinase SH2 and active site regions. *J. Biol. Chem.*, *276*, 9446–9451.
54. Biscardi, J. S., Tice, D. A., Parsons, S. J. (1999). c-Src, receptor tyrosine kinases, and human cancer. *Adv. Cancer Res.*, *76*, 61–119.
55. Jiang, G., Zhang, B. B. (2002). PI3-kinase and its up- and down-stream modulators as potential targets for the treatment of type II diabetes. *Front. Biosci.*, *7*, 903–907.
56. Frame, S., Cohen, P. (2001). GSK3 takes centre stage more than 20 years after its discovery. *Biochem. J.*, *359*(Pt. 1), 1–16.
57. Vlahos, C. J., McDowell, S. A., Clerk, A. (2003). Kinases as therapeutic targets for heart failure. *Nat. Rev. Drug Discov.*, *2*, 99–113.
58. Mucke, H. A. (2003). CEP-1347 (Cephalon). *IDrugs*, *6*, 377–383.
59. Lee, J. C., Kumar, S., Griswold, D. E., Underwood, D. C., Votta, B. J., Adams, J. L. (2000). Inhibition of p38 MAP kinase as a therapeutic strategy. *Immunopharmacology*, *47*, 185–201.

60. Traxler, P. (2003). Tyrosine kinases as targets in cancer therapy—Successes and failures. *Expert Opin. Ther. Patents*, 7, 215–234.
61. Gray, N., Detivaud, L., Doerig, C., Meijer, L. (1999). ATP-site directed inhibitors of cyclin-dependent kinases. *Curr. Med. Chem.*, 6, 859–875.
62. Collett, M. S., Erikson, R. L. (1978). Protein kinase activity associated with the avian sarcoma virus src gene product. *Proc. Natl. Acad. Sci. USA*, 75, 2021–2024.
63. Klint, P., Claesson-Welsh, L. (1999). Signal transduction by fibroblast growth factor receptors. *Front. Biosci.*, 4, D165–D177.
64. Dominguez, M., Wasserman, J. D., Freeman, M. (1998). Multiple functions of the EGF receptor in *Drosophila* eye development. *Curr. Biol.*, 8, 1039–1048.
65. Sieg, D. J., Hauck, C. R., Ilic, D., Klingbeil, C. K., Schaefer, E., Damsky, C. H., Schlaepfer, D. D. (2000). FAK integrates growth-factor and integrin signals to promote cell migration. *Nat. Cell Biol.*, 2, 249–256.
66. Casaccia-Bonnefil, P., Gu, C., Chao, M. V. (1999). Neurotrophins in cell survival/death decisions. The functional roles of glial cells in health and disease. *Adv. Exper. Med. Biol.*, 468, 275–282.
67. Van der Geer, P., Hunter, T., Lindberg, R. A. (1994). Receptor protein-tyrosine kinases and their signal transduction pathways. *Annu. Rev. Cell. Biol.*, 10, 251–337.
68. Carpenter, G., Cohen, S. (1979). Epidermal growth factor. *Annu. Rev. Biochem.*, 48, 193–216.
69. Carpenter, G. (1987). Receptors for epidermal growth factor and other polypeptide mitogens. *Annu. Rev. Biochem.*, 56, 881–914.
70. Quirion, R., Araujo, D., Nair, N. P., Chabot, J. G. (1988). Visualization of growth factor receptor sites in rat forebrain. *Synapse*, 2, 212–218.
71. Klapper, L. N., Kirschbaum, M. H., Sela, M., Yarden, Y. (2000). Biochemical and clinical implications of the ErbB/HER signaling network of growth factor receptors. *Adv. Cancer Res.*, 77, 25–79.
72. Olayioye, M. A., Neve, R. M., Lane, H. A., Hynes, N. E. (2000). The ErbB signaling network: Receptor heterodimerization in development and cancer. *EMBO J.*, 19, 3159–3167.
73. Sporn, M. B., Roberts, A. B. (1985). Autocrine growth factors and cancer, *Nature*, 313, 745–747.
74. Salomon, D. S., Brandt, R., Ciardiello, F., Normanno, N. (1995). Epidermal growth factor-related peptides and their receptors in human malignancies. *Crit. Rev. Oncol. Hematol.*, 19, 183–232.
75. Strawn, L. M., McMahon, G., App, H., Schreck, R., Kuchler, W. R., Longhi, M. P., Hui, T. H., Tang, C., Levitzki, A., Gazit, A., Chen, I., Keri, G., Orfi, L., Risau, W., Flamme, I., Ullrich, A., Hirth, K. P., Shawver, L. K. (1996). Flk-1 as a target for tumor growth inhibition. *Cancer Res.*, 56, 3540–3545.
76. Verheul, H. M., Jorna, A. S., Hoekman, K., Broxterman, H. J., Gebbink, M. F., Pinedo, H. M. (2000). Vascular endothelial growth factor-stimulated endothelial cells promote adhesion and activation of platelets. *Blood*, 96, 4216–4221.
77. Matsuzaki, K., Yoshitake, Y., Matuo, Y., Sasaki, H., Nishikawa, K. (1989). Monocolonal antibodies against heparin-binding growth factor II/basic fibroblast growth factor that block its biological activity: Invalidity of the antibodies for tumor angiogenesis. *Proc. Natl. Acad. Sci. USA*, 86, 9911–9915.

78. Dennis, P. A., Rifkin, D. B. (1990). Studies on the role of basic fibroblast growth factor in vivo: Inability of neutralizing antibodies to block tumor growth. *J. Cell Physiol.*, *144*, 84–98.
79. Ferrara, N., Hillan, K. J., Gerber, H.-P., Novotony, W. (2004). Discover and development of Bevacizumab, an anti-VEGF antibody for treating cancer. *Nat. Rev. Drug Dis.*, *3*, 391–400.
80. Abu-Duhier, F. M., Goodeve, A. C., Wilson, G. A., Care, R. S., Peake, I. R., Reilly, J. T. (2001). Identification of novel FLT-3 Asp835 mutations in adult acute myeloid leukaemia. *Br. J. Haematol.*, *113*, 983–988.
81. Laird, A. D., Christensen, J. G., Li, G., Carver, J., Smith, K., Xin, X., Moss, K. G., Louie, S. G., Mendel, D. B., Cherrington, J. M. (2002). SU6668 inhibits Flk-1/KDR and PDGFRbeta in vivo, resulting in rapid apoptosis of tumor vasculature and tumor regression in mice. *FASEB J.*, *16*, 681–690.
82. Mauro, M. J., Druker, B. J. (2001). STI571: Targeting BCR-ABL as therapy for CML. *Oncologists*, *6*, 233–238.
83. Sawyers, C. L. (1999). Chronic myeloid leukemia. *N. Engl. J. Med.*, *340*, 1330–1340.
84. Warmuth, M., Damoiseaux, R., Liu, Y., Fabbro, D., Gray, N. (2003). Src family kinases: Potential targets for the treatment of human cancer and leukemia. *Curr. Pharm. Des.*, *9*, 2043–2059.
85. Susa, M., Teti, A. (2000). Tyrosine kinase Src inhibitors: Potential therapeutic application. *Drugs News Perspect.*, *13*, 169–175.
86. Weber, T. K., Steele, G., Summerhayes, I. C. (1992). Differential pp60c-src activity in well and poorly differentiated human colon carcinomas and cell lines. *J. Clin. Invest.*, *90*, 815–821.
87. Ottenholf-Kalff, A. E., Rijksen, G., van Beurden, E. A., Hennipman, A., Michels, A. A., Staal, G. E. (1992). Characterization of protein tyrosine kinase from human breast cancer. *Cancer Res.*, *49*, 4773–4778.
88. Avizienyte, E., Wyke, A. W., Jones, R. J., McLean, G. W., Westhoff, M. A., Brunton, V. G., Frame, M. C. (2002). Src-induced de-regulation of E-cadherin in colon cancer cells requires integrin signaling. *Nat. Cell Biol.*, *4*, 632–638.
89. Thomas, S. M., Brugge, J. S. (1997). Cellular functions regulated by Src family kinases. *Annu. Rev. Cell Devel. Biol.*, *13*, 513–609.
90. Shakespeare, W. C., Metcalfe, C. A. 3rd, Wang, Y., Sundaramoorthi, R., Keenan, T., Weigele, M., Bohacek, R. S., Delgarne, D. C., Sawyer, T. K. (2003). Novel bone-targeted Src tyrosine kinase inhibitor drug discovery. *Curr. Clin. Drug. Discov. Devel.*, *6*, 729–741.
91. Lowe, C., Yoneda, T., Boyce, B. F., Chen, H., Mundy, G. R., Soriano, P. (1993). Osteopetrosis in Src-deficient mice is due to an autonomous defect of osteoclasts. *Proc. Natt. Acad. Sci. USA*, *90*, 4485–4489.
92. Boyce, B. F., Yoneda, T., Lowe, C., Soriano, P., Mundy, G. R. (1992). Requirement of pp60c-Src expression for osteoclasts to form ruffled borders and resorb bone in mice. *J. Clin. Invest.*, *90*, 1622–1627.
93. Soriano, P., Montgomery, C., Geske, R., Bradley, A. (1991). Targeted disruption of the c-src proto-oncogene leads to osteopetrosis in mice. *Cell*, *64*, 693–702.
94. Marzia, M., Sims, N. A., Voit, S., Migliaccio, S., Taranta, A., Bernardini, S., Faraggiana, T., Yoneda, T., Mundy, G. R., Boyce, B. F., Baron R., Teti. A. (2000). *J. Cell Biol.*, *151*, 311–320.

95. Kousteni, S., Bellido, T., Plotkin, L. I., O'Brien, C. A., Bodenner, D. L., Han, L., Han, K., DiGregorio, G. B., Katzenellenbogen, J. A., Katzenellenbogen, B. S., Roberson, P. K., Weinstein, R. S., Jilka, R. L., Manolagas, S. C. (2001). Nongenotropic, sex-nonspecific signaling through the estrogen or androgen receptors: Dissociation from transcriptional activity. *Cell*, *104*, 719–730.

96. Abu-Amer, Y., Ross, F. P., Schlesinger, P., Tondravi, M. M., Teitelbaum, S. L. (1997). Substrate recognition by osteoclast precursors induces C-src/microtubule association. *J. Cell Biol.*, *137*, 247–258.

97. Wong, B. R., Besser, D., Kim, N., Arron, J. R., Vologodskaia, M., Hanafusa, H., Choi, Y. (1999). TRANCE, a TNF family member, activates Akt/PKB through a signaling complex involving TRAF6 and c-Src. *Mol. Cell*, *4*, 1041–1049.

98. Tanaka, S., Amling, M., Neff, L., Peyman, A., Uhlmann, E., Levy J. B., Baron, R. (1996). c-Cbl is downstream of c-Src in a signalling pathway necessary for bone resorption. *Nature*, *383*, 528–531.

99. Jeschke, M., Brandi, M.-L., Susa, M. (1998). Expression of Src family kinases and their putative substrates in the human preosteoclastic cell line FLG 29.1. *J. Bone Miner. Res.*, *13*, 1880–1889.

100. Lowell, C. A., Berton, G. (1999). Integrin signal transduction in myeloid leukocytes. *J. Leukoc. Biol.*, *65*, 313–320.

101. Missbach, M., Jeschke, M., Feyen, J., Muller, K., Glatt, M., Green, J., Susa, M. (1999). A novel inhibitor of the tyrosine kinase Src suppresses phosphorylation of its major cellular substrates and reduces bone resorption in vitro and in rodent models in vivo. *Bone*, *24*, 437–449.

102. Shakespeare, W., Yang, M., Bohacek, R., Cerasoli, F., Stebbins, K., Sundaramoorthi, R., Azimioara, M., Vu, C., Pradeepan, S., Metcalf, C., Haraldson, C., Merry, T., Dalgarno, D., Narula, S., Hatada, M., Lu, X., van Schravendijk, M. R., Adams, S., Violette, S., Smith, J., Guan, W., Bartlett, C., Herson, J., Iuliucci, J., Weigele, M., Sawyer, T. (2000). Structure-based design of an osteoclast-selective, nonpeptide src homology 2 inhibitor with in vivo antiresorptive activity. *Proc. Natl. Acad. Sci. USA*, *97*, 9373–9378.

103. Luttrell, D. K., Lee, A., Lansing, T. J., Crosby, R. M., Jung, K. D., Willard, D., Luther, M., Rodriguez, M., Berman, J., Gilmer, T. M. (1994). Involvement of pp60c-src with two major signaling pathways in human breast cancer. *Proc. Natl. Acad. Sci. USA*, *91*, 83–87.

104. Frame, M. (2002). C-Src in cancer: Deregulation and consequences for cell behavior. *Biochim. Biophys. Acta*, *1602*, 114–130.

105. Kyriakis, J. M., Avruch, J. (2001). Mammalian mitogen-activated protein kinase signal transduction pathways activated by stress and inflammation. *Physiol. Rev.*, *81*, 807–869.

106. Pombo, C. P., Bonventre, J. V., Avruch, J., Woodgett, J. R., Kyriakis, J. M., Force, T. (1994). The stress-activated protein kinases (SAPKs) are major c-Jun aminoterminal kinases activated by ischemia and reperfusion. *J. Biol. Chem.*, *269*, 26546–26551.

107. Pearson, G., Robinson, F., Beers, G. T., Xu, B. E., Karandikar, M., Berman, K., Cobb, M. H. (2001). Mitogen-activated protein (MAP) kinase pathways: Regulation and physiological functions. *Endocr. Rev.*, *22*, 153–183.

108. Cobb, M. H., Goldsmith, E. J. (1995). How MAP kinases are regulated. *J. Biol. Chem., 270*, 14843–14864.

109. Minden, A., Karin, M. (1997). Regulation and function of the JNK subgroup of MAP kinases. *Biochem. Biophys. Acta, 1333*, F85–F104.

110. Hunter, T. (1995). When is a lipid kinase not a lipid kinase? When it is a protein kinase. *Cell, 80*, 225–236.

111. Chen, Z., Gibson, T. B., Robinson, F., Silvestro, L., Pearson, G., Xu, B., Wright, A., Vanderbilt, C., Cobb, M. H. (2001). MAP kinases. *Chem. Rev., 101*, 2449–2476.

112. Marshall, C. J. (1995). Specificity of receptor tyrosine kinase signaling: Transient versus sustained extracellular signal-regulated kinase activation. *Cell, 80*, 179–185.

113. Marshall, C. J. (1994). MAP kinase kinase kinase, MAP kinase kinase and MAP kinase. *Curr. Opin. Genet. Devel., 4*, 82–89.

114. Sugden, P. H., Clerk, A. (1997). Regulation of the ERK subgroup of MAP kinase cascades through G protein-coupled receptors. *Cell Signal, 9*, 337–351.

115. Cobb, M. H., Boulton, T. G., Robbins, D. J. (1991). Extracellular signal-regulated kinases: ERKs in progress. *Cell Regul., 2*, 965–978.

116. Ono, K., Han, J. (2000). The p38 signal transduction pathway: Activation and function. *Cell Signal, 12*, 1–13.

117. Mazzucchelli, C., Brambilla, R. (2000). Ras-related and MAPK signalling in neuronal plasticity and memory formation. *Cell Mol. Life Sci., 57*, 604–611.

118. Harper, J. W., Adams, P. D. (2001). Cyclin-dependent kinases. *Chem. Rev., 101*, 2511–2526.

119. Bramson, H. N., Corona, J., Davis, S. T., Dickerson, S. H., Edelstein, M., Frye, S. V., Gampe, R. T. Jr., Harris, P. A., Hassell, A., Holmes, W. D., Hunter, R. N., Lackey, K. E., Lovejoy, B., Luzzio, M. J., Montana, V., Rocque, W. J., Rusnak, D., Shewchuk, L., Veal, J. M., Walker, D. H., Kuyper, L. F. (2002). Oxindole-based inhibitors of cyclin-dependent kinase 2 (CDK2): Design, synthesis, enzymatic activities, and X-ray crystallographic analysis. *J. Med. Chem., 44*, 4339–4358.

120. Sherr, C. J. (1996). Cancer cell cycles. *Science, 274*, 1672–1677.

121. Frizelle, S. P., Grim, J., Zhou, J., Gupta, P., Curiel, D. T., Geradts, J., Kratzke, R. A. (1998). Re-expression of p16INK4a in mesothelioma cells results in cell cycle arrest, cell death, tumor suppression and tumor regression. *Oncogene, 16*, 3087–3095.

122. Barnes, D. M., Gillet, C. E. (1998). Cyclin D1 in breast cancer. *Breast Cancer Res. Treat., 52*, 1–15.

123. Roussel. M. F. (1999). The INK4 family of cell cycle inhibitors in cancer. *Oncogene, 18*, 5311–5317.

124. Nevins, J. R. (2001). The Rb/E2F pathway and cancer. *Hum. Mol. Genet., 10*, 699–703.

125. Tsao, H., Benoit, E., Sober, A. J., Thiele, C., Haluska, F. G. (1998). Novel mutations in the p16/CDKN2A binding region of the cyclin-dependent kinase-4 gene. *Cancer Res., 58*, 109–113.

126. Goekjian, P. G., Jirousek, M. R. (1999). Protein kinase C in the treatment of disease: Signal transduction pathways, inhibitors, and agents in development. *Curr. Med. Chem., 6*, 877–903.

127. Dekker, L. V., Parker, P. (1994). Protein kinase C—A question of specificity. *Trends Biochem. Sci.*, *19*, 73–77.
128. Tanaka, C., Saito, N. (1992). Localization of subspecies of protein kinase C in the mammalian central nervous system. *Neurochem. Int.*, *21*, 499–512.
129. Blobe, G. C., Obeid, L. M., Hannun, Y. A. (1994). Regulation of protein kinase C and role in cancer biology. *Cancer Metastasis Rev.*, *13*, 724–730.
130. Ron, D., Kazanietz, M. G. (1999). New insights into the regulation of protein kinase C and novel phorbol esters. *FASEB J.*, *27*, 1658–1676.
131. Bullock, W. H., Magnuson, S. R., Choi, S., Gunn, D. E., Rudolph. J. (2002). Prospects for kinase activity modulators in the treatment of diabetes and diabetic complications. *Curr. Top. Med. Chem.*, *2*, 915–938.
132. Way, K. J., Katai, N., King, G. L. (2001). Protein kinase C and the development of diabetic vascular complications. *Diabetes Med.*, *18*, 945–959.
133. Takenaga, K., Takahashi, K. (1996). Effects of 12-*O*-tetradecanoylphorbol-13-acetate on adhesiveness and lung-colonising ability of Lewis lung carcinoma cells. *Cancer Res.*, *46*, 375–380.
134. Schwartz, G. K., Juang, K., Kelsen, D., Albino, A. P. (1993). Protein kinase C: A novel target for inhibiting gastric cancer cell invasion. *J. Natl. Cancer Inst.*, *85*, 402–407.
135. O'Brian, C. A., Vogel, V. G., Singletary, S. E., Ward, N. E. (1989). Elevated protein kinase C expression in human breast tumour biopsies relative to normal breast tissue. *Cancer Res.*, *49*, 3215–3217.
136. Gokmen-Polar, Y., Murray, N., Velasco, M. A., Gatlica, Z., Fields, A. P. (2001). Elevated protein kinase C βII is an early promotive event in colon carcinogenesis. *Cancer Res.*, *61*, 1375–1381.
137. Robinson-White, A., Stratakis, C. (2002). Protein kinase A signaling: "Cross-talk" with other pathways in endocrine cells. *Ann. NY Acad. Sci.*, *968*, 256–270.
138. Rajnicek, A., Britland, S., McCaig, C. (1997). Contact guidance of CNS neurites on grooved quartz: Influence of groove dimensions, neuronal age and cell type. *J. Cell. Sci.*, *110*, 2905–2913.
139. Rajnicek, A., McCaig, C. (1997). Guidance of CNS growth cones by substratum grooves and ridges: Effects of inhibitors of the cytoskeleton, calcium channels and signal transduction pathways. *J. Cell. Sci.*, *110*, 2915–2924.
140. Hu, E., Lee, D. (2003). Rho kinase inhibitors as potential therapeutic agents for cardiovascular diseases. *Curr. Opin. Investig. Drugs*, *4*, 1065–1075.
141. Riento, K., Ridley A. J. (2003). Rocks: Multifunctional kinases in cell behaviour. *Nat. Rev. Mol. Cell Biol.*, *4*, 446–456.
142. Katso, R., Okkenhaug, K., Ahmadi, K., White, S., Timms, J., Waterfield, M. D. (2001). Cellular function of phosphoinositide 3-kinases: Implications for development, homeostasis, and cancer. *Annu. Rev. Cell Devel. Biol.*, *17*, 615–675.
143. McKenna, W. G., Muschel, R. J. (2003). Targeting tumor cells by enhancing radiation sensitivity. *Genes Chromosomes Cancer*, *38*, 330–338.
144. Woodgett, J. R. (1990). Molecular cloning and expression of glycogen synthase kinase-3/factor A. *EMBO J.*, *9*, 2431–2438.
145. Kim, H.-S., Skurk, C., Thomas, S. R., Bialik, A., Suhara, T., Kureishi, Y., Birnbaum, M., Keaney, J. F. Jr., Walsh. K. (2002). Regulation of angiogenesis by glycogen synthase kinase-3beta. *J. Biol. Chem.*, *277*, 41888–41896.

146. Kaidanovich, O., Eldar-Finkelman, H. (2002). The role of glycogen synthase kinase-3 in insulin resistance and Type 2 diabetes. *Exp. Opin. Ther. Targets*, *6*, 555–561.

147. Eldar-Finkelman, H. (2002). Glycogen synthase kinase 3: An emerging therapeutic target. *Trends Mol. Med.*, *8*, 126–132.

148. Wagman, A. S., Johnson, K. W., Bussiere, D. E. (2004). Discovery and development of GSK3 inhibitors for the treatment of type 2 diabetes. *Curr. Pharm. Des.*, *10*, 1105–1137.

149. Laird, A. D., Cherrington, J. M. (2003). Small molecule tyrosine kinase inhibitors: Clinical development of anticancer agents. *Expert Opin. Investig. Drugs*, *12*, 51–64.

150. Cohen, P. (2002). Protein kinases—The major drug targets of the twenty-first century? *Nat. Rev. Drug Discov.*, *1*, 309–315.

151. Druker, B. J. (2003). Imatinib alone and in combination for chronic myeloid leukemia. *Semin. Hematol.*, *40*, 50–58.

152. Wakeling, A. E. (2002). Epidermal growth factor receptor tyrosine kinase inhibitors. *Curr. Opin. Pharmacol.*, *2*, 382–387.

153. Marx, S. O., Marks, A. R. (2001). Bench to bedside: The development of rapamycin and its application to stent restenosis. *Circulation*, *104*, 852–855.

154. Slamon, D. J., Leyland-Jones, B., Shak, S., Fuchs, H., Paton, V., Bajamonde, A., Fleming, T., Eiermann, W., Wolter, J., Pegram, M., Baselga, J., Norton, L. (2001). Use of chemotherapy plus a monoclonal antibody against HER2 for metastatic breast cancer that overexpresses HER2. *N. Engl. J. Med.*, *344*, 783–792.

155. Mitchell, P. (2004). Erbitux diagnostic latest adjunct to cancer therapy. *Nat. Biotechnol.*, *22*, 363–364.

156. Roskoski, R. Jr. (2004). The ErbB/HER receptor protein-tyrosine kinases and cancer. *Biochem. Biophys. Res. Commun.*, *319*, 1–11.

157. Fabbro, D., Ruetz, S., Buchdunger, E., Cowan-Jacob, S. W., Fendrich, G., Liebetanz, J., Mestan, J., O'Reilly, T., Traxler, P., Chaudhuri, B., Fretz, H., Zimmermann, J., Meyer, T., Giorgio, C., Furet, G., Manley, P. W. (2002). Protein kinases as targets for anticancer agents: From inhibitors to useful drugs. *Pharmacol. Ther.*, *93*, 79–98.

158. Levitzki, A. (2003). Protein kinase inhibitors as a therapeutic modality. *Acc. Chem. Res.*, *36*, 462–469.

159. Force, T., Kuida, K., Namchuk, M., Parang, K., Kyriakis, J. M. (2004). Inhibitors of protein kinase signaling pathways. *Circulation*, *109*, 1196–1205.

160. Toledo, L. M., Lydon, N. B., Elbaum, D. (1999). The structure-based design of ATP-site directed protein kinase inhibitors. *Curr. Med. Chem.*, *6*, 775–805.

161. Traxler, P. (1998). Tyrosine kinase inhibitors in cancer treatment (part II). *Expert Opin. Ther. Patents*, *8*, 1599–1625.

162. Bridges, A. J. (2001). Chemical inhibitors of protein kinases. *Chem. Rev.*, *101*, 2541–2571.

163. Myers, M. R., He, W., Hulme, C. (1997). Inhibitors of tyrosine kinases involved in inflammation and autoimmune disease. *Curr. Pharm. Des.*, *3*, 473–502.

164. Dancey, J., Sausville, E. A. (2003). Issues and progress with protein kinase inhibitors for cancer treatment. *Nat. Drug Discov.*, *2*, 296–313.

165. Nam, N. H., Parang, K. (2003). Current targets for anticancer drug discovery. *Curr. Drug Targets*, *4*, 159–179.
166. Sawyer, T., Boyce, B., Dalgarno, D., Luliucci, J. (2001). Src inhibitors: Genomics to therapeutics. *Expert Opin. Investig. Drugs*, *10*, 1327–1344.
167. Orchard, S. (2002). Kinases as targets: Prospects for chronic therapy. *Curr. Opin. Drug Discov. Devel.*, *5*, 713–717.
168. Garcia-Echeverria, C., Traxler, P., Evans, D. B. (2000). ATP site-directed competitive and irreversible inhibitors of protein kinases. *Med. Res. Rev.*, *20*, 28–57.
169. Fischer, P. M. (2004). The design of drug candidate molecules as selective inhibitors of therapeutically relevant protein kinases. *Curr. Med. Chem.*, *11*, 1563–1583.
170. Fischer, O. M., Streit, S., Hart, S., Ullrich, A. (2003). Beyond Herceptin and Gleevec. *Curr. Opin. Chem. Biol.*, *7*, 490–495.
171. Sawyer, T. K. (2004). Cancer metastatis therapeutic targets and drug discovery: Emerging small-molecule protein kinase inhibitors. *Expert Opin. Invest. Drugs*, *13*, 1–19.
172. Wu, S. Y., McNae, I., Kontopidis, G., McClue, S. J., McInnes, C., Stewart, K. J., Wang, S., Zheleva, D. I., Marriage, H., Lane, D. P., Taylor, P., Fischer, P. M., Walkinshaw, M. D. (2003). Discovery of a novel family of CDK inhibitors with the program LIDAEUS: Structural basis for ligand-induced disordering of the activation loop. *Structure (Camb).*, *11*, 399–410.
173. Rossé, G., Séquin, U., Mett, H., Furet, P., Traxler, P., Fretz, H. (1997). Synthesis of modified tripeptides and tetrapeptides as potential bisubstrate inhibitors of the epidermal growth factor receptor protein tyrosine kinase. *Helv. Chim. Acta*, *80*, 653–670.
174. Yuan, C. J., Jakes, S., Elliot, S., Graves, D. (1990). A rationale for the design of an inhibitor of tyrosine kinase. *J. Biol. Chem.*, *265*, 16205–16209.
175. Cushman, M., Chinnasamy, P., Chakraborti, A. K., Jurayj, J., Geahlen, R. L., Haugwitz, R. D. (1990). Synthesis of [beta-(4-pyridyl-1-oxide)-L-alanine4]-angiotensin I as a potential suicide substrate for protein-tyrosine kinases. *Int. J. Pept. Protein Res.*, *36*, 538–543.
176. Bohmer, F. D., Kaagyozov, L., Uecker, A., Serve, H., Botzki, A., Mahboobi, S., Dove, S. (2003). A single amino acid exchange inverts susceptibility of related receptor tyrosine kinases for the ATP-site inhibitor STI-571. *J. Biol. Chem.*, *14*, 5148–5155.
177. Huse, M., Kuriyan, J. (2002). The conformational plasticity of protein kinases. *Cell*, *109*, 275–282.
178. Davies, S. P., Reddy, H., Caivano, M., Cohen, P. (2000). Specificity and mechanism of action of some commonly used protein kinase inhibitors. *Biochem. J.*, *351*, 95–105.
179. Bain, J., McLauchlan, H., Elliott, M., Cohen, P. (2003). The specificities of protein kinase inhibitors: An update. *Biochem. J.*, *371*, 199–204.
180. Lawrence, D. S., Niu, J. (1998). Protein kinase inhibitors: The tyrosine-specific protein kinases. *Pharmacol. Ther.*, *77*, 81–114.
181. Topaly, J., Zeller, W. J., Fruehauf, S. (2001). Synergistic activity of the new ABL-specific tyrosine kinase inhibitor STI571 and chemotherapeutic drugs on BCR-ABL-positive chronic myelogenous leukemia cells. *Leukemia*, *15*, 342–347.

182. Thiesing, J. T., Ohno-Jones, S., Kolibaba, K. S., Druker, B. J. (2000). Efficacy of STI571, an abl tyrosine kinase inhibitor, in conjunction with other antileukemic agents against bcr-abl-positive cells. *Blood*, 96, 3195–3199.
183. Bogoyevitch, M. A., Boehm, I., Oakley, A., Ketterman, A. J., Barr, R. K. (2004). Targeting the JNK MAPK cascade for inhibition: Basic science and therapeutic potential. *Biochim. Biophys. Acta*, 1697, 89–101.
184. Shen, G. X. (2003). Selective protein kinase C inhibitors and their applications. *Curr. Drug Targets Cardiovasc. Haematol. Disord.*, 3, 301–307.
185. Kim, H., Sasaki, T., Maeda, K., Koya, D., Kashiwagi, A., Yasuda, H. (2003). Protein kinase Cbeta selective inhibitor LY333531 attenuates diabetic hyperalgesia through ameliorating cGMP level of dorsal root ganglion neurons. *Diabetes*, 52, 2102–2109.
186. Showalter, H. D., Kraker, A. J. (1997). Small molecule inhibitors of the platelet-derived growth factor receptor, the fibroblast growth factor receptor, and Src family tyrosine kinases. *Pharmacol. Ther.*, 76, 55–71.
187. Druker, B. J., Lydon, N. B. (2000). Lessons learned from the development of an abl tyrosine kinase inhibitor for chronic myelogenous leukemia. *J. Clin. Invest.*, 105, 3–7.
188. Graziani, Y., Erikson, E., Erikson, R. L. (1983). The effect of quercetin on the phosphorylation activity of the Rous sarcoma virus transforming gene product in vitro and in vivo. *Eur. J. Biochem.*, 135, 583–589.
189. Hagiwara, M., Inoue, S., Tanaka, T., Nunoki, K., Ito, M., Hidaka, H. (1988). Differential effects of flavonoids as inhibitors of tyrosine protein kinases and serine/threonine protein kinases. *Biochem. Pharmacol.*, 37, 2987–2992.
190. Fischer, P. M., Lane, D. P. (2000). Inhibitors of cyclin-dependent kinases as anti-cancer therapeutics. *Curr. Med. Chem.*, 7, 1213–1245.
191. Sedlacek, H. H. (2001). Mechanisms of action of flavopiridol. *Crit. Rev. Oncol. Hematol.*, 38, 139–170.
192. Akiyama, T., Ishida, J., Nakagawa, S., Ogawara, H., Watanabe, S. I., Itoh, N., Shibuy, M., Fukami, Y. (1987). Genistein, a specific inhibitor of tyrosine-specific protein kinases. *J. Biol. Chem.*, 262, 5592–5595.
193. Nishio, K., Miura, K., Ohira, T., Heiko, Y., Saijo, N. (1994). Genistein, a tyrosine kinase inhibitor, decreased the affinity of p56lck to beta-chain of interleukin-2 receptor in human natural killer (NK)-rich cells and decreased NK-mediated cytotoxicity. *Proc. Soc. Exp. Biol. Med.*, 207, 227–233.
194. Dudley, D. T., Pang, L., Decker, S. J., Bridges, A. J., Saltiel, A. R. (1995). A synthetic inhibitor of the mitogen-activated protein kinase cascade. *Proc. Natl. Acad. Sci. USA*, 92, 7686–7689.
195. Pang, L., Sawada, T., Decker, S. J., Saltiel, A. R. (1995). Inhibition of MAP kinase kinase blocks the differentiation of PC-12 cells induced by nerve growth factor. *J. Biol. Chem.*, 270, 13585–13588.
196. Waters, S. B., Holt, K. H., Ross, S. E., Syu, L. J., Guan, K. L., Saltiel, A. R., Koretzky, G. A., Pessin, J. E. (1995). Desensitization of Ras activation by a feedback disassociation of the SOS-Grb2 complex. *J. Biol. Chem.*, 270, 20883–20886.
197. Langlois, W. J., Sasaoka, T., Saltiel, A. R., Olefsky, J. M. (1995). Negative feedback regulation and desensitization of insulin- and epidermal growth factor-stimulated p21ras activation. *J. Biol. Chem.*, 270, 25320–25323.

198. Alessi, D. R., Cuenda, A., Cohen, P., Dudley, D. T., Saltiel, A. R. (1995). PD 098059 is a specific inhibitor of the activation of mitogen-activated protein kinase kinase in vitro and in vivo. *J. Biol. Chem.*, 270, 27489–27494.
199. Kültz, D., Madhany, S., Burg, M. B. (1998). Hyperosmolality causes growth arrest of murine kidney cells. Induction of GADD45 and GADD153 by osmosensing via stress-activated protein kinase 2. *J. Biol. Chem.*, 273, 13645–13651.
200. Kemp, B. E., Graves, D. J., Benjamini, E., Krebs, E. G. (1977). Role of multiple basic residues in determining the substrate specificity of cyclic AMP-dependent protein kinase. *J. Biol. Chem.*, 252, 4888–4894.
201. Kemp, B. E., Pearson, R. B., House, C. M. (1991). Pseudosubstrate-based peptide inhibitors. *Methods Enzymol.*, 201, 287–304.
202. Ablooglu, A. J., Till, J. H., Kim, K., Parang, K., Cole, P. A., Hubbard, S. R., Kohanski, R. A. (2000). Probing the catalytic mechanism of the insulin receptor kinase with a tetrafluorotyrosine-containing peptide substrate. *J. Biol. Chem.*, 275, 30394–30398.
203. Niu, J., Lawrence, D. S. (1997). Nonphosphorylatable tyrosine surrogates. Implications for protein kinase inhibitor design. *J. Biol. Chem.*, 272, 1493–1499.
204. Pargellis, C., Tong, L., Churchill, L., Cirillo, P. F., Gilmore, T., Graham, A. G., Grob, P. M., Hickey, E. R., Moss, N., Pav, S., Regan, J. (2002). Inhibition of p38 MAP kinase utilizing a novel allosteric binding site. *Nat. Struct. Biol.*, 9, 268–272.
205. Nagar, B., Bornmann, W. G., Pellicena, P., Schindler, T., Veach, D. R., Miller, W. T., Clarkson, B., Kuriyan, J. (2002). Crystal structures of the kinase domain of c-Abl in complex with the small molecule inhibitors PD173955 and imatinib (STI-571). *Cancer Res.*, 62, 4236–4243.
206. Vu, C. B. (2000). Recent advances in the design and synthesis of SH2 inhibitors of Src, Grb2 and ZAP-70. *Curr. Med. Chem.*, 7, 1081–1100.
207. Vu, C. B., Corpuz, E. G., Pradeepan, S. G., Violette, S., Bartlett, C., Sawyer, T. K. (1999). Nonpeptidic SH2 inhibitors of the tyrosine kinase ZAP-70. *Bioorg. Med. Chem. Lett.*, 9, 3009–3014.
208. Muller, G. (2000). Peptidomimetics SH2 domain antagonists for targeting signal transduction. *Topics Curr. Chem.*, 211, 17–59.
209. Vu, C. B., Corpuz, E. G., Merry, T. J., Pradeepan, S. G., Bartlett, C., Bohacek, R. S., Botfield, M. C., Eyermann, C. J., Lynch, B. A., MacNeil, I. A., Ram, M. K., van Schravendijk, M. R., Violette, S., Sawyer, T. K. (1999). Discovery of potent and selective SH2 inhibitors of the tyrosine kinase ZAP-70. *J. Med. Chem.*, 42, 4088–4098.
210. Cody, W. L., Lin, Z., Panek, R. L., Rose, D. W., Rubin, J. R. (2000). Progress in the development of inhibitors of SH2 domains. *Curr. Pharm. Des.*, 6, 59–98.
211. Broom, A. D. (1989). Rational design of enzyme inhibitors: Multisubstrate analog inhibitors. *J. Med. Chem.*, 32, 2–7.
212. Page, M. I., Jencks, W. P. (1971). Entropic contributions to rate accelerations in enzymic and intramolecular reactions and the chelate effect. *Proc. Natl. Acad. Sci. USA*, 68, 1678–1683.
213. Bruice, T. C., Benkovic, S. J. (2000). Chemical basis for enzyme catalysis. *Biochemistry*, 30, 6267–6274.

214. Parang, K., Cole, P. A. (2002). Bisubstrate inhibitors of protein kinases. *Pharmacol. Ther.*, *93*, 1–13.
215. Kruse, C. H., Holden, K. G., Pritchard, M. L., Field, J. A., Rieman, D. J., Greig, R. G., Poste, G. (1988). Synthesis and evaluation of multisubstrate inhibitors of an oncogene-encoded tyrosine-specific protein kinase. *J. Med. Chem.*, *31*, 1762–1767.
216. Ricouart, A., Gesquiere, J. C., Tartar, A., Sergheraert, C. (1991). Design of potent protein-kinase inhibitors using the bisubstrate approach. *J. Med. Chem.*, *34*, 73–78.
217. Traxler, P. M., Wacker, O., Bach, H. L., Geissler, J. F., Kump, W., Meyer, T., Regenass, U., Roesel, J. L., Lydon, N. (1991). Sulfonylbenzoyl-nitrostyrenes: Potential bisubstrate type inhibitors of the EGF-receptor tyrosine protein kinase. *J. Med. Chem.*, *34*, 2328–2337.
218. Uri, A., Järlebark, L., von Kügelgen, I., Schöberg, T., Undén, A., Heibronn, E. (1994). A new class of compounds, peptide derivatives of adenosine 5′-carboxylic acid, includes inhibitors of ATP receptor-mediated responses. *Biorg. Med. Chem.*, *2*, 1099–1105.
219. Uri, A., Raidaru, G., Subbi, J., Padari, K., Pooga, M. (2002). Identification of the ability of highly charges nanomolar inhibitors of protein kinases to cross plasma membranes and carry a protein into cells. *Bioorg. Med. Chem. Lett.*, *12*, 2117–2120.
220. Loog, M., Uri, A., Raidaru, G., Järv, J., Ek, P. (2000). Bi-substrate analog ligands for affinity chromatography of protein kinases. *FEBS Lett.*, *480*, 244–248.
221. Medzihradszky, D., Chen, S. L., Kenyon, G. L., Gibson, B. (1994). Solid-phase synthesis of adenosine phosphopeptides as potential bisubstrate inhibitors of protein kinases. *J. Am. Chem. Soc.*, *116*, 9413–9419.
222. Profit, A. A., Lee, T. R., Lawrence, D. S. (1999). Bivalent inhibitors of protein tyrosine kinases. *J. Am. Chem. Soc.*, *121*, 280–283.
223. Cowburn, D., Zheng, J., Xu, Q., Barany, G. (1995). Enhanced affinities and specificities of consolidated ligands for the Src homology SH3 and SH2 domains of Abelson protein-tyrosine kinase. *J. Biol. Chem.*, *270*, 26738–26741.
224. Parang, K., Till, J. H., Ablooglu, A. J., Kohanski, R. A., Hubbard, S. R., Cole, P. A. (2001). Mechanism-based design of a protein kinase inhibitor. *Nat. Struct. Biol.*, *8*, 37–41.
225. Tanoue, T., Nishida, E. (2002). Docking interactions in the mitogen-activated protein kinase cascades. *Pharmacol. Ther.*, *93*, 193–202.
226. Chang, C. I., Xu, B. E., Akella, R., Cobb, M. H., Goldsmith, E. J. (2002). Crystal structures of MAP kinase p38 complexed to the docking sites on its nuclear substrate MEF2A and activator MKK3b. *Mol. Cell.*, *9*, 1241–1249.
227. Ho, D. T., Bardwell, A. J., Abdollahi, M., Bardwell, L. (2003). A docking site in MKK4 mediates high affinity binding to JNK MAPKs and competes with similar docking sites in JNK substrates. *J. Biol. Chem.*, *278*, 32662–32672.
228. Lee, S., Lin, X., Nam, N. H., Parang, K., Sun, G. (2003). Determination of the substrate-docking site of protein tyrosine kinase C-terminal Src kinase. *Proc. Natl. Acad. Sci. USA*, *100*, 14707–14712.
229. Denny, W. A. (2002). Irreversible inhibitors of the erbB family of protein tyrosine kinases. *Pharmacol. Ther.*, *93*, 253–261.

230. Allen, L. F., Eiseman, I. A., Fry, D. W., Lenehan, P. F. (2003). CI-1033, an irreversible pan-erbB receptor inhibitor and its potential application for the treatment of breast cancer. *Semin. Oncol.*, *30*(5, Suppl. 16), 65–78.

231. Wissner, A., Overbeek, E., Reich, M. F., Floyd, M. B., Johnson, B. D., Mamuya, N., Rosfjord, E. C., Discafani, C., Davis, R., Shi, X., Rabindran, S. K., Gruber, B. C., Ye, F., Hallett, W. A., Nilakantan, R., Shen, R., Wang, Y. F., Greenberger, L. M., Tsou, H. R. (2003). Synthesis and structure-activity relationships of 6,7-disubstituted 4-anilinoquinoline-3-carbonitriles. The design of an orally active, irreversible inhibitor of the tyrosine kinase activity of the epidermal growth factor receptor (EGFR) and the human epidermal growth factor receptor-2 (HER-2). *J. Med. Chem.*, *46*, 49–63.

232. Sreedhar, A. S., Soti, C., Csermely, P. (2004). Inhibition of Hsp90: A new strategy for inhibiting protein kinases. *Biochim. Biophys. Acta*, *1697*, 233–242.

233. Neckers. L. (2002). Hsp90 inhibitors as novel cancer chemotherapeutic agents. *Trends Mol. Med.*, *8*, S55–S61.

234. Nimmanapalli, R., O'Bryan, E., Kuhn, D., Yamaguchi, H., Wang, H. G., Bhalla, K. N. (2003). Regulation of 17-AAG-induced apoptosis: Role of Bcl-2, Bcl-xL, and Bax downstream of 17-AAG-mediated downregulation of Akt, Raf-1 and Src kinases. *Blood*, *102*, 269–275.

235. Masui, H., Kawamoto, T., Sato, J. D., Wolf, B., Sato, G., Mendelsohn, J. (1984). Growth inhibition of human tumor cells in athymic mice by anti-epidermal growth factor receptor antibodies. *Cancer Res.*, *44*, 1002–1007.

236. Slamon, D. J., Clark G. M., Wong, S. G., Levin, W. J., Ullrich, A., McGuire, W. L. (1987). Human breast cancer: Correlation of relapse and survival with amplification of the HER-2/neu oncogene. *Science*, *235*, 177–182.

237. Cho, H.-S., Mason, K., Ramyar, K. X., Stanley, A. M., Gabelli, S. B., Denney, Jr., D. W., Leahy, D. J. (2003). Structure of the extracellular region of HER2 alone and in complex with the Herceptin Fab. *Nature*, *421*, 756–760.

238. Morin, M. J. (2000). From oncogene to drug: Development of small molecule tyrosine kinase inhibitors as anti-tumor and anti-angiogenic agents. *Oncogene*, *19*, 6574–6583.

239. Cohen, M., H., Williams, G. A., Sridhara, R., Chen, G., Pazdur, R. (2003). FDA drug approval summary: Gefitinib (ZD1839) (Iressa) tablets. *Oncologist*, *8*, 303–306.

240. Blagosklonny, M. V., Darzynkiewicz, Z. (2003). Why Iressa failed: Toward novel use of kinase inhibitors (outlook). *Cancer Biol. Ther.*, *2*, 137–140.

241. Dancey, J. E., Freidlin, B. (2003). Targeting epidermal growth factor receptor—Are we missing the mark? *Lancet*, *362*, 62–64.

242. Azam, M., Latek, R. R., Daley, G. O. (2003). Mechanisms of autoinhibition and STI-571/Imatinib resistance revealed by mutagenesis of BCR-ABL. *Cell*, *112*, 831–843.

243. Hidalgo, M., Siu, L. L., Nemunaitis, J., Rizzo, J., Hammond, L. A., Takimoto, C., Eckhardt, S. G., Tolcher, A., Britten, C. D., Denis, L., Ferrante, K., Von Hoff, D. D., Silberman, S., Rowinsky, E. K. (2001). Phase I and pharmacologic study of OSI-774, an epidermal growth factor receptor tyrosine kinase inhibitor, in patients with advanced solid malignancies. *J. Clin. Oncol.*, *19*, 3267–3279.

244. Stamos, J., Sliwkowski, M. X., Eigenbrot, C. (2002). Structure of the epidermal growth factor receptor kinase domain alone and in complex with a 4-anilinoquinazoline inhibitor. *J. Biol. Chem.*, 277, 46265–46272.

245. Traxler, P., Bold, G., Buchdunger, E., Caravtti, G., Furet, P., Manley, P., O'Reilly, T., Wood, J., Zimmermann, J. (2001). Tyrosine kinase inhibitors: From rational design to clinical trials. *Med. Res. Rev.*, 21, 499–512.

246. Tiseo, M., Loprevite, M., Ardizzoni, A. (2004). Epidermal growth factor receptor inhibitors: A new prospective in the treatment of lung cancer. *Curr. Med. Chem. Anti-Cancer Agents*, 4, 139–148.

247. Folkman, J. (1971). Tumor angiogenesis: Therapeutic implications. *N. Engl. J. Med.*, 285, 1182–1186.

248. Giles, F. J. (2002). The emerging role of angiogenesis inhibitors in hematologic malignancies. *Oncol. (Huntington)*, 16, 23–29.

249. Giles, F., Cooper, M. A., Silverman, L., Karp, J. E., Lancet, J. E., Zangari, M., Shami, P. J., Khan, K. D., Hannah, A. L., Cherrington, J. M., Thomas, D. A., Garcia-Manero, G., Albitar, M., Kantarjian, H. M., Stopeck, A. T. (2003). Phase II study of SU5416—A small-molecule, vascular endothelial growth factor tyrosine kinase receptor inhibitor—in patients with refractory myeloproliferative diseases. *Cancer*, 97, 1920–1928.

250. Hasselbalch, H. C. (2003). SU6668 in idiopathic myelofibrosis—A rational therapeutic approach targeting several tyrosine kinases of importance for the myeloproliferation and the development of bone marrow fibrosis and angiogenesis. *Med. Hypoth.*, 61, 244–247.

251. Laird, A. D., Vajkoczy, P., Shawver, L. K., Thurnher, A., Liang, C., Mohammadi, M., Schlessinger, J., Ullrich, A., Hubbard, S. R., Blake, R. A., Fong, T. A., Strawn, L. M., Sun, L., Tang, C., Hawtin, R., Tang, F., Shenoy, N., Hirth, K. P., McMahon, G., Cherrington. (2000). SU6668 is a potent antiangiogenic and antitumor agent that induces regression of established tumors. *Cancer Res.*, 60, 4152–4160.

252. Breier, G., Blum, S., Peli, J., Groot, M., Wild, C., Risau, W., Reichmann, E. (2002). Transforming growth factor-beta and Ras regulate the VEGF/VEGF-receptor system during tumor angiogenesis. *Int. J. Cancer*, 97, 142–148.

253. Ferrara, N. (2000). VEGF: An update on biological and therapeutic aspects. *Curr. Opin. Biotechnol.*, 11, 617–624.

254. Ferrara, N. (2000). Vascular endothelial growth factor and the regulation of angiogenesis. *Recent Prog. Horm. Res.*, 55, 15–35.

255. Ferrara, N., Gerber, H. P. (2001). The role of vascular endothelial growth factor in angiogenesis. *Acta Haematol.*, 106, 148–156.

256. Kolch, W., Martiny-Baron, G., Kieser, A., Marme, D. (1995). Regulation of the expression of the VEGF/VPS and its receptors: Role in tumor angiogenesis. *Breast Cancer Res. Treat.*, 36, 139–155.

257. Bold, G., Altmann, K. H., Frei, J., Lang, M., Manley, P. W., Traxler, P., Wietfeld, B., Bruggen, J., Buchdunger, E., Cozens, R., Ferrari, S., Furet, P., Hofmann, F., Martiny-Baron, G., Mestan, J., Rosel, J., Sills, M., Stover, D., Acemoglu, F., Boss, E., Emmenegger, R., Lasser, L., Masso, E., Roth, R., Schlachter, C., Vetterli, W. (2000). New anilinophthalazines as potent and orally well absorbed inhibitors of

the VEGF receptor tyrosine kinases useful as antagonists of tumor-driven angiogenesis. *J. Med. Chem.*, *43*, 2310–2323.

258. Bold, G., Altmann, K. H., Frei, J., Lang, M., Manley, P. W., Traxler, P., Wietfeld, B., Bruggen, J., Buchdunger, E., Cozens, R., Ferrari, S., Furet, P., Hofmann, F., Martiny-Baron, G., Mestan, J., Rosel, J., Sills, M., Stover, D., Acemoglu, F., Boss, E., Emmenegger, R., Lasser, L., Masso, E., Roth, R., Schlachter, C., Vetterli, W. (2000). New anilinophthalazines as potent and orally well absorbed inhibitors of the VEGF receptor tyrosine kinases useful as antagonists of tumor-driven angiogenesis. *J. Med. Chem.*, *43*, 3200–3213.

259. Wood, J. M., Bold, G., Buchdunger, E., Cozens, R., Ferrari, S., Frei, J., Hofmann, F., Mestan, J., Mett, H., O'Reilly, T., Persohn, E., Rosel, J., Schnell, C., Stover, D., Theuer, A., Towbin, H., Wenger, F., Woods-Cook, K., Menrad, A., Siemeister, G., Schirner, M., Thierauch, K. H., Schneider, M. R., Drevs, J., Martiny-Baron, G., Totzke, F. (2000). PTK787/ZK 222584, a novel and potent inhibitor of vascular endothelial growth factor receptor tyrosine kinases, impairs vascular endothelial growth factor-induced responses and tumor growth after oral administration. *Cancer Res.*, *60*, 2178–2189.

260. Smith, J. K., Mamoon, N. M., Duhe, R. J. (2004). Emerging roles of targeted small molecule protein-tyrosine kinase inhibitors in cancer therapy. *Oncol. Res.*, *14*, 175–225.

261. Mohammadi, M., Froum, S., Hamby, J. M., Schroeder, M. C., Panek, R. L., Lu, G. H., Eliseenkova, A. V., Green, D., Schlessinger, J., Hubbard, S. R. (1998). Crystal structure of an angiogenesis inhibitor bound to the FGF receptor tyrosine kinase domain. *EMBO J.*, *17*, 5896–5904.

262. Fong, T. A., Shawver, L. K., Sun, L., Tang, C., App, H., Powell, T. J., Kim, Y. H., Schreck, R., Wang, X., Risau, W., Ullrich, A., Hirth, K. P., McMahon, G. (1999). SU5416 is a potent and selective inhibitor of the vascular endothelial growth factor receptor (Flk-1/KDR) that inhibits tyrosine kinase catalysis, tumor vascularization, and growth of multiple tumor types. *Cancer Res.*, *59*, 99–106.

263. Druker, B. J., Tamura, S., Buchdunger, E., Ohno, S., Segal, G. M., Fanning, S., Zimmermann, J., Lydon, N. B. (1996). Effects of a selective inhibitor of the Abl tyrosine kinase on the growth of Bcr-Abl positive cells. *Nat. Med.*, *2*, 561–566.

264. Barnes, D. J., Melo, J. V. (2003). Management of chronic myeloid leukemia: Targets for molecular therapy. *Semin. Hematol.*, *40*, 34–49.

265. Gorre, M. E., Mohammed, M., Ellwood, K., Hsu, N., Paquette, R., Rao, P. N., Sawyers, C. L. (2001). Clinical resistance to STI-1571 cancer therapy caused by BCR-ABL gene mutation or amplification. *Science*, *293*, 876–880.

266. Roumiantsev, S., Shah, N. P., Gorre, M. E., Nicoll, J., Brasher, B. B., Sawyers, C. L., Van Etten, R. A. (2000). Clinical resistance to the kinase inhibitor STI-571 in chronic myeloid leukemia by mutation of Tyr-253 in the Abl kinase domain P-loop. *Proc. Natl. Acad. Sci. USA*, *99*, 10700–10705.

267. Buchdunger, E., Cioffi, C. L., Law, N., Stiver, D., Ohno-Jones, S., Druker, B. J., Lyndon, N. B. (2000). Abl protein-tyrosine kinase inhibitor ST1571 inhibits in vitro signal transduction mediated by c-Kit and platelet-derived growth factor receptors. *J. Pharmacol. Exp. Ther.*, *295*, 139–145.

268. Okuda, K., Weisberg, E., Gilliland, D. G., Griffin, J. D. (2001). ARG tyrosine kinase activity is inhibited by ST1571. *Blood*, 97, 2440–2448.

269. van Oosterom, A. T., Judson, I., Verweij, J., Stroobants, S., Donato di Paola, E., Dimitrijevic, S., Martens, M., Webb, A., Sciot, R., Van Glabbeke, M., Silberman, S., Nielsen, O. S. European Organisation for Research and Treatment of Cancer Soft Tissue and Bone Sarcoma Group. (2001). Safety and efficacy of imatinib (STI571) in metastatic gastrointestinal stromal tumours: A phase I study. *Lancet*, 358, 1421–1423.

270. Blanke, C. D., Eisenberg, B. L., Heinrich, M. C. (2001). Gastrointestinal stromal tumors. *Curr. Treat. Options Oncol.*, 2, 485–491.

271. Heinrich, M. C., Griffith, D. J., Druker, B. J., Wait, C. L., Ott, K. A., Zigler, A. J. (2000). Inhibition of c-kit receptor tyrosine kinase activity by STI 571, a selective tyrosine kinase inhibitor. *Blood*, 96, 925–932.

272. Nagar, B., Hantschel, O., Young, M. A., Scheffzek, K., Veach, D., Bornmann, W., Clarkson, B., Superti-Furga, G., Kuriyan, J. (2003). Structural basis for the autoinhibition of c-Abl tyrosine kinase. *Cell*, 112, 859–871.

273. Noble, M. E., Endicott, J. A., Johnson, L. N. (2004). Protein kinase inhibitors: Insights into drug design from structure. *Science*, 303, 1800–1805.

274. Blencke, S., Zech, B., Engkvist, O., Greff, Z., Orfi, L., Keri, G., Ullrich, A., Daub, H. (2004). Characterization of a conserved structural determinant controlling protein kinase sensitivity to selective inhibitors. *Chem. Biol.*, 11, 691–701.

275. Blencke, S., Ullrich, A., Daub, H. (2003). Mutation of threonine 766 in the epidermal growth factor receptor reveals a hotspot for resistance formation against selective tyrosine kinase inhibitors. *J. Biol. Chem.*, 278, 15435–15440.

276. Wisniewski, D., Lambek, C. L., Liu, C., Strife, A., Veach, D. R., Nagar, B., Young, M. A., Schindler, T., Bornmann, W. G., Bertino, J. R., Kuriyan, J., Clarkson, B. (2002). Characterization of potent inhibitors of the Bcr-Abl and the c-kit receptor tyrosine kinases. *Cancer Res.*, 62, 4244–4255.

277. Dalgarno, D. C., Botfield, M. C., Rickles, R. J. (1998). SH3 domains and drug design: Ligands, structure, and biological function. *Biopolymers*, 43, 383–400.

278. Kuriyan, J., Cowburn, D. (1997). Molecular peptide recognition domains in eukaryotic signaling. *Annu. Rev. Biophys. Biomol. Struct.*, 26, 259–288.

279. Gilmer, T., Rodriguez, M., Jordan, S., Crosby, R., Alligood, K., Green, M., Kimery, M., Wagner, C., Kinder, D., Charifson, P. (1994). Peptide inhibitors of src SH3-SH2-phosphoprotein interactions. *J. Biol. Chem.*, 269, 31711–31719.

280. Waksman, G., Kominos, D., Robertson, S. C., Pant, N., Baltimore, D., Birge, R. B., Cowburn, D., Hanfusa, H., Majer, B. J., Overduin, M., Resh, M. D., Rios, C. B., Silverman, L., Kuriyan, J. (1992). Crystal structure of the phosphotyrosine recognition domain SH2 of v-src complexed with tyrosine-phosphorylated peptides. *Nature*, 358, 646–653.

281. Waksman, G., Shoelson, S. E., Pant, N., Cowburn, D., Kuriyan, J. (1993). Binding of a high affinity phosphotyrosyl peptide to the Src SH2 domain: Crystal structures of the complexed and peptide-free forms. *Cell*, 72, 779–789.

282. Eck, M. J., Shoelson, S. E., Harrison, S. C. (1993). Recognition of a high-affinity phosphotyrosyl peptide by the Src homology-2 domain of p56lck. *Nature*, 362, 87–91.

283. Grucza, R. A., Bradshaw, J. M., Futterer, K., Waksman, G. (1999). SH2 domains: From structure to energetics, a dual approach to the study of structure-function relationships. *Med. Res. Rev.*, *19*, 273–299.

284. Bradshaw, J. M., Mitaxov, V., Waksman, G. (1999). Investigation of phosphotyrosine recognition by the SH2 domain of the Src kinase. *J. Mol. Biol.*, *293*, 971–985.

285. Park, J., Fu, H., Pei, D. (2003). Peptidyl aldehydes as reversible covalent inhibitors of Src homology 2 domains. *Biochemistry*, *42*, 5159–5167.

286. Marseigne, I., Roques, B. P. (1988). Synthesis of new amino acids mimicking sulfated and phosphorylated tyrosine residues. *J. Org. Chem.*, *53*, 3621–3624.

287. Burke, T. R., Jr., Russ, P., Lim, B. (1991). Preparation of 4-[bis(tert-butoxy)phosphorylmethyl]-*N*-(fluoren-9-ylmethoxycarbonyl)-DL-phenylalanine. A hydrolytically stable analog of *O*-phosphotyrosine potentially suitable for peptide synthesis. *Synthesis*, *11*, 1019–1020.

288. Shoelson, S. E., Chatterjee, S., Chaudhuri, M., Burke, T. R., Jr. (1991). Solid-phase synthesis of nonhydrolyzable phosphotyrosyl peptide analogs with Nα-Fmoc-(*O*,*O*-di-tert-butyl)phosphono-*p*-methylphenylalanine. *Tet. Lett.*, *32*, 6061–6064.

289. Smyth, M. S., Ford, H., Jr., Burke, T. R. Jr. (1992). A general method for the preparation of benzylic α,α-difluorophosphonic acids; non-hydrolyzable mimetics of phosphotyrosine. *Tetrahedr. Lett.*, *33*, 4137–4140.

290. Burke, T. R., Jr., Smyth, M. S., Nomizu, M., Otaka, A., Roller, P. R. (1993). Preparation of fluoro- and hydroxy-4-(phosphonomethyl)-D,L-phenylalanine suitably protected for solid-phase synthesis of peptides containing hydrolytically stable analogs of *O*-phosphotyrosine. *J. Org. Chem.*, *58*, 1336–1340.

291. Burke, T. R., Jr, Smyth, M. S., Otaka, A., Nomizu, M., Roller, P. P., Wolf, G., Case, R., Shoelson, S. E. (1994). Nonhydrolyzable phosphotyrosyl mimetics for the preparation of phosphatase-resistant SH2 domain inhibitors. *Biochemistry*, *33*, 6490–6494.

292. Yao, Z. J., King, C. R., Cao, T., Kelley, J., Milne, G. W. A., Voigt, J. H., Burke, T. R. (1999). Potent inhibition of Grb2 SH2 domain binding by non-phosphate-containing ligands. *J. Med. Chem.*, *42*, 25–35.

293. Shahripour, A., Plummer, M. S., Lunney, E. A., Para, K. S., Stankovic, C. J., Rubin, J. R., Humblet, C., Fergus, J. H., Marks, J. S., Herrera, R., Hubbell, S. E., Saltiel, A. R., Sawyer, T. K. (1996). Novel phosphotyrosine mimetics in the design of peptide ligands for pp60(src) SH2 domain. *Bioorg. Med. Chem. Lett.*, *6*, 1209–1214.

294. Sundaramoorthi, R., Kawahata, N., Yang, M. G., Shakespeare, W. C., Metcalf, C. A. III, Wang, Y., Merry, T., Eyermann, C. J., Bohacek, R. S., Narula, S., Dalgarno, D. C., Sawyer, T. K. (2003). Structure-based design of novel nonpeptide inhibitors of the Src SH2 domain: Phosphotyrosine mimetics exploiting multifunctional group replacement chemistry. *Biopolymers*, *71*, 717–729.

295. Ye, B., Akamatsu, M., Shoelson, S. E., Wolf, G., Giorgetti-Peraldi, S., Yan, X., Roller, P. P., Burke, T. R., Jr. (1995). L-*O*-(2-Malonyl)tyrosine: A new phosphotyrosyl mimetic for the preparation of Src homology 2 domain inhibitory peptides. *J. Med. Chem.*, *38*, 4270–4275.

296. Burke, T. R., Yao, Z. J., Zhao, H., Milne, G. W. A., Wu, L., Zhang, Z. Y., Voigt, J. H. (1998). Enantioselective synthesis of nonphosphorus-containing phosphotyrosyl mimetics and their use in the preparation of tyrosine phosphatase inhibitory peptides. *Tetrahedron*, *54*, 9981–9994.

297. Gao, Y., Burke, T. R. (2000). Stereoselective preparation of L-4-(2′-malonyl)phenylalanine suitably protected for Fmoc-based synthesis of potent signal transduction inhibitory ligands. *Synlett*, *1*, 134–136.

298. Tong, L., Warren, T. C., Lukas, S., Schembri-King, J., Betageri, R., Proudfoot, J. R., Jakes, S. (1998). Carboxymethyl-phenylalanine as a replacement for phosphotyrosine in SH2 domain binding. *J. Biol. Chem.*, *273*, 20238–20242.

299. Nam, N. H., Pitts, R., Sun, G., Sardari, S., Tiemo, A., Xie, M., Yan, B., Parang, K. (2004). Design of tetrapeptide ligands as inhibitors of the Src SH2 domain. *Bioorg. Med. Chem.*, *12*, 779–787.

300. Nam, N. H., Ye, G., Sun, G., Parang, K. (2004). Conformationally constrained peptide analogues of pTyr-Glu-Glu-Ile as inhibitors of the Src SH2 domain binding. *J. Med. Chem.*, *47*, 3131–3141.

301. Alfaro-Lopez, J., Yuan, W., Phan, B., Kamath, J., Lou, Q., Lam, K. S., Hruby, V. J. (1998). Discovery of a novel series of potent and selective substrate-based inhibitors of p60^{c-src} protein tyrosine kinase: Conformational and topographical constraints in peptide design. *J. Med. Chem.*, *41*, 2252–2260.

302. Lou, Q., Leftwich, M. E., McKay, R. T., Salmon, S. E., Rychetsky, L., Lam, K. S. (1997). Potent pseudosubstrate-based peptide inhibitors for p60^{c-src} protein tyrosine kinase. *Cancer Res.*, *57*, 1877–1881.

303. Kamath, J. R., Liu, R., Enstrom, A. M., Lou, Q., Lam, K. S. (2003). Development and characterization of potent and specific peptide inhibitors of p60^{c-src} protein tyrosine kinase using pseudosubstrate-based inhibitor design approach. *J. Peptide Res.*, *62*, 260–268.

304. Metcalf, C. A. III, van Schravendijk, M. R., Dalgarno, D. C., Sawyer, T. K. (2002). Targeting protein kinases for bone disease: Discovery and development of Src inhibitors. *Curr. Pharm. Des.*, *8*, 2049–2075.

305. Sundaramoorthi, R., Shakespeare, W. C., Keenan, T. P., Metcalf, III C. A., Wang, Y., Mani, U., Taylor, M., Liu, S., Bohacek, R. S., Narula, S. S., Dalgarno, D. C., Van Schravandijk, M. R., Violette, S. M., Liou, S., Adams, S., Ram, M. K., Keats, J. A., Weigele, M., Sawyer, T. K. (2003). Bone-targeted Src kinase inhibitors: Novel pyrrolo- and pyrazolopyrimidine analogues. *Bioorg. Med. Chem. Lett.*, *13*, 3063–3066.

306. Sawyer, T. K., Bohacek, R. S., Eyermann, C. J., Kawahata, N., Metcalf, III, C. A., Shakespeare, W. C., Sundaramoorthi, R., Wang, Y., Yang, M. G. (2002). *Mini-Rev. Med. Chem.*, *2*, 475–489.

307. Uehara, Y., Hori, M., Takeuchi, T., Umezawa, H. (1986). Phenotypic change from transformed to normal induced by benzoquinonoid ansamycins accompanies inactivation of p60src in rat kidney cells infected with Rous sarcoma virus. *Mol. Cell. Biol.*, *6*, 2198–2206.

308. Murakami, Y., Mizuno, S., Hori, M., Uehara, Y. (1988). Reversal of transformed phenotypes by herbimycin A in *src* oncogene expressed rat fibroblasts. *Cancer Res.*, *48*, 1587–1590.

309. Kwon, H. J., Yoshida, M., Fukui, Y., Horinouchi, S., Beppu, T. (1992). Potent and specific inhibition of p60 v-*src* protein kinase both *in vivo* and *in vitro* by radicicol. *Cancer Res.*, 52, 6926–6930.

310. Bishop, A. C., Shokat, K. M. (1999). Acquisition of inhibitor-sensitive protein kinases through protein design. *Pharmacol. Ther.*, 82, 337–346.

311. Blake, R. A., Broome, M. A., Liu, X., Wu, J., Gishizky, M., Sun, L., Courtneidge, S. A. (2000). SU6656, a selective src family kinase inhibitor, used to probe growth factor signaling. *Mol. Cell Biol.*, 20, 9018–9027.

312. Tatoon, L., Molrey, G. M., Chopra, R., Khwaja, A. (2003). The Src-selective kinase inhibitor PP1 also inhibits kit and Bcr-Abl tyrosine kinases. *J. Biol. Chem.*, 278, 4847–4853.

313. Dorsey, J. F., Jove, R., Kraker, A. J., Wu, J. (2000). The pyrido[2,3-*d*-]pyrimidine derivative PD180970 inhibits p210Bcr-Abl tyrosine kinase and induces apoptosis of K562 leukemic cells. *Cancer Res.*, 60, 3127–3131.

314. Faltynek, C. R., Schroeder, J., Mauvais, P., Miller, D., Wang, S., Murphy, D., Lehr, R., Kelley, M., Maycock, A., Michne, W., Miski, M., Thunberg, A. L. (1995). Damnacanthal is a highly potent, selective inhibitor of p56lck tyrosine kinase activity. *Biochemistry*, 34, 12404–12410.

315. Huang, C. K., Wu, F. Y., Ai, Y. X. (1995). Polyhydroxylated 3-(*N*-phenyl) carbamoyl-2-iminochromene derivatives as potent inhibitors of tyrosine kinase p60^{c-src}. *Bioorg. Med. Chem. Lett.*, 5, 2423–2428.

316. Vermeulen, K. D., Van Bockstaele, R., Berneman, Z. N. (2003). The cell cycle: A review of regulation, deregulation and therapeutic targets in cancer. *Cell Prolif.*, 36, 131–149.

317. Senderowicz, A. M. (2002). Cyclin-dependent kinases as targets for cancer therapy. *Cancer Chemother. Biol. Response Modif.*, 20, 169–196.

318. Dai, Y., Grant, S. (2003). Cyclin-dependent kinase inhibitors. *Curr. Opin. Pharmacol.*, 3, 362–370.

319. Anchan, R. M., Reh, T. A., Angello, J., Balliet, A., Walker, M. (1991). EGF and TGF-alpha stimulate retinal neuroepithelial cell proliferation in vitro. *Neuron*, 6, 923–936.

320. Han, J., Lee J. D., Bibbs, L., Ulevitch, R. J. (1994). A MAP kinase targeted by endotoxin and hyperosmolarity in mammalian cells. *Science*, 265, 808–811.

321. Jiang, Y., Chen, C., Li, Z., Guo, W., Gegner, J. A., Lin, S., Han, J. (1996). Characterization of the structure and function of a new mitogen-activated protein kinase (p38beta). *J. Biol. Chem.*, 271, 17920–17926.

322. Li, Z., Jiang, Y., Ulevitch, R. J., Han, J. (1996). The primary structure of p38 gamma: A new member of p38 group of MAP kinases. *Biochem. Biophys. Res. Commun.*, 228, 334–340.

323. Gallagher, T. F., Seibel, G. L., Kassis, S., Laydon, J. T., Blumenthal, M. J., Lee, J. C., Lee, D., Boehm, J. C., Fier-Thompson, S. M., Abt, J. W., Soreson, M. E., Smietana, J. M., Hall, R. F., Garigipati, R. S., Bender, P. E., Erhard, K. F., Krog, A. J., Hofmann, G. A., Sheldrake, P. L., McDonnell, P. C., Kumar, S., Young, P. R., Adams, J. L. (1997). Regulation of stress-induced cytokine production by pyridinylimidazoles; inhibition of CSBP kinase. *Bioorg. Med. Chem.*, 5, 49–64.

REFERENCES

324. Adams, J. L., Boehm, J. C., Kassis, S., Gorycki, P. D., Webb, E. F., Hall, R., Sorenson, M., Lee, J. C., Ayrton, A., Griswold, D. E., Gallagher, T. F. (1998). Pyrimidinylimidazole inhibitors of CSBP/p38 kinase demonstrating decreased inhibition of hepatic cytochrome P450 enzymes. *Bioorg. Med. Chem. Lett.*, 8, 3111–3116.
325. de Laszlo, S. E., Visco, D., Agarwal, L., Chang, L., Chin, J., Croft, G., Forsyth, A., Fletcher, D., Frantz, B., Hacker, C., Hanlon, W., Harper, C., Kostura, M., Li, B., Luell, S., MacCoss, M., Mantlo, N., O'Neill, E. A., Orevillo, C., Pang, M., Parsons, J., Rolando, A., Sahly, Y., Sidler, K., O'Keefe, S. J. (1998). Pyrroles and other heterocycles as inhibitors of p38 kinase. *Bioorg. Med. Chem. Lett.*, 8, 2689–2694.
326. Revesz, L., Di Padova, F. E., Buhl, T., Feifel, R., Gram, H., Hiestand, P., Manning, U., Zimmerlin, A. G. (2000). SAR of 4-hydroxypiperidine and hydroxyalkyl substituted heterocycles as novel p38 map kinase inhibitors. *Bioorg. Med. Chem. Lett.*, 10, 1261–1264.
327. Fijen, J. W., Zijlstra, J. G., De Boer, P., Spanjersberg, R., Tervaert, J. W., Van Der Werf, T. S., Ligtenberg, J. J., Tulleken, J. E. (2001). Suppression of the clinical and cytokine response to endotoxin by RWJ-67657, a p38 mitogen-activated proteinkinase inhibitor, in healthy human volunteers. *Clin. Exp. Immunol.*, 124, 16–20.
328. Wadsworth, S. A., Cavender, D. E., Beers, S. A., Lalan, P., Schafer, P. H., Malloy, E. A., Wu, W., Fahmy, B., Olini, G. C., Davis, J. E., Pellegrino-Gensey, J. L., Wachter, M. P., Siekierka, J. J. (1999). RWJ 67657, a potent, orally active inhibitor of p38 mitogen-activated protein kinase. *J. Pharmacol. Exp. Ther.*, 291, 680–687.
329. Collis, A. J., Foster, M. L., Halley, F., Maslen, C., McLay, I. M., Page, K. M., Redford, E. J., Souness, J. E., Wilsher, N. E. (2001). RPR203494 a pyrimidine analogue of the p38 inhibitor RPR200765A with an improved in vitro potency. *Bioorg. Med. Chem. Lett.*, 11, 693–696.
330. Mclay, L. M., Halley, F., Souness, J. E., McKenna, J., Benning, V., Birrell, M., Burton, B., Belvisi, M., Collis, A., Constan, A., Foster, M., Hele, D., Jayyosi, Z., Kelley, M., Maslen, C., Miller, G., Ouldelhkim, M. C., Page, K., Phipps, S., Pollock, K., Porter, B., Ratcliffe, A. J., Redford, E. J., Webber, S., Slater, B., Thybaud, V., Wilsher, N. (2001). The discovery of RPR 200765A, a p38 MAP kinase inhibitor displaying a good oral antiarthritic efficacy. *Bioorg. Med. Chem.*, 9, 537–554.
331. Ge, B., Gram, H., Di Padova, F., Huang, B., New, L., Ulevitch, R. J., Luo, Y., Han, J. (2002). MAPKK-independent activation of p38alpha mediated by TAB1-dependent autophosphorylation of p38alpha. *Science*, 295, 1291–1294.
332. Lee, J. C., Laydon, J. T., McDonnell, P. C., Gallagher, T. F., Kumar, S., Green, D., McNulty, D., Blumenthal, M. J., Heys, J. R., Landvatter, S. W., Strickler, J. E., McLaughlin, M. M., Siemens, I. R., Fisher, S. M., Livi, G. P., White, J. R., Adams, J. L., Young, P. R. (1994). A protein kinase involved in the regulation of inflammatory cytokine biosynthesis. *Nature*, 372, 739–746.
333. Purves, T., Middlemas, A., Agthong, S., Jude, E. B., Boulton, A. J., Fernyhough, P., Tomlinson, D. R. (2001). A role for mitogen-activated protein kinases in the etiology of diabetic neuropathy. *FASEB J.*, 15, 2508–2514.
334. Igarashi, M., Wakasaki, H., Takahara, N., Ishii, H., Jiang, Z. Y., Yamauchi, T., Kuboki, K., Meier, M., Rhodes, C. J., King, G. L. (1999). Glucose or diabetes activates p38 mitogen-activated protein kinase via different pathways. *J. Clin. Invest.*, 103, 185–195.

335. Manson, M. M., Holloway, K. A., Howells, L. M., Hudson, E. A., Plummer, S. M., Squires, M. S., Prigent, S. A. (2000). Modulation of signal-transduction pathways by chemopreventive agents. *Biochem. Soc. Trans.*, 28, 7–12.

336. Ding, Q., Wang, Q., Evers, B. M. (2001). Alterations of MAPK activities associated with intestinal cell differentiation. *Biochem. Biophys. Soc. Trans.*, 284, 282–288.

337. Andreakos, E. (2003). Targeting cytokines in autoimmunity: New approaches, new promise. *Expert Opin. Biol. Ther.*, 3, 435–447.

338. Branger, J., van den Blink, B., Weijer, S., Gupta, A., van Deventer, S. J., Hack, C. E., Peppelenbosch, M. P., van der Poll, T. (2003). Inhibition of coagulation, fibrinolysis, and endothelial cell activation by a p38 mitogen-activated protein kinase inhibitor during human endotoxemia. *Blood*, 101, 4446–4448.

339. Jackson, P. F., Bullington, J. L. (2002). Pyridinylimidazole based p38 MAP kinase inhibitors. *Curr. Topics Med. Chem.*, 2, 1011–1020.

340. Cirillo, P. F., Pargellis, C., Regan, J. (2002). The non-diaryl heterocycle classes of p38 MAP kinase inhibitors. *Curr. Topics Med. Chem.*, 2, 1021–1035.

341. Regan, J., Capolino, A., Cirillo, P. F., Gilmore, T., Graham, A. G., Hickey, E., Kroe, R. R., Madwed, J., Moriak, M., Nelson, R., Pargellis, C. A., Swinamer, A., Torcellini, C., Tsang, M., Moss, N. (2003). Structure-activity relationships of the p38alpha MAP kinase inhibitor 1-(5-tert-butyl-2-*p*-tolyl-2*H*-pyrazol-3-yl)-3-[4-(2-morpholin-4-yl-ethoxy)naphthalen-1-yl]urea (BIRB 796). *J. Med. Chem.*, 46, 4676–4686.

342. Barone, F. C., Irving, E. A., Ray, A. M., Lee, J. C., Kassis, S., Kumar, S., Badger, A. M., Legos, J. J., Erhardt, J. A., Ohlstein, E. H., Hunter, A. J., Harrison, D. C., Philpott, K., Smith, B. R., Adams, J. L., Parsons, A. A. (2001). Inhibition of p38 mitogen-activated protein kinase provides neuroprotection in cerebral focal ischemia. *Med. Res. Rev.*, 21, 129–145.

343. Hardy, L. W., Malikayil, A. (December 2003). *The Impact of Structure-Guided Drug Design on Clinical Agents.* www.currentdrugdiscovery.com, 15.

344. Cuenda, A., Rouse, J., Doza, Y. N., Meier, R., Cohen, P., Gallagher, T. F., Young, P. R., Lee, J. C. (1995). SB 203580 is a specific inhibitor of a MAP kinase homologue which is stimulated by cellular stresses and interleukin-1. *FEBS Lett.*, 364, 229–233.

345. Gao, F., Yue, T. L., Shi, D. W., Christopher, T. A., Lopez, B. L., Ohlstein, E. H., Barone, F. C., Ma, X. L. (2002). p38 MAPK inhibition reduces myocardial reperfusion injury via inhibition of endothelial adhesion molecule expression and blockade of PMN accumulation. *Cardiovasc. Res.*, 53, 414–422.

346. Harper, S. J., LoGrasso, P. (2001). Inhibitors of the JNK signaling pathway. *Drugs Future*, 26, 957–973.

347. Bennett, B., Sasaki, D., Murray, B., O'leary, E., Sakata, S., Xu, W., Leisten, J. C., Motiwala, A., Pierce, S., Satoh, Y., Bhagwat, S. S., Manning, A. M., Anderson, D. W. (2001). SP600125, an anthrapyrazolone inhibitor of Jun N-terminal kinase. *Proc. Natl. Acad. Sci. USA*, 98, 13681–13686.

348. Schnabl, B., Bradham, C. A., Bennett, B. L., Manning, A. M., Stefanovic, B., Brenner, D. A. (2001). TAK1/JNK and p38 have opposite effects on rat hepatic stellate cells. *Hepatology*, 34, 953–963.

349. Goss, G. G., Jiang, L., Vandorpe, D. H., Kieller, D., Chernova, M. N., Robertson, M., Alper, S. L. (2001). Role of JNK in hypertonic activation of Cl⁻-dependent Na⁺/H⁺ exchange in *Xenopus* oocytes. *Am. J. Physiol.*, *281*, 1978–1990.

350. Rubinfeld, H., Seger, R. (2004). The ERK cascade as a prototype of MAPK signaling pathways. *Methods Mol. Biol.*, *250*, 1–28.

351. Brown, E. J., Beal, P. A., Keith, C. T., Chen, J., Shin, T. B., Schreiber, S. L. (1995). Control of p70 s6 kinase by kinase activity of FRAP in vivo. *Nature*, *377*, 441–446.

352. Kimball, S. D., Webster, K. R. (2001). Cell cycle kinases and checkpoint regulation in cancer. In A. M. Doherty (Ed.), *Annual Reports in Medicinal Chemistry*. Academic, San Diego, Vol. 36, pp. 139–148.

353. Webster K. R., Kimball S. D. (2000). Novel drugs targeting the cell cycle. *Emerg. Drugs*, *5*, 45–59.

354. Ikuta, M., Kamata, K., Fukasawa, K., Honma, T., Machida, T., Hirai, H., Suzuki-Takahashi, I., Hayama, T., Nishimura, S. (2001). Crystallographic approach to identification of cyclin-dependent kinase 4 (CDK4)-specific inhibitors by using CDK4 mimic CDK2 protein. *J. Biol. Chem.*, *276*, 27548–27554.

355. Carini, D. J., Kaltenbach, R. F., Liu, J., Benfield, P. A., Boylan, J., Boisclair, M., Brizuela, L., Burton, C. R., Cox, S., Grafstrom, R., Harrison, B. A., Harrison, K., Akamike, E., Markwalder, J. A., Nakano, Y., Seitz, S. P., Sharp, D. M., Trainor, G. L., Sielecki, T. M. (2001). Identification of selective inhibitors of cyclin dependent kinase 4. *Bioorg. Med. Chem. Lett.*, *11*, 2209–2211.

356. Honma, T., Hayashi, K., Aoyama, T., Hashimoto, N., Machida, T., Fukasawa, K., Iwama, T., Ikeura, C., Ikuta, M., Suzuki-Takahashi, I., Iwasawa, Y., Hayama, T., Nishimura, S., Morishima, H. (2001). Structure-based generation of a new class of potent Cdk4 inhibitors: New de novo design strategy and library design. *J. Med. Chem.*, *44*, 4615–4627.

357. Honma, T., Yoshizumi, T., Hashimoto, N., Hayashi, K., Kawanishi, N., Fukasawa, K., Takaki, T., Ikeura, C., Ikuta, M., Suzuki-Takahashi, I., Hayama, T., Nishimura, S., Morishima, H. (2001). A novel approach for the development of selective Cdk4 inhibitors: Library design based on locations of Cdk4 specific amino acid residues. *J. Med. Chem.*, *44*, 4628–4640.

358. Fry, D. W., Bedford, D. C., Harvey, P. H., Fritsch, A., Keller, P. R., Wu, Z., Dobrusin, E., Leopold, W. R., Fattaey, A., Garrett, M. D. (2001). Cell cycle and biochemical effects of PD 0183812. A potent inhibitor of the cyclin D-dependent kinases CDK4 and CDK6. *J. Biol. Chem.*, *276*, 16617–16623.

359. Soni, R., Muller, L., Furet, P., Schoepfer, J., Stephan, C., Zumstein-Mecker, S., Fretz, H., Chaudhuri, B. (2000). Inhibition of cyclin-dependent kinase 4 (Cdk4) by fascaplysin, a marine natural product. *Biochem. Biophys. Res. Commun.*, *275*, 877–884.

360. Soni, R., O'Reilly, T., Furet, P., Muller, L., Stephan, C., Zumstein-Mecker, S., Fretz, H., Fabbro, D., Chaudhuri, B. J. (2001). Selective in vivo and in vitro effects of a small molecule inhibitor of cyclin-dependent kinase 4. *Natl. Cancer Inst.*, *93*, 436–446.

361. Knockaert, M., Greengard, P., Meijer, L. (2002). Pharmacological inhibitors of cyclin-dependent kinases. *Trends Pharmacol. Sci.*, *23*, 417–425.

362. Jeong, H. W., Kim, M. R., Son, K. H., Han, M. Y., Ha, J. H., Garnier, M., Meijer, L., Kwon, B. M. (2000). Cinnamaldehydes inhibit cyclin dependent kinase 4/cyclin D1. *Bioorg. Med. Chem. Lett.*, *10*, 1819–1822.

363. Meijer, L., Leclerc, S., Leost, M. (1999). Properties and potential-applications of chemical inhibitors of cyclin-dependent kinases. *Pharmacol. Ther.*, *82*, 279–284.

364. Fischer, P. M., Gianella-Borradori, A. (2003). CDK inhibitors in clinical development for the treatment of cancer. *Exp. Opin. Invest. Drugs*, *12*, 955–970.

365. Zhai, S., Senderowicz, A. M., Sausville, E. A., Figg, W. D. (2002). Flavopiridol, a novel cyclin-dependent kinase inhibitor, in clinical development. *Ann. Pharmacother.*, *36*, 905–911.

366. Leclerc, S., Garnier, M., Hoessel, R., Marko, D., Bibb, J. A., Snyder, G. L., Greengard, P., Biernat, J., Wu, Y. Z., Mandelkow, E. M., Eisenbrand, G., Meijer, L. (2001). Indirubins inhibit glycogen synthase kinase-3 beta and CDK5/p25, two protein kinases involved in abnormal tau phosphorylation in Alzheimer's disease. A property common to most cyclin-dependent kinase inhibitors? *J. Biol. Chem.*, *276*, 251–260.

367. Meijer, L., Raymond, E. (2003). Roscovitine and other purines as kinase inhibitors. From starfish oocytes to clinical trials. *Acc. Chem. Res.*, *36*, 417–425.

368. Vesely, J., Havlicek, L., Strnad, M., Blow, J. J., Donella-Deana, A., Pinna, L., Letham, D. S., Kato, J., Detivaud, L., Leclerc, S. (1994). Inhibition of cyclin-dependent kinases by purine analogues. *Eur. J. Biochem.*, *224*, 771–786.

369. Meijer, L., Borgne, A., Mulner, O., Chong, J. P., Blow, J. J., Inagaki, N., Inagaki, M., Delcros, J. G., Moulinoux, J. P. (1997). Biochemical and cellular effects of Roscovitine, a potent and selective inhibitor of the cyclin-dependent kinases cdc2, cdk2 and cdk5. *Eur. J. Biochem.*, *243*, 527–536.

370. Gray, N. S., Wodicka, L., Thunnissen, A. W. H., Norman, T. C., Kwon, S., Espinoza, F. H., Morgan, D. O., Barnes, G., LeClerc, S., Meijer, L. (1998). Exploiting chemical libraries, structure and genomics in the search for kinase inhibitors. *Science*, *281*, 533–538.

371. Senderowicz, A. M. (2003). Novel small molecule cyclin-dependent kinases modulators in human clinical trials. *Cancer Biol. Ther.*, *2*, S84–95.

372. Goekjian, P. G., Jirousek, M. R. (2001). Protein kinase C inhibitors as novel anticancer drugs. *Exp. Opin. Invest. Drugs*, *10*, 2117–2140.

373. Sorbera, L. A., Silvestre, J., Rabasseda, X., Castaner, J. (2000). LY-333531 mesylate hydrate: Symptomatic antidiabetic; protein kinase C inhibitor. *Drugs Future*, *25*, 1017–1026.

374. Way, K. J., Chou, E., King, G. L. (2000). Identification of PKC-isoform-specific biological actions using pharmacological approaches. *Trends Pharmacol. Sci.*, *21*, 181–187.

375. Propper, D. J., McDonald, A. C., Man, A., Thavasu, P., Balkwill, F., Braybrooke, J. P., Caponigro, F., Graf, P., Dutreix, C., Blackie, R., Kaye, S. B., Ganesan, T. S., Talbot, D. C., Harris, A. L., Twelves, C. (2001). Phase I and pharmacokinetic study of PKC412 an inhibitor of protein kinase C. *J. Clin. Oncol.*, *19*, 1485–1492.

376. Fuse, E., Tanii, K., Kurata, N., Kobayashi, H., Shimada, Y., Tamura, T., Tanigawara, Y., Lush, R. D., Headlee, D., Figg, W. D., Arbuck, S. G., Senderowicz, A. M.,

Sauseville, E. A., Akinaga, S., Kuwabara, T., Kobayashi, S. (1998). Unpredicted clinical pharmacology of UCN-01 caused by specific binding to human alpha1-acid glycoprotein. *Cancer Res.*, *58*, 3248–3255.

377. Fuse, E., Tanii, K., Takai, K., Asanome, K., Kurata, N., Kobayashi, H., Kuwabara, T., Kobayashi, S., Sugiyama, Y. (1999). Altered pharmacokinetics of a novel anticancer drug, UCN-01, caused by specific high affinity binding to alpha1-acid glycoprotein in humans. *Cancer Res.*, *58*, 1054–1060.

378. Schaufelberger, D. E., Koleck, M. P., Beutler, J. A., Vatakis, A. M., Atvardo, A. B., Andrews, L. V., Muschick, G. M., Forenza, S. (1991). The large scale isolation of bryostatin 1 from *Bugula nerutina* following current good manufacturing practices. *J. Nat. Prod.*, *54*, 1265–1270.

379. Smith, J. B., Smith, L., Pettit, G. R. (1985). Bryostatins: Potent, new mitogens that mimic phorbol ester tumour promoters. *Biochem. Biophys. Res. Commun.*, *132*, 939–945.

380. Szallasi, Z., Smith, C. B., Pettit, G. R., Blumberg, P. M. (1994). Differential regulation of protein kinase C isozymes by bryostatin 1 and phorbol 12-myristate 13-acetate in NIH 3t3. broblasts. *J. Biol. Chem.*, *269*, 2118–2124.

381. Hocevar, B. A., Fields, A. P. (1991). Selective translocation of βII protein kinase C to the nucleus of human promyelocytic (HL-60) leukemia cells. *J. Cell. Biol.*, *266*, 28–33.

382. Neumunaitis, J., Holmlund, J. T., Kraynak, M., Richards, D., Bruce, J., Ognoskie, N., Kwoh, T. J., Greary, R., Dorr, A., Von Hoff, D., Eckhardt, S. G. (1999). Phase I evaluation of ISIS 3521, an antisense oligodeoxynucleotide to protein kinase C-α, in patients with advanced cancer. *J. Clin. Oncol.*, *17*, 3586–3595.

383. Breitenlechner, C., Gassel, M., Hidaka, H., Kinzel, V., Huber, R., Engh, R. A., Bossemeyer, D. (2003). Protein kinase A in complex with Rho-kinase inhibitors Y-27632, Fasudil, and H-1152P: Structural basis of selectivity. *Struct. (Cambr.)*, *11*, 1595–1607.

384. Walker, E. H., Pacold, M. E., Perisic, O., Stephens, L., Hawkins, P. T., Wymann, M. P., Williams, R. L. (2000). Structural determinants of phosphoinositide 3-kinase inhibition by wortmannin, LY294002, quercetin, myricetin, and staurosporine. *Mol. Cell.*, *6*, 909–919.

385. Lakkakorpi, P. T., Wesolowski, G., Zimolo, Z., Rodan, G. A., Rodan, S. B. (1997). Phosphatidylinositol 3-kinase association with the osteoclast cytoskeleton, and its involvement in osteoclast attachment and spreading. *Exp. Cell Res.*, *237*, 296–306.

386. Hunag, S., Houghton, P. J. (2002). Inhibitors of mammalian target of rapamycin as novel antitumor agents: From bench to clinic. *Curr. Opin. Invest. Drugs*, *3*, 295–304.

387. Podsypanina, K., Lee, R. T., Politis, C., Hennessy, I., Crane, A., Puc, J., Neshat, M., Wang, H., Yang, L., Gibbons, J., Frost, P., Dreisbach, V., Blenis, J., Gaciong, Z., Fisher, P., Sawyers, C., Hedrick-Ellenson, L., Parsons, R. (2001). An inhibitor of mTOR reduces neoplasia and normalizes p70/S6 kinase activity in Pten+/− mice. *Proc. Natl. Acad. Sci. USA*, *98*, 10314–10319.

27

RNA-BASED THERAPIES

STEVE PASCOLO
CureVac, The RNA people GmbH, Paul Ehrlich Strasse 15, 72076 Tuebingen, Germany

27.1	INTRODUCTION: RNA, THE SUPERLATIVE BIOMOLECULE	1260
	From the Isolation of Nuclein to the Chemical and Structural Characterization of Nuclei Acids	1260
	Natural Functions of RNA in Cells	1262
	Potentially Therapeutic Features of RNA	1263
	Advantages of RNA-Based Therapies	1264
	Natural and Modified RNA	1264
27.2	DRUGS BASED ON THE CAPACITY OF RNA TO FORM DEFINED THREE-DIMENSIONAL STRUCTURES: APTAMERS	1266
	Definition of Aptamers	1266
	RNA Mimetics Suitable for Aptamer Function	1270
	Preclinical and Clinical Utilization of Aptamers	1271
	Lead Company: Eyetech Pharmaceutical, Inc.	1271
27.3	SYNTHETIC RNA FOR GENE REGULATION: ANTISENSE, RIBOZYMES, AND SIRNA	1273
	Antisense and Ribozymes	1273
	Small Inhibitory RNA (siRNA)	1278
	Delivery of RNA Molecules for Suppression of Gene Expression	1282
27.4	RNA FOR PROTEIN PRODUCTION: mRNA	1283
	Definition of mRNA	1283
	RNA Mimetics Suitable for mRNA Function	1284
	Preclinical and Clinical Utilization of mRNA	1285
	Lead Company: MERIX Biosciences	1288
27.5	RNA FOR IMMUNOSTIMULATION: DOUBLE-STRANDED AND STABILIZED RNA	1288
	Definition of Immunostimulating RNAs	1288
	Preclinical and Clinical Utilization of Immunostimulating RNA	1289
	Lead Company: Hemispherx Biopharma, Inc.	1290

Drug Discovery Handbook, by Shayne Cox Gad
Copyright © 2005 by John Wiley & Sons, Inc.

27.6	PRODUCTION OF RNA AND LEGAL ISSUES	1290
	Manufacturing of RNA	1290
	Regulatory Aspects (Quality Management)	1294
27.7	CONCLUSION AND PERSPECTIVES: RNA-BASED THERAPIES OF THE FUTURE	1299
	References	1300

27.1 INTRODUCTION: RNA, THE SUPERLATIVE BIOMOLECULE

From the Isolation of Nuclein to the Chemical and Structural Characterization of Nucleic Acids

Living organisms are composed of four different types of organic molecules: sugars, lipids, proteins, and nucleic acids. Nucleic acids were first isolated from the nucleus of human pus cells by Friedrich Miescher (1844–1895) in the kitchen of Tübingen's castle (Germany) in 1868. Friedrich Miescher called this substance *nuclein*. He showed that it exists in all type of cells and found that when prepared from salmon sperm, it was associated to a peptide that he named *protamine*. Further fractionation of nuclein allowed the derivation of a pure product by Richard Altmann (1852–1900) in 1889. Altman named this product *nucleic acid*. Meanwhile, Albrecht Kossel (1853–1927, winner of the Nobel prize for Medicine in 1910) used chemical hydrolysis of nuclein to discover nucleotides (1885–1901): adenine (A) and guanine (G) are purines; thymidine (T), Cytosine (C), and uracile (U) are pyrimidines. Kossel demonstrated that the nucleotides contain pentose. The existence of two different nucleic acids was seen as early as 1910. One (prepared from yeast and plants) was found to be sensitive to alkaline lysis and contain A, C, G, and U (Fig. 27.1a), the other (prepared from thymus), was alkali resistant. Phoebus Levene (1869–1940) a student of Kossel, managed in 1929 to degrade thymus-derived nuclein using enzymes and demonstrated that it consists of A, C, G, and T, a demethylated version of U. Levene and Walter Abraham Jacobs (1883–1967) showed that the pentose of the ACGT nucleic acid, as opposed to that of ACGU nucleic acid, is a 2′-deoxyribose. The ACGT nucleic acid was named deoxyribonuclei-cacid (DNA) and the ACGU nucleic acid ribonucleicacid (RNA). Research by Tornbjörn Caspersson (1911–1998) and Jean Brachet (1909–1998) proved that DNA is contained in chromosomes located in the nucleus and RNA is found principally in the cytosole. In 1953 Jim Watson (1928–) and Francis Crick (1916–2004) demonstrated that nucleotides are complementary: A pairs with T or U and G pairs with C (Fig. 27.1a). Together with Maurice Wilkins (1917–) they received the Nobel prize in 1962 for their resolution of the structure of DNA. By the end of the 1950s, the characteristics of DNA and RNA were well described. Both nucleic acids consist of four nucleotides but have different subcellular localization. Base composition is U

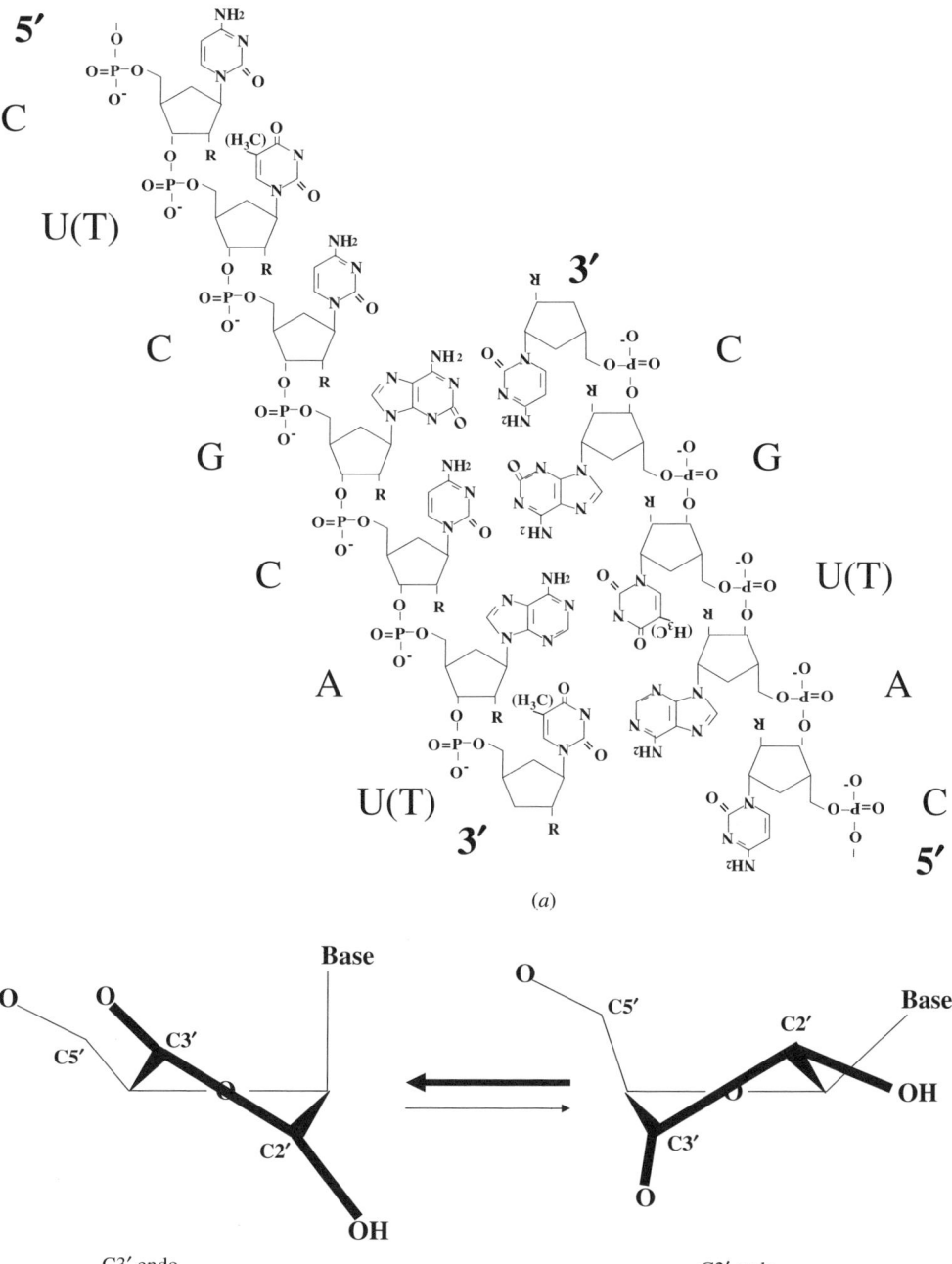

Figure 27.1 Structure of nucleic acids. (*a*) Representation of two hybridized nucleic acid strands: one of the sequence $^{5'}$CUCGCAU$^{3'}$ ($^{5'}$CTCGCAT$^{3'}$ for DNA) and one of the sequence $^{5'}$CAUGC$^{3'}$ ($^{5'}$CATGC$^{3'}$ for DNA). R is H for DNA and OH for RNA. (*b*) Structure of the pentose in nucleic acids. Due to steric effects and electronic bonds, ribonucleotides contain dominantly the C3'-endo form of the ribose.

for RNA and T for DNA (Fig. 27.1a). Chemical characteristics are deoxyribose in DNA, hydroxyribose in RNA, which is consequently sensitive to high pH (Fig. 27.1a). Isomeric structures are furane ring in equilibrium between the 2'-endo and 3'-endo for DNA, although the 3'-endo form is favored in RNA (Fig. 27.1b). Three-dimensional structures are DNA is usually a double-stranded molecule with two antiparallel strands associated by interstrand base pairing, whereas RNA usually consists of one strand where intramolecular base pairing dictates the secondary and tertiary structure of the molecule. It was only at the beginning of the 1960s that the hypothesized roles of RNA molecules in gene expression were being experimentally documented.

Natural Functions of RNA in Cells

At the end of the 1930s, by observing that RNA concentrations are higher in rapidly dividing cells, Casperson and Brachet presented the first experimental evidence that cellular RNA was involved in protein synthesis. The RNA-rich granules (now known as ribosomes and containing ribosomic RNA, rRNA) observed in the cytosole by Albert Claude were hypothesized by Brachet in 1946 to participate in protein synthesis. This theory was confirmed in the 1950s by experimental data coming from electron microscopy [1] and pulse-chase [2] experiments. In 1958, Crick formulated the central dogma of molecular biology [3]: DNA is transcribed into RNA, which is translated into protein. In 1961, Brenner [4] as well as Gros and co-workers [5] described unstable RNA molecules (now known as messenger RNA, mRNA), which carry genetic information from DNA in the nucleus to ribosomes in the cytosole. At the same time, another intermediate in protein synthesis was identified in the form of an aminoacyl-RNA (transfer RNA, tRNA) [6–8]. Thus, just a few years after the publication of "The Central Dogma of Molecular Biology" by Crick [3], all molecules mediating protein synthesis were identified: mRNA, rRNA, and tRNA are different forms of RNA molecules necessary in the process of gene expression.

At the end of the 1960s, another activity of RNA was discovered: double-stranded RNA structures activate the immune system [9]. More than 10 years later, a third function of RNA apart from coding–decoding genetic information and immunostimulation was reported: RNA can catalyze reactions. The discovery by Guerrier-Takada and co-workers [10] and Cech and Bass [11] of catalytic RNA (ribozymes) was acknowledged with a Nobel prize (Chemistry, 1989). Ribozymes can be used to specifically degrade mRNA, thus shutting down the expression of a defined gene. They can be seen as a catalytic version of antisense RNA [12] for inhibition of gene expression. The recent discovery of small inhibitory RNA (siRNA) is a new breakthrough in the world of molecular biology and molecular medicine [13]. Similarly to RNA antisense and ribozymes, siRNAs are molecules that naturally evolved to specifically suppress the synthesis of a defined protein, and they can be used as genetic tools to manipulate gene expression.

On reviewing RNA's natural functions, it appears that RNA is the only organic molecule that has the ability to recapitulate all biological activities necessary for life: containment of genetic information, mediation and regulation of protein synthesis, scaffolding of tri-dimensional structures and enzymatic activities. Accordingly, an "RNA world" was suggested as the start of life 4.2 billion years ago. At that time, self-replicating RNA molecules or RNA complexes may have been the basis of life. They could have evolved toward organisms that contained a new type of macromolecule such as DNA (an RNA derivative that resists basic pH), lipids, and proteins that can potentate the genetic, structural, and enzymatic activities of RNA. During this evolution, RNA remained the central player in all biological processes: RNA is the genome of many viruses [human immunodeficiency virus (HIV), human T-cell lymphotropic virus (HTLV), hepatitis C virus (HCV), Semliki Forest virus, etc.]. In cells it is the obligate transient copy of genes that are sent out of the nucleus (mRNA); it is the main constituent of the whole translation machinery (tRNA and rRNA); it regulates DNA replication (RNA primers for the production of Okazaki fragments); it modulates gene expression (ribozymes, siRNA, antisenses RNA); it is used as a matrix to build new DNA sequences (RNA contained in telomerase complexes); it provides energy (mainly adenine and guanine in their triphosphate form) and signals infection (double-stranded or stabilized RNA).

In addition to these characterized activities, we are still discovering new functions of RNA molecules. For example, microRNAs (miRNA) that are structurally (short ca. 21 bases double-stranded RNA) and functionally (post-transcriptional inhibition of gene expression) similar to siRNA but transcribed from the host instead of being brought by a pathogen are now found to affect the genome [14–16]. In particular, chromatin structure and DNA methylation at the locus of a gene targeted by miRNA is modified in a way that transcription of the precise gene is inhibited. This discovery together with the demonstration that RNA serves as a matrix for DNA synthesis in chromosome's telomers challenge Cricks' central dogma: RNA is not only a product of the genome but also a feedback modulator of DNA. This brings interesting perspectives such as RNA-directed evolution, whereby outside signals would be materialized into special RNA structures (encoded by the host) capable of "educating" the genome (modify genes or gene expression). Such an RNA-mediated evolution of genomes would be a new function of RNA molecules.

Potentially Therapeutic Features of RNA

All the natural functions of RNA have been explored as the basis for therapeutics. The capacity of RNA to form defined three-dimensional structures is exploited in aptamers (Section 27.2), its functions in gene regulation are tested for the specific control of tumor growth or virus infections (Section 27.3), its protein-encoding feature is used for gene therapy and vaccination (Section 27.4), and its natural capacity to activate the immune system is used in adjuvant

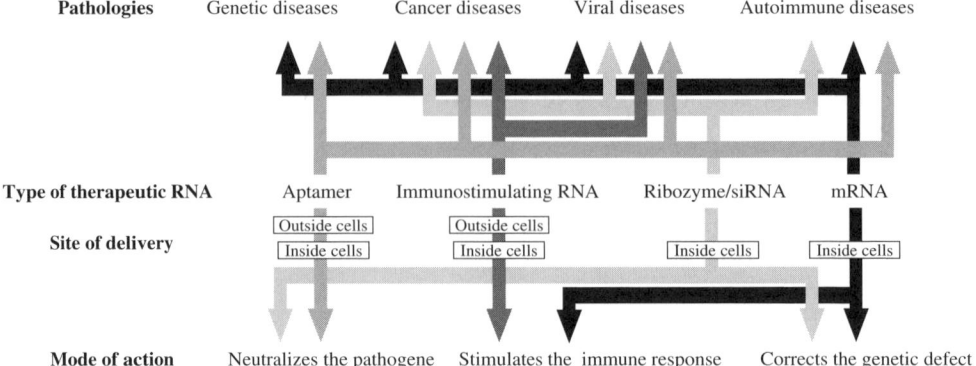

Figure 27.2 Applications of RNA-based therapies. The four types of therapeutic RNA are listed; their mode and site of action is shown. They can neutralize the pathogene (inhibit virus replication, block the activity of a protein, interfere with the growth of tumor cells), stimulate the specific immune response (trigger an antitumor, antiviral, or antibacteria immune response as well as suppressing allergy or other types of autoimmunity), or correct defects due to an inherited genetic mutation (mRNA coding for the wild-type protein in the context of diseases such as cystic fibrosis or siRNA destroying an mRNA coding for an aberrant protein that has a dominant phenotype).

formulations (Section 27.5). As shown in Figure 27.2, aptamers and mRNA approaches can virtually address all kinds of diseases, including pathologies resulting from inherited genetic defects. Gene interference mediated by siRNA and ribozymes as well as immunostimulation by RNA should prove to be especially useful in fights against infectious diseases, allergies, and cancer.

Advantages of RNA-Based Therapies

The main advantages of RNA as an active pharmaceutical ingredient (API) are:

1. Its production and conservation are easy (see Section 27.6).
2. Its half-life can be exactly determined through chemical modifications (see discussion of natural and modified RNA).
3. As opposed to DNA or proteins, RNA is not immunogenic (no specific immune response against RNA was reported).
4. Through its versatile activities, RNA can theoretically be an API against any disease: for correction of pathological genetic defects, prevention or cure of infections, treatment of tumors, therapy against degenerative diseases, and control of allergies.

Natural and Modified RNA

Chemically, RNA is a robust molecule at neutral or acidic pH (the oxygen on the 2' position of the ribose makes it sensitive to basic pH). As opposed to most proteins, lipids, or DNA, RNA molecules can be heated to more than

INTRODUCTION: RNA, THE SUPERLATIVE BIOMOLECULE

95°C, frost-thawed, lyophilized, and resuspended in water solutions as many times as necessary for production, storage, and use.

The main weakness of RNA resides in the ubiquitous presence of ribonucleases (RNases). Many species of RNases are produced by prokaryotic or eukaryotic life [17]. Some RNases are secreted outside the cells; some are resident in the cytosol. The function of secreted RNases is not fully understood but may be a mechanism to prevent infection by pathogens such as virus-containing RNA genomes (influenza, Sindbis, HIV, etc.). Alternatively, it was hypothesized that neighboring cells can communicate through the secretion and uptake of RNA. In this context, ubiquitous RNases are seen as controllers of RNA-based intercell communication. In any case, RNases are abundant in blood, in and on the skin, in cells, as well as in any extracellular spaces, thus rendering RNA-based treatments inefficient compared to those based on DNA.

The principal remedy against RNase activity is the modification of the therapeutic RNA backbone (Fig. 27.3). Several chemical modifications made in synthetic RNA were shown to increase the molecule's half-life in RNase-contaminated milieus. Mainly, replacement of the 2' oxygen by a fluorine atom, an ammonium (NH_2) or azido (N_3) group, replacement of the 2'-OH by O-alkyl (methyl, ethyl, propyl, methoxy-ethoxy) or O-allyl groups, replacement of an oxygen atom in the phosphorus group by a sulfur atom (phosphorothioate backbone), or replacement of the 3'OH by NH (phosphoramidate backbone) are standard chemical modifications that render RNA molecules resistant toward RNases [18]. These modified compounds are grouped under the term *RNA mimetics*. Chemical modifications can interfere with the

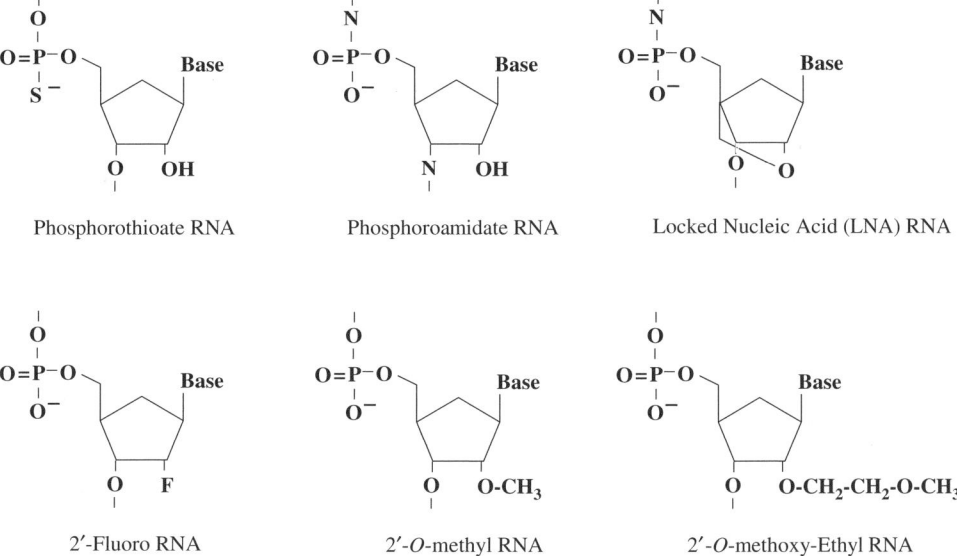

Figure 27.3 Description of the most common modifications introduced in RNA (especially synthetic oligonucleotides) designed for therapeutic use.

expected structural, genetic, immunomodulating or gene expression control activity of the RNA (see discussions of RNA mimetics through out this chapter). Partial chemical modification of the therapeutic RNA where only some residues are modified usually results in a good compromise between enhanced resistance toward RNases and conservation of biological activity.

Although for some of the modifications (2′ fluoro, 2′ amino, 2′ azido), the substitution of the 2′ oxygen by other atoms or chemical groups semantically turns the fully modified RNA mimetics into DNA, the designation of all these derivative as RNA molecules is justified by three characteristics:

1. Sensitivity to RNases that usually persists in spite of being severely reduced.
2. The presence of uridine as one of the bases.
3. The dominant C3′-endo structure of the pentose. The stabilized C3′-endo structure in all RNA and RNA derivatives provides them with an increased base-pairing thermostability in comparison to DNA. Consequently, RNA–RNA or RNA–DNA hybrids are thermodynamically more stable than DNA–DNA hybrids. Recently, a method to fix the optimal C3′-endo isomeric structure of the pentose was reported: A methylene bridge connects the 2′ oxygen to the 4′ carbon of the sugar [19]. Such structurally fixed furanose rings provide the RNA oligonucleotide (termed LNA for locked nucleic acid) with not only extreme resistance toward RNase but also an increased thermostability of base pairing: the melting temperature of LNA–target RNA duplexes is several degrees higher than the melting temperature of natural RNA–RNA duplexes.

The capacity of RNA mimetics to outscore natural RNA in therapeutic uses depends on the influence of the modification(s) on the RNA's function (whether aptamers, ribozymes, siRNA, immunostimulating, or mRNA activity is required), stability, membrane permeation, pharmacokinetic properties, and biodistribution. The design of an optimal RNA mimetic requires intense research efforts but has already resulted in the release of one medicament and several promising new potential drugs that are in clinical trials. The next section will present the different technologies behind RNA-based therapeutics, the optimization of RNA mimetics for each application, the results of in vitro and in vivo preclinical or clinical studies, and a look at one of the leading company that supports the development and commercialization of each therapeutic RNA compound.

27.2 DRUGS BASED ON THE CAPACITY OF RNA TO FORM DEFINED THREE-DIMENSIONAL STRUCTURES: APTAMERS

Definition of Aptamers

Specific interactions between RNA and proteins have been recognized. They largely depend on the secondary and tertiary structure of the partially double-stranded nucleic acid. As exemplified by tRNAs (see review by Wittenhagen and Kelley [20]), small alterations in the RNA structure can abrogate selective association to proteins and hinder the RNA's functions. Many RNA sequences, which fold into defined structures, especially stem-loop structures (a double-stranded stem that holds a single-stranded loop of as little as four and up to several hundreds nucleotides), control important biochemical processes. For example, stability or translation of mRNA by stem-loop structures in the mRNA [21], assembling of ribosomes after precise three-dimensional folding of the ribosomal RNAs [22] or replication of HIV that is dependent on stem-loop structures in some precise parts of the HIV RNA genome [23]. The idea to use RNA molecules, that is, three-dimensional structures generated by the refolding of an RNA sequence into stem-loop domains, as highly selective ligands, functionally similar to antibodies, was turned into a process of evolution (variation, selection, and replication) called SELEX (systematic evolution of ligands by exponential enrichment) by Tuerk and Gold [24]. At the same time, Ellington and Szostak [25] coined the name *aptamer* (from the Latin *aptus*: to fit) for such in vitro selected ligands composed exclusively of nucleic acids. An aptamer is an oligonucleotide that folds into a highly specific and stable three-dimensional structure due to intrastrand hybridizations that defined several stem loops structures of different sizes. The characterization of an aptamer proceeds through two steps represented in Figure 27.4.

Step 1: Selection From a library of random, ca. 30 to 40-mere sequences flanked by defined ca. 20-nucleotide sequences (one being also a promoter recognized by a specific RNA polymerase; Fig. 27.4*a*), oligonucleotides that can adapt a structure that allows them to bind to an immobilized substrate (a protein, a nucleic acid, or any type of chemical compound as well as entire cells [26]) are retained (Fig. 27.4*b*). The RNA oligonucleotides that cannot bind to the immobilized substrate are discarded by washing the solid support with standard salt solutions such as phosphate-buffered saline (PBS). Selectively bound oligonucleotides are eluted from the solid support by using detergents and/or heating and/or chaotropic solutions or releases of the ligand from the support.

Step 2: Amplification The eluted oligonucleotides are amplified by reverse transcription polymerase chain reaction (RT-PCR) thanks to the flanking defined sequences. Subsequently, they are transcribed into RNA and reselected on the immobilized ligand.

Several (usually 10 to 20) rounds of selection/amplification result in enrichment of the random mix with RNA that specifically interact with the immobilized ligand. At the end, these aptamers are reverse transcribed, cloned into a plasmid, and sequenced. Usually, most molecules are found to have related sequences or at least a common core sequence. This core sequence can be produced by in vitro transcription of an engineered plasmid or by chemical synthesis. Its specific binding to the ligand used for selection can be verified and

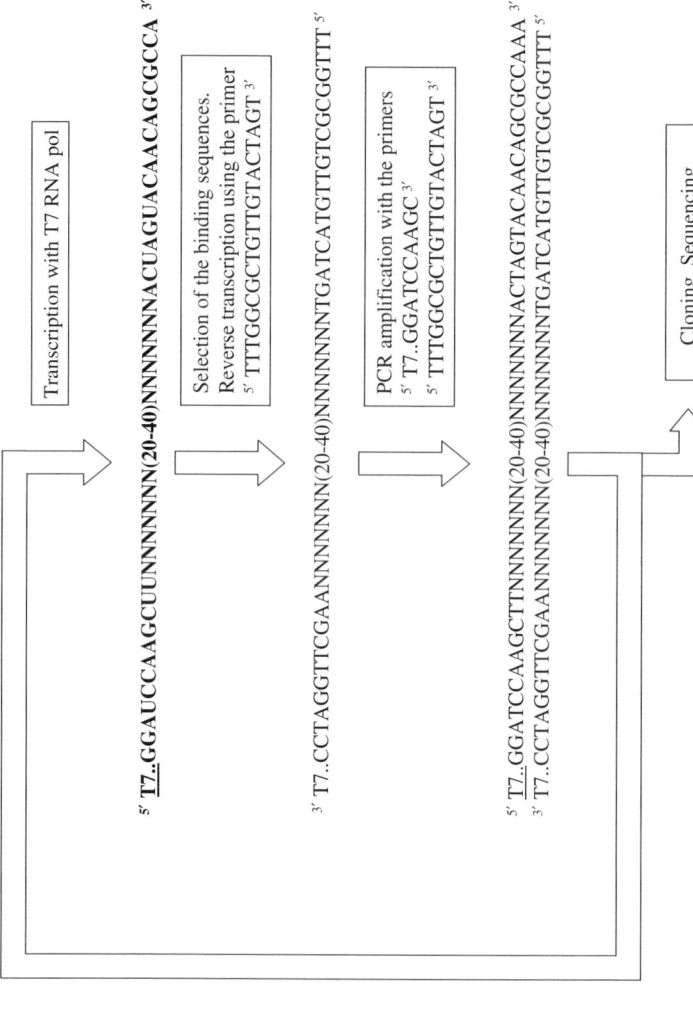

Figure 27.4 Generation of aptamers. A synthetic oligonucleotide containing a T7 promoter sequence (underlined) followed by a random sequence (from 20 to 40 nucleotides in length usually) and a defined ca. 20 residues sequence [shown in bold in (*a*)] is coincubated with an immobilized ligand [shown in (*b*)]. The sequences that can adopt a configuration that allows them to bind to the ligand are retained. They can be eluted, reverse transcribed, and PCR amplified [as shown in (*a*)] before being transcribed. The resulting RNA molecules can be reselected on the immobilized ligand. After several rounds (usually 10 to 20) the eluted, reverse-transcribed, and PCR-amplified molecules can be cloned and sequenced. Synthetic RNA corresponding to these sequences can be produced, stabilized by RNA modifications, and studied in vitro (affinity to their ligand, definition of the minimal active sequences) and *in vivo* (efficiency, biodistribution, half-life).

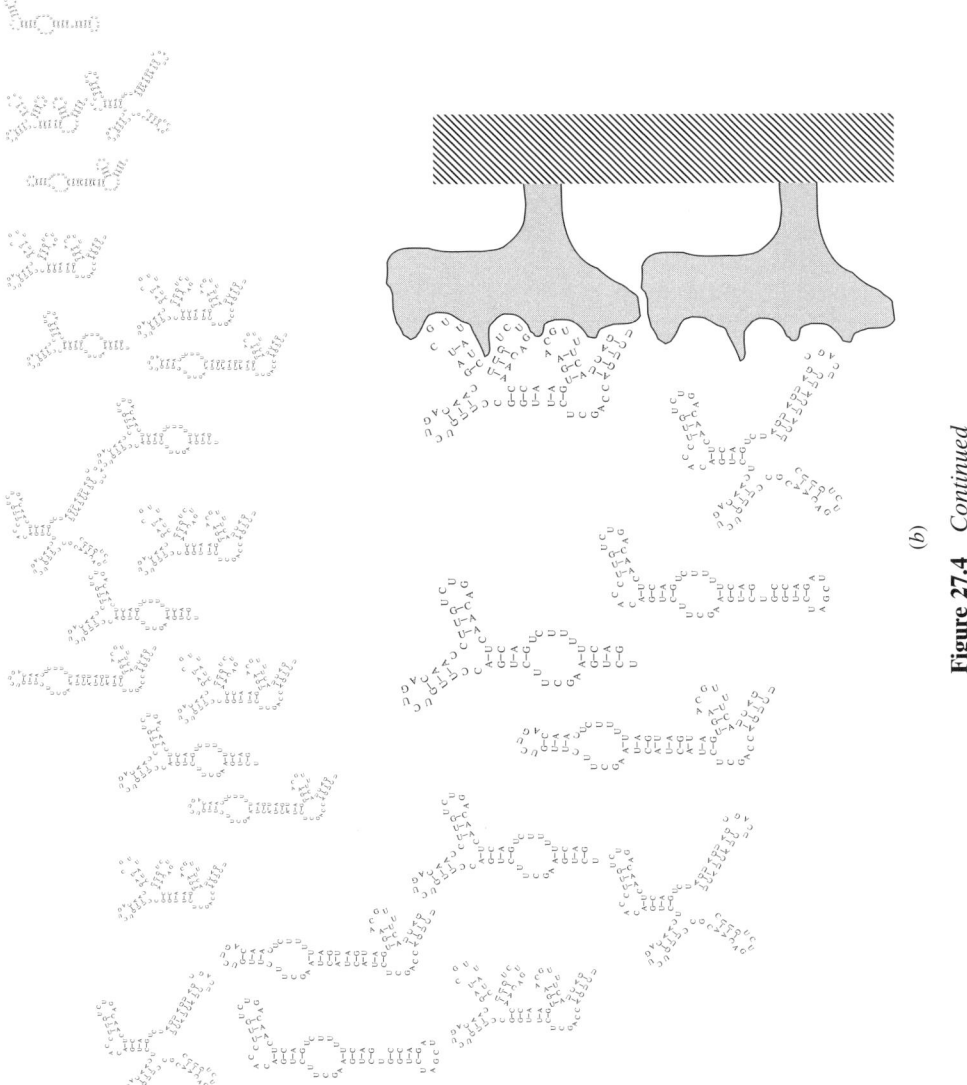

Figure 27.4 *Continued*

further studied by standard biochemical tests [e.g., enzyme-linked immunosorbent assay (ELISA) or Biacore]. Aptamers have affinities for their cognate "antigen" equal or superior to those obtained with antibodies (K_d values in the subnanomolar range). Since the whole process is done in vitro starting from a synthetic library of about 10^{14} different oligonucleotide sequences, aptamers can recognize any kind of ligands, whereas the isolation of high-affinity antibodies may be hindered by tolerance to self (especially for proteins conserved through evolution such as heat shock proteins). The possibility to produce large amounts of aptamers by chemical synthesis is another attractive characteristic of this technology in comparison to antibodies.

Aptamers can be DNA or RNA oligonucleotides. However, thanks to the following properties of RNA:

1. Its natural catabolism
2. The possibility to control its half-life in different physiologic milieus through the introduction of chemical modifications
3. Its lack of immunogenicity
4. The fact that it cannot modify the host genomes
5. The lack of systemic effects after in vivo application (as opposed to DNA oligonucleotides especially if they contain the so-called CpG motifs)

RNA aptamers are the safest version for therapeutic use. A large number of aptamers specific for a broad range of defined ligands have been isolated and characterized. Their inventory is presented in the aptamer database: http://aptamer.icmb.utexas.edu/ [27].

RNA Mimetics Suitable for Aptamer Function

Usually, the SELEX process is performed with 2′fluoro pyrimidine RNA. The T7 RNA polymerase used during the amplification step is permissive to the integration of 2′fluoro UTP and 2′fluoro CTP [28]. Since most natural RNases recognize pyrimidines [29], the 2′fluoropyrimidines RNA are more resistant than natural RNA and are consequently preferred both for the in vitro selection and for direct in vivo application. The modifications used during the selection process may be required not only for the stability but also for the high-affinity binding of the aptamer to its ligand [30]. Indeed, if the 2′fluoros are involved in the determination of the precise three-dimensional structure of the RNA, their replacement with OH or other groups may modify the functional characteristics of the aptamer. Nevertheless, it was shown in several cases that on top or instead of the 2′fluoro substitutions, some modifications such as a phosphoramidate backbone [31], 2′-amino or 2′-O-methyl [32] substitutions as well as LNA nucleotides [33] enhance the stability of aptamers without reducing their function. Increasing the resistance of aptamers toward nucleases by any one of or a combination of the possible modifications is an empirical process and must be individually tested as it may result in a decrease or an increase of the affinity of RNA to its target molecule.

Preclinical and Clinical Utilization of Aptamers

Potentially therapeutic aptamers have been described in diverse medical fields such as vascular biology, immunology, brain disorders, and infectious diseases [34, 35]. The selected molecular target of the aptamer may be outside the cells, for example, an extracellular receptor, a hormone, or a growth factor. In these cases, no transfection reagents are needed for aptamer applications. Of special interest for therapeutics and diagnostic use are the characterized aptamers that recognize HIV gp120 [36], thrombin [37], CTLA-4 [38], the surface protein from trypanosome [39], or prion (the pathogenic form of the prp protein: prp^{sc}) [40]. Each of these has direct pharmaceutical potential.

The first step toward the preclinical utilization of an aptamer is the definition of the minimal active sequence. Due to criteria of high-quality production, aptamers should not be longer than 40 bases (oligonucleotides longer than 40 bases are difficult to produce and to separate from N-1 contaminants). The second step is defining the chemical modifications that will provide the optimal stabilization with no loss of function.

With such minimal and stabilized oligonucleotides, preclinical tests can be conducted to test for efficiency in vivo. Several aptamers passed these tests and are now in human phase I and II trials. One aptamer, the anti-VEGF (vascular endothelial growth factor) pegylated aptamer (Macugen), went successfully through a phase III clinical trial and is now approved as a drug. VEGF known for inducing blood vessel permeability is an extracelluar growth factor required for neovascularisation. VEGF exists in 5 different isoforms. One of them, isoform 165, plays an important role in the abnormal blood vessel growth and leakage associated with several pathologies including the eye disease subfoveal choroidal neovascularization (CNV) secondary to age-related macular degeneration (AMD). This disease can lead to vision loss. In animal models, anti-VEGF agents have inhibited blood vessel formation and leakage. In humans, VEGF concentrations in the eyes correlates with wet AMD or diabetic macular edema (DME). Both AMD and DME are diseases of the retina that can lead to blindness. A 2'-fluoropyrimidine RNA aptamer capable of binding and of specifically inhibiting (blocks binding to its receptor) VEGF isoform 165 was characterized [41]. In a pegylated version [addition of polyethylene glycol (PEG) increases the size of the therapeutic RNA, preventing clearance of the molecule from the injected site], it showed long bioavailability (several weeks after intravitreal injection, [42]), good safety profile, and efficacy in human clinical trials [43, 44]. This anti-VEGF aptamer, called Macugen, is the first RNA oligonucleotide licensed as a drug.

Lead Company: Eyetech Pharmaceuticals, Inc.

Eyetech Pharmaceuticals, Inc. is a biopharmaceutical company that specialized in the development and commercialization of novel therapeutics to treat diseases of the eye (see www.eyetech.com). Eyetech is traded on the NASDAQ National Market System under the symbol EYET. Its initial public offering of 6,500,000 shares at a price of $21.00 per share was in January 2004. Today Eyetech is a company with over 150 employees with locations in New York, New Jersey and Massachusetts.

Macugen™ (Pegaptanib sodium injection), the anti-VEGF pegylated aptamer, Eyetech's lead product, is BEING developed for diseases of the eye together with Pfizer. By docking to and blocking the function of VEGF, it interferes with the development of abnormal vessel growth and leakage seen in neovascular (wet) age-related macular degeneration (AMD) and diabetic retinopathy.

In the United States, it is estimated that as many as 15 million people suffer from some form of AMD and that there are more than 1.6 million cases of wet AMD. Approximately 200,000 new cases of wet AMD arise each year in the United States. Although wet AMD represents approximately 10 per cent of all AMD cases, it is responsible for 90% of the vision loss associated with AMD. A majority of wet AMD patients experience severe vision loss in the affected eye within months to two years after diagnosis of the disease. Because AMD generally affects adults over 50 years of age, it is expected that the incidence of AMD will increase significantly as the baby boom generation ages and overall life expectancy increases.

Macugen™ has also been studied in patients with DME (diabetic macular edema). In the United States, there are approximately 500,000 people suffering from DME, with approximately 75,000 new cases each year. It is expected that the incidence of DME in the United States will increase as the number of people with diabetes increases. Because the existing treatments for both wet AMD and DME have significant limitations, there is a significant unmet medical need for a new therapy for these diseases.

In 2001, Eyetech initiated two Phase II/III pivotal clinical trials of Macugen for the treatment of wet AMD. It involved 117 medical centers and enrolled patients with subfoveal wet AMD: 578 patients in the North American trial and 612 patients in the international trial. Patients received one intravitreal injection of macugen EVERY SIX WEEKS. The primary efficacy endpoint in these trials was the proportion of patients losing less than 15 letters, or three lines, of visual acuity on the eye chart from baseline after 54 weeks. Based on the analysis of the data from the two trials, the primary efficacy endpoint was met. Macugen was approved as a drug against AMD in 2004.

Meanwhile, a phase II placebo controlled trial was done in the context of DME. Three doses were tested: 0.3 mg, 1 mg and 3 mg every six weeks for at least twelve weeks and endpoints were measured after 30 weeks. Results showed statistical significance for all doses, with the highest efficacy for the 0.3 mg dose, similarily to the AMD trial.

IN ADDITION TO AMD and DME, Macugen™ may prove to be efficient in the treatment of retinal vein occlusion (RVO), a condition that is also characterized by high VEGF levels, abnormal blood vessel growth and blood vessel leakage. RVO occurs when the circulation of a retinal vein becomes obstructed, causing blood vessel bleeding and leakage in the retina. Laser therapy is sometimes used to treat this condition, but with limited efficacy. A phase II trial with Macugen™ started in May 2004.

27.3 SYNTHETIC RNA FOR GENE REGULATION: ANTISENSE, RIBOZYMES, AND SIRNA

Antisense and Ribozymes

Definition of Antisense and Ribozymes The first RNA molecules found to be capable of suppressing gene expression were antisense RNA [45]. They consist of a short linear antiparallel copy of the target mRNA. Through hybridization to the mRNA, antisense RNA specifically block translation [5]. This process is not reversible: One antisense RNA inhibits the translation of one mRNA molecule. For an efficient suppression of gene expression an excess of the antisense molecules compared to the targeted mRNA in the cytosol is necessary [46, 47]. Systemic delivery of synthetic antisense RNA oligonucleotides is not expected to result in a high concentration of the therapeutic molecule in the cytosol of all target cells. Thus, it would probably not be efficient in vivo for complete gene suppression. The antisense approach best suits local delivery and utilization of the more stable DNA oligonucleotides [48]. As far as RNA-based therapies are concerned, efficient in vivo suppression of gene expression can be obtained with the catalytic version of antisense RNA: the ribozyme. This type of molecule can specifically catalyze the cleavage or religation of a target RNA and, as opposed to antisense molecules, are recycled. Ribozymes were discovered in the early 1980s by Cech and Bass [11]. Their characterization arose from the studies of a self-splicing intron from *Tetrahymena thermophilia* [49]. Further studies by Altman's group also pointed to the existence of natural ribozymes capable of specifically cleaving target RNA [10]. Since these initial results, ribozymes were found in several organisms (eukaryotes, viruses, bacteria) in different forms [50]: hammerhead, hairpin, Neurospora Varkud satellite, hepatitis delta virus, self-splicing group I and group II introns, and Rnase P-ribozymes. They were shown to be capable of performing different kinds of activities such as transesterification or hydrolysis. Furthermore, in vitro selected ribozyme can catalyze the synthesis of nucleotides [51] or the creation of amide bonds [52]. The attempts toward development of therapeutic ribozymes are mainly based on two forms of naturally occurring ribozymes: hammerheads and hairpins. Hammerhead ribozymes were isolated from viroid RNA [53]. Their enzymatic core is ca. 40 nucleotides in length and consists of two substrate binding arms and a conserved catalytic domain (Fig. 27.6). Hammerhead ribozymes can cleave the phosphodiester bond of a target RNA after a UA, UU, or UC dinucleotide sequence (AUC and GUC triplets are optimal [54]). Hairpin ribozymes were discovered in tobacco ringspot virus satellite RNA. They are usually designed to cleave the phosphodiester bond located 5′ to a GUC sequence in the target RNA [55]. The catalytic part of hairpin ribozymes contains a domain

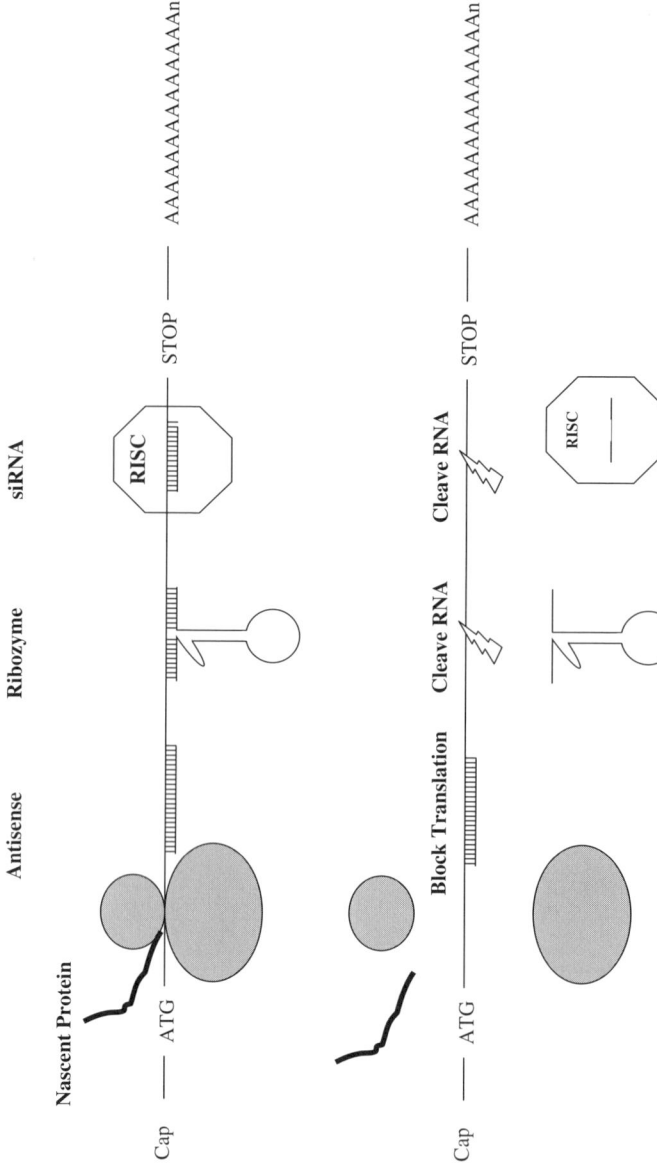

Figure 27.5 Strategies of gene suppression by RNA molecules: Antisense, ribozymes and siRNA can suppress gene expression by blocking (antisense) or cutting (ribozyme and siRNA) specifically a messenger RNA. Although antisense RNA and ribozymes act autonomously, siRNA require the housekeeping RISC complex in order to perform its catalytic activity.

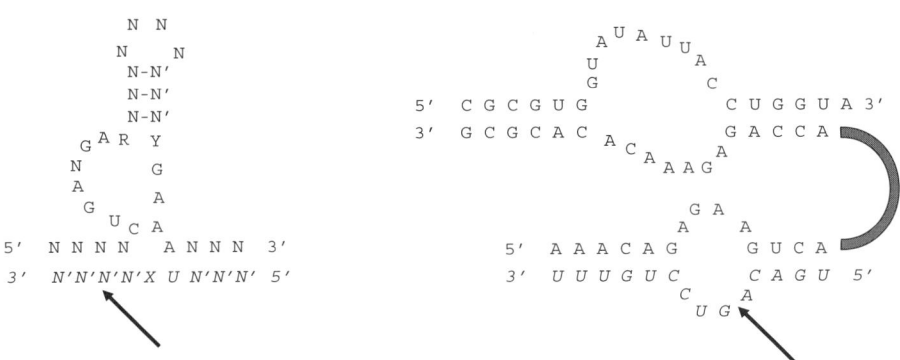

Figure 27.6 Representation of hammerhead and hairpin ribozymes. N is any nucleotide, N is a nucleotide complementary to N, X is any nucleotide but G, Y is a pyrimidine, R is a purine nucleotide complementary to Y. *The arrow represents the site of cleavage on the target RNA.* Adapted from Jen and Gewirtz (2000).

composed of a loop (two single-stranded RNA sequences of several bases) flanked by double-stranded structures and a single-stranded sequence that will recognize the target RNA (Fig. 27.6). The complex between hairpin and target RNA should also be structured in a loop flanked by double-stranded sequences. The cleavage happens in the single-stranded sequence of the loop in the target RNA. The single-stranded parts of hairpins (contained in the loops) are very conserved and necessary for the ribozyme's activity [56]. Both hairpin and hammerhead ribozymes require metal ions for their catalytic activity. As far as cleavage activity is required, hammerheads are preferred to hairpins because they are naturally prone to be more efficient in hydrolysis than transesterification (religation), whereas hairpin ribozymes are as efficient for both reactions. This is one of the reasons why the ribozymes that are being designed, improved, produced, and actually used in therapeutic settings are hammerheads.

RNA Mimetics Suitable for Hammerhead Ribozyme Function In order to be protected from quick degradation by ubiquitous RNases, synthetic ribozymes must be modified. A great variety of available modified nucleotides (listed in Natural and Modified RNA and described in Fig. 27.3) were tested for hammerhead ribozymes function [57–59]. Several modifications at the single-stranded catalytic site of the hammerhead inhibit its function. On the contrary, nuclease-resistant nucleotides such as 2′-fluoro, 2′-amino, 2′ alkyl, or allyl derivatives can be used at all positions involved in the recognition of the targeted sequence. There, a phosphorothioate backbone can bring further stability. In addition, the 3′ extremity of the molecule can be occupied by a modified nucleotide that will prevent degradation by 3′–5′ exonucleases. Through these modifications, the serum half-life of modified hammerhead ribozymes could be increased to more than 10 days as compared to less than 1 min for the unmodified version.

Preclinical and Clinical Utilizations of Hammerhead Ribozymes The first step toward the clinical use of hammerhead ribozymes is the characterization of the optimal accessible target sequence stretch in the mRNA to be cleaved. Messenger RNA or virus RNA genomes fold into complex, heterogeneous, and relatively unpredictable structures. Moreover, they associate with many proteins. Thus, a small part of the targeted RNA is exposed for recognition by ribozymes. To identify these sequences, a combination of computer prediction and RNAse H-sensitivity assay was developed [60]: Synthetic DNA oligonucleotides are incubated with the targeted mRNA; then, RNAse H is added. It will cleave the RNA only if the DNA probe could hybridize to it. Thus, the study of the integrity of the RNA will indicate if the sequence complementary to the DNA probe is accessible (RNA is cleaved) or not accessible (RNA is not cleaved). Only accessible sequences are considered as potential targets for ribozymes. Alternately, some random approaches were developed where optimal ribozymes are selected from libraries of active ribozymes (containing a hammerhead core catalytic sequence) having randomized substrate recognition arms [61–63]. Once the optimal target sequence is identified, the best RNA-stabilizing strategy is tested: At permissive positions (all positions involved in the recognition of the target RNA), 2′ modified nucleotides replace the natural nucleotides and a phosphorothioate backbone is included. Such RNase-resistant hammerhead ribozymes can to some extent spontaneously enter cells and show gene-suppressing activity in vitro in cell culture and in vivo after local [64] or systemic delivery [65]. In 1998, the first synthetic ribozyme ANGIOZYME entered a human clinical trial. It is a chemically stabilized RNA oligonucleotide capable of specifically cleaving the mRNA coding the high-affinity receptor for the vascular endothelial growth factor (VEGFR-1) [65]. Through blockade of angiogenesis by downregulation of VEGFR-1 expression, the ribozyme is expected to reduce tumor growth. Indeed, ANGIOZYME was shown to inhibit tumor growth in several mouse models. Daily intravenous or subcutaneous application resulted in detectable plasma levels of the synthetic RNA ribozyme [66]. Subsequently, it was tested in human phase I trials and proven to be nontoxic at doses up to $300\,mg/m^2$ administered intravenous or subcutaneous. ANGIOZYME was further evaluated in a phase II monotherapy trial involving breast cancer patients. Forty-five late-stage metastatic cancer patients with progressive disease were treated with doses of $100\,mg/m^2$ ANGIOZYME by daily subcutaneous injections. Most patients continued to progress as shown by computed tomography (CT) scans 6 weeks after the first injection. Nevertheless, the activity of the ribozyme could be recorded by a significant decrease of the amount of soluble VEGFR-1 detected in the blood at week 6 (soluble VEGFR-1 is encoded by an alternatively spliced mRNA derived from VEGFR-1 encoding premessenger RNA; both forms of the mRNA are recognized by ANGIOZYME). A phase II trial was conducted in which Angiozyme was used in combination with a standard chemotherapy (irinotecan/5-fluorouracil/leucovorin: IFL). Eighty-three colorectal cancer patients with metastatic disease were enrolled. They received daily subcutaneous injections of 100 mg/m^2 of ANGIOZYME in combination with IFL. Again, reduction of soluble VEGFR-1 in the blood was detectable after several weeks of injection. But, as far as the criteria "time to progression" is concerned, there was no clear difference between the group of the ANGIOZYME + IFL-treated patients compared to historical groups of IFL-treated patients. Nevertheless, since no control group (IFL alone) was included in the phase II study, definitive conclusions cannot be drawn. Another trial in tumor patients was performed with a hammerhead ribozyme called HERZYME and designed to inhibit the expression of the human epidermal growth factor receptor type 2 (Her2). This receptor is overexpressed in many carcinomas and drives cell proliferation. Its downregulation due to the ribozyme's activity should result in diminished cell growth and in attenuation of tumor progression. HERZYME was tested in a phase I dose-escalation study in patients with Her2 overexpressing cancers (subcutaneous delivery). HERZYME was well tolerated with no significant toxicity, but no biologic activity was established at the doses tested. The specific efficiency of ANGIOZYME and HERZYME needs to be further evaluated in randomized trials (with placebo or chemotherapy-alone control groups). The third hammerhead ribozyme that entered clinical trial is called HEPTAZYME. It was designed to cleave the hepatitis C RNA and reduce viral titers. The HCV virus is associated to liver cirrhosis and hepatocellular carcinoma. In a phase II clinical trial, HEPTAZYME has demonstrated a reduction in serum HCV RNA levels in 10 percent of treated patients. The Heptazyme trial had to be stopped due to some toxic side effects reported in one animal during toxicology studies. As of this writing there is no clinical evidence of the effectiveness of synthetic ribozymes in the context of any human disease.

One of the Leading Companies: Ribozyme Pharmaceuticals (now SIRNA Therapeutics) The potential of synthetic ribozymes as drugs has been developed and exploited by Ribozyme Pharmaceuticals (Boulder, Colorado). The company has expertise in oligonucleotide process development and manufacturing. Ribozyme Pharmaceuticals has established strategic corporate relationships with Archemix Corporation, atugen AG, Chiron Corporation, Fujirebio, and Geron Corporation. Ribozyme Pharmaceuticals was listed on the NASDAq under "RZYM" (now "RNAI"). It has performed studies on toxicology, biodistribution, biostability, and pharmacokinetic of modified synthetic ribozymes. Human clinical trials were sponsored by Ribozyme Pharmaceutical in the context of metastatic cancer (the hammerhead ribozymes ANGIOZYME and HERZYME, which target the mRNA coding VEGF receptor and epidermal growth factor, respectively) and hepatitis C infections (the hammerhead ribozyme HEPTAZYME).

The breakthrough discovery of siRNA [67, 68] led to a reorientation of Ribozyme Pharmaceuticals' scientists toward the utilization of this new RNA tool that is very efficient in gene suppression. Ribozyme Pharmaceuticals is now SIRNA Therapeutics (since 2003, "RNAI" on the NASDAq) and focuses primarily on the development and commercialization of siRNA-based treatments. A large part of the intellectual property of SIRNA Therapeutics is based on the patents and expertise from Ribozymes Pharmaceuticals in the areas of target validation, RNA synthesis, modifications, and delivery. SIRNA Therapeutics uses this expertise to develop a new class of siRNA. The company is in research and/or preclinical development of siRNA-based drugs capable of treating diseases such as AMD, HCV infection, and cancer (see *www.sirna.com*). In the field of oncology, SIRNA Therapeutics announced a collaboration with Eli Lilly & Company (Lilly) to jointly investigate chemically modified siRNAs against specific oncology targets provided by Lilly.

Small Inhibitory RNA (siRNA)

Definition of siRNA While engineering transgenic petunia to develop a more intense purple color than the wild variety, the Jorgensen group [68a] obtained genetically modified organisms (GMO) that developed white flowers! To explain these surprising results, they hypothesized a mechanism of posttranscriptional gene silencing. Eight years later, Fire et al. described in *Caenorhabditis elegans* the potential of double-stranded RNA molecules to specifically induce degradation of homologous single-stranded endogenous mRNA in a process that they named RNA interference (RNAi [69]). At the turn of the millennium, Elbashir and co-workers [67, 68] made a breakthrough discovery: RNAi can be obtained by small (21 to 23 mer) double-stranded synthetic RNA homologous to the target sequence. These small duplex must have 2 bases overhanging at their 3′ ends. They are called siRNA (for small inhibitory RNA) [67, 68]. The generalization of these observations in all tested organisms led to a revolution in fundamental research and molecular medicine [70, 71].

The advantages of siRNA over long double-stranded RNA molecules is that they:

1. Can be synthesized chemically.
2. Diffuse in the organism.
3. Do not induce activation of the interferon response pathway [72] (activation of the protein kinase PKR and of the 2′,5′-oligoadenylate synthetase that cause inhibition of translation by phosporylation of the translation initiation factor eIF2a and degradation of mRNA, respectively).

As opposed to ribozymes, RNAi requires housekeeping proteins to perform gene suppression: the type III RNase termed Dicer and the RISC (RNA-induced silencing complex, Fig. 27.6). Dicer chops long double-stranded RNA into ca. 21-mer double-stranded RNA that possess two nucleotides overhang at both strands 3' ends. In an adenosine triphosphate (ATP)-dependent process, RISC incorporates one strand from these ca. 21-mer double-stranded RNA. With that RNA probe, activated RISC recognizes and cleaves RNA molecules that can pair with the oligonucleotide. Using this mechanism, virtually any RNA can be specifically destroyed. Eukaryotes probably developed this enzymatic system for the recognition of double-stranded RNA (characteristic of several viruses) and destruction of homologous single-stranded RNA. Accordingly, siRNA can be regarded as the natural mediators of a cell immunity dedicated to protect the genome against pathogens (viruses or transposons).

Similarly to ribozymes and antisense technologies, the identification of an optimal target site in the mRNA to be cleaved (optimal sequence of the siRNA) is a prerequisite to further development of RNAi by siRNA. Although guidelines and websites are available (see, e.g., http://jura.wi.mit.edu [73] or EMBOSS at *http://bioweb.pasteur.fr/seqanal/interfaces/sirna.html*), the characterization of the most active siRNA for the selective degradation of a target gene requires the synthesis (enzymatically or chemically) and in vitro testing of several duplexes. Each duplex is introduced into target cells (usually tumor cells that can easily be transfected) that express the gene of interest and RNAi is measured by molecular biology tools (the amount of the targeted mRNA can be quantified by RT-PCR, Northern blot, or gene microarrays) and/or by biochemistry methods (quantification of the protein translated from the target mRNA by enzymatic tests when the activity of the target protein can be directly recorded or by ELISA, Western blots, etc.).

The sequence of a siRNA will dictate its effectiveness and specificity but will also dictate its sensitivity to RNases. The half-life can vary from minutes to days. Experimental measurements are the only way to determine the siRNA's half-life in physiologic milieus before further clinical development.

As far as therapeutic interventions are concerned, RNAi technology may translate into antivirus [HIV, human papilloma virus (HPV), hepatitis B virus (HBV), HCV] or antipathology-related proteins (e.g., oncogene, immunosuppressing agents, multidrug resistance transporters) siRNA-based drugs [74]. Although a very promising approach, the use of siRNA in modern medicine faces limits that must be solved:

1. Off target gene inhibition
2. The induction of stress by the double-stranded RNA structures
3. Inefficient intracellular penetration

The first point is crucial since the specificity of siRNA may not be as high as initially expected. Study of the global gene expression in siRNA transfected cells using gene microarrays indicated that a large number of nontargeted genes are affected by the siRNA. Actually, a complementarity between the siRNA and the target mRNA of as little as 11 continuous nucleotides is sufficient to trigger recognition and cleavage by the RISC complex [75]. Besides, only one of the two strands of the exogenous siRNA (the antisense as compared to the mRNA sequence) should theoretically be used, but both strands may practically be incorporated into the RISC, thus increasing the possible cross-recognition of irrelevant mRNA. A better knowledge of the fundamental mechanisms of RISC activity, especially in the choice of the captured siRNA strand, may allow a reduction of "off-target" genes affected by siRNA approaches.

The induction of stress was revealed by Semizarov et al. [76]. Here again, gene microarray analysis revealed the expression of several stress-related genes after delivery of siRNA, especially at high siRNA concentrations. Concerning the third point, it was shown that some siRNA can activate antigen presenting cells similarly to long double-stranded RNA [77, 78]. This phenomenon is mediated by Toll-like receptors [78a].

Altogether, the characterization of the optimal siRNA that will specifically and efficiently destroy the target mRNA with minimal off-target activity, highest stability, good biodistribution, and no side effects (immunostimulation) requires intensive experimental in vitro and in vivo testing. Several of these critical parameters are taken into account in the freely available siRNA prediction algorhythm available at *http://jura.wi.mit.edu/siRNAext/* [73], but in vivo stability, pharmacokinetics, and biodistributiuon must be tested experimentally for each individual siRNA duplex. Concerning the intracellular delivery of exogenous siRNA, refer to Delivery of RNA Molecules for Suppression of Gene Expression below.

RNA Mimetics Suitable for siRNA's Function Contrary to single-stranded RNA such as ribozymes, the short double-stranded siRNA are usually remarkably stable in physiologic milieus. Nevertheless, very large variations in siRNA half-lives (from several minutes to days) can be attributed to the sequence of the two oligonucleotides. In the case where further enhancement of the half-life of the optimal siRNA sequence deduced from in vitro experiments is required, some of the chemical modifications described in the discussion of natural and modified RNA can be introduced into the therapeutic siRNA. First, deoxy nucleotides can replace the two ribonucleotides in the 3' overhangs of each siRNA strand [67, 68]. Modifications such as allylation or alkylation of some [79] but not all [80] 2' OH as well as 2'fluoro substitutions and LNA nucleotides [81, 82] are tolerated with minimal or no decrease of siRNA function. Similarly, phosphorothioate backbones can be used in siRNA strands [81].

Up until now, the tested chemical modifications showed a relatively low increase in the efficiency of siRNA over time in transfected cells as well as in vivo [82, 83]. Thus, chemical modifications may not be required in the context of siRNA therapies. Nevertheless, together with the incorporation of RNase resistance data into siRNA sequence prediction programs, the usage of some chemical modifications at key positions may lead to the definition of siRNA with half-lives of several days or weeks in physiologic milieus. These improvements will result in a reduced dosage and frequency of application of siRNA for optimal activity in future therapeutic settings.

Preclinical Evaluation of siRNA Several siRNA with a potential therapeutic activity were characterized. For antitumor therapies were identified siRNA that target growth factors (VEGF) [83a] and antiapoptosis molecules (Bcl-2) or siRNA that targets the mRNA coding for oncogens such as the fusion protein Bcr-Abl [83b], the constitutively activated mutated Ras proteins or mutants of P53, which dominantly inhibit the tumor suppressing activity of wild type P53 in ca. 50 percent of human malignancies [83c, 83d]. In the context of infectious diseases, were characterized siRNA specific for viral genes (HIV, HCV, HBV). Although most reports are limited to the definition of the optimal siRNA sequence using in vitro testing in cell culture systems, several investigations went further to the biodistribution and efficiency studies in preclinical setups in mice. Concerning tissue distribution and pharmacokinetics, Braasch et al. [83] used radioactive iodine to label the siRNA. After intravenous injection of naked siRNA in mice, the molecule (the radioactivity) localizes in the liver and kidney (peak at 5 min postinjection). Later, a signal is found in the bladder suggesting quick elimination though urine. Lower amounts of radioactivity are found in lung, spleen, and heart. Close to no signal can be detected in the brain. Most siRNA seem to be eliminated from the mouse body in less than 24 h but some detectable amounts of radioactivity remain up to 72 h. These data suggest that (1) when a continuous inhibition of gene expression is required, injection of siRNA should be repeated every 24 h, and (2) as far as systemic delivery is concerned, siRNA-based therapies may be mostly relevant when the target gene is expressed in liver or kidneys. For delivery to other organs, encapsulating siRNA in specific delivery vehicles and/or local delivery could be envisioned. In agreement with this biodistribution data, Zender et al. [84] showed in LacZ transgenic mice that systemic application (intravenous delivery) of siRNA designed to inhibit β-galactosidase expression, induces a three- to fourfold reduction in β-galactosidase activity in liver. When using siRNA capable of cleaving the mRNA coding caspase 8 (a necessary mediator of apoptotic cell death), the authors could induce protection of hepatocytes against Fas-mediated apoptosis. In this mouse model, acute liver failure (ALF), similar to the disease observed in patients suffering from viral hepatitis, is induced by injection of activators of Fas (either antibodies against Fas or adenovirus expressing Fas-ligand). Histopathology studies demonstrated that the anti-caspase-8 siRNA could prevent Fas-induced apoptosis in hepatocytes. Song et al. [85] obtained similar protection against induced ALF by the in vivo delivery of siRNA designed to destroy the Fas-encoding mRNA. Besides, Sorensen et al. [86] also showed that systemic delivery of siRNA resulted in selective RNAi. They studied the potential of siRNA to decrease in vivo the production of tumor necrosis factor (TNF-α) (a major player in sepsis). Their study demonstrates that intraperitoneal injection of siRNA specific for the TNF-α encoding mRNA could induce a reduction of TNF-α production after exposure to lipopolysaccharide (LPS). Moreover, mice injected with these anti-TNF-α siRNA were protected against a lethal dose of LPS.

The results of in vivo studies summarized here provide "proof of concept" that open the door for pharmaceutical use of siRNA in humans. No human clinical trial results are yet reported, the technology being too new (described in 2001) for human studies to be completed. From the results obtained in the mouse models, it can be expected that new drugs with siRNA as the active component will prove to be safe and efficient in humans.

One of the Leading Company: Alnylam Pharmaceuticals Three companies were created to explore and develop siRNA: Ribopharma AG in Germany (created in 2000; located in Kulmbach, Bavaria, Germany), SIRNA (created in 2003 from Ribozyme Pharmaceuticals; located in Boulder, Colorado), and Alnylam Pharmaceuticals (created in 2002, Cambridge, Massachusetts). Ribopharma AG and Alnylam Pharmaceuticals fused in July 2003 to become Alnylam Pharmaceuticals. One of the founders of Alnylam Pharmaceuticals is Thomas Tuschl (Laboratory of RNA Molecular Biology, Rockefeller University), who discovered the structure of mammalian inhibitory RNA as a researcher at the Max-Planck-Institute for Biophysical Chemistry, in Gottingen, Germany. Alnylam Pharemaceuticals owns the first patent issued on the use of short double-stranded RNAs (siRNAs) to trigger RNAi. Alnylam Pharmaceuticals develops siRNA-based therapeutics for local (brain and eye diseases such as Parkinson's disease and wet AMD, respectively) or systemic (diabetes, obesity, autoimmune diseases, and cancer diseases) applications. No human clinical trials are disclosed on the web page from Alnylam Pharmaceuticals, www.alnylam.com, but such studies can be expected to start soon.

Delivery of RNA Molecules for Suppression of Gene Expression

Whether antisense RNA or catalytic RNA are used, the intracellular delivery of the oligonucleotides must be achieved in vivo. The resistance of cell membranes to penetration of nucleic acids limits the efficiency of such therapies. Still, naked stabilized RNA were shown to spontaneously enter cells in the context of ribozyme studies [65, 66]. In mice, a method called hydrodynamic injection consists in a fast (several seconds) intravenous delivery of a large volume (up to 1 mL) of solution containing the therapeutic RNA resulted in efficient delivery into hepatocytes [84, 85, 87, 88]. Of course, such an application method cannot be used in humans. A feasible method is the use of a delivery vehicle such as cationic liposomes [78, 89]. In mice, liposomal formulations of siRNA were associated with the clinical manifestations of the therapeutic siRNA activity [84, 86]. The problem is that most liposomal formulations are toxic. Other vehicles such as nanoparticles or microparticles may prove to be useful for efficient and safe delivery of siRNA. All these vehicles may be targeted to a precise cell type through coating with specific ligands [90]. Cell penetrating peptides (CPP) are another tool to deliver molecules intracellularly. CPP can spontaneously cross lipidic membranes and enter alone or with an associated cargo (protein or nucleic acid) in the cytosole. They are derived from diverse proteins such as Antennapedia or HIV TAT proteins [91]. Electrostatic interaction between an siRNA and a CPP was reported to result in efficient and specific RNAi in vitro [92]. The optimal (eventually selective) and nontoxic vehicles for siRNA is undoubtly the most needed tool for testing and developing efficient siRNA-based drugs. Many experts in this field as well as the pharmaceutical industry and biotechnology companies have advanced products that may fulfill the need for efficient delivery of siRNA. The characterization of their structure (homogeneity, stability, storage), their safety and efficiency profiles in vivo (preclinical studies, toxicology tests) as well as the possibility to produce great amounts of them at high quality good manufacturing practices (GMP) are many of the critical issues that must be addressed before new siRNA-based drugs are tested in human trials.

27.4 RNA FOR PROTEIN PRODUCTION: mRNA

Definition of mRNA

In eukaryotes, genes are transcribed in the nucleus. This process generates pre-messenger RNA (pre-mRNA). Pre-mRNA are heavily modified: A cap structure (7-methyl guanosine triphosphate) is added to the 5' end; some parts of the pre-mRNA (introns) are spliced out; some nucleotides of the pre-mRNA are changed by a process called editing, and a poly-A tail of usually several hundred adenine residues is added (Fig. 27.7). Such mature mRNA are between a minimum of ca. 200 up to several tens of thousands of bases in length. They exit the nucleus through nuclear pores. Their 5' cap structure is recognized by eukaryotic initiation factors (eIFs) and their 3' poly-A tail is recognized by the poly-A binding protein. Interactions of the 5' end and 3' end riboproteic complexes induce the beginning of translation by ribosomes. First, the small subunit of the ribosome will be recruited by the eIFs. It scans the mRNA until it reaches a AUG start codon (located in a Kozak surrounding: A/GNNNATGG) [93]. At this site, the large subunit of the ribosome associates to the small subunit and translation begins. Protein elongation continues until a stop codon (AAU, UGA, or UAG) [94]. For therapeutic use, mature messenger RNA is being produced by in vitro transcription (Fig. 27.7) thanks to RNA polymerase: usually T7, T3, or SP6 RNA polymerase derived from bacteriophages. See Manufacturing of RNA in GMP conditions under Section 27.6 for more details on mRNA production. When introduced into the cytosole of a cell, such in vitro produced mature mRNA can readily be translated into proteins. The natural catabolism of mRNA guarantees that the foreign nucleic acid will be eliminated from the cell's cytosole. The foreign protein translated from the recombinant mRNA will be expressed transiently. In a natural situation (endogenous production) intracellular mRNA has a half-life ranging from minutes (mRNA coding for proteins that control the cell cycle, e.g.) up

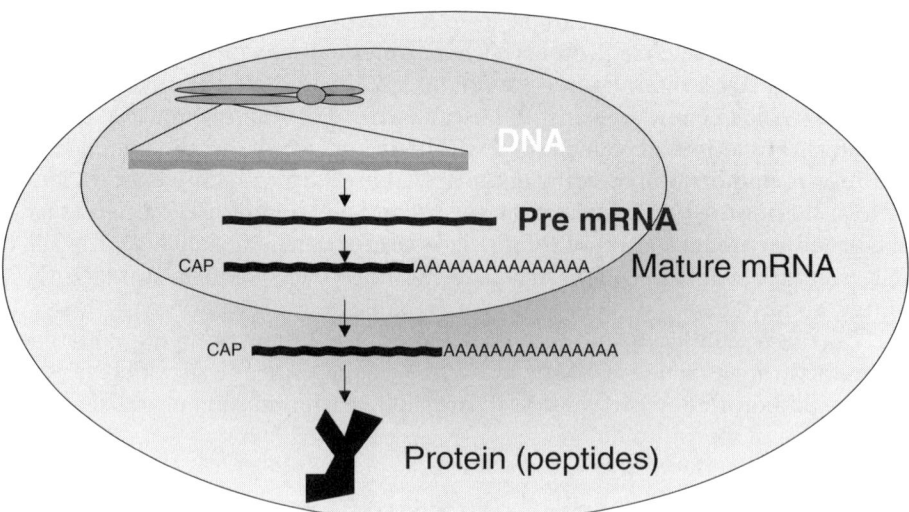

Figure 27.7 Production and structure of mRNA in vivo. Genes contained in the chromosomes located in the nucleus are transcribed into pre-mRNA. This molecule is spliced (excision of introns), capped at the 5′ end and polyadenylated at the 3′ end. Its sequence may be edited. The mature mRNA leaves the nucleus through nuclear pores and its translated in proteins until its degradation by specific RNases (not shown).

to weeks (mRNA coding globin proteins in terminally differentiated red blood cells) [95]. Several nucleotide sequences are known to dictate the transcript's half-life [96]. Most of them are located in the 3′ untranslated region (3′ UTR) of the mRNA. For example, the A and U rich elements (AURES) signal to the cell machinery in normal conditions that the mRNA should be quickly destroyed [97]. On the other hand, pyrimidine-rich elements such as those recognized by the so-called α complex (a multiprotein structure expressed ubiquitously) allow the mRNA to be very stable [98]. The deletion of destabilization sequences such as AURES and the addition of stabilization sequences such as globin UTRs at the 3′ end of the cDNA coding for the therapeutic proteins enables in vitro production of recombinant mRNA with enhanced stability [99].

RNA Mimetics Suitable for mRNA Function

Production of messenger RNA in vitro for clinical use depends on the activity of available purified RNA polymerases. Since only few modified NTPs are substrates for the RNA polymerases usually used for in vitro transcription (T7, SP6, and T3 RNA polymerases), the variety of chemical modifications that can be incorporated into therapeutic mRNA is very reduced in comparison to the variety of chemical modification available for synthetic RNA made by chemical synthesis. NTPs with modifications such as fluoro, azido, or amino substitutions of the 2' OH as well as NTPs with phosphorothioate modification can be used to synthesize mRNA. These, however, decrease the quality and quantity of the mRNA recovered by in vitro transcription [28] and impair translation in vivo (Probst et al., submitted). Until now, there was no NTP modifications that were shown to be permissive to RNA polymerases, to bring additional stability to the mRNA in regard of RNAse-mediated degradation and to be permissive to translation by the ribosomes in cells. Consequently, the mRNA currently used in therapies is nonmodified RNA.

Preclinical and Clinical Utilization of mRNA

A breakthrough discovery was reported by Wolf et al. in 1990: Direct injection of naked mRNA (as well as plasmid DNA) into mouse muscle led to the production of the protein encoded by the foreign nucleic acid [100]. This indicates that somehow, and in spite of ubiquitous RNases, some cells can take up the foreign genetic information and use it for protein production in vivo. Because mRNA will be only transiently expressed, its utilization for genetic complementation (expression of a protein that is not produced from the genome because of inherited mutations) would require a frequent (daily) delivery of the drug. This aspect of mRNA-based therapies was not yet addressed since more stable vectors such as plasmid DNA or recombinant viruses seemed more appropriate. Nevertheless, safety concerns in the field of DNA-based therapies lead to the reevaluation of mRNA-based therapies for in vivo gene complementation: Theoretically, a frequent uptake of a safe "gene therapy" drug is preferable than a single uptake of a therapeutic molecule that may lead to severe side effects. Indeed, DNA or recombinant viruses may persist for some hours in some recipients and for years in others and may even cause cancer (integration of the foreign genetic information at the site of an oncogen or a tumor suppressor gene), autoimmunity (induction of anti-DNA antibodies triggering diseases such as lupus erythematosus), or modification of the human specie (integration of the foreign genetic information into the patient's germ cell genomes). Thus, a frequent application of adequately protected and delivered mRNA may become a safe alternative to DNA-based gene therapies in the field of genetic complementation.

Meanwhile, starting in the early 1990s, the potential of mRNA for transient expression of an antigen (a protein against which an immune response would be therapeutically beneficial) was investigated (for a recent review, see Pascolo [101]). First, Martinon et al. showed in mice that injection of liposome-encapsulated mRNA coding for the influenza virus nucleoprotein could trigger an anti-influenza immune response [102]. Conry et al. [102a] confirmed and extended these results by using mRNA as the active component of an antitumor immunotherapy (a vaccine whereby immune cells become competent to specifically kill tumor cells). In their experiments, mice that received intramuscular injections of naked mRNA coding for the tumor antigen carcinoembryonic antigen (CEA) were resistant to the transplantation of CEA-expressing tumor cells. In these vaccination studies, delivery of mRNA was shown to stimulate all effector cells from the adaptive immune response: cytotoxic (CD8$^+$) T-cells capable of direct recognition and killing of target (tumor or infected) cells, B-cells that can secrete specific antibodies neutralizing the pathogenic agent, and helper (CD4$^+$) T-cells that orchestrate the immune response, enhance the activity of all immune cells, and ensure persistence of long-term memory effector cells. In 2000, two reports confirmed these results and moreover introduced tumor mRNA libraries as a tool for antitumor vaccination [102b, 102c]. In these studies, the whole tumor mRNA was injected into mice for vaccination. Such an approach could result in customized therapies where an antitumor vaccine is to be prepared from patient's tumors. Since every tumor expresses a characteristic set of immunogenic proteins specific for the tumor cells that include mutated (P53, Ras, e.g.) and recombined (Bcr-Abl, e.g.) proteins as well as the promiscuous "tumor antigens" (tissue-specific genes such as gp100, tyrosinase, or Melan-A, developmental genes, and overexpressed genes or oncogens), an autologous antitumor vaccine may be the most efficient strategy for immunotherapies. In mice, such vaccines trigger an immune response that specifically recognizes and destroys tumor cells. It prevents the dissemination of a primary tumor (blocking the formation of metastasis) and eventually results in the destruction of existing malignancies by immune cells (regression of the tumor disease). Recently, Carralot et al. showed that it is possible to tune the type of immune response triggered by mRNA-based vaccines to render it optimal for antitumor and antivirus immunotherapies [103]. In this study, granulocyte monocytes colony stimulation factor (GM-CSF), a cytokine used in the clinic for the stimulation of the growth of certain immune cells, is shown to shift the natural immunity induced by mRNA injection from a Th2 type (prone to the production of antibodies against the antigen encoded by the mRNA) to a Th1 type (prone to the proliferation of cytotoxic T-cells that can specifically kill target cells expressing the protein encoded by the mRNA vaccine). In anti-viral or antitumor immunotherapies, the mRNA plus GM-CSF form of mRNA vaccines should be used since mostly cytotoxic T-cells are required for prevention or treatment of the pathologies.

There are several other methods of mRNA-based vaccination aside from direct injection of simple mRNA. One consists in ballistic delivery where the mRNA is loaded on gold particle and delivered to the skin of patients with a gene gun [104, 105]. Cartouches containing the gold particles coated with mRNA (mRNA is precipitated on the gold microbeads by ethanol before being air-dried) are placed in the barrel of the gene gun. Pressing on the trigger while the gun is pointed to naked skin (ca. 10 cm from the skin) will result in a short release of high-pressured gas through the cartouche (the gene gun being connected to a helium bottle) that will bring the gold microbeads to the skin with high velocity. The beads will penetrate the epidermis and the dermis. Resident cells including antigen-presenting cells (APCs) such as Langerhans cells will take up the mRNA-coated beads, translate the nucleic acid, and expose the protein to the immune system to trigger an immune response. The cost of gold particles and of the gene gun as well as the difficult preclinical setting (mice must be well shaved before being vaccinated with gene gun) have probably limited the utilization of such a promising needleless vaccination technology (no human clinical studies based on gene gun delivery of mRNA vaccines has been reported).

Another technology is the injection of autoreplicative mRNA [106–110]. Thanks to an internal ribosome entry sequence (IRES), these mRNA are bicistronic. They code for two proteins: One protein is the antigen and the other one is an RNA replicase derived from RNA viruses such as the α viruses Sindbis or Semliki Forest. After application, the bicistronic mRNA will be taken up by some cells and will be transcribed into the antigen and the replicase. Due to the presence of sequences specifically recognized by the RNA replicase in the bicistronic RNA, the replicase will multiply the foreign mRNA, thus generating a large number of proteins: antigen and replicase. This system guarantees that the antigen will be expressed for a long time and in high amounts. The uncontrolled persistence of the autoreplicative mRNA in the host and the potential of such mRNA to recombine with RNA viruses (creating new viruses) are safety concerns that may limit the use of autoreplicative mRNA in human vaccination settings.

Instead of direct intraskin delivery of mRNA for vaccination, Boczkowski and co-workers introduced a method that consists of in vitro delivery of the mRNA into autologous dendritic cells (DC) [111]. DCs are the most powerful APCs for triggering an immune response. Messenger RNA-transfected DCs produce transiently the antigen encoded by the foreign nucleic acid and present it in an optimal context to the immune system. DCs can be derived in vitro from blood using a 1-week-long cell culture protocol whereby monocytes differentiates into DCs thanks to the cytokines GM-CSF and IL-4 [112]. Afterwards, the DCs are transfected with mRNA (using either simple co-incubation or active delivery of mRNA by liposomes or electroporation) and reinfused into the organism (usually by intradermal or subcutaneous injections). Such a method triggers in mice a strong immune response against the protein encoded by the transgenic mRNA [111]. It is the most popular mRNA-based vaccine strategy with more than 40 original publications documenting its efficiency in different models (tumors and infectious disease, see review Pascolo [101]). It was transferred from the laboratory to the clinic (phase I/II studies) with the development of facilities where DCs can be cultured in GMP conditions. Tumor patients received their own in vitro generated DCs that were transfected with mRNA coding for prostate specific antigen (PSA) in patients suffering from metastatic prostate carcinoma [113], coding for CEA in patients suffering from metastatic colon cancer [114], or coding for the whole set of proteins from autologous metastasis in patients suffering from metastatic renal cell carcinoma [115]. In all three reports, the expected antitumor T-cell response could be detected in most patients and some clinical benefits were recorded. The objective efficacy of the method as an antitumor or antivirus (HIV, HCV) treatment must be evaluated in large placebo-controlled trials.

Lead Company: MERIX Biosciences (now Argos Therapeutics)

MERIX Bioscience, Inc. (http://www.merixbio.com/index.html) was created in 1998 because of the scientific results of Prof. Gilboa. MERIX Bioscience is headquartered in Durham, North Carolina. According to its intellectual property MERIX Bioscience's therapeutic products are based on the production and utilization of mRNA-transfected dendritic cells as immunomodulators. This technology disclosed in the cornerstone article published in 1996 [111] consists of the in vitro derivation of DCs from patient's blood, transfection of such DCs by defined mRNA (coding for one tumor antigen) or mRNA libraries (derived from the patient's tumor), and reinjection of the DCs. Such a treatment was shown in mice to trigger T-cells that recognize and destroy specifically tumor cells. Three investigator-sponsored phase I/II human clinical trials indicated that the method owned by MERIX Bioscience is feasible, nontoxic and, efficient in triggering immunity against the antigens encoded by the transfected mRNA [113–115]. The impact of such a treatment on the tumor disease where clinical parameters such as "time to progression" or "time to death" are compared to control groups must be evaluated in phase IIb or III trials. MERIX Bioscience has now filed its first company-sponsored investigational new drug application with the FDA. RNA-loaded DC vaccine will be manufactured at the company's cGMP-approved vaccine manufacturing facility in Durham, North Carolina. Using mRNA-transfected DCs, MERIX Bioscience develops also treatments against infectious diseases, autoimmune disorders, and rejections of transplants.

27.5 RNA FOR IMMUNOSTIMULATION: DOUBLE-STRANDED AND STABILIZED RNA

Definition of Immunostimulating RNAs

There exist two types of immunostimulating RNA: (1) the double-stranded RNA molecules longer than 40 base pairs and (2) the stabilized single-stranded RNA that can be as short as 20 nucleotides. In both cases, the RNA mimics structures specific from pathogens: double-stranded RNA produced during virus replication (α virus or influenza virus, e.g.) [9] or stabilized single-stranded RNA similar to those found in viruses with RNA genomes (RNA molecules condensed and stabilized by cationic compounds such as histones) [116]. Toll-like receptors (TLR) are a family of molecules conserved through evolution. They signal infection or "danger" to the immune system [117]. Mammals kept a repertoire of TLR capable of recognizing bacterial proteins (flagellin is recognized by TLR5), bacterial LPS (recognized by TLR-2 and TLR-4), bacterial and viral nonmethylated DNA (the CpG motifs recognized by TLR-9), or RNA from infectious agents: double-stranded or stabilized single-stranded RNA recognized by TLR3 or TLR-7 and TLR-8, respectively. These receptors give different activation signals that result in different effector functions. For example, double-stranded RNA (TLR-3) induces a strong upregulation of costimulation molecules (e.g., CD86) at the surface of the responding cells (TLR-3$^+$) but low cytokine production; meanwhile stabilized single-stranded RNA (TLR-7 and -8) induces low upregulation of costimulation molecules but high cytokine secretion by responding cells. The stabilized mRNA can be mimicked by phosphorothioate RNA oligonucleotides for the stimulation of TLR-7 or -8 [118, 119]. Thus, chemically synthesized double-stranded RNA (usually a polyinosine molecule hybridized with a polycytosine nucleotide: poly I:C) or single-stranded phopshorothioate RNA as well as enzymatically produced mRNA complexed to cationic compounds are reagents available to trigger the natural mechanisms of immunostimulation.

Preclinical and Clinical Utilization of Immunostimulating RNA

Double-stranded RNA in the form of polyinosinic-polycytidylic acid (rIn X rCn or poly I:C) were shown to be potential antitumor drugs in animal models during the 1970s [120]. When applied locally at the site of tumor transplantation in rats, they could retard or suppress tumor growth. This phenomenon is probably a consequence of local immunostimulation that results in the triggering of an antitumor immunity. Unfortunately, poly I:C showed low efficacy and high toxicity when used in humans. A derived modified compound was found to have lower toxicity and retained antitumor immunotherapeutic effect in the 1980s [121]. It consists of a mismatch-interrupted double-stranded RNA. Mismatches were introduced in the double-stranded structure using the hybridization between a polyinosinic strand and a polycitidylic strand interrupted every 12 C by uracils [rIn X r(C12,U)n]. This modified double-stranded RNA was further developed under the name Ampligen. Intravenous injections of Ampligen (10 to 80 mg per dose, twice weekly) were well tolerated in treated tumor patients. Some stabilization of the tumor disease as well as regressions were observed [122].

The immunostimulating effects and inherent anti-HIV activity [123] of Ampligen were also evaluated as therapy in acquired immunodeficiency syndrome (AIDS) patients. The AIDS patients received higher doses than the tumor patients: 200 mg intravenous twice a week. Again, no severe side effects were recorded. Virus load was decreased and immune response increased during the treatment. When used in combination with a standard anti-HIV drug (zidovudine) in a placebo-controlled trial, Ampligen (up to 700 mg intravenous twice weekly) showed a synergistic benefit: Patients receiving zidovudine with Ampligen showed a reduced tendency in CD4+ cell loss over time compared to the placebo group that received zidovidine alone [124]. Other anti-HIV drugs such as abacavir, zalcitabine, didanosine, stavudine, efavirenz, indinavir, nelfinavir, and amprenavir can also work synergistically with Ampligen in reducing HIV-related pathologies [125].

Since double-stranded RNA and stabilized RNA stimulate different receptors and consequently trigger different responses, it would be useful to compare the therapeutic activities of Ampligen to phosphorothioate oligonucleotides or condensed mRNA in the context of local (injection in tumors) or systemic (adjuvant activity) applications. Meanwhile the development of DC-based therapies where DCs needs to be matured in vitro by contact to a danger signal before being reinjected offers a new therapeutic development for RNA-based adjuvant. Maturation of DCs through TLR-3, -7, or -8 may result in the triggering of the optimal maturation process that would allow the therapeutic DCs to secrete the required cytokines, express the adequate costimulation molecules, and acquire the necessary homing receptors that direct the DCs in lymph nodes. For such application, Ampligen is being tested [126].

Lead Company: Hemispherx Biopharma, Inc.

Hemispherx Biopharma, Inc. (http://www.hemispherx.netl) is based in Philadelphia, Pennsylvania. It develops drugs for treatment of viral and immune diseases. Hemispherx Biopharma, Inc. produces Ampligen under GMP conditions and evaluates its capacity as an adjuvant for anti-HIV drugs (a phase IIb trial is ongoing in United States) or as a therapeutic against chronic fatigue syndrome (CFS).

27.6 PRODUCTION OF RNA AND LEGAL ISSUES

Manufacturing of RNA

Oligonucleotides (<80 Nucleotides) Short RNA oligonucleotides (less than ca. 80 bases) to be used as aptamers, antisense, ribozymes, siRNA, or adjuvants are produced by chemical synthesis. This process guarantees that the production is well controlled and free of any biological products. It can be upscaled to provide the amounts of oligonucleotide necessary for all phases of clinical trials and commercialization (with the existing suppliers, up to a ton of oligonucleotide per year can be produced).

Several methods of synthesis exist. Each is based on the use of building blocks that consist of nucleotides protected on reactive groups: the 5' and 3' groups as well as the 2' OH (Fig. 27.8a). The most popular method of chemical synthesis is the "phosphoramidite" method [127]. The building blocks are nucleotides with an acyl protection of the amine function of the base (for A, C, and G), a dimethoxytrityl (DMT) group to protect the 5'-hydroxyl and a *t*-butyldimethylsilyl or a 2'-*O*-triisopropylsilyloxymethyl (TOM) group to protect the 2' OH. The 3' position has a *O*-phosphoramidite group. The chemical synthesis goes from 3' to 5' (reverse compared to the enzymatic synthesis). The 3' residue of the oligonucleotide is attached on a solid support by its 3' end. Its 5' OH group will make a link with the 3' phosphite group of an added building block in the presence of an appropriate activator (mostly tetrazole derivatives). The 5' protecting group of the added nucleotide will be

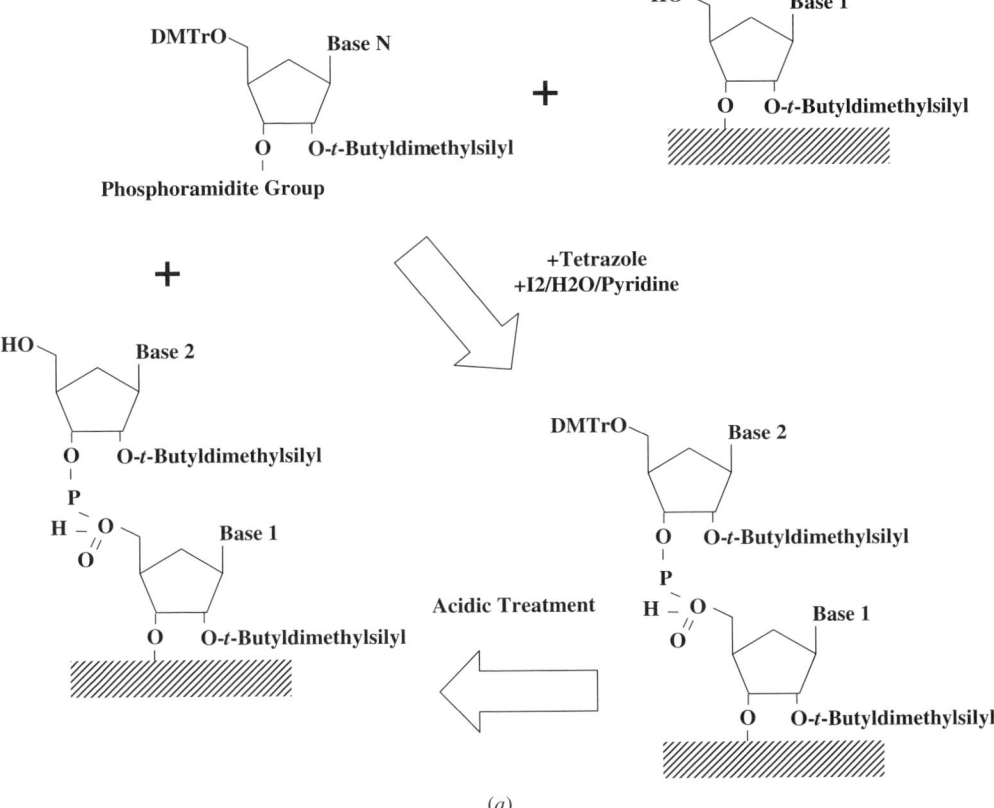

Figure 27.8 Production of RNA. (*a*) Chemical synthesis (phosphoramidite method). The 3′ residue of the oligonucleotide is attached on a solid support by its 3′ end. Its 5′ OH group is linked with the 3′ phosphite group of an added building block in the presence of an appropriate activator (mostly tetrazole derivatives). An acidic treatment will release the DMT group and prepare the nascent oligonucleotide to receive the next base. At the end of the synthesis, the oligonucleotide is released from the solid support by basic treatment. The 2′ protection groups are then removed by a treatment with tetrabutylammonium fluoride or triethylamine trihydrofluoride. The fully deprotected oligonucleotide can be purified by HPLC. (*b*) Production of long mRNA by in vitro transcription (e.g., with the T7 RNA polymerase). The recombinant plasmid template is purified from bacteria, linearized, and in vitro transcribed to generate mRNA. The template DNA is eliminated by treatment with DNase. The final mRNA product can be purified by selective precipitation with LiCl or by anion exchange chromatography or by selection on size through the PUREmessenger technology.

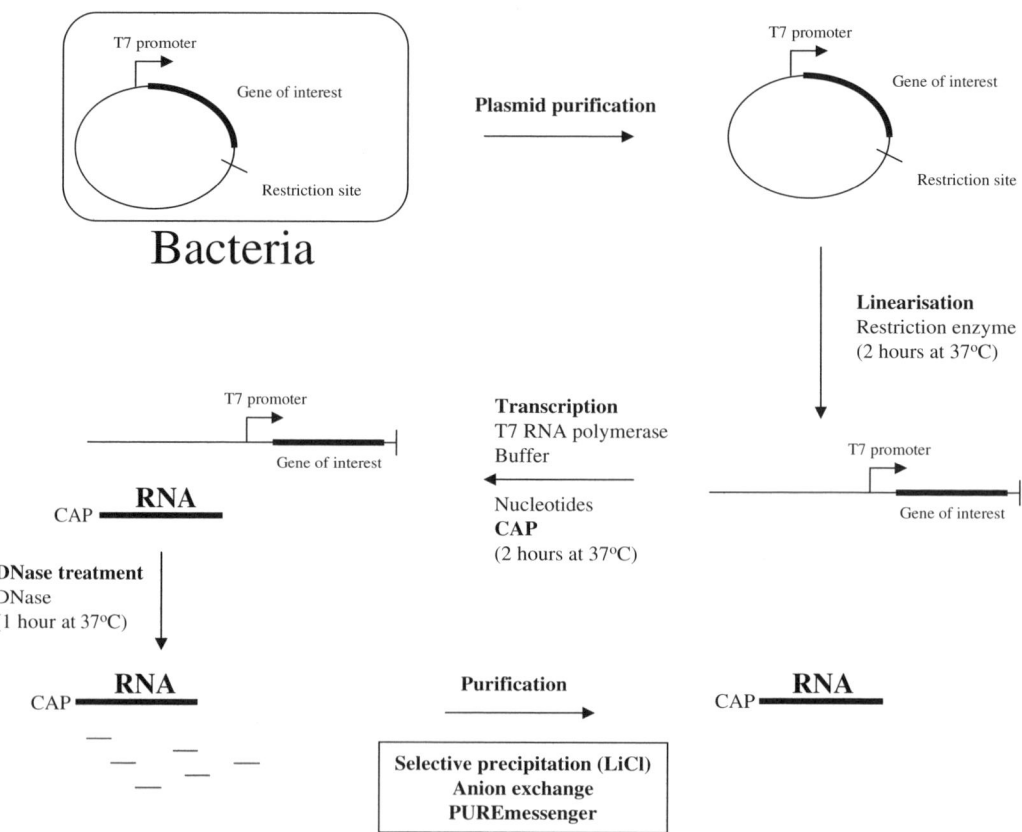

Figure 27.8 *Continued*

cleaved by acidic treatment. The oligonucleotide is ready to be elongated by another nucleotide building group thanks to the same chain of reactions. The full-length oligonucleotide can be released from the solid support by basic treatment with a mixture of ammonia and methylamine (AMA), which will also deprotect the amine function of the A, C, and G residues. Finally, the soluble oligonucleotide is treated by tetrabutylammonium fluoride or triethylamine trihydrofluoride to release the 2′ protections group.

A recently described method uses a new class of silyl ethers to protect the 5′-hydroxyl in combination with an acid-labile orthoester protecting group on the 2′-hydroxyl (2′-ACE). This set of protecting groups is used with standard phosphoramidite technology [128].

The fully deprotected oligonucleotide can be purified by adequate high-performance liquid chromatography (HPLC) (anion exchange chromatography or ion pairing reversed-phase chromatography).

Many companies are offering customized synthesis of RNA oligonucleotides, but only few offer GMP-certified oligonucleotides: Avecia

(Manchester, United Kingdom; www.avecia.com) can provide hundreds of kilograms of GMP-quality, RNA oligonucleotides per year; Girindus AG (Kuensebeck, Germany; www.girindus.com), CSS (Craigavon, United Kingdom; www.css-almac.com), Lonza Ltd (Basel, Switzerland; www.lonza.com), CureVac (Tuebingen, Germany; www.curevac.com), Ambion (Austin, Texas, www.ambion.com), and Transgenomic (Omaha, Nebraska; www.transgenomic.com), among others, can also produce GMP Gligonucleotides.

Long RNA (>80 Nucleotides) Messenger RNAs are always longer than 80 bases (the poly A tail being already at least 30 nucleotides). Their production requires an enzymatic synthesis. To achieve this, a genetically engineered plasmid DNA is used as a matrix. It contains a promoter recognized by a bacteriophage RNA polymerase (SP6, T7, or T3) located in front of the gene of interest. A poly-A sequence follows the gene (it should be of at least 30 A residues for the resulting mRNA to be recognized by the poly-A-binding protein necessary for translation). A unique restriction site located after the poly-A tail is used to linearize the plasmid DNA template. Through transcription this matrix generates a large amount of mRNA of a defined size (runoff transcription that stops at the site where the DNA is linearized). On the average, a DNA template is transcribed more than 100 times during a 1-h transcription reaction. The nucleotide mix in the in vitro transcription reaction contains an excess of cap nucleotide analog (N7-Me-guanosine-5′-triphosphate-5′-guanosine synthesized by chemistry) over GTP (usually a fourfold molar excess). Since the transcripts start with a G, there are statistically 80 percent of the transcripts that will start with a cap (the rest will start with a canonical G and will not be translated). DNase treatment of the transcription reaction will result in the destruction of the DNA template. The mRNA can be separated from nucleotides, enzymes, and DNA fragments by selective precipitation (e.g., with LiCl) or chromatography (e.g. anion exchange). Only two companies offer customized production of mRNA at laboratory quality or GMP quality: Ambion, (www.ambion.com) in the United States (Austin, Texas) and CureVac (www.curevac.com) in Europe (Tuebingen, Germany). These companies can generate GMP-certified batches of grams of mRNA. The homogeneity of an mRNA batch is challenged by the fact that, as is the case for oligonucleotides made by chemistry, the enzymatic production can generate shorter or longer contaminant products:

1. Short mRNA corresponds either to abortive transcriptions or to the start of transcription from a cryptic promoter (these alternative transcripts may be antisense and interfere with the biological activity of the mRNA).
2. Long mRNA that are made from traces of nonlinearized plasmids or from an upstream or antisense cryptic promoter. Here again the contaminants may affect the biological activity of the mRNA.

To solve this problem, the purification of mRNA is refined at CureVac by an innovative chromatography step called PUREmessenger. This technology not only removes all traces of contaminants (proteins, DNA fragments, and nucleotides) but also allows for the isolation of mRNA according to size. Thus shorter or longer by-products are eliminated. The so-called PUREmessenger are the highest quality in vitro transcripts available. Unexpectedly, together with their purity, their specific activity is increased. As shown in Figure 27.9, when transfected into eukaryotic cells, PUREmessenger mRNA is better translated than mRNA purified by classical methods such as precipitations, which do not sort mRNA on size.

Regulatory Aspects (Quality Management)

RNA-based therapies, whether they use oligonucleotides or long messenger RNA are not classified as "gene therapies." This makes RNA-based therapies easier to implement in the clinic compared to therapeutic formulations where recombinant plasmids, viruses, or bacteria as well as DNA-transfected cells are delivered to patients.

No formal guidance from the FDA (in the United States) or the European Medicines Agency (EMEA, in Europe) is available for the production and qualification of RNA. Just one publication addresses these points and provides hints for GMP production of oligonucleotides [129]. In general, the quality of the GMP RNA must be specified by the sponsor after dialog with the producer and in agreement with the relevant authorities.

Aside from the usual final quality controls performed on any pharmaceutical ingredient (sterility, residual solvents, heavy metals, endotoxin content, pH, appearance), two particularly delicate characteristics of the final RNA batch must be measured: purity and identity. These two aspects are discussed below. Additionally, a biological activity test should be implemented when possible (e.g., testing the activity of an aptamer or of a siRNA duplex in vitro using biochemical tools or cell culture methods).

Purity Measuring the purity of the RNA includes two levels of analysis that must answer two different questions. First, are there any products other than RNA in the preparation (endotoxin, microorganisms, proteins, lipids, DNA)? Second, is my RNA contaminated with some other forms of RNA: shorter, longer RNA, or RNA that does not have the expected modifications (e.g., Cap structure, 2′ substitutions, phosphorothioate backbone)? The first point is easily addressed with standard tests performed on any active pharmaceutical ingredient (API). The tests are available from many companies and are compatible with the chemical nature of RNA. The second point is much more difficult to quantify. The analytical methods available will more or less detect the contaminating RNA. Because RNA molecules are complex structures, a certain amount of RNA contaminants will remain whatever purification methods are used. The apparent quality of a RNA batch depends on the tests

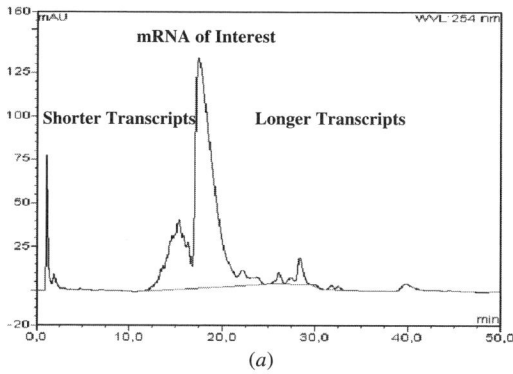

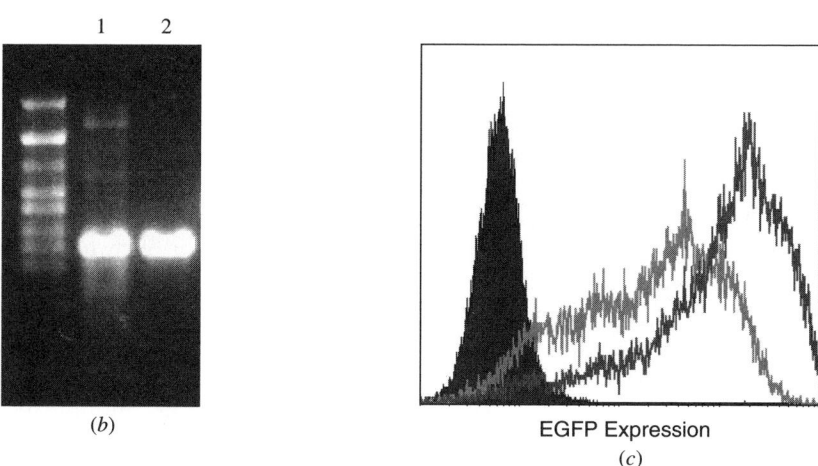

Figure 27.9 PUREmessenger. In (*a*) is shown the chromatography profile of an in vitro transcript coding for EGFP (raw product). The same product is shown on a gel electrophoresis analysis in (*b*), track 1. The mRNA contained in the main peak of the chromatograph was recovered and analyzed on the gel electrophoresis shown in (*b*), track 2. This PUREmessenger mRNA is free of smaller and longer transcripts. The primary transcripts shown in track 1 and the PUREmesseger product shown in track 2 were transfected by electroporation in BHK cells. Eighteen hours later, the expression of EGFP was measured by FACS analysis. The results are shown in (*c*). The filled blue line is the negative control: Mock-transfected cells; the gray line reports the fluorescence of BHK cells transfected with the raw product; and the darker line reports the fluorescence of BHK cells transfected with the PUREmessenger product. PUREmessenger mRNA are free of any contaminants, and they are more efficient than nonpurified mRNA for protein expression.

performed, the methods used, their sensitivity, and the degree of tolerance. This qualification will follow rules mainly defined together by the customer, the provider, and the relevant legal authorities. The limits of purity should not be too stringent in order to produce large amount of therapeutic RNA for reasonable costs.

Oligonucleotides As far as purity of an oligonucleotide batch is concerned, mass spectrometry analysis is a gold standard. The mass of the product(s) contained in the preparation are identified and, ideally, the only detectable mass is the one of the desired RNA oligonucleotide. However, a certain amount of contaminants will always be present. HPLC preparations can remove most of the shorter or longer contaminating oligonucleotides but not all. For example, N-1 oligonucleotides are usually still detectable (N-2 and N+1 oligonucleotides can also be present). In collaboration with the authorities in charge of API in his country, the sponsor who uses the oligonucleotide in the clinic needs to set limits of impurities that can be contained in his GMP-quality oligonucleotide. Ideally, the contaminants detected by mass spectrometry should all be characterized (sequence, structure). Additional analytical methods such as polyacrylamide gel electrophoresis (PAGE) analysis, capillary electrophoresis, HPLC profile, and nuclear magnetic resonance can also be performed. The more methods, the more contaminants will be detected. For each technique, the sponsor must set limits of purity that are acceptable. Purity of up to 99 percent (HPLC profile) can be reached but requires a very stringent purification (high loss of material). Usually purities higher than 96 percent (HPLC profile) are accepted for clinical tests. This limit can be set lower for modified oligonucleotides. For example, phosphorothioate backbone modifications made during synthesis by oxidation after each coupling step are not performed at 100 percent. Consequently, some nucleotide bonds will remain phosphodiester. These contaminants are difficult to eliminate by HPLC quantitatively and will be detected by mass spectrometry. Thus, depending on the length and number of bonds that must be modified, the final oligonucleotide batch may contain more than 5 percent of oligonucleotides that miss one or several phosphorothioate bonds. Moreover, isomeric forms of the molecules coexist. For example, in a phosphorothioate oligonucleotide, a mixture of S_p and R_p isomers are formed (no method to chemically produce chirally pure phosphorothioate oligonucleotides is available). This isomeric heterogeneity may give complex HPLC profiles where the biologically active product is contained in several peaks. The regulatory authorities are usually well aware of these limits in the production and purification of oligonucleotides. A close collaboration between the sponsor, the manufacturer, and the authorities is necessary to fix the set of analytical methods and the degree of acceptance of impurities in the oligonucleotide preparation dedicated to clinical use.

Messenger RNA Since mRNA is produced with biological tools and requires DNA as a template, the first purity controls to be done on a mRNA batch together with standard tests performed on any API (sterility, residual solvents, heavy metals, endotoxin content, pH, appearance) are the detection of contaminating DNA or proteins. The detection of DNA can be done by PCR or real-time PCR with primers specific for bacterial genomic DNA (the enzymes are produced from bacteria and the DNA matrix for transcription is usually produced from bacteria), plasmid DNA (when transcription is performed using a plasmid template), or PCR products (when the transcription is made from a PCR template, e.g., in the case of libraries). No amplicon should be detectable with 30 cycles of PCR performed on 10 ng of the mRNA batch. The detection of residual proteins can be made using a standard Bradford assay.

In addition, the purity of a defined mRNA (a single defined sequence) should be demonstrated by formaldehyde agarose gel electrophoresis analysis or ion-pair reverse-phase HPLC [130]. If a certain amount of in vitro transcribed product (more than 2 µg) is run in a gel with 0.5-cm slots, a certain amount of longer or shorter transcripts will usually be detectable [visualized by ultraviolet (UV) illumination of the ethidium bromide stained gel]. They should be characterized (purified, reverse transcribed, and sequenced) or, better, discarded during production using a purification method such as PUREmessenger from CureVac (Tuebingen, Germany). The sponsor, together with the producer and the relevant authorities, must fix the limit of detection of these longer and shorter transcript contaminants.

When mRNA libraries are needed (antitumor vaccination [115]), purity controls may be limited to the standard tests: sterility, residual solvents, heavy metals, endotoxin content, pH, appearance, together with detection of contaminating DNA and proteins.

Identity The main hurdle in GMP RNA production is to prove the identity of the molecules contained in the final batch. A general identity test can be simply asking "Is there RNA in the preparation?" The answer lies in the sensitivity of the product to purified RNases. An incubation of RNA molecules with RNases should result in the complete disappearance of the nucleic acid as evidenced by the relevant analytical method (mass spectrometry, PAGE or formaldehyde/agarose gels, depending on the size of the RNA). Modified RNA that are very resistant to RNases may, however, not be suitable for such an identity test.

In addition to this general test, further precise identity tests may be required by the authorities especially in phase III clinical trials and thereafter. They are discussed bellow.

Oligonucleotides As far as the sequence of the oligonucleotides is concerned, mass spectrometry can be used. It gives, of course, a very good indication of product identity: The measured mass of the dominant product should be exactly the predicted mass of the oligonucleotide. It does not, however, prove the identity of the RNA: Two RNA with the same A, C, G, and U content but different sequences will have the same mass. The sequencing of the oligonucleotide can be achieved by partial degradation of the molecule with exonucleases and mass spectrometry analysis of the resulting fragments. The mass difference between the initial oligonucleotide and the smaller degradation fragments (N-1, N-2, etc.) will indicate precisely which base is at each position. Performing this analysis separately with a 5'-exonuclease (phosphodiesterase II from bovine spleen) and a 3'-exonuclease (phosphodiesterase I from *Crotalus adamanteus*) allows confirmation of the entire oligonucleotide sequence in the batch. Again, such analysis may be difficult when the oligonucleotide contains chemical modifications that render it resistant to RNase degradation. For oligonucleotides with phosphorothioate backbones and/or DNA–RNA chimers similar enzymatic degradation methods followed by mass spectrometry analysis can be evolved to confirm the oligonucleotide's sequence and structure.

Fragmentation without enzymatic degradation can be performed by electrospray tandem mass spectrometry. The amount of fragments obtained with such a method makes the analysis difficult but may result in the confirmation of the oligonucleotide's sequence in a single experimental setup.

Manufacturers of GMP oligonucleotides have developed and qualified customized analytical methods. In conclusion, thanks to a combination of enzymatic tests and mass spectrometry measurements enabling a precise analysis of oligonucleotide's sequence and structure, it is now possible to confirm its identity. Since each oligonucleotide has a specific structure (sequence, modifications, DNA/RNA chimera, etc.), customized identity tests are required.

Messenger RNA Formaldehyde agarose gel electrophoresis analysis that is used to check the purity of the mRNA batch will also give information as to the identity of the mRNA. Compared to a defined RNA ladder (that must be qualified), the dominant product should have the expected size. Estimation of the size of an mRNA on a gel electrophoresis has an imprecision of ca. 10 percent of the length: a 1-kb fragment will be recognized as an RNA longer than 900 bases and shorter than 1100 bases. In this example, if the product is less than 900 bases or longer than 1100 bases, the gel electrophoresis analysis will allow the producer to recognize a problem in mRNA identity. Due to its relatively low precision, this analysis hints at the correct identity of the mRNA but provides no proof. A more precise identity test is possible after reverse transcription of the mRNA followed by PCR amplification with primer designed to recognize specifically the expected cDNA. This test will prove that at least the expected RNA sequence was present in the mRNA batch. It will, however, not demonstrate that all RNA molecules in the batch have the expected sequence. The cloning of the reverse transcribed product into a plasmid, followed by sequencing of randomly picked clones will demonstrate that virtually all RNA molecules contained in the batch have the expected sequence. This test is available but difficult to qualify at a GMP level. Nevertheless, it can be used if the regulatory authorities request an accurate identity check of the mRNA molecules contained in a GMP batch.

As far as mRNA libraries are concerned, no sequence identity test can be performed. On a formaldehyde/agarose gel electrophoresis the library should appear as a smear of fragments from 400 to 6000 bases with a maximum intensity around 1.5 kb (running 3 μg of mRNA in a 0.5-cm broad slot). Eventually, functional studies of the library can be undertaken: reverse transcription, cloning, and sequencing to confirm that most mRNA contained in the library is full-length mRNA or in vitro translation followed by Western blot analysis with antibodies specific for some housekeeping proteins to demonstrate that known proteins are translated from the mRNA batch. Again the relevance/cost ratio of these analyses must be set by the customer together with the local authorities.

A specific test necessary to qualify mRNA is the quantification of molecules that contain the 5′ cap structure. Theoretically, from the in vitro transcription setup, 80 percent of the molecules should contain the cap (usual cap:GTP ration is 4:1). Functionally, this structure is extremely important: The mRNA would not be translated into protein without the 5′ cap structure. Consequently, a test quantifying the percentage of capped molecule is required to qualify a GMP mRNA batch. To achieve this, some methods (corporate know-how, not publicly available) have been developed and validated by the two providers of GMP mRNA, Ambion (Austin, Texas) and CureVac (Tuebingen, Germany).

To conclude on the identity tests for mRNA, several methods exist to partially or accurately verify the sequence and functionality of the mRNA. Here again, the methods, the limit of sensitivity, and accuracy must be decided by the customer and the relevant authorities.

27.7 CONCLUSION AND PERSPECTIVES: RNA-BASED THERAPIES OF THE FUTURE

Aptamers, ribozymes, siRNA, immunostimulating RNA, and mRNA are as many tools that have proven in pre clinical and clinical studies to be effective against a wide variety of diseases. Production of these molecules is available at large scale and in GMP quality. The first RNA oligonucleotide that reached the market (Macugen, Eyetech) has proven the pharmaceutical potential of RNA and has paved the way for qualification and large-scale production of RNA.

Most diseases with unmet medical needs can be addressed by one or several therapies based on RNA. Cancer-related diseases, for example, may be treated by one or a combination of RNA therapies. It could consist in an antitumor vaccine (based on mRNA together with immunostimulating RNA) and/or specific inhibitors of oncogen production (ribozymes or siRNA specifically blocking the expression of growth factors) and/or aptamers that would recognize tumor cells and induce their neutralization or destruction. Similarly, the spread of infectious diseases such as AIDS or malaria may be prevented (mRNA-based vaccines) or treated (mRNA-based vaccines plus eventually aptamers and siRNA or ribozymes) with RNA-based therapeutics.

Although the 1990s were dominated by the development of DNA-based therapies (plasmid DNA, recombinant viruses, transfected bacteria, or cells), the future clearly will rely on the safe and versatile pharmaceutical use of RNA. As the source of life 4.2 billion years ago, RNA is now regarded as a base for future therapies dedicated to improving and saving lives.

REFERENCES

1. Palade, G. E. (1955). A small particulate component of the cytoplasm. *J. Biophys. Biochem. Cytol.*, *1*, 59–68.
2. Littlefield, J. W., Keller, E. B., Gross, J., Zamecnik, P. C. (1955). Studies on cytoplasmic ribonucleoprotein particles from the liver of the rat. *J. Biol. Chem.*, *217*, 111–123.
3. Crick, F. (1970). Central dogma of molecular biology. *Nature*, *227*, 561–563.
4. Brenner, S. J. F. M. M. (1961). An unstable intermediate carrying information from genes to ribosome for protein synthesis. *Nature*, *190*, 576–581.
5. Gros, F., Gilbert, W., Hiatt, H. H., Attardi, G., Spahr, P. F., Watson, J. D. (1961). Molecular and biological characterization of messenger RNA. *Cold Spring Harb. Symp. Quant. Biol.*, *26*, 111–132.
6. Hoagland, M. B., Stephenson, M. L., Scott, J. F., Hecht, L. I., Zamecnik, P. C. (1958). A soluble ribonucleic acid intermediate in protein synthesis. *J. Biol. Chem.*, *231*, 241–257.
7. Feldmann, H., Zachau, H. G. (1964). Chemical evidence for the 3′-linkage of amino acids to s-RNA+. *Biochem. Biophys. Res. Commun.*, *15*, 13–17.
8. Chapeville, F., Lipmann, F., Von Ehrenstein, G., Weisblum, B., Ray, W. J., Jr., Benzer, S. (1962). On the role of soluble ribonucleic acid in coding for amino acids. *Proc. Natl. Acad. Sci. USA*, *48*, 1086–1092.
9. Absher, M., Stinebring, W. R. (1969). Toxic properties of a synthetic double-stranded RNA. Endotoxin-like properties of poly I. poly C, an interferon stimulator. *Nature*, *223*, 715–717.
10. Guerrier-Takada, C., Gardiner, K., Marsh, T., Pace, N., Altman, S. (1983). The RNA moiety of ribonuclease P is the catalytic subunit of the enzyme. *Cell*, *35*, 849–857.
11. Cech, T. R., Bass, B. L. (1986). Biological catalysis by RNA. *Annu. Rev. Biochem.*, *55*, 599–629.
12. Delihas, N. (1995). Regulation of gene expression by trans-encoded antisense RNAs. *Mol. Microbiol.*, *15*, 411–414.
13. Tuschl, T., Borkhardt, A. (2002). Small interfering RNAs: A revolutionary tool for the analysis of gene function and gene therapy. *Mol. Interv.*, *2*, 158–167.
14. Matzke, M., Matzke, A. J. (2003). RNAi extends its reach. *Science*, *301*, 1060–1061.

15. Schramke, V., Allshire, R. (2003). Hairpin RNAs and retrotransposon LTRs effect RNAi and chromatin-based gene silencing. *Science*, *301*, 1069–1074.
16. Yao, M. C., Fuller, P., Xi, X. (2003). Programmed DNA deletion as an RNA-guided system of genome defense. *Science*, *300*, 1581–1584.
17. Deutscher, M. P., (1993). Ribonuclease multiplicity, diversity, and complexity. *J. Biol. Chem.*, *268*, 13011–13014.
18. Egli, M., Gryaznov, S. M. (2000). Synthetic oligonucleotides as RNA mimetics: 2′-Modified Rnas and N3′ → > P5′ phosphoramidates. *Cell Mol. Life Sci.*, *57*, 1440–1456.
19. Koshkin, A. A., Wengel, J. (1998). Synthesis of novel 2′,3′-linked bicyclic thymine ribonucleosides. *J. Org. Chem.*, *63*, 2778–2781.
20. Wittenhagen, L. M., Kelley, S. O. (2003). Impact of disease-related mitochondrial mutations on tRNA structure and function. *Trends Biochem. Sci.*, *28*, 605–611.
21. Andrake, M., Guild, N., Hsu, T., Gold, L., Tuerk, C., Karam, J. (1988). DNA polymerase of bacteriophage T4 is an autogenous translational repressor. *Proc. Natl. Acad. Sci. USA*, *85*, 7942–7946.
22. Powers, T., Daubresse, G., Noller, H. F. (1993). Dynamics of in vitro assembly of 16 S rRNA into 30 S ribosomal subunits. *J. Mol. Biol.*, *232*, 362–374.
23. Sullenger, B. A., Gallardo, H. F., Ungers, G. E., Gilboa, E. (1990). Overexpression of TAR sequences renders cells resistant to human immunodeficiency virus replication. *Cell*, *63*, 601–608.
24. Tuerk, C., Gold, L. (1990). Systematic evolution of ligands by exponential enrichment: RNA ligands to bacteriophage T4 DNA polymerase. *Science*, *249*, 505–510.
25. Ellington, A. D., Szostak, J. W. (1990). In vitro selection of RNA molecules that bind specific ligands. *Nature*, *346*, 818–822.
26. Daniels, D. A., Chen, H., Hicke, B. J., Swiderek, K. M., Gold, L. (2003). A tenascin-C aptamer identified by tumor cell SELEX: Systematic evolution of ligands by exponential enrichment. *Proc. Natl. Acad. Sci. USA*, *100*, 15416–15421.
27. Lee, J. F., Hesselberth, J. R., Meyers, L. A., Ellington, A. D. (2004). Aptamer database. *Nucleic Acids Res.*, *32 Database issue*, D95–100.
28. Aurup, H., Williams, D. M., Eckstein, F. (1992). 2′-Fluoro- and 2′-amino-2′-deoxynucleoside 5′-triphosphates as substrates for T7 RNA polymerase. *Biochemistry*, *31*, 9636–9641.
29. Melbye, S. W., Brant, B. A., Freedberg, I. W. (1977). Epidermal nucleases. III. The ribonucleases of human epidermis. *Br. J. Dermatol.*, *97*, 355–364.
30. Lorger, M., Engstler, M., Homann, M., Goringer, H. U. (2003). Targeting the variable surface of African trypanosomes with variant surface glycoprotein-specific, serum-stable RNA aptamers. *Eukaryot. Cell*, *2*, 84–94.
31. Darfeuille, F., Arzumanov, A., Gryaznov, S., Gait, M. J., Di Primo, C., Toulme, J. J. (2002). Loop-loop interaction of HIV-1 TAR RNA with N3′ → P5′ deoxyphosphoramidate aptamers inhibits in vitro Tat-mediated transcription. *Proc. Natl. Acad. Sci. USA*, *99*, 9709–9714.
32. Darfeuille, F., Arzumanov, A., Gait, M. J., Di Primo, C., Toulme, J. J. (2002). 2′-O-methyl-RNA hairpins generate loop-loop complexes and selectively inhibit HIV-1 Tat-mediated transcription. *Biochemistry*, *41*, 12186–12192.

33. Darfeuille, F., Hansen, J. B., Orum, H., Di Primo, C., Toulme, J. J. (2004). LNA/DNA chimeric oligomers mimic RNA aptamers targeted to the TAR RNA element of HIV-1. *Nucleic Acids Res.*, *32*, 3101–3107.

34. Hicke, B. J., Stephens, A. W. (2000). Escort aptamers: A delivery service for diagnosis and therapy. *J. Clin. Invest.*, *106*, 923–928.

35. White, R. R., Sullenger, B. A., Rusconi, C. P. (2000). Developing aptamers into therapeutics. *J. Clin. Invest.*, *106*, 929–934.

36. Sayer, N., Ibrahim, J., Turner, K., Tahiri-Alaoui, A., James, W. (2002). Structural characterization of a 2′ F-RNA aptamer that binds a HIV-1 SU glycoprotein, gp120. *Biochem. Biophys. Res. Commun.*, *293*, 924–931.

37. Bock, L. C., Griffin, L. C., Latham, J. A., Vermaas, E. H., Toole, J. J. (1992). Selection of single-stranded DNA molecules that bind and inhibit human thrombin. *Nature*, *355*, 564–566.

38. Santulli-Marotto, S., Nair, S. K., Rusconi, C., Sullenger, B., Gilboa, E. (2003). Multivalent RNA aptamers that inhibit CTLA-4 and enhance tumor immunity. *Cancer Res.*, *63*, 7483–7489.

39. Homann, M., Goringer, H. U. (1999). Combinatorial selection of high affinity RNA ligands to live African trypanosomes. *Nucleic Acids Res.*, *27*, 2006–2014.

40. Sayer, N. M., Cubin, M., Rhie, A., Bullock, M., Tahiri-Alaoui, A., James, W. (2004). Structural determinants of conformationally selective, prion-binding aptamers. *J. Biol. Chem.*, *279*, 13102–13109.

41. Ruckman, J., Green, L. S., Beeson, J., Waugh, S., Gillette, W. L., Henninger, D. D., Claesson-Welsh, L., Janjic, N. (1998). 2′-Fluoropyrimidine RNA-based aptamers to the 165-amino acid form of vascular endothelial growth factor (VEGF165). Inhibition of receptor binding and VEGF-induced vascular permeability through interactions requiring the exon 7-encoded domain. *J. Biol. Chem.*, *273*, 20556–20567.

42. Drolet, D. W., Nelson, J., Tucker, C. E., Zack, P. M., Nixon, K., Bolin, R., Judkins, M. B., Farmer, J. A., Wolf, J. L., Gill, S. C., Bendele, R. A. (2000). Pharmacokinetics and safety of an anti-vascular endothelial growth factor aptamer (NX1838) following injection into the vitreous humor of rhesus monkeys. *Pharm. Res.*, *17*, 1503–1510.

43. (2003). Anti-vascular endothelial growth factor therapy for subfoveal choroidal neovascularization secondary to age-related macular degeneration: Phase II study results. *Ophthalmology*, *110*, 979–986.

44. Csaky, K. (2003). Anti-vascular endothelial growth factor therapy for neovascular age-related macular degeneration: Promises and pitfalls. *Ophthalmology*, *110*, 879–881.

45. Pestka, S., Daugherty, B. L., Jung, V., Hotta, K., Pestka, R. K. (1984). Anti-mRNA: specific inhibition of translation of single mRNA molecules. *Proc. Natl. Acad. Sci. USA*, *81*, 7525–7528.

46. Coleman, J., Green, P. J., Inouye, M. (1992). The use of RNAs complementary to specific mRNAs to regulate the expression of individual bacterial genes. *Biotechnology*, *24*, 253–260.

47. Izant, J. G., Weintraub, H. (1985). Constitutive and conditional suppression of exogenous and endogenous genes by anti-sense RNA. *Science*, 229, 345–352.
48. Kurreck, J. (2003). Antisense technologies. Improvement through novel chemical modifications. *Eur. J. Biochem.*, 270, 1628–1644.
49. Kruger, K., Grabowski, P. J., Zaug, A. J., Sands, J., Gottschling, D. E., Cech, T. R. (1982). Self-splicing RNA: Autoexcision and autocyclization of the ribosomal RNA intervening sequence of Tetrahymena. *Cell*, 31, 147–157.
50. Doudna, J. A., Cech, T. R. (2002). The chemical repertoire of natural ribozymes. *Nature*, 418, 222–228.
51. Unrau, P. J., Bartel, D. P. (1998). RNA-catalysed nucleotide synthesis. *Nature*, 395, 260–263.
52. Lohse, P. A., Szostak, J. W. (1996). Ribozyme-catalysed amino-acid transfer reactions. *Nature*, 381, 442–444.
53. Haseloff, J., Gerlach, W. L. (1988). Simple RNA enzymes with new and highly specific endoribonuclease activities. *Nature*, 334, 585–591.
54. Eckstein, F. (1996). The hammerhead ribozyme. *Biochem. Soc. Trans.*, 24, 601–604.
55. Anderson, P., Monforte, J., Tritz, R., Nesbitt, S., Hearst, J., Hampel, A. (1994). Mutagenesis of the hairpin ribozyme. *Nucleic Acids Res.*, 22, 1096–1100.
56. Earnshaw, D. J., Gait, M. J. (1997). Progress toward the structure and therapeutic use of the hairpin ribozyme. *Antisense Nucleic Acid Drug Devel.*, 7, 403–411.
57. Beigelman, L., McSwiggen, J. A., Draper, K. G., Gonzalez, C., Jensen, K., Karpeisky, A. M., Modak, A. S., Matulic-Adamic, J., DiRenzo, A. B., Haeberli, P. (1995). Chemical modification of hammerhead ribozymes. Catalytic activity and nuclease resistance. *J. Biol. Chem.*, 270, 25702–25708.
58. Usman, N., Blatt, L. M. (2000). Nuclease-resistant synthetic ribozymes: Developing a new class of therapeutics. *J. Clin. Invest.*, 106, 1197–1202.
59. Pieken, W. A., Olsen, D. B., Aurup, H., Williams, D. M., Heidenreich, O., Benseler, F., Eckstein, F. (1991). Structure-function relationship of hammerhead ribozymes as probed by 2′-modifications. *Nucleic Acids Symp. Ser.*, 51–53.
60. Scherr, M., Rossi, J. J., Sczakiel, G., Patzel, V. (2000). RNA accessibility prediction: A theoretical approach is consistent with experimental studies in cell extracts. *Nucleic Acids Res.*, 28, 2455–2461.
61. Bramlage, B., Luzi, E., Eckstein, F. (2000). HIV-1 LTR as a target for synthetic ribozyme-mediated inhibition of gene expression: Site selection and inhibition in cell culture. *Nucleic Acids Res.*, 28, 4059–4067.
62. Pierce, M. L., Ruffner, D. E. (1998). Construction of a directed hammerhead ribozyme library: Towards the identification of optimal target sites for antisense-mediated gene inhibition. *Nucleic Acids Res.*, 26, 5093–5101.
63. Rhoades, K., Wong-Staal, F. (2003). Inverse genomics as a powerful tool to identify novel targets for the treatment of neurodegenerative diseases. *Mech. Ageing Devel.*, 124, 125–132.
64. Flory, C. M., Pavco, P. A., Jarvis, T. C., Lesch, M. E., Wincott, F. E., Beigelman, L., Hunt, S. W., III, Schrier, D. J. (1996). Nuclease-resistant ribozymes decrease stromelysin mRNA levels in rabbit synovium following exogenous delivery to the knee joint. *Proc. Natl. Acad. Sci. USA*, 93, 754–758.

65. Pavco, P. A., Bouhana, K. S., Gallegos, A. M., Agrawal, A., Blanchard, K. S., Grimm, S. L., Jensen, K. L., Andrews, L. E., Wincott, F. E., Pitot, P. A., Tressler, R. J., Cushman, C., Reynolds, M. A., Parry, T. J. (2000). Antitumor and antimetastatic activity of ribozymes targeting the messenger RNA of vascular endothelial growth factor receptors. *Clin. Cancer Res.*, *6*, 2094–2103.

66. Sandberg, J. A., Bouhana, K. S., Gallegos, A. M., Agrawal, A. B., Grimm, S. L., Wincott, F. E., Reynolds, M. A., Pavco, P. A., Parry, T. J. (1999). Pharmacokinetics of an antiangiogenic ribozyme (ANGIOZYME) in the mouse. *Antisense Nucleic Acid Drug Devel.*, *9*, 271–277.

67. Elbashir, S. M., Harborth, J., Lendeckel, W., Yalcin, A., Weber, K., Tuschl, T. (2001). Duplexes of 21-nucleotide RNAs mediate RNA interference in cultured mammalian cells. *Nature*, *411*, 494–498.

68. Elbashir, S. M., Lendeckel, W., Tuschl, T. (2001). RNA interference is mediated by 21- and 22-nucleotide RNAs. *Genes Devel.*, *15*, 188–200.

68a. Jorgensen, R. (1990). Altered gene expression in plants due to trans interactions between homologous genes. *Trends Biotech*, *8*, 340–344.

69. Fire, A., Xu, S., Montgomery, M. K., Kostas, S. A., Driver, S. E., Mello, C. C. (1998). Potent and specific genetic interference by double-stranded RNA in *Caenorhabditis elegans*. *Nature*, *391*, 806–811.

70. Caplen, N. J., Parrish, S., Imani, F., Fire, A., Morgan, R. A. (2001). Specific inhibition of gene expression by small double-stranded RNAs in invertebrate and vertebrate systems. *Proc. Natl. Acad. Sci. USA*, *98*, 9742–9747.

71. Couzin, J. (2002). Breakthrough of the year. Small RNAs make big splash. *Science*, *298*, 2296–2297.

72. Stark, G. R., Kerr, I. M., Williams, B. R., Silverman, R. H., Schreiber, R. D. (1998). How cells respond to interferons. *Annu. Rev. Biochem.*, *67*, 227–264.

73. Yuan, B., Latek, R., Hossbach, M., Tuschl, T., Lewitter, F. (2004). siRNA Selection Server: An automated siRNA oligonucleotide prediction server. *Nucleic Acids Res.*, *32*, W130–W134.

74. Sioud, M. (2004). Therapeutic siRNAs. *Trends Pharmacol. Sci.*, *25*, 22–28.

75. Jackson, A. L., Bartz, S. R., Schelter, J., Kobayashi, S. V., Burchard, J., Mao, M., Li, B., Cavet, G., Linsley, P. S. (2003). Expression profiling reveals off-target gene regulation by RNAi. *Nat. Biotechnol.*, *21*, 635–637.

76. Semizarov, D., Frost, L., Sarthy, A., Kroeger, P., Halbert, D. N., Fesik, S. W. (2003). Specificity of short interfering RNA determined through gene expression signatures. *Proc. Natl. Acad. Sci. USA*, *100*, 6347–6352.

77. Sioud, M., Sorensen, D. R. (2003). Cationic liposome-mediated delivery of siRNAs in adult mice. *Biochem. Biophys. Res. Commun.*, *312*, 1220–1225.

78. Sioud, M., Sorensen, D. R. (2004). Systemic delivery of synthetic siRNAs. *Methods Mol. Biol.*, *252*, 515–522.

78a. Hornung, V., Guenthner-Biller, M., Bourquin, C., Ablasser, A., Schlee, M., Uematsu, S., Noronha, A., Manoharan, M., Akira, S., de Fougerolles, A., Endres, S., Hartmann, G. (2005). Sequence-specific potent induction of IFN-alpha by short interfering RNA in plasmacytoid dendritic cells through TLR7. *Nat. Med.*, *11*, 263–270.

REFERENCES

79. Amarzguioui, M., Holen, T., Babaie, E., Prydz, H. (2003). Tolerance for mutations and chemical modifications in a siRNA. *Nucleic Acids Res.*, *31*, 589–595.
80. Elbashir, S. M., Martinez, J., Patkaniowska, A., Lendeckel, W., Tuschl, T. (2001). Functional anatomy of siRNAs for mediating efficient RNAi in *Drosophila melanogaster* embryo lysate. *EMBO J.*, *20*, 6877–6888.
81. Braasch, D. A., Jensen, S., Liu, Y., Kaur, K., Arar, K., White, M. A., Corey, D. R. (2003). RNA interference in mammalian cells by chemically-modified RNA. *Biochemistry*, *42*, 7967–7975.
82. Layzer, J. M., McCaffrey, A. P., Tanner, A. K., Huang, Z., Kay, M. A., Sullenger, B. A. (2004). In vivo activity of nuclease-resistant siRNAs. *RNA*, *10*, 766–771.
83. Braasch, D. A., Paroo, Z., Constantinescu, A., Ren, G., Oz, O. K., Mason, R. P., Corey, D. R. (2004). Biodistribution of phosphodiester and phosphorothioate siRNA. *Bioorg. Med. Chem. Lett.*, *14*, 1139–1143.
83a. Zhang, et al. (2003). Vector based RNAi, a novel tool for isoform-specific knockdown of VEGF and anti-angiogenesis gene therapy of cancer. *Biochem. Biophys. Res. Commun. 303*, 1169–1178.
84b. Scherr, et al. (2003). Specific inhibition of Bcr-abl gene expression by small interfering RNA. *Blood*, *101*, 1566–1569.
84c. Bullock and Fersht (2001). Rescuing the function of mutant p53. *Nat. Rev. Cancer*, *1*, 68–76.
84d. Martinez (2002). Synthetic small inhibiting RNAs: Efficient tools to inactivate oncogenic mutations and restore p53 pathways. *Proc. Natl. Acad. Sci. USA*, *99*, 14849–14854.
84. Zender, L., Hutker, S., Liedtke, C., Tillmann, H. L., Zender, S., Mundt, B., Waltemathe, M., Gosling, T., Flemming, P., Malek, N. P., Trautwein, C., Manns, M. P., Kuhnel, F., Kubicka, S. (2003). Caspase 8 small interfering RNA prevents acute liver failure in mice. *Proc. Natl. Acad. Sci. USA*, *100*, 7797–7802.
85. Song, E., Lee, S. K., Wang, J., Ince, N., Ouyang, N., Min, J., Chen, J., Shankar, P., Lieberman, J. (2003). RNA interference targeting Fas protects mice from fulminant hepatitis. *Nat. Med.*, *9*, 347–351.
86. Sorensen, D. R., Leirdal, M., Sioud, M. (2003). Gene silencing by systemic delivery of synthetic siRNAs in adult mice. *J. Mol. Biol.*, *327*, 761–766.
87. Lewis, D. L., Hagstrom, J. E., Loomis, A. G., Wolff, J. A., Herweijer, H. (2002). Efficient delivery of siRNA for inhibition of gene expression in postnatal mice. *Nat. Genet.*, *32*, 107–108.
88. McCaffrey, A. P., Meuse, L., Pham, T. T., Conklin, D. S., Hannon, G. J., Kay, M. A. (2002). RNA interference in adult mice. *Nature*, *418*, 38–39.
89. Templeton, N. S. (2002). Cationic liposome-mediated gene delivery in vivo. *Biosci. Rep.*, *22*, 283–295.
90. Liang, L., Liu, D. P., Liang, C. C. (2002). Optimizing the delivery systems of chimeric RNA. DNA oligonucleotides. *Eur. J. Biochem.*, *269*, 5753–5758.
91. Jarver, P., Langel, U. (2004). The use of cell-penetrating peptides as a tool for gene regulation. *Drug Discov. Today*, *9*, 395–402.

92. Simeoni, F., Morris, M. C., Heitz, F., Divita, G. (2003). Insight into the mechanism of the peptide-based gene delivery system MPG: Implications for delivery of siRNA into mammalian cells. *Nucleic Acids Res.*, *31*, 2717–2724.

93. Kozak, M. (1978). How do eucaryotic ribosomes select initiation regions in messenger RNA? *Cell*, *15*, 1109–1123.

94. Kozak, M. (1999). Initiation of translation in prokaryotes and eukaryotes. *Gene*, *234*, 187–208.

95. Ross, J. (1995). mRNA stability in mammalian cells. *Microbiol. Rev.*, *59*, 423–450.

96. Wilusz, C. J., Wormington, M., Peltz, S. W. (2001). The cap-to-tail guide to mRNA turnover. *Nat. Rev. Mol. Cell Biol.*, *2*, 237–246.

97. Chen, C. Y., Shyu, A. B. (1995). AU-rich elements: Characterization and importance in mRNA degradation. *Trends Biochem. Sci.*, *20*, 465–470.

98. Holcik, M., Liebhaber, S. A. (1997). Four highly stable eukaryotic mRNAs assemble 3' untranslated region RNA-protein complexes sharing cis and trans components. *Proc. Natl. Acad. Sci. USA*, *94*, 2410–2414.

99. Malone, R. W., Felgner, P. L., Verma, I. M. (1989). Cationic liposome-mediated RNA transfection. *Proc. Natl. Acad. Sci. USA*, *86*, 6077–6081.

100. Wolff, J. A., Malone, R. W., Williams, P., Chong, W., Acsadi, G., Jani, A., Felgner, P. L. (1990). Direct gene transfer into mouse muscle in vivo. *Science*, *247*, 1465–1468.

101. Pascolo, S. (2004). Messenger RNA-based vaccines. *Exp. Opin. Biol. Ther.*, *4*, 1285–1294.

102. Martinon, F., Krishnan, S., Lenzen, G., Magne, R., Gomard, E., Guillet, J. G., Levy, J. P., Meulien, P. (1993). Induction of virus-specific cytotoxic T lymphocytes in vivo by liposome-entrapped mRNA. *Eur. J. Immunol.*, *23*, 1719–1722.

102a. Conry, R. M., LoBuglio, A. F., Wright, M., Sumerel, L., Pike, M. J., Johanning, F., Benjamin, R., Lu, D., Curiel, D. T. (1995). Characterization of a messenger RNA polynucleotide vaccine vector. *Cancer Res.*, *55*, 1397–400.

102b. Hoerr, I., Obst, R., Rammensee, H. G., Jung, G. (2000). In vivo application of RNA leads to induction of specific cytotoxic T lymphocytes and antibodies. *Eur. J. Immunol.*, *30*, 1–7.

102c. Granstein, R. D., Ding, W., Ozawa, H. (2000). Induction of anti-tumor immunity with epidermal cells pulsed with tumor-derived RNA or intradermal administration of RNA. *J. Invest. Dermatol.*, *114*, 632–636.

103. Carralot, J. P., Probst, J., Hoerr, I., Scheel, B., Teufel, R., Jung, G., Rammensee, H., Pascolo, S. (2004). Polarization of the immunity induced by direct injection of naked globin UTR-stabilized mRNA vaccines. *Cell Mol. Life Sci.*

104. Qiu, P., Ziegelhoffer, P., Sun, J., Yang, N. S. (1996). Gene gun delivery of mRNA in situ results in efficient transgene expression and genetic immunization. *Gene Ther.*, *3*, 262–268.

105. Vassilev, V. B., Gil, L. H., Donis, R. O. (2001). Microparticle-mediated RNA immunization against bovine viral diarrhea virus. *Vaccine*, *19*, 2012–2019.

106. Mandl, C. W., Aberle, J. H., Aberle, S. W., Holzmann, H., Allison, S. L., Heinz, F. X. (1998). In vitro-synthesized infectious RNA as an attenuated live vaccine in a flavivirus model. *Nat. Med.*, *4*, 1438–1440.

107. Racanelli, V., Behrens, S. E., Aliberti, J., Rehermann, B. (2004). Dendritic cells transfected with cytopathic self-replicating RNA induce crosspriming of CD8(+) T cells and antiviral immunity. *Immunity*, 20, 47–58.

108. Schirmacher, V., Forg, P., Dalemans, W., Chlichlia, K., Zeng, Y., Fournier, P., von Hoegen, P. (2000). Intra-pinna anti-tumor vaccination with self-replicating infectious RNA or with DNA encoding a model tumor antigen and a cytokine. *Gene Ther.*, 7, 1137–1147.

109. Ying, H., Zaks, T. Z., Wang, R. F., Irvine, K. R., Kammula, U. S., Marincola, F. M., Leitner, W. W., Restifo, N. P. (1999). Cancer therapy using a self-replicating RNA vaccine. *Nat. Med.*, 5, 823–827.

110. Zhou, X., Berglund, P., Rhodes, G., Parker, S. E., Jondal, M., Liljestrom, P. (1994). Self-replicating Semliki Forest virus RNA as recombinant vaccine. *Vaccine*, 12, 1510–1514.

111. Boczkowski, D., Nair, S. K., Snyder, D., Gilboa, E. (1996). Dendritic cells pulsed with RNA are potent antigen-presenting cells in vitro and in vivo. *J. Exp. Med.*, 184, 465–472.

112. Sallusto, F., Lanzavecchia, A. (1994). Efficient presentation of soluble antigen by cultured human dendritic cells is maintained by granulocyte/macrophage colony-stimulating factor plus interleukin 4 and downregulated by tumor necrosis factor alpha. *J. Exp. Med.*, 179, 1109–1118.

113. Heiser, A., Coleman, D., Dannull, J., Yancey, D., Maurice, M. A., Lallas, C. D., Dahm, P., Niedzwiecki, D., Gilboa, E., Vieweg, J. (2002). Autologous dendritic cells transfected with prostate-specific antigen RNA stimulate CTL responses against metastatic prostate tumors. *J. Clin. Invest.*, 109, 409–417.

114. Morse, M. A., Nair, S. K., Mosca, P. J., Hobeika, A. C., Clay, T. M., Deng, Y., Boczkowski, D., Proia, A., Neidzwiecki, D., Clavien, P. A., Hurwitz, H. I., Schlom, J., Gilboa, E., Lyerly, H. K. (2003). Immunotherapy with autologous, human dendritic cells transfected with carcinoembryonic antigen mRNA. *Cancer Invest.*, 21, 341–349.

115. Su, Z., Dannull, J., Heiser, A., Yancey, D., Pruitt, S., Madden, J., Coleman, D., Niedzwiecki, D., Gilboa, E., Vieweg, J. (2003). Immunological and clinical responses in metastatic renal cancer patients vaccinated with tumor RNA-transfected dendritic cells. *Cancer Res.*, 63, 2127–2133.

116. Diebold, S. S., Kaisho, T., Hemmi, H., Akira, S., Reis E. S. (2004). Innate antiviral responses by means of TLR7-mediated recognition of single-stranded RNA. *Science*, 303, 1529–1531.

117. Sieling, P. A., Modlin, R. L. (2002). Toll-like receptors: Mammalian "taste receptors" for a smorgasbord of microbial invaders. *Curr. Opin. Microbiol.*, 5, 70–75.

118. Heil, F., Hemmi, H., Hochrein, H., Ampenberger, F., Kirschning, C., Akira, S., Lipford, G., Wagner, H., Bauer, S. (2004). Species-specific recognition of single-stranded RNA via toll-like receptor 7 and 8. *Science*, 303, 1526–1529.

119. Scheel, B., Braedel, S., Probst, J., Carralot, J. P., Wagner, H., Schild, H., Jung, G., Rammensee, H. G., Pascolo, S. (2004). Immunostimulating capacities of stabilized RNA molecules. *Eur. J. Immunol.*, 34, 537–547.

120. Pimm, M. V., Embleton, M. J., Baldwin, R. W. (1976). Treatment of transplanted rat tumours with double-stranded RNA (BRL 5907). I. Influence of systemic and local administration. *Br. J. Cancer*, *33*, 154–165.
121. Carter, W. A., Strayer, D. R., Hubbell, H. R., Brodsky, I. (1985). Preclinical studies with Ampligen (mismatched double-stranded RNA). *J. Biol. Response Mod.*, *4*, 495–502.
122. Brodsky, I., Strayer, D. R., Krueger, L. J., Carter, W. A. (1985). Clinical studies with ampligen (mismatched double-stranded RNA). *J. Biol. Response Mod.*, *4*, 669–675.
123. Schroder, H. C., Kelve, M., Schacke, H., Pfleiderer, W., Charubala, R., Suhadolnik, R. J., Muller, W. E. (1994). Inhibition of DNA topoisomerase I activity by 2′, 5′-oligoadenylates and mismatched double-stranded RNA in uninfected and HIV-1-infected H9 cells. *Chem. Biol. Interact.*, *90*, 169–183.
124. Thompson, K. A., Strayer, D. R., Salvato, P. D., Thompson, C. E., Klimas, N., Molavi, A., Hamill, A. K., Zheng, Z., Ventura, D., Carter, W. A. (1996). Results of a double-blind placebo-controlled study of the double-stranded RNA drug polyI:polyC12U in the treatment of HIV infection. *Eur. J. Clin. Microbiol. Infect. Dis.*, *15*, 580–587.
125. Essey, R. J., McDougall, B. R., Robinson, W. E., Jr. (2001). Mismatched double-stranded RNA (polyI-polyC(12)U) is synergistic with multiple anti-HIV drugs and is active against drug-sensitive and drug-resistant HIV-1 in vitro. *Antiviral Res.*, *51*, 189–202.
126. Adams, M., Navabi, H., Jasani, B., Man, S., Fiander, A., Evans, A. S., Donninger, C., Mason, M. (2003). Dendritic cell (DC) based therapy for cervical cancer: Use of DC pulsed with tumour lysate and matured with a novel synthetic clinically non-toxic double stranded RNA analogue poly [I]:poly [C(12)U] (Ampligen R). *Vaccine*, *21*, 787–790.
127. Scaringe, S. A., Francklyn, C., Usman, N. (1990). Chemical synthesis of biologically active oligoribonucleotides using beta-cyanoethyl protected ribonucleoside phosphoramidites. *Nucleic Acids Res.*, *18*, 5433–5441.
128. Scaringe, S. A., Wincott, F. E., Caruthers, M. H. (1998). Novel RNA synthesis method using 5′-silyl-2′-orthoester protecting groups. *J. Am. Chem. Soc.*, *120*, 11820–11821.
129. Kambhampati, R. V., Chiu, Y. Y., Chen, C. W., Blumenstein, J. J. (1993). Regulatory concerns for the chemistry, manufacturing, and controls of oligonucleotide therapeutics for use in clinical studies. *Antisense Res. Devel.*, *3*, 405–410.
130. Azarani, A., Hecker, K. H. (2001). RNA analysis by ion-pair reversed-phase high performance liquid chromatography. *Nucleic Acids Res.*, *29*, E7.

28

NOVEL IMAGING AGENTS FOR MOLECULAR MR IMAGING OF CANCER

DMITRI ARTEMOV AND ZAVER M. BHUJWALLA

The Johns Hopkins University School of Medicine
Baltimore, Maryland

28.1	INTRODUCTION	1310
28.2	CONTRAST AGENTS AND MECHANISM OF CONTRAST ENHANCEMENT IN MRI	1311
	T1 Contrast Agents	1311
	T2 Contrast Agents	1312
	Relaxation Properties of Main Classes of CA	1313
28.3	MRI OF TUMOR VASCULATURE AND VASCULAR TARGETS	1313
	Low-Molecular-Weight Contrast Agents (GdDTPA)	1315
	High-Molecular-Weight Contrast Agents	1316
	Molecular Imaging of Tumor Neovasculature	1316
28.4	MRI OF TUMOR CELL SURFACE RECEPTORS	1318
	MR Molecular Imaging of Isolated Cells	1318
	In Vivo MR Molecular Imaging of Receptors in Cancer	1320
28.5	MRI OF INTRACELLULAR TARGETS	1325
	Contrast Agents for Long-Term Labeling of Target Cells (MRI Cell Trafficking)	1326
	Specific Contrast Agents for Intracellular Targets	1327
28.6	MOLECULAR IMAGING WITH ACTIVATED MR CONTRAST AGENTS	1328
	Enzymatic Activation of CA	1328
	CEST and PARACEST Contrast Agents	1329
28.7	MOLECULAR MRI AND MR SPECTROSCOPY	1329
	CSI Molecular Imaging with Reporter Molecules	1330

Drug Discovery Handbook, by Shayne Cox Gad
Copyright © 2005 by John Wiley & Sons, Inc.

| 28.8 | DISCUSSION | 1331 |
| | References | 1333 |

28.1 INTRODUCTION

The response of solid tumors to therapy is traditionally measured by changes in tumor volume over the course of treatment. Because of its high spatial resolution and contrast in soft tissues, magnetic resonance imaging (MRI) provides an informative noninvasive approach to measuring tumor volumes. Additionally, nonspecific low-molecular-weight paramagnetic contrast agents (such as GdDTPA) can be used to derive perfusion characteristics of the lesion.

Novel therapeutic strategies often deviate from traditional cytotoxic chemotherapy as they do not directly kill tumor cells. As a result, tumor volume alone is no longer the ultimate response parameter. For example, one novel paradigm in tumor treatment is antiangiogenic therapy that targets tumor neovasculature. In this case treatment response can be best assessed from changes in tumor vascular parameters. Standard morphological MRI cannot detect these functional physiological parameters, and the application of dynamic MRI with appropriate contrast agents is required to determine the status of tumor vascularization. For example, dynamic MRI using a high-molecular-weight contrast agent enables quantitative measurements of tumor blood volume and vascular permeability surface area product and has been demonstrated to detect tumor response to antiangiogenic therapy [1]. Changes in the concentration of tumor metabolites detected by magnetic resonance spectroscopy (MRS) can also be an early marker of tumor response to conventional chemotherapy prior to changes in tumor volume [2, 3]. In vivo MRS might be a method of choice to detect early apoptotic events [4, 5] as well as tumor response to noncytotoxic (cytostatic) chemotherapy. In the latter case the lack of the tumor response to the therapy can be promptly addressed by switching to a more aggressive form of treatment.

Target-specific tumor therapy (based, e.g., on humanized monoclonal antibodies) is another important development in cancer treatment. The identification of a subpopulation of tumors that express therapeutic targets (such as cell surface receptors) and the monitoring of their status during treatment can help to optimize the therapeutic regimen and to improve cure. MRI with highly specific contrast agents (CA) can help to image these targets noninvasively with high spatial and temporal resolution. In the next section we present a brief overview of available contrast agents and the general mechanisms of contrast enhancement in MRI.

28.2 CONTRAST AGENTS AND MECHANISMS OF CONTRAST ENHANCEMENT IN MRI

The intensity of MR images typically depends upon the concentration of resonating nuclei (protons) in the sample and on their relaxation parameters. Relaxation processes that contribute to the measured signal can be described as T1, T2, and T2* relaxation [6]. Relaxation contrast agents that modify these relaxation rates are classified as "T1 agents" or "T2 (T2*) agents," respectively. T1 contrast agents provide a positive contrast or increased intensities in T1-weighted MR images. T2 or T2* contrast agents provide a negative contrast or reduced signals in T2 or T2* weighted MR images, respectively. To simplify our discussion we have combined T2 and T2* contrast agents as one group, discussing them separately only when necessary as a critical parameter for the method. The parameter relaxivity is used to define the effectiveness of a T1 or T2 relaxation CA. For CA based on paramagnetic metals, it is defined as a reciprocal of changes in T1 or T2 relaxation time per unit concentration of the metal $(mM \cdot s)^{-1}$. T1 and T2 relaxation processes are based on similar molecular mechanisms and most CA affect both T1 and T2 relaxation times. The classification of CA as T1 or T2 agents is therefore conditional and simply means that a particular CA under a certain set of experimental conditions will predominantly alter the T1 or T2 relaxation time.

T1 Contrast Agents

The important advantages of T1 relaxation agents for in vivo application are (1) the long precontrast tissue T1 relaxation time of about 1.7 s and (2) the positive contrast (or signal enhancement) generated by the agent in MR images. The majority of T1 contrast agents currently used both in the clinic and in animal studies are chelate complexes of paramagnetic metal ions such as Gd(III), Mn(II), or Fe(III). Several stable complexes of Gd(III) including linear GdDTPA (diethylenetriaminepentaacetic acid) and cyclic GdDOTA (1,4,7,10-tetraazacyclododecane-1,4,7,10-tetraacetic acid) chelates are extracellular low-molecular-weight compounds with high T1 relaxivity and are the only class of paramagnetic CA approved for clinical use. Typical examples of low-molecular-weight Gd-based agents include Gadoteridol, Gd-HP-DO3A (Prohance), Gadopentate dimeglumine, Gd-DTPA (Magnevist), Gadoversetamide (Optimark), and Gadodiamide, Gd-DTPA-BMA (Omniscan).

To improve the T1 relaxivity and to prolong the plasma lifetime of the agent, multiple Gd chelates can be attached to a single polymeric backbone molecule. The resulting complexes have a longer circulation time and increased relaxivity that is advantageous for a range of clinical applications, although the potential toxicity and induction of immune response to the polymeric carrier remain to be determined. Several gadolinium-based macromolecular imaging platforms have been designed and tested in preclinical models. Protein–Gd complexes include albumin–Gd conjugates, which are classic prototypes for intravascular imaging agents [7, 8]; poly-L-lysine–Gd [9, 10], avidin–Gd [11], as well as direct mAb–Gd conjugates [12, 13]. Polyamidoamine (PAMAM) dendrimers of different generations have also been used as a carrier for multiple Gd groups [14, 15]. For targeted intravascular imaging, several classes of very large molecular weight agents labeled with a high number of Gd atoms have been developed including cross-linked liposomes and nanoparticle emulsions [16, 17]. The typical molecular weight and the size of polymer Gd chelates varies from about 80 kD for protein-based agents to several million daltons for polymerized liposomes and emulsions with molecular diameters up to 200 nm. T1 relaxivity of macromolecular Gd complexes strongly depend on the B_0 magnetic field as well as on the rotational and tumbling correlation times of the complex. For targeted CA the correlation time can change significantly if the complex is immobilized by specific binding to the target receptor. The reported T1 relaxivity of the macromolecular CA ranges from 12 $(mM \cdot s)^{-1}$ for 200-nm nanoparticle emulsion at $B_0 = 4.7T$ [16] to 36 $(mM \cdot s)^{-1}$ for generation-nine PAMAM dendrimers [14] at 20 MHz and 80 $(mM \cdot s)^{-1}$ for MS-325 bound to albumin [18]. The relaxivity of a macromolecular Gd polychelate per unit concentration of the polymer increases linearly with the number of gadoliniums, as long as two Gd ions are separated by a distance that is much longer then a characteristic diffusion radius of water molecules. For a typical contact interaction between a water molecule and a metal ion of 100 ps and water diffusion coefficient of $10^{-5} cm^2/s$, the minimal distance between adjacent Gd ions should be larger than about 100 Å. If this condition is fulfilled, then the T1 relaxivity of these agents (per unit concentration of the metal) should not depend on the number of attached gadolinium ions. Indeed the measured T1 relaxivity of albumin GdDTPA complexes with a labeling ratio changing in the range of 9 to 18 Gd per albumin globule remains close to 12 $(mM \cdot s)^{-1}$ at 10.7 MHz [8].

T2 Contrast Agents

Contrast agents that increase T1 relaxation and accordingly reduce T1 relaxation time also change T2 relaxation as T2 is always shorter than T1. In fact all T1 agents discussed in the previous section can also act as T2 CA. However, because in a typical in vivo situation the inherent T2 relaxation time is significantly shorter (by an order of magnitude) than the corresponding T1 relaxation time, these agents induced dramatically more significant changes in T1 relaxation of the sample.

T2 contrast agents typically possess higher magnetic moments in comparison with T1 CA, and, because of their larger molecular size, they have correspondingly longer correlation times τ_c. This high magnetic moment makes them efficient at reducing T2 relaxation time and the relatively long τ_c reduces their effects on T1 relaxation time. T2 MR contrast agents usually consist of a superparamagnetic iron-oxide core that generates a macroscopic magnetic moment of the aligned electronic moments (so-called Curie spin [19]). This magnetic moment gives rise to strong local gradients of magnetic field that efficiently dephase nearby protons in an area much larger than the size of the magnetic particle [19]. Inhomogeneous line broadening induced by these agents also results in a significant shortening of T2* relaxation time. However, because of the very short inherent T2* in biological systems at high magnetic fields B_0, in the majority of in vivo applications T2 contrast enhancement is measured. Smaller superparamagnetic CA can also serve as efficient T1 contrast agents at low B_0 of about 1.5T, where τ_c^{-1} of the CA molecule is close to the proton magnetic resonance frequency.

The magnetic core of T2 CA usually consists of monocrystalline (MION) or polycrystalline (SPIO) iron oxide with a diameter of 5 and 30 nm, respectively. To improve the stability and biocompatibility of the complex, the core is coated with a polymer such as dextran or other polysaccharide resulting in a total particle diameter of 17 to 50 nm. Dextran-coated cross-linked iron oxide nanoparticles (CLIO) can be easily modified by functionalizing surface groups in the dextran coating [20]. Alternatively, the magnetic core can be embedded into a unilamellar liposome vesicle (magnetoliposomes [21]). Magnetodendrimers are CA with a superparamagnetic iron oxide core encapsulated within a dendrimer superstructure [22, 23].

An important feature of the iron-oxide-based CA is their high T2 relaxivity, which depends upon the size of the superparamagnetic core. The T2 relaxivity of the ultrasmall monocrystalline superparamagnetic iron oxide (MION-46L) with a diameter of the magnetic core of 4 nm (2046 Fe atoms) is close to 20 $(mM \cdot s)^{-1}$ at 1.5T magnetic field and 25°C [24]. SPIO particles with iron core diameter of 16 nm have the typical T2 relaxivity at 1.5T of about 240 $(mM \cdot s)^{-1}$. For comparison the T1 relaxivity of these compounds at 1.5T magnetic field is below 10 $(mM \cdot s)^{-1}$ [25].

Relaxation Properties of Main Classes of CA

Relaxation and molecular properties of different MRI agents suitable as platforms for development of targeted CA for MR molecular imaging are summarized in Table 28.1. MR molecular imaging presents unique challenges to researchers because of the inherently low sensitivity of MR detection (Gd concentrations in tens of micromolars) and the typically low number of imaging targets per cell. To design CA with the highest relaxivity per CA molecule, it is necessary either to use a large iron oxide core or large polymer carriers complexed with multiple Gd groups. The large molecular size prevents the efficient delivery of the agent to the imaging site because of slow extravasation and restricted diffusion of the large CA through the tumor interstitium. This problem may be less important for MRI of tumors with highly permeable vasculature [26] and is not relevant for imaging of molecular targets in the lumen of blood vessels such as the vascular endothelium [16, 17].

28.3 MRI OF TUMOR VASCULATURE AND VASCULAR TARGETS

Mechanistic approaches to characterizing the tumor vasculature include measurements of the following vascular parameters: tumor blood flow, tumor

TABLE 28.1 Typical Properties of Platforms for Targeted MR Contrast Agents

CA Class	Representative Compound	Number of Metal Atoms	T1 Relaxivity $(mM \cdot s)^{-1}$	T2 Relaxivity $(mM \cdot s)^{-1}$	Magnetic Field B_0 (T)	Molecular Diameter (nm)	Molecular Weight (Da)
Low molecular weight	GdDTPA Magnevist®	1 [Gd^{3+}]	4.5	5.7	1.5	—	743
Protein carriers	Albumin	19 [Gd^{3+}]	11.5	12.4	0.25	8	~80,000
	Poly-L-lysine	65 [Gd^{3+}]	13	15	0.47	—	~52,000
Dendrimers	PAMAM Generation 5	96–128 [Gd^{3+}]	22.5	27	1.5	6–8	~60,000
	Gadomer-17	24 [Gd^{3+}]	13		1.5	5–6	17,500 (35,000 apparent size)
Iron oxide nanoparticles	MION-46	2064 [Fe] 4.6 nm iron core	7	20	1.5	20	~775,000
	SPIO Feridex®	~6.5 nm Fe_2O_3 core	24	107	0.47	70–140	Megadalton range
Magnetodendrimers		9.5 nm Fe_2O_3 core	12	225	1.5	40–50	Megadalton range
Liposomes	ACPL (antibody-conjugated paramagnetic liposome)	~50,000 [Gd^{3+}]	12	11	1.5	200	Megadalton range
Nanoparticle emulsions	Gd-perfluorocarbon nanoparticles	74,000 [Gd^{3+}]	34	50	1.5	250	Megadalton range

vascular volume, and vascular permeability. A traditional approach to quantifying tumor vasculature is by histological evaluation of tumor microvessel density (MVD) [27]. Histological methods, however, do not provide information about the functionality of the vasculature that is known to be partly, permanently, or intermittently collapsed in tumors [28, 29]. Dynamic contrast-enhanced MRI provides an alternative noninvasive approach to determine some of these parameters, and we will discuss several applications of dynamic MRI in the following section. Imaging of vascular endothelium receptors that are unique for the tumor neovasculature such as selectins or integrins is yet another area where MRI with targeted contrast agents was successfully used in preclinical tumor models. This molecular MRI approach may facilitate early detection of the malignant growth and/or identify molecular targets for therapeutic intervention.

Low-Molecular-Weight Contrast Agents (GdDTPA)

Dynamic MRI studies of tumors with low-molecular-weight CA were performed in several tumor types including breast [30], brain [31], and uterine tumors [32]. GdDTPA is the most widely used Gd-based paramagnetic contrast agent. In the following sections we will use this generic name for general low-molecular-weight CA. The dynamic parameters derived from tracer kinetic analyses of these experiments, together with the assumptions made, have been discussed in several excellent studies [33, 34]. Generally, these agents are not freely diffusible; therefore, it is not possible to quantify tumor blood flow based on the two-compartment Kety model [35]. GdDTPA contrast agents rapidly extravasate from the leaky tumor vasculature, and the first-order kinetic constant defined as a transfer coefficient between the blood plasma compartment and the tumor interstitium is therefore affected by insufficient tumor perfusion (flow-limited regime) and generally depends on both PS (vascular permeability surface area product) and blood flow [36].

In bolus tracking experiments GdDTPA is injected at high doses, and the difference in magnetic susceptibility between blood vessels and extravascular space results in local magnetic field gradients or inhomogeneity broadening effects. These local gradients generate T2 or T2* contrast in spin echo or gradient echo imaging, respectively. Kinetic analysis of the dynamic susceptibility contrast in bolus tracking experiments provides a means to determine regional blood volume (rBV). While this approach is feasible for measurements of cerebral blood volume [37], its application to tumors is limited by the leakage of the low-molecular-weight CA across highly permeable tumor vasculature. Increased concentration of the CA in the tumor interstitium tends to attenuate local susceptibility gradients, and the experimental data have to be corrected for the leakage effects [38]. T2 and T2* dynamic MRI is also sensitive to the caliber of tumor vasculature, and appropriate experiments can be used to characterize micro- and macrovasculature in the tumor [37, 39–42].

High-Molecular-Weight Contrast Agents

A generic high-molecular-weight contrast agent is one in which serum albumin is conjugated to multiple GdDTPA groups using free amines on the surface of the polymer globule [43]. This CA is characterized by a long plasma half-life (about 2 h), and after initial equilibration period (about 3 min in mice) its blood concentrations remain constant for at least 40 min after an intravenous bolus injection [1]. The CA slowly leaks across the tumor vasculature, and tissue concentrations of CA increase linearly with time. Simple linear kinetic analysis of albumin–GdDTPA concentration as a function of time is sufficient to quantify tissue blood volume (from extrapolation of kinetic curve to time $t = 0$ min) and permeability surface area product, PS, from the slope of the line [7]. Typically albumin–GdDTPA tissue concentrations are determined from changes in T1 relaxation time under assumption of fast exchange of water moleculaes between blood, interstitium, and cellular compartments. Intermediate to slow water exchange can result in underestimation of vascular volume and PS [44].

High-molecular-weight CA extravasate from the leaky tumor vasculature at a relatively slow rate as compared to small molecular weight CA such as GdDTPA. Therefore, their in vivo kinetic is (1) sensitive to a wide range of PS present in solid tumors and (2) can be modeled using a simple linear kinetic model [45]. The utility of this method was demonstrated for imaging of breast and prostate cancer models preselected for different metastatic potential. More metastatic tumors were generally characterized by increased tumor vascular volume and PS [7]. We recently also investigated the effect of the antiangiogenic agent TNP-470 on tumor vascular characteristics using MRI [1]. TNP-470 is a fumagillin derivative and its antiangiogenic effects are thought to be due to inhibition of endothelial cell proliferation [46].

Representative vascular volume and permeability data obtained for a pair of volume-matched control and treated tumors are shown in Figures 28.1 and 28.2. Treatment with TNP-470 resulted in a significant decrease in the permeable and vascular regions within the tumors. However a "compensatory" increase in levels of vascular volume and permeability was also apparent in some regions of the treated tumors and demonstrate the critical importance of imaging techniques that provide spatial information in determining the effects of antiangiogenic therapy.

Molecular Imaging of Tumor Neovasculature

Tumor growth depends on the establishment of new vasculature recruited from the host through the process of angiogenesis [47, 48]. This neovasculature is characterized by the expression of specific molecular markers such as $\alpha_v\beta_3$ integrins. In adults these markers are mostly localized to the site of active angiogenesis and therefore can be used as targets for imaging tumor neovascularization with low background from quiescent mature vasculature [49].

One successful strategy for MRI of vascular molecular epitopes is based on using high-affinity functionalized MR contrast agents directed against a specific target. Selective retention of the CA by the target and clearance of the

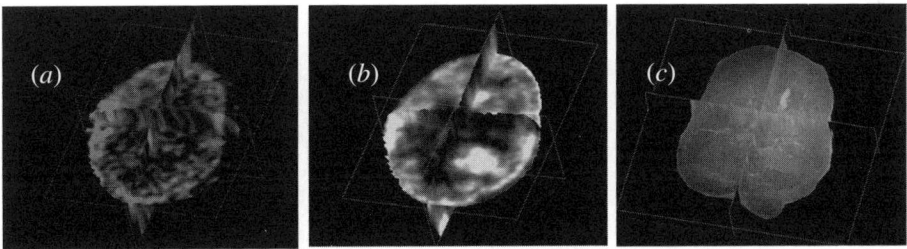

Figure 28.1 Triplanar views of three-dimensional reconstructed maps of (*a*) vascular volume, (*b*) permeability, and (*c*) hematoxylin and eosin-stained histological sections for a control MatLyLu tumor (volume 405 mm^3).

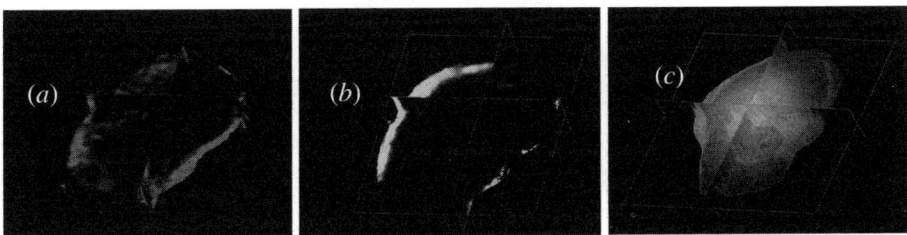

Figure 28.2 Triplanar views of three-dimensional reconstructed maps of (*a*) vascular volume, (*b*) permeability, and (*c*) hematoxyline and eosin-stained histological sections for a MatLyLu tumor treated with TNP-470, 30 mg/kg every second day, total dose 90 mg/kg (volume 395 mm^3).

unbound "free" agent generates measurable contrast in the image. Highly specific targeting can be achieved by attaching the imaging probe to monoclonal antibodies (mAb) or mAb fragments, adapter proteins, and/or synthetic polypeptides with high specific binding to the target. The intrinsically low sensitivity of MRI and low concentrations of molecular targets require using CA with the highest available relaxivity. The accessibility of molecular targets in the tumor vasculature allows the use of large contrast agents with molecular size of up to 1 μm. Molecules of CA are delivered within the bloodstream and reach and bind the target site without extravasation and/or diffusion across tissue barriers.

E-selectin, a proinflammatory marker of endothelial cells, is expressed in proliferating endothelium and can be a marker of tumor neovasculature [50]. An MR molecular imaging agent specific for E-selectins was developed by Kang et al. [50] and is a molecular conjugate between CLIO iron oxide particle and mAb F(ab)$_2$ fragment specific for the protein. MRI of endothelial cells in culture demonstrated strong retention of the contrast agent by the cells stimulated to express E-selectins [50].

$\alpha_v\beta_3$ receptors are expressed in the vascular angiogenic endothelium [16, 17]. $\alpha_v\beta_3$ receptors were imaged in vivo in the neovasculature of a rabbit tumor model with paramagnetic polymerized liposomes [17] and in the rabbit bFGF-

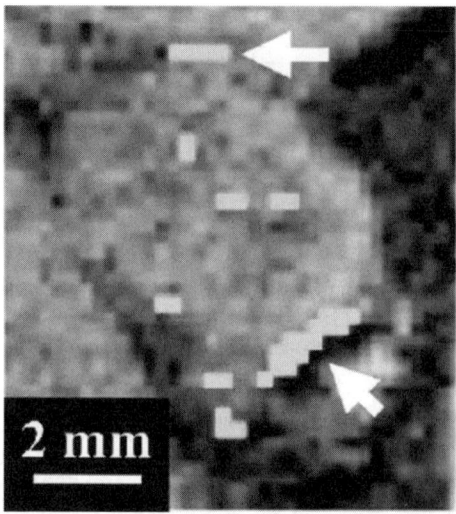

Figure 28.3 Enlarged section of T1-weighted MR image showing implanted Vx-2 tumor. Yellow overlay indicates MRI signal enhancement 120 min postinjection of $\alpha_v\beta_3$-targeted nanoparticles. MRI enhancement was predominantly, although not exclusively, asymmetrically distributed along the tumor periphery proximal to blood vessels and tissue fascial interfaces (white arrows). From [51].

induced corneal micropocket model using Gd-perfluorocarbon nanoparticles [16]. These imaging nanoparticles were targeted to the receptor by mAb covalently bound or attached via biotin-avidin linkers, respectively. More recently $\alpha_v\beta_3$ receptors were imaged in Vx-2 rabbit tumor model by targeted Gd-perfluorocarbon paramagnetic nanoparticles functionalized with a $\alpha_v\beta_3$-integrin peptidomimetic antagonist selected for high-affinity binding to the receptor [51]. MRI was performed on a standard clinical 1,5T scanner and demonstrated potential feasibility of the method for clinical translation (Fig. 28.3).

28.4 MRI OF TUMOR CELL SURFACE RECEPTORS

MR Molecular Imaging of Isolated Cells

Isolated cells in culture can be used as a test bed for identification of the potential tumor targets for imaging and/or therapy and for the development of surface receptor specific MRI contrast agents. Additionally, cell surface receptor CA systems can be used for long-term labeling of the cells for cell tracking MR application as an alternative to loading of cells with intracellular CA (discussed in next section). Caution should be exercised in the extrapolation of imaging results obtained in isolated cell to in vivo studies. Contrast agent pharmacokinetics, delivery properties, and stability in vivo can be determin-

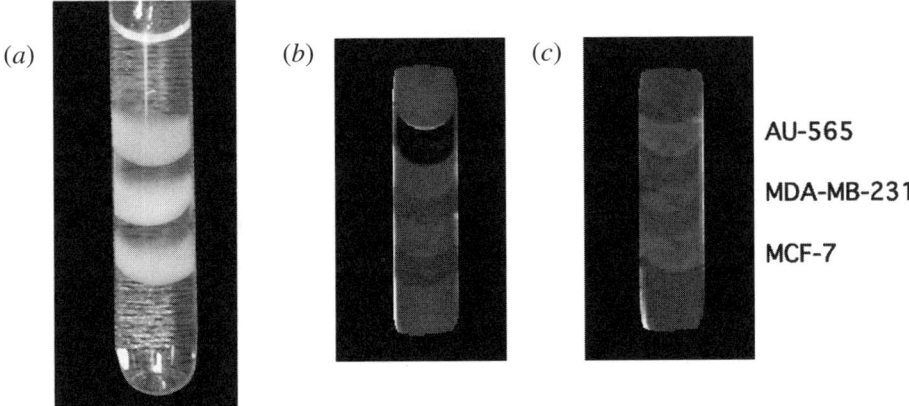

Figure 28.4 MRI of breast cancer cells. (*a–c*) Demonstrate the layout and MR images of cell samples consisting of layers of AU-565, MDA-MB-231, and MCF-7 cells embedded in agarose gel in a 5-mm NMR tube. Cells were pretargeted with biotinylated Herceptin and a nonspecific biotinylated mAb (negative control), and probed with streptavidin SPIO microbeads. T_2 maps of the cell samples were reconstructed from eight T_2-weighted images acquired with RD of 8s and TE in the range 20 to 250 ms. A T_2 map of a cell sample probed with Herceptin is shown in (*b*) and the control cell sample treated with a nonspecific biotinylated mAb is shown in (*c*). Adapted from [55].

ing factors for imaging efficiency in animal models and eventually in clinical applications.

In cultured cancer cells the surface receptors are entirely accessible for labeling with large molecular probes as in the case of endothelial cell receptors discussed above. The most efficient probes, as seen from Table 28.1, are iron-oxide-based CA as they provide the highest T2 relaxivity per mole of the compound. This group of CA includes SPIO, MION, CLIO, and magnetodendrimer nanoparticles. Chemical conjugation of the superparamagnetic iron oxide particle with mAb or mAb fragments provides selective binding of the CA to the target receptor on the cell surface. Original reports on using functionalized SPIO nanoparticles as MR receptor imaging agent were published in the 1990s [52, 53]. A multistep protocol based on biotinylated primary mAb, streptavidin, and biotinylated dextran-coated SPIO nanoparticles was proposed for imaging of lymphocytes with the mAb directed against the lymphocyte common antigen [54]. A two-step labeling method was also used to image HER-2/neu receptors expressed on the surface of malignant breast cancer cells using biotinylated Herceptin mAb and streptavidin-conjugated SPIO nanoparticles [55]. The results presented in Figure 28.4 demonstrate a significant negative T2 contrast in the HER-2/neu expressing cell layers. The highest contrast was detected in AU-565 cells expressing 2.7×10^6 receptors/cell and the lowest in MDA-MB-231 cells with 4×10^4 receptors/cell. The experimental data suggest a linear dependence between 1/T2 relaxation rate and concentration of the target sites for the SPIO-based CA [55].

Human engineered tranferrin receptor (ETR) was imaged in cancer cells using a conjugate of transferrin and MION nanoparticles [56, 57]. The CA was accumulated in ETR+ expressing cancer cells by internalization through receptor-mediated endocytosis mechanism.

A significant advantage of the iron-oxide-based CA is their high T2/T2* relaxivity that produces strong negative MR contrast already at nanomolar concentrations of the CA and provides sufficient sensitivity to detect cell surface receptors expressed at relatively low levels ($<10^4$ receptors/cell). However, the relatively large molecular size (>30 nm) of the superparamagnetic iron-oxide-based CA is a potential problem for in vivo applications, as discussed in the next section.

In Vivo MR Molecular Imaging of Receptors in Cancer

In this section we present a concise review of MRI results obtained in vivo in preclinical systems. These results may be considered as preliminary steps toward translation of the methods to clinical use. When translating these applications to the bedside, in contrast to nuclear imaging, molecular MRI encounters problems. These are primarily associated with safety issues due to potential toxicity and the triggering of the immune response by high doses of targeted CA, which need to be administered to produce sufficient contrast. On the other hand results obtained in animal systems do have intrinsic importance. They can provide insights in tumor biology, the tumor microenvironment and response to various stresses that essentially require in vivo longitudinal studies. These results can also be important for the rational design of novel anticancer treatments.

Gd-Based Contrast Agents We start our discussion with early attempts of in vivo MRI of various cell surface molecular targets in solid tumors using approaches similar to standard nuclear imaging protocols with radiolabeled mAb. To this end monoclonal antibodies were labeled with GdDTPA chelates, and reports on the use of the GdDTPA-mAb appear in the literature since 1985 [13, 58, 59]. Generally, these experiments produced discouraging results due to the low sensitivity of MR detection and limited contrast produced by insufficient concentration of the CA [58]. Many of these early studies also did not include appropriate control experiments. In comparison to nuclear imaging where nanomolar concentrations of immunoradioisotopes can be routinely detected in tumors, MRI detection requires concentrations of Gd above approximately 10 µM. Not more than 10 to 15 Gd complexes can be attached per antibody without significant reduction in the binding affinity [10]. Therefore, building a sufficient concentration of Gd at the target site is problematic if direct labeling of mAb with Gd is used. In a more recent study, Shahbazi-Gahrouei et al. reported on the ex vivo relaxometry of human MM-138 melanoma xenografts probed in vivo with antimelanoma GdDTPA-mAb [12]. These studies were performed with ex vivo MR at high magnetic field, with a high molar ratio of Gd to mAb. It is not clear that similar specific signal enhancement can be achieved in a typical in vivo MR setup.

The most straightforward way to increase the number of GD groups per mAb is to conjugate the antibody to a polymer carrier that is decorated by multiple Gd groups via an appropriate molecular linker. Antimucin mAb conjugated to poly-L-lysine-GdDTPA with a labeling ratio of 65 Gd ions per molecule were developed for imaging of mucinlike protein expressed in many types of gastrointestinal carcinomas [10]. The molecular weight of the conjugate was about 200 kDa, which may restrict its delivery to solid tumors. Konda et al. reported the development of a folate-conjugated dendrimer complexed with GdDTPA [60]. A fourth-generation PAMAM was used as the imaging platform for targeted MRI, and initial results were obtained for human folate receptor expressing ovarian tumor xenografts grown in nude mice [60].

Iron-Oxide-Based Contrast Agent Two quite different modes of application of iron-oxide-based CA were proposed for in vivo MRI. In the first variant iron oxide nanoparticles (MION) were conjugated to mAb or other high-affinity moieties to label cell surface receptors while mostly remaining in the extracellular compartment. In a study by Weissleder et al. human polyclonal IgG was used to target MION to sites of inflammation [61]. A similar approach was used for molecular MRI of apoptosis in EL4 solid tumor models exposed to a chemotherapeutic agent [62]. SPIO particles were conjugated to the C2 domain of the protein synaptotagmin, which binds with high affinity to phosphatidylserine residues that translocate to the outer leaflet of the plasma membrane in apoptotic cells. Efficient delivery of the CA to these sites can be explained by significantly increased vascular permeability in the sites of inflammation or in treated tumors. Targeting peptides conjugated to CLIO particles also labeled with fluorophores were used to image underglycosylated MUC-1 tumor antigen [63]. Combined T2 MRI and near-infrared fluorescence imaging demonstrated a specific accumulation of the contrast in the tumor that expressed the antigen as shown in Figure 28.5.

The second strategy for generating specific MR contrast is to load cells with iron oxide agent using a specific plasma membrane transporter system with an appropriate substrate. Transferrin-MION contrast agent was demonstrated in vivo in engineered transferring receptor (ETR) expressing 9L glioma cells [56]. The transgenic cells internalized up to 8×10^6 of the TF-targeted CA within an hour and MION-loaded cancer cells generated strong negative T2* contrast in vivo in gradient echo MR images. Images were obtained at 1.5T 24 h after intravenous injection of 3 mg of Tf-MION to a nude mouse with two ETR^- and ETR^+ 9L gliosarcoma tumors growing on opposite flanks of the animal. No differential contrast was detected in the T1 weighted image, and strong negative contrast was detected in the T2*-weighted image of the ETR^+ tumor. The efficient delivery of 17-nm MION particles to the tumor was most

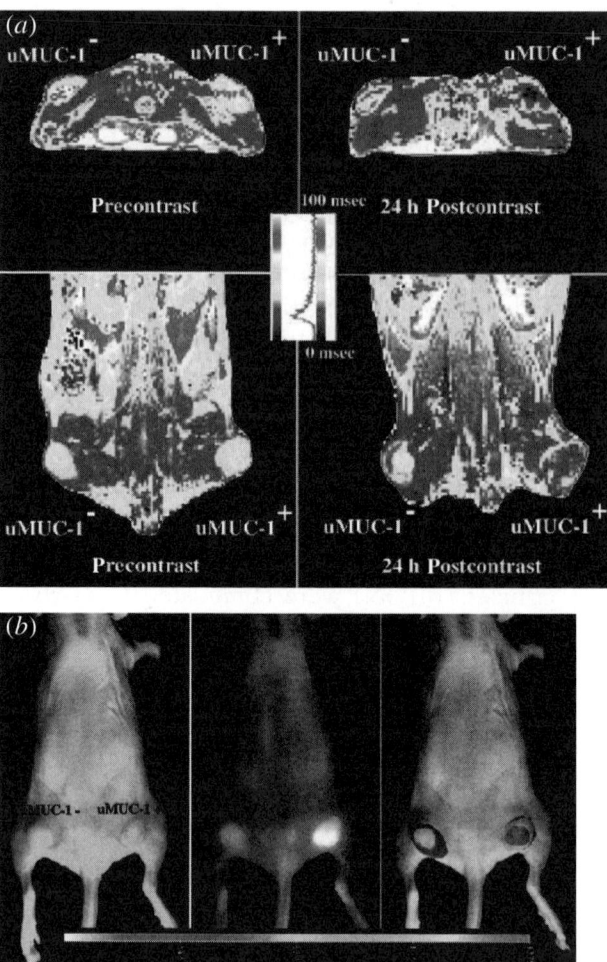

Figure 28.5 (*a*) Representative T2 maps of the animals bearing underglycosylated mucin-1 antigen (uMUC-1)-negative (U87) and uMUC-1-positive (LS174T) tumors. Transverse (*top*) and coronal (*bottom*) images showed a significant (52%; $P = 0.0001$) decrease in signal intensity in uMUC-1-positive tumors 24 h after administration of the CLIO-EPPT probe. (*b*) White light (*left*), NIRF (*middle*) images, and a color-coded map (*right*) of mice bearing bilateral underglycosylated mucin-1 antigen (uMUC-1)-negative (U87) and uMUC-1-positive (LS174T) tumors. NIRF imaging was performed immediately after the MRI session. Adapted from [63].

likely a result of the favorable pharmacokinetics of the agent due to its long circulation time and hyperpermeable tumor vasculature.

Multistep Targeting and Prelabeling Concept Tumor neovasculature is hyperpermeable in regions and permits extravasation of large molecules [64], which was demonstrated in several studies with macromolecular CA (above 20 nm molecular size), such as MION and SPIO [56, 62]. Solid-tumor vascularization is heterogeneous, and regions of high vascular permeability are often spatially distinct from regions of high vascular volume [7]. It was also shown that tumor perfusion is highly inhomogeneous and may have an intermittent blow flow [29]. Consequently, the delivery of high-molecular-weight imaging and therapeutic agents to tumor areas with low vascularity/flow/permeability can be significantly restricted. There is still a concern that extravasation and diffusion barriers for these nanocomplexes can significantly reduce their usefulness in viable regions of solid tumors following systemic administration [26]. Pharmacokinetics and delivery properties of targeted CA can be significantly improved by splitting the large targeted imaging/therapeutic probe to several functional components that can be administered and delivered to the tumor targets independently. This prelabeling concept is extensively used in nuclear imaging to increase the target/background ratio and to match the pharmacokinetics of radionuclides to their half-lifetimes [65].

HER-2/*neu* receptors were imaged in vivo using a two-step prelabeling strategy based on the avidin/biotin system [11]. Briefly, receptors in tumor xenografts were prelabeled with biotinylated mAb. After a 12-h interval, during which mAb were attached to the imaging targets and unbound mAb cleared from the systemic circulation, GdDTPA-labeled avidin was injected intravenously (IV). T1-weighted MRI was performed at 4.7T at different time points after the injection of avidin–GdDTPA and demonstrated positive T1 contrast in breast cancer tumor xenografts that overexpresssed the HER-2/*neu* receptor. The molecular weight of the individual components was 160 kDa for mAb and approximately 70 kDa for GdDTPA avidin, which corresponds to a molecular size smaller than 10 nm and provided efficient delivery of the CA to the interstitium of solid tumors in this model system. To further improve this approach we developed a three-component targeting system that included biotinylated primary mAb, purified avidin that served as a linker and mAb clearance agent, and biotinylated Gd-based macromolecule CA [66]. Avidin was injected 24 h after administration of the primary mAb and served two separate functions: (1) It provided selective labeling of cells with the attached biotinylated mAb and (2) avidin chase of circulating biotinylated mAb provided their rapid clearance from the blood, which significantly reduced background contrast [67]. In comparison to avidin–GdDTPA conjugates, this protocol also allows the use of optimized biotinylated paramagnetic agents with large number of Gd groups to generate contrast in MRI. An example of MRI of human HER-2/*neu* expressing BT-474 xenografts in SCID (severe combined immunodeficient) mouse is shown in Figure 28.6. Briefly, animals

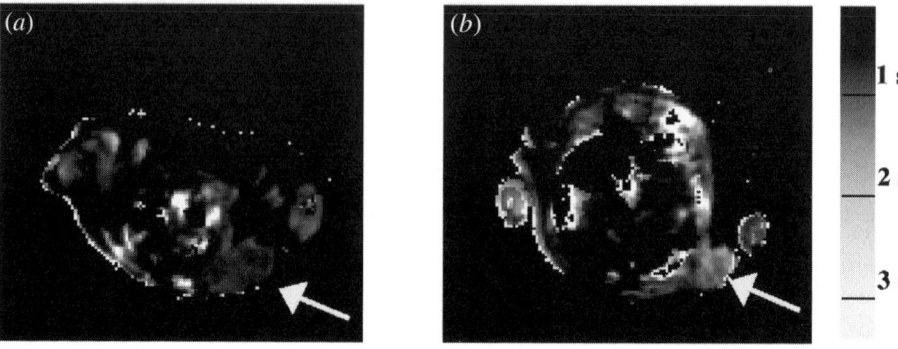

Figure 28.6 T1 maps of BT-474 tumors imaged with (*a*) biotinylated and (*b*) nonbiotinylated Herceptin. Tumors are indicated by arrows.

were injected IV with anti-HER-2/*neu* humanized mAb (Herceptin, Genentech) at a dose of 0.5 mg per animal. Biotinylated mAb was used for the experimental group and nonbiotinylated mAb for the control group of animals. Avidin (0.5 mg per animal) was administered IV 24 h after mAb, followed after 4 h by IV infusion of biotin-albumin-(GdDTPA)$_{20}$ contrast agent (0.5 mg/g). T1 relaxation maps were reconstructed from saturation recovery spin-echo imaging for several time points after administration of the contrast for the control and the experimental group and selective retention of the CA, and accordingly shorter T1 relaxation times were detected in BT-474 tumors treated with biotinylated Herceptin as shown in Figure 28.6.

Signal Amplification Concept The imaging techniques discussed so far rely on imaging of cell receptors expressed in large numbers (over a million) per cell. In many biologically relevant situations, expression levels of the receptor are significantly lower. It applies both to endogenous receptors that can be markers of a disease and to molecular reporter systems linked to the expression of extracellular receptors that can be used as molecular beacons for activation of relevant genes due to changes in the signaling and/or pharmacological interventions [68]. While MRI may not compare favorably with nuclear medicine for detecting cell surface receptors, its advantage arises from the ability to combine detection of these receptors with the array of functional imaging capabilities of MR methods. To achieve sufficient contrast enhancement from a low number of available imaging targets, signal amplification is necessary. Generally, signal amplification takes place if we can generate multiple contrast generating groups per single imaging target.

An increase in the number of CA molecules attached to a single extracellular target can be achieved by using the avidin–biotin system described above [69] as each avidin/streptavidin molecule can bind four biotins, and theoretically multiple labeling steps can give rise to treelike structures formed by the biotinylated CA. The relaxivity of individual CA units can be significantly improved by using polymer platforms labeled with multiple Gd chelates such as poly-L-lysine or dendrimer based CA [70]. Iron-oxide-based CA in principle can provide higher relaxivity in comparison with paramagnetic CA. Their large molecular size, however, can interfere with contrast delivery, and negative T2 contrast may be more difficult to detect. Recently, a novel frequency shift technique for detection of positive T2* contrast was developed that may provide an advantage for in vivo applications of SPIO base agents [71]. Another stratergy for signal amplification relies on loading target cells with CA molecules via active transport through the target receptor. Experiments with ETR receptor and MION nanoparticles conjugated to human holotransferrin is a good illustration of this approach [56]. The possible effects of the internalized CA on cell homeostasis and functions is one of the potential problems of the method.

The most effective approach to signal enhancement is to couple the targeting mechanism to a specific enzyme in such a way that the generated product acts as imaging or therapeutic CA. T1 relaxivity of a CA depends on the relation between the correlation time τ_c of the agent and B_0 magnetic field used for imaging. Enzymatic polymerization of small Gd complexes to a large-molecular-weight polymer product results in significant increase in τ_c and correspondingly significant increase in T1 relaxivity. Polymerization of hydroxyphenol-modified GdDOTA monomers was proposed as a sensitive probe for MRI of nanomolar concentrations of oxidoreductases (peroxidase) [72]. Low-relaxivity [3.75 (mM·s)$^{-1}$] monomeric substrates oxidized by the peroxidase-reduced peroxide produce activated molecules (A^*) that self-polymerase to form high-relaxivity [11.5 (mM·s)$^{-1}$] magnetic oligomers. This method was used for MRI of E-selectin expressed on the surface of endothelial cells using sandwich constructs of anti-E-selectin F(ab')$_2$ conjugated to peroxidase through DIG–anti-DIG Ab linkers [72]. It is still unclear if the method can be implemented in vivo; however, this strategy is definitely a significant step toward improving the performance of specific CA for molecular MR studies.

28.5 MRI OF INTRACELLULAR TARGETS

In previous sections we discussed various MR contrast agents suitable for molecular imaging of cellular epitopes using specific probes targeted against extracellular domain of cell surface receptors. In this chapter we will focus on intracellular MR contrast agents designed to penetrate the cell plasma membrane and accumulate in the cytoplasm and/or label specific intracellular targets. We will discuss two broad groups of intracellular CA that have recently been developed for MR imaging. The first group includes nonspecific intracellular agents primarily used for a long-term cell labeling for in vivo cell tracking. Typical examples of their application includes MR trafficking of inoculated stem cells and MRI of the docking of armed T cells to target sites. The second group consists of targeted intracellular contrast agents that have high affinity to molecular epitopes or can be activated by the action of cell enzymes.

Contrast Agents for Long-Term Labeling of Target Cells (MRI Cell Trafficking)

A typical experiment for MR cell trafficking includes in vitro labeling of cell population with an appropriate CA, administration of the labeled cells in vivo using systemic or local injection, and longitudinal MRI of the cell migration and accumulation to the target site(s). For a detailed discussion of MR tracking of magnetically labeled cells, we refer readers to excellent reviews by Bulte et al. [73–76].

Isolated cells labeled with iron oxide nanoparticles by phago- and/or endo/pinocytocis can be detected by MR at iron concentrations as low as 17 ng per 10^6 cells or 8.5×10^4 particles/cell (9L glioma cells loaded with MION particles [77]). MRI of a single T cell loaded with SPIO particles was demonstrated by Dodd et al. [78]. Typically, several million SPIO or USPIO particles must be loaded into the cell to provide sufficient contrast for MR detection. Detection of a cell labeled with just a single micron-size iron oxide particle was recently demonstrated in vitro [79]. One potential benefit of this approach is that the label is not diluted by subsequent cell division, and daughter cells can be detected for a long time postimplantation. T2*-weighted gradient-echo MRI that is sensitive to local disturbances in the magnetic field produced by the concentrated SPIO particles is usually used for detection of MR contrast in vitro in isolated cells. The benefit of T2* MRI is that "long-range" effects affect T2* relaxation of distant protons within an area much larger than the size of the nanoparticle, thus significantly amplifying sensitivity of the method [54, 78].

To improve internalization of the contrast agent, magnetic nanoparticles are typically mixed with standard cell transfecting agents such as poly-L-lysine [76]. Magnetodendrimers were proposed as an efficient cell-labeling MR agent. Stem cells loaded with magnetodendrimers (9 to 14 pg iron/cell) were successfully imaged both in vitro and in vivo for as long as 6 weeks after transplantation [23]. To further improve cell uptake properties, CA nanoparticles can be linked to the human immunodeficiency virus (HIV) viral transport Tat peptide that can channel the cargo molecules across the cell plasma membrane [80]. Efficient cellular uptake of Tat–CLIO as well as Tat–macrocyclic gadolinium chelates was recently demonstrated in in vitro and in in vivo systems [81, 82]. This approach is especially useful for loading of cells with low endocytotic/pinocytotic activity such as lymphocytes. Tat–CLIO nanoparticles were used to load cytotoxic T lymphocytes (CTL) ex vivo and to track their docking to the antigen-expressing tumor xenografts in vivo by T2-weighted MRI [83]. Repetitive noninvasive detection of the adoptively transferred cell targeting

Specific Contrast Agents for Intracellular Targets

Targeting intracellular epitopes is a major challenge for MR molecular imaging. Although many of potentially important biological targets are essentially intracellular such as enzymes, signaling proteins, deoxyribonucleic acid (DNA) and messenger ribonucleic acid (mRNA), most of the development in the area of MRI contrast agents was focused on extracellular CA. Indeed, in a traditional approach similar to the one used in nuclear imaging to specifically probe intracellular targets, contrast agent molecules need to freely diffuse across the cellular membrane and selectively accumulate in cells that express the marker. Large CA such as iron oxide nanoparticles primarily enter cells through endocytosis, and their retention in the cytoplasm is not target specific. To enable small gadolinium complexes to enter the cytoplasm and to provide a specific recognition of an intracellular target, a construct consisting of an amphiphilic transmembrane carrier peptide, Gd chelate, and c-Myc mRNA specific antisense peptide nuclei acid (PNA) [84] was created as shown in Figure 28.7. Specific uptake of the c-Myc-targeted CA was detected in rat

Figure 28.7 Spatial representation of one possible configuration of the c-myc-specific Gd3 complex. The transport peptide unit (green) is given as a ribbon representation of the peptide backbone. The heavy atoms of the address peptide (PNA; light brown), the disulfide bridge between the two peptide units and the two Lys-residues (cyan) connecting the Gd^{3+} complex to the PNA are displayed in a "ball-and-stick" representation. An atom color code (green, carbon; red, oxygen; blue, nitrogen) is used for the disulfide bridge and Gd^{3+} complex. *Magenta*, the van der Waals spheres of the Gd^{3+} ion. White, the hydrogen atoms of the two H_2O molecules in complex with the Gd^{3+}. The representation was generated using the INSIGHT II software package. From [84].

prostate adenocarcinoma model that expresses c-Myc. The structure of the gadolinium complex is not detailed, and the stability of the compound in vivo requires further evaluation.

28.6 MOLECULAR IMAGING WITH ACTIVATED MR CONTRAST AGENTS

The major idea that drives the development of activated MR probes is to gain the ability to control relaxation properties of the CA by chemical modification of its structure or by certain external action. A broad range of potential applications of such CA can be likened to the use of the cleavable optical probe technology, where fluorescence of the probe can be controlled by changing the intermolecular distance between the fluorescent and the quenching group [85, 86].

Enzymatic Activation of CA

One strategy to design target-specific MR contrast agents is to use a specific enzymatic reaction to activate the CA by increasing its relaxivity. The approach was originally reported by Louie et al. [87]. They designed "smart" relaxation agents in which the access of water to the first coordination sphere of a chelated paramagnetic ion is blocked with a substrate, galactopyranose, that can be removed by enzymatic cleavage by a traditional reporter gene, β-galactosidase. Following the cleavage of the blocking sugar group, the paramagnetic ion can interact directly with water protons to reduce their T1 by the inner sphere (contact) relaxation with the corresponding increase in the MR signal. Analogous approaches can be used to design CA with polypeptide blocking chains to probe protease activity. The enzyme β-galactosidase is expressed in the cell cytoplasm, therefore, the CA should be either microinjected to the cell(s) [87] or delivered to the cells by means of a certain cellular transport system such as amphiphilic Tat-type membrane translocation peptides [82]. This method can provide important information about expression patterns of endogenous enzymes or exogenous reporter genes expressed under control of a promoter region from the target gene.

A significant increase in T2 relaxivity is observed upon hybridization of the individual monodisperse iron oxide nanoparticles at the target site [88, 89]. This novel method of contrast enhancement was successfully used to study the processes of DNA cleavage or hybridization in vitro in cell-free systems [88, 90]. Cleavage of synthetic CLIO–oligonucleotide complexes with BamHI and DpnI restrictases significantly prolonged T2 relaxation time of the samples. MRI probes for detection activity of protease enzymes was developed using a similar approach [91].

CEST and PARACEST Contrast Agents

This new class of MR contrast agent is designed so that selective irradiation of an exchangeable proton(s) of these CA with an radio-frequency (RF) field induces a drop in the MR signal of bulk water because of transfer of the saturated magnetization from the irradiated protons to water protons by chemical exchange. Originally proposed by the group of Balaban et al. [92], CEST (chemical exchange saturation transfer) contrast was demonstrated for exchangeable amide protons of proteins [93], imino, and hydroxyl protons [94]. A significant increase in the sensitivity of detection can be gained if a polymer with a large number of equivalent exchangeable groups is used. In experiments with polyuridine (2000 uridine units) the sensitivity gain of more than 5000 per imino-proton was demonstrated, and the CA could be detected at concentration as low as 5μM [94]. The efficiency of CEST contrast increases as the frequency offset, $\Delta\omega$, between resonance frequency of the exchangeable group and water protons increases. The frequency difference can be significantly increased from several parts per million to more than 20ppm by attaching a paramagnetic shift reagent to the CEST agent. Experiments with a model poly-L-Arginine/Tm(HDOTP)$^{4-}$ system lowered the detection limit to 2.8μM in a phantom at 7T magnetic field [95]. This modification of the method was named PARACEST for paramagnetic enhanced CEST. The major potential advantages of the CEST/PARACEST methods to generate MR contrast are: (1) the ability to "turn" the contrast on and off by applying saturating RF field at the resonance frequency of the exchangeable group; (2) high sensitivity that can be achieved by using polymer probes and/or PARACEST agents with large $\Delta\omega$ shifts and fast exchange rates; (3) certain types of CEST agents such as poly-amino acids can be expressed in vivo by target cells with an appropriate reporter vector; and (4) it is possible to design PARACEST probes with different $\Delta\omega$ values and image them independently using RF saturation field with the appropriate frequency offset.

The CEST method was used in vivo for endogenous amide protons of proteins and peptides to detect amide proton transfer (APT) in rat brain tumors [96]. A typical example of APT imaging is shown in Figure 28.8.

28.7 MOLECULAR MRI AND MR SPECTROSCOPY

Generally, MR spectroscopic imaging [chemical shift imaging (CSI), [97]] can be considered a true molecular imaging modality as it provides selective images of distribution of specific chemical compounds such as metabolites, exogenous substances and drugs.

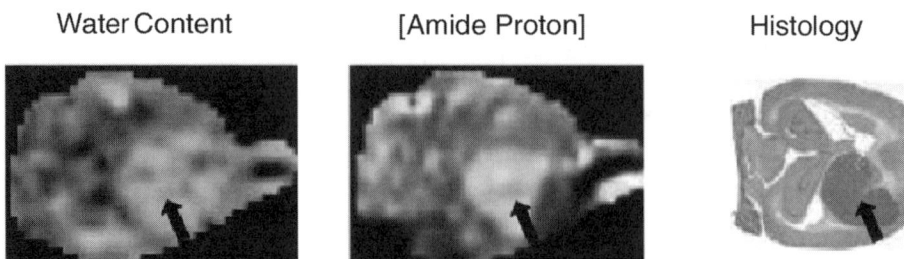

Figure 28.8 Comparison of APT images with MR images and histology for a rat brain tumor. The APT-weighted image was obtained by subtracting the S_{sat}/S_0 images acquired at frequency-labeling offsets of ±3.5 ppm (16 scans). The tumor is visible in all of the MR images (arrow), but its contour is much clearer in the APT images. Adapted from [96].

CSI Molecular Imaging with Reporter Molecules

In the following section we will focus on molecular aspects of imaging probes. These substances are typically designed in such a way that modification of the chemical structure of the probe induces a significant shift in the resonance frequency that can be detected by CSI. In comparison with contrast-enhanced MRI, where micromolar concentrations of CA can be detected, the low sensitivity of MR spectroscopy limits the detectable range of CSI agent concentrations at millimolar levels for protons and 19F MRS. For other nuclei such as 31P and 13C significantly higher concentrations are required.

A T1 relaxation-activated contrast agent to detect the enzyme β-galactosidase, a widely used reporter enzyme, was discussed in the previous section [87]. MR spectroscopy can also be used to assess the function of the enzyme by measuring changes in the chemical shift of a molecular probe induced by its enzymatic cleavage. 4-Fluoro-2-nitrophenyl-*b*-D-galactopyranoside was used as a prototype reporter molecule for β-galactosidase [98]. Upon the cleavage of the sugar by the enzyme, the chemical shift of 19F nucleus on the nitrophenyl ring changes by 4 to 8 ppm depending on the pH of the environment. A relatively high sensitivity of 19F MRS that approaches proton MRS, and the absence of background signals can render this method applicable to in vivo applications.

Expression of creatine kinase as a model marker gene was detected by in vivo MRS in mouse liver. Mice were injected in the tail vein with an adenoviral vector carrying the murine creatine kinase (*CK-B*) gene that induced expression of the gene in the mouse liver. Animals were given exogenous creatine, and 31P spectroscopy of the liver was performed in vivo. Production of phosphocreatine was only detected in the liver of mice injected with the *CK-B*. No spatial selection was used for MR spectroscopy apart from placing the RF surface coil onto exposed liver and shielding the surrounding tissue with a copper shield [99]. Generally the spatial resolution of 31P CSI is on the order of about 10 cm^3, which can be a serious limitation for practical applications of this technique.

28.8 DISCUSSION

Magnetic resonance imaging provides a very high spatial resolution that is invaluable for morphological reference in small animals. Dynamic MRI with small molecular weight and/or nonspecific high-molecular-weight intravascular agents enables measurements of perfusion parameters that can be indicators of angiogenesis and also of tumor response to therapy. Significant effort has been concentrated on the development of targeted MR probes for truly specific molecular MRI that can differentiate between molecular targets on the cellular and subcellular levels. In spite of significant progress, several issues should be addressed to enable routine use of these imaging technologies in mainstream radiology. Some of these are outlined here.

The low sensitivity of MRI is a serious obstacle for imaging sparse molecular targets that are present at micromolar concentrations in vivo. Since the sensitivity of MR spectroscopy is in the millimolar range, efficient mechanisms of signal enhancement such as enzyme-coupled reactions have to be used to enable in vivo application of MRS for detection of molecular targets. For contrast-enhanced MRI highly efficient targeted imaging agents have to be used. For an optimized MR relaxation contrast agent the efficiency or relaxivity of the agent generally increases with the molecular size of the agent. Therefore very large CA, such as polymerized liposomes or nanoparticles, are very efficient probes for imaging vascular targets. On the other hand, to efficiently target solid tumors, the CA molecules should traverse several diffusion barriers to reach the imaging target. Tumors are characterized by highly permeable nonmature vasculature with a pore cutoff size in the range between 380 and 780 nm [26]. However, as discussed above, the delivery pattern in tumors is very nonuniform, and highly permeable areas within a tumor are often interlaced with areas of low vascular permeability and inefficient delivery [100]. Albumin–GdDTPA, blood-pool contrast agent with a molecular weight of 80 kDa leaks into the tumor interstitium in several tumor models, but not efficiently in well-vascularized viable regions [7]. Generation-4 ^{153}Gd-folate dendrimer with a molecular weight in the range of 70 kDa demonstrated high accumulation in the high-affinity folate receptor-positive tumors [101]. On the other hand, generation-6 and higher dendrimers (MW of 240 kDa and above) are retained within the tumor vasculature [102]. Therefore, for the effective delivery of a CA to the nonvascular imaging target, its molecular weight should be restricted at the level of about 100 kDa, which corresponds to generation-4 or -5 dendrimers [14, 103] and protein carriers such as bovine serum albumin (BSA) with multiple modification sites.

Stability and in vivo pharmacokinetics of the probe molecules are important issues for the design of an efficient targeted CA. Paramagnetic CA generally use stable Gd complexes, as their biodegradation results in a release of toxic Gd^{3+} ions [104, 105]. Linear and macrocyclic chelates (DTPA and DOTA) can be used to form stable Gd complexes, and their derivatives, suitable for modification of macromolecules with Gd chelates, are commercially available (Macrocyclics, Dallas, Texas). CA delivery to cellular targets also depends on the plasma kinetics of the CA. Contrast agents with longer circulation time generally have favorable kinetics at the target site. In vivo CA clearance depends upon molecular size and surface characteristics of the agent [26], and several strategies can be used to prolong its plasma half-life time. Coating the iron core with a dextran shell results in longer circulation of iron oxide particles (LCDIO) [106]; incorporation of hydrophilic polymers such as polyethyleneglycol (PEG) to the coat reduces binding of plasma proteins and reduces macrophage uptake of the agent [21, 107]. Modification of the dendrimer surface with anionic groups reduces clearance and cytotoxicity of dendrimers in vivo [108], although both anionic and cationic surfaces showed significant liver accumulation [101].

Due to the low sensitivity of MRI to generate imaging contrast, it is necessary to build maximum local concentrations of the targeted CA at the cellular target site. Therefore, practically all available molecular targets are going to be occupied by the CA molecules. If the targets are functional cell surface receptors, then decorating them with CA molecules can interfere with normal cell signaling and may induce significant physiological response. For instance, using Herceptin as a targeting agent for HER-2/*neu* receptors can produce a therapeutic effect in HER-2/*neu* expressing tumors [109].

To provide efficient selective targeting, highly specific bispecific probes are required that can (1) recognize the molecular epitopes and (2) provide high-affinity binding sites for the CA molecule(s). Monoclonal antibodies are traditionally used for in vivo targeting. However, probes with a smaller molecular size such as mAb fragments, minibodies, diabodies, or peptides can provide better in vivo performance due to the improved delivery properties in solid tissues [69]. Biotin–avidin (streptavidin) linker system provides very high affinity for prelabeling applications but has limited applications in vivo mostly due to the induction of the host immune response against these foreign proteins [69].

For multistep labeling experiments with pretargeting of the epitopes with specific probes [110] and developing of MR contrast with CA that recognize these probes, special attention should be focused on optimized timing of imaging experiment. The available time window for a specific marker (such as mAb) on the cell surface can be limited by dissociation and internalization of the marker by endocytosis. Thus the reported half-life time of a radiolabeled mAb HER-2/*neu* receptor complex on the cell surface can be as short as 6 h [111]. On the other hand the circulating probes should be completely cleared from the system before administration of the contrast agent to provide low background in the images. Therefore, to maximize target–background contrast, a compromise between the clearance of circulating mAb and the availability of the receptor-bound mAb on the cell surface is required. An alternative approach may include removal of circulating targeting probes by a chasing agent [67].

In terms of optimizing MRI protocols, different requirements for T1 and T2 contrast agents need to be considered. Iron-oxide-based contrast agents such as SPIO or MION nanopartricles generally provide very high T2 relaxivity; however, their delivery to the target site can be significantly restricted. Another problem with this class of agent is the negative contrast (or reduced signal intensity) produced in MR images. Recently, a novel method for MR acquisition that provides positive signal enhancement with T2 CA was proposed to improve the in vivo performance of these agents [71]. T1 contrast agents can be designed to have an optimal molecular size [112], and they produce intrinsic positive contrast in T1-weighted MRI. Modern fast acquisition MR techniques enable measuring of quantitative T1 maps [7, 113] to quantify the contrast enhancement and concentration of the molecular targets, respectively.

The rapid development of novel contrast agents will greatly improve our ability to follow modulation in vascular perfusion and molecular composition in preclinical animal models of human cancer using high-resolution MR imaging. In comparison with the competing optical and nuclear imaging modalities such as fluorescence and bioluminescence imaging, SPECT, and PET imaging, MRI provides high-resolution three-dimensional information with excellent morphological registration and does not require radioactive exposure. Clinical applications of molecular target-specific MR agents on the other side are less apparent. Major problems include high concentration of the agent required for the measurable contrast and related potential toxicity and immune response effects. The lower magnetic field used for clinical MRI also contributes to the reduced sensitivity of detection that is of paramount importance for molecular MRI. Significant progress has been made in clinical tissue-specific MR contrast agents that provide better delineation of the diseased organs [114]. However, full-scale clinical applications of molecular targeted MRI in cancer detection and therapy may not be immediately available.

Acknowledgments We thank Dr. Jeff Bulte for help with preparation of the chapter. This work was supported by P50 CA103175.

REFERENCES

1. Bhujwalla, Z. M., Artemov, D., Natarajan, K., Solaiyappan, M., Kollars, P., Kristjansen, P. E. (2003). Reduction of vascular and permeable regions in solid tumors detected by macromolecular contrast magnetic resonance imaging after treatment with antiangiogenic agent TNP-470. *Clin. Cancer Res.*, 9, 355–362.

2. Bhujwalla, Z. M., Blackband, S. J., Wehrle, J. P., Grossman, S., Eller, S., Glickson, J. D. (1990). Metabolic heterogeneity in RIF-1 tumours detected in vivo by 31P NMR spectroscopy. *NMR Biomed.*, *3*, 233–238.

3. Aboagye, E. O., Artemov, D., Senter, P., Bhujwalla, Z. M. (1998). Intratumoral conversion of 5-fluorocytosine to 5-fluorouracil by monoclonal antibody-cytosine deaminase conjugates: Noninvasive detection of prodrug activation by magnetic resonance spectroscopy and spectroscopic imaging. *Cancer Res.*, *58*, 4075–4078.

4. Hakumaki, J. M., Brindle, K. M. (2003). Techniques: Visualizing apoptosis using nuclear magnetic resonance. *Trends Pharmacol. Sci.*, *24*, 146–149.

5. Hakumaki, J. M., Kauppinen, R. A. (2000). 1H NMR visible lipids in the life and death of cells. *Trends Biochem. Sci.*, *25*, 357–362.

6. Mansfield, P., Morris, P. G., Ordidge, R. J., Pykett, I. L., Bangert, V., Coupland, R. E. (1980). Human whole body imaging and detection of breast tumours by n.m.r. *Philos. Trans. R. Soc. Lond. B Biol. Sci.*, *289*, 503–510.

7. Bhujwalla, Z. M., Artemov, D., Natarajan, K., Ackerstaff, E., Solaiyappan, M. (2000). Vascular differences detected by MRI for metastatic versus nonmetastatic breast and prostate cancer xenografts. *Neoplasia*, *3*, 1–11.

8. Schmiedl, U., Ogan, M., Paajanen, H., Marotti, M., Crooks, L. E., Brito, A. C., Brasch, R. C. (1987). Albumin labeled with Gd-DTPA as an intravascular, blood pool-enhancing agent for MR imaging: Biodistribution and imaging studies. *Radiology*, *162*, 205–210.

9. Bogdanov, A. A., Jr., Weissleder, R., Frank, H. W., Bogdanova, A. V., Nossif, N., Schaffer, B. K., Tsai, E., Papisov, M. I., Brady, T. J. (1993). A new macromolecule as a contrast agent for MR angiography: Preparation, properties, and animal studies. *Radiology*, *187*, 701–706.

10. Gohr-Rosenthal, S., Schmitt-Willich, H., Ebert, W., Conrad, J. (1993). The demonstration of human tumors on nude mice using gadolinium-labelled monoclonal antibodies for magnetic resonance imaging. *Invest. Radiol.*, *28*, 789–795.

11. Artemov, D., Mori, N., Ravi, R., Bhujwalla, Z. M. (2003). Magnetic resonance molecular imaging of the Her-2/neu receptor. *Cancer Res.*, *63*, 2723–2727.

12. Shahbazi-Gahrouei, D., Williams, M., Rizvi, S., Allen, B. J. (2001). In vivo studies of Gd-DTPA-monoclonal antibody and gd-porphyrins: Potential magnetic resonance imaging contrast agents for melanoma. *J. Magn. Reson. Imag.*, *14*, 169–174.

13. Matsumura, A., Shibata, Y., Nakagawa, K., Nose, T. (1994). MRI contrast enhancement by Gd-DTPA-monoclonal antibody in 9L glioma rats. *Acta Neurochir. Suppl.*, *60*, 356–358.

14. Bryant, L. H., Jr., Brechbiel, M. W., Wu, C., Bulte, J. W., Herynek, V., Frank, J. A. (1999). Synthesis and relaxometry of high-generation ($G = 5, 7, 9$, and 10) PAMAM dendrimer-DOTA-gadolinium chelates. *J. Magn. Reson. Imag.*, *9*, 348–352.

15. Kobayashi, H., Sato, N., Hiraga, A., Saga, T., Nakamoto, Y., Ueda, H., Konishi, J., Togashi, K., Brechbiel, M. W. (2001). 3D-micro-MR angiography of mice using macromolecular MR contrast agents with polyamidoamine dendrimer core with reference to their pharmacokinetic properties. *Magn. Reson. Med.*, *45*, 454–460.

16. Anderson, S. A., Rader, R. K., Westlin, W. F., Null, C., Jackson, D., Lanza, G. M., Wickline, S. A., Kotyk, J. J. (2000). Magnetic resonance contrast enhancement of

neovasculature with alpha(v)beta(3)-targeted nanoparticles. *Magn. Reson. Med.*, *44*, 433–439.

17. Sipkins, D. A., Cheresh, D. A., Kazemi, M. R., Nevin, L. M., Bednarski, M. D., Li, K. C. (1998). Detection of tumor angiogenesis in vivo by alphaVbeta3-targeted magnetic resonance imaging. *Nat. Med.*, *4*, 623–626.
18. Caravan, P., Cloutier, N. J., Greenfield, M. T., McDermid, S. A., Dunham, S. U., Bulte, J. W., Amedio, J. C., Jr., Looby, R. J., Supkowski, R. M., Horrocks, W. D., Jr., McMurry, T. J., Lauffer, R. B. (2002). The interaction of MS-325 with human serum albumin and its effect on proton relaxation rates. *J. Am. Chem. Soc.*, *124*, 3152–3162.
19. Gillis, P., Roch, A., Brooks, R. A. (1999). Corrected equations for susceptibility-induced T2-shortening. *J. Magn. Reson.*, *137*, 402–407.
20. Hogemann, D., Josephson, L., Weissleder, R., Basilion, J. P. (2000). Improvement of MRI probes to allow efficient detection of gene expression. *Bioconjug. Chem.*, *11*, 941–946.
21. Bulte, J. W., de Cuyper, M., Despres, D., Frank, J. A. (1999). Short- vs. long-circulating magnetoliposomes as bone marrow-seeking MR contrast agents. *J. Magn. Reson. Imag.*, *9*, 329–335.
22. Strable, E., Bulte, J. W., Moskowitz, B. M., Vivekanandan, K., Allen, M., Douglas, T. (2001). Synthesis and characterization of soluble iron oxide-dendrimer composites. *Chem. Mater.*, *13*, 2201–2209.
23. Bulte, J. W., Douglas, T., Witwer, B., Zhang, S. C., Strable, E., Lewis, B. K., Zywicke, H., Miller, B., van Gelderen, P., Moskowitz, B. M., Duncan, I. D., Frank, J. A. (2001). Magnetodendrimers allow endosomal magnetic labeling and in vivo tracking of stem cells. *Nat. Biotechnol.*, *19*, 1141–1147.
24. Bulte, J. W., Brooks, R. A., Moskowitz, B. M., Bryant, L. H., Jr., Frank, J. A. (1999). Relaxometry and magnetometry of the MR contrast agent MION-46L. *Magn. Reson. Med.*, *42*, 379–384.
25. Bulte, J. W., Bryant, L. H., Jr., Frank, J. A. (2003). In L. E. Feinendegen, W. W. Shreeve, W. C. Eckelman, Y.-W. Bahk, and H. N. Wagner, Jr. (Eds.), *Molecular Nuclear Medicine—The Challenge of Genomics and Proteomics to Clinical Practice*. Springer, pp. 721–740.
26. Brigger, I., Dubernet, C., Couvreur, P. (2002). Nanoparticles in cancer therapy and diagnosis. *Adv. Drug Deliv. Rev.*, *54*, 631–651.
27. Weidner, N. (1993). Tumor angiogenesis: Review of current applications in tumor prognostication. *Semin. Diagn. Pathol.*, *10*, 302–313.
28. Durand, R. E., Aquino-Parsons, C. (2001). Clinical relevance of intermittent tumour blood flow. *Acta Oncol.*, *40*, 929–936.
29. Durand, R. E. (2001). Intermittent blood flow in solid tumours—An underappreciated source of "drug resistance". *Cancer Metastasis Rev.*, *20*, 57–61.
30. Furman-Haran, E., Margalit, R., Grobgeld, D., Degani, H. (1996). Dynamic contrast-enhanced magnetic resonance imaging reveals stress-induced angiogenesis in MCF7 human breast tumors. *Proc. Natl. Acad. Sci. USA*, *93*, 6247–6251.
31. Aronen, H. J., Cohen, M. S., Belliveau, J. W., Fordham, J. A., Rosen, B. R. (1993). Ultrafast imaging of brain tumors. *Top. Magn. Reson. Imag.*, *5*, 14–24.

32. Hawighorst, H., Knapstein, P. G., Weikel, W., Knopp, M. V., Zuna, I., Knof, A., Brix, G., Schaeffer, U., Wilkens, C., Schoenberg, S. O., Essig, M., Vaupel, P., van Kaick, G. (1997). Angiogenesis of uterine cervical carcinoma: Characterization by pharmacokinetic magnetic resonance parameters and histological microvessel density with correlation to lymphatic involvement. *Cancer Res.*, 57, 4777–4786.
33. Tofts, P. S., Brix, G., Buckley, D. L., Evelhoch, J. L., Henderson, E., Knopp, M. V., Larsson, H. B., Lee, T. Y., Mayr, N. A., Parker, G. J., Port, R. E., Taylor, J., Weisskoff, R. M. (1999). Estimating kinetic parameters from dynamic contrast-enhanced T(1)-weighted MRI of a diffusable tracer: Standardized quantities and symbols. *J. Magn. Reson. Imag.*, 10, 223–232.
34. Tofts, P. S. (1997). Modeling tracer kinetics in dynamic Gd-DTPA MR imaging. *J. Magn. Reson. Imaging.*, 7, 91–101.
35. Kety, S. S. (1965). Measurement of local contribution within the brain by means of inert, diffusible tracers; examination of the theory, assumptions and possible sources of error. *Acta Neurol. Scand. Suppl.*, 14, 20–23.
36. Larsson, H. B., Stubgaard, M., Frederiksen, J. L., Jensen, M., Henriksen, O., Paulson, O. B. (1990). Quantitation of blood-brain barrier defect by magnetic resonance imaging and gadolinium-DTPA in patients with multiple sclerosis and brain tumors. *Magn. Reson. Med.*, 16, 117–131.
37. Weisskoff, R. M., Chesler, D., Boxerman, J. L., Rosen, B. R. (1993). Pitfalls in MR measurement of tissue blood flow with intravascular tracers: Which mean transit time? *Magn. Reson. Med.*, 29, 553–558.
38. Donahue, K. M., Krouwer, H. G., Rand, S. D., Pathak, A. P., Marszalkowski, C. S., Censky, S. C., Prost, R. W. (2000). Utility of simultaneously acquired gradient-echo and spin-echo cerebral blood volume and morphology maps in brain tumor patients. *Magn. Reson. Med.*, 43, 845–853.
39. Pathak, A. P., Schmainda, K. M., Ward, B. D., Linderman, J. R., Rebro, K. J., Greene, A. S. (2001). MR-derived cerebral blood volume maps: Issues regarding histological validation and assessment of tumor angiogenesis. *Magn. Reson. Med.*, 46, 735–747.
40. Boxerman, J. L., Hamberg, L. M., Rosen, B. R., Weisskoff, R. M. (1995). MR contrast due to intravascular magnetic susceptibility perturbations. *Magn. Reson. Med.*, 34, 555–566.
41. Deane, B. R., Lantos, P. L. (1981). The vasculature of experimental brain tumours. Part 1. A sequential light and electron microscope study of angiogenesis. *J. Neurol. Sci.*, 49, 55–66.
42. Deane, B. R., Lantos, P. L. (1981). The vasculature of experimental brain tumours. Part 2. A quantitative assessment of morphological abnormalities. *J. Neurol. Sci.*, 49, 67–77.
43. Hnatowich, D. J., Layne, W. W., Childs, R. L. (1982). The preparation and labeling of DTPA-coupled albumin. *Int. J. Appl. Radiat. Isot.*, 33, 327–332.
44. Donahue, K. M., Weisskoff, R. M., Chesler, D. A., Kwong, K. K., Bogdanov, A. A., Jr., Mandeville, J. B., Rosen, B. R. (1996). Improving MR quantification of regional blood volume with intravascular T1 contrast agents: Accuracy, precision, and water exchange. *Magn. Reson. Med.*, 36, 858–867.
45. Bhujwalla, Z. M., Artemov, D., Glockner, J. (1999). Tumor angiogenesis, vascularization, and contrast-enhanced magnetic resonance imaging. *Top. Magn. Reson. Imag.*, 10, 92–103.

46. Teicher, B. A., Emi, Y., Kakeji, Y., Northey, D. (1996). TNP-470/minocycline/cytotoxic therapy: A systems approach to cancer therapy. *Eur. J. Cancer*, *32A*, 2461–2466.
47. Folkman, J. (1985). Tumor angiogenesis. *Adv. Cancer Res.*, *43*, 175–203.
48. Folkman, J. (1971). Tumor angiogenesis: Therapeutic implications. *N. Engl. J. Med.*, *285*, 1182–1186.
49. Brooks, P. C., Clark, R. A., Cheresh, D. A. (1994). Requirement of vascular integrin alpha v beta 3 for angiogenesis. *Science*, *264*, 569–571.
50. Kang, H. W., Josephson, L., Petrovsky, A., Weissleder, R., Bogdanov, A., Jr. (2002). Magnetic resonance imaging of inducible E-selectin expression in human endothelial cell culture. *Bioconjug. Chem.*, *13*, 122–127.
51. Winter, P. M., Caruthers, S. D., Kassner, A., Harris, T. D., Chinen, L. K., Allen, J. S., Lacy, E. K., Zhang, H., Robertson, J. D., Wickline, S. A., Lanza, G. M. (2003). Molecular imaging of angiogenesis in nascent Vx-2 rabbit tumors using a novel alpha(nu)beta3-targeted nanoparticle and 1.5 tesla magnetic resonance imaging. *Cancer Res.*, *63*, 5838–5843.
52. Josephson, L., Groman, E. V., Menz, E., Lewis, J. M., Bengele, H. (1990). A functionalized superparamagnetic iron oxide colloid as a receptor directed MR contrast agent. *Magn. Reson. Imag.*, *8*, 637–646.
53. Weissleder, R., Reimer, P., Lee, A. S., Wittenberg, J., Brady, T. J. (1990). MR receptor imaging: Ultrasmall iron oxide particles targeted to asialoglycoprotein receptors. *AJR Am. J. Roentgenol.*, *155*, 1161–1167.
54. Bulte, J. W., Hoekstra, Y., Kamman, R. L., Magin, R. L., Webb, A. G., Briggs, R. W., Go, K. G., Hulstaert, C. E., Miltenyi, S., The, T. H. (1992). Specific MR imaging of human lymphocytes by monoclonal antibody-guided dextran-magnetite particles. *Magn. Reson. Med.*, *25*, 148–157.
55. Artemov, D., Mori, N., Okollie, B., Bhujwalla, Z. M. (2003). MR molecular imaging of the Her-2/neu receptor in breast cancer cells using targeted iron oxide nanoparticles. *Magn. Reson. Med.*, *49*, 403–408.
56. Weissleder, R., Moore, A., Mahmood, U., Bhorade, R., Benveniste, H., Chiocca, E. A., Basilion, J. P. (2000). In vivo magnetic resonance imaging of transgene expression. *Nat. Med.*, *6*, 351–355.
57. Moore, A., Basilion, J. P., Chiocca, E. A., Weissleder, R. (1998). Measuring transferrin receptor gene expression by NMR imaging. *Biochim. Biophys. Acta*, *1402*, 239–249.
58. Anderson-Berg, W. T., Strand, M., Lempert, T. E., Rosenbaum, A. E., Joseph, P. M. (1986). Nuclear magnetic resonance and gamma camera tumor imaging using gadolinium-labeled monoclonal antibodies. *J. Nucl. Med.*, *27*, 829–833.
59. Unger, E. C., Totty, W. G., Neufeld, D. M., Otsuka, F. L., Murphy, W. A., Welch, M. S., Connett, J. M., Philpott, G. W. (1985). Magnetic resonance imaging using gadolinium labeled monoclonal antibody. *Invest. Radiol.*, *20*, 693–700.
60. Konda, S. D., Aref, M., Brechbiel, M., Wiener, E. C. (2000). Development of a tumor-targeting MR contrast agent using the high-affinity folate receptor: Work in progress. *Invest. Radiol.*, *35*, 50–57.
61. Weissleder, R., Lee, A. S., Fischman, A. J., Reimer, P., Shen, T., Wilkinson, R., Callahan, R. J., Brady, T. J. (1991). Polyclonal human immunoglobulin G labeled with polymeric iron oxide: Antibody MR imaging. *Radiology*, *181*, 245–249.

62. Zhao, M., Beauregard, D. A., Loizou, L., Davletov, B., Brindle, K. M. (2001). Noninvasive detection of apoptosis using magnetic resonance imaging and a targeted contrast agent. *Nat. Med.*, 7, 1241–1244.

63. Moore, A., Medarova, Z., Potthast, A., Dai, G. (2004). In vivo targeting of underglycosylated MUC-1 tumor antigen using a multimodal imaging probe. *Cancer Res.*, 64, 1821–1827.

64. Monsky, W. L., Fukumura, D., Gohongi, T., Ancukiewcz, M., Weich, H. A., Torchilin, V. P., Yuan, F., Jain, R. K. (1999). Augmentation of transvascular transport of macromolecules and nanoparticles in tumors using vascular endothelial growth factor. *Cancer Res.*, 59, 4129–4135.

65. Paganelli, G., Grana, C., Chinol, M., Cremonesi, M., De Cicco, C., De Braud, F., Robertson, C., Zurrida, S., Casadio, C., Zoboli, S., Siccardi, A. G., Veronesi, U. (1999). Antibody-guided three-step therapy for high grade glioma with yttrium-90 biotin. *Eur. J. Nucl. Med.*, 26, 348–357.

66. Artemov, D., Foss, C., Okollie, B., Bhujwalla, Z. M. (2004). In *The Society for Molecular Imaging, the Third Annual Meeting*, St. louis. p. 386.

67. Dafni, H., Israely, T., Bhujwalla, Z. M., Benjamin, L. E., Neeman, M. (2002). Overexpression of vascular endothelial growth factor 165 drives peritumor interstitial convection and induces lymphatic drain: Magnetic resonance imaging, confocal microscopy, and histological tracking of triple-labeled albumin. *Cancer Res.*, 62, 6731–6739.

68. Ichikawa, T., Hogemann, D., Saeki, Y., Tyminski, E., Terada, K., Weissleder, R., Chiocca, E. A., Basilion, J. P. (2002). MRI of transgene expression: Correlation to therapeutic gene expression. *Neoplasia*, 4, 523–530.

69. Goldenberg, D. M., Chang, C. H., Sharkey, R. M., Rossi, E. A., Karacay, H., McBride, W., Hansen, H. J., Chatal, J. F., Barbet, J. (2003). Radioimmunotherapy: Is avidin-biotin pretargeting the preferred choice among pretargeting methods? *Eur. J. Nucl. Med. Mol. Imag.*, 30, 777–780.

70. Kobayashi, H., Sato, N., Kawamoto, S., Saga, T., Hiraga, A., Haque, T. L., Ishimori, T., Konishi, J., Togashi, K., Brechbiel, M. W. (2001). Comparison of the macromolecular MR contrast agents with ethylenediamine-core versus ammonia-core generation-6 polyamidoamine dendrimer. *Bioconjug. Chem.*, 12, 100–107.

71. Seppenwoolde, J. H., Viergever, M. A., Bakker, C. J. (2003). Passive tracking exploiting local signal conservation: The white marker phenomenon. *Magn. Reson. Med.*, 50, 784–790.

72. Bogdanov, A., Jr., Matuszewski, L., Bremer, C., Petrovsky, A., Weissleder, R. (2002). Oligomerization of paramagnetic substrates result in signal amplification and can be used for MR imaging of molecular targets. *Mol. Imag.*, 1, 16–23.

73. Walter, G. A., Cahill, K. S., Huard, J., Feng, H., Douglas, T., Sweeney, H. L., Bulte, J. W. (2004). Noninvasive monitoring of stem cell transfer for muscle disorders. *Magn. Reson. Med.*, 51, 273–277.

74. Karmarkar, P. V., Kraitchman, D. L., Izbudak, I., Hofmann, L. V., Amado, L. C., Fritzges, D., Young, R., Pittenger, M., Bulte, J. W., Atalar, E. (2004). MR-trackable intramyocardial injection catheter. *Magn. Reson. Med.*, 51, 1163–1172.

75. Lee, I. H., Bulte, J. W., Schweinhardt, P., Douglas, T., Trifunovski, A., Hofstetter, C., Olson, L., Spenger, C. (2004). In vivo magnetic resonance tracking of

olfactory ensheathing glia grafted into the rat spinal cord. *Exp. Neurol.*, *187*, 509–516.

76. Bulte, J. W., Arbab, A. S., Douglas, T., Frank, J. A. (2004). Preparation of magnetically labeled cells for cell tracking by magnetic resonance imaging. *Methods Enzymol.*, *386*, 275–299.
77. Weissleder, R., Cheng, H. C., Bogdanova, A., Bogdanov, A. (1997). Magnetically labeled cells can be detected by MR imaging. *J. Magn. Reson. Imag.*, *7*, 258–263.
78. Dodd, S. J., Williams, M., Suhan, J. P., Williams, D. S., Koretsky, A. P., Ho, C. (1999). Detection of single mammalian cells by high-resolution magnetic resonance imaging. *Biophys. J.*, *76*, 103–109.
79. Shapiro, E. M., Skrtic, S., Sharer, K., Hill, J. M., Dunbar, C. E., Koretsky, A. P. (2004). MRI detection of single particles for cellular imaging. *Proc. Natl. Acad. Sci. USA*, *101*, 10901–10906.
80. Gammon, S. T., Villalobos, V. M., Prior, J. L., Sharma, V., Piwnica-Worms, D. (2003). Quantitative analysis of permeation peptide complexes labeled with technetium-99m: Chiral and sequence-specific effects on net cell uptake. *Bioconjug. Chem.*, *14*, 368–376.
81. Prantner, A. M., Sharma, V., Garbow, J. R., Piwnica-Worms, D. (2003). Synthesis and characterization of a Gd-DOTA-D-permeation peptide for magnetic resonance relaxation enhancement of intracellular targets. *Mol. Imag.*, *2*, 333–341.
82. Bhorade, R., Weissleder, R., Nakakoshi, T., Moore, A., Tung, C. H. (2000). Macrocyclic chelators with paramagnetic cations are internalized into mammalian cells via a HIV-tat derived membrane translocation peptide. *Bioconjug. Chem.*, *11*, 301–305.
83. Kircher, M. F., Mahmood, U., King, R. S., Weissleder, R., Josephson, L. (2003). A multimodal nanoparticle for preoperative magnetic resonance imaging and intraoperative optical brain tumor delineation. *Cancer Res.*, *63*, 8122–8125.
84. Heckl, S., Pipkorn, R., Waldeck, W., Spring, H., Jenne, J., von der Lieth, C. W., Corban-Wilhelm, H., Debus, J., Braun, K. (2003). Intracellular visualization of prostate cancer using magnetic resonance imaging. *Cancer Res.*, *63*, 4766–4772.
85. Mahmood, U., Tung, C. H., Bogdanov, A., Jr., Weissleder, R. (1999). Near-infrared optical imaging of protease activity for tumor detection. *Radiology*, *213*, 866–870.
86. Weissleder, R., Tung, C. H., Mahmood, U., Bogdanov, A., Jr. (1999). In vivo imaging of tumors with protease-activated near-infrared fluorescent probes. *Nat. Biotechnol.*, *17*, 375–378.
87. Louie, A. Y., Huber, M. M., Ahrens, E. T., Rothbacher, U., Moats, R., Jacobs, R. E., Fraser, S. E., Meade, T. J. (2000). In vivo visualization of gene expression using magnetic resonance imaging. *Nat. Biotechnol.*, *18*, 321–325.
88. Perez, J. M., O'Loughin, T., Simeone, F. J., Weissleder, R., Josephson, L. (2002). DNA-based magnetic nanoparticle assembly acts as a magnetic relaxation nanoswitch allowing screening of DNA-cleaving agents. *J. Am. Chem. Soc.*, *124*, 2856–2857.
89. Bremer, C., Weissleder, R. (2001). In vivo imaging of gene expression. *Acad. Radiol.*, *8*, 15–23.

90. Perez, J. M., Josephson, L., O'Loughlin, T., Hogemann, D., Weissleder, R. (2002). Magnetic relaxation switches capable of sensing molecular interactions. *Nat. Biotechnol.*, *20*, 816–820.

91. Zhao, M., Kircher, M. F., Josephson, L., Weissleder, R. (2002). Differential conjugation of tat peptide to superparamagnetic nanoparticles and its effect on cellular uptake. *Bioconjug. Chem.*, *13*, 840–844.

92. Guivel-Scharen, V., Sinnwell, T., Wolff, S. D., Balaban, R. S. (1998). Detection of proton chemical exchange between metabolites and water in biological tissues. *J. Magn. Reson.*, *133*, 36–45.

93. Zhou, J., Payen, J. F., Wilson, D. A., Traystman, R. J., van Zijl, P. C. (2003). Using the amide proton signals of intracellular proteins and peptides to detect pH effects in MRI. *Nat. Med.*, *9*, 1085–1090.

94. Snoussi, K., Bulte, J. W., Gueron, M., Van Zijl, P. C. (2003). Sensitive CEST agents based on nucleic acid imino proton exchange: Detection of poly(rU) and of a dendrimer-poly(rU) model for nucleic acid delivery and pharmacology. *Magn. Reson. Med.*, *49*, 998–1005.

95. Aime, S., Delli Castelli, D., Terreno, E. (2003). Supramolecular adducts between poly-L-arginine and [TmIIIdotp]: A route to sensitivity-enhanced magnetic resonance imaging-chemical exchange saturation transfer agents. *Angew Chem. Int. Ed. Engl.*, *42*, 4527–4529.

96. Zhou, J., Lal, B., Wilson, D. A., Laterra, J., van Zijl, P. C. (2003). Amide proton transfer (APT) contrast for imaging of brain tumors. *Magn. Reson. Med.*, *50*, 1120–1126.

97. Brown, T. R., Kincaid, B. M., Ugurbil, K. (1982). NMR chemical shift imaging in three dimensions. *Proc. Natl. Acad. Sci. USA*, *79*, 3523–3526.

98. Cui, W., Otten, P., Li, Y., Koeneman, K. S., Yu, J., Mason, R. P. (2004). Novel NMR approach to assessing gene transfection: 4-Fluoro-2-nitrophenyl-beta-D-galactopyranoside as a prototype reporter molecule for beta-galactosidase. *Magn. Reson. Med.*, *51*, 616–620.

99. Auricchio, A., Zhou, R., Wilson, J. M., Glickson, J. D. (2001). In vivo detection of gene expression in liver by 31P nuclear magnetic resonance spectroscopy employing creatine kinase as a marker gene. *Proc. Natl. Acad. Sci. USA*, *98*, 5205–5210.

100. Alexiou, C., Arnold, W., Hulin, P., Klein, R. J., Renz, H., Parak, F. G., Bergemann, C., Lubbe, A. S. (2001). Magnetic mitoxantrone nanoparticle detection by histology, X-ray and MRI after magnetic tumor targeting. *J. Magn. Magn. Mater.*, *225*, 187–193.

101. Konda, S. D., Wang, S., Brechbiel, M., Wiener, E. C. (2002). Biodistribution of a 153 Gd-folate dendrimer, generation = 4, in mice with folate-receptor positive and negative ovarian tumor xenografts. *Invest. Radiol.*, *37*, 199–204.

102. Kobayashi, H., Kawamoto, S., Saga, T., Sato, N., Hiraga, A., Konishi, J., Togashi, K., Brechbiel, M. W. (2001). Micro-MR angiography of normal and intratumoral vessels in mice using dedicated intravascular MR contrast agents with high generation of polyamidoamine dendrimer core: Reference to pharmacokinetic properties of dendrimer-based MR contrast agents. *J. Magn. Reson. Imag.*, *14*, 705–713.

103. Kobayashi, H., Kawamoto, S., Saga, T., Sato, N., Hiraga, A., Ishimori, T., Akita, Y., Mamede, M. H., Konishi, J., Togashi, K., Brechbiel, M. W. (2001). Novel liver

macromolecular MR contrast agent with a polypropylenimine diaminobutyl dendrimer core: Comparison to the vascular MR contrast agent with the polyamidoamine dendrimer core. *Magn. Reson. Med.*, 46, 795–802.

104. Spencer, A. W. S., Harpur, E. (1998). Gadolinium chloride toxicity in the mouse. *Hum. Exp. Toxicol.*, 17, 633–637.

105. Runge, V. M., Dickey, K. M., Williams, N. M., Peng, X. (2002). Local tissue toxicity in response to extravascular extravasation of magnetic resonance contrast media. *Invest. Radiol.*, 37, 393–398.

106. Moore, A., Marecos, E., Bogdanov, A., Jr., Weissleder, R. (2000). Tumoral distribution of long-circulating dextran-coated iron oxide nanoparticles in a rodent model. *Radiology*, 214, 568–574.

107. Oussoren, C., Storm, G. (1997). Lymphatic uptake and biodistribution of liposomes after subcutaneous injection: III. Influence of surface modification with poly(ethyleneglycol). *Pharm. Res.*, 14, 1479–1484.

108. Malik, N., Wiwattanapatapee, R., Klopsch, R., Lorenz, K., Frey, H., Weener, J. W., Meijer, E. W., Paulus, W., Duncan, R. (2000). Dendrimers: Relationship between structure and biocompatibility in vitro, and preliminary studies on the biodistribution of 125I-labelled polyamidoamine dendrimers in vivo. *J. Control Release*, 65, 133–148.

109. Pegram, M. D., Lipton, A., Hayes, D. F., Weber, B. L., Baselga, J. M., Tripathy, D., Baly, D., Baughman, S. A., Twaddell, T., Glaspy, J. A., Slamon, D. J. (1998). Phase II study of receptor-enhanced chemosensitivity using recombinant humanized anti-p185HER2/neu monoclonal antibody plus cisplatin in patients with HER2/neu-overexpressing metastatic breast cancer refractory to chemotherapy treatment. *J. Clin. Oncol.*, 16, 2659–2671.

110. Rosebrough, S. F. (1996). Two-step immunological approaches for imaging and therapy. *Q. J. Nucl. Med.*, 40, 234–251.

111. Zalutsky, M. R., Xu, F. J., Yu, Y., Foulon, C. F., Zhao, X. G., Slade, S. K., Affleck, D. J., Bast, R. C., Jr. (1999). Radioiodinated antibody targeting of the HER-2/neu oncoprotein: Effects of labeling method on cellular processing and tissue distribution. *Nucl. Med. Biol.*, 26, 781–790.

112. Kobayashi, H., Brechbiel, M. W. (2003). Dendrimer-based macromolecular MRI contrast agents: Characteristics and application. *Mol. Imag.*, 2, 1–10.

113. Nekolla, S., Gneiting, T., Syha, J., Deichmann, R., Haase, A. (1992). T1 maps by K-space reduced snapshot-FLASH MRI. *J. Comput. Assist. Tomogr.*, 16, 327–332.

114. Weinmann, H. J., Ebert, W., Misselwitz, B., Schmitt-Willich, H. (2003). Tissue-specific MR contrast agents. *Eur. J. Radiol.*, 46, 33–44.

29

TARGETS AND APPROACHES FOR CANCER DRUG DISCOVERY

SUSAN L. MOOBERRY
Southwest Foundation for Biomedical Research
San Antonio, Texas

29.1	INTRODUCTION	1344
	Advent of Targeted Therapy	1344
	Evolution of the Drug Discovery Process	1345
	Focus and Definitions	1345
29.2	SIGNAL TRANSDUCTION TARGETS	1346
	HER2	1347
	The EGF Receptor Pathway	1347
	Insulin-Like Growth Factor Signaling	1349
	Bcr/Abl, c-Kit, and Platelet-Derived Growth Factor	1350
	Mammalian Target of Rapamycin (mTOR)	1350
	Methods of Drug Discovery	1351
29.3	TARGETING TUMOR VASCULATURE	1352
	Inhibitors of Angiogenesis	1353
	Antivascular Agents	1356
	Methods of Drug Discovery	1356
29.4	DNA TARGETS	1357
	DNA Methylation	1357
	Histone Deacetylase Inhibitors	1358
	Telomerase Inhibitors	1359
29.5	ANTIMITOTICS	1360
	Microtubule-Interacting Agents	1360
	Kinesin Inhibitors	1361
	Aurora Kinase	1361
	Methods of Drug Discovery	1362

Drug Discovery Handbook, by Shayne Cox Gad
Copyright © 2005 by John Wiley & Sons, Inc.

29.6	PROMISING NEW TARGETS	1363
	Proteasome Inhibition	1363
	Heat Shock Protein 90 (Hsp90)	1364
29.7	SOURCES OF SMALL-MOLECULE DRUG LEADS	1365
	Natural Products	1365
	Synthetic Chemical Libraries	1366
29.8	RESOURCES FOR DRUG DISCOVERY	1367
	References	1368

29.1 INTRODUCTION

Advent of Targeted Therapy

Although significant advances in early diagnosis and treatment of cancer have been made in the past decade, there remains a compelling need for more effective therapies for the treatment of many different types of cancer. The identification of more specific and thereby less toxic therapies for these tumors will provide significant advantages for cancer control. With the sequencing of the human genome and identification of numerous oncogenes and tumor suppressor genes, the level of understanding of the molecular mechanisms underlying cancer has never been greater. These advances in understanding the molecular etiology of cancer led to efforts to discover and develop target-specific therapies that offer the promise of more specificity toward the key biological processes that are abnormal in cancer. Importantly, some but not all of the targeted therapies have proven efficacious in clinical trials. Single-target therapy was the initial aim of targeted therapy, but the genetic instability intrinsic to tumors causes multiple oncogenic abnormalities. Broader targeting of multiple molecular alterations that cause cancer may provide for optimum therapy for most tumors.

Along with an enhanced understanding of the molecular events that lead to tumor initiation and expansion, the discovery of the complex roles of stromal and infiltrating immune cells in tumor proliferation, neovascularization, and metastasis offers new avenues for therapeutic intervention. Angiogenesis is critical for both tumor expansion and metastasis, and the tumor vasculature is a promising new therapeutic target. The hypoxic tumor environment additionally provides therapeutic opportunities. Comprehension of many of the molecular driving forces behind cancer has given us better insights into the high degree of cross-talk between numerous signaling pathways between the tumor and its surrounding microenvironment. Elucidation of the interrelationships of these signaling networks facilitates the selection of rational drug combinations with complementary actions.

INTRODUCTION

Genetic instability is a hallmark of cancer, and tumors are highly heterogeneous, providing one of the many challenges in cancer treatment. Individual cells within a tumor have different growth and survival characteristics as well as different metastatic potentials. Effective targeting of the particular processes abnormal to each tumor is critical for optimal therapy. Together with development of targeted therapies, accurate predictive diagnostic tools are necessary to identify the molecular defects of tumors to allow for individualized targeted therapy.

Evolution of the Drug Discovery Process

Early therapies for cancer were identified in part by serendipity combined with astute observations by scientists and clinicians. Exposure of servicemen to nitrogen mustard as part of a secret government chemical warfare program during World War II led to the realization that these agents caused leucopoenia and bone marrow suppression. This knowledge was used in the first clinical tests of alkylating agents in the treatment of cancer. The observation that folate supplementation of children with leukemia caused leukemic cell proliferation led to the development of the antifolates and other antimetabolites. Extracts of the periwinkle plant were evaluated in rodents for antidiabetic properties where they caused aplasia of bone marrow and glanulocytopenia. The active constituents were isolated as the vinca alkaloids, a class of compounds with broad anticancer effects. The identification of drugs that worked and testing of these agents against a variety of different tumor types to find those that were sensitive characterized early cancer drug discovery.

Cancer drug discovery now is tied to advances in understanding the molecular causes of aberrant growth, resistance to immune surveillance and abnormal apoptotic processes, all of which occur during carcinogenesis. Today the majority of cancer drug discovery is related to identifying inhibitors of specific events that are aberrant in cancer. In this new era for cancer drug discovery, 40 percent of the compounds in clinical development are directed against novel targets, and it is estimated that 70 percent of targets for cancer drug discovery are novel, with no history of utility [1]. The prognosis for the treatment of cancer has never been better, and targeted therapy holds the key for more specific and effective therapy. Although cures may not be achieved, control of cancer as a chronic disease may be attainable.

Focus and Definitions

The goal of this chapter is to provide an introduction to the field of cancer drug discovery. It focuses on a subset of current targets, and a brief overview of some approaches for identification of leads, and sources of materials for drug discovery efforts. The list of targets is not all inclusive but covers many validated cellular targets and several promising targets that may yield effective anticancer pharmaceuticals. The lack of inclusion of a particular target in this introductory chapter should not in any way reduce enthusiasm for these drug targets for anticancer drug design. The chapter is focused on initial drug discovery of small-molecule inhibitors and not on the equally complex processes of lead optimization, preclinical pharmacology, in vivo testing, and clinical development. Additionally, some experimental approaches for the identification of active molecules are described. High-throughput screening (HTS) is of paramount importance in identifying new drug leads for cancer, and the details of assay design and approachs for HTS are covered in detail in other chapters of this text, so methods related to HTS are not covered in this chapter. Humanized monoclonal antibodies are discussed because they provide important proof-of-principle of the effectiveness of targeted therapy. Humanized monoclonal antibodies provide clinical utility for treating cancer, but the design of these therapies is not addressed. Gene therapy, antisense oligonucleotides, and small, interfering ribonucleic acid (siRNA) are additional viable avenues for anticancer therapies, but they are not a focus of this introductory chapter.

It is important that the cancer drug discovery community use a common language that designates the various stages of cancer drug development. These definitions were proposed years ago by the Developmental Therapeutics Branch of the National Cancer Institute [2] and are presented here to allow consistency and to acquaint those new in the field to appropriate terminology. *Cytotoxic* is used to describe activity against cells in vitro. This term does not imply any selectivity for cancerous versus normal cells. *Antitumor* is activity against a tumor in an in vivo model, and *anticancer* is used only to designate efficacy of a drug against cancer in humans following clinical testing.

29.2 SIGNAL TRANSDUCTION TARGETS

The goal of targeted therapy is to discover and develop target-specific therapies that offer the promise of more specificity toward the key biological processes that are abnormal in cancer. Abberant cellular signaling is a hallmark of cancer, with dysfunctions in receptor tyrosine kinases, intracellular tyrosine kinases, and protein phosphatases that result in increased cellular proliferation, survival, invasion, and angiogenesis. Identification of the role of specific signal transduction pathways in the carcinogenic process has led to efforts to specifically target these pathways for anticancer therapy. The most successful to date has been the targeting of Bcr/Abl that drives the oncogenic lesions in chronic myelogenous leukemia (CML). Dramatic responses are observed in the treatment of this disease with imatinib (Gleevec). Relapses occurred during therapy and whether this treatment will provide cures has yet to be seen. What has been brought to light with targeted therapy of signaling pathways is that in many cases multiple signaling pathways drive the oncogenic lesions and rational and individualized combinations of inhibitors may be needed for optimal therapy. Targeting signaling pathways for therapeutic benefit may be more complex than originally anticipated.

HER2

The first successful rationally developed targeted therapy for cancer was the inhibition of HER2, also known as ErbB2, a member of the epidermal growth factor (EGF) family of receptor tyrosine kinases. The oncogenic effects of *HER2/neu* were demonstrated when blockade of the receptor inhibited tumorigenesis [3]. These experiments provided valuable information about HER2 as a target, as well as the therapeutic possibilities of blockade by monoclonal antibodies. Amplification of the *HER2/neu* gene in breast cancers causes overexpression of the HER2 protein that drives breast cancer cell proliferation. Overexpression of HER2 leads to resistance to therapy and a poorer disease prognosis. It is estimated that 30 percent of all invasive breast cancers have amplified *HER2/neu*, and therapeutic targeting could have significant therapeutic benefits to those with the genetic signature of amplified *HER2/neu*. These findings led to the development of a humanized monoclonal antibody to HER2, trastuzumab (Herceptin). This antibody binds to the HER2 receptor and inhibits downstream signaling leading to growth inhibition as well as sensitization to cytotoxic chemotherapeutic drugs. The benefits of targeting HER2 are evident as a single agent and in combination with chemotherapy. With the clinical successes of the humanized monoclonal antibody trastuzumab, work is ongoing to identify small-molecule inhibitors of HER2. Importantly, the development of trastuzumab provided the first proof-of-principle that targeting the specific molecular defects of cancer is a viable approach for anticancer drug development.

The EGF Receptor Pathway

The therapeutic value of targeting one member of the EGF pathway, HER2 has been demonstrated and EGF receptor (EGFR) (HER1), a second member of this receptor tyrosine kinase family, is being evaluated as a drug target for multiple types of cancer. Activation of the EGFR through ligand binding initiates a cascade of effects, which are depicted graphically in Figure 29.1. This signaling pathway activates the mitogen activated protein kinase (MAPK), phosphoinositide-3 kinase (PI3K)/Akt and the Jak/Stat pathways. These signaling cascades are implicated in cell proliferation, growth, survival, apoptosis, migration, and angiogenesis [4–6]. Dysfunction of the EGFR occurs in many different types of cancer including lung, breast, head and neck, colon, pancreatic, ovarian, and prostate cancers. Over 33 percent of solid tumors express EGFR, and it is often associated with poorer prognosis [6]. Activation of EGFR can occur due to overexpression of the receptor, expression of a constitutively active receptor, and/or amplified levels of its ligands, EGF or transforming growth factor α (TGF-α). Genetic alterations of EGFR, including

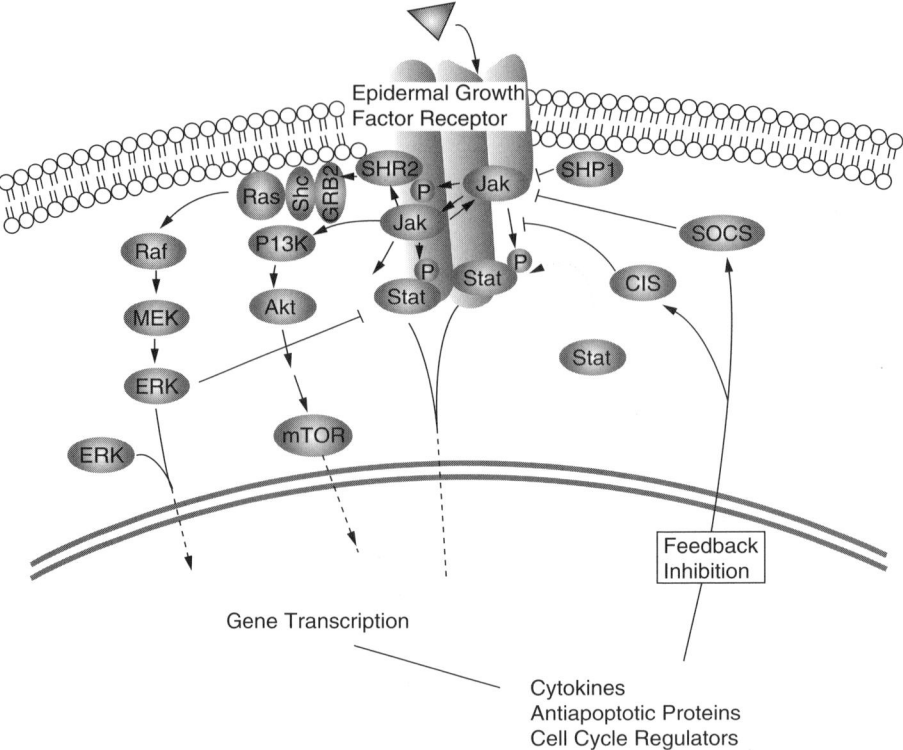

Figure 29.1 The EGFR signaling pathways. After ligand activation, the EGFR phosphorylates and activates the Ras-Raf-MAP kinase PI3K/Akt, and Jak/Stat pathways. This in turn results in activation of transcription factors and modulation of the cell cycle, growth, apoptosis, and angiogenic processes. From El-Rayes and LoRusso [4].

gene amplification and mutations leading to a constitutively active kinase, contribute to the pathogenesis of many tumors [6].

The prevalence of activated EGFR signaling in cancer led to several different targeting strategies including blockade of the receptor with monoclonal antibodies, analogous to the approach used with HER2, inhibition of the tyrosine kinase domain of the receptor, and use of antisense oligonucleotides against the receptor. Cetuximab (Eributex) is a humanized anti-EGFR antibody that binds to the EGFR causing dimerization and downregulation of MAPK, PI3K, and Jak/Stat signaling, which inhibits proliferation, cell cycle progression, metastasis, and angiogenesis [4, 6]. Cetuximab is well tolerated with limiting toxicities of allergic reactions and acneform rash, and it is being used for the treatment of advanced metastatic colorectal cancer [4, 6].

A second approach to limit EGFR's effects is through small-molecule tyrosine kinase inhibitors. Inhibition of the adenosine triphosphate (ATP) binding site has been the primary site of focus. Four different categories of inhibitors have been developed: EGFR-specific reversible inhibitors, EGFR-specific irreversible inhibitors, and reversible and irreversible pan HER inhibitors that have actions against several HER family members [4, 6]. Gefitinib (Iressa) is a reversible inhibitor of the EGFR ATP binding site. It is well tolerated, but only modest clinical responses were seen. Although the majority of non-small-cell lung cancer patients did not respond to inhibition of EGFR by gefitinib, a subgroup of 10 to 15 percent exhibited dramatic responses. Profiling of these tumors indicated that they had somatic activating mutations in the catalytic domain of the EGFR [7, 8]. The selectivity of gefitinib for these activating mutations had not been anticipated from preclinical studies. The mutations did not affect proliferation via the MAPK pathway, but they activated the Akt and Stat pathways and caused resistance to apoptosis, heightened survival, and drug resistance [9].

Antisense technologies are also being evaluated for the ability to decrease EGFR expression. What is clear though is that optimal use of these EGFR-targeted therapies may not be as apparent as first anticipated. The fact that the clinical indications for cetuximab and gefitinib do not overlap suggests that there is much more to learn about optimal use of these targeted therapies. The use of dual or pan-targeted therapies against multiple EGF family members, for example, EGFR and HER2, may also provide superior antitumor effects. Promiscuous inhibitors with broader specificity for multiple receptor kinases may be more active in tumors driven by abnormalities in several signaling pathways. Targeting several points in the signaling pathway may also yield better results.

Insulin-Like Growth Factor Signaling

Another family of receptor tyrosine kinases, the insulin-like growth factor (IGF) family, is implicated in the proliferation and survival of cancer cells. Activation of this signaling cascade in cancer may be due to amplified levels of the insulin-like growth factors IGF-1 and IGF-2 or its receptor IGF-1R. Similar to the signaling cascade initiated by EGFR, activation of IGF-1R stimulates the PI3K and MAP kinase pathways. Individuals with higher circulating levels of IGF-1 are at greater risk for lung, breast, prostate, colon, and bladder cancer [10]. Reduction of IGF-1 levels in murine models of cancer reduced tumor burden and incidence of metastasis [11]. In cell culture systems and in animal models, inhibition of IGF-1R expression or activity inhibited cell proliferation, colony formation, and tumor growth [12]. Efforts to identify specific inhibitors of this receptor tyrosine kinase have been successful. Inhibition of IGF-1R prevents downstream activation of both the PI3K and MAP kinase pathways and, ultimately, reduction in tumor growth [12]. Whether inhibition of IGF pathways will result in therapeutic benefit remains to be seen, but targeting this signaling cascade alone, or in combination with other signaling inhibitors or cytotoxic drugs, may lead to anticancer effects.

Bcr/Abl, c-Kit, and Platelet-Derived Growth Factor

The molecular driving force in CML is the Philadelphia chromosome (Ph$^+$), the name given to a reciprocal translocation between chromosomes 9 and 22. In this translocation the *Bcr* gene replaces the first exon of the *Abl* gene forming a chimeric oncoprotein Bcr/Abl. Bcr/Abl is deregulated and has constitutively active tyrosine kinase activities that leads to the phosphorylation of numerous cellular proteins including PI3K. Because this tumor is caused by an activating translocation in Bcr/Abl, this represented an optimal test of targeted therapies. The Bcr/Abl inhibitor imatinib (Gleevec), which binds and inhibits the ATP-binding site of the kinase, provided striking antitumor effects against CML and represents, to date, the most successful use of targeted therapy. In phase II studies using Ph$^+$ patients, a response rate of 95 percent was measured [5]. Imatinib is not specific for Bcr/Abl. It also has potent activity against c-Kit and the platelet-derived growth factor (PDGF) family of receptors. Activating mutations in the receptor tyrosine kinase c-Kit are found in gastrointestinal stromal tumors (GIST), and in some acute myeloid leukemias, small-cell lung cancer, neuroblastoma, and breast cancers [5]. Imatinib was highly effective against GIST, and it may have antitumor actions in other tumor types with activated c-Kit. The PDGF family is implicated in the pathogenesis of glioblastoma with over expression of both the receptor and the ligand, PDGF. Evidence suggests that PDGF autocrine signaling loops are involved in melanoma, neuroendocrine tumors, and tumors of the prostate, stomach, pancreas, and lung [5]. Targeting Bcr/Abl, c-Kit, and PDGF with imatinib provides utility against multiple types of cancer, and data suggest that the efficacy of imatinib may be due to its ability to target multiple tyrosine kinases [13].

Mammalian Target of Rapamycin (mTOR)

While significant efforts are directed toward inhibiting the receptor tyrosine kinases that drive proliferation and survival, several downstream effectors of these receptors are also emerging as cancer drug targets. Focusing on central junction points for multiple signaling pathways may provide broader antitumor activity. The serine/threonine kinase mTOR (mammalian target of rapamycin) fits this criterion as it is a convergence point for several signaling cascades (Fig. 29.1). mTOR is phosphorylated by PI3K/Akt in response to growth factors, the estrogen receptor, and oncogenic Ras [14–16]. Through a PI3K/Akt-independent pathway, mTOR senses nutritional status and controls cell cycle progression through G_1 [15]. Evidence has emerged that mTOR is a central controller of cell growth and proliferation mediated through its multiple effects on protein synthesis [14–16]. mTOR, as its name implies, was identified because it is the target of the bacterial secondary metabolite rapamycin. Inhibition of mTOR causes little change in global protein synthesis, but significant changes in the cellular levels of specific proteins including c-myc and cyclin D1, which play key roles in cell cycle progression, were observed [14].

Further cell cycle changes are mediated by the ability of rapamycin to stabilize the cyclin-depenent kinase inhibitor p27 (Kip 1) leading to inhibition of cyclin-depenent kinase 2 (Cdk2). Rapamycin has effects on the cell survival pathways where it causes activation of the stress kinase ASK1. In p53 mutant cells, sustained ASK1 phosphorylation causes cell death [14].

Defects in multiple upstream proteins can activite mTOR in cancer (Fig. 29.1). These include activated receptor tyrosine kinases and activating *Ras* mutations that occur in approximately 30 percent of epithelial-derived tumors [15]. Amplifications of the genes for PI3K and Akt and loss-of-function mutations in the tumor suppressor phosphatase and tensin homologue (PTEN), which inhibits PI3K, occur in many tumors [14, 15]. Mutations in eukaryotic translation initiation factor 4E (eIF4E), a downstream effector of mTOR, cause malignant transformations [14, 15]. The central position of mTOR in the signaling abnormalities that define cancer makes it a logical anticancer drug target. Rapamycin is an effective inhibitor of mTOR, but its inherent instability and poor solubility prevented its development for cancer therapy. Rapamycin analogs were designed and synthesized to overcome these limitations and were evaluated clinically. The results suggest low toxicity and some clinical responses as well as disease stabilization [16]. Optimization of scheduling and patient selection may provide additional opportunities to evaluate the effectiveness of inhibiting mTOR as a therapeutic approach. Identification of a predictive biomarker would allow correlation of drug levels with clinical response [14].

In summary, targeting signaling pathways implicated in the molecular etiology of cancer holds promise for more specific and less toxic therapy for neoplastic diseases. Many hurdles exist to fully implement the hope of this approach. There is significant cross-talk and redundancy of signaling pathways that control cell growth and survival. This, together with the inherent genetic instability of cancer, leads to questions of how best to use these molecular-targeted therapies. Rational combinations of these signaling blockers represent one approach, as is the testing of agents with broader target specificity inhibiting multiple receptor tyrosine kinases. A small molecule inhibitor that targets both EGFR and HER2 provides greater activity against HER2 positive breast cancer cells than HER2 inhibition alone [17]. Additionally, there is a need to understand the characteristics of tumors that are responsive to these targeted therapies and how biomarkers can be identified for appropriate patient selection. Successful use of these targeting agents requires integrated diagnosis and therapy.

Methods of Drug Discovery

The screening assays used to identify inhibitors of signaling cascade were initially based on biochemical kinase assays. With the availability of purified proteins and labeled peptide substrates, many different assay technologies are available to screen for inhibitors of protein kinases. New screening technologies are being developed at a rapid pace and most use a fluorescent-based readout based on the ability of the kinase to phosphorylate an exogenous fluorescent peptide substrate. HTS utilizes techniques such as enzyme-linked immunosorbent assay (ELISA), phospho-specific protein-specific antibodies, scintillation proximity assays, biotin-labeled substrate capture assays, fluorescence polymerization (FP), and time-resolved fluorescence resonance energy transfer (TR-FRET) [18–20]. IMAP® (immobilized metal ion affinity-based fluorescence polarization) is a new technology utilizing the ability of the fluorescently labeled phosphorylated substrate to bind to immobilized metal complexes on nanoparticles [20].

Every new assay should be carefully evaluated to ensure that it is robust, provides consistent and reliable results, and that it is predictive, selective, and sensitive. The assays need to have sufficient throughput to sample a wide variety of chemistry to identify molecules with activity against the target. It is important that new screening technologies retain the sensitivities and reliability of older generations of screening technologies. HTS (96- and 384-well format) have been used, and with improvements in liquid handling capabilities, ultrahigh-throughput (uHTS) (1536+ well formats) using ultra low volumes (1 µL) are now available. Sensitivity for uHTS is provided by TR-FRET and FP detection systems [21]. Microarrayed compound screening is an ultrahigh throughout screening method (uHTS) method that eliminates the need for complex liquid handling systems. The reagents that would be added to a well of a microplate are incorporated into agarose gels that are layered over the compounds arranged on polystyrene sheets [22]. This provides capabilities to screen at a density of 8640 reactions on one sheet.

The ATP-binding site of the kinases is a primary target for identification of inhibitors and most of the protein kinase inhibitors that have been evaluated clinically are directed to this site. With advances in the modeling of the catalytic site of these kinases, virtual screening using computer modeling of the binding site and compound structures is a useful approach. Certain kinases adopt different conformations when activated by cofactors. This should be considered in the design of the binding site for optimal anticancer efficacy and specificity.

29.3 TARGETING TUMOR VASCULATURE

Tumor growth is dependent upon the generation of a functional blood supply to provide nutrients and remove metabolic waste. Without neovascularization a tumor is limited in size to 1 to 2 mm in diameter [23]. Tumor vasculature can form by a number of different processes including sprouting from existing vessels and recruitment of circulating endothelial cell precursors. Unlike normal vasculature, tumor vasculature is highly abnormal and poorly differentiated, with tortuous dilated vessels [24]. Targeting either the process of new blood vessel development, angiogenesis, or the existing tumor vasculature, with antivascular therapy, are valid approaches to control and limit cancer and metastasis. In addition to targeting blood vessel formation, the development of a functional lymphatic system is critical to tumor metastasis. Inhibition of this process, called lymphangiogenesis, together with antiangiogenic therapy may also provide opportunities for better antitumor activities [25].

Inhibitors of Angiogenesis

The promise of controlling tumor growth by preventing angiogenesis was proposed over 30 years ago by Judah Folkman [26]. During tumor development, there exists an imbalance between proangiogenic factors and antiangiogenic factors that leads to rapid blood vessel growth (Fig. 29.2). This promotion of tumor angiogenesis is called the angiogenic switch and is critical for cancer progression and metastasis [27]. Stimulation of growth factor secretion by the tumor in response to oncogenic stimulation or hypoxia can unbalance the equilibrium and cause rapid neovascularization. This angiogenic switch occurs as an early event of tumorigenesis when quiescent microscopic tumors change into rapidly expanding aggressive tumors. The search for agents that could specifically inhibit tumor angiogenesis provides an attractive target for anticancer therapy. In the adult, with a few exceptions, the normal vasculature is quiescent with balance of positive and negative regulatory factors. Antiangiogenic therapy is expected to allow more specific targeting of abnormal events. Targeting genetically stable endothelial cells might have additional advantages in that it may circumvent drug resistance mechanisms that occur in tumor cells due to genetic instability. However, recent data suggest that the genetic abnormalities of tumor tissue may also be found in tumor endothelial cells [28]. The apparent heterogeneous nature of tumor endothelial cells might need to be considered in the future. Although antiangiogenic therapy was conceived decades ago, the first antiangiogenic anticancer agent, bevacizumab (Avastin)

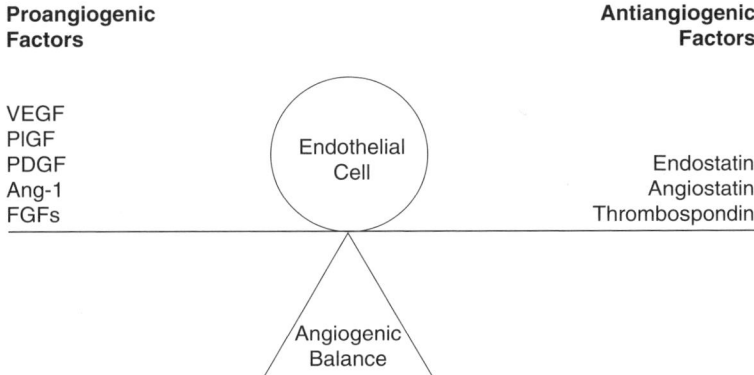

Figure 29.2 Angiogenic balance. In normal endothelial cells the proangiogenic factors are balanced by antiangiogenic factors. Disruption of the balance can lead to pathogenic conditions.

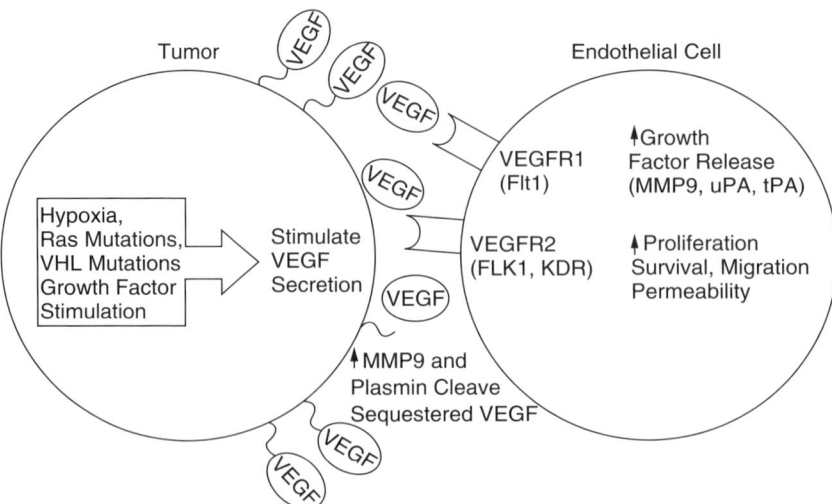

Figure 29.3 The VEGF pathway. Antiangiogenic approaches that target the tumor-driven VEGF pathway include blocking VEGF with antibodies, trapping VEGF with decoy receptors, blocking VEGFR with antibodies, and inhibiting VEGFR receptor tyrosine kinase activity.

was approved by the Food and Drug Administration (FDA) in early 2004 for use in colorectal cancer.

Angiogenesis is a complex process providing many avenues for control (Fig. 29.3). Key signaling events leading to the angiogenic switch include growth factors and receptors of the vascular endothelial growth factor (VEGF) family, basic fibroblast growth factor (bFGF), acidic fibroblast growth factor (aFGF), the platelet-derived growth factor (PDGF) family, and protease inhibitors. Inhibition of these may provide avenues to inhibit the angiogenic switch. Endogenous antiangiogenic signals are also present. The endogenous inhibitors endostatin and angiostatin have been tested for antiangiogenic actions in tumor models where they showed antitumor and antiangiogenic effects, yet anticancer effects have not yet been achieved.

The VEGF Axis The VEGF axis is critically important to normal embryonic and adult angiogenesis as well as to pathological blood vessel development. Loss of a single allele for VEGF, or the genes for VEGFR1 (Flt-1) or VEGFR2 (Flk-1, *KDR*) results in embryonic lethality. Although both normal and pathological angiogenesis involve the coordinated effort of many growth factors, the VEGF signaling cascade appears to be the rate-limiting step of new blood vessel formation [29]. The VEGF receptor ligands are a family of proteins, VEGF-A-E, and placental growth factor (PlGF). VEGF-A is the predominant VEGF with activity in angiogenesis, and its activities predominate in both normal and tumor angiogenesis [29]. The human VEGF gene contains eight exons and alternative splicing leads to the generation of several different

VEGF-A isoforms [30]. The isoforms differ in their heparin binding affinities, the smaller, acidic $VEGF_{121}$ having no affinity for heparin and the larger basic isoforms, $VEGF_{189}$ and $VEGF_{206}$, bind readily to extracellular matrix-associated heparin. The intermediate isoform, $VEGF_{165}$, has optimal biological activity and an intermediate capacity for heparin binding [30]. Tumor cells can directly secrete active VEGF or active isoforms of VEGF can be released from heparin in the extracellular matrix by the actions of proteases including plasmin and matrix metalloprotease 9 (MMP9) (Fig. 29.3). VEGF-C and D are implicated in lymphangiogenesis but are not thought to play a critical role in tumor angiogenesis. In addition to several different ligands, at least three different receptor types for the VEGF family have been identified (Fig. 29.3). The receptors of VEGF, VEGFR1–3, are receptor tyrosine kinases with varied cellular effects and localization. Currently, evidence suggests that the VEGFR2 receptor (Flt-1, KDR) plays the primary role in endothelial cell permeability, proliferation, migration, and release of endothelial stem cells from bone marrow, events critical for tumor angiogenesis [31, 32]. VEGFR1 is also present on endothelial cells, and activation can lead to induction of growth factor release and release of proteolytic enzymes including MMP9, urokinase-type plasminogen activator [33], and tissue type-plasminogen activator (tPA) [30, 31, 34]. An alternately spliced soluble isoform of VEGFR1 is implicated in blocking VEGF actions. Receptor cross-talk between VEGFR1 when activated by PlGF and VEGFR2 has been implicated in amplification of VEGF angiogenic signaling [31, 32]. VEGFR3 is involved primarily in lymphangiogenesis.

The FDA approval of the first antiangiogenic anticancer therapy, bevacizumab, a humanized monoclonal antibody to VEGF-A, highlights the importance of the VEGF pathway in tumor angiogenesis. Several different approaches to inhibit the tumor-driven VEGF pathway are active targets for cancer drug discovery. They include blockade of VEGF, with, for example, a monoclonal antibody, trapping VEGF with decoy receptors, antisense inhibition of VEGF, antibodies to VEGFR, and inhibition of VEGFR tyrosine kinase activity by small molecular protein kinase inhibitors [35].

HIF-1α as a Therapeutic Target The transcription factor hypoxia-inducible factor-1 (HIF-1) is a new target for antiangiogenic therapy and antitumor therapy because it occupies a central point in the responses of tissues to hypoxia. Tumors, by virtue of aberrant blood supply and high interstitial pressure, are hypoxic, and this causes activation of adaptive survival mechanisms mediated by the transcription factor HIF-1. HIF-1 binds to hypoxia response elements (HRE) leading to the transcription of numerous proteins involved in angiogenesis, cell survival and invasion, and drug resistance. Tumors maintain high levels of HIF-1 and in many tumor types, high expression of HIF-1 is associated with a worse prognosis, validating this protein as a therapeutic target for cancer. HIF-1 is a heterodimer composed of an α and a β subunit. Both subunits are constitutively transcribed, but the cellular levels of the α subunit are tightly controlled, providing the rate-limiting step in HIF-1 expression. Under normal oxygen conditions HIF-1α is rapidly degraded by the ubiquitin–proteasome pathway; however, under hypoxic conditions the protein is stabilized and binds to the β subunit to form the active transcription factor HIF-1. In addition to hypoxia, the expression of HIF-1α can be induced by oncogenes including *Ras, Src, HER2/neu*, and *mTOR* and by loss of tumor suppressor genes including *p53, PTEN*, and *VHL*. Growth factor stimulation of receptor tyrosine kinases and reactive oxygen species also stabilize HIF-1α. Over 60 genes are transactivated by HIF-1, including the genes for VEGF and VEGFR. Targeting HIF-1 may provide broad antiangiogenic and antitumor effects because of its ability to inhibit several different pathways of angiogeneic stimulation.

Antivascular Agents

With the focus on tumor blood vessels as a target for cancer therapy, some compounds with antivascular effects have been identified. Antivascular agents differ from antiangiogenic agents in that they target existing tumor vasculature, while antiangiogenics inhibit the process of neovascularization. Vascular disrupting agents cause rapid, often within minutes, and substantial disruption of the aberrant tumor vasculature, and subsequent loss of circulation leading to tumor ischemia and necrosis [36]. Vascular disrupting agents in clinical testing include multiple-tubulin targeting drugs such as combretastatin A4 and flavonoids that appear to act by inducing the localized release of TNFα from activated macrophages in the tumor [36]. The antivascular approach, while distinct from antiangiogenic therapies, appears promising. Combinations of antiangiogenic and antivascular targeting may be used to reduce initial tumor burden and to inhibit further tumor growth.

Methods of Drug Discovery

Diverse mechanisms are used in the identification of antiangiogenic agents. The VEGF axis represents many different avenues for drug discovery, and it has been validated as a therapeutic target for cancer. The efficacy of bevacizumab, which binds to VEGF ligand, has proven to be an effective approach to inhibiting VEGF signaling. HTS is used to identify inhibitors of the VEGF receptors tyrosine kinase activity. A high-throughput TR-FRET assay targeting VEGFR-2 measures the autophosphorylation of the receptor that occurs upon its activation [19]. The assay is homogenous in that no washing steps are required and is amenable to automation. Mechanism-based HTS to identify inhibitors of HIF-1 could address the ability of inhibitors to repress its transcriptional activation of HREs. To date most inhibitors of HIF-1 have been identified by mechanistic studies of known inhibitors [37].

The use of endothelial cells as a model has advantages in that a cell-based screen can identify new targets. Inhibition of endothelial cell proliferation, invasion through basement membrane material [38], and the formation of capillary-like tubules on Matrigel® matrices are assays used to predict antiangiogenic effects. Reagents are available commercially to evaluate endothelial cell migration and invasion in a 96-well plate format. Endothelial cells from different vascular beds can be obtained commercially, but it is important to note that these normal cells have a finite life span. An immortalized tumor endothelial cell line has been proposed as a good model of tumor endothelial cells [39]. Assays are used to confirm that inhibitors identified in HTS or cell-based screens actually have activity against neovascularization, an in vivo process involving not only the endothelial cells but their stromal environment as well. Several different models are available to assess the ability of compounds to inhibit angiogenesis. They include embryonic zebrafish [40], chorioallantoic membrane of the chick embryo [41], and growth of vessels into a subcutaneous matrigel plug [42]. Pathological evaluations of tumor samples are used to assess antiangiogenic effects concurrent with antitumor measurements. New imaging techniques with magnetic resonance imaging (MRI) and ultrasound allow real-time evaluation of both angiogenic and antivascular effects [43, 44].

29.4 DNA TARGETS

Nitrogen mustard, a deoxyribonucleic acid (DNA) alkylating agent, led the era of modern cancer chemotherapy. DNA binding agents remain some of the most effective anticancer agents available. Cross-linking of DNA and adduct formation by alkylating agents prevents DNA replication and normal repair functions culminating in apoptosis. Significant efforts over the course of many years led to optimization of alkylating agents including platinum-based compounds. Antimetabolites prevent the synthesis of essential DNA and RNA precursors, which then causes inhibition of DNA and protein synthesis leading ultimately to apoptosis. Topoisomerases are enzymes critical for maintaining normal DNA topography. Topoisomerase inhibitors bind to both DNA and the topoisomerase enzyme leading to DNA strand breaks, which accumulate and cause cell death. Drugs that target DNA by diverse mechanisms play a major role in cancer therapy. New DNA-based targets have emerged in the past few years that have the hope of wider therapeutic windows and better specificity for cancer cells as compared to normal cells.

DNA Methylation

Mutations in genes critical to cellular growth and survival can cause and promote cancer. Efforts are underway to target some of the specific genetic changes that occur during cancer. In addition to these genetic changes, epigenetic changes in DNA are also implicated in carcinogenesis and maintenance of the tumorigenic phenotype. Gene silencing that occurs through changes in chromatin structure may be reversible, providing a target for therapeutic intervention. Gene expression is controlled by many factors, including accessibility of the gene for transcription. DNA methylation patterns play a key role in gene expression. Actively transcribed genes have low methylation status, while areas that are silenced are hypermethylated. DNA methytransferases (DNMT) are enzymes that catalyze the addition of a methyl group to cytosine residues. Increased methylation in the cytosine-rich CpG promoter regions causes gene silencing. Hypermethylation of CpG islands is implicated in the silencing of a number of tumor suppression genes [45, 46]. DNA methylation appears to play a central role in cancer and yet global hypomethylation results in genetic instability [45, 46]. The ability to revert cancer cells back to a normal methylation pattern may hold promise for fixing the underlying events of cancer, yet care will need to be taken to prevent global changes in DNA methylation. Three DNMTs have been identified and specificity may hold the key to solving this conundrum. Inhibition of DNMT1 by a cytosine analog or by siRNA reverted transcriptional silencing of tumor suppressor genes [46]. Antisense oligonucleotides directed against DNMT1 are effective antitumor agents and a small-molecule inhibitor of DNMT1 is being evaluated clinically. It will be critically important to affect the abnormal DNA methylation without changing global methylation [47].

Histone Deacetylase Inhibitors

In addition to DNA methylation, histone acetylation is intricately involved in epigenetic control of gene transcription [46, 48]. Histones are the structural units that organize and condense chromatin. A nucleosome contains about 150 base pairs of DNA that is wrapped around a core of histone proteins. The amino terminus tails of histones contain lysine residues that can be acetylated by histone acetytransferases (HAT). Acetylation of these lysines leads to relaxation of the tight chromatin packaging in that region facilitating access to transcription factors and leading to transcriptional activation [48]. Histone deacetylases (HDAC) remove the acetyl group initiating a more compact chromatin structure and resultant transcriptional repression.

Histone acetylation is a dynamic process controlled by the balance between the actions of HATs and HDACs [48]. HDACs also have direct interactions with transcription factors including p53 and e2F [48]. In cancer, mutations in HAT and recruitment of HDAC, via in part by DNA methylation, are associated with transcriptional silencing of tumor suppressor genes [48, 49]. The ability to reverse the abnormal transcriptional repression is an attractive therapeutic target, and numerous HDAC inhibitors are being evaluated in preclinical and clinical models [48, 49]. Several diverse chemical classes of compounds inhibit HDAC. Trichostatin A, a microbial natural product was found to induce differentiation of leukemia cells, and additional studies showed that these effects were mediated through HDAC inhibition [48]. Other cellular effects of HDAC inhibitors include growth suppression, cell cycle arrest, induction of the cyclin-dependent kinase inhibitor p21, and initiation of apoptosis [49]. The effects of HDAC inhibitors appear to be cell-type dependent, and concerns have been raised about the ability of some of these inhibitors to increase genetic instability and telomerase (hTERT) expression [49]. Several HDAC inhibitors have been evaluated clinically. They appear to be well tolerated, and early results show some clinical responses that are tied with inhibition the target [48, 49]. Combinations of HDAC and DNMT inhibitors are being evaluated because targeting epigenetic silencing by two different approaches may provide optimal reversal of aberrant gene repression.

Telomerase Inhibitors

Telomeres are tandem nucleotide repeats found on the ends of chromosomes. A T-loop structure is found at the very end of the telomeres, and it differentiates the ends of the chromosomes from a chromosome break. These specialized structures prevent homologous recombination and nonhomologous end joining between chromosomes [50]. DNA polymerase cannot replicate to the very end of the chromosome because of the overhang needed for lagging strand synthesis. With progressive cell divisions telomeres are lost and aging cells contain shorter telomeres. Cellular replicative senescence occurs when the telomeres shorten to a certain threshold where they signal cell cycle arrest [51]. If this signal is lost and division continues, then chromosome ends fuse, leading to initiation of apoptosis. A normal functioning telomere prevents uncontrolled growth and cell immortality, and this is thought to help prevent cancer. Unfortunately, cancer cells escape replicative senescence and lose this control mechanism. The ability of cancer cells to escape this control is one of the hallmarks of cancer [50, 51]. Cancer cells reactivate telomerase, an enzyme that adds telomeric repeats to the ends of the chromosomes. Telomerase is composed of the enzymatic reverse transcriptase subunit, hTERT, and an RNA nucleotide template, hTR [51].

Activated telomerase is detected in almost 100 percent of cancer cell lines and in 80 percent of human tumors, and telomere-deficient mice have a low incidence of tumors [52]. Telomerase is a logical target for cancer therapeutics because cancer cells depend on active telomerase to escape replicative senescence. Telomerase inhibitors may be fairly specific because most normal cells do not express telomerase. Telomerase inhibitors are being developed. BIBR1523 is a specific and potent telomerase inhibitor that blocks cell proliferation and causes shortening of telomeres but does not initiate cancer cell death [52]. Consistent with its cytostatic actions, it did not cause tumor regression in tumor xenograft models. Concerns about the side effects of telomerase therapy have been expressed. Telomerase is expressed in hematopoetic cells and in germ cells and cells lining the intestinal crypts. Inhibition of telomerase may cause some of the side effects associated with other DNA-interacting anticancer drugs due to lack of specificity [50–52]. Whether telomerase will prove to be a valid target for cancer therapy remains to be seen.

29.5 ANTIMITOTICS

Antimitotics are one of the oldest and most effective classes of anticancer drugs. They are used in combination with other cytotoxic chemotherapeutics in the curative therapy of childhood and adult leukemias and testicular cancer. They are also effective in the adjuvant setting, providing activity against a broad range of solid tumors and hematological malignancies. The majority of antimitotics target tubulin, but newer antimitotics that target different components of the mitotic apparatus appear to be excellent therapeutic targets for cancer.

Microtubule-Targeting Agents

Tubulin is a validated anticancer target. Tubulin-targeting drugs are the one class of anticancer drugs that are equally effective against p53 mutant and p53 null tumors, and considering the fact that over half of human tumors contain abnormal p53, this is significant [53]. The history of the use of chemically diverse tubulin-targeting drug shows that structure cannot predict clinical utility or dose-limiting toxicities, and new tubulin-targeting agents may be expected to have different spectra of clinical activities. Microtubule-disrupting agents can be divided into two groups, microtubule stabilizers and microtubule depolymerizers. Microtubule stabilizers shift the equilibrium of cellular tubulin toward the polymerized state, and dramatic increases in the number of cellular microtubules can be observed in cells treated with these compounds. In contrast, microtubule depolymerizers cause a loss of cellular microtubules and a shift of cellular tubulin toward the cytosolic pool. At the lowest effective concentrations both classes of drugs inhibit microtubule dynamics and lead to interruption of mitosis and, ultimately, apoptosis [54]. Cancer cells are much more susceptible to tubulin-targeting drugs, and the molecular cause of this is not yet known. Identification of the nature of the susceptibility of cancer cells for microtubule-targeting agents may help in the identification of new mitotic targets for cancer therapy.

Taxol was the first microtubule stabilizer identified, and it represents one of the most effective drugs for cancer therapy to be identified in the last 20 years. Other microtubule stabilizers including taxane derivatives, the epothilones, and discodermolide are in clinical trials and the preliminary results look promising. Notably, they appear to be active in taxane-refractory tumors, even though they bind to the same binding site on tubulin. All the stabilizers tested clinically bind to tubulin on sites that overlap the Taxol binding site. A new group of stabilizers, represented by laulimalide and peloriside A, bind to a distinct, nonoverlapping site or sites on tubulin [55]. Whether stabilizers that do not interact with the Taxol binding site have anticancer actions is not yet known.

The vinca alkaloids are the prototypical microtubule depolymerizers. Several different vincas, vincristine, vinblastine, and vinorelbine, are used clinically, and newer vincas, including vinflunine, are in clinical trials. Although structurally very similar, they have some nonoverlapping indications and to some extent differing limiting toxicities. These drugs all bind to the vinca domain of tubulin. Although colchicine is not used to treat cancer, other compounds that share a binding site with colchicine are in clinical trials. There is a third binding site on tubulin that is occupied by microtubule depolymerizers; it is the peptide binding site within the vinca domain. Agents that bind within this site noncompetitively displace the vinca alkaloids. A number of drugs that bind within this site, including dolastatin 10 and cryptophycin 52, were unsuitable for the treatment for cancer. Whether these limitations are due to the binding site specifically or to the drugs themselves is not known. New tubulin binding agents may be expected to provide advantages, such as the ability to circumvent drug resistance mechanisms.

Kinesin Inhibitors

Until recently the only antimitotic drugs were those that bound to tubulin. The identification of monastrol, an Eg5 (KSP) inhibitor, led the way for the non-tubulin-targeting antimitotics [56, 57]. Eg5 is a mitotic kinesin that is active early in mitosis to separate the centrosomes to form a bipolar mitotic spindle. Inhibition of centrosome separation causes the formation of a monopolar mitotic spindle, which is dysfunctional and leads to mitotic arrest followed by initiation of apoptosis. Eg5 is preferentially expressed in mitotically active cells and this should provide some selectivity. It is not expressed in neural tissues, and thus Eg5-targeting agents should circumvent the peripheral neuropathies that occur with tubulin-targeting antimitotics. Eg5 is attractive as a anticancer target. Optimization of the lead monastrol led to the identification of CK0106023 [58]. CK0106023 is a potent inhibitor of Eg5 and cellular proliferation. Murine antitumor trials indicated that CK0106023 has tumor growth inhibitory activities comparable to Taxol, yet at the lowest effective concentration, it did not cause body weight loss that occurs with Taxol, indicating less toxicity [58]. Monopolar spindles in tumor sections confirmed Eg5 targeting during the in vivo trial. These data suggest that Eg5 inhibitors might have promising actions against human tumors.

Aurora Kinases

The Aurora kinases are a family of three serine/threonine kinases and two, Aurora A and B, play critical roles in mitosis and cytokinesis. Aurora A is localized at the centrosomes and mitotic spindles, and it has been referred to as the polar Aurora [59]. Aurora A is involved in centrosome separation and spindle assembly. Aurora B is called the equatorial Aurora because it is localized at the centromere and represents a chromosomal passenger protein [59]. Aurora B plays important roles in chromosome bi-orientation, cytokinesis, and perhaps chromosome condensation. It is also a critically important part of the spindle-assembly checkpoint that functions to prevent abnormal mitosis [59]. The diverse and central roles of Aurora A and B in cell division make these kinases encouraging targets for antimitotic effects. The Auroras are all implicated in cancer progression, and overexpression of each has been identified in cancer cell lines [59]. The gene encoding Aurora A will initiate the malignant phenotype when elevated, and gene amplification may occur in some tumors [59–61]. High expression levels of Aurora A are associated with a chromosomal instability, more aggressive disease, and poorer prognosis [60]. Elevated expression of Aurora B causes cytokinesis and chromosome segregation abnormalities leading ultimately to aneuploidy [59]. Targeting the Aurora kinases may directly target a functional abnormality that contributes to the pathogenesis of cancer.

Specific inhibitors of Aurora A have been identified. VX-680 is a potent and selective inhibitor of all three Aurora kinases, and it directed against the ATP-binding site of the kinase domain [62]. VX-680 has a 30-fold selectivity for Aurora A as compared to Aurora B. It has potent antiproliferative activities against a number of colorectal, leukemia, and lymphoma cell lines, and inhibits colony growth formation [62]. Mechanistically, VX-680 causes accumulation of cells with tetraploid DNA content, which then die by apoptosis [62]. These results suggest that it does not inhibit mitosis but cytokinesis. In xenograft models of AML, VX-680 had excellent antitumor effects and it caused tumor regression in pancreatic and colon tumor models [62]. Other Aurora kinase inhibitors have been identified [63], and it will be interesting to monitor whether Aurora kinases represent a new target for anticancer therapy. Mitosis is a complex and intricately controlled cellular event that is critical to tumor growth. Additional antimitotic targets are expected to be identified [61].

Methods in Drug Discovery

Cell-based assays constitute the primary screening methods to identify new microtubule depolymerizers and stabilizers. Many antimitotic drugs are concentrated in the cytosol at hundreds of times the concentration in the media, and this provides exquisite sensitivity. Purified tubulin-based assays are valuable as secondary confirmation screens but are limited use for primary screening because of lack of sensitivity. Cell-based screens include microscopic, flow cytometry, and cell-morphology-based screens. A phenotypic screen that simply evaluates the effects of compounds or natural product extracts on interphase smooth muscle cells led to the identification of two new classes of microtubule stabilizers, the laulimalides [64] and taccalonolides [65], and multiple depolymerizers, including the cryptaphycins [66] and symplostatins [67]. The assay is reliable and sensitive, but has only moderately high throughput. Use of automated microscopy combined with data algorithms to detect changes in cellular microtubule density make this assay amenable to HTS.

A flow-cytometry-based method uses a mitosis-specific antigen to quantify cancer cells in mitosis [68]. It is sensitive and has advantages over other flow-based technologies in that only a single mitosis-specific parameter is measured. This assay provides moderate throughout, but with a 96-well flow cytometry format it could be used for HTS. This assay has been used to identify a number of different new microtubule depolymerizing agents, and it has an advantage over the other two assays in that it can identify tubulin-independent antimitotics [68].

The ability of dibutyryl-cyclic AMP (db-cAMP) to cause glioma cells to round up due to tubulin polymerization is the basis for a third screening assay for tubulin-interacting agents [69]. In the presence of microtubule depolymerizers, dc-cAMP fails to initiate the morphology change. The morphology changes cause the cells to round up and lose adhesion and they can be easily removed by aspiration. Cells remaining can be quantified by any number of techniques. This assay is sensitive and amenable to HTS. Unlike the other assays it cannot detect microtubule stabilizers or antimitotics that work via a tubulin-independent mechanism [69].

High-throughput screening based on kinase activity of the mitotic kinases will be valuable approaches to identify specific inhibitors of the Aurora kinases. These techniques have already proved to be successful [63]. With the identification of other mitotic targets these types of assays will be invaluable. Modeling of the ATP binding site may also provide utility for optimization of Aurora kinase inhibitor leads.

29.6 NEW THERAPEUTIC TARGETS

Proteosome Inhibition

Precise control of the activities of cellular signaling proteins is important for maintenance of homeostatic balance. Manipulation of protein transcription/translation as well as degradation provides opportunities to alter cellular protein levels. Protein ubiquitination followed by digestion by the proteosome is particularly important for the control of cell cycle regulatory proteins [70]. The rationale of targeting the proteosome is to revert the cellular levels of amplified proteins that contribute to the malignant phenotype to normal levels. The transcription factor nuclear factor-κB (NF-κB) is believed to play a key role in cancer [71]. NF-κB is involved in many cellular actions that define malignancy, including proliferation, invasion, angiogenesis and metastasis [72]. NF-κB is a central transcription factor in inflammation. The chemopreventive actions of anti-inflammatory agents may be mediated through inhibition of NF-κB [72]. NF-κB is constitutively active in most cancer cell lines and in human cancer tissues. NF-κB matures from its precursor p105 form to its active p50 form by the actions of the proteosome. The activity of NF-κB is controlled by its associations with its inhibitory binding partner IκB. The degradation of IκB by the proteosome releases its inhibition of NF-κB. Inhibiting the proteosome can reduce the cellular levels of NFκB, a central prosurvival transcription factor, by two mechanisms, inhibiting NFκB maturation and preventing IκB degradation. The proteosome inhibitor bortezomib (Velcade, PS-341) is a reversible inhibitor of the 26S proteosome [71]. In vitro it has antiproliferative and apoptosis-inducing activity against drug-sensitive and drug-resistant multiple myeloma cells [73]. In clinical trials bortezomib was active against multiple myeloma, providing complete responses in 10 percent of patients and partial responses in 18 percent in patients with relapsed, refractory disease [71]. Preliminary indications suggest that inhibition of the proteosome may provide a new, effective anticancer target.

Heat Shock Protein 90 (Hsp90)

Heat shock protein 90 is a new target for cancer therapy and early clinical trial results of a Hsp90 inhibitor, 17-allylamino-17-desmethoxygeldanamycin (17AAG), suggest that it may have promise as an anticancer target. Hsp90 is a molecular chaperone that together with its co-chaperones forms a complex that is involved in the maturation and folding of nascent polypeptide chains into the tertiary structures required for proper activities. Disruption of the maturation process leads to ubiquitination and subsequent degradation of the proteins by the proteosome. Several of the client proteins of Hsp90 are implicated in cancer cell growth and survival. Client proteins of Hsp90 implicated in carcinogenesis include the tyrosine kinases v-Src, Akt, Raf-1, Bcr-Abl, and HER2, and the transcription factors HIF-1α, mutant p53, and steroid receptors [74, 75]. Recent data shed additional light on the selectivity of Hsp90 inhibitors toward malignant cells. During oncogenic transformation it appears that the HS90 protein complex becomes activated in a superchaperone form complexed with it co-chaperones [76]. In this configuration Hsp90 has high ATPase activity and a higher binding affinity for 17-AAG, which binds to the amino-terminal ATP binding site. Targeting Hsp90 is an approach to inhibit the machinery that allows a cancer cell to maintain its multifaceted and ever-changing growth and survival properties.

Inhibitors of Hsp90, which target the N-terminal ATPase site, have been developed and one, 17AAG is in clinical trials. There are numerous problems with 17AAG including limited solubility and oral bioavailability, but it has allowed the test of proof-of-principle of targeting Hsp90 for anticancer therapy. New inhibitors of the N-terminal ATP binding site have been identified [77] and other mechanisms of interfering with Hsp90 may also be effective [78].

29.7 SOURCES OF SMALL-MOLECULE DRUG LEADS

With the advent of exciting new targets for cancer drug discovery and numerous sensitive and specific high-throughput screens, it is critically important to evaluate a wide range of chemistry to optimize drug discovery. There are advantages of both natural product sources and synthetic sources for chemical libraries. There are strong rationales for using not just one source but both for optimal chemical diversity.

Natural Products

The chemical diversity found in nature in microorganisms, plants, and marine species has proved historically to be the best source for lead compounds for therapies for human diseases [79]. The majority of drugs that are in use for the treatment of cancer are natural products or are based on natural products (Fig. 29.4). An evaluation of all the cancer drugs approved by the FDA from the 1940s through 2002 showed that 40 percent of the anticancer drugs were natural products or natural product derivatives. Another 20 percent of the drugs represent synthetic compounds where the pharmacophore was defined by a natural product, and 30 percent were strictly synthetic [79]. Natural products have played an important role in the treatment of cancer over the

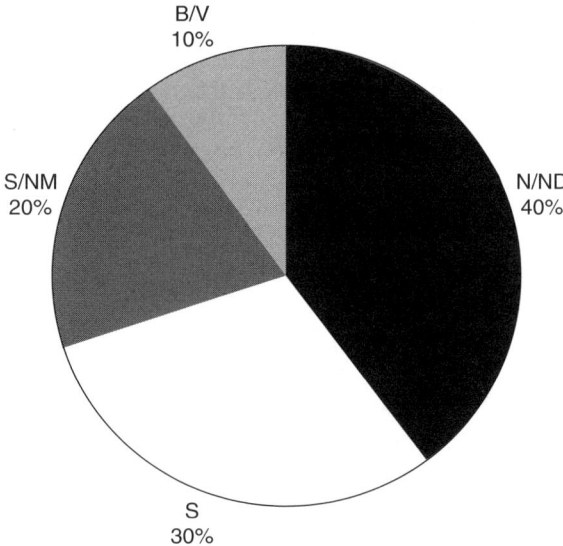

Figure 29.4 Cancer drugs, 1940s–2002. Sources of drugs approved for anticancer use: natural and natural derivatives (N/ND), synthetics (S), synthetics based on natural product mimetic or pharmacophore (S/NM), and biological or vaccine (B/V). Adapted from Newman et al. [79].

past 50 years and they continue to be useful for new targets. With the disappointment of leads obtained from the first efforts of combinatorial chemistry, there is a resurgence of natural products in drug discovery. Use of natural products has been streamlined to allow for the shorter timelines of modern drug development. Many groups, including some in industry are utilizing prefractionated natural product samples, peak libraries, and are evaluating these in HTS screening assays. The use of these enriched fractions allows for accelerated lead identification.

One of the limitations of natural products can be obtaining enough material for preclinical and clinical testing. With the great strides that have been made in synthetic organic chemistry, many complex natural products can now be synthesized. This approach can often provide a source for drugs for preclinical and clinical testing, but importantly, synthetic approaches provide the opportunity for designed analogs for lead optimization. While many natural products themselves have found utility in the treatment of cancer, analogs of the natural products often have superior activities, as the source organism is not expected to have optimized the compounds for human pharmacokinetics and anticancer actions. Clearly, the complex chemicals from nature cannot yet be duplicated in synthetic chemical libraries. Natural product peak libraries are available commercially but as yet do not have the diversity that are found in academic and corporate libraries.

Supply of natural products is always a concern and has limited the clinical development of many promising lead compounds. Advances in the understanding of the genetics of complex biosynthetic pathways have been utilized in bioengineering [80]. Biosynthetic gene clusters in bacterial expression systems have been engineered to provide a renewable source of rare natural products efficiently. Manipulation of biosynthetic gene clusters provides the opportunity for metabolic engineering of a wide diversity of new chemical entities [81]. The chemical diversity of natural products as sources for new drug leads should not be overlooked and can complement large synthetic chemical libraries.

Synthetic Chemical Libraries

Synthetic chemical libraries have many advantages over natural product libraries in that they represent structures easily amenable to chemical synthesis. Large chemical libraries, 100,000+ are available commercially for drug discovery screening. They are available in formats amenable to HTS and provide good chemical diversity at low cost. Many of the libraries are optimized for chemical diversity and for pharmaceutical characteristics. These libraries, together with HTS have been very important in the identification of new, targeted leads.

One of the most promising approaches for new drug discovery is combinatorial chemistry. Combinatorial chemistry is a relatively new technology for creating molecules en masse, compared with conventional one-molecule-at-a-time medicinal chemistry. The technology evolved from solid-phase peptide synthesis. Originally used to create polypeptide and oligonucleotide libraries, the field has moved toward libraries of small organic molecules with potential utility as drugs. Molecular diversity can be introduced into many different molecules being prepared in parallel while tethered by a reversible linker to an insoluble polymeric support. Although initially limited to peptide molecules, the method has been expanded to include carbohydrates.

Validation of the library of the compounds produced in a combinatorial synthesis has proven crucial to success. Early practitioners failed to recognize that not all reactions worked as predicted, leading to unpredicted by-products and incomplete reactions, leading to "dirty" libraries that created irreproducible effects. The utility of the combinatorial approach for creation of fundamentally different structures is hotly contested. It is clear, however, that for lead optimization by analog synthesis, combinatorial chemistry is working and is making major contributions.

29.8 RESOURCES FOR DRUG DISCOVERY

The approaches and targets for cancer drug discovery are changing at a rapid pace, and, as mentioned in the introduction, significant efforts are being placed on cancer drug discovery of targets that have yet to be validated. It can be anticipated that some targets will lead to successful pharmaceuticals for cancer treatment while others, for a number of reasons, will fail. Keeping current on the trends in cancer drug discovery is critically important.

A wealth of information about drug discovery in general and specifics about new screening technologies has become available in the past few years. A number of excellent new journals are published with the aims of helping the scientific community keep informed about the trends of drug discovery and in specific assay techniques. The Society of Biomolecular Screening was initiated in 1994 with the goal of "Discovery through Community." The society's journal, *The Journal of Biomolecular Screening* is published on a monthly basis and provides a variety of information on specific screening approaches and technologies. National and regional meetings are sponsored by the society. The inaugural issue of *Current Drug Discovery Technologies* was published in January of 2004. The aims and scope of the journal are to provide timely information on new approaches for all aspects of modern drug discovery. *Nature Reviews Drug Discovery*, which began publication in 2002, offers an overview of the drug discovery process, review articles for new therapeutic targets, as well as articles on new tools for drug discovery. These new resources allow rapid communications of approaches and solutions for various aspects of drug discovery.

The focus of this chapter is on initial drug discovery, but turning lead identification into a clinically useful anticancer therapy requires rational drug development, which is even more complex than initial drug discovery. A new reference, *Molecular Cancer Therapeutics: Strategies for Drug Discovery and Development* [82], was published this year and provides an excellent source for information on various aspects of cancer drug development.

Acknowledgments I apologize to the many scientists who have made important contributions to cancer drug discovery and whose work and approaches were not cited due to space considerations. The goal of this chapter is to provide an introduction to the field of oncology drug discovery, and, due to the large number of exciting new targets for drug discovery, many valid and promising targets have not been discussed. Special thanks are extended to Ms. April Hopstetter and Ms. Evelyn Jackson for their excellent help with this chapter; to Drs. Pat LoRusso and Basel El-Rayes for use of their EGFR-signaling pathway figure; and to Dr. David Newman for permission to adapt his figure on the sources of anticancer drugs.

REFERENCES

1. Booth, B., Glassman, R., Ma, P. (2003). Oncology's trials. *Nat. Rev. Drug Discov. 2*, 609–610.
2. Suffness, M., Pezzuto, J. M. (1991). Assays related to cancer drug discovery. In K. Hostettmann (Ed.), *Assays for Bioactivity*. Academic, New York, pp. 71–133.
3. Drebin, J. A., Link, V. C., Weinberg, R. A., Greene, M. I. (1986). Inhibition of tumor growth by a monoclonal antibody reactive with an oncogene-encoded tumor antigen. *Proc. Natl. Acad. Sci. USA, 83*, 9129–9133.
4. El-Rayes, B. F., LoRusso, P. M. (2004). Targeting the epidermal growth factor receptor. *Br. J. Cancer, 91*, 418–424.
5. Shawver, L. K., Slamon, D., Ullrich, A. (2002). Smart drugs: Tyrosine kinase inhibitors in cancer therapy. *Cancer Cell, 1*, 117–123.
6. Laskin, J. J., Sandler, A. B. (2004). Epidermal growth factor receptor: A promising target in solid tumours. *Cancer Treat. Rev., 30*, 1–17.
7. Lynch, T. J., Bell, D. W., Sordella, R., Gurubhagavatula, S., Okimoto, R. A., Brannigan, B. W., Harris, P. L., Haserlat, S. M., Supko, J. G., Haluska, F. G., Louis, D. N., Christiani, D. C., Settleman, J., Haber, D. A. (2004). Activating mutations in the epidermal growth factor receptor underlying responsiveness of non-small-cell lung cancer to gefitinib. *N. Engl. J. Med., 350*, 2129–2139.
8. Paez, J. G., Janne, P. A., Lee, J. C., Tracy, S., Greulich, H., Gabriel, S., Herman, P., Kaye, F. J., Lindeman, N., Boggon, T. J., Naoki, K., Sasaki, H., Fujii, Y., Eck, M. J., Sellers, W. R., Johnson, B. E., Meyerson, M. (2004). EGFR mutations in lung cancer: Correlation with clinical response to gefitinib therapy. *Science, 304*, 1497–1500.
9. Sordella, R., Bell, D. W., Haber, D. A., Settleman, J. (2004). Gefitinib-sensitizing EGFR mutations in lung cancer activate anti-apoptotic pathways. *Science, 305*, 1163–1167.
10. Chan, J. M., Stampfer, M. J., Giovannucci, E., Gann, P. H., Ma, J., Wilkinson, P., Hennekens, C. H., Pollak, M. (1998). Plasma insulin-like growth factor-I and prostate cancer risk: A prospective study. *Science, 279*, 563–566.

11. Wu, Y., Cui, K., Miyoshi, K., Hennighausen, L., Green, J. E., Setser, J., LeRoith, D., Yakar, S. (2003). Reduced circulating insulin-like growth factor I levels delay the onset of chemically and genetically induced mammary tumors. *Cancer Res.*, *63*, 4384–4388.
12. LeRoith, D., Helman, L. (2004). The new kid on the block(ade) of the IGF-1 receptor. *Cancer Cell*, *5*, 201–202.
13. Wong, S., McLaughlin, J., Cheng, D., Zhang, C., Shokat, K. M., Witte, O. N. (2004). Sole BCR-ABL inhibition is insufficient to eliminate all myeloproliferative disorder cell populations. *Proc. Natl. Acad. Sci. USA*. *101*, 17456–17461.
14. Bjornsti, M. A., Houghton, P. J. (2004). The TOR pathway: A target for cancer therapy. *Nat. Rev. Cancer*, *4*, 335–348.
15. Hay, N., Sonenberg, N. (2004). Upstream and downstream of mTOR. *Genes Dev.*, *18*, 1926–1945.
16. Sawyers, C. L. (2003). Will mTOR inhibitors make it as cancer drugs? *Cancer Cell*, *4*, 343–348.
17. Slamon, D. J. (2004). The FUTURE of ErbB-1 and ErbB-2 pathway inhibition in breast cancer: Targeting multiple receptors. *Oncologist*, *9*(Suppl. 3), 1–3.
18. Mallari, R., Swearingen, E., Liu, W., Ow, A., Young, S. W., Huang, S. G. (2003). A generic high-throughput screening assay for kinases: Protein kinase a as an example. *J. Biomol. Screen.*, *8*, 198–204.
19. Moshinsky, D. J., Ruslim, L., Blake, R. A., Tang, F. (2003). A widely applicable, high-throughput TR-FRET assay for the measurement of kinase autophosphorylation: VEGFR-2 as a prototype. *J. Biomol. Screen.*, *8*, 447–452.
20. Gaudet, E. A., Huang, K. S., Zhang, Y., Huang, W., Mark, D., Sportsman, J. R. (2003). A homogeneous fluorescence polarization assay adaptable for a range of protein serine/threonine and tyrosine kinases. *J. Biomol. Screen.*, *8*, 164–175.
21. Newman, M., Josiah, S. (2004). Utilization of fluorescence polarization and time resolved fluorescence resonance energy transfer assay formats for SAR studies: Src kinase as a model system. *J. Biomol. Screen.*, *9*, 525–532.
22. Anderson, S. N., Cool, B. L., Kifle, L., Chiou, W., Egan, D. A., Barrett, L. W., Richardson, P. L., Frevert, E. U., Warrior, U., Kofron, J. L., Burns, D. J. (2004). Microarrayed compound screening (microARCS) to identify activators and inhibitors of AMP-activated protein kinase. *J. Biomol. Screen.*, *9*, 112–121.
23. Folkman, J. (1990). What is the evidence that tumors are angiogenesis dependent? *J. Natl. Cancer Inst.*, *82*, 4–6.
24. Carmeliet, P., Jain, R. K. (2000). Angiogenesis in cancer and other diseases. *Nature*, *407*, 249–257.
25. Scavelli, C., Weber, E., Agliano, M., Cirulli, T., Nico, B., Vacca, A., Ribatti, D. (2004). Lymphatics at the crossroads of angiogenesis and lymphangiogenesis. *J. Anat.*, *204*, 433–449.
26. Folkman, J. (1971). Tumor angiogenesis: Therapeutic implications. *N. Engl. J. Med.*, *285*, 1182–1186.
27. Hanahan, D., Folkman, J. (1996). Patterns and emerging mechanisms of the angiogenic switch during tumorigenesis. *Cell*, *86*, 353–364.
28. Streubel, B., Chott, A., Huber, D., Exner, M., Jager, U., Wagner, O., Schwarzinger, I. (2004). Lymphoma-specific genetic aberrations in microvascular endothelial cells in B-cell lymphomas. *N. Engl. J. Med.*, *351*, 250–259.

29. Ferrara, N., Hillan, K. J., Gerber, H. P., Novotny, W. (2004). Discovery and development of bevacizumab, an anti-VEGF antibody for treating cancer. *Nat. Rev. Drug. Discov.*, *3*, 391–400.
30. Ferrara, N., Gerber, H. P., LeCouter, J. (2003). The biology of VEGF and its receptors. *Nat. Med.*, *9*, 669–676.
31. Ferrara, N. (2004). Vascular endothelial growth factor: Basic science and clinical progress. *Endocr. Rev.*, *25*, 581–611.
32. McDonald, D. M., Teicher, B. A., Stetler-Stevenson, W., Ng, S. S., Figg, W. D., Folkman, J., Hanahan, D., Auerbach, R., O'Reilly, M., Herbst, R., Cheresh, D., Gordon, M., Eggermont, A., Libutti, S. K. (2004). Report from the society for biological therapy and vascular biology faculty of the NCI workshop on angiogenesis monitoring. *J. Immunother.*, *27*, 161–175.
33. Hiratsuka, S., Nakamura, K., Iwai, S., Murakami, M., Itoh, T., Kijima, H., Shipley, J. M., Senior, R. M., Shibuya, M. (2002). MMP9 induction by vascular endothelial growth factor receptor-1 is involved in lung-specific metastasis. *Cancer Cell*, *2*, 289–300.
34. Ferrara, N. (2004). Vascular endothelial growth factor as a target for anticancer therapy. *Oncologist*, *9*(Suppl. 1), 2–10.
35. Tandle, A., Libutti, S. K., Libutti, S. K. (2003). Antiangiogenic therapy: Targeting vascular endothelial growth factors and its receptors. *Clin. Adv. Hematol. Oncol.*, *1*, 41–48.
36. Siemann, D. W., Chaplin, D. J., Horsman, M. R. (2004). Vascular-targeting therapies for treatment of malignant disease. *Cancer*, *100*, 2491–2499.
37. Powis, G., Kirkpatrick, L. (2004). Hypoxia inducible factor-1alpha as a cancer drug target. *Mol. Cancer Ther.*, *3*, 647–654.
38. Albini, A. (1998). Tumor and endothelial cell invasion of basement membranes. The matrigel chemoinvasion assay as a tool for dissecting molecular mechanisms. *Pathol. Oncol. Res.*, *4*, 230–241.
39. Walter-Yohrling, J., Morgenbesser, S., Rouleau, C., Bagley, R., Callahan, M., Weber, W., Teicher, B. A. (2004). Murine endothelial cell lines as models of tumor endothelial cells. *Clin. Cancer Res.*, *10*, 2179–2189.
40. Lawson, N. D., Weinstein, B. M. (2002). In vivo imaging of embryonic vascular development using transgenic zebrafish. *Dev. Biol.*, *248*, 307–318.
41. Ausprunk, D. H., Knighton, D. R., Folkman, J. (1975). Vascularization of normal and neoplastic tissues grafted to the chick chorioallantois. Role of host and preexisting graft blood vessels. *Am. J. Pathol.*, *79*, 597–628.
42. Passaniti, A., Taylor, R. M., Pili, R., Guo, Y., Long, P. V., Haney, J. A., Pauly, R. R., Grant, D. S., Martin, G. R. (1992). A simple, quantitative method for assessing angiogenesis and antiangiogenic agents using reconstituted basement membrane, heparin, and fibroblast growth factor. *Lab. Invest.*, *67*, 519–528.
43. Tozer, G. M. (2003). Measuring tumour vascular response to antivascular and antiangiogenic drugs. *Br. J. Radiol.*, *76*(Suppl. 1), S23–25.
44. Cosgrove, D. (2003). Angiogenesis imaging—ultrasound. *Br. J. Radiol.*, *76*(Suppl. 1), S43–49.
45. Hsieh, C. L., Jones, P. A. (2003). Meddling with methylation. *Nat. Cell. Biol.*, *5*, 502–504.

46. Szyf, M., Pakneshan, P., Rabbani, S. A. (2004). DNA demethylation and cancer: Therapeutic implications. *Cancer Lett.*, *211*, 133–143.
47. Brown, R., Plumb, J. A. (2004). Demethylation of DNA by decitabine in cancer chemotherapy. *Expert Rev. Anticancer Ther.*, *4*, 501–510.
48. Somech, R., Izraeli, S., Simon, A. J. (2004). Histone deacetylase inhibitors—a new tool to treat cancer. *Cancer Treat. Rev.*, *30*, 461–472.
49. Villar-Garea, A., Esteller, M. (2004). Histone deacetylase inhibitors: Understanding a new wave of anticancer agents. *Int. J. Cancer*, *112*, 171–178.
50. Shay, J. W. (2003). Telomerase therapeutics: Telomeres recognized as a DNA damage signal: Commentary re: K. Kraemer et al., antisense-mediated hTERT inhibition specifically reduces the growth of human bladder cancer cells. *Clin. Cancer Res.*, *9*, 3521–3525.
51. Shay, J. W., Wright, W. E. (2002). Telomerase: A target for cancer therapeutics. *Cancer Cell*, *2*, 257–265.
52. Parkinson, E. K. (2003). Telomerase as a novel and potentially selective target for cancer chemotherapy. *Ann. Med.*, *35*, 466–475.
53. O'Connor, P. M., Jackman, J., Bae, I., Myers, T. G., Fan, S., Mutoh, M., Scudiero, D. A., Monks, A., Sausville, E. A., Weinstein, J. N., Friend, S., Fornace, A. J., Jr. Kohn, K. W. (1997). Characterization of the p53 tumor suppressor pathway in cell lines of the National Cancer Institute anticancer drug screen and correlations with the growth-inhibitory potency of 123 anticancer agents. *Cancer Res.*, *57*, 4285–4300.
54. Wilson, L., Jordan, M. A. (1995). Microtubule dynamics: Taking aim at a moving target. *Chem. Biol.*, *2*, 569–573.
55. Pryor, D. E., O'Brate, A., Bilcer, G., Diaz, J. F., Wang, Y., Kabaki, M., Jung, M. K., Andreu, J. M., Ghosh, A. K., Giannakakou, P., Hamel, E. (2002). The microtubule stabilizing agent laulimalide does not bind in the taxoid site, kills cells resistant to paclitaxel and epothilones, and may not require its epoxide moiety for activity. *Biochemistry*, *41*, 9109–9115.
56. Mayer, T. U., Kapoor, T. M., Haggarty, S. J., King, R. W., Schreiber, S. L., Mitchison, T. J. (1999). Small molecule inhibitor of mitotic spindle bipolarity identified in a phenotype-based screen. *Science*, *286*, 971–974.
57. Maliga, Z., Kapoor, T. M., Mitchison, T. J. (2002). Evidence that monastrol is an allosteric inhibitor of the mitotic kinesin Eg5. *Chem. Biol.*, *9*, 989–996.
58. Sakowicz, R., Finer, J. T., Beraud, C., Crompton, A., Lewis, E., Fritsch, A., Lee, Y., Mak, J., Moody, R., Turincio, R., Chabala, J. C., Gonzales, P., Roth, S., Weitman, S., Wood, K. W. (2004). Antitumor activity of a kinesin inhibitor. *Cancer Res.*, *64*, 3276–3280.
59. Carmena, M., Earnshaw, W. C. (2003). The cellular geography of aurora kinases. *Nat. Rev. Mol. Cell. Biol.*, *4*, 842–854.
60. Meraldi, P., Honda, R., Nigg, E. A. (2004). Aurora kinases link chromosome segregation and cell division to cancer susceptibility. *Curr. Opin. Genet. Dev.*, *14*, 29–36.
61. Sausville, E. A. (2004). Aurora kinases dawn as cancer drug targets. *Nat. Med.*, *10*, 234–235.
62. Harrington, E. A., Bebbington, D., Moore, J., Rasmussen, R. K., Ajose-Adeogun, A. O., Nakayama, T., Graham, J. A., Demur, C., Hercend, T., Diu-Hercend, A., Su, M., Golec, J. M., Miller, K. M. (2004). VX-680, a potent and selective small-

molecule inhibitor of the Aurora kinases, suppresses tumor growth in vivo. *Nat. Med.*, *10*, 262–267.

63. Ditchfield, C., Johnson, V. L., Tighe, A., Ellston, R., Haworth, C., Johnson, T., Mortlock, A., Keen, N., Taylor, S. S. (2003). Aurora B couples chromosome alignment with anaphase by targeting BubR1, Mad2, and Cenp-E to kinetochores. *J. Cell. Biol.*, *161*, 267–280.

64. Mooberry, S. L., Tien, G., Hernandez, A. H., Plubrukarn, A., Davidson, B. S. (1999). Laulimalide and isolaulimalide, new paclitaxel-like microtubule-stabilizing agents. *Cancer Res.*, *59*, 653–660.

65. Tinley, T. L., Randall-Hlubek, D. A., Leal, R. M., Jackson, E. M., Cessac, J. W., Quada, J. C., Jr., Hemscheidt, T. K., Mooberry, S. L. (2003). Taccalonolides E and A: Plant-derived steroids with microtubule-stabilizing activity. *Cancer Res.*, *63*, 3211–3220.

66. Smith, C. D., Zhang, X., Mooberry, S. L., Patterson, G. M., Moore, R. E. (1994). Cryptophycin: A new antimicrotubule agent active against drug-resistant cells. *Cancer Res.*, *54*, 3779–3784.

67. Mooberry, S. L., Leal, R. M., Tinley, T. L., Luesch, H., Moore, R. E., Corbett, T. H. (2003). The molecular pharmacology of symplostatin 1: A new antimitotic dolastatin 10 analog. *Int. J. Cancer*, *104*, 512–521.

68. Anderson, H. J., de Jong, G., Vincent, I., Roberge, M. (1998). Flow cytometry of mitotic cells. *Exp. Cell. Res.*, *238*, 498–502.

69. Kokoshka, J. M., Ireland, C. M., Barrows, L. R. (1996). Cell-based screen for identification of inhibitors of tubulin polymerization. *J. Nat. Prod.*, *59*, 1179–1182.

70. King, R. W., Deshaies, R. J., Peters, J. M., Kirschner, M. W. (1996). How proteolysis drives the cell cycle. *Science*, *274*, 1652–1659.

71. Paramore, A., Frantz, S. (2003). Bortezomib. *Nat. Rev. Drug Discov.*, *2*, 611–612.

72. Aggarwal, B. B. (2004). Nuclear factor-kappaB: The enemy within. *Cancer Cell*, *6*, 203–208.

73. Hideshima, T., Richardson, P., Chauhan, D., Palombella, V. J., Elliott, P. J., Adams, J., Anderson, K. C. (2001). The proteasome inhibitor PS-341 inhibits growth, induces apoptosis, and overcomes drug resistance in human multiple myeloma cells. *Cancer Res.*, *61*, 3071–3076.

74. Neckers, L. (2002). Hsp90 inhibitors as novel cancer chemotherapeutic agents. *Trends Mol. Med.*, *8*, S55–61.

75. Isaacs, J. S., Xu, W., Neckers, L. (2003). Heat shock protein 90 as a molecular target for cancer therapeutics. *Cancer Cell*, *3*, 213–217.

76. Kamal, A., Thao, L., Sensintaffar, J., Zhang, L., Boehm, M. F., Fritz, L. C., Burrows, F. J. (2003). A high-affinity conformation of Hsp90 confers tumour selectivity on Hsp90 inhibitors. *Nature*, *425*, 407–410.

77. Vilenchik, M., Solit, D., Basso, A., Huezo, H., Lucas, B., He, H., Rosen, N., Spampinato, C., Modrich, P., Chiosis, G. (2004). Targeting wide-range oncogenic transformation via PU24FCl, a specific inhibitor of tumor Hsp90. *Chem. Biol.*, *11*, 787–797.

78. Chiosis, G., Vilenchik, M., Kim, J., Solit, D. (2004). Hsp90: The vulnerable chaperone. *Drug. Discov. Today*, *9*, 881–888.

79. Newman, D. J., Cragg, G. M., Snader, K. M. (2003). Natural products as sources of new drugs over the period 1981–2002. *J. Nat. Prod.*, *66*, 1022–1037.
80. Kerwin, S. (2002). Toward bioengineering anticancer drugs. *Chem. Biol.*, *9*, 956–958.
81. Sherman, D. H. (2002). New enzymes for "warheads." *Nat. Biotechnol.*, *20*, 984–985.
82. Prendergast, G. C. (Ed.) (2004). *Molecular Cancer Therapeutics: Strategies for Drug Discovery and Development.* Wiley, New York.

INDEX

A. archeri, research and development, 18
Abacavir, reverse transcriptase inhibitors, 1156–1158
Absorption, distribution, metabolism, and excretion (ADME) evaluation:
 drug development, 3
 quantitative structure-activity relationships, 228–229
Absorption, distribution, metabolism, excretion, and toxicology (ADME/Tox):
 high-throughput screening, 196
 systems biology and:
  basic principles, 126–127
  drug development applications, 156–159
Absorption mechanisms, prodrug development, 752–766
 carrier-mediated transport, 754–756
 dermal drug delivery, 756–760
  topical drugs, 757–759
  transdermal delivery, 759–760

nasal drug delivery, 762–766
 ehanced brain delivery, 766
 lipophilic prodrugs, 764–766
 peptide prodrugs, 766
 transport-mediated absorption, 766
 water-soluble prodrugs, 763–764
ophthalmic drug delivery, 760–762
oral drug delivery, 752–756
 lipophilic drugs, 753–754
Acceptance criteria:
 potency precision estimates, 674
 replication-experiment studies, 683–687
Access software, laboratory data management, 289
Accuracy parameters, high-throughput screening automation, 580–581
ACD/Labs software, NMR spectral information management, 278–279
Acetic acid measurement, chronic gastric ulcer, 1015–1016

Drug Discovery Handbook, by Shayne Cox Gad
Copyright © 2005 by John Wiley & Sons, Inc.

Acetylcholine binding protein (AChBP), GABA$_A$ receptor complex, receptor and pharmacophore models, 811–814
   benzodiazepine binding sites, 821–823
Acetylcholine esterase (AChE), click chemistry, 996–997
A-5021 compound, antiviral therapy, DNA polymerase inhibitors, 1153–1155
Acquired immunodeficiency syndrome (AIDS), human cytomegalovirus (HCMV), 1114
Acta Crystallographica D-Biological Crystallography, 389
Activated protein C (APC), clinical trials, 946–950
Activation domain (AD):
   protein kinases, 1198–1199
   protein-protein interactions:
      mammalian two-hybrid systems, 506
      protein linkage maps, 507–509
      three-hybrid systems, 505–506
      yeast two-hybrid systems, 487–489
      target and bail vectors, 490–494
ActiveSheet function, Excell software, 302–303
Activity screening:
   criteria for, 4–9
   natural products, 43–47
Acute gastric lesions, animal models, 1015
Acute gastric mucosal injury, animal models, 1014–1015
Acute laryngotracheobronchities (croup), respiratory viruses and, 1116
Acute lymphoblastic leukemia (ALL), Bcr-Abl protein tyrosine kinase, 1201
Acyclic nucleoside phosphonates, antiviral therapy, 1158–1161
Acyclovir:
   DNA polymerase inhibitors, 1148–1155
   prodrug structures, nasal drug delivery, 764–766
   respiratory viral infection therapy, 1120–1121
N-acyl prodrugs, bioreversible prodrug structures, 751
Acyloxyalkyl prodrugs, bioconversion mechanisms, 743–744
N-α-acyloxymethyl*N*-phosphoroxymethyl derivatives, bioreversible prodrug structures, 751–752
Adefovir:
   acyclic nucleoside phosphonates, 1159–1161
   DNA polymerase inhibitors, 1148–1150
   lipophilic characteristics, 754
S-Adenosylhomocysteine hydrolase (SAH) inhibitors, antiviral therapy, 1173–1177
Adenosine analogs, *S*-adenosylhomocysteine hydrolase inhibitors, 1175–1177
Adenosine diphosphate (ADP):
   coupled bioluminescent assays, high-throughput screening, 696–697
   protein kinase phosphorylation, 1195–1196
Adenosine 5′-triphosphate (ATP):
   coupled bioluminescent assays:
      backward reactions, 721–722
      firefly luciferase, ATP/luciferase reactions, 695–696
      kinases, 698–699
      high-throughput screening, 696–697
   DeathTRAK cycotoxicity assay, relase assays, 704
   protein kinases:
      binding site inhibitors, 1210–1212
      bisubstate analog inhibitors, 1214
      phosphorylation mechanisms, 1195–1196
      signal transduction, 1193–1194
      structural features, 1196–1198
   Src family kinase inhibitors, binding sites, 1225
   targeted cancer therapy, binding site inhibitors, 1348–1349

Adenoviridae, characteristics of, 1113
ADH1 promoter, protein-protein interactions, yeast two-hybrid systems, 494
Adrenal corticosteroids, 1060–1066
  aldosterone, 1064–1065
  corticosteroid synthesis inhibitors, 1063–1064
  glucocorticoids, 1060–1063
  mineralcorticoids, 1064–1065
  antagonists, 1065–1066
Adrenal function, septic shock and, glucocorticoid efficacy, 943–946
β2-Adrenergic receptor (β2AR), high-throughput flow cytometry, drug discovery targets:
  GPCR signaling pathways, 190
  intracellular arrestin regulation, 199–200
  proof of principle, 202
  soluble and bead-based assemblies, 204–207
Adrenocorticotropic hormone (ACTH) syndrome, corticotropin-releasing factor therapy, 1042
Advanced computational analysis, microarray data, 139–142
  biological networks, 141–142
  linear pathways to complex networks, 139–140
  literature sources, 140–141
Affinity constant (K_a), crystal-based drug discovery, 386–388
Affinity data:
  antiviral therapy, DNA polymerase inhibitors, 1150–1155
  biomolecular screening assays, 669
  defined, 36
Affinity tagging, protein-protein interactions, mass spectroscopy, 531–533
Agarose microarrays, protein-protein interactions, protein microarray supports, 524–526
Age-related macular degeneration (AMD), RNA mimetics, aptamer function, 1271–1273

Agonists, general principles, 34
Agranulocytosis, thionamide drugs, 1055–1056
Aldosterone, characteristics and therapeutic applications, 1064–1065
Alkyl ester prodrugs, basic properties, 741–743
Allosteric modulators:
  $GABA_A$ receptor complex, 814–821
  $GABA_B$ receptors, 827
  general principles, 35
  metabotropic glutamate receptors:
    noncompetitive receptor antagonists, 860–862
    positive modulators, 862
    protein kinase inhibitors, 12123
Alnylam Pharmaceuticals, small inhibitory RNA production, 1282
α-fetoprotein (AFP), cancer cell proteomics, molecular diagnosis, 74–75
α-substitutions, metabotropic glutamate competitive receptor antagonists, 859–860
Alzheimer's disease (AD), GABA and glutamate receptor ligands, 800–801
AMCA-DNA, metal-enhanced fluorescence, enhanced energy transfer, silver surfaces, 645–649
Amides:
  bioconversion mechanisms, 744
  bioreversible prodrug structures, 750–752
    N-acyl prodrugs, 751
    N-α-acyloxymethyl$N$-phosphoroxymethyl derivatives, 751–752
    N-mannich bases, 751
Amines, bioreversible prodrug structures, 750–752
  N-acyl prodrugs, 751
  N-α-acyloxymethyl$N$-phosphoroxymethyl derivatives, 751–752
  N-mannich bases, 751

Amino acid esters, water-soluble prodrugs, 748–749, 769
Amino acid sequencing, macromolecule crystallographic models, 422–425
α-Aminoamide moiety, prodrug development, peptide formulations, stability improvement, 781–782
2-Amino-3-(3-hydroxy-5-methyl-4-isoxazolyl)propionic acid (AMPA), glutamatic acid receptors, 829–830
  competitive/noncompetitive receptor antagonists, 840–842
  modulatory agents, 842–843
  receptor agonists, 837–840
1-Amino-3-propylethoxy silane (APS) coating, metal-enhanced fluorescence:
  laser-deposited silver, 618–619
  ratiometric surface sensing, 649–653
  silver colloid films, 611–615
Aminoglutethimide:
  aromatase inhibitors and, 1070–1071
  corticosteroid synthesis inhibition, 1064
5-Aminosalicylic acid (5-ASA), colon targeting, 778–779
Amorpha fruitcosa, research and development, 17
AMPA-kines, modulatory functions, 842–843
Ampicillin prodrug derivatives, lipophilic characteristics, oral drug delivery, 753–754
Amplification, three-dimensional RNA structures, 1270
Anabolic actions:
  parathyroid hormone, 1058
  steroids, 1067–1068
Analytical chemistry assays:
  drug discovery, 462–464
  formulation development, 470–471
Androgens, structure and applications, 1066–1068
  anabolic steroids, 1067–1068
  antiandrogens, 1068

receptors, 1066–1067
5α-reductase inhibitors, 1068
Angelica sinensis, polysaccharide extracts, 1024
Angiogenesis, targeted cancer therapy inhibitors, tumor vasculature, 1353–1356
  HIF-1α target, 1355–1356
  VEGF axis, 1354–1355
Angiongenesis, herbal therapeutic agents, 1021–1022
Angiotensin-converting enzyme (ACE):
  aklyl and aryl ester prodrugs, 741–743
  structure-based drug design, 437–438
  vertebrate sources, 31
Angular dependence:
  metal-enhanced fluorescence, directional emission, 656–660
  x-ray diffraction, protein structures, 396–397
    electron clouds and thermal motion, 398–399
Animal models:
  gastrointestinal disease:
    acute gastric lesion assessment, 1015
    acute gastric mucosal injury, 1014–1015
    chronic gastric ulcer, 1015–1016
    colorectal cancer, 1020–1021
      chemical induction model, 1020–1021
      genetically modified model, 1020
      inflammation-associated model, 1020
    duodenal ulcers, acute and chronic, 1016
    gastrointestinal cancer, 1019–1021
      chemical induction model, 1019–1020
      human gastric cancer cell implantation, 1019
    herbal medicines, 1021–1028
      biological processes, 1021–1022
      licorice, 1025–1027
      phenolic compounds and terpenoids, 1023–1028
      curcumin, 1023–1024

polysaccharides, 1022–1024
  mushroom/seaweed sources, 1022–1023
  plant sources, 1023
  terpenoid sources, 1027–1028
inflammatory bowel disease, 1016–1019
  colitis severity assessment, 1018–1019
  dextran sulfuate sodium, 1017–1018
  dinitrobenzene sulfonic acid/ethanol, 1017
  trinitrobenzene sulfonic acid/ethanol, 1017
respiratory viruses, 1126–1127
Anisotropic silver nanostructures, metal-enhanced fluorescence, 616–618
Anisotropy, defined, 80
Anomalous diffraction techniques, protein crystallization, 388
Anomalous dispersion techniques, x-ray crystal structure determination, 412–416
Antagonists, general principles, 34–35
Antalarmin, corticotropin-releasing factor antagonist, 1042
Anterior pituitary hormones, 1047–1049
  follicle-stimulating hormone, 1048–1049
  growth hormone (somatotropin), 1047–1048
  luteinizing hormone, 1049
Antiandrogens, structure and applications, 1068
Antiarrhythmic agents:
  ATP-sensitive potassium channel antagonists:
    nonselective antagonists, 915–917
    proarrhythmic effects, 923–924
    selective antagonists, 917–923
      HMR 1402 cardioselective compound, 921–923
      HMR 1883 cardioselective compound, 919–921
    ventricular arryhthmias, 914–924

extracellular potassium:
  myocardial ischemia effects, 911–913
  ventricular rhythm, 913–914
  research background, 909–911
Antibacterial agents, marine sources, 32–33
Antibiotics:
  comparative structural connectivity spectral analysis, ^{13}C-^{15}N heteronuclear connectivity matrix, cephalosporins, 271–275
  dual mode cycotoxicity/proliferation assays, DeathTRAK cycotoxicity assay, 707–709
  natural product drug development, 24, 29
Antibodies:
  bioanalytical chemistry assays, 474–476
  dual mode cycotoxicity/proliferation assays, DeathTRAK cycotoxicity assay, 705–712
  protein-protein interactions:
    microarrays, 519, 522
    phage display technology, 513–515
  tumor marker detection, 75–76
Antibody-directed enzyme prodrug therapy (ADEPT), tumor targeting, 772–775
Anticancer agents. See also Tumor targeting
  cell-based assays, antimitotic compounds, 362–363
  natural product development, 19–22
    information sources, 29
    marine sources, 32–33
    plant sources, 30
  prodrug tumor targeting, 770–775
  protein kinases, 1199
    activation, 1198–1199
    classification, 1194–1195
    human disease and, 1199
    inhibitor design:
      allosteric inhibitors, 1213
      alternative design strategies, 1214–1215

ATP binding site inhibitors, 1210–1212
  potency, 1212
  selectivity, 1210–1212
basic principles, 1205
bisubstrate analog inhibitors, 1213–1214
clinical trials, 1206–1209, 1215–1232
  future research issues, 1232
  receptor tyrosine kinases, 1216–1221
SH2 domain inhibitors, 1213
substrate binding site inhibitors, 1212–1213
molecular drug action, 40–41
protein phosphorylation mechanisms, 1195–1196
serine/threonine kinases, 1202–1205
  cyclin-dependent kinases, 1203–1204
    inhibitors, 1229
  glycogen synthase kinase 3, 1205
    inhibitors, 1231–1232
  inhibitors, 1225–1232
  MAP kinases, 1202–1203
    inhibitors, 1226–1229
  phosphoinositide 3-kinase, 1204–1025
    inhibitors, 1231
  protein kinase C and A, 1204
    inhibitors, 1229–1231
  Rho kinase, 1204
    inhibitors, 1231
signal transduction phosphorylation, 1193–1194
structural features, 1196–1198
structure-based drug design, 440–441
therapeutic applications, 1199–1205
  protein tyrosine kinases, clinical trials, 1206–1209
tyrosine kinases, 1199–1202
  nonreceptor tyrosine kinases, 1201–1202
    Bcr-Abl, 1201
      inhibitors, 1221–1222
    inhibitors, 1221–1225

  Src kinases, 1201–1202
    inhibitors, 1222–1225
  receptor tyrosine kinases, 1200–1201
    EGF receptors, 1200
    extracellular domain inhibition, 1216–1217
    intracellular domain inhibition, 1217–1219
    inhibitors, 1216–1221
    PDGF receptors, 1201
      inhibitors, 1221
    VEGF/bEGF receptors, 1200–1201
      inhibitors, 1219–1221
targeting techniques:
  antimimotics, 1360–1363
    aurora kinases, 1361–1362
    cell-based assays, 1362–1363
    kinesin inhibitors, 1361
    microtubule-targeting agents, 1360–1361
  Bcr/Abl, c-Kit, and PDGF, 1350
  DNA targets, 1357–1360
    DNA methylation, 1357–1358
    histone deacetylase inhibitors, 1358–1359
    telomerase inhibitors, 1359–1360
  EGF receptor pathway, 1347–1349
  evolution of, 1344–1346
  heat shock protein 90, 1364
  HER2/Neu, 1347
  information resources for, 1367–1368
  insulin-like growth factor signaling, 1349
  mammalian target of rapamycin (mTOR), 1350–1351
  proteosome inhibition, 1363–1364
  screening assays, 1351–1352
  signal transduction targets, 1346–1352
  small-molecule drug sources, 1365–1367
    natural products, 1365–1366
    synthetic chemical libraries, 1366–1367

tumor vasculature, 1352–1357
  angiogenesis inhibitors, 1353–1356
    HIF-1α target, 1355–1356
    VEGF axis, 1354–1355
  antivascular agents, 1356
  high-throughput techniques, 1356–1357
  Taxol case study, 57–59
Anti-coagulants, sepsis management, 946–948
  clinical trials, 946
  coagulation pathways, 946
  efficacy factors, 946–948
  future research, 949–950
Anti-CXR2 antibodies, respiratory syncytial virus therapy, 1122
Anti-$G_{i\alpha 2}$ antibody, high-throughput flow cytometry characterization, 204–207
Antimimotic compounds, targeted cancer therapy, 1360–1363
  aurora kinases, 1361–1362
  cell-based assays, 1362–1363
  kinesin inhibitors, 1361
  microtubule-targeting agents, 1360–1361
Antimitotic compounds, cell-based assays, 362–363
Antiobesity agents, structure and applications, 1083–1086
  orlistat, 1085–1086
Anti-proliferative compounds, multiplexed screening and identification, 367–370
Antisauvagine-30, corticotropin-releasing factor antagonist, 1042
Antisense molecules:
  oligonucleotides:
    antiviral agents, 1125–1126
    protein kinase A and C inhibition, 1231
  synthetic RNA, 1273–1278
    definitions, 1273–1274
    hammerhead ribozymes:
      mimetics, 1274–1276

preclinical/clinical utilization, 1276–1277
small inhibitory RNA, 1277–1278
targeted cancer therapy, EGFR signal transduction, 1349
Antistructures, x-ray crystallography-based drug discovery, 441–442
Antitarget proteins, x-ray crystallography-based drug discovery, 441–442
Antithrombin-IIII (AT-III), sepsis management, 946–948
Antivascular agents, targeted cancer therapy, tumor vasculature, 1356
Antiviral therapy:
  design strategies:
    S-adenosyl-L-homocysteine hydrolase inhibitors, 1173–1177
    DNA/RNA synthesis inhibitors, 1147–1163
      acyclic nucleoside phosphonates, 1158–1161
      DNA polymerase inhibitorsr, 1147–1152
      reverse transcriptase inhibitors, 1155–1158
      RNA synthesis processes, 1161–116
    IMP dehydrogenase inhibitors, 1171–1173
    neuraminidase inhibitors, 1169–1171
    research background, 1135–1140
    viral adsorption inhibitors, 1140–1141
    viral protease inhibitors, 1163–1169
    virus-cell fusion inhibitors, 1141–1147
  respiratory viruses:
    current treatments, 1120–1121
    development strategies, 1122–1126
    infections, 1121–11222
Apo-crystals, drug discovery applications, 386–388
Apoptosis, herbal therapeutic agents, 1021–1022

Apoptotic index, TUNEL assay, multiplex dose curve analysis, 364–366
Apparent partition coefficient, ophthalmic drug delivery, prodrug drug development, 760–762
Aptamers:
cancer cell proteomics:
fluorescence anisotropy probe, proetin analysis, 80–81
fluorescence energy transfer, PDGF analysis, 76–80
molecular diagnosis, 74–75
multiple tumor markers, whole-cell-based selection, 81–83
research background, 73–74
tumor marker detection, 75–76
molecular beacon aptamer, platelet-derived growth factor analysis, fluorescence resonance energy transfer, 77–80
peptide aptamers, protein-protein interaction inhibition, 539–540
three-dimensional RNA structures, 1267–1273
amplification, 1270
definitions, 1267–1269
Eyetech Pharmaceutical development, 1272–1273
mimetics, 1270–1271
preclinical and clinical utilization, 1271–1272
selection, 1267
Aqueous solubility:
bioreversible prodrugs:
aklyl and aryl ester prodrugs, 742–743
hydroxyl groups, 748–749
prodrug development:
improvement protocols, 766–769
hemiester structures, 769
phosphates, 767–769
etoposides, 767–768
fosphenytoin, 768–769
nasal drug delivery, 763–764

peptide formulation and delivery, 780
Arab medicine, history of, 13–14
Area under the curve (AUC) evaluation:
drug development, 3
pharmacokinetic assays, 476–477
Argand diagrams:
heavy atom replacement, x-ray crystal structure determination, 409–412
x-ray crystal structure determination, anomalous dispersion, 413–416
Arithmetic mean, coupled luminescent assays, 719–720
Aromatase enzyme:
comparative structural connectivity spectral analysis binding model, 261–264
comparative structurally assigned spectra analysis model, 240, 242–244
estrogen inhibitors, 1070–1071
Array approach, protein linkage maps, protein-protein interactions, yeast two-hybrid systems, 507–509
Arrestins, high-throughput flow cytometry, drug discovery targets:
FPR assemblies, 197–198
FPR colocalization in cells, 198–199
G-protein-coupled receptor constructs, 196–197
G-protein-coupled receptor signaling pathways, 188–190
intracellular FPR traffic regulation, 199–200
Arrhythmias. See Antiarrhythmic agents; Ventricular arrhythmias
Artemisinin, natural product development, 20–22
Artificial neural networks, comparative structural connectivity spectra analysis model, corticosterone binding, 259–261
Aryl ester prodrugs, basic properties, 741–743

Aryl hydrocarbon receptor binding:
  comparative structural connectivity spectra analysis models, 264–271
  comparative structurally assigned spectra analysis, 244–245
Aspartyl enzymes, viral protease inhibitors, 1163–1169
Aspirin, natural ingredients in, 14
Assays. *See also* specific assays
  biomolecular screening, concentration-response models and outcomes, 669–671
  cell-based assays:
    CellCard preparation, 360
    compound addition and assay protocol, 360–361
    high-throughput screening platforms, 566–567
      design criteria, 595–597
      fluorescent methods, 595
      format innovations, 597–598
    historical background, 354–356
    image analysis and data visualization, 361–362
    mixing and dispensing protocols, 360
    multiplexing principles, 357–359
      antiproliferation screening, cell type selective compounds, 367–370
      TUNEL assay, 363–366
    reading protocols, 361
    selective antimitotic compound identification and profiling, 362–363
    tissue culture, 360
  coupled luminescent assays:
    abbreviations and acronyms, 725–726
    assay specifications, 718–720
      cost issues, 720
      operational characteristics, 719
      sensitivity, 719
      speed and linear response, 718
      Z values, 719–720
    backward reactions, 721–722
    chemiluminescent assays, 693–694
    coupled bioluminescent assay methods, 694–700
      bacterial luciferases, 694
      firefly luciferase, 694–700
        ATP/luciferase reactions:
          advantages of, 695–696
          drug discovery, 695
          high-throughput screening, 696–697
          kinases, 698–699
          limitations and considerations, 697–698
          phosphodiesterase, 699–700
          phosphorolysis, 699
    coupled enzyme assays, 692–693
    cytotoxicity measurement, release assays, 700–701
    DeathTRAK cycotoxicity assay:
      basic properties, 701–702
      biomass measurement (competitive comparison), 704–711
        ATP release assays, 704
        dual-mode cytotoxicity/proliferation, 705–712
        gram-negative bacteria, 707–709
        gram-positive bacteria, 709–711
        metabolic assays, 704–705
      cytotoxic process measurements, 702–704
      drug discovery applications, 702–711
    fluorescence *vs.* luminance, 691
    as fluorescent assay alternatives, 693
    formulations, 722–725
      lytic formulations, 725
      PhosTRAK formulations, 723–724
      reaction master mixes and dilution buffers, 722–723
    phosphatase activity measurement, 701
    PhosTRAK phosphatase assay, 712–718
      applications, 713–714
      basic principles, 713

complexity and thermodynamic
issues, 716–717
diverse results, 715–716
future research issues, 717–718
phosphatase activity
measurement, 712–713
research background, 690–691
scintillation proximity assays, 691–692
developability assays:
basic principles, 466–468
bioanalytical chemistry assays, 474–476
drug delivery assays, 471–474
drug metabolism assays, 477–478
formulation development, 469–471
pharmacokinetic assays, 476–477
pharmacology evaluations, 468–469
toxicology assays, 478–481
in vitro toxicology, 479–480
in vivo toxicology, 480–481
discovery assays:
analytical chemistry assays, 462–464
basic princples, 460
in vitro pharmacology assays, 460–462
in vivo pharmacology assays, 464–465
efficacy indicators in, 4–9
high-throughput screening procedures:
detection systems, 582
development of, 561–563, 566–567
plate density and reaction volumes, 570–572
quality control, 569–570
metal-enhanced fluorescence:
drug discovery applications, 660–662
hybridization assays, enhanced DNA/RNA detection, 630–637
solution-based assays, 653–654
natural product activity, 43–47
statistical parameters, 668–669
targeted cancer therapy:
antimimotics, 1362–1363
signal transduction, 1351–1352
Asymmetric units, x-ray diffraction, Fourier synthesis, 396

Atomic force microscopy (AFM):
metal-enhanced fluorescence, silver colloid films, 611–615
protein-protein interactions, protein microarray detection, 528
ATP-sensitive potassium channel antagonists:
extracellular potassium:
myocardial ischemia effects, 911–913
ventricular rhythm, 913–914
nonselective antagonists, 915–917
proarrhythmic effects, 923–924
research background, 909–911
selective antagonists, 917–923
HMR 1402 cardioselective compound, 921–923
HMR 1883 cardioselective compound, 919–921
ventricular arryhthmias, 914–924
Aurora kinases, targeted cancer therapy, 1361–1362
Automated high-throughput screening (HTS):
infrastructure for, 582–584
technology and process, 570–586
accuracy and precision, 580–581
assay detection, 582
compound management, 572–578
close-loop screening, 578
composition, 572–573
nanoliter despensing, 576–578
plate format storage, 575
replication, 575–576
wet or dry storage, 573–575
plate design, 570–572
Auxotrophic markers, protein-protein interactions, yeast two-hybrid systems, 485–489
Azidothymidine (AZT):
brain targeting, chemical drug delivery system principle, 776
DNA polymerase inhibitors, 1148–1150
phosphate group prodrug development, 744–746
reverse transcriptase inhibitors, 1156–1158
Azo-bond derivatives, colon targeting, 778

Backbone chain length:
  metabotropic glutamate receptor ligands, 855–857
  natural and modified RNA, 1265–1266
Background emission, metal-enhanced fluorescence (MEF), enhanced DNA/RNA detection, 631–637
Back-propagation algorithm, comparative structural connectivity spectra analysis model, corticosterone binding, 259–261
"Backward" kinase reactions, coupled bioluminescent assays:
  ATP/luciferase reactions, 699
  firefly luciferase, 721–722
Baclofen, $GABA_B$ receptor agonists, 825
Bacterial display systems, protein-protein interactions, 515
Bacterial host, phage display technologies, protein-protein interactions, 514–515
Bacterial luciferases, coupled bioluminescent assays, 694
B42AD protein, protein-protein interactions, yeast two-hybrid systems, target and bail vectors, 490–494
Bait systems, protein-protein interactions:
  protein linkage maps, 507–509
  yeast two-hybrid systems, 485–489
  selection criteria, 489–494
  two-bait systems, 502
BALB cell line, platelet-derived growth factor analysis, fluorescence resonance energy transfer, 78–80
Bambuterol, prodrug development, prolonged drug action technologies, 770–771
Barbiturates, $GABA_A$ receptor complex, 819
Barcode readout, split-pool synthesis resins, 979–980

Basic Local Alignment Search Tool (BLAST), macromolecule crystallographic models, 422–425
Basidiomycetes, polysaccharide extracts, 1023–1024
Batch crystallization, format for, 380–381
"Baton" building, macromolecule crystallographic models, 424–425
Bayesian clustering, gene expression data, systems biology, 138–139
Bayes's theorem, protein structural models, refinement mathematics, 426–432
Bcr-Abl protein tyrosine kinase:
  nonreceptor inhibitors, 1201, 1221–1222
  targeted cancer therapy, signal transduction pathways, 1346, 1350
Bead-based assemblies, high-throughput flow cytometry, 202–207
  β2AR, 205–206
  cell cycle protein display, 206–207
  FPR assemblies, 204–205
Benzodiazepines:
  $GABA_A$ receptor complex:
    binding sites, 814–821
    receptor/pharmacophore models, 821–823
  glutamate receptors, AMPA receptor antagonists, 842
  parallel library synthesis, 972
  prodrug development, nasal drug delivery, 764
Betulinic acid, natural product development, 20–22
Bevacizumab:
  targeted cancer therapy, angiogenesis inhibitors, 1355
  VEGFR inhibitor, 1219
B factors:
  macromolecule crystallographic models, 420–422
  protein structural models, 428–432
Bicalutamide, antiandrogen properties, 1068

Bicuculline, GABA$_A$ receptor agonists and antagonists, 808–811
Bi-functional prodrug, defined, 736–737
Biguanides, diabetes therapy, 1077
Bijvoet pair of reflections, x-ray crystal structure determination, anomalous dispersion, 413–416
Binary encoding, split-pool synthesis, 977–979
Binary kernel discrimination, virtual screening, 113–114
Binary quantitative structure-activity relationships (QSARs), virtual screening, probabilistic methods, 113–114
BIND database:
  contents and location, 127
  protein-protein interactions, 533–536
Binding site homology, metabotropic glutamate receptor structure, 865–868
Binning spectral data algorithms, CoSCoSA modeling, 279–280
Bioanalytical chemistry assays, drug delivery, 474–476
Bioassays:
  high-throughput screening, 586–598
    assay formats, 587–588
    cell-based assay design, 595–598
    design criteria, 591
      ligand-receptor binding interactions, 592–594
      mechanistic issues, 591–592
    fluorescent methods, 594–595
    reagent production, 588–591
  historical background, 354–356
  natural product activity screening, product isolation, 45–47
Biocarta database, contents and location, 127–128
BIochemical assessment, colitis severity, 1018–1019
Biochemical combinatorial chemistry, natural product activity screening, 44–47
Bioconversion mechanisms, carboxyl prodrugs, 741–744

BioCyc database, contents and location, 128
Bioisosteres, metabotropic glutamate receptor ligands, 857
Biological effect modeling, NMR spectral information, 231
Biological Macromolecule Crystallization Database (BMCD), crystallization data, 389
Biological markers:
  developability assays, 469
  in vitro toxicology assays, 480
Biological networks:
  condition-specific molecular/functional networks, 163–164
  gaucoma drug development, 148–152
  microarray data, linear pathways, 139–140
  systems biology interpretations, 141–142
Biologics Control Act, 340
Bioluminescence, coupled assay methods, 694–700
  bacterial luciferases, 694
  firefly luciferase, 694–700
    ATP/luciferase reactions:
      advantages of, 695–696
      drug discovery, 695
      high-throughput screening, 696–697
      kinases, 698–699
      limitations and considerations, 697–698
      phosphodiesterase, 699–700
      phosphorolysis, 699
Biomass measurement, DeathTRAK cycotoxicity assay, 704–711
  ATP release assays, 704
  dual-mode cytotoxicity/proliferation, 705–712
  gram-negative bacteria, 707–709
  gram-positive bacteria, 709–711
  metabolic assays, 704–705
Biomolecular screening assay, concentration-response models and outcomes, 669–671
  affinity models and parameter estimates, 669

muscarinic M_1 receptor GTPγ^{35}S assay, 670–671
Biopanning cycle, protein-protein interactions, phage display technology, 513–515
Biopharmaceutical classification scheme (BSC), prodrug development, 739–740
Biopolymers, solid-phase synthesis, 965–970
Bioreductive enzyme expression, prodrug tumor targeting, 771–774
Bioreversible prodrug structures, 740–752
  amines and amides, 750–752
    N-acyl prodrugs, 751
    N-α-acyloxymethylN-phosphoroxymethyl derivatives, 751–752
    N-mannich bases, 751
  carboxyl groups, 741–744
    acyloxyalkyl prodrugs, 743–744
    aklyl and aryl esters, 741–743
    amides, 744
  hydroxyl groups, 747–750
    lactones, 750
    lipophilic prodrugs, 749–750
    steric hindrance, promoieties for, 750
    water-soluble prodrugs, 748–749
  peptide backbone cyclization, 782–783
  phosphate groups, 744–746
  phosphonate groups, 747
Biotinylated GPCR tail peptides, high-throughput flow cytometry characterization, 204–207
BIRB 796:
  allosteric modulation, 1213
  P38 MAPK inhibitors, 1228
Bis-azamacrocycles, antiviral therapy, virus-cell fusion inhibitors, 1144–1147
Bisphosphonates, bone metabolism, 1058–1059
Bisubstrate analogs, protein kinase inhibitors, 1213–1214

Bit string representations, molecular fingerprints, virtual screening, 102–106
Blind controls, screening procedures, 6–7
Blood-based diseases, natural product development for, 17–18
Blood-brain barrier (BBB):
  brain targeting prodrug development, 775–778
  drug delivery assays, 472–474
  GABA receptor ligands:
    GABA$_B$ receptor agonists, 825
    GABA$_B$ receptor antagonists, 825–827
  reuptake inhibition, 805–806
Blood collection protocols, regulatory guidelines, 338–339
Boltzmann's distribution, comparative structural connectivity spectra analysis, aromatase enzyme binding models, 263–264
Bonding interactions, macromolecule crystallographic models, 422
Bone metabolism, endocrine hormone agents, 1057–1060
  bisphosphonates, 1058–1059
  calcimimetics, 1059–1060
  calcitonin, 1058
  mithramycin, 1059
  parathyroid hormone, 1057–1058
Book sources, natural product drug development, 25–27
Bothrops jaracaca, drug development, 31
Bragg's law, x-ray diffraction, protein structures, 396–397
Brain targeting, prodrug development, 775–778
  carrier-mediated transport, 776–778
  chemical drug delivery system, 776
  lipophilic prodrugs, 775–776
Breast cancer:
  estrogen receptors, 1071
  MRI imaging, tumor cell surface receptors, 1319–1325
  systems biology-based drug development for, 151, 153–156

targeted cancer therapy, HER2/Neu signal transduction, 1347
BRENDA database, contents and location, 128
"Bridging" enzymes, coupled bioluminescent assays, firefly luciferase, ATP/luciferase reactions, 695–696
Bright-field imaging, metal-enhanced fluorescence, silver fractal-lilke structures, 626–630
Brivudin (BVDU), antiviral therapy, DNA polymerase inhibitors, 1148–1155
Bronchiolitis, viral agents for, 1117
Bryostatins, protein kinase A and C inhibition, 1231
Bulpuerum falcatum, polysaccharide extracts, 1024
Bunyaviridae, characteristics of, 1112–1113

Caco-2 model:
  bioreversible prodrug structures, peptide membrane permiation, 782–783
  drug delivery assays, 472–474
Caenorhabditis elegans, protein-protein interactions, 508–509
Calcimimetics, bone metabolism, 1059–1060
Calcineurin, PhosTRAK phosphatase assay, 715–717
Calcitonin, bone metabolism, 1058
Calcium channels:
  $GABA_B$ receptors, allosteric modulation, 827
  protein kinase C and A, 1204
Calcium homeostasis, parathyroid hormone action, 1057–1058
Calcium sensing receptor (CaSR), calcimimetics, 1059–1060
Cancer. *See also* Anticancer agents; Tumor targeting
  breast cancer:
    estrogen receptors, 1071
    MRI imaging, tumor cell surface receptors, 1319–1325
  systems biology-based drug development for, 151, 153–156
  targeted cancer therapy, HER2/Neu signal transduction, 1347
  colorectal cancer, animal models, 1020–1021
  gastrointestinal cancer, animal models, 1019–1021
    chemical induction model, 1019–1020
    human gastric cancer cell implantation, 1019
  natural product drug development, 19–22
Cancer cell proteomics, molecular aptamers:
  fluorescence anisotropy probe, proetin analysis, 80–81
  fluorescence energy transfer, PDGF analysis, 76–80
  molecular diagnosis, 74–75
  multiple tumor markers, whole-cell-based selection, 81–83
  research background, 73–74
  tumor marker detection, 75–76
Capacity, screening for, 5–7
Captopril, structure-based drug design, 437–438
Carbenicillin, prodrug development, oral drug delivery, 756
Carbimazole, therapeutic applications, 1055–1056
Carbobenzoxy group, biopolymer solid-phase synthesis, 966–970
^{13}C-^{15}N heteronuclear connectivity matrix, comparative structural connectivity spectral analysis, 271–278
  antibiotic cephalosporins, 271–275
  four-dimensional matrix:
    configurational entropy, 276–278
    molecular dynamics, 275–276
^{13}C NMR spectroscopy:
  comparative structural connectivity spectral analysis:
    aromatase enzyme binding model, 261–264

aryl hydrocarbon receptor binding, 264–271
^{13}C-^{15}N heteronuclear connectivity matrix, antibiotic cephalosporins, 271–275
estrogen receptor binding compounds, 251–256
future applications, 281
molecular structural analysis, 246–271
PD-ANN corticosterone binding model, 260–261
three-dimensional corticosterone binding, 257–259
comparative structurally assigned spectra analysis:
aromatase enzyme model, 240, 242–244
aryl hydrocarbon receptor binding, 244–245
corticosterone binding model, 240–242
molecular modeling, 239–240
distance spectra, 235–236
predicted spectra, 236
spectral information content analysis, 229–231
simulated vs. experimental spectra, 232
Carbonic anhydrase inhibitors, structure-based drug design, 438–440
Carboxyl groups, bioreversible prodrug structures, 741–744
acyloxyalkyl prodrugs, 743–744
aklyl and aryl esters, 741–743
amides, 744
peptide membrane permiation, 782–783
Carcinoembryonic antigen (CEA):
cancer cell proteomics, molecular diagnosis, 74–75
mRNA protein production, 1286–1288
Cardiac sarcolemmal ATP-sensitive potassium channel antagonists:
extracellular potassium:
myocardial ischemia effects, 911–913
ventricular rhythm, 913–914

nonselective antagonists, 915–917
proarrhythmic effects, 923–924
research background, 909–911
selective antagonists, 917–923
HMR 1402 cardioselective compound, 921–923
HMR 1883 cardioselective compound, 919–921
ventricular arryhthmias, 914–924
Cardiovascular agents:
corticotropin-releasing factor and related peptides, 1041
thyroid hormone analogs and receptors, 1053–1055
Carrier-mediated transport, prodrug structures:
brain targeting properties, 776–778
oral drug delivery, 754–756
Casein, PhosTRAK phosphatase assay, 715–716
Caspofungin, reserach and development, 16
Cathelicidins, vertebrate sources, 31
CAT reporter gene, protein-protein interactions, mammalian two-hybrid systems, 506
CCR5 compound, antiviral therapy, virus-cell fusion inhibitors, 1144–1147
Cell adhesion molecules, herbal terpenoid inhibition, 1027–1028
Cell-based assays:
CellCard preparation, 360
compound addition and assay protocol, 360–361
high-throughput screening platforms, 566–567
bioassays, 595
design criteria, 595–597
format innovations, 597–598
overview of, 587–588
reagent production, 588–591
historical background, 354–356
image analysis and data visualization, 361–362
mixing and dispensing protocols, 360
multiplexing principles, 357–359

antiproliferation screening, cell type selective compounds, 367–370
TUNEL assay, 363–366
reading protocols, 361
selective antimitotic compound identification and profiling, 362–363
targeted cancer therapy, antimimotics, 1362–1363
tissue culture, 360
in vitro pharmacology assays, 461–462
in vivo pharmacology assays, 464–465
Cell-based partitioning, virtual screening, 108–109
CellCard system, multiple cell line screening, 356–357
antimitotic cellular arrays case study, 362–363
card preparation, 360
compound addition and assay protocol, 360–361
experiment design and plate layout, 359–360
image analysis and data visualization, 361–362
key features of, 370
multiplexed antiproliferation screen, 367–370
multiplexing cell-based assays, 357–359
96-well mixing and dispensing format, 360
reading protocol, 361
tissue culture, 360
TUNEL assay case study, 363
multiplex dose curve analysis, 363–366
Cell cycle control:
breast cancer drug development, 155–156
natural product development for, 19–22
protein domain display, 206–207
RNA function in, 1262–1263
Cell-free expression systems, protein-protein interactions, protein microarray reagents, 524
Cell illustrator software, biological pathway analysis, 143

Cell line cultures:
high-throughput screening bioassays, 590–591
cell-based assay design, 596–597
platelet-derived growth factor analysis, fluorescence resonance energy transfer, 78–80
Cell penetrating peptides (CPP), gene expression suppression, 1283
Cell proliferation, herbal therapeutic agents, 1021–1022
Cell sorting, microfluidic mixing, high-throughput flow cytometry, 208–209
Cells property, Excell object entities, 300–303
Cell type selective compounds, multiplexed antiproliferation screen, 367–370
Cellular proteins, viral replication, 1123–1124
Central nervous system ischemia, glutamate receptor antagonists, 800
CEP-7055 compound, VEGFR inhibitor, 1220–1221
Cephalosporins, ^{13}C-^{15}N heteronuclear connectivity matrix,, comparative structural connectivity spectral analysis model, 271–275
Cerebyx, N-α-acyloxymethyl$N$-phosphoroxymethyl derivatives, 751
Charge-coupled device (CCD):
detector systems:
protein-protein interactions, protein microarrays, 527–528
x-ray diffraction, synchrotron data collection, 404–405
high-throughput screening platforms:
assay detection, 583
miniaturization, 566
ChemBank:
high-throughput flow cytometry, 195
virtual screening, 213
Chemical delivery system (CDS):
brain targeting prodrugs, 776
defined, 737–738

INDEX

Chemical derivatization, anomalous diffraction, 388
Chemical diversity:
  high-throughput flow cytometry, informatics techniques, 194–195
  high-throughput screening, compound management, 572–573
Chemical exchange saturation transfer (CEST) contrast agent, MRI molecular imaging, 1329
Chemical induction model:
  colorectal cancer, 1020–1021
  gastrointestinal cancer, 1019–1020
Chemical shift information:
  ^{13}C-^{15}N heteronuclear connectivity matrix, antibiotic cephalosporins, comparative structural connectivity spectral analysis, 274–275
  molecular MRI and, 1329–1331
  NMR spectral data, 232–233
Chemical signals, molecular drug actions, 37–39
Chemical spaces, virtual screening, 91–92
Chemiluminescent assays, coupled luminescent assays *vs.*, 693–694
Chemokines:
  antiviral agents, virus-cell fusion inhibitors, 1142–1147
  respiratory viral infections, 1121–1122
Chemotherapeutic agents:
  antiviral natural product development techniques, 21–22
  natural product development, 19–22
Chinese herbal medicine:
  estimates on development of, 14
  history of, 12–13
Chiral chromatography, analytical chemistry assays, 463–464
Chlorpropamide, vasopressin receptor agonists/antagonists, 1051
Cholecystokinin (CCK), antiobesity agents, 1084–1085
Cholesterol drugs, thyroid hormone analogs and receptors, 1054–1055
Chorionic gonadotropin, luteinizing hormone receptors, 1049
Chromatographic techniques. *See also* specific techniques
  analytical chemistry assays, 463–464
  bioanalytical chemistry assays, 474–476
  natural product structural identification, 51–53
  split-pool synthesis, binary encoding, 977–979
Chronic myelogenous leukema (CML):
  Bcr-Abl protein tyrosine kinase, 1201
  targeted cancer therapy, signal transduction pathways, 1346, 1350
Cidofovir:
  acyclic nucleoside phosphonates, 1159–1161
  DNA polymerase inhibitors, 1148–1150
C-jun N-terminal kinases (JNKs):
  inhibitors, 1228
  therapeutic targeting, 1202–1205
c-Kit gene, targeted cancer therapy, signal transduction, 1350
Cladosiphon fucoidan, polysaccharide extracts, 1024
Clavanins, marine sources, 32–33
Click chemistry, combinatorial techniques, 992–997
Clinical syndromes, respiratory viruses, 1114–1118
  acute laryngotracheobronchitis, 1116
  bronchitis, 1117
  common cold, 1114, 1116
  laryngitis, 1116
  pharyngitis, 1116
  pneumonia, 1117–1118
  tracheitis and tracheobronchitis, 1116
Clinical trials. *See also* Preclinical studies
  glucocorticoids, 943–945
  hammerhead ribozymes, 1276–1277
  immunostimulating RNA, 1290
  mediator-specific anti-inflammatory agents, 939–941
  mRNA function, protein production, 1285–1288
  natural product development, 57–58
  prodrug requirements, 738–740

protein kinases inhibitors, 1206–1209, 1215–1232
RNA mimetics, aptamer function, 1271–1272
Clipboard function, Excel data management, 304–306
Closed-loop screening, high-throughput screening compounds, 578
Clozapine, $GABA_A$ receptor complex, 819
Cluster analysis:
 breast cancer drug development, 155–156
 gene expression data, systems biology, 138–139
 virtual screening, 106–107
Coagulation drugs, insect sources, 30–31
Coagulation pathways, sepsis pathophysiology, 946
Coating materials, high-throughput screening procedures, plate design, 571–572
Co-crystallization techniques, crystal-based drug discovery, 386–388
Code execution control, Visual Basic Editor, 295–297
Code of Federal Regulations, drug discovery guidelines, 343–344
Cognia Molecular System, biological pathway analysis, 145
Collaborative research, natural product development, 62
Colon targeting, prodrug development for, 778
Colorectal cancer, animal models, 1020–1021
Combinatorial chemistry. *See also* Biochemical combinatorial chemistry
 drug development based on, 8–9
 high-throughput methods, 983–985
 libraries:
  design issues, 1001–1004
  on-bead screening, 974
  parallel synthesis, 971–972
  split-pool synthesis, 973–974
   binary encoding, 977–979
   infrared coded resins, 979–980

radio-frequency tagging, 980–981
recursive deconvolution, 976
tagging methods, 977
virtual screening sources, 90
natural product synthesis, 54–55
 biosynthesis, 60–62
 future research issues, 60–62
photolithographic synthetic methods, 981–983
research background, 962–970
solid-phase biopolymer synthesis, 965–970
tag-free methodology, 985–1001
 click and in situ click chemistry, 992–997
 dynamic chemistory and allied methods, 985–992
 macromolecular ligands, thiol-based in situ assembly, 997–998
 structure-activity relationship, nuclear magnetic resonance, 998–1000
targeted cancer therapy, small-molecule inhibitors, synthetic chemicallibraries, 1366–1367
theoretical origins, 963–965
Comma-delimited data, Excel clipboard format, 305–306
Commercially Available Data Base version 0.8:
 high-throughput flow cytometry, 195
 virtual screening:
  generation and maintenance, 213
  high-performance computing applications, 213–216
Common cold, respiratory viruses and, 1114, 1116
Comparative mean field analysis (CoMFA):
 estrogen receptor binding compounds, comparative structural connectivity spectral analysis, 256
 simulated *vs.* experimental NMR spectra, 232
Comparative spectral analysis (CoSA):
 aromatase enzyme binding models, 264
 biological effect modeling, 231–232

INDEX
1393

simulated *vs.* experimental NMR spectra, 232
structurally assigned chemical shift information, 232–233
Comparative structural connectivity spectral analysis (CoSCoSA):
  binning spectral data algorithms, 279–280
  molecular structural models, 245–271
    aromatase enzyme model, 261–264
    aryl hydrocarbon receptor binding model, 264–271
    corticosterone binding:
      artificial neural network model, 259–261
      three-dimensional models, 256–259
    two-dimensional model of estrogen receptor binding compounds, 247–256
  multidimensional NMR spectra, 234–236
Comparative structurally assigned spectral analysis (CoSASA):
  aromatase enzyme model, 240, 242–244
  aryl hydrocarbon receptor binding, 244–245
  CoSCoSA model comparisons, 264–271
  corticosterone binding model, 240–242
  molecular model flowchart, 239–240
  NMR chemical shift information management, 233
Competitive receptor antagonists:
  AMPA receptors, 840–842, 850
  metabotropic glutamate receptor ligands, 857–860
  α-substitutions, 859–860
  phenylglycines, 857–859
Complementary DNA (cDNA):
  gene expression profiling, systems biology, 132–133
  protein-protein interactions, yeast two-hybrid systems, 485–489
  library sources, 496–501
Complement-mediated lysis, DeathTRAK cycotoxicity assay, 702–704

Compliance strategies, risk assessment, 344–349
Compound management:
  CellCard system, addition and assay protocol, 360–361
  high-throughput screening, 560–563
    automated systems technology, 572–578
    close-loop screening, 578
    composition, 572–573
    nanoliter despensing, 576–578
    plate format storage, 575
    replication, 575–576
    wet or dry storage, 573–575
  high-throughput screening procedures, compound flow, 567–568
  virtual screening, 89–90
Computational analysis, protein-protein interactions, 533–536
Computed tomography (CT), cancer cell proteomics, 74
Computer-assisted structure elucidation (CASE), natural product development, 51–53
Concanamycin F., research and development, 18–19
Concavalin A ligand, dynamic combinatorial chemistry, 990–992
Concentration-response data:
  biochemical assays, 668–669
  biomolecular screening assays, 669–671
Configurational entropy, comparative structural connectivity spectral analysis, four-dimensional connectivity matrix, 276–278
Conformationally constraint analogs, metabotropic glutamate receptor ligands, 853–855
Connectivity matrix construction:
  ^{13}C-^{15}N heteronuclear connectivity matrix, antibiotic cephalosporins, comparative structural connectivity spectral analysis, 271–275

comparative structural connectivity spectral analysis:
configurational entropy estimation, 276–278
molecular dynamics of compounds, 275–276
multidimensional NMR spectra, 233–236
Conotoxins, marine sources, 32–33
Constitutive androstane receptor (CAR), ADME/Tox drug development, 157–159
Consumer protection, FDA role in, 341–343
Content analysis:
high-throughput flow cytometry, 193–194
nuclear magnetic resonance spectral information, 229–231
Contraception, estrogen receptors, 1071
Contrast agents (CA), magnetic resonance imaging (MRI), mechanisms and enhancement, 1311–1314
enzymatic activation, 1328
intracellular targeting, long-term labeling, 1326–1328
low-molecular-weight agents, vasculature and vascular targets, 1315
relaxation properties, 1313–1314
signal amplification, 1324–1325
T1 contrast agents, 1311–1312
T2 contrast agents, 1312–1313
Coomassie brilliant blue staining:
protein crystallization, 382–383
protein-protein interactions, mass spectrometry techniques, 530
Copper anode X-ray source, x-ray diffraction data:
home laboratory, 402–403
synchrotron data, 404–405
Coriolus versicolor, polysaccharide extracts, 1023–1024
Coronaviridae, structural properties, 1109
Correlation spectroscopy (COSY):
comparative structural connectivity spectral analysis:

aromatase enzyme binding model, 261–264
aryl hydrocarbon receptor binding, 264–271
^{13}C-^{15}N heteronuclear connectivity matrix, antibiotic cephalosporins, 273–275
corticosterone binding, 256–259
estrogen receptor binding compounds, 247–256
molecular structural analysis, 246–271
PD-ANN corticosterone binding, 260–261
multidimensional NMR spectra, 234–236
Corticosteroids:
adrenal corticosteroids, 1060–1066
aldosterone, 1064–1065
corticosteroid synthesis inhibitors, 1063–1064
glucocorticoids, 1060–1063
mineralcorticoids, 1064–1065
antagonists, 1065–1066
synthesis inhibition, 1063–1064
Corticosterone binding:
comparative structural connectivity spectra analysis model, 240–242
artificial neural network model, 259–261
comparative structural connectivity spectral analysis model, three-dimensional models, 256–259
Corticotropin-releasing factor (CRF), analogs and antagonists, 1040–1042
Cost issues in drug discovery:
coupled luminescent assays, 720
natural product research and development, 15–22
respiratory viral infection, 1107
systems biology and, 124–125
Coupled enzyme assays, basic principles, 692–693
Coupled luminescent assays:
abbreviations and acronyms, 725–726
assay specifications, 718–720

INDEX

cost issues, 720
operational characteristics, 719
sensitivity, 719
speed and linear response, 718
Z values, 719–720
backward reactions, 721–722
bioluminescent assay methods, 694–700
  bacterial luciferases, 694
  firefly luciferase, 694–700
    ATP/luciferase reactions:
      advantages of, 695–696
      drug discovery, 695
      high-throughput screening, 696–697
    kinases, 698–699
    limitations and considerations, 697–698
    phosphodiesterase, 699–700
    phosphorolysis, 699
chemiluminescent assays, 693–694
coupled enzyme assays, 692–693
cytotoxicity measurement, release assays, 700–701
DeathTRAK cycotoxicity assay:
  basic properties, 701–702
  biomass measurement (competitive comparison), 704–711
    ATP release assays, 704
    dual-mode cytotoxicity/proliferation, 705–712
      gram-negative bacteria, 707–709
      gram-positive bacteria, 709–711
    metabolic assays, 704–705
  cytotoxic process measurements, 702–704
  drug discovery applications, 702–711
fluorescence vs. luminance, 691
as fluorescent assay alternatives, 693
formulations, 722–725
  lytic formulations, 725
  PhosTRAK formulations, 723–724
  reaction master mixes and dilution buffers, 722–723
phosphatase activity measurement, 701

PhosTRAK phosphatase assay, 712–718
  applications, 713–714
  basic principles, 713
  complexity and thermodynamic issues, 716–717
  diverse results, 715–716
  formulations, 723–724
  future research issues, 717–718
  phosphatase activity measurement, 712–713
  research background, 690–691
  scintillation proximity assays, 691–692
COX-1 inhibition:
  herbal sources for, 1027–1028
  prodrug development, side effects reduction, 784
CP-154,526 antagonist, corticotropin-releasing factor, 1042
CREB binding protein (CBP), protein-protein interactions, split-hybrid inhibition system, 538–539
Cremophor EL, Taxol development and, 59
CRFSoft, HyperCyt system, 210
Crohn's disease (CD), animal models, 1016–1019
  trinitrobenzene sulfonic acid/ethanol-induce colitis, 1017
Cross-linked iron oxide nanoparticles (CLIO), magnetic resonance imaging (MRI), mechanisms and enhancement:
  T2 contrast agents, 1313
  tumor cell surface receptors, 1319–1325
  tumor neovasculature, 1317–1318
Cross-validated variance coefficients:
  comparative structural connectivity spectra analysis, aromatase enzyme binding models, 263–264
  comparative structural connectivity spectra analysis model, PD-ANN corticosterone binding model, 260–261
Croup, respiratory viruses and, 1116

Cryopreservation techniques, protein crystallization, 385–386
Cryostats, x-ray diffraction data, home laboratories, 403
Cryptotheca crypta, drug development, 31–32
Crystal-based drug discovery, protein targeting, 386–388
Crystallization. *See also* X-ray crystallography
  defined, 376–377
  formats for, 380–381
  ionotropic glutamate receptors, 846–851
  metabotropic glutamate receptor structure, 863–868
  protein kinase structures, 1196–1198
  vendor sources, 378–379
Crystallographic fragment screening, crystallographic structural models, 444–445
Crystallographic screening:
  crystallographic structural models, 443–444
  limitations of, 449–450
CrystalMiner software, protein crystallization, 384
CRYSTOOL software, protein crystallization formulation, 385
CSNDB database, contents and location, 128
CSV files, Excel data file management, 307
C-terminal ubiquitin moiety (Cub), protein-protein interactions, 504
Cucumaria japonica, research and development, 17
Curcumin-base medicinal agents, gastrointestinal disease, animal models, 1024–1025
Curie spin, magnetic resonance imaging (MRI), mechanisms and enhancement, T2 contrast agents, 1312–1313
Cushing's syndrome/Cushing's disease, corticotropin-releasing factor therapy, 1041–1042

CV-N protein, natural product drug development, 27
Cyanobacterium, natural product drug development, 27
Cyclic adenosine 5'-monophosphate (cAMP) response element binding protein (CREB), protein-protein interactions, split-hybrid inhibition system, 538–539
Cyclic electrophiles, click chemistry, 993–997
Cyclin-dependent kinases (CDKs):
  inhibitors, 1229
  natural product development, 19–22
  RNA synthesis inhibition, 1162–1163
  therapeutic targeting, 1203–1204
Cycloaddition reactions, click chemistry, 995–997
Cyclohexenylguanine, antiviral therapy, DNA polymerase inhibitors, 1153–1155
Cyclopentane derivatives, neuraminidase inhibitors, 1171
Cyclosporine A:
  prodrug development, peptide formulation and delivery, 780
  research and development, 18–19
Cy3/Cy5-DNA, metal-enhanced fluorescence (MEF), 637–639
CYH2 gene, protein-protein interaction inhibition, reverse two-hybrid systems, 537–538
CYN-154806 antagonist, somatostatin receptors, 1046
Cysteamine derivatives, acute/chronic duodenal ulcers, 1016
Cytochrome P450 (CYP) enzymes:
  ADME/Tox drug development, 157–159
  antistructure drug discovery, 441–442
  corticosteroid synthesis inhibition, 1063–1064
  drug metabolism assays, 477–478
Cytokines:
  corticotropin-releasing factor, 1041
  P38 MAPK inhibitors, 1226–1228

INDEX

respiratory viral infections, 1121–1122
thyroid hormone analogs and receptors, 1054–1055
Cytotoxicity assays:
coupled luminescent assays, 700–701
DeathTRAK cycotoxicity assay:
basic properties, 701–702
biomass measurement (competitive comparison), 704–711
ATP release assays, 704
dual-mode cytotoxicity/proliferation, 705–712
gram-negative bacteria, 707–709
gram-positive bacteria, 709–711
metabolic assays, 704–705
cytotoxic process measurements, 702–704
drug discovery applications, 702–711
Cytotoxic T lymphocytes (CTL), DeathTRAK cycotoxicity assay, 702–704

DABCYL quencher, aptamer-based fluorescence resonance energy transfer analysis, platelet-derived growth factor, 77–80
Database sources:
Excel Visual Basic for Applications, external database imports, 319–327
dabatase characteristics, 319–320
data imports, 322–326
ODBC access issues, 320–322
high-throughput flow cytometry, 194–195
macromolecule crystallographic models, 422–425
protein crystallization, 388–389
protein-protein interactions, computational methods, 533–536
systems biology, 127–131
future applications, 159–164
virtual screening, generation and maintenance, 212–213
Data-generating model, statistical properties, 678–679
Data management:
high-throughput screening:
automated management, 584–586
hit selection, 585–586
IC_{50} analysis, 585
quality control and data capture, 584–585
laboratory data, Excel software:
clipboard function, 304–306
opening and saving data files, 306–307
research background, 287–289
saving text and CSV files, 307
text import wizard, 306–307
Visual Basic for Applications, 290–303
automated report generation, 327–334
heat map display, 331–334
worksheet formatting, 328–330
data file reading and writing, 307–319
file access types, 309
file selection, 308–309
opening and closing sequential access files, 309–311
sequential access files, 311–315
event-driven programming, 297–299
external database imports, 319–327
dabatase characteristics, 319–320
data imports, 322–326
ODBC access issues, 320–322
fundamental principles, 291–303
language primer, 293–297
macro recording, 290–291
object entities, 299–303
Oracle ODBC connection, 327
string data manipulation and parsing, 315–319
Visual Basic Editor elements, 292–293

x-ray diffraction, 399–402
   home laboratory collection, 402–403
   synchrotron diffraction data, 403–405
Data phasing methods, x-ray diffraction, protein structures, molecular replacement, 420
Data Source Names (DSN), Excel Visual Basic for Applications, 322–326
   data import, 325–327
Daylight fingerprints, virtual screening, 103
DeathTRAK cycotoxicity assay:
   backward reactions, 721–722
   basic properties, 701–702
   biomass measurement (competitive comparison), 704–711
      ATP release assays, 704
      dual-mode cytotoxicity/proliferation, 705–712
         gram-negative bacteria, 707–709
         gram-positive bacteria, 709–711
      metabolic assays, 704–705
   cytotoxic process measurements, 702–704
   drug discovery applications, 702–711
   operational characteristics, 719
   sensitivity limits, 719
   standard cocktail formulations, 722–723
Decision-tree algorithms, virtual screening:
   multidomain clustering, 107
   partitioning techniques, 110–112
Defensive compounds, combinatorial libraries, 963–965
Densensitization, drug tolerance mechanisms, 42–43
Density skeletonization calculations, macromolecule crystallographic models, 423–425
Deoxyribonucleic acid (DNA):
   cancer cell proteomics, 74
   cell-based assays, 355–356
   multiple tumor markers, whole-cell-based aptamers, 82–83

sequence variations, systems biology, 132
targeted cancer therapy, 1357–1360
   DNA methylation, 1357–1358
   histone deacetylase inhibitors, 1358–1359
   telomerase inhibitors, 1359–1360
viral synthesis inhibition:
   acyclic nucleoside phosphonates, 1158–1161
   DNA polymerase inhibitors, 1147–1155
   reverse transcriptase inhibition, 1155–1158
Dermal delivery routes, drug delivery assays, 473–474
Dermal drug delivery, prodrug absorption mechanisms, 756–760
   topical drugs, 757–759
   transdermal delivery, 759–760
Desmopressin-1-deamino, 8-D-arginine vasopressin-dDAVP (DDAVP), vasopressin receptor agonists/antagonists, 1051–1052
Detection devices:
   high-throughput screening procedures, assay detection, 582
   protein-protein interactions:
      mass spectrometry techniques, 530
      protein microarrays, 527–528
Deterministic algorithms, natural product structural identification, 53
Developability assays:
   applications for, 481–482
   basic principles, 466–468
   bioanalytical chemistry assays, 474–476
   drug delivery assays, 471–474
   drug metabolism assays, 477–478
   formulation development, 469–471
   pharmacokinetic assays, 476–477
   pharmacology evaluations, 468–469
   toxicology assays, 478–481
      in vitro toxicology, 479–480
      in vivo toxicology, 480–481
Development phase, natural products, 55–58

Dexamethasone, gaucoma drug
 development, 147–151
Dextran-SNAFL-2 probes, metal-
 enhanced fluorescence,
 ratiometric surface sensing,
 649–653
Dextran sulfate sodium, inflammatory
 bowel disease, animal
 models, 1017–1018
DHA bead competitive assay, β2-
 adrenergic receptor (β2AR),
 205–207
Diabetes mellitus:
 antiobesity agents, 1083–1086
 endocrine hormone therapy,
   1073–1083
  biguanides, 1077
  exendin-4, 1082–1083
  glucagon, 1083
  glucagon-like peptide-1, 1082
  α-glucosidase inhibitors, 1081–1082
  glucose-dependent insulinotropic
   polypeptide, 1082
  insulin, 1073–1075
  meglitinide analogs, 1080–1081
  sulfonylureas, 1079–1080
  thiazolidinediones, 1077–1079
 natural product development for,
   17–22
 ventricular arrhythmias, nonselective
  ATP-sensitive potassium
  channel antagonists, 916–
  917
Diabetic macular edema (DME), RNA
 mimetics, aptamer function,
 1271–1273
Diagnostic testing, viral respiratory
 infection, 1117–1120
Dialysis-based protein crystallization,
 381
Didemnins, research and development,
 18–19
Dideoxynucleoside analogs, reverse
 transcriptase inhibitors,
 1156–1158
Diffraction theory. See also X-ray
 crystallography
 Bragg's law and angular dependence,
  396–397

electron clouds and thermal motion,
  398–399
 x-ray crystallization of protein
  structures, 389–390
  Fourier synthesis, 392–396
  macromolecule crystals, 390–392
Difhydrofolate reductase (DHFR),
 protein-protein interactions,
 504
Diflunisal, antistructure drug discovery,
 442
Dihydrotrigonelline, brain targeting
 prodrug development, 776
3,5-Diiodothyropropionic acid (DITPA),
 thyroid hormone analogs and
 receptors, 1054–1055
Dilution buffers, coupled luminescent
 assays, 722–725
Dimension reduction, cell-based
 partitioning, virtual
 screening, 108–109
Dimer topology, metabotropic glutamate
 receptor structure, 863–868
1,2-Dimethylhydrazine (DMH),
 colorectal cancer models,
 1020–1021
Dimethyl sulfoxide (DMSO), high-
 throughput screening
 compounds:
 nanoliter dispensing, 576–578
 replication protocols, 575–576
 wet or dry storage, 573–575
DiMSim, biological pathway analysis,
 143
Dinitrobenzene sulfonic acid (DNBS)/
 ethanol, inflammatory bowel
 disease, animal models,
 1017
DIP database, contents and location,
 128
Diphosphate 1,2-diacyl glycerol
 derivatives, phosphate group
 prodrug, 746
Dipivalyl epinephrine, ophthalmic drug
 delivery, prodrug drug
 development, 761–762
Direct fluorescent antibody (DFA), viral
 respiratory infection
 diagnosis, 1119–1120

Directional emission, metal-enhanced fluorescence, 654–660
Direct response, molecular drug actions, 37–39
Direct space partitioning, virtual screening, 109–110
Discoderma dissoluta, research and development, 18–19
Discovery assays:
  analytical chemistry assays, 462–464
  applications for, 481–482
  basic princples, 460
  in vitro pharmacology assays, 460–462
  in vivo pharmacology assays, 464–465
Discovery phase, natural products, 55–58
Disease-seeking drugs:
  mechanism-based drug development, 8
  in vitro pharmacology assays, 461–462
Display vector, protein-protein interactions, peptide display reagents, 517
Dissociated glucocorticoids, development of, 1062–1063
Dissociative molecules, protein-protein interactions, inhibition, with reverse two-hybrid systems, 537–538
Distance functions, molecular similarity, virtual screening, 94
Distance variables, comparative structural connectivity spectral analysis, estrogen receptor binding compounds, 252–256
"Distinctive Name Proviso" (FDA), 341–342
Diversification step, split-pool synthesis, 973–974
Diversity-oriented Organic Synthesis (DOS):
  combinatorial chemistry libraries, 1002–1004
  high-throughput flow cytometry, 195
DNA binding domain (DBD), protein-protein interactions:
  genetic selection systems and inhibition of, 540–541
  mammalian two-hybrid systems, 506
  protein linkage maps, 507–509

three-hybrid systems, 505–506
yeast two-hybrid systems, 485–489
target and bail vectors, 490–494
DNA methylation, targeted cancer therapy, 1357–1358
Docking algorithms:
  FPR case study, 217
  protein structural models, 433–434
  virtual screening, 212
Documentation protocols, protein crystallization, 384
Domain-closed binding sites, metabotropic glutamate receptor structure, 865–868
Donor-acceptor distances, metal-enhanced fluorescence, enhanced energy transfer, silver surfaces, 646–649
Dose curve analysis:
  glucocorticoid efficacy, 944–946
  TUNEL assay, 363–366
Dose-reponse curves:
  drug tolerance mechanisms, 42–43
  platelet-derived growth factor analysis, fluorescence resonance energy transfer, 78–80
  potency measurements, 36
Dose selection criteria, screening procedures, 6–7
Dot-blot filter arrays, protein-protein interactions, 522–523
Dot notation, Excell object entities, 300–303
Double mass spectrometry (MS/MS), natural product activity screening, 46–47
Double-prodrug concept, defined, 735–737
Double-stranded RNA (dsRNA), immunostimulation therapy, 1288–1290
  definitions, 1288–1289
  hemispherx Biopharma, 1290
  preclinical/clinical utilization, 1289–1290
Do-While example, Visual Basic Editor task, 296
Do-While Loop example, Visual Basic Editor task, 296

INDEX **1401**

"Drop out" compounds, reasons for, 1–2
Drosophila melanogaster, protein-protein interactions, 508–509
Drug action:
  general principles, 34–37
  molecular aspects, 37–42
    enzyme receptors, 40–41
    G-protein-coupled systems, 40
    ion channels, 39
    receptors, 37–39
    target selection, 41–42
    transcription factors, 41
  prodrug development, prolonged duration designs, 769–770
Drug delivery assays, drug development, 471–474
Drug discovery and development:
  antiviral agents, process for, 1135–1140
  attrition rates in, 1–2
  cancer drug targeting:
    antimimotics, 1360–1363
      aurora kinases, 1361–1362
      cell-based assays, 1362–1363
      kinesin inhibitors, 1361
      microtubule-targeting agents, 1360–1361
    Bcr/Abl, c-Kit, and PDGF, 1350
    DNA targets, 1357–1360
      DNA methylation, 1357–1358
      histone deacetylase inhibitors, 1358–1359
      telomerase inhibitors, 1359–1360
    EGF receptor pathway, 1347–1349
    evolution of, 1344–1346
    heat shock protein 90, 1364
    HER2/Neu, 1347
    information resources for, 1367–1368
    insulin-like growth factor signaling, 1349
    mammalian target of rapamycin (mTOR), 1350–1351
    proteosome inhibition, 1363–1364
    screening assays, 1351–1352
    signal transduction targets, 1346–1352

    small-molecule drug sources, 1365–1367
      natural products, 1365–1366
      synthetic chemical libraries, 1366–1367
    tumor vasculature, 1352–1357
      angiogenesis inhibitors, 1353–1356
      HIF-1α target, 1355–1356
      VEGF axis, 1354–1355
      antivascular agents, 1356
      high-throughput techniques, 1356–1357
  coupled bioluminescent assays, ATP/luciferase reactions, 695
  DeathTRAK cycotoxicity assay, 702–711
  high-throughput screening procedures, 560–563
    workflow, 567–569
  metal-enhanced fluorescence and, 660–662
  modern paradigm for, 734–735
  phases of, 457–460
  prodrug design and application:
    bioreversible structures, 740–752
      amines and amides, 750–752
        N-acyl prodrugs, 751
        N-α-acyloxymethyl$N$-phosphoroxymethyl derivatives, 751–752
        N-mannich bases, 751
      carboxyl groups, 741–744
        acyloxyalkyl prodrugs, 743–744
        aklyl and aryl esters, 741–743
        amides, 744
      hydroxyl groups, 747–750
        lactones, 750
        lipophilic prodrugs, 749–750
        steric hindrance, promoieties for, 750
        water-soluble prodrugs, 748–749
      phosphate groups, 744–746
      phosphonate groups, 747
    clinically useful requirements, 738–740
    definitions and concepts, 735–738
      chemical delivery system, 737–738
      double-prodrug concept, 735–737
      soft drug concept, 737

pharmaceutical applications, 752–785
 aqueous solubility improvement, 766–769
  hemiester structures, 769
  phosphates, 767–769
  etoposides, 767–768
  fosphenytoin, 768–769
 drug absorption mechanisms, 752–766
  carrier-mediated transport, 754–756
  dermal drug delivery, 756–760
   topical drugs, 757–759
   transdermal delivery, 759–760
  nasal drug delivery, 762–766
   ehanced brain delivery, 766
   lipophilic prodrugs, 764–766
   peptide prodrugs, 766
   transport-mediated absorption, 766
   water-soluble prodrugs, 763–764
  ophthalmic drug delivery, 760–762
  oral drug delivery, 752–756
   lipophilic drugs, 753–754
 drug targeting improvements, 770–779
  brain targeting, 775–778
   carrier-mediated transport, 776–778
   chemical drug delivery system, 776
   lipophilic prodrugs, 775–776
  colon targeting, 778
  kidney targeting, 778–779
  tumor targeting, 770–775
   antibody-directed enzyme prodrug therapy, 774
   gene-directed enzyme prodrug therapy, 774–775
   hypoxia-selective drugs, 772–774
 duration of action, 769–770
 macromolecular promoieties, 784–785

peptide formulation and delivery, 779–783
 aqueous solubility improvement, 780
 enzyme stability, 780–782
 membrane permeation improvement, 782–783
 side effects reduction, 783–784
respiratory antiviral agents, 1124–1126
Drug likeness, virtual screening, 114
Drug metabolism (DM) assays, drug development, 477–478
Drug-seeking diseases, therapeutic area drug development, 7
Drug tolerance, natural product pharmacodynamics, 42–43
ds-Fl-DNA-SH, metal-enhanced fluorescence (MEF), enhanced DNA/RNA detection, 631–637
Dual mode cycotoxicity/proliferation assays, DeathTRAK cycotoxicity assay, 705–712
 gram-negative bacteria, 707–709
 gram-positive bacteria, 709–711
Dual reporter systems, protein-protein interactions, yeast two-hybrid systems, 495
Duodenal ulcers, animal models, 1014–1015
 acute/chronic, 1016
Dynamic combinatorial chemistry (DCC), tag-free techniques, 985–992
Dynamic combinatorial libraries (DCLs), tag-free techniques, 985–992
Dynamic mapping of consensus positions, virtual screening, 110–111

E. coli display, protein-protein interactions, 515
E. coli phosphatase, PhosTRAK phosphatase assay, 715–716
Ebers Papyrus, 12–13

E-CELL software, systems biology and, 126–127
Echinocandin-type lipopeptide (FR901379), reserach and development, 16
EcoCyc, contents and location, 128
EDGE database, gene expression profiles, 158–159
Effective dose response (ED_{50}) values:
　biomolecular screening assays:
　　affinity models and parameter estimates, 669
　　muscarinic M_1 receptor GTPγ^{35}S assay, 670–671
　developability assays, 468–469
　$GABA_B$ receptors, allosteric modulation, 827
　potency precision estimates, 676–678
　TUNEL assay, multiplex dose curve analysis, 364–366
　in vivo pharmacology assays, 465
Efficacy indicators:
　anti-coagulant agents, 946–948
　general principles, 37
　mediator-specific anti-inflammatory agents, 940–943
　in screening techniques, 4–9
Electrochemically-deposited silver, metal-enhanced fluorescence, 619–622
Electron clouds, x-ray diffraction, protein structure, 398–399
Electron density mapping:
　macromolecule crystallographic models, de novo model construction, 422–425
　x-ray diffraction, protein structures:
　　anomalous dispersion, 415–416
　　crystallographic screening, 443–444
　　electron clouds and thermal motion, 399
　　molecular replacement, 418–420
　　multiple isomorphous replacement, 407–412
Electroplated silver, metal-enhanced fluorescence, 622
Electrostatic forces, antiviral therapy, DNA polymerase inhibitors, 1152–1155

Emission spectra, metal-enhanced fluorescence, enhanced energy transfer, silver surfaces, 645–649
EMP database, contents and location, 128
Encapsulation, Excell object entities, 299–303
Endocrine hormones:
　adrenal corticosteroids, 1060–1066
　aldosterone, 1064–1065
　corticosteroid synthesis inhibitors, 1063–1064
　glucocorticoids, 1060–1063
　mineralcorticoids, 1064–1065
　antagonists, 1065–1066
　androgens, 1066–1068
　anabolic steroids, 1067–1068
　antiandrogens, 1068
　receptors, 1066–1067
　5α-reductase inhibitors, 1068
　anterior pituitary hormones, 1047–1049
　follicle-stimulating hormone, 1048–1049
　growth hormone (somatotropin), 1047–1048
　luteinizing hormone, 1049
　antiobesity agents, 1083–1086
　orlistat, 1085–1086
　bone metabolism agents, 1057–1060
　bisphosphonates, 1058–1059
　calcimimetics, 1059–1060
　calcitonin, 1058
　mithramycin, 1059
　parathyroid hormone, 1057–1058
　diabetes mellitus therapy, 1073–1083
　biguanides, 1077
　exendin-4, 1082–1083
　glucagon, 1083
　glucagon-like peptide-1, 1082
　α-glucosidase inhibitors, 1081–1082
　glucose-dependent insulinotropic polypeptide, 1082
　insulin, 1073–1075
　meglitinide analogs, 1080–1081
　sulfonylureas, 1079–1080
　thiazolidinediones, 1077–1079

1404   INDEX

disruptors, reserach and development, 17
estrogens, 1069–1072
  agonists, 1070
  antagonists, 1070
  aromatase inhibitors, 1070–1071
  progestins, 1071–1072
  receptors, 1069
  selective estrogen receptor modulators, 1070
hypothalamic hormone analogs and antagonists, 1039–1046
  corticotropin-releasing hormone, 1040–1042
  gonadotrophin-releasing hormone, 1043–1045
  growth-hormone-releasing hormone, 1043
  somatostatin, 1045–1046
  thyrotropin-releasing hormone, 1039–1040
neurohypophyseal hormones, 1049–1053
  oxytocin, 1052–1053
  vasopressin, 1049–1052
thyroid hormones, 1053–1057
  analogs, 1054–1055
  iodide, 1055
  potassium perchlorate, 1056–1057
  therapeutic analogs, 1054
  thionamide drugs, 1055–1056
  thyroxine/tri-iodothyronine receptors, 1053–1054
Endopeptidases, prodrug development, peptide formulations, stability improvement, 780–782
Enfuviritide, virus-cell fusion inhibitors, 1146–1147
Enhanced brain delivery, prodrug structures, 766
Enhanced DNA labeling, metal-enhanced fluorescence (MEF), 637–639
Enhanced DNA/RNA detection, metal-enhanced fluorescence (MEF), 630–637
Environmental degradation, natural product development, 61

Environmental sources, natural product development and discovery, 23–24, 29–33
  insects, 30–31
  marine organisms, 31–33
  microbes, 24, 29
  plants, 29–30
  vertebrates, 31
Enzyme fragment complementation, high-throughput screening cellular assays, 597–5989
Enzyme-linked immunosorbent assays (ELISA), high-throughput screening formats, 563–565
Enzyme-linked lectin assay (ELLA), dynamic combinatorial chemistry, 990–992
Enzymes. *See also* specific enzymes
  chemical shift molecular imaging, reporter molecules, 1330–1331
  contrast agent activation, MRI molecular imaging, 1328
  coupled enzyme assays, 692–693
    PhosTRAK phosphatase assay, 712–718
  gastrointestinal disease models, acute gastric lesions, 1015
  high-throughput screening, bioassay design, 591–592
  molecular drug action:
    receptors, 40–41
    target enzymes, 41–42
  natural product synthesis, 54–55
  prodrug development:
    bioreversible prodrug structures, 740–752
    amines and amides, 750–752
      N-acyl prodrugs, 751
      N-α-acyloxymethyl*N*-phosphoroxymethyl derivatives, 751–752
      N-mannich bases, 751
    carboxyl groups, 741–744
      acyloxyalkyl prodrugs, 743–744
      aklyl and aryl esters, 741–743
      amides, 744

INDEX   1405

hydroxyl groups, 747–750
  lactones, 750
  lipophilic prodrugs, 749–750
  steric hindrance, promoieties
    for, 750
  water-soluble prodrugs, 748–
    749
  phosphate groups, 744–746
  phosphonate groups, 747
  peptide formulations, stability
    improvement, 780–782
  tumor targeting, 770–775
  viral replication, 1122–1124
EO-9 derivative, hypoxia-selective
  prodrugs, tumor targeting,
  773–774
Epidermal growth factor receptor
  (EGFR):
  protein-protein interactions, 505–506
  targeted cancer therapy, signal
    transduction, 1347–1349
  tyrosine kinases, 1200
    extracellular domain inhibition,
      1216–1217
    intracellular domain inhibition,
      1217–1219
Epipedobates tricolor, drug development
  from, 31
Epstein-Barr virus, natural product
  development for, 17–18
Erbitux, receptor tyrosine kinase
  inhibitors, 1216–1217
ERGO database, contents and location,
  128
E-selectin, MRI tumor targeting, tumor
  neovasculature, 1317–1318
Estradiol:
  chemical drug delivery system, brain
    targeting prodrugs, 776
  estrogen agonists, 1070
Estrogens:
  agonists, 1070
  antagonists, 1070
  aromatase inhibitors, 1070–1071
  progestins, 1071–1072
  receptors, 1069
    binding compounds, comparative
      structural connectivity
      spectral analysis, 247–256

selective estrogen receptor
  modulators, 1070
structure and applications, 1069–1072
Etoposide phosphate, prodrug
  development, aqueous
  solubility, 767–768
Euclidian distance calculations, virtual
  screening, topological and
  shape representation, 100
Eukaryotic display systems, protein-
  protein interactions, 515–516
Eulerian angles, x-ray diffraction, protein
  structures, molecular
  replacement, 417–420
Euphorbia lateriflora, drug development,
  30
Event-driven programming, Visual Basic
  for Applications, laboratory
  data management, 297–299
Evolutionary process:
  dynamic combinatorial chemistry
    (DCC), 985–992
  natural product research and
    development, 15–22
Excel software, laboratory data
  management:
  clipboard function, 304–306
  opening and saving data files, 306–307
  research background, 287–289
  saving text and CSV files, 307
  text import wizard, 306–307
  Visual Basic for Applications, 290–303
    automated report generation,
      327–334
      heat map display, 331–334
      worksheet formatting, 328–330
    data file reading and writing,
      307–319
      file access types, 309
      file selection, 308–309
      opening and closing sequential
        access files, 309–311
      sequential access files, 311–315
    event-driven programming, 297–299
    external database imports, 319–327
      dabatase characteristics, 319–320
      data imports, 322–326
      ODBC access issues, 320–322
    fundamental principles, 291–303

language primer, 293–297
macro recording, 290–291
object entities, 299–303
Oracle ODBC connection, 327
string data manipulation and parsing, 315–319
Visual Basic Editor elements, 292–293
Excitatory neurotransmitters, glutamate receptor ligands, 829–868
　AMPA receptor ligands, 837–843
　　agonists, 837–840
　　competitive/noncompetitive antagonists, 840–842
　　modulatory agents, 842–843
　ionotropic receptor structure, 846–852
　KA receptor agonists and antagonists, 843–846
　metabotropic receptor ligands, 851, 853–868
　　agonists, 853–857
　　　backbone chain length extension, 855–857
　　　bioisosteres, 857
　　　conformationally constraint analogs, 853–855
　　allosteric modulation, 860–862
　　　noncompetitivve antagonists, 860–862
　　　positive modulators, 862
　　competitive antagonists, 857–860
　　　α-substitutions, 859–860
　　　phenylglycines, 857–859
　　noncompetitive antagonists, 860–862
　　putative drug targeting, 862–863
　　structural properties, 863–868
　NMDA receptor ligands, 830–837
　　competitive receptor antagonists, 832–833
　　glycine co-agonist site, 835–836
　　receptor agonists, 831–832
　　subtype-selective ligands, therapeutic potential, 836–837
　　uncompetitive/noncompetitive receptor antagonists, 833–835
　receptor classification, 829–830

Exendin-4, diabetes therapy, 1082–1083
Exo-THPO, GABA reuptake inhibition, 806
Experiment Management software, CellCard system, 359–360
Expert Protein Analysis System (ExPASy), macromolecule crystallographic models, 422–425
Exponential enrichment (SELEX) process, aptamer sequencing, tumor marker detection, 75–76
Expressed Sequencing Tags (EST) data, biological pathway analysis, 142–146
External test set predictions, estrogen receptor binding compounds, comparative structural connectivity spectral analysis, 254–256
Extracellular domains:
　EGFR receptor inhibitor targeting, 1216–1217
　metabotropic glutamate receptor structure, 864–868
Extracellular potassium, myocardial ischemia, 911–913
Extracellular signal-regulated kinases (ERK1/2):
　inhibitors, 1228–1229
　therapeutic targeting, 1202–1205
Extraction procedures, natural product purification, 49–51
Eyetech Pharmaceutical, Inc., RNA mimetics, aptamer function, 1272–1273

"Fail early/fail often" principle, drug discovery and development, 459–460
False negatives, protein-protein interactions, yeast two-hybrid systems, 489
False positives, protein-protein interactions, yeast two-hybrid systems, 488–489
Fasudil, Rho kinase inhibition, 1231

FCSQuery software, HyperCyt system, 209–210
Fibroblast growth factor receptors (FGFR), tyrosine kinases, 1200–1201
inhibitors, 1219–1221
File access functions, Excel software, 309
Filter disk techniques, parallel library synthesis, 971–972
Firefly luciferase, coupled bioluminescent assays, 694–700
  ATP/luciferase reactions:
    advantages of, 695–696
    drug discovery, 695
  high-throughput screening, 696–697
  kinases, 698–699
  limitations and considerations, 697–698
  phosphodiesterase, 699–700
  phosphorolysis, 699
FITC-HSA, metal-enhanced fluorescence:
  overlabled proteins as ultrabright probes, 639–642
  silver fractal-like structures, 625–630
FK506 binding protein (FKBP), structure-activity relationship, 999–1001
FLAG tags, high-throughput flow cytometry, soluble and bead-based assemblies, 203–207
Flavone derivatives, $GABA_A$ receptor complex, receptor and pharmacophore models, benzodiazepine binding sites, 822–823
Flavopiridol:
  CDK inhibition, 1229
  protein kinase inhibition, ATP binding site inhibitors, 1211–1212
FLIPR imaging system, high-throughput bioassays, 594–597
FliTrx Random Display Lybrary system, protein-protein interactions, peptide display reagents, 517

Flow cytometry. See High-throughput flow cytometry
  targeted cancer therapy, antimimotics, 1363
Fludrocortisone, characteristics and therapeutic applications, 1065
Fluorescein-conjugated peptides, fMLFK-FITC, FPR ligand screening, 210–211
Fluorescein probes, metal-enhanced fluorescence (MEF), 639–642
Fluorescence anisotropy (FA) probes, molecular aptamers, 80–81
Fluorescence correlation spectroscopy (FCS), high-throughput screening bioassays, 594–595
Fluorescence imaging plate reader (FLIPR), cell-based assays, 354–356
Fluorescence in situ hybridization (FISH), metal-enhanced fluorescence, enhanced energy transfer, silver surfaces, 648–649
Fluorescence polarization (FP), coupled luminescent assays, 691
Fluorescence resonance energy transfer (FRET):
  aptamer-based analysis, platelet-derived growth factor, 76–80
  coupled luminescent assays, 691
  high-throughput screening bioassays, 594–595
Fluorescent assays, coupled luminescent assays vs., 693
Fluorescent labeling, protein-protein interactions, protein microarrays, 527–528
Fluorescent techniques, high-throughput screening, 594–595
  cell-based bioassays, 595
5-Fluorocytosine (5-FU), gene-directed enzyme prodrug therapy, 775
Fluorophore-metal interactions, metal-enhanced fluorescence, 604–609
  anisotropic silver nanostructures, 615–618

Fluorophore molecules:
  aptamer-based fluorescence resonance energy transfer analysis, platelet-derived growth factor, 77–80
  metal-enhanced fluorescence:
    directional emission, 655–660
    enhanced DNA/RNA detection, 631–637
    multiphonon excitation, 643–645
Fluoxetine, $GABA_A$ receptor complex, 819
Fluphenazine, prodrug development, prolonged drug action technologies, 770
Flutamide, antiandrogen properties, 1068
Follicle-stimulating hormone (FSH):
  drug actions and applications, 1048–1049
  gonadotrophin-releasing hormone regulation, 1043–1045
Food and Drug Administration (FDA):
  origins of, 339–343
  risk assessment regulations, 345–349
Formula property, Excell software, 303
Formulation development assays, drug development, 469–471
Formulations:
  coupled luminescent assays, 722–725
  prodrug development, peptide delivery, 779–783
Formulation screens, protein crystallization, 384–385
Formyl peptide receptor (FPR):
  high-throughput flow cytometry, drug discovery targets:
    arrestin assemblies, 197–198
    cell-based colocalization with arrestin, 198–199
    GPCR signaling pathways, 190
    intracellular FPR traffic, arrestin regulation, 199–200
    ligand screening assay, 210–211
    soluble and bead-based assemblies, 204–207
    tail assembly, protein domain display, 201–202
    virtual screening case study with, 216–217

For-Next Loop, Visual Basic Editor task, 296–297
Förster distance value, metal-enhanced fluorescence, enhanced energy transfer, silver surfaces, 645–649
Förster-like quenching:
  metal-enhanced fluorescence, overlabled proteins as ultrabright probes, 639–642
  metal-fluorophore interactions, 604–609
Fosphenytoin, prodrug development:
  aqueous solubility, 768–769
  side effects reduction, 783–785
Four-dimensional connectivity matrix, comparative structural connectivity spectral analysis:
  configurational entropy estimation, 276–278
  molecular dynamics of compounds, 275–276
Four-dimensional quantitative structure-activity relationships (4D-QSARs), virtual screening, 95–96, 113
Fourier coefficients, x-ray diffraction, protein structures, molecular replacement, 419–420
Fourier synthesis:
  x-ray crystallography, 392–396
    Bragg's law and angular momentum, 397
    X-ray crystal structure determination, phase problem, 405
Fourier transform, x-ray diffraction, protein structures:
  electron clouds and thermal motion, 399
  multiple isomorphous replacement, 407–412
Fourier transformation mass spectrometry (FT-MS), metabolomic/metabonomic analysis, 137–138
Four-parameter logistic (4PL) model:
  biomolecular screening assays, 669
  muscarinic M_1 receptor $GTP\gamma^{35}S$ assay, 674–676

FPR-G$_{i\alpha2}$ fusion protein, high-throughput flow cytometry characterization, 200–201
Fractal-like structures, metal-enhanced fluorescence, silver nanoparticles, 625–630
Fractionation-driven bioassays, natural product development:
　activity screening, 45–47
　purification and isolation, 49–51
Fragment-based shape-matching algorithm, virtual screening, topological and shape representation, 100
Fragment screening, defined, 375
Fragment tethering, crystallographic structural models, site-directed leads, 445–446
Friedel's law, x-ray crystal structure determination, anomalous dispersion, 412–416
F test, comparative structural connectivity spectral analysis, corticosterone binding, 257–259
Fucosyltransferase (Fuc-T), click chemistry, 993–997
Functionally related enzyme clusters (FREC), biological networks, 162–164
Functional protein microarrays, protein-protein interactions, 519
Function-oriented synthesis, combinatorial chemistry libraries, 1002–1004
Fungal disease, natural product development for, 16–22
Fusion proteins:
　antiviral agents, 1125–1126
　crystallization factors, 382–383
　protein-protein interactions:
　　mammalian two-hybrid systems, 506
　　peptide display reagents, 516–517
　　protein linkage maps, 507–509
　　three-hybrid systems, 505–506
　　yeast two-hybrid systems, 487–489
　　　target and bail vectors, 490–494
　virus-cell fusion inhibitors, 1141–1147

F-180 vasopressin analog, receptor agonists/antagonists, 1050
GABA receptor ligands:
　inhibitory neurotransmitter mechanisms, 802–829
　　biosynthesis and metabolism, 803–804
　　classification and therapeutic targets, 802–803
　　GABA$_A$ receptor complex, 807–823
　　　agonists and antagonists, 808–811
　　　benzodiazepine-allosteric binding sites, 814–821
　　　pharmacophore models, 821–823
　　　receptor and pharmacophore models, 811–814
　　GABA$_B$ receptor complex, 823–827
　　　agonists, 825
　　　allosteric modulators, 827
　　　antagonists, 825–827
　　GABA$_C$ receptor agonists and antagonists, 828–829
　　metabolism inhibitors, 803–804
　　reuptake mechanisms, 804–806
　therapeutic applications:
　　Alzheimer's disease, 800–801
　　central nervous system ischemia, 800
　　disease conditions, 801–802
　　neurological diseases, 799
　　research background, 798–799
GABA-T enzyme:
　GABA biosynthesis, 803
　GABA metabolism inhibition, 803–804
Gadolinium-based contrast agents:
　magnetic resonance imaging (MRI), mechanisms and enhancement:
　　limitations, 1331–1333
　　T1 contrast agents, 1311–1312
　　in vivo MR molecular imaging, tumor cell surface receptors, 1320–1321

GAL4 hybrid protein, protein-protein interactions, 486–489
  host yeast strains, 495
  inhibition, with reverse two-hybrid systems, 537–538
  library selection, 501
  mammalian two-hybrid systems, 506
  three-hybrid system, 506
  yeast two-hybrid systems, target and bail vectors, 490–494
Ganciclovir (GCV):
  gene-directed enzyme prodrug therapy, 775
  respiratory viral infection therapy, 1120–1121
Gas chromatography-mass spectrometry, metabolomic analysis, 137–138
Gastric ulcers, animal models, 1015–1016
Gastrointestinal cancer, animal models, 1019–1021
  chemical induction model, 1019–1020
  human gastric cancer cell implantation, 1019
Gastrointestinal disease, animal models:
  acute gastric lesion assessment, 1015
  acute gastric mucosal injury, 1014–1015
  chronic gastric ulcer, 1015–1016
  colorectal cancer, 1020–1021
    chemical induction model, 1020–1021
    genetically modified model, 1020
    inflammation-associated model, 1020
  duodenal ulcers, acute and chronic, 1016
  gastrointestinal cancer, 1019–1021
    chemical induction model, 1019–1020
    human gastric cancer cell implantation, 1019
  herbal medicines, 1021–1028
    biological processes, 1021–1022
    licorice, 1025–1027
    phenolic compounds and terpenoids, 1023–1028
      curcumin, 1023–1024
    polysaccharides, 1022–1024
      mushroom/seaweed sources, 1022–1023
      plant sources, 1023
      terpenoid sources, 1027–1028
  inflammatory bowel disease, 1016–1019
    colitis severity assessment, 1018–1019
    dextran sulfate sodium, 1017–1018
    dinitrobenzene sulfonic acid/ethanol, 1017
    trinitrobenzene sulfonic acid/ethanol, 1017
GdDTPA contrast agent:
  MRI tumor targeting, vasculature and vascular targets, 1315
  in vivo MR molecular imaging, tumor cell surface receptors, 1320–1321
Gene-directed enzyme prodrug therapy (GDEPT), tumor targeting, 772–775
Gene expression:
  synthetic RNA molecules and suppression of, 1282–1283
  systems biology, 132–135
    ADME/Tox drug development, 157–159
    breast cancer drug development, 151, 153–156
    clustering approaches, 138–139
    gaucoma drug development, 148–151
    microarrays, 132–133
    serial analysis, 133–135
Gene network signatures, systems biology, breast cancer drug development, 155–156
Gene Ontology (GO) classification, systems biology pathways, 160–164
Gene-prodrug activation therapy (GPAT), tumor targeting, 772–775
GeNet database, contents and location, 129
Genetically modified models, colorectal cancer, 1020

GenMAPP database, contents and
    location, 128–129
Genome-wide protein-protein
    interactions, survey of,
    508–510
Genomics:
  drug development based on, 8
  natural product activity screening, 44–
    47
  structural genomics:
    crystallographic structural models,
      446–447
    defined, 376
  systems biology and, 124–125
Geometric refinement statistics, protein
    structural models validation,
    432–433
Gilbenclamide, ventricular arrhythmias,
    915–917
Gingko biloba, research and
    development of, 17
Glass surface geometry, metal-enhanced
    fluorescence, 609–630
  anisotropic silver nanostructures, 615–
    618
  electrochemically deposited silver,
    619–622
  electroplated silver, 622
  laser-deposited silver, 618–619
  roughened silver electrodes, 622–624
  silver colloid films, 610–615
  silver fractal-like structures, glass
    substrates, 624–630
  silver island films, 610
Glaucoma drug development, systems
    biology techniques, 146–152
Glitazones, diabetes therapy, 1077–1079
Global network architecture, biological
    networks, 141–142
Global screening procedures, protein-
    protein interactions, yeast
    two-hybrid systems, 508–509
Glucagon, diabetes therapy, 1083
Glucagon-like peptide-1 (GLP-1),
    diabetes therapy, 1082
Glucans, plant sources, 1022–1024
Glucocorticoid response elements
    (GREs), glucocorticoid
    receptors, 1061

Glucocorticoids:
  anti-inflammatory and adverse effects,
    1062–1063
  nongenomic activities, 1062
  receptors, 1060–1063
  sepsis management, 943–946
    clinical trials, 943–944
    dose-dependent effects, 943, 945
    future research, 949
    mediator-specific anti-inflammatory
      agents *vs.*, 945–946
    structure and therapeutic applications,
      1060–1063
Glucose-dependent insulinotropic
    polypeptide (GIP), diabetes
    therapy, 1082
Glucose-induced insulin secretion,
    diabetes mellitus therapy,
    1074
α-Glucosidase inhibitors, diabetes
    therapy, 1081–1082
Glu decarboxylase (GAD), GABA
    biosynthesis, 803–804
Glu homologs:
  AMPA receptor agonists, 839–840
  AMPA receptor antagonists, 840–842
  kainic acid receptor agonists and
    antagonists, 844–846
Glutamate receptor ligands:
  excitatory neurotransmitters, 829–868
    AMPA receptor ligands, 837–843
      agonists, 837–840
      competitive/noncompetitive
        antagonists, 840–842
      modulatory agents, 842–843
    glutamic acid receptor classification,
      829–830
    ionotropic receptor structure,
      846–852
    KA receptor agonists and
      antagonists, 843–846
    metabotropic receptor ligands, 851,
      853–868
      agonists, 853–857
        backbone chain length
          extension, 855–857
        bioisosteres, 857
        conformationally constraint
          analogs, 853–855

allosteric modulation, 860–862
  noncompetitivve antagonists, 860–862
  positive modulators, 862
competitive antagonists, 857–860
  α-substitutions, 859–860
  phenylglycines, 857–859
  noncompetitive antagonists, 860–862
  putative drug targeting, 862–863
  structural properties, 863–868
NMDA receptor ligands, 830–837
  competitive receptor antagonists, 832–833
  glycine co-agonist site, 835–836
  receptor agonists, 831–832
  subtype-selective ligands, therapeutic potential, 836–837
  uncompetitive/noncompetitive receptor antagonists, 833–835
therapeutic applications:
  Alzheimer's disease, 800–801
  central nervous system ischemia, 800
  disease conditions, 801–802
  neurological diseases, 799
  research background, 798–799
Glutamic acid receptors, classification, 829–830
Glutathione S-transferases (GSTs):
  protein-protein interactions:
    affinity tagging and mass spectroscopy, 531–533
    tag systems, 523
  in vitro toxicology assays, 480
1,3-β-Glycan, reserach and development, 16
Glyceraldehyde-3-phosphate dehydrogenase (G3PDH):
  backward reactions, 721–722
  coupled luminescent assay speed and linear response, 718
  cytotoxicity measurements, coupled luminescent assays, 700–701
DeathTRAK cycotoxicity assay:
  basic properties, 701–702
  biomass measurement, 704

3-(Glycidoxypropyl) trimethoxysilane (GOPTES), metal-enhanced fluorescence, ratiometric surface sensing, 649–653
Glycine co-agonist site, NMDA receptor antagonists, 835–836
Glycogen synthase kinase 3 (GSK3):
  inhibitors, 1231–1232
  therapeutic targeting, 1205
Glycoprotein gp120, virus-cell fusion inhibitors, 1142–1147
Glycyrrhizin medicinal compounds, gastrointestinal disease, animal models, 1026–1027
Gold-coated glass surfaces, protein-protein interactions, protein microarray supports, 525–526
Gonadotrophin-releasing hormone (GnRH), analogs and antagonists, 1043–1045
Goniometer, x-ray diffraction data collection, 399–402
Good Laboratory Practices (GLP):
  natural product development, 56–58
  toxicology assays, 479–481
Good Manufacturing Practices (GMP), natural product development, 56–58
G-protein-coupled receptor kinases (GRKs):
  drug tolerance mechanisms, 42–43
  high-throughput flow cytometry, drug discovery targets, 188–190
G-protein-coupled receptors (GPCR):
  agonist/antagonist discrimination, 194
  calcitonin, 1058
  coupled bioluminescent assays, 694
  drug tolerance mechanisms, 42–43
  $GABA_B$ receptors, 823–827
  glucagon, 1083
  gonadotrophin-releasing hormone agonists, 1043–1045
  high-throughput flow cytometry, drug discovery applications, 187–196
    content volume, 193–194
    diversity analysis, 190–192
    FPR-arrestin colocalization in cells, 198–199

FPR-G$_{i\alpha2}$ fusion protein
  characterization, 200–201
FPR tail assembly, 201–202
HyperCyt system, 192–193, 196
informatics technology, 194–195
protein constructs, 196–197
screening techniques, 195–196
signaling pathways, 187–190
soluble and bead-based assemblies,
  202–207
soluble receptors with G proteins,
  197
high-throughput screening platforms,
  593–594
luteinizing hormone, 1049
molecular drug action, 40
structural studies of, 450
thyrotropin-releasing hormone
  receptors, 1040
vasopressin, 1050
virtual screening case study of,
  216–217
G proteins, high-throughput flow
  cytometry:
bead assembly, 205–207
βγ subunits, 203–207
soluble GPCRs, 197
Gram-negative bacteria, dual mode
  cycotoxicity/proliferation
  assays, DeathTRAK
  cycotoxicity assay, 707–709
Gram-positive bacteria, dual mode
  cycotoxicity/proliferation
  assays, DeathTRAK
  cycotoxicity assay, 709–
  711
Graphpad/Prism software, biomolecular
  screening assays:
affinity models and parameter
  estimates, 669
muscarinic M$_1$ receptor GTPγ^{35} S
  assay, 670–671
Graves' disease:
thionamide drugs, 1055–1056
thyroid hormone analogs and
  receptors, iodide, 1055
Growth factors:
cancer cell proteomics, molecular
  diagnosis, 74–75

thyroid hormone analogs and
  receptors, 1054–1055
Growth hormone (GH), drug actions
  and applications, 1047–1048
Growth-hormone-releasing hormone
  (GHRH), analogs and
  antagonists, 1043
Guanine quadruplex DNA ligands,
  dynamic combinatorial
  chemistry, 989–992
Guanosine analogs:
antiviral therapy, DNA polymerase
  inhibitors, 1152–1155
IMP dehydrogenase inhibitors,
  1173
Guanosine 5′-diphosphate (GDP):
click chemistry, 993–997
G-protein-coupled systems, molecular
  drug action, 40
Guanosine 5′-diphosphate (GDP)-
  guanosine 5′-triphosphate
  (GDP-GTP) exchange factor
  (GEF), protein-protein
  interactions, Sos recruitment
  system (SRS), 502–503
Guanosine 5′-triphosphate, G-protein-
  coupled systems, molecular
  drug action, 40
Guanyl cyclase, molecular drug action,
  enzyme receptors, 41
Guanylguanidine, diabetes therapy,
  1077
Guvacine, GABA reuptake inhibition,
  805–806

Hammerhead ribozymes:
preclinical/clinical utilizations,
  1276–1277
RNA mimetics, 1274–1276
Hantaviruses, characteristics, 1112–
  1113
Haystack storage system, high-
  throughput screening,
  compound storage, 574–
  576
Heat map display, Excell reports,
  331–334
Heat shock protein 90, targeted cancer
  therapy, 1364

Heavy atom replacement methods:
x-ray crystal structure determination, 406
anomalous dispersion, 412–416
isomorphous replacement, 407–412
x-ray diffraction, protein structures, molecular replacement, 417–420
Heavy metal compounds, anomalous diffraction experiments, 388
Helicase-RNA primase, biomolecular screening assays, RNA synthesis inhibition, 1155
Helicobacter pylori:
plant-base medicines, 1024
polyphenol medicinal compounds, 1026–1027
Hemiester prodrugs, aqueous solubility, 769
Hemispherx Biopharma, Inc., immunostimulating RNA, 1290
Heparin, clinical trials, 949–950
Hepatitis C virus (HCV), antiviral therapy, 1121
IMP dehydrogenase inhibitors, 1173
protease inhibitors, 1168–1169
RNA synthesis inhibition, 1162–1163
Herbal medicines. *See also* Natural products
gastrointestinal disease, animal models, 1021–1028
biological processes, 1021–1022
licorice, 1025–1027
phenolic compounds and terpenoids, 1023–1028
curcumin, 1023–1024
polysaccharides, 1022–1024
mushroom/seaweed sources, 1022–1023
plant sources, 1023
terpenoid sources, 1027–1028
Herceptin, protein kinase inhibitor targeting, 1210
HER2/Neu, targeted cancer therapy, signal transduction pathways, 1347

Herpes simplex virus (HSV), antiviral therapy:
DNA polymerase inhibitors, 1147–1155
evolution of, 1135–1140
S-adenosylhomocysteine hydrolase inhibitors, 1177
virus adsorption inhibitors, 1140–1141
virus-cell fusion inhibitors, 1147
Herpesviridae:
antiviral therapy:
DNA polymerase inhibitors, 1147–1155
protease inhibitors, 1168–1169
characteristics of, 1113–1114
Heteronuclear multiple quantum coherence (HMQC) experiments, multidimensional NMR spectra, 235–236
Heteronuclear single quantum correlation (HSQC) experiments, multidimensional NMR spectra, 235–236
Hierarchical clustering:
gene expression data, systems biology, 138–139
virtual screening, 107
Hierarchically ordered spherical description of environment (HOSE), predicted spectra, 236
High content screening (HCS), cellular assay development, 598
Highly active antiretroviral therapy (HAART), HIV protease inhibitors, 1168–1169
High-molecular-weight contrast agents, MRI tumor targeting, vasculature and vascular targets, 1316
High-performance computing, virtual screening and, 213–216
High-throughput flow cytometry:
FPR ligand screening assay, 210–211
future research, 217

INDEX
1415

G-protein-coupled receptor drug
 discovery, 187–196
 content volume, 193–194
 diversity analysis, 190–192
 HyperCyt system, 192–193, 196
 informatics technology, 194–195
 screening techniques, 195–196
 signaling pathways, 187–190
high content techniques, HyperCyt
 applications, 207
microfluidic mixing:
 HyperCyt cell sorting, 208–209
 sample carryover, 209
 soluble compounds, 207–208
protein expression and
 characterization, 196–202
 arrestin-regulated intracellular FPR
 traffic, 199–200
 β2AR proof of principle, 202
 FPR assemblies with arrestin,
 197–198
 FPR colocalization with arrestin,
 198–199
 FPR-$G_{i\alpha2}$ fusion protein, 200–201
 FPR tail assembly, 201–202
 GPCR constructs, 196–197
 G protein-soluble GPCR
 reconstitution, 197
research background, 186–187
software development, 209–210
soluble and bead-based assemblies,
 202–207
 β2AR, 205–206
 cell cycle protein display, 206–207
 FPR assemblies, 204–205
High-throughput protein production,
 protein-protein interactions,
 protein microarray reagents,
 523–524
High-throughput screening (HTS):
 automated systems technology,
 570–586
 accuracy and precision, 580–581
 assay detection, 582
 compound management, 572–578
 close-loop screening, 578
 composition, 572–573
 nanoliter despensing, 576–578
 plate format storage, 575

replication, 575–576
 wet or dry storage, 573–575
data analysis, 584–586
 data capture and quality control,
 584–585
 hit selection, 585–586
 IC_{50} analysis, 585
infrastructure, 582–584
liquid handling, 578–580
plate design, 570–572
reagent management, 584
scheduling software, 583–584
bioassay technology, 586–598
 assay formats, 587–588
 cell-based assay design, 595–598
 design criteria, 591
 ligand-receptor binding
 interactions, 592–594
 mechanistic issues, 591–592
 fluorescent methods, 594–595
 reagent production, 588–591
cell-based assays, 354–356
combinatorial synthesis, 983–985
compound sources, 89–90
coupled bioluminescent assays, 696–697
crystallographic structural models, in
 silico screening vs., 443
current applications, 560–563
drug discovery and development and,
 458–460, 560–563
flow cytometry and, 195–196
metal-enhanced fluorescence:
 acronyms and symbols, 663
 drug discovery applications, 660–662
 future research issues, 662–663
 metal-fluorophore interactions, 604–609
 metal substrates, 609–630
 anisotropic silver nanostructures,
 615–618
 electrochemically deposited silver,
 619–622
 electroplated silver, 622
 laser-deposited silver, 618–619
 roughened silver electrodes, 622–624
 silver colloid films, 610–615

silver fractal-like structures, glass substrates, 624–630
silver island films, 610
plate well sensing applications, 630–660
  directional emission, 654–660
  enhanced DNA labels, 637–639
  enhanced DNA/RNA detection, 630–637
  enhanced energy transfer, silver surfaces, 645–649
  multiphoton excitation and metallic nanoparticles, 643–645
  overlabeled proteins, ultrabright probes as, 639–642
  ratiometric surface sensing, 649–653
  solution-based assays, 653–654
multidomain clustering and, 107
natural product development:
  activity screening, 46–47
  future applications, 60–62
  research background, 15–22
platform evolution and integration, 563–570
  cellular assays, 566–567
  drug discovery process, 567–569
  instrumentation and screening formats, 563–565
  miniaturization, 565–566
  quality control, 569–570
systems biology, 131–142
  DNA sequence variations, 132
  gene expression profiling, 132–135
    clustering approaches, 138–139
    microarrays, 132–133
    serial analysis, 133–135
  metabolomics and metabonomics, 137–138
  microarray data, advanced computational analysis, 139–142
    biological networks, 141–142
    linear pathways to complex networks, 139–140
    literature sources, 140–141

proteomics, 135–137
  protein arrays, 136–137
  yeast two hybrid system, 136
targeted cancer therapy, 1346
  antimimotics, 1363
  tumor vasculature, 1356–1357
  virtual screening vs., 89, 114–116
His6-tagged FPR, high-throughput flow cytometry, drug discovery targets, 203–207
Histone deacetylase inhibitors, targeted cancer therapy, 1358–1359
Histopathological assessment, colitis severity, 1018
Hit selection criteria, automated high-throughput screening, 585–586
HMR 1402 cardioselective ATP-potassium channel antagonists, 921–923
HMR 1883 cardioselective ATP-potassium channel antagonists, 919–921
HMR 1098 sodium salt, sulfonylurea agents, 919–912
Home laboratory systems, x-ray diffraction data collection, 402–403
Homeostasis, molecular drug actions, 37–39
Homogeneous assays:
  DeathTRAK cycotoxicity assay:
    basic properties, 701–702
    biomass measurement (competitive comparison), 704–711
      ATP release assays, 704
      dual-mode cytotoxicity/proliferation, 705–712
      gram-negative bacteria, 707–709
      gram-positive bacteria, 709–711
      metabolic assays, 704–705
    cytotoxic process measurements, 702–704
    drug discovery applications, 702–711
    high-throughput screening formats, 565

INDEX **1417**

cell-based assays, 566–567
homogeneous time-resolved fluorescence, 594–595
Homogeneous time-resolved fluorescence (HTRF), high-throughput screening bioassays, 594–595
Homologous protein families, macromolecule crystallographic models, 422–425
Hormone replacement therapy, estrogen receptor agonists, 1071
Hormones. *See* Endocrine hormones; specific hormones
Horseradish peroxidase (HRP), coupled enzyme assays, 692–693
Host yeast strains, protein-protein interactions, yeast two-hybrid systems, 494–496
  new developments in, 501–502
HPRD database, contents and location, 129
HTB cell line, platelet-derived growth factor analysis, fluorescence resonance energy transfer, 78–80
Human Cyc database, contents and location, 129
Human cytomegalovirus (HCMV):
  characteristics, 1114
  respiratory viral infection therapy, 1120–1121
Human epidermal growth factor receptors:
  cancer cell proteomics, molecular diagnosis, 74–75
  receptor tyrosine kinases, 1200
Human gastric cancer cells, animal implantation model, 1019
Human genome project, natural product research and development, 15–22
Human immunodeficiency virus (HIV):
  antiviral therapy, 1120–1121
    acyclic nucleoside phosphonates, 1158–1161
    evolution of, 1135–1140
    protease inhibitors, 1163–1169

  reverse transcriptase inhibitors, 1155–1158
  RNA synthesis inhibition, 1161–1163
  virus adsorption inhibitors, 1140–1141
  virus-cell fusion inhibitors, 1142–1147
biopolymer solid-phase synthesis, 969–970
FDA guidelines concerning, 343
life cycle, 1136–1137
natural product development for, 16–17, 20–22
  information sources, 27
  marine sources, 32–33
  plant sources, 29–30
phosphate prodrug development, 744–746
protease inhibitors:
  antiviral therapies, 1163–1169
  structure-based drug design, 438–440
protein therapeutics, 442
Human metapneumovirus (HMPV), characteristics of, 1111
Human serum albumin (HSA):
antistructure drug discovery, 442
metal-enhanced fluorescence:
  anisotropic silver nanostructures, 616–618
  electrochemically-deposited silver, 619–622
  electroplating silver, 622
  laser-deposited silver, 618–619
  roughened silver electrodes, 622–625
  silver colloid films, 611–615
  silver fractal-like structures, 626–630
  silver island films, 610
Human trabecular meshwork (HTM) cells, gaucoma drug development, 148–151
Hybridization assays, metal-enhanced fluorescence (MEF), enhanced DNA/RNA detection, 630–637
Hydrochloric acid, acute/chronic duodenal ulcers, 1016

Hydrogen atoms, protein structural models, 428–432
Hydrogen bonding:
dynamic combinatorial chemistry, 988–992
protein structural models validation, 433
Hydrophilic corticosteroid prodrugs, water-solubility, 749
Hydrostatic pressure (HP) mechanisms, gaucoma drug development, 147–151
Hydroxyl groups, bioreversible prodrug structures, 747–750
lactones, 750
lipophilic prodrugs, 749–750
steric hindrance, promoieties for, 750
water-soluble prodrugs, 748–749
Hydroxypropoxymethylphosphonate adenine derivative (HPMPA), acyclic nucleoside phosphonates, 1160–1161
HyperCyt system:
content volume management, 193–194, 207
FPR case study using, 217
microfluidic mixing:
cell sorting, 208–209
soluble compound analysis, 207–208
sample delivery, 192–193
software development, 209–210
technology of, 196
Hyperthyroidism, thyroid hormone analogs and receptors, 1055
Hypoglycemia, sulfonylurea agents, 1080
Hypotension, corticotropin-releasing factor and related peptides, 1041
Hypothalamic hormones:
analogs and antagonists, 1039–1046
corticotropin-releasing hormone, 1040–1042
gonadotrophin-releasing hormone, 1043–1045
growth-hormone-releasing hormone, 1043

somatostatin, 1045–1046
thyrotropin-releasing hormone, 1039–1040
antiobesity agents, 1084–1085
Hypoxia-inducible factor-1 (HIF-1), targeted cancer therapy, angiogenesis inhibition, 1355–1356
Hypoxia-selective prodrugs, tumor targeting, 770–774

IC_{50} curve analysis:
biomolecular screening assays, affinity models and parameter estimates, 669
high-throughput screening, automated procedures, 585
protein kinase inhibition, ATP binding site inhibitors, 1211–1212
ICH guidelines, in vivo toxicology assays, 480–481
Iconix Drug Matrix, gene expression profiles, 157–159
Identity, RNA production, 1298–1299
If-Then-Else task, Visual Basic Editor, 296–297
Imaging techniques. See specific techniques
Imidazole-4-acetic acid (IAA), $GABA_A$ receptor agonists and antagonists, 809–811
Immediate interactions algorithm, gaucoma drug development, 148–151
Immobilized pH gradients (IPGs), protein-protein interactions, spectrometric reagents and actions, 529
Immunoglobulin M (IgM), viral respiratory infection diagnosis, 1119–1120
Immunostimulation therapy, double-stranded and stablilized RNA, 1288–1290
definitions, 1288–1289
hemispherx Biopharma, 1290
preclinical/clinical utilization, 1289–1290

INDEX

Immunosuppressive agents:
  insect sources, 30–31
  natural product development, 18–19
IMP dehydrogenase inhibitors, antiviral therapy, 1171–1173
Incubation media, multiplexing cell-based assays, 357–359
Indirect response, molecular drug actions, 37–39
Indirubin, CDK inhibition, 1229
Indocyanine green (ICG), metal-enhanced fluorescence:
  anisotropic silver nanostructures, 616–618
  electrochemically-deposited silver, 619–622
  electroplating silver, 622
  laser-deposited silver, 618–619
  metal substrates, 610
  roughened silver electrodes, 622–625
  silver colloid films, 610–615
  silver island films, 610
Infectious disease:
  natural product development for, 20–22
  respiratory viruses:
    adenoviridae, 1113
    animal models, 1126–1127
    antiviral therapy:
      current treatments, 1120–1121
      development strategies, 1122–1126
      infections, 1121–11222
    bunyaviridae, 1112–1113
    clinical syndromes, 1114–1118
      acute laryngotracheobronchitis, 1116
      bronchitis, 1117
      common cold, 1114, 1116
      laryngitis, 1116
      pharyngitis, 1116
      pneumonia, 1117–1118
      tracheitis and tracheobronchitis, 1116
    coronaviridae, 1109
    diagnosis, 1118–1120
    future research issues, 1127–1128
    herpesviridae, 1113–1114
    infection control, 1107–1108

orthomyxoviridae, 1111–1112
    paramyxoviridae, 1109–1111
    piconaviridae, 1108
    public health and economic impact, 1107
    seasonal patterns, 1106
    specific pathogens, 1108–1114
    structure and classification, 1106
Inflammation-associated models, colorectal cancer, 1020
Inflammatory bowel disease (IBD), animal models, 1016–1019
  colitis severity assessment, 1018–1019
  dextran sulfate sodium, 1017–1018
  dinitrobenzene sulfonic acid/ethanol, 1017
  trinitrobenzene sulfonic acid/ethanol, 1017
Inflammatory mediators:
  dissociated glucocorticoids, 1062–1063
  sepsis pathophysiology, 939
Influenza virus:
  neuraminidase inhibitors, 1169–1171
  orthomyxoviridae, 1111–1112
  seasonal patterns, 1106
Informatics:
  high-throughput flow cytometry, 194–195
  high-throughput screening, accuracy and precision measurements, 580–581
Information retrieval services, natural product development and discovery, 24
Infrared coded resins, split-pool synthesis, 979–980
Ingenuity pathways analysis, biological pathway analysis, 143
Inhibitors:
  antiviral therapy:
    S-adenosyl-L-homocysteine hydrolase inhibitors, 1173–1177
    DNA/RNA synthesis inhibitors, 1147–1163
      acyclic nucleoside phosphonates, 1158–1161
      DNA polymerase inhibitorsr, 1147–1152

1420 INDEX

reverse transcriptase inhibitors, 1155–1158
RNA synthesis processes, 1161–116
IMP dehydrogenase inhibitors, 1171–1173
neuraminidase inhibitors, 1169–1171
research background, 1135–1140
viral adsorption inhibitors, 1140–1141
viral protease inhibitors, 1163–1169
virus-cell fusion inhibitors, 1141–1147
protein kinases:
allosteric inhibitors, 1213
alternative design strategies, 1214–1215
ATP binding site inhibitors, 1210–1212
potency, 1212
selectivity, 1210–1212
basic principles, 1205
bisubstrate analog inhibitors, 1213–1214
clinical trials, 1206–1209, 1215–1232
future research issues, 1232
serine/threonine kinases, 1225–1232
SH2 domain inhibitors, 1213
substrate binding site inhibitors, 1212–1213
tyrosine kinases:
nonreceptor tyrosine kinases, 1221–1225
receptor tyrosine kinases, 1216–1221
protein-protein interactions:
microarrays and small molecules, 541–542
phage display, 541
in vivo genetic selection systems, 540–541
yeast two-hybrid system variants, 536–540
peptide aptamers, 539–540
reverse two-hybrid systems, 537–538
split-hybrid system, 538–539

targeted cancer therapy:
angiogenesis, 1353–1356
histone deacetylase, 1358–1359
kinesin antimimotics, 1361
proteosome inhibition, 1363–1364
small-molecule inhibitors, 1365–1367
telomerase, 1359–1360
Inhibitory neurotransmitters, GABA receptor ligand inhibition, 802–829
biosynthesis and metabolism, 803–804
classification and therapeutic targets, 802–803
$GABA_A$ receptor complex, 807–823
agonists and antagonists, 808–811
benzodiazepine-allosteric binding sites, 814–821
pharmacophore models, 821–823
receptor and pharmacophore models, 811–814
$GABA_B$ receptor complex, 823–827
agonists, 825
allosteric modulators, 827
antagonists, 825–827
$GABA_C$ receptor agonists and antagonists, 828–829
metabolism inhibitors, 803–804
reuptake mechanisms, 804–806
Insects, drug development using, 30–31
In silico screening, crystallographic structural models, 442–443
In situ assembly, macromolecular ligands, 997–998
In situ click chemistry, combinatorial techniques, 992–997
In situ target-induced selection and amplification, dynamic combinatorial chemistry, 987–992
Instrumentation, high-throughput screening platforms, 563–565
Insulin:
diabetes mellitus therapy, 1073–1075
mechanism of action, 1075–1076
prodrug development, prolonged drug action technologies, 770

INDEX   1421

Insulin-like growth factors (IGFs):
    growth hormone mediation, 1047–1048
    targeted cancer therapy, signal transduction, 1349
Insulin receptor substrate (IRS) proteins, mechanism of action, 1076
Integrase, RNA synthesis inhibition, 1161–1163
Intellectual property rights, natural product development, 61–62
Intensity decays, metal-enhanced fluorescence, metal-fluorophore interactions, 607–609
Interaction arrays, protein-protein interactions, 522–523
Interaction suppression, protein-protein interactions, yeast two-hybrid systems, 506–507
Interferon-α:
    respiratory viral infections, 1121–1122
    ribivarin and, 1121
Interlaboratory comparisons, potency precision estimates, 676–678
Interleukin (IL)-1:
    respiratory viral infections, 1121–1122
    sepsis pathophysiology, 939
Internal ribosome entry sequence (IRES), mRNA protein production, 1287–1288
International Cooperative Biodiversity Group (ICBG) program, natural product development, 62
Internet sites, natural product development and discovery, 23, 58
Intracellular domains, EGFR receptor inhibitor targeting, 1217–1221
Intracellular phosphorylation:
    acyclic nucleoside phosphonates, 1161
    antiviral therapy, DNA polymerase inhibitors, 1148–1155

Intracellular targeting, magnetic resonance imaging, 1325–1328
    long-term cell labeling contrast agents, 1326–1327
    specific contrast agents, 1327–1328
Intraocular pressure (IOP), glaucoma drug therapy and, 146–151
Inverse agonists, general principles, 35
Investigational New Drug (IND):
    drug development phase for, 1–3
    firefly luciferase, ATP/luciferase reactions, phosphodiesterases, 699–700
    natural product development, 56–58
In vitro techniques:
    drug delivery assays, 472–474
    drug development and, 1–2
    drug metabolism assays, 477–478
    mRNA protein production, 1287–1288
    natural product development, P-450 studies, 55–58
    pharmacology assays, 460–462
        bioanalytical chemistry assays and, 475–476
    potency principles, 36
    protein-protein interactions, display technologies, 516
    toxicology assays, 479–480
In vivo techniques:
    drug metabolism assays, 477–478
    genetic selection systems, protein-protein interaction inhibition, 540–541
    magnetic resonance imaging, tumor cell surface receptors, 1320–1325
        Gd-based contrast agents, 1320–1321
        iron-oxide-based contrast agent, 1321–1323
        multistep targeting and prelabeling, 1323–1324
        signal amplification, 1324–1325
    molecular imaging MRI, 1331–1332
    pharmacology assays, 464–465
    toxicology assays, 480–481

Iodide transport:
  thyroid hormone analogs and receptors, 1055
  thyroid hormones, 1053–1057
Ion channels:
  high-throughput screening cellular assays, 597–598
  ionotropic glutamate receptors, 851–853
  molecular drug actions, 39
Ionizable/polar promoiety, prodrug derivatization, aqueous solubility improvement, 766–769
Ionotropic amino acid receptors:
  glutamate receptor ligands, central nervous system ischemia, 800
  research background, 798–799
Ionotropic glutamate receptors, structural properties, 846–853
Iron-oxide contrast agents
  magnetic resonance imaging (MRI), mechanisms and enhancement, 1313
    intracellular targeting, long-term labeling, 1326–1328
    signal amplification, 1325
    tumor cell surface receptors, 1319–1325
      in vivo MR molecular imaging, 1321–1323
Irori Corporation scanning device, split-pool synthesis resins, 980–981
Isoguvacine, $GABA_A$ receptor agonists and antagonists, 810–811
Isolation techniques, natural product development, 47–51
Isomorphous replacement, x-ray crystal structure determination, 406–412
Iterative single-wavelength anomalous scattering (ISAS), x-ray diffraction, protein structures, 416
ITO-coated glass electrodes, metal-enhanced fluorescence, electroplated silver, 622, 624

Jablonski diagrams, metal-fluorophore interactions, 604–605
Jarvis-Patrick clustering, virtual screening, 107
Jod-Basedow effect, thyroid hormone analogs and receptors, 1055

Kainic acid (KA):
  glutamatic acid receptor classification, 829–830
    AMPA receptor agonists, 838–840
    receptor agonists and antagonists, 843–846
Kazlausku destruction/amplification, dynamic combinatorial chemistry, 991–992
KB cytotoxicity assay, taxol research, 58–59
KEGG database, contents and location, 129
Ketoconazole, corticosteroid synthesis inhibition, 1064
Kidney targeting, prodrug development, 778–779
Kinases, coupled luminescent assays, firefly luciferase, 698–699
Kinesin inhibitors, antimimotic targeted cancer therapy, 1361
Knowledge-based methods, virtual screening, docking and scoring software, 212
Kohonen maps, gene expression data, systems biology, 138–139
K-value clustering, gene expression data, systems biology, 138–139

Laboratory data management:
  Excel software:
    clipboard function, 304–306
    opening and saving data files, 306–307
    research background, 287–289
    saving text and CSV files, 307
    text import wizard, 306–307

INDEX **1423**

Visual Basic for Applications, 290–303
  automated report generation, 327–334
    heat map display, 331–334
    worksheet formatting, 328–330
  data file reading and writing, 307–319
    file access types, 309
    file selection, 308–309
    opening and closing sequential access files, 309–311
    sequential access files, 311–315
  event-driven programming, 297–299
  external database imports, 319–327
    dabatase characteristics, 319–320
    data imports, 322–326
    ODBC access issues, 320–322
  fundamental principles, 291–303
  language primer, 293–297
  macro recording, 290–291
  object entities, 299–303
  Oracle ODBC connection, 327
  string data manipulation and parsing, 315–319
  Visual Basic Editor elements, 292–293
x-ray diffraction data, home laboratories, 402–403
lac repressor, protein-protein interactions:
  bacterial display systems, 515
  genetic selection systems and inhibition of, 540–541
Lactate dehydrogenase, severe acute respiratory syndrome, 0
β-lactam antibiotics, prodrug development:
  bioconversion mechanisms, 743–744
  oral drug delivery, 756
Lactones, hydroxyl prodrugs, 750
λ phage display, protein-protein interactions, 514–515
  genetic selection systems and inhibition of, 540–541

Lamivudine, reverse transcriptase inhibitors, 1156–1158
Lanreotide, somatostatin receptors, 1046
Large neutral amino acids transport (LAT), prodrug structures, brain targeting properties, 777–778
Laryngitis, respiratory viruses and, 1116
Laser-deposited silver, metal-enhanced fluorescence, 618–619
Latent hits, virtual screening limitations, 97
Lead discovery, high-throughput screening procedures, 562–563
Lead optimization:
  bioanalytical chemistry assays, 474–476
  drug delivery assays, 472–474
  drug discovery and development and, 458–460
  high-throughput screening *vs.*, compound flow, 567–568
  HIV protease inhibitors, 1166–1169
  protein structural models:
    crystallographic screening, 443–444
    structure-based drug design, 436–438
  in vitro toxicology assays, 479–480
LeadScope program, virtual screening, two-dimensional substructure searching, 99
Least squares refinement, protein structural models, 425–432
Leave-*N*-out (LNO) cross validation, pattern recognition models, 238–239
Leave-one-out (LOO) cross validation:
  comparative structural connectivity spectral analysis, aryl hydrocarbon receptor binding, 271
  comparative structurally assigned spectra analysis, aromatase enzyme model, 240, 242–244
  pattern recognition models, 238–239

Legal issues, RNA production,
1290–1300
  manufacturing protocols, 1290–1294
    long RNA, 1291–1294
    oligonucleotides, 1290–1291
  regulatory aspects, 1294–1300
    identity, 1298–1299
      messenger RNA, 1299–1300
    mRNA, 1297–1298
    oligonucleosides, 1296–1297
    oligonucleotides, 1298–1299
    purity, 1296
Lentinus edodes, polysaccharide extracts, 1023
Lethal dose (LD_{50}):
  dual mode cycotoxicity/proliferation assays, DeathTRAK cycotoxicity assay, 705–712
  in vivo toxicology assays, 480–481
Leukemia, natural product development for, 17–18
Leukocyte antigen-related phosphatase (LARP), PhosTRAK phosphatase assay, 715–716
Levodopa, prodrug structures, enhanced brain delivery, 766
Lex-A protein, protein-protein interactions:
  library selection, 501
  three-hybrid system, 506
  yeast two-hybrid systems, target and bail vectors, 490–494
Library design:
  combinatorial chemistry:
    design issues, 1001–1004
    on-bead screening, 974
    parallel synthesis, 971–972
    split-pool synthesis, 973–974
      binary encoding, 977–979
      infrared coded resins, 979–980
      radio-frequency tagging, 980–981
      recursive deconvolution, 976
      tagging methods, 977
  protein-protein interactions:
    screening approach, 508
    yeast two-hybrid systems, 495–501
  targeted cancer therapy, small-molecule inhibitors, synthetic chemicallibraries, 1366–1367

Licorice-based medicinal agents, gastrointestinal disease, animal models, 1025–1027
Ligand-based virtual screening, basic principles, 90
Ligand binding:
  protein crystallization, 386–388
    limitations of, 448–449
  receptor mechanisms, 38
  structure-based drug design, 436–438
Ligand-gated ion channels, molecular drug actions, 39
Ligand-growing algorithms, protein structural models, 433–434
Ligand-receptor binding, high-throughput screening bioassays, 592–594
Likelihood estimation, protein structural models, 426–432
Limit of detection (LOD):
  bioanalytical chemistry assays, 475–476
  cytotoxicity measurements, coupled luminescent assays, 700–701
  DeathTRAK cycotoxicity assay, 719
  PhosTRAK phosphatase assay, 719
Limits of agreement (LsA):
  acceptance criteria properties, 683–687
  estimated outcomes properties, 682–683
  potency precision estimates:
    acceptance criteria, 674
    replication-experiment studies, 672–673
Linear pathway analysis, microarray data, 139–140
Linear response, coupled luminescent assays, 718
Linear techniques, pattern recognition models, 238
Lipinski's rules of five, drug delivery assays, 472–474
Lipopeptidolactone (FR901469), reserach and development, 16
Lipophilic characteristics:
  drug delivery assays, 472–474
  GABA receptor ligands, reuptake inhibition, 805–806

glucocorticoid receptors, 1060–1063
prodrug structures:
    brain targeting prodrugs, 775–776
    hydroxyl prodrugs, 749–750
    nasal drug delivery, 764–766
    oral drug delivery, 753–754
    phosphonate group prodrugs, 747
    prolonged duration of action designs, 769–770
Lipoxygenase (LOX) inhibition, polyphenol medicinal compounds, 1026–1027
Liquid chromatography-mass spectrometry (LC-MS):
    metabolomic analysis, 137–138
    protein-protein interactions, 530
Liquid handling protocols, high-throughput screening automation, 578–580
Literature sources:
    microarray data, 140–141
    natural product development, 22–28
Long RNA, legal manufacturing protocols, 1291–1294
Long terminal repeat (LTR) promoter, RNA synthesis inhibition, 1161–1163
Long-term labeling, MRI cell trafficking, intracellular targeting, 1326–1327
Low-molecular-weight contrast agents, MRI tumor targeting, vasculature and vascular targets, 1315
Low-molecular-weight natural products, activity screening, 45–47
Luminance properties, coupled luminescent assays, 691
    fluorescence assays *vs.*, 693
Luminescence. *See* specific luminescence techniques
Luteinizing hormone (LH):
    drug action and applications, 1049
    gonadotrophin-releasing hormone regulation, 1043–1045
Lymphopenia, severe acute respiratory syndrome, 0
Lymphostin, research and development, 19

Lytic formulations, coupled luminescent assays, 725

MACCS keys, virtual screening, two-dimensional molecular fingerprints, 102–103
Macrobeads, split-pool synthesis, 984–985
Macroimagers, cell-based assays, 354–356
Macromolecules:
    combinatorial chemistry, thiol-based in situ ligand assembly, 997–998
    crystallographic models, 420–422
    de novo model construction, 422–425
    formulation development assays, 470–471
    prodrug development, 784–785
    x-ray diffraction, 390–392
Macro recording:
    heat map display, 331–334
    Visual Basic for Applications (Excel software), 290–291
Macugen, RNA mimetics, aptamer function, 1271–1273
Magnetic resonance imaging (MRI):
    cancer research, 1310
    contrast agents and enhancement, 1311–1313
        relaxation properties, 1313–1314
        T1 contrast agents, 1311–1312
        T2 contrast agents, 1312–1313
    intracellular targets, 1325–1328
        long-term cell labeling contrast agents, 1326–1327
        specific contrast agents, 1327–1328
    limitations and future research, 1331–1333
    molecular imaging, activated contrast agents, 1328–1329
        CEST/PARACEST contrast agents, 1329
        enzymatic activation, 1328
    molecular MRI and MR spectroscopy, 1329–1331
        reporter molecules, CSI imaging, 1330–1331

tumor cell surface receptors, 1318–1325
  isolated cell molecular imaging, 1318–1320
  in vivo molecular imaging, 1320–1325
    Gd-based contrast agents, 1320–1321
    iron-oxide-based contrast agent, 1321–1323
    multistep targeting and prelabeling, 1323–1324
    signal amplification, 1324–1325
  tumor vasculature and vascular targets, 1313, 1315–1318
    high-molecular-weight contrast agents, 1316
    low-molecular-weight contrast agents, 1315
    moleular imaging, 1316–1318
Magnetic resonance spectroscopy (MRS):
  molecular MRI and, 1329–1331
  tumor therapy vs. MRI, 1310
Magnolia officinalis, research and development, 17
Mammalian AdoHcy hydrolase, S-adenosylhomocysteine hydrolase inhibitors, 1174–1177
Mammalian peptide display system, protein-protein interactions, 515–516
Mammalian target of rapamycin (mTOR), targeted cancer therapy, signal transduction, 1350–1351
Mammalian two-hybrid systems, protein-protein interactions, 506
N-mannich bases, bioreversible prodrug structures, 751
Mapacalcine, marine sources, 32–33
Marine organisms, as drug sources, 31–33
Mass spectrometry (MS):
  analytical chemistry assays, 463–464
  protein-protein interactions, 528–533
    affinity tagging, 531–533
    LC-MS approaches, 530
    protein detection and image analysis, 530
    protein identification, 530–531
    reagents, 528–533
    two-dimensional polyacrylamid gel electrophoresis, 528–529
Matrix approach, protein linkage maps, protein-protein interactions, yeast two-hybrid systems, 507–509
Maximum likelihood refinement, protein structural models, 425–432
Maximum tolerated dose (MTD), toxicology assays, 478–481
M13 bacteriophage, protein-protein interactions, peptide display reagents, 517
Mean ratio (MR):
  acceptance criteria properties, 683–687
  estimated outcomes, 680–683
  potency precision estimates, replication-experiment studies, 672–673
Measles, paramyxovirus, 1110–1111
Mechanism-based drug development:
  disease-seeking drugs, 8
  natural product reserach, 60–62
Mechanistic enzymology, high-throughput screening, bioassay design, 591–592
Median partitioning, virtual screening, 109–110
Mediator-specific anti-inflammatory therapies, sepsis management:
  clinical trials, 939–940
  efficacy measurements, 940–943
  future research, 948–949
  glucocorticoids vs., 945–946
  inflammatory mediators, 939
  research background, 938–939
Medicinal chemistry-based drug development, defined, 8
MEDLINE abstracts:
  biological pathway analysis, 145–146
  microarray data, 140–141
MedScan system, microarray data, 140–141

Meglitinide analogs, diabetes therapy, 1080–1081
Melting point reduction, prodrug derivatization, aqueous solubility improvement, 766–769
Membrane-anchoring (MA) domain, protein-protein interactions, Sos recruitment system (SRS), 502–503
Membrane permeation, prodrug development, peptide formulations, 782–783
Membrane proteins, crystallization, 381–382
Membrane transporters, ophthalmic drug delivery, prodrug drug development, 761–762
MERIX BIosciences, mRNA protein production, 1288
Messenger RNA (mRNA):
 gaucoma drug development, 147–151
 gene expression profiling, systems biology, 133–134
 hammerhead ribozymes, 1276–1277
 identity techniques, 1299–1300
 protein production, 1283–1288
  definitions, 1283–1284
  MERIX Biosciences, 1288
  mimetics, 1285
  preclinical/clinical utilization, 1285–1288
  purity controls, 1297–1298
 S-adenosylhomocysteine hydrolase inhibitors, 1173–1177
 small inhibitory RNA (siRNA) synthesis, 1279–1280
Metabolic assays, DeathTRAK cycotoxicity assay, biomass measurement, 704–705
Metabolic pathways:
 diabetes mellitus therapy, 1073
 GABA biosynthesis, 803–804
 systems biology and, 126–127
  microarray data, 139–140
Metabolite isolation, drug metabolism assays, 478
Metabolomics, systems biology applications, 137–138
Metabonomics, systems biology applications, 137–138
Metabotropic glutamate receptor ligands, excitatory neurotransmitters, 851, 853–868
 agonists, 853–857
  backbone chain length extension, 855–857
  bioisosteres, 857
  conformationally constraint analogs, 853–855
 allosteric modulation, 860–862
  noncompetitivve antagonists, 860–862
  positive modulators, 862
 competitive antagonists, 857–860
  α-substitutions, 859–860
  phenylglycines, 857–859
 noncompetitive antagonists, 860–862
 putative drug targeting, 862–863
 structural properties, 863–868
MetaCore:
 biological pathway analysis, 143–144
 systems biology, future applications, 161–164
Metal-enhanced fluorescence (MEF), high-throughput screening:
 acronyms and symbols, 663
 drug discovery applications, 660–662
 future research issues, 662–663
 metal-fluorophore interactions, 604–609
 metal substrates, 609–630
  anisotropic silver nanostructures, 615–618
  electrochemically deposited silver, 619–622
  electroplated silver, 622
  laser-deposited silver, 618–619
  roughened silver electrodes, 622–624
  silver colloid films, 610–615
  silver fractal-like structures, glass substrates, 624–630
  silver island films, 610
 plate well sensing applications, 630–660
  directional emission, 654–660
  enhanced DNA labels, 637–639

enhanced DNA/RNA detection, 630–637
enhanced energy transfer, silver surfaces, 645–649
multiphoton excitation and metallic nanoparticles, 643–645
overlabeled proteins, ultrabright probes as, 639–642
ratiometric surface sensing, 649–653
solution-based assays, 653–654
Metal-fluorophore interactions, metal-enhanced fluorescence, 604–609
Metallic nanoparticles, metal-enhanced fluorescence, multiphonon excitation, 643–645
Metallic spheroids, metal-enhanced fluorescence, anisotropic silver nanostructures, 615–618
Metallic surfaces, metal-fluorophore interactions, 604–609
Metal substrates, metal-enhanced fluorescence, 609–630
  anisotropic silver nanostructures, 615–618
  electrochemically deposited silver, 619–622
  electroplated silver, 622
  laser-deposited silver, 618–619
  roughened silver electrodes, 622–624
  silver colloid films, 610–615
  silver fractal-like structures, glass substrates, 624–630
  silver island films, 610
Metaregression analysis, mediator-specific anti-inflammatory agents, efficacy assesment, 941–943
Metformin, diabetes therapy, 1077
N-Methyl-N'-nitro-N-nitrosoguanidine (MNNG):
  colorectal cancer models, 1020–1021
  gastrointestinal cancer cell models, 1020
N-Methyl-N-nitrosourea (MNU), gastrointestinal cancer cell models, 1020

Micafungin, reserach and development, 16
Microarrays:
  photolithographic synthesis, 981–983
  protein microarrays, protein-protein interactions, 517–528
    antibody microarrays, 519, 522
    detection methods, 527–528
    functional protein, 519
    high-throughput protein production, 523–524
    interaction arrays, 522–523
    peptide microarrays, 522
    ProteinChip technology, 518
    protein delivery systems, 526–527
    reagents, 523–528
    small-molecule inhibition, 541–542
    supports, 524–526
    surface plasmon resonance arrays, 522
  protein-protein interactions, 517–528
    antibody microarrays, 519, 522
    detection methods, 527–528
    functional protein, 519
    high-throughput protein production, 523–524
    interaction arrays, 522–523
    peptide microarrays, 522
    ProteinChip technology, 518
    protein delivery systems, 526–527
    reagents, 523–528
    supports, 524–526
    surface plasmon resonance arrays, 522
  systems biology, 124–125
    ADME/Tox drug development, 158–159
    advanced computational analysis, 139–142
      biological networks, 141–142
      linear pathways to complex networks, 139–140
      literature sources, 140–141
    breast cancer drug development, 153–156
    gaucoma drug development, 148–152
    gene expression profiling, 132–133

Microbes, natural product drug
  development, 24, 29
  information sources, 29
Microfluidic mixing, high-throughput
    flow cytometry:
  HyperCyt cell sorting, 208–209
  sample carryover, 209
  soluble compounds, 207–208
Microimaging technologies, cell-based
  assays, 355–356
Microorganisms, natural product drug
  development, 27, 29
MicroRNAs (miRNA), cell function,
  1263
Microsoft Query application, Excel
    Visual Basic for
    Applications, data import,
    322–326
Microtubule-targeting agents,
    antimimotic targeted cancer
    therapy, 1360–1361
Microvessel density (MVD), magnetic
    resonance imaging (MRI),
    mechanisms and
    enhancement, vasculature
    and vascular targets, 1315
Miller indices:
  x-ray crystal structure
      determination:
    anomalous dispersion, 412–416
    multiple isomorphous replacement,
      407–412
    structure factor, 405–406
    x-ray diffraction data collection, 400–
      402
  x-ray diffraction patterns:
    Bragg's law, 397
    macromolecule crystals, 392
Mimetics:
  mRNA function, protein production,
    1285
  RNA-based therapies:
    aptamer function, 1270–1271
    natural and modified RNA, 1266
    small inhibitory RNA (siRNA)
      synthesis, 1280–1281
    synthetic RNA, hammerhead
      ribozyme function, 1274–
      1276

Mineralocorticoids, characteristics and
    therapeutic applications,
    1064–1065
Miniaturization:
  high-throughput screening platforms,
    565–566
  photolithographic synthesis, 981–
    983
  split-pool synthesis, radio-frequency
    tagging, 980–982
Minifingerprint techniques, virtual
  screening, 103
Minimum significant ratio (MSR):
  acceptance criteria properties, 683–
    687
  concentration-response assays, 668–
    669
  potency precision estimates:
    acceptance criteria, 674
    fou-parameter logistic (4PL)
      model, 674–676
    model parameter and outcome
      estimates, 673–674
    replication-experiment studies, 672–
      673
  statistical properties:
    data-generating model, 678–679
    estimated outcomes, 679–683
MINT database, protein-protein
    interactions, 533–536
MIPS database, microarray data, 141
Mithramycin, bone metabolism, 1059
Mitogen-activated protein kinases
    (MAPK):
  inhibitors, 1225–1232
    ERK compounds, 1228–1229
    JNK compounds, 1228
    P38 MAP compounds, 1226–
      1228
  therapeutic targeting, 1202–1203
Mitomycin C, hypoxia-selective
    prodrugs, tumor targeting,
    773–774
Mitotane, corticosteroid synthesis
  inhibition, 1064
Mixing techniques:
  CellCard system, 360
  high-throughput flow cytometry,
    192–193

Modeling pathways, systems biology, 125–127
Model parameters:
comparative structural connectivity spectral analysis:
aryl hydrocarbon receptor binding, 267–271
corticosterone binding, 258–259
protein structural models, 431–432
replication-experiment studies, 673–674
Modulatory agents, AMPA receptors, 842–843
Molecular aspects of drug action, 37–42
enzyme receptors, 40–41
G-protein-coupled systems, 40
ion channels, 39
natural product activity screening, 46–47
receptors, 37–39
target selection, 41–42
transcription factors, 41
Molecular beacon aptamer (MBA), platelet-derived growth factor analysis, fluorescence resonance energy transfer, 77–80
Molecular biology, combinatorial chemistry and, 963–965
Molecular chaperones, protein kinase inhibitor design, 1215
Molecular descriptors:
quantitative structure-activity relationships, 95–96
virtual screening, 91–92
Molecular diagnosis, cancer cell proteomics, 74–75
Molecular diversity:
high-throughput flow cytometry, drug discovery targets, 190–192
virtual screening, 94
Molecular dynamics, comparative structural connectivity spectral analysis:
configurational entropy, 276–278
four-dimensional connectivity matrix, 275–276

Molecular fingerprints, virtual screening, 102–106
profiling and scaling, 104–106
three-dimensional pharmacophores, 103–104
two-dimensional, 102–103
Molecular geometry, protein structural models, refinement of, 427–432
Molecular imaging, MRI tumor targeting:
activated contrast agents, 1328–1329
CEST/PARACEST contrast agents, 1329
enzymatic activation, 1328
tumor cell surface receptors, 1318–1325
tumor neovasculature, 1316–1318
Molecular networks, systems biology interpretations, 141–142
condition-specific molecular/functional networks, 163–164
Molecular prototypes, natural product research and development, 14–22
Molecular replacement (MR), x-ray diffraction, protein structures, 416–420
Molecular similarity concept, virtual screening, 92–94
quantification, 97–98
Molecular structural models, comparative structural connectivity spectral analysis, 245–271
aromatase enzyme model, 261–264
aryl hydrocarbon receptor binding model, 264–271
corticosterone binding:
artificial neural network model, 259–261
three-dimensional models, 256–259
two-dimensional model of estrogen receptor binding compounds, 247–256
Molecular weight (MW), analytical chemistry assays, 463–464

Monocrystalline iron oxide (MION), magnetic resonance imaging (MRI), mechanisms and enhancement:
  T2 contrast agents, 1313
  tumor cell surface receptors, 1319–1325
    in vivo MR molecular imaging, 1321–1323
Monovalent systems, protein-protein interactions, phage display technology, 512–515
Monte Carlo simulation:
  biological networks, 142
    future applications, 161–164
  estimated outcomes, statistical properties, 679–683
  macromolecule crystallographic models, 425
Morphological assessment, colitis severity, 1018
Mozenavir:
  protease inhibitor properties, 1166–1169
  structure-based drug design, 439–440
MPW database, contents and location, 129
Multidimensional nuclear magnetic resonance, quantitative spectrometric data-activity relationships, 233–236
Multidomain clustering, virtual screening, 107
Multiphonon excitation, metal-enhanced fluorescence, 643–645
Multiple cell line screening:
  CellCard technology, 356–357
    antimitotic cellular arrays case study, 362–363
    card preparation, 360
    compound addition and assay protocol, 360–361
    experiment design and plate layout, 359–360
    image analysis and data visualization, 361–362
    key features of, 370
    multiplexed antiproliferation screen, 367–370

multiplexing cell-based assays, 357–359
    96-well mixing and dispensing format, 360
    reading protocol, 361
    tissue culture, 360
    TUNEL assay case study, 363
      multiplex dose curve analysis, 363–366
  research background, 354–356
Multiple isomorphous replacement (MIR), x-ray crystal structure determination, 406, 410–412
  anomalous dispersion, 412–416
Multiple isomorphous replacement with anomalous dispersion (MIRAS), x-ray crystal structure determination, 415–416
Multiple linear regression (MLR) statistical techniques:
  comparative structural connectivity spectral analysis:
    aryl hydrocarbon receptor binding, 266–271
    ^{13}C-^{15}N heteronuclear connectivity matrix, antibiotic cephalosporins, 273–275
    estrogen receptor binding compounds, 251–256
  comparative structurally assigned spectra analysis:
    aromatase enzyme model, 240, 242–244
    aryl hydrocarbon receptor binding, 244–245
  quantitative spectrometrid data-activity relationships, 232
Multiple-wavelength anomalous dispersion (MAD), x-ray crystal structure determination, 415–416
Multiplexing techniques:
  cell-based assays:
    antimitotic compounds, 362–363
    antiproliferation screening, cell type selective compounds, 367–370
    basic principles, 357–359

high-throughput flow cytometry, drug discovery targets, diversity analysis, 191–192
TUNEL assay, dose curve analysis, 363–366
Multistep targeting techniques:
molecular MRI imaging, limitations, 1332–1333
tumor cell surface receptors, in vivo MR molecular imaging, 1323–1324
Muscarinic M_1 receptor GTPγ^{35} S assay:
biomolecular screening assay, concentration-response models and outcomes, 670–671
example data, 674–676
Muscimol, GABA$_A$ receptor complex:
agonists and antagonists, 808–811
pharmacophore models, 812–814
Mushrooms, polysaccharide agents from, 1022–1023
Mycena pura, research and development, 17–18
Mycophenolic acid, IMP dehydrogenase inhibitors, 1172–1173
Mycoplasma genitalium, systems biology and, 126–127
Mycoplasma pneumonia, bronchiolitis, 1117
Mycoses, natural product development for, 16–22
Myeloperoxidase (MPO), colitis severity assessment, 1018–1019
Myocardial ischemia, extracellular potassium effects, 911–914
Myocilin gene expression, gaucoma drug development, 148–151

Nabumetone, bioconversion, 741
N-acetylneuraminic acid (NANA), influenza virus inhibition, 1169–1171
NanoKan benzopyran library, radio-frequency tagging, 980–982
Nanoliter dispensing, high-throughput screening compounds, 576–578

Nanorods, metal-enhanced fluorescence, anisotropic silver nanostructures, 615–618
Nanowell structures, protein-protein interactions, protein microarray supports, 525–526
Naproxen, prodrug development, topical drug delivery, 757–759
Nasal drug delivery, prodrug development, 762–766
ehanced brain delivery, 766
lipophilic prodrugs, 764–766
peptide prodrugs, 766
transport-mediated absorption, 766
water-soluble prodrugs, 763–764
Nateglinide, diabetes therapy, 1081
Native American medicine, history of, 13–14
Natural killer cells, herbal terpenoid inhibition, 1027–1028
Natural products. *See also* Herbal medicines
activity screening, 43–47
defined, 12
development and discovery, 22–33, 55–59
environmental sources, 23–24, 29–33
insects, 30–31
marine organisms, 31–33
microbes, 24, 29
plants, 29–30
vertebrates, 31
historical background, 58–59
literature sources, 22–28
regulatory guidelines and nonclinical development, 55–58
future prospects, 60–63
history and background, 12–14
isolation and purification, 47–51
pharmacodynamics, 33–43
drug tolerance, 42–43
general principles, 34–37
molecular aspects, 37–42
enzyme receptors, 40–41
G-protein-coupled systems, 40
ion channels, 39
receptors, 37–39

INDEX

target selection, 41–42
transcription factors, 41
protein targets, 33–34
research and development update, 14–22
structure identification, 51–53
synthesis, 53–55
targeted cancer therapy, small-molecule inhibitors, 1365–1366
Natural Products Repository (NPR), natural product development, 62
NCBI database, serial analysis of gene expression depository, 134–135
Negative accuracy, screening for, 5–7
"Negative-signal" characteristics, coupled luminescent assays, firefly luciferase, 698–699
Nei Ching anthology, 12
Neighborhood behavior, molecular similarity quantification, 98
Neovasculature, MRI tumor targeting, molecular imaging, 1316–1318
Network analysis algorithms, future applications, 161–164
Network motifs, biological networks, 142
Neural networks, virtual screening, drug likeness studies, 114
Neuraminidase inhibitors:
  antiviral therapy, 1169–1171
  influenza virus, 1112
  oseltamivir, antiviral agents, 1124–1126
Neurohypophyseal hormones, drug action and therapeutic applications, 1049–1053
  oxytocin, 1052–1053
  vasopressin, 1049–1052
Neurological drugs, GABA and glutamate receptor ligands, 799
Neuropeptide Y, antiobesity agents, 1084–1085
Neuropsychiatric disorders, natural product development for, 17
Neurotoxins, insect sources, 30–31

Neurotransmitters:
  GABA receptor ligand inhibition, 802–829
    biosynthesis and metabolism, 803–804
    classification and therapeutic targets, 802–803
    $GABA_A$ receptor complex, 807–823
      agonists and antagonists, 808–811
      benzodiazepine-allosteric binding sites, 814–821
      pharmacophore models, 821–823
      receptor and pharmacophore models, 811–814
    $GABA_B$ receptor complex, 823–827
      agonists, 825
      allosteric modulators, 827
      antagonists, 825–827
    $GABA_C$ receptor agonists and antagonists, 828–829
    metabolism inhibitors, 803–804
    reuptake mechanisms, 804–806
  glutamate receptor ligands excitation, 829–868
    AMPA receptor ligands, 837–843
      agonists, 837–840
      competitive/noncompetitive antagonists, 840–842
      modulatory agents, 842–843
    ionotropic receptor structure, 846–852
    KA receptor agonists and antagonists, 843–846
    metabotropic receptor ligands, 851, 853–868
      agonists, 853–857
        backbone chain length extension, 855–857
        bioisosteres, 857
        conformationally constraint analogs, 853–855
      allosteric modulation, 860–862
      noncompetitivve antagonists, 860–862
      positive modulators, 862
      competitive antagonists, 857–860
        α-substitutions, 859–860
        phenylglycines, 857–859

noncompetitive antagonists, 860–862
  putative drug targeting, 862–863
  structural properties, 863–868
NMDA receptor ligands, 830–837
  competitive receptor antagonists, 832–833
  glycine co-agonist site, 835–836
  receptor agonists, 831–832
  subtype-selective ligands, therapeutic potential, 836–837
  uncompetitive/noncompetitive receptor antagonists, 833–835
  receptor classification, 829–830
Newbutonia vellutina, drug development, 30
New Drug Application (NDA), natural product development, 57–58
NewSheet coding, Excell object entities, 301–303
Nidula candida, research and development, 17–18
NIEHS database, ADME/Tox drug development, 158–159
Nilutamide, antiandrogen properties, 1068
96-well platform design:
  drug delivery assays, 473–474
  high-throughput screening, 564–565
  multiple cell line screening, 360
  protein-protein interactions, protein microarray delivery systems, 526–527
Nipecotic acid, GABA reuptake inhibition, 805–806
Niphatevirin, marine sources, 32–33
^{15}N-labeled protein, combinatorial chemistry, structure-activity relationship, 999–1001
2-Nitroimidazoles, hypoxia-selective prodrugs, tumor targeting, 773–774
N-methyl-D-aspartic acid (NMDA):
  glutamate receptor ligands, 830–837
  competitive receptor antagonists, 832–833
  glycine co-agonist site, 835–836
  receptor agonists, 831–832
  subtype-selective ligands, therapeutic potential, 836–837
  uncompetitive/noncompetitive receptor antagonists, 833–835
  glutamatic acid receptors, 829–830
Nonclinical development, natural products, 55–58
Noncompetitive receptor antagonists:
  AMPA receptors, 840–842
  kainic acid receptors, 845–846
  metabotropic glutamate receptors, allosteric modulation, 860–862
  NMDA receptors, 833–834
  subtype-selective NMDA ligands, 836–837
Noncrystallographic symmetry (NCS) restraints, protein structural models, 429–432
Nongenomic activity, glucocorticoid receptors, 1062
Nonhierarchical clustering, virtual screening, 107
Nonlabile metal complexes, dynamic combinatorial chemistry, 988–922
Nonlinear techniques, pattern recognition models, 238
Nonnucleoside reverse transcriptase inhibitors (NNRTIs), antiviral therapy, 1158
HIV protease inhibitors, 1168–1169
Nonreceptor tyrosine kinase inhibitors, therapeutic targeting, 1201–1202, 1221–1225
  Bcr-Abl, 1201, 1221–1222
  Src kinases, 1201–1202, 1222–1225
Nonspecific effects, natural product pharmacodynamics, 33–34
Nonsteroidal anti-inflammatory drugs (NSAIDs), prodrug development:
  side effects reduction, 784
  topical drug delivery, 757–759
Nonstructural proteins, antiviral therapy:
  protease inhibitors, 1168–1169
  RNA synthesis inhibition, 1162–1163

No-observable-adverse-effect-level (NOAEL):
  developability assays, 468–469
  toxicology assays, 478–481
Normalized log relative binding activity, comparative structural connectivity spectral analysis, estrogen receptor binding compounds, 252–256
Normalizing factor, protein structural models, refinement mathematics, 426–432
Nostoc ellipsosporum, natural product drug development, 27
Novel chemical entity (NCE):
  discovery and development assessment, research background, 457–460
  discovery assays, 460
N-terminal ubiquitin moiety (NubG), protein-protein interactions, 504
Nuclear factor-κB, targeted cancer therapy, proteosome inhibition, 1363–1364
Nuclear magnetic resonance (NMR):
  analytical chemistry assays, 463–464
  combinatorial chemistry, structure-activity relationship, 998–1001
  high-throughput screening and, 195–196
  metabolomic/metabonomic analysis, 137–138
  natural product development:
    isolation and purification, 49–51
    tandem assays, activity screening, 46–47
  spectral information management:
    binning algorithm development, 279–280
    biological effect modeling, 231
    carbon-nitrogen connectivity matrix, 271–278
      four-dimensional matrix construction:
        configurational entropy estimation, 276–278
        molecular dynamics of compounds, 275–276
      heteronuclear CoSCoSA antibiotic cephalosporin model, 271–275
    comparative modeling approaches, 231–232
    content analysis, 229–231
    CoSASA models:
      aromatase enzyme, 240, 242–244
      aryl hydrocarbon receptor binding, 244–245
      corticosterone binding, 240–241
    CoSCoSA molecular structure incorporation, 245–271
      aromatase enzyme model, 261–264
      aryl hydrocarbon receptor binding model, 264–271
      corticosterone binding:
        artificial neural network model, 259–261
        three-dimensional models, 256–259
      two-dimensional model of estrogen receptor binding compounds, 247–256
    future research issues, 280–281
    linear and nonlinear aspects, 238
    LOO/LNO model validation, 238–239
    molecular structure integration, 239–245
    multidimensional techniques, 233–236
    pattern recognition and model development, 236–237
    predicted spectra, 236
    research background, 228–236
    simulated *vs.* experimental modeling, 232
    software product development, 278–279
    structurally assigned chemical shift information, 232–233
Nuclear Overhauser effect spectroscopy (NOESY), multidimensional spectra, 235–236

Nuclear receptors, ADME/Tox drug development, 157–159
Nucleic acid detection (NAD):
 nuclein isolation, 1260–1262
 viral respiratory infection diagnosis, 1119–1120
Nuclein isolation, nucleic acid characterization, 1260–1262
Nucleosides:
 antiviral therapy:
  DNA polymerase inhibitors, 1148–1155
  HIV protease inhibitors, 1168–1169
  reverse transcriptase inhibitors, 1156–1158
  virus-cell fusion inhibitors, 1147–1148
 natural product development, 20–22

Object entities, Excel program, 299–303
Occupancy parameter, macromolecule crystallographic models, 420–422
Octreotide, somatostatin receptors, 1046
Odds ratio of survival:
 anticoagulant agents, 946–948
 mediator-specific anti-inflammatory agents, 940–943
OE-MPI code, virtual screening, high-performance computing applications, 214–216
Oligomycin F, research and development, 18–19
Oligonucleosides, RNA production, regulatory guidelines, 1296–1297
Oligonucleotides:
 antiviral agents, 1125–1126
 biopolymer solid-phase synthesis, 967–970
 identity of, 1298–1299
 legal manufacturing protocols, 1290–1291
 purity controls, 1298–1299
 split-pool synthesis libraries, 974–981

On-bead screening, split-pool synthesis, 974–975
 binary encoding, 977–979
 infrared coded resins, 979–980
 tagging methods, 977
On-chip probe synthesis, protein-protein interactions, protein microarray reagents, 524
One-dimensional ^{13}C NMR spectroscopy:
 biological effect modeling, 231–232
 comparative structural connectivity spectral analysis, molecular structural analysis, 245–271
One-dimensional quantitative spectrometric data-activity relationships:
 biological effect modeling, 231
 chemical shift information, 232–233
ONH astrocytes, gaucoma drug development, 146–151
Open DataBase Connectivity (ODBC), Excel Visual Basic for Applications:
 access protocols, 320–322
 data import, 322–326
 Oracle connection, 327
Open reading frames (ORFs), protein-protein interactions, yeast two-hybrid systems:
 library approach, 508–509
 protein linkage maps, 507–509
Ophthalmic drug delivery, prodrug drug development, 760–762
Oracle software:
 Excel Visual Basic for Applications, ODBC connection, 327
 laboratory data management, 289
Oral drug delivery:
 prodrug development, absorption mechanisms, 752–756
  carrier-mediated transport, 754–756
  lipophilic drugs, 753–754
 zanamivir, neuraminidase inhibitors, 1171
Orlistate, structure and properties, 1085–1086

Orthomyxoviridae, characteristics of, 1111–1112
Oseltamivir:
 neuraminidase inhibitors, 1171
 respiratory viral infection therapy, 1120
OSI-774, EGFR intracellular domain targeting, 1217–1218
Osteoporosis, bone metabolism agents, 1057–1060
 biophosphonates, 1058–1059
 calcimimetics, 1059–1060
 calcitonin, 1058
 mithramycin, 1059
 parathyroid hormone, 1057–1058
Osteoprotegrin (OPG), bone metabolism, 1057–1058
Outcome estimates:
 replication-experiment studies, 673–674
 statistical properties, 679–683
Overlabeled proteins, metal-enhanced fluorescence (MEF), 639–642
OvrMSR:
 acceptance criteria properties, 683–687
 estimated outcomes, statistical properties, 679–683
 statistical properties, data-generating model, 678–679
Oxytocin, drug action and therapeutic applications, 1052–1053

Paclitaxel. *See* Taxol
Palivizumab, respiratory syncytial virus therapy, 1110
Panax ginseng, drug development from, 29–30
Pancreatic polypeptide, antiobesity agents, 1084–1085
Panning procedure, protein-protein interactions, phage display technology, 512–515
Paradigm Genetics, gene expression profiles, 158
Parainfluenza viruses, characteristics of, 1110–1111

Parallel-distributed-artificial neural networks (PD-ANN), comparative structural connectivity spectra analysis model, corticosterone binding, 259–261
Parallel library synthesis:
 combinatorial chemistry, 971–972
 on-bead screening, 974
Paramagnetic enhanced chemical exchange saturation transfer (PARACEST) contrast agent, MRI molecular imaging, 1329
Parameter estimates, biomolecular screening assays, 669
Paramyxoviridae, structural properties, 1109–1111
Parasiticde, natural products for, 30
Parathyroid hormone (PTH), bone metabolism, 1057–1058
Parsing strings, Excel VBA function, 315–319
Partial agonists, general principles, 35
Partial least-squares discriminant analysis (PLS-DA), biological effect modeling, 231–232
Particle carryover, high-throughput flow cytometry, 192–193
Partitioning methods, virtual screening, 108–112
 cell-based partitioning, 108–109
 decision tree algorithms, 110–112
 direct space methods, 109–110
 performance evaluation, 115–116
PathArt, biological pathway analysis, 144
PathDB, biological pathway analysis, 145
Pathway Assist, biological pathway analysis, 145
Pattern recognition model development, NMR spectral information management, 236–237
Patterson function:
 x-ray crystal structure determination:
  anomalous dispersion, 413–416
  multiple isomorphous replacement, 407–412

x-ray diffraction, protein structures, molecular replacement, 417–420
Peak plasma levels (C_{max}), drug development, 3
Pegvisomant, growth hormone receptor antagonist, 1048
Peptide aptamers, protein-protein interaction inhibition, 539–540
Peptide chains, macromolecule crystallographic models, 423–425
Peptide display technologies, protein-protein interactions, 509, 512–517
  bacterial display, 515
  eukaryotic display systems, 515–516
  phage display, 512–515
  reagents, 516–517
  in vitro display, 516
Peptide mass fingerprinting (PMF), protein-protein interactions:
  affinity tagging and mass spectroscopy, 532–533
  mass spectrometry, 530–531
Peptide microarrays, protein-protein interactions, 522
Peptide nucleic acid (PNA), protein-protein interactions, microarray inhibitions, 542
Peptides:
  antiobesity agents, 1084–1085
  biopolymer solid-phase synthesis, 965–970
  corticotropin-releasing factor, 1041
  gonadotrophin-releasing hormone antagonists, 1044–1045
  gucagon-like peptide-1, diabetes therapy, 1082
  natural product research and development, 14–15
    marine sources, 32–33
    vertebrate sources, 31
  prodrug development:
    formulation improvements, 779–783

aqueous solubility improvement, 780
enzyme stability, 780–782
membrane permeation improvement, 782–783
nasal delivery, 766
somatostatin receptor antagonists, 1046
Performance evaluation:
  high-throughput screening automation, 580–581
  virtual screening, 114–116
Periodicals, natural product development sources, 28
Peroxisome proliferator activated receptors (PPAR):
  herbal sources for, 1027–1028
  thiazolidinedione receptors, 1078
pG5CAT, protein-protein interactions, mammalian two-hybrid systems, 506
Phaclofen, $GABA_B$ receptor antagonists, 825–827
Phage display, protein-protein interactions, 512–515
  inhibitors, 541
Pharmacodynamics, natural products, 33–43
  drug tolerance, 42–43
  general principles, 34–37
  molecular aspects, 37–42
    enzyme receptors, 40–41
    G-protein-coupled systems, 40
    ion channels, 39
    receptors, 37–39
    target selection, 41–42
    transcription factors, 41
  protein targets, 33–34
Pharmacokinetic assays:
  drug development, 476–477
  molecular imaging MRI, 1331–1332
Pharmacokinetic/metabolism (PKM), drug development, 3
Pharmacology assays:
  developability assays, 468–469
  in vitro techniques, 460–462
  in vivo techniques, 464–465

Pharmacophores:
  antiviral agents, 1141–1143
  GABA$_A$ receptor complex models, 811–814
    benzodiazepine binding sites, 821–823
    virtual screening, 100–102
      three-dimensional fingerprints, 103–104
Pharyngitis, respiratory viruses and, 1116
Phase problem, X-ray crystal structure determination, 405
Phenolic compounds, gastrointestinal disease, animal models, 1023–1028
  curcumin, 1023–1024
Phenylglyceines, metabotropic glutamate competitive receptor antagonists, 857–859
Philadelphia chromosome, Bcr-Abl protein tyrosine kinase, 1201
Phosphatase activity measurements:
  cytotoxicity measurements, coupled luminescent assays, 701
  PhosTRAK phosphatase assay, 712–713
    enzyme characteristics, 713–714
Phosphates, prodrug development, 744–746
  aqueous solubility, 748–749
    etoposide phosphate, 767–768
    fosphenytoin, 768–769
Phosphodiesterases, firefly luciferase, ATP/luciferase reactions, 699–700
Phosphoglycerokinase (PGK), DeathTRAK cycotoxicity assay, 701–702
Phosphoinositide 3-kinase (PI3K):
  inhibitors, 1231
  therapeutic targeting, 1204–1205
Phosphonate groups, prodrug development, 747
  acyclic nucleoside phosphonates, 1158–1161
Phosphonic acid moiety, oral drug delivery, lipophilic characteristics of prodrugs, 754

Phosphoroylsis, coupled bioluminescent assays, firefly luciferase, ATP/luciferase reactions, 699
Phosphorylation:
  acyclic nucleoside phosphonates, 1159–1161
  molecular drug action, enzyme receptors, 40–41
  protein kinases:
    mechanistic features, 1195–1196
    signal transduction pathways, 1193–1194
Phosphotyrosine levels, high-throughput screening formats, 564–565
[^{32}P]-adenosine triphosphate (ATP), high-throughput screening formats, 564–565
PhosTRAK phosphatase assay, 712–718
  applications, 713–714
  basic principles, 713
  complexity and thermodynamic issues, 716–717
  diverse results, 715–716
  future research issues, 717–718
  operational characteristics, 719
  phosphatase activity measurement, 712–713
  sensitivity limits, 719
  standard cocktail formulations, 723–724
Photolithographic synthetic techniques, combinatorial chemistry, 981–983
Photomultiplier tubes (PMT), high-throughput screening procedures:
  assay detection, 583
  plate design, 571–572
Photostability, metal-enhanced fluorescence (MEF):
  enhanced DNA labeling, 637–639
  enhanced DNA/RNA detection, 631–637
Phyllospongia foliascens, research and development, 18
Phytolacca americana, drug development from, 29–30

Picornaviridae:
acute laryngotracheobronchitis, 1116
antiviral therapy, 1121
bronchiolitis, 1117
structural properties, 1108
therapeutic agents for, 1125
Pin-based spotting, combinatorial synthesis, 938–985
4-PIOL, $GABA_A$ receptor complex:
agonists and antagonists, 809–811
pharmacophore models, 812–814
Piperidine-4-sulfonic acid, $GABA_A$ receptor agonists and antagonists, 809–811
PKI116 compounds, EGFR intracellular domain targeting, 1219
P13K inhibitors, structure and properties, 1231
PKR database, contents and location, 129
Plant-based medicines. *See also* Herbal medicines
drug development techniques, 29–30
herbal therapeutic agents, polysaccharides, 1022–1024
mushroom/seaweed sources, 1022–1023
history of, 12–14
research opportunities and, 15–22
Plasmid reagents, protein-protein interactions, yeast two-hybrid systems, 489–501
Platelet-derived growth factor (PDGF):
aptamer-based analysis:
fluorescence anistropy probe, 80–81
fluorescence energy transfer, 76–80
targeted cancer therapy, signal transduction, 1350
tyrosine kinases, 1201
inhibitors, 1219–1221
Plate systems, high-throughput screening:
compound storage, 575
design criteria, 570–572
metal-enhanced fluorescence (MEF), 630–660
directional emission, 654–660
enhanced DNA labels, 637–639
enhanced DNA/RNA detection, 630–637
enhanced energy transfer, silver surfaces, 645–649
multiphoton excitation and metallic nanoparticles, 643–645
overlabeled proteins, ultrabright probes as, 639–642
ratiometric surface sensing, 649–653
solution-based assays, 653–654
miniaturization, 565–566
Platform technology, high-throughput screening, 563–570
cellular assays, 566–567
drug discovery process, 567–569
instrumentation and screening formats, 563–565
miniaturization, 565–566
quality control, 569–570
Pleconaril, antiviral agents, 1125
Pleurotus ostreatus, natural product drug development, 29
P38 MAPK inhibitors, structure and properties, 1226–1228
Pneumocystis carinii, natural product development for, 20–22
Pneumonia, viral agents, 1117–1118
Podophyllum peltatum, drug development, 30
Polak-Ribiere conjugate gradient energy minimization, comparative structural connectivity spectral analysis, configurational entropy, 277–278
Polyacrylamid gel, protein-protein interactions, protein microarray supports, 524–526
Polyamidoamine (PAMAM) dendrimers, magnetic resonance imaging (MRI), mechanisms and enhancement, T1 contrast agents, 1312
Polyanionic compounds, virus adsorption inhibitors, 1140–1141
Polychlorinated biphenyls (PCBs), comparative structural connectivity spectral analysis, 264–271

Polychlorinated dibenzo-difurans (PCDD), comparative structural connectivity spectral analysis, 264–271
Polychlorinated dibenzo-*p*-dioxins (PCDD):
comparative structural connectivity spectral analysis, 264–271
comparative structurally assigned spectra analysis, 244–245
Polycrystalline iron oxide (SPIO), magnetic resonance imaging (MRI), mechanisms and enhancement:
T2 contrast agents, 1313
tumor cell surface receptors, 1319–1325
Polydimethylsiloxane (PDMS), protein-protein interactions, functional protein microarrays, 519
Polymerase chain reaction (PCR):
multiple tumor markers, whole-cell-based aptamers, 82–83
protein crystallization, 383
protein-protein interactions:
protein linkage maps, 507–509
in vitro display technologies, 516
serial analysis of gene expression, 134–135
viral respiratory infection diagnosis, 1119–1120
Polymorphisms, systems biology, DNA sequence variation, 132
Polyoxometalate (PM523), antiviral agents, 1124–1126
Polypeptides:
glucose-dependent insulinotropic polypeptide, 1082
insect sources for, 30–31
Polyphenol-based herbal medicines, gastrointestinal disease, animal models, 1026–1027
Polysaccharides, herbal therapeutic agents, 1022–1024
mushroom/seaweed sources, 1022–1023
plant sources, 1023

Polystyrene-co-divinylbenzene, biopolymer solid-phase synthesis, 966–970
Polyvalent display, protein-protein interactions, 512–515
Positive accuracy, screening for, 5–7
Positive control, coupled luminescent assays, 719–720
Positron emission tomography (PET), cancer cell proteomics, 74
Posterior probability, protein structural models, 426–432
Postsource decay (PSD), protein-protein interactions, peptide mass fingerprinting, 531
Posttranslational modifications (PTMs), proteomics, systems biology applications, 135–136
Potassium channels, ATP-sensitive potassium channel antagonists:
myocardial ischemia effects, 911–913
nonselective antagonists, 915–917
proarrhythmic effects, 923–924
research background, 909–911
selective antagonists, 917–923
HMR 1402 cardioselective compound, 921–923
HMR 1883 cardioselective compound, 919–921
ventricular arryhthmias, 914–924
ventricular rhythm, 913–914
Potassium perchlorate, structure and therapeutic application, 1056–1057
Potency:
general principles, 36–37
protein kinase inhibition, 1212
virtual screening limitations and, 98–99
Potency precision estimates, 671–678
acceptance criteria, 674
example data, 674–676
interlaboratory comparisons, 676–678
model parameter and outcome estimates, 673–674

replication-experiment protocol, 671–672
  outcomes, 672–673
  statistical model, 672
Power law principles, biological networks, 141–142
Precipitation agents, protein crystallization, 385
Precision parameters, high-throughput screening automation, 580–581
Preclinical studies. *See also* Clinical trials
  developability assays, 466–468
  hammerhead ribozymes, 1276–1277
  immunostimulating RNA, 1289–1290
  mRNA function, protein production, 1285–1288
  natural product development, 56–58
  RNA mimetics, aptamer function, 1271–1272
  small inhibitory RNA, 1281–1282
  Taxol case study, 59
Predicted binding models, comparative structural connectivity spectral analysis:
  aromatase enzyme, 262–264
  aryl hydrocarbon receptor binding, 266–271
  corticosterone binding, 258–259
Predicted distance dependencies, metal-fluorophore interactions, 604–605
Pregnane X-receptor (PXR), ADME/Tox drug development, 157–159
Prelabeling techniques, tumor cell surface receptors, in vivo MR molecular imaging, 1323–1324
Prestwick Chemical Library (PCL), FPR case study using, 217–218
Prey structures, protein-protein interactions:
  protein linkage maps, 507–509
  yeast two-hybrid systems, 485–489

Principal component (PC) of variation: comparative structural connectivity spectral analysis, corticosterone binding, 257–259
NMR spectral information content, 230–231
Prior probability, protein structural models, refinement mathematics, 426–432
Proarrhythmic effects, ATP-sensitive potassium agonists, 923–925
Probabilistic methods, quantitative structure-activity relationships, virtual screening, 113–114
Prodrugs:
  bioreversible structures, 740–752
    amines and amides, 750–752
      N-acyl prodrugs, 751
      N-α-acyloxymethyl$N$-phosphoroxymethyl derivatives, 751–752
      N-mannich bases, 751
    carboxyl groups, 741–744
      acyloxyalkyl prodrugs, 743–744
      aklyl and aryl esters, 741–743
      amides, 744
    hydroxyl groups, 747–750
      lactones, 750
      lipophilic prodrugs, 749–750
      steric hindrance, promoieties for, 750
      water-soluble prodrugs, 748–749
    phosphate groups, 744–746
    phosphonate groups, 747
  clinically useful requirements, 738–740
  definitions and concepts, 735–738
    chemical delivery system, 737–738
    double-prodrug concept, 735–737
    soft drug concept, 737
  pharmaceutical applications, 752–785
    aqueous solubility improvement, 766–769
      hemiester structures, 769
      phosphates, 767–769
        etoposides, 767–768
        fosphenytoin, 768–769

INDEX 1443

drug absorption mechanisms, 752–766
  carrier-mediated transport, 754–756
  dermal drug delivery, 756–760
    topical drugs, 757–759
    transdermal delivery, 759–760
  nasal drug delivery, 762–766
    ehanced brain delivery, 766
    lipophilic prodrugs, 764–766
    peptide prodrugs, 766
    transport-mediated absorption, 766
    water-soluble prodrugs, 763–764
  ophthalmic drug delivery, 760–762
  oral drug delivery, 752–756
    lipophilic drugs, 753–754
drug targeting improvements, 770–779
  brain targeting, 775–778
    carrier-mediated transport, 776–778
    chemical drug delivery system, 776
    lipophilic prodrugs, 775–776
  colon targeting, 778
  kidney targeting, 778–779
  tumor targeting, 770–775
    antibody-directed enzyme prodrug therapy, 774
    gene-directed enzyme prodrug therapy, 774–775
    hypoxia-selective drugs, 772–774
  duration of action, 769–770
  macromolecular promoieties, 784–785
  peptide formulation and delivery, 779–783
    aqueous solubility improvement, 780
    enzyme stability, 780–782
    membrane permeation improvement, 782–783
    side effects reduction, 783–784
  research background, 734–735
Profiling techniques, molecular fingerprints, virtual screening, 104–106

Progesterone, structure and properties, 1071–1072
Progestins, structure and properties, 1071–1072
Proliferation assays, DeathTRAK cycotoxicity assay:
  basic properties, 701–702
  biomass measurement (competitive comparison), 704–711
    ATP release assays, 704
    dual-mode cytotoxicity/proliferation, 705–712
      gram-negative bacteria, 707–709
      gram-positive bacteria, 709–711
      metabolic assays, 704–705
  cytotoxic process measurements, 702–704
  drug discovery applications, 702–711
Promoter genes, protein-protein interactions, yeast two-hybrid systems:
  library selection, 501
  target and bail vectors, 494
"Proof of concept' paradigm, small inhibitory RNA, 1282
Propofol, GABA$_A$ receptor complex, 819
Propranolol, transdermal drug delivery, prodrug development, 759–760
Propylthiouracil, therapeutic applications, 1055–1056
Pro-soft drug, defined, 737
Prostaglandin analogs, ophthalmic drug delivery, prodrug drug development, 761–762
Prostate-specific antigen (PSA), cancer cell proteomics, molecular diagnosis, 74–75
Protamine, nuclein isolation, 1260–1262
Protease inhibitors:
  antiviral therapy, 1163–1169
  biopolymer solid-phase synthesis, 969–970
  prodrug development, peptide formulation and delivery, 780
  structure-based drug design, 438–440

Protease trap system, protein-protein interactions, yeast two-hybrid systems, 507
Protein analysis:
  aptamer-based fluorescence anisotropy probe, 80–81
  database sources, 130–131
  formulation development assays, 470–471
  multiple tumor markers, whole-cell-based aptamers, 81–83
  systems biology applications, 136–137
ProteinChip technology, protein-protein interactions, protein microarrays, 518
Protein-compound complexes, x-ray diffraction, protein structures, molecular replacement, 419–430
Protein Data Bank (PBD):
  crystallization data sources, 388–389
  macromolecule crystallographic models, 420–422
  protein structural models:
    limitations of, 450
    validation, 432–433
Protein delivery systems, protein-protein interactions, protein microarrays, 526–527
Protein domain display, high-throughput flow cytometry, drug discovery targets:
  cell cycle proteins, 206–207
  FPR assembly characterization of, 204–205
  FPR tail assembly, 201–202
  soluble and bead-based assemblies, 202–207
Protein expression and characterization:
  high-throughput flow cytometry, 196–202
    arrestin-regulated intracellular FPR traffic, 199–200
    β2AR proof of principle, 202
    FPR assemblies with arrestin, 197–198
    FPR colocalization with arrestin, 198–199
    FPR-$G_{i\alpha2}$ fusion protein, 200–201
    FPR tail assembly, 201–202
    GPCR constructs, 196–197
    G protein-soluble GPCR reconstitution, 197
x-ray crystallography:
  anomalous dispersion, 412–416
    Friedel's law, 412
    structure determination, 413–416
  Bragg's law and angular dependence, 396–397
  crystallization trials, 376–389
    anomalous diffraction derivatization, 388
    cryopreservation for, 385–386
    data sources, 388–389
    demonstration experiment, 389
    drug discovery and, 386–388
    factors in, 382–383
    formats, 380–381
    formulation screens, 384–385
    membrane protein, 381–382
    observation and documentation, 384
    research background, 377–379
  data collection, 399–402
  de novo model construction, 422–425
  diffraction principles, 389–390
  drug discovery, 434–447
    antistructures, 441–442
    crystallographic screening, 443–444
    fragment screening, 444–445
    interface, 433–434
    lead optimization, structure-based drug design, 436–441
      HIV protease inhibitors, 438–440
      protein kinases, 440–441
    in silico screening, 442–443
    site-directed leads, fragment tethering, 445–446
    structural genomics, 446–447
    therapeutics, 442
  electron clouds and thermal motion, 398–399
  Fourier synthesis, 392–396
  heavy atom replacement, 406

INDEX   1445

home laboratory data collection, 402–403
limitations in drug discovery, 447–4508
macromolecule crystals, 390–392, 420–422
model preparation, 432–433
model refinement and analysis, 425–432
molecular replacement, 416–420
multiple isomorphous replacement, 406–412
phase problem, 405
research background, 374–376
structure factor, 405–406
synchrotron diffraction data collection, 403–404
Protein identification, protein-protein interactions, mass spectrometry, 530–531
Protein in situ array (PISA), protein-protein interactions, protein microarray reagents, 524
Protein kinase A and C:
  inhibitors, 1229–1231
  therapeutic targeting, 1204
Protein kinases (PKs):
  activation, 1198–1199
  classification, 1194–1195
  human disease and, 1199
  inhibitor design:
    allosteric inhibitors, 1213
    alternative design strategies, 1214–1215
    ATP binding site inhibitors, 1210–1212
      potency, 1212
      selectivity, 1210–1212
    basic principles, 1205
    bisubstrate analog inhibitors, 1213–1214
    clinical trials, 1206–1209, 1215–1232
    future research issues, 1232
    receptor tyrosine kinases, 1216–1221
    SH2 domain inhibitors, 1213
    substrate binding site inhibitors, 1212–1213
  molecular drug action, 40–41

protein phosphorylation mechanisms, 1195–1196
serine/threonine kinases, 1202–1205
  cyclin-dependent kinases, 1203–1204
    inhibitors, 1229
  glycogen synthase kinase 3, 1205
    inhibitors, 1231–1232
  inhibitors, 1225–1232
  MAP kinases, 1202–1203
    inhibitors, 1226–1229
  phosphoinositide 3-kinase, 1204–1025
    inhibitors, 1231
  protein kinase C and A, 1204
    inhibitors, 1229–1231
  Rho kinase, 1204
    inhibitors, 1231
signal transduction phosphorylation, 1193–1194
structural features, 1196–1198
structure-based drug design, 440–441
therapeutic applications, 1199–1205
tyrosine kinases, 1199–1202
  nonreceptor tyrosine kinases, 1201–1202
    Bcr-Abl, 1201
      inhibitors, 1221–1222
    inhibitors, 1221–1225
    Src kinases, 1201–1202
      inhibitors, 1222–1225
  receptor tyrosine kinases, 1200–1201
    EGF receptors, 1200
      extracellular domain inhibition, 1216–1217
      intracellular domain inhibition, 1217–1219
      inhibitors, 1216–1221
    PDGF receptors, 1201
      inhibitors, 1221
    VEGF/bEGF receptors, 1200–1201
      inhibitors, 1219–1221
Protein-ligand complexes, protein structural models, 433–434
Protein linkage maps (PLMs), protein-protein interactions, yeast two-hybrid systems, 507–509

Protein microarrays, protein-protein
   interactions, 517–528
  antibody microarrays, 519, 522
  detection methods, 527–528
  functional protein, 519
  high-throughput protein production,
   523–524
  interaction arrays, 522–523
  peptide microarrays, 522
  ProteinChip technology, 518
  protein delivery systems, 526–527
  reagents, 523–528
  small-molecule inhibition, 541–542
  supports, 524–526
  surface plasmon resonance arrays, 522
Protein-protein interactions:
  computational analysis, 533–536
  inhibitors:
    microarrays and small molecules,
     541–542
    phage display, 541
    in vivo genetic selection systems,
     540–541
    yeast two-hybrid system variants,
     536–540
      peptide aptamers, 539–540
      reverse two-hybrid systems,
       537–538
      split-hybrid system, 538–539
  peptide display technologies, 509,
   512–517
    bacterial display, 515
    eukaryotic display systems, 515–516
    phage display, 512–515
    reagents, 516–517
    in vitro display, 516
  protein mass spectrometry, 528–533
    affinity tagging, 531–533
    LC-MS approaches, 530
    protein detection and image
     analysis, 530
    protein identification, 530–531
    reagents, 528–533
    two-dimensional polyacrylamid gel
     electrophoresis, 528–529
  protein microarrays, 517–528
    antibody microarrays, 519, 522
    detection methods, 527–528
    functional protein, 519

  high-throughput protein production,
   523–524
  interaction arrays, 522–523
  peptide microarrays, 522
  ProteinChip technology, 518
  protein delivery systems, 526–527
  reagents, 523–528
  supports, 524–526
  surface plasmon resonance arrays,
   522
 research background, 483–484
 yeast two-hybrid systems, 484–511
  basic principles, 485–489
  classical system improvements,
   501–502
  interaction suppression, 506–507
  mammalian two-hybrid systems, 506
  new applications for, 506–511
  new developments in, 501–506
  protease trap, 507
  protein three-hybrid systems,
   503–506
  reagents, 489–501
    library choice, 495–501
    reporter genes and host yeast
     strains, 494–495
    target and bait vectors, 489–494
  Sos recruitment system, 502–503
  two-bait system, 502
  USPS system, 503
  whole genome applications, 507–509
Protein structural models, refinement
   process, 432
Protein targets, natural product
   pharmacodynamics, 33–34
Protein therapeutics, x-ray
   crystallographic studies, 442
Protein three-hybrid systems, protein-
   protein interactions, 504–506
Protein tyrosine kinases (PTKs),
   therapeutic targeting,
   1199–1202
  nonreceptor tyrosine kinases,
   1201–1202
  Bcr-Abl, 1201
    inhibitors, 1221–1222
  inhibitors, 1221–1225
  Src kinases, 1201–1202
    inhibitors, 1222–1225

receptor tyrosine kinases, 1200–1201
  EGF receptors, 1200
    extracellular domain inhibition, 1216–1217
    intracellular domain inhibition, 1217–1219
    inhibitors, 1216–1221
  PDGF receptors, 1201
    inhibitors, 1221
  VEGF/bEGF receptors, 1200–1201
    inhibitors, 1219–1221
  targeted cancer therapy, 1348–1349
Proteoglycans, plant sources, 1023–1024
Proteomics:
  cancer cells, molecular aptamers:
    fluorescence anisotropy probe, proetin analysis, 80–81
    fluorescence energy transfer, PDGF analysis, 76–80
    molecular diagnosis, 74–75
    multiple tumor markers, whole-cell-based selection, 81–83
    research background, 73–74
    tumor marker detection, 75–76
  drug development based on, 9
  macromolecule crystallographic models, 422–425
  protein-protein interactions, microarray technologies, 523
  systems biology applications, 135–137
    protein arrays, 136–137
    yeast two hybrid system, 136
Proteosome inhibition, targeted cancer therapy, 1363–1364
Protonation state, protein structural models validation, 433
Proximal causality principle:
  CFR Part II guidelines, 343–344
  drug discovery regulations, 349–351
  regulatory guidelines, 338–339
Pseudolaric acids, herbal sources for, 1027–1028
Pseudolarix kaempferi, terpenoid herbal compounds, 1027–1028
PTK787/ZK222584, VEGFR inhibitor, 1220
PTP1B, in silico screening, 443
PubGene, biological pathway analysis, 145

Public health impact:
  respiratory viral infections, 1107
  viral respiratory infection control, 1107–1108
Pulmonary delivery routes, drug delivery assays, 473–474
PUREmessenger, mRNA quality control, 1294–1298
Purgative, natural products as, 30
Purification techniques:
  natural product development, 47–51
  protein crystallization, 382–383
  protein-protein interactions, affinity tagging and mass spectroscopy, 531–533
  RNA production, quality control, 1296–1299
    messenger RNA, 1299–1300
    oligonucleotides, 1298–1299
Pyridoxal-5′-phosphate (PLP), GABA biosynthesis, 803–804

QT intervals, ATP-sensitive potassium channels, 910
Quality control (QC):
  bioanalytical chemistry assays, sample quality, 475–476
  high-throughput screening, 569–570
    data capture, 584–585
  RNA production, 1294–1300
Quantitative spectrometric data-activity relationships (QSDARs):
  comparative structural connectivity spectra analysis:
    aromatase enzyme binding models, 263–264
    binning algorithms, 279–280
    configurational entropy, 277–278
  information management, nuclear magnetic resonance:
    binning algorithm development, 279–280
    biological effect modeling, 231
    carbon-nitrogen connectivity matrix, 271–278
    four-dimensional matrix construction:

configurational entropy estimation, 276–278
molecular dynamics of compounds, 275–276
heteronuclear CoSCoSA antibiotic cephalosporin model, 271–275
comparative modeling approaches, 231–232
content analysis, 229–231
CoSASA models:
aromatase enzyme, 240, 242–244
aryl hydrocarbon receptor binding, 244–245
corticosterone binding, 240–241
CoSCoSA molecular structure incorporation, 245–271
aromatase enzyme model, 261–264
aryl hydrocarbon receptor binding model, 264–271
corticosterone binding:
artificial neural network model, 259–261
three-dimensional models, 256–259
two-dimensional model of estrogen receptor binding compounds, 247–256
future research issues, 280–281
linear and nonlinear aspects, 238
LOO/LNO model validation, 238–239
molecular structure integration, 239–245
multidimensional techniques, 233–236
pattern recognition and model development, 236–237
predicted spectra, 236
research background, 228–236
simulated vs. experimental modeling, 232
software product development, 278–279
structurally assigned chemical shift information, 232–233

Quantitative structure-activity relationships (QSARs):
ADME/Tox drug development, 157–159
comparative structural connectivity spectral analysis:
aromatase enzyme binding models, 261–264
configurational entropy, 277–278
corticosterone binding, 258–259
comparative structurally assigned spectra analysis, aromatase enzyme model, 243–244
current applications, 228–229
pharmacophore analysis, 102
screening techniques, 5–7
virtual screening:
basic principles, 95–96
four-dimensional models, 113
future issues, 116
probabilistic methods, 113–114
three-dimensional models, 112–113
two-dimensional models, 112
Quassinoside glycoside, natural product development, 20–22
Quencher molecules, aptamer-based fluorescence resonance energy transfer analysis, platelet-derived growth factor, 77–80
Quinones, hypoxia-selective prodrugs, tumor targeting, 772–774

Radioactive decay engineering (RDE), metal-fluorophore interactions, 604–609
Radio-frequency tagging, split-pool synthesis resins, 980–981
Radiolabeled ligands, high-throughput screening bioassays, 592–594
Raloxifene, structure and properties, 1070
Ramachandran plot, protein structural models validation, 432–433
Randomized Aldactone Evaluation Study (RALES), aldosterone antagonists, 1066
Range object, Excell software, 301–303

RANKL compound, bone metabolism, 1057–1058
RANTES chemokines, viral respiratory disease, 1211–122
Ras genes:
  MAPK protein kinase targeting, 1202–1205
  protein-protein interactions, Sos recruitment system (SRS), 503
Ratiometric surface sensing, metal-enhanced fluorescence, 649–653
RB6145, hypoxia-selective prodrugs, tumor targeting, 773–774
R568 compound, calcimimetics properties, 1060
Reaction master mixes, coupled luminescent assays, 722–725
READIT assay, firefly luciferase, ATP/luciferase reactions, phosphoroylsis, 699
Reagents:
  high-throughput screening:
    automated management, 584
    bioassay reagent production, 588–591
  protein-protein interactions:
    peptide display technologies, 516–517
    protein microarrays, 523–528
    yeast two-hybrid systems, 489–501
      library choice, 495–501
      reporter genes and host yeast strains, 494–495
      target and bait vectors, 489–494
Real-time polymerase chain reaction (RT-PCR), three-dimensional RNA amplification, 1270
Receptor agonists:
  ATP-sensitive potassium channels, proarrhythmic effects, 923–925
  estrogens, 1070
  follicle-stimulating hormone, 1048–1049
  GABA$_A$ receptor complex, 808–811
  GABA$_B$ receptor complex, 825
  GABA$_C$ complex, 828–829

glutamate receptors:
  AMPA receptor agonists, 837–840, 850
  NMDA receptors, 831–832
    glycine co-agonist site, 835–836
  gonadotrophin-releasing hormone, 1043–1044
  kainic acid receptors, 843–846
  metabotropic glutamate receptor ligands, 853–857
    backbone chain length extension, 855–857
    bioisosteres, 857
    conformationally constrained analogs, 853–855
  oxytocin mediation, 1052–1053
  progesterone, 1072
  somatostatin, 1045–1046
  thiazolidinediones, 1078
  vasopressin mediation, 1050–1051
Receptor antagonists:
  ATP-sensitive potassium channel antagonists:
    nonselective antagonists, 915–917
    selective antagonists, 917–923
      HMR 1402 cardioselective compound, 921–923
      HMR 1883 cardioselective compound, 919–921
  corticotropin-releasing factor, 1042
  estrogens, 1070
  follicle-stimulating hormone, 1048–1049
  GABA$_A$ receptor complex, 808–811
  GABA$_B$ receptor complex, 825–827
  GABA$_C$ complex, 828–829
  glucocorticoids, 1063
  glutamate receptors:
    competitive/noncompetitive AMPA receptors, 840–842
    NMDA receptors, 832–833
      glycine co-agonist site, 835–836
    uncompetitive and noncompetitive antagonists, 833–835
  gonadotrophin-releasing hormone, 1044–1045
  kainic acid receptors, 843–846

metabotropic glutamate receptor ligands, 857–862
  competitive antagonists, 857–860
    α-substitutions, 859–860
    phenylglycines, 857–859
  noncompetitive antagonists, allosteric modulators, 860–862
mineralocorticoids, 1065–1066
oxytocin mediation, 1052–1053
progesterone, 1072
somatostatin, 1046
somatotropin, 1048
vasopressin mediation, 1050–1051
Receptors:
  aldosterone, 1065
  androgens, 1066–1067
  antiobesity agents, 1084–1085
  calcitonin, 1058
  corticotropin-releasing factor, 1040–1042
  estrogen, 1069
  follicle-stimulating hormone, 1048–1049
  $GABA_A$ receptor complex models, 811–814
    benzodiazepine sites, 821–823
  $GABA_B$ receptor complex, 823–827
    agonists, 825
    allosteric modulators, 827
    antagonists, 825–827
  glucagon, 1083
  glucocorticoid, 1060–1063
  insulin, 1075–1076
  mineralocorticoids, 1065
  molecular drug action, 37–39
    enzymes, 40–41
    ion channels, 39
  oxytocin mediation, 1052
  parathyroid hormone, 1058
  progesterone, 1071–1072
  somatostatin, 1045
  sulfonylureas, 1079–1080
  thiazolidinediones, 1078
  thyroid hormone mediation, 1053–1054
  thyrotropin-releasing hormone, 1040
  vasopressin mediation, 1050

Receptor tyrosine kinases (RTKs), therapeutic targeting, 1200–1201
  EGF receptors, 1200
    extracellular domain inhibition, 1216–1217
    intracellular domain inhibition, 1217–1219
    inhibitors, 1216–1221
  PDGF receptors, 1201
    inhibitors, 1221
  VEGF/bEGF receptors, 1200–1201
    inhibitors, 1219–1221
Reciprocal lattice, x-ray diffraction, Fourier synthesis, 396
Recombinant human granulocyte colony-stimulating factor (rhG-CSF), thionamide drugs, 1056
Recombinant proteins:
  crystallization factors, 382–383
  natural product research and development, 14–22
Recursive deconvolution, split-pool synthesis, 976
Recursive partitioning, virtual screening, decision tree algorithms, 110–112
Redman's screening characteristics, 5–7
5α-Reductase inhibitors, structure and applications, 1068
Refinement process, macromolecule crystallographic models, 425–432
Reflections, x-ray defraction spots, 396–397
Regression-based methods, virtual screening, docking and scoring software, 212
Regulators of G-protein signaling (RGS) proteins, high-throughput flow cytometry, drug discovery targets, 188–190
Regulatory guidelines:
  CFR Part 11 guidelines, 343–344
  FDA origins, 339–343
  future issues, 349–351

INDEX

natural product development, 55–58
research background, 337–339
risk assessment, 344–349
RNA production, 1294–1300
   identity, 1298–1299
      messenger RNA, 1299–1300
   mRNA, 1297–1298
   oligonucleosides, 1296–1297
   oligonucleotides, 1298–1299
   purity, 1296
Rejection rules, acceptance criteria properties, 685–687
Relational databases, Excel Visual Basic for Applications, 319–320
Relative binding activity (RBA):
   comparative structural connectivity spectral analysis, configurational entropy, 277–278
   estrogen receptor binding compounds, comparative structural connectivity spectral analysis, 247–256
Relative growth parameter, TUNEL assay, multiplex dose curve analysis, 364–366
Relaxation properties, contrast agents, magnetic resonance imaging, 1313
Relenza, respiratory viral infection therapy, 1120
Remote-docking mechanisms, protein kinase inhibitor design, 1214–1215
Renal-specific drug delivery, prodrug development, 778–779
Repaglinide, diabetes therapy, 1081
Replication-experiment studies:
   acceptance criteria, 683–687
   data-generating model, 678–679
   estimated outcomes, 679–683
   high-throughput screening compounds, 575–576
   potency precision estimates, 671–672
   outcomes, 672–673
   statistical model, 672
   statistical properties of estimates and acceptance criteria, 678–687

Reporter genes:
   high-throughput screening cellular assays, 597–598
   protein-protein interactions, yeast two-hybrid systems, 489
   host yeast strains, 494–496
   reagents, 489–501
Reporter molecules, chemical shift molecular imaging, 1330–1331
Report generation, Excel software, 327–334
Reproducibility, screening for, 5–7
Research Consortium for Structural Biology, macromolecule crystallographic models, 422
Resin beads, split-pool synthesis, 973–974
   infrared coding, 979–980
   radio-frequency tagging, 980–981
Resolution limit, x-ray diffraction data collection, 402
Resonance energy transfer (RET), metal-enhanced fluorescence:
   drug discovery applications, 661–662
   silver surfaces, 645–649
Respiratory syncytial virus immune globulin intravenous (RSV-IGIV) agent, 1110–1111
Respiratory syncytial virus (RSV):
   acute laryngotracheobronchitis, 1116
   animal models, 1126–1127
   bronchiolitis, 1117
   immunologic and nonimmunologic factors, 1121–1122
   paramyxovirus, 1110–1111
   pneumonia and, 1118
   seasonal patterns, 1106
Respiratory viruses:
   adenoviridae, 1113
   animal models, 1126–1127
   antiviral therapy:
      current treatments, 1120–1121
      development strategies, 1122–1126
      infections, 1121–11222
   bunyaviridae, 1112–1113
   clinical syndromes, 1114–1118
      acute laryngotracheobronchitis, 1116
      bronchitis, 1117
      common cold, 1114, 1116

laryngitis, 1116
pharyngitis, 1116
pneumonia, 1117–1118
tracheitis and tracheobronchitis, 1116
coronaviridae, 1109
future research issues, 1127–1128
herpesviridae, 1113–1114
infection control, 1107–1108
infection diagnosis, 1118–1120
orthomyxoviridae, 1111–1112
paramyxoviridae, 1109–1111
piconaviridae, 1108
public health and economic impact, 1107
seasonal patterns, 1106
specific pathogens, 1108–1114
structure and classification, 1106
Retigabine, $GABA_A$ receptor complex, 819
Reuptake mechanisms, GABA receptor ligand, 804–806
Reverse Kretschmann (RK) configuration, metal-enhanced fluorescence, directional emission, 656–660
Reverse-phase high-pressure liquid chromatography (RP-HPLC), protein-protein interactions, 530
Reverse three-hybrid systems, protein-protein interaction inhibition, 536–537
Reverse transcriptase inhibitors, antiviral therapy, 1155–1158
Reverse two-hybrid systems, protein-protein interaction inhibition, 536–538
Rhinoviruses:
  animal models, 1126–1127
  genomic structure, 1108
  protease inhibitor therapy, 1168–1169
  therapeutic agents for, 1125
Rhodamine B, metal-enhanced fluorescence:
  metal-fluorophore interactions, 607–609

multiphonon excitation, 643–645
silver island films, 642
Rhodopsin, high-throughput flow cytometry, drug discovery targets, G-protein-coupled receptor signaling pathways, 188–190
Rho kinase:
  inhibitors, 1231
  therapeutic targeting, 1204
Ribivarin:
  IMP dehydrogenase inhibitors, 1172–1173
  respiratory viral infection therapy, 1120–1121
Ribonucleases (Rnases), natural/modified RNA, 1265–1266
Ribonucleic acid (RNA):
  antiviral drug development, 1120–1121
    helicase-RNA primase inhibition, 1155
    IMP dehydrogenase inhibitors, 1173
    synthesis inhibitors, 1161–1163
  aptamer three-dimensional structures, 1267–1273
    amplification, 1270
    definitions, 1267–1269
    Eyetech Pharmaceutical development, 1272–1273
    mimetics, 1270–1271
    preclinical and clinical utilization, 1271–1272
  future research issues, 1300
  immunostimulation therapy, 1288–1290
    definitions, 1288–1289
    hemispherx Biopharma, 1290
    preclinical/clinical utilization, 1289–1290
  legal issues in production of, 1290–1300
  manufacturing protocols, 1290–1294
    long RNA, 1291–1294
    oligonucleotides, 1290–1291

INDEX 1453

regulatory aspects, 1294–1300
  identity, 1298–1299
    messenger RNA, 1299–1300
  mRNA, 1297–1298
  oligonucleosides, 1296–1297
  oligonucleotides, 1298–1299
  purity, 1296
mRNA protein production, 1283–1288
  definitions, 1283–1284
  MERIX Biosciences, 1288
  mimetics, 1285
  preclinical/clinical utilization, 1285–1288
multiple tumor markers, whole-cell-based aptamers, 82–83
respiratory viruses:
  Bunyaviridae, 1112–1113
  Coronaviridae, 1109
  Orthomyxoviridae, 1111–1112
  Paramyxoviridae, 1109–1111
  Picornaviridae, 1108
structural and chemical properties, 1260–1267
  active pharmaceutical ingredients, 1264
  cellular functions, 1262–1263
  natural and modified forms, 1265–1267
  nuclein isolation, 1260–1262
  therapeutic applications, 1263–1264
synthetic compounds:
  antisense and ribozymes, 1273–1278
    definitions, 1273–1274
    hammerhead ribozymes:
      mimetics, 1274–1276
      preclinical/clinical utilization, 1276–1277
    small inhibitory RNA, 1277–1278
  gene expression suppression, 1282–1283
  small inhibitory RNA, 1278–1282
    definitions, 1278–1280
    mimetics, 1280–1283
    preclinical evaluation, 1281–1282
systems biology and, 124–125
viral synthesis inhibition, process inhibitors, 1161–1163

Ribozyme Pharmaceuticals, small inhibitory RNA, 1277–1278
Ribozymes, synthetic RNA, 1273–1278
  definitions, 1273–1274
  hammerhead ribozymes:
    mimetics, 1274–1276
    preclinical/clinical utilization, 1276–1277
  small inhibitory RNA, 1277–1278
Ringworm, natural product development for, 30
Risk assessment, regulatory guidelines, 339, 344–349
RNA interference (RNA-i)-based agents:
  antiviral agents, 1125–1126
  viral respiratory infections, 1122–1124
Robotic systems:
  automated high-throughput screening, 582–584
  sample mounter, x-ray diffraction, synchrotron data collection, 405
Rolling circle amplification (RCA), protein-protein interactions, protein microarrays, 527–528
Rose Bengal, metal-enhanced fluorescence, metal-fluorophore interactions, 607–609
Rotational search methods, x-ray diffraction, protein structures, molecular replacement, 417–420
Roughened silver electrodes, metal-enhanced fluorescence, 622–625
Rozites caperata, natural product development, 29
R121919 prototype, corticotropin-releasing factor antagonist, 1042
RSU1069, hypoxia-selective prodrugs, tumor targeting, 773–774
R_{sym}, x-ray diffraction data collection, 400–402

RU 486, structure and properties, 1072
Rubidium quantum yields, metal-enhanced fluorescence, metal-fluorophore interactions, 608–609
"Rule of three," crystallographic fragment screening, 445

Saccharomyces cerevisiae, protein-protein interactions:
  eukaryotic display systems, 515–516
  hybrid inhibition systems, 536–539
  yeast two-hybrid systems, 487–489, 508–509
Saclofen, $GABA_B$ receptor antagonists, 825–827
Safety assessment:
  natural product development, 56–58
  screening procedures, 6–7
  in vivo toxicology assays, 480–481
Salicylaldimine monomers, dynamic combinatorial chemistry, 988–992
Salting in/out process, protein crystallization, 377–379
Sample carryover process, microfluidic mixing, high-throughput flow cytometry, 209
Sample delivery techniques, high-throughput flow cytometry, drug discovery targets, 192–193
Sample size, screening procedures, 6–7
Saponaria officinalis, drug development from, 29–30
Saquinavir, protease inhibitor properties, 1164–1169
SB203580, P38 MAPK inhibitors, 1228
Scaling factors, molecular fingerprints, virtual screening, 104–106
SCH-C antagonist, virus-cell fusion inhibitors, 1144–1147
Scheduling software, automated high-throughput screening, 583–584
Schizophrenia, glutamate transmitter systems, 801–802
Scintillation proximity assay (SPA):
  coupled luminescent assays, 691–692
  limitations and considerations, 697–698
  high-throughput screening platforms, 565–566
  bioassay development, 593–594
Scoring programs, virtual screening, 212
Screening techniques. *See also* Activity screening
  characteristics of, 5–7
  crystallographic structural models:
    crystallographic fragment screening, 444–445
    crystallographic screening, 443–444
    in silico screening, 442–443
  drug discovery and interpretation of, 3–9
  FPR ligands, 210–211
  fragment screening, defined, 375
  high-throughput flow cytometry, 195–196
  high-throughput screening formats, 563–565
  multiple cell lines:
    CellCard technology, 356–357
    antimitotic cellular arrays case study, 362–363
    card preparation, 360
    compound addition and assay protocol, 360–361
    experiment design and plate layout, 359–360
    image analysis and data visualization, 361–362
    key features of, 370
    multiplexed antiproliferation screen, 367–370
    multiplexing cell-based assays, 357–359
    96-well mixing and dispensing format, 360
    reading protocol, 361
    tissue culture, 360
    TUNEL assay case study, 363
    multiplex dose curve analysis, 363–366
    research background, 354–356

INDEX 1455

natural product activity, 43–47
respiratory viruses, 1123–1124
targeted cancer therapy, signal transduction, 1351–1352
x-ray crystallography, defined, 375
Seaweed, polysaccharide agents from, 1022–1023
Second-messenger kinases, G-protein-coupled receptors (GPCR), high-throughput flow cytometry, drug discovery targets, 188–190
Second messenger response:
  high-throughput screening formats, cell-based assays, 567
  molecular drug action, enzyme receptors, 41
  molecular drug actions, 37–39
Selective androgen receptor modulators (SARMs), anabolic action, 1067–1068
Selective estrogen receptor modulator (SERM):
  action of, 1069–1070
  therapeutic applications, 1071
Selectivity:
  general principles, 36
  protein kinase inhibition, ATP binding site inhibitors, 1210–1212
  virtual screening limitations and, 98–99
Selenium, x-ray diffraction, protein structures, anomalous dispersion, 416
SELEX production, RNA mimetics, aptamer function, 1270–1271
Self-assembled monolayer (SAM), protein-protein interactions, protein microarray supports, 525–526
Self-organizing maps (SOM), gene expression data, systems biology, 138–139
Self-quenching mechanisms, metal-enhanced fluorescence, overlabled proteins as ultrabright probes, 639–642

Self-regulation:
  drug discovery guidelines, 349–351
  protocols, 339
Self-selected components, dynamic combinatorial chemistry, 986–992
Self-treatment guides, history of, 13–14
Seminaftofluoresceins (SNAFLs), metal-enhanced fluorescence, ratiometric surface sensing, 650–653
Seminaftorhodafluors (SNARFs), metal-enhanced fluorescence, ratiometric surface sensing, 650–653
Sensitivity:
  coupled luminescent assay speed and linear response, 719
  molecular imaging MRI, 133–1332
  PhosTRAK phosphatase assay, 719
  screening for, 5–7
SENTRA database, contents and location, 129
Separation/chemical techniques, natural product development:
  activity screening, 45–47
  isolation and purification, 49–51
Sepsis management:
  anti-coagulants, 946–948
    clinical trials, 946
    coagulation pathways, 946
    efficacy factors, 946–948
  glucocorticoids, 943–946
    clinical trials, 943–944
    dose-dependent effects, 943, 945
    mediator-specific anti-inflammatory agents vs., 945–946
  mediator-specific anti-inflammatory therapies:
    clinical trials, 939–940
    efficacy measurements, 940–943
    glucocorticoids, 943–946
      clinical trials, 943–944
      dose-dependent effects, 943, 945
    inflammatory mediators, 939
    research background, 938–939
Sequence assignment, macromolecule crystallographic models, 424–425

Sequential access files, Excel software:
opening and closing, 309–311
reading and writing, 311–315
Serial analysis of gene expression (SAGE), systems biology, 133–135
Serine proteases, HSV antiviral therapy, 1168–1169
Serine/threonine kinases:
 targeted cancer therapy:
  aurora kinases, 1361–1362
  mammalian target of rapamycin, 1350–1351
  therapeutic targeting, 1202–1205
  cyclin-dependent kinases, 1203–1204
   inhibitors, 1229
  glycogen synthase kinase 3, 1205
   inhibitors, 1231–1232
  inhibitors, 1225–1232
  MAP kinases, 1202–1203
   inhibitors, 1226–1229
  phosphoinositide 3-kinase, 1204–1025
   inhibitors, 1231
  protein kinase C and A, 1204
   inhibitors, 1229–1231
  Rho kinase, 1204
   inhibitors, 1231
Sesquiterpene lactone, natural product development, 20–22
Severe acute respiratory syndrome (SARS):
 antiviral drug development, 1125–1126
 coronavirus agents, 1109
 pneumonia and, 1117–1118
 replication system, 1124
Shape representation, virtual screening, 99–100
Side-chain modification, prodrug development, peptide formulations, stability improvement, 782
Side effects reduction, prodrug development, 783–785
Signal amplification, in vivo MR molecular imaging, tumor cell surface receptors, 1324–1325

Signaling pathways:
 G-protein-coupled receptors (GPCR), high-throughput flow cytometry, drug discovery targets, 187–190
 microarray data, 139–140
Signal-to-noise ratio, x-ray diffraction data collection, 402
Signal transduction pathways:
 protein kinase pohsphorylation, 1193–1194
 activation mechanisms, 1198–1199
 targeted cancer therapy, 1346–1352
  Bcr/Abl, c-Kit, and PDGF, 1350
  EGF receptor pathway, 1347–1349
  HER2/Neu inhibition, 1347
  insulin-like growth factor signaling, 1349
  mammalian target of rapamycin, 1350–1351
  screening assays, 1351–1352
Silver nanoparticles, metal-enhanced fluorescence:
 anisotropic silver nanostructures, 615–618
 electrochemically-deposited silver, 619–622
 electroplating techniques, 622
 enhanced energy transfer, 645–649
 fractal-like structures on glass substrates, 625–630
 laser-deposited silver, 618–619
 metal-fluorophore interactions, 607–609
 roughened silver electrodes, 622–625
 silver colloid films, 610–615
 silver island films, 610
  overlabled proteins as ultrabright probes, 640–642
 ratiometric surface sensing, 650–653
Silver staining, protein-protein interactions, mass spectrometry techniques, 530
Similarity coefficients, molecular similarity, virtual screening, 94
Similarity paradox, virtual screening limitations, 97

INDEX 1457

Similar property principle:
  molecular similarity, virtual screening, 93–94
  virtual screening limitations and paradox of, 96–97
Single isomorphous replacement with anomalous dispersion (SIRAS), x-ray crystal structure determination, 414–416
Single-nucleotide polymorphism (SNPs), systems biology, DNA sequence variation, 132
Single-wavelength anomalous dispersion (SAD), x-ray crystal structure determination, 415–416
Single-well compound profiling, multiplexed antiproliferation screening, 367–370
Sin Nombre virus (SNV), characteristics, 1112–1113
Site-directed lead discovery by fragment tethering:
  crystallographic structural models, 445–446
  defined, 375–376
Site-specific drug delivery. *See also* Targeting mechanisms; Tumor targeting
  prodrug development, 770–779
    brain targeting, 775–778
      carrier-mediated transport, 776–778
      chemical drug delivery system, 776
      lipophilic prodrugs, 775–776
    colon targeting, 778
    kidney targeting, 778–779
    tumor targeting, 770–775
      antibody-directed enzyme prodrug therapy, 774
      gene-directed enzyme prodrug therapy, 774–775
      hypoxia-selective drugs, 772–774
Size exclusion chromatography (SEC), protein crystallization, 382–383

Skeletonization calculations, macromolecule crystallographic models, 423–425
Small inhibitory RNA (siRNA):
  Alnylam Pharmaceuticals manufacture of, 1282
  definitions, 1278–1280
  gene expression suppression, 1282–1283
  preclinical evaluation, 1281–1282
  RNA mimetics, 1280–1281
  synthetic manufacturing, 1277–1278
Small-molecule inhibitors:
  protein-protein interactions, 537–538
  microarrays, 541–542
  targeted cancer therapy, 1365–1367
    natural products, 1365–1366
    synthetic chemical libraries, 1366–1367
Soaking compounds, crystal-based drug discovery, 386–388
Sodium dodecyl sulfate polyacrylamide gel electrophoresis (SDS-PAGE):
  analytical chemistry assays, 463–464
  protein crystallization, 382–383
Soft drug concept, defined, 737
Software programs. *See also* specific programs
  automated high-throughput screening, scheduling software, 583–584
  biological pathway analysis, 142–146
  high-throughput flow cytometry, 209–210
  natural product structural identification, 52–53
  NMR spectral information management, 278–279
  pathway analysis, future development, 159–164
  virtual screening:
    docking and scoring software, 212
    high-performance computing, 214–216

Solid-phase synthesis:
　combinatorial chemistry and, 963–970
　　biopolymers, 965–970
　　click and in situ click chemistry, 992–997
　　organic synthesis, 970
Soluble assemblies, high-throughput flow cytometry, 202–207
　β2AR, 205–206
　cell cycle protein display, 206–207
　FPR assemblies, 204–205
　microfluidic mixing, 207–208
Solution-based assays, metal-enhanced fluorescence, 653–654
Solvent flattening, x-ray diffraction, protein structures, anomalous dispersion, 415–416
Solvent X-ray scattering, protein structural models, 429–432
Somatostatin, analogs and antagonists, 1045–1046
Somatotropin, drug actions and applications, 1047–1048
Sos recruitment system (SRS), protein-protein interactions:
　development of, 502–503
　yeast two-hybrid systems, 488–489
SPAD database, contents and location, 129
Sparse matrix design, protein crystallization formulation, 384–385
Spatial coordinates, macromolecule crystallographic models, 420–422
Specificity:
　general principles, 35
　molecular drug actions, receptors, 38–39
　natural product pharmacodynamics, protein targets, 33–34
　protein-protein interactions, antibody microarrays, 522
　screening for, 5–7
Spectrometric data-activity relationships (SDAR):
　NMR spectral information content, 229–231
　pattern recognition techniques, 237

Spectroscopic analysis:
　information management, nuclear magnetic resonance:
　　binning algorithm development, 279–280
　　biological effect modeling, 231
　　carbon-nitrogen connectivity matrix, 271–278
　　four-dimensional matrix construction:
　　　configurational entropy estimation, 276–278
　　　molecular dynamics of compounds, 275–276
　　　heteronuclear CoSCoSA antibiotic cephalosporin model, 271–275
　　comparative modeling approaches, 231–232
　　content analysis, 229–231
　　CoSASA models:
　　　aromatase enzyme, 240, 242–244
　　　aryl hydrocarbon receptor binding, 244–245
　　　corticosterone binding, 240–241
　　CoSCoSA molecular structure incorporation, 245–271
　　　aromatase enzyme model, 261–264
　　　aryl hydrocarbon receptor binding model, 264–271
　　　corticosterone binding:
　　　　artificial neural network model, 259–261
　　　　three-dimensional models, 256–259
　　　two-dimensional model of estrogen receptor binding compounds, 247–256
　　future research issues, 280–281
　　linear and nonlinear aspects, 238
　　LOO/LNO model validation, 238–239
　　molecular structure integration, 239–245
　　multidimensional techniques, 233–236
　　pattern recognition and model development, 236–237

predicted spectra, 236
research background, 228–236
simulated vs. experimental modeling, 232
software product development, 278–279
structurally assigned chemical shift information, 232–233
natural product structural identification, 52–53
proteomics, systems biology applications, 135–136
Speed limitations, coupled luminescent assays, 718
Spironolactone, aldosterone antagonists, 1066
Split-hybrid system, protein-protein interaction inhibition, 536, 538–539
Split-pool synthesis:
combinatorial chemistry libraries, 973–974
binary encoding, 977–979
infrared coded resins, 979–980
radio-frequency tagging, 980–981
recursive deconvolution, 976
tagging methods, 977
macrobeads, 984–985
SPOT technology, protein-protein interactions, protein microarray reagents, 524
Spotted arrays, combinatorial synthesis, 983–985
Square wave analysis, x-ray diffraction, Fourier synthesis, 393–396
SR121463, vasopressin receptor agonists/antagonists, 1051
SRA880 antagonist, somatostatin receptors, 1046
Src family kinases:
nonreceptor tyrosine kinase inhibitors, 1201–1202
domain inhibitors, 1222–1224
substrate binding sites, 1224–1225
protein kinase structures, 1197–1198
Src homology 2 (SH2), protein kinase structures, 1197–1198
domain inhibitors, 1213

ss-Fl-DNA, metal-enhanced fluorescence (MEF), enhanced DNA/RNA detection, 631–637
SSR149415, vasopressin receptor agonists/antagonists, 1051
Stability-indicating assay, drug development, 470–471
Stabilized RNA, immunostimulation therapy, 1288–1290
definitions, 1288–1289
hemispherx Biopharma, 1290
preclinical/clinical utilization, 1289–1290
Standard deviation (STDEV), coupled luminescent assays, 719–720
Statistical modeling:
acceptance criteria properties, 683–687
assay systems, 4–9
comparative structural connectivity spectral analysis:
aryl hydrocarbon receptor binding, 266–271
estrogen receptor binding compounds, 247–256
estimates and acceptance criteria, 678–687
acceptance criteria properties, 683–687
data-generating model, 678–679
estimated outcome properties, 679–683
high-throughput screening bioassays, 588
potency precision estimates, replication-experiment studies, 672
Staurosporine, protein kinase A and C inhibition, 1229–1231
Stavudine-5′(phenyl methoxy-L-alaninyl) phosphate, prodrug development, 746
Stereoisomeric structures, molecular drug actions, receptors, 38–39
Steric hindrans, hydroxyl prodrugs, 750
Steroids, anabolic:
actions and applications, 1067–1068
inhibitors, 1070–1071

STKE database, contents and location, 129
Storage protocols, high-throughput screening compounds:
  closed-loop screening, 578
  nanoliter dispensing, 576–578
  plate formats, 575
  replication, 575–576
  wet or dry storage, 573–575
Streptomyces hygroscopicus ascomyceticus, research and development, 18
Streptomyces osterogriseus, research and development, 18–19
Streptomyces platensis, research and development, 18
String data manipulation and parsing, Excel VBA function, 315–319
Structural diversity, combinatorial chemistry libraries, 1002–1004
Structural genomics:
  crystallographic structural models, 446–447
  defined, 376
Structure-activity relationship (SAR):
  antiviral agents:
    reverse transcriptase inhibitors, 1157–1158
    virus-cell fusion inhibitors, 1144–1147
  biochemical assays, 354–356
  combinatorial chemistry:
    nuclear magnetic resonance, 998–1001
    thiol-based in situ macromolecular ligand assembly, 998
  crystallographic screening, 444
  current applications, 228–229
  drug discovery and development and, 458–460
  $GABA_A$ receptor complex, receptor and pharmacophore models, 811–814
  high-throughput screening, compound management, 573
  natural product development, 61–62

Structure-based compound libraries, defined, 376
Structure-based drug design (SBDD):
  defined, 376
  HIV protease inhibitors, 438–440
  lead optimization, 436–438
  limitations of, 449–450
  protein kinases, 440–441
  work flow for, 434–436
Structured Query Language (SQL), Excel Visual Basic for Applications, data import, 325–327
Structure factor, X-ray crystal structure determination, 405–406
Structure identification, natural product development, 51–53
S-T segment changes, ATP-sensitive potassium channels, 913
Study design, failures in, 6–7
Subroutines, Excel software:
  GetSaveAsFilename, 308–309
  Visual Basic Editor task, 293–297
Substrate binding sites:
  protein kinase inhibitors, 1212–1213
  Src family kinase inhibitors, 1224–1224
Substructure analysis, virtual screening, two-dimensional substructure searching, 99
Subtype-selective NMDA ligands, glutamate receptors, therapeutic applications, 836–837
Subunit proteins, molecular drug action, ion channels, 39
SU5416 compound, VEGFR inhibitor, 1219
SU6668 compounds:
  PDGFR inhibitor, 1221
  VEGFR inhibitor, 1219
"Suicide gene therapy," tumor targeting, 772–775
Sulfinpyrazone, virtual screening, 217
Sulfonylurea agents:
  diabetes therapy, 1079–1080
  HMR 1883 cardioselective ATP-potassium channel antagonists, 919–921

ventricular arrhythmias:
    nonselective ATP-sensitive potassium channel antagonists, 916–917
    selective ATP-sensitive potassium channel antagonists, 917–923
Sum-squared difference, protein structural models, 425–432
Surface-enhanced laser/desorption ionization (SELDI):
    multiple tumor marker detection, 81–83
    protein-protein interactions, protein microarrays, 527–528
Surface-enhanced Raman scattering (SERS), metal-enhanced fluorescence:
    metal substrates, 610
    roughened silver electrodes, 625
    silver fractal-lilke structures, 626–630
Surface plasmon-coupled emission (SPCE), metal-enhanced fluorescence, directional emission, 655–660
Surface plasmon resonance-biomolecular interaction analysis (SPR-BIA), protein-protein interactions, 522
Surface plasmon resonance (SPR) arrays, protein-protein interactions, 522
    protein microarray detection, 527–528
Surface supports, protein-protein interactions, protein microarray reagents, 524–526
SwissProt database, contents and location, 129
"Switch" two-hybrid system, protein-protein interactions, 487–489
Symmetry-related reflections, x-ray diffraction data collection, 400–402
Synchrotron diffraction data, protein crystals, 403–405
Synergy screen, protein crystallization formulation, 384–385

Synthesis procedures:
    corticosteroid synthesis inhibition, 1063–1064
    natural product development, 53–55
Synthetic mimics, virtual screening and, 114–116
Synthetic RNA:
    antisense and ribozymes, 1273–1278
    definitions, 1273–1274
    hammerhead ribozymes:
        mimetics, 1274–1276
        preclinical/clinical utilization, 1276–1277
    small inhibitory RNA, 1277–1278
    gene expression suppression, 1282–1283
    small inhibitory RNA, 1278–1282
    definitions, 1278–1280
    mimetics, 1280–1283
    preclinical evaluation, 1281–1282
Systems biology:
    biological pathway analysis, resources for, 142–146
    database sources, 127–131
    definitions, 124–125
    drug discovery applications, 146–159
        ADME/Tox systems, 156–159
        breast cancer, 151–156
        glaucoma, 146–151
    future applications, 164–165
    high-throughput data, 131–142
        DNA sequence variations, 132
        gene expression profiling, 132–135
            clustering approaches, 138–139
            microarrays, 132–133
            serial analysis, 133–135
        metabolomics and metabonomics, 137–138
        microarray data, advanced computational analysis, 139–142
            biological networks, 141–142
            linear pathways to complex networks, 139–140
            literature sources, 140–141
        proteomics, 135–137
            protein arrays, 136–137
            yeast two hybrid system, 136

modeling pathways, cells, whole organs, and diseases, 125–127
pathway analysis software development, 159–164
condition-specific molecular and functional networks, 163–164

Tag-free methodology, combinatorial chemistry, 985–1001
click and in situ click chemistry, 992–997
dynamic chemistry and allied methods, 985–992
macromolecular ligands, thiol-based in situ assembly, 997–998
Tag fusions, recombinant protein crystallization, 382–383
Tagging methods:
affinity tagging, protein-protein interactions, mass spectroscopy, 531–533
split-pool synthesis, 977
binary encoding, 977–979
radio-frequency tagging, 980–981
TAK-779 compound, virus-cell fusion inhibitors, 1144–1147
Tamiflu, respiratory viral infection therapy, 1120, 1124–1126
Tamoxifen, estrogen receptors, 1069–1070
Tandem affinity purification (TAP), protein-protein interactions, affinity tagging and mass spectroscopy, 532–533
Tandem assays, natural product activity screening, 45–47
Tanimoto coefficient (Tc), molecular similarity quantification, 98
Target-based virtual screening, basic principles, 90
TAR-GET database, contents and location, 129–130
Target-directed selection, dynamic combinatorial chemistry, 985–992

Targeting mechanisms. *See also* Tumor targeting
cancer drug discovery:
antimimotics, 1360–1363
aurora kinases, 1361–1362
cell-based assays, 1362–1363
kinesin inhibitors, 1361
microtubule-targeting agents, 1360–1361
Bcr/Abl, c-Kit, and PDGF, 1350
DNA targets, 1357–1360
DNA methylation, 1357–1358
histone deacetylase inhibitors, 1358–1359
telomerase inhibitors, 1359–1360
EGF receptor pathway, 1347–1349
evolution of, 1344–1346
heat shock protein 90, 1364
HER2/Neu, 1347
information resources for, 1367–1368
insulin-like growth factor signaling, 1349
mammalian target of rapamycin (mTOR), 1350–1351
proteosome inhibition, 1363–1364
screening assays, 1351–1352
signal transduction targets, 1346–1352
small-molecule drug sources, 1365–1367
natural products, 1365–1366
synthetic chemical libraries, 1366–1367
tumor vasculature, 1352–1357
angiogenesis inhibitors, 1353–1356
HIF-1α target, 1355–1356
VEGF axis, 1354–1355
antivascular agents, 1356
high-throughput techniques, 1356–1357
metabotropic glutamate receptors, 862–863
prodrug development, 770–779
brain targeting, 775–778
carrier-mediated transport, 776–778

chemical drug delivery system, 776
lipophilic prodrugs, 775–776
colon targeting, 778
kidney targeting, 778–779
tumor targeting, 770–775
antibody-directed enzyme prodrug therapy, 774
gene-directed enzyme prodrug therapy, 774–775
hypoxia-selective drugs, 772–774
protein kinases, 1199–1205
protein tyrosine kinases, 1199–1202
serine/threonine kinases, 1202–1205
Target vectors, protein-protein interactions, yeast two-hybrid systems, 489–494
Taxol:
antimimotic targeted cancer therapy, 1360–1361
historical perspective on, 58–59
hydroxyl prodrugs, steric hindrance inhibition, 750
research and development, 18–19
supply crisis involving, 59, 61–62
toxicology studies, 57
Taylor expansion, protein structural models, refinement mathematics, 426–432
T-Cell Protein Tyrosine Phosphatase (TCPTP), PhosTRAK phosphatase assay, 715–716
T1 contrast agents, magnetic resonance imaging (MRI), mechanisms and enhancement, 1311–1312
T2 contrast agents, magnetic resonance imaging (MRI), mechanisms and enhancement, 1311–1312
"Tea bag" parallel synthesis, early research, 971–972
Technique-based drug development. *See also* specific techniques
defined, 8–9
Technology platforms, high-throughput screening procedures, 561–563
Telomerase inhibitors, targeted cancer therapy, 1359–1360

Tenofovir, acyclic nucleoside phosphonates, 1159–1161
Terminal Uracil Nick-End Label (TUNEL) assay:
CellCard system, 363
multiplexed dose curve analysis, 363–366
Terpenoids, gastrointestinal disease, animal models, 1023–1028
Testosterone:
anabolic action, 1067–1068
prodrug development, nasal drug delivery, 763–764
Tethering technique, thiol-based in situ macromolecular ligand assembly, 997–998
Tetracycline repressor (TetR), protein-protein interactions, split-hybrid inhibition system, 538–539
Tetraethylorthosilicate (TEOS), metal-enhanced fluorescence, solution-based assays, 653–654
Text Import Wizard, Excel data file management, 306–307
Therapeutic area drug development, drug-seeking diseases, 7
Therapeutic index (TI), developability assays, 468–469
Therapeutic ratio (TR), developability assays, 468–469
Thermal factor, protein structural models, 430–432
Thermal motion:
macromolecule crystallographic models, 420–422
x-ray diffraction, protein structure, 398–399
Thermodynamics, PhosTRAK phosphatase assay, 716–717
Thiazolidinediones (TZDs), diabetes therapy, 1077–1079
Thionamide drugs, thyroid hormone mediation, 1055–1056
THIP bicyclic analog, $GABA_A$ receptor complex:
agonists and antagonists, 808–811
pharmacophore models, 812–814

Three-dimensional arrays, protein-protein interactions, protein microarray supports, 524–526
Three-dimensional connectivity matrix: biological binding activity models, 280–281
multidimensional NMR spectra, 233–236
Three-dimensional pharmacophores, virtual screening, 100–102
molecular fingerprints, 103–104
Three-dimensional quantitative spectrometric data activity relationships (3D-QSDAR), corticosteroid binding, CoSCoSA model, 256–259
Three-dimensional quantitative structure-activity relationships (3D-QSARs):
comparative structural connectivity spectra analysis, aromatase enzyme binding models, 263–264
current applications, 228–229
two-dimensional quantitative spectrometric data-activity relationships, comparison, 231
virtual screening, 95–96, 112–113, 115–116
Three-dimensional techniques, virtual screening, 104, 106
Thrombin inhibitors, prodrug development, carrier-mediated transport, 754–756
Through-space models, comparative structural connectivity spectral analysis, aryl hydrocarbon receptor binding, 267–271
Thymidine kinase (TK), antiviral therapy, DNA polymerase inhibitors, 1148–1155
Thymidylate synthase (TM), thiol-based in situ macromolecular ligand assembly, 997–998

Thyroid hormones, structure and therapeutic application, 1053–1057
analogs, 1054–1055
iodide, 1055
potassium perchlorate, 1056–1057
therapeutic analogs, 1054
thionamide drugs, 1055–1056
thyroxine/tri-iodothyronine receptors, 1053–1054
Thyroid peroxidases (TPO):
iodide mediation, 1055
thionamide drugs, 1055–1056
Thyrotrophin-releasing hormone receptors (TRH-R), analogs and antagonists, 1040
Thyrotrophin-releasing hormone (TRH):
analogs and antagonists, 1039–1040
prodrug development, peptide formulations, stability improvement, 782
Thyroxine (T_4), structure and therapeutic application, 1053–1057
Tiagabine, GABA reuptake inhibition, 805–806
Tibetan medicine, history of, 13
Tiered screening/testing, drug development and, 2–3
Time-dependent hybridization, metal-enhanced fluorescence (MEF), enhanced DNA/RNA detection, 631–637
Time-resolved fluorimetry (TRF), protein-protein interactions, protein microarray detection, 527
Time/space constraints, protein-protein interactions, yeast two-hybrid systems, 488–489
Timing issues, natural product purification and isolation, 51
Tirapazamide (TPZ), hypoxia-selective prodrugs, tumor targeting, 773–774
Tissue culture, CellCard system, 360
Tissue factor pathway inhiibitor (TFPI), sepsis management, 946–948

TLS groups, protein structural models, 428–432
Tolbutamide, diabetes therapy, 1079
Topical drug delivery, prodrug development, 757–759
Topological representation, virtual screening, 99–100
Topological segregation, split-pool synthesis, reaction functionality, 979
Toxicity mechanisms. *See also* Cytotoxicity assays
   protein-protein interactions, phage display technology, 514–515
Toxicology studies:
   developability assays, 478–481
   in vitro toxicology, 479–480
   in vivo toxicology, 480–481
   natural product development, 56–58
   pharmacokinetic assays, 477
   Taxol case study, 57–59
Tracheitis, viral agents for, 1116
Tracheobronchitis:
   influenza virus, 1112
   viral agents for, 1116
Training sets, pattern recognition techniques, 237
Transcription activation:
   glucocorticoid receptors, 1061–1062
   protein-protein interactions, yeast two-hybrid systems, 487–489
   RNA synthesis inhibition, 1162–1163
Transcription factors, molecular drug action, 41
Transdermal drug delivery, prodrug development, 759–760
Transiently transfected cell populations, high-throughput screening bioassays, 590–591
Transmembrane helices:
   FPR models, 216–217
   protein structures, 382
Transmembrane receptors, protein-protein interactions, yeast two-hybrid systems, 486–489
Transmission electron microscopy (TEM), metal-enhanced fluorescence, solution-based assays, 653–654

TRANSPATH database, contents and location, 130
Transport-mediated absorption, prodrug structures, 766
Triazole synthesis, click chemistry, 993–997
Tribrid systems, protein-protein interactions:
   kinase tribid system, 506
   yeast two-hybrid systems, 488–489
Trichoderma polysporum, research and development, 18–19
Trichoderma viride, natural product development, 29
Trichothecium roseum, research and development, 17
Trididemnum solidum, research and development, 18–19
Tri-iodothyronine (T_3), structure and therapeutic application, 1053–1057
Trilostane, corticosteroid synthesis inhibition, 1064
"Trimer-of-hairpins" structure, virus-cell fusion inhibitors, 1144–1147
Trinitrobenzene sulfonic acid (TNBS)/ethanol-induce colitis, inflammatory bowel disease, animal models, 1017
Tripterygium wilfordii species, terpenoid herbal compounds, 1027–1028
Tripterygium winfordii, research and development, 19
Triptolide, research and development, 19
Triterpenoid, natural product development, 20–22
T7Select vector, protein-protein interactions, peptide display reagents, 517
Tubulin-targeting drugs, antimimotic targeted cancer therapy, 1360–1361
Tumor cell surface receptors, magnetic resonance imaging, 1318–1325
   isolated cell molecular imaging, 1318–1320

in vivo molecular imaging, 1320–1325
  Gd-based contrast agents, 1320–1321
  iron-oxide-based contrast agent, 1321–1323
  multistep targeting and prelabeling, 1323–1324
  signal amplification, 1324–1325
Tumor markers, cancer cell proteomics:
  aptamer-based detection, 75–76
  molecular diagnosis, 74–75
  multiple markers, whole-cell-based aptamer selection, 81–83
Tumor necrosis factor (TNF)-α:
  herbal terpenoid inhibition, 1027–1028
  respiratory viral infections, 1121–1122
  sepsis pathophysiology, 939
Tumor targeting:
  magnetic resonance imaging, 1310
    contrast agents and enhancement, 1311–1313
      relaxation properties, 1313–1314
      T1 contrast agents, 1311–1312
      T2 contrast agents, 1312–1313
    intracellular targets, 1325–1328
      long-term cell labeling contrast agents, 1326–1327
      specific contrast agents, 1327–1328
    limitations and future research, 1331–1333
    molecular imaging, activated contrast agents, 1328–1329
      CEST/PARACEST contrast agents, 1329
      enzymatic activation, 1328
    molecular MRI and MR spectroscopy, 1329–1331
      reporter molecules, CSI imaging, 1330–1331
    tumor cell surface receptors, 1318–1325
      isolated cell molecular imaging, 1318–1320
      in vivo molecular imaging, 1320–1325
        Gd-based contrast agents, 1320–1321
        iron-oxide-based contrast agent, 1321–1323
        multistep targeting and prelabeling, 1323–1324
        signal amplification, 1324–1325
    tumor vasculature and vascular targets, 1313, 1315–1318
      high-molecular-weight contrast agents, 1316
      low-molecular-weight contrast agents, 1315
      moleular imaging, 1316–1318
  prodrug development, 770–775
    antibody-directed enzyme prodrug therapy, 774
    gene-directed enzyme prodrug therapy, 774–775
    hypoxia-selective drugs, 772–774
Two-dimensional chemical structure representation, protein structural models, 428–432
Two-dimensional ^{13}C NMR spectroscopy, comparative structural connectivity spectral analysis, molecular structural analysis, 245–271
Two-dimensional gel electrophoresis (2-DE), proteomics, systems biology applications, 135–136
Two-dimensional polyacrylamide gel electrophoresis (2D PAGE), protein-protein interactions, spectrometric reagents and actions, 528–529
Two-dimensional quantitative spectrometric data-activity relationships (2DSDAR), biological effect modeling, 231
Two-dimensional quantitative structure-activity relationships (2D-QSARs), virtual screening, 95–96, 112
Two-dimensional substructure searching, virtual screening, 99
Two-dimensional techniques, virtual screening, 104, 106
Tyrosine kinases. *See also* Protein tyrosine kinases
  high-throughput screening formats, 564–565

INDEX

Ubiquitin-based split-protein sensor (USPS), protein-protein interactions, 504
  yeast two-hybrid systems, 488–489
Ubiquitin-specific proteases (UBPs), protein-protein interactions, 504
Ulcerative colitis (UC):
  animal models, 1016–1019
    severity assessment, 1018–1019
    trinitrobenzene sulfonic acid/ethanol-induce colitis, 1017
    dinitrobenzene sulfonic acid/ethanol, 1017
Ulcers:
  animal models:
    chronic gastric, 1015–1016
    duodenal, 1014–1015
    acute/chronic, 1016
  herbal therapeutic agents, 1021–1028
    biological processes, 1021–1022
    licorice, 1025–1027
    phenolic compounds and terpenoids, 1023–1028
      curcumin, 1023–1024
    polysaccharides, 1022–1024
      mushroom/seaweed sources, 1022–1023
      plant sources, 1023
    terpenoid sources, 1027–1028
Ultrabright probes, metal-enhanced fluorescence (MEF), overlabeled proteins, 639–642
Ultraviolet (UV) detection, analytical chemistry assays, 463–464
Ulvans lactuc, polysaccharide extracts, 1023–1024
UMBBD database, contents and location, 130
Uncompetitive receptor antagonists, NMDA receptors, 833–835
Unformatted Text format, Excel data management, 304–306
Unigene database, contents and location, 130

Upper limit of quantification (ULQ), bioanalytical chemistry assays, 475–476
Uracil compounds, AMPA receptor agonists, 838–840
URA3 gene, protein-protein interactions, inhibition, with reverse two-hybrid systems, 537–538
Urocortin, drug actions of, 1041

Vaccines:
  current research on, 1135–1140
  influenza virus, 1112
Valacyclovir, prodrug development, carrier-mediated transport, 756
Validation procedures, protein structural models, 432–433
Valinomycin, biopolymer solid-phase synthesis, 967–970
van der Waals forces, antiviral therapy, DNA polymerase inhibitors, 1152–1155
Vapor diffusion, crystallization format, 380–381
Varicella-zoster virus (VZV):
  acyclovir therapy, 1120–1121
  characteristics, 1113–1114
  DNA polymerase inhibitors, 1148–1155
Vascular endothelilal growth factor receptors (VEGFR):
  hammerhead ribozymes, 1276–1277
  receptor tyrosine kinases, 1200–1201
  inhibitors, 1219–1221
Vascular endothelilal growth factor (VEGF):
  RNA mimetics, aptamer function, 1271–1272
  targeted cancer therapy, angiogenesis inhibitors, 1354–1355
Vasculature and vascular targets:
  MRI tumor targeting, 1313, 1315–1318
    high-molecular-weight contrast agents, 1315
    low-molecular-weight contrast agents, 1315

neovasculature mmolecular imaging, 1316–1318
targeted cancer therapy, 1352–1357
  angiogenesis inhibitors, 1353–1356
    HIF-1α target, 1355–1356
    VEGF axis, 1354–1355
  antivascular agents, 1356
  high-throughput techniques, 1356–1357
Vasopressin, drug action and therapeutic applications, 1049–1052
Vasopressors, sepsis management, glucocorticoids vs. mediator-specific anti-inflammatory agents, 945–946
Venereal disease, natural product development, 30
Venom sources, drug development using, 30–31
Ventricular arrhythmias:
  ATP-sensitive potassium channel antagonists, 914–924
  channel agonists, proarrhythmic effects, 923–924
  nonselective antagonists, 915–917
  selective antagonists, 917–924
    HMR 1402 cardioselective antagonist, 921–923
    HMR 1883 cardioselective antagonist, 919–921
  extracellular potassium effects, 911–913
Vertebrates, drug development from, 31
Vibrational properties, protein structural models, 428–432
Vigabatrin, GABA metabolism inhibition, 803–804
Vinca alkaloids, antimimotic targeted cancer therapy, 1361
Viral disease. See also Antiviral therapy
  natural product development for, 20–22
  respiratory clinical syndromes, 1114–1118
    acute laryngotracheobronchitis, 1116
    bronchitis, 1117
    common cold, 1114, 1116
    laryngitis, 1116
    pharyngitis, 1116
    pneumonia, 1117–1118
    tracheitis and tracheobronchitis, 1116
  respiratory infection, diagnosis, 1118–1120
Viral DNA synthesis:
  acyclic nucleoside phosphonate inhibitors, 1158–1161
  polymerase inhibitors, 1147–1155
  reverse transcriptase inhibition, 1155–1158
Viral genome products, natural product development techniques, 21–22
"Virtual cell" construction, systems biology and, 126–127
Virtual screening (VS), 210
  chemical database generation and maintenance, 212–213
  clustering techniques, 106–107
  compound sources, 89–90
  docking and scoring, 212
  drug likeness, 114
  evolution of, 88
  FPR case study, 216–218
  future research issues, 116
  high-performance computing, 213–216
  high-throughput screening vs., 89, 210
  ligand- vs. target-based techniques, 90
  limitations of, 96–99
    molecular similarity quantification, 97–98
    potency and selectivity issues, 98–99
    similar property principle, 96–97
  molecular descriptors and chemical spaces, 91–92
  molecular fingerprints, 102–106
    profiling and scaling, 104–106
    three-dimensional pharmacophores, 103–104
    two-dimensional, 102–103
  molecular similarity concept, 92–94
  partitioning methods, 108–112
    cell-based partitioning, 108–109
    decision tree algorithms, 110–112
    direct space methods, 109–110

INDEX

performance evaluation, 114–116
quantitative structure-activity relationship analysis:
  basic principles, 95–96
  four-dimensional models, 113
  probabilistic methods, 113–114
  three-dimensional models, 112–113
  two-dimensional models, 112
three-dimensional pharmacophores, 100–102
topological and shape representations, 99–100
two-dimensional substructure searching, 99
two-dimensional vs. three-dimensional methods, 104, 106
Virus-directed enzyme prodrug therapy (VDEPT), tumor targeting, 772–775
Viruses:
  adsorption inhibitors, 1140–1141
  life cycle, 1136–1137
  neuraminidase inhibitors, 1169–1171
  protease inhibitors, 1163–1169
  structure and properties, 1106
  virus-cell fusion inhibitors, 1141–1147
Visual Basic Editor (VBE):
  basic elements, 292–293
  language primer, 293–297
  macro recording, 290–291
Visual Basic for Applications (Excel software), laboratory data management, 290–303
  automated report generation, 327–334
    heat map display, 331–334
    worksheet formatting, 328–330
  data file reading and writing, 307–319
    file access types, 309
    file selection, 308–309
    opening and closing sequential access files, 309–311
    sequential access files, 311–315
  event-driven programming, 297–299
  external database imports, 319–327
    dabatase characteristics, 319–320
    data imports, 322–326
    ODBC access issues, 320–322
  fundamental principles, 291–303

language primer, 293–297
macro recording, 290–291
object entities, 299–303
Oracle ODBC connection, 327
string data manipulation and parsing, 315–319
Visual Basic Editor elements, 292–293
VLX Biological Modeler, biological pathway analysis, 145
Volvariella volvacea, natural product drug development, 29

Waller's test data, estrogen receptor binding compounds, comparative structural connectivity spectral analysis, 254–256
Ward's agglomerative clustering, virtual screening, 107
Water-soluble prodrugs. *See* Aqueous solubility
Wave scattering properties, x-ray crystallization of protein structures, 390
  Fourier synthesis, 392–396
Weighting factor, protein structural models, 430–432
Well platforms:
  high-throughput screening:
    accuracy and precision measurements, 580–581
    compound storage, 574–575, 575
    design criteria, 570–572
    metal-enhanced fluorescence (MEF), 630–660
      directional emission, 654–660
      enhanced DNA labels, 637–639
      enhanced DNA/RNA detection, 630–637
      enhanced energy transfer, silver surfaces, 645–649
      multiphoton excitation and metallic nanoparticles, 643–645
      overlabeled proteins, ultrabright probes as, 639–642

ratiometric surface sensing, 649–653
solution-based assays, 653–654
miniaturization, 565–566
96-well platform design:
drug delivery assays, 473–474
high-throughput screening, 564–565
multiple cell line screening, 360
protein-protein interactions, protein microarray delivery systems, 526–527
Whole-cell-based aptamers, multiple tumor markers, 81–83
Whole genome approach, protein-protein interactions, yeast two-hybrid systems, 507–509
WIT2 database, contents and location, 130
Wolff-Chaikoff effect, thyroid hormone analogs and receptors, 1055
WOMBAT (WOrld of Molecular BioAcTivity) database, virtual screening, 212–213
Workbook events, Visual Basic for Applications:
laboratory data management, 298–299
ODBC protocol, 321–322
Worksheet objects, Excell object entities, 300–303
formatting protocols, 328–330
World Health Organization (WHO), natural product research and development and, 14
Wortmanin, structure and properties, 1231

X-ray crystallography, protein structures:
anomalous dispersion, 412–416
Friedel's law, 412
structure determination, 413–416
Bragg's law and angular dependence, 396–397
crystallization trials, 376–389
anomalous diffraction derivatization, 388
cryopreservation for, 385–386
data sources, 388–389
demonstration experiment, 389
drug discovery and, 386–388
factors in, 382–383
formats, 380–381
formulation screens, 384–385
membrane protein, 381–382
observation and documentation, 384
research background, 377–379
data collection, 399–402
de novo model construction, 422–425
diffraction principles, 389–390
drug discovery applications, 434–447
antistructures, 441–442
crystallographic screening, 443–444
fragment screening, 444–445
lead optimization, structure-based drug design, 436–441
HIV protease inhibitors, 438–440
protein kinases, 440–441
in silico screening, 442–443
site-directed leads, fragment tethering, 445–446
structural genomics, 446–447
therapeutics, 442
drug discovery interface, 433–434
electron clouds and thermal motion, 398–399
Fourier synthesis, 392–396
heavy atom replacement, 406
home laboratory data collection, 402–403
limitations in drug discovery, 447–4508
macromolecule crystals, 390–392, 420–422
model analysis and preparation, 432–433
model refinement and analysis, 425–432
molecular replacement, 416–420
multiple isomorphous replacement, 406–412
phase problem, 405
research background, 374–376
structure factor, 405–406
synchrotron diffraction data collection, 403–404
X-ray detector, x-ray diffraction data collection, 399–402
X-ray source, x-ray diffraction data collection, 399–402

Yeast two-hybrid systems:
  protein-protein interactions, 484–511
    affinity tagging and mass spectroscopy vs., 533
    basic principles, 485–489
    classical system improvements, 501–502
    inhibitors, 536–540
      peptide aptamers, 539–540
      reverse two-hybrid systems, 537–538
      split-hybrid system, 538–539
    interaction suppression, 506–507
    mammalian two-hybrid systems, 506
    new applications for, 506–511
    new developments in, 501–506
    protease trap, 507
    protein three-hybrid systems, 503–506
    reagents, 489–501
      library choice, 495–501
      reporter genes and host yeast strains, 494–495
      target and bait vectors, 489–494
    Sos recruitment system, 502–503
    two-bait system, 502
    USPS system, 503
    whole genome applications, 507–509

proteomics, systems biology applications, 136
YOYO-1-DNA, metal-enhanced fluorescence (MEF), enhanced DNA labeling, 638–639

Zanamivir:
  antiviral agents, 1124–1126
  neuraminidase inhibitors, 1171
  respiratory viral infection therapy, 1120
ZD1839, EGFR intracellular domain targeting, 1217–1218
Zidovudine (AZT):
  brain targeting, chemical drug delivery system principle, 776
  DNA polymerase inhibitors, 1148–1150
  phosphate group prodrug development, 744–746
  reverse transcriptase inhibitors, 1156–1158
Zinc complexes, dynamic combinatorial chemistry, 988–992
Z values, coupled luminescent assays, 719–720

Ref.
RM
301.25
.D784
2005

Drug discovery handbook.

47878

$170.00

	DATE		

BAKER & TAYLOR

SOUTH UNIVERSITY
709 MALL BLVD.
SAVANNAH, GA 31406